Book companion website:

CAD: **cad.eecs.umich.edu**

CIRCUIT ANALYSIS AND DESIGN

Fawwaz T. Ulaby
The University of Michigan

Michel M. Maharbiz
The University of California, Berkeley

Cynthia M. Furse
The University of Utah

This book is published by Michigan Publishing under an agreement with the authors. It is made available free of charge in electronic form to any student or instructor interested in the subject matter.

Published in the United States of America by Michigan Publishing
Manufactured in the United States of America

ISBN 978-1-60785-483-8 (hardcover)
ISBN 978-1-60785-484-5 (electronic)

To an academic, writing a book is an endeavor of love.

We dedicate this book to Jean, Anissa, and Katie.

Brief Contents

Contents

List of Technology Briefs

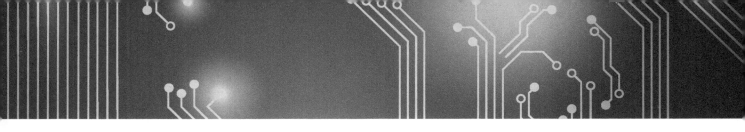

Preface

Welcome to *Circuit Analysis and Design*.

As the foundational course in the majority of electrical and computer engineering curricula, an *electric circuits* course should serve four vital objectives:

(1) It should introduce the fundamental principles of circuit *analysis* and equip the student with the skills necessary to analyze any planar, linear circuit, including those driven by dc or ac sources, or by more complicated waveforms such as pulses and exponentials.

(2) It should start the student on the journey of *circuit design*.

(3) It should guide the student into the seemingly magical world of *domain transformations*—such as the Laplace and Fourier transforms, not only as circuit analysis tools, but also as mathematical languages that are "spoken" by many fields of science and engineering.

(4) It should expand the student's technical horizon by introducing him/her to some of the many allied fields of science and technology.

This book aims to accomplish exactly those objectives. Among its distinctive features are:

Technology Briefs: The book contains 32 Technology Briefs, each providing an overview of a topic that every electrical and computer engineering professional should become familiar with. Electronic displays, data storage media, sensors and actuators, supercapacitors, and 3-D imaging are typical of the topics shared with the reader. The Briefs are presented at a technical level intended to introduce the student to how the concepts in the chapter are applied in real-world applications and to interest the reader in pursuing the subject further on his/her own. Technology Briefs cover applications in *circuits*, *medicine*, the *physical world*, *optics*, *signals and systems*, and more.

Application Notes: Most chapters include a section focused on how certain devices or circuits might be used in practical applications. Examples include *power supplies*, *CMOS inverters* in *computer processors*, *signal modulators*, and several others.

Multisim and MathScript: Multisim is a SPICE circuit simulator available from National Instruments (see cad.eecs.umich.edu for details). Multisim is highlighted through many end-of-chapter demonstrations. The student is strongly encouraged to take advantage of this rich resource. The Math-Script software can perform matrix inversion and many other calculations, much like the MathWorks, Inc. MATLAB® software.

myDAQ: The myDAQ board does not come with this e-book, but it can be purchased directly from National Instruments.

The myDAQ is a convenient, portable measurement tool that turns a PC into a basic electrical engineering lab with a DVM, analog and digital power supplies, function generator, oscilloscope, Bode plot analyzer, and diode analyzer. A written myDAQ tutorial is available in Appendix F and online video tutorials are available at http://www.ni.com/mydaq. The book contains 53 *integrative* end-of-chapter problems, each intended to be solved analytically, by Multisim using software simulation, and by constructing the circuit and measuring its currents and voltages using myDAQ. ***The three-way complementary approach is an exceedingly valuable learning experience.***

Acknowledgments

A science or engineering textbook is the product of an integrated effort by many professionals. Invariably, the authors receive far more of the credit than they deserve, for if it were not for the creative talents of so many others, the book would never have been possible, much less a success. We are indebted to many students and colleagues, most notably the following individuals:

Richard Carnes: For his meticulous typing of the manuscript, careful drafting of its figures, and overall stewardship of the project. Richard imparted the same combination of precision and passion to the manuscript as he always does when playing Chopin on the piano.

Joe Steinmeyer: For testing the Multisim problems contained in the text and single-handedly developing all of the Multisim modules on the DVD-ROMs. Shortly thereafter, Joe went to MIT at which he completed a Ph.D. in electrical engineering.

Professor Ed Doering: For developing a comprehensive tutorial that includes 36 circuit problems, each of which is solved analytically, with Multisim, and with myDAQ. In addition, he created instructive video tutorials on how to use a variety of computer-based instruments, including the multimeter, oscilloscope, waveform generator, and Bode analyzer.

Nathan Sawicki: For developing a tutorial (Appendix F) on myDAQ and how to build circuits using it.

Rose Anderson: For developing an elegant cover design and a printable InDesign version of the book.

For their reviews of the overall manuscript and for offering many constructive criticisms, we are grateful to Professors Fred Terry and Jamie Phillips of the University of Michigan, Keith Holbert of Arizona State University, Ahmad Safaai-Jazi of Virginia Polytechnic Institute and State University, Robin Strickland of the University of Arizona, and Frank Merat of Case Western Reserve University. The manuscript was also scrutinized by a highly discerning group of University of Michigan graduate students: Mike Benson, Fikadu Dagefu, Scott Rudolph, and Jane Whitcomb. Multisim sections were reviewed by Peter Ledochowitsch.

Many of the 818 end-of-chapter problems were solved and checked by students from the University of Michigan and the University of California at Berkeley. They include Holly Chiang, David Hiskens, Tonmoy Monsoor, Zachary Hargeaves, James Dunn, Christopher Lo, Chris Buonocore, and Randolf Tjandra. We thank them for their contributions.

We enjoyed writing this book, and we hope you enjoy learning from it.

Fawwaz Ulaby, Michel Maharbiz, and Cynthia Furse

Photo Credits

Page 4 (left) Dorling Kindersley/Getty Images; (right) © Bettmann/CORBIS; Chuck Eby
Page 5 (left) Chuck Eby; John Jenkins, sparkmuseum.com; IEEE History Center; History San José; (right) LC-USZ62-39702, Library of Congress; History San José; © Bettmann/ CORBIS
Page 6 (left) MIT Museum; © Bettmann/ CORBIS; Emilio Segre Visual Archives/American Institute of Physics/Science Photo Library; (right) Emilio Segre Visual Archives/American Institute of Physics/Science Photo Library
Page 7 (left) Courtesy of Dr. Steve Reyer; Courtesy of Texas Instruments Incorporated; NASA; Digital Equipment Corporation; (right) used with permission of SRI International; Courtesy of Texas Instruments Incorporated
Page 8 (left) Courtesy of IBM; Courtesy of Palra Inc., US Robotics, Inc.
Page 10 Miguel Rodriguez
Page 11 Courtesy Office of Basic Energy Sciences, Office of Science, U.S. Department of Energy
Page 14 From "When will computer hardware match the human brain?" by Hans Moravec, *Journal of Transhumanism*, Vol. 1, 1998. Used with permission
Page 32 National Geographic
Page 57 Pacific Northwest National Laboratory
Page 57 Courtesy of Central Japan Railway Company; Courtesy General Electric Healthcare
Page 58 GE Healthcare
Page 71 Courtesy of Khalil Najafi, University of Michigan
Page 91 Soomi Park
Page 126 *Geophysical Research Letters*
Page 127 Wendy Pyper
Page 136 Courtesy of Veljko Milanovic
Page 139 Courtesy of International Business Machines Corporation
Page 203 ZeptoBars
Page 226 ZYPEX, Inc.
Page 228 CST MICROWAVE STUDIOS
Page 229 Courtesy of Prof. Ken Wise, University of Michigan
Page 265 *Science*, August 18, 2006, Vol. 313 (#5789). Reprinted with permission of AAAS
Page 293 © Steve Allen/Brand X/Corbis
Page 302 Balluff
Page 303 STMicroelectronics
Page 339 (left to right) Analog Devices; Courtesy of Prof. Khalil Najafi, University of Michigan
Page 340 Analog Devices
Page 363 (left to right) Cochlear Americas and MED-EL
Page 364 (left to right) John Wyatt and Medtronic
Page 365 Todd Kuiken, Spine-health.com, and Orthomedical
Page 424 NIST
Page 477 © Reuters/CORBIS
Page 478 Suljo
Page 480 (top to bottom) Advance Dermatology Pocono Medical Care, Inc.; NASA/SDO
Page 533 Agoora.co.uk
Page 534 (top to bottom) Wordpress.com; SunglassWarehouse.com
Page 546 Martin Logan
Page 587 Intermountain Power Project, IECACA
Page 608 Image from WebPath, courtesy of Edward C. Klatt MD
Page 610 Emilee Minalga, Robb Merrill
Page 648 NASA
Page 649 NASA
Page 649 ROBYN BECK/AFP/Getty Images
Page 696 Aaron Chevalier and *Nature* (Nov. 24, 2005)
Page 702 Deka Corp., UC Berkeley, EPFL
Page 714 Courtesy of Renaldi Winoto

CHAPTER 1

Circuit Terminology

Contents

Objectives

Learn to:

- Differentiate between active and passive devices; analysis and synthesis; device, circuit, and system; and dc and ac.
- Point to important milestones in the history of electrical and computer engineering.
- Relate electric charge to current; voltage to energy; power to current and voltage; and apply the passive sign convention.
- Describe the properties of dependent and independent sources.
- Describe the operation of SPST and SPDT switches.

The iPhone is a perfect example of an *integrated electronic architecture* composed of a large number of interconnected circuits. Learning a new language starts with the alphabet. This chapter introduces the terms and conventions used in the *language of electronics*.

Overview

Electrical engineering is an exciting field through which we interface with the world using electrical signals. In this chapter you will learn about the basis of electrical engineering—voltage and current—where they come from, what they mean, and how to measure them. The chapter provides you the nomenclature and symbols to draw and represent electric circuits. You will also learn your first circuit analysis tool, Ohm's law, which describes the relationship between voltage, current, and resistance. In the first section of this chapter, enjoy electrical engineering's innovative past, and in the micro-nano Technology Brief, imagine the things you could do with it in the future. As you explore this chapter and start to pick up the tools you need in your engineering career, imagine an application that particularly interests you, and how these concepts and ideas apply to that application.

Figure 1-1: Cell phone.

Cell-Phone Circuit Architecture

Electronic circuits are contained in just about every gadget we use in daily living. In fact, electronic sensors, computers, and displays are at the operational heart of most major industries, from agricultural production and transportation to healthcare and entertainment. The ubiquitous cell phone (Fig. 1-1), which has become practically indispensable, is a perfect example of an integrated electronic architecture made up of a large number of interconnected circuits. It includes a two-way antenna (for transmission and reception), a diplexer (which facilitates the simultaneous transmission and reception through the antenna), a microprocessor for computing and control, and circuits with many other types of functions (Fig. 1-2). Factors such as compatibility among the various circuits and proper electrical connections between them are critically important to the overall operation and integrity of the cell phone.

Usually, we approach electronic analysis and design through a hierarchical arrangement where we refer to the overall entity as a *system*, its subsystems as *circuits*, and the individual circuit elements as *devices* or components. Thus, we may regard the cell phone as a system (which is part of a much larger communication system); its audio-frequency amplifier, for example, as a circuit, and the resistors, integrated circuits (ICs), and other constituents of the amplifier as devices. In actuality, an IC is a fairly complex circuit in its own right, but its input/output functionality is such that usually it can be represented by a relatively simple equivalent circuit, thereby allowing us to treat it like a device. Generally, we refer to devices that do not require an external power source in order to operate as *passive devices*; these include resistors, capacitors, and inductors. In contrast, an *active device* (such as a transistor or an IC) cannot function without a power source.

This book is about *electric circuits*. A student once asked: "What is the difference between an *electric* circuit and an *electronic* circuit? Are they the same or different?" Strictly speaking, both refer to the flow of electric charge carried by electrons, but historically, the term "electric" preceded "electronic," and over time the two terms have come to signify different things:

► An *electric circuit* is one composed of passive devices, in addition to voltage and current sources, and possibly some types of switches. In contrast, the term *electronic* has become synonymous with transistors and other active devices. ◄

The study of electric circuits usually precedes and sets the stage for the study of electronic circuits, and even though a course on electric circuits usually does not deal with the internal operation of an active device, it does incorporate active devices in circuit examples by representing them in terms of equivalent circuits.

An *electric circuit*, as defined by *Webster's English Dictionary*, is a "complete or partial path over which current may flow." The path may be confined to a physical structure (such as a metal wire connecting two components), or it may be an unbounded channel carrying electrons through it. An example of the latter is when a lightning bolt strikes the ground, creating an electric current between a highly charged atmospheric cloud and the earth's surface.

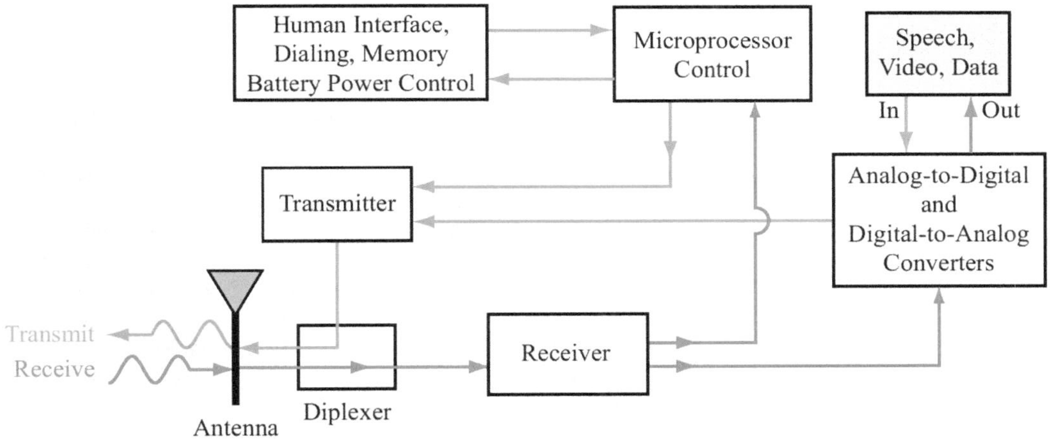

Figure 1-2: Basic cell-phone block diagram. Each block consists of multiple circuits that together provide the required functionality.

Electrical engineering design is about how we use and control voltages and currents to do the things we want to do. To interface with the real world, sensors are the electrical tools that convert real world inputs—like heat, sound, light, pressure, user inputs like button presses or touch screen, motion, etc.—into voltages and currents. We then manipulate these input voltages and currents using various circuits. We may amplify them if they are too small, switch them on or off, change their frequency (filter, oscillate, modulate them), or convert them into a digital signal a computer circuit can further analyze. In the end, we want to have an output voltage or current we can use to interface back to the real world—turn on a light, buzzer, alarm, motor/actuator, or control a cell phone, car airplane, robot, medical device, etc. Electrical engineers design both the ***input/output*** (I/O) systems as well as the control and actuation circuits, and often the software and algorithms as well. Electrical engineering is about "what you can do to a voltage" and how to use it to do something important in the real world.

The study of electric circuits consists of two complementary tasks: ***analysis*** and ***synthesis*** (Fig. 1-3). Through analysis, we develop an understanding of "how" a given circuit works. If we think of a circuit as having an input—a ***stimulus***—and an output—a ***response***, the tools we use in circuit analysis allow us to mathematically relate the output response to the input stimulus, enabling us to analytically and graphically "observe" the behavior of the output as we vary the relevant parameters of the input. An example might be a specific amplifier circuit, in which case the objective of circuit analysis might be to establish how the output voltage varies as a function of the input voltage over the full operational range of the amplifier parameters. By analyzing the operation of each circuit in a system containing multiple circuits, we can characterize the operation of the overall system.

As a process, synthesis is the reverse of analysis. In engineering, we tend to use the term ***design*** as a synonym for synthesis. The design process usually starts by defining the operational specifications that a gadget or system should meet, and then we work backwards (relative to the analysis process) to develop circuits that will satisfy those specifications. In analysis, we are dealing with a single circuit with a specific set of operational characteristics. We may employ different analysis tools and techniques, but the circuit is unique, and so are its operational characteristics. That is not necessarily the case for synthesis; the design process may lead to multiple

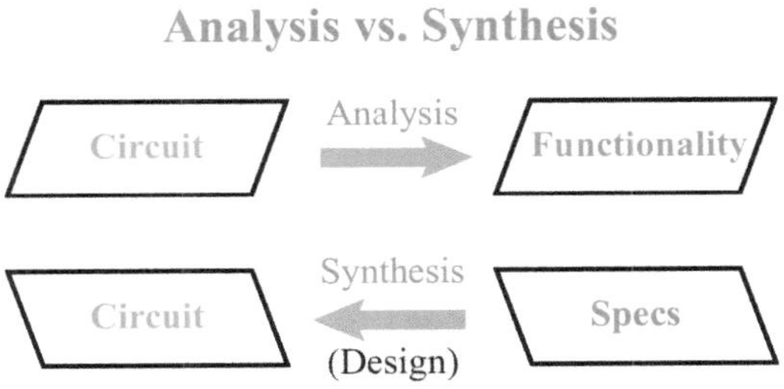

Figure 1-3: The functionality of a circuit is discerned by applying the tools of circuit analysis. The reverse process, namely the realization of a circuit whose functionality meets a set of specifications, is called circuit synthesis or design.

circuit realizations—each one of which exhibits or satisfies the desired specifications.

Given the complementary natures of analysis and synthesis, it stands to reason that developing proficiency with the tools of circuit analysis is a necessary prerequisite to becoming a successful design engineer. This textbook is intended to provide you with a solid foundation of the primary set of tools and mathematical techniques commonly used to analyze both *direct current (dc)* and *alternating current (ac)* circuits, as well as circuits driven by pulses and other types of waveforms. A dc circuit is one in which voltage and current sources are constant as a function of time, whereas in ac circuits, sources vary sinusoidally with time. Even though this is not a book on circuit design, design problems occasionally are introduced into the discussion as a way to illustrate how the analysis and synthesis processes complement each other.

> **Concept Question 1-1:** What are the differences between a device, a circuit, and a system? (See CAD)

> **Concept Question 1-2:** What is the difference between analysis and synthesis? (See CAD)

1-1 Historical Timeline

We live today in the age of electronics. No field of science or technology has had as profound an influence in shaping the operational infrastructure of modern society as has the field of electronics. Our computers and communication systems are at the nexus of every major industry. Even though no single event marks the beginning of a discipline, electrical engineering became a recognized profession sometime in the late 1800s (see chronology). *Alexander Graham Bell* invented the telephone (1876); *Thomas Edison* perfected his incandescent light bulb (1880) and built an electrical distribution system in a small area in New York City; *Heinrich Hertz* generated radio waves (1887); and *Guglielmo Marconi* demonstrated radio telegraphy (1901). The next 50 years witnessed numerous developments, including radio communication, TV broadcasting, and radar for civilian and military applications—all supported by electronic circuitry that relied entirely on vacuum tubes. The invention of the *transistor* in 1947 and the development of the *integrated circuit* (IC) shortly thereafter (1958) transformed the field of electronics by setting it on an exponentially changing course towards "smaller, faster, and cheaper."

Computer engineering is a relatively young discipline. The first *all-electronic computer*, the ENIAC, was built and demonstrated in 1945, but computers did not become available for business applications until the late 1960s and for personal use until the introduction of Apple I in 1976. Over the past 20 years, not only have computer and communication technologies expanded at a truly impressive rate (see Technology Brief 1), but more importantly, it is the seamless integration of the two technologies that has made so many business and personal applications possible.

Generating a comprehensive chronology of the events and discoveries that have led to today's technologies is beyond the scope of this book, but ignoring the subject altogether would be a disservice to both the reader and the subject of electric circuits. The abbreviated chronology presented on the next few pages represents our compromise solution.

Chronology: **Major Discoveries, Inventions, and Developments in Electrical and Computer Engineering**

ca. 1100 BC **Abacus**: the earliest known calculating device.

ca. 900 BC **Magnetite**: According to legend, a shepherd in northern Greece, **Magnus**, experienced a pull on the iron nails in his sandals by the black rock he was standing on. The rock later became known as magnetite [a form of iron with permanent magnetism].

ca. 600 BC **Static electricity**: Greek philosopher **Thales** described how amber, after being rubbed with cat fur, can pick up feathers.

1600 **Electric**: The term was coined by **William Gilbert** (English) after the Greek word for amber (*elektron*). He observed that a compass needle points north to south, indicating the Earth acts as a bar magnet.

1614 **Logarithm**: developed by **John Napier** (Scottish).

1642 **First adding machine**: built by **Blaise Pascal** (French) using multiple dials.

1733 **Electric charge**: **Charles François du Fay** (French) discovers that charges are of two forms and that like charges repel and unlike charges attract.

1745 **Capacitor**: **Pieter van Musschenbroek** (Dutch) invented the Leyden jar, the first electrical capacitor.

1800 **First electric battery**: developed by **Alessandro Volta** (Italian).

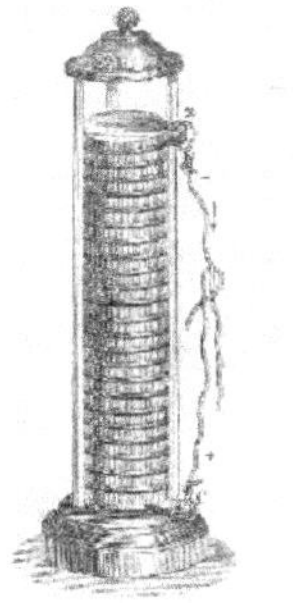

1827 **Ohm's law**: formulated by **Georg Simon Ohm** (German), relating electric potential to current and resistance.

1827 **Inductance**: introduced by **Joseph Henry** (American), who built one of the earliest electric motors. He also assisted Samuel Morse in the development of the telegraph.

1837 **Telegraph**: concept patented by **Samuel Morse** (American), who used a code of dots and dashes to represent letters and numbers.

1843 **Computer algorithm**: original concept and plan attributed to **Ada Byron Lovelace** (British), the daughter of poet Lord Byron. The "Ada" software language was developed in 1979 by the U.S. Department of Defense in her honor.

1876 **Telephone**: invented by **Alexander Graham Bell** (Scottish-American): the rotary dial became available in 1890, and by 1900, telephone systems were installed in many communities.

1879 **Incandescent light bulb**: demonstrated by **Thomas Edison** (American), and in 1880, his power distribution system provided dc power to 59 customers in New York City.

1887 **Radiowaves**: **Heinrich Hertz** (German) built a system that could generate electromagnetic waves (at radio frequencies) and detect them.

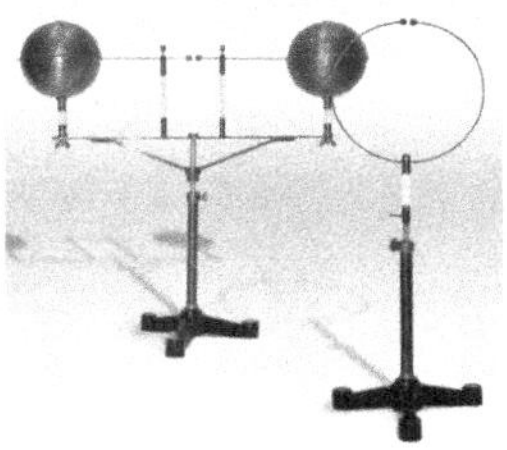

Courtesy of John Jenkins (sparkmuseum.com)

1888 **ac motor**: invented by **Nikola Tesla** (Croatian-American).

1893 **Magnetic sound recorder**: invented by **Valdemar Poulsen** (Danish) using steel wire as recording medium.

1895 X-rays: discovered by **Wilhelm Röntgen** (German). One of his first X-ray images was of the bones in his wife's hands. [1901 Nobel prize in physics.]

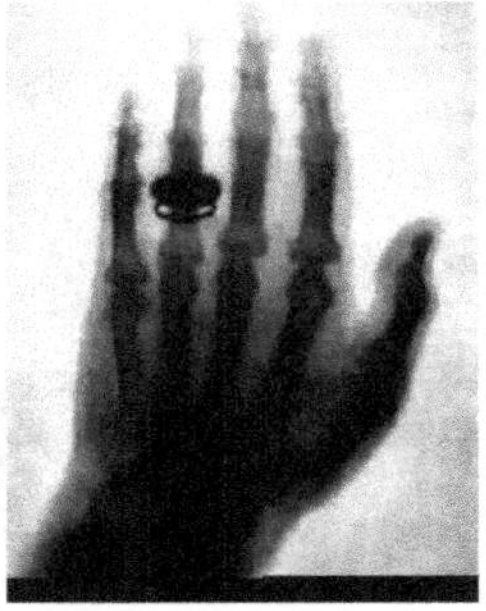

1896 Radio wireless transmission: patented by **Guglielmo Marconi** (Italian). In 1901, he demonstrated radio telegraphy across the Atlantic Ocean. [1909 Nobel prize in physics, shared with Karl Braun (German).]

1897 Cathode ray tube (CRT): invented by **Karl Braun** (German). [1909 Nobel prize, shared with Marconi.]

1897 Electron: discovered by **Joseph John Thomson** (English), who also measured its charge-to-mass ratio. [1906 Nobel prize in physics.]

1902 Amplitude modulation: invented by **Reginald Fessenden** (American) for telephone transmission. In 1906, he introduced AM radio broadcasting of speech and music on Christmas Eve.

1904 Diode vacuum tube: patented by **John Fleming** (British).

1907 Triode tube amplifier: developed by **Lee De Forest** (American) for wireless telegraphy, setting the stage for long-distance phone service, radio, and television.

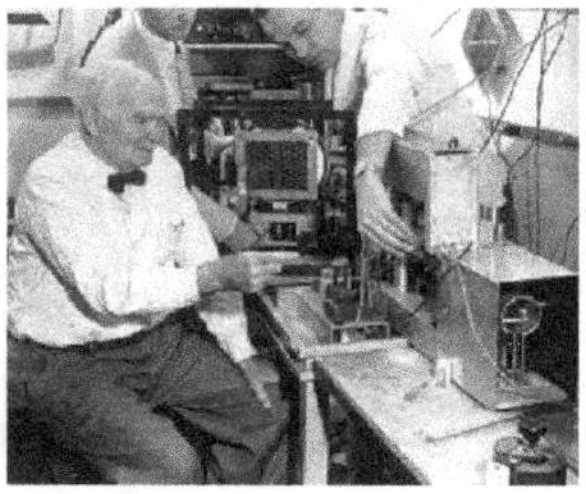

1917 Superheterodyne and frequency modulation (FM): invented by **Edwin Howard Armstrong** (American), providing superior sound quality of radio transmissions over AM radio.

1920 Commercial radio broadcasting: **Westinghouse Corporation** established radio station KDKA in Pittsburgh, Pennsylvania.

1923 Television: invented by **Vladimir Zworykin** (Russian-American). In 1926, **John Baird** (Scottish) transmitted TV images over telephone wires from London to Glasgow. Regular TV broadcasting began in Germany (1935), England (1936), and the United States (1939).

1926 Transatlantic telephone service established between London and New York.

1930 Analog computer: developed by **Vannevar Bush** (American) for solving differential equations.

1935 Anti-glare glass: developed by **Katharine Blodgett** by transferring thin monomolecular coatings to glass.

1935 Radar: invented by **Robert Watson-Watt** (Scottish).

1944 Computer compiler: One of the earliest compilers was designed by **Grace Hopper** for Harvard's Mark I computer. She retired as a rear admiral in the U.S. Navy in 1986.

1945 ENIAC: The first all-electronic computer was developed by **John Mauchly** and **J. Presper Eckert** (both American).

1947 Transistor: invented by **William Shockley**, **Walter Brattain**, and **John Bardeen** (all Americans) at Bell Labs. [1956 Nobel prize in physics.]

1948 Modern communication: **Claude Shannon** (American) published his *Mathematical Theory of Communication*, which formed the foundation of information theory, coding, cryptography, and other related fields.

1950 Floppy disk: invented by **Yoshiro Nakama** (Japanese) as a magnetic medium for storing data.

1954 First AM transistor radio: introduced by **Texas Instruments**.

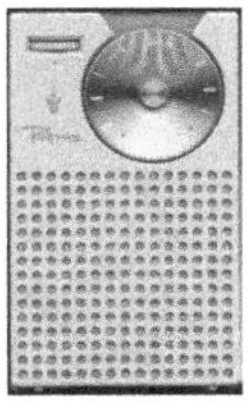

Courtesy of Dr. Steve Reyer

1955 Optical fiber: demonstrated by **Narinder Kapany** (Indian-American) as a low-loss, light-transmission medium.

1956 FORTRAN: developed by **John Backus** (American), the first major programming language.

```
C     FORTRAN PROGRAM FOR
PRINTING A TABLE OF CUBES
      DO 5 I = 1, 64
      ICUBE = I * I * I
      PRINT 2, I, ICUBE
    2 FORMAT (1H , I3,I7)
    5 CONTINUE
      STOP
```

1958 Laser: concept developed by **Charles Townes** and **Arthur Schawlow** (both Americans). [Townes shared 1964 Nobel prize in physics with Aleksandr Prokhorov and Nicolay Bazov (both Soviets).] In 1960 **Theodore Maiman** (American) built the first working model of a laser.

1958 Modem: developed by **Bell Labs**.

1958 Integrated circuit (IC): **Jack Kilby** (American) built the first IC on germanium, and independently, **Robert Noyce** (American) built the first IC on silicon.

1960 Echo: The first passive communication satellite was launched and successfully reflected radio signals back to Earth. In 1962, the first communication satellite, Telstar, was placed in geosynchronous orbit.

1960 Microcomputer: introduced by **Digital Equipment Corporation** as the PDP-1, which was followed with the PDP-8 in 1965.

1961 Thick-film resistor: one of 28 electronic devices patented by **Otis Boykin** (African-American).

1962 MOSFET: invented by **Steven Hofstein** and **Frederic Heiman** (both American), which became the workhorse of computer microprocessors.

1964 IBM's 360 mainframe: became the standard computer for major businesses.

1965 BASIC computer language: developed by **John Kemeny** and **Thomas Kurtz** (both American).

```
PRINT
FOR Counter = 1 TO Items
  PRINT USING "##."; Counter;
   LOCATE , ItemColumn
   PRINT Item$(Counter);
   LOCATE , PriceColumn
 PRINT Price$(Counter)
NEXT Counter
```

1965 Programmable digital computer: developed by **Konrad Zuse** (German) using binary arithmetic and electric relays.

1968 Word processor: demonstrated by **Douglas Engelbart** (American), followed by the mouse pointing device and the use of a Windows-like operating system.

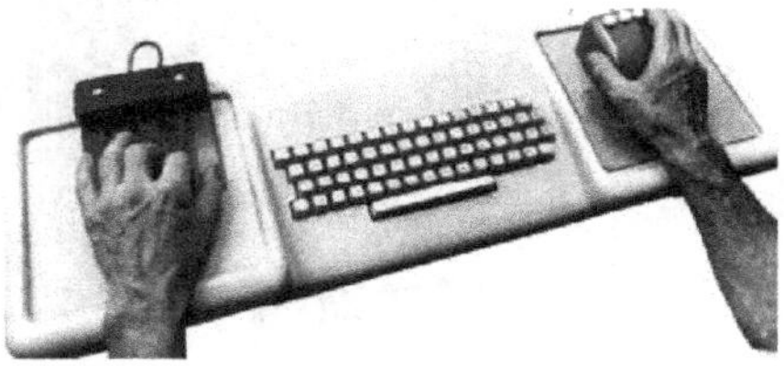

1969 ARPANET: established by the U.S. Department of Defense, which later evolved into the Internet.

1970 CD-ROM: patented by **James Russell** (American), as the first system capable of digital-to-optical recording and playback.

1971 Pocket calculator: introduced by **Texas Instruments**.

Courtesy of Texas Instruments

1971 Intel 4004 four-bit microprocessor: capable of executing 60,000 operations per second.

1972 Computerized axial tomography scanner (CAT scan: developed by **Godfrey Hounsfield** (British) and **Alan Cormack** (South African–American) as a diagnostic tool. [1979 Nobel Prize in physiology or medicine.]

1976 Laser printer: introduced by IBM.

1976 Apple I: sold by **Apple Computer** in kit form, followed by the fully assembled Apple II in 1977, and the Macintosh in 1984.

1979 First cellular telephone network: built in Japan:

- 1983 cellular phone networks started in the United States.
- 1990 electronic beepers became common.
- 1995 cell phones became widely available.

1980 MS-DOS computer disk operating system: introduced by **Microsoft**: Windows marketed in 1985.

1981 PC: introduced by **IBM**.

1984 Internet became operational worldwide.

1988 First transatlantic optical fiber cable: installed between the U.S. and Europe.

1988 Touchpad: invented by **George Gerpheide** (American).

1989 World Wide Web: invented by **Tim Berners-Lee** (British) by introducing a networking hypertext system.

1996 Hotmail: launched by **Sabeer Bhatia** (Indian-American) and **Jack Smith** (American) as the first webmail service.

1997 Palm Pilot: became widely available.

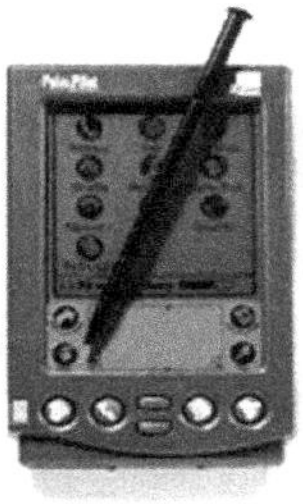

2007 White LED: invented by **Shuji Nakamura** (Japanese) in the 1990s. It promises to replace Edison's lightbulb in most lighting applications.

2007 iPhone: released by Apple.

2009 Cloud computing: went mainstream.

2011 Humans vs. supercomputer: IBM's Watson supercomputer beat the top two human contestants of *Jeopardy!* for a $1M prize.

2011 Text messages: 8×10^{12} (8 trillion) text messages sent worldwide.

2014 Mobile subscribers: Approximately 96% of the world population is a mobile phone subscriber (7 billion people).

Concept Question 1-3: What do you consider to be the most important electrical engineering milestone that is missing from this historical timeline? (See CAD)

1-2 Units, Dimensions, and Notation

The standard system used in today's scientific literature to express the units of physical quantities is the *International System of Units* (SI), abbreviated after its French name *Système Internationale*. Time is a fundamental *dimension*, and the second is the *unit* by which it is expressed relative to a specific reference standard. The SI configuration is based on the seven *fundamental dimensions* listed in Table 1-1, and their units are called *fundamental SI units*. All other dimensions, such as velocity, force, current, and voltage, are regarded as *secondary* because their units are based on and can be expressed in terms of the seven fundamental units. For example, electric current is measured in amps, which is an abbreviation for coulombs/second. Appendix A provides a list of the quantities used in this book, together with their symbols and units.

Table 1-1: **Fundamental and electrical SI units.**

Dimension	Unit	Symbol
Fundamental:		
Length	meter	m
Mass	kilogram	kg
Time	second	s
Electric charge	coulomb	C
Temperature	kelvin	K
Amount of substance	mole	mol
Luminous intensity	candela	cd
Electrical:		
Current	ampere	A
Voltage	volt	V
Resistance	ohm	Ω
Capacitance	farad	F
Inductance	henry	H
Power	watt	W
Frequency	hertz	Hz

In science and engineering, a set of *prefixes* commonly are used to denote multiples and submultiples of units. These prefixes, ranging in value between 10^{-18} and 10^{18}, are listed in Table 1-2. An electric current of 3×10^{-6} A, for example, may be written as 3 μA. The physical quantities we discuss in this book (such as voltage and current) may be constant in time or may vary with time.

Table 1-2: **Multiple and submultiple prefixes.**

Prefix	Symbol	Magnitude
exa	E	10^{18}
peta	P	10^{15}
tera	T	10^{12}
giga	G	10^{9}
mega	M	10^{6}
kilo	k	10^{3}
milli	m	10^{-3}
micro	μ	10^{-6}
nano	n	10^{-9}
pico	p	10^{-12}
femto	f	10^{-15}
atto	a	10^{-18}

Technology Brief 1
Micro- and Nanotechnology

Scale of Things

Our ability as humans to shape and control the environment around us has improved steadily over time, most dramatically in the past 100 years. The degree of control is reflected in the *scale* (size) at which objects can be constructed, which is governed by the tools available for constructing them. This refers to the construction of both very large and very small objects. Early tools—such as flint, stone, and metal hunting gear—were on the order of tens of centimeters. Over time, we were able to build ever-smaller and ever-larger tools. The world's largest antenna[*] is the radio telescope at the Arecibo observatory in Puerto Rico (Fig. TF1-1). The dish is 305 m (1000 ft) in diameter and 50 m deep and covers nearly 20 acres. It is built from nearly 40,000 perforated 1 m $\times$ 2 m aluminum plates. On the other end of the size spectrum, some of the smallest antennas today are nanocrescent antennas that are under 100 nm long. These are built by sputtering aluminum against glass beads and then removing the beads to expose crescent-shaped antennas (Fig. TF1-2).

Miniaturization continues to move forward: the first hydraulic valves, for example, were a few meters in length (ca. 400 BCE); the first toilet valve was tens of centimeters in size (ca. 1596); and by comparison, the largest dimension in a modern microfluidic valve used in biomedical analysis-chips is less than 100 μm!

[*] http://www.naic.edu/general/

Figure TF1-1: Arecibo radio telescope.

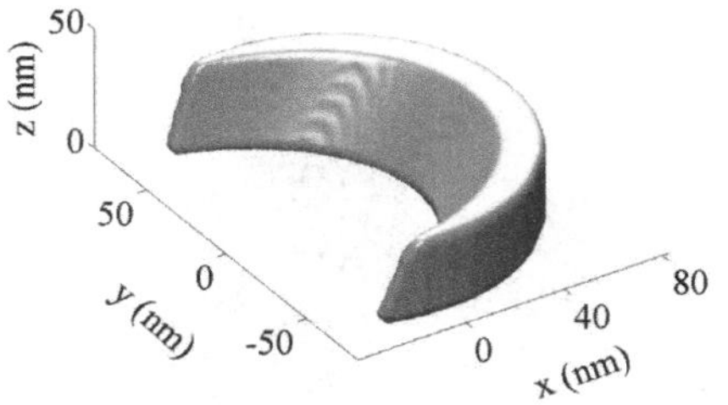

Figure TF1-2: Nano-crescent antenna for use in the ultraviolet range (320 nm to 370 nm wavelength). (Credit: Miguel Rodriguez.)

The chart in Fig. TF1-3 displays examples of manmade and natural things whose dimensions fall in the range between 0.1 nm (10^{-10} m) and 1 cm, which encompasses both micrometer (1 $\mu\text{m} = 10^{-6}$ m) and nanometer (1 nm $= 10^{-9}$ m) ranges. *Microtechnology*, which refers to our ability to manipulate matter at a precision of 1 μm or better, became possible in the 1960s, ushering in an electronics revolution that led to the realization of the laptop computer and the ubiquitous cell phone. It then took another 30 years to improve the manufacturing precision down to the nanometer scale (*nanotechnology*), promising the development of new materials and devices with applications in electronics, medicine, energy, and construction.

Moore's Law

With the invention of the semiconductor transistor in 1947 and the subsequent development of the *integrated circuit* in 1959, it became possible to build thousands (now trillions) of electronic components onto a single substrate or *chip*. The 4004 microprocessor chip (ca. 1971) had 2250 transistors and could execute 60,000 instructions per second; each transistor had a "gate" on the order of 10 μm (10^{-5} m). In comparison, the 2006 Intel Core had 151 million transistors with each transistor gate measuring 65 nm (6.5×10^{-8} m); it could

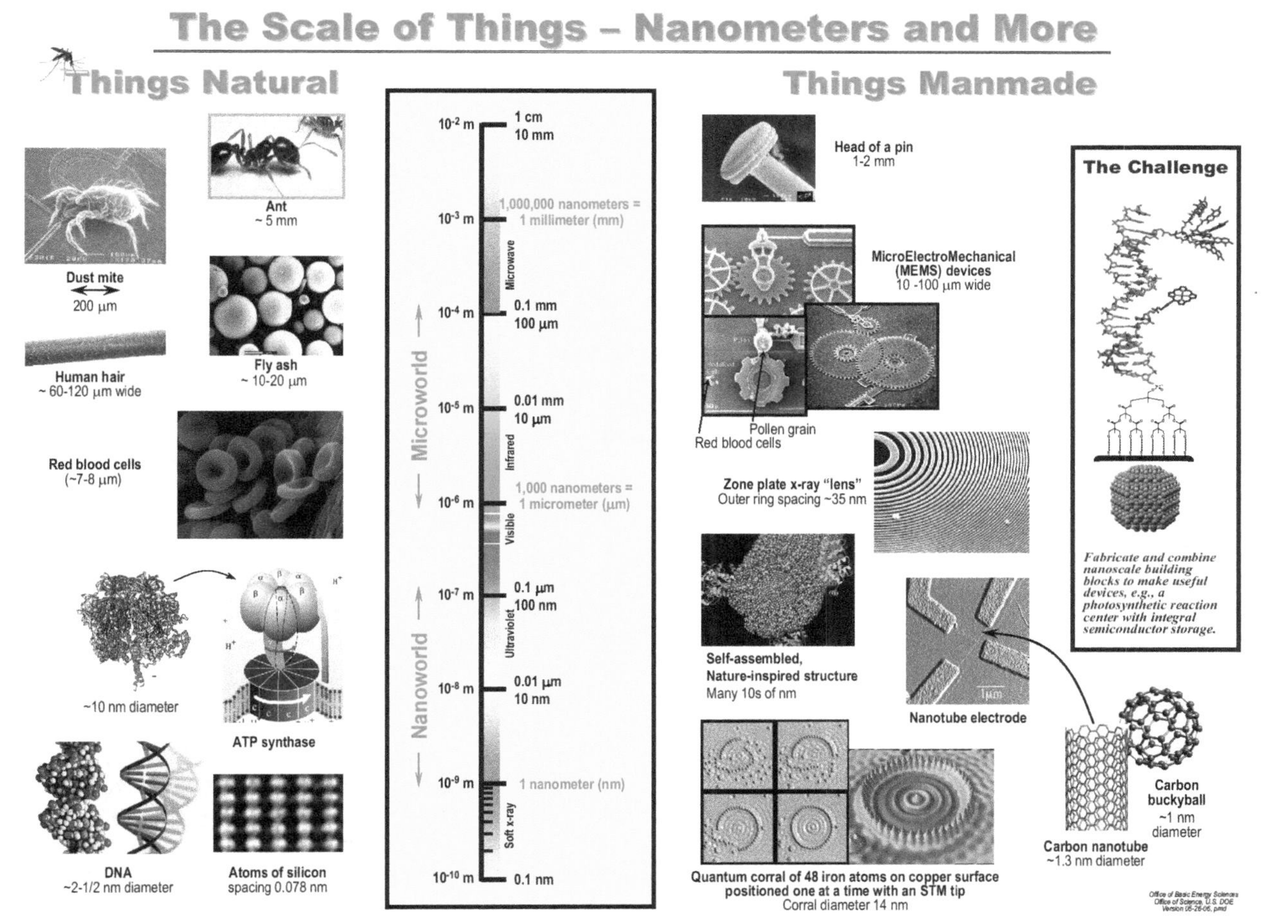

Figure TF1-3: The scale of natural and man-made objects, sized from nanometers to centimeters. (Courtesy of U.S. Department of Energy.)

perform 27 billion instructions per second. The 2011 Intel Core i7 "Gulftown" processors have 1.17 *billion* transistors and can perform ~ 150 billion instructions per second. In recent years, the extreme miniaturization of transistors (the smallest transistor gate in an i7 Core is ~ 32 nanometers wide!) has led to a number of design innovations and trade-offs at the processor level, as devices begin to approach the physical limits of classic semiconductor devices. Among these, the difficulty of dissipating the heat generated by a billion transistors has led to the emergence of ***multicore processors***; these devices distribute the work (and heat) between more than one processor operating simultaneously on the same chip (2 processors on the same chip are called a ***dual core***, 4 processors are called a ***quad core***, etc.). This type of architecture requires additional components to manage computation between processors and has led to the development of new software paradigms to deal with the parallelism inherent in such devices.

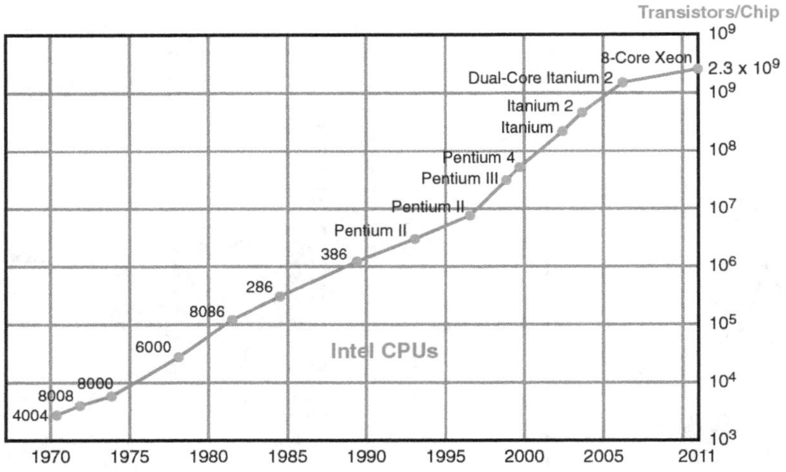

Figure TF1-4: Moore's Law predicts that the number of transistors per processor doubles every two years.

Moore's Law and Scaling

In 1965, Gordon Moore, co-founder of Intel, predicted that the number of transistors in the minimum-cost processor would double every two years (initially, he had guessed they would double every year). Amazingly, this prediction has proven true of semiconductor processors for 40 years, as demonstrated by Fig. TF1-4.

In order to understand Moore's Law, we have to understand the basics of how transistors are used in computers. Computers carry all of their information (numbers, letters, sounds, etc.) in coded strings of electrical signals that are either "on" or "off." Each "on" or "off" signal is called a ***bit***, and 8 bits in a row are called a ***byte***. Two bytes are a ***word***, and (when representing numbers) they provide 16-bit precision. Four bytes give 32-bit precision. These bits can be added, subtracted, moved around, etc., by switching each bit individually on or off, so a computer processor can be thought of as a big network of (trillions of) switches. Transistors are the basic switches in computers. We will learn more about them in Chapter 3, but for now, the important thing to know is that they can act as very tiny, very fast, very low power switches. Trillions of transistors are built directly onto a single silicon wafer (read more about how in Technology Brief 7), producing ***very-large-scale integrated*** (VLSI) circuits or ***chips***. Transistors are characterized by their ***feature size***, which is the smallest line width that can be drawn in that VLSI manufacturing process. Larger transistors are used for handling more current (such as in the power distribution system for the chip). Smaller transistors are used where speed and efficiency are critical. The 22 nm processes available today can make lines and features $\sim$22 nm in dimension. They produce transistors that are about 100 nm on a side, switched on and off over 100 billion times a second (it would take you over 2000 years to flip a light switch that many times),[†] and can do about 751 billion operations per watt.[‡] Even smaller, 5 nm transistors are expected to become commercially viable by 2020. The VLSI design engineer uses ***computer-aided design*** (CAD) tools to design chips by combining transistors into larger subsystems (such as logic gates that add/multiply/etc.), choosing the smallest, fastest transistors that can be used for every part of the circuit.

The following questions then arise: How small can we go? What is the fundamental limit to shrinking down the size of a transistor? As we ponder this, we immediately observe that we likely cannot make a transistor smaller than the diameter of one silicon or metal atom (i.e., $\sim$0.2 to 0.8 nm). But is there a limit prior to this? Well, as we shrink transistors down to the point that they are

---

[†] http://download.intel.com/newsroom/kits/22nm/pdfs/22nm_Fun_Facts.pdf

[‡] https://newsroom.intel.com/servlet/JiveServlet/previewBody/2834-102-1-5130/Intel%20at%20VLSI%20Fact%20Sheet.pdf

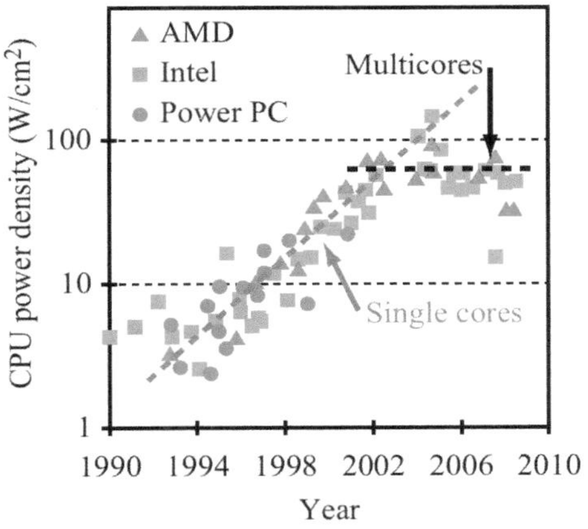

	Light Bulb	Integrated Circuit
Power dissipation	100 W	50 W
Surface area	106 cm^2 (bulb surface area)	1.5 cm^2 (die area)
Heat flux	0.9 W/cm^2	33.3 W/cm^2

Figure TF1-5: (a) Heat power density generated by consumer processors (From "Energy Dissipation and Transport in Nanoscale Devices" by E. Pop, Nano Research, V3, 2010, (b) heat generation by a light bulb and a typical processor.

made of just one or a few atomic layers ($\sim$1 to 5 nm), we run into issues related to the stochastic nature of quantum physics. At these scales, the random motion of electrons between both physical space and energy levels becomes significant with respect to the size of the transistor, and we start to get spurious or random signals in the circuit. There are even more subtle problems related to the statistics of yield. If a certain piece of a transistor contained only 10 atoms, a deviation of just one atom in the device (to a 9-atom or an 11-atom transistor) represents a huge change in the device properties! This would make it increasingly difficult to economically fabricate chips with hundreds of millions of transistors. Additionally, there is an interesting issue of heat generation: Like any dissipative device, each transistor gives off a small amount of heat. But when you add up the heat produced by more than 1 billion transistors, you get a very large number! Figure TF1-5 compares the power density (due to heat) produced by different processors over time. The heat generated by single core processors increased exponentially until the mid-2000s when power densities began approaching 100 W/cm^2 (in comparison, a nuclear reactor produces about 200 W/cm^2!). The inability to practically dissipate that much heat led, in part, to the development of multicore processors and a leveling off of heat generation for consumer processors.

None of these issues are insurmountable. Challenges simply spur creative people to come up with innovative solutions. Many of these problems will be solved, and in the process, provide engineers (like you) with jobs and opportunities. But, more importantly, the minimum feature size of a processor is not the end goal of innovation: it is the means to it. Innovation seeks simply to make *increasingly powerful* computation, not smaller feature sizes. Hence, the move towards multicore processors. By sharing the workload among various processors (called ***distributed computing***) we increase processor performance while using less energy, generating less heat, and without needing to run at warp speed. So it seems, as we approach ever-smaller features, we simply will creatively transition into new physical technologies and also new computational techniques. As Gordon Moore himself said, "It will not be like we hit a brick wall and stop."

Scaling Trends and Nanotechnology

It is an observable fact that each generation of tools enables the construction of an even newer, smaller, more powerful generation of tools. This is true not just of mechanical devices, but electronic ones as well. Today's high-power processors could not have been designed, much less tested, without the use of previous processors that were employed to draw and simulate the next generation. Two observations can be made in this regard. First, we now have the technology to build tools

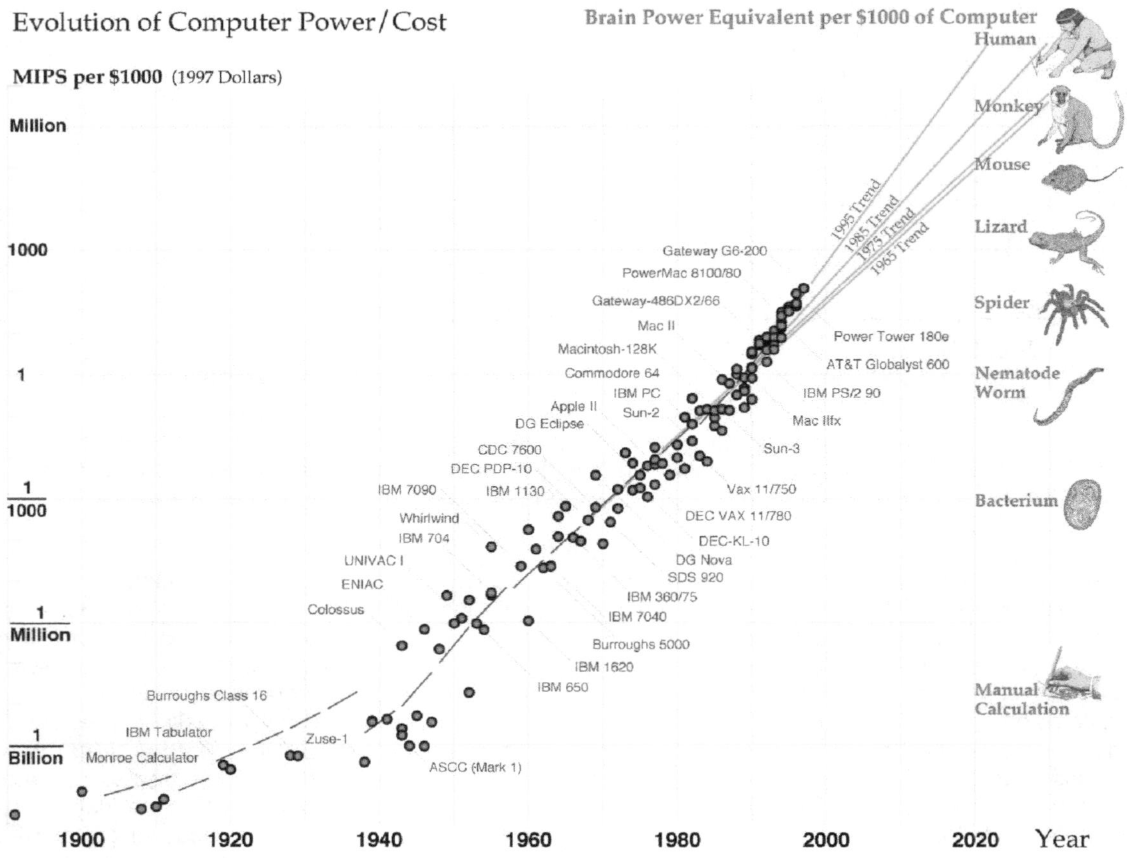

Figure TF1-6: Time plot of computer processing power in MIPS per $1000. (From "When will computer hardware match the human brain?" by Hans Moravec, *Journal of Transhumanism*, Vol. 1, 1998.)

to manipulate the environment at atomic resolution. At least one generation of micro-scale techniques (ranging from *microelectromechanical systems*—or MEMS—to micro-chemical devices) has been developed that, while useful in themselves, are also enabling the construction of newer, nano-scale devices. These newer devices range from 5 nm transistors to femtoliter (10^{-15}) microfluidic devices that can manipulate single protein molecules. At these scales, the lines between mechanics, electronics, and chemistry begin to blur! It is to these ever-increasing interdisciplinary innovations that the term *nanotechnology* rightfully belongs. Second, the rate at which these innovations are occurring seems to be increasing exponentially! (Consider Fig. TF1-6 and note that the y axis is logarithmic and the plots are very close to straight lines.) Keeping up with rapidly changing technology is one of the exciting and challenging aspects of an engineering career. Electrical engineers use the Institute of Electrical and Electronic Engineers (IEEE) to find professional publications, workshops, and conferences to provide lifelong learning opportunities to stay current and creative (see IEEE.org).

As a general rule, we use:

- A lowercase letter, such as i for current, to represent the general case:

  i *may or may not be time-varying*

- A lowercase letter followed with (t) to emphasize time:

  $i(t)$ *is a time-varying quantity*

- An uppercase letter if the quantity is not time-varying; thus:

  I *is of constant value (dc quantity)*

- A letter printed in boldface to denote that:

  I *has a specific meaning, such as a vector, a matrix, the phasor counterpart of $i(t)$, or the Laplace or Fourier transform of $i(t)$*

Exercise 1-1: Convert the following quantities to scientific notation: (a) 52 mV, (b) 0.3 MV, (c) 136 nA, and (d) 0.05 Gbits/s.

Answer: (a) 5.2×10^{-2} V, (b) 3×10^{5} V, (c) 1.36×10^{-7} A, and (d) 5×10^{7} bits/s. (See CAD)

Exercise 1-2: Convert the following quantities to a prefix format such that the number preceding the prefix is between 1 and 999: (a) 8.32×10^{7} Hz, (b) 1.67×10^{-8} m, (c) 9.79×10^{-16} g, (d) 4.48×10^{13}V, and (e) 762 bits/s.

Answer: (a) 83.2 MHz, (b) 16.7 nm, (c) 979 ag, (d) 44.8 TV, and (e) 762 bits/s. (See CAD)

Exercise 1-3: Simplify the following operations into a single number, expressed in prefix format: (a) $A = 10\ \mu\text{V} + 2.3$ mV, (b) $B = 4\text{THz} - 230$ GHz, (c) $C = 3\ \text{mm}/60\ \mu\text{m}$.

Answer: (a) $A = 2.31$ mV, (b) $B = 3.77$ THz, (c) $C = 50$. (See CAD)

1-3 Circuit Representation

When we design circuits, we first think of what we want the circuit to do (its *functional block diagram*), then we design circuits to do this (a *circuit diagram*). We then select and lay out the components in the circuit (*PCB layout*) and build it. Let's consider a capacitive-touch sensor such as the touch screen on the iphone. The circuit includes a flat conducting plate, two ICs, one diode (⊣▶⊢), and several resistors and capacitors. When the plate is touched by a finger, the capacitance introduced by the finger causes the output voltage to rise above a preset threshold, signifying the fact that the plate has been touched. The voltage rise can then be used to trigger a follow-up circuit such as a light-emitting diode (LED). Figure 1-4 contains four parts: (a) a block diagram of a circuit designed as a capacitor-touch-sensor, (b) a *circuit diagram* representing the circuit's electrical configuration, (c) the circuit's *printed-circuit-board (PCB)* layout, and (d) a photograph of the circuit with all of its components.

The PCB layout shown in part (c) of Fig. 1-4 displays the intended locations of the circuit elements and the *printed conducting lines* needed to connect the elements to each other. These lines are used in lieu of wires. The diagram in part (d) is the symbolic representation of the physical circuit. In this particular representation the resistors are drawn as rectangular boxes instead of the more familiar symbol —⋀⋀⋀—. Designing the PCB layout and the circuit's physical architecture is an important step in the production process, but it is outside the scope of this book. Our prime interest is to help the reader understand *how circuits work*, and to use that understanding to *design circuits* to perform functions of interest. Accordingly, circuit diagrams will be regarded as *true representations* of the many circuits and systems we discuss in this and the following chapters.

1-3.1 Circuit Elements

Table 1-3 provides a partial list of the symbols used in this book to represent circuit elements in circuit diagrams.

By way of an example, the diagram in Fig. 1-5 contains the following elements:

- A 12 V ac source, denoted by the symbol (±∼). An ac source varies sinusoidally with time (such as a 60 Hz wall outlet).
- A 6 V dc source, denoted by the symbol +|−. A dc source is constant in time (such as a battery).
- Six resistors, all denoted by the symbol —⋀⋀⋀—
- One capacitor, denoted by the symbol —| |—
- One inductor, denoted by the symbol —⌒⌒⌒—

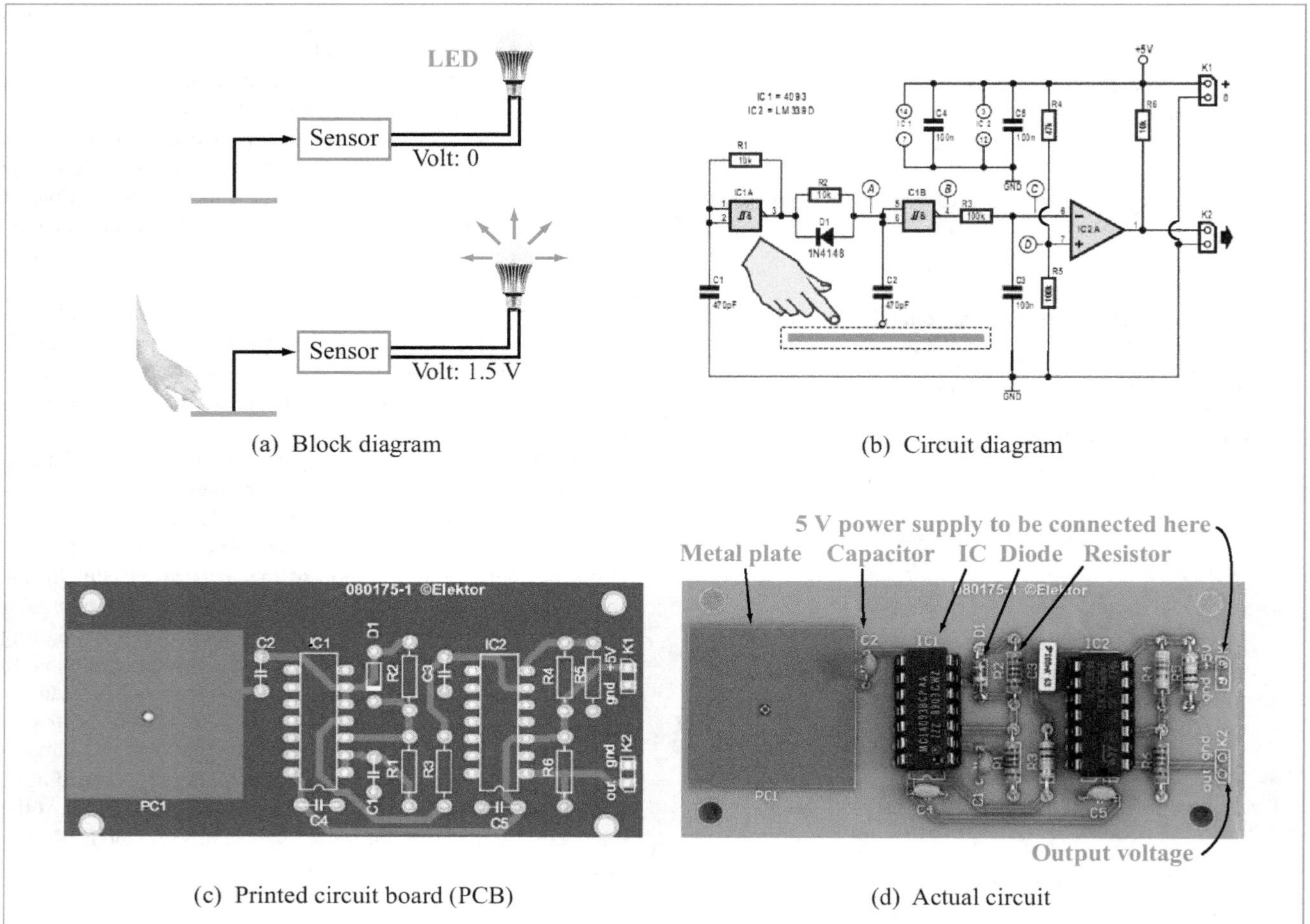

Figure 1-4: (a) Block diagram, (b) circuit diagram, (c) printed-circuit-board (PCB) layout, (d) photograph of a touch-sensor circuit.

- An important integrated circuit known as an *operational amplifier* (or *op amp* for short), denoted by a triangular symbol (the internal circuit of the op amp is not shown).

1-3.2 Circuit Architecture

The vocabulary commonly used to describe the architecture of an electric circuit includes a number of important terms. Short, but precise, definitions follow.

- *Node*: electrical conductor(s) or wires that connect two or more circuit elements. The node is not just a point, but includes the entire set of wires between two or more circuit elements. Nodes are color-coded in Fig. 1-5. For example, node N_1 is red, N_2 is green, and N_3 is orange. The dot at N_1 is typically used to emphasize that the wires are actually connected together. *All conductors in a node always have the same voltage.*

- *Ordinary node*: an electrical connection point that connects only two elements, such as all the yellow nodes in Fig. 1-5.

- *Extraordinary node*: node connected to three or more elements. Figure 1-5 contains four extraordinary nodes, denoted N_1 through N_4, of which N_4 has been selected as a reference voltage node, often referred to as the *ground* node. When two points with no element between them are connected by a conducting wire, they are regarded as the

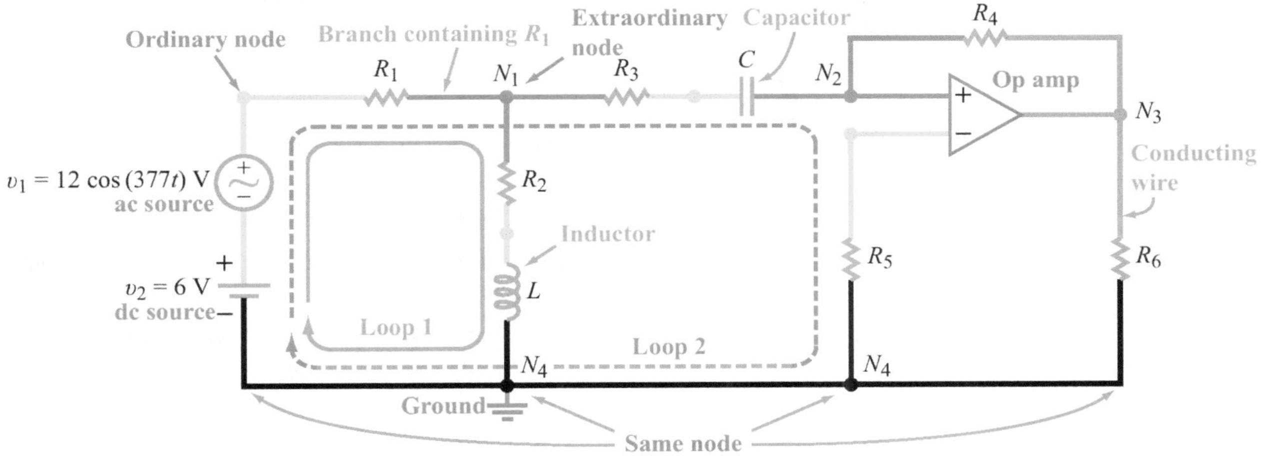

Figure 1-5: Diagram representing a circuit that contains dc and ac sources, passive elements (six resistors, one capacitor, and one inductor), and one active element (operational amplifier). Ordinary nodes are in yellow, extraordinary nodes in other colors, and the ground node in black.

same node. Hence, all of the black wires together located at the bottom of the circuit in Fig. 1-5 make up node N_4.

- *Branch*: the trace between two consecutive nodes containing one and only one element between them.
- *Path*: any continuous sequence of branches, *provided that no one node is encountered more than once*. The path between nodes N_1 and N_2 consists of two branches, one containing R_3 and another containing C.
- *Loop*: a closed path in which the start and end node is one and the same. Figure 1-5 contains several loops, of which two are shown explicitly.
- *Mesh*: a loop that encloses no other loop. In Fig. 1-5, Loop 1 is a mesh, but Loop 2 is not.
- *In series*: *path in which elements share the same current.* As you move along a series path you encounter only ordinary nodes. Elements on these paths are in series. In Fig. 1-6(a), the two light bulbs are in series because the same current flows through both of them. Also, in Fig. 1-5, the two sources and R_1 are all in series, as are R_2 and L, and R_3 and C.

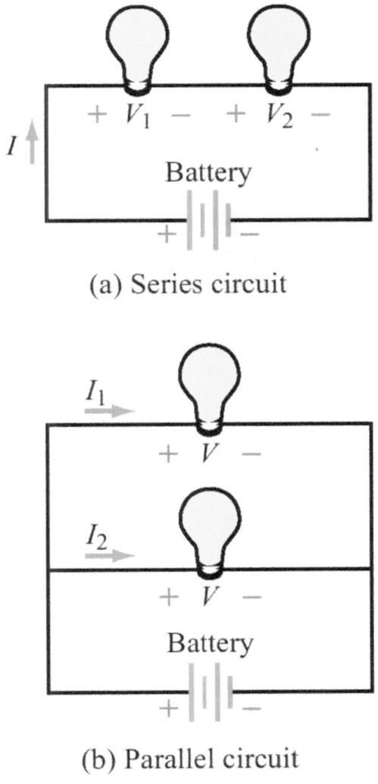

Figure 1-6: Two light bulbs connected (a) in series and (b) in parallel.

Table 1-3: **Symbols for common circuit elements.**

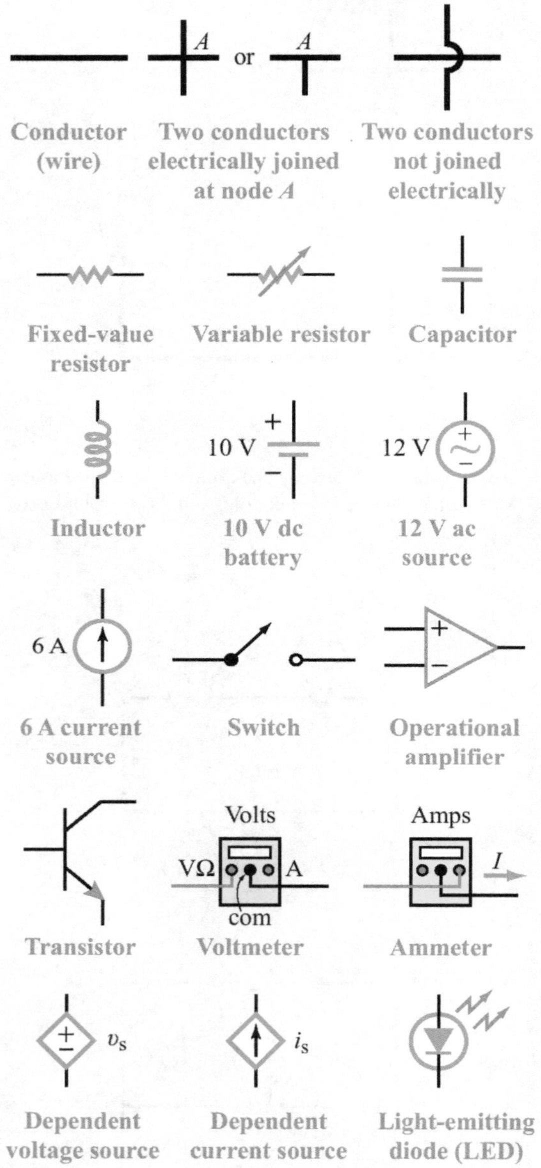

- *In parallel*: *path in which elements share the same voltage*, which means they share the same pair of nodes. In Fig. 1-6(b), the two bulbs are in parallel because they share the same battery voltage across them. In Fig. 1-5 the series combination ($\upsilon_1 - \upsilon_2 - R_1$) is in parallel with the series combination ($R_2 - L$).

Table 1-4: **Circuit terminology.**

Node: An electrical connection between two or more elements.

Ordinary node: An electrical connection node that connects to only two elements.

Extraordinary node: An electrical connection node that connects to three or more elements.

Branch: Trace between two consecutive nodes with only one element between them.

Path: Continuous sequence of branches with no node encountered more than once.

Extraordinary path: Path between two adjacent extraordinary nodes.

Loop: Closed path with the same start and end node.

Independent loop: Loop containing one or more branches not contained in any other independent loop.

Mesh: Loop that encloses no other loops.

In series: Elements that share the same current. They have only ordinary nodes between them.

In parallel: Elements that share the same voltage. They share two extraordinary nodes.

A summary of circuit terminology is given in Table 1-4.

Example 1-1: In Series and In Parallel

(a) For the circuit in Fig. 1-7(a):

(1) Which current is the same as I_2?

(2) Under what circumstance would $I_1 = I_2$?

(b) For the circuit in Fig. 1-7(b):

(1) Which node voltages are at the same voltage as node 4?

(2) Which node voltages are the same as the ground voltage?

(c) Which elements, or combinations of elements, in the circuits of Fig. 1-7 are connected in series and which are connected in parallel?

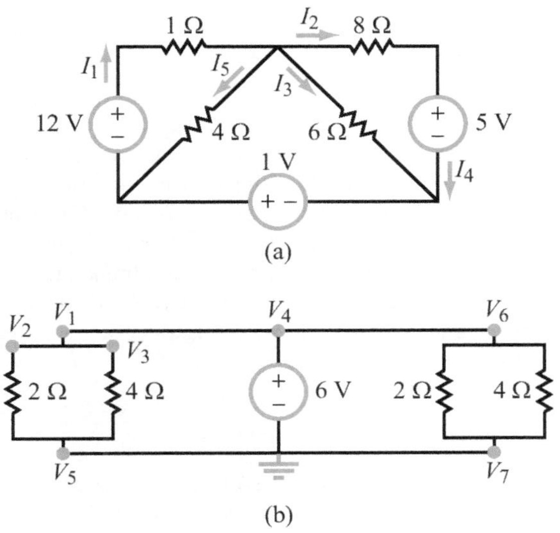

Figure 1-7: Circuits for Example 1-1.

Solution: (a) Two currents are the same if they flow in the same branch and in the same direction. Hence:

(1) $I_2 = I_4$.

(2) $I_1 = I_2$ only if $I_3 = I_5 = 0$.

(b) Two nodes are electrically the same if the only connection between them is a short circuit. Hence:

(1) $V_1 = V_2 = V_3 = V_4 = V_6$, relative to the ground node. Hence, all five nodes are electrically the same.

(2) Nodes V_5 and V_7 are the same as the ground node.

(c) Two or more elements are connected electrically in series if the same current flows through all of them, and they are connected in parallel if they share the same nodes.

Circuit in Fig. 1-7(a):

In series: 8 Ω resistor and 5 V voltage source (call it combination 1).

In series: 1 Ω resistor and 12 V voltage source (call it combination 2).

In parallel: 6 Ω resistor and combination 1. (Call this combination 3.)

In parallel: 4 Ω resistor and combination 2. (Call this combination 4.)

Also, combination 3, combination 4, and the 1 V source are all in series.

Circuit in Fig. 1-7(b):

In series: none.

In parallel: all five elements.

1-3.3 Planar Circuits

▶ A circuit is ***planar*** if it is possible to draw it on a two-dimensional plane without having any two of its branches cross over or under one another (Fig. 1-8). ◀

If such a crossing is unavoidable, then the circuit is ***nonplanar***. This concept becomes particularly important when we construct circuit boards (see Fig. 1-4) or layers on an integrated circuit. To clarify what we mean, we start by examining the circuit in Fig. 1-8(a). An initial examination of the circuit topology might suggest that the circuit is nonplanar because the branches containing resistors R_3 and R_4 appear to cross one another without having physical contact between them (absence of a solid dot at crossover point). However, if we redraw the branch containing R_4 on the outside, as shown in configuration (b) of Fig. 1-8, we would then conclude that the circuit is planar after all, and that is so because *it is possible* to draw it in a single plane without crossovers. In contrast, the circuit in Fig. 1-8(c) is indeed nonplanar because no matter how we might try to redraw it, it will always include at least one crossover of branches.

▶ Circuits in this book will be presumed to be planar. ◀

Concept Question 1-4: What is the difference between the symbol for a dc voltage source and that for an ac source? (See CAD)

Concept Question 1-5: What differentiates an extraordinary node from an ordinary node? A loop from a mesh? (See CAD)

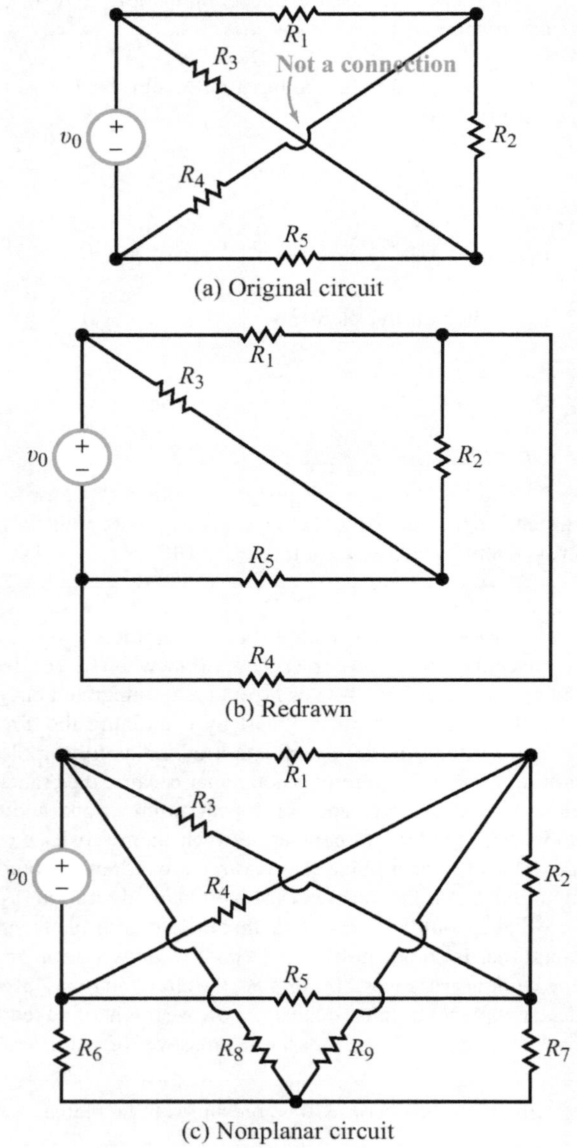

Figure 1-8: The branches containing R_3 and R_4 in (a) *appear to cross* over one another, but redrawing the circuit as in (b) avoids the crossover, thereby demonstrating that the circuit is planar.

> **Concept Question 1-6:** Color-code all of the nodes in Fig. 1-8(b), using Fig. 1-5 as a model. (See CAD)

1-4 Electric Charge and Current

1-4.1 Charge

At the atomic scale, all matter contains a mixture of neutrons, positively charged protons, and negatively charged electrons. The nature of the force induced by electric charge was established by the French scientist Charles Augustin de Coulomb (1736–1806) during the latter part of the 18th century. This was followed by a series of experiments on electricity and magnetism over the next 100 years, culminating in J. J. Thompson's discovery of the electron in 1897. Through these and more recent investigations, we can ascribe to electric charge the following fundamental properties:

> 1. Charge can be either positive or negative.
> 2. The fundamental (smallest) quantity of charge is that of a single electron or proton. Its magnitude usually is denoted by the letter e.
> 3. According to the law of conservation of charge, the (net) charge in a closed region can neither be created nor destroyed.
> 4. Two like charges repel one another, whereas two charges of opposite polarity attract.

The unit for charge is the coulomb (C) and the magnitude of e is

$$e = 1.6 \times 10^{-19} \quad \text{(C)}. \tag{1.1}$$

The symbol commonly used to represent charge is q. The charge of a single proton is $q_p = e$, and that of an electron, which is equal in magnitude but opposite in polarity, is $q_e = -e$. It is important to note that the term *charge* implies "net charge," which is equal to the combined charge of all protons present in any given region of space minus the combined charge of all electrons in that region. Hence, *charge is always an integral multiple of* e.

The actions by charges attracting or repelling each other are responsible for the movement of charge from one location to another, thereby constituting an ***electric current***. Consider the simple circuit in Fig. 1-9 depicting a battery of voltage V connected across a resistor R using metal wires. The arrangement gives rise to an electric current I given by ***Ohm's law*** (which is discussed in more detail in Chapter 2):

$$I = \frac{V}{R}. \tag{1.2}$$

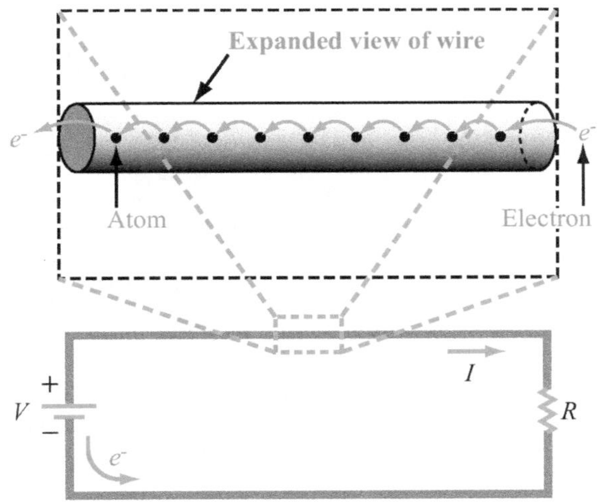

Figure 1-9: The current flowing in the wire is due to electron transport through a drift process, as illustrated by the magnified structure of the wire.

As shown in Fig. 1-9:

> ▶ The current flows from the positive (+) terminal of the battery to its negative (−) terminal, along the path *external* to the battery. ◀

Through chemical or other means, the battery generates a supply of electrons at its negatively labeled terminal by ionizing some of the molecules of its constituent material. A convenient model for characterizing the functionality of a battery is to regard the internal path between its terminals as unavailable for the flow of charge, forcing the electrons to flow from the (−) terminal, through the external path, and towards the (+) terminal to achieve neutrality. It is important to note that:

> ▶ The direction of electric current I is defined to be the same as the direction of flow that positive charges would follow, which is opposite to the direction of flow of electrons e^-. ◀

Even though we talk about electrons *flowing* through the wires and the resistor, in reality the process is a *drift* movement rather than free-flow. The wire material consists of atoms with loosely attached electrons. The positive polarity of the (+) terminal exerts an attractive force on the electrons of the hitherto neutral atoms adjacent to that terminal, causing some

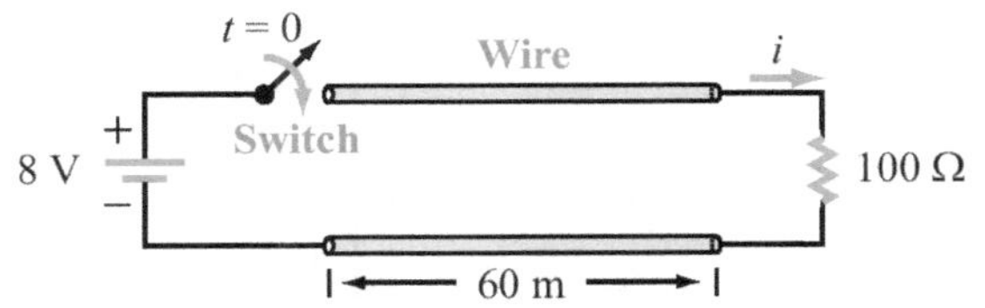

Figure 1-10: After closing the switch, it takes only 0.2 μs to observe a current in the resistor.

of the loosely attached electrons to detach and jump to the (+) terminal. The atoms that have lost those electrons now become positively charged (*ionized*), thereby attracting electrons from their neighbors and compelling them to detach from their hosts and to attach themselves to the ionized atoms instead. This process continues throughout the wire segment (between the (+) battery terminal and the resistor), into the longitudinal path of the resistor, and finally through the wire segment between the resistor and the (−) terminal. The net result is that the (−) terminal loses an electron and the (+) terminal gains one, making it *appear* as if the very same electron that left the (−) terminal actually flowed through the wires and the resistor and finally appeared at the (+) terminal. It is as if the path itself were not involved in the electron transfer, which is not the case.

The process of sequential migration of electrons from one atom to the next is called *electron drift*, and it is this process that gives rise to the flow of *conduction current* through a circuit. To illustrate how important this process is in terms of the electronic transmission of information, let us examine the elementary transmission experiment represented by the circuit shown in Fig. 1-10. The circuit consists of an 8-volt battery and a switch on one end, a resistor on the other end, and a 60 m long two-wire transmission line in between. The wires are made of copper, and they have a circular cross section with a 2 mm diameter. After closing the switch, a current starts to flow through the circuit. It is instructive to compare two velocities associated with the consequence of closing the switch, namely the actual (physical) *drift velocity* of the electrons inside the copper wires and the *transmission velocity* (of the information announcing that the switch has been closed) between the battery and the resistor. For the specified parameters of the circuit shown in Fig. 1-10, the electron drift velocity—which is the actual physical velocity of the electrons along the wire—can be calculated readily and shown to be on the order of only 10^{-4} m/s. Hence, it would take about 1 million seconds (~ 10 days) for an electron to physically travel over a distance of 120 m. In contrast, the time delay between closing the switch at the sending end and observing a response at the receiving end (in the form of current flow through the resistor) is extremely

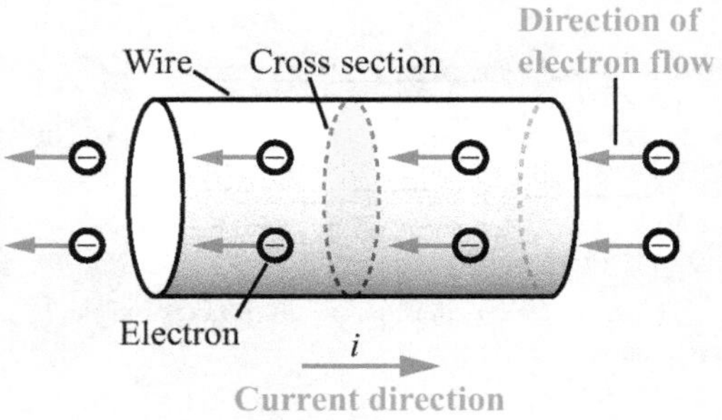

Figure 1-11: Direction of (positive) current flow through a conductor is opposite that of electrons.

short ($\approx 0.2\ \mu$s). This is because the transmission velocity is on the order of the velocity of light $c = 3 \times 10^8$ m/s. Thus:

▶ The rate at which information can be transmitted electronically using conducting wires is about 12 orders of magnitude faster than the actual transport velocity of the electrons flowing through those wires! ◀

This fact is at the heart of what makes electronic communication systems viable.

1-4.2 Current

Moving charge gives rise to current.

▶ ***Electric current*** is defined as the time rate of transfer of electric charge across a specified cross section. ◀

For the wire segment depicted in Fig. 1-11, the current i flowing through it is equal to the amount of charge dq that crosses the wire's cross section over an infinitesimal time duration dt, given as

$$i = \frac{dq}{dt} \quad \text{(A)}, \tag{1.3}$$

and the unit for current is the ampere (A). In general, both positive and negative charges may flow across the hypothetical interface, and the flow may occur in both directions.

▶ By convention, the direction of i is defined to be the direction of the net flow of (net) charge (positive minus negative). ◀

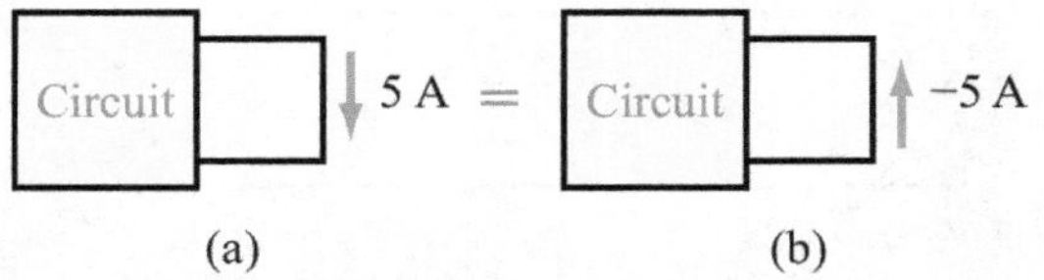

Figure 1-12: A current of 5 A flowing "downward" is the same as −5 A flowing "upward" through the wire.

The circuit segment denoted with an arrow in Fig. 1-12(a) signifies that a current of 5 A is flowing through that wire segment in the direction of the arrow. The same information about the current magnitude and direction may be displayed as in Fig. 1-12(b), where the arrow points in the opposite direction and the current is expressed as −5 A.

When a battery is connected to a circuit, the resultant current that flows through it usually is constant in time (Fig. 1-13(a))—at least over the time duration of interest—in which case we refer to it as a ***direct current*** or ***dc*** for short. In contrast, the currents flowing in household systems (as well as in many

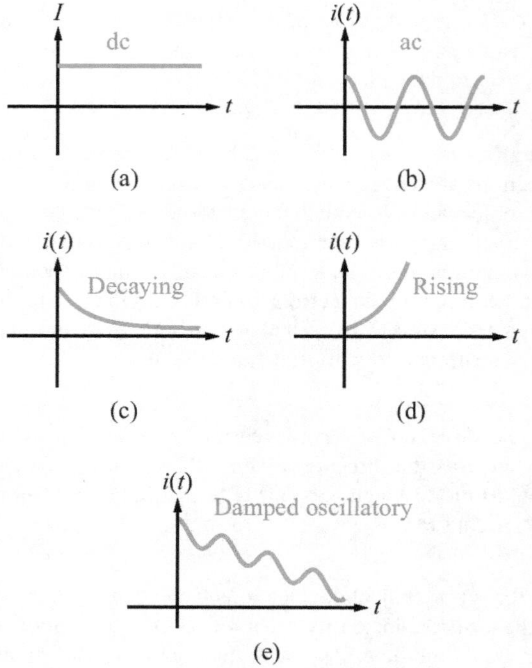

Figure 1-13: Graphical illustrations of various types of current variations with time.

electrical systems) are called *alternating currents* or simply *ac*, because they vary sinusoidally with time (Fig. 1-13(b)). Other time variations also may occur in circuits, such as exponential rises and decays (Fig. 1-13(c) and (d)), exponentially damped oscillations (Fig. 1-13(e)), and many others.

> ▶ As a reminder, we use uppercase letters, such as V and I, to denote dc quantities (with no time variation), and lowercase letters, such as υ and i, to denote the general case, which may be either dc or ac. ◀

Even though in the overwhelming majority of cases the current flowing through a material is dominated by the movement of electrons (as opposed to positively charged ions), it is advisable to *start thinking of the current in terms of positive charge*, primarily to avoid having to keep track of the fact that current direction is defined to be in opposition to the direction of flow of negative charges.

Example 1-2: Charge Transfer

In terms of the current $i(t)$ flowing past a reference cross section in a wire:

(a) Develop an expression for the *cumulative charge* $q(t)$ that has been transferred past that cross section up to time t. Apply the result to the exponential current displayed in Fig. 1-14(a), which is given by

$$i(t) = \begin{cases} 0 & \text{for } t < 0, \\ 6e^{-0.2t} \text{ A} & \text{for } t \geq 0. \end{cases} \tag{1.4}$$

(b) Develop an expression for the *net charge* $\Delta Q(t_1, t_2)$ that flowed through the cross section between times t_1 and t_2, and then compute ΔQ for $t_1 = 1$ s and $t_2 = 2$ s.

Solution: (a) We start by rewriting Eq. (1.3) in the form:

$$dq = i\ dt.$$

Then by integrating both sides over the limits $-\infty$ to t, we have

$$\int_{-\infty}^{t} dq = \int_{-\infty}^{t} i\ dt,$$

which yields

$$q(t) - q(-\infty) = \int_{-\infty}^{t} i\ dt, \tag{1.5}$$

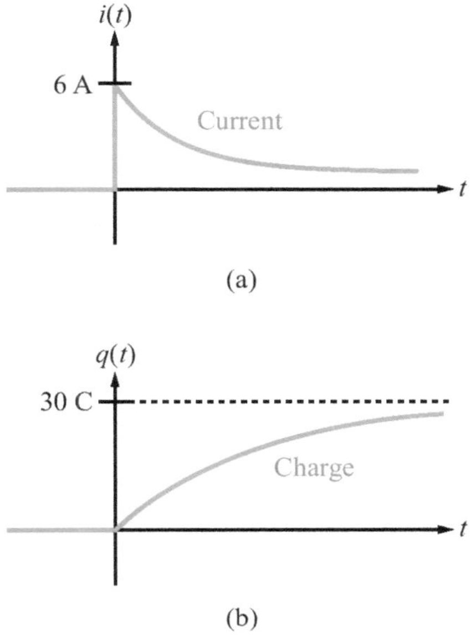

Figure 1-14: The current $i(t)$ displayed in (a) generates the cumulative charge $q(t)$ displayed in (b).

where $q(-\infty)$ represents the charge that was transferred through the wire "at the beginning of time." We choose $-\infty$ as a reference limit in the integration, because it allows us to set $q(-\infty) = 0$, implying that no charge had been transferred prior to that point in time. Hence, Eq. (1.5) becomes

$$q(t) = \int_{-\infty}^{t} i\ dt \qquad \text{(C)}. \tag{1.6}$$

For $i(t)$ as given by Eq. (1.4), $i(t) = 0$ for $t < 0$. Upon changing the lower integration limit to zero and inserting the expression for $i(t)$ in Eq. (1.6), the integration leads to

$$q(t) = \int_{0}^{t} 6e^{-0.2t}\ dt = \frac{-6}{0.2}\ e^{-0.2t}\Big|_0^t = 30[1 - e^{-0.2t}] \text{ C}.$$

A plot of $q(t)$ versus t is displayed in Fig. 1-14(b). The cumulative charge that would transfer after a long period of time is obtained by setting $t = +\infty$, which yields $q(+\infty) = 30$ C.

(b) The cumulative charge that has flowed through the cross section up to time t_1 is $q(t_1)$, and a similar definition applies to $q(t_2)$. Hence, the ***net charge*** that flowed through the cross section over the time interval between t_1 and t_2 is

$$\Delta Q(t_1, t_2) = q(t_2) - q(t_1) = \int_{-\infty}^{t_2} i\,dt - \int_{-\infty}^{t_1} i\,dt = \int_{t_1}^{t_2} i\,dt.$$

For $t_1 = 1$ s, $t_2 = 2$ s, and $i(t)$ as given by Eq. (1.4),

$$\Delta Q(1, 2) = \int_1^2 6e^{-0.2t}\,dt = \left.\frac{6e^{-0.2t}}{-0.2}\right|_1^2$$
$$= -30(e^{-0.4} - e^{-0.2}) = 4.45 \text{ C}.$$

Example 1-3: Current

The charge flowing past a certain location in a wire is given by

$$q(t) = \begin{cases} 0 & \text{for } t < 0, \\ 5te^{-0.1t} \text{ C} & \text{for } t \geq 0. \end{cases}$$

Determine (a) the current at $t = 0$ and (b) the instant at which $q(t)$ is a maximum and the corresponding value of q.

Solution: (a) Application of Eq. (1.3) yields

$$i = \frac{dq}{dt} = \frac{d}{dt}(5te^{-0.1t}) = 5e^{-0.1t} - 0.5te^{-0.1t}$$
$$= (5 - 0.5t)e^{-0.1t} \text{ A}.$$

Setting $t = 0$ in the expression gives $i(0) = 5$ A.

Note that $i \neq 0$, even though $q(t) = 0$ at $t = 0$.

(b) To determine the value of t at which $q(t)$ is a maximum, we find dq/dt and then set it equal to zero:

$$\frac{dq}{dt} = (5 - 0.5t)e^{-0.1t} = 0,$$

which is satisfied when

$$5 - 0.5t = 0 \quad \text{or} \quad t = 10 \text{ s},$$

as well as when

$$e^{-0.1t} = 0 \quad \text{or} \quad t = \infty.$$

The first value ($t = 10$ s) corresponds to a maximum and the second value ($t = \infty$) corresponds to a minimum (which can be verified either by graphing $q(t)$ or by taking the second derivative of $q(t)$ and evaluating it at $t = 10$ s and at $t = \infty$).

At $t = 10$ s,

$$q(10) = 5 \times 10e^{-0.1 \times 10} = 50e^{-1} = 18.4 \text{ C}.$$

Concept Question 1-7: What are the four fundamental properties of electric charge? (See CAD)

Concept Question 1-8: Is the direction of electric current in a wire defined to be the same as or opposite to the direction of flow of electrons? (See CAD)

Concept Question 1-9: How does electron drift lead to the conduction of electric current? (See CAD)

Exercise 1-4: If the current flowing through a given resistor in a circuit is given by $i(t) = 5[1 - e^{-2t}]$ A for $t \geq 0$, determine the total amount of charge that passed through the resistor between $t = 0$ and $t = 0.2$ s.

Answer: $\Delta Q(0, 0.2) = 0.18$ C. (See CAD)

Exercise 1-5: If $q(t)$ has the waveform shown in Fig. E1.5, determine the corresponding current waveform.

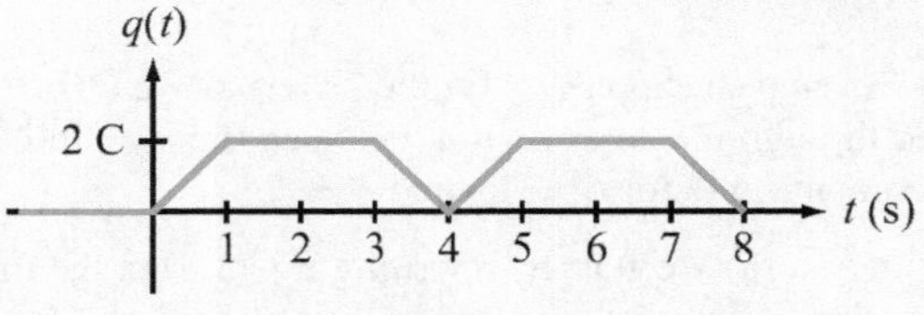

Figure E1.5

Answer:

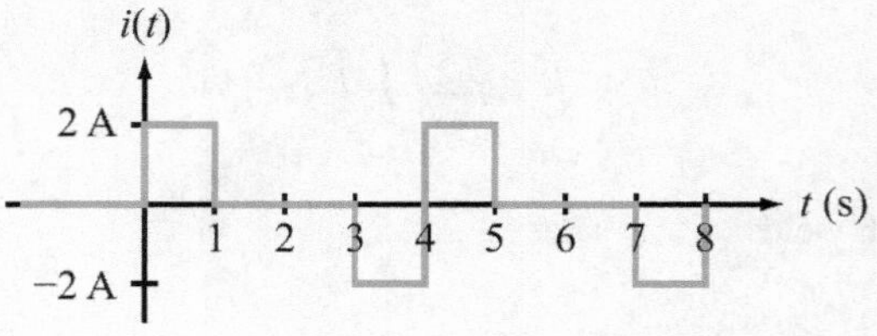

(See CAD)

1-5 Voltage and Power

1-5.1 Voltage

The two primary quantities used in circuit analysis are electrical current and voltage. Current is associated with the movement (flow) of electric charge and voltage is associated with the displacement or concentration of that charge. Before we offer a formal definition for voltage, let us consider a water analogy. Suppose we were to take a very small (differential) amount of water of mass dm from ground level at elevation $z = b$ and raise it (pump it up) to an elevation $z = a$ to fill a water tank, as depicted in Fig. 1-15(a). Doing so requires the expenditure of kinetic energy dw, which is gained by mass dm in the form of gravitational potential energy. [Were we to open a valve to allow the water to flow back down (under the force of gravity), the water would expend its potential energy by converting it into kinetic energy as it flows downward.] At height a, mass dm has potential energy dw relative to the ground surface. Accordingly, we can define a "gravitational voltage" V_{ab} as

$$V_{ab} = \frac{dw}{dm} . \qquad (1.7a)$$

Thus, V_{ab} is a measure of the potential energy change dw, per differential mass dm, between heights a and b.

Next, we consider the electrical voltage associated with the electrical force of attraction between charges of opposite polarity. Let us examine the energy implications of polarizing a hitherto neutral material, thereby establishing opposite electrical polarities on its two ends. Suppose we have a piece of material (such as a resistor) to which we connect two short wires and label their end points a and b, as shown in Fig. 1-15(b). At each point, we have two small metal plates, the combination of which constitutes a capacitor. Starting out with an electrically neutral structure, assume that we are able to detach an electron from one of the atoms at point a and move it to point b. Moving a negative charge from the (remaining) positively charged atom against the attraction force between them requires the expenditure of a certain amount of energy. Voltage is a measure of this expenditure of energy relative to the amount of charge involved, and it always involves two spatial locations:

▶ Voltage often is denoted $\boldsymbol{v_{ab}}$ to emphasize the fact that it is the voltage difference *between* points $\boldsymbol{a}$ and $\boldsymbol{b}$. ◀

The two points may be two locations in a circuit or any two points in space.

Against this background, we now offer the following formal definition for voltage:

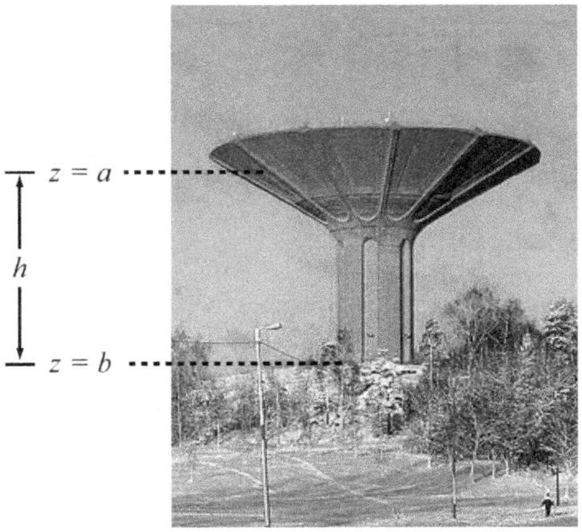

(a) Raising water from ground level at b to height a

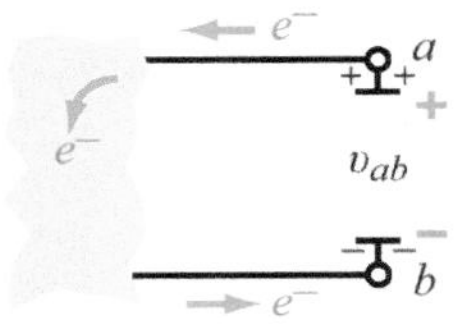

(b) Moving charge from a to b

Figure 1-15: Moving charge dq through the material in (b) is analogous to raising mass dm in (a).

▶ The voltage v_{ab} between location $\boldsymbol{a}$ and location $\boldsymbol{b}$ is the ratio of $\boldsymbol{dw}$ to $\boldsymbol{dq}$, where $\boldsymbol{dw}$ is the energy in joules (J) required to move (positive) charge $\boldsymbol{dq}$ from $\boldsymbol{b}$ to $\boldsymbol{a}$ (or negative charge from a to b). ◀

That is,

$$v_{ab} = \frac{dw}{dq}, \qquad (1.7b)$$

and the unit for voltage is the volt (V), named after the inventor of the first battery, Alessandro Volta (1745–1827). Voltage also is called *potential difference*. In terms of that terminology, if v_{ab} has a positive value, it means that point a is at a potential higher than that of point b. Accordingly, points a and b in Fig. 1-15(b) are denoted with (+) and (−) signs, respectively. If $v_{ab} = 5$ V, we often use the terminology: "The *voltage rise* from b to a is 5 V," or "The *voltage drop* from a to b is 5 V."

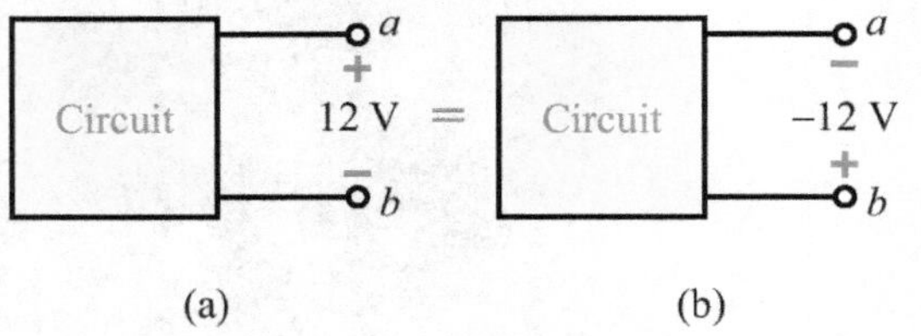

Figure 1-16: In (a), with the (+) designation at node a, $V_{ab} = 12$ V. In (b), with the (+) designation at node b, $V_{ba} = -12$ V, which is equivalent to $V_{ab} = 12$ V. [That is, $V_{ab} = -V_{ba}$.]

Just as 5 A of current flowing from a to b in a circuit conveys the same information as −5 A flowing in the opposite direction, a similar analogy applies to voltage. Thus, the two representations in Fig. 1-16 convey the same information with regard to the voltage between terminals a and b. Also, the terms *dc* and *ac* defined earlier for current apply to voltage as well. *A constant voltage is called a dc voltage and a sinusoidally time-varying voltage is called an ac voltage.*

Ground

Let us look again at the water and circuit analogies in Fig. 1-15. We originally considered only a very localized potential difference as we pumped the water up into the tank. But its total potential energy (the ability to create water pressure in your shower!) is different if this tank is on a hill or in a valley. In order to design a water system for a city, we have to define some location to be the real "ground" point from which all other heights are measured. For convenience, this is typically the lowest elevation in the terrain.

Similarly, for the electrical system we originally considered only a very localized potential difference as we moved electric charge from one plate of the capacitor to the other. But the potential of such a capacitor (the ability to turn on a light bulb) depends not only on how much energy it has, but also on how and where it is connected in the rest of the circuit. In order to design an electrical system, we have to define some location to be the real "ground" location from which all other voltages are calculated. For convenience, this is typically the lowest voltage in the system. For mobile systems, this is usually the chassis or metal structure (called *chassis ground*), and for buildings and fixed systems, this is typically the *Earth ground* (usually physical rods or poles are buried in the dirt near the structure).

Since by definition voltage is not an absolute quantity but rather the difference in electric potential between two locations, it is often convenient to select a reference point in the circuit,

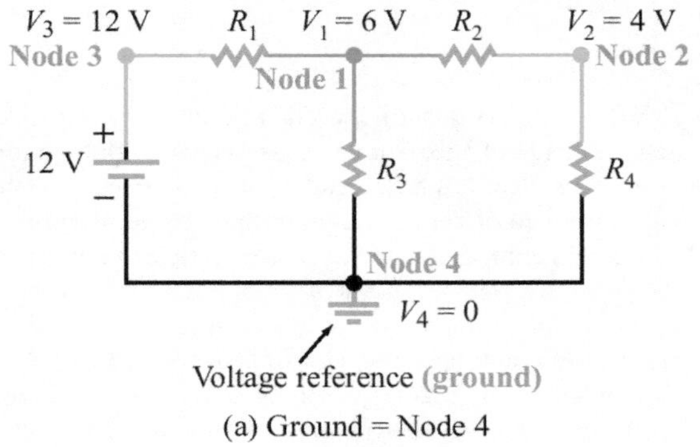

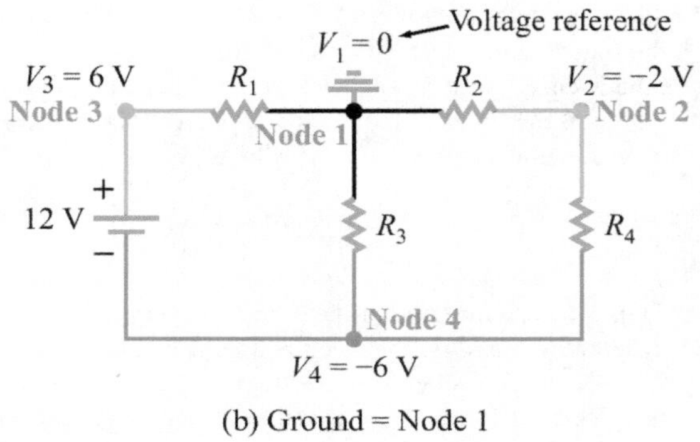

Figure 1-17: Ground is any point in the circuit selected to serve as a reference point for all points in the circuit.

label it *ground*, and then define the voltage at any point in the circuit with respect to that ground point. Thus, when we say that the voltage V_1 at node 1 in Fig. 1-17(a) is 6 V, we mean that the potential difference between node 1 and the ground reference point (node 4) is 6 V, which is equivalent to having assigned the ground node a voltage of zero. Also, since $V_1 = 6$ V and $V_2 = 4$ V, it follows that

$$V_{12} = V_1 - V_2 = 6 - 4 = 2 \text{ V}.$$

The voltage at node 3 is $V_3 = 12$ V, relative to node 4. This is because nodes 3 and 4 are separated by a 12 V voltage source with its (+) terminal next to node 3 and (−) terminal next to node 4.

Had we chosen a node other than node 4 as our ground node, node voltages V_1 to V_4 would have had entirely different values (see Example 1-4). The takeaway message is:

▶ *Node voltages* are defined relative to a specific reference (ground) node whose voltage is assigned a voltage of zero. If a different node is selected as ground, the values of the node voltages may change to reflect the fact that the reference node has changed. ◀

Voltage difference is defined between any two nodes. It is often denoted with two subscripts, as in $V_{12} = V_1 - V_2$, where V_1 and V_2 are the voltages at nodes 1 and 2, with both defined to a common reference (ground).

Example 1-4: Node Voltages

In Fig. 1-17(a), node 4 was selected as the ground node. Suppose node 1 is selected as the ground node instead, as shown in Fig. 1-17(b). Use the information in Fig. 1-17(a) to determine node voltages V_2 to V_4 when defined relative to V_1 at node 1.

Solution: In the circuit of Fig. 1-17(a), V_2 is 2 V lower in level than V_1 (4 V compared to 6 V). Hence, in the new configuration in Fig. 1-17(b), V_2 will still be 2 V lower than V_1, and since $V_1 = 0$, it follows that $V_2 = -2$ V. Similarly, $V_3 = 6$ V and $V_4 = -6$ V.

To summarize:

	node 4 = ground	node 1 = ground
$V_1 =$	6 V	0
$V_2 =$	4 V	−2 V
$V_3 =$	12 V	6 V
$V_4 =$	0 V	−6 V

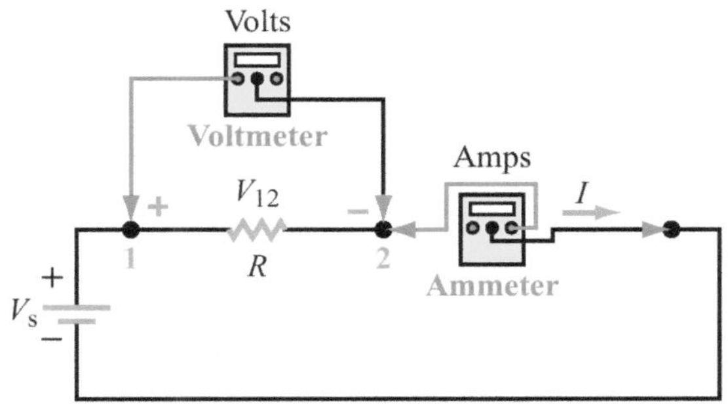

(a) Voltmeter and ammeter connections

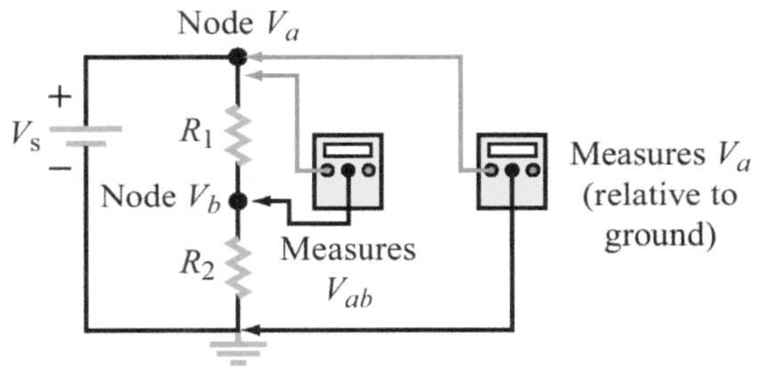

(b) Voltmeters connected to measure voltage difference V_{ab} and node voltage V_a (relative to ground)

Figure 1-18: An ideal voltmeter measures the voltage difference between two points (such as nodes 1 and 2 in (a)) without interfering with the circuit (i.e., no current runs through the voltmeter). Similarly, an ideal ammeter measures the current magnitude and direction with no voltage drop across itself. In (b), one voltmeter is used to measure voltage difference V_{ab} and another to measure node voltage V_a. Note the polarity of the meters. The red leads are connected to the + terminals of the voltages or currents, and the black leads are connected to the − terminals of the voltages or currents. For the voltmeter, the red port on the left is (+) and the black port in the center is (−), and for the ammeter the red port on the right is the (+).

When a circuit is constructed in a laboratory, the chassis often is used as the common ground point—in which case it is called *chassis ground*. As discussed later in Section 10-1, in a household electrical network, outlets are connected to three wires—one of which is called *Earth ground* because it is connected to the physical ground next to the house.

Measuring voltage and current

The voltmeter is the standard instrument used to measure the *voltage difference* between two points in a circuit. To measure V_{12} in the circuit of Fig. 1-18, we connect the (+) red terminal of the voltmeter to terminal 1 and the (−) **black** terminal to terminal 2 in parallel with V_{12}. To measure a *node voltage*, we connect the (+) red terminal to the node and the (−) black terminal to the ground node. Connecting the voltmeter to the circuit does not require any changes to the circuit, and *in the ideal case, the presence of the voltmeter has no effect on any of the voltages and currents associated with the circuit.* In reality, the voltmeter has to extract some current from the circuit in order to perform the voltage measurement, but the voltmeter is designed such that the amount of extracted current is so small as to have a negligible effect on the circuit.

To measure the current flowing through a wire, it is necessary to *insert* an ammeter *in series* in that path, as illustrated by Fig. 1-18(a). The ammeter is connected so that positive current

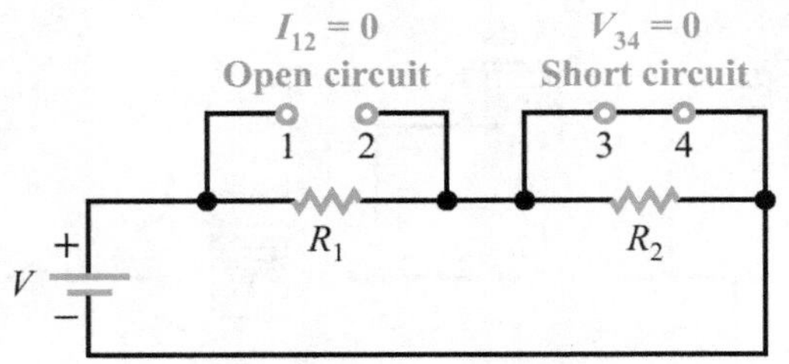

Figure 1-19: Open circuit between terminals 1 and 2, and short circuit between terminals 3 and 4.

flows from the (+) red lead to the (−) black lead. *The voltage drop across an ideal ammeter is zero.*

Open and short circuits

> ▶ An ***open circuit*** refers to the condition of path discontinuity (infinite resistance) between two points. No current can flow through an open circuit, regardless of the voltage across it. ◀

The path between terminals 1 and 2 in Fig. 1-19 is an open circuit.

> ▶ In contrast, a ***short circuit*** constitutes the condition of complete path continuity (with zero electrical resistance) between two points, such as between terminals 3 and 4 in Fig. 1-19. ◀

> ▶ No voltage drop occurs across a short circuit, regardless of the magnitude of the current flowing through it. ◀

Switches

Switches come in many varieties, depending on the intended function. They can be manual (such as an ordinary household light switch) or electrically controlled by a voltage or current (such as a circuit breaker). The simple ON/OFF switch depicted in Fig. 1-20(a) is known as a *single-pole single-throw (SPST)* switch. The ON (closed) position acts like a short circuit, allowing current to flow while extracting no voltage drop across the switch's terminals; the OFF (open) position acts like an open circuit. The specific time $t = t_0$ denoted below or above the switch (Fig. 1-20(a)) refers to the time t_0 at which it opens or closes.

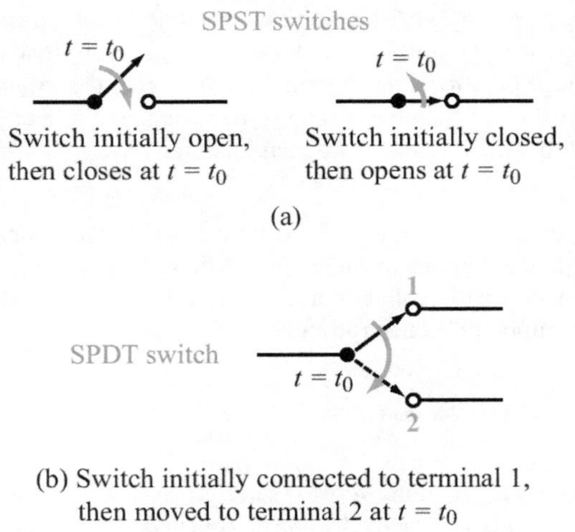

Figure 1-20: (a) Single-pole single-throw (SPST) and (b) single-pole double-throw (SPDT) switches.

If the purpose of the switch is to combine two switching functions so as to connect a common terminal to either of two other terminals, then we need to use the ***single-pole double-throw*** (SPDT) switch illustrated in Fig. 1-20(b). Before $t = t_0$, the common terminal is connected to terminal 1; then at $t = t_0$, that connection ceases (becomes open), and it is replaced with a connection between the common terminal and terminal 2.

1-5.2 Power

The circuit shown in Fig. 1-21(a) consists of a battery and a light bulb connected by an SPST switch in the open position. No current flows through the open circuit, but the battery has a voltage V_{bat} across it, due to the excess positive and negative charges it has at its two terminals. After the switch is closed at $t = 5$ s, as indicated in Fig. 1-21(b), a current I will flow through the circuit along the indicated direction. The battery's excess positive charges flow from its positive terminal downward through the light bulb towards the battery's negative terminal, and (since current direction is defined to coincide with the direction of flow of positive charge) the current direction is as indicated in the figure.

The consequences of current flow through the circuit are: (1) The battery acts as a supplier of power and (2) The light bulb acts as a recipient of power, which gets absorbed by its filament, causing it to heat up and glow, resulting in the conversion of electrical power into light and heat.

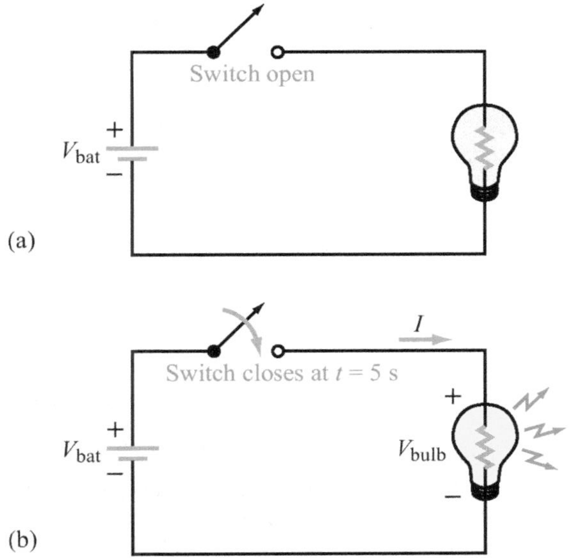

Figure 1-21: Current flow through a resistor (light-bulb filament) after closing the witch.

► A power supply, such as a battery, offers a *voltage rise* across it as we follow the current from the terminal at which it enters (denoted with a (−) sign) to the terminal from which it leaves (denoted with a (+) sign). In contrast, a power recipient (such as a light bulb) exhibits a *voltage drop* across its corresponding terminals. This set of assignments of voltage polarities relative to the direction of current flow for devices generating power versus those consuming power is known as the *passive sign convention* (Fig. 1-22). We will adhere to it throughout the book. ◄

Our next task is to establish an expression for the power p delivered to or received by an electrical device. By definition, *power is the time rate of change of energy*,

$$p = \frac{dw}{dt} \qquad \text{(W)}, \tag{1.8}$$

and its unit is the watt (W), named after the Scottish engineer and inventor James Watt (1736–1819), who is credited with the development of the steam engine from an embryonic stage into a viable and efficient source of power. Using Eqs. (1.3) and (1.7b), we can rewrite Eq. (1.8) as

$$p = \frac{dw}{dt} = \frac{dw}{dq} \cdot \frac{dq}{dt}$$

Passive Sign Convention

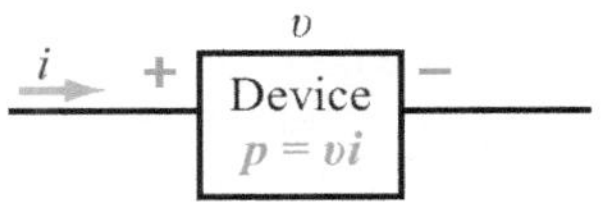

$p > 0$ power delivered to device
$p < 0$ power supplied by device

Note that i direction is defined as entering (+) side of υ.

Figure 1-22: Passive sign convention.

or simply

$$p = \upsilon i \qquad \text{(W)}. \tag{1.9}$$

Consistent with the passive sign convention:

► The power delivered to a device is equal to the voltage across it multiplied by the current entering through its (+) voltage terminal. ◄

For example, a 100 W light bulb in a 120 V household electrical system draws 0.83 A of current.

If the algebraic value of p is negative, then the device is a supplier of energy. For an isolated electric circuit composed of multiple elements, the *law of conservation of power* requires that the algebraic sum of power for the entire circuit be always zero. That is, for a circuit with n elements,

$$\sum_{k=1}^{n} p_k = 0, \tag{1.10}$$

which means that the total power supplied by the circuit always must equal the total power absorbed by it.

Power supplies are sometimes assigned ratings to describe their capacities to deliver energy. A battery may be rated as having an output capacity of 200 *ampere-hours* (Ah) at 9 volts, which means that it can deliver a current I over a period of time T (measured in hours) such that $IT = 200$ Ah, and it can do so while maintaining a voltage of 9 V. Alternatively, its output capacity may be expressed as 1.8 *kilowatt-hours* (kWh), which represents the total amount of energy it can supply, namely $W = VIT$ (with T in hours).

Technology Brief 2
Voltage: How Big Is Big?

Electrical voltage plays a central role in all of our electrical circuits, our bodies, and many other effects seen in the natural world. Table TT2-1 gives some perspective on really little and really big voltages.

Table TT2-1: **A wide range of voltage levels.**

Bird standing on a power line (foot to foot)	10 mV
Neuron action potential	55 mV
Cardiac action potential	100 mV
AA battery	1.5 V
TTL digital logic gates	5 V
Residential electricity (US)	110 V / 220 V
High voltage lines	110 kV +
Static electricity	20 to 25 kV
Lightning	1 billion volts

Big Voltages: Lightning

Lightning begins with clouds and the water cycle. Storm clouds have tremendous amounts of turbulent air (updrafts and downdrafts). This results in a thunderhead, a cumulonimbus cloud that has the typical vertical shape we all associate with a storm coming on. These clouds can build quite suddenly from otherwise mild skies, thus bringing on the classic afternoon thunderstorm. Freezing and collisions of the water particles in the cloud break some of the electrons away from the particles, making the storm clouds positively charged at the top and negatively charged at the bottom (Fig. TF2-1). This creates a voltage difference, similar to a battery, with values around a billion volts!

Like a battery, these charges cannot just travel through the air, because air is a good insulator. Normally, a wire or other metal conductor would be needed in order to carry the current from a battery. Not so with lightning. The separation of charges (voltage difference) creates an ***electric field***. *When the electric field is high*

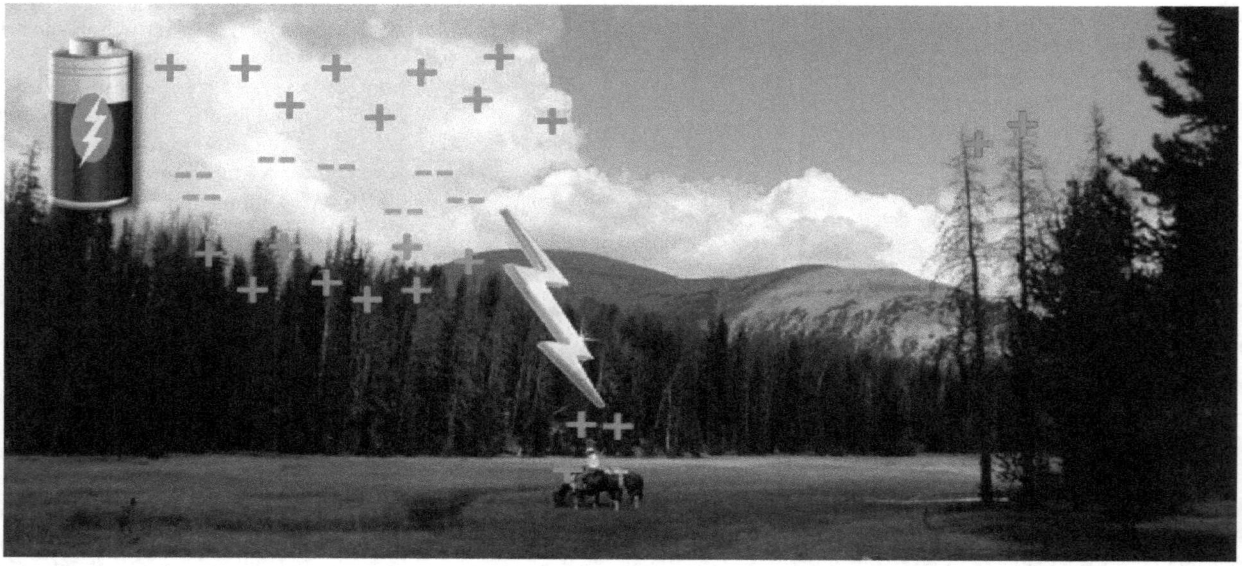

Figure TF2-1: Turbulent air causes negative charges to build up on the bottom of cumulonimbus clouds, separated from the positive charges on the top. The negative charges attract positive charges from the Earth, which move to the top of tall objects. A lightning strike can occur between the negative cloud and positive Earth charges.

enough (around 3 MV/m), the air breaks down and partially ionizes. This means it changes from an insulator (that cannot conduct electricity) to a conductor (that can). The air breakdown creates ozone, and the "fresh air" smell associated with lightning storms. The path of ionized air is called a step leader. The negative charges on the bottom of the cloud begin drawing positive charge towards the Earth's surface. The positive charges are pulled as close to the negative cloud charges as possible. They concentrate on the tops of things that are tall, like trees, golfers, farmers on their tractors, and hikers in the mountains. These positive charges create streamers, reaching towards the negative cloud charges. When a positive streamer and a negative step leader meet, they can form a complete path (like a wire) for lightning to travel from the cloud to the ground (other types of lightning follow a slightly different process). Silently, the lightning strike occurs.

But the ionized air is only a partial conductor. When the current of lightning passes through the resistive air, the air heats up and expands so much and so quickly that it causes a shock wave that produces a sound wave to radiate away from the strike path. That's thunder.

What should you do if a lightning storm approaches? First, go indoors if you can, and stay away from water lines and electrical appliances. Unplug sensitive electronics. Lightning may strike the building, but the currents will pass through the walls or the electrical system, to ground. If you are outdoors, avoid high places, move off the ridges and into draws and lowlands.

Also stay away from high, pointy things (such as tall trees, flag poles, and raised golf clubs). Objects that are pointy will concentrate the charge (and create a stronger streamer) than things that are smooth and rounded. Lightning rods use this principle to protect buildings and structures. The lightning rod produces a much stronger streamer than the rest of the building, so it is more likely to be struck. The current from the lightning bolt can then (hopefully safely) go down the cable to a ground rod buried under the building. Figure TF2-2 shows an example on the old rock church at Sleepy Hollow. Every chimney and the weather vane on the steeple has a separate lightning rod and cable. People and animals also make good lightning rods. We are about 2/3 salt water, which is a pretty good conductor, and we are tall and pointy, similar to a lightning rod. Thus, people (and other animals) are very capable of sending up positive streamers that attract negative step leaders. Consider your profile if you are golfing, hiking, horseback riding, riding on a tractor or mower. In all cases, you are the tallest thing around. Golfers and farmers on tractors have some of the highest incidences of lightning strikes. So, avoid being a lightning rod. Avoid being the tallest thing around.

Figure TF2-2: Lightning rod and grounding cable on Old Rock Church at Sleepy Hollow, New York. The lightning rod attracts the strike by concentrating charges at its tip. The cable shunts the current to ground, carrying it on the outside of the (rock) church, rather than on the inside where materials (wood, plaster, etc.) are more flammable. The cable is large enough in diameter to carry the current without burning, although it will still be hot to the touch after a lightning strike.

Figure TF2-3: Radial, dendritic pattern of scorched grass caused by lightning strike of golf course pin flag. [From *National Geographic*, Colton, 1950.]

The most common cause of lightning injury is not a direct strike, but the ground current. When lightning strikes, it brings negative charges from the cloud down to the positively charged Earth. It then spreads those charges until all of the negative lightning charges are combined with positive Earth charges. Some of the charge spreads over the surface of the ground. (See for example the pattern on the ground by the golf flag in Fig. TF2-3.) Some current also penetrates deeper into the Earth. The charges spreading on the surface of the Earth are called ground currents, and they are real currents that can cause injury.

Electrical Safety

Electrical safety is a function of the current that goes through your body. From Ohm's Law we know that $I = V/R$, so the current depends on the voltage and resistance. The voltage depends on the source (see Table TT2-1). The resistance depends on how you connect to the voltage source—did you touch it with a dry finger, a sweaty shoulder, or were you walking across a wet field when lightning produced a ground current? Were you wearing rubber soled tennis shoes or golf shoes with metal cleats?

The minimum current a human can feel (the ***threshold of sensation***) depends on the frequency and whether the current is ac, dc, or pulsed. Most people can feel 5 mA at dc or 1 mA at household 60 Hz ac. This is generally considered benign, although most people are not comfortable with the sensation. You will feel a mildly painful current if you briefly touch a 9 V battery to your tongue. A more dangerous condition occurs around 10 mA when the muscles lock up and cannot release an electrified object. This is the "let go threshold" and is a criterion in electrical regulations for ***shock hazard***. Additional risk is associated with sensitive organs, particularly those that are controlled by electrical signals such as the heart and brain. As little as 10 μV applied directly to the heart can cause ***fibrillation***. Typical voltages used to deliberately pace the heart with internal defibrillators or pace makers are -100 to 35 mV. You might have noticed a change in units from current to voltage in this description. Some disciplines use voltage, others use current, mainly due to what they find easiest to measure. We know they are related via Ohm's law, although more information is always needed to define the resistance and the specific conditions under which it is assumed, calculated, or measured. The ANSI/IEEE Standard 80-1986 uses 1 kΩ for the body resistance. Adding dry shoes and standing on dry ground, the total resistance is 5–10 kΩ.

Current flow requires two contact points (a node where the current enters the body and a node where it leaves). The resistance R is made up of a combination of series and parallel resistances between these two nodes. For example, in the case of lightning-induced ground current, the current will typically enter one foot,

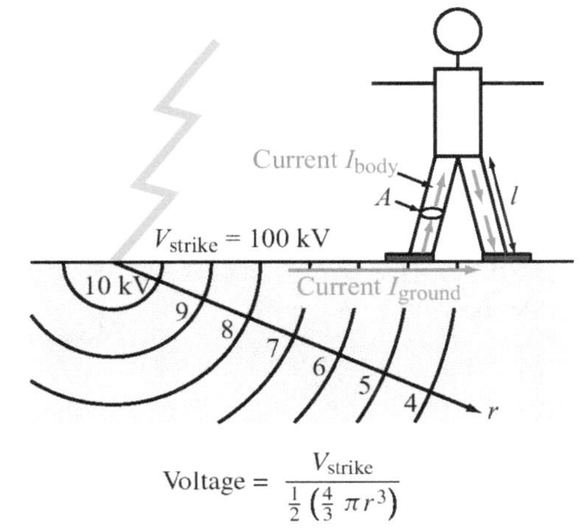

Figure TF2-4: Current path from a lightning strike.

travel through the body, and exit through the other foot. The total resistance will be the sum of resistance from one shoe (R_{shoe}), the series and parallel resistances as the current travels through the body to the other foot (R_{body}), and the resistance of the other shoe (R_{shoe}). The total resistance $R = R_{shoe} + R_{body} + R_{shoe}$ (see Fig. TF2-4). There is another resistance here too, the resistance through the ground, which is parallel to R, and it is controlled by soil type and moisture content.

The resistance between the source of the current and the body is often called the ***contact resistance*** (in this case, it is R_{shoe}). In applications where you want to maximize the current in the body or other object (such as reading the voltages from the heart with an electromyogram (EMG)), you want to minimize the contact resistance. This is often done by using large, conducting electrodes to connect to the body, and placing conductive gels between the electrode and the body. In applications where you want to minimize the current in the body (such as protection from electric shock), you want to maximize the contact resistance. This can be done by minimizing the surface area of the body in contact with the current source and making sure the contact area is dry and insulating (for instance wearing rubber-soled shoes).

Electrical engineers protect people, buildings, circuits, etc., in several ways. Preventing contact between the source and a person or animal can be done with locked buildings and fences, warning signs, and insulators as simple as rubber handles on tools and fiberglass (rather than aluminum) ladders. Circuit protection devices such as circuit breakers and fuses limit the current by tripping (opening the circuit up) if the current exceeds their maximum rating. In circuit breakers, a bimetal junction heats up when current passes through the element. One metal heats up faster than the other, bending/breaking away and disconnecting the circuit. Fuses use a thin metal filament that burns away when its current rating is exceeded, opening the circuit. Current limiting resistors in series with other circuit elements such as potentiometers prevent the resistance from going to zero, thereby preventing large currents. Current limiting devices are effective within moderate ranges of voltage, but very high voltages such as lightning can simply "jump the gaps" even when the circuit is opened up. Rather than trying to simply "stop" the current, protection from very high currents typically relies on shunting the current away from more sensitive circuits, sending it straight to ground. The lightning rod/cable system is one example of this. The cable is a short circuit straight to ground and is sized large enough to carry these very large currents without melting. Other lightning protection circuits use bypass capacitors or various types of filters in parallel with the circuit being protected.

Example 1-5: Conservation of Power

For each of the two circuits shown in Fig. 1-23, determine how much power is being delivered to each device and whether it is a power supplier or recipient.

Solution: (a) For the circuit in Fig. 1-23(a), the current entering the (+) terminal of the device is 0.2 A. Hence, the power P (where we use an uppercase letter because both the current and voltage are dc) is:

$$P = VI = 12 \times 0.2 = 2.4 \text{ W},$$

and since $P > 0$, the device is a recipient of power. As we know, the law of conservation of power requires that if the device receives 2.4 W of power, then the battery has to deliver exactly that same amount of power. For the battery, the current entering its (+) terminal is -0.2 A (because 0.2 A of current is shown leaving that terminal), so according to the passive sign convention, the power that would be absorbed by the battery (had it been a passive device) is

$$P_{\text{bat}} = 12(-0.2) = -2.4 \text{ W}.$$

The fact that P_{bat} is negative is confirmation that the battery is indeed a supplier of power.

(b) For device 1 in Fig. 1-23(b), the current entering its (+) terminal is 3 A. Hence,

$$P_1 = V_1 I_1 = 18 \times 3 = 54 \text{ W},$$

and the device is a power recipient.

For device 2,

$$P_2 = V_2 I_2 = (-6) \times 3 = -18 \text{ W},$$

and the device is a supplier of power (because P_2 is negative).

By way of confirmation, the power associated with the battery is

$$P_{\text{bat}} = 12(-3) = -36 \text{ W},$$

thereby satisfying the law of conservation of power, which requires the net power of the overall circuit to be exactly zero.

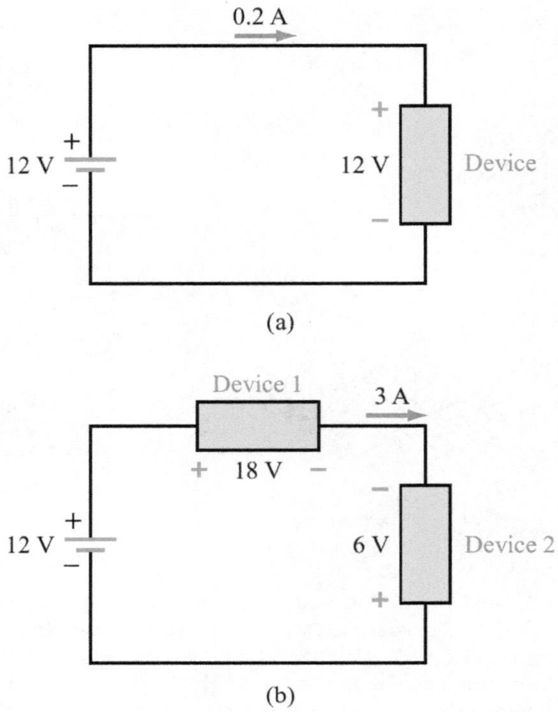

Figure 1-23: Circuits for Example 1-5.

Example 1-6: Energy Consumption

A resistor connected to a 100 V dc power supply was consuming 20 W of power until the switch was turned off, after which the voltage decayed exponentially to zero. If $t = 0$ is defined as the time at which the switch was turned to the *off* position and if the subsequent voltage variation is given by

$$v(t) = 100e^{-2t} \text{ V} \qquad \text{for } t \geq 0$$

(where t is in seconds), determine the total amount of energy consumed by the resistor after the switch was turned off.

Solution: Before $t = 0$, the current flowing through the resistor was $I = P/V = 20/100 = 0.2$ A. Hence, the resistance R of the resistor is

$$R = \frac{V}{I} = \frac{100}{0.2} = 500 \ \Omega,$$

and the current variation after the switch was turned off is

$$i(t) = \frac{v(t)}{R} = 0.2e^{-2t} \text{ A} \qquad \text{for } t \geq 0.$$

The instantaneous power is

$$p(t) = \upsilon(t) \cdot i(t) = (100e^{-2t})(0.2e^{-2t}) = 20e^{-4t} \text{ W}.$$

We note that the power decays at a rate (e^{-4t}) much faster than the rate for current and voltage (e^{-2t}). The total energy dissipated in the resistor after engaging the switch is obtained by integrating $p(t)$ from $t = 0$ to infinity (the integral equation form of Eq. (1.8)), namely

$$W = \int_0^\infty p(t)\, dt = \int_0^\infty 20e^{-4t}\, dt = -\frac{20}{4} e^{-4t}\Big|_0^\infty = 5 \text{ J}.$$

Concept Question 1-10: Explain how node voltage relates to voltage difference. To what do the (+) and (−) leads of the voltmeter connect to in each case? (See CAD)

Exercise 1-6: If a positive current is flowing through a resistor from its terminal a to its terminal b, is υ_{ab} positive or negative?

Answer: $\upsilon_{ab} > 0$. (See CAD)

Exercise 1-7: A certain device has a voltage difference of 5 V across it. If 2 A of current is flowing through it from its (−) voltage terminal to its (+) terminal, is the device a power supplier or a power recipient, and how much energy does it supply or receive in 1 hour?

Answer: $P = VI = 5(-2) = -10$ W. Hence, the device is a power supplier. Since $p(t) =$ (not time-varying), $|W| = |P|\,\Delta t = 36$ kJ. (See CAD)

Exercise 1-8: A car radio draws 0.5 A of dc current when connected to a 12 V battery. How long does it take for the radio to consume 1.44 kJ?

Answer: 4 minutes. (See CAD)

1-6 Circuit Elements

Electronic circuits used in functional systems employ a wide range of circuit elements, including transistors and integrated circuits. The operation of most electronic circuits and devices—no matter how complex—can be modeled (represented) in terms of an *equivalent circuit* composed of *basic elements* with idealized characteristics. The equivalent circuit offers a circuit behavior that closely resembles the behavior of the actual electronic circuit or device over a certain range of specified conditions, such as the range of input signal level or output load resistance. The set of basic elements commonly used in circuit analysis include voltage and current sources; passive elements (which include resistors, capacitors, and inductors); and various types of switches. The basic attributes of switches were covered in Section 1-5.1. The nomenclature and current–voltage relationships associated with the other two groups are the subject of this section.

1-6.1 i–υ Relationship

The relationship between the current flowing through a device and the voltage across it defines the fundamental operation of that device. As was stated earlier, Ohm's law states that the current i entering into the (+) terminal of the voltage υ across a resistor is given by

$$i = \frac{\upsilon}{R}.$$

This is called the ***i–υ relationship*** for the resistor. We note that the resistor exhibits a ***linear i–υ relationship***, meaning that i and υ always vary in a proportional manner, as shown in Fig. 1-24(a), so long as R remains constant. A circuit composed exclusively of elements with linear i–υ responses is called a ***linear circuit***. The linearity property of a circuit is an underlying requirement for the various circuit analysis techniques presented in this and future chapters. Diodes and transistors exhibit nonlinear i–υ relationships. To apply the analysis techniques specific to linear circuits to circuits containing nonlinear devices, we can represent those devices in terms of linear subcircuits that contain ***dependent sources***. The concept of a dependent source and how it is used is introduced in Section 1-6.4.

1-6.2 Independent Voltage Source

An ***ideal, independent voltage source*** provides a specified voltage across its terminals, independent of the type of load or circuit connected to it (so long as it is not a short circuit). Hence, for a voltage source with a specified voltage V_s, its i–υ relationship is given by

$$\upsilon = V_s \qquad \text{for any } i \neq \infty.$$

The i–υ profile of an ideal voltage source is a vertical line, as illustrated in Fig. 1-24(b).

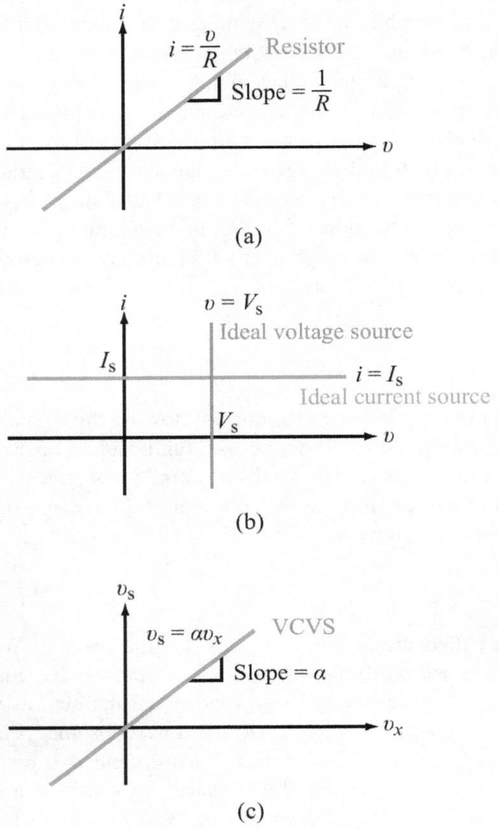

Figure 1-24: i–v relationships for (a) an ideal resistor, (b) ideal, independent current and voltage sources, and (c) a dependent, voltage-controlled voltage source (VCVS).

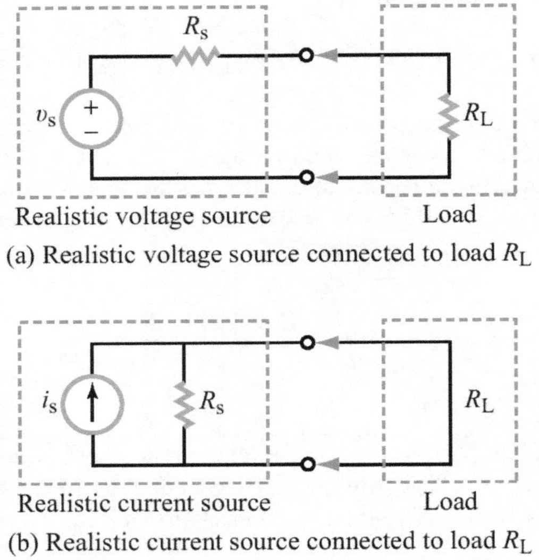

Figure 1-25: (a) A realistic voltage source has a nonzero series resistance R_s, which can be replaced with a short circuit if R_s is much smaller than the load resistance R_L. (b) A realistic current source has a nonzero parallel resistance R_s, which can be replaced with an open circuit if $R_s \gg R_L$.

The circuit symbol used for independent sources is a circle, as shown in Table 1-5, although for dc voltage sources the traditional "battery" symbol is used as well. A household electrical outlet connected through an electrical power-distribution network to a hydroelectric- or nuclear-power generating station provides continuous power at an approximately constant voltage level. Hence, it may be classified appropriately as an independent voltage source. On a shorter time scale, a flashlight's 9-volt battery may be regarded as an independent voltage source, but only until its stored charge has been used up by the light bulb. Thus, strictly speaking, a battery is a storage device (not a generator), but we tend to treat it as a generator so long as it acts like a constant voltage source.

In reality, no sources can provide the performance specifications ascribed to ideal sources. If a 5 V voltage source is connected across a short circuit, for example, we run into a serious problem of ambiguity. From the standpoint of the source, the voltage is 5 V, but by definition, the voltage across the short circuit is zero. How can it be both zero and 5 V simultaneously? The answer resides in the fact that our description of the ideal voltage source breaks down in this situation. Most often, in such cases, the circuit malfunctions as well. Short-circuiting a battery will draw more current than the battery is intended to provide, thereby overheating it, damaging it, and possibly causing a fire or explosion.

► A more realistic model for a voltage source includes an internal series resistor, as shown in Fig. 1-25(a). ◄

A ***real voltage source*** (which may have an elaborate circuit configuration) behaves like a combination of an ***equivalent, ideal voltage source*** v_s in series with an ***equivalent resistance*** R_s. Usually, the voltage source is designed such that its series resistance R_s is much smaller than the resistance values of the types of loads it is intended to energize. Under such a condition, R_s becomes inconsequential and can be ignored, in which case the realistic voltage source behaves approximately the same as an ideal voltage source.

Table 1-5: **Voltage and current sources.**

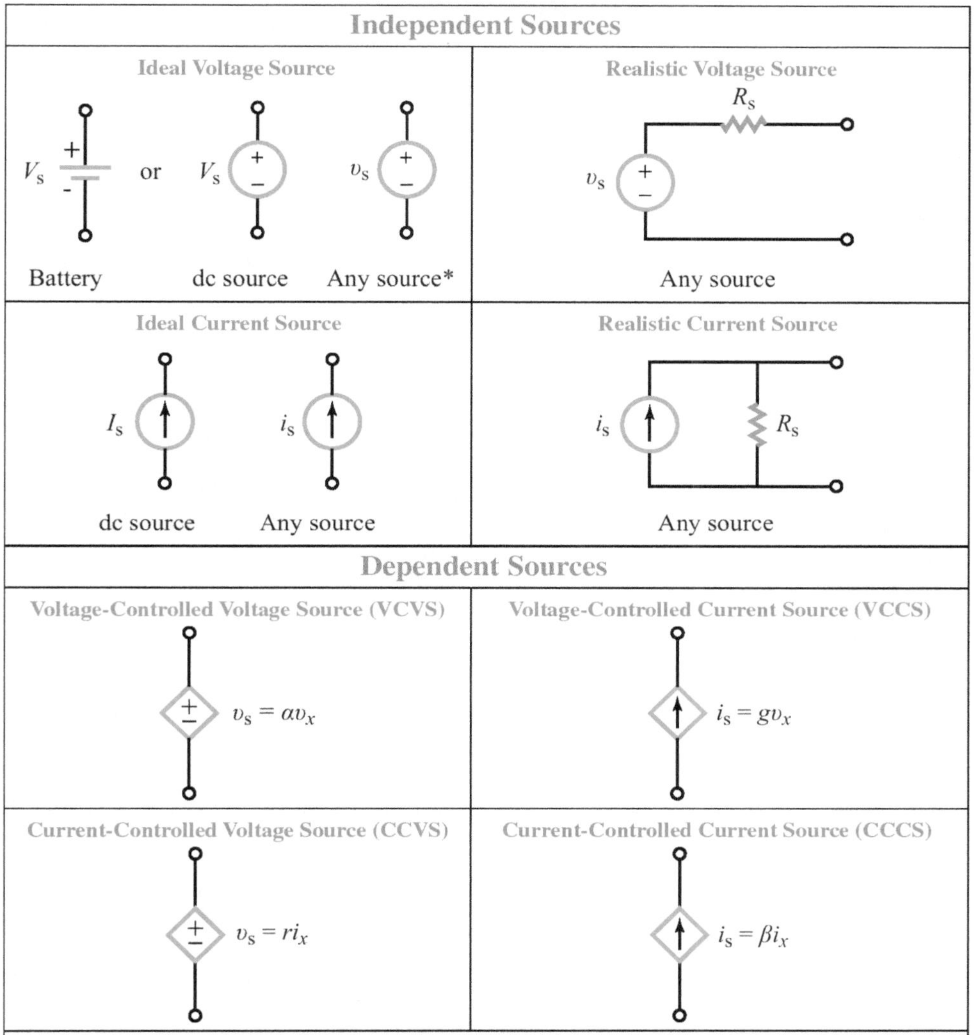
Independent Sources
Ideal Voltage Source
V_s
+
-
or
V_s
v_s
Battery
dc source
Any source*
Realistic Voltage Source
R_s
v_s
Any source
Ideal Current Source
I_s
i_s
dc source
Any source
Realistic Current Source
i_s
R_s
Any source
Dependent Sources
Voltage-Controlled Voltage Source (VCVS)
$v_s = \alpha v_x$
Voltage-Controlled Current Source (VCCS)
$i_s = g v_x$
Current-Controlled Voltage Source (CCVS)
$v_s = r i_x$
Current-Controlled Current Source (CCCS)
$i_s = \beta i_x$

Note: α, g, r, and β are constants; v_x and i_x are a specific voltage and a specific current elsewhere in the circuit. *Lowercase v and i represent voltage and current sources that may or may not be time-varying, whereas uppercase V and I denote dc sources.

1-6.3 Independent Current Source

Based on our common experience with stand-alone chemical batteries used in cars, flashlights, and other systems, we readily accept the notion of an electric circuit acting like a voltage source by providing at its output terminals a specified voltage level V_s. By contrast, there is no such thing as a "current battery," one that provides a constant current to flow through the load connected to its terminals. Nevertheless, we can build an electric circuit that behaves like a current source. An *ideal, independent current source* provides a specified current flowing through it, regardless of the voltage across it (except when connected to an open circuit). Its i–v relationship is

$$i = I_s \qquad \text{for any } v_s \neq \infty.$$

The i–υ profile of an ideal current source is a horizontal line, as shown in Fig. 1-24(b). A current source may be built from a voltage source with a current limiter, so long as the voltage source can supply the desired current independently of the attached load.

In the same way that a realistic voltage source consists of an ideal voltage source in series with a resistor R_s, a *realistic current source* consists of an ideal current source i_s in parallel with a resistor R_s (Fig. 1-25(b)). In a well-designed current source, R_s is very large, thereby extracting from the current source a very small fraction in comparison to the current that flows into the load.

Example 1-7: AA Battery

The circuit shown in Fig. 1-26(a) represents an AA battery, with voltage V_s and internal resistance R_s, connected to a light bulb represented by a load resistance $R_L = 10\ \Omega$. If $V_s = 1.5$ V and independent of ambient temperature, and R_s is as profiled in Fig. 1-26(b), use the voltage division equation (which will be derived later in Chapter 2) given by

$$V_L = \left(\frac{R_L}{R_s + R_L}\right) V_s$$

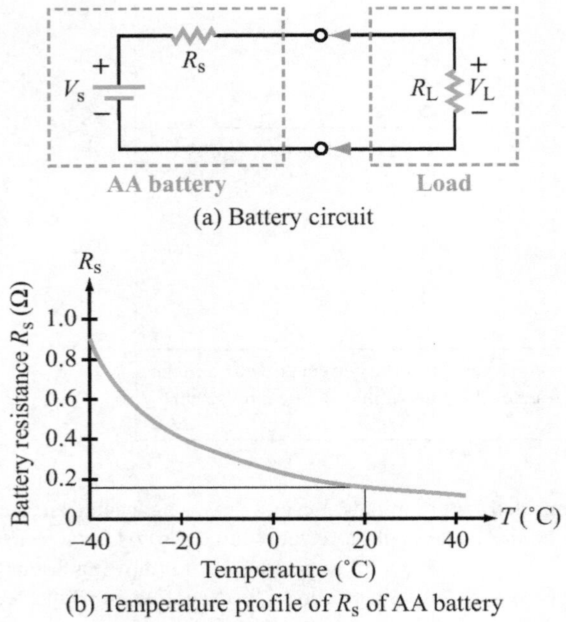

Figure 1-26: Circuit and temperature profile of battery's R_s of Example 1-7.

to determine V_L at (a) room temperature (20 °C) and (b) in Antarctica at −40 °C.

Solution: (a) From the plot in Fig. 1-26(b), $R_s \approx 0.15\ \Omega$ at $T = 20$ °C. Hence,

$$V_L = \left(\frac{10}{0.15 + 10}\right) \times 1.5 = 1.4778 \text{ V},$$

which is within 1.5% of $V_s = 1.5$ V.

(b) At $T = -40$ °C, $R_s = 0.9\ \Omega$. Hence,

$$V_L = \left(\frac{10}{0.9 + 10}\right) \times 1.5 = 1.376 \text{ V}.$$

In this case, ignoring R_s altogether would lead to an error of about 8%. At low temperatures, batteries are less efficient and often cease to provide the desired voltage and current, as anyone whose car battery has "died" on a cold winter day has discovered.

1-6.4 Dependent Sources

As alluded to in the opening paragraph of Section 1-6, we often use equivalent circuits to model the behavior of transistors and other electronic devices. The ability to represent complicated devices by equivalent circuits composed of basic elements greatly facilitates not only the circuit analysis process but the design process as well. Such circuit models incorporate the relationships between various parts of the device through the use of a set of *artificial* sources known as *dependent sources*. The voltage level of a *dependent voltage source* is defined in terms of a specific voltage or current elsewhere in the circuit. An example of circuit equivalence is illustrated in Fig. 1-27. In part (a) of the figure, we have a Model 741 *operational amplifier* (*op amp*), denoted by the triangular circuit symbol, used in a simple amplifier circuit intended to provide a voltage amplification factor of −2; that is, the output voltage $\upsilon_0 = -2\upsilon_s$, where υ_s is the input signal voltage. The op amp, which we will examine later in Chapter 4, is an electronic device with a complex architecture composed of transistors, resistors, capacitors, and diodes, but in practice, its circuit behavior can be represented by a rather simple circuit consisting of two resistors (input resistor R_i and output resistor R_o) and a dependent voltage source, as shown in Fig. 1-27(b). The voltage υ_2 on the right-hand side of the circuit in Fig. 1-27(b) is given by $\upsilon_2 = A\upsilon_i$, where A is a very large constant ($> 10^4$) and υ_i

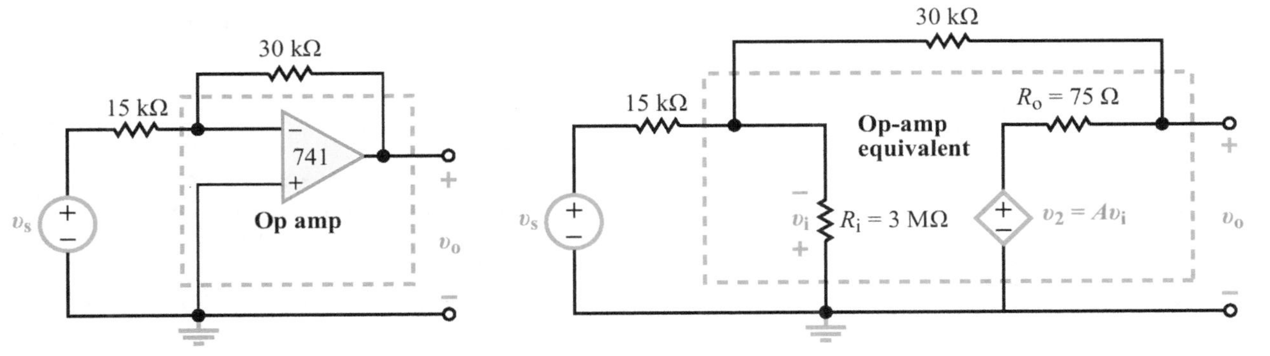

(a) Op-amp circuit (b) Equivalent circuit with dependent voltage source

Figure 1-27: An operational amplifier is a complex device, but its circuit behavior can be represented in terms of a simple equivalent circuit that includes a dependent voltage source.

is the voltage across the resistor R_i located on the left-hand side of the equivalent circuit. In this case, the magnitude of v_2 always depends on the magnitude of v_i, which depends in turn on the input signal voltage v_s and on the values chosen for some of the resistors in the circuit. Since the controlling quantity v_i is a voltage, v_2 is called a *voltage-controlled voltage source* (VCVS). Had the controlling quantity been a current source, the dependent source would have been called a *current-controlled voltage source* (CCVS) instead. A parallel analogy exists for voltage-controlled and current-controlled current sources.

▶ The characteristic symbol for a dependent source is the diamond (Table 1-5). ◀

Proportionality constant α in Table 1-5 relates voltage to voltage. Hence, it is dimensionless, as is β, since it relates current to current. Constants g and r have units of (A/V) and (V/A), respectively. Because dependent sources are characterized by linear relationships, so are their i–v profiles. An example is shown in Fig. 1-24(c) for the VCVS.

Example 1-8: Dependent Source

Find the magnitude of the voltage V_1 of the dependent source in Fig. 1-28. What type of source is it?

Solution: Since V_1 depends on current I_1, it is a current-controlled voltage source with a coefficient of 4 V/A.

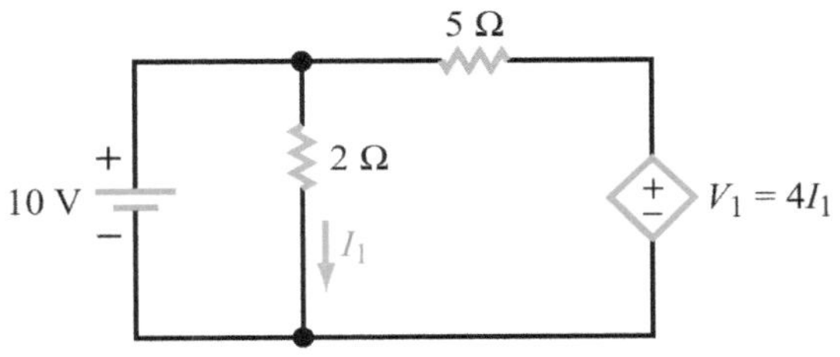

Figure 1-28: Circuit for Example 1-8.

The 10 V dc voltage is connected across the 2 Ω resistor. Hence, the current I_1 along the designated direction is

$$I_1 = \frac{10}{2} = 5\ \text{A}.$$

Consequently,

$$V_1 = 4I_1 = 4 \times 5 = 20\ \text{V}.$$

Example 1-9: Switches

The circuit in Fig. 1-29 contains one SPDT switch that changes position at $t = 0$, one SPST switch that opens at $t = 0$, and one SPST switch that closes at $t = 5$ s. Generate circuit diagrams that include only those elements that have current flowing through them for (a) $t < 0$, (b) $0 \le t < 5$ s, and (c) $t \ge 5$ s.

Solution: See Fig. 1-30.

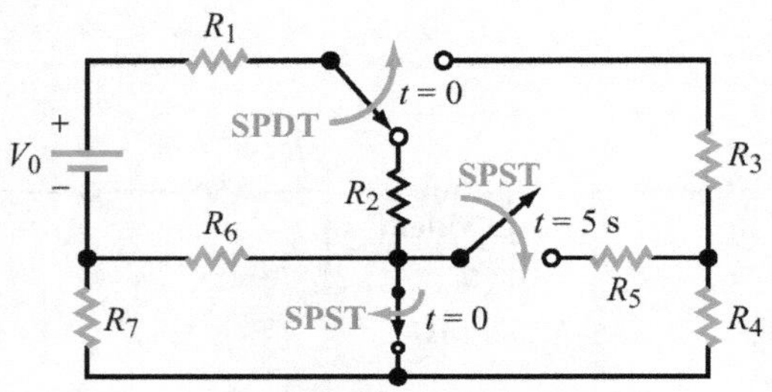

Figure 1-29: Circuit for Example 1-9.

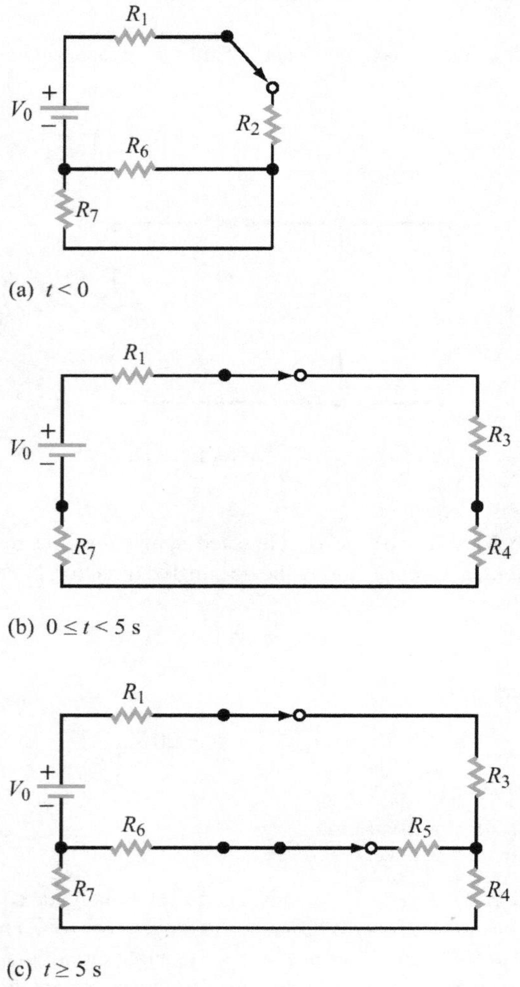

Figure 1-30: Solutions for circuit in Fig. 1-29.

Concept Question 1-11: What is the difference between an SPST switch and an SPDT switch? (See CAD)

Concept Question 1-12: What is the difference between an independent voltage source and a dependent voltage source? Is a dependent voltage source a real source of power? (See CAD)

Concept Question 1-13: What is an "equivalent-circuit" model? How is it used? (See CAD)

Exercise 1-9: Find I_x from the diagram in Fig. E1.9.

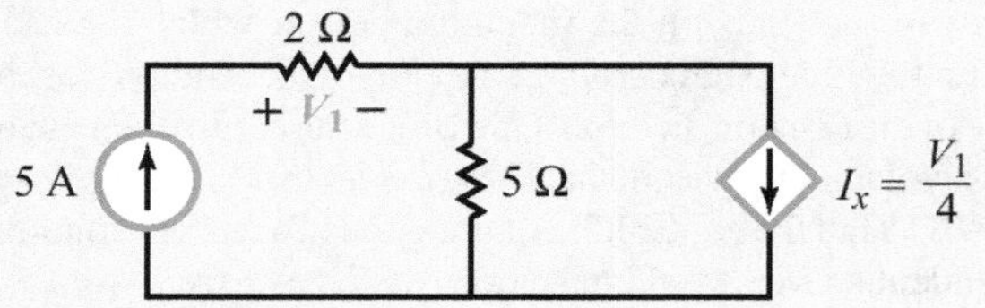

Figure E1.9

Answer: $I_x = 2.5$ A. (See CAD)

Exercise 1-10: In the circuit of Fig. E1.10, find I at (a) $t < 0$ and (b) $t > 0$.

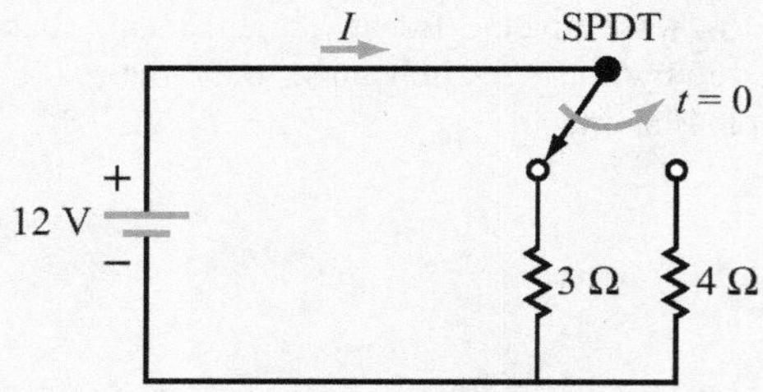

Figure E1.10

Answer: (a) $I = 4$ A, (b) $I = 3$ A. (See CAD)

Summary

Concepts

- Active devices (such as transistors and ICs) require an external power source to operate; in contrast, passive devices (resistors, capacitors, and inductors) do not.
- Analysis and synthesis (design) are complementary processes.
- Current is related to charge by $i = dq/dt$; voltage between locations a and b is $\upsilon_{ab} = dw/dq$, where dw is the work (energy) required to move dq from b to a; and power $p = \upsilon i$.
- Passive sign convention assigns i direction as entering the (+) side of υ; if $p > 0$, the device is recipient (consumer) of power, and if $p < 0$, it is a supplier of power.
- Node voltage refers to the voltage difference between the node and ground by selecting $V_{\text{ground}} = 0$.
- Independent voltage and current sources are real sources of energy; dependent sources are artificial representations used in modeling the nonlinear behavior of active devices (transistors and integrated circuits) in terms of an equivalent linear circuit.

Mathematical and Physical Models

Ohm's law $\quad i = \upsilon/R$

Current $\quad i = dq/dt$

Direction of i = direction of flow of (+) charge

Charge transfer $\quad q(t) = \int_{-\infty}^{t} i\,dt$

$$\Delta Q = q(t_2) - q(t_1) = \int_{t_1}^{t_2} i\,dt$$

Voltage = potential energy difference per unit charge

Passive sign convention

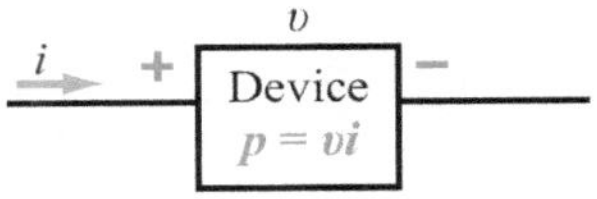

$p > 0$ power delivered to device

$p < 0$ power supplied by device

Note that i direction is defined as entering (+) side of υ.

Energy $\quad w = \int_{0}^{\infty} p(t)\,dt$

Important Terms

Provide definitions or explain the meaning of the following terms:

ac
active device
Alexander Graham Bell
all-electronic computer
alternating current
ampere-hours
analysis
basic elements
branch
chassis ground
circuit
circuit diagram
conduction current
cumulative charge
current-controlled voltage source
dc
dependent voltage source
dependent source
design
device
dimension
direct current
drift
drift velocity
Earth ground
electric circuit
electric current
electron drift
electronic
electronic circuit
equivalent circuit
equivalent, ideal voltage source
equivalent resistance
external
extraordinary node
functional block diagram
fundamental dimension
fundamental SI unit
ground
Guglielmo Marconi

Important Terms (continued)

Heinrich Hertz
i–υ relationship
ideal, independent current source
ideal, independent voltage source
in parallel
in series
input/output
integrated circuit
International System of Units
ionized
kilowatt-hours
law of conservation of power
linear circuit
linear i–υ relationship
loop
mesh
net charge
nonplanar
Ohm's law
op amp
open circuit
operational amplifier
ordinary node
passive device
passive sign convention
path
PCB layout
planar
potential difference
prefix
printed circuit board
printed conducting lines
real voltage source
realistic current source
response
secondary dimension
short circuit
single-pole single-throw
single-pole double-throw
stimulus
synthesis
system
Thomas Edison
transistor
transmission velocity
unit
voltage-controlled voltage source
voltage drop
voltage rise

PROBLEMS

Sections 1-2 to 1-4: Dimensions, Charge, and Current

1.1 Use appropriate multiple and submultiple prefixes to express the following quantities:

(a) 3,620 watts (W)
*(b) 0.000004 amps (A)
(c) 5.2×10^{-6} ohms (Ω)
*(d) 3.9×10^{11} volts (V)
(e) 0.02 meters (m)
(f) 32×10^{5} volts (V)

1.2 Use appropriate multiple and submultiple prefixes to express the following quantities:

(a) 4.71×10^{-8} seconds (s)
(b) 10.3×10^{8} watts (W)
(c) 0.00000000321 amps (A)
(d) 0.1 meters (m)
(e) 8,760,000 volts (V)
(f) 3.16×10^{-16} hertz (Hz)

*Answer(s) available in Appendix G.

1.3 Convert:

(a) 16.3 m to mm
(b) 16.3 m to km
*(c) 4×10^{-6} μF (microfarad) to pF (picofarad)
(d) 2.3 ns to μs
(e) 3.6×10^{7} V to MV
(f) 0.03 mA (milliamp) to μA

1.4 Convert:

(a) 4.2 m to μm
(b) 3 hours to μseconds
(c) 4.2 m to km
(d) 173 nm to m
(e) 173 nm to μm
(f) 12 pF (picofarad) to F (farad)

1.5 For the circuit in Fig. P1.5:

(a) Identify and label all distinct nodes.
(b) Which of those nodes are extraordinary nodes?
(c) Identify all combinations of 2 or more circuit elements that are connected in series.
(d) Identify pairs of circuit elements that are connected in parallel.

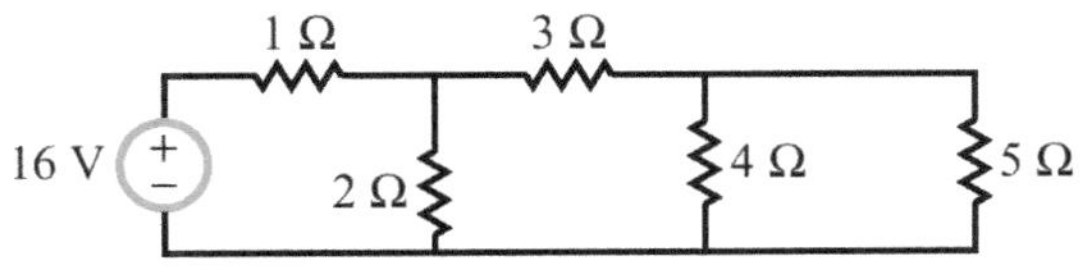

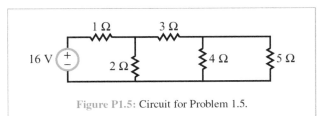

Figure P1.5: Circuit for Problem 1.5.

1.6 For the circuit in Fig. P1.6:

(a) Identify and label all distinct nodes.

(b) Which of those nodes are extraordinary nodes?

(c) Identify all combinations of 2 or more circuit elements that are connected in series.

(d) Identify pairs of circuit elements that are connected in parallel.

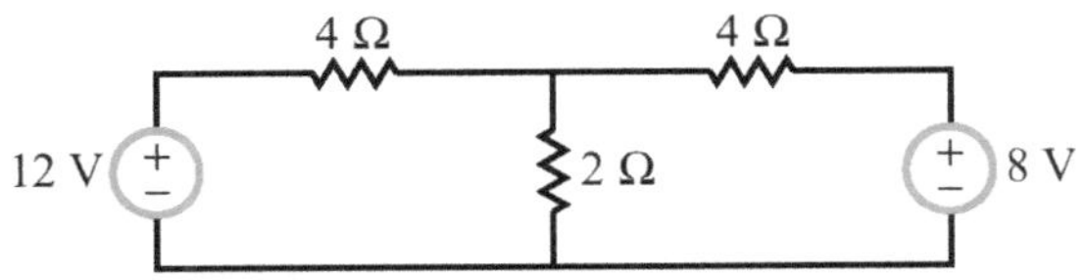

Figure P1.6: Circuit for Problem 1.6.

1.7 For the circuit in Fig. P1.7:

(a) Identify and label all distinct nodes.

(b) Which of those nodes are extraordinary nodes?

(c) Identify all combinations of 2 or more circuit elements that are connected in series.

(d) Identify pairs of circuit elements that are connected in parallel.

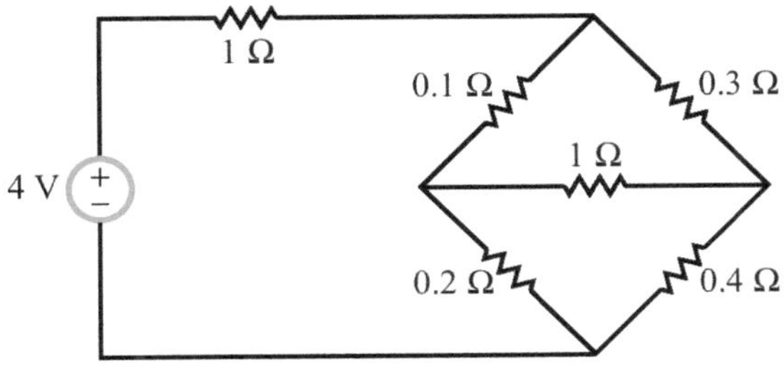

Figure P1.7: Circuit for Problem 1.7.

1.8 For the circuit in Fig. P1.8:

(a) Identify and label all distinct nodes.

(b) Which of those nodes are extraordinary nodes?

(c) Identify all combinations of 2 or more circuit elements that are connected in series.

(d) Identify pairs of circuit elements that are connected in parallel.

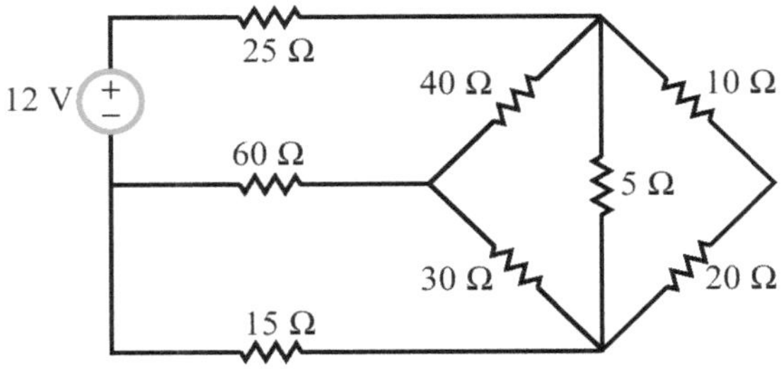

Figure P1.8: Circuit for Problem 1.8.

1.9 For the circuit in Fig. P1.9:

(a) Identify and label all distinct nodes.

(b) Which of those nodes are extraordinary nodes?

(c) Identify all combinations of 2 or more circuit elements that are connected in series.

(d) Identify pairs of circuit elements that are connected in parallel.

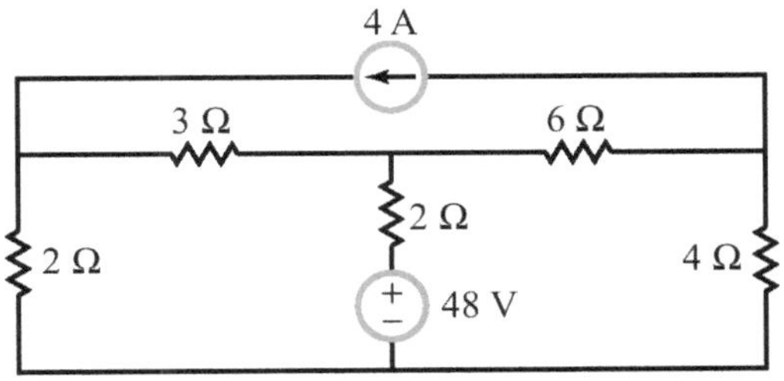

Figure P1.9: Circuit for Problem 1.9.

1.10 For the circuit in Fig. P1.10:

(a) Identify and label all distinct nodes.

(b) Which of those nodes are extraordinary nodes?

(c) Identify all combinations of 2 or more circuit elements that are connected in series.

(d) Identify pairs of circuit elements that are connected in parallel.

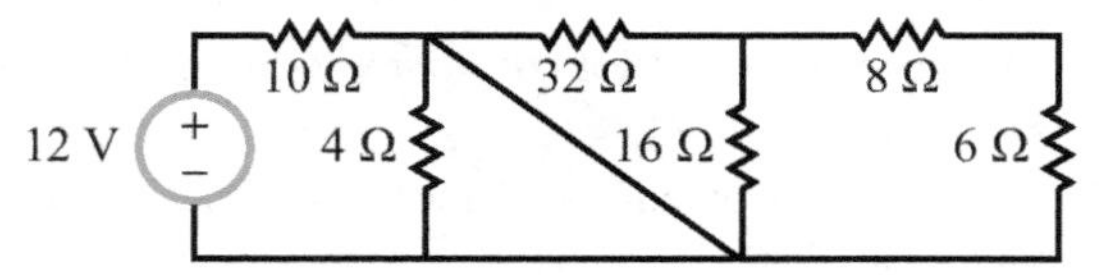

Figure P1.10: Circuit for Problem 1.10.

1.11 For the circuit in Fig. P1.11:

(a) Identify and label all distinct nodes.

(b) Which of those nodes are extraordinary nodes?

(c) Identify all combinations of 2 or more circuit elements that are connected in series.

(d) Identify pairs of circuit elements that are connected in parallel.

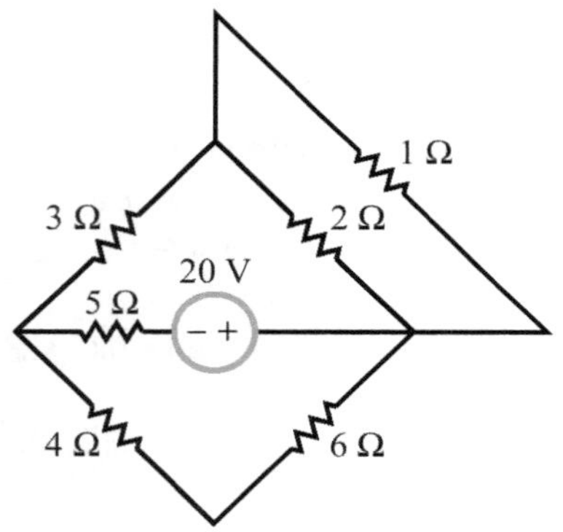

Figure P1.11: Circuit for Problem 1.11.

1.12 The total charge contained in a certain region of space is -1 C. If that region contains only electrons, how many does it contain?

*__1.13__ A certain cross section lies in the x–y plane. If 3×10^{20} electrons go through the cross section in the z direction in 4 seconds, and simultaneously 1.5×10^{20} protons go through the same cross section in the negative z direction, what is the magnitude and direction of the current flowing through the cross section?

1.14 Determine the current $i(t)$ flowing through a resistor if the cumulative charge that has flowed through it up to time t is given by

(a) $q(t) = 3.6t$ mC

(b) $q(t) = 5\sin(377t)$ μC

*(c) $q(t) = 0.3[1 - e^{-0.4t}]$ pC

(d) $q(t) = 0.2t\sin(120\pi t)$ nC

1.15 Determine the current $i(t)$ flowing through a certain device if the cumulative charge that has flowed through it up to time t is given by

(a) $q(t) = -0.45t^3$ μC

(b) $q(t) = 12\sin^2(800\pi t)$ mC

(c) $q(t) = -3.2\sin(377t)\cos(377t)$ pC

*(d) $q(t) = 1.7t[1 - e^{-1.2t}]$ nC

1.16 Determine the net charge ΔQ that flowed through a resistor over the specified time interval for each of the following currents:

(a) $i(t) = 0.36$ A, from $t = 0$ to $t = 3$ s

*(b) $i(t) = [40t + 8]$ mA, from $t = 1$ s to $t = 12$ s

(c) $i(t) = 5\sin(4\pi t)$ nA, from $t = 0$ to $t = 0.05$ s

(d) $i(t) = 12e^{-0.3t}$ mA, from $t = 0$ to $t = \infty$

1.17 Determine the net charge ΔQ that flowed through a certain device over the specified time intervals for each of the following currents:

(a) $i(t) = [3t + 6t^3]$ mA, from $t = 0$ to $t = 4$ s

*(b) $i(t) = 4\sin(40\pi t)\cos(40\pi t)$ μA, from $t = 0$ to $t = 0.05$ s

(c) $i(t) = [4e^{-t} - 3e^{-2t}]$ A, from $t = 0$ to $t = \infty$

(d) $i(t) = 12e^{-3t}\cos(40\pi t)$ nA, from $t = 0$ to $t = 0.05$ s

1.18 If the current flowing through a wire is given by $i(t) = 3e^{-0.1t}$ mA, how many electrons pass through the wire's cross section over the time interval from $t = 0$ to $t = 0.3$ ms?

1.19 The cumulative charge in mC that entered a certain device is given by

$$q(t) = \begin{cases} 0 & \text{for } t < 0, \\ 5t & \text{for } 0 \le t \le 10 \text{ s}, \\ 60 - t & \text{for } 10 \text{ s} \le t \le 60 \text{ s} \end{cases}$$

(a) Plot $q(t)$ versus t from $t = 0$ to $t = 60$ s.

(b) Plot the corresponding current $i(t)$ entering the device.

*__1.20__ A steady flow resulted in 3×10^{15} electrons entering a device in 0.1 ms. What is the current?

1.21 Given that the current in (mA) flowing through a wire is given by:

$$i(t) = \begin{cases} 0 & \text{for } t < 0 \\ 6t & \text{for } 0 \leq t \leq 5 \text{ s} \\ 30e^{-0.6(t-5)} & \text{for } t \geq 5 \text{ s}, \end{cases}$$

(a) Sketch $i(t)$ versus t.

(b) Sketch $q(t)$ versus t.

1.22 The plot in Fig. P1.22 displays the cumulative amount of charge $q(t)$ that has entered a certain device up to time t. What is the current at

(a) $t = 1$ s

*(b) $t = 3$ s

(c) $t = 6$ s

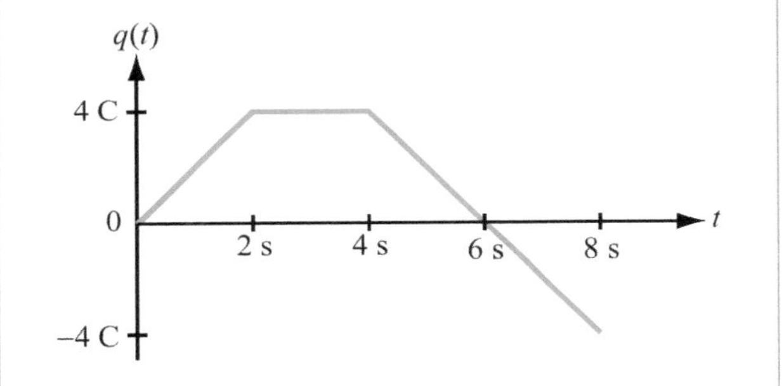

Figure P1.22: $q(t)$ for Problem 1.22.

1.23 The plot in Fig. P1.23 displays the cumulative amount of charge $q(t)$ that has exited a certain device up to time t. What is the current at

*(a) $t = 2$ s

(b) $t = 6$ s

(c) $t = 12$ s

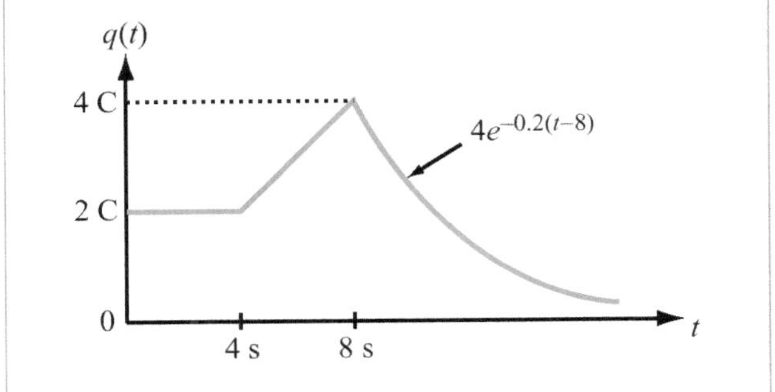

Figure P1.23: $q(t)$ for Problem 1.23.

1.24 The plot in Fig. P1.24 displays the cumulative charge $q(t)$ that has entered a certain device up to time t. Sketch a plot of the corresponding current $i(t)$.

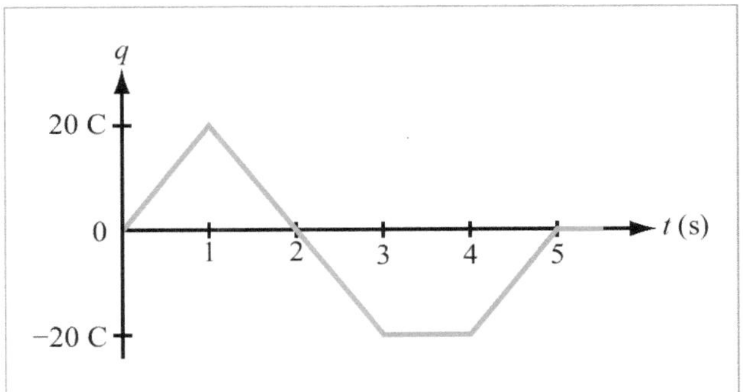

Figure P1.24: $q(t)$ for Problem 1.24.

Sections 1-5 and 1-6: Voltage, Power, and Circuit Elements

1.25 In the circuit of Fig. P1.25, node V_1 was selected as the ground node.

*(a) What is the voltage at node V_2?

(b) What is the voltage difference $V_{32} = V_3 - V_2$?

(c) What are the voltages at nodes 1, 3, 4, and 5 if node 2 is selected as the ground node instead of node 1?

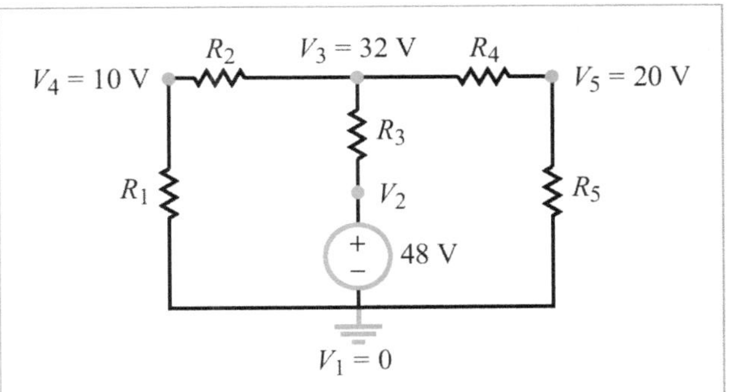

Figure P1.25: Circuit for Problem 1.25.

1.26 In the circuit of Fig. P1.26, node V_1 was selected as the ground node.

*(a) What is the voltage difference across R_6?

(b) What are the voltages at nodes 1, 3, and 4 if node 2 is selected as the ground node instead of node 1?

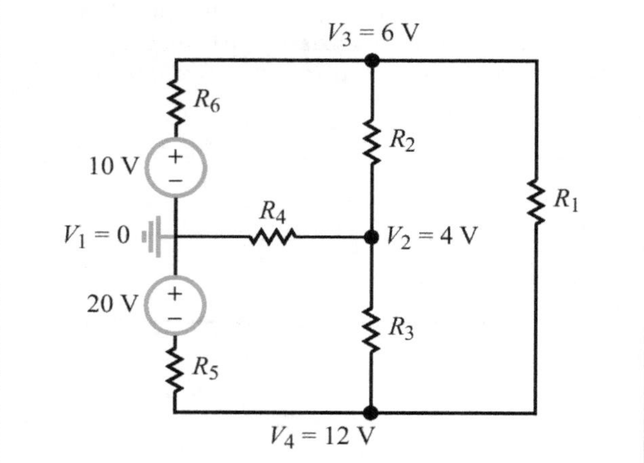

Figure P1.26: Circuit for Problem 1.26.

1.27 For each of the eight devices in the circuit of Fig. P1.27, determine whether the device is a supplier or a recipient of power and how much power it is supplying or receiving.

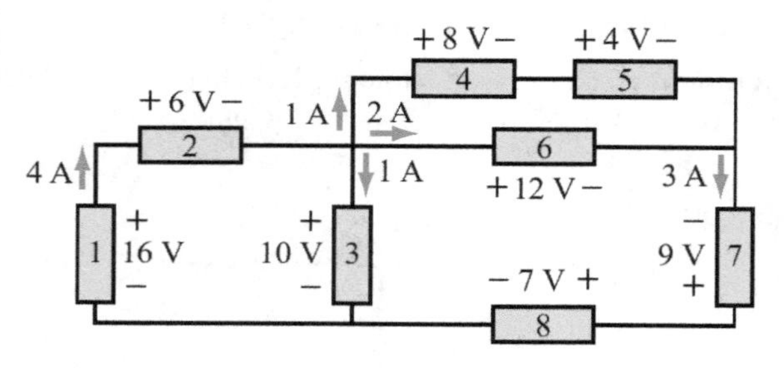

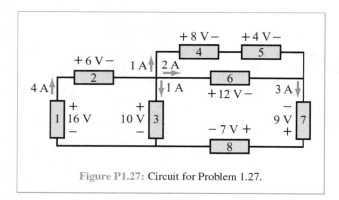

Figure P1.27: Circuit for Problem 1.27.

1.28 For each of the seven devices in the circuit of Fig. P1.28, determine whether the device is a supplier or a recipient of power and how much power it is supplying or receiving.

*__1.29__ An electric oven operates at 120 V. If its power rating is 0.6 kW, what amount of current does it draw, and how much energy does it consume in 12 minutes of operation?

1.30 A 9 V flashlight battery has a rating of 1.8 kWh. If the bulb draws a current of 100 mA when lit; determine the following:

(a) For how long will the flashlight provide illumination?

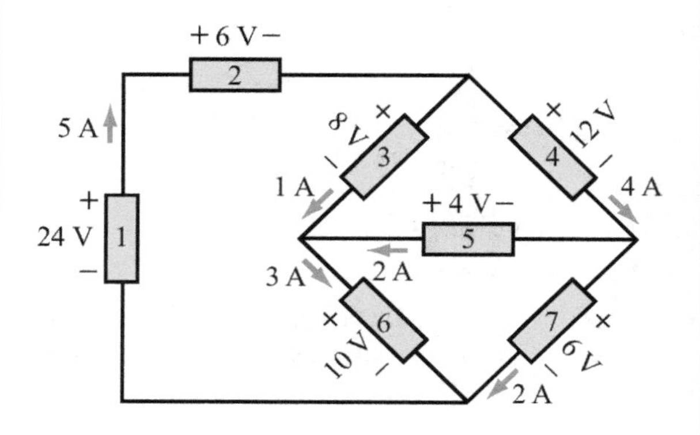

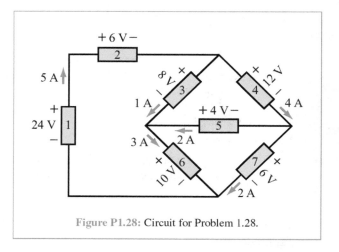

Figure P1.28: Circuit for Problem 1.28.

(b) How much energy in joules is contained in the battery?

(c) What is the battery's rating in ampere-hours?

1.31 The voltage across and current through a certain device are given by

$$v(t) = 5\cos(4\pi t) \text{ V}, \qquad i(t) = 0.1\cos(4\pi t) \text{ A}.$$

Determine:

*(a) The instantaneous power $p(t)$ at $t = 0$ and $t = 0.25$ s.

(b) The average power p_{av}, defined as the average value of $p(t)$ over a full time period of the cosine function (0 to 0.5 s).

1.32 The voltage across and current through a certain device are given by

$$v(t) = 100(1 - e^{-0.2t}) \text{ V}, \qquad i(t) = 30e^{-0.2t} \text{ mA}.$$

Determine:

(a) The instantaneous power $p(t)$ at $t = 0$ and $t = 3$ s.

(b) The cumulative energy delivered to the device from $t = 0$ to $t = \infty$.

1.33 The voltage across a device and the current through it are shown graphically in Fig. P1.33. Sketch the corresponding power delivered to the device and calculate the energy absorbed by it.

1.34 The voltage across a device and the current through it are shown graphically in Fig. P1.34. Sketch the corresponding power delivered to the device and calculate the energy absorbed by it.

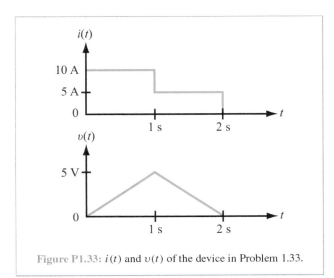

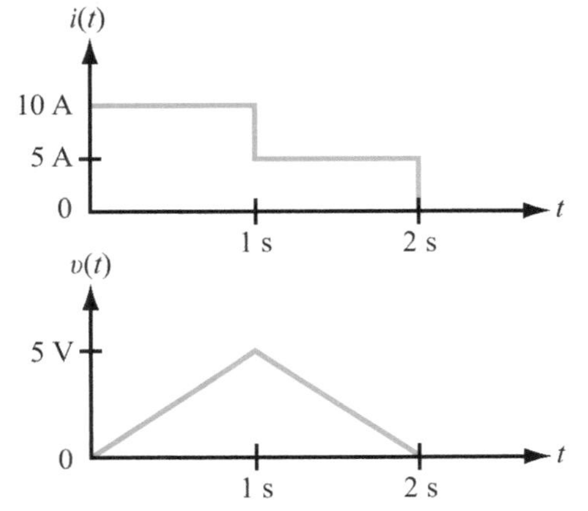

Figure P1.33: $i(t)$ and $\upsilon(t)$ of the device in Problem 1.33.

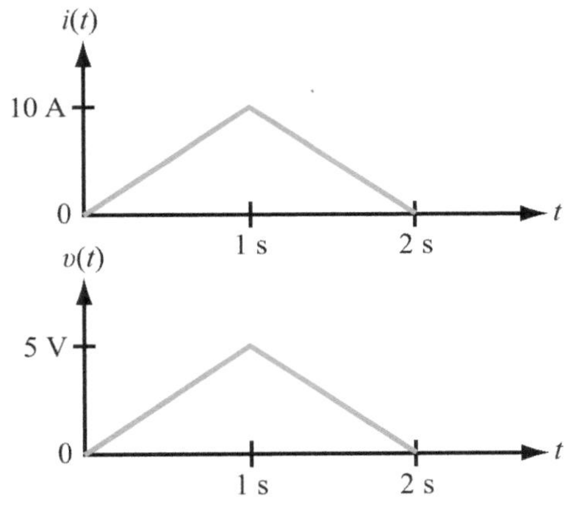

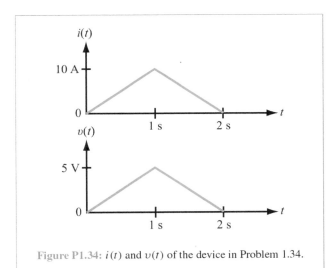

Figure P1.34: $i(t)$ and $\upsilon(t)$ of the device in Problem 1.34.

1.35 The voltage across a device and the current through it are shown graphically in Fig. P1.35. Sketch the corresponding power delivered to the device and calculate the energy absorbed by it.

*__1.36__ After $t = 0$, the current entering the positive terminal of a flashlight bulb is given by

$$i(t) = 2(1 - e^{-10t}) \qquad \text{(A)},$$

and the voltage across the bulb is $\upsilon(t) = 12e^{-10t}$ (V). Determine the maximum power level delivered to the flashlight.

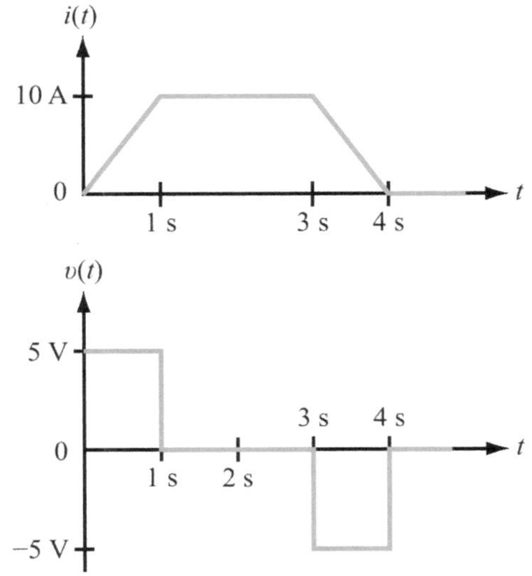

Figure P1.35: $i(t)$ and $\upsilon(t)$ of the device in Problem 1.35.

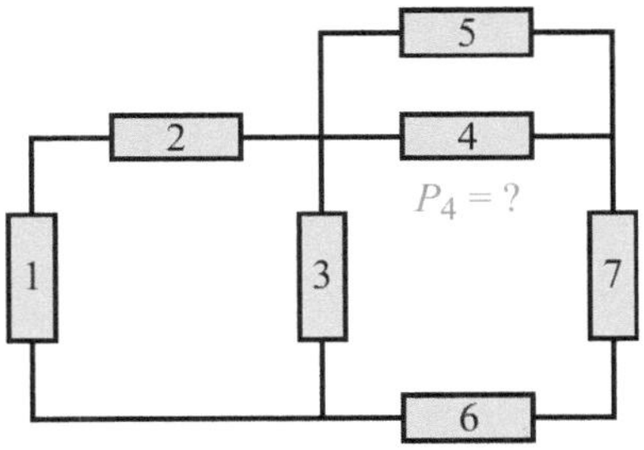

Figure P1.37: Circuit for Problem 1.37.

1.37 Apply the law of conservation of power to determine the amount of power delivered to device 4 in the circuit of Fig. P1.37, given that that the amounts of power delivered to the other devices are: $p_1 = -100$ W, $p_2 = 30$ W, $p_3 = 22$ W, $p_5 = 67$ W, $p_6 = -201$ W, and $p_7 = 120$ W.

*__1.38__ Determine V_y in the circuit of Fig. P1.38.

1.39 Determine V, the voltage of the dependent voltage source in the circuit of Fig. P1.39.

*__1.40__ Determine V_z in the circuit of Fig. P1.40.

1.41 Determine I_x in the circuit of Fig. P1.41.

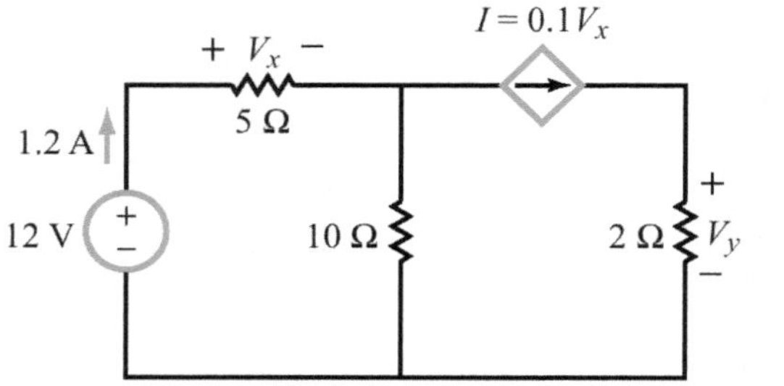

Figure P1.38: Circuit for Problem 1.38.

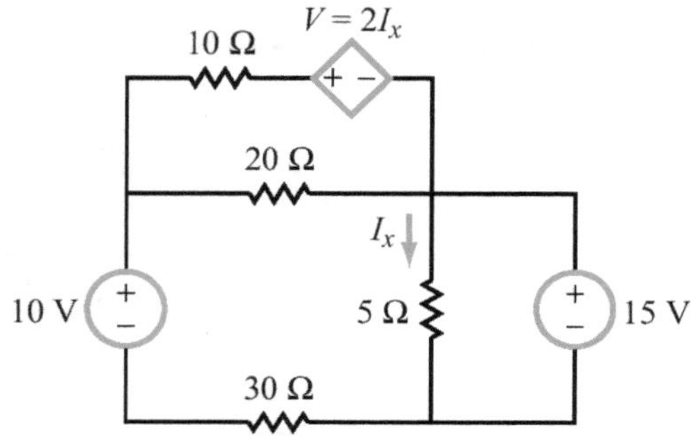

Figure P1.39: Circuit for Problem 1.39.

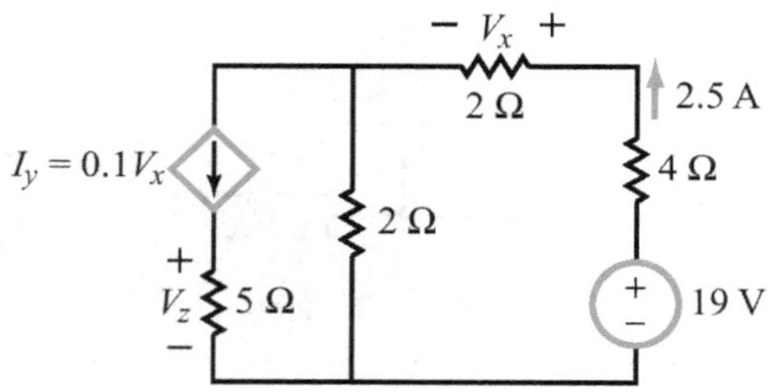

Figure P1.40: Circuit for Problem 1.40.

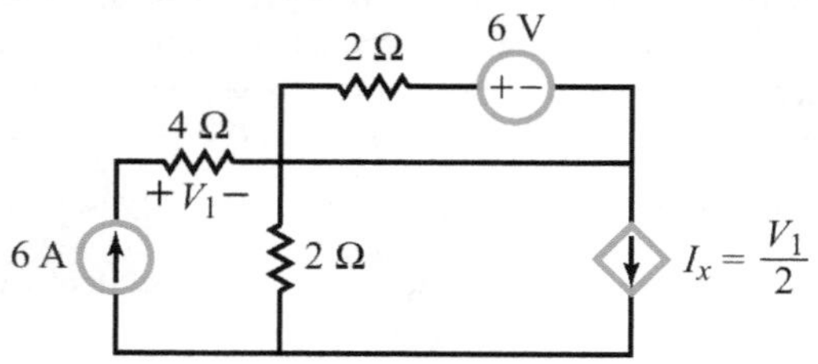

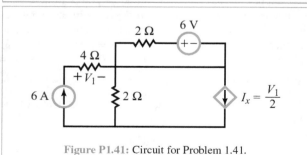

Figure P1.41: Circuit for Problem 1.41.

1.42 For the circuit in Fig. P1.42, generate circuit diagrams that include only those elements that have current flowing through them for

(a) $t < 0$

(b) $0 < t < 2$ s

(c) $t > 2$ s

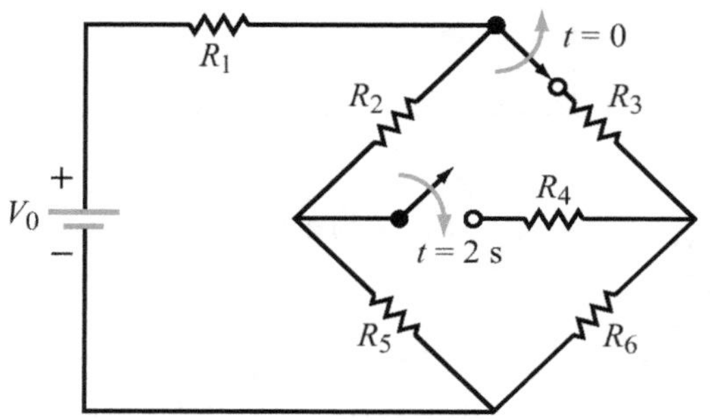

Figure P1.42: Circuit for Problem 1.42.

1.43 For the circuit in Fig. P1.43, generate circuit diagrams that include only those elements that have current flowing through them for

(a) $t < 0$

(b) $0 < t < 2$ s

(c) $t > 2$ s

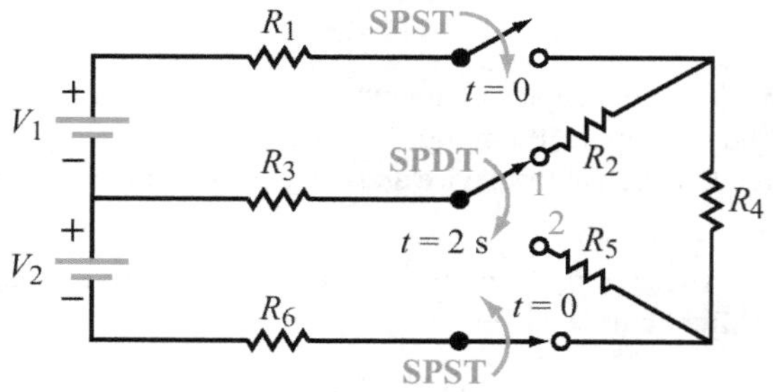

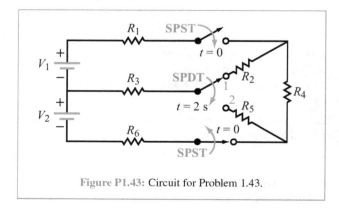

Figure P1.43: Circuit for Problem 1.43.

1.44 The switch in the circuit of Fig. P1.44 closes at $t = 0$. Which elements are in series and which are in parallel at (a) $t < 0$ and (b) $t > 0$?

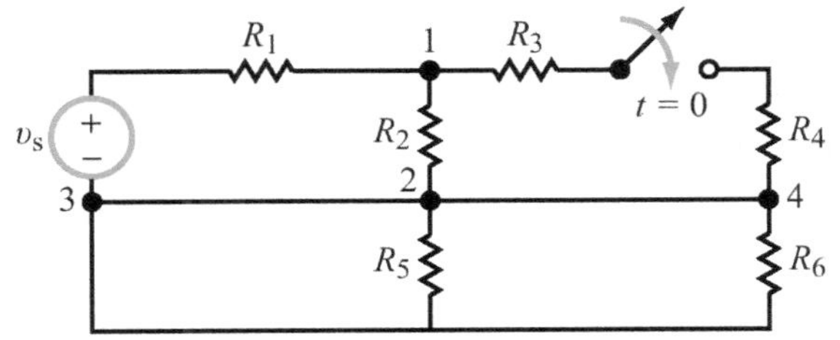

Figure P1.44: Circuit for Problem 1.44.

Potpourri Questions

1.45 What aspect of electrical engineering particularly interests you? Check out http://spectrum.ieee.org/ to learn more.

1.46 Will the prediction of Moore's Law continue to hold true indefinitely? If not, why not?

1.47 Provide a definition of what the term "nanotechnology" means to you.

1.48 What is the typical voltage level associated with lightning? With a bird standing on a power line (foot to foot)?

CHAPTER 2

Resistive Circuits

Contents

Objectives

Learn to:

- Apply Ohm's law and explain the basic properties of piezoresistivity and superconductivity.
- State Kirchhoff's current and voltage laws; apply them to resistive circuits.
- Define circuit equivalency, combine resistors in series and in parallel, and apply voltage and current division.
- Apply source transformation between voltage and current sources and Y–Δ circuits.
- Describe the operation of the Wheatstone-bridge circuit and how it is used to measure small deviations.
- Use Multisim and myDAQ to analyze simple circuits.

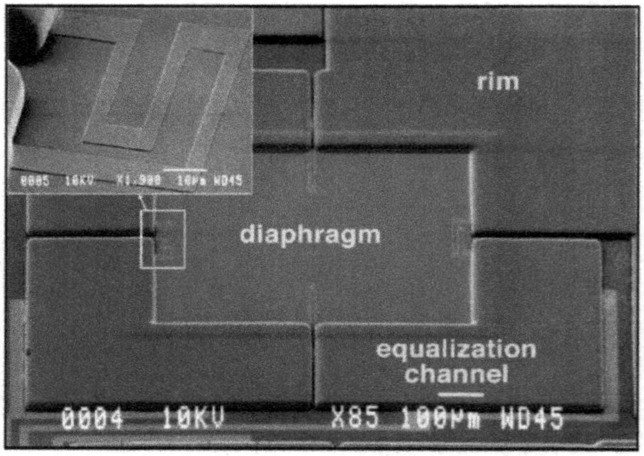

Microfabricated pressure sensor

The basic laws of *circuit theory* are used to develop fluency in analyzing *resistive circuits* and characterizing their performance.

Overview

The study of any field of inquiry starts with nomenclature: defining the terms specific to that field. That is exactly what we did in the preceding chapter. We introduced and defined electric current, voltage, power, open and closed circuits, and dependent and independent voltage and current sources, among others. Now, we are ready to acquire our first set of circuit-analysis tools, which will enable us to analyze a variety of different types of circuits. We limit our discussion to *resistive circuits*, namely those circuits containing only sources and resistors. (In future chapters, we will extend these tools to circuits containing capacitors, inductors, and other elements.) Our new toolbox includes three simple, yet powerful laws—Ohm's law and Kirchhoff's voltage and current laws—and several circuit simplification and transformation techniques. You will learn how to divide the voltage (using voltage dividers) and current (using current dividers), how to combine resistors in series and parallel combinations, how to analyze resistive sensors using Wheatstone bridges, how to use diodes to control the direction of a current, plus how to use a light-emitting diode (LED) as a visual output, warning light, etc. You will also learn how to use *Multisim* to simulate and analyze your circuits, and how to build a circuit on a circuit board and measure its properties using your computer via the *NI myDAQ*.

2-1 Ohm's Law

▶ The *conductivity* σ of a material is a measure of how easily electrons can drift through the material when an external voltage is applied across it. *Resistivity* (ρ) is the inverse ($1/\sigma$) of conductivity. ◀

Materials are classified as *conductors* (primarily metals), *semiconductors*, or *dielectrics* (insulators) according to the magnitudes of their conductivities. Tabulated values of σ expressed in units of siemens per meter (S/m) are given in Table 2-1 for a select group of materials. The siemen is the inverse of the ohm, $S = 1/\Omega$, and the inverse of σ is called the *resistivity* ρ,

$$\rho = \frac{1}{\sigma} \quad (\Omega\text{-m}), \tag{2.1}$$

which is a measure of how well a material *impedes* the flow of current through it. The conductivity of most metals is on the order of 10^7 S/m, which is 17 or more orders of magnitude greater than the conductivity of typical insulators. Common semiconductors, such as silicon and germanium, fall in the in-between range on the conductivity scale.

Table 2-1: Conductivity and resistivity of some common materials at 20 °C.

Material	Conductivity σ (S/m)	Resistivity ρ (Ω-m)
Conductors		
Silver	6.17×10^7	1.62×10^{-8}
Copper	5.81×10^7	1.72×10^{-8}
Gold	4.10×10^7	2.44×10^{-8}
Aluminum	3.82×10^7	2.62×10^{-8}
Iron	1.03×10^7	9.71×10^{-8}
Mercury (liquid)	1.04×10^6	9.58×10^{-7}
Semiconductors		
Carbon (graphite)	7.14×10^4	1.40×10^{-5}
Pure germanium	2.13	0.47
Pure silicon	4.35×10^{-4}	2.30×10^3
Insulators		
Paper	$\sim 10^{-10}$	$\sim 10^{10}$
Glass	$\sim 10^{-12}$	$\sim 10^{12}$
Teflon	$\sim 3.3 \times 10^{-13}$	$\sim 3 \times 10^{12}$
Porcelain	$\sim 10^{-14}$	$\sim 10^{14}$
Mica	$\sim 10^{-15}$	$\sim 10^{15}$
Polystyrene	$\sim 10^{-16}$	$\sim 10^{16}$
Fused quartz	$\sim 10^{-17}$	$\sim 10^{17}$
Common materials		
Distilled water	5.5×10^{-6}	1.8×10^5
Drinking water	$\sim 5 \times 10^{-3}$	~ 200
Sea water	4.8	0.2
Graphite	1.4×10^{-5}	71.4×10^3
Rubber	1×10^{-13}	1×10^{13}
Biological tissues		
Blood	~ 1.5	~ 0.67
Muscle	~ 1.5	~ 0.67
Fat	~ 0.1	10

The values of σ and ρ given in Table 2-1 are specific to room temperature at 20 °C. In general, the conductivity of a metal increases with decreasing temperature. At very low temperatures (in the neighborhood of absolute zero), some conductors become *superconductors*, because their conductivities become practically infinite and their corresponding resistivities approach zero. To learn more about superconductivity, refer to Technology Brief 3.

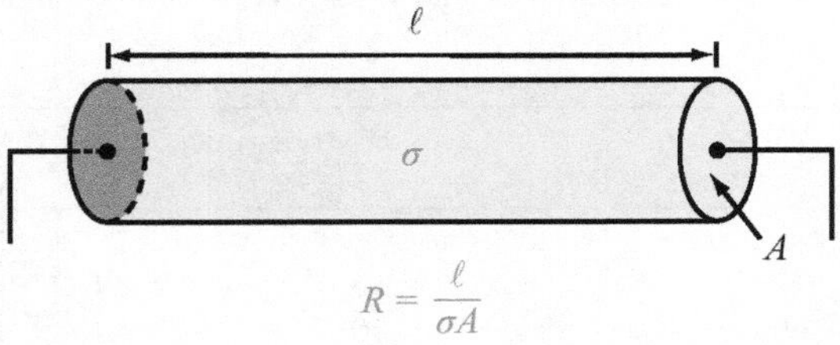

Figure 2-1: Longitudinal resistor of conductivity σ, length ℓ, and cross-sectional area A.

2-1.1 Resistance

▶ The *resistance* R of a device incorporates two factors: (a) the inherent bulk property of its material to conduct (or impede) current, represented by the conductivity σ (or resistivity ρ), and (b) the shape and size of the device. ◀

For a longitudinal resistor (Fig. 2-1), R is given by

$$R = \frac{\ell}{\sigma A} = \rho \frac{\ell}{A} \qquad (\Omega), \tag{2.2}$$

where ℓ is the length of the device and A is its cross-sectional area. In addition to its direct dependence on the resistivity ρ, R is directly proportional to ℓ, which is the length of the path that the current has to flow through, and inversely proportional to A, because the larger A is, the more electrons can drift through the material.

Every element of an electric circuit has a certain resistance associated with it. This even includes the wires used to connect devices to each other, but we usually treat them like zero-resistance segments because their resistances are so much smaller than the resistances of the other devices in the circuit. To illustrate with an example, let us consider a 10 cm long segment of one of the wire sizes commonly found in circuit boards, such as the AWG-18 copper wire. According to Table 2-2, which lists the diameter d for various wire sizes as specified by the *American Wire Gauge* (AWG) system, the AWG-18 wire has a diameter $d = 1$ mm. Using the values specified for ℓ and d and the value for ρ of copper given in Table 2-1, we have

$$R = \rho \frac{\ell}{A} = \rho \frac{\ell}{\pi (d/2)^2} = 1.72 \times 10^{-8} \times \frac{0.1}{\pi (0.5 \times 10^{-3})^2}$$
$$= 2.2 \times 10^{-3}\ \Omega = 2.2\ \text{m}\Omega.$$

Table 2-2: **Diameter d of wires, according to the American Wire Gauge (AWG) system.**

AWG Size Designation	Diameter d (mm)
0	8.3
2	6.5
4	5.2
6	4.1
10	2.6
14	1.6
18	1.0
20	0.8

▶ Thus, R of a 10 cm long AWG-18 copper wire is on the order of milliohms. If the wire segment connects to circuit elements with resistances of ohms or larger, ignoring the resistance of the wire would have no significant impact on the overall behavior of the circuit. ◀

The preceding justification should be treated with some degree of caution. While it is true that a piece of wire may be treated like a short circuit in the majority of circuit configurations, there are certain situations for which such an assumption may not be valid. One obvious example is when the wire is very long, as in the case of a kilometers long electric power-transmission cable. Another is when very thin wires or channels with micron-size diameters are used in microfabricated circuits.

Resistive elements used in electronic circuits are fabricated in many different sizes and shapes to suit the intended application and requisite circuit architecture. Discrete resistors usually are cylindrical in shape and made of a carbon composite. Hybrid and miniaturized circuits use film-shaped metal or carbon resistors. In integrated circuits, resistive elements are fabricated through a diffusion process (see Technology Brief 7).

Figure 2-2 displays photographs of three types of resistors, amongst which the tubular-shaped resistor is the most familiar. Resistors are generally marked with a banded color code to denote the resistor's specifications:

(a) 4-Band color code: b_1 b_2 b_4 b_5

Note that a wider spacing exists between b_4 and b_5 than between the earlier bands. The resistor value is given by

$$R = (b_1 b_2) \times 10^{b_4} \pm b_5,$$

with the values of b_1, b_2, b_4, and b_5 specified by the color code shown in Fig. 2-2. For example,

▮ ▮ ▮ ▮ $= 25 \times 10^0 \pm 10\% = 25 \pm 10\%\ \Omega$.

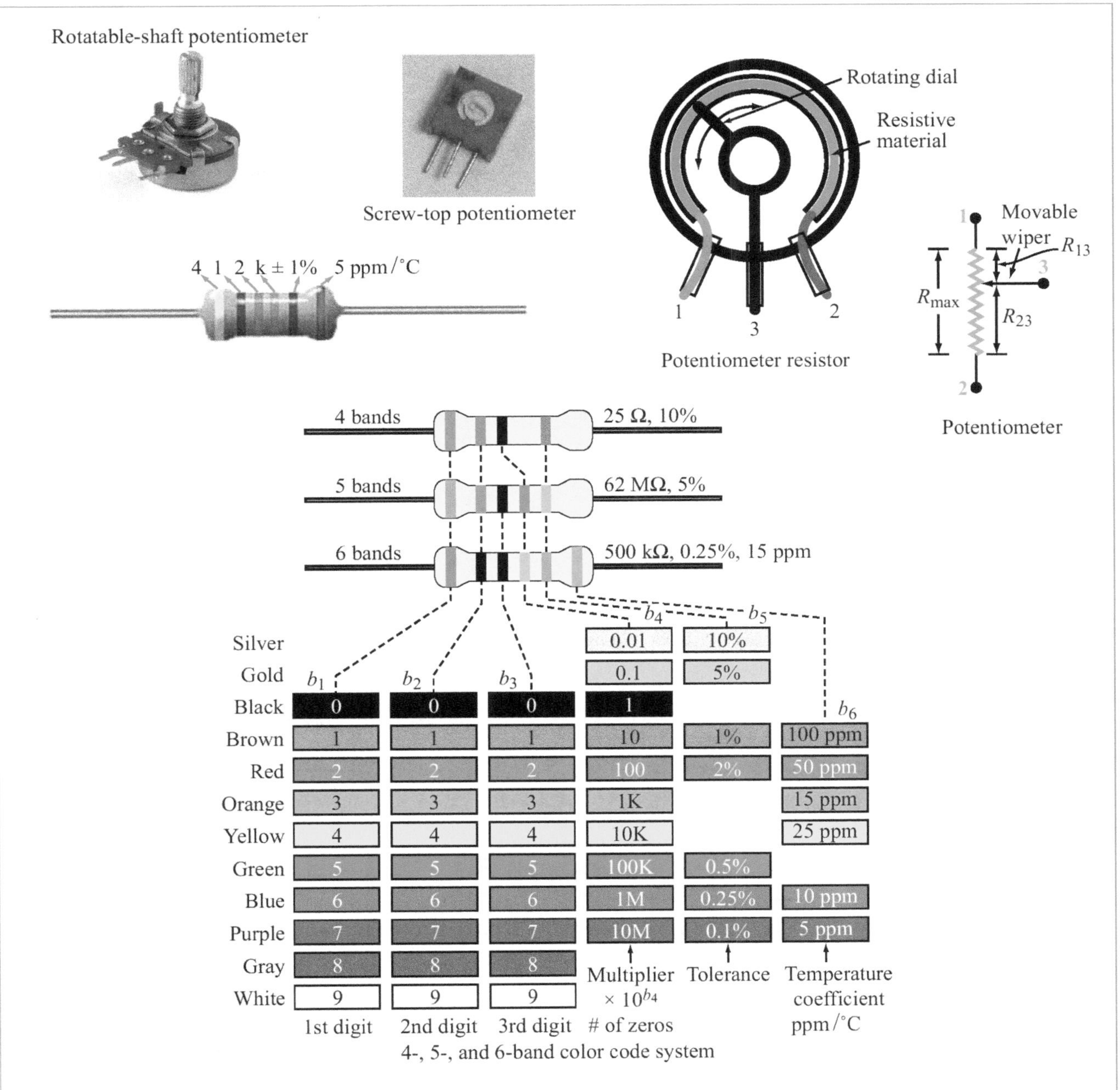

Figure 2-2: Various types of resistors. Tubular-shaped resistors usually are color-coded by 4-, 5-, or 6-band systems.

(b) 5-Band color code: b_1 b_2 b_3 b_4 b_5

In this case

$$R = (b_1 b_2 b_3) \times 10^{b_4} \pm b_5.$$

(c) 6-Band color code: b_1 b_2 b_3 b_4 b_5 b_6

This code adds one more piece of information in the form of b_6 which denotes the temperature coefficient of the resistor, measured in parts-per-million/ °C.

Table 2-3: Common resistor terminology.

Thermistor	R sensitive to temperature
Piezoresistor	R sensitive to pressure
Light-dependent R (LDR)	R sensitive to light intensity
Rheostat	2-terminal variable resistor
Potentiometer	3-terminal variable resistor

For some metal oxides, the resistivity ρ exhibits a strong sensitivity to temperature. A resistor manufactured of such materials is called a *thermistor* (Table 2-3), and it is used for temperature measurement, temperature compensation, and related applications. Another interesting type of resistor is the *piezoresistor*, which is used as a pressure sensor in many household appliances, automotive systems, and biomedical devices. More coverage on resistive sensors is available in Technology Brief 4.

Certain applications, such as volume adjustment on a radio, may call for the use of a resistor with *variable resistance*. The rheostat and the potentiometer are two standard types of variable resistors in common use. The *rheostat* [Fig. 2-3(a)] is a two-terminal device with one of its terminals connected to one end of a resistive track and the other terminal connected to a movable wiper. Movement of the wiper across the resistive track, through rotation of a shaft, can change the resistance between the two terminals from (theoretically) zero resistance to the full resistance value of the track. Thus, if the total resistance of the track is R_{max}, the rheostat can provide any resistance between zero and R_{max}.

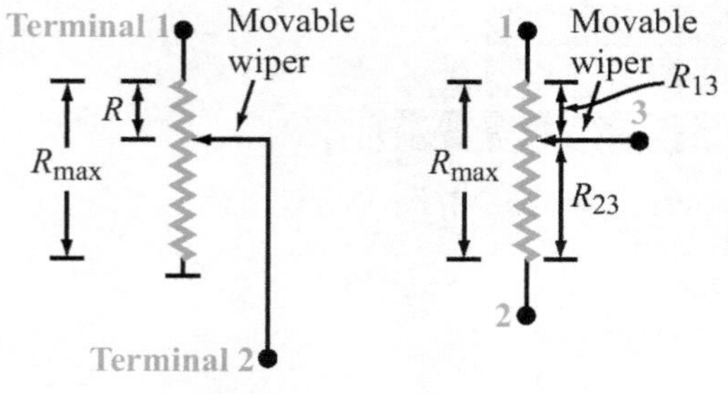

(a) Rheostat (b) Potentiometer

Figure 2-3: (a) A rheostat is used to set the resistance between terminals 1 and 2 at any value between zero and R_{max}; (b) the wiper in a potentiometer divides the resistance R_{max} among R_{13} and R_{23}.

The *potentiometer* is a three-terminal device. Terminals 1 and 2 in Fig. 2-3(b) are connected to the two ends of the track (with total resistance R_{max}) and terminal 3 is connected to a movable wiper. When terminal 3 is at the end next to terminal 1, the resistance between terminals 1 and 3 is zero and that between terminals 2 and 3 is R_{max}. Moving terminal 3 away from terminal 1 increases the resistance between terminals 1 and 3 and decreases the resistance between terminals 2 and 3. A potentiometer can be used as a rheostat by connecting to only terminals 1 and 3.

2-1.2 i–υ Characteristics of Ideal Resistor

Based on the results of his experiments on the nature of conduction in circuits, German physicist Georg Simon Ohm (1787–1854) formulated in 1826 the i–υ relationship for a resistor, which has become known as *Ohm's law*. He discovered that the voltage υ across a resistor is directly proportional to the current i flowing through it, namely

$$\upsilon = iR, \tag{2.3}$$

with the resistance R being the proportionality factor.

▶ In compliance with the passive sign convention, current enters a resistor at the "+" side of the voltage across it.

$$\upsilon_{ab} = \upsilon_a - \upsilon_b$$

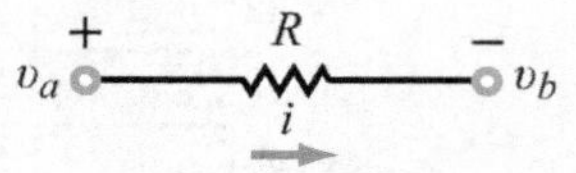

$$i = \frac{\upsilon_a - \upsilon_b}{R}$$ ◀

An ideal *linear resistor* is one whose resistance R is constant and independent of the magnitude of the current flowing through it, in which case its *i–υ response* is a straight line (Fig. 2-4(a)). In practice, the i–υ response of a real linear resistor is indeed approximately linear, as illustrated in Fig. 2-4(b), so long as i remains within the *linear region* defined by $-i_{max}$ to i_{max}. The slope of the curve is the resistance R. Outside this range, the response deviates from the straight-line model. When we use Ohm's law as expressed by Eq. (2.3), we tacitly assume that the resistor is being used in its linear range of operation.

Some resistive devices exhibit highly nonlinear i–υ characteristics. These include diode elements and light-bulb filaments, among others. Unless noted otherwise, the common use of the term *resistor* in circuit analysis and design usually refers to the linear resistor exclusively.

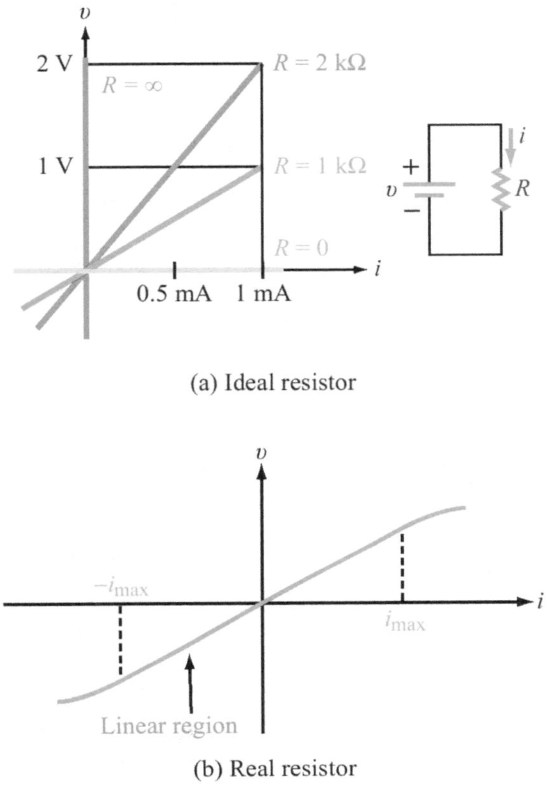

Figure 2-4: i–v responses of ideal and real resistors.

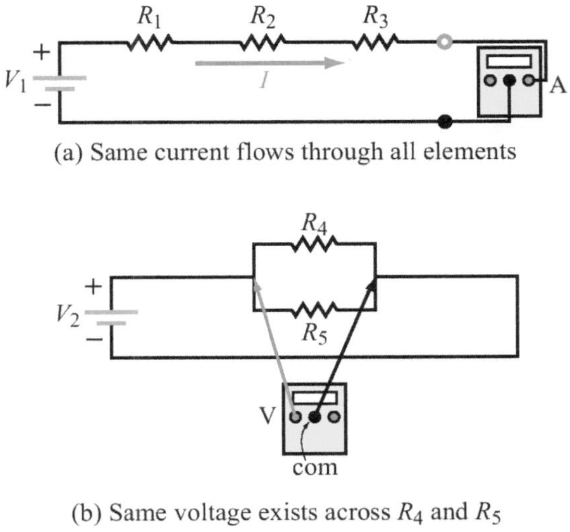
R_1
R_2
R_3
+
V_1
−
I
A
(a) Same current flows through all elements
R_4
+
V_2
−
R_5
V
com
(b) Same voltage exists across R_4 and R_5

Figure 2-5: In-series and in-parallel connections.

The flow of current in a resistor leads to power dissipation in the form of heat (or the combination of heat and light in the case of a light bulb's filament). Using Eq. (2.3) in Eq. (1.9) provides the following expression for the power p dissipated in a resistor:

$$p = iv = i^2R = \frac{v^2}{R} \quad \text{(W)}. \tag{2.4}$$

▶ The *power rating* of a resistor defines the maximum continuous power level that the resistor can dissipate without getting damaged. Excessive heat can cause melting, smoke, and even fire. ◀

For electric circuits with a fixed voltage (such as a 120 V for a house), the power rating refers to the maximum current limit. Current-limiting devices, such as fuses and circuit breakers, are used to protect against dangerous overloading of circuits.

2-1.3 In-Series and In-Parallel Connections

Recall from Chapter 1 that two or more elements are considered to be connected in series if the same current flows through all of them. This is indeed the case for voltage source V_1 and the resistors shown in Fig. 2-5(a). For two or more elements to be in parallel, they have to share the same voltage, which is the case for R_4 and R_5 in Fig. 2-5(b).

Example 2-1: Series Connection Resistances for a dc Motor

A 12 V car battery is connected via a 6 m long, twin-wire cable to a dc motor that drives the wiper blade on the rear window. The cable is copper AWG-10 and the motor exhibits to the rest of the circuit an equivalent resistance $R_m = 2\ \Omega$. Determine: (a) the resistance of the cable and (b) the fraction of the power contributed by the battery that gets delivered to the motor.

Solution: The circuit described in the problem statement is represented by Fig. 2-6.

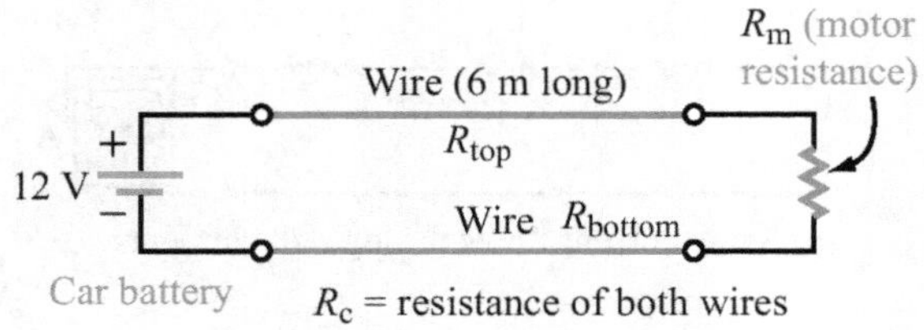

Figure 2-6: Circuit for Example 2-1.

(a) We need to include both the top wire and the bottom wire, as each represents a resistor through which the current flows, and therefore contributes to the resistive losses of the circuit. With $\ell = 12$ m (total for twin wires), $\rho = 1.72 \times 10^{-8}$ Ω-m for copper, $A = \pi(d/2)^2$, and $d = 2.6$ mm for AWG-10, the cable resistance R_c is

$$R_c = \rho \frac{\ell}{A} = 1.72 \times 10^{-8} \times \frac{12}{\pi(1.3 \times 10^{-3})^2} = 0.04\ \Omega.$$

(b) The total resistance in the circuit is equal to the sum of the cable and motor resistances. [In a later section, we will learn that the resistance of two resistors connected in series is simply equal to the sum of their resistances.] Hence,

$$R = R_c + R_m = 0.04 + 2 = 2.04\ \Omega.$$

Consequently, the current flowing through the circuit is

$$I = \frac{V}{R} = \frac{12}{2.04} = 5.88\ \text{A},$$

and the power contributed by the battery P and the power delivered to the motor P_m are:

$$P = IV = 5.88 \times 12 = 70.56\ \text{W}$$

and

$$P_m = I^2 R_m = (5.88)^2 \times 2 = 69.15\ \text{W},$$

and the fraction of P delivered to the load (motor) is

$$\text{Fraction} = \frac{P_m}{P} = \frac{69.15}{70.56} = 0.98 \text{ or } 98 \text{ percent.}$$

Thus, 2 percent of the power is dissipated in the cable.

> **Concept Question 2-1:** If the terminals of the battery in Fig. 2-6 were corroded, how would that change the problem and the results? (See CAD)

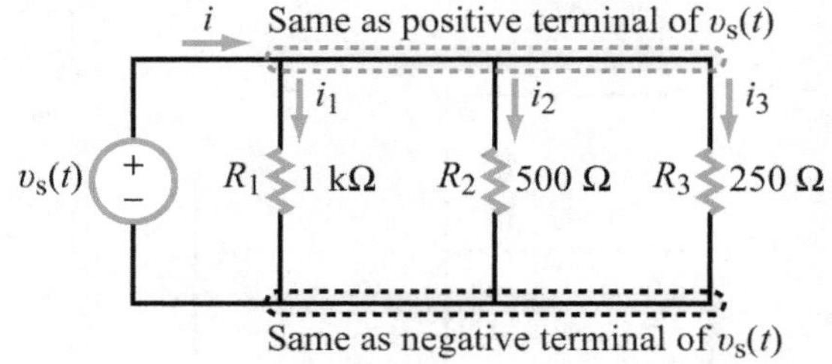

Figure 2-7: Circuit for Example 2-2.

Example 2-2: Parallel Loads

Three loads—a 1 kΩ light bulb, a 500 Ω computer, and a 250 Ω TV, each represented by a resistor, are connected in parallel to a household ac voltage source as shown in Fig. 2-7. The source is cosinusoidal in time at a frequency of 60 Hz and its amplitude is 170 V. Hence, it can be described as

$$v_s(t) = 170\cos(2\pi \times 60t) = 170\cos(377t)\ \text{V}.$$

Determine the currents supplied by the source to the three loads.

Solution: All three loads share the same positive terminal (node) of $v_s(t)$ on one end and the same negative terminal (node) on the other. Consequently, application of Ohm's law leads to

$$i_1(t) = \frac{v_s(t)}{R_1} = \frac{170}{10^3}\cos(377t) = 0.17\cos(377t)\ \text{A},$$

$$i_2(t) = \frac{v_s(t)}{R_2} = \frac{170}{500}\cos(377t) = 0.34\cos(377t)\ \text{A},$$

$$i_3(t) = \frac{v_s(t)}{R_3} = \frac{170}{250}\cos(377t) = 0.68\cos(377t)\ \text{A}.$$

As we see in the next section, the current i supplied by the source is the sum of the three load currents,

$$i(t) = i_1 + i_2 + i_3 = 1.19\cos(377t)\ \text{A}.$$

> **Concept Question 2-2:** How does the magnitude of the conductivity of a metal, such as copper, compare with that of a typical insulator, such as mica? (See CAD)

> **Concept Question 2-3:** What is piezoresistivity, and how is it used? (See CAD)

Technology Brief 3
Superconductivity

When an electric voltage is applied across two points in a conductor, such as copper or silver, current flows between them. The relationship between the voltage difference V and the current I is given by Ohm's law, $V = IR$, where R is the resistance of the conducting material between the two points. It is helpful to visualize the electric current as a fluid of electrons flowing through a dense forest of sturdy metal atoms, called the *lattice*. Under the influence of the electric field (induced by the applied voltage), the electrons can attain very high instantaneous velocities, but their overall forward progress is impeded by the frequent collisions with the lattice atoms. Every time an electron collides and bounces off an atom, some of that electron's kinetic energy is transferred to the atom, causing the atom to vibrate—which heats up the material—and causing the electron to slow down. The resistance R is a measure of how much of an obstacle the resistor poses to the flow of current, as well as a measure of how much heat it generates for a given current. The power dissipated in R is I^2R if I is a dc current, and it is $\frac{1}{2}\, I^2R$ if the current is ac with an amplitude I.

Can a conductor ever have zero resistance? The answer is most definitely yes! In 1911, the Dutch physicist Heike Kamerlingh Onnes developed a refrigeration technique so powerful that it could cool helium down low enough to condense it into liquid form at 4.2 K (0 kelvin $= -273$ °C). Into his new liquid helium container, he immersed (among other things) mercury; he soon discovered that the resistance of a solid piece of mercury at 4.2 K was *zero*! The phenomenon, which was completely unexpected and not predicted by classical physics, was coined *superconductivity*. According to quantum physics, many materials experience an abrupt change in behavior (called a *phase transition*) when cooled below a certain *critical temperature* T_C.

Superconductors have some amazing properties. The current in a superconductor can persist with no external voltage applied. Even more interesting, currents have been observed to persist in superconductors for many years without decaying. When a magnet is brought close to the surface of a superconductor, the currents induced by the magnetic field are mirrored exactly by the superconductor (because the superconductor's resistance is zero), and consequently the magnet is repelled (Fig. TF3-1). This property has been used to demonstrate magnetic levitation and is the basis of some super-fast *maglev trains* (Fig. TF3-2) being used around the world. The same phenomenon is used in the *Magnetic Resonance Imaging* (MRI) machines that hospitals use to perform 3-D scans of organs and tissues (Fig. TF3-3) and in *Superconducting Quantum Interference Devices* (SQUIDs) to examine brain activity at high resolution.

Figure TF3-1: The Meissner effect, or strong diamagnetism, seen between a high-temperature superconductor and a rare earth magnet. (Courtesy of Pacific Northwest National Laboratory.)

Figure TF3-2: Maglev train. (Courtesy of Central Japan Railway Company.)

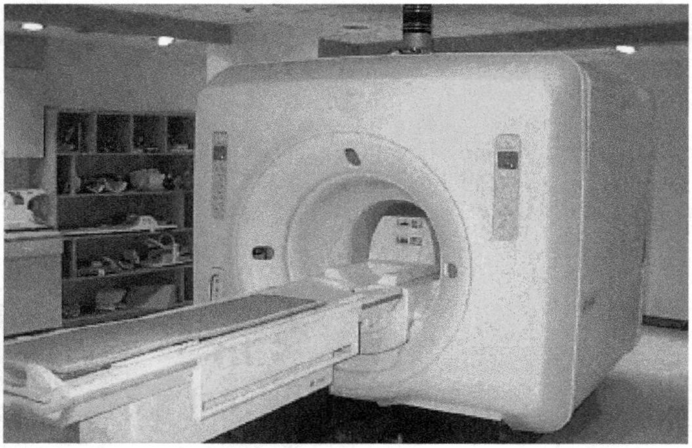

Figure TF3-3: Magnetic Resonance Imaging machine. (Courtesy GE Healthcare.)

Superconductivity is one of the last frontiers in solid-state physics (see Table TT3-1). Even though the physics of low-temperature superconductors (like mercury, lead, niobium nitride, and others) is now fairly well understood, a different class of ***high-temperature superconductors*** still defies complete theoretical explanation. This class of materials was discovered in 1986 when Alex Müller and Georg Bednorz, at IBM Research Laboratory in Switzerland, created a ceramic compound that superconducted at 30 K. This discovery was followed by the discovery of other ceramics with even higher T_C values; the now-famous YBCO ceramic discovered at the University of Alabama-Huntsville (1987) has a T_C of 92 K, and the world record holder is a group of mercury-cuprate compounds with a T_C of 138 K (1993). New superconducting materials and conditions are still being found; carbon nanotubes, for example, were recently shown to have a T_C of 15 K (Hong Kong University, 2001). Are there higher-temperature superconductors? What theory will explain this higher-temperature phenomenon? Can so-called *room-temperature* superconductors exist? For engineers (like you) the challenges are just beginning: How can these materials be made into useful circuits, devices, and machines? What new designs will emerge? The race is on!

Table TT3-1: Critical temperatures.

Critical Temperature T_c [K]	Material	Type
138	$HgBa_2Ca_2Cu_3O_x$	Copper-oxide superconductors
138	$Bi_2Sr_2Ca_2Cu_3O_{10}$ (BSCCO)	
92	$YBa_2Cu_3O_7$ (YBCO)	
55	SmFeAs	Iron-based superconductors
41	CeFeAs	
26	LaFeAs	
18	Nb_3Sn	Metallic low-temperature superconductors
10	NbTi	
9.2	Nb	
4.2	Hg (mercury)	

Concept Question 2-4: What is meant by the *linear region* of a resistor? Is it related to its power rating? (See CAD)

Exercise 2-1: A cylindrical resistor made of carbon has a power rating of 2 W. If its length is 10 cm and its circular cross section has a diameter of 1 mm, what is the maximum current that can flow through the resistor without damaging it?

Answer: 1.06 A. (See CAD)

Exercise 2-2: A rectangular bar made of aluminum has a current of 3 A flowing through it along its length. If its length is 2.5 m and its square cross section has 1 cm sides, how much power is dissipated in the bar at 20 °C?

Answer: 5.9 mW. (See CAD)

2-1.4 *i–υ* Characteristics of LEDs

A resistor is a bidirectional device, meaning that current can flow through it in either direction. This is because it is constructed of the same material along the dimension between its two terminals. In contrast, a diode allows current to flow in only one direction. It is built of two sections of different semiconductor materials, denoted p and n in Fig. 2-8(a). The p-type material has excess positive charges and the n-type material has excess negative charges. When connected to a voltage source, the diode acts like a resistor in one direction, but like an open circuit in the other. Specifically:

(a) *Reverse bias*: If the voltage V_D applied across the diode is negative (relative to its own terminals), as shown in Fig. 2-8(b), no current flows through it, which is equivalent to having infinite resistance. That is, the diode behaves like an open circuit.

(b) *Forward bias*: If the voltage V_D is positive, as in Fig. 2-8(c), current will flow through the diode, but the relationship between I and V_D is not a constant. For a

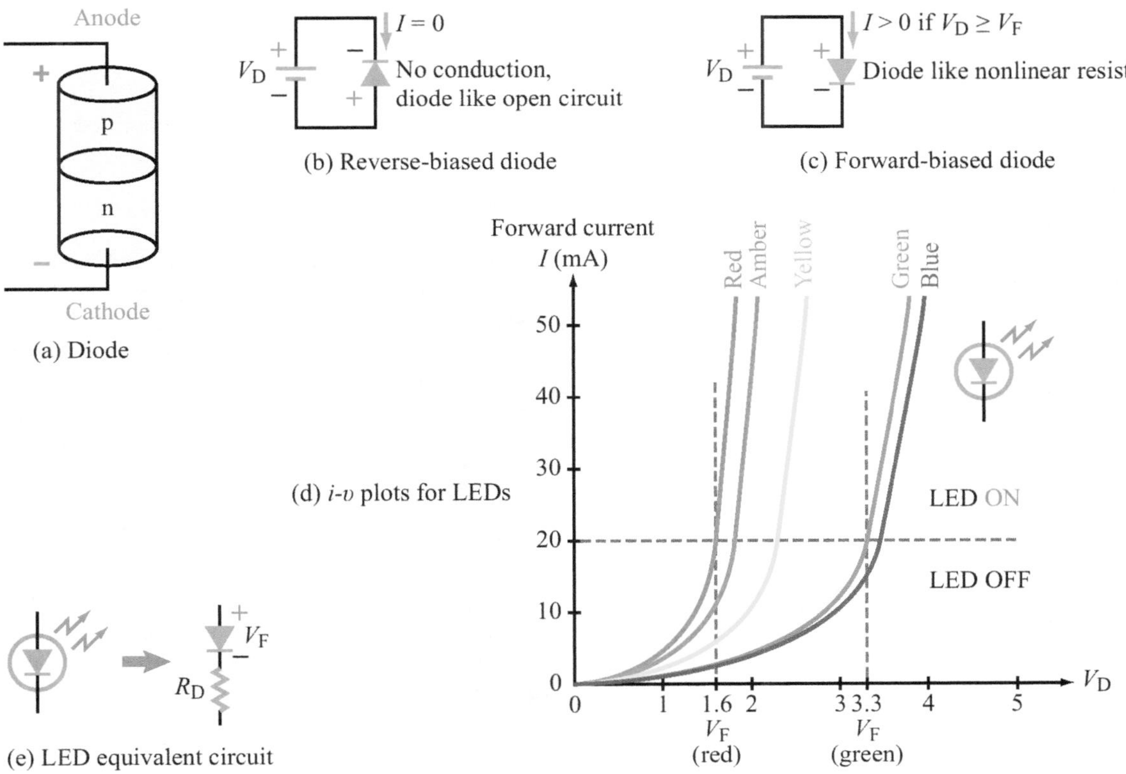

Figure 2-8: p-n junction diode (a) configuration, (b) reverse biased, (c) forward biased, (d) typical *i-υ* plots for LEDs, and (e) LED equivalent circuit.

resistor, $V_D/I = R$ and R is a constant, but for a diode the relationship between V_D and I is more complicated. However, its i–υ relationship can be approximated by

$$I = aV_D^2 \qquad (V_D > 0),$$

where a is a constant that depends on the semiconductor material used to build the LED.

A light-emitting diode is a special kind of diode in that it emits light if I exceeds a certain threshold. Figure 2-8(d) displays plots of I versus V_D for five LEDs of different colors. The color of light emitted by an LED depends on the semiconductor compounds from which it is constructed. The voltage at which the diode becomes approximately linear is the forward bias voltage V_F, and it becomes part of the diode model shown in Fig. 2-8(e). For the typical family of LEDs shown in Fig. 2-8(d), the current I has to exceed 20 mA in order for the LED to fully light up. This current threshold has a corresponding voltage threshold called the *forward voltage* V_F. Below this threshold, the diode conducts little or no current and is considered "OFF" (although it does generate a small amount of light). For the red LED, for example, $V_F = 1.6$ V, and the current flowing through the LED at that voltage is exactly 20 mA. Higher values of V_F are required to cause the LEDs of the other colors to emit light.

When we analyze a circuit containing an LED, the LED can be modeled as an ideal diode with a voltage drop of V_F in series with a small internal resistance R_D, as shown in Fig. 2-8(e). We can determine the approximate LED resistance R_D from the slope of the linear section (above V_F) of the i–υ curve in Fig. 2-8(d); i.e., $R_D \approx \Delta V_D/\Delta I$.

Exercise 2-3: A certain type of diode exhibits a nonlinear relationship between υ—the voltage across it—and i—the current entering into its (+) voltage terminal. Over its operational voltage range (0 to 1 V), the current is given by

$$i = 0.5\upsilon^2 \qquad \text{for } 0 \le \upsilon \le 1 \text{ V}.$$

Determine how the diode's effective resistance varies with υ and calculate its value at $\upsilon = 0$, 0.01 V, 0.1 V, 0.5 V, and 1 V.

Answer: $R = \dfrac{2}{\upsilon}$, (See CAD)

υ	R
0	∞
0.01 V	200 Ω
0.1 V	20 Ω
0.5 V	4 Ω
1 V	2 Ω

2-1.5 Conductance

The reciprocal of resistance is called *conductance*,

$$G = \frac{1}{R} \qquad \text{(S)}, \tag{2.5}$$

and its unit is Ω^{-1}, which is called the *siemen* (S, or sometimes called "mho"). In terms of G, Ohm's law can be rewritten in the form

$$i = \frac{\upsilon}{R} = G\upsilon, \tag{2.6}$$

and the expression for power becomes

$$p = i\upsilon = G\upsilon^2 \qquad \text{(W)}. \tag{2.7}$$

Since $G = 1/R$, what is the point in dealing with both G and R? The answer is: convenience. In some circuit solutions it is easier to work with R for all resistors in the circuit, whereas in other circuit configurations (especially those in parallel) it may be easier to work with conductances instead.

2-2 Kirchhoff's Laws

Circuit theory—encompassing both analysis and synthesis—is built upon a foundation comprised of a small number of fundamental laws. Among the cornerstones are *Kirchhoff's current and voltage laws*. Kirchhoff's laws, which constitute the subject of this section, were introduced by the German physicist Gustav Robert Kirchhoff (1824–1887) in 1847, some 21 years after a fellow German, Georg Simon Ohm, developed his famous law.

2-2.1 Kirchhoff's Current Law (KCL)

As defined earlier, a node is a connection point for two or more branches. As such, it is not a real circuit element, and therefore it cannot generate, store, or consume electric charge. This assertion, which follows from the *law of conservation of charge*, forms the basis of *Kirchhoff's current law* (KCL), which states that:

▶ The algebraic sum of the currents entering a node must always be zero. ◀

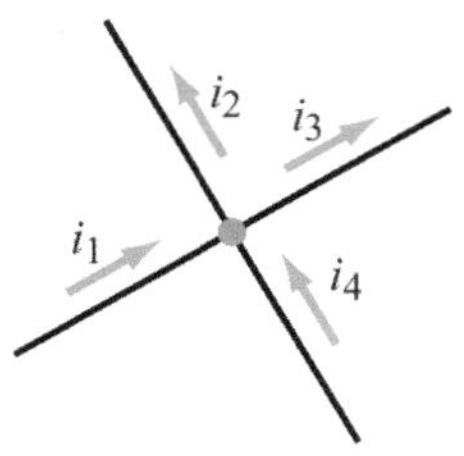

Figure 2-9: Currents at a node.

Mathematically, KCL can be expressed by the compact form:

$$\sum_{n=1}^{N} i_n = 0 \qquad \text{(KCL)}, \tag{2.8}$$

where N is the total number of branches connected to the node, and i_n is the nth current.

▶ A common convention is to assign a positive "+" sign to a current if it is entering the node and a negative "−" sign if it is leaving it. ◀

For the node in Fig. 2-9, the sum of currents entering the node is

$$i_1 - i_2 - i_3 + i_4 = 0, \tag{2.9}$$

where currents i_1 and i_4 were assigned positive signs because they are labeled in the figure as entering the node, and i_2 and i_3 were assigned negative signs because they are leaving the node.

▶ Alternatively, the sum of currents leaving a node is zero, in which case we assign a "+" to a current leaving the node and a "−" to a current entering it. ◀

Either convention is equally valid so long as it is applied consistently to all currents entering and leaving the node.

By moving i_2 and i_3 to the right-hand side of Eq. (2.9), we obtain the alternative form of KCL, namely

$$i_1 + i_4 = i_2 + i_3, \tag{2.10}$$

which states that:

▶ The total current entering a node must be equal to the total current leaving it. ◀

How do we know which way a current is flowing in a circuit? Often, we do not. So, we guess by assigning a direction to each current, and then applying Kirchhoff's laws to compute the currents. If the value for a particular current is a positive number, then our guess was correct, but if it is a negative number, then the direction of the current is opposite the one we assigned it.

Example 2-3: KCL Equations

Write the KCL equations at nodes 1 through 5 in the circuit of Fig. 2-10.

Solution:

At node 1:	$-I_1 - I_3 + I_5 = 0$
At node 2:	$I_1 - I_2 + 2 = 0$
At node 3:	$-2 - I_4 + I_6 = 0$
At node 4:	$-5 - I_5 - I_6 = 0$
At node 5:	$I_3 + I_4 + I_2 + 5 = 0$

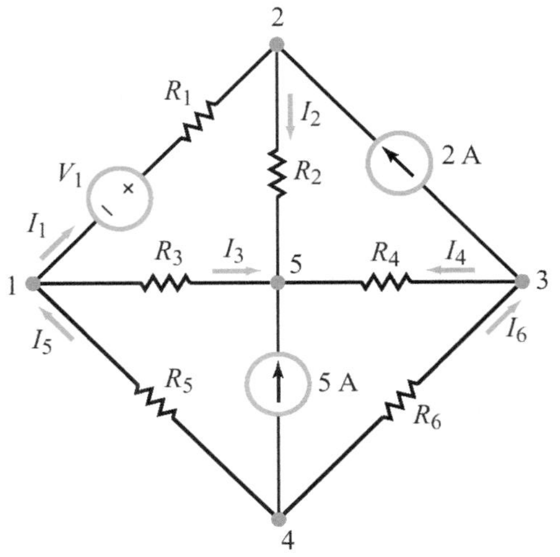

Figure 2-10: Circuit for Example 2-3.

Example 2-4: Applying KCL

If V_4, the voltage across the 4 Ω resistor in Fig. 2-11, is 8 V, determine I_1 and I_2.

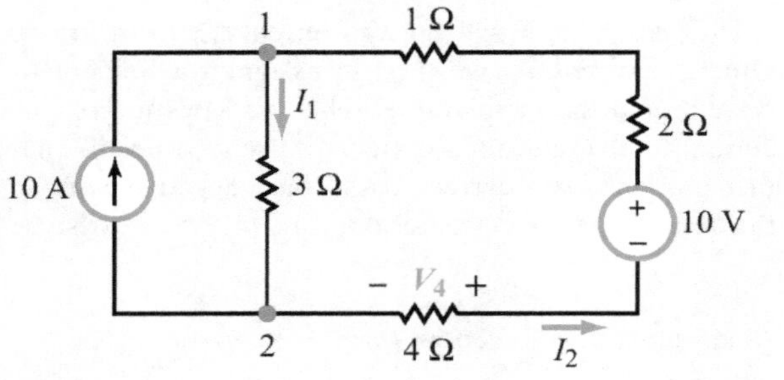

Figure 2-11: Circuit for Example 2-4.

Solution: The designated direction of I_2 is such that it enters the negative (−) terminal of V_4, whereas according to Ohm's law, the current should enter through the positive (+) terminal of the voltage across a resistor. Hence, in the present case, we should include a negative sign in the relationship between I_2 and V_4, namely

$$I_2 = -\frac{V_4}{4} = -\frac{8}{4} = -2 \text{ A}.$$

Thus, the true direction of the current flowing through the 4 Ω resistor is opposite of that of I_2.

Using the KCL convention that defines a current as positive if it is leaving a node and negative if it is entering it, at node 2:

$$10 - I_1 + I_2 = 0,$$

which leads to

$$I_1 = 10 + I_2 = 10 - 2 = 8 \text{ A}.$$

2-2.2 Kirchhoff's Voltage Law (KVL)

The voltage across an element represents the amount of energy expended in moving positive charge from the negative terminal to the positive terminal, thereby establishing a potential energy difference between those terminals. The *law of conservation of energy* mandates that if we move electric charge around a closed loop, starting and ending at exactly the same location, the net gain or loss of energy must be zero. Since voltage is a surrogate for potential energy:

> ▶ The algebraic sum of the voltages around a closed loop must always be zero. ◀

This statement defines *Kirchhoff's voltage law* (KVL). In equation form, KVL is given by

$$\sum_{n=1}^{N} \upsilon_n = 0 \qquad \text{(KVL)}, \tag{2.11}$$

where N is the total number of branches in the loop and υ_n is the nth voltage across the nth branch. Application of Eq. (2.11) requires the specification of a sign convention to use with it. Of those used in circuit analysis, the sign convention we chose to use in this book consists of two steps.

Sign Convention

- Add up the voltages in a systematic clockwise movement around the loop.
- Assign a positive sign to the voltage across an element if the (+) side of that voltage is encountered first, and assign a negative sign if the (−) side is encountered first.

Hence, for the loop in Fig. 2-12, starting at the negative terminal of the 4 V voltage source, application of Eq. (2.11) yields

$$-4 + V_1 - V_2 - 6 + V_3 - V_4 = 0. \tag{2.12}$$

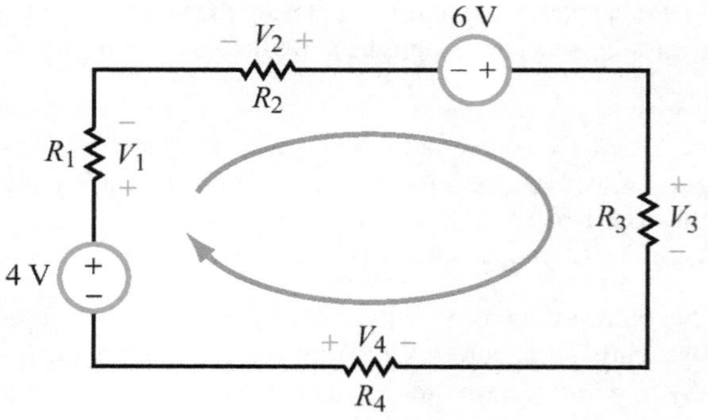

Figure 2-12: One-loop circuit.

Table 2-4: Equally valid, multiple statements of Kirchhoff's Current Law (KCL) and Kirchhoff's Voltage Law (KVL).

KCL	• Sum of all currents entering a node = 0 [i = "+" if entering; i = "−" if leaving] • Sum of all currents leaving a node = 0 [i = "+" if leaving; i = "−" if entering] • Total of currents entering = Total of currents leaving
KVL	• Sum of voltages around closed loop = 0 [υ = "+" if + side encountered first in clockwise direction] • Total voltage rise = Total voltage drop

▶ An alternative statement of KVL is that *the total voltage rise around a closed loop must equal the total voltage drop around the loop.* ◀

Recalling that a voltage rise is realized by moving from the (−) voltage terminal to the (+) terminal across the element, and voltage drop is the converse of that, the clockwise movement around the loop in Fig. 2-12 gives

$$4 + V_2 + 6 + V_4 = V_1 + V_3, \tag{2.13}$$

which mathematically conveys the same information contained in Eq. (2.12).

Table 2-4 provides a summary of KCL and KVL statements.

Concept Question 2-5: Explain why KCL is (in essence) a statement of the law of conservation of charge. (See CAD)

Concept Question 2-6: Explain why KVL is a statement of conservation of energy. What sign convention is used with KVL? (See CAD)

Example 2-5: Applying KVL

Determine the value of current I in the circuit of Fig. 2-13(a).

Solution: For the specified direction of I, we designate voltages V_1, V_2, and V_3 across the three resistors, as shown in Fig. 2-13(b). In each case, the positive polarity of the voltage across a resistor is placed at the terminal at which the current enters the resistor.

Starting at the negative terminal of the 12 V voltage source and moving clockwise around the loop, KVL gives

$$-12 + V_1 + V_2 + V_3 = 0.$$

By Ohm's law, $V_1 = 10I$, $V_2 = 20I$, and $V_3 = 30I$. Hence,

$$-12 + 10I + 20I + 30I = 0,$$

which leads to

$$60I = 12,$$

or

$$I = \frac{12}{60} = 0.2 \text{ A}.$$

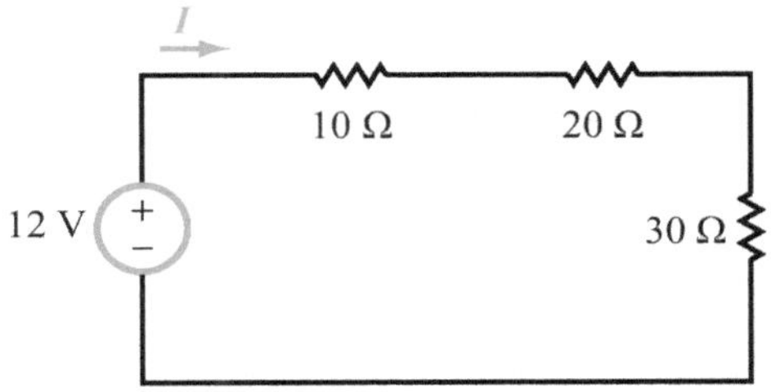

(a) Circuit for Example 2-5

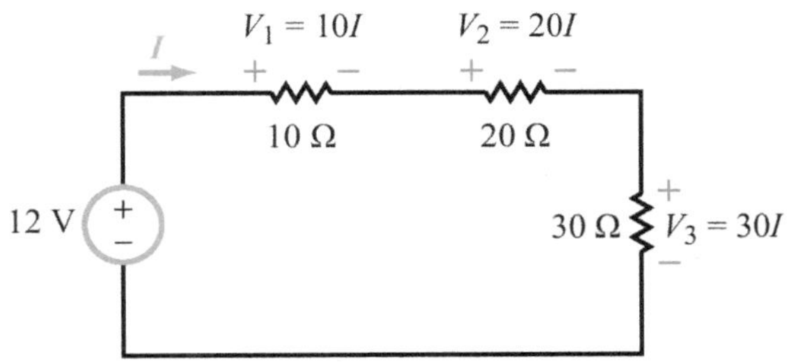

(b) After labeling voltages across resistors

Figure 2-13: Circuit for Example 2-5 before and after labeling voltages across the three resistors with polarities consistent with Ohm's law.

KCL/KVL Solution Recipe

- Use KCL, KVL, and Ohm's law to develop as many independent equations as the number of unknowns (N).

  (a) Write as many KVL loop equations as you can, picking up at least one additional circuit element for each loop. Let M be the number of such loop equations. Exclude loops that go through current sources.

  (b) Write $(N - M)$ KCL equations, making sure each node picks up an additional current.

- Write the equations in standard form (see Eq. (B.2) in Appendix B).
- Cast the standard-form equations in matrix form, as in Eqs. (B.19) and (B.20) of Appendix B.
- Apply matrix inversion to compute the values of the circuit unknowns (Appendix B).

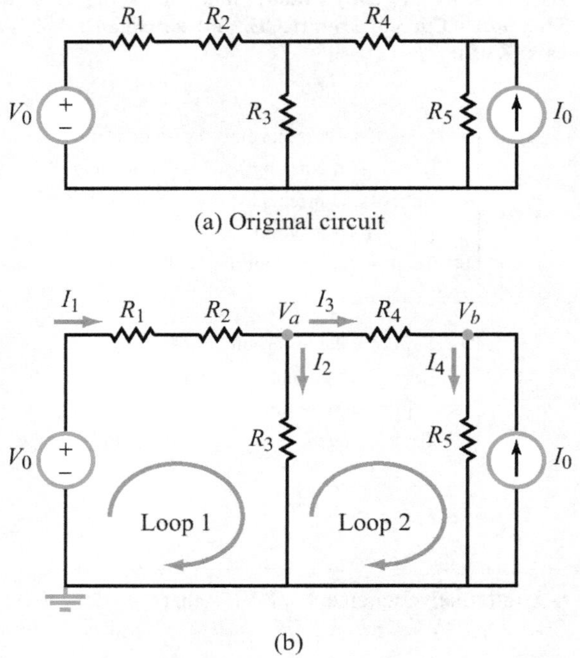

Figure 2-14: Circuit for Example 2-6.

Example 2-6: Matrix Inversion of KVL/KCL Equations

For the circuit in Fig. 2-14(a): (a) identify all N unknown branch currents and assign them preliminary directions, (b) develop M KVL loop equations through all possible elements (while excluding loops containing current sources), (c) develop $(N - M)$ KCL node equations, (d) arrange the equations in matrix form, (e) solve by matrix reduction to find the unknown branch currents, (f) determine the power dissipated in R_5, and (g) find the voltages of all extraordinary nodes relative to the negative terminal of the voltage source. The element values are: $V_0 = 10$ V, $I_0 = 0.8$ A, $R_1 = 2\ \Omega$, $R_2 = 3\ \Omega$, $R_3 = 5\ \Omega$, $R_4 = 10\ \Omega$, and $R_5 = 2.5\ \Omega$.

Solution:

(a) Identify unknown currents

Excluding the branch containing I_0 (since we know that the current in that branch is $I_0 = 0.8$ A), we have 4 unknown branch currents, which we denote I_1 to I_4 in Fig. 2-14(b). Also, with the negative terminal of the voltage source denoted as the voltage reference (ground), we have two extraordinary nodes, with designated voltages V_a and V_b.

(b) KVL equations

The circuit contains two independent loops that do not contain the current source I_0. The associated KVL equations are:

$$-V_0 + I_1R_1 + I_1R_2 + R_3I_2 = 0 \qquad \textbf{(Loop 1)},$$

$$-I_2R_3 + I_3R_4 + I_4R_5 = 0 \qquad \textbf{(Loop 2)}.$$

Alternatively, we can replace either of the two loop equations with the KVL equation for the perimeter loop that includes both of them, namely the loop that starts at the ground node, then goes clockwise through V_0, R_1, R_2, R_4, and R_5, and back to the ground node. Either approach leads to the same final result.

(c) KCL equations

We have two extraordinary nodes (in addition to the ground node). We designate their voltages as shown in Fig. 2-14(b). With current defined as positive when entering a node, their KCL equations are

$$I_1 - I_2 - I_3 = 0 \qquad \textbf{(Node } a\textbf{)},$$

$$I_3 - I_4 + I_0 = 0 \qquad \textbf{(Node } b\textbf{)}.$$

(d) Arrange equations in matrix form

$$\underbrace{\begin{bmatrix} (R_1+R_2) & R_3 & 0 & 0 \\ 0 & -R_3 & R_4 & R_5 \\ 1 & -1 & -1 & 0 \\ 0 & 0 & 1 & -1 \end{bmatrix}}_{\mathbf{A}} \underbrace{\begin{bmatrix} I_1 \\ I_2 \\ I_3 \\ I_4 \end{bmatrix}}_{\mathbf{I}} = \underbrace{\begin{bmatrix} V_0 \\ 0 \\ 0 \\ -I_0 \end{bmatrix}}_{\mathbf{B}}.$$

This is in the form

$$\mathbf{AI} = \mathbf{B}.$$

(e) Matrix inversion

After replacing the sources and resistors with their specified numerical values, matrix reduction, per MATLAB, MathScript, or the procedure outlined in Appendix B-2, leads to

$$I_1 = 1.1 \text{ A}, \qquad I_2 = 0.9 \text{ A},$$
$$I_3 = 0.2 \text{ A}, \qquad I_4 = 1 \text{ A}, .$$

(f) Power in R_5

$$P = I_4^2 R_5 = 1^2 \times 2.5 = 2.5 \text{ W}.$$

(g) Node voltages

$$V_a = I_2 R_3 = 0.9 \times 5 = 4.5 \text{ V},$$
$$V_b = I_4 R_5 = 1 \times 2.5 = 2.5 \text{ V}.$$

Example 2-7: Two-Source Circuit

Determine V_{ab} in the circuit of Fig. 2-15(a).

Solution: The circuit contains two independent loops and two extraordinary nodes, which we label node 1 and node 2 in Fig. 2-15(b). At extraordinary node 1, we assign currents I_1, I_2, and I_3. Their directions are chosen arbitrarily; for I_1, for example, if the solution yields a positive value, then the direction we assigned it is indeed the correct direction, and if the solution yields a negative value, then its true direction is the opposite of what we had assigned it.

Once the directions of I_1 to I_3 are specified at node 1, continuity of current automatically specifies their directions at node 2, as shown in Fig. 2-15(b). Since we have 3 unknowns (I_1, I_2, and I_3), we need $N = 3$ equations.

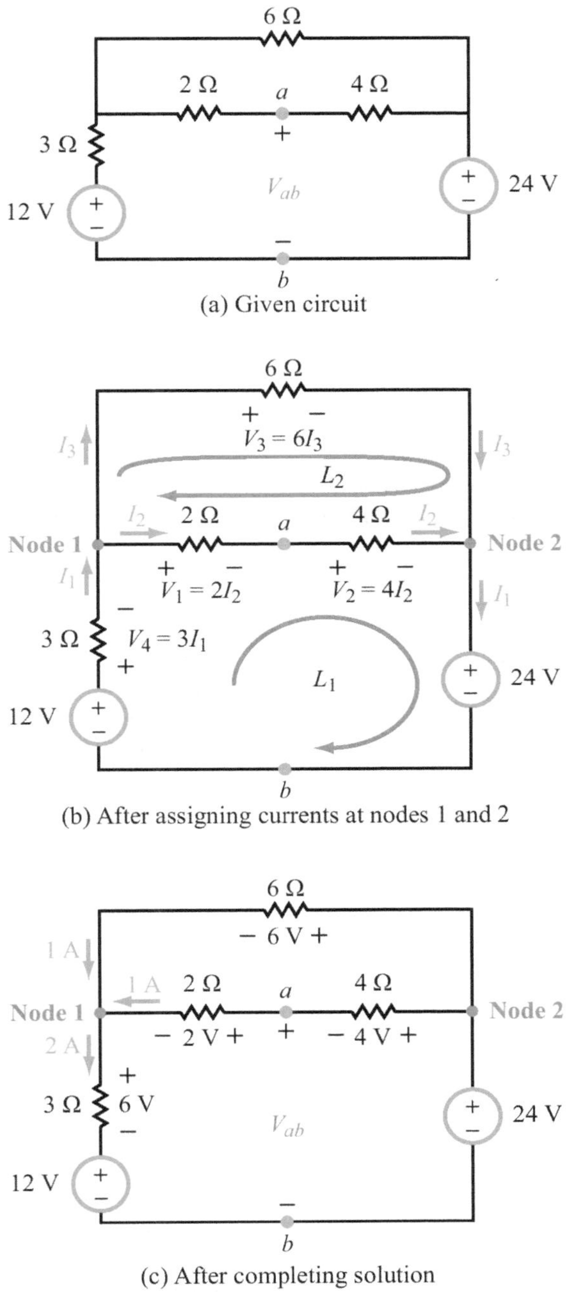

Figure 2-15: Circuit for Example 2-7.

In terms of the labeled voltages, application of KVL around the two loops gives

$$-12 + V_4 + V_1 + V_2 + 24 = 0, \qquad \text{(KVL for Loop 1)} \qquad (2.14a)$$

$$V_3 - V_2 - V_1 = 0. \qquad \text{(KVL for Loop 2)} \qquad (2.14b)$$

Using Ohm's law for the four resistors, the two KVL equations become

$$-12 + 3I_1 + 2I_2 + 4I_2 + 24 = 0, \qquad \text{(KVL for Loop 1)} \qquad (2.15a)$$

$$6I_3 - 4I_2 - 2I_2 = 0. \qquad \text{(KVL for Loop 2)} \qquad (2.15b)$$

The two simultaneous equations contain three unknowns, namely I_1 to I_3. A third equation is supplied by KCL at node 1 or node 2:

$$I_1 = I_2 + I_3. \qquad \text{(KCL @ node 1 or 2)} \qquad (2.16)$$

Equations (2.15a), (2.15b), and (2.16) constitute 3 equations in 3 unknowns. We can solve for I_1 to I_3 either by the substitution method or by matrix inversion (Appendix B). To apply the latter, we need to cast the three equations in standard form:

$$\begin{aligned} 3I_1 + 6I_2 \qquad &= -12, \\ -6I_2 + 6I_3 &= 0, \\ I_1 - I_2 - I_3 &= 0. \end{aligned}$$

In matrix form:

$$\begin{bmatrix} 3 & 6 & 0 \\ 0 & -6 & 6 \\ 1 & -1 & -1 \end{bmatrix} \begin{bmatrix} I_1 \\ I_2 \\ I_3 \end{bmatrix} = \begin{bmatrix} -12 \\ 0 \\ 0 \end{bmatrix}.$$

Matrix inversion, as outlined in Appendix B, leads to

$$I_1 = -2\text{ A}, \qquad I_2 = -1\text{ A}, \qquad I_3 = -1\text{ A}.$$

Hence, the true directions of the three currents are exactly opposite those we supposed , and so are the polarities of the voltages across the resistors. Incorporating both the calculated magnitudes and signs of I_1 to I_3 leads to the diagram shown in Fig. 2-15(c). To calculate V_{ab}, we start at node b and move clockwise towards node a in loop 1, while keeping track of voltage rises and drops. From node b to the (+) terminal of the 12 V source is a voltage rise of 12 V, from there to node 1 is a voltage rise of 6 V, and from node 1 to node b is a third voltage rise of 2 V. Hence

$$V_{ab} = 12 + 6 + 2 = 20\text{ V}.$$

Alternatively, we can calculate V_{ab} by moving from node b to node a counterclockwise through node 2. In that case

$$V_{ab} = 24 - 4 = 20\text{ V},$$

which is identical to the earlier result.

Example 2-8: Circuit with Dependent Source

The circuit in Fig. 2-16 includes a current-dependent voltage source. Apply KVL and KCL to determine the amount of power consumed by the 12 Ω resistor.

Solution: We start by assigning currents I_2 and I_3 at node 1, and using those currents to designate the voltages across the three resistors. Note that in all cases, the designated (+) side of

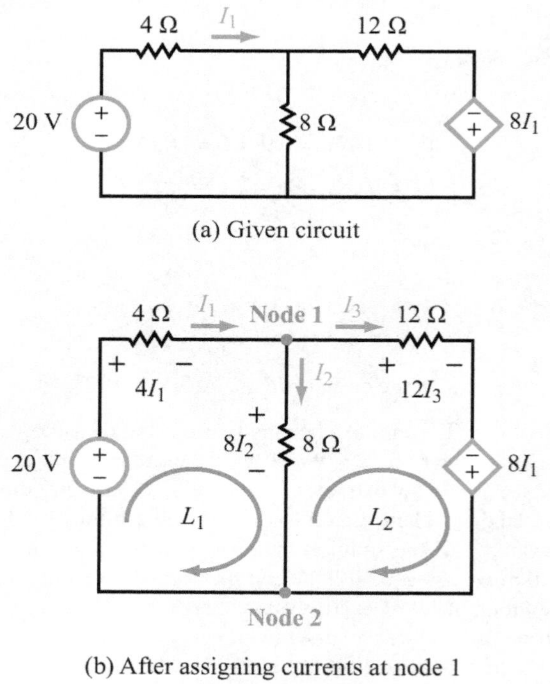

Figure 2-16: Circuit for Example 2-8.

the voltage corresponds to the terminal at which the current is entering.

For loops 1 and 2, KVL gives

$$-20 + 4I_1 + 8I_2 = 0, \qquad \text{(KVL for Loop 1)}$$
$$-8I_2 + 12I_3 - 8I_1 = 0. \qquad \text{(KVL for Loop 2)}$$

Note that there is another loop in the circuit, namely the perimeter loop around the whole circuit, but if we write a KVL equation for that loop, it would not provide an equation independent of the other loop equations because it would not include any circuit element not already included in loops L_1 and L_2.

At node 1, KCL states that

$$I_1 = I_2 + I_3.$$

The combination of the three equations in unknowns I_1, I_2, and I_3 leads to the solution

$$I_1 = \frac{25}{7} \text{ A},$$
$$I_2 = \frac{5}{7} \text{ A},$$
$$I_3 = \frac{20}{7} \text{ A}.$$

Hence, the power dissipated in the 12 Ω resistor is

$$P = I_3^2 R = \left(\frac{20}{7}\right)^2 \times 12 = 97.96 \text{ W}.$$

Exercise 2-4: If $I_1 = 3$ A in Fig. E2.4, what is I_2?

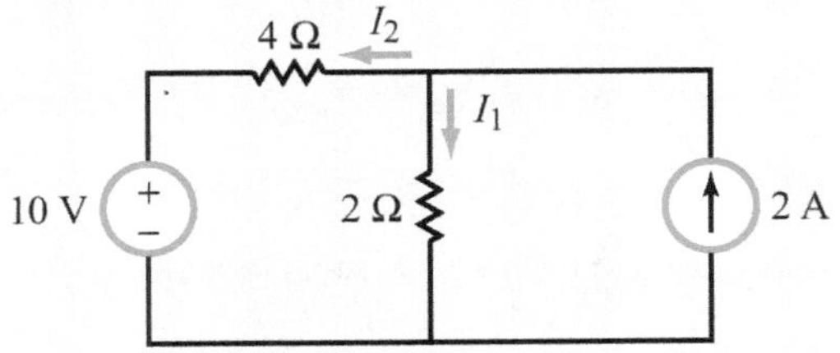

Figure E2.4

Answer: $I_2 = -1$ A. (See CAD)

Exercise 2-5: Apply KCL and KVL to find I_1 and I_2 in Fig. E2.5.

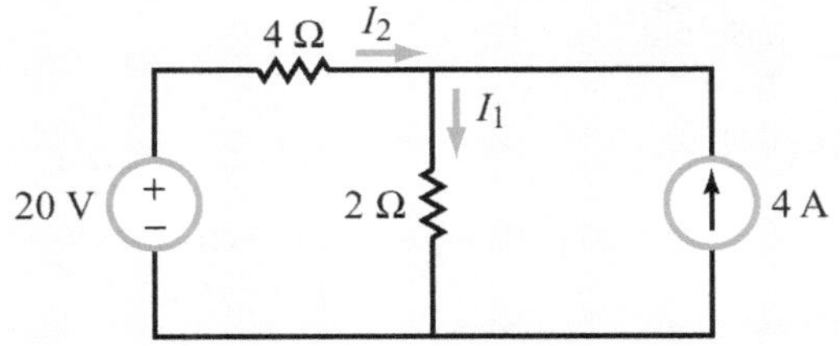

Figure E2.5

Answer: $I_1 = 6$ A, $I_2 = 2$ A. (See CAD)

Exercise 2-6: Determine I_x in the circuit of Fig. E2.6.

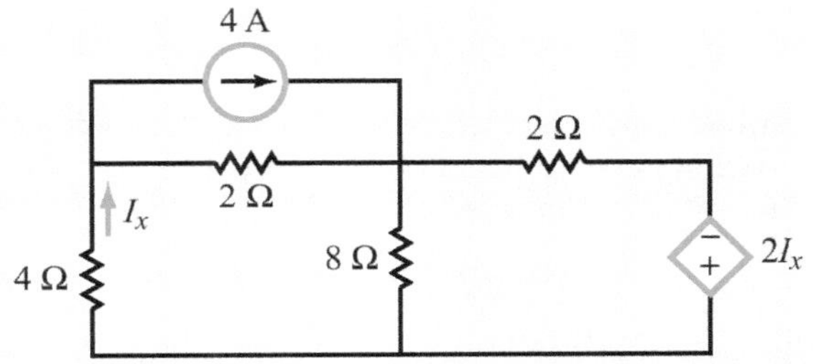

Figure E2.6

Answer: $I_x = 1.33$ A. (See CAD)

2-3 Equivalent Circuits

Even though Kirchhoff's current and voltage laws can be used to write down the requisite number of node and loop equations that are necessary to solve for all of the voltages and currents in a circuit, it is often easier to determine a certain unknown voltage or current by first simplifying the other parts of the circuit. The simplification process involves the use of *circuit equivalence*, wherein a circuit segment connected between two nodes (such as the *original* circuit segment connected between nodes 1 and 2 in Fig. 2-17) is replaced with another, simpler, circuit whose behavior is such that the voltage difference $(\upsilon_1 - \upsilon_2)$ between the two nodes—as well as the currents entering into them (or exiting from them)—remain unchanged. That is:

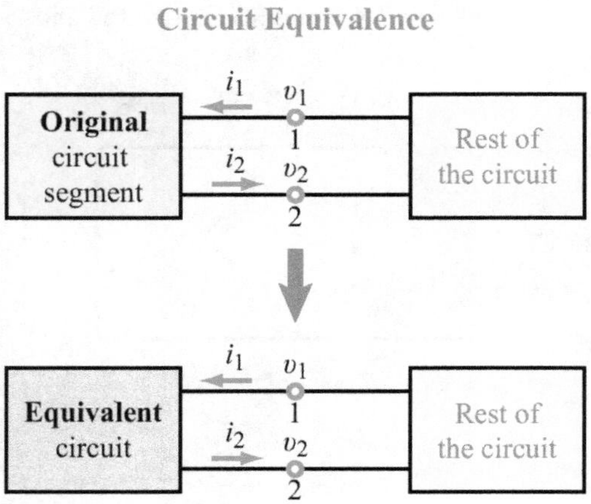

Figure 2-17: Circuit equivalence requires that the equivalent circuit exhibit the same i–v characteristic as the original circuit.

▶ Two circuits connected between a pair of nodes are considered to be equivalent—as seen by the rest of the circuit—if they exhibit identical i–v characteristics at those nodes. ◀

To the rest of the circuit, the original and equivalent circuit segments *appear* identical. The equivalent-circuit technique can be applied on the source side of a circuit, as well as on the load side.

We now will examine several types of equivalent circuits and then provide an overall summary at the conclusion of this section.

2-3.1 Resistors in Series

Consider the single-loop circuit of Fig. 2-18(a) in which a voltage source v_s is connected *in series* with five resistors. The KVL equation is given by

$$-v_s + R_1 i_s + R_2 i_s + R_3 i_s + R_4 i_s + R_5 i_s = 0, \quad (2.17)$$

which can be rewritten as

$$\begin{aligned} v_s &= R_1 i_s + R_2 i_s + R_3 i_s + R_4 i_s + R_5 i_s \\ &= (R_1 + R_2 + R_3 + R_4 + R_5) i_s = R_{eq} i_s, \end{aligned} \quad (2.18)$$

where R_{eq} is an ***equivalent resistor*** whose resistance is equal to the sum of the five in-series resistances,

$$R_{eq} = R_1 + R_2 + R_3 + R_4 + R_5. \quad (2.19)$$

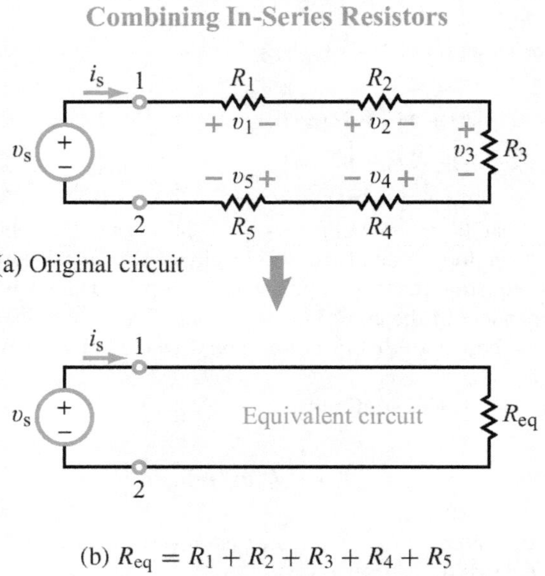

Figure 2-18: In a single-loop circuit, R_{eq} is equal to the sum of the resistors.

From the standpoint of the source voltage v_s and the current i_s it supplies, the circuit in Fig. 2-18(a) is *equivalent* to that in Fig. 2-18(b). That is,

$$i_s = \frac{v_s}{R_{eq}}. \quad (2.20)$$

▶ Multiple resistors connected in series (experiencing the same current) can be combined into a single equivalent resistor R_{eq} whose resistance is equal to the sum of all of their individual resistances. ◀

Mathematically,

$$R_{eq} = \sum_{i=1}^{N} R_i \qquad \text{(resistors in series)}, \quad (2.21)$$

where N is the total number of resistors in the group.

Voltage division

For resistor R_2 in Fig. 2-18(a), the voltage across it is given by

$$v_2 = R_2 i_s = \left(\frac{R_2}{R_{eq}}\right) v_s. \quad (2.22a)$$

Similar expressions apply to the other resistors, wherein the voltage across a resistor is equal to υ_s multiplied by the ratio of its own resistance to the sum total R_{eq}. Thus, *the single-loop circuit, in effect, divides the source voltage among the series resistors.*

▶ The voltage across any individual resistor R_i in a series circuit is a proportionate fraction (R_i/R_{eq}) of the voltage across the entire group

$$\upsilon_i = \left(\frac{R_i}{R_{eq}}\right)\upsilon_s \qquad \textbf{(voltage division).} \qquad (2.22b)$$

◀

Example 2-9: The Voltage Divider

The term ***voltage divider*** is used commonly in reference to a circuit of the type shown in Fig. 2-19, whose purpose is to supply a secondary load circuit a specific voltage υ_2 that is smaller than the available source voltage υ_s. In other words, the goal is to scale υ_s down to υ_2. If $\upsilon_s = 100$ V, choose appropriate values for R_1 and R_2 so that $\upsilon_2 = 60$ V.

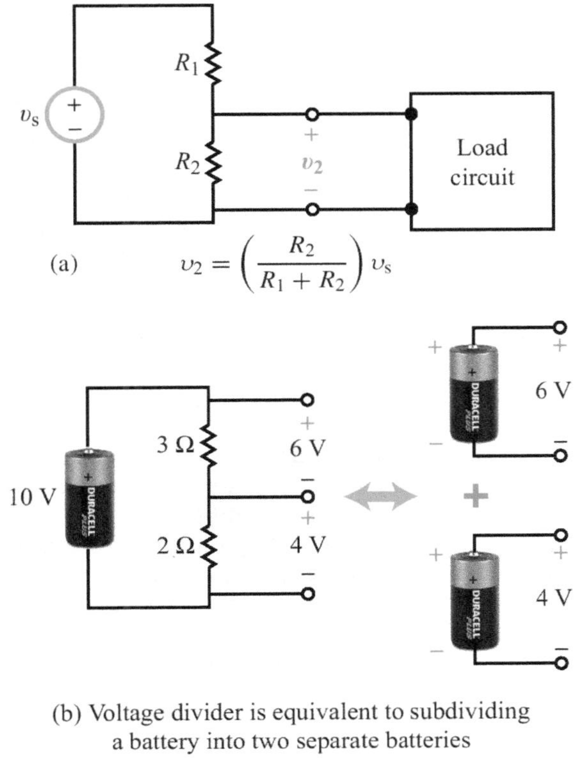

Figure 2-19: Voltage dividers are important tools in circuit analysis and design.

Solution: In view of Eq. (2.22a), application of the voltage-division property gives

$$\upsilon_2 = \left(\frac{R_2}{R_1 + R_2}\right)\upsilon_s.$$

To obtain the desired division, we require

$$\frac{R_2}{R_1 + R_2} = \frac{\upsilon_2}{\upsilon_s} = \frac{60}{100} = 0.6,$$

which can be satisfied through an infinite combination of choices of R_1 and R_2. Hence, we arbitrarily choose

$$R_1 = 2\ \Omega \qquad \text{and} \qquad R_2 = 3\ \Omega.$$

Note that the circuit in Fig. 2-19(b) will provide approximately the indicated voltages to a load circuit, so long as the resistance of the load circuit is very large compared with the resistance of R_2. Otherwise, the load circuit would draw current, thereby "loading down" the source circuit and changing V_2.

2-3.2 Sources in Series

Figure 2-20 contains a single-loop circuit composed of a voltage source, a resistor, and two current sources, all connected in series. One of the current sources indicates that the current flowing through it is 4 A in magnitude and clockwise in direction, while the other current source indicates that the

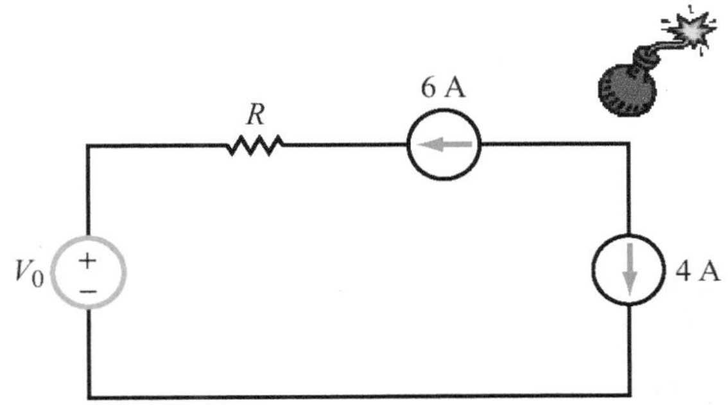

Figure 2-20: Unrealizable circuit; two current sources with different magnitudes or directions cannot be connected in series.

Technology Brief 4
Resistive Sensors

Resistive sensors can convert many physical parameters in our environment into a resistance that varies with temperature, light, pressure, moisture, chemical composition, sound, or other inputs. This variable resistance will then change the voltage or current in a circuit, which can be further manipulated in an electrical system to produce a desired output (turning on a warning light or buzzer, adjusting a valve, or otherwise control the pressure/light/heat/sound automatically). When a system measures a parameter (e.g., temperature) in order to control that parameter, the process is called a ***feedback loop***. Sensors are a very important part of a feedback system.

So how do resistive sensors work? The resistance R of a semiconductor accounts for the reduction in the electrons' velocities due to collisions with the much larger atoms of the conducting material (see Technology Brief 3). The question is: What happens to R if we disturb the atoms of the conductor by applying an external, nonelectrical ***stimulus***, such as heating or cooling it, stretching or compressing it, or shining light on it? Through proper choice of materials, we can *modulate* (change) the magnitude of R in response to such external stimuli.

Piezoresistive Sensors (Pressure, Bending, Force, etc.)

In 1856, Lord Kelvin discovered that applying a mechanical load on a bar of metal changed its resistance. Over the next 150 years, both theoretical and practical advances made it possible to describe the physics behind this effect in both conductors and semiconductors. The phenomenon is referred to as the ***piezoresistive effect*** (Fig. TF4-1) and is used in many practical devices to convert a mechanical signal into an electrical one. Such sensors (Fig. TF4-2) are called ***strain gauges***. Piezoresistive sensors are used in a wide variety of consumer applications, including writing styluses for tablets (some high-precision styluses are resistive and others are capacitive—which we will learn about in Chapter 5), robot toy "skins" that sense force, microscale gas-pressure sensors, and micromachined accelerometers that sense acceleration. They all use piezoresistors in electrical circuits to generate a signal from a mechanical stimulus.

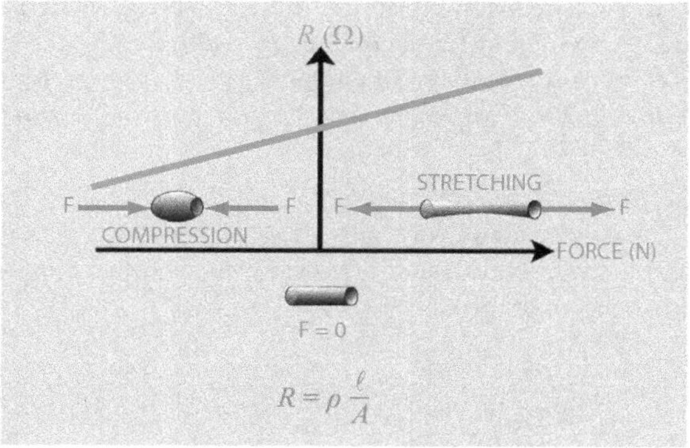

Figure TF4-1: Piezoresistance varies with applied force. The word "piezein" means "to press" in Greek.

In its simplest form, a resistance change ΔR occurs when a mechanical pressure P (N/m^2) is applied along the axis of the resistor (Fig. TF4-1)

$$\Delta R = R_0 \alpha P,$$

where R_0 is the unstressed resistance and α is known as the ***piezoresistive coefficient*** (m^2/N). The piezoresistive coefficient is a material property, and for crystalline materials (such as silicon), the piezoresistive coefficient also varies depending on the direction of the applied pressure (relative to the crystal planes of the material). When the horizontal and vertical components are different the material is called ***anisotropic***. The total resistance of a piezoresistor under stress is therefore given by

$$R = R_0 + \Delta R = R_0(1 + \alpha P).$$

The pressure P, which usually is called the ***mechanical stress*** or ***mechanical load***, is equal to F/A, where F is the force acting on the piezoresistor and A is the cross-sectional area it is acting on. The sign of P is defined as positive for a compressive force and negative for a stretching force. The piezoresistive coefficient α usually has a negative value, so the product αP leads to a decrease in R for compression and an increase for stretching.

Thermistor Sensors

Changes in temperature also can lead to changes in the resistance of a piece of conductor or semiconductor;

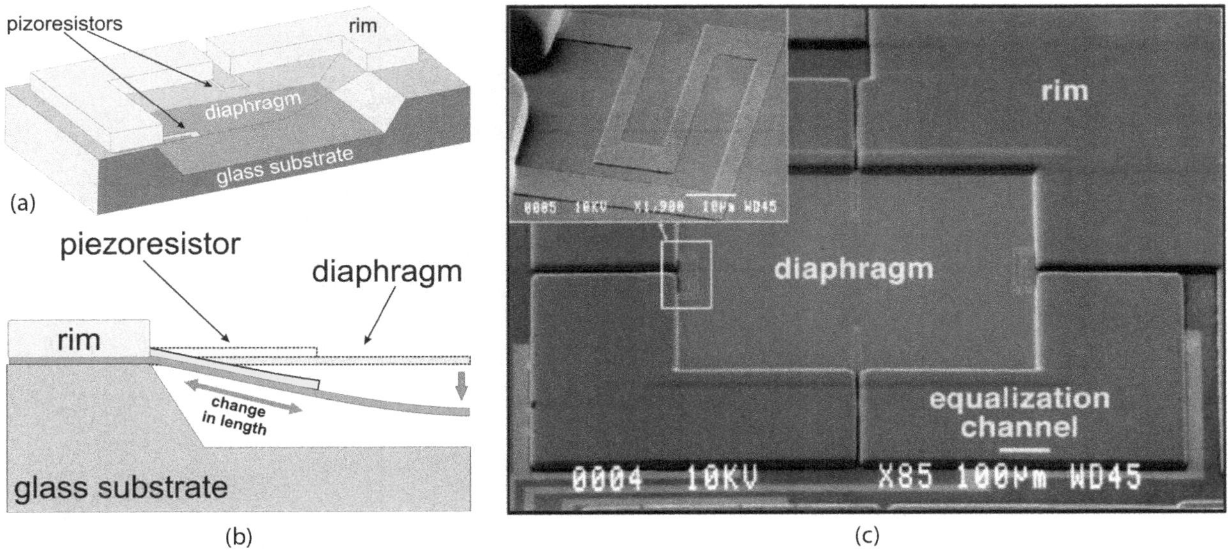

Figure TF4-2: A microfabricated pressure sensor utilizing piezoresistors as sensors. (a) A thin diaphragm (blue) is suspended over a depression etched into a glass substrate (grey). Serpentine piezoresistors (yellow) are patterned onto the membrane. (b) Differences in pressure between between the ambient and the gas in the depression will move the membrane. When this happens, the resistors stretch (or compress), changing their resistance as explained in the text. (c) A false color scanning electron micrograph of an actual microfabricated pressure sensor. Note the piezoresistors (yellow) patterned along the four sides of the diaphragm and the white, 100 μm scale bar. (Courtesy of Khalil Najafi, University of Michigan.)

when used as a sensor, such an element is called a ***thermistor***. As a simple approximation, the change in resistance can be modeled as

$$\Delta R = k\,\Delta T,$$

where ΔT is the temperature change (in degrees C) and k is the first-order temperature coefficient of resistance ($\Omega/°$C). Thermistors are classified according to whether k is negative or positive (i.e., if an increase in temperature decreases or increases the resistance). This approximation works only for small temperature changes; for larger swings, higher-order terms must be included in the equation. Resistors used in electrical circuits that are not intended to be used as sensors are manufactured from materials with the lowest k possible, since circuit designers do not want their resistors changing during operation. In contrast, materials with high values of k are desirable for sensing temperature variations. Care must be taken, however, to incorporate into the sensor response the self-heating effect that occurs due to having a current passing through the resistor itself (as in the flow sensor shown in Fig. TF4-3).

Thermistors are used routinely in modern thermostats, cell phones, automotive and industrial applications, weather monitoring, and battery-pack chargers (to prevent batteries from overheating). Thermistors also have found niche applications (Fig. TF4-3) in low-temperature sensing and as fuse replacements (for thermistors with large, positive k values). In the case of current-limiting fuse replacements, a large enough current self-heats the thermistor, and the resistance increases. There is a threshold current above which the thermistor cannot be cooled off by its environment; as it continues to get hotter, the resistance continues to increase, which in turn, causes even more self-heating. This "runaway" effect rapidly shuts current off almost entirely. Thermistors are specified based on their linear range where resistance varies linearly with the temperature, and a wide variety of options are available.

Moisture and Chemical Sensors

Resistive sensors can also be built with two electrodes measuring the material between them. A simple moisture

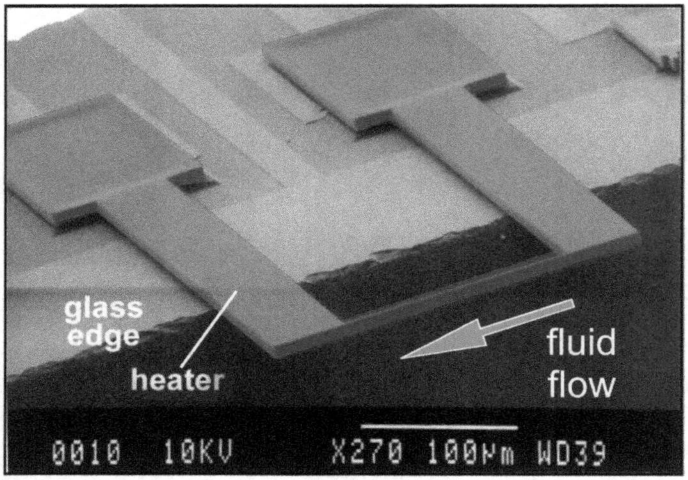

Figure TF4-3: This micromachined anemometer (flow meter) is a thermistor that measures fluid velocity. The resistor (red) serves as both a heater and a thermistor. During operation, a voltage across the resistor produces a current ($I = V/R$) which heats the resistor (recall the heat power, $P = V * I$). As fluid flows by the resistor (blue), the flow draws away heat. Since increasing the flow increases the cooling of the resistor and temperature changes the resistance, the flow can be inferred from the thermistor. (Courtesy of Khalil Najafi, University of Michigan.)

sensor you can build yourself consists of two electrodes with an absorbing material between them (Fig. TF4-4). Just draw two thick pencil (graphite) lines on paper, clip to them with alligator clips, and measure the resistance with your myDAQ. Then drip water between the two lines, so that it makes contact between them. The resistance will immediately drop in magnitude.

In a similar approach, resistive sensors can sometimes be used to determine chemical composition of a liquid material. The resistivity of the material depends strongly on the number of dissolved or loose ions in the material (see Table 2-1). Deionized water has high resistivity, drinking water has moderate resistivity, and sea water has low resistivity. Placing two electrodes into a container of fluid, or running fluid over two electrodes in a microfluidic system can be used to measure the resistivity of the material and hence its chemical composition. This is often used as a simple way to monitor the purity of drinking water.

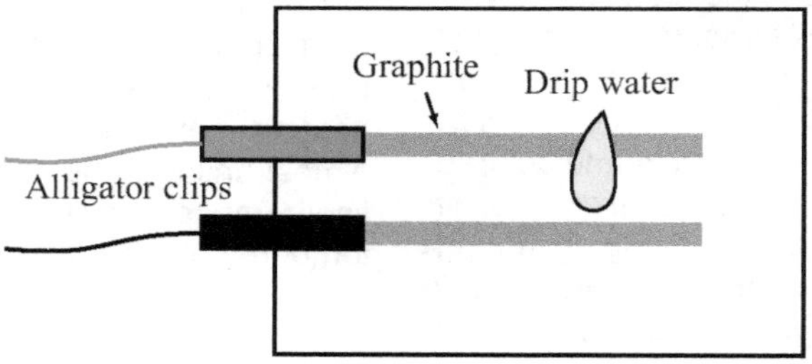

Figure TF4-4: Increased ions (from dissolved solids, for example) increase the conductivity (reduce resistivity), which can be measured by an ohmmeter.

current is 6 A in magnitude and counterclockwise in direction. Continuity of current flow mandates that the current flowing through the loop be exactly the same in both magnitude and direction at every location over the full extent of the loop. So our dilemma is: Is the current 4 A, 6 A, or the difference between the two? It is none of those guesses. The true answer is that the circuit is *unrealizable*, meaning that it is not possible to construct a circuit with two current sources of different magnitudes or different directions that are connected in series. The problem with the circuit of Fig. 2-20 has to do with our representation of ideal current sources. As was stated in Section 1-6.2 and described in Table 1-5, a real current source can be modeled as the parallel combination of an ideal current source and a shunt resistor R_s. Usually, R_s is very large, so very little current flows through it in comparison with the current flowing through the other part of the circuit, in which case it can be deleted without much consequence. In the present case, however, had such shunt resistors been included in the circuit of Fig. 2-20, the dilemma would not have arisen. The lesson we should learn from this discussion is that when we idealize current sources by deleting their parallel resistors, we should never connect them in series in circuit diagrams.

▶ Ideal current sources cannot be added in series. ◀

Whereas current sources cannot be connected in series, voltage sources can. In fact, it follows from KVL that from the standpoint of an external load resistor R_L connected between nodes 1 and 2, the circuit in Fig. 2-21(a) can be simplified into the equivalent circuit of Fig. 2-21(b) with

$$\upsilon_{eq} = \upsilon_1 - \upsilon_2 + \upsilon_3 \tag{2.23}$$

and

$$R_{eq} = R_1 + R_2. \tag{2.24}$$

Thus:

▶ Multiple voltage sources connected in series can be combined into an equivalent voltage source whose voltage is equal to the algebraic sum of the voltages of the individual sources. ◀

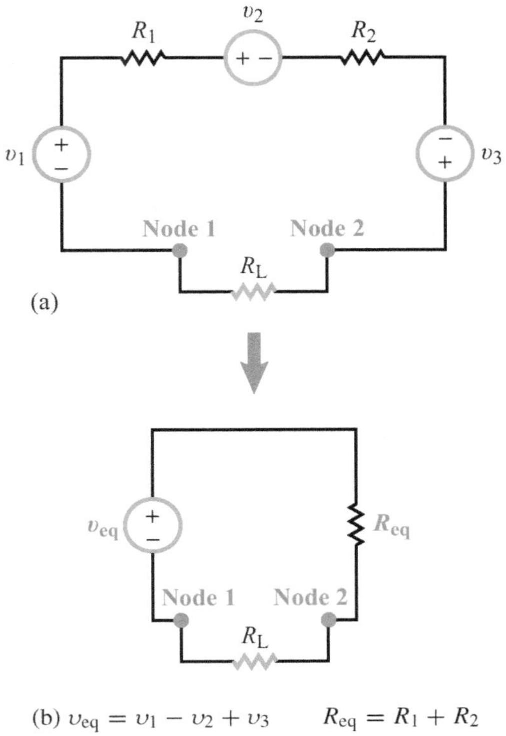

Figure 2-21: In-series voltage sources can be added together algebraically.

2-3.3 Resistors and Sources in Parallel

When multiple resistors are connected in series, they all share the same current, but each has its own individual voltage across it. The converse is true for multiple resistors connected in parallel: the three resistors in Fig. 2-22(a) experience the same voltage across all of them, namely υ_s, but each carries its own individual current. The current supplied by the source is *divided* among the branches containing the three resistors. Thus,

$$i_s = i_1 + i_2 + i_3. \tag{2.25}$$

Application of Ohm's law provides

$$i_1 = \frac{\upsilon_s}{R_1}, \qquad i_2 = \frac{\upsilon_s}{R_2}, \qquad \text{and} \qquad i_3 = \frac{\upsilon_s}{R_3}, \tag{2.26}$$

which when used in Eq. (2.25) leads to

$$i_s = \frac{\upsilon_s}{R_1} + \frac{\upsilon_s}{R_2} + \frac{\upsilon_s}{R_3}. \tag{2.27}$$

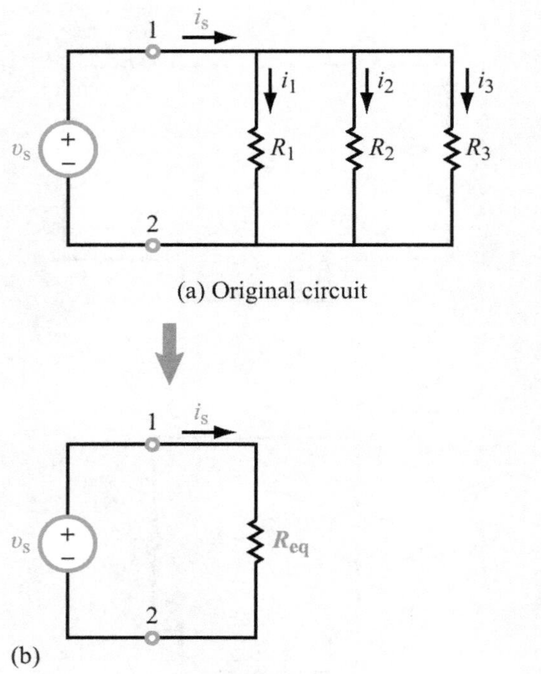

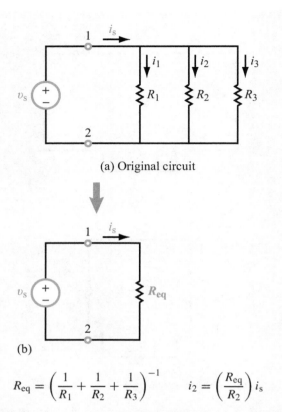

$$R_{eq} = \left(\frac{1}{R_1} + \frac{1}{R_2} + \frac{1}{R_3}\right)^{-1} \qquad i_2 = \left(\frac{R_{eq}}{R_2}\right) i_s$$

Figure 2-22: Voltage source connected to a parallel combination of three resistors.

We wish to replace the parallel combination of the three resistors with a single equivalent resistor R_{eq}, as depicted in Fig. 2-22(b), such that the current i_s remains unchanged. For the equivalent circuit,

$$i_s = \frac{v_s}{R_{eq}}. \tag{2.28}$$

If the two circuits in Fig. 2-22 are to function the same, as regards the source, then i_s as given by Eq. (2.27) for the original circuit should be equal to the expression for i_s given by Eq. (2.28) for the equivalent circuit. Thus,

$$\frac{v_s}{R_{eq}} = \frac{v_s}{R_1} + \frac{v_s}{R_2} + \frac{v_s}{R_3}, \tag{2.29}$$

from which we conclude that

$$\frac{1}{R_{eq}} = \frac{1}{R_1} + \frac{1}{R_2} + \frac{1}{R_3}. \tag{2.30}$$

This result can be generalized to any N resistors connected in parallel

$$\frac{1}{R_{eq}} = \sum_{i=1}^{N} \frac{1}{R_i} \qquad \textbf{(resistors in parallel).} \tag{2.31}$$

Current division

> ▶ Multiple resistors connected in parallel divide the input current among them. ◀

For R_2 in Fig. 2-22(a),

$$i_2 = \frac{v_s}{R_2} = \left(\frac{R_{eq}}{R_2}\right) i_s. \tag{2.32}$$

By extension, for a *current divider* composed of N in-parallel resistors, the current flowing through R_i is a proportionate fraction (R_{eq}/R_i) of the input current.

It is useful to note that the equivalent resistance for a parallel combination of two resistors R_1 and R_2 (Fig. 2-23) is given by

$$R_{eq} = \frac{R_1 R_2}{R_1 + R_2}. \tag{2.33}$$

> ▶ As a short-hand notation, we will sometimes denote such a parallel combination $R_1 \parallel R_2$. We also denote the series combination of R_1 and R_2 as $(R_1 + R_2)$. ◀

As was noted earlier in Section 2-1.5, *the inverse of the resistance R is the conductance G;* $G = 1/R$. For N conductances

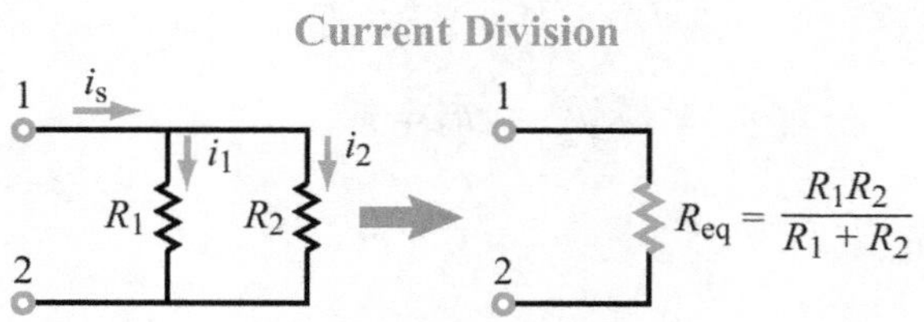

$$i_1 = \left(\frac{R_2}{R_1 + R_2}\right) i_s \qquad i_2 = \left(\frac{R_1}{R_1 + R_2}\right) i_s$$

Figure 2-23: Equivalent circuit for two resistors in parallel.

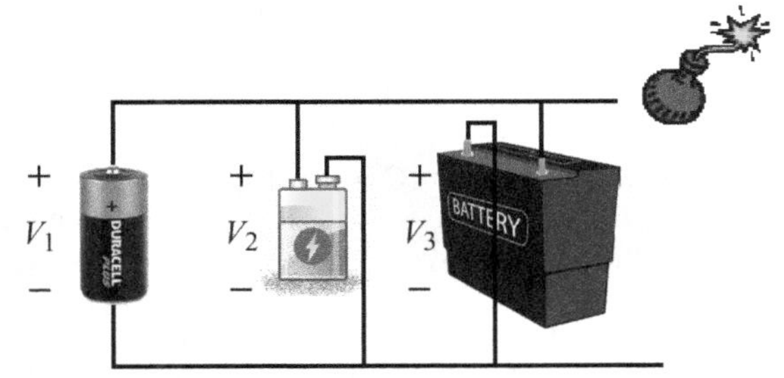

Figure 2-24: This is an unrealizable circuit unless all voltage sources have identical voltages and polarities; that is, $V_1 = V_2 = V_3$.

connected in parallel, Eq. (2.31) assumes the form of a linear sum

$$G_{eq} = \sum_{i=1}^{N} G_i \quad \text{(conductances in parallel).} \qquad (2.34)$$

Two resistors always can be combined together, whether they are connected in series (sharing the same current) or in parallel (sharing the same voltage). Two voltage sources can be combined when connected in series, but they cannot be connected in parallel, unless they have identical voltages (Fig. 2-24). Two current sources can be combined when connected in parallel (as illustrated by Fig. 2-25), but they cannot be connected in series.

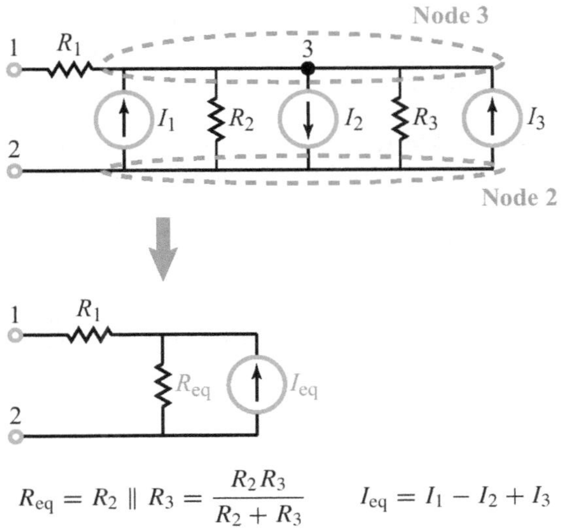

Figure 2-25: Adding current sources connected in parallel.

Example 2-10: Current Division Using Conductance

For the circuit in Fig. 2-26:

(a) Relate I_3 to I_0 and resistances R_1 to R_3.

(b) Relate I_3 to I_0 and conductances G_1 to G_3, where $G_i = 1/R_i$.

Solution: (a) Application of the expressions given in Fig. 2-22 leads to

$$I_3 = \left(\frac{R_{eq}}{R_3}\right) I_0,$$

with

$$R_{eq} = \left(\frac{1}{R_1} + \frac{1}{R_2} + \frac{1}{R_3}\right)^{-1} = \left(\frac{R_2R_3 + R_1R_3 + R_1R_2}{R_1R_2R_3}\right)^{-1}$$
$$= \left(\frac{R_1R_2R_3}{R_2R_3 + R_1R_3 + R_1R_2}\right).$$

Hence,

$$I_3 = \left(\frac{R_{eq}}{R_3}\right) I_0 = \left(\frac{R_1R_2}{R_2R_3 + R_1R_3 + R_1R_2}\right) I_0.$$

(b) Rewriting the expressions for I_3 and R_{eq} in terms of conductances gives

$$I_3 = \left(\frac{G_3}{G_{eq}}\right) I_0,$$

with

$$G_{eq} = \frac{1}{R_{eq}} = \left(\frac{1}{R_1} + \frac{1}{R_2} + \frac{1}{R_3}\right) = G_1 + G_2 + G_3.$$

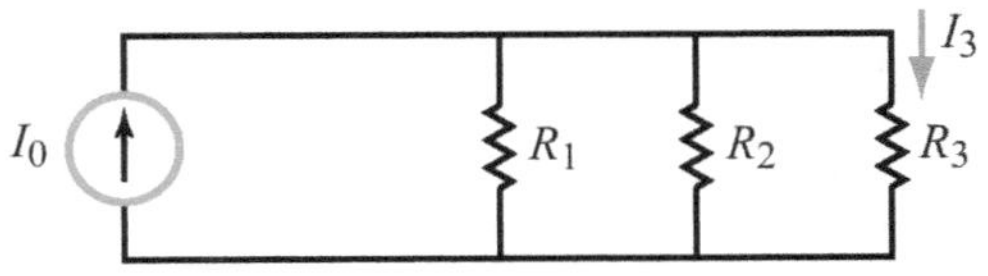

Figure 2-26: Circuit of Example 2-10.

Hence,

$$I_3 = \left(\frac{G_3}{G_1 + G_2 + G_3} \right) I_0.$$

Current division using conductances assumes the same functional form as voltage division using resistances (Eq. (2.22b)).

Example 2-11: Realizable and Unrealizable Circuits

Given that the voltage difference between any two nodes in a circuit has to be unique (cannot have multiple values simultaneously), and that the current in any given branch also is unique, determine which of the three circuits in Fig. 2-27 are realizable and which are unrealizable.

Solution: (a) Circuit of Fig. 2-27(a): Circuit is not realizable.

From the perspective of the ideal voltage source V_s, the voltage difference between nodes A and B is V_s, but according to the dependent source the voltage is $2V_s$.

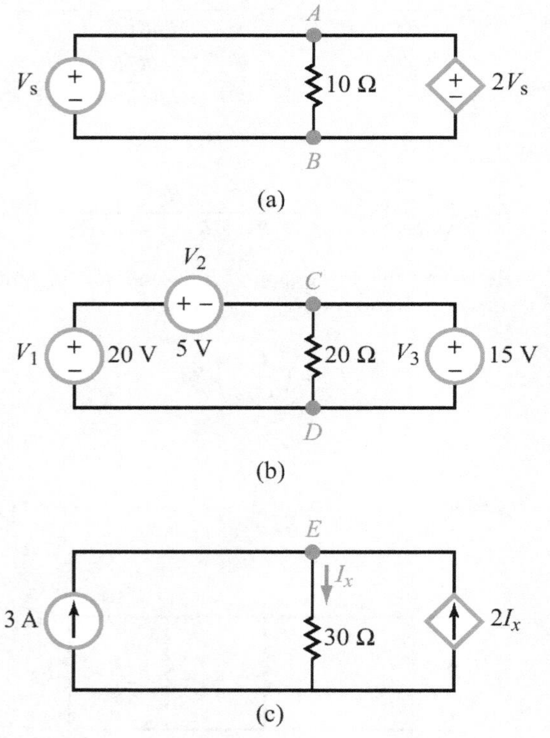

Figure 2-27: Circuits of Example 2-11.

(b) Circuit of Fig. 2-27(b): Circuit is realizable.

From the standpoint of the two voltage sources to the left of nodes CD,

$$V_{CD} = V_1 + V_2 = 20 - 5 = 15 \text{ V}.$$

Also connected across nodes CD is voltage source V_3, but its voltage is exactly 15 V. Two voltage sources can be connected in parallel if they have the same voltage.

(c) Circuit of Fig. 2-27(c): Circuit is realizable.

KCL at node E requires that the sum of the three currents entering the node be zero. Hence,

$$3 + 2I_x - I_x = 0,$$

which leads to

$$I_x = -3 \text{ A}.$$

This means that the direction of I_x is upwards and the dependent current source has a downward-moving current of 6 A.

Example 2-12: Equivalent-Circuit Solution

Use the equivalent-resistance approach to determine V_2, I_1, I_2, and I_3 in the circuit of Fig. 2-28(a).

Solution: In the circuit of Fig. 2-28(a), the part of the circuit connected to the voltage source is equivalent to a resistor $R_{eq} = R_1 + [(R_3 + R_4) \parallel (R_2 + R_5)]$. Hence, our first step is to combine the 2 Ω and 4 Ω in-series resistances into a 6 Ω resistance and to combine the two 6 Ω in-parallel resistances into a 3 Ω resistance (by applying Eq. (2.33)). The simplifications lead to the circuit in Fig. 2-28(b). Next, we calculate the parallel combination of the 3 Ω and 6 Ω resistors, $(3 \parallel 6)$, again using Eq. (2.33), to get $(3 \times 6)/(3 + 6) = 18/9 = 2$ Ω. The new equivalent circuit is displayed in Fig. 2-28(c), from which we deduce that

$$I_1 = \frac{24}{10 + 2} = 2 \text{ A}$$

and

$$V_2 = 2I_1 = 2 \times 2 = 4 \text{ V}.$$

Returning to Fig. 2-28(b), we apply Ohm's law to find I_2 and I_3.

$$I_2 = \frac{V_2}{3} = \frac{4}{3} = 1.33 \text{ A},$$

and

$$I_3 = \frac{V_2}{6} = \frac{4}{6} = 0.67 \text{ A}.$$

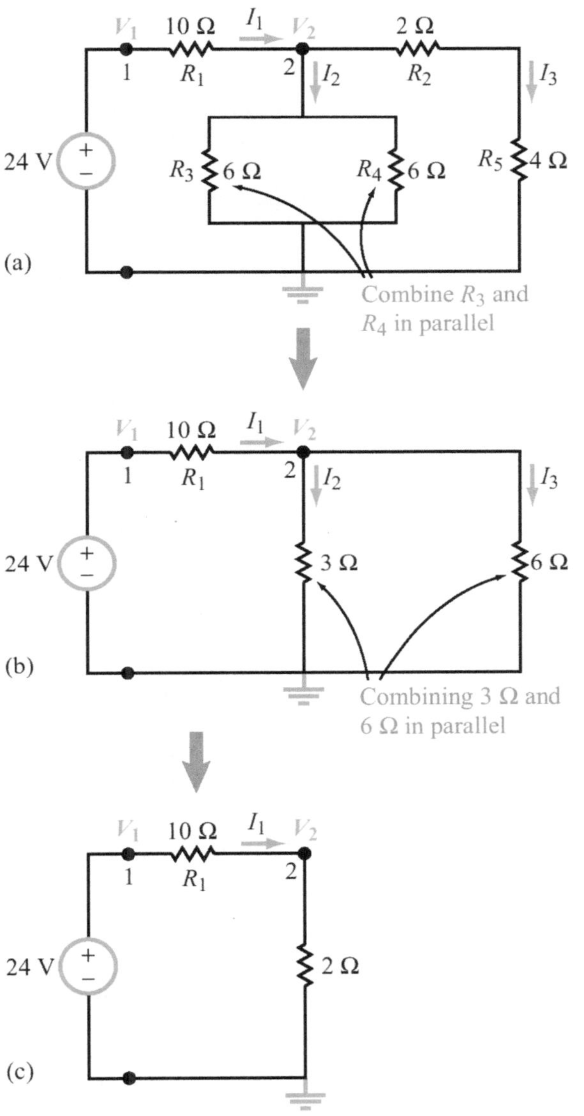

Figure 2-28: Example 2-12. (a) Original circuit, (b) after combining R_3 and R_4 in parallel and combining R_2 and R_5 in series, and (c) after combining the 3 Ω and 6 Ω resistances in parallel.

> **Concept Question 2-7:** What conditions must be satisfied in order for two circuits to be considered *equivalent*? (See CAD)

> **Concept Question 2-8:** What is a voltage divider and what is a current divider? (See CAD)

> **Concept Question 2-9:** What is the i–υ relationship for a conductance G? (See CAD)

Exercise 2-7: Apply resistance combining to simplify the circuit of Fig. E2.7 in order to find I. All resistor values are in ohms.

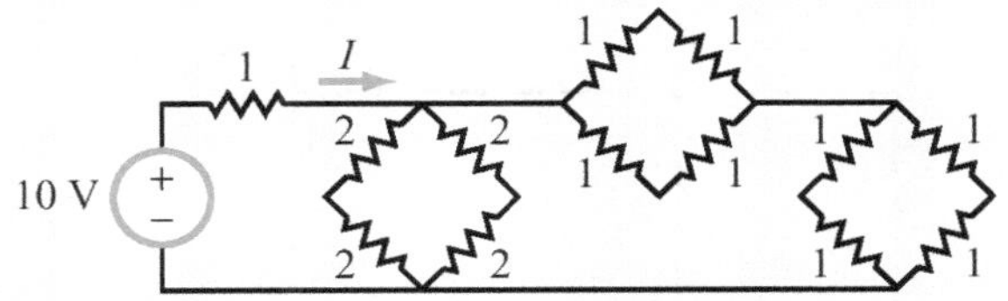

Figure E2.7

Answer: $I = 5$ A. (See CAD)

2-3.4 Source Transformation

We now will demonstrate how a realistic voltage source composed of an ideal voltage source in series with a resistor can be exchanged for a realistic current source composed of an ideal current source in parallel with a shunt resistor, or vice versa. The two circuits are shown in parts (a) and (b) of Fig. 2-29. Exchanging the one source for the other requires that they be equivalent—from the vantage point of the external circuit.

> ▶ A voltage-source circuit and a current-source circuit are considered equivalent and interchangeable if they deliver the same input current i and voltage υ_{12} to the external circuit. ◀

For the voltage-source circuit, application of KVL gives

$$-\upsilon_s + iR_1 + \upsilon_{12} = 0, \tag{2.35}$$

from which we obtain the following expression for i:

$$i = \frac{\upsilon_s}{R_1} - \frac{\upsilon_{12}}{R_1}. \tag{2.36}$$

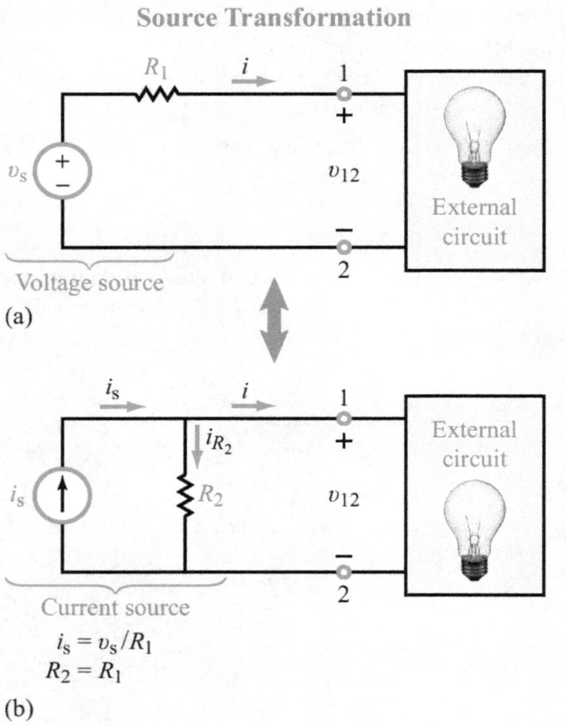

Figure 2-29: Realistic voltage and current sources connected to an external circuit. Equivalence requires that $i_s = v_s/R_1$ and $R_2 = R_1$.

Application of KCL to the current-source circuit gives

$$i = i_s - i_{R_2} = i_s - \frac{v_{12}}{R_2}, \tag{2.37}$$

where we used Ohm's law to relate i_{R_2} to v_{12}. Equivalence of Eqs. (2.36) and (2.37) is satisfied for all values of i and v_{12} if and only if:

$$R_1 = R_2 \tag{2.38a}$$

and

$$i_s = \frac{v_s}{R_1}. \tag{2.38b}$$

In summary:

> ► A voltage source v_s in series with a source resistance R_s is equivalent to the combination of a current source $i_s = v_s/R_s$, in parallel with a shunt resistance R_s. The direction of the equivalent current source is the same as the direction from (−) to (+) terminals of the voltage source. ◄

This equivalence is called *source transformation* because it allows us to replace a realistic voltage source with a realistic current source, or vice versa.

A summary of in-series and in-parallel equivalent circuits involving sources and resistors is available in Table 2-5.

Example 2-13: Source Transformation

Determine the current I in the circuit of Fig. 2-30(a).

Solution: It is best to avoid transformations that would involve the 3 Ω resistor with the unknown current I. Hence, we will apply multiple source-transformation steps, moving from the left end of the circuit towards the 3 Ω resistor.

Step 1: Current to voltage transformation allows us to convert the combination (I_{s_1}, R_{s_1}) to a voltage source

$$V_{s_1} = I_{s_1} R_{s_1} = 16 \times 2 = 32 \text{ V},$$

in series with R_{s_1}.

Step 2: Combining R_{s_1} in series with the 6 Ω resistor results in

$$R_{s_2} = 2 + 6 = 8 \ \Omega.$$

Hence, the new input source becomes (V_{s_1}, R_{s_2}).

Step 3: Convert (V_{s_1}, R_{s_2}) back into a current source

$$I_{s_2} = V_{s_1}/R_{s_2} = 32/8 = 4 \text{ A},$$

in parallel with R_{s_2}.

Step 4: Combine $R_{s_2} = 8$ Ω in parallel with the other 8 Ω resistor (8 ∥ 8) to obtain an equivalent resistance $R_{s_3} = 4$ Ω.

Table 2-5: **Equivalent circuits.**

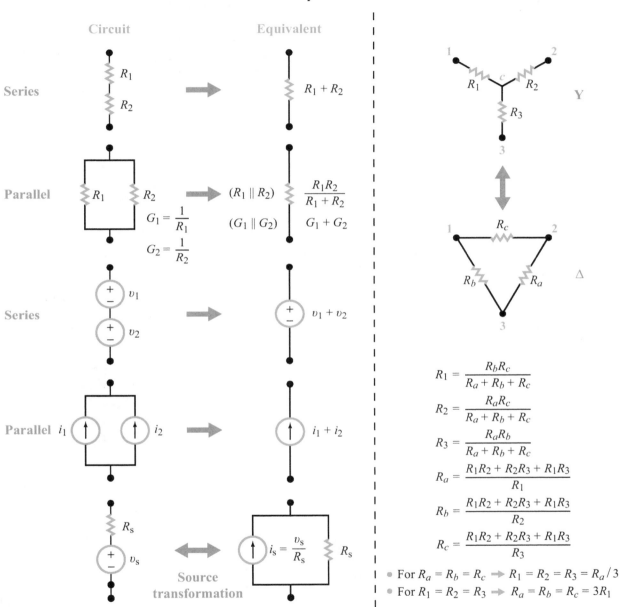

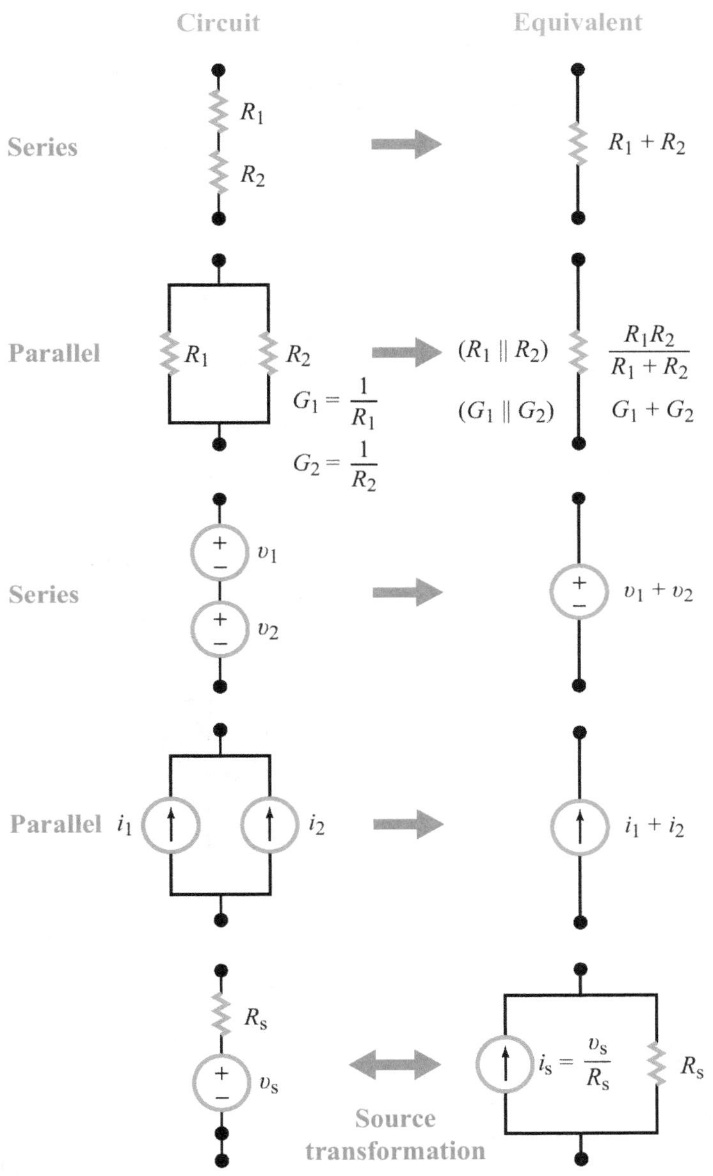

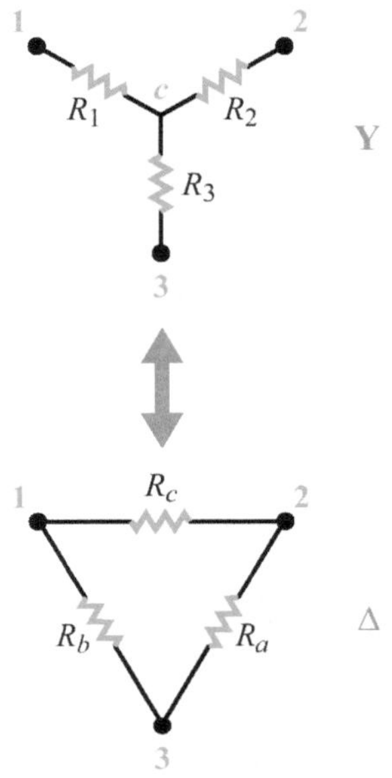

$$R_1 = \frac{R_bR_c}{R_a + R_b + R_c}$$

$$R_2 = \frac{R_aR_c}{R_a + R_b + R_c}$$

$$R_3 = \frac{R_aR_b}{R_a + R_b + R_c}$$

$$R_a = \frac{R_1R_2 + R_2R_3 + R_1R_3}{R_1}$$

$$R_b = \frac{R_1R_2 + R_2R_3 + R_1R_3}{R_2}$$

$$R_c = \frac{R_1R_2 + R_2R_3 + R_1R_3}{R_3}$$

- For $R_a = R_b = R_c \rightarrow R_1 = R_2 = R_3 = R_a/3$
- For $R_1 = R_2 = R_3 \rightarrow R_a = R_b = R_c = 3R_1$

Step 5: Convert again to a voltage source

$$V_{s_2} = I_{s_2} R_{s_3} = 4 \times 4 = 16 \text{ V},$$

in series with R_{s_3}.

For the single loop realized in the final step,

$$I = \frac{V_{s_2}}{4 + 1 + 3} = \frac{16}{8} = 2 \text{ A}.$$

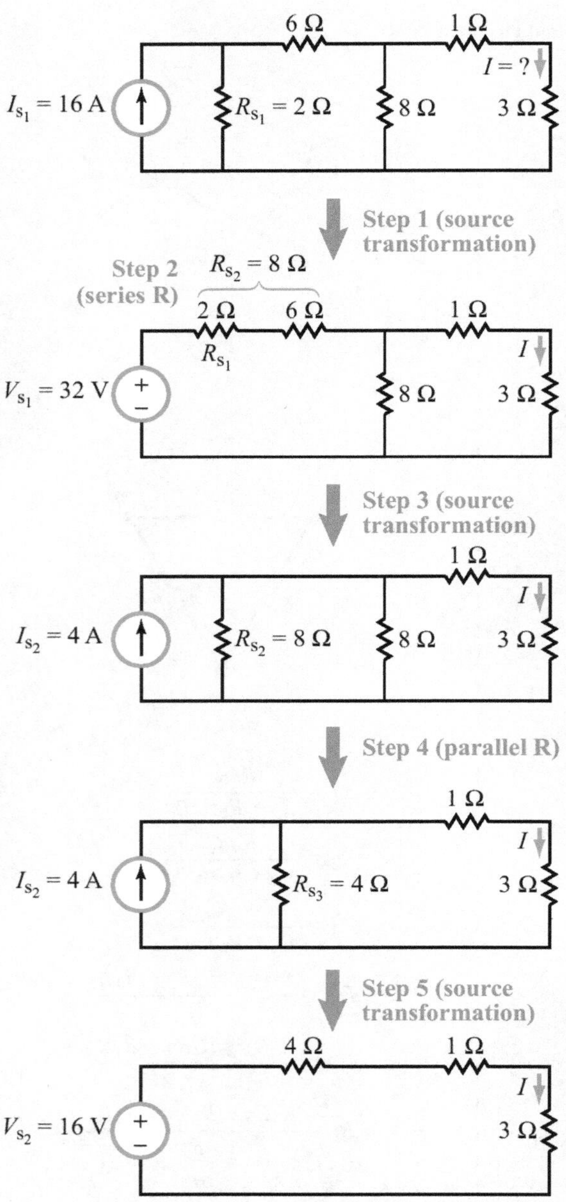

Figure 2-30: Example 2-13 circuit evolution.

Example 2-14: Finding V_{ab}

While keeping the load resistor R_L in the top circuit of Fig. 2-31 intact, apply source transformations until the circuit simplifies to a current divider, then determine V_{ab} for $R_L = 10\ \Omega$.

Solution:

Step 1: Convert the 2 A current source in parallel with the 20 Ω resistor into a 40 V voltage source in series with a 20 Ω resistor.

Step 2: Combine the two in-series 20 Ω and 40 Ω resistances into a 60 Ω resistance, and combine the 40 V and 16 V sources into a single 24 V source.

Step 3: Convert each voltage source (together with its in-series resistance) into a current source with a resistance in parallel.

Step 4: Combine the two in-parallel resistances and the two in-parallel current sources.

Step 5: For $R_L = 10\ \Omega$, current division yields

$$I = \frac{20}{10 + 20} \times 3 = 2\text{ A},$$

and the associated voltage across R_L is

$$V_{ab} = 10I = 20\text{ V}.$$

Exercise 2-8: Apply source transformation to the circuit in Fig. E2.8 to find I.

Answer: $I = 4$ A. (See CAD)

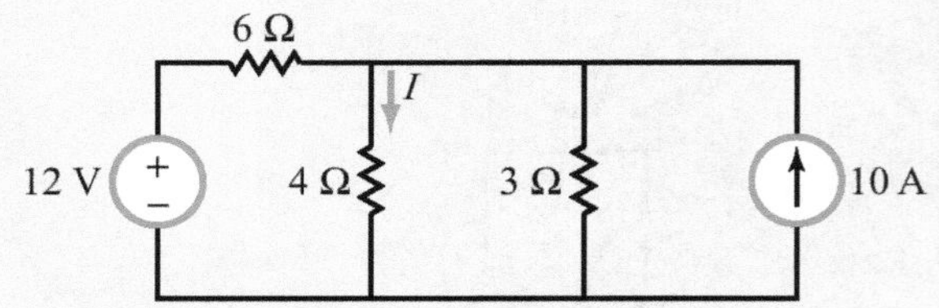

Figure E2.8

2-4 Wye–Delta (Y–Δ) Transformation

In principle, it always is possible to simplify the behavior of a resistive circuit when measured across any two nodes—no matter how complex its topology—down to a simple equivalent circuit composed of an equivalent voltage source in series with an equivalent resistor. The preceding sections offered us tools for combining resistors together whenever they are connected in series or in parallel, as well as for combining in-series voltage sources and in-parallel current sources. Sometimes, however, we may encounter circuit topologies that cannot be simplified using those tools because their resistors are connected neither in series nor in parallel. A case in point is the circuit in Fig. 2-32, in which no two resistors share the same current or voltage. This

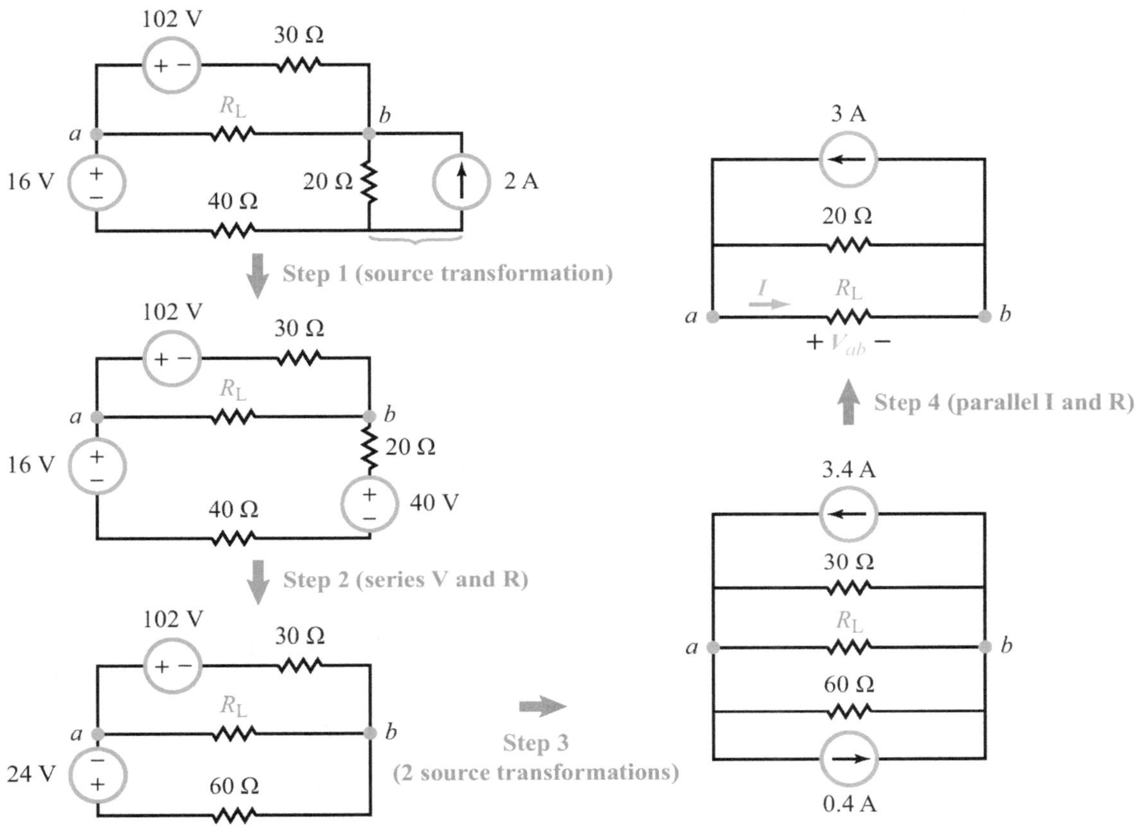

Figure 2-31: Circuit evolution for Example 2-14.

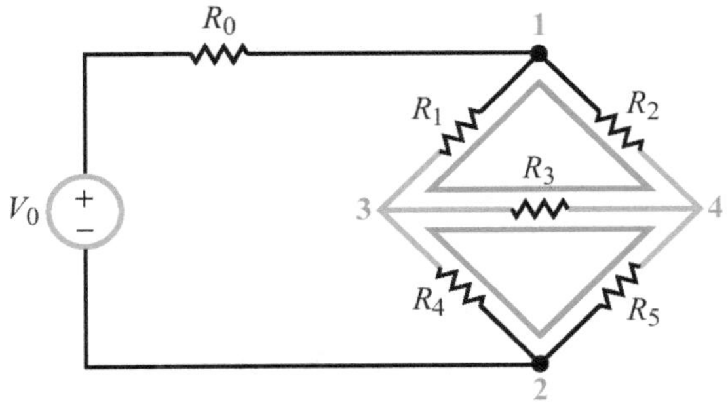

Figure 2-32: No two resistors of this circuit share the same current (connected in series) or voltage (connected in parallel).

section introduces a new circuit-simplification tool—known as the *Wye–Delta (Y–Δ) transformation*—for dealing specifically with such a circuit arrangement.

To that end, let us start by considering the Y and Δ circuit segments shown in Fig. 2-33(a) and (b), respectively. Let us assume that the same external circuit is connected to the Y and Δ circuits at nodes 1, 2, and 3. Our task is to develop a set of transformation relations between the resistor set (R_1, R_2, R_3) of the Y circuit and the resistor set (R_a, R_b, R_c) of the Δ circuit that will allow us to replace the Y circuit with the Δ circuit (or vice versa) without affecting the terminal characteristics (currents and voltages) at nodes 1, 2, and 3. That is, from the standpoint of the external circuit, the Y and Δ circuits should behave equivalently.

The standard procedure employed in deriving the transformation relations is to (a) set one node as an open circuit (i.e.,

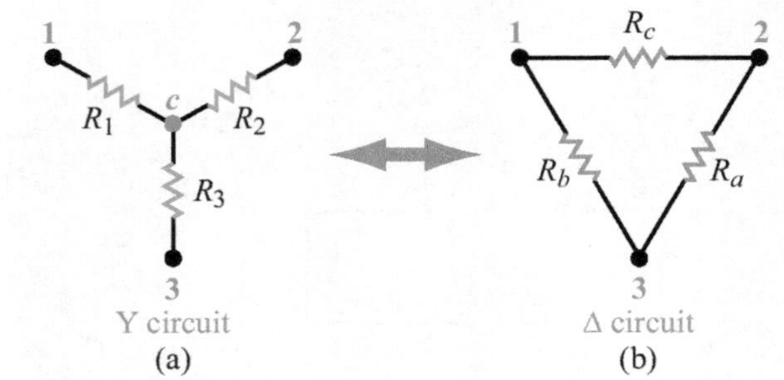

Figure 2-33: Y–Δ equivalent circuits.

not connected to an external circuit), (b) derive an expression for the resistance between the other two nodes (as if a voltage source were connected between them) of the Y circuit, (c) follow the same procedure for the Δ circuit, and then (d) equate the expressions obtained in steps (b) and (c). For example, with node 3 open-circuited, the Y circuit reduces to just two in-series resistors R_1 and R_2, in which case the resistance between nodes 1 and 2 is simply

$$R_{12} = R_1 + R_2 \qquad \text{(Y-circuit)}. \tag{2.39}$$

Repeating the procedure for the Δ circuit (again with node 3 not connected to the external circuit) leads to a configuration between nodes 1 and 2 consisting of R_c in parallel with the series combination of R_a and R_b. Hence,

$$R_{12} = \frac{R_c(R_a + R_b)}{R_a + R_b + R_c} \qquad (\Delta\text{-circuit}). \tag{2.40}$$

Upon equating the expressions for R_{12} given by Eqs. (2.39) and (2.40), we have

$$R_1 + R_2 = \frac{R_c(R_a + R_b)}{R_a + R_b + R_c}. \tag{2.41a}$$

When applied to the other two combinations of nodes, the foregoing procedure leads to:

$$R_2 + R_3 = \frac{R_a(R_b + R_c)}{R_a + R_b + R_c} \tag{2.41b}$$

and

$$R_1 + R_3 = \frac{R_b(R_a + R_c)}{R_a + R_b + R_c}. \tag{2.41c}$$

2-4.1 Δ →Y Transformation

Solution of the preceding set of equations provides the following expressions for R_1, R_2, and R_3:

$$R_1 = \frac{R_b R_c}{R_a + R_b + R_c} \tag{2.42a}$$

$$R_2 = \frac{R_a R_c}{R_a + R_b + R_c} \tag{2.42b}$$

$$R_3 = \frac{R_a R_b}{R_a + R_b + R_c} \tag{2.42c}$$

Note the symmetry associated with the form of these expressions:

► R_1 of the Y circuit, which is connected to node 1, is given by an expression (Eq. (2.42a)) whose numerator is the product of the two resistors connected to node 1 in the Δ circuit, namely R_b and R_c. The same form of symmetry applies to R_2 and R_3. ◄

The transformation represented by the three parts of Eq. (2.42) enables us to replace the Δ circuit with a Y circuit without having any impact on the external circuit.

2-4.2 Y→ Δ Transformation

When applied in the reverse direction, from Y to Δ, the associated transformation relations are given by the following expressions.

$$R_a = \frac{R_1R_2 + R_2R_3 + R_1R_3}{R_1} \tag{2.43a}$$

$$R_b = \frac{R_1R_2 + R_2R_3 + R_1R_3}{R_2} \tag{2.43b}$$

$$R_c = \frac{R_1R_2 + R_2R_3 + R_1R_3}{R_3} \tag{2.43c}$$

For this transformation, the symmetry is as follows:

▶ R_a of the Δ circuit, which is connected between nodes 2 and 3, is given by an expression (Eq. (2.43a)) whose denominator is R_1, the resistor connected to node 1 of the Y circuit. This form of symmetry also applies to R_b and R_c. ◀

When we started our examination of the Y–Δ transformation, we referred to Fig. 2-32. Returning to that figure, we note that the circuit contains two obvious Δ circuits, namely R_1–R_2–R_3 and R_3–R_5–R_4, as well as two not-so-obvious Y circuits: R_1–R_3–R_4 and R_2–R_3–R_5. To demonstrate that those two combinations are indeed Y circuits, we have redrawn the circuit in the form shown in Fig. 2-34(a) where we stretched nodes 1 and 2 from single points into two horizontal lines. Electrically, we did not change the circuit whatsoever.

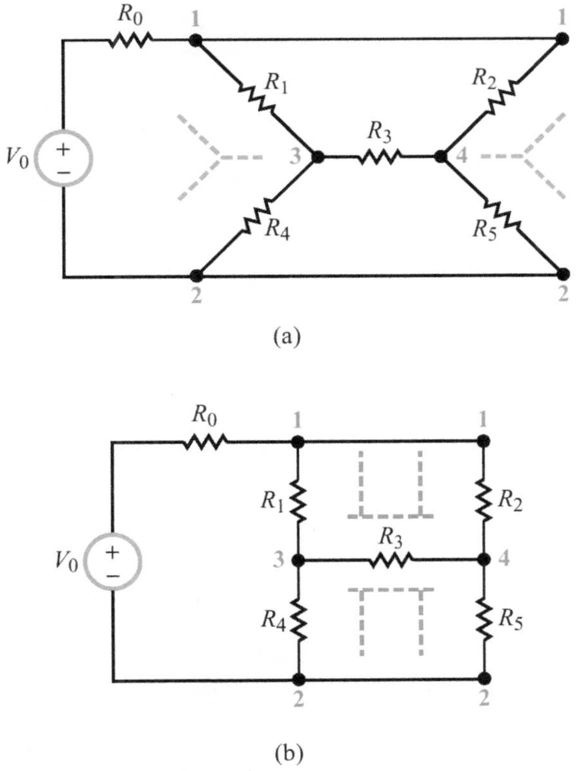

Figure 2-34: Redrawing the circuit of Fig. 2-32 to resemble (a) Y and (b) T and Π subcircuits.

Figure 2-34(b) depicts another rendition of the same circuit. In this case, the Y circuit given by R_1–R_3–R_4 resembles a sideways T rather than a Y, and the Δ circuit given by R_1–R_3–R_2 resembles a Π. Hence, it is not surprising that the Y–Δ transformation is oftentimes called the T–Π transformation. It is instructive to note that the shape in which a circuit is drawn is irrelevant electrically; what does matter is how the branches are connected to the nodes.

2-4.3 Balanced Circuits

If the resistors of the Δ circuit are all equal, the circuit is said to be *balanced* (because the three resistors will have equal voltages across them and equal currents through them), as a result of which the Y circuit will also be balanced and will have equal resistors given by

$$R_1 = R_2 = R_3 = \frac{R_a}{3} \quad (\text{if } R_a = R_b = R_c), \tag{2.44a}$$

and conversely

$$R_a = R_b = R_c = 3R_1 \quad (\text{if } R_1 = R_2 = R_3). \tag{2.44b}$$

Example 2-15: Applying Y–Δ Transformation

Simplify the circuit in Fig. 2-35(a) by applying the Y–Δ transformation so as to determine the current I.

Solution: Noting the symmetry rules associated with the transformation, the Δ circuit connected to nodes 1, 3, and 4 can be replaced with a Y circuit, as shown in Fig. 2-35(b), with resistances

$$R_1 = \frac{24 \times 36}{24 + 36 + 12} = 12\ \Omega,$$

$$R_2 = \frac{24 \times 12}{24 + 36 + 12} = 4\ \Omega,$$

and

$$R_3 = \frac{36 \times 12}{24 + 36 + 12} = 6\ \Omega.$$

Next, we add the 4 Ω and 20 Ω resistors in series, obtaining 24 Ω for the right branch of the trapezoid. Similarly, the left branch combines into 12 Ω and the two in-parallel branches reduce to a resistance equal to $(24 \times 12)/(24 + 12) = 8\ \Omega$. When added

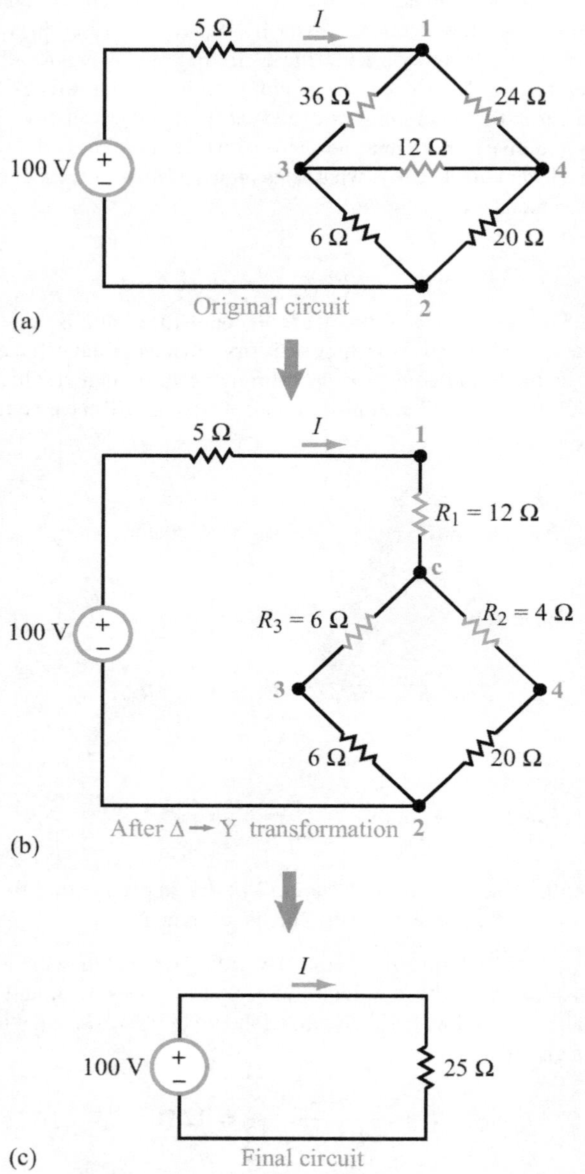

Figure 2-35: Example 2-15 circuit evolution.

to the 5 Ω and 12 Ω in-series resistances, this leads to the final circuit in Fig. 2-35(c). Hence,

$$I = \frac{100}{25} = 4\text{ A}.$$

Concept Question 2-10: When is the Y–Δ transformation used? Describe the inherent symmetry between the resistance values of the Y circuit and those of the Δ circuit. (See CAD)

Concept Question 2-11: How are the elements of a balanced Y circuit related to those of its equivalent Δ circuit? (See CAD)

Exercise 2-9: For each of the circuits shown in Fig. E2.9, determine the equivalent resistance between terminals (a, b).

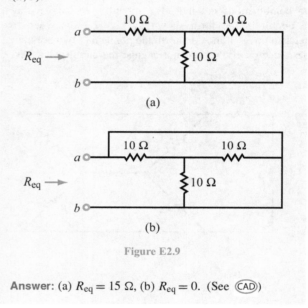

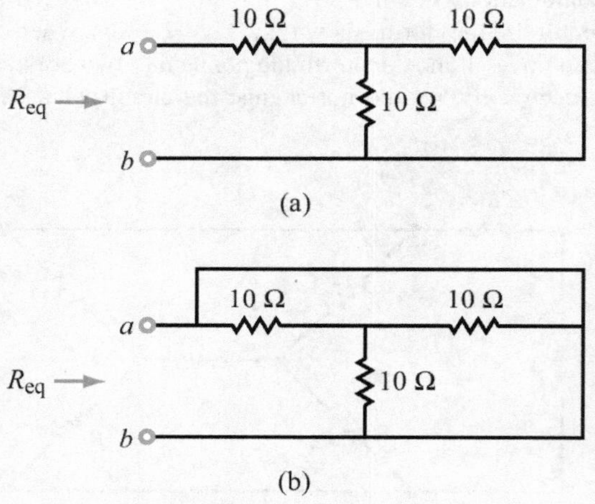

Figure E2.9

Answer: (a) $R_{\text{eq}} = 15\ \Omega$, (b) $R_{\text{eq}} = 0$. (See CAD)

2-5 The Wheatstone Bridge

Developed initially by Samuel Christie (1784–1865) in 1833 as an accurate ohmmeter for measuring resistance, the Wheatstone bridge subsequently was popularized by Sir Charles Wheatstone (1802–1875), who used it in a variety of practical applications. Today, the Wheatstone-bridge circuit is integral to numerous sensing devices, including strain gauges, force and torque sensors, and inertial gyros. The reader is referred to Technology Brief 3 for an illustrative example.

The Wheatstone-bridge circuit shown in Fig. 2-36 consists of four resistors: two fixed resistors (R_1 and R_2) of known values,

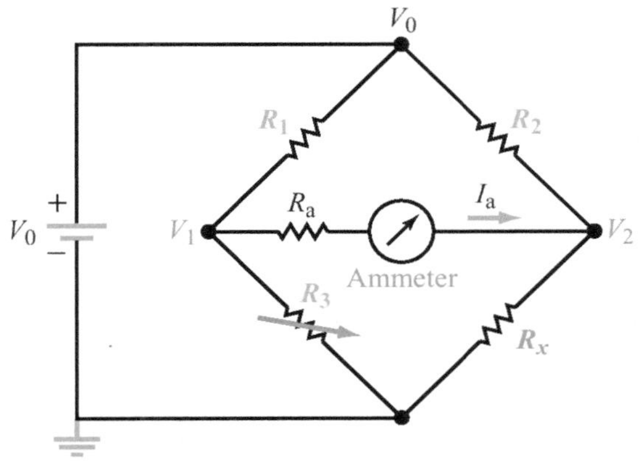

Figure 2-36: Wheatstone-bridge circuit containing an adjustable variable resistor R_3 and an unknown resistor R_x. When R_3 is adjusted to make $I_a = 0$, R_x is determined from $R_x = (R_2/R_1)R_3$.

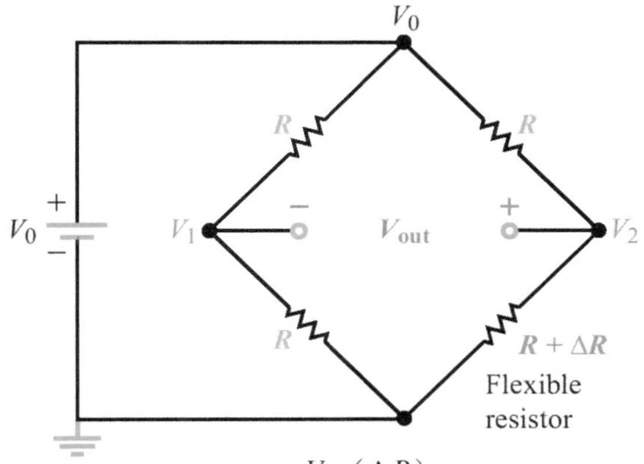

$$V_{\text{out}} \approx \frac{V_0}{4}\left(\frac{\Delta R}{R}\right)$$

Figure 2-37: Circuit for Wheatstone-bridge sensor.

an adjustable resistor R_3 whose value also is known, and a resistor R_x of unknown resistance. A dc voltage source V_0 is connected between the top node and ground, and an ammeter is connected between nodes 1 and 2. The standard procedure for determining R_x starts by adjusting R_3 so as to make $I_a = 0$.

▶ The absence of current flow between nodes 1 and 2, called the ***balanced condition***, implies that $V_1 = V_2$. ◀

From voltage division, $V_1 = R_3V_0/(R_1 + R_3)$, and $V_2 = R_xV_0/(R_2 + R_x)$. Hence,

$$\frac{R_3V_0}{R_1 + R_3} = \frac{R_xV_0}{R_2 + R_x}. \tag{2.45}$$

A balanced bridge also implies that the voltages across R_1 and R_2 are equal,

$$\frac{R_1V_0}{R_1 + R_3} = \frac{R_2V_0}{R_2 + R_x}. \tag{2.46}$$

Dividing Eq. (2.45) by Eq. (2.46) leads to

$$\frac{R_3}{R_1} = \frac{R_x}{R_2},$$

from which we have

$$R_x = \left(\frac{R_2}{R_1}\right)R_3 \qquad \text{(balanced condition)}. \tag{2.47}$$

Example 2-16: Wheatstone-Bridge Sensor

A special version of the Wheatstone bridge (Fig. 2-37) is configured specifically for *measuring small deviations from a reference condition*. An example of a reference condition might be a highway bridge with no load on it. A strain gauge employing a high-sensitivity flexible resistor can measure the small deflection in the bridge surface caused by the weight (force) of a car or truck when present on it. As the force deflects the surface of the bridge to which the resistor is attached, the resistor stretches in length, causing its resistance to increase from a nominal value R (under no stress) to $R + \Delta R$. The other three resistors in the Wheatstone-bridge circuit are all identical and equal to R. Thus, when no vehicles are present on the bridge, the circuit is in the balanced condition.

Develop an approximate expression for V_{out} (the output voltage between nodes 1 and 2) for $\Delta R/R \ll 1$.

Solution: Voltage division gives

$$V_1 = \frac{V_0R}{R + R} = \frac{V_0}{2}$$

and

$$V_2 = \frac{V_0(R + \Delta R)}{R + (R + \Delta R)} = \frac{V_0(R + \Delta R)}{2R + \Delta R}.$$

Hence,

$$\begin{aligned} V_{\text{out}} = V_2 - V_1 &= \frac{V_0(R + \Delta R)}{2R + \Delta R} - \frac{V_0}{2} \\ &= \frac{2V_0(R + \Delta R) - V_0(2R + \Delta R)}{2(2R + \Delta R)} \\ &= \frac{V_0\,\Delta R}{4R + 2\,\Delta R} = \frac{V_0\,\Delta R}{4R(1 + \Delta R/2R)}. \end{aligned}$$

Since $\Delta R/R \ll 1$, ignoring the second term in the denominator would incur negligible error. Such an approximation leads to

$$V_{\text{out}} \approx \frac{V_0}{4}\left(\frac{\Delta R}{R}\right), \qquad (2.48)$$

providing a simple linear relationship between the change in resistance ΔR and the output voltage V_{out}.

Concept Question 2-12: What is a Wheatstone bridge used for? (See CAD)

Concept Question 2-13: What is the *balanced condition* in a Wheatstone bridge? (See CAD)

Exercise 2-10: If in the sensor circuit of Fig. 2-37, $V_0 = 4$ V and the smallest value of V_{out} that can be measured reliably is 1 μV, what is the corresponding accuracy with which $(\Delta R/R)$ can be measured?

Answer: 10^{-6} or 1 part in a million. (See CAD)

2-6 Application Note: Linear versus Nonlinear *i–υ* Relationships

Ideal resistors and voltage and current sources are all considered linear elements; the relationship between the current and the voltage across any one of them is described by a straight line. The *i–υ* relationships plotted in Fig. 2-38 for the current source, the voltage source, and the resistor have slopes of 0, ∞, and $1/R$, respectively.

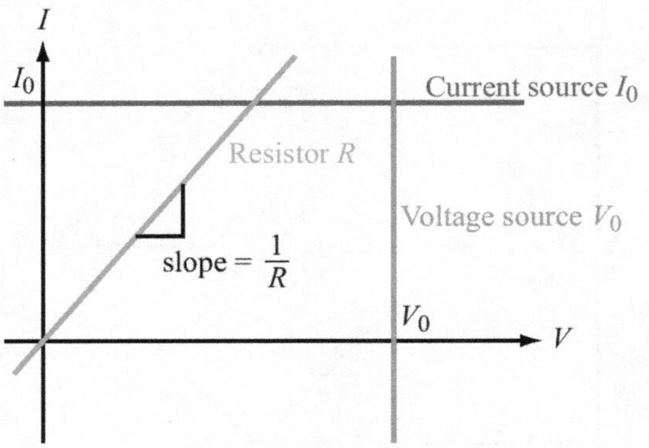

Figure 2-38: I–V relationships for a resistor R, an ideal voltage source V_0, and an ideal current source I_0.

2-6.1 The Fuse: A Simple Nonlinear Element

Many very useful circuit elements do not have linear *i–υ* relationships. Consider Fig. 2-39(a). A realistic voltage source is connected to a load R_L at terminals (a, b). Note that the resistance value of the source resistor R_s is much smaller than that of the load (1 Ω versus 1 kΩ). It is typical of a well-designed voltage source to have a small source resistor so as to minimize the voltage drop across it. The switch simulates an accidental short circuit. Application of KVL to the loop in Fig. 2-39(a) (with the switch in the open position) leads to

$$I_s = \frac{V_s}{R_s + R_L} = \frac{100}{1 + 1000} \approx 0.1 \text{ A} \qquad \text{(switch open).}$$

If, *accidentally*, a short circuit were to be introduced across terminal (a, b), which is represented schematically by the closing of the SPST switch, the current I_s will flow entirely through the short circuit, resulting in

$$I_s = \frac{V_s}{R_s} = 100 \text{ A!} \qquad \text{(switch closed).}$$

This is a very large current. Many household wires would begin to overheat and melt off their insulation at such high currents.

It is precisely for this reason that the *fuse* (and later, the *breaker*) came into heavy use in power-distribution circuits [Fig. 2-39(b)]. The *i–υ* curve for a fuse, shown in [Fig. 2-39(c)], is decidedly nonlinear: Above a certain current level, the fuse will cease to allow more current to pass through it, acting like a current limiter. The physical device contains a small metal wire that is designed to melt away at a specific current level

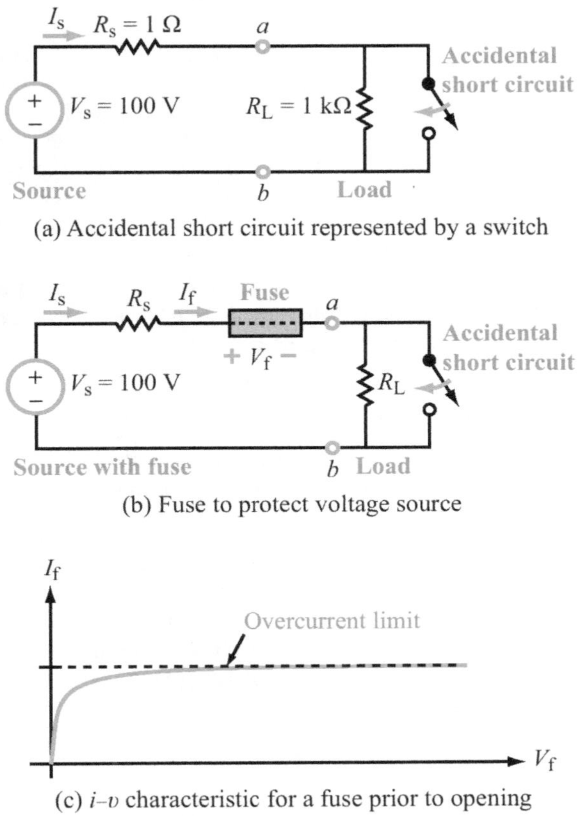

Figure 2-39: Use of a fuse to protect a voltage source.

(called its ***overcurrent***), thereby becoming an open circuit and preventing large currents from flowing through the circuit. Note that Fig. 2-39(b) does not explain the fuse's time-dependent behavior; it describes the fuse's behavior only until the moment at which the current exceeds the overcurrent. After that, the fuse just looks like an open circuit.

Fuses also are rated for several other important characteristics such as how fast they can respond. Ultra-fast fuses can trip in milli- to micro-seconds. Another important attribute is the maximum voltage it can sustain across its terminals. Note that in Fig. 2-39(b), once the fuse assumes the role of an open circuit, the voltage across it becomes V_s. If this voltage is too high, arcing and sparks might develop between the terminals (we know from physics that a large-enough voltage in air will break down the air molecules, causing them to conduct and generate a bright spark). Clearly, that is an important rating factor to keep in mind when selecting a fuse.

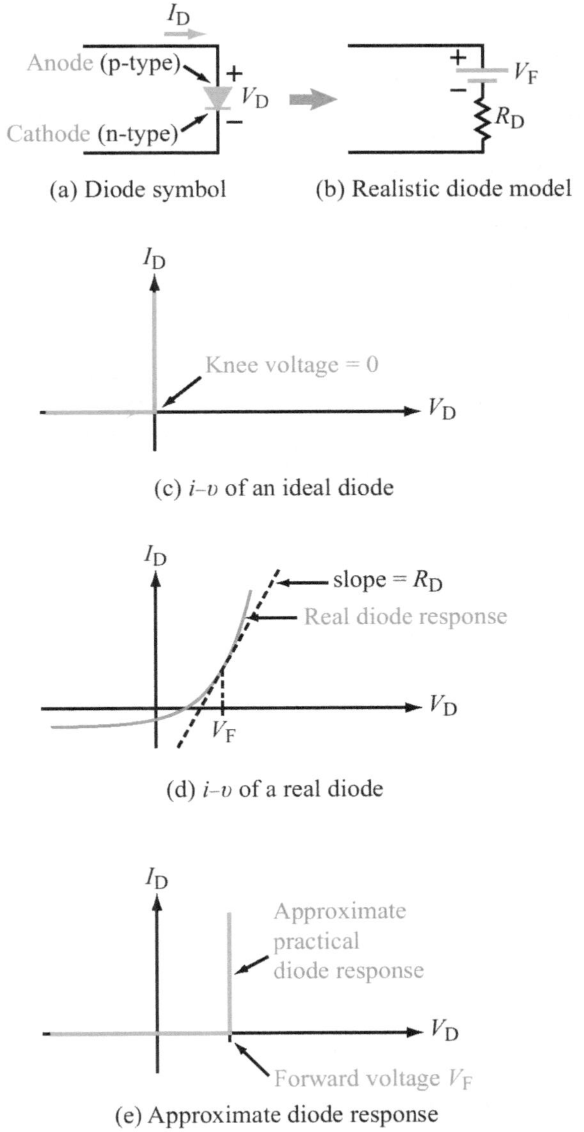

Figure 2-40: pn-junction diode schematic symbol and i–υ characteristics.

2-6.2 The Diode: A Solid-State Nonlinear Element

The ***diode*** is a mainstay of solid-state circuits. Its circuit schematic symbol is shown in Fig. 2-40(a) with V_D as the voltage across the diode, defined such that the (+) side is at the anode terminal of the diode and the (−) side at its cathode terminal. There are many types of diodes, including the basic *pn-junction diode*, the Zener and Schottky diodes, and the ubiquitous *light-emitting diode* (LED) used in consumer electronics. A brief introduction of the LED was made earlier in Section 2-1.4, and a more detailed overview of its operation is provided in Technology Brief 5. For the present, we will limit our discussion to the pn-junction diode, commonly referred to simply as *the diode*. The *pn* diode consists of a *p-type* semiconductor placed in contact with an *n-type* semiconductor, thereby forming a *junction*. The p-type material is so named because the impurities that have been added to its bulk material result in a crystalline structure in which the available charged carriers are predominantly *positive* charges. The opposite is true for the n-type material; different types of impurities are added to the bulk material, as a result of which the predominant carriers are *negative* charges (electrons). In the absence of a voltage across the diode, the two sets of carriers diffuse away from each other at the edge of the junction, generating an associated built-in potential barrier (voltage), called the *forward-bias voltage* or *offset voltage* V_F.

The main use of the diode is as a *one-way valve* for current. Figure 2-40(c) displays the i–υ relationship for an *ideal diode*, which conveys the following behavior:

> ▶ Current can flow through the diode from the (+) terminal to the (−) terminal unimpeded, regardless of its magnitude, but it cannot flow in the opposite direction. ◀

In other words, an ideal diode looks like a short circuit for positive values of V_D and like an open circuit for negative values of V_D. These two states are called *forward bias* and *reverse bias*, respectively. When a positive-bias voltage exceeding V_F is applied to the diode, the potential barrier is counteracted, allowing the flow of current from p to n (which includes positive charges flowing in that direction as well as negative charges flowing in the opposite direction). On the other hand, if a negative-bias voltage is applied to the diode, it adds to the potential barrier, further restricting the flow of charges across the barrier and resulting in no current flow from n to p.

The voltage level at which the diode switches from reverse bias to forward bias is called the *knee voltage* or *forward-bias voltage*. For the ideal diode, $V_F = 0$ and the knee is at $V_D = 0$, which means that the forward-bias segment of its i–υ characteristic is aligned perfectly along the I_D axis, as shown in Fig. 2-40(c).

Real diodes differ from the ideal diode model in two important respects: (1) the knee in the curve is not at $V_D = 0$, and (2) the diode does not behave exactly like a perfect short circuit when in forward bias nor like a perfect open circuit when in reverse bias. Figure 2-40(d) shows a realistic diode i–υ curve. Note how nonlinear a real diode really is! For many electrical engineering applications, however, the nonlinearities are not so important, and the approximate ideal-like diode model shown in Fig. 2-40(e) is quite sufficient. The only difference between the ideal diode model of Fig. 2-40(c) and the approximate diode model of Fig. 2-40(e) is that in the latter the transition from reverse to forward bias occurs at a non-zero, positive value of V_D, namely the *forward-bias voltage* V_F. For a silicon pn-junction diode, a typical value of V_F is 0.7 V and a realistic model is shown in Fig. 2-40(b). A typical value of R_D is 10–20 Ω. We always should remember that V_F is a property of the diode itself, not of the circuit it is a part of.

Example 2-17: Diode Circuit

The circuit in Fig. 2-41 contains a diode with $V_F = 0.7$ V. Determine I_D, assuming R_D to be negligibly small.

Solution: Initially, we do not know whether the diode is forward biased or reverse biased. We will first assume it is forward biased in order to compute I_D. Then, if it turns out that I_D is positive, our assumption will have been validated, but if I_D is negative, we will conclude that the diode is reverse biased and no current flows through the circuit.

Application of KVL around the loop gives

$$-V_s + I_D R + V_D = 0.$$

If the diode is forward biased, $V_D = 0.7$ V, which leads to

$$I_D = \frac{V_s - V_D}{R} = \frac{5 - 0.7}{100} = 43 \text{ mA}.$$

The positive sign of I_D confirms our assumption that the diode is indeed forward biased.

As an interesting aside, one could use this circuit to control the current through a light-emitting diode (LED). As explained in Technology Brief 5, the amount of light emitted by an LED (i.e., how bright it appears) is proportional directly to the current I_D passing through it when it is forward biased. By using the circuit

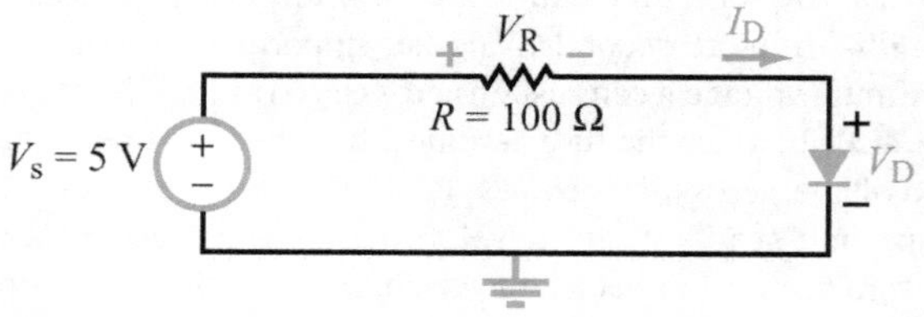

Figure 2-41: Diode circuit of Example 2-17.

in Fig. 2-41 and choosing an appropriate value for R, we can build a circuit that forward biases an LED and controls its brightness.

Example 2-18: Square-Wave Waveform

The circuit in Fig. 2-42 contains two diodes, both with $V_F = 0.7$ V. The waveform of the voltage source consists of a single cycle of a square wave. Generate plots for $i_1(t)$ and $i_2(t)$. Ignore R_D for both diodes.

Solution: Again, we will use the diode model of Fig. 2-40(b). From the analysis of Example 2-17, we concluded that if the voltage across a series combination of a diode and a resistor exceeds V_F of the diode (with the + polarity of the voltage coinciding with the + side of the diode), current will flow through the series combination, but if the voltage is negative, no current will flow through the diode.

For the first half of the source voltage cycle, υ_a, the voltage across the series combination (D_1, R_1) is positive at 6 V. Hence,

$$i_1(t) = \frac{\upsilon_a - 0.7}{R_1} = \frac{6 - 0.7}{53} = 0.1 \text{ A} \qquad \text{for } 0 \le t \le 1 \text{ s}.$$

But for series combination (D_2, R_2), no current will flow through diode D_2 because the polarity of υ_a is opposite of that of the diode. Hence,

$$i_2(t) = 0 \qquad \text{for } 0 \le t \le 1 \text{ s}.$$

The opposite behavior occurs during the second half of the cycle of $\upsilon_s(t)$, diode D_2 will conduct current through it, but diode D_1 will not. Hence,

$$i_1(t) = 0 \qquad \text{for } 1 \le t \le 2 \text{ s},$$

$$i_2(t) = \frac{6 - 0.7}{R_2} = \frac{6 - 0.7}{106} = 0.05\text{A} \qquad \text{for } 1 \le t \le 2 \text{ s}.$$

The combined results are displayed in Fig. 2-42(c).

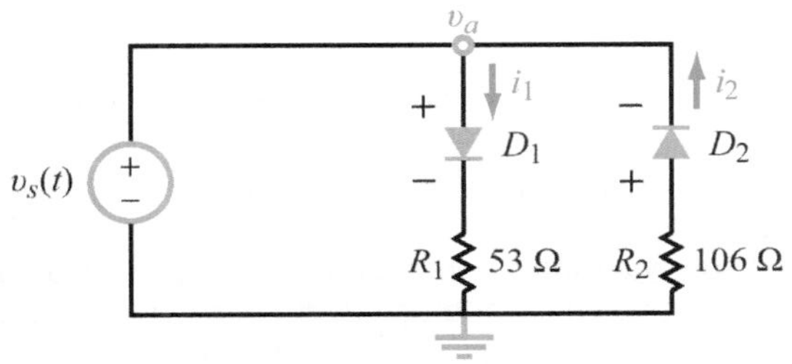

(a) Diode circuit

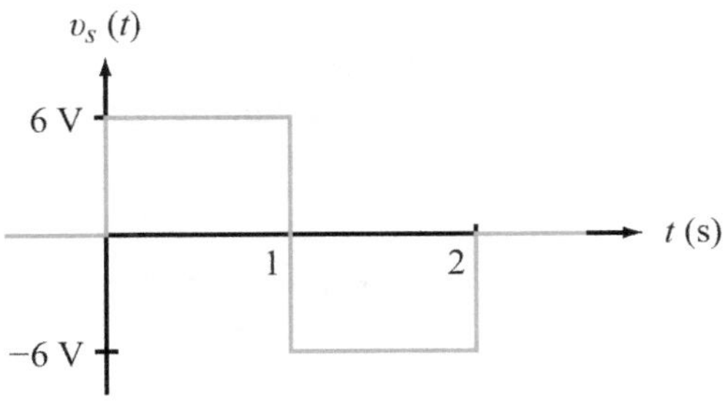

(b) Source voltage waveform

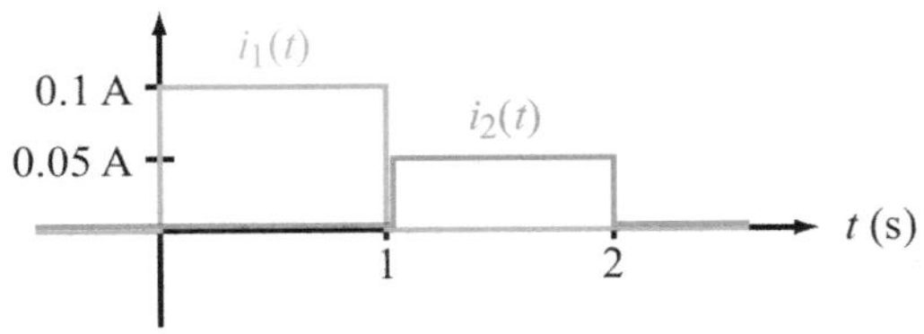

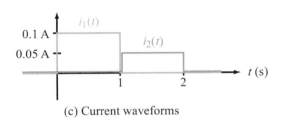

(c) Current waveforms

Figure 2-42: Diode circuit and waveforms of Example 2-18.

> **Concept Question 2-14:** What is the *overcurrent* of a fuse? (See CAD)

> **Concept Question 2-15:** Why does a pn-junction diode have a non-zero forward-bias voltage V_F? (See CAD)

Exercise 2-11: Determine I in the two circuits of Fig. E2.11. Assume $V_F = 0.7$ V for all diodes.

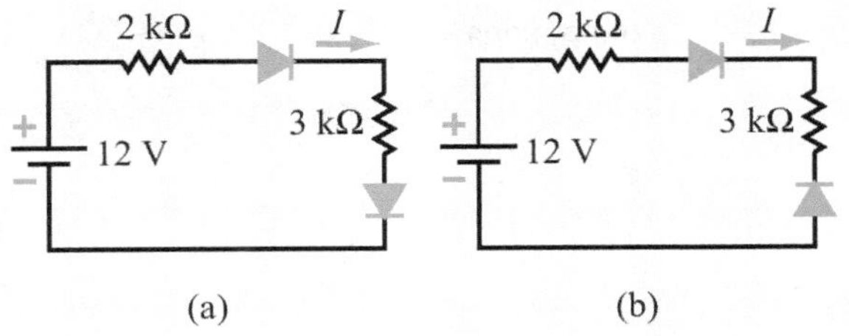

Figure E2.11

Answer: (a) $I = 2.12$ mA, (b) $I = 0$. (See CAD)

Technology Brief 5
Light-Emitting Diodes (LEDs)

How LEDs Are Made

LEDs are a specific type of the much larger family of semiconductor diodes, whose basic behavior we discussed earlier in Section 2-6. When a voltage is applied in the forward-biased direction across an LED, current flows and photons are emitted (Fig. TF5-1). This occurs because as electrons surge through the diode material, they recombine with charge carriers in the material and release energy in the form of photons (quanta of light). The energy of the emitted photon (and hence the wavelength/color) depends on the type of material used to make the diode. For example, a diode made of indium gallium aluminum phosphide (InGaAlP) emits red light, while a diode made from gallium nitride (GaN) emits bluish light. Extensive research over many decades has yielded materials that can emit photons at practically any wavelength from the infrared through ultraviolet (Fig. TF5-2). Various "tricks" have also been employed to modify the emitted light after emission. To make white light diodes, for example, certain blue light diodes can be coated with crystal powders which convert the blue light into a broad-spectrum "white" light. Other coatings such as ***quantum dots*** are still the subject of today's research. In a traditional package, the LED transmits light in a hemispherical pattern, but numerous other light-focused packaging methods are available that can focus the light in virtually any way imaginable. LEDs can be focused using highly reflective coatings to intensify their light for higher power applications.

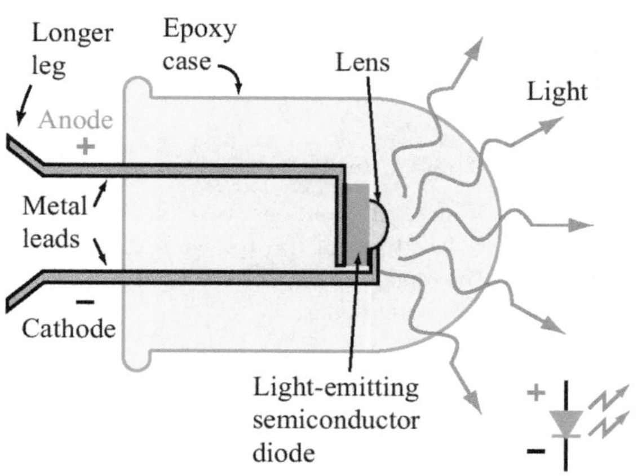

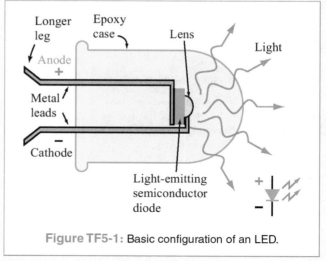

Figure TF5-1: Basic configuration of an LED.

In addition to semiconductor LEDs, a newer class of devices called ***organic light emitting diodes*** (OLEDs) are the subject of intense research efforts. OLEDs operate in a manner that is analogous to conventional LEDs, but are made from organic molecules (often polymers). Because OLEDs are lighter weight than conventional LEDs and can be made to be flexible, they have the potential to revolutionize handheld and lightweight displays, such as those used in phones, PDAs and flexible screens. Imagine a flexible contact lens that could allow you to see a heads-up display or augmented reality!

LED Advantages

LEDs have several major attributes that have made them a key element of many applications. First, they can be

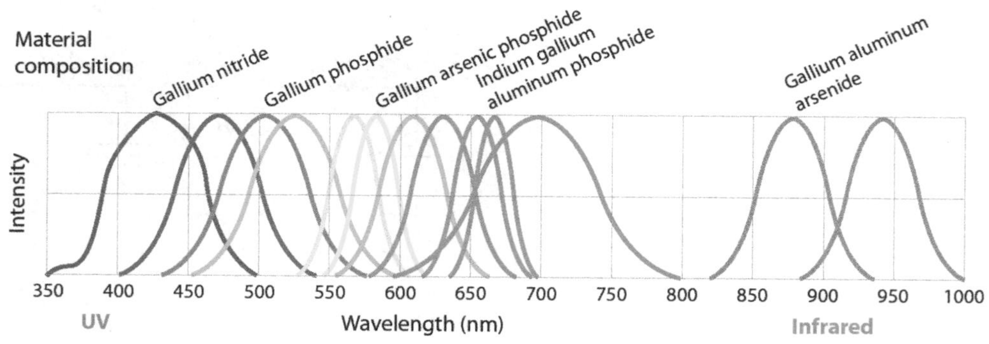

Figure TF5-2: Emission spectra of LEDs made of different material composites.

Figure TF5-3: LED-lit building.

Figure TF5-4: LED eyelashes can be worn in many colors, and can be made to turn on or off with a tip of the head. (Credit: Soomi Park.)

produced in a wide variety of wavelengths from infrared through ultraviolet. Targeted or broad spectra can also be produced, making them applicable to virtually any optical application. Second, they are energy efficient. An incandescent lightbulb uses 80% of its energy for heat and 20% to produce light. LEDs use only about 20% of their energy for heat and 80% for light. This also makes them cool, requiring less energy to remove the excess heat. Third, they are manufactured in a huge array of colors, sizes, shapes, designs, and more. They are affordable (not yet less expensive than incandescent bulbs in the initial purchase price, but definitely less expensive over the lifetime of the bulb). Fourth, they last longer (often > 100k hours) than incandescent bulbs, which is particularly important in hard-to-reach applications. Fifth, they can be integrated directly into semiconductor circuits, printed circuit boards, and light-focusing packages. Various combinations of these advantages are key to the following broad range of applications of LEDs.

LEDs for Lighting

In an era where energy efficiency matters financially, environmentally, and practically, LEDs have become a popular mainstay in home and office lighting, street lighting and consumer products from home appliances and toys to high-efficiency tail lights for cars and flashlights. Of growing importance is the replacement of traditional incandescent bulbs with LEDs in homes and buildings (Fig. TF5-3), because of their energy efficiency. But lighting is more than just enabling us to see at night. LEDs can be used in horticulture to efficiently target ideal wavelengths for plant growth, and exposing produce to certain wavelengths of light can help it ripen on demand, or can extend its ripened shelf life. UV LEDS are being explored to enhance development of polyphenol, which are believed to have antioxidant qualities, in growth of green, leafy vegetables. LEDs provide high visibility bike lights, safety vests, tennis shoes, and more. They are also used artistically for decoration and advertising on buildings and signs, woven into clothes often augmented by plastic fiber optic threads (e.g., Philips Research Lumalive textiles), or even worn with LED eyelashes (see Fig. TF5-4)!

LEDs for Medical Applications

LEDs are used for a variety of medical applications. One particularly important application is the pulse oximeter (Fig. TF5-5), which measures blood oxygen level and pulse rate. Oxygenated blood absorbs light at 660 nm (red light), whereas deoxygenated blood absorbs light at 940 nm (infrared). Pulse oximeters use two LEDs, one at 600 nm and another at 940 nm, which are arranged to transmit through a translucent section of the body such as the finger or ear lobe. Two associated light collecting sensors are placed on the opposite side to measure the amount of each wavelength that is transmitted through the body. The ratio of the red and infrared light indicates how much oxygen is in the blood. To insure that the received light signals are actually from the blood, the measurement is made over several seconds (several pulses), focusing in on the pulsing blood rather than the static surrounding tissues.

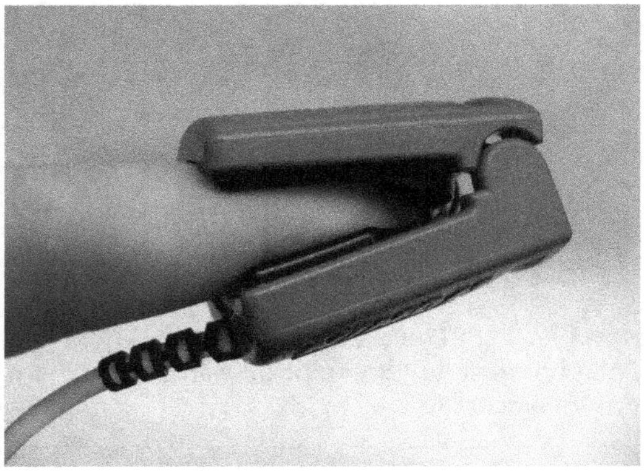

Figure TF5-5: Pulse oximeter used to measure blood oxygen content.

Figure TF5-6: Large LED display.

LEDs are also used to treat many superficial (skin) conditions. Red light in the range of 600–950 nm can be used to treat acne, rosacea, and wrinkles. The red light works by stimulating the mitochondria in the skin to make older cells behave like younger cells. Blue-light therapy in the 405–420 nm range is used for acne treatments and "anti-aging" skin therapies because of its ability to stimulate collagen in the skin. Green to yellow light (532–595 nm) can reduce skin redness (rosacea). Combining LED light sources with topical drug treatments that are ***photoactivated*** may be used to treat a variety of skin conditions including skin cancer and pre-cancer.

LEDs are also used extensively in dentistry. Blue LEDs can be used to cure (harden) polymer composite materials used for fillings. The rate at which the filling material cures is proportional to the power carried by the LED light, so high power LEDs are used to speed up the curing process.

Ultraviolet (UV) LEDs

The UV range provides a wealth of applications, and low-cost high-power UV LEDs are enabling many of these applications. Inks (printing), adhesives and coatings are often cured with LEDs in the UV range (primarily 395 nm, 385 nm or 365 nm). UV LED flashlights are used to detect fraudulent identification (at the airport, for example) and currency. UV-LEDs are used extensively in forensic analysis and drug discovery. In the lower UV spectral range (100–280 nm) LEDs sterilize air and water by breaking up the DNA and RNA of microorganisms and preventing their reproduction. For example, 275 nm is believed to be the most effective wavelength for eradicating pathogens such as E-coli in water. LEDs in this range are also used for spectroscopic and fluorescence measurements and for chemical and biological detectors.

LED Displays

LEDs, with their wide range of colors, efficiency, low cost, flexibility, low profile and light weight, are ideal for both handheld displays and much larger displays (such as billboards and signage, as shown in Fig. TF5-6). Some LED displays use edge lighting where LEDs shine light across the screen (allowing the display to be thinner than traditional screens but not improving picture quality). Others use RGB LEDs. These LEDs use a common anode but have separate cathodes for red, green and blue LEDs (making the composite a 4-pin LED). They can be made to generate light with almost any color, depending on the voltages applied across the combination of RGB pins. This greatly enhances picture color. RGB LEDs can also be dimmed independently and instantly (giving a more dynamic picture, especially great "black" levels for dark scenes). The flexibility and bendability of OLEDs promise new, creative options for the next generation of TVs and smart phones—can you imagine rolling your TV up like a poster and carrying it with you anywhere? Or wearing it? Or ...?

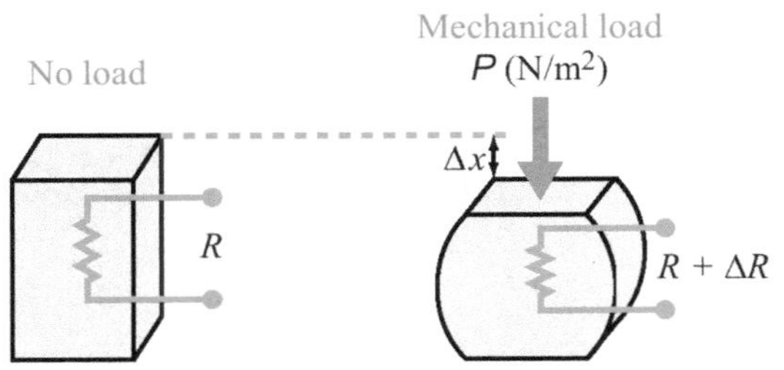

Figure 2-43: The resistance of a piezoresistor changes when mechanical stress is applied.

2-6.3 Piezoresistor Circuit

According to Technology Brief 4, if we apply a force on a resistor along its axis (Fig. 2-43), the resistance changes from R_0, which is the resistance with no stress (pressure) applied, to R as

$$R = R_0 + \Delta R, \tag{2.49}$$

and the deviation ΔR is given by

$$\Delta R = R_0 \alpha P, \tag{2.50}$$

where α is a property of the material that the resistor is made of and is called its *piezoresistive coefficient*, and P is the *mechanical stress* applied to the resistor. The unit for P is newtons/m^2 (N/m^2) and the unit for α is the inverse of that. Compression decreases the length of the resistor and increases its cross section, so in view of Eq. (2.2), which states that the resistance of a longitudinal resistor is given by $R = \rho\ell/A$, the consequence of a compressive force—namely reduction in ℓ and increase in A—leads to a reduction in the magnitude of R.

► Hence, for compression, ΔR is negative, requiring that α in Eq. (2.50) be defined as a negative quantity. ◄

If a piezoresistor is integrated into a Wheatstone-bridge circuit (as in Fig. 2-37), such that all three other resistors are given by R_0, the expression for the voltage output given by Eq. (2.48) becomes

$$V_{\text{out}} = \frac{V_0}{4}\left(\frac{\Delta R}{R_0}\right) = \frac{V_0}{4}\,\alpha P. \tag{2.51}$$

Since V_0 and α are both constants, the linear relationship between the applied stress P and the output voltage V_{out} makes the piezoresistor a natural sensor for detecting or measuring mechanical stress. However, we should examine the sensitivity of such a sensor. As a reference, a finger can apply about 50 N of force across an area of 1 cm^2 (10^{-4} m^2), which is equivalent to a pressure $P = 5 \times 10^5$ N/m^2. If the piezoresistor is made of silicon with $\alpha = -1 \times 10^{-9}$ m^2/N and if the dc source in the Wheatstone bridge is $V_0 = 1$ V, Eq. (2.51) yields the result that $V_{\text{out}} = -125\ \mu$V, which is not impossible to measure but quite small nevertheless. How then are such pressure sensors used?

The answer is simple: We need a mechanism to amplify the signal. We can do so electronically by feeding V_{out} into a high-gain amplifier, or we can amplify the mechanical pressure itself before applying it to the piezoresistor. The latter approach can be realized by constructing the piezoresistor into a cantilever structure, as shown in Fig. 2-44 (a cantilever is a fancy name for a "diving board" with one end fixed and the other free). Deflection of the cantilever tip induces stress at the base of the cantilever near the attachment point. If properly designed, the cantilever—which usually is made of silicon or metal—can amplify the applied stress by several orders of magnitude, as we see in the following example.

Example 2-19: A Realistic Piezoresistor Sensor

When a force F is applied on the tip of a cantilever of width W, thickness H, and length L (as shown in Fig. 2-44) the corresponding stress exerted on the piezoresistor attached to the cantilever base is given by

$$P = \frac{FL}{WH^2}. \tag{2.52}$$

Determine the output voltage of a Wheatstone-bridge circuit if $F = 50$ N, $V_0 = 1$ V, the piezoresistor is made of silicon, and the cantilever dimensions are $W = 0.5$ cm, $H = 0.5$ mm, and $L = 1$ cm.

Solution: Combining Eqs. (2.51) and (2.52) gives

$$\begin{aligned} V_{\text{out}} &= \frac{V_0}{4}\,\alpha \cdot \frac{FL}{WH^2} \\ &= \frac{1}{4} \times (-1 \times 10^{-9}) \times \frac{50 \times 10^{-2}}{(5 \times 10^{-3}) \times (5 \times 10^{-4})^2} \\ &= -0.1 \text{ V}. \end{aligned}$$

The integrated piezoresistor–cantilever arrangement generates an output voltage whose magnitude is on the order of 800 times greater than that generated by pressing on the resistor directly!

Concept Question 2-16: Does compression along the current direction increase or decrease the resistance? Why? (See CAD)

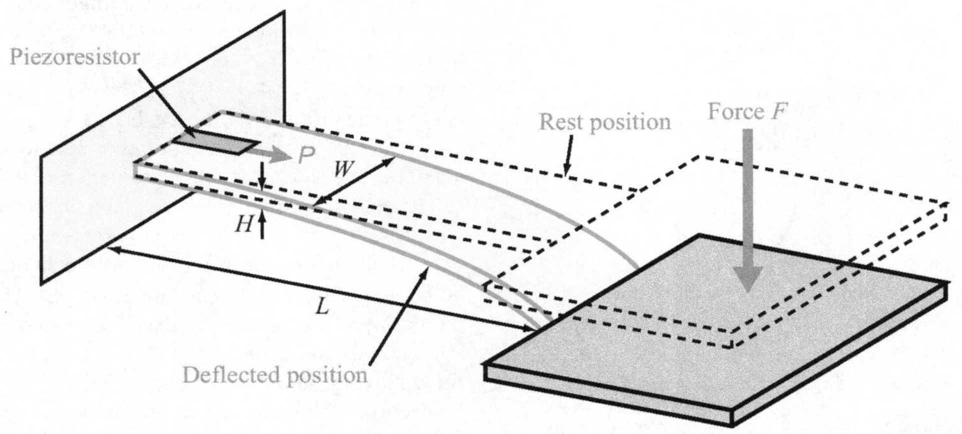

Figure 2-44: A cantilever structure with integrated piezoresistor at the base.

Concept Question 2-17: Why are piezoresistors placed at the base of cantilevers and other deflecting structures? (See CAD)

Exercise 2-12: What would the output voltage associated with the circuit of Example 2-18 change to, if the cantilever thickness is reduced by a factor of 2?

Answer: $V_{\text{out}} = -0.4$ V. (See CAD)

2-7 Introducing Multisim

Multisim 13 is the latest edition of National Instrument's ***SPICE*** simulator software. SPICE, originally short for Simulation Program with Integrated Circuit Emphasis, was developed by Larry Nagel at the University of California, Berkeley, in the early 1970s. It since has inspired and been used in many academic and commercial software packages to simulate analog, digital, and mixed-signal circuits. Modern SPICE simulators like Multisim are indispensable in integrated circuit design; ICs are so complex that they cannot be built and tested on a breadboard ahead of production (see Technology Brief 7). With SPICE, you can draw a circuit from a library of components, specify how the components are connected, and ask the program to solve for all voltages and currents at any point in time. Modern SPICE packages like Multisim include very intuitive graphic user interface (GUI) tools that make both circuit design and analysis very easy. Multisim allows the user to simulate a laboratory experience on his/her computer ahead of actually working with real components.

In this section, you will learn how to:

- Set up and analyze a simple dc circuit in Multisim.
- Use the Measurement Probe tool to quickly solve for voltages and currents.
- Use the Analysis tools for more comprehensive solutions.

We will return to these concepts and learn to apply many other analysis tools throughout the book. Appendix C provides an introduction to the Multisim Tutorial available on the book website http://c3.eecs.umich.edu/. The Tutorial is a useful reference if you have never used Multisim before. When defining menu selections starting from the main window, the format Menu → Sub-Menu1 → Sub-Menu2 will be used.

2-7.1 Drawing the Circuit

After installing and running Multisim, you will be presented with the ***basic user interface*** window, also referred to as the ***circuit window*** or the ***schematic capture window*** (see Multisim Tutorial on the book website). Here, we will draw our circuits much as if we were drawing them on paper.

Placing resistors in the circuit

Components in Multisim are organized into a hierarchy going in a descending general order from Database → Group → Family → Component. Every component that you use in Multisim will fit into this hierarchy somewhere.

Place → Component opens the Select a Component window. (Ctrl-W is the shortcut key for the place-component command. Multisim has many shortcut keys,

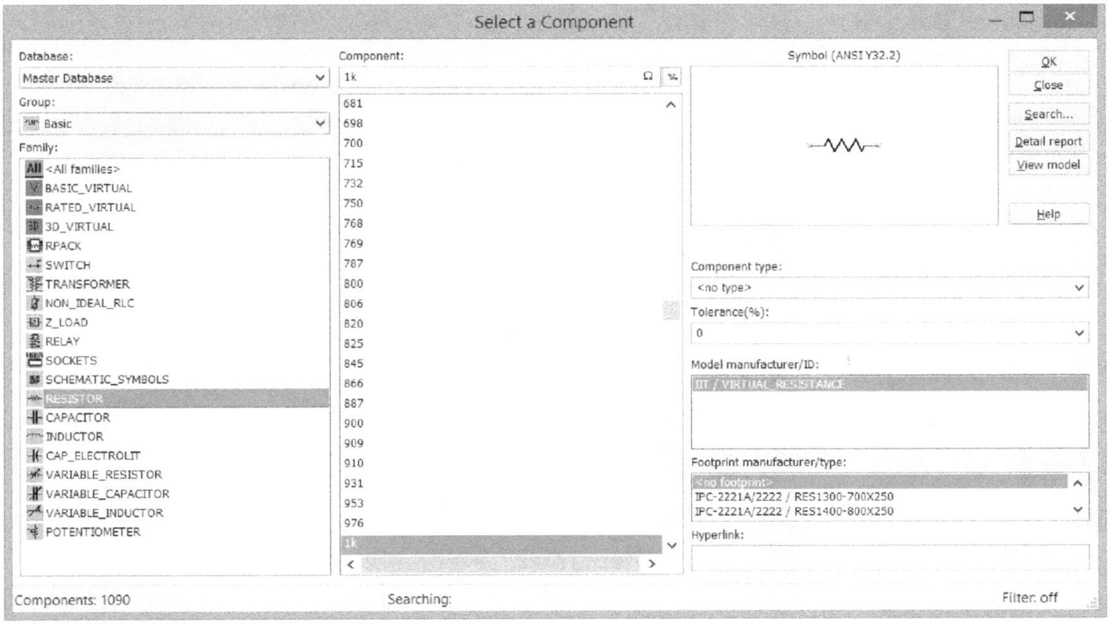

Figure 2-45: Multisim screen for selecting and placing a resistor.

and it will be worthwhile for you to learn some of the basic ones to improve your efficiency in creating and testing circuits.)

Choose Database: Master Database and Group: Basic in the pulldown menus.

Now select Family: RESISTOR.

You should see a long list of resistor values under Component and the schematic symbol for a resistor (Fig. 2-45). Note that the Family menu contains other components like inductors, capacitors, potentiometers, and many more. We will use these in later chapters.

Scroll down and select a 1k value (the units are in ohms) and then click OK. You should see a resistor in the capture window. Before clicking in the window, Ctrl-R allows you to rotate the resistor in the window. Rotate the resistor such that it is vertical and then click anywhere on the window to place it. Repeat this operation; this time place a vertical 100-ohm resistor directly below the first one (as in Fig. 2-46). How to connect them together will be described shortly. Once you are finished placing components, click Close to return to the schematic capture window.

Note that the components have symbolic names (R1 and R2) and values displayed next to them (1k and 100). Also, by double-clicking on a specific component, you can access many details of the component model and its values. For now, it is sufficient to know that the Resistance value can be altered at any time through the Value menu.

Placing an independent voltage source

Just as you did with the resistors, open up the Select a Component window.

Choose Database: Master Database and Group: Sources in the pulldown menus.

Select Family: POWER_SOURCES.

Under Component select DC_POWER and click OK.

Place the part somewhere to the left of the two resistors (Fig. 2-46).

Once placed, close the component window, then double-click on component V1. Under the Value tab, change the Voltage to 10 V. Click OK.

Wiring components together

Place → Wire allows you to use your mouse to wire components together with click-and-drag motions (Ctrl-Q is the shortcut key for the wire command). You

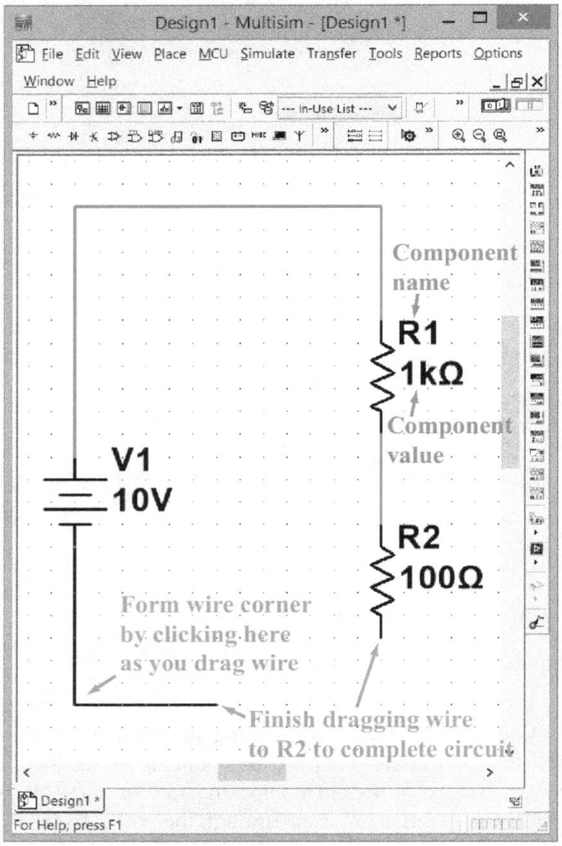

Figure 2-46: Adding a voltage source and completing the circuit.

can also enable the wire tool automatically by moving the cursor very close to a component node; you should see the mouse pointer change into a black circle with a cross-hair.

Click on one of the nodes of the dc source with the wire tool activated (you should see the mouse pointer change from a black cross to a black circle with a cross hair when you hover it over a node). Additional clicks anywhere in the schematic window will make corners in the wire. Double-clicking will terminate the wire. Additionally, when not already dragging a wire, double-clicking on any blank spot of the schematic will generate a wire based at the origin of clicking.

Wire the components as shown in Fig. 2-46. Add a GROUND reference point as shown in Fig. 2-47. The Ground can be found in the Component list of POWER_SOURCES. We now have a resistive divider.

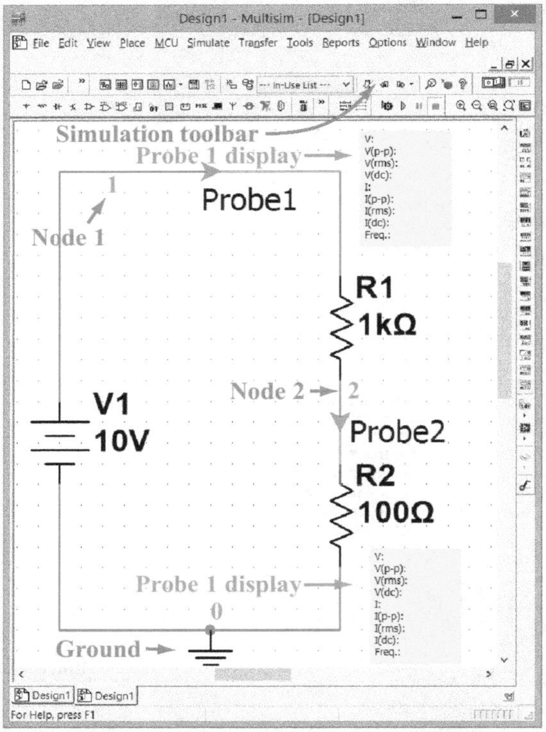

Figure 2-47: Executing a simulation.

2-7.2 Solving the Circuit

In Multisim, there are two broad ways in which to solve a circuit. The first, called *Interactive Simulation*, allows you to utilize virtual instruments (such as ohmmeters, oscilloscopes, and function generators) to measure aspects of a circuit in a time-based environment. It is best to think of the Interactive Simulation as a simulated "in-lab" experience. Just as in real life, time proceeds in the Interactive Simulation as you analyze the circuit (although the rate at which time proceeds is heavily dependent on your computer's processor speed and the resolution of the simulation). The Interactive Simulation is started using the F5 key, the  button, or the  toggle switch. The simulation is paused using the F6 key, the  button, or the  button. The simulation is terminated using either the  button or the  toggle switch.

The other main way in which to solve a circuit in Multisim is through *Analyses*. These simulations display their outputs not in instruments, but rather in the *Grapher* window (which may produce tables in some instances). These simulations are run for controlled amounts of time or over controlled sweeps of specific variables or other aspects of the circuit. For example, a dc sweep simulates the values of a specified voltage or current in the circuit over a defined range of dc input values.

Each of the methods described has its own advantages and disadvantages, and in fact, both varieties can perform many of the same simulations, albeit with different advantages. The choice of method to be used for a given circuit really comes down to your preferences, which will be formed as you gain more experience with Multisim.

For the circuit in Fig. 2-47, we wish to solve for the voltages at every node and the currents running through every branch. As you will often see in Multisim, the solution can be obtained using either the Interactive Simulation or through one of the Analyses. We will demonstrate both approaches.

Interactive simulation

Selecting Simulate → Instruments → Measurement Probe allows you to drag and place a measurement probe onto any node in the circuit. (Note that the Instruments menu contains many common types of equipment used in an electronics laboratory.) The Measurement Probe constantly reports both the current running through the branch to which it is assigned and the voltage at that node. Place two probes into the circuit as shown in Fig. 2-47. When placed, by default, the probes should be pointing in the direction shown in Fig. 2-47. If they are not, you can reverse a probe's direction by right-clicking on it and pressing Reverse Probe Direction. Once the probes are in place, you must run the simulation using the commands for Interactive Simulations.

As expected, the current running through both wires is the same since the circuit has only one loop.

$$I = \frac{V_1}{R_1 + R_2} = \frac{10}{1000 + 100} = 9.09 \text{ mA}.$$

The voltage at node 1 is 10 V, as defined by the source. Application of voltage division (Fig. 2-19) gives

$$V_2 = \left(\frac{R_2}{R_1 + R_2}\right) V_1 = \left(\frac{100}{1100}\right) 10 = 0.909 \text{ V}.$$

DC operating point analysis

The circuit also can be solved using Simulate → Analyses → DC Operating Point. This method is more convenient than the Interactive Simulation when solving circuits with many nodes. After opening this window, you can specify which voltages and currents you want solved. [The Interactive Simulation mode must be stopped, not just paused, in order for the DC Operating Point Analysis mode to work.] Under the Output tab, select the two node voltages and the branch current in the Variables in Circuit window. Make sure the Variables in Circuit pull-down menu is set to All Variables. Once selected, click Add and they will appear in the Selected variables for analysis window. Once you have selected all of the variables for which you want solutions, simply click Simulate. Multisim then solves the entire circuit and opens a window showing the values of the selected voltages and currents (Fig. 2-48).

2-7.3 Dependent Sources

Multisim provides both defined dependent sources (voltage-controlled current, current-controlled current, etc.) and a generic dependent source whose definition can be entered as a mathematical equation. We will use this second type in the following example.

Step 1: The dependent sources are established as follows: Place → Component opens the Select a Component window.

Choose Database: Master Database and Group: Sources in the pulldown menus.

Select Family: CONTROLLED_VOLTAGE or CONTROLLED_CURRENT.

Under Component, select ABM_VOLTAGE or ABM_CURRENT and click OK.

The value of ABM sources (which stands for Analog Behavioral Modeling) can be set directly with mathematical expressions using any variables in the circuit. For information on the variable nomenclature, which may be somewhat confusing, see the Multisim Tutorial on the book website.

Step 2: Using what you learned in Section 2-7.1, draw the circuit shown in Fig. 2-49 (including the probe at node 2).

Step 3: Double-click the ABM_CURRENT source. Under the value tab, enter: 3*V(2). The expression V(2) refers to the voltage at node 2. This effectively defines this source as a voltage-controlled current source. Note that when making the circuit, if the node numbering in your circuit differs from that in the example (e.g., if nodes 1 and 2 are switched), then take care to keep track of the differences so that you will use the proper node voltage when writing the equation. To edit or change node labels, double-click any wire to open the Net Window. Under Net name enter the label you like for that node.

To write the expression for I1 next to the current source, go to Place → Text, and then type in the expression at a location near

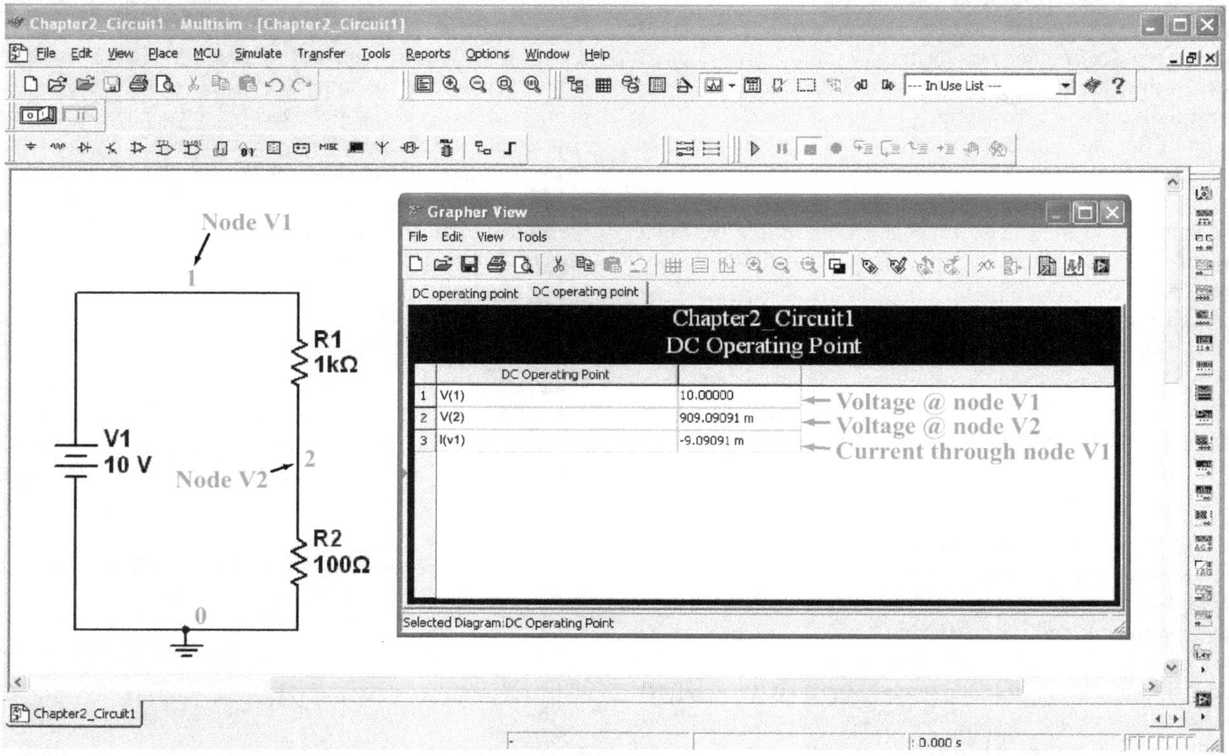

Figure 2-48: Solution window.

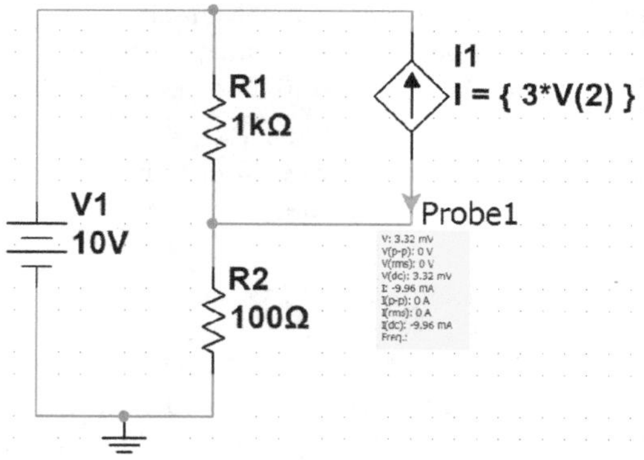

Figure 2-49: Creating a dependent source.

the current source. [Ctrl-T is the shortcut key for the place-text command.]

Referencing currents in arbitrary branches

Now let us analyze the circuit using the DC Operating Point Analysis. Our goal is to solve for the voltages at every node and the current running through each branch. Remove the probe from the circuit if you still have it in there by clicking on it so it is highlighted and pressing the Delete key.

To perform a DC operating point analysis, just as we did earlier in Section 2-7.2, go to Simulate → Analyses → DC Operating Point and transfer all available variables into the Selected variables for analysis window. You should notice that the only variables available are V(1), V(2), and I(v1); if Probe 1 is still connected to your circuit, you should also see I(Probe 1) and V(Probe 1). Where are the other currents, such as the current flowing through R1, the current through R2, or even the current coming out of the dependent source? In Multisim and most SPICE software in general, you can only measure/manipulate currents through a Voltage Source (there are some exceptions, but we will ignore them for now). This is why the current through V1, denoted I(v1), is available but the currents through the other components are not. A simple trick, however, to obtain these currents is to add a 0 V dc source into the branches where you want to measure current. Do this to your circuit, so that it ends up looking like that shown in Fig. 2-50.

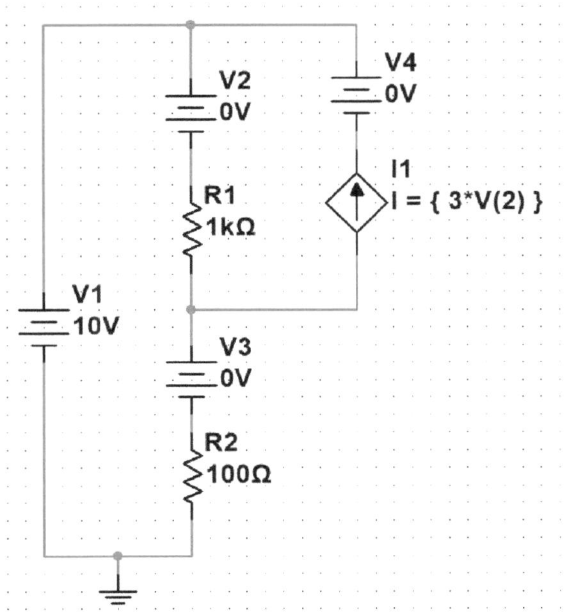

Figure 2-50: Circuit from Fig. 2-49 adapted to read out the currents through R1, R2, and the dependent source.

You will notice that there are new nodes in the circuit now, but since V2, V3, and V4 are 0 V sources, V(3) = V(4) = V(1) and V(5) = V(2).

Go back to the DC Operating Point Analysis window and under the Variables in Circuit window there should now be four currents [I(v1), I(v2), I(v3), and I(v4)] and the five voltages. Highlight all four currents as well as V(1) and V(2) and click Add and then click OK. This will bring up the Grapher window with the results of the analysis.

Note that when we analyze the currents through the branches, the current through a voltage source is defined as going *into* the positive terminal. For example, in source V1, this corresponds to the current flowing *from* Node 1 into V1 and then *out* of V1 to Node 0.

Concept Question 2-18: In Multisim, how are components placed and wired into circuits? (See CAD)

Concept Question 2-19: How do you obtain and visualize the circuit solution? (See CAD)

Exercise 2-13: The circuit in Fig. E2.13 is called a resistive bridge. How does $V_x = (V_3 - V_2)$ vary with the value of potentiometer R_1?

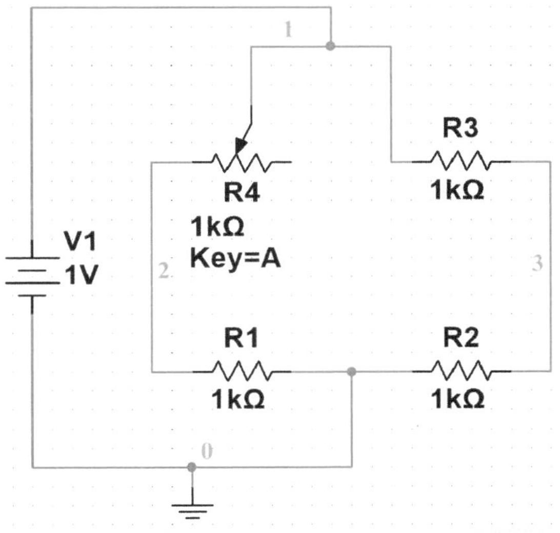

Figure E2.13

Answer: (See CAD)

Exercise 2-14: Simulate the circuit shown in Fig. E2.14 and solve it for the voltage across R_3. The magnitude of the dependent current source is $V_1/100$.

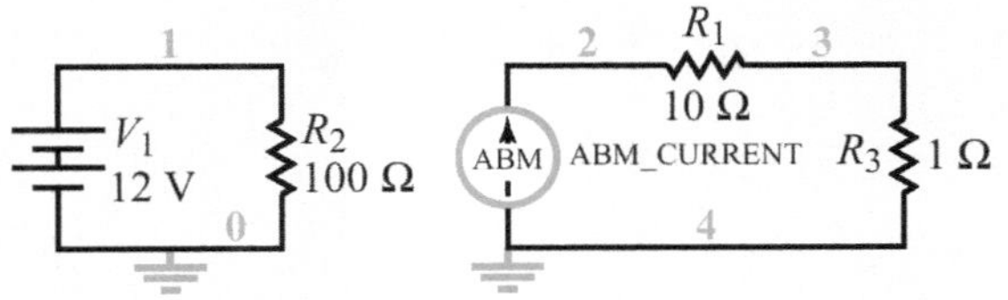

Figure E2.14

Answer: (See CAD)

Summary

Concepts

- As described by Ohm's law, the i–υ relationship of a resistor is linear over a specific range ($-i_{max}$ to $+i_{max}$); however, R may vary with temperature (thermistors), pressure (piezoresistors), and light intensity (LDR).
- Kirchhoff's current and voltage laws form the foundation of circuit analysis and synthesis.
- Two circuits are considered equivalent if they exhibit identical i–υ characteristics relative to an external circuit.
- Source transformation allows us to represent a real voltage source by an equivalent real current source, and vice versa.
- A Y circuit configuration can be transformed into a Δ configuration, and vice versa.
- The Wheatstone bridge is a circuit used to measure resistance, as well as to detect small deviations (from a reference condition), as in strain gauges and other types of sensors.
- Nonlinear resistive elements include the light bulb, the fuse, the diode, and the light-emitting diode (LED).
- Multisim is a software simulation program capable of simulating electric circuits and analyzing their behavior.
- A diode is a one-way valve for current. An LED is a diode that also emits light.

Mathematical and Physical Models

Linear resistor

$$R = \rho \ell / A$$
$$p = i^2 R$$

Kirchhoff current law (KCL)

$$\sum_{n=1}^{N} i_n = 0$$

i_n = current entering node n

Kirchhoff voltage law (KVL)

$$\sum_{n=1}^{N} \upsilon_n = 0$$

υ_n = voltage across branch n

Resistor combinations

In series $$R_{eq} = \sum_{i=1}^{N} R_i$$

In parallel $$\frac{1}{R_{eq}} = \sum_{i=1}^{N} \frac{1}{R_i}$$

or $$G_{eq} = \sum_{i=1}^{N} G_i$$

Voltage division

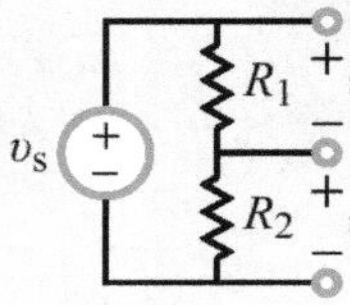

$$\upsilon_1 = \left(\frac{R_1}{R_1 + R_2}\right) \upsilon_s$$
$$\upsilon_2 = \left(\frac{R_2}{R_1 + R_2}\right) \upsilon_s$$

Current division

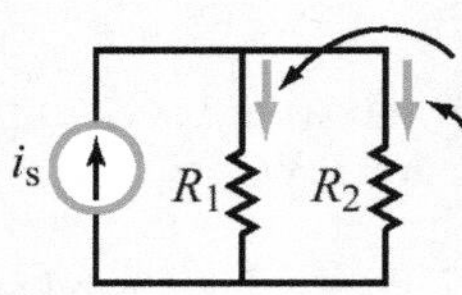

$$i_1 = \left(\frac{R_2}{R_1 + R_2}\right) i_s = \frac{G_1}{G_{eq}} i_s$$
$$i_2 = \left(\frac{R_1}{R_1 + R_2}\right) i_s = \frac{G_2}{G_{eq}} i_s$$

Source transformation

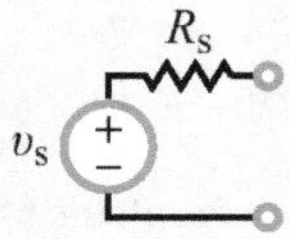

$\Leftrightarrow$ $$i_s = \frac{\upsilon_s}{R_s}$$

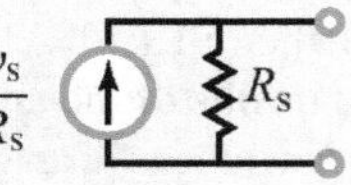

Y–Δ transformation Table 2-5

Wheatstone bridge (Fig. 2-37) $$\upsilon_{out} \approx \frac{V_0}{4}\left(\frac{\Delta R}{R}\right)$$

Important Terms

Provide definitions or explain the meaning of the following terms:

American Wire Gauge
ammeter
Analyses
balanced
balanced condition
basic user interface
breaker
circuit equivalence
circuit window
conductance
conductivity
conductor
current divider
dielectric
diode
equivalent resistor
forward bias
forward-bias voltage
forward voltage
fuse
Grapher
i–υ response
ideal diode
impede
in series
Interactive Simulation
Kirchhoff's current law (KCL)
Kirchhoff's voltage law (KVL)
knee voltage
law of conservation of charge
law of conservation of energy
light-emitting diode
linear region
linear resistor
mechanical stress
Multisim
n-type
negative
NI myDAQ
offset voltage
Ohm's law
one-way valve
overcurrent
p-type
piezoresistive coefficient
piezoresistor
pn-junction diode
positive
potentiometer
power rating
resistance
resistive circuit
resistivity
reverse bias
rheostat
schematic capture window
semiconductor
siemen
source transformation
superconductor
SPICE
thermistor
variable resistance
voltage divider
Wye–Delta (Y–Δ) transformation

PROBLEMS

Section 2-1: Ohm's Law

*2.1 An AWG-14 copper wire has a resistance of 17.1 Ω at 20 °C. How long is it?

2.2 A 3 km long AWG-6 metallic wire has a resistance of approximately 6 Ω at 20 °C. What material is it made of?

2.3 A thin-film resistor made of germanium is 2 mm in length and its rectangular cross section is 0.2 mm × 1 mm, as shown in Fig. P2.3. Determine the resistance that an ohmmeter would measure if connected across its:

(a) Top and bottom surfaces

*(b) Front and back surfaces

(c) Right and left surfaces

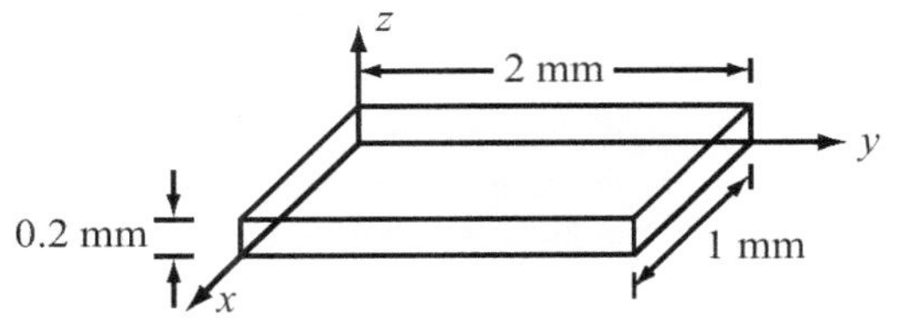

Figure P2.3: Film resistor of Problem 2.3.

2.4 A resistor of length ℓ consists of a hollow cylinder of radius a surrounded by a layer of carbon that extends from $r = a$ to $r = b$, as shown in Fig. P2.4.

(a) Develop an expression for the resistance R.

(b) Calculate R at 20 °C for $a = 2$ cm, $b = 3$ cm and $\ell = 10$ cm.

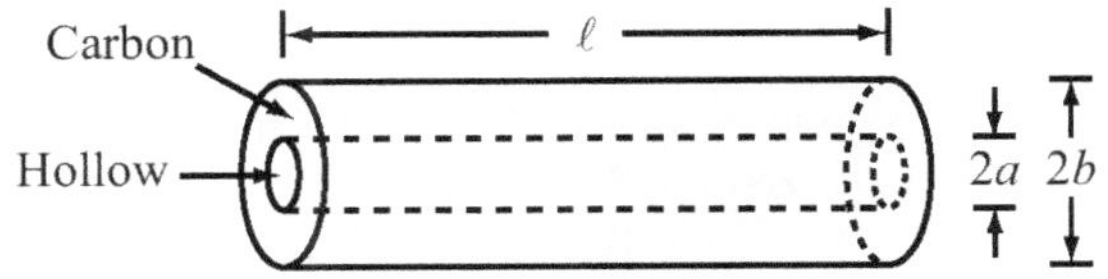

Figure P2.4: Carbon resistor for Problem 2.4.

2.5 A standard model used to describe the variation of resistance with temperature T is given by

$$R = R_0(1 + \alpha T),$$

where R is the resistance at temperature T (measured in °C), R_0 is the resistance at $T = 0$ °C, and α is a

*Answer(s) available in Appendix G.

temperature coefficient. For copper, $\alpha = 4 \times 10^{-3}$ °C^{-1}. At what temperature is the resistance greater than R_0 by 1 percent?

2.6 A light bulb has a filament whose resistance is characterized by a temperature coefficient $\alpha = 6 \times 10^{-3}$ °C^{-1} (see resistance model given in Problem 2.5). The bulb is connected to a 100 V household voltage source via a switch. After turning on the switch, the temperature of the filament increases rapidly from the initial room temperature of 20 °C to an operating temperature of 1800 °C. When it reaches its operating temperature, it consumes 80 W of power.

(a) Determine the filament resistance at 1800 °C.

(b) Determine the filament resistance at room temperature.

(c) Determine the current that the filament draws at room temperature and also at 1800 °C.

(d) If the filament deteriorates when the current through it approaches 10 A, is the damage done to the filament greater when it is first turned on or later when it arrives at its operating temperature?

*2.7 A 110 V heating element in a stove can boil a standard-size pot of water in 1.2 minutes, consuming a total of 136 kJ of energy. Determine the resistance of the heating element and the current flowing through it.

2.8 A certain copper wire has a resistance R characterized by the model given in Problem 2.5 with $\alpha = 4 \times 10^{-3}$ °C^{-1}. If $R = 60\ \Omega$ at 20 °C and the wire is used in a circuit that cannot tolerate an increase in the magnitude of R by more than 10 percent over its value at 20 °C, what would be the highest temperature at which the circuit can be operated within its tolerance limits?

Section 2-2: Kirchhoff's Laws

2.9 The circuit shown in Fig. P2.9 includes two identical potentiometers with per-length resistance of 20 Ω/cm. Determine I_a and I_b.

2.10 Determine V_L in the circuit of Fig. P2.10.

*2.11 Select the value of R in the circuit of Fig. P2.11 so that $V_L = 9$ V.

2.12 A high-voltage direct-current generating station delivers 10 MW of power at 250 kV to a city, as depicted in Fig. P2.12. The city is represented by resistance R_L and each of the two wires of the transmission line between the generating station and the city is represented by resistance R_{TL}. The distance between the two locations is 2000 km and the transmission lines are made of 10 cm diameter copper wire. Determine (a) how much power is consumed by the transmission line and (b)

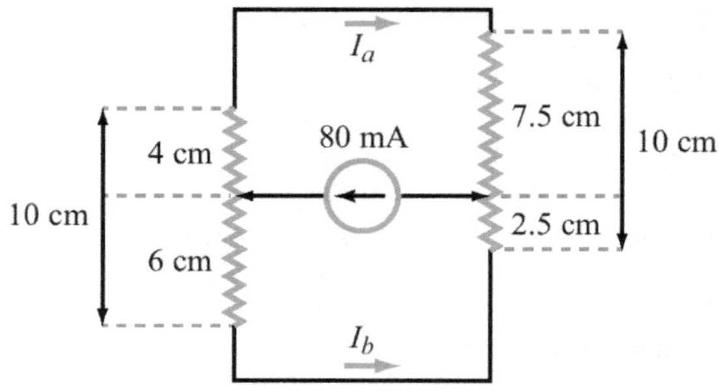

Figure P2.9: Circuit for Problem 2.9.

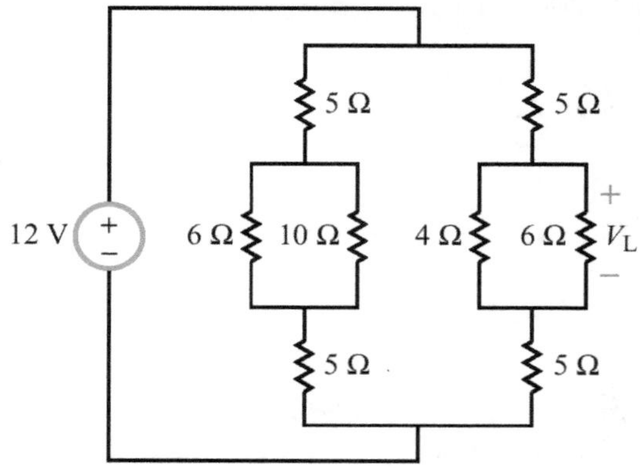

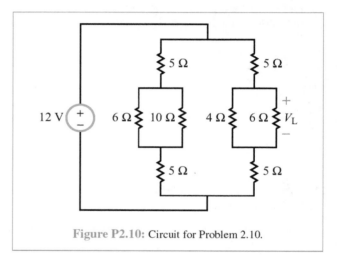

Figure P2.10: Circuit for Problem 2.10.

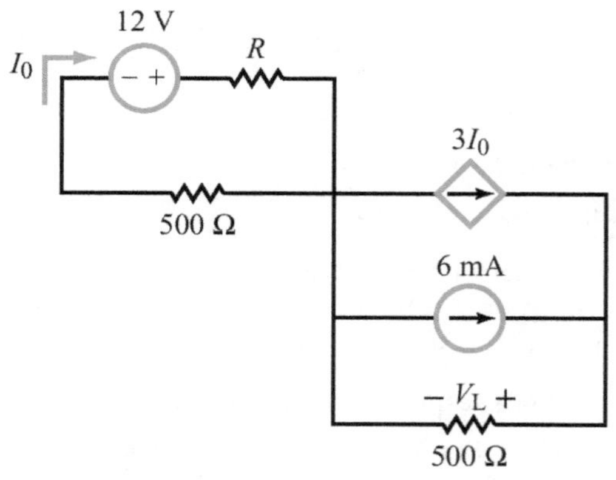

Figure P2.11: Circuit for Problem 2.11.

what fraction of the power generated by the generating station is used by the city.

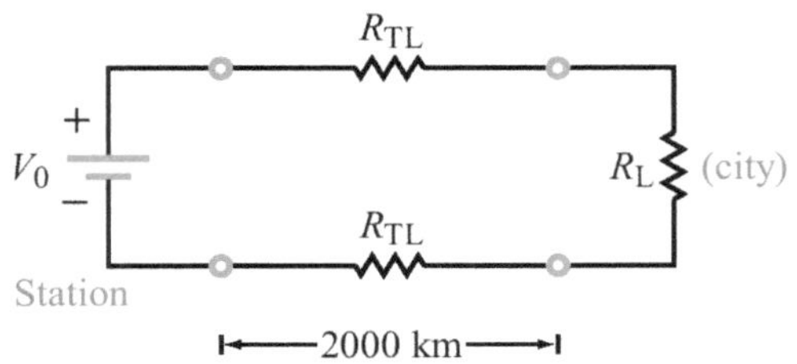

Figure P2.12: Diagram for Problem 2.12.

2.13 Determine the current I in the circuit of Fig. P2.13 given that $I_0 = 0$.

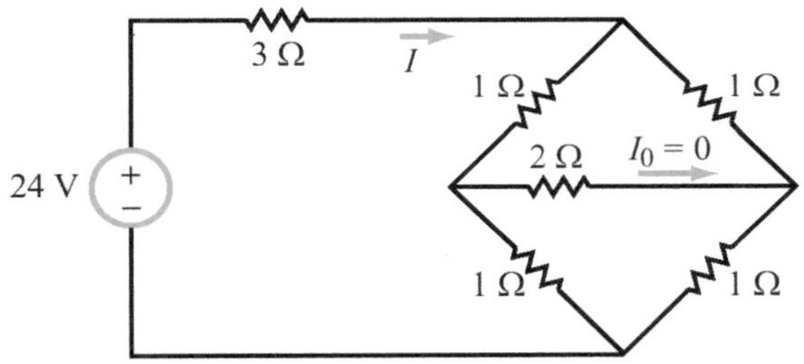

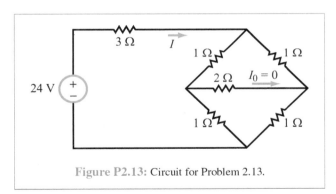

Figure P2.13: Circuit for Problem 2.13.

2.14 Determine currents I_1 to I_3 in the circuit of Fig. P2.14.

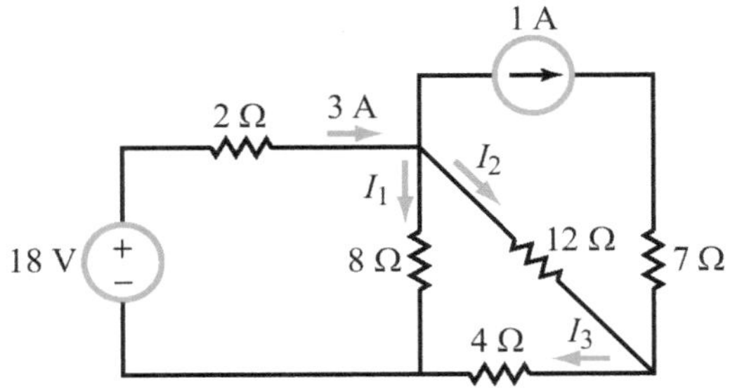

Figure P2.14: Circuit for Problem 2.14.

***2.15** Determine I_x in the circuit of Fig. P2.15.

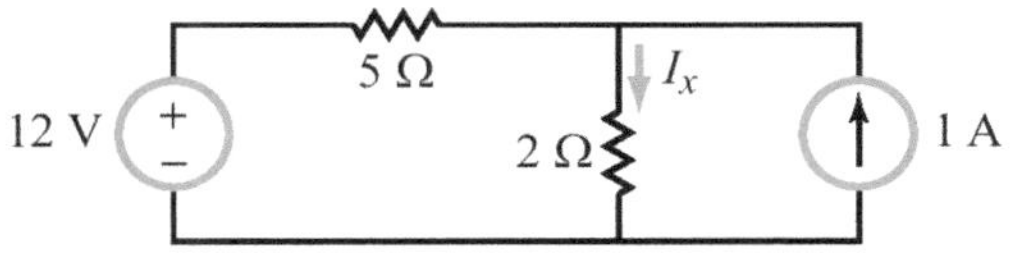

Figure P2.15: Circuit for Problem 2.15.

2.16 Determine currents I_1 to I_4 in the circuit of Fig. P2.16.

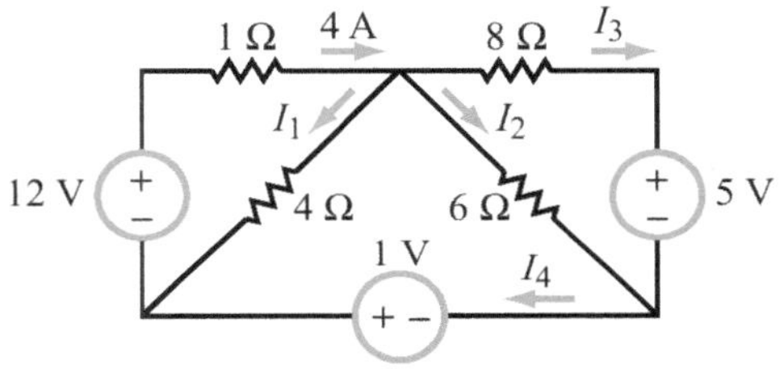

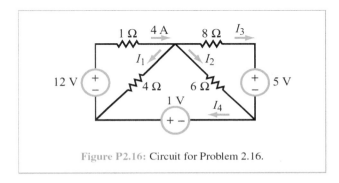

Figure P2.16: Circuit for Problem 2.16.

***2.17** Determine currents I_1 to I_4 in the circuit of Fig. P2.17.

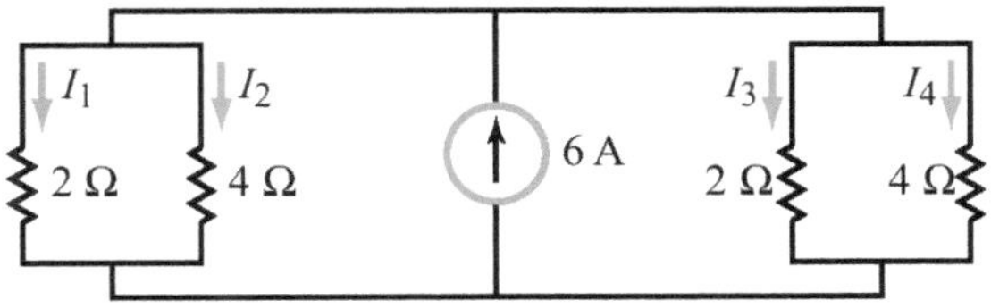

Figure P2.17: Circuit for Problem 2.17.

2.18 Determine the amount of power dissipated in the 3 kΩ resistor in the circuit of Fig. P2.18.

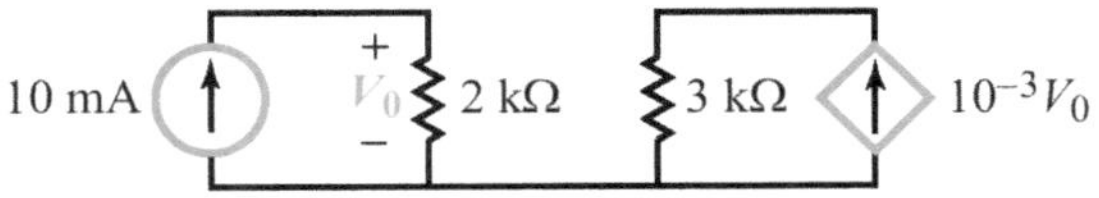

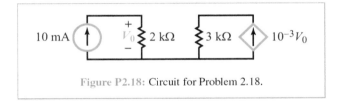

Figure P2.18: Circuit for Problem 2.18.

*2.19 Determine I_x and I_y in the circuit of Fig. P2.19.

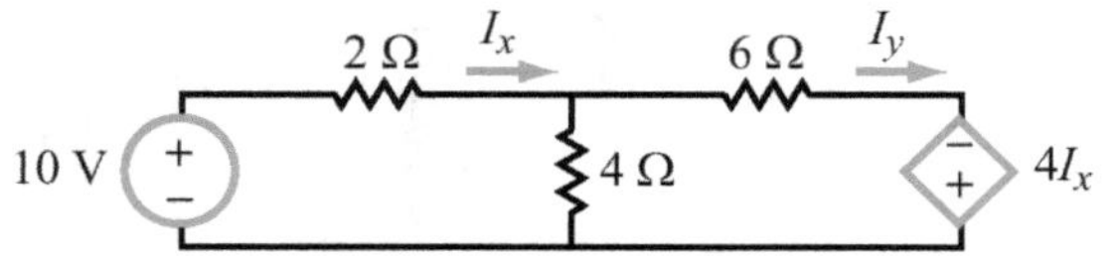

Figure P2.19: Circuit for Problem 2.19.

2.20 Find V_{ab} in the circuit of Fig. P2.20.

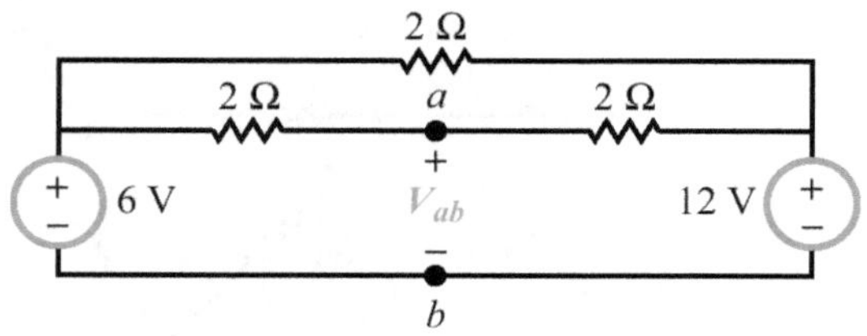

Figure P2.20: Circuit for Problem 2.20.

2.21 Find I_1 to I_3 in the circuit of Fig. P2.21.

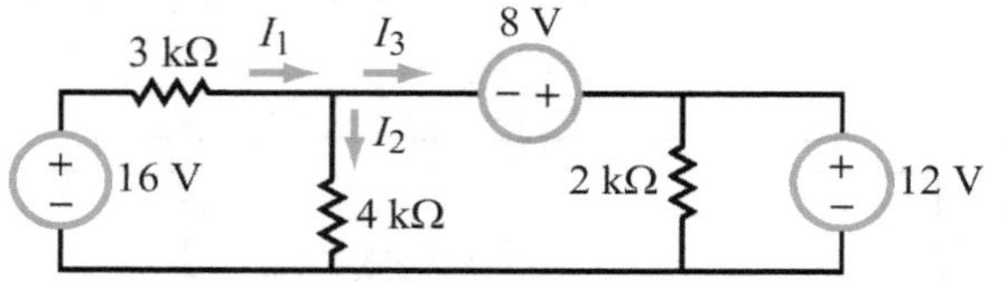

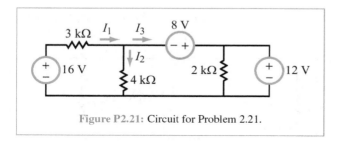

Figure P2.21: Circuit for Problem 2.21.

2.22 Find I in the circuit of Fig. P2.22.

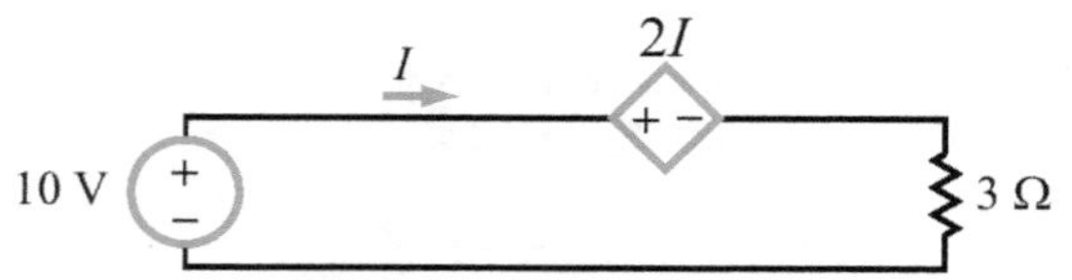

Figure P2.22: Circuit for Problem 2.22.

*2.23 Determine the amount of power supplied by the independent current source in the circuit of Fig. P2.23.

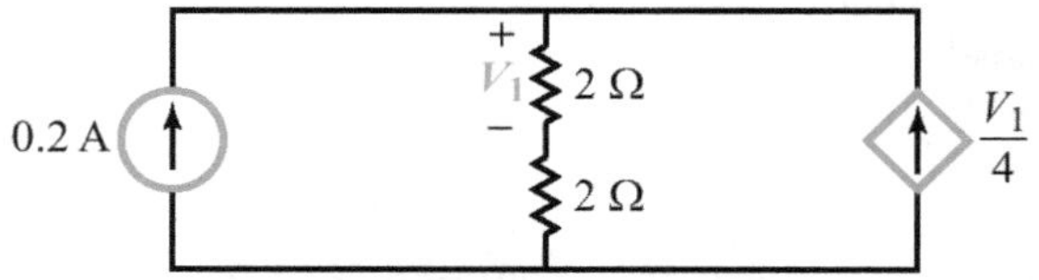

Figure P2.23: Circuit for Problem 2.23.

2.24 Given that in the circuit of Fig. P2.24, $I_1 = 4$ A, $I_2 = 1$ A, and $I_3 = 1$ A, determine node voltages V_1, V_2, and V_3.

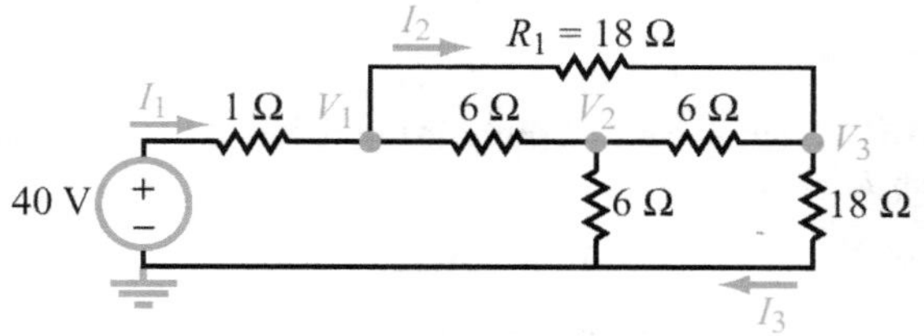

Figure P2.24: Circuit for Problem 2.24.

*2.25 After assigning node V_4 in the circuit of Fig. P2.25 as the ground node, determine node voltages V_1, V_2, and V_3.

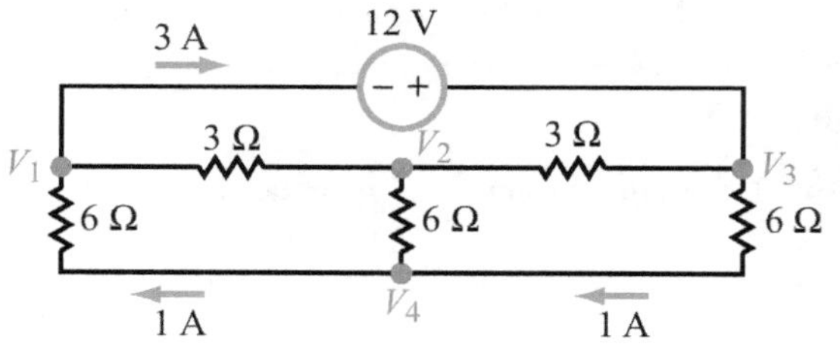

Figure P2.25: Circuit for Problems 2.25 and 2.26.

2.26 After assigning node V_1 in the circuit of Fig. P2.25 as the ground node, determine node voltages V_2, V_3, and V_4.

2.27 In the circuit of Fig. P2.27, $I_1 = 42/81$ A, $I_2 = 42/81$ A, and $I_3 = 24/81$ A. Determine node voltages V_2, V_3, and V_4 after assigning node V_1 as the ground node.

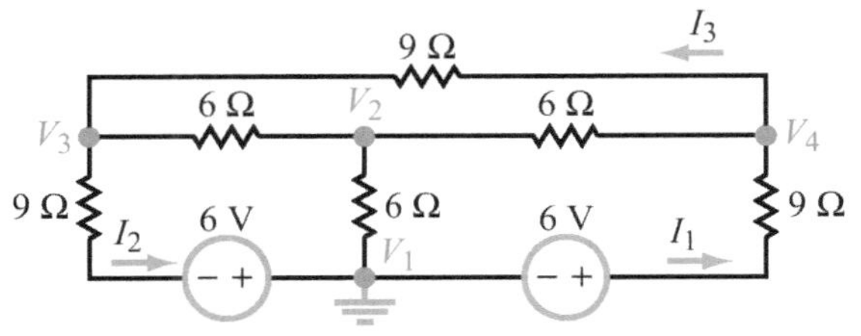

Figure P2.27: Circuit for Problem 2.27.

2.28 The independent source in Fig. P2.28 supplies 48 W of power. Determine I_2.

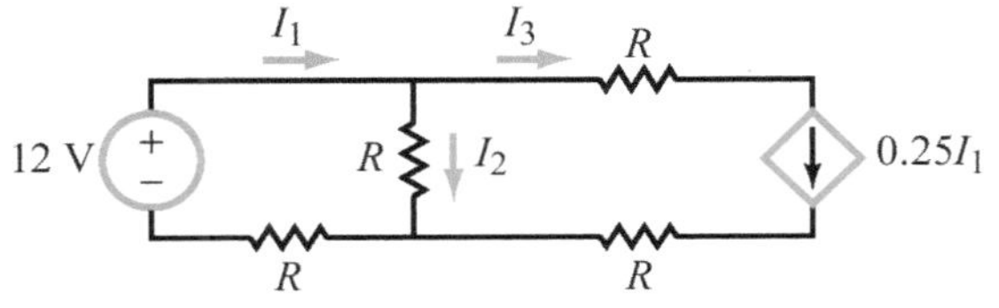

Figure P2.28: Circuit for Problem 2.28.

Section 2-3: Equivalent Circuits

*__2.29__ Given that $I_1 = 1$ A in the circuit of Fig. P2.29, determine I_0.

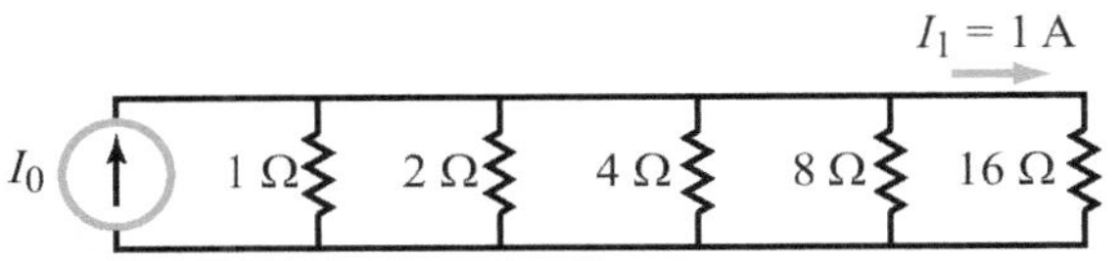

Figure P2.29: Circuit for Problem 2.29.

2.30 What should R be in the circuit of Fig. P2.30 so that $R_{eq} = 4\ \Omega$?

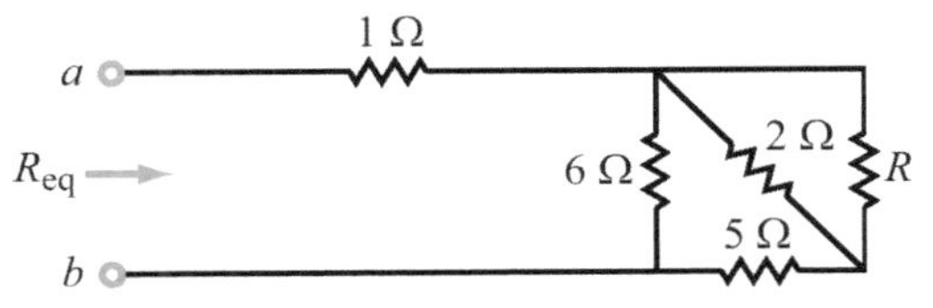

Figure P2.30: Circuit for Problem 2.30.

2.31 Find I_0 in the circuit of Fig. P2.31.

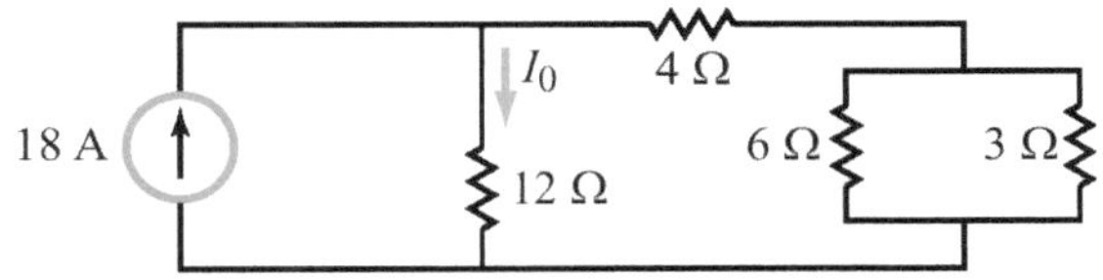

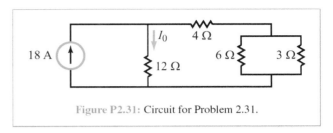

Figure P2.31: Circuit for Problem 2.31.

2.32 For the circuit in Fig. P2.32, find I_x for $t < 0$ and $t > 0$.

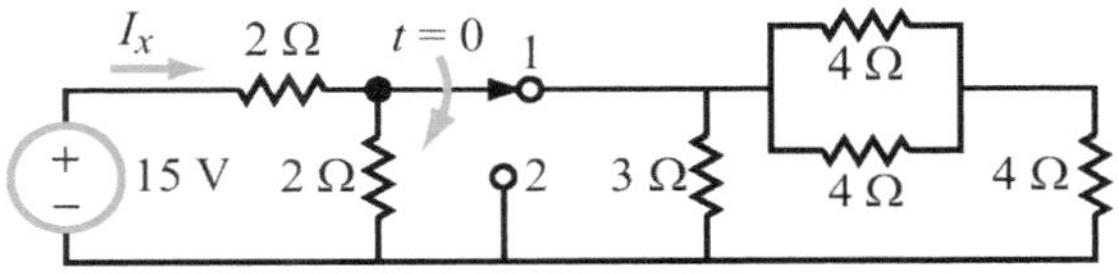

Figure P2.32: Circuit with SPDT switch for Problem 2.32.

2.33 Determine R_{eq} at terminals (a, b) in the circuit of Fig. P2.33.

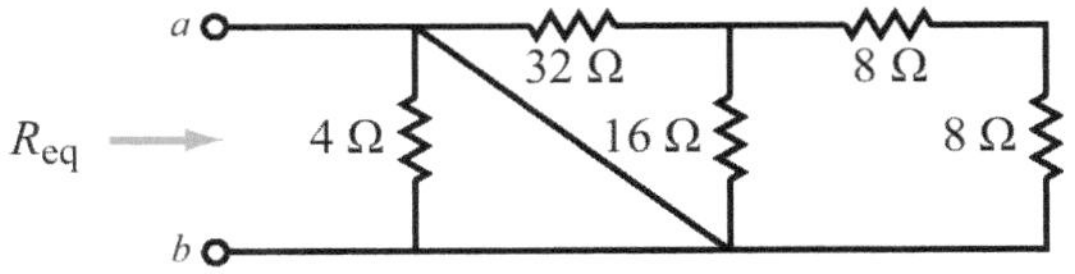

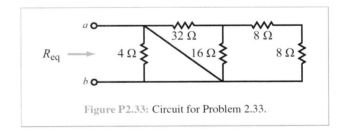

Figure P2.33: Circuit for Problem 2.33.

*__2.34__ Select R in the circuit of Fig. P2.34 so that $V_L = 5$ V.

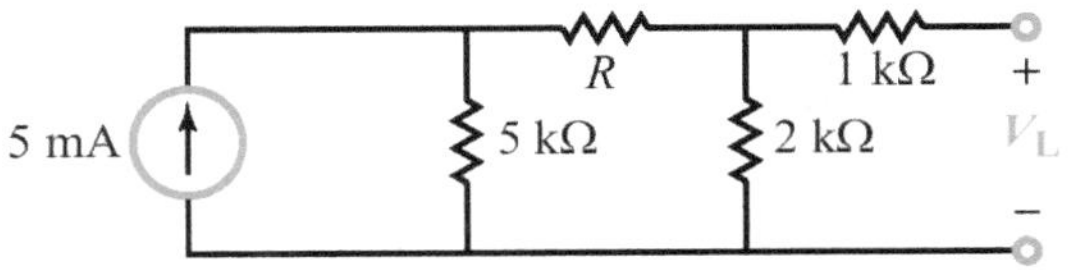

Figure P2.34: Circuit for Problem 2.34.

2.35 If $R = 12\ \Omega$ in the circuit of Fig. P2.35, find I.

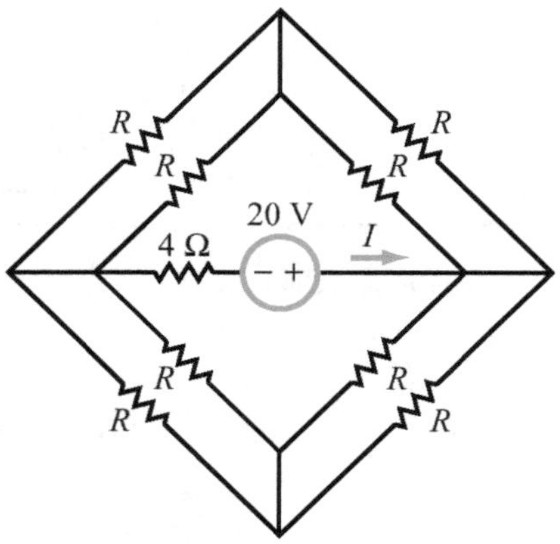

Figure P2.35: Circuit for Problem 2.35.

***2.36** Use resistance reduction and source transformation to find V_x in the circuit of Fig. P2.36. All resistance values are in ohms.

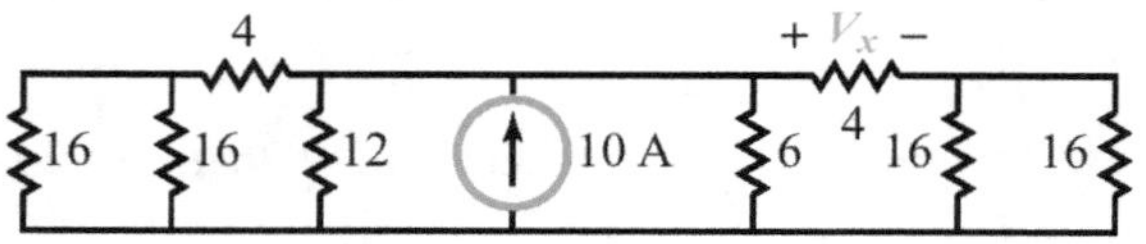

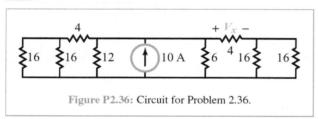

Figure P2.36: Circuit for Problem 2.36.

2.37 Determine A if $V_{out}/V_s = 9$ in the circuit of Fig. P2.37.

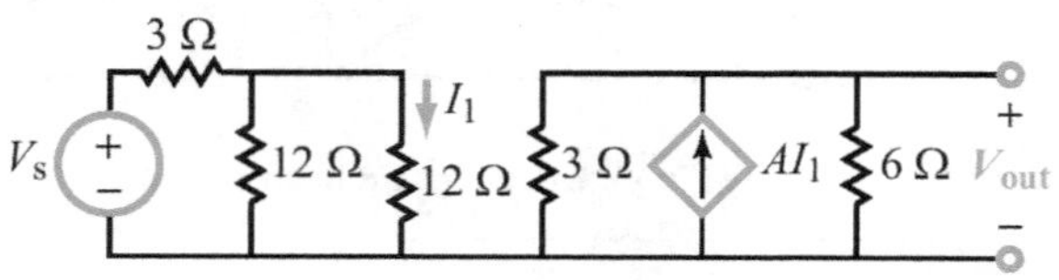

Figure P2.37: Circuit for Problem 2.37.

***2.38** For the circuit in Fig. P2.38, find R_{eq} at terminals (a, b).

Figure P2.38: Circuit for Problems 2.38 and 2.39.

2.39 Find R_{eq} at terminals (c, d) in the circuit of Fig. P2.38.

2.40 Simplify the circuit to the right of terminals (a, b) in Fig. P2.40 to find R_{eq}, and then determine the amount of power supplied by the voltage source. All resistances are in ohms.

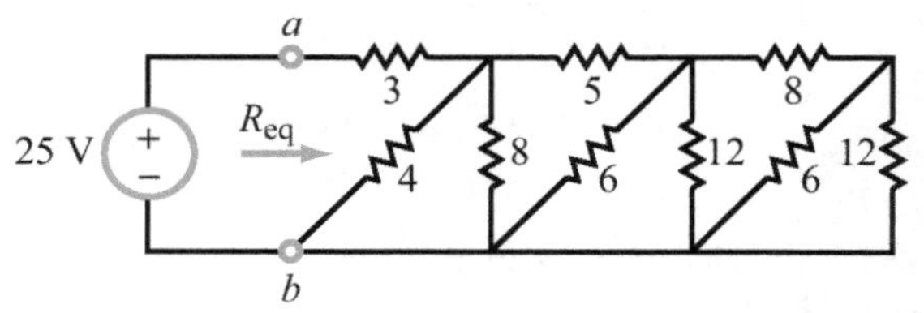

Figure P2.40: Circuit for Problem 2.40.

2.41 For the circuit in Fig. P2.41, determine R_{eq} at

*(a) Terminals (a, b)

(b) Terminals (a, c)

(c) Terminals (a, d)

(d) Terminals (a, f)

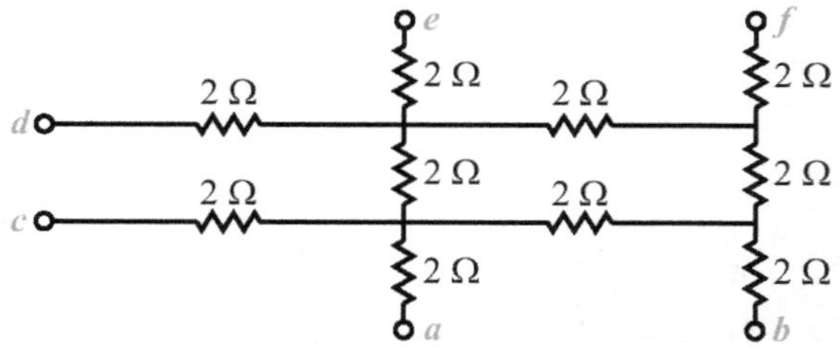

Figure P2.41: Circuit for Problem 2.41.

2.42 Find R_{eq} for the circuit in Fig. P2.42. All resistances are in ohms.

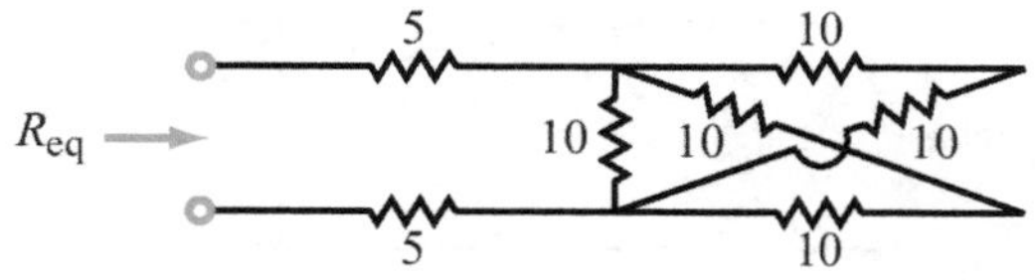

Figure P2.42: Circuit for Problem 2.42.

2.43 Apply voltage and current division to determine V_0 in the circuit of Fig. P2.43 given that $V_{out} = 0.2$ V.

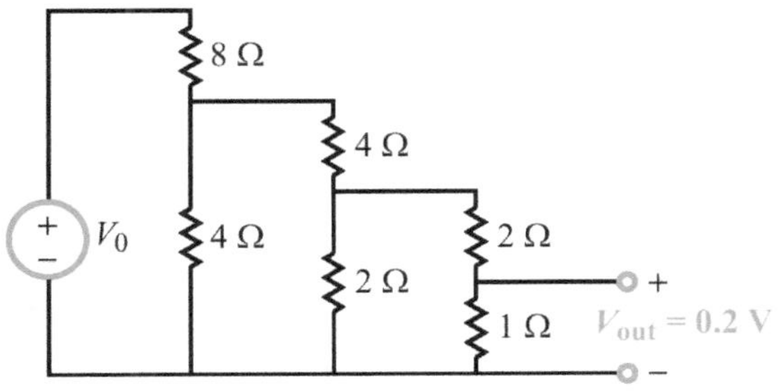

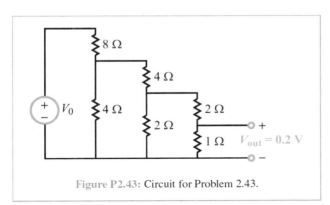

Figure P2.43: Circuit for Problem 2.43.

*2.44 Apply source transformations and resistance reductions to simplify the circuit to the left of nodes (a, b) in Fig. P2.44 into a single voltage source and a resistor. Then, determine I.

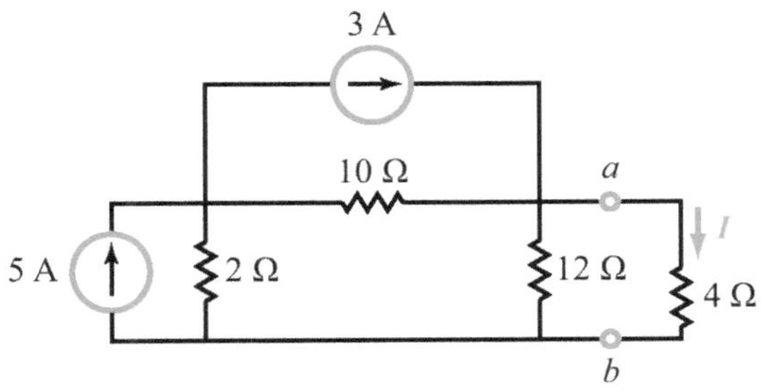

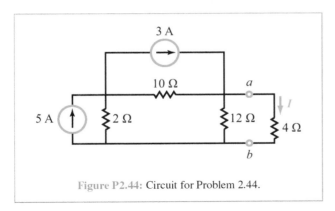

Figure P2.44: Circuit for Problem 2.44.

2.45 Determine the open-circuit voltage V_{oc} across terminals (a, b) in Fig. P2.45.

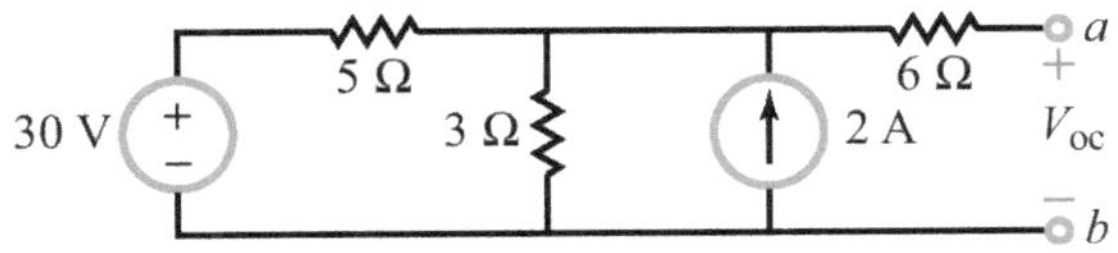

Figure P2.45: Circuit for Problem 2.45.

2.46 Use circuit transformations to determine I in the circuit of Fig. P2.46.

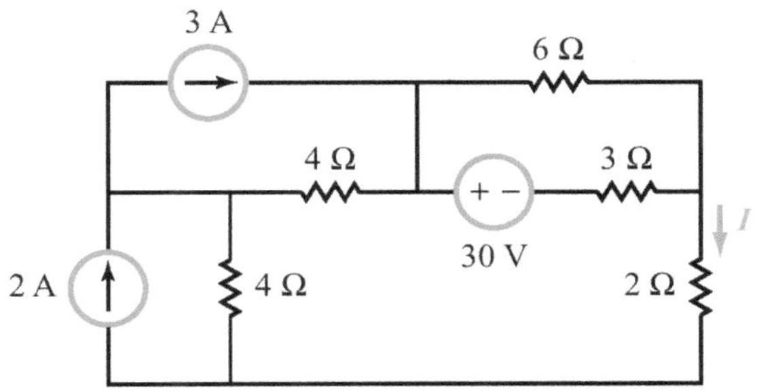

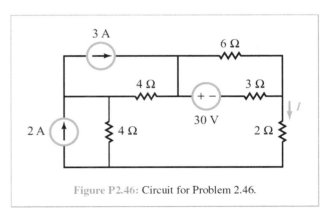

Figure P2.46: Circuit for Problem 2.46.

2.47 Determine currents I_1 to I_4 in the circuit of Fig. P2.47.

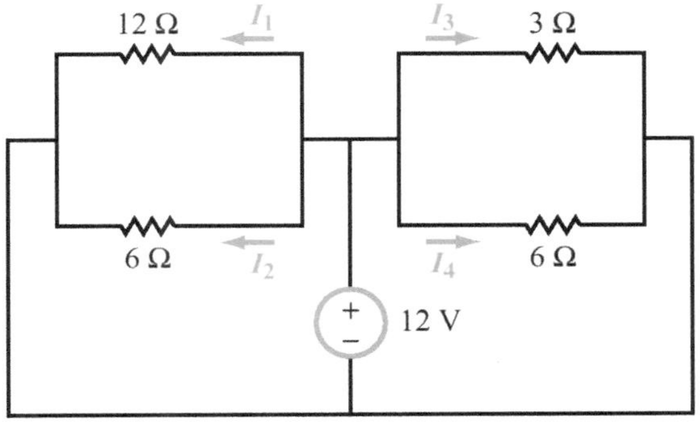

Figure P2.47: Circuit for Problems 2.47 and 2.48.

2.48 Replace the 12 V source in the circuit of Fig. P2.47 with a 4 A current source pointing upwards. Then, determine currents I_1 to I_4.

*2.49 Determine current I in the circuit of Fig. P2.49.

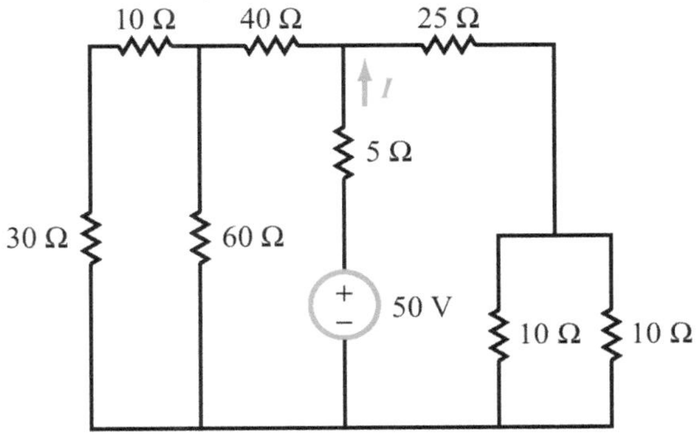

Figure P2.49: Circuit for Problem 2.49.

2.50 Determine the equivalent resistance R_{eq} at terminals (a, b) in the circuit of Fig. P2.50.

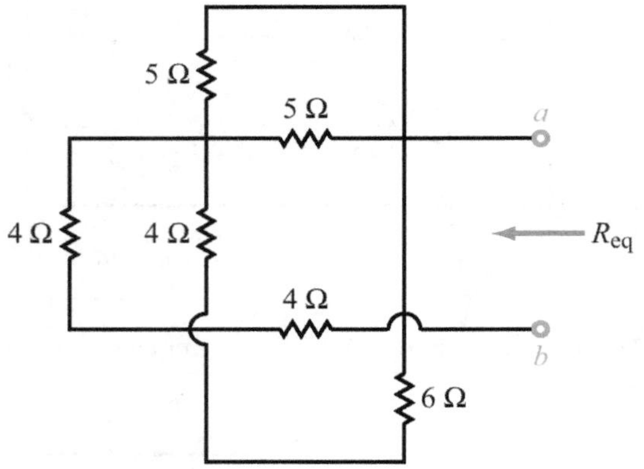

Figure P2.50: Circuit for Problem 2.50.

*2.51 Determine current I in the circuit of Fig. P2.51.

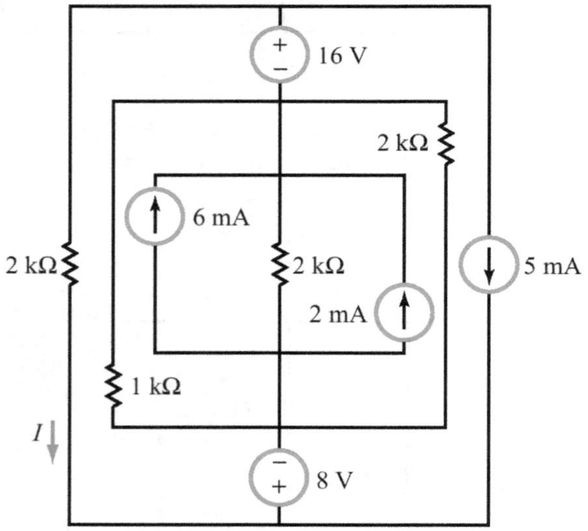

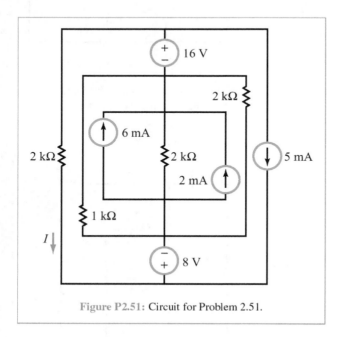

Figure P2.51: Circuit for Problem 2.51.

2.52 Determine voltage V_a in the circuit of Fig. P2.52.

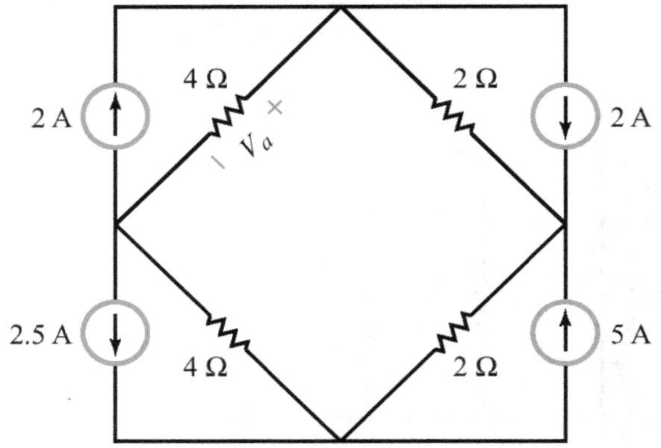

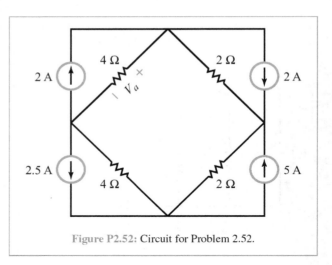

Figure P2.52: Circuit for Problem 2.52.

Sections 2-4 and 2-5: Y–Δ and Wheatstone Bridge

2.53 Convert the circuit in Fig. P2.53(a) from a Δ to a Y configuration.

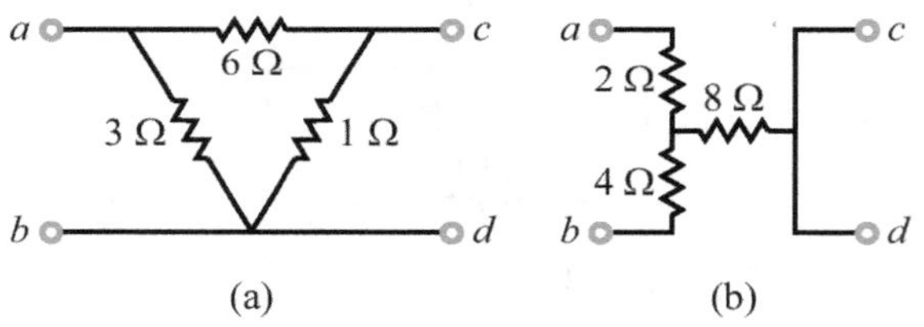

Figure P2.53: Circuit for Problems 2.53 and 2.54.

2.54 Convert the circuit in Fig. P2.53(b) from a T to a Π configuration.

*2.55 Find the power supplied by the generator in Fig. P2.55.

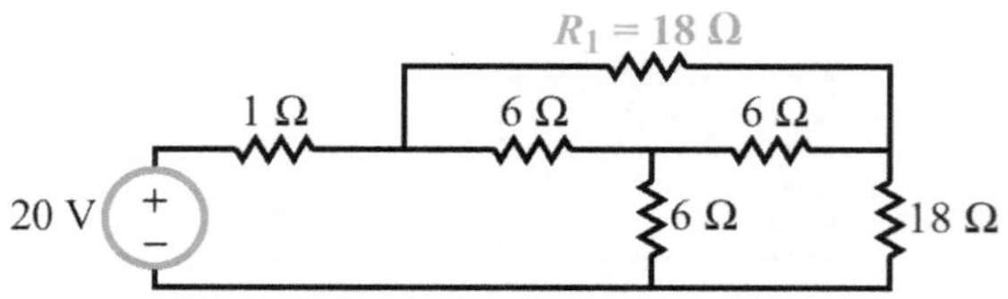

Figure P2.55: Circuit for Problems 2.55 and 2.56.

2.56 Repeat Problem 2.55 after replacing R_1 with a short circuit.

2.57 Find I in the circuit of Fig. P2.57.

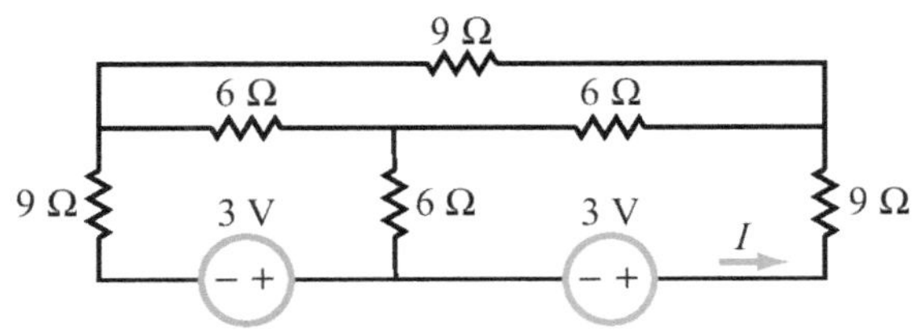

Figure P2.57: Circuit for Problem 2.57.

2.58 Find the power supplied by the voltage source in Fig. P2.58.

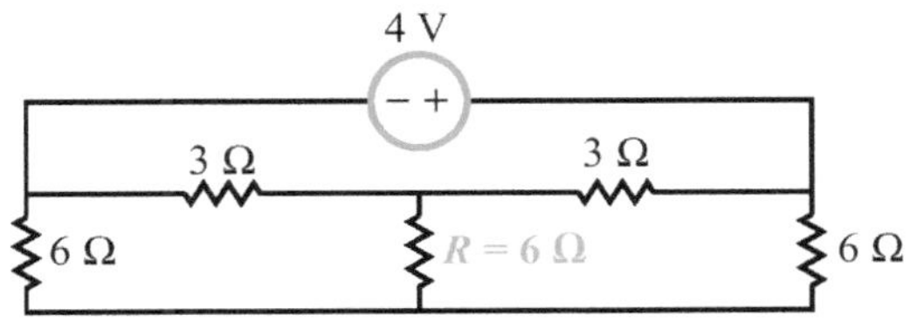

Figure P2.58: Circuit for Problems 2.58 and 2.59.

*__2.59__ Repeat Problem 2.58 after replacing R with a short circuit.

2.60 Find I in the circuit of Fig. P2.60. All resistances are in ohms.

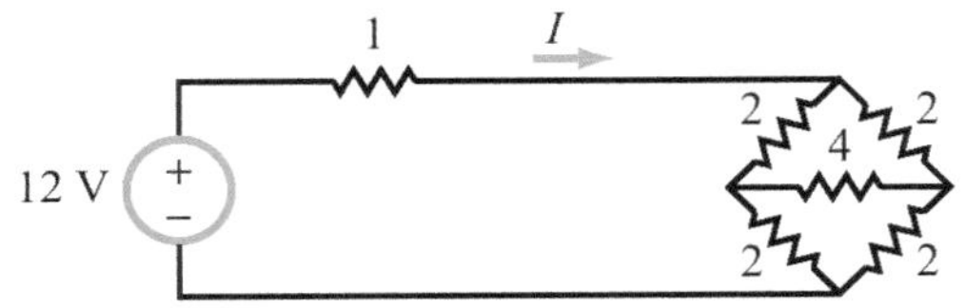

Figure P2.60: Circuit for Problem 2.60.

*__2.61__ Find R_{eq} for the circuit in Fig. P2.61.

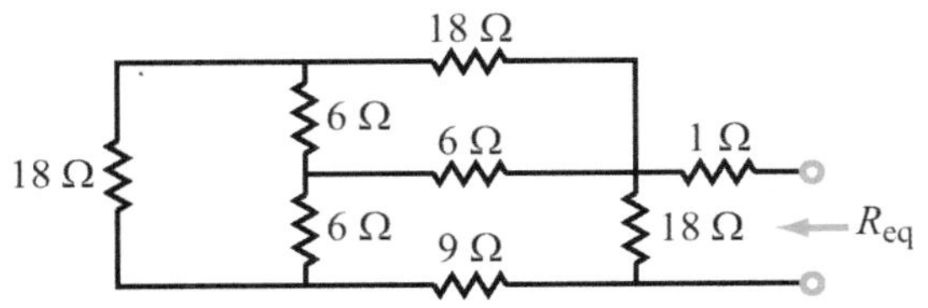

Figure P2.61: Circuit for Problem 2.61.

2.62 Find R_{eq} at terminals (a, b) in Fig. P2.62 if

(a) Terminal c is connected to terminal d by a short circuit

(b) Terminal e is connected to terminal f by a short circuit

(c) Terminal c is connected to terminal e by a short circuit

All resistance values are in ohms.

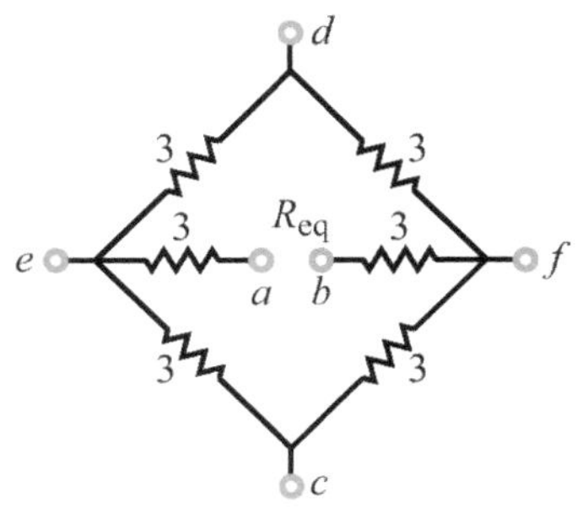

Figure P2.62: Circuit for Problem 2.62.

2.63 For the Wheatstone-bridge circuit of Fig. 2-36, solve the following problems.

*(a) If $R_1 = 1\ \Omega$, $R_2 = 2\ \Omega$, and $R_x = 3\ \Omega$, to what value should R_3 be adjusted so as to achieve a balanced condition?

(b) If $V_0 = 6$ V, $R_a = 0.1\ \Omega$, and R_x were then to deviate by a small amount to $R_x = 3.01\ \Omega$, what would be the reading on the ammeter?

2.64 If $V_0 = 10$ V in the Wheatstone-bridge circuit of Fig. 2-37 and the minimum voltage V_{out} that a voltmeter can read is 1 mV, what is the smallest resistance fraction $(\Delta R/R)$ that can be measured by the circuit?

2.65 Suppose the cantilever system shown in Fig. 2-44 is used in the Wheatstone-bridge sensor of Fig. 2-37 with $V_0 = 2$ V, $\alpha = -1 \times 10^{-9}$ m^2/N, $L = 0.5$ cm, $W = 0.2$ cm, and $H = 0.2$ mm. If the measured voltage is $V_{out} = -2$ V, what is the force applied to the cantilever?

*__2.66__ A touch sensor based on a piezoresistor built into a micromechanical cantilever made of silicon is connected in a Wheatstone-bridge configuration with a $V_0 = 1$ V. If $L = 1.44$ cm and $W = 1$ cm, what should the thickness H be so that the touch sensor registers a voltage magnitude of 10 mV when the touch pressure is 10 N?

Section 2-6: i–υ Relationships

*__2.67__ Determine I_1 and I_2 in the circuit of Fig. P2.67. Assume $V_F = 0.7$ V for both diodes.

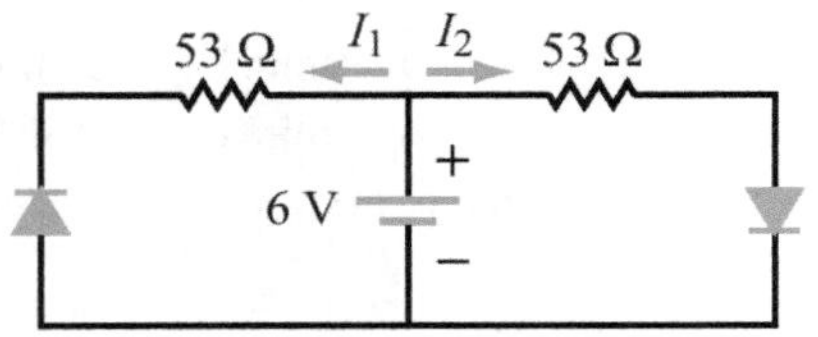

Figure P2.67: Circuit for Problem 2.67.

2.68 Determine V_1 in the circuit of Fig. P2.68. Assume $V_F = 0.7$ V for all diodes.

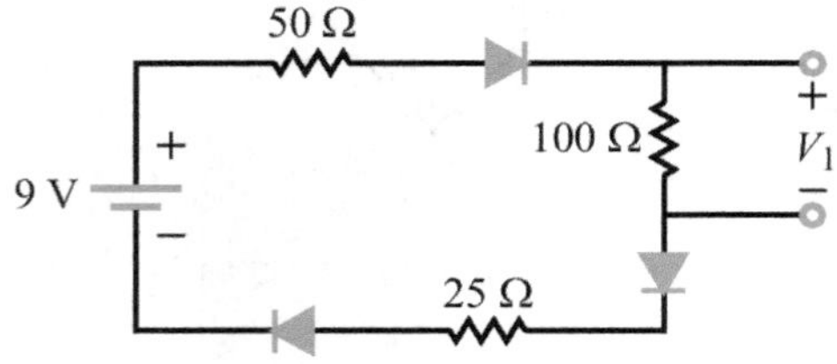

Figure P2.68: Circuit for Problem 2.68.

2.69 If the voltage source in the circuit of Fig. P2.69 generates a single square wave with an amplitude of 2 V, generate a plot for v_{out} for the same time period.

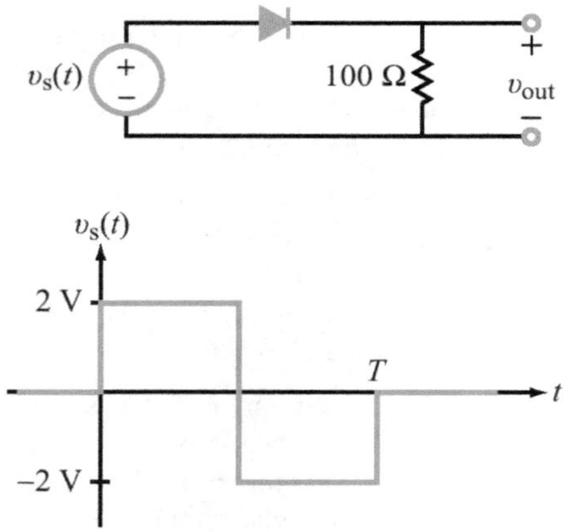

Figure P2.69: Circuit and voltage waveform for Problem 2.69.

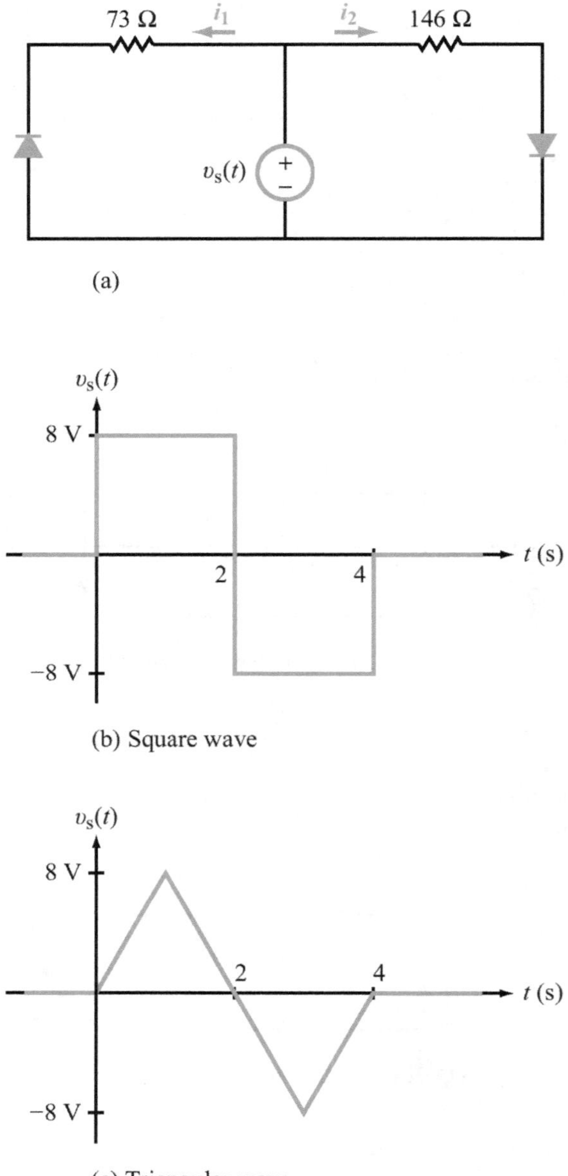

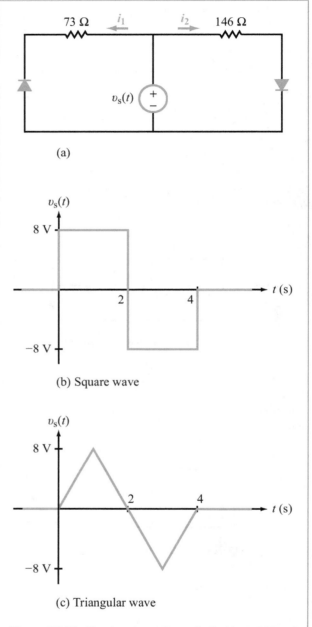

Figure P2.70: Circuit and waveforms for Problems 2.70 and 2.71.

2.70 If the voltage source in the circuit of Fig. P2.70(a) generates the single square waveform shown in Fig. P2.70(b), generate plots for $i_1(t)$ and $i_2(t)$.

2.71 If the voltage source in the circuit of Fig. P2.70(a) generates the single triangular waveform shown in Fig. P2.70(c), generate plots for $i_1(t)$ and $i_2(t)$.

2.72 The circuit shown in Fig. P2.72 is used to control a red LED. The LED is designed to turn on when the resistance R of the rheostat is 50 Ω or lower. Use the information contained in Fig. 2-8(d) to determine the value of the constant resistor R_0.

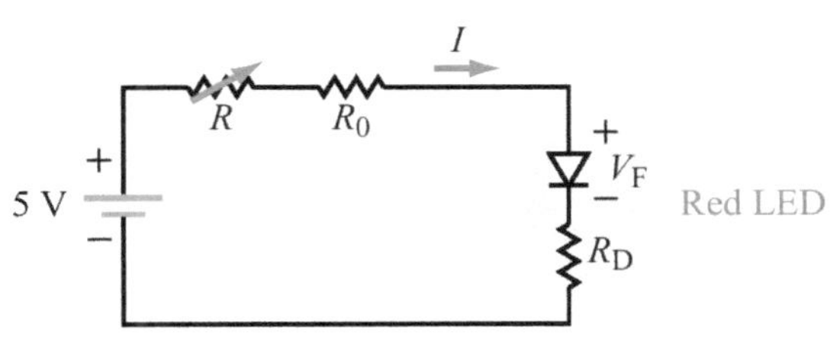

Figure P2.72: Circuit for Problem 2.72.

Section 2-7: Multisim

2.73 Use the DC Operating Point Analysis in Multisim to solve for voltage V_{out} in the circuit of Fig. P2.73. Solve for V_{out} by hand and compare with the value generated by Multisim. See the solution for Exercise 2.14 (on CAD) for how to incorporate circuit variables into algebraic expressions.

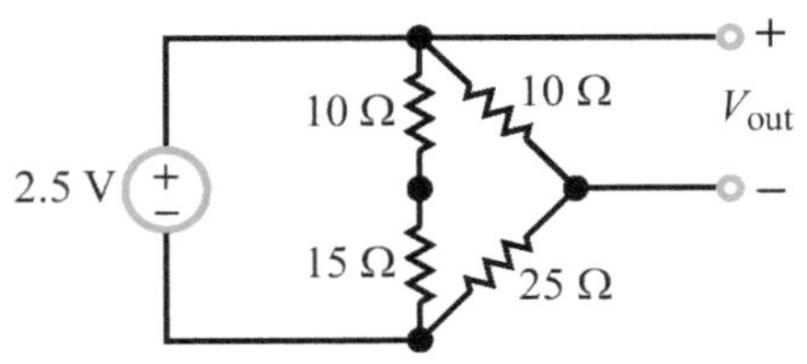

Figure P2.73: Circuit for Problem 2.73.

2.74 Find the ratio V_{out}/V_{in} for the circuit in Fig. P2.74 using DC Operating Point Analysis in Multisim. See the Multisim Tutorial included on the book website on how to reference currents in ABM sources [you should not just type in I(V1)].

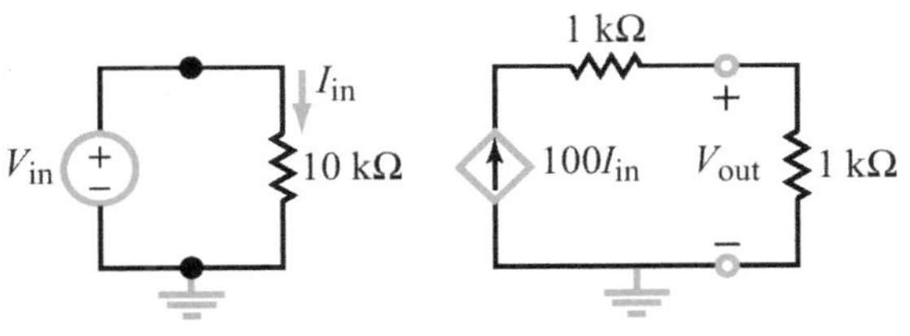

Figure P2.74: Circuit for Problem 2.74.

2.75 Use DC Operating Point Analysis in Multisim to solve for all six labeled resistor currents in the circuit of Fig. P2.75.

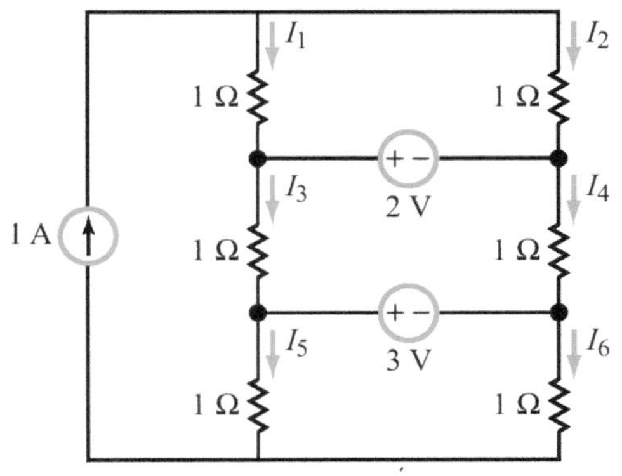

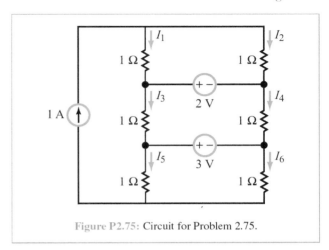

Figure P2.75: Circuit for Problem 2.75.

2.76 Find the voltages across R_1, R_2, and R_3 in the circuit of Fig. P2.76 using the DC Operating Point Analysis tool in Multisim.

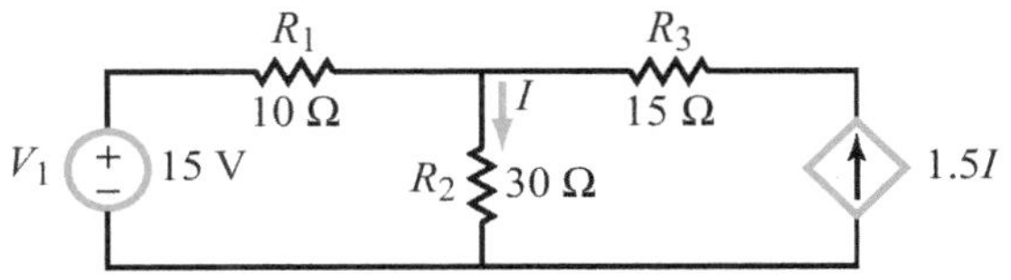

Figure P2.76: Circuit for Problem 2.76.

2.77 Find the equivalent resistance looking into the terminals of the circuit in Fig. P2.77 using a test voltage source and current probes in the Interactive Simulation in Multisim. Compare the answer you get to what you obtain from series and parallel combining of resistors carried out by hand.

Potpourri Questions

2.78 What is a superconducting material and what happens when its physical temperature is below or above its critical temperature? How is superconductivity used in practice?

2.79 What is a piezoresistor? How is it used? Resistors are also used as chemical sensors. Explain how.

2.80 What determines the color of the light emitted by an LED? Why are LEDs economical to use?

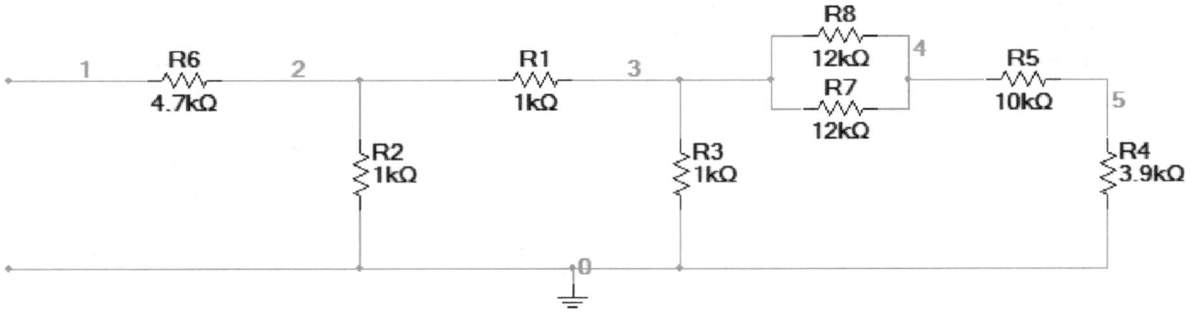

Figure P2.77: Circuit for Problem 2.77.

Integrative Problems: Analytical / Multisim / myDAQ

To master the material in this chapter, solve the following problems using three complementary approaches: (a) analytically, (b) with Multisim, and (c) by constructing the circuit and using the *myDAQ* interface unit to measure quantities of interest via your computer. [myDAQ tutorials and videos are available in Appendix F and on CAD.]

m2.1 Kirchhoff's Laws: Determine currents I_1 to I_3 and the voltage V_1 in the circuit of Fig. m2.1 with component values $I_{src} = 1.8$ mA, $V_{src} = 9.0$ V, $R_1 = 2.2$ kΩ, $R_2 = 3.3$ kΩ, and $R_3 = 1.0$ kΩ.

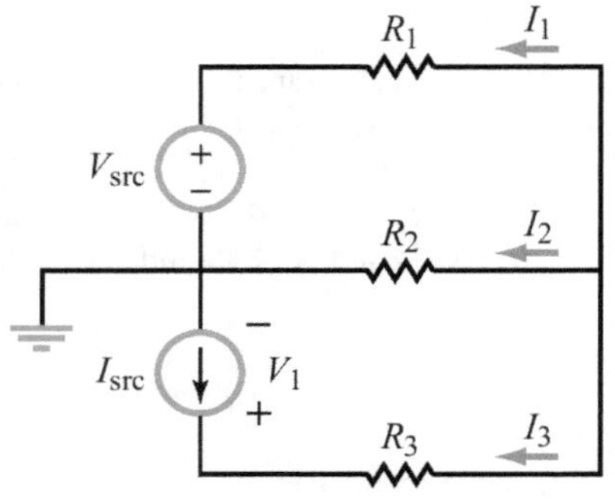

Figure m2.1 Circuit for Problem m2.1.

m2.2 Equivalent Resistance: Find the equivalent resistance between the following terminal pairs in the circuit of Fig. m2.2 under the stated conditions:

(a) *a*-*b* with the other terminals unconnected,

(b) *a*-*d* with the other terminals unconnected,

(c) *b*-*c* with a wire connecting terminals *a* and *d*, and

(d) *a*-*d* with a wire connecting terminals *b* and *c*.

Use these component values: $R_1 = 10$ kΩ, $R_2 = 33$ kΩ, $R_3 = 15$ kΩ, $R_4 = 47$ kΩ, and $R_5 = 22$ kΩ.

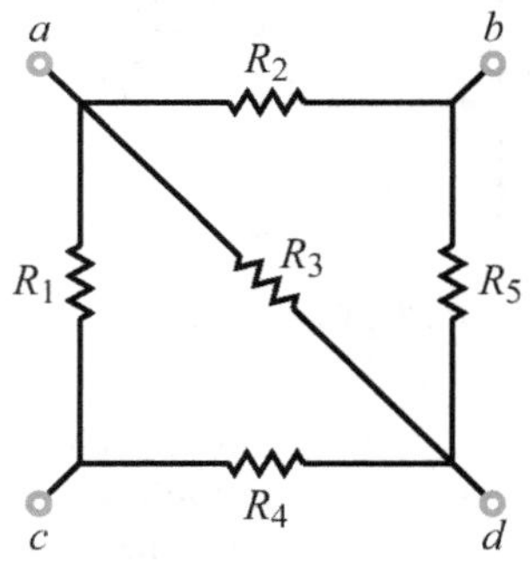

Figure m2.2 Circuit for Problem m2.2.

m2.3 Current and Voltage Dividers: Apply the concepts of voltage dividers, current dividers, and equivalent resistance to find the currents I_1 to I_3 and the voltages V_1 to V_3 in the circuit of Fig. m2.3. Use these component values: $V_{src} = 12$ V, $R_1 = 1.0$ kΩ, $R_2 = 10$ kΩ, $R_3 = 1.5$ kΩ, $R_4 = 2.2$ kΩ, $R_5 = 4.7$ kΩ, and $R_6 = 3.3$ kΩ.

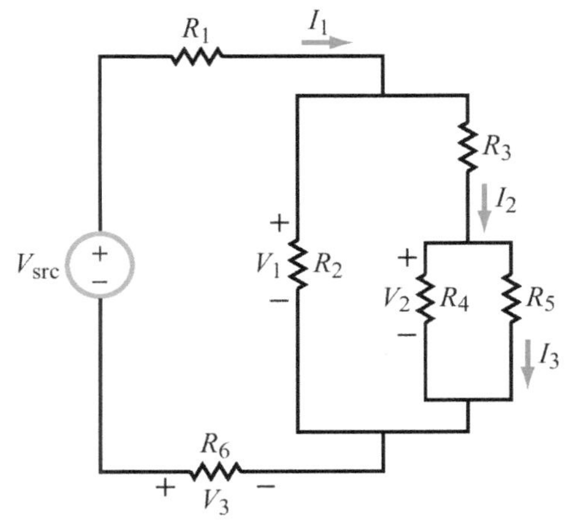

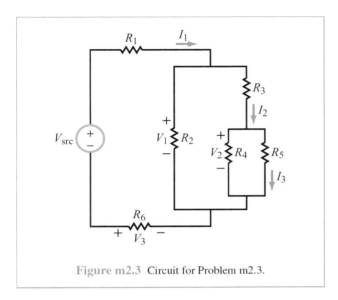

Figure m2.3 Circuit for Problem m2.3.

m2.4 Wye-Delta Transformation: Find (a) the currents I_1 and I_2 in the circuit of Fig. m2.4 and (b) the power delivered by each of the two voltage sources. Use these component values: $V_1 = 15$ V, $V_2 = 15$ V, $R_1 = 3.3$ kΩ, $R_2 = 1.5$ kΩ, $R_3 = 4.7$ kΩ, $R_4 = 5.6$ kΩ, $R_5 = 1.0$ kΩ, and $R_6 = 2.2$ kΩ.

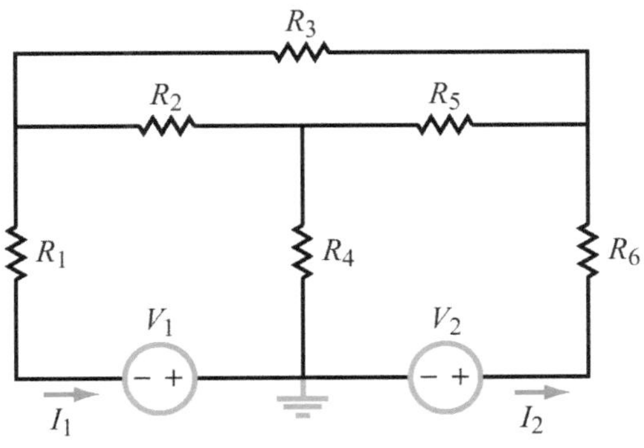

Figure m2.4 Circuit for Problem m2.4.

m2.5 Kirchoff's Laws and Equivalent Resistance: In the circuit of Fig. m2.5:

(a) Find the voltage drop across the 46 kΩ resistor.

(b) What is the equivalent resistance seen by the 15 V source?

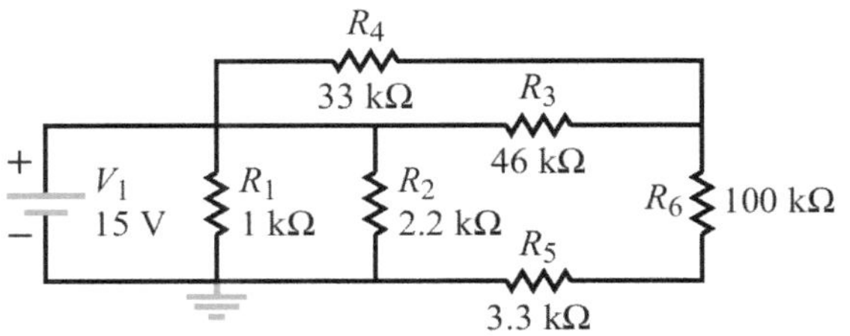

Figure m2.5 Circuit for Problem m2.5.

m2.6 Multiple Sources: To create multiple sources,use the AO 0 and AO 1 ports simultaneously for the myDAQ portion of this problem. Use the Arbitrary waveform generator to create the 3 V and 5 V sources.

(a) Find currents I_1 and I_2 in the circuit of Fig. m2.6. For the myDAQ portion, make sure to measure current correctly or you could blow the myDAQ's fuse.

(b) Find the voltage drop across the 47 kΩ resistor.

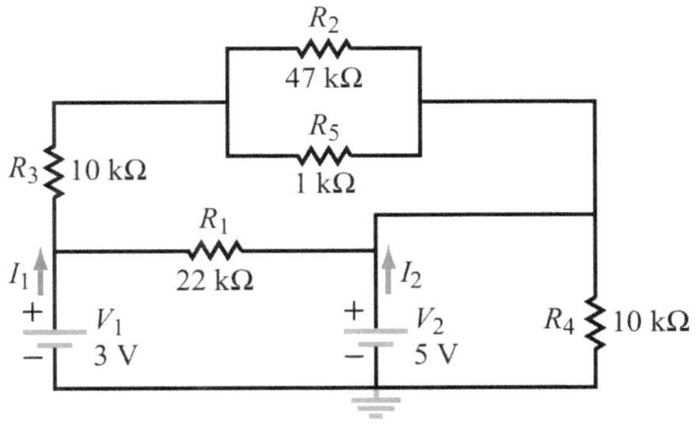

Figure m2.6 Circuit for Problem m2.6.

m2.7 Current Source: This problem is relatively straightforward to solve by hand and with Multisim. However, to create the myDAQ version of the circuit in Fig. m2.7, you will need to use an LM371 Regulator with a 100 Ω connected between V_{out} and V_{adj}. For more information, consult Appendix F or look up the specification of the LM371-LZ regulator.

(a) Determine the voltage drop across each 1 kΩ resistor.

(b) Determine the current through the 3.3 kΩ resistor.

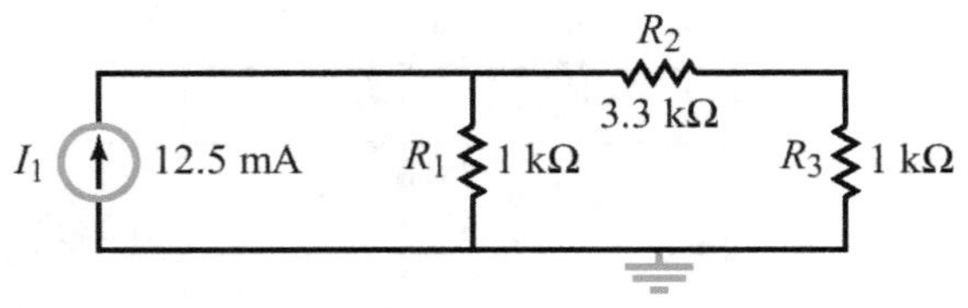

Figure m2.7 Circuit for Problem m2.7.

m2.8 Equivalent Resistance: Determine the equivalent resistance of the circuit in Fig. m2.8 as seen at terminals (1, 2).

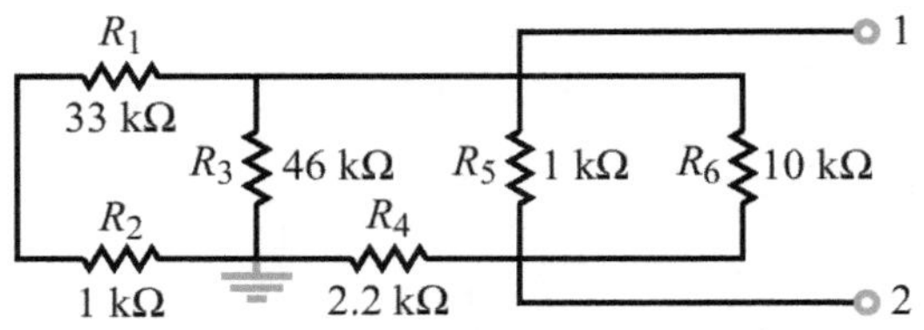

Figure m2.8 Circuit for Problem m2.8.

CHAPTER 3

Analysis Techniques

Contents

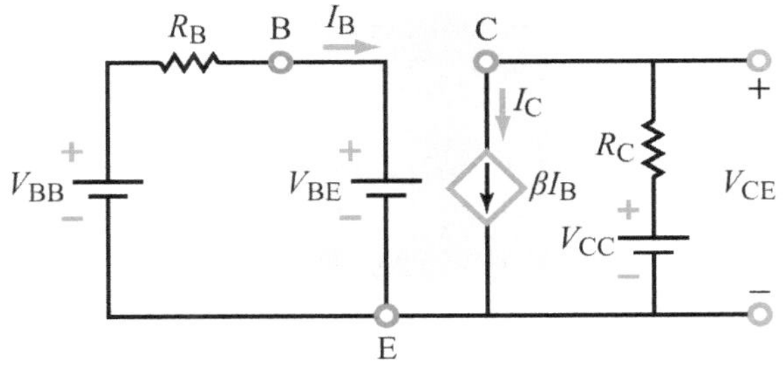

Transistor equivalent circuit

The basic laws of Chapter 2 are used in the present chapter to *develop standard solution methods* that can be applied to analyze any linear circuit, no matter how complex.

Objectives

Learn to:

- Apply the node-voltage and mesh-current methods to analyze an electric circuit of any configuration, so long as it is linear and planar.
- Apply the by-inspection methods to circuits that satisfy certain conditions.
- Use the source-superposition method to evaluate the sensitivity of a circuit to the various sources in the circuit.
- Determine the Thévenin and Norton equivalent circuits of any input circuit and use them to evaluate the response of an external load (or an output circuit) to the input circuit.
- Establish the conditions for maximum transfer of current, voltage, and power from an input circuit to an external load.
- Learn the basic properties of the bipolar junction transistor.

Overview

By applying the circuit-analysis skills we developed in the preceding chapter, we now extend our capability further so we may tackle any linear, planar circuit—no matter how complex. Node-voltage and mesh-current equations will be cast into a systematic structure in Sections 3-2 through 3-4, so we may take advantage of standard methods for solving linear, simultaneous equations, either by the use of determinants and matrices (Appendix B) or the execution of computer simulation packages such as MATLAB or MathScript (Appendix E). The nodal and mesh analysis techniques are followed with treatments of two special tools: the source superposition method and the Thévenin/Norton equivalent-circuit method. These methods allow us to break any complex electrical system into smaller, manageable subcircuits for analysis. With these tools, you are ready to analyze pretty much any circuit you may encounter for the rest of your career. We will also introduce you to semiconductor manufacturing and the relationships between analog and digital signals.

3-1 Linear Circuits

A circuit is a system with inputs and outputs; its inputs are the independent voltage and current sources that energize the circuit, and its outputs are all of the currents flowing through and voltages across all of the passive elements of the circuit. By *passive* element, we mean that it does not generate energy of its own. A resistor is a perfect example of a passive element. By comparison, an *active* element requires an external power supply in order to function. Examples of active elements include transistors (such as the BJT described in Section 3-9) and operational amplifiers (Chapter 4).

A *linear circuit* is a circuit composed entirely of independent sources and *linear elements*. An element is linear if it is passive and exhibits a linear i–υ relationship. For a resistor R, for example,

$$\upsilon = Ri. \tag{3.1}$$

A circuit element, or an entire circuit, is nonlinear if its i–υ relationship is not linear. The LED (Section 2-1.4) is an example of a nonlinear device.

3-1.1 Homogeneity Property

If i through resistor R is increased by a factor K, so will υ. This proportional increase of i and υ by the same factor is called the *homogeneity* (or *scaling*) property of a linear element.

3-1.2 Superposition Principle

If current i_1 can give rise to voltage $\upsilon_1 = Ri_1$, and another current i_2 can give rise to voltage $\upsilon_2 = Ri_2$, then the *simultaneous* presence of both currents gives rise to

$$\upsilon = R(i_1 + i_2) = Ri_1 + Ri_2 = \upsilon_1 + \upsilon_2. \tag{3.2}$$

Thus, the output (υ) due to the two inputs (i_1 and i_2) is equal to the sum of the two outputs (υ_1 and υ_2) had each input been introduced separately. This is a statement of the *superposition principle* (also known as the *additivity property*). We will use this principle in Section 3-5 to simplify our analysis for circuits containing multiple sources.

3-1.3 Linear and Nonlinear Elements

Linear elements

By virtue of its linear i–υ relationship, the resistor is an obvious candidate for the list of linear circuit elements, which includes:

- Resistors
- Capacitors
- Inductors
- Linear dependent sources

The i–υ relationship for a capacitor, which we will learn more about in Chapter 5, is given by

$$i = C\,\frac{d\upsilon}{dt}. \tag{3.3a}$$

If we multiply both sides by a factor K, we get

$$Ki = KC\,\frac{d\upsilon}{dt} = C\,\frac{d}{dt}\,(K\upsilon). \tag{3.3b}$$

Hence, increasing υ by a factor K leads to an increase in i by the same factor, which implies that the d/dt differentiation operator has no bearing on the homogeneity property linking i to υ. The time derivative does not impact the additivity property either.

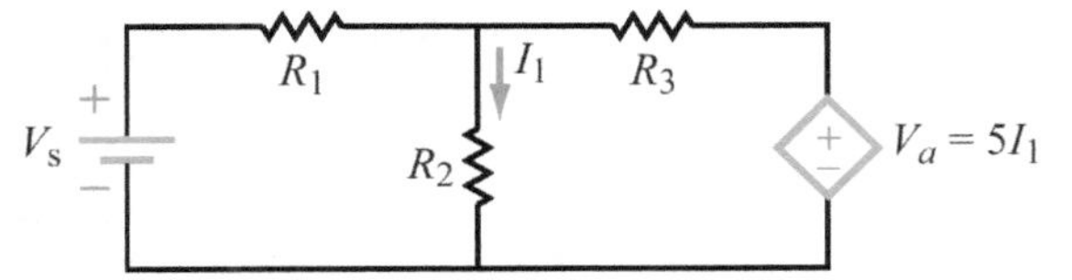

Figure 3-1: Circuit with dependent source $V_a = 5I_1$.

▶ Since the capacitor is a passive element and obeys both the homogeneity and additivity (superposition) properties, it is classified as a linear circuit element. A similar argument applies to the inductor, for which $\upsilon = L\, di/dt$. ◀

Next we consider dependent sources, which were first introduced in Section 1-6.4. Dependent sources are *artificial sources* (because they do not generate energy of their own) used in *equivalent* linear circuits intended to *model* the approximate behavior of nonlinear circuits and elements like transistors and operational amplifiers. Let us consider the simple circuit shown in Fig. 3-1, which includes an independent voltage source V_s and a dependent voltage source V_a. The magnitude of V_a depends on I_1, which, in turn, depends on the real source V_s. If $V_s = 0$, no currents would flow in the circuit, so I_1 would be zero, and so would V_a.

▶ Hence, dependent source V_a is a passive element, and since it is also directly proportional to I_1 (raised to first order), V_a is classified as a linear element. The same is true for a dependent voltage source whose magnitude is linearly related to a voltage elsewhere in the circuit (instead of to a current), as well as for dependent current sources that depend linearly on a voltage or current elsewhere in the circuit. ◀

Nonlinear elements

The circuit analysis techniques developed in this book apply primarily to linear circuits, and yet many devices—such as diodes, transistors, and integrated circuits—exhibit nonlinear i–υ relationships. Consequently:

(1) The analysis techniques do not directly apply to circuits containing such nonlinear elements.

(2) However, it is often possible to replace nonlinear elements with equivalent circuits containing linear elements, including dependent sources, and then use them to obtain approximate, but fairly accurate results, provided certain conditions are satisfied. Examples of equivalent circuits will be presented in Section 3-8 for the bipolar junction transistor (BJT) and in Chapter 4 for the operational amplifier and the CMOS transistor.

3-1.4 Advantages of Linear Circuits

The linearity properties of a linear circuit allow us to use certain analysis techniques that would be otherwise not applicable had the circuit contained one or more nonlinear elements (unless they can be adequately represented by equivalent linear circuits). Through the application of such analysis techniques, which include the Thévenin and superposition methods presented later in Sections 3-5 and 3-6, we can simplify the analysis (and design) of a complex circuit considerably.

3-2 Node-Voltage Method

3-2.1 General Procedure

According to Kirchhoff's current law (KCL), the algebraic sum of all currents entering any node in an electric circuit is equal to zero. Built on that principle, the node-voltage analysis method provides a systematic and efficient procedure for determining all of the currents and voltages in a circuit. This determination is realized through the solution of a system of linear, simultaneous equations in which the unknown variables are the voltages at the extraordinary nodes in the circuit. As a reminder, in Section 1-3 we defined *an extraordinary node as a node connected to three or more elements*. For a circuit containing n_{ex} extraordinary nodes, implementation of the node-voltage method consists of three basic steps:

Solution Procedure: Node Voltage

Step 1: Identify all extraordinary nodes, select one of them as a reference node (ground), and then assign node voltages to the remaining $(n_{ex} - 1)$ extraordinary nodes.

Step 2: At each of the $(n_{ex} - 1)$ extraordinary nodes, apply the form of KCL requiring the sum of all currents *leaving* a node to be zero (see KCL template).

Step 3: Solve the $(n_{ex} - 1)$ independent simultaneous equations to determine the unknown node voltages (see Appendix B).

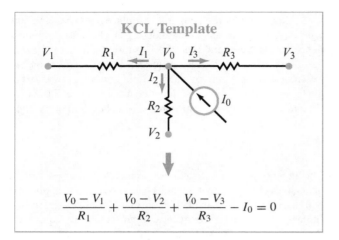

KCL Template

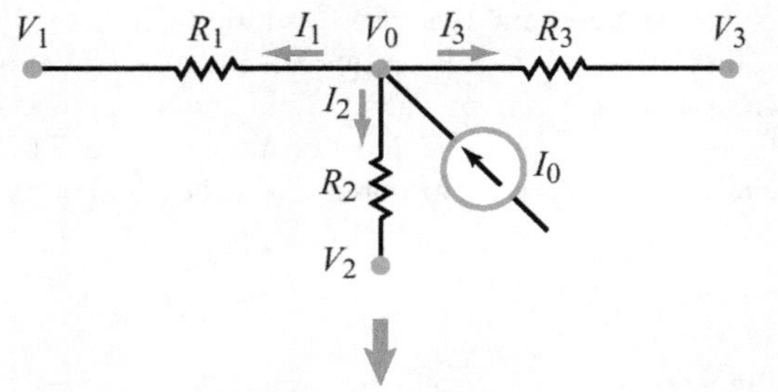

$$\frac{V_0 - V_1}{R_1} + \frac{V_0 - V_2}{R_2} + \frac{V_0 - V_3}{R_3} - I_0 = 0$$

Once the node voltages have been determined, all currents through branches and voltages across elements can be calculated readily.

Example 3-1: Circuit with Two Sources

For the circuit in Fig. 3-2, (a) identify all extraordinary nodes and select one of them as the ground node, (b) develop node-voltage equations at the remaining extraordinary nodes, (c) solve for the node voltages, and then (d) calculate the power consumed by R_5. The element values are $V_0 = 10$ V, $I_0 = 0.8$ A, $R_1 = 5\ \Omega$, $R_2 = 2\ \Omega$, $R_3 = 3\ \Omega$, $R_4 = 10\ \Omega$, and $R_5 = 2.5\ \Omega$.

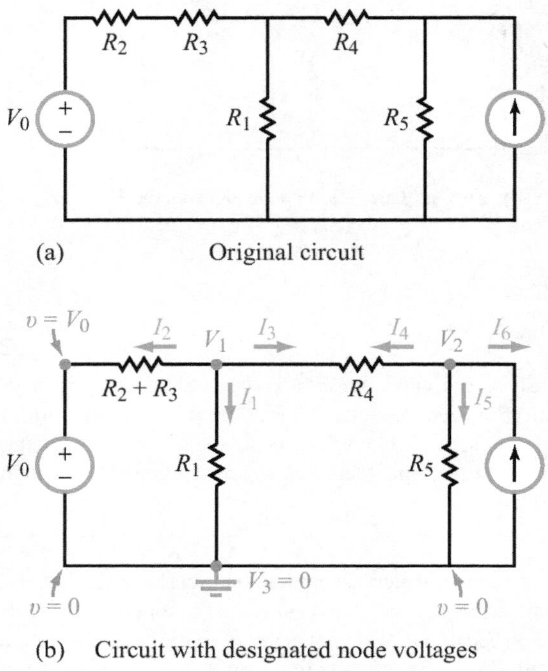

Figure 3-2: Circuit for Example 3-1.

Solution:

(a) **Identify extraordinary nodes and assign node voltages**

The circuit has three extraordinary nodes, labeled as shown in Fig. 3-2(b). Node 3 is selected as the ground node and its voltage is labeled $V_3 = 0$. Nodes 1 and 2 are assigned (unknown) voltages V_1 and V_2, with both defined relative to $V_3 = 0$.

(b) **Apply KCL at nodes 1 and 2**

At each non-ground extraordinary node, we designate currents and we choose their directions as *leaving* the node. We realize that $I_3 = -I_4$, for example, but for the sake of consistency we treat each node the same by designating a current leaving it through every branch connected to it.

Node 1:

$$I_1 + I_2 + I_3 = 0. \tag{3.4}$$

Unless we already know the value of a current (such as I_0 entering node V_2), we should express it in terms of the node voltages connected to the branch through which it is flowing. We do so by applying Ohm's law, while reminding ourselves that the convention we adopted for the current direction is that it flows through a resistor from the (+) voltage terminal to the (−) terminal. Hence:

▶ The current leaving a node is equal to the voltage at that node, minus the voltage at the node to which the current is going, and divided by the resistance.

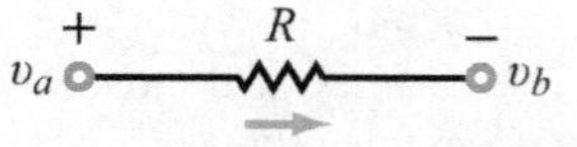

$$i = \frac{\upsilon_a - \upsilon_b}{R}$$

◀

Consequently, I_1 flowing through R_1 is given by

$$I_1 = \frac{V_1 - 0}{R_1} = \frac{V_1}{R_1}. \quad (3.5a)$$

Similarly,

$$I_3 = \frac{V_1 - V_2}{R_4}. \quad (3.5b)$$

The voltage across the in-series resistances $(R_2 + R_3)$ is $(V_1 - V_0)$, where V_0 is the node voltage at the positive terminal of the voltage source. Hence, I_2 is given by

$$I_2 = \frac{V_1 - V_0}{R_2 + R_3}. \quad (3.5c)$$

Inserting Eqs. (3.5a) through (3.5c) into Eq. (3.4) gives

$$\frac{V_1}{R_1} + \frac{V_1 - V_0}{R_2 + R_3} + \frac{V_1 - V_2}{R_4} = 0 \quad \text{(node 1 Voltage Eq.).} \quad (3.6)$$

Node 2:

$$I_4 + I_5 + I_6 = 0,$$

or equivalently,

$$\frac{V_2 - V_1}{R_4} + \frac{V_2}{R_5} - I_0 = 0 \quad \text{(node 2 Voltage Eq.),} \quad (3.7)$$

where we incorporated the fact that $I_6 = -I_0$, as required by the current source.

We note that by designating all current directions at a node as leaving that node:

> ▶ The node-voltage expression for any node (such as node 1 or node 2) always has V of that node preceded with a plus (+) sign. Also, the node voltages of the other nodes are preceded with negative (−) signs. ◀

Thus, V_1 in Eq. (3.6)—which is specific to node 1—has a positive sign wherever it appears in that equation, whereas V_2 and V_3 always have negative signs if they appear in that equation. Conversely, in the node-2 equation given by Eq. (3.7), V_2 is always preceded by a (+) sign and V_1 is preceded by a (−) sign.

(c) **Solve simultaneous equations**

As a prelude to solving Eqs. (3.6) and (3.7) to determine the unknown voltages V_1 to V_3, we need to reorganize them into a standard system of equations as

$$\left(\frac{1}{R_1} + \frac{1}{R_2 + R_3} + \frac{1}{R_4}\right) V_1 - \left(\frac{1}{R_4}\right) V_2 = \frac{V_0}{R_2 + R_3}, \quad (3.8a)$$

and

$$-\left(\frac{1}{R_4}\right) V_1 + \left(\frac{1}{R_4} + \frac{1}{R_5}\right) V_2 = I_0. \quad (3.8b)$$

These are equivalent to

$$a_{11} V_1 + a_{12} V_2 = b_1, \quad (3.9a)$$

and

$$a_{21} V_1 + a_{22} V_2 = b_2, \quad (3.9b)$$

with

$$a_{11} = \left(\frac{1}{R_1} + \frac{1}{R_2 + R_3} + \frac{1}{R_4}\right) = \frac{1}{5} + \frac{1}{2+3} + \frac{1}{10} = 0.5,$$

$$a_{12} = -\frac{1}{R_4} = -\frac{1}{10} = -0.1,$$

$$a_{21} = -\frac{1}{R_4} = -0.1,$$

$$a_{22} = \left(\frac{1}{R_4} + \frac{1}{R_5}\right) = \frac{1}{10} + \frac{1}{2.5} = 0.5,$$

$$b_1 = \frac{V_0}{R_2 + R_3} = \frac{10}{2+3} = 2,$$

and

$$b_2 = I_0 = 0.8.$$

Inserting these values in Eq. (3.9) gives

$$0.5V_1 - 0.1V_2 = 2,$$
$$-0.1V_1 + 0.5V_2 = 0.8.$$

The system of two equations is now amenable for solution by Cramer's rule or matrix inversion (as illustrated in Appendix B) either manually or by using MATLAB or MathScript software (Appendix E). The solution leads to

$$V_1 = 4.5 \text{ V}, \qquad V_2 = 2.5 \text{ V}.$$

(d) **Determine power in R_5**

The current flowing through R_5 in Fig. 3-2(b) is

$$I_5 = \frac{V_2}{R_5} = \frac{2.5}{2.5} = 1 \text{ A},$$

and the power dissipated in R_5 is

$$P = I_5^2 R_5 = (1)^2 \times 2.5 = 2.5 \text{ W}.$$

Concept Question 3-1: The node-*voltage* method relies on the application of Kirchhoff's *current* law. Explain. (See CAD)

Concept Question 3-2: Why does a circuit with n_{ex} extraordinary nodes require only $(n_{ex} - 1)$ node-voltage equations to analyze it? (See CAD)

Exercise 3-1: Apply nodal analysis to determine the current I in the circuit of Fig. E3.1.

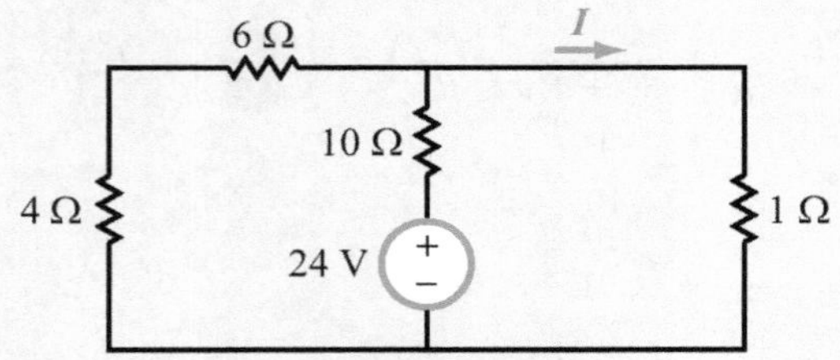

Figure E3.1

Answer: $I = 2$ A. (See CAD)

3-2.2 Circuits Containing Dependent Sources

When a circuit contains dependent sources, the node-voltage analysis method remains applicable, as does the solution procedure outlined in the preceding subsection. However, each dependent source defines a relationship between its own magnitude and some current or voltage elsewhere in the circuit, and that relationship needs to be incorporated into the solution.

Example 3-2: Dependent Current Source

The circuit of Fig. 3-3 contains a current-controlled current source (CCCS) whose magnitude I_x is governed by the current flowing through the 6 Ω resistor in the direction shown. Determine I_x.

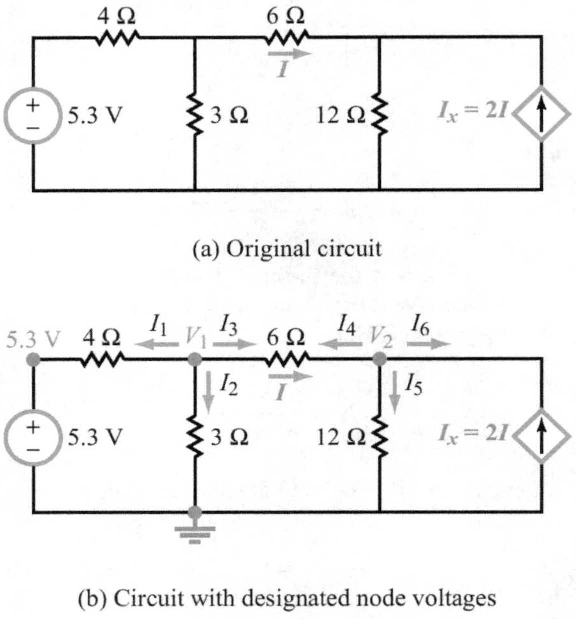

Figure 3-3: Example 3-2.

Solution: Following the standard procedure outlined earlier, we start by selecting a ground node and assigning node voltages to the other extraordinary nodes in the circuit, as shown in Fig. 3-3(b). We also designate currents with their directions out of the nodes for all branches connected to nodes 1 and 2.

Next, we write down the node-voltage equations for nodes 1 and 2 as

$$\frac{V_1 - 5.3}{4} + \frac{V_1}{3} + \frac{V_1 - V_2}{6} = 0 \qquad \text{(node 1)},$$

and

$$\frac{V_2 - V_1}{6} + \frac{V_2}{12} - I_x = 0 \qquad \text{(node 2)}.$$

In the equation for node 1, the three terms represent I_1 to I_3, each expressed as a voltage difference divided by a resistance. The same is true for node 2 except that I_6 is replaced with $(-I_x)$.

We have three unknowns (V_1, V_2, and I_x), but only two equations, so we need to express I_x in terms of the unknown variables, V_1 and V_2. The dependent source I_x is given in terms of I, which in turn is dependent on the voltage difference

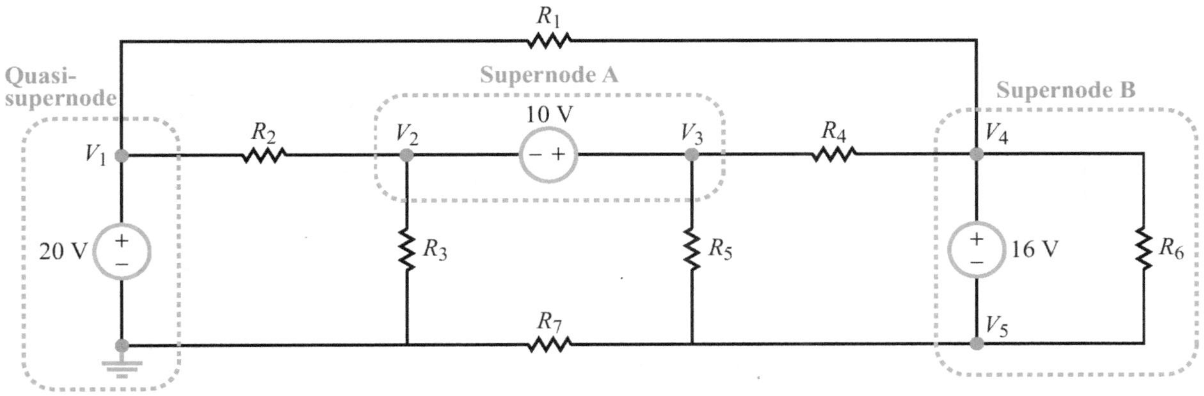

Figure 3-4: Circuit containing two supernodes and one quasi-supernode.

between V_1 and V_2. That is,

$$I_x = 2I = 2\frac{(V_1 - V_2)}{6} = \frac{V_1 - V_2}{3}.$$

This is effectively Ohm's law for I_x. Upon substituting this expression for I_x into the second of the node-voltage equations and rearranging its terms, we end up with

$$9V_1 - 2V_2 = 15.9 \qquad \text{(node 1)}$$

and

$$-6V_1 + 7V_2 = 0 \qquad \text{(node 2)}.$$

Simultaneous solution of the two equations gives $V_1 = 2.18$ V and $V_2 = 1.87$ V. Hence,

$$I_x = \frac{V_1 - V_2}{3} = \frac{2.18 - 1.87}{3} = 0.1 \text{ A}.$$

Exercise 3-2: Apply nodal analysis to find V_a in the circuit of Fig. E3.2.

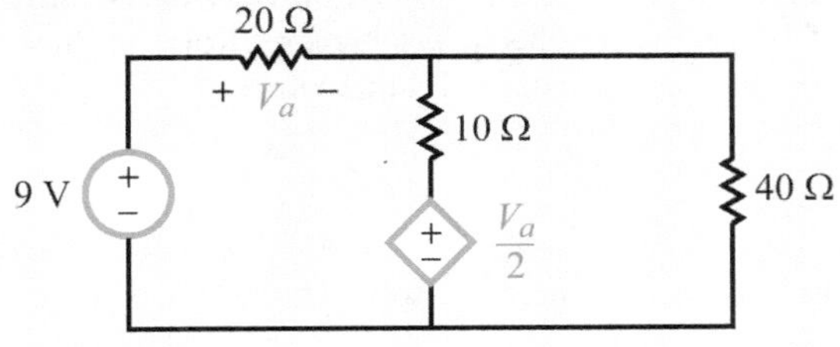

Figure E3.2

Answer: $V_a = 5$ V. (See CAD)

3-2.3 Supernodes

Occasionally, a circuit may contain a solitary voltage source nestled between two extraordinary nodes, with no other elements in series with it between those nodes. Such an arrangement is called a *supernode*. Examples of supernodes are shown in Fig. 3-4. Formally:

> ▶ A *supernode* is the combination of two extraordinary nodes (excluding the reference node) between which a voltage source exists. The voltage source may be of the independent or dependent type, and the voltage source may include elements in parallel with it (such as R_4 in parallel with the 16-V source of supernode B in Fig. 3-4) but not in series with it. If one of the two nodes of a supernode is a reference (ground) node, it is called a *quasi-supernode*. ◀

For a quasi-supernode, the only relevant information we need is that the voltage of the non-reference node is equal to the voltage magnitude of the voltage source. Thus, $V_1 = 20$ V in Fig. 3-4.

The complication caused by a supernode is that we can no longer apply Ohm's law to define the current through a resistor between two extraordinary nodes, because we now have a voltage source between the two nodes instead of a resistor. Hence, we need to treat the supernode in a special way.

To explain the properties of a supernode and how we use it, let us analyze supernode A, all on its own. In Fig. 3-5(a),

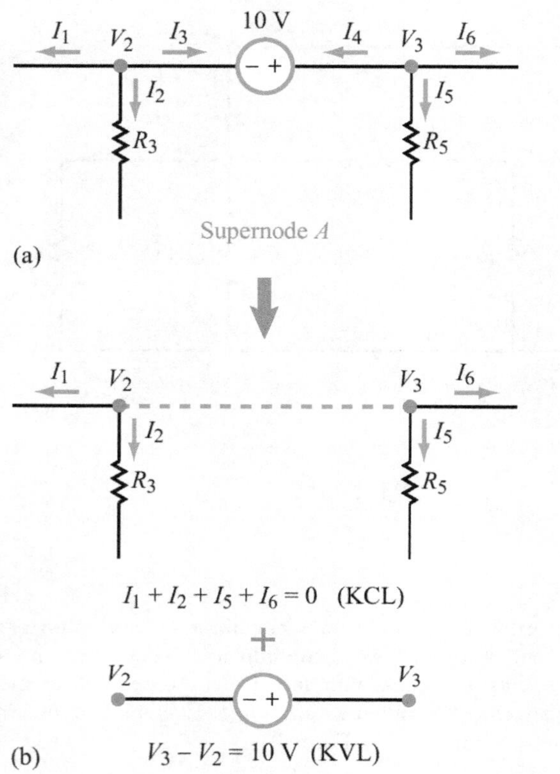

Figure 3-5: A supernode composed of nodes V_2 and V_3 can be represented as a single node, in terms of summing currents flowing out of them, plus an auxiliary equation that defines the voltage difference between V_3 and V_2.

we show currents I_1 to I_3 leaving node 2 and currents I_4 to I_6 leaving node 3. KCL requires that

$$I_1 + I_2 + I_3 = 0 \qquad \text{(node } V_2\text{)}, \qquad (3.10a)$$

and

$$I_4 + I_5 + I_6 = 0 \qquad \text{(node } V_3\text{)}. \qquad (3.10b)$$

Adding the two equations together and recognizing that $I_3 = -I_4$ leads to

$$I_1 + I_2 + I_5 + I_6 = 0 \qquad \text{(supernode A)}, \qquad (3.11)$$

which constitutes the four currents leaving supernode A. The implication of Eq. (3.11) is that we can treat nodes 2 and 3 as a combined single node, connected by a dashed line (Fig. 3-5(b)), but we also should acknowledge the fact that

$$V_3 - V_2 = 10 \text{ V} \qquad \text{(supernode A auxiliary equation)},$$

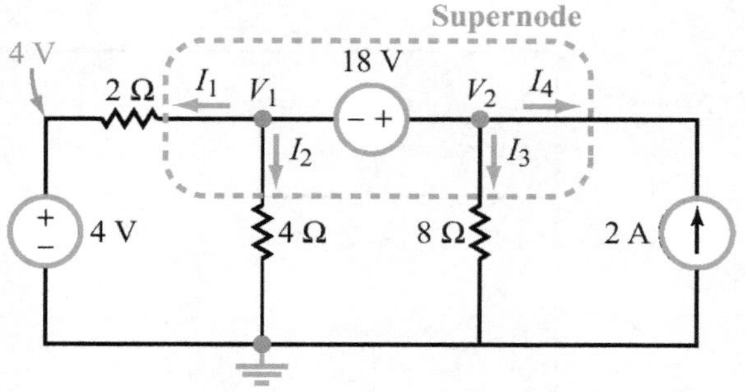

Figure 3-6: Circuit for Example 3-3.

which is a much simpler equation than the typical node-voltage equation.

Supernode Attributes

(1) At a supernode, Kirchhoff's current law (KCL) can be applied to the combination of the two nodes as if they are a single node, but the two nodes retain their own identities.

(2) Kirchhoff's voltage law (KVL) is used to express the voltage difference between the two nodes in terms of the voltage of the source between them. This provides the ***supernode auxiliary equation***.

(3) If a supernode contains a resistor in parallel with the voltage source, the resistor exercises no influence on the currents and voltages in the other parts of the circuit, and therefore, it may be ignored altogether.

(4) For a quasi-supernode, the node-voltage of the non-reference node is equal to the voltage magnitude of the source.

In the circuit of Fig. 3-4, the voltage difference between nodes 4 and 5 is specified by the 16 V source, regardless of the value of R_6 (so long as R_6 is not a short circuit).

Example 3-3: Circuit with a Supernode

Use the supernode concept to solve for the node voltages in Fig. 3-6.

Solution: The combination of nodes 1 and 2 constitutes a supernode, with an associated node-voltage equation given by

$$I_1 + I_2 + I_3 + I_4 = 0$$

or, equivalently,

$$\frac{V_1 - 4}{2} + \frac{V_1}{4} + \frac{V_2}{8} - 2 = 0,$$

which may be simplified to

$$6V_1 + V_2 = 32.$$

Additionally, the supernode KVL equation is

$$V_2 - V_1 = 18.$$

Simultaneous solution of the two equations yields

$$V_1 = 2\text{ V}, \qquad V_2 = 20\text{ V}.$$

> **Concept Question 3-3:** What impact does the presence of a dependent source have on the implementation of the node-voltage method? (See CAD)

> **Concept Question 3-4:** What is a supernode? How is it treated in nodal analysis? (See CAD)

Exercise 3-3: Apply the supernode concept to determine I in the circuit of Fig. E3.3.

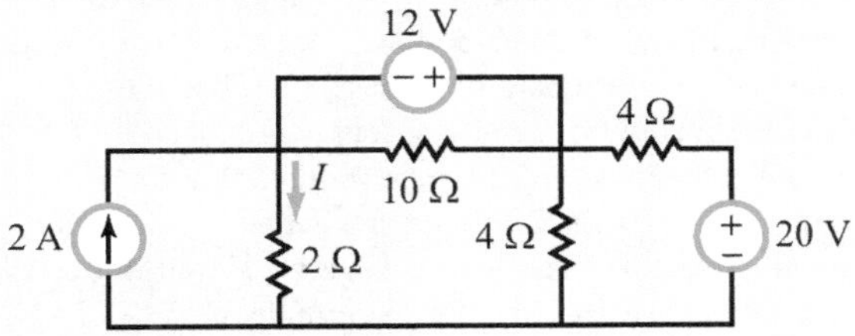

Figure E3.3

Answer: $I = 0.5$ A. (See CAD)

3-3 Mesh-Current Method

3-3.1 General Procedure

A *mesh* was defined in Section 1-3 as a loop that encloses no other loop. The current associated with a mesh is called its *mesh current*. The circuit in Fig. 3-7 contains two meshes with

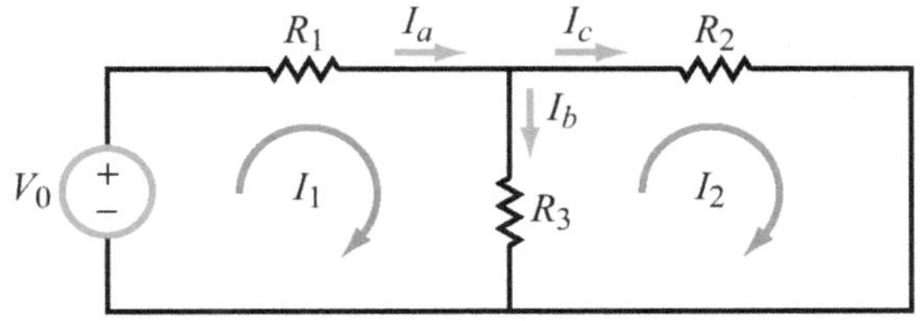

Figure 3-7: Circuit containing two meshes with mesh currents I_1 and I_2.

mesh currents I_1 and I_2. *A mesh current may be thought of as the current flowing through the branches of that mesh, with no regard for the currents in neighboring meshes.* That does not mean, however, that the mesh current is the same as the actual currents flowing through the elements of that mesh. For an element that belongs to only one mesh, such as R_1 in Fig. 3-7, the current through it is indeed identical to the current in mesh 1. That is,

$$I_a = I_1.$$

On the other hand, if an element is shared by two meshes, as is the case for R_3, the true branch current through it is the combination of the two branch currents:

$$I_b = I_1 - I_2.$$

Current I_1 is assigned a positive sign because its direction through R_3 is the same as that of I_b, but I_2 is assigned a negative sign because it flows "upward" through R_3. The mesh-current analysis method is based on the application of KVL to all of the meshes in the circuit. The solution procedure, which is analogous with that discussed earlier in Section 3-2 for the node-voltage method, consists of the following steps:

> **Solution Procedure: Mesh Current**
>
> **Step 1:** Identify all meshes and assign each of them an unknown mesh current. For convenience, define all mesh currents to be clockwise in direction.
>
> **Step 2:** Apply Kirchhoff's voltage law (KVL) to each mesh.
>
> **Step 3:** Solve the resultant simultaneous equations to determine the mesh currents (see Appendix B).

For the circuit in Fig. 3-7, application of KVL to mesh 1, starting at the bottom left-hand corner and moving clockwise around the loop, gives

$$-V_0 + I_1 R_1 + (I_1 - I_2) R_3 = 0 \quad \text{(mesh 1)}, \quad (3.12)$$

where for each term we assigned a (+) or (−) sign to it depending on which of its voltage terminals is encountered first. Also, for a resistor, current flows into the (+) terminal of the voltage across it. For mesh 2,

$$(I_2 - I_1) R_3 + I_2 R_2 = 0 \quad \text{(mesh 2)}. \quad (3.13)$$

The two simultaneous equations can be rearranged by collecting coefficients of I_1 and I_2 as

$$\underbrace{(R_1 + R_3)}_{\substack{\text{Sum of resistances} \\ \text{in mesh 1} \\ + \text{ sign}}} I_1 - \underbrace{I_2 R_3}_{\substack{\text{Resistance shared} \\ \text{by meshes 1 and 2} \\ - \text{ sign}}} = \underbrace{V_0}_{\substack{\text{Voltage source} \\ \text{in mesh 1}}} \quad \text{(mesh 1)},$$

(3.14a)

and

$$-\underbrace{R_3 I_1}_{\substack{\text{Resistance shared} \\ \text{by meshes 1 and 2} \\ - \text{ sign}}} + \underbrace{(R_2 + R_3)}_{\substack{\text{Sum of resistances} \\ \text{in mesh 2} \\ + \text{ sign}}} I_2 = 0 \quad \text{(mesh 2)}.$$

(3.14b)

Note the built-in symmetry reflected by the structure of Eqs. (3.14a and b). For mesh 1, the coefficient of I_1 in Eq. (3.14a) is the sum of all of the resistors contained in mesh 1, and the coefficient of I_2 contains the resistor that mesh 1 shares with mesh 2. Furthermore, the coefficients of I_1 and I_2 have opposite signs. The same pattern applies for mesh 2 in Eq. (3.14b); the coefficient of I_2 contains all of the resistors of mesh 2, and the coefficient of I_1 contains the resistor shared by the two meshes. The magnitude of the voltage source in mesh 1 (namely, V_0) appears on the right-hand side of Eq. (3.14a), with its polarity defined as positive if I_1 flows through it from its negative to positive terminals. This structural pattern allows us to write the mesh-current equations directly, as discussed in more detail later in Section 3-4.

Example 3-4: Circuit with Three Meshes

Use mesh analysis to (a) obtain mesh-current equations for the circuit in Fig. 3-8 and then (b) determine the current in R_4, given that $V_0 = 18$ V, $R_1 = 6\ \Omega$, $R_2 = R_3 = 2\ \Omega$, and $R_4 = R_5 = R_6 = 4\ \Omega$.

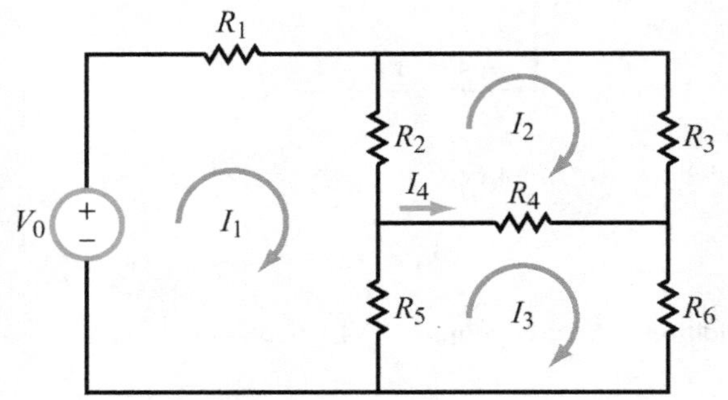

Figure 3-8: Circuit for Example 3-4.

Solution: (a) Applying the symmetry pattern inherent in the structure of the mesh-current equations, we have

$$(R_1 + R_2 + R_5) I_1 - R_2 I_2 - R_5 I_3 = V_0 \quad \text{(mesh 1)}, \quad (3.15a)$$

$$-R_2 I_1 + (R_2 + R_3 + R_4) I_2 - R_4 I_3 = 0 \quad \text{(mesh 2)}, \quad (3.15b)$$

and

$$-R_5 I_1 - R_4 I_2 + (R_4 + R_5 + R_6) I_3 = 0 \quad \text{(mesh 3)}. \quad (3.15c)$$

We note that in Eq. (3.15a) the coefficient of I_1 is positive and is composed of the sum of all resistors in mesh 1 and the coefficients of I_2 and I_3 are negative and include the resistors that meshes 2 and 3 share with mesh 1, respectively. An equivalent pattern pertains to Eqs. (3.15b and c).

If the mesh contains a voltage source, its magnitude appears on the right-hand side of the mesh equation and it is assigned a positive sign if it is a voltage rise when moving clockwise around the mesh. It is assigned a negative sign if it is a voltage drop. In the case of mesh 1 in the circuit of Fig. 3-8, V_0 is a voltage rise, so it appears on the right-hand side of Eq. (3.15a) with a positive sign.

(b) For the specified values of V_0 and the six resistors, the three parts of Eq. (3.15) become

$$12 I_1 - 2 I_2 - 4 I_3 = 18,$$
$$-2 I_1 + 8 I_2 - 4 I_3 = 0,$$
$$-4 I_1 - 4 I_2 + 12 I_3 = 0,$$

and solution of the simultaneous equations leads to

$$I_1 = 2\ \text{A}, \qquad I_2 = 1\ \text{A}, \qquad I_3 = 1\ \text{A}.$$

The current through R_4 is

$$I_4 = I_3 - I_2 = 1 - 1 = 0.$$

Given that the circuit is a Wheatstone bridge (Section 2-5) operated under the balanced condition ($R_2 R_6 = R_3 R_5$), the result $I_4 = 0$ is exactly what we should have expected.

Exercise 3-4: Apply mesh analysis to determine I in the circuit of Fig. E3.4.

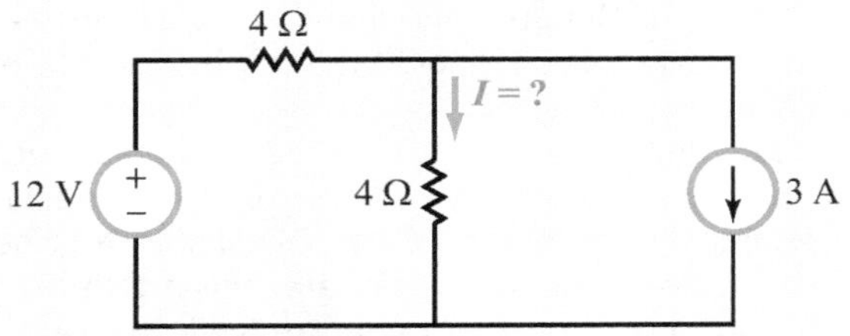

Figure E3.4

Answer: $I = 0$. (See CAD)

3-3.2 Circuit with Dependent Sources

The presence of a dependent source in a circuit does not alter the basic procedure of the mesh-current method, but it requires the addition of a supplemental equation expressing the relationship between the dependent source and the other parts of the circuit.

Example 3-5: Dependent Current Source

Use mesh-current analysis to determine the magnitude of the dependent source I_x in Fig. 3-9.

Solution: For the meshes with mesh currents I_1 and I_2,

$$(1 + 2)I_1 - 2I_2 - I_3 = 10 \quad \text{(mesh 1)}, \qquad (3.16a)$$

and

$$-2I_1 + (2 + 1 + 3)I_2 - I_3 = 0 \quad \text{(mesh 2)}. \qquad (3.16b)$$

For mesh 3, we do not need to write a mesh-current equation, because I_3 is specified by the current source as

$$I_3 = I_x = 4V_1.$$

The voltage V_1 across the 2 Ω resistor is given by

$$V_1 = 2(I_1 - I_2).$$

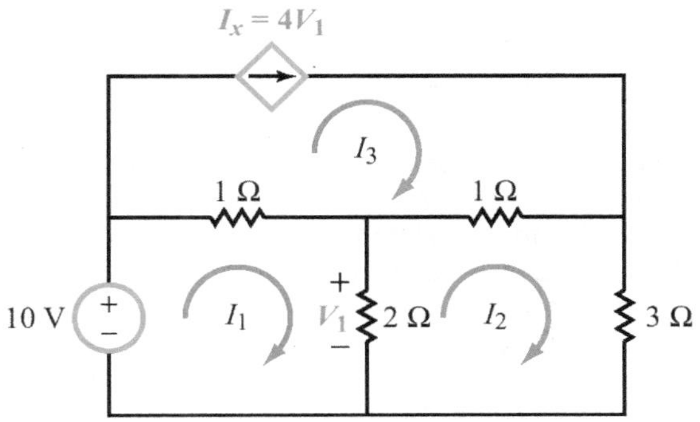

Figure 3-9: Mesh-current solution for a circuit containing a dependent source (Example 3-5).

Hence,

$$I_3 = 4V_1 = 8(I_1 - I_2). \qquad (3.17)$$

After inserting Eq. (3.17) into Eqs. (3.16a and b) and collecting terms in I_1 and I_2, we end up with

$$-5I_1 + 6I_2 = 10,$$
$$-10I_1 + 14I_2 = 0.$$

Solution of this pair of simultaneous equations gives

$$I_1 = -14 \text{ A}, \qquad I_2 = -10 \text{ A}.$$

Hence,

$$I_x = 8(I_1 - I_2) = 8(-14 + 10) = -32 \text{ A}.$$

Exercise 3-5: Determine the current I in the circuit of Fig. E3.5.

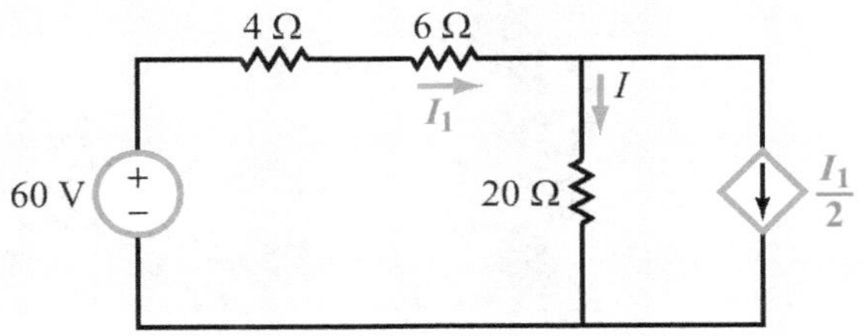

Figure E3.5

Answer: $I = 1.5$ A. (See CAD)

Technology Brief 6
Measurement of Electrical Properties of Sea Ice

Climate change is often first measured by the decrease of our polar ice caps. This sea ice is a unique and vibrant type of ice; the fresh water freezes first, leaving pockets of more and more briny (salty) water, that eventually freezes only when the temperature gets below its eutectic point around −21 °C. A combination of gravity and freeze-thaw cycles elongates these tiny brine pockets (initially sub-mm in size), and many of them start linking together to form fluidic channels (which eventually expand to become a full centimeter or more in diameter), from the top of the ice all the way through one or two meters of ice to the sea below the ice pack (Fig. TF6-1). In this columnar type of sea ice, which is prevalent in the Arctic, there is a critical brine volume fraction of about 5%, called the ***percolation threshold***, above which there are large-scale connected channels or pathways through which fluid can flow, and below which the sea ice is effectively impermeable. For a typical bulk sea-ice salinity of 5 parts per thousand, this brine volume fraction corresponds to a critical temperature of about −5 °C. This *on-off switch* for fluid flow is known as the ***rule of fives***. The brine channels can moderate the formation of melt ponds (Fig. TF6-2) by quickly draining them and returning the ice to its more reflective white coloring.

This brine percolation threshold has been quantified through measurements of the electrical resistivity of the ice, as well as X-ray computed tomography and measurements of the fluid permeability. Salty brine pockets are very conductive, and the surrounding ice is a near insulator. As the brine pockets join into channels, the overall conductivity of the ice increases substantially by providing a conducting path for current in pretty much the same way it provides a path for the water to percolate (drain) through. Conductivity, then, is highly correlated with the percolation threshold and can be used to help us study melt-pond formation.

The electrical properties of the ice are measured by drilling out a 9 cm cylindrical core of ice, measuring its resistance using a model very similar to that seen in Fig. 2-1. Stainless steel nails are driven into the ice core (drilling holes for them first, to avoid cracking the core) to make the electrical connection to the ice. But this method has a problem. It is very hard to get a consistent electrical connection between the nail and the ice. This contact resistance is very much a part of the circuit, and it varies with each connection. A circuit model of this resistance measurement is shown below. The total resistance is the series combination of the two (variable) contact resistances and the resistance of the

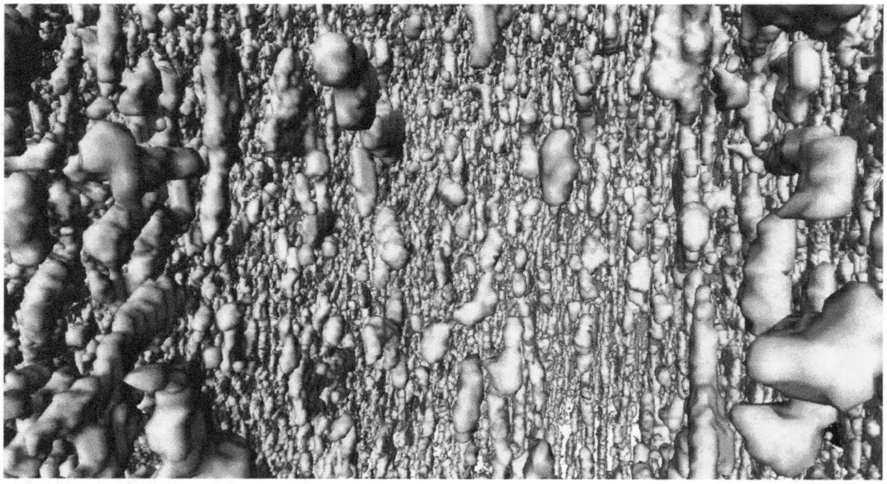

Figure TF6-1: X-ray CT images (approximately 1 cm across) of the brine microstructure of sea ice. The brine volume fraction is 5.7%, and the temperature is −8 °C. Channels are beginning to form but are not fully connected yet. (From Golden et al., *Geophys. Res. Letters*, 2007.)

Figure TF6-2: As ice melts, the liquid water collects in depressions on the surface and deepens them, forming these melt ponds in the Arctic. These fresh water ponds are separated from the salty sea below and around it, until breaks in the ice merge the two.

ice. Without being able to better control the contact resistance, $R_{\text{sea ice}}$ cannot be accurately measured.

To solve this problem, rather than doing a simple 2-wire resistance measurement as shown in Fig. TF6-3, a 4-wire measurement system can be used as shown in Fig. TF6-4. This system employs both an ammeter and a voltmeter (which are combined into the single yellow AEMC resistance meter shown in Fig. TF6-5). Two wires are used to connect the ammeter in series with the resistances, and two are used to connect the voltmeter in parallel with $R_{\text{sea ice}}$ (hence, 4 wires). We do not need to know the driving voltage or the contact resistances in order to accurately measure $R_{\text{sea ice}}$ with this method.

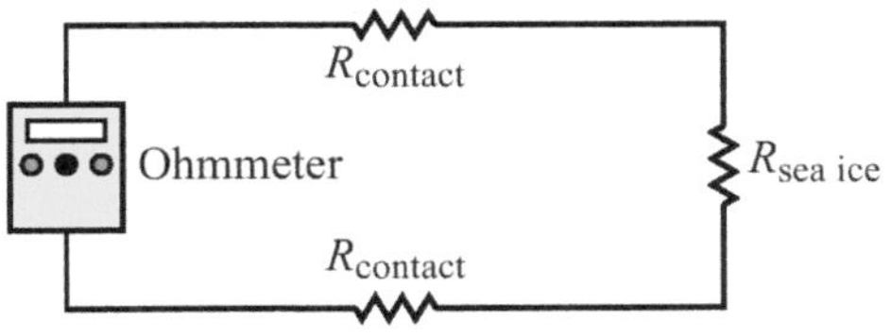

Ohmmeter indicates $R_{\text{contact}} + R_{\text{sea ice}} + R_{\text{contact}}$

Figure TF6-3: Simple 2-wire resistance measurement circuit.

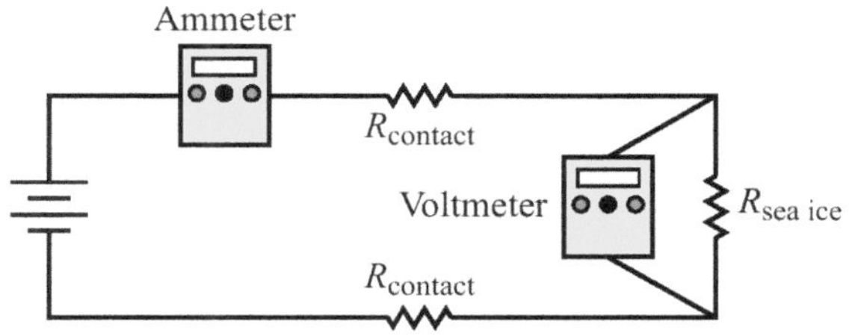

$$R_{\text{subject}} = \frac{\text{Voltmeter indication}}{\text{Ammeter indication}}$$

Figure TF6-4: 4-wire measurement circuit.

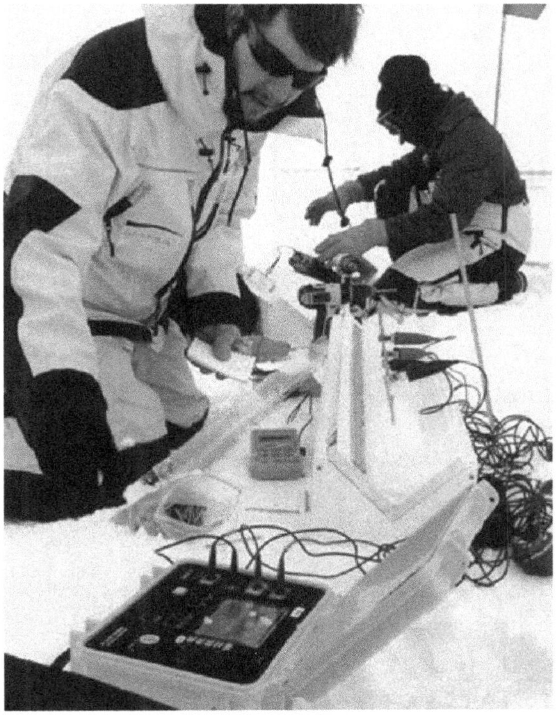

Figure TF6-5: University of Utah mathematics Ph.D. student Christian Sampson measures the electrical conductivity of a sea-ice core during the Sea Ice Physics and Ecosystem eXperiment in 2012. Electrical clamps are attached to nails inserted along the length of the ice core. (© Wendy Pyper/Australian Antarctic Division.)

3-3.3 Supermeshes

▶ Two adjoining meshes that share a current source constitute a supermesh. The current source may be of the independent or dependent type, and it may include a resistor in series with it, but not in parallel. ◀

The presence of a supermesh in a circuit, such as the one shown in Fig. 3-10(a), simplifies the solution by (a) combining the two mesh-current equations into one and (b) adding a simpler, auxiliary equation that relates the current of the source to the mesh currents of the two meshes.

In Fig. 3-10(b), the current source of the supermesh has been removed (as has the series resistor R_4) and replaced with a dashed line. The dashed line is a reminder to relate I_0 to the mesh currents, namely

$$I_0 = I_2 - I_3 \qquad \text{(auxiliary eq.).} \tag{3.18}$$

The mesh-current equations for mesh 1 and the joint combination of meshes 2 and 3 are

$$(R_1 + R_2 + R_5)I_1 - R_2 I_2 - R_5 I_3 = V_0 \qquad \text{(mesh 1)}, \tag{3.19}$$

and

$$-(R_2 + R_5)I_1 + (R_2 + R_3)I_2 + (R_5 + R_6)I_3 = 0 \qquad \text{(supermesh)}. \tag{3.20}$$

The two mesh-current equations, together with the auxiliary equation given by Eq. (3.18), are sufficient to solve for the three mesh currents.

It is instructive to note that the series resistor R_4 played no role in the solution. This is because the current through it is specified by I_0, regardless of the magnitude of R_4 (so long as it is not an open circuit).

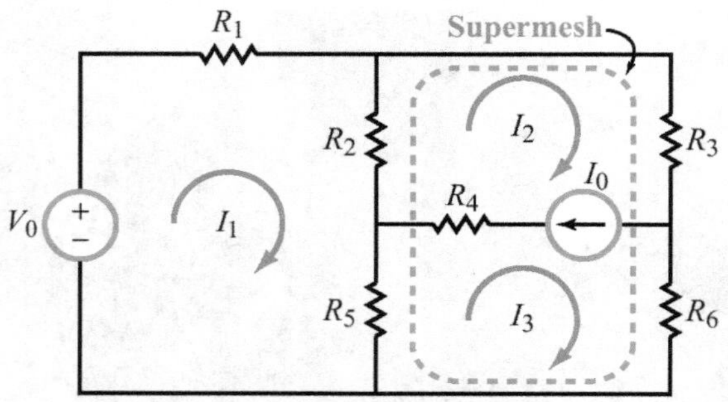

(a) Two adjoining meshes sharing a current source constitute a supermesh.

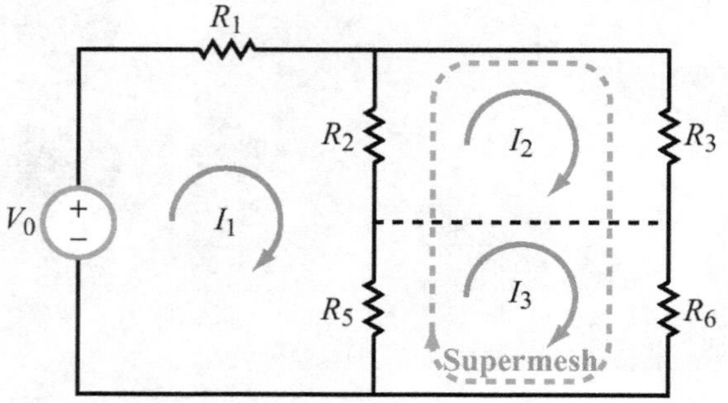

(b) Meshes 2 and 3 can be combined into a single supermesh equation, plus an auxiliary equation $I_0 = I_2 - I_3$.

Figure 3-10: Concept of a supermesh.

Example 3-6: Circuit with a Supermesh

For the circuit in Fig. 3-11(a), determine (a) the mesh currents and (b) the power supplied by each of the two sources.

Solution: (a) Meshes 3 and 4 share a current source, thereby forming a supermesh. Figure 3-11(b) shows the circuit redrawn such that meshes 3 and 4 can be combined into a single supermesh equation. Consequently, the mesh-current equations for mesh 1, mesh 2, and supermesh 3 and 4 respectively, are

$$(10 + 2 + 4)I_1 - 2I_2 - 4I_3 = 6 \qquad \text{(mesh 1)}, \tag{3.21a}$$

$$-2I_1 + (2 + 2 + 2)I_2 - 2I_4 = 0 \qquad \text{(mesh 2)}, \tag{3.21b}$$

and

$$-4I_1 - 2I_2 + 4I_3 + (2 + 4)I_4 = 0 \qquad \text{(supermesh)}. \tag{3.21c}$$

The auxiliary equation associated with the current source is given by

$$I_4 - I_3 = 3 \qquad \text{(auxiliary equation)}. \tag{3.22}$$

Inserting Eq. (3.22) to eliminate I_4 in Eqs. (3.21b and c) leads to

$$16I_1 - 2I_2 - 4I_3 = 6,$$

$$-2I_1 + 6I_2 - 2I_3 = 6,$$

$$-4I_1 - 2I_2 + 10I_3 = -18.$$

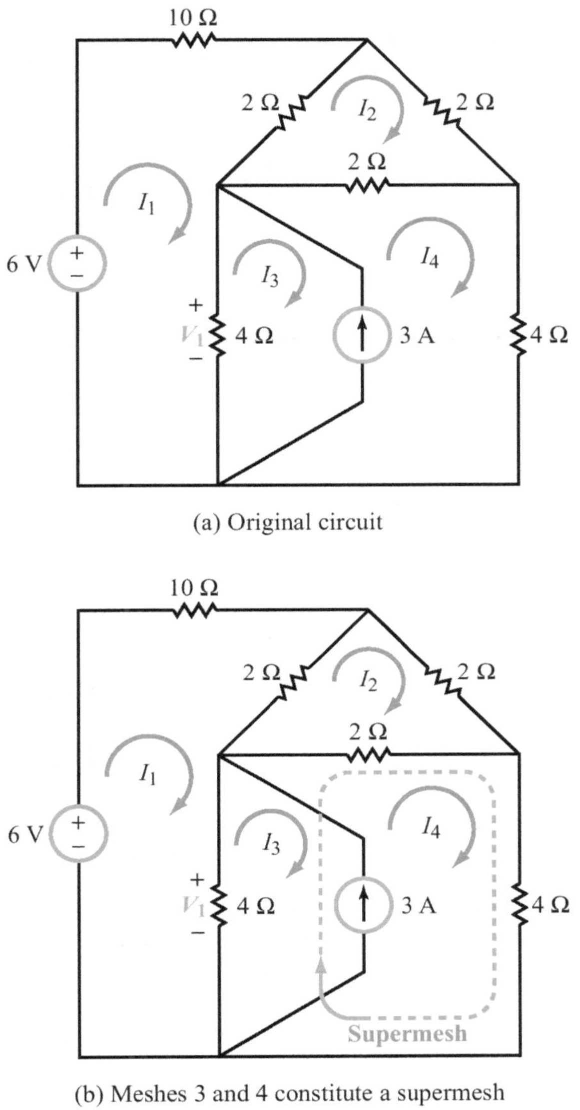

Figure 3-11: Using the supermesh concept to simplify solution of the circuit in Example 3-6.

Solution of the three simultaneous equations gives

$$I_1 = 0, \qquad I_2 = \frac{3}{7} \text{ A},$$

$$I_3 = -\frac{12}{7} \text{ A}, \qquad I_4 = \frac{9}{7} \text{ A}.$$

(b) Since $I_1 = 0$, the power supplied by the 6 V source is

$$P_1 = 6I_1 = 0.$$

To calculate the power supplied by the 3 A current source, we need to know the voltage V_1 across it, which is also the voltage across the 4 Ω resistor given as

$$V_1 = 4(I_1 - I_3) = 4\left(0 - \left(-\frac{12}{7}\right)\right) = \frac{48}{7} \text{ V}.$$

Hence,

$$P_2 = 3V_1 = 3 \times \frac{48}{7} = 20.6 \text{ W}.$$

Thus, all of the power is supplied by the 3 A source alone and is dissipated in the circuit resistances, except for the 10 Ω resistance (because the current through it is $I_1 = 0$).

Concept Question 3-5: How does the presence of a dependent source in the circuit influence the implementation procedure of the mesh-current method? (See CAD)

Concept Question 3-6: What is a supermesh, and how is it used in mesh analysis? (See CAD)

Exercise 3-6: Apply mesh analysis to determine I in the circuit of Fig. E3.6.

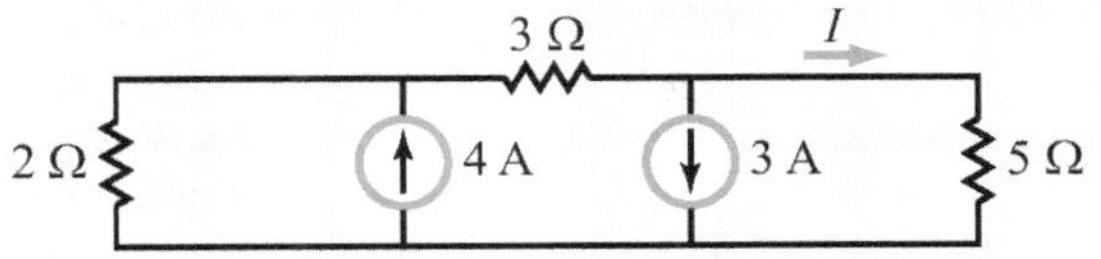

Figure E3.6

Answer: $I = -0.7$ A. (See CAD)

3-4 By-Inspection Methods

The node-voltage and mesh-current methods can be used to analyze any planar circuit, including those containing dependent sources. The solution process relies on the application of KCL and KVL to generate the requisite number of equations necessary to solve for the unknown currents and voltages.

► For circuits that *contain only independent sources*, their KCL and KVL equations exhibit standard patterns, allowing us to write them down by direct *inspection* of the circuit. The method of *nodal analysis by inspection* is easy to implement, but it requires that all sources in the circuit be independent current sources. Similarly, *mesh analysis by inspection* requires that all sources be independent voltage sources. ◄

If a circuit contains a mixture of independent current and voltage sources, implementation of the by-inspection methods will require a prerequisite step in which current sources are converted to voltage sources, or vice versa, so as to secure the requirement that all sources exclusively are current sources or voltage sources. The conversion process can be realized with the help of the source-transformation technique of Section 2-3.4.

3-4.1 Nodal Analysis by Inspection

Even though it is common practice to characterize the i–υ relationship of a resistor in terms of its resistance R, it is more convenient in some cases to work in terms of its conductance $G = 1/R$ and to apply the form of Ohm's law given by

$$I = \frac{V}{R} = GV.$$

The node-voltage by-inspection method is one such case.

We shortly will demonstrate the method for the general case of a circuit composed of n (nonreference) extraordinary nodes. As noted earlier, *applicability of the method is limited to circuits with independent current sources*. By way of introducing the method, let us consider the simple circuit of Fig. 3-12(a), whose resistances have been relabeled in terms of conductances in Fig. 3-12(b). In a circuit diagram, the value next to the symbol of a resistor may be designated in ohms (Ω) or siemens (S), with the former referring to the value of its resistance R and the latter referring to the value of its conductance G. Both designations convey the same information about the resistor.

The circuit has two extraordinary nodes. According to the node-voltage by-inspection method, the circuit is characterized by two node-voltage equations given by

$$G_{11}V_1 + G_{12}V_2 = I_{t_1}, \tag{3.23a}$$

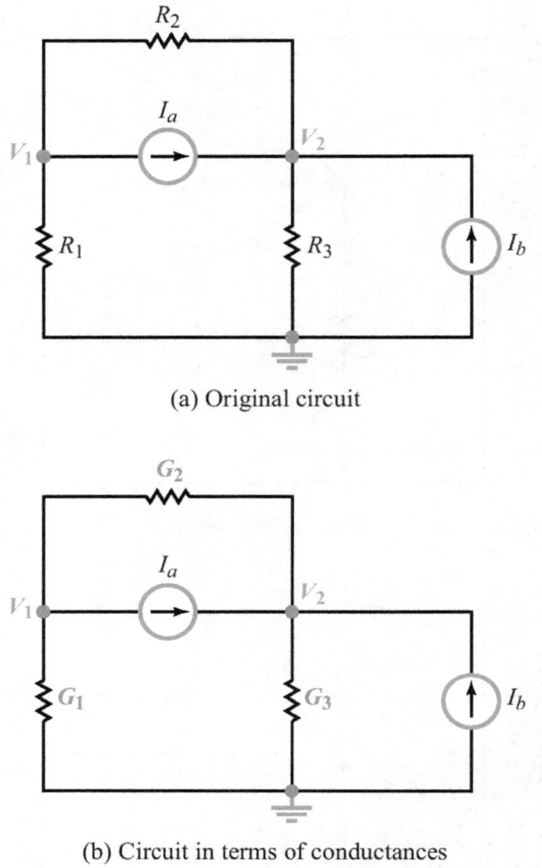

Figure 3-12: Application of the nodal-analysis by-inspection method is facilitated by replacing resistors with conductances.

and

$$G_{21}V_1 + G_{22}V_2 = I_{t_2}, \tag{3.23b}$$

where

G_{11} and G_{22} = sum of all conductances connected to nodes 1 and 2, respectively

$G_{12} = G_{21}$ = *negative* of the sum of all conductances connected between nodes 1 and 2

I_{t_1} and I_{t_2} = total of all independent current sources *entering* nodes 1 and 2, respectively (a negative sign applies to a current source leaving a node).

Application of these definitions to Fig. 3-12(b) gives

$$G_{11} = G_1 + G_2,$$
$$G_{22} = G_2 + G_3,$$
$$G_{12} = G_{21} = -G_2,$$
$$I_{t_1} = -I_a,$$

and

$$I_{t_2} = I_a + I_b.$$

Hence,

$$(G_1 + G_2)V_1 - G_2V_2 = -I_a \quad (3.24a)$$

and

$$-G_2V_1 + (G_2 + G_3)V_2 = I_a + I_b. \quad (3.24b)$$

It is a straightforward task to ascertain that Eqs. (3.24a and b) are indeed the correct node-voltage equations for the circuit in Fig. 3-12(b).

Generalizing to the n-node case, the node-voltage equations can be cast in matrix form as

$$\begin{bmatrix} G_{11} & G_{12} & \cdots & G_{1n} \\ G_{21} & G_{22} & \cdots & G_{2n} \\ \vdots & & & \\ G_{n1} & G_{n2} & \cdots & G_{nn} \end{bmatrix} \begin{bmatrix} V_1 \\ V_2 \\ \vdots \\ V_n \end{bmatrix} = \begin{bmatrix} I_{t_1} \\ I_{t_2} \\ \vdots \\ I_{t_n} \end{bmatrix}, \quad (3.25)$$

and abbreviated as

$$\mathbf{GV} = \mathbf{I}_t, \quad (3.26)$$

where $\mathbf{G}$ is the *conductance matrix* of the circuit, $\mathbf{V}$ is an *unknown voltage vector* representing the node voltages, and $\mathbf{I}_t$ is the *source vector*. The elements of these matrices are defined as

G_{kk} = sum of all conductances connected to node k

$G_{k\ell} = G_{\ell k}$ = *negative* of conductance(s) connecting nodes k and ℓ, with $k \neq \ell$ ($G_{k\ell} = 0$ if no conductance connects nodes k and ℓ directly)

V_k = voltage at node k

I_{t_k} = total of current sources *entering* node k (a negative sign applies to a current source leaving the node).

Solution of Eq. (3.26) for the elements of vector $\mathbf{V}$ can be obtained through matrix inversion (Appendix B) or the application of MATLAB or MathScript (Appendix E).

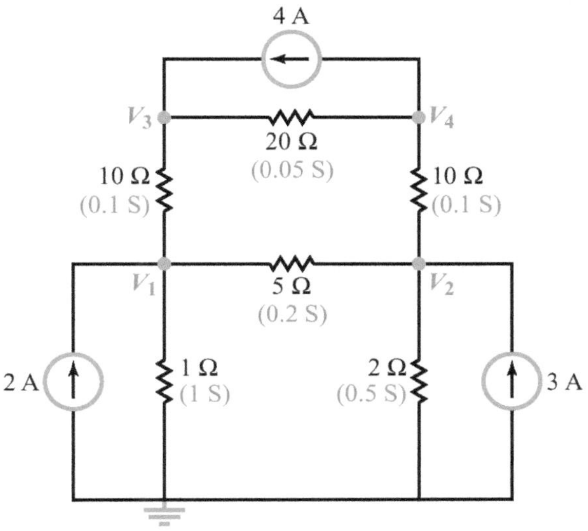

Figure 3-13: Circuit for Example 3-7.

Example 3-7: Four-Node Circuit

Obtain the node-voltage matrix equation for the circuit in Fig. 3-13 by inspection.

Solution: At node 1,

$$G_{11} = \frac{1}{1} + \frac{1}{5} + \frac{1}{10} = 1.3 \text{ S}.$$

Similarly, at nodes 2, 3, and 4,

$$G_{22} = \frac{1}{5} + \frac{1}{2} + \frac{1}{10} = 0.8 \text{ S},$$

$$G_{33} = \frac{1}{10} + \frac{1}{20} = 0.15 \text{ S},$$

and

$$G_{44} = \frac{1}{10} + \frac{1}{20} = 0.15 \text{ S}.$$

The off-diagonal elements of the matrix are

$$G_{12} = G_{21} = -\frac{1}{5} = -0.2 \text{ S},$$

$$G_{13} = G_{31} = -\frac{1}{10} = -0.1 \text{ S},$$

$$G_{14} = G_{41} = 0,$$

$$G_{23} = G_{32} = 0,$$

$$G_{24} = G_{42} = -\frac{1}{10} = -0.1 \text{ S},$$

and

$$G_{34} = G_{43} = -\frac{1}{20} = -0.05 \text{ S}.$$

The total currents entering nodes 1 to 4 are

$$I_{t_1} = 2 \text{ A},$$

$$I_{t_2} = 3 \text{ A},$$

$$I_{t_3} = 4 \text{ A},$$

and

$$I_{t_4} = -4 \text{ A}.$$

Hence, the node-voltage matrix equation is given by

$$\begin{bmatrix} 1.3 & -0.2 & -0.1 & 0 \\ -0.2 & 0.8 & 0 & -0.1 \\ -0.1 & 0 & 0.15 & -0.05 \\ 0 & -0.1 & -0.05 & 0.15 \end{bmatrix} \begin{bmatrix} V_1 \\ V_2 \\ V_3 \\ V_4 \end{bmatrix} = \begin{bmatrix} 2 \\ 3 \\ 4 \\ -4 \end{bmatrix},$$

Solution by matrix inversion or MATLAB or MathScript software gives

$$V_1 = 3.73 \text{ V},$$

$$V_2 = 2.54 \text{ V},$$

$$V_3 = 23.43 \text{ V},$$

$$V_4 = -17.16 \text{ V}.$$

Exercise 3-7: Apply the node-analysis by-inspection method to generate the node-voltage matrix for the circuit in Fig. E3.7.

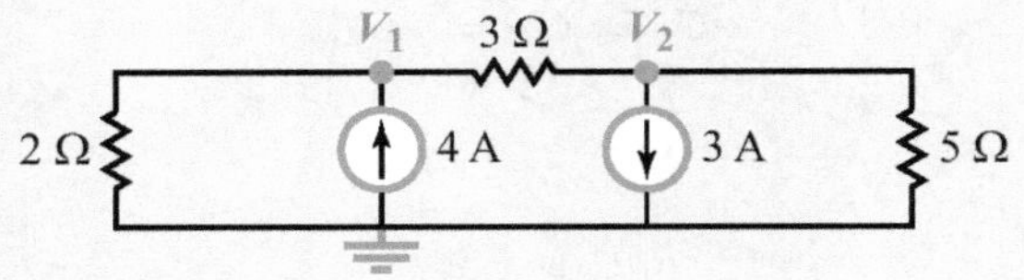

Figure E3.7

Answer:

$$\begin{bmatrix} \frac{5}{6} & -\frac{1}{3} \\ -\frac{1}{3} & \frac{8}{15} \end{bmatrix} \begin{bmatrix} V_1 \\ V_2 \end{bmatrix} = \begin{bmatrix} 4 \\ -3 \end{bmatrix}.$$

(See CAD)

3-4.2 Mesh Analysis by Inspection

By analogy with the node-voltage by-inspection method, for a circuit containing only independent voltage sources, its n mesh-current equations can be cast in matrix form as

$$\mathbf{RI} = \mathbf{V}_t, \tag{3.27}$$

where $\mathbf{R}$ is the *resistance matrix* of the circuit, $\mathbf{I}$ is a vector representing the unknown mesh currents, and $\mathbf{V}$ is the *source vector*. Equation (3.27) is an abbreviation for

$$\begin{bmatrix} R_{11} & R_{12} & \cdots & R_{1n} \\ R_{21} & R_{22} & \cdots & R_{2n} \\ \vdots & & & \\ R_{n1} & R_{n2} & \cdots & R_{nn} \end{bmatrix} \begin{bmatrix} I_1 \\ I_2 \\ \vdots \\ I_n \end{bmatrix} = \begin{bmatrix} V_{t_1} \\ V_{t_2} \\ \vdots \\ V_{t_n} \end{bmatrix}, \tag{3.28}$$

where

R_{kk} = sum of all resistances in mesh k,
$R_{k\ell} = R_{\ell k}$ = *negative* of the sum of all resistances shared between meshes k and ℓ (with $k \neq \ell$) ($R_{k\ell} = 0$ if meshes k and ℓ do not share a resistor).
I_k = current of mesh k
V_{t_k} = total of all independent voltage sources in mesh k, with positive assigned to a voltage rise when moving around the mesh in a clockwise direction.

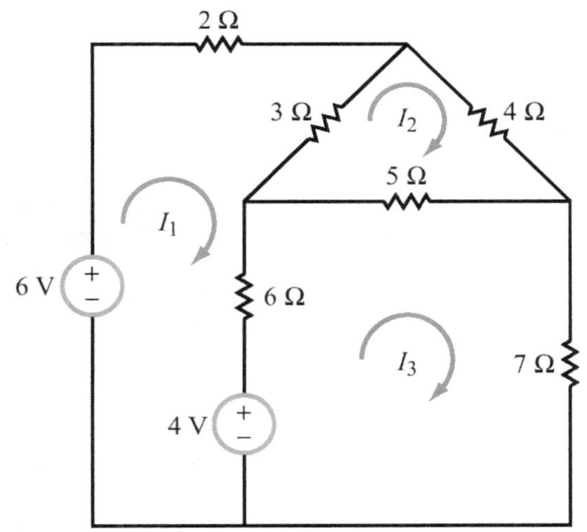

Figure 3-14: Three-mesh circuit of Example 3-8.

Example 3-8: Three-Mesh Circuit

Obtain the mesh-current matrix equation for the circuit in Fig. 3-14, by inspection.

Solution: Application of the definitions for the elements of the matrix **R** and vector $\mathbf{V}_t$ leads to

$$\begin{bmatrix} (2+3+6) & -3 & -6 \\ -3 & (3+4+5) & -5 \\ -6 & -5 & (5+6+7) \end{bmatrix} \begin{bmatrix} I_1 \\ I_2 \\ I_3 \end{bmatrix} = \begin{bmatrix} 6-4 \\ 0 \\ 4 \end{bmatrix},$$

which simplifies to

$$\begin{bmatrix} 11 & -3 & -6 \\ -3 & 12 & -5 \\ -6 & -5 & 18 \end{bmatrix} \begin{bmatrix} I_1 \\ I_2 \\ I_3 \end{bmatrix} = \begin{bmatrix} 2 \\ 0 \\ 4 \end{bmatrix}.$$

Solution of the matrix equation gives $I_1 = 0.55$ A, $I_2 = 0.35$ A, and $I_3 = 0.50$ A.

Concept Question 3-7: Are the by-inspection methods applicable to (a) circuits containing a mixture of independent voltage and current sources or (b) circuits containing a mixture of independent and dependent voltage sources? (See CAD)

Concept Question 3-8: If the circuit contains a mixture of real voltage and current sources, what step should be taken to prepare the circuit for application of one of the two by-inspection methods? (See CAD)

Exercise 3-8: Use the by-inspection method to generate the mesh-current matrix for the circuit in Fig. E3.8.

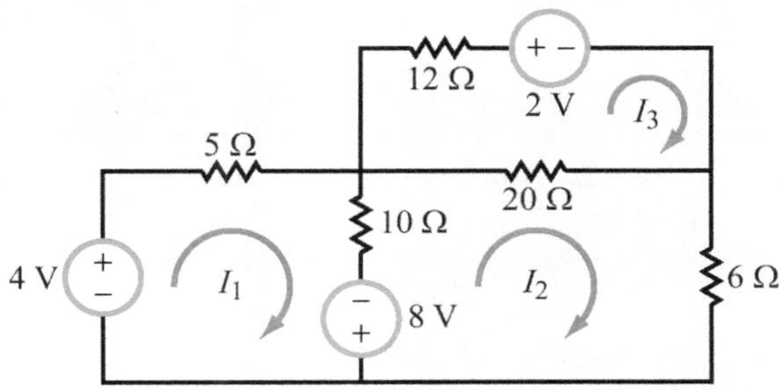

Figure E3.8

Answer:

$$\begin{bmatrix} 15 & -10 & 0 \\ -10 & 36 & -20 \\ 0 & -20 & 32 \end{bmatrix} \begin{bmatrix} I_1 \\ I_2 \\ I_3 \end{bmatrix} = \begin{bmatrix} 12 \\ -8 \\ -2 \end{bmatrix}$$

(See CAD)

3-5 Linear Circuits and Source Superposition

▶ A system is said to be linear if its output response is directly proportional to the excitation at its input. ◀

In the case of a resistive circuit, the input excitation consists of the combination of all independent voltage and current sources in the circuit, and the output response consists of the set of all voltages across all passive elements in the circuit (namely, the resistors), or all currents through them. As noted in Section 3-1, *circuits with ideal elements (including those containing capacitors and inductors) satisfy the linearity property, and therefore qualify as linear systems.* A linear system obeys the superposition principle (Section 3-1.2), which for a linear circuit translates into:

▶ If a circuit contains more than one independent source, the voltage (or current) response of any element in the circuit is equal to the algebraic sum (*superposition*) of the individual responses associated with the individual independent sources, as if each had been acting alone. ◀

Thus, for a circuit with n independent voltage or current sources labeled as sources 1 to n, the voltage υ across a given passive circuit element is given by

$$\upsilon = \upsilon_1 + \upsilon_2 + \cdots + \upsilon_n, \tag{3.29}$$

where υ_k is the response when all sources have been set to zero, except for source k. A similar expression applies to the current i through the circuit,

$$i = i_1 + i_2 + \cdots + i_n. \tag{3.30}$$

The superposition principle can be used to find υ (or i) by executing the following steps:

Solution Procedure: Source Superposition

Step 1: Set all independent sources equal to zero (by replacing voltage sources with short circuits and current sources with open circuits), except for source 1.

Step 2: Apply node-voltage, mesh-current, or any other convenient analysis technique to solve for the response υ_1 due to source 1 acting alone.

Step 3: Repeat the process for sources 2 through n, calculating in each case the response due to that one source acting alone.

Step 4: Use Eq. (3.29) to determine the total response υ.

Alternatively, the procedure can be used to find currents i_1 to i_n and then to add them up algebraically to find the total current i using Eq. (3.30).

Because it entails solving a circuit multiple times, the source-superposition method may not seem attractive, particularly for analyzing circuits with many sources. However, it is a useful tool in both analysis and design for evaluating the sensitivity of a response (such as the current in a load resistor) to specific sources in the circuit.

▶ Whereas the source-superposition method is applicable for calculating voltage and current, it is not applicable for power (see Example 3-9). ◀

Example 3-9: Circuit Analysis by Source Superposition

(a) Use source superposition to determine the current I in the circuit of Fig. 3-15. (b) Determine the amount of power dissipated in the 10 Ω resistor due to each source acting alone and due to both sources acting simultaneously.

Solution: (a) The circuit contains two sources, I_0 and V_0. We start by transforming the circuit into the sum of two new circuits (one with I_0 alone and another with V_0 alone), as shown in parts (b) and (c) of Fig. 3-15, respectively. The current through R_2 due to I_0 alone is labeled I_1, and that due to V_0 alone is labeled I_2.

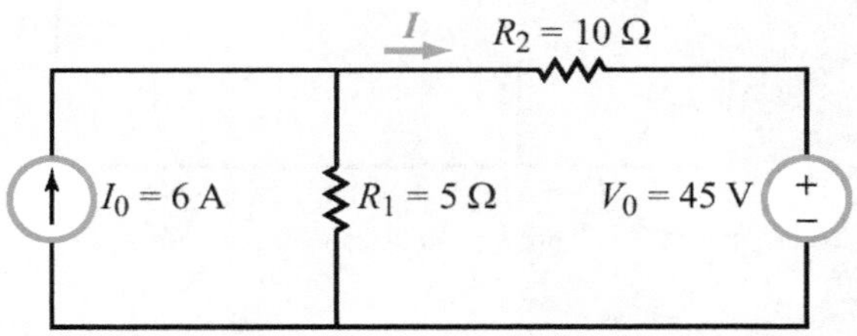

(a) Original circuit

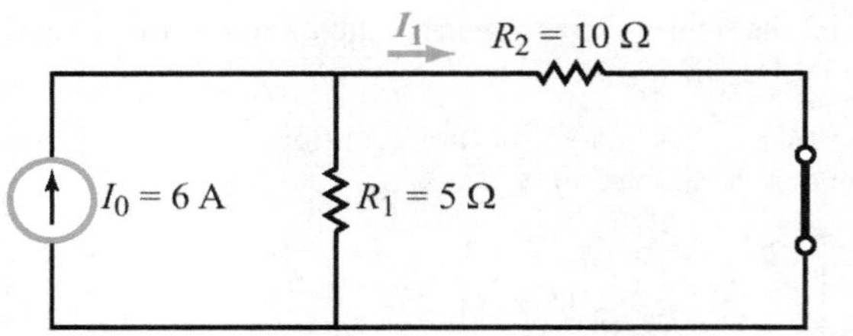

(b) Source I_0 alone generates I_1
[Eliminating a voltage source = replacing it with short circuit]

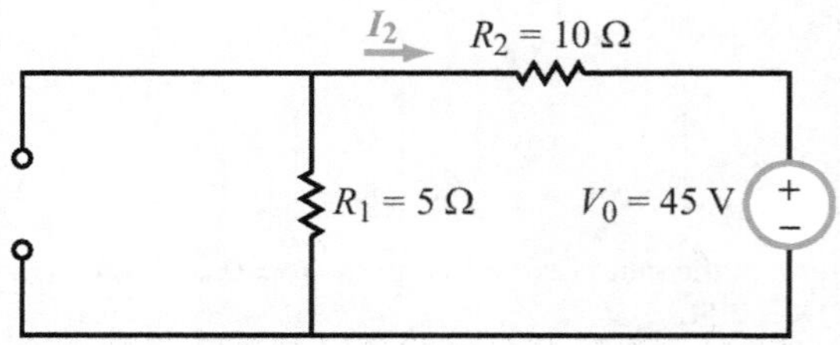

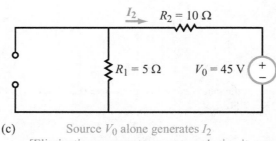

(c) Source V_0 alone generates I_2
[Eliminating a current source = replacing it with open circuit]

Figure 3-15: Application of the source-superposition method to the circuit of Example 3-9.

Circuit with current source alone

Setting $V_0 = 0$ means replacing the voltage source with a short circuit, as shown in Fig. 3-15(b). By current division,

$$I_1 = \left(\frac{R_1}{R_1 + R_2}\right) I_0 = \left(\frac{5}{5 + 10}\right) 6 = 2 \text{ A}.$$

Circuit with current source alone

Setting $I_0 = 0$ means replacing the current source with an open circuit, as shown in Fig. 3-15(c). Application of KVL leads to

$$I_2 = -\left(\frac{V_0}{R_1 + R_2}\right) = \frac{-45}{5 + 10} = -3 \text{ A}.$$

Hence,

$$I = I_1 + I_2 = 2 - 3 = -1 \text{ A}.$$

(b) The amounts of power dissipated in the 10 Ω resistor due to I_1 alone, I_2 alone, and the total current I are, respectively;

$$P_1 = I_1^2 R = 2^2 \times 10 = 40 \text{ W},$$

$$P_2 = I_2^2 R = (-3)^2 \times 10 = 90 \text{ W},$$

and

$$P = I^2 R = 1^2 \times 10 = 10 \text{ W}.$$

Note that $P \neq P_1 + P_2$, because the linearity property does not apply to power.

Example 3-10: Superposition for Dependent-Source Circuit

Apply the superposition principle to the circuit shown in Fig. 3-16(a) to determine V_x.

Solution: The circuit in Fig. 3-16(a) contains two independent voltage sources. Our task is to determine voltage V_x across the 6 Ω resistor. The superposition method allows us to represent the original circuit by two new circuits, one containing the 6 V source while excluding the 2 V source, and another with the opposite arrangement. The first circuit generates V_{x_1} across the 6 Ω resistor and the second circuit generates V_{x_2}. The unknown voltage V_x is the sum of the two.

Circuit with 6 V source alone

At node V_{x_1} in the circuit of Fig. 3-16(b), KCL gives

$$\frac{V_{x_1} - 6}{2} + \frac{V_{x_1}}{9} - 2V_{x_1} + \frac{V_{x_1}}{6} = 0,$$

which leads to

$$V_{x_1} = -2.45 \text{ V}.$$

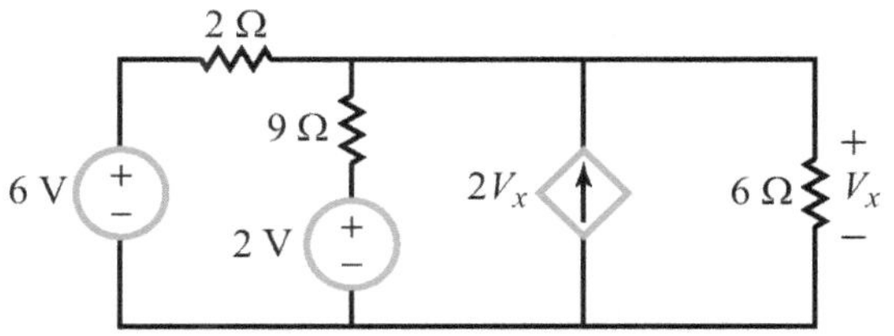

(a) Original circuit

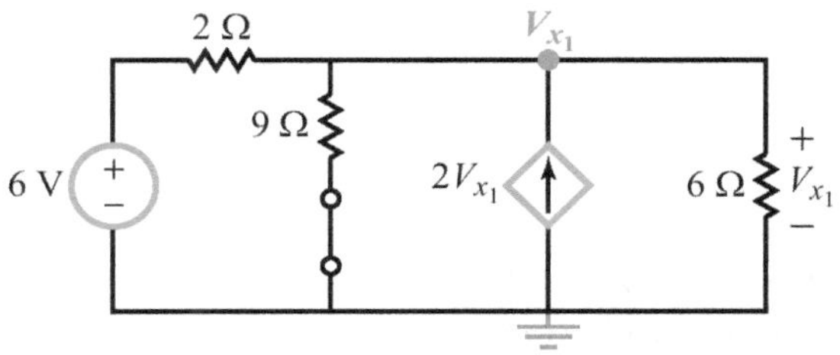

(b) The 6 V source acting alone generates voltage V_{x_1}

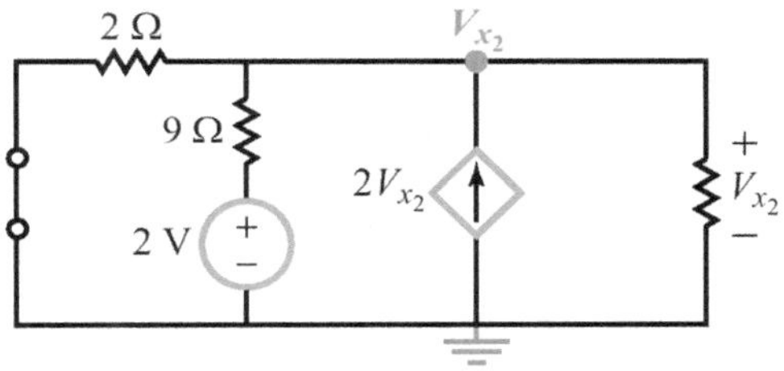

(c) The 2 V source acting alone generates voltage V_{x_2}

Figure 3-16: Application of superposition to the circuit of Example 3-10.

Circuit with 2 V source alone

At node V_{x_2} in the circuit of Fig. 3-16(c), KCL gives

$$\frac{V_{x2}}{2} + \frac{V_{x2} - 2}{9} - 2V_{x_2} + \frac{V_{x2}}{6} = 0,$$

which leads to

$$V_{x_2} = -0.18 \text{ V}.$$

Hence,

$$V_x = V_{x_1} + V_{x_2} = -2.45 - 0.18 = -2.63 \text{ V}.$$

Technology Brief 7
Integrated Circuit Fabrication Process

Do you ever wonder how the processor in your computer was actually fabricated? How is it that engineers can put hundreds of millions of transistors into one device that measures only a few centimeters on a side (and with so few errors) so the devices actually function as expected?

Devices such as modern computer processors and semiconductor memories fall into a class known as ***integrated circuits*** (IC). They are so named because all of the components in the circuit (and their "wires") are fabricated simultaneously (integrated) onto a circuit during the manufacturing process. This is in contrast to circuits where each component is fabricated separately and then soldered or wired together onto a common board (such as those you probably build in your lab classes). Integrated circuits were first demonstrated independently by Jack Kilby at Texas Instruments and Robert Noyce at Fairchild Semiconductor in the late 1950s. Once developed, the ability to easily manufacture components and their connections with good quality control meant that circuits with thousands (then millions, then billions) of components could be designed and built reliably.

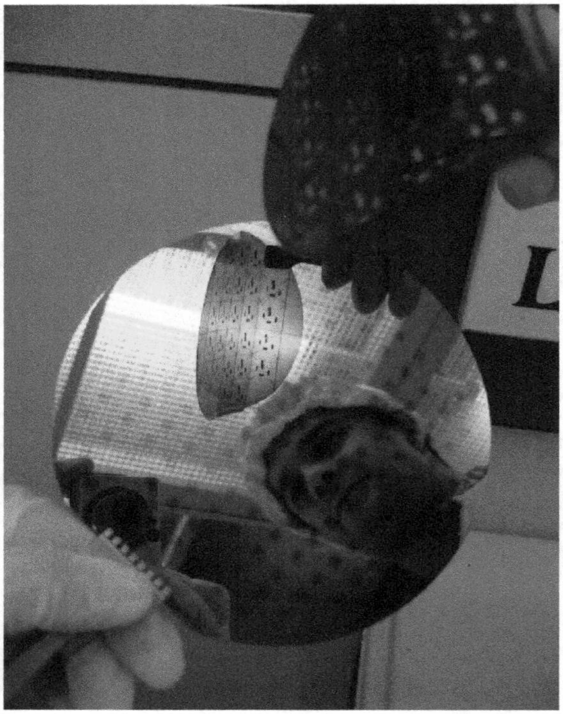

Figure TF7-1: A single 4-inch silicon wafer. Note the wafer's mirror-like surface. (Courtesy of Veljko Milanovic.)

Semiconductor Processing Basics

All mainstream semiconductor integrated-circuit processes start with a thin slice of silicon, known as a ***substrate*** or ***wafer***. This wafer is circular and ranges from 4 to 18 inches in diameter and is approximately 1 mm thick (hence its name). Each wafer is cut from a single crystal of the element silicon and polished to its final thickness with atomic smoothness (Fig. TF7-1). Most circuit designs (like your processor) fit into a few square centimeters of silicon area; each self-contained area is known as a ***die***. After fabrication, the wafer is cut to produce independent, rectangular dies often known as ***chips***, which are then packaged with plastic covers and metal pins or other external connections to produce the final component you buy at the store.

A specific sequence or ***process*** of chemical and mechanical modifications is performed on certain areas of the wafer. Although complex processes employ a variety of techniques, a basic IC process will employ one of the following three modifications to the wafer:

- ***Implantation***: Atoms or molecules are driven into (implanted in) the silicon wafer, changing its electronic properties (Fig. TF7-2(a)).

- ***Deposition***: Materials such as metals, insulators, or semiconductors are deposited in thin layers (like spray painting) onto the wafer (Fig. TF7-2(b)).

- ***Etching***: Material is removed from the wafer through chemical reactions or mechanical motion (Fig. TF7-2(c)).

Lithography

When building an IC, we need to perform different modifications to different areas of the wafer. We may want to etch some areas and add metal to others, for example. The method by which we define which areas will be modified is known as ***lithography***.

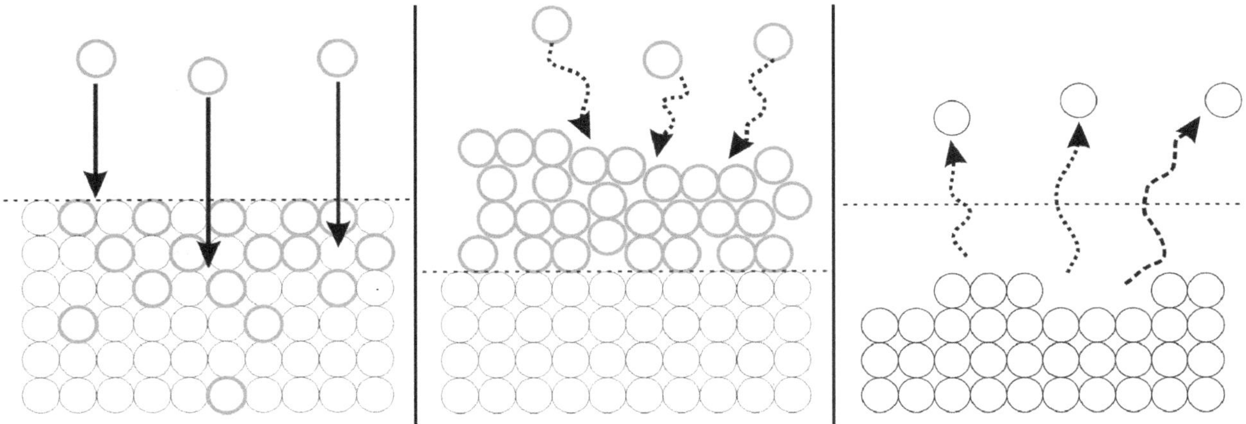

(a) *Implantation:* High-energy ions are driven into the silicon. Most become lodged in the first few nanometers, with decreasing concentration away from the surface. In this example, boron (an electron donor) is implanted into a silicon substrate to make a p-type material.

(b) *Deposition:* Atoms (or molecules) impact the surface but do not have the energy required to penetrate the surface. They accumulate on the surface in *thin films*. In this example, aluminum is deposited in a conductive film onto the silicon.

(a) *Etching:* Chemical, mechanical, or high-energy plasma methods are used to remove silicon (or other material) from the surface. In this example, silicon is etched away from the substrate.

Figure TF7-2: Cross-section of basic fabrication processes. The dashed line in each drawing indicates the original surface of the wafer.

Lithography has evolved much over the last 40 years and will continue to do so. Modern lithography employs all of the basic principles described below, but uses complex computation, specialized materials, and optical devices to achieve the very high resolutions required to reach modern feature sizes.

At its heart, lithography is simply a stencil process. In an old-fashioned stencil process, when a plastic sheet with cut-out letters or numbers is laid on a flat surface and painted, the cutout areas will be painted and the rest will be masked. Once the stencil is removed, the design left behind consists of only the painted areas with clean edges and a uniform surface. The total surface area of the IC depends on the number and complexity of the circuit elements on the IC, and on the minimum feature size, which is 10 nm (10^{-8} m) today. With that in mind, consider Fig. TF7-3. Given a flat wafer, we first apply a thin coating of liquid polymer known as ***photoresist*** (PR). This layer usually is several hundred nanometers thick and is applied by placing a drop in the center of the wafer and then spinning the wafer very fast (1000 to 5000 rpm) so that the drop spreads out evenly over the surface. Once coated, the PR is heated (usually between 60 to 100 °C) in a process known as ***baking***; this allows the PR to solidify slightly to a plastic-like consistency. This layer is then exposed to ultraviolet (UV) light, the bonds that hold the PR molecules together are "chopped" up; this makes it easy to wash away the UV-exposed areas (some varieties of PR behave in exactly the opposite manner: UV light makes the PR very strong or cross-linked, but we will ignore that technique here). In lithography, UV light is focused through a glass plate with patterns on it; this is known as ***exposure***. These patterns act as a "light stencil" for the PR. Wherever UV light hits the PR, that area subsequently can be washed away in a process called ***development***. After development, the PR film remains behind with holes in the exposed and washed areas.

How is this helpful? Let's look at how the modifications presented earlier can be masked with PR to produce patterned effects (Fig. TF7-4). In each case, we first use lithography to pattern areas onto the wafer (Fig. TF7-4(a)) then we perform one of our three

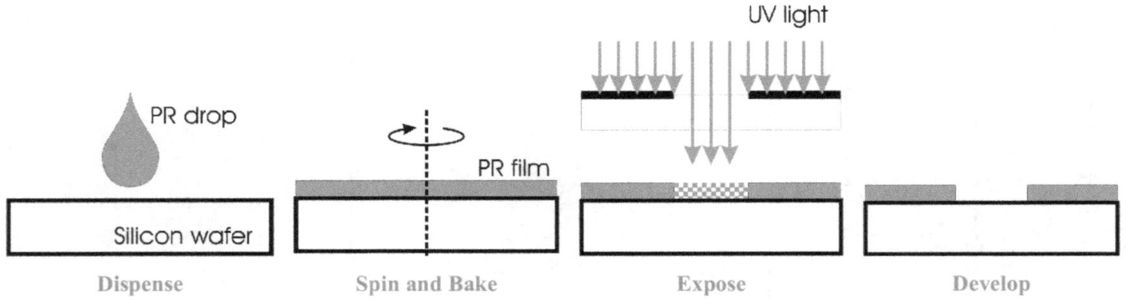

Figure TF7-3: Basic lithography steps.

processes (Fig. TF7-4(b)), and finally, we use a strong solvent such as acetone (nail polish remover) to completely wash away the PR (Fig. TF7-4(c)). The PR allows us to implant, deposit, or etch only in defined areas.

Fabricating a Diode

In Section 2-6, we discussed the functional performance of the diode as a circuit component. Here, we will examine briefly how a diode is fabricated. Similar but more complex multi-step processes are used to make transistors and integrated circuits. Conceptually, the simplest diode is made from two slabs of silicon—each implanted with different atoms—pressed together such that they share a boundary (Fig. TF7-5). The n and p areas are pieces of silicon that have been implanted with atoms (known as impurities) that change the number of electrons capable of flowing freely through the silicon. This changes the semiconducting properties of the silicon and creates an electrically active boundary

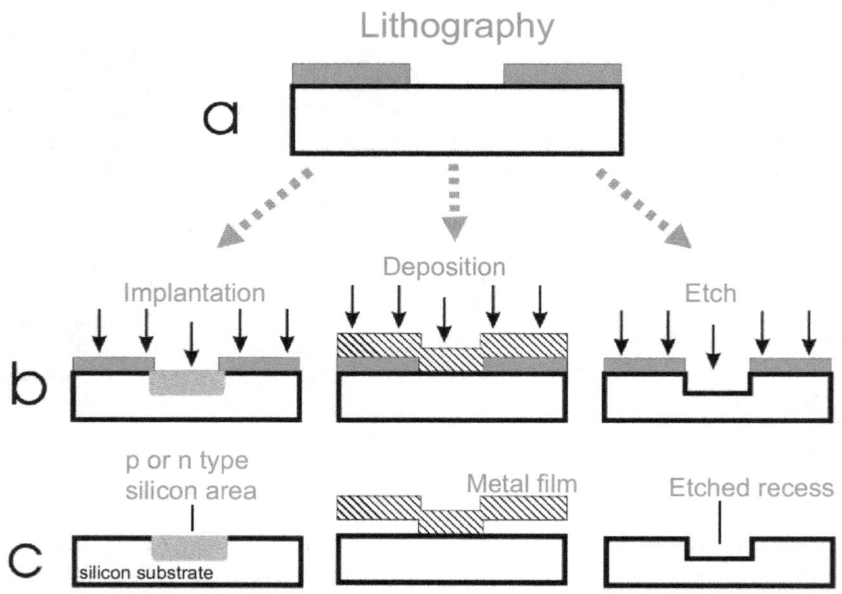

Figure TF7-4: Lithography used to pattern implantation areas, deposit metal features, and etch areas.

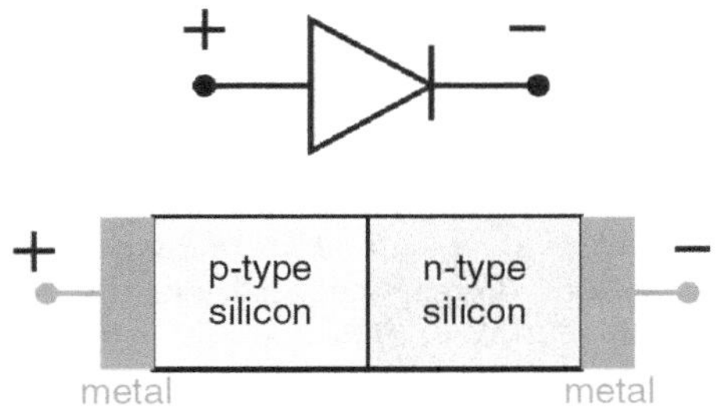

Figure TF7-5: The basic diode (top) circuit symbol and (bottom) conceptual depiction of the physical structure.

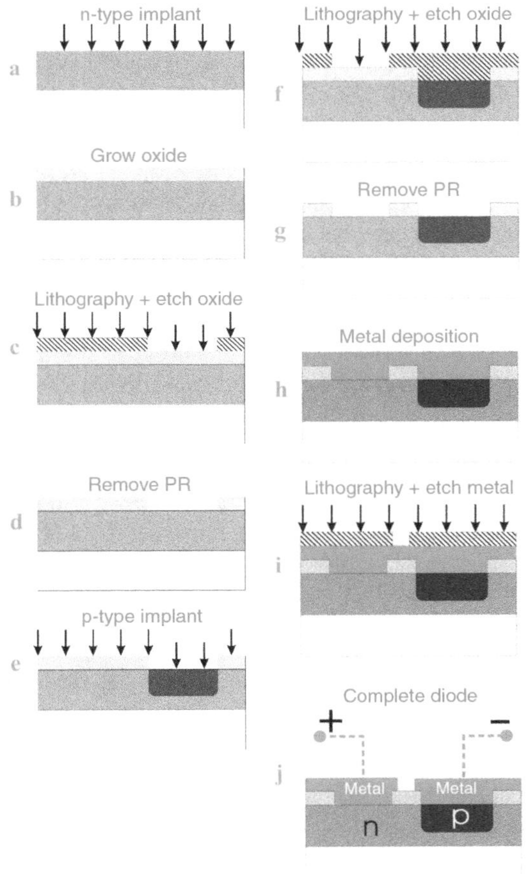

Figure TF7-6: A simple pn-junction diode fabrication process.

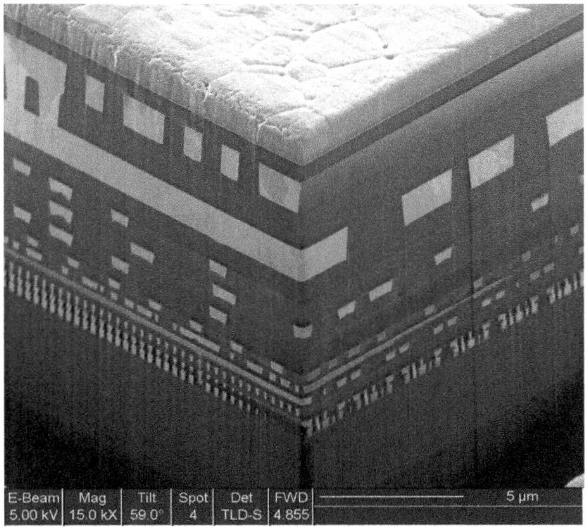

Figure TF7-7: Colorized scanning electron-microscope cross section of a 64-bit high-performance microprocessor chip built in IBM's 90 nm Server-Class CMOS technology. Note that several metal interconnect levels are used (metal lines are orange, insulator is green); the transistors lie below this metal on the silicon wafer itself (dark blue). (Courtesy of International Business Machines Corporation.)

(called a ***junction***) between the n and the p areas of silicon. If a forward-biased voltage is applied, it is as if the p charges move towards the n side, allowing current to flow, even though no actual p or n atoms move in the diode. When both the n and p pieces of silicon are connected to metal wires, this two-terminal device exhibits the diode i–υ curve shown in Fig. 2-40(c).

Figure TF7-6 shows the process for making a single diode. Only one step needs further definition: ***oxidation***. During oxidation, the silicon wafer is heated to > 1000 °C in an oxygen atmosphere. At this temperature, the oxygen atoms and the silicon react and form a layer of SiO_2 on the surface (this layer is often called an ***oxide layer***). SiO_2 is a type of glass and is used as an insulator.

Wires are made by depositing metal layers on top of the device; these are called ***interconnects***. Modern ICs have 6 to 7 such interconnect layers (Fig. TF7-7). These layers are used to make electrical connections between all of the various components in the IC in the same way that macroscopic wires are used to link components on a breadboard.

Concept Question 3-9: Explain why the linearity property of electric circuits is an underlying requirement for the application of the source-superposition method. (See CAD)

Concept Question 3-10: How is the superposition method used as a sensitivity tool in circuit analysis and design? (See CAD)

Concept Question 3-11: Is the source-superposition method applicable to power? In other words, if source 1 alone supplies power P_1 to a certain device and source 2 alone supplies power P_2 to the same device, will the two sources acting simultaneously supply power $P_1 + P_2$ to the device? (See CAD)

Exercise 3-9: Apply the source-superposition method to determine the current I in the circuit of Fig. E3.9.

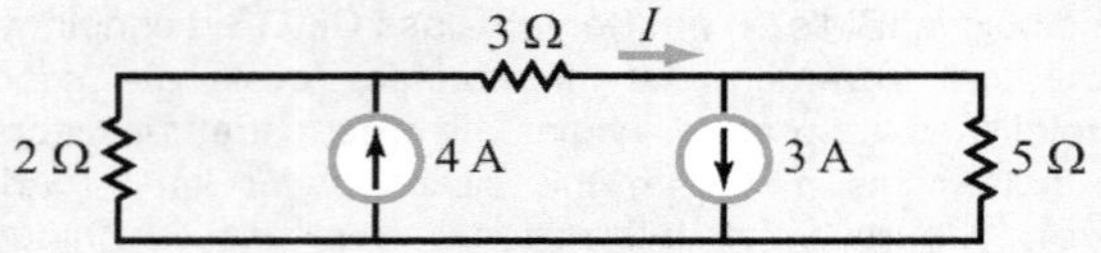

Figure E3.9

Answer: $I = 2.3$ A. (See CAD)

Exercise 3-10: Apply source superposition to determine V_{out} in the circuit of Fig. E3.10.

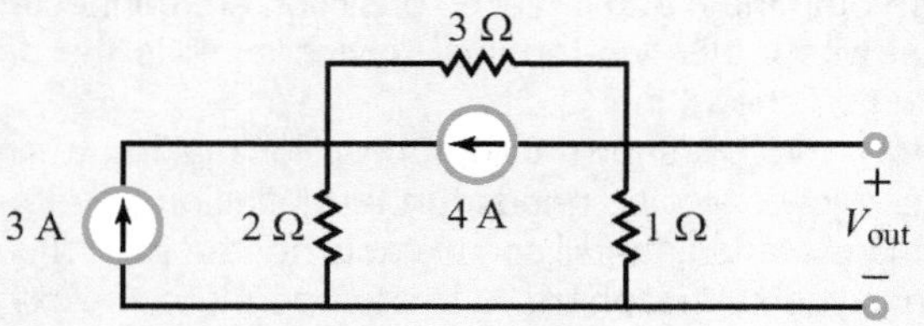

Figure E3.10

Answer: $V_{out} = -1$ V. (See CAD)

3-6 Thévenin and Norton Equivalent Circuits

As depicted by the block diagram shown in Fig. 3-17, a generic *cell-phone circuit* consists of several individual circuits, including amplifiers, oscillators, analog-to-digital (A/D) and digital-to-analog (D/A) converters, an antenna, a diplexer that allows the antenna to be used for both transmission and reception, a microprocessor, and other auxiliary circuits. Many of these circuits are quite complex and may contain a large number of active and passive elements, in both discrete and integrated form. So the question one might ask is: *How does an engineer approach an analysis or design task involving such a complex architecture?*

Dealing with the entire circuit all at once would be next to impossible, not only because of its daunting complexity, but also because the individual circuits call for engineers with different specializations.

Fortunately, we have a straightforward answer to the question, namely that each circuit gets modeled as a "black box," or *block*, with specified input and output (I/O) terminal characteristics allowing the engineer working with a particular circuit to treat the other circuits connected to it in terms of only those (I/O) characteristics without much regard to the details of their internal architectures. For an amplifier, for example, its overall specifications might include voltage gain and frequency bandwidth, among other attributes, but its terminal characteristics refer to how it would "appear" from the perspective of other circuits.

Conversely, from the amplifier's perspective, other circuits are specified in terms of how they appear to the amplifier. Figure 3-18 illustrates the concept from the perspective of the radio-frequency (RF) low-noise amplifier in the receive channel of the cell-phone circuit. The combination of the antenna and diplexer (including the input signal picked up by the antenna) is represented at the input side of the amplifier by an equivalent circuit composed of a voltage source υ_s in series with an *impedance* $\mathbf{Z}_s$. Impedance (which we shall introduce in a later chapter) is the ac-equivalent of resistance in dc circuits. At the output side of the amplifier, the mixer (whose function is to shift the center frequency of the input signal from 834 MHz down to 70 MHz) is represented by a *load impedance* $\mathbf{Z}_L$. Thus, the output terminal characteristics of the antenna/diplexer combination become the input source to the amplifier, and the input impedance of the mixer becomes the load to which the amplifier is connected.

▶ Isolating the amplifier, while keeping it in the context of its input and output neighbors, facilitates both the analysis and design processes. ◀

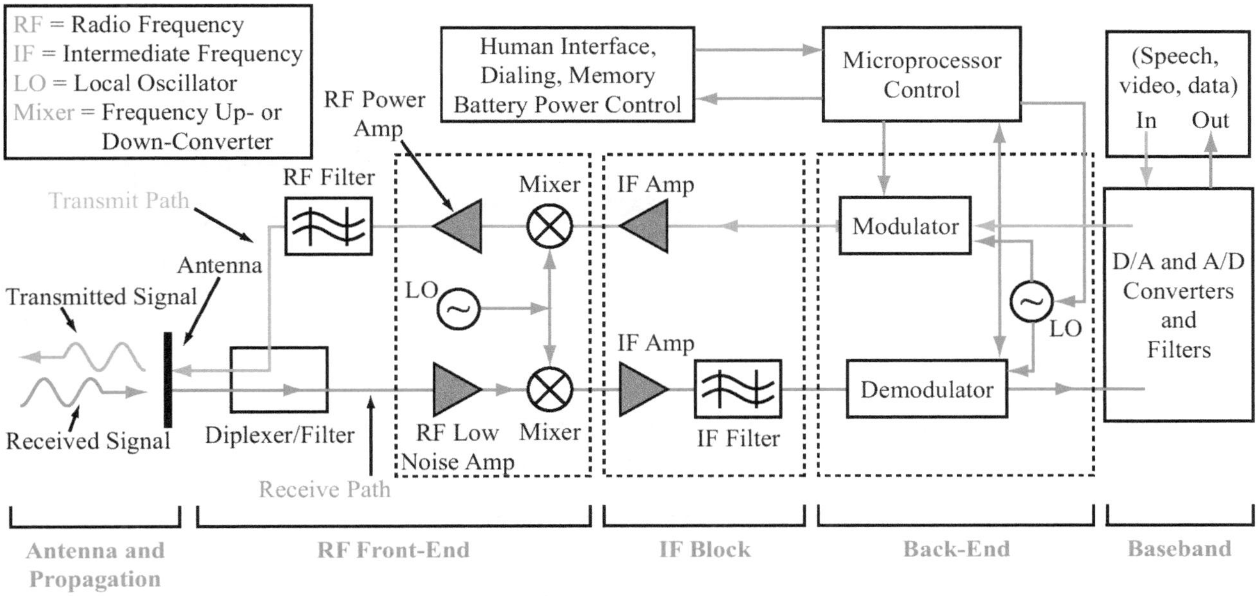

Figure 3-17: Cell-phone block diagram.

3-6.1 Input and Output Resistances

Example 1: Household wiring

Our homes are powered by some kind of electrical generation plant (coal-powered, hydroelectric, etc.) that produces high voltage, which is run to our city on high-voltage transmission lines, split into smaller voltages at various substations, and eventually delivered to the breakers or fuse boxes of our homes (Chapter 10). This is a rather complex system with many parts, so we prefer not to analyze the entire system every time we consider a change in a household electrical circuit. We can represent the entire power distribution system as a voltage source (in this case, 110 V) in series with a small source resistance R_s that represents the losses in the transmission lines and connections, as shown in the simplified ***block diagram*** in Fig. 3-19. Even though the source is ac, we will treat it as if it were a dc source with $V_s = 110$ V.

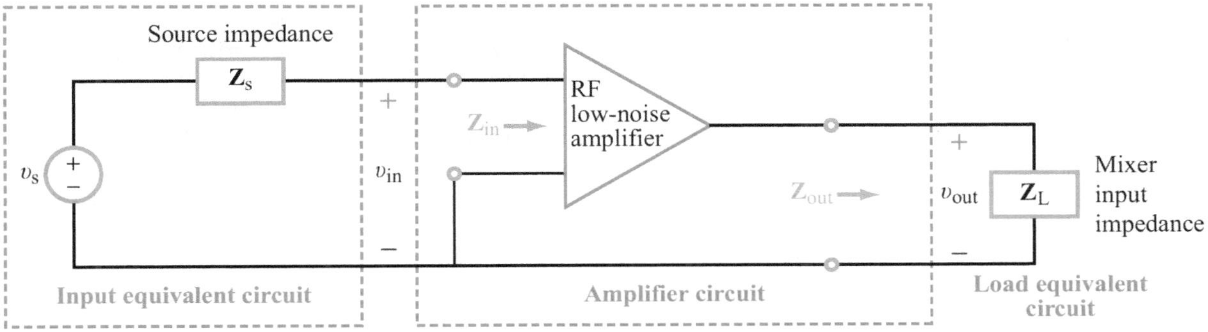

Figure 3-18: Input and output circuits as seen from the perspective of a radio-frequency amplifier circuit.

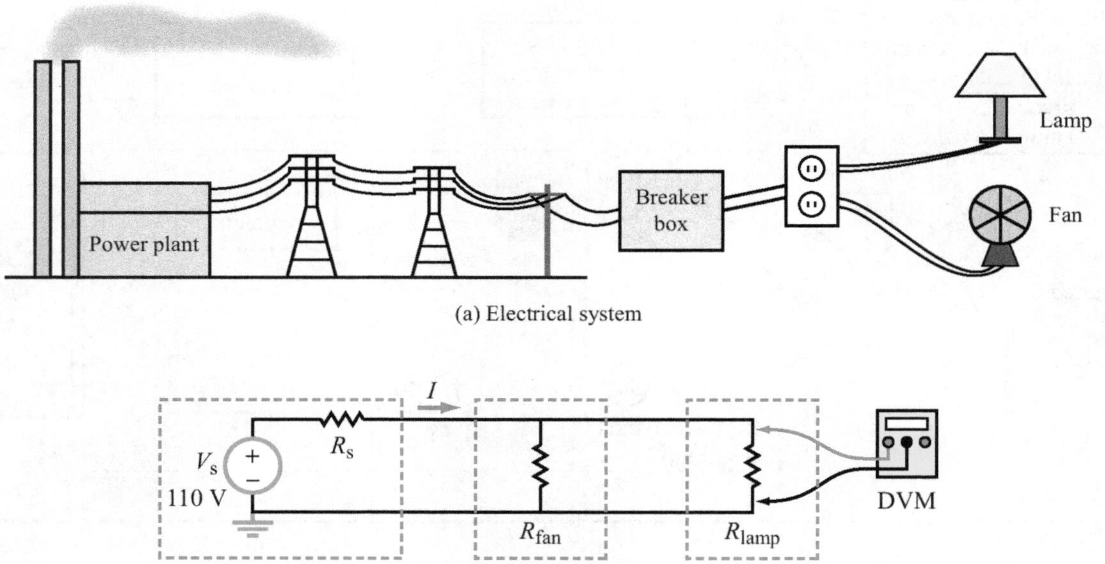

(b) Equivalent circuit

Figure 3-19: (a) Power distribution system driving a fan and a lamp in a house, and (b) block diagram of the source (power distribution system), fan, lamp, and a voltmeter measuring the voltage in the outlet.

Every device we plug in (such as the fan and lamp) is in parallel with the power source block. The lamp is just a switch and a light bulb, which we might even simplify further by ignoring the switch and assuming it is always on, thus giving us very simply a resistor R_{lamp} in the block diagram in Fig. 3-19(b). The fan, on the other hand, is a little more complicated because it includes a motor and a switch that controls various speeds, but we can still represent it by a parallel resistor R_{fan}. If $V_s = 110$ V, $R_s = 10\ \Omega$, and $R_{fan} = R_{lamp} = 100\ \Omega$, what is the current drawn from the source?

The parallel combination of R_{fan} and R_{lamp} is

$$R_{\parallel} = R_{fan} \parallel R_{lamp} = 100 \parallel 100 = 50\ \Omega.$$

The total resistance connected to V_s is the series sum of $R_{\parallel}$ and R_s:

$$R_{total} = R_{\parallel} + R_s = 50 + 10 = 60\ \Omega.$$

Hence, by Ohm's law,

$$I = \frac{V_s}{R_{total}} = \frac{110}{60} = 1.83\ \text{A}.$$

What is the voltage across the fan and lamp?

Application of voltage division gives

$$V_{fan} = V_{lamp} = V_s\,\frac{R_{\parallel}}{R_{total}} = \frac{110 \times 50}{60} = 91.67\ \text{V}.$$

This is measurably less than the 110 V of V_s. The voltage reduction is called *loading* the circuit, and it occurs when the series source *input resistance* and the *load resistance* ($R_{fan} \parallel R_{lamp}$) are on the same order of magnitude, or if the source resistance is larger than the load resistance. If too many appliances are plugged into the outlet, all of their resistances combine in parallel, thereby reducing the total load resistance, drawing more current, and loading down the circuit (reducing the voltage across the devices). Eventually, the devices will no longer function properly (if the voltage gets too low) or the circuit breaker creates an open circuit if the current gets too high.

This example illustrates the concept of input and output resistances and why they matter. The input resistance is what is seen looking into a block "from the left," and the output resistance is what is seen looking in "from the right." For the voltage source, we do not really have an input resistance, and its output resistance is R_s. For the fan and lamp, the input and output resistances are both R_{fan} and R_{lamp}, respectively. If we have a small output resistance looking into (connected to) a large input resistance of a load (such as the fan/lamp circuit), the load will not draw down the voltage (load the circuit) very much. In fact, if the input resistance of the load is high enough, we can even ignore it in the analysis of the circuit because the voltage across it is essentially the same as the source voltage. On the other hand, if the output and input resistances are similar in magnitude, the load will indeed draw down the voltage (and load the circuit). The load clearly has an impact on the circuit, and we cannot ignore it in the analysis of how the circuit works. And if the output resistance of the source is larger than the input resistance of the load, the voltage of the load will be significantly reduced (loaded).

Example 2: Voltmeter

Voltmeters are deliberately designed with high input resistance ($\geq$ 2 MΩ) so that they do not affect the circuit being measured. Consider, for example, measuring the voltage (or resistance) across the fan/lamp circuit in Fig. 3-19(b). The resistance of the fan and lamp in parallel is 50 Ω. If the voltmeter (DVM) has an input resistance of 2 MΩ, the fan/lamp/DVM circuit has a total resistance of 49.999 Ω, a change of less than 0.01%. Another way to think of this is that the DVM will draw very little current through it, because of its high input resistance. If the DVM is used to measure resistance instead of voltage, its input resistance also is high, and would have minimal effect on the circuit being measured. In contrast, the input resistance of an ammeter is very small (about 1 $\mu\Omega$), much smaller than the fan/lamp combination.

Summary: What have we learned from these examples?

- Input resistance (looking toward the load) and output resistance (looking toward the source) are important parameters of the circuit.

- If the input resistance of the load is very high compared with the output resistance of the rest of the circuit (such as the case with the voltmeter), that part of the circuit (the load) can basically be ignored when we analyze the other parts of the circuit. In fact, this means that these blocks can be designed and analyzed *individually*. We call them ***independent***, ***uncoupled***, or ***decoupled***. Being able to design and analyze blocks of a circuit individually is such a powerful concept that we often deliberately build circuits to have high input resistance. Circuits with high input resistance draw minimal current.

- If the input resistance of the load is low (or similar) compared with the output resistance of the input circuit, significant current is drawn into the load circuit. This may *load* the source circuit and reduce the voltage at the load. Also, the circuits can no longer be analyzed individually; they are coupled and must be analyzed together.

3-6.2 Thévenin's Theorem

> ► Our ability to develop equivalent-circuit representations is made possible (in part) by a pair of theorems of fundamental significance known as Thévenin's and Norton's theorems. ◄

Most electrical systems are quite complex, so that each subsystem (such as the filter, demodulator, amplifier, etc., in Fig. 3-17) is designed independently, and often by different engineers and even by different companies. We established in the preceding subsection that in order to design subsystems independently, it is necessary that each has a high input resistance. This is often not feasible, however, so we need another approach to designing cascaded circuits and then combining them together into a larger system. The Thévenin and Norton concepts described in this section help us do that. They are very powerful techniques used extensively in electrical engineering design. In practice, the system engineering team determines what blocks are needed for the system, lays out the block diagram, and specifies the input voltages and currents, and input and output resistances for each block of the circuit. Design teams then create circuits for each block, and test them independently using the input and output resistances/voltages/currents for their neighboring subsystems in the test protocol. The integration team puts the subsystems together, often with the mechanical parts of the system as well, and then tests the overall system as an integrated unit to insure that its performance meets the design specifications.

In the 1880s, a French engineer named Léon Thévenin introduced the concept known today as ***Thévenin's theorem***, which asserts:

► A linear circuit can be represented at its output terminals by an equivalent circuit consisting of a series combination of a voltage source υ_{Th} and a resistor R_{Th}, where υ_{Th} is the open-circuit voltage at those terminals (no load) and R_{Th} is the equivalent resistance between the same terminals when all independent sources in the circuit have been deactivated. R_{Th} is the output resistance of the Thévenin circuit. ◄

A pictorial representation of Thévenin's theorem is shown in Fig. 3-20, where the actual circuit in part (a) has been replaced with the Thévenin equivalent circuit in part (b). The implication of this model is that when the circuit is connected to a load resistor R_{L}, the current i_{L} running through it will be identical for both the actual circuit and the equivalent circuit. This equivalence holds true for any value of R_{L}, from zero (short circuit) to ∞ (open circuit). *Thus, from the standpoint of the load, the two circuits are indistinguishable.*

Even though the present discussion pertains to dc currents, the Thévenin concept extends to ac circuits as well. We will revisit the concept in a future chapter for circuits containing capacitors and inductors.

3-6.3 Finding υ_{Th}

Thévenin equivalency means that from the standpoint of the load R_{L}, the two circuits in Fig. 3-20 are indistinguishable. For

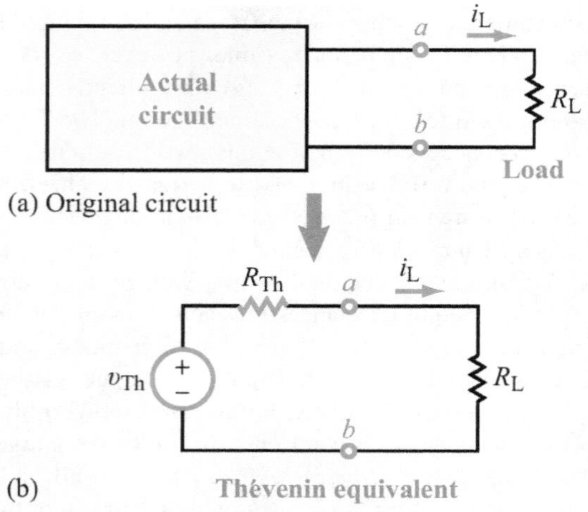

Figure 3-20: A circuit can be represented in terms of a Thévenin equivalent comprising a voltage source υ_{Th} in series with a resistance R_{Th}.

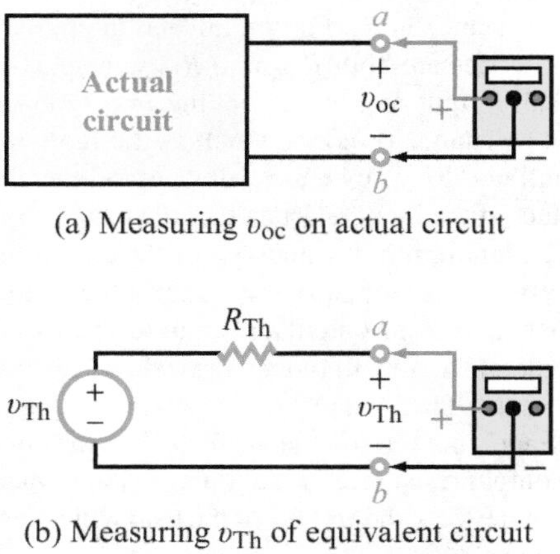

Figure 3-21: Equivalency means that υ_{Th} of the Thévenin equivalent circuit is equal to the open-circuit voltage for the actual circuit.

any value we assign to R_{L}, both circuits generate the same i_{L}. Hence, if we disconnect R_{L} altogether from both circuits and then measure the voltage across terminals (a, b), we should measure the same voltage. The scenario is depicted in Fig. 3-21. In part (a), a voltmeter would measure the open-circuit voltage υ_{oc} of the actual circuit, and in part (b) the voltmeter would measure υ_{Th} (since there is no voltage drop across R_{Th}). We are effectively measuring the output voltage of our blackbox. Equivalency requires that

$$\upsilon_{\text{Th}} \text{ (of Thévenin equivalent)} = \upsilon_{\text{oc}} \text{ (of actual circuit)}. \tag{3.31}$$

The procedure is equally valid for circuits with or without dependent sources. For a circuit with no independent sources, $\upsilon_{\text{Th}} = 0$.

3-6.4 Finding R_{Th}—Short-Circuit Method

Multiple methods are available for finding the Thévenin resistance R_{Th}. We start with the short-circuit method. From Fig. 3-20(b),

$$i_{\text{L}} = \frac{\upsilon_{\text{Th}}}{R_{\text{Th}} + R_{\text{L}}}. \tag{3.32}$$

If $R_{\text{L}} = 0$ (short-circuit load), we call i_{L} the short-circuit current i_{sc}, which would be given by

$$i_{\text{sc}} = \frac{\upsilon_{\text{Th}}}{R_{\text{Th}}}. \tag{3.33}$$

Open-Circuit / Short-Circuit Method

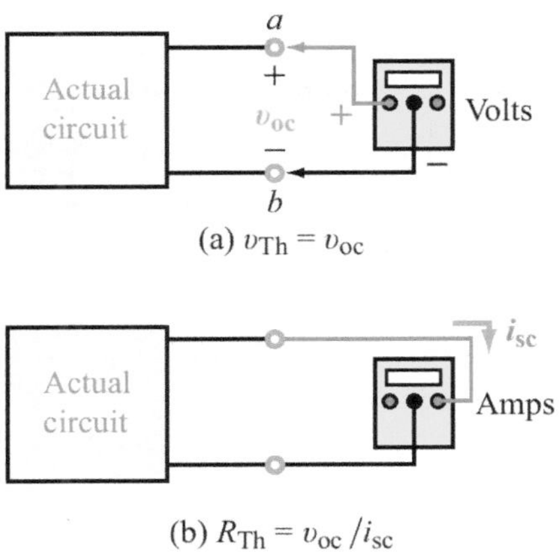

Figure 3-22: Thévenin voltage is equal to the open-circuit voltage and Thévenin resistance is equal to the ratio of v_{oc} to i_{sc}, where i_{sc} is the short-circuit current between the output terminals.

By analyzing the circuit configuration in Fig. 3-22(b) to find i_{sc} or, measuring i_{sc} with an ammeter, we can apply Eq. (3.33) to find R_{Th},

$$R_{Th} = \frac{v_{Th}}{i_{sc}}. \tag{3.34}$$

The only potential problem with this type of measurement is that when short-circuiting the source circuit, the current threshold of the ammeter may be exceeded (if the output resistance of the source circuit is very small).

This method is applicable to any circuit with at least one independent source, regardless of whether or not it contains dependent sources.

> ▶ The Thévenin voltage v_{Th} is obtained by removing the load R_L (replacing it with an open circuit), and then measuring or computing the open-circuit voltage at the same terminals. The short-circuit current i_{sc} is obtained by replacing the load with a short circuit and then measuring or computing the short-circuit current flowing through it (Fig. 3-22). ◀

Example 3-11: Open Circuit / Short Circuit Method

The input circuit to the left of terminals (a, b) in Fig. 3-23(a) is connected to a variable load resistor R_L. Determine (a) the Thévenin equivalent of the circuit to the left of terminals (a, b) and (b) use it to find the value of R_L that will cause the magnitude of the current through it to be 0.5 A.

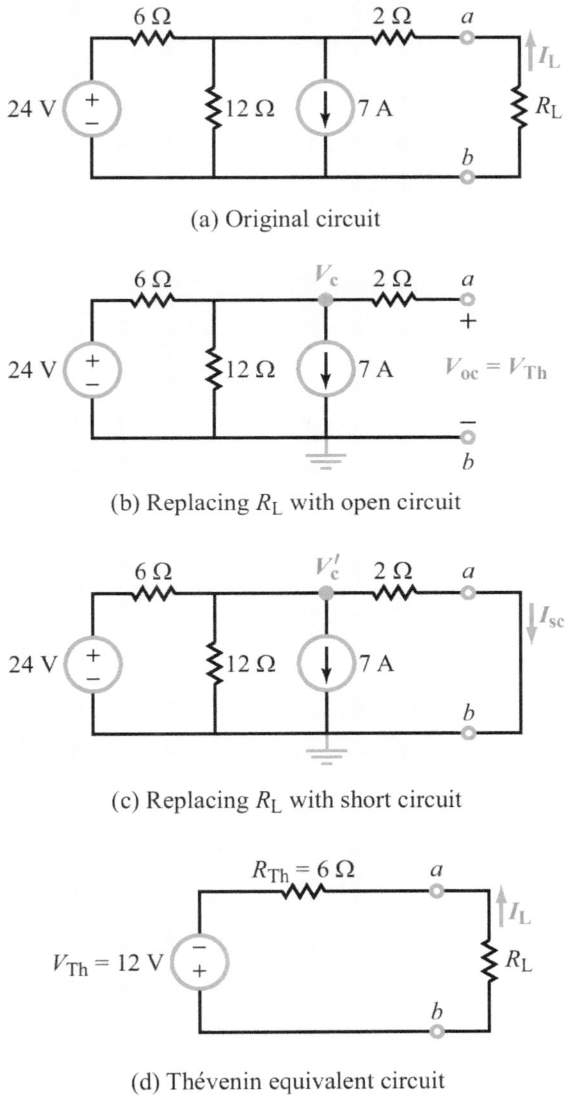

Figure 3-23: Applying open circuit/short circuit method to find the Thévenin equivalent for the circuit of Example 3-10.

Solution: (a) With R_L replaced with an open circuit in Fig. 3-23(b), V_{Th} is the open-circuit voltage between terminals (a, b). Since no current flows through the 2 Ω resistor, $V_{Th} = V_c$ at node c. The node-voltage equation at node c is

$$\frac{V_c - 24}{6} + \frac{V_c}{12} + 7 = 0,$$

which leads to $V_c = -12$ V. Hence,

$$V_{Th} = -12 \text{ V}.$$

Next, we replace R_L with a short circuit (Fig. 3-23(c)) and repeat the process to find V_c':

$$\frac{V_c' - 24}{6} + \frac{V_c'}{12} + 7 + \frac{V_c'}{2} = 0,$$

whose solution gives $V_c' = -4$ V, and

$$I_{sc} = \frac{V_c'}{2} = -\frac{4}{2} = -2 \text{ A}.$$

Hence,

$$R_{Th} = \frac{V_{Th}}{I_{sc}} = \frac{-12}{-2} = 6\ \Omega,$$

and the Thévenin equivalent circuit is shown in Fig. 3-23(d).

(b) In view of Fig. 3-23(d), for I_L to be 0.5 A, it is necessary that

$$I_L = \frac{12}{6 + R_L} = 0.5 \text{ A}$$

or

$$R_L = 18\ \Omega.$$

Exercise 3-11: Determine the Thévenin-equivalent circuit at terminals (a, b) in Fig. E3.11.

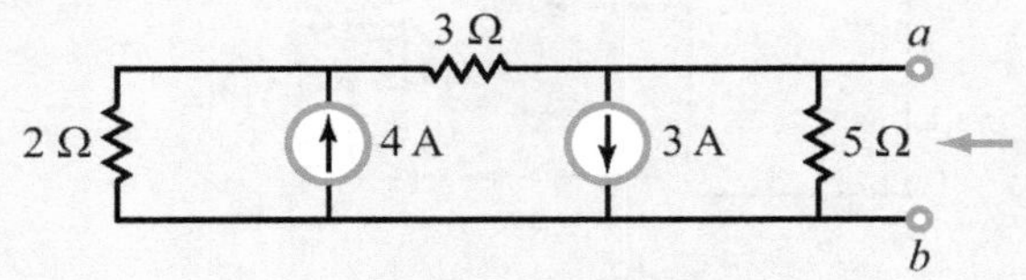

Figure E3.11

Answer: $V_{Th} = -3.5$ V, $I_{sc} = -1.4$ A, $R_{Th} = 2.5\ \Omega$. (See CAD)

3-6.5 Finding R_{Th}—Equivalent Resistance Method

If the circuit does not contain dependent sources, R_{Th} can be determined by deactivating all sources (replacing voltage sources with short circuits and current sources with open circuits) and then simplifying the circuit down to a single equivalent resistance between its output terminals, as portrayed by Fig. 3-24. In that case,

$$R_{Th} = R_{eq}. \tag{3.35}$$

This method does not apply to circuits that contain dependent sources.

Example 3-12: Thévenin Resistance

Find R_{Th} at terminals (a, b) for the circuit in Fig. 3-25(a).

Solution: Since the circuit has no dependent sources, we can apply the equivalent-resistance method. We start by

Equivalent-Resistance Method

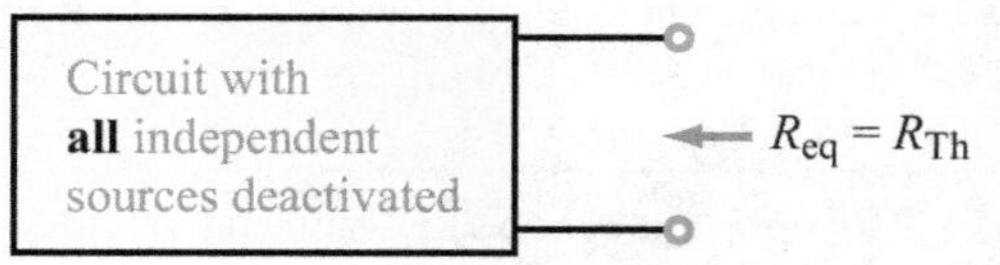

Figure 3-24: For a circuit that does not contain dependent sources, R_{Th} can be determined by deactivating all sources (replacing voltage sources with short circuits and current sources with open circuits) and then simplifying the circuit down to a single resistance R_{eq}.

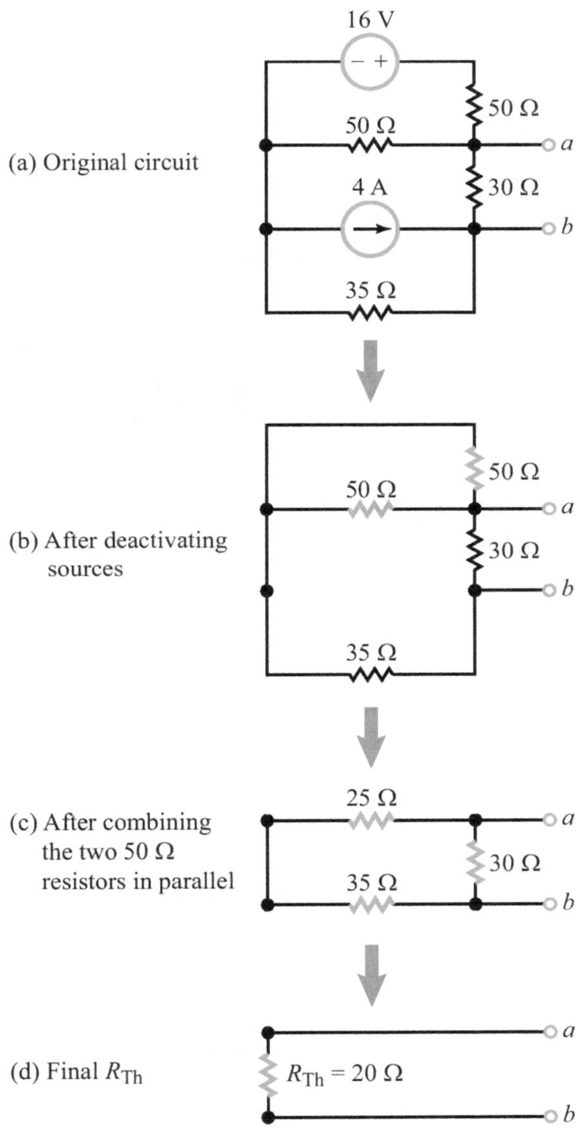

Figure 3-25: After deactivation of sources, systematic simplification leads to R_{Th} (Example 3-12).

deactivating all of the sources (as shown in Fig. 3-25(b)) where we replaced the voltage source with a short circuit and the current source with an open circuit. After (a) combining the two 50 Ω resistors in parallel, (b) combining their 25 Ω combination in series with the 35 Ω resistance, and (c) finally combining the resultant 60 Ω with the 30 Ω resistance in parallel, we obtain

$$R_{\text{Th}} = 20\ \Omega.$$

Exercise 3-12: Find the Thévenin equivalent of the circuit to the left of terminals (a, b) in Fig. E3.12, and then determine the current I.

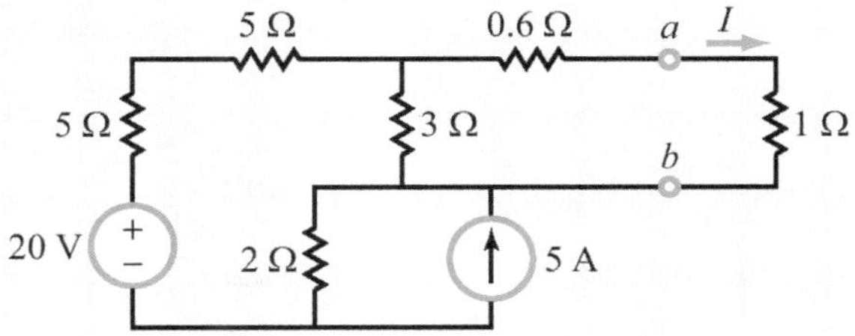

Figure E3.12

Answer:

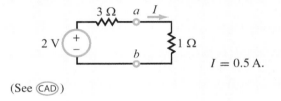

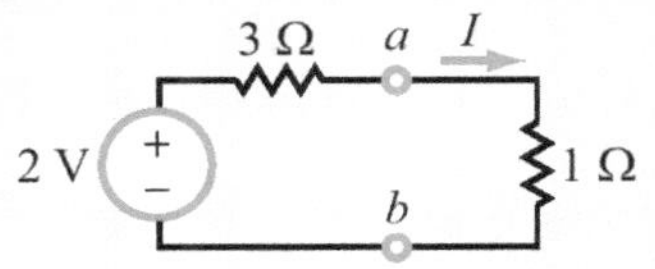

$I = 0.5$ A.

(See CAD)

3-6.6 Finding R_{Th}—External-Source Method

The equivalent-resistance method described previously does not apply to circuits containing dependent sources. Hence, an alternative variation is called for. Independent sources again are deactivated (but dependent sources are left alone) and an external voltage source υ_{ex} is introduced to excite the circuit, as shown in Fig. 3-26. After analyzing the circuit to determine the current i_{ex}, R_{Th} is found by applying

$$R_{\text{Th}} = \frac{\upsilon_{\text{ex}}}{i_{\text{ex}}}. \qquad (3.36)$$

Since i_{ex} is caused by υ_{ex}, it is directly proportional to it. Hence, we may choose any value for υ_{ex}, such as $\upsilon_{\text{ex}} = 1$ V, as long as we use the same value both in Fig. 3-26 when analyzing the circuit to find i_{ex} and in applying Eq. (3.36) to compute R_{Th}.

Example 3-13: Circuit with Dependent Source

Find the Thévenin equivalent circuit at terminals (a, b) for the circuit in Fig. 3-27(a) by applying the combination of open-circuit-voltage and external-source methods.

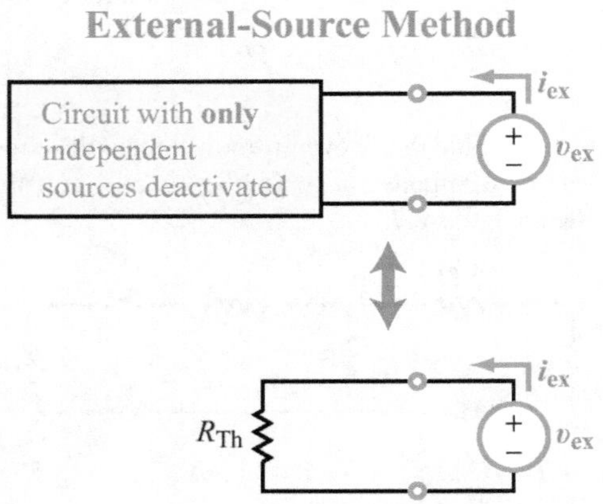

Figure 3-26: If a circuit contains both dependent and independent sources, R_{Th} can be determined by (a) deactivating only independent sources (by replacing independent voltage sources with short circuits and independent current sources with open circuits), (b) adding an external source v_{ex}, and then (c) solving the circuit to determine i_{ex}. The solution is $R_{Th} = v_{ex}/i_{ex}$.

Solution: The KVL equation for mesh current I_1 in Fig. 3-27(a) is given by

$$-33 + 6I_1 + 2I_1 + 3I_x = 0.$$

Recognizing that $I_x = I_1$, solution of the preceding equation leads to

$$I_1 = 3 \text{ A}.$$

Since there is no voltage drop across the 4 Ω resistor (because no current is flowing through it),

$$V_{Th} = V_{ab} = 2I_1 + 3I_x = 5I_1 = 15 \text{ V}.$$

To find R_{Th} using the external-source method, we deactivate the 33 V voltage source and we add an external voltage source V_{ex}, as shown in Fig. 3-27(b). Our task is to obtain an expression for I_{ex} in terms of V_{ex}. In Fig. 3-27(b) we have two mesh currents, which we have labeled I'_1 and I'_2. Their equations are given by

$$6I'_1 + 2(I'_1 - I'_2) + 3I_x = 0,$$

$$-3I_x + 2(I'_2 - I'_1) + 4I'_2 + V_{ex} = 0.$$

After replacing I_x with I'_1 and solving the two simultaneous equations, we obtain

$$I'_1 = -\frac{1}{28} V_{ex},$$

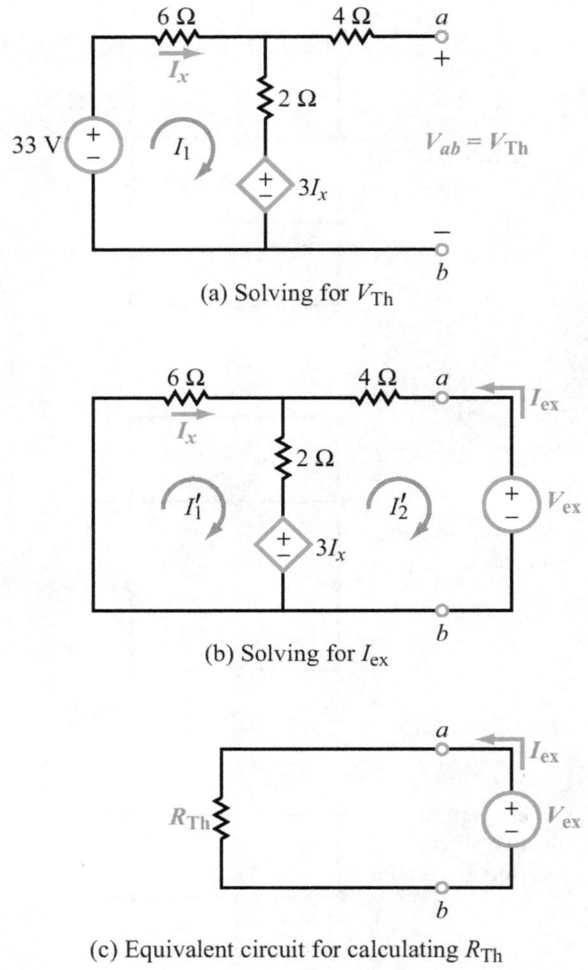

Figure 3-27: Solution of the open-circuit voltage gives $V_{ab} = V_{Th} = 15$ V. Use of the external-voltage method leads to $R_{Th} = 56/11$ Ω (Example 3-13).

and

$$I'_2 = -\frac{11}{56} V_{ex}.$$

For the equivalent circuit shown in Fig. 3-27(c),

$$R_{Th} = \frac{V_{ex}}{I_{ex}}.$$

In terms of our solution, $I_{ex} = -I'_2$. Hence,

$$R_{Th} = -\frac{V_{ex}}{I'_2} = \frac{56}{11} \ \Omega.$$

Table 3-1: **Properties of Thévenin/Norton analysis techniques.**

To Determine	Method	Can Circuit Contain Dependent Sources?	Relationship
υ_{Th}	Open-circuit υ	Yes	$\upsilon_{\text{Th}} = \upsilon_{\text{oc}}$
υ_{Th}	Short-circuit i (if R_{Th} is known)	Yes	$\upsilon_{\text{Th}} = R_{\text{Th}} i_{\text{sc}}$
R_{Th}	Open/short	Yes	$R_{\text{Th}} = \upsilon_{\text{oc}}/i_{\text{sc}}$
R_{Th}	Equivalent R	No	$R_{\text{Th}} = R_{\text{eq}}$
R_{Th}	External source	Yes	$R_{\text{Th}} = \upsilon_{\text{ex}}/i_{\text{ex}}$
$i_{\text{N}} = \upsilon_{\text{Th}}/R_{\text{Th}}$; $R_{\text{N}} = R_{\text{Th}}$			

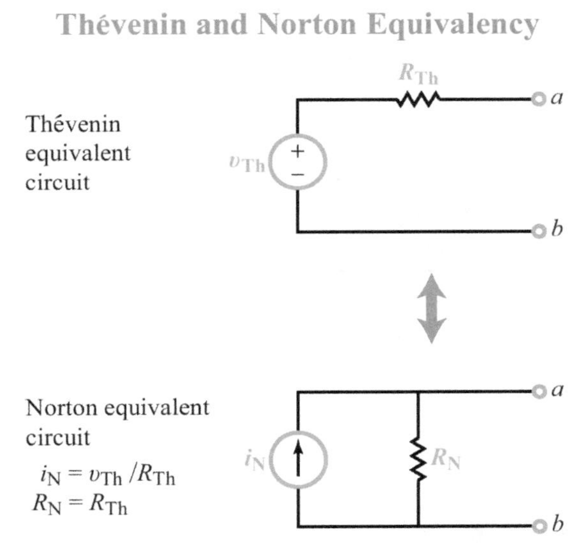

Figure 3-28: Equivalence between Thévenin and Norton equivalent circuits, consistent with the source transformation method of Section 2-3.4.

3-6.7 Norton's Theorem

A corollary of Thévenin's theorem, Norton's theorem states that a linear circuit can be represented at its output terminals by an equivalent circuit composed of a parallel combination of a current source i_{N} and a resistor R_{N}. Application of source transformation (Section 2-3.4) on the Thévenin equivalent circuit shown in Fig. 3-28 leads to the straightforward conclusion that i_{N} and R_{N} of the Norton equivalent circuit are given by

$$i_{\text{N}} = \frac{\upsilon_{\text{Th}}}{R_{\text{Th}}} \tag{3.37a}$$

and

$$R_{\text{N}} = R_{\text{Th}}. \tag{3.37b}$$

Table 3-1 provides a summary of the various methods available for finding the elements of the Thévenin and Norton equivalent circuits.

> **Concept Question 3-12:** Why is the Thévenin-equivalent circuit method such a powerful tool when analyzing a complex circuit, such as that of a cell phone? (See CAD)

3-6.8 Analyzing Cascaded Systems

Let us go back to the simple household circuit of Fig. 3-19(b) and redraw it in Fig. 3-29(a) as a series combination of blocks: the voltage source consisting of V_{s} and associated resistance R_{s}, the fan, the lamp, and the DVM. We intend to use the circuit to demonstrate how the Thévenin technique is used in practice to analyze much more complicated circuits. Our goal is to determine the voltage measured by the DVM.

Blocks 1 and 2

We start with the combination of the first two blocks, namely the source and the fan, after disconnecting everything to the right of terminals (c, d) from the circuit. The Thévenin voltage between terminals (c, d) in Fig. 3-29(b) is labeled V_{cd} and is given by

$$V_{cd} = \frac{V_{\text{s}} R_{\text{fan}}}{R_{\text{s}} + R_{\text{fan}}} = \frac{110 \times 100}{10 + 100} = 100 \text{ V}.$$

The Thévenin resistance of the circuit at terminals (c, d) in Fig. 3-29(b) is the parallel combination of R_{s} and R_{fan}:

$$R_{cd} = R_{\text{s}} \parallel R_{\text{fan}} = 10 \parallel 100 = 9.09 \ \Omega.$$

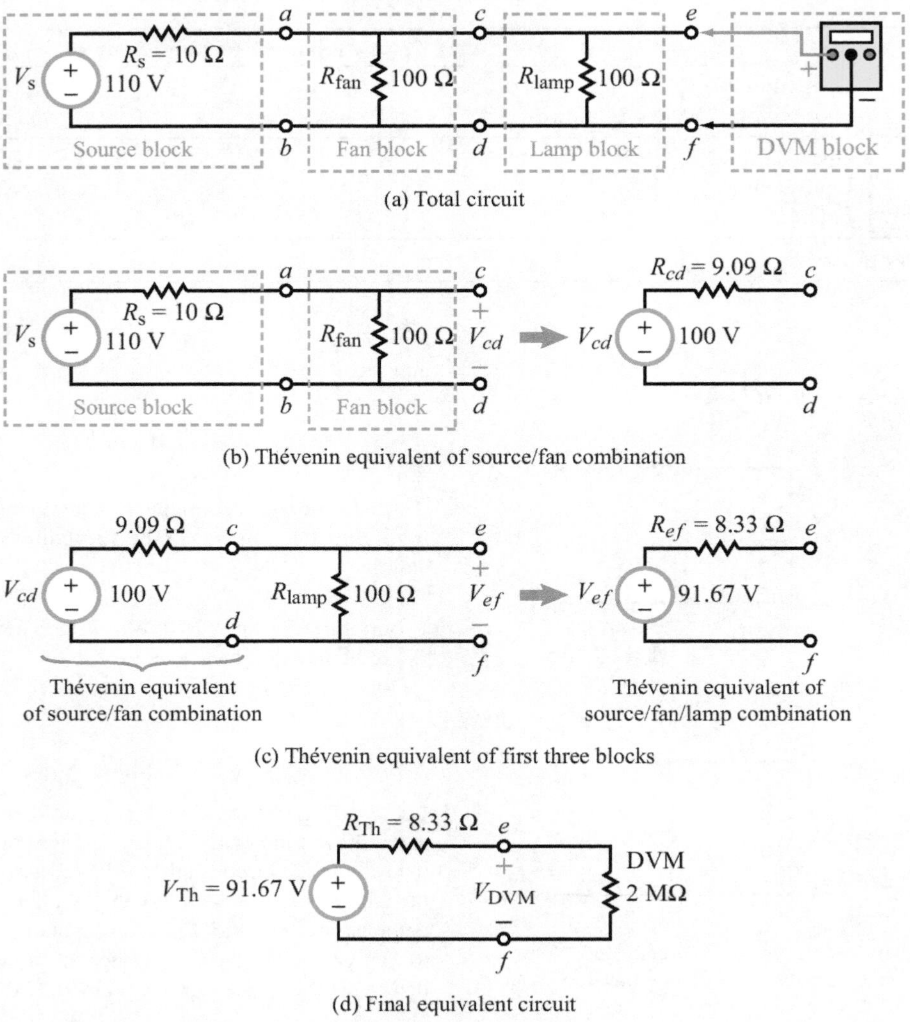

Figure 3-29: Repeated application of Thévenin-equivalent circuit technique.

Blocks 1, 2, and 3

Next, we repeat the Thévenin technique at terminals (e, f) by combining the lamp with the two earlier blocks. The Thévenin voltage at terminals (e, f) in Fig. 3-29(c) is labeled V_{ef} and is given by

$$V_{ef} = \frac{100 \times 100}{9.09 + 100} = 91.67 \text{ V},$$

and the Thévenin resistance, R_{ef}, is

$$R_{ef} = 100 \parallel 9.09 = 8.33\ \Omega.$$

Blocks 1–4

In part (d) of Fig. 3-29, we show the Thévenin equivalent of all blocks to the left of the DVM connected to the DVM at terminals (e, f). Voltage division leads to

$$V_{DVM} = \frac{91.67 \times 2 \times 10^6}{8.33 + 2 \times 10^6} \approx 91.67 \text{ V}.$$

This is the same answer we would have obtained had we analyzed the entire circuit at once using KCL/KVL. For this simple circuit, the multiple application of the Thévenin equivalent technique is obviously unwarranted, but when dealing with complex circuits comprising multiple subsections, the Thévenin technique is not only desirable, but also the only practical way to analyze and design circuits.

> **Concept Question 3-13:** Section 3-6 offers three different approaches for finding R_{Th}. Which ones apply to circuits containing dependent sources? (See CAD)

Exercise 3-13: Find the Norton equivalent at terminals (a, b) of the circuit in Fig. E3.13.

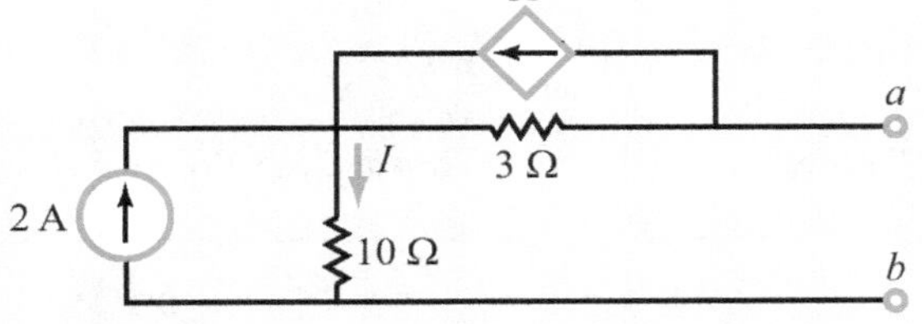

Figure E3.13

Answer:

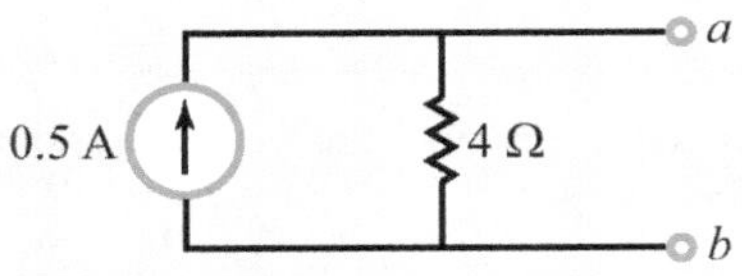

(See CAD)

3-7 Comparison of Analysis Methods

In this and the preceding chapter, we presented several different methods for analyzing electric circuits. Which method is *best*? Which one is the easiest to implement and why? The answer depends on the circuit configuration and the intended application. Table 3-2 provides a summary of the key attributes of the three circuit-analysis laws (Ohm's law, KCL, and KVL) and the analysis methods covered thus far. If the circuit contains no dependent sources and the goal is to determine the currents and voltages in the circuit, the two by-inspection methods provide a straightforward solution approach. When dependent sources are present, the node voltage and mesh current methods are always applicable. For cascaded circuits, the Thévenin (and Norton) equivalent-circuit technique is invariably the preferred choice.

3-8 Maximum Power Transfer

Suppose an *active* linear circuit is connected to a *passive* linear circuit, as depicted by Fig. 3-30(a). An active circuit contains at least one independent source, whereas a passive circuit may contain dependent sources, but no independent sources. For convenience, we shall refer to them as the *source and load circuits*, respectively. For certain applications, it is desirable to maximize the magnitude of the current i_L that flows from the source circuit to the load circuit, while other applications may call for maximizing the voltage υ_L at the input to the load circuit, or maximizing the power p_L that gets transferred from the source to the load. Given a specified source circuit, how, then, does one approach the design of the load circuit so as to achieve these different goals?

The solution to the problem posed by our question is facilitated by the equivalence offered by Thévenin's theorem. We demonstrated in the preceding section that any active, linear circuit always can be represented by an equivalent circuit composed of a Thévenin voltage υ_{Th} connected in series with a Thévenin resistance R_{Th}. In the case of the passive load circuit, its equivalent circuit consists of only a Thévenin resistance. To avoid confusion between the two circuits, we denote υ_{Th} and R_{Th} of the source circuit as υ_s and R_s, and we denote R_{Th} of the load circuit as R_L, as shown in Fig. 3-30(b). The current i_L

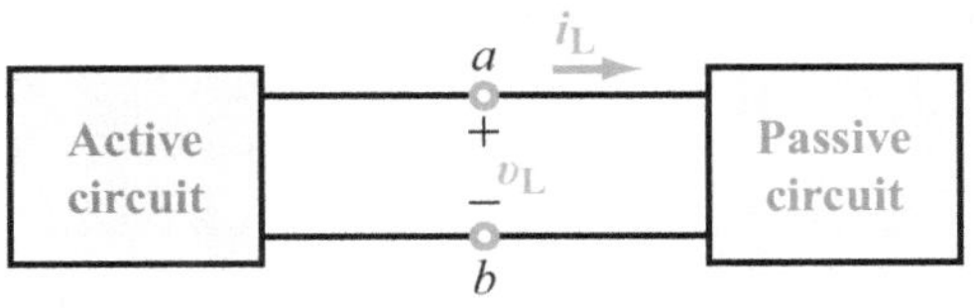

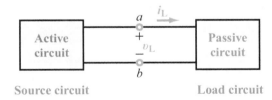

(a) Source and load circuits

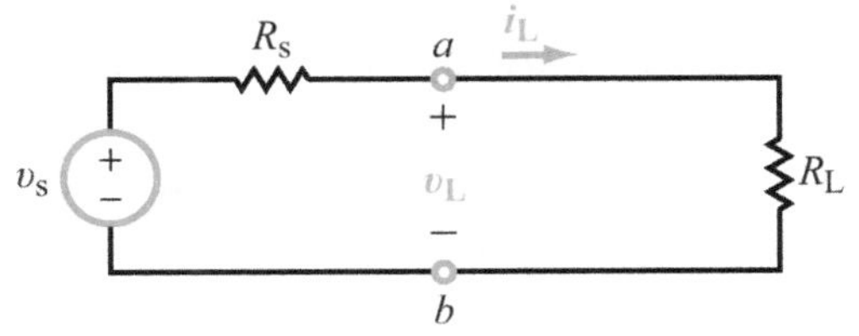

(b) Replacing source and load circuits with their Thévenin equivalents

Figure 3-30: To analyze the transfer of voltage, current, and power from the source circuit to the load circuit, we first replace them with their Thévenin equivalents.

Table 3-2: **Summary of circuit analysis methods.**

Method	Common Use
Ohm's law	Relates V, I, R. Used with all other methods to convert $V \Leftrightarrow I$.
R, G in series and ‖	Combine to simplify circuits. R in series adds, and is most often used. G in ‖ adds, so may be used when much of the circuit is in parallel.
Y-Δ or Π-T	Convert resistive networks that are not in series or in ‖ into forms that can often be combined in series or in ‖. Also simplifies analysis of bridge circuits.
Voltage/current dividers	Common circuit configurations used for many applications, as well as handy analysis tools. Dividers can also be used as combiners when used "backwards."
Kirchhoff's laws (KVL/KCL)	Solve for branch currents. Often used to derive other methods.
Node-voltage method	Solves for node voltages. Probably the most commonly used method because (1) node voltages are easy to measure, and (2) there are usually fewer nodes than branches and therefore fewer unknowns (smaller matrix) than KVL/KCL.
Mesh-current method	Solves for mesh currents. Fewer unknowns than KVL/KCL, approximately the same number of unknowns as node voltage method. Less commonly used, because mesh currents seem less intuitive, but useful when combining additional blocks in cascade.
Node-voltage by-inspection method	Quick, simplified way of analyzing circuits. Very commonly used for quick analysis in practice. Limited to circuits containing only independent current sources.
Mesh-current by-inspection method	Quick, simplified way of analyzing circuits. Very commonly used for quick analysis in practice. Limited to circuits containing only independent voltage sources.
Superposition	Simplifies circuits with multiple sources. Commonly used for both calculation and measurement.
Source transformation	Simplifies circuits with multiple sources. Commonly used for both calculation/design and measurement/test applications.
Thévenin and Norton equivalents	Very often used to simplify circuits in both calculation and measurement applications. Also used to analyze cascaded systems. Thévenin is the more commonly used form, but Norton is often handy for analyzing parallel circuits. Source transformation allows easy conversion between Thévenin and Norton.
Input/output resistance ($R_{\text{in}}/R_{\text{out}}$)	Commonly used to evaluate when cascaded circuits can be analyzed individually or when full circuit analysis or a buffer is needed.

and associated voltage υ_{L} are given by Ohm's law as

$$i_{\text{L}} = \frac{\upsilon_{\text{s}}}{R_{\text{s}} + R_{\text{L}}},$$

and by voltage division:

$$\upsilon_{\text{L}} = \frac{\upsilon_{\text{s}} R_{\text{L}}}{R_{\text{s}} + R_{\text{L}}}.$$

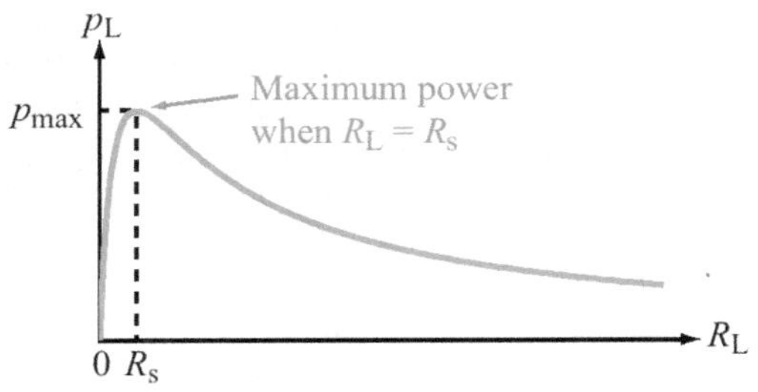

Figure 3-31: Variation of power p_L dissipated in the load R_L, as a function of R_L.

If the source-circuit parameters υ_s and R_s are fixed and the intent is to *transfer* maximum current to the load circuit, then R_L should be zero (short circuit). For a real circuit with a functional purpose, the circuit will need to receive some energy in order to function. Hence, R_L cannot be exactly zero, but it can be made to be very small in comparison with R_s. Thus, to maximize current transfer, the load circuit should be designed such that

$$R_L \ll R_s \quad \textbf{(maximum current transfer).} \tag{3.39}$$

Based on Eq. (3.38), the opposite is true for maximum voltage transfer, namely

$$R_L \gg R_s \quad \textbf{(maximum voltage transfer).} \tag{3.40}$$

The situation for power transfer calls for maximizing the product of i_L and υ_L,

$$p_L = i_L \upsilon_L = \frac{\upsilon_s^2 R_L}{(R_s + R_L)^2}. \tag{3.41}$$

The expression given by Eq. (3.41) is a nonlinear function of R_L. The power p_L goes to zero as R_L approaches either end of its range (0 and ∞), as illustrated by the plot in Fig. 3-31, and it is at a maximum when

$$R_L = R_s \quad \textbf{(maximum power transfer).} \tag{3.42}$$

▶ This equality is referred to as ***matching*** the source to the load. ◀

The proof of Eq. (3.42) is given in Example 3-14. Use of $R_L = R_s$ in Eq. (3.41) leads to

$$p_{max} = \frac{\upsilon_s^2 R_L}{(R_L + R_L)^2} = \frac{\upsilon_s^2}{4R_L}, \tag{3.43}$$

which represents 50 percent of the total power generated by the equivalent input source υ_s. The other 50 percent is dissipated in R_s.

Example 3-14: Maximum Power Transfer

Prove that p_L, as given by Eq. (3.41), is at a maximum when $R_L = R_s$.

Solution: To find the value of R_L at which the expression for p_L is at a maximum, we differentiate the expression with respect to R_L and then set the result equal to zero. That is,

$$\frac{dp_L}{dR_L} = \frac{d}{dR_L}\left[\frac{\upsilon_s^2 R_L}{(R_s + R_L)^2}\right] = \upsilon_s^2\left[\frac{1}{(R_s + R_L)^2} - \frac{2R_L}{(R_s + R_L)^3}\right] = 0.$$

A few simple steps of algebra lead to

$$R_L = R_s.$$

Concept Question 3-14: Under what conditions is the power transferred from a power source to a load resistor a maximum? When is the voltage a maximum? When is the current a maximum? (See CAD)

Concept Question 3-15: Of the power generated by an input circuit, what is the maximum fraction that can be transferred to an external load? (See CAD)

Technology Brief 8
Digital and Analog

Most of electrical engineering depends on the manipulation of voltages and currents. The real world interfaces with our circuits through sensors (such as the resistive sensors in Technology Brief 4), and we interface back to the real world through user interfaces (such as turning on an LED in Technology Brief 5). In between these ***transducers*** are *circuits*! In the physical world, most signals of interest are ***analog signals***; that is, they vary continuously with time and can take on any value between their possible minimum and maximum values. When electrical sensors transduce these signals into changes in voltage or current, the electrical signals produced are thus also analog. Analog electrical signals can be transduced from sound (using a microphone), mechanical vibration (using a piezoelectric vibration sensor), light or images (using sensor arrays in a camera), temperature (using a thermistor), and many other sources.

All of the circuits we have examined so far are analog circuits. The voltages (and currents) present in these circuits can take on any value between a maximum and a minimum (typically set by the power source). By contrast, a ***digital*** signal can only assume a few discrete values. Most digital systems are binary, which is to say they can only assume two such values, usually called "0" and "1" (alternatively, "on" and "off"). The exact voltages which represent the two logic states depend on the type of digital logic used; for example, many modern digital processors represent "0" with 0 V and "1" with 1.2 V.

Because any single digital line can only assume two values, many such lines can be used to represent a range of numbers. Consider, for example, Table TT8-1: three digital lines (or ***bits***) are used to encode 8 different numbers within a given range. In the same way that base-10 numbers can encode 10^N different values with N discrete numbers in the range 0–9 (e.g., two base 10 numbers can encode 0–99), 2^N different values can be encoded by N binary bits. Eight such bits make up a ***byte*** (e.g., the value 01101111 is a byte). Two bytes (16 bits) are a ***word***. Standard encoding schemes exist for representing commonly used data. For example, letters, carriage returns, and other typographics can be represented using the 7-bit American Standard Code for Information Interchange (***ASCII***, pronounced "ask-ee"). Table TT8-2 gives these codes for capital letters. Many such standards exist (ranging from the data encoding format for, say, Blu-ray data to data transmission across ATM networks).

When representing floating point numbers (such as −2.3), the computer must encode the sign (−1), the number and the exponent. The precision to which a number can be represented depends on how many bits are used. Four words (32 bits) are considered single precision, and 64 bits are double precision. The first bit is the sign (1 = negative), and the next 8 bits are the exponent. The remaining 23 bits (single precision) or 55 bits (double precision) are used to represent the number. This means that the floating point representation of the number has a certain predictable ***round-off error***, and when the computer adds, subtracts, multiplies, etc., this error is also present in the calculations. Usually it is too small to be noticed, but in some cases $(2 - 1.9999\ldots \neq 0)$ it can cause unexpected problems in computer programs.

We commonly convert back and forth between analog and digital voltages. Almost all analog signals are converted to digital signals for storage (e.g., images), wireless transmission (your voice in a cell phone call), and performing mathematical functions (in your calculator). This is done with an ***analog-to-digital converter*** (ADC). Sometimes the digital signal must be converted back to

Table TT8-1: Three-bit counting scheme.

Bits		Number
000	=	0
001	=	1
010	=	2
011	=	3
100	=	4
101	=	5
110	=	6
111	=	7

Table TT8-2: **ASCII characters for capital letters.**

0100 0001 = A	0100 1110 = N
0100 0010 = B	0101 1111 = O
0100 0011 = C	0101 0000 = P
0100 0100 = D	0101 0001 = Q
0100 0101 = E	0101 0010 = R
0100 0110 = F	0101 0011 = S
0100 0111 = G	0101 0100 = T
0100 1000 = H	0101 0101 = U
0100 1001 = I	0101 0110 = V
0100 1010 = J	0101 0111 = W
0100 1011 = K	0101 1000 = X
0100 1100 = L	0101 1001 = Y
0100 1101 = M	0101 1010 = Z

analog (so your friend can hear your voice on his cell phone). This is done with a ***digital-to-analog converter*** (DAC).

The analog voltage in Fig. TF8-1 can be converted to digital using an ADC to sample it, find the closest step that matches the signal, and convert the value of that step to a digital value. The number of steps (controlled by the number of bits in the ADC) controls the precision of the ADC. Figure TF8-1 shows a very coarse 3-bit ADC that can represent 8 levels. The difference between the actual analog signal and the level that can be represented with the ADC is called the ***quantization error***.

One of the strengths of digital representations of data is that manipulations of this data (mathematical operations, storage, etc.) can be carried out efficiently with switching networks. These are circuits of components wherein each component can only produce one of two voltage values. Transistors, in particular MOSFETS (see Chapter 4), are particularly well-suited to act as switches in these circuits; modern integrated circuits contain on the order of a billion MOSFETS arranged into circuits to manipulate digital data. Importantly, most modern integrated circuits contain both analog and digital circuits and are known as ***mixed-signal circuits*** (see Section 13-9). Using built-in ADC and

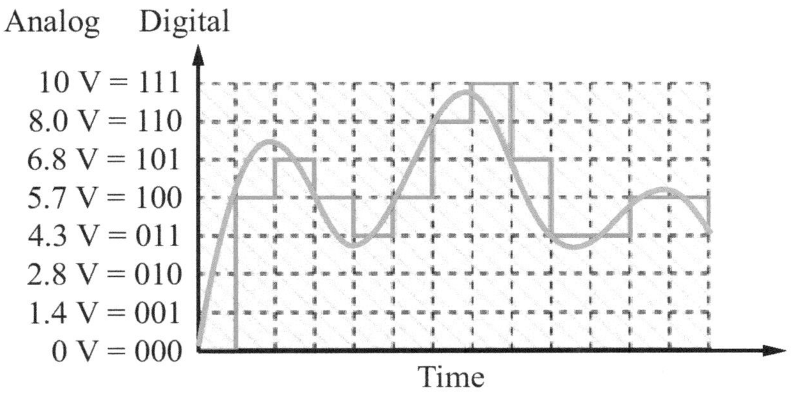

Figure TF8-1: Three-bit digital representation of a continuous signal.

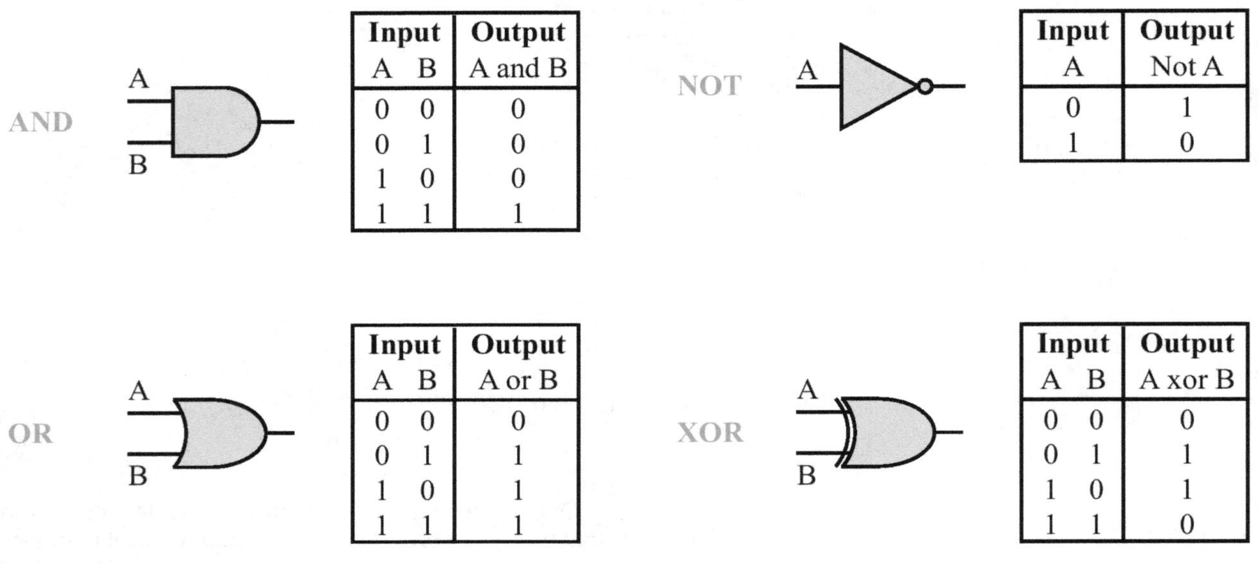

Input A	Input B	Output A and B
0	0	0
0	1	0
1	0	0
1	1	1

Input A	Output Not A
0	1
1	0

Input A	Input B	Output A or B
0	0	0
0	1	1
1	0	1
1	1	1

Input A	Input B	Output A xor B
0	0	0
0	1	1
1	0	1
1	1	0

Figure TF8-2: Logic gates.

DAC circuits, data is moved from the analog to the digital domain within a single chip.

But sometimes we use only a few gates for simple control operations or prototyping. Each gate takes two digital signals (which can be either a 0 or 1) as input, and outputs a different digital signal (based on these inputs). Figure TF8-2 shows a few of these common logic gates. An ***AND*** gate outputs a 1 if both input A AND input B are 1. An ***OR*** gate outputs a 1 if either input A OR input B are 1. A ***NOT*** gate outputs a 1 if input A is NOT a 1; i.e., it inverts the input. An exclusive OR gate, called an ***XOR*** gate, outputs a 1 if one and only one of input A OR input B is 1.

One way to prototype with logic gates is to use a chip that plugs into your protoboard (see Appendix F). Figure TF8-3 shows an example of a quad AND package. Each pin on the chip is numbered 1–14 and plugs into a separate node (row) on the protoboard. Logic gates are active devices, which means they require an external power supply, so V_{cc} is plugged into pin 14, and GND into pin 7.

Interfacing from the real world to a computer most often involves an analog sensor (such as a thermistor for measuring temperature), a level-shifter (amplifier or de-amplifier or comparator that converts the analog output voltage to digital levels), and then a logic circuit to act upon the output (turn a switch to a heater on or off, for instance). When interfacing back to the real world, the digital signal may be applied in digital form, or may need to be converted back to an analog signal (to drive speakers for voice and music, or precision control of an engine air intake, for example).

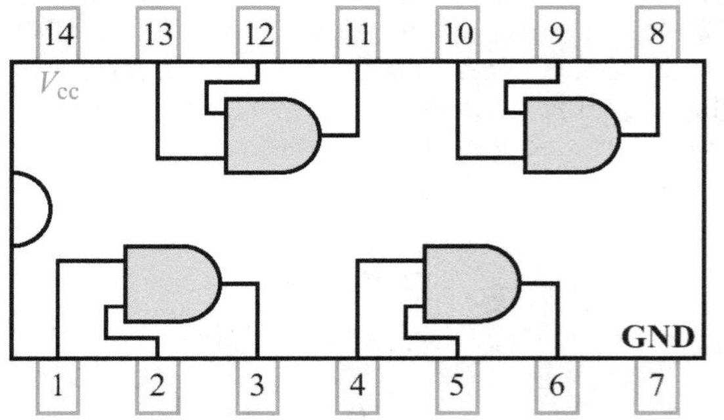

7408 Quad 2 Input AND

Figure TF8-3: Quad AND package.

Exercise 3-14: The bridge circuit of Fig. E3.14 is connected to a load R_L between terminals (a, b). Choose R_L such that maximum power is delivered to R_L. If $R = 3\ \Omega$, how much power is delivered to R_L?

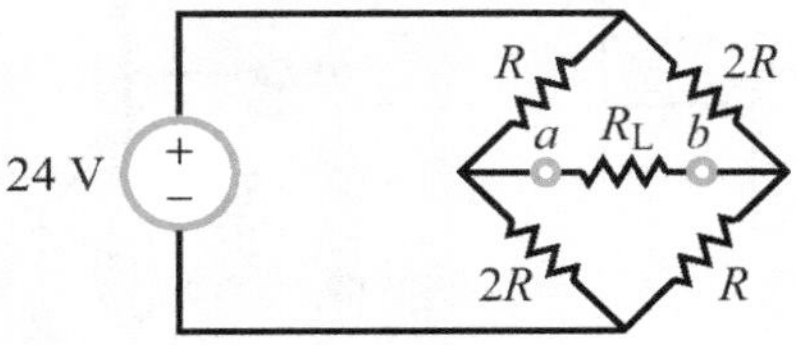

Figure E3.14

Answer: $R_L = 4R/3 = 4\ \Omega$, $p_{max} = 4$ W. (See CAD)

Example 3-15: Bridge Circuit

In the bridge circuit shown in Fig. 3-32(a), choose R_L so that the power delivered to it is a maximum. How much power will that be?

Solution: After temporarily removing R_L from the circuit, we proceed to find the Thévenin equivalent of the circuit at terminals (a, b).

Open-Circuit Voltage: In the circuit shown in Fig. 3-32(b), we designate the bottom node of the bridge as ground and the top node as voltage V_1. Application of KCL at node V_1 gives

$$\frac{V_1 - 16}{5} + \frac{V_1}{2+4} + \frac{V_1}{2+4} = 0,$$

which leads to

$$V_1 = 6\ \text{V}.$$

Voltage division gives

$$V_a = \left(\frac{4}{2+4}\right) V_1 = 4\ \text{V},$$

$$V_b = \left(\frac{2}{2+4}\right) V_1 = 2\ \text{V}.$$

Hence,

$$V_{Th} = V_{oc} = V_a - V_b = 4 - 2 = 2\ \text{V}.$$

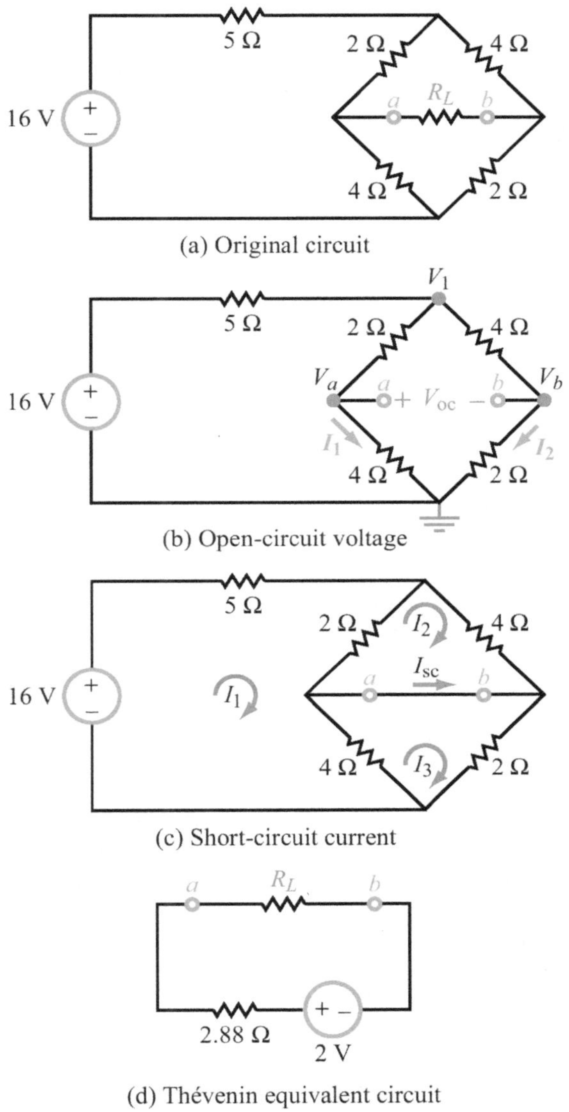

Figure 3-32: Evolution of the circuit of Example 3-15.

Short-Circuit Current: In the circuit configuration shown in Fig. 3-32(c), terminals (a, b) are connected by a short circuit. Application of the mesh-analysis by-inspection method (Section 3-4.2) leads to the matrix equation

$$\begin{bmatrix} 11 & -2 & -4 \\ -2 & 6 & 0 \\ -4 & 0 & 6 \end{bmatrix} \begin{bmatrix} I_1 \\ I_2 \\ I_3 \end{bmatrix} = \begin{bmatrix} 16 \\ 0 \\ 0 \end{bmatrix}.$$

Matrix inversion by MATLAB or MathScript yields

$$I_1 = \frac{96}{46}\text{ A}, \qquad I_2 = \frac{32}{46}\text{ A}, \qquad I_3 = \frac{64}{46}\text{ A}.$$

Hence, the short-circuit current is

$$I_{sc} = I_3 - I_2 = \frac{64}{46} - \frac{32}{46} = \frac{32}{46} = 0.7,$$

and

$$R_{Th} = \frac{V_{oc}}{I_{sc}} = \frac{2}{0.7} = 2.88\ \Omega.$$

The Thévenin equivalent circuit is shown in Fig. 3-32(d). Power transfer to R_L is a maximum when

$$R_L = R_{Th} = 2.88\ \Omega,$$

and according to Eq. (3.45),

$$p_{max} = \frac{\upsilon_s^2}{4R_L} = \frac{(2)^2}{4 \times 2.88} = 0.35\text{ W}.$$

3-9 Application Note: Bipolar Junction Transistor (BJT)

With the exception of the SPDT switch, all of the elements we have discussed thus far have been two-terminal devices, each characterized by a single i–υ relationship. These include resistors, voltage and current sources, as well as the pn-junction diode of Section 2-6.2. The potentiometer (Fig. 2-3(b)) may appear to be like a three-terminal device, but in reality it is no more than two resistors—each with its own pair of terminals. This section introduces a true three-terminal device, the *bipolar junction transistor* (BJT).

The BJT is a three-layer semiconductor structure commonly made of silicon. Other compounds sometimes are used for special-purpose applications (such as for operation at microwave and optical frequencies), but for the present, we will limit our examination to silicon-based transistors and their uses in dc circuits. The three terminals of a BJT are called the *emitter*, *collector*, and *base*, and each is made of either a p-type (silicon with positive charge carriers) or n-type (silicon with negative charge carriers) semiconductor material. The emitter and collector are made of the same material—either p-type or n-type—and the base is made of the other material. Thus, the BJT can be constructed to have either a *pnp configuration* or an *npn configuration*, as shown in the diagrams of Fig. 3-33. The geometries and fabrication details of real transistors are far more elaborate than the simple diagrams suggest, but the basic idea that the BJT consists of three alternating layers of p- and n-type material is quite sufficient from the standpoint of its external electrical behavior.

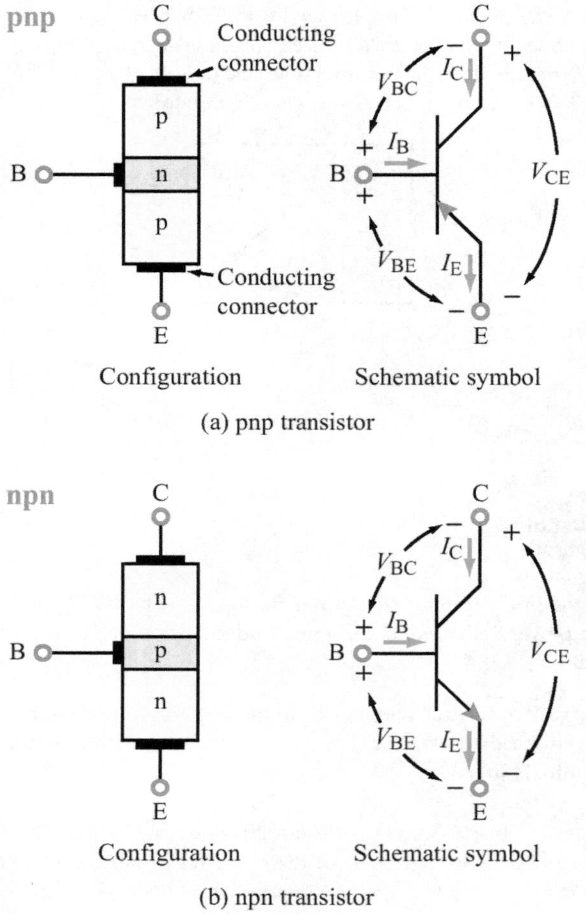

Figure 3-33: Configurations and symbols for (a) pnp and (b) npn transistors.

Figure 3-33 also shows schematic symbols used for the pnp and npn transistors. The center terminal is always the base. One of the three leads includes an arrow. *The lead containing the arrow identifies the emitter terminal* and whether the transistor is a pnp or npn. The arrow always points towards an n-type material, so *in the pnp transistor, the arrow points towards the base, whereas in an npn transistor, the arrow points away from the base.*

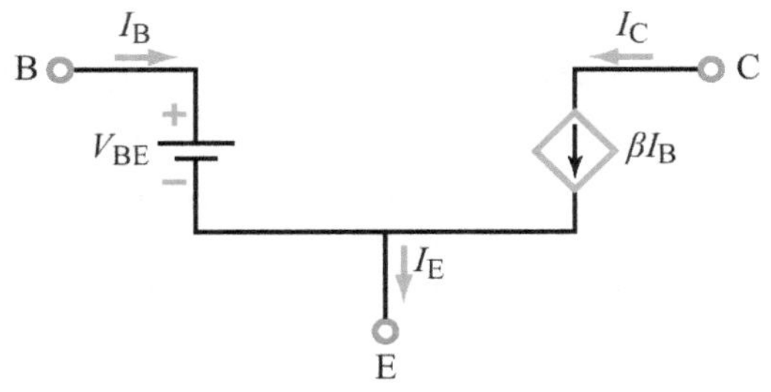

Figure 3-34: dc equivalent model for the npn transistor. The equivalent dc source $V_{BE} \approx 0.7$ V.

The directions of the terminal currents shown in Fig. 3-33 are defined such that the base and collector currents I_B and I_C, respectively, flow into the transistor, and the emitter current I_E flows out of it. KCL requires that

$$I_E = I_B + I_C. \quad (3.44)$$

Under normal operating conditions, I_E has the largest magnitude of the three currents, and I_B is much smaller than either I_C or I_E. The transistor can operate under both dc and ac conditions, but we will limit our present discussion to dc circuits. For simplicity, we will consider only the npn common-emitter configuration. Accordingly, we can describe the operation of the npn transistor by the dc equivalent model shown in Fig. 3-34. The circuit contains a constant dc voltage source V_{BE} and a dependent current-controlled current source that relates I_C to I_B by

$$I_C = \beta I_B, \quad (3.45)$$

where β is a transistor parameter called the *common-emitter current gain*. This is a perfect example of how a nonlinear element can be modeled in terms of a linear circuit containing a dependent source. Under normal operation, $V_{BE} \approx 0.7$ V, and β may assume values in the range between 30 and 1000, depending on its specific design configuration. The V_{BE} source in Fig. 3-34 models a built-in voltage drop that arises within the transistor at the interface of p-type and n-type regions; it is not a true independent source as it can never supply power. Transistors never supply power, they modify the flow of power through them in interesting and useful ways. To operate in its active mode, the transistor requires that certain dc voltages be applied at its base and collector terminals. We shall refer to these voltages as V_{BB} and V_{CC}, respectively.

Example 3-16: BJT Amplifier Circuit

Apply the equivalent-circuit model with $V_{BE} \approx 0.7$ V and $\beta = 200$ to determine I_B, I_C, and V_{CE} in the circuit of Fig. 3-35(a). Assume that $V_{BB} = 2$ V, $V_{CC} = 10$ V, $R_B = 26$ kΩ, and $R_C = 200$ Ω.

Solution: Upon replacing the npn transistor with its equivalent circuit, we end up with the circuit shown in Fig. 3-35(b). In the left-hand loop, KVL gives

$$-V_{BB} + R_B I_B + V_{BE} = 0,$$

which leads to

$$I_B = \frac{V_{BB} - V_{BE}}{R_B} = \frac{2 - 0.7}{26 \times 10^3} = 5 \times 10^{-5} \text{ A} = 50\ \mu\text{A}.$$

Given that $\beta = 200$,

$$I_C = \beta I_B = 200 \times 50 \times 10^{-6} = 10 \text{ mA}$$

and

$$V_{CE} = V_{CC} - I_C R_C = 10 - 10^{-2} \times 200 = 8 \text{ V},$$

which is a 4-times amplification of source V_{BB}.

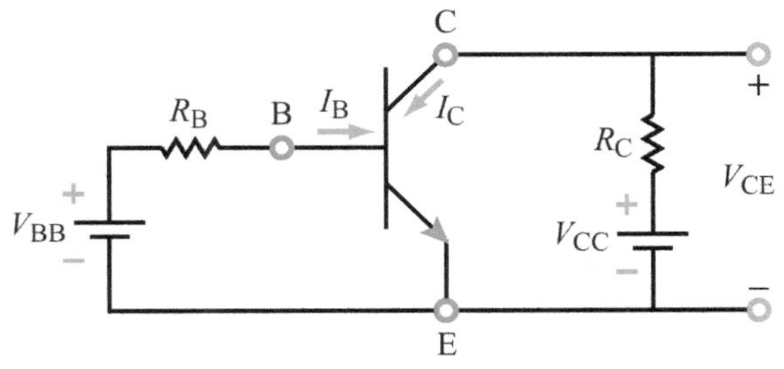

(a) Transistor circuit

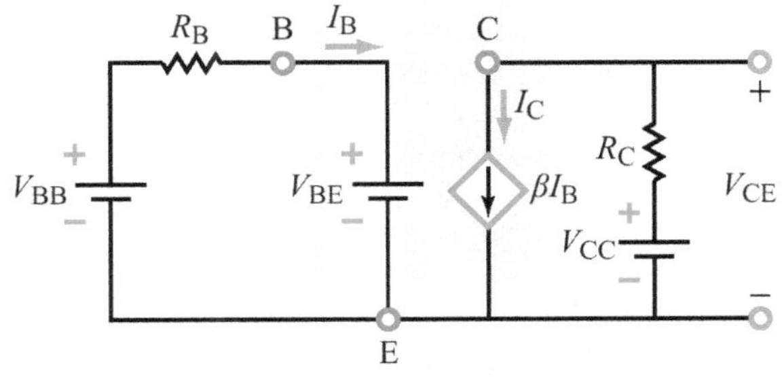

(b) Equivalent circuit

Figure 3-35: Circuit for Example 3-16.

Example 3-17: Digital-Inverter Circuit

Digital logic deals with two states, "0" and "1" (or equivalently "low" and "high"). A digital-inverter circuit provides one of the logic operations performed by a computer processor, namely to invert the state of an input bit from low to high or from high to low. Demonstrate that the transistor circuit shown in Fig. 3-36 functions as a digital inverter by plotting its output voltage V_{out} versus the input voltage V_{in}. A bit is assumed to be in state 0 (low) if its voltage is between 0 and 0.5 V and in state 1 (high) if its voltage is greater than 4 V. Assume that the equivalent model given by Fig. 3-34 is applicable (with $\beta = 20$) with the following qualifications: neither I_B nor V_{out} can have negative values, so if the analysis using the equivalent-circuit model generates a negative value for either one of them, it should be replaced with zero.

Solution: The equivalent circuit shown in Fig. 3-36(b) provides the following expressions:

$$I_B = \frac{V_{\text{in}} - 0.7}{20k}, \tag{3.46}$$

$$I_C = \beta I_B = 200 I_B, \tag{3.47}$$

and

$$V_{\text{out}} = V_{\text{CC}} - I_C R_C. \tag{3.48}$$

Combining the three equations leads to

$$V_{\text{out}} = V_{\text{CC}} - \frac{\beta R_C}{R_B}(V_{\text{in}} - 0.7) = 12 - 10 V_{\text{in}} \quad \text{(V)}. \tag{3.49}$$

Since V_{out} is linearly related to V_{in}, the plot would be a straight line, as shown in Fig. 3-36(c), but we also have to incorporate the provisions that I_B cannot be negative (which occurs when $V_{\text{in}} < 0.7$ V), and V_{out} cannot be negative (which occurs when $V_{\text{in}} = 1.2$ V). The resultant transfer function clearly satisfies the digital inverter requirements:

Input: **Low**   Output: **High**
If $V_{\text{in}} < 0.5$ V ➡ $V_{\text{out}} = 5$ V,

-

Input: **High**   Output: **Low**
If $V_{\text{in}} > 1.2$ V ➡ $V_{\text{out}} = 0$.

(a) Inverter circuit

(b) Equivalent circuit

(c) V_{out} versus V_{in}

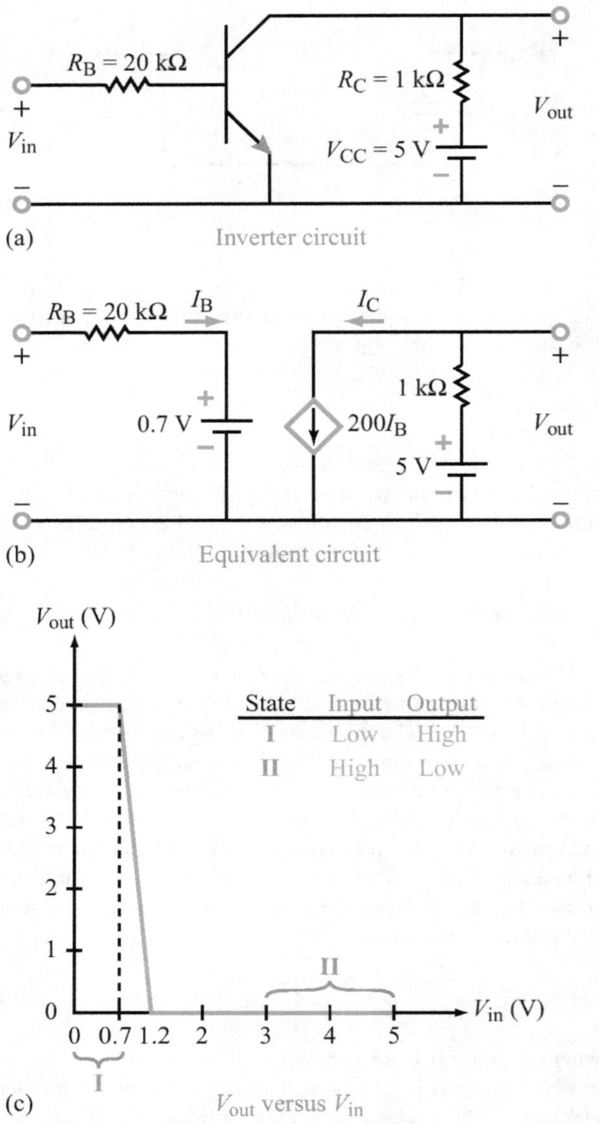

Figure 3-36: Circuit for Example 3-17.

Concept Question 3-16: How is the collector current related to the base current in a BJT? (See CAD)

Concept Question 3-17: What is a digital inverter? How are its input and output voltages related to one another? (See CAD)

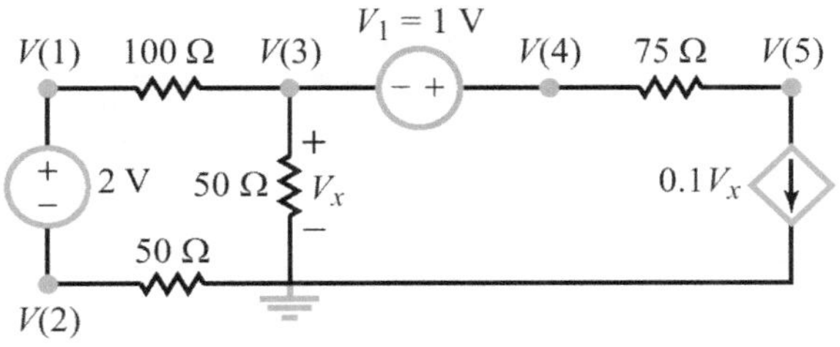

(a) Six-node circuit

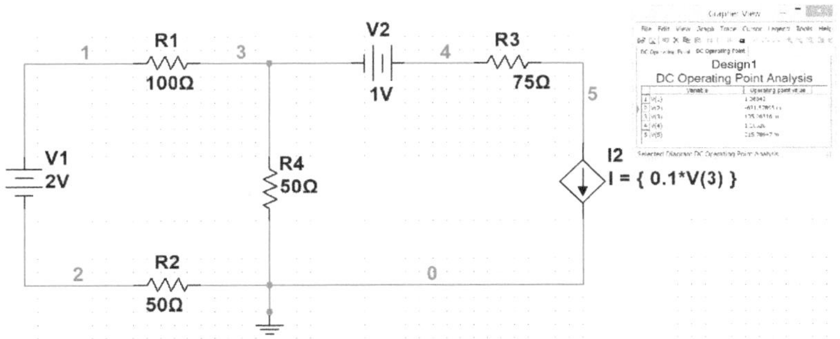

(b) Multisim circuit and solution

Figure 3-37: Circuit analysis with Multisim.

Exercise 3-15: Determine I_B, V_{out_1}, and V_{out_2} in the transistor circuit of Fig. E3.15, given that $V_{BE} = 0.7$ V and $\beta = 200$.

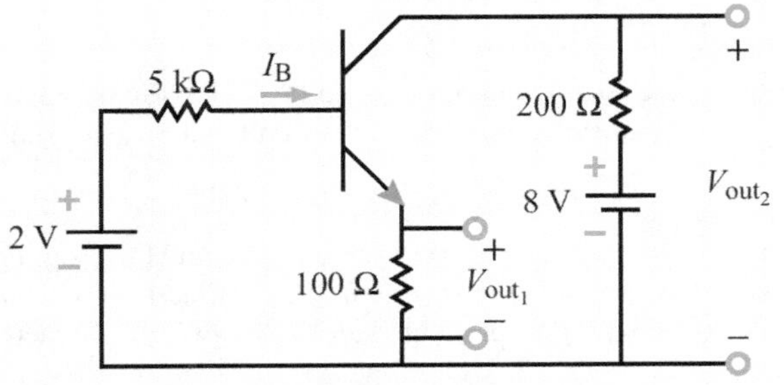

Figure E3.15

Answer: $I_B = 51.79$ μA, $V_{out_1} = 1.04$ V, $V_{out_2} = 5.93$ V. (See CAD)

3-10 Nodal Analysis with Multisim

Multisim is a particularly useful tool for analyzing circuits with many nodes. Consider the six-node circuit shown in Fig. 3-37(a), in which the voltages and currents are designated in accordance with the Multisim notation system. In Multisim, V1 refers to the voltage of source 1 and V(1) refers to the voltage at node 1. Application of nodal analysis would generate five equations with five unknowns, V(1) to V(5), whose solution would require the use of matrix algebra or several steps of elimination of variables. [For this simple two-loop circuit, mesh analysis is much easier to apply, as it involves only two mesh equations and one auxiliary equation for the dependent current source, but the objective of the present section is to illustrate how Multisim can be used for circuits involving a large number of nodes.] When drawn in Multisim, the circuit appears in the form shown in Fig. 3-37(b). Application of either Measurement Probes or DC Operating Point Analysis generates the values of V(1) to V(5) listed in the inset of Fig. 3-37(b).

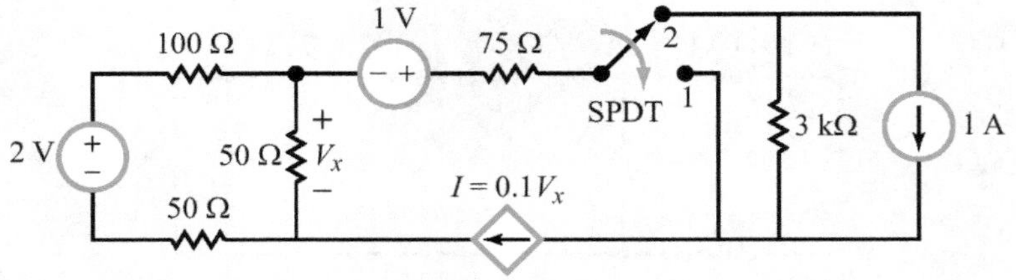

(a) Circuit with SPDT switch

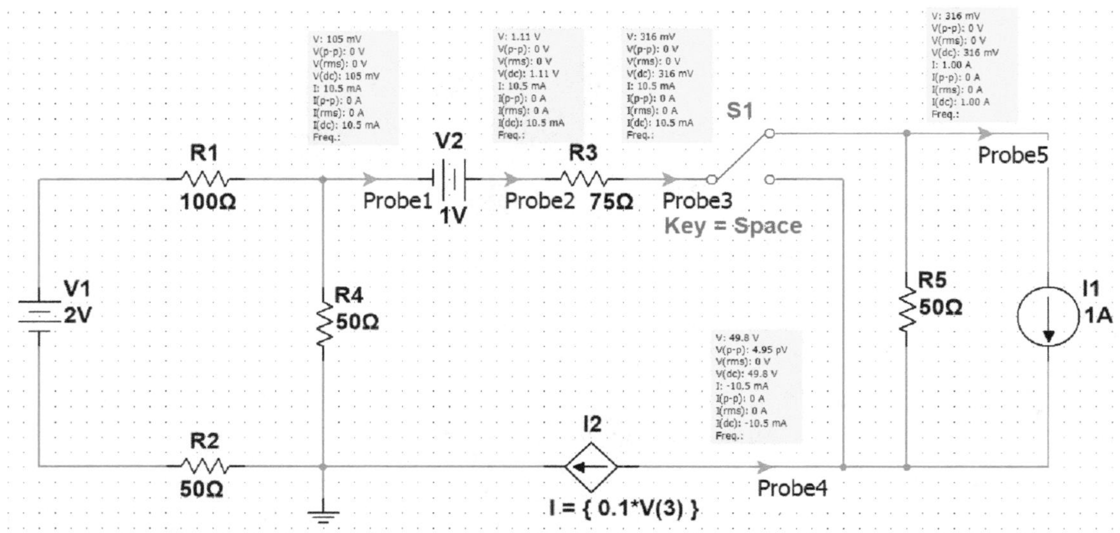

(b) Multisim configuration

Figure 3-38: (a) Circuit with a switch, and (b) its Multisim representation.

For circuits containing more than four or five nodes, analyzing the circuit *by hand* becomes unwieldy. Moreover, some circuits may contain time-varying sources or elements. Consider, for example, the circuit in Fig. 3-38(a), which is a replica of the circuit in Fig. 3-37 except for the addition of an SPDT switch. [In Multisim, the switch can be *toggled* between positions 1 and 2 using the space bar on your computer.] When connected to position 1, the state of the circuit is identical to that in Fig. 3-37, but when the SPDT switch is moved to position 2, the new circuit configuration includes two additional elements and one extra node.

The circuit drawn in Multisim is shown in Fig. 3-38(b). The SPDT is available in the Select a Component window under the Basic group in the SWITCH family. Measurement Probes were added to nodes 4, 5, and 6. Using the Interactive Simulation feature of Multisim, the circuit can be analyzed in each of its two states by pressing F5 (or the button or toggle switch) to start the simulation, and then toggling the switch by pressing the space bar. This live-action switching capability is why this particular tool is known as Interactive Simulation.

In the Multisim section of Chapter 2, we examined how the DC Operating Point Analysis tool can be used to determine differences between node voltages. In addition to basic subtraction, there are many operators that you can apply to variables (or combinations of variables) to obtain the desired quantities. [See the Multisim Tutorial on the book website http://c3.eecs.umich.edu for a list of the basic operators].

We will now use variable manipulation in the DC Operating Point Analysis to calculate the power dissipated or supplied in each component in the circuit in Fig. 3-37(a). To calculate the power for each component, we need to know both the current through and voltage across each component. However, for many devices, Multisim can calculate the power automatically. Open up the DC Operating Point Analysis window. Notice that for the voltage sources and resistors, Multisim allows you to select to solve for the power, using the P() notation. You can also ask Multisim to solve for expressions which use the available variables. In the output tab enter equations via the Add Expression... button. We'll enter an expression for the power across the controlled source this way using the expression V(5)*I(BI2). Click OK after entering any expressions. [Remember proper sign notation and current direction.] The equations for power should be

Source V1:	(V(4)-V(3))*I(v1)
Source V2:	(V(1)-V(2))*I(v2)
Source I1:	-V(5)*I(v1)
Resistor R1:	(V(3)-V(1))*I(v2)
Resistor R2:	V(3)*I(v3)
Resistor R3:	(V(5)-V(4))*I(v1)
Resistor R4:	V(2)*I(v2)

Note: Remember that these variable names apply to the circuit shown in Fig. 3-39(a). If your circuit has a different numbering for nodes or voltage sources, your equations will differ in number accordingly.

Once these equations are entered, the Selected Variables for Analysis field should resemble that in Fig. 3-39(b). To obtain the values, press the Simulate button. The results should agree with those shown in Fig. 3-39(c).

Knowing how to write equations such as these in Multisim is very important, because many other Analyses which you will encounter later in the book utilize identical syntax to that used for the DC Operating Point Analysis.

> **Concept Question 3-18:** What is the difference between the Measurement Probe tool and the DC Operating Point Analysis? (See CAD)

Exercise 3-16: Use Multisim to calculate the voltage at node 3 in Fig. 3-38(b) when the SPDT switch is connected to position 2.

Answer: (See CAD)

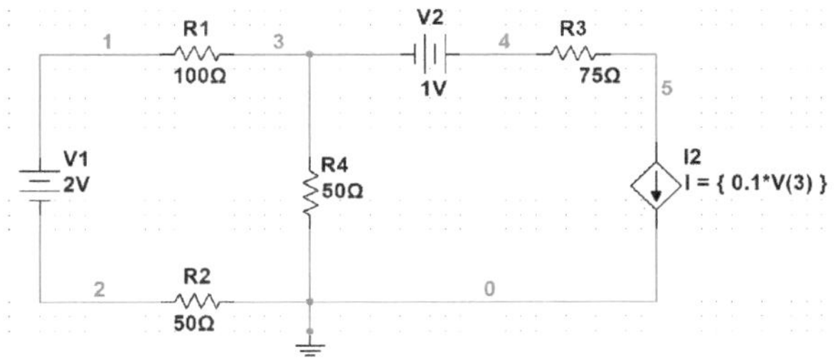

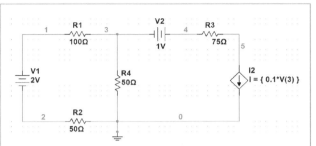

(a) Multisim circuit of Fig. 3-32(a) ready for power calculations

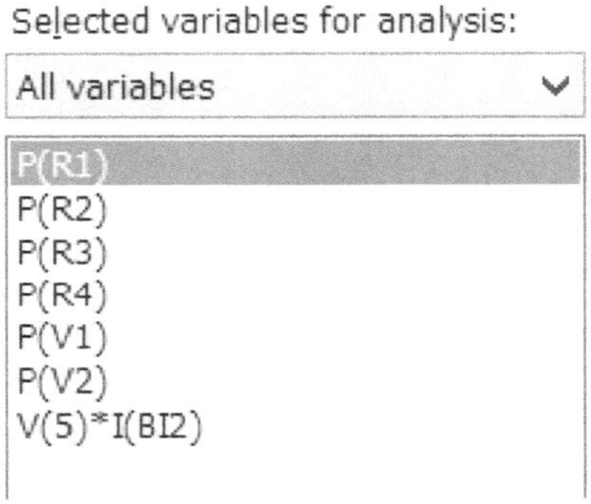

(b) Selected variables for analysis visible in DC Operating Point Analysis window

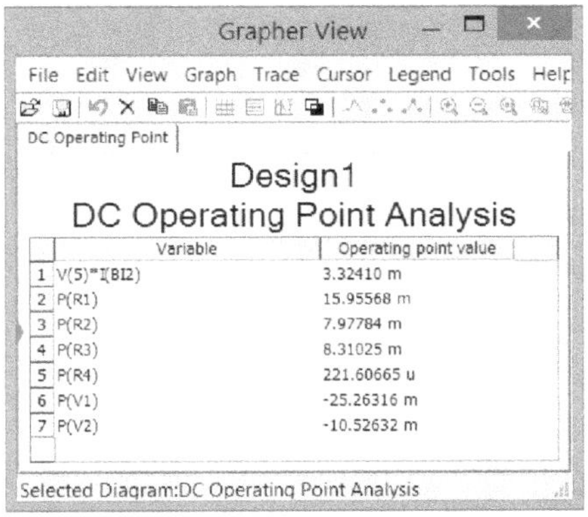

(c) Output of simulations (remember that all values are in watts)

Figure 3-39: Multisim procedure for calculating power consumed (or generated) by the seven elements in the circuit of Fig. 3-37(a).

Summary

Concepts

- After designating one of the extraordinary nodes in a circuit as reference (ground), KCL at the remaining extraordinary nodes provides the requisite number of equations for determining the voltages at those nodes.
- Two extraordinary nodes connected by a solitary voltage source constitute a supernode. The two nodes can be treated as a single node, augmented by an auxiliary relation specifying the voltage difference between them.
- By assigning a mesh current to each independent loop, application of KVL leads to the requisite number of equations for determining the unknown mesh currents.
- Two adjoining loops sharing a branch containing a solitary current source constitute a supermesh. The two loops can be treated as a single loop, augmented by an auxiliary relation specifying the relationship between their mesh currents..
- A circuit containing no dependent sources and only current sources can be analyzed by the node-voltage by-inspection method.
- Similarly, a circuit containing no dependent sources and only voltage sources can be analyzed the mesh-current by-inspection method.
- Thévenin's (Norton's) theorem states that a linear circuit can be represented by an equivalent circuit composed of a voltage source (current source) in series (in parallel) with a resistor.
- Thévenin and Norton equivalent circuits are powerful tools for analyzing and designing complex, cascaded circuits.
- The power transferred by an input circuit to an external load is at a maximum when the load resistance is equal to the Thévenin resistance of the input circuit. The fraction of the power thus transferred is 50 percent of the power supplied by the generator.
- Multisim is a useful tool for simulating the behavior of a circuit and examining its sensitivity to specific variables of interest.

Mathematical and Physical Models

Node-voltage method

$\sum$ of all current leaving a node $= 0$
[current entering a node is $(-)$]

Mesh-current method

$\sum$ of all voltages around a loop $= 0$
[passive sign convention applied to mesh currents in clockwise direction]

Nodal analysis by inspection $\mathbf{GV} = \mathbf{I}_t$

Mesh analysis by inspection $\mathbf{RI} = \mathbf{V}_t$

Thévenin equivalent circuit
$\upsilon_{Th} = \upsilon_{oc}$
$R_{Th} = \upsilon_{oc}/i_{sc}$

Norton equivalent circuit
$i_N = i_{sc}$
$R_N = R_{Th}$

Maximum power transfer
$R_L = R_s$
$$p_{max} = \frac{\upsilon_s^2}{4R_L}$$

Important Terms

Provide definitions or explain the meaning of the following terms:

active
additivity property
artificial sources
base
bipolar junction transistor (BJT)
block
bridge circuit
by-inspection method
cell-phone circuit
collector
common collector amplifier
common-emitter amplifier
common-emitter current gain
conductance matrix
current mirror
decoupled
digital inverter
emitter

Important Terms (continued)

extraordinary node
homogeneity
impedance
independent
linear circuit
linear elements
load circuit
load impedance
loading
matching
maximum power transfer
mesh
mesh analysis by inspection
mesh current
nodal analysis by inspection
node-voltage method
Norton's theorem
npn configuration
passive
pnp configuration
quasi-supernode
resistance matrix
scaling
source circuit
source superposition
source vector
supermesh
supernode
superposition principle
Thévenin's theorem
Thévenin's voltage
Thévenin's resistance
uncoupled
voltage vector

PROBLEMS

Section 3-2: Node-Voltage Method

*__3.1__ Apply nodal analysis to find the node voltage V in the circuit of Fig. P3.1. Use the information to determine the current I.

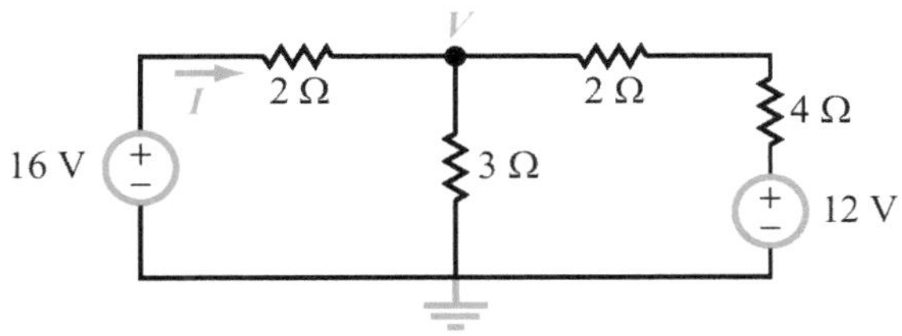

Figure P3.1: Circuit for Problem 3.1.

3.2 Apply nodal analysis to determine V_x in the circuit of Fig. P3.2.

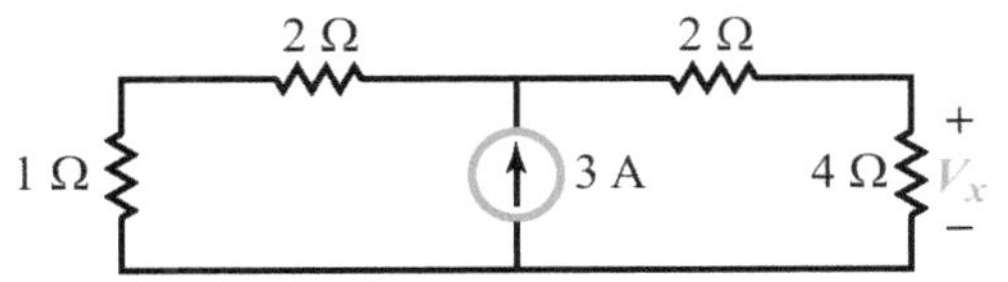

Figure P3.2: Circuit for Problem 3.2.

*__3.3__ Use nodal analysis to determine the current I_x and amount of power supplied by the voltage source in the circuit of Fig. P3.3.

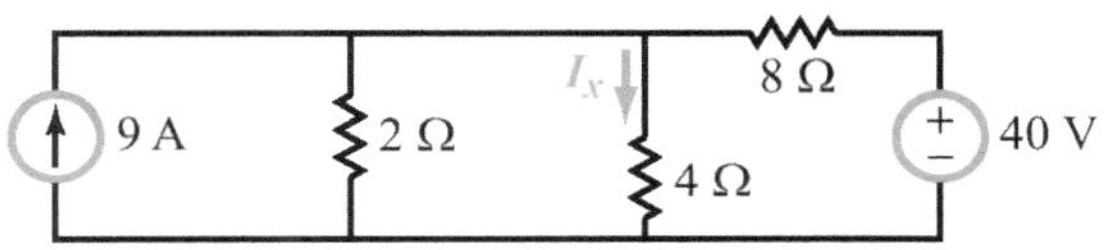

Figure P3.3: Circuit for Problem 3.3.

3.4 For the circuit in Fig. P3.4:

(a) Apply nodal analysis to find node voltages V_1 and V_2.

(b) Determine the voltage V_R and current I.

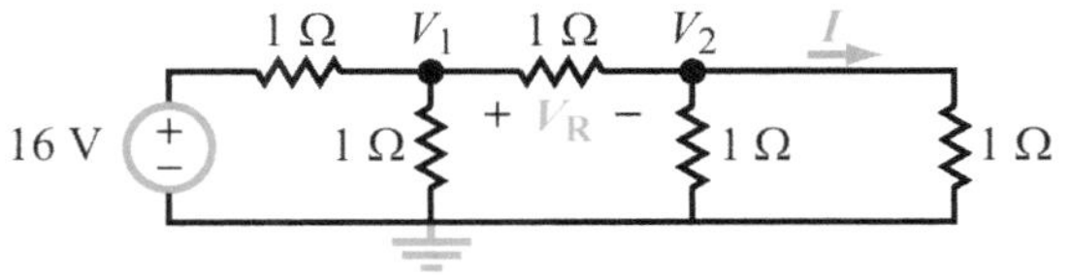

Figure P3.4: Circuit for Problem 3.4.

*__3.5__ Apply nodal analysis to determine the voltage V_R in the circuit of Fig. P3.5.

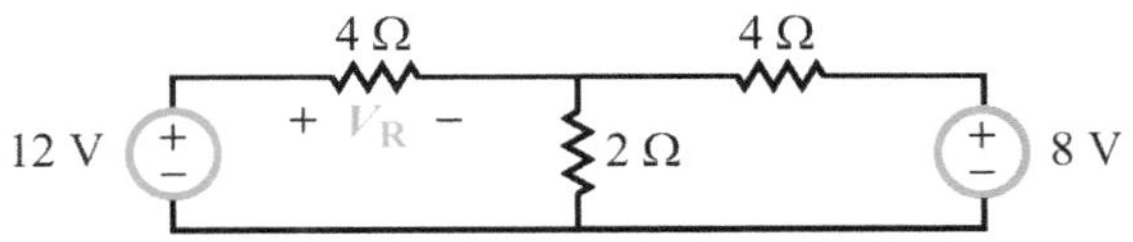

Figure P3.5: Circuit for Problem 3.5.

*Answer(s) available in Appendix G.

3.6 Use the nodal-analysis method to find V_1 and V_2 in the circuit of Fig. P3.6, and then apply that to determine I_x.

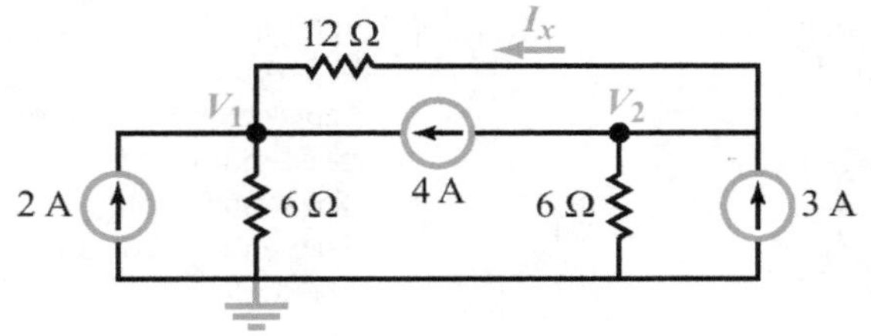

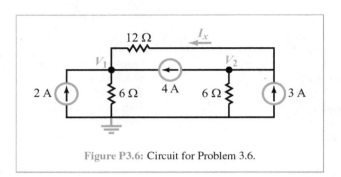

Figure P3.6: Circuit for Problem 3.6.

*3.7 Find I_x in the circuit for Fig. P3.7.

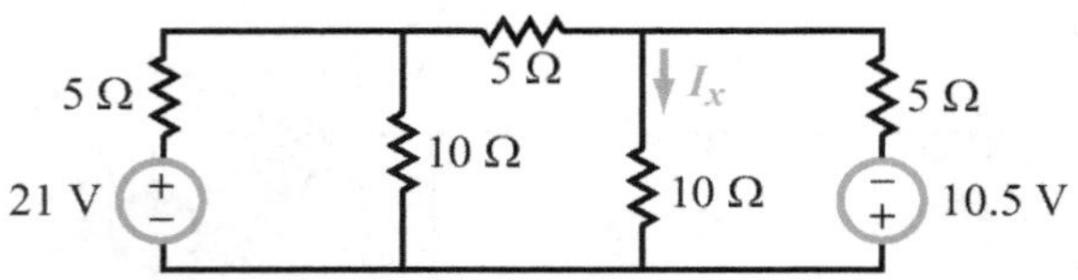

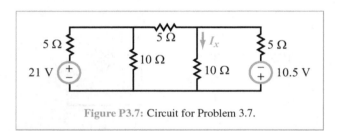

Figure P3.7: Circuit for Problem 3.7.

3.8 For the circuit in Fig. P3.8:

(a) Determine I.

(b) Determine the amount of power supplied by the voltage source.

(c) How much influence does the 4 A source have on the circuit to the left of the 3 A source?

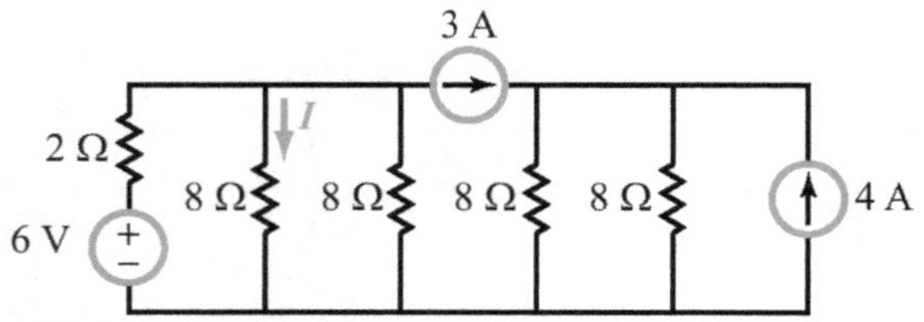

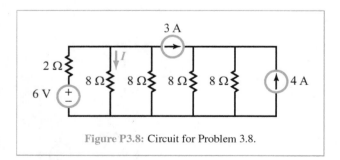

Figure P3.8: Circuit for Problem 3.8.

3.9 Apply nodal analysis to find node voltages V_1 to V_3 in the circuit of Fig. P3.9 and then determine I_x.

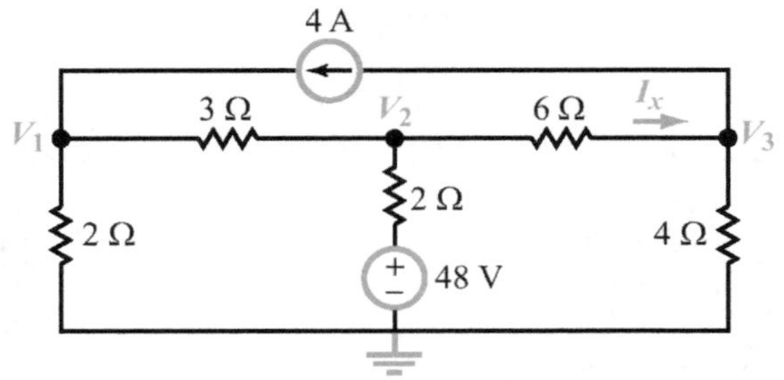

Figure P3.9: Circuit for Problem 3.9.

3.10 The circuit in Fig. P3.10 contains a dependent current source. Determine the voltage V_x.

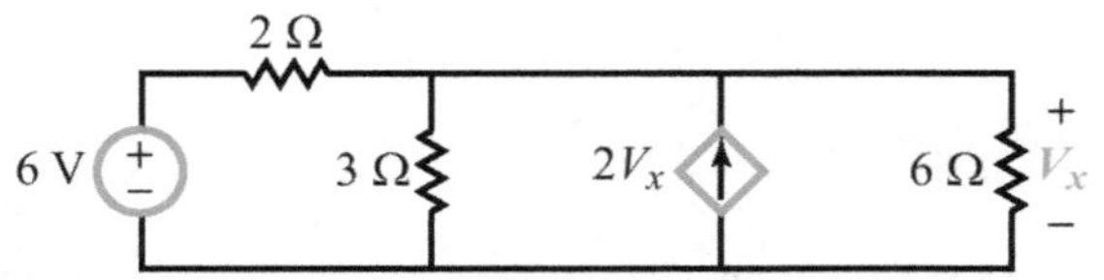

Figure P3.10: Circuit for Problem 3.10.

*3.11 Determine the power supplied by the independent voltage source in the circuit of Fig. P3.11.

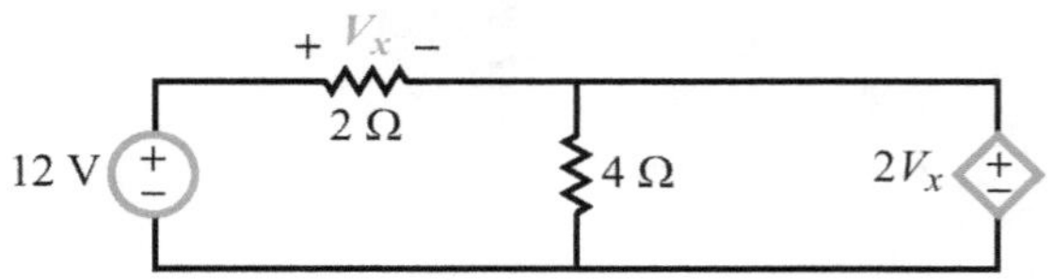

Figure P3.11: Circuit for Problem 3.11.

3.12 The magnitude of the dependent current source in the circuit of Fig. P3.12 depends on the current I_x flowing through the 10 Ω resistor. Determine I_x.

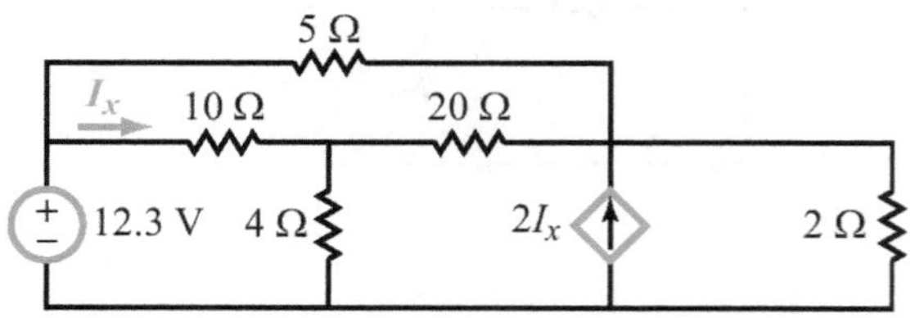

Figure P3.12: Circuit for Problems 3.12 and 3.13.

*3.13 Repeat Problem 3.12 after replacing the 5 Ω resistor in Fig. P3.12 with a short circuit.

3.14 Apply nodal analysis to find the current I_x in the circuit of Fig. P3.14.

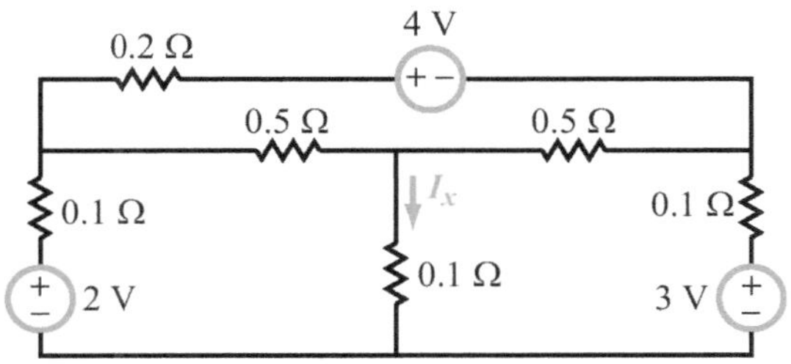

Figure P3.14: Circuit for Problem 3.14.

*3.15 Use the supernode concept to find the current I_x in the circuit of Fig. P3.15.

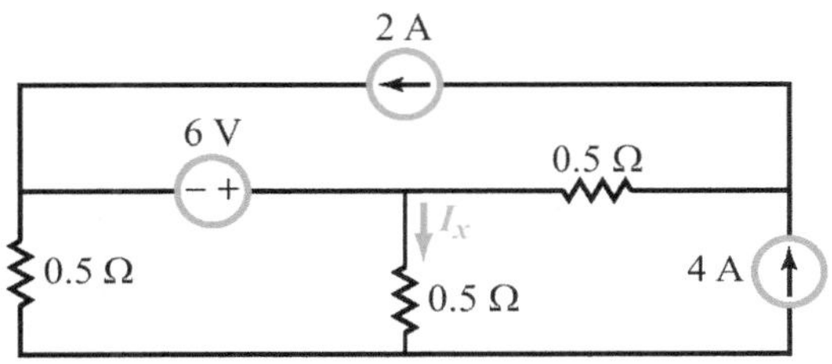

Figure P3.15: Circuit for Problem 3.15.

3.16 Apply the supernode technique to determine V_x in the circuit of Fig. P3.16.

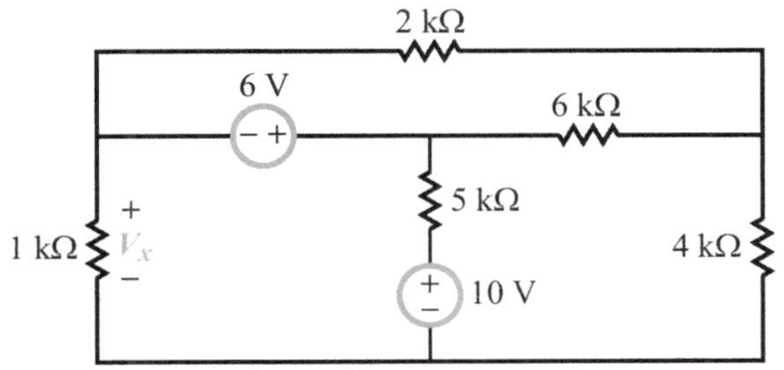

Figure P3.16: Circuit for Problem 3.16.

*3.17 Determine V_x in the circuit of Fig. P3.17.

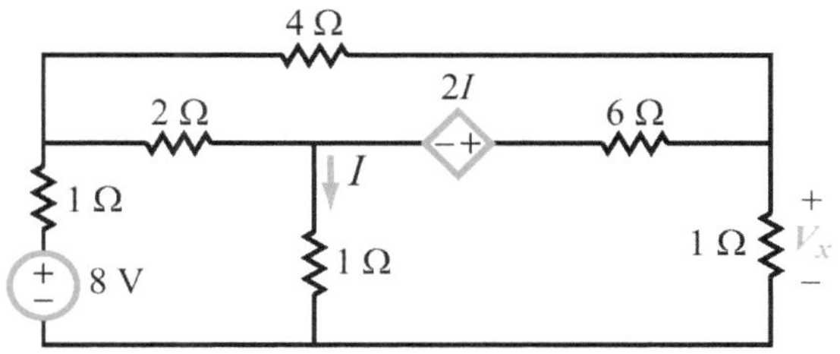

Figure P3.17: Circuit for Problems 3.17 and 3.18.

3.18 Repeat Problem 3.17 after replacing the 2 Ω resistor in Fig. P3.17 with a short circuit.

3.19 For the circuit shown in Fig. P3.19:

(a) Determine R_{eq} between terminals (a, b).

(b) Determine the current I using the result of (a).

(c) Apply nodal analysis to the original circuit to determine the node voltages and then use them to determine I. Compare the result with the answer of part (b).

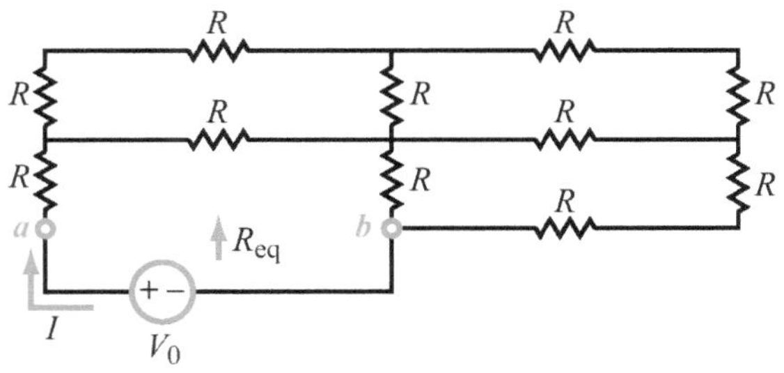

Figure P3.19: Circuit for Problem 3.19.

*3.20 For the circuit in Fig. P3.20, determine the current I_x.

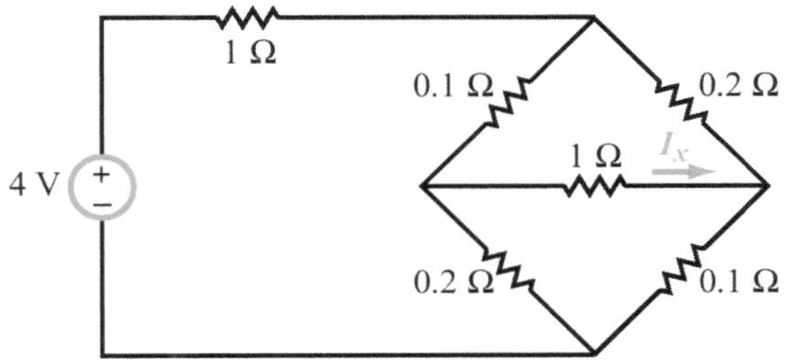

Figure P3.20: Circuit for Problem 3.20.

3.21 Apply nodal analysis to determine V_x in the circuit of Fig. P3.21.

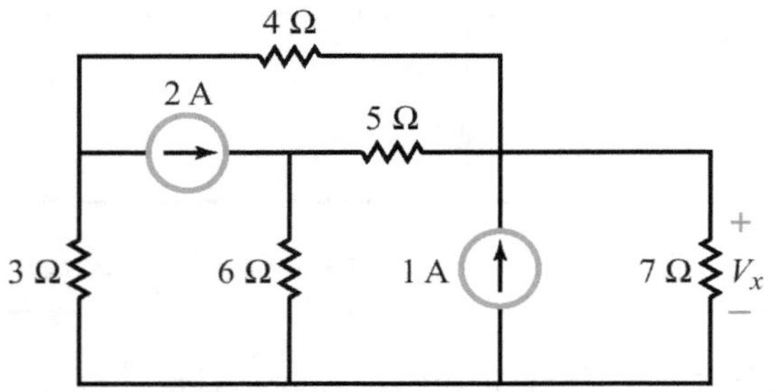

Figure P3.21: Circuit for Problem 3.21.

3.22 Apply nodal analysis to determine V_L in the circuit of Fig. P3.22.

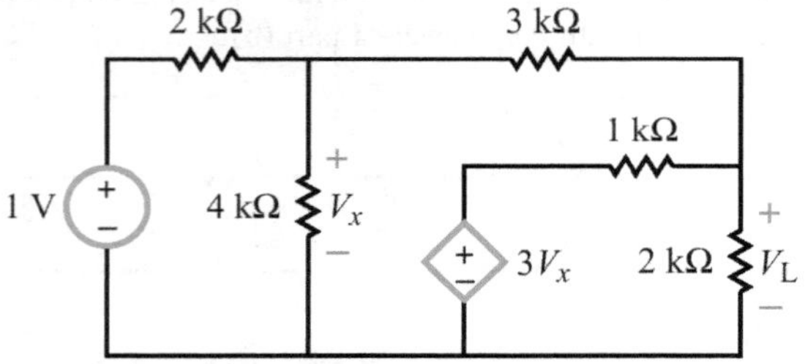

Figure P3.22: Circuit for Problem 3.22.

*__3.23__ Apply nodal analysis to determine V_x in the circuit of Fig. P3.23.

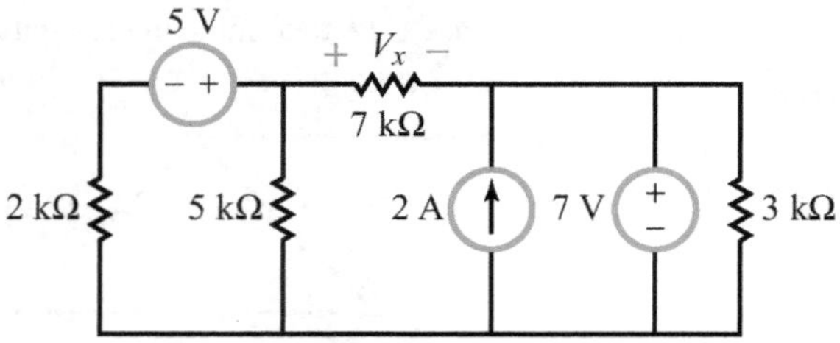

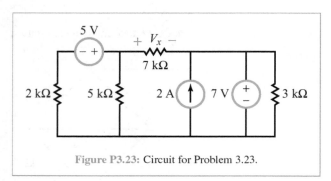

Figure P3.23: Circuit for Problem 3.23.

3.24 Apply nodal analysis to determine V_x in the circuit of Fig. P3.24.

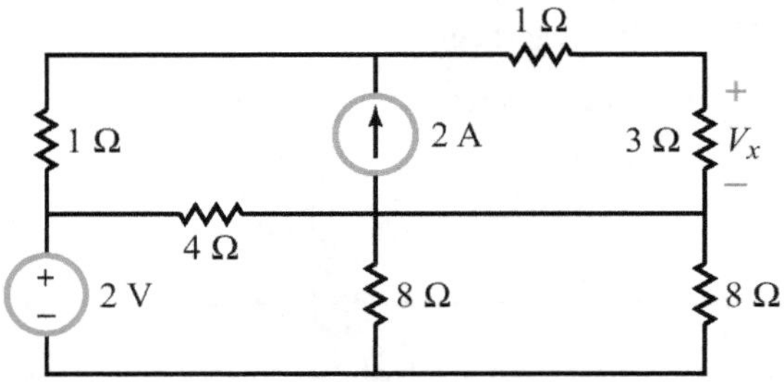

Figure P3.24: Circuit for Problem 3.24.

3.25 Apply nodal analysis to determine V_a, V_b, and V_c in the circuit of Fig. P3.25.

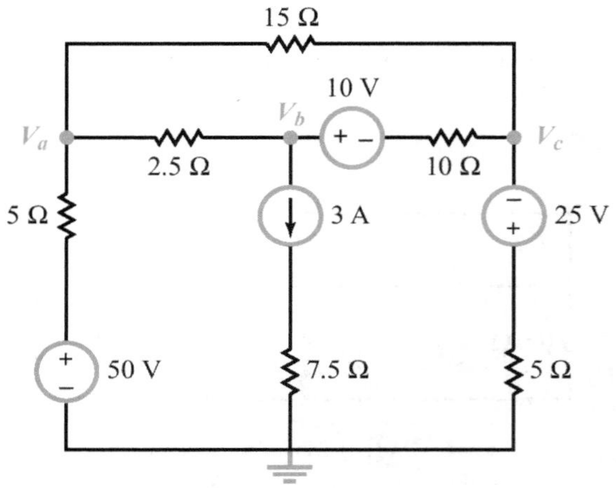

Figure P3.25: Circuit for Problem 3.25.

Section 3-3: Mesh-Current Method

*__3.26__ Apply mesh analysis to find the mesh currents in the circuit of Fig. P3.26. Use the information to determine the voltage V.

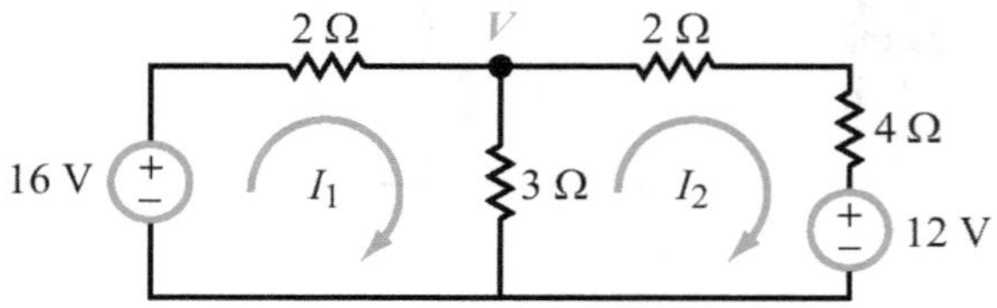

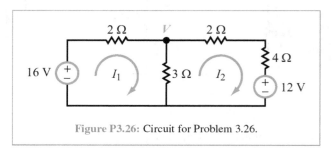

Figure P3.26: Circuit for Problem 3.26.

3.27 Use mesh analysis to determine the amount of power supplied by the voltage source in the circuit of Fig. P3.27.

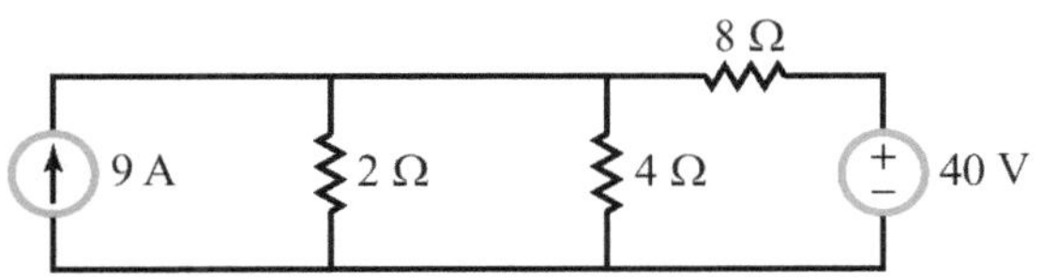

Figure P3.27: Circuit for Problem 3.27.

*__3.28__ Determine V in the circuit of Fig. P3.28 using mesh analysis.

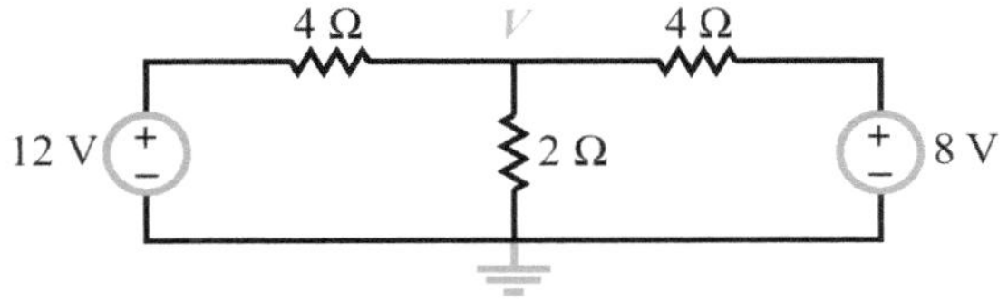

Figure P3.28: Circuit for Problem 3.28.

3.29 Apply mesh analysis to find I in the circuit of Fig. P3.29.

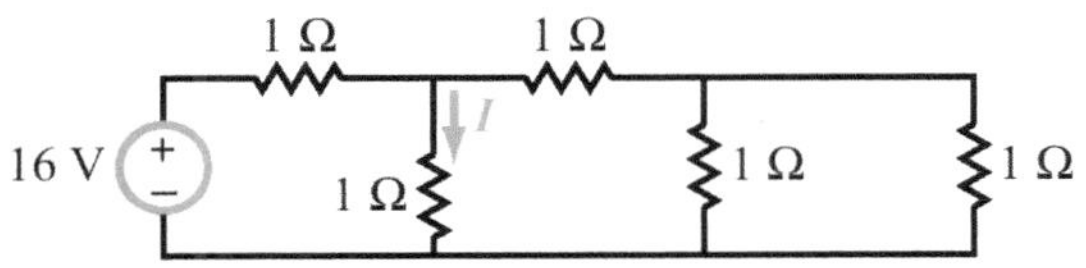

Figure P3.29: Circuit for Problem 3.29.

*__3.30__ Apply mesh analysis to find I_x in the circuit of Fig. P3.30.

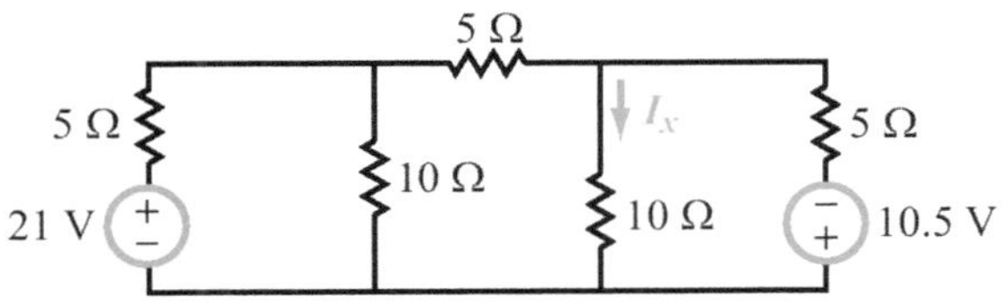

Figure P3.30: Circuit for Problem 3.30.

3.31 Apply mesh analysis to determine the amount of power supplied by the voltage source in Fig. P3.31.

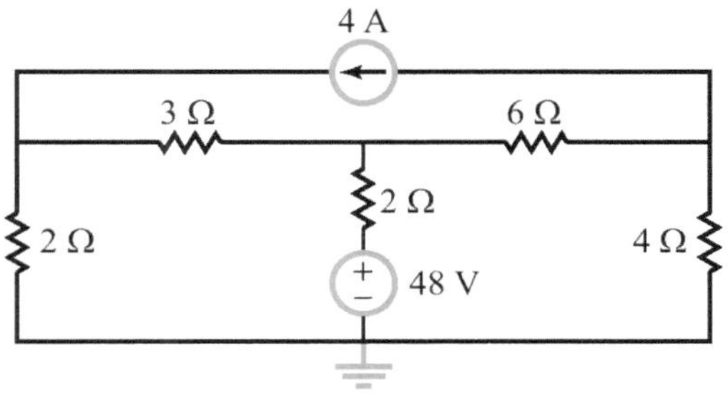

Figure P3.31: Circuit for Problem 3.31.

*__3.32__ Use the supermesh concept to solve for V_x in the circuit of Fig. P3.32.

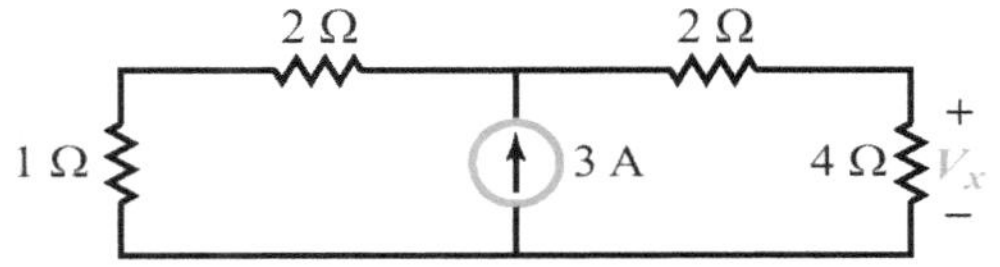

Figure P3.32: Circuit for Problem 3.32.

3.33 Use the supermesh concept to solve for I_x in the circuit of Fig. P3.33.

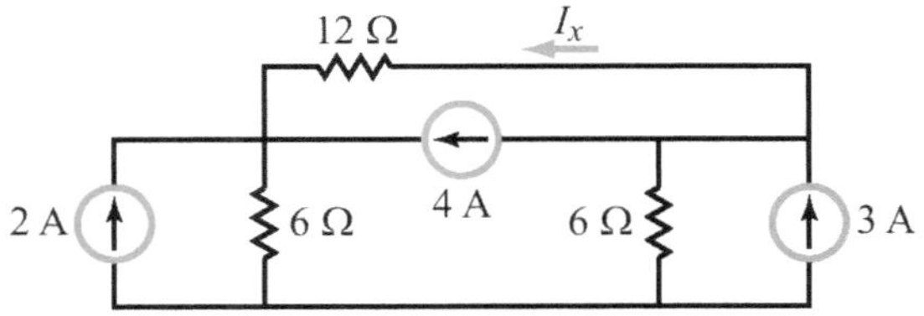

Figure P3.33: Circuit for Problem 3.33.

3.34 Apply mesh analysis to the circuit in Fig. P3.34 to determine V_x.

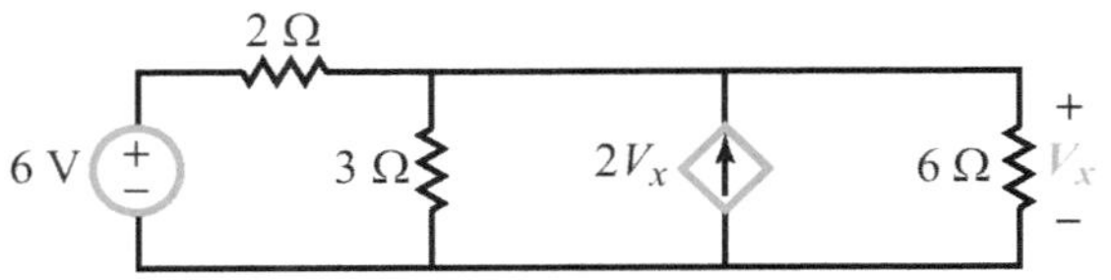

Figure P3.34: Circuit for Problem 3.34.

3.35 Determine the amount of power supplied by the independent voltage source in Fig. P3.35 by applying the mesh-analysis method.

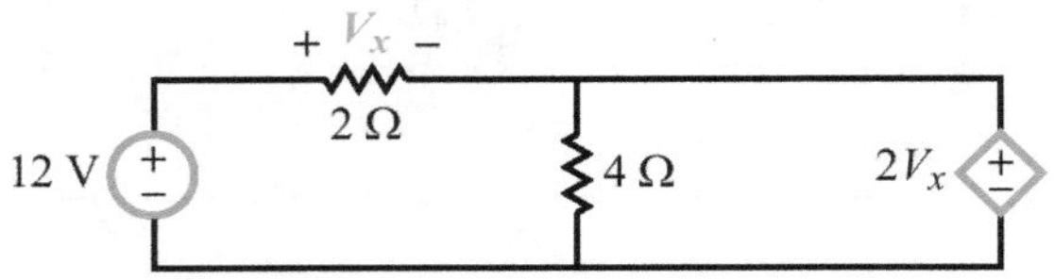

Figure P3.35: Circuit for Problem 3.35.

*__3.36__ Use mesh analysis to find I_x in the circuit of Fig. P3.36.

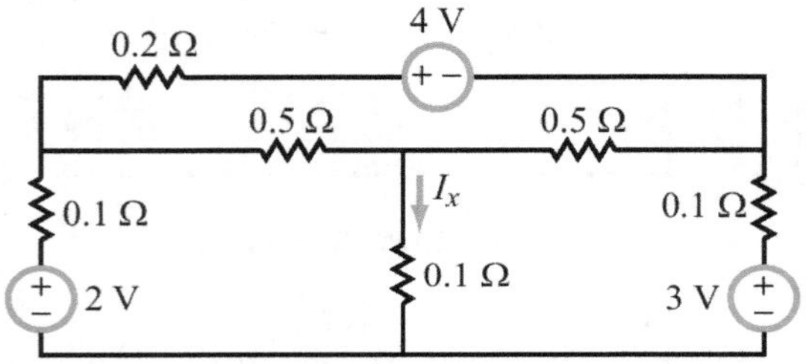

Figure P3.36: Circuit for Problem 3.36.

3.37 The circuit in Fig. P3.37 includes a dependent current source. Apply mesh analysis to determine I_x.

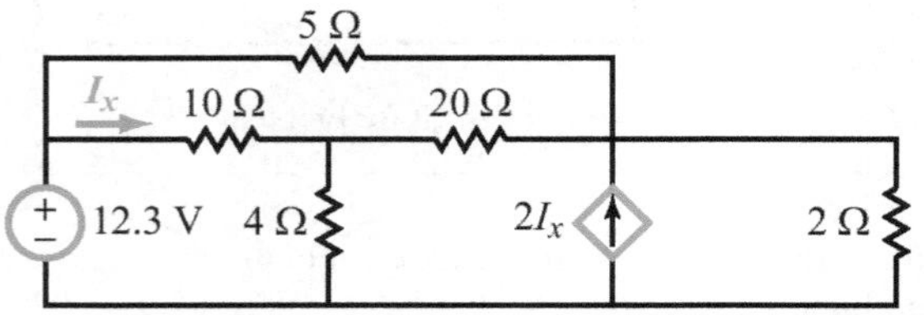

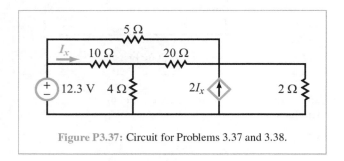

Figure P3.37: Circuit for Problems 3.37 and 3.38.

3.38 Repeat Problem 3.37 after replacing the 5 Ω resistor in Fig. P3.37 with a short circuit.

*__3.39__ Apply mesh analysis to the circuit of Fig. P3.39 to determine I_x.

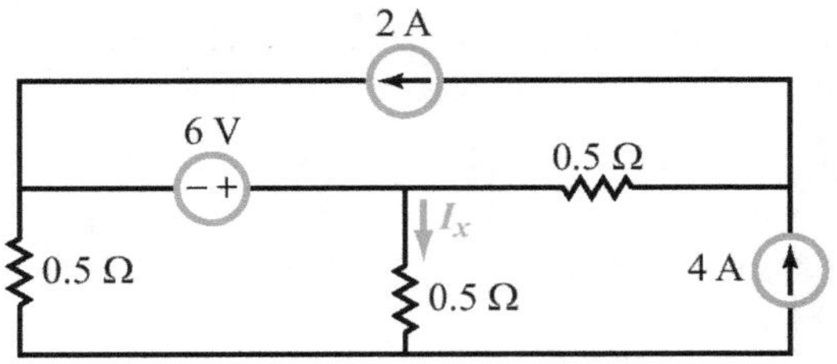

Figure P3.39: Circuit for Problem 3.39.

3.40 Determine V_x in the circuit of Fig. P3.40.

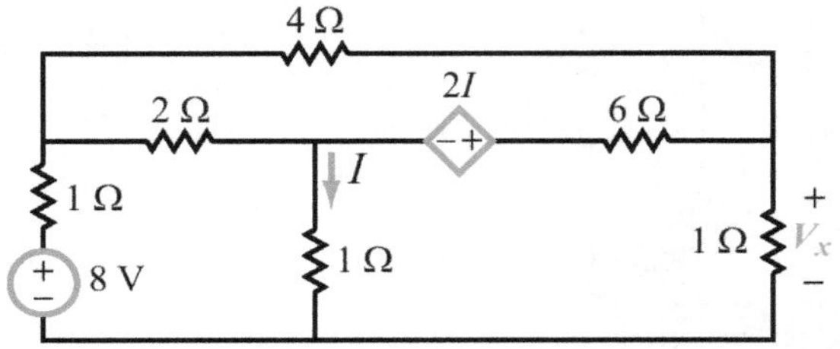

Figure P3.40: Circuit for Problems 3.40 and 3.42.

3.41 Apply the supermesh technique to find V_x in the circuit of Fig. P3.41.

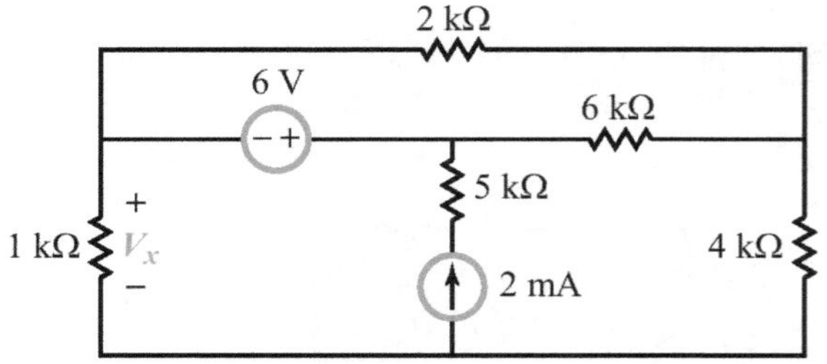

Figure P3.41: Circuit for Problem 3.41.

*__3.42__ Repeat Problem 3.40 after replacing the 2 Ω resistor in Fig. P3.40 with a short circuit.

3.43 Apply mesh analysis to the circuit of Fig. P3.43 to find I_x.

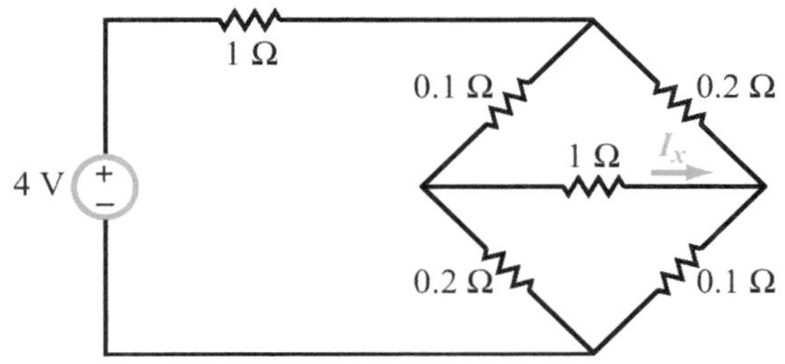

Figure P3.43: Circuit for Problem 3.43.

3.44 Determine I_0 in Fig. P3.44 through mesh analysis.

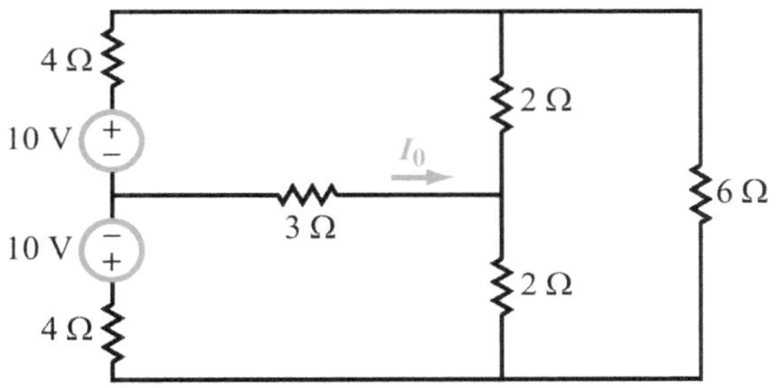

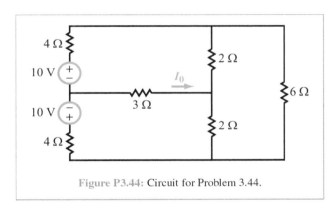

Figure P3.44: Circuit for Problem 3.44.

*__3.45__ Use an analysis method of your choice to determine I_0 in the circuit of Fig. P3.45.

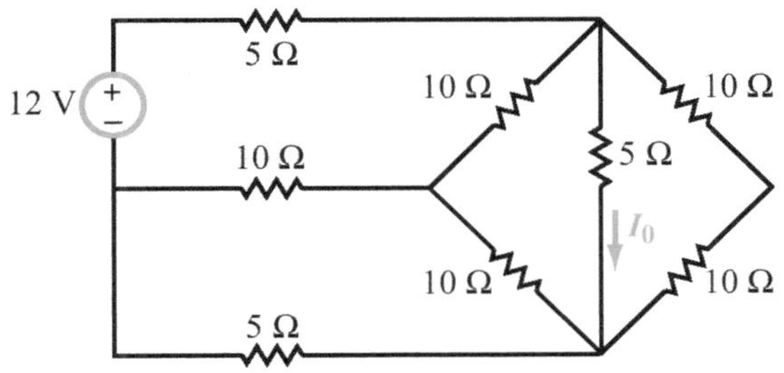

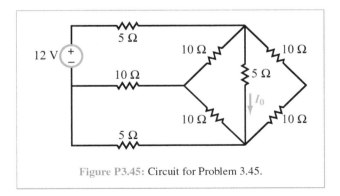

Figure P3.45: Circuit for Problem 3.45.

3.46 Simplify the circuit in Fig. P3.46 as much as possible using source transformation and resistance combining, and then apply mesh analysis to determine I_x.

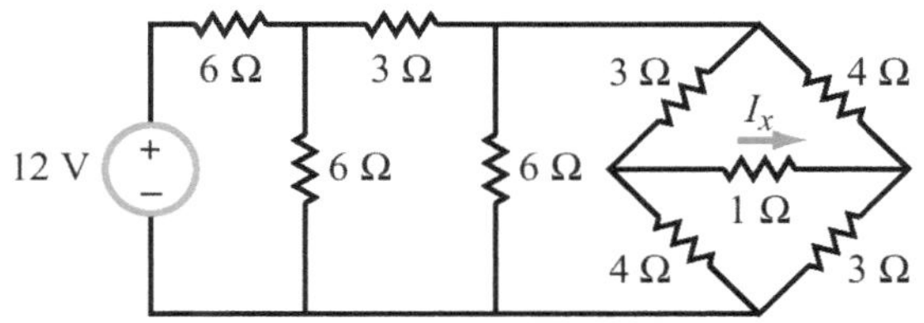

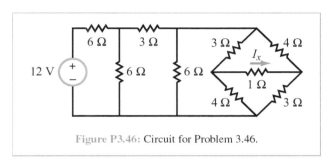

Figure P3.46: Circuit for Problem 3.46.

3.47 Apply mesh analysis to determine I_0 in the circuit in Fig. P3.47.

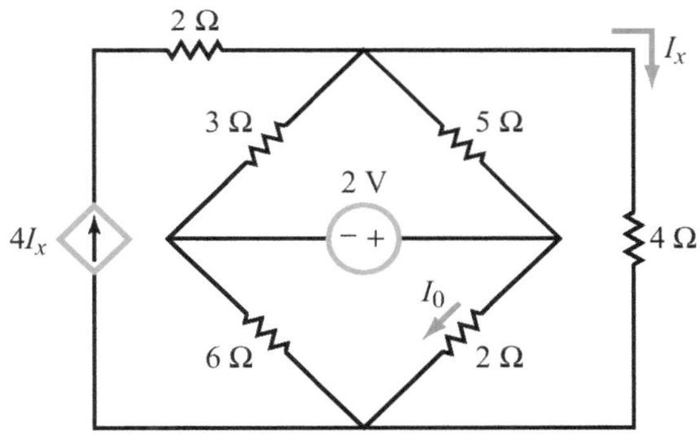

Figure P3.47: Circuit for Problem 3.47.

*__3.48__ Apply mesh analysis to determine I_x in the circuit in Fig. P3.48.

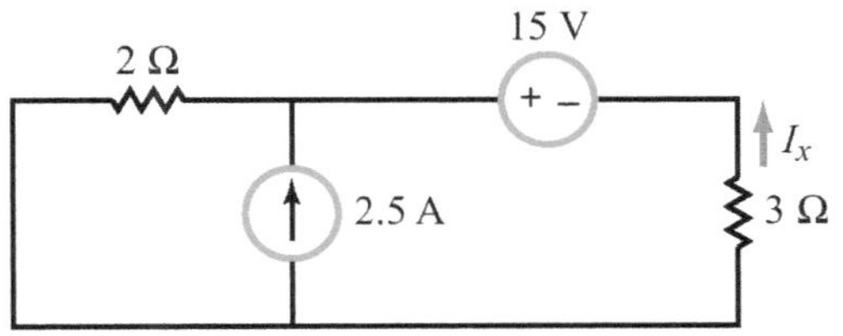

Figure P3.48: Circuit for Problem 3.48.

3.49 Apply mesh analysis to determine I_x in the circuit in Fig. P3.49.

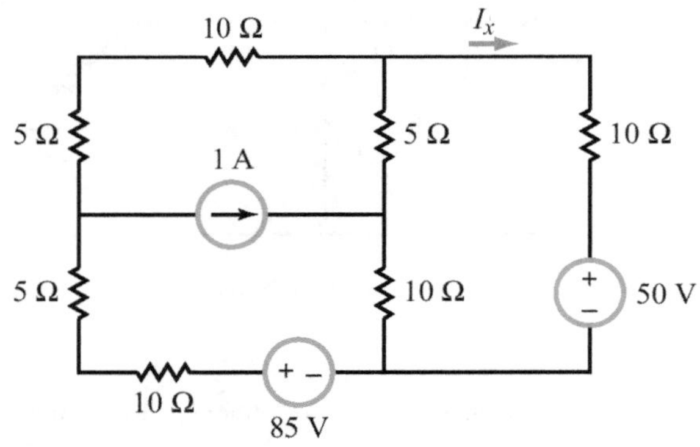

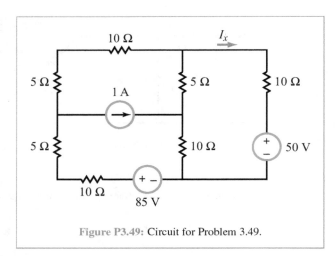

Figure P3.49: Circuit for Problem 3.49.

3.50 Apply mesh analysis to determine V_x in the circuit in Fig. P3.50.

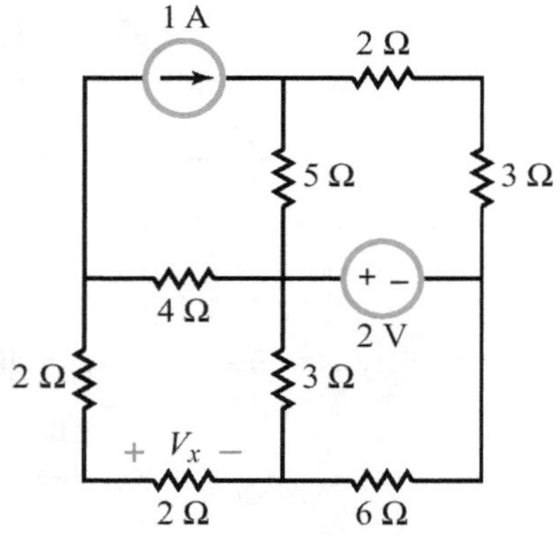

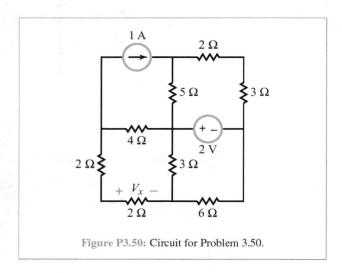

Figure P3.50: Circuit for Problem 3.50.

3.51 Consider the circuit shown in Fig. P3.51.

(a) How many extraordinary nodes does it have?

(b) How many independent meshes does it have?

(c) The values of how many of those mesh currents can be determined immediately from the circuit?

(d) Apply mesh analysis to find I'.

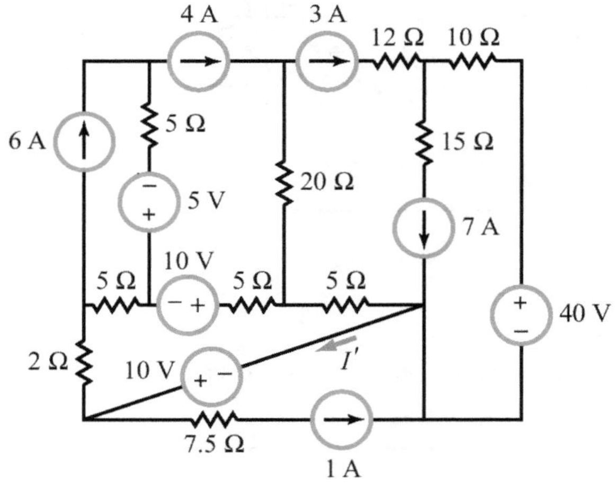

Figure P3.51: Circuit for Problem 3.51.

Sections 3-4 and 3-5: By-Inspection and Superposition Methods

*3.52 Apply the by-inspection method to develop a node-voltage matrix equation for the circuit in Fig. P3.52 and then use MATLAB or MathScript software to solve for V_1 and V_2.

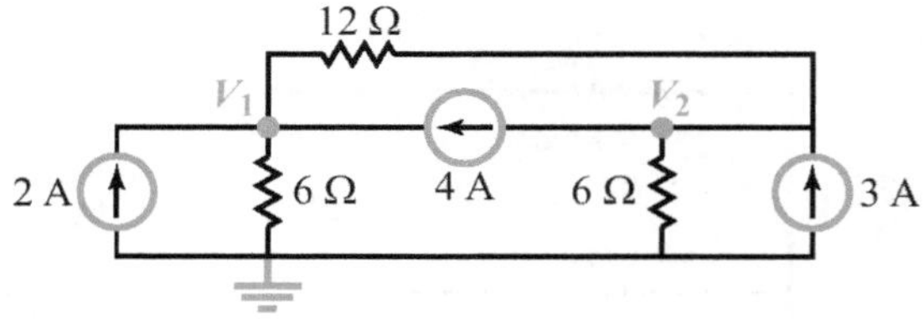

Figure P3.52: Circuit for Problem 3.52.

3.53 Use the by-inspection method to establish a node-voltage matrix equation for the circuit in Fig. P3.53. Solve the matrix equation by MATLAB or MathScript software to find V_1 to V_4.

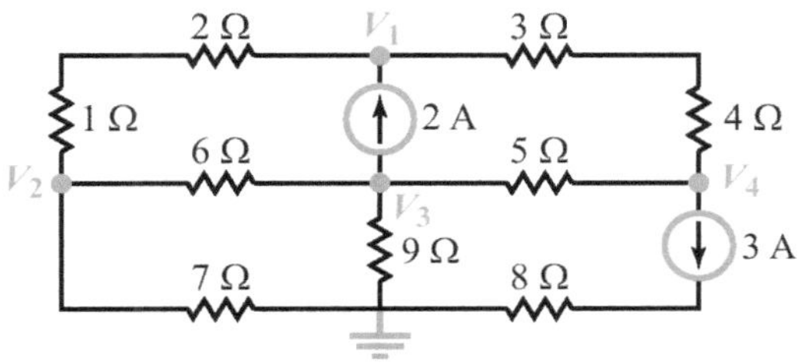

Figure P3.53: Circuit for Problem 3.53.

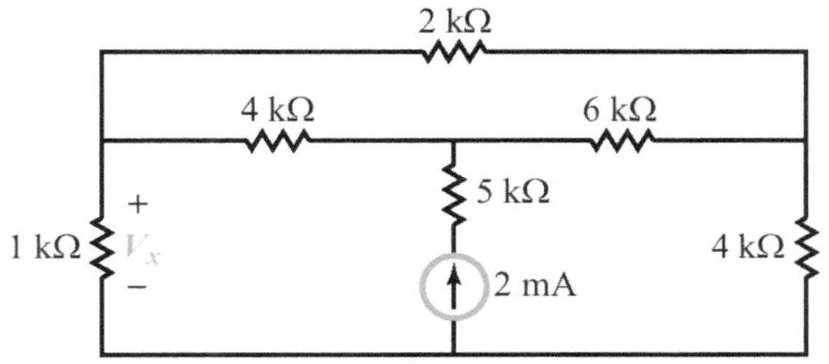

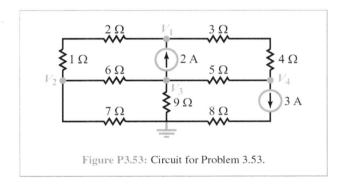

Figure P3.56: Circuit for Problem 3.56.

3.54 Develop a mesh-current matrix equation for the circuit in Fig. P3.54 by applying the by-inspection method. Solve for I_1 to I_3.

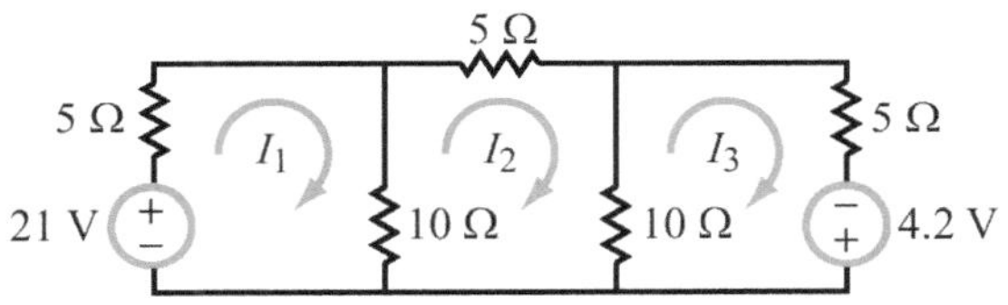

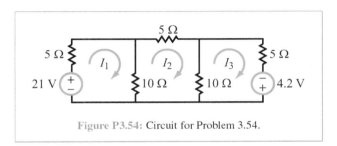

Figure P3.54: Circuit for Problem 3.54.

3.55 Find I_0 in the circuit of Fig. P3.55 by developing a mesh-current matrix equation and then solving it using MATLAB or MathScript software.

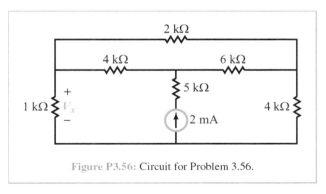

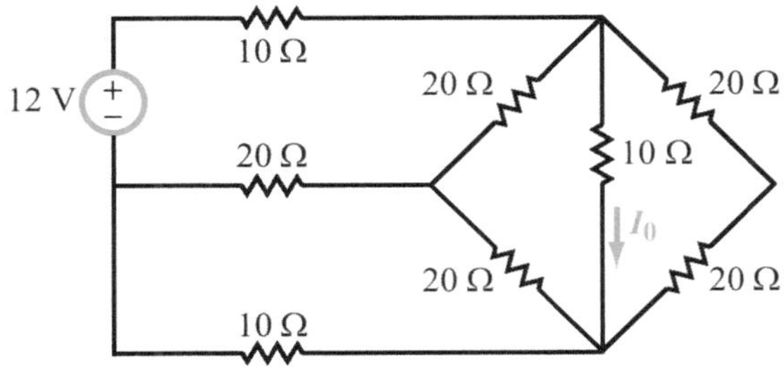

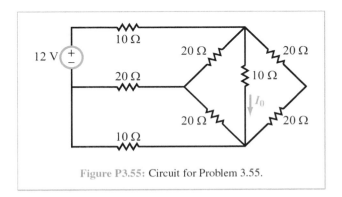

Figure P3.55: Circuit for Problem 3.55.

*3.56 Apply the by-inspection method to derive a node-voltage matrix equation for the circuit in Fig. P3.56 and then solve it using MATLAB or MathScript software to find V_x.

3.57 Use the by-inspection method to establish the mesh-current matrix equation for the circuit in Fig. P3.57 and then solve the equation to determine V_{out}.

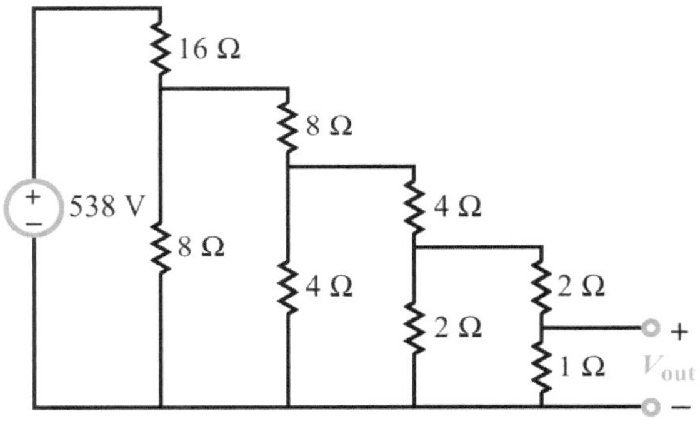

Figure P3.57: Circuit for Problem 3.57.

*3.58 Develop a node-voltage matrix equation for the circuit in Fig. P3.58. Solve it to determine I.

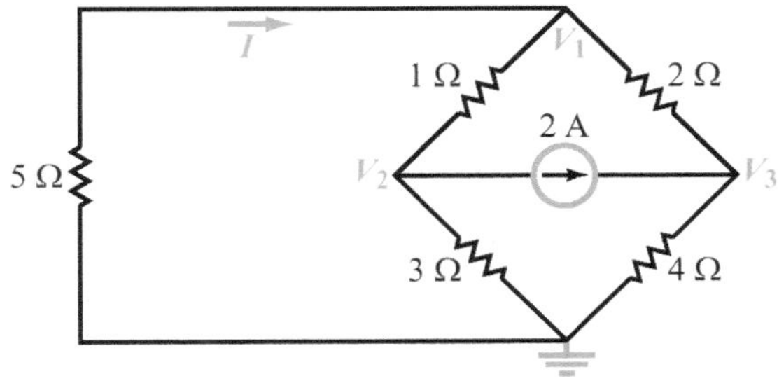

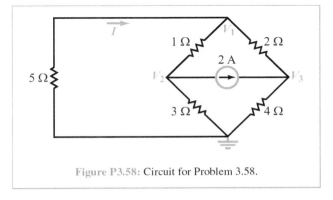

Figure P3.58: Circuit for Problem 3.58.

3.59 Determine the amount of power supplied by the voltage source in Fig. P3.59 by establishing and then solving the mesh-current matrix equation of the circuit.

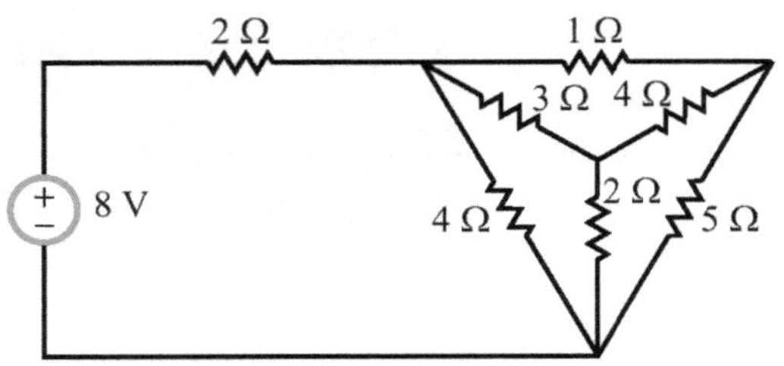

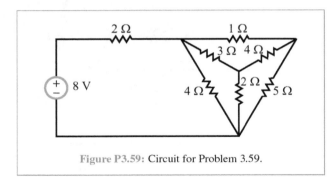

Figure P3.59: Circuit for Problem 3.59.

3.60 Determine the current I_x in the circuit of Fig. P3.60 by applying the source-superposition method. Call I'_x the component of I_x due to the voltage source alone, and I''_x the component due to the current source alone. Show that $I_x = I'_x + I''_x$ is the same as the answer to Problem 3.9.

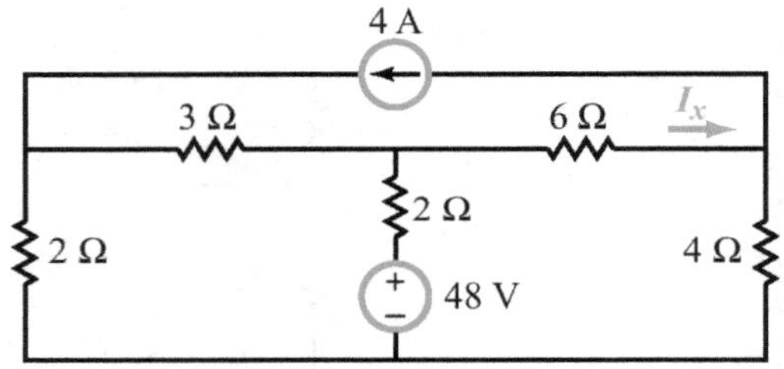

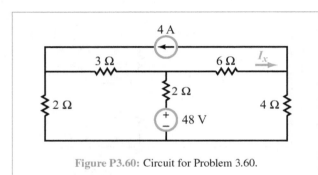

Figure P3.60: Circuit for Problem 3.60.

3.61 Apply the source-superposition method to the circuit in Fig. P3.61 to determine:

(a) I'_x, the component of I_x due to the voltage source alone

(b) I''_x, the component of I_x due to the current source alone

(c) The total current $I_x = I'_x + I''_x$

(d) P', the power dissipated in the 4 Ω resistor due to I'_x

(e) P'', the power dissipated in the 4 Ω resistor due to I''_x

(f) P, the power dissipated in the 4 Ω resistor due to the total current I. Is $P = P' + P''$? If not, why not?

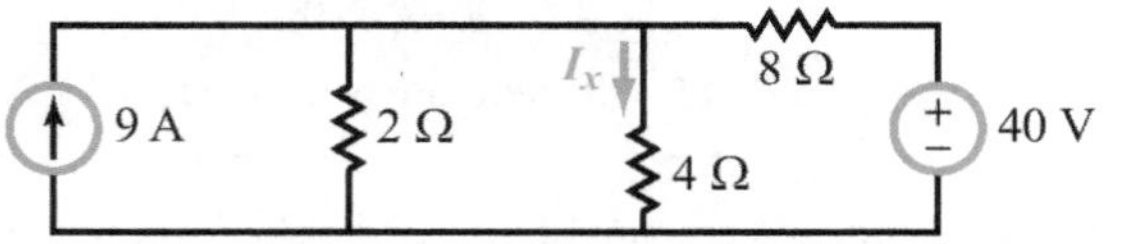

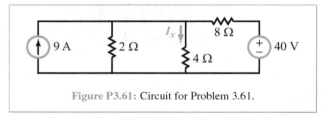

Figure P3.61: Circuit for Problem 3.61.

*__3.62__ Perform necessary source transformations and then use the mesh analysis by-inspection method to determine V_x in the circuit of Fig. P3.62.

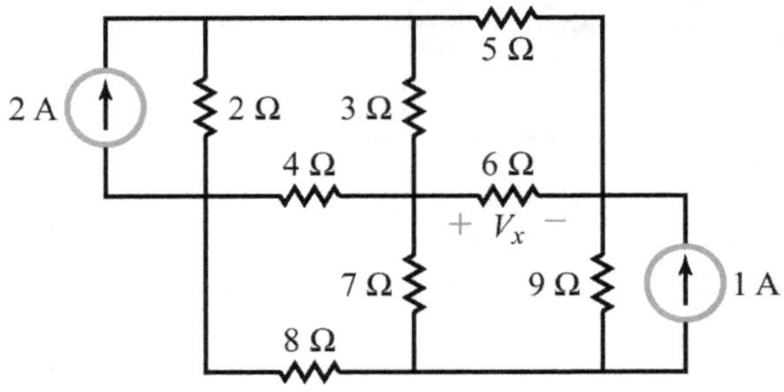

Figure P3.62: Circuit for Problem 3.62.

3.63 Apply the source-superposition method to the circuit in Fig. P3.63 to determine:

(a) V'_x, the component of V_x due to the 1 A current source alone.

(b) V''_x, the component of V_x due to the 10 V voltage source alone.

(c) V'''_x, the component of V_x due to the 3 A current source alone.

(d) The total voltage $V_x = V'_x + V''_x + V'''_x$.

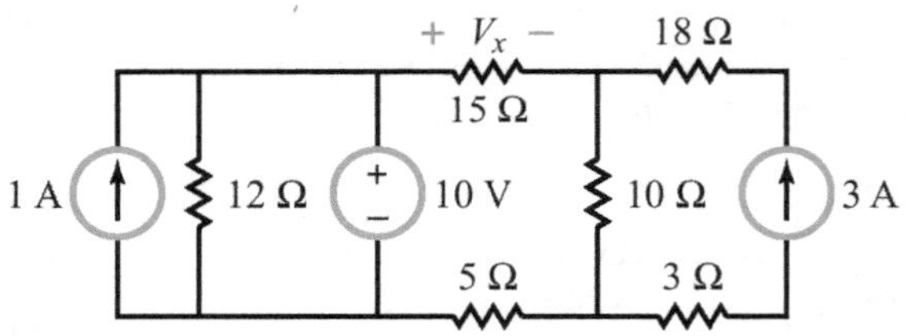

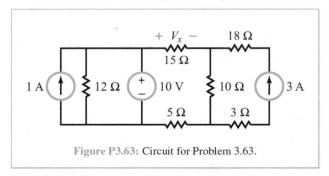

Figure P3.63: Circuit for Problem 3.63.

Section 3-6: Thévenin and Norton Equivalents

*__3.64__ Find the Thévenin equivalent circuit at terminals (a, b) for the circuit in Fig. P3.64.

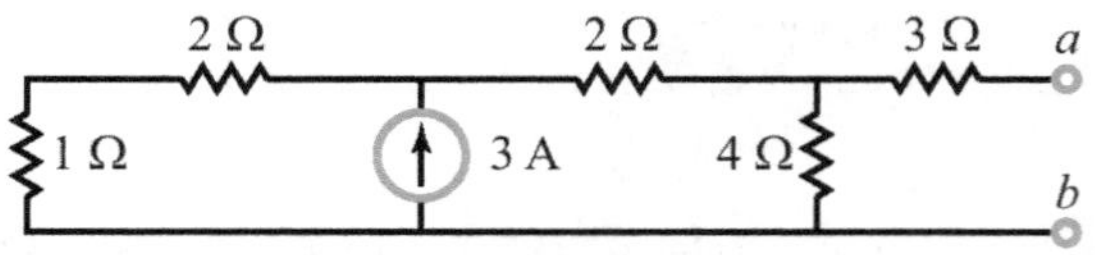

Figure P3.64: Circuit for Problem 3.64.

3.65 Find the Thévenin equivalent circuit at terminals (a, b) for the circuit in Fig. P3.65.

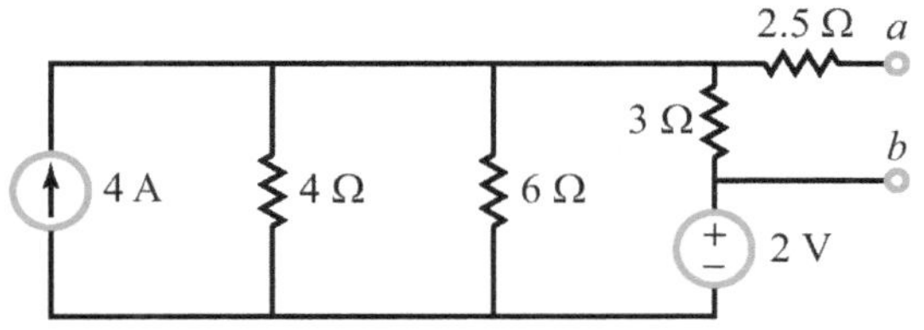

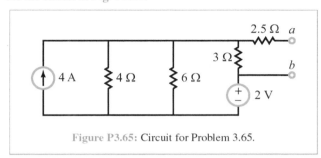

Figure P3.65: Circuit for Problem 3.65.

3.66 The circuit in Fig. P3.66 is to be connected to a load resistor R_L between terminals (a, b).

(a) Find the Thévenin equivalent circuit at terminals (a, b).

(b) Choose R_L so that the current flowing through it is 0.5 A.

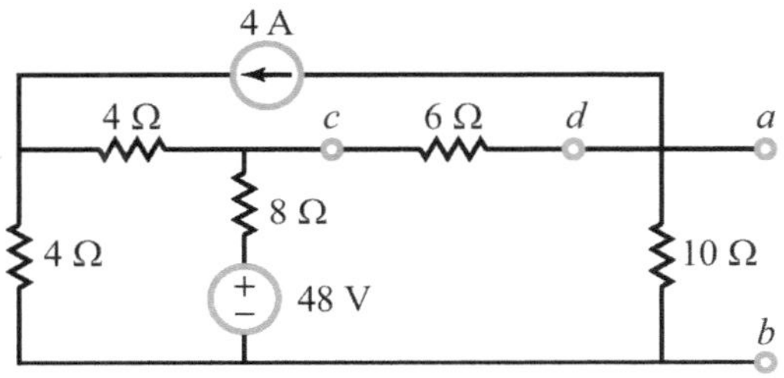

Figure P3.66: Circuit for Problems 3.66 and 3.67.

3.67 For the circuit in Fig. P3.66, find the Thévenin equivalent circuit as seen by the 6 Ω resistor connected between terminals (c, d) as if the 6 Ω resistor is a load resistor connected to (but external to) the circuit. Determine the current flowing through that resistor.

*__3.68__ Find the Thévenin equivalent circuit at terminals (a, b) for the circuit in Fig. P3.68.

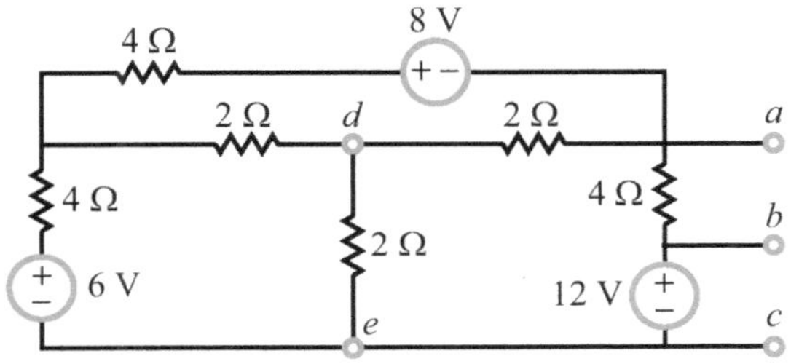

Figure P3.68: Circuit for Problems 3.68 through 3.70.

3.69 Repeat Problem 3.68 for terminals (a, c).

3.70 Repeat Problem 3.68 for terminals (d, e) as seen by the 2 Ω resistor between them (as if it were a load resistor external to the circuit).

3.71 Find the Thévenin equivalent circuit at terminals (a, b) of the circuit in Fig. P3.71.

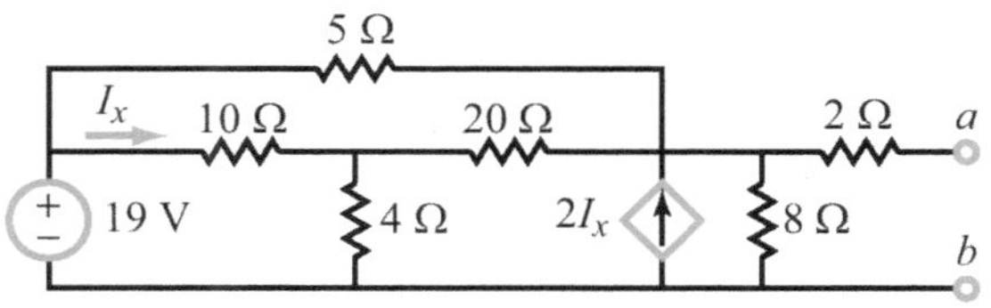

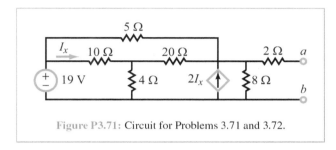

Figure P3.71: Circuit for Problems 3.71 and 3.72.

*__3.72__ Find the Norton equivalent circuit of the circuit in Fig. P3.71 after increasing the magnitude of the voltage source to 38 V.

3.73 Find the Norton equivalent circuit at terminals (a, b) for the circuit in Fig. P3.73.

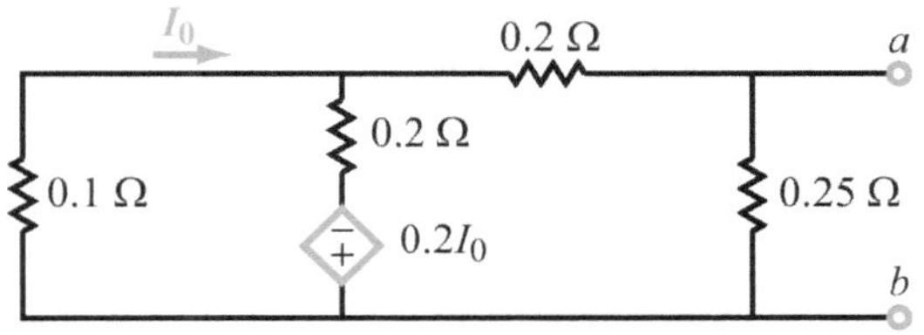

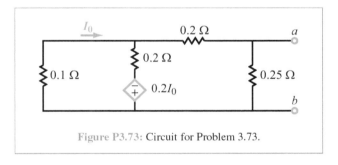

Figure P3.73: Circuit for Problem 3.73.

*__3.74__ Find the Norton equivalent circuit at terminals (a, b) of the circuit in Fig. P3.74.

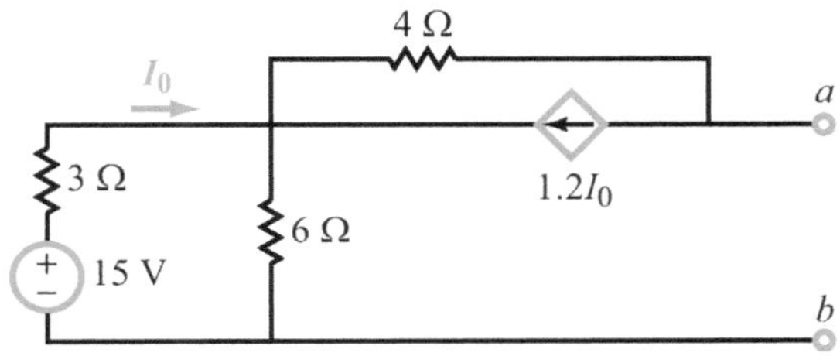

Figure P3.74: Circuit for Problems 3.74 and 3.75.

3.75 Repeat Problem 3.74 after replacing the 6 Ω resistor with an open circuit.

3.76 Find the Norton equivalent circuit at terminals (a, b) of the circuit in Fig. P3.76.

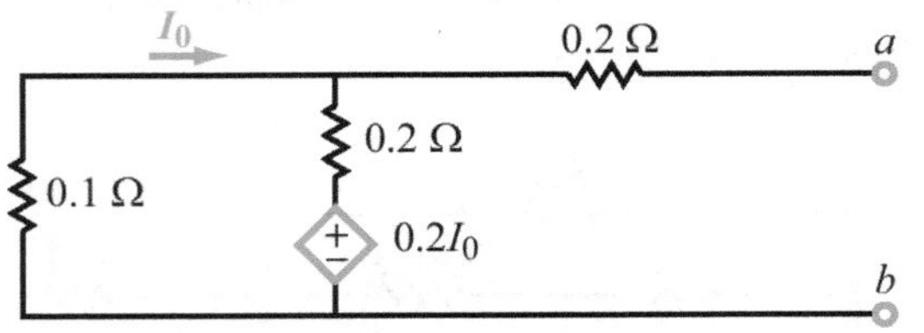

Figure P3.76: Circuit for Problems 3.76.

*__3.77__ Obtain the Thévenin equivalent circuit at terminals (a, b) in Fig. P3.77.

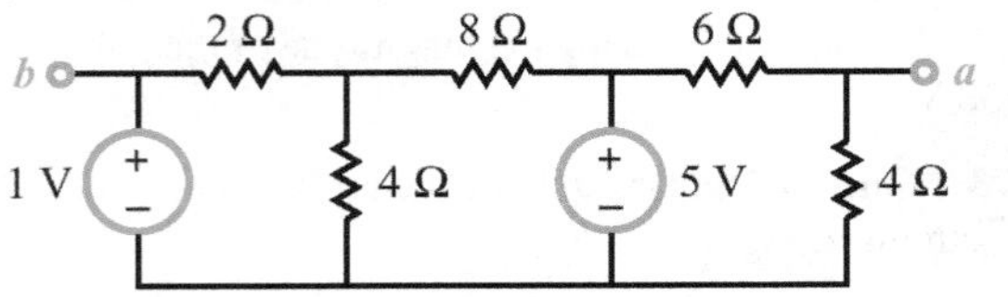

Figure P3.77: Circuit for Problem 3.77.

3.78 Obtain the Thévenin equivalent of the circuit to the left of terminals (a, b) in Fig. P3.78. Use your result to compute the power dissipated in the 0.4 Ω load resistor.

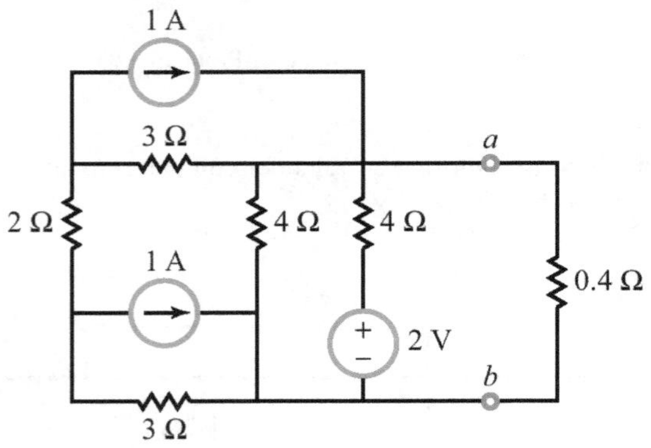

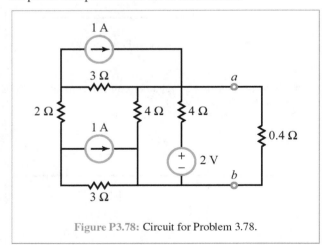

Figure P3.78: Circuit for Problem 3.78.

3.79 Obtain the Thévenin equivalent of the circuit in Fig. P3.79 at terminals (a, b).

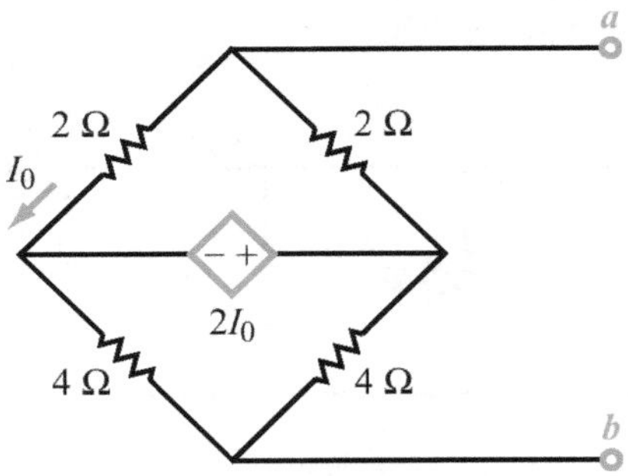

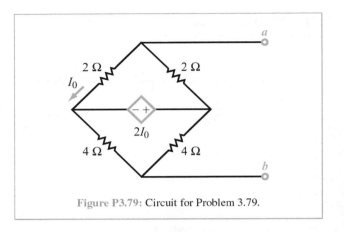

Figure P3.79: Circuit for Problem 3.79.

*__3.80__ Obtain the Thévenin equivalent of the circuit in Fig. P3.80 at terminals (a, b).

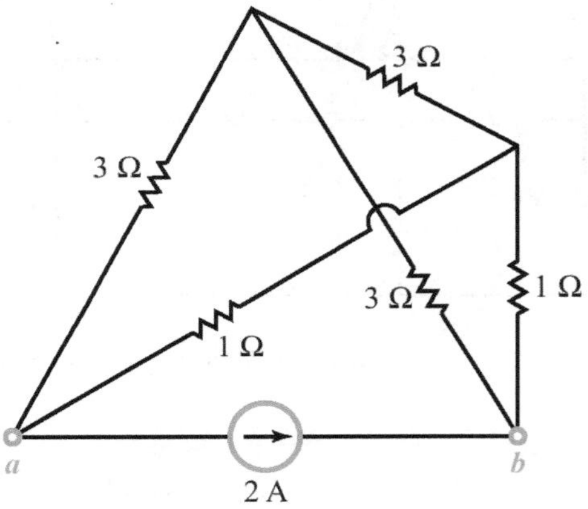

Figure P3.80: Circuit for Problem 3.80.

Section 3-8: Maximum Power Transfer

3.81 What value of the load resistor R_L will extract the maximum amount of power from the circuit in Fig. P3.81, and how much power will that be?

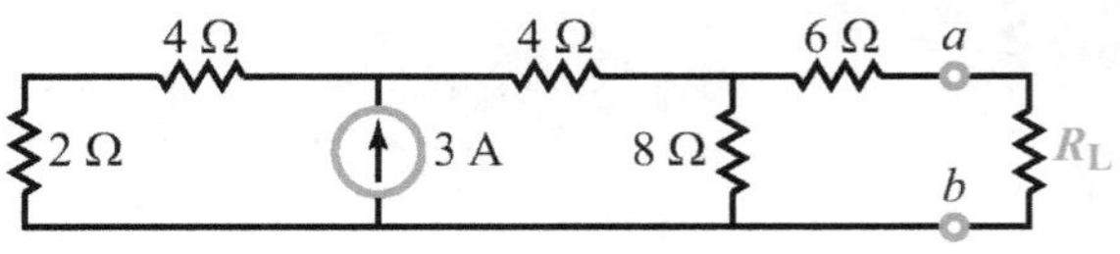

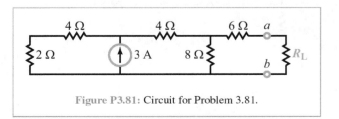

Figure P3.81: Circuit for Problem 3.81.

3.82 For the circuit in Fig. P3.82, choose the value of R_L so that the power dissipated in it is a maximum.

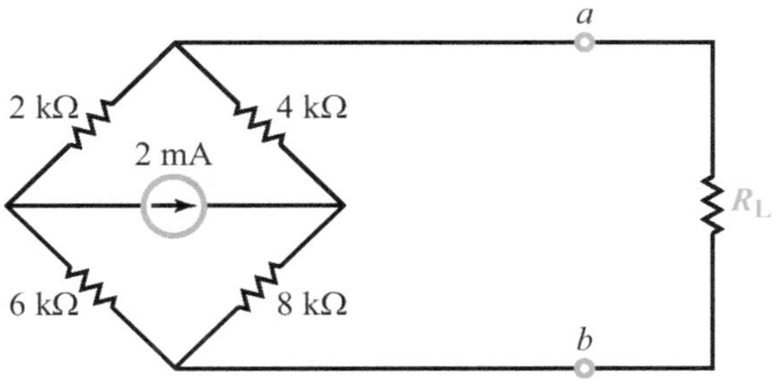

Figure P3.82: Circuit for Problem 3.82.

*__3.83__ Determine the maximum power that can be extracted by the load resistor from the circuit in Fig. P3.83.

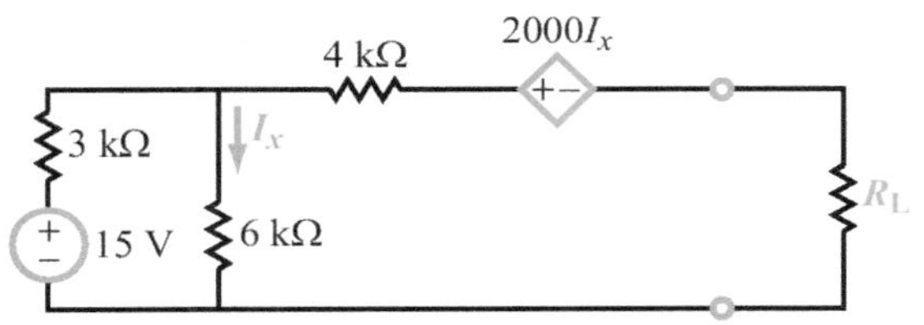

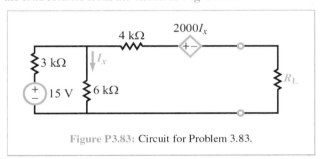

Figure P3.83: Circuit for Problem 3.83.

3.84 Figure P3.84 depicts a 0-to-10 kΩ potentiometer as a variable load resistor R_L connected to a circuit of an unknown architecture. When the wiper position on the potentiometer was adjusted such that $R_L = 1.2$ kΩ, the current through it was measured to be 3 mA, and when the wiper was lowered so that $R_L = 2$ kΩ, the current decreased to 2.5 mA. Determine the value of R_L that would extract maximum power from the circuit.

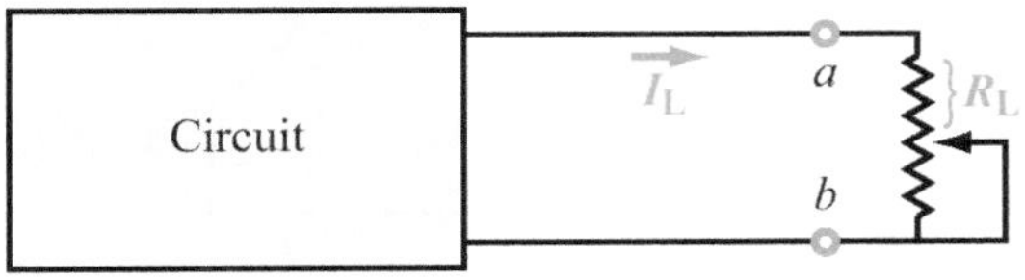

Figure P3.84: Circuit for Problem 3.84.

*__3.85__ The circuit shown in Fig. P3.85 is connected to a variable load R_L through a resistor R_s. Choose R_s so that I_L never exceeds 4 mA, regardless of the value of R_L. Given that choice, what is the maximum power that R_L can extract from the circuit?

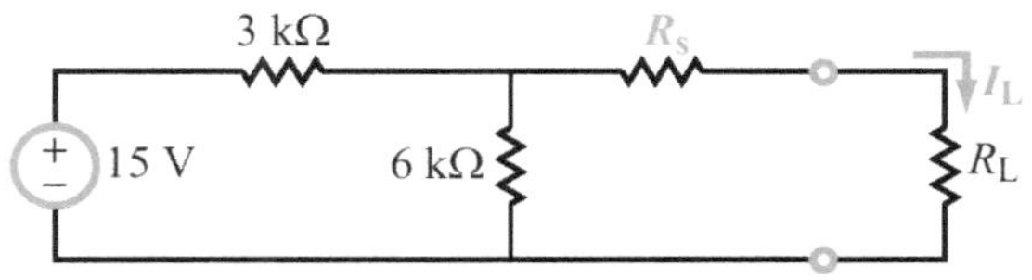

Figure P3.85: Circuit for Problem 3.85.

3.86 In the circuit shown in Fig. P3.86, a potentiometer is connected across the load resistor R_L. The total resistance of the potentiometer is $R = R_1 + R_2 = 5$ kΩ.

(a) Obtain an expression for the power P_L dissipated in R_L for any value of R_1.

(b) Plot P_L versus R_1 over the full range made possible by the potentiometer's wiper.

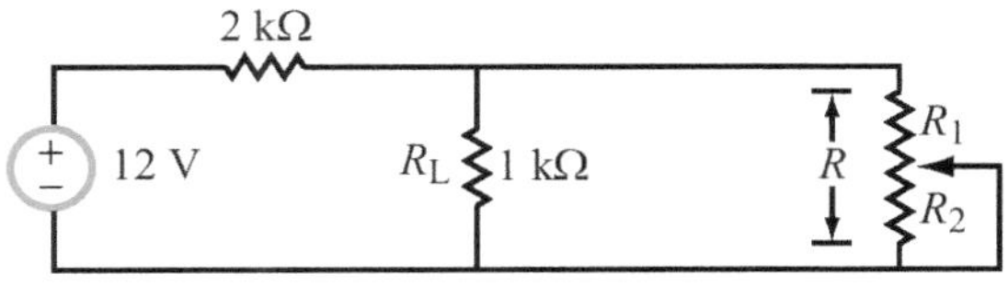

Figure P3.86: Circuit for Problem 3.86.

3.87 Determine the maximum power extractable from the circuit in Fig. P3.87 by the load resistor R_L.

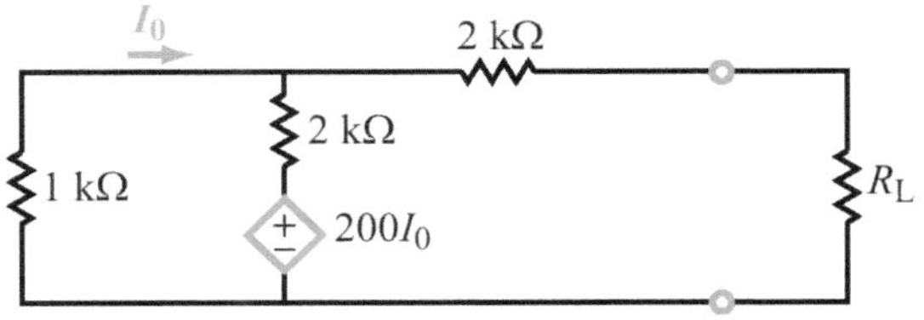

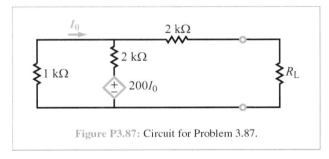

Figure P3.87: Circuit for Problem 3.87.

3.88 In the circuit Fig. P3.88, what value of R_s would result in maximum power transfer to the 10 Ω load resistor?

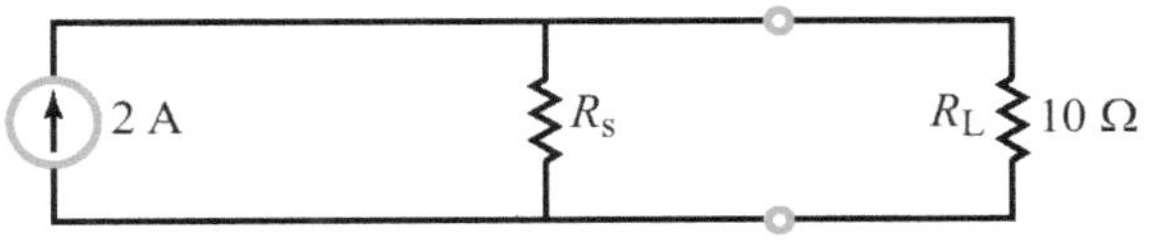

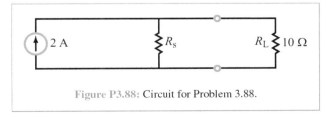

Figure P3.88: Circuit for Problem 3.88.

Section 3-9: Bipolar Junction Transistor

*3.89 The two-transistor circuit in Fig. P3.89 is known as a *current mirror*. It is useful because the current I_0 controls the current I_{REF} regardless of external connections to the circuit. In other words, this circuit behaves like a current-controlled current source. Assume both transistors are the same size such that $I_{B_1} = I_{B_2}$. Find the relationship between I_0 and I_{REF}. (*Hint:* You do not need to know what is connected above or below the transistors. Nodal analysis will suffice.)

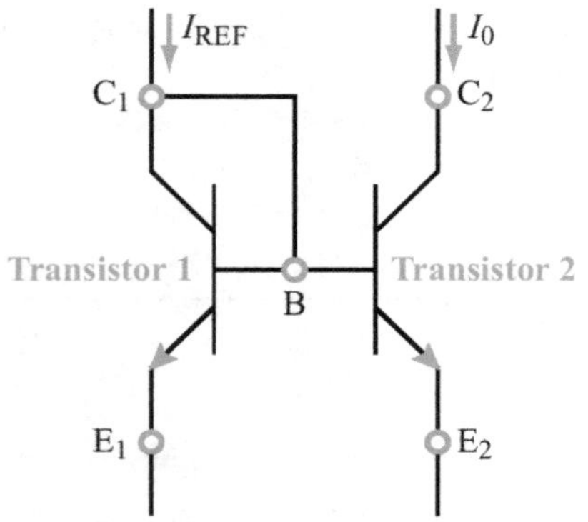

Figure P3.89: A simple current mirror (Problem 3.89).

3.90 The circuit in Fig. P3.90 is a BJT *common collector amplifier*. Obtain expressions for both the voltage gain ($A_V = V_{out}/V_{in}$) and the current gain ($A_I = I_{out}/I_{in}$). Assume $V_{in} \gg V_{BE}$.

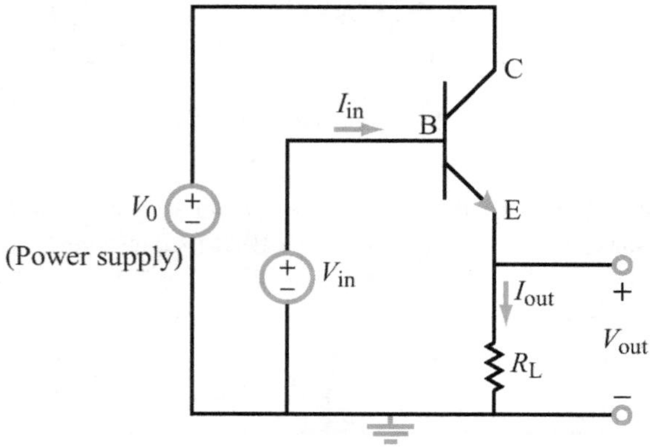

Figure P3.90: Circuit for Problem 3.90.

3.91 The circuit in Fig. P3.91 is identical to the circuit in Fig. P3.90, except that the voltage source V_{in} is more realistic in that it has an associated resistance R_{in}. Find both the voltage gain ($A_V = V_{out}/V_{in}$) and the current gain ($A_I = I_{out}/I_{in}$). Assume $V_{in} \gg V_{BE}$.

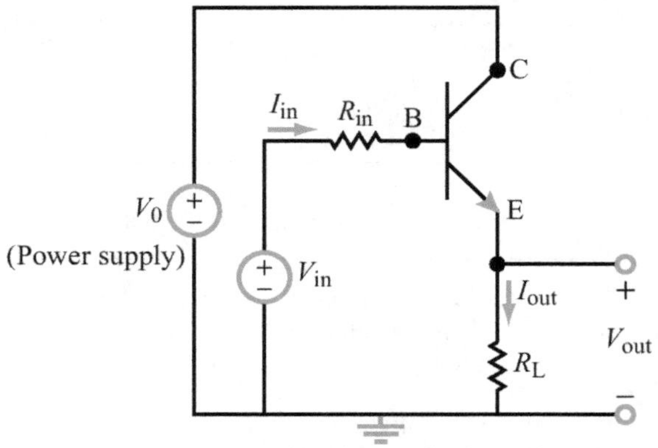

Figure P3.91: Circuit for Problem 3.91.

3.92 The circuit in Fig. P3.92 is a BJT *common-emitter amplifier*. Find V_{out} as a function of V_{in}.

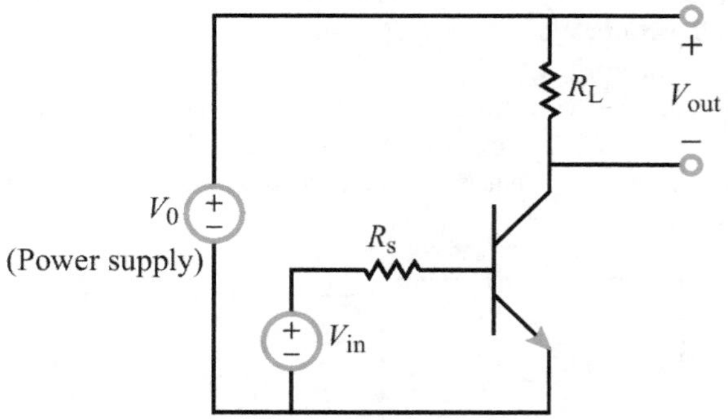

Figure P3.92: Circuit for Problem 3.92.

*3.93 Obtain an expression for V_{out} in terms of V_{in} for the common emitter-amplifier circuit in Fig. P3.93. Assume $V_{in} \gg V_{BE}$.

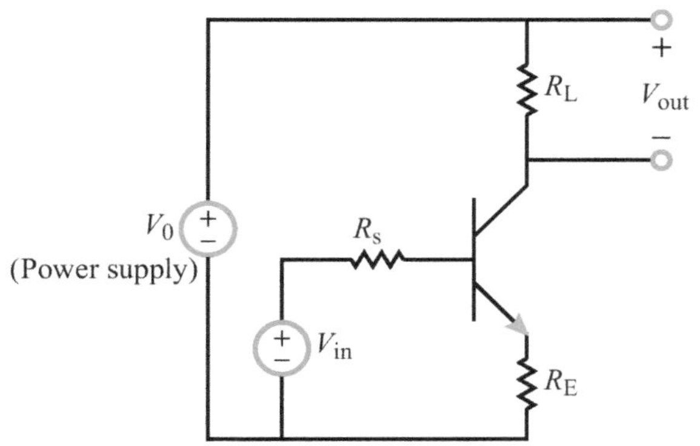

Figure P3.93: Circuit for Problem 3.93.

Section 3-10: Multisim Analysis

3.94 Using Multisim, draw the circuit in Fig. P3.94 and solve for voltages V_1 and V_2.

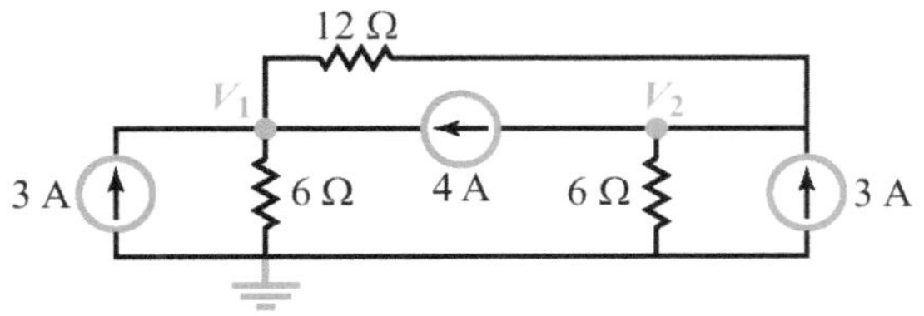

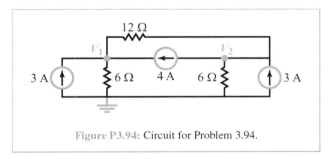

Figure P3.94: Circuit for Problem 3.94.

3.95 The circuit in Problem 3.55 was solved using MATLAB or MathScript software. It can be solved just as easily using Multisim. Using Multisim, draw the circuit in Fig. P3.55 and solve for all node voltages and the current I_0.

3.96 Using Multisim, draw the circuit in Fig. P3.96 and solve for V_x.

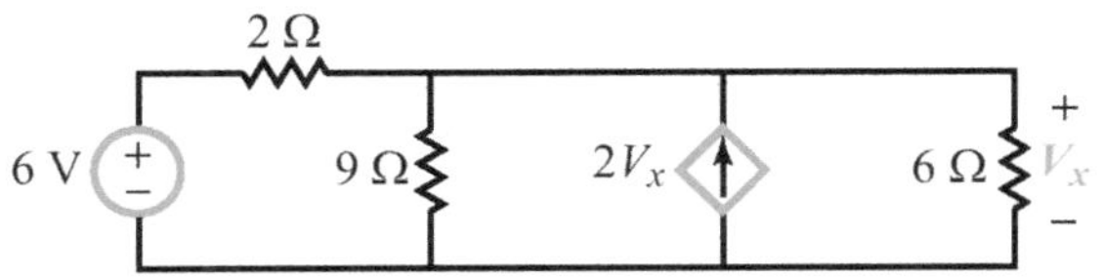

Figure P3.96: Circuit for Problem 3.96.

3.97 Use Multisim to draw the circuit in Fig. P3.97 and solve for V_x.

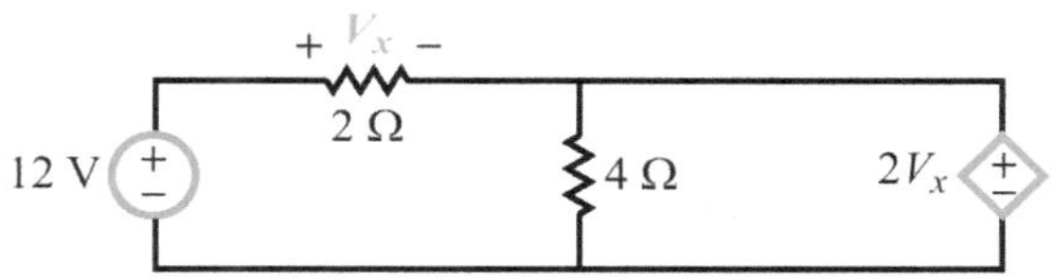

Figure P3.97: Circuit for Problem 3.97.

3.98 Use the DC Operating Point Analysis in Multisim to find the power dissipated or supplied by each component in the circuit in Fig. P3.98 and show that the sum of all powers is zero.

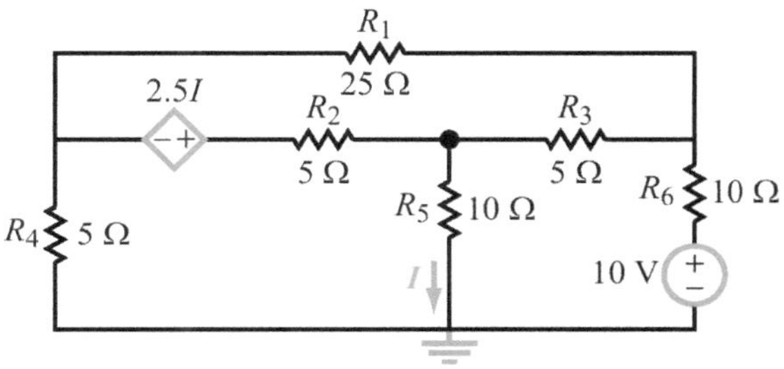

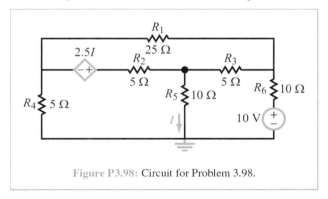

Figure P3.98: Circuit for Problem 3.98.

3.99 Simulate the circuit found in Fig. P3.99 with a 10 Ω resistor placed across the terminals (a, b). Then either by hand or by using tools in Multisim (see Multsim Demo 3.3), find the Thévenin and Norton equivalent circuits and simulate both of those circuits in Multisim with 10 Ω resistors across their output terminals. Show that the voltage drop across and current through the 10 Ω load resistor is the same in all three simulations.

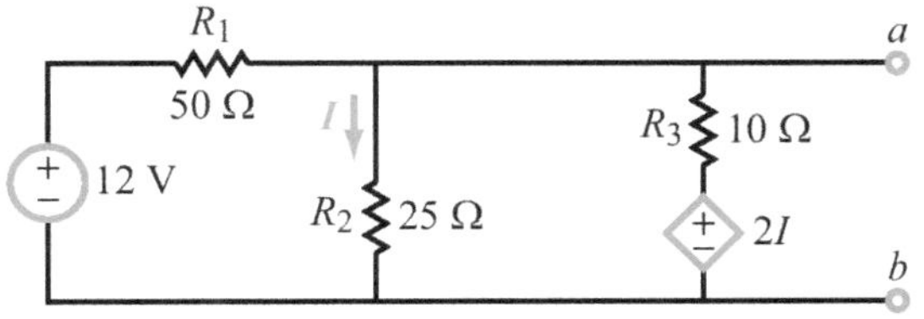

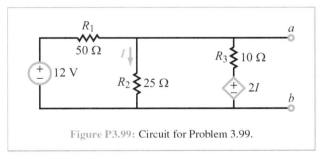

Figure P3.99: Circuit for Problem 3.99.

Potpourri Questions

3.100 Why is it of interest to measure the conductivity of sea ice?

3.101 In integrated circuit fabrication, what is a wafer? A die? A chip?

3.102 How is lithography related to feature size in IC fabrication? Why are ICs fabricated under super-clean conditions?

3.103 What is a *bit* in a digital signal? A *byte*? A *word*? What does the acronym ASCII stand for?

Integrative Problems: Analytical / Multisim / myDAQ

To master the material in this chapter, solve the following problems using three complementary approaches: (a) analytically, (b) with Multisim, and (c) by constructing the circuit and using the myDAQ interface unit to measure quantities of interest via your computer. [myDAQ tutorials and videos are available on (CAD).]

m3.1 Node-Voltage Method: Apply the node-voltage method to determine node voltages V_1 to V_4 for the circuit of Fig. m3.1. From these results determine which resistor dissipates the most power and which resistor dissipates the least power, and report these two values of power. Use these component values: $I_{\text{src1}} = 3.79$ mA, $I_{\text{src2}} = 1.84$ mA, $V_{\text{src}} = 4.00$ V, $R_1 = 3.3$ kΩ, $R_2 = 2.2$ kΩ, $R_3 = 1.0$ kΩ, and $R_4 = 4.7$ kΩ.

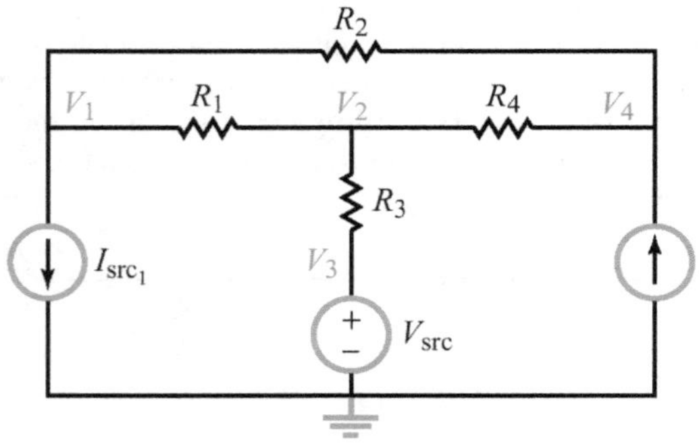

Figure m3.1 Circuit for Problem m3.1.

m3.2 Mesh-Current Method: Apply the mesh-current method to determine mesh currents I_1 to I_4 in the circuit of Fig. m3.2. From these results determine V_1, the voltage across the current source. Use these component values: $I_{\text{src1}} = 12.5$ mA, $V_{\text{src}} = 15$ V, $R_1 = 5.6$ kΩ, $R_2 = 2.2$ kΩ, $R_3 = 3.3$ kΩ, and $R_4 = 4.7$ kΩ.

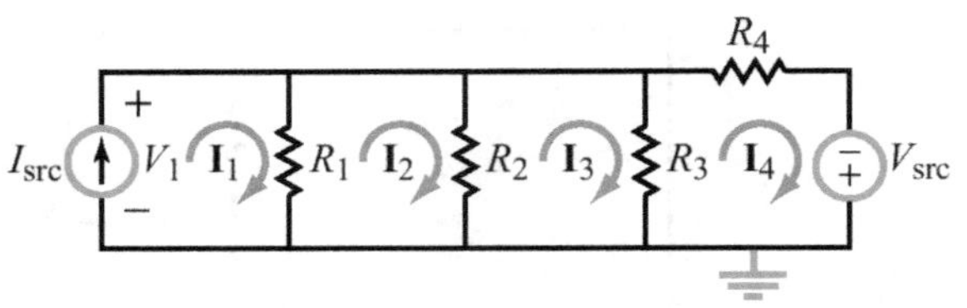

Figure m3.2 Circuit for Problem m3.2.

m3.3 Superposition: In the circuit of Fig. m3.3:

(a) Solve for I_a and V_b using nodal analysis.

(b) Solve for I_a and V_b using superposition. *Hint:* Solve for I_a and V_b with one source on at a time.

(c) Determine I_a and V_b using any method.

Use these component values: $I_1 = 1.84$ mA, $V_2 = 3.0$ V, $R_1 = 1.0$ kΩ, $R_2 = 2.2$ kΩ, and $R_3 = 4.7$ kΩ.

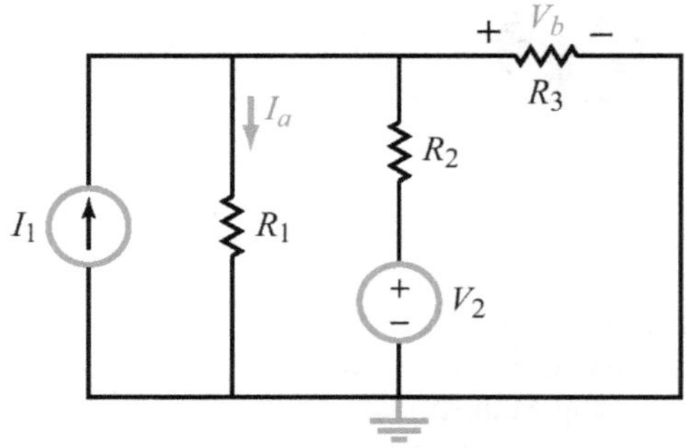

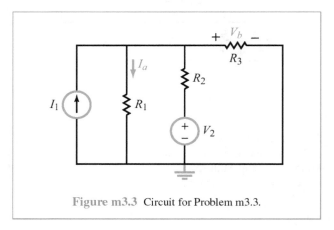

Figure m3.3 Circuit for Problem m3.3.

m3.4 Thévenin Equivalents and Maximum Power Transfer: In the circuit of Fig. m3.4, find the Thévenin equivalent of the circuit at terminals (a, b) as would be seen by a load resistor R_L. Specifically:

(a) Determine the open-circuit voltage V_{oc} that appears at terminals (a, b).

(b) Determine the short-circuit current I_{sc} that flows through a wire connecting terminals (a, b) together.

(c) Determine the Thévenin resistance.

(d) Determine the maximum power P_{max} that could be delivered by this circuit.

Use these component values: $V_{src} = 10$ V, $R_1 = 680\ \Omega$, $R_2 = 3.3\ k\Omega$, $R_3 = 4.7\ k\Omega$, and $R_4 = 1.0\ k\Omega$.

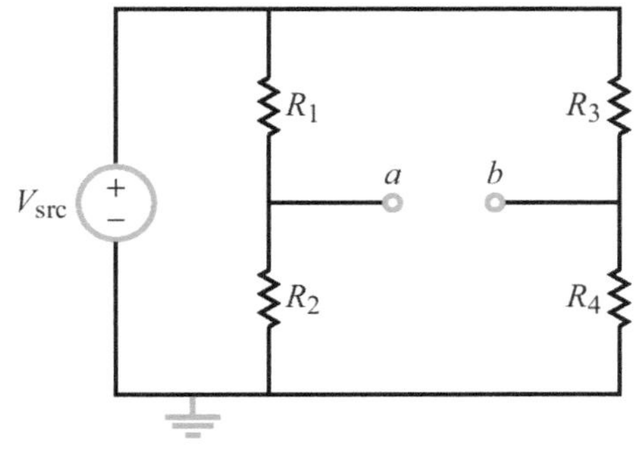

Figure m3.4 Circuit for Problem m3.4.

m3.5 Power Dissipation: For the circuit shown in Fig. m3.5:

(a) Find the combined total power generated by the two current sources analytically and with Multisim. Do not build this circuit (there is no myDAQ portion for part (a)).

(b) Use source transformations to reduce the current sources in Fig. m2.5 into a single voltage source. Now, build this circuit and measure the total power dissipated by all four resistors. *Hint:* To create the voltage source, use the myDAQ arbitrary waveform generator.

(c) Is the power found in part (a) the same as in part (b)?

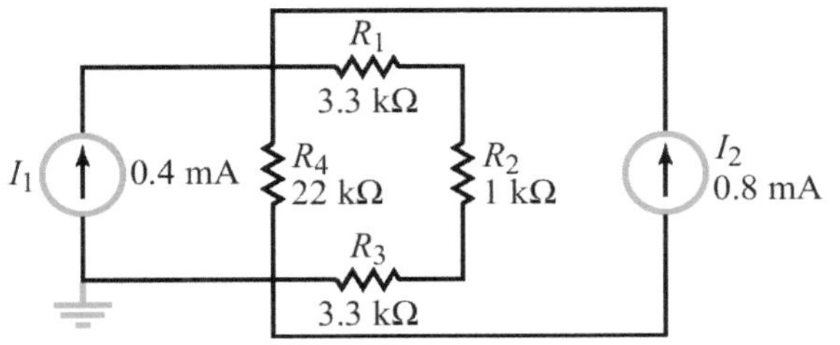

Figure m3.5 Circuit for Problem m3.5.

m3.6 Thévenin Equivalents: For the circuit in Fig. m3.6:

(a) Find the open circuit voltage between nodes 1 and 2.

(b) Add a short circuit between nodes 1 and 2, and then find the short circuit current between them. Use this information to calculate the Thévenin resistance.

(c) Turn off the 4 V and 8 V sources. Verify the Thévenin resistance from part (b) by measuring the equivalent resistance between terminals 1 and 2 (using Multisim and myDAQ).

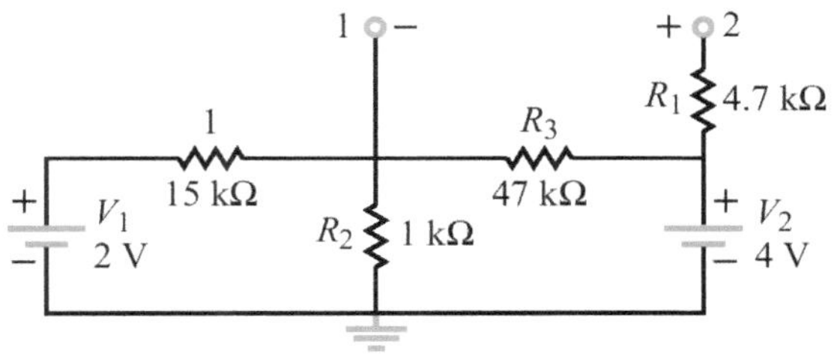

Figure m3.6 Circuit for Problem m3.6.

m3.7 Power Dissipation with Current Source: Creating an ideal current source with the myDAQ requires a current regulator. For the myDAQ portion of this problem use the LM371 and a 220 Ω resistor to create the current source in Fig. m3.7.

(a) Determine the power generated by the current source. For the myDAQ portion of this problem, be sure to measure the current through the LM371 regulator.

(b) Determine the total power dissipated by all other circuit elements. Compare your answer to the result obtained in part (a).

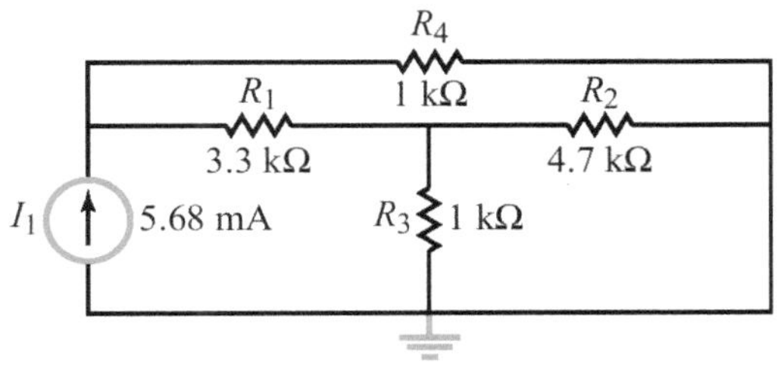

Figure m3.7 Circuit for Problem m3.7.

m3.8 Thévenin Equivalent with Current Source: Creating an ideal current source with the myDAQ requires a current regulator. For the myDAQ portion of this problem, use the LM371 and a 1 kΩ resistor to create the current source in Fig. m3.8.

(a) Determine the open circuit voltage.

(b) Determine the short circuit current between the output terminals.

(c) Determine the Thévenin resistance for the circuit.

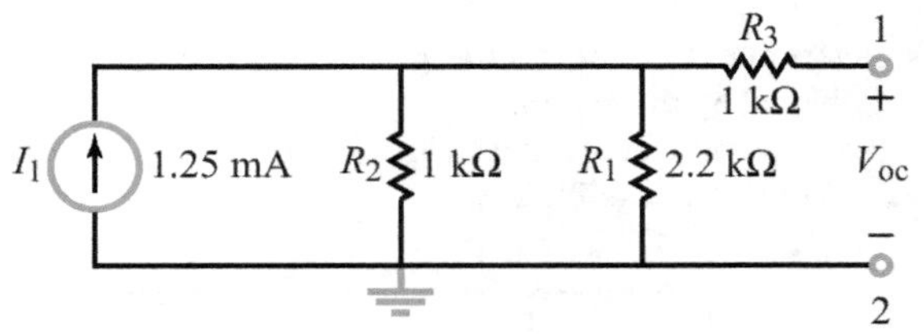

Figure m3.8 Circuit for Problem m3.8.

CHAPTER 4

Operational Amplifiers

Contents

The introduction of the *operational amplifier* chip in the 1960s has led to the development of a wide array of *signal processing circuits*, enabling the creation of an ever-increasing number of *electronic applications*.

Objectives

Learn to:

- Describe the basic properties of an op amp and state the constraints of the ideal op-amp model.
- Explain the role of negative feedback and the trade-off between circuit gain and dynamic range.
- Analyze and design inverting amplifiers, summing amplifiers, difference amplifiers, and voltage followers.
- Combine multiple op-amp circuits together to perform signal processing operations.
- Analyze and design high-gain, high-sensitivity instrumentation amplifiers.
- Design an n-bit digital-to-analog converter.
- Use the MOSFET in analog and digital circuits.
- Apply Multisim to analyze and simulate circuits that include op amps.

Overview

Since its first realization by Bob Widlar in 1963 and then its introduction by Fairchild Semiconductor in 1968, the *operational amplifier*, or *op amp* for short, has become the workhorse of many signal-processing circuits. It acquired the adjective *operational* because it is a versatile device capable not only of amplifying a signal but also inverting it (reversing its polarity), integrating it, or differentiating it. When multiple signals are connected to its input, the op amp can perform additional mathematical operations—including addition and subtraction. Consequently, op-amp circuits often are cascaded together in various arrangements to support a variety of different applications. In this chapter, we explore several op-amp circuit configurations, including amplifiers, summers that add multiple signals together, and digital-to-analog converters that convert signals from digital format to analog.

4-1 Op-Amp Characteristics

The internal architecture of an op-amp circuit consists of many interconnected transistors, diodes, resistors and capacitors (Fig. 4-1), all fabricated on a chip of silicon. Despite its internal complexity, however, *an op amp can be modeled in terms of a relatively simple equivalent circuit that exhibits a linear input-output response.* This equivalence allows us to apply the tools we developed in the preceding chapters to analyze (as well as design) a large array of op-amp circuits and to do so with relative ease.

4-1.1 Nomenclature

Commercially available op amps are fabricated in encapsulated packages of various shapes. A typical example is the eight-pin *DIP configuration* shown in Fig. 4-2(a) [DIP stands for dual-in-line package]. The pin diagram for the op amp is shown in Fig. 4-2(b), and its circuit symbol (the triangle) is displayed in Fig. 4-2(c). Of the eight pins (terminals) only five need to be connected to an outside circuit in order for the op amp to function (the remaining three are used for specialized applications). The op amp has two input voltage terminals (υ_p and υ_n) and one output voltage terminal (υ_o).

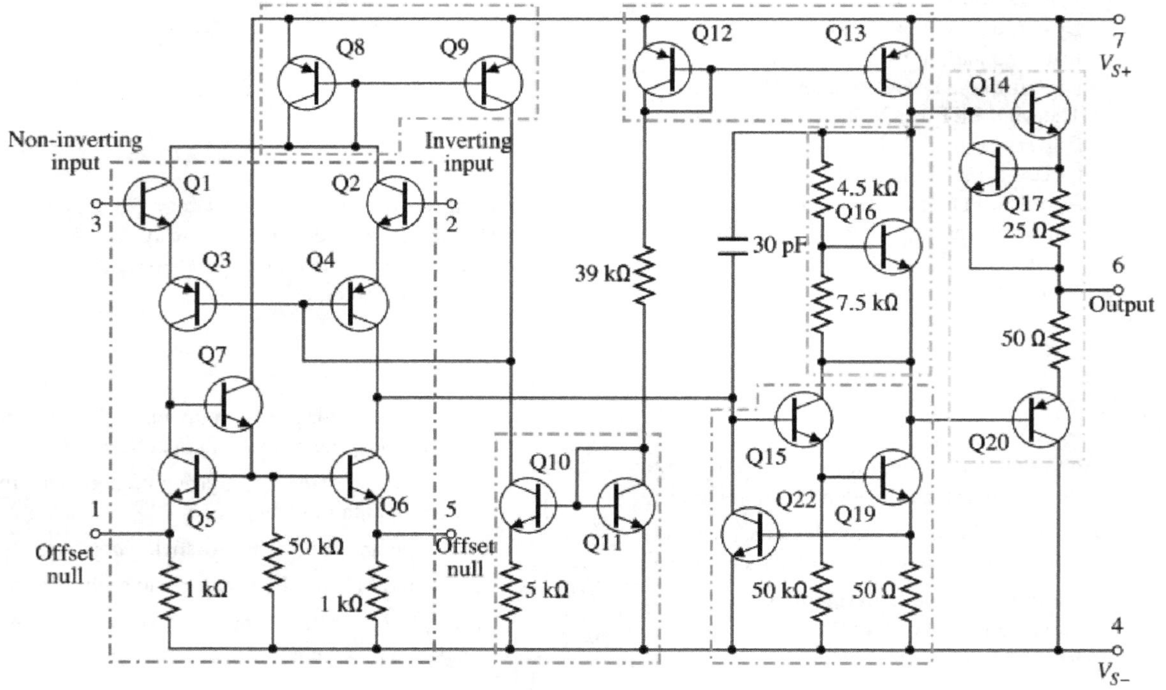

Figure 4-1: The circuit diagram of the Model 741 op amp consists of 20 transistors, several resistors, and one capacitor.

Op-Amp Pin Designation

Pin 2	inverting (or *negative*) input voltage, υ_n
Pin 3	noninverting (or *positive*) input voltage, υ_p
Pin 4	negative (−) terminal of power supply V_{cc}
Pin 7	positive (+) terminal of power supply V_{cc}
Pin 6	output voltage, υ_o

▶ The terms used to describe pins 3 and 2 as *noninverting* and *inverting* are associated with the property of the op amp that its output voltage υ_o is directly proportional to both the noninverting input voltage υ_p and the negative of the inverting input voltage υ_n. ◀

Kirchhoff's current law applies to any volume of space, including an op amp. Hence, for the five terminals connected to the op amp, KCL mandates that

$$i_o = i_p + i_n + i_+ + i_-, \tag{4.1}$$

where i_p, i_n, and i_o may be constant (dc) or time-varying currents. Currents i_+ and i_- are dc currents generated by the dc power supply V_{cc}.

▶ From here on forward, *we will ignore the pins connected to V_{cc} when we draw circuit diagrams involving op amps, because so long as the op amp is operated in its linear region, V_{cc} will have no bearing on the operation of the circuit.* ◀

Hence, the op-amp triangle often is drawn with only three terminals, as shown in Fig. 4-2(d). Moreover, voltages υ_p, υ_n, and υ_o are defined relative to a common reference or ground.

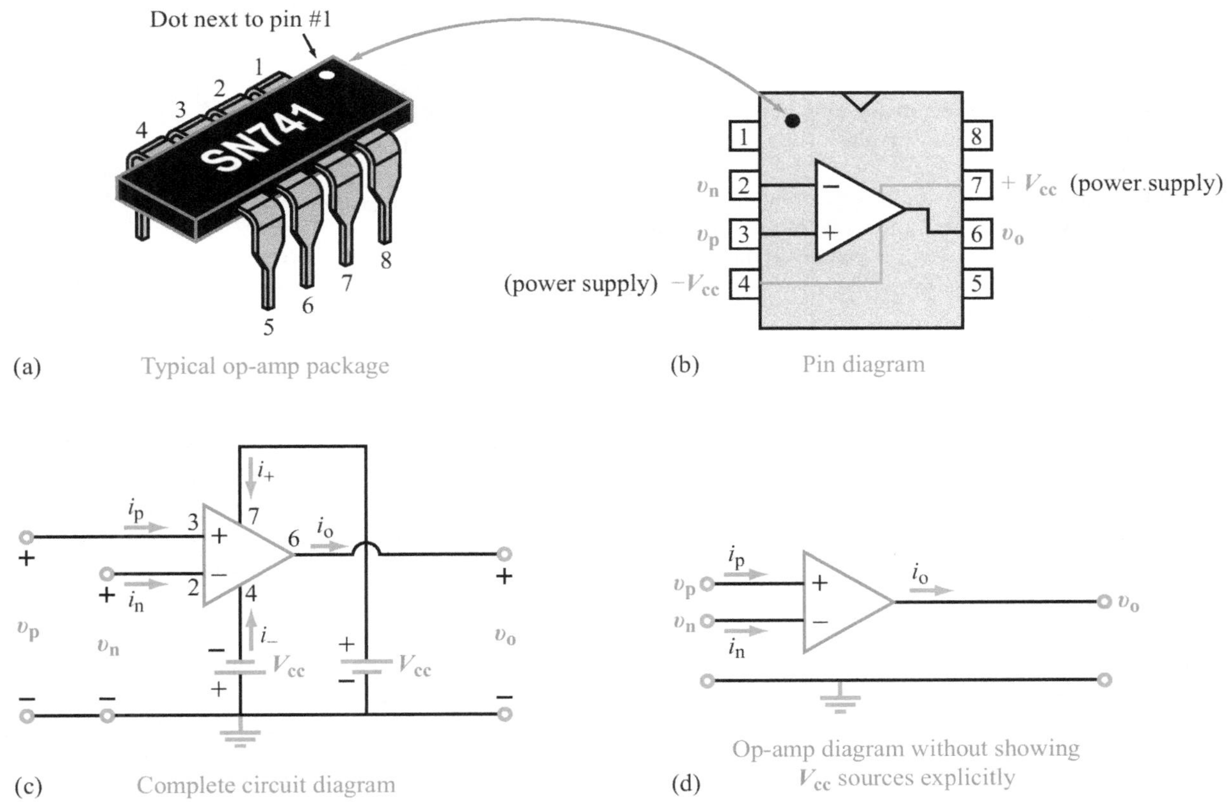

Figure 4-2: Operational amplifier.

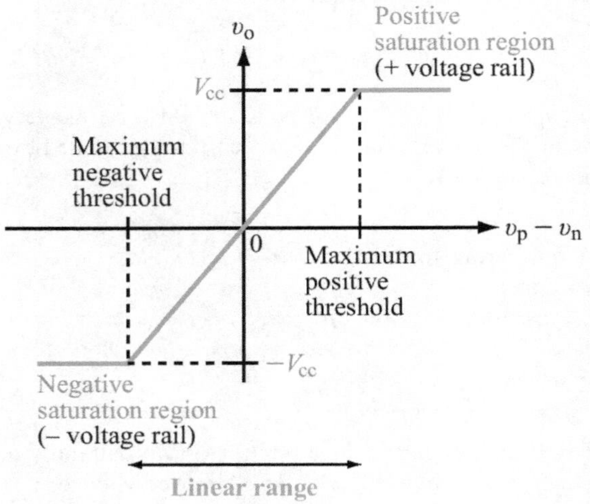

Figure 4-3: Op-amp transfer characteristics. The linear range extends between $v_o = -V_{cc}$ and $+V_{cc}$. The slope of the line is the op-amp gain A

The (+) and (−) labels printed on the op-amp triangle simply denote the noninverting and inverting pins of the op amp not the polarities of v_p or v_n.

Ignoring the pins associated with the power-supply voltage V_{cc} does not mean we can ignore currents i_+ and i_-. To avoid making the mistake of writing a KCL equation on the basis of the simplified diagram given in Fig. 4-2(d), we explicitly state that fact by writing

$$i_o \neq i_p + i_n. \tag{4.2}$$

4-1.2 Transfer Characteristics

The output voltage v_o of the op amp depends on the difference $(v_p - v_n)$ at the input side. The plot shown in Fig. 4-3, which depicts the input-output voltage-transfer characteristic of the op amp, is divided into three regions of operation, denoted the *negative saturation*, *linear*, and *positive saturation* regions. In the linear region, the output voltage v_o is related to the input voltages v_p and v_n by

$$v_o = A(v_p - v_n), \tag{4.3}$$

where A is called the *op-amp gain*, or the *open-loop gain*. The output voltage can be either positive or negative depending on whether v_p is larger than v_n or the other way around. Strictly speaking, this relationship is valid only when the op amp is not connected to an external circuit on the output side (open loop), but as will become clearer in future sections, it continues to hold (approximately) if the output circuit satisfies certain conditions (has high enough input resistance so as not to load the circuit). The open-loop gain is specific to the op-amp device itself, in contrast with the *circuit gain* or *closed-loop gain* G, which defines the gain of the entire circuit. Thus, if v_s is the signal voltage of the circuit connected at the input side of the op-amp circuit (Fig. 4-4), and v_L is the voltage across the load connected at its output side, then

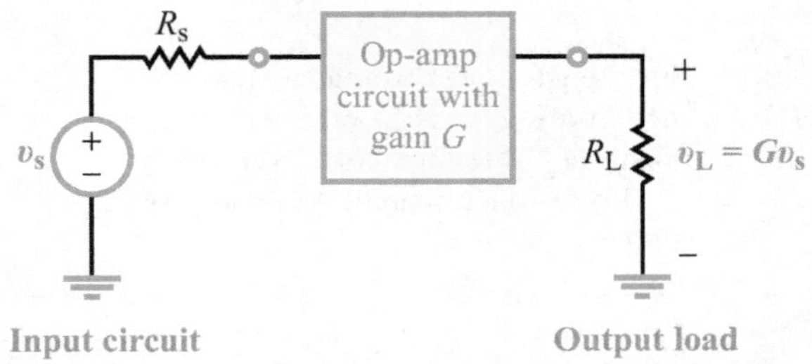

Figure 4-4: Circuit gain G is the ratio of the output voltage v_L to the signal input voltage v_s.

$$v_L = Gv_s. \tag{4.4}$$

According to Eq. (4.3), v_o is related linearly to the difference between v_p and v_n or to either one of them if the other is held constant. Excluding circuits that contain magnetically coupled transformers, in a regular circuit no voltage can exceed the net voltage level of the power supply.

> ▶ The maximum value that v_o can attain is $|V_{cc}|$. The op amp goes into a saturation mode if $|A(v_p - v_n)| > |V_{cc}|$, which can occur on both the negative and positive sides of the linear region. ◀

As we will discuss shortly, the op-amp gain A is typically on the order of 10^5 or greater, and the supply voltage is on the order of volts or tens of volts. In the linear region, v_o is bounded between $-V_{cc}$ and $+V_{cc}$, which means that $(v_p - v_n)$ is bounded between $-V_{cc}/A$ and $+V_{cc}/A$. For $V_{cc} = 10$ V and $A = 10^6$, the operating range of $(v_p - v_n)$ is -10 μV to $+10$ μV. So a basic op-amp configuration is able to amplify only very small voltages, but the configuration can be modified so as to amplify a wider range of voltages (Section 4-2). Even in such cases, however, the maximum output voltage is V_{cc} and the minimum is $-V_{cc}$. These are called the *voltage rails*. It is important to keep this in mind as we deal with circuits containing operational amplifiers.

4-1.3 Op-Amp Switch

An op amp is an active device. Hence, to operate, it needs to be connected to a power supply that can provide the necessary voltages. Specifically, the op amp requires a positive supply voltage V_{cc} at pin 7 and a negative supply voltage $-V_{cc}$ at pin 4. The magnitude of V_{cc} is specified by the manufacturer. For some models, the positive and negative supply voltages need not be of the same magnitude, but most often they are. Hence, our default assumption in all future considerations of op-amp circuits is that the dc supply voltages connected to pins 4 and 7 are equal in magnitude and opposite in polarity. Among various op-amp models, V_{cc} typically is between 5 and 24 V.

As noted earlier in connection with Fig. 4-4, if $(\upsilon_p - \upsilon_n)$ exceeds a certain maximum positive threshold, the output voltage υ_o saturates at V_{cc}, and if $(\upsilon_p - \upsilon_n)$ is negative (because the voltage connected to υ_p is smaller than that connected to υ_n) and its magnitude exceeds a maximum negative threshold, then υ_o saturates at $-V_{cc}$. This op-amp behavior can be used to operate the op amp like an electronic switch, either as an ON/OFF switch, or as a switch to activate one device versus another. An example is illustrated by the circuit in Fig. 4-5. At the input side, the positive terminal is connected to a dc voltage V_p that can be set at either +2 V or −2 V, and the negative input terminal is connected to ground. At the output side, the op amp is connected to the parallel combination of two LEDs, one that can emit red light and another that can emit green light. The two LEDs are arranged in opposite directions, so that when V_0 is positive and sufficiently large to cause a current to flow through the red LED, it lights up, but the green LED will neither conduct nor emit green light because it is reverse biased relative to V_0. This is the scenario depicted in Fig. 4-5(b); the input $V_p = +2$ V (and $V_n = 0$) causes the output to saturate at $V_0 = V_{cc} = 12$ V (the vertical flag with $V_{cc} = 12$ V is used to denote that this LED uses a $V_{cc} = 12$ V), which is quite sufficient to cause the red LED to conduct. When V_p is switched to −2 V, as in the scenario depicted in Fig. 4-5(c), the output saturates at $V_0 = -12$ V, in which case the green LED starts to conduct and emit green light and the red LED stops conducting altogether. Thus, switching the input of the op amp between +2 V and −2 V causes the two LEDs to alternate roles between active and inactive.

4-1.4 Equivalent-Circuit Model in Linear Region

When operated in its linear region, the op-amp input-output behavior can be modeled in terms of the equivalent linear circuit shown in Fig. 4-6. The equivalent circuit consists of a voltage-controlled voltage source of magnitude $A(\upsilon_p - \upsilon_n)$, an input resistance R_i, and an output resistance R_o. Table 4-1 lists the

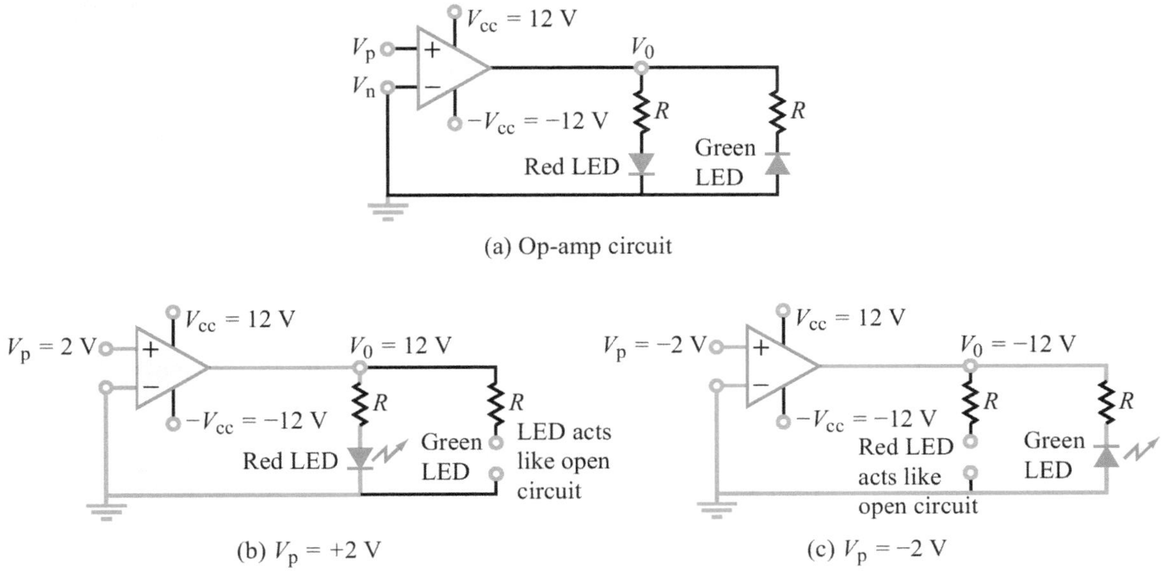

Figure 4-5: Op amp operated as a switch. The $\pm V_{cc}$ flags indicate the dc supply voltages connected to pins 7 and 4.

Table 4-1: Characteristics and typical ranges of op-amp parameters. The rightmost column represents the values assumed by the ideal op-amp model.

Op-Amp Characteristics	Parameter	Typical Range	Ideal Op Amp
• Linear input-output response	Open-loop gain A	10^4 to 10^8 (V/V)	∞
• High input resistance	Input resistance R_i	10^6 to 10^{13} Ω	∞ Ω
• Low output resistance	Output resistance R_o	1 to 100 Ω	0 Ω
• Very high gain	Supply voltage V_{cc}	5 to 24 V	As specified by manufacturer

typical range of values that each of these op-amp parameters may assume. Based on these values, we note that an op amp is characterized by:

(1) *High input resistance* R_i: at least 1 MΩ, which is highly desirable from the standpoint of voltage transfer from an input circuit (as discussed previously in Section 3-7).

(2) *Low output resistance* R_o: which is desirable from the standpoint of transfering the op-amp's output voltage to a load circuit.

(3) *High open loop voltage gain* A: which is the key, as we see later, to allowing us to further simplify the equivalent circuit into an "ideal" op-amp model with infinite gain.

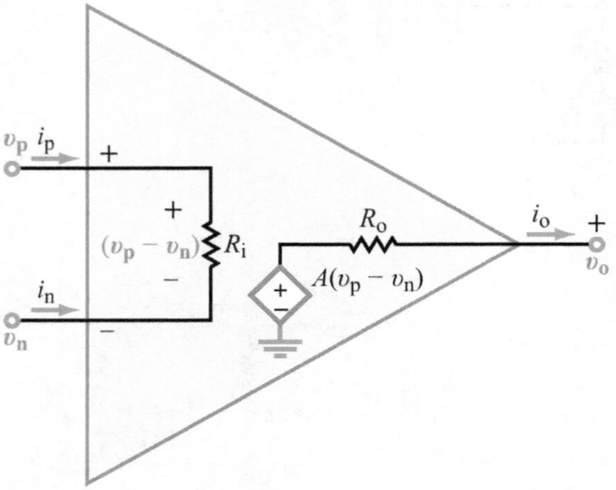

Figure 4-6: Equivalent circuit model for an op amp operating in the linear range ($\upsilon_o \le |V_{cc}|$). Voltages υ_p, υ_n, and υ_o are referenced to ground.

Example 4-1: Noninverting Amplifier

The circuit shown in Fig. 4-7 uses an op amp to amplify the input signal voltage υ_s. The circuit uses *feedback* to connect the op-amp output (at node a) to the inverting input terminal υ_n through a resistor R_1. Obtain an expression for the circuit gain $G = \upsilon_o/\upsilon_s$, and then evaluate it for $V_{cc} = 10$ V, $A = 10^6$, $R_i = 10$ MΩ, $R_o = 10$ Ω, $R_1 = 80$ kΩ, and $R_2 = 20$ kΩ.

Solution: For reference purposes, we label the output as terminal a and the node from which a current is fed back into the op amp as terminal b. The current i_3 flowing from terminal b to terminal a is the same as the current i_4 flowing from terminal a towards R_o. (The presence of the voltmeter used to measure υ_o has no impact on the operation of the circuit because of the very high input resistance of the voltmeter.) When expressed in terms of node voltages, the equality $i_3 = i_4$ gives

$$\frac{\upsilon_n - \upsilon_o}{R_1} = \frac{\upsilon_o - A(\upsilon_p - \upsilon_n)}{R_o} \qquad \text{(node } a\text{)}. \tag{4.5}$$

At node b, KCL gives $i_1 + i_2 + i_3 = 0$, or

$$\frac{\upsilon_n - \upsilon_p}{R_i} + \frac{\upsilon_n}{R_2} + \frac{\upsilon_n - \upsilon_o}{R_1} = 0. \qquad \text{(node } b\text{)}. \tag{4.6}$$

Additionally,

$$\upsilon_p = \upsilon_s. \tag{4.7}$$

Solution of these simultaneous equations leads to the following expression for the circuit gain G:

$$G = \frac{\upsilon_o}{\upsilon_s} = \frac{[AR_i(R_1 + R_2) + R_2R_o]}{AR_2R_i + R_o(R_2 + R_i) + R_1R_2 + R_i(R_1 + R_2)}. \tag{4.8}$$

For $V_{cc} = 10$ V, $A = 10^6$, $R_i = 10^7$ Ω, $R_o = 10$ Ω, $R_1 = 80$ kΩ, and $R_2 = 20$ kΩ,

$$G = \frac{\upsilon_o}{\upsilon_s} = 4.999975 \approx 5.0. \tag{4.9}$$

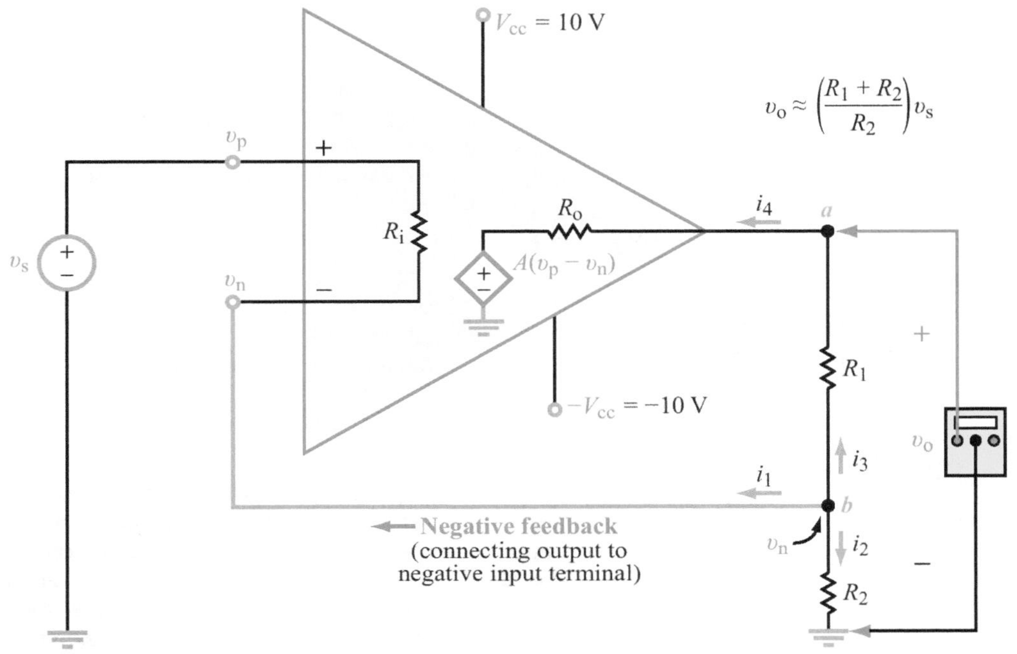

Figure 4-7: Noninverting amplifier circuit of Example 4-1.

In the expression for G, the two parameters A and R_i are several orders of magnitude larger than all of the others. Also, R_o is in series with R_1, which is 8000 times larger. Hence, we would incur minimal error if we let $A \to \infty$, $R_i \to \infty$, and $R_o \to 0$, in which case the expression for G reduces to

$$G \approx \frac{R_1 + R_2}{R_2} \qquad \text{(ideal op-amp model)}. \qquad (4.10)$$

This approximation, based on the ideal op-amp model that will be introduced in Section 4-3, gives

$$G = \frac{80\ \text{k}\Omega + 20\ \text{k}\Omega}{20\ \text{k}\Omega} = 5.$$

Concept Question 4-1: How is the linear range of an op amp defined? (See CAD)

Concept Question 4-2: What is the difference between the op-amp gain A and the circuit gain G? (See CAD)

Concept Question 4-3: How is an op amp used as a switch? (See CAD)

Concept Question 4-4: An op amp is characterized by three important input-output attributes. What are they? (See CAD)

Exercise 4-1: In the circuit of Example 4-1 shown in Fig. 4-7, insert a series resistance R_s between v_s and v_p and then repeat the solution to obtain an expression for G. Evaluate G for $R_s = 10\ \Omega$ and use the same values listed in Example 4-1 for the other quantities. What impact does the insertion of R_s have on the magnitude of G?

Answer:

$$G = \frac{[A(R_i + R_s)(R_1 + R_2) + R_2 R_o]}{[AR_2(R_i + R_s) + R_o(R_2 + R_i + R_s) + R_1 R_2 + (R_i + R_s)(R_1 + R_2)]}$$
$$= 4.999977 \qquad \text{(negligible impact)}.$$

(See CAD)

Technology Brief 9
Display Technologies

From cuneiform-marked clay balls to the abacus to today's digital projection technology, advances in visual displays have accompanied almost every major leap in information technology. While the earliest "modern" computers relied on cathode ray tubes (CRT) to project interactive images, today's computers can access a wide variety of displays ranging from plasma screens and LED arrays to digital micromirror projectors, electronic ink, and virtual reality interfaces. In this Technology Brief, we will review the major technologies currently available for two-dimensional visual displays.

Cathode Ray Tube (CRT)

The earliest computers relied on the same technology that made the television possible. In a CRT television or monitor (Fig. TF9-1), an ***electron gun*** is placed behind a positively charged glass screen, and a negatively charged electrode (the ***cathode***) is mounted at the input of the electron gun.

- During operation, the cathode emits streams of electrons into the electron gun.

- The emitted electron stream is steered onto different parts of the positively charged screen by the electron gun; the direction of the electron stream is controlled by the electric field of the deflecting coils through which the beam passes.

- The screen is composed of thousands of tiny dots of phosphorescent material arranged in a two-dimensional array. Every time an electron hits a phosphor dot, it glows a specific color (red, blue, or green). A pixel on the screen is composed of phosphors of these three colors.

- In order to make an image appear to move on the screen, the electron gun constantly steers the electron stream onto different phosphors, lighting them up faster than the eye can detect the changes, and thus, the images appear to move. In modern color CRT displays, three electron guns shoot different electron streams for the three colors.

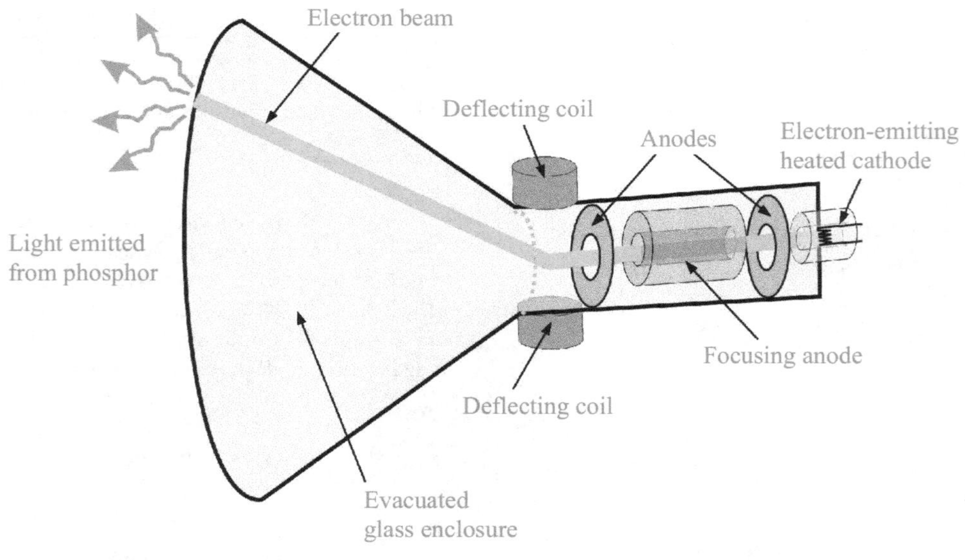

Figure TF9-1: Schematic of CRT operation.

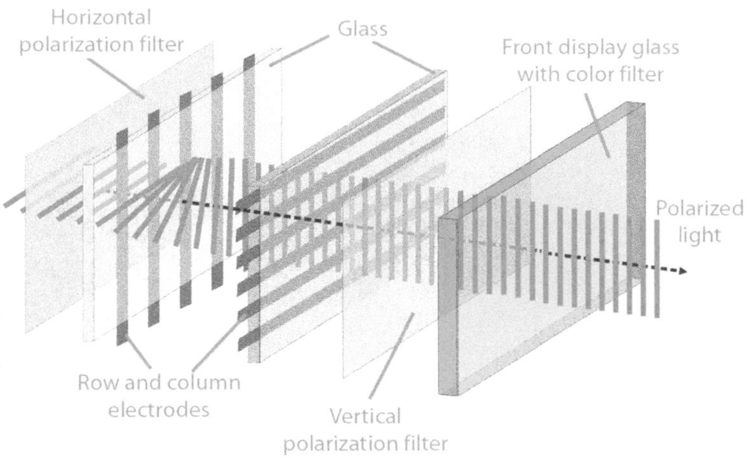

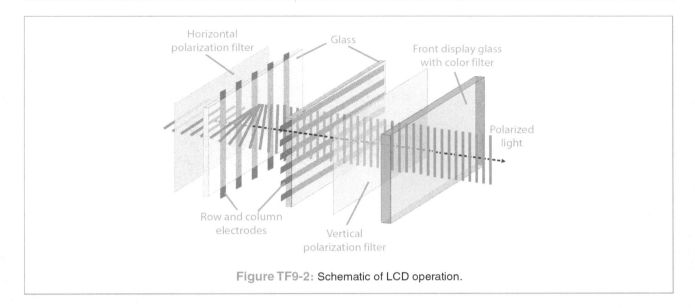

Figure TF9-2: Schematic of LCD operation.

The basic concept behind CRT was explored in the early 2000s in the development of ***field emission displays*** (FED), which used a thin film of atomically sharp electron emitter tips to generate electrons. The electrons emitted by the film collide with phosphor elements just as in the traditional CRT. The primary advantage of this type of "flat-panel" display is that it can provide a wider viewing angle (i.e., one can look at an FED screen at a sharp angle and still see a good image) than possible with conventional LCD or LED technology (discussed next).

Liquid Crystal Displays (LCD)

LCDs are used in digital clocks, cellular phones, desktop and laptop computers, and some televisions and other electronic systems. They offer a decided advantage over other display technologies (such as cathode ray tubes) in that they are lighter and thinner and consume a lot less power to operate. LCD technology relies on special electrical and optical properties of a class of materials known as ***liquid crystals***, first discovered in the 1880s by botanist Friedrich Reinitzer. In the basic LCD display, light shines through a thin stack of layers as shown in Fig. TF9-2.

- Each stack consists of layers in the following order (starting from the viewer's eye): color filter, vertical (or horizontal) polarizer filter, glass plate with transparent electrodes, liquid crystal layer, second glass plate with transparent electrodes, horizontal (or vertical) polarizer filter.
- Light is shone from behind the stack (called the ***backlight***). As light crosses through the layer stack, it is polarized along one direction by the first filter.
- If no voltage is applied on any of the electrodes, the liquid crystal molecules align the filtered light so that it can pass through the second filter.
- Once through the second filter, it crosses the color filter (which allows only one color of light through) and the viewer sees light of that color.
- If a voltage is applied between the electrodes on the glass plates (which are on either side of the liquid crystal), the induced electric field causes the liquid crystal molecules to rotate. Once rotated, the crystals no longer align the light coming through the first filter so that it can pass through the second filter plate.
- If light cannot cross, the area with the applied voltage looks dark. This is precisely how simple hand-held calculator displays work; usually the bright background is made dark every time a character is displayed.

Modern monitors, laptops, phones, and tablets use a version of the LCD called ***thin-film transistor*** (TFT) LCD; these also are known as ***active matrix*** displays. In TFT

LCDs, several thin films are deposited on one of the glass substrates and patterned into transistors. Each color component of a pixel has its own microscale transistor that controls the voltage across the liquid crystal; since the transistors only take up a tiny portion of the pixel area, they effectively are invisible. Thus, each pixel has its own electrode driver built directly into it. This specific feature enabled the construction of the flat high-resolution screens now in common use (and made the CRT display increasingly obsolete). Since LCD displays also weigh considerably less than a CRT tube, they enabled the emergence of laptop computers in the 1980s. Early laptops used large, heavy monochrome LCDs; most of today's mobile devices use active-matrix displays.

Light-Emitting Diode (LED) Displays

A different but very popular display technology employs tiny light-emitting diodes (LED) in large pixel arrays on flat screens (see Technology Brief 5 on LEDs). Each pixel in an LED display is composed of three LEDs (one each of red, green, and blue). Whenever a current is made to pass through a particular LED, it emits light at its particular color. In this way, displays can be made flatter (i.e., the LED circuitry takes up less room than an electron gun or LCD) and larger (since making large, flat LED arrays technically is less challenging than giant CRT tubes or LCD displays). Unlike LCDs, LED displays do not need a backlight to function and easily can be made multicolor.

Modern LED research is focused mostly on flexible and ***organic LEDs*** (OLEDs), which are made from polymer light-emitting materials and can be fabricated on flexible substrates (such as an overhead transparency). Flexible displays of this type have been demonstrated by several groups around the world.

Plasma Displays

Plasma displays have been around since 1964 when invented at the University of Illinois. While attractive due to their low profile, large viewing angle, brightness, and large screen size, they largely were displaced in the 1980s in the consumer market by LCD displays for manufacturing-cost reasons. In the late 1990s, plasma displays became popular for ***high-definition television*** (HDTV) systems.

Each pixel in a plasma display contains one or more microscale pocket(s) of trapped noble gas (usually neon or xenon); electrodes patterned on a glass substrate are placed in front and behind each pocket of gas (Fig. TF9-3).

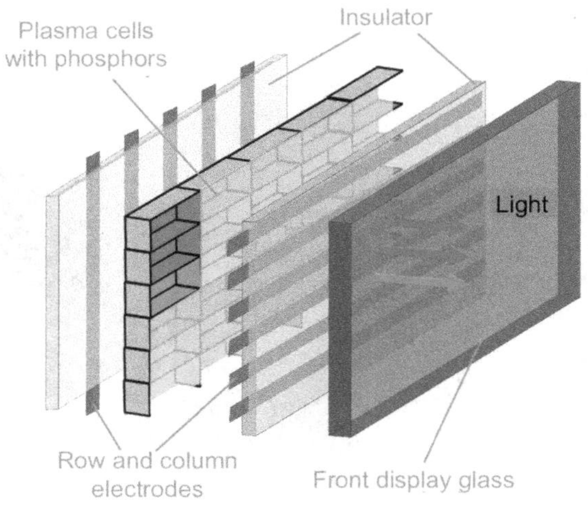

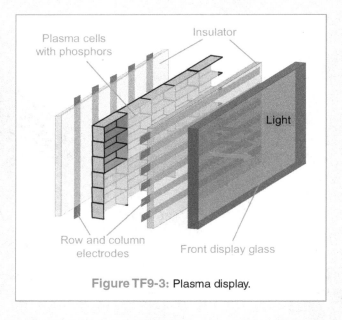

Figure TF9-3: Plasma display.

The back of one of the glass plates is coated with light-emitting phosphors. When a sufficient voltage is applied across the electrodes, a large electric field is generated across the noble gas, and a plasma (ionized gas) is ignited. The plasma emits ultraviolet light which impacts the phosphors; when impacted with UV light, the phosphors emit light of a certain color (blue, green, or red). In this way, each pocket can generate one color.

Electronic Ink

Electronic ink, ***e-paper***, or ***e-ink*** are all names for a set of display technologies made to look like paper with ink on it. In all cases, the display is very thin (almost as thin as real paper), does not use a backlight (ambient light is reflected off the display, just like real paper), and little to no power is consumed when the image is kept constant. The first version of e-paper was invented in the 1970s at Xerox, but it was not until the 1990s that a commercially viable version was developed at MIT. A number of electronic ink technologies are in production or in development.

- Most common electronic ink technologies trap a thin layer of oil between two layers of glass or plastic onto which have been patterned transparent electrodes. The total stack is usually less than a tenth of a millimeter.

- Within the oil are suspended charged particles. In some versions, the oil is colored.

Table TT9-1: A comparison of some characteristics of common display technologies; see also http://en.wikipedia.org/wiki/Comparison_of_CRT,_LCD,_Plasma,_and OLED.

Pros	Cons
Cathode Ray Tube (CRT)	
• Good dynamic range (~15,000 : 1) • Very little distortion • Excellent viewing angle • No inherent pixels	• Large and heavy, limiting maximum practical size • High power consumption and heat generation • Burn-in possible • Produces noticeable flicker at low refresh rates • Minimum size for color limited to 7″ diagonal • Can contain lead, barium, and cadmium, which are toxic
Plasma Displays	
• Excellent contrast ratios (~1,000,000 : 1) • Sub-millisecond response time • Near zero distortion • Excellent viewing angle • Very scalable (easier than other technologies to make large displays)	• Large minimum pixel pitch; suitable for larger displays • High power consumption than LCD • Limited color depth since plasma pixels can only be turned on or off, no grading of emission • Image burn-in possible
Organic Light-Emitting Diode (LED) Displays	
• Excellent viewing angle • Very light • Very fast, so no image distortion during fast motion • Excellent color quality because no backlight is used	• Limited lifetime of organic materials (but progress in this area is rapid) • Burn-in possible • More expensive than other technologies (ca. 2012)
Liquid Crystal Displays (LCD)	
• Small and light • Lower power consumption than plasma or CRT • No geometric distortion • Can be made in almost any size or shape • Liquid crystal has no inherent resolution limit	• Limited viewing angle • Slower response than plasma or CRT can cause image distortion during fast motion • Slow response at low temperatures • Requires a backlight, which can vary across screen
Digital Light Projection (DLP) Displays	
• No burn-in • Cheaper than LCD or plasma displays • DLPs with LED and laser sources do not need light source replacement very often • Excellent for very large screens (theaters) due to possibility of using multiple color sources (color depth) and no inherent size limitation to hardware	• Requires light source replacement • Reduced viewing angle compared with CRT, plasma, and LCD • Some viewers perceive the colors in the projection, producing a rainbow effect
Electronic Ink Displays	
• Very low power consumption • Works with reflected light; excellent for viewing in bright light • Lightweight • Flexible and bendable	• Slow, consumer units not yet suitable for fast video • Ghost images persist without refresh • Color displays are still under development

- Applying a potential across the electrodes on either side of the oil suspension attracts the charged particles to either the top or bottom substrates (depending on the polarity). Some displays use white particles in black fluid. Thus, when the white particles move to the top, they block the black fluid and the display appears white. When they move to the bottom, the display appears dark. Some displays use

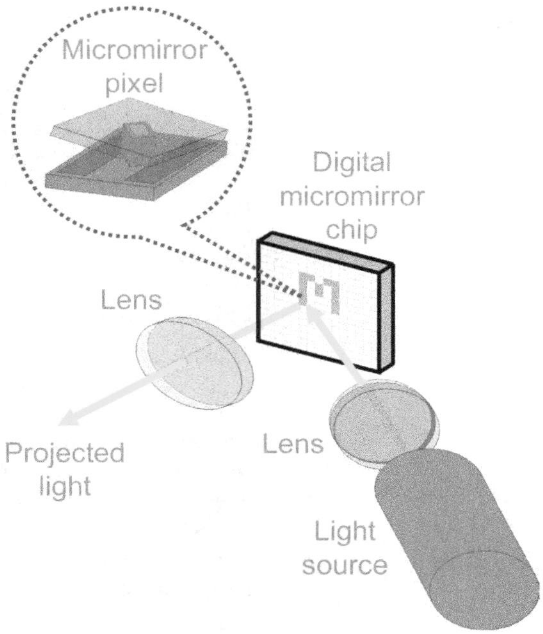

Figure TF9-4: A typical digital light processor (DLP) arrangement includes a light source, lenses, and a micromirror array that steers the light to create projected pixels.

a combination of black and white particles to achieve the same effect.

Digital Light Processing (DLP)

Digital light processing (DLP) is the name given to a technology that uses arrays of individual, micro-mechanical mirrors to manipulate light at each pixel position. Invented in 1987 by Dr. Hornbeck at Texas Instruments, this technology has revolutionized projection technology; many of today's digital projectors are made possible by DLP chips. DLP also was used heavily in large, rear-projection televisions.

- A basic DLP consists of an array of metal micromirrors, each about 100 micrometers on a side (Fig. TF9-4(inset)). One micromirror corresponds to one pixel on a digital image.
- Each micromirror is mounted on micromechanical hinges and can be tilted towards or away from a light source several thousand times per second!
- The mirrors are used to reflect light from a light source (housed within the television or projector case) and through a lens to project it either from behind a screen (as is the case in rear-projection televisions) or onto a flat surface (in the case of projectors), as in (Fig. TF9-4). If a micromirror is tilted away from the light source, that pixel on the projected image becomes dark (since the mirror is not passing the light onto the lens).
- If it is tilted towards the light source, the pixel lights up. By varying the relative time a given mirror is in each position, grey values can be generated as well.
- Color can be added by using multiple light sources and either one chip (with a filter wheel) or three chips. The three-chip color DLP used in high-resolution cinema systems can purportedly generate 35 trillion different colors!

4-2 Negative Feedback

▶ *Feedback* refers to taking a part of the output signal and feeding it back into the input. It is called *positive feedback* if it increases the intensity of the input signal, and it is called *negative feedback* if it decreases it. In negative feedback, the output terminal is connected to the υ_n terminal, either directly or through a resistor. ◀

Positive feedback causes the op amp to saturate, thereby forcing its output voltage υ_o to become equal to its supply voltage V_{cc}. This behavior is used to advantage in certain types of applications but they are outside the scope of this book. Negative feedback, on the other hand, is an essential ingredient of all of the op-amp circuits covered in this and forthcoming chapters.

Why do some op-amp circuits need feedback and why negative feedback specifically? It seems counter-intuitive to want to decrease the input signal when the intent is to amplify it! We will answer this question by examining the circuit of Example 4-1 in some detail. To facilitate the discussion we have reproduced the circuit diagram (into a smaller version) and inserted it in Fig. 4-8(a).

When we say an op amp has a supply voltage V_{cc} of 10 V, we actually mean that a positive (10 V) dc voltage source is connected to pin 7 of its package and another, negative (−10 V) source is connected to its pin 4 (Fig. 4-2(b)). The op-amp circuit cannot generate an output voltage υ_o that exceeds its supply voltage. Hence, υ_o is bounded to $\pm V_{cc}$ which means

$$|\upsilon_o| \leq V_{cc},$$

or equivalently,

$$-V_{cc} \leq \upsilon_o \leq V_{cc}. \tag{4.11}$$

Thus, the *linear dynamic range* of υ_o extends from $-V_{cc}$ to $+V_{cc}$.

According to Example 4-1, υ_o is related to the signal voltage υ_s by

$$\upsilon_o = G\upsilon_s, \tag{4.12}$$

with

$$G \approx \frac{R_1 + R_2}{R_2}. \tag{4.13}$$

Inserting Eq. (4.12) into Eq. (4.11) gives

$$|G\upsilon_s| \leq V_{cc}, \tag{4.14}$$

or

$$\frac{-V_{cc}}{G} \leq \upsilon_s \leq \frac{V_{cc}}{G}, \tag{4.15}$$

which states that the linear dynamic range of υ_s is inversely proportional to the circuit gain G.

(a) Unity Gain: If $R_2 = \infty$ (open circuit between node b and ground in the circuit of Fig. 4-8(a)), Eq. (4.13) gives $G \approx 1$. The corresponding dynamic range of υ_s extends from $-V_{cc}$ to $+V_{cc}$, the same as the output. The input-output transfer plot relating υ_o to υ_s is displayed in green in Fig. 4-8(b).

(b) Modest Gain: If we choose $R_1/R_2 = 4$, Eq. (4.13) gives $G = 5$, and the dynamic range of υ_s now extends from $-(10/5) = -2$V to +2 V. Thus, the gain is higher than the unity-gain case by a factor of 5, but the dynamic range of υ_s is narrower by the same factor.

(c) Maximum Gain: If R_1 is removed (replaced with an open circuit between nodes a and b) and R_2 is set equal to zero (short circuit), no feedback will take place in the circuit of Fig. 4-8(a). Use of the exact expression for G given by Eq. (4.8) leads to $G = A$. Since $A = 10^6$, the absence of feedback provides a huge gain, but operationally υ_s becomes limited to a very narrow range extending from -10 μV to $+10$ μV.

▶ Application of negative feedback offers a trade-off between circuit gain and dynamic range for the input voltage. ◀

Concept Question 4-5: Why is negative feedback used in op-amp circuits? (See CAD)

Concept Question 4-6: How large is the circuit gain G in the absence of feedback? How large is it with 100 percent feedback (equivalent to setting $R_1 = 0$ in the circuit of Fig. 4-8(a))? (See CAD)

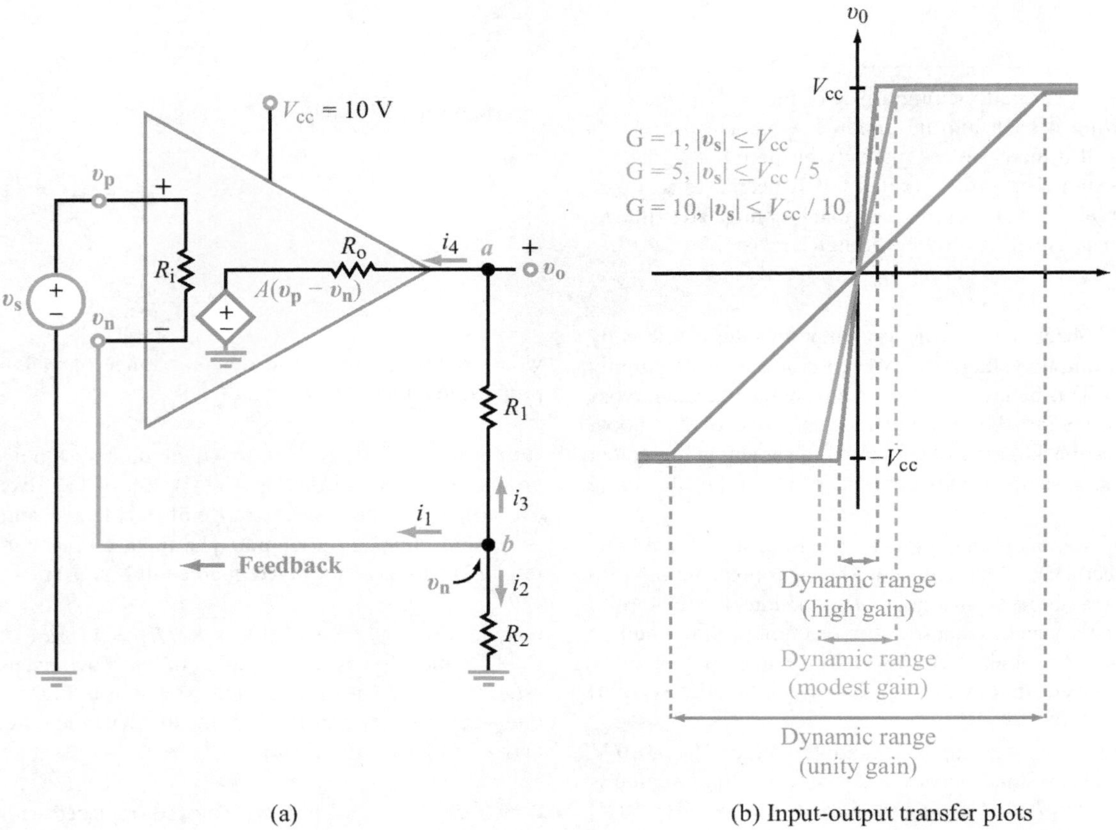

Figure 4-8: Trade-off between gain and dynamic range.

Exercise 4-2: To evaluate the trade-off between the circuit gain G and the linear dynamic range of v_s, apply Eq. (4.8) to find the magnitude of G and then determine the corresponding dynamic range of v_s for each of the following values of R_2: 0 (no feedback), 800 Ω, 8.8 kΩ, 40 kΩ, 80 kΩ, and 1 MΩ. Except for R_2, all other quantities remain unchanged.

Answer:

R_2	G	v_s Range
0	10^6	$-10\ \mu$V to $+10\ \mu$V
800 Ω	101	−99 mV to +99 mV
8.8 kΩ	10.1	−0.99 V to +0.99 V
40 kΩ	3	−3.3 V to +3.3 V
80 kΩ	2	−5 V to +5 V
1 MΩ	1.08	−9.26 V to +9.26 V

(See CAD)

4-3 Ideal Op-Amp Model

We noted in Section 4-1 that the op amp has a very large input resistance R_i on the order of 10^7 Ω, a relatively small output resistance R_o on the order of 1–100 Ω, and an open-loop gain $A \approx 10^6$. Usually, the series resistances of the input circuit connected to terminals v_p and v_n are several orders of magnitude smaller than R_i. Consequently, not only will very little current flow through the input circuit, but also the voltage drop across the input-circuit resistors will be negligibly small in comparison with the voltage drop across R_i. These considerations allow us to simplify the equivalent circuit of the op amp by replacing it with the ideal op-amp circuit model shown in Fig. 4-9, in which R_i has been replaced with an open circuit. An open circuit between terminals v_p and v_n implies

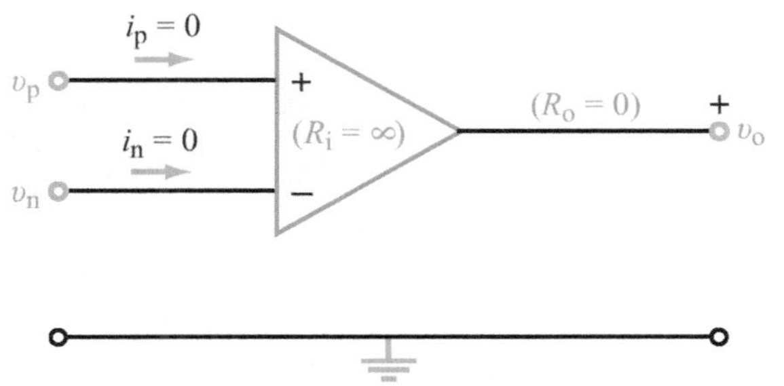

Figure 4-9: Ideal op-amp model.

the following *ideal op-amp current constraint*:

$$i_p = i_n = 0 \qquad \text{(ideal op-amp model).} \tag{4.16}$$

In reality, i_p and i_n are very small but not identically zero; for if they were, there would be no amplification through the op amp. Nevertheless, the current condition given by Eq. (4.16) will prove quite useful.

Similarly, at the output side, *if the load resistor connected in series with R_o is several orders of magnitude larger than R_o, then R_o can be ignored by setting it equal to zero.* Finally, in the ideal op-amp model, the large open-loop gain A is made infinite—the consequence of which is that

$$v_p - v_n = \frac{v_o}{A} \to 0 \qquad \text{as } A \to \infty.$$

Hence, we obtain the *ideal op-amp voltage constraint*

$$v_p = v_n \qquad \text{(ideal op-amp model).} \tag{4.17}$$

In reality v_p and v_n are not exactly equal, but very close to being equal, and only when negative feedback is in use. Nevertheless, setting $v_p = v_n$ leads to highly accurate results when relating the output to the input. In summary:

> ▶ The ideal op-amp model characterizes the op amp in terms of an equivalent circuit in which $R_i = \infty$, $R_o = 0$, and $A = \infty$. ◀

The operative consequences are given by Eqs. (4.16) and (4.17) and in Table 4-2.

Table 4-2: **Characteristics of the ideal op-amp model.**

Ideal Op Amp	
• Current constraint	$i_p = i_n = 0$
• Voltage constraint	$v_p = v_n$
• $A = \infty$ $\quad R_i = \infty$	$R_o = 0$

Noninverting Amplifier

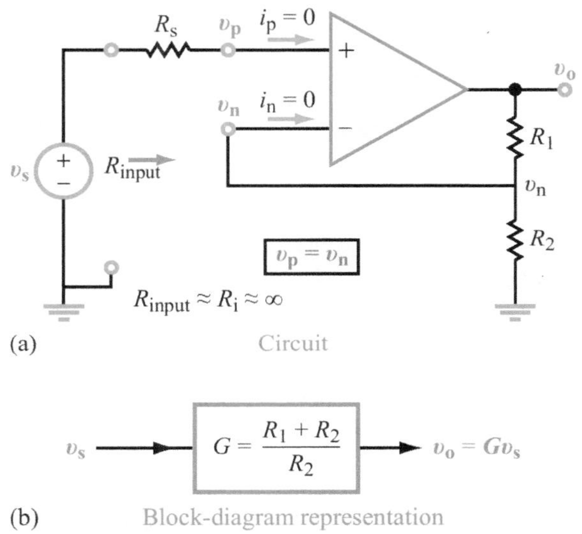

Figure 4-10: Noninverting amplifier circuit: (a) using ideal op-amp model and (b) equivalent block-diagram representation.

To illustrate the utility of the ideal op-amp model, let us re-examine the circuit we analyzed earlier in Example 4-1, but we will do so this time using the ideal model. The new circuit, as shown in Fig. 4-10, includes a source resistance R_s, but because the op amp draws no current ($i_p = 0$), there is no voltage drop across R_s. Hence,

$$v_p = v_s, \tag{4.18}$$

and on the output side, v_o and v_n are related through voltage division by

$$v_o = \left(\frac{R_1 + R_2}{R_2}\right) v_n. \tag{4.19}$$

Using these two equations, in conjunction with $v_p = v_n$ (from Eq. (4.17)), we end up with the following result for the circuit gain G:

$$G = \frac{v_o}{v_s} = \left(\frac{R_1 + R_2}{R_2}\right), \quad (4.20)$$

which is identical to Eq. (4.10).

► The *input resistance* of the noninverting amplifier circuit shown in Fig. 4-10 is the Thévenin resistance of the op-amp circuit as seen by the input source v_s. Because $i_p = 0$, it is easy to show that $R_{input} = R_i \approx \infty$, where R_i is the input resistance of the op amp (typically on the order of $10^9\ \Omega$). ◄

► From here on forward, we use the ideal op-amp model exclusively. ◄

Concept Question 4-7: What are the current and voltage constraints of the ideal op amp? (See CAD)

Concept Question 4-8: What are the values of the input and output resistances of the ideal op amp? (See CAD)

Concept Question 4-9: In the ideal op-amp model, R_o is set equal to zero. To satisfy such an approximation, does the load resistance need to be much larger or much smaller than R_o? Explain. (See CAD)

Exercise 4-3: Consider the noninverting amplifier circuit of Fig. 4-10(a) under the conditions of the ideal op-amp model. Assume $V_{cc} = 10$ V. Determine the value of G and the corresponding dynamic range of v_s for each of the following values of R_1/R_2: 0, 1, 9, 99, 10^3, 10^6.

Answer:

R_1/R_2	G	v_s Range
0	1	−10 V to +10 V
1	2	−5 V to +5 V
9	10	−1 V to +1 V
99	100	−0.1 V to +0.1 V
1000	∼ 1000	−10 mV to +10 mV (approx.)
10^6	∼ 10^6	−10 μV to +10 μV (approx.)

(See CAD)

4-4 Inverting Amplifier

► In an *inverting amplifier* op-amp circuit, the input source is connected to terminal v_n (instead of to terminal v_p) through an *input source resistance* R_s, and terminal v_p is connected to ground. ◄

Feedback from the output continues to be applied at v_n (through a *feedback resistance* R_f), as shown in Fig. 4-11. It is called an *inverting* amplifier because (as we will see shortly) the circuit gain G is negative.

To relate the output voltage v_o to the input signal voltage v_s, we start by writing down the node-voltage equation at terminal v_n as

$$i_1 + i_2 + i_n = 0 \quad (4.21)$$

or

$$\frac{v_n - v_s}{R_s} + \frac{v_n - v_o}{R_f} + i_n = 0. \quad (4.22)$$

Upon invoking the op-amp current constraint given by Eq. (4.16), namely $i_n = 0$, and the voltage constraint $v_n = v_p$,

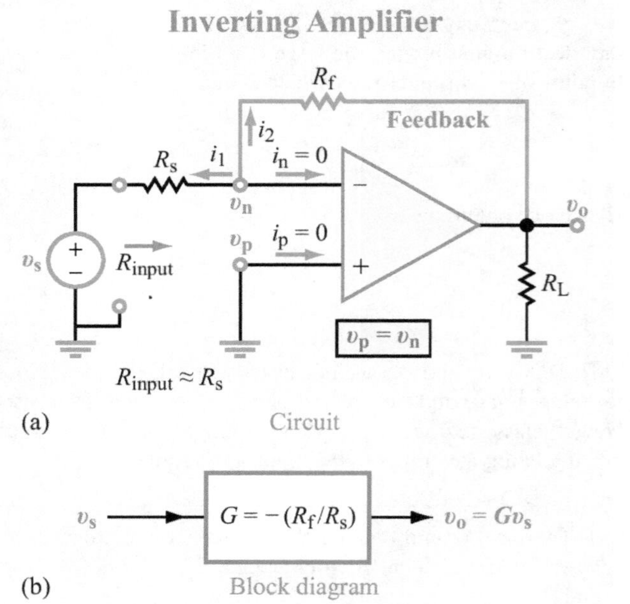

Figure 4-11: Inverting amplifier circuit and its block-diagram equivalent.

as well as recognizing that $\upsilon_p = 0$ (because terminal υ_p is connected to ground), we obtain the relationship

$$\upsilon_o = -\left(\frac{R_f}{R_s}\right)\upsilon_s. \tag{4.23}$$

The circuit voltage gain of the inverting amplifier therefore is given by

$$G = \frac{\upsilon_o}{\upsilon_s} = -\left(\frac{R_f}{R_s}\right). \tag{4.24}$$

▶ In addition to amplifying υ_s by the ratio (R_f/R_s), the inverting amplifier also reverses the polarity of υ_s. ◀

▶ υ_o is independent of the magnitude of the load resistance R_L, so long as R_L is much larger than the op-amp output resistance R_o (which is an implicit assumption of the ideal op-amp model). ◀

Because $\upsilon_n = 0$, a Thévenin analysis of the circuit in Fig. 4-11(a) would reveal that the *input resistance* of the inverting amplifier circuit (as seen by source υ_s) is $R_{input} = R_{Th} = R_s$.

▶ **Caution:** Under the ideal op-amp model, it is not possible to compute i_o, the current that flows into the op amp from output terminal υ_o. Hence, it is inappropriate to apply KCL at that terminal because additional current can be delivered by the supply voltage sources V_{cc} and $-V_{cc}$. ◀

Example 4-2: Amplifier with Input Current Source

For the circuit shown in Fig. 4-12(a): (a) obtain an expression for the input-output transfer function $K_t = \upsilon_o/i_s$ and evaluate it for $R_1 = 1$ kΩ, $R_2 = 2$ kΩ, $R_f = 30$ kΩ, and $R_L = 10$ kΩ; and (b) determine the linear dynamic range of i_s if $V_{cc} = 20$ V.

Solution: (a) Application of the source transformation method converts the combination of i_s and R_2 into a voltage source $\upsilon_s = i_s R_2$, in series with a resistance R_2. Upon combining R_2 in series with R_1, we obtain the new circuit shown in Fig. 4-12(b), which is identical in form to the inverting amplifier circuit of Fig. 4-11, except that now the source

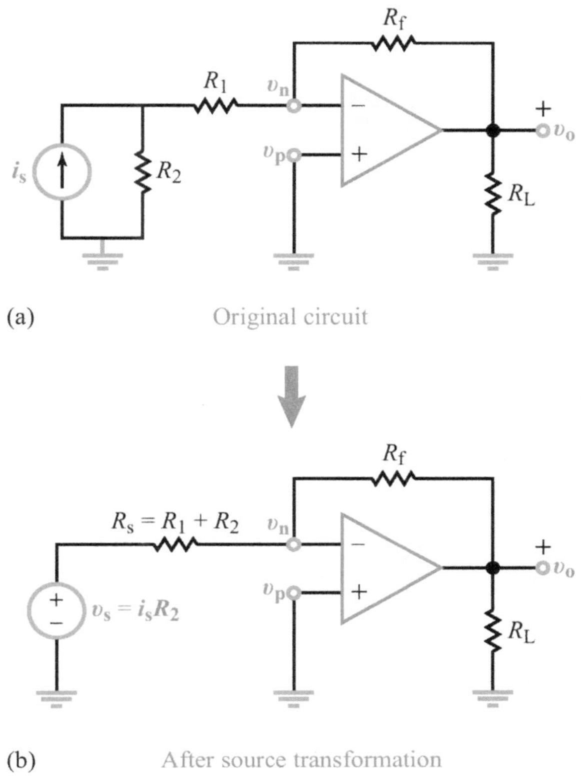

Figure 4-12: Inverting amplifier circuit of Example 4-2.

resistance is $R_s = (R_1 + R_2)$. Hence, application of Eq. (4.23) gives

$$\upsilon_o = -\left(\frac{R_f}{R_1 + R_2}\right)\upsilon_s = -\left(\frac{R_f}{R_1 + R_2}\right)R_2 i_s, \tag{4.25}$$

from which we obtain the transfer function

$$K_t = \frac{\upsilon_o}{i_s} = -\frac{R_f R_2}{R_1 + R_2}. \tag{4.26}$$

For $R_1 = 1$ kΩ, $R_2 = 2$ kΩ, and $R_f = 30$ kΩ,

$$K_t = \frac{\upsilon_o}{i_s} = -2 \times 10^4 \qquad \text{(V/A)}.$$

(b) From the expression for K_t,

$$i_s = -\frac{\upsilon_o}{2 \times 10^4},$$

and since $|v_o|$ is bounded by $V_{cc} = 20$ V, the linear range for i_s is bounded by

$$|i_s| = \left|\frac{V_{cc}}{2 \times 10^4}\right| = \left|\frac{20}{2 \times 10^4}\right| = 1 \text{ mA}.$$

Thus, the linear range of i_s extends from -1 mA to $+1$ mA.

Concept Question 4-10: How does feedback control the gain of the inverting-amplifier circuit? (See CAD)

Concept Question 4-11: The expression given by Eq. (4.24) states that the gain of the inverting amplifier is independent of the magnitude of R_L. Would the expression remain valid if $R_L = 0$? Explain. (See CAD)

Exercise 4-4: The input to an inverting-amplifier circuit consists of $v_s = 0.2$ V and $R_s = 10$ Ω. If $V_{cc} = 12$ V, what is the maximum value that R_f can assume before saturating the op amp?

Answer: $G_{max} = -60$, $R_f = 600$ Ω. (See CAD)

4-5 Inverting Summing Amplifier

By connecting multiple sources in parallel at terminal v_n of the inverting amplifier, the circuit becomes an *adder* (or more precisely a *scaled inverting adder*), as depicted by the block diagram of Fig. 4-13(d). After we demonstrate how such a circuit (usually called an *inverting summing amplifier*) works for two input voltages v_1 and v_2, we will extend it to multiple sources. There are many applications where we may want to scale and add multiple voltages together, such as combining or averaging results from several sensors.

For the circuit shown in Fig. 4-13(a), our goal is to relate the output voltage v_o to v_1 and v_2. To do so, we apply the source-transformation technique so as to cast the input circuit in the form of a single voltage source v_s in series with a source resistance R_s. The steps involved in the transformation are illustrated in Fig. 4-13(b) and (c). Voltage to current transformation gives $i_{s_1} = v_1/R_1$ and $i_{s_2} = v_2/R_2$, which can be combined together into a single current source as

$$i_s = i_{s_1} + i_{s_2} = \frac{v_1}{R_1} + \frac{v_2}{R_2} = \frac{v_1 R_2 + v_2 R_1}{R_1 R_2}. \quad (4.27)$$

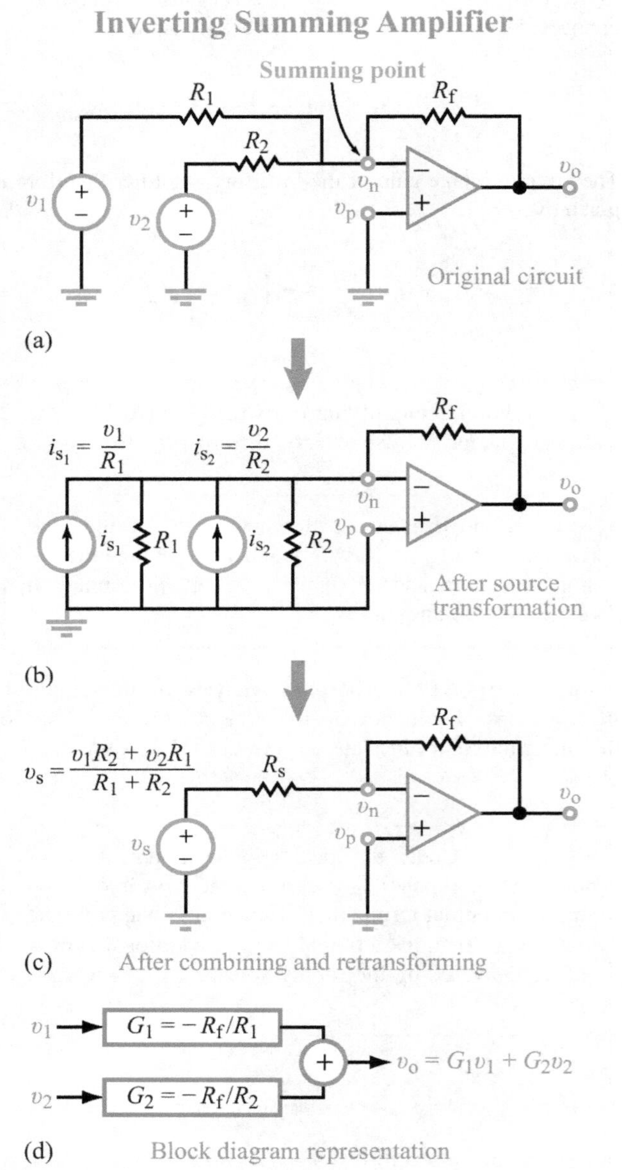

Figure 4-13: Inverting summing amplifier.

Similarly, the two parallel resistors add up to

$$R_s = \frac{R_1 R_2}{R_1 + R_2}. \quad (4.28)$$

If we transform (i_s, R_s) into a voltage source (υ_s, R_s), we get

$$\upsilon_s = i_s R_s = \left(\frac{\upsilon_1 R_2 + \upsilon_2 R_1}{R_1 R_2}\right) \frac{R_1 R_2}{R_1 + R_2} = \frac{\upsilon_1 R_2 + \upsilon_2 R_1}{R_1 + R_2}. \tag{4.29}$$

The circuit in Fig. 4-13(c) is identical in form to that of the inverting amplifier of Fig. 4-11. Hence, by applying the input-output voltage relationship given by Eq. (4.23), we have

$$\upsilon_o = -\left(\frac{R_f}{R_s}\right)\upsilon_s = -\frac{R_f}{\left(\frac{R_1 R_2}{R_1 + R_2}\right)}\left(\frac{\upsilon_1 R_2 + \upsilon_2 R_1}{R_1 + R_2}\right)$$
$$= -\left(\frac{R_f}{R_1}\right)\upsilon_1 - \left(\frac{R_f}{R_2}\right)\upsilon_2. \tag{4.30}$$

This expression for υ_o can be written in the form

$$\upsilon_o = G_1\upsilon_1 + G_2\upsilon_2, \tag{4.31}$$

where $G_1 = -(R_f/R_1)$ is the (negative) gain applied to source voltage υ_1, and $G_2 = -(R_f/R_2)$ is the gain applied to υ_2. Thus:

> ▶ The summing amplifier scales υ_1 by negative gain G_1 and υ_2 by negative gain G_2 and adds them together. ◀

4-5.1 Special Cases

For the special case where $R_1 = R_2 = R$,

$$\upsilon_o = -\left(\frac{R_f}{R}\right)(\upsilon_1 + \upsilon_2) \qquad \left(\begin{matrix}\text{equal gain} \\ R_1 = R_2 = R\end{matrix}\right), \tag{4.32}$$

and if additionally $R_f = R_1 = R_2$, then $G_1 = G_2 = -1$. In this case, the summing amplifier becomes an inverted adder with

$$\upsilon_o = -(\upsilon_1 + \upsilon_2) \qquad \left(\begin{matrix}\text{inverted adder} \\ R_1 = R_2 = R_f\end{matrix}\right). \tag{4.33}$$

Generalizing to the case where the input consists of n input voltage sources υ_1 to υ_n (and associated source resistances R_1 to R_n, respectively), all connected in parallel at the same summing point (terminal υ_n), the output voltage becomes

$$\upsilon_o = \left(-\frac{R_f}{R_1}\right)\upsilon_1 + \left(-\frac{R_f}{R_2}\right)\upsilon_2 + \cdots + \left(-\frac{R_f}{R_n}\right)\upsilon_n. \tag{4.34}$$

Example 4-3: Summing Circuit

Use inverting amplifiers to design a circuit that performs the operation

$$\upsilon_o = 4\upsilon_1 + 7\upsilon_2.$$

Solution: The desired circuit has to amplify υ_1 by a factor of 4, amplify υ_2 by a factor of 7, and add the two together. A summing amplifier can do that, but it also inverts the sum. Hence, we will need to use a two-stage cascaded circuit with the first stage providing the desired operation within a "−" sign and then follow it up with an inverting amplifier with a gain of (−1). The two-stage circuit is shown in Fig. 4-14.

For the first stage, we need to select values for R_1, R_2, and R_{f_1} such that

$$\frac{R_{f_1}}{R_1} = 4 \qquad \text{and} \qquad \frac{R_{f_1}}{R_2} = 7.$$

Since we have only two constraints, we can satisfy the specified ratios with an infinite number of combinations. Arbitrarily, we choose $R_{f_1} = 56$ kΩ, which then specifies the other resistors as

$$R_1 = 14\text{ k}\Omega \qquad \text{and} \qquad R_2 = 8\text{ k}\Omega.$$

For the second stage, a gain of (−1) requires that

$$\frac{R_{f_2}}{R_{s_2}} = 1.$$

Arbitrarily, we choose $R_{f_2} = R_{s_2} = 20$ kΩ.

4-5.2 Noninverting Summer

To perform the summing operation, the solution offered in Example 4-3 employed two inverting amplifier circuits—one

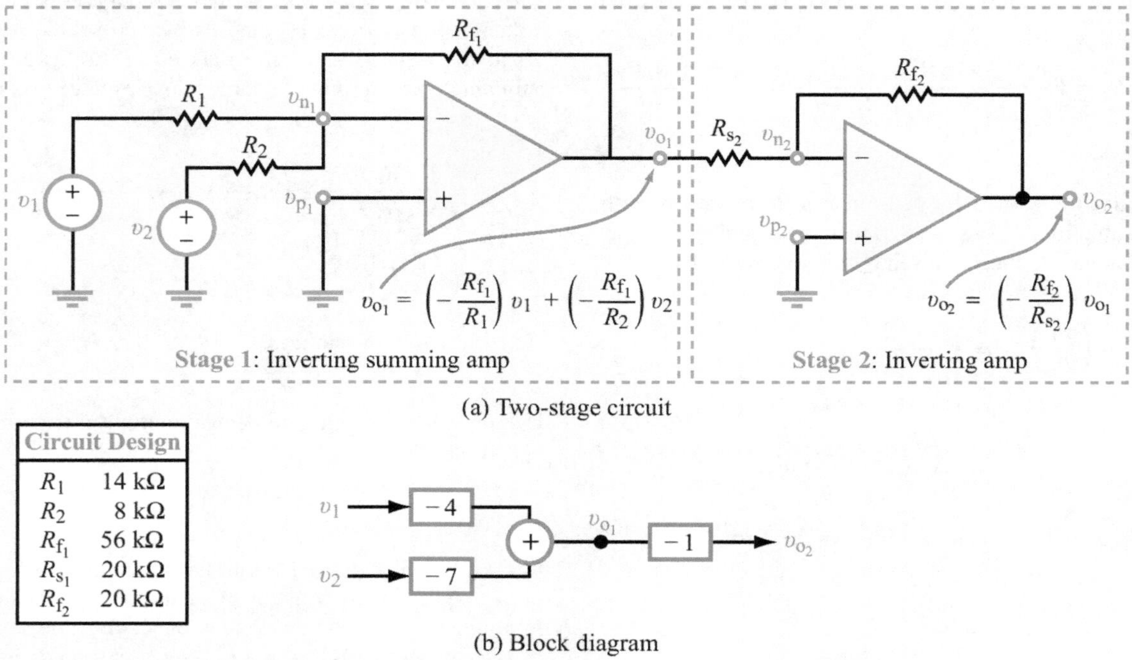

Circuit Design	
R_1	14 kΩ
R_2	8 kΩ
R_{f_1}	56 kΩ
R_{s_1}	20 kΩ
R_{f_2}	20 kΩ

Figure 4-14: Two-stage circuit realization of $v_o = 4v_1 + 7v_2$.

to perform an inverted sum, and a second one to provide multiplication by (-1). Alternatively, the same result can be achieved by using a single op amp in a noninverting amplifier circuit, as shown in Fig. 4-15.

From our analysis in Section 4-3, we established that the output voltage v_o of the noninverting amplifier circuit is related to v_p by

$$\frac{v_o}{v_p} = G = \frac{R_1 + R_2}{R_2}. \tag{4.35}$$

For the circuit in Fig. 4-15, in view of the ideal op-amp constraint that the op amp draws no current ($i_p = 0$), it is a straightforward task to show that

$$v_p = \frac{v_1 R_{s_2} + v_2 R_{s_1}}{R_{s_1} + R_{s_2}}. \tag{4.36}$$

Combining Eqs. (4.35) and (4.36) leads to

$$v_o = G\left[\left(\frac{R_{s_2}}{R_{s_2} + R_{s_1}}\right) v_1 + \left(\frac{R_{s_1}}{R_{s_1} + R_{s_2}}\right) v_2\right]. \tag{4.37}$$

To realize a coefficient of 4 for v_1 and a coefficient of 7 for v_2, it is necessary that

$$\frac{G R_{s_2}}{R_{s_1} + R_{s_2}} = 4$$

and

$$\frac{G R_{s_1}}{R_{s_1} + R_{s_2}} = 7.$$

A possible solution that satisfies these two constraints is $R_{s_1} = 7$ kΩ, $R_{s_2} = 4$ kΩ, and $G = 11$. Furthermore, the specified value of G can be satisfied by choosing $R_1 = 50$ kΩ and $R_2 = 5$ kΩ.

4-5.3 Multiple Ways of Building a System

There are often several different choices for how to implement a linear equation such as $v_o = 4v_1 + 7v_2$ (Example 4-3) with op-amp circuits. Here are a few options:

(a) $v_o = (4v_1) + (7v_2)$: Multiply v_1 by 4 (noninverting amplifier with a gain of 4) and v_2 by 7 (noninverting amplifier with a gain of 7), and then add them together (noninverting summer with a gain of 1).

Technology Brief 10
Computer Memory Circuits

The storage of information in electronically addressable devices is one of the hallmarks of all modern computer systems. Among these devices are a class of storage media, collectively called ***solid-state*** or ***semiconductor memories***, which store information by changing the state of an electronic circuit. The state of the circuit usually has two possibilities (0 or 1) and is termed a ***bit*** (see Technology Brief 8). Values in memories are represented by a string of ***binary*** bits; a 5-bit sequence $[V_1 V_2 V_3 V_4 V_5]$, for example, can be used to represent any integer decimal value between 0 and 31. How do computers store these bits? Many types of technologies have emerged over the last 40 years, so in this Brief, we will highlight some of the principal technologies in use today or under development. It is worth noting that memory devices usually store these values in arrays. For example, a small memory might store sixteen different 16-bit numbers; this memory usually would be referred to as a 16 $\times$ 16 block or a 256-bit memory. Of course, modern multi-gigabyte computer memories use thousands of much larger blocks to store very large numbers of bits (Fig. TF10-1).

Read-Only Memories (ROMs)

One of the oldest, still-employed, memory architectures is the ***read-only memory*** (ROM). The ROM is so termed because it can only be "written" once, and after that it can only be read. ROMs usually are used to store information that will not need to be changed (such as certain startup information on your computer or a short bit of code always used by an integrated circuit in your camera). Each bit in the ROM is held by a single MOSFET transistor.

Consider the circuit in Fig. TF10-2(a), which operates much like the circuit in Fig. 4-25. The MOSFET has three voltages, all referenced to ground. For convenience, the input voltage is labeled V_{READ} and the output voltage is labeled V_{BIT}. The third voltage, V_{DD}, is the voltage of the dc power supply connected to the drain terminal via a resistor R. If $V_{\text{READ}} \ll V_{\text{DD}}$, then the output registers a voltage $V_{\text{BIT}} = V_{\text{DD}}$ denoting the binary state "1," but if $V_{\text{READ}} \geq V_{\text{DD}}$, then the output terminal shorts to ground, generating $V_{\text{BIT}} = 0$ denoting the binary state "0." But how does this translate into a permanent memory on a chip? Let us examine the 4-bit ROM diagrammed in Fig. TF10-2(b). In this case, some bits simply do not have transistors; V_{BIT2}, for example, is permanently connected to V_{DD} via a resistor. This may seem trivial,

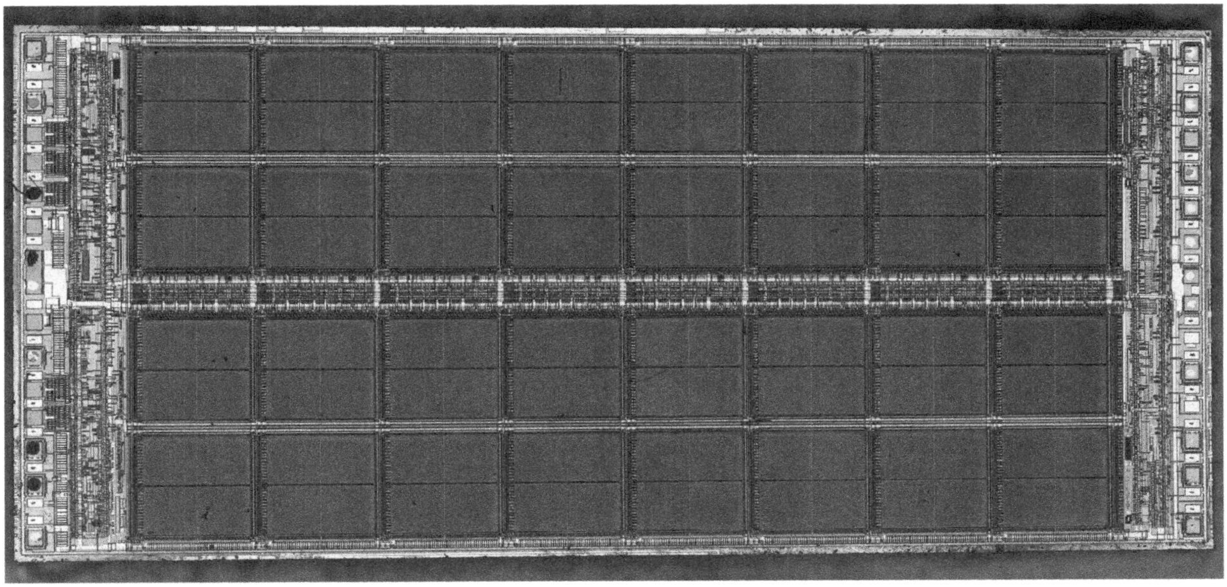

Figure TF10-1: Integrated circuit die photo of a Micron MT4C1024 2^{20}-bit DRAM chip. Die size is 8.662 mm $\times$ 3.969 mm. (Courtesy of ZeptoBars.)

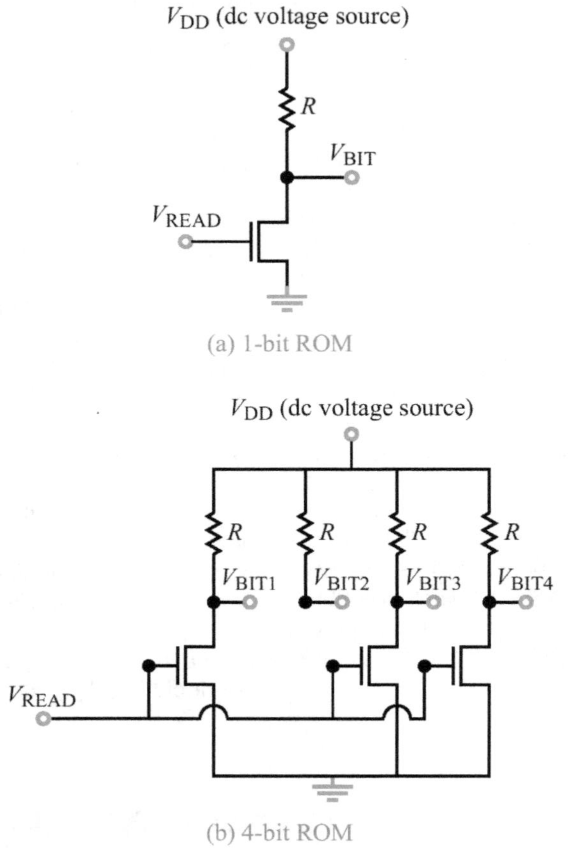

Figure TF10-2: (a) 1-bit ROM that uses a MOSFET transistor, and (b) 4-bit ROM configured to store the sequence [0100], whose decimal value is 4.

but this specific 4-bit memory configuration always stores the value [0100]. In this same way, thousands of such components can be strung together in rows and columns in $N \times N$ arrays. As long as a power supply of voltage V_{DD} is connected to the circuit, the memory will report its contents to an external circuit as [0100]. Importantly, even if you remove power altogether, the values are not lost; as soon as you add power back to the chip, the same values appear again (i.e., you would have to break the chip to make it forget what it is storing!). Because of the permanency of this data, these memories also often are called ***nonvolatile memories*** (NVM).

Random-Access Memories (RAMs)

RAMs are a class of memories that can be read to and written from constantly. RAMs generally fall into two categories: ***static RAMs*** and ***dynamic RAMs*** (DRAMs). Because RAMs lose the state of their bits if the power is removed, they are termed ***volatile memories***. Static RAMs not only can be read from and written to, but also do not forget their state as long as power is supplied. These circuits also are composed of transistors, but each single bit in a modern static RAM consists of four transistors wired up in a bi-stable circuit (the explanation of which we will leave to your intermediate digital components classes!). Dynamic RAMs, on the other hand, are illustrated more easily. Dynamic RAMs usually hold more bits per area than static RAMs, but they need to be refreshed constantly (even when power is supplied continuously to the chip).

Figure TF10-3 shows a simple one-transistor dynamic RAM. Again, we will treat the transistor as we did in

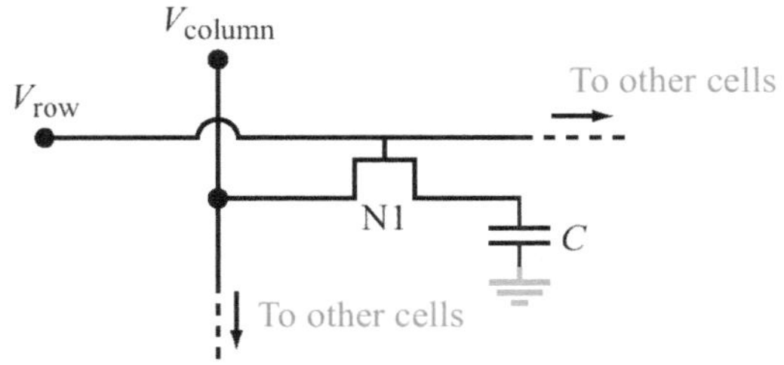

Figure TF10-3: 1-bit DRAM cell.

Section 4-11. Note that if we make $V_{ROW} > V_{DD}$, then the transistor will conduct and the capacitor C will start charging to whatever value we select for V_{COLUMN}. When writing a bit, V_{COLUMN} usually is set at either 0 (GND) or 1 (V_{DD}). We can calculate how long this charging-up process will require, because we know the value of C and the transistor's current gain g (see Section 5-7). When the capacitor is charged to V_{DD}, a value of 1 is stored in the DRAM. Had we applied instead a value of zero volts to V_{COLUMN}, the transistor would have discharged to ground (instead of charged to V_{DD}) and the bit would have a value of 0. However, note that unlike the ROM, the state of the bit is not "hardwired." That is, if even tiny leakage currents were to flow through the transistor when it is not on (that is, when $V_{ROW} < V_{DD}$), then charge will constantly leak away and the voltage of the transistor will drop slowly with time. After a short time (on the order of a few milliseconds in the dynamic RAM in your computer), the capacitor will have irrecoverably lost its value. How is that mitigated? Well, it turns out that a modern memory will read and then re-write every one of its (several billion) bits every 64 milliseconds to keep them refreshed! Because each bit is so simple (one transistor and one capacitor), it is possible to manufacture DRAMs with very high memory densities (which is why 1-Gbit DRAMs are now available in packages of reasonable size). Other variations of DRAMs also exist whose architectures deviate slightly from the previous model—at either the transistor or system level. ***Synchronous Graphics RAM*** (SGRAM), for example, is a DRAM modified for use with graphics adaptors; ***Double Data Rate 4 RAM*** (DDR4RAM) is a fourth-generation enhancement over DRAM which allows for faster clock speeds and lower operating voltages.

Advanced Memories

Several substantially different technologies are emerging that likely will change the market landscape—just as Flash memories revolutionized portable memory (like your USB memory stick). Apart from the drive to increase storage density and access speed, one of the principal drivers in today's memory research is the development of non-volatile memories that do not degrade over time (unlike Flash).

The ***Ferroelectric RAM*** (FeRAM) is the first of these technologies to enter mainstream production; FeRAM replaces the capacitor in DRAM (Fig. TF10-3) with a ferroelectric capacitor that can hold the binary state even with power removed. While FeRAM can be faster than Flash memories, FeRAM densities are still much smaller than modern Flash (and Flash densities continue to increase rapidly). FeRAM currently is used in niche applications where the increased speed is important. ***Magnetoresistive RAM*** (MRAM) is another emerging technology, currently commercialized by Everspin Technologies (spun out from Freescale Semiconductor), which relies on magnetic plates to store bits of data. In MRAM, each cell is composed of two ferromagnetic plates separated by an insulator. The storage and retrieval of bits occurs by manipulation of the magnetic polarization of the plates with associated circuits. Like FeRAM, MRAM currently is overshadowed by Flash memories, but improvements in density, speed, and fabrication methods may make it a viable alternative in the mainstream consumer market in the future. Even more speculative is the idea of using single carbon nanotubes to store binary bits by changing their configuration electronically; this technology is currently known as ***Nano RAM*** (NRAM).

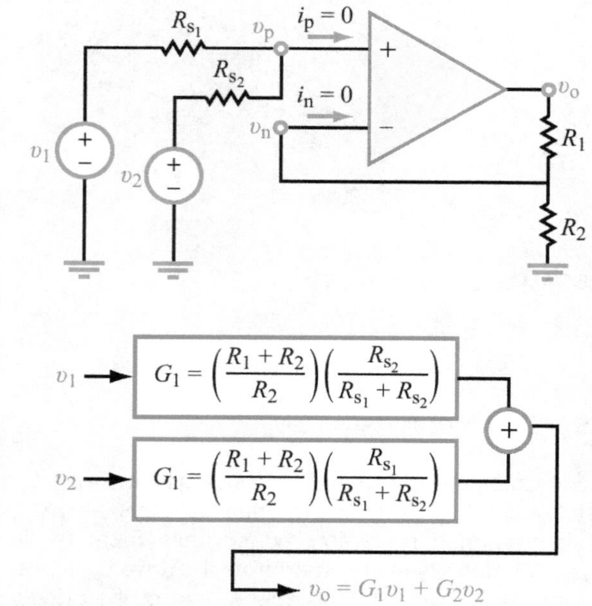

Figure 4-15: Noninverting summer.

(b) $v_o = (-4v_1 - 7v_2)(-1)$: Multiply v_1 by -4 and v_2 by -7 and add them together (inverting summing amplifier with gains of -4 and -7), and then multiply the result by -1 (inverting amplifier with a gain of -1).

(c) $v_o = (4v_1 + 7v_2)$: Multiply v_1 by 4 and v_2 by 7 and add them together (noninverting summing amplifier with gains of 4 and 7).

(d) $v_o = [(2v_1)+(3.5v_2)]\times 2$: Multiply v_1 by 2 (noninverting amplifier with a gain of 2) and v_2 by 3.5 (noninverting amplifier with a gain of 3.5), and then add them (noninverting summer with a gain of 2).

Why might you choose one of these systems over another? There are several reasons:

- To minimize the number of op amps (option c)
- To meet gain limitations. An inverting amplifier can have a gain of less than 1, but a noninverting amplifier cannot.
- To avoid saturation. The output voltage of any individual stage is limited by its V_{cc}. The order in which multiplication/summation is done must keep each individual stage from exceeding $+/-V_{cc}$.
- Sensitivity when adding large and small values. Care is typically taken to add values that are similar in magnitude, so amplification is typically done prior to summation if two values have significantly different magnitudes.
- Other considerations ...

Concept Question 4-12: What type of op-amp circuits (inverting, noninverting, and others) might one use to perform the operation $v_o = G_1v_1 + G_2v_2$ with G_1 and G_2 both positive? (See CAD)

Concept Question 4-13: What is an inverting adder? (See CAD)

Exercise 4-5: The circuit shown in Fig. 4-14(a) is to be used to perform the operation

$$v_o = 3v_1 + 6v_2.$$

If $R_1 = 1.2\ \text{k}\Omega$, $R_{s_2} = 2\ \text{k}\Omega$, and $R_{f_2} = 4\ \text{k}\Omega$, select values for R_2 and R_{f_1} so as to realize the desired result.

Answer: $R_{f_1} = 1.8\ \text{k}\Omega$, $R_2 = 600\ \Omega$. (See CAD)

4-6 Difference Amplifier

When an input signal v_2 is connected to terminal v_p of a noninverting amplifier circuit, the output is a scaled version of v_2. A similar outcome is generated by an inverting amplifier circuit when an input voltage v_1 is connected to the op amp's v_n terminal, except that in addition to scaling v_1 its polarity is reversed as well. The *difference amplifier* circuit combines these two functions to perform *subtraction*.

In the difference-amplifier circuit of Fig. 4-16(a), the input signals are v_1 and v_2, R_2 is the feedback resistance, R_1 is the source resistance of v_1, and resistances R_3 and R_4 serve to control the scaling factor (gain) of v_2. To obtain an expression that relates the output voltage v_o to the inputs v_1 and v_2, we apply KCL at nodes v_n and v_p. At v_n, $i_1 + i_2 + i_n = 0$, which is equivalent to

$$\frac{v_n - v_1}{R_1} + \frac{v_n - v_o}{R_2} + i_n = 0 \qquad \text{(node } v_n\text{)}. \qquad (4.38)$$

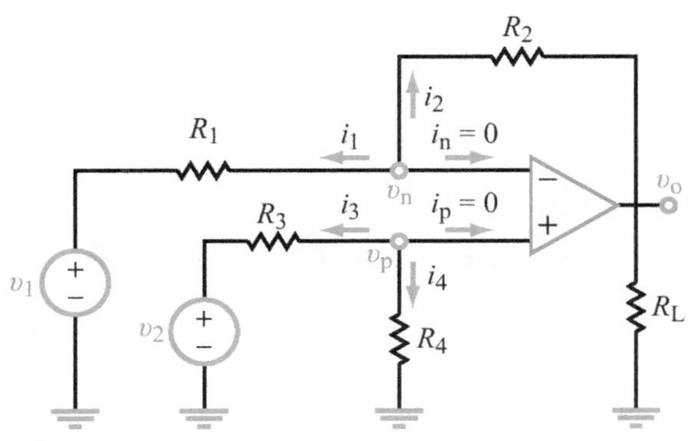

(a) Difference circuit

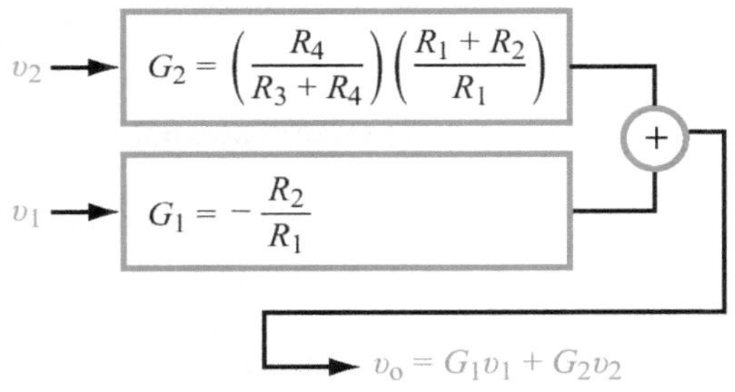

(b) Block diagram

Figure 4-16: Difference-amplifier circuit.

At v_p, $i_3 + i_4 + i_p = 0$, or

$$\frac{v_p - v_2}{R_3} + \frac{v_p}{R_4} + i_p = 0 \qquad \text{(node } v_p\text{)}. \tag{4.39}$$

Upon imposing the ideal op-amp constraints $i_p = i_n = 0$ and $v_p = v_n$, we end up with

$$v_o = \left[\left(\frac{R_4}{R_3+R_4}\right)\left(\frac{R_1+R_2}{R_1}\right)\right] v_2 - \left(\frac{R_2}{R_1}\right) v_1, \tag{4.40}$$

which can be cast in the form

$$v_o = G_2 v_2 + G_1 v_1, \tag{4.41}$$

where the scale factors (gains) are given by

$$G_2 = \left(\frac{R_4}{R_3+R_4}\right)\left(\frac{R_1+R_2}{R_1}\right) \tag{4.42a}$$

and

$$G_1 = -\left(\frac{R_2}{R_1}\right). \tag{4.42b}$$

According to Fig. 4-16(b) which is a block-diagram representation of the difference amplifier circuit:

▶ The difference amplifier scales v_2 by positive gain G_2, v_1 by negative gain G_1 and adds them together. ◀

For the difference amplifier to function as a subtraction circuit with equal gain, its resistors have to be interrelated by

$$R_2 R_3 = R_1 R_4, \tag{4.43}$$

in which case Eq. (4.41) reduces to

$$v_o = \left(\frac{R_2}{R_1}\right)(v_2 - v_1) \qquad \textbf{(equal gain)}. \tag{4.44}$$

Exact subtraction with no scaling requires that $R_1 = R_2$.

Exercise 4-6: The difference-amplifier circuit of Fig. 4-16 is used to realize the operation

$$v_o = (6v_2 - 2) \text{ V}.$$

Given that $R_3 = 5$ kΩ, $R_4 = 6$ kΩ, and $R_2 = 20$ kΩ, specify values for v_1 and R_1.

Answer: $v_1 = 0.2$ V, $R_1 = 2$ kΩ. (See CAD)

4-7 Voltage Follower/Buffer

4-7.1 No Buffer

In electronic circuits, we often need to incorporate the functionality of a relatively simple (but important) circuit that serves to isolate the input source from variations in the load resistance R_L. Such a circuit is called a *voltage follower*, *buffer*, or *unity gain amplifier*. To appreciate the utility of the voltage follower, let us first examine the circuit shown in Fig. 4-17(a). A source input circuit represented by its Thévenin equivalent (υ_s, R_s), is connected to a load R_L. The output voltage is

$$\upsilon_o = \frac{\upsilon_s R_L}{R_s + R_L} \quad \text{(without voltage follower)}, \qquad (4.45)$$

which obviously is dependent on both R_s and R_L, so if the load resistance R_L changes, so will the output voltage υ_o.

4-7.2 With Op-Amp Buffer

In contrast, when the op-amp voltage follower circuit shown in Fig. 4-17(b) is inserted in between the source circuit and the load, the output voltage becomes completely independent of both R_s and R_L. Because $i_p = 0$, it follows that $\upsilon_p = \upsilon_s$. Furthermore, in view of the op-amp constraint $\upsilon_p = \upsilon_n$ and because the output node is connected directly to υ_n, it follows that

$$\upsilon_o = \upsilon_p = \upsilon_s \quad \text{(with voltage follower)}, \qquad (4.46)$$

and this is true regardless of the values of R_s and R_L (excluding R_s = open circuit and/or R_L = short circuit, either of which would invalidate the entire circuit). Thus:

▶ The output of the voltage follower ***follows*** the input signal while remaining immune to changes in R_L because it has a high input resistance and low output resistance. ◀

A circuit that offers this type of protection is often called a *buffer*.

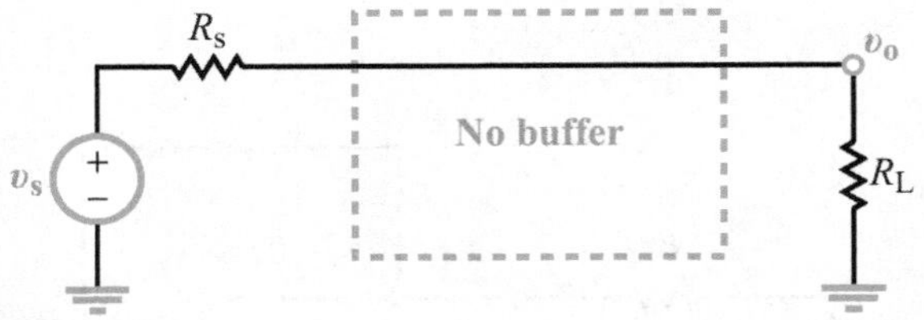

(a) Source circuit connected directly to a load

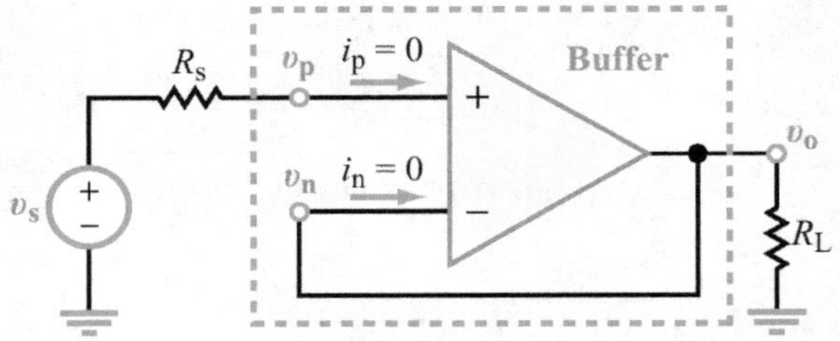

(b) Source circuit separated by a buffer

Figure 4-17: The voltage follower provides no voltage gain ($\upsilon_o = \upsilon_s$), but it insulates the input circuit from the load.

▶ When designing and building a multistage circuit, designers usually insert buffers between adjacent stages, which allows them to design each stage separately and then cascade them all together with buffers in between them. ◀

4-7.3 Input-Output Resistance

When is a buffer needed? Consider again the circuit in Fig. 4-17(a). Let us examine υ_o for various values of R_s and R_L.

R_s (kΩ)	R_L (Ω)	υ_o (V)	% change	Buffer needed?
1	100	0.09	91%	Yes
1	1	0.5	50%	Yes
1	10	0.91	9%	Probably
1	100	0.99	1%	No

If $R_s < R_L$, or even if $R_s \approx R_L$, there is a substantial difference between υ_o and υ_s. This is *overloading* the circuit, which we typically just call *loading*. Substantial current is drawn from the source, and the voltage is decreased as a result. To prevent this, a buffer is needed. But if $R_s \ll R_L$, the change is minimal, and the circuit does not require a buffer.

An additional interesting aspect of buffering has to do with where the current is coming from and where it is going to in the circuit. In Fig. 4-17(a), the current is coming from the source and going to the load. Excess current is being drawn, and the circuit is (over)loaded, thus reducing the output voltage υ_o. In Fig. 4-17(b), the current is *not* coming from the source, but it is going to the load. Where is it coming from? The answer is that it is coming from the output of the buffer, extracted from the power supply voltage V_{cc} that powers the op amp in the buffer.

Concept Question 4-14: What is the function of a voltage follower, and why is it called a "buffer"? (See CAD)

Concept Question 4-15: How much voltage gain is provided by the voltage follower? (See CAD)

Exercise 4-7: Express υ_o in terms of υ_1, υ_2, and υ_3 for the circuit in Fig. E4.7.

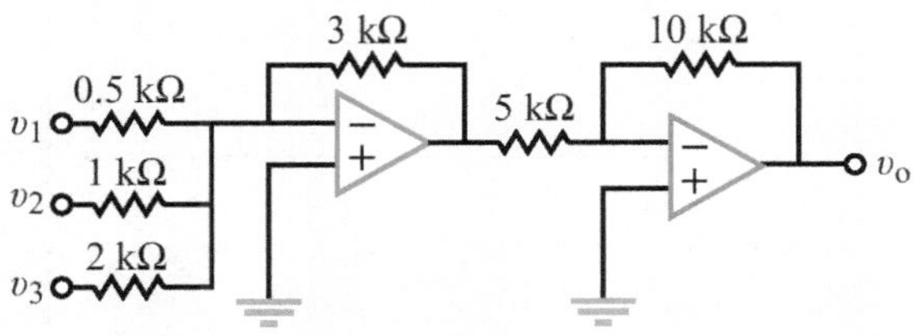

Figure E4.7

Answer: $\upsilon_o = 12\upsilon_1 + 6\upsilon_2 + 3\upsilon_3$. (See CAD)

4-8 Op-Amp Signal-Processing Circuits

Table 4-3 provides a summary of the op-amp circuits we have considered thus far, together with their functional characteristics in the form of block-diagram representations. These circuits can be used in various combinations to realize specific signal-processing operations. We note that the input-output transfer functions are independent of the load resistance R_L that may be connected between the output terminal υ_o and ground. In the case of the noninverting amplifier, the transfer function is also independent of the source resistance R_s.

▶ When cascading multiple stages of op-amp circuits in series, care must be exercised to ensure that none of the op amps is driven into saturation by the cumulative gain of the multiple stages. ◀

When analyzing circuits that involve op amps, whether in configurations similar to or different from those we encountered so far in this chapter, the basic rules to remember are as follows:

Basic Rules of Op-Amp Circuits

(1) KCL and KVL always apply everywhere in the circuit, but KCL is inapplicable at the output node when applying the ideal op-amp model. All other circuit-analysis tools can be applied to op-amp circuits.

(2) The op amp will operate in the linear range so long as $|\upsilon_o| < |V_{cc}|$.

(3) The ideal op-amp model assumes that the source resistance R_s (connected to terminals υ_p or υ_n) is much smaller than the op-amp input resistance R_i (which usually is no less than 10 MΩ), and the load resistance R_L is much larger than the op-amp output resistance R_o (which is on the order of tens of ohms).

(4) The ideal op-amp constraints are $i_p = i_n = 0$ and $\upsilon_p = \upsilon_n$.

Example 4-4: Block-Diagram Representation

Generate a block-diagram representation for the circuit shown in Fig. 4-18(a).

Solution: The first op amp is an inverting amplifier (Table 4-3(b)) with a dc input voltage $\upsilon_1 = 0.42$ V. Its circuit gain G_i (with the subscript added to denote "inverting amp") is

$$G_i = -\frac{30K}{10K} = -3,$$

Table 4-3: Summary of op-amp circuits.

Op-Amp Circuit	Block Diagram
(a) 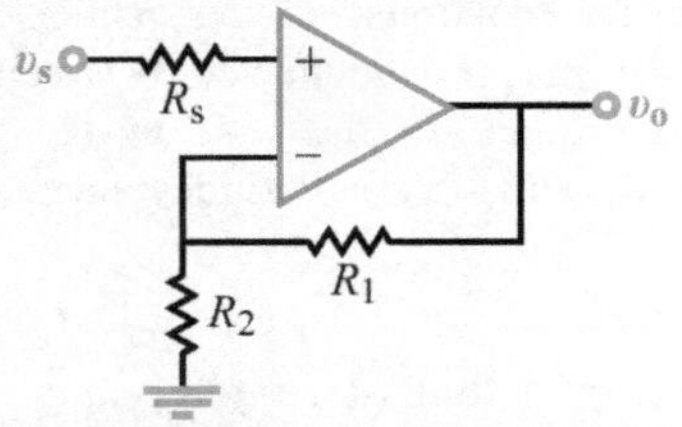	$v_s \rightarrow \boxed{G = \frac{R_1 + R_2}{R_2}} \rightarrow v_o = Gv_s$ **Noninverting Amp** (v_o independent of R_s)
(b) 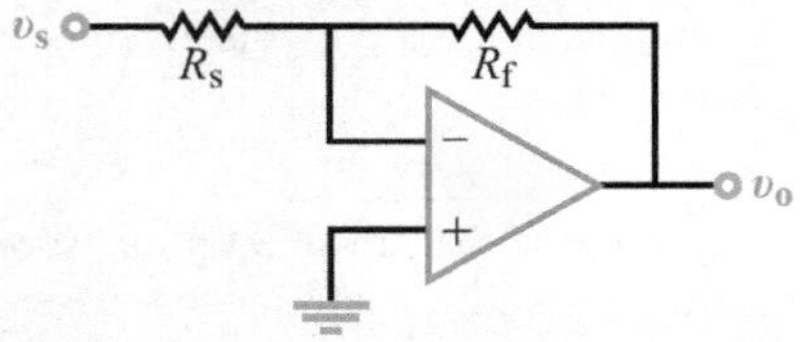	$v_s \rightarrow \boxed{G = -\frac{R_f}{R_s}} \rightarrow v_o = Gv_s$ **Inverting Amp**
(c) 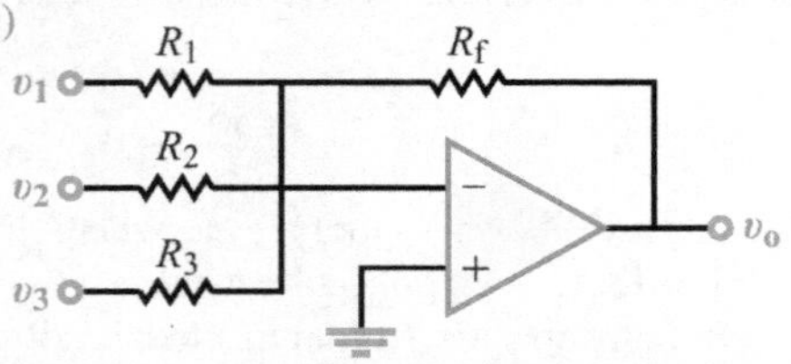	$v_1 \rightarrow \boxed{G_1 = -R_f/R_1}$, $v_2 \rightarrow \boxed{G_2 = -R_f/R_2}$, $v_3 \rightarrow \boxed{G_3 = -R_f/R_3}$ $\rightarrow (+) \rightarrow v_o = G_1v_1 + G_2v_2 + G_3v_3$ **Inverting Summing Amp**
(d) 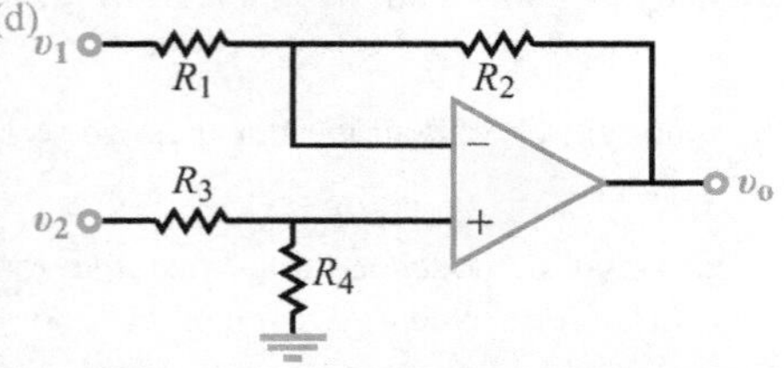	$v_1 \rightarrow \boxed{G_1 = -\frac{R_2}{R_1}}$, $v_2 \rightarrow \boxed{G_2 = \left(\frac{R_1 + R_2}{R_1}\right)\left(\frac{R_4}{R_3 + R_4}\right)}$ $\rightarrow (+) \rightarrow v_o = G_1v_1 + G_2v_2$ **Subtracting Amp**
(e) 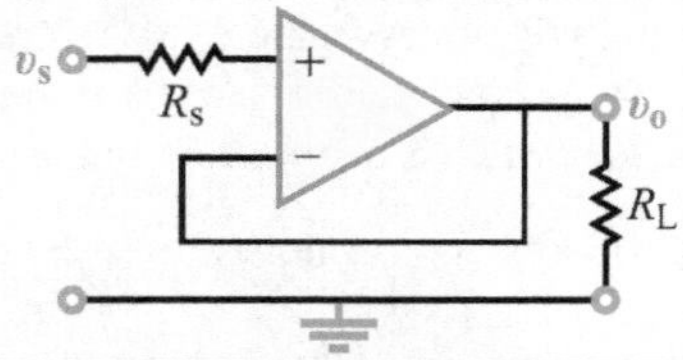	$v_s \rightarrow \boxed{G = 1} \rightarrow v_o = v_s$ **Voltage Follower / Buffer** (v_o independent of R_s and R_L)
(f) 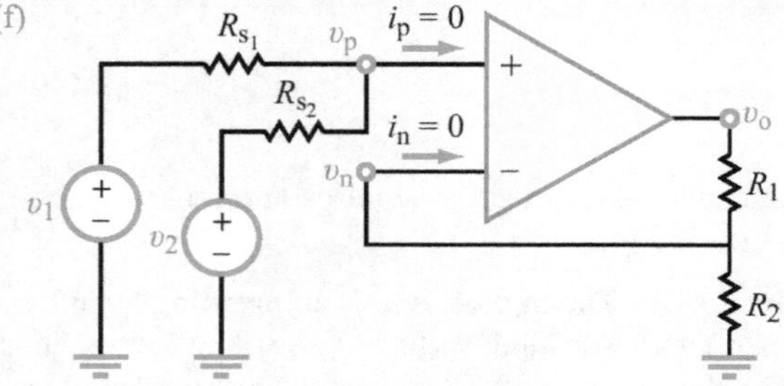	$v_1 \rightarrow \boxed{G_1 = \left(\frac{R_1 + R_2}{R_2}\right)\left(\frac{R_{s_2}}{R_{s_1} + R_{s_2}}\right)}$, $v_2 \rightarrow \boxed{G_2 = \left(\frac{R_1 + R_2}{R_2}\right)\left(\frac{R_{s_1}}{R_{s_1} + R_{s_2}}\right)}$ $\rightarrow (+) \rightarrow v_o = G_1v_1 + G_2v_2$ **Noninverting Summing Amp**

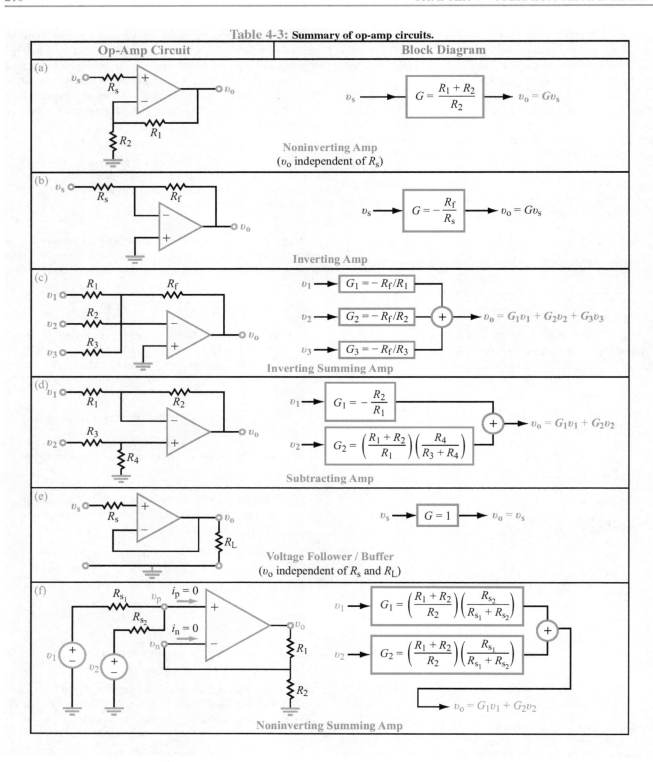

and its output is

$$v_{o_1} = G_1 v_1 = -3(0.42) = -1.26 \text{ V}.$$

The second op amp is a difference amplifier. Using Table 4-3(d), the gains of its positive and negative channels are

$$G_2 = \left(\frac{R_4}{R_3 + R_4}\right)\left(\frac{R_1 + R_2}{R_1}\right) = \left(\frac{2K}{1K + 2K}\right)\left(\frac{10K + 20K}{10K}\right) = 2$$

and

$$G_1 = -\frac{R_2}{R_1} = -\frac{20K}{10K} = -2.$$

Hence,

$$v_o = G_2 v_2 + G_1 v_{o_1} = 2v_2 - 2(-1.26) = (2v_2 + 2.52) \text{ V}.$$

Example 4-5: Elevation Sensor

A hand-held elevation sensor uses a pair of capacitors separated by a flexible metallic membrane (Fig. 4-19(a)) to measure the height h above sea level. The lower chamber in Fig. 4-19(a) is sealed, and its pressure is P_0, which is the standard atmospheric pressure at sea level. The pressure in the upper chamber, which is open to the outside air, is P. When at sea level, $P = P_0$, so the membrane assumes a flat shape and the two capacitances are equal. Since atmospheric pressure decreases with elevation, a rise in altitude results in a change in the pressure P in the upper chamber, causing the membrane to bend upwards (Fig. 4-19(b)), thereby changing the capacitances of the two capacitors. The sensor measures a voltage v_s that is proportional to the change in capacitance.

Based on measurements of v_s as a function of h, the data was found to exhibit an approximately linear variation given by

$$v_s = 2 + 0.2h \qquad \text{(V)}, \tag{4.47}$$

where h is in km. The sensor is designed to operate over the range $0 \leq h \leq 10$ km. Design a circuit whose output voltage v_o (in volts) is an exact indicator of the height h (in km).

Solution: Based on the given information, the sensor voltage v_s will serve as the input to the circuit we are asked to design, and the output v_o will represent the height elevation h. We therefore need a circuit that can perform the operation

$$v_o = h = \frac{1}{0.2}\, v_s - \frac{2}{0.2} = 5v_s - 10, \tag{4.48}$$

where we have inverted Eq. (4.47) to solve for h in terms of v_s. The functional form of Eq. (4.48) indicates that we have

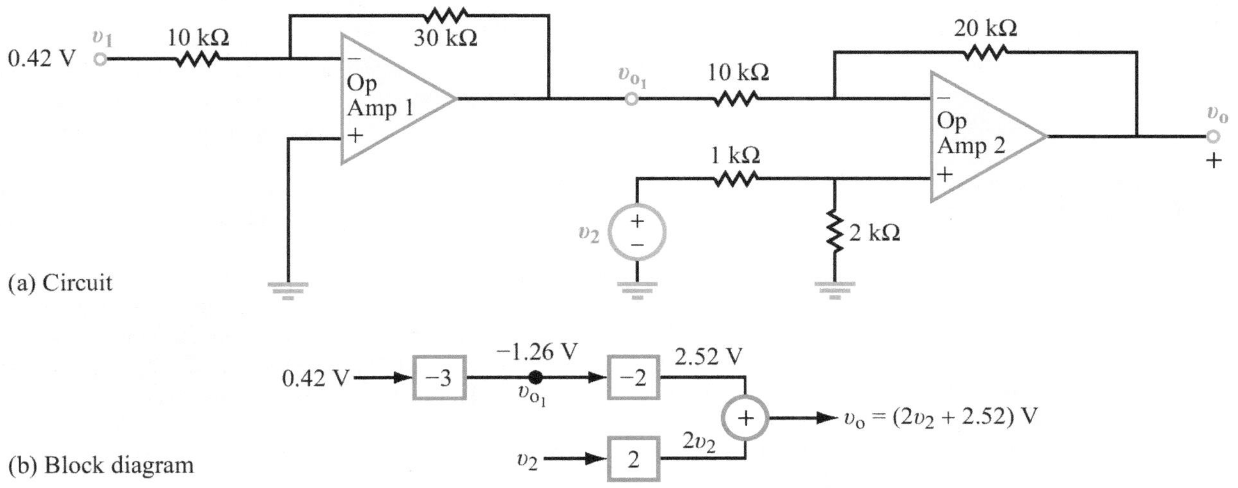

Figure 4-18: Block-diagram representation (Example 4-4).

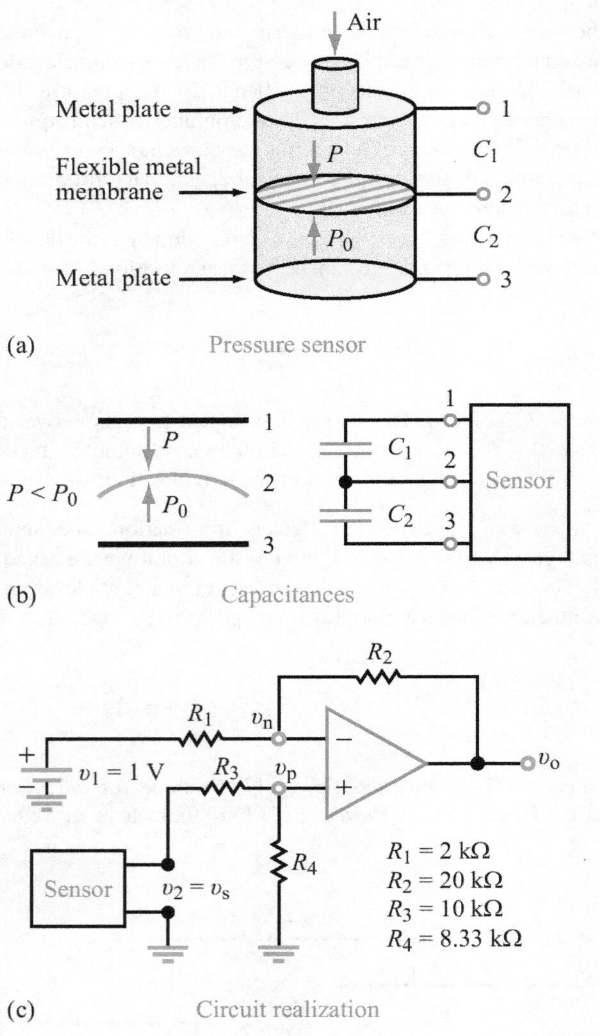

Figure 4-19: Design of a circuit for the pressure sensor of Example 4-5 with P_0 = pressure at sea level and P = pressure at height h.

only one active (variable) input, namely υ_s, which we need to amplify by a factor of 5, but we also need to subtract 10 V from it. There are multiple circuit configurations that can achieve the desired operation, including the subtractor circuit shown in Table 4-3(d) and in Fig. 4-19(c). According to Eq. (4.40), the output of the difference amplifier is given by

$$\upsilon_o = \left[\left(\frac{R_4}{R_3 + R_4}\right)\left(\frac{R_1 + R_2}{R_1}\right)\right]\upsilon_2 - \left(\frac{R_2}{R_1}\right)\upsilon_1. \quad (4.49)$$

Equation (4.49) can be made to implement Eq. (4.48) if we select the following

(a) $\upsilon_s = \upsilon_2$

(b) υ_1 as a dc voltage source such that $(R_2/R_1)\upsilon_1 = 10$ V, which can be satisfied by arbitrarily selecting $\upsilon_1 = 1$ V and $(R_2/R_1) = 10$

(c) values for R_1 through R_4 that simultaneously satisfy the conditions

$$\frac{R_2}{R_1} = 10 \qquad \text{and} \qquad \left(\frac{R_4}{R_3 + R_4}\right)\left(\frac{R_1 + R_2}{R_1}\right) = 5.$$

A possible set of values that meets these conditions is

$$R_1 = 2\ \text{k}\Omega, \qquad R_2 = 20\ \text{k}\Omega,$$
$$R_3 = 10\ \text{k}\Omega, \qquad R_4 = 8.33\ \text{k}\Omega.$$

Before we conclude the design, we should check to make sure that the op amp will operate in its linear range over the full range of operation of the sensor. According to Eq. (4.47), as h varies from zero to 10 km, υ_s varies from 2 V to 4 V. The corresponding range of variation of υ_o, from Eq. (4.48), is from zero to 10 V. Hence, we should choose an op amp designed to function with a dc supply voltage V_{cc} that exceeds 10 V.

Example 4-6: Circuit with Multiple Op Amps

Relate the output voltage υ_o to the input voltages υ_1 and υ_2 of the circuit in Fig. 4-20.

Solution: By comparing the circuit connections surrounding the four op amps with those given in Table 4-3, we recognize op amps 1 and 2 as noninverting amplifiers (sources υ_1 and υ_2 are connected to + input terminals), op amp 3 as an inverting amplifier with a gain of -1 (equal input and feedback resistors R_4), and op amp 4 as an inverting summing amplifier (Table 4-3(b)) with equal gain (same input resistances R_6 at summing point).

We start by examining the pair of input op amps. Because they are not among the standard configurations in Table 4-3, we will use KVL/KCL to evaluate them. For op amp 1, $\upsilon_{p_1} = \upsilon_1$ and $\upsilon_{p_1} = \upsilon_{n_1}$ (op-amp voltage constraint). Hence,

$$\upsilon_a = \upsilon_{n_1} = \upsilon_1.$$

Similarly, for op amp 2,

$$\upsilon_b = \upsilon_{n_2} = \upsilon_2.$$

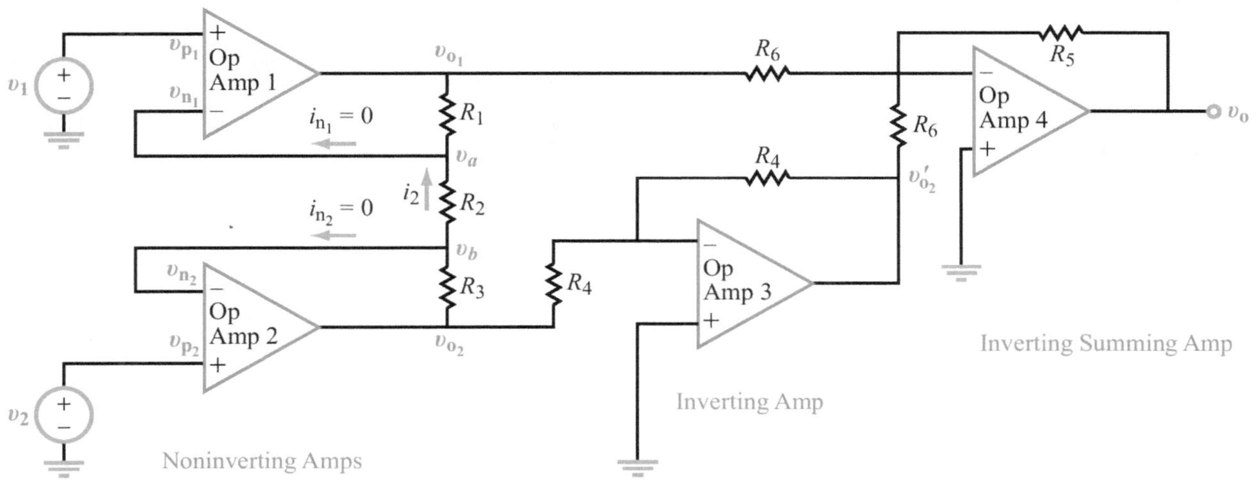

Figure 4-20: Example 4-6.

Since $i_{n_1} = i_{n_2} = 0$ (op-amp current constraint),

$$i_2 = \frac{\upsilon_b - \upsilon_a}{R_2} = \frac{\upsilon_2 - \upsilon_1}{R_2},$$

and

$$\begin{aligned} \upsilon_{o_2} - \upsilon_{o_1} &= i_2(R_1 + R_2 + R_3) \\ &= \left(\frac{R_1 + R_2 + R_3}{R_2}\right)(\upsilon_2 - \upsilon_1). \end{aligned} \tag{4.50}$$

Op amp 3 is a standard inverting amplifier, so we can use Table 4-3(c) to obtain

$$\upsilon'_{o_2} = -\left(\frac{R_4}{R_4}\right)\upsilon_{o_2} = -\upsilon_{o_2}.$$

Op amp 4 is an inverting summing amplifier (Table 4-3(c)) with output

$$\begin{aligned} \upsilon_o &= -\frac{R_5}{R_6}(\upsilon_{o_1} + \upsilon'_{o_2}) \\ &= -\frac{R_5}{R_6}(\upsilon_{o_1} - \upsilon_{o_2}) \\ &= \frac{R_5}{R_6}(\upsilon_{o_2} - \upsilon_{o_1}) = R_5\left(\frac{R_1 + R_2 + R_3}{R_6 R_2}\right)(\upsilon_2 - \upsilon_1). \end{aligned} \tag{4.51}$$

Example 4-7: Interesting Op-Amp Circuit

Generate a plot for i_L at the output side of the circuit shown in Fig. 4-21(a) versus υ_s, covering the full linear range of υ_s.

Solution: This circuit is not one of the standard op-amp configurations in Table 4-3, so we need to analyze it using KVL/KCL. At node υ_n, KCL gives

$$\frac{\upsilon_n}{2k} + \frac{\upsilon_n - \upsilon_o}{6k} = 0,$$

which leads to

$$\upsilon_o = 4\upsilon_n.$$

At node υ_p, KCL gives

$$\frac{\upsilon_p - (\upsilon_s - 0.5)}{2k} = 0,$$

which leads to

$$\upsilon_p = \upsilon_s + 0.5.$$

By imposing the op-amp constraint $\upsilon_p = \upsilon_n$, we have

$$\upsilon_o = 4\upsilon_n = 4(\upsilon_s + 0.5) = 4\upsilon_s + 2.$$

At the output side,

$$i_L = \frac{v_o - 4}{1k} = \frac{4v_s + 2 - 4}{1k} = (4v_s - 2) \text{ mA}.$$

For $v_o = V_{cc} = 10$ V,

$$10 = 4v_s + 2, \qquad \text{or } v_s = 2 \text{ V},$$

and for $v_o = -V_{cc} = -10$ V,

$$-10 = 4v_s + 2, \qquad \text{or } v_s = -3 \text{ V}.$$

Hence, linear range of v_s is

$$-3 \text{ V} \leq v_s \leq 2 \text{ V} \qquad \text{(linear range)}.$$

Figure 4-21(b) displays a plot of i_L versus v_s over the latter's linear range. Note that the linear range is not symmetrical.

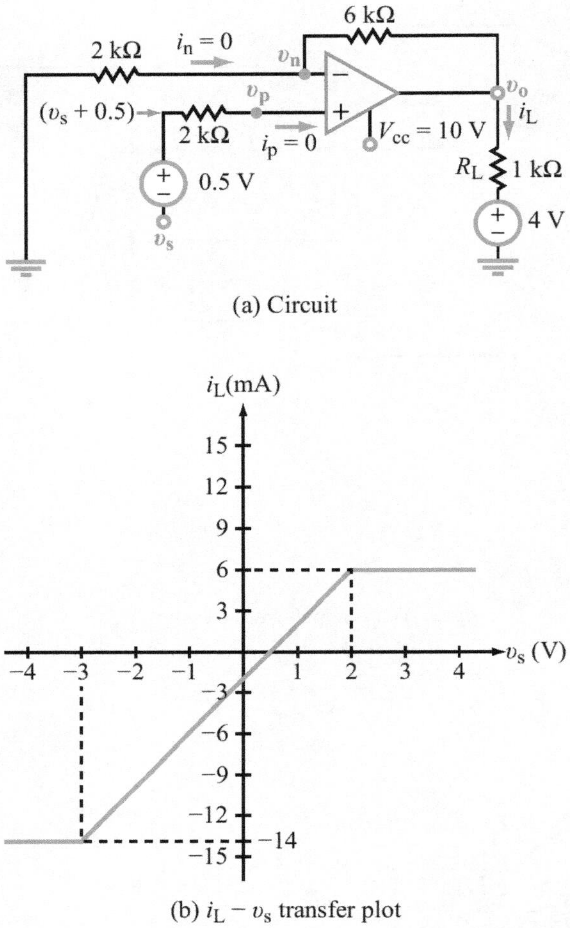

Figure 4-21: Circuit for Example 4-7.

4-9 Instrumentation Amplifier

▶ An electric *sensor* is a circuit used to measure a physical quantity, such as distance, motion, temperature, pressure, or humidity. *In some applications, the intent is not to measure the magnitude of a certain quantity, but rather to sense small deviations from a nominal value.* ◀

For example, if the temperature in a room is to be maintained at 20 °C, the functional goal of the temperature sensor is to measure the difference between the room temperature T and the reference temperature $T_0 = 20$ °C and then to activate an air conditioning or heating unit if the deviation exceeds a certain prespecified threshold. Let us assume the threshold is 0.1 °C. Instead of requiring the sensor to be able to measure T with an absolute accuracy of no less than 0.1 °C, an alternative approach would be to design the sensor to measure $\Delta v = v_2 - v_1$, where v_2 is the voltage output of a thermocouple circuit responding to the room temperature T and v_1 is the voltage corresponding to what a calibrated thermocouple would measure when $T_0 = 20$ °C. Thus, the sensor is designed to measure the deviation of T from T_0, rather than T itself, with an absolute accuracy of no less than 0.1 °C. The advantage of such an approach is that the signal is now Δv, which is more than two orders of magnitude smaller than v_2. A circuit with a precision of 10 percent is not good enough for measuring v_2, but it is plenty good for measuring Δv.

To appreciate the advantage of the differential measurement approach over the direct measurement approach, consider the two system configurations represented in Fig. 4-22.

(a) Direct Measurement Approach

In the configuration depicted in Fig. 4-22(a), input voltage v_2 represents the voltage across a thermistor used to measure the temperature T in a house. The voltage is related to T by

$$v_2 = 0.01T,$$

with T in °C. The application circuit has a gain of 100 and a measurement precision of ±1% of the amplified output. Thus,

$$v_o = (100 \pm 1)v_2 = (100 \pm 1) \times 0.01T = T \pm 0.01T.$$

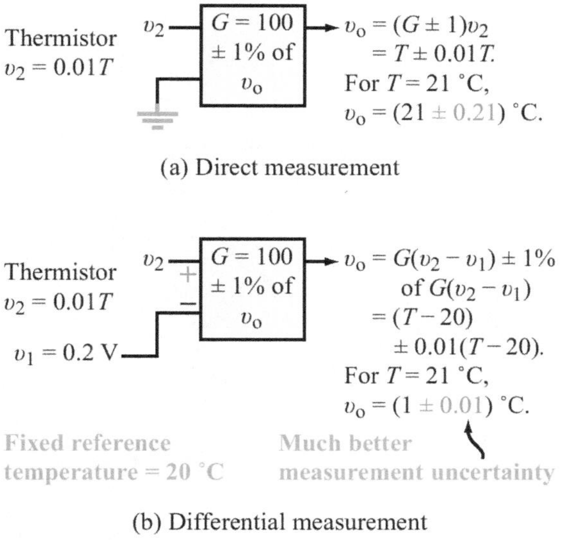

Figure 4-22: Comparison of direct and differential measurement uncertainties.

If $T = 21$ °C, the output registers 21 °C, and the associated precision is 0.21 °C.

(b) Differential Measurement Approach

The differential system in Fig. 4-22(b) also uses v_2 to measure T, but it also uses a fixed voltage v_1 at the negative terminal, with v_1 set at the desired reference temperature of 20 °C. Hence, $v_1 = 0.2$ V. The differential output is given by

$$\begin{aligned} v_o &= 100(v_2 - v_1) \pm (v_2 - v_1) \\ &= 100(v_2 - 0.2) \pm (v_2 - 0.2) \\ &= 100(0.01T - 0.2) \pm (0.01T - 0.2) \\ &= (T - 20) \pm 0.01(T - 20). \end{aligned}$$

If $T = 21$ °C,

$$v_o = (1 \pm 0.01) \text{ °C}.$$

In the differential system, v_o measures the deviation from the reference temperature of 20 °C, which is the same information provided by the direct measurement system, but with an associated precision on the order of 20 times better (± 0.01 °C compared with ± 0.21 °C for the direct measurement system).

▶ The instrumentation amplifier is perfectly suited for detecting and amplifying a small signal deviation when superimposed on one or the other of two much larger (and otherwise identical) signals. ◀

An instrumentation amplifier consists of three op amps, as shown in Fig. 4-23. The circuit configuration for the first two is the same as the one we examined earlier in connection with Example 4-6. According to Eq. (4.50), the voltage difference between the outputs of op amps 1 and 2 is

$$v_{o_2} - v_{o_1} = \left(\frac{R_1 + R_2 + R_3}{R_2}\right)(v_2 - v_1) = G_1(v_2 - v_1), \tag{4.52}$$

where G_1 is the circuit gain of the first stage (which includes op amps 1 and 2) and is given by

$$G_1 = \frac{R_1 + R_2 + R_3}{R_2}. \tag{4.53}$$

The third op amp is a difference amplifier that amplifies $(v_{o_2} - v_{o_1})$ by a gain factor G_2 given by

$$G_2 = \frac{R_4}{R_5}. \tag{4.54}$$

Hence,

$$v_o = G_2 G_1(v_2 - v_1) = \left(\frac{R_4}{R_5}\right)\left(\frac{R_1 + R_2 + R_3}{R_2}\right)(v_2 - v_1). \tag{4.55}$$

To simplify the circuit, and improve precision, all resistors—with the exception of R_2—often are chosen to be identical in design and construction, thereby minimizing deviations between their resistances. If we set $R_1 = R_3 = R_4 = R_5 = R$ in Eq. (4.55), the expression for v_o reduces to

$$v_o = \left(1 + \frac{2R}{R_2}\right)(v_2 - v_1). \tag{4.56}$$

In that case, R_2 becomes the *gain-control resistance* of the circuit; its value (relative to R) sets the gain. If the expected signal deviation $(v_2 - v_1)$ is on the order of microvolts to millivolts, the instrumentation amplifier is designed to have an overall gain that would amplify the signal to the order of volts.

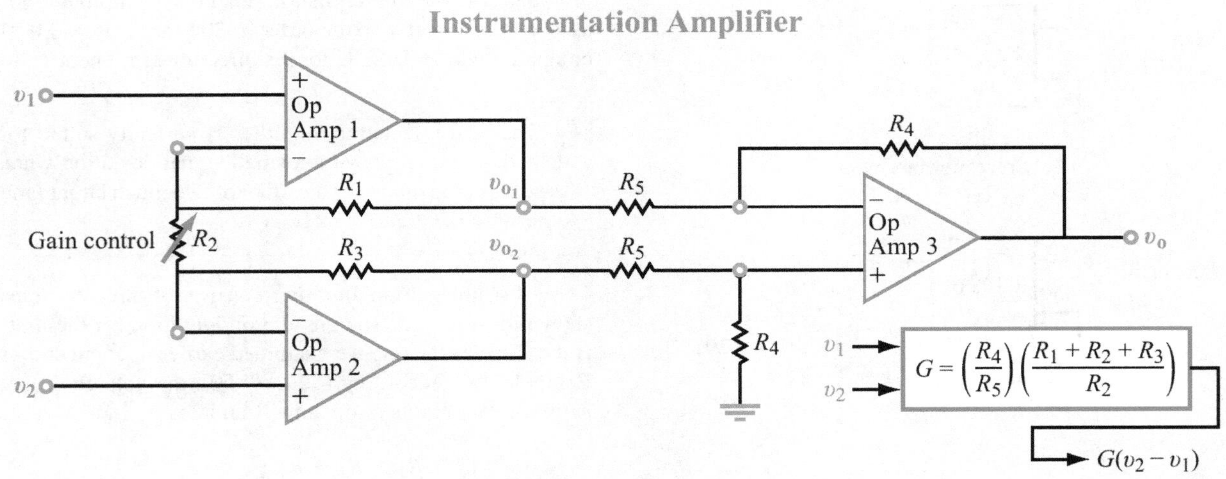

Figure 4-23: Instrumentation-amplifier circuit.

▶ *The instrumentation amplifier is a high-sensitivity, high-gain, deviation sensor.* Several semiconductor manufacturers offer instrumentation-amplifier circuits in the form of integrated packages. ◀

Concept Question 4-16: When designing a multistage op-amp circuit, what should the design engineer do to insure that none of the op amps is driven into saturation? (See CAD)

Concept Question 4-17: If the goal is to measure small deviations between a pair of input signals, what is the advantage of using an instrumentation amplifier over using a difference amplifier? (See CAD)

Exercise 4-8: To monitor brain activity, an instrumentation-amplifier sensor uses a pair of needle-like probes inserted at different locations in the brain to measure the voltage difference between them. If the circuit is of the type shown in Fig. 4-23 with $R_1 = R_3 = R_4 = R_5 = R = 50\ \text{k}\Omega$, $V_{cc} = 12$ V, and the maximum magnitude of the voltage difference that the brain is likely to exhibit is 3 mV, what should R_2 be to maximize the sensitivity of the brain sensor?

Answer: $R_2 = 25\ \Omega$. (See CAD)

4-10 Digital-to-Analog Converters (DAC)

▶ A *digital-to-analog converter* (DAC) is a circuit that transforms a digital sequence presented to its input into an analog output voltage whose magnitude is proportional to the decimal value of the input signal. ◀

An n-bit digital signal is described by the sequence $[V_1 V_2 V_3 \ldots V_n]$, where V_1 is called the *most significant bit* (MSB) and V_n is the *least significant bit* (LSB). Voltages V_1 through V_n can each assume only two possible states—either a 0 or a 1. When a bit is in the 1 state, its decimal value is 2^m, where m depends on the location of that bit in the sequence. For the most significant bit (V_1), its decimal value is $2^{(n-1)}$; for V_2 it is $2^{(n-2)}$; and so on. The decimal value of the least significant bit is $2^{n-n} = 2^0 = 1$, when that bit is in state 1. Any bit in state 0 has a decimal value of 0. Table 4-4 illustrates the correspondence between the binary sequences of a 4-bit digital signal and their decimal values. The binary sequences start at [0000] and end at [1111], representing 16 decimal values extending from 0 to 15 and inclusive of both ends. To do so, the DAC in Fig. 4-24 has to sum V_1 to V_n after weighting each by a factor equal to its decimal value. Thus, for a 4-bit digital sequence, for example, the output voltage of the DAC has to be related to the input by

$$\begin{aligned} V_{\text{out}} &= G(2^{4-1}V_1 + 2^{4-2}V_2 + 2^{4-3}V_3 + 2^{4-4}V_4) \\ &= G(8V_1 + 4V_2 + 2V_3 + V_4), \end{aligned} \tag{4.57}$$

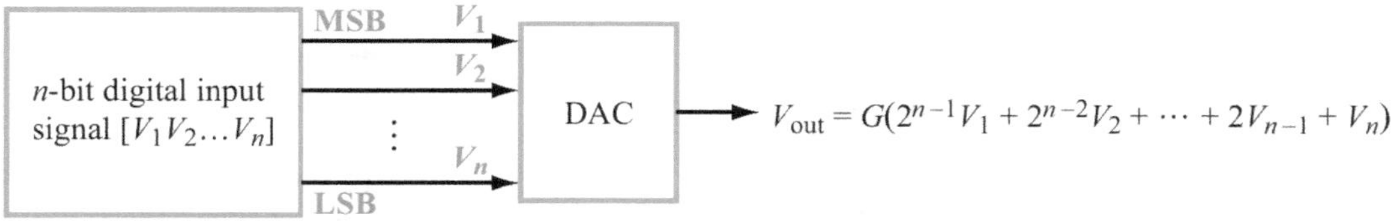

Figure 4-24: A digital-to-analog converter transforms a digital signal into an analog voltage proportional to the decimal value of the digital sequence.

Table 4-4: Correspondence between binary sequence and decimal value for a 4-bit digital signal and output of a DAC with $G = -0.5$.

$V_1V_2V_3V_4$	Decimal Value	DAC Output (V)
0000	0	0
0001	1	−0.5
0010	2	−1
0011	3	−1.5
0100	4	−2
0101	5	−2.5
0110	6	−3
0111	7	−3.5
1000	8	−4
1001	9	−4.5
1010	10	−5
1011	11	−5.5
1100	12	−6
1101	13	−6.5
1110	14	−7
1111	15	−7.5

where G is a scale factor that has no influence on the relative weights of the four terms. The magnitude of G is selected to suit the range of the output voltage. If the input is a 3-bit sequence whose range of decimal values extends from 0 to 7, one might design the circuit so that $G = 1$, because in that case, the maximum output voltage is 7 V, which is below V_{cc} for most op amps. For digital signals with longer sequences, G needs to be smaller than 1 in order to avoid saturating the op amp.

The weighted-sum operation of a DAC can be realized by many different signal-processing circuits. A rather straightforward implementation is shown in Fig. 4-25, where an inverting summer (Table 4-3(c)) uses the ratios of R_f to the individual resistances to realize the necessary weights, and the positions of the switches determine the 0/1 states of the 4 bits. Reference to either Table 4-3(c) or Eq. (4.34) yields

$$\begin{aligned} V_{out} &= -\frac{R_f}{R} V_1 - \frac{R_f}{2R} V_2 - \frac{R_f}{4R} V_3 - \frac{R_f}{8R} V_4 \\ &= \frac{-R_f}{8R}(8V_1 + 4V_2 + 2V_3 + V_4), \end{aligned} \tag{4.58}$$

which satisfies the relative weights given in Eq. (4.57). Also, in this case,

$$G = -\frac{R_f}{8R}. \tag{4.59}$$

For $[V_1 V_2 V_3 V_4] = [1111]$, $V_{out} = 15G$. By selecting $G = -0.5$ (corresponding to $R_f = 4R$), the output will vary from 0 to −7.5.

Example 4-8: *R*–2*R* Ladder

The circuit in Fig. 4-26(a) offers an alternative approach to realizing digital-to-analog conversion of a 4-bit signal. It is called an R–$2R$ ladder, because all of the resistors of its input circuit have values of R or $2R$, thereby limiting the input resistance seen by the dc source to a 2 : 1 range no matter how many bits are contained in the digital sequence. This is in contrast with the DAC of Fig. 4-25, whose input-resistance range is dependent on the number of bits; 8 : 1 for a 4-bit converter, and 128 : 1 for an 8-bit converter. Additionally, circuit performance and precision depend on resistor tolerance and are superior when fewer groups of resistors are involved in the input circuit. Resistors fabricated in the same production process are likely to exhibit less variability among them than resistors fabricated by different processes.

Show that the R–$2R$ ladder in Fig. 4-26(a) does indeed provide the appropriate weighting for a 4-bit DAC. If $R = 2$ kΩ and $V_{cc} = 10$ V, what is the maximum realistic value that R_f can have?

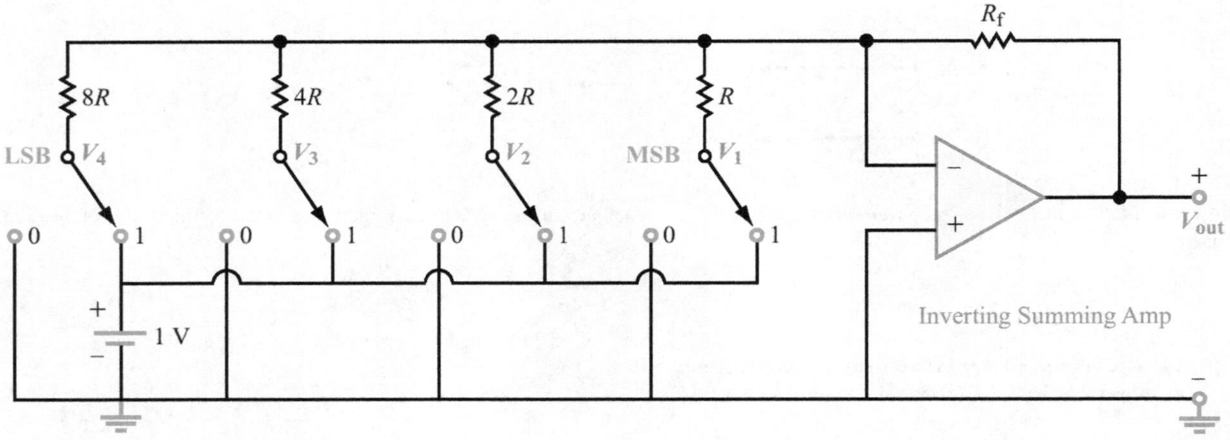

Figure 4-25: Circuit implementation of a DAC.

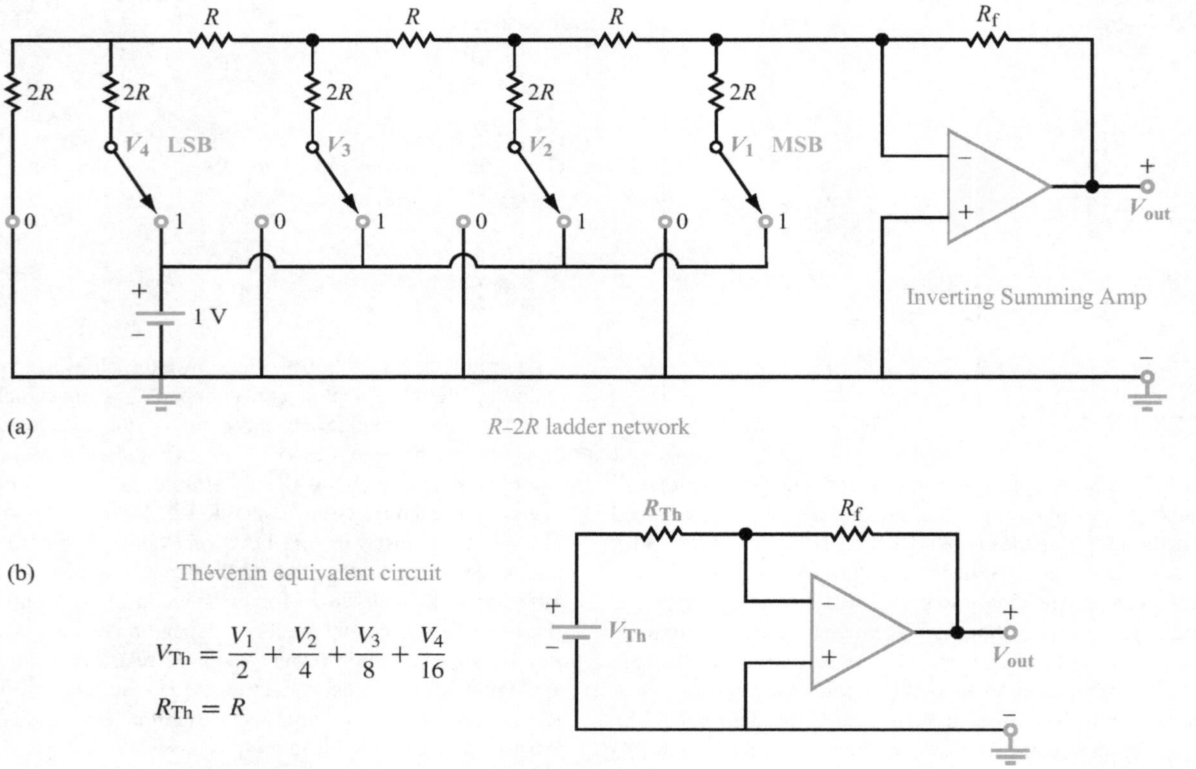

$$V_{Th} = \frac{V_1}{2} + \frac{V_2}{4} + \frac{V_3}{8} + \frac{V_4}{16}$$

$$R_{Th} = R$$

Figure 4-26: R–$2R$ ladder digital-to-analog converter.

Solution: Even though we know that (depending on the positions of the switches) V_1 to V_4 can each assume only 2 binary values, namely 0 or 1 V, let us treat V_1 to V_4 as dc power supplies and apply multiple iterations of voltage-current transformations (starting on the left with the LSB) to arrive at the Thévenin equivalent circuit at the input side of the op amp. The result of such a transformation process is shown in Fig. 4-26(b), in which

$$V_{Th} = \frac{V_1}{2} + \frac{V_2}{4} + \frac{V_3}{8} + \frac{V_4}{16} \tag{4.60a}$$

and

$$R_{Th} = R. \tag{4.60b}$$

Consequently,

$$\begin{aligned} V_{out} &= -\frac{R_f}{R_{Th}} V_{Th} \\ &= -\frac{R_f}{R}\left(\frac{V_1}{2} + \frac{V_2}{4} + \frac{V_3}{8} + \frac{V_4}{16}\right) \\ &= -\frac{R_f}{16R}(8V_1 + 4V_2 + 2V_3 + V_4). \end{aligned} \tag{4.61}$$

The voltage $|V_{out}|$ is a maximum when $[V_1 V_2 V_3 V_4] = [1111]$, in which case

$$V_{out} = -\frac{15}{16}\frac{R_f}{R}.$$

To insure that $|V_{out}|$ does not exceed $|V_{cc}| = 10$ V as well as to provide a safety margin of 2 V it is necessary that

$$8 \geq \frac{15}{16}\frac{R_f}{2k},$$

which gives $R_f \leq 17.1$ kΩ.

Concept Question 4-18: In a digital-to-analog converter, what dictates the maximum value that R_f can assume? (See CAD)

Concept Question 4-19: What is the advantage of the R–$2R$ ladder (Fig. 4-26) over the traditional DAC (Fig. 4-25)? (See CAD)

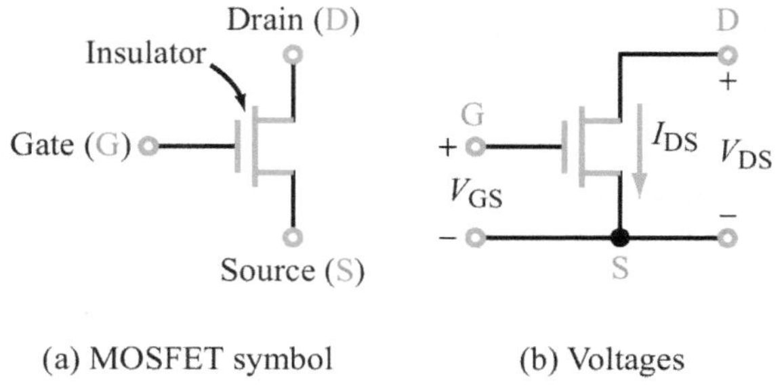

Figure 4-27: MOSFET symbol and voltage designations.

Exercise 4-9: A 3-bit DAC uses an R–$2R$ ladder design with $R = 3$ kΩ and $R_f = 24$ kΩ. If $V_{cc} = 10$ V, write an expression for V_{out} and evaluate it for $[V_1 V_2 V_3] = [111]$.

Answer:

$$V_{out} = -\frac{R_f}{8R}(4V_1 + 2V_2 + V_3) = -(4V_1 + 2V_2 + V_3).$$

For $[V_1 V_2 V_3] = [111]$, $V_{out} = -7$ V, whose magnitude is smaller than $V_{cc} = 10$ V. (See CAD)

4-11 The MOSFET as a Voltage-Controlled Current Source

In earlier sections, we demonstrated how op amps can be used to build buffers and amplifiers. We now examine how to realize the same outcome using MOSFETs. The simplest model of a *MOSFET*, which stands for *metal-oxide semiconductor field-effect transistor*, is shown in Fig. 4-27(a). The vast majority of commercial computer processors are built with MOSFETs; as mentioned in Technology Brief 1 on nanotechnology, a 2010 Intel Core processor contains over 1 billion independent MOSFETs. A MOSFET has three terminals: the *gate* (G), the *source* (S), and the *drain* (D). Actually, it has a fourth terminal, namely its body (B), but we will ignore it for now because for many applications it is simply connected to the ground terminal. The circuit symbol for the MOSFET may look somewhat unusual, but it is actually a stylized depiction of the physical cross section of a real MOSFET. In a real MOSFET, the gate consists of a very thin layer (< 500 nm thick) of a conducting material adjacent to an even thinner layer (< 100 nm) of insulator. The insulator in turn is placed directly on the surface of a relatively large slab of semiconductor material, usually referred to as "the chip" in everyday conversation (usually made of silicon 0.5 to 1.5 mm thick). The drain and the source sections are fabricated into this semiconductor chip on either side of the gate.

▶ Because the gate G is separated from the rest of the transistor by the thin insulating layer, no dc current can flow from G to either D or S. ◀

Nonetheless, it turns out that the voltage difference between terminals G and S is key to the operation of the MOSFET.

Using terminal S as a reference in Fig. 4-27(b), we denote V_{DS} and V_{GS} as the voltages at terminals D and G, respectively. We also denote the current that flows through the MOSFET from D to S as I_{DS}. This simplification is justified by the assumption that no current flows through the gate node to either the drain or source node. The operation of the MOSFET can be analyzed by placing it in the simple circuit shown in Fig. 4-28(a), in which V_{DD} is a dc power supply voltage usually set at a level close to but not greater than, the maximum rated value of V_{DS} for the specific MOSFET model under consideration. The resistance R_D is external to the MOSFET, and its role will be discussed later. The input voltage is synonymous with V_{GS} and the output voltage is synonymous with V_{DS},

$$V_{in} = V_{GS}, \quad \text{and} \quad V_{out} = V_{DS}. \tag{4.62}$$

Moreover V_{out} is related to V_{DD} by

$$V_{out} = V_{DD} - I_{DS} R_D. \tag{4.63}$$

Since current cannot flow from G to either D or S, the only current that can flow through the MOSFET is I_{DS}. The dependence of I_{DS} on V_{GS} and V_{DS} is shown for a typical MOSFET in Fig. 4-28(b) in the form of characteristic curves displaying the response of I_{DS} to V_{DS} at specific values of V_{GS}. We observe that if V_{DS} is greater than a certain *saturation threshold value* V_{SAT}, the curves assume approximately constant levels, and that these levels are approximately proportional to V_{GS}. These observations allow us to characterize the MOSFET in terms of the simple, equivalent circuit model shown in Fig. 4-28(c), which consists of a single dependent current source given by

$$I_{DS} = g V_{GS}, \tag{4.64}$$

where g is a *MOSFET gain constant*. The characteristic curves associated with this model, which is valid only if V_{DS} exceeds V_{SAT}, are shown in Fig. 4-28(d).

Even though this equivalent circuit is very simple and more sophisticated models usually are required, it nevertheless serves as a useful approximate model for introducing some common uses of MOSFETs. In real MOSFETs, the relationship between I_{DS} and V_{GS} at saturation is not strictly linear. How linear the relationship is depends (in part) on the size of the transistor. Modern sub-micron transistors used in digital processors exhibit a linear relationship between I_{DS} and V_{GS} at saturation, whereas larger MOSFETs used for power switching may behave nonlinearly. For our purposes, the simplification denoted by Eq. (4.64) will suffice.

4-11.1 Digital Inverter

We now will use the model given by Eq. (4.64) to demonstrate how the MOSFET can function as a *digital inverter* by generating an output state of "0" when the input state is "1," and vice versa. Combining Eqs. (4.62) to (4.64) gives

$$V_{out} = V_{DD} - g R_D V_{in}. \tag{4.65}$$

The constant g is a MOSFET parameter, so if we choose R_D such that $g R_D \approx 1$, Eq. (4.65) simplifies to

$$\frac{V_{out}}{V_{DD}} = 1 - \frac{V_{in}}{V_{DD}}. \tag{4.66}$$

In a digital inverter, we are interested in output responses to only two input states. According to Eq. (4.66):

$$\text{If } \frac{V_{in}}{V_{DD}} = 1, \quad \Rightarrow \quad \frac{V_{out}}{V_{DD}} = 0, \tag{4.67a}$$

and

$$\text{if } \frac{V_{in}}{V_{DD}} = 0, \quad \Rightarrow \quad \frac{V_{out}}{V_{DD}} = 1. \tag{4.67b}$$

Hence, the MOSFET circuit in Fig. 4-28(a) behaves like a digital inverter, provided the model given by Eq. (4.64) holds true and requiring that V_{DS} exceeds V_{SAT}. In a real circuit,

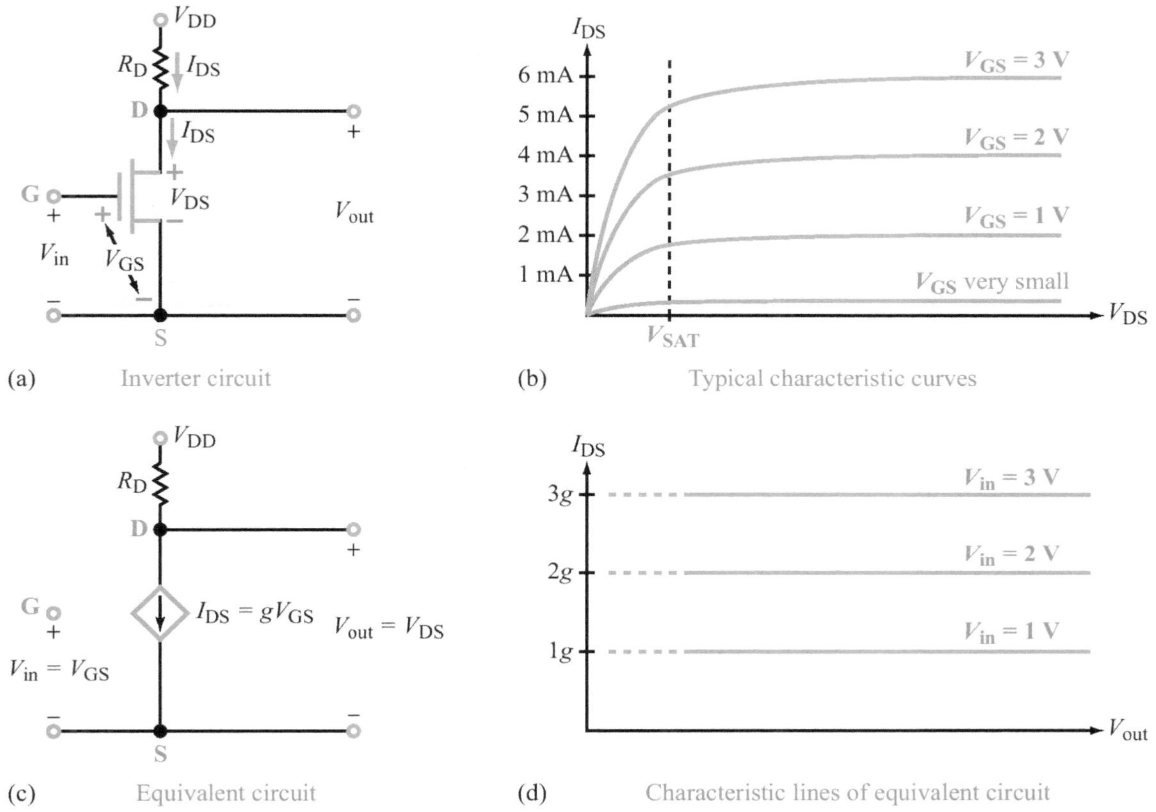

Figure 4-28: MOSFET (a) circuit, (b) characteristic curves, (c) equivalent circuit, and (d) associated characteristic lines.

V_{in} and V_{out} are not given by the simple results indicated by Eq. (4.67), but each can be categorized easily into high and low voltage values to satisfy the functionality of a digital inverter.

4-11.2 NMOS versus PMOS Transistors

The MOSFET circuit of Fig. 4-28(a) actually is called an n-channel MOSFET or *NMOS* for short. Its operation is limited to the first quadrant in Fig. 4-28(d), where both I_{DS} and V_{DS} can assume positive values only. A second type of MOSFET called *PMOS* (p-channel MOSFET) is designed and fabricated to operate in the third quadrant, corresponding to negative values for I_{DS} and V_{DS}, as illustrated in Fig. 4-29. To distinguish between the two types, the symbol for PMOS includes a small open circle at terminal G.

The NMOS inverter circuit of Fig. 4-28(a) provides the correct functionality required from a digital inverter, but it suffers from a serious power-dissipation problem. Let us consider the power consumed by R_{D} under realistic conditions:

Input State 0:

$$\frac{V_{\text{in}}}{V_{\text{DD}}} = 0 \quad \Rightarrow \quad I_{\text{DS}} \approx 0 \quad \Rightarrow \quad P_{R_{\text{D}}} = I_{\text{DS}}^2 R_{\text{D}} \approx 0 \tag{4.68a}$$

Input State 1:

$$\frac{V_{\text{in}}}{V_{\text{DD}}} = 1 \quad \Rightarrow \quad I_{\text{DS}} = \frac{V_{\text{DD}}}{R_{\text{D}}} \quad \Rightarrow \quad P_{R_{\text{D}}} = \frac{V_{\text{DD}}^2}{R_{\text{D}}}. \tag{4.68b}$$

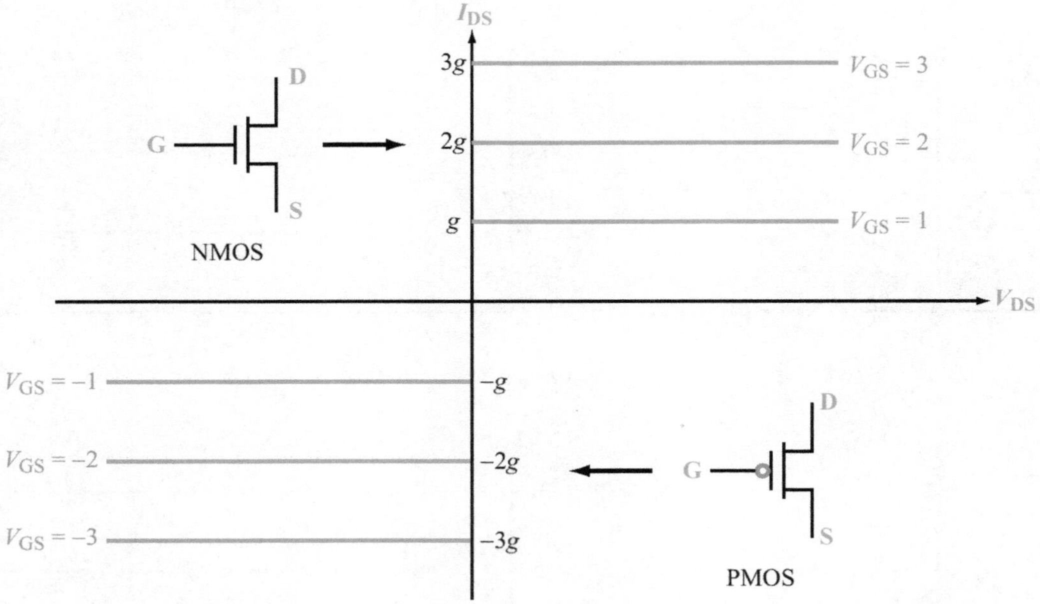

Figure 4-29: Complementary characteristic curves for NMOS and PMOS.

Heat dissipation in R_D is practically zero for input state 0, but for input state 1, it is equal to V_{DD}^2/R_D. The value of V_{DD}, which is dictated by the MOSFET specifications, is typically on the order of volts, and R_D can be made very large—on the order of kΩ or tens of kΩ. If R_D is much larger than that, I_{DS} becomes too small for the MOSFET to function as an inverter. For V_{DD} on the order of 1 V and R_D on the order of 10 kΩ, P_{R_D} for an individual NMOS is on the order of 100 μW. This amount of heat generation is trivial for a single transistor, but when we consider that a typical computer processor contains on the order of 10^9 transistors, all confined to a relatively small volume of space, the total amount of heat that would be generated by such an NMOS-based processor would likely *burn a hole through the computer!* To address this heat-dissipation problem, a new technology was introduced in the 1980s called ***CMOS***, which stands for ***complementary MOS***.

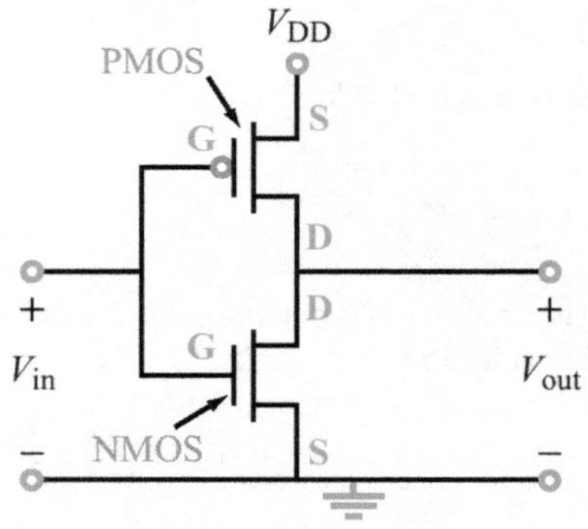

Figure 4-30: CMOS inverter.

▶ CMOS has revolutionized the microprocessor industry and led to the rise of the x86 family of PC processors. ◀

CMOS is a configuration that attaches an NMOS to a PMOS at their drain terminals, as shown in Fig. 4-30. The CMOS inverter provides the same functionality as the simpler NMOS inverter, but it has the distinct advantage in that it dissipates

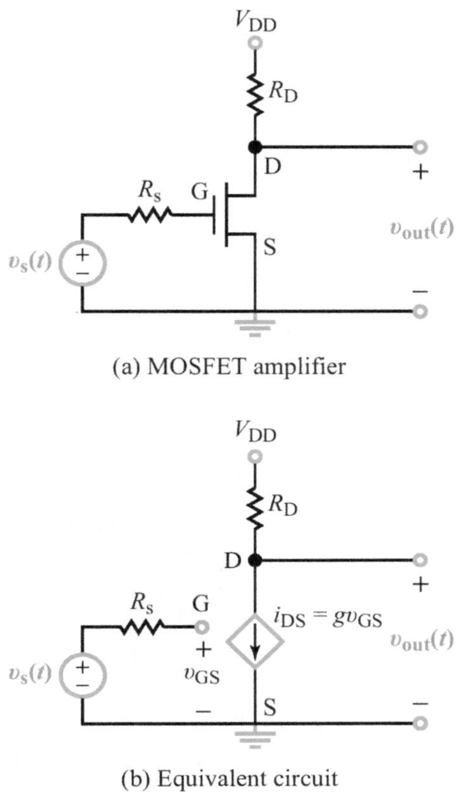

Figure 4-31: MOSFET amplifier circuit for Example 4-9.

negligible power for *both* input states. The significance of the inverter is in the role it plays as a basic building block for more complicated logic circuits, such as those that perform AND and OR operations.

4-11.3 MOSFETs in Analog Circuits

In addition to their use in digital circuits, MOSFETs also can be used in analog circuits as buffers and amplifiers, as demonstrated by Examples 4-9 and 4-10. As we discussed earlier in Section 4-7, a buffer is a circuit that insulates the input voltage from variations in the load resistance.

Example 4-9: MOSFET Amplifier

The circuit shown in Fig. 4-31(a) is known as a common-source amplifier and uses a MOSFET with a dc drain voltage $V_{DD} = 10$ V and a drain resistance $R_D = 1$ kΩ. The input signal $v_s(t)$ is an ac voltage with a dc-bias given by

$$v_s(t) = [500 + 40\cos 300t] \qquad (\mu\text{V}).$$

Note that the amplitude of the input ac signal is several orders of magnitude smaller than that of the dc voltage V_{DD}. Apply the MOSFET equivalent model with $g = 10$ A/V to obtain an expression for $v_{out}(t)$.

Solution: Upon replacing the MOSFET with its equivalent circuit, we end up with the circuit in Fig. 4-31(b). At the input side, because no current flows through R_s, it follows that

$$v_{GS}(t) = v_s(t),$$

and at the output side,

$$\begin{aligned} v_{out}(t) &= V_{DD} - i_{DS} R_D \\ &= V_{DD} - g R_D v_{GS}(t) \\ &= V_{DD} - g R_D v_s(t). \end{aligned}$$

We observe that the output voltage consists of a constant dc component (namely V_{DD}) and an ac component that is directly proportional to the input signal $v_s(t)$. For the element values specified in the problem,

$$\begin{aligned} v_{out}(t) &= 10 - 10 \times 10^3 \times (500 + 40\cos 300t) \times 10^{-6} \\ &= 5 - 0.4\cos 300t \quad \text{V}. \end{aligned}$$

The 5 V dc component is simply a level shift superimposed on which is a cosinusoidal signal that is identical to the input signal but is inverted and amplified by an ac gain of 10^4 (from 40 μV to 0.4 V).

Example 4-10: MOSFET Buffer

The circuit in Fig. 4-32(a) consists of a real voltage source (v_s, R_s) connected directly to a load resistor R_L. In contrast, the circuit in Fig. 4-32(b) uses a common-drain MOSFET circuit inbetween the source and the load to *buffer* (insulate) the source from the load. Let us define the source as being buffered from the load if the output voltage across the load is equal to at

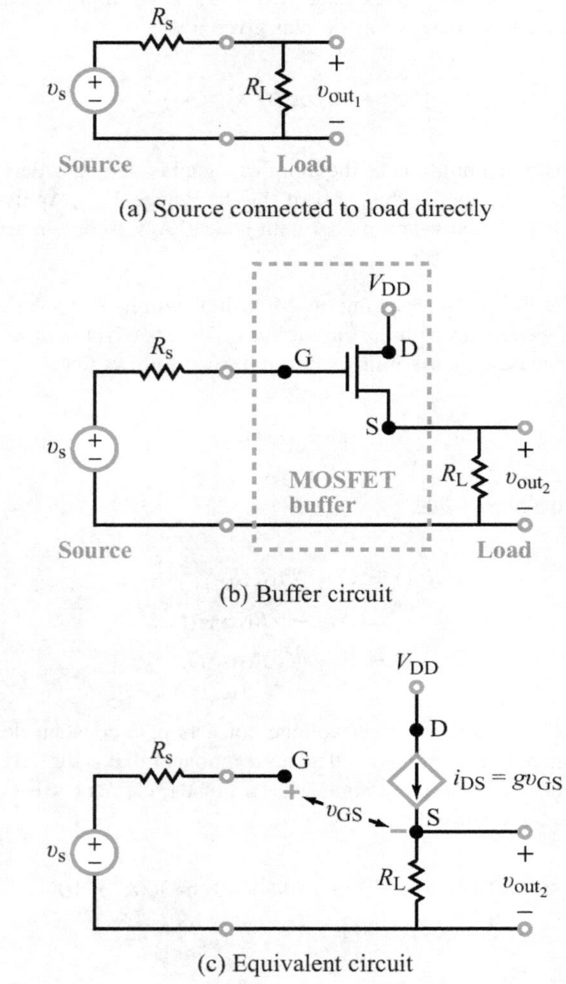

Figure 4-32: Buffer circuit for Example 4-10.

least 99 percent of v_s. For each circuit, determine the condition on R_L that will satisfy this criterion. Assume $R_s = 100\ \Omega$ and the MOSFET gain factor $g = 10$ A/V.

Solution:

(a) No-Buffer Circuit

For the circuit in Fig. 4-32(a),

$$v_{out_1} = \frac{v_s R_L}{R_s + R_L}.$$

In order for $v_{out_1}/v_s \geq 0.99$, it is necessary that

$$\frac{R_L}{R_s} \geq 99$$

or

$$R_L \geq 9.9\text{ k}\Omega \qquad (\text{for } R_s = 100\ \Omega).$$

(b) With MOSFET Buffer

For the circuit in Fig. 4-32(c), in which the MOSFET has been replaced with its equivalent circuit, KVL gives

$$-v_s + v_{GS} + v_{out_2} = 0.$$

Also,

$$v_{out_2} = I_{DS} R_L = g R_L v_{GS}.$$

Simultaneous solution of the two equations gives

$$v_{out_2} = \left(\frac{gR_L}{1 + gR_L}\right) v_s.$$

With $g = 10$ A/V and in order for v_{out_2} to be no less than $0.99 v_s$, it is necessary that

$$R_L \geq 9.9\ \Omega,$$

which is three orders of magnitude smaller than the requirement for the unbuffered circuit.

> **Concept Question 4-20:** What is the major advantage of a CMOS over an NMOS circuit as a digital inverter? (See CAD)

> **Concept Question 4-21:** When a MOSFET is used in a buffer circuit, $v_{out} \approx v_s$, where v_s is the input signal voltage. So, why is it used? (See CAD)

Exercise 4-10: In the circuit of Example 4-9, what value of R_D will give the highest possible ac gain while keeping $v_{out}(t)$ always positive?

Answer: $R_D = 1.85$ kΩ. (See CAD)

Exercise 4-11: Repeat Example 4-10, but require that v_{out} be at least 99.9 percent of v_s. What should R_L be (a) without the buffer and (b) with the buffer?

Answer: (a) $R_L \geq 99.9$ kΩ, (b) $R_L \geq 99.9\ \Omega$. (See CAD)

Technology Brief 11
Circuit Simulation Software

In Chapters 2 and 3 we examined all of the common methods used for analyzing linear electric circuits. In practice, these are used for designing and analyzing the many building blocks that make up larger circuits, or for obtaining approximate solutions for how more complex circuits function. In Technology Brief 1, we noted that ***very large scale integrated circuits*** (VLSI) have experienced exponential scaling for almost 50 years, so some of today's electrical networks may include as many as 100 billion transistors! The standard circuit analysis methods available to us are accurate and applicable, but it takes a great deal of computer automation to apply them to a 100 billion–transistor network. The ***Multisim*** circuit analysis software provides an excellent start towards modeling the behavior of complex circuits. Accordingly, Multisim will be the first of two computer-based tools we will explore in this Technology Brief. Whereas Multisim is an excellent tool, it treats a circuit as a 2-D configuration, which does not account for thermal effects associated with heat generation by the circuit elements, nor possible capacitive or inductive cross-coupling of voltages between elements (through the air or insulator medium between them). To account for these effects, we need to use a sophisticated ***3-D computer simulation tool***. This is the subject of the second part of this Technology Brief.

Multisim Software

(1) Using Simulation Tools to Calculate and Understand

Engineers use ***electronic design automation*** (EDA) tools, such as Multisim, to understand the function of a circuit and calculate its response. Consider the simple example shown in Fig. TF11-1(a), and let us assume we need to determine what voltage V_r would be measured by the voltmeter shown in the circuit. In this case, because the circuit is very simple, we can analyze it by hand or we can implement it and solve it by Multisim (Fig. TF11-1(b)). But if the circuit has more than five nodes, the by-hand approach becomes tedious, and the Multisim option becomes far more practical.

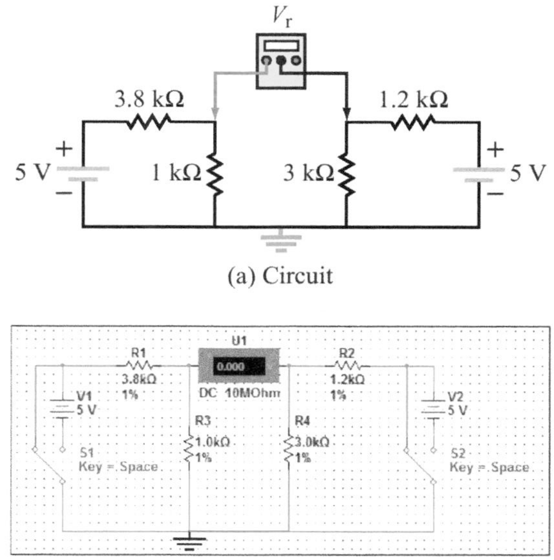

(a) Circuit

(b) Multisim layout

Figure TF11-1: Two-source circuit and Multisim representation using switches to switch one or both voltage sources on or off.

(2) Using Simulation Tools to Lay Out a Circuit

Once a circuit has been designed, we can either build it on a protoboard or, alternatively, we can have a circuit board built for it and then solder the parts to the board to create the circuit. ***Printed circuit board*** (PCB) layout tools help us plan the circuit layout and routing architecture, which often are multiple layers deep, as in the circuit of Fig. TF11-2.

When using silicon chips, for example, these designs involve hundreds, millions, or trillions of components arranged in one or more layers, and carrying thousands of simultaneous signals throughout the circuit, all acting together to obtain the desired voltage and/or current output of the circuit. Classic EDA tools (such as Multisim) begin with a ***graphical user interface*** (GUI) that allows users to specify what type of circuit elements (sources, resistors, switches, etc.) are needed and how they are connected together. Circuits made up of several elements can often be grouped or bundled together and stored in ***libraries*** for later reuse. Often, libraries of complex

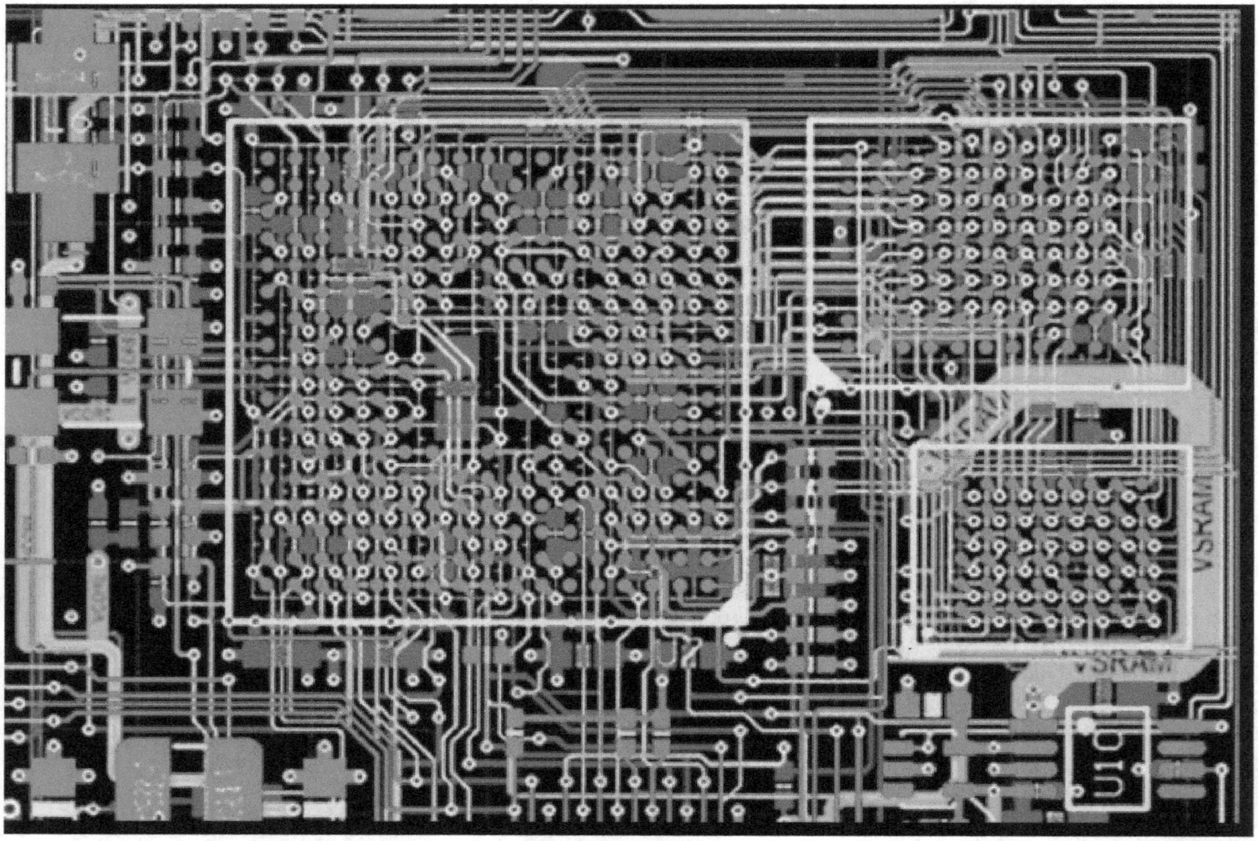

Figure TF11-2: Multilayer PCB layout, with each layer assigned a different color. Holes and solder pads are planned for each chip and component attached to the board, and multilayer routing built into the circuit board connects them all together. (Courtesy of ZYPEX Inc.)

circuits (such as the core of a computer processor) are shared or purchased to reduce engineering design time. For circuits whose design can be expressed as either logical rules or a desired logical function—primarily digital circuits—modern software tools transform circuit design into an exercise in writing code. In essence, programs can be written in ***hardware description languages*** (HDL), which define the structure and/or operation of digital circuits. The program is then executed and a circuit description suitable for manufacture, or instantiation into a field-programmable gate array (FPGA), is synthesized. Programming in HDLs is similar to assembly language or C coding, although major differences exist. Most modern complex digital circuits are designed, simulated, and synthesized with the aid of HDL tools.

Once the elements and their connections are defined, they are then modeled with either more or less detail (by specifying tolerance levels or other relevant parameters) depending on the level of accuracy needed. Simulation results are only as good as the circuit model and input parameters, so this is a very important consideration when using EDA software. The more detailed the model, the more accurate the results can be expected to be, but also the longer it takes the simulation to run. Consider, for example, the ideal and the more realistic models for voltage and current sources listed in Table 1-5. The realistic source models are certainly more accurate than the ideal models, but even the "realistic" models are approximate, because they neglect nonlinearities, stray capacitance and inductance, and potential feedback loops within the sources. For many applications, the ideal model is sufficient, for others the first-order realistic model (including a resistor) is sufficient, but for others, a more detailed nonlinear model is required. How do you, the engineer, know what model to use? The intuition and knowledge gained from working with the common circuit analysis tools from Chapters 2 and 3 help you determine when you may or may not need a more realistic model. Often, we will first try a simplified model, and then one that is slightly more realistic. If there is minimal change, we do not go on to a more complex model, but if there is substantial change, we may try more and more realistic models (each requiring more time and memory for the software to run), until the result converges and we are satisfied that we have modeled the real system at hand.

Now let's consider VLSI circuits involving trillions of transistors. Even with relatively simple models of the transistor (such as the BJT in Section 3-9 or MOSFET in Section 4-11), there are still more unknowns than we generally care to wait for the computer to solve. In this case, two simplifications are essential. First, we must break the circuit down into ***functional blocks***, so we can design each block individually and cascade or connect the blocks together. We have already seen simple examples of doing this using the Thévenin equivalent circuit technique. Thévenin is also used this way in much larger circuits, including VLSI designs. Second, we must simplify the models we use for each circuit element. Fortunately (or perhaps necessarily!) the largest circuits electrical engineers design are digital circuits, for which we can use the simplest models of all. We can assume that all voltages are either ***high/on*** (digital 1) or ***low/off*** (digital 0). This flexibility in the voltages allows us to use much simpler models. The transistor, for example, can be modeled as just a switch (on or off), or just as a resistor that is switched in or out of the circuit. Assuming all voltages are either on or off is the simplest assumption. We also can model them as on/off or in transition between on and off. The transition (which is actually a ***bouncy switch***) can be modeled as a linear slope from low to high or high to low. The length of this slope is the ***rise time*** of the transition, and the faster the rise time, the faster the circuit can send data.

3-D Modeling Tools

Model-based EDA tools define how a circuit is supposed to function electrically, but sometimes effects not included in the models come into play to make the circuit malfunction. Two of these that are particularly relevant are associated with thermal problems and coupling problems. We know that resistors and other devices are designed with specific power ratings. The power rating is related to the size and material the resistors are made of and their ability to withstand the heat generated by current moving through them. If we start pushing all of the elements of the circuit to their maximum capability, their interactions (hot chips next to other hot chips) may make the most vulnerable of these parts fail. But how do we determine which parts are the most vulnerable, and what solution can we offer to mitigate the heat problem? 3-D simulation tools help us to identify these potential problems or (all too often) diagnose them when they occur. The 3-D simulation process starts with the physical model of a given part, such as the high-speed IC package shown in Fig. TF11-3(a). The spatial distributions of electrical voltage and current are then modeled for part or all of the package, as shown in Fig. TF11-3(b). The current density at a given location is representative of what the temperature will be at that location. If overheating were to occur, it would most likely occur at the points with the highest current. More detailed thermal modeling can include the effects of heat sinks, fans, and other cooling effects. The voltage is used to calculate coupling between nearby electrical signals (such as two adjacent legs of this package).

Another interesting circuit simulation is shown in Fig. TF11-4, which displays the amount of power radiated by a crescent antenna.

So WHY Should You Learn the Circuit Analysis Methods Introduced in This Book?

Having learned how to apply the various circuit analysis tools covered in this book thus far, you may wonder why you need to learn so many different methods when they all can give you the same result. And now that you have read this Technology Brief and seen that you can use a computer to analyze circuits, you may wonder why you need to learn these analytical methods at all!

While it is true that automated tools are essential for testing circuits used in practical applications, it is equally true that the success of the design process is highly coupled to one's understanding of the fundamental

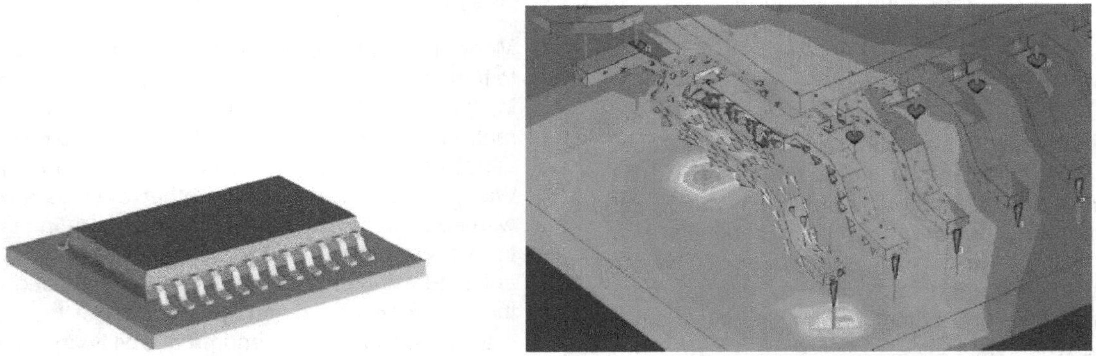

(a) Physical package (b) Current density contour

Figure TF11-3: High-speed IC package and contour and vector plot of the current density flowing through it at 5 GHz. The brighter/redder colors show higher current density (A/m^2) (which also results in higher temperature) than the darker/bluer colors. The arrows show the direction in which the current is flowing, and the size of the arrow is also proportional to the magnitude of the current density. (Courtesy: CST MICROWAVE STUDIO© IC Package Simulation.)

concepts in circuit analysis and design. Designing a new circuit to address a specified application is a creative endeavor that relies on one's past experience and fluency in circuit behavior and performance. Once an initial circuit configuration has been developed, computer simulation tools are then used to fine-tune the design and optimize the circuit performance.

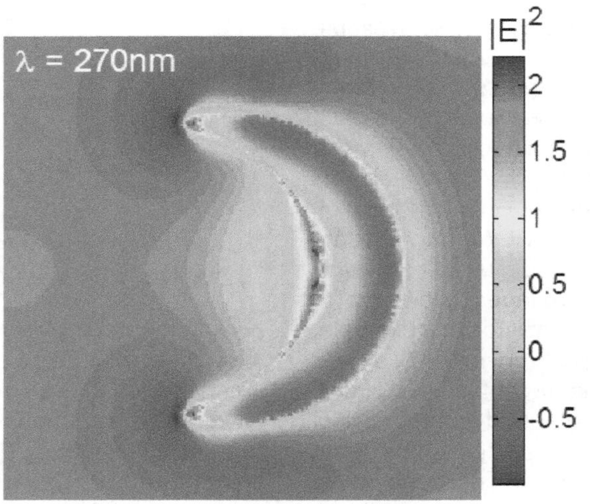

Figure TF11-4: This 3D electromagnetic simulation was used to evaluate the fields (in this case the square of the electric field, which is proportional to power) in the nanocrescent antenna shown in Technology Brief 1. We can see the strong fields at the tips (because charge congregates there), and also in the center. (Credit: Miguel Rodriguez.)

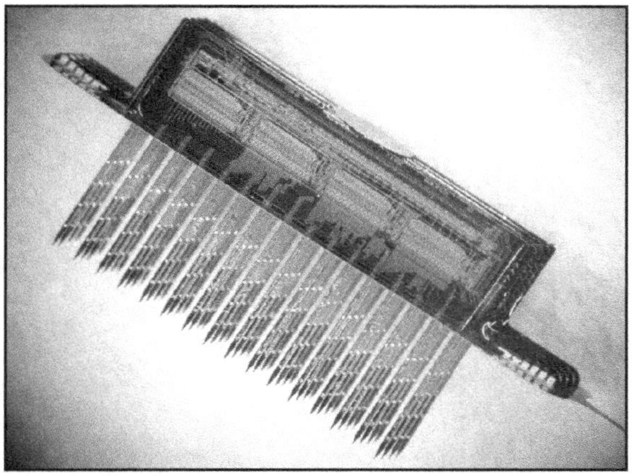

Figure 4-33: Three-dimensional neural probe (5 mm × 5 mm × 3 mm). (Courtesy of Prof. Ken Wise and Gayatri Perlin, University of Michigan.)

4-12 Application Note: Neural Probes

The human brain is composed, in part, of interconnected networks of individual, information-processing cells known as *neurons*. There are about one trillion (10^{12}) neurons in the human brain with each neuron having on average 7000 connections to other neurons. Although the working of the neural system is well beyond the scope of this book, it is important to note that when a neuron transmits information, it causes a change in the concentrations of various ions in its vicinity. This movement of ions gives rise to an electric current through the neuron's membrane which in turn generates a change in potential (voltage) between various parts of the cell and its surroundings. Thus, when a given neuron fires, a small ($\sim$ 100 mV) but detectable potential drop develops between the cell and its surroundings.

Over the past few decades, various types of devices were built for measuring this electrical phenomenon in neurons. In recent years, however, the field has achieved phenomenal success due in part to the successful development of *neural probes* (also known as *neural interfaces*) with very high sensitivities. An example of a 3-dimensional probe is shown in Fig. 4-33. It consists of a 2-D array of very thin probes—each instrumented with a sensor at each of several locations along its length. With such a probe, it is now possible to measure the *action potentials* of firing neurons at a large number of brain locations simultaneously. Modern neural interface systems also have been developed to stimulate or change the electrical state of specific neurons, thereby affecting their operation in the brain. These types of devices not only offer the potential of unraveling aspects of brain development and operation, but they also are beginning to see use in clinical applications for the treatment of chronic neurological disorders, such as Parkinson's disease (see Technology Briefs 17 and 32 on neural stimulation and computer-brain interfaces, respectively).

Because these voltage signals are so small, on-board amplification, noise-removal, and analog-to-digital circuitry are needed to process the signal from the brain to the recording device.

Example 4-11: Neural Probe

The neural probe shown in Fig. 4-34 consists of a long shank at the end of which lie two metal electrodes. This shank is inserted a short distance into the brain and the signal coming from these electrodes is recorded. For simplicity, we will model the brain activity between the two probes just like a realistic voltage source V_s in series with a resistance R_s. The source produces inverted pulses with -100 mV amplitudes. Note that neither V_a nor V_b are grounded relative to the ground level of the circuit. The neural signal needs to be inverted and amplified so that it can be presented to an analog-to-digital converter (ADC) which only operates in the 0 to 5 V range. Design the amplifier circuit.

Solution: The input signal is represented by the difference between V_a and V_b, and since neither of those terminals is grounded, some sort of differential amplifier is the logical choice for the intended application.

The amplifier should invert the input signal and amplify it into the 0 to 5 V range required by the ADC. Given these constraints, we propose to use the op-amp instrumentation amplifier circuit of Fig. 4-23 with V_a as input υ_1 and V_b as input υ_2. The amplifier output is proportional to $(\upsilon_2 - \upsilon_1)$, so the choice of connections we made will realize the inversion requirement automatically. According to Eq. (4.56), if we choose the circuit resistors such that $R_1 = R_3 = R_4 = R_5 = R$, the output voltage is given by

$$\upsilon_o = \left(1 + \frac{2R}{R_2}\right)(\upsilon_2 - \upsilon_1)$$

$$= \left(1 + \frac{2R}{R_2}\right)(V_b - V_a) = -\left(1 + \frac{2R}{R_2}\right)(V_a - V_b).$$

To amplify $(V_a - V_b)$ from -100 mV to $+5$ V, the ratio (R/R_2) should be chosen such that

$$5 = -\left(1 + \frac{2R}{R_2}\right) \times (-100 \times 10^{-3})$$

or, equivalently,

$$\frac{R}{R_2} = 24.5.$$

If we set $R = 100$ kΩ, then R_2 should be 4.08 kΩ. This will yield a 5 V pulse to the ADC every time a -100 mV pulse is generated by the neuron.

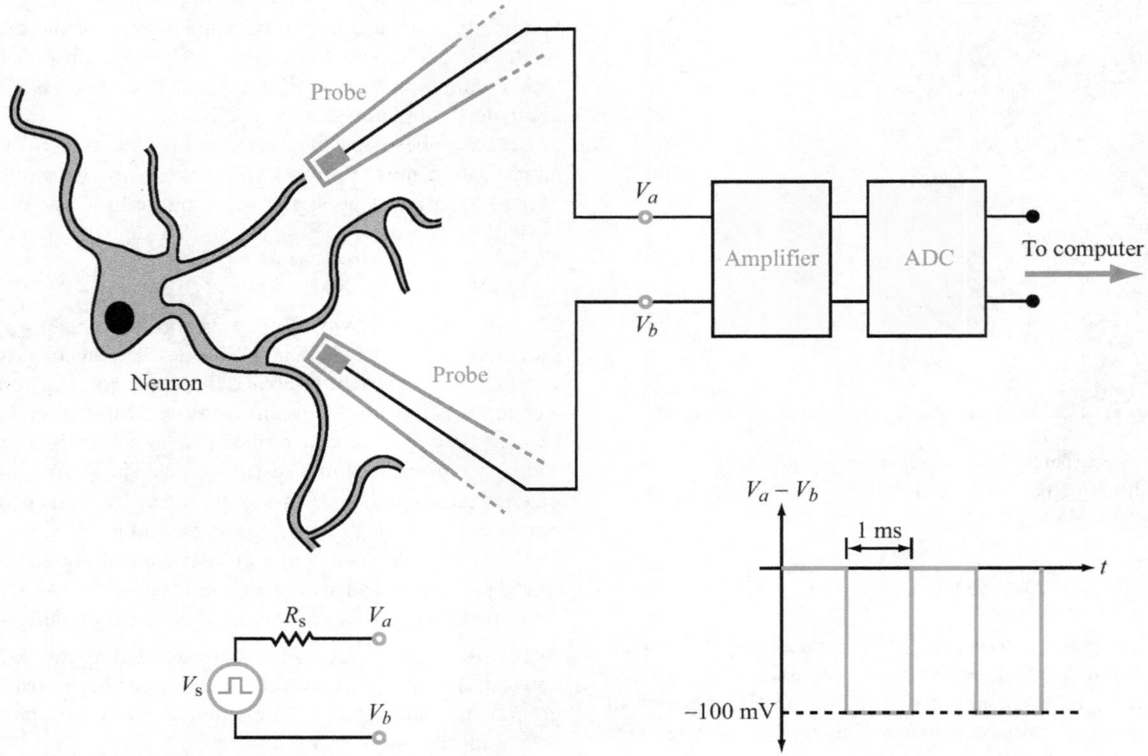

Figure 4-34: Neural-probe circuit for Example 4-11.

4-13 Multisim Analysis

One of the most attractive features of Multisim is its interactive-simulation mode, which we began to utilize in Sections 2-7 and 3-8. The simulation mode allows you to connect virtual test instruments to your circuit and to operate them in real time as Multisim simulates the circuit behavior. In this section, we will explore this feature with an op-amp circuit and two MOSFET circuits.

4-13.1 Op Amps and Virtual Instruments

The circuit shown in Fig. 4-35 uses a resistive Wheatstone bridge (Section 2-5) to detect the change of resistance induced in a sensor modeled as a variable resistor (see Technology Brief 4 on resistive sensors). The output of the bridge is fed into a pair of voltage followers and then into a differential amplifier. The circuit can be constructed and tested in Multisim using the components listed in Table 4-5. The resistance value of the potentiometer component is adjustable with a keystroke (the default is the key "a" to change the resistance in one direction and the default key combination Shift-a to change the resistance in the opposite direction) or by using the mouse slider under the component. In order to observe how changes in the potentiometer cause changes in the output, we need to connect the output to an *oscilloscope*. Multisim provides several oscilloscopes to choose from, including a generic instrument and virtual versions of commercial oscilloscopes made by Agilent and Tektronix. For starters, it is easiest to use the generic instrument by selecting Simulate → Instruments → Oscilloscope, or by selecting and dragging an oscilloscope from the instrument dock. Figure 4-36 shows the complete circuit drawn in Multisim. The power supplies for the op amps can be found under Components → Sources → POWER SOURCES → VDD (or VSS). Once placed, double-click the VDD (or VSS) component, select the values tab and set the voltage to 15 V for VDD and −15 V for VSS. Once the circuit is complete, you can begin the simulation by pressing F5 (or Simulate → Run) and pause it by pressing F6. Double-click on the oscilloscope element in the schematic to bring up the

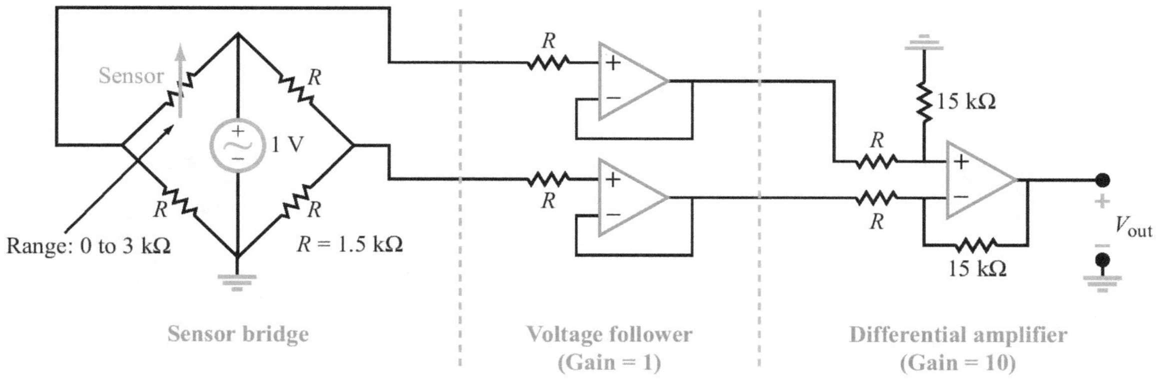

Figure 4-35: Wheatstone-bridge op-amp circuit.

oscilloscope's screen. The output voltage should be visible as Channel A in the oscilloscope window. In order to get a good view of the trace, you might need to adjust both its timebase and voltage scale using the controls found at the bottom of the Oscilloscope window. Observe the change in the amplitude of the output by shifting the resistance value of the sensor potentiometer.

With Multisim, you can modify different parts of the circuit and observe the consequent changes in behavior. Make sure to stop your simulation (not just pause it) before changing components or wiring.

Concept Question 4-22: What types of Multisim instruments are available for testing a circuit? (See CAD)

Concept Question 4-23: Explain what the timebase is on the oscilloscope. (See CAD)

Exercise 4-12: Why are the voltage followers necessary in the circuit of Fig. 4-36? Remove them from the Multisim circuit and connect the resistive bridge directly to the two inputs of the differential amplifier. How does the output vary with the potentiometer setting?

Answer: (See CAD)

4-13.2 The Digital Inverter

The MOSFET inverter introduced in Section 4-11.2 provides a good opportunity to explore the difference between steady-state and time-dependent analysis techniques. Consider again the MOSFET digital inverter of Fig. 4-30. When analyzing this type of logic gate, we usually are interested in both the response of the output voltage to a change in input voltage and in how fast the gate generates the output voltage in response to a change in input voltage. Both types of analyses are possible with Multisim.

Table 4-5: List of Multisim components for the circuit in Fig. 4-35.

Component	Group	Family	Quantity	Description
1.5 k	Basic	Resistor	7	1.5 kΩ resistor
15 k	Basic	Resistor	2	15 kΩ resistor
3 k	Basic	Variable resistor	1	3 kΩ resistor
OP_AMP_5T_VIRTUAL	Analog	Analog_Virtual	3	Ideal op amp with 5 terminals
AC_POWER	Sources	Power_Sources	1	1 V ac source, 60 Hz
VDD	Sources	Power_Sources	1	15 V supply
VSS	Sources	Power_Sources	1	−15 V supply

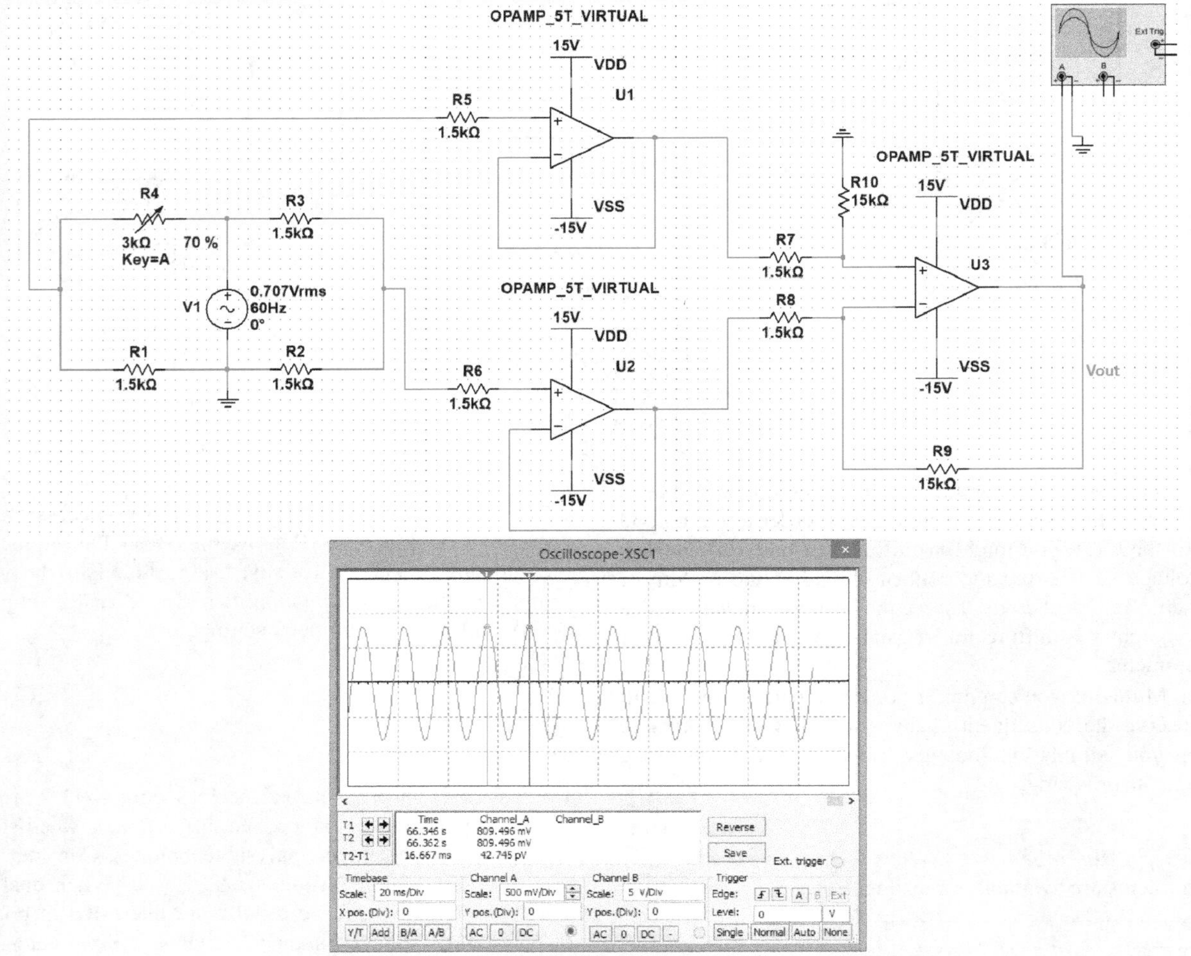

Figure 4-36: Multisim window of the circuit of Fig. 4-35. The oscilloscope trace shows the 60 Hz waveform of the output voltage. Had the voltage source been a dc source, the oscilloscope trace would have been a horizontal line.

Figure 4-37 shows a MOSFET inverter circuit in Multisim. To draw this circuit, you need the components listed in Table 4-6.

Transient Analysis

We can use a function generator (Simulate → Instruments → Function Generator) to observe the inverter output as a function of time. Double-click on the function generator to bring up its control window. Set the function generator to Square Wave mode with a frequency of 1 kHz, amplitude of 2.5 V, and an offset of 1.25 V. This will generate a 0–2.5 V square-wave input. The input and output can be plotted separately as a function of time using Simulate → Analyses → Transient Analysis. Whereas in Interactive Simulation the course of time is open ended (by default it is limited to a duration of 1×10^{30} s), when using Transient Analysis we can define the start and stop times. Maintain the start time at 0 s, set the final time to 0.005 s, and under the Output tab select the input voltage V(1) as the voltage to plot. Click Simulate. The input voltage is plotted as a function of time, as in Fig. 4-38(a). Repeat the simulation after removing V(1) and adding V(2) under the Output tab.

Table 4-6: **Components for the circuit in** Fig. 4-37.

Component	Group	Family	Quantity	Description
MOS_N	Transistors	Transistors_VIRTUAL	1	3-terminal N-MOSFET
MOS_P	Transistors	Transistors_VIRTUAL	1	3-terminal P-MOSFET
VDD	Sources	Power Sources	1	2.5 V supply
GND	Sources	Power Sources	2	Ground node

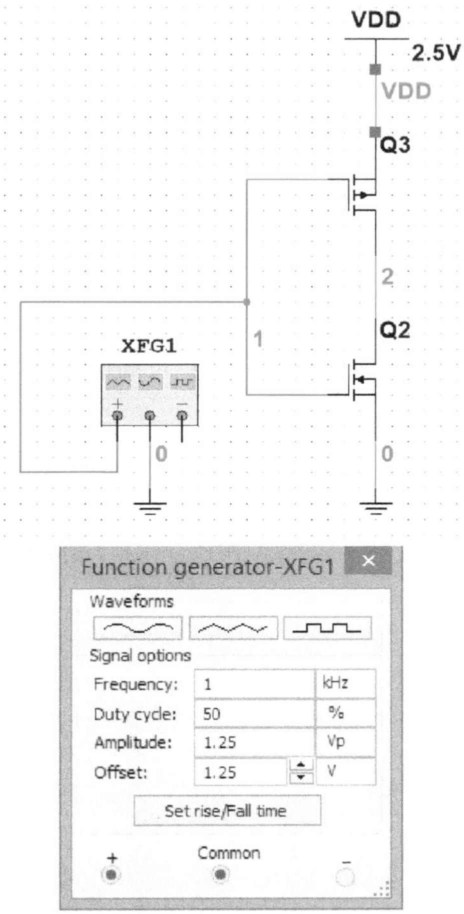

Figure 4-37: Multisim equivalent of the MOSFET circuit of Fig. 4-30.

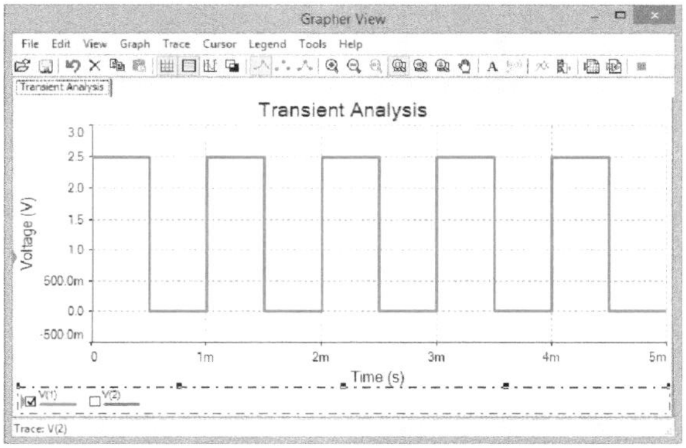

(a) Input voltage

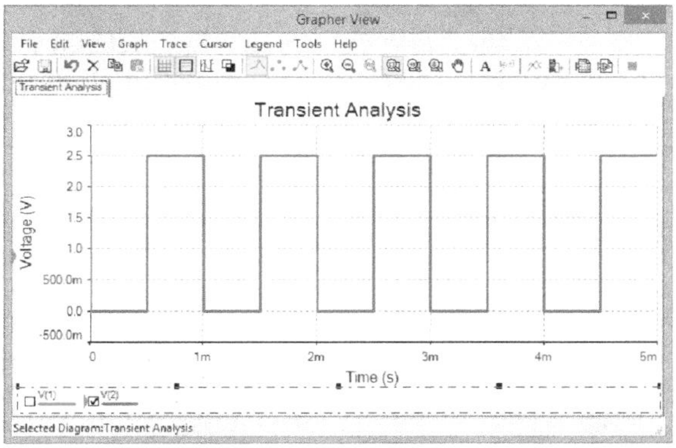

(b) Output voltage

Figure 4-38: Input and output voltages V(1) and V(2) in the circuit of Fig. 4-37 as a function of time.

Figure 4-38(b) shows the output voltage as a function of time. The input and output plots are essentially mirror images of one another.

Steady-State Analysis

In order to analyze the steady-state output behavior, we first must remove the function generator and replace it with a

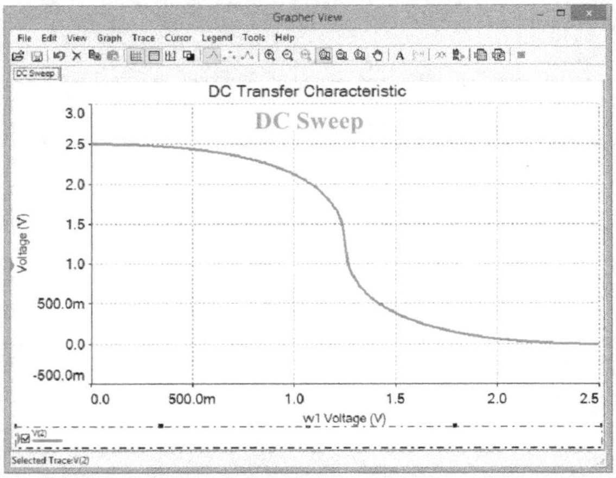

Figure 4-39: Output response of the MOSFET inverter circuit of Fig. 4-37 as a function of the amplitude of the input voltage.

dc voltage source. The actual voltage value of the source is unimportant. Once wired, select Simulate → Analyses → DC Sweep. This analysis is similar to the DC Operating Point Analysis, but it sweeps through a range of voltages at a node of your choice and solves for the resultant steady-state voltage (or current) at any other node you select. In this way, you can generate and plot input-output relationships for circuits and components.

Choose the source name vv1 as the input and enter 0 V, 2.5 V, and 0.01 V for the start, stop, and increment values, respectively. Under the Output tab, select the output voltage V(2) as the voltage to plot. Click Simulate. Figure 4-39 shows that the output displays the expected inverter behavior: an input in the 0 to 500 mV range generates an output of ~ 2.5 V; conversely, when the input is in the range between 2 and 2.5 V, the circuit generates an output voltage of ~ 0 V. In between, we see a gradual transition zone.

Concept Question 4-24: How do the DC Operating Point Analysis, Transient Analysis, and DC Sweep analyses differ? (See CAD)

Concept Question 4-25: How many types of waveforms can the generic function-generator instrument provide? (See CAD)

Exercise 4-13: The *IV Analyzer* is another useful Multisim instrument for analyzing circuit performance. To demonstrate its utility, let us use it to generate characteristic curves for an NMOS transistor similar to those in Fig. 4-28(b). Figure E4.13(a) shows an NMOS connected to an IV Analyzer. The instrument sweeps through a range of gate (G) voltages and generates a current-versus-voltage (IV) plot between the drain (D) and source (S) for each gate voltage. Show that the display of the IV analyzer is the same as that shown in Fig. E4.13(b).

Answer: (See CAD)

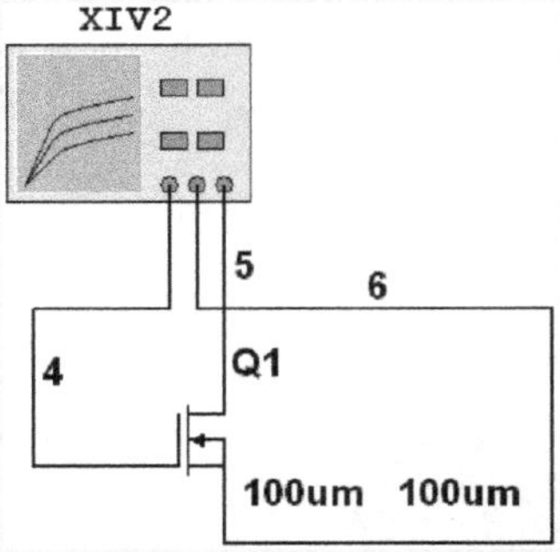

(a)

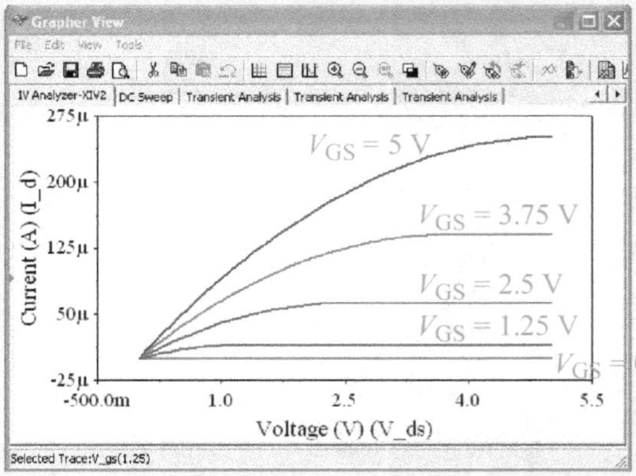

(b)

Figure E4.13 (a) Circuit schematic and (b) IV analyzer traces for I_{DS} versus V_{DS} at selected values of V_{GS}.

Summary

Concepts

- Despite its complex circuit architecture, the op amp can be modeled in terms of a relatively simple, linear equivalent circuit.
- The ideal op amp has infinite gain A, infinite input resistance R_i, and zero output resistance R_o.
- Through resistive feedback connections between its output and its two inputs, the op amp can be made to amplify, sum, and subtract multiple input signals.
- Multistage op-amp circuits can be configured to support a variety of signal-processing functions.
- Cascaded circuit blocks can be analyzed or designed individually and then combined together if R_o of the first circuit is much smaller than R_i of the second circuit.
- Buffers are used to increase R_i of the followup circuit.
- The instrumentation amplifier is a high-gain, high-sensitivity detector of small signals, making it particularly suitable for sensing deviations from reference conditions.
- Multisim can accommodate op-amp circuits and simulate their input-output responses.

Mathematical and Physical Models

Ideal op amp $\upsilon_p = \upsilon_n$
$i_p = i_n = 0$

Noninverting amp* $G = \dfrac{\upsilon_o}{\upsilon_s} = \dfrac{R_1 + R_2}{R_2}$

Inverting amp* $G = \dfrac{\upsilon_o}{\upsilon_s} = -\left(\dfrac{R_f}{R_s}\right)$

Summing amp* $\upsilon_o = -R_f\left(\dfrac{\upsilon_1}{R_1} + \dfrac{\upsilon_2}{R_2}\right)$

Difference amp* $\upsilon_o = G_2\upsilon_2 + G_1\upsilon_2$

Voltage follower* $\upsilon_o = \upsilon_s$

Instrumentation amp $\upsilon_o = \left(1 + \dfrac{2R}{R_2}\right)(\upsilon_2 - \upsilon_1)$
(with gain-control resistor R_2)

MOSFET $V_{out} = V_{DD} - gR_D V_{in}$

*See Table 4-3.

Important Terms

Provide definitions or explain the meaning of the following terms:

action potential
ADC
adder
bit
buffer
circuit gain
closed-loop gain
CMOS
complementary MOS
current constraint
difference amplifier
digital inverter
digital-to-analog converter
DIP configuration
drain
dynamic range
feedback
feedback resistance
gain-control resistance
gate
ideal op-amp current constraint
ideal op-amp voltage constraint
input resistance
input source resistance
instrumentation amplifier
inverter
inverting
inverting adder
inverting amplifier
inverting input
inverting summing amplifier
IV Analyzer
least significant bit
linear
linear dynamic range
loading
metal-oxide semiconductor field-effect transistor
MOSFET
MOSFET gain constant
most significant bit
negative feedback
negative saturation
neural interface
neural probe
neuron
NMOS
noninverting amplifier
noninverting
noninverting input
noninverting summing amplifier
oscilloscope
op amp
op-amp gain
open-loop gain
operational amplifier
output resistance
overloading
percent clipping
PMOS
positive feedback
positive saturation
R–$2R$ ladder
saturation threshold value
scaled inverting adder
sensor
signal-processing circuit
source
subtraction
summing amplifier
unity gain amplifier
voltage constraint
voltage follower
voltage rails

PROBLEMS

Sections 4-1 and 4-2: Op-Amp Characteristics and Negative Feedback

*4.1 An op amp with an open-loop gain of 10^6 and $V_{cc} = 12$ V has an inverting-input voltage of 20 μV and a noninverting-input voltage of 10 μV. What is its output voltage?

4.2 An op amp with an open-loop gain of 6×10^5 and $V_{cc} = 10$ V has an output voltage of 3 V. If the voltage at the inverting input is -1 μV, what is the magnitude of the noninverting-input voltage?

*4.3 What is the output voltage for an op amp whose noninverting input is connected to ground and its inverting-input voltage is 4 mV? Assume that the op-amp open-loop gain is 2×10^5 and its supply voltage is $V_{cc} = 10$ V.

4.4 With its noninverting-input voltage at 10 μV, the output voltage of an op amp is -15 V. If $A = 5 \times 10^5$ and $V_{cc} = 15$ V, can you determine the magnitude of the inverting-input voltage? If not, can you determine its possible range?

4.5 For the op-amp circuit shown in Fig. P4.5:

(a) Use the model given in Fig. 4-6 to develop an expression for the current gain $G_i = i_L/i_s$.

(b) Simplify the expression by applying the ideal op-amp model (taking $A \to \infty$, $R_i \to \infty$, and $R_o \to 0$).

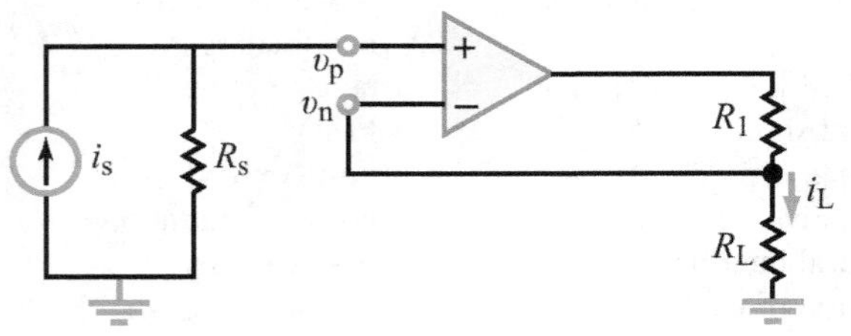

Figure P4.5: Circuit for Problem 4.5.

4.6 The inverting-amplifier circuit shown in Fig. P4.6 uses a resistor R_f to provide feedback from the output terminal to the inverting-input terminal.

(a) Use the equivalent-circuit model of Fig. 4-6 to obtain an expression for the closed-loop gain $G = \upsilon_o/\upsilon_s$ in terms of R_s, R_i, R_o, R_L, R_f, and A.

(b) Determine the value of G for $R_s = 10\ \Omega$, $R_i = 10$ MΩ, $R_f = 1$ kΩ, $R_o = 50\ \Omega$, $R_L = 1$ kΩ, and $A = 10^6$.

(c) Simplify the expression for G obtained in (a) by letting $A \to \infty$, $R_i \to \infty$, and $R_o \to 0$ (ideal op-amp model).

*(d) Evaluate the approximate expression obtained in (c) and compare the result with the value obtained in (b).

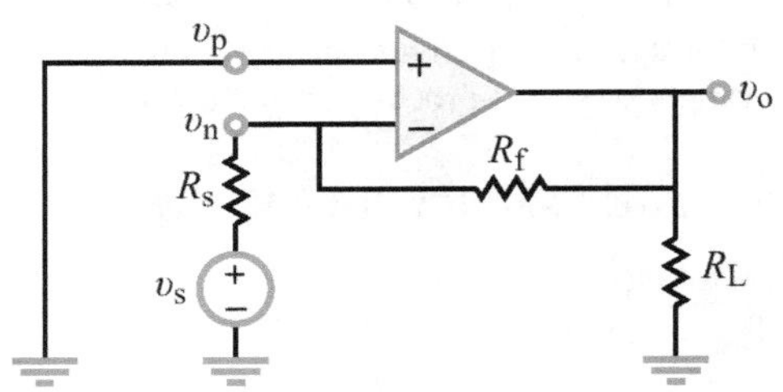

Figure P4.6: Circuit for Problem 4.6.

4.7 For the circuit in Fig. P4.7:

(a) Use the op-amp equivalent-circuit model to develop an expression for $G = \upsilon_o/\upsilon_s$.

(b) Simplify the expression by applying the ideal op-amp model parameters, namely $A \to \infty$, $R_i \to \infty$, and $R_o \to 0$.

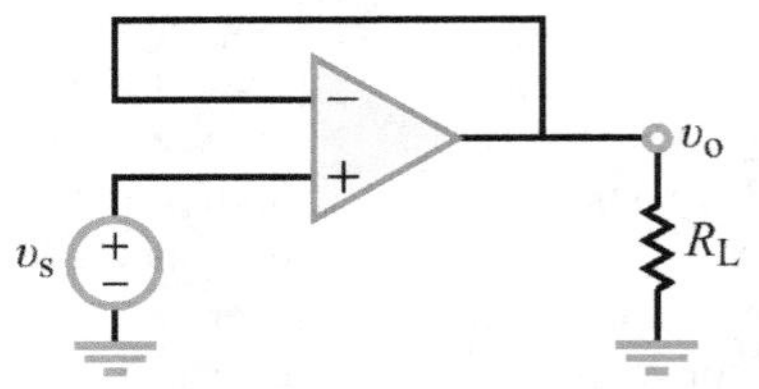

Figure P4.7: Circuit for Problem 4.7.

4.8 The op-amp circuit shown in Fig. P4.8 has a constant dc voltage of 6 V at the noninverting input. The inverting input is the sum of two voltage sources consisting of a 6 V dc source and a small time-varying signal υ_s.

*Answer(s) available in Appendix G.

(a) Use the op-amp equivalent-circuit model given in Fig. 4-6 to develop an expression for v_o.

(b) Simplify the expression by applying the ideal op-amp model, which lets $A \to \infty$, $R_i \to \infty$, and $R_o \to 0$.

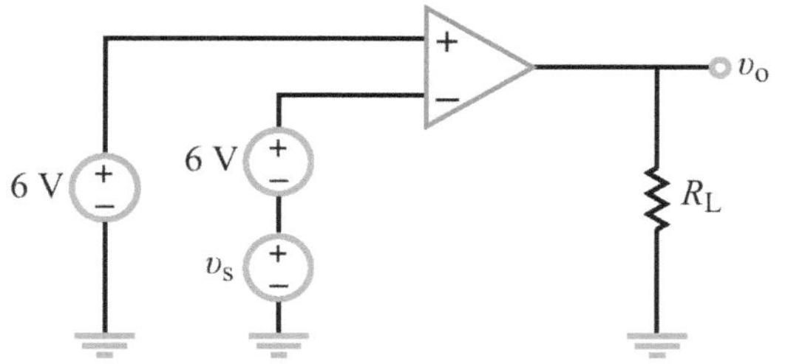

Figure P4.8: Circuit for Problem 4.8.

Sections 4-3 and 4-4: Ideal Op Amp and Inverting Amp

Assume all op amps to be ideal from here on forward.

*4.9 The supply voltage of the op amp in the circuit of Fig. P4.9 is 16 V. If $R_L = 3$ kΩ, assign a resistance value to R_f so that the circuit would deliver 75 mW of power to R_L.

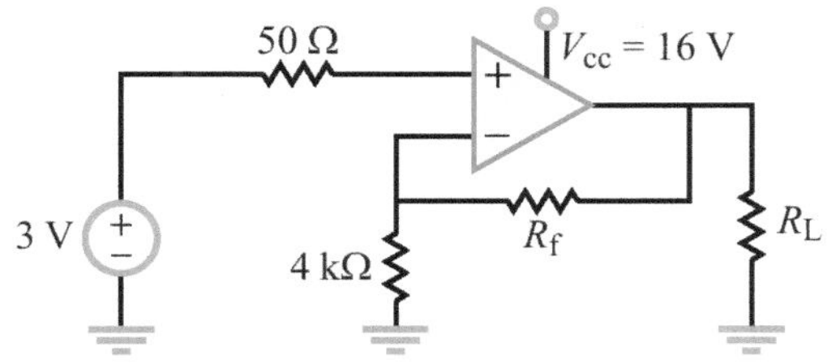

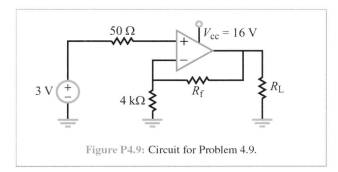

Figure P4.9: Circuit for Problem 4.9.

4.10 In the circuit of Fig. P4.10, a bridge circuit is connected at the input side of an inverting op-amp circuit.

(a) Obtain the Thévenin equivalent at terminals (a, b) for the bridge circuit.

(b) Use the result in (a) to obtain an expression for $G = v_o/v_s$.

(c) Evaluate G for $R_1 = R_4 = 100$ Ω, $R_2 = R_3 = 101$ Ω, and $R_f = 100$ kΩ.

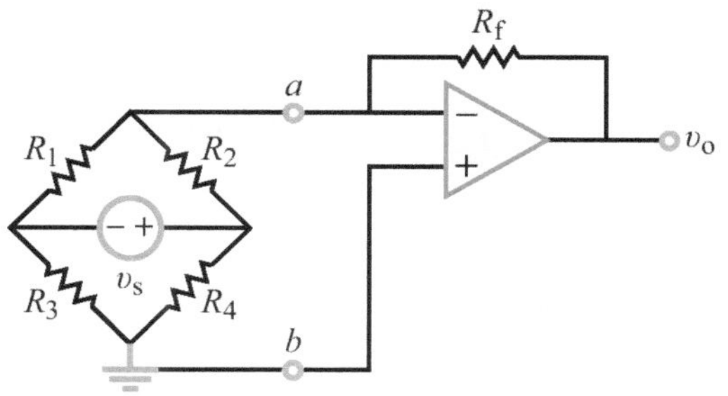

Figure P4.10: Circuit for Problem 4.10.

4.11 Determine the output voltage for the circuit in Fig. P4.11 and specify the linear range for v_s, given that $V_{cc} = 15$ V and $V_0 = 0$.

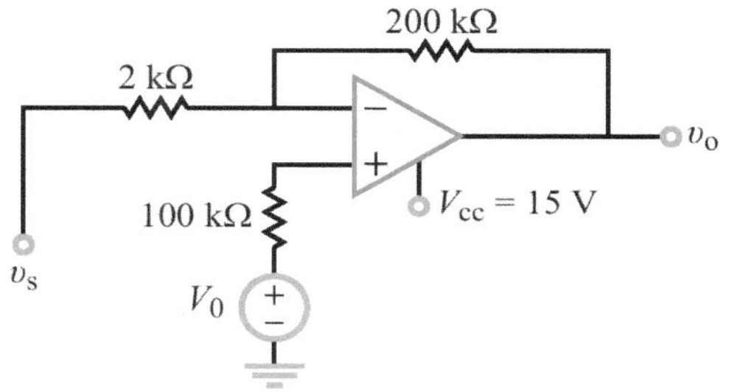

Figure P4.11: Circuit for Problems 4.11 and 4.12.

4.12 Repeat Problem 4.11 for $V_0 = 0.1$ V.

*4.13 Obtain an expression for the voltage gain $G = v_o/v_s$ for the circuit in Fig. P4.13.

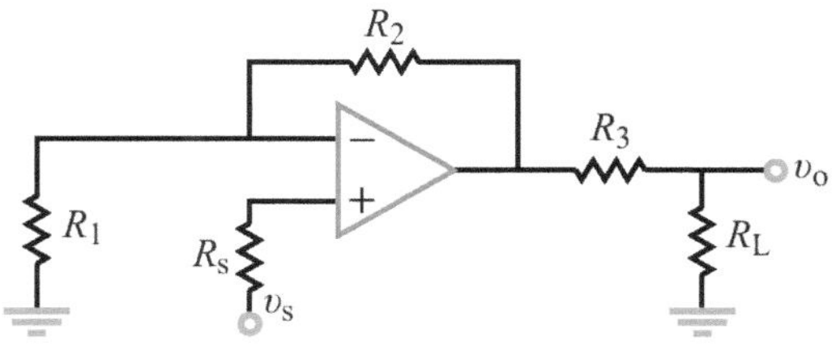

Figure P4.13: Circuit for Problem 4.13.

4.14 For the op-amp circuit shown in Fig. P4.14:

(a) Obtain an expression for the current gain $G_i = i_L/i_s$.

*(b) If $R_L = 12$ kΩ, choose R_f so that $G_i = -15$.

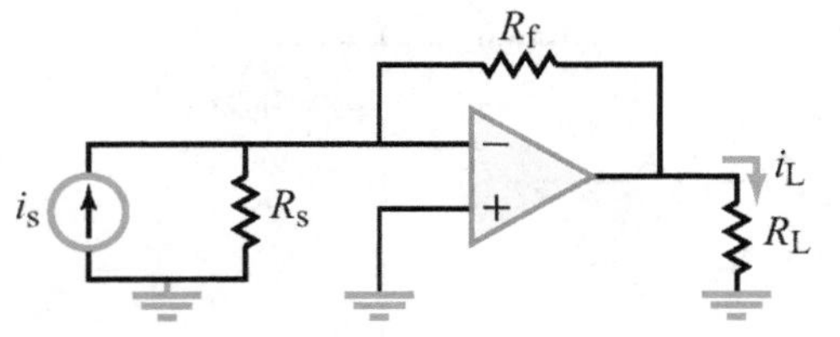

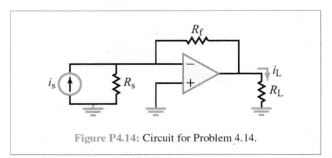

Figure P4.14: Circuit for Problem 4.14.

4.15 Determine the gain $G = v_L/v_s$ for the circuit in Fig. P4.15 and specify the linear range of v_s for $R_L = 4$ kΩ.

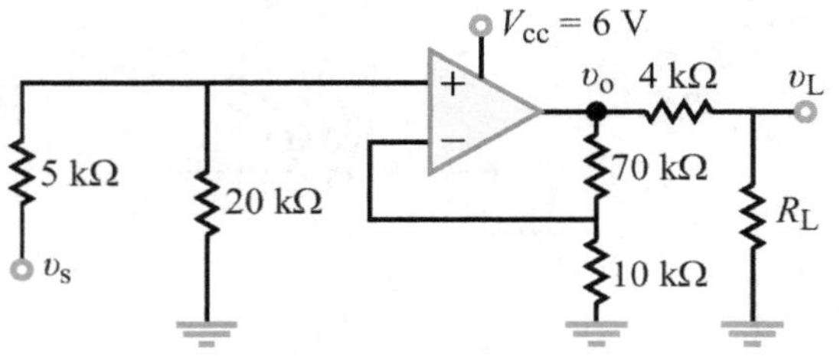

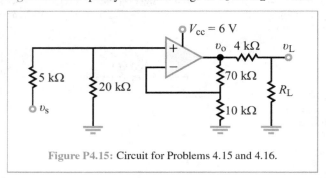

Figure P4.15: Circuit for Problems 4.15 and 4.16.

***4.16** For the circuit of Fig. P4.15, what should the resistance value of R_L be so as to have maximum transfer of power into it?

4.17 Determine v_o across the 10 kΩ resistor in the circuit of Fig. P4.17.

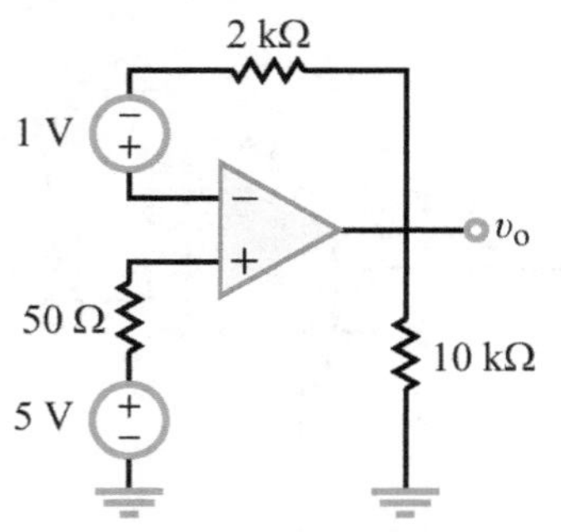

Figure P4.17: Circuit for Problem 4.17.

4.18 Evaluate $G = v_o/v_s$ for the circuit in Fig. P4.18, and specify the linear range of v_s. Assume $R_f = 2400$ Ω.

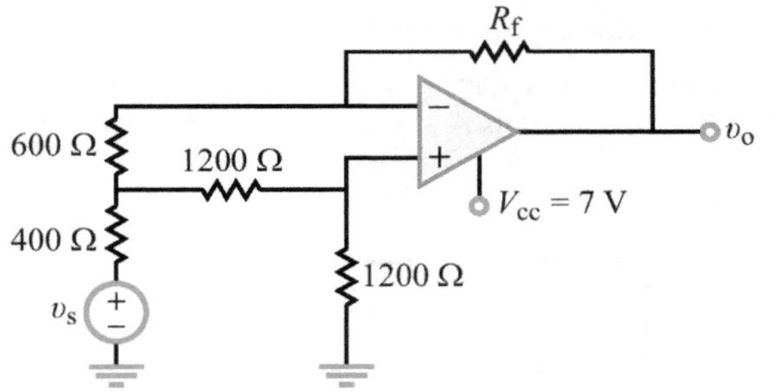

Figure P4.18: Circuit for Problems 4.18 and 4.19.

***4.19** Repeat Problem 4.18 for $R_f = 0$.

4.20 Determine the linear range of the source v_s in the circuit of Fig. P4.20.

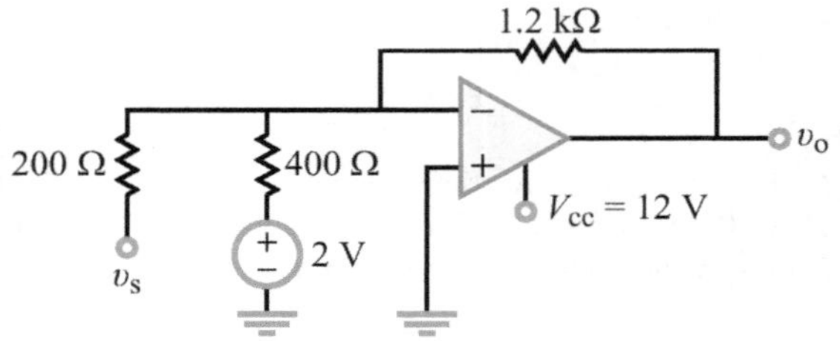

Figure P4.20: Circuit for Problems 4.20 and 4.21.

***4.21** Repeat Problem 4.20 after replacing the 2 V dc source in Fig. P4.20 with a short circuit.

4.22 The circuit in Fig. P4.22 uses a potentiometer whose total resistance is $R = 10$ kΩ with the upper section being βR and the bottom section $(1-\beta)R$. The stylus can change β from 0 to 0.9. Obtain an expression for $G = v_o/v_s$ in terms of β and evaluate the range of G (as β is varied over its own allowable range).

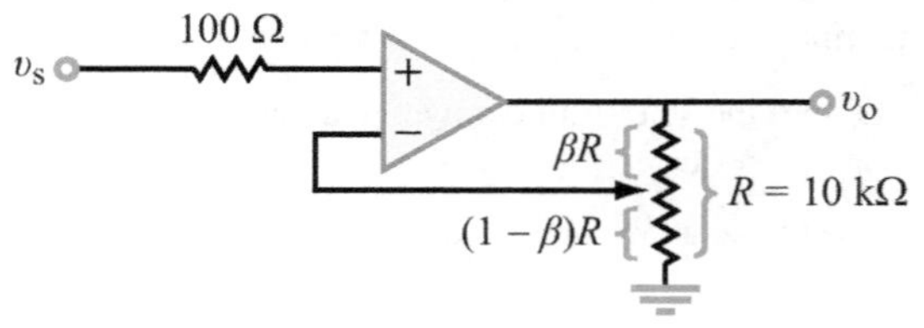

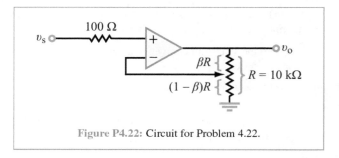

Figure P4.22: Circuit for Problem 4.22.

4.23 For the circuit in Fig. P4.23, obtain an expression for voltage gain $G = \upsilon_o/\upsilon_s$.

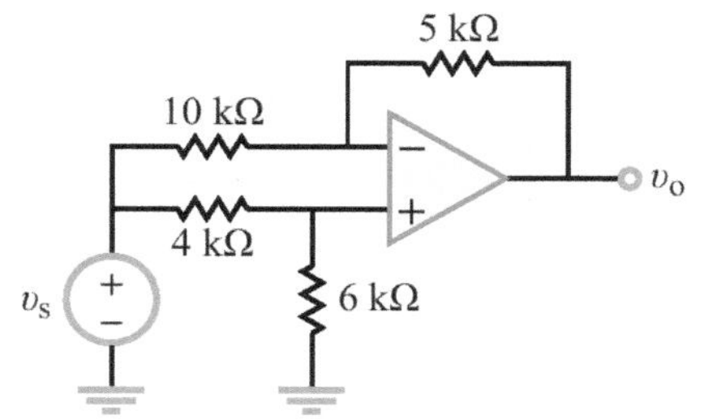

Figure P4.23: Circuit for Problem 4.23.

***4.24** Find the value of υ_o in the circuit in Fig. P4.24.

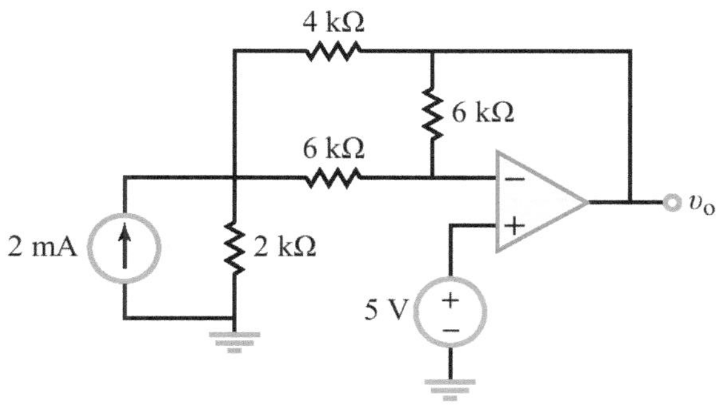

Figure P4.24: Circuit for Problem 4.24.

4.25 Determine the linear range of υ_s for the circuit in Fig. P4.25.

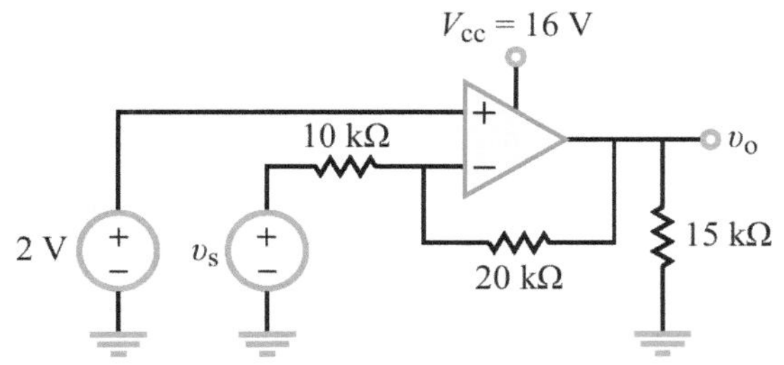

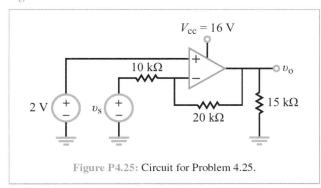

Figure P4.25: Circuit for Problem 4.25.

Sections 4-5 and 4-6: Summing and Difference Amplifiers

4.26 If $R_2 = 4$ kΩ, select values for R_{s_1}, R_{s_2}, and R_1 in the circuit of Fig. 4-15 so that $\upsilon_o = 3\upsilon_1 + 5\upsilon_2$.

4.27 Design an op-amp circuit that performs an averaging operation of five inputs υ_1 to υ_5.

4.28 For the circuit in Fig. P4.28, generate a plot for υ_L as a function of υ_s over the full linear range of υ_s.

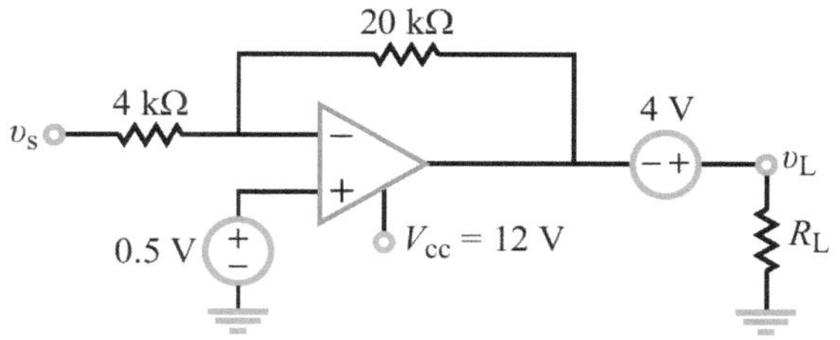

Figure P4.28: Circuit for Problem 4.28.

4.29 Relate υ_o in the circuit of Fig. P4.29 to υ_s and specify the linear range of υ_s. Assume $V_0 = 0$.

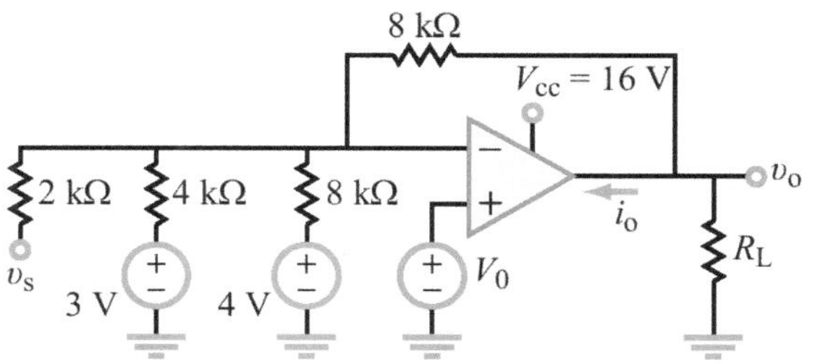

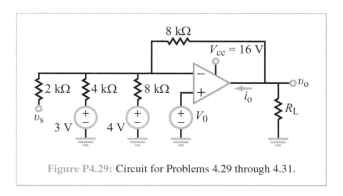

Figure P4.29: Circuit for Problems 4.29 through 4.31.

***4.30** Repeat Problem 4.29 for $V_0 = 6$ V.

4.31 Determine the current i_o flowing into the op-amp of the circuit in Fig. P4.29 under the conditions $\upsilon_s = 0.5$ V, $V_0 = 0$, and $R_L = 10$ kΩ.

4.32 Design a circuit containing a single op amp that can perform the operation $\upsilon_o = 3 \times 10^4(i_2 - i_1)$, where i_2 and i_1 are input current sources.

4.33 Design a circuit that can perform the operation $\upsilon_o = 3\upsilon_1 + 4\upsilon_2 - 5\upsilon_3 - 8\upsilon_4$, where υ_1 to υ_4 are input voltage signals.

4.34 Relate υ_o in the circuit of Fig. P4.34 to υ_1, υ_2, and υ_3.

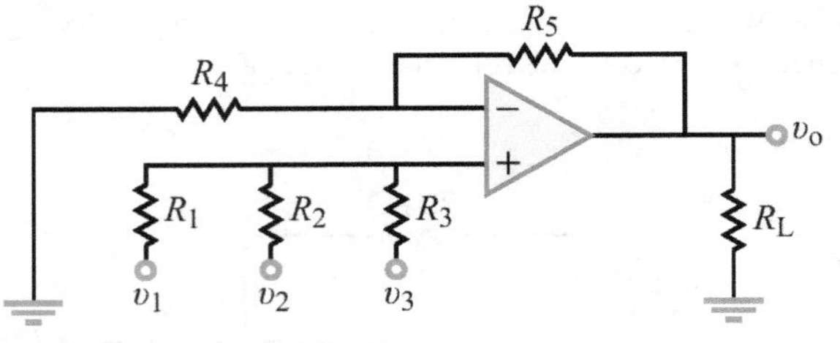

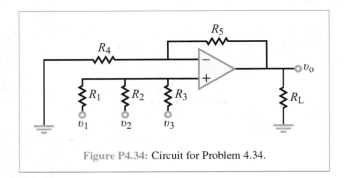

Figure P4.34: Circuit for Problem 4.34.

*__4.35__ For the circuit in Fig. P4.35, obtain an expression for υ_o in terms of υ_1, υ_2, and the four resistors. Evaluate υ_o if $\upsilon_1 = 0.1$ V, $\upsilon_2 = 0.5$ V, $R_1 = 100$ Ω, $R_2 = 200$ Ω, $R_3 = 2.4$ kΩ, and $R_4 = 1.2$ kΩ.

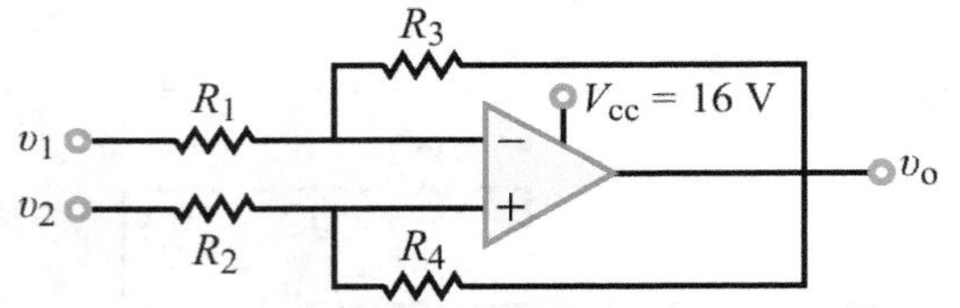

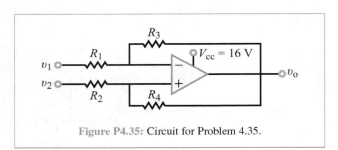

Figure P4.35: Circuit for Problem 4.35.

4.36 Find the value of υ_o in the circuit in Fig. P4.36.

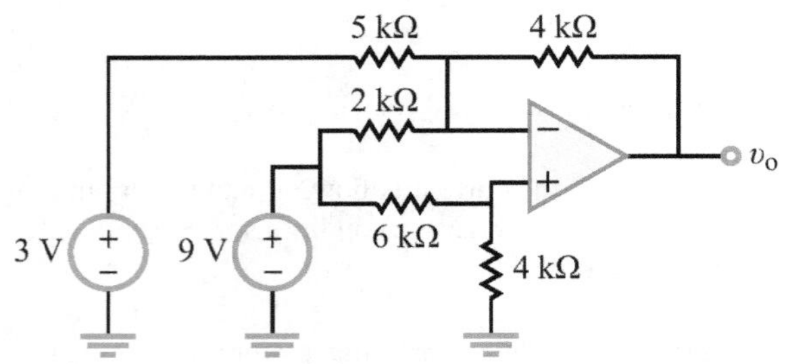

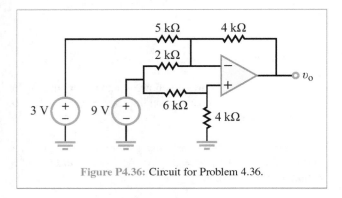

Figure P4.36: Circuit for Problem 4.36.

4.37 Find the range of R_f for which the op amp in the circuit of Fig. P4.37 does not saturate.

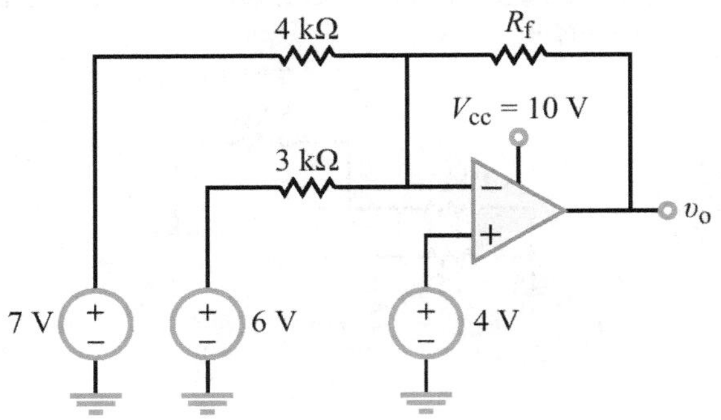

Figure P4.37: Circuit for Problem 4.37.

*__4.38__ Determine υ_o and the power dissipated in R_L in the circuit of Fig. P4.38.

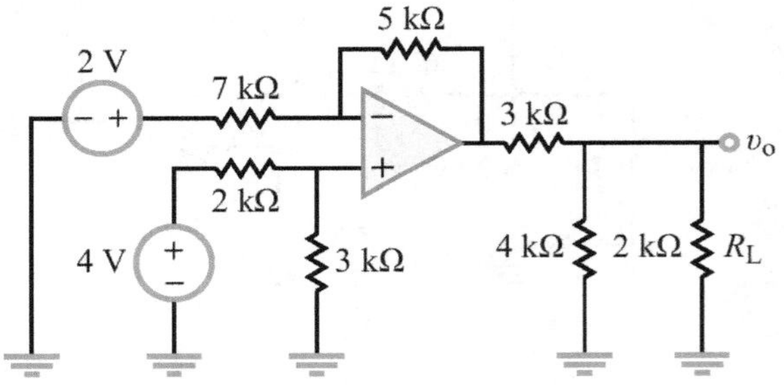

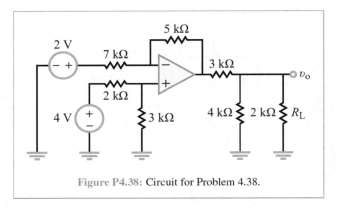

Figure P4.38: Circuit for Problem 4.38.

4.39 The circuit in Fig. P4.39 contains two single-pole single-throw switches, S_1 and S_2. Determine the closed-circuit gain $G = \upsilon_o/\upsilon_s$ for each of the four possible closed/open switch combinations.

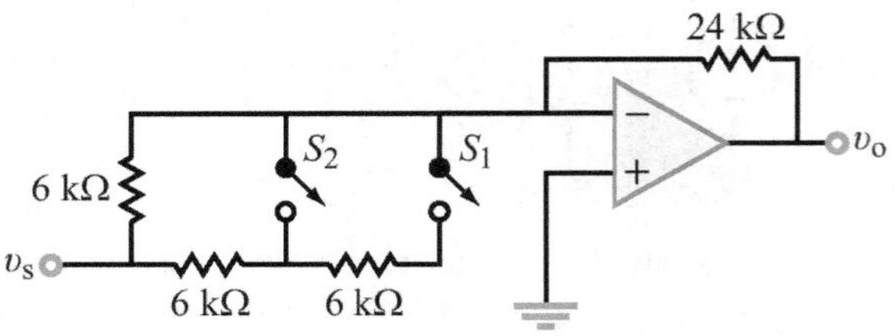

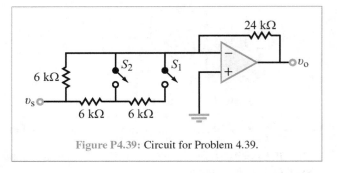

Figure P4.39: Circuit for Problem 4.39.

Section 4-8: Op-Amp Signal-Processing Circuits

4.40 Develop a block-diagram representation for the circuit in Fig. P4.40 for $v_{s_2} = v_{s_3} = 0$ and

*(a) R_1 = open circuit

(b) $R_1 = 10$ kΩ.

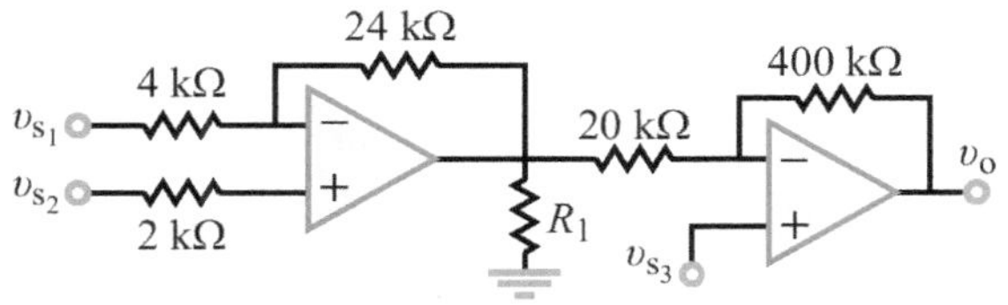

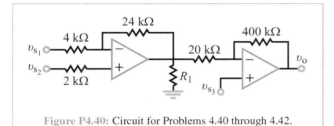

Figure P4.40: Circuit for Problems 4.40 through 4.42.

4.41 Develop a block-diagram representation for the circuit in Fig. P4.40 for $v_{s_3} = 0$ and $R_1 = \infty$.

4.42 Develop a block-diagram representation for the circuit in Fig. P4.40 for $v_{s_2} = 0$ and $R_1 = \infty$.

4.43 For the circuit in Fig. P4.43:

(a) Develop a block-diagram representation with R_L as a variable parameter.

(b) Specify the linear range of v_s.

(c) Determine v_o for $v_s = 0.3$ V and $R_L = 10$ kΩ.

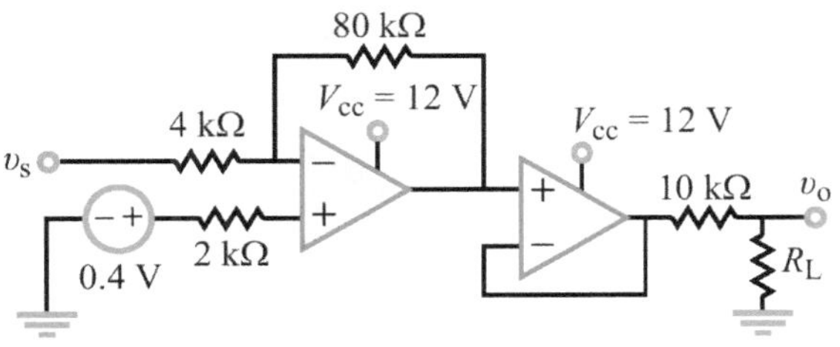

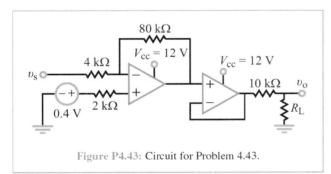

Figure P4.43: Circuit for Problem 4.43.

4.44 Design an op-amp circuit that can perform the operation $v_o = 12v_{s_1} + 3v_{s_2}$, while simultaneously presenting an input resistance of 50 kΩ on the input side for source v_{s_1} and an input resistance of 25 kΩ on the input side for source v_{s_2}.

4.45 Design an op-amp circuit that can perform the operation $v_o = 4v_{s_1} - 3v_{s_2}$, while simultaneously presenting an input resistance of 10 kΩ on the input side for source v_{s_1} and an input resistance of 5 kΩ on the input side for source v_{s_2}.

*__4.46__ Relate v_o in the circuit of Fig. P4.46 to v_s.

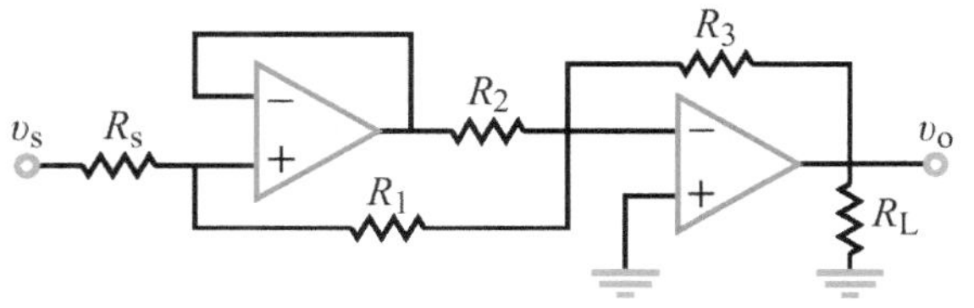

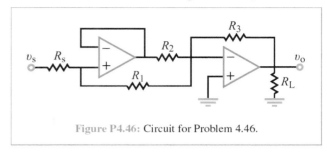

Figure P4.46: Circuit for Problem 4.46.

4.47 In the circuit of Fig. P4.47, op amp 1 receives feedback at its input from its own output as well as from the output of op amp 2. Relate v_o to v_s.

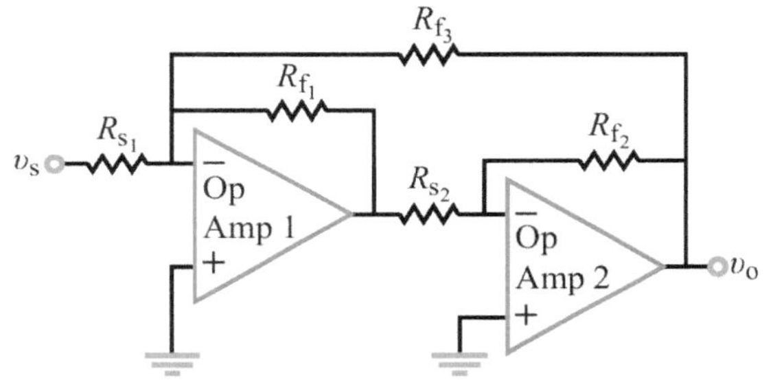

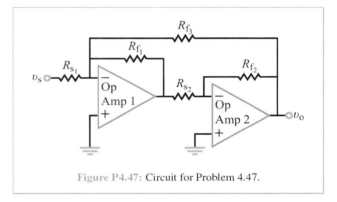

Figure P4.47: Circuit for Problem 4.47.

4.48 Relate v_o in the circuit of Fig. P4.48 to v_1 and v_2.

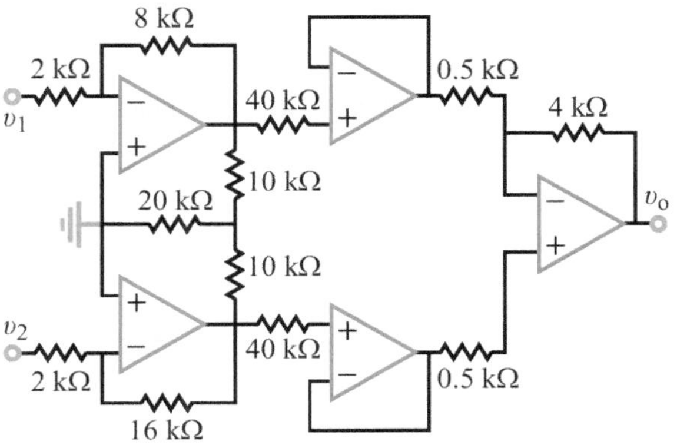

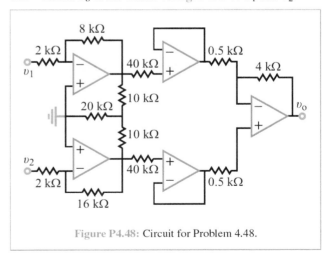

Figure P4.48: Circuit for Problem 4.48.

4.49 Design an op-amp circuit that can perform the operation $i_o = (30i_1 - 8i_2 + 0.6)$ A where i_1 and i_2 are input current sources.

*4.50 Relate the output voltage v_o in Fig. P4.50 to v_s.

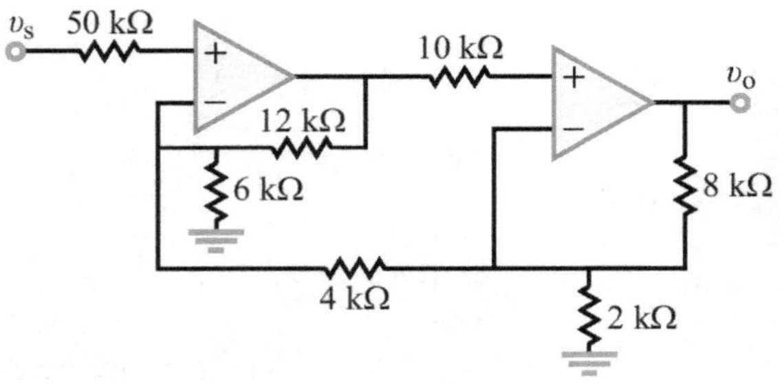

Figure P4.50: Circuit for Problem 4.50.

4.51 Solve for v_o in terms of v_s for the circuit in Fig. P4.51.

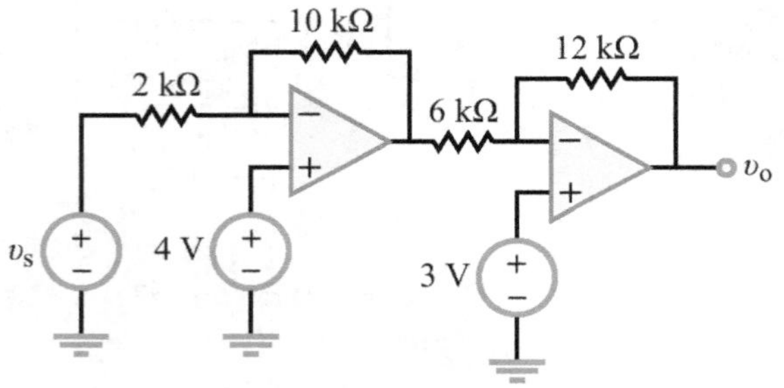

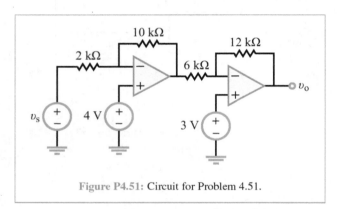

Figure P4.51: Circuit for Problem 4.51.

*4.52 Find the value of v_o in the circuit in Fig. P4.52.

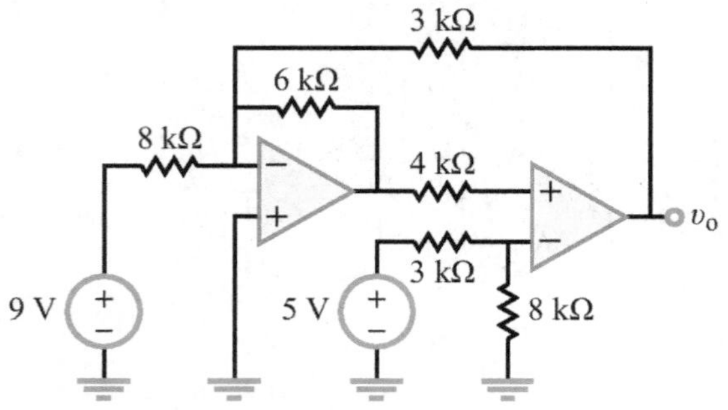

Figure P4.52: Circuit for Problem 4.52.

4.53 Solve for v_o in the circuit in Fig. P4.53.

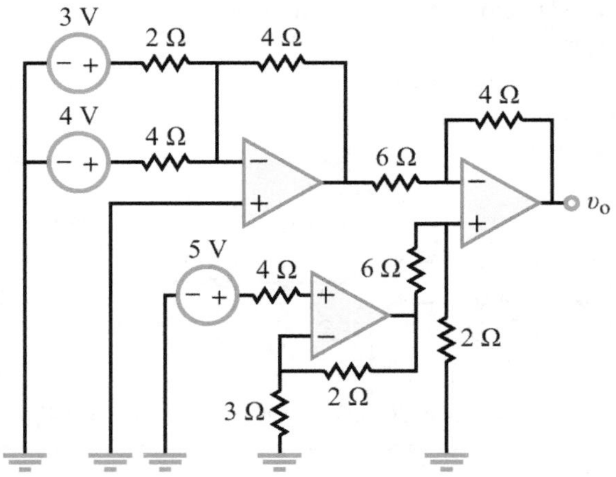

Figure P4.53: Circuit for Problem 4.53.

*4.54 If $v_o = -3$ V, what is the value of v_s in the circuit in Fig. P4.54?

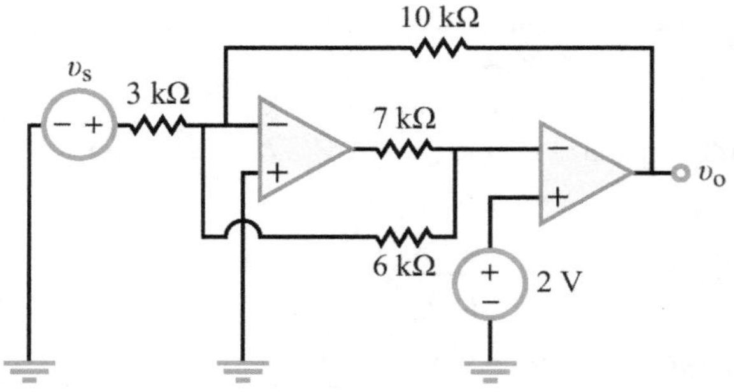

Figure P4.54: Circuit for Problem 4.54.

Sections 4-9 and 4-10: Instrumentation Amp and D/A Converter

4.55 The instrumentation-amplifier circuit shown in Fig. 4-23 is used to measure the voltage differential $\Delta v = v_2 - v_1$. If the range of variation of Δv is from -10 to $+10$ mV and $R_1 = R_3 = R_4 = R_5 = 100$ kΩ, choose R_2 so that the corresponding range of v_o is from -5 to $+5$ V.

*4.56 An instrumentation amplifier with $R_1 = R_3 = 10$ kΩ, $R_4 = 1$ MΩ, and $R_5 = 1$ kΩ uses a potentiometer for the gain-control resistor R_2. If the potentiometer resistance can be varied between 10 and 100 Ω, what is the corresponding variation of the circuit gain $G = \upsilon_o/(\upsilon_2 - \upsilon_1)$?

4.57 Design a five-bit DAC using a circuit configuration similar to that in Fig. 4-25.

4.58 Design a six-bit DAC using a R–$2R$ ladder configuration.

Section 4-11: MOSFET

4.59 In Example 4-9, we analyzed a common-source amplifier without a load resistance. Consider the amplifier in Fig. P4.59; it is identical to the circuit in Fig. 4-31, except that we have added a load resistor R_L. Obtain an expression for υ_{out} as a function of υ_s.

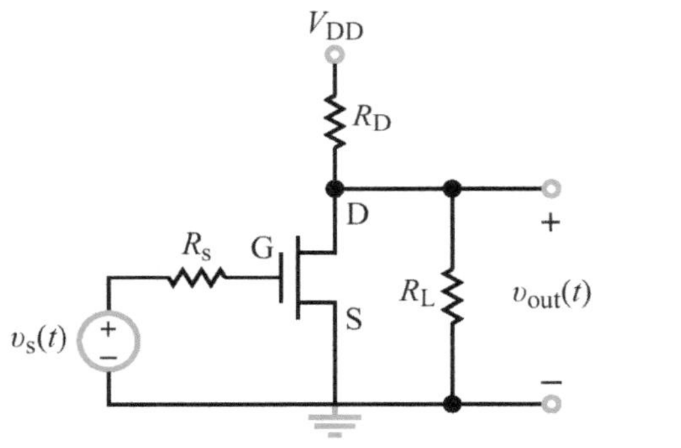

Figure P4.59: MOSFET circuit for Problem 4.59.

*4.60 Determine $\upsilon_{out}(t)$ as a function of $\upsilon_s(t)$ for the circuit in Fig. P4.60. Assume $V_{DD} = 2.5$ V.

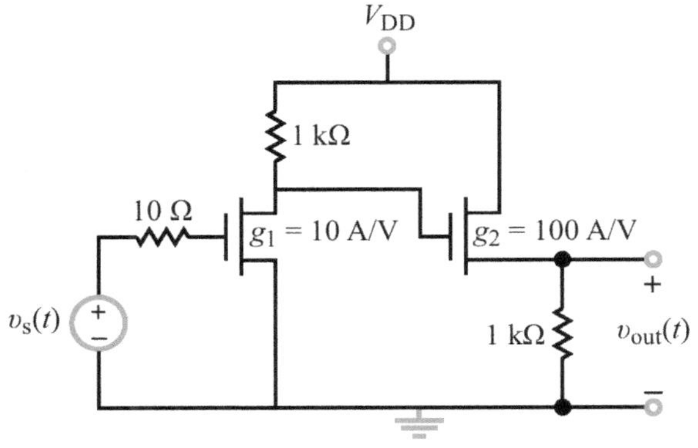

Figure P4.60: Two-MOSFET circuit for Problem 4.60.

4.61 In Problem 3.73 of Chapter 3, we analyzed a current mirror circuit containing BJTs. Current mirror circuits also can be designed using MOSFETs, as shown in Fig. P4.61. Determine the relationship between I_0 and I_{REF}.

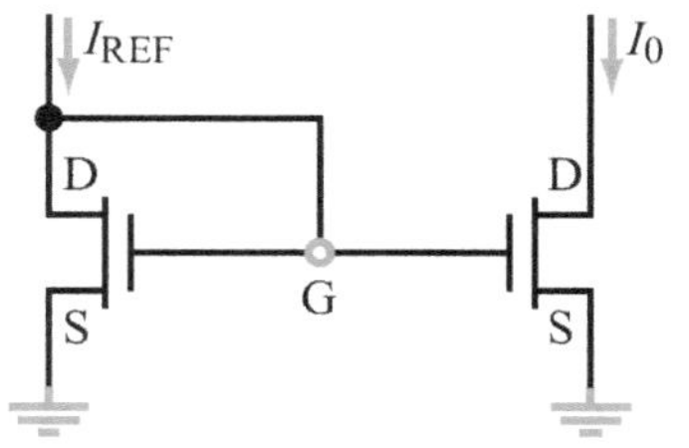

Figure P4.61: Circuit for Problem 4.61.

Section 4-13: Multisim Analysis

4.62 Draw a noninverting amplifier (Fig. 4-7) with a gain of 2 in Multisim. Show that the circuit works as expected by connecting a 1 V pulse source and plotting both the input and the output voltages using the Grapher Tool and Transient Analysis. Use the 3-terminal virtual op-amp component.

4.63 Draw an inverting amplifier (Fig. 4-11) with a gain of −3.5 in Multisim. Show that the circuit works as expected by connecting a 1 V dc voltage source and solving the circuit using the DC Operating Point analysis. Use the 3-terminal virtual op-amp component.

4.64 In Multisim, draw a summing amplifier that adds the values of four different dc voltage sources, each with an inverting gain of 4. Use the DC Operating Point analysis tool to verify the circuit performance.

4.65 In Multisim, draw a noninverting summing amplifier that adds the values of three different dc voltage sources V_1, V_2, and V_3 with gains of 1, 2, and 5, respectively. Apply the DC Operating Point Solution tool to demonstrate that the circuit functions as specified.

4.66 Draw the op-amp circuit shown in Fig. P4.66 in Multisim, provide a DC Operating Point Analysis solution that demonstrates its operation, and state what function the circuit performs.

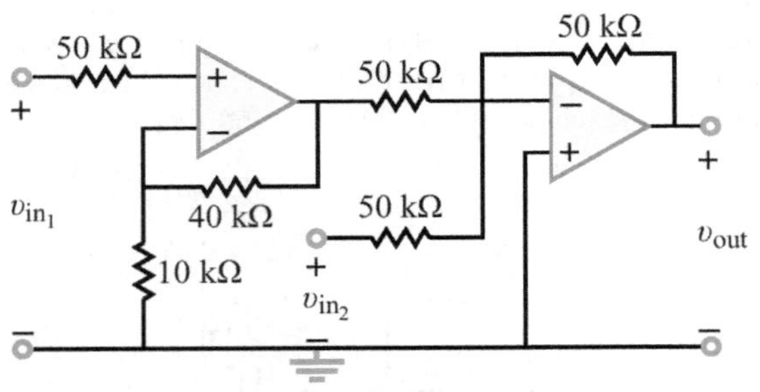

Figure P4.66: Circuit for Problem 4.66.

4.67 Construct the noninverting amplifier circuit shown in Fig. P4.67 in Multisim. Set the value of R to 50 kΩ and then perform a DC Sweep analysis of the input voltage from −5 to +5 V. Plot the Output. Now change the value of R to 80 kΩ and repeat the DC Sweep analysis. Compare the two plots either side by side or by overlapping them using the Overlay Traces button on the Grapher toolbar. (Use the three-terminal virtual op amp for the simulation.)

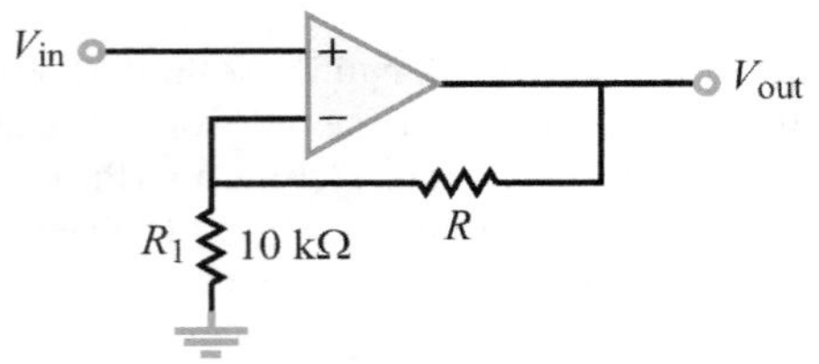

Figure P4.67: Circuit for Problem 4.67.

4.68 Until the 1970s, much research was carried out on analog computers (as distinguished from the digital computers found everywhere today). In fact, analog computers were one of the originally intended users of operational amplifiers. Op amps easily can be incorporated to perform many mathematical operations.

Using the basic op-amp circuits shown in this chapter, construct a circuit that expresses the following algebraic equation in voltage:

$$v = 2x - 3.5y + 0.2z,$$

where v is the output voltage and x, y, and z are three input voltages. Once you have the circuit designed, build it in Multisim and demonstrate that the circuit behaves appropriately by giving it the following inputs: $x = 1.2$, $y = 0.4$, and $z = 0.9$.

Potpourri Questions

4.69 Based on the information provided in Table TT9-1 of Technology Brief 9, which types of display technologies are best suited for a large football stadium? A home TV? A cell phone screen?

4.70 What are the limitations of today's computer memory circuits (ROM and RAM), and what emerging technologies are becoming available to improve them?

4.71 Circuit analysis and design can be performed analytically by applying the techniques covered in this book, or they can be performed by computer simulation. Are these competing or complementary approaches? Explain.

Integrative Problems: Analytical / Multisim / myDAQ

To master the material in this chapter, solve the following problems using three complementary approaches: (a) analytically, (b) with Multisim, and (c) by constructing the circuit and using the myDAQ interface unit to measure quantities of interest via your computer. [myDAQ tutorials and videos are available on CAD.]

m4.1 Ideal Op-Amp Model:

(a) Determine a general expression for v_{out} in terms of the resistor values and i_s for the circuit of Fig. m4.1 (no Multisim or myDAQ for this part).

(b) Find V_{out} for these specific component values: $R_1 = 3.3$ kΩ, $R_2 = 4.7$ kΩ, $R_3 = 1.0$ kΩ, and $I_s = 1.84$ mA.

(c) Replace R_2 with a potentiometer. Use myDAQ and the potentiometer to determine V_{out} for each of the following values of R_2: 2.5 kΩ, 10 kΩ, and 25 kΩ.

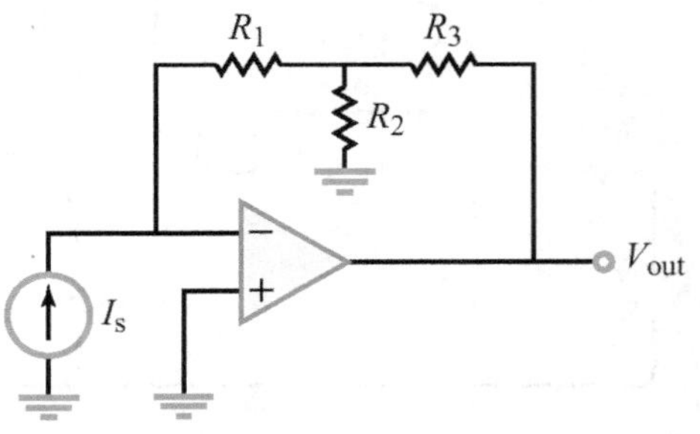

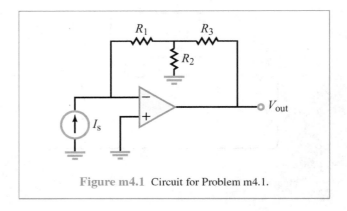

Figure m4.1 Circuit for Problem m4.1.

m4.2 Noninverting Amplifier: The circuit in Fig. m4.2 uses a potentiometer whose total resistance is R_1. The movable stylus on terminal 2 creates two variable resistors: βR_1 between terminals 1 and 2 and $(1-\beta)R_1$ between terminals 2 and 3. The movable stylus varies β over the range $0 \le \beta \le 1$.

(a) Obtain on expression for $G = \upsilon_o/\upsilon_s$ in terms of β.

(b) Calculate the amplifier gain for $\beta = 0.0$, $\beta = 0.5$, and $\beta = 1.0$ with component values $R_1 = 10$ kΩ and $R_2 = 1.5$ kΩ.

(c) Let υ_s be a 100 Hz sinusoidal signal with a 1 V peak value. Plot υ_o and υ_s to scale for $\beta = 0.0$, $\beta = 0.5$, and $\beta = 1.0$.

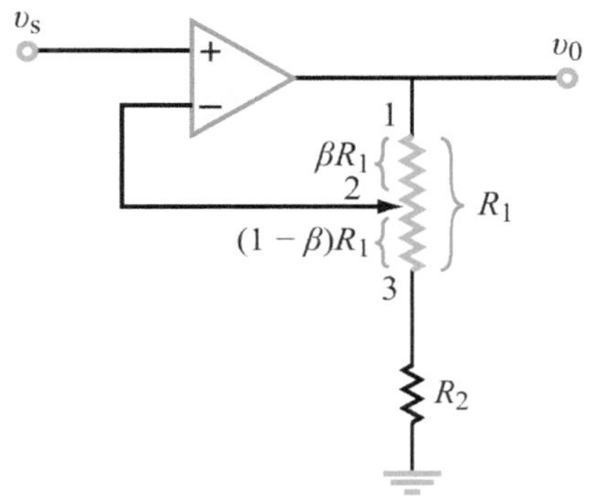

Figure m4.2 Circuit for Problem m4.2.

m4.3 Summing Amplifier:

(a) Design an op-amp summing circuit that performs the operation $\upsilon_o = -(2.14\upsilon_1 + 1.00\upsilon_2 + 0.47\upsilon_3)$. Use not more than four standard-value resistors with values between 10 kΩ and 100 kΩ. Refer to the resistor parts list in Appendix A of the myDAQ tutorial on the EM.

(b) Draw the output waveform υ_o for the input waveforms υ_1 and υ_2 shown in Fig. m4.3 and $\upsilon_3 = 4.7$ V.

(c) State the minimum and maximum values of υ_o.

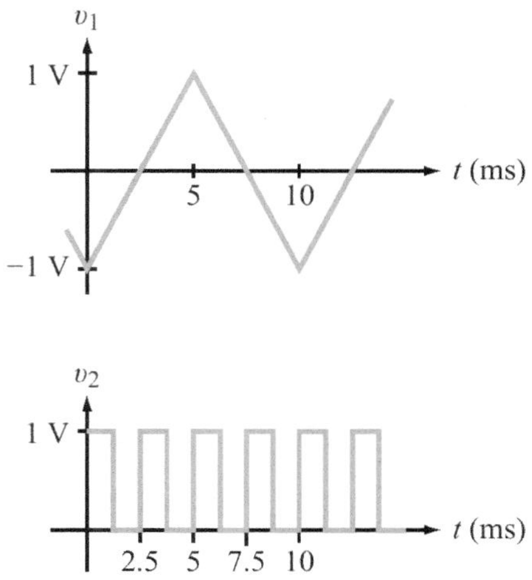

Figure m4.3 Input waveforms for Problem m4.3.

m4.4 Signal Processing Circuits:

(a) Design a two-stage signal processor to serve as a "distortion box" for an electric guitar. The first-stage amplifier applies a variable gain magnitude in the range 13.3 to 23.3 while the second-stage amplifier attenuates the signal by 13.3, i.e., the second-stage amplifier has a fixed gain of 1/13.3. Note that when the first-stage amplifier gain is 13.3 the overall distortion box gain is unity. The distortion effect relies on intentionally driving the first-stage amplifier into saturation (also called "clipping") when its gain is higher than 13.3. Use a 10 kΩ potentiometer and standard-value resistors in the range 1.0 kΩ and 100 kΩ; see the resistor parts list in Appendix A of the myDAQ tutorial on CAD. You may combine two standard-value resistors in series to achieve the required amplifier gains.

(b) Derive a general formula for *percent clipping* of a unit-amplitude sinusoidal test signal; this is the percent of time during one period in which the signal is clipped. The formula includes the peak sinusoidal voltage V_p that would appear at the output of the first-stage amplifier with saturation ignored and the actual maximum value V_s due to saturation.

(c) Apply your general formula to calculate percent clipping of a 1 V peak amplitude sinusoidal signal for the potentiometer dial in three positions: fully counter-clockwise (no distortion), midscale (moderate distortion), and fully clockwise (maximum distortion). Assume the op-amp outputs saturate at ±13.5 V.

(d) Apply a 1 V peak amplitude sinusoidal signal with 100 Hz frequency to the distortion box input and plot its output for the potentiometer dial in the same three positions as above. State the maximum and minimum values of the distortion box output.

m4.5 Multiple Op-Amp Stages: Determine V_{out} in each of the two circuits in Fig. m4.5.

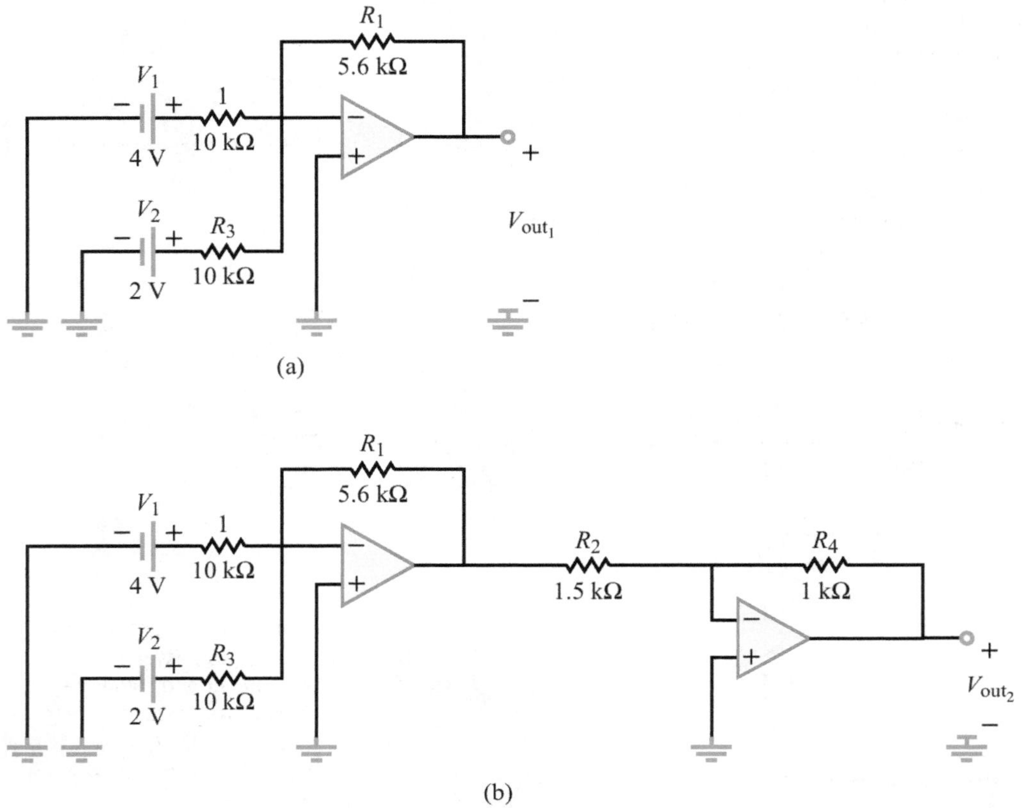

Figure m4.5 Circuits for Problem m4.5.

m4.6 The Importance of Voltage Followers: Suppose you are asked to design a circuit to power a certain gadget and the only source available to you is the 15 V source from your NI myDAQ. Your boss tells you that in order for the gadget to operate properly, its input voltage should be 10.3 V. Moreover, you are told that the input equivalent load resistance of the gadget is exceedingly high (greater than 10 MΩ). To generate the required 10.3 V source, you used the voltage divider shown in Fig. m4.6.

(a) Confirm that the voltage divider provides an output voltage of 10.3 V.

(b) It turns out that the information given to you about the load resistance is in error; the true load resistance of the gadget is 10 kΩ, not 10 MΩ, and the required input current is 1.03 mA. Reevaluate your circuit in light of the new information. What is the input voltage for the gadget and what is the input current?

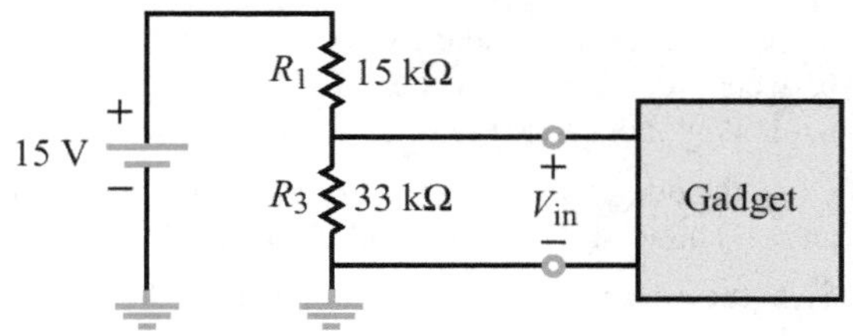

Figure m4.6 Circuit for Problem m4.6.

(c) To fix the problem, you decide to use a voltage follower. Design a voltage follower in conjunction with your voltage divider from part (a) to achieve a 1.03 mA current through the 10 kΩ load resistor.

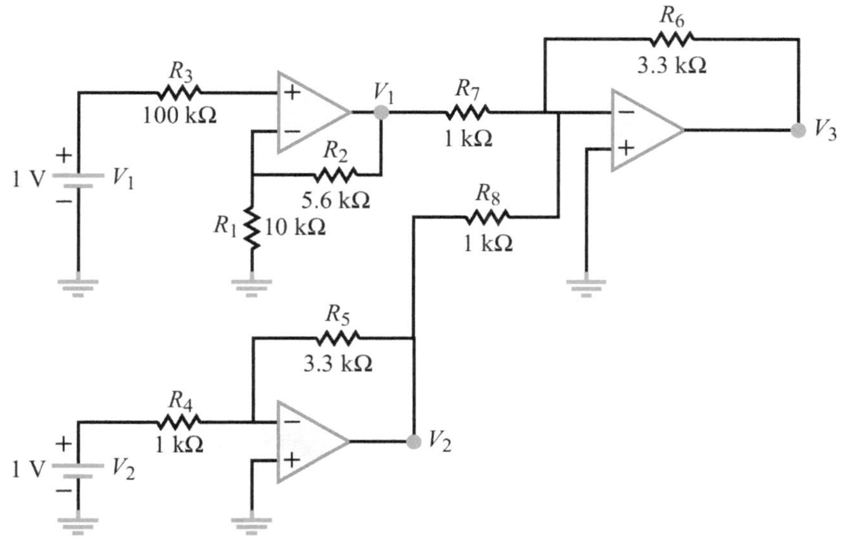

Figure m4.7 Circuit for Problem m4.7.

m4.7 Cascaded Op Amps: Find the voltage at each of the three op-amp outputs in the circuit of Fig. m4.7.

CHAPTER 5

RC and RL First-Order Circuits

Contents

Objectives

Learn to:

- Use mathematical functions to describe several types of nonperiodic waveforms.
- Define the electrical properties of a capacitor, including its i-υ relationship and energy equation.
- Combine multiple capacitors when connected in series or in parallel.
- Define the electrical properties of an inductor, including its i-υ relationship and energy equation.
- Combine multiple inductors when connected in series or in parallel.
- Analyze the transient responses of RC and RL circuits.
- Design RC op-amp circuits to perform differentiation and integration and related operations.
- Apply Multisim to analyze RC and RL circuits.

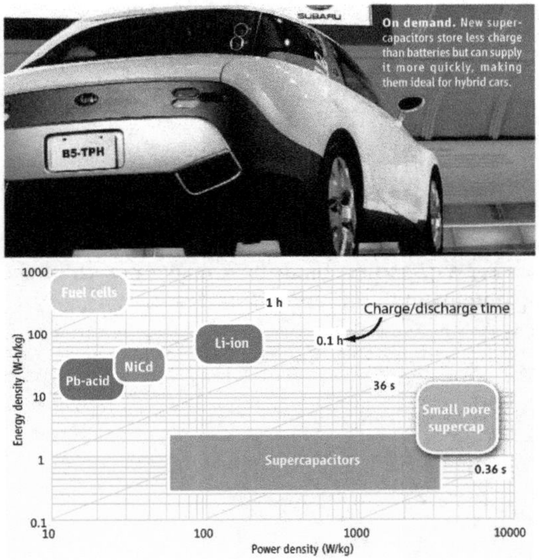

Capacitors (C) and inductors (L) are ***energy storage devices***, in contrast with resistors, which are energy dissipation devices. This chapter examines the behavior of ***RC and RL circuits***, to be followed in Chapter 6 with an examination of RLC circuits.

Overview

A resistor is characterized by a linear i–υ relationship, namely $\upsilon = iR$, which does not involve time explicitly. When we apply Kirchhoff's current and voltage laws to resistive circuits, we end up with one or more simultaneous *linear* equations. The process of solving a set of linear equations is relatively straightforward and does not involve time explicitly, but if i varies with time, so will υ, in a linearly proportionate manner, and the character of the time variation remains the same for both. Hence, even when a certain voltage or current source in the circuit varies with time, we solve the resistive circuit using *static* formulas that do not depend on time rather than dynamic formulas that do, because the time variation is merely a scale change. Another important feature of resistive elements is that they consume electrical energy by converting it into heat.

Resistive circuits are used to change the relationship between υ and i, divide voltages and currents, and (with the addition of op amps) amplify, add, subtract, and compare voltages. Resistive sensors allow us to convert properties of the physical world—light, heat, sound, moisture, pressure, etc.—to voltage and current values that we can use in our circuits.

► Capacitors and inductors represent a contrasting (yet complimentary) class of electrical devices. Not only is time t (or more precisely d/dt) at the heart of how capacitors and inductors function, but they also differ from resistors in that they do not dissipate energy. They can store energy and then release it—but not consume it. ◄

The addition of capacitors and inductors to circuits containing time-varying sources opens the door to *dynamic circuits* with a wide range of practical applications. Because capacitors and inductors store energy, they can be used to smooth out or average time-varying voltages or currents, select or filter out different frequencies, and delay circuit responses. Capacitive sensors can also be used to measure proximity, touch, pressure, moisture, vibration, and more. Both capacitors and inductors also are found as unintended parasitics in all circuits. The dynamic, time-varying responses of capacitors and inductors provide a new and important set of tools for controlling voltage and current. The dynamic response of a circuit to a certain voltage or current source depends on both the architecture of the circuit and the waveform characterizing the time variation of that source. In general, the response consists of a *transient component* and a *steady-state component*.

► The *transient response* represents the initial reaction immediately after a sudden change, such as closing or opening a switch to connect a source to the circuit. This is also called the *early time response*. ◄

Most (but not all) electronic circuits are designed such that the transient response usually dies out or reaches an approximately constant level within a fraction of a second after the introduction of the external excitation. An example of a transient response is when the energy stored in a capacitor is transferred into the flashbulb of a camera. Figure 5-1 shows examples of two typical circuit responses. In part (a), the external excitation is a dc voltage source, and the displayed response represents the current flowing through a certain capacitor in the circuit, starting when the switch is closed. This is much like the camera flash example. The current levels labeled i_0 and i_∞ denote the values exhibited by the transient response at the onset of the change (closing the switch at $t = 0$) and a long time afterward (at $t = \infty$), respectively. They are called the *initial* and *final* (or steady state) values of $i(t)$. For this example, the steady state current is $i_\infty = 0$.

Our second example displays in Fig. 5-1(b) the response of another circuit to turning on a sinusoidally time-varying source. The combination of the ac source and switch action initially elicit a transient response that quickly transitions into a steady-state response. This steady state ac case belongs to a class of external excitations and circuit responses called *periodic waveforms* (which repeat periodically). In contrast, a dc waveform is *nonperiodic* (it does not repeat). As we shall see later, the tools of circuit analysis and design lend themselves to different mathematical approaches when dealing with periodic versus nonperiodic waveforms. We will first examine the behavior of circuits excited by nonperiodic external excitations in this and the following chapter, before we pursue the treatment of periodic ac circuits starting in Chapter 7.

Section 5-1 introduces some of the nonperiodic waveforms commonly used in electric circuits, followed in Sections 5-2 and 5-3 with presentations of the circuit properties of capacitors and inductors, respectively. Our treatment of the circuit response to nonperiodic excitations is divided into two segments. The first, covered in Sections 5-4 through 5-6 of this chapter, deals with *first-order circuits*, so named because their Kirchhoff voltage and current equations are characterized by first-order differential equations.

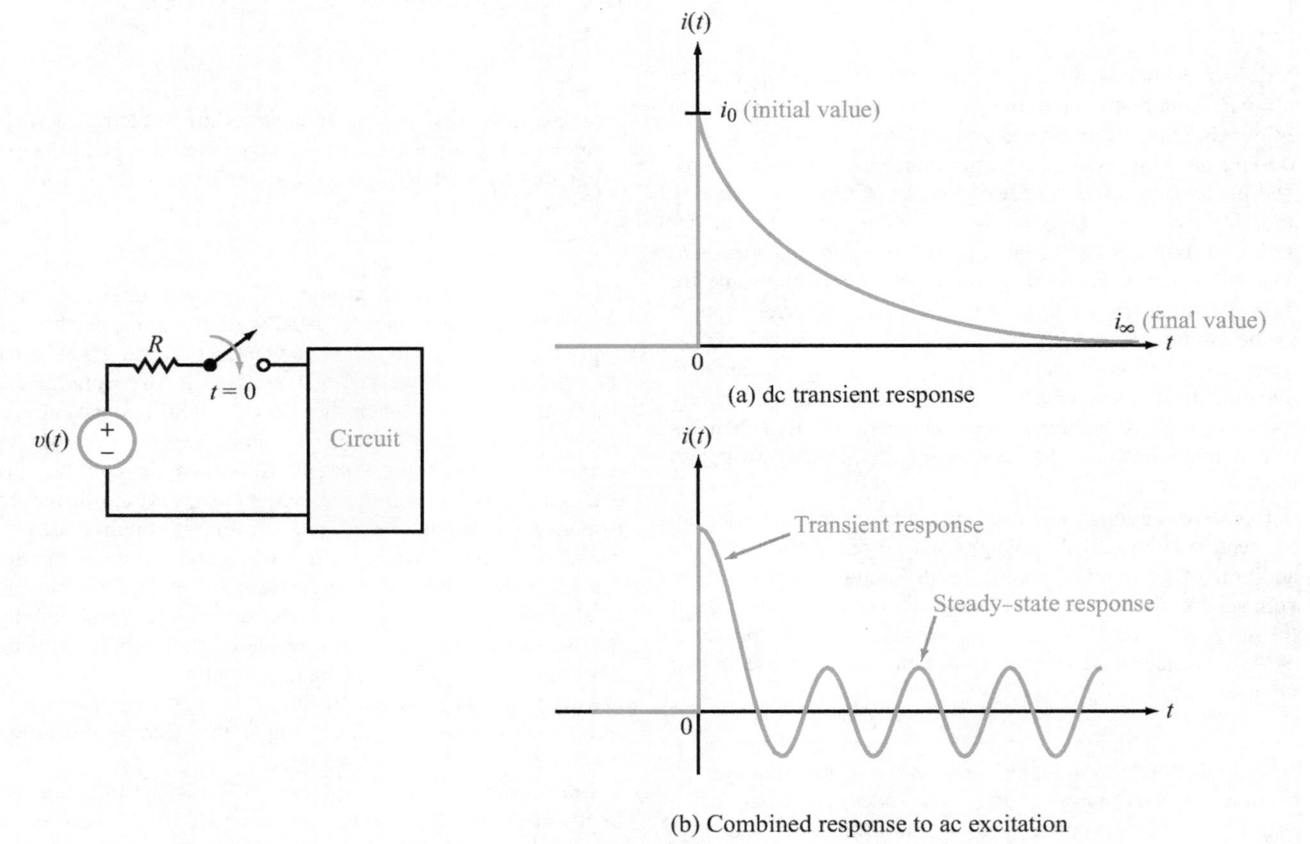

Figure 5-1: Circuit response to (a) dc source $\upsilon(t) = V_0$ and (b) ac source $\upsilon(t) = V_0 \cos \omega t$.

▶ First-order circuits include ***RC circuits***—composed of sources, resistors, and a single capacitor (or multiple capacitors that can be combined into a single equivalent capacitor)—and ***RL circuits***, but not circuits containing capacitors and inductors simultaneously. ◀

RLC circuits, which give rise to second-order differential equations, are the subject of Chapter 6.

Concept Question 5-1: What is the difference between the transient and steady-state components of the circuit response? (See CAD)

Concept Question 5-2: Why do we study the circuit response to dc and ac sources separately? (See CAD)

5-1 Nonperiodic Waveforms

Among the multitudes of possible nonperiodic waveforms, the step, ramp, pulse, and exponential waveforms are encountered most frequently in electrical circuits. In this section, we review the geometrical properties and corresponding mathematical expressions associated with each of these four waveforms, as well as introduce some of the connections between them.

5-1.1 Step-Function Waveform

The waveform $\upsilon(t)$ shown in Fig. 5-2(a) is an (ideal) ***step function***: it is equal to zero for $t < 0$, at $t = 0$ it makes a

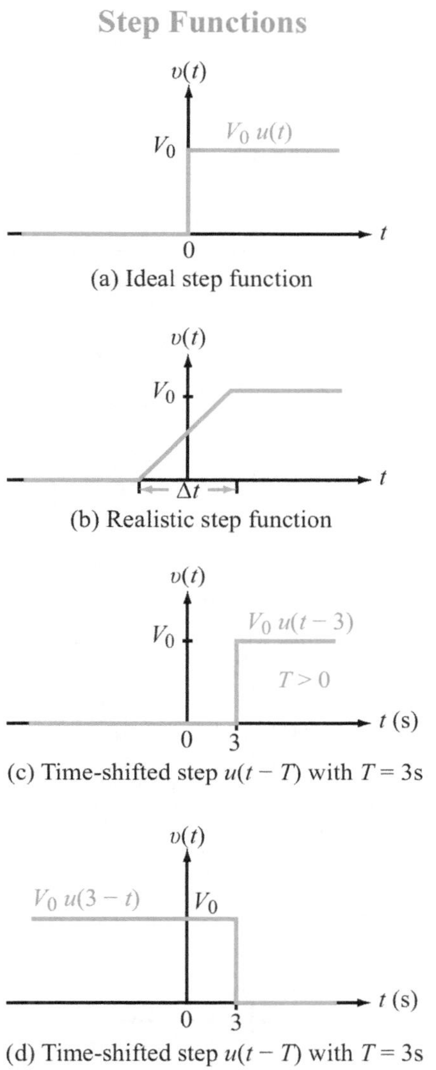

Figure 5-2: Step functions: (a) ideal step function, (b) realistic step function with transition duration Δt, (c) time-shifted step function $V_0\, u(t-3)$, (d) time-shifted step function $V_0\, u(3-t)$.

discontinuous jump to V_0, and from there on forward it remains at V_0. The process represents an ideal switch that turns on a dc voltage at $t = 0$. Mathematically, it can be described as

$$\upsilon(t) = V_0\, u(t), \tag{5.1}$$

where $u(t)$ is known as the *unit step function* and is defined as

$$u(t) = \begin{cases} 0 & \text{for } t < 0, \\ 1 & \text{for } t > 0. \end{cases} \tag{5.2}$$

In reality, it is not possible to turn on a switch with an (ideal) step function, because that would require changing the value of $\upsilon(t)$ from 0 to V_0 in zero time. A more realistic shape of the step function is illustrated in Fig. 5-2(b); the discontinuous jump is replaced with a ramp waveform with rise time Δt, providing a smooth voltage turn-on.

If $\upsilon(t)$ transitions between its two levels at a time other than zero, such as at $t = T$, it is written as

$$\upsilon(t) = V_0\, u(t-T) = \begin{cases} 0 & \text{for } t < T, \\ V_0 & \text{for } t > T. \end{cases} \tag{5.3}$$

► $u(t-T)$ is called the *time-shifted step function*, which is defined to be zero when its argument $(t-T)$ is less than zero and 1 when its argument is greater than zero. Thus, $u(t-T) = 1$ for $t > T$. ◄

By the same definition, $u(T-t)$ is zero when $T - t < 0$ (which is true when $t > T$), and 1 when $T - t > 0$ (which is true when $t < T$). Figure 5-2(c) and (d) display step-function waveforms for $V_0\, u(t-3)$ and $V_0\, u(3-t)$, respectively. We often use combinations of step functions to represent voltage sources turning on and off.

An example of a step function is when a switch is closed so as to connect a voltage source to a circuit, as shown in Fig. 5-3(a). When writing KCL and KVL equations for circuits that include switches, the switching action (closing or opening) can be represented mathematically by step functions. In Fig. 5-3(a), closing the switch at $t = 3$ s is represented by $u(t-3)$, whereas disconnecting the source by opening the switch in Fig. 5-3(b) is represented by $u(3-t)$.

If the time associated with closing the switch is very short in comparison with the time scale of interest, then it may be acceptable to approximate the switch closing by an ideal step function. On the other hand, if we are interested in analyzing the circuit response at a sampling rate whose interval is shorter than or comparable with the transition interval associated with closing the switch, then it may be necessary to use a more realistic, continuous, step function to represent the switch action.

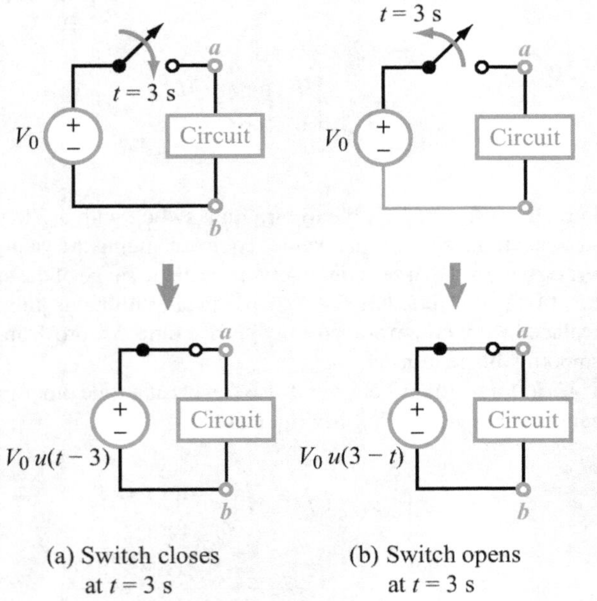

Figure 5-3: Connecting/disconnecting a voltage source to/from a circuit via a switch can be represented mathematically by a step function.

5-1.2 Ramp-Function Waveform

A waveform that varies linearly with time, starting at a specific time $t = T$, is called a ***time-shifted ramp function*** and is denoted by $r(t-T)$. If $T = 0$, it simply is called a ***ramp function*** and is denoted by $r(t)$. Formally, $r(t-T)$ is defined as

$$r(t-T) = \begin{cases} 0 & \text{for } t \leq T, \\ (t-T) & \text{for } t \geq T. \end{cases} \quad (5.4)$$

Plots of $\upsilon(t) = r(t-T)$ are displayed in Fig. 5-4(a) and (b) for $T = -1$ s and $T = 2$ s, respectively. A voltage $\upsilon(t)$ that ramps up at 3 V per second, starting at $t = 1$ s, is shown graphically in Fig. 5-4(c). Mathematically, $\upsilon(t)$ can be expressed as

$$\upsilon(t) = 3r(t-1) \qquad \text{V}. \quad (5.5)$$

If the coefficient of $r(t-T)$ is negative, $\upsilon(t)$ would exhibit a negative slope, as illustrated by Fig. 5-4(d) for $\upsilon(t) = -2r(t+1)$.

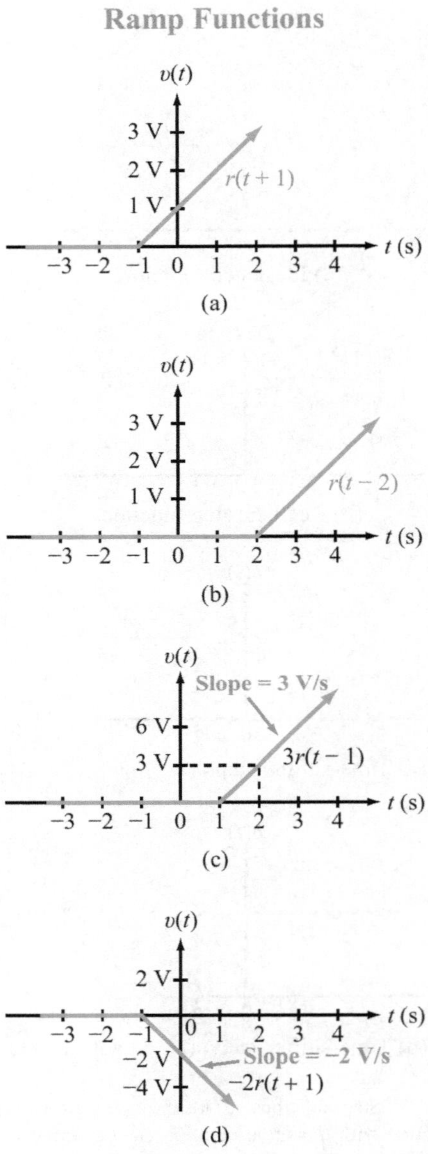

Figure 5-4: Time-shifted ramp functions.

A unit ramp function is related to the unit step function by

$$r(t) = \int_{-\infty}^{t} u(t)\, dt = t\, u(t), \quad (5.6)$$

and for the case where the ramping action starts at $t = T$,

$$r(t - T) = \int_{-\infty}^{t} u(t - T)\, dt = (t - T)\, u(t - T). \qquad (5.7)$$

Example 5-1: Realistic Step Waveform

Generate an expression to describe the waveform shown in Fig. 5-5(a). Note that the time scale is in ms.

Solution: The voltage $\upsilon(t)$ can be synthesized as the sum of two time-shifted ramp functions (Fig. 5-5(b)): one with a positive slope of 3 V/s and a ramp start-up time $T = -2$ ms and a second ramp function that starts at $T = +2$ ms but its slope is -3 V/s. Thus,

$$\upsilon(t) = \upsilon_1(t) + \upsilon_2(t) = 3r(t + 2 \text{ ms}) - 3r(t - 2 \text{ ms}) \qquad \text{V}.$$

In view of Eq. (5.7), $\upsilon(t)$ also can be expressed in terms of time-shifted step functions as

$$\upsilon(t) = 3(t + 2 \text{ ms})\, u(t + 2 \text{ ms}) - 3(t - 2 \text{ ms})\, u(t - 2 \text{ ms}) \qquad \text{V}.$$

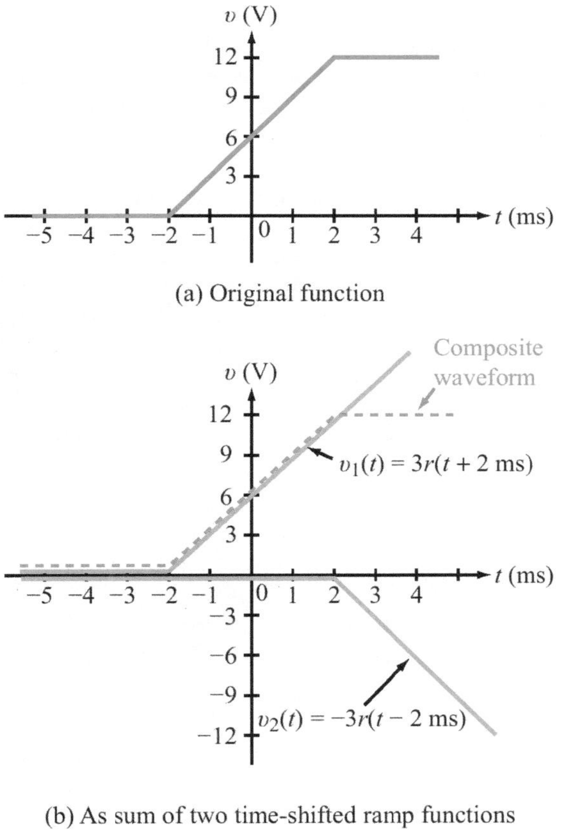

Figure 5-5: Step waveform of Example 5-1.

5-1.3 Pulse waveform

The diagram in Fig. 5-6(a) depicts a SPDT switch that moves from position 1 to position 2 at $t = 1$ s, connects a dc voltage source to an electric circuit, and then returns to position 1 at $t = 5$ s. From the standpoint of the circuit, the switch actions constitute the introduction of a *rectangular pulse* of voltage V_0, as illustrated in Fig. 5-6(b). A pulse also may be triangular or Gaussian in shape or may assume other forms, but in all cases, it usually is assumed that a pulse rises from some specified base level up to a peak value, remains constant for a while, and then declines back to its original base level.

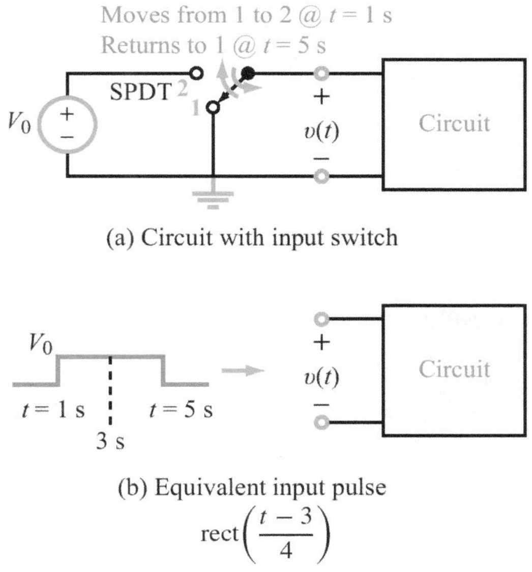

Figure 5-6: Connecting a switch to a dc source at $t = 1$ s and then returning it to ground at $t = 5$ s constitutes a voltage pulse centered at $T = 3$ s and of duration $\tau = 4$ s.

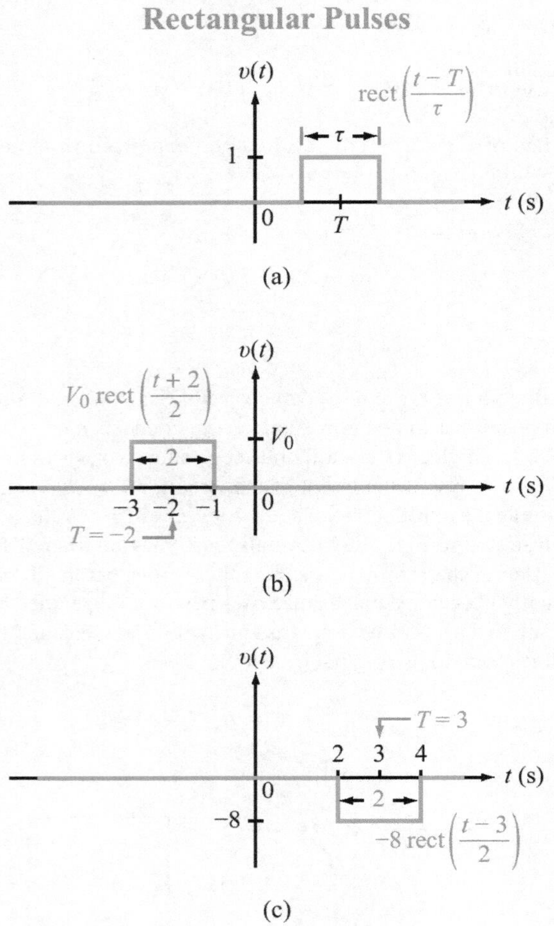

Figure 5-7: Rectangular pulses.

A rectangular pulse can be described in terms of the *unit rectangular function* $\mathrm{rect}[(t-T)/\tau]$, which is characterized by two parameters: location of the center of the pulse T and the *duration of the pulse* τ, as shown in Fig. 5-7. Its mathematical definition is given by

$$\mathrm{rect}\left(\frac{t-T}{\tau}\right) = \begin{cases} 0 & \text{for } t < (T-\tau/2), \\ 1 & \text{for } (T-\tau/2) \le t \le (T+\tau/2), \\ 0 & \text{for } t > (T+\tau/2). \end{cases} \tag{5.8}$$

A rectangular pulse can be constructed out of two time-shifted step functions: one that causes the rise in level and another (delayed in time) that cancels the first one. The details are given in Example 5-2.

Example 5-2: Pulses

Construct expressions for (a) the rectangular pulse shown in Fig. 5-8(a) and (b) the trapezoidal pulse shown in Fig. 5-8(b) in terms of step and ramp functions.

Solution: (a) From Fig. 5-8(a), it is evident that the amplitude of the rectangular pulse is 4 V and its duration is 2 s, extending from $T_1 = 2$ s to $T_2 = 4$ s. Hence, with its center at 3 s and its duration equal to 2 s,

$$\upsilon_a(t) = 4\,\mathrm{rect}\left(\frac{t-3}{2}\right) \quad \text{V}. \tag{5.9}$$

The sequential addition of two time-shifted step functions, $\upsilon_1(t)$ at $t = 2$ s and $\upsilon_2(t)$ at $t = 4$ s, as demonstrated graphically in Fig. 5-8(c), accomplishes the task of synthesizing the rectangle function in terms of two step functions. Specifically,

$$\upsilon_a(t) = \upsilon_1(t) + \upsilon_2(t) = 4[u(t-2) - u(t-4)] \quad \text{V}. \tag{5.10}$$

(b) The trapezoidal pulse consists of three segments, a ramp with a positive slope that starts at $t = 0$ and ends at $t = 1$ s, followed by a plateau that extends to $t = 3$ s, and finally, a ramp with a negative slope that ends at 4 s. Building on the experience gained from Example 5-1, we can synthesize the trapezoidal pulse in terms of four ramp functions. The process, which is illustrated graphically in Fig. 5-8(d), leads to

$$\begin{aligned}\upsilon_b(t) &= \upsilon_1(t) + \upsilon_2(t) + \upsilon_3(t) + \upsilon_4(t) \\ &= 5[r(t) - r(t-1) - r(t-3) + r(t-4)] \quad \text{V}.\end{aligned} \tag{5.11}$$

Equivalently, using the relationship between the ramp and step functions given by Eq. (5.7), $\upsilon_b(t)$ can be expressed as

$$\begin{aligned}\upsilon_b(t) = 5[&t\,u(t) - (t-1)\,u(t-1) \\ &- (t-3)\,u(t-3) + (t-4)\,u(t-4)] \quad \text{V}.\end{aligned} \tag{5.12}$$

There are often multiple ways for representing waveforms of these types, all of which should lead to the same result in the end.

Waveform Synthesis

(a) Rectangular pulse

(b) Trapezoidal pulse

(c) $v_a(t) = 4u(t-2) - 4u(t-4)$

(d) $v_b(t) = v_1(t) + v_2(t) + v_3(t) + v_4(t)$

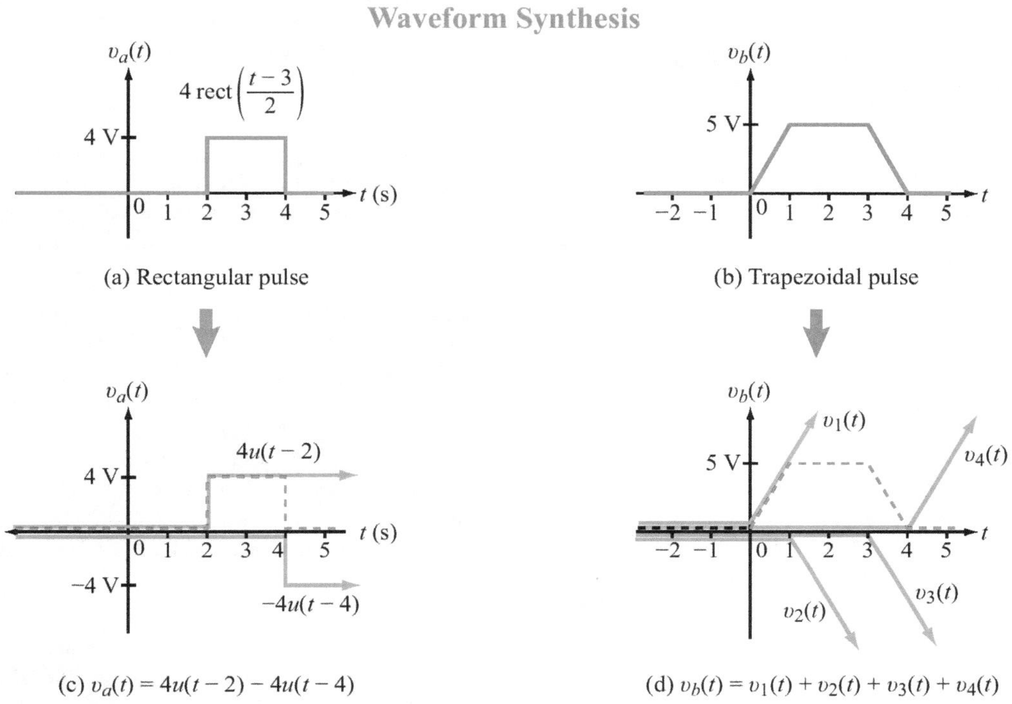

Figure 5-8: Rectangular and trapezoidal pulses of Example 5-2.

Concept Question 5-3: What determines the slope of a ramp waveform? (See CAD)

Concept Question 5-4: How are the ramp and rectangle functions related to the step function? (See CAD)

Concept Question 5-5: A unit step function $u(t)$ is equivalent to closing an SPST switch at $t = 0$. What is $u(-t)$ equivalent to? (See CAD)

Exercise 5-1: Express the waveforms shown in Fig. E5.1 in terms of unit step functions.

Answer:
(a) $v(t) = 10\, u(t) - 20\, u(t-2) + 10\, u(t-4)$,
(b) $v(t) = 2.5\, r(t) - 10\, u(t-2) - 2.5\, r(t-4)$.
(See CAD)

(a)

(b)

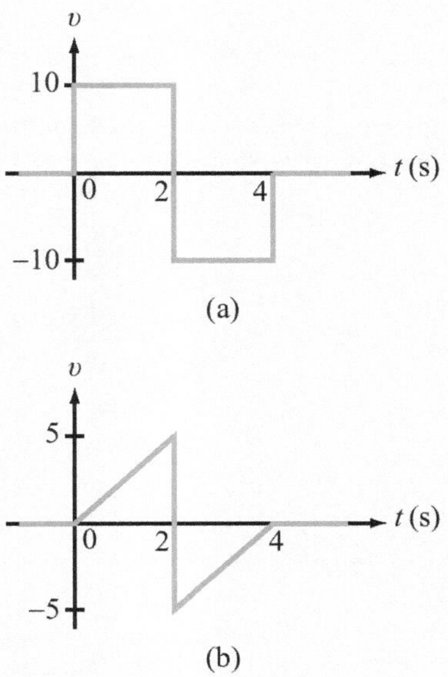

Figure E5.1

Exercise 5-2: How is $u(t)$ related to $u(-t)$?

Answer: They are mirror images of one another (with respect to the y-axis). (See CAD)

Exercise 5-3: Consider the SPDT switch in Fig. 5-6(a). Assume that it started out at position 2, was moved to position 1 at $t = 1$ s, and then moved back to position 2 at $t = 5$ s. This is the reverse of the sequence shown in Fig. 5-6(a). Express $\upsilon(t)$ in terms of (a) units step functions and (b) the rectangle function.

Answer: (a) $\upsilon(t) = V_0[u(1-t) + u(t-5)]$, (b) $\upsilon(t) = V_0\left[1 - \text{rect}\left(\frac{t-3}{4}\right)\right]$. (See CAD)

5-1.4 Exponential waveform

The ***exponential function*** is a particularly useful tool for characterizing fast-rising and fast-decaying waveforms, which, as we will see in later sections, are related to the transient responses of RC and RL circuits. The (positive) exponential function given by

$$\upsilon_\text{p}(t) = e^{t/\tau} \tag{5.13}$$

is shown graphically in Fig. 5-9 for a positive value of the ***time constant*** τ. The figure also includes a plot of the ***negative exponential function***, where

$$\upsilon_\text{n}(t) = e^{-t/\tau}. \tag{5.14}$$

When $t = \tau$, $\upsilon_\text{n} = e^{-1} = 0.37$. Thus, if a certain quantity (such as a voltage or current) is said to decay exponentially with time, it means that after τ seconds its amplitude decreases to $1/e$ or 37 percent of its initial value. Symmetrically, $\upsilon_\text{p} = e^{-1} = 0.37$ when $t = -\tau$.

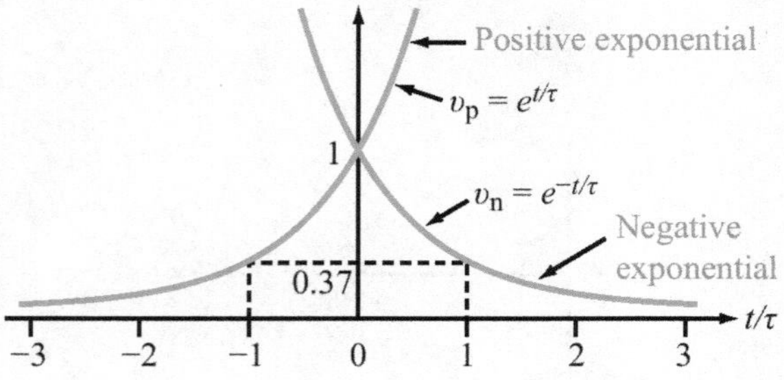

Figure 5-9: By $t = \tau$, the exponential function $e^{-t/\tau}$ has decayed to 37 percent of its original value at $t = 0$.

▶ An exponential function with a short time constant rises or decays faster than an exponential function with a longer time constant, as illustrated by the plots in Fig. 5-10(a). ◀

Replacing t in the exponential with $(t-T)$ shifts the exponential curve to the right if T has a positive value and to the left if T is negative (Fig. 5-10(b)). In Fig. 5-10(c), the range of the exponential function has been limited to $t > 0$ by multiplying $e^{-t/\tau}$ by $u(t)$, and in Fig. 5-10(d) the function $\upsilon(t) = V_0(1 - e^{-t/\tau})\, u(t)$ is used to describe a waveform that builds up as a function of time towards a saturation value V_0.

Table 5-1 provides a summary of common waveform shapes and their equivalent expressions.

Concept Question 5-6: If the time constant of a negative exponential function is doubled in value, will the corresponding waveform decay faster or slower? (See CAD)

Concept Question 5-7: What is the approximate shape of the waveform described by the function $(1 - e^{-|t|})$? (See CAD)

Exercise 5-4: The radioactive decay equation for a certain material is given by $n(t) = n_0 e^{-t/\tau}$, where n_0 is the initial count at $t = 0$. If $\tau = 2 \times 10^8$ s, how long is its half-life? [Half-life $t_{1/2}$ is the time it takes a material to decay to 50 percent of its initial value.]

Answer: $t_{1/2} = 1.386 \times 10^8$ s $= 4$ years, 144 days, 12 hours, 10 minutes, 36 s. (See CAD)

Exercise 5-5: If the current $i(t)$ through a resistor R decays exponentially with a time constant τ, what is the value of the power dissipated in the resistor at $t = \tau$, compared with its value at $t = 0$?

Answer: $p(t) = i^2R = I_0^2R(e^{-t/\tau})^2 = I_0^2Re^{-2t/\tau}$, $p(\tau)/p(0) = e^{-2} = 0.135$ or 13.5 percent. (See CAD)

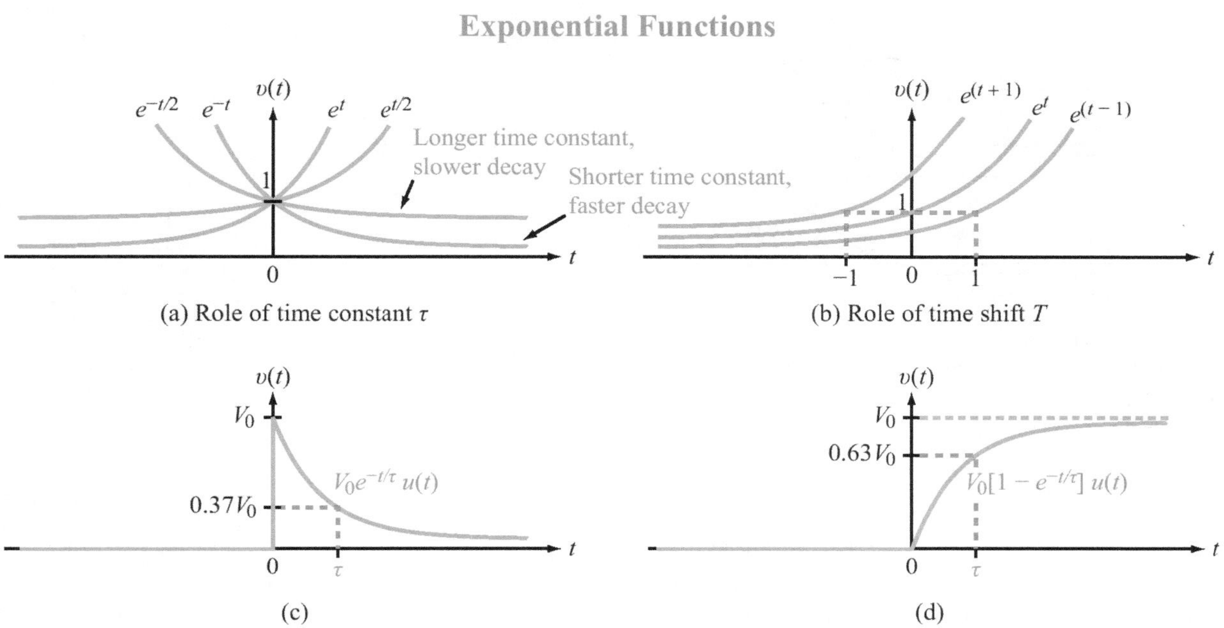

Figure 5-10: Properties of the exponential function.

Table 5-1: Common nonperiodic waveforms.

waveform	Expression	General Shape
Step	$u(t-T) = \begin{cases} 0 & \text{for } t < T \\ 1 & \text{for } t > T \end{cases}$	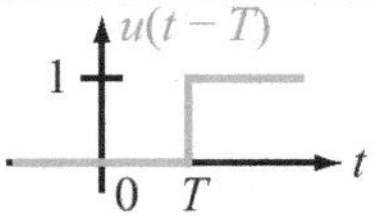
Ramp	$r(t-T) = (t-T)\,u(t-T)$	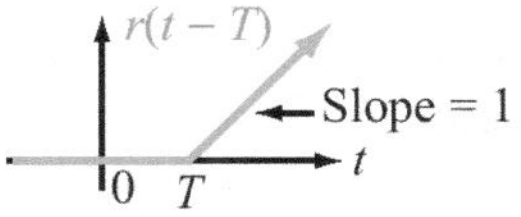
Rectangle	$\text{rect}\left(\frac{t-T}{\tau}\right) = u(t-T_1) - u(t-T_2)$ $T_1 = T - \frac{\tau}{2}; \quad T_2 = T + \frac{\tau}{2}$	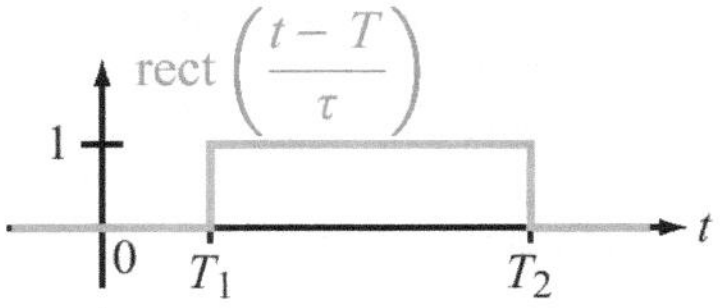
Exponential	$\exp[-(t-T)/\tau]\,u(t-T)$	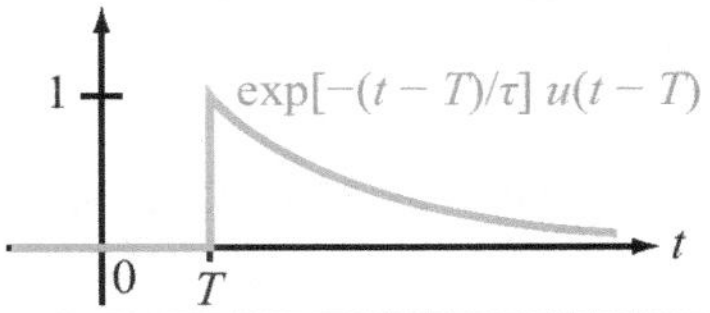

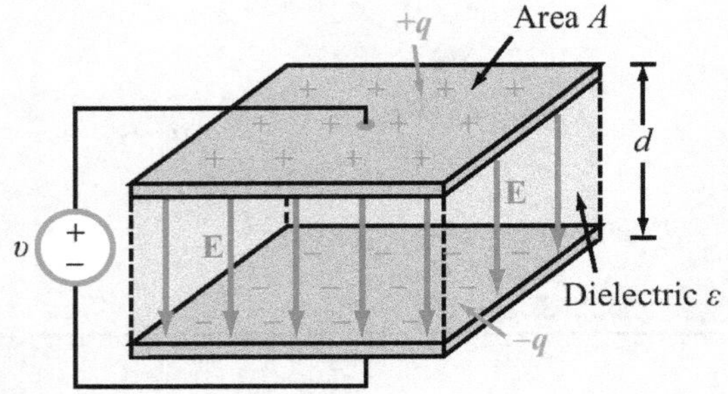

Figure 5-11: Parallel-plate capacitor with plates of area A, separated by a distance d, and filled with an insulating dielectric material of permittivity ε.

Table 5-2: Relative electrical permittivity of common insulators: $\varepsilon_r = \varepsilon/\varepsilon_0$ and $\varepsilon_0 = 8.854 \times 10^{-12}$ F/m.

Material	Relative Permittivity ε_r
Air (at sea level)	1.0006
Teflon	2.1
Polystyrene	2.6
Paper	2–4
Glass	4.5–10
Quartz	3.8–5
Bakelite	5
Mica	5.4–6
Porcelain	5.7

5-2 Capacitors

▶ When separated by an insulating medium, any two conducting bodies (regardless of their shapes and sizes) form a *capacitor*. A capacitor can store electric charge. ◀

The *parallel-plate capacitor* shown in Fig. 5-11 represents a simple configuration in which two identical conducting plates (each of area A) are separated by a distance d containing an insulating (dielectric) material of *electrical permittivity* ε. The permittivity of a material is usually referenced to that of free space, namely $\varepsilon_0 = 8.85 \times 10^{-12}$ farads/m (F/m). Hence, the *relative permittivity* of a material is defined as

$$\varepsilon_r = \frac{\varepsilon}{\varepsilon_0}. \tag{5.15}$$

When a dielectric material is subjected to an electric field, its atoms become partially polarized; i.e., the atom is rearranged into positive and negative domains.. The *electric field* E induced in the space between the conducting plates is the result of the voltage υ applied across the plates. The *electrical susceptibility* χ_e of a material is a measure of how susceptible that material is to electrical polarization. The permittivity ε and susceptibility χ_e are related by

$$\varepsilon = \varepsilon_0(1 + \chi_e). \tag{5.16}$$

In view of Eq. (5.15), the relative permittivity ε_r is given by

$$\varepsilon_r = \frac{\varepsilon}{\varepsilon_0} = 1 + \chi_e. \tag{5.17}$$

Free space contains no atoms; hence, its $\chi_e = 0$ and $\varepsilon_r = 1$. For air at sea level, $\varepsilon_r = 1.0006 \approx 1.0$. Table 5-2 provides typical values of ε_r for common types of insulators.

Returning to the parallel-plate capacitor, if a voltage source is connected across the two plates, as shown in Fig. 5-11, charge of equal and opposite polarity is transferred to the conducting surfaces. The plate connected to the (+) terminal of the voltage source will accumulate charge $+q$, and charge $-q$ will accumulate on the other plate. The charges induce a nearly uniform electric field E in the dielectric medium, given by

$$E = \frac{q}{\varepsilon A}, \tag{5.18}$$

with the direction of E being from the plate with $+q$ to the plate with $-q$. Moreover, E, whose unit is V/m, is related to the voltage υ through

$$E = \frac{\upsilon}{d} \qquad \text{(V/m)} \qquad \text{(parallel-plate capacitor)}. \tag{5.19}$$

▶ For any capacitor, its *capacitance* C, measured in farads (F), is defined as the amount of charge q that its positive-polarity plate holds, normalized to the applied voltage responsible for that charge accumulation. ◀

Thus,

$$C = \frac{q}{\upsilon} \qquad \text{(F)} \qquad \text{(any capacitor)}. \tag{5.20}$$

For the parallel-plate capacitor, combining Eqs. (5.18) and (5.19) leads to $q = \varepsilon A \upsilon / d$. Upon inserting this expression for q in Eq. (5.20), we have

$$C = \frac{\varepsilon A}{d} \qquad \text{(parallel-plate capacitor)}. \tag{5.21}$$

Even though the expression given by Eq. (5.21) is specific to the parallel-plate capacitor, the general tenor of the expression holds true for other geometrical configurations as well. In general, the capacitance C of any two-conductor system increases with the area of the conducting surfaces, decreases with the separation between them, and is directly proportional to ε of the insulating material. For example, the capacitance of a *coaxial capacitor* consisting of two concentric conducting cylinders of radii a and b (Fig. 5-12(a)) and separated by a dielectric material of permittivity ε is given by

$$C = \frac{2\pi \varepsilon \ell}{\ln(b/a)} \qquad \text{(coaxial capacitor)}, \tag{5.22}$$

where ℓ is the length of the capacitor and $\ln(b/a)$ is the natural logarithm of (b/a). The spacing between the cylinders is $(b-a)$; reducing this spacing, while holding b constant, requires reducing the ratio (b/a), which reduces the value of $\ln(b/a)$, thereby increasing the magnitude of C.

The *mica capacitor* shown in Fig. 5-12(b) consists of a stack of conducting plates, interleaved by sheets of mica (dielectric). The *plastic-foil capacitor* in Fig. 5-12(c) is constructed by rolling flexible conducting foils (separated by a plastic layer) into a spindle-like configuration. Small capacitors used in microcircuits typically have capacitances in the picofarad (10^{-12} F) to microfarad (10^{-6} F) range. Large capacitors used in power-transmission substations may have capacitors in the range of millifarads (10^{-3} F). Using thin-film polymers for the dielectric insulator and carbon nanotubes for the electrodes (terminals), a new type of capacitor (sometimes called a *supercapacitor* or *nanocapacitor*) was developed in the 1990s with the express goal of significantly increasing the amount of charge that the conductors can hold (at a specified voltage level). Such capacitors have capacitance values that are several orders of magnitude greater than conventional capacitors of comparable size. The new fabrication techniques have not only expanded the versatility of capacitors in electronic circuits, but they have also introduced the use of supercapacitors as energy-storage devices in many electronic applications (see Technology Brief 12: Supercapacitors).

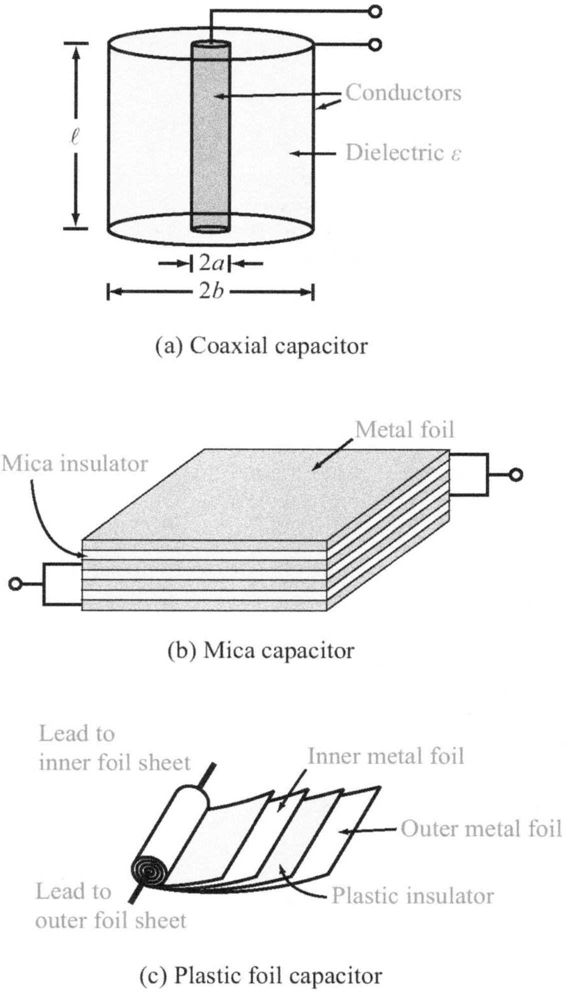

Figure 5-12: Various types of capacitors.

5-2.1 Electrical Properties of Capacitors

According to Eq. (5.20), $q = C\upsilon$. Application of the standard definition for current (Eq. (1.3) provides the expression for the current i through a capacitor as

$$i = \frac{dq}{dt} = C\,\frac{d\upsilon}{dt}, \tag{5.23}$$

where the direction of i and the polarity of υ are defined in accordance with the passive sign convention (Fig. 5-13).

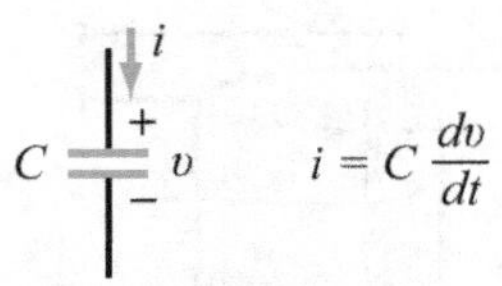

Figure 5-13: Passive sign convention for capacitor: if current i is entering the (+) voltage terminal across the capacitor, then power is getting transferred into the capacitor. Conversely, if i is leaving the (+) terminal, then power is getting released from the capacitor.

The i–υ relationship expressed by Eq. (5.23) conveys a very important condition, namely:

▶ The voltage across a capacitor cannot change instantaneously, but the current can. ◀

This assertion is supported by the observation that if υ were to change values in zero time, $d\upsilon/dt$ would be infinite, as a result of which the current i would be also infinite. Since i cannot be infinite, υ cannot change instantaneously.

Another attribute of Eq. (5.23) relates to the behavior of a capacitor under dc conditions (constant voltage across it). Since $d\upsilon/dt = 0$ for a dc voltage, it follows that $i = 0$. Such a behavior is characteristic of an open circuit, through which no current flows even when a non-zero voltage exists across it. Thus:

▶ Under dc conditions, a capacitor behaves like an open circuit. ◀

To express $\upsilon(t)$ in terms of $i(t)$, we replace t with a dummy variable t' and integrate both sides of Eq. (5.23) from $t' = t_0$ to $t' = t$,

$$\int_{t_0}^{t} \left(\frac{d\upsilon}{dt'}\right) dt' = \frac{1}{C}\int_{t_0}^{t} i\ dt', \tag{5.24}$$

where t_0 is the initial reference point in time at which the initial condition $\upsilon(t_0)$ is known. Since the integral of the derivative of a function is the function itself, integrating the left-hand side and rearranging terms leads to

$$\upsilon(t) = \upsilon(t_0) + \frac{1}{C}\int_{t_0}^{t} i\ dt'. \tag{5.25}$$

In view of $dq = i\ dt$, we recognize that the integral $\int_{t_0}^{t} i\ dt$ represents the amount of charge accumulation on the capacitor at time t. If we are dealing with a capacitor that had no charge on it until a switch was closed or a signal was injected into the circuit and if we conveniently set our time reference such that the signal injection commenced at $t_0 = 0$, then Eq. (5.25) simplifies to

$$\upsilon(t) = \frac{1}{C}\int_{0}^{t} i\ dt' \tag{5.26}$$

(capacitor uncharged before $t = 0$).

Charging up a capacitor creates an electric field in the dielectric medium between the capacitor's conductors. The electric field becomes the mechanism for storage of electrical energy in that medium. The stored energy can be released by discharging the capacitor. Thus, a capacitor can store energy and release previously stored energy but cannot dissipate energy.

The instantaneous power $p(t)$ transferring into or out of a capacitor is given by

$$p(t) = \upsilon i = C\upsilon\,\frac{d\upsilon}{dt} \qquad \text{(W)}, \tag{5.27}$$

where i is defined as entering the capacitor at its positive voltage terminal (Fig. 5-13).

▶ If the magnitude of $p(t)$ is positive, then by the passive sign convention, the capacitor is receiving power (charging up), and if $p(t)$ is negative, it is delivering power (discharging). ◀

Energy is the integral of the product of power and time. Hence, the amount of energy stored in the capacitor at any time t is equal to the time integral of $p(t)$ from $-\infty$ (at which time the capacitor was uncharged) to t and is given by

$$\begin{aligned} w(t) &= \int_{-\infty}^{t} p\ dt' = C\int_{-\infty}^{t} \left(\upsilon\,\frac{d\upsilon}{dt'}\right) dt' \\ &= C\int_{-\infty}^{t} \left[\frac{d}{dt'}\left(\frac{1}{2}\upsilon^2\right)\right] dt', \end{aligned} \tag{5.28}$$

which yields

$$w(t) = \frac{1}{2} C\, \upsilon^2(t) \qquad \text{(J)}. \qquad (5.29)$$

We note that since the capacitor had no charge at $t = -\infty$, then its voltage also was zero at $t = -\infty$.

Equation (5.29) states that:

> ▶ The electrical energy stored in a capacitor at a given instant in time depends on the voltage across the capacitor at that instant, without regard to prior history. ◀

This stored energy is akin to potential energy in a physical system.

Example 5-3: Capacitor Response to Voltage Waveform

The voltage waveform shown in Fig. 5-14(a) was applied across a 0.6 μF capacitor. Determine the corresponding waveforms for (a) the current $i(t)$, (b) the power $p(t)$, and (c) the energy stored in the capacitor $w(t)$.

Solution: (a) We start by establishing a suitable expression for the waveform of $\upsilon(t)$, shown in Fig. 5-14(a), in terms of ramp functions. Noting that the ramp starts at $t = 0$ and has a slope of $10/2 = 5$ V/s, $\upsilon(t)$ can be written as

$$\upsilon(t) = 5r(t) - 5r(t-2) - 5r(t-4) + 5r(t-5) \qquad \text{V}.$$

Recalling that according to Eq. (5.7),

$$r(t-T) = (t-T)\, u(t-T),$$

the expression for $\upsilon(t)$ corresponds to

$$\upsilon(t) = \begin{cases} 0 & \text{for } t \le 0, \\ 5t \text{ V} & \text{for } 0 \le t \le 2 \text{ s}, \\ 10 \text{ V} & \text{for } 2 \text{ s} \le t \le 4 \text{ s}, \\ (-5t+30) \text{ V} & \text{for } 4 \text{ s} \le t \le 5 \text{ s}, \\ 5 \text{ V} & \text{for } t \ge 5 \text{ s}. \end{cases} \qquad (5.30)$$

Application of Eq. (5.23), while recalling that the derivative is the same as the slope of a line or curve, gives:

$$i(t) = C\,\frac{d\upsilon}{dt} = \begin{cases} 0 & \text{for } t \le 0, \\ 3\ \mu\text{A} & \text{for } 0 \le t \le 2 \text{ s}, \\ 0 & \text{for } 2 \text{ s} \le t \le 4 \text{ s}, \\ -3\ \mu\text{A} & \text{for } 4 \text{ s} \le t \le 5 \text{ s}, \\ 0 & \text{for } t \ge 5 \text{ s}. \end{cases} \qquad (5.31)$$

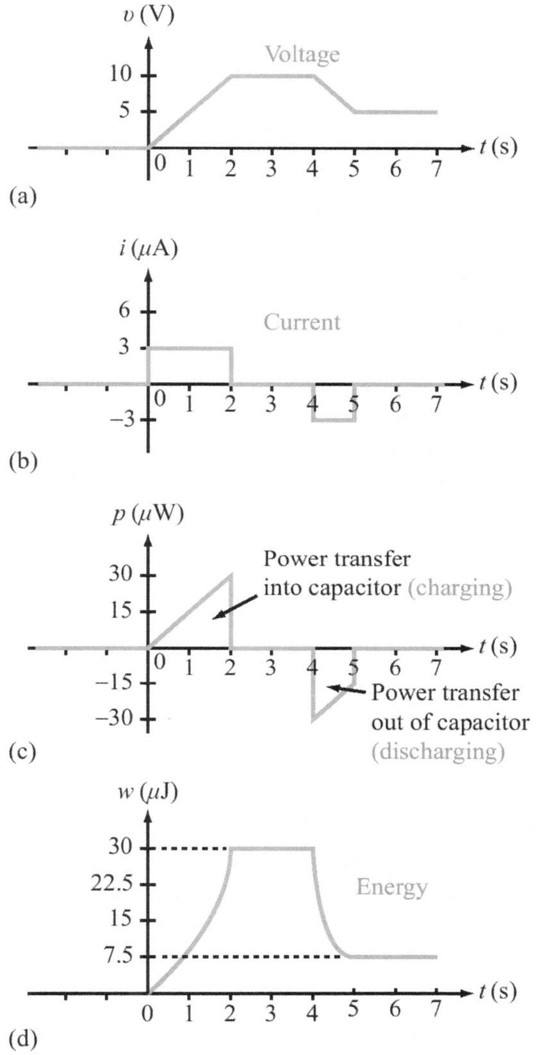

Figure 5-14: Example 5-3 waveforms for i, υ, p, and w.

A plot of the current waveform is displayed in Fig. 5-14(b). We note that $i(t) > 0$ when $\upsilon(t)$ has a positive slope, and $i(t) < 0$ when $\upsilon(t)$ has a negative slope.

(b) The power $p(t)$, which is equal to the product of Eqs. (5.30) and (5.31), is shown in Fig. 5-14(c).

(c) We can calculate the stored energy $w(t)$ either by integrating $p(t)$—which is graphically equivalent to computing the area under the curve—or by applying Eq. (5.29). In either case, we end up with the plot displayed in Fig. 5-14(d).

We note that after $t = 5$ s, the current is zero, the voltage is constant, the power getting transferred into the capacitor is zero (because $i = 0$), and the stored energy remains unchanged at 7.5 μJ.

Let us examine the energy transfer process from the standpoint of the current and voltage. Between $t = 0$ and 2 s, a constant positive current flows to the capacitor, causing the deposition of positive charge on one side of the capacitor and a net increase of negative charge by the same amount on the other side of the capacitor. The increase in charge leads to a linear increase in voltage. By Eq. (5.29), increasing the voltage leads to a quadratic increase in stored energy, as shown in Fig. 5-14 during the time span between 0 and 2 s.

Between 2 and 4 s, $i = 0$ and υ is a constant. Hence, the stored energy remains unchanged. Then, between 4 and 5 s, the current reverses direction, which entails repatriating some of the positive charges back to their original location. Consequently, υ decreases and so does the stored energy, until $t = 5$ s. Beyond that time, the remaining charge stays in place, the voltage remains constant at 5 V, and the corresponding 7.5 μJ of energy stored in the capacitor remains in that state until some future action.

Example 5-4: RC Circuit under dc Conditions

Determine voltages υ_1 and υ_2 across capacitors C_1 and C_2 in the circuit of Fig. 5-15(a). Assume that the circuit has been in its present (charged) condition for a long time.

Solution: "Long time" implies steady state. Under steady-state dc conditions, no current flows through a capacitor. Replacing capacitors C_1 and C_2 with open circuits, as in Fig. 5-15(b), allows us to apply KCL at node V as

$$\frac{V - 20}{20 \times 10^3} + \frac{V}{(30 + 50) \times 10^3} = 0,$$

which gives $V = 16$ V. Hence,

$$V_1 = V = 16 \text{ V}.$$

Through voltage division, V_2 across the 50 kΩ resistor is given by

$$V_2 = \frac{V \times 50\text{k}}{(30 + 50)\text{k}} = \frac{16 \times 50}{80} = 10 \text{ V}.$$

Concept Question 5-8: Explain why a capacitor behaves like an open circuit under dc conditions. (See CAD)

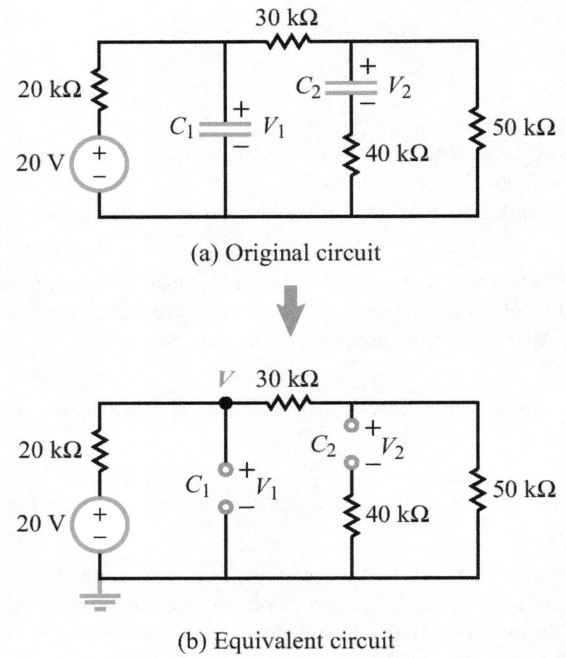

Figure 5-15: Under dc conditions, capacitors behave like open circuits.

Concept Question 5-9: The voltage across a capacitor cannot change instantaneously. Can the current change instantaneously, and why? (See CAD)

Concept Question 5-10: For the capacitor, can $p(t)$ be negative? Can $w(t)$ be negative? Explain. (See CAD)

Exercise 5-6: It is desired to build a parallel-plate capacitor capable of storing 1 mJ of energy when the voltage across it is 1 V. If the capacitor plates are 2 cm × 2 cm each and its insulating material is Teflon, what should the separation d be? Is such a capacitor practical?

Answer: $d = 3.72 \times 10^{-12}$ m. No, it is not practical to build a capacitor with such a small d, because it is about two orders of magnitude smaller than the typical spacing between two adjacent atoms in a solid material.
(See CAD)

Exercise 5-7: Instead of specifying A and calculating the spacing d needed to meet the 1 mJ requirement in Exercise 5-6, suppose we specify d as 1 μm and then calculate A. How large would A have to be?

Answer: $A = 10.4$ m × 10.4 m, equally impractical! (See CAD)

Exercise 5-8: Determine the current i in the circuit of Fig. E5.8, under dc conditions.

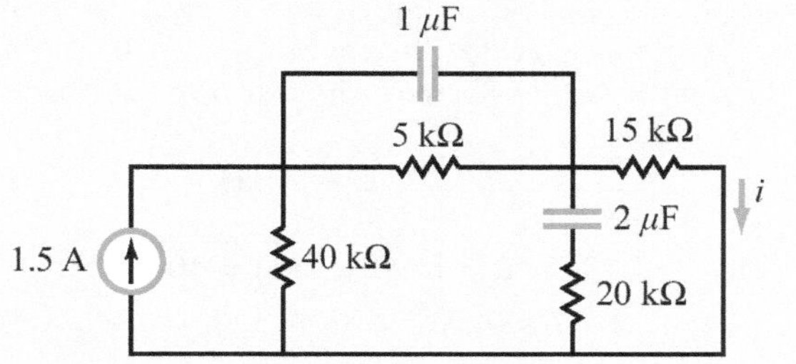

Figure E5.8

Answer: $i = 1$ A. (See CAD)

5-2.2 Series and Parallel Combinations of Capacitors

In Chapter 2, we established that multiple resistors connected in series are equivalent to a single resistor whose resistance is equal to the algebraic sum of the resistances of the individual resistors. This equivalence relationship does not hold true for capacitors. In fact, we will shortly determine that:

> ▶ The equivalence relationship for capacitors connected in series is similar in form to the relationship for resistors connected in parallel, and vice versa. ◀

Capacitors in series

Consider the three capacitors shown in Fig. 5-16. They share the same current i_s, and are therefore in series. Current is related to their individual voltages by

$$i_s = C_1 \frac{dv_1}{dt} = C_2 \frac{dv_2}{dt} = C_3 \frac{dv_3}{dt}. \tag{5.32}$$

Also,

$$v_s = v_1 + v_2 + v_3. \tag{5.33}$$

Combining In-Series Capacitors

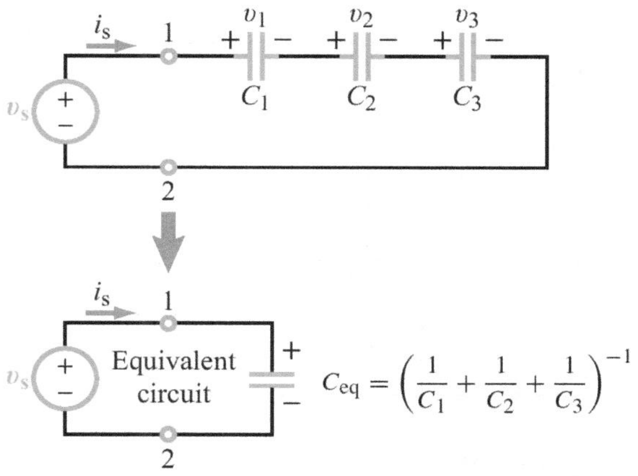

Figure 5-16: Capacitors in series.

We wish to relate C_{eq} of the equivalent circuit to C_1, C_2, and C_3, subject to the requirement that the actual circuit and its equivalent exhibit identical i–v characteristics at terminals (1, 2). For the equivalent circuit,

$$\begin{aligned} i_s &= C_{eq} \frac{dv_s}{dt} = C_{eq} \left(\frac{dv_1}{dt} + \frac{dv_2}{dt} + \frac{dv_3}{dt} \right) \\ &= C_{eq} \left(\frac{i_s}{C_1} + \frac{i_s}{C_2} + \frac{i_s}{C_3} \right), \end{aligned} \tag{5.34}$$

which leads to

$$\frac{1}{C_{eq}} = \frac{1}{C_1} + \frac{1}{C_2} + \frac{1}{C_3}. \tag{5.35}$$

Generalizing to the case of N capacitors in series,

$$\frac{1}{C_{eq}} = \sum_{i=1}^{N} \frac{1}{C_i} = \frac{1}{C_1} + \frac{1}{C_2} + \cdots + \frac{1}{C_N} \tag{5.36}$$

(capacitors in series).

Additionally, if at reference time t_0 the capacitors had initial voltages $v_1(t_0)$ to $v_N(t_0)$, the initial voltage of the equivalent capacitor is

$$v_{eq}(t_0) = \sum_{i=1}^{N} v_i(t_0). \tag{5.37}$$

Combining In-Parallel Capacitors

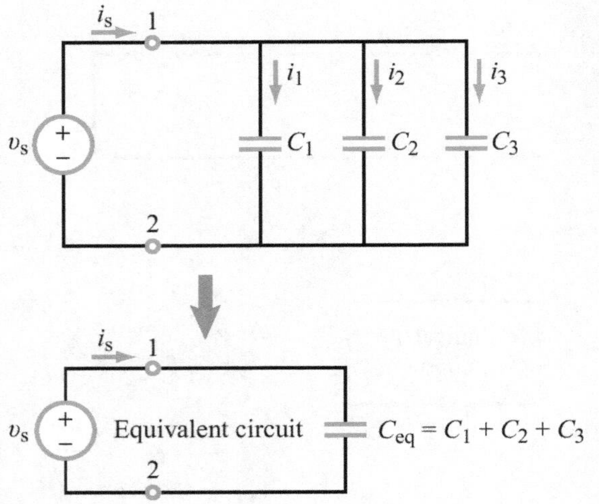

Figure 5-17: Capacitors in parallel.

Capacitors in parallel

The three capacitors shown in Fig. 5-17 share the same voltage v_s and are therefore connected in parallel. The source current i_s is equal to the sum of their currents,

$$i_s = i_1 + i_2 + i_3 = C_1 \frac{dv_s}{dt} + C_2 \frac{dv_s}{dt} + C_3 \frac{dv_s}{dt}. \quad (5.38)$$

For the equivalent circuit with equivalent capacitor C_{eq},

$$i_s = C_{eq} \frac{dv_s}{dt}. \quad (5.39)$$

Equating the expressions given by Eqs. (5.38) and (5.39) leads to

$$C_{eq} = C_1 + C_2 + C_3, \quad (5.40)$$

which can be generalized to N capacitors in parallel as

$$C_{eq} = \sum_{i=1}^{N} C_i \quad \textbf{(capacitors in parallel)}. \quad (5.41)$$

Since the capacitors are connected in parallel, they shared the same voltage $v(t_0)$ at reference time t_0. Hence, for the equivalent capacitor

$$v_{eq}(t_0) = v(t_0). \quad (5.42)$$

Example 5-5: Equivalent Circuit

Reduce the circuit of Fig. 5-18(a) into the simplest equivalent configuration.

Solution: Resistors are combined independently of capacitors. For the resistors, we first combine R_2 and R_3 in parallel, and then add the result to R_1 in series, noting that interchanging the locations of two elements connected in series is perfectly permissible, as such an action has no influence on either the current flowing through them or the voltages across them. A similar procedure can be followed for the capacitors, but we have to keep in mind that the equivalence relationships for resistors and capacitors are the reciprocal of one another:

$$R_2 \parallel R_3 = \frac{R_2 R_3}{R_2 + R_3} = \frac{3\text{k} \times 6\text{k}}{3\text{k} + 6\text{k}} = 2\text{ k}\Omega.$$

$$R_{eq} = R_1 + 2\text{ k}\Omega = 8\text{ k}\Omega + 2\text{ k}\Omega = 10\text{ k}\Omega,$$

$$C_2 \parallel C_3 = C_2 + C_3 = 1\ \mu\text{F} + 5\ \mu\text{F} = 6\ \mu\text{F},$$

$$C_{eq} = \frac{C_1 \times 6 \times 10^{-6}}{C_1 + 6 \times 10^{-6}} = \left(\frac{12 \times 6}{12 + 6}\right) \times 10^{-6} = 4\ \mu\text{F}.$$

The equivalent circuit is shown Fig. 5-18(b).

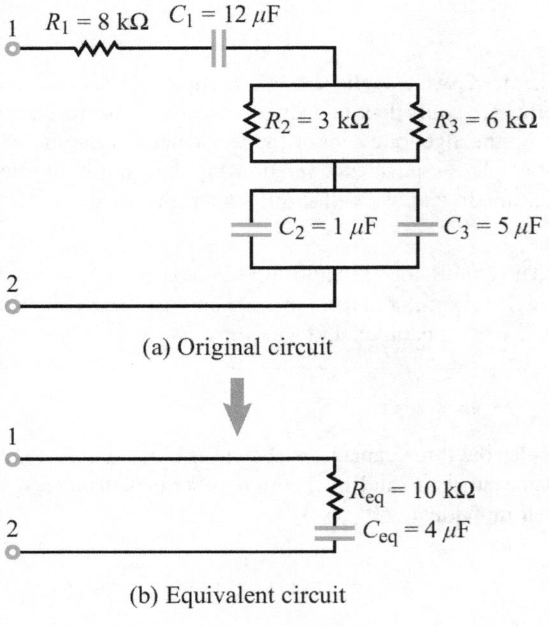

Figure 5-18: Circuit for Example 5-5.

Technology Brief 12
Supercapacitors

As shown in Section 5-2.1, the energy (in joules) stored in a capacitor is given by $w = \frac{1}{2}CV^2$, where C is the capacitance and V is the voltage across it. Why then do we not charge capacitors by applying a voltage across them and then use them instead of batteries in support of everyday gadgets and systems? To help answer this question, we refer the reader to Fig. TF12-1, whose axes represent two critical attributes of storage devices. It is the combination (intersection) of these attributes that determines the type of applications best suited for each of the various energy devices displayed in the figure.

Energy density W' is a measure of how much energy a device or material can store per unit weight. That is, $W' = w/m$, where m is the mass of the capacitor in kilograms. [Alternatively, energy density can be defined in terms of volume (instead of weight) for applications where minimizing the volume of the energy source is more important than minimizing its weight.] Even though the formal SI unit for energy density is (J/kg), a more common unit is the watt-hour/kg (Wh/kg) with 1 Wh = 3600 J. The second dimension in Fig. TF12-1 is the ***power density*** P' (W/kg), which is a measure of how fast energy can be added to or removed from an energy-storage device (also per unit weight). Power is defined as energy per unit time as $P' = dW'/dt$.

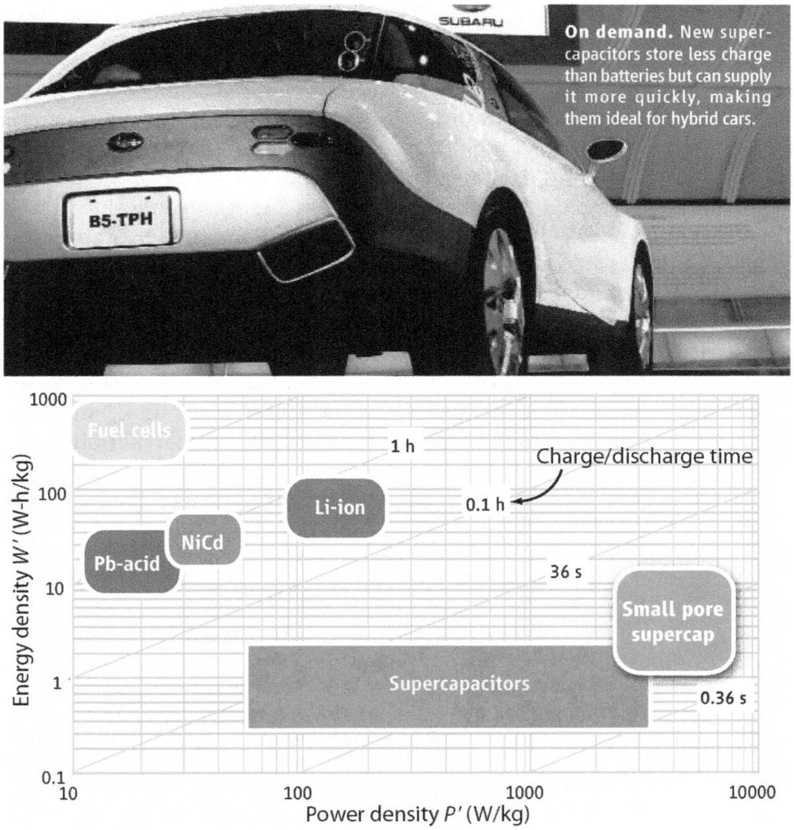

Figure TF12-1: Energy and power densities of modern energy-storage technologies. Even though supercapacitors store less charge than batteries, they can discharge their energy more quickly, making them more suitable for hybrid cars. (*Science*, Vol. 313, p. 902.)

Table TT12-1: Comparison of a conventional capacitor, supercapacitor, and lithium battery size and mass required to hold ~ 1 megajoule (MJ) of energy (300 watt-hours). 1 MJ of energy will power a laptop with an average consumption of 50 W for 6 hours. Note from the first column that a lithium ion battery might hold 1000 times more energy than a conventional capacitor for reasonable voltages (< 50 V).

Sample device	Specific Energy [Watt hours/ kg]	Specific Energy [MJ / kg]	Energy Density [MJ / liter]	Volume required to hold 1 MJ [liter]	Weight required to hold 1 MJ [kg]
Conventional capacitor	0.01 – 0.1	$4\text{x}10^{-5}$-$4\text{x}10^{-4}$	$6\text{x}10^{-5}$-$6\text{x}10^{-4}$	17000-1700	25000 - 2500
Supercapacitor	1 - 10	0.004 – 0.04	0.006 - 0.06	166 – 16	250 – 25
Lithium ion battery	100 - 250	0.36 - 0.9	1 - 2	1 – 0.5	2.8 – 1.1

According to Fig. TF12-1, fuel cells can store large amounts of energy, but they can deliver that energy only relatively slowly (several hours). In contrast, conventional capacitors can store only small amounts of energy—several orders of magnitude less than fuel cells—but it is possible to charge or discharge a capacitor in just a few seconds—or even a fraction of a second. Batteries occupy the region in-between fuel cells and conventional capacitors; they can store more energy per unit weight than the ordinary capacitor by about three orders of magnitude, and they can release their energy faster than fuel cells by about a factor of 10. Thus, capacitors are partly superior to other energy devices because they can accomodate very fast rates of energy transfer, but the amount of energy that can be "packed into" a capacitor is limited by its size and weight. To appreciate what that means, let us examine the relation

$$w = \frac{1}{2} CV^2.$$

To increase w, we need to increase either C or V. We can develop an intuitive feel for this if we compare how large a storage element would have to be to hold 1 MJ (~ 300 watt-hours). From Table TT12-1, we can see that a conventional capacitor would have to be thousands of liters in size (and weigh thousands of kilograms), whereas a supercapacitor or a battery would be considerably smaller.

For a parallel-plate capacitor, $C = \varepsilon A/d$, where ε is the permittivity of the material between the plates, A is the area of each of the two plates, and d is the separation between them. The material between the plates should be a good insulator, and for most such insulators, the value of ε is in the range between ε_0 (permittivity of vacuum) and $6\varepsilon_0$ (for mica), so the choice of material can at best increase C by a factor of 6. Making A larger increases both the volume and weight of the capacitor. In fact, since the mass m of the plates is proportional directly to A, the energy density $W' = w/m$ is independent of A. That leaves d as the only remaining variable. Reducing d will indeed increase C, but such a course will run into two serious obstacles: (a) to avoid voltage breakdown (arcing), V has to be reduced along with d such that V/d remains lower than the breakdown value of the insulator; (b) eventually d approaches subatomic dimensions, making it infeasible to construct such a capacitor. Increasing V also increases the energy stored (by V^2) but here, too, we run into problems with breakdown. Another serious limitation of the capacitor as an energy storage device is that its voltage does not remain constant as energy is transferred to and from it.

Supercapacitor Technology

A new generation of capacitor technologies, termed ***supercapacitors*** or ***ultracapacitors***, is narrowing the gap between capacitors and batteries. These capacitors can have sufficiently high energy densities to approach within 10 percent of battery storage densities, and additional improvements may increase this even more. Importantly, supercapacitors can absorb or release energy much faster than a chemical battery of identical volume. This helps immensely during recharging. Moreover, most batteries can be recharged only a few hundred times before they are degraded completely; supercapacitors can be charged and discharged millions

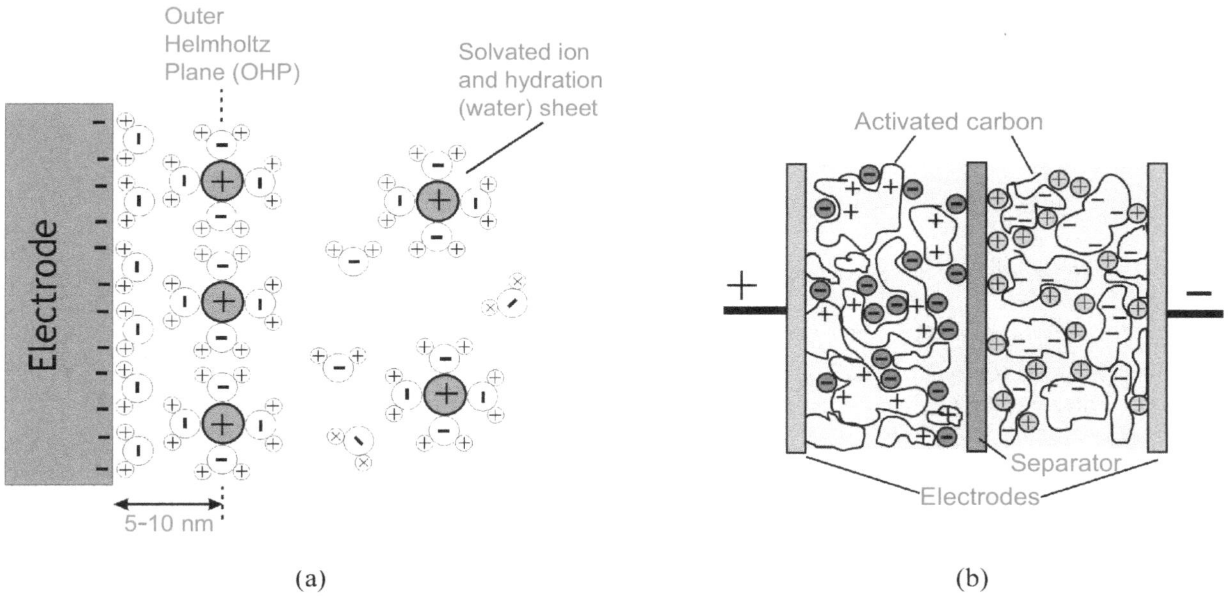

Figure TF12-2: (a) Conceptual illustration of the water double layer at a charged metal surface; (b) conceptual illustration of an electrochemical capacitor.

of times before they wear out. Supercapacitors also have a much smaller environmental footprint than conventional chemical batteries, making them particularly attractive for green energy solutions.

History and Design

Supercapacitors are a special class of capacitor known as an ***electrochemical capacitor***. This should not be confused with the term ***electrolytic capacitor***, which is a term applied to a specific variety of the conventional capacitor. Electrochemical capacitors work by making use of a special property of water solutions (and some polymers and gels). When a metal electrode is immersed in water and a potential is applied, the water molecules (and any dissolved ions) immediately align themselves to the charges present at the surface of the metal electrode, as illustrated in Fig. TF12-2(a). This rearrangement generates a thin layer of organized water molecules (and ions), called a ***double layer***, that extends over the entire surface of the metal. The very high charge density, separated by a tiny distance on the order of a few nanometers, effectively looks like a capacitor (and a very large one: capacitive densities on the order of $\sim 10\ \mu\text{F/cm}^2$ are common for water solutions). This phenomenon has been known to physicists and chemists since the work of von Helmholtz in 1853, and later Guoy, Chapman, and Stern in the early 20th century. In order to make capacitors useful for commercial applications, several technological innovations were required. Principal among these were various methods for increasing the total surface area that forms the double layer. The first working capacitor based on the electrochemical double layer (patented by General Electric in 1957) used very porous conductive carbon. Modern electrochemical capacitors employ ***carbon aerogels***, and more recently ***carbon nanotubes*** have been shown to effectively increase the total double layer area (Fig. TF12-2(b)).

Supercapacitors are beginning to see commercial use in applications ranging from transportation to low-power consumer electronics. Several bus lines around the world now run with buses powered with supercapacitors; train systems are also in development. Supercapacitors intended for small portable electronics (like your MP3 player) are in the pipeline as well!

Example 5-6: Voltage Division

Figure 5-19(a) contains two resistors R_1 and R_2 connected in series to a voltage source υ_s. In Chapter 2, we demonstrated that the voltage υ_s is divided among the two resistors and, for example, υ_1 is given by

$$\upsilon_1 = \left(\frac{R_1}{R_1 + R_2}\right) \upsilon_s. \quad (5.43)$$

Derive the equivalent voltage-division equation for the series capacitors C_1 and C_2 in Fig. 5-19(b). Assume that the capacitors had no charge on them before they were connected to υ_s.

Solution: From the standpoint of the source υ_s, it "sees" an equivalent, single capacitor C given by the series combination of C_1 and C_2, namely

$$C = \frac{C_1 C_2}{C_1 + C_2}. \quad (5.44)$$

The voltage across C is υ_s. The law of conservation of energy requires that the energy that would be stored in the equivalent capacitor C be equal to the sum of the energies stored in C_1 and C_2. Hence, application of Eq. (5.29) gives

$$\frac{1}{2}\, C\upsilon_s^2 = \frac{1}{2}\, C_1\upsilon_1^2 + \frac{1}{2}\, C_2\upsilon_2^2. \quad (5.45)$$

Upon replacing C with the expression given by Eq. (5.44) and replacing the source voltage with $\upsilon_s = \upsilon_1 + \upsilon_2$, we have

$$\frac{1}{2}\left(\frac{C_1 C_2}{C_1 + C_2}\right)(\upsilon_1 + \upsilon_2)^2 = \frac{1}{2}\, C_1\upsilon_1^2 + \frac{1}{2}\, C_2\upsilon_2^2, \quad (5.46)$$

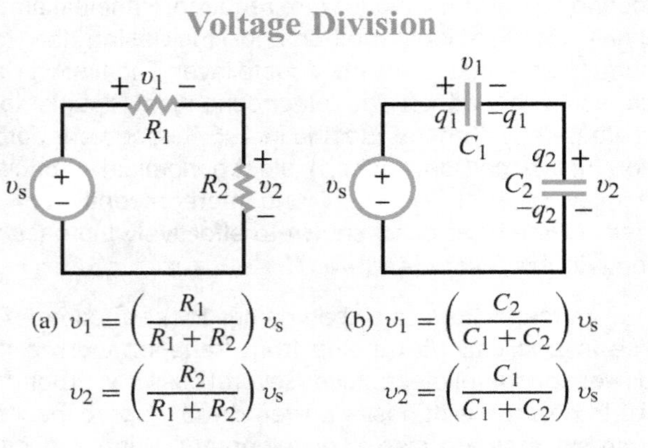

Figure 5-19: Voltage-division rules for (a) in-series resistors and (b) in-series capacitors.

which reduces to

$$C_1\upsilon_1 = C_2\upsilon_2. \quad (5.47)$$

Using $\upsilon_2 = \upsilon_s - \upsilon_1$ in Eq. (5.47) leads to

$$C_1\upsilon_1 = C_2(\upsilon_s - \upsilon_1)$$

or

$$\upsilon_1 = \left(\frac{C_2}{C_1 + C_2}\right) \upsilon_s. \quad (5.48)$$

We note that in the voltage-division equation for resistors, υ_1 is directly proportional to R_1, whereas in the capacitor case, υ_1 is directly proportional to C_2 (instead of to C_1). Additionally, in view of the relationship given by Eq. (5.47), application of the basic definition for capacitance, namely $C = q/\upsilon$, leads to

$$q_1 = q_2. \quad (5.49)$$

This result is exactly what one would expect when viewing the circuit from the perspective of the voltage source υ_s.

Concept Question 5-11: Compare the voltage-division equation for two capacitors in series with that for two resistors in series. Are they identical or different in form? (See CAD)

Concept Question 5-12: Two capacitors are connected in series between terminals (a, b) in a certain circuit with capacitor 1 next to terminal a and capacitor 2 next to terminal b. How does the magnitude and polarity of charge q_1 on the plate (of capacitor 1) near terminal a compare with charge q_2 on the plate (of capacitor 2) near terminal b? (See CAD)

Exercise 5-9: Determine C_{eq} and $\upsilon_{eq}(0)$ at terminals (a, b) for the circuit in Fig. E5.9 given that $C_1 = 6\ \mu$F, $C_2 = 4\ \mu$F, $C_3 = 8\ \mu$F, and the initial voltages on the three capacitors are $\upsilon_1(0) = 5$ V and $\upsilon_2(0) = \upsilon_3(0) = 10$ V, respectively.

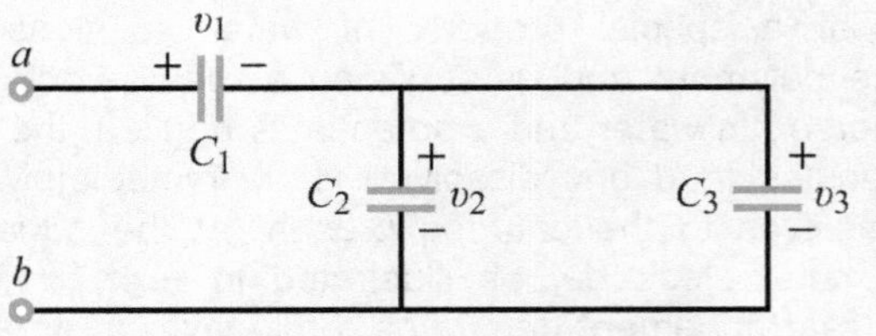

Figure E5.9

Answer: $C_{eq} = 4\ \mu$F, $\upsilon_{eq}(0) = 15$ V. (See CAD)

Exercise 5-10: Suppose the circuit of Fig. E5.9 is connected to a dc voltage source $V_0 = 12$ V. Assuming that the capacitors had no charge before they were connected to the voltage source, determine υ_1 and υ_2 given that $C_1 = 6\ \mu$F, $C_2 = 4\ \mu$F, and $C_3 = 8\ \mu$F.

Answer: $\upsilon_1 = 8$ V, $\upsilon_2 = 4$ V. (See CAD)

5-3 Inductors

Any current-carrying conductor, whether straight or coiled, forms an inductor. A current produces a magnetic field, which stores energy that can be released later in the form of another current. Also, since every wire acts like an inductor, we have small amounts of stray inductance in every circuit. Fortunately, this can be ignored except at extremely high frequencies (microwave band).

Inductors exhibit a number of useful properties, including magnetic coupling and electromagnetic induction. They are employed in microphones and loudspeakers, magnetic relays and sensors, theft detection devices, and motors and generators, and they provide wireless power transmission and data communication (albeit over relatively short distances).

▶ Capacitors and inductors constitute a canonical pair of devices. Whereas a capacitor can store energy through the electric field induced by the voltage imposed across its terminals, an inductor can store magnetic energy through the magnetic field induced by the current flowing through its wires. ◀

The i–υ relationship for a capacitor is $i = C\ d\upsilon/dt$; the converse is true for an inductor with $\upsilon = L\ di/dt$. As we will see in Chapter 7, the capacitor acts like an open circuit to low-frequency signals and like a short circuit to high-frequency signals; the exact opposite behavior is exhibited by the inductor.

A typical example of an inductor is the *solenoid* configuration shown in Fig. 5-20. The solenoid consists of multiple turns of wire wound in a helical geometry around a cylindrical core. The core may be air filled or may contain a magnetic material (typically iron) with *magnetic permeability* μ. If the wire carries a current $i(t)$ and the turns are closely spaced, the solenoid produces a relatively uniform *magnetic field* B within its interior region.

Magnetic-flux linkage Λ is defined as the total magnetic flux linking (passing through) a coil or a given circuit. For a solenoid with N turns carrying a current i,

$$\Lambda = \left(\frac{\mu N^2 S}{\ell}\right) i \qquad \text{(Wb)}, \tag{5.50}$$

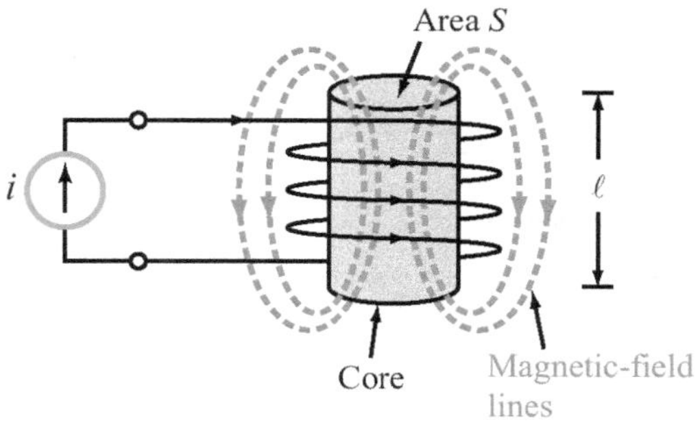

Figure 5-20: The inductance of a solenoid of length ℓ and cross-sectional area S is $L = \mu N^2 S/\ell$, where N is the number of turns and μ is the magnetic permeability of the core material.

where ℓ is the length of the solenoid and S is its cross-sectional area. The unit for Λ is the weber (Wb), named after the German scientist Wilhelm Weber (1804–1891).

Self-inductance refers to the magnetic-flux linkage of a coil (or circuit) with itself, in contrast with *mutual inductance*, which refers to magnetic-flux linkage in a coil due to the magnetic field generated by another coil (or circuit). Usually, when the term *inductance* is used, the intended reference is to self-inductance. Mutual inductance is covered in Chapter 11.

The (self) inductance of any conducting system is defined as the ratio of Λ to the current i responsible for generating it, given as

$$L = \frac{\Lambda}{i} \qquad \text{(H)}, \tag{5.51}$$

and its unit is the henry (H), so named to honor the American inventor Joseph Henry (1797–1878). Using the expression for Λ given by Eq. (5.50), we have

$$L = \frac{\mu N^2 S}{\ell} \qquad \textbf{(solenoid)}. \tag{5.52}$$

The inductance L is directly proportional to μ, the magnetic permeability of the core material. The relative magnetic permeability μ_r is defined as

$$\mu_r = \frac{\mu}{\mu_0}, \tag{5.53}$$

where $\mu_0 \approx 4\pi \times 10^{-7}$ (H/m) is the magnetic permeability of free space.

Table 5-3: **Relative magnetic permeability of materials, $\mu_r = \mu/\mu_0$ and $\mu_0 = 4\pi \times 10^{-7}$ H/m.**

Material	Relative Permeability μ_r
All Dielectrics and Non-Ferromagnetic Metals	≈ 1.0
Ferromagnetic Metals	
Cobalt	250
Nickel	600
Mild steel	2,000
Iron (pure)	4,000–5,000
Silicon iron	7,000
Mumetal	$\sim 100,000$
Purified iron	$\sim 200,000$

▶ Except for ferromagnetic materials, $\mu_r \approx 1$ for all dielectrics and conductors. According to Table 5-3, μ_r of ferromagnetic materials (which include iron, nickel, and cobalt) can be as much as five orders of magnitude larger than that of other materials. Consequently, L of an *iron-core solenoid* is about 5000 times that of an *air-core solenoid* of the same size and shape. ◀

Air-core inductors have relatively low inductances, on the order of 10 μH or smaller. Consequently, they are used mostly in high-frequency circuits, such as those designed to support AM and FM radio, cell phones, TV, and similar types of transmitters and receivers. *Ferrite-core inductors* have the inductance-size advantage over air-core inductors, but they have the disadvantage that the ferrite material is subject to hysteresis effects, and they tend to be larger and heavier than their air-core counterparts. One of the consequences of magnetic hysteresis is that the inductance L becomes a function of the current flowing through it. Magnetic hysteresis is outside the scope of this book; hence, *we will always assume that an inductor is an ideal linear device and its inductance is constant and independent of the current flowing through it.*

In modern circuit design and manufacturing, it is highly desirable to contain circuit size down to the smallest dimensions possible. To that end, it is advantageous to use planar integrated-circuit (IC) devices whenever possible. It is relatively easy to manufacture resistors and capacitors in a planar IC format and to do so for a wide range of resistance and capacitance values, but the same is not true for inductors. even though inductors can be manufactured in planar form, as illustrated by the coil shown in Fig. 5-21, their inductance values are too small for most circuit applications, necessitating the use of the more bulky, discrete form instead.

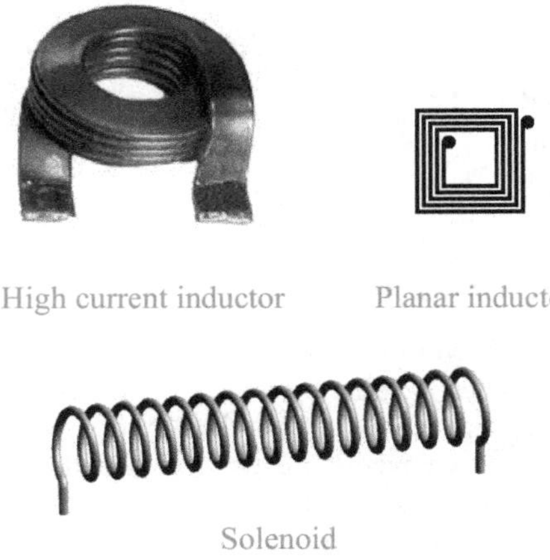

Figure 5-21: Various types of inductors.

5-3.1 Electrical Properties

According to Faraday's law, if the magnetic-flux linkage in an inductor (or circuit) changes with time, it induces a voltage υ across the inductor's terminals given by

$$\upsilon = \frac{d\Lambda}{dt}. \tag{5.54}$$

In view of Eq. (5.51),

$$\upsilon = \frac{d}{dt}(Li) = L\,\frac{di}{dt}. \tag{5.55}$$

This i–υ relationship adheres to the passive sign convention introduced earlier for resistors and capacitors. If the direction of i is into the $(+)$ voltage terminal of the inductor (Fig. 5-22), then the inductor is receiving power. Also, the same logic that led us earlier to the conclusion that the voltage across a capacitor cannot change instantaneously leads us now to the conclusion:

▶ The current through an inductor cannot change instantaneously, but the voltage can. ◀

(Otherwise, the voltage across it would become infinite.) The implication of this restriction is that *when a current source connected to an inductor is disconnected by a switch, the current*

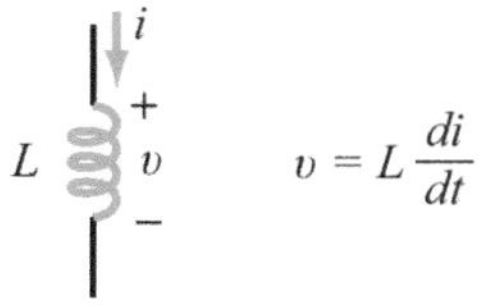

Figure 5-22: Passive sign convention for an inductor.

continues to flow for a short amount of time through the air between the switch terminals, manifesting itself in the form of a spark! In large power systems, current must always be ramped up and down slowly to avoid this problem.

When we discussed the capacitor's i–υ relationship given by Eq. (5.23), we noted that under dc conditions a capacitor acts like an open circuit. In contrast, Eq. (5.55) asserts that:

> ▶ Under dc conditions, an inductor acts like a short circuit. ◀

To express $i(t)$ in terms of $\upsilon(t)$, we duplicate the procedure we followed earlier in connection with the capacitor, which for the inductor leads to

$$i(t) = i(t_0) + \frac{1}{L}\int_{t_0}^{t} \upsilon \, dt', \qquad (5.56)$$

where t_0 is an initial reference point in time.

The power delivered to the inductor is given by

$$p(t) = \upsilon i = Li \, \frac{di}{dt}, \qquad (5.57)$$

and as with the resistor and the capacitor, the sign of p determines whether the inductor is receiving power ($p > 0$) or delivering it ($p < 0$). The accumulation of power over time constitutes the storage of energy. The magnetic energy stored in an inductor is

$$w(t) = \int_{-\infty}^{t} p \, dt' = \int_{-\infty}^{t} \left(Li \, \frac{di}{dt'}\right) dt', \qquad (5.58)$$

which yields

$$w(t) = \frac{1}{2} \, L \, i^2(t) \qquad \text{(J)}, \qquad (5.59)$$

where it is presumed that at $t = -\infty$ no current was flowing through the inductor. Note the analogy with the capacitor for which $w(t) = \frac{1}{2} \, C \, \upsilon^2(t)$.

> ▶ The magnetic energy stored in an inductor at a given instant in time depends on the current flowing through the inductor at that instant—without regard to prior history. ◀

Example 5-7: Inductor Response to Current Waveform

Upon closing the switch at $t = 0$ in the circuit of Fig. 5-23(a), the voltage source generates a current waveform through the circuit given by

$$i(t) = 10e^{-0.8t} \sin(\pi t/2) \text{ A}, \qquad (\text{for } t \geq 0).$$

(a) Plot the waveform $i(t)$ versus t and determine the locations of its first maximum, first minimum, and their corresponding amplitudes.

(b) given that $L = 50$ mH, obtain an expression for $\upsilon(t)$ across the inductor and plot its waveform.

(c) Generate a plot of the power $p(t)$ delivered to the inductor.

Solution: (a) The waveform of $i(t)$ is shown in Fig. 5-23(b). To determine the locations of its maxima and minima, we take the derivative of $i(t)$ and equate it to zero, which leads to

$$-0.8 \times 10e^{-0.8t} \sin(\pi t/2) + \left(\frac{\pi}{2}\right) \times 10e^{-0.8t} \cos\left(\frac{\pi t}{2}\right) = 0,$$

which in turn simplifies to

$$\tan\left(\frac{\pi t}{2}\right) = \frac{\pi}{1.6}.$$

Its solution is

$$\frac{\pi t}{2} = 1.1 + n\pi \qquad (\text{for } n = 0, 1, 2, \ldots).$$

For $n = 0$, $t = 0.7$ s, which is the location in time of the first maximum of $i(t)$. The next solution, corresponding to $n = 1$, gives the location of the first minimum of $i(t)$ at 2.7 s. The amplitudes of $i(t)$ at these locations are

$$i_{\text{max}} = i(t = 0.7 \text{ s}) = 10e^{-0.8 \times 0.7} \sin(\pi \times 0.7/2) = 5.09 \text{ A}$$

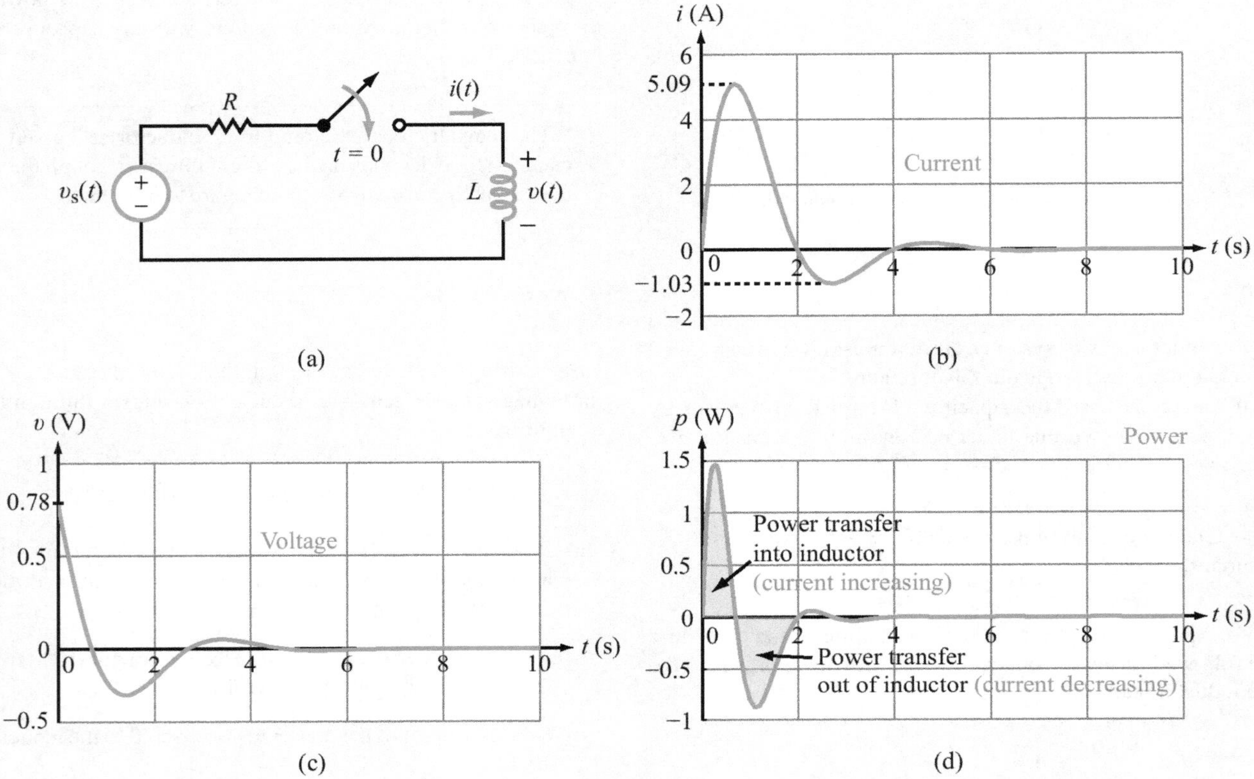

Figure 5-23: Circuit for Example 5-7.

and

$$i_{\min} = i(t = 2.7 \text{ s}) = 10e^{-0.8\times 2.7} \sin(\pi \times 2.7/2) = -1.03 \text{ A}.$$

(b)

$$\begin{aligned}
\upsilon(t) &= L \frac{di}{dt} \\
&= L \frac{d}{dt}[10e^{-0.8t} \sin(\pi t/2)] \\
&= 50 \times 10^{-3} \cdot [-8e^{-0.8t} \sin(\pi t/2) \\
&\qquad + 5\pi e^{-0.8t} \cos(\pi t/2)] \\
&= [-0.4 \sin(\pi t/2) + 0.25\pi \cos(\pi t/2)]e^{-0.8t} \text{ V}.
\end{aligned}$$

The waveform of $\upsilon(t)$ is shown in Fig. 5-23(c).

(c)

$$\begin{aligned}
p(t) &= \upsilon(t)\, i(t) \\
&= [-0.4 \sin(\pi t/2) + 0.25\pi \cos(\pi t/2)]e^{-0.8t} \\
&\qquad \times 10e^{-0.8t} \sin(\pi t/2) \\
&= [-4 \sin^2(\pi t/2) + 2.5\pi \cos(\pi t/2) \sin(\pi t/2)] \\
&\qquad \times e^{-1.6t} \text{ W}.
\end{aligned}$$

The waveform of $p(t)$ shown in Fig. 5-23(d) includes both positive and negative values. During periods when $p(t) > 0$, magnetic energy is getting stored in the inductor. Conversely, when $p(t) < 0$, the inductor is releasing some of its previously stored energy.

Concept Question 5-13: What type of material exhibits a magnetic permeability higher than μ_0? (See CAD)

Concept Question 5-14: Can the voltage across an inductor change instantaneously? (See CAD)

Exercise 5-11: Calculate the inductance of a 20-turn air-core solenoid if its length is 4 cm and the radius of its circular cross section is 0.5 cm.

Answer: $L = 9.87 \times 10^{-7}$ H $= 0.987$ μH. (See CAD)

Exercise 5-12: Determine currents i_1 and i_2 in the circuit of Fig. E5.12, under dc conditions.

Answer: $i_1 = 0$, $i_2 = 6$ A. (See CAD)

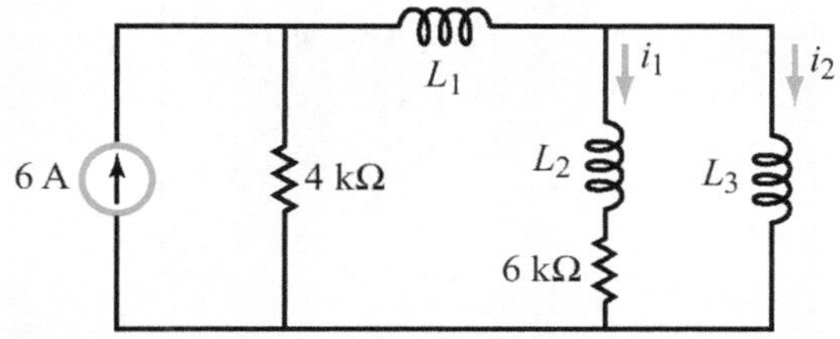

Figure E5.12

5-3.2 Series and Parallel Combinations of Inductors

▶ The rules for combining multiple inductors in series or in parallel are the same as those for resistors. ◀

Inductors in series

For the three inductors in series in Fig. 5-24,

$$\upsilon_s = \upsilon_1 + \upsilon_2 + \upsilon_3 = L_1 \frac{di_s}{dt} + L_2 \frac{di_s}{dt} + L_3 \frac{di_s}{dt}$$
$$= (L_1 + L_2 + L_3) \frac{di_s}{dt}, \qquad (5.60)$$

and for the equivalent circuit,

$$\upsilon_s = L_{eq} \frac{di_s}{dt}. \qquad (5.61)$$

Hence,

$$L_{eq} = L_1 + L_2 + L_3, \qquad (5.62)$$

and for N inductors in series,

$$L_{eq} = \sum_{i=1}^{N} L_i = L_1 + L_2 + \cdots + L_N$$
$$\textbf{(inductors in series)}. \qquad (5.63)$$

Combining In-Series Inductors

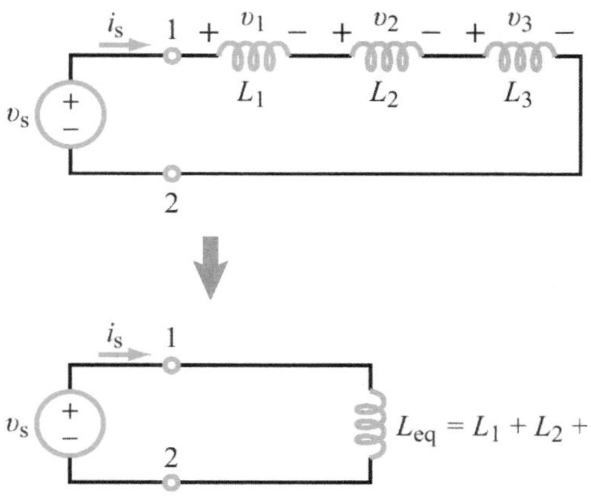

Figure 5-24: Inductors in series.

Combining In-Parallel Inductors

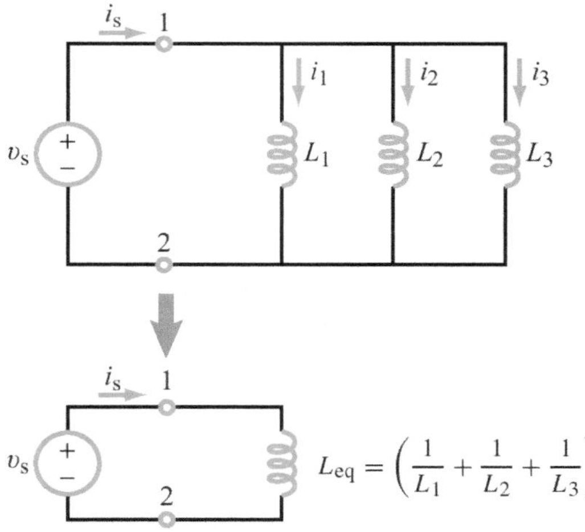

$$L_{eq} = \left(\frac{1}{L_1} + \frac{1}{L_2} + \frac{1}{L_3}\right)^{-1}$$

Figure 5-25: Inductors in parallel.

Inductors in parallel

A similar analysis for the currents in the parallel circuit of Fig. 5-25 leads to

$$\frac{1}{L_{eq}} = \frac{1}{L_1} + \frac{1}{L_2} + \frac{1}{L_3}. \qquad (5.64)$$

Table 5-4: **Basic properties of R, L, and C.**

Property	R	L	C
i–υ relation	$i = \dfrac{\upsilon}{R}$	$i = \dfrac{1}{L}\displaystyle\int_{t_0}^{t} \upsilon\, dt' + i(t_0)$	$i = C\,\dfrac{d\upsilon}{dt}$
υ-i relation	$\upsilon = iR$	$\upsilon = L\,\dfrac{di}{dt}$	$\upsilon = \dfrac{1}{C}\displaystyle\int_{t_0}^{t} i\, dt' + \upsilon(t_0)$
p (power transfer in)	$p = i^2R$	$p = Li\,\dfrac{di}{dt}$	$p = C\upsilon\,\dfrac{d\upsilon}{dt}$
w (stored energy)	0	$w = \dfrac{1}{2}Li^2$	$w = \dfrac{1}{2}C\upsilon^2$
Series combination	$R_{eq} = R_1 + R_2$	$L_{eq} = L_1 + L_2$	$\dfrac{1}{C_{eq}} = \dfrac{1}{C_1} + \dfrac{1}{C_2}$
Parallel combination	$\dfrac{1}{R_{eq}} = \dfrac{1}{R_1} + \dfrac{1}{R_2}$	$\dfrac{1}{L_{eq}} = \dfrac{1}{R_1} + \dfrac{1}{R_2}$	$C_{eq} = C_1 + C_2$
dc behavior	no change	short circuit	open circuit
Can υ change instantaneously?	yes	yes	no
Can i change instantaneously?	yes	no	yes

Generalizing to the case of N inductors,

$$\frac{1}{L_{eq}} = \sum_{i=1}^{N} \frac{1}{L_i} = \frac{1}{L_1} + \frac{1}{L_2} + \cdots + \frac{1}{L_N}. \qquad (5.65)$$

(inductors in parallel)

If $i_1(t_0)$ through $i_N(t_0)$ are the initial currents flowing through the parallel inductors L_1 to L_N at t_0, then the initial current $i_{eq}(t_0)$ that would be flowing through the equivalent inductor L_{eq} is given by

$$i_{eq}(t_0) = \sum_{j=1}^{N} i_j(t_0). \qquad (5.66)$$

A summary of the electrical properties of resistors, inductors and capacitors is available in Table 5-4.

Example 5-8: Energy Storage under dc Conditions

The circuit in Fig. 5-26(a) has been in its present state for a long time. Determine the amount of energy stored in the capacitors and inductors.

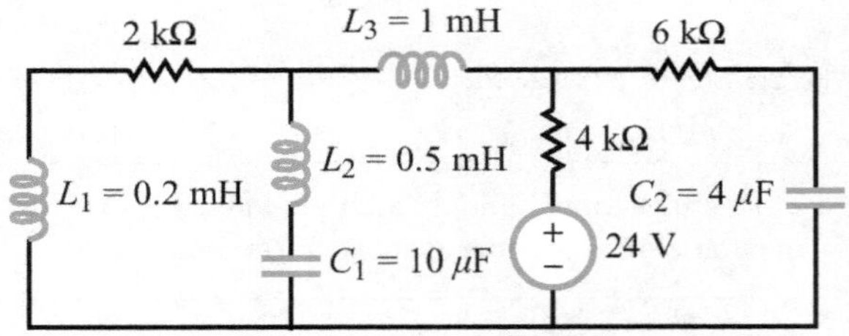

(a) Original circuit

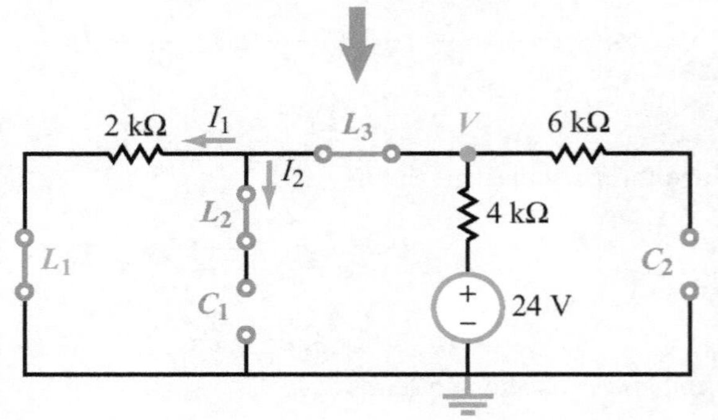

(b) Equivalent circuit under steady state conditions

Figure 5-26: Under steady-state dc conditions, capacitors act like open circuits, and inductors act like short circuits.

Solution: Our first step is to replace components with their dc equivalents (capacitors with open circuits and inductors with short circuits). The process leads to the circuit in Fig. 5-26(b), which can be solved using any of the analysis methods used previously with resistive circuits. Current I_1 then is given by

$$I_1 = \frac{24}{(2+4)\text{k}} = 4 \text{ mA},$$

and node voltage V is

$$V = 24 - (4 \times 10^{-3} \times 4 \times 10^3) = 8 \text{ V}.$$

Hence, the amounts of energy stored in C_1, C_2, L_1, L_2, and L_3 are

$$C_1: \quad W = \frac{1}{2} C_1 V^2 = \frac{1}{2} \times 10^{-5} \times 64 = 0.32 \text{ mJ},$$

$$C_2: \quad W = \frac{1}{2} C_2 V^2 = \frac{1}{2} \times 4 \times 10^{-6} \times 64 = 0.128 \text{ mJ},$$

$$L_1: \quad W = \frac{1}{2} L_1 I_1^2 = \frac{1}{2} \times 0.2 \times 10^{-3} \times (4 \times 10^{-3})^2 = 1.6 \text{ nJ},$$

$$L_2: \quad W = \frac{1}{2} L_2 I_2^2 = \frac{1}{2} \times 0.5 \times 10^{-3} \times (0) = 0,$$

and

$$L_3: \quad W = \frac{1}{2} L_3 I_1^2 = \frac{1}{2} \times 10^{-3} \times (4 \times 10^{-3})^2 = 8 \text{ nJ}.$$

Concept Question 5-15: How do the rules for adding inductors in series and in parallel compare with those for resistors and capacitors? (See CAD)

Concept Question 5-16: An inductor stores energy through the magnetic field B, but the equation for the energy stored in an inductor is $w = \frac{1}{2} Li^2$. Explain. (See CAD)

Exercise 5-13: Determine L_{eq} at terminals (a, b) in the circuit of Fig. E5.13.

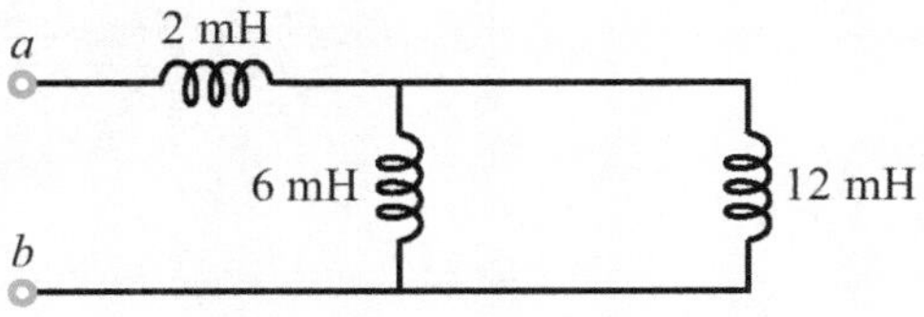

Figure E5.13

Answer: $L_{\text{eq}} = 6$ mH. (See CAD)

5-4 Response of the RC Circuit

The preceding sections described the behavior of capacitors and inductors under dc conditions (i.e., a static circuit with none of its voltages or currents varying with time). We now turn our attention to the time-varying (dynamic) conditions of these circuits.

From the standpoint of analysis and design, circuits containing capacitors and inductors are divided into three groups:

- **RC Circuits**: composed of sources (either constant or time-varying), capacitors, and resistors.
- **RL Circuits**: composed of sources (either constant or time-varying), inductors, and resistors.
- **RLC Circuits**: composed of any combination and any number of sources, capacitors, inductors, and resistors.

In this and succeeding sections of this chapter, we examine the responses of relatively simple RC and RL circuits to sudden changes, such as closing or opening a switch—or both sequentially—and we limit the sources to dc voltage and current sources. The RLC circuit response is addressed in Chapter 6, also for dc sources with switches. RLC circuits driven by ac sources are treated in Chapters 7–11, and RLC circuits driven by other types of sources are the subject of Chapters 12 and 13.

The circuit shown in Fig. 5-27 is called a *first-order RC circuit*; it contains a resistor and a capacitor, and its current and voltage responses are determined by solving a first-order differential equation. The name also applies to any other circuit containing sources, resistors, and capacitors—provided it can be reduced to the form of the generic RC circuit of Fig. 5-27 or its Norton equivalent. This can be realized by combining elements in series or in parallel, as well as through Y-Δ transformations. The voltage source exciting the circuit is a rectangular pulse of amplitude V_s and duration T_0, which includes both turn-on (charging) and turn-off (discharging) periods. The objective of the present section is to develop a

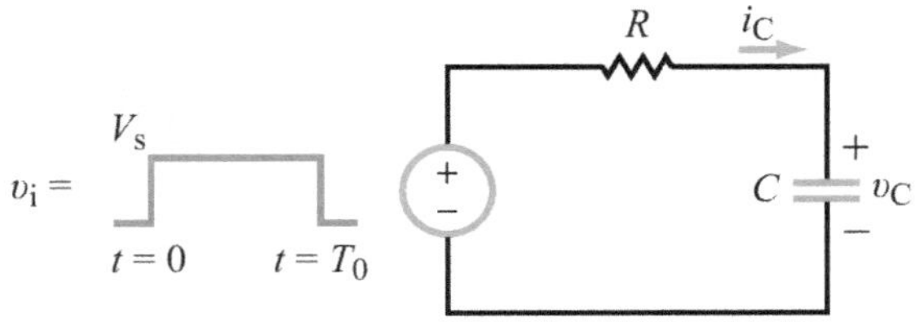

Figure 5-27: Generic first-order RC circuit.

methodology appropriate for RC circuits, so we may apply it to evaluate the circuit's response to the rectangular-pulse waveform or to other types of nonperiodic waveforms.

5-4.1 Natural Response of a Charged Capacitor

We begin by considering what is called the *natural response* of the circuit, which refers to the time variations of the voltages and currents in reaction to moving a switch that allows a fully charged capacitor to discharge its accumulated charge. This occurs at $t = T_0$ in Fig. 5-27. To that end, let us examine the more realistic circuit in Fig. 5-28(a). Until $t = 0$, the series RC circuit had been connected to dc voltage source V_s for a long time. At $t = 0$, the switch disconnects the RC circuit from the source and connects it to terminal 2. We seek to determine the voltage response of the capacitor $\upsilon(t)$ for $t \geq 0$.

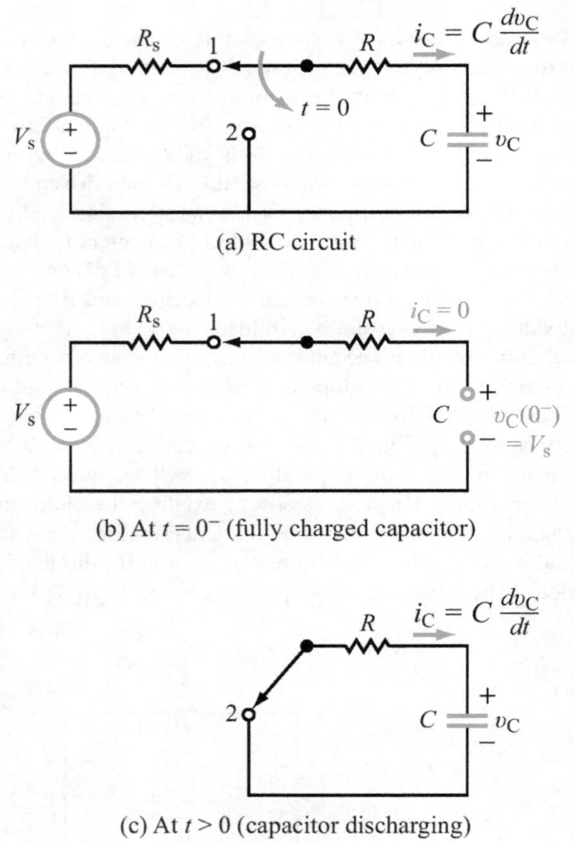

Figure 5-28: RC circuit with an initially charged capacitor that starts to discharge its energy after $t = 0$.

Before we start our solution, it is important to consider the implication of the information we are given about the state of the capacitor before and after moving the switch. For purposes of clarity, we define:

(a) $t = 0^-$ as the instant *just before the switch is moved* from terminal 1 to terminal 2, and

(b) $t = 0$ as the instant *just after it was moved; $t = 0$ is synonymous with $t = 0^+$.*

At $t = 0^-$, the circuit had been in the condition shown in Fig. 5-28(a) for a long time. As we noted earlier in Section 5-2.1, when a dc circuit is in a steady state, its capacitors act like open circuits. Consequently, the open circuit in Fig. 5-28(b), representing the state of the circuit at $t = 0^-$, allows no current to flow through the loop, and, therefore, there is no voltage drop across either of the two resistors. Hence, $\upsilon_C(0^-) = V_s$, and since the voltage across the capacitor cannot change instantaneously, it follows that $\upsilon_C(0)$, the voltage after moving the switch, is given by

$$\upsilon_C(0) = \upsilon_C(0^-) = V_s. \tag{5.67}$$

As we see shortly, we will need this piece of information for when we apply this initial condition to the solution of the differential equation of $\upsilon_C(t)$.

For $t \geq 0$, application of KVL to the loop in Fig. 5-28(c) gives

$$Ri_C + \upsilon_C = 0 \qquad (\text{for } t \geq 0), \tag{5.68}$$

where i_C is the current through and υ_C is the voltage across the capacitor. Since $i_C = C\, d\upsilon_C/dt$, Eq. (5.68) becomes

$$RC\,\frac{d\upsilon_C}{dt} + \upsilon_C = 0. \tag{5.69}$$

Upon dividing both terms by RC, Eq. (5.69) takes the form

$$\frac{d\upsilon_C}{dt} + a\upsilon_C = 0 \qquad \textbf{(source-free)}, \tag{5.70}$$

where

$$a = \frac{1}{RC}. \tag{5.71}$$

When arranging a differential equation in $\upsilon_C(t)$, it is customary to place all terms that involve $\upsilon_C(t)$ on the left-hand side of the equation and to place terms that do not involve $\upsilon_C(t)$ on the right-hand side. The term(s) on the right-hand side is (are) called the *forcing function*. For a circuit, the forcing function is related directly to the voltage and current sources in the circuit. Because the RC circuit in Fig. 5-28(c) does not contain any sources, Eq. (5.70) has a zero on its right-hand side and it is called (appropriately) a *source-free, first-order differential equation*.

> ▶ The solution of the source-free equation is called the *natural response* (discharging condition) of the circuit. ◀

The standard procedure for solving Eq. (5.70) starts by replacing t with dummy variable t' and multiplying both sides by $e^{at'}$,

$$\frac{d\upsilon_C}{dt'}\, e^{at'} + a\upsilon_C e^{at'} = 0. \qquad (5.72)$$

Next, we recognize that the sum of the two terms on the left-hand side is equal to the expansion of the differential of $(\upsilon_C e^{at'})$,

$$\frac{d}{dt'}(\upsilon_C e^{at'}) = \frac{d\upsilon_C}{dt'}\, e^{at'} + a\upsilon_C e^{at'}. \qquad (5.73)$$

Hence, Eq. (5.72) becomes

$$\frac{d}{dt'}(\upsilon_C e^{at'}) = 0. \qquad (5.74)$$

Integrating both sides, we have

$$\int_0^t \frac{d}{dt'}(\upsilon_C e^{at'})\, dt' = 0, \qquad (5.75)$$

where we have chosen the lower limit to be $t' = 0$ (because we are given specific information on the state of the circuit at that point in time). Performing the integration gives

$$\upsilon_C e^{at'}\Big|_0^t = 0$$

or

$$\upsilon_C(t)\, e^{at} - \upsilon_C(0) = 0. \qquad (5.76)$$

Solving for $\upsilon_C(t)$, we have

$$\upsilon_C(t) = \upsilon_C(0)\, e^{-at} = \upsilon_C(0)\, e^{-t/RC} \qquad (\text{for } t \geq 0), \quad (5.77)$$

where we used Eq. (5.71) for a and appended the inequality $t \geq 0$ to indicate that the expression given by Eq. (5.77) is valid only for $t \geq 0$.

The coefficient of t in the exponent is a critically important parameter, because it determines the temporal rate of $\upsilon_C(t)$. It is customary to rewrite Eq. (5.77) in the form

$$\upsilon_C(t) = \upsilon_C(0)\, e^{-t/\tau}, \qquad (5.78)$$

(natural response discharging),

with

$$\tau = RC \qquad (\text{s}), \qquad (5.79)$$

where τ is called the *time constant* of the circuit, and it is measured in seconds (s).

In view of the initial condition given by Eq. (5.67), namely $\upsilon_C(0) = V_s$, the expression for $\upsilon_C(t)$ becomes

$$\upsilon_C(t) = V_s e^{-t/\tau}\, u(t), \qquad (5.80)$$

where we inserted the unit step function $u(t)$ as a multiplication factor as a substitute for "for $t \geq 0$." The plot shown in Fig. 5-29(a) indicates that in response to the switch action, $\upsilon_C(t)$ decays exponentially with time from V_s at $t = 0$ down to its final value of zero as $t \to \infty$. The decay rate is dictated by the time constant τ. At $t = \tau$,

$$\upsilon_C(t = \tau) = V_s e^{-1} = 0.37 V_s, \qquad (5.81)$$

which means that at τ seconds after activating the switch, the voltage across the capacitor is down to 37 percent of its initial value. At $t = 2\tau$, it reaches 14 percent, and at $t = 5\tau$, it is less than 1 percent of its initial value. Hence, for all practical purposes, we can treat the circuit as having reached its final state when the switch has been in its new configuration for a time equal to or longer than 5τ.

> ▶ The magnitude of the time constant τ is a measure of how fast or how slowly a circuit responds to a sudden change. ◀

As we will see later in Section 5-7, the *clock speed* of a computer processor is, to first order, proportional to $1/\tau$. Hence, a *slow* circuit with $\tau = 1$ ms would have a clock speed on the order of 1 kHz, whereas a *fast* circuit with $\tau = 1$ ns can support clock speeds as high as 1 GHz.

The current $i_C(t)$ flowing through the capacitor is given by

$$i_C(t) = C\,\frac{d\upsilon_C}{dt} = C\,\frac{d}{dt}\,(V_s e^{-t/\tau})$$
$$= -C\,\frac{V_s}{\tau}\, e^{-t/\tau} \qquad (\text{for } t \geq 0), \qquad (5.82)$$

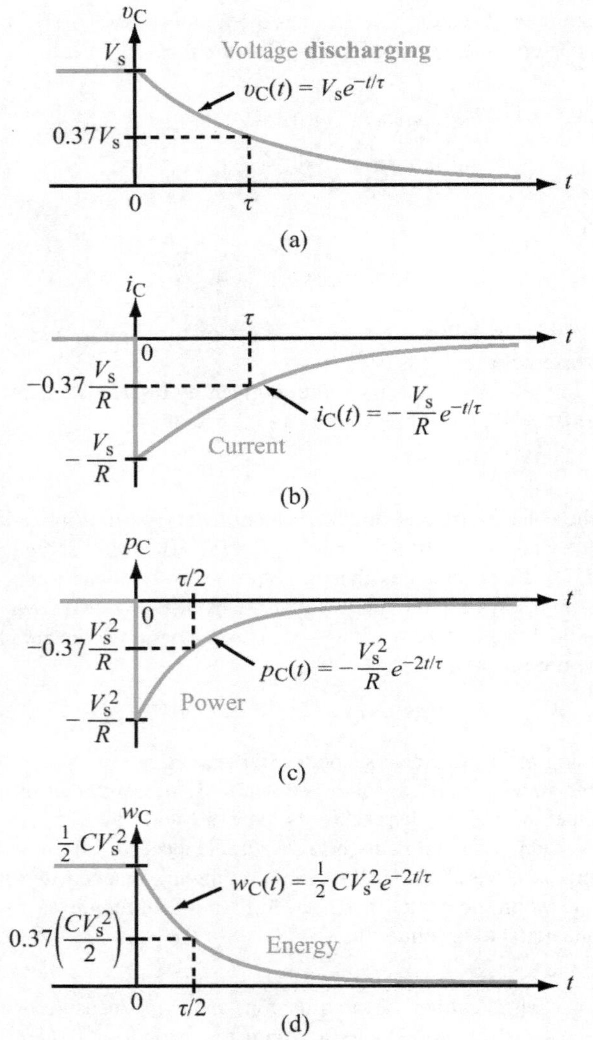

Figure 5-29: Response of the RC circuit in Fig. 5-28(a) to moving the SPDT switch to terminal 2.

which simplifies to

$$i_C(t) = -\frac{V_s}{R}\, e^{-t/\tau}\, u(t), \qquad (5.83)$$

(natural response discharging)

where, again, $u(t)$ is used to emphasize the fact that the expression is valid for only $t \geq 0$. The plot of $i_C(t)$ shown in Fig. 5-29(b) indicates that after closing the switch at $t = 0$, the current changes instantly to $(-V_s/R)$—as if the capacitor were a voltage source V_s—and then it decays exponentially down to zero. The negative sign of i signifies that it flows in a counterclockwise direction through the loop, consistent with the behavior of the capacitor as a voltage source.

Given $v_C(t)$ and $i_C(t)$, we can provide an expression for $p_C(t)$, the instantaneous power getting transferred to the capacitor, as

$$p_C(t) = i_C v_C = -\frac{V_s}{R}\, e^{-t/\tau} \times V_s e^{-t/\tau} = -\frac{V_s^2}{R}\, e^{-2t/\tau}\, u(t). \qquad (5.84)$$

Note that from the definition of $u(t)$ given by Eq. (5.2), $u(t) \cdot u(t) = u(t)$.

In general, power transfer is into a device if $p_C > 0$ and out of it if $p_C < 0$. Prior to $t = 0$, the capacitor had been connected to the voltage source for a long time. Hence, power already had flowed into the capacitor and was stored as electrical energy. The minus sign in Eq. (5.84) denotes that after $t = 0$ power flows out of the capacitor and gets dissipated in the resistor.

▶ The decay rate for $p_C(t)$ is $2/\tau$, which is twice as fast as that for $v_C(t)$ or $i_C(t)$. ◀

The amount of energy $w_C(t)$ contained in the medium between the capacitor's oppositely charged conducting plates can be calculated either by integrating $p_C(t)$ over time from 0 to t or by applying Eq. (5.29). The latter approach gives

$$w_C(t) = \frac{1}{2}\, C\, v_C^2(t) = \frac{CV_s^2}{2}\, e^{-2t/\tau}\, u(t). \qquad (5.85)$$

Parts (c) and (d) of Fig. 5-29 display the time waveforms of $p_C(t)$ and $w_C(t)$, respectively.

Concept Question 5-17: What specific characteristic defines a first-order circuit? (See CAD)

Concept Question 5-18: What does the time constant of an RC circuit represent? Would a larger capacitor discharge faster or more slowly than a small one? (See CAD)

Concept Question 5-19: For the natural response of an RC circuit, how does the decay rate for voltage compare with that for power? (See CAD)

Exercise 5-14: If in the circuit of Fig. E5.14 $\upsilon_C(0^-) = 24$ V, determine $\upsilon_C(t)$ for $t \geq 0$.

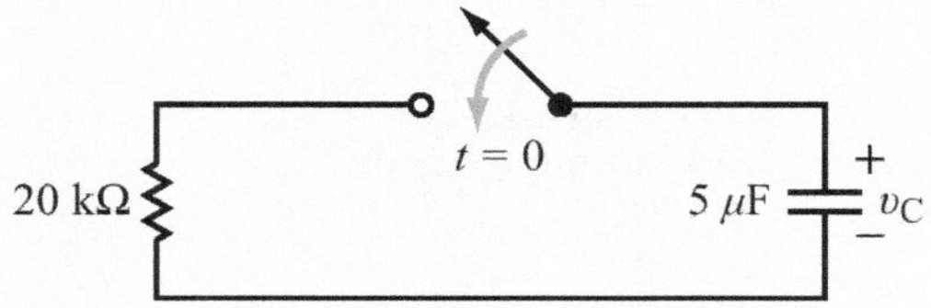

Figure E5.14

Answer: $\upsilon_C(t) = 24e^{-10t}$ V for $t \geq 0$. (See CAD)

5-4.2 General Form of the Step Response of the RC Circuit

When we use the term *circuit response*, we mean the reaction of a certain voltage or current in the circuit to change, such as the introduction of a new source, the elimination of a source, or some other change in the circuit configuration. Whenever possible, we usually designate $t = 0$ as the instant at which the change occurred and $t \geq 0$ as the time interval over which we seek the circuit response. In the general case, the capacitor may start with a voltage $\upsilon_C(0)$ at $t = 0$ (immediately after the sudden change) and may approach a value denoted $\upsilon_C(\infty)$ as $t \to \infty$. A circuit configuration that can represent such a scenario is the series RC circuit shown in Fig. 5-30(a). Prior to $t = 0$, the RC circuit is connected to a source V_{s_1}, and after $t = 0$, it is connected to a different source V_{s_2}. The circuit can be reduced to the following special cases:

- *Step response* (due to V_{s_2}) of an *uncharged capacitor* (if $V_{s_1} = 0$)
- *Step response* (due to V_{s_2}) of a *charged capacitor* (if $V_{s_1} \neq 0$)
- *Natural response* (if $V_{s_2} = 0$) of a *charged capacitor* ($V_{s_1} \neq 0$)

For obvious reasons, we excluded the trivial case where both V_{s_1} and V_{s_2} are zero, and we will now treat the general case where neither V_{s_1} nor V_{s_2} is zero.

At $t = 0^-$ (Fig. 5-30(b)), the capacitor has been in steady state for a long time. Hence, it acts like an open circuit. Consequently, $i_C(0^-) = 0$, and $\upsilon_C(0^-) = V_{s_1}$. Since υ_C across the capacitor cannot change in zero time, the (initial condition) voltage $\upsilon_C(0)$ after moving the switch to terminal 2 is

$$\upsilon_C(0) = \upsilon_C(0^-) = V_{s_1}. \tag{5.86}$$

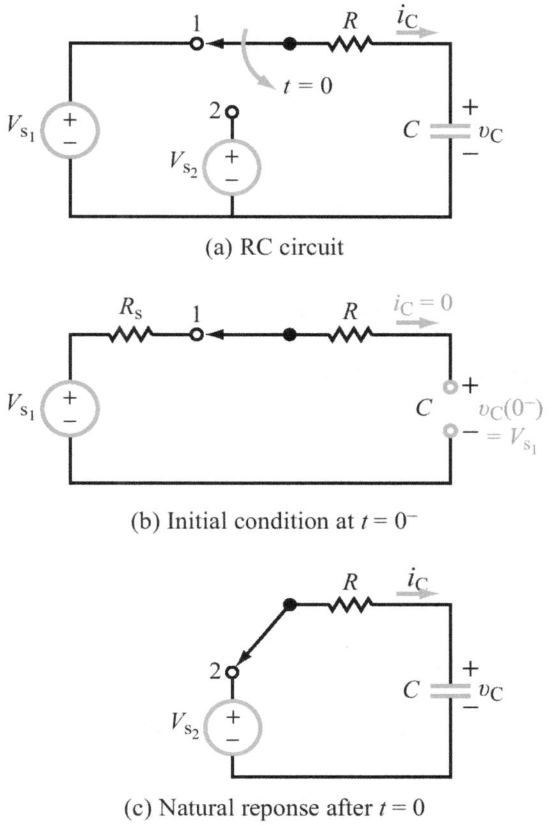

Figure 5-30: RC circuit switched from source V_{s_1} to source V_{s_2} at $t = 0$.

For $t \geq 0$, the (natural response) voltage equation for the loop in Fig. 5-30(c) is

$$-V_{s_2} + i_C R + \upsilon_C = 0. \tag{5.87}$$

Upon using $i_C = C\, d\upsilon_C/dt$ and rearranging its terms, Eq. (5.87) can be written in the differential-equation form

$$\frac{d\upsilon_C}{dt} + a\upsilon_C = b, \tag{5.88}$$

where

$$a = \frac{1}{RC} \quad \text{and} \quad b = \frac{V_{s_2}}{RC}. \tag{5.89}$$

We note that Eq. (5.88) is similar to Eq. (5.70), except that now we have a non-zero term on the right-hand side of the equation. Nevertheless, the method of solution remains the same. After replacing t with dummy variable t' and multiplying both sides of Eq. (5.88) by $e^{at'}$, we have

$$e^{at'} \frac{d\upsilon_C}{dt'} + a\upsilon_C e^{at'} = be^{at'}. \tag{5.90a}$$

In view of Eq. (5.73), Eq. (5.90a) can be rewritten as

$$\frac{d}{dt'}(\upsilon_C e^{at'}) = be^{at'}. \tag{5.90b}$$

Integrating both sides from $t' = 0$ to $t' = t$, namely

$$\int_0^t \frac{d}{dt'}(\upsilon_C e^{at'})\, dt' = \int_0^t be^{at'}\, dt' \tag{5.91}$$

gives

$$\upsilon_C e^{at'}\Big|_0^t = \frac{b}{a} e^{at'}\Big|_0^t. \tag{5.92}$$

Upon evaluating the functions at the two limits, we have

$$\upsilon_C(t)\, e^{at} - \upsilon_C(0) = \frac{b}{a} e^{at} - \frac{b}{a}, \tag{5.93}$$

and then solving for $\upsilon_C(t)$, we have

$$\upsilon_C(t) = \upsilon_C(0)\, e^{-at} + \frac{b}{a}(1 - e^{-at}). \tag{5.94}$$

As $t \to \infty$, $e^{-\infty} = 0$ and $\upsilon_C(t)$ reduces to the *final condition*

$$\upsilon_C(\infty) = \frac{b}{a} = V_{s_2}. \tag{5.95}$$

By reintroducing the time constant $\tau = RC = 1/a$ and replacing b/a with $\upsilon_C(\infty)$, we can rewrite Eq. (5.94) in the general form:

$$\upsilon_C(t) = \left\{\upsilon_C(\infty) + [\upsilon_C(0) - \upsilon_C(\infty)]e^{-t/\tau}\right\} u(t).$$

(series RC circuit with switch action at $t = 0$)

(5.96)

> ► The voltage response of any RC circuit is determined by three parameters: the initial voltage $\upsilon_C(0)$, the final voltage $\upsilon_C(\infty)$, and the time constant τ. ◄

For the specific circuit in Fig. 5-30(a), Eqs. (5.86) and (5.95) give $\upsilon_C(0) = V_{s_1}$ and $\upsilon_C(\infty) = V_{s_2}$. Hence,

$$\upsilon_C(t) = V_{s_2} + (V_{s_1} - V_{s_2})e^{-t/\tau}. \tag{5.97}$$

If the switch action causing the change in voltage across the capacitor occurs at time T_0 instead of at $t = 0$, Eq. (5.96) assumes the form

$$\upsilon_C(t) = \left\{\upsilon_C(\infty) + [\upsilon_C(T_0) - \upsilon_C(\infty)]e^{-(t-T_0)/\tau}\right\} \cdot u(t - T_0), \tag{5.98}$$

(series RC circuit with switch action at $t = T_0$)

where we have replaced t with $(t - T_0)$ on the right-hand side of Eq. (5.96). Now $\upsilon_C(T_0)$ is the initial voltage at $t = T_0$. For easy reference, this expression is made available in Table 5-5, along with expressions for three other types of circuits discussed in future sections.

Series RC Circuit Solution

1: If switch action is at $t = 0$, analyze circuit at $t = 0^-$ to determine initial conditions $\upsilon_C(0^-)$ and $i_C(0^-)$. Use this information to determine $\upsilon_C(0)$ and $i_C(0)$, at t immediately after the switch action. Remember that the voltage across a capacitor cannot change instantaneously (between $t = 0^-$ and $t = 0$), but the current can.

2: Analyze the circuit to determine $\upsilon_C(\infty)$, the voltage across the capacitor long after the switch action.

3: Determine the time constant $\tau = RC$.

4: Incorporate the information obtained in the previous three steps in Eq. (5.96):

$$\upsilon_C(t) = \left\{\upsilon_C(\infty) + [\upsilon_C(0) - \upsilon_C(\infty)]e^{-t/\tau}\right\} u(t).$$

5: If the switch action is at $t = T_0$ instead of $t = 0$, replace 0 with T_0 and use Eq. (5.98):

$$\upsilon_C(t) = \left\{\upsilon_C(\infty) + [\upsilon_C(T_0) - \upsilon_C(\infty)] \cdot e^{-(t-T_0)/\tau}\right\} \cdot u(t - T_0).$$

5-4.3 Thévenin Approach

For a circuit containing dc sources, resistors, switches and a single capacitor (or multiple capacitors that can be combined into a single equivalent capacitor), the voltage response across the capacitor, $v_C(t)$, can be calculated with relative ease by taking advantage of the Thévenin theorem. The procedure involves the following steps:

Thévenin Approach to RC Response

Step 1: If the circuit includes a single switch action (open, close, or move between two terminals) at $t = T_0$, analyze the circuit at $t = T_0^-$ (just before the switch action) to determine $v_C(T_0^-)$. When so doing, the capacitor should be replaced with an open circuit. Then set $v_C(T_0) = v_C(T_0^-)$, where $v_C(T_0)$ is the voltage across the capacitor *after* the switch action.

Step 2: For the circuit configuration at $t \geq T_0$ (after the switch action), obtain the Thévenin equivalent circuit as "seen" by the capacitor. Figure 5-31(a) depicts a general circuit (composed of possibly two subcircuits) connected to a capacitor C. After removing (temporarily) the capacitor and calculating V_{Th} and R_{Th} of the equivalent Thévenin circuit at terminals (a, b), reinstate the capacitor as in Fig. 5-31(b).

Step 3: The capacitor's voltage response is then given by

$$v_C(t) = \left\{ v_C(\infty) + [v_C(T_0) - v_C(\infty)]e^{-(t-T_0)/\tau} \right\} \cdot u(t - T_0),$$

with $v_C(\infty) = V_{Th}$, $v_C(T_0)$ as obtained in step 1, and $\tau = R_{Th}C$.

Step 4: If the circuit undergoes multiple switch actions, repeat the procedure for each time segment and use the property that the voltage across a capacitor cannot change instantaneously to match the responses at the boundaries between adjacent time segments.

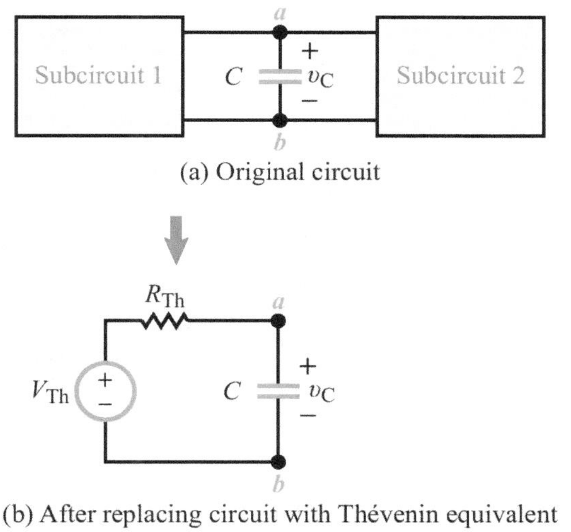

Figure 5-31: Replacing a resistive circuit with its Thévenin equivalent as seen by capacitor C.

Example 5-9: Thévenin Approach

The switch in the circuit of Fig. 5-32(a) had been in position 1 for a long time until it was moved to position 2 at $t = 0$. Determine $v_C(t)$ for $t \geq 0$.

Solution:

Step 1: Figure 5-32(b) depicts the state of the circuit at $t = 0^-$ (initial condition), with the capacitor represented by an open circuit. Because of the open circuit, $i = 0$ in the left-hand side of the circuit. Hence, no voltage drop occurs across the 3 kΩ resistor. Consequently, the voltage at node V_1, relative to the designated ground node, is

$$V_1 = 24 \text{ V}.$$

On the right-hand side of the circuit, the current source flows entirely through the 4 kΩ resistor, generating a node voltage

$$V_2 = 4.5 \times 10^{-3} \times 4 \times 10^3 = 18 \text{ V}.$$

Hence, the initial voltage is

$$v_C(0^-) = V_1 - V_2 = 24 - 18 = 6 \text{ V}.$$

Since the voltage across the capacitor cannot change instantaneously, it follows that

$$v_C(0) = v_C(0^-) = 6 \text{ V}.$$

Step 2: Figure 5-32(c) represents the state of the circuit after moving the switch to position 2 and removing the capacitor so as to calculate the elements of the Thévenin circuit at terminals (a, b). In step (d), conversion of the current source and 4 kΩ

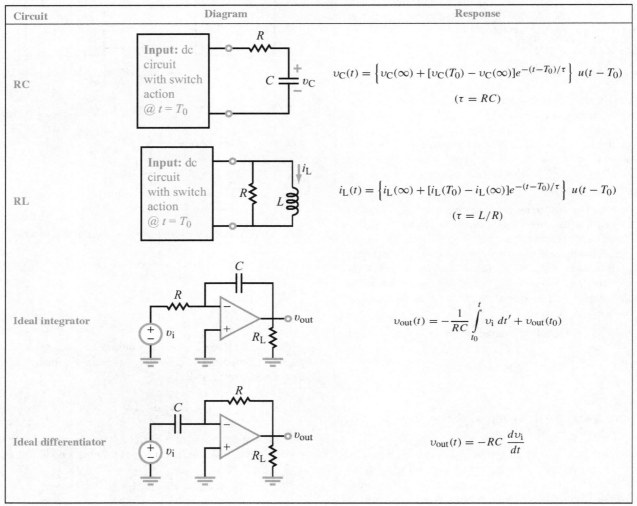

Table 5-5: Response forms of basic first-order circuits.

Circuit	Diagram	Response
RC	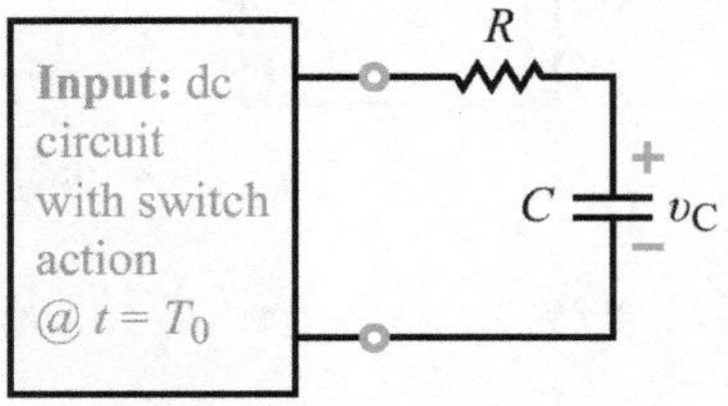	$v_C(t) = \left\{ v_C(\infty) + [v_C(T_0) - v_C(\infty)]e^{-(t-T_0)/\tau} \right\} u(t - T_0)$ $(\tau = RC)$
RL	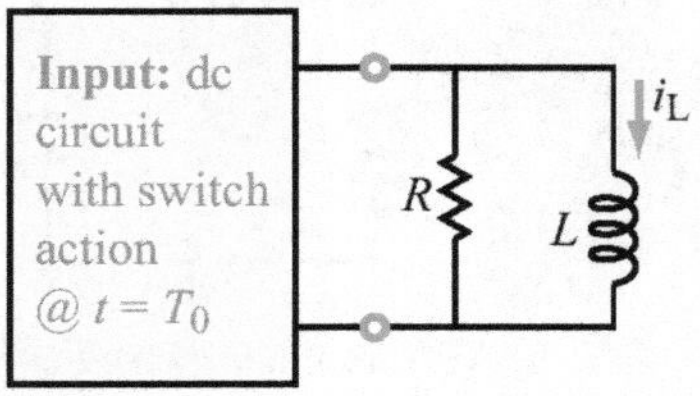	$i_L(t) = \left\{ i_L(\infty) + [i_L(T_0) - i_L(\infty)]e^{-(t-T_0)/\tau} \right\} u(t - T_0)$ $(\tau = L/R)$
Ideal integrator	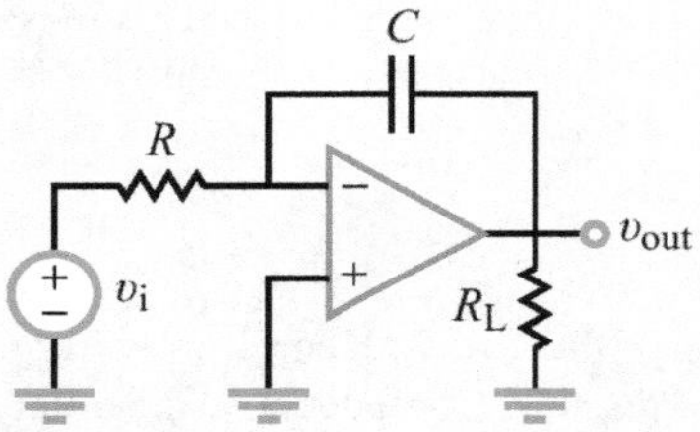	$v_{\text{out}}(t) = -\dfrac{1}{RC} \int_{t_0}^{t} v_i \, dt' + v_{\text{out}}(t_0)$
Ideal differentiator	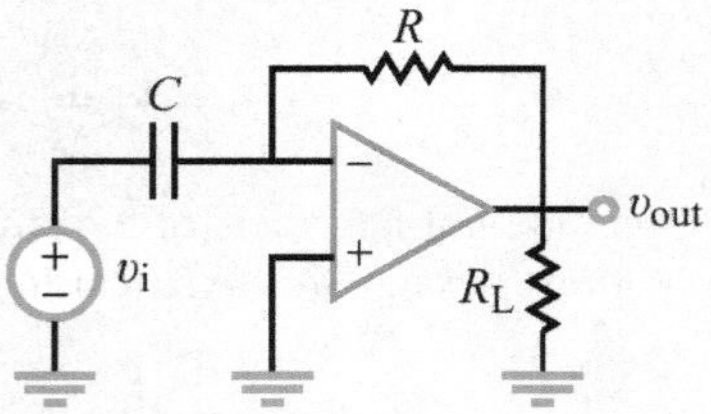	$v_{\text{out}}(t) = -RC \dfrac{dv_i}{dt}$

resistor into a voltage source in series with a resistor leads to

$$R_{\text{Th}} = 4 \text{ k}\Omega + 1 \text{ k}\Omega = 5 \text{ k}\Omega,$$

$$V_{\text{Th}} = -4.5 \times 10^{-3} \times 4 \times 10^{3} = -18 \text{ V}.$$

► Note that the polarity of the Thévenin voltage source has to be assigned to match that of v_C, the voltage across the capacitor. In the present case, the current to voltage transformation led to a voltage source with the opposite polarity to that defined for V_{Th}. Hence, $V_{\text{Th}} = -18$ V, not 18 V. ◄

Step 3: The capacitor is reinserted in part (e). With $v_C(0) = 6$ V, $v_C(\infty) = V_{\text{Th}} = -18$ V, and

$$\tau = R_{\text{Th}}C = 5 \times 10^{3} \times 100 \times 10^{-6} = 0.5 \text{ s},$$

we have

$$v_C(t) = \left\{ v_C(\infty) + [v_C(0) - v_C(\infty)]e^{-t/\tau} \right\} u(t)$$
$$= [-18 + 24e^{-2t}] \, u(t) \text{ V}.$$

This solution indicates that at $t = 0$, the initial voltage across the capacitor is $v_C(0) = -18 + 24 = 6$ V, which is consistent with the result obtained in step 1. After a long time t such that e^{-2t} approaches zero, $v_C(t)$ approaches -18 V, which is $v_C(\infty)$. In between, the capacitor discharges to zero and

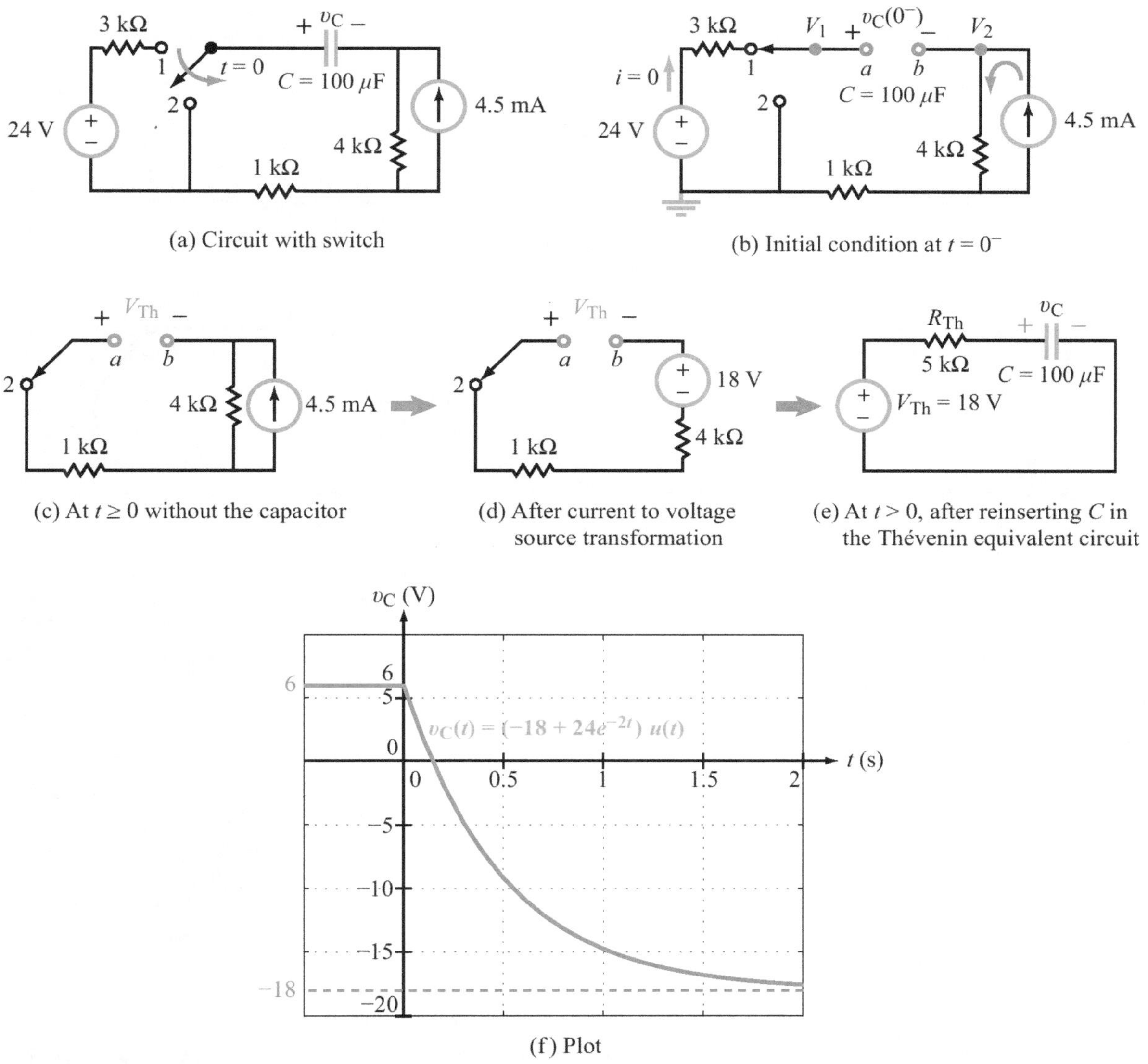

Figure 5-32: Circuit for Example 5-9.

then builds up charge again, but of opposite polarity. The time variation of $v_C(t)$ is displayed in Fig. 5-32(f).

Example 5-10: Switching between Two Sources

In the circuit of Fig. 5-33(a), the SPDT switch is moved from position 1 to position 2 after it had been in position 1 for a long time. Determine the voltage $v_C(t)$ for $t \geq 0$ if the switch is moved at (a) $t = 0$ or (b) $t = 3$ s.

Solution: (a) For $T_0 = 0$ and $t \geq 0$, the complete solution of $v_C(t)$ is given by Eq. (5.96) as

$$v_C(t) = \left\{v_C(\infty) + [v_C(0) - v_C(\infty)]e^{-t/\tau}\right\}\ u(t). \qquad (5.99)$$

We need to determine three quantities: the initial voltage $v_C(0)$, the final voltage $v_C(\infty)$, and the time constant τ.

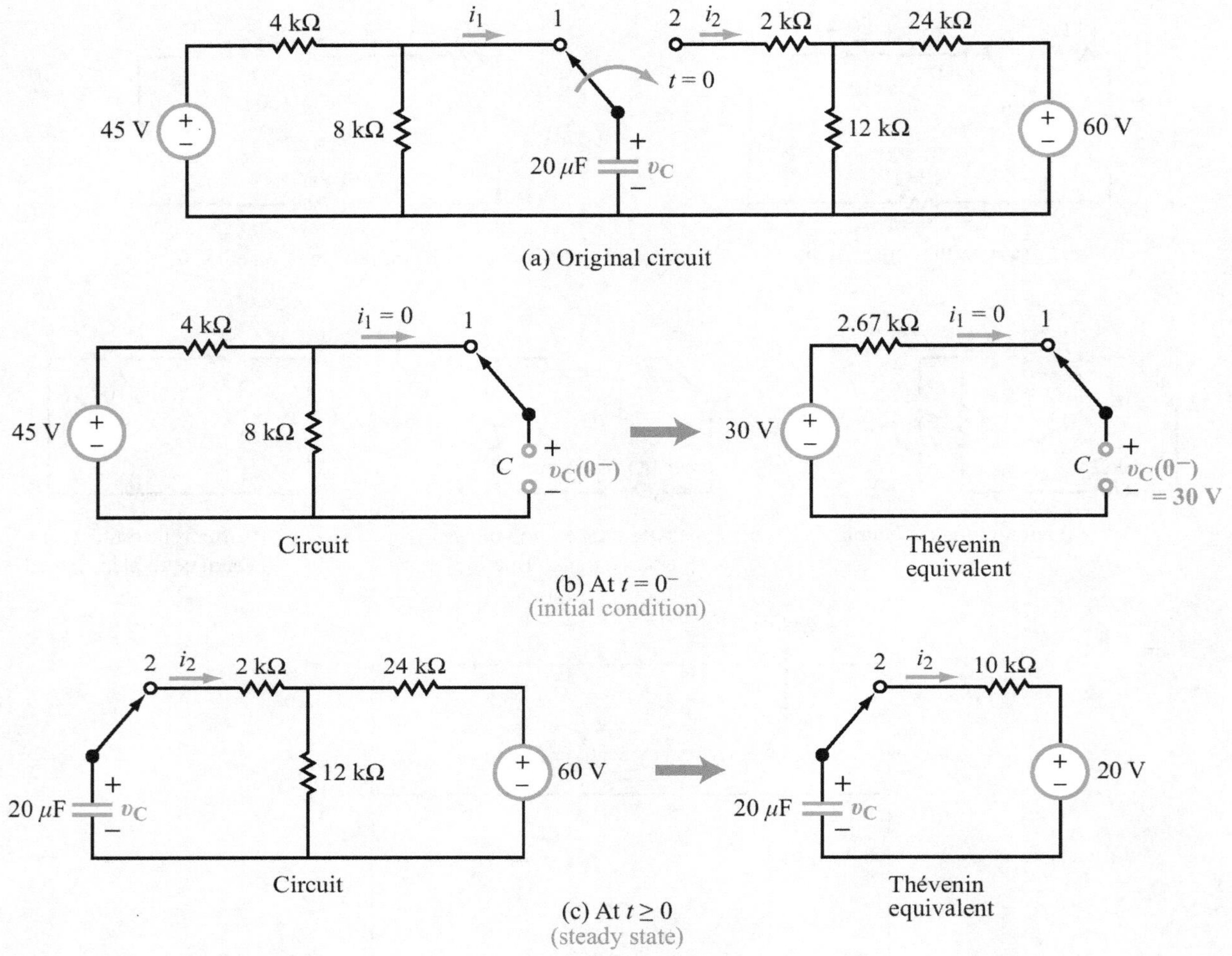

Figure 5-33: Circuit for Example 5-10 [part (a)].

The initial voltage is the voltage that existed across the capacitor before moving the switch. Since the switch had been in that position for a long time, we presume that the circuit in Fig. 5-33(b) had reached its steady-state condition long before the switch was moved. Hence, at $t = 0^-$ (just before moving the switch), the capacitor behaves like an open circuit. The voltage $v_C(0^-)$ across the capacitor is the same as that across the 8 kΩ resistor, and since $i_1 = 0$ at $t = 0^-$, application of voltage division yields

$$v_C(0^-) = \left(\frac{8\text{k}}{4\text{k} + 8\text{k}}\right) \times 45 = 30 \text{ V}.$$

Incidentally, we could have obtained the same result by transforming the circuit in Fig. 5-33(b) into its Thévenin equivalent.

Incorporating the constraint that the voltage across the capacitor cannot change instantaneously, it follows that

$$v_C(0) = v_C(0^-) = 30 \text{ V}.$$

Now we turn our attention to finding $v_C(\infty)$. After moving the switch to position 2 (Fig. 5-33(c)) and allowing the circuit sufficient time to reach its final state, the capacitor again will behave like on open circuit, which means that $i_2 = 0$ at $t = \infty$. Voltage division gives

$$v_C(\infty) = \left(\frac{12k}{12k + 24k}\right) \times 60 = 20 \text{ V}.$$

The time constant of the circuit to the right of terminal 2 is given by $\tau = RC$, with R being the Thévenin resistance of that circuit. After suppressing (short-circuiting) the 60 V source, we get

$$R = R_{Th} = 2 \text{ k}\Omega + 12 \text{ k}\Omega \parallel 24 \text{ k}\Omega$$
$$= 2 \text{ k}\Omega + \frac{12k \times 24k}{12k + 24k} = 10 \text{ k}\Omega.$$

Hence,

$$\tau = RC = 10 \times 10^3 \times 20 \times 10^{-6} = 0.2 \text{ s}.$$

Substituting the values we obtained for $v_C(0)$, $v_C(\infty)$, and τ in Eq. (5.99) leads to

$$v_C(t) = [(20 + 10e^{-5t})\, u(t)] \text{ V}.$$

(b) This is a repetition of the previous case except that now the switch action takes place at $T_0 = 3$ s. The applicable expression is given by Eq. (5.98),

$$v_C(t) = \left\{v_C(\infty) + [v_C(3) - v_C(\infty)]e^{-(t-3)/\tau}\right\} u(t-3).$$

Of course, $v_C(t) = 30$ V before $t = 3$ s. Hence, for the specified time duration $t \geq 0$,

$$v_C(t) = \begin{cases} 30 \text{ V} & \text{for } 0 \leq t \leq 3 \text{ s}, \\ [20 + 10e^{-5(t-3)}] \text{ V} & \text{for } t \geq 3 \text{ s}. \end{cases}$$

Example 5-11: Charge/Discharge Action

Given that the switch in Fig. 5-34 was moved to position 2 at $t = 0$ (after it had been in position 1 for a long time) and then returned to position 1 at $t = 10$ s, determine the voltage response $v_C(t)$ for $t \geq 0$ and evaluate it for $V_1 = 20$ V, $R_1 = 80$ kΩ, $R_2 = 20$ kΩ, and $C = 0.25$ mF.

(a) Actual circuit

(b) Circuit during $0 \leq t \leq 10$ s

(c) Circuit after $t = 10$ s

(d) Voltage response

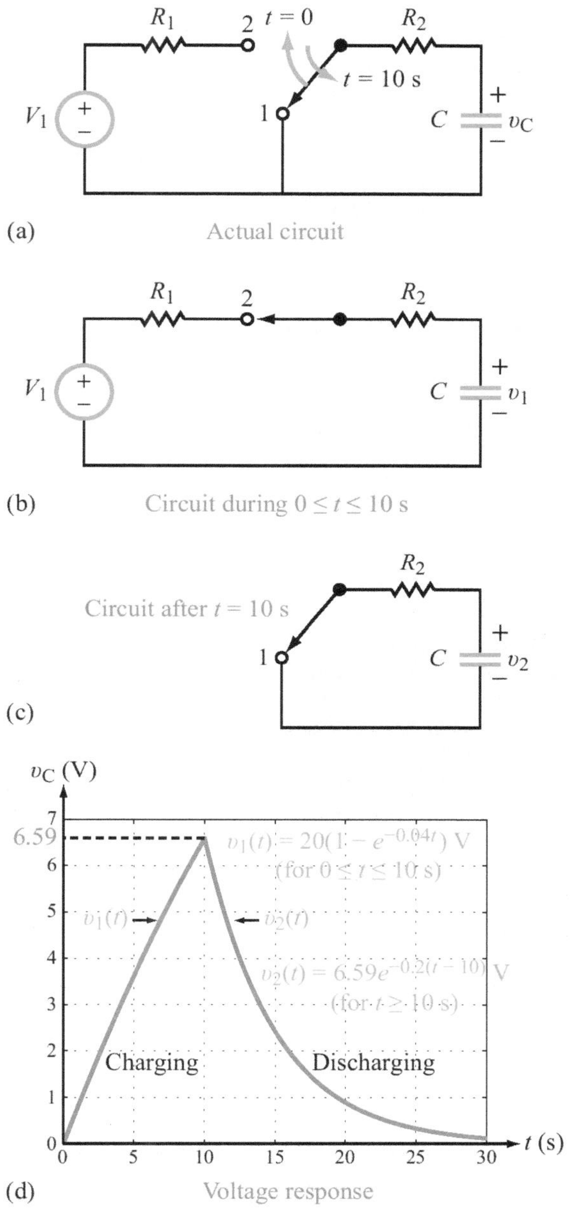

Figure 5-34: After having been in position 1 for a long time, the switch is moved to position 2 at $t = 0$ and then returned to position 1 at $t = 10$ s (Example 5-11).

Solution: We will divide our solution into two time segments: $v_C = v_1(t)$ for $0 \leq t \leq 10$ s and $v_C = v_2(t)$ for $t \geq 10$ s.

Time Segment 1: $0 \le t \le 10$ s

When the switch is in position 2 (Fig. 5-34(b)), the resistance of the circuit is $R = R_1 + R_2$. Hence, the time constant during this first time segment is

$$\tau_1 = (R_1 + R_2)C = (80 + 20) \times 10^3 \times 0.25 \times 10^{-3} = 25 \text{ s}.$$

Application of Eq. (5.96) with $\upsilon_1(0) = 0$ (the capacitor had no charge prior to $t = 0$), $\upsilon_1(\infty) = V_1 = 20$ V, and $\tau_1 = 25$ s leads to

$$\begin{aligned}\upsilon_1(t) &= \upsilon_1(\infty) + [\upsilon_1(0) - \upsilon_1(\infty)]e^{-t/\tau_1}\\ &= 20(1 - e^{-0.04t}) \quad \text{V} \quad (\text{for } 0 \le t \le 10 \text{ s}).\end{aligned}$$

Time Segment 2: $t \ge 10$ s

Voltage $\upsilon_2(t)$, corresponding to the second time segment (Fig. 5-34(c)), is given by Eq. (5.98) with a new time constant τ_2 as

$$\upsilon_2(t) = \upsilon_2(\infty) + [\upsilon_2(10) - \upsilon_2(\infty)]e^{-(t-10)/\tau_2}.$$

The new time constant is associated with the capacitor circuit after returning the switch to position 1,

$$\tau_2 = R_2 C = 20 \times 10^3 \times 0.25 \times 10^{-3} = 5 \text{ s}.$$

The initial voltage $\upsilon_2(10)$ is equal to the capacitor voltage υ_1 at the end of time segment 1, namely

$$\upsilon_2(10) = \upsilon_1(10) = 20(1 - e^{-0.04\times 10}) = 6.59 \text{ V}.$$

With no voltage source present in the R_2C circuit, the charged capacitor will dissipate its energy into R_2, exhibiting a ***natural response*** with a final voltage of $\upsilon_2(\infty) = 0$. Consequently,

$$\begin{aligned}\upsilon_2(t) &= \upsilon_2(10)\, e^{-(t-10)/\tau_2}\\ &= 6.59e^{-0.2(t-10)} \text{ V} \quad (\text{for } t \ge 10 \text{ s}).\end{aligned}$$

The complete time response of $\upsilon(t)$ is displayed in Fig. 5-34(d).

Example 5-12: RC-Circuit Response to Rectangular Pulse

Determine the voltage response of a previously uncharged RC circuit to a rectangular pulse $\upsilon_i(t)$ of amplitude V_s and duration T_0, as depicted in Fig. 5-35(a). Evaluate and plot the response for $R = 25$ kΩ, $C = 0.2$ mF, $V_s = 10$ V, and $T_0 = 4$ s.

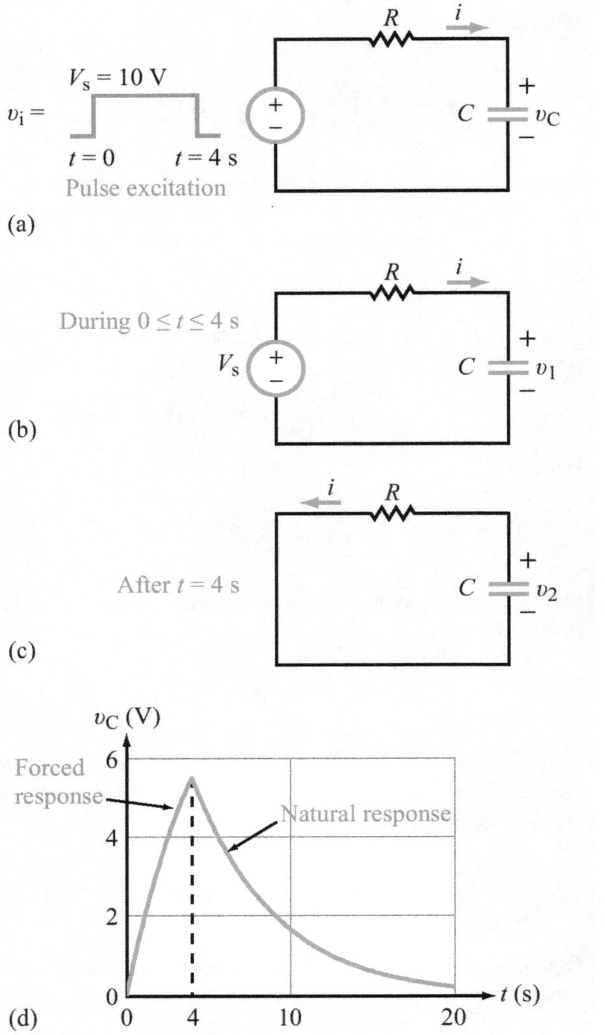

Figure 5-35: RC-circuit response to a 4 s long rectangular pulse.

Solution: According to Example 5-2, a rectangular pulse is equivalent to the sum of two step functions. Thus

$$\upsilon_i(t) = V_s[u(t - T_1) - u(t - T_2)],$$

where $u(t - T_1)$ accounts for the rise in level from 0 to 1 at $t = T_1$ and the second term (with negative amplitude) serves to counteract (cancel) the first term after $t = T_2$. For the present problem, $T_1 = 0$, and $T_2 = 4$ s. Hence, the input pulse can be written as

$$\upsilon_i(t) = V_s\, u(t) - V_s\, u(t - 4).$$

Since the circuit is linear, we can apply the superposition theorem to determine the capacitor response $\upsilon_C(t)$. Thus,

$$\upsilon_C(t) = \upsilon_1(t) + \upsilon_2(t),$$

where $\upsilon_1(t)$ is the response to $V_s\, u(t)$ acting alone and, similarly, $\upsilon_2(t)$ is the response to $-V_s\, u(t-4)$ also acting alone.

Response to $V_s\, u(t)$ alone

The response $\upsilon_1(t)$ is given by Eq. (5.96) with $\upsilon_1(0) = 0$, $\upsilon_1(\infty) = V_s$, and $\tau = RC$. Hence,

$$\begin{aligned}\upsilon_1(t) &= \upsilon_1(\infty) + [\upsilon_1(0) - \upsilon_1(\infty)]e^{-t/\tau}\\ &= V_s(1 - e^{-t/\tau}) \qquad (\text{for } t \geq 0).\end{aligned}$$

For $V_s = 10$ V and $\tau = RC = 25 \times 10^3 \times 0.2 \times 10^{-3} = 5$ s,

$$\upsilon_1(t) = 10(1 - e^{-0.2t}) \text{ V} \qquad (\text{for } t \geq 0).$$

Response to $-V_s\, u(t-4)$ alone

The second step function has an amplitude of $-V_s$ and is delayed in time by 4 s. Upon reversing the polarity of V_s and replacing t with $(t-4)$, we have

$$\upsilon_2(t) = -10[1 - e^{-0.2(t-4)}] \text{ V} \qquad (\text{for } t \geq 4 \text{ s}).$$

Total response

The total response for $t \geq 0$ therefore is given by

$$\begin{aligned}\upsilon_C(t) &= \upsilon_1(t) + \upsilon_2(t)\\ &= 10[1 - e^{-0.2t}] - 10[1 - e^{-0.2(t-4)}]\, u(t-4) \text{ V},\end{aligned} \tag{5.100}$$

where we introduced the time-shifted step function $u(t-4)$ to assert that the second term is zero for $t \leq 4$ s. The plot of $\upsilon_C(t)$ displayed in Fig. 5-35(d) shows that $\upsilon_C(t)$ builds up to a maximum of 5.5 V by the end of the pulse (at $t = 4$ s) and then decays exponentially back to zero thereafter. The build-up part is due to the external excitation and often is called the *forced response*. In contrast, during the time period after $t = 4$ s, $\upsilon_C(t)$ exhibits a *natural decay response* as the capacitor discharges its energy into the resistor. During this latter time segment, $i(t)$ flows in a counterclockwise direction.

Concept Question 5-20: What are the three quantities needed to establish $\upsilon_C(t)$ across a capacitor in an RC circuit? (See CAD)

Concept Question 5-21: If $V_{s_2} < V_{s_1}$ in the circuit of Fig. 5-30, what would you expect the direction of the current to be after the switch is moved from position 1 to 2? Analyze the process in terms of charge accumulation on the capacitor. (See CAD)

Exercise 5-15: Determine $\upsilon_1(t)$ and $\upsilon_2(t)$ for $t \geq 0$, given that in the circuit of Fig. E5.15 $C_1 = 6\ \mu$F, $C_2 = 3\ \mu$F, $R = 100$ kΩ, and neither capacitor had any charge prior to $t = 0$.

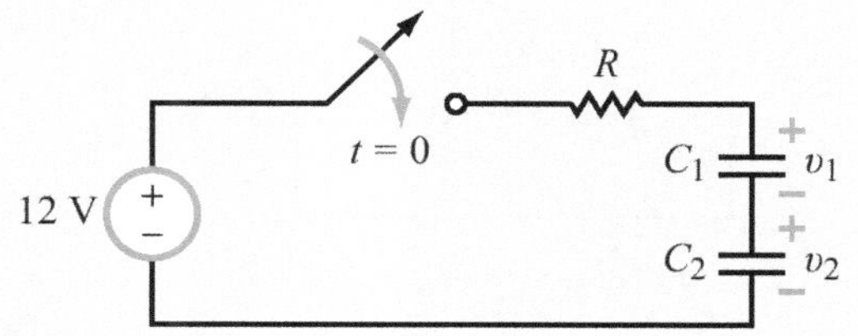

Figure E5.15

Answer: $\upsilon_1(t) = 4(1 - e^{-5t})$ V, for $t \geq 0$, $\upsilon_2(t) = 8(1 - e^{-5t})$ V, for $t \geq 0$. (See CAD)

5-5 Response of the RL Circuit

With series RC circuits, we developed a first-order differential equation for $\upsilon_C(t)$, the voltage across the capacitor, and then we solved it (subject to initial and final conditions) to obtain a complete expression for $\upsilon_C(t)$. By applying $i_C = C\, d\upsilon_C/dt$, $p_C = i_C \upsilon_C$, and $w_C = \frac{1}{2} C\upsilon_C^2$, we were able to determine the corresponding current passing through the capacitor, the power getting transferred to it, and the net energy stored in it. We now follow an analogous procedure for the parallel RL circuit, but our analysis will focus on the current $i(t)$ through the inductor, instead of on the voltage across it.

5-5.1 Natural Response of the RL Circuit

After having been in the closed position for a long time, the switch in the RL circuit of Fig. 5-36(a) was moved to position 2 at $t = 0$, thereby disconnecting the RL circuit from the current source I_s. What happens to the current i flowing through the inductor after the sudden change caused by moving the switch? That is, what is the waveform of $i_L(t)$ for $t \geq 0$? To answer this question, we first note that at $t = 0^-$ (just before moving the switch), the RL circuit can be represented by the circuit in Fig. 5-36(b), in which the inductor has been

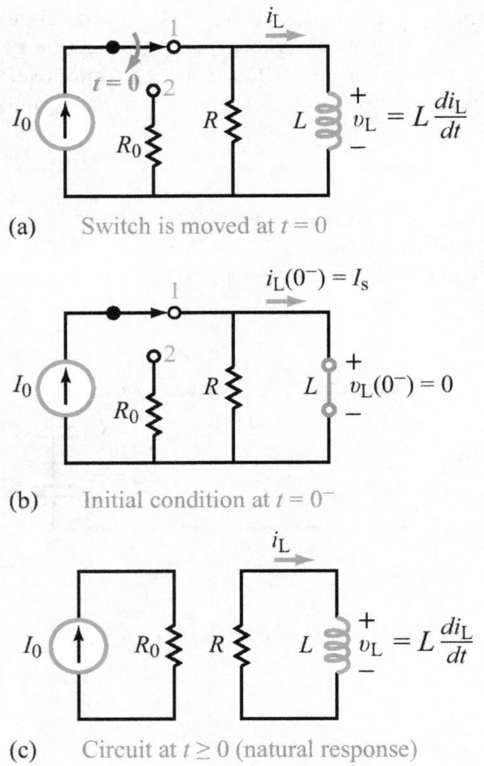

Figure 5-36: RL circuit disconnected from a current source at $t = 0$.

replaced with a short circuit. This is because under steady-state conditions i_L no longer changes with time, which leads to $v_L = L\,di_L/dt = 0$. We also know that the current will take the path of least resistance through the short circuit. A current source entering a node connected to another node via a parallel combination of a resistor R and a short circuit will flow entirely through the short circuit. Hence, $i_L(0^-) = I_s$. Moreover, since the current through an inductor cannot change instantaneously, the initial current at $t = 0$ (after moving the switch) has to be

$$i_L(0) = i_L(0^-) = I_s.$$

For the time period $t \geq 0$, the loop equation for the RL circuit in Fig. 5-36(c) is given by

$$Ri_L + L\,\frac{di_L}{dt} = 0,$$

which can be cast in the form

$$\frac{di_L}{dt} + ai_L = 0, \qquad (5.101)$$

where a is a temporary constant given by

$$a = \frac{R}{L}. \qquad (5.102)$$

The form of Eq. (5.101) is identical to that of Eq. (5.70) for the source-free RC circuit, except that now the variable is $i_L(t)$, whereas then it was $v_L(t)$. By analogy with the solution given by Eq. (5.78), our solution for $i_L(t)$ is given by

$$i_L(t) = i_L(0)\, e^{-t/\tau}\, u(t), \qquad (5.103)$$

(natural response discharging)

where for the RL circuit, the *time constant* is given by

$$\tau = \frac{1}{a} = \frac{L}{R}. \qquad (5.104)$$

5-5.2 General Form of the Step Response of the RL Circuit

To generalize our solution to the case where the RL circuit may contain sources both before and after the sudden change in the circuit configuration, we adopt the basic circuit shown in Fig. 5-37(a) in which two switches are moved simultaneously at $t = 0$ so as to switch the RL circuit from current source I_{s_1} to current source I_{s_2}. The initial state of the circuit at $t = 0^-$ (Fig. 5-37(b)) leads to the conclusion that

$$i_L(0) = i_L(0^-) = I_{s_1}.$$

The circuit in Fig. 5-37(c) represents the arrangement at $t \geq 0$. Application of KCL at the common node gives

$$-I_{s_2} + i_R + i_L = 0.$$

Since v is common to R and L, $i_R = v/R$, and by applying $v_L = L\,di_L/dt$, the KCL equation becomes

$$\frac{di_L}{dt} + ai_L = b, \qquad (5.105)$$

where a is as given previously by Eq. (5.102) and

$$b = aI_{s_2} = \frac{R}{L}\,I_{s_2}. \qquad (5.106)$$

Not surprisingly, Eq. (5.105) has the same form as Eq. (5.88) for the RC circuit and therefore exhibits a solution analogous

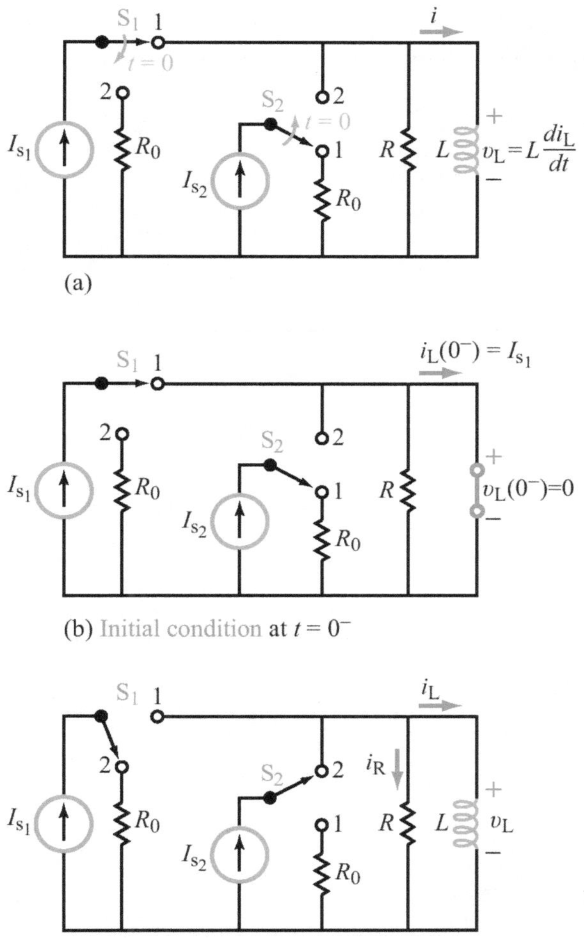

(a)

(b) Initial condition at $t = 0^-$

(c) At $t \geq 0$ (natural response)

Figure 5-37: RL circuit switched between two current sources at $t = 0$.

to the expression given by Eq. (5.96). Thus, the general form for the current through an inductor in an RL circuit is given by

$$i_L(t) = \left[i_L(\infty) + [i_L(0) - i_L(\infty)]e^{-t/\tau}\right] u(t), \quad (5.107)$$

(switch action at $t = 0$)

with *time constant* $\tau = L/R$. For the specific circuit in Fig. 5-37(a), $i_L(0) = I_{s_1}$ and $i_L(\infty) = I_{s_2}$.

If the sudden change in the circuit configuration happens at $t = T_0$ instead of at $t = 0$, the general expression for $i_L(t)$ becomes

$$i_L(t) = \left\{i_L(\infty) + [i_L(T_0) - i_L(\infty)]e^{-(t-T_0)/\tau}\right\} \cdot u(t - T_0),$$

(switch action at $t = T_0$) (5.108)

where $i_L(T_0)$ is the current at T_0. This expression is the analogue of Eq. (5.98) for the voltage across the capacitor.

Parallel RL Circuit Solution

1: If switch action is at $t = 0$, analyze circuit at $t = 0^-$ (by replacing L with a short circuit) to determine initial conditions $i_L(0^-)$ and $\upsilon_L(0^-)$. Use this information to determine $i_L(0)$ and $i_L(0)$, at t immediately after the switch action. Remember that the current through an inductor cannot change instantaneously (between $t = 0^-$ and $t = 0$), but the voltage can.

2: Analyze the circuit to determine $i_L(\infty)$, the current through the inductor long after the switch action.

3: Determine the time constant $\tau = L/R$.

4: Incorporate the information obtained in the previous three steps in Eq. (5.107):

$$i_L(t) = \left[i_L(\infty) + [i_L(0) - i_L(\infty)]e^{-t/\tau}\right] u(t).$$

5: If the switch action is at $t = T_0$ instead of $t = 0$, replace 0 with T_0 everywhere and use Eq. (5.108):

$$i_L(t) = \left\{i_L(\infty) + [i_L(T_0) - i_L(\infty)]e^{-(t-T_0)/\tau}\right\} u(t - T_0).$$

Example 5-13: Circuit with Two RL Branches

After having been in position 1 for a long time, the SPDT switch in Fig. 5-38(a) was moved to position 2 at $t = 0$. Determine i_1, i_2, and i_3 for $t \geq 0$, given that $V_s = 9.6$ V, $R_s = 4$ kΩ, $R_1 = 6$ kΩ, $R_2 = 12$ kΩ, $L_1 = 1.2$ H, and $L_2 = 0.36$ H.

Solution: We start by examining the initial state of the circuit before moving the switch. At $t = 0^-$, the inductors behave

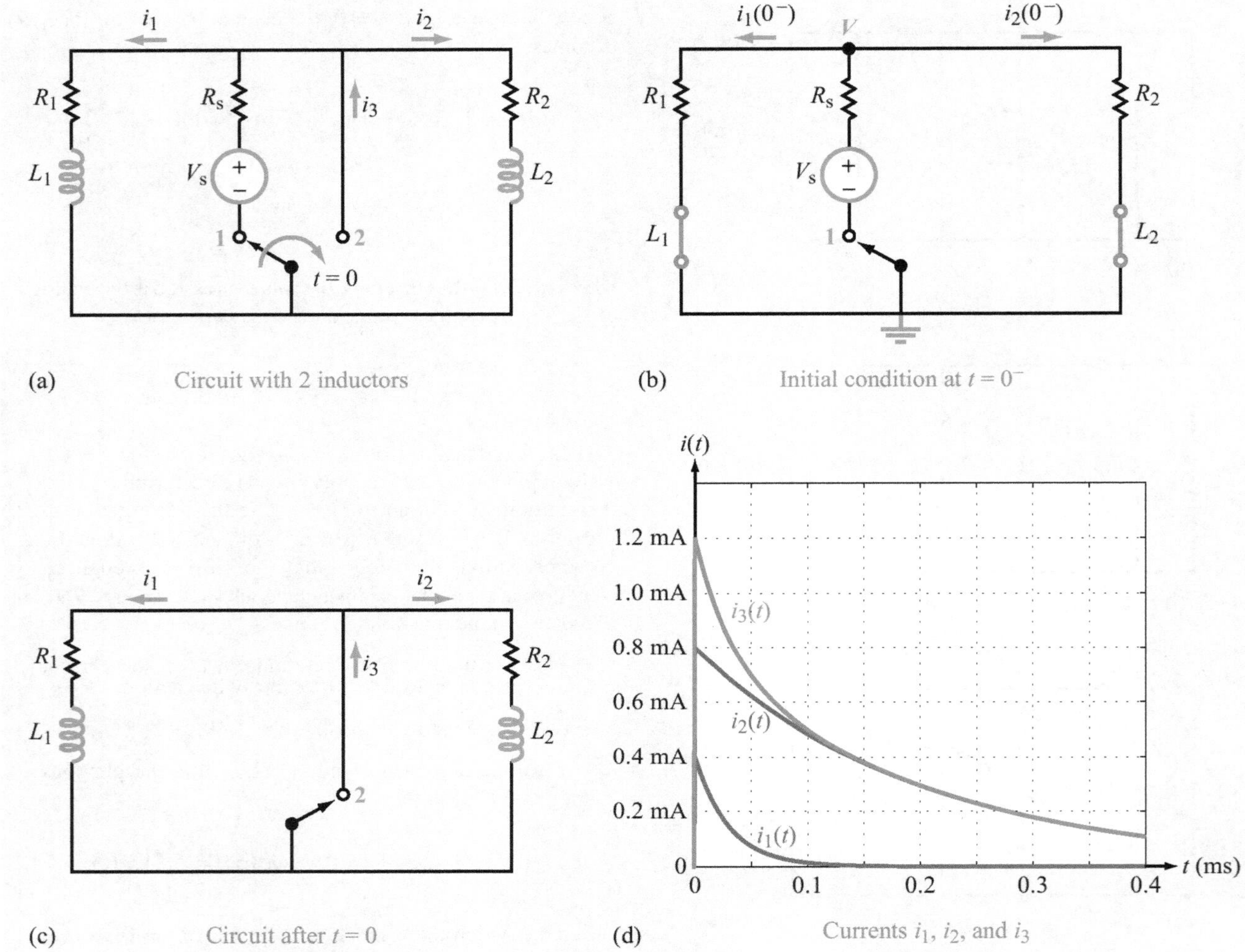

Figure 5-38: Circuit for Example 5-13.

like short circuits, resulting in the equivalent circuit shown in Fig. 5-38(b). Application of KCL to node V gives

$$\frac{V}{R_1} + \frac{V - V_s}{R_s} + \frac{V}{R_2} = 0,$$

whose solution is

$$V = \frac{R_1 R_2 V_s}{R_1 R_2 + R_1 R_s + R_2 R_s} = \frac{6 \times 12 \times 9.6}{6 \times 12 + 6 \times 4 + 12 \times 4} = 4.8 \text{ V}.$$

Hence, the initial currents $i_1(0)$ and $i_2(0)$ are given by

$$i_1(0) = i_1(0^-) = \frac{V}{R_1} = \frac{4.8}{6 \times 10^3} = 0.8 \text{ mA}$$

and

$$i_2(0) = i_2(0^-) = \frac{V}{R_2} = \frac{4.8}{12 \times 10^3} = 0.4 \text{ mA}.$$

The circuit in Fig. 5-38(c) represents the natural response circuit condition after $t = 0$. Even though we have two resistors and two inductors in the overall circuit, it can be treated as two independent RL circuits because each RL branch is connected across a short circuit. In both cases, the inductors will dissipate their magnetic energy (that they had stored prior to moving the switch) through their respective resistors. Hence, $i_1(\infty) = i_2(\infty) = 0$. The complete expressions for $i_1(t)$ and $i_2(t)$ for $t \geq 0$ then are given by

$$i_1(t) = [i_1(\infty) + [i_1(0) - i_1(\infty)]e^{-t/\tau_1}] = 0.8e^{-t/\tau_1} \, u(t) \quad \text{mA}$$

and

$$i_2(t) = [i_2(\infty) + [i_2(0) - i_2(\infty)]e^{-t/\tau_2}] = 0.4e^{-t/\tau_2} \, u(t) \quad \text{mA},$$

where τ_1 and τ_2 are the time constants of the two RL circuits, namely

$$\tau_1 = \frac{L_1}{R_1} = \frac{1.2}{6 \times 10^3} = 2 \times 10^{-4} \, \text{s}$$

and

$$\tau_2 = \frac{L_2}{R_2} = \frac{0.36}{12 \times 10^3} = 3 \times 10^{-5} \, \text{s}.$$

The current flowing through the short circuit is simply

$$i_3 = i_1 + i_2 = (0.8e^{-t/\tau_1} + 0.4e^{-t/\tau_2}) \, u(t) \quad \text{mA}.$$

Example 5-14: Response to a Triangle Excitation

The source voltage in the circuit of Fig. 5-39(a) generates a triangular ramp function that starts at $t = 0$, rises linearly to 12 V at $t = 3$ ms, and then drops abruptly down to zero. Additionally, $R = 250 \, \Omega$, $/ L = 0.5$ H, and no current was flowing through L prior to $t = 0$.

(a) Synthesize $v_s(t)$ in terms of unit step functions and plot it.

(b) Develop the differential equation for $i_L(t)$ for $t \geq 0$.

(c) solve the equation and plot $i_L(t)$ for $t \geq 0$.

Solution: (a) The waveform of $v_s(t)$ shown in Fig. 5-39(b) can be synthesized as the sum of two ramp functions:

$$v_s(t) = 4r(t) - 4r(t) \, u(t - 3 \text{ ms})$$
$$= 4t \, u(t) - 4t \, u(t) \, u(t - 3 \text{ ms})$$
$$= 4t \, u(t) - 4t \, u(t - 3 \text{ ms}) \quad \text{V}. \qquad (5.109)$$

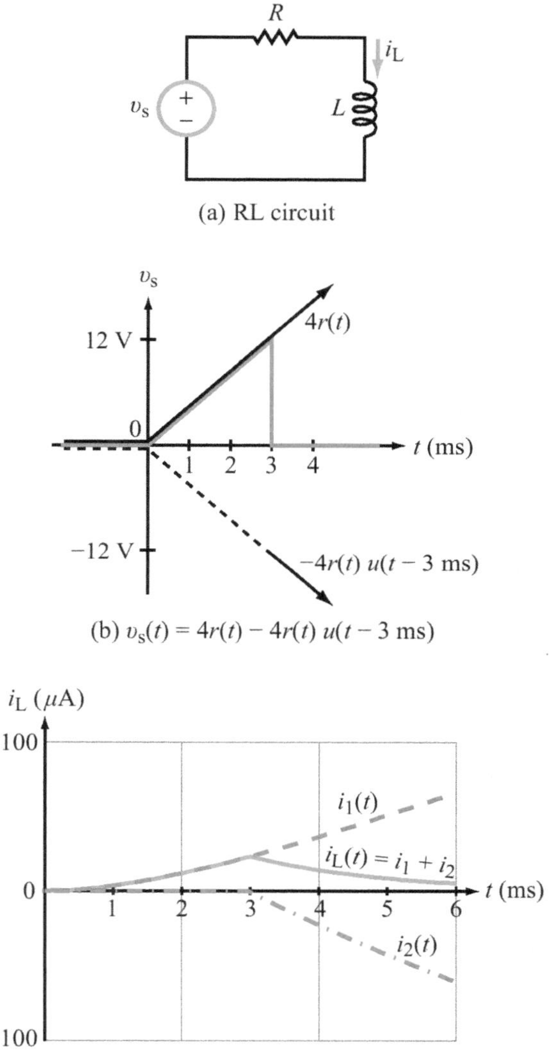

(a) RL circuit

(b) $v_s(t) = 4r(t) - 4r(t) \, u(t - 3 \text{ ms})$

(c) $i_L(t) = i_1(t) + i_2(t)$

Figure 5-39: Circuit and associated plot for Example 5-14.

(b) For $t \geq 0$, the KVL loop equation is given by

$$-v_s + Ri_L + L\frac{di_L}{dt} = 0,$$

which can be rearranged into the form

$$\frac{di_L}{dt} + ai_L = \frac{v_s}{L}, \qquad (5.110)$$

where $a = R/L$. Since $\upsilon_s(t)$ is composed of two components, we will write $i_L(t)$ as the sum of two components,

$$i_L(t) = i_1(t) + i_2(t), \tag{5.111}$$

where $i_1(t)$ is the solution of Eq. (5.110) with $\upsilon_s = 4t\ u(t)$ acting alone and $i_2(t)$ is the solution of Eq. (5.110) with $\upsilon_s = -4t\ u(t - 3\text{ ms})$ acting alone. That is,

$$\frac{di_1}{dt} + ai_1 = \frac{4t}{L} = bt \qquad \text{for } t \ge 0 \tag{5.112a}$$

and

$$\frac{di_2}{dt} + ai_2 = \frac{-4t}{L} = -bt \qquad \text{for } t \ge 3\text{ ms} \tag{5.112b}$$

with $b = 4/L$.

Current $i_1(t)$ alone

We start by multiplying both sides of Eq. (5.112a) by $e^{at'}$ and then integrating from 0 to t:

$$\int_0^t \left(e^{at'} \frac{di_1}{dt'} + ai_1e^{at'}\right) dt' = \int_0^t bt'e^{at'}\, dt'. \tag{5.113}$$

For the left-hand side,

$$\int_0^t \left[e^{at'} \frac{di_1}{dt'} + ai_1e^{at'}\right] dt' = \int_0^t \left[\frac{d}{dt'}(i_1e^{at'})\right] dt'$$
$$= i_1e^{at'}\Big|_0^t, \tag{5.114}$$

and for the right-hand side,

$$\int_0^t bt'e^{at'}\, dt' = \frac{b}{a^2}\, e^{at'}(at' - 1)\Big|_0^t. \tag{5.115}$$

In view of Eqs. (5.114) and (5.115), Eq. (5.113) becomes

$$i_1e^{at'}\Big|_0^t = \frac{b}{a^2}\, e^{at'}(at' - 1)\Big|_0^t, \tag{5.116}$$

which leads to

$$i_1(t)\, e^{at} - i_1(0) = \frac{b}{a_2}\, [e^{at}(at - 1) + 1]. \tag{5.117}$$

given that $i_1(0) = 0$, the expression for $i_1(t)$ becomes

$$i_1(t) = \frac{b}{a^2}\, [(at - 1) + e^{-at}] \qquad (\text{for } t \ge 0). \tag{5.118}$$

Current $i_2(t)$ alone

Equations (5.112a) and (5.112b) are identical in form, except for two important differences:

(1) The forcing function for $i_1(t)$ is bt whereas the forcing function for $i_2(t)$ is $-bt$.

(2) The temporal domain of applicability for $i_2(t)$ starts at $t = 3$ ms, instead of at $t = 0$.

Hence, Eq. (5.116) can be adapted to i_2 by replacing b with $-b$ and changing the lower limit of integration to 3 ms, which gives

$$i_2e^{at'}\Big|_{3\text{ ms}}^t = \frac{-b}{a^2}\, e^{at'}(at' - 1)\Big|_{3\text{ ms}}^t, \tag{5.119}$$

which leads to

$$i_2(t)\, e^{at} - i_2(3\text{ ms})\, e^{0.003a}$$
$$= -\frac{b}{a^2}[e^{at}(at - 1) - e^{0.003a}(0.003a - 1)]. \tag{5.120}$$

When we apply superposition, we apply the same initial condition to both RL circuits (corresponding to the two components of $\upsilon_s(t)$). Thus, $i_1(0) = i_2(3\text{ ms}) = 0$, and Eq. (5.120) simplifies to

$$i_2(t) = -\frac{b}{a^2}[(at - 1) - (0.003a - 1)e^{-a(t-0.003)}]$$
$$(\text{for } t \ge 3\text{ ms}). \tag{5.121}$$

Total solution for $i_L(t)$

For $R = 250\ \Omega$ and $L = 0.5$ H, $a = R/L = 500$, $b = 4/L = 8$, and

$$i_L(t) = \begin{cases} i_1(t) & \text{for } 0 \le t \le 3\text{ ms}, \\ i_1(t) + i_2(t) & \text{for } t \ge 3\text{ ms}, \end{cases}$$

$$= \begin{cases} 32[(500t - 1) + e^{-500t}]\ \ \mu\text{A} \\ \qquad \text{for } 0 \le t < 3\text{ ms}, \\ 103.7e^{-500t}\ \ \mu\text{A} \\ \qquad \text{for } t \ge 3\text{ ms}. \end{cases} \tag{5.122}$$

Figure 5-39(c) displays a plot of $i_L(t)$ versus t.

> **Concept Question 5-22:** Compare Eq. (5.96) with Eq. (5.107) to draw an analogy between RC and RL circuits. υ_C, R, and C of the RC circuit correspond to which parameters of the RL circuit? (See CAD)

Technology Brief 13
Hard Disk Drives (HDD)

Although invented in 1956, the ***hard disk drive*** (HDD) arguably is still the most commonly used data-storage device among nonvolatile storage media available today. It is the availability of vast amounts of relatively inexpensive hard-drive space that has made search engines, webmail, and online games possible. Over the past 40 years, improvements in HDD technology have led to huge increases in storage density, which are simultaneous with the significant reduction in physical size. The term ***hard disk*** or ***hard drive*** evolved from common usage as a means to distinguish these devices from flexible (***floppy***) disk drives.

HDD Operation

Hard drives make use of magnetic material to read and write data. A nonmagnetic disc ranging in diameter from 36 to 146 mm is coated with a thin film of magnetic material, such as an iron or cobalt alloy. When a strong magnetic field is applied across a small area of the disc, it causes the atoms in that area to align along the orientation of the field, providing the mechanism for writing bits of data onto the disc (Fig. TF13-1). Conversely, by detecting the aligned field, data can be read back from the disc. The hard drive is equipped with an arm that can be moved across the surface of the disc (Fig. TF13-2), and the disc itself is spun around to make all of the magnetic surface accessible to the writing or reading heads. The reading and writing elements are physically moved along the radius of the disk by using a magnet with a coil wrapped around it. When current is driven into the coil, it produces a magnetic force that moves the actuator. Because writing onto or reading from the magnetized surface can be performed very rapidly (fraction of a microsecond), hard drives are spun at very high speeds (5,000 to 15,000 rpm) when directed to record or retrieve information. Amazingly, hard-drive heads usually hover at a height of about 25 nm above the surface of the magnetic disc while the disc is spinning at such high speeds! The extremely small gap between the head and the disc is maintained by having the head "ride" on a thin cushion of air trapped between the head and the surface of the spinning disc. To prevent accidental scratches, the disc is coated with carbon- or Teflon-like materials.

Hard drives are packaged carefully to prevent dust and other airborne particles from interfering with the drive's operation. In combination with the air motion caused by the spinning disc, a very fine air filter is used to keep dust out while maintaining the air pressure necessary to cushion the spinning discs. Hard drives intended for operation at high altitudes (or low air pressure) are sealed hermetically so as to make them airtight.

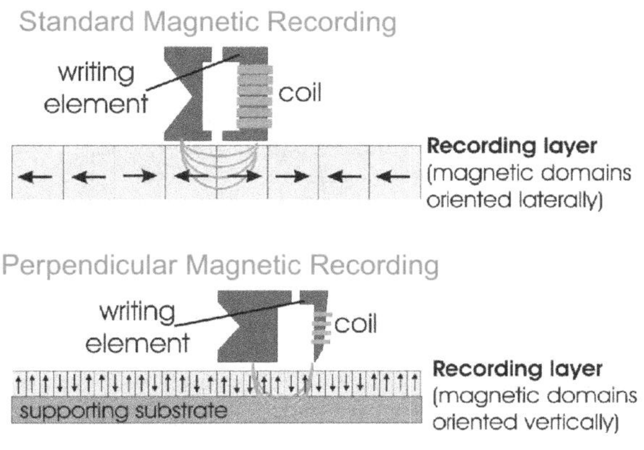

Figure TF13-1: Longitudinal and perpendicular writing techniques.

Modern Drive Technology

Early hard drives performed read and write operations by using an inductor coil placed at the tip of the head. When electric current is made to flow through the coil, the coil induces a magnetic field which in turn aligns the

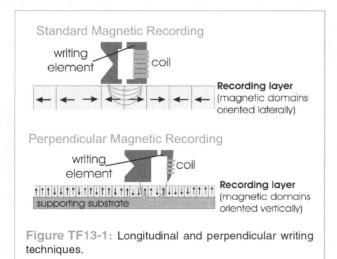

Figure TF13-2: Close-up of a disassembled hard drive showing the magnetic discs mounted on a spindle and an actuator arm. The head sits at the end of the arm and performs the read/write operations as the disc spins.

atoms of the magnetic material (i.e., a *write* operation). The same coil also is used to detect the presence of aligned atoms, thereby providing the *read* operation. The many major developments that shaped the evolution of read/write heads over the past 50 years have introduced two major differences between the modern hard-drive heads and the original models. Instead of using the same head for both reading and writing, separate heads are now used for the two operations. Furthermore, the writing operation is now carried out with a lithographically defined *thin-film* head, thereby reducing the *feature size* of the head by several orders of magnitude. The feature size is the area occupied by a single bit on the disc surface, which is determined in part by the size of the write head. Decreasing feature size leads to increased recording density. The read operation—housed separately next to the write head—uses a *magnetoresistive* material whose resistance changes when exposed to a magnetic field—even when the field intensity is exceedingly small. In modern hard drives, high magnetoresistive sensitivities are realized through the application of either the *giant magnetoresistance* (GMR) phenomena or the *tunneling magnetoresistance* (TMR) effect exhibited by certain materials. The 2007 Nobel prize in physics was awarded to Albert Fert and Peter Grünberg for their discovery of GMR. A consequence of the extremely small size of the magnetic bits (each bit in a 100-Gb/in^2 disc is about 40 nm long) is that temperature variations can lead to loss of information over time. One method developed to combat this issue is to use two magnetic layers separated by a thin (~ 1 nm) insulator, which increases the stability of the stored bit. Another recent innovation that is already in production involves the use of *perpendicular magnetic recording* (PMR) as illustrated in Fig. TF13-1. PMR makes it possible to align bits more compactly next to each other.

Recent Developments

A new wave of developments is pushing hard drives into the tens of terabytes. Already in commercial use is *shingled magnetic recording* (SMR). Conventional drives write bits in parallel rows Fig. TF13-3(a)), usually with a slight gap between them. Making the individual track width smaller is extremely difficult because, as mentioned above, very small magentic grains are not stable (or, conversely, to make very small grains stable makes them very hard to read/write with a magnetic head). The SMR solution (Fig. TF13-3(b)) is to lay bits down in overlapping tracks, exactly like roof shingles (where each shingle row sits slightly on top of one adjacent row and slightly below the other). The advantage is that the size of the track (and hence, the grain), does not change but the overall density increases. This works because a magnetic head can still read the state of the magentic grain even if it slightly overlapped with a nearby grain. The difficulty of this method is that the writing process slows down since every time we write to one of the overlapped rows, we must also rewrite the neighboring rows. The tracks are organized into bands (Fig. TF13-3(b)) and each band is thus rewritten as needed. Coordinating this write activity can be handled in firmware on the drive itself or in the computer's operating system (if it has the appropriate driver to handle such drives).

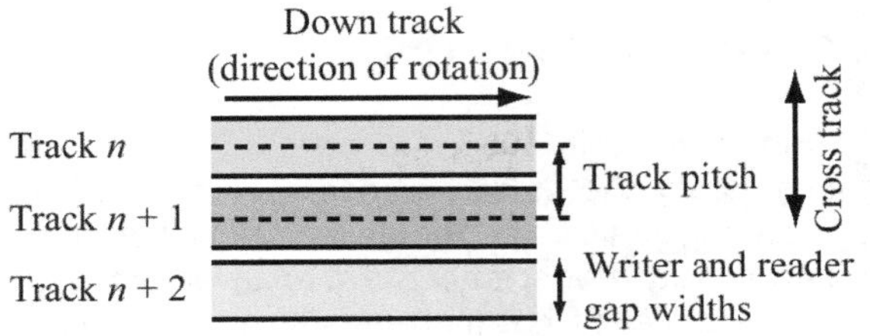

(a) Schematic of conventional magnetic recording

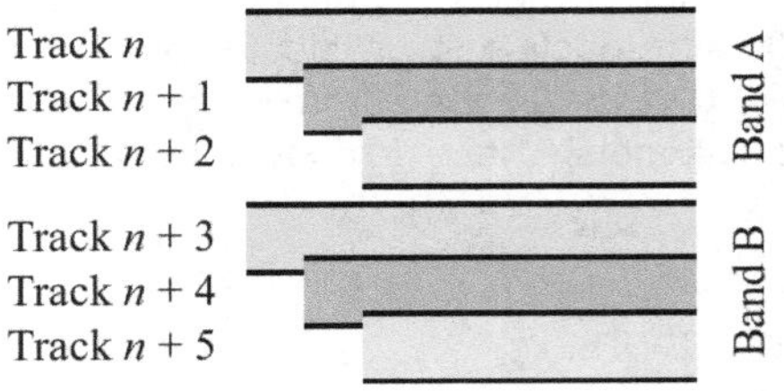

(b) Schematic of shingled magnetic recording

Figure TF13-3: Schematics of (a) conventional magnetic recording and (b) shingled magnetic recording with two 3-track bands.

A variety of other techniques (including the GMR heads discussed above) are being explored to increase areal density; in general, these focus on allowing smaller grains by making them harder to write magnetically (which makes them consequently more temperature stable). Among these are heat-assisted, microwave-assisted and patterning single-grain (or close to single-grain) isolated magnetic islands (instead of a continuous magnetic thin film); this is known as *bit-patterned media* (BPM). It is estimated that techniques such as these will enable densities on the order of 1–10 Tb/in^2 in the next decade.

Concept Question 5-23: Suppose the switch in the circuit of Fig. 5-36(a) had been open for a long time, and then it was closed suddenly. Will I_s initially flow through R or L? (See CAD)

Exercise 5-16: Determine $i_1(t)$ and $i_2(t)$ for $t \geq 0$ given that, in the circuit of Fig. E5.16, $L_1 = 6$ mH, $L_2 = 12$ mH, and $R = 2\ \Omega$. Assume

$$i_1(0^-) = i_2(0^-) = 0.$$

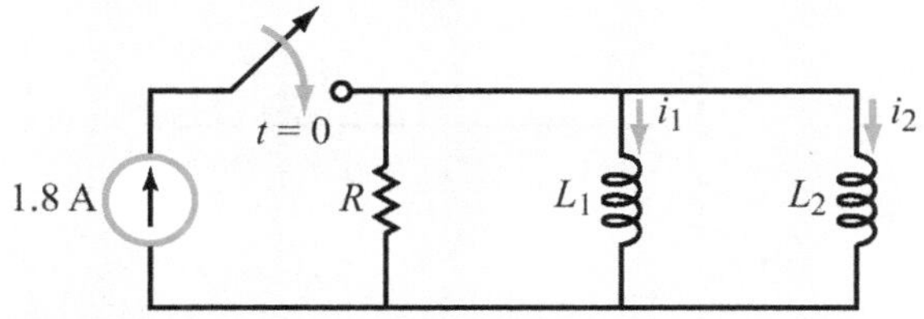

Figure E5.16

Answer: $i_1(t) = 1.2(1 - e^{-500t})\, u(t)$ A, $i_2(t) = 0.6(1 - e^{-500t})\, u(t)$ A. (See CAD)

5-6 RC Op-Amp Circuits

Adding capacitors and inductors to resistive circuits vastly expands their utility and versatility. In this section, we consider a few examples of circuits in which capacitors are used in conjunction with op amps to perform integration, differentiation, and related operations. Even though these specific functions also can be realized through the use of inductors, capacitors are usually the preferred option (whenever such a choice is possible) because of their smaller physical size and availability in planar form.

5-6.1 Ideal Op-Amp Integrator

The circuit shown in Fig. 5-40 resembles the standard inverting-amplifier circuit of Section 4-4, except that its feedback resistor R_f has been replaced with a capacitor C, converting it into an op-amp integrator. As we show shortly:

▶ The output voltage υ_{out} of the RC integrator circuit is directly proportional to the time integral of the input signal υ_i. ◀

RC Integrator

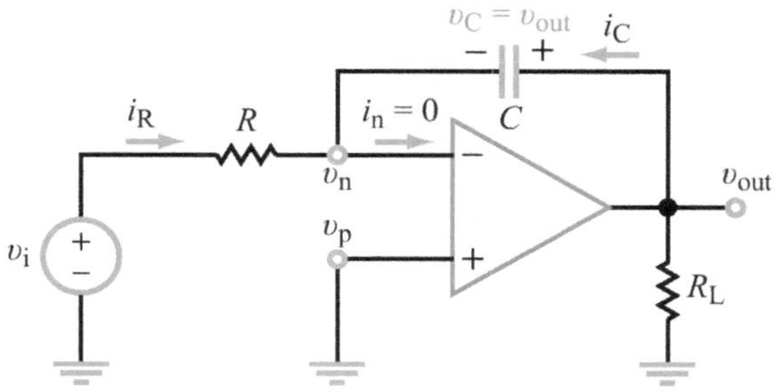

Figure 5-40: Integrator circuit.

The ideal op-amp model has two constraints. The voltage constraint states that $\upsilon_p = \upsilon_n$, and since $\upsilon_p = 0$ in the circuit of Fig. 5-40, it follows that $\upsilon_n = 0$. Hence, the current i_R flowing through R is given by

$$i_R = \frac{\upsilon_i}{R}. \tag{5.123}$$

Given that $\upsilon_n = 0$, the voltage υ_C across C is simply υ_{out}, and the current flowing through it is

$$i_C = C\,\frac{d\upsilon_{out}}{dt}. \tag{5.124}$$

At node υ_n,

$$i_R + i_C - i_n = 0. \tag{5.125}$$

In view of the second op-amp constraint, namely $i_n = i_p = 0$, it follows that

$$i_C = -i_R \tag{5.126}$$

or

$$\frac{d\upsilon_{out}}{dt} = -\frac{1}{RC}\,\upsilon_i. \tag{5.127}$$

Upon integrating both sides of Eq. (5.127) from an initial reference time t_0 to time t, we have

$$\int_{t_0}^{t} \left(\frac{d\upsilon_{out}}{dt'}\right) dt' = -\frac{1}{RC}\int_{t_0}^{t} \upsilon_i\, dt', \tag{5.128}$$

which leads to

$$\upsilon_{out}(t) = -\frac{1}{RC}\int_{t_0}^{t} \upsilon_i(t')\, dt' + \upsilon_{out}(t_0). \tag{5.129}$$

Time t_0 is the time at which the integration process begins, and $\upsilon_{out}(t_0)$ is the initial voltage across the capacitor at that instant in time. Thus, according to Eq. (5.129), the output voltage (which is also the voltage across the capacitor) is equal to whatever voltage existed across the capacitor at the start of the integration process, $\upsilon_{out}(t_0)$, incremented by an amount equal to the integrated value of the input voltage (from t_0 to present time t) and multiplied by a (negative) *scaling factor* $(-1/RC)$.

► Since the magnitude of the output voltage, $|\upsilon_{out}|$, cannot exceed the supply voltage V_{cc}, the values of R and C have to be chosen carefully so as to avoid saturating the op amp. ◄

If the time scale can be conveniently chosen such that the initial reference time $t_0 = 0$ and the capacitor was uncharged at that point in time (i.e., $\upsilon_{out}(0) = 0$), then Eq. (5.129) simplifies to

$$\upsilon_{out}(t) = -\frac{1}{RC}\int_0^t \upsilon_i(t')\,dt' \quad (\text{if } \upsilon_{out}(0) = 0). \tag{5.130}$$

Example 5-15: Square-Wave Input Signal

The square-wave signal shown in Fig. 5-41(a) is applied at the input of an ideal integrator circuit with an initial capacitor voltage of zero at $t = 0$. If $R = 200$ kΩ and $C = 2.5\ \mu$F, determine the waveform of the corresponding output voltage for an amp with (a) $V_{cc} = 14$ V and (b) $V_{cc} = 9$ V.

Solution: (a) The scaling factor is given by

$$-\frac{1}{RC} = -\frac{1}{2\times 10^5 \times 2.5\times 10^{-6}} = -2\ \text{s}^{-1}.$$

For the time period $0 \le t \le 2$ s (first half of the first cycle),

$$\upsilon_{out}(t) = -2\int_0^t \upsilon_i\,dt' = -2\int_0^t 3\,dt' = -6t \text{ V}$$

$$(0 \le t \le 2 \text{ s}),$$

which is represented by the first ramp function shown in Fig. 5-41(b). The polarity reversal of υ_i during the second half of the first cycle causes the energy that had been stored in the capacitor to be discharged, concluding the cycle with no net voltage across the capacitor. The process then is repeated during succeeding cycles.

We note that because $|\upsilon_{out}|$ never exceeds $|V_{cc}| = 14$ V, no saturation occurs in the op amp.

(b) For the op amp with $V_{cc} = 9$ V, the waveform shown in Fig. 5-41(c) is the same as that in Fig. 5-41(b), except that it is *clipped* at −9 V.

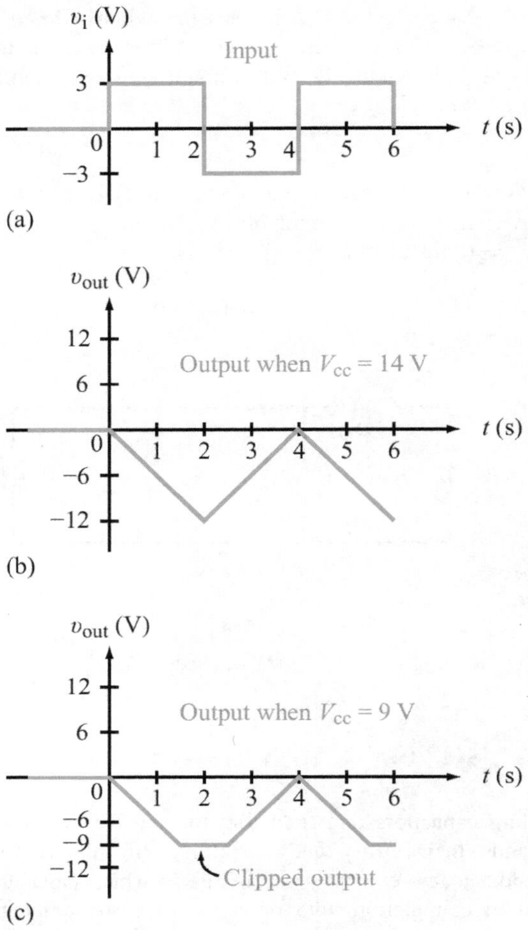

Figure 5-41: Example 5-15 (a) input signal, (b) output signal with no op-amp saturation, and (c) output signal with op-amp saturation at −9 V.

RC Differentiator

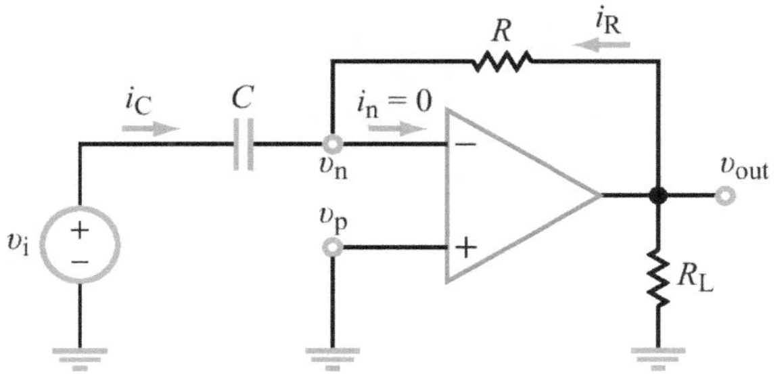

Figure 5-42: Differentiator circuit.

5-6.2 Ideal Op Amp Differentiator

The integrator circuit of Fig. 5-40 can be converted into the differentiator circuit of Fig. 5-42 by simply interchanging the locations of R and C. For the differentiator circuit, application of the voltage and current constraints leads to

$$i_C = C\,\frac{d\upsilon_i}{dt}, \qquad i_R = \frac{\upsilon_{out}}{R}, \qquad \text{and} \qquad i_C = -i_R.$$

Consequently,

$$\upsilon_{out} = -RC\,\frac{d\upsilon_i}{dt}, \tag{5.131}$$

which states that:

▶ The output voltage of the differentiator circuit is proportional directly to the time derivative of its input voltage υ_i, and the proportionality factor is $(-RC)$. The differentiator circuit performs the inverse function of that performed by the integrator circuit. ◀

5-6.3 Other Op-Amp Circuits

The relative ease with which we were able to develop input–output relationships for the ideal integrator and differentiator circuits is attributed (at least in part) to the relative simplicity of those circuits. Aside from the load resistor R_L (which exercised no influence on the solutions), the circuits in Figs. 5-40 and 5-42 consisted each of one resistor and one capacitor. Now, through two examples, we demonstrate ways to approach the analysis of RC op-amp circuits that may have more complicated architectures.

Example 5-16: Pulse Response of an Op-Amp Circuit

The op-amp circuit shown in Fig. 5-43(a) is subjected to an input pulse of amplitude $V_s = 2.4$ V and duration $T_0 = 0.3$ s. Determine and plot the output voltage $\upsilon_{out}(t)$ for $t \geq 0$, assuming that the capacitor was uncharged before $t = 0$.

Solution: One possible approach to solving the problem is to analyze the circuit twice—once for the duration of the pulse (0 to 0.3 s) and a second time for $t > 0.3$ s. An alternative approach is to synthesize the rectangular pulse as the sum of two step functions, to seek an independent solution for each step function, and then to add up the solutions (superposition). We will illustrate both methods.

(a) Method 1: **Two Time Segments**

Time Segment 1: $0 \leq t \leq 0.3$ s, and $\upsilon_i = V_s = 2.4$ V.

At node υ_n,

$$i_1 + i_2 + i_3 = 0,$$

or, using the node voltage method,

$$\frac{\upsilon_n - V_s}{R_1} + C\,\frac{d}{dt}(\upsilon_n - \upsilon_{out_1}) + \frac{\upsilon_n - \upsilon_{out_1}}{R_2} = 0,$$

where υ_{out_1} is the output voltage during time segment 1. Since $\upsilon_p = 0$, injection of the ideal op-amp voltage constraint $\upsilon_p = \upsilon_n$ leads to

$$C\,\frac{d\upsilon_{out_1}}{dt} + \frac{\upsilon_{out_1}}{R_2} = -\frac{V_s}{R_1},$$

which can be cast in the standard first-order differential-equation form given by

$$\frac{d\upsilon_{out_1}}{dt} + a\upsilon_{out_1} = b, \tag{5.132}$$

where

$$a = \frac{1}{R_2 C}, \qquad \text{and} \qquad b = -\frac{V_s}{R_1 C}.$$

Equation (5.132) is analogous to Eq. (5.88), so its solution is analogous to that given by Eq. (5.94), namely

$$\begin{aligned}\upsilon_{out_1}(t) &= \upsilon_{out_1}(0)\,e^{-at} + \frac{b}{a}(1 - e^{-at}) \\ &= \upsilon_{out_1}(0)\,e^{-t/\tau} - \frac{V_s R_2}{R_1}(1 - e^{-t/\tau}),\end{aligned} \tag{5.133}$$

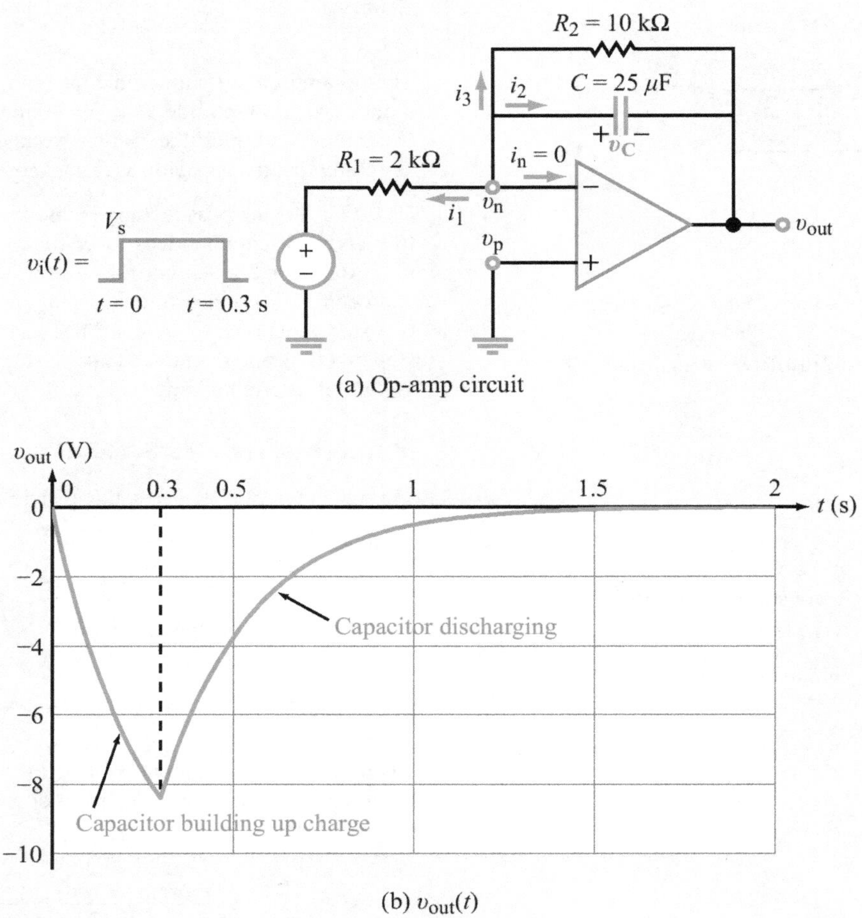

Figure 5-43: Op-amp circuit of Example 5-16.

where

$$\tau = \frac{1}{a} = R_2 C = 0.25 \text{ s}.$$

Given that $v_n = 0$, it is evident from the circuit in Fig. 5-43(a) that

$$v_{out_1} = -v_{C_1},$$

where v_{C_1} is the voltage across the capacitor during the first time segment. According to the problem statement, the initial condition $v_{C_1}(0^-) = 0$, and since the voltage across a capacitor cannot change instantaneously, it follows that

$$v_{out_1}(0) = -v_{C_1}(0) = -v_{C_1}(0^-) = 0.$$

Upon incorporating this piece of information into our solution, we have the natural response

$$v_{out_1}(t) = -\frac{V_s R_2}{R_1}(1 - e^{-t/\tau})$$
$$= -12(1 - e^{-4t}) \text{ V} \quad (\text{for } 0 \le t \le 0.3 \text{ s}). \quad (5.134)$$

Time Segment 2: $t > 0.3$ s, and $v_i = 0$.

The form of the solution for this time segment is the same as that given by Eq. (5.133) for the preceding time segment, except for three modifications:

(a) The input voltage is now zero, so we should set $V_s = 0$.

(b) The time variable t should be replaced with (t − 0.3 s) to reflect the fact that our starting (reference) time is $t = 0.3$ s, not $t = 0$.

(c) The initial voltage $\upsilon_{\text{out}_2}(0.3 \text{ s})$ is not zero (because the capacitor had been building up charge during the previous time segment).

Hence, for time segment 2, υ_{out_2} is given by

$$\upsilon_{\text{out}_2}(t) = \upsilon_{\text{out}_2}(0.3)\, e^{-4(t-0.3)} \qquad (\text{for } t > 0.3 \text{ s}).$$

The initial voltage $\upsilon_{\text{out}_2}(0.3)$ is equal to the voltage that existed during the previous time segment at $t = 0.3$ s. Hence,

$$\upsilon_{\text{out}_2}(0.3) = \upsilon_{\text{out}_1}(0.3) = -12(1 - e^{-4\times 0.3}) = -8.4 \text{ V}.$$

Hence,

$$\upsilon_{\text{out}_2}(t) = -8.4e^{-4(t-0.3 \text{ s})} \text{ V} \qquad (\text{for } t > 0.3 \text{ s}). \tag{5.135}$$

The combined output response to the input pulse is displayed in Fig. 5-43(b).

(b) Method 2: **Two Step Functions**

By modeling the rectangular pulse as

$$\upsilon_{\text{i}}(t) = V_{\text{s}}[u(t) - u(t - 0.3 \text{ s})], \tag{5.136}$$

we can develop a generic solution to a step-function input and then use it to find

$$\upsilon_{\text{out}}(t) = \upsilon_{\text{out}_a}(t) + \upsilon_{\text{out}_b}(t).$$

We will treat the two step functions as two independent sources, and we will apply the same initial-condition information to both cases; that is, when treating the case of the second step function, we do so as if the first step function had never existed.

To that end, the response of the first step function is given by Eq. (5.134) as

$$\upsilon_{\text{out}_a}(t) = -12(1 - e^{-4t})\, u(t) \text{ V} \qquad (\text{for } t \geq 0). \tag{5.137}$$

Similarly, after reversing the polarity of V_{s} and incorporating a time delay of 0.3 s,

$$\upsilon_{\text{out}_b}(t) = 12(1-e^{-4(t-0.3)})\, u(t-0.3) \text{ V} \qquad (\text{for } t \geq 0.3 \text{ s}). \tag{5.138}$$

In view of the definition of the step function, the complete solution is given by

$$\upsilon_{\text{out}}(t) = \upsilon_{\text{out}_a}(t) + \upsilon_{\text{out}_b}(t) = \begin{cases} \upsilon_{\text{out}_a}(t) & \text{for } 0 \leq t \leq 0.3 \text{ s} \\ \upsilon_{\text{out}_a}(t) + \upsilon_{\text{out}_b}(t) & \text{for } t > 0.3 \text{ s}. \end{cases} \tag{5.139}$$

It is a relatively straightforward exercise to demonstrate that the two methods do indeed provide the same solution.

Example 5-17: Op-Amp Circuit with Output Capacitor

Determine $\upsilon_C(t)$, the voltage across the capacitor in Fig. 5-44(a), given that $\upsilon_{\text{i}}(t) = 3u(t)$ V, the capacitor had no charge on it prior to $t = 0$, $R_1 = 1$ kΩ, $R_2 = 15$ kΩ, $R_3 = 30$ kΩ, $R_4 = 12$ kΩ, $R_5 = 24$ kΩ, and $C = 50\ \mu$F.

Solution: The capacitor is on the output (load) side of the op amp, so one possible approach to solving the problem is to

(a) temporarily replace the capacitor with an open circuit;

(b) determine the Thévenin equivalent circuit at terminals (a, b); and

(c) reinsert the capacitor as in Fig. 5-44(c) and analyze the circuit.

To that end, we start by relating υ_{out} to υ_{i}. Given that for the ideal op amp $\upsilon_{\text{n}} = \upsilon_{\text{p}}$ and $i_{\text{p}} = 0$, it follows that

$$\upsilon_{\text{n}} = \upsilon_{\text{p}} = \upsilon_{\text{i}}.$$

Moreover, since $i_{\text{n}} = 0$, υ_{n} and υ_{out} are related by a voltage divider between nodes c and d:

$$\upsilon_{\text{out}} = \left(\frac{R_2 + R_3}{R_2}\right)\upsilon_{\text{n}} = \left(\frac{R_2 + R_3}{R_2}\right)\upsilon_{\text{i}}.$$

With the capacitor removed, the Thévenin voltage across terminals (a, b) in Fig. 5-44(a) is equal to the voltage across R_5, which is related to υ_{out} by the voltage-division rule

$$\begin{aligned}\upsilon_{\text{Th}} &= \left(\frac{R_5}{R_4 + R_5}\right)\upsilon_{\text{out}} \\ &= \left(\frac{R_5}{R_4 + R_5}\right)\left(\frac{R_2 + R_3}{R_2}\right)\upsilon_{\text{i}} \\ &= \left(\frac{24}{12 + 24}\right)\left(\frac{15 + 30}{15}\right) \times 3 = 6u(t) \text{ V} \quad (\text{for } t \geq 0).\end{aligned}$$

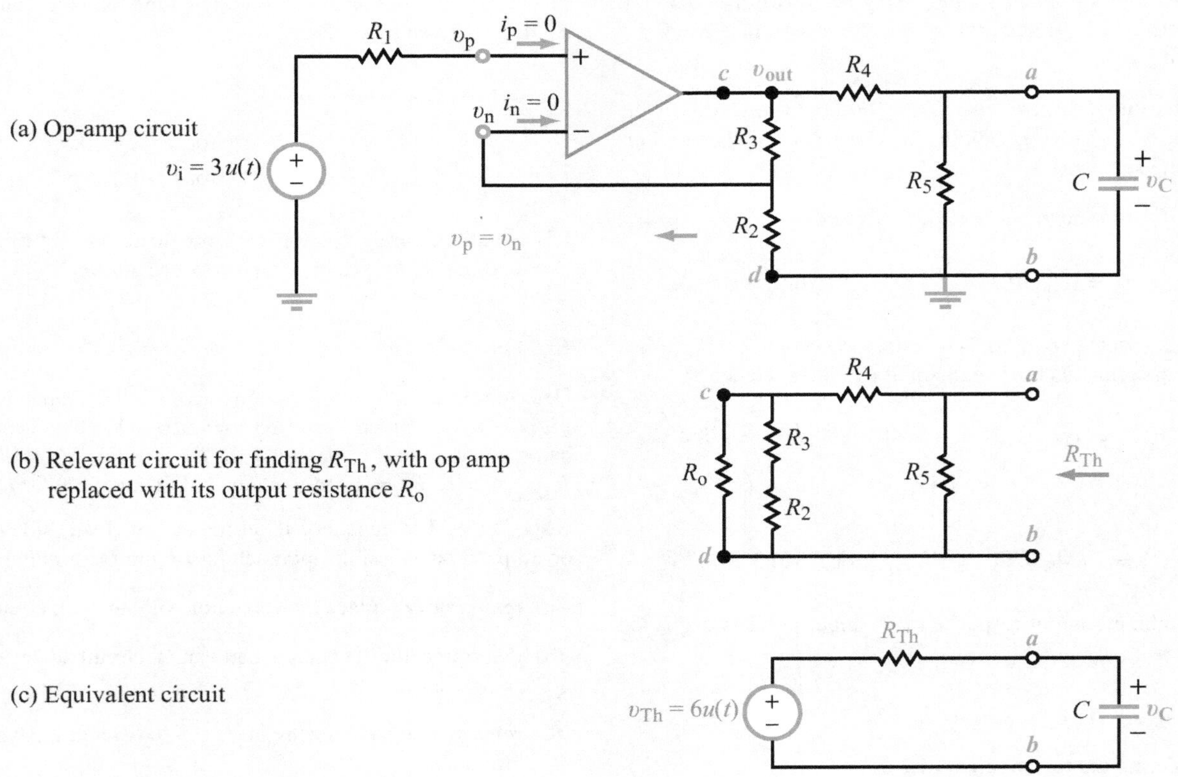

Figure 5-44: Circuit for Example 5-17.

Our next task is to determine the value of R_{Th}. To that end, we set $\upsilon_i = 0$. Consequently, $\upsilon_p - \upsilon_n = 0$, in which case the op-amp's equivalent circuit at terminals (c, d) consists of only its output resistance R_0. Figure 5-44(b) contains the relevant part of the overall circuit seen by terminals (a, b). For the real op amp, R_0 is on the order of 10 to 100 Ω, which is at least two orders of magnitude smaller than any of the other resistors in the circuit, lending justification to the ideal op-amp model which sets $R_0 = 0$ (thereby shorting out $(R_2 + R_3)$). Consequently,

$$R_{Th} = R_4 \parallel R_5 = \frac{R_4 R_5}{R_4 + R_5} = \frac{12 \times 24}{12 + 24} = 8 \text{ k}\Omega.$$

With υ_{Th} and R_{Th} known, we now have a circuit (Fig. 5-44(c)) that resembles the step-function circuit of Fig. 5-30(a). Its solution is given by Eq. (5.97) using $V_{s_1} = 0$ and $V_{s_2} = V_s$, namely

$$\upsilon_C(t) = V_s(1 - e^{-t/\tau}).$$

In the present case, $V_s = \upsilon_{Th} = 6$ V, and

$$\tau = R_{Th}C = 8 \times 10^3 \times 50 \times 10^{-6} = 0.4 \text{ s}.$$

The capacitor response is therefore given by

$$\upsilon_C(t) = 6(1 - e^{-2.5t})\ u(t) \quad \text{V}.$$

Example 5-18: Differential Equation Solver

Design an op-amp circuit whose output is the solution of the differential equation

$$\frac{d^2\upsilon}{dt^2} + 8\,\frac{d\upsilon}{dt} + 2\upsilon = 4\upsilon_s(t), \tag{5.140}$$

where $\upsilon_s(t)$ is a sinusoidal source given by

$$\upsilon_s(t) = 3\sin(200t)\ u(t).$$

Technology Brief 14
Capacitive Sensors

Capacitive sensors are used to convert information from the real world to a change in capacitance that can be detected by an electric circuit. Even though capacitors can assume many different shapes, the basic concepts can be easily explained using the shape and properties of the parallel plate capacitor, for which the capacitance C is given by

$$C = \frac{\varepsilon A}{d},$$

where ε is the permittivity of the material between the plates, A is the area of each plate, and d is the spacing between the plates. So, most capacitive sensors operate by measuring the change in one or more of these three basic parameters, in response to external physical stimuli. Let us examine each one of these three parameters separately and how it can be used to measure external stimuli.

Applications Based on Change in Permittivity ε

The electrical permittivity ε of a given material is an inherent property of that material; its value is dictated by the polarization behavior of that material's molecular structure, relative to the absence of polarizability (as in free space or vacuum). In free space, $\varepsilon = \varepsilon_0 = 8.854 \times 10^{-12}$ F/m, and for all other media, it is convenient to express the permittivity of a material relative to that for free space through the relative permittivity $\varepsilon_r = \varepsilon/\varepsilon_0$. Table TT14-1 provides a list for various types of materials. We note that for plastic, glass, and most ceramics, ε_r is in the range between 2 and 4, which makes them different (electrically) from air ($\varepsilon_r = 1$ for air), but not markedly so. In contrast, water-based materials—such as biological materials or parts of the body—have an ε_r in the range of 60–80, making them electrically very different from both air and dry materials. This means that their presence can be easily detected by a capacitive sensor, which is the basis of capacitive touchscreens, fluid and moisture meters, and some proximity meters.

Table TT14-1: Relative permittivity ε_r of common materials.[a]

$\varepsilon = \varepsilon_r\varepsilon_0$ and $\varepsilon_0 = 8.854 \times 10^{-12}$ F/m

Material	Relative Permittivity, ε_r
Vacuum	1
Air (at sea level)	1.0006
Low Permittivity Materials	
Styrofoam	1.03
Teflon	2.1
Petroleum oil	2.1
Wood (dry)	1.5–4
Paraffin	2.2
Polyethylene	2.25
Polystyrene	2.6
Paper	2–4
Rubber	2.2–4.1
Plexiglass	3.4
Glass	4.5–10
Quartz	3.8–5
Water	72–80
Biological Materials	40–70

[a]These are at room temperature (20 °C).

Capacitive Touch Buttons

An example of a capacitive touch sensor is shown in Fig. TF14-1. The capacitor has two conducting surfaces labeled ***sensor pad*** and ***ground hatch***. In general, the two conductors are separated either vertically or horizontally, and covered with a layer of glass or plastic. By applying a voltage source (supplied by the printed circuit board) between the conducting surfaces, electric field lines get established between them. When no finger (or a capacitive stylus) is present near the sensor pad, the electric field lines flow through the glass or plastic cover, but when in the proximity of a finger, the electric field lines pass partially through the finger, and since the finger has a relative permittivity comparable to that of water, its

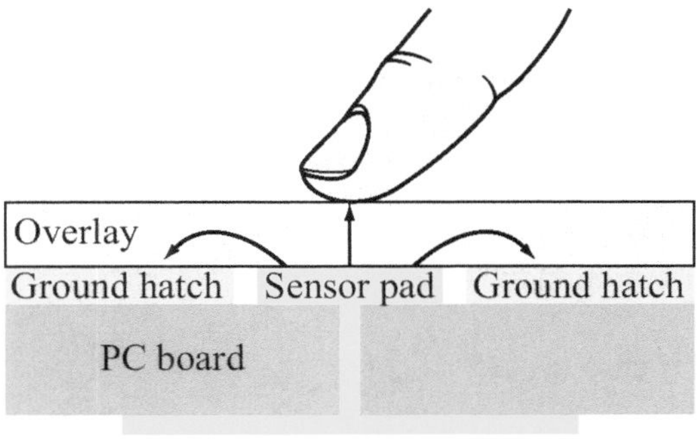

Figure TF14-1: A capacitive touch sensor uses the high permittivity of the finger to change the capacitance. The finger does not need to come in direct contact with the sensor in order to be detected.

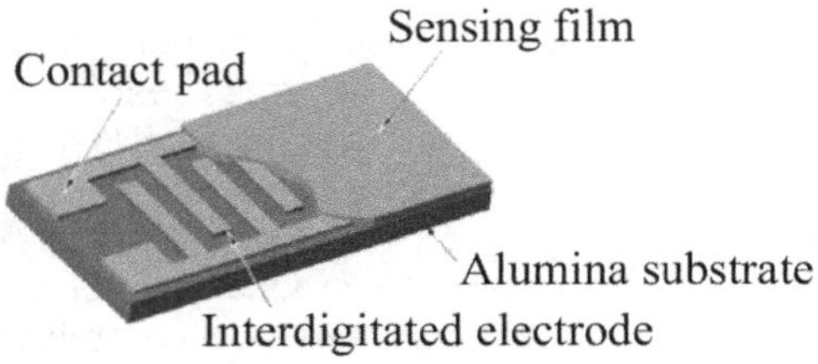

Figure TF14-2: Interdigitated humidity sensor. (Credit: Hygrometrix.)

proximity changes the overall capacitance of the circuit. The electric field starts on one of the conductors and ends on the other, basically making an arc between them. When the finger comes near either one or both of the two conductors, it changes this field (note the electric field arrow pointing straight up at the finger, which would not be there without the finger), and this in turn changes the capacitance. Another way to think about the process is in terms of the electric charge stored at the two conductors. The presence of the finger changes the effective permittivity of the medium through which the electric field lines flow, thereby changing the effective capacitance C. Since for any capacitor, $C = Q/V$—where Q is the charge on the conductor connected to the positive terminal of the voltage source and V is the voltage of the source—it follows that increasing C leads to an increase in Q (with V remaining constant). Hence, when the finger approaches the sensor pad, additional charge accumulates at the two conductors (with more $+Q$ at the sensor pad and a corresponding $-Q$ at the ground hatch).

Humidity Sensor

Another example of a capacitive sensor that also relies on measuring the change in permittivity is the humidity sensor featured in Fig. TF14-2. A sensing film absorbs moisture from the air, thereby changing the capacitance of the interdigitated line in proportion to the humidity in the air surrounding the sensor.

"Seeing" through Walls

The capacitive sensing technique also is used to "see" inside boxes, through walls, or through basically any low-conductivity low-permittivity material (paper, plastic, glass, etc.). An example is illustrated in Fig. TF14-3, in which a capacitive sensor on an assembly line is used to determine if a metal object is placed inside a box. The

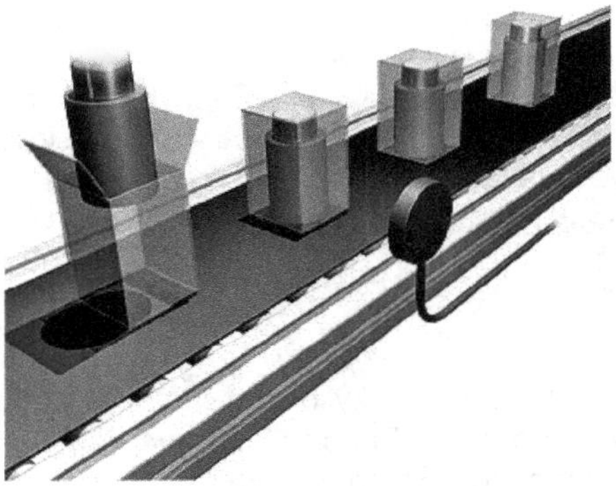

Figure TF14-3: Capacitive proximity sensors can "see" through low permittivity materials such as paper, cardboard, plastic, and glass and detect objects composed of a wide variety of materials including metals, fluids, etc. Here, a capacitive sensor detects the contents of a box. (Graphic courtesy of Balluff.)

object does not have to be metal, but its permittivity has to be significantly different from that of the paper or plastic enclosure. A similar application of capacitive sensors is to locate wooden studs through plaster walls.

Fluid Gauge

Capacitive sensors can serve as fluid gauges by measuring the height of a fluid in a tank or reservoir. Examples include gasoline and oil level gauges used in cars. If the tank is made of plastic or glass, metal strips on the outside of the tank can determine the height of the fluid without having to make contact with the fluid. This is very useful when the fluid is caustic or sterile. If the tank is metal, the strips must be placed inside. In either case, the sensor consists of two capacitors, one (C_2 in Fig. TF14-4) with metal plates separated by a reference fluid, and another (C_1) in which the fluid level is a variable. If the permittivity of the fluid is ε and the height of the fluid in the upper container in Fig. TF14-4 is h, the ratio of the two capacitances is given by

$$\frac{C_1}{C_2} = ah + b,$$

where a and b are known constants related to ε and the dimensions of the two capacitors. Hence, by measuring the two capacitances with an external circuit, the sensor provides a direct measurement of the fluid height h.

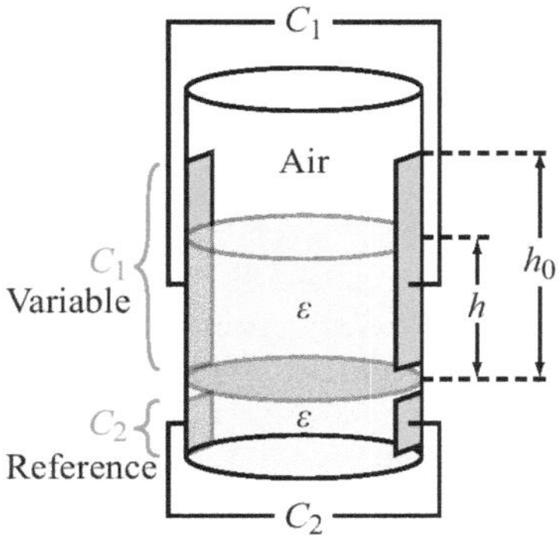

Figure TF14-4: Fluid height can be measured from the outside of a plastic or glass tank using a pair of parallel plate capacitors on the outside of the tank.

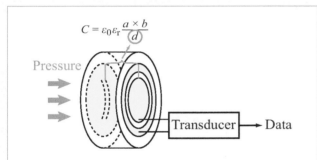

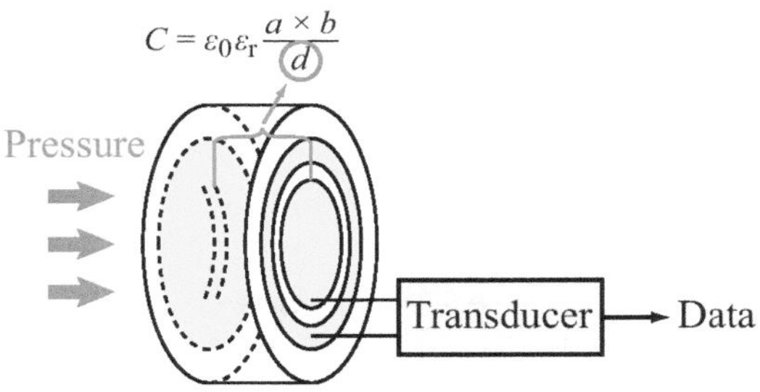

Figure TF14-5: Capactive transducer responding to pressure from a sound wave.

Applications Based on Change in Inter-Conductor Distance d

As noted earlier, the capacitance C is inversely proportional to the distance d between the two conductors. This dependence can be used to measure pressure, as illustrated by the diagram in Fig. TF14-5. We call such a sensor an electrical ***transducer*** because it converts one type of energy (mechanical) into another (electrical). The capacitor has one stationary conducting plate on the back side and a flexible conducting membrane on the side exposed to the incident pressure carried by an acoustic wave. The sound wave causes the membrane to vibrate, thereby changing the capacitance, which is measured and processed by an external circuit. This type of capacitive transducer is used in numerous industrial applications.

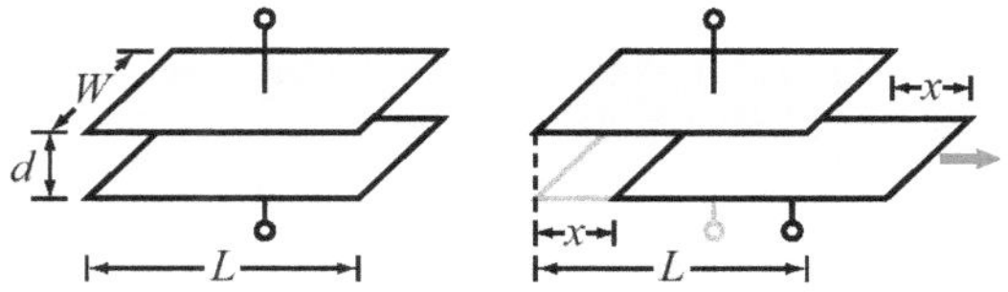

Figure TF14-6: Capacitance is proportional to overlap area $A = W(L - x)$, so when plates slide past each other the capacitance decreases in proportion to the shifted distance x.

Applications Based on Change in Area A

The change in the effective area common to the two conducting surfaces can also change the capacitance C. If one plate is slid past the other in Fig. TF14-6, the effective area A changes as a function of the shifted distance x. The capacitance is maximum when they are perfectly lined up, corresponding to $x = 0$, and changes approximately linearly as $(L - x)$. This can be used to align two objects, or to determine any other manual displacement in either one or two directions. The MEMS capacitive vibration sensor shown in Fig. TF14-7 uses two interdigital electrodes, one static and another moveable. When mounted in a car, for example, car acceleration or deceleration causes the moveable electrode to respond accordingly, which changes the capacitance between the two electrodes, thereby providing the means to measure acceleration. Such a sensor is called an ***accelerometer***.

Figure TF14-7: Microelectromechanical system (MEMS) vibration sensor using interdigitated static and movable electrodes. (Credit: STMicroelectronics.)

The step function $u(t)$ denotes that the source is connected to the circuit at $t = 0$. In your circuit, you may use a sinusoidal source of any amplitude and angular frequency.

Solution: Using op amps, multiple circuit configurations can be constructed to solve the given differential equation. One such configuration is shown in Fig. 5-45. If in Eq. (5.140) we denote $dv/dt = v'$ and $d^2v/dt^2 = v''$ and then solve for v, we have

$$\begin{aligned} v &= -\frac{1}{2}\, v'' - 4v' + 2v_s(t) \\ &= -\frac{1}{2}\, v'' - 4v' + 6\sin(200t)\; u(t). \end{aligned} \qquad (5.141)$$

One approach for designing this circuit is to realize that the output must be v, and somehow within the circuit we will also need v' and v''. We can design a differentiator with a gain of 1 and feed in the v (output), and then feed that into a second differentiator to get v''. The values of v', v'', and v_s can be combined by a weighted op-amp summer in which the gains can be adjusted to obtain the desired output v.

In Fig. 5-45, v is the output of op amp 4, as well as the input to op amp 1, which is a differentiator with a gain factor of $-RC = -1$ (the values of R and C are selected such that their product is 1). The output of op amp 1 is simply $-v'$. When followed by a second differentiator (op amp 2), we obtain v''. Op amp 3 serves as an inverter with gain of -1. Finally, op amp 4 is a summing amplifier that performs the sum of all three terms in Eq. (5.141). The values of the resistors preceding the summing point at the input to op amp 4 are selected to provide the correct weights, namely $(6R/12R) = 1/2$ for v'', $(6R/1.5R) = 4$ for v', and $(6R/R) = 6$ for the sinusoidal source. The switch serves to initiate the process at $t = 0$. Prior to that, $v = 0$. To avoid saturation, the supply voltage V_{cc} of each op amp should exceed the maximum possible voltage at its output.

If one were to construct the circuit and close the switch, the voltage $v(t)$ observed at the output of op amp 4 would be the same solution we would obtain were we to solve the differential equation analytically.

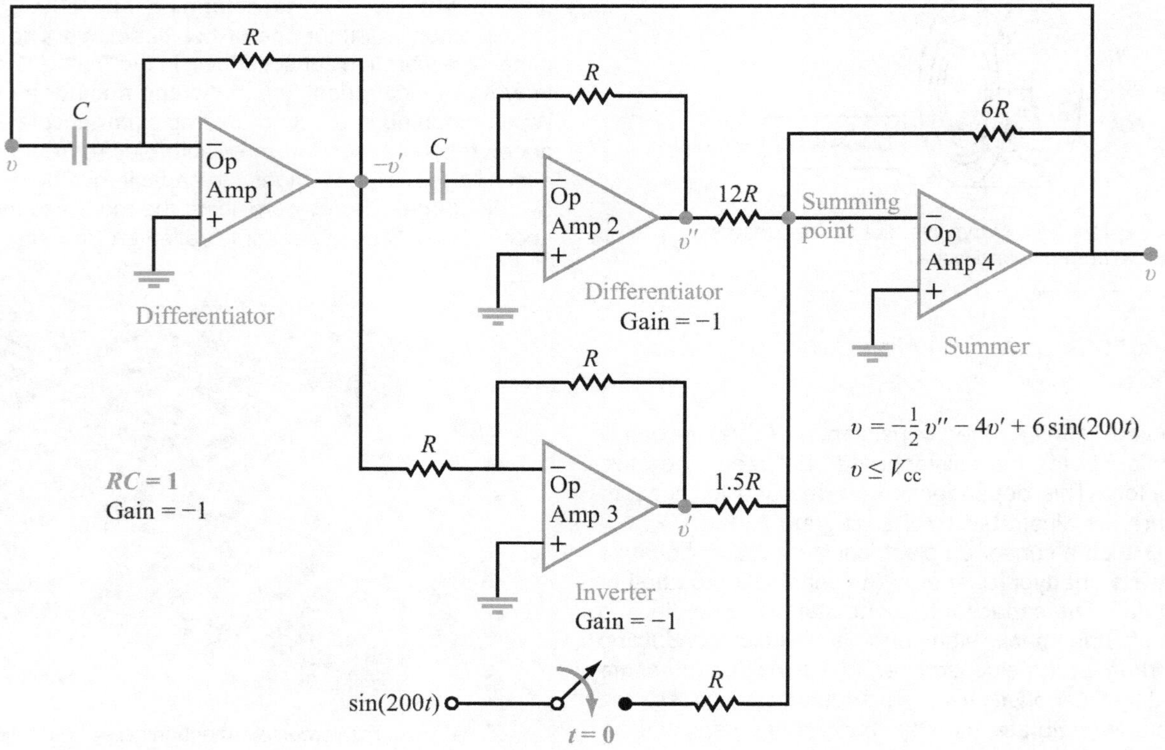

Figure 5-45: Op-amp circuit whose output $v(t)$ is a solution to $v'' + 8v' + 2v = 12\sin(200t)\; u(t)$.

Concept Question 5-24: What causes clipping of the waveform at the output of an op-amp integrator circuit? Can clipping occur at the output of a differentiator circuit? (See CAD)

Concept Question 5-25: If $\upsilon_s(t)$ is the input signal to a two-stage op-amp circuit with the first stage being an integrator with $R_1C_1 = 0.01$ and the second stage being a differentiator with $R_2C_2 = 0.01$, under what circumstances will the output waveform $\upsilon_{out}(t)$ be the same or different from $\upsilon_s(t)$? (See CAD)

Exercise 5-17: The input signal to an ideal integrator circuit with $RC = 2 \times 10^{-3}$ s and $V_{cc} = 15$ V is given by $\upsilon_s(t) = 2 \sin 100t$ V. What is $\upsilon_{out}(t)$?

Answer: $\upsilon_{out}(t) = 10[\cos(100t) - 1]$ V. (See CAD)

Exercise 5-18: Repeat Exercise 5-17 for a differentiator instead of an integrator.

Answer: $\upsilon_{out}(t) = -0.4 \cos 100t$ V. (See CAD)

5-7 Application Note: Parasitic Capacitance and Computer Processor Speed

As was noted in Section 4-11 and in Technology Brief 10, the primary computational element in modern computer processors is the CMOS transistor. How quickly a single logic gate is able to switch its output between logic states 0 and 1 determines how fast the entire processor can perform complex calculations. Figure 5-46(a) displays a sample of a digital sequence, perhaps at the output of a digital inverter. The individual pulses, each denoting a logic state of 0 or 1, are each of duration T. *If it were possible to switch between states instantaneously, the maximum number of pulses that can be sequenced per 1 second is $1/T$. We refer to this rate by several names, including the* ***pulse repetition frequency, switching frequency,*** *and* ***clock speed.*** *In the present case, we shall call it the switching frequency* and assign it the symbol f_s. That is,

$$f_s = \frac{1}{T} \qquad \text{(Hz).} \tag{5.142}$$

So if $T = 1$ ns, $f_s = 1/10^{-9} = 1$ GHz, and if we can make the pulse duration narrower, we can increase f_s accordingly.

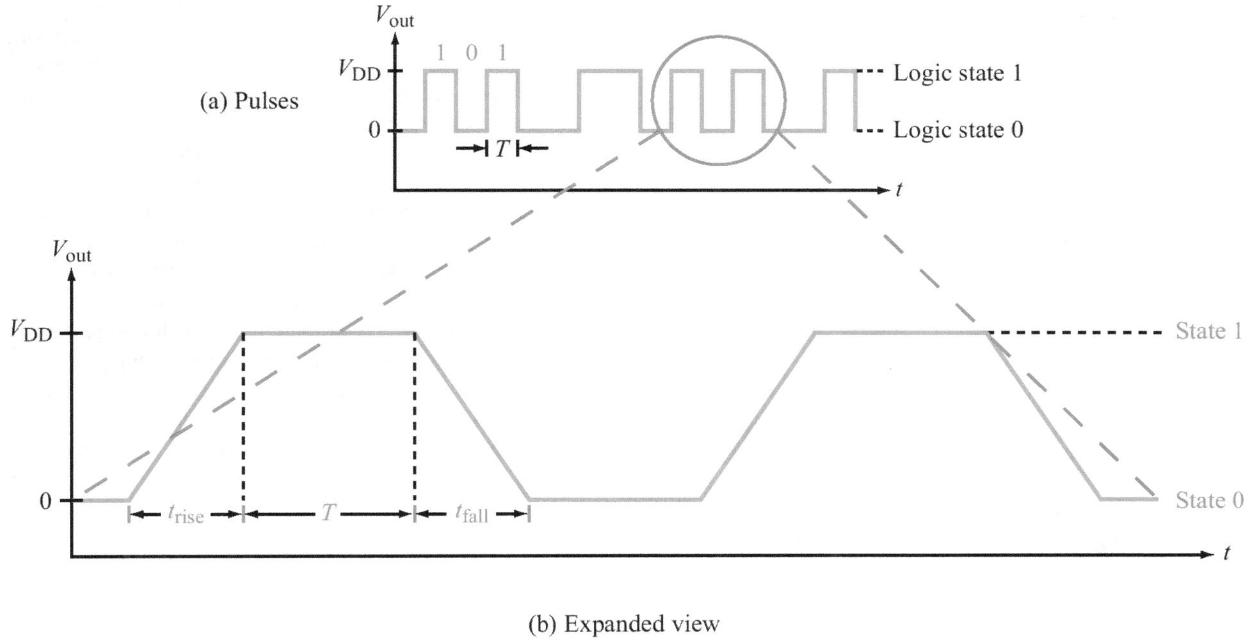

Figure 5-46: Pulse sequence.

Such a conclusion would be true if we can indeed arrange to have the logic circuit switch between states instantaneously, but it cannot. In Fig. 5-46(b), we show an expanded view of three pulses representing the sequence 101. We observe that the switching process is represented by ramp functions (rather than step functions) and it takes a finite amount of time for the voltage to change between a 0 state and a 1 state, which we shall call the *rise time* t_{rise}. Similarly, the *fall time* between states 1 and 0 is t_{fall}. [The linear rise and fall responses are actually artifacts of certain simplifying assumptions. In general, the responses involve exponentials, in which case it is more appropriate to define t_{rise} and t_{fall} as the durations between the 10 percent level and 90 percent level of the change in voltage.] The total time associated with a pulse is

$$T_{\text{total}} = T + t_{\text{rise}} + t_{\text{fall}} = T + 2t_{\text{rise}} \qquad (\text{if } t_{\text{rise}} = t_{\text{fall}}),$$

and the associated switching frequency is

$$f_s = \frac{1}{T_{\text{total}}} = \frac{1}{T + 2t_{\text{rise}}}.$$

Even if T can be reduced to zero, the maximum possible switching speed (without overlap between adjacent pulses) would be

$$f_s(\max) = \frac{1}{2t_{\text{rise}}}. \qquad (5.143)$$

As we shall see shortly, the switching times (t_{rise} and t_{fall}) are governed in part by the capacitances in the circuit. Consequently, capacitances play a major role in determining the ultimate switching speed of a digital circuit. In fact, capacitances also govern the switching speeds of the *wires*—often referred to as the *bus*—that connect the processor to the various other devices on a computer *motherboard*.

> ▶ Whereas the processor speed of a modern computer is in the GHz range, the *bus speed* usually is slower by a factor of 3 to 10. ◀

This is (in part) why a computer appears to slow down when the processor needs to access data through the bus. The following section will examine why this is so.

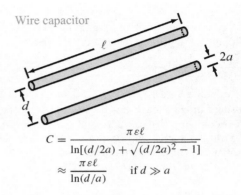

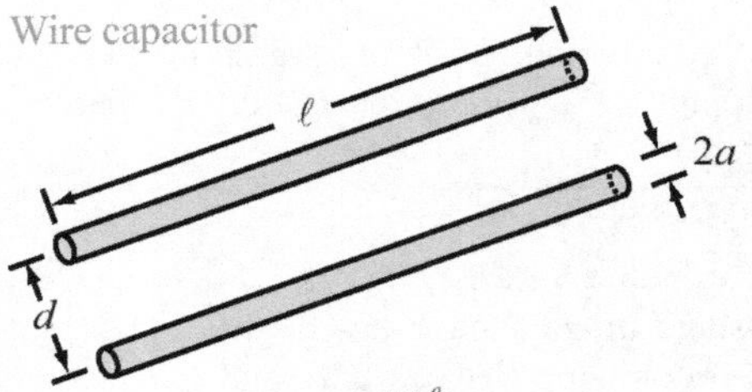

$$C = \frac{\pi\varepsilon\ell}{\ln[(d/2a) + \sqrt{(d/2a)^2 - 1}]} \approx \frac{\pi\varepsilon\ell}{\ln(d/a)} \quad \text{if } d \gg a$$

Figure 5-47: Capacitance of a two-wire configuration where ε is the permittivity of the material separating the wires.

5-7.1 Parasitic Capacitance

Functionally, any two conducting bodies separated by an insulating material (including air, plastic, and all non-conductors) form a capacitor. The capacitors we have considered thus far are the type designed and fabricated intentionally for use as components in circuits. In some situations, however, *unintentional* capacitance may exist in the circuit, in which case it usually is called *parasitic capacitance*. (Parasitic inductance also is present, but it is usually very small, so we will ignore it.) Consider, for example, the capacitance formed by two parallel wires running side by side on a circuit board. The capacitance of such a two-wire *transmission line* (Fig. 5-47) is proportional directly to the length of the wires ℓ and inversely proportional to a logarithmic function involving d, the spacing between the wires. Thus, C increases with ℓ and decreases with d. If the wires are sufficiently long, or sufficiently close to one another, or some combination of the two [as to result in a capacitance of significant magnitude relative to the other capacitances in the circuit] such a wire capacitor (the conductor traces between the different components in the circuit) can slow down the response time of the circuit. In a digital circuit, slower response time means slower switching speed. To explore this subject further, we now examine the impact of parasitic capacitance on the operation of a MOSFET.

5-7.2 CMOS Switching Speed

Recall from Section 4-11 that the gate node in a MOSFET is composed of a metal and a semiconductor separated by a thin layer of silicon dioxide that serves as a dielectric insulator. This geometry is somewhat similar to that of the parallel-plate capacitor of Fig. 5-11. Hence, during normal

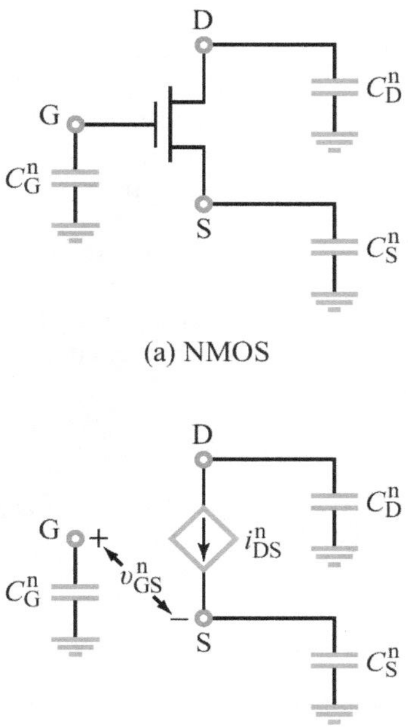

(a) NMOS

(b) Equivalent circuit

Figure 5-48: n-channel MOSFET (NMOS): (a) circuit symbol with added parasitic capacitances and (b) equivalent circuit. [In a PMOS, parasitic capacitances C_D^p and C_S^p should be shown connected to V_{DD} instead of to ground.]

operation, the gate (G) and the source (S) nodes form a capacitor between them, as do the gate and the drain (D) nodes. Other parasitic capacitances also exist in a MOSFET, mainly due to charges separated between the source and the large silicon chip and between the drain and the chip. For simplicity, the various parasitic capacitances can be lumped together into an equivalent model containing three capacitances (all connected to ground) from G, S, and D. As shown in Fig. 5-48, these capacitances are designated C_G^n, C_S^n, and C_D^n, respectively, with the superscript "n" denoting that the circuit configuration applies to the n-channel MOSFET (or NMOS for short) whose body node usually is connected to ground. In a p-channel MOSFET, the body node is connected to V_{DD}. Hence, the model for PMOS would show parasitic capacitances C_D^p and C_S^p connected to V_{DD}, instead of to ground.

Now we are ready to analyze the operation of a CMOS inverter in the presence of parasitic capacitances. The circuit in Fig. 5-49(a) is essentially the same CMOS circuit of Fig. 4-30, except with added parasitic capacitances. The capacitances

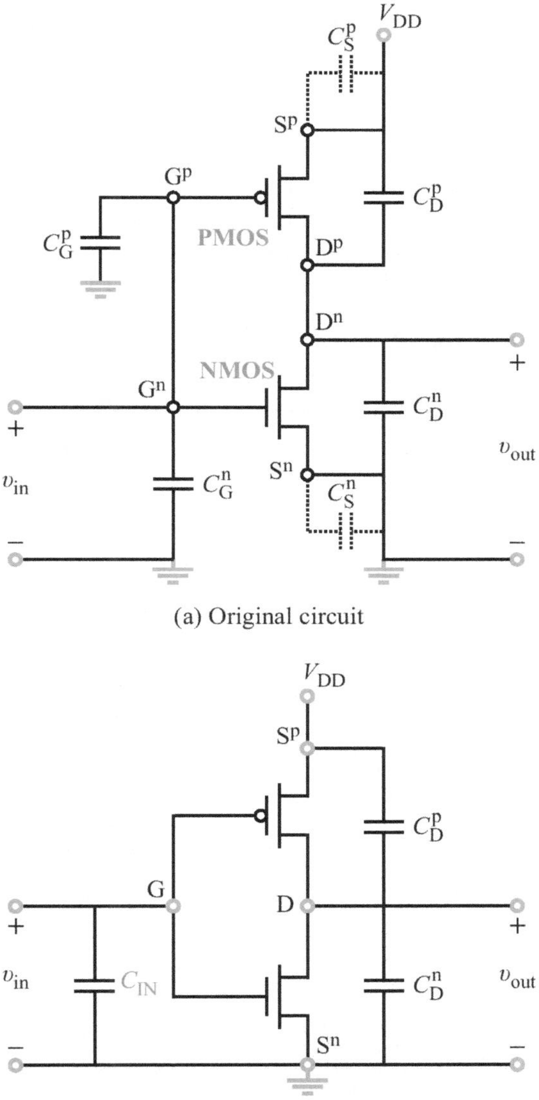

(a) Original circuit

(b) Simplified circuit

Figure 5-49: Common drain inverter circuit with parasitic capacitances. Superscripts "n" and "p" refer to the NMOS and PMOS transistors, respectively.

associated with the n-channel MOSFET are shown connected from terminals G^n, D^n, and S^n to ground. For the p-channel MOSFET, capacitance C_G^p is also connected to ground, but for the other two terminals, the capacitances are shown connected

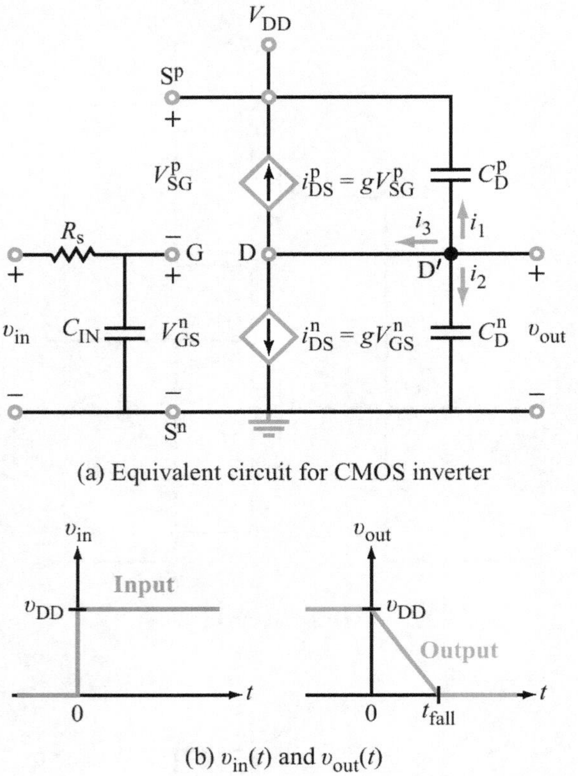

Figure 5-50: (a) Equivalent circuit for the CMOS inverter; (b) the response of $v_{out}(t)$ to v_{in} changing states from 0 to V_{DD} at $t = 0$.

to V_{DD}. The two MOSFETs share a common gate terminal at the input side and a common drain terminal at the output side. Terminal S^n of the NMOS is connected directly to ground, which renders capacitance C^n_S irrelevant. Terminal S^p of the PMOS is connected directly to V_{DD}, which similarly renders C^p_S irrelevant. Capacitances C^n_G and C^p_G both are connected from the common gate terminal to ground and therefore can be combined into an *equivalent capacitance* C_{IN}. Incorporating these simplifications leads to the circuit shown in Fig. 5-49(b).

Our next step is to determine the output response $v_{out}(t)$ to a sudden change of state at the input from $v_{in} = 0$ to $v_{in} = V_{DD}$. Let us assume that the change happens at $t = 0$ and that the circuit was already in a steady-state condition by then.

(a) Initial condition at $t = 0^-$:

The capacitances in Fig. 5-50(a) act like open circuits. Also, $v_{in} = 0$, which means that $V^n_{GS} = 0$ for the NMOS and $V^p_{SG} = V_{DD}$ for the PMOS. Under such circumstances,

$$i^n_{DS} = gV^n_{GS} = 0, \qquad \text{and} \qquad i^p_{DS} = gV^p_{SG} = gV_{DD}, \tag{5.144}$$

where g is the MOSFET gain constant. Furthermore, the PMOS behavior is such that, if V^p_{SG} approaches V_{DD}, the voltage V^p_{DS} across the dependent current source goes to zero. With i^n_{DS} not conducting and i^p_{DS} acting like a short circuit, it follows that the voltage across capacitor C^n_D is

$$v_{out}(0^-) = V_{DD}. \tag{5.145}$$

Since the voltage across a capacitor cannot change instantaneously,

$$v_{out}(0) = V_{DD}. \tag{5.146}$$

(b) At $t \geq 0$:

If v_{in} is a step function that changes from 0 to V_{DD} at $t = 0$, the following pair of responses will take place:

(a) At the input side in the circuit of Fig. 5-50(a), we have an isolated loop comprising v_{in}, R_s, and C_{IN}. In response to the change in v_{in}, capacitor C_{IN} will charge up to a final voltage V_{DD} at a rate governed by the time constant $\tau = R_sC_{IN}$. Through proper choice of R_s (very small), C_{IN} can charge up to V_{DD} so quickly (in comparison with the response time of the output) that it can be assumed that $V^n_{GS} = V_{DD}$ immediately after $t = 0$.

(b) At the output side, with $V^n_{GS} = V_{DD}$, it follows that $V^p_{SG} = 0$. Hence,

$$i^n_{DS} = gV_{DD}, \qquad \text{and} \qquad i^p_{DS} = gV^p_{SG} = 0. \tag{5.147}$$

At node D',

$$i_1 + i_2 + i_3 = 0, \tag{5.148}$$

and at node D,

$$i_3 = i^n_{DS} + i^p_{DS} = gV_{DD}. \tag{5.149}$$

Also,

$$i_1 = C^p_D \frac{d}{dt}(v_{out} - V_{DD}) = C^p_D \frac{d}{dt} v_{out}, \tag{5.150}$$

and

$$i_2 = C^n_D \frac{d}{dt} v_{out}. \tag{5.151}$$

Upon inserting the expressions given by Eqs. (5.149) through (5.151) into Eq. (5.148) and then rearranging terms, we have

$$\frac{d\upsilon_{\text{out}}}{dt} = \frac{-gV_{\text{DD}}}{C_{\text{D}}^{\text{n}} + C_{\text{D}}^{\text{p}}}. \tag{5.152}$$

Integrating both sides from 0 to t gives

$$\upsilon_{\text{out}}\Big|_0^t = \frac{-gV_{\text{DD}}}{C_{\text{D}}^{\text{n}} + C_{\text{D}}^{\text{p}}} \int_0^t dt, \tag{5.153}$$

which leads to

$$\upsilon_{\text{out}}(t) = \upsilon_{\text{out}}(0) - \left(\frac{gV_{\text{DD}}}{C_{\text{D}}^{\text{n}} + C_{\text{D}}^{\text{p}}}\right) t. \tag{5.154}$$

In view of Eq. (5.146), the expression for $\upsilon_{\text{out}}(t)$ becomes

$$\upsilon_{\text{out}}(t) = V_{\text{DD}}\left[1 - \left(\frac{g}{C_{\text{D}}^{\text{n}} + C_{\text{D}}^{\text{p}}}\right) t\right]. \tag{5.155}$$

Plots of $\upsilon_{\text{in}}(t)$ changing states from 0 to V_{DD} at $t = 0$ and of the corresponding response $\upsilon_{\text{out}}(t)$ are displayed in Fig. 5-50(b). We observe that t_{fall} is the time it takes for υ_{out} to change states from V_{DD} to zero. From Eq. (5.155), we deduce that

$$t_{\text{fall}} = \frac{C_{\text{D}}^{\text{n}} + C_{\text{D}}^{\text{p}}}{g}. \tag{5.156}$$

Example 5-19: Processor Speed

The input to a CMOS inverter consists of a sequence of bits, each 25 picoseconds in duration. Determine the maximum switching frequency at which the CMOS inverter can be operated without causing overlap between adjacent bits (pulses) under each of the following conditions: (a) parasitic capacitances totally ignored and (b) parasitic capacitances included. In both cases, $g = 10^{-5}$ A/V, and $C_{\text{D}}^{\text{n}} = C_{\text{D}}^{\text{p}} = 0.5$ fF.

Solution: (a) With $T = 25$ ps $= 25 \times 10^{-12}$ s and no capacitances to slow down the switching process, the maximum switching frequency is

$$f_{\text{s}} = \frac{1}{T} = \frac{1}{25 \times 10^{-12}} = 40 \text{ GHz}.$$

(b) From Eq. (5.156),

$$t_{\text{fall}} = \frac{C_{\text{D}}^{\text{n}} + C_{\text{D}}^{\text{p}}}{g} = \frac{(0.5 + 0.5) \times 10^{-15}}{10^{-5}} = 10^{-10} \text{ s}.$$

To determine t_{rise}, we have to repeat the solution that led to Eq. (5.156) but with υ_{in} starting in state 1 (i.e., $\upsilon_{\text{in}} = V_{\text{DD}}$) and switching to state 0 at $t = 0$. Such a process would lead to

$$\upsilon_{\text{out}}(t) = V_{\text{DD}}\left(\frac{g}{C_{\text{D}}^{\text{n}} + C_{\text{D}}^{\text{p}}}\right) t.$$

The time duration that it takes $\upsilon_{\text{out}}(t)$ to reach V_{DD} is

$$t_{\text{rise}} = \frac{C_{\text{D}}^{\text{n}} + C_{\text{D}}^{\text{p}}}{g} = t_{\text{fall}}.$$

Hence, in the presence of parasitic capacitances, Eq. (5.143) is applicable. Namely,

$$f_{\text{s}} = \frac{1}{T + 2t_{\text{rise}}} = \frac{1}{25 \times 10^{-12} + 2 \times 10^{-10}} = 4.44 \text{ GHz}.$$

In this example, the parasitic capacitances are responsible for slowing down the switching speed of the CMOS processor by about one order of magnitude.

In the preceding example, we essentially ignored the input capacitances of the CMOS. Since logic gates are strung along in series such that one gate's output is the next gate's input, input capacitances usually are lumped together with the previous gate's output capacitances. To properly incorporate the roles of both input and output parasitic capacitances, a more thorough treatment is needed than the first-order approximation we carried out in this section. Nevertheless, the approximation did succeed in making the point that at high switching rates parasitic capacitances are important and should not be ignored.

Concept Question 5-26: What is the rationale for adding parasitic capacitances to nodes G, D, and S in Fig. 5-48? (See CAD)

Concept Question 5-27: What determines the maximum switching frequency for a CMOS inverter? (See CAD)

Exercise 5-19: A CMOS inverter with $C_{\text{D}}^{\text{n}} + C_{\text{D}}^{\text{p}} = 20$ fF has a fall time of 1 ps. What is the value of its gain constant?

Answer: $g = 2 \times 10^{-2}$ A/V. (See CAD)

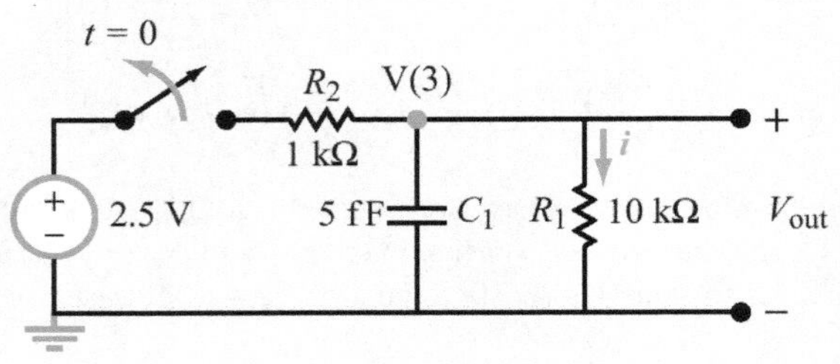

Figure 5-51: RC circuit with an SPST switch.

5-8 Analyzing Circuit Response with Multisim

5-8.1 Modeling Switches in Multisim

Determining the time-dependent behavior of large, complex circuits often is difficult to do and extremely time-consuming. Accordingly, designs of commercial circuits rely heavily on SPICE simulators for evaluating the response of a candidate circuit design before constructing the real version. In this section, we demonstrate how Multisim can be used to analyze the transient response of a circuit driven by a time-dependent source.

Because the first-order RC circuit is straightforward to analyze *by hand*, it makes for a useful example with which we can compare Multisim simulation results to hand calculations. Consider the circuit shown in Fig. 5-51, in which the switch is opened at $t = 0$ after it had been in the closed position for a long time. Hence, prior to $t = 0$, the circuit was in a steady state and the capacitor was fully charged with no current flowing through it (behaving like an open circuit). The voltage across the capacitor is designated V(3) (so as to match the Multisim circuit that we will be constructing soon) and is given by

$$\text{V}(3) = \frac{2.5 \times 10\text{ k}}{1\text{ k} + 10\text{ k}} = 2.27\text{ V} \quad (@\ t = 0^-).$$

Upon opening the switch, the capacitor will discharge through the 10 kΩ resistor with a time constant given by

$$\tau_{\text{discharge}} = R_1 C_1 = 10^4 \times 5 \times 10^{-15} = 50\text{ ps}.$$

Likewise, if the switch were to close at a later time after the circuit had fully discharged, the capacitor would again charge up to 2.27 V, but in this case, the time constant would be

$$\tau_{\text{charge}} = (R_1 \parallel R_2)C_2 = \frac{1\text{ k} \times 10\text{ k}}{11\text{ k}} \times 5 \times 10^{-15} = 4.54\text{ ps}.$$

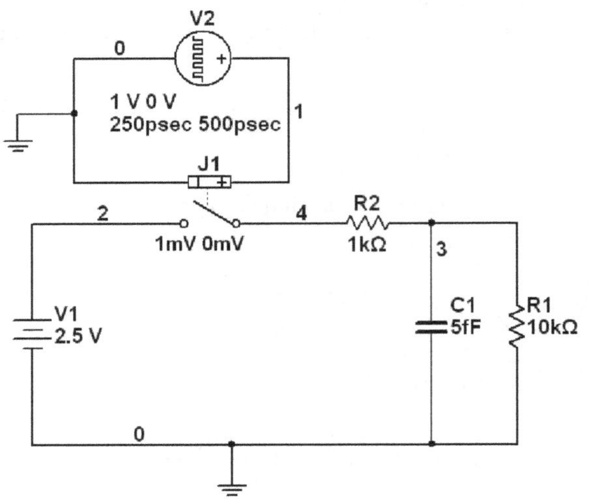

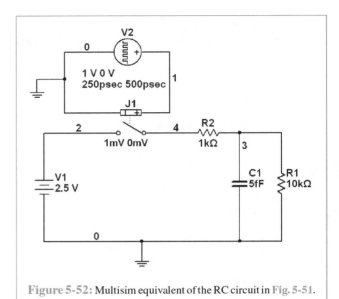

Figure 5-52: Multisim equivalent of the RC circuit in Fig. 5-51.

Thus, the charge-up response of the circuit is much faster (by about one order of magnitude) than its discharge response.

To demonstrate the transient behavior of the circuit with Multisim, we construct the circuit model shown in Fig. 5-52 using the component list given in Table 5-6. The only oddity in the circuit is the use of a Voltage-Controlled Switch and a Pulse Generator source to drive it. Multisim does not provide the user the option to use time-programmable switches, so in order to observe the circuit response to multiple opening and closing events of the switch, we use a voltage-controlled switch in combination with an appropriately configured pulse generator. The exact voltage amplitude of the pulse (V2 in Fig. 5-52) is not important (so long as it is larger than the 1 mV threshold of the switch), but the timing of the pulse is critically important, as we want to allow enough time between opening and closing events to observe the complete transient responses of the circuit. Since the longest time constant is 50 ps, double-click on the Pulse Generator and set the Pulse width at 250 ps and the Period at 500 ps so as to provide an adequate time window. Also set the Rise Time and Fall Time to 1 ps.

To analyze the behavior, we select Simulate → Analyses → Transient Analysis. Make sure to select an End Time equal to a few periods; 3 ns should suffice. (If you forget this, you may need to abort the simulation to prevent it from running for a long time since the default value is 0.001 s! To abort the simulation or any general Analyses which may be taking too long, go to Simulate → Analyses → Stop Analysis.) In the Output tab, select the non-ground node of the capacitor V(3) and the pulse voltage V(1) for time references. Figure 5-53 shows the

Table 5-6: Multisim component list for the circuit in Fig. 5-52.

Component	Group	Family	Quantity	Description
1 k	Basic	Resistor	1	1 kΩ resistor
10 k	Basic	Resistor	1	10 kΩ resistor
5 f	Basic	Capacitor	1	5 fF capacitor
VOLTAGE_CONTROLLED_SPST	Basic	Switch	1	Switch
DC_POWER	Sources	Power_Sources	1	2.5 V dc source
PULSE_VOLTAGE	Sources	Signal_Voltage_Source	1	Pulse-generating voltage source

output of the transient analysis. Enabling the Cursor tool in the Grapher window allows the user to read out the exact voltage and time values for any trace.

5-8.2 Modeling Time-Dependent Sources in Multisim

In the previous subsection, we examined how to create switches that toggle with time. What if we wanted to simulate the circuit shown in Fig. 5-54(a) and plot υ_C over a certain time duration? The circuit has three time-dependent sources, which would make adding switches and pulse generators rather complicated. Multisim allows us to create the time-dependent sources found in this circuit by using the ABM Voltage and Current sources.

In Multisim's ABM syntax, the step function $u(t)$ is represented by the stp(TIME) function. Also, to guard against Multisim calculating incorrect initial conditions prior to the step function, it is advisable to shift the step-function transition to occur 10 ms after the start of the simulation. Hence, we use the

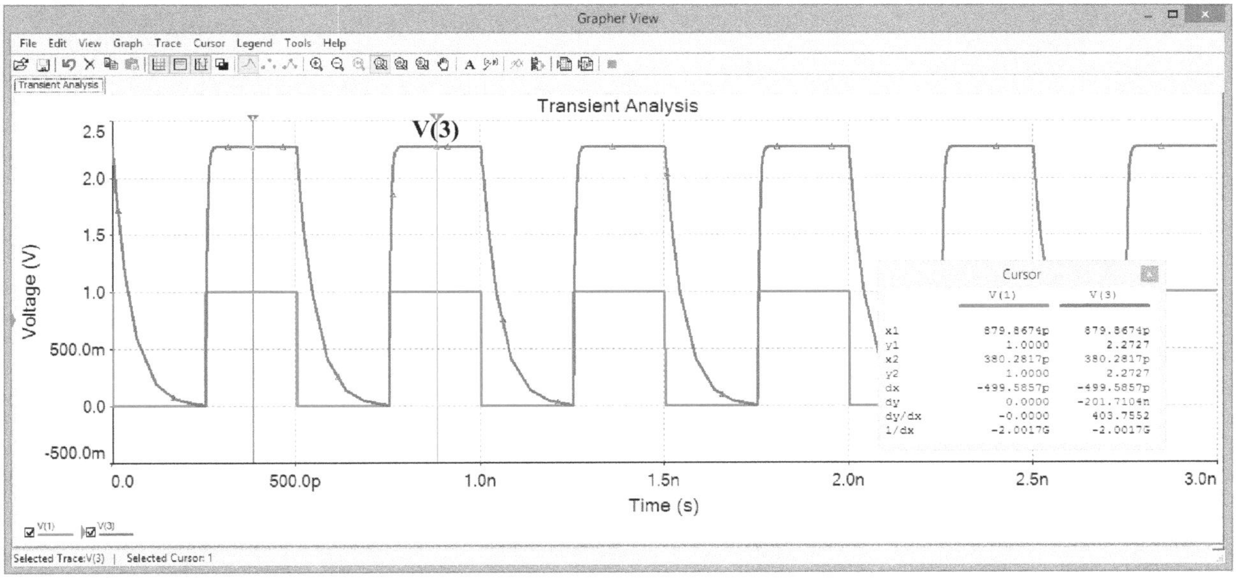

Figure 5-53: Transient response of the circuit in Fig. 5-52.

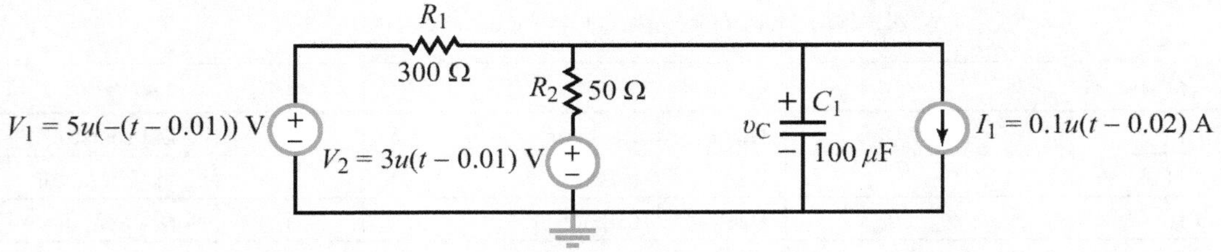

(a) Circuit with three time-dependent sources

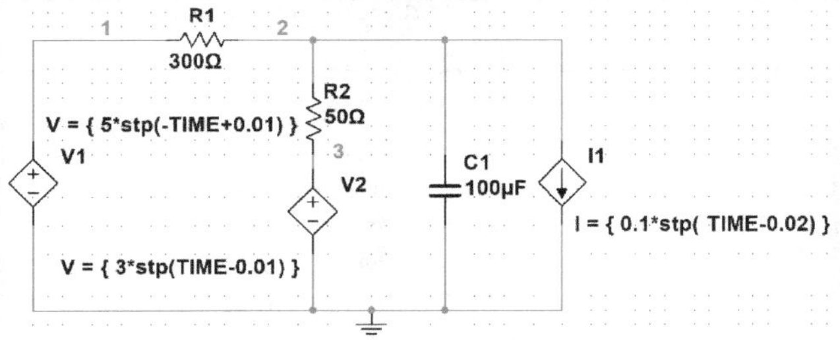

(b) Multisim circuit

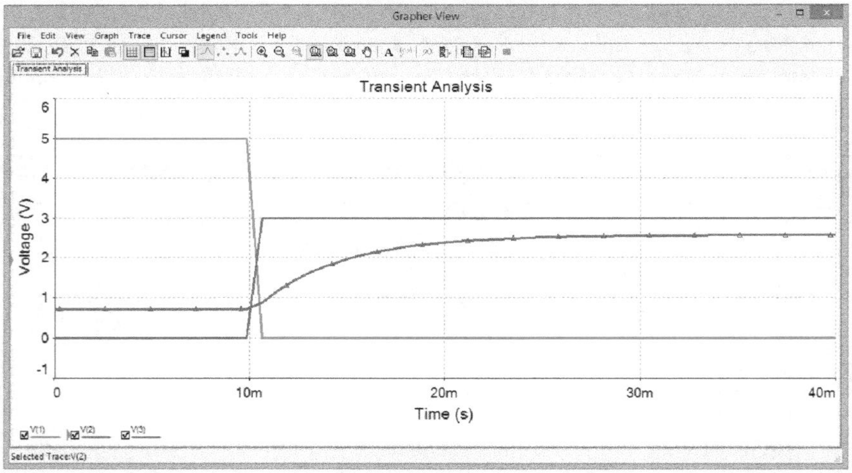

(c) Trace of $v_C(t)$

Figure 5-54: Multisim analysis of a circuit containing time-dependent sources.

following ABM expressions:

For V1 $= 5u(-(t-0.01))$ V: ➡ 5*stp(-TIME+0.01)

For V2 $= 3u(t-0.01)$ V: ➡ 3*stp(TIME-0.01)

For I1 $= 0.1u(t-0.02)$ A: ➡ 0.1*stp(TIME-0.02)

Once these expressions have been entered, go to Simulate → Analyses → Transient Analysis. Leave the Start Time at 0 s, and set the End Time to 0.04 s. Under the Output tab, select the voltages V(1), V(2), and V(3) and press Simulate. This generates the plots shown in Fig. 5-54(c).

Summary

Concepts

- The step, ramp, rectangle, and exponential functions can be used to characterize a variety of nonperiodic waveforms.
- A capacitor stores electrical energy when a voltage exists across it.
- An inductor stores magnetic energy when a current passes through it.
- Under dc conditions, a capacitor acts like an open circuit and an inductor acts like a short circuit.
- A series RC circuit excited by a dc source exhibits a voltage response (across the capacitor) characterized by an exponential function containing a time constant $\tau = RC$.
- A parallel RL circuit exhibits a current response (through the inductor) that has the same form as the voltage response of the series RC circuit, but for the RL circuit, $\tau = L/R$.
- The output voltage of the ideal op-amp RC integrator circuit is directly proportional to the time integral of the input signal.
- An integrator circuit becomes a differentiator circuit upon interchanging the locations of R and C.
- Parasitic capacitance is often the factor that ultimately limits the processor speed of a computer.
- Multisim allows us to evaluate the switching response of a circuit.

Mathematical and Physical Models

Unit step function

$$u(t) = \begin{cases} 0 & \text{for } t < 0 \\ 1 & \text{for } t > 0 \end{cases}$$

Time-shifted step function

$$u(t-T) = \begin{cases} 0 & \text{for } t < T \\ 1 & \text{for } t > T \end{cases}$$

Unit ramp function

$$r(t) = \begin{cases} 0 & \text{for } t \le 0 \\ t & \text{for } t \ge 0 \end{cases}$$

Time-shifted ramp function

$$r(t-T) = \begin{cases} 0 & \text{for } t \le T \\ (t-T) & \text{for } t \ge T \end{cases}$$

Unit rectangular function
(pulse center at $t = T$; pulse length $= \tau$)

$$\text{rect}\left[\frac{(t-T)}{\tau}\right] = \begin{cases} 0 & \text{for } t < (T - \tau/2), \\ 1 & \text{for } (T-\tau/2) \le t \le (T+\tau/2), \\ 0 & \text{for } t > (T+\tau/2). \end{cases}$$

Capacitor

$$i = C\,\frac{d\upsilon}{dt}$$

$$\upsilon(t) = \upsilon(t_0) + \frac{1}{C}\int_{t_0}^{t} i\,dt'$$

$$w = \tfrac{1}{2}\,C\upsilon^2 \qquad \text{(stored electrical energy)}$$

Parallel plate $C = \dfrac{\varepsilon A}{d}$

Inductor

$$\upsilon = L\,\frac{di}{dt}$$

$$i(t) = i(t_0) + \frac{1}{L}\int_{t_0}^{t} \upsilon\,dt'$$

$$w = \tfrac{1}{2}\,Li^2 \qquad \text{(stored magnetic energy)}$$

Solenoid $L = \dfrac{\mu N^2 S}{\ell}$

Series RC circuit response (sudden change at $t = 0$)

$$\upsilon_C(t) = \upsilon_C(\infty) + [\upsilon(0) - \upsilon(\infty)]e^{-t/\tau}$$
$$\tau = RC$$

Parallel RL circuit response (sudden change at $t = 0$)

$$i_L(t) = i_L(\infty) + [i_L(0) - i_L(\infty)]e^{-t/\tau}$$
$$\tau = L/R$$

Op-amp integrator

$$\upsilon_{out}(t) = -\frac{1}{RC}\int_{t_0}^{t} \upsilon_i\,dt' + \upsilon_{out}(t_0)$$

Op-amp differentiator

$$\upsilon_{out}(t) = -RC\,\frac{d\upsilon_i}{dt}$$

Important Terms Provide definitions or explain the meaning of the following terms:

air-core solenoid
bus
bus speed
capacitance
capacitor
charge/discharge
charged capacitor
circuit response
clip
clock speed
coaxial capacitor
dc condition
duration of the pulse
dynamic circuit
early time response
electric field
electrical permittivity
electrical susceptibility
equivalent capacitance
exponential function
ferrite-core inductor
final condition
final value
first-order circuit
first-order RC circuit
forced response
forcing function
inductance
initial value
iron-core solenoid
magnetic field
magnetic flux linkage
magnetic permeability
mica capacitor
motherboard
mutual inductance
nanocapacitor
natural decay response
natural response
negative exponential function
nonperiodic waveform
op-amp differentiator
op-amp integrator
parallel-plate capacitor
parasitic capacitance
periodic waveform
permeability
permittivity
plastic-foil capacitor
pulse repetition frequency
pulse waveform
ramp function
RC circuit
rectangle function
rectangular pulse
relative permittivity
rise time
RL circuit
scaling factor
self-inductance
solenoid
source-free
source-free, first-order differential equation
static
steady-state component
steady-state response
step function
step function response
supercapacitor
switching frequency (speed)
time constant
time-shifted ramp function
time-shifted step function
transient component
transient response
transmission line
uncharged capacitor
unit rectangular function
unit step function

PROBLEMS

Section 5-1: Nonperiodic Waveforms

5.1 Generate plots for each of the following step-function waveforms over the time span from −5 to +5 s.

(a) $\upsilon_1(t) = -6u(t+3)$

(b) $\upsilon_2(t) = 10u(t-4)$

(c) $\upsilon_3(t) = 4u(t+2) - 4u(t-2)$

(d) $\upsilon_4(t) = 8u(t-2) + 2u(t-4)$

(e) $\upsilon_5(t) = 8u(t-2) - 2u(t-4)$

5.2 Provide expressions in terms of step functions for the waveforms displayed in Fig. P5.2.

***5.3** A 10 V rectangular pulse with a duration of 5 μs starts at $t = 2$ μs. Provide an expression for the pulse in terms of step functions.

5.4 Generate plots for each of the following functions over the time span from −4 to +4 s.

(a) $\upsilon_1(t) = 5r(t+2) - 5r(t)$

(b) $\upsilon_2(t) = 5r(t+2) - 5r(t) - 10u(t)$

(c) $\upsilon_3(t) = 10 - 5r(t+2) + 5r(t)$

(d) $\upsilon_4(t) = 10 \text{ rect}\left(\frac{t+1}{2}\right) - 10 \text{ rect}\left(\frac{t-3}{2}\right)$

(e) $\upsilon_5(t) = 5 \text{ rect}\left(\frac{t-1}{2}\right) - 5 \text{ rect}\left(\frac{t-3}{2}\right)$

5.5 Provide expressions for the waveforms displayed in Fig. P5.5 in terms of ramp and step functions.

5.6 Provide plots for the following functions (over a time span and with a time scale that will appropriately display the shape of the associated waveform):

(a) $\upsilon_1(t) = 100e^{-2t}\, u(t)$

(b) $\upsilon_2(t) = -10e^{-0.1t}\, u(t)$

(c) $\upsilon_3(t) = -10e^{-0.1t}\, u(t-5)$

(d) $\upsilon_4(t) = 10(1 - e^{-10^3 t})\, u(t)$

(e) $\upsilon_5(t) = 10e^{-0.2(t-4)}\, u(t)$

(f) $\upsilon_6(t) = 10e^{-0.2(t-4)}\, u(t-4)$

*Answer(s) available in Appendix G.

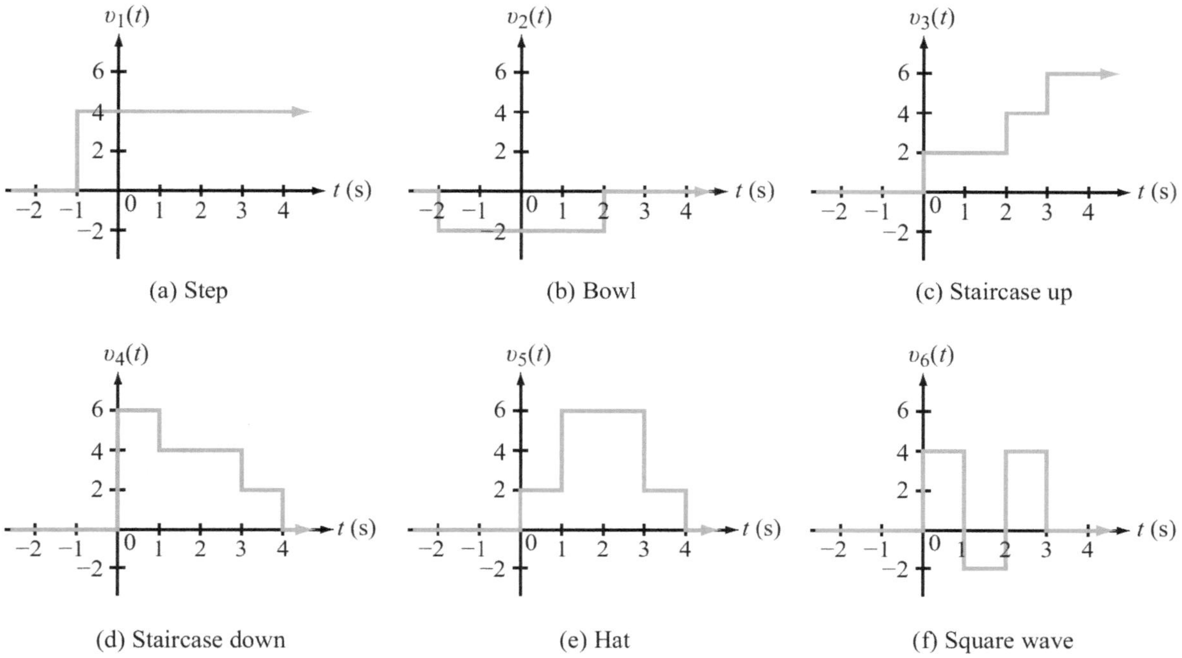

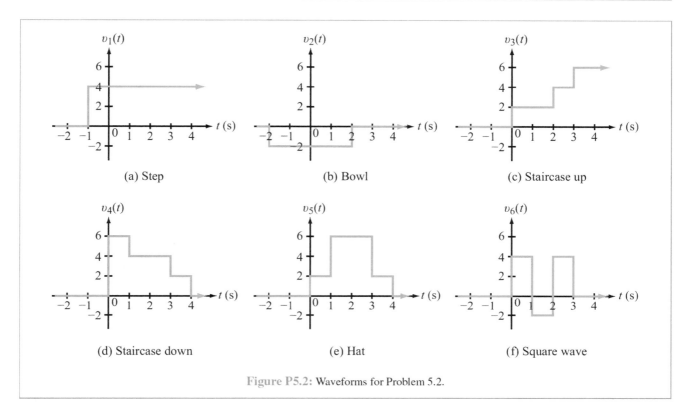

Figure P5.2: Waveforms for Problem 5.2.

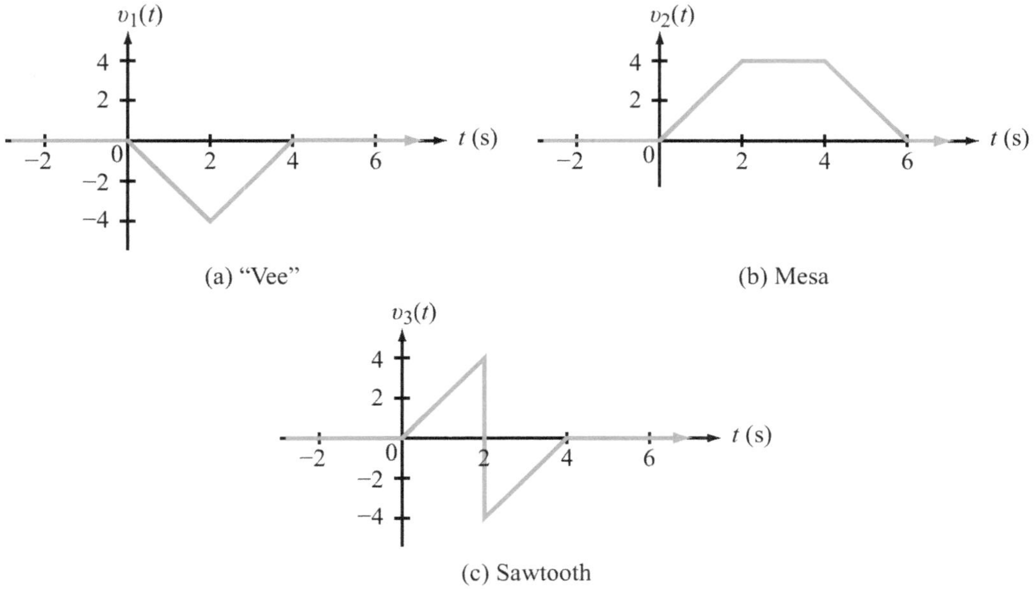

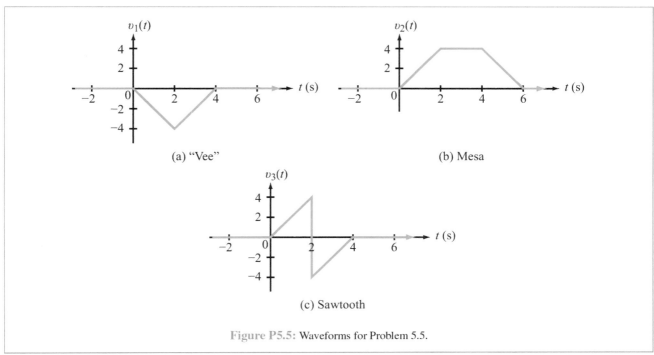

Figure P5.5: Waveforms for Problem 5.5.

*5.7 After opening a certain switch at $t = 0$ in a circuit containing a capacitor, the voltage across the capacitor started decaying exponentially with time. Measurements indicate that the voltage was 7.28 V at $t = 1$ s and 0.6 V at $t = 6$ s. Determine the initial voltage at $t = 0$ and the time constant of the voltage waveform.

Section 5-2: Capacitors

5.8 After plotting the voltage waveform, obtain expressions and generate plots for $i(t)$, $p(t)$, and $w(t)$ for a 0.2 mF capacitor. The voltage waveforms are given by

(a) $\upsilon_1(t) = 5r(t) - 5r(t-2)$ V

(b) $\upsilon_2(t) = 10u(-t) + 10u(t) - 5r(t-2) + 5r(t-4)$ V

(c) $\upsilon_3(t) = 15u(-t) + 15e^{-0.5t}\, u(t)$ V

(d) $\upsilon_4(t) = 15[1 - e^{-0.5t}]\, u(t)$ V

*5.9 In response to a change introduced by a switch at $t = 0$, the current flowing through a 100 μF capacitor, defined in accordance with the passive sign convention, was observed to be

$$i(t) = -0.4e^{-0.5t} \text{ mA} \qquad (\text{for } t > 0).$$

If the final energy stored in the capacitor (at $t = \infty$) is 0.2 mJ, determine $\upsilon(t)$ for $t \geq 0$.

5.10 The voltage $\upsilon(t)$ across a 20 μF capacitor is given by the waveform shown in Fig. P5.10.

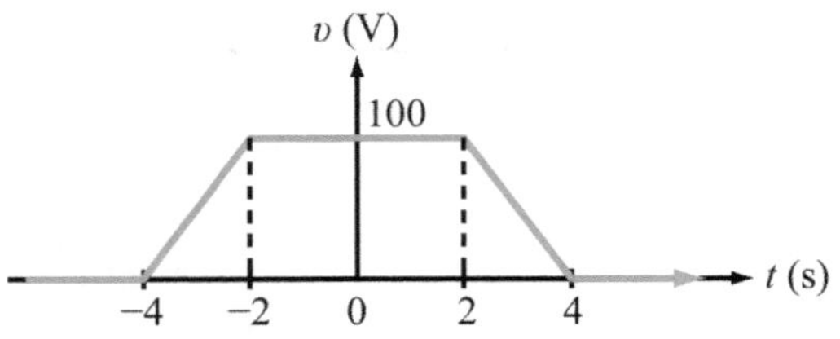

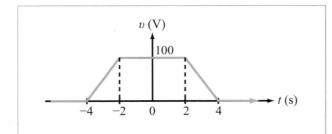

Figure P5.10: Waveform for Problems 5.10 and 5.11.

(a) Determine and plot the corresponding current $i(t)$.

(b) Specify the time interval(s) during which power transfers into the capacitor and that (those) during which it transfers out of the capacitor.

(c) At what instant in time is the power transfer into the capacitor a maximum? And at what instant is the power transfer out of the capacitor a maximum?

(d) What is the maximum amount of energy stored in the capacitor, and when does it occur?

5.11 Suppose the waveform shown in Fig. P5.10 is the current $i(t)$ through a 0.2 mF capacitor (rather than the voltage) and its peak value is 100 μA. given that the initial voltage on the capacitor was zero at $t = -4$ s, determine and plot $\upsilon(t)$.

5.12 The current through a 40 μF capacitor is given by a rectangular pulse as

$$i(t) = 40 \text{ rect}\left(\frac{t-1}{2}\right) \text{ mA}.$$

If the capacitor was initially uncharged, determine $\upsilon(t)$, $p(t)$, and $w(t)$.

5.13 The voltage across a 0.2 mF capacitor was 20 V until a switch in the circuit was opened at $t = 0$, causing the voltage to vary with time as

$$\upsilon(t) = (60 - 40e^{-5t}) \text{ V} \qquad (\text{for } t > 0).$$

(a) Did the switch action result in an instantaneous change in $\upsilon(t)$?

(b) Did the switch action result in an instantaneous change in the current $i(t)$?

(c) How much initial energy was stored in the capacitor at $t = 0$?

(d) How much final energy will be stored in the capacitor (at $t = \infty$)?

5.14 Determine voltages υ_1 to υ_4 in the circuit of Fig. P5.14 under dc conditions.

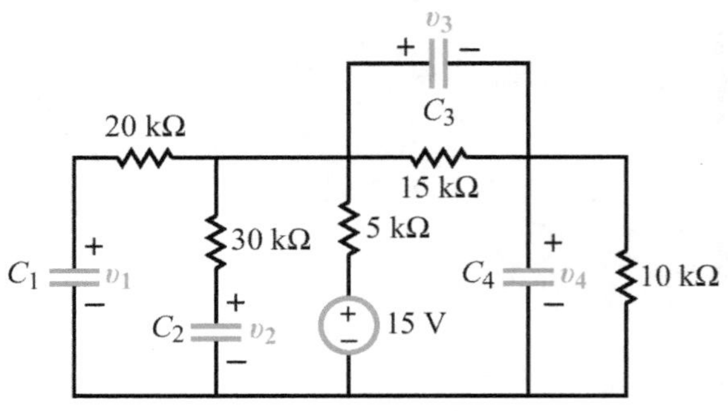

Figure P5.14: Circuit for Problem 5.14.

*5.15 Determine voltages v_1 to v_3 in the circuit of Fig. P5.15 under dc conditions.

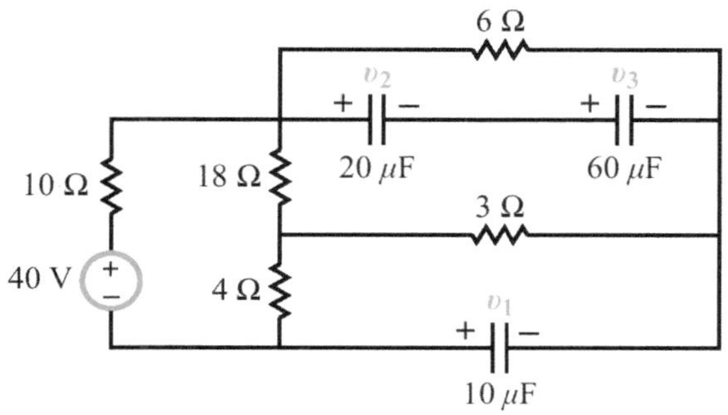

Figure P5.15: Circuit for Problem 5.15.

5.16 Determine the voltages across the two capacitors in the circuit of Fig. P5.16 under dc conditions.

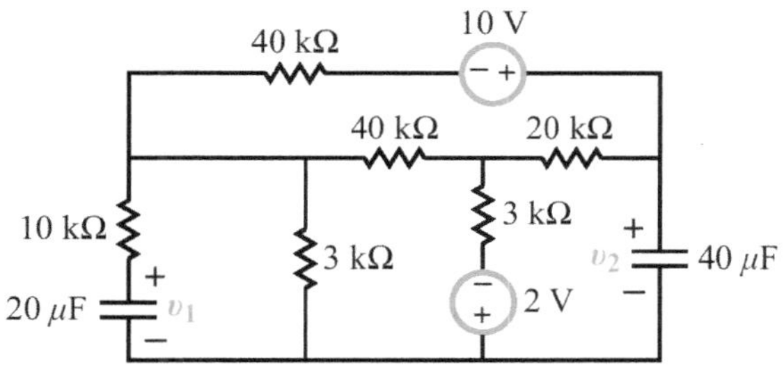

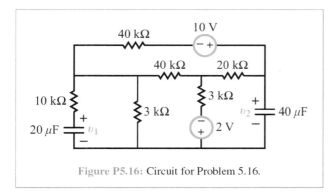

Figure P5.16: Circuit for Problem 5.16.

*5.17 Reduce the circuit in Fig. P5.17 into a single equivalent capacitor at terminals (a, b). Assume that all initial voltages are zero at $t = 0$.

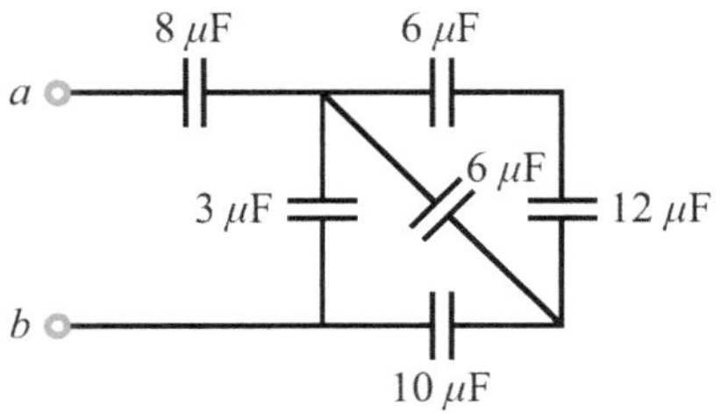

Figure P5.17: Circuit for Problems 5.17 and 5.21.

5.18 Reduce the circuit in Fig. P5.18 into a single equivalent capacitor at terminals (a, b). Assume that all initial voltages are zero at $t = 0$.

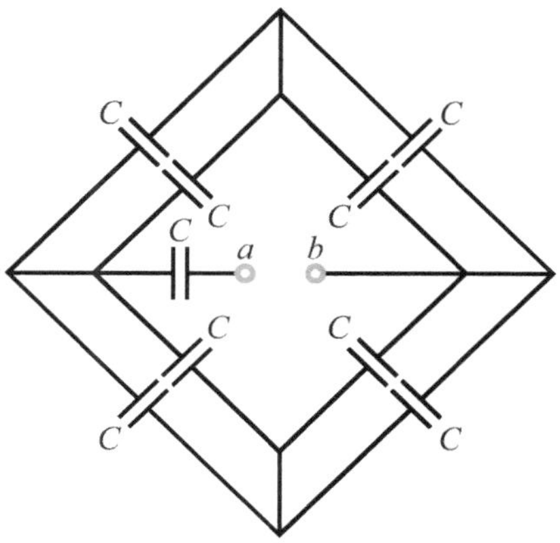

Figure P5.18: Circuit for Problem 5.18.

*5.19 For the circuit in Fig. P5.19, find C_{eq} at terminals (a, b). Assume all initial voltages to be zero.

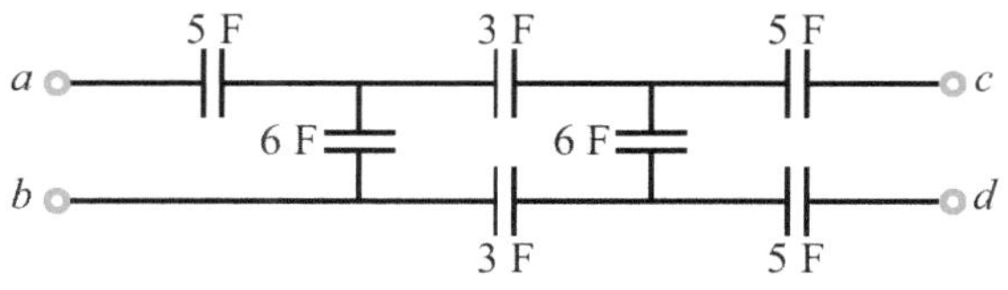

Figure P5.19: Circuit for Problems 5.19 and 5.20.

5.20 Find C_{eq} at terminals (c, d) in the circuit of Fig. P5.19.

*5.21 Assume that a 120 V dc source is connected at terminals (a, b) to the circuit in Fig. P5.17. Determine the voltages across all capacitors.

5.22 Determine (a) the amount of energy stored in each of the three capacitors shown in Fig. P5.22, (b) the equivalent capacitance at terminals (a, b), and (c) the amount of energy stored in the equivalent capacitor.

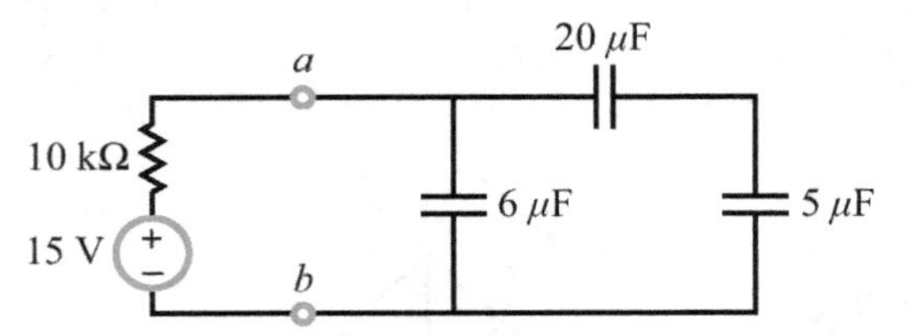

Figure P5.22: Circuit for Problem 5.22.

Section 5-3: Inductors

5.23 After plotting the current waveform, obtain expressions and generate plots for $\upsilon(t)$, $p(t)$, and $w(t)$ for a 0.5 mH inductor. The current waveforms are given by

(a) $i_1(t) = 0.2r(t-2) - 0.2r(t-4) - 0.2r(t-8) + 0.2r(t-10)$ A

(b) $i_2(t) = 2u(-t) + 2e^{-0.4t}\, u(t)$ A

(c) $i_3(t) = -4(1 - e^{-0.4t})\, u(t)$ A

5.24 The current $i(t)$ passing through a 0.1 mH inductor is given by the waveform shown in Fig. P5.24.

(a) Determine and plot the corresponding voltage $\upsilon(t)$ across the inductor.

(b) Specify the time interval(s) during which power is transferred into the inductor and that (those) during which power transfers out of the inductor. Also specify the amount of energy transferred in each case.

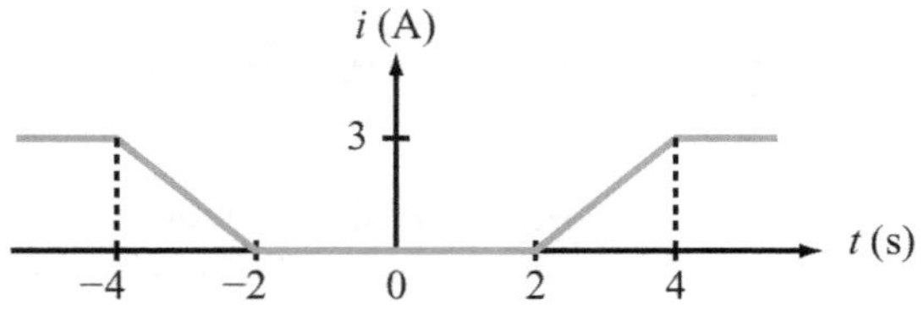

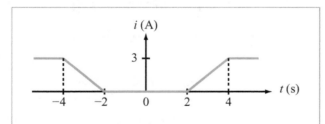

Figure P5.24: Current waveform for Problem 5.24.

***5.25** Activation of a switch at $t = 0$ in a certain circuit caused the voltage across a 20 mH inductor to exhibit the voltage response

$$\upsilon(t) = 4e^{-0.2t}\,\text{mV} \qquad (\text{for } t \geq 0).$$

Determine $i(t)$ for $t \geq 0$ given that the energy stored in the inductor at $t = \infty$ is 0.64 mJ.

5.26 The waveform shown in Fig. P5.26 represents the voltage across a 0.2 H inductor for $t \geq 0$. If the current flowing through the inductor is -20 mA at $t = 0$, determine the current $i(t)$ for $t \geq 0$.

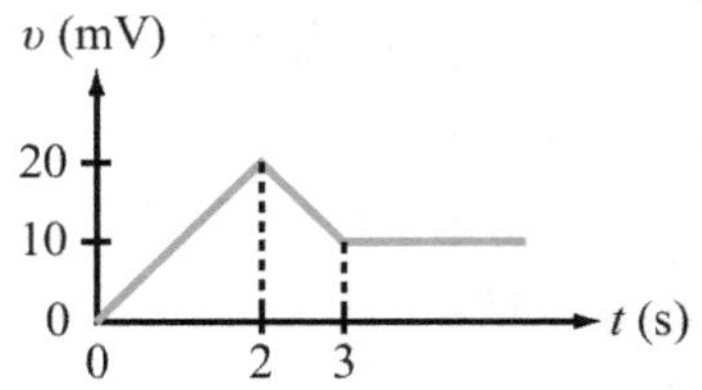

Figure P5.26: Voltage waveform for Problem 5.26.

5.27 The waveform shown in Fig. P5.27 represents the voltage across a 50 mH inductor. Determine the corresponding current waveform. Assume $i(0) = 0$.

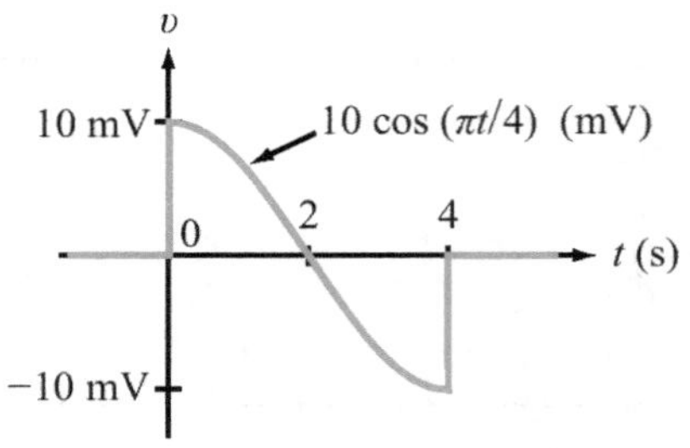

Figure P5.27: Voltage waveform for Problem 5.27.

5.28 For the circuit in Fig. P5.28, determine the voltage across C and the currents through L_1 and L_2 under dc conditions.

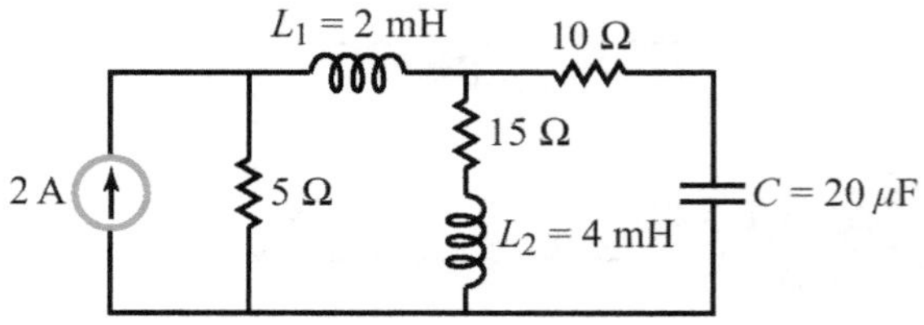

Figure P5.28: Circuit for Problem 5.28.

*5.29 For the circuit in Fig. P5.29, determine the voltages across C_1 and C_2 and the currents through L_1 and L_2 under dc conditions.

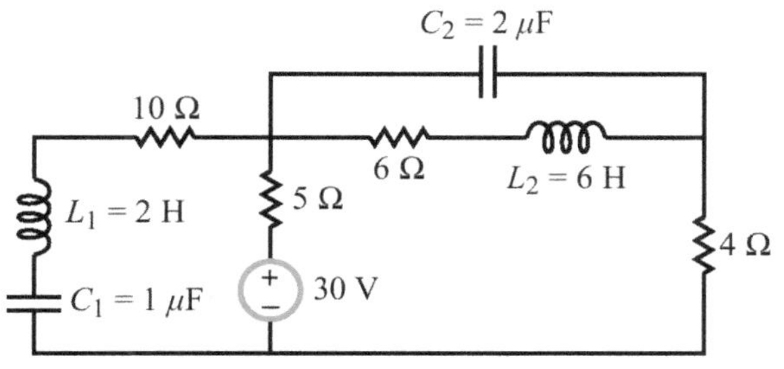

Figure P5.29: Circuit for Problem 5.29.

5.30 All elements in Fig. P5.30 are 10 mH inductors. Determine L_{eq}.

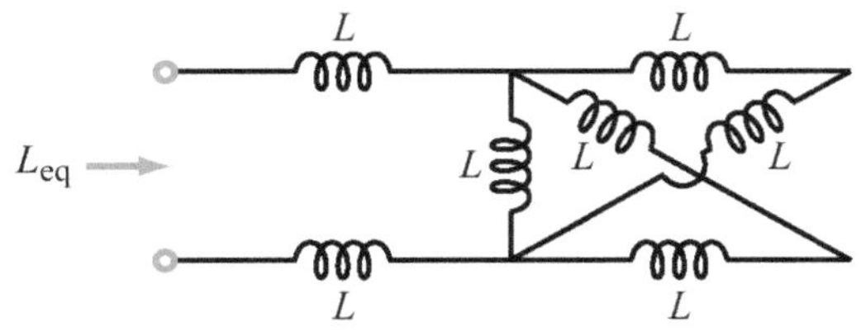

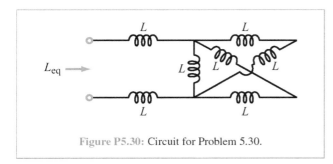

Figure P5.30: Circuit for Problem 5.30.

*5.31 The values of all inductors in the circuit of Fig. P5.31 are in millihenrys. Determine L_{eq}.

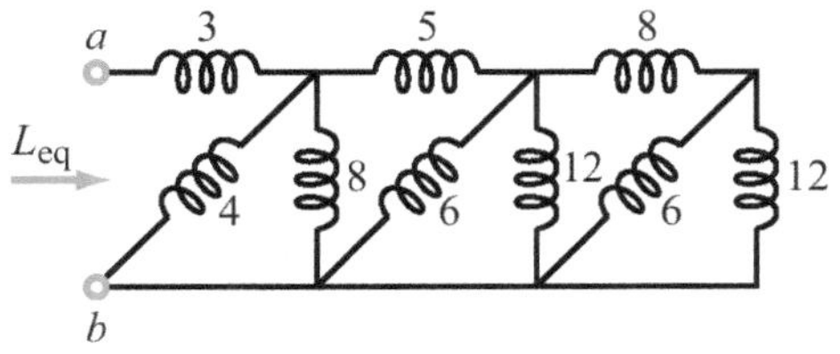

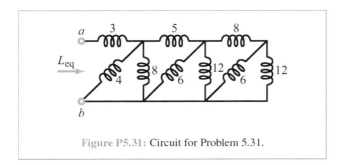

Figure P5.31: Circuit for Problem 5.31.

5.32 Determine L_{eq} at terminals (a, b) in the circuit of Fig. P5.32. All inductor values are in millihenrys.

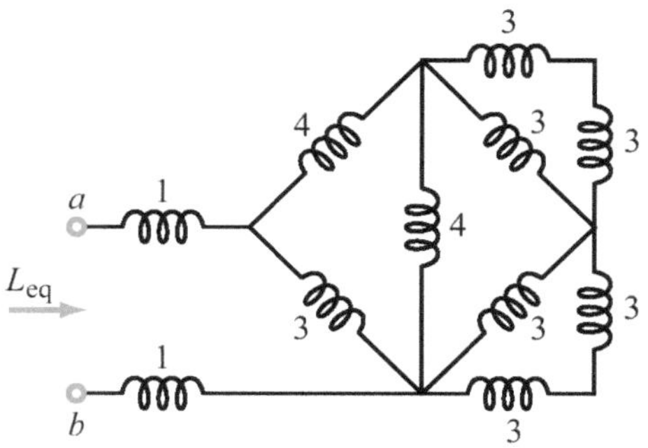

Figure P5.32: Circuit for Problem 5.32.

Section 5-4: Response of the RC Circuit

5.33 After having been in position 1 for a long time, the switch in the circuit of Fig. P5.33 was moved to position 2 at $t = 0$. Given that $V_0 = 12$ V, $R_1 = 30$ kΩ, $R_2 = 120$ kΩ, $R_3 = 60$ kΩ, and $C = 100\ \mu$F, determine:

(a) $i_C(0^-)$ and $\upsilon_C(0^-)$

(b) $i_C(0)$ and $\upsilon_C(0)$

(c) c $i_C(\infty)$ and $\upsilon_C(\infty)$

(d) $\upsilon_C(t)$ for $t \geq 0$

(e) $i_C(t)$ for $t \geq 0$

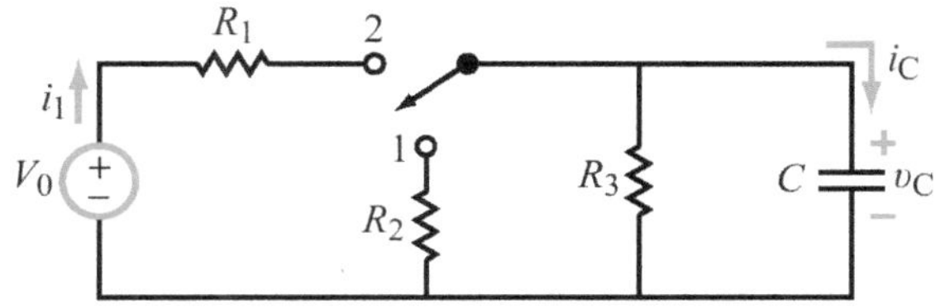

Figure P5.33: Circuit for Problems 5.33 and 5.34.

5.34 Repeat Problem 5.33, but with the switch having been in position 2 for a long time, and then moved to position 1 at $t = 0$.

5.35 The circuit in Fig. P5.35 contains two switches, both of which had been open for a long time before $t = 0$. Switch 1 closes at $t = 0$, and switch 2 follows suit at $t = 5$ s. Determine and plot $\upsilon_C(t)$ for $t \geq 0$ given that $V_0 = 24$ V, $R_1 = R_2 = 16$ kΩ, and $C = 250\ \mu$F. Assume $\upsilon_C(0) = 0$.

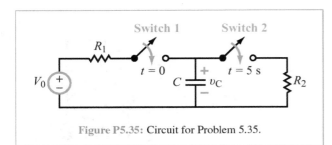

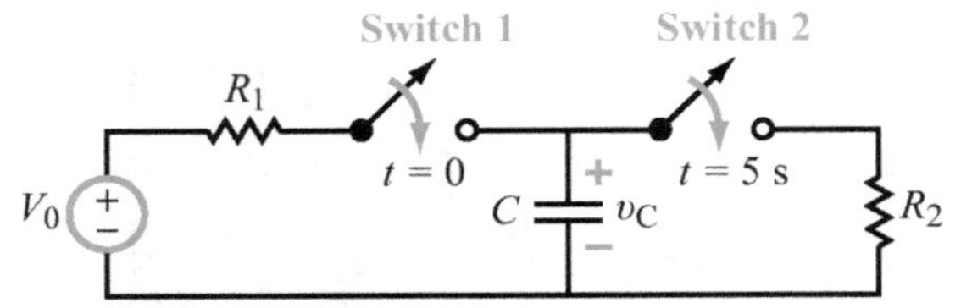

Figure P5.35: Circuit for Problem 5.35.

*5.36 The circuit in Fig. P5.36 was in steady state until the switch was moved from terminal 1 to terminal 2 at $t = 0$. Determine $v(t)$ for $t \geq 0$ given that $I_0 = 21$ mA, $R_1 = 2$ kΩ, $R_2 = 3$ kΩ, $R_3 = 4$ kΩ, and $C = 50$ μF.

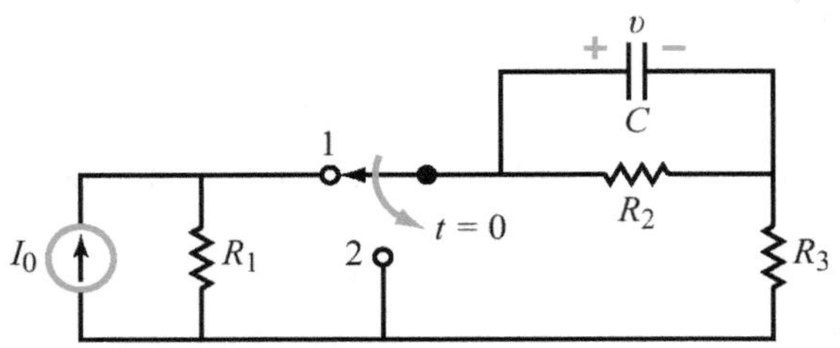

Figure P5.36: Circuit for Problem 5.36.

5.37 Prior to $t = 0$, capacitor C_1 in the circuit of Fig. P5.37 was uncharged. For $I_0 = 5$ mA, $R_1 = 2$ kΩ, $R_2 = 50$ kΩ, $C_1 = 3$ μ F, and $C_2 = 6$ μ F, determine:

(a) The equivalent circuit involving the capacitors for $t \geq 0$. Specify $v_1(0)$ and $v_2(0)$.

(b) $i(t)$ for $t \geq 0$.

(c) $v_1(t)$ and $v_2(t)$ for $t \geq 0$.

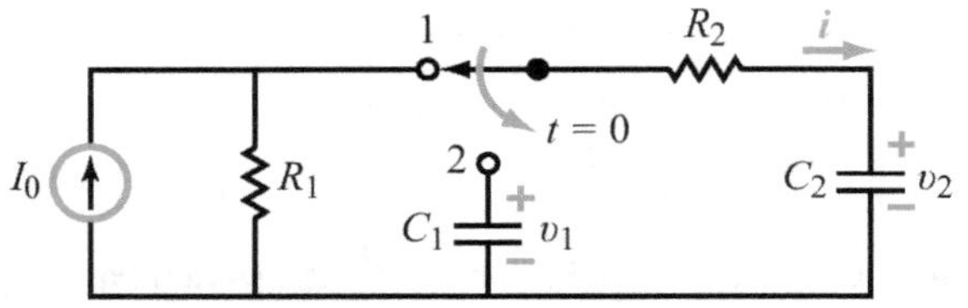

Figure P5.37: Circuit for Problem 5.37.

5.38 The switch in the circuit of Fig. P5.38 had been closed for a long time before it was opened at $t = 0$. Given that $V_s = 10$ V, $R_1 = 20$ kΩ, $R_2 = 100$ kΩ, $C_1 = 6$ μF, and $C_2 = 12$ μF, determine $i(t)$ for $t \geq 0$.

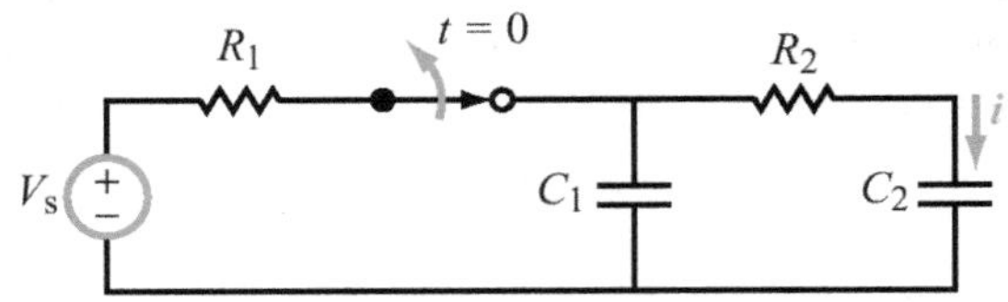

Figure P5.38: Circuit for Problem 5.38.

*5.39 The switch in the circuit of Fig. P5.39 had been in position 1 for a long time until it was moved to position 2 at $t = 0$. Determine $v(t)$ for $t \geq 0$, given that $I_0 = 6$ mA, $V_0 = 18$ V, $R_1 = R_2 = 4$ kΩ, and $C = 200$ μF.

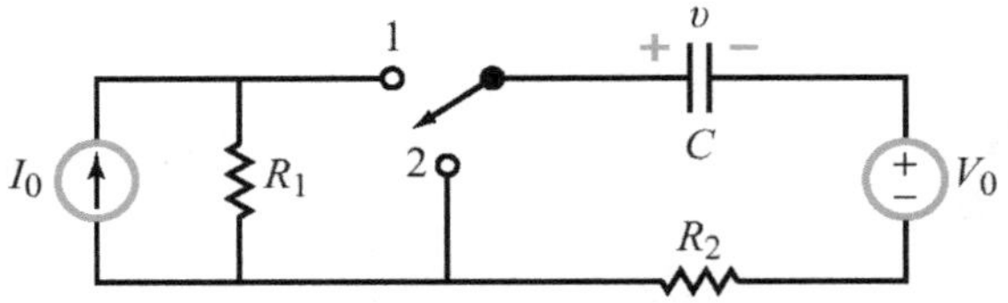

Figure P5.39: Circuit for Problems 5.39 and 5.40.

5.40 Repeat Problem 5.39, but reverse the switching sequence. [Switch starts in position 2 and is moved to position 1 at $t = 0$.]

5.41 Determine $i(t)$ for $t \geq 0$ where i is the current passing through R_3 in the circuit of Fig. P5.41. The element values are $v_s = 16$ V, $R_1 = R_2 = 2$ kΩ, $R_3 = 4$ kΩ, and $C = 25$ μF. Assume that the switch had been open for a long time prior to $t = 0$.

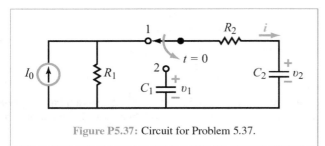

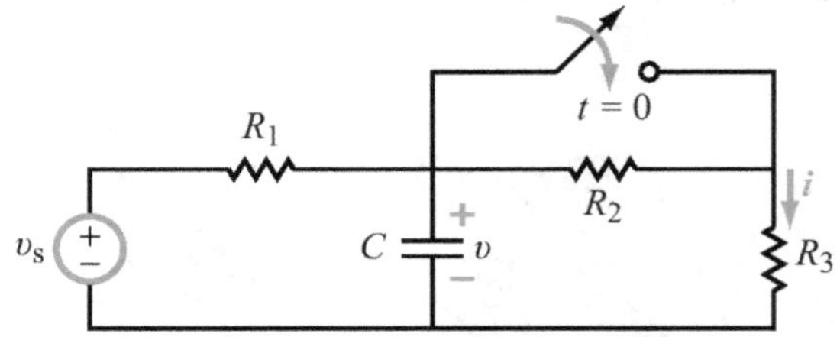

Figure P5.41: Circuit for Problems 5.41 to 5.43.

5.42 Repeat Problem 5.41, but start with the switch being closed prior to $t = 0$ and then opened at $t = 0$.

*5.43 Consider the circuit in Fig. P5.41, but without the switch. If the source v_s represents a 12 V, 100 ms long rectangular pulse that starts at $t = 0$ and the element values are $R_1 = 6$ kΩ, $R_2 = 2$ kΩ, $R_3 = 4$ kΩ, and $C = 15$ μF, determine the voltage response $\upsilon(t)$ for $t \geq 0$.

5.44 Given that in Fig. P5.44, $I_1 = 4$ mA, $I_2 = 6$ mA, $R_1 = 3$ kΩ, $R_2 = 6$ kΩ, and $C = 0.2$ mF, determine $\upsilon(t)$. Assume the switch was connected to terminal 1 for a long time before it was moved to terminal 2.

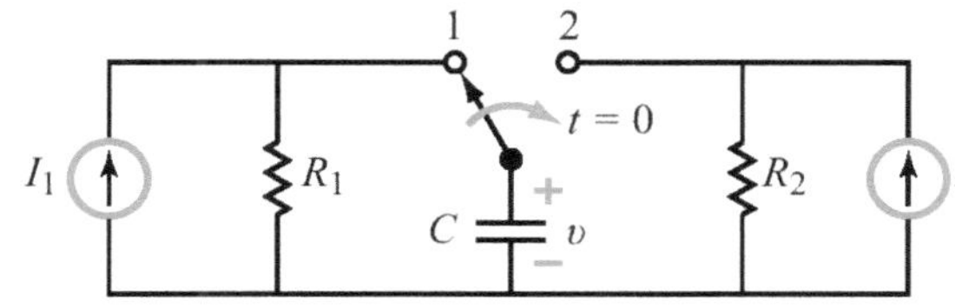

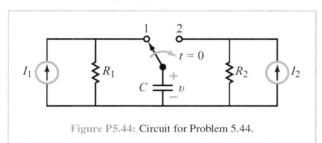

Figure P5.44: Circuit for Problem 5.44.

***5.45** Determine $\upsilon_C(t)$ in the circuit of Fig. P5.45 for $t \geq 0$, given that the switch had been closed for a long time prior to $t = 0$.

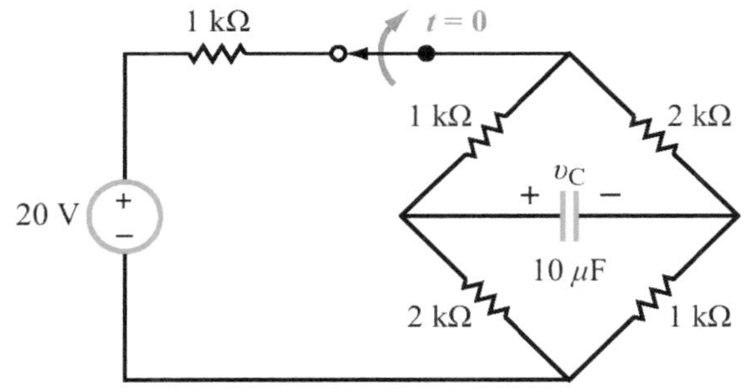

Figure P5.45: Circuit for Problem 5.45.

Section 5-5: Response of the RL Circuit

5.46 After having been in position 1 for a long time, the switch in the circuit of Fig. P5.46 was moved to position 2 at $t = 0$. Given that $V_0 = 12$ V, $R_1 = 30$ Ω, $R_2 = 120$ Ω, $R_3 = 60$ Ω, and $L = 0.2$ H, determine:

(a) $i_L(0^-)$ and $\upsilon_L(0^-)$

(b) $i_L(0)$ and $\upsilon_L(0)$

(c) $i_L(\infty)$ and $\upsilon_L(\infty)$

(d) $i_L(t)$ for $t \geq 0$

(e) $\upsilon_L(t)$ for $t \geq 0$

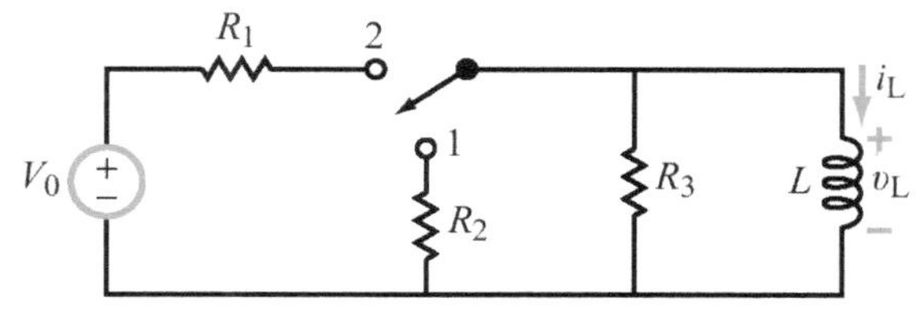

Figure P5.46: Circuit for Problems 5.46 and 5.47.

5.47 Repeat Problem 5.46, but with the switch having been in position 2 for a long time and then moved to position 1 at $t = 0$.

***5.48** Determine $i(t)$ for $t \geq 0$ given that the circuit in Fig. P5.48 had been in steady state for a long time prior to $t = 0$. Also, $I_0 = 5$ A, $R_1 = 2$ Ω, $R_2 = 10$ Ω, $R_3 = 3$ Ω, $R_4 = 7$ Ω, and $L = 0.15$ H.

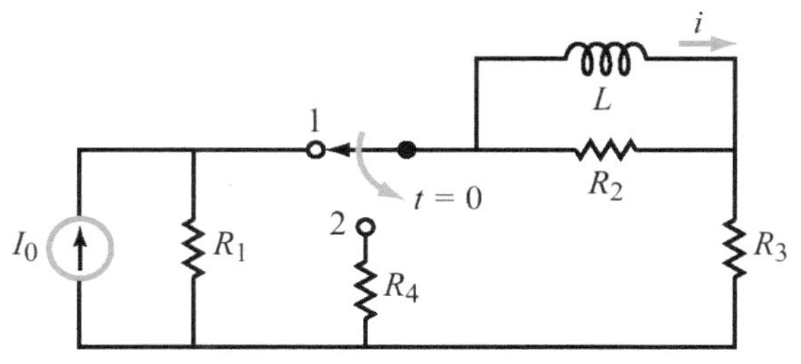

Figure P5.48: Circuit for Problem 5.48.

5.49 For the circuit in Fig. P5.49, determine $i_L(t)$ and plot it as a function of t for $t \geq 0$. The element values are $I_0 = 4$ A, $R_1 = 6$ Ω, $R_2 = 12$ Ω, and $L = 2$ H. Assume that $i_L = 0$ before $t = 0$.

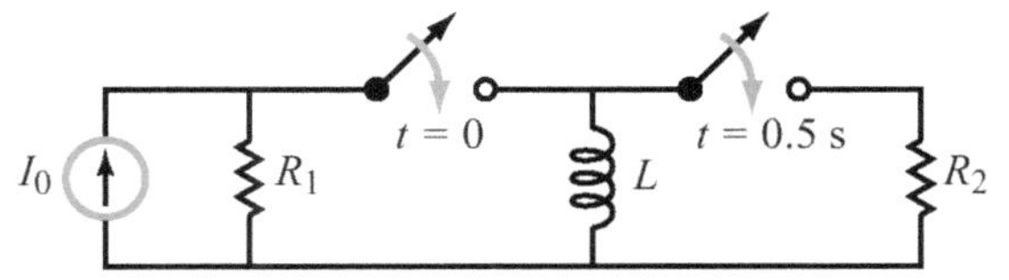

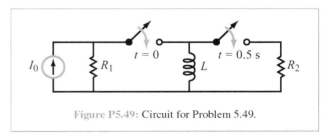

Figure P5.49: Circuit for Problem 5.49.

***5.50** After having been in position 1 for a long time, the switch in the circuit of Fig. P5.50 was moved to position 2 at $t = 0$. Determine $i_1(t)$ and $i_2(t)$ for $t \geq 0$, given that $I_0 = 6$ mA, $R_0 = 12$ Ω, $R_1 = 10$ Ω, $R_2 = 40$ Ω, $L_1 = 1$ H, and $L_2 = 2$ H.

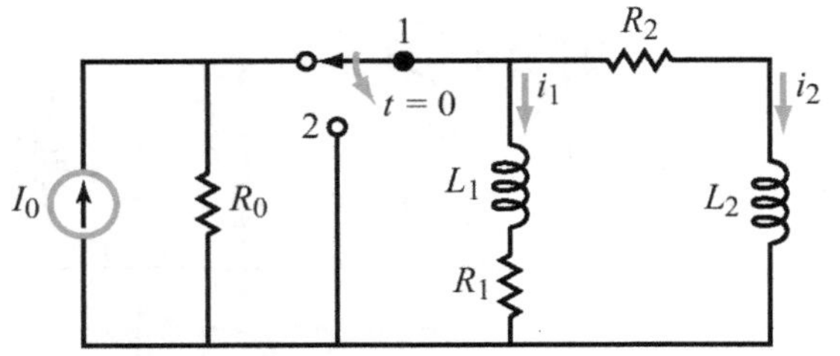

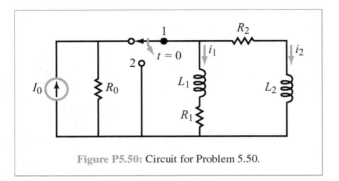

Figure P5.50: Circuit for Problem 5.50.

5.51 Derive an expression for $i_2(t)$ in the circuit of Fig. P5.51 in terms of the circuit variables, given that I_s is a dc current source and the switch was closed at $t = 0$ after it had been open for a long time.

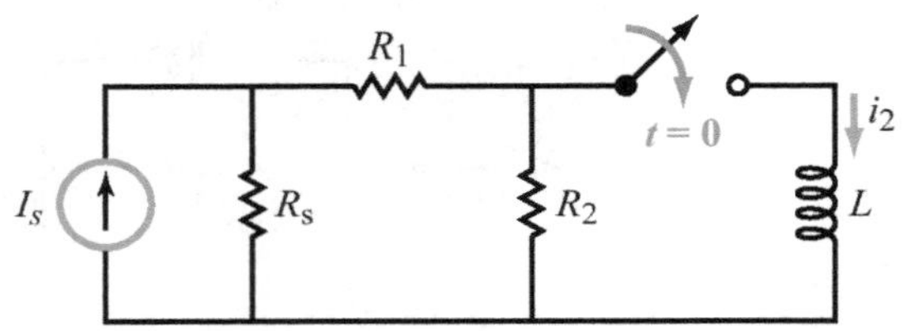

Figure P5.51: Circuit for Problem 5.51.

5.52 Determine $i_L(t)$ in the circuit of Fig. P5.52 for $t \geq 0$.

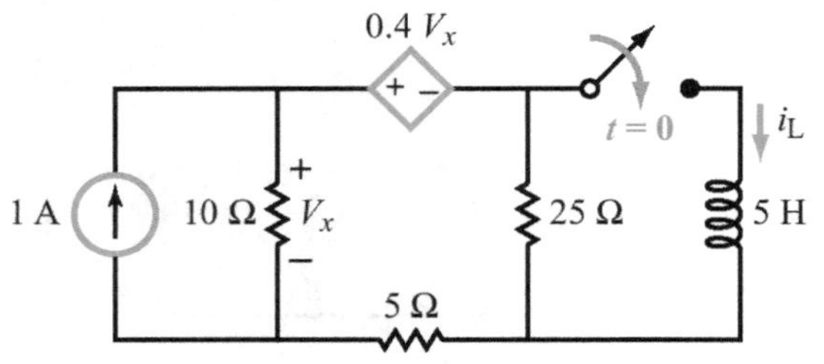

Figure P5.52: Circuit for Problem 5.52.

***5.53** In the circuit of Fig. P5.53(a), $R_1 = R_2 = 20\ \Omega$, $R_3 = 10\ \Omega$, and $L = 2.5$ H. Determine $i(t)$ for $t \geq 0$ given that $v_s(t)$ is the step function described in Fig. P5.53(b).

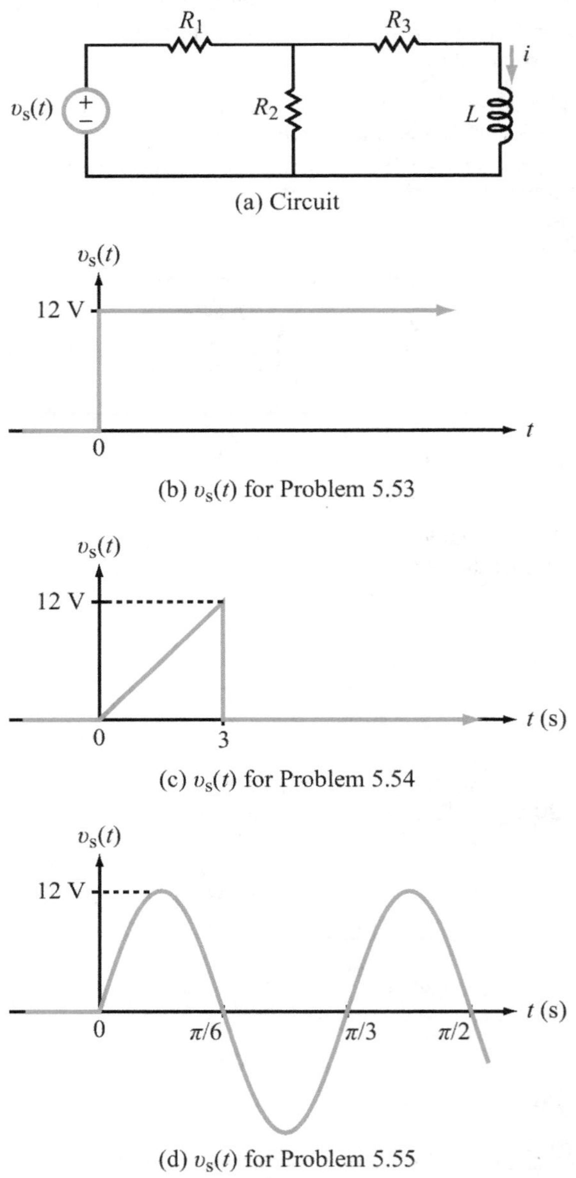

Figure P5.53: Circuit and excitation voltages for Problems 5.53 to 5.55.

5.54 Repeat Problem 5.53 for the triangular-source excitation given in Fig. P5.53(c).

$$\textit{Hint}: \quad \int x e^{ax}\, dx = \frac{e^{ax}}{a^2}(ax - 1).$$

5.55 Repeat Problem 5.53 for the sinusoidal-source excitation $\upsilon_s(t) = 12 \sin 6t$ V displayed in Fig. P5.53(d).

$$\textit{Hint}: \quad \int e^{ax} \sin bx \, dx = e^{ax} \frac{[a \sin bx - b \cos(bx)]}{a^2 + b^2}.$$

*__5.56__ The switch in the circuit of Fig. P5.56 was moved from position 1 to position 2 at $t = 0$, after it had been in position 1 for a long time. If $L = 80$ mH, determine $i(t)$ for $t \geq 0$.

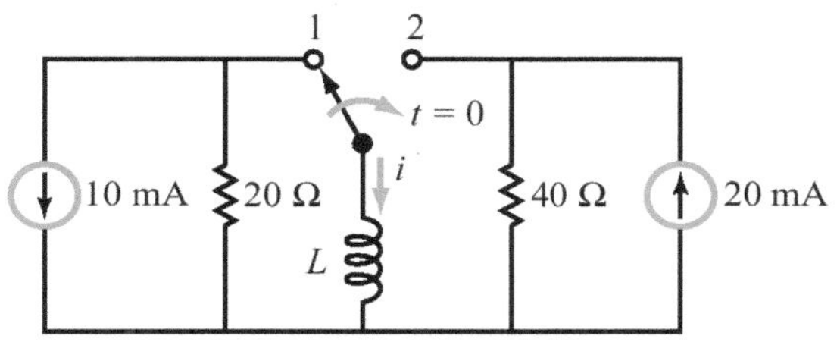

Figure P5.56: Circuit for Problems 5.56 and 5.57.

5.57 Repeat Problem 5.56, but with the switch having been in position 2 and then moved to position 1 at $t = 0$.

5.58 Determine $i(t)$ for $t \geq 0$ due to the rectangular-pulse excitation in the circuit of Fig. P5.58.

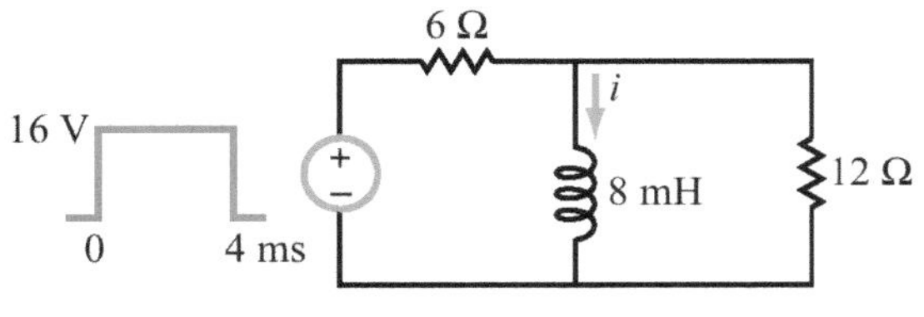

Figure P5.58: Circuit for Problem 5.58.

Section 5-6: RC Op-Amp Circuits

5.59 The input-voltage waveform shown in Fig. P5.59(a) is applied to the circuit in Fig. P5.59(b). Determine and plot the corresponding $\upsilon_{out}(t)$.

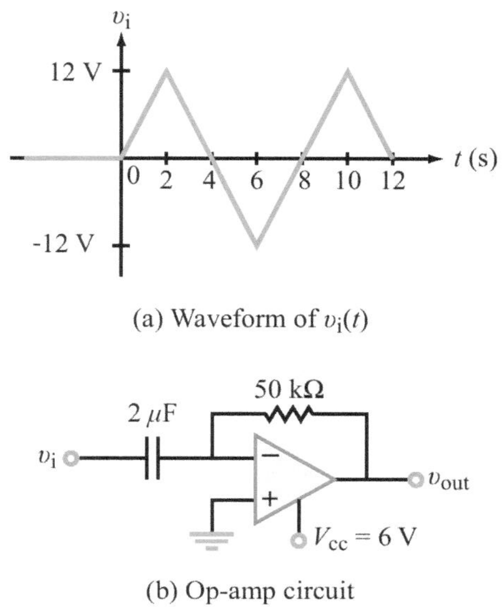

Figure P5.59: Waveform and circuit for Problem 5.59.

*__5.60__ Relate υ_{out} to υ_i in the circuit of Fig. P5.60.

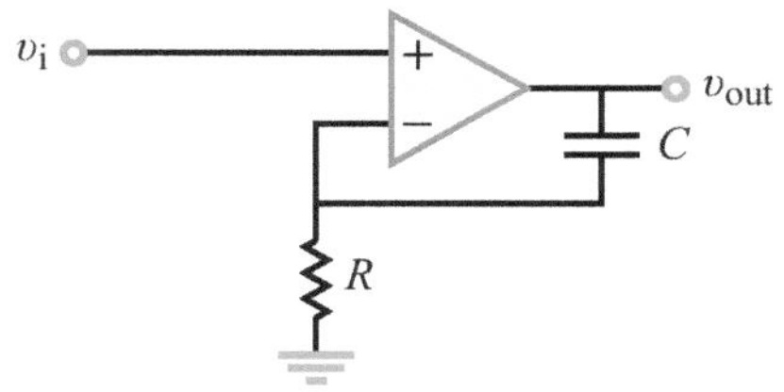

Figure P5.60: Circuit for Problem 5.60.

5.61 Develop the relationship between the output voltage υ_{out} and the input voltage υ_i for the circuit in Fig. P5.61.

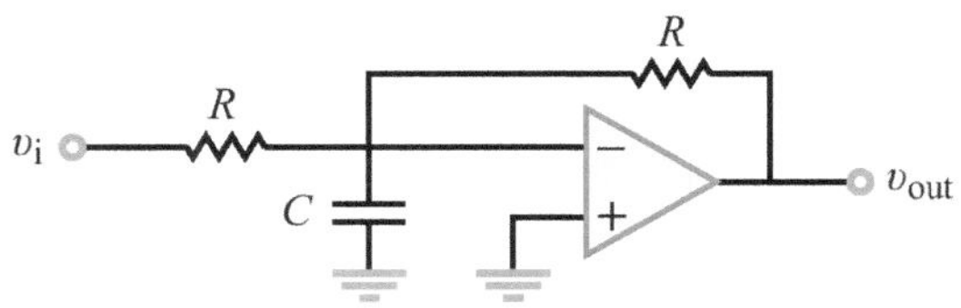

Figure P5.61: Circuit for Problem 5.61.

5.62 Relate υ_{out} to υ_i in the circuit of Fig. P5.62. Assume $\upsilon_C = 0$ at $t = 0$.

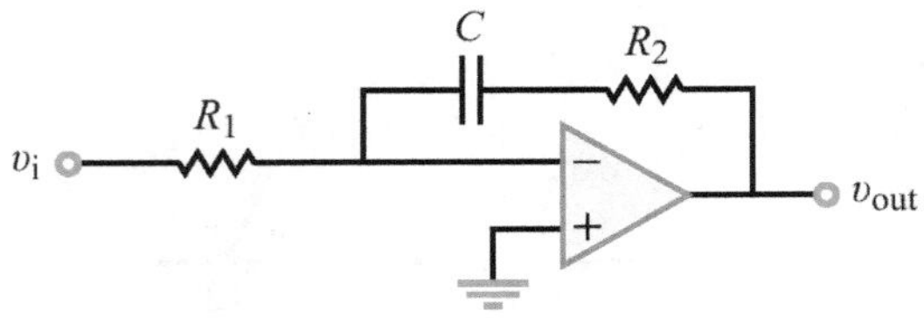

Figure P5.62: Circuit for Problem 5.62.

*5.63 Relate $i_{out}(t)$ to $\upsilon_i(t)$ in the circuit of Fig. P5.63. Evaluate it for $\upsilon_C(0) = 3$ V, $R = 10$ kΩ, $C = 50$ μF, and $\upsilon_i(t) = 9\,u(t)$ V.

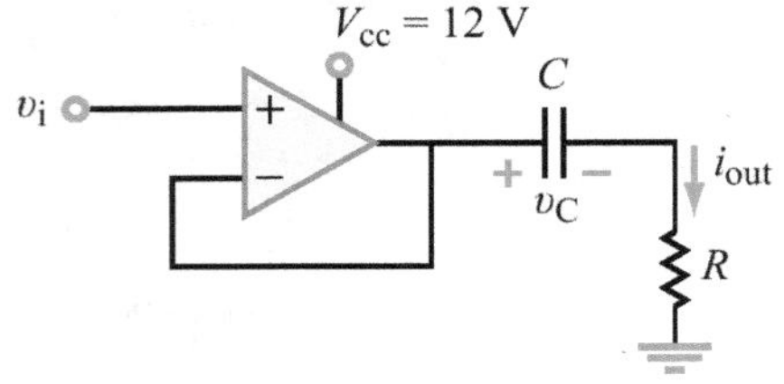

Figure P5.63: Circuit for Problem 5.63.

5.64 Determine $\upsilon_{out}(t)$ in the circuit of Fig. P5.64 for $t \geq 0$.

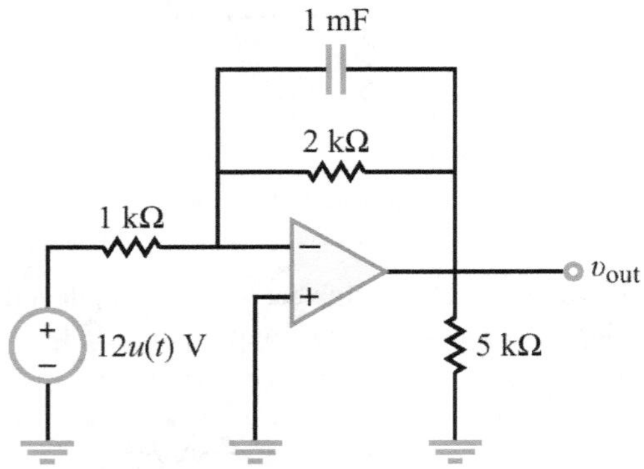

Figure P5.64: Circuit for Problem 5.64.

5.65 In the circuit of Fig. P5.65:

(a) Derive an expression for $\upsilon_{out}(t)$ for $t \geq 0$ in terms of R_1, R_2, R_3, C, and A.

*(b) Evaluate the expression for $R_1 = 1$ kΩ, $R_2 = 5$ kΩ, $R_3 = 2$ kΩ, $C = 0.25$ mF, and $A = 12$ V.

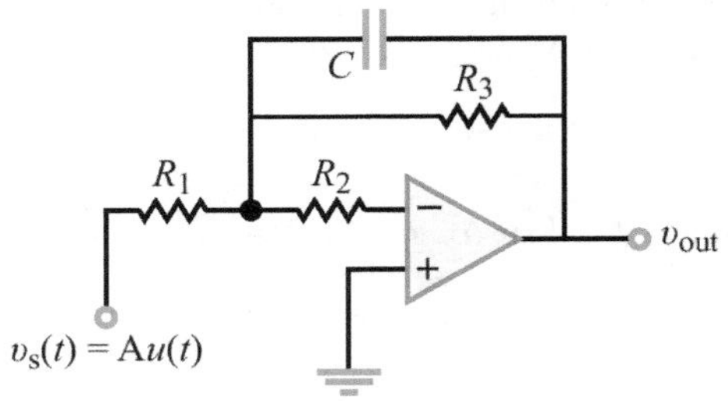

Figure P5.65: Circuit for Problem 5.65.

5.66 Design a single op-amp circuit with a 40 μF capacitor to generate a circuit output given by

$$\upsilon_{out}(t) = \int_0^t [6 - 2\upsilon_s(t')]\,dt' = 6t - 2\int_0^t \upsilon_s(t')\,dt' \qquad \text{(V)},$$

where $\upsilon_s(t)$ is any input voltage source that starts at $t = 0$.

5.67 Design a circuit that can perform the following relationship between its output and input voltages:

$$\upsilon_{out} = -100\int_0^t \upsilon_i\,dt,$$

with $\upsilon_{out}(0) = 0$ at $t = 0$. You are limited to one op-amp, one capacitor that does not exceed 0.1 F, and any resistor(s) of your choice.

5.68 The two-stage op-amp circuit in Fig. P5.68 is driven by an input step voltage given by $\upsilon_i(t) = 10\,u(t)$ mV. If $V_{cc} = 10$ V for both op amps and the two capacitors had no charge prior to $t = 0$, determine and plot:

*(a) $\upsilon_{out_1}(t)$ for $t \geq 0$;

(b) $\upsilon_{out_2}(t)$ for $t \geq 0$.

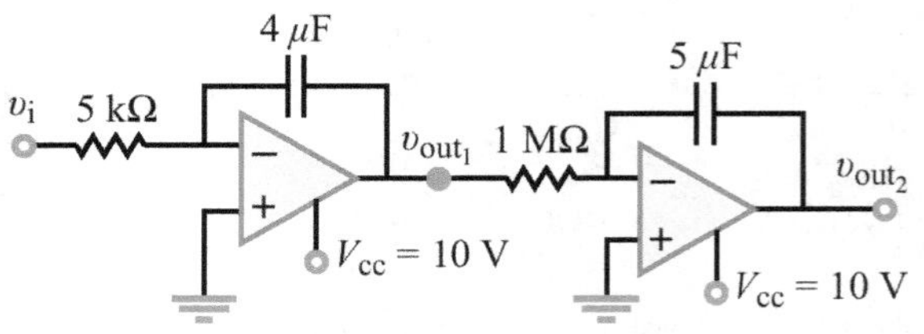

Figure P5.68: Op-amp circuit for Problem 5.68.

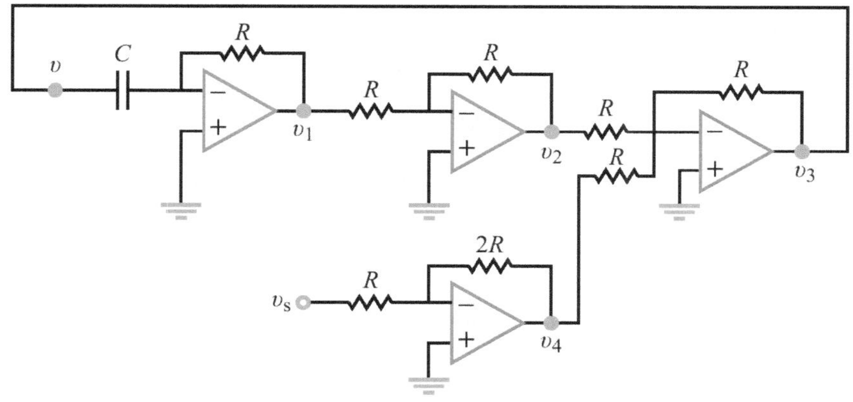

Figure P5.71: Circuit for Problem 5.71.

5.69 Design a single op-amp circuit that can perform the operation

$$v_{\text{out}} = -\int_0^t (5v_1 + 2v_2 + v_3)\, dt.$$

5.70 Design a single op-amp circuit that can perform the operation

$$i_{\text{out}} = -\int_0^t \left(\frac{v_1}{100} + \frac{v_2}{200} + \frac{v_3}{400}\right) dt.$$

5.71 Show that the op-amp circuit in Fig. P5.71 (in which $R = 10$ kΩ and $C = 20\ \mu$F) simulates the differential equation

$$\frac{dv}{dt} + 5v = 10v_s.$$

5.72 Design an op-amp circuit that can solve the differential equation

$$\frac{dv}{dt} + 0.2v = 4 \sin 10t$$

with $v(0) = 0$. *Hint*: See Problem 5.71.

Sections 5-7 and 5-8: Parasitic Capacitance and Multisim Analysis

*5.73 In real transistors, both the MOSFET gain g and parasitic capacitances C_D^n and C_D^p depend on the size of the transistor. Assuming the functional relationships

$$g = 10^6 W \qquad \text{and} \qquad C_D^n = C_D^p = (2.5 \times 10^3) W^2,$$

where W is the transistor width in meters, how small should W be in order for the CMOS inverter to have a fall time of 1 ns? [The width of modern digital MOSFETs varies between 40 nm and 4 μm.]

5.74 Draw and simulate in Multisim the circuit in Fig. 5-43(a) of Example 5-15. Using the Grapher tool, plot $v_{\text{out}}(t)$ for $t \geq 0$.

5.75 Consider the circuit in Fig. P5.75. Switch S1 begins in the closed position and opens at $t = 0$. Switch S2 begins in the open position and toggles between the open and closed positions every 250 ps. Model this circuit in Multisim and plot v_0 and v_1 as a function of time until all nodes are discharged below 1 mV.

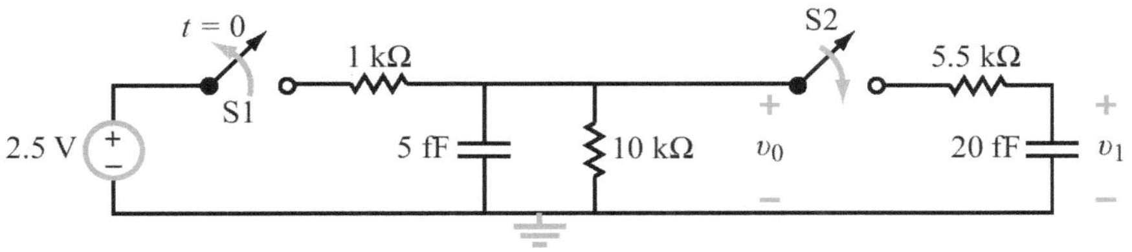

Figure P5.75: Circuit for Problem 5.75.

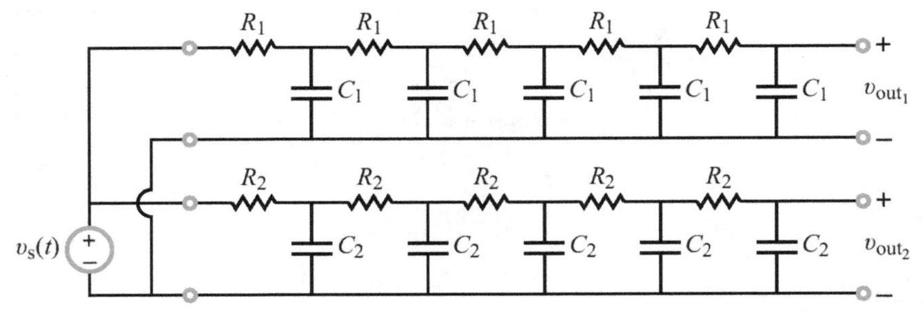

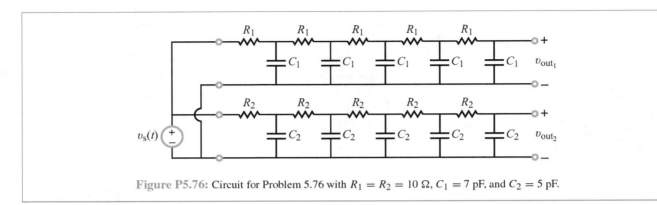

Figure P5.76: Circuit for Problem 5.76 with $R_1 = R_2 = 10\ \Omega$, $C_1 = 7$ pF, and $C_2 = 5$ pF.

5.76 A step voltage source $v_s(t)$ sends a signal down two transmission lines simultaneously (Fig. P5.76). In Multisim, the step voltage may be modeled as a 1 V square wave with a period of 10 ns. Model the circuit in Multisim and answer the following questions:

(a) If a detector registers a signal when the output voltage reaches 0.75 V, which signal arrives first?

(b) By how much?

Hint: When using cursors in the Grapher View, select a trace, then right-click on a cursor and select Set Y_Value, and enter 750 m. This will give you the exact time point at which that trace equals 0.75 V.

5.77 Consider the delta topology in Fig. P5.77. Use Multisim to generate response curves for v_a, v_b, and v_c. Apply Transient Analysis with TSTOP $= 3 \times 10^{-10}$ s.

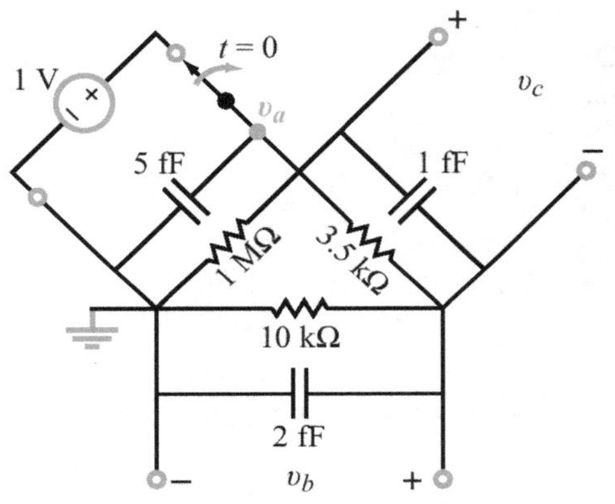

Figure P5.77: Circuit for Problem 5.77.

5.78 Use Multisim to generate a plot for current $i(t)$ in the circuit in Fig. P5.78 from 0 to 15 ms.

5.79 Construct the integrator circuit shown in Fig. P5.79, using a 3-terminal virtual op amp. Print the output corresponding to each of the following input signals:

(a) $v_{in}(t)$ is a 0-to-1 V square wave with a period of 1 ms and a 50 percent duty cycle. Plot the output from 0 to 10 ms.

(b) $v_{in}(t) = -0.2t$ V. Plot the output from 0 to 50 ms.

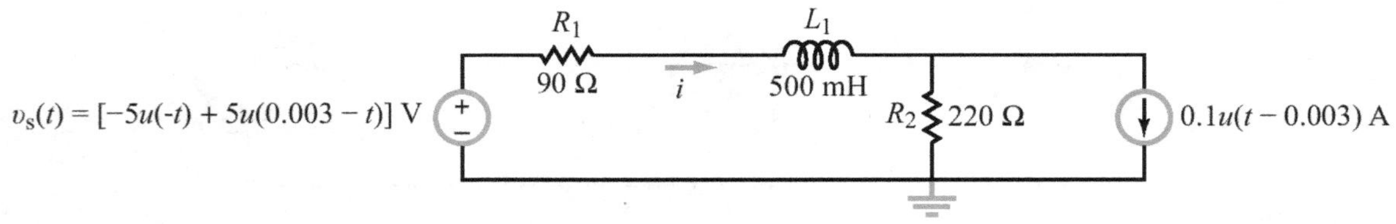

Figure P5.78: Circuit for Problem 5.78.

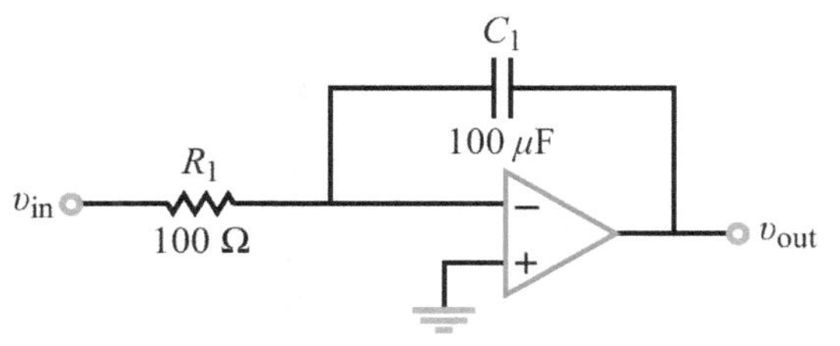

Figure P5.79: Circuit for Problem 5.79.

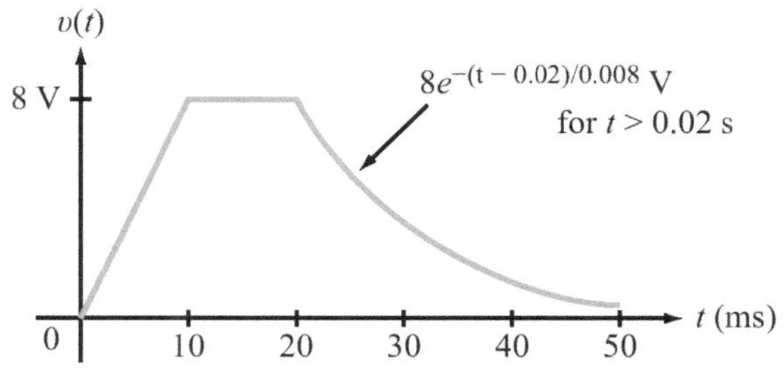

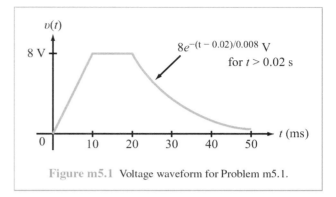

Figure m5.1 Voltage waveform for Problem m5.1.

Potpourri Questions

5.80 Calculate the plate area required to store 1 MJ of energy in a traditional air-filled parallel plate capacitor at a voltage of 10 V. Assume the plate separation to be 1 cm.

5.81 What are the advantages and disadvantages of supercapacitors relative to a lithium-ion battery?

5.82 Is the memory stored on a hard disk drive volatile or nonvolatile? What is the advantage of perpendicular magnetic recording over the standard recording method?

5.83 How does the proximity of a finger change the capacitance of a pixel in a touchscreen? How does the MEMS capacitor measure the acceleration of a moving vehicle?

Integrative Problems: Analytical / Multisim / myDAQ

To master the material in this chapter, solve the following problems using three complementary approaches: (a) analytically, (b) with Multisim, and (c) by constructing the circuit and using the myDAQ interface unit to measure quantities of interest via your computer. [myDAQ tutorials and videos are available on CAD.]

m5.1 Capacitors: The voltage $v(t)$ across a 10 μF capacitor is given by the waveform shown in Fig. m5.1.

(a) Determine the equation for the capacitor current $i(t)$ and plot it over the time period from 0 to 50 ms.

(b) Calculate the values of the capacitor current at times 0, 25, and 30 ms.

m5.2 Inductors: The voltage $v(t)$ across a 33 mH inductor is given by the sinusoidal pulse waveform shown in Fig. m5.2.

(a) Determine the equation for the inductor current $i(t)$ and plot it over the time period from 0 to 0.4 ms. Assume zero initial inductor current.

(b) Determine the time at which the inductor current reaches its maximum value.

(c) Calculate the total peak-to-peak range of inductor current; i.e., the maximum value minus the minimum value.

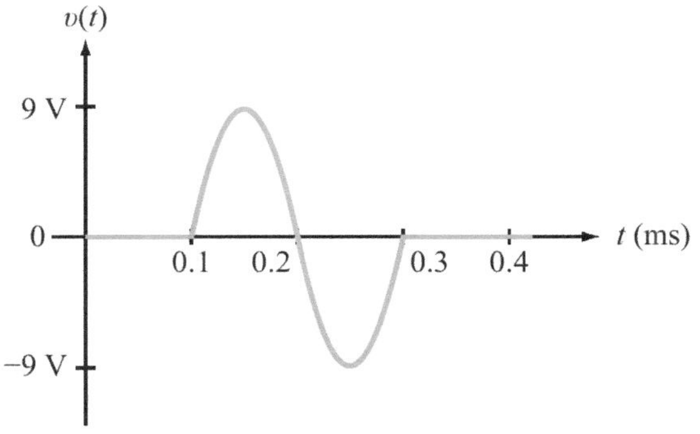

Figure m5.2 Voltage waveform for Problem m5.2.

m5.3 Response of the RC Circuit: Figure m5.3(a) shows a resistor-capacitor circuit with a pair of switches and Fig. m5.3(b) shows the switch opening-closing behavior as a function of time. The initial capacitor voltage is −9 V. Component values are $R_1 = 10$ kΩ, $R_2 = 3.3$ kΩ, $R_3 = 2.2$ kΩ, $C = 1.0$ μF, $V_1 = 9$ V, and $V_2 = -15$ V.

(a) Determine the equation that describes $v(t)$ over the time range 0 to 50 ms.

(b) Plot $v(t)$ over the time range 0 to 50 ms.

(c) Determine the values of $v(t)$ at the times 5, 15, 25, 35, and 45 ms.

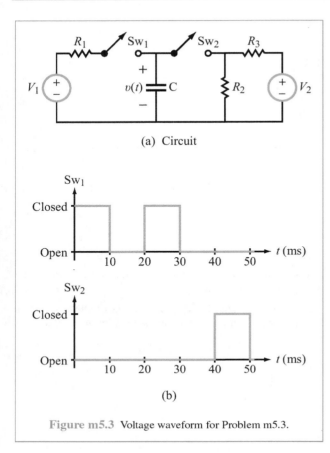

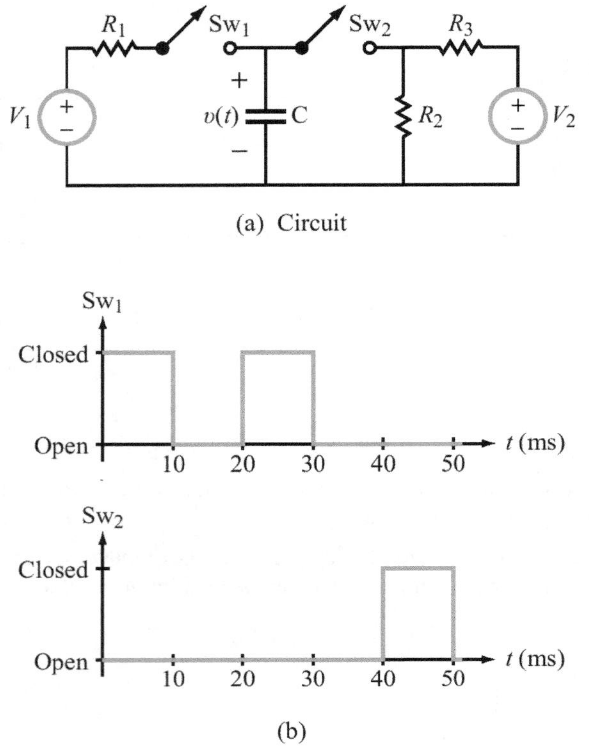

Figure m5.3 Voltage waveform for Problem m5.3.

m5.4 Response of the RL Circuit: The circuit of Fig. m5.4 demonstrates how an inductor can produce a high-voltage pulse across a load resistor R_{load} that is considerably higher than the circuit's power supply V_{batt}, a 1.5 V "AA" battery. High-voltage pulses drive photo flash bulbs, strobe lights, and cardiac defibrillators, as examples.

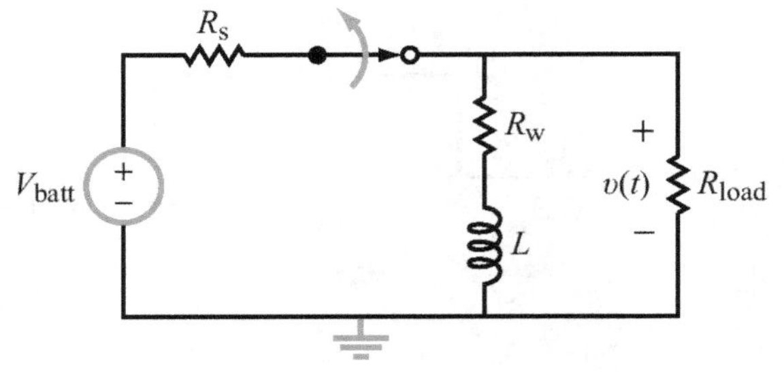

Figure m5.4 Circuit for Problem m5.4.

Resistor R_s models the finite resistance of an electronic analog switch and R_w models the finite winding resistance of the inductor. Component values are: $R_s = 16\ \Omega$, $R_w = 90\ \Omega$, $R_{load} = 680\ \Omega$, $L = 33$ mH, and $V_{batt} = 1.5$ V.

(a) Determine the load voltage υ after the switch had been closed for a long time.

(b) Determine the equation that describes $\upsilon(t)$ after the switch opens at time $t = 0$.

(c) Determine the magnitude of the peak value of $\upsilon(t)$. How many times larger is this value compared to the battery voltage V_{batt}?

(d) State the value of the circuit time constant τ with the switch open. Plot $\upsilon(t)$ over the time range $-\tau \leq t \leq 5\tau$.

m5.5 RC Differentiator: The circuit in Fig. m5.5 is a differentiator. Find $\upsilon_{out}(t)$, given that $\upsilon_s(t)$ is a 300 Hz sinusoid with an amplitude of 3 V. You will need to use the myDAQ's Function Generator and Oscilloscope for this problem.

m5.6 RC Integrator: The circuit in Fig. m5.6 is an RC integrator circuit. Find $\upsilon_{out}(t)$, given that $\upsilon_s(t)$ is a 100 Hz sinusoid with an amplitude of 5 V. You will need to use the myDAQ's Function Generator and Oscilloscope for this problem.

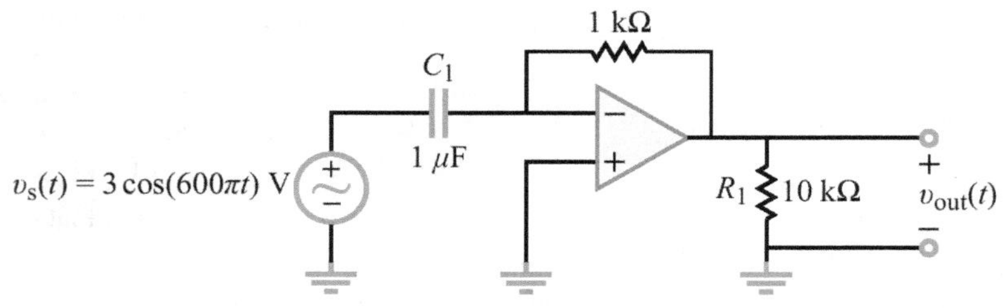

Figure m5.5 A differentiator circuit.

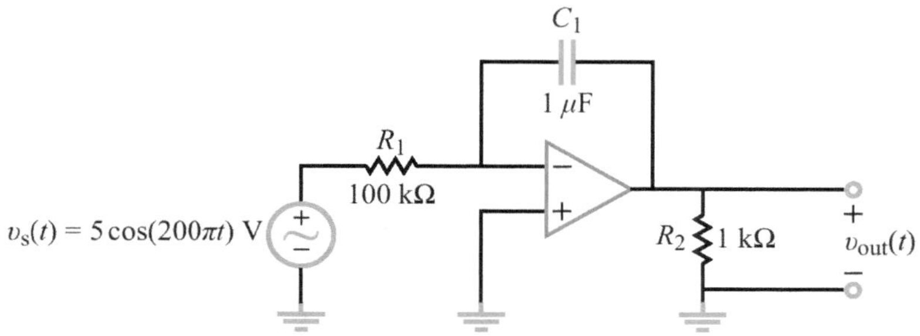

Figure m5.6 Circuit for Problem m5.6.

CHAPTER 6

RLC Circuits

Contents

Objectives

Learn to:

- Analyze series and parallel RLC circuits containing dc sources and switches.
- Analyze RC op-amp circuits.
- Understand RFID circuits.

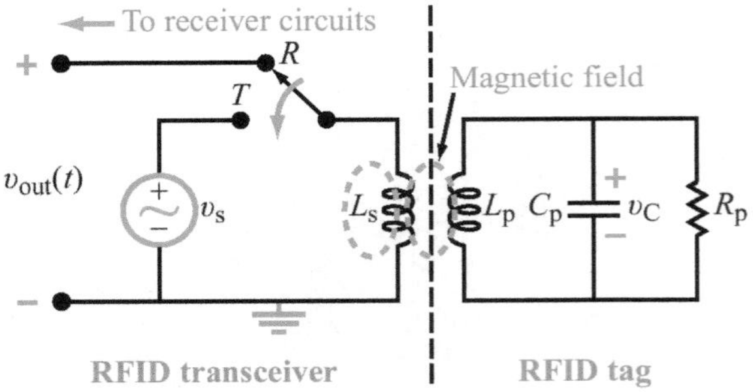

Overview

In this chapter we evaluate the operation of second-order RLC circuits—those with any combination of two inductors and/or capacitors—in response to dc sources (the response of RLC circuits to ac sources is covered in Chapter 7). These circuits are particularly interesting because they allow us to design oscillators and resonators for communication and wireless power transmission systems, or to create sensors that use the oscillation or resonance to detect capacitive (usually) or inductive (rarely) changes caused by environmental parameters (moisture, pressure, proximity, etc.). One particularly interesting example is wireless power transfer for radiofrequency ID (RFID) systems, as described in Section 6-9 and Technology Brief 16. Using two inductors and a capacitor, the current in one loop is converted into voltage in the capacitor, that can then be used to power the RFID circuit.

The currents and voltages of the first-order RC and RL circuits we examined in the preceding chapter were characterized by first-order differential equations. A key provision of a *first-order circuit* is that it is reducible to a single series or parallel circuit containing a single capacitor or a single inductor, in addition to sources and resistors. If a circuit contains two capacitors, as in Fig. 6-1(a), and if the circuit architecture is such that it is not possible to combine the two capacitors into a single in-series or in-parallel equivalent, then the circuit does not qualify as a first-order circuit. The two-capacitor circuit is a *second-order circuit* characterized by a second-order differential equation. The same is true for the two-inductor circuit in part (b) and for the series and parallel RLC circuits shown in parts (c) and (d) of the same figure.

▶ A second-order circuit may contain any combination of two energy-storage elements (2 capacitors, 2 inductors, or one of each), provided like-elements cannot be replaced with a single-element equivalent. ◀

In general, the order of a circuit, and hence the order of the differential equation describing any of its currents or voltages, is governed by the number of irreducible storage elements (capacitors and inductors) contained in the circuit. The complexity of the solution depends on the order of the differential equation and the character of the excitation source. In this chapter we examine the response of series and parallel RLC circuits to dc excitations, and we do so by solving their differential equations in the time domain. Time-domain solutions are reasonably tractable, so long as the forcing function is a dc source or a rectangular pulse, and the differential equation describing the voltages and currents in the circuit is not higher than second order. For more complicated circuits, a more robust method of solution is called for, such as the Laplace transform analysis technique introduced in Chapter 12, which is perfectly suited to deal with a wide range of circuits and any type of realistic forcing function, including pulses and sinusoids.

6-1 Initial and Final Conditions

The general form of the solution of the differential equation associated with a second-order circuit always includes a number of unknown constants. To determine the values of these constants, we usually match the solution to known values of the voltage or current under consideration. For a circuit where the solution we seek is for the time period following a sudden change (such as when a SPST switch is closed or opened, or when a SPDT switch is moved from one terminal to another)

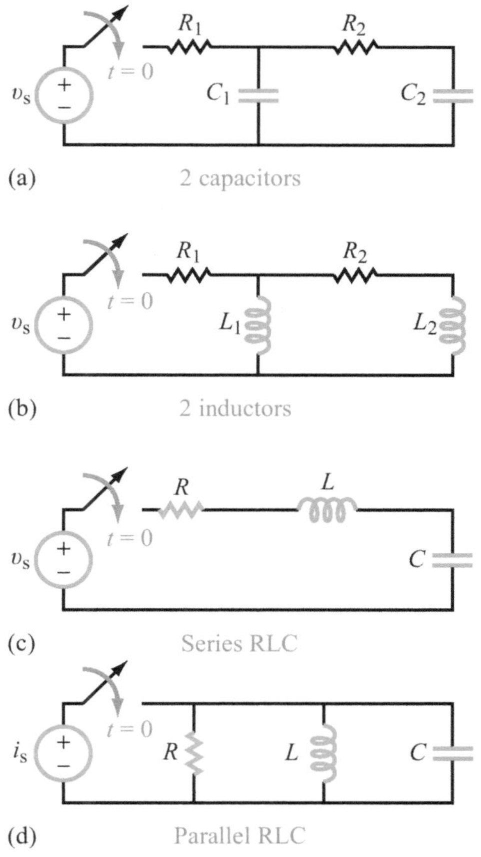

Figure 6-1: Examples of second-order circuits.

we can analyze the circuit conditions at the beginning and at the end of that time period and then use the results to match the solution of the differential equation. We call the process *invoking initial and final conditions*.

Analyzing a circuit in its initial and final states relies on the following fundamental properties:

▶
- The voltage v_C across a capacitor cannot change instantaneously, and neither can the current i_L through an inductor.
- In circuits containing dc sources, the steady state condition of the circuit (after all transients have died out) is such that no currents flow through capacitors and no voltages exist across inductors, allowing us to represent capacitors as open circuits and inductors as short circuits under steady state conditions. ◀

Example 6-1: Initial and Final Values

The circuit in Fig. 6-2(a) contains dc source V_s and a switch that had been in position 1 for a long time prior to $t = 0$. Determine: (a) initial values $v_C(0)$ and $i_L(0)$, (b) $i_C(0)$ and $v_L(0)$, and (c) final values $v_C(\infty)$ and $i_L(\infty)$.

Solution: (a) To determine $v_C(0)$ and $i_L(0)$, we analyze the circuit configuration at $t = 0^-$ (before moving the switch), whereas to determine $i_C(0)$ and $v_L(0)$, we analyze the circuit configuration at $t = 0$ (after moving the switch). At $t = 0^-$, the circuit is equivalent to the arrangement shown in Fig. 6-2(b), in which C has been replaced with an open circuit and L with a short circuit. Because the circuit contains no closed loops, no current flows anywhere in the circuit. With no voltage drop across R_1, it follows that

$$v_C(0^-) = V_s.$$

Also,

$$i_L(0^-) = 0.$$

Time-continuity of v_C and i_L mandates that after moving the switch to terminal 2:

$$v_C(0) = v_C(0^-) = V_s,$$
$$i_L(0) = i_L(0^-) = 0.$$

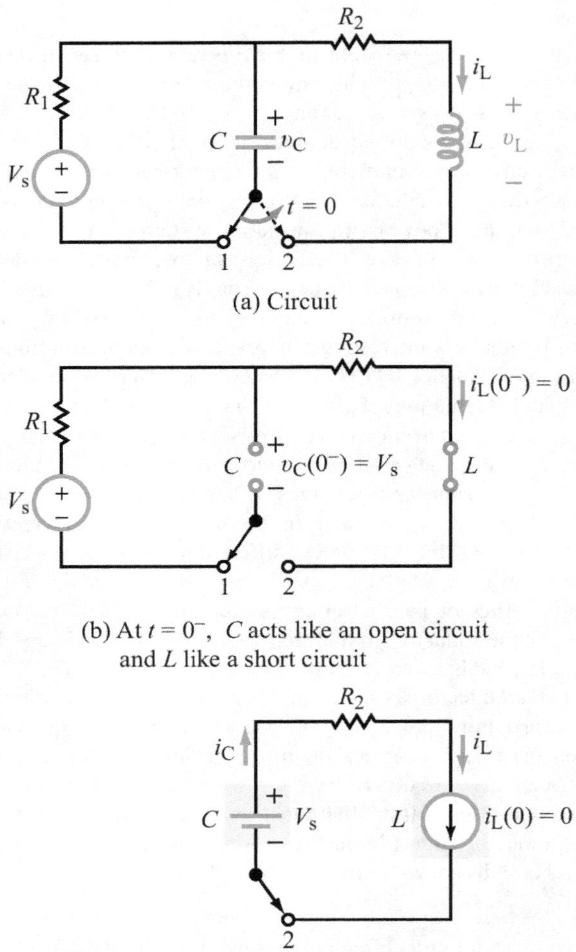

R_2
i_L
R_1
C
v_C
L
v_L
V_s
$t = 0$
1
2
(a) Circuit
R_2
$i_L(0^-) = 0$
R_1
C
$v_C(0^-) = V_s$
L
V_s
1
2
(b) At $t = 0^-$, C acts like an open circuit and L like a short circuit
R_2
i_C
i_L
C
V_s
L
$i_L(0) = 0$
2
(c) At $t = 0$, C acts like a voltage source and L like a current source with zero current

Figure 6-2: Circuit of Example 6-1.

(b) The circuit in Fig. 6-2(c) depicts the state of the circuit at $t = 0$ (after moving the switch). The capacitor behaves like a dc voltage source of magnitude V_s, and the inductor behaves like a dc current source with zero current, which is equivalent to an open circuit. Even though in general there is no requirement disallowing a sudden change in i_C, in this case $i_C = i_L$ and $i_L(0) = 0$.

Consequently,

$$i_C(0) = 0.$$

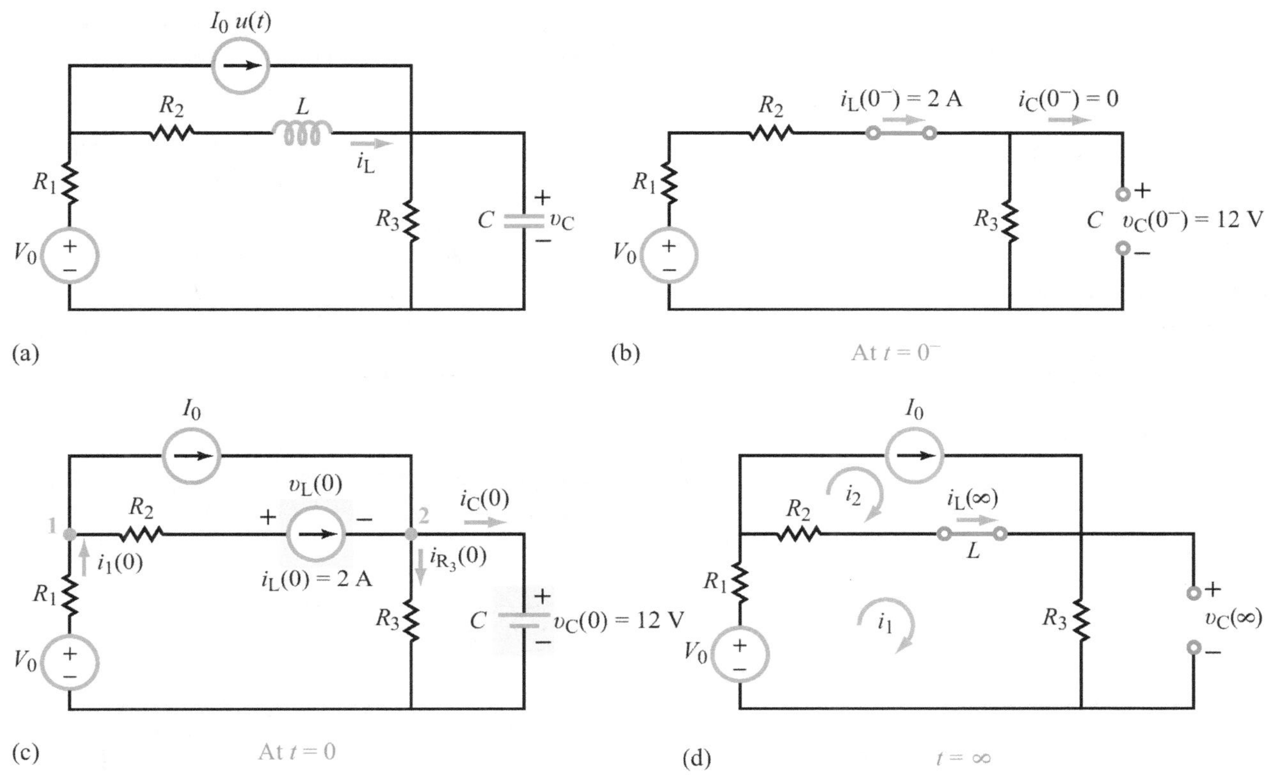

Figure 6-3: Circuit for Example 6-2.

With no voltage drop across R_2, the voltage across the inductor is

$$v_L(0) = v_C(0) = V_s.$$

(c) The analysis for v_C and i_L as $t \to \infty$ is totally straightforward; with no active sources remaining in the part of the circuit that contains L and C, all of the energy that may have been stored in L and C will have dissipated completely by $t = \infty$, rendering the circuit inactive. Hence,

$$v_C(\infty) = 0, \qquad i_L(\infty) = 0.$$

Example 6-2: Initial and Final Conditions

The circuit in Fig. 6-3(a) contains a dc voltage source and a step-function current source. The element values are $V_0 = 24$ V, $I_0 = 4$ A, $R_1 = 2\ \Omega$, $R_2 = 4\ \Omega$, $R_3 = 6\ \Omega$, $L = 0.2$ H, and $C = 8$ mF. Determine: (a) $v_C(0)$ and $i_L(0)$, (b) $i_C(0)$ and $v_L(0)$, and (c) $v_C(\infty)$ and $i_L(\infty)$.

Solution: (a) To find initial values of v_C and i_L at $t = 0$, we have to determine their values at $t = 0^-$, and then invoke the requirement that neither the voltage across a capacitor nor the current through an inductor can change in zero time. The state of the circuit at $t = 0^-$ is shown in Fig. 6-3(b), wherein the inductor has been replaced with a short circuit, the capacitor replaced with an open circuit, and the current source is absent altogether. Since $i_C(0^-) = 0$,

$$i_L(0^-) = \frac{V_0}{R_1 + R_2 + R_3} = 2\text{ A},$$

and

$$v_C(0^-) = i_L(0^-)\,R_3 = 12\text{ V}.$$

Hence,

$$i_L(0) = i_L(0^-) = 2\text{ A},$$

and

$$v_C(0) = v_C(0^-) = 12 \text{ V}.$$

(b) At $t = 0$, the state of the circuit is as shown in Fig. 6-3(c). Since

$$v_{R_3}(0) = v_C(0) = 12 \text{ V},$$

it follows that

$$i_{R_3}(0) = \frac{12}{6} = 2 \text{ A}.$$

We did this because we need $i_C(0)$. Application of KCL at node 2 leads to

$$i_C(0) = I_0 + i_L(0) - i_{R_3}(0) = 4 + 2 - 2 = 4 \text{ A}.$$

Next, we need to determine $v_L(0)$. At node 1,

$$i_1(0) = I_0 + i_L(0) = 4 + 2 = 6 \text{ A}.$$

By applying KVL around the lower left loop, we find that

$$v_L(0) = -8 \text{ V}.$$

(c) The state of the circuit at $t = \infty$ shown in Fig. 6-3(d) resembles that at $t = 0^-$, except that now we also have the current source I_0. The mesh equation for loop 1 is

$$-V_0 + R_1 i_1 + R_2(i_1 - i_2) + R_3 i_1 = 0,$$

and for loop 2,

$$i_2 = I_0 = 4 \text{ A}.$$

Solving for i_1 gives

$$i_1 = 3.33 \text{ A},$$

which leads to

$$i_L(\infty) = i_1 - I_0 = 3.33 - 4 = -0.67 \text{ A}$$

and

$$v_C(\infty) = i_1 R_3 = 3.33 \times 6 = 20 \text{ V}.$$

Concept Question 6-1: Determination of initial circuit conditions after a sudden change relies on two fundamental properties of capacitors and inductors. What are they? (See CAD)

Concept Question 6-2: Under dc steady state conditions, does a capacitor resemble an open circuit or a short circuit? What does an inductor resemble? (See CAD)

Concept Question 6-3: What role do initial and final values play in the solution of a circuit? (See CAD)

Exercise 6-1: For the circuit in Fig. E6.1, determine $v_C(0)$, $i_L(0)$, $v_L(0)$, $i_C(0)$, $v_C(\infty)$, and $i_L(\infty)$.

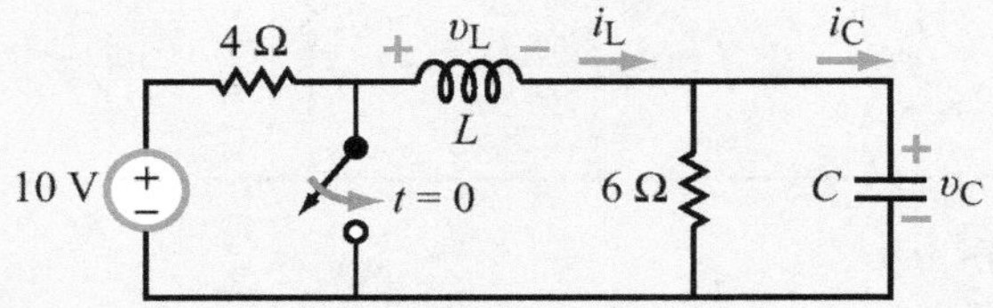

Figure E6.1

Answer: $v_C(0) = 6$ V, $i_L(0) = 1$ A, $v_L(0) = -6$ V, $i_C(0) = 0$, $v_C(\infty) = 0$, $i_L(\infty) = 0$. (See CAD)

Exercise 6-2: For the circuit in Fig. E6.2, determine $v_C(0)$, $i_L(0)$, $v_L(0)$, $i_C(0)$, $v_C(\infty)$, and $i_L(\infty)$.

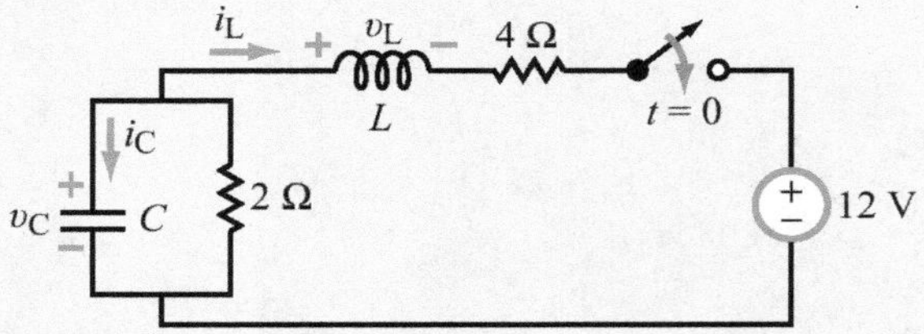

Figure E6.2

Answer: $v_C(0) = 0$, $i_L(0) = 0$, $v_L(0) = -12$ V, $i_C(0) = 0$, $v_C(\infty) = 4$ V, $i_L(\infty) = -2$ A. (See CAD)

6-2 Introducing the Series RLC Circuit

6-2.1 Charging-Up Mode

The circuit in part (a) of Fig. 6-4 depicts a scenario in which a series RLC circuit with no stored energy is connected to a dc voltage source V_s at $t = 0$. After closing the switch, charge supplied by the source starts to flow to the (+) voltage terminal of the capacitor, and continues to do so until the capacitor reaches the maximum voltage possible, namely V_s. Hence, our expectation is that $v_C(t)$ will start at zero at $t = 0$ and then build up to reach V_s as $t \to \infty$. The specific path it takes,

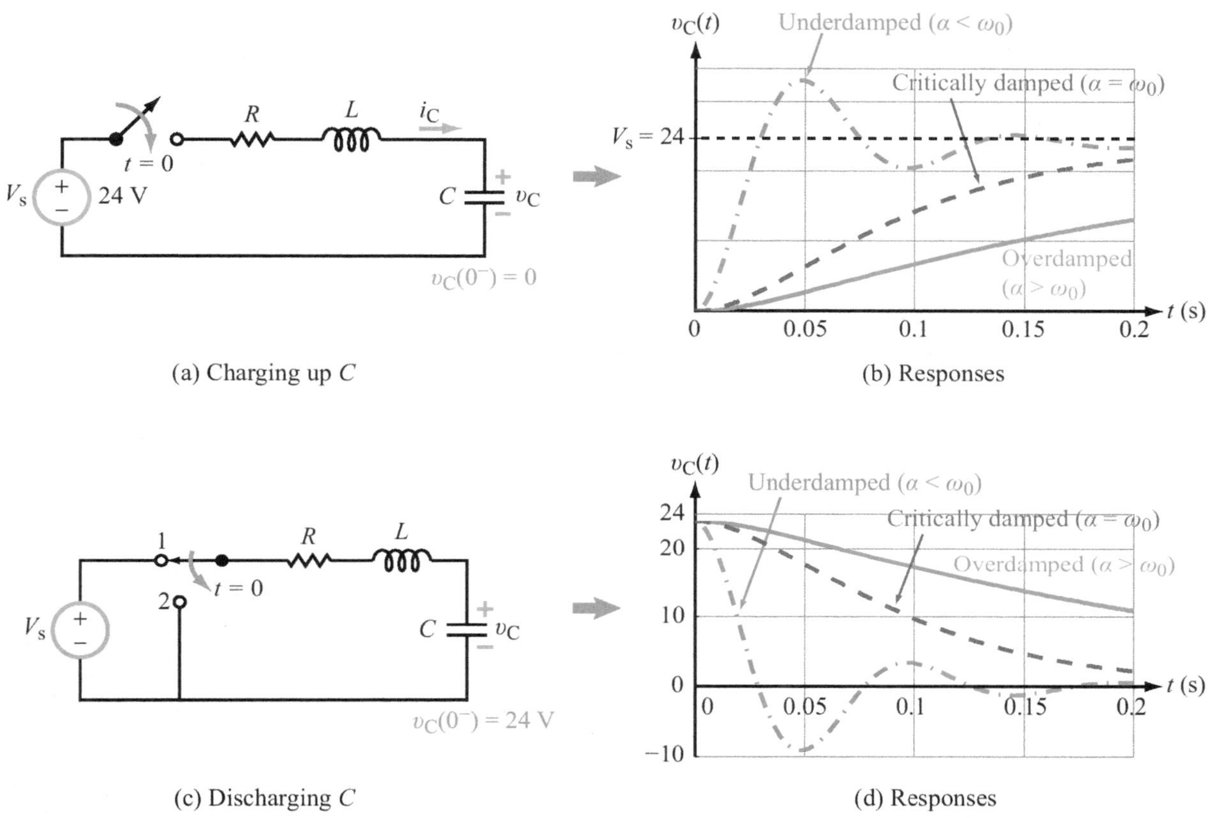

Figure 6-4: Illustrating the charge-up and discharge responses of a series RLC circuit with $V_s = 24$ V. In all cases $R = 12\ \Omega$ and $L = 0.3$ H, which specifies $\alpha = R/2L = 20$ Np/s. When $C = 0.01$ F, the response is overdamped, when $C = 8.33$ mF, the response is critically damped, and when $C = 0.72$ mF, the response is underdamped.

however, depends on the relative magnitudes of two important parameters. These are:

$$\textit{damping coefficient}\ \alpha = \frac{R}{2L} \quad \text{(Np/s)}, \tag{6.1a}$$

$$\textit{resonant frequency}\ \omega_0 = \frac{1}{\sqrt{LC}} \quad \text{(rad/s)}. \tag{6.1b}$$

(series RLC)

The parameter α is measured in nepers/second (Np/s) and ω_0 is an angular frequency, measured in radians per second (rad/s). The magnitudes of the two parameters are specified by the values chosen for R, L, and C.

Figure 6-4(b) displays three different response curves for $v_C(t)$, labeled as follows:

Overdamped response	$\alpha > \omega_0$,
Critically damped response	$\alpha = \omega_0$,
Underdamped response	$\alpha < \omega_0$.

The critically damped response represents the fastest smooth path for $v_C(t)$ between its initial and final values. In comparison, the overdamped response is slower than the underdamped response, which starts out faster but exhibits an oscillatory (*ringing*) behavior. The mathematical solutions for all three cases are presented in detail in forthcoming sections. The intent is to provide an overview of how $v_C(t)$ varies with time under these various scenarios.

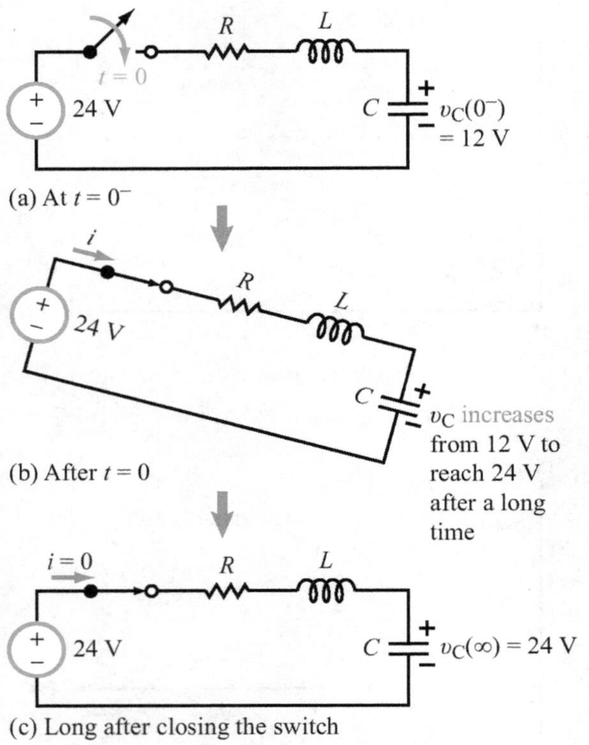

Figure 6-5: Connecting a series RLC circuit with a charged-up capacitor to a source with higher voltage.

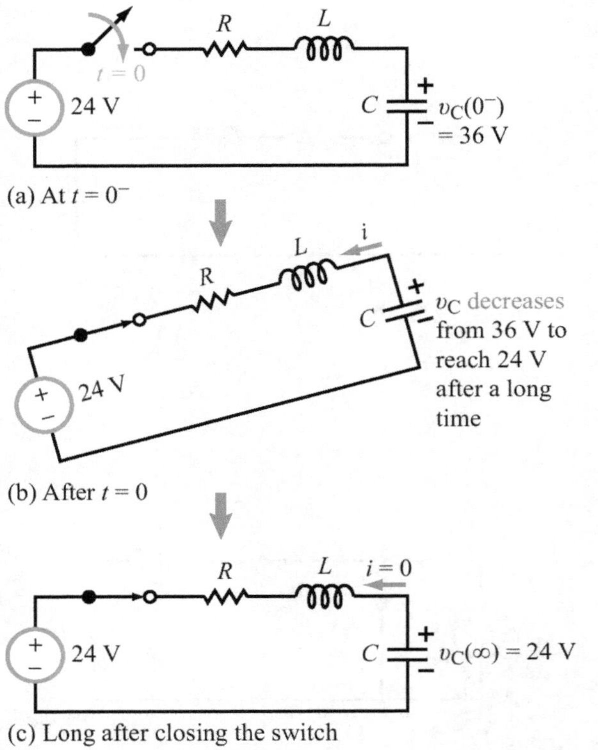

Figure 6-6: Connecting a series RLC circuit with a charged-up capacitor to a source with lower voltage.

6-2.2 Discharging Mode

If instead of starting out with an uncharged RLC circuit, we were to start with a fully charged capacitor, as depicted by the circuit in Fig. 6-4(c), and then discharge it by moving the SPDT switch from terminal 1 to terminal 2, the voltage $v_C(t)$ across the capacitor will decay from its initial value, V_s, to a final value of zero volts. The specific path between V_s and zero again depends on the value of α relative to that of ω_0, as shown in Fig. 6-4(d). In fact, the three responses of the discharging RLC circuit are essentially mirror images of those for the charging-up circuit; the initial and final conditions of the circuit in Fig. 6-4(a) are the converse of those for the circuit in Fig. 6-4(c). The capacitor voltage of the changing-up circuit starts at zero and concludes at 24 V, in contrast to the discharging circuit that starts at 24 V and concludes at zero.

Now let us consider an RLC circuit in which the capacitor has 12 V across it (due to some previous charging-up action), and then a switch is closed to connect the RLC segment to a source with $V_s = 24$ V, as shown in Fig. 6-5(a). After closing the switch (Fig. 6-5(b)), the situation is such that $V_s = 24$ V exceeds the initial voltage of 12 V across the capacitor. Consequently, charge will flow to the capacitor to build up its voltage, and will continue to do so until the capacitor reaches the maximum possible voltage, namely $V_s = 24$ V. When it reaches that state, the current goes to zero (Fig. 6-5(c)).

The scenario in Fig. 6-6 depicts a similar circuit, but one that starts with a capacitor whose initial voltage $v_C(0^-)$ is 36 V, which is higher than that of $V_s = 24$ V. In this case, the capacitor will start to discharge after closing the switch and then continue to discharge until it reaches 24 V. Thus, in both circuit scenarios, the capacitor will charge up or discharge down so as to equalize its voltage to that of the source, V_s. Recall that a short circuit is equivalent to a voltage source with $V_s = 0$. Hence, if we connect an RLC circuit with a charged-up capacitor to a short circuit, the capacitor will discharge down until it reaches a final voltage of zero, the same as the scenario depicted in Fig. 6-4(c).

Technology Brief 15
Micromechanical Sensors and Actuators

Energy is stored in many different forms in the world around us. The conversion of energy from one form to another is called *transduction*. Each of our five senses, for example, transduces a specific form of energy into electrochemical signals: tactile transducers on the skin convert mechanical and thermal energy; the eye converts electromagnetic energy; smell and taste receptors convert chemical energy; and our ears convert the mechanical energy of pressure waves. Any device, whether natural or man-made, that converts energy signals from one form to another is a *transducer*.

Most modern man-made systems are designed to manipulate signals (i.e., information) using electrical energy. Computation, communication, and storage of information are examples of functions performed mostly with electrical circuits. Most systems also perform a fourth class of signal manipulation: the transduction of energy from the environment into electrical signals that circuits can use in support of their intended application. If a transducer converts external signals into electrical signals, it is called a *sensor*. The charge-coupled device (CCD) chip on your camera is a sensor that converts electromagnetic energy (light) into electrical signals that can be processed, stored, and communicated by your camera circuits. Some transducers perform the reverse function, namely to convert a circuit's electrical signal into an environmental excitation. Such a transducer then is called an *actuator*. The components that deploy the airbag in your car are actuators: given the right signal from the car's microcontroller, the actuators convert electrical energy into mechanical energy and the airbag is released and inflated.

Microelectromechanical Systems (MEMS)

Micro- and nanofabrication technology have begun to revolutionize many aspects of sensor and actuator design. Humans increasingly are able to embed transducers at very fine scales into their environment. This is leading to big changes, as our computational elements are becoming increasingly aware of their environment. Shipping containers that track their own acceleration profiles, laptops that scan fingerprints for routine login, cars that detect collisions, and even office suites that modulate energy consumption based on human activity are all examples of this transduction revolution. In this technology brief, we will focus on a specific type of microscale transducers that lend themselves to direct integration with silicon ICs. Collectively, devices of this type are called *microelectromechanical systems* (MEMS) or *microsystems technologies* (MST); the two names are used interchangeably.

A Capacitive Sensor: The MEMS Accelerometer

According to Eq. (5.21), the capacitance C of a parallel plate capacitor varies directly with A, the effective area of overlap between its two conducting plates, and inversely with d, the spacing between the plates. By capitalizing on these two attributes, capacitors can be made into motion sensors that can measure velocity and acceleration along x, y, and z.

Figure TF15-1 illustrates two mechanisms for translating motion into a change of capacitance. The first generally is called the *gap-closing mode*, while the second one is called the *overlap mode*. In the gap-closing mode, A remains constant, but if a vertical force is applied onto the upper plate, causing it to be displaced from its nominal position at height d above the lower plate to a new position $(d-z)$, then the value of capacitance C_z will change in accordance with the expression given in Fig. TF15-1(a). The sensitivity of C_z to the vertical displacement is given by dC_z/dz.

The overlap mode (Fig. TF15-1(b)) is used to measure horizontal motion. If a horizontal force causes one of the plates to shift by a distance y from its nominal position (where nominal position corresponds to a 100 percent overlap), the decrease in effective overlap area will lead to a corresponding change in the magnitude of capacitance C_y. In this case, d remains constant, but the width of the overlapped areas changes from w to $(w-y)$. The expression for C_y given in Fig. TF15-1(b) is reasonably accurate (even though it ignores the effects of the *fringing electric field* between the edges of the two plates) so long as $y \ll w$. To measure and amplify changes in capacitance, the capacitor can be integrated into an appropriate op-amp circuit whose output voltage is proportional to C. As we shall see shortly, a combination of three capacitors, one to sense vertical motion and two to measure horizontal motion along orthogonal axes, can provide complete information on both the velocity and acceleration vectors associated with the applied force. The capacitor configurations shown in Fig. TF15-1 illustrate the basic concept of how a capacitor is used to measure motion, although more complex capacitor

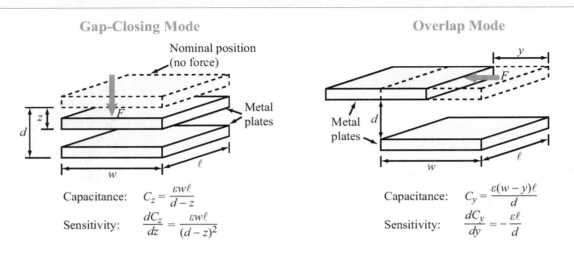

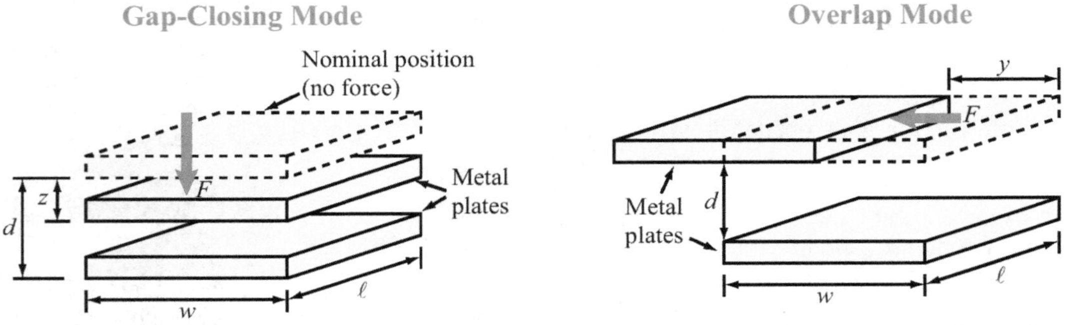

Capacitance: $C_z = \dfrac{\varepsilon w\ell}{d-z}$

Sensitivity: $\dfrac{dC_z}{dz} = \dfrac{\varepsilon w\ell}{(d-z)^2}$

Capacitance: $C_y = \dfrac{\varepsilon(w-y)\ell}{d}$

Sensitivity: $\dfrac{dC_y}{dy} = -\dfrac{\varepsilon\ell}{d}$

Figure TF15-1: Basic capacitive measurement modes. For (b), the expressions hold only for small displacements such that $y \ll w$.

geometries also are possible, particularly for sensing angular motion.

To convert the capacitor-accelerometer concept into a practical sensor—such as the automobile accelerometer that controls the release of the airbag—let us consider the arrangement shown in Fig. TF15-2(b). The lower plate is fixed to the body of the vehicle, and the upper plate sits on a plane at a height d above it. The upper plate is attached to the body of the vehicle through a spring with a *spring constant* k. When no horizontal force is acting on the upper plate, its position is such that it provides a 100 percent overlap with the lower plate, in which case the capacitance will be a maximum at $C_y = \varepsilon W\ell/d$. If the vehicle accelerates in the y-direction with acceleration a_y, the acceleration force F_{acc} will generate an opposing spring force F_{sp} of equal magnitude.

Equating the two forces leads to an expression relating the displacement y to the acceleration a_y, as shown in the figure. Furthermore, the capacitance C_y is directly proportional to the overlap area $\ell(w-y)$ and therefore is proportional to the acceleration a_y. Thus, by measuring C_y, the accelerometer determines the value of a_y. A similar overlap-mode capacitor attached to the vehicle along the x-direction can be used to measure a_x. Through a similar analysis for the gap-closing mode capacitor shown in Fig. TF15-2(a), we can arrive at a functional relationship that can be used to determine the vertical acceleration a_z by measuring capacitance C_z.

For example, if we designate the time when the ignition starts the engine as $t = 0$, we then can set the initial conditions on both the velocity u of the vehicle and its acceleration a as zero at $t = 0$. That is, $u(0) = a(0) = 0$. The capacitor accelerometers measure continuous-time waveforms $a_x(t)$, $a_y(t)$, and $a_z(t)$. Each waveform then can be used by an op-amp integrator circuit to calculate the corresponding velocity waveform. For u_x, for example,

$$u_x(t) = \int_0^t a_x(t)\, dt,$$

and similar expressions apply to u_y and u_z.

Commercial MEMS Accelerometers

Figure TF15-3 shows the Analog Devices ADXL202 accelerometer which uses the gap-closing mode to detect accelerations on a tiny micromechanical capacitor structure that works on the same principle described above, although slightly more complicated geometrically. Commercial accelerometers, such as this one, make use of negative feedback to prevent the plates from physically moving. When an acceleration force attempts to move the plate, an electric negative-feedback circuit applies a voltage across the plates to generate an electrical force between the plates that counteracts the acceleration force exactly, thereby preventing any motion by the plate. The magnitude of the applied voltage becomes a measure of the acceleration force that the capacitor plate is subjected to. Because of their small size and low power consumption, chip-based microfabricated silicon accelerometers

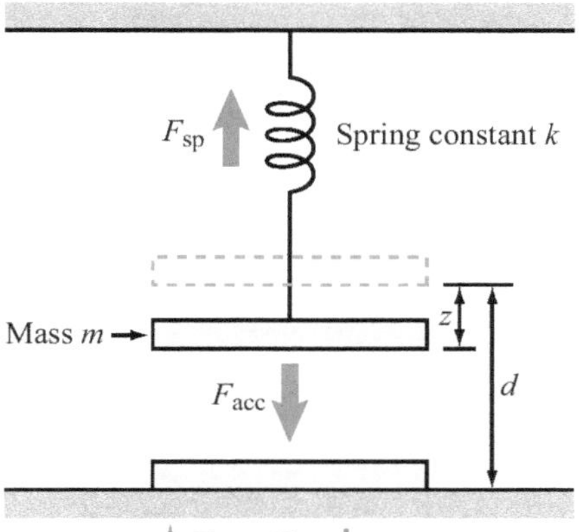

$$F_{sp} = F_{acc}$$
$$kz = ma_z$$
$$z = \frac{ma_z}{k}$$
$$C_z = \frac{\varepsilon w \ell}{\left(d - \frac{ma_z}{k}\right)}$$

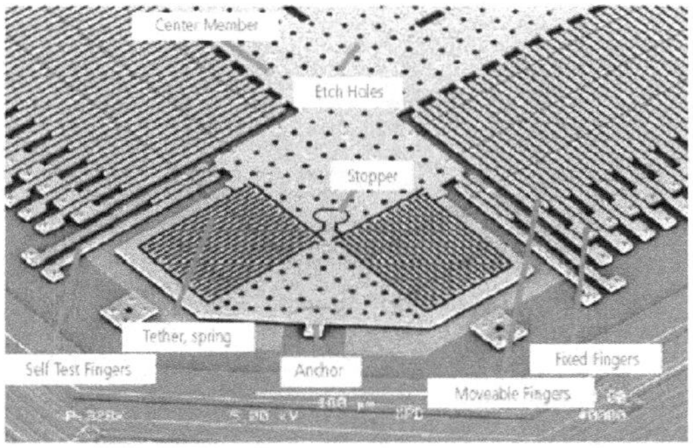

(a) The ADXL202 accelerometer employs many gap-closing capacitor sensors to detect acceleration. (Courtesy Analog Devices.)

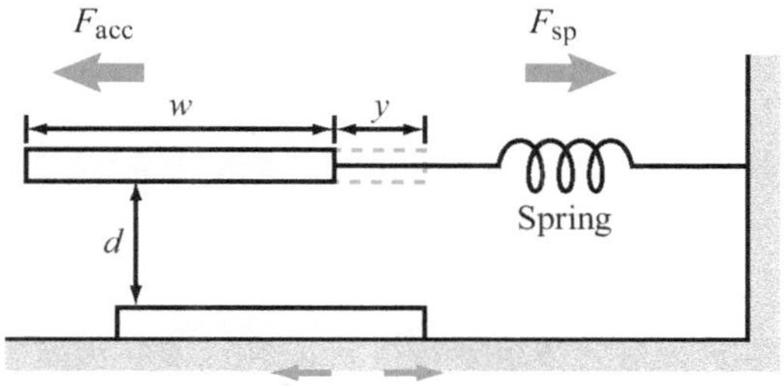

$$F_{acc} = F_{sp}$$
$$ma_y = ky$$
$$y = \frac{ma_y}{k}$$
$$C_y = \frac{\varepsilon \ell (w - y)}{d} = \frac{\varepsilon \ell \left(w - \frac{ma_y}{k}\right)}{d}$$

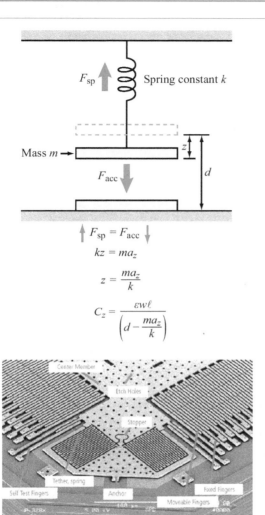

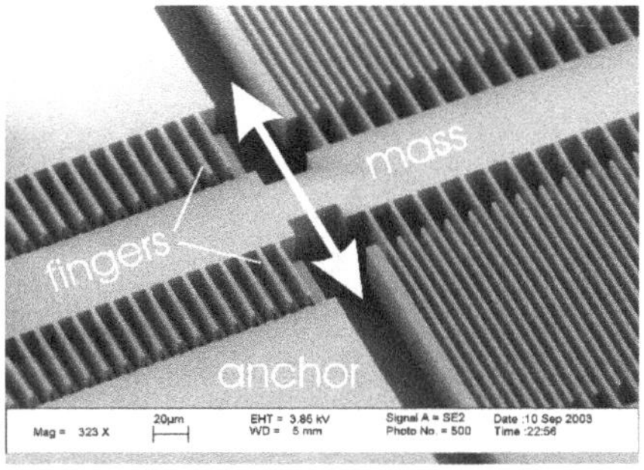

(b) A silicon sensor that uses overlap mode fingers. The white arrow shows the direction of motion of the moving mass and its fingers in relation to the fixed anchors. Note that the moving fingers move into and out of the fixed fingers on either side of the mass during motion. (Courtesy of the Adriatic Research Institute.)

Figure TF15-2: Adding a spring to a movable plate capacitor makes an accelerometer.

are used in most modern cars to activate the release mechanism of airbags. They also are used heavily in many toy applications to detect position, velocity and acceleration. The Nintendo Wii, for example, uses accelerometers in each remote to detect orientation and acceleration. Incidentally, a condenser microphone operates much like the device shown in Fig. TF15-2(a): as air pressure waves (sound) hit the spring-mounted plate, it moves and the change in capacitance can be read and recorded.

A Capacitive Actuator: MEMS Electrostatic Resonators

Not surprisingly, we can drive the devices discussed previously in reverse to obtain actuators. Consider again

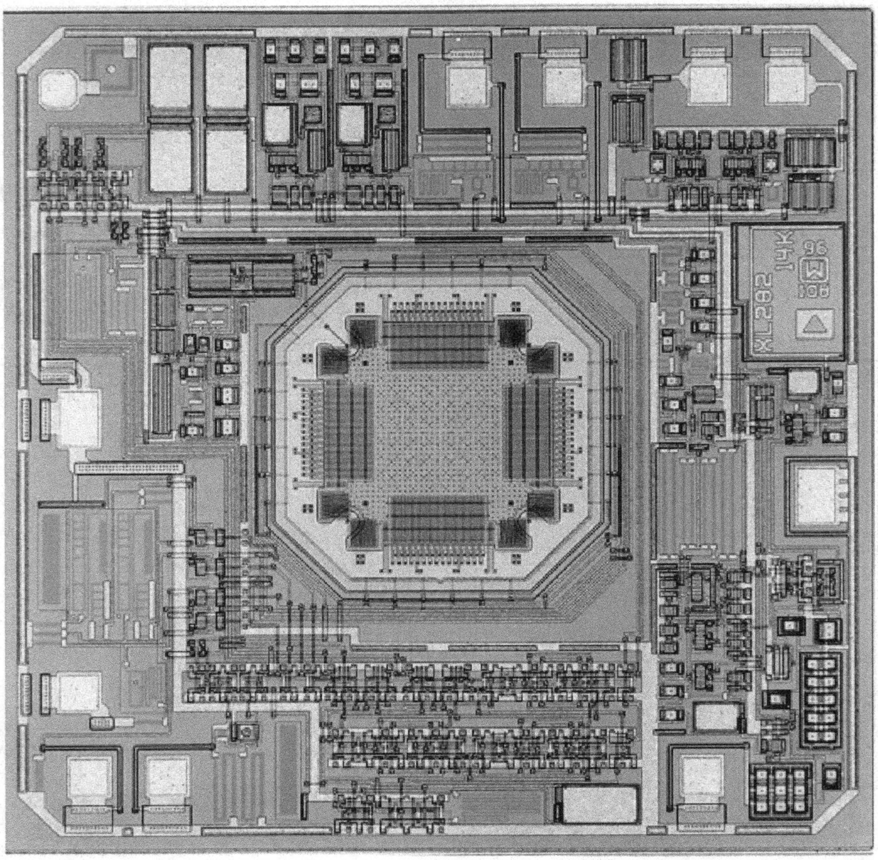

Figure TF15-3: The complete ADXL202 accelerometer chip. The center region holds the micromechanical sensor; the majority of the chip space is used for the electronic circuits that measure the capacitance change, provide feedback, convert the measurement into a digital signal, and perform self-tests. (Courtesy of Analog Devices.)

the configuration in Fig. TF15-2(a). If the device is not experiencing any external forces and we apply a voltage V across the two plates, an attractive force F will develop between the plates. This is because charges of opposite polarity on the two plates give rise to an electrostatic force between them. This, in fact, is true for all capacitors. In the case of our actuator, however, we replace the normally stiff, dielectric material with air (since air is itself a dielectric) and attach it to a spring as before. With this modification, an applied potential generates an electrostatic force that moves the plates.

This basic idea can be applied to a variety of applications. A classic application is the ***digital light projector*** (DLP) system that drives most digital projectors used today. In the DLP, hundreds of thousands of capacitor actuators are arranged in a 2-D array on a chip, with each actuator corresponding to a pixel on an image displayed by the projector. One capacitive plate of each pixel actuator (which is mirror smooth and can reflect light exceedingly well) is connected to the chip via a spring. In order to brighten or darken a pixel, a voltage is applied between the plates, causing the mirror to move into or out of the path of the projected light. These same devices have been used for many other applications, including microfluidic valves and tiny force sensors used to measure forces as small as a zeptonewton (1 zeptonewton = 10^{-21} newtons).

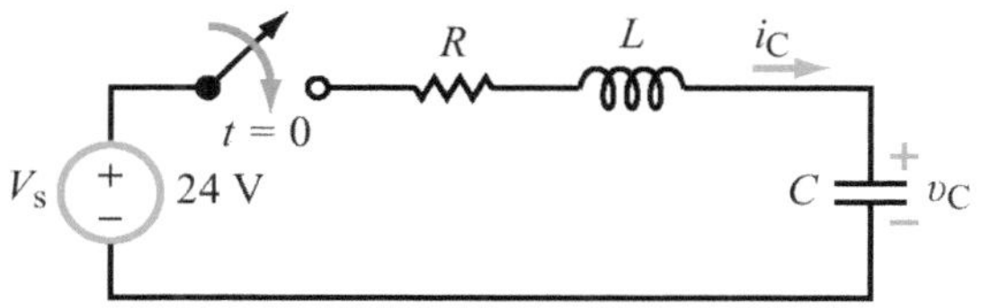

Figure 6-7: Series RLC circuit connected to a source V_s at $t = 0$. In general, the capacitor may have had an initial charge on it at $t = 0^-$, with a corresponding initial voltage $\upsilon_C(0^-)$.

6-3 Series RLC Overdamped Response ($\alpha > \omega_0$)

A key takeaway lesson from the qualitative description given in the preceding section is that after closing the switch in a series RLC circuit, the voltage across the capacitor will charge up or discharge down to equalize to the voltage across the source. In this section, we derive the differential equation for the series RLC circuit in Fig. 6-7 and then solve it to obtain an expression for $\upsilon_C(t)$ for $t \geq 0$, with $t = 0$ designated as the time *immediately after* the switch is closed.

As noted in the preceding section, the nature of the solution for $\upsilon_C(t)$ depends on how the magnitude of the damping coefficient α compares with that of the resonant frequency ω_0. The values of the two parameters are dictated by the values of R, L, and C, per the expressions in Eq. (6.1). In the present section, we consider the case corresponding to $\alpha > \omega_0$, which is called the *overdamped response*. The other two cases are treated in follow-up sections.

6-3.1 Differential Equation

For the circuit in Fig. 6-7, the KVL loop equation for $t \geq 0$ (after closing the switch) is

$$Ri_C + L\,\frac{di_C}{dt} + \upsilon_C = V_s \qquad (\text{for } t \geq 0), \tag{6.2}$$

where i_C and υ_C are the current through and voltage across the capacitor. The capacitor may or may not have had charge on it. If it had, we denote the value of the initial voltage across it $\upsilon_C(0)$, which is the same as $\upsilon_C(0^-)$, the voltage across it before closing the switch (since the voltage across a capacitor cannot change instantaneously).

By incorporating the relation

$$i_C = C\,\frac{d\upsilon_C}{dt}\,, \tag{6.3}$$

and rearranging terms, Eq. (6.2) becomes

$$\frac{d^2\upsilon_C}{dt^2} + \frac{R}{L}\,\frac{d\upsilon_C}{dt} + \frac{1}{LC}\,\upsilon_C = \frac{V_s}{LC}\,. \tag{6.4}$$

For convenience, we rewrite Eq. (6.4) in the abbreviated form

$$\upsilon''_C + a\upsilon'_C + b\upsilon_C = c, \tag{6.5}$$

where

$$a = \frac{R}{L}\,, \qquad b = \frac{1}{LC}\,, \qquad c = \frac{V_s}{LC}\,. \tag{6.6}$$

The second-order differential equation given by Eq. (6.5) is specific to the capacitor voltage of the series RLC circuit of Fig. 6-7, but the form of the equation is equally applicable to any current or voltage in any second-order circuit (although the values of the constants a, b, and c are different for different circuits). The same is true for the general form of the solution of the differential equation.

6-3.2 Solution of Differential Equation

The general solution of the second-order differential equation given by Eq. (6.5) consists of two components:

$$\upsilon_C(t) = \upsilon_{tr}(t) + \upsilon_{ss}(t), \tag{6.7}$$

where $\upsilon_{tr}(t)$ is the *transient* (also called *homogeneous* solution of Eq. (6.5) or the *natural response* of the RLC circuit) and $\upsilon_{ss}(t)$ is the *steady-state* solution (also called *particular* solution). The transient solution is the solution of Eq. (6.5) under source-free conditions; i.e., with $V_s = 0$, which means that $c = V_s/LC$ also is zero. Thus $\upsilon_{tr}(t)$ is the solution of

$$\upsilon''_{tr} + a\upsilon'_{tr} + b\upsilon_{tr} = 0 \qquad \textbf{(source-free)}. \tag{6.8}$$

The steady-state solution $\upsilon_{ss}(t)$ is related to the forcing function on the right-hand side of Eq. (6.5), and its functional form is similar to that of the forcing function. Since in the present case, the forcing function c is simply a constant, so is $\upsilon_{ss}(t)$. That is, $\upsilon_{ss}(t)$ is a non–time-varying constant υ_{ss} that will be determined later from initial and final conditions. Moreover, as we will see shortly, the transient component $\upsilon_{tr}(t)$ always goes to zero as $t \to \infty$ (that's why it is called *transient*). Hence, as $t \to \infty$, Eq. (6.7) reduces to

$$\upsilon_C(\infty) = \upsilon_{ss}, \tag{6.9}$$

in which case Eq. (6.7) can be rewritten as

$$\upsilon_C(t) = \upsilon_{tr}(t) + \upsilon_C(\infty). \tag{6.10}$$

Our remaining task is to determine $\upsilon_{tr}(t)$.

When differentiated, the exponential function e^{st} replicates itself (within a multiplying factor), so it is often offered as a candidate solution when solving homogeneous differential equations. Thus, we assume that

$$\upsilon_{\rm tr}(t) = Ae^{st}, \tag{6.11}$$

where A and s are constants to be determined later. To ascertain that Eq. (6.11) is indeed a viable solution of Eq. (6.8), we insert the proposed expression for $\upsilon_{\rm tr}(t)$ and its first and second derivatives in Eq. (6.8). The result is

$$s^2Ae^{st} + asAe^{st} + bAe^{st} = 0, \tag{6.12}$$

which simplifies to

$$s^2 + as + b = 0. \tag{6.13}$$

Hence, the proposed solution given by Eq. (6.11) is indeed an acceptable solution so long as Eq. (6.13) is satisfied.

The quadratic equation given by Eq. (6.13) is known as the *characteristic equation* of the differential equation. It has two roots:

$$s_1 = -\frac{a}{2} + \sqrt{\left(\frac{a}{2}\right)^2 - b}\,, \tag{6.14a}$$

$$s_2 = -\frac{a}{2} - \sqrt{\left(\frac{a}{2}\right)^2 - b}\,. \tag{6.14b}$$

Since the values of a and b are governed by the values of only the passive components in the circuit, so are the values of s_1 and s_2. Strictly speaking, the unit of s_1 and s_2 is 1/second, but it is customary to add the dimensionless neper to the units of quantities that appear in exponential functions. Hence, s_1 and s_2 are measured in *nepers/second* (Np/s).

The existence of two distinct roots implies that Eq. (6.8) has two viable solutions, one in terms of e^{s_1t} and another in terms of e^{s_2t}. Hence, we should generalize the form of our solution to

$$\upsilon_{\rm tr}(t) = A_1e^{s_1t} + A_2e^{s_2t} \qquad \text{for } t \geq 0, \tag{6.15}$$

where constants A_1 and A_2 are to be determined shortly.

Inserting Eq. (6.15) into Eq. (6.10) leads to

$$\upsilon_{\rm C}(t) = A_1e^{s_1t} + A_2e^{s_2t} + \upsilon_{\rm C}(\infty). \tag{6.16}$$

The exponential coefficients s_1 and s_2 are given by Eq. (6.14) in terms of constants a and b, both of which are defined in Eq. (6.6). By reintroducing the *damping coefficient* α and *resonant frequency* ω_0, which we defined earlier in Eq. (6.1), as

$$\alpha = \frac{R}{2L} = \frac{a}{2} \qquad \text{(Np/s)}, \tag{6.17a}$$

$$\omega_0 = \frac{1}{\sqrt{LC}} = b \qquad \text{(rad/s)}, \tag{6.17b}$$

the expressions given by Eq. (6.14) become

$$s_1 = -\alpha + \sqrt{\alpha^2 - \omega_0^2}\,, \tag{6.18a}$$

$$s_2 = -\alpha - \sqrt{\alpha^2 - \omega_0^2}\,, \tag{6.18b}$$

The solution in the present section pertains to the overdamped case corresponding to $\alpha > \omega_0$. Under this condition, both s_1 and s_2 are real, negative numbers. Consequently, as $t \to \infty$, the first two terms in Eq. (6.16) go to zero, just as we asserted earlier.

6-3.3 Invoking Initial Conditions

To determine the values of constants A_1 and A_2 in Eq. (6.16), we need to *invoke initial conditions*, which means that we need to use information available to us about the values of $\upsilon_{\rm C}$ and its time derivative $\upsilon'_{\rm C}$, both at $t = 0$. Since

$$i_{\rm C}(t) = C\,\frac{d\upsilon_{\rm C}}{dt} = C\,\upsilon'(t), \tag{6.19}$$

the second requirement is equivalent to needing to know $i_{\rm C}(0)$.

At $t = 0$, Eq. (6.16) simplifies to

$$\upsilon_{\rm C}(0) = A_1 + A_2 + \upsilon_{\rm C}(\infty), \tag{6.20}$$

and

$$\begin{aligned} i_{\rm C}(0) = C\,\frac{d\upsilon_{\rm C}}{dt}\bigg|_{t=0} &= C(s_1A_1e^{s_1t} + s_2A_2e^{s_2t})\big|_{t=0} \\ &= C(s_1A_1 + s_2A_2). \end{aligned} \tag{6.21}$$

Simultaneous solution of Eqs. (6.20) and (6.21) for A_1 and A_2 gives

$$A_1 = \frac{\frac{1}{C}\,i_{\rm C}(0) - s_2[\upsilon_{\rm C}(0) - \upsilon_{\rm C}(\infty)]}{s_1 - s_2}\,, \tag{6.22a}$$

$$A_2 = \frac{\frac{1}{C}\,i_{\rm C}(0) - s_1[\upsilon_{\rm C}(0) - \upsilon_{\rm C}(\infty)]}{s_2 - s_1}\,. \tag{6.22b}$$

This concludes the general solution for the overdamped response. A summary of relevant expressions is available in Table 6-1.

Table 6-1: Step response of RLC circuits for $t \geq 0$.

Series RLC	Parallel RLC
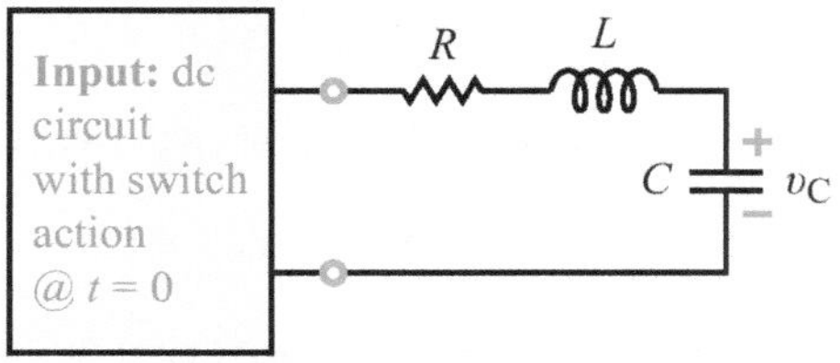	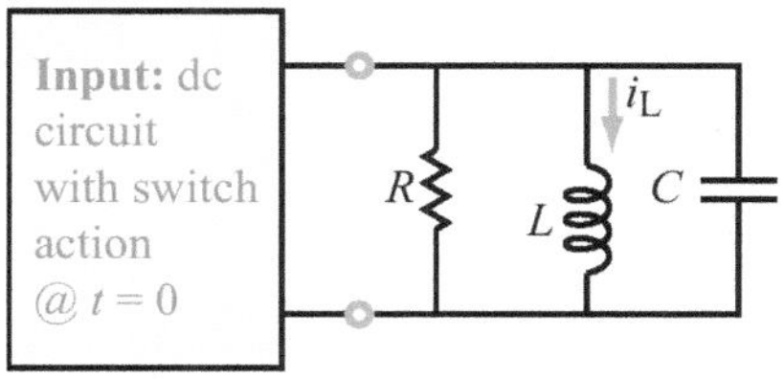
Total Response	**Total Response**
Overdamped ($\alpha > \omega_0$) $v_C(t) = A_1 e^{s_1 t} + A_2 e^{s_2 t} + v_C(\infty)$ $A_1 = \dfrac{\frac{1}{C}\, i_C(0) - s_2[v_C(0) - v_C(\infty)]}{s_1 - s_2}$ $A_2 = \left[\dfrac{\frac{1}{C}\, i_C(0) - s_1[v_C(0) - v_C(\infty)]}{s_2 - s_1}\right]$	**Overdamped** ($\alpha > \omega_0$) $i_L(t) = A_1 e^{s_1 t} + A_2 e^{s_2 t} + i_L(\infty)$ $A_1 = \dfrac{\frac{1}{L}\, v_L(0) - s_2[i_L(0) - i_L(\infty)]}{s_1 - s_2}$ $A_2 = \left[\dfrac{\frac{1}{L}\, v_L(0) - s_1[i_L(0) - i_L(\infty)]}{s_2 - s_1}\right]$
Critically Damped ($\alpha = \omega_0$) $v_C(t) = (B_1 + B_2 t)e^{-\alpha t} + v_C(\infty)$ $B_1 = v_C(0) - v_C(\infty)$ $B_2 = \frac{1}{C}\, i_C(0) + \alpha[v_C(0) - v_C(\infty)]$	**Critically Damped** ($\alpha = \omega_0$) $i_L(t) = (B_1 + B_2 t)e^{-\alpha t} + i_L(\infty)$ $B_1 = i_L(0) - i_L(\infty)$ $B_2 = \frac{1}{L}\, v_L(0) + \alpha[i_L(0) - i_L(\infty)]$
Underdamped ($\alpha < \omega_0$) $v_C(t) = e^{-\alpha t}(D_1 \cos\omega_d t + D_2 \sin\omega_d t) + v_C(\infty)$ $D_1 = v_C(0) - v_C(\infty)$ $D_2 = \dfrac{\frac{1}{C}\, i_C(0) + \alpha[v_C(0) - v_C(\infty)]}{\omega_d}$	**Underdamped** ($\alpha < \omega_0$) $i_L(t) = e^{-\alpha t}(D_1 \cos\omega_d t + D_2 \sin\omega_d t) + i_L(\infty)$ $D_1 = i_L(0) - i_L(\infty)$ $D_2 = \dfrac{\frac{1}{L}\, v_L(0) + \alpha[i_L(0) - i_L(\infty)]}{\omega_d}$
Auxiliary Relations	
$\alpha = \begin{cases} \dfrac{R}{2L} & \text{Series RLC} \\ \dfrac{1}{2RC} & \text{Parallel RLC} \end{cases}$	$\omega_0 = \dfrac{1}{\sqrt{LC}}$ $\omega_d = \sqrt{\omega_0^2 - \alpha^2}$
$s_1 = -\alpha + \sqrt{\alpha^2 - \omega_0^2}$	$s_2 = -\alpha - \sqrt{\alpha^2 - \omega_0^2}$

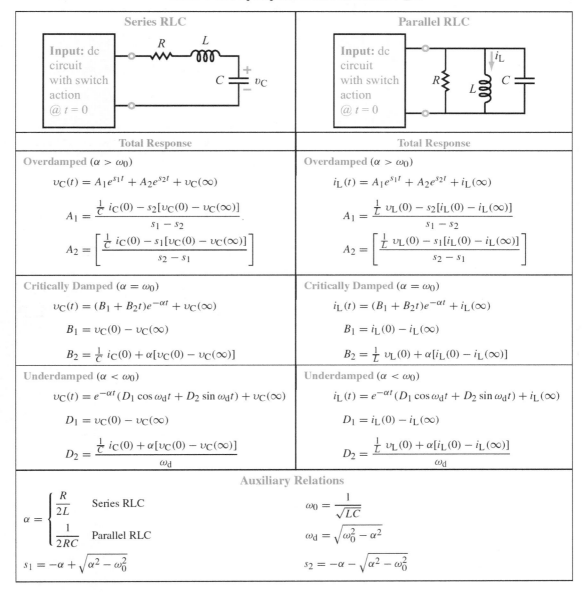

Example 6-3: Charging Up Capacitor with No Prior Charge

Given that in the circuit of Fig. 6-8(a), $V_s = 16$ V, $R = 64\ \Omega$, $L = 0.8$ H, and $C = 2$ mF, determine $v_C(t)$ and $i_C(t)$ for $t \geq 0$. The capacitor had no charge prior to $t = 0$.

Solution: We begin by establishing the damping condition of the circuit. From the definitions for α and ω_0 given by Eq. (6.17), we have

$$\alpha = \frac{R}{2L} = \frac{64}{2 \times 0.8} = 40 \text{ Np/s},$$

$$\omega_0 = \frac{1}{\sqrt{LC}} = \frac{1}{\sqrt{0.8 \times 2 \times 10^{-3}}} = 25 \text{ rad/s}.$$

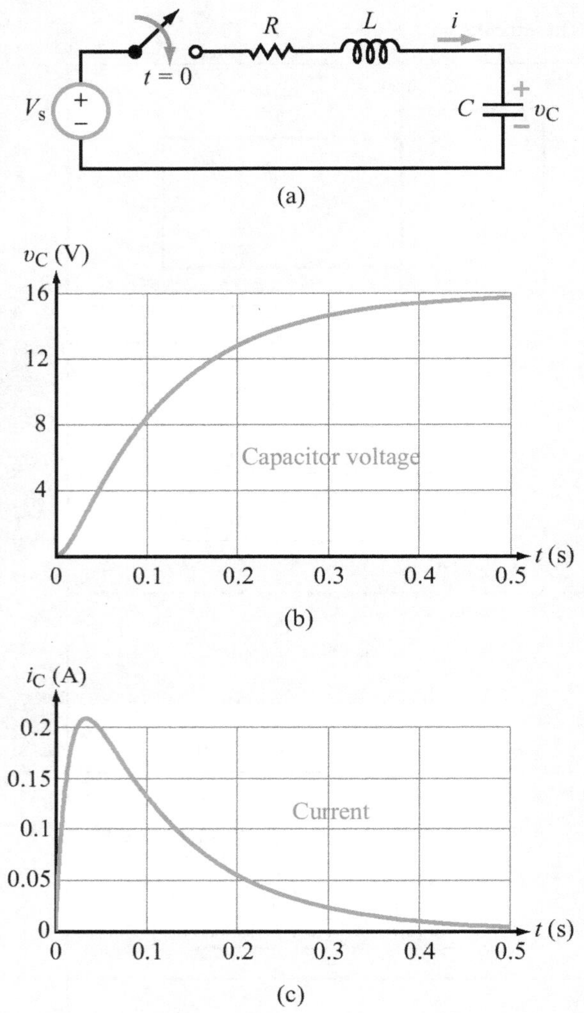

Figure 6-8: Example 6-3: (a) circuit, (b) $\upsilon_C(t)$, and (c) $i_C(t)$.

Hence, $\alpha > \omega_0$, which means that the circuit will exhibit an overdamped response after the switch is closed. The applicable expression for $\upsilon_C(t)$ is given by Eq. (6.16),

$$\upsilon_C(t) = [A_1 e^{s_1 t} + A_2 e^{s_2 t} + \upsilon_C(\infty)].$$

From Eq. (6.18),

$$\begin{aligned} s_1 &= -\alpha + \sqrt{\alpha^2 - \omega_0^2} \\ &= -40 + \sqrt{40^2 - 25^2} = -8.8 \text{ Np/s}, \\ s_2 &= -\alpha - \sqrt{\alpha^2 - \omega_0^2} = -71.2 \text{ Np/s}. \end{aligned}$$

As $t \to \infty$, the circuit reaches steady state and the capacitor becomes like an open circuit, allowing no current to flow through the circuit. Consequently,

$$\upsilon_C(\infty) = V_s = 16 \text{ V}.$$

At $t = 0^-$, the capacitor was uncharged. Hence,

$$\upsilon_C(0) = \upsilon_C(0^-) = 0.$$

Prior to $t = 0$, there was no current in the circuit, and since the current through L (which is also the current through C) cannot change instantaneously, it follows that

$$i_C(0) = i_L(0) = i_L(0^-) = 0.$$

From Eq. (6.22), A_1 and A_2 are given by

$$\begin{aligned} A_1 &= \frac{\frac{1}{C}\, i_C(0) - s_2[\upsilon_C(0) - \upsilon_C(\infty)]}{s_1 - s_2} \\ &= \frac{0 + 71.2(0 - 16)}{-8.8 + 71.2} = -18.25 \text{ V}, \\ A_2 &= -\left[\frac{\frac{1}{C}\, i_C(0) - s_1[\upsilon_C(0) - \upsilon_C(\infty)]}{s_1 - s_2}\right] \\ &= -\left[\frac{0 + 8.8(0 - 16)}{-8.8 + 71.2}\right] = 2.25 \text{ V}. \end{aligned}$$

The total response $\upsilon_C(t)$ is then given by

$$\upsilon_C(t) = [-18.25e^{-8.8t} + 2.25e^{-71.2t} + 16] \text{ V} \quad (\text{for } t \geq 0),$$

and the associated current is

$$\begin{aligned} i_C(t) &= C\,\frac{d\upsilon_C}{dt} \\ &= 2 \times 10^{-3}[18.25 \times 8.8e^{-8.8t} - 2.25 \times 71.2e^{-71.2t}] \\ &= 0.32(e^{-8.8t} - e^{-71.2t}) \text{ A} \quad (\text{for } t \geq 0). \end{aligned}$$

The waveforms of $\upsilon_C(t)$ and $i_C(t)$ are displayed in Figs. 6-8(b) and (c), respectively.

Example 6-4: RLC Circuit with a Current Source

Determine $\upsilon_C(t)$ in the circuit of Fig. 6-9(a), given that $I_s = 2$ A, $R_s = 10\ \Omega$, $R_1 = 1.81\ \Omega$, $R_2 = 0.2\ \Omega$, $L = 5$ mH,

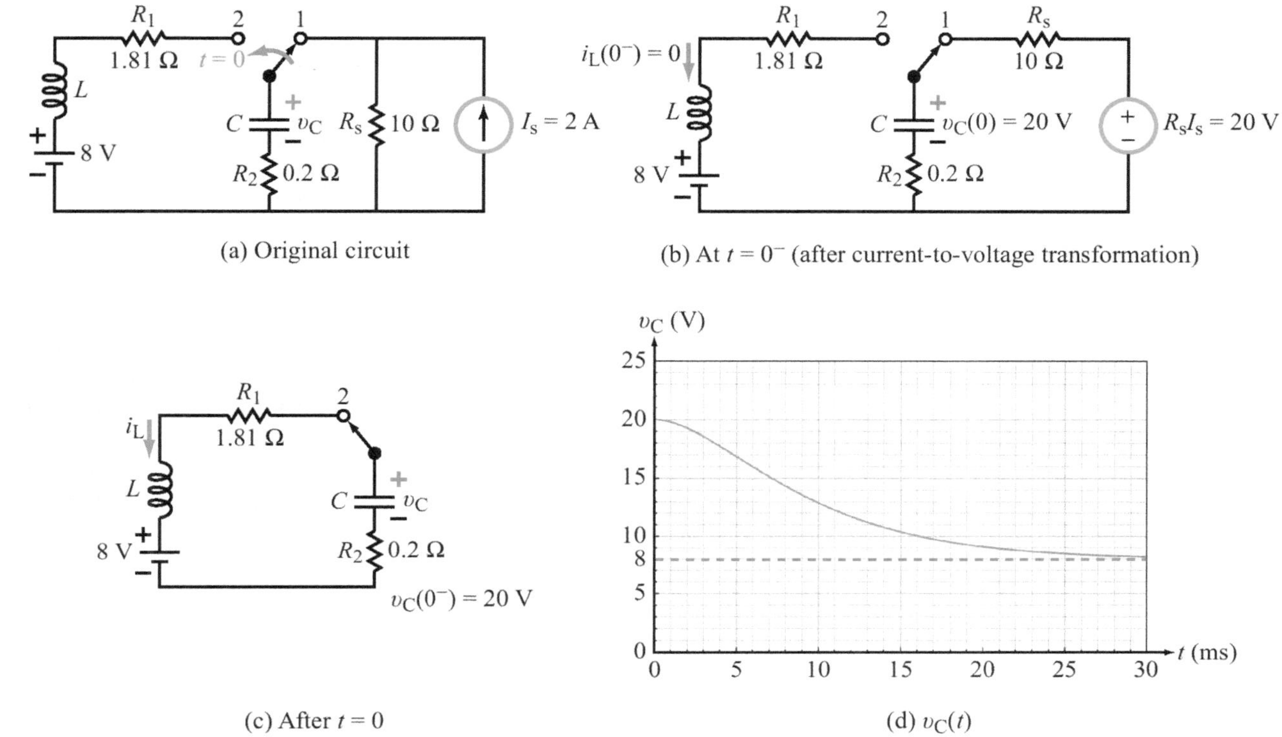

Figure 6-9: Circuit for Example 6-4.

and $C = 5$ mF. Assume that the circuit had been in the condition shown in Fig. 6-9(a) for a long time prior to $t = 0$.

Solution:

At $t = 0^-$: Figure 6-9(b) depicts the state of the circuit at $t = 0^-$, but after making a current source to voltage source transformation. The replacement voltage source is 20 V. Since the circuit had been in steady state for a long time, the capacitor behaves like an open circuit with

$$\upsilon_C(0^-) = 20 \text{ V}.$$

We also note that in the left-hand part of the circuit, no current can flow, mandating that

$$i_L(0^-) = 0.$$

At $t \geq 0$: After moving the switch to terminal 2, the capacitor becomes part of a new circuit composed of a combination of $R = R_1 + R_2 = 1.81 + 0.2 = 2.01\ \Omega$, $L = 5$ mH, and $C = 5$ mF, all connected in series with an 8 V source (Fig. 6-9(c)). The current through C is the same as the current through L, and since the current through an inductor cannot charge instantaneously, it follows that

$$i_C(0) = i_L(0) = i_L(0^-) = 0.$$

For the capacitor,

$$\upsilon_C(0) = \upsilon_C(0^-) = 20 \text{ V}.$$

Also, as t approaches ∞, $\upsilon_C(t)$ approaches the voltage of the 8 V source. Hence,

$$\upsilon_C(\infty) = 8 \text{ V}.$$

The parameters α and ω_0 are given by

$$\alpha = \frac{R_1 + R_2}{2L} = \frac{2.01}{2 \times 5 \times 10^{-3}} = 201 \text{ Np/s},$$

$$\omega_0 = \frac{1}{\sqrt{LC}} = \frac{1}{\sqrt{5 \times 10^{-3} \times 5 \times 10^{-3}}} = 200 \text{ rad/s}.$$

Since $\alpha > \omega_0$, the response is overdamped and given by Eq. (6.16),

$$v_C(t) = A_1 e^{s_1 t} + A_2 e^{s_2 t} + v_C(\infty),$$

with

$$\begin{aligned} s_1 &= -\alpha + \sqrt{\alpha^2 - \omega_0^2} \\ &= -201 + \sqrt{(201)^2 - (200)^2} = -181 \text{ Np/s}, \\ s_2 &= -\alpha - \sqrt{\alpha^2 - \omega_0^2} = -221 \text{ Np/s}, \\ A_1 &= \frac{\frac{1}{C}\, i_C(0) - s_2[v_C(0) - v_C(\infty)]}{s_1 - s_2} \\ &= \frac{0 + 221[20 - 8]}{-181 + 221} = 66.3, \\ A_2 &= \frac{\frac{1}{C}\, i_C(0) - s_1[v_C(0) - v_C(\infty)]}{s_2 - s_1} \\ &= \frac{0 + 181[20 - 8]}{-221 + 181} = -54.3. \end{aligned}$$

Inserting the values of s_1, s_2, A_1, A_2, and $v_C(\infty)$ in Eq. (6.16) leads to

$$v_C(t) = (66.3 e^{-181t} - 54.3 e^{-221t} + 8) \text{ V} \qquad \text{for } t \geq 0.$$

Figure 6-9(d) displays the time response of $v_C(t)$.

Exercise 6-3: After interchanging the locations of L and C in Fig. 6-9(a), repeat Example 6-4 to determine $v_C(t)$ across C.

Answer: $v(t) = 9.8(e^{-221t} - e^{-181t})$ V. (See CAD)

6-4 Series RLC Critically Damped Response ($\alpha = \omega_0$)

▶ The critically damped response is the fastest response the circuit can exhibit, without oscillation, between initial and final conditions. ◀

When

$$R = 2\sqrt{\frac{L}{C}} \qquad \text{(critically damped)}, \tag{6.23}$$

$\alpha = \omega_0$, and according to Eq. (6.18),

$$s_1 = s_2 = -\alpha. \tag{6.24}$$

Repeated roots are problematic because Eq. (6.16) becomes

$$\begin{aligned} v_C(t) &= A_1 e^{-\alpha t} + A_2 e^{-\alpha t} + v_C(\infty) \\ &= (A_1 + A_2) e^{-\alpha t} + v_C(\infty) = (A_3) e^{-\alpha t} + v_C(\infty), \end{aligned} \tag{6.25}$$

where $A_3 = A_1 + A_2$. A solution containing a single constant (A_3) cannot simultaneously satisfy the initial conditions on both the voltage across the capacitor and the current through the inductor.

For this *critically damped* case, we introduce two new constants, B_1 and B_2, and we adopt the modified form

$$\begin{aligned} v_C(t) &= B_1 e^{-\alpha t} + B_2 t e^{-\alpha t} + v_C(\infty) \\ &= (B_1 + B_2 t) e^{-\alpha t} + v_C(\infty) \\ &\quad (\text{for } t \geq 0) \qquad \textbf{(critically damped)}, \end{aligned} \tag{6.26}$$

which contains a term with $e^{-\alpha t}$ and a second term with $(te^{-\alpha t})$. It is a relatively straightforward task to show that the expression given by Eq. (6.26) is indeed a valid solution of the differential equation given by Eq. (6.4). When doing so, however, we need to keep in mind that under the critically damped condition, R, L, and C are interrelated by Eq. (6.23), and $v_C(\infty) = V_s$.

The constants B_1 and B_2 are governed by the initial conditions on v_C and i_c. Thus, at $t = 0$, Eq. (6.26) provides

$$v_C(0) = B_1 + v_C(\infty), \tag{6.27a}$$

$$\begin{aligned} i_C(0) &= C \left. \frac{dv_C}{dt} \right|_{t=0} \\ &= C\, (-\alpha B_1 - \alpha B_2 t + B_2) e^{-\alpha t} \Big|_{t=0} \\ &= C(-\alpha B_1 + B_2). \end{aligned} \tag{6.27b}$$

Simultaneous solution of Eqs. (6.27a and b) leads to

$$B_1 = v_C(0) - v_C(\infty), \tag{6.28a}$$

$$B_2 = \frac{1}{C}\, i_C(0) + \alpha[v_C(0) - v_C(\infty)]. \tag{6.28b}$$

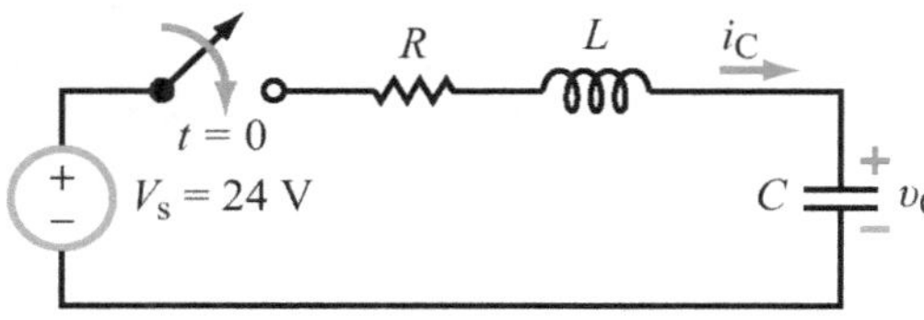

(a)

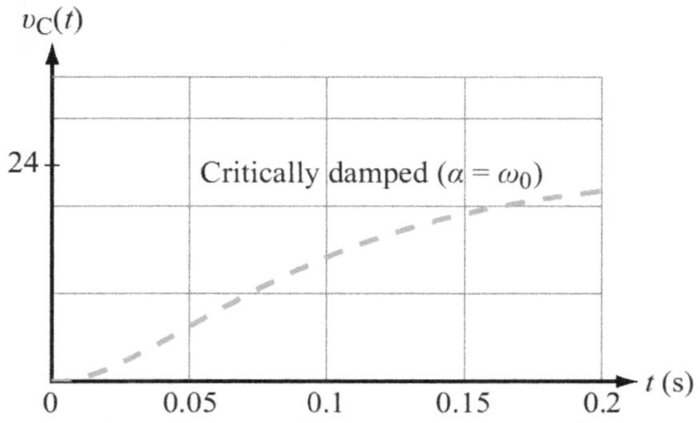

(b) $v_C(t)$

Figure 6-10: Circuit response for Example 6-5.

Example 6-5: Critically Damped Response

Evaluate the response of the circuit in Fig. 6-10(a) for $t \geq 0$, given that the capacitor had no charge prior to $t = 0$ and $V_s = 24$ V, $R = 12\ \Omega$, $L = 0.3$ H, and $C = 8.33$ mF.

Solution: The parameters α and ω_0 are given by

$$\alpha = \frac{R}{2L} = \frac{12}{2 \times 0.3} = 20 \text{ Np/s},$$

$$\omega_0 = \frac{1}{\sqrt{LC}} = \frac{1}{\sqrt{0.3 \times 8.33 \times 10^{-3}}} = 20 \text{ rad/s}.$$

Hence, because $\alpha = \omega_0$, the response is critically damped and given by Eq. (6.26) as

$$v(t) = (B_1 + B_2 t)e^{-20t} + v_C(\infty).$$

The initial conditions at $t = 0$ are

$$v_C(0) = 0 \quad \text{and} \quad i_C(0) = 0,$$

and the final condition on v_C is

$$v_C(\infty) = V_s = 24 \text{ V}.$$

Application of these initial and final conditions to Eq. (6.28) leads to

$$B_1 = v_C(0) - v_C(\infty) = -24 \text{ V},$$

$$B_2 = \frac{1}{C}\, i_C(0) + \alpha[v_C(0) - v_C(\infty)]$$

$$= 0 + 20[0 - 24] = -480.$$

Hence,

$$v_C(t) = (B_1 + B_2 t)e^{-\alpha t} + v_C(\infty)$$

$$= [-(24 + 480t)e^{-20t} + 24] \text{ V}, \qquad \text{for } t \geq 0.$$

The response is plotted in Fig. 6-10(b).

Exercise 6-4: The switch in Fig. E6.4 is moved to position 2 after it had been in position 1 for a long time. Determine: (a) $v_C(0)$ and $i_C(0)$, and (b) $i_C(t)$ for $t \geq 0$.

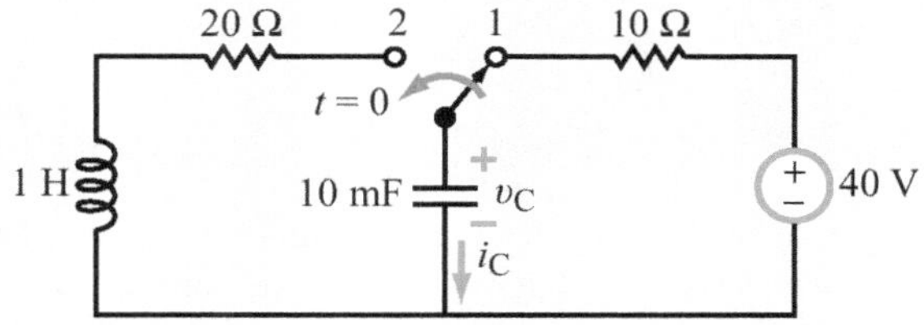

Figure E6.4

Answer: (a) $v_C(0) = 40$ V, $i_C(0) = 0$.
(b) $i_C(t) = [-40te^{-10t}]$ A. (See CAD)

Exercise 6-5: The circuit in Fig. E6.5 is a replica of the circuit in Fig. E6.4, but with the capacitor and inductor interchanged in location. Determine: (a) $i_L(0)$ and $v_L(0)$, and (b) $i_L(t)$ for $t \geq 0$.

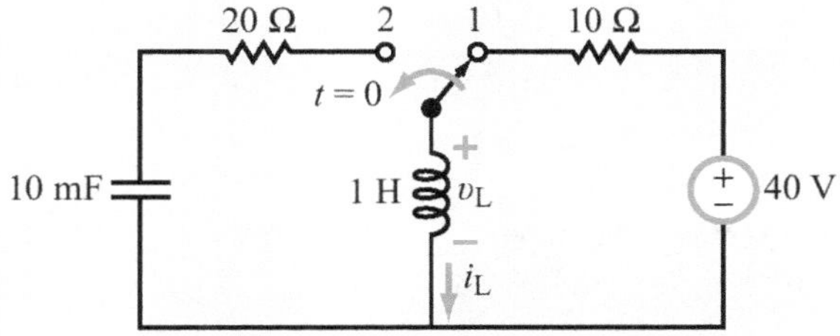

Figure E6.5

Answer: (a) $i_L(0) = 4$ A, $v_L(0) = -80$ V.
(b) $i_L(t) = [4(1 - 10t)e^{-10t}]$ A. (See CAD)

6-5 Series RLC Underdamped Response ($\alpha < \omega_0$)

If $\alpha < \omega_0$, corresponding to

$$R < 2\sqrt{\frac{L}{C}} \quad \textbf{(underdamped)}, \tag{6.29}$$

we introduce the ***damped natural frequency*** ω_d defined as

$$\omega_d^2 = \omega_0^2 - \alpha^2. \tag{6.30}$$

Since $\alpha < \omega_0$, it follows that $\omega_d > 0$. In terms of ω_d, the expressions for the roots s_1 and s_2 given by Eq. (6.18) become

$$s_1 = -\alpha + \sqrt{\alpha^2 - \omega_0^2} = -\alpha + \sqrt{-\omega_d^2} = -\alpha + j\omega_d, \tag{6.31a}$$

$$s_2 = -\alpha - \sqrt{\alpha^2 - \omega_0^2} = -\alpha - j\omega_d, \tag{6.31b}$$

where $j = \sqrt{-1}$. The fact that s_1 and s_2 are complex conjugates of one another will prove central to the form of the solution. Inserting the expressions for s_1 and s_2 into Eq. (6.16) gives

$$\upsilon_C(t) = A_1 e^{-\alpha t} e^{j\omega_d t} + A_2 e^{-\alpha t} e^{-j\omega_d t} + \upsilon_C(\infty). \tag{6.32}$$

The Euler identity

$$e^{\pm j\theta} = \cos\theta \pm j\sin\theta \tag{6.33}$$

allows us to expand Eq. (6.32) as follows:

$$\begin{aligned}\upsilon_C(t) &= A_1 e^{-\alpha t}(\cos\omega_d t + j\sin\omega_d t) \\ &\quad + A_2 e^{-\alpha t}(\cos\omega_d t - j\sin\omega_d t) + \upsilon_C(\infty) \\ &= e^{-\alpha t}[(A_1 + A_2)\cos\omega_d t \\ &\quad + j(A_1 - A_2)\sin\omega_d t] + \upsilon_C(\infty).\end{aligned} \tag{6.34}$$

Next, by introducing a new pair of constants, $D_1 = A_1 + A_2$ and $D_2 = j(A_1 - A_2)$, we have

$$\upsilon_C(t) = e^{-\alpha t}[D_1 \cos\omega_d t + D_2 \sin\omega_d t] + \upsilon_C(\infty)$$
$$(\text{for } t \geq 0) \quad \textbf{(underdamped)}. \tag{6.35}$$

The negative exponential $e^{-\alpha t}$ signifies that $\upsilon(t)$ has a damped waveform with a ***time constant*** $\tau = 1/\alpha$, and the sine and cosine terms signify that $\upsilon_C(t)$ is oscillatory with an angular frequency ω_d and a corresponding ***time period***

$$T = \frac{2\pi}{\omega_d}. \tag{6.36}$$

Since ω_d is a measure of the oscillation associated with the damped natural response of the circuit, it is only appropriate that it be called the "damped natural frequency" of the circuit.

Invoking initial conditions on the expression given by Eq. (6.35) leads to

$$D_1 = \upsilon_C(0) - \upsilon_C(\infty), \tag{6.37a}$$

$$D_2 = \frac{\frac{1}{C}\, i_C(0) + \alpha[\upsilon_C(0) - \upsilon_C(\infty)]}{\omega_d}. \tag{6.37b}$$

The oscillatory behavior of the underdamped response is illustrated by Example 6-6.

Example 6-6: Underdamped Response

Determine $\upsilon_C(t)$ for the circuit in Fig. 6-11, given that $V_s = 24$ V, $R = 12\ \Omega$, $L = 0.3$ H, and $C = 0.72$ mF. The circuit had been in steady state prior to moving the switch at $t = 0$.

Solution: For the specified values of R, L, and C,

$$\alpha = \frac{R}{2L} = \frac{12}{2 \times 0.3} = 20 \text{ Np/s}$$

and

$$\omega_0 = \frac{1}{\sqrt{LC}} = \frac{1}{\sqrt{0.3 \times 0.72 \times 10^{-3}}} = 68 \text{ rad/s}.$$

Since $\alpha < \omega_0$, the voltage response is underdamped and given by Eq. (6.35) as

$$\upsilon_C(t) = e^{-\alpha t}[D_1 \cos\omega_d t + D_2 \sin\omega_d t] + \upsilon_C(\infty),$$

with

$$\omega_d = \sqrt{\omega_0^2 - \alpha^2} = \sqrt{(68)^2 - (20)^2} = 65 \text{ rad/s}.$$

Prior to $t = 0$, the circuit was in steady state, which means that the capacitor was fully charged at $V_s = 24$ V and acting like an open circuit. Hence, $\upsilon_C(0^-) = 24$ V and $i_C(0^-) = 0$.

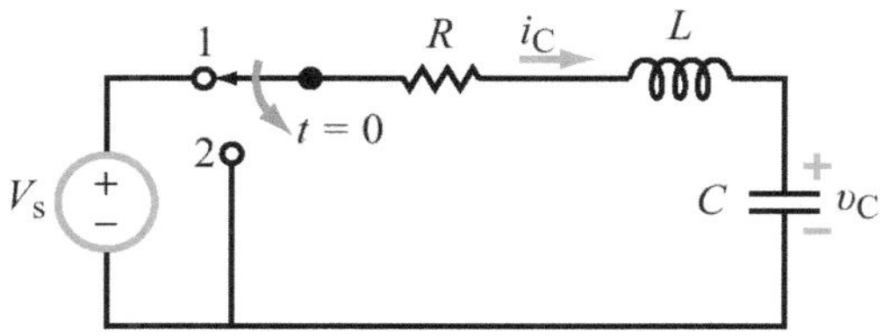

(a)

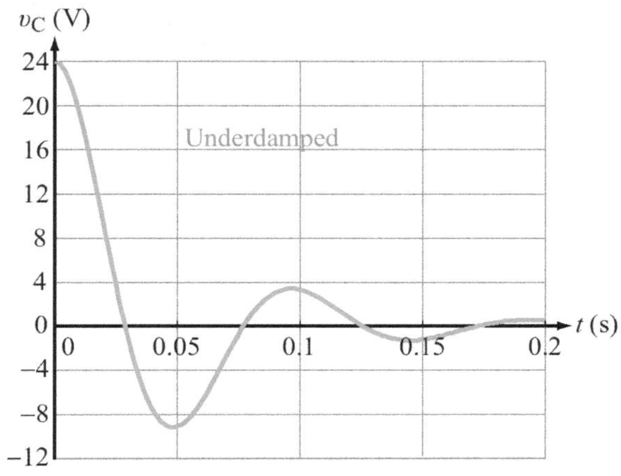

(b)

Figure 6-11: Example 6-6 (a) circuit and (b) $v_C(t)$.

Since both v_C across C and i_L through L cannot change instantaneously,

$$v_C(0) = 24 \text{ V},$$

$$i_C(0) = i_L(0) = i_L(0^-) = 0.$$

After $t = 0$, the closed RLC circuit will no longer have any active sources, allowing the capacitor to dissipate all its energy in the resistor. Hence, as $t \to \infty$, $v_C(\infty) = 0$. Using these initial and final values in the appropriate expressions for D_1 and D_2 in Eq. (6.37) leads to $D_1 = 24$ V, $D_2 = 7.4$ V, and

$$v_C(t) = e^{-20t}[24 \cos 65t + 7.4 \sin 65t] \text{ V}, \qquad \text{for } t \geq 0.$$

Figure 6-11(b) shows a time plot of $v_C(t)$, which exhibits an exponential decay (due to e^{-20t}) in combination with the oscillatory behavior associated with the sine and cosine functions.

Concept Question 6-4: What specific feature distinguishes the waveform of the underdamped response from those of the overdamped and critically damped responses? (See CAD)

Concept Question 6-5: Why is ω_d called the damping frequency? (See CAD)

Exercise 6-6: Repeat Example 6-4 after replacing the 8 V source with a short circuit and changing the value of R_1 to 1.7 Ω.

Answer:

$$v(t) = e^{-190t}(20 \cos 62.45t + 60.85 \sin 62.45t) \text{ V}.$$

(See CAD)

6-6 Summary of the Series RLC Circuit Response

6-6.1 Switch Action at $t = 0$

The left-hand column of Table 6-1 provides the general expressions for $v_C(t)$ for each of the three damping conditions associated with the series RLC circuit. The table also includes expressions for the constants in those expressions in terms of the initial and final values of v_C and the initial value of i_C. In all three cases, the starting point is to compute the values of α and ω_0, then their relative values determines the applicable damping condition.

6-6.2 Switch Action at $t = T_0$

If the sudden change in the circuit occurs at $t = T_0$, instead of at $t = 0$, the only changes that need to be made are:

(1) t should be replaced with $(t - T_0)$ everywhere on the right-hand side of all equations in Table 6-1.

(2) $v_C(0)$ and $i_C(0)$ should be replaced with $v_C(T_0)$ and $i_C(T_0)$, respectively, in the expressions for the constants in Table 6-1.

Example 6-7: Rectangular-Pulse Excitation

The switch in the circuit of Fig. 6-12(a) was in position 1 for a long time before it was moved to position 2 at $t = 0$, and

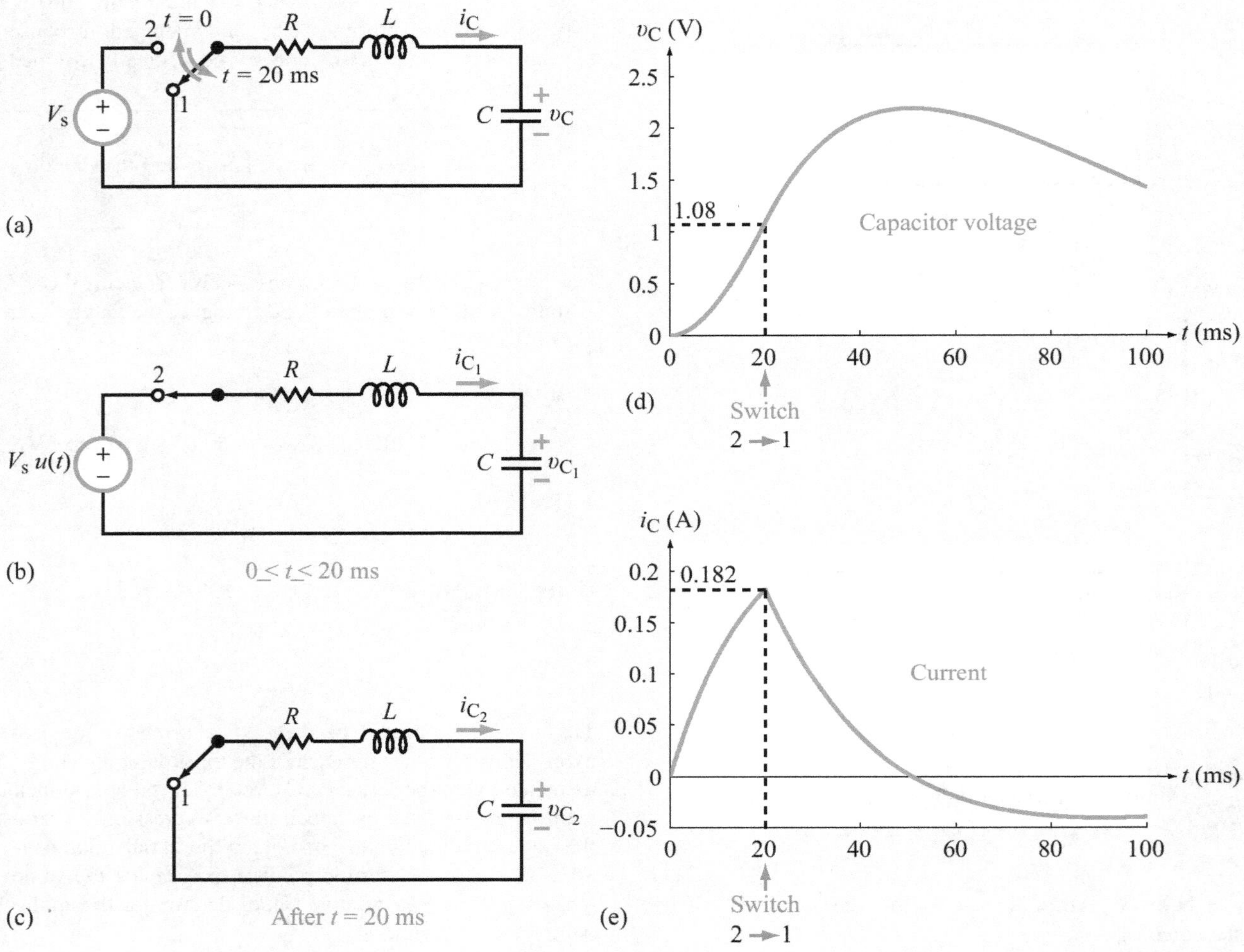

Figure 6-12: Example 6-7 with $V_s = 12$ V, $R = 40\ \Omega$, $L = 0.8$ H, and $C = 2$ mF.

then back to position 1 at $t = 20$ ms. If $V_s = 12$ V, $R = 40\ \Omega$, $L = 0.8$ H, and $C = 2$ mF, determine the waveforms of $\upsilon_C(t)$ and $i(t)$ for $t \geq 0$.

Solution: From Eq. (6.17),

$$\alpha = \frac{R}{2L} = \frac{40}{2 \times 0.8} = 25 \text{ Np/s},$$

$$\omega_0 = \frac{1}{\sqrt{LC}} = \frac{1}{\sqrt{0.8 \times 2 \times 10^{-3}}} = 25 \text{ rad/s}.$$

Since $\alpha = \omega_0$, the circuit will exhibit a critically damped response. We will divide the solution into two time segments.

Time Segment 1: $0 \leq t \leq 20$ ms.

The general expression for the critically damped response of the series RLC circuit is given by Eq. (6.26) as

$$\upsilon_{C_1}(t) = (B_1 + B_2 t)e^{-\alpha t} + \upsilon_1(\infty). \tag{6.38}$$

Even though we know that the switch will be moved back to position 1 at $t = 20$ ms, when we evaluate the constants in Eq. (6.38) for Time Segment 1, we do so as if the state of the circuit shown in Fig. 6-12(b) is to remain the same until $t = \infty$. Since the circuit is "unaware" of the change that will be taking place at $t = 20$ ms, its reaction to the change at $t = 0$ presumes that the new condition of the circuit will continue indefinitely. Hence, the voltage across the capacitor at $t = \infty$ would have been

$$\upsilon_{C_1}(\infty) = V_s = 12 \text{ V}. \tag{6.39}$$

At $t = 0^-$, the RLC circuit contains no active sources, so both $\upsilon_1(0^-)$ and $i_1(0^-)$ are zero. Moreover, since neither the voltage across C nor the current through L can change instantaneously, it follows that

$$\upsilon_{C_1}(0) = \upsilon_{C_1}(0^-) = 0,$$
$$i_{C_1}(0) = i_{C_1}(0^-) = 0.$$

Application of the expressions for B_1 and B_2 available in Table 6-1 gives

$$B_1 = \upsilon_C(0) - \upsilon_C(\infty) = 0 - 12 = -12 \text{ V}, \tag{6.40a}$$

$$B_2 = \frac{1}{C}\, i_{C_1}(0) + \alpha[\upsilon_{C_1}(0) - \upsilon_{C_1}(\infty)]$$
$$= 0 + 25[0 - 12] = -300 \text{ V/s}. \tag{6.40b}$$

Consequently, $\upsilon_{C_1}(t)$ is given by

$$\upsilon_{C_1}(t) = 12 - (12 + 300t)e^{-25t} \text{ V}, \qquad \textbf{for } 0 \le t \le 20 \textbf{ ms.} \tag{6.41}$$

The associated current is

$$i_{C_1}(t) = C\,\frac{d\upsilon_{C_1}}{dt} = 2 \times 10^{-3}\,\frac{d}{dt}[12 - (12 + 300t)e^{-25t}]$$
$$= 15te^{-25t} \text{ A}, \quad (\text{for } 0 \le t \le 20 \text{ ms}). \tag{6.42}$$

Time Segment 2: $t \ge 20$ ms.

After moving the switch back to position 1 at $t = 20$ ms, the circuit no longer has any active sources, and yet it is part of a closed circuit (Fig. 6-12(c)), allowing the capacitor and inductor to dissipate their stored energies through the resistor. Hence, at $t = \infty$,

$$\upsilon_{C_2}(\infty) = 0.$$

Upon shifting t by 0.02 s, the expression for $\upsilon_{C_2}(t)$ assumes the form

$$\upsilon_{C_2}(t) = [B_3 + B_4(t - 0.02)]e^{-25(t-0.02)} \text{ V} \qquad \text{for } t \ge 20 \text{ ms}, \tag{6.43}$$

where constants B_3 and B_4 are so labeled to avoid confusion with B_1 and B_2 of the earlier time segment. The associated current is

$$iC_2(t) = C\,\frac{d\upsilon_{C_2}}{dt}$$
$$= 2 \times 10^{-3}\,\frac{d}{dt}\{[B_3 + B_4(t - 0.02)]e^{-25(t-0.02)}\}$$
$$= [(2B_4 - 50B_3) - 50B_4(t - 0.02)]$$
$$\cdot e^{-25(t-0.02)} \times 10^{-3} \text{ A} \qquad \textbf{for } t \ge 20 \textbf{ ms.} \tag{6.44}$$

Across the juncture between time segment 1 and time segment 2, neither the voltage can change (as mandated by the capacitor) nor can the current (as mandated by the inductor). Thus,

$$\upsilon_{C_1}(t = 20 \text{ ms}) = \upsilon_{C_2}(t = 20 \text{ ms}), \tag{6.45a}$$
$$i_{C_1}(t = 20 \text{ ms}) = i_{C_2}(t = 20 \text{ ms}). \tag{6.45b}$$

Application of Eqs. (6.45a and b) to the expressions given by Eqs. (6.41) to (6.44) gives

$$12 - (12 + 300 \times 0.02)e^{-25\times0.02} = B_3,$$
$$15 \times 0.02e^{-25\times0.02} = (2B_4 - 50B_3) \times 10^{-3},$$

whose joint solution leads to

$$B_3 = 1.08 \text{ V}, \qquad B_4 = 118.04 \text{ V/s}.$$

Consequently,

$$\upsilon_{C_2}(t) = [1.08 + 118.04(t - 0.02)]e^{-25(t-0.02)} \text{ V} \qquad \text{for } t \ge 20 \text{ ms} \tag{6.46a}$$

and

$$i_{C_2}(t) = [0.182 - 5.90(t - 0.02)]e^{-25(t-0.02)} \text{ A} \qquad \text{for } t \ge 20 \text{ ms}. \tag{6.46b}$$

The waveforms of $\upsilon_C(t)$ and $i_C(t)$ are displayed in Figs. 6-12(d) and (e), respectively.

Example 6-8: Two-Source Circuit

The switch in the circuit of Fig. 6-13(a) was opened at $t = 0$, after it had been closed for a long time. If $V_{s_1} = 20$ V, $V_{s_2} = 24$ V, $R_1 = 40\ \Omega$, $R_2 = R_3 = 20\ \Omega$, $R_4 = 10\ \Omega$, $L = 0.8$ H, and $C = 2$ mF, determine $\upsilon_C(t)$ for $t \ge 0$.

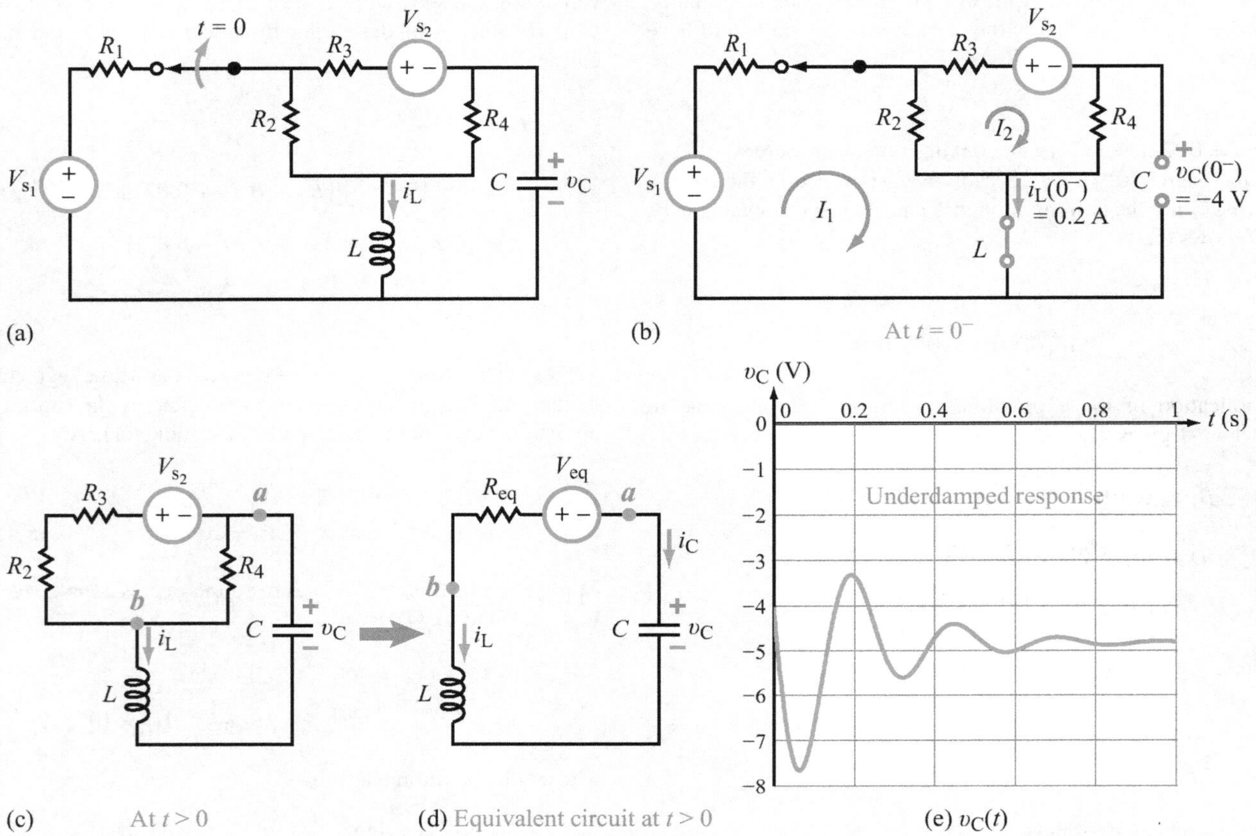

Figure 6-13: Circuit for Example 6-8.

Solution: Consider the state of the circuit at $t = 0^-$ (before opening the switch), as depicted by Fig. 6-13(b). The mesh current equations for the indicated loops are

$$-V_{s_1} + R_1 I_1 + R_2(I_1 - I_2) = 0,$$
$$R_2(I_2 - I_1) + R_3 I_2 + V_{s_2} + R_4 I_2 = 0.$$

After substituting the given values for the sources and the resistors, simultaneous solution of the two equations leads to

$$I_1 = 0.2\text{ A}, \qquad I_2 = -0.4\text{ A}.$$

Hence,

$$v_C(0^-) = I_2 R_4 = -0.4 \times 10 = -4\text{ V}, \quad (6.47a)$$
$$i_L(0^-) = I_1 = 0.2\text{ A}. \quad (6.47b)$$

Next, we consider Fig. 6-13(c), which depicts the circuit configuration at $t > 0$ (after opening the switch). To simplify the analysis, we use source transformation to convert the circuit into its Thévenin equivalent, as shown in Fig. 6-13(d), where

$$R_{eq} = (R_2 + R_3) \parallel R_4 = \frac{(R_2 + R_3)R_4}{R_2 + R_3 + R_4} = 8\ \Omega,$$
$$V_{eq} = \frac{V_{s_2}}{R_2 + R_3} \times R_{eq} = 4.8\text{ V}.$$

Now we are ready to analyze the series RLC circuit of Fig. 6-13(d). To that end, we compute α and ω_0:

$$\alpha = \frac{R_{eq}}{2L} = \frac{8}{2 \times 0.8} = 5\text{ Np/s},$$
$$\omega_0 = \frac{1}{\sqrt{LC}} = \frac{1}{\sqrt{0.8 \times 2 \times 10^{-3}}} = 25\text{ rad/s}.$$

Since $\alpha < \omega_0$, the capacitor voltage υ_C will exhibit an underdamped oscillatory response of the form given by Eq. (6.35) as

$$\upsilon_C(t) = \{e^{-\alpha t}[D_1 \cos \omega_d t + D_2 \sin \omega_d t]\} + \upsilon_C(\infty), \quad (6.48)$$

where

$$\omega_d = \sqrt{\omega_0^2 - \alpha^2} = \sqrt{25^2 - 5^2} = 24.5 \text{ rad/s}.$$

It is evident from the circuit in Fig. 6-13(d) that

$$\upsilon_C(\infty) = -V_{eq} = -4.8 \text{ V}.$$

To determine D_1 and D_2, we apply Eq. (6.37) with $\upsilon_C(0) = -4$ V, $i_C(0) = -i_L(0) = -0.2$ A, and $\upsilon_C(\infty) = -4.8$ V,

$$D_1 = \upsilon_C(0) - \upsilon_C(\infty) = -4 + 4.8 = 0.8 \text{ V}, \quad (6.49a)$$

$$D_2 = \frac{\frac{1}{C}\, i_C(0) + \alpha[\upsilon_C(0) - \upsilon_C(\infty)]}{\omega_d} = \frac{-100 + 5[-4 + 4.8]}{24.5} = -3.92 \text{ V}. \quad (6.49b)$$

With all unknown quantities accounted for,

$$\upsilon_C(t) = \{-4.8 + e^{-5t}[0.8 \cos 24.5t - 3.92 \sin 24.5t]\} \text{ V}, \quad \text{for } t \geq 0. \quad (6.50)$$

The waveform of $\upsilon_C(t)$ is displayed in Fig. 6-13(e).

6-7 The Parallel RLC Circuit

Having completed our examination of the series RLC circuit [Fig. 6-14(a)], we now turn our attention to the parallel RLC circuit shown in Fig. 6-14(b). As we will see shortly, the current $i_L(t)$ flowing through the inductor in the parallel RLC circuit is characterized by a second-order differential equation identical in form to that for the voltage $\upsilon_C(t)$ across the capacitor of the series RLC circuit. Accordingly, we will take advantage of this correspondence between the series and parallel RLC circuits by adapting the solutions we obtained in the preceding section for the series circuit to the solutions we seek in this section for the parallel circuit.

Application of KCL to the circuit in Fig. 6-14(b) gives

$$i_R + i_L + i_C = I_s \qquad \text{for } t \geq 0. \quad (6.51)$$

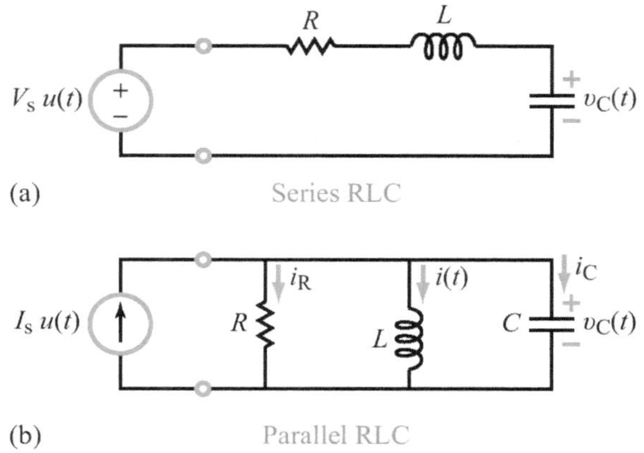

Figure 6-14: The differential equation for $\upsilon_C(t)$ of the series RLC circuit shown in (a) is identical in form to that of the current $i_L(t)$ in the parallel RLC circuit in (b).

When expressed in terms of $\upsilon_C(t)$, the voltage common to all three passive elements, Eq. (6.51) becomes

$$\frac{\upsilon_C}{R} + i_L + C\,\frac{d\upsilon_C}{dt} = I_s. \quad (6.52)$$

Using $\upsilon_C = \upsilon_L = L\, di_L/dt$, and rearranging terms, leads to

$$\frac{d^2 i_L}{dt^2} + \frac{1}{RC}\,\frac{di_L}{dt} + \frac{1}{LC}\, i_L = \frac{I_s}{LC}, \quad (6.53)$$

which can be rewritten in the abbreviated form

$$i_L'' + a_2 i_L' + b_2 i_L = c_2, \quad (6.54)$$

where

$$a_2 = \frac{1}{RC}, \qquad b_2 = \frac{1}{LC}, \qquad c_2 = \frac{I_s}{LC}. \quad (6.55)$$

Comparison of Eq. (6.54) with Eq. (6.5) for the capacitor voltage of the series RLC circuit reveals that the two differential equations are identical in form, albeit the constant coefficients have different expressions in the two cases. The overdamped, underdamped, and critically damped expressions for $i_L(t)$ are given in Table 6-1.

Quantities s_1, s_2, ω_0, and ω_d retain the same expressions given earlier, but α is now given by

$$\alpha = \frac{1}{2RC} \qquad \textbf{(parallel RLC)}. \quad (6.56)$$

Parallel RLC

Overdamped ($\alpha > \omega_0$)

$$i_L(t) = [A_1 e^{s_1 t} + A_2 e^{s_2 t} + i_L(\infty)], \qquad \text{(for } t \geq 0\text{)} \qquad (6.57a)$$

Critically damped ($\alpha = \omega_0$)

$$i_L(t) = [(B_1 + B_2 t)e^{-\alpha t} + i_L(\infty)], \qquad \text{(for } t \geq 0\text{)} \qquad (6.57b)$$

Underdamped ($\alpha < \omega_0$)

$$i_L(t) = [e^{-\alpha t}(D_1 \cos \omega_d t + D_2 \sin \omega_d t) + i_L(\infty)], \qquad \text{(for } t \geq 0\text{)} \qquad (6.57c)$$

*More details in Table 6-1.

Example 6-9: Parallel RLC Circuit

Determine $i_L(t)$ in the circuit of Fig. 6-15(a) for $t \geq 0$, given that $I_s = 0.5$ A, $V_0 = 12$ V, $R_1 = 60$ Ω, $R_2 = 30$ Ω, $L = 0.2$ H, and $C = 500$ μF.

Solution: The circuit in Fig. 6-15(b) represents the steady state condition of the circuit at $t = 0^-$ (prior to moving the switch). Under constant conditions, C acts like an open circuit and L acts like a short circuit. Given that I_s flows entirely through the short circuit representing the inductor, it follows that

$$i_L(0^-) = I_s = 0.5 \text{ A},$$

$$\upsilon_C(0^-) = 0.$$

Since i_L through an inductor cannot change instantaneously, nor can υ_C across a capacitor, these conditions are equally applicable at $t = 0$. Consequently,

$$i_L(0) = i_L(0^-) = 0.5 \text{ A},$$

and

$$\upsilon_L(0) = \upsilon_C(0) = 0.$$

After moving the switch ($t > 0$), the circuit assumes the configuration shown in Fig. 6-15(c). After application of source transformation, current source I_0' and the equivalent resistance R' in Fig. 6-15(d) are given by

$$I_0' = \frac{V_0}{R_1} = \frac{12}{60} = 0.2 \text{ A},$$

$$R' = R_1 \parallel R_2 = \frac{R_1 R_2}{R_1 + R_2} = 20\Omega.$$

For the parallel RLC circuit in Fig. 6-15(d), the expressions for α and ω_0 are given by

$$\alpha = \frac{1}{2R'C} = \frac{1}{2 \times 20 \times 500 \times 10^{-6}} = 50 \text{ Np/s},$$

$$\omega_0 = \frac{1}{\sqrt{LC}} = \frac{1}{\sqrt{0.2 \times 500 \times 10^{-6}}} = 100 \text{ rad/s}.$$

Since $\alpha < \omega_0$, the circuit will exhibit an underdamped response with a damped natural frequency ω_d given in Table 6-1 as

$$\omega_d = \sqrt{\omega_0^2 - \alpha^2} = \sqrt{100^2 - 50^2} = 86.6 \text{ rad/s}.$$

From Table 6-1, the expression for $i_L(t)$ is given by

$$i_L(t) = [e^{-\alpha t}(D_1 \cos \omega_d t + D_2 \sin \omega_d t) + i_L(\infty)]$$
$$\text{for } t \geq 0.$$

At $t = \infty$, the inductor behaves like a short circuit, forcing I_0' to flow through it exclusively. Hence,

$$i_L(\infty) = I_0' = 0.2 \text{ A}.$$

The only remaining unknowns are D_1 and D_2, which we determine by applying the expressions given in Table 6-1, namely

$$D_1 = i_L(0) - i_L(\infty) = (0.5 - 0.2) \text{ A} = 0.3 \text{ A},$$

and

$$D_2 = \frac{\frac{1}{L}\,\upsilon_L(0) + \alpha[i_L(0) - i_L(\infty)]}{\omega_d}$$
$$= \frac{0 + 50(0.5 - 0.2)}{86.6} = 0.17 \text{ A}.$$

The final expression for $i_L(t)$ is then given by

$$i_L(t) = [0.2 + e^{-50t}(0.3 \cos 86.6t + 0.17 \sin 86.6t)] \text{ A},$$
$$\text{for } t \geq 0,$$

and its plot is displayed in Fig. 6-15(e).

Exercise 6-7: Determine the initial and final values for i_L in the circuit of Fig. E6.7 on the following page and provide an expression for $i_L(t)$.

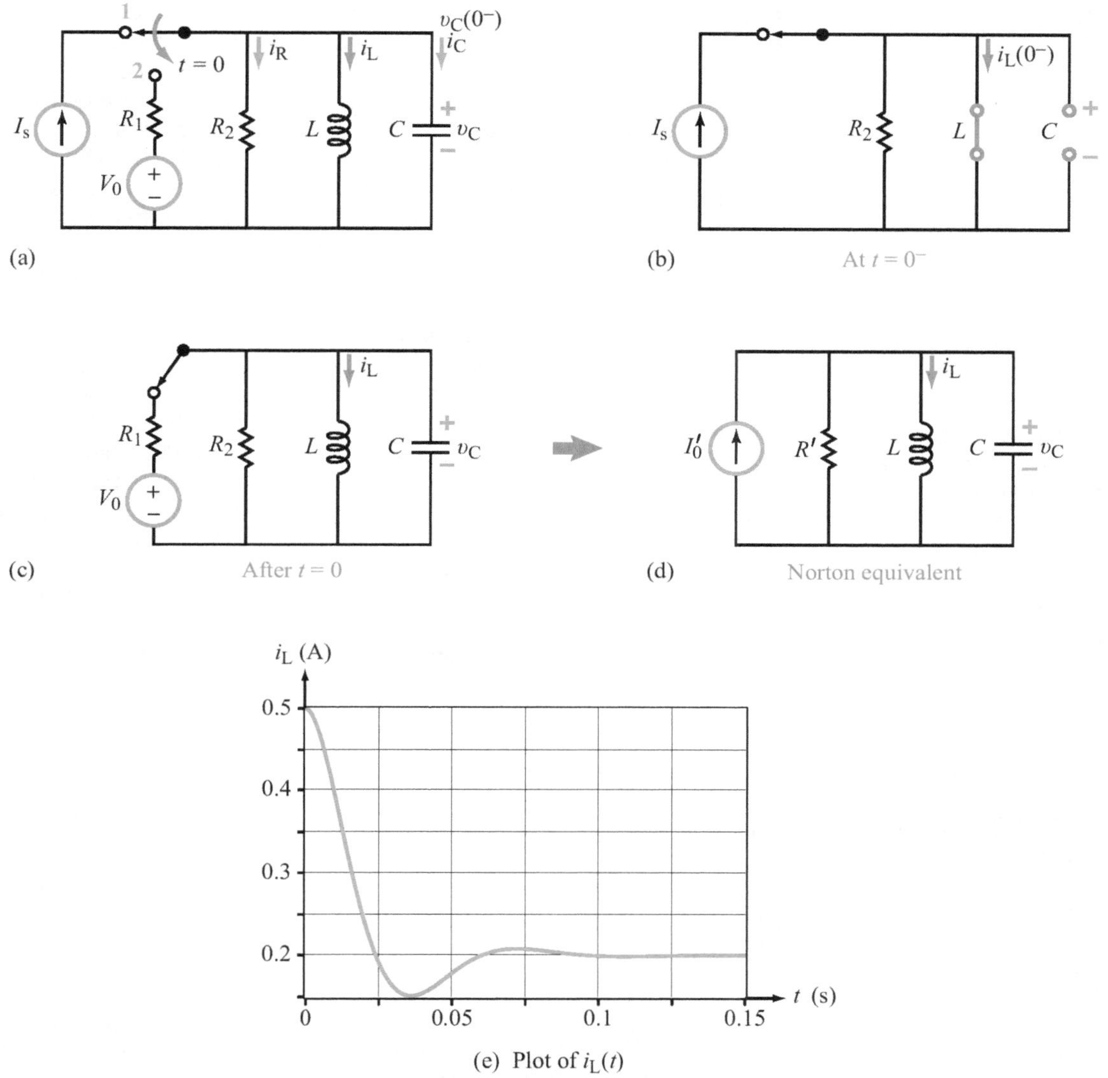

Figure 6-15: Circuit for Example 6-9.

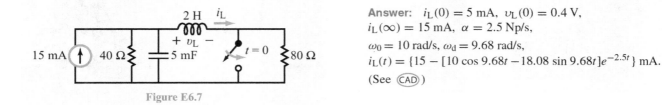

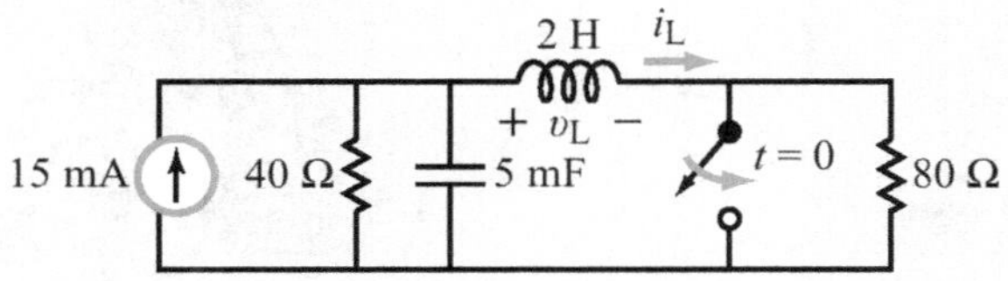

Figure E6.7

Answer: $i_L(0) = 5$ mA, $v_L(0) = 0.4$ V, $i_L(\infty) = 15$ mA, $\alpha = 2.5$ Np/s, $\omega_0 = 10$ rad/s, $\omega_d = 9.68$ rad/s, $i_L(t) = \{15 - [10\cos 9.68t - 18.08 \sin 9.68t]e^{-2.5t}\}$ mA. (See CAD)

Technology Brief 16
RFID Tags and Antenna Design

RFID Applications

Radio-frequency identification (RFID) uses electromagnetic fields to transfer identifying information from a small electrical ID circuit to an external receiver. These are commonly used for identifying or tracking animals, packages and goods, smart cards, tags, etc. (Fig. TF16-1). RFID circuits are injected in pets to help identify and return lost or stolen animals, attached via ear tags to livestock to identify their whereabouts and activities (how much time they spend eating or drinking), attached to athletes via wrist bands to track and verify their progress in a race, affixed to consumer goods and packaging to track, locate, and maintain inventory, and prevent theft. RFID tags can be based on either static, unchanging data (such as the ID number for a dog or cat), or their data can be changed by either an internal circuit (monitoring and reporting temperature of a refrigerated shipping container, for instance) or an external circuit (such as marking the last time a box was inspected).

When combined with other circuits, the information provided by RFID tags can be used in a myriad of ways. For instance, credit-card sized RFID tags attached to valuable art or other one-of-a-kind objects contain a unique ID number, as well as circuits detecting tilt and vibration. This information is continuously transmitted to receivers on the ceiling of a museum to create a security system that constantly monitors their location and status, and generates alarms if they are moved. RFID tags permanently installed in new guitars can help track them throughout their lives, and those installed in vintage guitars can help prevent fraud and theft. RFID tags are in most access-monitoring cards today, and can uniquely identify a person and his/her time of entry and exit. If other items are also tracked (sensitive documents for instance), an RFID reader can also identify what he/she is carrying and can generate an alarm if documents are leaving a room (or books leaving a library) that shouldn't be. RFID tags can be used in numerous medical applications to identify a person and identify and track the drugs or treatments he/she receives.

RFID and bar code scanners can be used for similar applications, but work in very different ways. Bar code scanners require direct visual access for a laser to read

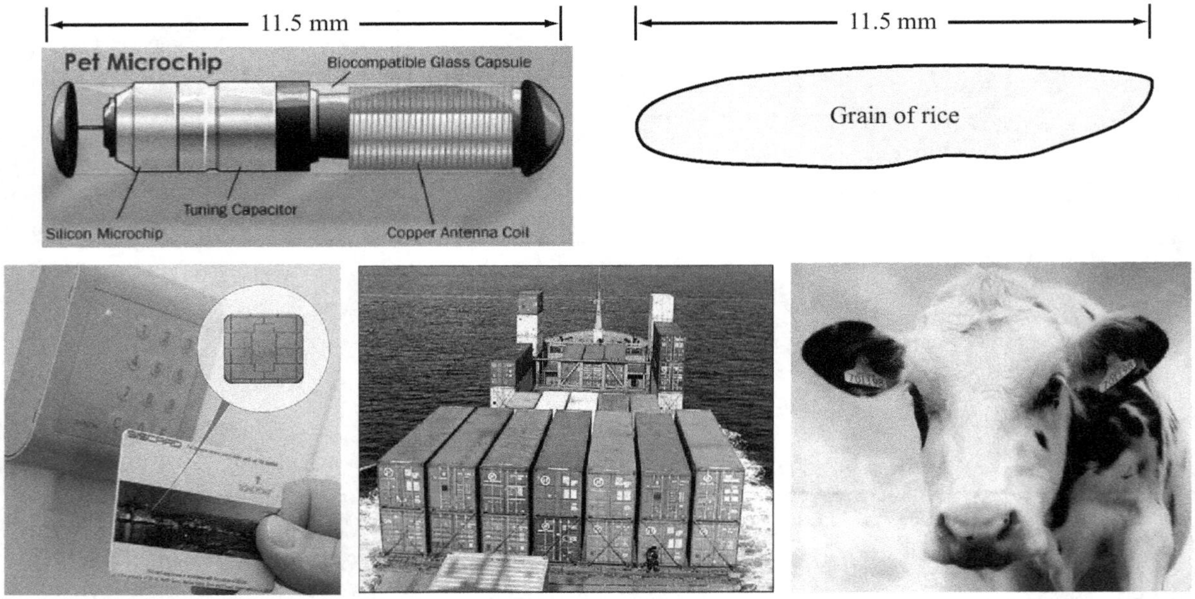

Figure TF16-1: RFID examples.

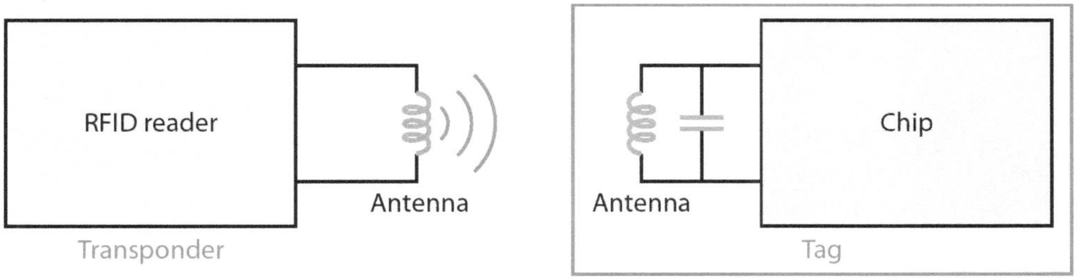

Figure TF16-2: RFID system.

the bar code. RFID circuits can be out of sight (inside a pet or package) as long as the wireless electromagnetic signal can penetrate the external packaging. Bar codes are *read only*. RFID systems can be *read only* or *read-write*. Bar codes are printed directly on packaging, or stickers affixed to packaging. RFID systems require an external antenna and a (tiny) computer chip. The antenna can be printed, but the chip must be somehow affixed. The entire system is often implemented in a sticker or card. Bar codes are essentially free (printed), whereas RFID tags cost 15 US cents and up.

RFID Operation

In a *passive RFID* system, an external transponder transmits a wireless signal to the RFID circuit (Fig. TF16-2), which "wakes up" and receives power from the signal through inductive coupling or other power harvesting methods. It then transmits its coded ID information back to the transponder, through the inductive link. The advantage of passive RFID systems is that they can be very small, not much bigger than a grain of rice, and can last for decades without maintenance as they do not require an internal battery to power the circuit. But the transponder must be within a short distance (less than 1 m) of the RFID circuit in order to receive the ID information. *Active RFID* systems have a battery to power the internal RFID circuit and can therefore transmit much further, up to 200 m.

RFID systems consist of an RFID transceiver with a sinusoidal source and (typically) a loop antenna, through which the current flows, creating a magnetic field. The magnetic field is part of an electromagnetic wave that travels a short distance through the air to the RFID tag. The RFID tag has another (typically) loop or loop-like antenna to receive the magnetic field and convert it back to a current, and an RF circuit to convert it to a small voltage that can be used to power the data circuit in the chip. Frequencies used for RFID and some of their applications are listed in Table TT16-1.

RFID Antennas

Two examples of RFID antennas are shown in Fig. TF16-3. Both are printed 2-D antennas containing an inductor, in either a coiled design as in part (a) or in a "squiggly" design (yes, it really is called a squiggle tag),

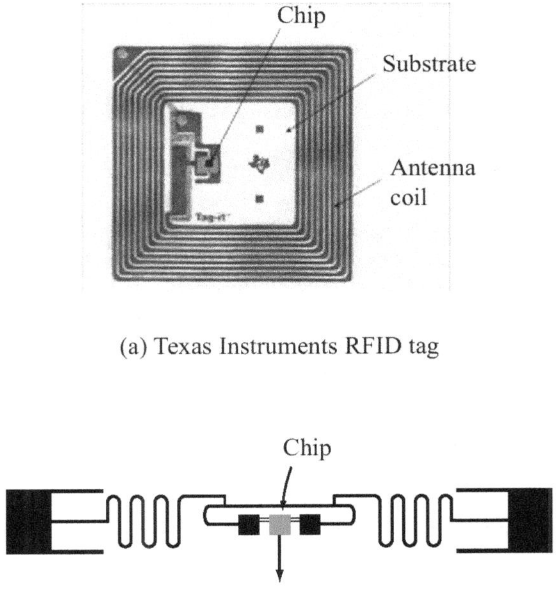

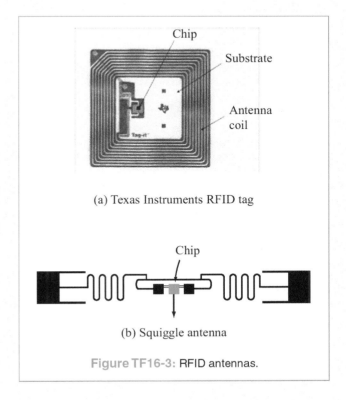

(a) Texas Instruments RFID tag

(b) Squiggle antenna

Figure TF16-3: RFID antennas.

Table TT16-1: **RFID frequency bands.**

Band	Regulations	Range	Data Speed	Remarks	Approximate Tag Cost in Volume (2006) US$
120–150 kHz (LF)	Unregulated	10 cm	Low	Animal identification, factory data collection	$1
13.56 MHz (HF)	(ISM) band worldwide	10 cm – 1 m	Low to moderate	Smart cards (MIFARE, ISO/IEC 14443)	$0.50
433 MHz (UHF)	Short-range devices	1–100 m	Moderate	Defense applications, with active tags	$5
865–868 MHz (Europe), 902–928 MHz (North America) UHF	ISM band	1–12 m	Moderate to high	EAN, various standards	$0.15 (passive tags)
2450–5800 MHz (microwave)	ISM band	1–2 m	High	802.11 (WLAN), Bluetooth standards	$25 (active tag)
3.1–10 GHz (microwave)	Ultra wide band	1 to 200 m	High	Requires semi-active or active tags	$5

which is often printed on a sticker label for consumer products.

Antenna design is a subspecialty of electrical engineering. Antenna designers consider ways to either convert current and voltage to electric and magnetic fields in the air (for wireless transmission) or to collect those fields in the air and convert them back into currents and voltages. In general, the same antenna can be used to receive and transmit the RFID signals. Antenna performance is governed by the shape of the antenna and its size relative to the wavelength λ of the electromagnetic (EM) wave it radiates or intercepts. The wavelength, in turn, is related to the signal frequency f by $\lambda = c/f$, where c is the velocity of light in vacuum. Hence, the size of an antenna usually is chosen to match the EM frequency that the RFID is intended to use. The ratio of electric to magnetic field is called the impedance of the antenna, and it needs to be matched to the same ratio of voltage and current that are produced or received by the circuit (the impedance of the circuit). The impedance of the circuit is controlled by the capacitors, resistors, inductors, and other elements at the input or output of the circuit. The impedance of the antenna is controlled by its shape and size. Coils tend to be more inductive, which means their impedance is more like an inductor (has a positive imaginary part). Antennas shaped like plates tend to be more capacitive (having a negative imaginary part). Most antennas are a combination of inductive and capacitive, and can be modeled in circuit analysis as circuits containing both inductors and capacitors. Circuit elements are called ***lumped elements*** because their capacitance, inductance, and resistance are built from individual components, whereas an antenna is a ***distributed element*** whose capacitance, inductance, and resistance are spatially distributed along the length of the antenna. Taking all of these design factors into account at once is fairly daunting, so computer software is used extensively in antenna design, leading to creative designs such as the squiggle antenna and beyond. Antenna designers sometimes say they are "painting with copper" to describe the creative artistry of their field.

Exercise 6-8: In the parallel RLC circuit shown in Fig. 6-14(b), how much energy will be stored in L and C at $t = \infty$?

Answer: $w_L = \frac{1}{2} L I_s^2$, $w_C = 0$. (See CAD)

6-8 General Solution for Any Second-Order Circuit with dc Sources

According to the material covered in the preceding sections, series and parallel RLC circuit share a common set of characteristics. An RLC circuit is characterized by a resonant frequency ω_0 and a damping coefficient α, and when driven by a sudden dc excitation, the circuit exhibits a response that decays exponentially as $e^{-\alpha t}$, and it may or may not contain an oscillatory variation, depending on whether ω_0 is or is not larger than α in magnitude, respectively. These characteristics arise from the interplay between energy storage and energy dissipation. During the operation of the RLC circuit, energy is exchanged between the two storage elements—the capacitor and the inductor—through the resistor. Dissipation is governed by $e^{-\alpha t}$, which we can redefine as $e^{-t/\tau}$, with

$$\tau = \frac{1}{\alpha} \quad \text{(s)}. \tag{6.58}$$

In this alternative form, the decay rate is specified by the time constant τ. If τ is short (rapid decay) in comparison with the duration of a single oscillation period T, where $T = 2\pi/\omega_d$, it means that energy burns away too quickly to generate an oscillation. This is the overdamped case. On the other hand, if τ is sufficiently long (slow decay) in comparison with T, energy will move back and forth between L and C, generating an oscillation. With every cycle, however, the resistance will burn off some of the remaining energy, resulting in an underdamped response that decays and oscillates simultaneously. If $R = 0$, the circuit will oscillate forever at the resonant frequency ω_0 (see Exercise 6-9).

Building on the experience we gained from our examination of the series and parallel RLC circuits, we now extend the method of solution to any second-order circuit, including those containing op amps. For a circuit containing only dc sources (or no independent sources at all), we seek to find the circuit response $x(t)$ for $t \geq 0$, where $x(t)$ is a voltage or current of interest in the circuit, and $t = 0$ is the instant at which the circuit experiences a sudden change (usually caused by a switch). To that end, we propose the following solution outline:

Step 1: Develop a second-order differential equation for $x(t)$, for $t \geq 0$. Express the equation in the general form

$$x'' + ax' + bx = c, \tag{6.59}$$

where a, b, and c are constants.

Step 2: Determine the values of α and ω_0:

$$\alpha = \frac{a}{2}, \qquad \omega_0 = \sqrt{b}\,. \tag{6.60}$$

Step 3: Determine whether the response $x(t)$ is overdamped, critically damped, or underdamped, and write down the expression corresponding to that case from the following general solution:

General Solution

Overdamped ($\alpha > \omega_0$)

$$x(t) = [A_1 e^{s_1 t} + A_2 e^{s_2 t} + x(\infty)], \qquad (\text{for } t \geq 0) \tag{6.61a}$$

Critically Damped ($\alpha = \omega_0$)

$$x(t) = [(B_1 + B_2 t)e^{-\alpha t} + x(\infty)], \qquad (\text{for } t \geq 0) \tag{6.61b}$$

Underdamped ($\alpha < \omega_0$)

$$x(t) = [e^{-\alpha t}(D_1 \cos \omega_d t + D_2 \sin \omega_d t) + x(\infty)], \qquad (\text{for } t \geq 0) \tag{6.61c}$$

where

$$s_1 = -\alpha + \sqrt{\alpha^2 - \omega_0^2}\,, \tag{6.62a}$$

$$s_2 = -\alpha - \sqrt{\alpha^2 - \omega_0^2}\,, \tag{6.62b}$$

$$\omega_d = \sqrt{\omega_0^2 - \alpha^2}\,. \tag{6.62c}$$

▶ The three expressions given by Eq. (6.61) represent the circuit response to a sudden change that occurs at $t = 0$. Had the sudden change occurred at $t = T_0$ instead, the expressions would continue to apply, but t will need to be replaced with $(t - T_0)$ everywhere on the right-hand side (only) of those expressions. ◀

Table 6-2: **General solution for second-order circuits for $t \geq 0$.**

$x(t)$ = unknown variable (voltage or current)	
Differential equation:	$x'' + ax' + bx = c$
Initial conditions:	$x(0)$ and $x'(0)$
Final condition:	$x(\infty) = \frac{c}{b}$
$\alpha = \frac{a}{2}$	$\omega_0 = \sqrt{b}$
Overdamped Response $\alpha > \omega_0$	
$x(t) = [A_1 e^{s_1 t} + A_2 e^{s_2 t} + x(\infty)]\, u(t)$	
$s_1 = -\alpha + \sqrt{\alpha^2 - \omega_0^2}$	$s_2 = -\alpha - \sqrt{\alpha^2 - \omega_0^2}$
$A_1 = \frac{x'(0) - s_2[x(0) - x(\infty)]}{s_1 - s_2}$	$A_2 = -\left[\frac{x'(0) - s_1[x(0) - x(\infty)]}{s_1 - s_2}\right]$
Critically Damped $\alpha = \omega_0$	
$x(t) = [(B_1 + B_2 t)e^{-\alpha t} + x(\infty)]\, u(t)$	
$B_1 = x(0) - x(\infty)$	$B_2 = x'(0) + \alpha[x(0) - x(\infty)]$
Underdamped $\alpha < \omega_0$	
$x(t) = [D_1 \cos\omega_d t + D_2 \sin\omega_d t + x(\infty)]e^{-\alpha t}\, u(t)$	
$D_1 = x(0) - x(\infty)$	$D_2 = \frac{x'(0) + \alpha[x(0) - x(\infty)]}{\omega_d}$
$\omega_d = \sqrt{\omega_0^2 - \alpha^2}$	

Step 4: Evaluate the circuit to determine $x(\infty)$ at $t = \infty$. Alternatively, we can use

$$x(\infty) = \frac{c}{b} . \tag{6.63}$$

Step 5: Apply initial conditions for $x(t)$ and $x'(t)$ at $t = 0$ (or at $t = T_0$ if the sudden change occurred at T_0) to determine the remaining unknown constants.

This procedure is highlighted in Table 6-2 and demonstrated through Examples 6-10 to 6-12.

Example 6-10: RLC Circuit with a Short-Circuit Switch

The switch in the circuit of Fig. 6-16(a) had been open for a long time before it was closed at $t = 0$. Determine $i_L(t)$ for $t \geq 0$. The circuit elements have the following values: $V_0 = 24$ V, $R_1 = 4\ \Omega$, $R_2 = 8\ \Omega$, $R_3 = 12\ \Omega$, $L = 2$ H, and $C = 0.2$ F.

Solution: Figures 6-16(b), (c), and (d) depict the state of the circuit at $t = 0^-$, $t \geq 0$, and $t = \infty$, respectively.

Step 1: **Obtain differential equation for $i_L(t)$**

After closing the switch, node 1 gets connected to node 2 and R_2 becomes inconsequential to the rest of the circuit because it is connected in parallel with a short circuit. At node 2 of the circuit in Fig. 6-16(c), KCL gives

$$-i_1 + i_L + i_C = 0. \tag{6.64}$$

In terms of the node voltage υ_C,

$$-i_1 = \frac{\upsilon_C - V_0}{R_1} , \tag{6.65a}$$

$$i_C = C\,\frac{d\upsilon_C}{dt} . \tag{6.65b}$$

Hence,

$$\frac{\upsilon_C}{R_1} + i_L + C\,\frac{d\upsilon_C}{dt} = \frac{V_0}{R_1} . \tag{6.66}$$

The voltage υ_C is equal to the sum of the voltages across L and R_3,

$$\upsilon_C = L\,\frac{di_L}{dt} + i_L R_3. \tag{6.67}$$

6-8 GENERAL SOLUTION FOR ANY SECOND-ORDER CIRCUIT WITH DC SOURCES

(a) Circuit with switch

(b) At $t = 0^-$: $i_L(0^-) = V_0/(R_1 + R_2 + R_3) = 1$ A, and $v_C(0^-) = i_L(0^-) R_3 = 12$ V.

(c) At $t \geq 0$

(d) At $t = \infty$: $i_L(\infty) = V_0/(R_1 + R_3) = 1.5$ A.

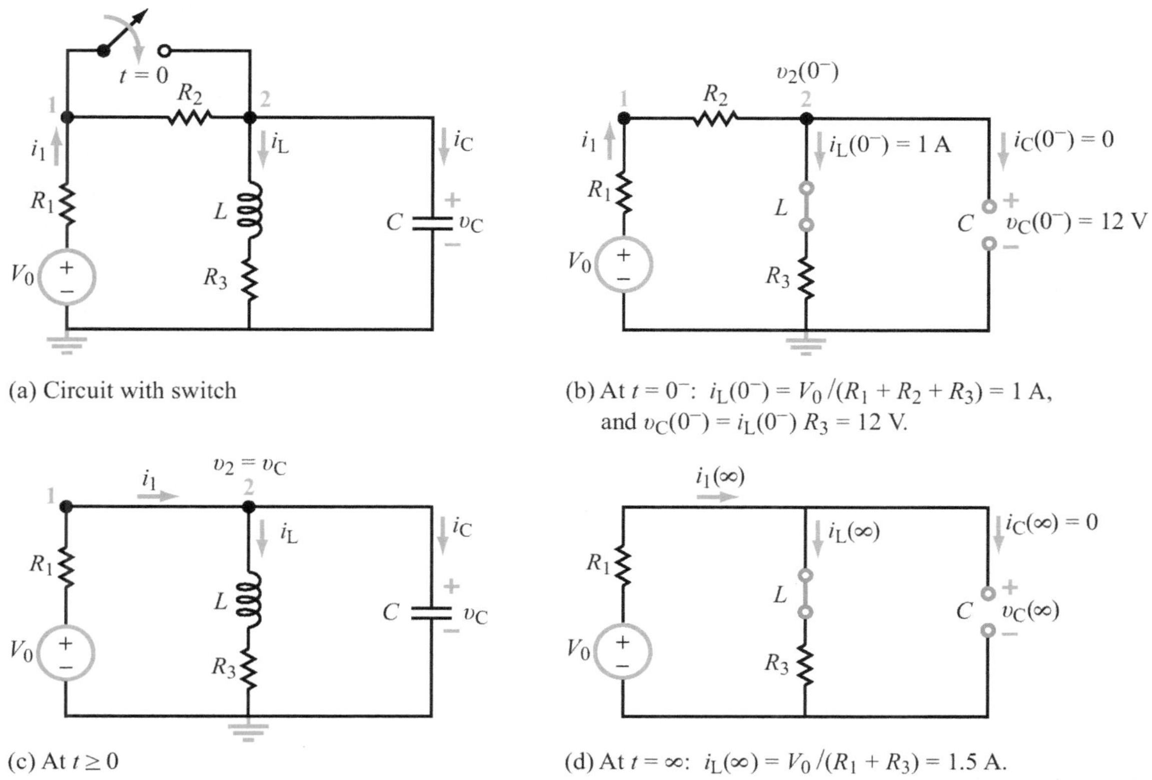

Figure 6-16: Circuit for Example 6-10.

Substituting Eq. (6.67) in Eq. (6.66) leads to

$$\frac{1}{R_1}\left(L\frac{di_L}{dt} + i_L R_3\right) + i_L + C\frac{d}{dt}\left(L\frac{di_L}{dt} + i_L R_3\right) = \frac{V_0}{R_1}.$$
(6.68)

After carrying out the differentiation in the third term and rearranging terms, we have

$$\frac{d^2 i_L}{dt^2} + \left(\frac{L + R_1 R_3 C}{R_1 LC}\right)\frac{di_L}{dt} + \left(\frac{R_1 + R_3}{R_1 LC}\right) i_L = \frac{V_0}{R_1 LC}.$$
(6.69)

For convenience, we rewrite Eq. (6.69) in the compact form

$$i_L'' + a i_L' + b i_L = c,$$
(6.70)

where

$$a = \frac{L + R_1 R_3 C}{R_1 LC} = \frac{2 + 4 \times 12 \times 0.2}{4 \times 2 \times 0.2} = 7.25,$$
(6.71a)

$$b = \frac{R_1 + R_3}{R_1 LC} = \frac{4 + 12}{4 \times 2 \times 0.2} = 10,$$
(6.71b)

$$c = \frac{V_0}{R_1 LC} = \frac{24}{4 \times 2 \times 0.2} = 15.$$
(6.71c)

Step 2: **Determine α and ω_0**

$$\alpha = \frac{a}{2} = \frac{7.25}{2} = 3.625$$
(6.72a)

and

$$\omega_0 = \sqrt{b} = \sqrt{10} = 3.162.$$
(6.72b)

Step 3: **Determine damping condition and select appropriate expression**

Since $\alpha > \omega_0$, the response is overdamped, and

$$i_L(t) = A_1 e^{s_1 t} + A_2 e^{s_2 t} + i_L(\infty) \qquad (6.73)$$

with

$$s_1 = -\alpha + \sqrt{\alpha^2 - \omega_0^2} = -1.85 \text{ Np/s} \qquad (6.74a)$$

and

$$s_2 = -\alpha - \sqrt{\alpha^2 - \omega_0^2} = -5.40 \text{ Np/s}. \qquad (6.74b)$$

Step 4: **Determine $i_L(\infty)$**

From the circuit in Fig. 6-16(d), $i_C = 0$ (open-circuit capacitor) and

$$i_L(\infty) = \frac{V_0}{R_1 + R_3} = \frac{24}{4 + 12} = 1.5 \text{ A}. \qquad (6.75)$$

Step 5: **Invoke initial conditions**

With C acting like an open circuit at $t = 0^-$ (Fig. 6-16(b)),

$$I_L(0^-) = i_1(0^-) = \frac{V_0}{R_1 + R_2 + R_3} = 1 \text{ A}.$$

Since i_L cannot change in zero time,

$$i_L(0) = i_L(0^-) = 1 \text{ A}. \qquad (6.76)$$

We need one additional relationship involving A_1 and A_2, which can be provided by the initial condition on i'_L. From the circuit in Fig. 6-16(b) at $t = 0^-$, we have

$$\upsilon_C(0^-) = i_L(0^-)\, R_3 = 1 \times 12 = 12 \text{ V}. \qquad (6.77)$$

As we transition from $t = 0^-$ (before closing the switch) to $t = 0$ (after closing the switch), neither i_L nor υ_C can change, which means that the voltage $\upsilon_2(0)$ at node 2 will continue to be 12 V and the current i_L through R_3 will continue to be 1 A. Hence, the voltage $\upsilon_L(0)$ has to be

$$\upsilon_L(0) = \upsilon_2(0) - i_L(0)\, R_3 = 12 - 1 \times 12 = 0. \qquad (6.78)$$

Since $\upsilon_L = L\, di_L/dt$, it follows that

$$i'_L(0) = 0. \qquad (6.79)$$

The expressions for A_1 and A_2 in Table 6-2 are given in terms of x, the variable associated with the second-order differential equation. In the present case, our differential equation is given by Eq. (6.70), with $i_L(t)$ as the unknown variable. Hence, by setting $x = i_L$ in the expressions for A_1 and A_2, we have

$$A_1 = \frac{i'_L(0) - s_2[i_L(0) - i_L(\infty)]}{s_1 - s_2} = \frac{0 + 5.4(1 - 1.5)}{-1.85 + 5.4} = -0.76 \text{ A} \qquad (6.80a)$$

and

$$A_2 = -\left[\frac{i'_L(0) - s_1[i_L(0) - i_L(\infty)]}{s_1 - s_2}\right] \qquad (6.80b)$$

$$= -\left[\frac{0 + 1.85(1 - 1.5)}{-1.85 + 5.4}\right] = 0.26 \text{ A}, \qquad (6.80c)$$

and the final solution is then given by

$$i_L(t) = [1.5 - 0.76 e^{-1.85t} + 0.26 e^{-5.4t}] \text{ A} \quad \text{for } t \geq 0. \qquad (6.81)$$

Exercise 6-9: Develop an expression for $i_C(t)$ in the circuit of Fig. E6.9 for $t \geq 0$.

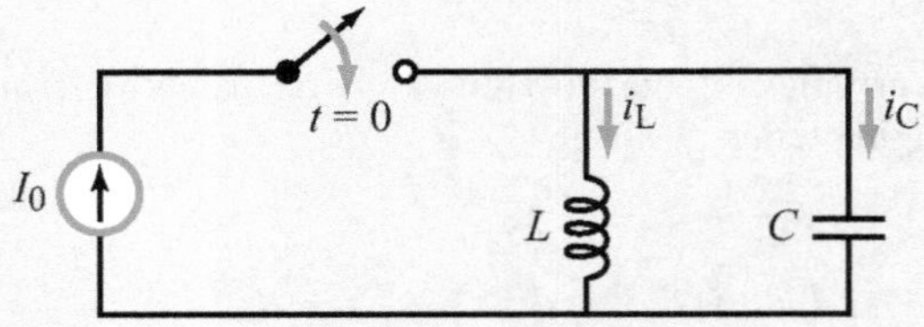

Figure E6.9

Answer: $i_C(t) = I_0 \cos \omega_0 t$, with $\omega_0 = 1/\sqrt{LC}$. This is an LC *oscillator* circuit in which dc energy provided by the current source is converted into ac energy in the LC circuit. (See CAD)

Technology Brief 17
Neural Stimulation and Recording

Section 4-12 introduced neural probes and how they can be used to measure voltage at specific locations in the brain. They can also be used to stimulate neurons to control movement, sight, hearing, touch, smell, emotion, and more. Neural stimulation and recording begin with a neural probe such as the three dimensional neural probe shown in Fig. 4-30 or the spiral-shaped cochlear implant electrodes shown in Fig. TF17-1. Each electrode is meant to stimulate one or more nearby neurons.

The electrodes are surgically inserted in proximity to the neurons of interest, and connected onto an electrical stimulation device that sends carefully designed electrical pulses into the extracellular fluid around them (for neural stimulation), or connected to an electrical receiver (that reads signals from them in the case of neural recording). There are many different devices, both commercially available and in research applications, that utilize neural stimulation or recording. These bioelectronics are one of the most exciting and rapidly advancing areas of electrical engineering. Several examples of these devices are given below.

Cochlear Implant

In the *cochlear implant* shown in Fig. TF17-2, the ear drum and stapes (inner bones of the ear) are replaced by a microphone and electrical circuitry. The sounds are picked up by the microphone mounted behind the ear, processed or coded (using electrical circuitry) into electrical pulses associated with the sounds, and then transmitted through the skin via inductive coupling or direct connection to the electrodes. The electrodes place these signals directly onto the auditory nerves, which then send the signals to the brain, which "hears" the sound. If the auditory nerve is not functional, an *auditory brainstem implant* is used instead, wherein electrodes directly stimulate the cochlear nucleus complex in the lower brain stem.

Artificial Eye Retina

The *artificial retina*, or *cortical implant*, replaces damaged eye structures with an external camera, a wireless link (shown as the two orange inductive coils in Fig. TF17-3), and an electrode array that stimulates the optic nerve in the back of the eye. Another alternative is to bypass the optical nerve and stimulate the visual cortex of the brain directly. The resolution of sight depends on the number of electrodes, as shown in Fig. TF17-4.

Brain Stimulation

The *deep brain stimulation* (DBS) or *cognitive prosthesis* shown in Fig. TF17-5 is used to stimulate

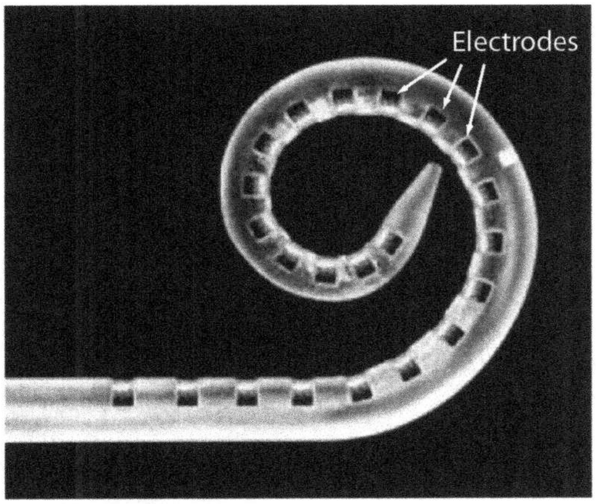

Figure TF17-1: Preformed spiral electrode for cochlear implant. (Courtesy of Cochlear Americas, © 2015 Cochlear Americas.)

1. Sounds are picked up by the microphone.
2. The signal is then "coded" (turned into a special pattern of electrical pulses).
3. These pulses are sent to the coil and are then transmitted across the skin to the implant.
4. The implant sends a pattern of electrical pulses to the electrodes in the cochlea.
5. The auditory nerve picks up these electrical pulses and sends them to the brain. The brain recognizes these signals as sound.

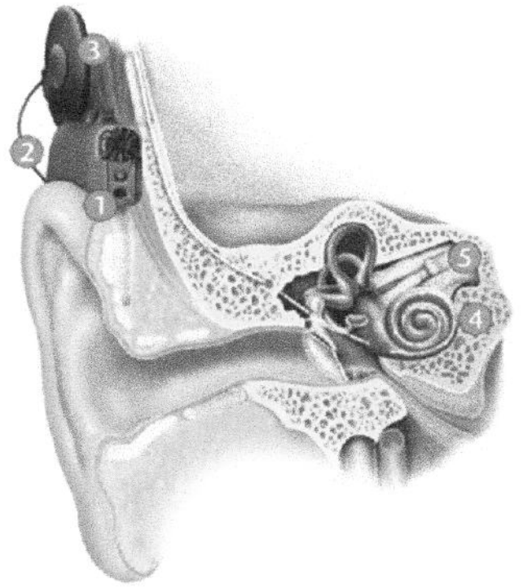

Figure TF17-2: A cochlear implant stimulates the auditory nerves to help deaf people hear. (Courtesy MED-EL.)

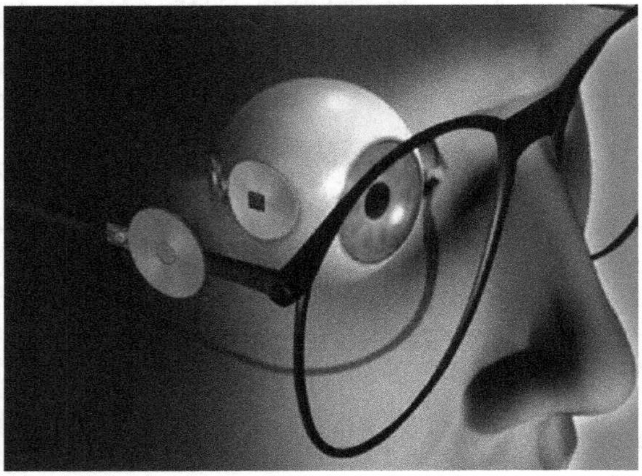

Figure TF17-3: Artificial retina simulates the optic nerve to help blind people see. (Credit: John Wyatt.)

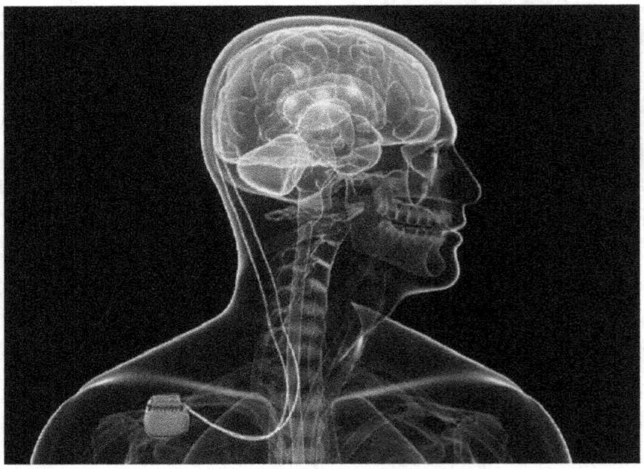

Figure TF17-5: Deep brain stimulation (DBS) is used to treat depression and tremors associated with Parkinson's disease. (Credit: Medtronic.)

nerves deep within the brain. This has been used to reduce tremors due to Parkinson's disease and to relieve some types of depression, and it has been proposed for treating a number of other psychological and physiological disorders. The development of applications for direct stimulation of the brain is often preceded by neural recording, to help researchers better understand the natural electrical signals in the body.

Sensory and Motor Prostheses

Several designs of ***sensory/motor prostheses*** are being developed to help patients with spinal cord injuries, damaged or amputated limbs, loss of bladder control, and other physical impairments. If only the nerve connections are damaged, these may be replaced by neural recording (to receive signals) and stimulation devices (to transmit them), thus returning some level of motion control. If a limb is entirely gone, it can be replaced by an artificial limb, controlled by neural recording and stimulation (Fig. TF17-6). An interesting phenomenon associated with these and many other types of neural prosthetics is that the plasticity of the brain often allows the user to learn and train the brain and body to see, hear, touch, and move based on the adapted machine-brain interface from the neural signals.

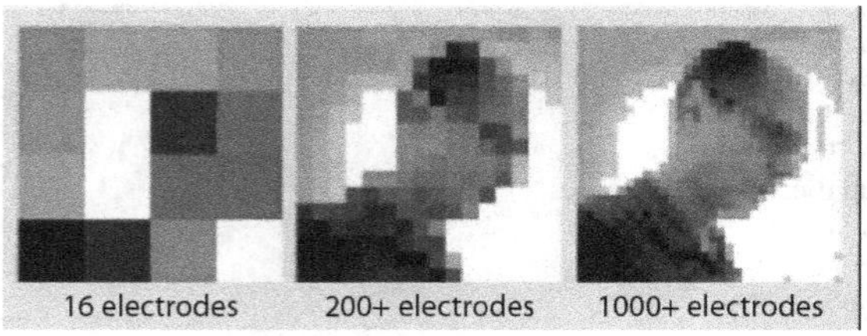

Figure TF17-4: Vision resolution expected with various numbers of sight-stimulating electrodes.

Pain Control

Another application of both internal and external electrical stimulation is in control of pain. Basically, the pain signals are masked by a stimulation-induced tingling known as ***paresthesia***. Internal devices used to induce paresthesia include the ***spinal cord stimulator*** (SCS) shown in Fig. TF17-7 and external devices include ***pulsed electromagnetic field*** (PEMF) stimulators. External devices use one of two methods for directing the pulsed energy to the location of the pain. One method involves inductive coupling (using coils external to the body), and the other involves the use of two electrodes on either side of the region, transmitting current from one electrode through the body region to the other electrode (Fig. TF17-8). PEMF devices have also been used to improve bone and soft tissue healing.

Emerging technology in neural prostheses and other body-machine interfaces has already provided life improvements for many. This technology is still in its infancy,

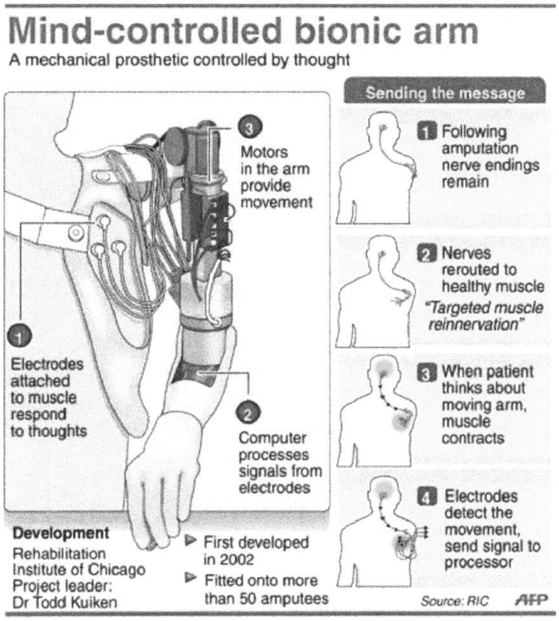

Figure TF17-6: Mind-controlled bionic arm uses both neural recording and neural stimulation within the brain and at the attachment site of the artificial limb. (Credit: Todd Kuiken, MD, Center for Bionic Medicine.)

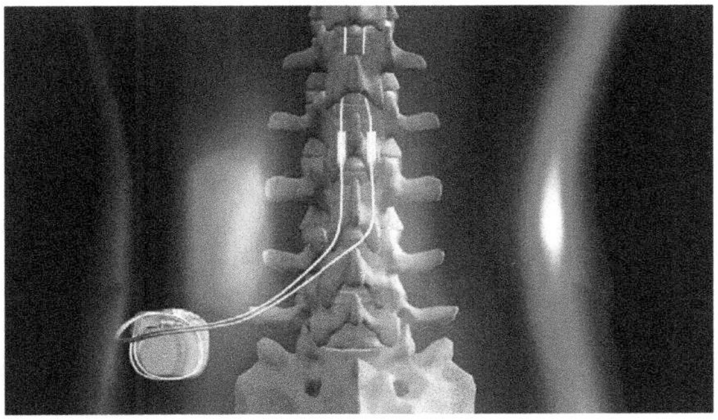

Figure TF17-7: Spinal cord stimulator (SCS). (Credit: Spine-health.com.)

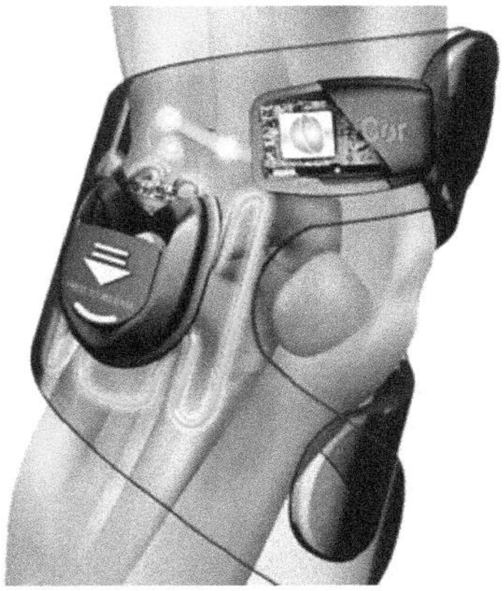

Figure TF17-8: Wearable pulsed electromagnetic field (PEMF) pain-control device for the knee. (Credit: Orthomedical.)

and many interesting challenges remain. How to create a full-function, long-term biocompatible implant small enough to be placed directly into the eye, brain, spine, bladder, brain and other organs, with battery life and/or power harvesting to support its operation, but with heat and power low enough not to damage the critical neurons it is connected to, surgically placing it correctly every time for every patient, with easy ways to get information to and from the device ... there are enough challenges to keep engineers engaged for decades to come!

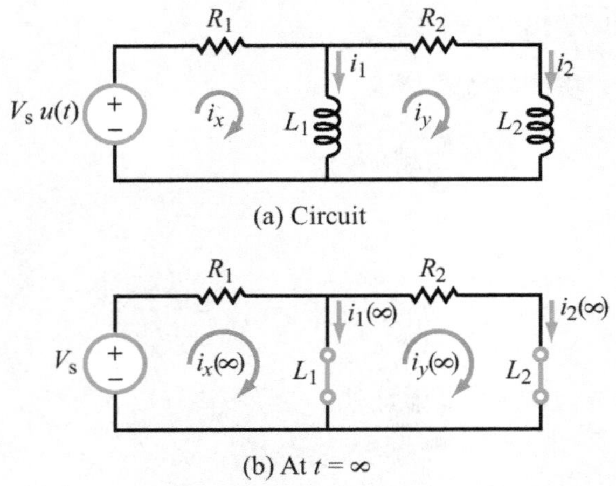

Figure 6-17: Circuit for Example 6-11.

Example 6-11: Two-Inductor Circuit

Determine $i_1(t)$ and $i_2(t)$ in the circuit of Fig. 6-17 for $t \geq 0$. The component values are $V_s = 1.4$ V, $R_1 = 0.4\ \Omega$, $R_2 = 0.3\ \Omega$, $L_1 = 0.1$ H, and $L_2 = 0.2$ H.

Solution: We designate i_x and i_y as the mesh currents in the two loops, as shown. We will analyze the circuit in terms of i_x and i_y and then use the solutions to determine i_1 and i_2.

For $t \geq 0$, the mesh equations are given by:

$$-V_s + R_1 i_x + L_1 \frac{d}{dt}(i_x - i_y) = 0, \qquad (i_x \text{ loop})$$

$$L_1 \frac{d}{dt}(i_y - i_x) + R_2 i_y + L_2 \frac{di_y}{dt} = 0, \qquad (i_y \text{ loop})$$

which can be rearranged and rewritten in the form

$$R_1 i_x + L_1 i'_x - L_1 i'_y = V_s, \qquad (i_x \text{ loop}) \qquad (6.82)$$

$$-L_1 i'_x + R_2 i_y + (L_1 + L_2) i'_y = 0. \qquad (i_y \text{ loop}) \qquad (6.83)$$

Step 1: **Develop a differential equation in i_x alone**

Take the time derivative of all terms in the i_y-loop equation:

$$-L_1 i''_x + R_2 i'_y + (L_1 + L_2) i''_y = 0. \qquad (6.84)$$

To convert Eq. (6.84) into a differential equation in i_x alone, we need to develop expressions for i'_y and i''_y in terms of i_x and its derivatives. By isolating i'_y in Eq. (6.82), we have

$$i'_y = \frac{R_1}{L_1} i_x + i'_x - \frac{V_s}{L_1}. \qquad (6.85)$$

To obtain an expression for i''_y, we simply take the derivative of Eq. (6.85),

$$i''_y = \frac{R_1}{L_1} i'_x + i''_x. \qquad (6.86)$$

After inserting Eqs. (6.85) and (6.86) into Eq. (6.84) and rearranging terms, we have

$$i''_x + \left[\frac{(R_1 + R_2)L_1 + R_1 L_2}{L_1 L_2}\right] i'_x + \left(\frac{R_1 R_2}{L_1 L_2}\right) i_x = \frac{R_2 V_s}{L_1 L_2}, \qquad (6.87)$$

which can be rewritten in the compact form

$$i''_x + a i'_x + b i_x = c, \qquad (6.88)$$

where

$$a = \frac{(R_1 + R_2)L_1 + R_1 L_2}{L_1 L_2} = 7.5,$$

$$b = \frac{R_1 R_2}{L_1 L_2} = 6, \qquad c = \frac{R_2 V_s}{L_1 L_2} = 21.$$

Step 2: **Evaluate α, ω_0, s_1, and s_2**

$$\alpha = \frac{a}{2} = \frac{7.5}{2} = 3.75 \text{ Np/s}, \qquad (6.89a)$$

$$\omega_0 = \sqrt{b} = \sqrt{6} = 2.45 \text{ rad/s}, \qquad (6.89b)$$

$$s_1 = -\alpha + \sqrt{\alpha^2 + \omega_0^2}$$
$$= -3.75 + \sqrt{(3.75)^2 - 6} = -0.91 \text{ Np/s}, \qquad (6.89c)$$

and

$$s_2 = -3.75 - \sqrt{(3.75)^2 - 6} = -6.6 \text{ Np/s}. \qquad (6.89d)$$

Step 3: **Write expression for $i_x(t)$**

Since $\alpha > \omega_0$, i_x will exhibit an overdamped response given by

$$i_x(t) = [i_x(\infty) + A_1 e^{s_1 t} + A_2 e^{s_2 t}]$$
$$= [i_x(\infty) + A_1 e^{-0.91t} + A_2 e^{-6.6t}]. \qquad (6.90)$$

Step 4: **Evaluate final condition**

At $t = \infty$, the inductors in the circuit behave like short circuits (Fig. 6-17(b)), in which case the current generated by V_s will flow entirely through L_1. Hence,

$$i_x(\infty) = \frac{V_s}{R_1} = \frac{1.4}{0.4} = 3.5 \text{ A} \qquad (6.91a)$$

and

$$i_y(\infty) = 0. \qquad (6.91b)$$

The expression for $i_x(t)$ becomes

$$i_x(t) = 3.5 + A_1 e^{-0.91t} + A_2 e^{-6.6t}. \qquad (6.92)$$

Step 5: **Invoke initial conditions**

Before $t = 0$, the circuit contained no sources. Hence,

$$i_1(0) = i_1(0^-) = 0 \tag{6.93a}$$

and

$$i_2(0) = i_2(0^-) = 0, \tag{6.93b}$$

which implies that

$$i_x(0) = i_x(0^-) = 0 \tag{6.94}$$

and

$$i_y(0) = i_y(0^-) = 0. \tag{6.95}$$

At $t = 0$, with no currents flowing through either loop, the voltages across L_1 and L_2 are both equal to V_s. That is,

$$i_1'(0) = \frac{1}{L_1}\, \upsilon_{L_1}(0) = \frac{V_s}{L_1} \tag{6.96a}$$

and

$$i_2'(0) = \frac{1}{L_2}\, \upsilon_{L_2}(0) = \frac{V_s}{L_2}, \tag{6.96b}$$

Consequently,

$$i_x'(0) = i_1'(0) + i_2'(0) = \frac{V_s}{L_1} + \frac{V_s}{L_2} = 21. \tag{6.97}$$

Now that we know the values of $i_x(0)$, $i_x'(0)$, and $i_x(\infty)$, we can apply the general expressions for A_1 and A_2 in Table 6-2 to get

$$A_1 = \frac{i_x'(0) - s_2[i_x(0) - i_x(\infty)]}{s_1 - s_2} = \frac{21 + 6.6(0 - 3.5)}{-0.91 + 6.6} = -0.36 \text{ A}$$

and

$$A_2 = -\left[\frac{i_x'(0) - s_1[i_x(0) - i_x(\infty)]}{s_1 - s_2}\right] = -\left[\frac{21 + 0.91(0 - 3.5)}{-0.91 + 6.6}\right] = -3.14 \text{ A}.$$

The final expression for $i_x(t)$ is then given by

$$i_x(t) = [3.5 - 0.36e^{-0.91t} - 3.14e^{-6.6t}] \text{ A}. \tag{6.98}$$

Repetition of steps 1–4 for i_y requires that we start by taking the time derivative of the i_x-loop equation (Eq. (6.82)) and then using the i_y-loop equation (Eq. (6.83)) to generate expressions for i_x' and i_x''. The procedure leads to

$$i_y(t) = 1.23(e^{-0.91t} - e^{-6.6t}) \text{ A}. \tag{6.99}$$

Finally, the solutions for $i_1(t)$ and $i_2(t)$ are:

$$i_1(t) = i_x(t) - i_y(t) = [3.5 - 1.59e^{-0.91t} - 1.91e^{-6.6t}] \text{ A} \tag{6.100a}$$

and

$$i_2(t) = i_y(t) = 1.23(e^{-0.91t} - e^{-6.6t}) \text{ A} \tag{6.100b}$$

(for $t \geq 0$)

Exercise 6-10: For the circuit in Fig. E6.10, determine $i_C(t)$ for $t \geq 0$.

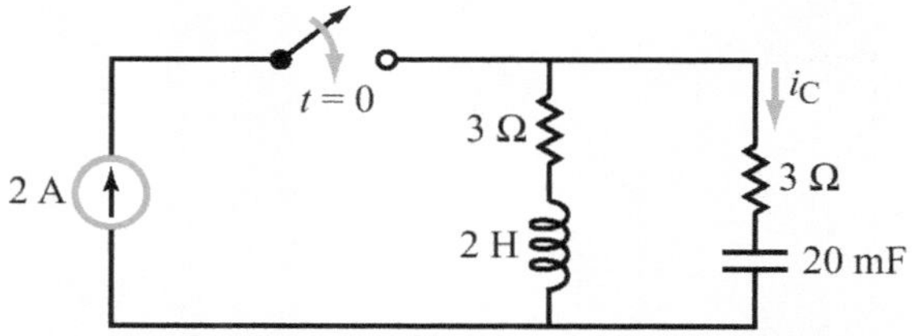

Figure E6.10

Answer: $i_C(t) = 2e^{-1.5t} \cos 4.77t$ A. (See CAD)

Example 6-12: Second-Order Op-Amp Circuit

Determine $i_L(t)$ in the op-amp circuit of Fig. 6-18(a) for $t \geq 0$. Assume $V_s = 1$ mV, $R_1 = 10$ kΩ, $R_2 = 1$ MΩ, $R_3 = 100$ Ω, $L = 5$ H and $C = 1$ μF.

Solution: KCL at node υ_n gives

$$i_1 + i_n + i_2 + i_3 = 0,$$

or equivalently,

$$\frac{\upsilon_n - V_s}{R_1} + i_n + \frac{\upsilon_n - \upsilon_{out}}{R_2} + C\,\frac{d}{dt}(\upsilon_n - \upsilon_{out}) = 0. \tag{6.101}$$

Since $\upsilon_n = \upsilon_p = 0$, $i_n = 0$, and

$$\upsilon_{out} = R_3 i_L + L\,\frac{di_L}{dt}, \tag{6.102}$$

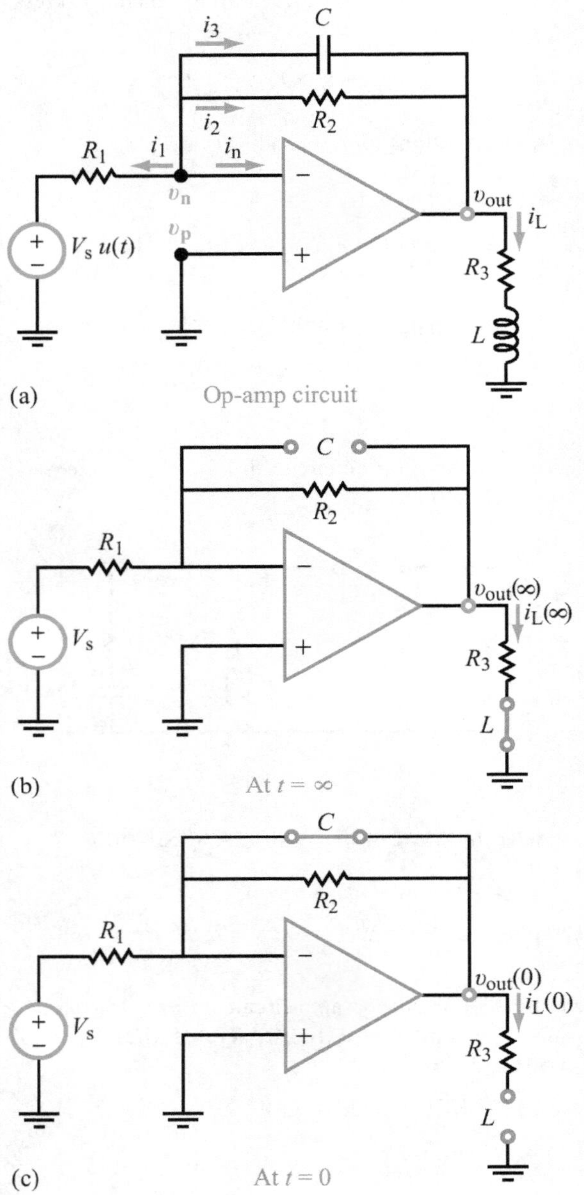

Figure 6-18: Op-amp circuit of Example 6-12.

Equation (6.101) becomes

$$\frac{R_3}{R_2}\, i_{\rm L} + \left(\frac{L}{R_2} + R_3 C\right) \frac{di_{\rm L}}{dt} + LC\, \frac{d^2 i_{\rm L}}{dt^2} = -\frac{V_{\rm s}}{R_1}. \quad (6.103)$$

Rearranging, we have

$$i_{\rm L}'' + a i_{\rm L}' + b i_{\rm L} = c, \quad (6.104)$$

where

$$a = \frac{L + R_2 R_3 C}{R_2 LC} = 21,$$

$$b = \frac{R_3}{R_2 LC} = 20,$$

and

$$c = \frac{-V_{\rm s}}{R_1 LC} = -0.02.$$

The damping behavior of $i_{\rm L}$ is determined by how the magnitude of α compares with that of ω_0:

$$\alpha = \frac{a}{2} = 10.5 \text{ Np/s},$$

$$\omega_0 = \sqrt{b} = \sqrt{20} = 4.47 \text{ rad/s}.$$

Since $\alpha > \omega_0$, $i_{\rm L}$ will exhibit an overdamped response given by

$$i_{\rm L}(t) = [A_1 e^{s_1 t} + A_2 e^{s_2 t} + i_{\rm L}(\infty)]\, u(t),$$

with

$$s_1 = -\alpha + \sqrt{\alpha^2 - \omega_0^2} = -1.0,$$

$$s_2 = -\alpha - \sqrt{\alpha^2 - \omega_0^2} = -20.$$

At $t = \infty$, the circuit assumes the equivalent configuration shown in Fig. 6-18(b), which is an inverting amplifier with an output voltage

$$\upsilon_{\rm out}(\infty) = -\frac{R_2}{R_1}\, V_{\rm s}.$$

Hence,

$$i_{\rm L}(\infty) = \frac{\upsilon_{\rm out}(\infty)}{R_3} = -\frac{R_2 V_{\rm s}}{R_1 R_3} = -1 \text{ mA}.$$

The expression for $i_{\rm L}(t)$ becomes

$$i_{\rm L}(t) = [A_1 e^{-t} + A_2 e^{-20t} - 10^{-3}]. \quad (6.105)$$

To determine the values of A_1 and A_2, we examine initial conditions for $i_{\rm L}$ and $i_{\rm L}'$. At $t = 0^-$, there were no active sources in the circuit, and since i_L cannot change instantaneously, it follows that

$$i_L(0) = i_L(0^-) = 0,$$

which means that the inductor behaves like an open circuit at $t = 0$, as depicted in Fig. 6-18(c). Also, since the voltage v_C across the capacitor was zero before $t = 0$, it has to remain at zero at $t = 0$, which is why it has been replaced with a short circuit in Fig. 6-18(c). Consequently, $v_{out}(0) = 0$, $v_L(0) = 0$, and

$$i'_L(0) = \frac{1}{L}\, v_L(0) = 0.$$

From Table 6-2, with $x = i_L$,

$$\begin{aligned} A_1 &= \frac{i'_L(0) - s_2[i_L(0) - i_L(\infty)]}{s_1 - s_2} \\ &= \frac{0 + 20(0+1)}{-1+20} \times 10^{-3} = 1.05 \text{ mA} \end{aligned} \tag{6.106}$$

and

$$\begin{aligned} A_2 &= -\left[\frac{i'_L(0) - s_1[i_L(0) - i_L(\infty)]}{s_1 - s_2}\right] \\ &= -\left[\frac{0 + 1(0+1)}{-1+20}\right] \times 10^{-3} = -0.053 \text{ mA}. \end{aligned} \tag{6.107}$$

The final expression for $i_L(t)$ is then given by

$$i_L(t) = [1.05e^{-t} - 0.053e^{-20t} - 1] \text{ mA}, \qquad \text{for } t \geq 0.$$

Concept Question 6-6: A circuit contains two capacitors and three inductors, in addition to resistors and sources. Under what circumstance is it a second-order circuit? (See CAD)

Concept Question 6-7: Suppose $a = 0$ in Eq. (6.59). What type of response will $x(t)$ have in that case? (See CAD)

6-9 Multisim Analysis of Circuit Response

Understanding the behavior of even a simple RLC circuit is sometimes a challenging task for electrical and computer engineering students. In reaction to a sudden change, a circuit gives rise to voltage and current variations that depend on the circuit topology, the initial conditions of its components, and the values of those components. In this section, we describe how to use Multisim to analyze the response of the series RLC circuit we discussed in earlier sections. The procedure is intended to demonstrate the steps one would follow to analyze any circuit with Multisim. As an example of a real-world application of the RLC-circuit response, we will then examine how such a circuit is used in RFID (radio frequency identification) technology.

6-9.1 The Series RLC Circuit

Using the now (hopefully) familiar schematic tools, draw a series RLC circuit, including a switch, in the Multisim Schematic Capture window. Use the parts and component values listed in Table 6-3, and add an oscilloscope as shown in Fig. 6-19. The scope is used for both L_1 and C_1, so that we may compare the voltages across them on the same screen. Make sure that before starting the interactive simulation, the initial condition of the switch is in position 2, so that the dc voltage source is not connected directly to the RLC circuit. Upon starting the simulation, you should see no voltage across any of the three components. After hitting the space bar to move the switch (Fig. 6-20), $v_L(t)$ will initially jump in level to 1 V and then exhibit an underdamped oscillatory response as a function of time. In contrast, $v_C(t)$ will exhibit an oscillatory behavior that will dampen out with time to assume a final value of 1 V.

A note on the Interactive Simulation settings is appropriate here. When you run an Interactive Simulation, Multisim numerically solves for the solution to the circuit at successive points in time. The resolution of this time step can be modified under Simulate → Interactive Simulation Settings. Both the *maximum time step* (TMAX) and the *initial time step* can be changed. Normally, there is no reason to do this and Multisim's defaults will work well. However, when using the virtual instruments, sometimes time points are generated too quickly and this makes it difficult for the user to observe the behavior, or conversely the resolution may be too small so that

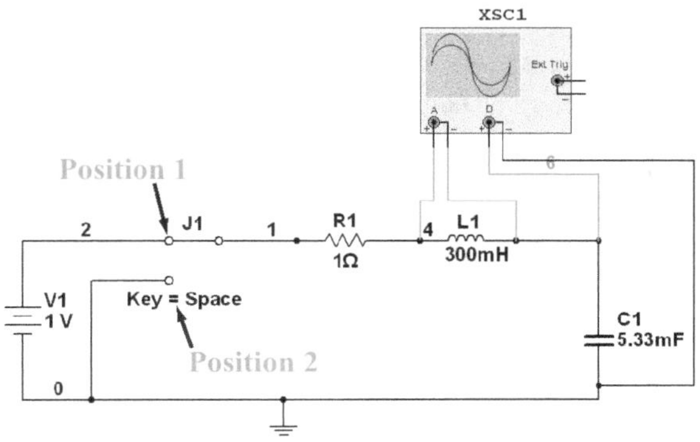

Figure 6-19: Multisim screen with RLC circuit.

Table 6-3: **Component values for the circuit in Fig. 6-19.**

Component	Group	Family	Quantity	Description
1	Basic	Resistor	1	1 Ω resistor
300 m	Basic	Inductor	1	300 mH inductor
5.33 m	Basic	Capacitor	1	5.33 mF capacitor
SPDT	Basic	Switch	1	Single-pole double-throw (SPDT) switch
DC_POWER	Sources	Power_Sources	1	1 V dc source

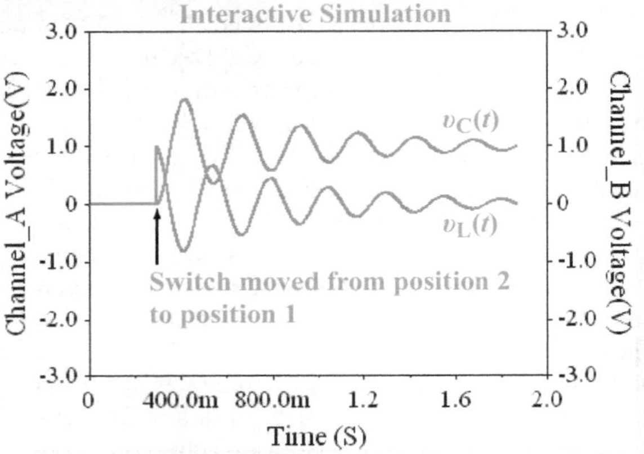

Figure 6-20: Voltage responses to moving the switch in the RLC circuit from position 2 to position 1.

the progression of time in the Interactive Simulation becomes annoyingly slow. When generating the traces in Fig. 6-20, for example, it may be difficult to see the damped behavior directly on the scope window because it scrolls by too fast. In that case, it can be helpful to reduce both the maximum and initial time steps (10–100 × reduction usually works fine). This forces the computer to simulate more data points and slows it down, allowing you to see the trace appear more slowly. The drawback of this tweak is that you also use up more memory (and filespace).

Exercise 6-11: Given the component values in the Multisim circuit of Fig. 6-19, what are the values of ω_0 and α for the circuit response?

Answer: (See CAD)

Exercise 6-12: Is the natural response for the circuit in Fig. 6-19 over-, under-, or critically damped? You can determine this both graphically (from the oscilloscope) and mathematically, by comparing ω_0 and α.

Answer: (See CAD)

Exercise 6-13: Modify the value of R in the circuit of Fig. 6-19 so as to obtain a critically damped response.

Answer: (See CAD)

6-9.2 RFID Circuit

Radio frequency identification (RFID) circuits are fast becoming ubiquitous in many mass consumer applications, ranging from tracking parcels and shipments to "smart" ID badges (see Technology Brief 16). Most systems in use today rely on a transceiver (usually handheld) that can remotely interrogate one or more RFID tags (ranging in size from a few millimeters to a few centimeters). Some tags reply with only a serial number, while others are connected to miniature sensors and return values for temperature, humidity, acceleration, position, etc. The key to the widespread success of these RFID tags is that they do not require batteries to operate! If the transceiver is in close proximity to the tag (usually within a fraction of a meter), the radio-frequency power it transmits is sufficient to activate the RFID tag. The RFID tag uses an RLC circuit to harvest this power and communicate back to the transceiver (Fig. 6-21). The essential elements of the RFID communication system are shown in the circuit of Fig. 6-22. [An actual RFID circuit is more sophisticated, but the basic principle

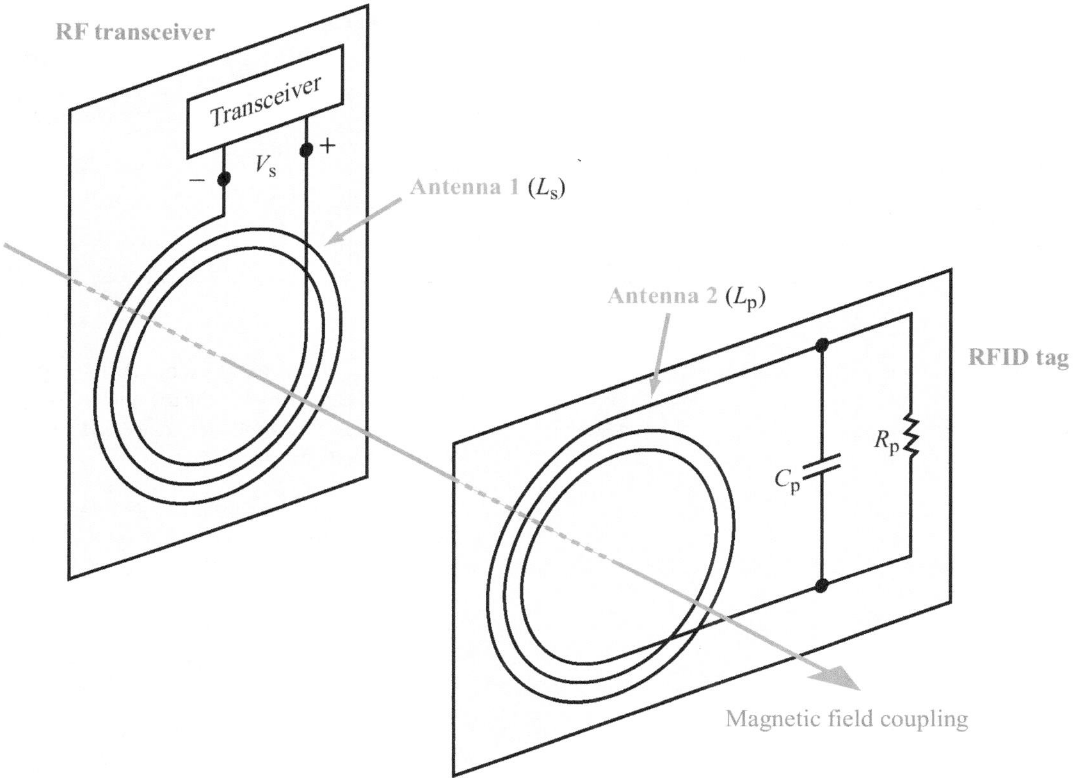

Figure 6-21: Illustration of an RFID transceiver in close proximity to an RFID tag. Note that the RFID tag will only couple to the transceiver when the two inductors are aligned along the magnetic field (shown in blue).

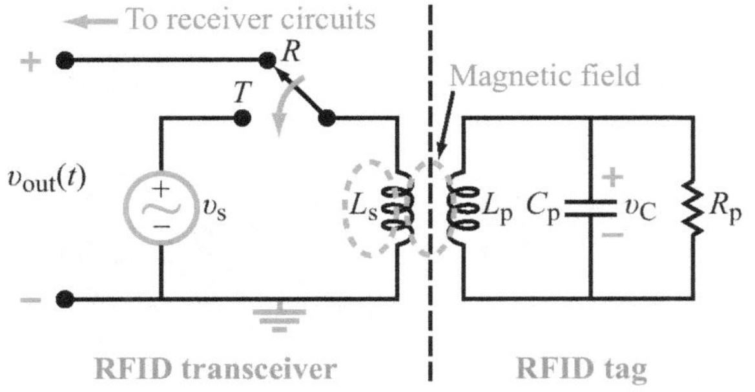

Figure 6-22: Basic elements of the RFID.

of operation is the same.] In transmit mode—with the SPDT switch connected to terminal T—the transceiver circuit consists of a ac voltage source, υ_s, connected in series with inductor L_s. By moving the switch to terminal R, the transceiver circuit becomes a receiver with output voltage $\upsilon_{out}(t)$. In transmit mode, υ_s generates a current through L_s, which induces a magnetic field around it. If inductor L_p of the RFID tag is close to L_s, the magnetic field generated by L_s will induce a current through L_p. This current becomes the power source in the RFID-tag circuit, and the mechanism for building up the voltage across C_p to some maximum value V_C.

When the switch is moved from transmit mode to receive mode, υ_s stops delivering power to L_s. The current through L_p, however, cannot change to zero instantaneously. The RLC circuit will react to the sudden change with an oscillatory underdamped response characterized by a damped natural frequency ω_d, whose value is governed by the choice of values for R_p, L_p, and C_p of the RFID tag. This oscillation frequency becomes part of the ID of that particular tag. In the same way

Table 6-4: **Parts for the Multisim circuit in Fig. 6-23.**

Component	Group	Family	Quantity	Description
TS_IDEAL	Basic	Transformer	1	1 mH:1 mH ideal transformer
1 k	Basic	Resistor	1	1 kΩ resistor
1 μ	Basic	Capacitor	1	1 μF capacitor
SPDT	Basic	Switch	1	SPDT switch
AC_CURRENT	Sources	Signal_Current_Source	1	1 mA, 5.033 kHz

that magnetic coupling served to transfer power from L_s to L_p during the transmit mode, it also serves to transfer information in the opposite direction—from L_p to L_s—during the receive mode. Since

$$\upsilon_{out}(t) = L_s \frac{di_{L_s}}{dt},$$

the output voltage recorded after moving the switch to receive mode provides the *reply* by the RFID tag to the earlier excitation introduced by υ_s during the transmit mode. [Real RFID transceivers transmit a few bits of data by superimposing digital bits onto the oscillations.]

To illustrate the operation of the RFID tag, we can simulate the process in Multisim. Using the parts listed in Table 6-4, we can build the circuit shown in the Multisim window of Fig. 6-23.

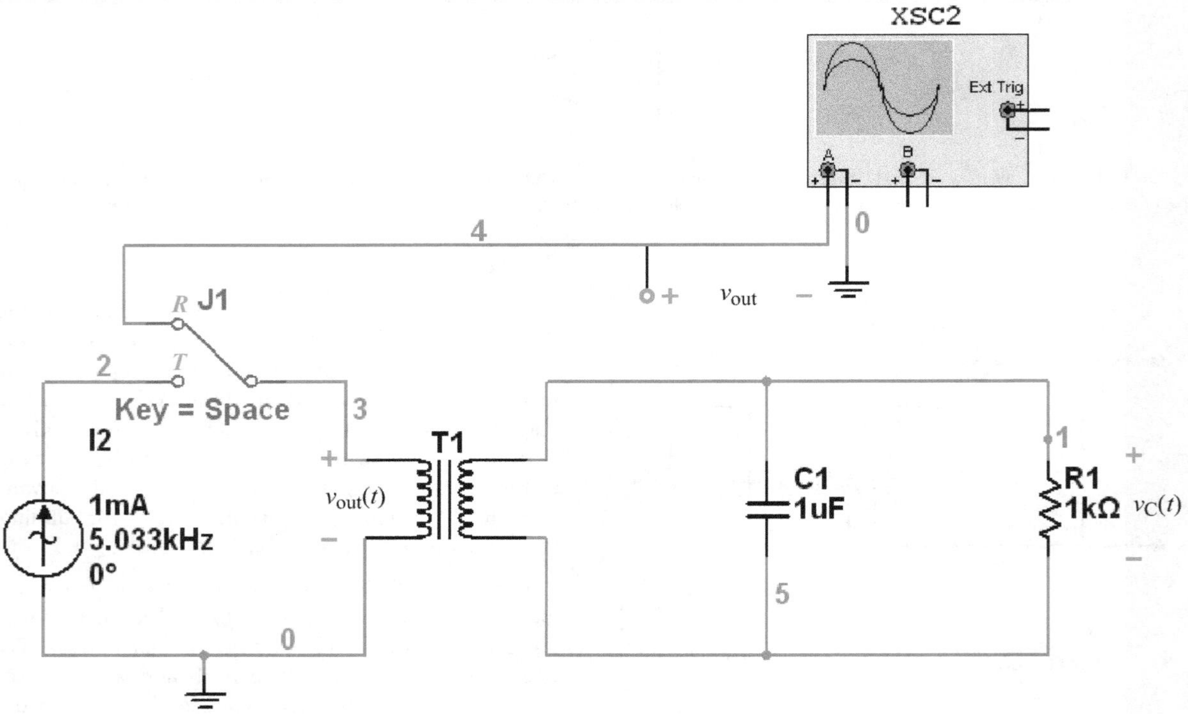

Figure 6-23: Multisim rendition of RFID circuit.

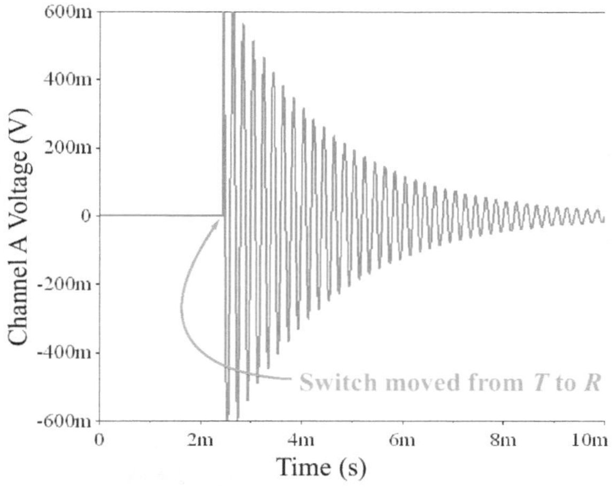

Figure 6-24: Oscilloscope trace for RFID receive channel $v_{out}(t)$ after moving the switch from T to R.

To simulate magnetic coupling between inductors L_s and L_p, we use transformer T_1, which represents two closely coupled inductors sharing a common magnetic field. In Multisim we set the inductance of each of the two transformer units to 1 mH and the coupling coefficient to 1. The circuit uses an oscilloscope to monitor $v_{out}(t)$. The oscilloscope trace is displayed in Fig. 6-24. Note that when the switch is moved from transmit to receive mode, $v_{out}(t)$ exhibits an immediate response that then decays exponentially with time. You may also want to plot $v_C(t)$ and $i_C(t)$ to examine the voltage and current experienced by the RFID tag itself during transmit and receive periods.

Concept Question 6-8: How does the transmitter in the RFID system transfer power to the RLC circuit? (See CAD)

Concept Question 6-9: How does the transceiver elicit a reply from the RFID tag? (See CAD)

Exercise 6-14: Calculate ω_0, α, and ω_d for the RLC circuit in Fig. 6-23. How do ω_0 and ω_d compare with the angular frequency of the current source? This result, as we will learn later when we study resonant circuits in Chapter 9, is not at all by coincidence.

Answer: (See CAD)

Exercise 6-15: Ideally, we would like the response of the RFID tag to take a very long time to decay down to zero, so as to contain as many digital bits as possible. What determines the decay time? Change the values of some of the components in Fig. 6-23 so as to decrease the damping coefficient by a factor of 2.

Answer: (See CAD)

Summary

Concepts

- Under dc steady state conditions, a capacitor behaves like an open circuit and an inductor behaves like a short circuit.
- Second-order circuits include series and parallel RLC circuits, as well as any circuit containing two passive, energy storage elements (capacitors and inductors).
- The response of a second-order circuit (containing dc sources) to a sudden change consists of a transient component, which decays to zero as $t \to \infty$, and a steady state component that has a constant value.
- The transient response may be overdamped, critically damped, or underdamped, depending on the values of the circuit elements.
- The general solution for second-order circuits is applicable to circuits containing op-amps.
- Multisim can be used to simulate the response of any second-order circuit.

Mathematical and Physical Models

Step response of series and parallel RLC circuits (See Table 6-1)

General Solution for Second Order Circuits:
(see details in Table 6-3)

Differential equation $x'' + ax' + bx = c$

General Solution for Second Order Circuits (cont'd.):

Overdamped Response $(\alpha > \omega_0)$
$$x(t) = [x(\infty) + A_1 e^{s_1 t} + A_2 e^{s_2 t}]\, u(t)$$

Critically Damped Response $(\alpha > \omega_0)$
$$x(t) = [x(\infty) + (B_1 + B_2 t)e^{-\alpha t}]\, u(t)$$

Underdamped Response $(\alpha > \omega_0)$
$$x(t) = [x(\infty) + [D_1 \cos \omega_d t + D_2 \sin \omega_d t]e^{-\alpha t}\; u(t)$$

Important Terms

Provide definitions or explain the meaning of the following terms:

characteristic equation
critically damped
critically damped response
damped natural frequency
damping coefficient
initial condition
initial time step
final condition
first-order circuit
homogeneous
homogeneous solution
invoke initial and final conditions
MEMS
maximum time step
natural response
nepers/second
oscillator
overdamped response
particular
particular solution
resonant frequency
radio frequency identification
RFID
second-order circuit
steady-state
steady-state response
time constant
time period
transient
transient response
underdamped response

PROBLEMS

Section 6-1: Initial and Final Conditions

*6.1 The SPST switch in the circuit of Fig. P6.1 closes at $t = 0$ after it had been open for a long time. Draw the configurations that appropriately represent the state of the circuit at $t = 0^-$, $t = 0$, and $t = \infty$ and use them to determine (a) $v_C(0)$ and $i_L(0)$, (b) $i_C(0)$ and $v_L(0)$, and (c) $v_C(\infty)$ and $i_L(\infty)$.

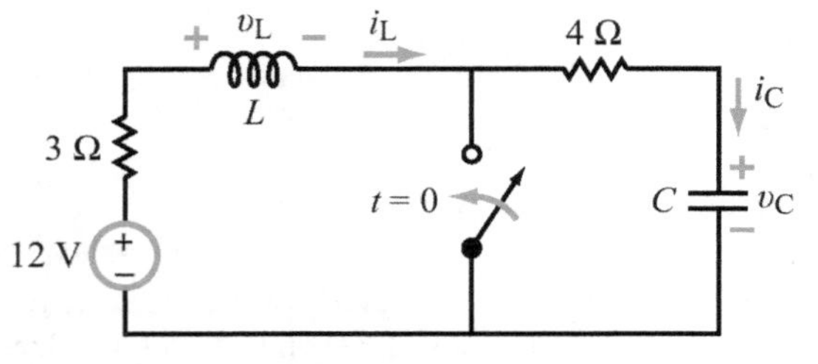

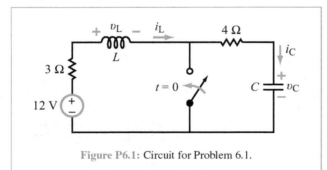

Figure P6.1: Circuit for Problem 6.1.

6.2 The SPST switch in the circuit of Fig. P6.2 opens at $t = 0$, after it had been closed for a long time. Draw the configurations that appropriately represent the state of the circuit at $t = 0^-$, $t = 0$, and $t = \infty$ and use them to determine (a) $v_C(0)$ and $i_L(0)$, (b) $i_C(0)$ and $v_L(0)$, and (c) $v_C(\infty)$ and $i_L(\infty)$.

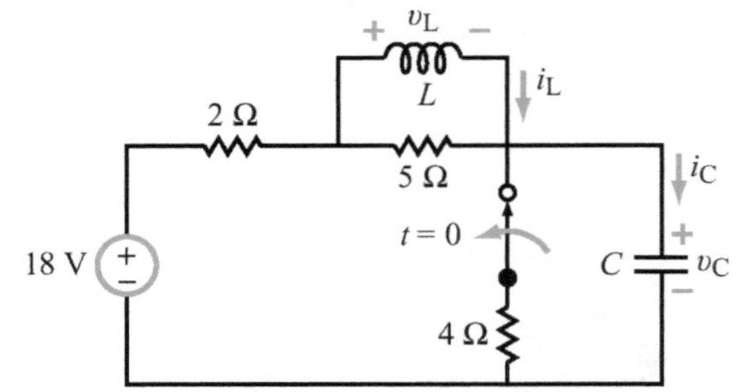

Figure P6.2: Circuit for Problem 6.2.

6.3 The SPST switch in the circuit of Fig. P6.3 opens at $t = 0$, after it had been closed for a long time. Draw the configurations that appropriately represent the state of the circuit at $t = 0^-$, $t = 0$, and $t = \infty$ and use them to determine

*(a) $v_C(0)$ and $i_L(0)$,

(b) $i_C(0)$ and $v_L(0)$, and

*Answer(s) available in Appendix G.

(c) $v_C(\infty)$ and $i_L(\infty)$.

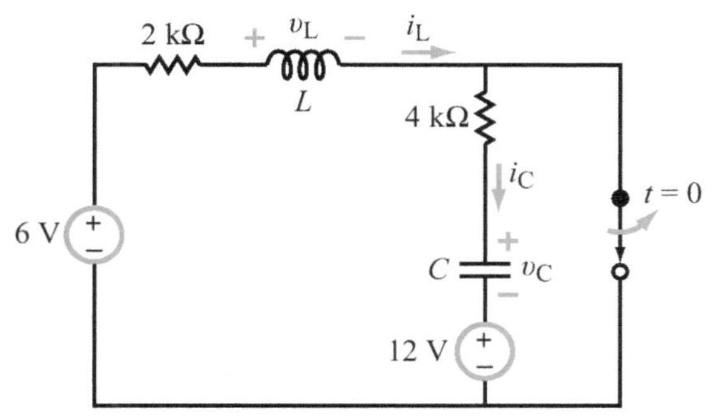

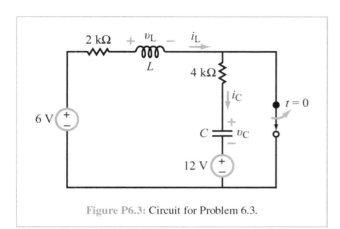

Figure P6.3: Circuit for Problem 6.3.

6.4 The SPST switch in the circuit of Fig. P6.4 opens at $t = 0$, after it had been closed for a long time. Draw the configurations that appropriately represent the state of the circuit at $t = 0^-$, $t = 0$, and $t = \infty$ and use them to determine (a) $v_C(0)$ and $i_L(0)$, (b) $i_C(0)$ and $v_L(0)$, and (c) $v_C(\infty)$ and $i_L(\infty)$.

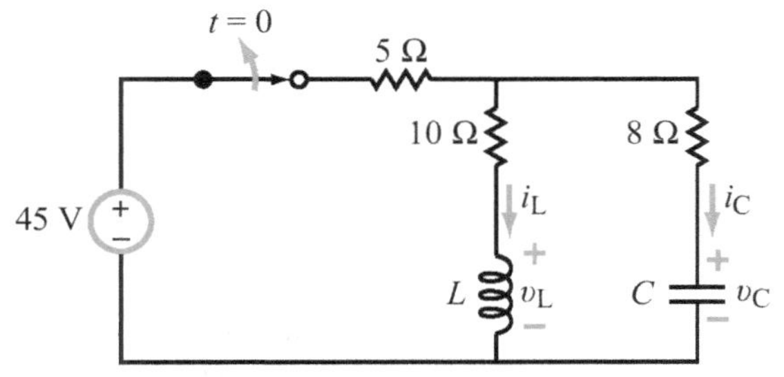

Figure P6.4: Circuit for Problem 6.4.

6.5 The SPST switch in the circuit of Fig. P6.5 closes at $t = 0$, after it had been opened for a long time. Draw the configurations that appropriately represent the state of the circuit at $t = 0^-$, $t = 0$, and $t = \infty$ and use them to determine (a) $v_C(0)$ and $i_L(0)$, (b) $i_C(0)$ and $v_L(0)$, and (c) $v_C(\infty)$ and $i_L(\infty)$.

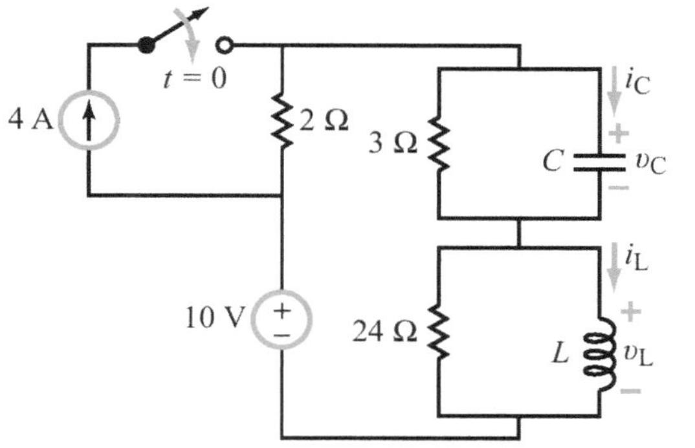

Figure P6.5: Circuit for Problems 6.5 and 6.6.

6.6 Repeat Problem 6.5, but start with a closed switch that opens at $t = 0$.

***6.7** For the circuit in Fig. P6.7, determine $i_1(0)$ and $i_2(0)$.

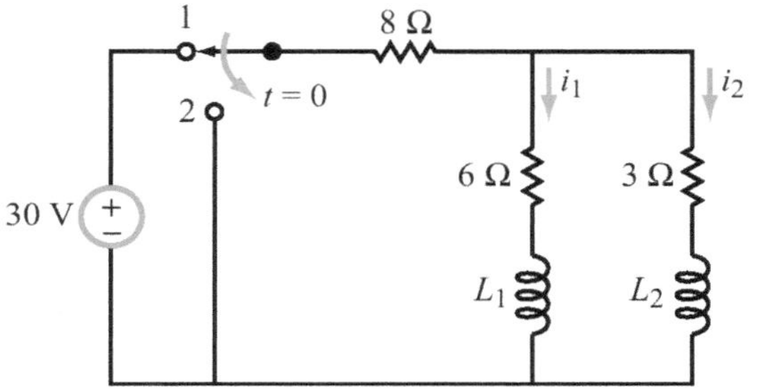

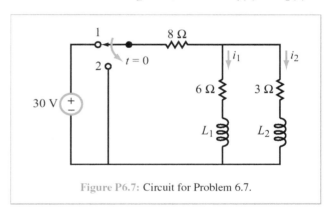

Figure P6.7: Circuit for Problem 6.7.

6.8 For the circuit of Fig. P6.8, determine (a) $i_{C_1}(0)$, $i_{R_1}(0)$, $i_{C_2}(0)$, and $i_{R_2}(0)$ and (b) $v_{C_1}(\infty)$ and $v_{C_2}(\infty)$.

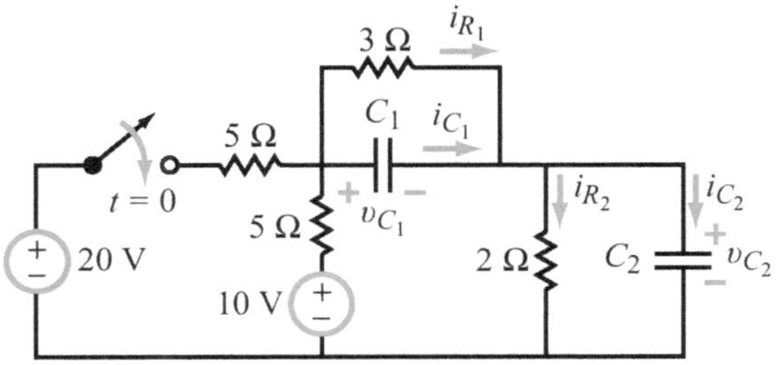

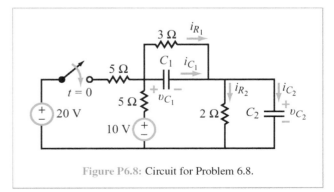

Figure P6.8: Circuit for Problem 6.8.

6.9 For the circuit in Fig. P6.9:

(a) Draw the configurations that appropriately represent the state of the circuit at $t = 0^-$, $t = 0$, and $t = \infty$.

(b) Use the configurations to determine $i_L(0^-)$, $\upsilon_C(0^-)$, $i_L(0)$, $\upsilon_C(0)$, $i_L(\infty)$, and $\upsilon_C(\infty)$.

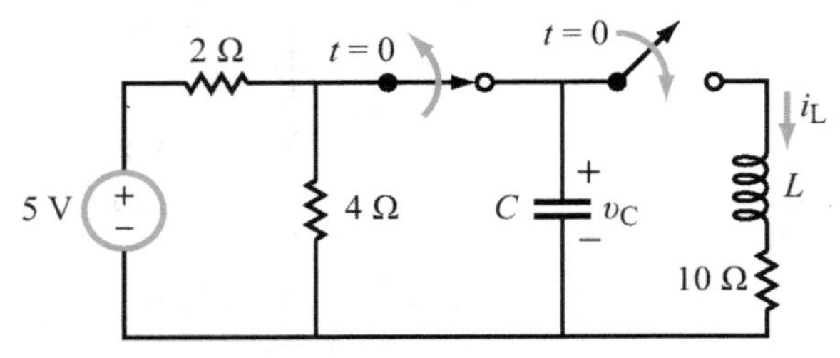

Figure P6.9: Circuit for Problem 6.9.

*6.10 For the circuit in Fig. P6.10, determine $i_C(0)$, $\upsilon_C(0)$, $i_R(0)$, $\upsilon_R(0)$, $i_L(0)$, $\upsilon_L(0)$, $\upsilon_L(\infty)$, $i_R(\infty)$, $\upsilon_C(\infty)$, and $i_L(\infty)$.

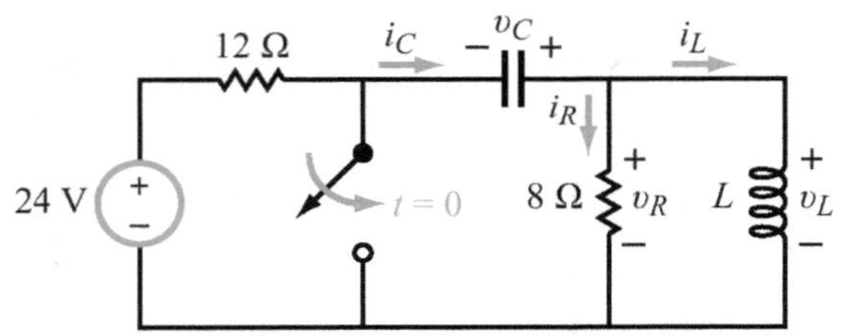

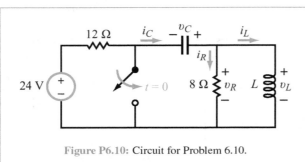

Figure P6.10: Circuit for Problem 6.10.

6.11 For the circuit in Fig. P6.11, find $i_1(0^-)$, $i_2(0)$, $\upsilon_C(0)$, and $i_3(\infty)$.

Sections 6-2 to 6-6: Series RLC Circuit

*6.12 Determine $\upsilon_C(t)$ in the circuit of Fig. P6.12 and plot its waveform for $t \geq 0$, given that $V_0 = 12$ V, $R_1 = 0.4\ \Omega$, $R_2 = 1.2\ \Omega$, $L = 0.1$ H, and $C = 0.4$ F. Use a time scale that appropriately captures the shape of the waveform in your plot.

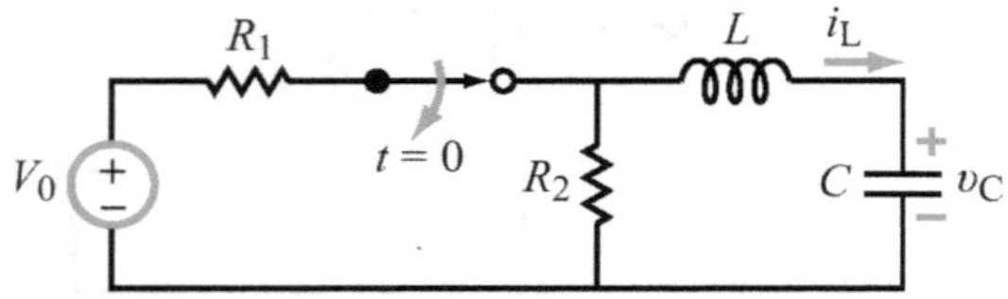

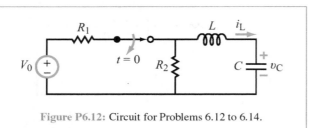

Figure P6.12: Circuit for Problems 6.12 to 6.14.

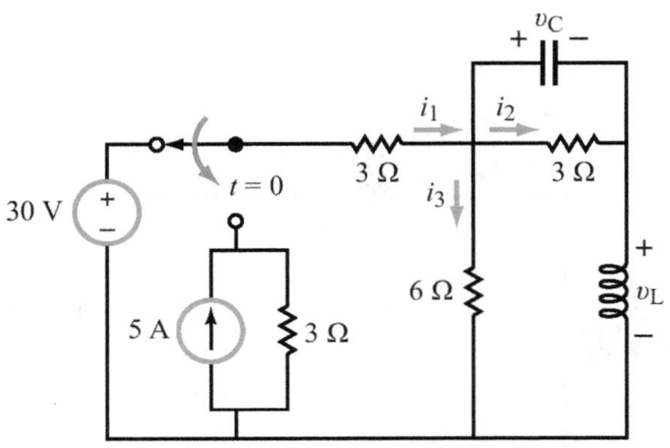

Figure P6.11: Circuit for Problem 6.11.

6.13 Determine $i_L(t)$ in the circuit of Fig. P6.12 and plot its waveform for $t \geq 0$, given that $V_0 = 12$ V, $R_1 = 0.4\ \Omega$, $R_2 = 1.2\ \Omega$, $L = 0.1$ H, and $C = 0.1$ F. Use a time scale that appropriately captures the shape of the waveform in your plot.

*6.14 In the circuit of Fig. P6.12, $V_0 = 12$ V, $R_1 = 0.4\ \Omega$, $R_2 = 1.2\ \Omega$, and $L = 0.1$ H. What should the value of C be in order for $i_L(t)$ to exhibit a critically damped response? Provide an expression for $i_L(t)$ and plot its waveform for $t \geq 0$.

6.15 The voltage υ in a certain circuit is described by the differential equation

$$3\upsilon'' + 24\upsilon' + 75\upsilon = 0.$$

(a) Determine the values of α and ω_0.

(b) What type of damping is exhibited by $\upsilon(t)$?

*6.16 In the circuit of Fig. P6.16, the switch is moved from position 1 to position 2 at $t = 0$. Provide an expression for $\upsilon_C(t)$ for $t \geq 0$.

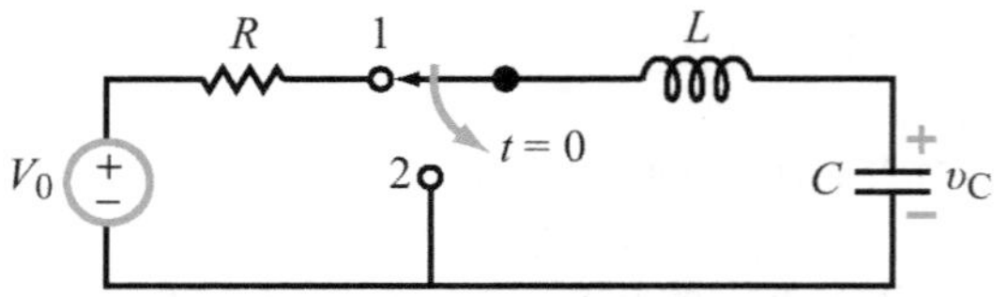

Figure P6.16: Circuit for Problem 6.16.

6.17 A series RLC circuit exhibits the following voltage and current responses:

$$v_C(t) = (6\cos 4t - 3\sin 4t)e^{-2t}\ u(t)\ \text{V},$$

$$i_C(t) = -(0.24\cos 4t + 0.18\sin 4t)e^{-2t}\ u(t)\ \text{A}.$$

Determine α, ω_0, R, L, and C.

***6.18** Determine $i_C(t)$ in the circuit of Fig. P6.18 for $t \geq 0$.

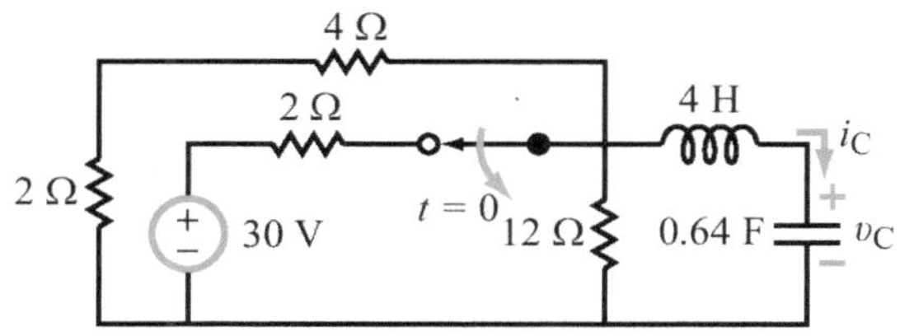

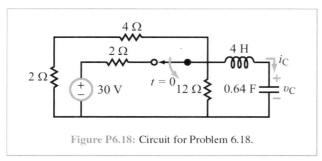

Figure P6.18: Circuit for Problem 6.18.

6.19 Determine $v_C(t)$ in the circuit of Fig. 6.19 for $t \geq 0$.

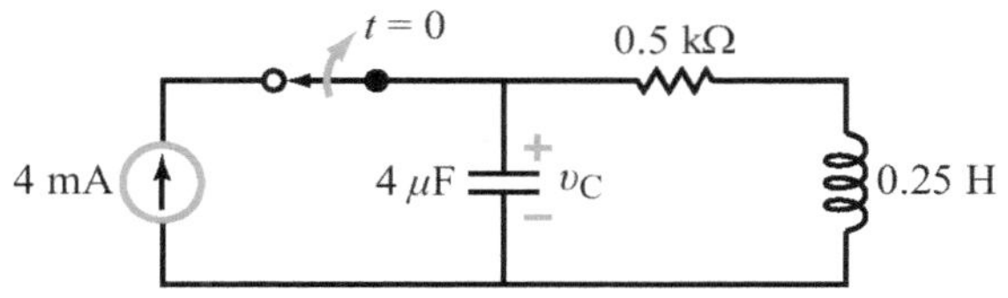

Figure P6.19: Circuit for Problem 6.19.

6.20 Determine $i_C(t)$ in the circuit of Fig. 6.20 for $t \geq 0$.

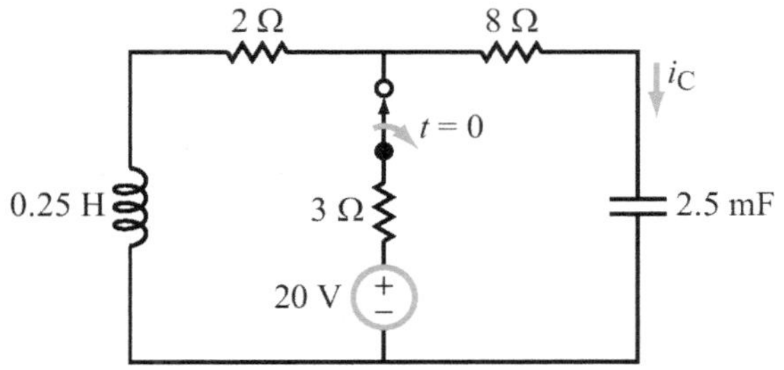

Figure P6.20: Circuit for Problem 6.20.

6.21 The circuit in Fig. 6-4(c) exhibits the response

$$v(t) = (12 + 36t)e^{-3t}\ \text{V}, \quad (\text{for } t \geq 0).$$

If $R = 12\ \Omega$, determine the values of V_s, L, and C.

***6.22** Determine $i_C(t)$ in the circuit of Fig. 6.22 and plot its waveform for $t \geq 0$.

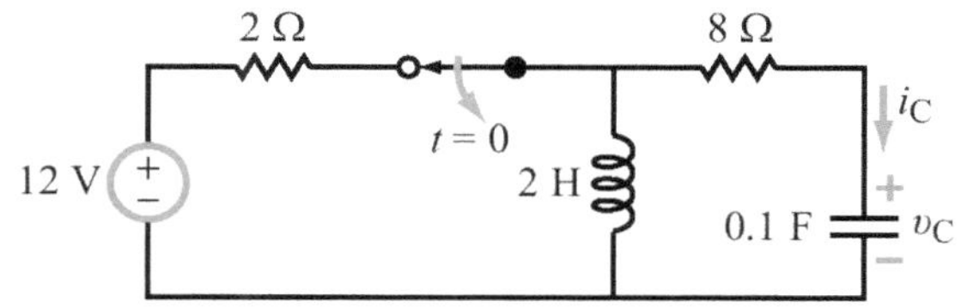

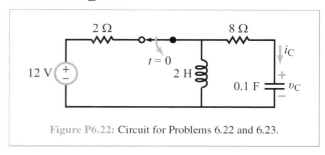

Figure P6.22: Circuit for Problems 6.22 and 6.23.

6.23 Repeat Problem 6.22, retaining the same values for all elements in the circuit except C. Choose the value of C so that the response of $i_C(t)$ is critically damped.

6.24 Determine $i_C(t)$ in the circuit of Fig. 6.24 and plot its waveform for $t \geq 0$, given that $L = 0.05$ H. Use a time scale that appropriately captures the shape of the waveform in your plot.

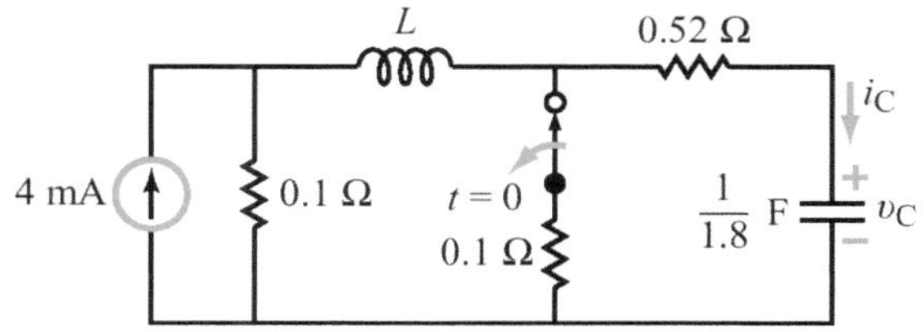

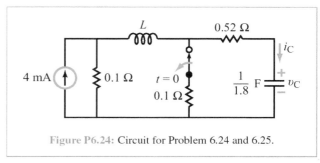

Figure P6.24: Circuit for Problem 6.24 and 6.25.

***6.25** Choose the value of the inductor in the circuit of Fig. 6.24 so that v_C exhibits a critically damped response and determine $v_C(t)$ for $t \geq 0$.

6.26 Determine $i_C(t)$ in the circuit of Fig. 6.26 and plot its waveform for $t \geq 0$, given that $V_s = 24$ V, $R_1 = 2\ \Omega$, $R_2 = 4\ \Omega$, $L = 0.4$ H, and $C = \frac{10}{24}$ F.

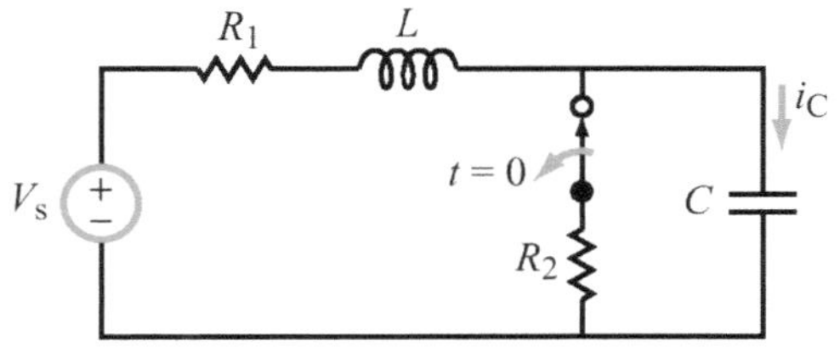

Figure P6.26: Circuit for Problems 6.26 and 6.27.

6.27 Repeat Problem 6.26 with the elements retaining their values, except change C to $\frac{10}{29}$ F.

6.28 In the circuit of Fig. 6.28:

*(a) What is the value of $v_C(\infty)$?

(b) How long does it take after $t = 0$ for v_C to reach 0.99 of its final value? [*Hint:* After solving for $v_C(t)$, step through values of t over the range $2 \le t \le 2.5$ to determine the value that satisfies the stated condition.]

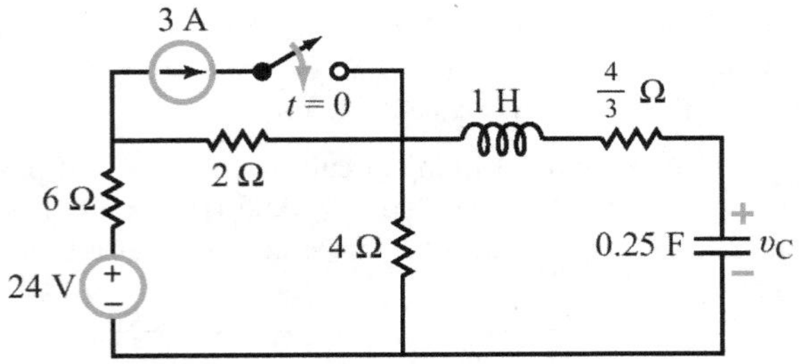

Figure P6.28: Circuit for Problem 6.28.

***6.29** Choose the value of C in the circuit of Fig. 6.29 so that $v_C(t)$ has a critically damped response for $t \ge 0$. Plot the waveform of $v_C(t)$.

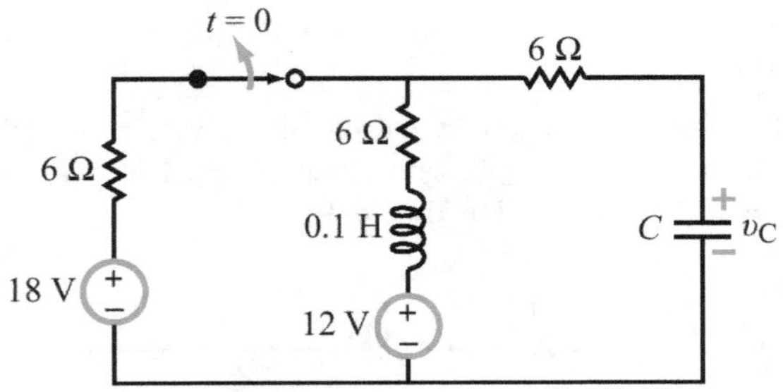

Figure P6.29: Circuit for Problem 6.29.

6.30 Determine $i_L(t)$ in the circuit of Fig. 6.30 and plot its waveform for $t \ge 0$.

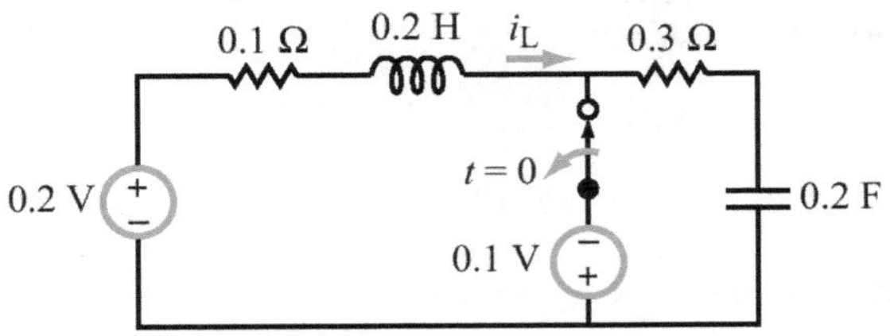

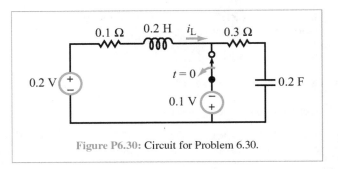

Figure P6.30: Circuit for Problem 6.30.

6.31 Determine $i_C(t)$ and $i_L(t)$ in the circuit of Fig. 6.31 for $t \ge 0$.

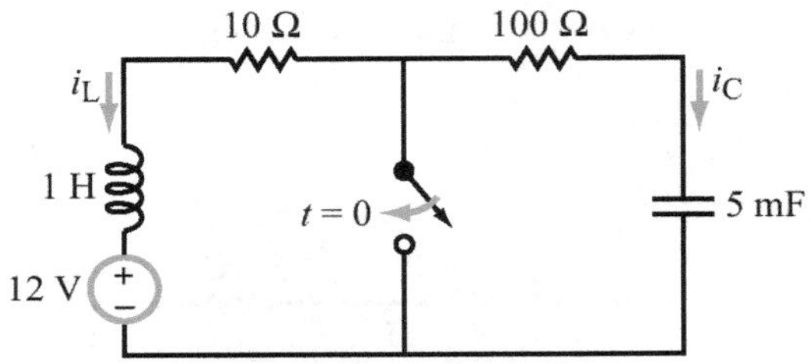

Figure P6.31: Circuit for Problem 6.31.

***6.32** For the circuit in Fig. P6.32, assume that before $t = 0$, the circuit had been in that state for a long time. Find $v_C(t)$ and $i_L(t)$ for $t \ge 0$.

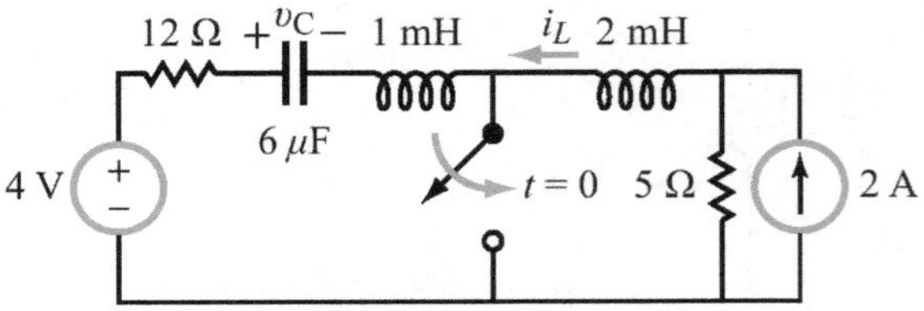

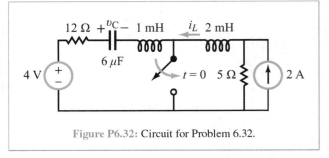

Figure P6.32: Circuit for Problem 6.32.

6.33 Find $v_C(t)$ for $t \ge 0$ in the circuit in Fig. P6.33.

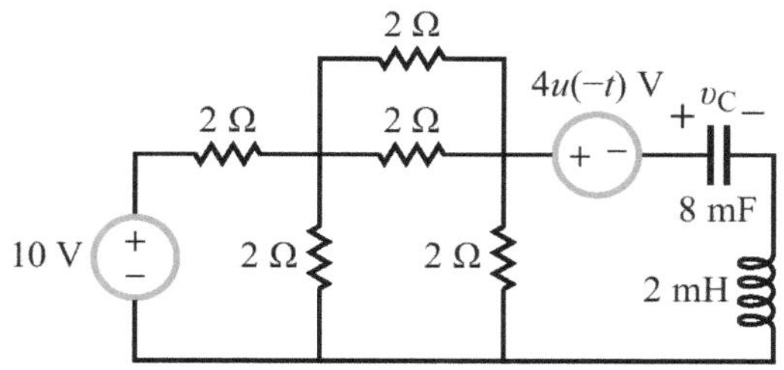

Figure P6.33: Circuit for Problem 6.33.

6.34 For the circuit in Fig. P6.34, determine:

(a) $v_C(0)$.

(b) α, ω_0, and the type of response you expect $v_C(t)$ to exhibit.

(c) $i_C(t)$ for $t \geq 0$.

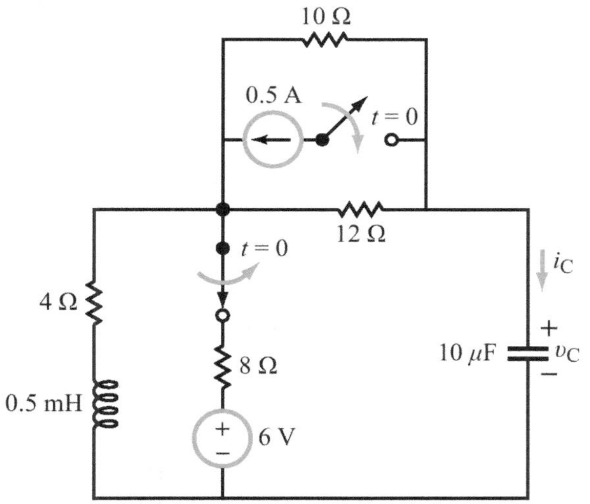

Figure P6.34: Circuit for Problem 6.34.

*6.35 For the circuit in Fig. P6.35, find $v_C(t)$ for $t \geq 0$.

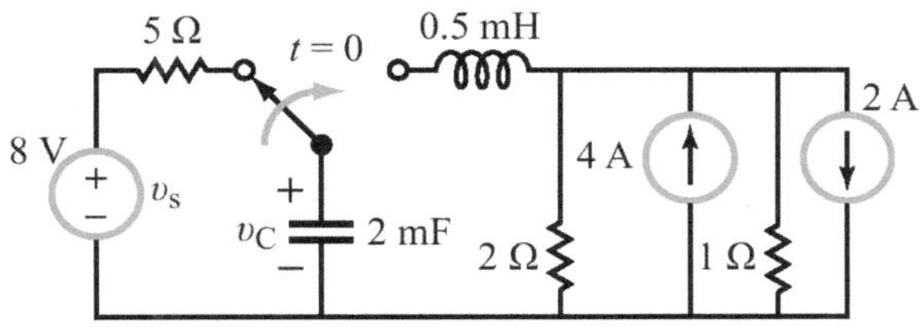

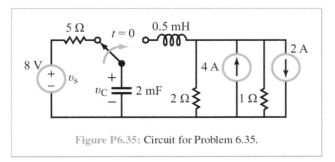

Figure P6.35: Circuit for Problem 6.35.

Section 6-7: Parallel RLC Circuit

6.36 Determine $i_L(t)$ and $i_C(t)$ in the circuit of Fig. 6.36 and plot both waveforms for $t \geq 0$. The SPDT switch was moved from position 1 to position 2 at $t = 0$.

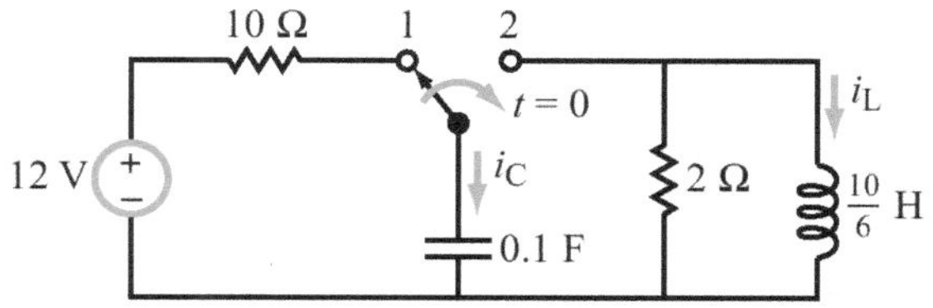

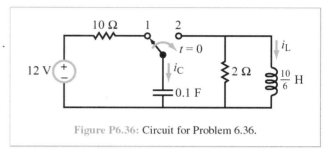

Figure P6.36: Circuit for Problem 6.36.

6.37 Determine $i_L(t)$ in the circuit of Fig. 6.37 and plot its waveform for $t \geq 0$.

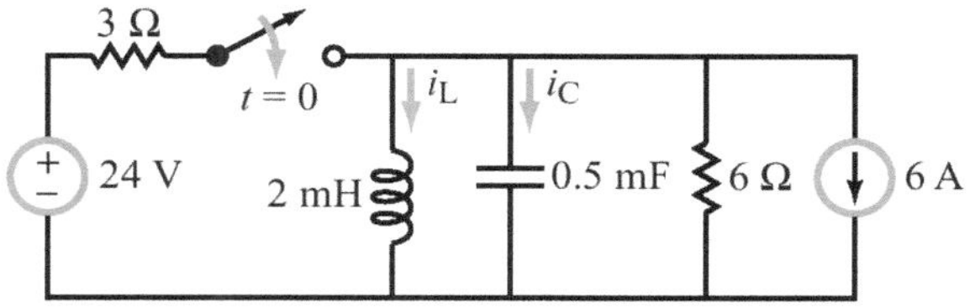

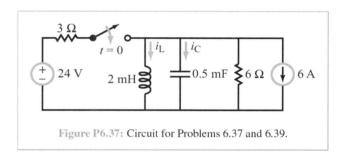

Figure P6.37: Circuit for Problems 6.37 and 6.39.

*6.38 Determine $i_L(t)$ in the circuit of Fig. 6.38 and plot its waveform for $t \geq 0$. The capacitor had no charge on it prior to $t = 0$.

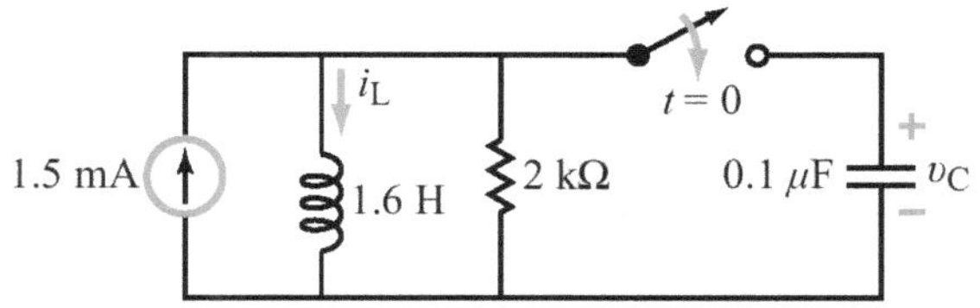

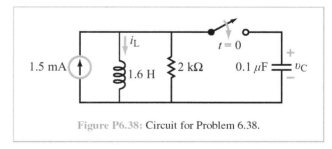

Figure P6.38: Circuit for Problem 6.38.

6.39 Determine $i_C(t)$ in the circuit of Fig. 6.37 for $t \geq 0$.

*6.40 Determine $i_L(t)$ in the circuit of Fig. 6.40 and plot its waveform for $t \geq 0$.

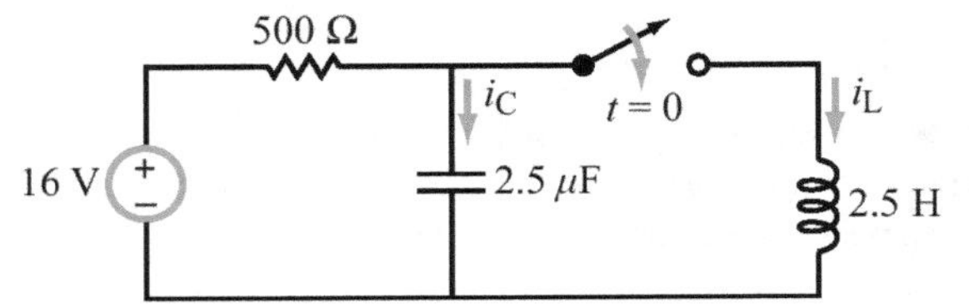

Figure P6.40: Circuit for Problems 6.40 and 6.41.

6.41 Determine $i_C(t)$ in the circuit of Fig. 6.40 and plot its waveform for $t \geq 0$.

6.42 Determine $i_L(t)$ in the circuit of Fig. 6.42 and plot its waveform for $t \geq 0$.

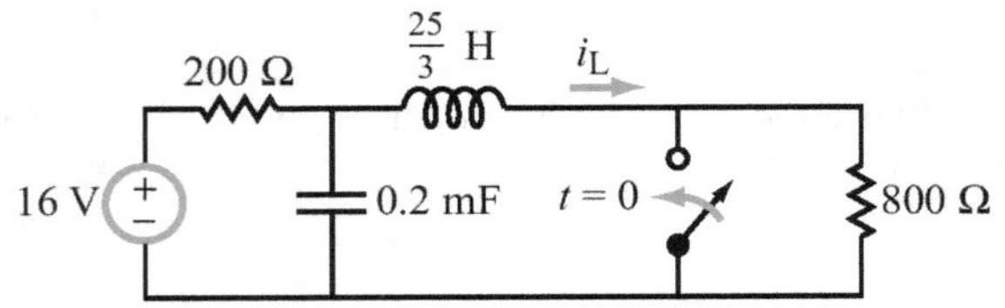

Figure P6.42: Circuit for Problem 6.42.

*__6.43__ For the circuit of Fig. 6.43, determine:

(a) $i_L(t)$ for $t \geq 0$

(b) The amount of energy stored in the capacitor at $t = \infty$.

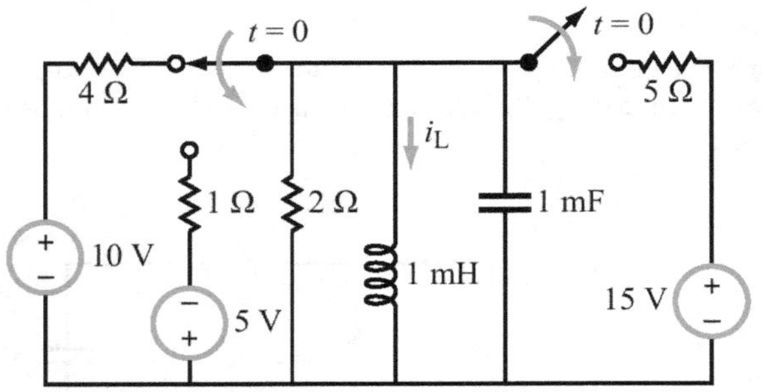

Figure P6.43: Circuit for Problem 6.43.

6.44 Assume that the circuit in Fig. P6.44 had been in that state for a long time prior to $t = 0$.

(a) Determine the value of C for which $i_L(t)$ exhibits the fastest smooth response.

(b) Use the value of C found in part (a) to find $i_L(t)$ for $t \geq 0$.

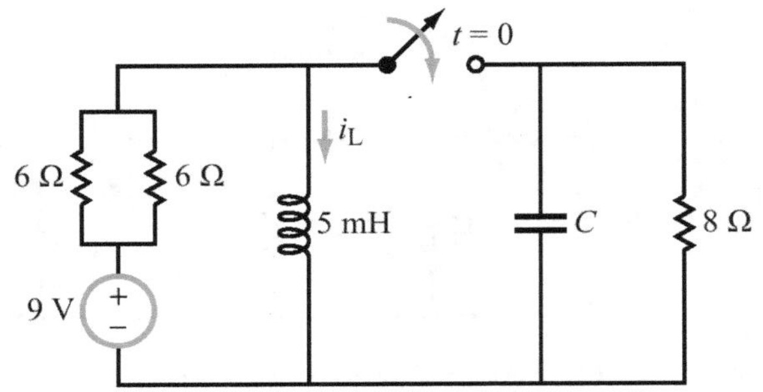

Figure P6.44: Circuit for Problem 6.44.

6.45 For the circuit in Fig. P6.45:

(a) Determine $v_C(t)$ for $t \geq 0$.

(b) Determine the time at which the inductor has maximum energy stored in it and calculate the amount of that maximum energy.

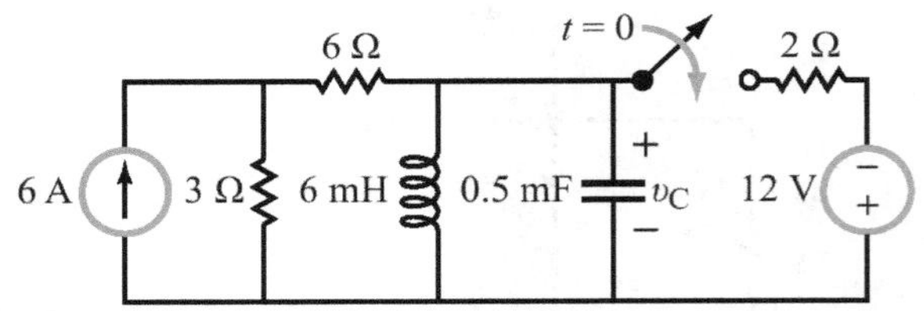

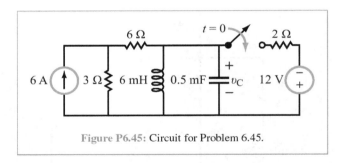

Figure P6.45: Circuit for Problem 6.45.

*__6.46__ In the circuit in Fig. P6.46, $v_s = 20$ V.

(a) Determine $i_L(t)$ for $t \geq 0$.

(b) If the source is changed to $v_s(t) = e^{-2t}\,u(t)$, can you still use the solution method in part (a) to find $i_L(t)$? If not, why not?

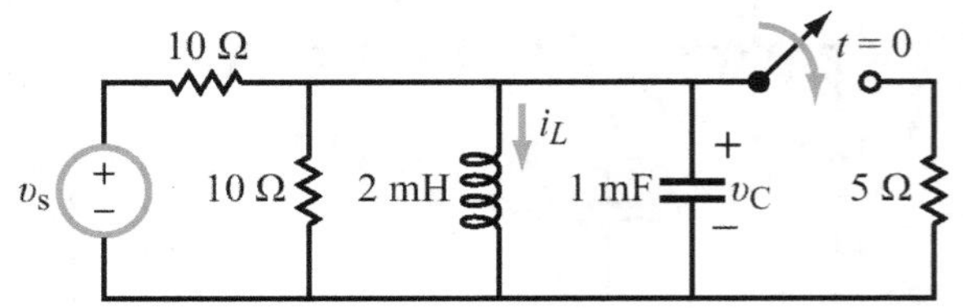

Figure P6.46: Circuit for Problem 6.46.

Section 6-8: General Solution

6.47 The switch in the circuit of Fig. P6.47 was closed at $t = 0$ and then reopened at $t = 1$ ms. Determine $i_L(t)$ and $v_C(t)$ for $t \geq 0$. Assume the capacitor had no charge prior to $t = 0$.

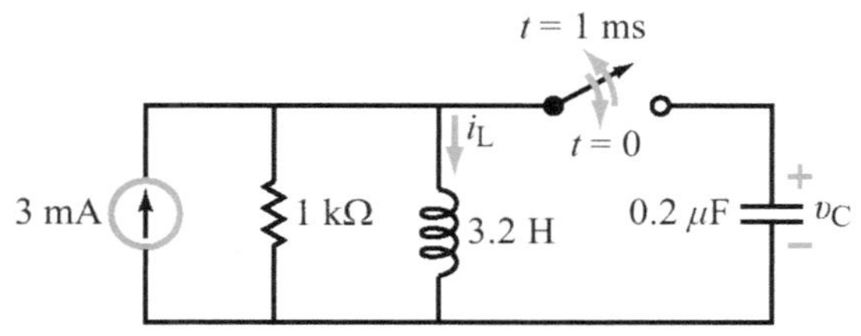

Figure P6.47: Circuit for Problem 6.47.

*6.48 After closing the switch in the circuit of Fig. P6.48 at $t = 0$, it was reopened at $t = 1$ ms. Determine $i_C(t)$ and plot its waveform for $t \geq 0$. Assume no energy was stored in either L or C prior to $t = 0$.

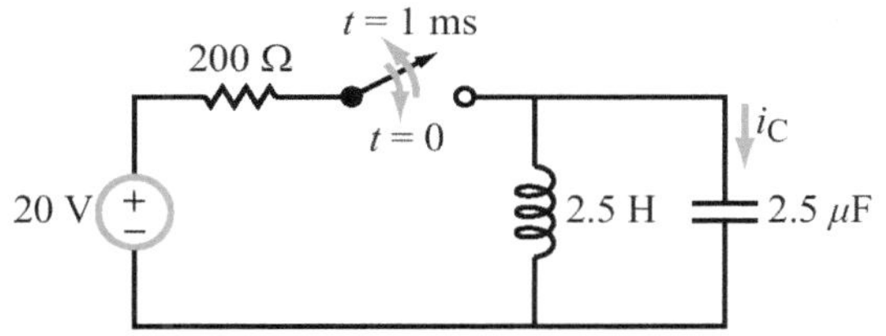

Figure P6.48: Circuit for Problem 6.48.

6.49 Determine the current responses $i_L(t)$ and $i_C(t)$ to a rectangular-current pulse as shown in Fig. P6.49, given that $I_s = 10$ mA and $R = 499.99\ \Omega$. Plot the waveforms of $i_L(t)$, $i_C(t)$, and $i_s(t)$ on the same scale.

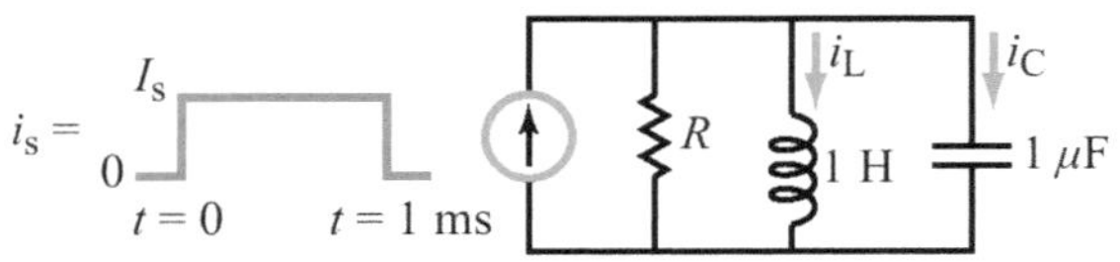

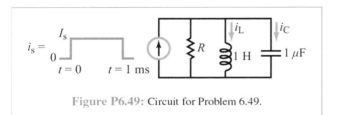

Figure P6.49: Circuit for Problem 6.49.

*6.50 The voltage in a certain circuit is described by the differential equation

$$v'' + 5v' + 6v = 144 \qquad (\text{for } t \geq 0).$$

Determine $v(t)$ for $t \geq 0$ given that $v(0) = 16$ V and $v'(0) = 9.6$ V/s.

6.51 The current in a certain circuit is described by the differential equation

$$i'' + \sqrt{24}\, i' + 6i = 18 \qquad (\text{for } t \geq 0).$$

Determine $i(t)$ for $t \geq 0$ given that $i(0) = -2$ A and $i'(0) = 8\sqrt{6}$ A/s.

6.52 For the circuit in Fig. P6.52:

(a) Determine $i_L(0)$ and $v_L(0)$.

(b) Derive the differential equation for $i_L(t)$ for $t \geq 0$.

*(c) Solve the differential equation and obtain an explicit expression for $i_L(t)$, given that $V_s = 12$ V, $R_s = 3\ \Omega$, $R_1 = 0.5\ \Omega$, $R_2 = 1\ \Omega$, $L = 2$ H, and $C = 2$ F.

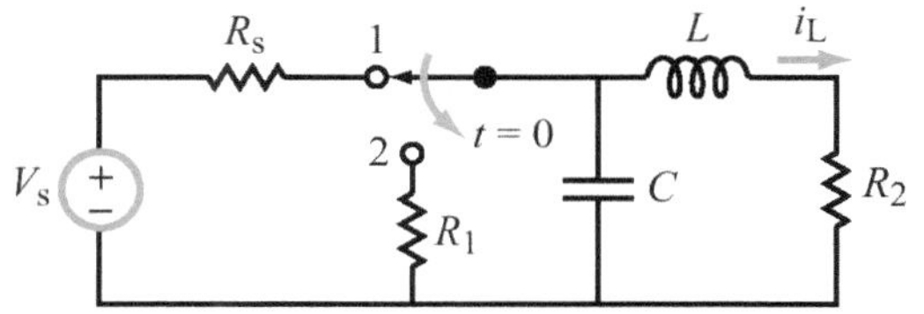

Figure P6.52: Circuit for Problem 6.52.

6.53 Develop a differential equation for $i_L(t)$ in the circuit of Fig. P6.53. Solve it to determine $i_L(t)$ for $t \geq 0$ subject to the following element values: $I_s = 36\ \mu$A, $R_s = 100$ kΩ, $R = 100\ \Omega$, $L = 10$ mH, and $C = 10\ \mu$F.

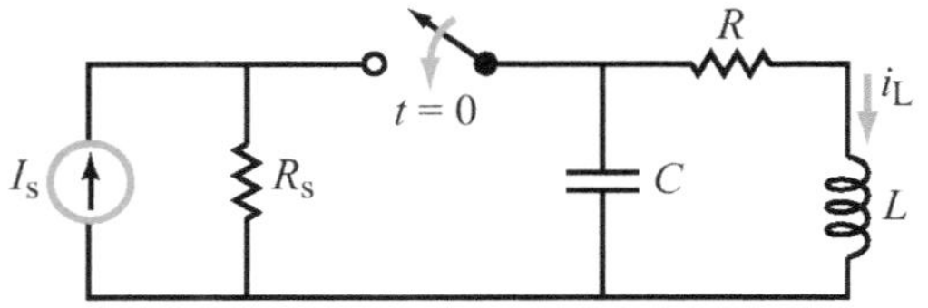

Figure P6.53: Circuit for Problem 6.53.

*6.54 Develop a differential equation for v_C in the circuit of Fig. P6.54. Solve it to determine $v_C(t)$ for $t \geq 0$. The element values are $I_s = 0.2$ A, $R_s = 30\ \Omega$, $R_1 = 10\ \Omega$, $R_2 = 20\ \Omega$, $R_3 = 20\ \Omega$, $L = 4$ H, and $C = 5$ mF.

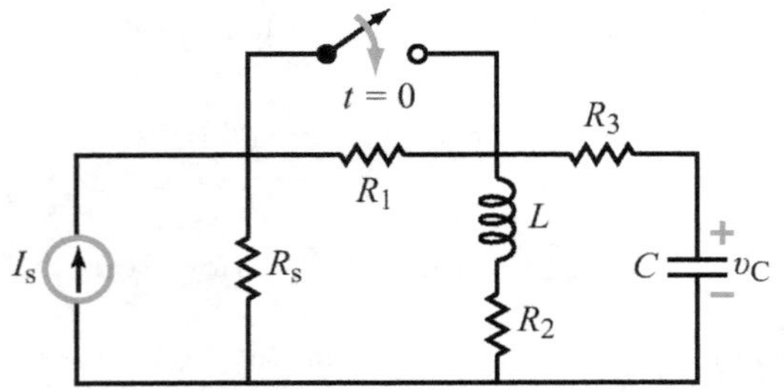

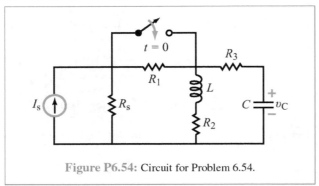

Figure P6.54: Circuit for Problem 6.54.

6.55 Develop a differential equation for i_L in the circuit of Fig. P6.55. Solve it for $t \geq 0$. The switch was closed at $t = 0$ and then reopened at $t = 0.5$ s, and the element values are $V_s = 18$ V, $R_s = 1\ \Omega$, $R_1 = 5\ \Omega$, $R_2 = 2\ \Omega$, $L = 2$ H, and $C = \frac{1}{17}$ F.

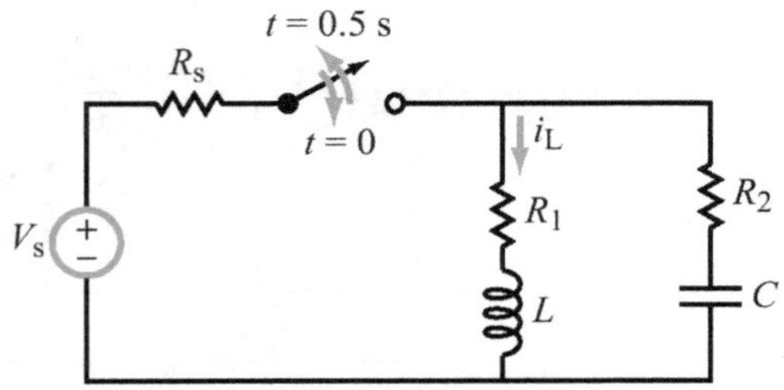

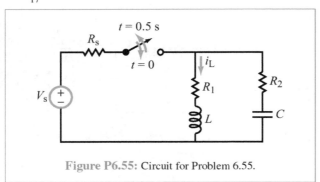

Figure P6.55: Circuit for Problem 6.55.

*6.56 Determine i_2 in the circuit of Fig. P6.56 for $t \geq 0$, given that $V_s = 10$ V, $R_s = 0.1$ MΩ, $R = 1$ MΩ, $C_1 = 1\ \mu$F, and $C_2 = 2\ \mu$F.

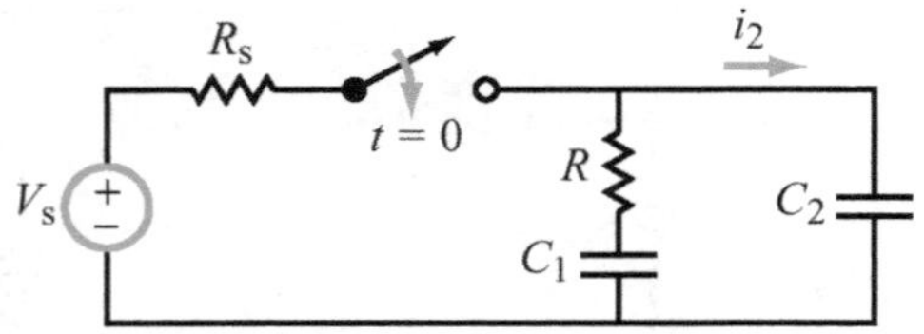

Figure P6.56: Circuit for Problems 6.56 and 6.57.

6.57 Repeat Problem 6.56, but this time assume that the switch had been closed for a long time and then opened at $t = 0$.

*6.58 The op-amp circuit shown in Fig. P6.58 is called a multiple-feedback bandpass filter. If $v_{in} = A\ u(t)$, determine $v_{out}(t)$ for $t \geq 0$ for $A = 6$ V, $R_1 = 10$ kΩ, $R_2 = 5$ kΩ, $R_f = 50$ kΩ, and $C_1 = C_2 = 1\ \mu$F.

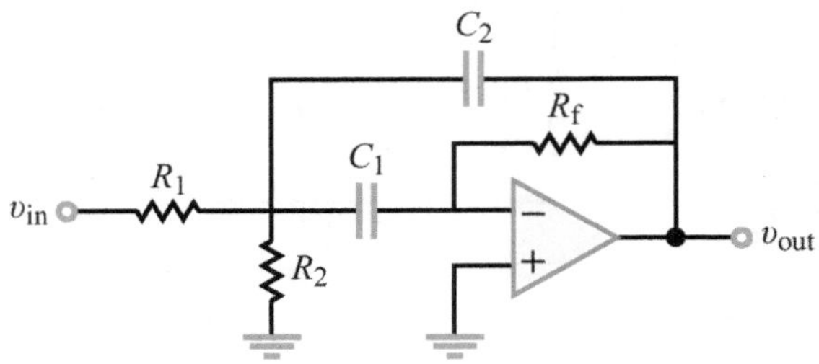

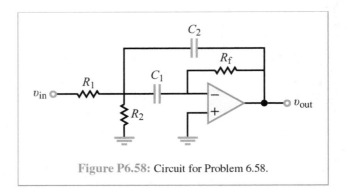

Figure P6.58: Circuit for Problem 6.58.

6.59 The op-amp circuit shown in Fig. P6.59 is called a two-pole low-pass filter. If $v_{in} = A\ u(t)$, determine $v_{out}(t)$ for $t \geq 0$ for $A = 2$ V, $R_1 = 5$ kΩ, $R_2 = 10$ kΩ, $R_3 = 12$ kΩ, $R_4 = 20$ kΩ, $C_1 = 100\ \mu$F, and $C_2 = 200\ \mu$F.

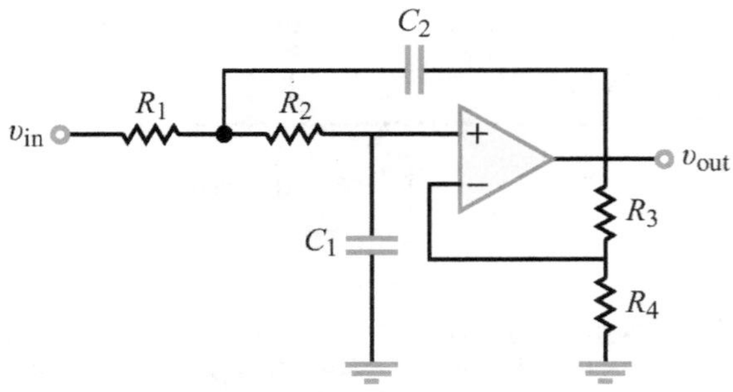

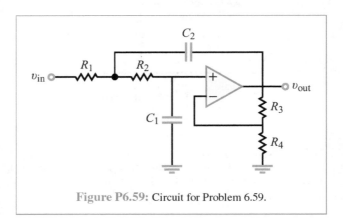

Figure P6.59: Circuit for Problem 6.59.

Section 6-9: Multisim

6.60 Using Multisim, draw a series RLC circuit with $V_s = 24$ V, $R = 12\ \Omega$, $L = 300$ mH, and $C = 10$ mF. Use the Transient Analysis tool to obtain a plot of $v_C(t)$ for $0 < t < 0.2$ s.

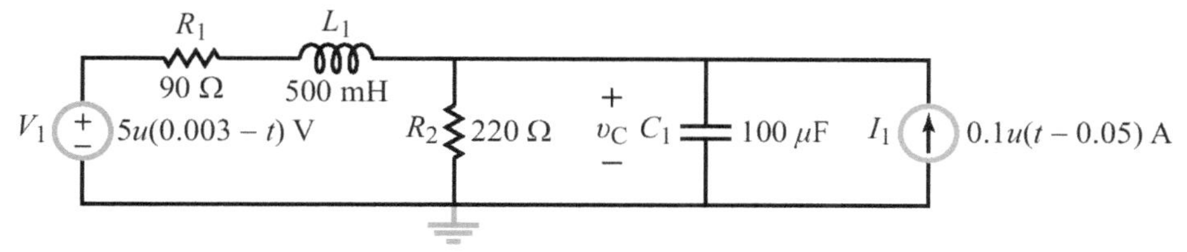

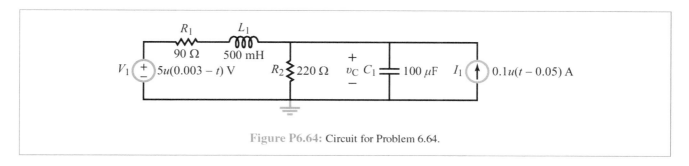

Figure P6.64: Circuit for Problem 6.64.

6.61 Using Multisim, draw the circuit in Fig. E6.4 of Exercise 6.4. Use the Transient Analysis tool to obtain a plot of $i_C(t)$ for $0 < t < 1$ s.

6.62 Using Multisim, draw the circuit in Fig. E6.4 of Exercise 6.4. Use the Transient Analysis tool to obtain three plots of $i_C(t)$ for (a) an underdamped response, (b) a critically damped response, and (c) an overdamped response. To obtain the three desired responses, adjust the value of the 20 Ω resistor as needed.

6.63 Adjust the values of the source and the components in Fig. 6-23 such that the RLC circuit is excited and oscillates at a frequency of 1 MHz and the oscillation envelope decays to 10 percent of its initial value after 12 oscillations once the circuit is switched to "listen" mode.

6.64 Build the circuit shown in Fig. P6.64 in Multisim and then plot the voltage $v_C(t)$ from 0 to 200 ms using Transient Analysis.

6.65 Build the active second-order circuit shown in Fig. P6.65, plot the signal v_{out} from 0 to 5 ms, and note how long it takes before the amplitude of the oscillations drops below 1 V. Change the value of R_2 to 100 kΩ and repeat the simulation. (You may need to readjust your timescale.)

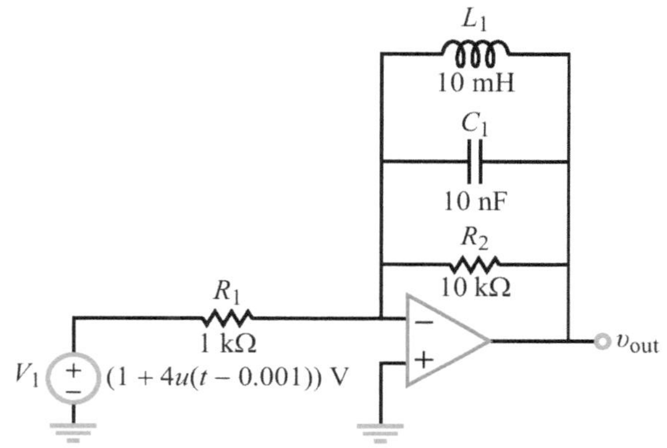

Figure P6.65: Circuit for Problem 6.65.

Potpourri Questions

6.66 How are transducers and actuators related?

6.67 How does a capacitive accelerometer work?

6.68 What are the differences between a passive RFID tag and an active RFID tag?

6.69 RFID tags operate at several frequency bands. How does the data speed change as the frequency is increased from the LF band to the microwave band?

6.70 Describe how electrical stimulation is used in a cochlear implant, in motor prostheses, and in reducing tremors in patients with Parkinson's disease.

Integrative Problems: Analytical / Multisim / myDAQ

To master the material in this chapter, solve the following problems using three complementary approaches: (a) analytically, (b) with Multisim, and (c) by constructing the circuit and using the myDAQ interface unit to measure quantities of interest via your computer. [myDAQ tutorials and videos are available on CAD.]

m6.1 Initial and Final Conditions: The SPST switch in the circuit of Fig. m6.1 opens at $t = 0$, after it had been closed for a long time. Draw the circuit configurations that appropriately represent the state of the circuit at $t = 0^-$, $t = 0$, and $t = \infty$ and use them to determine:

(a) $v_C(0)$, $i_C(0)$ and $v_C(\infty)$, and

(b) $i_L(0)$, $v_L(0)$ and $i_L(\infty)$.

Component values are: $R_1 = 680$ Ω, $R_2 = 100$ Ω, $R_3 = 100$ Ω, switch resistance $R_{sw} = 10$ Ω, wire resistance $R_w = 10$ Ω, $L = 3.3$ mH, $C = 0.1$ μF, and $V_s = 4.7$ V.

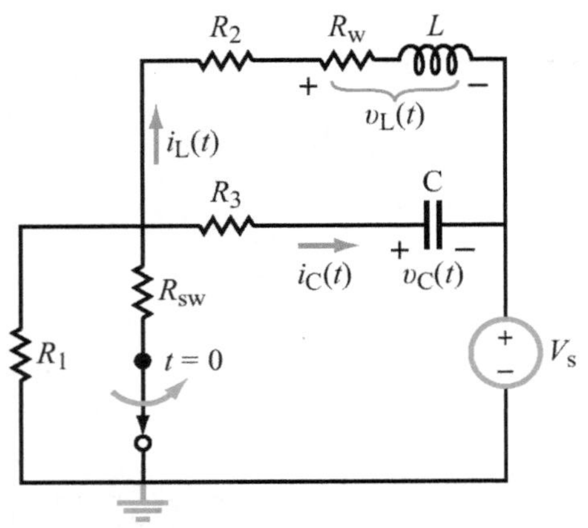

Figure m6.1 Circuit for Problem m6.1.

m6.2 Natural Response of the Series RLC Circuit: The SPST switch in the circuit of Fig. m6.2 opens at $t = 0$, after it had been closed for a long time.

(a) Determine $v_C(t)$ for $t \geq 0$.

(b) Plot $v_C(t)$ over the time range $0 \leq t \leq 1$ ms with a plotting tool such as MathScript or MATLAB.

(c) Determine the following numerical values; use either the equation $v_C(t)$ or take cursor measurements from the plot you created in the previous step:

- Initial voltage v_C,
- $v_C(0)$,
- Maximum value of v_C,
- Damped oscillation frequency $f_d = \omega_d/2\pi$ in Hz, and
- Damping coefficient α.

Use these component values: $R_1 = 220\ \Omega$, $R_2 = 330\ \Omega$, $L = 33$ mH, $C = 0.01\ \mu$F, and $V_{src} = 3.0$ V.

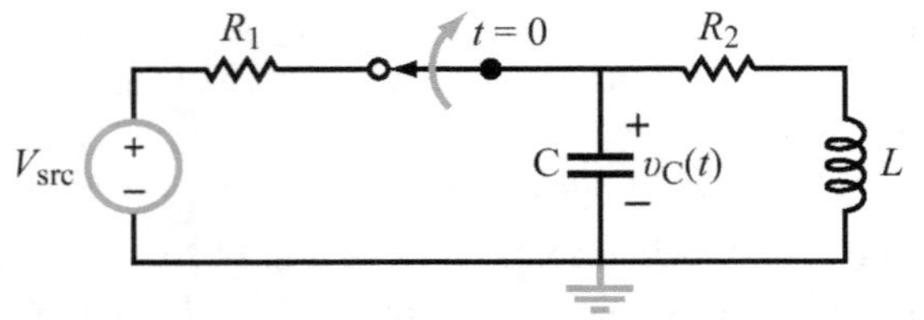

Figure m6.2 Circuit for Problem m6.2.

m6.3 Three-Resistor Circuit: Determine $v(t)$ of the circuit shown in Fig. m6.3 for $t \geq 0$, given that the switch is opened at $t = 0$, after having been closed for a long time. Use the following component values: $V_{src} = 8$ V, $R_1 = 470\ \Omega$, $R_2 = 100\ \Omega$, $R_w = 90\ \Omega$, $C = 1.0\ \mu$F, and $L = 33$ mH.

(a) Plot $v(t)$ from 0 to 5 ms using a tool such as MathScript or MATLAB. Include a hard copy of the script used to create the plot.

(b) Determine the following values for $v(t)$:

- Initial value $v(0)$,
- Final value of $v(t)$,
- Minimum value of $v(t)$, and
- Time to reach the minimum value of $v(t)$.

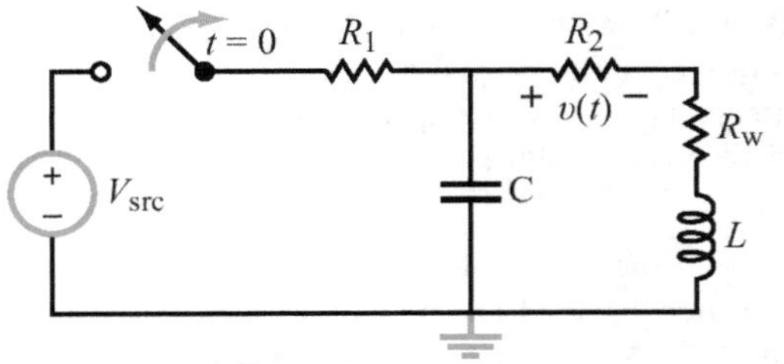

Figure m6.3 Circuit for Problem m6.3.

CHAPTER 7

ac Analysis

Contents

Objectives

Learn to:

- Transform time-varying sinusoidal functions to the phasor domain and vice versa.
- Analyze any linear circuit in the phasor domain.
- Determine the impedance of any passive element, or the combination of elements connected in series or in parallel.
- Perform source transformations, current division and voltage division, and determine Thévenin and Norton equivalent circuits, all in the phasor domain.
- Apply nodal analysis, mesh analysis, and other analysis techniques, all in the phasor domain.
- Design simple RC phase-shift circuits.
- Design a dc power-supply circuit.
- Use Multisim to analyze ac circuits

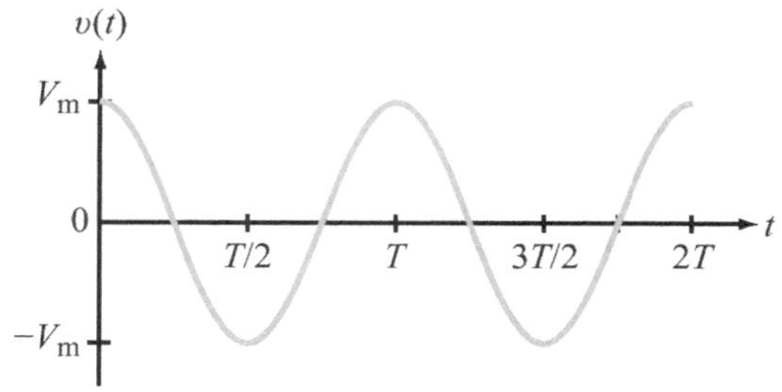

Electric circuits whose currents and voltages vary sinusoidally with time—called *alternating current (ac)* circuits—are at the heart of most analog applications. This chapter and the next four are dedicated to ac circuits.

Overview

From solar illumination to radio and cell-phone transmissions, we are surrounded by electromagnetic (EM) waves all of the time. *EM waves are composed of sinusoidally varying electric and magnetic fields*, and the fundamental parameter that distinguishes one EM wave from another is the wave's frequency f (or equivalently, its wavelength $\lambda = c/f$, where $c = 3 \times 10^8$ m/s is the velocity of light in a vacuum). The frequency of red light, for example, is 4.3×10^{14} Hz, and one of the frequencies assigned to cell-phone traffic is 1,900 MHz (1.9×10^9 Hz). Both are EM waves—and so are X-rays, infrared waves, and microwaves—but they oscillate sinusoidally at different frequencies and interact with matter differently (see Technology Brief 20 on the Electromagnetic Spectrum).

The term "*ac*" (*alternating current*) is associated with electric circuits whose currents and voltages vary sinusoidally with time, just like EM waves. In fact, ac circuits and EM waves are not only similar, but they also are connected directly: when flowing in a conductor, an ac current with an oscillation frequency f radiates EM waves of the same frequency. The radiated waves can couple signals from one part of the circuit to another through the air space they share or the insulating regions between them. The coupling may serve as an intentional means of communication, as in the case of *radio frequency identification* (RFID) circuits, or it may introduce unwelcome signals that interfere with the intended operation of the circuit. Mitigation of such undesirable consequences is part of a subdiscipline of electrical engineering called *electromagnetic compatibility*.

This and the next four chapters will be devoted to the study of ac circuits, which are far more prevalent than dc circuits and offer a much broader array of practical applications. In our study, we will assume that all currents and voltages are confined to the discrete elements in the circuit and to the connections between them, allowing us to ignore EM-compatibility issues altogether.

In Chapter 12, we will learn how to use the Laplace transform technique to determine the response of a circuit to any source with any realistic waveform, including ac sources. In general, the solution consists of two components, a transient component—in response to sudden changes, such as the opening or closing of switches—and a steady state component that mimics the time variation of the source. If (a) all the sources in the circuit are ac sources and (b) our interest is in only the steady state component (because the transient component decays to approximately zero within a short time after connecting the circuit to the ac source), we can use the *phasor domain technique* (instead of the Laplace transform technique) to analyze the circuit, because it is mathematically simpler and easier to implement. In fact, the phasor domain technique is a special case of the Laplace transform technique.

> ▶ The *phasor domain technique*—also known as the *frequency domain technique*—applies to ac circuits only, and provides a solution of only the steady state component of the total solution. ◀

7-1 Sinusoidal Signals

The voltage between two points in a circuit (or the current flowing through a branch) is said to have a *sinusoidal waveform* if its time variation is given by a sinusoidal function. *The term sinusoid includes both sine and cosine functions.* For example, the expression

$$\upsilon(t) = V_{\mathrm{m}} \cos \omega t \qquad (7.1)$$

describes a sinusoidal voltage $\upsilon(t)$ that has an *amplitude* V_{m} and an *angular frequency* ω. The amplitude defines the maximum or *peak value* that $\upsilon(t)$ can reach, and $-V_{\mathrm{m}}$ is its lowest negative value. The *argument* of the cosine function, ωt, is measured either in degrees or in radians, with

$$\pi \text{ (rad)} \approx 3.1416 \text{ (rad)} = 180^\circ. \qquad (7.2)$$

Since ωt is measured in radians, the unit for ω is (rad/s). Figure 7-1(a) displays a plot of $\upsilon(t)$ as a function of ωt. The familiar cosine function starts at its maximum value (at $\omega t = 0$), decreases to zero at $\omega t = \pi/2$, goes into negative territory for half of a cycle, and completes its first cycle at $\omega t = 2\pi$. Occasionally, we may want to display a sinusoidal signal as a function of t, instead of ωt. We note that the angular frequency ω is related to the *oscillation frequency* (or simply the *frequency*) f of the signal by

$$\omega = 2\pi f \qquad \text{(rad/s)}, \qquad (7.3)$$

with f measured in hertz (Hz), which is equivalent to cycles/second. A sinusoidal voltage with a frequency of 100 Hz makes 100 oscillations in 1 s, each of duration $1/100 = 0.01$ s. The duration of a cycle is its *period* T. Thus,

$$T = \frac{1}{f} \qquad \text{(s)}. \qquad (7.4)$$

By combining Eqs. (7.1), (7.3), and (7.4), $\upsilon(t)$ can be rewritten as

$$\upsilon(t) = V_{\mathrm{m}} \cos \frac{2\pi t}{T}, \qquad (7.5)$$

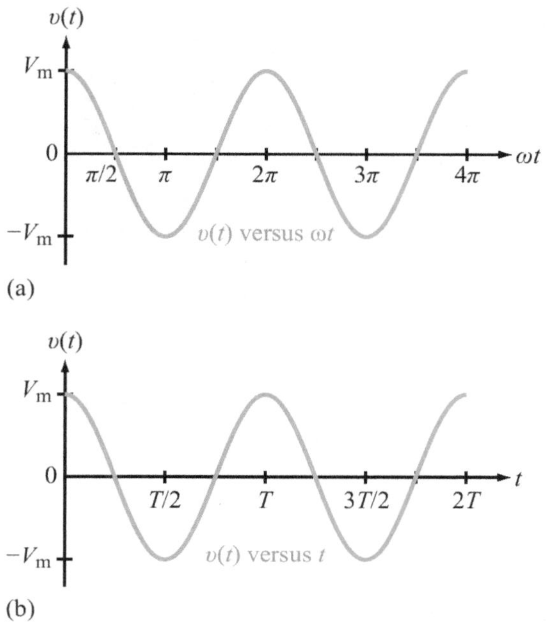

Figure 7-1: The function $\upsilon(t) = V_m \cos \omega t$ plotted as a function of (a) ωt and (b) t.

which is displayed in Fig. 7-1(b) as a function of t. We observe that the cyclical pattern of the waveform repeats itself every T seconds. That is,

$$\upsilon(t) = \upsilon(t + nT) \tag{7.6}$$

for any integer value of n.

Sinusoidal waveforms can be expressed in terms of either sine or cosine functions.

> ▶ To avoid confusion, we adopt the cosine form as our reference standard throughout this and followup chapters. ◀

This means that we will always express voltages and currents in terms of cosine functions, so *if a voltage (or current) waveform is given in terms of a sine function, we should first convert it to a cosine form with a positive amplitude before proceeding with our circuit analysis.* Conversion from sine to cosine form is realized through the application of Eq. (7.7a) of Table 7-1. For example,

$$\begin{aligned} i(t) &= 6\sin(\omega t + 30^\circ) \\ &= 6\cos(\omega t + 30^\circ - 90^\circ) = 6\cos(\omega t - 60^\circ). \end{aligned} \tag{7.8}$$

Table 7-1: Useful trigonometric identities (additional relations are given in Appendix D).

$\sin x = \pm\cos(x \mp 90^\circ)$	(7.7a)
$\cos x = \pm\sin(x \pm 90^\circ)$	(7.7b)
$\sin x = -\sin(x \pm 180^\circ)$	(7.7c)
$\cos x = -\cos(x \pm 180^\circ)$	(7.7d)
$\sin(-x) = -\sin x$	(7.7e)
$\cos(-x) = \cos x$	(7.7f)
$\sin(x \pm y) = \sin x \cos y \pm \cos x \sin y$	(7.7g)
$\cos(x \pm y) = \cos x \cos y \mp \sin x \sin y$	(7.7h)
$2\sin x \sin y = \cos(x - y) - \cos(x + y)$	(7.7i)
$2\sin x \cos y = \sin(x + y) + \sin(x - y)$	(7.7j)
$2\cos x \cos y = \cos(x + y) + \cos(x - y)$	(7.7k)

In addition to ωt, the argument of the cosine function contains a constant angle of -60°. A ***cosine-referenced*** sinusoidal function generally takes the form

$$\upsilon(t) = V_m \cos(\omega t + \phi), \tag{7.9}$$

where ϕ is called its ***phase angle***. For $i(t)$ of Eq. (7.8), $\phi = -60^\circ$.

The angle ϕ may assume any positive or negative value, but we usually add or subtract multiples of 2π radians (or equivalently, multiples of 360°) so that the remainder is between -180° and $+180^\circ$. The magnitude and sign (+ or −) of ϕ determine, respectively, by how much and in what direction the waveform of $\upsilon(t)$ is shifted along the time axis, relative to the reference waveform corresponding to $\upsilon(t)$ with $\phi = 0$. Figure 7-2 displays three waveforms:

$$\upsilon_1(t) = V_m \cos\left(\frac{2\pi t}{T} - \frac{\pi}{4}\right) \quad \text{(lags by } \pi/4), \tag{7.10a}$$

$$\upsilon_2(t) = V_m \cos\frac{2\pi t}{T} \quad \textbf{(reference waveform with } \phi = 0), \tag{7.10b}$$

$$\upsilon_3(t) = V_m \cos\left(\frac{2\pi t}{T} + \frac{\pi}{4}\right) \quad \text{(leads by } \pi/4). \tag{7.10c}$$

We observe that waveform $\upsilon_3(t)$, which is shifted backwards in time relative to the reference waveform $\upsilon_2(t)$, attains its peak value before $\upsilon_2(t)$ does. Consequently, waveform $\upsilon_3(t)$ is said to ***lead*** $\upsilon_2(t)$ by a ***phase lead*** of $\pi/4$. Similarly, waveform $\upsilon_1(t)$ ***lags*** $\upsilon_2(t)$ by a ***phase lag*** of $\pi/4$. A cosine function with a negative phase angle ϕ takes longer to reach a specified

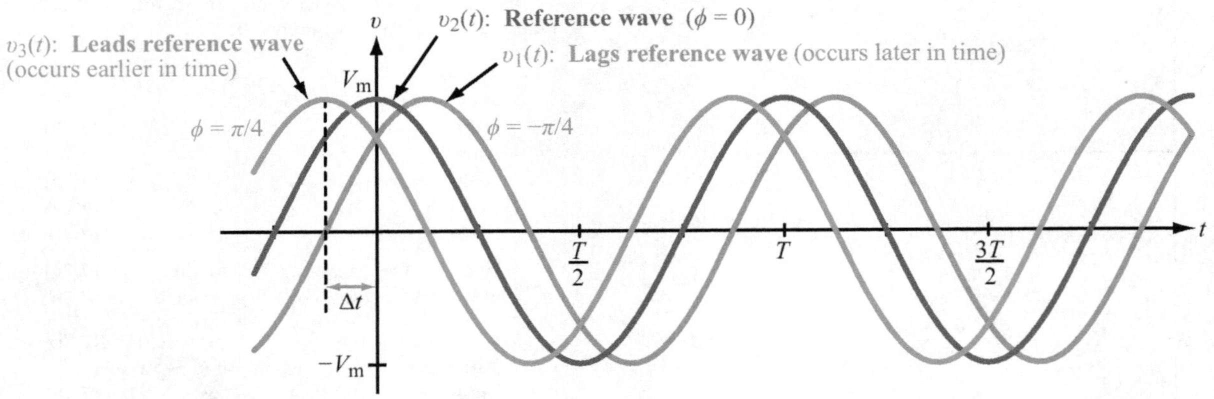

Figure 7-2: Plots of $\upsilon(t) = V_m \cos[(2\pi t/T) + \phi]$ for three different values of ϕ.

reference level (such as the peak value) than it takes the zero-phase angle function to reach that level, signifying a phase lag. When ϕ is positive, it signifies a phase lead. A phase angle of 2π corresponds to a *time shift* along the time axis equal to one full period T. Proportionately, a phase angle of ϕ (in radians) corresponds to a *time shift* Δt given by

$$\Delta t = \left(\frac{\phi}{2\pi}\right) T. \qquad (7.11)$$

We generalize our discussion of phase lead and lag by stating that:

► Given two sinusoidal functions with the same angular frequency ω, and both expressed in standard cosine form as

$$\upsilon_1(t) = V_1 \cos(\omega + \phi_1)$$

and

$$\upsilon_2(t) = V_2 \cos(\omega + \phi_2),$$

the relevant terminology is:

υ_2 **leads** υ_1 **by** $(\phi_2 - \phi_1)$,

υ_2 **lags** υ_1 **by** $(\phi_1 - \phi_2)$,

υ_1 and υ_2 are **in phase** if $\phi_2 = \phi_1$,

υ_1 and υ_2 are **in phase-opposition** if $\phi_2 = \phi_1 \pm 180°$.
(out of phase)

◄

Example 7-1: Voltage Waveform

A sampling oscilloscope is used to measure a voltage signal $\upsilon(t)$. The measurements reveal that $\upsilon(t)$ is periodic with an amplitude of 10 V, its maxima are separated by 20 ms, and one of its maxima occurs at $t = 1.2$ ms. Determine the functional form of $\upsilon(t)$.

Solution: Given that $V_m = 10$ V and

$$T = 20 \text{ ms} = 2 \times 10^{-2} \text{ s},$$

$\upsilon(t)$ is given by

$$\upsilon(t) = 10\cos\left(\frac{2\pi t}{2 \times 10^{-2}} + \phi\right) = 10\cos(100\pi t + \phi) \text{ V}.$$

Application of $\upsilon(t = 1.2 \text{ ms}) = 10$ V gives

$$10 = 10\cos(100\pi \times 1.2 \times 10^{-3} + \phi),$$

which requires the argument of the cosine to be a multiple of 2π,

$$0.12\pi + \phi = 2n\pi, \qquad n = 0, \pm 1, \pm 2, \ldots$$

The smallest value of ϕ in the range $[-180°,\ 180°]$ that satisfies the preceding equation corresponds to $n = 0$, and is given by

$$\phi = -0.12\pi = -21.6°.$$

Hence,

$$\upsilon(t) = 10\cos(100\pi t - 21.6°) \text{ V}.$$

Example 7-2: Phase Lead / Lag

Given the current waveforms

$$i_1(t) = -8\cos(\omega t - 30^\circ) \text{ A}$$

and

$$i_2(t) = 12\sin(\omega t + 45^\circ) \text{ A},$$

does $i_1(t)$ lead $i_2(t)$, or the other way around, and by how much?

Solution: Standard cosine format requires that the sinusoidal functions be cosines and that the amplitudes have positive values. Application of Eq. (7.7d) of Table 7-1 allows us to remove the negative sign preceding the amplitude of $i_1(t)$,

$$\begin{aligned} i_1(t) = -8\cos(\omega t - 30^\circ) &= 8\cos(\omega t - 30^\circ + 180^\circ) \\ &= 8\cos(\omega t + 150^\circ) \text{ A}. \end{aligned}$$

Application of Eq. (7.7a) to $i_2(t)$ leads to

$$\begin{aligned} i_2(t) = 12\sin(\omega t + 45^\circ) &= 12\cos(\omega t + 45^\circ - 90^\circ) \\ &= 12\cos(\omega t - 45^\circ) \text{ A}. \end{aligned}$$

Hence, $\phi_1 = 150^\circ$, $\phi_2 = -45^\circ$, and

$$\Delta\phi = \phi_2 - \phi_1 = -195^\circ.$$

The concept of phase lead/lag requires that $\Delta\phi$ be within the range $[-180^\circ, 180^\circ]$. Addition of 360° to $\Delta\phi$ converts it to 165°, which means that i_2 leads i_1 by 165°.

Concept Question 7-1: A sinusoidal waveform is characterized by three parameters. What are they, and what does each one of them specify? (See CAD)

Concept Question 7-2: Waveforms $\upsilon_1(t)$ and $\upsilon_2(t)$ have the same angular frequency, but $\upsilon_1(t)$ leads $\upsilon_2(t)$. Will the peak value of $\upsilon_1(t)$ occur sooner or later than that of $\upsilon_2(t)$? Explain. (See CAD)

Exercise 7-1: Provide an expression for a 100 V, 60 Hz voltage that exhibits a minimum at $t = 0$.

Answer: $\upsilon(t) = 100\cos(120\pi t + 180^\circ)$ V. (See CAD)

Exercise 7-2: Given two current waveforms:

$$i_1(t) = 3\cos\omega t$$

and

$$i_2(t) = 3\sin(\omega t + 36^\circ),$$

does $i_2(t)$ lead or lag $i_1(t)$, and by what phase angle?

Answer: $i_2(t)$ lags $i_1(t)$ by 54°. (See CAD)

7-2 Review of Complex Algebra

This section provides a review of complex algebra, in preparation for the introduction of the phasor domain technique in Section 7-3.

A *complex number* $\mathbf{z}$ may be written in the *rectangular form*

$$\mathbf{z} = x + jy, \tag{7.12}$$

where x and y are the *real* ($\Re\mathfrak{e}$) and *imaginary* ($\Im\mathfrak{m}$) parts of $\mathbf{z}$, respectively, and $j = \sqrt{-1}$. That is,

$$x = \Re\mathfrak{e}(\mathbf{z}), \qquad y = \Im\mathfrak{m}(\mathbf{z}). \tag{7.13}$$

Alternatively, $\mathbf{z}$ may be written in *polar form* as

$$\mathbf{z} = |\mathbf{z}|e^{j\theta} = |\mathbf{z}|\underline{/\theta} \tag{7.14}$$

where $|\mathbf{z}|$ is the magnitude of $\mathbf{z}$, θ is its phase angle, and the form $\underline{/\theta}$ is a useful shorthand representation commonly used in numerical calculations. A phase angle may be expressed in degrees, as in $\theta = 30^\circ$, or in radians, as in $\theta = 0.52$ rad.

By applying *Euler's identity*,

$$e^{j\theta} = \cos\theta + j\sin\theta, \tag{7.15}$$

we can convert $\mathbf{z}$ from polar form, as in Eq. (7.14), into rectangular form, as in Eq. (7.12)),

$$\mathbf{z} = |\mathbf{z}|e^{j\theta} = |\mathbf{z}|\cos\theta + j|\mathbf{z}|\sin\theta, \tag{7.16}$$

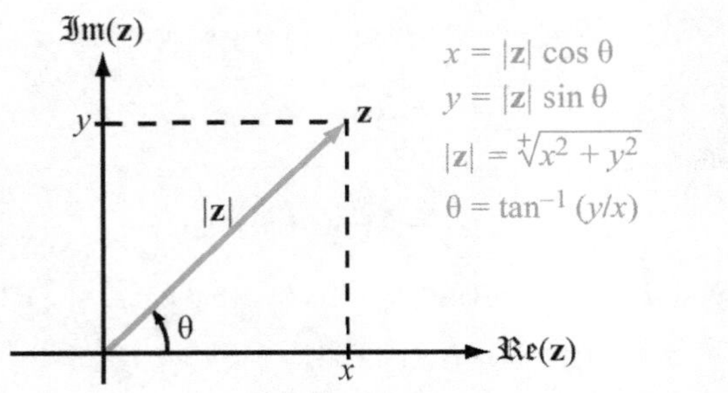

Figure 7-3: Relation between rectangular and polar representations of a complex number $\mathbf{z} = x + jy = |\mathbf{z}|e^{j\theta}$.

which leads to the relations

$$x = |\mathbf{z}|\cos\theta, \qquad y = |\mathbf{z}|\sin\theta, \tag{7.17}$$

$$|\mathbf{z}| = \sqrt{x^2 + y^2}, \qquad \theta = \tan^{-1}(y/x). \tag{7.18}$$

The two forms of $\mathbf{z}$ are illustrated graphically in Fig. 7-3. Because in the complex plane, a complex number assumes the form of a vector, it is represented by a bold letter in this book.

When using Eq. (7.18), care should be taken to ensure that θ is in the proper quadrant by noting the signs of x and y individually, as illustrated in Fig. 7-4. Complex numbers $\mathbf{z}_2$ and $\mathbf{z}_4$ point in opposite directions and their phase angles θ_2 and θ_4 differ by 180°, despite the fact that (y/x) has the same value in both cases. Also note that, since $|\mathbf{z}|$ is a positive quantity, only the positive root in Eq. (7.18) is applicable.

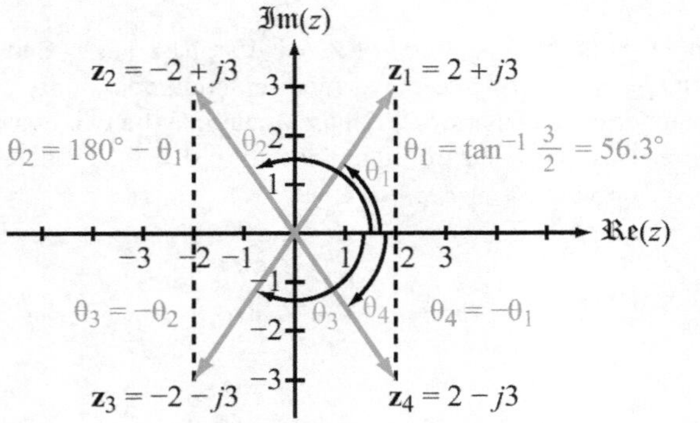

Figure 7-4: Complex numbers $\mathbf{z}_1$ to $\mathbf{z}_4$ have the same magnitude $|\mathbf{z}| = \sqrt{2^2 + 3^2} = 3.61$, but their polar angles depend on the polarities of their real and imaginary components.

The *complex conjugate* of $\mathbf{z}$, denoted with a star superscript (or asterisk), is obtained by replacing j (wherever it appears) with $-j$, so that

$$\mathbf{z}^* = (x + jy)^* = x - jy = |\mathbf{z}|e^{-j\theta} = |\mathbf{z}|\underline{/-\theta}. \tag{7.19}$$

The magnitude $|\mathbf{z}|$ is equal to the positive square root of the product of $\mathbf{z}$ and its complex conjugate:

$$|\mathbf{z}| = \sqrt{\mathbf{z}\,\mathbf{z}^*}. \tag{7.20}$$

We now highlight some of the properties of complex algebra that we will likely encounter in future sections.

Equality: If two complex numbers $\mathbf{z}_1$ and $\mathbf{z}_2$ are given by

$$\mathbf{z}_1 = x_1 + jy_1 = |\mathbf{z}_1|e^{j\theta_1}, \tag{7.21a}$$

$$\mathbf{z}_2 = x_2 + jy_2 = |\mathbf{z}_2|e^{j\theta_2}, \tag{7.21b}$$

then $\mathbf{z}_1 = \mathbf{z}_2$ if and only if (*iff*) $x_1 = x_2$ and $y_1 = y_2$ or, equivalently, $|\mathbf{z}_1| = |\mathbf{z}_2|$ and $\theta_1 = \theta_2$.

Addition:

$$\mathbf{z}_1 + \mathbf{z}_2 = (x_1 + x_2) + j(y_1 + y_2). \tag{7.22}$$

Multiplication:

$$\begin{aligned}\mathbf{z}_1\mathbf{z}_2 &= (x_1 + jy_1)(x_2 + jy_2)\\ &= (x_1x_2 - y_1y_2) + j(x_1y_2 + x_2y_1),\end{aligned} \tag{7.23a}$$

or

$$\begin{aligned}\mathbf{z}_1\mathbf{z}_2 &= |\mathbf{z}_1|e^{j\theta_1} \cdot |\mathbf{z}_2|e^{j\theta_2}\\ &= |\mathbf{z}_1||\mathbf{z}_2|e^{j(\theta_1+\theta_2)}\\ &= |\mathbf{z}_1||\mathbf{z}_2|[\cos(\theta_1 + \theta_2) + j\sin(\theta_1 + \theta_2)].\end{aligned} \tag{7.23b}$$

Division: For $\mathbf{z}_2 \neq 0$,

$$\begin{aligned}\frac{\mathbf{z}_1}{\mathbf{z}_2} = \frac{x_1 + jy_1}{x_2 + jy_2} &= \frac{(x_1 + jy_1)}{(x_2 + jy_2)} \cdot \frac{(x_2 - jy_2)}{(x_2 - jy_2)}\\ &= \frac{(x_1x_2 + y_1y_2) + j(x_2y_1 - x_1y_2)}{x_2^2 + y_2^2},\end{aligned} \tag{7.24a}$$

Table 7-2: **Properties of complex numbers.**

Euler's Identity: $e^{j\theta} = \cos\theta + j\sin\theta$	
$\sin\theta = \dfrac{e^{j\theta} - e^{-j\theta}}{2j}$	$\cos\theta = \dfrac{e^{j\theta} + e^{-j\theta}}{2}$
$\mathbf{z} = x + jy = \lvert\mathbf{z}\rvert e^{j\theta}$	$\mathbf{z}^* = x - jy = \lvert\mathbf{z}\rvert e^{-j\theta}$
$x = \mathfrak{Re}(\mathbf{z}) = \lvert\mathbf{z}\rvert\cos\theta$	$\lvert\mathbf{z}\rvert = \sqrt{\mathbf{z}\mathbf{z}^*} = \sqrt{x^2 + y^2}$
$y = \mathfrak{Im}(\mathbf{z}) = \lvert\mathbf{z}\rvert\sin\theta$	$\theta = \begin{cases} \tan^{-1}(y/x) & \text{if } x > 0, \\ \tan^{-1}(y/x) \pm \pi & \text{if } x < 0, \\ \pi/2 & \text{if } x = 0 \text{ and } y > 0, \\ -\pi/2 & \text{if } x = 0 \text{ and } y < 0. \end{cases}$
$\mathbf{z}^n = \lvert\mathbf{z}\rvert^n e^{jn\theta}$	$\mathbf{z}^{1/2} = \pm\lvert\mathbf{z}\rvert^{1/2} e^{j\theta/2}$
$\mathbf{z}_1 = x_1 + jy_1$	$\mathbf{z}_2 = x_2 + jy_2$
$\mathbf{z}_1 = \mathbf{z}_2$ iff $x_1 = x_2$ and $y_1 = y_2$	$\mathbf{z}_1 + \mathbf{z}_2 = (x_1 + x_2) + j(y_1 + y_2)$
$\mathbf{z}_1\mathbf{z}_2 = \lvert\mathbf{z}_1\rvert\lvert\mathbf{z}_2\rvert e^{j(\theta_1+\theta_2)}$	$\dfrac{\mathbf{z}_1}{\mathbf{z}_2} = \dfrac{\lvert\mathbf{z}_1\rvert}{\lvert\mathbf{z}_2\rvert} e^{j(\theta_1-\theta_2)}$
$-1 = e^{j\pi} = e^{-j\pi} = 1\angle{\pm 180^\circ}$	
$j = e^{j\pi/2} = 1\angle{90^\circ}$	$-j = e^{-j\pi/2} = 1\angle{-90^\circ}$
$\sqrt{j} = \pm e^{j\pi/4} = \pm\dfrac{(1+j)}{\sqrt{2}}$	$\sqrt{-j} = \pm e^{-j\pi/4} = \pm\dfrac{(1-j)}{\sqrt{2}}$

or

$$\frac{\mathbf{z}_1}{\mathbf{z}_2} = \frac{\lvert\mathbf{z}_1\rvert e^{j\theta_1}}{\lvert\mathbf{z}_2\rvert e^{j\theta_2}} = \frac{\lvert\mathbf{z}_1\rvert}{\lvert\mathbf{z}_2\rvert} e^{j(\theta_1-\theta_2)}$$
$$= \frac{\lvert\mathbf{z}_1\rvert}{\lvert\mathbf{z}_2\rvert}[\cos(\theta_1 - \theta_2) + j\sin(\theta_1 - \theta_2)]. \tag{7.24b}$$

Powers: For any positive integer n,

$$\mathbf{z}^n = (\lvert\mathbf{z}\rvert e^{j\theta})^n$$
$$= \lvert\mathbf{z}\rvert^n e^{jn\theta} = \lvert\mathbf{z}\rvert^n(\cos n\theta + j\sin n\theta), \tag{7.25}$$
$$\mathbf{z}^{1/2} = \pm\lvert\mathbf{z}\rvert^{1/2} e^{j\theta/2}$$
$$= \pm\lvert\mathbf{z}\rvert^{1/2}[\cos(\theta/2) + j\sin(\theta/2)]. \tag{7.26}$$

Useful Relations:

$$-1 = e^{j\pi} = e^{-j\pi} = 1\angle{180^\circ}, \tag{7.27a}$$
$$j = e^{j\pi/2} = 1\angle{90^\circ}, \tag{7.27b}$$
$$-j = -e^{j\pi/2} = e^{-j\pi/2} = 1\angle{-90^\circ}, \tag{7.27c}$$
$$\sqrt{j} = (e^{j\pi/2})^{1/2} = \pm e^{j\pi/4} = \frac{\pm(1+j)}{\sqrt{2}}, \tag{7.27d}$$
$$\sqrt{-j} = \pm e^{-j\pi/4} = \frac{\pm(1-j)}{\sqrt{2}}. \tag{7.27e}$$

For quick reference, the preceding properties of complex numbers are summarized in Table 7-2. Note that if a complex number is given by $(a + jb)$ and $b = 1$, it can be written either as $(a + j1)$ or simply as $(a + j)$. Thus, j is synonymous with $j1$.

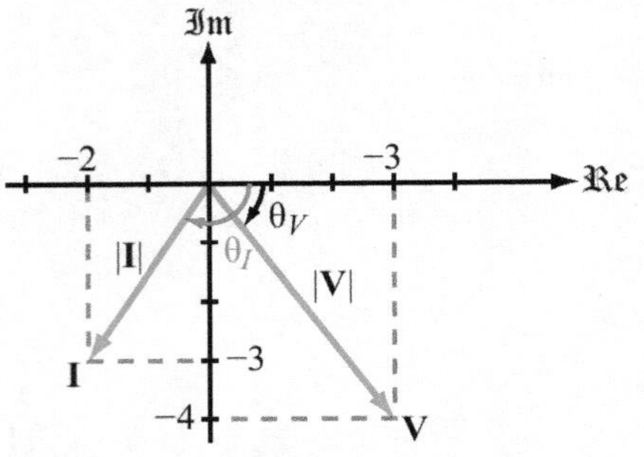

Figure 7-5: Complex numbers **V** and **I** in the complex plane (Example 7-3).

Example 7-3: Working with Complex Numbers

Given two complex numbers

$$\mathbf{V} = 3 - j4,$$
$$\mathbf{I} = -(2 + j3).$$

(a) Express **V** and **I** in polar form, and find (b) **VI**, (c) **VI***, (d) **V**/**I**, and (e) $\sqrt{\mathbf{I}}$.

Solution: (a)

$$|\mathbf{V}| = \sqrt{\mathbf{VV}^*} = \sqrt{(3 - j4)(3 + j4)} = \sqrt{9 + 16} = 5,$$
$$\theta_V = \tan^{-1}(-4/3) = -53.1°,$$
$$\mathbf{V} = |\mathbf{V}|e^{j\theta_V} = 5e^{-j53.1°} = \underline{5\angle -53.1°},$$
$$|\mathbf{I}| = \sqrt{2^2 + 3^2} = \sqrt{13} = 3.61.$$

Since $\mathbf{I} = (-2 - j3)$ is in the third quadrant in the complex plane (Fig. 7-5),

$$\theta_\mathbf{I} = -180° + \tan^{-1}\left(\tfrac{3}{2}\right) = -123.7°,$$
$$\mathbf{I} = \underline{3.61\angle -123.7°}.$$

Alternatively, whenever the real part of a complex number is negative, we can factor out a (-1) multiplier and then use Eq. (7.27a) to replace it with a phase angle of either $+180°$ or $-180°$, as needed. In the case of **I**, the process is as follows:

$$\mathbf{I} = -2 - j3 = -(2 + j3) = e^{\pm j180°} \cdot \sqrt{2^2 + 3^2}\; e^{j\tan^{-1}(3/2)}$$
$$= 3.61e^{j57.3°} e^{\pm j180°}.$$

Since our preference is to end up with a phase angle within the range between $-180°$ and $+180°$, we will choose $-180°$. Hence,

$$\mathbf{I} = 3.61e^{-j123.7°}.$$

(b)

$$\mathbf{VI} = (5\angle -53.1°)(3.61\angle -123.7°)$$
$$= (5 \times 3.61)\angle(-53.1° - 123.7°) = \underline{18.05\angle -176.8°}.$$

(c)

$$\mathbf{VI}^* = 5e^{-j53.1°} \times 3.61e^{j123.7°} = \underline{18.05e^{j70.6°}}.$$

(d)

$$\frac{\mathbf{V}}{\mathbf{I}} = \frac{5e^{-j53.1°}}{3.61e^{-j123.7°}} = \underline{1.39e^{j70.6°}}.$$

(e)

$$\sqrt{\mathbf{I}} = \sqrt{3.61e^{-j123.7°}}$$
$$= \pm\sqrt{3.61}\; e^{-j123.7°/2} = \underline{\pm 1.90e^{-j61.85°}}.$$

Concept Question 7-3: If **Z** is a complex number that lies in the first quadrant in the complex plane, its complex conjugate **Z*** will lie in which quadrant? (See CAD)

Concept Question 7-4: If two complex numbers have the same magnitude, are they necessarily equal to each other? (See CAD)

Exercise 7-3: Express the following complex functions in polar form:

$$\mathbf{z}_1 = (4 - j3)^2,$$
$$\mathbf{z}_2 = (4 - j3)^{1/2}.$$

Answer: $\mathbf{z}_1 = \underline{25\angle -73.7°}$, $\mathbf{z}_2 = \underline{\pm\sqrt{5}\angle -18.4°}$. (See CAD)

Exercise 7-4: Show that $\sqrt{2j} = \pm(1 + j)$. (See CAD)

Technology Brief 18
Touchscreens and Active Digitizers

Touchscreen is the common name given to a wide variety of technologies that allow computer displays to directly sense information from the user. In older systems, this usually meant the display could detect and pinpoint where a user touched the screen surface; newer systems can detect multiple touch locations as well as the associated touch pressures simultaneously, with very high resolution. This has led to a surge of applications in mobile computing, cell phones, personal digital assistants (PDA), and consumer appliances. Interactive touchscreens which detect multiple touches and interact with styli are now commonly used in phones, tablet computers and e-readers.

Numerous technologies have been developed since the invention of the electronic touch interface in 1971 by Samuel C. Hurst. Some of the earlier technologies were susceptible to dust, damage from repeat use, and poor transparency. These issues largely have been resolved over the years (even for older technologies) as experience and advanced material selection have led to improved devices. With the explosion of consumer interest in portable, interactive electronics, newer technologies have emerged that are more suitable for these applications. Figure TF18-1 summarizes the general categories of touchscreens in use today. Historically, touchscreens were manufactured separately from displays and added as an extra layer of the display. More recently, display companies have begun to manufacture sensing technology directly into the displays; some of the newer technologies reflect this.

Resistive

Resistive touchscreens are perhaps the simplest to understand. A thin, flexible membrane is separated from a plastic base by insulating spacers. Both the thin membrane and the plastic base are coated on the inside with a transparent conductive film (indium tin oxide (ITO) often is used). When the membrane is touched, the two conductive surfaces come into contact. Detector circuits at the edges of the screen can detect this change in resistance between the two membranes and pinpoint the location on the *X–Y* plane. Older designs of this type were susceptible to membrane damage (from repeated flexing) and suffered from poor transparency.

Capacitive

Older capacitive touchscreens employ a single thin, transparent conductive film (usually indium tin oxide (ITO)) on a plastic or glass base. The conductive film is coated with another thin, transparent insulator for protection. Since the human body stores charge, a finger tip moved close to the surface of the film effectively forms a capacitor where the film acts as one of the plates and the finger as the other. The protective coating and the air form the intervening dielectric insulator. This capacitive coupling changes how a current flowing across the film surface is distributed; by placing electrodes at the screen corners and applying an ac electric signal, the location of the finger capacitance can be calculated precisely. One variant of this idea is to divide the sensing area into many smaller squares (just like pixels on the display) and to sense the change in capacitance across each of them continuously and independently; this is commonly known as *self-capacitance sensing*. A newer development, found in many modern portable devices, is the use of *mutual capacitance sensing* touchscreens, which employ two sets of conductive lines, each on a different layer. On one layer, the lines might run horizontally, while on another layer below the first the lines run vertically. At each point of overlap between the lines on the two layers, a parallel plate capacitor is formed. If there are M lines on the top layer and N lines on the bottom, there will be $M \times N$ such nodes. Whenever a finger moves near a node, the capacitance of the node changes. By monitoring the capacitance of each node continuously, the touchscreen can detect when touches occur and where. The principal advantages of a touchscreen of this type are its ability to detect many simultaneous touches and its ability to detect very light ones. Capacitive technologies are much more resistant to wear and tear (since they are not flexed) than resistive touchscreen and are somewhat more transparent (> 85 percent transparency) since they can have fewer films and avoid air gaps. These types of screens can be used to detect metal objects as well, so pens with conductive tips can be used on writing interfaces.

Not all capacitive touch systems are integrated with screens; a number of interactive media technologies developed over the last 15 years integrate the touch sensing technology into furniture, household objects, or even countertops and overlay a display using nearby projection equipment. Some interactive tables operate this way. A completely different way to detect touch relies on the measurement of acoustic energy on or near the touchscreen. There are several ways to make use of

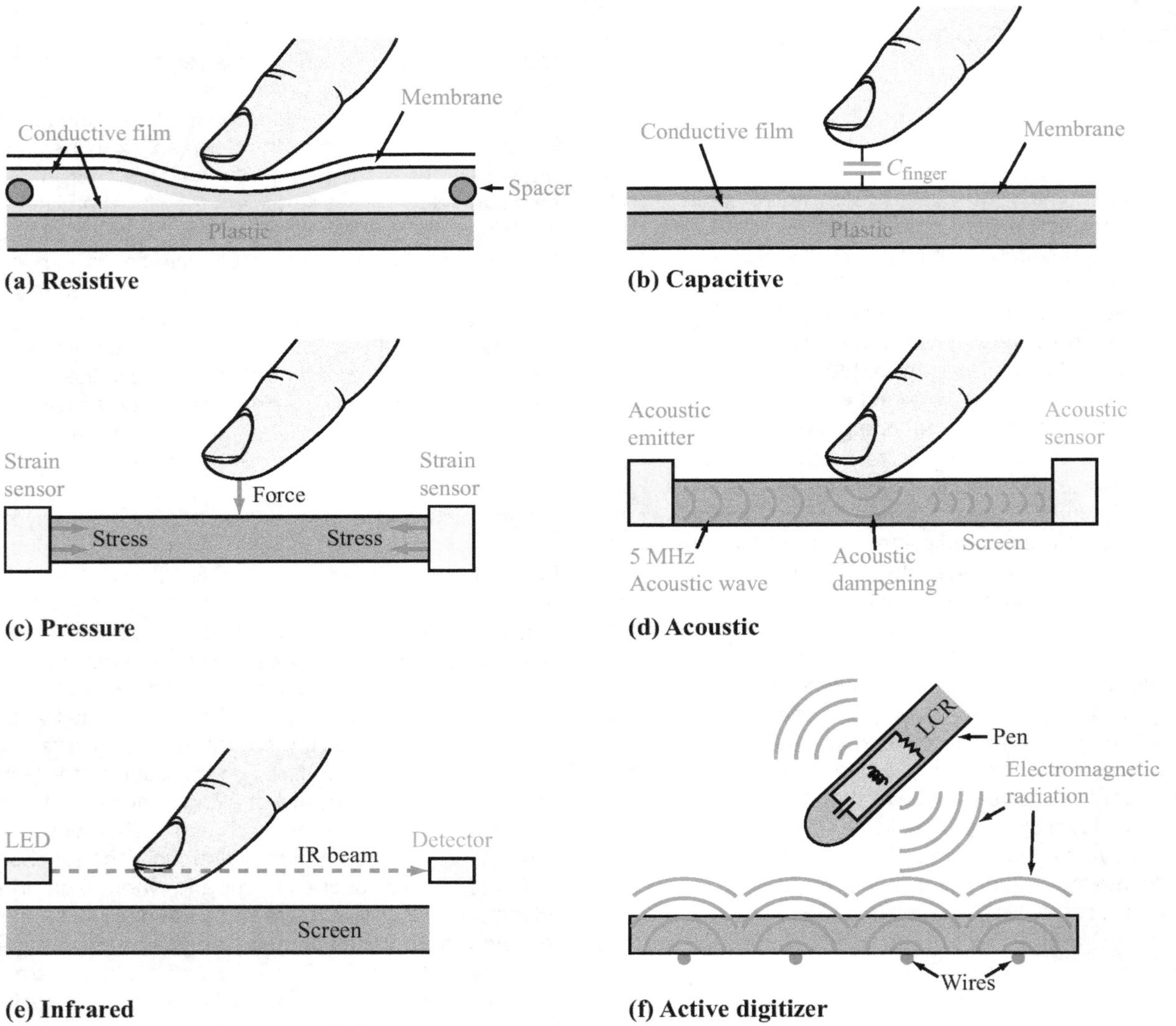

Figure TF18-1: Touchscreen technologies: (a) resistive, (b) capacitive, (c) pressure/strain sensor, (d) acoustic, (e) infrared, and (f) active digitizer.

acoustic energy to measure touch. One implementation relies on transmission of high-frequency acoustic energy across the surface of the display material.

Pressure

Touch also can be detected mechanically. Pressure sensors can be placed at the corners of the display screen or even the entire display assembly, so whenever the screen is depressed, the four corners will experience different stresses depending on the (X, Y) position of the pressure point. Pressure screens benefit from high resistance to wear and tear and no losses in transparency (since there is no need to add layers over the display screen).

Acoustic

A completely different way to detect touch relies on the transmission of high-frequency acoustic energy across the surface of the display material. Bursts of 5 MHz tones are launched by acoustic actuators from two corners of the screen. Acoustic reflectors all along the edges of the screen re-direct the incoming waves to the sensors. Any time an object comes into contact with the screen, it dampens or absorbs some fraction of the energy traveling across the material. The exact (X, Y) position can be calculated from the energy hitting the acoustic sensors. The contact force can be calculated as well, because the acoustic energy is dampened more or less depending on how hard the screen is pressed.

Another approach is to listen, with very sensitive acoustic transducers (i.e., microphones) to the characteristic pressure signal (e.g., sound) made in the touchscreen material when it is touched. By placing several transducers around the edge of the screen, the system can determine if a touch occurred and where. One drawback is that motionless fingers cannot be detected. However, this does provide an advantage in that resting objects (i.e., your cheek) do not trigger the screen. This method is sometimes known as ***acoustic pulse recognition***.

Infrared

One of the oldest and least used technologies is the infrared touchscreen. This technology relies on infrared emitters (usually infrared diodes) aligned along two adjoining edges of the screen and infrared detectors aligned across from the emitters at the other two edges. The position of a touch event can be determined through a process based on which light paths are interrupted. The detection of multiple simultaneous touch events is possible. Infrared screens are somewhat bulky, prone to damage or interference from dust and debris, and need special modifications to work in daylight. They largely have been displaced by newer technologies.

Electromagnetic Resonance

Another technology in widespread use is the electromagnetic resonance detection scheme used by many tablet PCs. Strictly speaking, many tablet PC screens are not touchscreens; they are called ***active digitizers*** because they can detect the presence and location of the tablet pen as it approaches the screen (even without contact). In this scheme, a very thin wire grid is integrated within the display screen (which usually is a flat-profile LCD display). The pen itself contains a simple RLC resonator (see Section 6-1) with no power supply. The wire grid alternates between two modes (transmit and receive) every $\sim$ 20 milliseconds. The grid essentially acts as an antenna. During the transmit mode, an ac signal is applied to the grid and part of that signal is emitted into the air around the display. As the pen approaches the grid, some energy from the grid travels across to the pen's resonator which begins to oscillate. In receive mode, the grid is used to "listen" for ac signals at the resonator frequency; if those signals are present, the grid can pinpoint where they are across the screen. A tuning fork provides a good analogy. Imagine a surface vibrating at a musical note; if a tuning fork designed to vibrate at that note comes very close to that surface, it will begin to oscillate at the same frequency. Even if we were to stop the surface vibrations, the tuning fork will continue to make a sound for a little while longer (as the resonance dies down). In a similar way, the laptop screen continuously transmits a signal and listens for the pen's electromagnetic resonance. Functions (such as buttons and pressure information) can be added to the pen by having the buttons change the capacitance value of the LCR when pressed; in this way, the resonance frequency will shift (see Section 6-2), and the shift can be detected by the grid and interpreted as a button press.

Increased Integration

Mobile devices have largely driven the development of advanced touch technologies in the last few years. Given the constant pressure to miniaturize and integrate, a number of companies have or are developing integrated touch and display systems. Unlike the earlier-generation technologies, the display and the touch sensor are not manufactured separately and then integrated during assembly. Rather, the touch sensor conductors (in the case of capacitive sensing) are designed into the very display itself, either into the conductive traces in/on the pixels of the display or immediately over them. In other designs, light-sensing pixels are manufactured into each display pixel of a display, giving the display not only the ability to produce images but also to sense nearby objects that occlude light landing on the sensing pixels. Even the integrated circuits are increasingly being integrated; earlier-generation systems relied on stand-alone ***touch controller IC*** chips that managed the sensor information and communicated it to the application processor in the mobile devices. There is a push to integrate this functionality into some phone processors directly.

7-3 Phasor Domain

In this chapter, we explore how currents and voltages defined in the time domain are transformed into their counterparts in the phasor domain (also called the *frequency domain*), and why such a transformation facilitates the analysis of ac circuits.

The KVL and KCL equations characterizing an ac circuit containing capacitors and inductors take the form of integro-differential equations with forcing functions (representing the real sources in the circuit) that vary sinusoidally with time. The phasor technique allows us to transform the equations from the time domain to the phasor domain, as a result of which the integro-differential equations get converted into linear equations with no sinusoidal functions. After solving for the desired variable—such as a particular voltage or current—in the phasor domain, conversion back to the time domain provides the same solution that we would have obtained had we solved the integro-differential equations entirely in the time domain. The procedure involves multiple steps, but it avoids the complexity of solving differential equations containing sinusoidal functions.

7-3.1 Time-Domain/Phasor-Domain Correspondence

Transformation from the time domain to the phasor domain entails transforming all time-dependent quantities in the circuit, which in effect transforms the entire circuit from the time domain to an equivalent circuit in the phasor domain. The quantities involved in the transformation include all currents and voltages, all sources, and all capacitors and inductors. The values of capacitors and inductors do not change per se, but their i–υ relationships undergo a transformation because they involve differentiation or integration with respect to t.

Any cosinusoidally *time-varying function* $x(t)$, representing a voltage or a current, can be expressed in the form

$$x(t) = \Re\mathfrak{e}[\underbrace{\mathbf{X}}_{\text{phasor}} e^{j\omega t}], \tag{7.28}$$

where $\mathbf{X}$ is a *time-independent* function called the *phasor counterpart* of $x(t)$. Thus, $x(t)$ is defined in the time domain, while its counterpart $\mathbf{X}$ is defined in the phasor domain.

> ▶ To distinguish phasor quantities from their time-domain counterparts, phasors are always represented by **bold** letters in this book. ◀

In general, the phasor-domain quantity $\mathbf{X}$ is complex, consisting of a magnitude $|\mathbf{X}|$ and a phase angle ϕ,

$$\mathbf{X} = |\mathbf{X}|e^{j\phi}. \tag{7.29}$$

Using this expression in Eq. (7.28) gives

$$x(t) = \Re\mathfrak{e}[|\mathbf{X}|e^{j\phi}e^{j\omega t}] = \Re\mathfrak{e}[|\mathbf{X}|e^{j(\omega t+\phi)}] = |\mathbf{X}|\cos(\omega t+\phi). \tag{7.30}$$

Application of the $\Re\mathfrak{e}$ operator allows us to transform a function from the phasor domain to the time domain. The reverse operation, namely to specify the phasor-domain equivalent of a time function, can be ascertained by comparing the two sides of Eq. (7.30). Thus, for a voltage $\upsilon(t)$ with phasor counterpart $\mathbf{V}$, the correspondence between the two domains is as follows:

Time Domain		Phasor Domain	
$\upsilon(t) = V_0 \cos\omega t$	$\longleftrightarrow$	$\mathbf{V} = V_0$	(7.31a)
$\upsilon(t) = V_0 \cos(\omega t + \phi)$	$\longleftrightarrow$	$\mathbf{V} = V_0 e^{j\phi}$.	(7.31b)

If $\phi = -\pi/2$,

$$\upsilon(t) = V_0 \cos(\omega t - \pi/2) \longleftrightarrow \mathbf{V} = V_0 e^{-j\pi/2}. \tag{7.32}$$

Since $\cos(\omega t - \pi/2) = \cos(\pi/2 - \omega t) = \sin\omega t$ and

$$e^{-j\pi/2} = \cos(\pi/2) - j\sin(\pi/2) = -j,$$

Eq. (7.32) reduces to

$$\upsilon(t) = V_0 \sin\omega t \longleftrightarrow \mathbf{V} = -jV_0, \tag{7.33}$$

which can be generalized to

$$\upsilon(t) = V_0 \sin(\omega t + \phi) \longleftrightarrow \mathbf{V} = V_0 e^{j(\phi-\pi/2)}. \tag{7.34}$$

Occasionally, voltage and current time functions may encounter differentiation or integration. For example, consider a current $i(t)$ with a corresponding phasor $\mathbf{I}$,

$$i(t) = \Re\mathfrak{e}[\mathbf{I}e^{j\omega t}], \tag{7.35}$$

where $\mathbf{I}$ may be complex but, by definition, not a function of time. The derivative di/dt is given by

$$\frac{di}{dt} = \frac{d}{dt}[\Re\mathfrak{e}(\mathbf{I}e^{j\omega t})] = \Re\mathfrak{e}\left[\frac{d}{dt}(\mathbf{I}e^{j\omega t})\right] = \Re\mathfrak{e}[\underbrace{j\omega\mathbf{I}}_{\text{phasor of } di/dt} e^{j\omega t}], \tag{7.36}$$

where in the second step we interchanged the order of the two operators, $\Re\mathfrak{e}$ and d/dt, which is justified by the fact that the two operators are independent of one another, meaning that *taking the real part* of a quantity has no influence on taking its time derivative, and vice versa. We surmise from Eq. (7.36) that

$$\frac{di}{dt} \leftrightarrow j\omega\mathbf{I}, \quad (7.37)$$

or:

▶ Differentiation of a time function $i(t)$ in the time domain is equivalent to multiplication of its phasor counterpart **I** by $j\omega$ in the phasor domain. ◀

Similarly,

$$\int i\,dt = \int \mathfrak{Re}[\mathbf{I}e^{j\omega t}]\,dt$$
$$= \mathfrak{Re}\left[\int \mathbf{I}e^{j\omega t}\,dt\right] = \mathfrak{Re}\left[\underbrace{\frac{\mathbf{I}}{j\omega}}_{\text{phasor of } \int i\,dt} e^{j\omega t}\right], \quad (7.38)$$

or

$$\int i\,dt \leftrightarrow \frac{\mathbf{I}}{j\omega}, \quad (7.39)$$

which states that:

▶ Integration of $i(t)$ in the time domain is equivalent to dividing its phasor **I** by $j\omega$ in the phasor domain. ◀

Table 7-3 provides a summary of some time functions and their phasor-domain counterparts.

Table 7-3: Time-domain sinusoidal functions $x(t)$ and their cosine-reference phasor-domain counterparts X, where $x(t) = \mathfrak{Re}\,[\mathbf{X}e^{j\omega t}]$.

$x(t)$		$\mathbf{X}$
$A\cos\omega t$	$\leftrightarrow$	A
$A\cos(\omega t+\phi)$	$\leftrightarrow$	$Ae^{j\phi}$
$-A\cos(\omega t+\phi)$	$\leftrightarrow$	$Ae^{j(\phi\pm\pi)}$
$A\sin\omega t$	$\leftrightarrow$	$Ae^{-j\pi/2} = -jA$
$A\sin(\omega t+\phi)$	$\leftrightarrow$	$Ae^{j(\phi-\pi/2)}$
$-A\sin(\omega t+\phi)$	$\leftrightarrow$	$Ae^{j(\phi+\pi/2)}$
$\frac{d}{dt}(x(t))$	$\leftrightarrow$	$j\omega\mathbf{X}$
$\frac{d}{dt}[A\cos(\omega t+\phi)]$	$\leftrightarrow$	$j\omega Ae^{j\phi}$
$\int x(t)\,dt$	$\leftrightarrow$	$\frac{1}{j\omega}\mathbf{X}$
$\int A\cos(\omega t+\phi)\,dt$	$\leftrightarrow$	$\frac{1}{j\omega}Ae^{j\phi}$

7-3.2 Impedance of Circuit Elements

Resistors

The υ–i relationship for a resistor R is

$$\upsilon_\text{R} = Ri_\text{R}. \quad (7.40)$$

If i_R is a sinusoidal function of t, the same is true for υ_R. The time-domain quantities υ_R and i_R are related to their phasor-domain counterparts by

$$\upsilon_\text{R} = \mathfrak{Re}[\mathbf{V}_\text{R}e^{j\omega t}] \quad (7.41a)$$

and

$$i_\text{R} = \mathfrak{Re}[\mathbf{I}_\text{R}e^{j\omega t}]. \quad (7.41b)$$

Inserting these expressions into Eq. (7.40) gives

$$\mathfrak{Re}[\mathbf{V}_\text{R}e^{j\omega t}] = R\,\mathfrak{Re}[\mathbf{I}_\text{R}e^{j\omega t}] = \mathfrak{Re}[R\mathbf{I}_\text{R}e^{j\omega t}]. \quad (7.42)$$

Upon combining both sides under the same real-part ($\mathfrak{Re}$) operator, we have

$$\mathfrak{Re}[(\mathbf{V}_\text{R} - R\mathbf{I}_\text{R})e^{j\omega t}] = 0. \quad (7.43a)$$

Through a somewhat similar treatment that uses a sine reference—rather than a cosine reference—to define sinusoidal functions, we can obtain the result

$$\mathfrak{Im}[(\mathbf{V}_\text{R} - R\mathbf{I}_\text{R})e^{j\omega t}] = 0, \quad (7.43b)$$

which, for the sake of expediency, we simply state without taking the steps to prove it. In view of Eqs. (7.43a) and (7.43b), both the real and imaginary components of the quantity inside the square bracket are zero. Hence, the quantity itself is zero, and since $e^{j\omega t} \neq 0$, it follows that

$$\mathbf{V}_\text{R} - R\mathbf{I}_\text{R} = 0. \quad (7.44)$$

In the phasor domain:

▶ The *impedance* **Z** of a circuit element is defined as the ratio of the phasor voltage across it to the phasor current entering through its plus (+) terminal, ◀

$$\mathbf{Z} = \frac{\mathbf{V}}{\mathbf{I}} \quad (\Omega), \tag{7.45}$$

and the unit of **Z** is the ohm (Ω). For a resistor, Eq. (7.44) gives

$$\mathbf{Z}_R = \frac{\mathbf{V}_R}{\mathbf{I}_R} = R. \tag{7.46}$$

Thus, for a resistor the impedance is entirely real, and the form of the υ–i relationship is the same in both the time and phasor domains.

Inductors

In the time domain, the voltage υ_L across an inductor L is related to i_L by

$$\upsilon_L = L\,\frac{di_L}{dt}. \tag{7.47}$$

Phasors $\mathbf{V}_L$ and $\mathbf{I}_L$ are related to their time-domain counterparts by

$$\upsilon_L = \mathfrak{Re}[\mathbf{V}_L e^{j\omega t}] \tag{7.48a}$$

and

$$i_L = \mathfrak{Re}[\mathbf{I}_L e^{j\omega t}]. \tag{7.48b}$$

Consequently,

$$\mathfrak{Re}[\mathbf{V}_L e^{j\omega t}] = L\,\frac{d}{dt}[\mathfrak{Re}(\mathbf{I}_L e^{j\omega t})] = \mathfrak{Re}[j\omega L\mathbf{I}_L e^{j\omega t}], \tag{7.49}$$

which leads to

$$\mathbf{V}_L = j\omega L\mathbf{I}_L. \tag{7.50}$$

Hence, the impedance of an inductor L is

$$\mathbf{Z}_L = \frac{\mathbf{V}_L}{\mathbf{I}_L} = j\omega L. \tag{7.51}$$

According to Eq. (7.51), $\mathbf{Z}_L$ is positive and entirely imaginary (no real component); $\mathbf{Z}_L \to 0$ as $\omega \to 0$ (dc); and $\mathbf{Z}_L \to \infty$ as $\omega \to \infty$. Consequently:

▶ In the phasor domain, an inductor behaves like a short circuit at dc and like an open circuit at very high frequencies. ◀

Capacitors

Since for a capacitor

$$i_C = C\,\frac{d\upsilon_C}{dt}, \tag{7.52}$$

it follows that in the phasor domain,

$$\mathbf{I}_C = j\omega C\mathbf{V}_C \tag{7.53}$$

and the impedance of a capacitor C is

$$\mathbf{Z}_C = \frac{\mathbf{V}_C}{\mathbf{I}_C} = \frac{1}{j\omega C}. \tag{7.54}$$

Because $\mathbf{Z}_L$ and $\mathbf{Z}_C$ are, respectively, directly and inversely proportional to ω, $\mathbf{Z}_L$ and $\mathbf{Z}_C$ assume inverse roles as ω approaches zero and ∞.

▶ In the phasor domain, a capacitor behaves like an open circuit at dc and like a short circuit at very high frequencies. ◀

We note that the impedance of a resistor is purely real, that of an inductor is purely imaginary and positive, and that of a capacitor is purely imaginary and negative (because $1/j\omega C = -j/\omega C$). Table 7-4 provides a summary of the υ–i properties for R, L, and C.

Example 7-4: Phasor Quantities

Determine the phasor-domain counterparts of the following quantities:

(a) $\upsilon_1(t) = 10\cos(2 \times 10^4 t + 53^\circ)$ V,

(b) $\upsilon_2(t) = -6\sin(3 \times 10^3 t - 15^\circ)$ V,

(c) $L = 0.4$ mH at 1 kHz,

(d) $C = 2\ \mu$F at 1 MHz.

Table 7-4: **Summary of υ–i properties for R, L, and C.**

Property	R	L	C
υ–i	$\upsilon = Ri$	$\upsilon = L\,\dfrac{di}{dt}$	$i = C\,\dfrac{d\upsilon}{dt}$
V–**I**	$\mathbf{V} = R\mathbf{I}$	$\mathbf{V} = j\omega L\mathbf{I}$	$\mathbf{V} = \dfrac{\mathbf{I}}{j\omega C}$
Z	R	$j\omega L$	$\dfrac{1}{j\omega C}$
dc equivalent	R	Short circuit	Open circuit
High-frequency equivalent	R	Open circuit	Short circuit
Frequency response	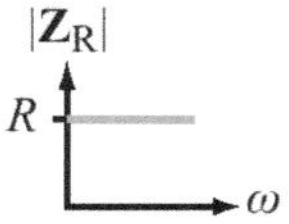	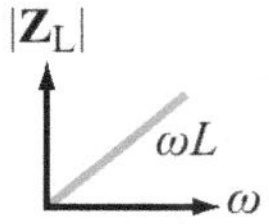	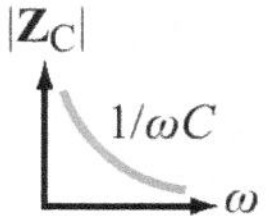

Solution: (a) Since $\upsilon_1(t)$ is already in cosine format,

$$\mathbf{V}_1 = 10e^{j53^\circ} = 10\angle 53^\circ \text{ V}.$$

(b) To determine the phasor $\mathbf{V}_2$ corresponding to $\upsilon_2(t)$, we should either convert the expression for $\upsilon_2(t)$ to standard cosine format or apply the transformation for a sine function given in Table 7-3. We choose the first option,

$$\begin{aligned}\upsilon_2(t) &= -6\sin(3\times 10^3 t - 15^\circ)\\ &= -6\cos(3\times 10^3 t - 15^\circ - 90^\circ)\\ &= -6\cos(3\times 10^3 t - 105^\circ) \text{ V}.\end{aligned}$$

To convert the amplitude from -6 to $+6$, we use Eq. (7.7d) of Table 7-1, namely

$$-\cos(x) = \cos(x \pm 180^\circ).$$

We can either add or subtract 180° from the argument of the cosine. Since the argument has a negative phase angle (-105°), it is more convenient to add 180°. Hence,

$$\begin{aligned}\upsilon_2(t) &= 6\cos(3\times 10^3 t - 105^\circ + 180^\circ)\\ &= 6\cos(3\times 10^3 t + 75^\circ) \text{ V},\end{aligned}$$

and

$$\mathbf{V}_2 = 6e^{j75^\circ} = 6\angle 75^\circ \text{ V}.$$

(c)

$$\mathbf{Z}_\mathrm{L} = j\omega L = j2\pi \times 10^3 \times 0.4 \times 10^{-3} = j2.5\ \Omega.$$

(d)

$$\mathbf{Z}_\mathrm{C} = \frac{-j}{\omega C} = \frac{-j}{2\pi \times 10^6 \times 2 \times 10^{-6}} = -j0.08\ \Omega.$$

Concept Question 7-5: Why is the phasor domain useful for analyzing ac circuits? (See CAD)

Concept Question 7-6: Differentiation in the time domain corresponds to what mathematical operation in the phasor domain? (See CAD)

Concept Question 7-7: The unit for inductance is the henry (H). What is the unit for the impedance $\mathbf{Z}_\mathrm{L}$ of an inductor? (See CAD)

Concept Question 7-8: What type of circuit is equivalent to the behavior of (a) an inductor at dc and (b) a capacitor at very high frequencies? (See CAD)

Exercise 7-5: Determine the phasor counterparts of the following waveforms:

(a) $i_1(t) = 2\sin(6 \times 10^3 t - 30°)$ A,

(b) $i_2(t) = -4\sin(1000t + 136°)$ A

Answer: (a) $\mathbf{I}_1 = 2\angle{-120°}$ A, (b) $\mathbf{I}_2 = 4\angle{-134°}$ A. (See CAD)

Exercise 7-6: Obtain the time-domain waveforms (in standard cosine format) corresponding to the following phasors at angular frequency $\omega = 3 \times 10^4$ rad/s:

(a) $\mathbf{V}_1 = (-3 + j4)$ V

(b) $\mathbf{V}_2 = (3 - j4)$ V

Answer: (a) $\upsilon_1(t) = 5\cos(3 \times 10^4 t + 126.87°)$ V, (b) $\upsilon_2(t) = 5\cos(3 \times 10^4 t - 53.13°)$ V. (See CAD)

Exercise 7-7: At $\omega = 10^6$ rad/s, the phasor voltage across and current through a certain element are given by $\mathbf{V} = 4\angle{-20°}$ V and $\mathbf{I} = 2\angle{70°}$ A. What type of element is it?

Answer: Capacitor with $C = 0.5\ \mu$F. (See CAD)

7-4 Phasor-Domain Analysis

In the time domain, Kirchhoff's voltage law states that the algebraic sum of all voltages υ_1 to υ_n around a closed path containing n elements is zero,

$$\upsilon_1(t) + \upsilon_2(t) + \cdots + \upsilon_n(t) = 0. \qquad (7.55)$$

If $\mathbf{V}_1$ to $\mathbf{V}_n$ are respectively the phasor-domain counterparts of υ_1 to υ_n, then

$$\mathfrak{Re}[\mathbf{V}_1 e^{j\omega t}] + \mathfrak{Re}[\mathbf{V}_2 e^{j\omega t}] + \cdots + \mathfrak{Re}[\mathbf{V}_n e^{j\omega t}] = 0, \qquad (7.56)$$

or equivalently,

$$\mathfrak{Re}[(\mathbf{V}_1 + \mathbf{V}_2 + \cdots + \mathbf{V}_n)e^{j\omega t}] = 0. \qquad (7.57)$$

Since $e^{j\omega t} \neq 0$, it follows that

$$\mathfrak{Re}[\mathbf{V}_1 + \mathbf{V}_2 + \cdots + \mathbf{V}_n] = 0. \qquad (7.58a)$$

Had we used a sine convention—instead of a cosine convention—we would have arrived at the result

$$\mathfrak{Im}[\mathbf{V}_1 + \mathbf{V}_2 + \cdots + \mathbf{V}_n] = 0. \qquad (7.58b)$$

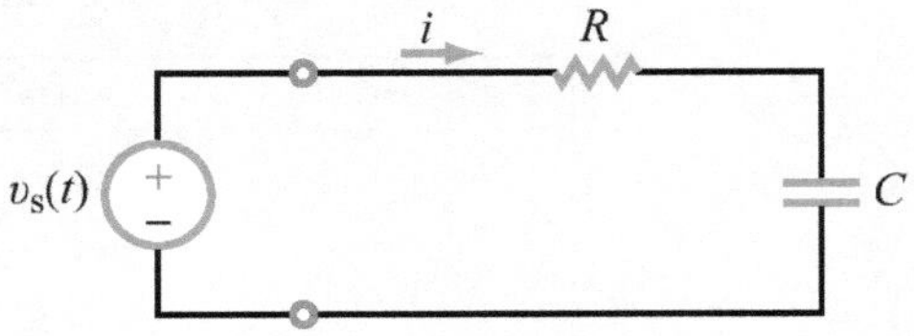

Figure 7-6: RC circuit connected to an ac source.

The combination of Eqs. (7.58a)(a) and (b) asserts that

$$\mathbf{V}_1 + \mathbf{V}_2 + \cdots + \mathbf{V}_n = 0, \qquad (7.58c)$$

which states that *KVL is equally applicable in the phasor domain.*

Similarly, KCL at a node leads to

$$\mathbf{I}_1 + \mathbf{I}_2 + \cdots + \mathbf{I}_n = 0, \qquad (7.59)$$

where $\mathbf{I}_1$ to $\mathbf{I}_n$ are the phasor counterparts of i_1 to i_n.

▶ The fact that KCL and KVL are valid in the phasor domain is highly significant, because it implies that the analysis tools we developed earlier on the basis of these two laws also are valid in the phasor domain. These include the nodal and mesh analysis methods, the Thévenin and Norton techniques, and several others. ◀

Revisiting these tools and learning to apply them to ac circuits is the subject of future sections in this chapter. However, we will now introduce the basic elements of the phasor analysis process through a simple example.

The phasor analysis method consists of five steps. To assist us in presenting it, we use the RC circuit shown in Fig. 7-6. The voltage source is given by

$$\upsilon_s = 12\sin(\omega t - 45°) \text{ V}, \qquad (7.60)$$

with $\omega = 10^3$ rad/s, $R = \sqrt{3}$ kΩ, and $C = 1\ \mu$F. Application of KVL generates the following loop equation:

$$Ri + \frac{1}{C}\int i\, dt = \upsilon_s \qquad \text{(time domain)}. \qquad (7.61)$$

Our goal is to obtain a solution for $i(t)$. In general, $i(t)$ consists of a transient response, obtained by solving Eq. (7.61) with υ_s set equal to zero (as we had done previously in Chapters 5 and 6), and a steady-state response that involves the sinusoidal function $v_s(t)$. Our interest at present is in only the sinusoidal response, which we can obtain by solving Eq. (7.61) in the time domain, but the method of solution is somewhat cumbersome—even for such a simple circuit—on account of the sinusoidal voltage source. Alternatively, we can obtain the desired solution by applying the phasor technique, which avoids dealing with sine and cosine functions altogether.

Step 1: **Adopt cosine reference**

All voltages and currents with known sinusoidal functions should be expressed in the standard cosine format (Section 7-1). For our RC circuit, $v_s(t)$ is the only time-varying quantity with an explicit expression, and since $v_s(t)$ is given in terms of a sine function, we need to convert it into a cosine by applying Eq. (7.7a) of Table 7-1:

$$\begin{aligned} v_s(t) &= 12\sin(\omega t - 45^\circ) \\ &= 12\cos(\omega t - 45^\circ - 90^\circ) = 12\cos(\omega t - 135^\circ) \text{ V.} \end{aligned} \tag{7.62}$$

In accordance with Table 7-3, the phasor equivalent of $v_s(t)$ is

$$\mathbf{V}_s = 12e^{-j135^\circ} \text{ V.} \tag{7.63}$$

Step 2: **Transform circuit to phasor domain**

The current $i(t)$ in Eq. (7.61) is related to its phasor counterpart $\mathbf{I}$ by

$$i(t) = \mathfrak{Re}[\mathbf{I}e^{j\omega t}]. \tag{7.64}$$

As yet, we do not have an explicit expression for either $i(t)$ or $\mathbf{I}$, but we will obtain those expressions later on in Steps 4 and 5. Step 2 in Fig. 7-7 shows the RC circuit in the phasor domain, with loop current $\mathbf{I}$, impedance $\mathbf{Z}_R = R$ representing the resistance and impedance $\mathbf{Z}_C = 1/j\omega C$ representing the capacitor. The voltage source is represented by its phasor $\mathbf{V}_s$.

Step 3: **Cast KCL and/or KVL equations in phasor domain**

For the circuit in Step 2 of Fig. 7-7, its loop equation is given by

$$\mathbf{Z}_R\mathbf{I} + \mathbf{Z}_C\mathbf{I} = \mathbf{V}_s, \tag{7.65}$$

which is equivalent to

$$\left(R + \frac{1}{j\omega C}\right)\mathbf{I} = 12e^{-j135^\circ}. \tag{7.66}$$

This equation also could have been obtained by transforming Eq. (7.61) from the time domain to the phasor domain, which entails replacing i with $\mathbf{I}$, $\int i\,dt$ with $\mathbf{I}/j\omega$, and v_s with $\mathbf{V}_s$.

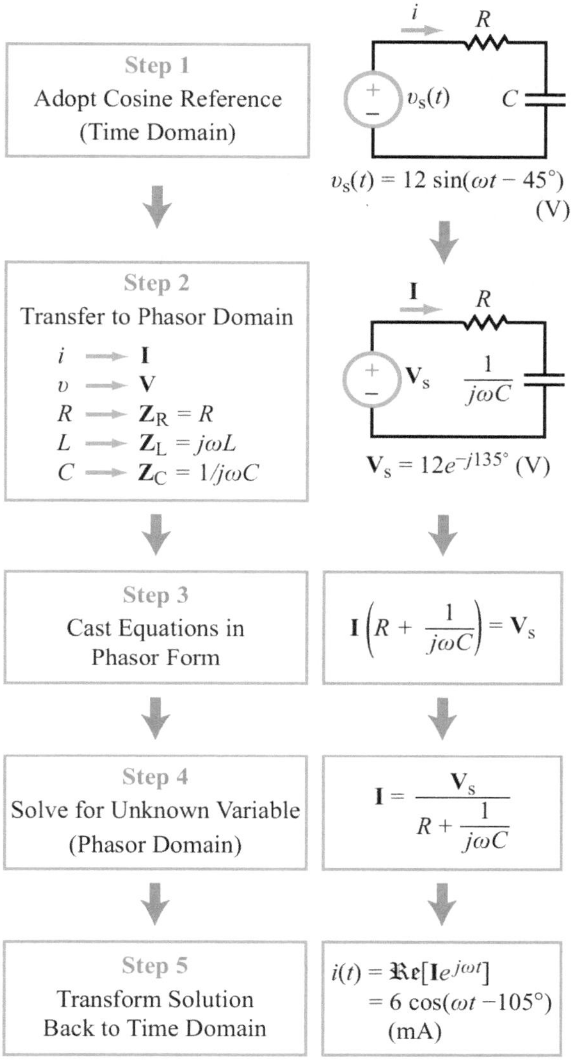

Figure 7-7: Five-step procedure for analyzing ac circuits using the phasor-domain technique.

Step 4: **Solve for unknown variable**

Solving Eq. (7.66) for $\mathbf{I}$ gives

$$\mathbf{I} = \frac{12e^{-j135^\circ}}{R + \frac{1}{j\omega C}} = \frac{j12\omega Ce^{-j135^\circ}}{1 + j\omega RC}. \tag{7.67}$$

Using the specified values, namely $R = \sqrt{3}\,\text{k}\Omega$, $C = 1\,\mu\text{F}$, and $\omega = 10^3$ rad/s, Eq. (7.67) becomes

$$\mathbf{I} = \frac{j12 \times 10^3 \times 10^{-6} e^{-j135^\circ}}{1 + j10^3 \times \sqrt{3} \times 10^3 \times 10^{-6}} = \frac{j12e^{-j135^\circ}}{1 + j\sqrt{3}} \text{ mA}.$$

In preparation for the next step, we should convert the expression for **I** into polar form ($Ae^{j\theta}$, where A is a positive real number) because it is easier to multiply or divide two complex numbers using the polar form. To that end, we should replace j in the numerator with $e^{j\pi/2}$ and convert the denominator into polar form:

$$1 + j\sqrt{3} = \sqrt{1+3}\,e^{j\phi} = 2e^{j\phi},$$

where

$$\phi = \tan^{-1}\left(\frac{\sqrt{3}}{1}\right) = 60^\circ.$$

Hence,

$$\mathbf{I} = \frac{12e^{-j135^\circ} \cdot e^{j90^\circ}}{2e^{j60^\circ}} = 6e^{j(-135^\circ + 90^\circ - 60^\circ)} = 6e^{-j105^\circ} \text{ mA}.$$

Step 5: **Transform solution back to time domain**

To return to the time domain, we apply the fundamental relation between a sinusoidal function and its phasor counterpart, namely

$$i(t) = \mathfrak{Re}[\mathbf{I}e^{j\omega t}] = \mathfrak{Re}[6e^{-j105^\circ}e^{j\omega t}] = 6\cos(\omega t - 105^\circ) \text{ mA}.$$

This concludes our demonstration of the five-step procedure of the phasor-domain analysis technique. The procedure is equally applicable for solving any linear ac circuit.

Example 7-5: RL Circuit

The voltage source of the circuit shown in Fig. 7-8(a) is given by

$$\upsilon_s(t) = 15\sin(4 \times 10^4 t - 30^\circ) \text{ V}.$$

Also, $R = 3\ \Omega$ and $L = 0.1$ mH. Obtain an expression for the voltage across the inductor.

Solution:

Step 1: Convert $\upsilon_s(t)$ to the cosine reference.

$$\begin{aligned}\upsilon_s(t) &= 15\sin(4 \times 10^4 t - 30^\circ)\\ &= 15\cos(4 \times 10^4 t - 30^\circ - 90^\circ)\\ &= 15\cos(4 \times 10^4 t - 120^\circ) \text{ V},\end{aligned}$$

and its corresponding phasor $\mathbf{V}_s$ is given by

$$\mathbf{V}_s = 15e^{-j120^\circ} \text{ V}.$$

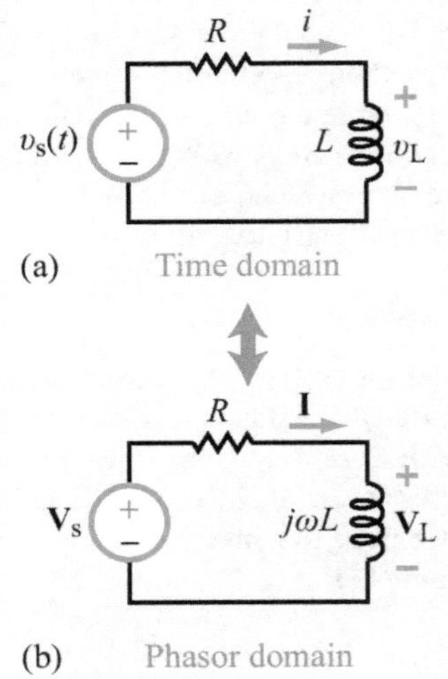

Figure 7-8: RL circuit of Example 7-5.

Step 2: Transform circuit to the phasor domain.

Phasor-domain circuit is shown in Fig. 7-8(b), in which R remains R, L becomes $j\omega L$, $i(t)$ becomes **I**, and $\upsilon_s(t)$ becomes $\mathbf{V}_s$.

Step 3: Cast KVL in phasor domain.

$$R\mathbf{I} + j\omega L\mathbf{I} = \mathbf{V}_s.$$

Step 4: Solve for unknown variable.

$$\begin{aligned}\mathbf{I} = \frac{\mathbf{V}_s}{R + j\omega L} &= \frac{15e^{-j120^\circ}}{3 + j4 \times 10^4 \times 10^{-4}}\\ &= \frac{15e^{-j120^\circ}}{3 + j4} = \frac{15e^{-j120^\circ}}{5e^{j53.1^\circ}} = 3e^{-j173.1^\circ} \text{ A}.\end{aligned}$$

The phasor voltage across the inductor is related to **I** by

$$\begin{aligned}\mathbf{V}_L = j\omega L\mathbf{I} &= j4 \times 10^4 \times 10^{-4} \times 3e^{-j173.1^\circ}\\ &= j12e^{-j173.1^\circ}\\ &= 12e^{-j173.1^\circ} \cdot e^{j90^\circ} = 12e^{-j83.1^\circ} \text{ V},\end{aligned}$$

where we replaced j with e^{j90°.

Step 5: Transform solution to the time domain.

The corresponding time-domain voltage is obtained by multiplying $\mathbf{V}_L$ by $e^{j\omega t}$ and then taking the real part:

$$\begin{aligned} \upsilon_L(t) &= \mathfrak{Re}[\mathbf{V}_L e^{j\omega t}] = \mathfrak{Re}[12e^{-j83.1^\circ} e^{j4\times10^4 t}] \\ &= 12\cos(4\times10^4 t - 83.1^\circ)\text{ V}. \end{aligned}$$

Exercise 7-8: Repeat the analysis of the circuit in Example 7-4 for $\upsilon_s(t) = 20\cos(2\times10^3 t + 60^\circ)$ V, $R = 6\ \Omega$, and $L = 4$ mH.

Answer: $\upsilon_L(t) = 16\cos(2\times10^3 t + 96.9^\circ)$ V. (See CAD)

7-5 Impedance Transformations

Voltage division, current division, and the Y–Δ transformation are among the many analysis tools we developed in Chapter 2 in connection with circuits composed solely of sources and resistors. All of these tools are based on two fundamental laws: KCL and KVL. Having established in the preceding section that KCL and KVL also are valid in the phasor domain, it follows that these simplification and transformation techniques can be used in the phasor domain as well. The fundamental difference between the two cases is that in Chapter 2 we dealt with resistors, and with voltages and currents expressed in the time domain, whereas in the phasor domain the circuit quantities are impedances and phasors. Thus, once an ac circuit has been transformed into the phasor domain, we can apply the same techniques of Chapters 2 and 3, but we do so using complex algebra.

In this and the next section, we illustrate how impedance and source transformations are executed in the phasor domain. Before we do so, however, we should expand our definition of impedance to encompass more than the impedance of a single element. The three passive elements, R, L, and C, are measured in ohms, henrys, and farads. *Their corresponding impedances* $\mathbf{Z}_R$, $\mathbf{Z}_L$, *and* $\mathbf{Z}_C$ *are all measured in ohms,* and are given by

$$\mathbf{Z}_R = R, \qquad \mathbf{Z}_L = j\omega L, \qquad \mathbf{Z}_C = \frac{-j}{\omega C}. \tag{7.68}$$

Consider the three series combinations shown in Fig. 7-9. Application of KVL to the circuits on the left-hand side and to their counterparts leads to

$$\mathbf{Z}_1 = \mathbf{Z}_{R_1} + \mathbf{Z}_{L_1} = R_1 + j\omega L_1,$$

$$\mathbf{Z}_2 = \mathbf{Z}_{R_2} + \mathbf{Z}_{C_2} = R_2 - \frac{j}{\omega C_2},$$

and

$$\mathbf{Z}_3 = \mathbf{Z}_{L_3} + \mathbf{Z}_{C_3} = j\left(\omega L_3 - \frac{1}{\omega C_3}\right).$$

From these three simple examples, we observe that an impedance $\mathbf{Z}$ is, in general, a complex quantity composed of a real part and an imaginary part. We usually use the symbol R to represent its real part and we call it its *resistance*, and we use the symbol X to represent its imaginary part and we call it its *reactance*. Thus,

$$\mathbf{Z} = R + jX. \tag{7.69}$$

Impedances $\mathbf{Z}_1$ and $\mathbf{Z}_2$ have reactances with opposite polarities. When X is positive, as in $\mathbf{Z}_1$, we call $\mathbf{Z}$ an *inductive impedance*, and when X is negative, we call it a *capacitive impedance*. Impedance $\mathbf{Z}_2$ is capacitive. Impedance $\mathbf{Z}_3$ is purely imaginary, and it may be inductive or capacitive depending on how the magnitude of ωL compares with that of $1/\omega C$.

Occasionally, we may need to express $\mathbf{Z}$ in polar form

$$\mathbf{Z} = |\mathbf{Z}|e^{j\theta}, \tag{7.70}$$

where its magnitude $|\mathbf{Z}|$ and phase angle θ are related to components R and X of the rectangular form by

$$|\mathbf{Z}| = \sqrt[+]{R^2 + X^2}, \qquad \text{and} \qquad \theta = \tan^{-1}\left(\frac{X}{R}\right). \tag{7.71}$$

The inverse relationships are given by

$$R = \mathfrak{Re}[\mathbf{Z}] = \mathfrak{Re}[|\mathbf{Z}|e^{j\theta}] = |\mathbf{Z}|\cos\theta \tag{7.72a}$$

and

$$X = \mathfrak{Im}[\mathbf{Z}] = \mathfrak{Im}[|\mathbf{Z}|e^{j\theta}] = |\mathbf{Z}|\sin\theta. \tag{7.72b}$$

In Chapter 2, we defined the conductance G as the reciprocal of R, namely $G = 1/R$. The phasor analogue of G is the *admittance* $\mathbf{Y}$, defined as

$$\mathbf{Y} = \frac{1}{\mathbf{Z}} = G + jB, \tag{7.73}$$

where $G = \mathfrak{Re}[\mathbf{Y}]$ is called the *conductance* of $\mathbf{Y}$ and $B = \mathfrak{Im}[\mathbf{Y}]$ is called its *susceptance*. The unit for $\mathbf{Y}$, G, and B is the siemen (S).

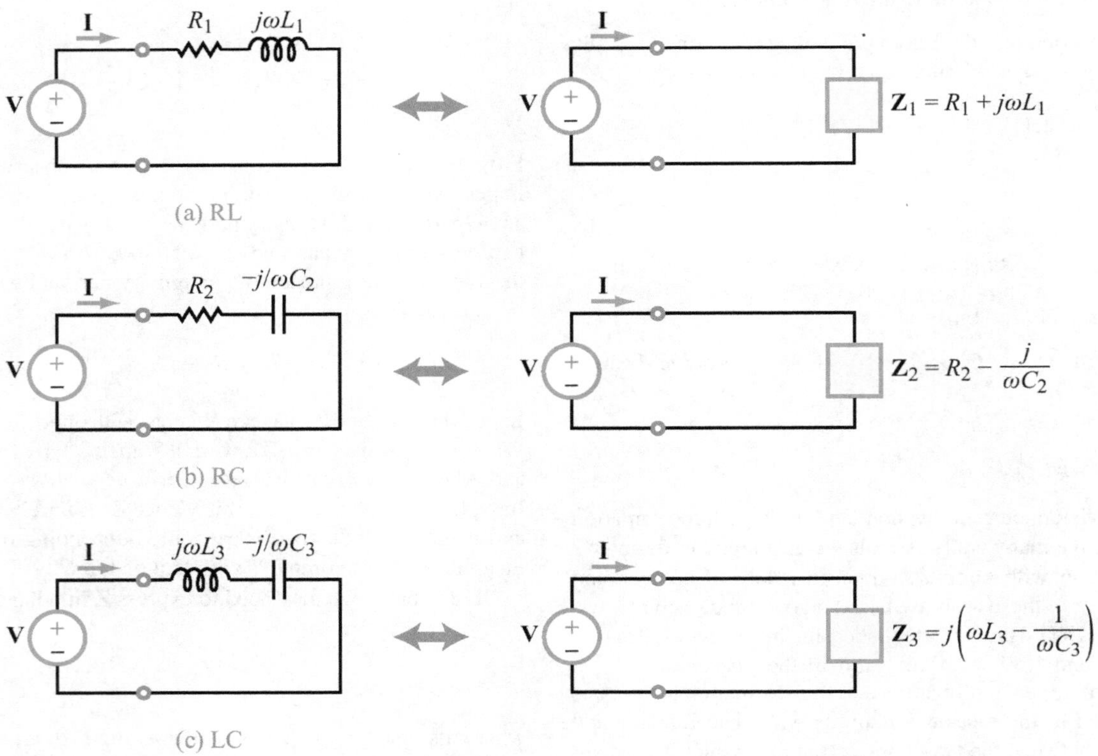

Figure 7-9: Three different, two-element, series combinations.

7-5.1 Impedances in Series and in Parallel

The three in-series examples of Fig. 7-9 consisted each of only two impedances. By extension, we can assert that:

> ▶ N impedances connected in series (sharing the same phasor current) can be combined into a single equivalent impedance $\mathbf{Z}_{eq}$ whose value is equal to the algebraic sum of the individual impedances. ◀

$$\mathbf{Z}_{eq} = \sum_{i=1}^{N} \mathbf{Z}_i \qquad \text{(impedances in series).} \qquad (7.74)$$

The phasor voltage across any individual impedance $\mathbf{Z}_i$ is a proportionate fraction $(\mathbf{Z}_i/\mathbf{Z}_{eq})$ of the phasor voltage across the entire group.

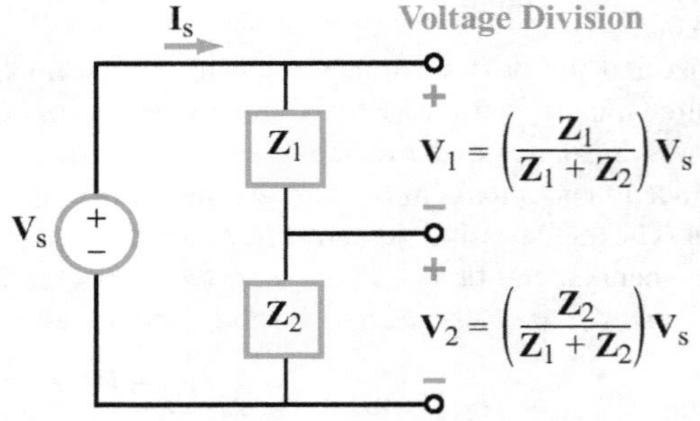

Figure 7-10: Voltage division among two impedances in series.

This is a statement of ***voltage division***, which for the two-impedance circuit of Fig. 7-10, assumes the form

$$\mathbf{V}_1 = \left(\frac{\mathbf{Z}_1}{\mathbf{Z}_1 + \mathbf{Z}_2}\right)\mathbf{V}_s, \qquad \mathbf{V}_2 = \left(\frac{\mathbf{Z}_2}{\mathbf{Z}_1 + \mathbf{Z}_2}\right)\mathbf{V}_s. \qquad (7.75)$$

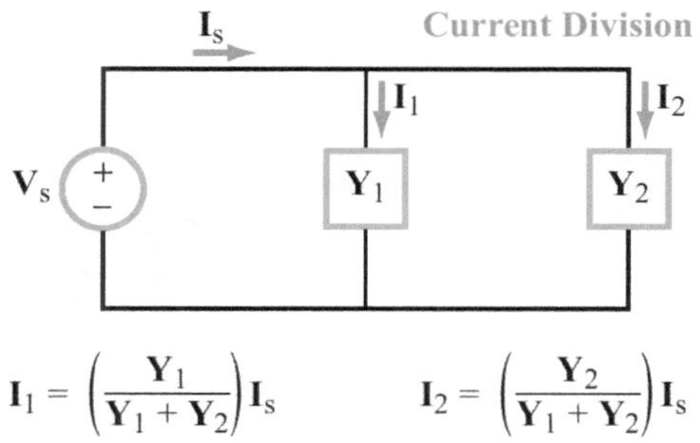

Figure 7-11: Current division among two admittances in parallel.

Admittance $\mathbf{Y}$ is the inverse of impedance $\mathbf{Z}$. That is, $\mathbf{Y} = 1/\mathbf{Z}$. Hence,

> ▶ N admittances connected in parallel between a pair of nodes, all sharing the same voltage, can be combined into a single, equivalent admittance $\mathbf{Y}_{eq}$, whose value is equal to the algebraic sum of the individual admittances. ◀

$$\mathbf{Y}_{eq} = \sum_{i=1}^{N} \mathbf{Y}_i \qquad \textbf{(admittances in parallel)} \qquad (7.76a)$$

or, equivalently,

$$\mathbf{Z}_{eq} = \left[\sum_{i=1}^{N} \frac{1}{\mathbf{Z}_i}\right]^{-1}. \qquad (7.76b)$$

The phasor current flowing through any individual admittance $\mathbf{Y}_i$ is a proportionate fraction $(\mathbf{Y}_i/\mathbf{Y}_{eq})$ of the phasor current flowing through the entire group.

The ***current division*** analogue of Eq. (7.75), defining how current splits up among two admittances connected in parallel (Fig. 7-11), is

$$\mathbf{I}_1 = \left(\frac{\mathbf{Y}_1}{\mathbf{Y}_1 + \mathbf{Y}_2}\right)\mathbf{I}_s, \qquad \mathbf{I}_2 = \left(\frac{\mathbf{Y}_2}{\mathbf{Y}_1 + \mathbf{Y}_2}\right)\mathbf{I}_s. \qquad (7.77)$$

Since $\mathbf{Z}_1 = 1/\mathbf{Y}_1$ and $\mathbf{Z}_2 = 1/\mathbf{Y}_2$, Eq. (7.77) can be rewritten in terms of impedances as

$$\mathbf{I}_1 = \left(\frac{\mathbf{Z}_2}{\mathbf{Z}_1 + \mathbf{Z}_2}\right)\mathbf{I}_s, \qquad \mathbf{I}_2 = \left(\frac{\mathbf{Z}_1}{\mathbf{Z}_1 + \mathbf{Z}_2}\right)\mathbf{I}_s. \qquad (7.78)$$

Example 7-6: Input Impedance

The circuit in Fig. 7-12(a) is connected to a source given by

$$\upsilon_s(t) = 16\cos 10^6 t \text{ V}.$$

Determine (a) the input impedance of the circuit, given that $R_1 = 2$ kΩ, $R_2 = 4$ kΩ, $L = 3$ mH, and $C = 1$ nF, and (b) the voltage $\upsilon_2(t)$ across R_2.

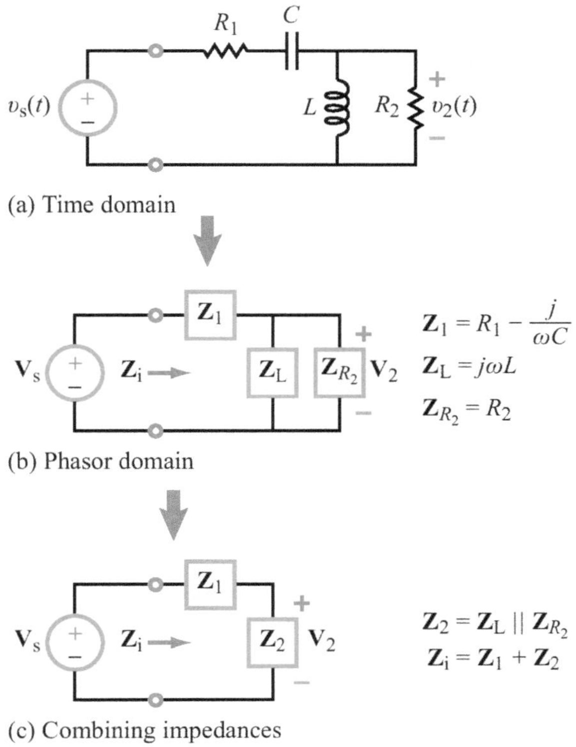

Figure 7-12: Circuit for Example 7-6.

Solution: (a) The phasor-domain equivalent circuit is shown in Fig. 7-12(b), where

$$\mathbf{V}_s = 16,$$

$$\mathbf{Z}_1 = R_1 - \frac{j}{\omega C} = 2 \times 10^3 - \frac{j}{10^6 \times 10^{-9}} = (2 - j1)\ \text{k}\Omega,$$

$$\mathbf{Z}_L = j\omega L = j \times 10^6 \times 3 \times 10^{-3} = j3\ \text{k}\Omega,$$

and

$$\mathbf{Z}_{R_2} = R_2 = 4\ \text{k}\Omega.$$

The parallel combination of $\mathbf{Z}_L$ and $\mathbf{Z}_{R_2}$ is denoted $\mathbf{Z}_2$ in Fig. 7-12(c), and it is given by

$$\mathbf{Z}_2 = \mathbf{Z}_L \parallel \mathbf{Z}_{R_2} = \frac{\mathbf{Z}_L\mathbf{Z}_{R_2}}{\mathbf{Z}_L + \mathbf{Z}_{R_2}} = \frac{j3 \times 10^3 \times 4 \times 10^3}{(4 + j3) \times 10^3} = \frac{j12 \times 10^3}{4 + j3}.$$

A useful "trick" for converting the expression for $\mathbf{Z}_2$ into the form $(a + jb)$ is to multiply the numerator and denominator by the complex conjugate of the denominator:

$$\mathbf{Z}_2 = \frac{j12 \times 10^3}{4 + j3} \times \frac{4 - j3}{4 - j3} = \frac{36 + j48}{16 + 9} \times 10^3 = (1.44 + j1.92)\ \text{k}\Omega.$$

The input impedance $\mathbf{Z}_i$ is equal to the sum of $\mathbf{Z}_1$ and $\mathbf{Z}_2$,

$$\mathbf{Z}_i = \mathbf{Z}_1 + \mathbf{Z}_2 = (2 - j1 + 1.44 + j1.92) \times 10^3 = (3.44 + j0.92)\ \text{k}\Omega.$$

(b) By voltage division,

$$\mathbf{V}_2 = \frac{\mathbf{Z}_2\mathbf{V}_s}{\mathbf{Z}_1 + \mathbf{Z}_2} = \frac{(1.44 + j1.92) \times 10^3 \times 16}{(3.44 + j0.92) \times 10^3} = 10.8e^{j38.2^\circ}\ \text{V}.$$

Transforming $\mathbf{V}_2$ to its time-domain counterpart leads to

$$\upsilon_2(t) = \mathfrak{Re}[\mathbf{V}_2 e^{j\omega t}] = \mathfrak{Re}[10.8e^{j38.2^\circ} e^{j10^6 t}] = 10.8\cos(10^6 t + 38.2^\circ)\ \text{V}.$$

Example 7-7: Current Division

The circuit in Fig. 7-13(a) is connected to a source

$$\upsilon_s(t) = 4\sin(10^7 t + 15^\circ)\ \text{V}.$$

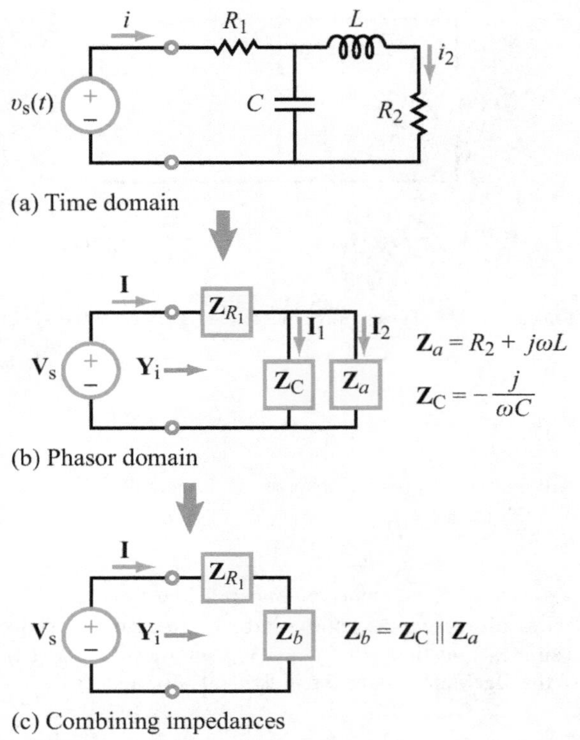

Figure 7-13: Circuit for Example 7-7.

Determine (a) the input admittance $\mathbf{Y}_i$, given that $R_1 = 10\ \Omega$, $R_2 = 30\ \Omega$, $L = 2\ \mu$H, and $C = 10$ nF, and (b) the current $i_2(t)$ flowing through R_2.

Solution: (a) We start by converting $\upsilon_s(t)$ to cosine format:

$$\upsilon_s(t) = 4\sin(10^7 t + 15^\circ) = 4\cos(10^7 t + 15^\circ - 90^\circ) = 4\cos(10^7 t - 75^\circ)\ \text{V}.$$

The corresponding phasor voltage is

$$\mathbf{V}_s = 4e^{-j75^\circ}\ \text{V},$$

and the impedances shown in Fig. 7-13(b) are given by

$$\mathbf{Z}_{R_1} = R_1 = 10\ \Omega,$$

$$\mathbf{Z}_C = \frac{-j}{\omega C} = \frac{-j}{10^7 \times 10^{-8}} = -j10\ \Omega,$$

and

$$\mathbf{Z}_a = R_2 + j\omega L = 30 + j10^7 \times 2 \times 10^{-6} = (30 + j20)\ \Omega.$$

In Fig. 7-13(c), $\mathbf{Z}_b$ represents the parallel combination of $\mathbf{Z}_C$ and $\mathbf{Z}_a$,

$$\begin{aligned}\mathbf{Z}_b &= \mathbf{Z}_C \parallel \mathbf{Z}_a \\ &= \frac{(-j10)(30+j20)}{-j10+30+j20} \\ &= \frac{20-j30}{3+j1} = \frac{(20-j30)}{(3+j1)}\frac{(3-j1)}{(3-j1)} = (3-j11)\ \Omega.\end{aligned}$$

The input impedance is

$$\mathbf{Z}_i = \mathbf{Z}_{R_1} + \mathbf{Z}_b = 10 + 3 - j11 = (13 - j11)\ \Omega,$$

and its reciprocal is

$$\begin{aligned}\mathbf{Y}_i = \frac{1}{\mathbf{Z}_i} &= \frac{1}{13-j11} \times \frac{13+j11}{13+j11} \\ &= \frac{13+j11}{169+121} \\ &= (4.5+j3.8)\times 10^{-2} = 5.89\times 10^{-2}e^{-j40.2^\circ}\ \text{S}.\end{aligned}$$

(b) The current $\mathbf{I}$ is given by

$$\mathbf{I} = \mathbf{V}_s\mathbf{Y}_i = (4e^{-j75^\circ})(5.89\times 10^{-2}e^{-j40.2^\circ}) = 0.235e^{-j34.8^\circ}\ \text{A}.$$

By current division in Fig. 7-13(b),

$$\begin{aligned}\mathbf{I}_2 &= \frac{\mathbf{Z}_C}{\mathbf{Z}_a + \mathbf{Z}_C}\,\mathbf{I} \\ &= \frac{-j10}{30+j20-j10} \times 0.235e^{-j34.8^\circ} \\ &= \frac{2.35e^{-j34.8^\circ}\cdot e^{-j90^\circ}}{31.6e^{j18.4^\circ}} = 7.4\times 10^{-2}e^{-j143.2^\circ}\ \text{A}.\end{aligned}$$

The corresponding current in the time domain is

$$\begin{aligned}i_2(t) &= \mathfrak{Re}[\mathbf{I}_2e^{j\omega t}] = \mathfrak{Re}[7.4\times 10^{-2}e^{-j143.2^\circ}e^{j10^7t}] \\ &= 7.4\times 10^{-2}\cos(10^7t - 143.2^\circ)\ \text{A}.\end{aligned}$$

Concept Question 7-9: The rule for adding the capacitances of two in-series capacitors is different from that for adding the resistances of two in-series resistors, but the rule for adding the impedances of those two in-series capacitors is the same as the rule for adding two in-series resistors. Does this pose a contradiction? Explain. (See CAD)

Concept Question 7-10: Is it possible to construct a circuit composed solely of capacitors and inductors such that the impedance of the overall combination has a non-zero real part? Explain. (See CAD)

Exercise 7-9: Determine the input impedance at $\omega = 10^5$ rad/s for each of the circuits in Fig. E7.9.

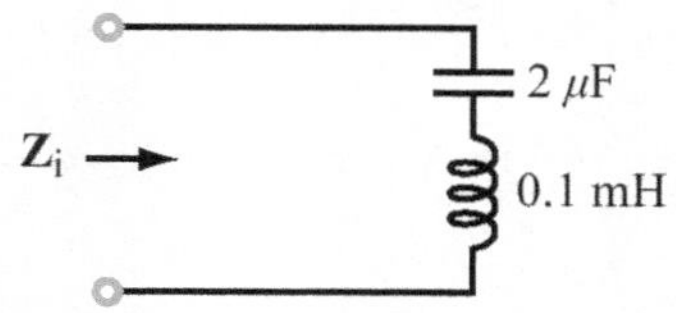

(a)

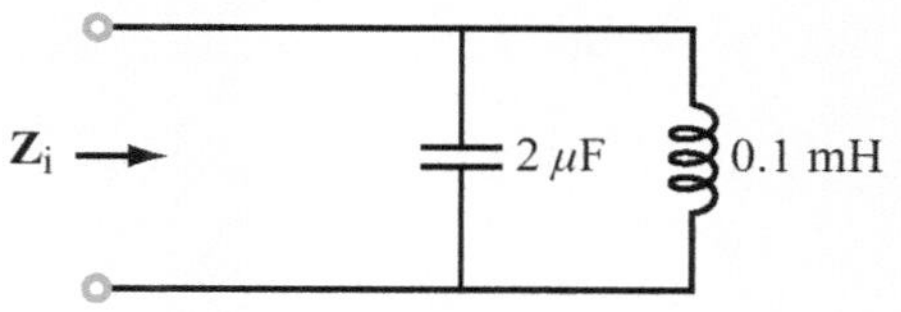

(b)

Figure E7.9

Answer: (a) $\mathbf{Z}_i = j5\ \Omega$, (b) $\mathbf{Z}_i = -j10\ \Omega$. (See CAD)

7-5.2 Y–Δ Transformation

The Y–Δ transformation outlined in Section 2-4 allows us to replace a Y circuit connected to three nodes with a Δ circuit, or vice versa, without altering the voltages at the three nodes or the currents entering them. The same principle applies to impedances, as do the relationships between impedances $\mathbf{Z}_1$ to $\mathbf{Z}_3$ of the Y circuit (Fig. 7-14) and impedances $\mathbf{Z}_a$ to $\mathbf{Z}_c$ of the Δ circuit.

Δ →Y transformation:

$$\mathbf{Z}_1 = \frac{\mathbf{Z}_b\mathbf{Z}_c}{\mathbf{Z}_a + \mathbf{Z}_b + \mathbf{Z}_c}, \tag{7.79a}$$

$$\mathbf{Z}_2 = \frac{\mathbf{Z}_a\mathbf{Z}_c}{\mathbf{Z}_a + \mathbf{Z}_b + \mathbf{Z}_c}, \tag{7.79b}$$

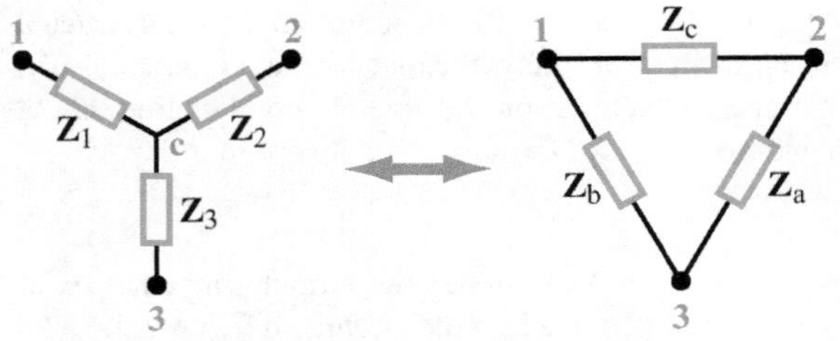

Figure 7-14: Y–Δ equivalent circuits.

$$\mathbf{Z}_3 = \frac{\mathbf{Z}_a\mathbf{Z}_b}{\mathbf{Z}_a + \mathbf{Z}_b + \mathbf{Z}_c}. \qquad (7.79c)$$

Y→ Δ transformation:

$$\mathbf{Z}_a = \frac{\mathbf{Z}_1\mathbf{Z}_2 + \mathbf{Z}_2\mathbf{Z}_3 + \mathbf{Z}_1\mathbf{Z}_3}{\mathbf{Z}_1}, \qquad (7.80a)$$

$$\mathbf{Z}_b = \frac{\mathbf{Z}_1\mathbf{Z}_2 + \mathbf{Z}_2\mathbf{Z}_3 + \mathbf{Z}_1\mathbf{Z}_3}{\mathbf{Z}_2}, \qquad (7.80b)$$

$$\mathbf{Z}_c = \frac{\mathbf{Z}_1\mathbf{Z}_2 + \mathbf{Z}_2\mathbf{Z}_3 + \mathbf{Z}_1\mathbf{Z}_3}{\mathbf{Z}_3}. \qquad (7.80c)$$

Balanced circuits:

If the Y circuit is balanced (all of its impedances are equal), so will be the Δ circuit, and vice versa. Accordingly:

$$\mathbf{Z}_1 = \mathbf{Z}_2 = \mathbf{Z}_3 = \frac{\mathbf{Z}_a}{3}, \quad \text{if } \mathbf{Z}_a = \mathbf{Z}_b = \mathbf{Z}_c, \qquad (7.81a)$$

$$\mathbf{Z}_a = \mathbf{Z}_b = \mathbf{Z}_c = 3\mathbf{Z}_1, \quad \text{if } \mathbf{Z}_1 = \mathbf{Z}_2 = \mathbf{Z}_3. \qquad (7.81b)$$

Example 7-8: Applying Y–Δ Transformation

(a) Simplify the circuit in Fig. 7-15(a) by applying the Y–Δ transformation so as to determine the current **I**. (b) Determine the corresponding $i(t)$, given that the oscillation frequency of the voltage source is 1 MHz.

Solution: (a) The Δ circuit connected to nodes 1, 3, and 4 can be replaced with a Y circuit, as shown in Fig. 7-15(b), with impedances

$$\begin{aligned}\mathbf{Z}_1 &= \frac{\mathbf{Z}_b\mathbf{Z}_c}{\mathbf{Z}_a + \mathbf{Z}_b + \mathbf{Z}_c} \\ &= \frac{-j6 \times 12}{24 - j12 - j6 + 12} = \frac{-j72}{36 - j18} = (0.8 - j1.6)\ \Omega, \\ \mathbf{Z}_2 &= \frac{\mathbf{Z}_a\mathbf{Z}_c}{\mathbf{Z}_a + \mathbf{Z}_b + \mathbf{Z}_c} = \frac{(24 - j12) \times 12}{36 - j18} = 8\ \Omega,\end{aligned}$$

and

$$\mathbf{Z}_3 = \frac{\mathbf{Z}_b\mathbf{Z}_a}{\mathbf{Z}_a + \mathbf{Z}_b + \mathbf{Z}_c} = \frac{-j6(24 - j12)}{36 - j18} = -j4\ \Omega.$$

In Fig. 7-15(c), $\mathbf{Z}_f$ represents the series combination of $\mathbf{Z}_3$ and $\mathbf{Z}_d$,

$$\mathbf{Z}_f = \mathbf{Z}_3 + \mathbf{Z}_d = -j4 + j2 = -j2\ \Omega.$$

Similarly,

$$\mathbf{Z}_g = \mathbf{Z}_2 + \mathbf{Z}_e = (8 + j6)\ \Omega.$$

Impedances $\mathbf{Z}_f$ and $\mathbf{Z}_g$ are connected in parallel, and their combination is in series with $\mathbf{Z}_0$ and $\mathbf{Z}_1$. Hence,

$$\begin{aligned}\mathbf{I} &= \frac{\mathbf{V}_s}{\mathbf{Z}_0 + \mathbf{Z}_1 + (\mathbf{Z}_f \parallel \mathbf{Z}_g)} \\ &= \frac{16e^{j30^\circ}}{2.4 + (0.8 - j1.6) + \frac{-j2 \times (8 + j6)}{-j2 + 8 + j6}}.\end{aligned}$$

After a few steps of complex algebra, we obtain the result

$$\mathbf{I} = 3.06\underline{/76.55^\circ}\ \text{A}.$$

(b)

$$\begin{aligned}i(t) &= \mathfrak{Re}[\mathbf{I}e^{j\omega t}] = \mathfrak{Re}[3.06e^{j76.55^\circ}e^{j2\pi \times 10^6 t}] \\ &= 3.06\cos(2\pi \times 10^6 t + 76.55^\circ)\ \text{A}.\end{aligned}$$

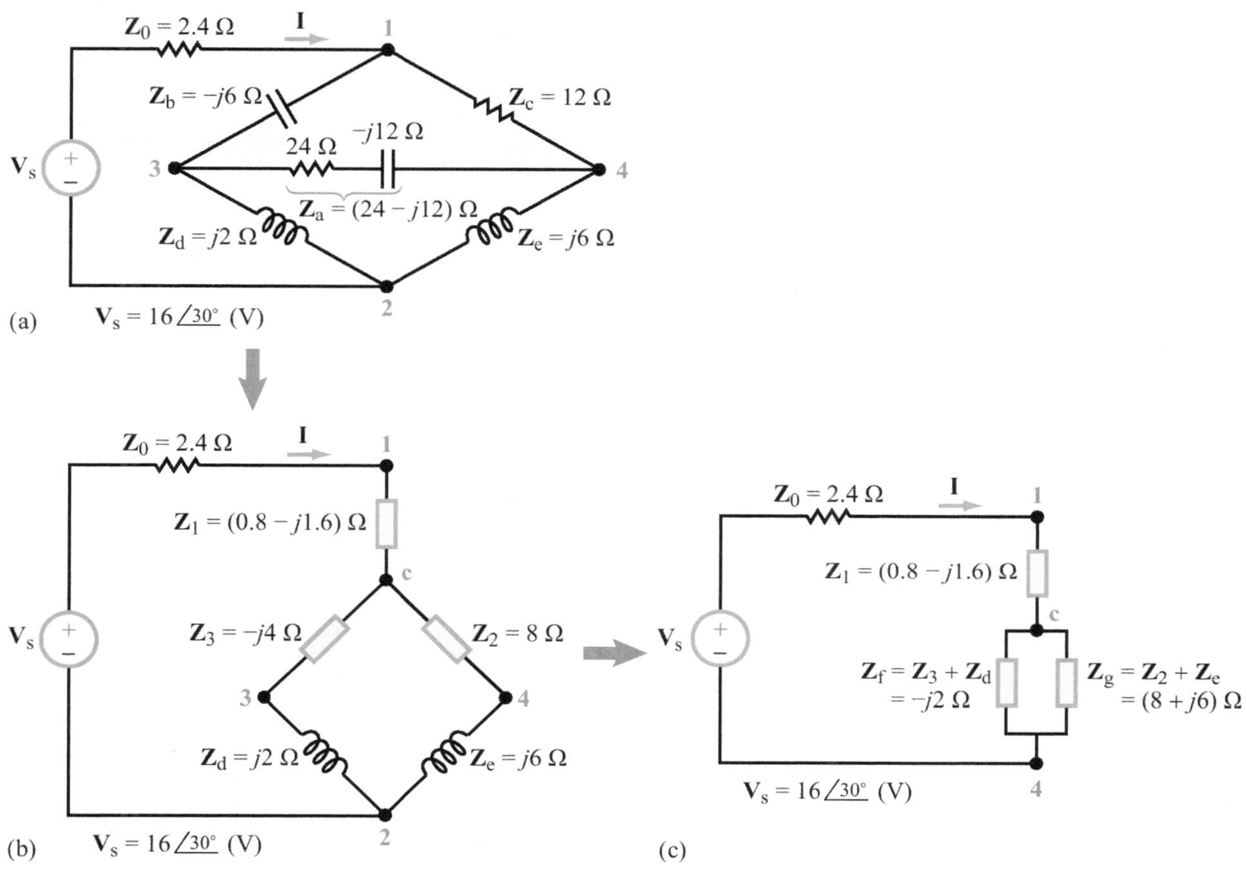

Figure 7-15: Example 7-8 circuit evolution.

Exercise 7-10: Convert the Y-impedance circuit in Fig. E7.10 into a Δ-impedance circuit.

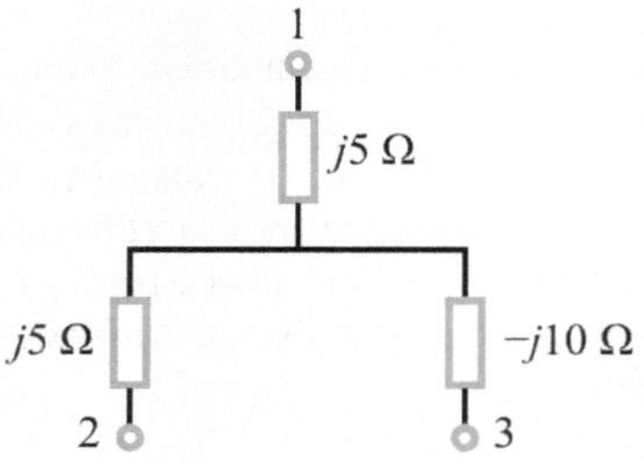

Figure E7.10

Answer:

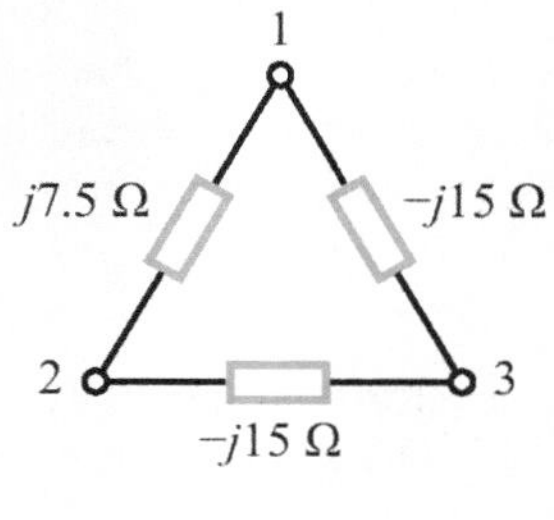

(See CAD)

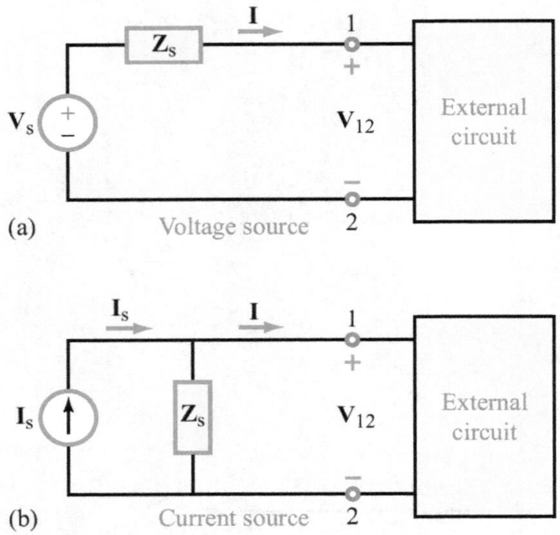

Figure 7-16: Source-transformation equivalency.

7-6 Equivalent Circuits

Having examined in the preceding section how phasor-domain circuits can be simplified by applying impedance transformations, we now extend our review of the rules of circuit equivalency to circuits containing voltage and current sources.

7-6.1 Source Transformation

Section 2-3.4 provides an outline of the ***source-transformation principle*** as it applies to resistive circuits. Its phasor-domain analogue is diagrammed in Fig. 7-16 from the vantage point of the external circuit.

▶ A voltage source $\mathbf{V}_s$ in series with a source impedance $\mathbf{Z}_s$ is equivalent to the combination of a current source $\mathbf{I}_s = \mathbf{V}_s/\mathbf{Z}_s$, in parallel with a shunt impedance $\mathbf{Z}_s$. The direction of $\mathbf{I}_s$ is the same as the arrow from the (−) terminal to the (+) terminal of $\mathbf{V}_s$. ◀

Equivalence implies that both input circuits would deliver the same current $\mathbf{I}$ and voltage $\mathbf{V}_{12}$ to the external circuit.

7-6.2 Thévenin Equivalent Circuit

When restated for the phasor domain, ***Thévenin's theorem*** of Section 3-5.1 becomes:

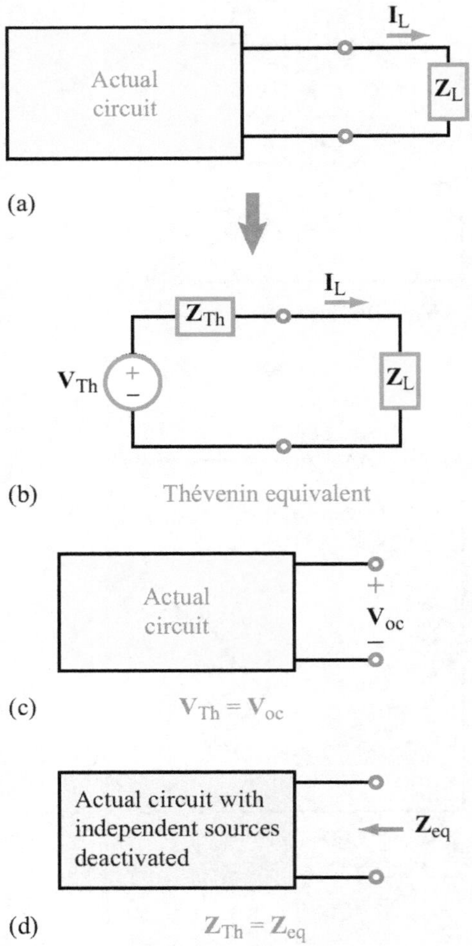

Figure 7-17: Thévenin-equivalent method for a circuit with no dependent sources.

▶ A linear circuit can be represented at its output terminals by an equivalent circuit consisting of a series combination of a voltage source $\mathbf{V}_{Th}$ and an impedance $\mathbf{Z}_{Th}$, where $\mathbf{V}_{Th}$ is the open-circuit voltage at those terminals (no load) and $\mathbf{Z}_{Th}$ is the equivalent impedance between the same terminals when all independent sources in the circuit have been deactivated. ◀

Equivalence implies that if a load $\mathbf{Z}_L$ is connected at the output terminals of any actual circuit (as portrayed in Fig. 7-17(a)) thereby inducing a current $\mathbf{I}_L$ to flow through it, the Thévenin

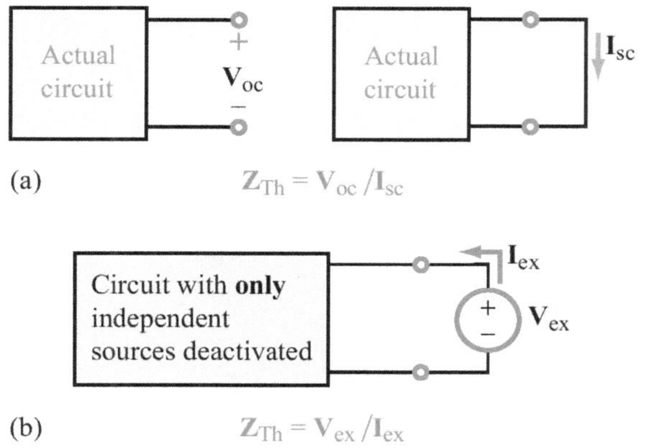

Figure 7-18: The (a) open-circuit/short-circuit method and (b) the external-source method are both suitable for determining $\mathbf{Z}_{Th}$, whether or not the circuit contains dependent sources.

equivalent circuit (Fig. 7-17(b)) would deliver the same current $\mathbf{I}_L$ when connected to the same load impedance $\mathbf{Z}_L$. For the equivalence to hold, the voltage $\mathbf{V}_{Th}$ and impedance $\mathbf{Z}_{Th}$ of the Thévenin circuit have to be related to the actual circuit by (Figs. 7-17(c) and (d)):

$$\mathbf{V}_{Th} = \mathbf{V}_{oc} \tag{7.82a}$$

and

$$\mathbf{Z}_{Th} = \mathbf{Z}_{eq}. \tag{7.82b}$$

Application of Eq. (7.82a) to determine $\mathbf{V}_{Th}$ by calculating or measuring the open-circuit voltage $\mathbf{V}_{oc}$ is always a valid approach, whether or not the actual circuit contains dependent sources. That is not so for Eq. (7.82b). The equivalent-impedance method cannot be used to determine $\mathbf{Z}_{Th}$ if the circuit contains dependent sources. Alternative approaches include the following.

Open-circuit / short-circuit method

$$\mathbf{Z}_{Th} = \frac{\mathbf{V}_{oc}}{\mathbf{I}_{sc}}, \tag{7.83}$$

where $\mathbf{I}_{sc}$ is the short-circuit current at the circuit's output terminals (Fig. 7-18(a)).

External-source method

$$\mathbf{Z}_{Th} = \frac{\mathbf{V}_{ex}}{\mathbf{I}_{ex}}, \tag{7.84}$$

where $\mathbf{I}_{ex}$ is the current generated by an external source $\mathbf{V}_{ex}$ connected at the circuit's terminals (as shown in Fig. 7-18(b)) after deactivating all independent sources in the circuit.

For the sake of completeness, we should remind the reader that a Thévenin equivalent circuit always can be transformed into a Norton equivalent circuit—or vice versa—by applying the source-transformation method of Section 7-6.1.

Example 7-9: Thévenin Circuit

The circuit shown in Fig. 7-19(a) contains a sinusoidal source given by

$$\upsilon_s(t) = 10\cos 10^5 t \text{ V}.$$

Determine the Thévenin equivalent circuit at terminals (a, b).

Solution:

Step 1: The phasor counterpart of $\upsilon_s(t)$ is

$$\mathbf{V}_s = 10 \text{ V}.$$

Figure 7-19(b) displays the circuit in the phasor domain, in addition to having replaced the series combination $(\mathbf{V}_s, R_s)$ with the parallel combination $(\mathbf{I}_s, R_s)$, where

$$\mathbf{I}_s = \frac{\mathbf{V}_s}{R_s} = \frac{10}{5} = 2 \text{ A}.$$

Step 2: Combining R_s with $\mathbf{Z}_1$ in parallel gives

$$\mathbf{Z}'_1 = R_s \parallel \mathbf{Z}_1 = \frac{5(6+j8)}{5+6+j8} = (3.51 + j1.08)\ \Omega.$$

Step 3: Converting back to a voltage source in series with $\mathbf{Z}'_1$ leads to the circuit in Fig. 7-19(d), with

$$\mathbf{V}'_s = \mathbf{I}_s\mathbf{Z}'_1 = 2(3.51 + j1.08) = (7.02 + j2.16) \text{ V}.$$

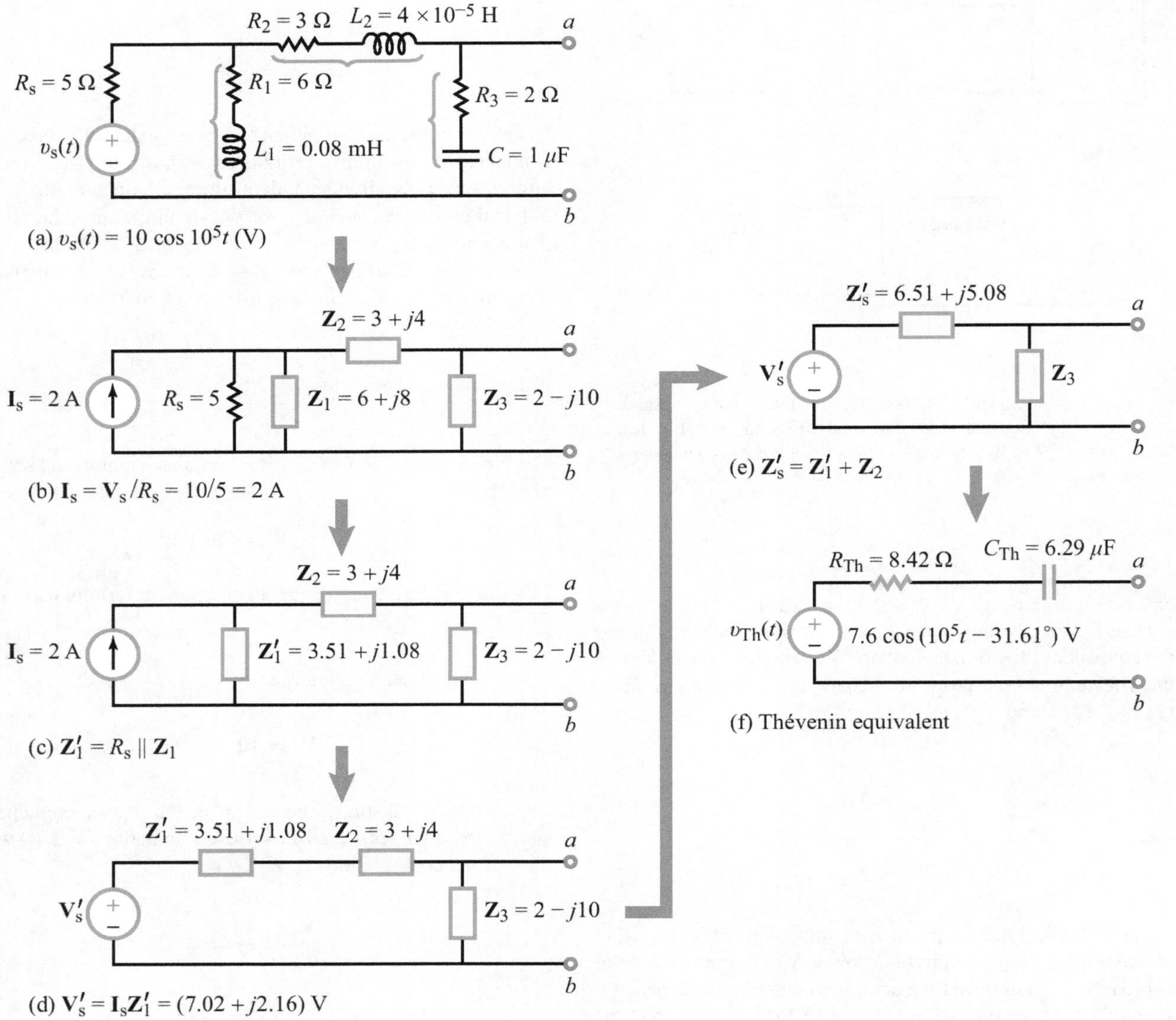

Figure 7-19: Using source transformation to simplify the circuit of Example 7-9. (All impedances are in ohms.)

Step 4: Combining $\mathbf{Z}_1'$ with $\mathbf{Z}_2$ in series leads to the circuit in Fig. 7-19(e), where

$$\mathbf{Z}_s' = \mathbf{Z}_1' + \mathbf{Z}_2 = (3.51 + j1.08) + (3 + j4) = (6.51 + j5.08)\ \Omega.$$

Step 5: Application of voltage division provides

$$\mathbf{V}_{Th} = \mathbf{V}_{oc} = \frac{\mathbf{V}_s' \mathbf{Z}_3}{\mathbf{Z}_s' + \mathbf{Z}_3} = \frac{(7.02 + j2.16)(2 - j10)}{(6.51 + j5.08) + (2 - j10)} = 7.6\angle{-31.61°}\ \text{V}.$$

Hence,

$$v_{\text{Th}}(t) = \Re\mathfrak{e}[\mathbf{V}_{\text{Th}} e^{j\omega t}] = \Re\mathfrak{e}[7.6 e^{-j31.61^\circ} e^{j10^5 t}]$$
$$= 7.6\cos(10^5 t - 31.61^\circ) \text{ V}.$$

Step 6: Suppressing the source $\mathbf{V}'_s$ in Fig. 7-19(e) reduces the circuit at terminals (a, b) to $\mathbf{Z}'_s$ in parallel with $\mathbf{Z}_3$, leading to

$$\mathbf{Z}_{\text{Th}} = \mathbf{Z}'_s \parallel \mathbf{Z}_3$$
$$= \frac{(6.51 + j5.08)(2 - j10)}{(6.51 + j5.08) + (2 - j10)} = (8.42 - j1.59)\ \Omega.$$

Step 7: The impedance $\mathbf{Z}_{\text{Th}}$ is capacitive because the sign of the imaginary component is negative. Hence, it is equivalent to

$$\mathbf{Z}_{\text{Th}} = R_{\text{Th}} - \frac{j}{\omega C_{\text{Th}}}.$$

Matching the two expressions gives

$$R_{\text{Th}} = 8.42\ \Omega, \qquad C_{\text{Th}} = \frac{1}{1.59\omega} = 6.29\ \mu\text{F}.$$

The time-domain Thévenin equivalent circuit is shown in Fig. 7-19(f).

Concept Question 7-11: In the phasor domain, is the Thévenin equivalent method valid for circuits containing dependent sources? If yes, what methods are amenable to finding $\mathbf{Z}_{\text{Th}}$ of such circuits? (See CAD)

Concept Question 7-12: If $\mathbf{Z}_{\text{Th}}$ of a certain circuit is purely imaginary, what would be your expectation about whether or not the circuit contains resistors? (See CAD)

Exercise 7-11: Determine $\mathbf{V}_{\text{Th}}$ and $\mathbf{Z}_{\text{Th}}$ for the circuit in Fig. E7.11 at terminals (a, b).

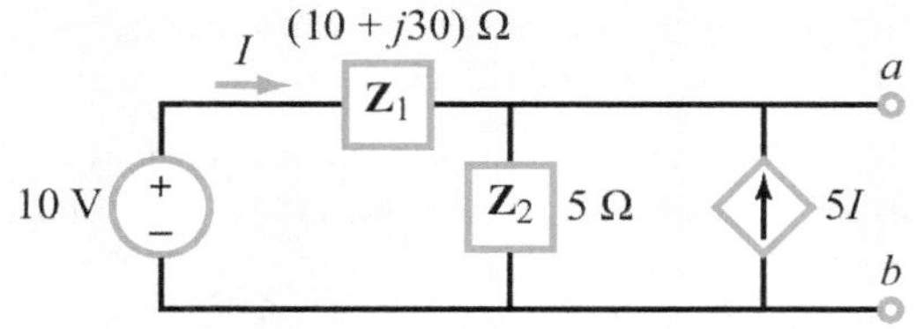

Figure E7.11

Answer: $\mathbf{V}_{\text{Th}} = 6\angle{-36.9^\circ}$ V, $\mathbf{Z}_{\text{Th}} = (2.6 + j1.8)\ \Omega$. (See CAD)

7-7 Phasor Diagrams

Consider the following sinusoidal signal $v_s(t)$ and its phasor counterpart $\mathbf{V}_s$:

$$v_s(t) = V_0\cos(\omega t + \phi) \iff \mathbf{V}_s = V_0\angle\phi. \tag{7.85}$$

The time-domain voltage $v_s(t)$ is characterized by three attributes: the amplitude V_0, the angular frequency ω, and the phase angle ϕ. In contrast, its counterpart in the phasor domain $\mathbf{V}_s$ is specified by only two attributes, V_0 and ϕ. This may suggest that ω becomes irrelevant when we analyze a circuit in the phasor domain, but that certainly is not true if the circuit contains capacitors and/or inductors. Whereas ω does not appear explicitly in the expressions for phasor currents and voltages, it is integral to the definitions of the capacitor impedance $\mathbf{Z}_C$ and inductor impedance $\mathbf{Z}_L$, which in turn define the **I**–**V** relationships for those two elements as

$$\mathbf{Z}_C = \frac{\mathbf{V}_C}{\mathbf{I}_C} = \frac{1}{j\omega C} = \frac{1}{\omega C}\angle{-90^\circ} \tag{7.86a}$$

and

$$\mathbf{Z}_L = \frac{\mathbf{V}_L}{\mathbf{I}_L} = j\omega L = \omega L\angle 90^\circ. \tag{7.86b}$$

In fact, the value of ω (relative to the values of L of C) can drastically change the behavior of a circuit:

▶ At dc, $\mathbf{Z}_C \to \infty$ (open circuit) and $\mathbf{Z}_L \to 0$ (short circuit); and conversely, as $\omega \to \infty$, $\mathbf{Z}_C \to 0$ and $\mathbf{Z}_L \to \infty$. ◀

A *phasor diagram* is a useful graphical tool for examining the relationships among the various currents and voltages in a circuit. Before considering multielement circuits, however, we will start by examining the phasor diagrams for R, L and C, individually. Figure 7-20 displays the phasor diagrams for **I** and **V** for all three elements, with **V** chosen as a reference by selecting its phase angle to be zero. Each phasor quantity is displayed in the complex plane in terms of its magnitude and phase angle. For the resistor, $\mathbf{V}_R$ and $\mathbf{I}_R$ always line up along the same direction because they are always in-phase. Since $\mathbf{V}_R$ was chosen to be purely real, so is $\mathbf{I}_R$.

Next, we consider the capacitor. In view of Eq. (7.86a),

$$\mathbf{I}_C = \frac{\mathbf{V}_C}{\mathbf{Z}_C} = j\omega C\mathbf{V}_C = \omega C\mathbf{V}_C\angle 90^\circ, \tag{7.87a}$$

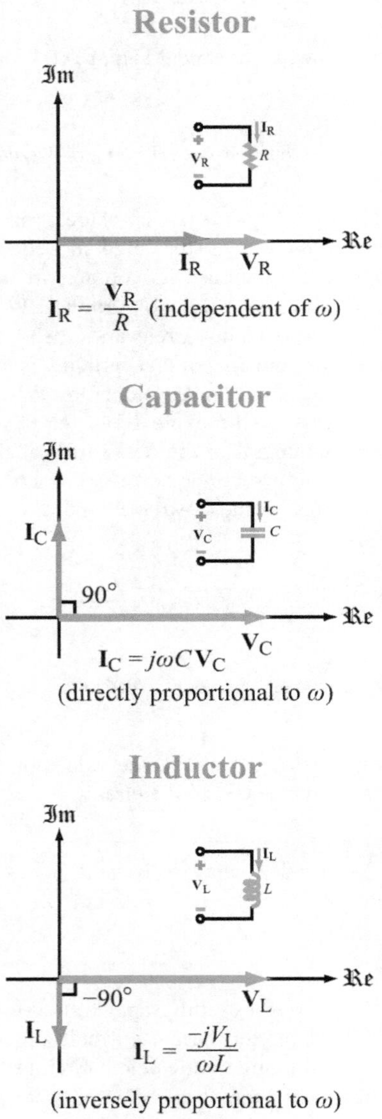

Figure 7-20: Phasor diagrams for R, L, and C.

which *positions the vector* $\mathbf{I}_C$ *ahead of* $\mathbf{V}_C$ *by* 90°. Hence:

> ▶ $\mathbf{I}_C$ leads $\mathbf{V}_C$ by 90°. ◀

For the inductor,

$$\mathbf{I}_L = \frac{\mathbf{V}_L}{j\omega L} = \frac{-j\mathbf{V}_L}{\omega L} = \frac{\mathbf{V}_L}{\omega L}\angle{-90^\circ}. \tag{7.87b}$$

Consequently,

> ▶ $\mathbf{I}_L$ lags $\mathbf{V}_L$ by 90°. ◀

For individual elements, the relationship between **I** and **V** is straightforward; given the position of either one of them in the complex plane, we can place the other one in accordance with the phase-angle shift appropriate to that element.

> ▶ For a multielement circuit, we can draw either a ***relative phasor diagram*** or an ***absolute phasor diagram***. For the relative phasor diagram, we usually choose a specific current or voltage and designate it as our reference phasor by arbitrarily assigning it a phase angle of 0°. ◀

The goal then is to use the phasor diagram to examine the relationships between and among the various currents and voltages in the circuit—which includes their magnitudes and *relative* phase angles—rather than to establish their absolute phase angles. In principle, it does not matter much which specific phasor voltage or current is selected as the reference, but in practice, we usually choose a phasor current or voltage that is common to lots of elements in the circuit. By way of illustration, Example 7-10 examines a series RLC circuit by displaying its phasor diagram twice, once using the current flowing through the loop as reference, and a second time with the voltage source as reference. The former results in a *relative phasor diagram*, whereas the latter results in an *absolute phasor diagram*.

Example 7-10: Relative versus Absolute Phasor Diagrams

The circuit in Fig. 7-21(a) is driven by a voltage source given by

$$\upsilon_s(t) = 20\cos(500t + 30^\circ) \text{ V}.$$

Generate: (a) a relative phasor diagram by selecting the phasor current **I** as a reference, and (b) an absolute phasor diagram.

Solution: Figure 7-21(b) displays the phasor-domain circuit with its RLC elements represented by their respective impedances.

(a) Relative Phasor Diagram

Selecting **I** as the reference phasor means that we assign it an unknown magnitude I_0 and a phase angle of 0°:

$$\mathbf{I} = I_0\angle{0^\circ}.$$

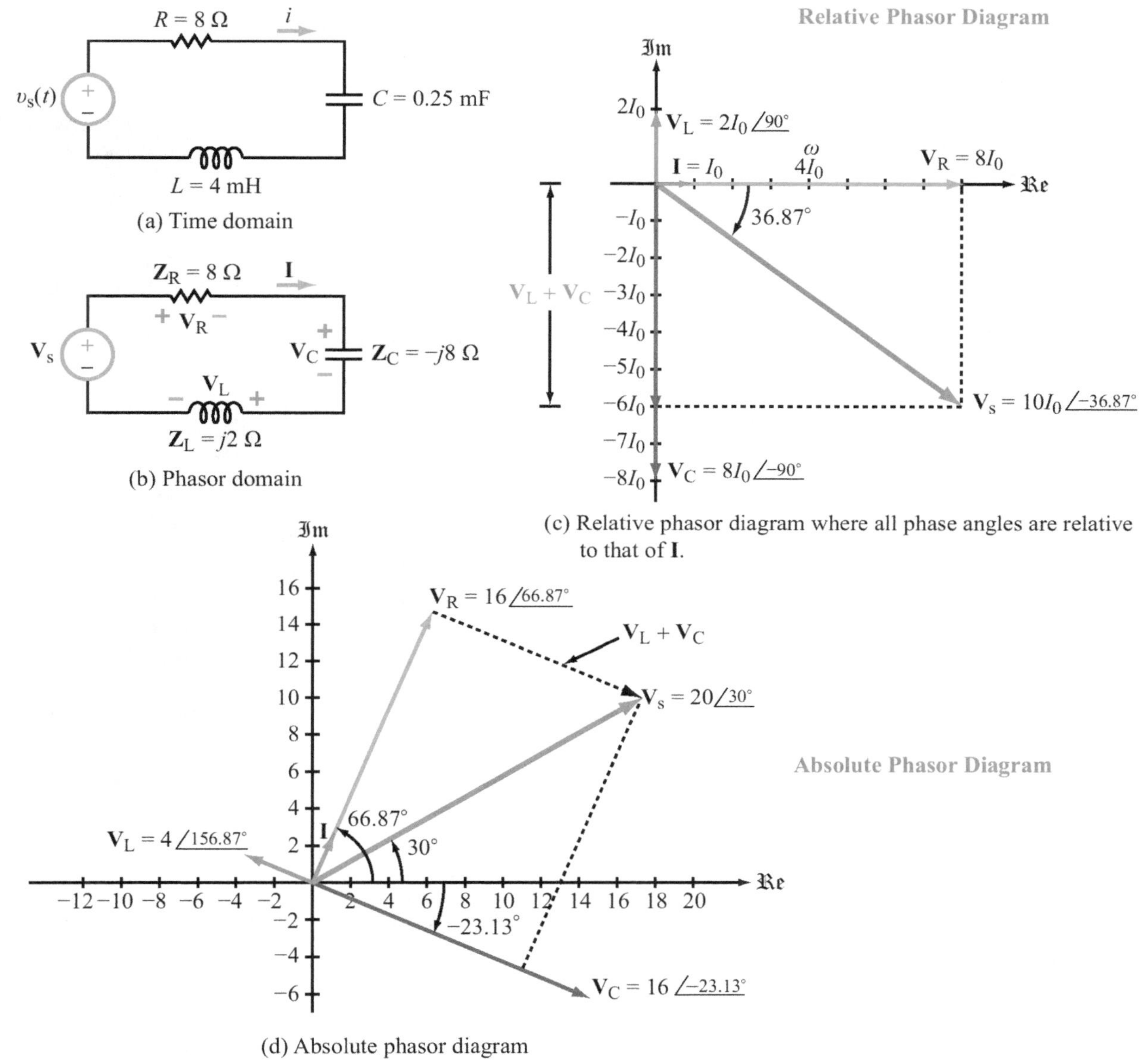

Figure 7-21: Circuit and phasor diagrams for Example 7-10. The true phase angle of **I** is 66.87°, so if the relative phasor diagram in (c) were to be rotated counterclockwise by that angle and the scale adjusted to incorporate the fact $I_0 = 2$, the diagram would coincide with the absolute phasor diagram in (d).

Because the true phase angle of **I** actually may not be zero, the vectors we will draw in the complex plane of the relative phasor diagram all will be shifted in orientation by exactly the same amount (namely by the true phase angle of **I**) so even though they may not have the correct orientations, they all will bear the correct *relative* orientations to one another.

We deduce from the functional form of $\upsilon_s(t)$ that $\omega = 500$ rad/s. In terms of $\mathbf{I}$, the voltages across R, C, and L are

$$\mathbf{V}_R = R\mathbf{I} = 8I_0\angle 0^\circ,$$

$$\mathbf{V}_C = \frac{\mathbf{I}}{j\omega C} = \frac{-jI_0}{500 \times 2.5 \times 10^{-4}} = -j8I_0 = 8I_0\angle -90^\circ,$$

and

$$\mathbf{V}_L = j\omega L\mathbf{I} = j500 \times 4 \times 10^{-3} I_0 = j2I_0 = 2I_0\angle 90^\circ,$$

and the sum of all three gives

$$\begin{aligned}\mathbf{V}_s &= \mathbf{V}_R + \mathbf{V}_C + \mathbf{V}_L \\ &= 8I_0 - j8I_0 + j2I_0 \\ &= (8 - j6)I_0 = \sqrt{8^2 + 6^2}\, I_0 e^{j\phi} = 10I_0\angle\phi,\end{aligned}$$

with

$$\phi = -\tan^{-1}\frac{6}{8} = -36.87^\circ.$$

Figure 7-21(c) displays the relative phasor diagram of the RLC circuit with $\mathbf{I}$ as a reference; the magnitudes of $\mathbf{V}_R$, $\mathbf{V}_C$, $\mathbf{V}_L$, and $\mathbf{V}_s$ are all measured in units of I_0, and their orientations are relative to that of $\mathbf{I}$.

(b) Absolute Phasor Diagram

The phasor counterpart of $\upsilon_s(t)$ is

$$\mathbf{V}_s = 20\angle 30^\circ \text{ V},$$

and the application of KVL around the loop leads to

$$\begin{aligned}\mathbf{I} &= \frac{\mathbf{V}_s}{R + j\omega L - \dfrac{j}{\omega C}} \\ &= \frac{20e^{j30^\circ}}{8 + j2 - j8} = \frac{20e^{j30^\circ}}{8 - j6} = \frac{20e^{j30^\circ}}{10e^{-j36.87^\circ}} = 2e^{j66.87^\circ} \text{ A},\end{aligned}$$

which states that the *true* phase angle of $\mathbf{I}$ is 66.87°. Given $\mathbf{I}$, we easily can calculate $\mathbf{V}_R$, $\mathbf{V}_C$, and $\mathbf{V}_L$. The phasor diagram shown in Fig. 7-21(d) is identical to that in Fig. 7-21(c), except that all vectors have been rotated in a counterclockwise direction by 66.87°.

Concept Question 7-13: For a capacitor, what is the phase angle of its phasor current, relative to that of its phasor voltage? (See CAD)

Concept Question 7-14: What is the difference between a *relative* phasor diagram and an *absolute* phasor diagram? (See CAD)

Exercise 7-12: Establish the relative phasor diagram for the circuit in Fig. E7.12 with $\mathbf{V}$ as the reference phasor.

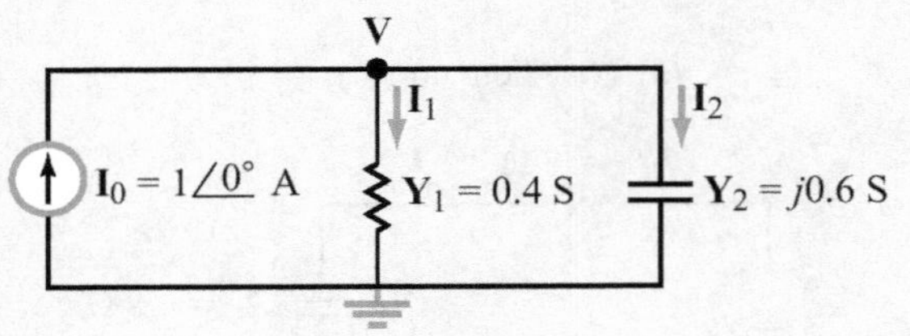

Figure E7.12

Answer:

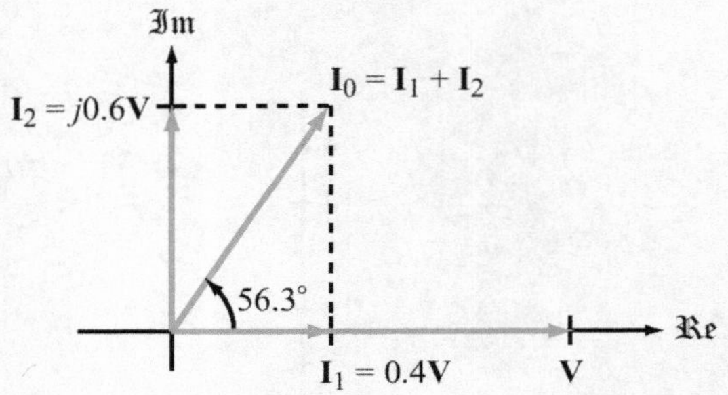

(See CAD)

7-8 Phase-Shift Circuits

In certain communication and signal-processing applications, we often need to shift the phase of an ac signal by adding (or subtracting) a phase angle of a specified value, ϕ. Thus, if the input voltage in Fig. 7-22 is

$$\upsilon_{in}(t) = V_1 \cos\omega t, \tag{7.88}$$

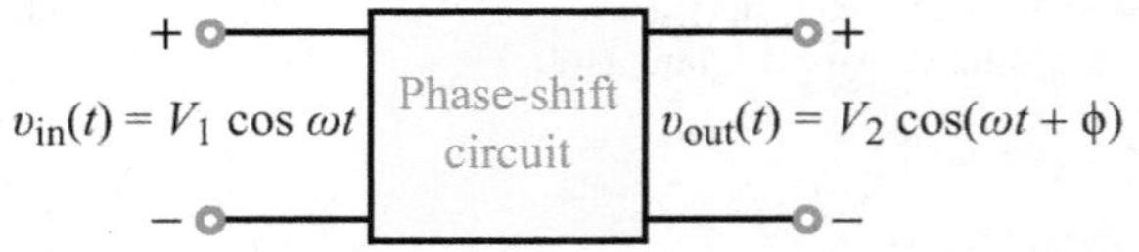

Figure 7-22: The phase-shift circuit changes the phase of the input signal by ϕ.

the function of the *phase-shift circuit* is to provide an output voltage given by

$$v_{out}(t) = V_2 \cos(\omega t + \phi). \quad (7.89)$$

The amplitude V_2 of the output voltage is related to V_1 (the amplitude of the input voltage) and to the configuration of the phase-shift circuit. RC circuits can be designed as phase shifters, with any specified positive or negative value of ϕ:

$$\begin{cases} v_{out} \text{ leads } v_{in} & \text{if } 0 \le \phi \le 180°, \\ v_{out} \text{ lags } v_{in} & \text{if } -180° \le \phi \le 0. \end{cases}$$

To illustrate the process, let us consider the simple RC circuit shown in Fig. 7-23(a). The input signal is given by

$$v_{in}(t) = 10 \cos 10^6 t \quad \text{V},$$

and the element values are $R = 2\ \Omega$ and $C = 0.2\ \mu$F. At $\omega = 10^6$ rad/s, the capacitor impedance is

$$\mathbf{Z}_C = \frac{-j}{\omega C} = \frac{-j}{10^6 \times 0.2 \times 10^{-6}} = -j5\ \Omega.$$

By voltage division in the phasor domain (Fig. 7-23(b)),

$$\mathbf{V}_{out1} = \frac{\mathbf{V}_{in} R}{R - \dfrac{j}{\omega C}} = \frac{\omega RC}{\sqrt{1 + \omega^2 R^2 C^2}}\, \mathbf{V}_{in} \underline{/\phi_1}, \quad (7.90a)$$

$$\mathbf{V}_{out2} = \frac{\mathbf{V}_{in}\left(\dfrac{-j}{\omega C}\right)}{R - \dfrac{j}{\omega C}} = \frac{1}{\sqrt{1 + \omega^2 R^2 C^2}}\, \mathbf{V}_{in} \underline{/\phi_2}, \quad (7.90b)$$

and the phase angles ϕ_1 and ϕ_2 are given by

$$\phi_1 = \tan^{-1}\left(\frac{1}{\omega RC}\right) \quad (7.91a)$$

and

$$\phi_2 = \phi_1 - 90° = \tan^{-1}\left(\frac{1}{\omega RC}\right) - 90°. \quad (7.91b)$$

For $\omega = 10^6$ rad/s, $R = 2\ \Omega$, $C = 0.2\ \mu$F, and $\mathbf{V}_{in} = 10$ V,

$$\mathbf{V}_{out1} = 3.71 \underline{/68.2°} = (1.38 + j3.45) \text{ V}$$

and

$$\mathbf{V}_{out2} = 9.28 \underline{/-21.8°} = (8.62 - j3.45) \text{ V}.$$

The phase angle ϕ_1 associated with $\mathbf{V}_{out1}$ is 68.2°, and the angle ϕ_2 associated with $\mathbf{V}_{out2}$ is −21.8°. As shown in the complex plane of Fig. 7-23(c), the angular separation between $\mathbf{V}_{out1}$ and $\mathbf{V}_{out2}$ is exactly 90°. Also, if we were to add $\mathbf{V}_{out1}$ and $\mathbf{V}_{out2}$ in the complex plane, their imaginary parts would cancel out and their real parts would add up to 10 V (the amplitude of $\mathbf{V}_{in}$).

In the time domain,

$$v_{out1}(t) = \mathfrak{Re}[\mathbf{V}_{out1} e^{j\omega t}] = 3.716 \cos(10^6 t + 68.2°) \text{ V} \quad (7.92)$$

and

$$v_{out2}(t) = \mathfrak{Re}[\mathbf{V}_{out2} e^{j\omega t}] = 9.285 \cos(10^6 t - 21.8°) \text{ V}. \quad (7.93)$$

Figure 7-23(a) provides a comparison of the waveform of the input signal $v_{in}(t)$ with that of $v_{out2}(t)$, the voltage across the capacitor. We note that because v_{out2} lags v_{in}, it always crosses the time axis later than v_{in} by a time delay Δt. If we denote t_0 as the time when $v_{in}(t)$ crosses the time axis and t_2 as the time when $v_{out2}(t)$ does, then

$$\omega t_0 = 10^6 t_0 = \frac{\pi}{2}$$

and

$$\omega t_2 + \phi_2 = 10^6 t_2 + \phi_2 = \frac{\pi}{2},$$

with

$$\phi_2 = -21.8° \times \left(\frac{\pi}{180°}\right) = -0.38 \text{ radians}.$$

Now that all quantities are in the same units, we can determine the time delay from

$$\Delta t_2 = t_2 - t_0 = -\phi_2 \times 10^{-6} = -(-0.38) \times 10^{-6} = 0.38\ \mu\text{s}.$$

By the same argument, v_{out1} leads v_{in} by 68.2°, and it crosses the time axis *sooner* than does $v_{in}(t)$ by

$$\Delta t_1 = 68.2° \times \frac{\pi}{180°} \times 10^{-6} = 1.19\ \mu\text{s}.$$

From the foregoing analysis, we conclude that for the simple RC circuit, we can use v_{out1} as our output if we want to add

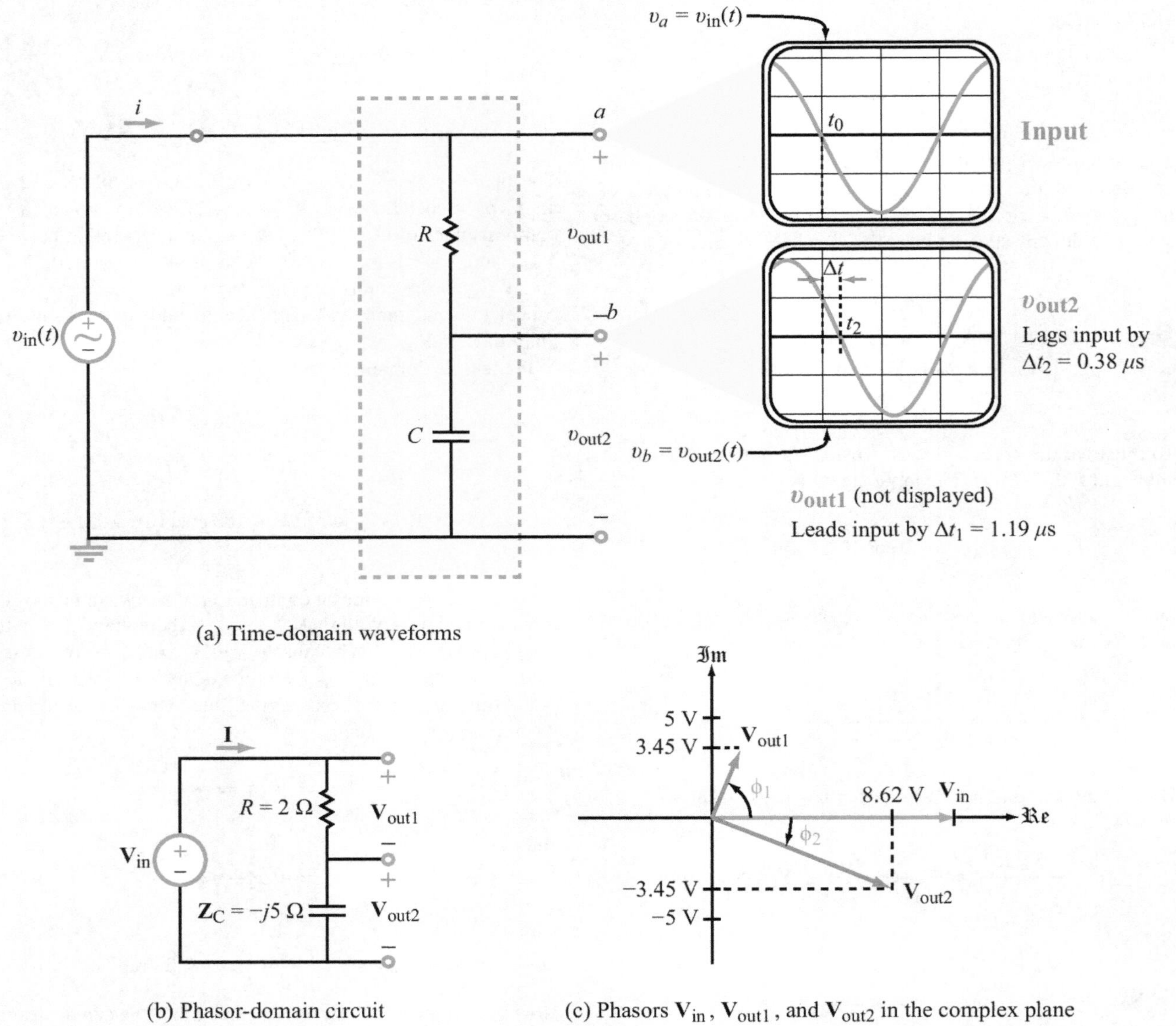

Figure 7-23: RC phase-shift circuit: the phase of v_{out1} (across R) leads the phase of $v_{in}(t)$, whereas the phase of v_{out2} (across C) lags the phase of $v_{in}(t)$.

a positive phase angle to the input v_{in}, and we can use v_{out2} as our output if we want to add a negative phase angle to v_{in}. Moreover, by adjusting the values of R and C (at a specific value of ω), we can change ϕ_1 to any value between 0 and 90°, and similarly, we can change ϕ_2 to any value between 0 and −90° (but not independently); as was noted earlier in connection with Fig. 7-23(c), the absolute values of ϕ_1 and ϕ_2 always add up to 90°. Another consideration that we should be aware of is that the magnitudes of v_{out1} and v_{out2} are linked to the magnitudes of ϕ_1 and ϕ_2 through the choices we make for R, C, and ω. For example, as ϕ_1 approaches 90°, v_{out1} approaches zero, so we can indeed phase-shift the input signal by an angle close to 90°,

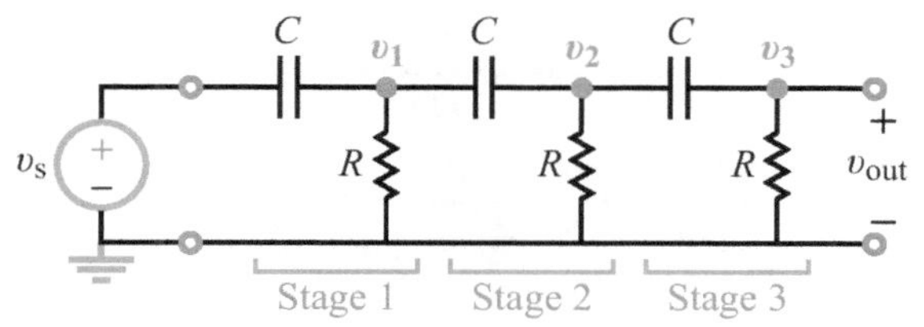

Figure 7-24: Three-stage, cascaded, RC phase-shifter (Example 7-11).

but the magnitude of the output signal will be too small to be useful. To overcome this limitation or to introduce phase-shift angles greater than 90°, we can use circuits with more than two elements, such as the cascaded circuit of Example 7-11.

▶ To generate a phase lead at the output, the cascading arrangement should be as that shown in Fig. 7-24, but to generate a phase lag, the locations of R and C should be interchanged. ◀

Example 7-11: Cascaded Phase-Shifter

The circuit in Fig. 7-24 uses a 3-stage cascaded phase-shifter to produce an output signal $v_{out}(t)$ whose phase is 120° ahead of the input signal $v_s(t)$. If $\omega = 10^3$ (rad/s) and $C = 1\ \mu$F, determine R and the ratio of the amplitude of v_{out} to that of v_s.

Solution: Application of nodal analysis at nodes $\mathbf{V}_1$ and $\mathbf{V}_2$ in the phasor domain gives

$$\frac{\mathbf{V}_1 - \mathbf{V}_s}{\mathbf{Z}_C} + \frac{\mathbf{V}_1}{R} + \frac{\mathbf{V}_1 - \mathbf{V}_2}{\mathbf{Z}_C} = 0 \quad (7.94)$$

and

$$\frac{\mathbf{V}_2 - \mathbf{V}_1}{\mathbf{Z}_C} + \frac{\mathbf{V}_2}{R} + \frac{\mathbf{V}_2}{R + \mathbf{Z}_C} = 0, \quad (7.95)$$

where $\mathbf{Z}_C = 1/j\omega C$. Moreover, through voltage division, $\mathbf{V}_3$ is related to $\mathbf{V}_2$ by

$$\mathbf{V}_3 = \left(\frac{R}{R + \mathbf{Z}_C}\right)\mathbf{V}_2. \quad (7.96)$$

Simultaneous solution of Eqs. (7.94) and (7.95), followed by several steps of algebra, leads to the expressions

$$\frac{\mathbf{V}_1}{\mathbf{V}_s} = \frac{x[(x^2-1) - j3x]}{(x^3 - 5x) + j(1 - 6x^2)}, \quad (7.97)$$

$$\frac{\mathbf{V}_2}{\mathbf{V}_s} = \frac{x^2(x - j1)}{(x^3 - 5x) + j(1 - 6x^2)}, \quad (7.98)$$

and

$$\frac{\mathbf{V}_3}{\mathbf{V}_s} = \frac{x^3}{(x^3 - 5x) + j(1 - 6x^2)}, \quad (7.99)$$

where

$$x = \omega RC. \quad (7.100)$$

The magnitude and phase of $\mathbf{V}_3$ (both relative to those of $\mathbf{V}_s$) are

$$\left|\frac{\mathbf{V}_3}{\mathbf{V}_s}\right| = \frac{x^3}{[(x^3 - 5x)^2 + (1 - 6x^2)^2]^{1/2}}, \quad (7.101a)$$

and

$$\phi_3 = -\tan^{-1}\left(\frac{1 - 6x^2}{x^3 - 5x}\right). \quad (7.101b)$$

To satisfy the stated requirement, we set $\phi_3 = 120°$ and solve for x:

$$\tan 120° = -1.732 = -\left(\frac{1 - 6x^2}{x^3 - 5x}\right),$$

which leads to

$$x = 1.1815. \quad (7.102)$$

Given that $\omega = 10^3$ rad/s and $C = 1\ \mu$F, it follows that

$$R = \frac{x}{\omega C} = \frac{1.1815}{10^3 \times 10^{-6}} = 1.1815\ \text{k}\Omega \approx 1.2\ \text{k}\Omega.$$

With $x = 1.1815$, Eq. (7.101a) gives

$$\left|\frac{\mathbf{V}_3}{\mathbf{V}_s}\right| = 0.194.$$

Note that:

- The use of multiple stages allowed us to shift the phase by more than 90°.
- However, the magnitude of the output voltage is about 20% of that of the input.

Concept Question 7-15: Describe the function of a phase-shift circuit in terms of time delay or time advance of the waveform. (See CAD)

Concept Question 7-16: When is it necessary to use multiple stages to achieve the desired phase shift? (See CAD)

Exercise 7-13: Repeat Example 7-11, but use only two stages of RC phase shifters.

Answer: $R \approx 2.2$ kΩ; $|\mathbf{V}_{\text{out}}/\mathbf{V}_{\text{s}}| = 0.63$. (See CAD)

Exercise 7-14: Design a two-stage RC phase shifter that provides a phase shift of negative 120° at $\omega = 10^4$ rad/s. Assume $C = 1\ \mu$F.

Answer: $R \approx 220\ \Omega$. (See CAD)

7-9 Phasor-Domain Analysis Techniques

The analysis techniques introduced in Chapter 3 in connection with resistive circuits are all equally applicable for analyzing ac circuits in the phasor domain. The only fundamental difference is that after transferring the circuit from the time domain to the phasor domain, the operations conducted in the phasor domain involve the use of complex algebra, as opposed to just real numbers. Otherwise, the circuit laws and methods of solution are identical.

At this stage, instead of repeating the details of these various techniques, a more effective approach is to illustrate their implementation procedures through concrete examples. Examples 7-12 through 7-16 are designed to do just that.

Example 7-12: Nodal Analysis

Apply the nodal-analysis method to determine $i_{\text{L}}(t)$ in the circuit of Fig. 7-25(a). The sources are given by:

$$\upsilon_{\text{s}_1}(t) = 12 \cos 10^3 t \text{ V},$$

$$\upsilon_{\text{s}_2}(t) = 6 \sin 10^3 t \text{ V}.$$

Solution: We first demonstrate how to solve this problem using the standard nodal-analysis method (Section 3-2), and then we solve it again by applying the by-inspection method (Section 3-4).

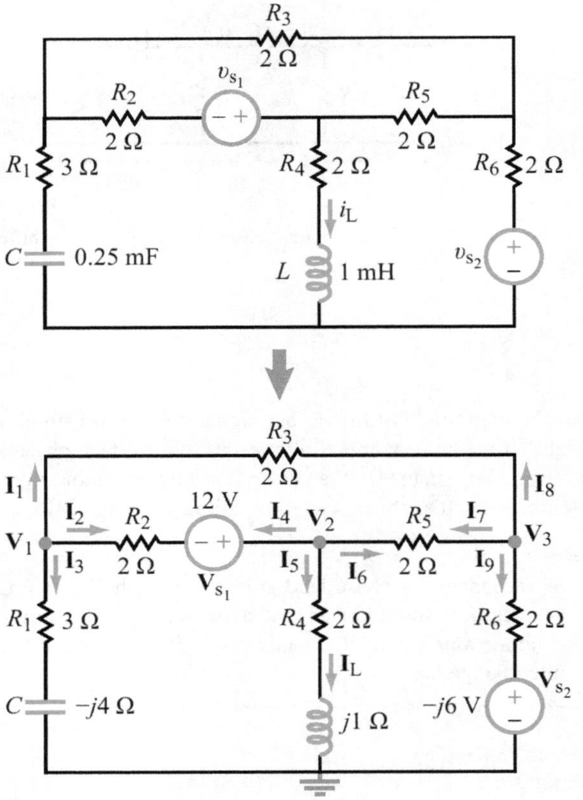

Figure 7-25: Circuit for Example 7-12 in (a) the time domain and (b) the phasor domain.

Nodal-analysis method

Our first step is to transform the given circuit to the phasor domain. Accordingly,

$$\mathbf{Z}_{\text{C}} = \frac{1}{j\omega C} = \frac{-j}{10^3 \times 0.25 \times 10^{-3}} = -j4\ \Omega,$$

$$\mathbf{Z}_{\text{L}} = j\omega L = j\,10^3 \times 10^{-3} = j1\ \Omega,$$

$$\upsilon_{\text{s}_1} = 12 \cos 10^3 t \quad \longleftrightarrow \quad \mathbf{V}_{\text{s}_1} = 12 \text{ V},$$

and

$$\upsilon_{\text{s}_2} = 6 \sin 10^3 t \quad \longleftrightarrow \quad \mathbf{V}_{\text{s}_2} = -j6 \text{ V},$$

where for $\mathbf{V}_{\text{s}_2}$ we used the property given in Table 7-2, namely that the phasor counterpart of $\sin \omega t$ is $-j$. Using these values, we generate the phasor-domain circuit given in Fig. 7-25(b) in which we selected one of the extraordinary nodes as a ground node and assigned phasor voltages $\mathbf{V}_1$ to $\mathbf{V}_3$ to the other three.

Our plan is to write the voltage node equations at nodes 1 to 3, solve them simultaneously to find $\mathbf{V}_1$ to $\mathbf{V}_3$, and then use the value of $\mathbf{V}_2$ to obtain $\mathbf{I}_\mathrm{L}$. The final step will involve transforming $\mathbf{I}_\mathrm{L}$ to the time domain to obtain $i_\mathrm{L}(t)$.

At node 1, KCL requires that

$$\mathbf{I}_1 + \mathbf{I}_2 + \mathbf{I}_3 = 0. \tag{7.103}$$

In terms of node voltages $\mathbf{V}_1$ to $\mathbf{V}_3$,

$$\mathbf{I}_1 = \frac{\mathbf{V}_1 - \mathbf{V}_3}{R_3} = \frac{\mathbf{V}_1 - \mathbf{V}_3}{2},$$

$$\mathbf{I}_2 = \frac{\mathbf{V}_1 - \mathbf{V}_2 + \mathbf{V}_{\mathrm{s}_1}}{R_2} = \frac{\mathbf{V}_1 - \mathbf{V}_2 + 12}{2},$$

and

$$\mathbf{I}_3 = \frac{\mathbf{V}_1}{R_1 + \mathbf{Z}_\mathrm{C}} = \frac{\mathbf{V}_1}{3 - j4}.$$

Inserting the expressions for $\mathbf{I}_1$ to $\mathbf{I}_3$ in Eq. (7.103) and then rearranging the terms leads to

$$\left(\frac{1}{2} + \frac{1}{2} + \frac{1}{3 - j4}\right)\mathbf{V}_1 - \frac{1}{2}\mathbf{V}_2 - \frac{1}{2}\mathbf{V}_3 = -6. \tag{7.104}$$

The coefficient of $\mathbf{V}_1$ can be simplified as follows:

$$\begin{aligned}\frac{1}{2} + \frac{1}{2} + \frac{1}{3 - j4} &= 1 + \frac{1}{3 - j4} \\ &= \frac{3 - j4 + 1}{3 - j4} \\ &= \frac{4 - j4}{3 - j4} \times \frac{3 + j4}{3 + j4} \\ &= \frac{(12 + 16) + j(16 - 12)}{9 + 16} = 1.12 + j0.16.\end{aligned} \tag{7.105}$$

Inserting Eq. (7.105) in Eq. (7.104) and multiplying all terms by 2 leads to the following simplified algebraic equation for node 1:

$$(2.24 + j0.32)\mathbf{V}_1 - \mathbf{V}_2 - \mathbf{V}_3 = -12 \quad \textbf{(node 1)}. \tag{7.106}$$

Similarly, at node 2,

$$\frac{\mathbf{V}_2 - \mathbf{V}_1 - 12}{2} + \frac{\mathbf{V}_2}{2 + j1} + \frac{\mathbf{V}_2 - \mathbf{V}_3}{2} = 0,$$

which can be simplified to

$$-\mathbf{V}_1 + (2.8 - j0.4)\mathbf{V}_2 - \mathbf{V}_3 = 12 \quad \textbf{(node 2)}, \tag{7.107}$$

and at node 3,

$$\frac{\mathbf{V}_3 - \mathbf{V}_2}{2} + \frac{\mathbf{V}_3 - \mathbf{V}_1}{2} + \frac{\mathbf{V}_3 + j6}{2} = 0,$$

or

$$-\mathbf{V}_1 - \mathbf{V}_2 + 3\mathbf{V}_3 = -j6 \quad \textbf{(node 3)}. \tag{7.108}$$

Equations (7.106) to (7.108) now are ready to be cast in matrix form:

$$\begin{bmatrix} (2.24 + j0.32) & -1 & -1 \\ -1 & (2.8 - j0.4) & -1 \\ -1 & -1 & 3 \end{bmatrix} \begin{bmatrix} \mathbf{V}_1 \\ \mathbf{V}_2 \\ \mathbf{V}_3 \end{bmatrix} = \begin{bmatrix} -12 \\ 12 \\ -j6 \end{bmatrix}. \tag{7.109}$$

Matrix inversion, either manually or by MATLAB or MathScript software, provides the solution:

$$\mathbf{V}_1 = -(4.72 + j0.88) \text{ V}, \tag{7.110a}$$

$$\mathbf{V}_2 = (2.46 - j0.89) \text{ V}, \tag{7.110b}$$

and

$$\mathbf{V}_3 = -(0.76 + j2.59) \text{ V}. \tag{7.110c}$$

Hence,

$$\mathbf{I}_\mathrm{L} = \frac{\mathbf{V}_2}{2 + j1} = \frac{2.46 - j0.89}{2 + j1} = 0.81 - j0.85 = 1.17e^{-j46.5^\circ} \text{ A},$$

and its corresponding time-domain counterpart is

$$\begin{aligned} i_\mathrm{L}(t) &= \mathfrak{Re}[\mathbf{I}_\mathrm{L} e^{j1000t}] \\ &= \mathfrak{Re}[1.17e^{-j46.4^\circ} e^{j1000t}] = 1.17\cos(1000t - 46.5^\circ) \text{ A}. \end{aligned} \tag{7.111}$$

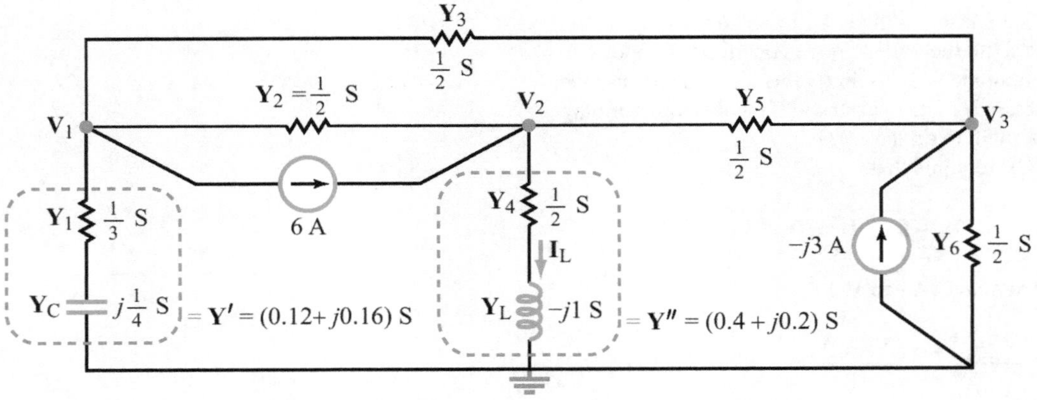

Figure 7-26: Equivalent of the circuit in Fig. 7-25, after source transformation of voltage sources into current sources and replacement of passive elements with their equivalent admittances.

By-inspection method

Implementation of the nodal-analysis by-inspection method requires that the circuit contain no dependent sources and that all independent sources in the circuit be current sources. The first condition is valid for the circuit in Fig. 7-25(b), but the second one is not. However, both voltage sources in Fig. 7-25(b) have in-series resistors associated with them, so we easily can transform them into current sources. The resultant circuit is shown in Fig. 7-26, in which not only have the voltage sources been replaced with equivalent current sources, but all impedances have also been replaced with their equivalent admittances ($\mathbf{Y} = 1/\mathbf{Z}$). For the 3-node case, the phasor-domain equivalent of Eq. (3.25) is given by

$$\begin{bmatrix} \mathbf{Y}_{11} & \mathbf{Y}_{12} & \mathbf{Y}_{13} \\ \mathbf{Y}_{21} & \mathbf{Y}_{22} & \mathbf{Y}_{23} \\ \mathbf{Y}_{31} & \mathbf{Y}_{32} & \mathbf{Y}_{33} \end{bmatrix} \begin{bmatrix} \mathbf{V}_1 \\ \mathbf{V}_2 \\ \mathbf{V}_3 \end{bmatrix} = \begin{bmatrix} \mathbf{I}_{t_1} \\ \mathbf{I}_{t_2} \\ \mathbf{I}_{t_3} \end{bmatrix}, \qquad (7.112)$$

where

$\mathbf{Y}_{kk} =$ sum of all admittances connected to node k

$\mathbf{Y}_{k\ell} = \mathbf{Y}_{\ell k} =$ *negative* of admittance(s) connecting nodes k and ℓ, with $k \neq \ell$

$\mathbf{V}_k =$ unknown phasor voltage at node k

$\mathbf{I}_{t_k} =$ total of phasor current sources entering node k (a negative sign applies to a current source leaving the node).

For the circuit in Fig. 7-26,

$$\mathbf{Y}_{11} = \mathbf{Y}' + \mathbf{Y}_2 + \mathbf{Y}_3 = (\mathbf{Y}' + 0.5 + 0.5) \text{ S}, \qquad (7.113)$$

where $\mathbf{Y}'$ is the sum of $\mathbf{Y}_1$ and $\mathbf{Y}_C$. The rule for adding two in-series admittances is the same as that for adding two in-parallel impedances:

$$\mathbf{Y}' = \mathbf{Y}_1 \parallel \mathbf{Y}_C = \frac{\frac{1}{3} \times j\frac{1}{4}}{\frac{1}{3} + j\frac{1}{4}} = (0.12 + j0.16) \text{ S}.$$

Hence,

$$\mathbf{Y}_{11} = (1.12 + j0.16) \text{ S}.$$

Similarly,

$$\mathbf{Y}_{22} = \mathbf{Y}'' + 0.5 + 0.5$$

$$= (\mathbf{Y}_4 \parallel \mathbf{Y}_L) + 1 = \frac{0.5 \times (-j1)}{0.5 - j1} + 1 = (1.4 - j0.2) \text{ S},$$

$$\mathbf{Y}_{33} = 0.5 + 0.5 + 0.5 = 1.5 \text{ S}.$$

Also, $\mathbf{Y}_{12} = \mathbf{Y}_{21} = \mathbf{Y}_{13} = \mathbf{Y}_{31} = \mathbf{Y}_{23} = \mathbf{Y}_{32} = -0.5$ S, $\mathbf{I}_{t_1} = -6$ A, $\mathbf{I}_{t_2} = 6$ A, and $\mathbf{I}_{t_3} = -j3$ A. Entering the values of all of these quantities in Eq. (7.112) gives

$$\begin{bmatrix} (1.12 + j0.16) & -0.5 & -0.5 \\ -0.5 & (1.4 - j0.2) & -0.5 \\ -0.5 & -0.5 & 1.5 \end{bmatrix} \begin{bmatrix} \mathbf{V}_1 \\ \mathbf{V}_2 \\ \mathbf{V}_3 \end{bmatrix} = \begin{bmatrix} -6 \\ 6 \\ -j3 \end{bmatrix}. \qquad (7.114)$$

Multiplication of both sides of Eq. (7.114) by a factor of 2 would produce exactly the matrix equation given by Eq. (7.109), as expected. Consequently, the final expression for $i_L(t)$ is identical to that given by Eq. (7.111).

Technology Brief 19
Crystal Oscillators

Circuits that produce well-defined ac oscillations are fundamental to many applications: frequency generators for radio transmitters, filters for radio receivers, and processor clocks, among many. An ***oscillator*** is a circuit that takes a dc input and produces an ac output at a desired frequency. Temperature stability, long lifetime, and little frequency drift over time are important considerations when designing oscillators.

A circuit consisting of an inductor and a capacitor will resonate at a specific natural frequency $\omega_0 = 1/\sqrt{LC}$. In such a circuit, energy is stored in the capacitor's electric field and the inductor's magnetic field. Once energy is introduced into the circuit (for example, by applying an initial voltage to the capacitor), it will begin to flow back and forth (oscillate) between the two components; this constant conversion gives rise to oscillations in voltage and current at the resonant frequency. In an ideal circuit with no dissipation (no resistor), the oscillations will continue at this one frequency forever.

Making oscillating circuits from individual inductor and capacitor components, however, is relatively impractical and yields devices with poor reproducibility, high temperature drift (i.e., the resonant frequency changes with the temperature surrounding the circuit), and poor overall lifetime. Since the early part of the 20th century, resonators have been made in a completely different way, namely by using tiny, mechanically resonating pieces of quartz glass.

Quartz Crystals and Piezoelectricity

In 1880, the Curie brothers demonstrated that certain crystals—such as *quartz*, topaz, and tourmaline—become electrically polarized when subjected to mechanical stress. That is, such a crystal exhibits a voltage across it if compressed, and a voltage of opposite polarity if stretched. The converse property, namely that if a voltage is applied across a crystal it will change its shape (compress or stretch), was predicted a year later by Gabriel Lippman (who received the 1908 Nobel Prize in physics for producing the first color photographic plate). Collectively, these bidirectional properties of crystals are known as ***piezoelectricity***. Piezoelectric crystals are used in microphones to convert mechanical vibrations of the crystal surface, caused by acoustic waves, into electrical signals, and the converse is used in loudspeakers. Piezoelectricity can also be applied to make a quartz crystal resonate. If a voltage of the proper polarity is applied across one of the principal axes of the crystal, it will shrink along the direction of that axis. Upon removing the voltage, the crystal will try to restore its shape to its original unstressed state by stretching itself, but its stored compression energy is sufficient to allow it to stretch beyond the unstressed state, thereby generating a voltage whose polarity is opposite of that of the original voltage that was used to compress it. This induced voltage will cause it to shrink, and the process will continue back and forth until the energy initially introduced by the external voltage source is totally dissipated. The behavior of the crystal is akin to an underdamped RLC circuit.

In addition to crystals, some metals and ceramics are also used for making oscillators. Because the resonant frequency can be chosen by specifying the type of material and its shape, such oscillators are easy to manufacture in large quantities, and their oscillation frequencies can be designed with a high degree of precision. Moreover, quartz crystals have good temperature performance, which means that they can be used in many applications without the need for temperature compensation, including in clocks, radios, and cellphones.

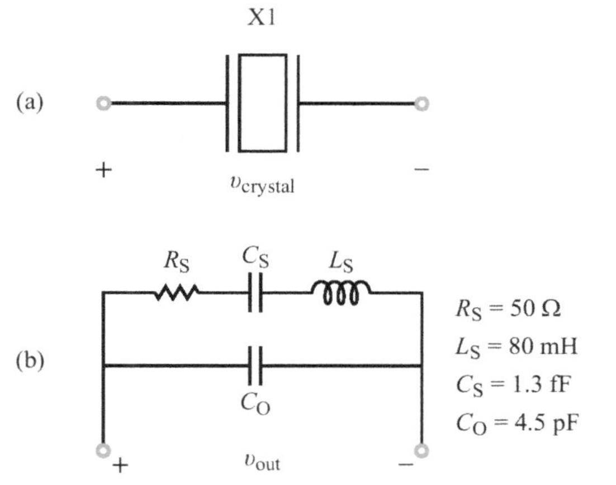

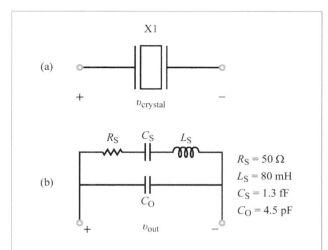

Figure TF19-1: (a) Quartz crystal circuit symbol and (b) equivalent circuit. Values given are for a 5 MHz crystal.

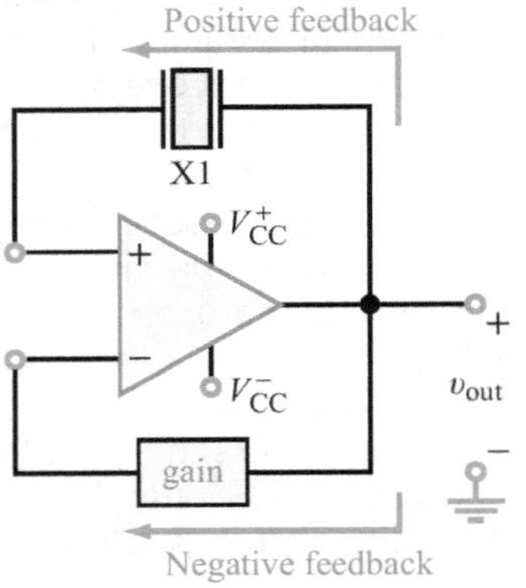

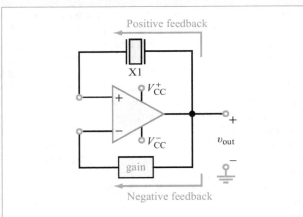

Figure TF19-2: Schematic block diagram of an oscillator circuit. An oscillator is wired into the positive feedback path, while a negative feedback path is used to control gain.

Crystal Equivalent Circuit and Oscillator Design

The electrical behavior of a quartz crystal can be modeled as a series RLC circuit (L_S, C_S, R_S) in parallel with a shunt capacitor (C_O). The RLC circuit models the fundamental oscillator behavior with dissipation. The shunt capacitor is mostly due to the capacitance between the two plates that actuate the quartz crystal. Figure TF19-1 shows the circuit symbol, the equivalent circuit with sample values for a commercial 12 MHz crystal along with expressions and values for the resonant frequencies and Q.

The crystal is, of course, not sufficient to produce a continuous oscillating waveform; we need to excite the circuit and keep it running. A common way to do this is to insert the crystal in the positive feedback path of an amplifier (Fig. TF19-2). The amplifier, of course, is supplied with dc power (V_{CC}^+ and V_{CC}^-). Note that no input signal is applied to the circuit. Initially, the output generates no oscillations; however, any noise at v_{out} that is at the resonant frequency of X1 will be fed back to the input and amplified. This positive feedback will quickly ramp up the output so that it is oscillating at the resonant frequency of the crystal. A negative feedback loop is also commonly used to control the overall gain and prevent the circuit from clipping the signal against the op amp's supply voltages V_{CC}^+ and V_{CC}^-.

In order to oscillate continuously, a circuit must meet the following two ***Barkhausen criteria***: (1) The gain of the circuit must be greater than 1. (This makes sense, for otherwise the signal will neither get amplified nor establish a resonating condition.) (2) The phase shift from the input to the output, then across the feedback loop to the input must be 0. (This also makes sense, since if there is non-zero phase shift, the signals will destructively interfere and the oscillator will not be able to start up.)

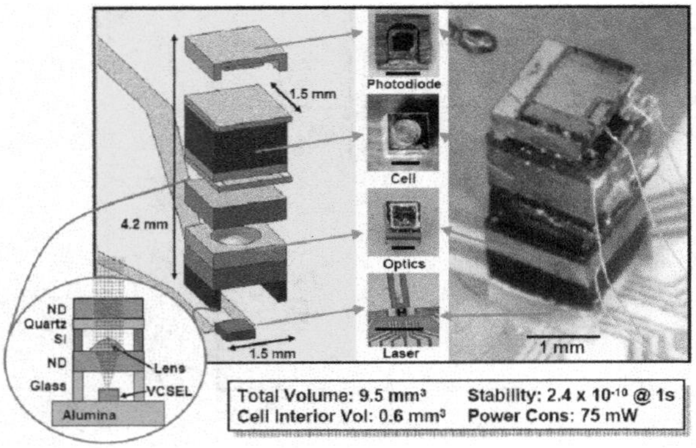

Figure TF19-3: Schematic (left) and photo (right) of a tiny atomic physics package used in a chip-scale atomic clock. (Courtesy of Clark Nguyen, U.C. Berkeley, and John Kitching, National Institutes of Standards and Technology.

Advances in Resonators and Clocks

As good as quartz resonators are, even the best among them will drift in frequency by 0.01 ppm per year as a result of aging of the crystal. If the oscillator is being used to keep time (as in your digital watch), this dictates how many seconds (or fractions thereof) the clock will lose per year. Put differently, this drift puts a hard limit on how long a clock can run without calibration. The same phenomenon limits how well independent clocks can stay synchronized with each other. Atomic clocks provide an extra level of precision by basing their oscillations on atomic transitions; these clocks are accurate to about 10^{-9} seconds per day. Recently, a chip-scale version of an atomic clock (Fig. TF19-3) was demonstrated by the ***National Institute for Standards and Technology*** (NIST); it consumes 75 mW and was the size of a grain of rice (10 mm^3). Other recent efforts for making oscillators for communication have focused on replacing the quartz crystal with a type of micromechanical resonator.

Exercise 7-15: Write down the node-voltage matrix equation for the circuit in Fig. E7.15.

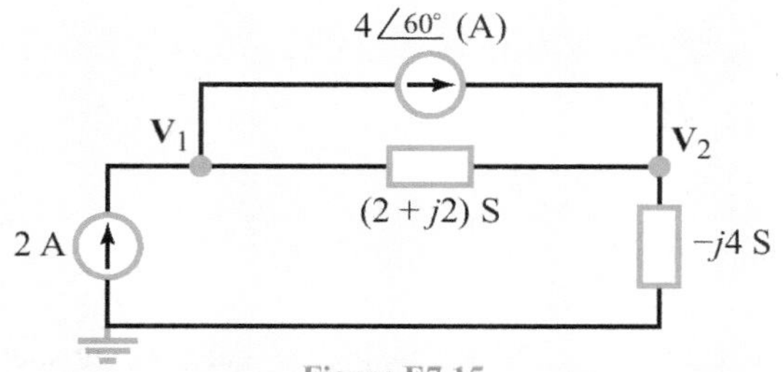

Figure E7.15

Answer:

$$\begin{bmatrix} (2+j2) & -(2+j2) \\ -(2+j2) & (2-j2) \end{bmatrix} \begin{bmatrix} \mathbf{V}_1 \\ \mathbf{V}_2 \end{bmatrix} = \begin{bmatrix} 2-4e^{j60^\circ} \\ 4e^{j60^\circ} \end{bmatrix}.$$

(See CAD)

Example 7-13: Circuit with a Supernode

The circuit in Fig. 7-27, which is already in the phasor domain, contains two independent voltage sources, both oscillating at an angular frequency $\omega = 2 \times 10^3$ rad/s, and both characterized by a phase angle of 0°. Determine $i_L(t)$.

Solution: Because nodes $\mathbf{V}_1$ and $\mathbf{V}_2$ are connected by a voltage source, their combination constitutes a supernode. When we apply KCL to a supernode, we simply sum all the currents leaving both of its nodes as if the two nodes are one,

$$\mathbf{I}_1 + \mathbf{I}_2 + \mathbf{I}_3 + \mathbf{I}_4 = 0,$$

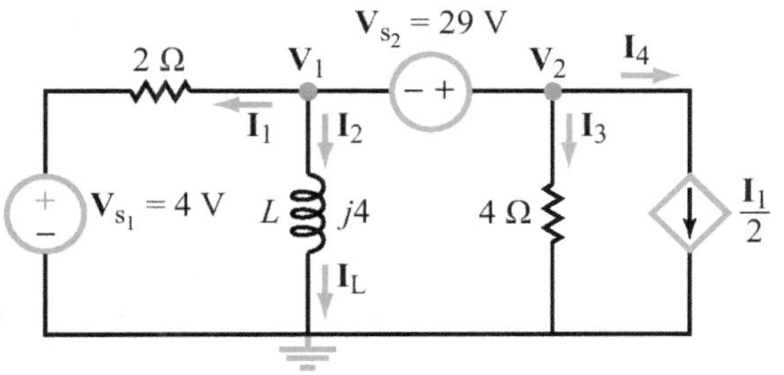

Figure 7-27: Phasor-domain circuit containing a supernode and a dependent source (Example 7-13).

or

$$\frac{\mathbf{V}_1 - 4}{2} + \frac{\mathbf{V}_1}{j4} + \frac{\mathbf{V}_2}{4} + \frac{\mathbf{I}_1}{2} = 0, \qquad (7.115)$$

and we also incorporate the auxiliary equation relating the two nodes, namely

$$\mathbf{V}_2 - \mathbf{V}_1 = 29. \qquad (7.116)$$

From the circuit, the current $\mathbf{I}_1$ in Eq. (7.115) is given by

$$\mathbf{I}_1 = \frac{\mathbf{V}_1 - 4}{2}. \qquad (7.117)$$

Using Eqs. (7.116) and (7.117) in Eq. (7.115) and then solving for $\mathbf{V}_1$ leads to

$$\mathbf{V}_1 = -(4 + j1) \text{ V},$$

which in turn gives

$$\mathbf{I}_L = \frac{\mathbf{V}_1}{j4} = -\frac{(4+j1)}{j4} = (-0.25 + j1) = 1.03\angle 104^\circ \text{ A}.$$

With $\omega = 2 \times 10^3$ rad/s, the inductor current in the time domain is given by

$$\begin{aligned} i_L(t) &= \Re\mathfrak{e}[\mathbf{I}_L e^{j\omega t}] = \Re\mathfrak{e}[1.03 e^{j104^\circ} e^{j2\times 10^3 t}] \\ &= 1.03 \cos(2 \times 10^3 t + 104^\circ) \text{ A}. \end{aligned}$$

Example 7-14: Mesh Analysis

Apply the mesh-analysis method to determine $i_L(t)$ in the circuit of Fig. 7-28, given that $\omega = 1000$ rad/s.

Solution: The circuit shown in Fig. 7-28 has mesh currents $\mathbf{I}_1$ to $\mathbf{I}_3$. Since the circuit has no dependent sources and no independent current sources, it is suitable for application of the mesh-analysis by-inspection method. For a three-loop circuit, the phasor-domain parallel of Eq. (3.28) assumes the form:

$$\begin{bmatrix} \mathbf{Z}_{11} & \mathbf{Z}_{12} & \mathbf{Z}_{13} \\ \mathbf{Z}_{21} & \mathbf{Z}_{22} & \mathbf{Z}_{23} \\ \mathbf{Z}_{31} & \mathbf{Z}_{32} & \mathbf{Z}_{33} \end{bmatrix} \begin{bmatrix} \mathbf{I}_1 \\ \mathbf{I}_2 \\ \mathbf{I}_3 \end{bmatrix} = \begin{bmatrix} \mathbf{V}_{t_1} \\ \mathbf{V}_{t_2} \\ \mathbf{V}_{t_3} \end{bmatrix}, \qquad (7.118)$$

where

$\mathbf{Z}_{kk}$ = sum of all impedances in loop k

$\mathbf{Z}_{k\ell}$ = $\mathbf{Z}_{\ell k}$ = *negative* of impedance(s) shared by loop k and ℓ, with $k \neq \ell$

$\mathbf{I}_k$ = unknown phasor current of loop k

$\mathbf{V}_{t_k}$ = total of phasor voltage sources contained in loop k, with the polarity defined as positive if $\mathbf{I}_k$ flows from (−) to (+) through the source.

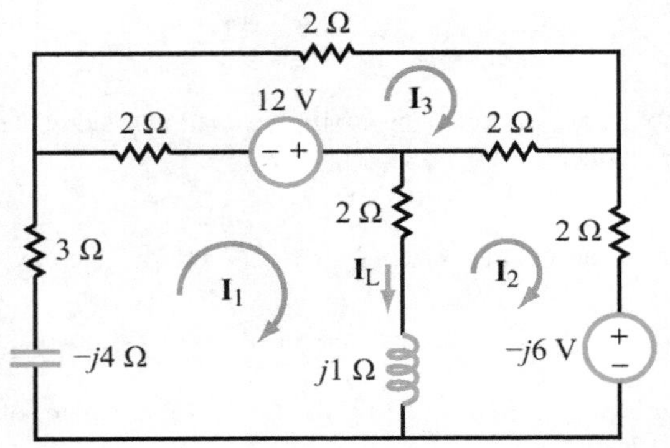

Figure 7-28: Circuit for Example 7-14.

In view of these definitions, the matrix equation for the circuit in Fig. 7-28 is given by

$$\begin{bmatrix} (7-j3) & -(2+j1) & -2 \\ -(2+j1) & (6+j1) & -2 \\ -2 & -2 & 6 \end{bmatrix} \begin{bmatrix} \mathbf{I}_1 \\ \mathbf{I}_2 \\ \mathbf{I}_3 \end{bmatrix} = \begin{bmatrix} 12 \\ j6 \\ -12 \end{bmatrix}. \quad (7.119)$$

Matrix inversion leads to

$$\mathbf{I}_1 = (0.43 + j0.86) \text{ A},$$

$$\mathbf{I}_2 = (-0.38 + j1.71) \text{ A},$$

and

$$\mathbf{I}_3 = (-1.98 + j0.86) \text{ A}.$$

The current $\mathbf{I}_L$ through the inductor is given by

$$\begin{aligned} \mathbf{I}_L = \mathbf{I}_1 - \mathbf{I}_2 &= (0.43 + j0.86) - (-0.38 + j1.71) \\ &= 0.81 - j0.85 = 1.17e^{-j46.5^\circ} \text{ A}, \quad (7.120) \end{aligned}$$

and its time-domain counterpart is

$$\begin{aligned} i_L(t) &= \mathfrak{Re}[\mathbf{I}_L e^{j\omega t}] = \mathfrak{Re}[1.17e^{-j46.5^\circ} e^{j1000t}] \\ &= 1.17\cos(1000t - 46.5^\circ) \text{ A}. \quad (7.121) \end{aligned}$$

Exercise 7-16: Write down the mesh-current matrix equation for the circuit in Fig. E7.16.

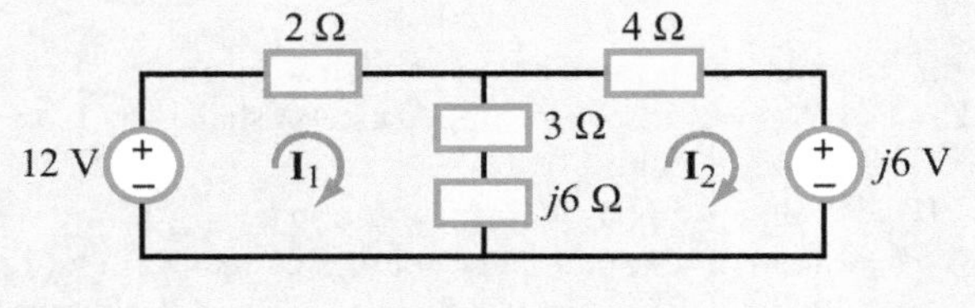

Figure E7.16

Answer:

$$\begin{bmatrix} (5+j6) & -(3+j6) \\ -(3+j6) & (7+j6) \end{bmatrix} \begin{bmatrix} \mathbf{I}_1 \\ \mathbf{I}_2 \end{bmatrix} = \begin{bmatrix} 12 \\ -j6 \end{bmatrix}.$$

(See CAD)

Example 7-15: Source Superposition

The circuit in Fig. 7-29(a) contains two independent sources. Apply the source-superposition method to demonstrate that $\mathbf{I}_L$ is given by the same expression obtained in Example 7-14, namely Eq. (7.120).

Solution: With the source-superposition method, we activate one independent source at a time.

Source 1 Alone: In part (b) of Fig. 7-29, only the 12 V source is active, and the other source has been replaced with a short circuit. The loop currents are designated $\mathbf{I}'_1$ through $\mathbf{I}'_3$, and the corresponding current through the inductor is $\mathbf{I}'_L$. Application of the mesh-current by-inspection method gives the matrix equation

$$\begin{bmatrix} (7-j3) & -(2+j1) & -2 \\ -(2+j1) & (6+j1) & -2 \\ -2 & -2 & 6 \end{bmatrix} \begin{bmatrix} \mathbf{I}'_1 \\ \mathbf{I}'_2 \\ \mathbf{I}'_3 \end{bmatrix} = \begin{bmatrix} 12 \\ 0 \\ -12 \end{bmatrix}, \quad (7.122)$$

whose inversion leads to

$$\mathbf{I}'_1 = (0.79 + j0.52) \text{ A},$$

$$\mathbf{I}'_2 = (-0.36 + j0.48) \text{ A},$$

and

$$\mathbf{I}'_3 = (-1.86 + j0.33) \text{ A}.$$

Hence,

$$\begin{aligned} \mathbf{I}'_L = \mathbf{I}'_1 - \mathbf{I}'_2 &= (0.79 + j0.52) - (-0.36 + j0.48) \\ &= (1.15 + j0.04) \text{ A}. \quad (7.123) \end{aligned}$$

Source 2 Alone: Deactivation of the 12 V source and reactivation of the $-6j$ V source produces the circuit shown in part (c) of Fig. 7-29. Now the loop currents are $\mathbf{I}''_1$, $\mathbf{I}''_2$, and $\mathbf{I}''_3$, and their matrix equation is

$$\begin{bmatrix} (7-j3) & -(2+j1) & -2 \\ -(2+j1) & (6+j1) & -2 \\ -2 & -2 & 6 \end{bmatrix} \begin{bmatrix} \mathbf{I}''_1 \\ \mathbf{I}''_2 \\ \mathbf{I}''_3 \end{bmatrix} = \begin{bmatrix} 0 \\ j6 \\ 0 \end{bmatrix}. \quad (7.124)$$

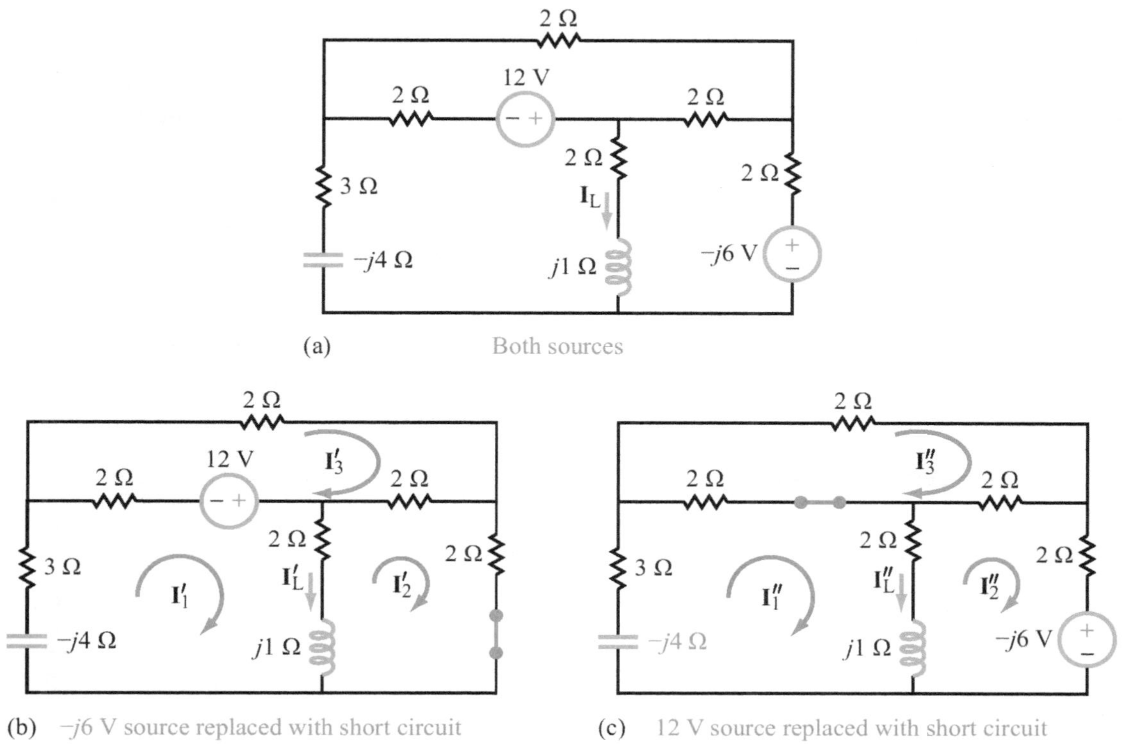

Figure 7-29: Demonstration of the source-superposition technique (Example 7-15).

The solution of Eq. (7.124) is

$$\mathbf{I}_1'' = (-0.36 + j0.34)\text{ A},$$
$$\mathbf{I}_2'' = (-0.02 + j1.23)\text{ A},$$
$$\mathbf{I}_3'' = (-0.13 + j0.53)\text{ A},$$

and

$$\mathbf{I}_L'' = \mathbf{I}_1'' - \mathbf{I}_2'' = -0.36 + j0.34 - (-0.02 + j1.23) = (-0.34 - j0.89)\text{ A}.$$

Total Superposition Solution: Given $\mathbf{I}_L'$ due to source 1 alone and $\mathbf{I}_L''$ due to source 2 alone, the total current due to both sources simultaneously is

$$\mathbf{I}_L = \mathbf{I}_L' + \mathbf{I}_L'' = (1.15 + j0.04) + (-0.34 - j0.89) = (0.81 - j0.85)\text{ A}, \quad (7.125)$$

which is identical to the expression given by Eq. (7.120).

Example 7-16: Thévenin Approach

For the circuit of Fig. 7-30, (a) obtain its Thévenin equivalent at terminals (a, b), as if the inductor were an external load, and (b) then use the Thévenin circuit to determine $\mathbf{I}_L$.

Solution: (a) We will apply the open-circuit/short-circuit method to determine the values of $\mathbf{V}_{Th}$ and $\mathbf{Z}_{Th}$ of the Thévenin equivalent circuit.

Open-Circuit Voltage: With the inductor replaced with an open circuit in Fig. 7-30(b), the matrix equation for loop currents $\mathbf{I}_1$ and $\mathbf{I}_2$ is

$$\begin{bmatrix} (9 - j4) & -4 \\ -4 & 6 \end{bmatrix} \begin{bmatrix} \mathbf{I}_1 \\ \mathbf{I}_2 \end{bmatrix} = \begin{bmatrix} 12 + j6 \\ -12 \end{bmatrix}, \quad (7.126)$$

and its inversion gives

$$\mathbf{I}_1 = (0.02 + j0.96)\text{ A} \quad \text{and} \quad \mathbf{I}_2 = (-1.98 + j0.64)\text{ A}.$$

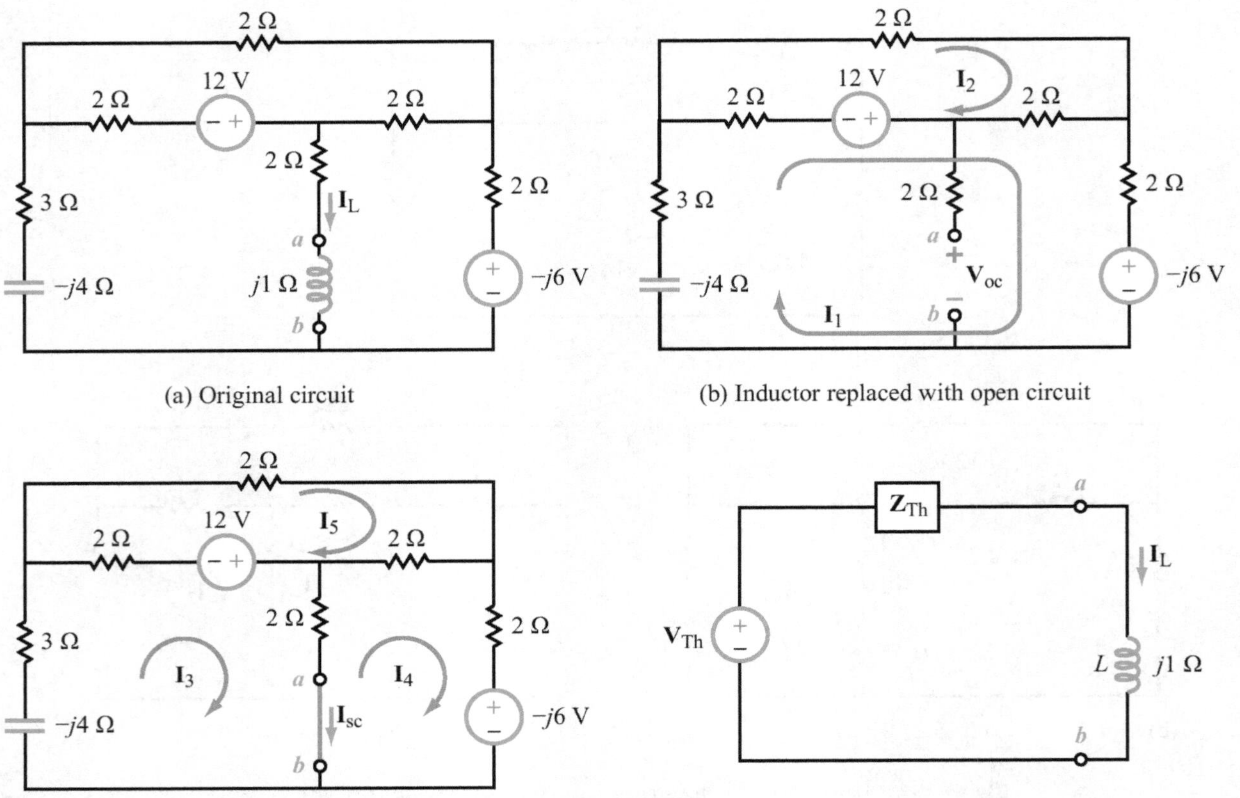

(a) Original circuit

(b) Inductor replaced with open circuit

(c) Inductor replaced with short circuit

(d) Thévenin circuit connected to inductor

Figure 7-30: After determining the open-circuit voltage in part (b) and the short-circuit current in part (c), the Thévenin equivalent circuit is connected to the inductor to determine $\mathbf{I}_L$.

With $\mathbf{I}_1$ and $\mathbf{I}_2$ known, application of KVL around the loop containing the $-j6$ V source leads to

$$\begin{aligned}\mathbf{V}_{Th} = \mathbf{V}_{oc} &= 2(\mathbf{I}_1 - \mathbf{I}_2) + 2\mathbf{I}_1 - j6 \\ &= 4I_1 - 2I_2 - j6 = (4.06 - j3.44) \text{ V}. \end{aligned} \quad (7.127)$$

Short-Circuit Current: In part (c) of Fig. 7-30, the inductor has been replaced with a short circuit. The matrix equation for loop currents $\mathbf{I}_3$ to $\mathbf{I}_5$ is given by

$$\begin{bmatrix} (7-j4) & -2 & -2 \\ -2 & 6 & -2 \\ -2 & -2 & 6 \end{bmatrix} \begin{bmatrix} \mathbf{I}_3 \\ \mathbf{I}_4 \\ \mathbf{I}_5 \end{bmatrix} = \begin{bmatrix} 12 \\ j6 \\ -12 \end{bmatrix}. \quad (7.128)$$

Solution of Eq. (7.128) gives

$$\mathbf{I}_3 = (0.44 + j0.95) \text{ A},$$

$$\mathbf{I}_4 = (-0.53 + j1.60) \text{ A},$$

and

$$\mathbf{I}_5 = (-2.03 + j0.85) \text{ A},$$

from which we have

$$\begin{aligned}\mathbf{I}_{sc} = \mathbf{I}_3 - \mathbf{I}_4 &= (0.44 + j0.95) - (-0.53 + j1.60) \\ &= (0.97 - j0.65) \text{ A}. \end{aligned} \quad (7.129)$$

Given $\mathbf{V}_{oc}$ and $\mathbf{I}_{sc}$, it follows that

$$\mathbf{Z}_{Th} = \frac{\mathbf{V}_{oc}}{\mathbf{I}_{sc}} = \frac{4.06 - j3.44}{0.97 - j0.65} = (4.53 - j0.51) \ \Omega. \quad (7.130)$$

(b) Having established $\mathbf{V}_{\text{Th}}$ and $\mathbf{Z}_{\text{Th}}$, we now connect the Thévenin equivalent circuit to the inductor at terminals (a, b), as shown in Fig. 7-30(d). The current $\mathbf{I}_{\text{L}}$ is simply

$$\mathbf{I}_{\text{L}} = \frac{\mathbf{V}_{\text{Th}}}{\mathbf{Z}_{\text{Th}} + j1} = \frac{4.06 - j3.44}{4.53 - j0.51 + j1} = (0.80 - j0.85) \text{ A}. \tag{7.131}$$

7-10 ac Op-Amp Circuits

Question 1: Are op amps used in ac circuits?

Answer 1: Yes.

Question 2: Is the ideal op-amp model applicable to ac circuits?

Answer 2: The ideal op-amp model is based on the assumption that the open-loop gain A is very large ($> 10^4$), which is true at dc and low frequencies, but not necessarily so at high frequencies. The range of frequencies over which A is large depends on the specific op-amp design. As we shall see later on in this section, when the standard LM741 op amp is used in an inverting amplifier circuit, the ideal op-amp model is applicable for ac circuits so long as the frequency is less than about 1 kHz. For operations at higher frequencies, other models should be used instead, so the selection of a particular op-amp model for a particular application (such as amplification and processing of video signals) becomes an important consideration.

To explain what we mean by the answer to the second question, let us start with a quick review of the op-amp models introduced earlier in Chapter 4. The operation of the op amp can be represented by the equivalent circuit shown in Fig. 7-31(a). The model parameters include large input resistance R_{i} on the order of megaohms, small output resistance R_{o} on the order of 50 Ω, and an open-loop gain A. At dc, A is very large, on the order of 10^5 or greater. These attributes allowed us to adopt the ideal op-amp model in which we set $R_{\text{i}} \approx \infty$, $R_{\text{o}} \approx 0$, and $A \approx \infty$. By invoking these approximations, we obtained the two constraints:

$$\upsilon_{\text{p}} = \upsilon_{\text{n}}$$

and

$$i_{\text{p}} = i_{\text{n}} = 0.$$

The use of these constraints served to significantly simplify the analysis of op-amp circuits containing dc sources. An important underlying assumption is that A is very large. Whereas this assumption is certainly valid for dc, it is not necessarily so at ac.

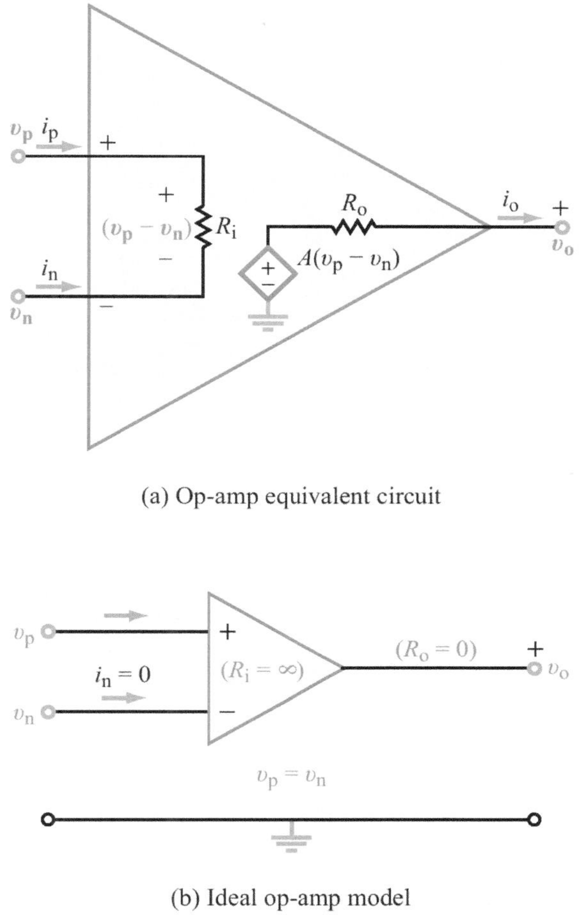

Figure 7-31: Op-amp (a) equivalent circuit (for both dc and ac) and (b) ideal model (for dc, and ac at low frequencies).

Figure 7-32 displays a typical plot of the open-loop gain A as a function of the oscillator frequency f for the LM741 op amp. At dc, the gain (denoted A_0) is indeed very large (10^5), but A decreases rapidly with increasing frequency. The gain spectrum of an op amp is characterized by three important parameters:

(a) the dc gain A_0: the value of A at $f = 0$ Hz.

(b) the corner frequency f_{c}: the frequency at which $A = A_0/\sqrt{2} = 0.707A_0$.

(c) the unity gain frequency f_{u}: the frequency at which $A = 1$.

For the op-amp gain displayed in Fig. 7-32, $A_0 = 10^5$, $f_{\text{c}} = 10$ Hz, and $f_{\text{u}} = 1$ MHz. The ideal op-amp model assumes

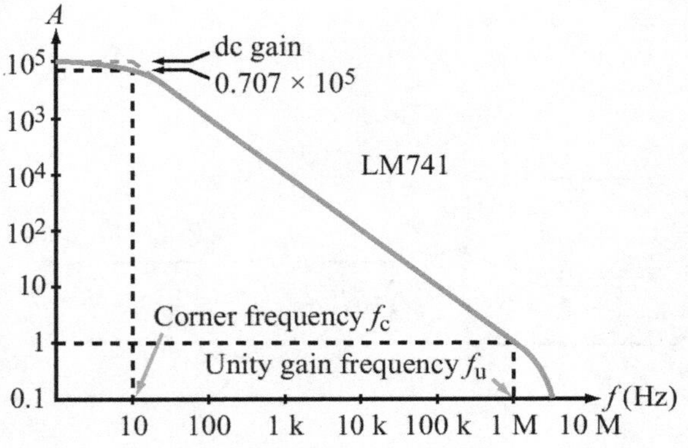

Figure 7-32: Open-loop gain A versus frequency for the LM741 op amp.

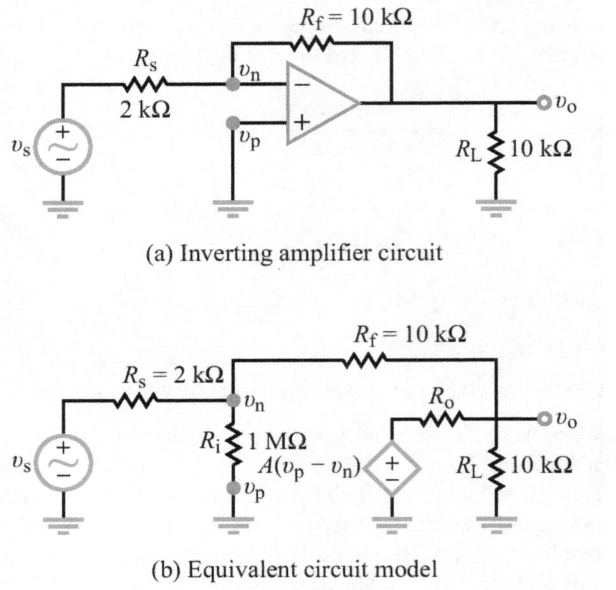

Figure 7-33: Inverting amplifier.

that A is very large, which is a valid assumption at dc and at frequencies as high as 10 kHz, but it is certainly not valid at much higher frequencies.

What are the implications of a nonuniform spectrum for A (i.e., A not a constant as a function of f)? We answer the question through Example 7-17.

Example 7-17: Audio and Video Amplifier

The objective of this example is to establish whether or not the inverting amplifier circuit shown in Fig. 7-33(a) is suitable for amplifying (a) audio signals with spectra extending to 1 kHz and video signals with spectra extending to 1 MHz. The op-amp gain spectrum is given in Fig. 7-32, and the input and output resistances are $R_i = 1$ MΩ, and $R_o = 50$ Ω.

Solution: Since A is not uniformly high at all the frequencies under consideration, we should compute the circuit gain $G = v_o/v_s$ using the op-amp equivalent circuit model, and then compare it with the value obtained using the ideal model. We will perform the comparison at multiple frequencies between dc and 1 MHz.

Ideal op-amp model

From Eq. (4.24),

$$G_{\text{ideal}} = \frac{v_o}{v_s} = -\frac{R_f}{R_s} = -\frac{10\text{ k}}{2\text{ k}} = -5. \tag{7.132}$$

Equivalent circuit model

The node equations at nodes v_n and v_o in Fig. 7-33(b) are given by:

$$\frac{v_n - v_s}{R_s} + \frac{v_n}{R_i} + \frac{v_n - v_o}{R_f} = 0, \tag{7.133}$$

$$\frac{v_o - v_n}{R_f} + \frac{v_o - A(v_p - v_n)}{R_o} + \frac{v_o}{R_L} = 0. \tag{7.134}$$

After setting $v_p = 0$ (because the positive input terminal is connected to the ground terminal) and solving the two equations simultaneously to obtain an expression for the circuit gain, we have

$$G = \frac{v_o}{v_s} = \frac{R_f}{R_s} \cdot \left[\frac{R_s R_i (R_o - A R_f)}{(R_L R_o + R_f R_L + R_f R_o)(R_i R_f + R_s R_f + R_s R_i) - R_s R_i (R_o - A R_f)}\right] \tag{7.135}$$

Using the values $R_i = 10^6$ Ω, $R_s = 2 \times 10^3$ Ω, $R_f = 10^4$ Ω, $R_o = 50$ Ω, $R_L = 10$ kΩ, and the value of A from Fig. 7-32, we obtain the results summarized in Table 7-5.

(a) Audio Signal: Based on the gain data listed in Table 7-5, an audio signal consisting of frequencies extending between

Table 7-5: **Inverting amplifier gain G as a function of oscillation frequency f. $G_{\text{ideal}} = -5$**

f (Hz)	A	G	Error
0 (dc)	10^5	−4.997	0.06%
100	10^4	−4.970	0.6%
1 k	10^3	−4.714	5.7%
10 k	10^2	−3.111	37.8%
100 k	10	−0.707	85.9%
1 M	1	−0.081	98.4%

The error is defined as

$$\% \text{ error} = \left(\frac{G_{\text{ideal}} - G}{G_{\text{ideal}}}\right) \times 100.$$

dc and 1 kHz would experience relatively minimal distortion because they would all be amplified by a factor of about −5, within a maximum variation of 5.7% (at 1 kHz).

(b) Video Signal: Because the video signal extends to 1 MHz and the op-amp circuit does not provide good amplification at frequencies above 10 kHz, the output signal will be highly distorted. Hence, to amplify video signals with minimal distortion, it is necessary to use an op amp with a corner frequency (Fig. 7-32) as high as 1 MHz or higher.

7-11 Op-Amp Phase Shifter

In Section 7-8, we examined how an RC circuit can be used as a phase shifter with an output voltage having the same angular frequency ω of the input voltage, but whose phase angle is increased or decreased (shifted) by a desired amount. That is, if the input is

$$v_{\text{in}}(t) = V_1 \cos \omega t, \tag{7.136a}$$

the phase-shifted output is

$$v_{\text{out}}(t) = V_2 \cos(\omega t + \phi). \tag{7.136b}$$

As was shown earlier in Section 7-8, an RC circuit can indeed realize the desired phase shift, but at a cost in amplitude. The amplitude of the output voltage, V_2, is smaller than V_1, and the degree of reduction depends on ϕ and the number of RC stages used in the phase shifter.

An op-amp circuit can serve as a phase shifter, without necessarily sacrificing a reduction in amplitude. Consider the circuit in Fig. 7-34(a). It is an inverting amplifier with complex source and feedback impedances:

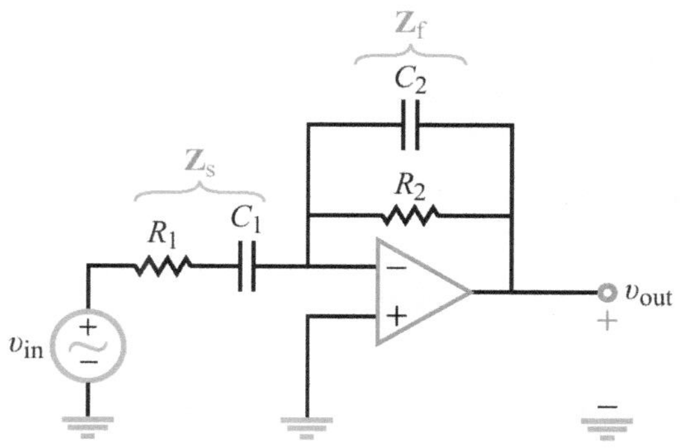

Figure 7-34: Inverting amplifier as a phase-shift circuit.

$$\mathbf{Z}_s = R_1 + \frac{1}{j\omega C_1} = \frac{j\omega R_1 C_1 + 1}{j\omega C_1}, \tag{7.137a}$$

$$\mathbf{Z}_f = R_2 \parallel \frac{1}{j\omega C_2} = \frac{R_2/j\omega C_2}{R_2 + 1/j\omega C_2} = \frac{R_2}{1 + j\omega R_2 C_2}. \tag{7.137b}$$

The circuit gain is

$$\mathbf{G} = \frac{\mathbf{V}_{\text{out}}}{\mathbf{V}_{\text{in}}} = -\frac{\mathbf{Z}_f}{\mathbf{Z}_s} = \frac{-j\omega R_2 C_1}{(1 + j\omega R_1 C_1)(1 + j\omega R_2 C_2)}, \tag{7.138}$$

which can be expressed as

$$\mathbf{G} = |\mathbf{G}| e^{j\phi}, \tag{7.139}$$

with

$$|\mathbf{G}| = \frac{\omega R_2 C_1}{[(1 + \omega^2 R_1^2 C_1^2)(1 + \omega^2 R_2^2 C_2^2)]^{1/2}}, \tag{7.140a}$$

$$\phi = 270^\circ - \tan^{-1}(\omega R_1 C_1) - \tan^{-1}(\omega R_2 C_2), \tag{7.140b}$$

where 270° is the phase angle corresponding to $(-j)$ in the numerator of Eq. (7.138). In the time domain,

$$v_{\text{out}}(t) = |\mathbf{G}| V_1 \cos(\omega t + \phi). \tag{7.141}$$

Through judicious choice of the values of R_1, R_2, C_1, and C_2, it should be possible to design a phase shifter that provides the desired value of ϕ, with $|G| \geq 1$. The process is illustrated by Example 7-18.

Example 7-18: Op-Amp Phase Shifter

Select values for R_1, R_2, C_1, and C_2 in the circuit of Fig. 7-34 so that $\phi = 120^\circ$ and $|G| = 2$ at $\omega = 500$ rad/s.

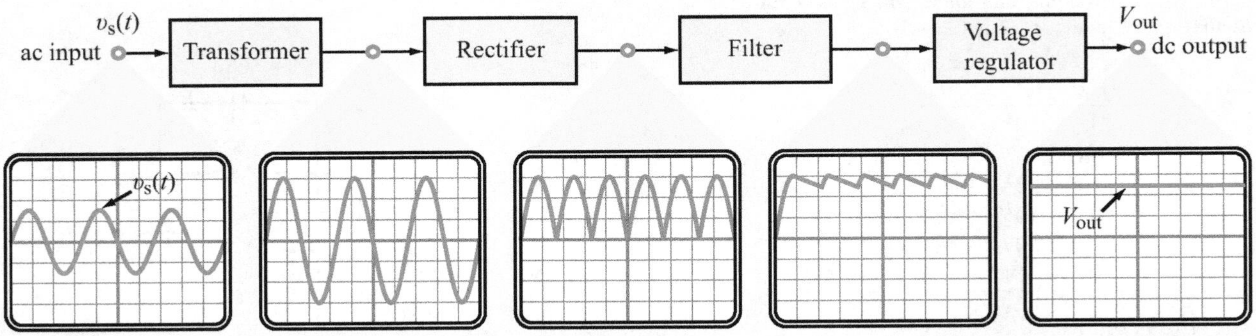

Figure 7-35: Block diagram of a basic dc power supply.

Solution: With 4 selectable parameters against only 2 specified parameters, the desired outcome can be realized through many different combinations of (R_1, R_2, C_1, C_2). Hence, we arbitrarily choose

$$R_1 = 2\text{ k}\Omega, \qquad C_1 = 3\ \mu\text{F},$$

which leads to

$$\omega R_1 C_1 = 500 \times 2 \times 10^3 \times 3 \times 10^{-6} = 3.$$

Using this value and $\phi = 120°$ in Eq. (7.140b) leads to

$$120° = 270° - \tan^{-1}(3) - \tan^{-1}(\omega R_2 C_2),$$

which simplifies to

$$\begin{aligned}\tan^{-1}(\omega R_2 C_2) &= 270° - 120° - \tan^{-1} 3\\ &= 150° - 71.57° = 78.43°.\end{aligned}$$

Hence,

$$\omega R_2 C_2 = \tan 78.43° = 4.89.$$

With $\omega R_1 C_1 = 3$ and $\omega R_2 C_2 = 4.89$, and $|\mathbf{G}| = 2$, solution of Eq. (7.140a) leads to

$$R_2 = 21\text{ k}\Omega,$$

and

$$C_2 = \frac{4.89}{\omega R_2} = \frac{4.89}{500 \times 21 \times 10^3} = 0.47\ \mu\text{F}.$$

7-12 Application Note: Power-Supply Circuits

Systems composed of one or more electronic circuits usually contain power-supply circuits that convert the ac power available from the wall outlet into dc power, thereby providing the internal dc voltages required for proper operation of the electronic circuits. Most dc power supplies consist of the four subsystems diagrammed in Fig. 7-35. The input is an ac voltage $v_s(t)$ of amplitude V_s and angular frequency ω, and the final output is a dc voltage V_{out}. Our plan in this section is to describe the operation of each of the intermediate stages, and then connect them all together.

7-12.1 Ideal Transformers

A transformer consists of two inductors called *windings*, that are in close proximity to each other but not connected electrically. The two windings are called the *primary* and the *secondary*, as shown in Fig. 7-36. Even though the two windings are isolated electrically—meaning that no current flows between them—when an ac voltage is applied to the primary, it creates a magnetic flux that permeates both windings through a common *core*, inducing an ac voltage in the secondary.

▶ The *transformer* gets its name from the fact that it is used to transform currents, voltages, and impedances between its primary and secondary circuits. ◀

The key parameter that determines the relationships between the primary and the secondary is the *turns ratio* $n = N_2/N_1$,

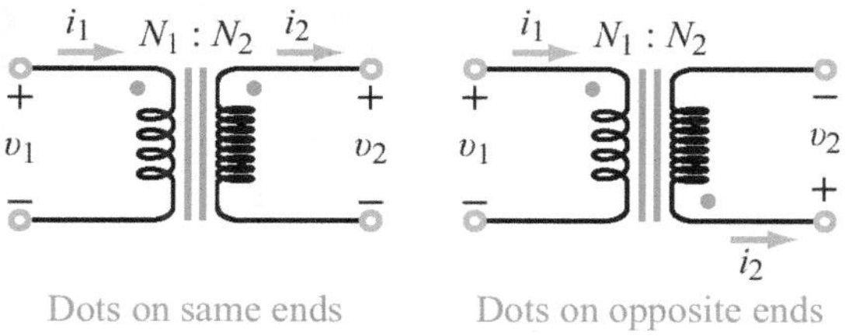

Figure 7-36: Schematic symbol for an ideal transformer. Note the reversal of the voltage polarity and current direction when the dot location at the secondary is moved from the top end of the coil to the bottom end. For both configurations:

$$\frac{v_2}{v_1} = \frac{N_2}{N_1} = n, \qquad \frac{i_2}{i_1} = \frac{N_1}{N_2} = \frac{1}{n}, \qquad \frac{p_2}{p_1} = \frac{v_2 i_2}{v_1 i_1} = 1$$

where N_1 is the number of turns in the primary coil and N_2 is the number of turns in the secondary. An additionally important attribute is the direction of the primary winding, relative to that of the secondary, around the common magnetic core. The relative directions determine the voltage polarity and current direction at the secondary, relative to those at the primary. To distinguish between the two cases, a dot usually is placed at one or the other end of each winding, as shown in Fig. 7-36. For the *ideal transformer*, voltage v_2 at the secondary side is related to voltage v_1 at the primary side by

$$\frac{v_2}{v_1} = \frac{N_2}{N_1} = n, \qquad (7.142)$$

where *the polarities of v_1 and v_2 are defined such that their (+) terminals are at the ends with the dots*. In an ideal transformer, no power is lost in the core, so all of the power supplied by a source to its primary coil is transferred to the load connected at its secondary side. Thus, $p_1 = p_2$, and since $p_1 = i_1 v_1$ and $p_2 = i_2 v_2$, it follows that

$$\frac{i_2}{i_1} = \frac{N_1}{N_2}, \qquad (7.143)$$

with i_1 *always defined in the direction towards the dot* on the primary side and i_2 *defined in the direction away from the dot* on the secondary side. The purpose of the dot designation is to indicate whether the windings in the primary and secondary coils curl in the same (clockwise or counterclockwise) direction or in opposite directions. The coil directions determine the direction of magnetic flux coupling between the two coils. More details are available in Chapter 11.

▶ If $N_2/N_1 > 1$, the transformer is called a *step-up transformer* because it transforms v_1 to a higher voltage, and if $N_2/N_1 < 1$, it is called a *step-down transformer*. ◀

Most office and household electronic gadgets (such as telephones, clocks, radios, and answering machines) require dc voltages that are on the order of volts (or at most a few tens of volts), which is much smaller than the voltage level available at the wall outlet. The transformer in such gadgets is invariably a step-down transformer.

As discussed in great detail in Chapter 11, the input-output relationships for a real transformer are more elaborate than those given by Eqs. (7.142) and (7.143) for the ideal transformer. Nevertheless, these simple relationships are reasonable first-order approximations and serve our current discussion quite adequately.

Concept Question 7-17: In a transformer, how are the voltage polarities and current directions defined relative to the dots on the primary and secondary windings? (See CAD)

Concept Question 7-18: For an ideal transformer, how is power p_2 related to power p_1? (See CAD)

7-12.2 Rectifiers

A rectifier is a diode circuit that converts an ac waveform into one that is either always positive or always negative, depending on the direction(s) of the diode(s). Power supplies usually use a *bridge rectifier*, but to appreciate how such a bridge functions, we will first consider the simple single-diode rectifier circuit shown in Fig. 7-37. As discussed in Section 2-6.2, a diode is modeled by a practical response that allows current to flow through it in the direction shown in Fig. 7-37 if and only if the voltage across it is greater than a threshold value known as the *forward-bias voltage* V_F. That is, for the circuit in Fig. 7-37, the output voltage across the load resistor is given by

$$v_{out} = \begin{cases} v_{in} - V_F & \text{if } v_{in} \geq V_F, \\ 0 & \text{if } v_{in} \leq V_F. \end{cases} \qquad (7.144)$$

For an ideal diode with $V_F = 0$, the output waveform is identical to the input waveform for the half cycles during which v_{in} is positive, and the output is zero when v_{in} is negative. In the case

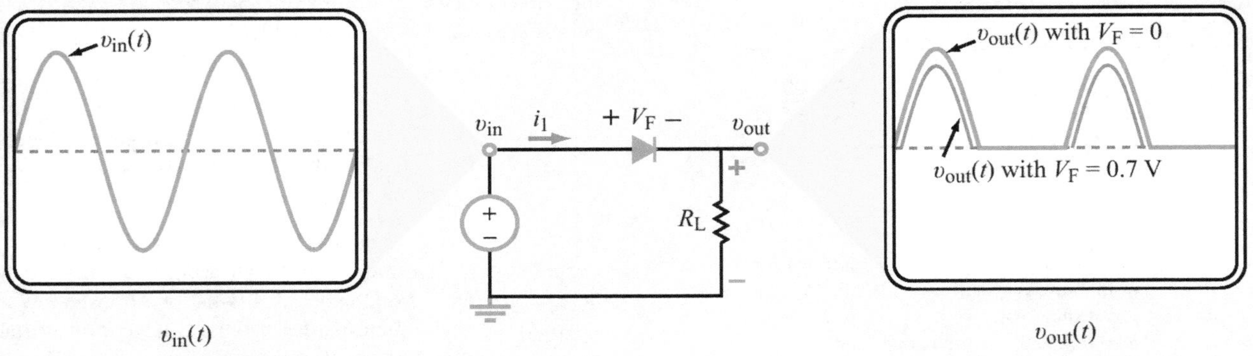

Figure 7-37: Half-wave rectifier circuit.

of a real diode with $V_F \approx 0.7$ V, the peak amplitude of the output is smaller than that of the input by 0.7 V. Because the output waveform essentially replicates only the positive half cycles of the input waveform (with a negative amplitude shift equal to V_F), the circuit of Fig. 7-37 is called a *half-wave rectifier*.

Next, we consider the *bridge-rectifier* circuit of Fig. 7-38. The bridge rectifier uses four diodes. During the positive half cycle of $v_{in}(t)$, two of the diodes conduct, and the other two are OFF. The reverse happens during the second half cycle, *but the direction of the current through R_L is the same during both half cycles*. Consequently, the output waveform essentially is equivalent to taking the absolute value of the input waveform (if V_F is so small relative to the peak value as to be neglected). Because a bridge rectifier acts on both halves of a cycle, it is often called a *full-wave rectifier*.

Exercise 7-17: Suppose the input voltage in the circuit of Fig. 7-38 is a 10 V amplitude square wave. What would the output look like?

Answer: 8.6-V dc. (See CAD)

7-12.3 Smoothing Filters

So far, we have examined two of the four subcircuits of the dc power supply. The transformer serves to adjust the amplitude of the ac signal to a level close to the desired dc voltage level of the final output. The bridge rectifier converts the ac signal into an all-positive waveform. Next, we need to reduce the variations of the full-wave rectified waveform to bring it to as close to a constant level as possible. We accomplish this by subjecting the full-wave rectified waveform to a smoothing (averaging) filter. This is realized by adding a capacitor C in parallel with the load resistor. The modified circuit is shown in Fig. 7-39(a), and the associated output waveform is displayed in Fig. 7-39(b). The capacitor is a storage device that goes through partial charging-up and discharging-down cycles. During the charging-up period, the *upswing time constant* of the circuit is given by

$$\tau_{up} = (2R_D \parallel R_L)C \approx 2R_D C \qquad \text{if } R_L \gg R_D, \tag{7.145}$$

where R_D is the diode resistance. Typically, R_D is on the order of ohms and R_L is on the order of kiloohms, so the approximation given by Eq. (7.145) is quite reasonable. In the absence of the capacitor in the circuit, R_D usually is ignored because it is in series with a much larger resistance, R_L. Adding a capacitor, however, creates an RC circuit in which R is the parallel combination of R_D and R_L, placing R_D in a controlling position.

During the discharging period, the diode turns off, and the capacitor discharges through R_L alone. Consequently, the *downswing time constant* involves R_L and C only,

$$\tau_{dn} = R_L C. \tag{7.146}$$

For a specified value of the diode resistance R_D, we can choose the values of R_L and C so that τ_{up} is short and τ_{dn} is long—both relative to the period of the rectified waveform—thereby realizing a fast response on the upswing part and a very slow response on the downswing part. In practice, it is possible to generate an approximately constant dc voltage with a ripple component on the order of 1 to 10 percent of its average value (Fig. 7-39(b)).

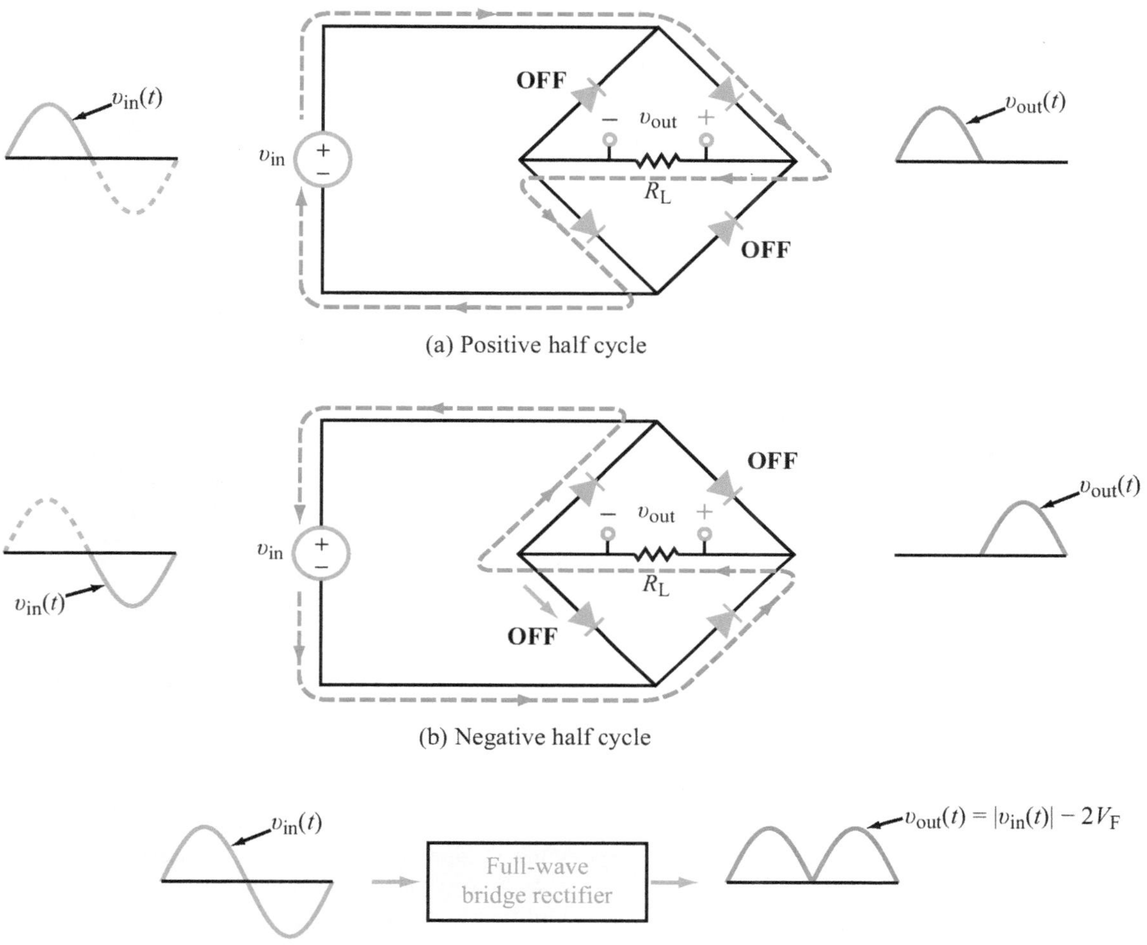

(a) Positive half cycle

(b) Negative half cycle

(c) Input-output response

Figure 7-38: Full-wave bridge rectifier. Current flows in the same direction through the load resistor for both half cycles.

Example 7-19: Filter Design

If the bridge rectifier circuit of Fig. 7-39(a) has a 60 Hz ac input signal, determine the values of R_L and C that would result in $\tau_{up} = T_{rect}/12$ and $\tau_{dn} = 12T_{rect}$, where T_{rect} is the period of the *rectified* waveform. Assume $R_D = 5\ \Omega$.

Solution: If the frequency of the original ac signal is 60 Hz, the frequency of the rectified waveform is 120 Hz. Hence, the period of the rectified waveform is

$$T_{rect} = \frac{1}{120} = 8.33 \text{ ms},$$

and the corresponding design specifications are

$$\tau_{up} = \frac{T_{rect}}{12} = 0.69 \text{ ms}, \quad \text{and} \quad \tau_{dn} = 12T_{rect} = 100 \text{ ms}.$$

Application of Eq. (7.145) leads to

$$\tau_{up} \approx 2R_D C$$

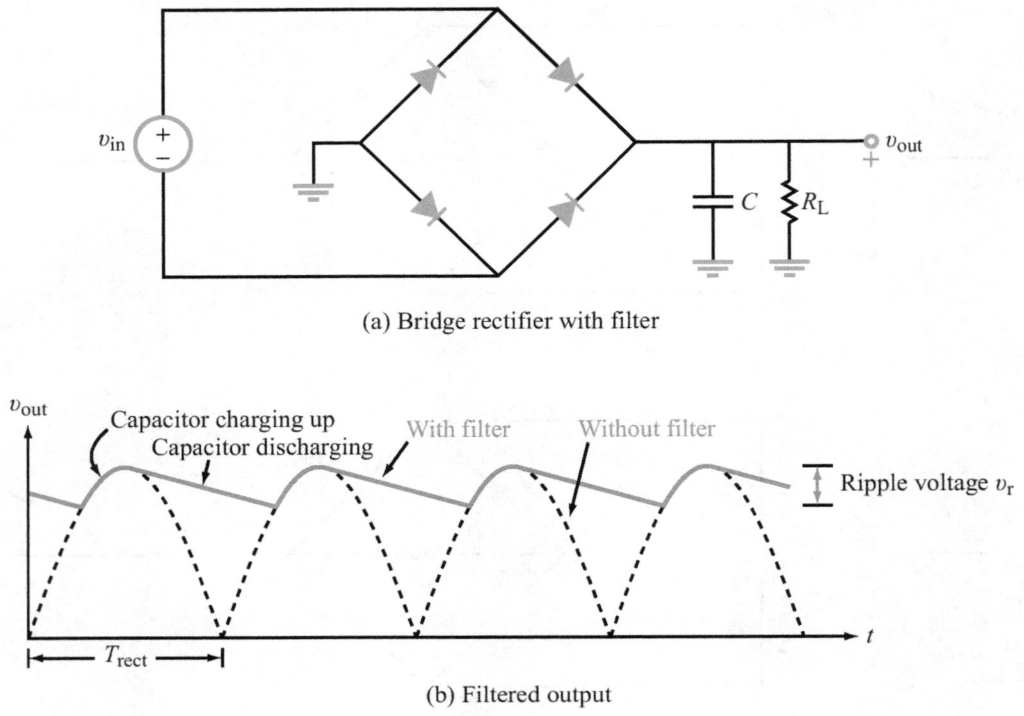

Figure 7-39: Smoothing filter reduces the variations of waveform $v_{\text{out}}(t)$.

or

$$C = \frac{\tau_{\text{up}}}{2R_D} = \frac{0.69 \times 10^{-3}}{2 \times 5} = 69 \ \mu\text{F}.$$

With the value of C known, application of Eq. (7.146) gives

$$R_L = \frac{\tau_{\text{dn}}}{C} = \frac{100 \times 10^{-3}}{69 \times 10^{-6}} = 1.45 \ \text{k}\Omega.$$

7-12.4 Voltage Regulator

The circuit shown in Fig. 7-40 includes all of the power-supply subcircuits we have discussed thus far, plus two additional elements, namely a series resistance R_s and a *zener diode*. When operated in reverse breakdown, the zener diode maintains the voltage across it at a constant level V_z—so long as the current i_z passing through it remains between certain limits. Since the diode is connected in parallel with R_L, the output voltage becomes equal to the *zener voltage* V_z, and the effective time constant of the smoothing filter becomes $\tau = R_s C$. It is worth noting that the addition of the zener diode reduces the *peak-to-peak ripple voltage* V_r (Fig. 7-39(b)) at the output of the RC filter by about an order of magnitude. An approximate expression for the peak-to-peak ripple voltage with the zener diode in place is given by

$$V_r = \frac{[(V_{s_1} - 1.4) - V_z]T_{\text{rect}}}{R_s C} \times \frac{(R_z \parallel R_L)}{R_s + (R_z \parallel R_L)}, \quad (7.147)$$

where V_{s_1} is the amplitude of the ac signal at the output of the transformer (Fig. 7-40), the factor 1.4 V accounts for the voltage drop across a pair of diodes in the rectifier, V_z is the manufacturer-rated zener voltage for the specific model used in the circuit, T_{rect} is the period of the rectified waveform, and R_z is the manufacturer specified value of the *zener-diode resistance*.

Example 7-20: Power-Supply Design

A power supply with the circuit configuration shown in Fig. 7-40 has the following specifications: the input voltage is 60 Hz with an rms amplitude $V_{\text{rms}} = 110$ V where $V_{\text{rms}} = V_s/\sqrt{2}$ (the rms value of a sinusoidal function is

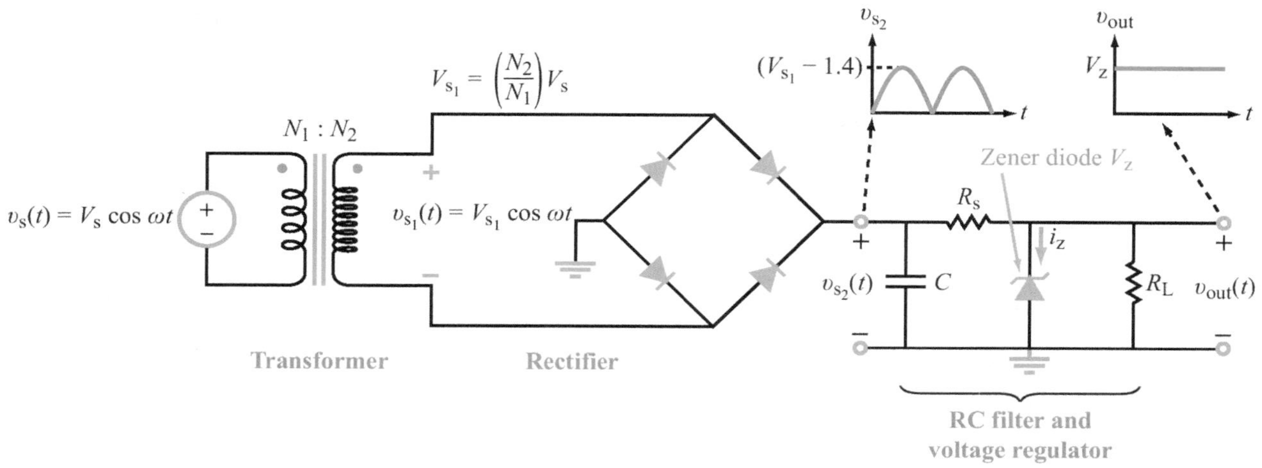

Figure 7-40: Complete power-supply circuit.

discussed in Chapter 8), $N_1/N_2 = 5$, $C = 2$ mF, $R_s = 50$ Ω, $R_L = 1$ kΩ, $V_z = 24$ V, and $R_z = 20$ Ω. Determine v_{out}, the ripple voltage, and the ripple fraction relative to v_{out}.

Solution: At the secondary side of the transformer,

$$v_{s_1}(t) = \left(\frac{N_2}{N_1}\right)(V_s \cos 377t)$$

$$= \frac{1}{5} \times 110\sqrt{2} \cos 377t = 31.11 \cos 377t \text{ V}.$$

Hence, $V_{s_1} = 31.11$ V, which is greater than the zener voltage $V_z = 24$ V.

Consequently, the zener diode will limit the output voltage at

$$v_{\text{out}} = V_z = 24 \text{ V}.$$

In Example 7-19, we established that $T_{\text{rect}} = 8.33$ ms. Also,

$$R_z \parallel R_L = \frac{20 \times 1000}{20 + 1000} = 19.6 \text{ Ω}.$$

Application of Eq. (7.147) gives

$$V_r = \frac{[(V_{s_1} - 1.4) - V_z]T_{\text{rect}}}{R_s C} \times \frac{(R_z \parallel R_L)}{R_s + (R_z \parallel R_L)}$$

$$= \frac{[(31.11 - 1.4) - 24]}{50 \times 2 \times 10^{-3}}(8.33 \times 10^{-3}) \times \frac{19.6}{50 + 19.6}$$

$$= 0.13 \text{ V (peak-to-peak)}.$$

Hence,

$$\text{ripple fraction} = \frac{(V_r/2)}{V_z} = \frac{0.13/2}{24} = 0.0027,$$

which represents a relative variation of less than ±0.3 percent.

7-13 Multisim Analysis of ac Circuits

Even though we usually treat the wires in a circuit as ideal short circuits, in reality a wire has a small but non-zero resistance. Also, as noted earlier in Section 5-7.1, when two wires are in close proximity to one another, they form a non-zero capacitor. A pair of parallel wires on a circuit board is modeled as a distributed transmission line with each small length segment ℓ represented by a series resistance R and a shunt capacitance C, as depicted by the circuit model shown in Fig. 7-41. For a parallel-wire segment of length ℓ, R and C are given by

$$R = \frac{2\ell}{\pi a^2 \sigma} \quad \begin{array}{l}\text{(low-frequency approximation)} \\ (a\sqrt{f\sigma} \leq 500),\end{array} \tag{7.148a}$$

or

$$R = \sqrt{\frac{\pi f \mu}{\sigma}}\left(\frac{\ell}{\pi a}\right) \quad \begin{array}{l}\text{(high-frequency approximation)} \\ (a\sqrt{f\sigma} \geq 1250),\end{array} \tag{7.148b}$$

and

$$C = \frac{\pi \varepsilon \ell}{\ln(d/a)} \quad \text{for } (d/2a)^2 \gg 1, \tag{7.148c}$$

where a is the wire radius, d is the separation between the wires, f is the frequency of the signal propagating along the wires, μ and σ are respectively the magnetic permeability and conductivity of the wire material, and ε is the permittivity of the material between the two wires. Note that R represents the

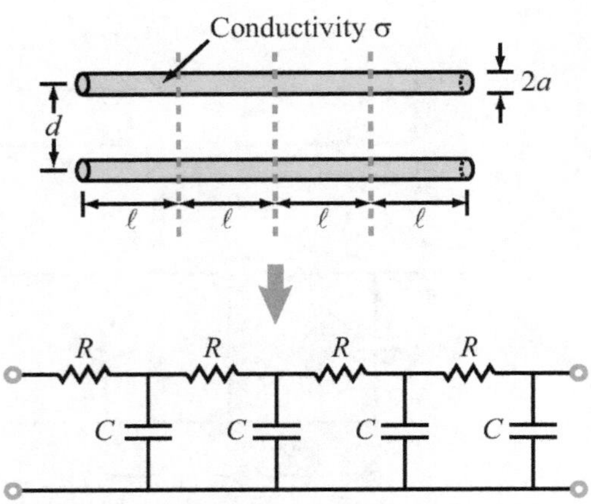

Figure 7-41: Distributed impedance model of two-wire transmission line.

resistance of both wires. There is actually a third distributed element to consider in the general case of a transmission line: the distributed inductance. This inductance is placed in series with the resistance R of each segment. It arises because current flowing through the transmission-line wires gives rise to a magnetic field around the wires and, hence, an inductance (as discussed in Section 5-3). However, modeling the behavior of a transmission line with all three components is rather complex. So, for the purposes of this section, we will ignore the inductance altogether so that we may illustrate the performance of an RC transmission line using Multisim. Keeping this in mind, the distributed model shown in Fig. 7-41 allows us to represent the wires by a series of cascaded RC circuits. For the model to faithfully represent the behavior of the real two-wire configuration, each RC stage should represent a physical length ℓ that is no longer than a fraction ($\approx$ 10 percent) of the distance that the signal travels during one period of the signal frequency. Thus, ℓ should be on the order of

$$\ell \le \frac{u_p T}{10} \approx \frac{c}{10f}, \tag{7.149}$$

where u_p is the signal velocity along the wires, which is on the order of the velocity of light by $c = 3 \times 10^8$ m/s, and the period T is related to the frequency f by $T = 1/f$. For example, if the signal frequency is 1 GHz ($= 10^9$ Hz), then ℓ should be on the order of

$$\ell \approx \frac{c}{10f} = \frac{3 \times 10^8}{10 \times 10^9} = 3 \text{ cm},$$

and if the total length of the parallel wires is $\ell_t = 15$ cm, then their transmission-line equivalent circuit should consist of n sections with

$$n = \frac{\ell_t}{\ell} = \frac{15 \text{ cm}}{3 \text{ cm}} = 5.$$

We will now use Multisim to simulate such a transmission line.

Example 7-21: Transmission-Line Simulation

A pair of parallel wires made of a conducting material with conductivity $\sigma = 1.9 \times 10^5$ S/m is used to carry a 1 GHz square-wave signal between two circuits on a circuit board. The wires are 15 cm in length and separated by 1 mm, and their radii are 0.1 mm. (a) Develop a transmission-line equivalent model for the wires and (b) use Multisim to evaluate the voltage response along the transmission line.

Solution: (a) With $\ell = 3$ cm (to satisfy Eq. (7.149)), application of Eqs. (7.148b and c) gives

$$\begin{aligned} R &= \sqrt{\frac{\pi f \mu}{\sigma}} \left(\frac{\ell}{\pi a}\right) \\ &= \sqrt{\frac{\pi \times 10^9 \times 4\pi \times 10^{-7}}{1.9 \times 10^5}} \left(\frac{3 \times 10^{-2}}{\pi \times 10^{-4}}\right) \\ &= 13.76\ \Omega \end{aligned}$$

and

$$\begin{aligned} C &= \frac{\pi \varepsilon \ell}{\ln(d/a)} \\ &= \frac{\pi \times (10^{-9}/36\pi) \times 3 \times 10^{-2}}{\ln(10)} \\ &= 3.6 \times 10^{-13} \text{ F} \\ &= 0.36 \text{ pF}. \end{aligned}$$

(b) To use Multisim, we need to select values for R and C—from the libraries of available values—that are approximately equal to those we calculated. The selected values are less critical to the simulation than the value of their product, because it is the product $RC = 13.76 \times 0.36 \times 10^{-12} \approx 5 \times 10^{-12}$ s that determines the time constant of the voltage response. Hence, we select

$$R = 10\ \Omega \qquad \text{and} \qquad C = 0.5 \text{ pF},$$

and we draw the 5-stage circuit shown in Fig. 7-42. The square wave is generated by a pulse generator that alternates between 0 and 1 V. Its pulses are 500 ps long and the pulse period is

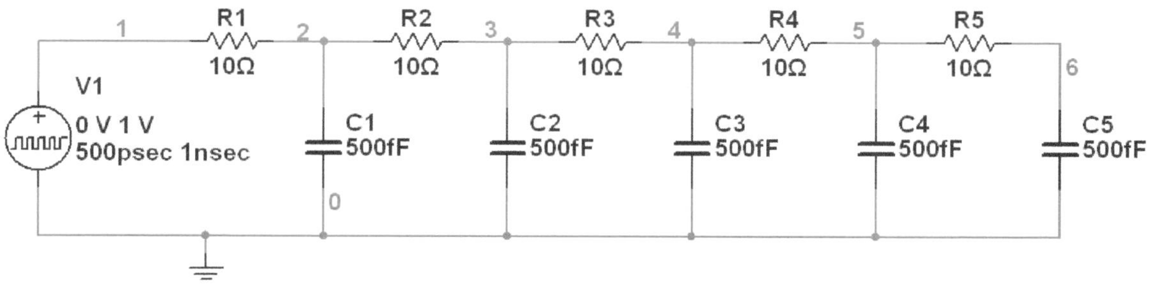

Figure 7-42: Transmission-line circuit in Multisim.

1000 ps (or equivalently, 1 ns, which is the period corresponding to a frequency $f = 1$ GHz). The Rise Time and Fall Time should be set to 1 ps. Figure 7-43 displays V(1) at node 1, which represents the pulse-generator voltage waveform, and the voltages at nodes 2, 3, 4, 5, and 6 corresponding to the outputs of the five RC stages.

During the charging-up period, it takes longer for the nodes further away from the pulse generator to reach the steady-state voltage of 1 V than it does for those closer to the generator. The same pattern applies during the discharge period. In addition to the parallel-wire configuration, the distributed transmission-line concept is equally applicable

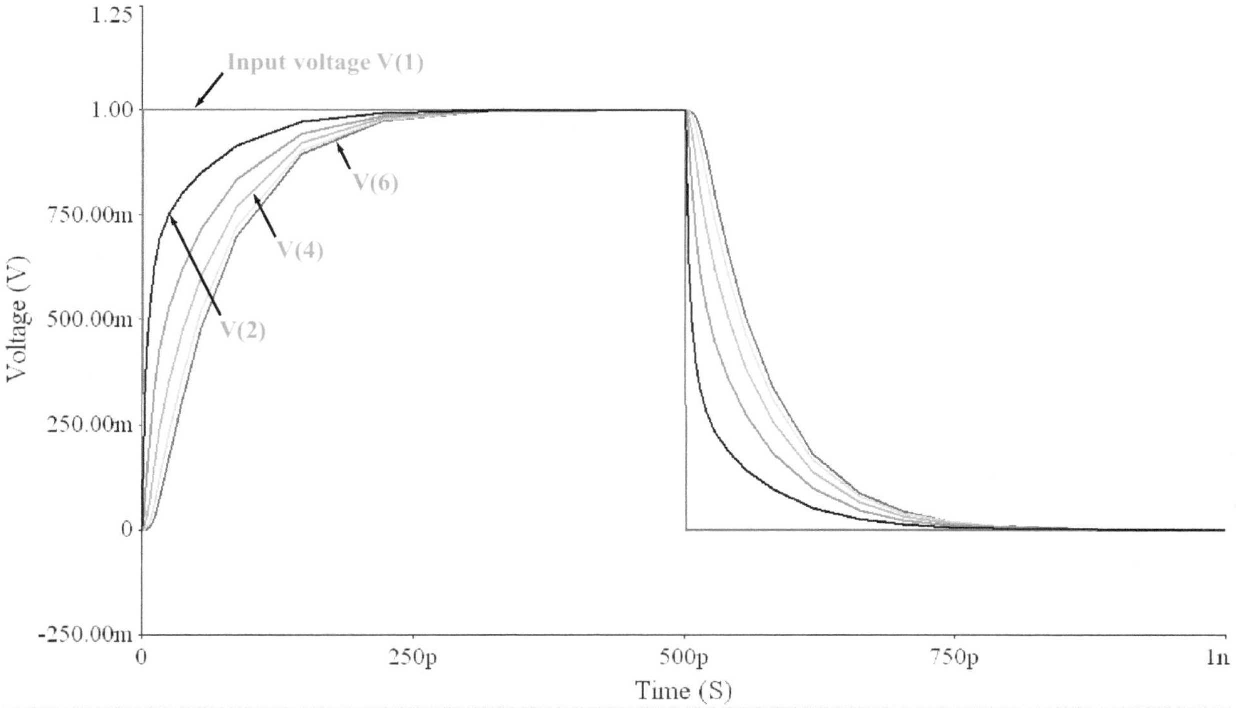

Figure 7-43: Multisim display of voltage waveforms at nodes 1, 2, 3, 4, 5, and 6.

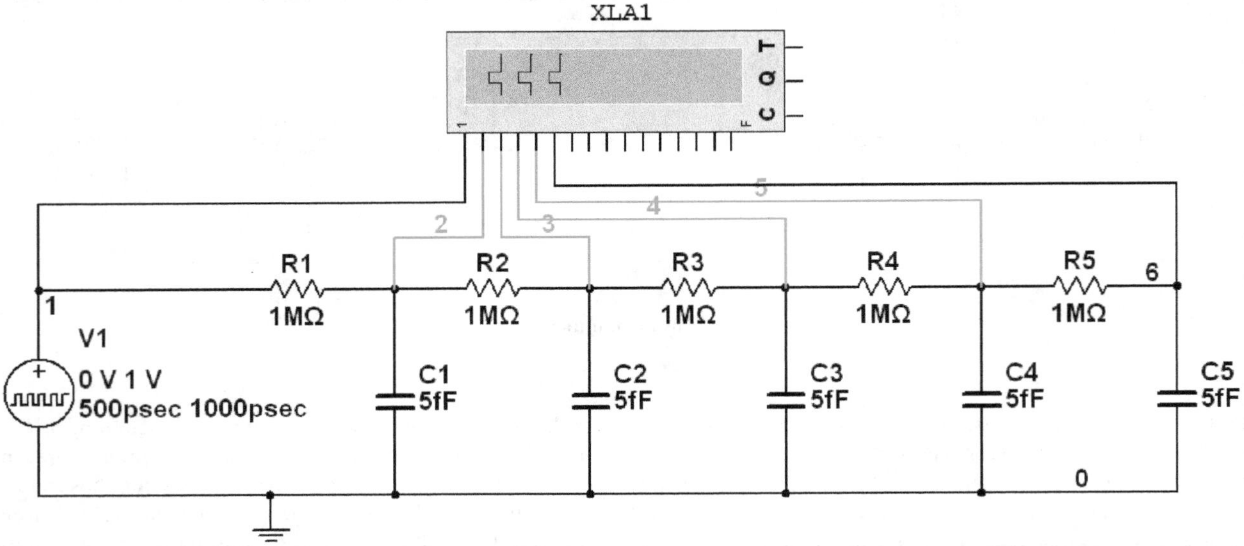

Figure 7-44: Using the Logic Analyzer to measure time delay in Multisim.

to other transmission media, including the shielded cable commonly used for the transmission of audio, video, and digital data between different circuits. If a digital signal with *logic* $0 = 0$ V and *logic* $1 = 1$ V is to be transmitted along a coaxial cable or some other transmission line, it may be of interest to simulate the process using Multisim to determine how long it takes to charge the different nodes along the line up to 1 V. This is also known as *propagating* the *logic 1* down the transmission line. The Logic Analyzer (Simulate → Instruments → Logic Analyzer) is used to visualize a large number of logic levels at once. (See the Multisim Tutorial for a detailed explanation on how to use the Logic Analyzer Instrument.) An example is shown in Fig. 7-44. The circuit uses 1 MΩ resistors, 5 fF capacitors, and a pulse generator. The pulse length is set at 500 ps and the pulse period at 1000 ps (= 1 ns). The circuit nodes are wired to the logic analyzer. In Fig. 7-45, we can observe how long it takes each node to charge up sufficiently to register as a *logic* 1. Note that the logic analyzer's cursor can be used to read out the exact time points.

Example 7-22: Measuring Phase Shift

Run a Transient Analysis on the Multisim circuit in Fig. 7-44 after replacing the pulse generator with a 1 V amplitude, 10 MHz ac source. The goal is to determine the phase of node 2, relative to the phase of node 1 (the voltage source). Select a Start Time of 2.7 μs and an End Time (TSTOP) of 3.0 μs, and set TSTEP and TMAX to 1e-10 seconds so as to generate smooth-looking curves. [We did not choose a Start Time of 0 s simply because it takes the circuit a few microseconds to reach its steady-state solution.]

Solution: Figure 7-46 shows the traces of selected nodes V(1), V(2), and V(6) on Grapher View. Clicking on the Show/Hide Cursors button enables the measurement cursor, which can be used to quantify the amplitude (vertical axis) and time (horizontal axis) for each curve. To measure the phase shift between nodes V(2) and V(1), two cursors are needed.

Step 1: Place cursor 1 slightly to the left of a maximum of the V(1) trace.

Step 2: Click on the trace for V(1) to select it. White triangles will appear on the V(1) trace.

Step 3: Right-click the cursor itself and select Go to next Y_Max=>. On row x1, at column V(1), the value in the table should be 2.7250 μs.

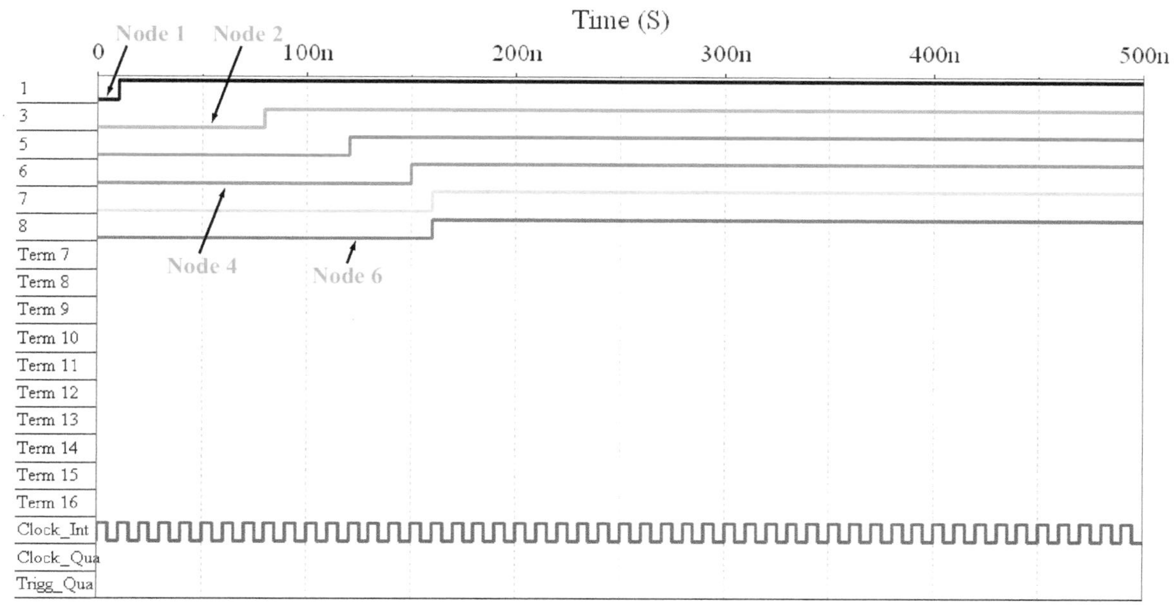

Figure 7-45: Logic Analyzer readout at nodes 1, 2, 3, 4, 5, and 6.

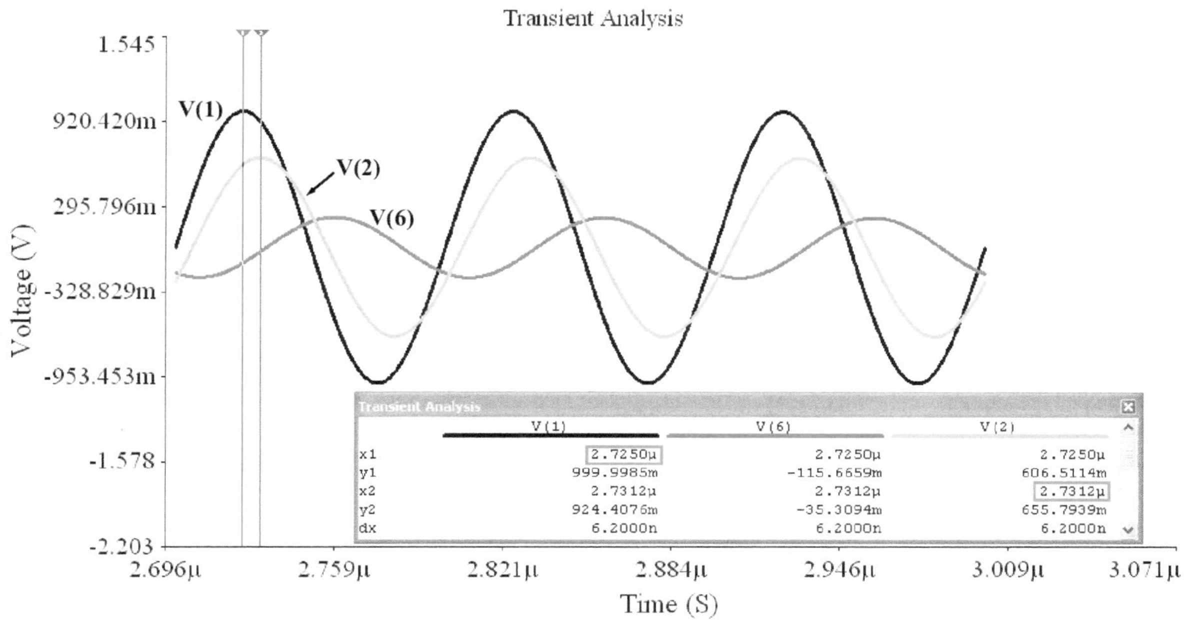

Figure 7-46: Multisim Grapher Plot of voltage nodes V(1), V(2), and V(6) in the circuit of Fig. 7-42.

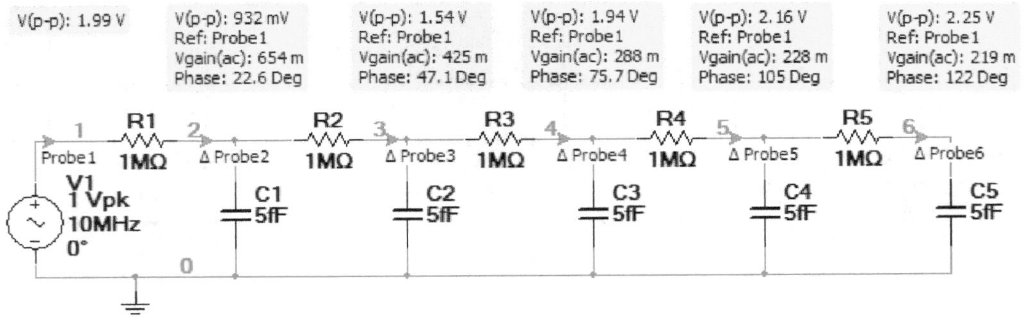

Figure 7-47: Using Measurement Probes to determine phase and amplitude of signal at various points on transmission line.

Step 4: Repeat the process using cursor 2 to select the nearby maximum of the V(2) trace. The entry in row x2, at column V(2), should be 2.7312 μs.

The time difference between the two values is

$$\Delta t = 2.7312\ \mu\text{s} - 2.7250\ \mu\text{s}$$
$$= 0.0062\ \mu\text{s}.$$

Given that $f = 10$ MHz, the period is

$$T = \frac{1}{f} = \frac{1}{10^7} = 10^{-7} = 0.1\ \mu\text{s}.$$

Application of Eq. (7.11) gives

$$\phi = 2\pi\left(\frac{\Delta t}{T}\right) = 360^\circ \times \left(\frac{0.0062}{0.1}\right) = 22.3^\circ.$$

We also can determine the ratio of the amplitude of V(2) to that of V(1). The ratio of y2 in column V(2) to y1 in column V(1) gives

$$\frac{\text{V(2)}}{\text{V(1)}} = \frac{0.656}{1} \approx 66 \text{ percent.}$$

Exercise 7-18: Determine the amplitude and phase of V(6) in the circuit of Example 7-22, relative to those of V(1).

Answer: (See CAD)

Additional method to measure amplitude and phase

Let us continue working with the transmission-line circuit of the previous two examples. Place a Measurement Probe (of the type we introduced in Chapters 2 and 3) at each of the appropriate nodes in the circuit. Double-click on the Probe, and under the Parameters tab, select the appropriate parameters so that only V(p-p), Vgain(ac), and Phase are printed in the Probe output. Additionally, with the exception of Probe 1 (located right above V1), at the top of the Probe Properties window, check Use reference probe, and select Probe 1. Note that "phase" here refers to the phase difference between the voltage at the specific probe and the reference probe. So if a particular signal is leading the reference node, then the phase will appear *negative*, and if a particular signal is lagging the reference node, then the phase will appear *positive*. This is the opposite of how we are taught to think of phase, so keep this at the front of your mind when using this approach.

Run the Interactive Simulation by pressing F5 (or any of the appropriate buttons or toggles, which you should know by now) and the result should resemble that shown in Fig. 7-47. We can see that the Phase at Node 2 is 22.6°, which of course is opposite to what we see in Fig. 7-46, where the signal at V(2) is *behind* V(1) by 22.3°. However, we must remember that the phase values are flipped in the Measurement Probe readings, so the values actually are in agreement. Additionally, we see in Fig. 7-47 that the Vgain(ac) at Node 2 is "654m" (which corresponds to 65.4 percent), which is very nearly in agreement with the value of 66 percent obtained in Example 7-22.

Summary

Concepts

- A sinusoidal waveform is characterized by three independent parameters: its amplitude A, its angular frequency ω, and its phase angle ϕ.
- Complex algebra is used extensively in the phasor domain to analyze ac circuits. Hence, it behooves every student taking a course in circuit analysis to become proficient in using complex numbers (by hand, with a scientific calculator, and with MATLAB/Mathworks).
- By transforming an ac circuit from the time domain to the phasor domain, its integro-differential equation gets transformed into a linear equation. After solving the linear equation, the solution is then transformed back to the time domain.
- Voltages and currents in the time domain have phasor counterparts in the phasor domain; resistors, capacitors, and inductors are transformed into impedances.
- The rules for combining impedances (when connected in series or in parallel) are the same as those for resistors in resistive circuits. The same is true for Y–Δ transformations.
- All of the techniques of circuit analysis are equally applicable in the phasor domain.
- A phase shifter is a circuit that can modify the phase angle of a sinusoidal waveform.
- An ac waveform can be converted into dc by subjecting it to a four-step process that includes a transformer, bridge rectifier, smoothing filter, and voltage regulator.
- Multisim is very useful for analyzing an ac circuit and evaluating its response as a function of frequency.

Mathematical and Physical Models

Trigonometric identities Table 7-1

Time domain/phasor domain correspondence Table 7-2

Impedance
$$\mathbf{Z}_{\rm R} = R$$
$$\mathbf{Z}_{\rm C} = 1/j\omega C$$
$$\mathbf{Z}_{\rm L} = j\omega L$$

Impedances in series
$$\mathbf{Z}_{\rm eq} = \sum_{i=1}^{N} \mathbf{Z}_i$$

Admittances in parallel
$$\mathbf{Y}_{\rm eq} = \sum_{i=1}^{N} \mathbf{Y}_i$$

Y–Δ transformation Section 7-4.2

Transformer
$$\frac{\upsilon_2}{\upsilon_1} = \frac{N_2}{N_1}$$
$$\frac{i_2}{i_1} = \frac{N_1}{N_2}$$

Wire resistance
$$R = \frac{2\ell}{\pi a^2 \sigma} \quad \text{for } (a\sqrt{f\sigma} \leq 500)$$
$$R = \sqrt{\frac{\pi f \mu}{\sigma}} \left(\frac{\ell}{\pi a}\right) \quad \text{for } (a\sqrt{f\sigma} \geq 1250)$$

Wire capacitor
$$C = \frac{\pi \varepsilon \ell}{\ln(d/a)} \quad \text{for } (d/2a)^2 \gg 1$$

Important Terms

Provide definitions or explain the meaning of the following terms:

absolute phasor diagram
ac
admittance
alternating current
amplitude
angular frequency
argument
bridge rectifier
capacitive impedance
complex conjugate
complex number
conductance
core
cosine-referenced
current division
downswing time constant
electromagnetic compatibility
Euler's identity
forward-bias voltage

Important Terms (continued)

frequency	peak value	reactance	Thévenin's theorem
frequency domain technique	period (of a cycle)	real	time-independent
full-wave rectifier	phase angle	relative phasor diagram	time shift
half-wave rectifier	phase lag	rectangular form	time-varying function
iff	phase lead	rectifier	transformer
ideal transformer	phase-shift circuit	resistance	true
imaginary	phase-shift oscillator	ripple	turns ratio
impedance	phasor counterpart	secondary winding	upswing time constant
inductive impedance	phasor diagram	sinusoidal waveform	voltage division
lag	phasor domain	source-transformation principle	voltage regulator
lead	phasor domain technique	step-down transformer	winding
oscillation frequency	polar form	step-up transformer	zener diode
peak-to-peak ripple voltage	primary winding	susceptance	zener-diode resistance
	radio frequency identification		zener voltage

PROBLEMS

Section 7-1: Sinusoidal Signals

*7.1 Express the sinusoidal waveform

$$\upsilon(t) = -4\sin(8\pi \times 10^3 t - 45°) \text{ V}$$

in standard cosine form and then determine its amplitude, frequency, period, and phase angle.

7.2 Express the current waveform

$$i(t) = -0.2\cos(6\pi \times 10^9 t + 60°) \text{ mA}$$

in standard cosine form and then determine the following:

(a) Its amplitude, frequency, and phase angle.

(b) $i(t)$ at $t = 0.1$ ns.

*7.3 A 4 kHz sinusoidal voltage waveform $\upsilon(t)$, with a 12 V amplitude, was observed to have a value of 6 V at $t = 1$ ms. Determine the functional form of $\upsilon(t)$.

7.4 Two waveforms, $\upsilon_1(t)$ and $\upsilon_2(t)$, have identical amplitudes and oscillate at the same frequency, but $\upsilon_2(t)$ lags $\upsilon_1(t)$ by a phase angle of 60°. If

$$\upsilon_1(t) = 4\cos(2\pi \times 10^3 t + 30°) \text{ V},$$

write the expression appropriate for $\upsilon_2(t)$ and plot both waveforms over the time span from -1 ms to $+1$ ms.

7.5 Waveforms $\upsilon_1(t)$ and $\upsilon_2(t)$ are given by:

$$\upsilon_1(t) = -4\sin(6\pi \times 10^4 t + 30°) \text{ V},$$
$$\upsilon_2(t) = 2\cos(6\pi \times 10^4 t - 30°) \text{ V}.$$

Does $\upsilon_2(t)$ lead or lag $\upsilon_1(t)$, and by what phase angle?

*7.6 A phase angle of 120° was added to a 3 MHz signal, causing its waveform to shift by Δt along the time axis. In what direction did it shift and by how much?

7.7 Provide an expression for a 24 V signal that exhibits adjacent minima at $t = 1.04$ ms and $t = 2.29$ ms.

7.8 A multiplier circuit has two input ports, designated υ_1 and υ_2, and one output port whose voltage υ_{out} is equal to the product of υ_1 and υ_2. Assume

$$\upsilon_1 = 10\cos 2\pi f_1 t \text{ V},$$
$$\upsilon_2 = 10\cos 2\pi f_2 t \text{ V}.$$

(a) Obtain an expression for υ_{out} in terms of the sum and difference frequencies, $f_s = f_1 + f_2$ and $f_d = f_1 - f_2$.

(b) Plot its waveform over the time interval [0, 2 s], given that $f_1 = 3$ Hz and $f_2 = 2$ Hz.

*7.9 Provide an expression for a 12 V signal that exhibits a maximum at $t = 2.5$ ms, followed by an adjacent minimum at $t = 12.5$ ms.

*Answer(s) available in Appendix G.

Section 7-2: Complex Algebra

7.10 Express the following complex numbers in polar form:

(a) $\mathbf{z}_1 = 3 + j4$

(b) $\mathbf{z}_2 = -6 + j8$

*(c) $\mathbf{z}_3 = -6 - j4$

(d) $\mathbf{z}_4 = j2$

*(e) $\mathbf{z}_5 = (2 + j)^2$

(f) $\mathbf{z}_6 = (3 - j2)^3$

(g) $\mathbf{z}_7 = (-1 + j)^{1/2}$

7.11 Express the following complex numbers in rectangular form:

(a) $\mathbf{z}_1 = 2e^{j\pi/6}$

(b) $\mathbf{z}_2 = -3e^{-j\pi/4}$

*(c) $\mathbf{z}_3 = \sqrt{3}\, e^{-j3\pi/4}$

(d) $\mathbf{z}_4 = -j^3$

(e) $\mathbf{z}_5 = -j^{-4}$

*(f) $\mathbf{z}_6 = (2 + j)^2$

(g) $\mathbf{z}_7 = (3 - j2)^3$

7.12 Complex numbers $\mathbf{z}_1$ and $\mathbf{z}_2$ are given by:

$$\mathbf{z}_1 = 6 - j4,$$
$$\mathbf{z}_2 = -2 + j1.$$

(a) Express $\mathbf{z}_1$ and $\mathbf{z}_2$ in polar form.

(b) Determine $|\mathbf{z}_1|$ by applying Eq. (7.20) to the given expression.

(c) Determine the product $\mathbf{z}_1\mathbf{z}_2$ in polar form.

(d) Determine the ratio $\mathbf{z}_1/\mathbf{z}_2$ in polar form.

(e) Determine $\mathbf{z}_1^2$ and compare it with $|\mathbf{z}_1|^2$.

(f) Determine $\mathbf{z}_1/(\mathbf{z}_1 - \mathbf{z}_2)$ in polar form.

7.13 For the complex number $\mathbf{z} = 1 + j$, show that

$$\mathbf{z}^2 - |\mathbf{z}|^2 = -2(1 - j).$$

7.14 If $\mathbf{z} = -8 + j6$, determine the following quantities:

(a) $|\mathbf{z}|^2$

*(b) $\mathbf{z}^2$, in polar form

(c) $1/\mathbf{z}$, in polar form

(d) $\mathbf{z}^{-3}$, in polar form

(e) $\mathfrak{Re}(1/\mathbf{z}^2)$

*(f) $\mathfrak{Im}(\mathbf{z}^*)$

(g) $\mathfrak{Im}[(\mathbf{z}^*)^2]$

(h) $\mathfrak{Re}[(\mathbf{z}^*)^{-1/2}]$

7.15 Complex numbers $\mathbf{z}_1$ and $\mathbf{z}_2$ are given by

$$\mathbf{z}_1 = 2\angle{-60^\circ},$$
$$\mathbf{z}_2 = 5\angle{45^\circ}.$$

Determine in polar form:

(a) $\mathbf{z}_1\mathbf{z}_2$

(b) $\mathbf{z}_1/\mathbf{z}_2$

*(c) $\mathbf{z}_1\mathbf{z}_2^*$

(d) $\mathbf{z}_1^2$

(e) $\sqrt{\mathbf{z}_2}$

(f) $\sqrt{\mathbf{z}_2^*}$

(g) $\mathbf{z}_1(\mathbf{z}_2 - \mathbf{z}_1)^*$

(h) $\mathbf{z}_2^*/(\mathbf{z}_1 + \mathbf{z}_2)$

7.16 Given $\mathbf{z} = 1.2 - j2.4$, determine the value of:

(a) $\ln \mathbf{z}$

*(b) $e^{\mathbf{z}}$

(c) $\ln(\mathbf{z}^*)$

(d) $\exp(\mathbf{z}^* + 1)$

7.17 Simplify the following expressions into the form $(a + jb)$, where a and b are real numbers:

(a) $\sqrt{j} + \sqrt{-j}$

(b) $\sqrt{j}\sqrt{-j}$

(c) $\dfrac{(1 + j)^2}{(1 - j)^2}$

7.18 Simplify the following expressions and express the result in polar form:

(a) $\mathbf{A} = \dfrac{5e^{-j30^\circ}}{2 + j3} - j4$

*(b) $\mathbf{B} = \dfrac{(-20\angle{45^\circ})(3 - j4)}{(2 - j)} + (2 + j)$

(c) $\mathbf{C} = \dfrac{j4}{(3 + j2) - 2(1 - j)} + \dfrac{1}{1 + j4}$

(d) $\mathbf{D} = \begin{vmatrix} (2 - j) & -(3 + j4) \\ -(3 + j4) & (2 + j) \end{vmatrix}$

(e) $\mathbf{E} = \begin{vmatrix} 5\angle{30^\circ} & -2\angle{45^\circ} \\ -2\angle{45^\circ} & 4\angle{60^\circ} \end{vmatrix}$

7.19 Calculate the following complex numbers and express the results in rectangular form:

(a) $3e^{j\pi/4} - j4$
(b) $(3 - j4)/(2\angle 45^\circ)$
(c) $[(7 - j5)/(3 + j2)] - 2e^{j3\pi/4}$

7.20 Calculate the following complex numbers and express the results in polar form:

(a) $2e^{j\pi/3} - 5e^{j\pi/2}$
*(b) $(8 - j3)(7 + j2)$
(c) $[(6 - j7)/(-2 + j9)] + 3e^{-j\pi/4}$

Sections 7-3 to 7-5: Phasor Domain and Impedance Transformations

7.21 Transform the following sinusoidal currents into their phasor counterparts:

(a) $i_1(t) = 10\sin(8t + 75^\circ)$ A
(b) $i_2(t) = -17\cos(9t - 25^\circ)$ A
(c) $i_3(t) = [8\cos(6t - 45^\circ) - 5\sin(6t)]$ A

7.22 Determine the phasor counterparts of the following sinusoidal functions:

(a) $\upsilon_1(t) = 4\cos(377t - 30^\circ)$ V
(b) b $\upsilon_2(t) = -2\sin(8\pi \times 10^4 t + 18^\circ)$ V
(c) $\upsilon_3(t) = 3\sin(1000t + 53^\circ) - 4\cos(1000t - 17^\circ)$ V

7.23 Determine the instantaneous time functions corresponding to the following phasors:

(a) $\mathbf{I}_1 = 6e^{j60^\circ}$ A at $f = 60$ Hz
(b) $\mathbf{I}_2 = -2e^{-j30^\circ}$ A at $f = 1$ kHz
*(c) $\mathbf{I}_3 = j3$ A at $f = 1$ MHz
(d) $\mathbf{I}_4 = -(3 + j4)$ A at $f = 10$ kHz
(e) $\mathbf{I}_5 = -4\angle{-120^\circ}$ A at $f = 3$ MHz

7.24 Show that the instantaneous time function corresponding to the phasor $\mathbf{V} = 4e^{j60^\circ} + 6e^{-j60^\circ}$ V is given by

$$\upsilon(t) = 5.29\cos(\omega t - 19.1^\circ) \text{ V}.$$

7.25 Determine the impedances of the following elements:

(a) $R = 1$ kΩ at 1 MHz
(b) $L = 30\ \mu$H at 1 MHz
*(c) $C = 50\ \mu$F at 1 kHz

7.26 The function $\upsilon(t)$ is the sum of two sinusoids,

$$\upsilon(t) = 4\cos(\omega t + 30^\circ) + 6\cos(\omega t + 60^\circ) \text{ V}. \quad (1)$$

(a) Apply the necessary trigonometric identities from Table 7-1 to show that

$$\upsilon(t) = 9.67\cos(\omega t + 48.1^\circ) \text{ V}. \quad (2)$$

(b) Transform the expression given by Eq. (1) to the phasor domain, simplify it into a single term, and then transform it back to the time domain to show that the result is identical to the expression given by Eq. (2).

7.27 Use phasors to simplify each of the following expressions into a single term [*Hint*: See Problem 7.26]:

(a) $\upsilon_1(t) = 12\cos(6t + 30^\circ) - 6\cos(6t - 45^\circ)$ V
*(b) $\upsilon_2(t) = -3\sin(1000t - 15^\circ) - 6\sin(1000t + 15^\circ) + 12\cos(1000t - 60^\circ)$ V
(c) $\upsilon_3(t) = 2\cos(377t + 60^\circ) - 2\cos(377t - 60^\circ)$ V
(d) $\upsilon_4(t) = 10\cos 800t + 10\sin 800t$ V

7.28 Simply the following expressions using phasors:

(a) $i_1(t) = 20\cos(\omega t - 30^\circ) + 16\cos(\omega t + 15^\circ)$ A
(b) $i_2(t) = 14\sin(\omega t + 45^\circ) - 17\cos(\omega t + 60^\circ)$ A
(c) $i_3(t) = 2\cos(5t) - 7\sin(5t)$ A

***7.29** The current source in the circuit of Fig. P7.18 is given by

$$i_s(t) = 12\cos(2\pi \times 10^4 t - 60^\circ) \text{ mA}.$$

Apply the phasor-domain analysis technique to determine $i_C(t)$, given that $R = 20\ \Omega$ and $C = 1\ \mu$F.

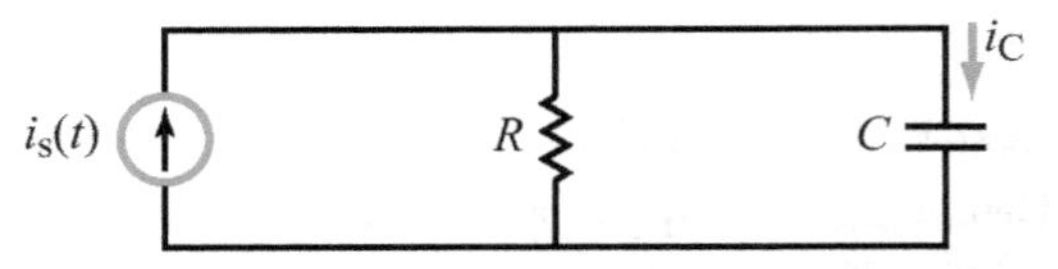

Figure P7.29: Circuit for Problems 7.29 and 7.30.

7.30 Repeat Problem 7.29, after replacing the capacitor with a 0.5 mH inductor and then calculating the current through it.

7.31 Find $i_s(t)$ in the circuit of Fig. P7.31, given that $\upsilon_s(t) = 15\cos(5 \times 10^4 t - 30^\circ)$ V, $R = 1$ kΩ, $L = 120$ mH, and $C = 5$ nF.

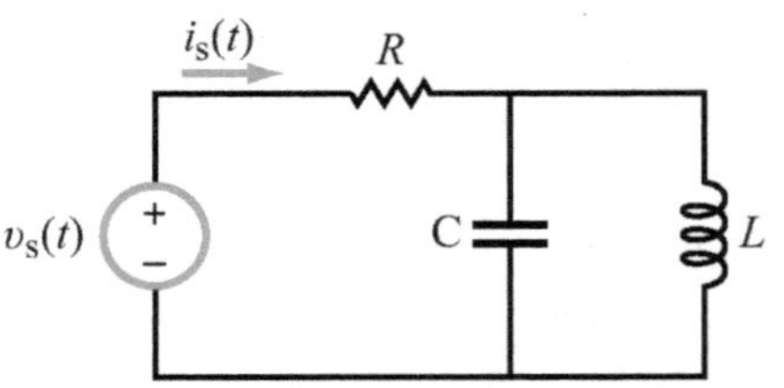

Figure P7.31: Circuit for Problem 7.31.

*7.32 Find voltage $v_{ab}(t)$ in the circuit of Fig. P7.32, given that $i_s(t) = 35\sin(300t - 15°)$ mA, $R = 80\ \Omega$, $L = 15$ mH, and $C = 200\ \mu$F.

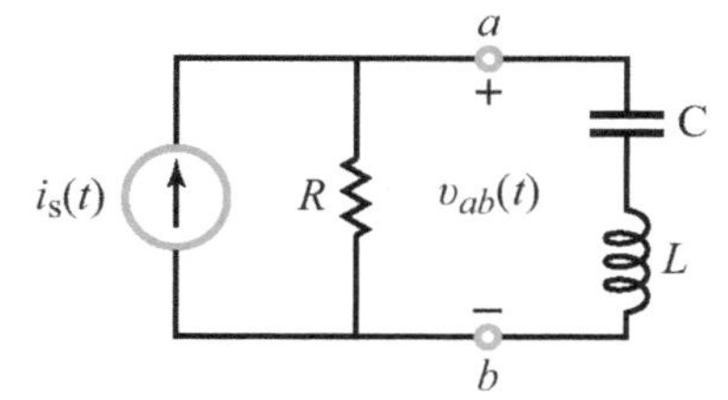

Figure P7.32: Circuit for Problem 7.32.

7.33 Find $i_a(t)$ in the circuit of Fig. P7.33, given that $v_s(t) = 40\sin(200t - 20°)$ V.

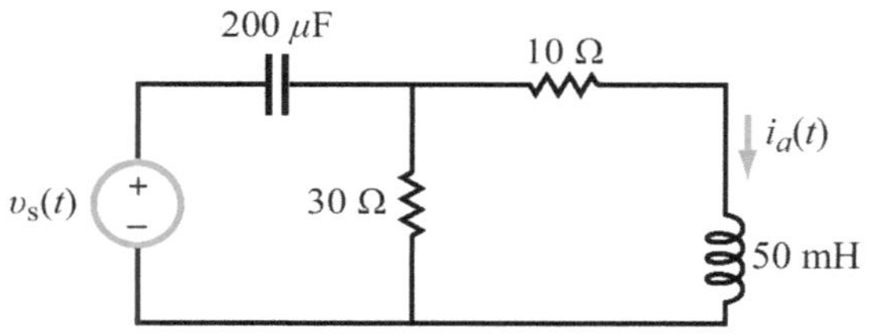

Figure P7.33: Circuit for Problem 7.33.

7.34 Determine the equivalent impedance:

(a) $\mathbf{Z}_1$ at 1000 Hz (Fig. P7.34(a))

(b) b $\mathbf{Z}_2$ at 500 Hz (Fig. P7.34(b))

(c) $\mathbf{Z}_3$ at $\omega = 10^6$ rad/s (Fig. P7.34(c))

(d) $\mathbf{Z}_4$ at $\omega = 10^5$ rad/s (Fig. P7.34(d))

(e) $\mathbf{Z}_5$ at $\omega = 2000$ rad/s (Fig. P7.34(e))

7.35 Find the input impedance **Z** of the circuit in Fig. P7.35.

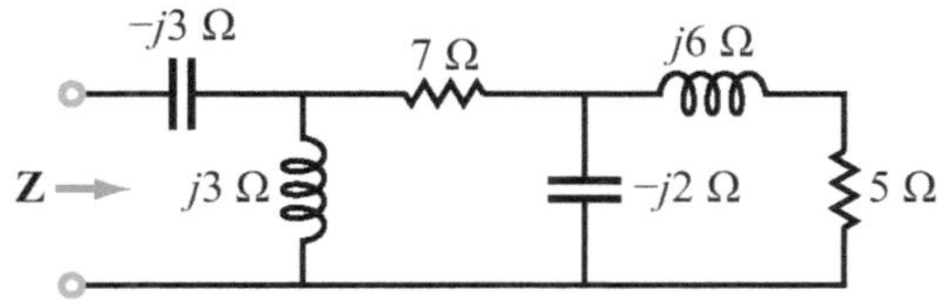

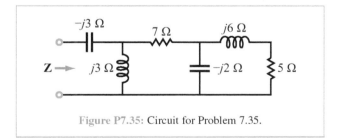

Figure P7.35: Circuit for Problem 7.35.

*7.36 Find the input impedance **Z** of the circuit in Fig. P7.36 at $\omega = 400$ rad/s.

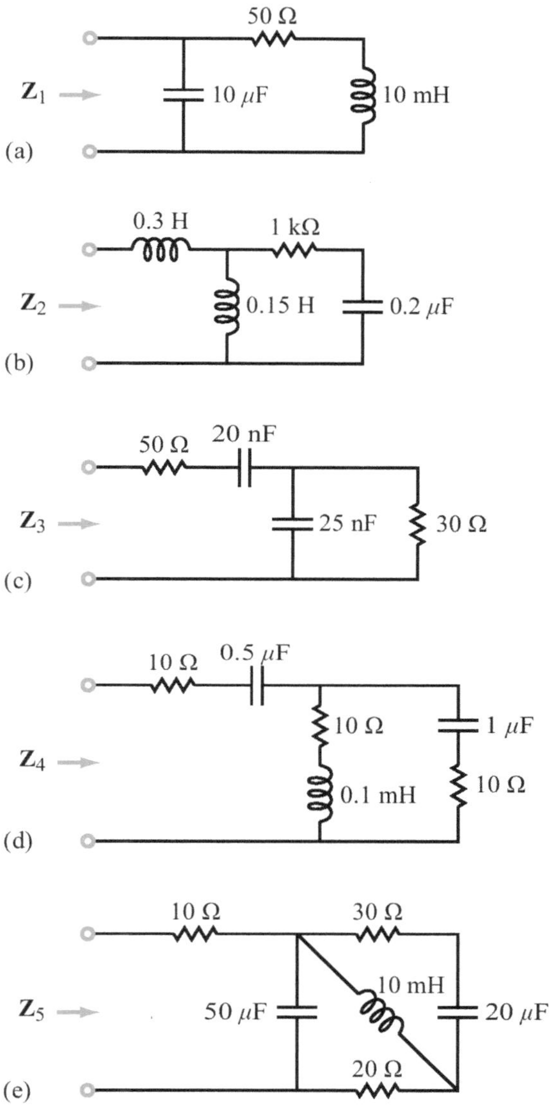

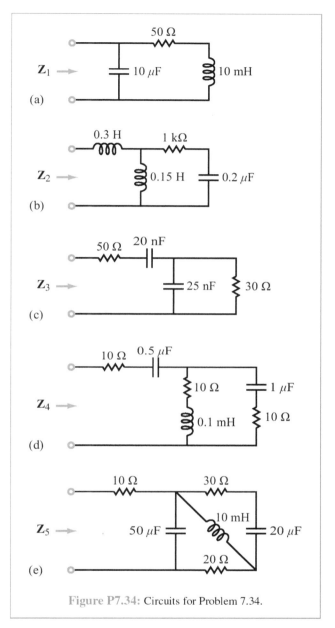

Figure P7.34: Circuits for Problem 7.34.

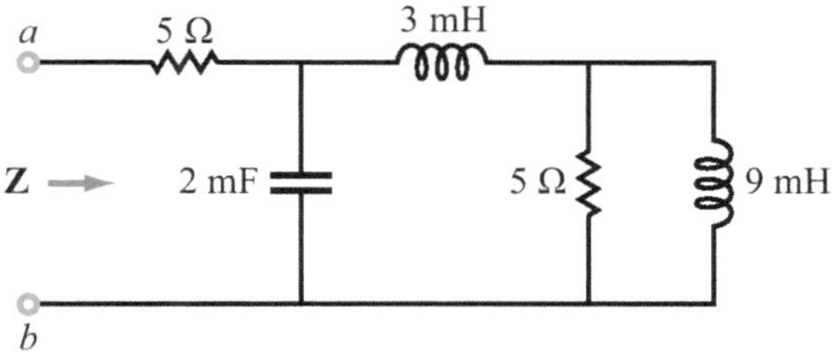

Figure P7.36: Circuit for Problem 7.36.

7.37 Find **Z** of the circuit in Fig. P7.37, given that

$$v_s(t) = 10\cos(377t + 15°) \text{ V}$$

and

$$i_s(t) = 3\sin(377t + 30°) \text{ A}.$$

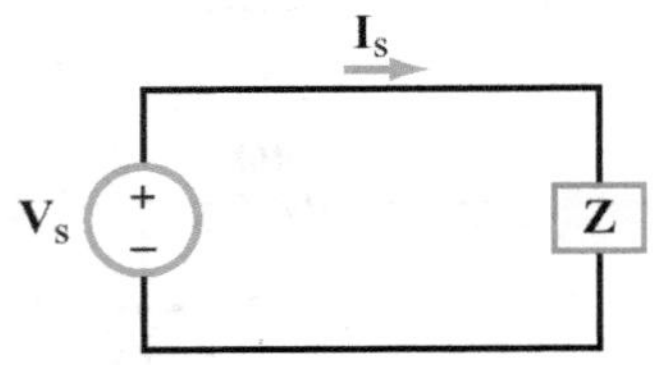

Figure P7.37: Circuit for Problem 7.37.

*7.38 Find $\mathbf{I}_R$ in the circuit of Fig. P7.38, given that $\mathbf{V}_s = 25$ V.

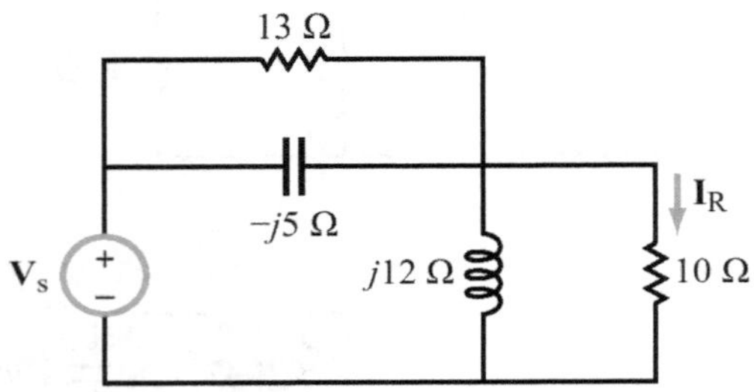

Figure P7.38: Circuit for Problem 7.38.

7.39 The voltage source in the circuit of Fig. P7.39 is given by

$$v_s(t) = 12\cos 10^4 t \text{ V}.$$

(a) Transform the circuit to the phasor domain and then determine the equivalent impedance **Z** at terminals (a, b). [*Hint*: Application of Δ–Y transformation should prove helpful.]

(b) Determine the phasor **I**, corresponding to $i(t)$.

(c) Determine $i(t)$.

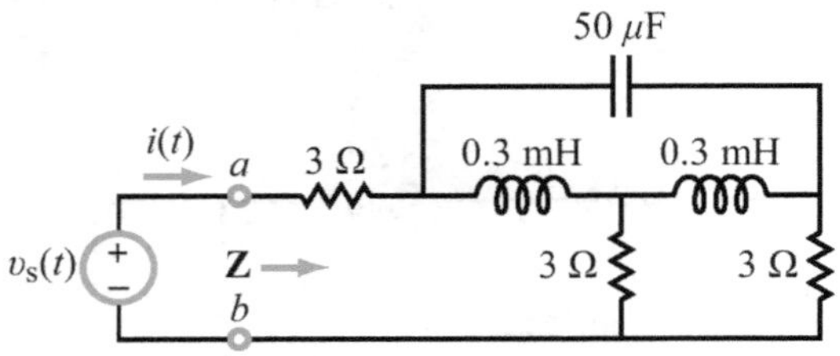

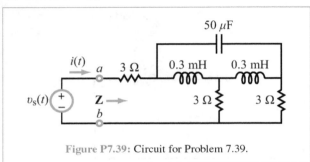

Figure P7.39: Circuit for Problem 7.39.

*7.40 The circuit in Fig. P7.40 is in the phasor domain. Determine the following:

(a) The equivalent input impedance **Z** at terminals (a, b).

(b) The phasor current **I**, given that $\mathbf{V}_s = 25\angle 45°$ V.

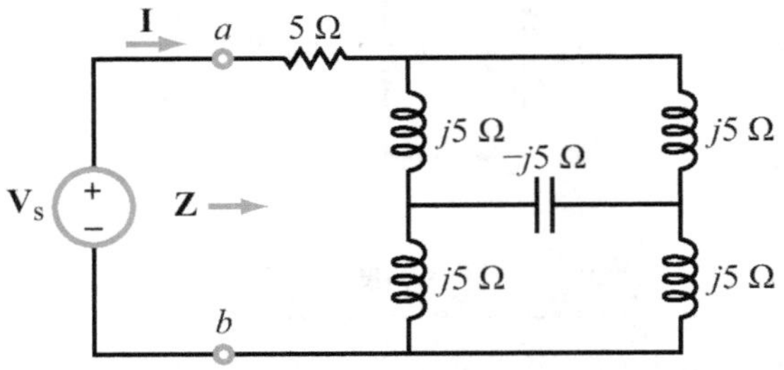

Figure P7.40: Circuit for Problem 7.40.

7.41 Use the phasor domain circuit in Fig. P7.41.

(a) Determine the value of $\mathbf{Z}_x$ that would make the input impedance **Z** purely real.

(b) Specify what type of element would be needed to realize that condition, and what its magnitude should be if $\omega = 6250$ rad/s.

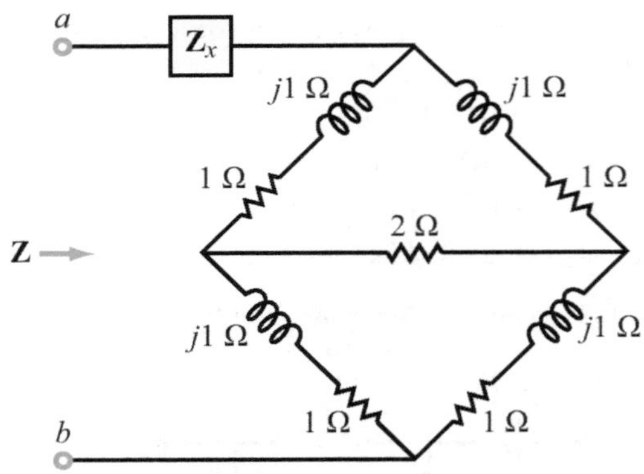

Figure P7.41: Circuit for Problem 7.41.

7.42 In response to an input signal voltage $v_s(t) = 24\cos 2000\pi t$, the input current in the circuit of Fig. P7.42 was measured as $i(t) = 6\cos(2000\pi t - 60°)$ mA. Determine the equivalent input impedance **Z** of the circuit.

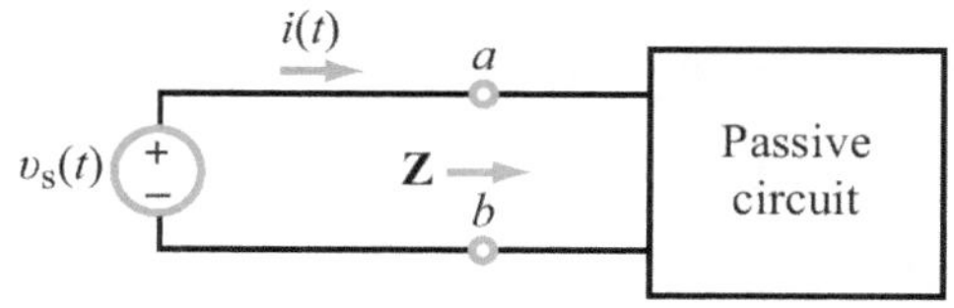

Figure P7.42: Configuration for Problem 7.42.

7.43 At $\omega = 400$ rad/s, the input impedance of the circuit in Fig. P7.43 is $\mathbf{Z} = (74 + j72)\ \Omega$. What is the value of L?

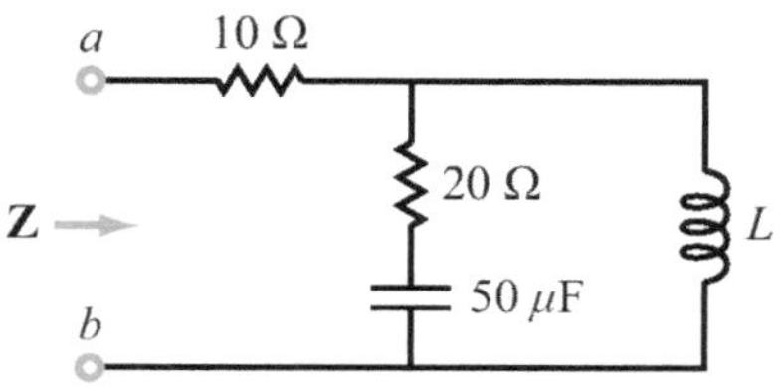

Figure P7.43: Circuit for Problem 7.43.

*__7.44__ In the circuit of Fig. P7.44, what should the value of L be at $\omega = 10^4$ rad/s so that $i(t)$ is in-phase with $\upsilon_s(t)$?

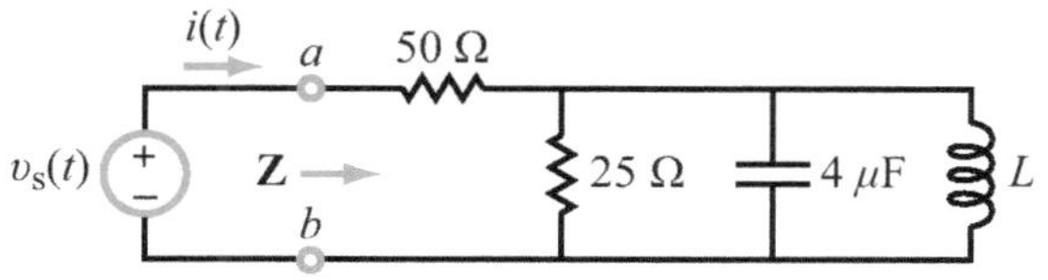

Figure P7.44: Circuit for Problem 7.44.

7.45 At what angular frequency ω is the current $i(t)$ in the circuit of Fig. P7.45 in-phase with the source voltage $\upsilon_s(t)$?

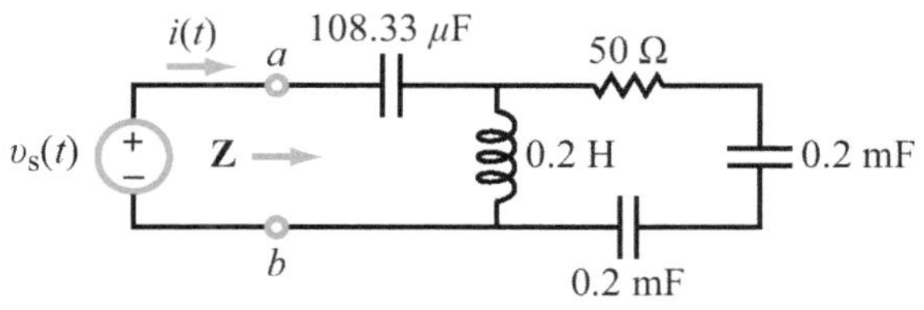

Figure P7.45: Circuit for Problem 7.45.

Sections 7-6: Equivalent Circuits

7.46 Your objective is to obtain a Thévenin equivalent for the circuit shown in Fig. P7.46, given that $i_s(t) = 3\cos 4 \times 10^4 t$ A. To that end:

(a) Transform the circuit to the phasor domain.

(b) Apply the source-transformation technique to obtain the Thévenin equivalent circuit at terminals (a, b).

(c) Transform the phasor-domain Thévenin circuit back to the time domain.

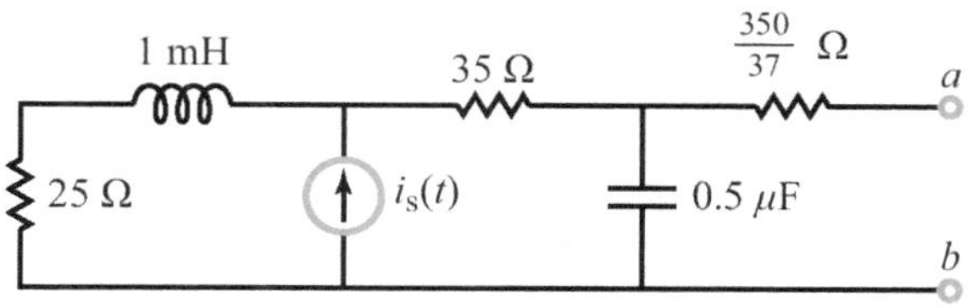

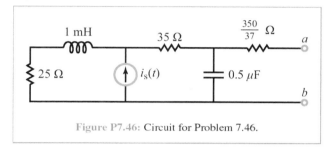

Figure P7.46: Circuit for Problem 7.46.

7.47 The input circuit shown in Fig. P7.47 contains two sources, given by

$$i_s(t) = 2\cos 10^3 t \text{ A},$$
$$\upsilon_s(t) = 8\sin 10^3 t \text{ V}.$$

This input circuit is to be connected to a load circuit that provides optimum performance when the impedance $\mathbf{Z}$ of the input circuit is purely real. The circuit includes a "matching" element whose type and magnitude should be chosen to realize that condition. What should those attributes be?

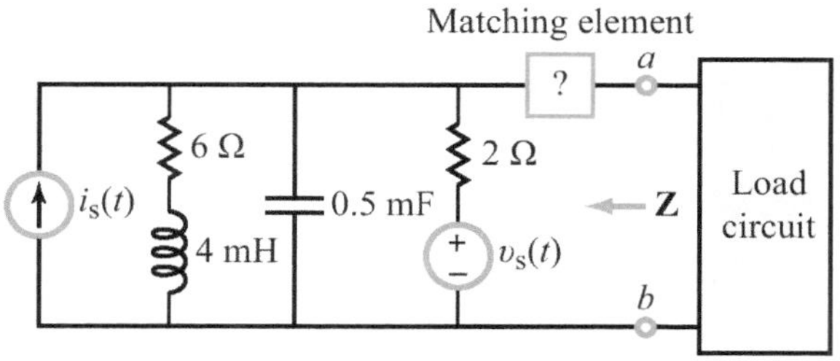

Figure P7.47: Circuit for Problem 7.47.

*__7.48__ Determine the Thévenin equivalent of the circuit in Fig. P7.48 at terminals (a, b), given that

$$\upsilon_s(t) = 12\cos 2500t \text{ V},$$
$$i_s(t) = 0.5\cos(2500t - 30°) \text{ A}.$$

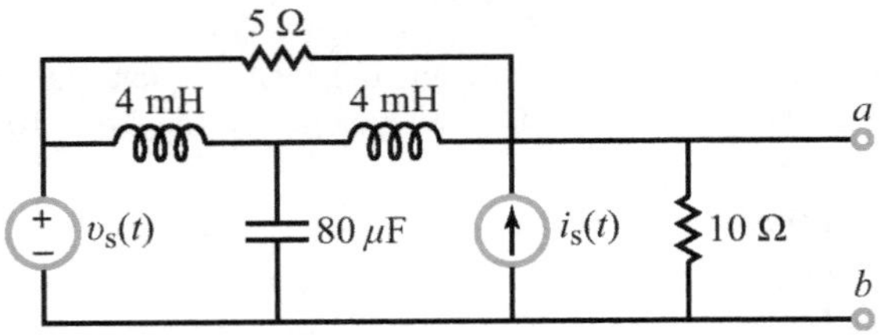

Figure P7.48: Circuit for Problem 7.48.

7.49 The circuit in Fig. P7.49 is in the phasor domain. Determine its Thévenin equivalent at terminals (a, b).

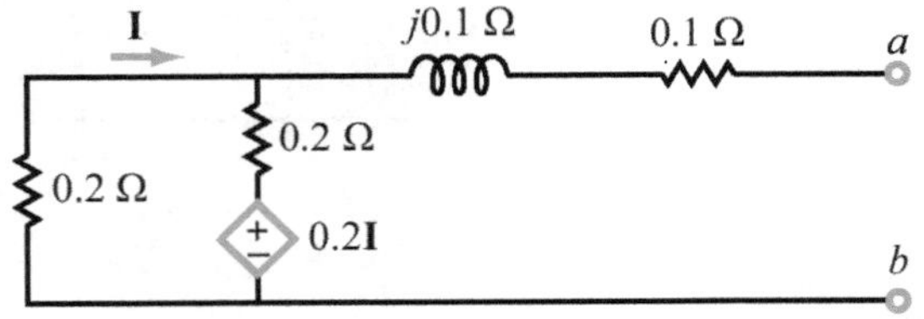

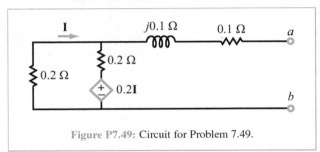

Figure P7.49: Circuit for Problem 7.49.

7.50 As we will learn in Chapter 8, to maximize the transfer of power from an input circuit to a load $\mathbf{Z}_L$, it is necessary to choose $\mathbf{Z}_L$ such that it is equal to the complex conjugate of the impedance of the input circuit. For the circuit in Fig. P7.50, such a condition translates into requiring $\mathbf{Z}_L = \mathbf{Z}^*_{Th}$. Determine $\mathbf{Z}_L$ such that it satisfies this condition.

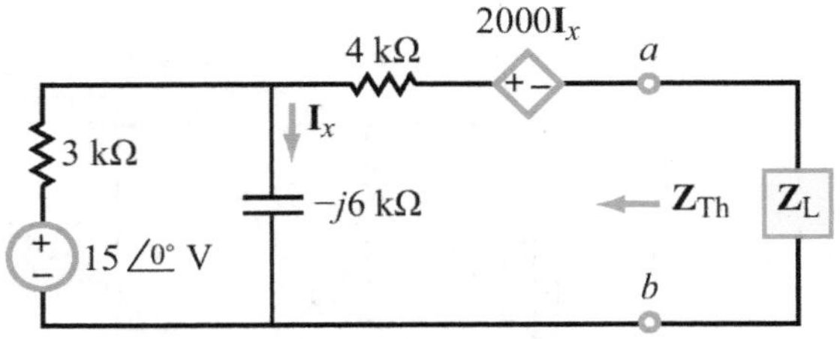

Figure P7.50: Circuit for Problem 7.50.

*__7.51__ The phasor current $\mathbf{I}_L$ in the circuit of Fig. P7.51 was measured to be

$$\mathbf{I}_L = \left(\frac{78}{41} + j\frac{36}{41}\right) \text{ mA}.$$

Determine $\mathbf{Z}_L$.

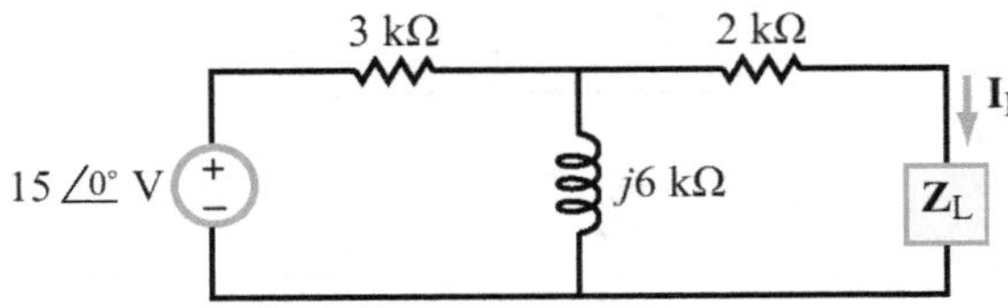

Figure P7.51: Circuit for Problem 7.51.

Sections 7-7 and 7-8: Phasor Diagrams and Phase Shifters

7.52 For the circuit in Fig. P7.52:

(a) Apply current division to express $\mathbf{I}_C$ and $\mathbf{I}_R$ in terms of $\mathbf{I}_s$.

(b) Using $\mathbf{I}_s$ as reference, generate a relative phasor diagram showing $\mathbf{I}_C$, $\mathbf{I}_R$, and $\mathbf{I}_s$ and demonstrate that the vector sum $\mathbf{I}_R + \mathbf{I}_C = \mathbf{I}_s$ is satisfied.

(c) Analyze the circuit to determine $\mathbf{I}_s$ and then generate the absolute phasor diagram with $\mathbf{I}_C$, $\mathbf{I}_R$, and $\mathbf{I}_s$ drawn according to their true phase angles.

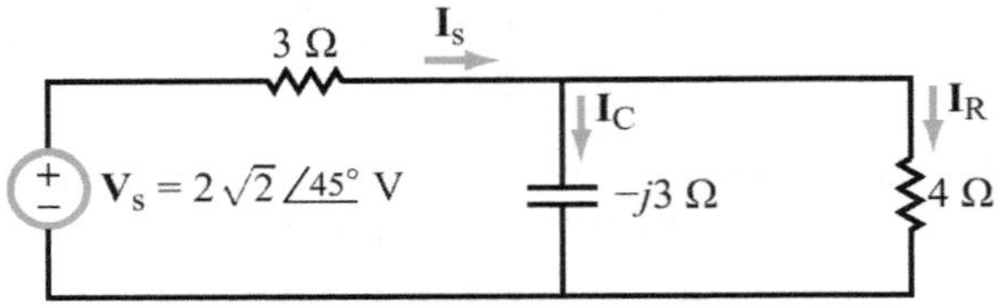

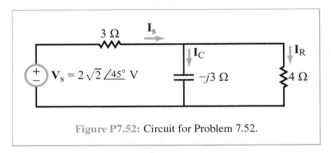

Figure P7.52: Circuit for Problem 7.52.

7.53 For the circuit in Fig. P7.53:

(a) Apply current division to express $\mathbf{I}_1$ and $\mathbf{I}_2$ in terms of $\mathbf{I}_s$.

(b) With $\mathbf{I}_s$ as reference, generate a relative phasor diagram showing that the vector sum $\mathbf{I}_1 + \mathbf{I}_2 = \mathbf{I}_s$ is indeed satisfied.

(c) Analyze the circuit to determine $\mathbf{I}_s$ and then generate the absolute phasor diagram for the three currents.

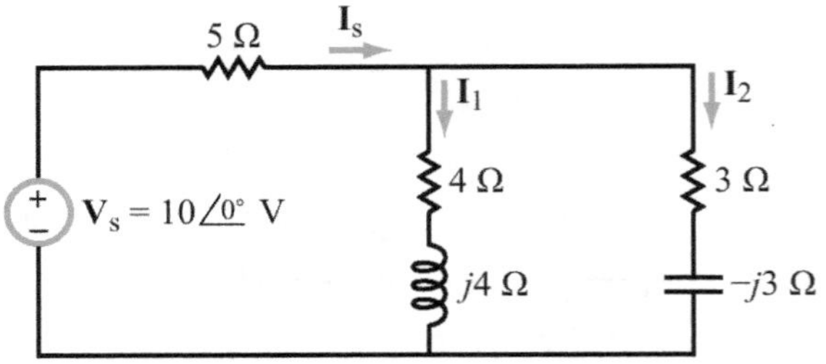

Figure P7.53: Circuit for Problem 7.53.

7.54 Design a two-stage 1 MHz RC phase-shift circuit whose output voltage is 120° behind that of the input signal. All capacitors are 1 nF each. Determine the values of the resistors and the amplitude ratio of the output voltage to that of the input.

7.55 A two-stage RC circuit provides a phase shift lead of 120°. What is the ratio of the output-voltage amplitude to that of the input?

*__7.56__ The element values of a single-stage phase-shift circuit are $R = 40\ \Omega$ and $C = 5\ \mu$F. At what frequency f is $\phi_1 = -\phi_2$, where ϕ_1 and ϕ_2 are the phase angles of the output voltages across R and C, respectively?

Section 7-9: Analysis Techniques

7.57 Apply nodal analysis in the phasor domain to determine $i_x(t)$ in the circuit of Fig. P7.57.

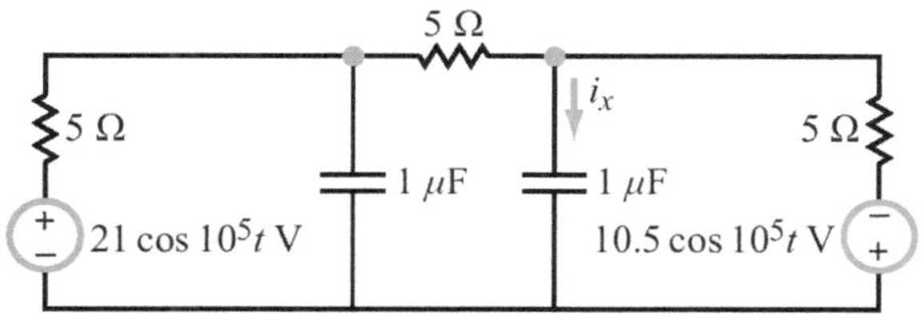

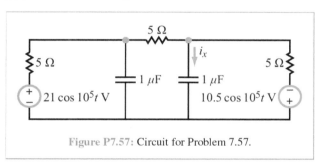

Figure P7.57: Circuit for Problem 7.57.

*__7.58__ Apply nodal analysis in the phasor domain to determine $i_C(t)$ in the circuit of Fig. P7.58.

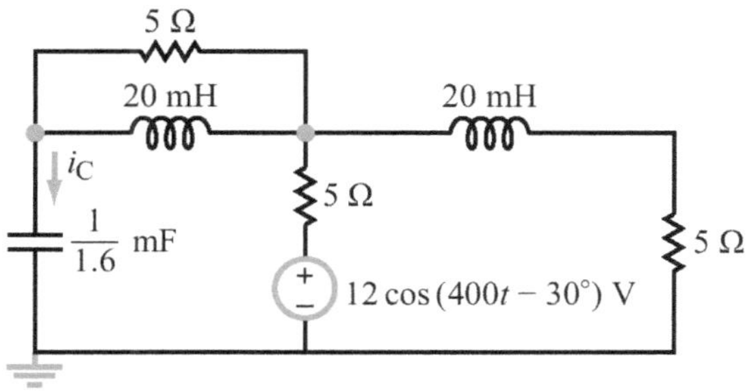

Figure P7.58: Circuit for Problem 7.58.

7.59 The circuit in Fig. P7.59 contains a supernode between nodes $\mathbf{V}_1$ and $\mathbf{V}_2$. Apply the supernode method to determine $\mathbf{V}_1$, $\mathbf{V}_2$, and $\mathbf{V}_3$, and then calculate $\mathbf{I}_C$.

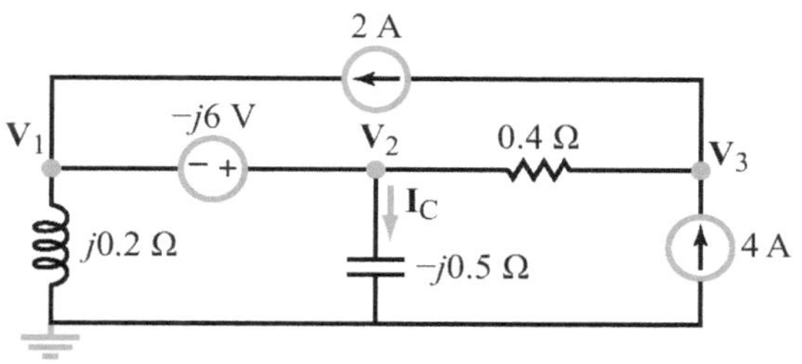

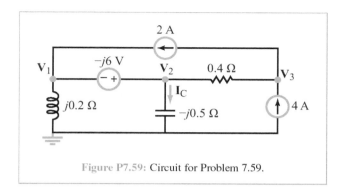

Figure P7.59: Circuit for Problem 7.59.

7.60 Apply the by-inspection method to develop a node-voltage matrix equation for the circuit in Fig. P7.60, and then use MATLAB or MathScript software to solve for $\mathbf{V}_1$ and $\mathbf{V}_2$.

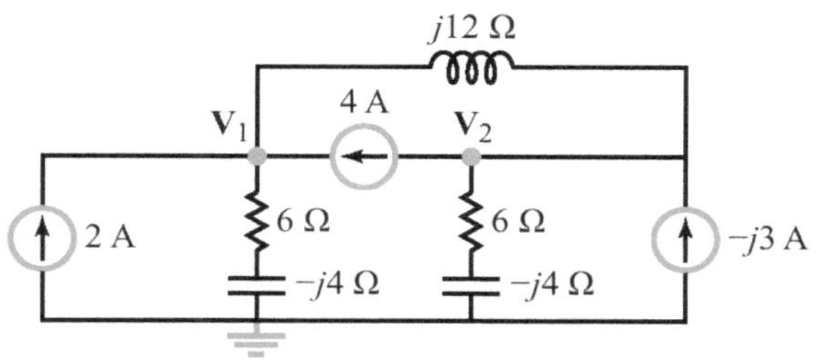

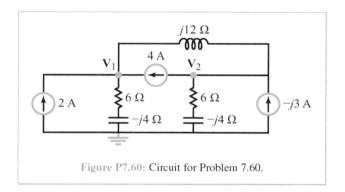

Figure P7.60: Circuit for Problem 7.60.

7.61 With $\mathbf{I}_s = 12\,\angle 20°$ V in the circuit of Fig. P7.61, apply the by-inspection method to develop a node-voltage matrix equation and then use MATLAB or MathScript software to solve for $\mathbf{I}_x$.

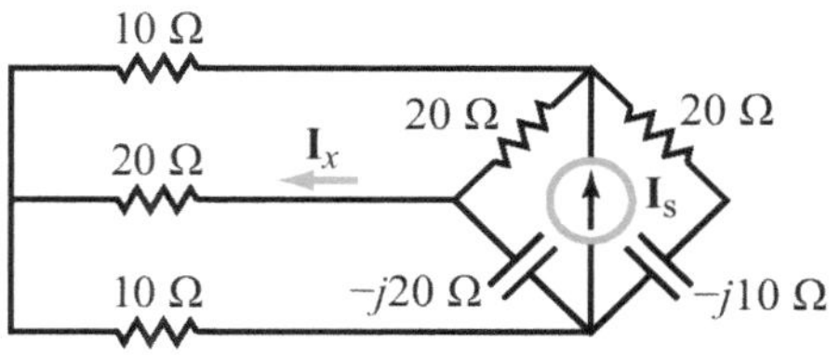

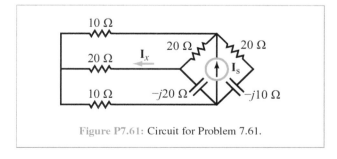

Figure P7.61: Circuit for Problem 7.61.

*__7.62__ Apply nodal analysis to determine $\mathbf{I}_C$ in the circuit of Fig. P7.62.

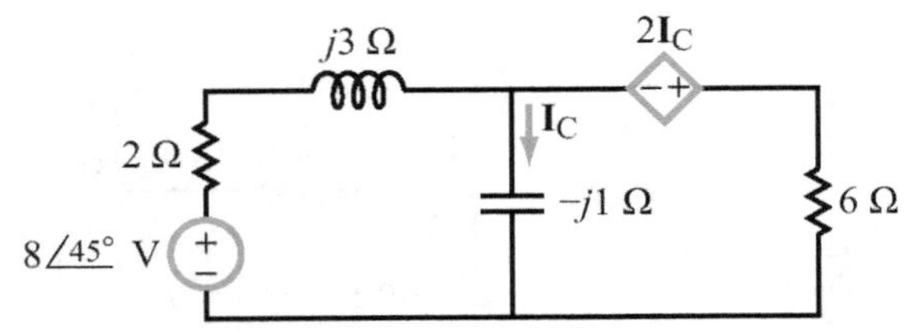

Figure P7.62: Circuits for Problems 7.62 and 7.63.

7.63 Apply mesh analysis to determine $\mathbf{I}_C$ in the circuit of Fig. P7.62.

7.64 Apply mesh analysis to determine $i_L(t)$ in the circuit of Fig. P7.64.

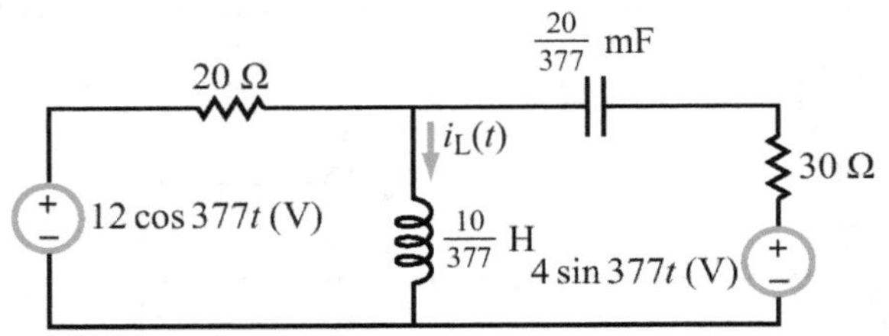

Figure P7.64: Circuit for Problem 7.64.

*__7.65__ Use mesh analysis to obtain an expression for the phasor $\mathbf{V}_{out}$ in the circuit of Fig. P7.65, in terms of $\mathbf{V}_s$ and R, given that $R = \omega L = 1/\omega C$.

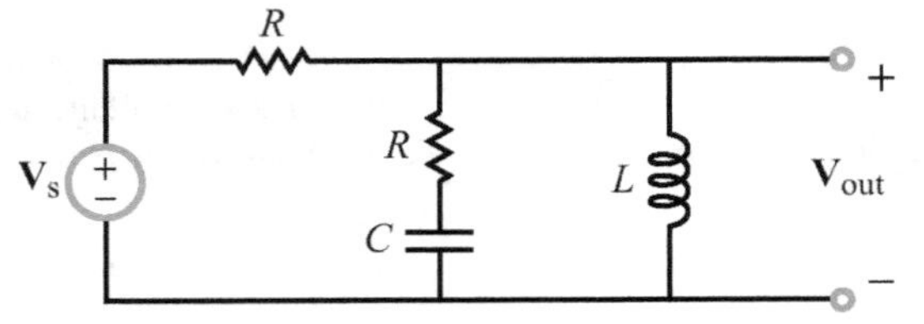

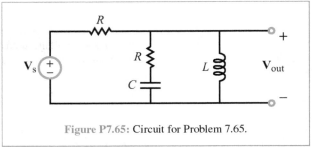

Figure P7.65: Circuit for Problem 7.65.

7.66 Apply the by-inspection method to develop a mesh-current matrix equation for the circuit in Fig. P7.66 and then use MATLAB or MathScript software to solve for $\mathbf{I}_1$, $\mathbf{I}_2$, and $\mathbf{I}_3$.

7.67 Use any analysis technique of your choice to determine $i_C(t)$ in the circuit of Fig. P7.67.

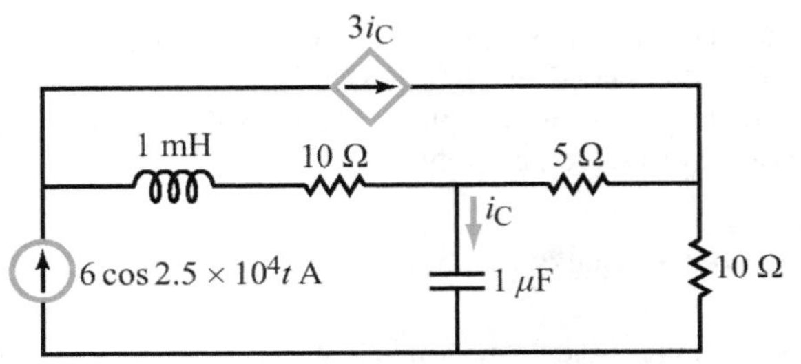

Figure P7.67: Circuit for Problem 7.67.

*__7.68__ Determine $i_x(t)$ in the circuit of Fig. P7.68, given that $v_s(t) = 6\cos 5 \times 10^5 t$ V.

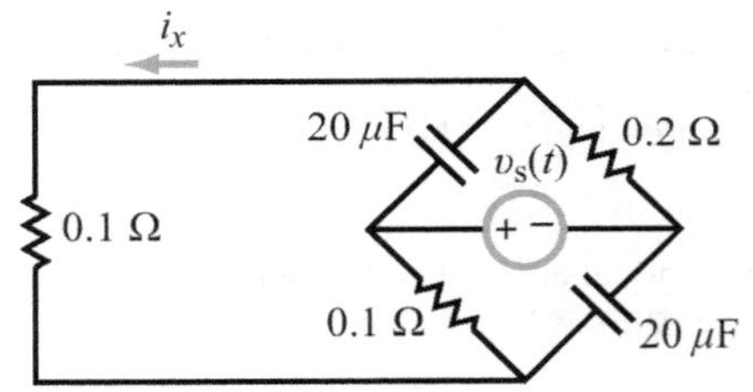

Figure P7.68: Circuit for Problem 7.68.

7.69 Find $\mathbf{I}_s$ in the circuit of Fig. P7.69, given that $\mathbf{V}_s = 8\angle 15^\circ$ V.

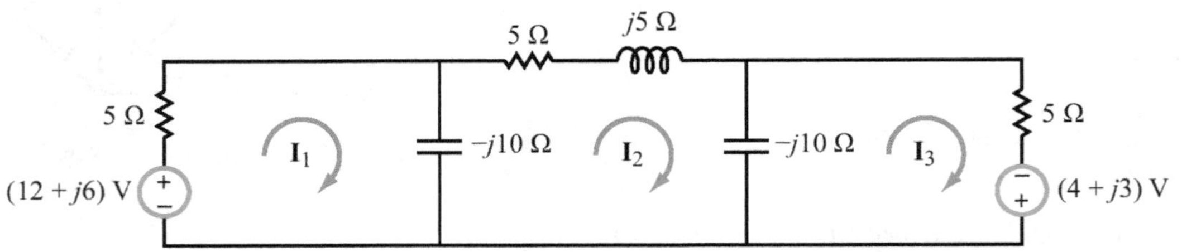

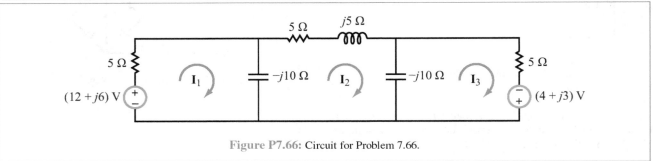

Figure P7.66: Circuit for Problem 7.66.

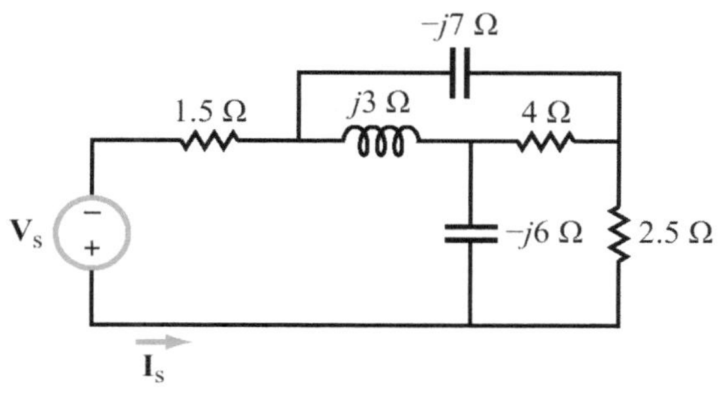

Figure P7.69: Circuit for Problem 7.69.

7.70 Find **Z** in the circuit of Fig. P7.70, given that $\mathbf{V}_s = 40$ V and $\mathbf{V}_a = 17.22e^{-j132.2°}$ V.

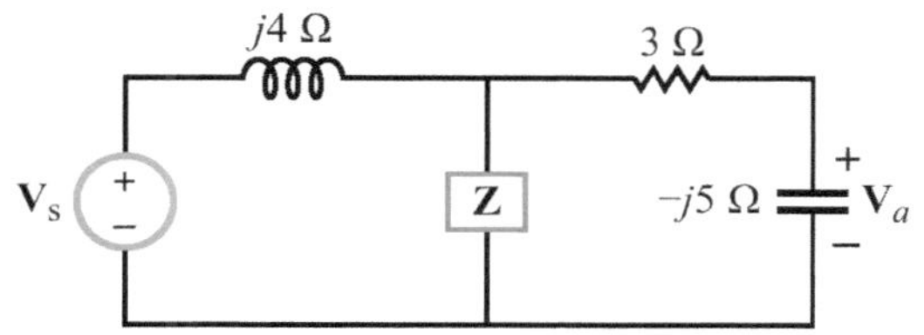

Figure P7.70: Circuit for Problem 7.70.

*__7.71__ Find the Thévenin equivalent at terminals (a, b) for the circuit in Fig. P7.71. The source is $\mathbf{V}_s = 10\angle 45°$ V.

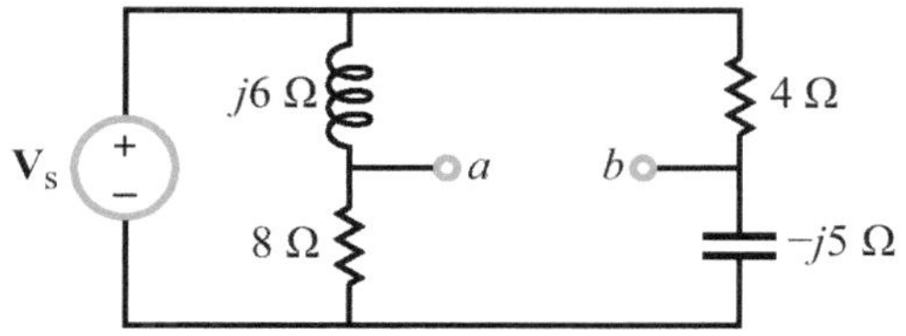

Figure P7.71: Circuit for Problem 7.71.

7.72 Find ω such that $\upsilon_a(t)$ and $i_s(t)$ in the circuit of Fig. P7.72 are in-phase. The element values are $R_1 = 5\ \Omega$, $R_2 = 3\ \Omega$, $L = 35$ mH, and $C = 7$ mF.

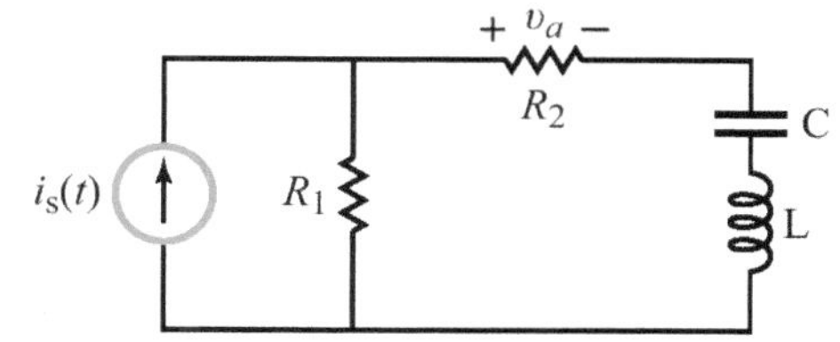

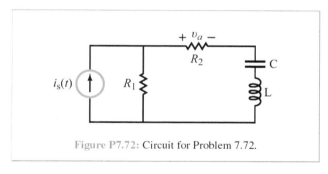

Figure P7.72: Circuit for Problem 7.72.

7.73 Find $i_s(t)$ in the circuit of Fig. P7.73, given that $\upsilon_s = 15\cos(\omega t)$, and:

(a) $\omega = 50$ rad/s

(b) $\omega = 75$ rad/s

(c) $\omega = 200$ rad/s

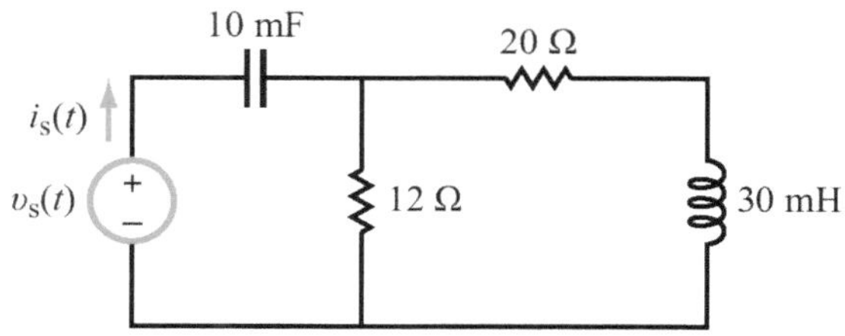

Figure P7.73: Circuit for Problem 7.73.

*__7.74__ Find $i_a(t)$ in the circuit of Fig. P7.74, given that $i_s(t) = 18\cos(35t + 75°)$ A.

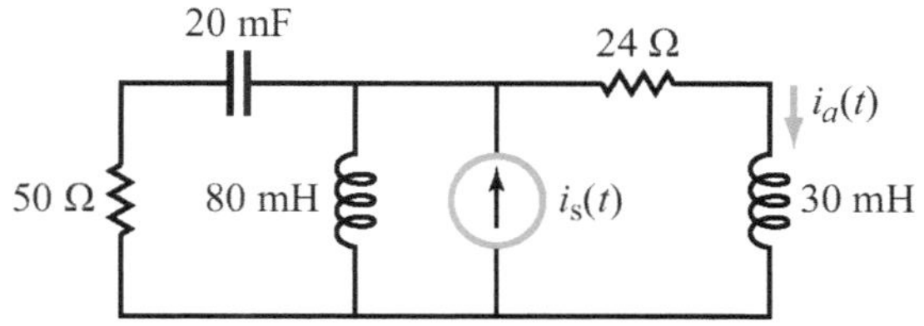

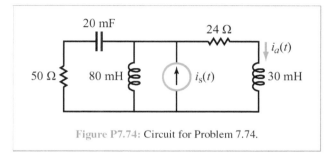

Figure P7.74: Circuit for Problem 7.74.

7.75 Find the value of ω at which $\upsilon_s(t)$ and $i_s(t)$ in the circuit of Fig. P7.75 are in phase.

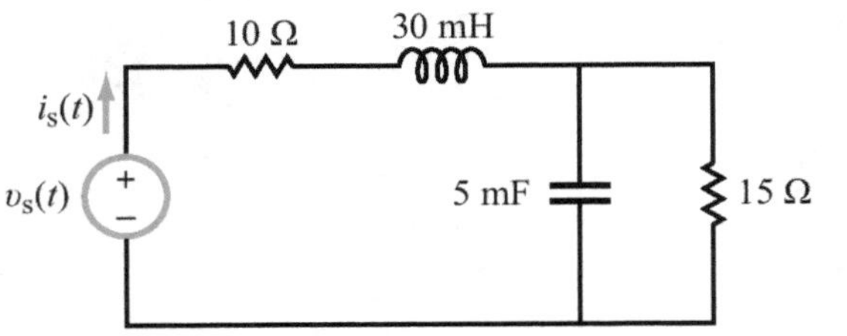

Figure P7.75: Circuit for Problem 7.75.

7.76 In the circuit of Fig. P7.76, find the value of ω, given that $\upsilon_s(t) = 20\cos(\omega t + 31.4°)$ V and $i_s(t) = 50\sin(\omega t + 80°)$ A.

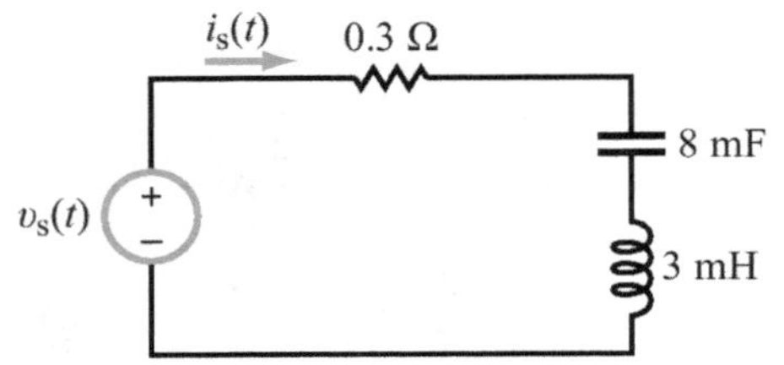

Figure P7.76: Circuit for Problem 7.76.

*__7.77__ Find $\mathbf{I}_a$ in the circuit of Fig. P7.77, given that $\mathbf{V}_s = 10$ V and $\mathbf{I}_s = 5\underline{/30°}$ A.

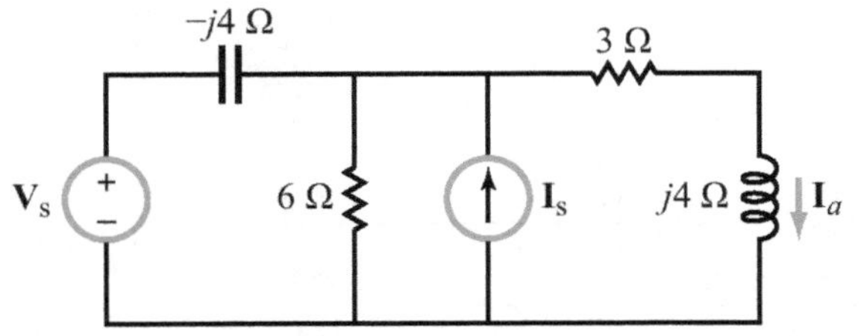

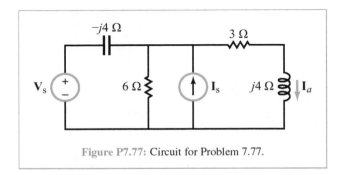

Figure P7.77: Circuit for Problem 7.77.

7.78 Use the superposition principle to solve for $\mathbf{V}_a$ in the circuit of Fig. P7.78.

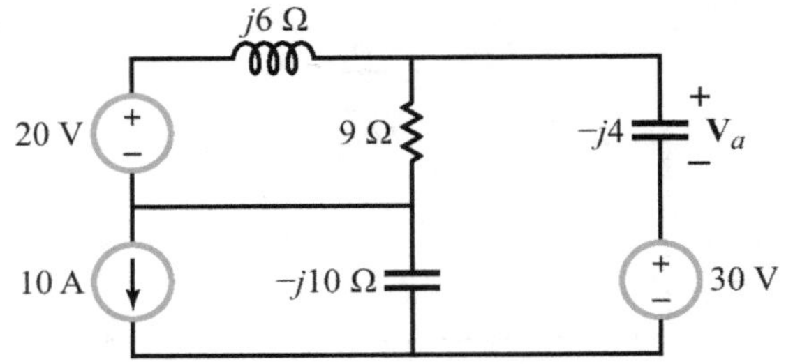

Figure P7.78: Circuit for Problem 7.78.

7.79 Find $\mathbf{V}_o$ in the circuit of Fig. P7.79.

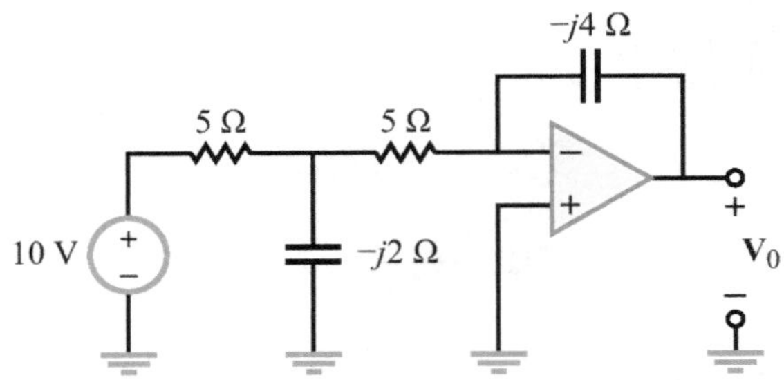

Figure P7.79: Circuit for Problem 7.79.

*__7.80__ The input signal in the op-amp circuit of Fig. P7.80 is given by

$$\upsilon_{in}(t) = V_0 \cos \omega t.$$

Assuming the op amp is operating within its linear range, obtain an expression for $\upsilon_{out}(t)$ by applying the phasor-domain technique and then evaluate it for $\omega RC = 1$.

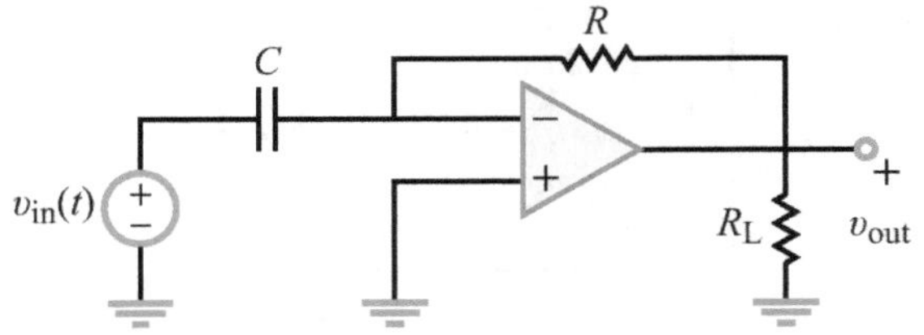

Figure P7.80: Op-amp circuit for Problem 7.80.

7.81 The input signal in the op-amp circuit of Fig. P7.81 is given by

$$\upsilon_{in}(t) = 0.5 \cos 2000t \text{ V}.$$

Obtain an expression for $v_{out}(t)$ and then evaluate it for $R_1 = 2$ kΩ, $R_2 = 10$ kΩ, and $C = 0.1$ μF.

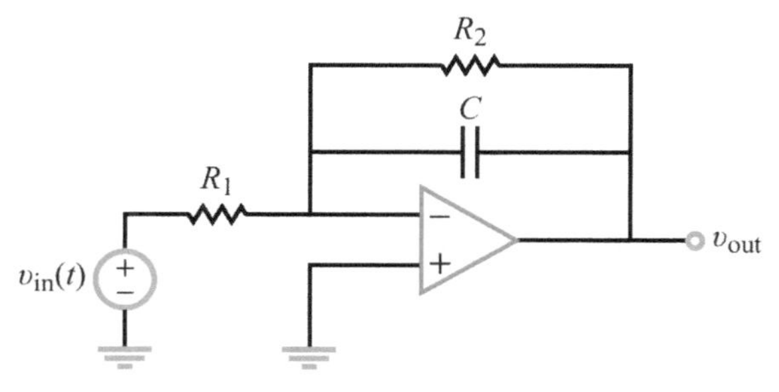

Figure P7.81: Op-amp circuit for Problem 7.81.

7.82 For $v_i(t) = V_0 \cos \omega t$, obtain an expression for $v_{out}(t)$ in the circuit of Fig. P7.82 and then evaluate it for $V_0 = 4$ V, $\omega = 400$ rad/s, $R = 5$ kΩ, and $C = 2.5$ μF.

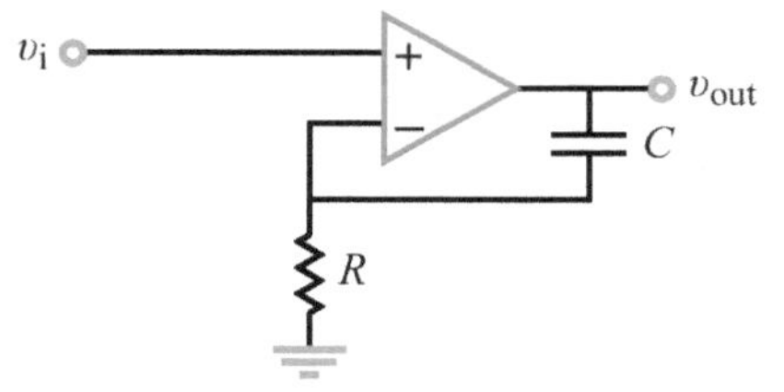

Figure P7.82: Circuit for Problem 7.82.

*__7.83__ For $v_i(t) = V_0 \cos \omega t$, obtain an expression for $v_{out}(t)$ in the circuit of Fig. P7.83 and then evaluate it for $V_0 = 2$ V, $\omega = 377$ rad/s, $R_1 = 2$ kΩ, $R_2 = 10$ kΩ, and $C = 0.5$ μF.

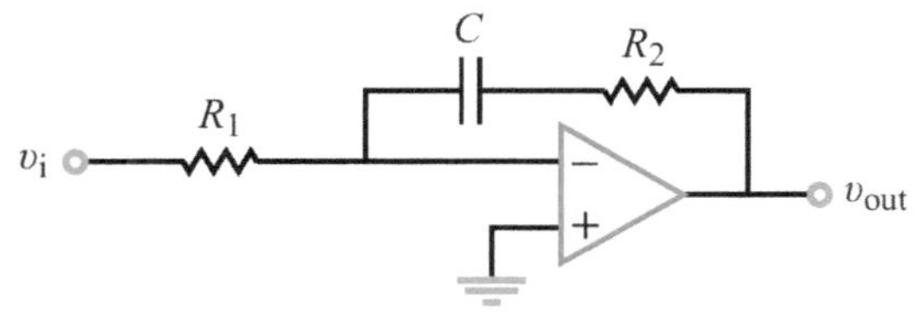

Figure P7.83: Circuit for Problem 7.83.

Section 7-12: Power-Supply Circuits

7.84 The signal voltage at the input of a half-wave rectifier circuit is given by $v_{in}(t) = A\cos(377t + 30°)$ V. Determine and plot the waveform of $v_{out}(t)$. Calculate the fraction of a full period over which $v_{out} = 0$ for each of the following values of A (assume $V_F = 0.7$ V):

(a) $A = 0.5$ V

(b) $A = 5$ V

7.85 A bridge rectifier is driven by a 1 kHz input signal with an amplitude of 10 V. The smoothing filter at the rectifier output uses a 1-μF capacitor in parallel with a load resistor R_L. If $R_D = 5$ Ω:

(a) What should R_L be so that $\tau_{dn}/\tau_{up} = 2500$?

(b) How does τ_{dn} compare with the period of the rectified waveform?

(c) What is the approximate peak value of the output waveform?

7.86 A power supply with the circuit configuration shown in Fig. 7-40 has the following specifications: $v_s = 24\cos(2\pi \times 10^3 t + 30°)$ V, $N_2/N_1 = 2$, $C = 0.1$ mF, $R_s = 50$ Ω, $R_L = 20$ kΩ, $R_z = 20$ Ω, and $V_z = 42$ V. Determine v_{out} and the peak-to-peak ripple voltage.

Section 7-13: Multisim Analysis

7.87 Use the Network Analyzer (see Appendix C) in Multisim to determine the equivalent impedance $\mathbf{Z}_{eq}$ of the circuit in Fig. P7.87. Using the Network Analyzer, plot $\mathbf{Z}_{eq}$ from 1 kHz to 1 MHz and provide a hand calculation demonstrating that the simulated results are correct.

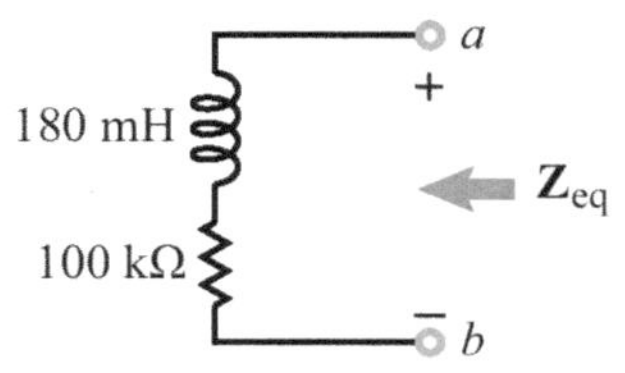

Figure P7.87: Circuit for Problem 7.87.

7.88 Use the Network Analyzer (see Appendix C) in Multisim to determine the equivalent impedance $\mathbf{Z}_{eq}$ of the circuit in Fig. P7.88. Using the Network Analyzer, plot the real and imaginary parts of $\mathbf{Z}_{eq}$ from 100 Hz to 100 kHz and provide a hand calculation demonstrating that the simulated results are correct.

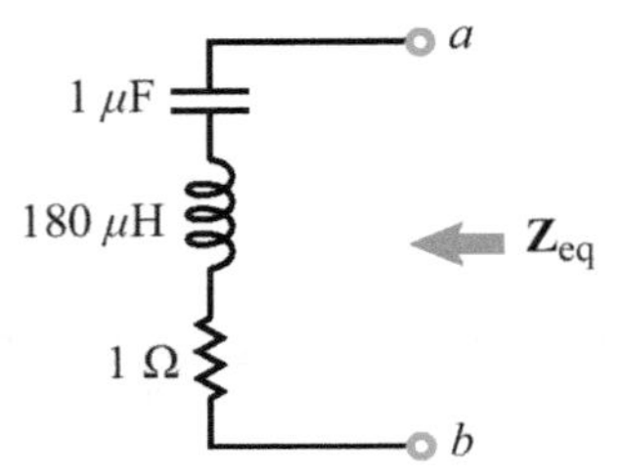

Figure P7.88: Circuit for Problem 7.88.

7.89 A 1 V, 100 MHz voltage source $\upsilon_s(t)$ sends a signal down two transmission lines simultaneously, as depicted by Fig. P7.89. Model this circuit in Multisim with $R_1 = R_2 = 10\ \Omega$, $C_1 = 7$ pF, and $C_2 = 5$ pF and answer the following questions.

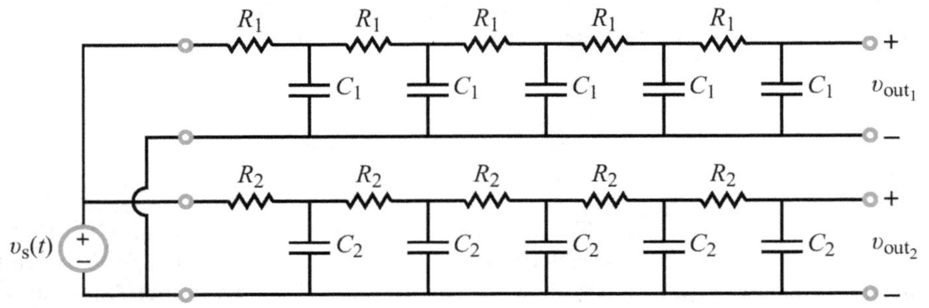

Figure P7.89: Circuit for Problem 7.89.

(a) What is the phase shift between $\upsilon_s(t)$ and the two output nodes, υ_{out_1} and υ_{out_2}?

(b) What is the amplitude ratio for $\upsilon_{out_1}/\upsilon_s$ and $\upsilon_{out_2}/\upsilon_s$?

7.90 Phase-shift circuits have many uses. They can be the fundamental component of an oscillator (a circuit which produces a repetitive electronic signal). The circuit shown in Fig. P7.90 is a *phase-shift oscillator*. While a detailed analysis is too complex for this text, Multisim allows us to easily create and analyze this circuit. Using the 3-terminal virtual op-amp component, construct the phase-shift oscillator shown in the figure and plot the output from 0 to 1.5 ms in Transient Analysis. Determine the frequency and amplitude of the oscillations as well as the DC offset. Note that you may need to decrease the maximum time step (TMAX) in order to get a clear plot.

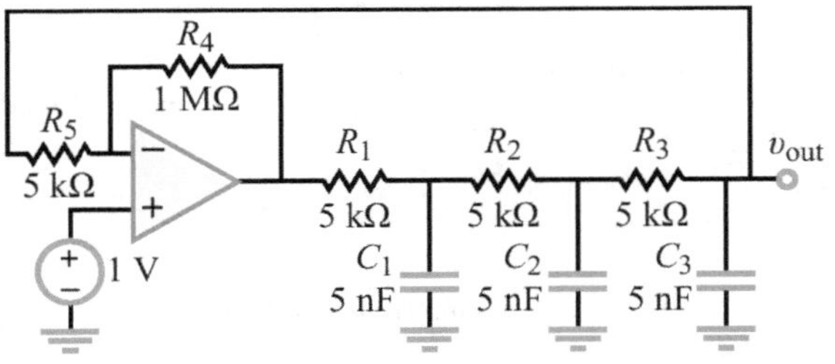

Figure P7.90: Circuit for Problem 7.90.

7.91 Using a Multisim tool or analysis of your choice, find the phase and magnitude of the voltage at each node in the circuit in Fig. P7.91.

Potpourri Questions

7.92 Select two from among the touchscreen sensing mechanisms depicted in Fig. TF18-1 of Technology Brief 18. Compare and contrast their advantages and limitations.

7.93 What is the "crystal" in a crystal oscillator? How is it related to piezoelectricity?

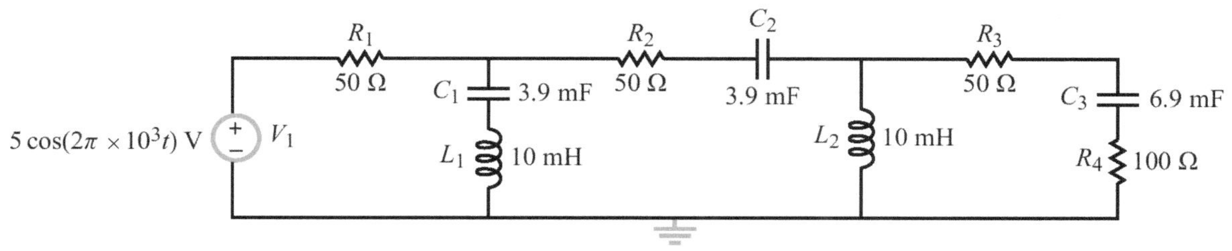

Figure P7.91: Circuit for Problem 7.91.

Integrative Problems: Analytical / Multisim / myDAQ

To master the material in this chapter, solve the following prob-lems using three complementary approaches: (a) analytically, (b) with Multisim, and (c) by constructing the circuit and using the myDAQ interface unit to measure quantities of interest via your computer. [myDAQ tutorials and videos are available on CAD.]

m7.1 Impedance Transformations: Determine the equivalent impedance **Z** looking into terminals (a, b) for the circuit of Fig. m7.1 at the following frequencies: 100 Hz, 500 Hz, 1000 Hz, and 2000 Hz. Report your results in polar form. Use these component values: $R_1 = 100\ \Omega$, $R_2 = 90\ \Omega$, $C = 1.0\ \mu$F, and $L = 33$ mH.

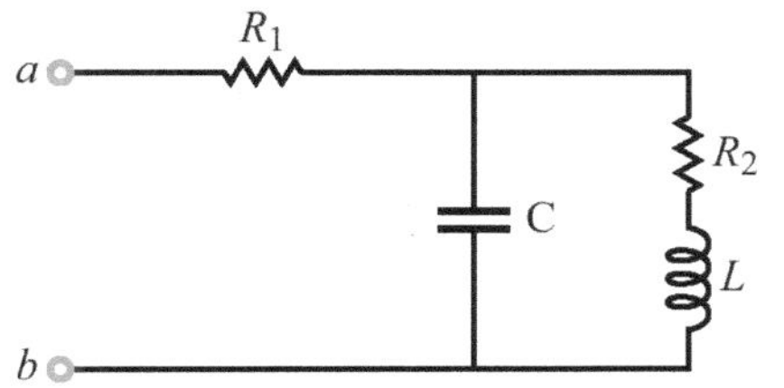

Figure m7.1 Circuit for Problem m7.1.

m7.2 Equivalent Circuits: For the circuit in Fig. m7.2:

(a) Determine the Thévenin equivalent circuit at terminals (a, b) using the open-circuit/short-circuit method. Show the Thévenin impedance as a resistor in series with a single reactive element (capacitor or inductor) and determine the values of all components in the equivalent circuit. The sinusoidal source is $\mathbf{V}_s = 3$ V and $f = 500$ Hz. Component values are: $R_1 = 90\ \Omega$, $R_2 = 100\ \Omega$, $C = 1.0\ \mu$F, and $L = 33$ mH.

(b) Repeat with the source frequency increased to 1100 Hz.

(c) Does the circuit seem to "change its personality" with different source frequencies? Explain your answer.

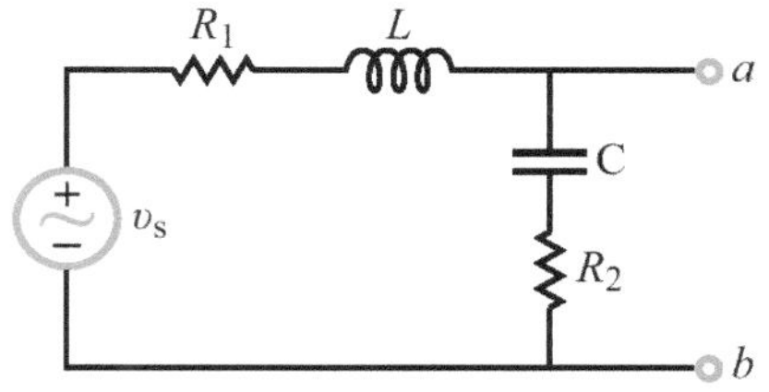

Figure m7.2 Circuit for Problem m7.2.

m7.3 Phase-Shift Circuits: Figure m7.3 shows a phase-shift circuit based on op amps.

(a) Write the general expression for the magnitude of $\mathbf{V}_{out}$ with frequency taken as a variable. *Hint:* View the circuit as the cascade of two standard op-amp circuits.

(b) Write the general expression for the phase of $\mathbf{V}_{out}$ with frequency taken as a variable.

(c) Set $C = 0.1\ \mu$F and set all resistors to 1.0 kΩ. Determine the frequency in Hz at which $\mathbf{V}_{out}$ and $\mathbf{V}_{in}$ share the same magnitude. What is the phase shift at this frequency?

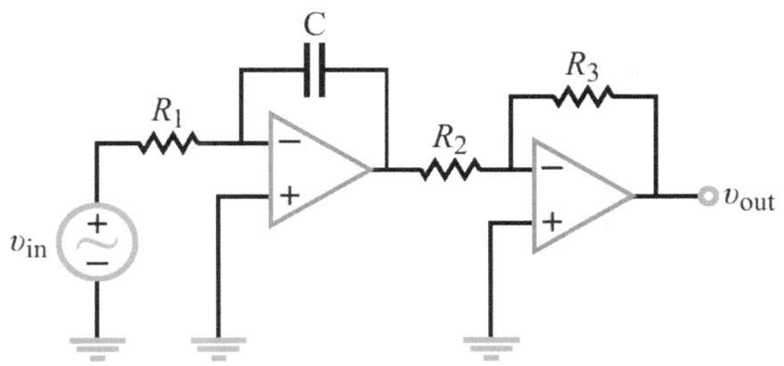

Figure m7.3 Circuit for Problem m7.3.

m7.4 Introduction to Bode Plots: Determine $\mathbf{V}_a$ for the circuit of Fig. m7.4 for $\mathbf{V}_1 = 1$ V at the following frequencies: 100 Hz, 500 Hz, and 2000 Hz. Use these component values: $R_1 = 4.7$ kΩ, $R_2 = 3.3$ kΩ, $R_3 = 2.2$ kΩ, $C_1 = 0.047\ \mu$F, and $C_2 = 0.1\ \mu$F.

For the myDAQ and Multisim portions of this question, use a Bode plotter to capture a representation of $\mathbf{V}_a$. The Bode plot provides the magnitude and phase of the gain ratio $\mathbf{V}_a/\mathbf{V}_1$ as a function of frequency. Use cursors to verify the gain and phase of $\mathbf{V}_a$ from the analytical portion.

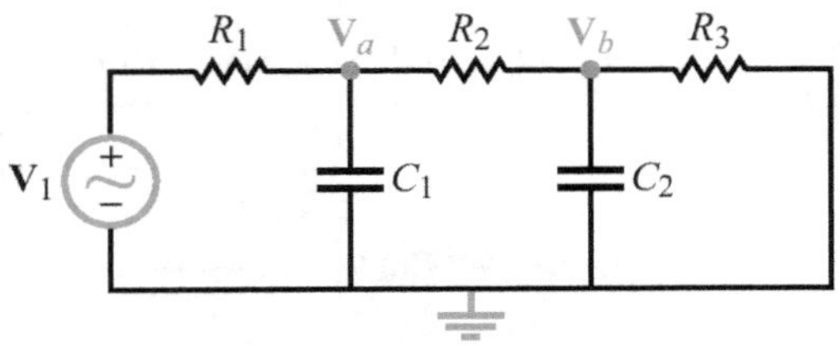

Figure m7.4 Circuit for Problem m7.4.

m7.5 Frequency Response: An ac circuit may respond differently at different frequencies. Find the peak-to-peak voltage across the 33 kΩ resistor in Fig. m7.5 at each of following frequencies:

(a) 1 kHz

(b) 2 kHz

(c) 20 kHz

The amplitude of the ac source is 1 V.

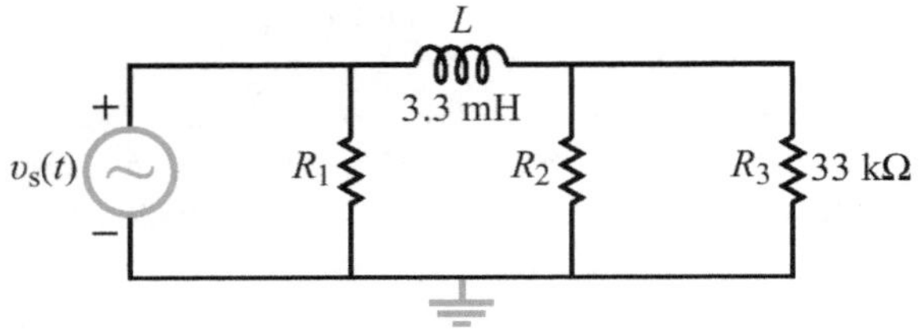

Figure m7.5 Circuit for Problem m7.5.

m7.6 Arbitrary Sources: Use the myDAQ's arbitrary waveform generator "ARB" and the AO 0 and AO 1 ports to create the circuit in Fig. m7.6. Then use Multisim to verify your answer (no analytical component to this problem).

- V1 is a square wave with LO value: 0, HI value: 1 V, period: 1 ms, and 50% duty cycle.
- V2 is a square wave with LO value: 0 V, HI value: 0.5 V, period: 0.5 ms, and 50% duty cycle.

(a) From the waveform of the voltage across R_4, determine the four distinct voltage levels (you will need to use the myDAQ's oscilloscope and Arbitrary Waveform Generator).

(b) Using your answer from part (a), find the four corresponding currents flowing through resistor R_4. Verify your answer using Multisim.

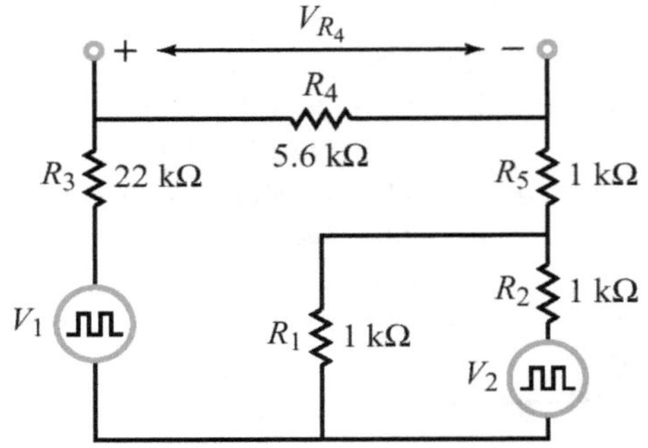

Figure m7.6 Circuit for Problem m7.6.

CHAPTER 8

ac Power

Contents

Objectives

Learn to:

- Calculate the average and rms value of a periodic waveform.
- Determine the complex power, average real power, and reactive power for any complex load with known input voltage or current.
- Determine the power factor for a complex load and evaluate the improvement realized by compensating the load through the addition of a shunt capacitor.
- Choose the load impedance so as to maximize the transfer of power from the input circuit to the load.
- Apply Multisim to measure power.

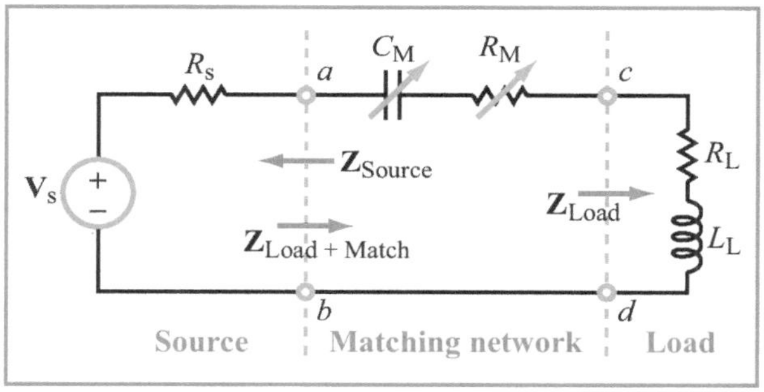

A *matching network* is a circuit used to optimize the transfer of *ac power* between a source and a load. This chapter provides the tools to analyze circuits from the perspective of the total *complex power* they consume (in their resistors) and store (in their capacitors and inductors).

Overview

The power absorbed by a resistor R when a current i passes through it is given by $p = i^2R$. If i is time varying, we usually designate it $i(t)$, call it the instantaneous current, and call the corresponding power the *instantaneous power*

$$p(t) = i^2(t)\,R \qquad \text{(W)}. \tag{8.1}$$

Usually, we are interested in the *average power* P consumed by a given circuit—or by a collection of circuits, as in an entire household—and since ac signals are periodic in time with an angular frequency ω and a time period $T = 2\pi/\omega$, we define P_{av} to be the average value of $p(t)$ over one (or more) complete period(s). For an ac current given by

$$i(t) = I_{\text{m}} \cos \omega t, \tag{8.2}$$

the average power consumed by a resistor R is

$$P_{\text{av}} = \tfrac{1}{2}\, I_{\text{m}}^2 R \qquad \text{(W)}. \tag{8.3}$$

This result, which we will derive in Section 8-2, is somewhat intriguing, primarily because P_{av} is independent of ω and its expression contains a factor of $1/2$. The explanations are fairly straightforward and will be covered later. The more important point we wish to make at this time is that had our intent been to discuss ac power in resistive circuits only, the discussion would not have required more than just a few pages and perhaps no more than one or two examples. Instead, we are devoting this entire chapter to ac power because real circuits contain more than just resistors; they contain capacitors and inductors, both of which cannot consume power but can store it and then release it.

The current through a resistor is always in phase with the voltage across it. This phase attribute is responsible, in part, for the functional form of the expression for P_{av} given by Eq. (8.3). The expression, however, generally is not valid when the load circuit contains reactive elements (capacitors and inductors) either alone or in combination with resistive elements. So, for the general case, we need to develop a formulation appropriate for any complex load—from the purely resistive to the purely reactive. That defines one of the objectives of the present chapter.

When we transform a circuit from the time domain to the phasor domain, voltages and currents are assigned phasor counterparts, and passive elements become impedances. What about power? Is there a phasor power $\mathbf{P}$, corresponding to $p(t)$? The answer is: Not exactly. We will introduce a quantity $\mathbf{S}$ which we will call *complex power*, but $\mathbf{S}$ is not the phasor counterpart of $p(t)$. In fact, we assign it the symbol $\mathbf{S}$ (rather than $\mathbf{P}$) to avoid the possible misinterpretation that it bears a one-to-one correspondence to $p(t)$. As we will see in Section 8-3, $\mathbf{S}$ consists of a real part and an imaginary part with the real part representing the real average power *consumed* by the circuit and the imaginary part representing the average power *stored* by the circuit.

Towards the end of Chapter 3, we posed the question: *When an input circuit is connected to a resistive load, under what condition(s) is the power transferred from the circuit to the load a maximum?* Through the application of Thévenin's theorem, we demonstrated that the transferred power is a maximum when the load resistance is equal to the Thévenin resistance of the input circuit. In the present chapter, we pose the question again, but we generalize the load to a complex load $\mathbf{Z}_{\text{L}} = R_{\text{L}} + jX_{\text{L}}$, composed of a resistive part R_{L} and a reactive part X_{L}. In view of the fact that $\mathbf{Z}_{\text{L}}$ consists of two parts, we should expect the answer to consist of two conditions (not just one) and it does. The details are given in Section 8-5.

8-1 Periodic Waveforms

Even though the focus of this chapter is on the ac power carried by sinusoidally time-varying signals, we will preface our examination by first reviewing some of the important properties shared by all periodic waveforms, including sinusoids.

Mathematically, a periodic waveform $x(t)$ with period T satisfies the *periodicity property*

$$x(t) = x(t + nT) \tag{8.4}$$

for any integer value of n. The periodicity property simply states that the waveform of $x(t)$ repeats itself every T seconds. Figure 8-1 displays the waveforms of three typical (and unrelated) periodic functions. In part (a), $\upsilon(t)$ is a sine wave; in (b) $i(t)$ is a sawtooth with a clipped top; and part (c) displays a function given by $p(t) = P_{\text{m}} \cos^2 \omega t$.

8-1.1 Instantaneous and Average Values

Each of the three waveforms shown in Fig. 8-1 describes the exact variation of its magnitude as a function of time. Consequently, the time function $\upsilon(t)$, for example, is referred to as the *instantaneous voltage*. Similarly, $i(t)$ is the *instantaneous current*, and $p(t)$ is the *instantaneous power*. Often times, however, we may be interested in specifying an attribute of the waveform that conveys useful information about it, and yet it is much simpler to use than the complete waveform. When ac circuits are concerned, two attributes of particular interest are the *average value* of the waveform and its *root-mean-square (rms) value*. The latter is introduced in the next

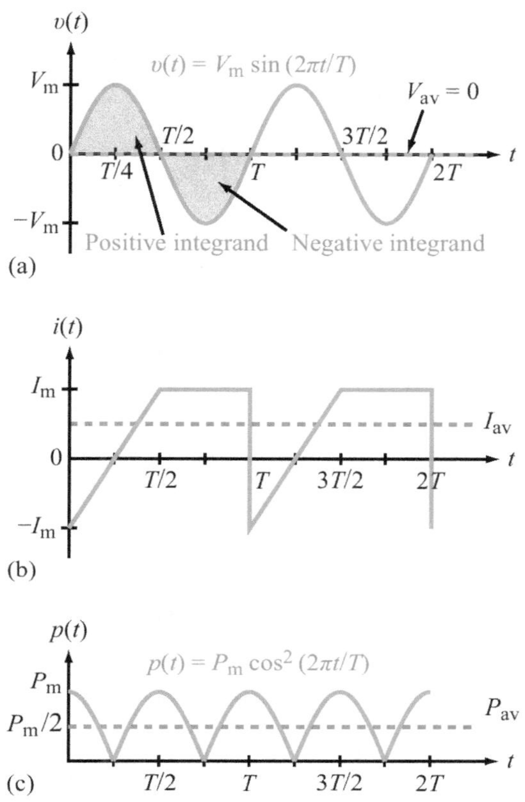

Figure 8-1: Examples of three periodic waveforms.

subsection, so for the present we pursue only the former. The average value of a periodic function $x(t)$ with period T is given by

$$X_{\text{av}} = \frac{1}{T}\int_0^T x(t)\,dt. \tag{8.5}$$

▶ We note that X_{av} is obtained by integrating $x(t)$ over a complete period T and then normalizing the integrated value by dividing it by T. The limits of integration are from 0 to T, but the definition is equally valid for any two limits so long as the upper limit is greater than the lower limit by exactly T (such as from T_0 to $T_0 + T$) or an integer multiple of T. ◀

The voltage waveform shown in Fig. 8-1(a) is given by

$$v(t) = V_{\text{m}} \sin \frac{2\pi t}{T}. \tag{8.6}$$

Application of Eq. (8.5) gives

$$\begin{aligned} V_{\text{av}} &= \frac{1}{T}\int_0^T V_{\text{m}} \sin \frac{2\pi t}{T}\,dt \\ &= \frac{V_{\text{m}}}{T}\left(-\frac{T}{2\pi}\right)\cos \frac{2\pi t}{T}\bigg|_0^T = -\frac{V_{\text{m}}}{2\pi}[1-1] = 0. \end{aligned} \tag{8.7}$$

The fact that the *average value of a sine wave is zero* is not at all surprising; it is clear from the characteristic symmetry of its waveform that the area under the curve (integrand) during the first half of any cycle is equal (but opposite in polarity) to the area under the curve during the second half of the cycle, so the net sum of the two is exactly zero. In contrast, the lack of symmetry in the waveforms of $i(t)$ and $p(t)$ in Figs. 8-1(b) and (c) (between that part of the waveform above the t axis and the part below it) is an obvious indicator that their average values are not only nonzero but also positive.

Example 8-1: Average Values

Determine the average values of the waveforms displayed in parts (b) and (c) of Fig. 8-1.

Solution: During the first half of the first cycle, $i(t)$ is described by a linear ramp of the form

$$i(t) = at + b \qquad \text{for } 0 \le t \le \frac{T}{2}.$$

Its slope is

$$a = \frac{2I_{\text{m}}}{T/2} = \frac{4I_{\text{m}}}{T},$$

and its intercept at $t = 0$ is

$$b = -I_{\text{m}}.$$

Hence,

$$i(t) = \begin{cases} [(4t/T) - 1]I_{\text{m}} & \text{for } 0 \le t \le T/2, \\ I_{\text{m}} & \text{for } T/2 \le t < T. \end{cases} \tag{8.8}$$

By Eq. (8.5), the average value of $i(t)$ is

$$I_{av} = \frac{1}{T}\int_0^T i(t)\,dt$$

$$= \frac{1}{T}\left[\int_0^{T/2}\left(\frac{4t}{T}-1\right) I_m\,dt + \int_{T/2}^{T} I_m\,dt\right] = \frac{I_m}{2}. \qquad (8.9)$$

We also can obtain the same result by adding up the areas bounded by one cycle of the waveform of $i(t)$ and then dividing the net total by T, as in

$$I_{av} = \frac{1}{T}\left[-\frac{1}{2}I_m \times \frac{T}{4} + \frac{1}{2}I_m \times \frac{T}{4} + I_m \times \frac{T}{2}\right] = \frac{I_m}{2}.$$

To determine P_{av} of $p(t)$, we apply Eq. (8.5) to the $\cos^2$ function:

$$P_{av} = \frac{1}{T}\int_0^T P_m \cos^2\left(\frac{2\pi t}{T}\right) dt.$$

The integration is facilitated by applying the trigonometric relation

$$\cos^2 x = \frac{1}{2} + \frac{1}{2}\cos 2x,$$

which leads to the final result

$$P_{av} = \frac{P_m}{2}.$$

We should take note for future reference of the fact that *the average value of* $\cos^2 \omega t$ *is* $1/2$. In fact, it is easy to show that

$$\frac{1}{T}\int_0^T \cos^2\left(\frac{2\pi n t}{T} + \phi_1\right) dt = \frac{1}{2},$$

and $\qquad (8.10)$

$$\frac{1}{T}\int_0^T \sin^2\left(\frac{2\pi n t}{T} + \phi_2\right) dt = \frac{1}{2},$$

for any values of ϕ_1 and ϕ_2.

> ▶ The average values of $\cos^2(n\omega t)$ and $\sin^2(n\omega t)$ are both $1/2$ for any integer values of n equal to or greater than 1, irrespective of whether or not their arguments are shifted by constant phase angles, so long as the averaging process is performed over a complete period $T = 2\pi/\omega$. That is, the average values of $\cos^2(n\omega t + \phi)$ and $\sin^2(n\omega t + \phi)$ also are $1/2$ for any constant value of ϕ. ◀

8-1.2 Root-Mean-Square (rms) Value

For a periodic current waveform $i(t)$ flowing through a resistor R, the average power absorbed by the resistor is

$$P_{av} = \frac{1}{T}\int_0^T p(t)\,dt = \frac{1}{T}\int_0^T i^2(t)\,R\,dt. \qquad (8.11)$$

> ▶ We would like to introduce a new attribute of $i(t)$, called its *effective value*, I_{eff}, defined such that the average power P_{av} delivered by $i(t)$ to resistor R is equivalent to what a dc current I_{eff} would deliver to R, namely $I_{eff}^2 R$. ◀

That is,

$$I_{eff}^2 R = P_{av} = \frac{1}{T}\int_0^T i^2(t)\,R\,dt. \qquad (8.12)$$

Solving for I_{eff} gives

$$I_{eff} = \sqrt{\frac{1}{T}\int_0^T i^2(t)\,dt} \qquad (8.13)$$

According to Eq. (8.13), I_{eff} is obtained by taking the square *root* of the *mean* (average value) of the *square* of $i(t)$. The three terms characterizing the operation are coupled together to form *root-mean-square* (*rms* for short) and I_{eff} is relabeled I_{rms}.

Even though the idea to define an effective or rms value is introduced in connection with a periodic current waveform, the definition is equally applicable to a periodic voltage waveform as well as to any other periodic waveform. For a periodic waveform $x(t)$, its rms value therefore is defined as

$$X_{rms} = X_{eff} = \sqrt{\frac{1}{T}\int_0^T x^2(t)\,dt}. \qquad (8.14)$$

When a multimeter is used to measure an ac voltage waveform, it records the rms value of the voltage. In contrast, when the waveform is displayed on an oscilloscope, the entire waveform is displayed.

Example 8-2: rms Values

Determine the rms values of (a) $\upsilon(t) = V_m \sin(2\pi t/T + \phi)$ and (b) $i(t)$ of Fig. 8-1(b).

Solution: (a) Application of Eq. (8.14) to $\upsilon(t)$ gives

$$V_{\rm rms} = \left[\frac{1}{T}\int_0^T V_{\rm m}^2 \sin^2\left(\frac{2\pi t}{T} + \phi\right) dt\right]^{1/2} \quad (8.15)$$

In view of Eq. (8.10),

$$V_{\rm rms} = \frac{V_{\rm m}}{\sqrt{2}}. \quad (8.16)$$

Hence:

▶ For any sinusoidal function, its rms value is equal to its maximum value (its amplitude) divided by $\sqrt{2}$. ◀

(b) From Eq. (8.8) of Example 8-1, $i(t)$ is given by

$$i(t) = \begin{cases} \left(\frac{4t}{T} - 1\right) I_{\rm m} & \text{for } 0 \le t \le \frac{T}{2}, \\ I_{\rm m} & \text{for } \frac{T}{2} \le t < T. \end{cases}$$

Its rms value therefore is given by

$$I_{\rm rms} = \left\{\frac{1}{T}\left[\int_0^{T/2} \left(\frac{4t}{T} - 1\right)^2 I_{\rm m}^2\, dt + \int_{T/2}^{T} I_{\rm m}^2\, dt\right]\right\}^{1/2},$$

which leads to

$$I_{\rm rms} = \frac{2I_{\rm m}}{\sqrt{6}} = 0.82 I_{\rm m}.$$

Concept Question 8-1: What is the average value of a sinusoidal waveform? What is its rms value? (See CAD)

Concept Question 8-2: Why is Eq. (8.10) true, irrespective of the values of ϕ_1 and ϕ_2? Explain in terms of a diagram. (See CAD)

Concept Question 8-3: What does rms stand for and how does it relate to its definition? (See CAD)

Exercise 8-1: Determine the average and rms values of the waveform $\upsilon(t) = 12 + 6\cos 400t$ V.

Answer: $V_{\rm av} = 12$ V, $V_{\rm rms} = 12.73$ V. (See CAD)

Exercise 8-2: Determine the average and rms value of the waveform

$$i(t) = 8\cos 377t - 4\sin(377t - 30°) \text{ A}.$$

Answer: $I_{\rm av} = 0$, $I_{\rm rms} = 7.48$ A. (See CAD)

8-2 Average Power

The circuit configuration shown in Fig. 8-2 consists of an active ac circuit supplying power to a passive load. The load circuit is not restricted in terms of either its architecture or the combination of resistors, capacitors, and inductors it may contain. The instantaneous voltage across the load is $\upsilon(t)$ and the corresponding instantaneous current flowing into it—whose direction is defined in accordance with the passive sign convention—is $i(t)$. Since this is an ac circuit, all of its currents and voltages oscillate sinusoidally at the same angular frequency ω. The general functional forms for $\upsilon(t)$ and $i(t)$ are given by

$$\upsilon(t) = V_{\rm m}\cos(\omega t + \phi_\upsilon), \quad (8.17a)$$

and

$$i(t) = I_{\rm m}\cos(\omega t + \phi_i), \quad (8.17b)$$

where $V_{\rm m}$ and $I_{\rm m}$ are the amplitudes of $\upsilon(t)$ and $i(t)$, and ϕ_υ and ϕ_i are their phase angles, respectively. Our objective is to relate the average power absorbed by the load $P_{\rm av}$ to the parameters of $\upsilon(t)$ and $i(t)$.

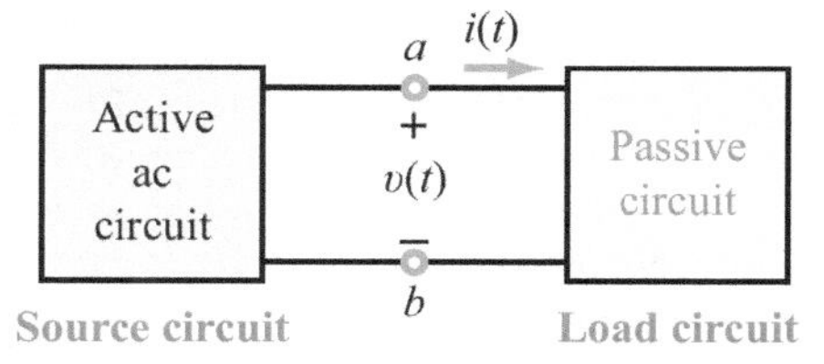

Figure 8-2: Passive load circuit connected to an input source at terminals (a, b).

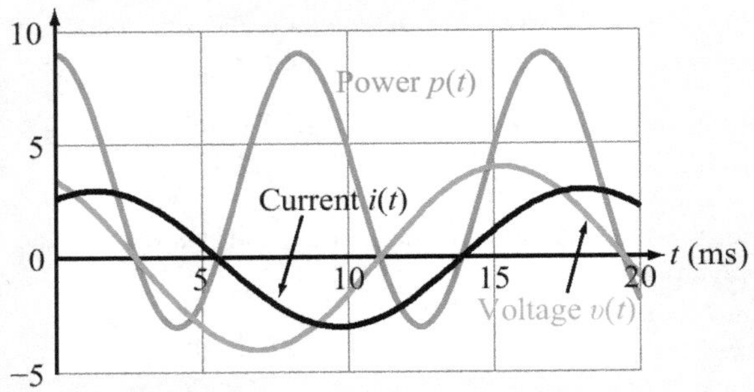

Figure 8-3: Waveforms for a 60 Hz circuit with $\upsilon(t) = 4\cos(377t + 30°)$ V, $i(t) = 3\cos(377t - 30°)$ A, and $p(t) = \upsilon(t)\, i(t)$. The waveform of $i(t)$ is shifted by 60° behind that of $\upsilon(t)$, and the oscillation frequency of $p(t)$ is twice that of $\upsilon(t)$ or $i(t)$.

The instantaneous power flowing into the load circuit is

$$p(t) = \upsilon(t)\, i(t) = V_m I_m \cos(\omega t + \phi_\upsilon)\cos(\omega t + \phi_i). \quad (8.18)$$

By applying the trigonometric identity

$$\cos x \cos y = \frac{1}{2}\cos(x - y) + \frac{1}{2}\cos(x + y), \quad (8.19)$$

$p(t)$ can be cast in the form

$$p(t) = \frac{V_m I_m}{2}\cos(\phi_\upsilon - \phi_i) + \frac{V_m I_m}{2}\cos(2\omega t + \phi_\upsilon + \phi_i). \quad (8.20)$$

Before proceeding to find the average value of $p(t)$, let us briefly examine the significance of the two terms of Eq. (8.20). The first term is a constant, as it contains no dependence on t, and the second term is sinusoidal, but its angular frequency is 2ω. Thus:

> ► $p(t)$ is the sum of a dc-like term and an ac term that oscillates at a frequency twice that of $i(t)$ and $\upsilon(t)$. ◄

This behavior is evident in the waveforms of $\upsilon(t)$, $i(t)$, and $p(t)$ displayed in Fig. 8-3. The angular frequency $\omega = 2\pi f$ corresponds to $f = 60$ Hz, and the phase angles were arbitrarily chosen as $\phi_\upsilon = 30°$ and $\phi_i = -30°$. The waveform patterns elicit the following observations:

(a) The voltage $\upsilon(t)$ oscillates symmetrically relative to the t axis, with a peak-to-peak variation extending from −4 V to +4 V. The current $i(t)$ exhibits a similar pattern between −3 A and +3 A.

(b) The waveforms of $\upsilon(t)$ and $i(t)$ are separated from each other by a time shift Δt, corresponding to the difference in phase angle between them. Given that $\phi_\upsilon = +30°$ and $\phi_i = -30°$, $\upsilon(t)$ leads $i(t)$ by 60°. A complete period

$$T = 1/f = 1/60 = 16.67 \text{ ms}$$

corresponds to a total phase angle of 360°. Hence, $\upsilon(t)$ leads $i(t)$ by

$$\Delta t = \left(\frac{60°}{360°}\right) \times 16.67 \text{ ms} = 2.78 \text{ ms}.$$

(c) The waveform of the power $p(t)$ is not symmetrical with respect to the t axis. It has a dc shift equal to the first term in Eq. (8.20). Also, it traces twice as many cycles per unit time, in comparison with the waveforms of $\upsilon(t)$ or $i(t)$.

Returning to the task at hand, we now apply Eq. (8.5) to the expression of $p(t)$ given by Eq. (8.20) to determine P_{av}, the average power delivered to the load:

$$\begin{aligned} P_{av} &= \frac{1}{T}\int_0^T p(t)\, dt \\ &= \frac{1}{T}\int_0^T \frac{V_m I_m}{2}[\cos(\phi_\upsilon - \phi_i) + \cos(2\omega t + \phi_\upsilon + \phi_i)]\, dt. \end{aligned} \quad (8.21)$$

For any sinusoidal function with $\omega = 2\pi/T$, it is fairly straightforward to show that for integer values of n equal to or greater than 1 and any constant angle θ

$$\frac{1}{T}\int_0^T \cos(n\omega t + \theta)\, dt = 0 \quad (n = 1, 2, \ldots). \quad (8.22)$$

Thus, the average value over a period $T = 2\pi/\omega$ of a sinusoidal function of angular frequency ω or integer multiple of ω is zero. In view of Eq. (8.22), the integral of the second term in Eq. (8.21) is zero. Consequently, the expression for P_{av} simplifies to

$$P_{av} = \frac{V_m I_m}{2}\cos(\phi_\upsilon - \phi_i) \quad \text{(W)}. \quad (8.23)$$

Technology Brief 20
The Electromagnetic Spectrum

Electromagnetic Energy

The sun's rays, the signal transmitted by a cell phone, and the radiation emitted by plutonium share a fundamental property: they all carry electromagnetic (EM) energy. It is an interesting and fundamental observation that this energy can be described both as a *wave* moving through space and as a *particle*. Neither model alone is sufficient to explain the phenomena we observe in the world around us. This correspondence, called the *wave-particle duality*, sparked scientific debate as far back as the 1600s, and it was not until the 20th century and the advent of quantum mechanics that this duality was fully incorporated into modern physics.

When we treat EM energy as a wave with alternating electric and magnetic fields, we ascribe to the wave a wavelength λ and an oscillation frequency f, whose product defines the velocity of the wave u as

$$u = f\lambda.$$

If the propagation medium is free space, then u is equal to c, which is the speed of light in vacuum at 3×10^8 m/s. Because of the wave-particle duality, when EM energy is regarded as a particle, each such particle will have the same velocity u as its wave counterpart and will carry energy E whose magnitude is specified by the frequency f through

$$E = hf,$$

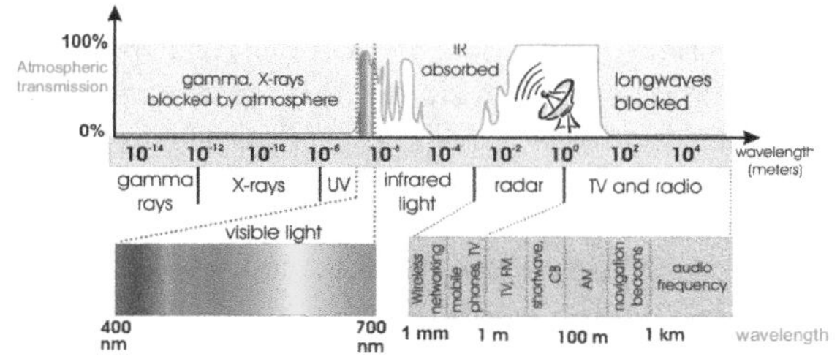

Figure TF20-1: The electromagnetic spectrum extends over a wide range of wavelengths—from gamma rays to radio waves. The atmosphere is transparent in the microwave and in selected windows in the visible and infrared.

where h is Planck's constant (6.6×10^{-34} J·s). In view of the direct link between E and f, we can refer to an EM particle (also called a *photon*) either by its energy E or by the frequency f of its wave counterpart. The higher the frequency is, the higher is the energy carried by a photon, but also the shorter is its wavelength λ.

The Spectrum

In terms of the wavelength λ, the EM spectrum extends across many orders of magnitude (Fig. TF20-1), from the radio region on one end to the gamma-ray region on the other. The degree to which an EM wave is absorbed or scattered as it travels through a medium depends on the types of constituents present in that medium and their sizes relative to λ of the wave. For Earth's atmosphere, the composition and relative distributions of its gases are responsible for the near total opacity of the atmosphere to EM waves across most of the EM spectrum, except for narrow "windows" in the visible, infrared, and radio spectral regions (Fig. TF20-1). It is precisely because EM waves with these wavelengths can propagate well through the atmosphere that human sight, thermal infrared imaging, and radio communication are possible through the air.

1. Cosmic Rays: Emitted by the decay of the nuclei of unstable elements and by cosmic, high-energy sources in the universe, cosmic rays—which include gamma, beta, and alpha radiation—are highly energetic particles that can be dangerous to organisms and destructive to matter. Earth emits gamma rays of its own, but at very weak levels.

2. X-Rays: Slightly lower energy radiation falls into the X-ray region; this radiation is energetic enough to be dangerous to organisms in large doses, but small doses are safe. More importantly, their relatively high energy allows them to traverse much farther into solid objects than lower frequency radiation (such as visible light). This phenomenon allows for modern medical radiology, in which X-rays are used to measure the opacity of the medium between the X-ray source and the detector or film. Thankfully, Earth's atmosphere efficiently screens the surface from high-energy radiation, such as cosmic rays and X-rays.

3. Ultraviolet Rays: The atmosphere is only partially opaque to ultraviolet (UV) waves, which border the visible

Table TT20-1: Some examples of radio frequency communication channels and their frequency bands.

Communication modality	Band name	Frequencies
Medium wave AM radio (US)	MF	520 – 1610 kHz, broken into 10 kHz channels
FM radio (US)	VHF	88 – 108 MHz, broken into 100 – 200 kHz channels
GPS L1 and L2	UHF	1575.42 MHz (L1) and 1227.60 MHz (L2)
802.11g wifi	ISM	2.4 – 2.5 GHz, broken into 13 overlapping 22 MHz channels
Bluetooth®	ISM	2400 – 2483 MHz, broken into 1 MHz channels
802.15.4 – ZigBee (US)	UHF	902 – 928 MHz, broken into 30 channels
802.15.4 – ZigBee (Asia)	UHF	2.4 GHz, broken into 16 channels

spectrum on the short-wavelength side. UV radiation is both useful in modern technology and potentially harmful to living things in high doses. Among its many uses, UV radiation is used routinely in electronic fabrication technology for erasing programmable memory chips, polymer processing, and even as a curing ink and adhesive. While UV's potential danger to human skin is well recognized, it is for the same reasons that UV lamps are used to sterilize hospital and laboratory equipment.

4. Visible Light Rays: The wavelength range of visible light extends from about 380 nm (violet color) to 740 nm (red/brown color), although the exact range varies from one human to another. Some species can see well into the infrared (IR) or the UV, so the definition of *visible* is completely anthropocentric. It is no coincidence that evolution led to the development of sight organs that are sensitive to precisely that part of the spectrum where atmospheric absorption is very low. In the visible spectrum, blue light is more susceptible to scattering by atmospheric particles than the longer wavelengths, which is why the sky appears blue to us.

5. Infrared Rays: The infrared (IR) region, straddled in between the visible spectrum and the radio region, is particularly useful for thermal applications. When an object is heated, the added energy increases the vibrations of its molecules. These molecular vibrations, in turn, release electromagnetic radiation at many frequencies. Within the range of our thermal environment, the peak of the radiated spectrum is in the IR region. This feature has led to the development of IR detectors and cameras for both civilian and military thermal-imaging applications. Nightvision systems use IR detector arrays to image a scene when the intensity of visible-wavelength light is insufficient for standard cameras. This is because material objects emit IR energy even in pitch-black darkness. Conversely, IR energy can be used to heat an object, because a good radiator of IR is also a good absorber. Additionally, IR beams are used extensively in short-distance communication, such as in the remote control of most modern TV sets and garage door openers.

6. Radio Waves: The frequency range of the radio spectrum extends from essentially dc (or zero frequency) to $f = 1$ THz $= 10^{12}$ Hz. It is subdivided into many bands with formal designations (Fig. TF20-1) such as VHF (30 to 300 MHz) and UHF (300 to 3000 MHz), and some of those bands combine together to form bands commonly known by historic designations, such as the microwave band (300 MHz to 30 GHz). All major free-space communication systems operate at frequencies in the radio region, including wireless local area networks (LANs), cell phones, satellite communication, and television and radio transmissions (Table TT20-1). Because the radio spectrum is used so heavily, spectrum allocation is controlled (often sold) by various national and international agencies that set standards for what types of devices are permitted to operate, within what frequency bands, and at what maximum-power transmission levels. Cell phones, for example, are allowed to transmit and receive in the 2.11 to 2.2 GHz band and in the 1.885 to 2.025 GHz band. Radio waves are the range where the classic antenna-to-antenna transmission occurs.

This is the dc term in Eq. (8.20). According to Eq. (8.16), the rms value of a sinusoidal voltage waveform is related to its amplitude by $V_{\rm rms} = V_{\rm m}/\sqrt{2}$, and a similar relationship holds for $i(t)$. Hence,

$$P_{\rm av} = V_{\rm rms} I_{\rm rms} \cos(\phi_\upsilon - \phi_i) \qquad \text{(W)}. \tag{8.24}$$

The quantity $(\phi_\upsilon - \phi_i)$ is called the *power factor angle* and plays a critical role with respect to $P_{\rm av}$. For a purely resistive load R, $\upsilon(t)$ and $i(t)$ are *in phase*, which means that $\phi_\upsilon = \phi_i$. Consequently,

$$P_{\rm av} = V_{\rm rms} I_{\rm rms} = \frac{V_{\rm rms}^2}{R} . \tag{8.25}$$

(purely resistive load)

The average power supplied to the load is exactly what we would expect for a resistor at dc, except that in the ac case, we substitute rms for dc values.

For a purely reactive load (capacitors and/or inductors, with no resistors), we established in Section 7-7 that $(\phi_\upsilon - \phi_i) = \pm 90°$, with the (+) sign corresponding to an inductive load (because $\upsilon_{\rm L}$ leads $i_{\rm L}$ by 90°) and the (−) sign corresponding to a capacitive load ($\upsilon_{\rm C}$ lags $i_{\rm C}$ by 90°). In either case,

$$P_{\rm av} = V_{\rm rms} I_{\rm rms} \cos 90° = 0. \tag{8.26}$$

(purely reactive load)

▶ A purely reactive load can store power and then release it, but the net average power it absorbs is zero. ◀

Concept Question 8-4: How is the rms value related to the amplitude of a sinusoidal signal? (See CAD)

Concept Question 8-5: How much average power is consumed by a reactive load? Explain. (See CAD)

Exercise 8-3: The voltage across and current through a certain load are given by

$$\upsilon(t) = 8\cos(754t - 30°) \text{ V}$$

and

$$i(t) = 0.2 \sin 754t \text{ A}.$$

What is the average power consumed by the load, and by how far in time is $i(t)$ shifted relative to $\upsilon(t)$?

Answer: $P_{\rm av} = 0.4$ W; $\Delta t = 1.39$ ms. (See CAD)

8-3 Complex Power

The correspondence between the instantaneous voltage $\upsilon(t)$ and instantaneous current $i(t)$ and their respective phasors (**V** and **I**) is embodied by the relationships

$$\upsilon(t) = V_{\rm m}\cos(\omega t + \phi_\upsilon) \quad \Longleftrightarrow \quad \mathbf{V} = V_{\rm m} e^{j\phi_\upsilon} \tag{8.27a}$$

and

$$i(t) = I_{\rm m}\cos(\omega t + \phi_i) \quad \Longleftrightarrow \quad \mathbf{I} = I_{\rm m} e^{j\phi_i} . \tag{8.27b}$$

In the time domain, in general it is not possible to combine all of the elements of a passive load circuit into a single equivalent element, but it is possible to do so in the phasor domain. A passive ac circuit always can be represented by an equivalent impedance **Z**, as shown in Fig. 8-4, and it has to satisfy the condition

$$\mathbf{Z} = \frac{\mathbf{V}}{\mathbf{I}} = \frac{V_{\rm m}}{I_{\rm m}} e^{j(\phi_\upsilon - \phi_i)} \qquad (\Omega), \tag{8.28a}$$

where **V** and **I** are the phasor voltage and current at its input terminals. Since in general

$$\mathbf{Z} = |\mathbf{Z}|\, e^{j\phi_z},$$

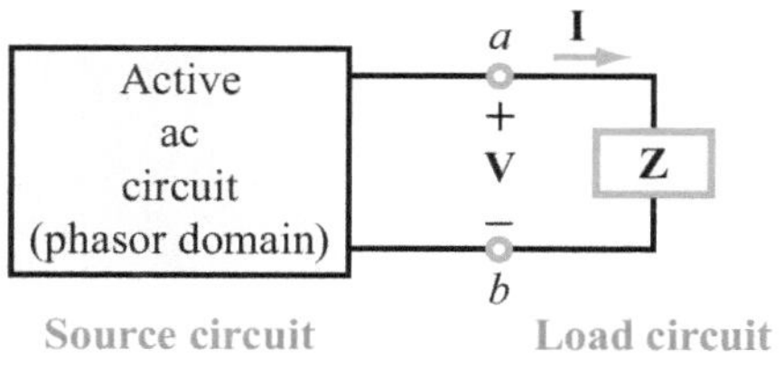

Figure 8-4: Source circuit connected to an impedance **Z** of a load circuit.

where ϕ_z is the phase angle of **Z**, it follows that

$$|\mathbf{Z}| = \frac{V_m}{I_m}, \qquad \phi_z = \phi_v - \phi_i. \tag{8.28b}$$

The *complex power* **S** is a phasor quantity defined in terms of **V** and **I**, but it is not simply the product of **V** and **I**. The definition of **S** is constructed such that the real part of **S** is exactly equal to P_{av}, the real average power absorbed by the load **Z**. To that end, **S** is defined as

$$\mathbf{S} = \frac{1}{2}\mathbf{V}\mathbf{I}^* \qquad \text{(VA)}, \tag{8.29}$$

where $\mathbf{I}^*$ is the complex conjugate of **I**, realized by replacing j with $-j$ everywhere in **I**. Upon inserting the expressions for **V** and **I** given by Eqs. (8.27a and b) into Eq. (8.29) (after replacing $j\phi_i$ with $-j\phi_i$ so as to convert **I** to $\mathbf{I}^*$), we obtain the result

$$\begin{aligned}\mathbf{S} &= \frac{1}{2}(V_m e^{j\phi_v})(I_m e^{-j\phi_i}) \\ &= \frac{1}{2}V_m I_m e^{j(\phi_v-\phi_i)} \\ &= \frac{1}{2}V_m I_m \cos(\phi_v-\phi_i) + j\frac{1}{2}V_m I_m \sin(\phi_v-\phi_i).\end{aligned} \tag{8.30}$$

For the sake of consistency, we introduce the rms phasor voltage and current as

$$\mathbf{V}_{rms} = \frac{\mathbf{V}}{\sqrt{2}} = \frac{V_m}{\sqrt{2}}e^{j\phi_v} \tag{8.31a}$$

and

$$\mathbf{I}_{rms} = \frac{\mathbf{I}}{\sqrt{2}} = \frac{I_m}{\sqrt{2}}e^{j\phi_i}, \tag{8.31b}$$

and we rewrite Eqs. (8.29) and (8.30) in terms of rms quantities as

$$\mathbf{S} = \mathbf{V}_{rms}\mathbf{I}^*_{rms} \qquad \text{(VA)}, \tag{8.32}$$

and

$$\mathbf{S} = V_{rms} I_{rms}\cos(\phi_v-\phi_i) + jV_{rms}I_{rms}\sin(\phi_v-\phi_i). \tag{8.33}$$

We note that the real part of **S** (first term) is equal to the expression for P_{av} given by Eq. (8.24). The second term is called the *reactive power* Q:

$$Q = V_{rms}I_{rms}\sin(\phi_v-\phi_i) \qquad \text{(VAR)}. \tag{8.34}$$

Hence,

$$\mathbf{S} = P_{av} + jQ \qquad \text{(VA)}, \tag{8.35}$$

and conversely,

$$P_{av} = \mathfrak{Re}[\mathbf{S}] \qquad \text{(average absorbed power)} \tag{8.36a}$$

and

$$Q = \mathfrak{Im}[\mathbf{S}] \qquad \text{(peak exchanged power)}. \tag{8.36b}$$

▶ Whereas P_{av} represents real dissipated power, Q represents the peak amount of power exchanged (back and forth) between the source circuit and the load circuit. ◀

During a single oscillation cycle of duration T:

$P_{av}T$ = energy dissipated in the load,
QT = energy transferred to the load and then returned to the source.

The three quantities—**S**, P_{av}, and Q—are each a product of a voltage and a current and therefore should be measured in watts.

▶ However, to help distinguish between them, only P_{av} retains the unit of watt, and the other two have been assigned artificially different units. **S** has been given the unit *volt-ampere* (VA) and Q the unit *volt-ampere reactive* (VAR). ◀

8-3.1 Complex Power for a Load

So far, we have expressed **S** in terms of **V** and **I**, but **V** and **I** are linked to one another through the impedance of the load circuit **Z** (Fig. 8-4). In general, **Z** has a real, resistive component R, and an imaginary, reactive component X:

$$\mathbf{Z} = R + jX.$$

We should recall from Chapter 7 that the reactive component is inductive if $X > 0$ and capacitive if $X < 0$. In terms of **Z**,

$$\mathbf{V} = \mathbf{Z}\,\mathbf{I}, \tag{8.37}$$

and the expression for **S** given by Eq. (8.29) becomes

$$\mathbf{S} = \frac{1}{2}\mathbf{V}\,\mathbf{I}^* = \frac{1}{2}\mathbf{Z}\,\mathbf{I}\,\mathbf{I}^* = \frac{1}{2}|\mathbf{I}|^2\,\mathbf{Z} = I^2_{rms}(R+jX), \tag{8.38}$$

From this, we deduce that

$$P_{av} = \mathfrak{Re}[\mathbf{S}] = \frac{1}{2}|\mathbf{I}|^2 R = I_{rms}^2 R \quad \text{(W)} \tag{8.39a}$$

and

$$Q = \mathfrak{Im}[\mathbf{S}] = \frac{1}{2}|\mathbf{I}|^2 X = I_{rms}^2 X \quad \text{(VAR)}. \tag{8.39b}$$

The relationship between $\mathbf{S}$ and its components P_{av} and Q is illustrated graphically in Fig. 8-5(a) for an impedance with an inductive component ($X > 0$). A similar illustration is contained in Fig. 8-5(b) for an impedance with a capacitive component ($X < 0$). The vector $\mathbf{S}$ lies in quadrant 1 ($0 < (\phi_v - \phi_i) \leq 90°$) if X is inductive and in quadrant 4 ($-90° \leq (\phi_v - \phi_i) < 0$) if X is capacitive. (If $\mathbf{S}$ were to lie in quadrants 2 or 3, P_{av} would be negative, indicating that the load is actually a source supplying power, not consuming it.)

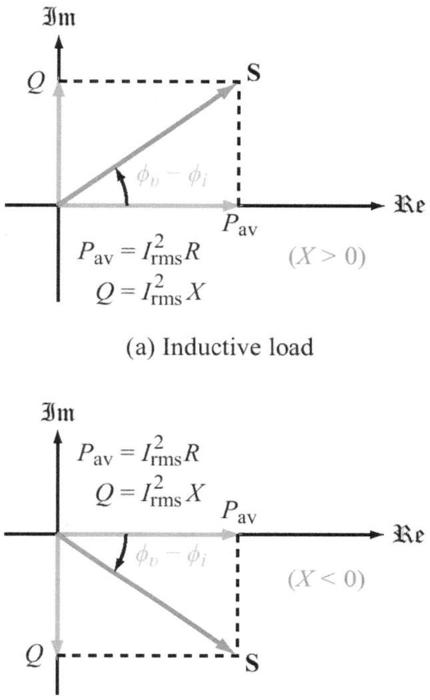

Figure 8-5: Complex power $\mathbf{S}$ lies in quadrant 1 for an inductive load and in quadrant 4 for a capacitive load.

8-3.2 Conservation of Complex Power

In a circuit containing n elements, energy conservation requires that the sum of the complex powers associated with all n elements be equal to zero:

$$\sum_{i=1}^{n} \mathbf{S}_i = 0.$$

Since $\mathbf{S}_i$ is complex, it follows that both the real and imaginary components of the sum have to individually be equal to zero, which, in view of Eq. (8.35), leads to

$$\sum_{i=1}^{n} P_{av_i} = 0, \qquad \sum_{i=1}^{n} Q_i = 0. \tag{8.40}$$

Keeping in mind that P_{av_i} has a positive (+) value if the ith element is a resistor and a negative (−) value if it is a generator of power, the first summation in Eq. (8.40) states that the power consumed by the resistors is equal to the (real) power generated by the sources in the circuit. Similarly, the summation over Q_i states that there is no net exchange of reactive power between the sources and the reactive elements in the circuit.

Example 8-3: RL Load

An input circuit consisting of a source $v_s = 10 \cos 10^5 t$ V in series with a source resistance $R_s = 100\ \Omega$ is connected to an RL load circuit, as shown in Fig. 8-6(a). If $R = 300\ \Omega$ and $L = 3$ mH, determine: $\mathbf{I}$, $\mathbf{S}$ into the RL load and ϕ_v of the voltage across the load.

Solution: From the expression for v_s, we deduce that $\mathbf{V}_s = 10$ V and $\omega = 10^5$ rad/s. Hence, the load impedance is

$$\mathbf{Z} = R + j\omega L = 300 + j10^5 \times 3 \times 10^{-3} = (300 + j300)\ \Omega.$$

The phasor current $\mathbf{I}$ corresponding to $i(t)$ is given by

$$\mathbf{I} = \frac{\mathbf{V}_s}{R_s + \mathbf{Z}} = \frac{10}{100 + 300 + j300}$$
$$= \frac{10}{400 + j300} = 20e^{-j36.87°} \text{ mA}.$$

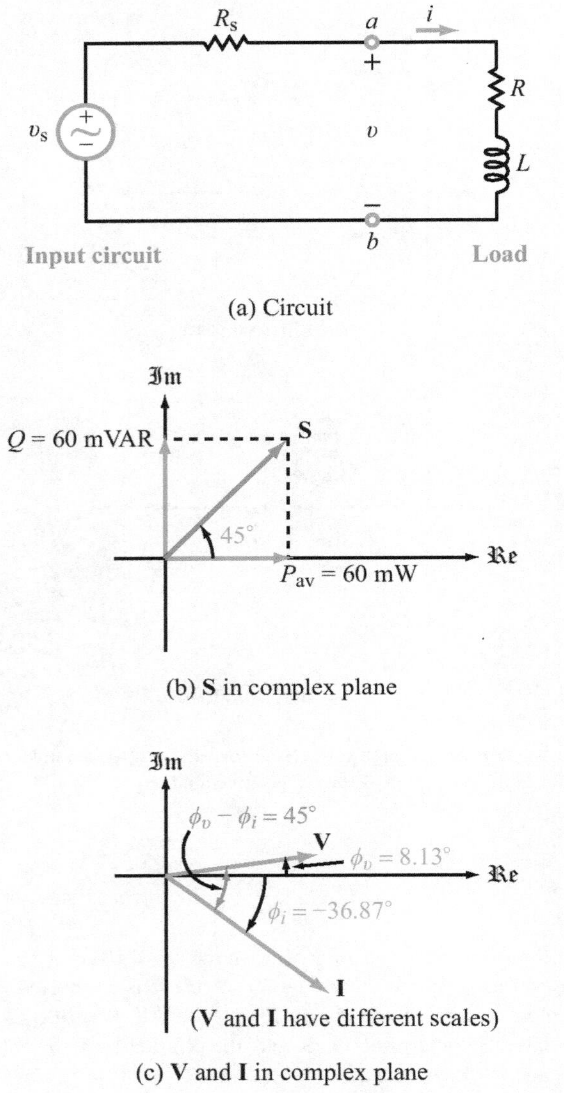

Figure 8-6: Example 8-3.

Given that $I_m = 20$ mA and $R = X = 300\ \Omega$,

$$P_{av} = I_{rms}^2 R = \frac{I_m^2 R}{2} = \frac{(20 \times 10^{-3})^2}{2} \times 300 = 60 \text{ mW}$$

and

$$Q = I_{rms}^2 X = \frac{(20 \times 10^{-3})^2}{2} \times 300 = 60 \text{ mVAR},$$

and their combination specifies **S** as

$$\mathbf{S} = 60 + j60 = 84.85e^{j45°} \text{ mVA}.$$

According to Eq. (8.30), the phase angle of **S** is equal to $\phi_v - \phi_i$. Hence,

$$45° = \phi_v - (-36.87°),$$

which yields

$$\phi_v = 8.13°.$$

We also can determine **V** independently by applying voltage division to the circuit,

$$\mathbf{V} = \frac{\mathbf{V}_s \mathbf{Z}}{R_s + \mathbf{Z}} = \frac{10(300 + j300)}{400 + j300} = 8.48e^{j8.13°} \text{ V},$$

which confirms the value we found earlier for ϕ_v. Figure 8-6(b) and (c) provide graphical renditions of **S**, **V**, and **I** in the complex plane.

Example 8-4: Capacitive Load

The current source $i_s(t) = 20\cos(10^3 t + 30°)$ mA and associated shunt resistance $R_s = 400\ \Omega$, as shown in Fig. 8-7(a), provide ac power to the load circuit to the right of terminals (a, b). If $R_1 = 200\ \Omega$, $R_2 = 2$ kΩ, and $C = 1\ \mu$F, determine: (a) **I**, **V**, **S**, P_{av}, and Q for the entire load circuit (to the right of terminals (a, b)), (b) $\mathbf{S}_C$ for the capacitor alone, and (c) $\mathbf{S}_s$ for the current source.

Solution: (a) In the phasor domain,

$$\mathbf{I}_s = 20e^{j30°} \text{ mA}$$

and

$$\mathbf{Z}_C = \frac{-j}{\omega C} = \frac{-j}{10^3 \times 10^{-6}} = -j1000\ \Omega,$$

and the impedance **Z** of the load circuit is

$$\mathbf{Z} = R_1 + R_2 \parallel \mathbf{Z}_C$$
$$= 200 + \frac{2000 \times (-j1000)}{2000 - j1000} = (600 - j800)\ \Omega.$$

Current division in the phasor-domain circuit of Fig. 8-7(b) yields

$$\mathbf{I} = \frac{\mathbf{I}_s R_s}{R_s + \mathbf{Z}} = \frac{20 \times 10^{-3} e^{j30°} \times 400}{400 + (600 - j800)} = 6.25e^{j68.66°} \text{ mA},$$

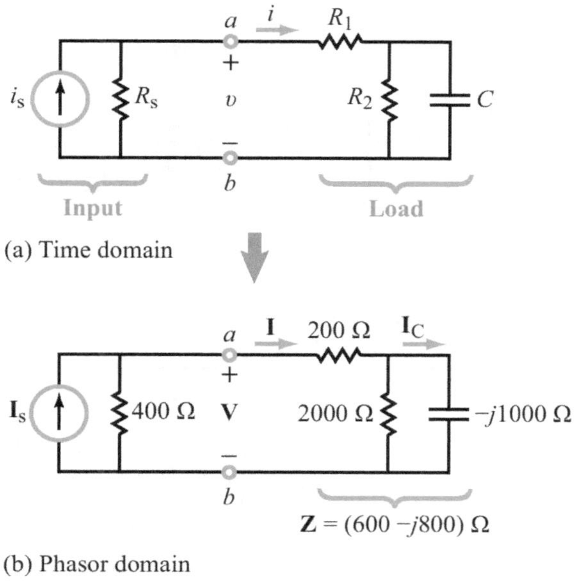

Figure 8-7: Circuit for Example 8-4.

and the phasor voltage at terminals (a, b) is

$$\mathbf{V} = \mathbf{IZ} = 6.25\times10^{-3}e^{j68.66}\times(600-j800) = 6.25e^{j15.53^\circ}\ \text{V}.$$

Given **I** and **V**, the complex power **S** is

$$\mathbf{S} = \frac{1}{2}\,\mathbf{VI}^* = \frac{1}{2}\times 6.25e^{j15.53^\circ}\times 6.25\times 10^{-3}e^{-j68.66^\circ}$$
$$= 19.53e^{-j53.13^\circ}\ \text{mVA},$$

with real and imaginary components given by

$$P_{\text{av}} = \mathfrak{Re}[\mathbf{S}] = 19.53\times 10^{-3}\cos(-53.13^\circ) = 11.72\ \text{mW}$$

and

$$Q = \mathfrak{Im}[\mathbf{S}] = 19.53\times 10^{-3}\sin(-53.13^\circ) = -15.62\ \text{mVAR}.$$

(b) The phasor current $\mathbf{I}_C$ flowing through C is related to **I** by

$$\mathbf{I}_C = \frac{R_2\mathbf{I}}{R_2+\mathbf{Z}_C} = \frac{2000\times 6.25\times 10^{-3}e^{j68.66^\circ}}{2000-j1000}$$
$$= 5.59e^{j95.23^\circ}\qquad (\text{mA}),$$

and the corresponding voltage $\mathbf{V}_C$ across the capacitor is

$$\mathbf{V}_C = \mathbf{I}_C\mathbf{Z}_C = 5.59e^{j95.23^\circ}\times 10^{-3}\times(-j1000)$$
$$= 5.59e^{j5.23^\circ}\ \text{V},$$

where we used the identity

$$-j = e^{-j90^\circ}.$$

The complex power associated with the capacitor is

$$\mathbf{S}_C = \frac{1}{2}\,\mathbf{V}_C\mathbf{I}_C^* = \frac{1}{2}\,5.59e^{j5.23^\circ}\times 5.59\times 10^{-3}e^{-j95.23^\circ}$$
$$= 15.62e^{-j90^\circ} = 0 - j15.62\ \text{mVA}.$$

As expected, the real part of $\mathbf{S}_C$ (representing the amount of power *dissipated* in the capacitor) is zero, and the imaginary part is exactly equal to Q of the overall load circuit (the capacitor is the only element in the load circuit capable of *exchanging* power back and forth with the input circuit).

(c) Recall that for any device, **S** represents the complex power transferred into the device, and it is defined such that the current direction through the device is from the (+) terminal to the (−) terminal of the voltage across it. For the current source $\mathbf{I}_s$, it flows through itself from the (−) terminal of **V** to the (+) terminal of **V**, in exact opposition to the definition of **S**. Hence,

$$\mathbf{S}_s = -\frac{1}{2}\,\mathbf{VI}_s^* = -\frac{1}{2}\times 6.25e^{j15.53^\circ}\times 20e^{-j30^\circ}\times 10^{-3}$$
$$= -62.5e^{-j14.47^\circ}\ \text{mVA}$$
$$= -62.5\cos(-14.47^\circ) - j62.5\sin(-14.47^\circ)$$
$$= (-60.52 + j15.62)\ \text{mVA}.$$

The real part of $\mathbf{S}_s$ represents the real average power generated by $\mathbf{I}_s$ and is equal in magnitude to the average power dissipated in the three resistors in the circuit. The imaginary part of $\mathbf{S}_s$ is equal in magnitude and opposite in sign to $\mathbf{S}_C$.

> **Concept Question 8-6:** What are the two components of the complex power **S**, what type of power do they represent, and what units are assigned to them? (See CAD)

> **Concept Question 8-7:** If **S** lies in quadrant 2 in the complex plane, what does that tell you about the load? (See CAD)

Exercise 8-4: The current flowing into a load is given by $i(t) = 2\cos 2500t$ A. If the load is known to consist of a series of two passive elements, and $\mathbf{S} = (10 - j8)$ VA, determine the identities of the elements and their values.

Answer: $R = 5\ \Omega$, $C = 100\ \mu$F. (See CAD)

Table 8-1: **Summary of power-related quantities.**

Time Domain ⟷ **Phasor Domain**

Time Domain		Phasor Domain
$\upsilon(t) = V_m \cos(\omega t + \phi_\upsilon)$	⟷	$\mathbf{V} = V_m e^{j\phi_\upsilon}$
$i(t) = I_m \cos(\omega t + \phi_i)$	⟷	$\mathbf{I} = I_m e^{j\phi_i}$
$V_{rms} = V_m/\sqrt{2}$	⟷	$\mathbf{V}_{rms} = V_{rms} e^{j\phi_\upsilon}$
$I_{rms} = I_m/\sqrt{2}$	⟷	$\mathbf{I}_{rms} = I_{rms} e^{j\phi_i}$

Complex Power

$$\mathbf{S} = \tfrac{1}{2}\,\mathbf{VI}^* = \mathbf{V}_{rms}\mathbf{I}^*_{rms} = P_{av} + jQ$$

Real Average Power

$$P_{av} = \mathfrak{Re}\,[\mathbf{S}] = V_{rms} I_{rms} \cos(\phi_\upsilon - \phi_i) = I^2_{rms} R = V^2_{rms} R/|\mathbf{Z}|^2$$

Reactive Power

$$Q = \mathfrak{Im}\,[\mathbf{S}] = V_{rms} I_{rms} \sin(\phi_\upsilon - \phi_i) = I^2_{rms} X = V^2_{rms} X/|\mathbf{Z}|^2$$

Apparent Power

$$S = |\mathbf{S}| = \sqrt{P^2_{av} + Q^2} = V_{rms} I_{rms} = I^2_{rms}|\mathbf{Z}| = V^2_{rms}/|\mathbf{Z}|$$

$$\mathbf{S} = S e^{j(\phi_\upsilon - \phi_i)} = S e^{j\phi_z}$$
$$\phi_z = \phi_\upsilon - \phi_i$$

Power Factor

$$pf = \frac{P_{av}}{S} = \cos(\phi_\upsilon - \phi_i) = \cos\phi_z$$

8-4 The Power Factor

Several power-related terms were introduced in the preceding two sections, including the complex power $\mathbf{S}$, the real average power P_{av}, and the reactive power Q. We plan to introduce two additional terms in this section, so lest this apparent profusion of terms contribute to any possible confusion, we have prepared a summary of all relevant terms and expressions in the form of Table 8-1. This is intended to provide the reader easy access to and greater clarity about the interrelationships among the various power quantities.

In terms of the complex quantities $\mathbf{V}$ and $\mathbf{I}$ (representing the phasor voltage across a load circuit and the associated current into it) the complex power $\mathbf{S}$ transferred to the load circuit (Fig. 8-8) is given by

$$\mathbf{S} = P_{av} + jQ, \tag{8.41}$$

with

$$P_{av} = V_{rms} I_{rms} \cos(\phi_\upsilon - \phi_i) \tag{8.42a}$$

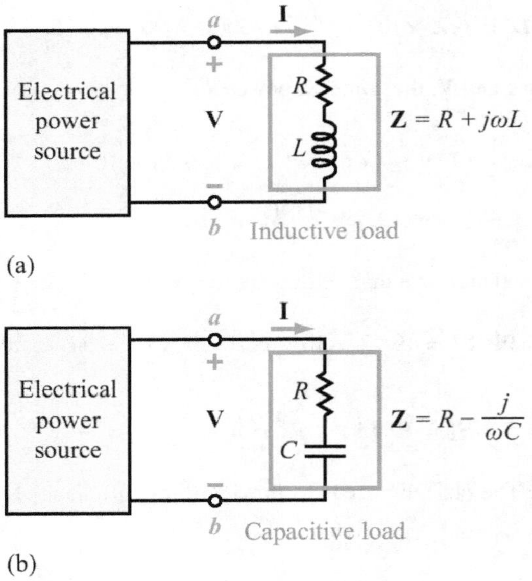

Figure 8-8: Inductive and capacitive loads connected to an electrical source.

Table 8-2: **Power factor leading and lagging relationships for a load $\mathbf{Z} = R + jX$.**

Load Type	$\phi_z = \phi_v - \phi_i$	I-V Relationship	*pf*
Purely Resistive ($X = 0$)	$\phi_z = 0$	**I** in phase with **V**	1
Inductive ($X > 0$)	$0 < \phi_z \leq 90°$	**I** lags **V**	lagging
Purely Inductive ($X > 0$ and $R = 0$)	$\phi_z = 90°$	**I** lags **V** by 90°	lagging
Capacitive ($X < 0$)	$-90° \leq \phi_z < 0$	**I** leads **V**	leading
Purely Capacitive ($X < 0$ and $R = 0$)	$\phi_z = -90°$	**I** leads **V** by 90°	leading

and

$$Q = V_{\text{rms}} I_{\text{rms}} \sin(\phi_v - \phi_i). \qquad (8.42b)$$

For reasons that we will discuss in the next subsection, the magnitude of **S** is called the *apparent power* S, and it is given by

$$S = |\mathbf{S}| = \sqrt{P_{\text{av}}^2 + Q^2} = V_{\text{rms}} I_{\text{rms}}, \qquad (8.43)$$

and the ratio of P_{av} to S is called the *power factor pf*, and is given by

$$pf = \frac{P_{\text{av}}}{S} = \cos(\phi_v - \phi_i). \qquad (8.44)$$

The argument of the cosine $(\phi_v - \phi_i)$ is called the *power factor angle*. Per Eq. (8.28b), this angle is equal to the phase angle of the load impedance ϕ_z:

$$\phi_z = \phi_v - \phi_i. \qquad (8.45)$$

In view of Eq. (8.45), the expression for the power factor can be rewritten as

$$pf = \cos\phi_z. \qquad (8.46)$$

Inductive load

An inductive load, such as a series RL circuit, has an impedance

$$\mathbf{Z}_{\text{ind}} = R + j\omega L. \qquad (8.47)$$

As both components of $\mathbf{Z}_{\text{ind}}$ are positive quantities, ϕ_z is positive. Since R cannot be negative, the range of ϕ_z is $0 \leq \phi_z \leq 90°$ with 0° corresponding to a purely resistive load and 90° corresponding to a purely inductive load.

Capacitive load

The equivalent circuit of a capacitive load is a series RC circuit with

$$\mathbf{Z}_{\text{cap}} = R - \frac{j}{\omega C}. \qquad (8.48)$$

Consequently, ϕ_z is negative and its range is $-90° \leq \phi_z \leq 0$, with $-90°$ corresponding to a purely capacitive load.

Because $\cos(-\theta) = \cos\theta$ for any angle θ between $-90°$ and $+90°$, the power factor (Eq. (8.46)) is insensitive to the sign of ϕ_z, and therefore, it cannot differentiate between an inductive load and a capacitive load. To qualify *pf* with such information:

▶ The load is said to have a *leading pf* or a *lagging pf*, depending on whether the current **I** leads or lags the voltage **V** (see Table 8-2). ◀

8-4.1 Power Factor Significance

Most industrial loads involve the use of large motors or other inductive machinery that require the supply of tens of kilowatts of power, typically at 440 V rms. Household appliances (such as refrigerators and air conditioners) also contain inductive coils, and most are designed to operate at either 110 V rms or 220 V rms. Thus, most loads to which an electrical source has to supply power have an RL equivalent circuit of the type shown in Fig. 8-8(a). From the perspective of an energy supplier (such as the electric power company) the load has two important attributes: S and P_{av}. The amount of power the company has to supply is S, but it can charge for only P_{av}, because P_{av} is the only real power consumed by the load. The company appears to supply S—hence, the name apparent power—but it gets paid for a fraction of that, and the power factor is that fraction. For two loads—one purely resistive with $\mathbf{Z}_1 = R$ and the second inductive with $\mathbf{Z}_2 = R + j\omega L$—with both requiring the same voltage $\mathbf{V}$ and consuming the same power P_{av}, the inductive load will require the transmission of a larger current to it than would the purely resistive load. This point is demonstrated numerically through Example 8-5.

Example 8-5: ac Motor

The equivalent circuit of a dishwasher motor is characterized by an impedance $\mathbf{Z} = (20 + j20)\ \Omega$. The household voltage is 110 V rms. Determine: (a) *pf*, *S*, and P_{av} and (b) the current that the electric company would have supplied to the motor had it been purely resistive and consumed the same amount of power.

Solution: (a) We will treat the phase of the voltage as our reference by setting it arbitrarily equal to zero. Thus,

$$\mathbf{V}_{rms} = V_{rms}\angle 0^\circ = 110\ \text{V}. \quad (8.49)$$

This is justified by the fact that none of the quantities of interest require knowledge of the values of ϕ_v and ϕ_i individually; it is the difference $(\phi_v - \phi_i)$ that counts. The corresponding current is

$$\mathbf{I}_{rms} = \frac{\mathbf{V}_{rms}}{\mathbf{Z}} = \frac{110}{20 + j20} = \frac{110}{20\sqrt{2}\, e^{j45^\circ}} = 3.9\angle -45^\circ\ \text{A},$$

from which we deduce that $I_{rms} = 3.9$ A and $\phi_z = 45^\circ$. The quantities of interest are then given by

$$S = V_{rms} I_{rms} = 110 \times 3.9 = 427.8\ \text{VA},$$

$$P_{av} = S\cos\phi_z = 429\cos 45^\circ = 302.5\ \text{W},$$

and

$$pf = \frac{P_{av}}{S} = 0.707. \quad (8.50)$$

(b) A purely resistive load that consumes 302.5 W at 110 V rms must have a current of

$$I_{rms} = \frac{P_{av}}{V_{rms}} = \frac{302.5}{110} = 2.75\ \text{A}. \quad (8.51)$$

For the same amount of consumed power, the power supplier has to provide 3.9 A to an inductive load with a power factor of 0.707, compared with only 2.75 A to a purely resistive load with $pf = 1$.

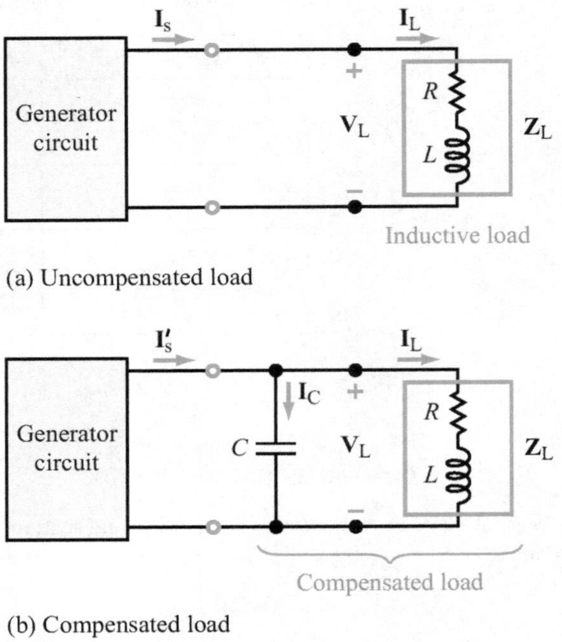

Figure 8-9: Adding a shunt capacitor across an inductive load reduces the current supplied by the generator.

8-4.2 Power Factor Compensation

Raising the power factor of an inductive load (such as an electric drill or a compressor) is highly desirable, not only for the energy supplier but also ultimately for its customers as well. Redesigning the load circuit itself to raise its power factor to a value closer to 1, however, may not be practical, primarily because its motor or other inductive components were presumably selected to meet certain operational specifications that may be incompatible with a higher power factor. This problem of partial incompatibility raises the following question: can we raise the *pf* of a load (as seen by the generator circuit) while keeping it the same as far as the inductive load itself is concerned? The answer is yes, and the solution is fairly straightforward: it entails adding a shunt capacitor across the inductive load, as shown in Fig. 8-9(b). Without the capacitor (Fig. 8-9(a)), the inductor load requires a voltage $\mathbf{V}_L$ across it and a current $\mathbf{I}_L$ through it. The source current $\mathbf{I}_s$ is equal to $\mathbf{I}_L$. The presence of the shunt capacitor does not change $\mathbf{V}_L$, and by virtue of the load impedance $\mathbf{Z}_L$, the current $\mathbf{I}_L = \mathbf{V}_L/\mathbf{Z}_L$ also remains unchanged. In other words, the capacitor exercises no influence on the inductive load, but it does change the overall load circuit as far as the generator is concerned. The new load

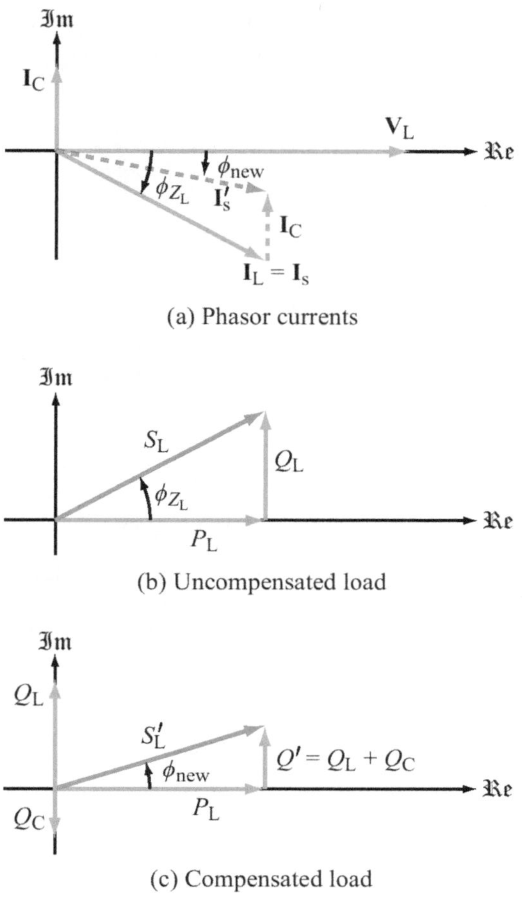

Figure 8-10: Comparison of source currents and *power factor triangles* for the compensated and uncompensated circuits.

circuit—which we will call the *compensated load circuit*—consists of the parallel combination of C and the original RL circuit. Because of the new current $\mathbf{I}_C$, the source current becomes

$$\mathbf{I}'_s = \mathbf{I}_L + \mathbf{I}_C. \quad (8.52)$$

Had C and the RL load been purely resistive, both $\mathbf{I}_C$ and $\mathbf{I}_L$ would have been real and of the same sign, resulting in a larger source current rather than smaller. Fortunately, $\mathbf{I}_C$ and $\mathbf{I}_L$ are phasor quantities, and their imaginary components have opposite polarities (actually, $\mathbf{I}_C$ is purely imaginary). With $\mathbf{V}_L$ chosen to serve as the phase reference, Fig. 8-10(a) illustrates how the vector sum of $\mathbf{I}_L$ (the current into the RL circuit) and $\mathbf{I}_C$ leads to a vector $\mathbf{I}'_s$, whose length (or equivalently, its magnitude) is shorter than the length it was before adding the capacitor. In terms of the power factor,

$$pf = \begin{cases} \cos\phi_{Z_L} & \text{for the RL circuit alone,} \\ \cos\phi_{new} & \text{for the compensated circuit,} \end{cases} \quad (8.53)$$

where ϕ_{new} is the phase angle between $\mathbf{I}'_s$ and $\mathbf{V}_L$ in the compensated load circuit.

Another approach to demonstrate how the addition of the capacitor improves the power factor is by comparing the *power factor triangle* of the RL circuit alone with that of the compensated load circuit that includes the capacitor. The two triangles are diagrammed in parts (b) and (c) of Fig. 8-10, in which P_L and Q_L represent the consumed and reactive powers associated with the RL load, and Q_C is associated with the capacitor C. The capacitor introduces reactive power Q_C, and since Q_C is negative, the net sum

$$Q' = Q_L + Q_C \quad (8.54)$$

is smaller than Q_L alone, thereby reducing the phase angle from ϕ_{Z_L} to ϕ_{new}, where

$$\phi_{new} = \tan^{-1}\left(\frac{Q'}{P_L}\right). \quad (8.55)$$

Example 8-6: *pf* Compensation

A 60 Hz electric generator supplies a 220 V rms to a load that consumes 200 kW at $pf = 0.8$ lagging. By adding a shunt capacitor C, the power factor of the overall circuit was improved to 0.95 lagging. Determine the value of C.

Solution: A power factor of 0.8 corresponds to a phase angle ϕ_{Z_L} given by

$$\phi_{Z_L} = \cos^{-1}(pf_1) = \cos^{-1}(0.8) = 36.87°.$$

The values of S_L and Q_L for the load alone are

$$S_L = \frac{P_L}{pf_1} = \frac{200 \times 10^3}{0.8} = 250 \text{ kVA}$$

and

$$Q_L = S_L \sin\phi_{Z_L} = 250 \sin 36.87° = 150 \text{ kVAR}.$$

The associated power triangle is shown in Fig. 8-11(a).

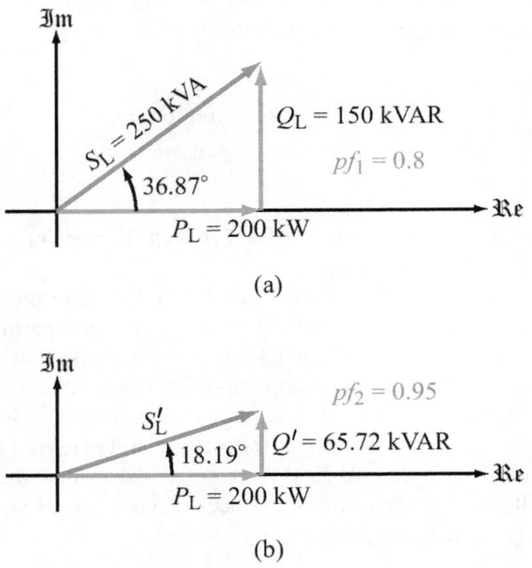

Figure 8-11: Power triangles for Example 8-6.

Addition of the capacitor changes the power factor to $pf_2 = 0.95$, with a corresponding angle as

$$\phi_{\text{new}} = \cos^{-1}(pf_2) = \cos^{-1}(0.95) = 18.19°.$$

The consumed power P_L does not change, but from Fig. 8-11(b), the new reactive power is now

$$Q' = 200 \tan \phi_{\text{new}} = 200 \tan 18.19° = 65.72 \text{ kVAR}.$$

Using the value of Q_L we determined earlier, the reactive power introduced by the capacitor is

$$Q_C = Q' - Q_L = (65.74 - 150) = -84.26 \text{ kVAR}.$$

With $\mathbf{Z}_C = 1/j\omega C$, the complex power of C is

$$\mathbf{S}_C = \mathbf{V}_{L_{\text{rms}}} \mathbf{I}^*_{C_{\text{rms}}} = \mathbf{V}_{L_{\text{rms}}} \frac{\mathbf{V}^*_{L_{\text{rms}}}}{\mathbf{Z}^*_C} = -j|\mathbf{V}_{L_{\text{rms}}}|^2 \omega C.$$

Hence, $P_C = 0$, and

$$Q_C = -|\mathbf{V}_{L_{\text{rms}}}|^2 \omega C.$$

Solving for C gives

$$C = \frac{-Q_C}{2\pi f V^2_{\text{rms}}} = \frac{84.26 \times 10^3}{2\pi \times 60 \times (220)^2} = 4.62 \text{ mF}.$$

Concept Question 8-8: Why is the power factor of a household appliance significant to an electric utility company? (See CAD)

Concept Question 8-9: What is pf compensation, and why is it used? (See CAD)

Exercise 8-5: At 60 Hz, the impedance of a RL load is $\mathbf{Z}_L = (50 + j50)\ \Omega$. (a) What is the value of the power factor of $\mathbf{Z}_L$. (b) What will be the new power factor if a capacitance $C = \frac{1}{12\pi}$ mF is added in parallel with the RL load?

Answer: (a) $pf_1 = 0.707$, (b) $pf_2 = 1$. (See CAD)

8-5 Maximum Power Transfer

Consider the network configuration shown in Fig. 8-12 in which an ac source circuit is represented by its Thévenin equivalent circuit, composed of a phasor voltage $\mathbf{V}_s$ and a source impedance

$$\mathbf{Z}_s = R_s + jX_s. \tag{8.56}$$

Similarly, the load is represented by its impedance $\mathbf{Z}_L$ with

$$\mathbf{Z}_L = R_L + jX_L. \tag{8.57}$$

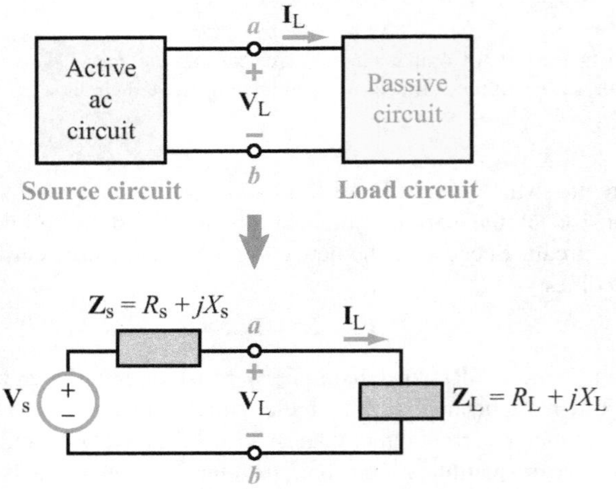

Figure 8-12: Replacing the source and load circuits with their respective Thévenin equivalents.

Technology Brief 21
Seeing without Light

When we think of optical technology, we most often think of visible light—the wavelengths we can see, that our eyes are sensitive to. These wavelengths range from about 750 nm (red) to 400 nm (violet), as shown in Fig. TF21-1. From $f = c/\lambda$, where $c = 3 \times 10^8$ m/s and λ is the wavelength in meters, the corresponding frequency range extends from 400 THz (1 THz $= 10^{12}$ Hz) for red light to 750 THz for violet light.

Our eyes are insensitive to electromagnetic waves whose frequencies are outside this range, but we can build sensors that are. *Infrared* (IR) frequencies (those below the visible spectrum) can be used for thermal imaging (sensing heat) and night vision (seeing in the dark). *Ultraviolet* (UV) frequencies (those above the visible spectrum) can be used for dermal (skin) imaging as well as numerous surface treatments (see Technology Brief 5 on LEDs).

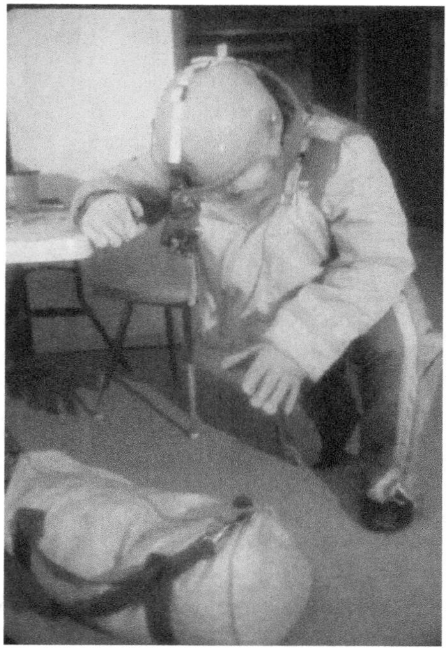

Figure TF21-2: A night-vision image taken with military-grade goggles.

Thermal — Infrared (IR) Imaging

Night-vision imaging is used for a wide variety of applications including imaging people for security and rescue (as seen in Figs. TF21-2 and TF21-3). Helicopters can fly over large regions, locating people and animals from their IR signatures. Firefighters can use IR goggles to see through smoke and find victims. Thermal imaging is also used for medical applications (inflammation warms injured body parts) including those involving animals and small children who cannot tell you "where it hurts" and

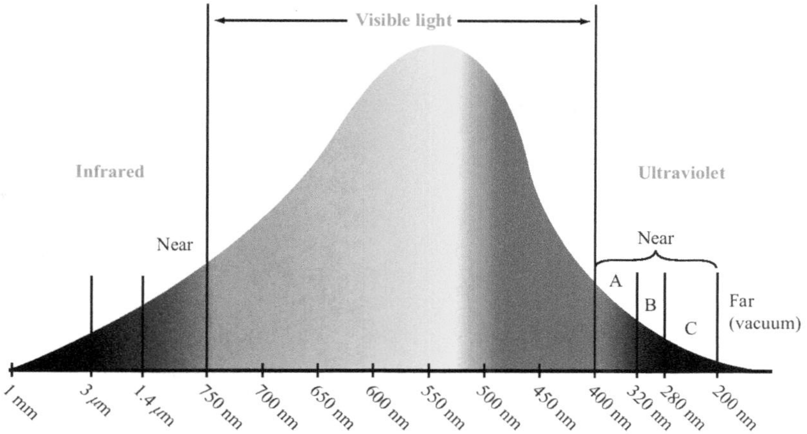

Figure TF21-1: Spectrum of visible light and its two neighbors, the infrared and the ultraviolet.

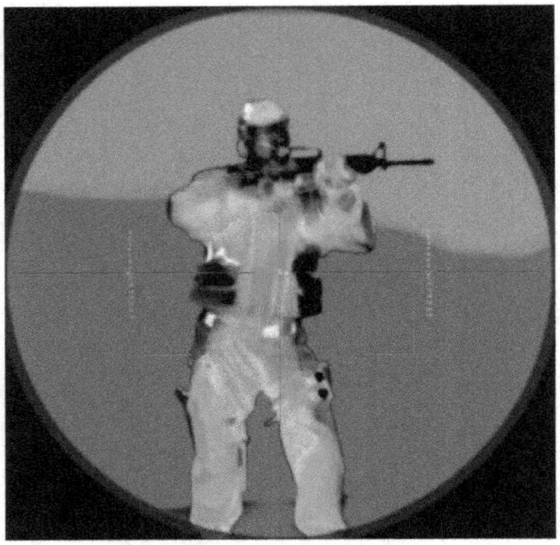

Figure TF21-3: A full-color thermal-infrared image of a soldier.

industrial and mechanical applications (damaged/failing/inefficient parts often heat up). This also is used to locate problems in electrical circuits at the board or chip level (Fig. TF21-4). Night vision is important for security, and is highly valued by outdoor enthusiasts as well (IR wildlife cameras can catch pictures of animals when they are most likely to be moving around at night). IR is used for things other than imaging, too, including motion detection and measuring body temperature.

Historically, two approaches have been pursued to "see in the dark": one that relies on measuring self-emitted *thermal energy* by the scene and another that focuses on *intensifying* the light reflected by the scene when illuminated by very weak sources, such as the moon or the stars. We will explore each of the two approaches briefly.

The visible spectrum extends from the violet (wavelength $\lambda \approx 0.38$ μm) to the red (≈ 0.78 μm). The spectral region next to the visible is the infrared (IR), and it is subdivided into the *near-IR* (≈ 0.7 to 1.3 μm), *mid-IR* (1.3 to 3 μm), and *thermal-IR* (3 to 30 μm). Infrared waves cannot be perceived by humans, because our eyes are not sensitive to EM waves outside of the visible spectrum. In the visible spectrum, we see or image a scene by detecting the light reflected by it, but in the thermal-IR region, we image a scene without an external source of energy, because the scene itself is the source. All material media emit electromagnetic energy all of the time—with hotter objects emitting more than cooler objects. The amount of energy emitted by an object and the shape of its emission spectrum depend on the object's

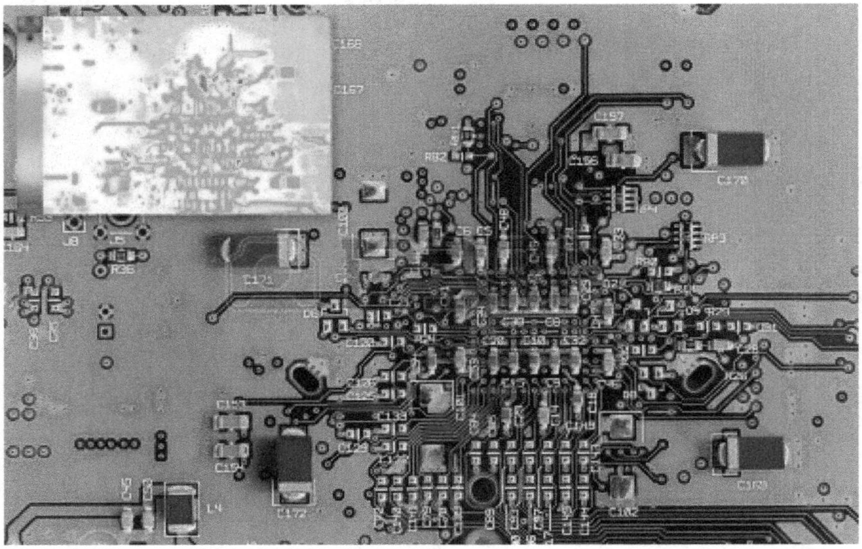

Figure TF21-4: Circuit board with superimposed IR image (inset) identifying (in red) high-temperature components or connections. (Credit: Suljo.)

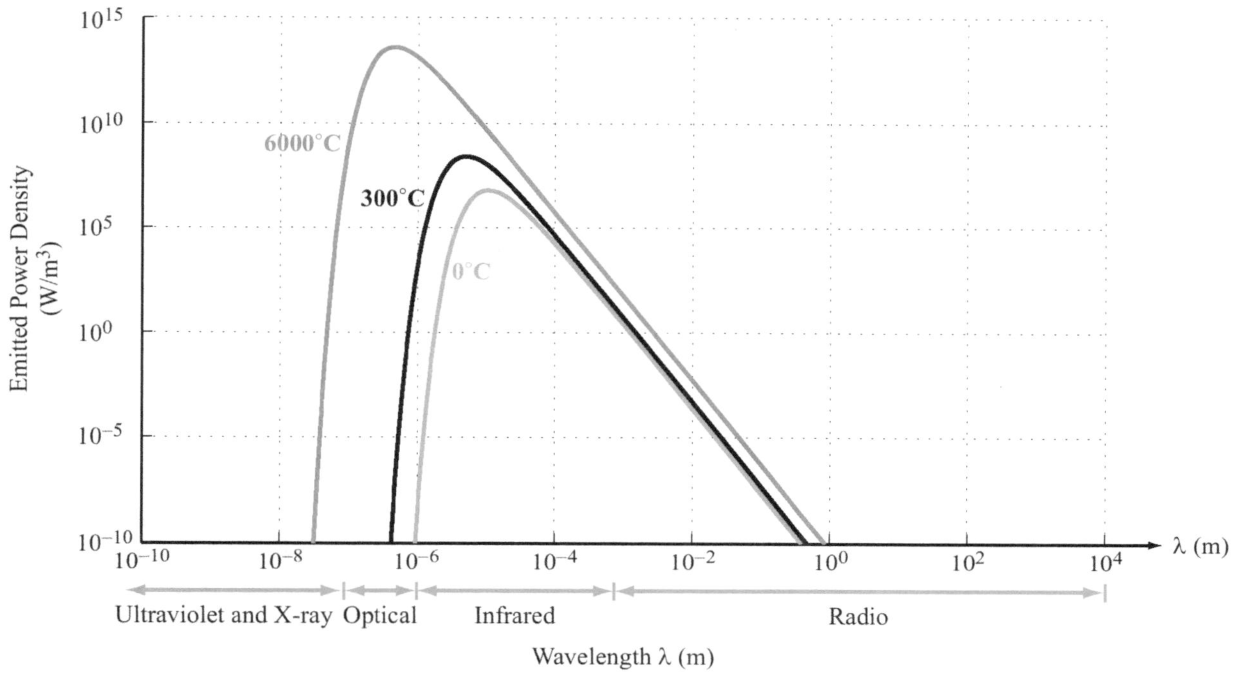

Figure TF21-5: Spectra of power density emitted by ideal blackbodies at 0 °C, 300 °C, and 6000 °C.

temperature and its material properties. Most of the emitted energy occurs over a relatively narrow spectral range, as illustrated in Fig. TF21-5, which is centered around a peak value that is highly temperature dependent. For a high-temperature object like the sun ($\approx$ 6000 °C), the peak value occurs at about 0.5 μm (red-orange color), whereas for a terrestrial object, the peak value occurs in the thermal-IR region.

Through a combination of lenses and a 2-D array of infrared detectors, the energy emitted by a scene can be focused onto the array, thereby generating an image of the scene. The images sometimes are displayed with a rainbow coloring—with hotter objects displayed in red and cooler objects in blue.

In the near- and mid-IR regions, the imaging process is based on reflection—just as in the visible. Interestingly, the sensor chips used in commercial digital cameras are sensitive not only to visible light but to near-IR energy as well. To avoid image blur caused by the IR energy, the camera lens usually is coated with an IR-blocking film that filters out the IR energy but passes visible light with near-perfect transmission. TV remote controls use near-IR signals to communicate with TV sets, so if an inexpensive digital camera with no IR-blocking coating is used to image an activated TV remote control in the dark, the image will show a bright spot at the tip of the remote control. Some cameras are now making use of this effect to offer IR-based night-vision recording. These cameras emit IR energy from LEDs mounted near the lens, so upon reflection by a nighttime scene, the digital camera is able to record an image "in the dark."

Image Intensifier

A second approach to nighttime imaging is to build sensors with much greater detection sensitivity than the human eye. Such sensors are called ***image intensifiers***. Greater sensitivity means that fewer photons are required in order to detect and register an input signal against the random "noise" in the receiver (or the brain in the case of vision). Some animals can see in the dark (but not in total darkness) because their eye receptors and neural networks require fewer numbers of photons than humans to generate an image under darker conditions. Image

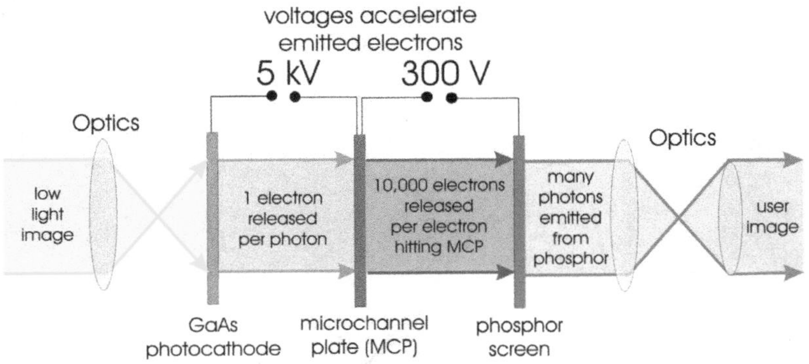

Figure TF21-6: Schematic of image intensifier assembly and operation.

intensifiers work by a simple principle (Fig. TF21-6). Incident photons (of which there are relatively few in a dark scene) are focused through lenses and onto a thin plate of *gallium arsenide* material. This material emits one electron every time a photon hits it. Importantly, these electrons are emitted at the locations where the photons hit the plate, preserving the shape of the light image. These photoelectrons then are accelerated by a high voltage ($\sim$ 5000 V) onto a *microchannel plate* (MCP). The MCP is a plate that emits 10,000 new electrons every time one electron impacts its surface. In essence, it is an amplifier with a current gain of 10,000. These secondary electrons again are accelerated—this time onto phosphors that glow when impacted with electrons. This works on the same principle as the cathode ray tube. The phosphors are arranged in arrays and form pixels on a display, allowing the image to be seen by the naked eye.

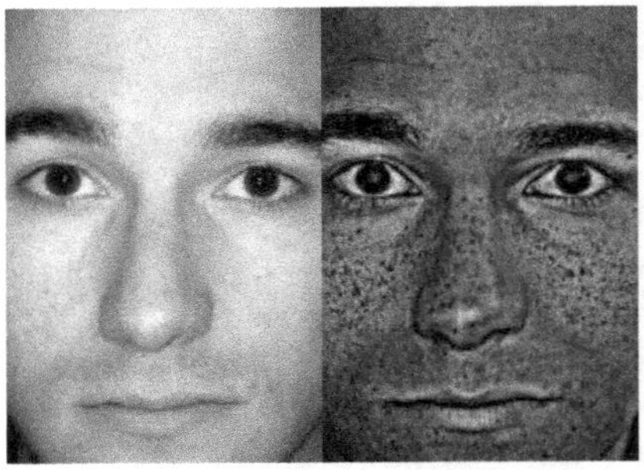

Visible light UV image

Figure TF21-7: Comparison of visible light and UV images. The latter shows skin damage. (Credit: Milford MD Advanced Dermatology Pocono Medical Care, Inc.)

Ultraviolet Imaging

On the other end of the spectrum, UV wavelengths range from 400–200 nm and beyond. UV can also be used to see things that are out of the visible spectrum, particularly skin or soft tissue damage, as shown in the picture of sun-damaged skin in Fig. TF21-7. Dark areas of the skin show where UV is absorbed and not reflected. This can be used for treatment planning, and also to show people the value of skin protection from the sun. UV is also used for numerous astrophysical observations, including the solar flare image shown in Fig. TF21-8. Much information in the universe is outside of the visible spectrum.

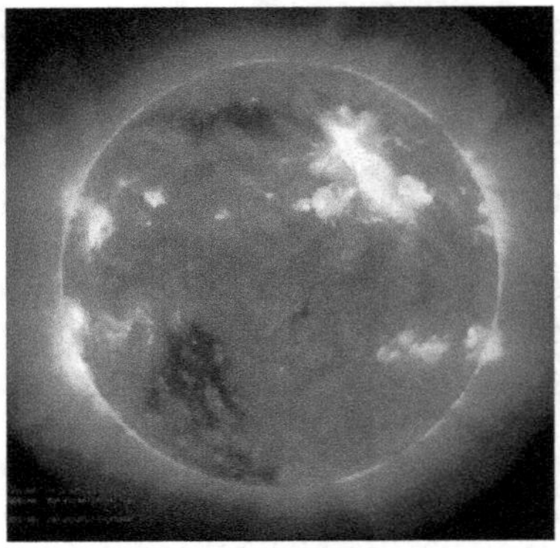

Figure TF21-8: Giant solar flare captured in UV light. (Courtesy NASA/SDO and the AIA, EVE, and HMI science team.)

In Section 3-6, we established that for a purely resistive circuit, the power transferred from the source circuit to the load is a maximum when $R_L = R_S$. The question we now pose is: What are the equivalent conditions for an ac circuit with complex impedances?

To answer the question, we start by writing down the expression for $\mathbf{I}_L$ (the current flowing into the load) namely

$$\mathbf{I}_L = \frac{\mathbf{V}_s}{\mathbf{Z}_s + \mathbf{Z}_L} = \frac{\mathbf{V}_s}{(R_s + R_L) + j(X_s + X_L)}. \tag{8.58}$$

From Eq. (8.39a), the average power transferred to (consumed by) the load is

$$\begin{aligned} P_{av} &= \frac{1}{2}|\mathbf{I}_L|^2 R_L = \frac{1}{2}\mathbf{I}_L \times \mathbf{I}_L^* R_L \\ &= \frac{1}{2}\frac{\mathbf{V}_s}{(R_s + R_L) + j(X_s + X_L)} \\ &\quad \times \frac{\mathbf{V}_s^*}{(R_s + R_L) - j(X_s + X_L)} \cdot R_L \\ &= \frac{1}{2}\frac{|\mathbf{V}_s|^2 R_L}{(R_s + R_L)^2 + (X_s + X_L)^2}. \end{aligned} \tag{8.59}$$

The load parameters R_L and X_L represent orthogonal dimensions in the complex plane. Hence, the values of R_L and X_L that maximize P_{av} can be obtained by performing independent maximization processes: one by setting $\partial P_{av}/\partial R_L = 0$ and another by setting $\partial P_{av}/\partial X_L = 0$. For R_L,

$$\frac{\partial P_{av}}{\partial R_L} = \frac{1}{2}|\mathbf{V}_s|^2 \left[\frac{(R_s + R_L)^2 + (X_s + X_L)^2 - 2R_L(R_s + R_L)}{[(R_s + R_L)^2 + (X_s + X_L)^2]^2}\right]. \tag{8.60}$$

The right-hand side of Eq. (8.60) is equal to zero if its numerator is equal to zero (because the other alternative, namely setting the denominator equal to infinity, produces a solution in which $\mathbf{Z}_L$ and $\mathbf{Z}_s$ are open circuits corresponding to no power transfer to the load). That is,

$$(R_s + R_L)^2 + (X_s + X_L)^2 - 2R_L(R_s + R_L) = 0,$$

which simplifies to

$$R_s^2 - R_L^2 + (X_s + X_L)^2 = 0. \tag{8.61}$$

Similarly, the partial derivative of P_{av} with respect to X_L is

$$\frac{\partial P_{av}}{\partial X_L} = \frac{1}{2}|\mathbf{V}_s|^2 R_L \left[\frac{-2(X_s + X_L)}{(R_s + R_L)^2 + (X_s + X_L)^2}\right], \tag{8.62}$$

which when set equal to zero yields

$$X_L = -X_s. \tag{8.63}$$

Incorporating Eq. (8.63) in Eq. (8.61) gives

$$R_L = R_s. \tag{8.64}$$

The conditions on X_L and R_L can be combined into

$$\mathbf{Z}_L = \mathbf{Z}_s^* \qquad \textbf{(maximum power transfer)}, \tag{8.65}$$

where $\mathbf{Z}_s^* = (R_s - jX_s)$ is the complex conjugate of $\mathbf{Z}_s$. When the condition represented by Eq. (8.65) is true, the load is said to be ***conjugate matched*** to the source.

According to the result encapsulated by Eq. (8.65):

> ▶ The average power transferred to (consumed by) an ac load is a maximum when its impedance $\mathbf{Z}_L$ is equal to $\mathbf{Z}_s^*$, which is the complex conjugate of the Thévenin impedance of the source circuit. ◀

Under the conditions of maximum power transfer represented by Eqs. (8.63) and (8.64), the expression for P_{av} given by Eq. (8.59) reduces to

$$P_{av}(\max) = \frac{1}{8}\frac{|\mathbf{V}_s|^2}{R_L}. \tag{8.66}$$

Example 8-7: Maximum Power

Determine the maximum amount of power that can be consumed by the load $\mathbf{Z}_L$ in the circuit of Fig. 8-13.

Solution: We start by determining the Thévenin equivalent of the circuit to the left of terminals (a, b). In Fig. 8-13(b), the load has been removed so as to calculate the open-circuit voltage. Voltage division yields

$$\mathbf{V}_s = \mathbf{V}_{oc} = \frac{(4 + j6)}{4 + 4 + j6} \times 24 = 17.31\angle 19.44^\circ \text{ V},$$

where $\mathbf{V}_s$ is the Thévenin voltage of the source circuit to the left of terminals (a, b). The Thévenin impedance of the source

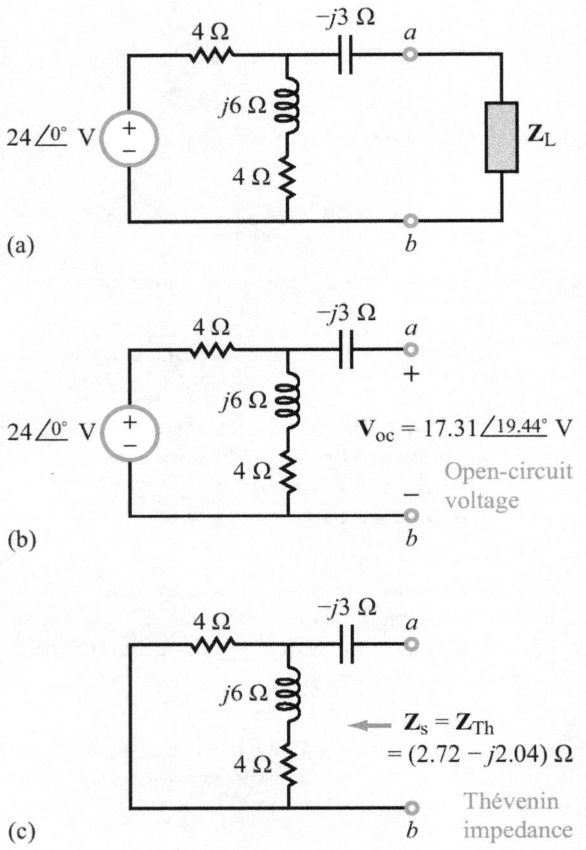

Figure 8-13: Circuit for Example 8-7.

circuit, $\mathbf{Z}_s$, is obtained by calculating the impedance at terminals (a, b), as shown in Fig. 8-13(c), after deactivating the 24 V voltage source,

$$\mathbf{Z}_s = 4 \parallel (4+j6) - j3 = \frac{4(4+j6)}{4+4+j6} - j3 = (2.72 - j2.04)\ \Omega.$$

For maximum transfer of power to the load, the load impedance should be

$$\mathbf{Z}_L = \mathbf{Z}_s^* = (2.72 + j2.04)\ \Omega,$$

and the corresponding value of P_{av} is

$$P_{av}(\max) = \frac{|\mathbf{V}_s|^2}{8R_L} = \frac{(17.31)^2}{8 \times 2.72} = 13.77\ \text{W}.$$

Concept Question 8-10: To achieve maximum transfer of power from a source circuit to a load, how should the impedance of the load be related to that of the source circuit? (See CAD)

Concept Question 8-11: Suppose that a certain passive circuit—containing resistors, capacitors, and inductors—is connected to a square-wave voltage source. What procedure would you use to analyze the voltages and currents in the circuit? (See CAD)

8-6 Measuring Power with Multisim

This section introduces Multisim power-measurement tools and demonstrates their ability through an interactive simulation of an *impedance-matching network*. In Section 8-5 we established that the amount of power transferred to a load from a source is at a maximum when the impedance of the load $\mathbf{Z}_{\text{Load}}$ is the complex conjugate of the source impedance $\mathbf{Z}_{\text{Source}}$. That is,

$$\mathbf{Z}_{\text{Load}} = \mathbf{Z}^*_{\text{Source}}. \tag{8.67}$$

Consider the circuit shown in Fig. 8-14. The circuit is supplied by a realistic source composed of an ideal voltage source $\mathbf{V}_s$ in series with a source resistance R_s. The load is a series RL circuit. In the phasor domain:

$$\mathbf{Z}_{\text{Source}} = R_s \quad \text{and} \quad \mathbf{Z}_{\text{Load}} = R_L + j\omega L_L. \tag{8.68}$$

For the general case where $L_L \neq 0$ and $R_s \neq R_L$, the load would not be matched to the source, and power transfer would not be a maximum. By inserting a matching network in between the source and the load and selecting the values of

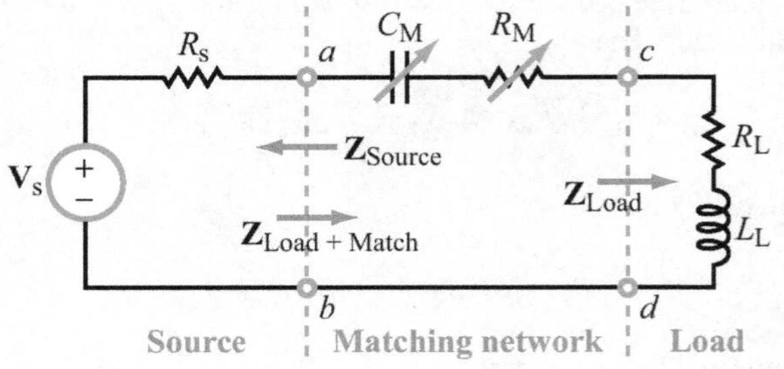

Figure 8-14: Matching network in between the source and the load.

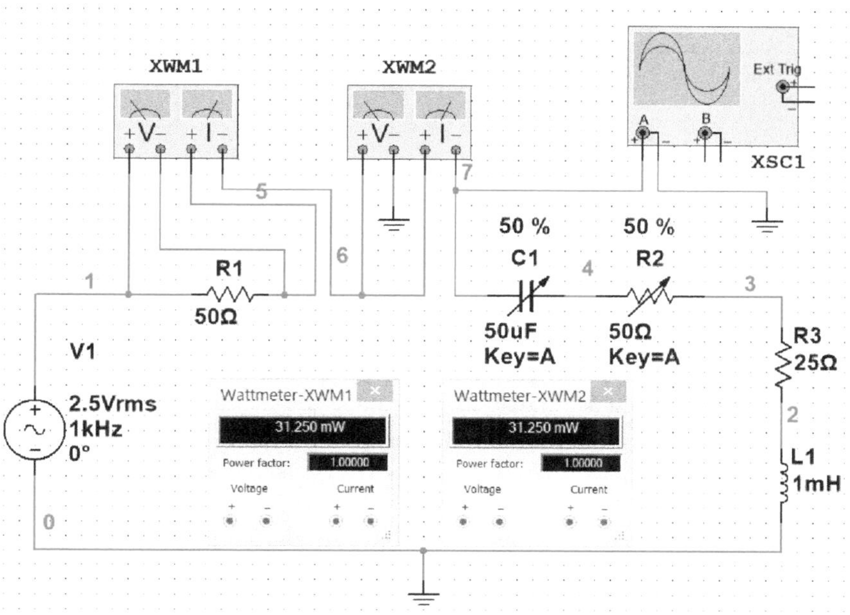

Figure 8-15: Multisim simulation of matching network (CM, RM) in between the source and the load, and wattmeter displays for maximum power transfer.

its components appropriately, we can match the source to the load, thereby realizing the maximum transfer of power from the source to the circuit segment to the right of terminals (a, b), which includes the matching network and the load. If the load has an inductor, the matching network should have a capacitor, and vice versa.

For the circuit to the right of terminals (a, b), which includes both the matching network and the load,

$$\mathbf{Z}_{\text{Load+Match}} = (R_M + R_L) + j\left(\omega L_L - \frac{1}{\omega C_M}\right). \tag{8.69}$$

For the maximum transfer of power at terminals (a, b) towards the load, it is necessary that

$$\mathbf{Z}_{\text{Source}} = \mathbf{Z}^*_{\text{Load+Match}}, \tag{8.70}$$

which can be satisfied by selecting R_M and C_M as

$$R_M = R_s - R_L \qquad \text{and} \qquad C_M = \frac{1}{\omega^2 L_L}, \tag{8.71}$$

provided $R_s \geq R_L$. Under these matched conditions, the impedance of the capacitor cancels out the impedance of L_L, and the source is matched to the combination of the matching network and load. This means that the power transferred from the source to this combination is a maximum, but it does not mean that the power transferred to the load alone is a maximum. In fact, if the values of R_s and R_L cannot be changed, power transfer to the load is a maximum when $R_M = 0$ and $C_M = 1/(\omega^2 L_L)$.

We also should note that the value of C_M required to achieve the matching condition is a function of ω. Thus, if the value of C_M is selected so as to match the circuit at a given frequency, the circuit will cease to remain matched if ω is changed to a significantly different value. To serve its intended function with significant flexibility, the matching network usually is configured to include a potentiometer and an adjustable capacitor, allowing for manual tuning of R_M and C_M to satisfy Eq. (8.71) at any specified value of ω (within a certain range).

The circuit in Fig. 8-14 can be simulated and analyzed by Multisim, as shown in Fig. 8-15. For variable components, you can choose which keys will shift the component values by double-clicking the component and selecting the desired key letter under Values → Key. Measurement instruments XWM1 and XWM2 are wattmeters configured to measure the average

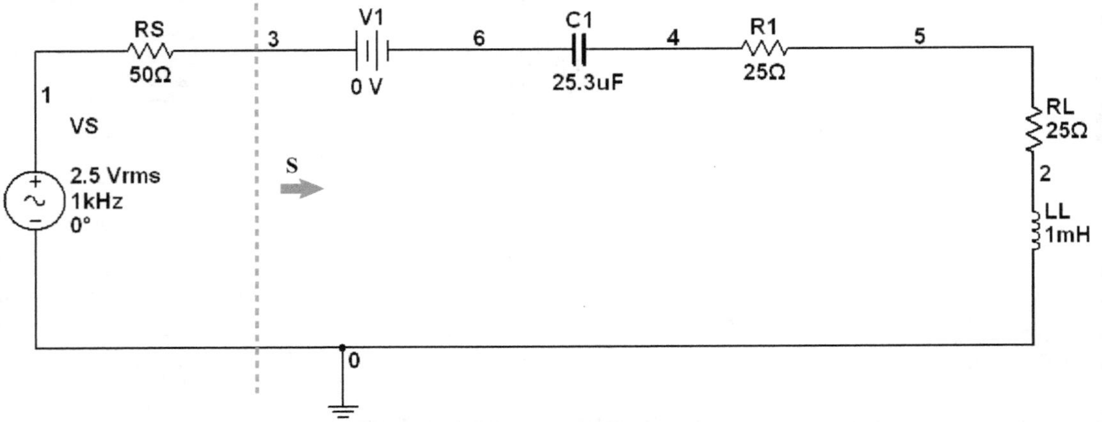

Figure 8-16: Multisim circuit without instruments.

power dissipated by a component or circuit:

$$P_{\mathrm{av}} = \frac{1}{2}\,\mathfrak{Re}[\mathbf{V}\mathbf{I}^*],$$

where **V** is the phasor voltage across the component or circuit and **I** is the phasor current flowing into its positive voltage terminal. In Fig. 8-15, XWM2 measures the current through R_s and the voltage across it, and XWM1 measures the voltage at node 7 (relative to the ground terminal) and the current through the loop at node 7. Thus, XWM2 measures the average power dissipated in R_s, and XWM1 measures the average power delivered by the source to the matching network and load combined. To match the load to the source in the circuit of Fig. 8-15, we should select

$$R_M = R_s - R_L = 50 - 25 = 25\ \Omega$$

and

$$C_M = \frac{1}{\omega^2 L_L} = \frac{1}{(2\pi \times 10^3)^2 \times 10^{-3}} = 25.33\ \mu\text{F}.$$

In Fig. 8-15, R_M is a 50 Ω potentiometer set at 50 percent of its maximum value (or 25 Ω), and C_M is a variable 50 μF capacitor, also set at 50 percent of its maximum value (which is very close to the required value of 25.33 μF). The wattmeter displays confirm that the average powers reported by XWM1 and XWM2 are indeed equal.

It is important to note that the wattmeter calculates the average power by measuring the voltage and current at a sampling rate specified by the Maximum Time Step (TMAX) in the Interactive Simulation Settings. The default value is 10^{-5} s, which means that the voltage and current are sampled at a time spacing of 10^{-5} s. At 1 kHz, the period is 10^{-3} s. Hence at a time spacing of 10^{-5} s each cycle gets sampled 100 times, which is quite adequate for generating a reliable measurement of the average power. At higher oscillation frequencies, however, the period is much shorter necessitating that TMAX be selected such that TMAX $\leq 10^{-2}/f$ where f is the oscillation frequency in Hz. Thus, at $f = 1$ MHz, for example, TMAX should be set at 10^{-8} s.

Another method for measuring average power in Multisim is to use the Analysis functions to plot the complex power across any section of a circuit. Figure 8-16 is a Multisim reproduction of the circuit in Fig. 8-15 but with no instruments and fixed-value components. Note that to perform the AC Analysis Simulation properly, the AC Analysis Magnitude value of the VS source must be changed to 2.5 ∗ sqrt(2) = 3.5355 V. We can plot the magnitude and phase of the complex power **S** across terminals (3,0) in Fig. 8-16 by performing AC Analysis in Multisim. Under Simulate → Analyses → AC Analysis, set FSTART to 1 Hz and FSTOP to 1 MHz. Make sure to include at least 10 points per decade to produce a good plot. Under Output, enter the following expression: 0.5*(real(I(v1)),-imag(I(v1)))*V(3). Note that this expression is equivalent to $\mathbf{S} = \frac{1}{2}\mathbf{I}^*\mathbf{V}$ (Eq. (8.29)). (The expression (real(X),-imag(X)) gives us the complex conjugate of any complex number X; we need to do this because Multisim does not have a complex conjugate function). Figure 8-17 shows a plot of the AC Analysis output. As expected, the phase of **S** goes to 0 at 1 kHz (since it is at this frequency that the inductor and capacitor reactances cancel each other out).

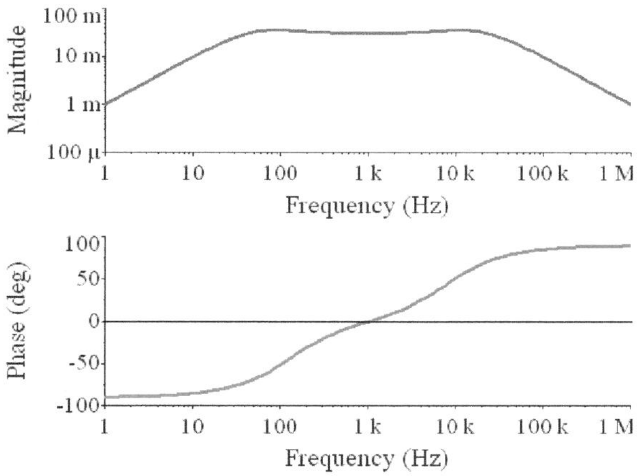

Figure 8-17: Spectral plots of the magnitude and phase of the complex power **S** at terminals (3,0) in Fig. 8-16.

Concept Question 8-12: How is power measured in Multisim? Why must all four terminals of the wattmeter be used to obtain a power measurement? (See CAD)

Concept Question 8-13: Assuming the values of $\mathbf{V}_s$, R_s, R_L, and L_L are fixed, what values of R_M and C_M lead to maximum transfer of power from the source to R_L? (See CAD)

Exercise 8-6: Use Multisim to simulate the circuit in Fig. 8-15. Connect Channel B of the oscilloscope across the voltage source $\mathbf{V}_s$. Vary C_M over its full range, noting the phase difference between the two channels of the oscilloscope at $C_M = 0$, $C_M = 25$ μF, and $C_M = 50$ μF.

Answer: (See CAD)

Summary

Concepts

- Even though the average values of the sinusoidal voltage across and current through a load are both zero, the average power consumed by the load is not zero, unless the load is purely reactive (no resistors).
- Power is characterized by several attributes, including the complex power **S**, the average power P_{av}, and reactive power Q.
- The power factor *pf* is the ratio of the average real power P_{av} consumed by the load to S (the magnitude of the complex power) which incorporates the reactive power Q through $S = [P_{av}^2 + Q^2]^{1/2}$.
- An RL load can be compensated by adding a shunt capacitor, causing its *pf* to increase, and in turn reducing the amount of current that has to be supplied by the electrical power source.
- The power transferred from an input source circuit with Thévenin impedance $\mathbf{Z}_s = R_s + jX_s$ to a complex load with impedance $\mathbf{Z}_L = R_L + jX_L$ is at a maximum when $\mathbf{Z}_L = \mathbf{Z}_s^*$.
- Multisim can be used to measure the magnitude and phase of complex power as a function of frequency.

Mathematical and Physical Models

Average value	$X_{av} = \dfrac{1}{T}\displaystyle\int_0^T x(t)\,dt$
rms value	$X_{rms} = X_{eff} = \sqrt{\dfrac{1}{T}\displaystyle\int_0^T x^2(t)\,dt}$
Average power	$P_{av} = V_{rms} I_{rms} \cos(\phi_v - \phi_i)$ (W)
Complex power	$\mathbf{S} = \frac{1}{2}\,\mathbf{V}\mathbf{I}^*$ (VA)
Reactive power	$Q = V_{rms} I_{rms} \sin(\phi_v - \phi_i)$ (VAR)
Power factor	$pf = \dfrac{P_{av}}{S} = \cos(\phi_v - \phi_i)$
Power factor lead or lag	Table 8-2
Maximum power transfer	$\mathbf{Z}_L = \mathbf{Z}_s^*$
Maximum power	$P_{av}(\max) = \dfrac{1}{8}\,\dfrac{|\mathbf{V}_s|^2}{R_L}$

Important Terms Provide definitions or explain the meaning of the following terms:

apparent power
average power
average value
compensated load
compensated load circuit
complex power
consume
effective value
impedance matching network
in phase
instantaneous current
instantaneous power
instantaneous voltage
lagging
leading
matched load
mean
periodicity property
power factor
power factor angle
power factor compensation
reactive power
root
root-mean-square (rms) value
square
store
VAR
volt-ampere
volt-ampere reactive

PROBLEMS

Section 8-1: Periodic Waveforms

*8.1 Determine (a) the average and (b) rms values of the periodic voltage waveform shown in Fig. P8.1.

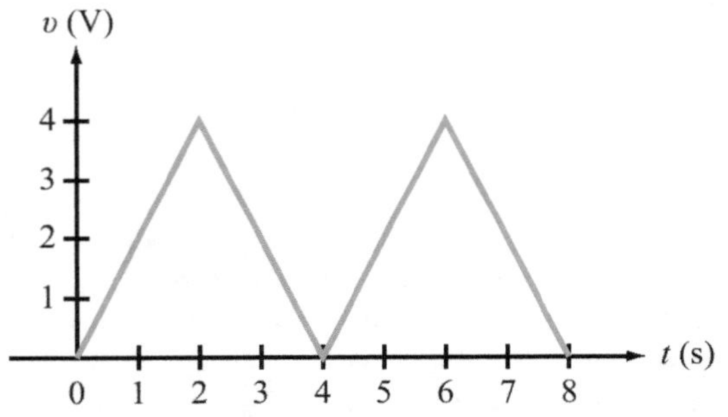

Figure P8.1: Waveform for Problem 8.1.

8.2 Determine (a) the average and (b) rms values of the periodic voltage waveform shown in Fig. P8.2.

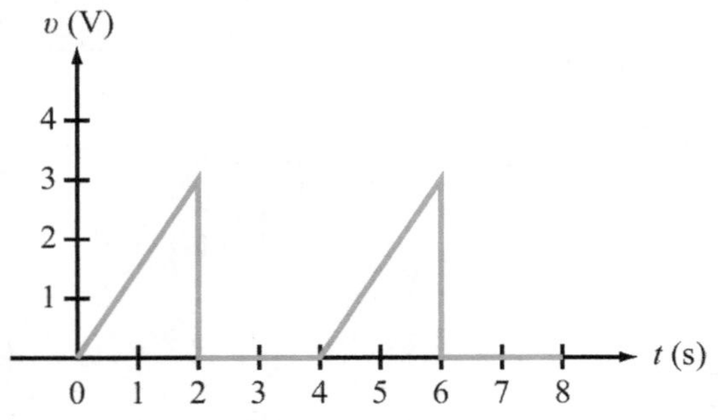

Figure P8.2: Waveform for Problem 8.2.

8.3 Determine (a) the average and (b) rms values of the periodic current waveform shown in Fig. P8.3.

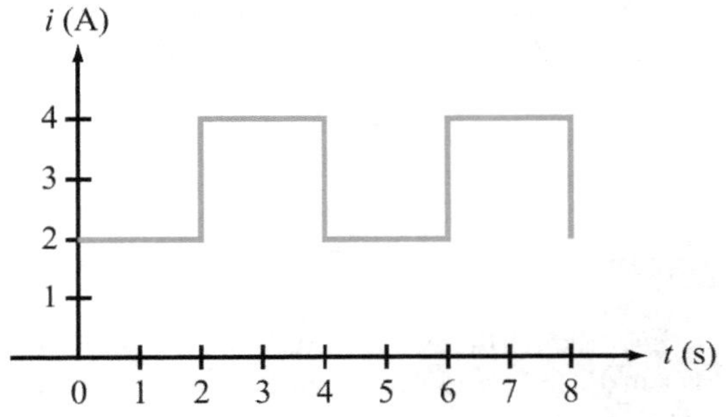

Figure P8.3: Waveform for Problem 8.3.

*8.4 Determine (a) the average and (b) rms values of the periodic current waveform shown in Fig. P8.4.

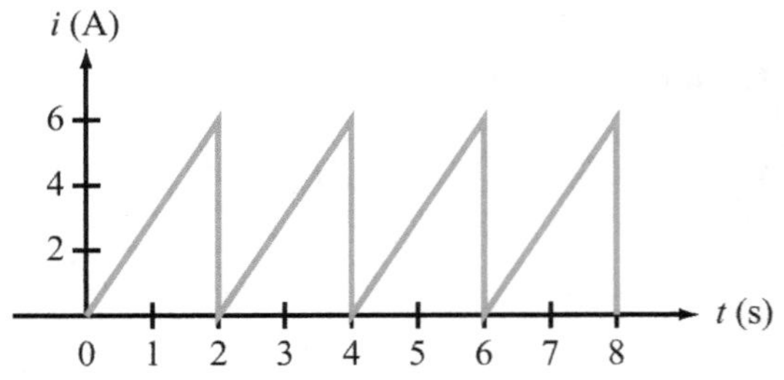

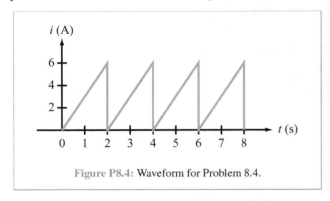

Figure P8.4: Waveform for Problem 8.4.

8.5 Determine (a) the average and (b) rms values of the periodic voltage waveform shown in Fig. P8.5.

*Answer(s) available in Appendix G.

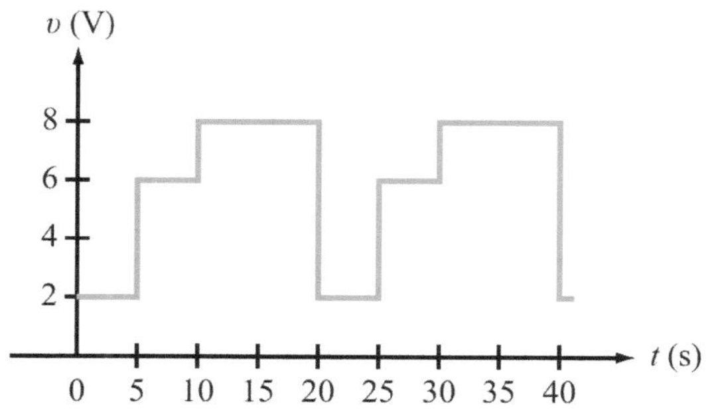

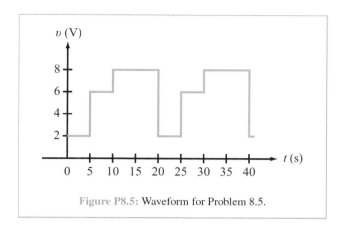

Figure P8.5: Waveform for Problem 8.5.

8.6 Determine (a) the average and (b) rms values of the periodic current waveform shown in Fig. P8.6.

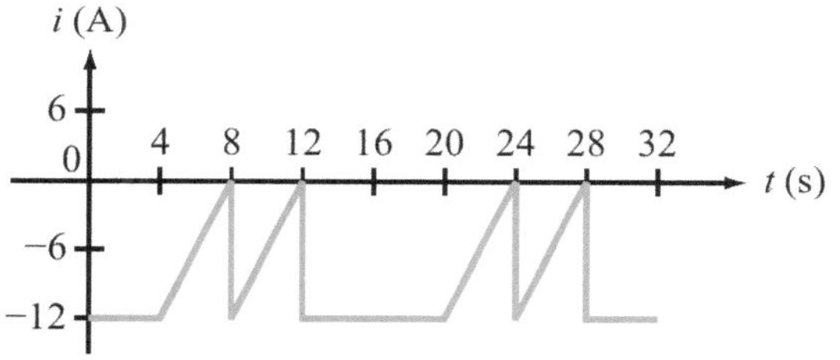

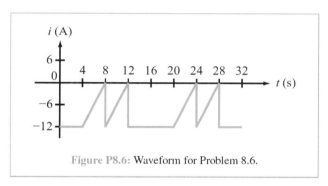

Figure P8.6: Waveform for Problem 8.6.

***8.7** Determine (a) the average and (b) rms values of the periodic voltage waveform shown in Fig. P8.7.

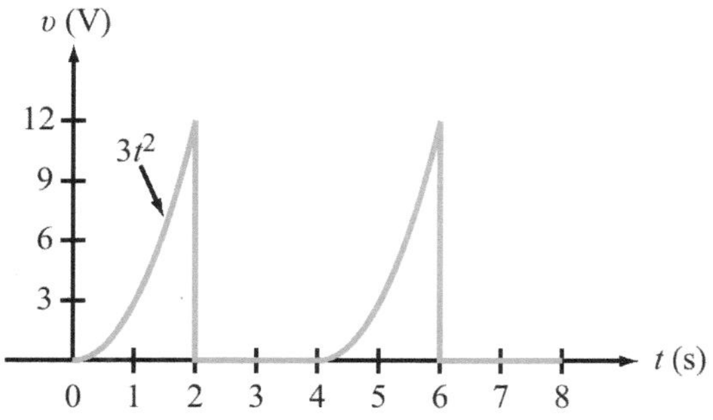

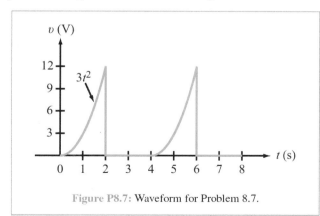

Figure P8.7: Waveform for Problem 8.7.

8.8 Determine (a) the average and (b) rms values of the periodic current waveform shown in Fig. P8.8.

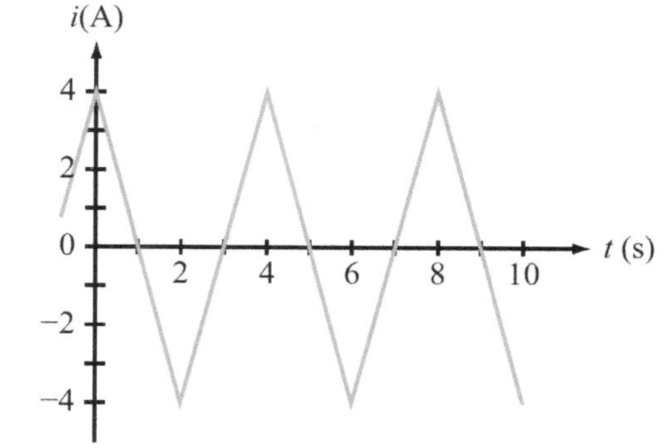

Figure P8.8: Waveform for Problem 8.8.

8.9 Determine (a) the average and (b) rms values of the periodic voltage waveform shown in Fig. P8.9.

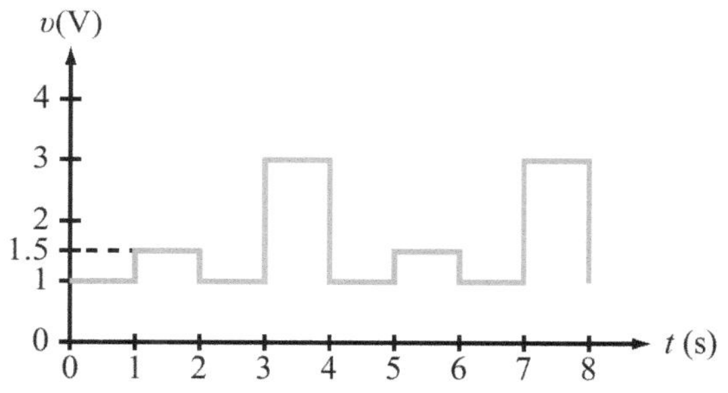

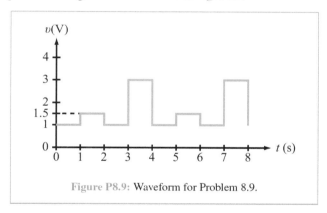

Figure P8.9: Waveform for Problem 8.9.

***8.10** Determine (a) the average and (b) rms values of the periodic voltage waveform shown in Fig. P8.10.

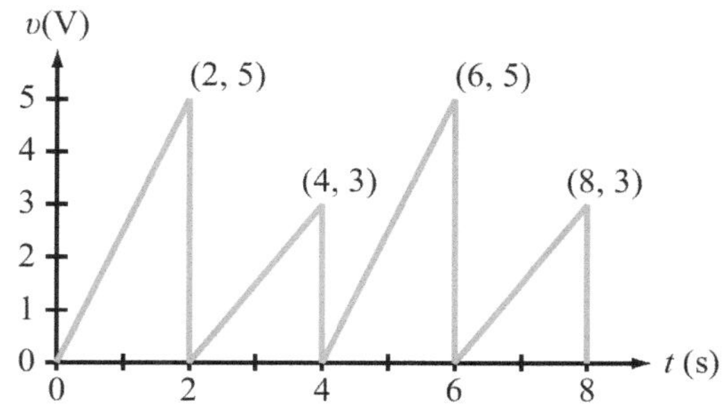

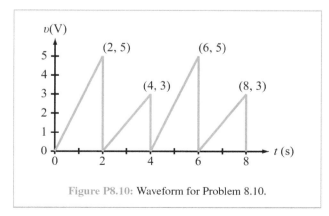

Figure P8.10: Waveform for Problem 8.10.

8.11 Determine (a) the average and (b) rms values of the periodic voltage waveform shown in Fig. P8.11.

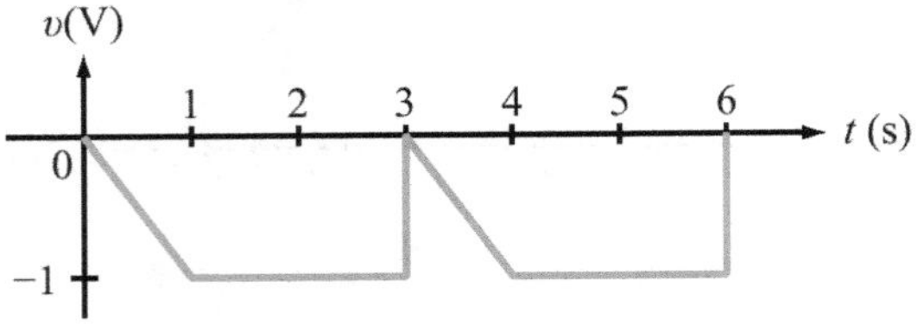

Figure P8.11: Waveform for Problem 8.11.

*__8.12__ Determine (a) the average and (b) rms values of the periodic voltage waveform shown in Fig. P8.12.

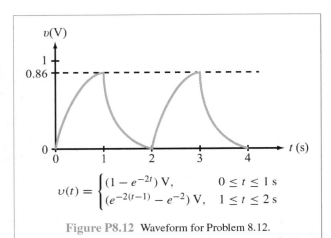

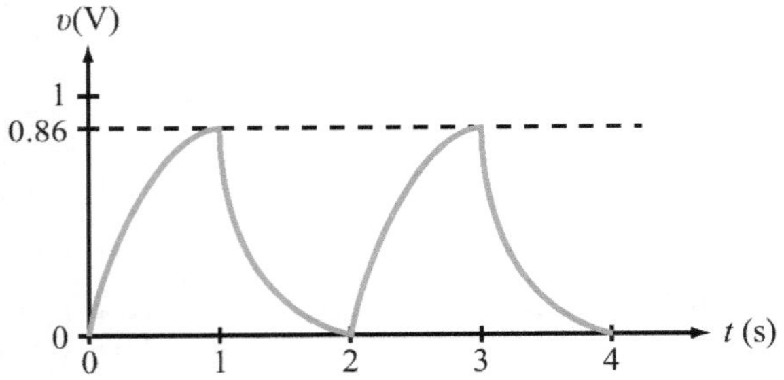

$$\upsilon(t) = \begin{cases} (1 - e^{-2t})\ \text{V}, & 0 \le t \le 1\ \text{s} \\ (e^{-2(t-1)} - e^{-2})\ \text{V}, & 1 \le t \le 2\ \text{s} \end{cases}$$

Figure P8.12 Waveform for Problem 8.12.

8.13 Determine (a) the average and (b) rms values of the periodic voltage waveform shown in Fig. P8.13.

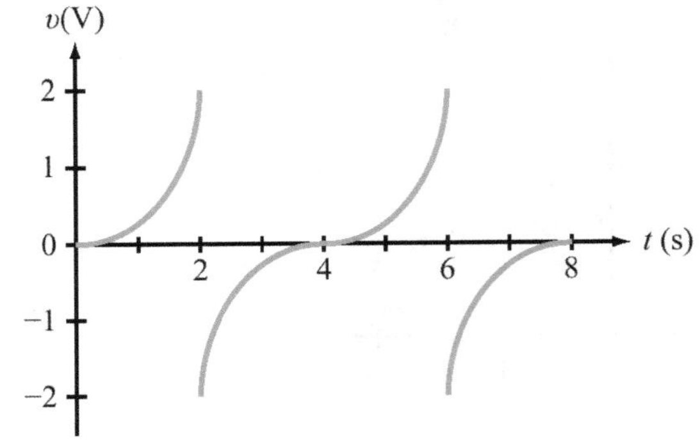

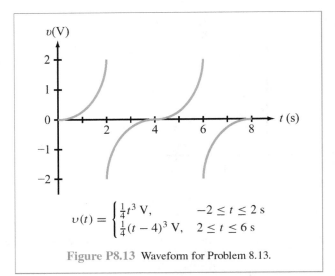

$$\upsilon(t) = \begin{cases} \frac{1}{4}t^3\ \text{V}, & -2 \le t \le 2\ \text{s} \\ \frac{1}{4}(t-4)^3\ \text{V}, & 2 \le t \le 6\ \text{s} \end{cases}$$

Figure P8.13 Waveform for Problem 8.13.

8.14 The current waveform shown in Fig. P8.8 dissipates average power at a rate of 3.2 kW when connected to a resistor. What is the value of the resistor?

8.15 Determine the average and rms values of the following periodic waveforms:

(a) $\upsilon_1(t) = 4\cos(60t - 30°)$ V.

(b) $i_1(t) = 2.5$ A.

*(c) $\upsilon_2(t) = 12 - \sin(2t + 45°)$ V.

(d) $i_2(t) = 4\sin(10t) - 6\sin^2(5t)$ A.

8.16 Determine the average and rms values of the following periodic waveforms:

(a) $\upsilon(t) = |12\cos(\omega t + \theta)|$ V

(b) $\upsilon(t) = 4 + 6\cos(2\pi f t + \phi)$ V

(c) $\upsilon(t) = 2\cos\omega t - 4\sin(\omega t + 30°)$ V

(d) $\upsilon(t) = 9\cos\omega t \sin(\omega t + 30°)$ V

Section 8-2 and 8-3: Average and Complex Power

8.17 Determine the complex power, apparent power, average power absorbed, reactive power, and power factor (including whether it is leading or lagging) for a load circuit whose voltage and current at its input terminals are given by:

(a) $\upsilon(t) = 100\cos(377t - 30°)$ V,
$i(t) = 2.5\cos(377t - 60°)$ A.

(b) $\upsilon(t) = 25\cos(2\pi \times 10^3 t + 40°)$ V,
$i(t) = 0.2\cos(2\pi \times 10^3 t - 10°)$ A.

*(c) $\mathbf{V}_{\text{rms}} = 110\angle 60°$ V, $\mathbf{I}_{\text{rms}} = 3\angle 45°$ A.

(d) $\mathbf{V}_{\text{rms}} = 440\angle 0°$ V, $\mathbf{I}_{\text{rms}} = 0.5\angle 75°$ A.

(e) $\mathbf{V}_{\text{rms}} = 12\angle 60°$ V, $\mathbf{I}_{\text{rms}} = 2\angle -30°$ A.

8.18 Determine the complex power, apparent power, average power absorbed, reactive power, and power factor (including whether it is leading or lagging) for a load circuit whose voltage and current at its input terminals are given by:

(a) $v(t) = 110\cos(60t - 60^\circ)$ V,
$i(t) = 4\cos(60t - 25^\circ)$ A.

(b) $v(t) = 12\cos(1000t + 30^\circ)$ V,
$i(t) = 0.4\cos(1000t - 15^\circ)$ A.

(c) $v(t) = -10\cos(3000t + 10^\circ)$ V,
$i(t) = 0.5\sin(3000t - 5^\circ)$ A.

(d) $\mathbf{V}_{\text{rms}} = 240\angle 0^\circ$ V, $\mathbf{I}_{\text{rms}} = 0.8\angle 50^\circ$ A.

(e) $\mathbf{V}_{\text{rms}} = 6\angle 20^\circ$ V, $\mathbf{I}_{\text{rms}} = 255\angle 15^\circ$ mA.

(f) $\mathbf{V}_{\text{rms}} = 18\angle 40^\circ$ V, $\mathbf{I}_{\text{rms}} = 1\angle -50^\circ$ A.

8.19 Determine the impedance of a load characterized by the following attributes:

(a) $\mathbf{S} = 1.2\angle 30^\circ$ kVA, $\mathbf{V}_{\text{rms}} = 40\angle 0^\circ$ V,

(b) $|\mathbf{S}| = 80$ VA, $Q = 26$ VAR, $\mathbf{I}_{\text{rms}} = 4\angle 45^\circ$ A.

(c) $\mathbf{V}_{\text{rms}} = 25\angle -15^\circ$ V, $\mathbf{I}_{\text{rms}} = 0.5\angle 35^\circ$ A.

8.20 In the circuit shown in Fig. P8.20, $v(t) = 40\cos(10^5 t)$ V, $R_1 = 100\ \Omega$, $R_2 = 500\ \Omega$, $C = 0.1\ \mu$F, and $L = 0.5$ mH. Determine the complex power for each passive element, and verify that conservation of energy is satisfied.

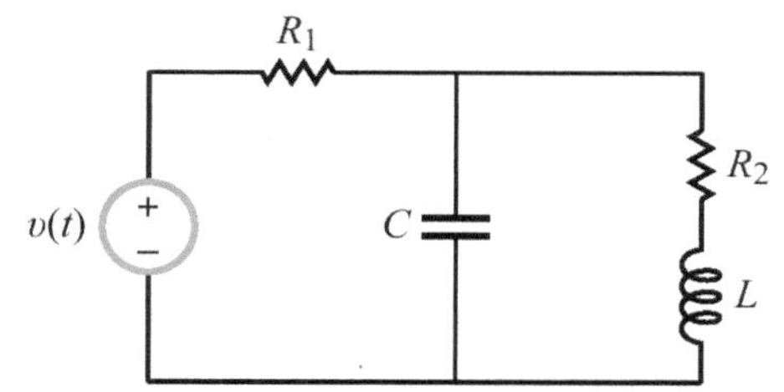

Figure P8.20: Circuit for Problem 8.20.

*8.21 In the circuit shown in Fig. P8.21, $v(t) = 12\cos(2000t)$ V, $R = 20\ \Omega$, and $C = 4.7\ \mu$F. Determine the complex power of the source. What is the power factor of the voltage source.

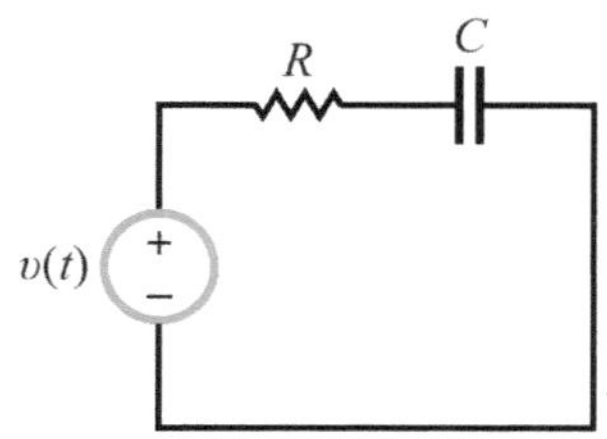

Figure P8.21: Circuit for Problem 8.21.

8.22 In the phasor-domain shown in Fig. P8.22, $\mathbf{V} = 120\angle 0^\circ$ V, $\mathbf{I} = 0.3\angle 30^\circ$ A, $\omega = 1000$ rad/s, $R_1 = 200\ \Omega$, $R_2 = 200\ \Omega$, $R_3 = 1.2$ kΩ, $L = 0.2$ H, and $C = 10\ \mu$F. Determine the complex power for each passive element, and verify that conservation of energy is satisfied.

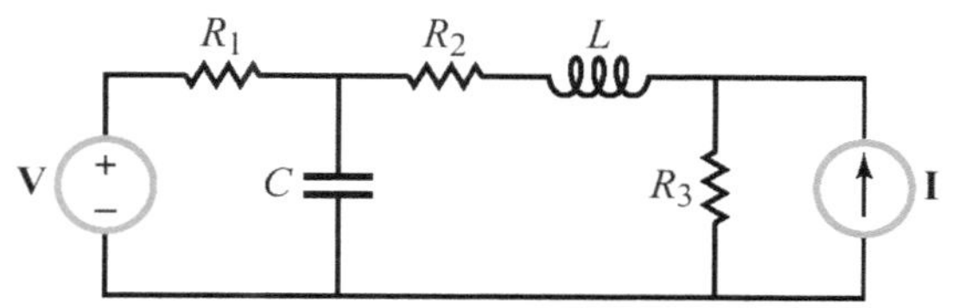

Figure P8.22: Circuit for Problem 8.22.

*8.23 In the circuit of Fig. P8.23, $v_s(t) = 60\cos 4000t$ V, $R_1 = 200\ \Omega$, $R_2 = 100\ \Omega$, and $C = 2.5\ \mu$F. Determine the average power absorbed by each passive element and the average power supplied by the source.

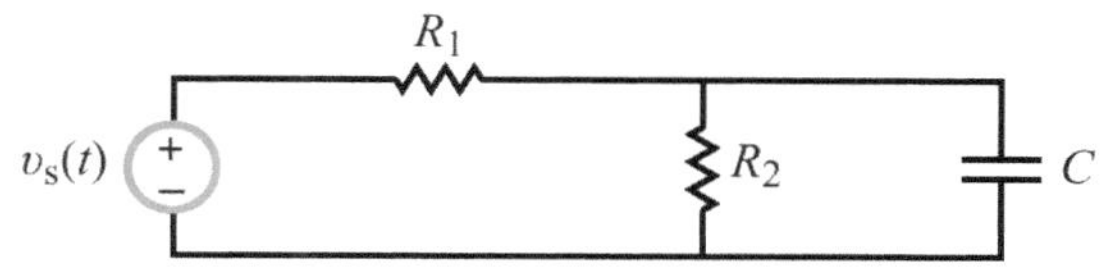

Figure P8.23: Circuit for Problem 8.23.

8.24 In the circuit of Fig. P8.24, $i_s(t) = 0.2\sin 10^5 t$ A, $R = 20\ \Omega$, $L = 0.1$ mH, and $C = 2\ \mu$F. Show that the sum of the complex powers for the three passive elements is equal to the complex power of the source.

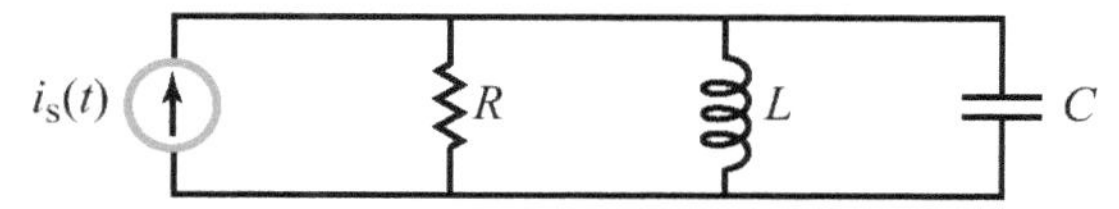

Figure P8.24: Circuit for Problem 8.24.

8.25 In the phasor-domain circuit of Fig. P8.25, $\mathbf{V}_s = 20$ V, $\mathbf{I}_s = 0.3\angle 30^\circ$ A, $R_1 = R_2 = 100\ \Omega$, $\mathbf{Z}_L = j50\ \Omega$, and $\mathbf{Z}_C = -j50\ \Omega$. Determine the complex power for each of the four passive elements and for each of the two sources. Verify that conservation of energy is satisfied.

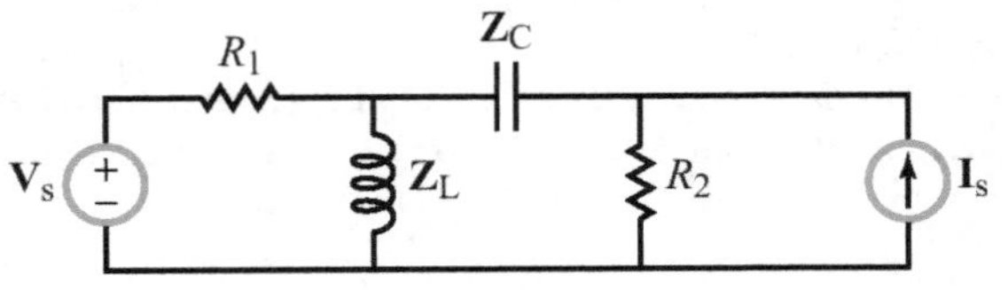

Figure P8.25: Circuit for Problem 8.25.

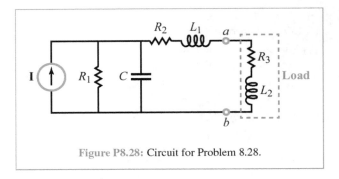

*8.26 Determine the average power dissipated in the load resistor R_L of the circuit in Fig. P8.26, given that $\mathbf{V}_s = 100$ V, $R_1 = 1$ kΩ, $R_2 = 0.5$ kΩ, $R_L = 2$ kΩ, $\mathbf{Z}_L = j0.8$ kΩ, and $\mathbf{Z}_C = -j4$ kΩ.

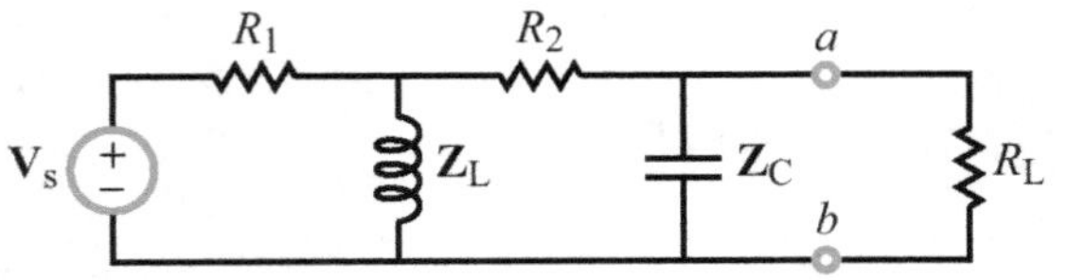

Figure P8.26: Circuit for Problem 8.26.

8.27 Determine **S** for the RL load in the circuit of Fig. P8.27, given that $\mathbf{I}_s = 4\angle 0°$ A, $R_1 = 10$ Ω, $R_2 = 5$ Ω, $\mathbf{Z}_C = -j20$ Ω, $R = 10$ Ω, and $\mathbf{Z}_L = j20$ Ω.

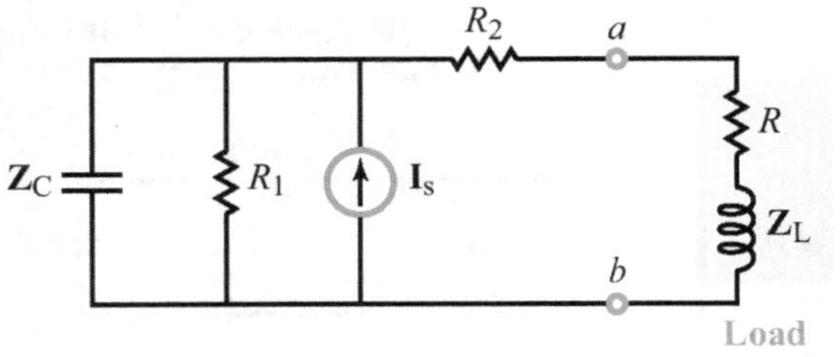

Figure P8.27: Circuit for Problem 8.27.

8.28 In the phasor-domain circuit shown in Fig. P8.28, $\mathbf{I} = 2\angle 0°$ A, $R_1 = 20$ Ω, $R_2 = 1$ Ω, $R_3 = 5$ Ω, $\mathbf{Z}_{L1} = j5$ Ω, $\mathbf{Z}_{L2} = j25$ Ω, and $\mathbf{Z}_C = -20$ Ω. Determine **S** of the load.

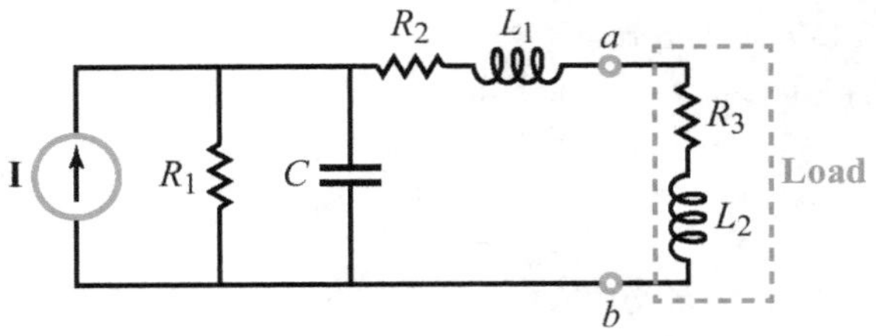

Figure P8.28: Circuit for Problem 8.28.

*8.29 In the phasor-domain circuit shown in Fig. P8.29, $\mathbf{V} = 100\angle 0°$ V, $R_1 = 1$ kΩ, $R_2 = 0.6$ kΩ, $R_L = 3$ kΩ, $\mathbf{Z}_L = j0.8$ kΩ, and $\mathbf{Z}_C = -j0.5$ kΩ. Determine the average power dissipated in R_L.

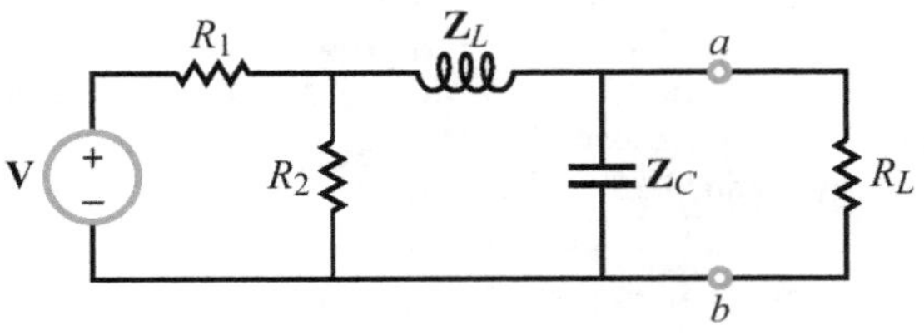

Figure P8.29: Circuit for Problem 8.29.

8.30 In the circuit shown in Fig. P8.30, $i(t) = 3\cos(1000t)$ A, $R_1 = 2$ kΩ, $R_2 = 560$ Ω, $R_L = 2$ kΩ, and $L = 0.4$ H. Determine the average power dissipated in R_L.

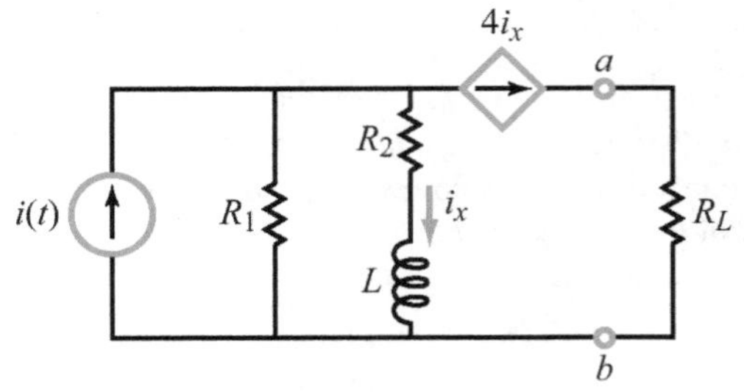

Figure P8.30: Circuit for Problem 8.30.

*8.31 In the phasor-domain circuit shown in Fig. P8.31, $\mathbf{V} = 15\angle 45°$ V, $R_1 = 5$ Ω, $R_2 = 2$ Ω, $\mathbf{Z}_C = -j1$ Ω, $\mathbf{Z}_L = j2$ Ω, and $\mathbf{Z}_{\text{load}} = 6 + j4$ Ω. Determine the complex power of the load.

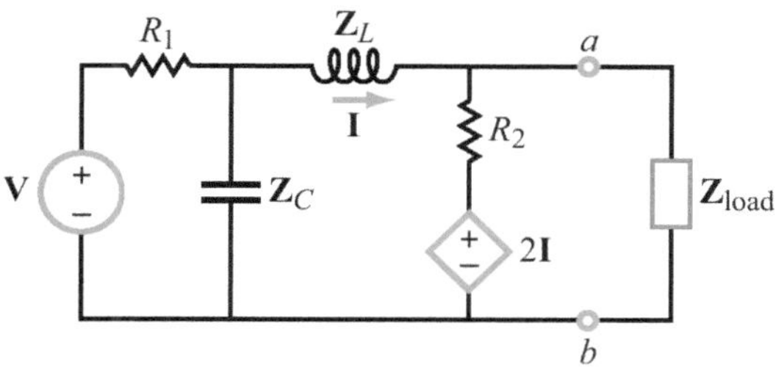

Figure P8.31: Circuit for Problem 8.31.

8.32 In the op-amp circuit shown in Fig. P8.32, $v_{\text{in}}(t) = 12\cos(1000t)$ V, $R = 10$ kΩ, $R_{\text{L}} = 5$ kΩ, and $C = 1$ μF. Determine the complex power for each of the passive elements in the circuit. Is conservation of energy satisfied?

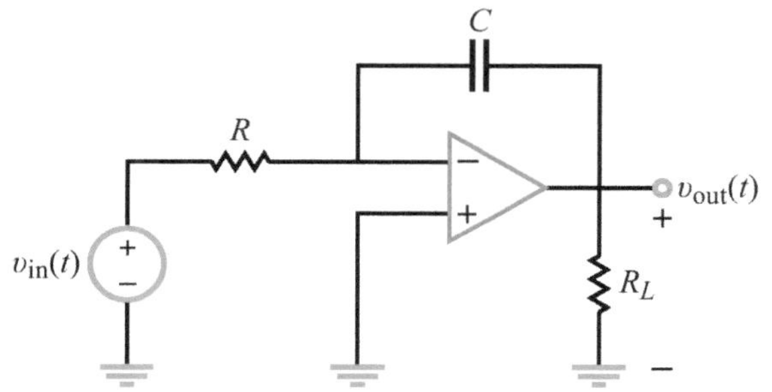

Figure P8.32: Op-amp circuit for Problem 8.32.

*8.33 In the phasor-domain op-amp circuit shown in Fig. P8.33, $\mathbf{V}_{\text{in}} = 2\angle 0^\circ$ V, $R_1 = 200$ Ω, $R_2 = 2.4$ kΩ, $R_{\text{L}} = 10$ kΩ, $\mathbf{Z}_{\text{C}} = -j500$ Ω, and $\mathbf{Z}_{\text{L}} = j1$ kΩ. Determine the average power delivered to R_{L}.

8.34 In the phasor-domain op-amp circuit shown in Fig. P8.34, $v(t) = 8\cos(200t)$ V, $R_1 = 5$ kΩ, $R_{\text{L}} = 2$ kΩ, $C_1 = 1$ μF, and $C_2 = 4.7$ μF. Determine the average power delivered to R_{L}.

8.35 Determine the power dissipated in R_{L} of the circuit in Fig. P8.35.

*8.36 Determine the power dissipated in R_{L} of the circuit in Fig. P8.36.

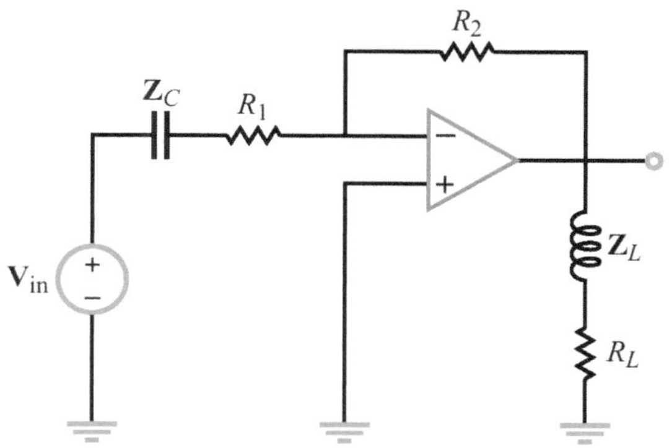

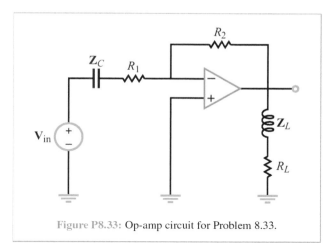

Figure P8.33: Op-amp circuit for Problem 8.33.

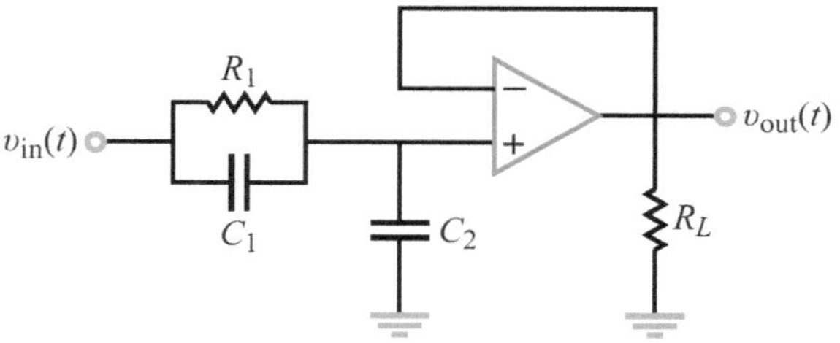

Figure P8.34: Op-amp circuit for Problem 8.34.

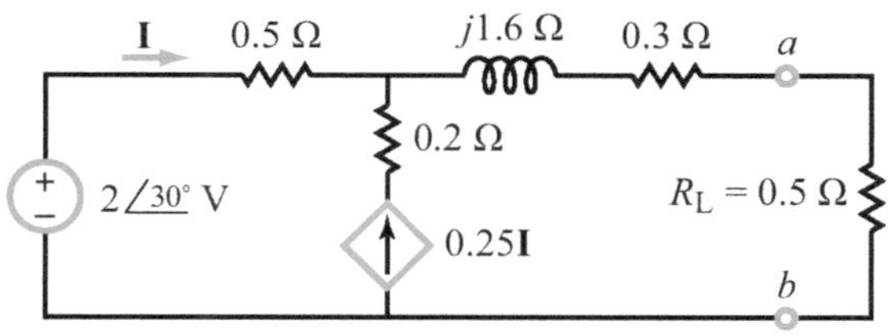

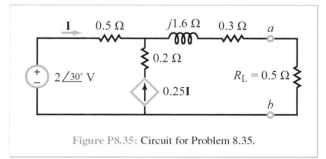

Figure P8.35: Circuit for Problem 8.35.

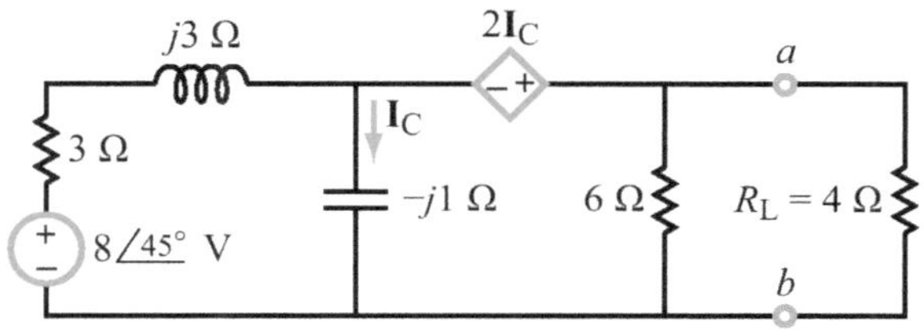

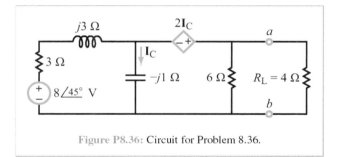

Figure P8.36: Circuit for Problem 8.36.

8.37 In the op-amp circuit of Fig. P8.37,

$$v_{\text{in}}(t) = V_0 \cos \omega t \text{ V},$$

with $V_0 = 10$ V, $\omega RC = 1$, and $R_L = 10$ kΩ. Determine the power delivered to R_L.

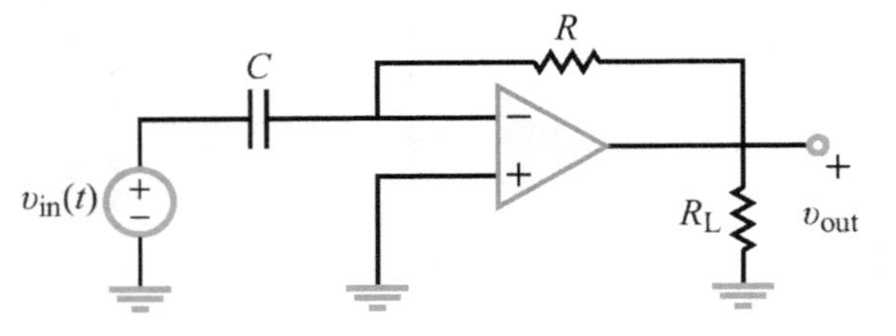

Figure P8.37: Op-amp circuit for Problem 8.37.

8.38 Determine the amount of power delivered to R_L in the circuit of Fig. P8.38, given that $v_{\text{in}}(t) = 0.5 \cos 2000t$ V, $R_1 = 1$ kΩ, $R_2 = 10$ kΩ, $C = 0.1$ μF, $R_L = 1$ kΩ, and $L = 0.2$ H.

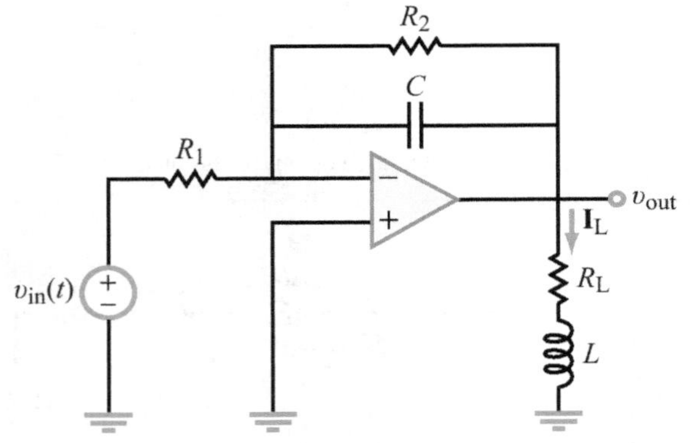

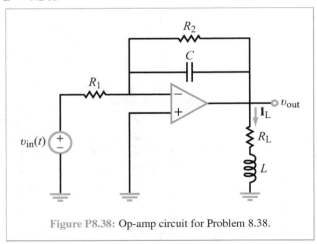

Figure P8.38: Op-amp circuit for Problem 8.38.

*__8.39__ Given that $v_s(t) = 2\cos 10^3 t$ V in the circuit of Fig. P8.39, determine the power delivered to R_L.

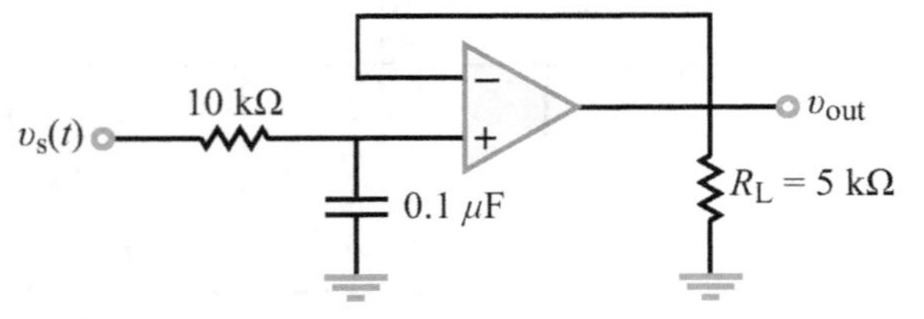

Figure P8.39: Circuit for Problem 8.39.

8.40 The apparent power entering a certain load **Z** is 250 VA at a power factor of 0.8 leading. If the rms phasor voltage of the source is 125 V at 1 MHz:

(a) Determine $\mathbf{I}_{\text{rms}}$ going into the load

(b) Determine **S** into the load

(c) Determine **Z**

(d) The equivalent impedance of the load circuit should be of the form $\mathbf{Z} = R + j\omega L$ or $\mathbf{Z} = R - j/\omega C$. Determine the value of L or C, whichever is applicable.

8.41 Voltage source $\mathbf{V}_s$ in the circuit of Fig. P8.41 supplies power to three load circuits with impedances $\mathbf{Z}_1$, $\mathbf{Z}_2$, and $\mathbf{Z}_3$. The following partial power information was deduced from measurements performed on the three load circuits:

Load $\mathbf{Z}_1$:	80 W at $pf = 0.8$ lagging
Load $\mathbf{Z}_2$:	60 VA at $pf = 0.7$ leading
Load $\mathbf{Z}_3$:	40 VA at $pf = 0.6$ leading

If $\mathbf{I}_{\text{rms}} = 0.4\angle 37^\circ$ A, determine:

(a) the rms value of $\mathbf{V}_s$ by applying the law of conservation of energy

(b) $\mathbf{Z}_1$, $\mathbf{Z}_2$, and $\mathbf{Z}_3$.

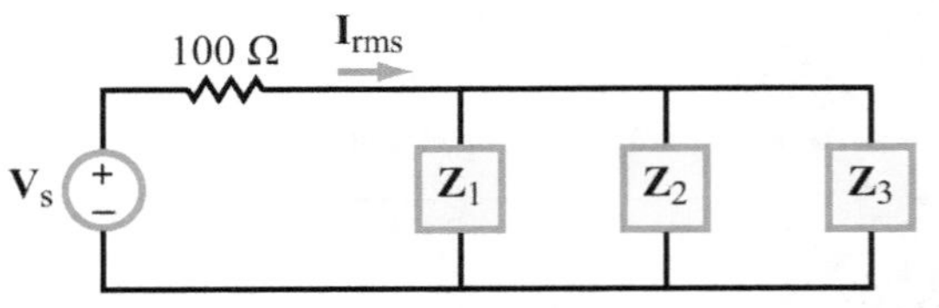

Figure P8.41: Circuit for Problem 8.41.

8.42 The apparent power entering a certain load **Z** is 120 VA at a power factor of 0.866 lagging. If the rms phasor voltage is 240 V at 60 Hz:

(a) Determine **S** into the load

(b) Determine $\mathbf{I}_{\text{rms}}$ going into the load

(c) Determine **Z**

(d) The equivalent impedance of the load circuit should be of the form $\mathbf{Z} = R + j\omega L$ or $\mathbf{Z} = R + 1/j\omega C$. Determine the value of L or C, whichever is applicable.

*8.43 In the circuit in Fig. P8.43, voltage source $\mathbf{V}_s$ supplies power to three load circuits with impedances $\mathbf{Z}_1$, $\mathbf{Z}_2$, and $\mathbf{Z}_3$. The following information was deduced from measurements performed on the three load circuits:

Load $\mathbf{Z}_1$:	100 VA at $pf = 0.6$ lagging
Load $\mathbf{Z}_2$:	70 VA at $pf = 0.75$ leading
Load $\mathbf{Z}_3$:	45 W at $pf = 0.95$ lagging

If $\mathbf{V}_s = 100\angle 0^\circ$ V, determine the equivalent impedance. Is it inductive or capacitive? Are $\mathbf{Z}_1$, $\mathbf{Z}_2$, and $\mathbf{Z}_3$ inductive or capacitive?

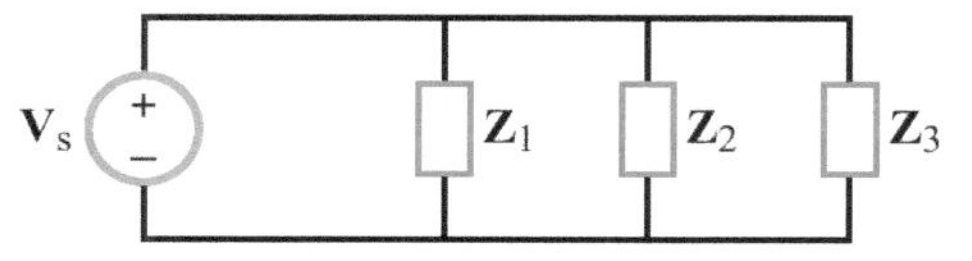

Figure P8.43: Circuit for Problem 8.43.

8.44 In the circuit shown in Fig. P8.44, voltage source $\mathbf{V}_s$ supplies power to three load circuits with impedances $\mathbf{Z}_1$, $\mathbf{Z}_2$, and $\mathbf{Z}_3$. The following information was deduced from measurements performed on the three load circuits.

Load $\mathbf{Z}_1$:	60 VA at $pf = 0.866$ lagging
Load $\mathbf{Z}_2$:	80 W at $pf = 0.750$ leading
Load $\mathbf{Z}_3$:	100 VAR at $pf = 0.600$ leading

If $\mathbf{I}_{rms} = 0.5\angle 45^\circ$ and $R = 100\ \Omega$, determine:

(a) the rms value of $\mathbf{V}_s$, by applying the law of conservation of energy

(b) $\mathbf{Z}_1$, $\mathbf{Z}_2$, and $\mathbf{Z}_3$.

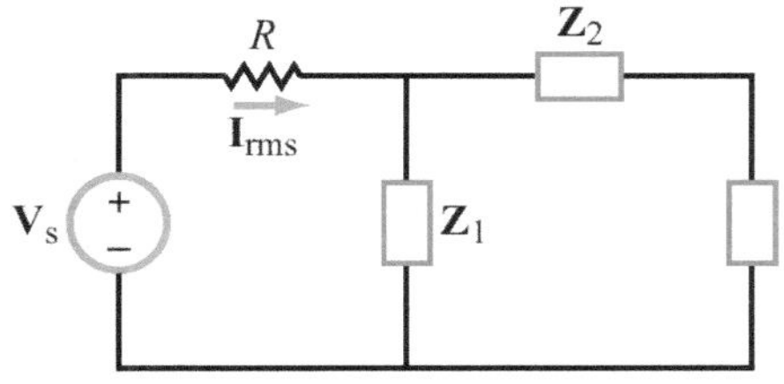

Figure P8.44: Circuit for Problem 8.44.

Section 8-4: Power Factor

*8.45 The RL load in Fig. P8.45 is compensated by adding the shunt capacitance C so that the power factor of the combined (compensated) circuit is exactly unity. How is C related to R, L, and ω in that case?

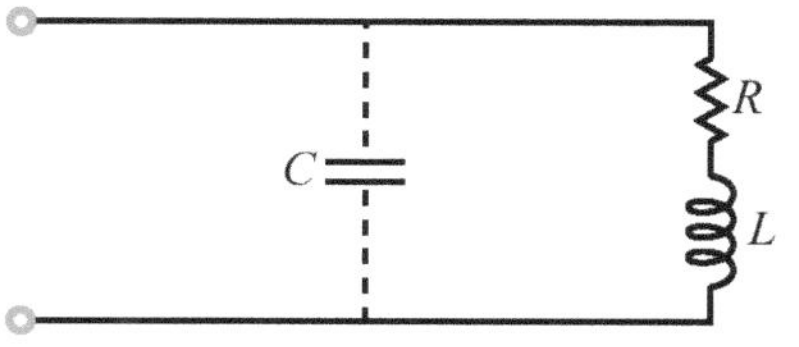

Figure P8.45: Circuit for Problem 8.45.

8.46 The generator circuit shown in Fig. P8.46 (see page 493) is connected to a distant load via a long coaxial transmission line. The overall circuit can be modeled as in Fig. P8.46(b), in which the transmission line is represented by an equivalent impedance $\mathbf{Z}_{line} = (5 + j2)\ \Omega$.

(a) Determine the power factor of voltage source $\mathbf{V}_s$.

(b) Specify the capacitance of a shunt capacitor C that would raise the power factor of the source to unity when connected between terminals (a, b). The source frequency is 1.5 kHz.

8.47 Source $\mathbf{V}_s$ in the circuit of Fig. P8.47 is connected to two industrial loads, with equivalent impedances $\mathbf{Z}_1$ and $\mathbf{Z}_2$, via two identical transmission lines, each characterized by an equivalent impedance $\mathbf{Z}_{line} = (0.5 + j0.3)\ \Omega$. If $\mathbf{Z}_1 = (8 + j12)\ \Omega$ and $\mathbf{Z}_2 = (6 + j3)\ \Omega$:

(a) Determine the power factors for $\mathbf{Z}_1$, $\mathbf{Z}_2$, and source $\mathbf{V}_s$.

(b) Specify the capacitance of a shunt capacitor C that would raise the power factor of the source to 0.95 when connected between terminals (a, b). The source frequency is 12 kHz.

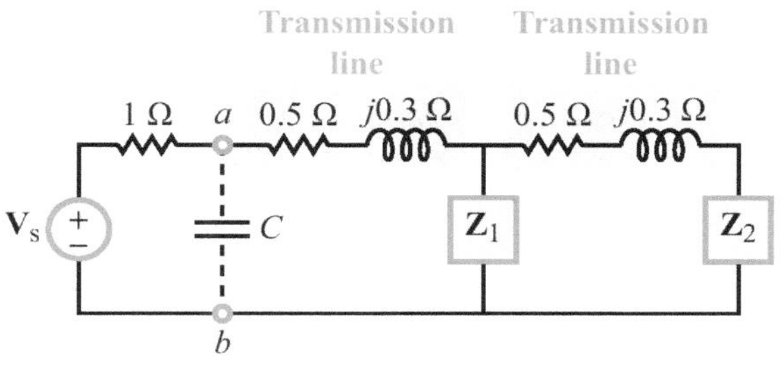

Figure P8.47: Circuit for Problem 8.47.

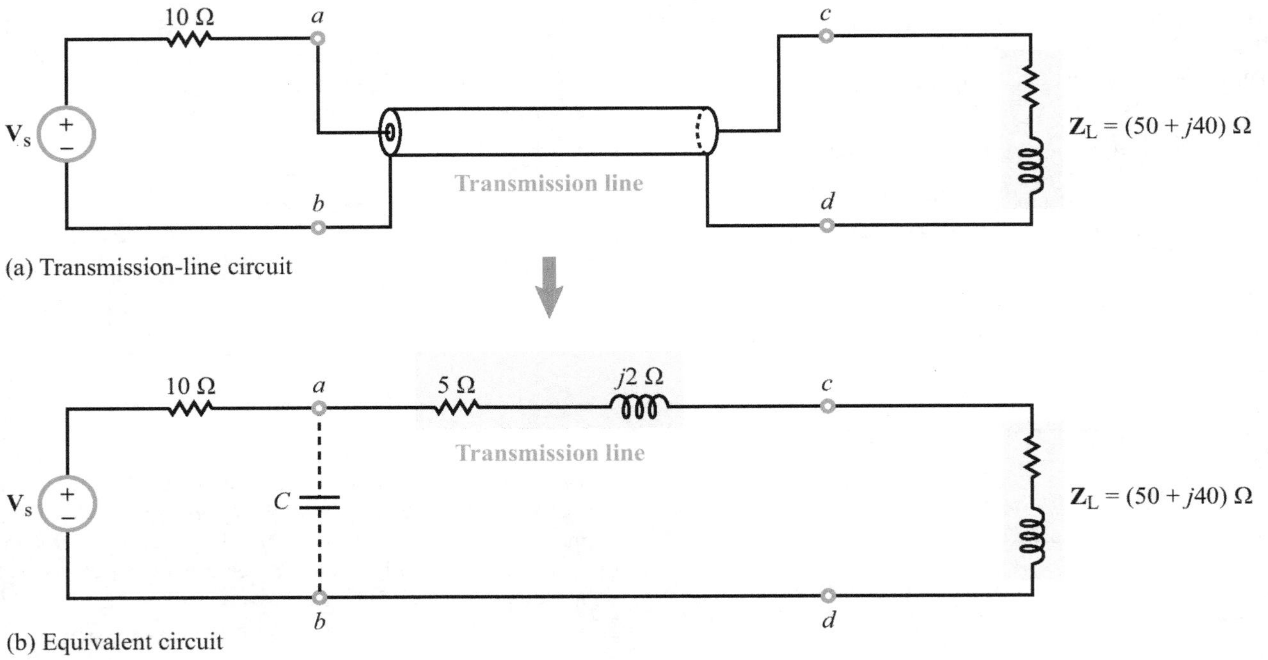

Figure P8.46: Circuit for Problem 8.46.

*8.48 Use the power information given for the circuit in Fig. P8.48 to determine:

(a) $\mathbf{Z}_1$ and $\mathbf{Z}_2$

(b) the rms value of $\mathbf{V}_s$.

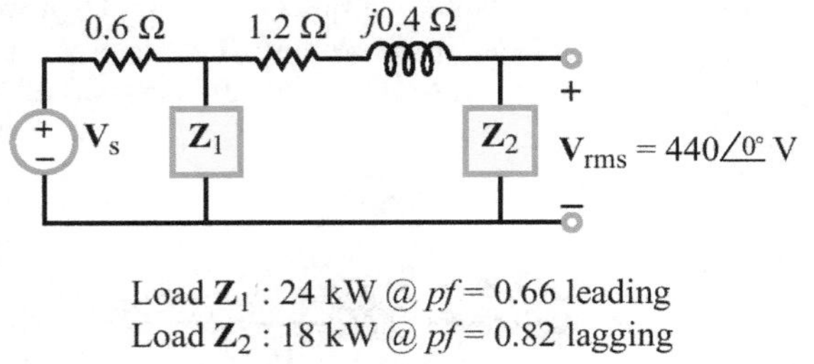

Figure P8.48: Circuit for Problem 8.48.

8.49 In the circuit shown in Fig. P8.49, a generator is connected to a load via a transmission line. Given that $R_s = 10$ Ω, $\mathbf{Z}_{\text{line}} = (4 + j2)$ Ω, and $\mathbf{Z}_{\text{load}} = (40 + j30)$ Ω:

(a) Determine the power factor of the load, the power factor of the transmission line, and the power factor of the voltage source.

(b) Specify the capacitance of a shunt capacitor C that would raise the power factor of the source to unity when connected between terminals (a, b). The source frequency is 60 Hz.

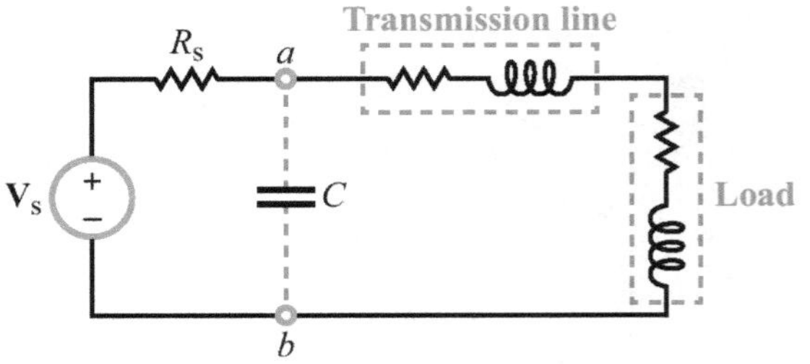

Figure P8.49: Circuit for Problem 8.49.

8.50 In the phasor-domain circuit shown in Fig. P8.50, $R_1 = 1.2$ kΩ, $R_2 = 5$ kΩ, $L_1 = 0.8$ H, $L_1 = 0.6$ H, $C = 2$ μF, and $\omega = 1000$ rad/s.

*(a) Determine the power factor of the voltage source **V**.

(b) Specify the capacitance of a shunt capacitor C_{shunt} that would raise the power factor at terminals (a, b) to 0.9 when connected between terminals (a, b).

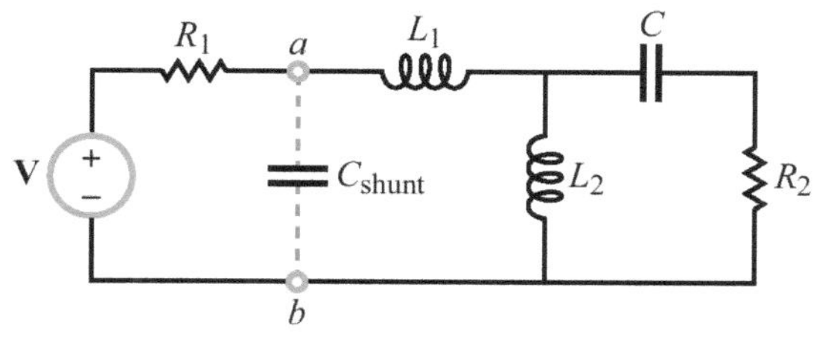

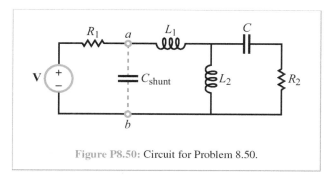

Figure P8.50: Circuit for Problem 8.50.

Section 8-5: Maximum Power Transfer

8.51 For the circuit in Fig. P8.51, choose the load impedance $\mathbf{Z}_L$ so that the power dissipated in it is a maximum. How much power will that be?

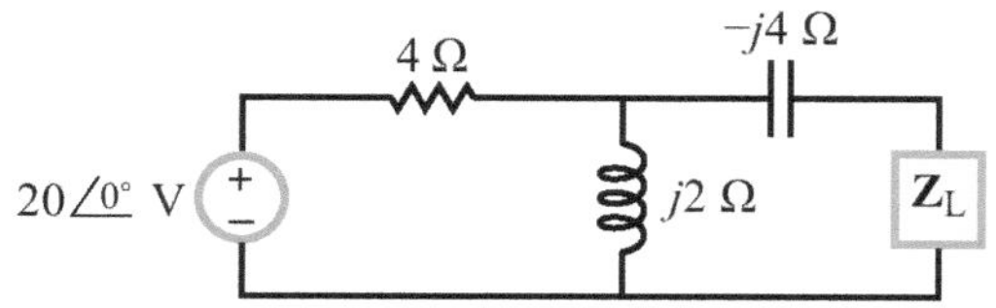

Figure P8.51: Circuit for Problem 8.51.

*8.52 For the circuit in Fig. P8.52, choose the load impedance $\mathbf{Z}_L$ so that the power dissipated in it is a maximum. How much power will that be?

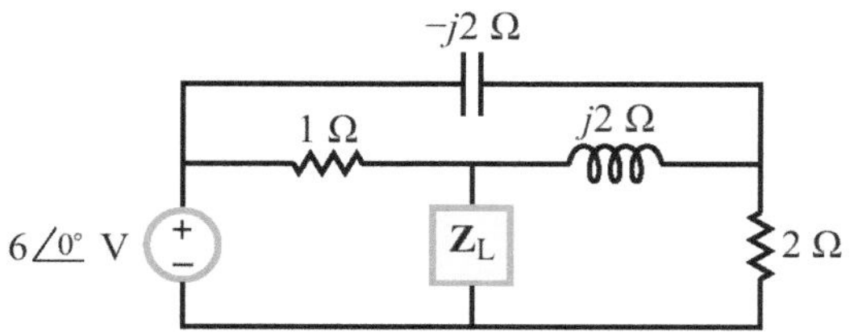

Figure P8.52: Circuit for Problem 8.52.

8.53 For the circuit in Fig. P8.53, choose the load impedance $\mathbf{Z}_L$ so that the power dissipated in it is a maximum. How much power will that be?

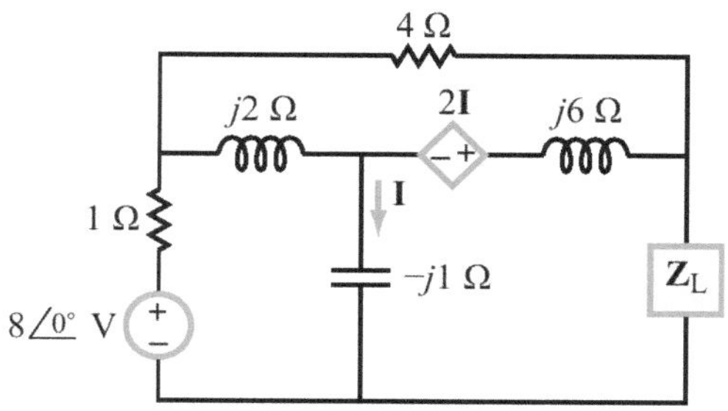

Figure P8.53: Circuit for Problem 8.53.

8.54 For the circuit in Fig. P8.54, choose the load impedance $\mathbf{Z}_L$ so that the power dissipated in it is a maximum. How much power will that be?

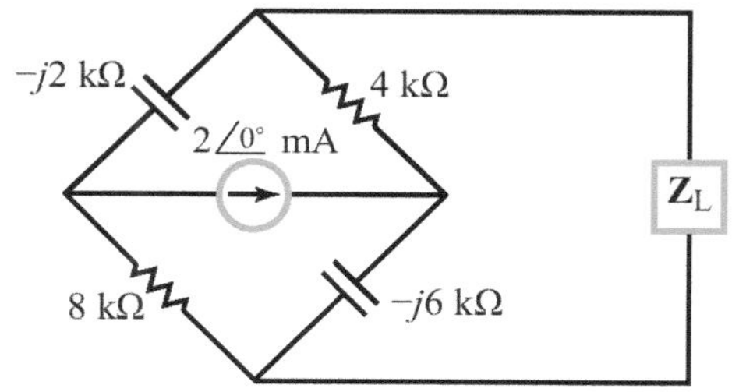

Figure P8.54: Circuit for Problem 8.54.

*8.55 For the circuit in Fig. P8.55, choose the load impedance $\mathbf{Z}_L$ so that the power dissipated in it is a maximum. How much power will that be?

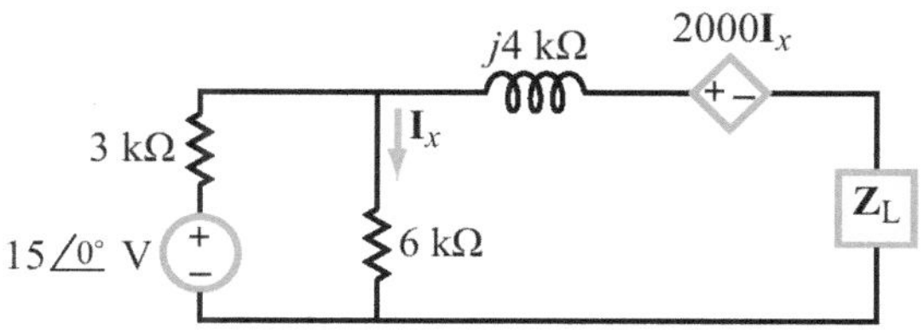

Figure P8.55: Circuit for Problem 8.55.

8.56 In the phasor-domain shown in Fig. P8.56, $\mathbf{V} = 20\underline{/0^\circ}$ V, $R_1 = 10\ \Omega$, $R_2 = 5\ \Omega$, $\mathbf{Z}_C = -j5\ \Omega$, and $\mathbf{Z}_L = j3\ \Omega$. Choose the load impedance $\mathbf{Z}_{\text{load}}$ so that the power dissipated in it is a maximum. How much power will that be?

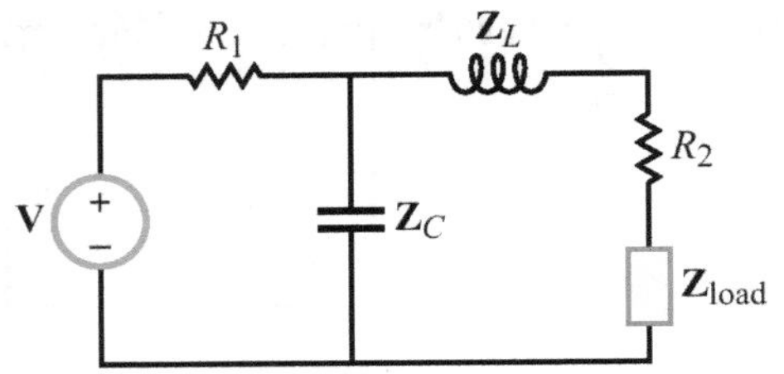

Figure P8.56: Circuit for Problem 8.56.

8.57 In the phasor-domain circuit shown in Fig. P8.57, $\mathbf{V} = 30\underline{/60^\circ}$ V, $R_1 = 5\ \Omega$, $R_2 = 20\ \Omega$, $R_3 = 10\ \Omega$, $\mathbf{Z}_C = -j4\ \Omega$, and $\mathbf{Z}_L = j6\ \Omega$. Choose the load impedance $\mathbf{Z}_{\text{load}}$ so that the power dissipated in it is a maximum. How much power will that be?

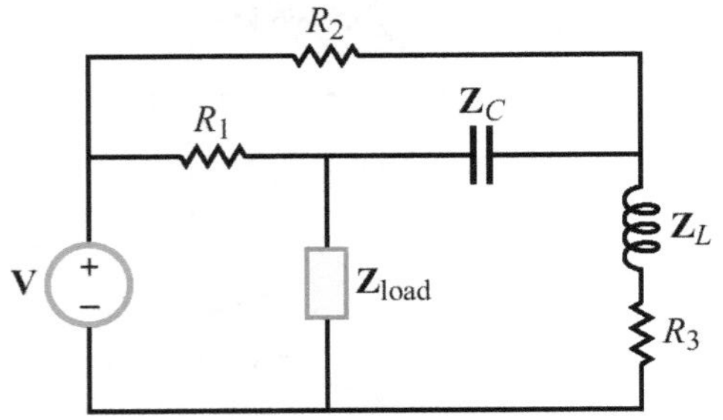

Figure P8.57: Circuit for Problem 8.57.

*__8.58__ In the phasor-domain circuit shown in Fig. P8.58, $\mathbf{V} = 6\underline{/0^\circ}$ V, $R_1 = 1\ \Omega$, $R_2 = 2\ \Omega$, $R_3 = 5\ \Omega$, $\mathbf{Z}_{L1} = j2\ \Omega$, $\mathbf{Z}_{L2} = j5\ \Omega$, and $\mathbf{Z}_C = -j6\ \Omega$. Choose the load impedance $\mathbf{Z}_{\text{load}}$ so that the power dissipated in it is a maximum. How much power will that be?

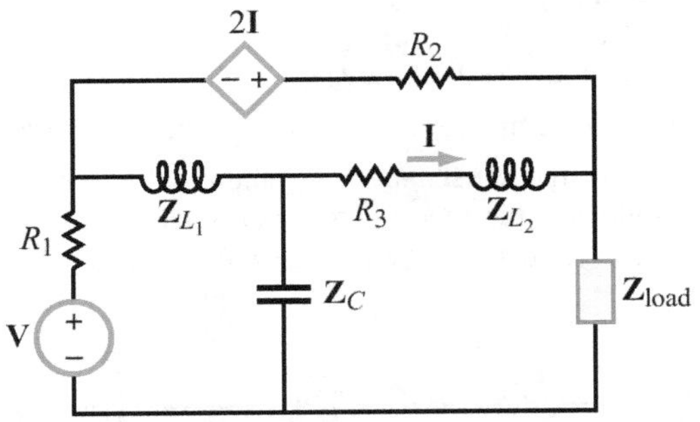

Figure P8.58: Circuit for Problem 8.58.

8.59 In the phasor-domain circuit shown in Fig. P8.59, $\mathbf{V} = 12\underline{/0^\circ}$ V, $R_1 = 2$ kΩ, $R_2 = 4$ kΩ, $\mathbf{Z}_{L1} = j2$ kΩ, and $\mathbf{Z}_{L2} = j5$ kΩ. Choose the load impedance $\mathbf{Z}_{\text{load}}$ so that the power dissipated in it is a maximum. How much power will that be?

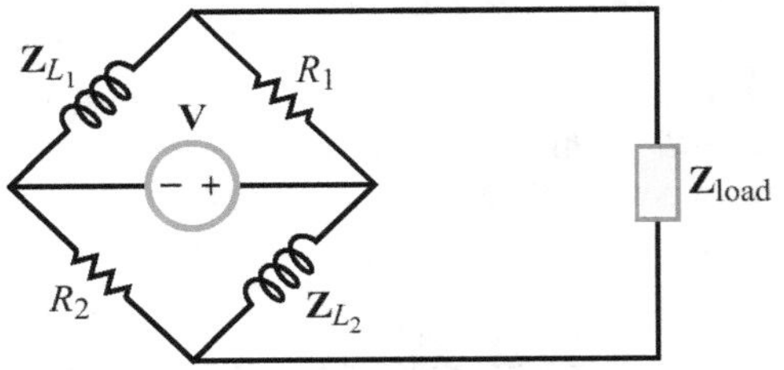

Figure P8.59: Circuit for Problem 8.59.

*__8.60__ In the phasor-domain circuit shown in Fig. P8.60, $\mathbf{I} = 2.5\underline{/0^\circ}$ mA, $R_1 = 10$ kΩ, $R_2 = 2.4$ kΩ, $R_3 = 10$ kΩ, and $\mathbf{Z}_C = -j4$ kΩ. Choose the load impedance $\mathbf{Z}_{\text{load}}$ so that the power dissipated in it is a maximum. How much power will that be?

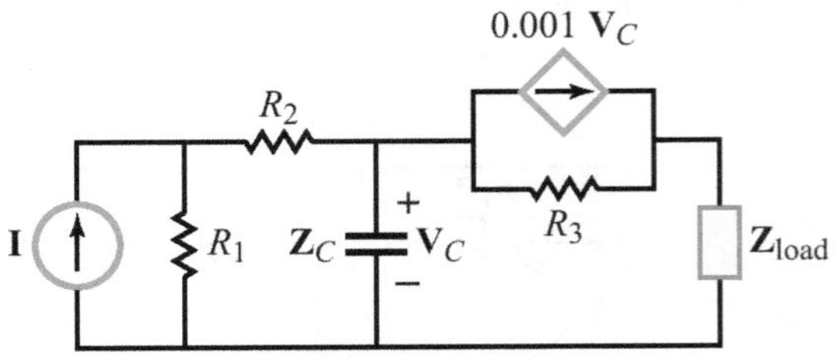

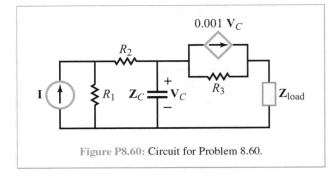

Figure P8.60: Circuit for Problem 8.60.

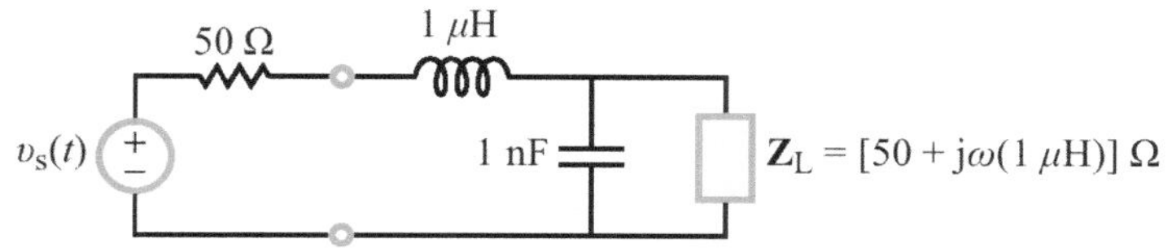

Figure P8.61: Circuit for Problem 8.61.

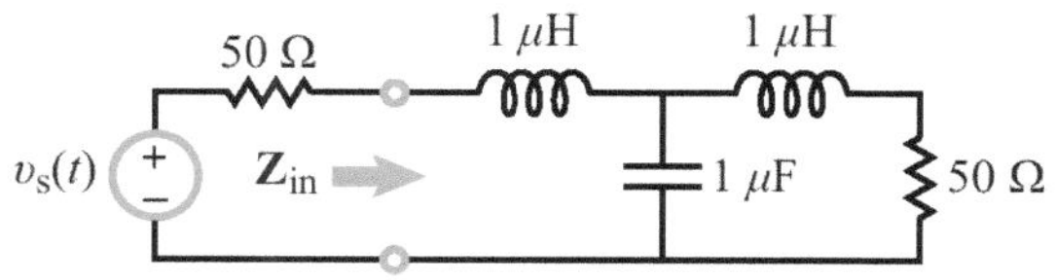

Figure P8.62: Circuit for Problems 8.62 and 8.65.

Section 8-6: Multisim

8.61 Model the circuit in Fig. P8.61 in Multisim and plot the complex power through the load $\mathbf{Z}_L$ as a function of frequency from 1 kHz to 1 GHz. Assume $v_s(t)$ has an amplitude of 1 V.

8.62 Model the circuit in Fig. P8.62 in Multisim and find the frequency at which the input impedance of the load circuit $\mathbf{Z}_{in}$ is purely real. Assume $v_s(t)$ has an amplitude of 1 V.

8.63 Model the circuit in Fig. P8.63 in Multisim and find the frequency at which the input impedance of the load circuit $\mathbf{Z}_{in}$ is purely real.

8.64 Model the circuit in Fig. P8.64 and use the wattmeter to determine the average power consumed by the load $\mathbf{Z}_L$. Also, perform an AC Analysis from 100 kHz to 1 GHz and show that the average power value given by the AC Analysis at 1 MHz matches the value provided by the wattmeter.

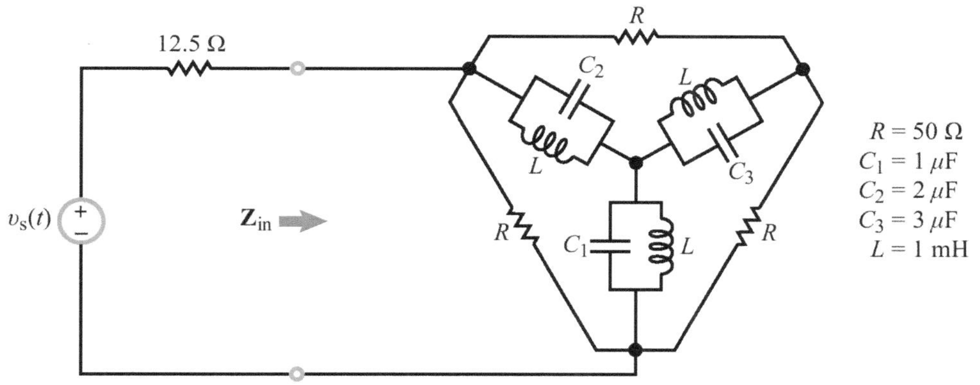

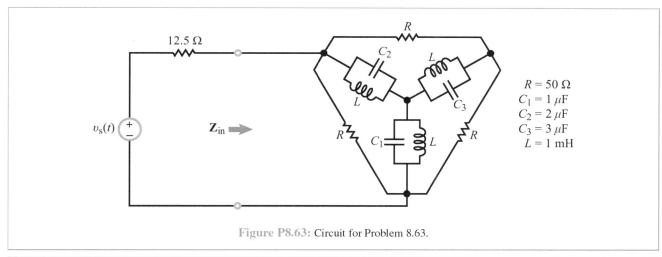

Figure P8.63: Circuit for Problem 8.63.

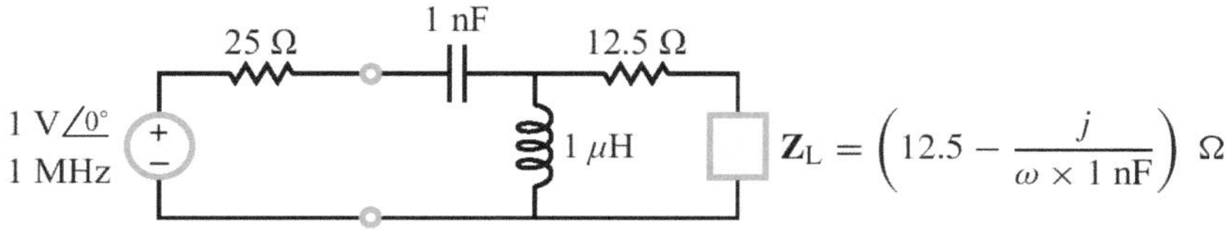

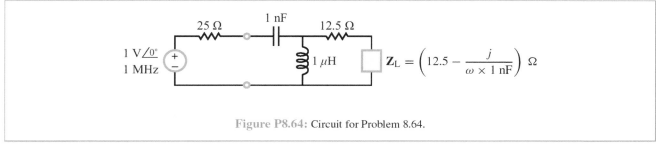

Figure P8.64: Circuit for Problem 8.64.

8.65 Plot the power factor and phase angle ϕ_z across the load $\mathbf{Z}_{in}$ in Fig. P8.62 using AC Analysis in Multisim from 1 kHz to 1 MHz. (See Multisim Demo 8.3 in the Tutorial for help on how to do this.)

Potpourri Questions

8.66 What is the wavelength range of visible light? The corresponding frequency range?

8.67 What is the frequency range assigned to Bluetooth communication?

8.68 What does "thermal" imaging refer to? How is ultraviolet light used in skin treatment?

Integrative Problems: Analytical / Multisim / myDAQ

To master the material in this chapter, solve the following problems using three complementary approaches: (a) analytically, (b) with Multisim, and (c) by constructing the circuit and using the myDAQ interface unit to measure quantities of interest via your computer. [myDAQ tutorials and videos are available on CAD.]

m8.1 Periodic Waveforms: Plot the half-wave rectifier output and full-wave rectifier output for each of the three standard waveforms shown in Fig. m8.1. Then:

(a) Determine the general expressions for the (1) average value and (2) rms value for each of the six rectified waveforms.

(b) Evaluate your expressions for average and rms values for $V_m = 10$ V and $T = 10$ ms.

m8.2 Average Power: The circuit shown in Fig. m8.2 operates in sinusoidal steady state at 1500 Hz. The voltage source amplitude is 3 V. Find the average power delivered by the source. Use these component values: $R = 100\ \Omega$, $C = 1.0\ \mu$F, and $L = 3.3$ mH.

m8.3 Complex Power: The circuit shown in Fig. m8.3 operates in sinusoidal steady state at 1000 Hz. The voltage source amplitude is 2.5 V. Component values are: $R = 100\ \Omega$, $C = 1.0\ \mu$F, and $L = 3.3$ mH.

(a) Find the complex power in rectangular format for each of the four circuit elements: $\mathbf{S}_{src}$, $\mathbf{S}_R$, $\mathbf{S}_L$, and $\mathbf{S}_C$.

(b) Demonstrate conservation of complex power with these four values.

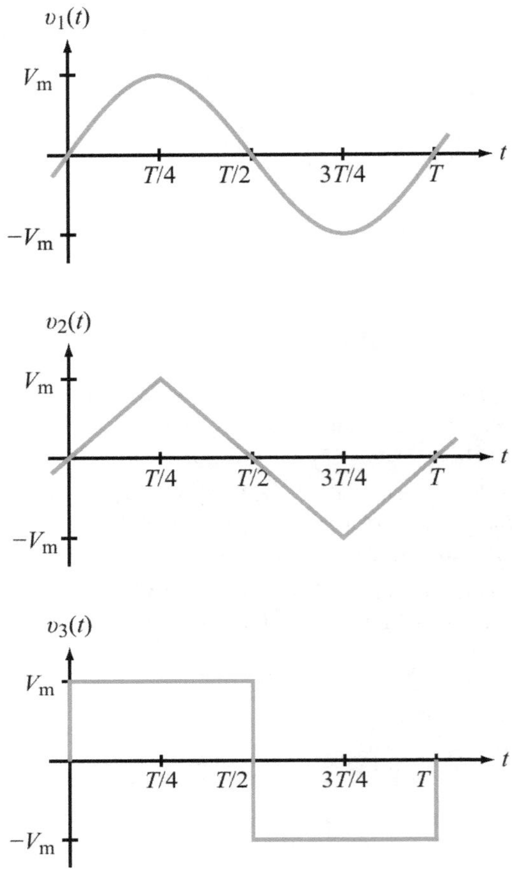

Figure m8.1 Circuit for Problem m8.1.

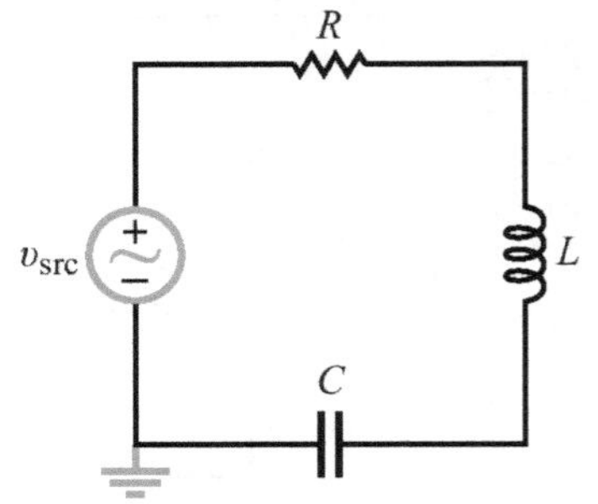

Figure m8.2 Circuit for Problem m8.2.

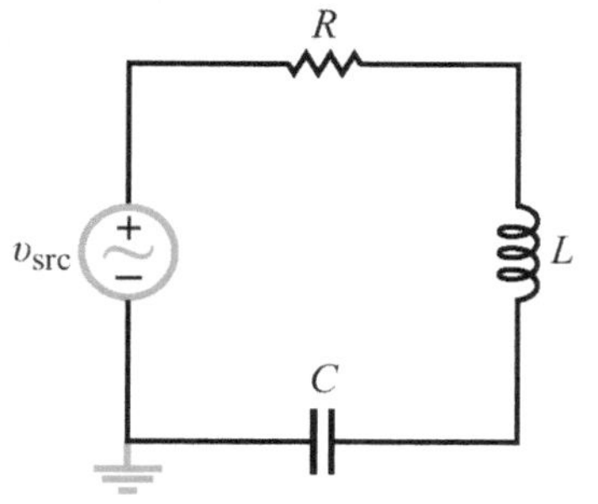

Figure m8.3 Circuit for Problem m8.3.

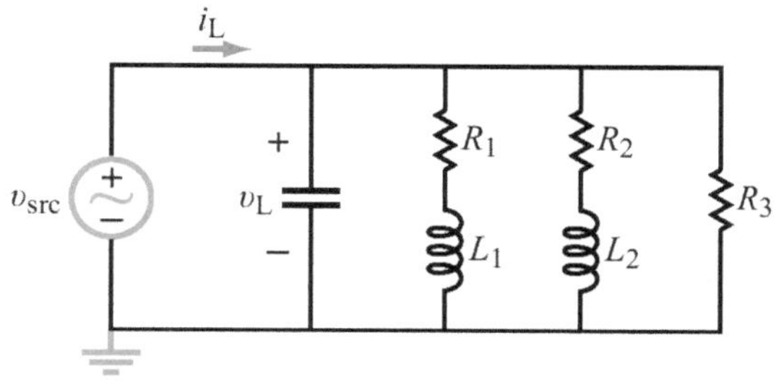

Figure m8.4 Circuit for Problem m8.4.

m8.4 The Power Factor: The circuit shown in Fig. m8.4 is a "scale model" of two industrial electric motors and a heating unit connected to a manufacturing plant power distribution network. The resistor/inductor combinations, R_1-L_1 and R_2-L_2, model the winding resistance and magnetic fields of the motors. Resistor R_3 models the heater coils. C represents the power factor compensation equipment—essentially a capacitor bank with high power capacity.

(a) Determine the power factor of the uncompensated load, and draw its power triangle to scale.

(b) Determine the value of the compensation capacitor C required to improve the load power factor to 0.90 lagging.

(c) Available power factor compensation capacitors include 0.1 μF, 1.0 μF, and 10 μF; the cost of compensation equipment increases with capacitance. Choose the least expensive compensation capacitor closest to C and then determine the power factor and power triangle (also drawn to scale) of the compensated load.

Component values are: $R_1 = 10\ \Omega$, $R_2 = 100\ \Omega$, $R_3 = 100\ \Omega$, $L_1 = 3.3$ mH, $L_2 = 33$ mH, and $V_{src} = 1$ V at 2500 Hz (actual industrial motors operate at hundreds of volts and 50 Hz to 60 Hz).

CHAPTER 9

Frequency Response of Circuits and Filters

Contents

Objectives

Learn to:

- Derive the transfer function of an ac circuit.
- Generate magnitude and phase spectral plots.
- Design first-order lowpass, highpass, bandpass, and bandreject filters.
- Generate Bode plots for any transfer function.
- Design active filters.
- Apply Multisim to generate spectral responses for passive and active circuits.

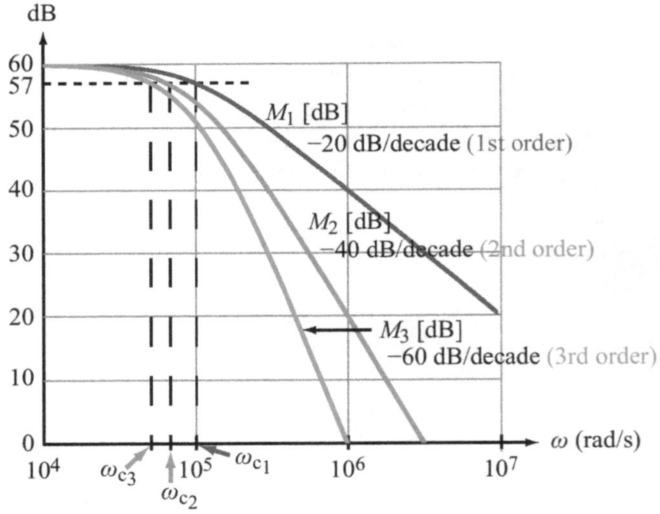

Frequency filters are used to suppress noise, remove interfering signals, and to channel multifrequency signals along their intended paths.

Overview

In Chapter 7 we learned how to analyze an ac circuit excited by an individual source at a particular frequency f. Often, the input signal is a superposition of many sinusoidal signals at different frequencies. A good analogue is sunlight incident on an eye's pupil. The light consists of many ac-like signals at many frequencies, extending from below the frequency of violet to beyond that of red. The pupil is like a receiver that detects the incident light, but it is also a *filter* because it detects only a portion of the spectrum incident upon it, while rejecting other bands such as the ultraviolet and infrared. Additional filtering can be effected through the use of tinted sunglasses, or similar optical filters. In digital photography, filtering can be performed in software during post-processing (using Photoshop™, for example) to enhance certain colors of interest over others. This chapter is about how to design RLC circuits that can *filter in* (pass through) the range of frequencies of interest (to a certain application, such as in a communication, imaging, or sensing system) and *filter out* (reject) the range of frequencies of signals that are either problematic or not of interest.

To avoid interference, every radio and TV transmission station is assigned a unique transmission frequency different from those assigned to other radio and TV stations in the area. At the receiver end, even though the antenna will intercept the signals transmitted by all sources within a certain distance, the receiver is able to select from among them the specific channel of interest, while rejecting all others. The selection process is based on the oscillation frequency of the desired signal, and it is realized by passing the intercepted signals through a narrow *bandpass filter* whose center frequency is aligned with the frequency of the desired channel. The bandpass filter is one of many different types of *frequency-selective circuits* employed in analog and digital communication networks to manage the traffic of signals between multiple sources and multiple recipients. The behavior of an ac circuit as a function of the angular frequency ω is called its *frequency response*. Building on the phasor-domain analysis tools we acquired in the preceding two chapters, we now are ready to develop and adopt a standard set of metrics and design methodologies for characterizing the frequency response of any resonant circuit and to apply them to various types of active and passive circuits.

9-1 The Transfer Function

The passive linear circuit represented by the block diagram in Fig. 9-1 has an input phasor voltage $\mathbf{V}_{\text{in}}$ applied at input terminals (a, b), causing an associated input phasor current $\mathbf{I}_{\text{in}}$ to flow into the circuit. In general, a corresponding set of phasors, $\mathbf{V}_{\text{out}}$ and $\mathbf{I}_{\text{out}}$, exist at output terminals (c, d). The voltage gain of the circuit is defined as

$$\text{Voltage gain:} \quad \mathbf{H}(\omega) = \frac{\mathbf{V}_{\text{out}}(\omega)}{\mathbf{V}_{\text{in}}(\omega)}, \tag{9.1}$$

where all quantities are written explicitly as functions of the angular frequency ω simply to emphasize the notion that ω will play a central role in our forthcoming discussions. If the circuit contains capacitors and inductors, $\mathbf{V}_{\text{out}}$ likely will be a function of ω, and in the general case $\mathbf{V}_{\text{in}}$ may vary with ω also. The phasor $\mathbf{H}(\omega)$ is called the *voltage transfer function* of the circuit and carries a connotation broader than just another name for voltage gain. In fact, $\mathbf{H}(\omega)$ can be defined to convey the relationship between any input excitation and any output response. For example, we may define other transfer functions for the circuit in Fig. 9-1, such as:

$$\text{Current gain:} \quad \mathbf{H}_I(\omega) = \frac{\mathbf{I}_{\text{out}}(\omega)}{\mathbf{I}_{\text{in}}(\omega)}, \tag{9.2a}$$

$$\text{Transfer impedance:} \quad \mathbf{H}_\mathbf{Z}(\omega) = \frac{\mathbf{V}_{\text{out}}(\omega)}{\mathbf{I}_{\text{in}}(\omega)}, \tag{9.2b}$$

and

$$\text{Transfer admittance:} \quad \mathbf{H}_\mathbf{Y}(\omega) = \frac{\mathbf{I}_{\text{out}}(\omega)}{\mathbf{V}_{\text{in}}(\omega)}. \tag{9.2c}$$

In any case, because $\mathbf{H}(\omega)$ always is defined as the ratio of an output quantity to an input quantity, we may think of it as equal to the output generated by the circuit in response to a *unity input* ($1\angle 0^\circ$).

As a complex quantity, the transfer function $\mathbf{H}(\omega)$ has a *magnitude*—to which we assign the symbol $M(\omega)$—and an associated *phase angle* $\phi(\omega)$,

$$\mathbf{H}(\omega) = M(\omega)\, e^{j\phi(\omega)}, \tag{9.3}$$

where by definition,

$$M(\omega) = |\mathbf{H}(\omega)| \quad \text{and} \quad \phi(\omega) = \tan^{-1}\left\{\frac{\mathfrak{Im}[\mathbf{H}(\omega)]}{\mathfrak{Re}[\mathbf{H}(\omega)]}\right\}. \tag{9.4}$$

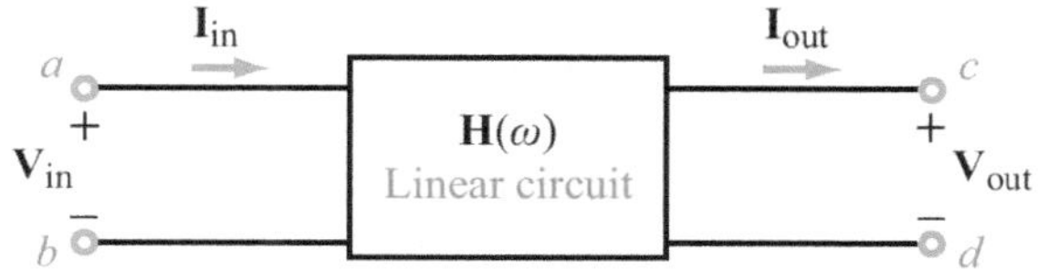

Figure 9-1: The voltage-gain transfer function is $\mathbf{H}(\omega) = \mathbf{V}_{\text{out}}(\omega)/\mathbf{V}_{\text{in}}(\omega)$

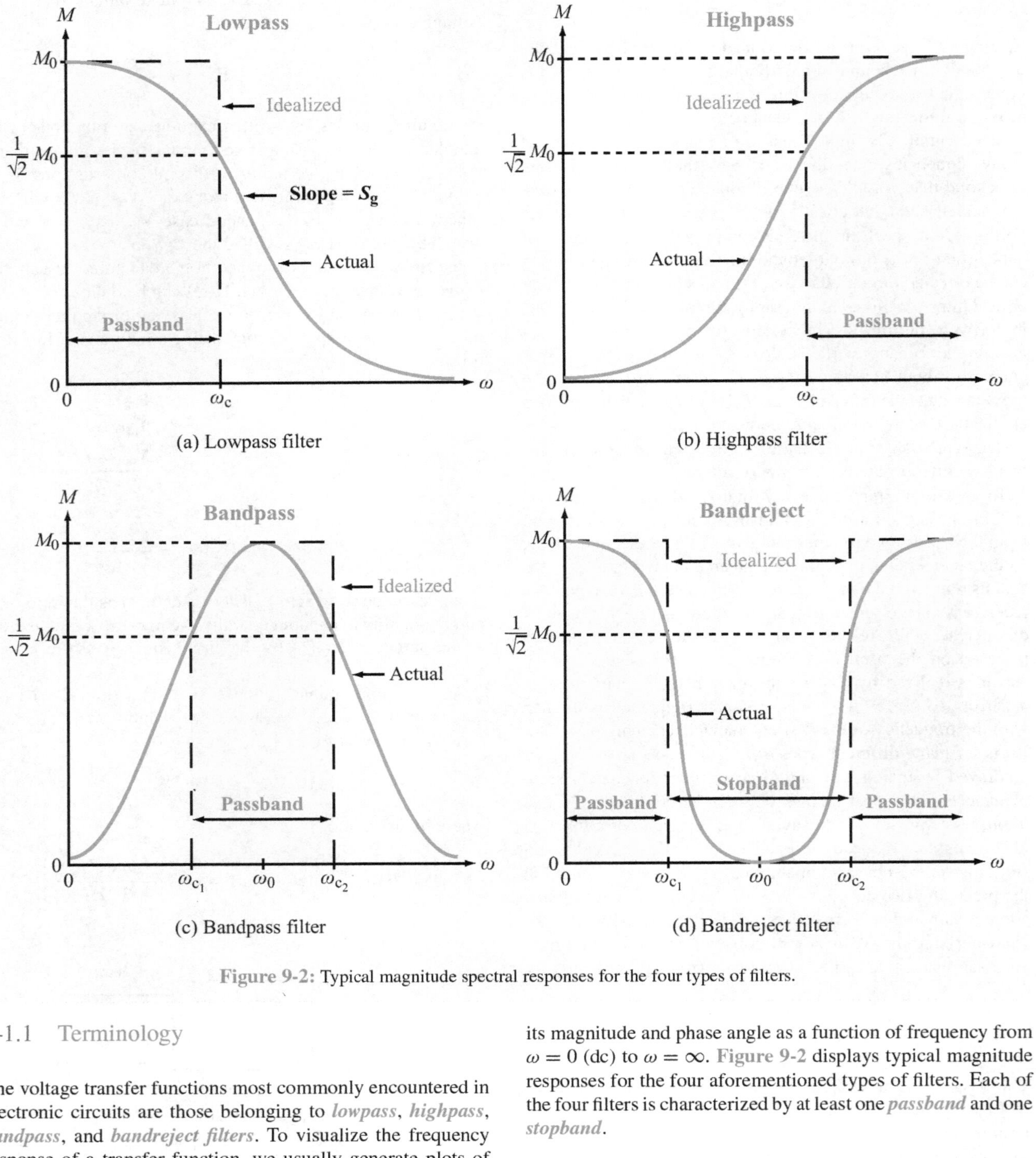

Figure 9-2: Typical magnitude spectral responses for the four types of filters.

9-1.1 Terminology

The voltage transfer functions most commonly encountered in electronic circuits are those belonging to ***lowpass***, ***highpass***, ***bandpass***, and ***bandreject filters***. To visualize the frequency response of a transfer function, we usually generate plots of its magnitude and phase angle as a function of frequency from $\omega = 0$ (dc) to $\omega = \infty$. Figure 9-2 displays typical magnitude responses for the four aforementioned types of filters. Each of the four filters is characterized by at least one ***passband*** and one ***stopband***.

▶ The lowpass filter allows low-frequency signals to pass through (essentially unimpeded) but blocks the transmission of high-frequency signals. ◀

The qualifiers *low* and *high* are relative to the ***corner frequency*** ω_c (Fig. 9-2(a)), which we shall define shortly.

▶ The highpass filter exhibits the opposite behavior, blocking low-frequency signals while allowing high frequencies to go through. ◀

The bandpass filter (Fig. 9-2(c)) is transparent to signals whose frequencies are within a certain range centered at ω_0, but cuts off both very high and very low frequencies. The response of the bandreject filter provides the opposite function to that of the bandpass filter; it is transparent to low- and high-frequency signals and opaque to intermediate-frequency signals.

We often use the term "frequency" for both the angular frequency ω and the oscillation frequency $f = \omega/2\pi$. Because the impedances of inductors and capacitors are given by $j\omega L$ and $1/j\omega C$, it is easier to analyze a circuit and plot its response as a function of ω, but if the circuit performance is specified in Hz, ω should be replaced with $2\pi f$ everywhere.

Gain factor M_0

All four spectral plots shown in Fig. 9-2 exhibit smooth patterns as a function of ω, and each has a peak value M_0 in its passband. If M_0 occurs at dc, as in the case of the lowpass filter, it is called the ***dc gain***; if it occurs at $\omega = \infty$, it is called the ***high-frequency gain***; and for the bandpass filter, it is called simply the ***gain factor***.

In some cases, the transfer function of a lowpass or highpass filter may exhibit a resonance behavior that manifests itself in the form of a peaking pattern in the neighborhood of the resonant frequency of the circuit, ω_0, as illustrated in Fig. 9-3. Obviously, the peak value at $\omega = \omega_0$ exceeds M_0, but we will continue to refer to M_0 as the dc gain of $M(\omega)$ because *M_0 is defined as the reference level in the passband of the transfer function*, whereas the behavior of $M(\omega)$ in the neighborhood of ω_0 is specific to that neighborhood.

Corner frequency ω_c

The corner frequency ω_c is defined as the angular frequency at which $M(\omega)$ is equal to $1/\sqrt{2}$ of the ***reference value*** M_0,

$$M(\omega_c) = \frac{M_0}{\sqrt{2}} = 0.707 M_0. \tag{9.5}$$

Since $M(\omega)$ is a voltage transfer function, $M^2(\omega)$ is the transfer function for power. The condition described by Eq. (9.5) is equivalent to

$$M^2(\omega_c) = \frac{M_0^2}{2} \quad \text{or} \quad P(\omega_c) = \frac{P_0}{2}. \tag{9.6}$$

Hence, ω_c also is called the ***half-power frequency***. The spectra of the lowpass and highpass filters shown in Fig. 9-2(a) and (b) have only one half-power frequency each, but the bandpass and bandreject responses have two half-power frequencies each, ω_{c_1} and ω_{c_2}.

Even though the *actual* frequency response of a filter is a gently varying curve, it usually is approximated to that of an equivalent ***idealized response***, as illustrated in Fig. 9-2. The idealized version for the lowpass filter has a rectangle-like envelope with a sudden transition at $\omega = \omega_c$. Accordingly, ω_c also is referred to as the ***cutoff frequency*** of the filter. This term also applies to the other three types of filters.

Bandwidth B

For lowpass and bandpass filters, the *bandwidth B is defined as the range of ω corresponding to the filter's idealized passband* (Fig. 9-2):

$$B = \omega_c \qquad \text{for lowpass filter,} \tag{9.7a}$$

$$B = \omega_{c_2} - \omega_{c_1} \qquad \text{for bandpass filter.} \tag{9.7b}$$

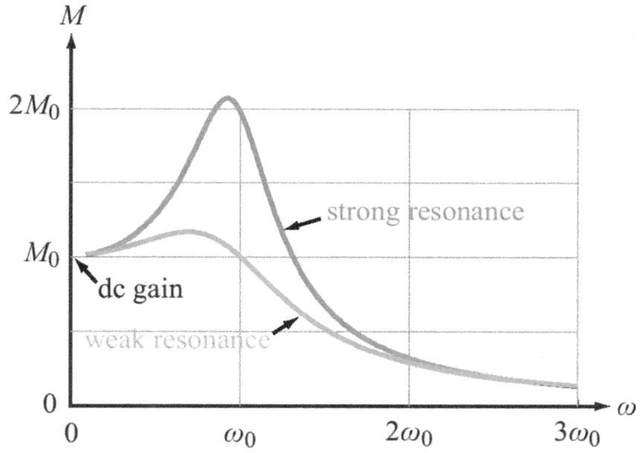

Figure 9-3: Resonant peak in the spectral response of a lowpass filter circuit.

Resonant frequency ω_0

▶ Resonance is a condition that occurs when the input impedance or input admittance of a circuit *containing reactive elements* is purely real, and the angular frequency at which it occurs is called the ***resonant frequency*** ω_0. ◀

Often (but not always) the transfer function $\mathbf{H}(\omega)$ also is purely real at $\omega = \omega_0$, and its magnitude is at its maximum or minimum value.

Let us consider the two circuits shown in Fig. 9-4. The input impedance of the RL circuit is simply

$$\mathbf{Z}_{\text{in}_1} = R + j\omega L. \tag{9.8}$$

Resonance corresponds to when the imaginary part of $\mathbf{Z}_{\text{in}_1}$ is zero, which occurs at $\omega = 0$. Hence, the resonant frequency of the RL circuit is

$$\omega_0 = 0 \qquad \text{(RL circuit)}. \tag{9.9}$$

When $\omega_0 = 0$ (dc) or ∞, the resonance is regarded as a *trivial resonance* because it occurs at the extreme ends of the spectrum. This usually happens when the circuit has either an inductor or a capacitor (but not both simultaneously). A circuit that exhibits only a trivial resonance, such as the RL circuit in Fig. 9-4(a), is not considered a resonator.

If the circuit contains at least one capacitor and at least one inductor, resonance can occur at intermediate values of ω. A case in point is the series RLC circuit shown in Fig. 9-4(b). Its input impedance is

$$\mathbf{Z}_{\text{in}_2} = R + j\left(\omega L - \frac{1}{\omega C}\right). \tag{9.10}$$

At resonance ($\omega = \omega_0$), the imaginary part of $\mathbf{Z}_{\text{in}_2}$ is equal to zero. Thus,

$$\omega_0 L - \frac{1}{\omega_0 C} = 0,$$

or

$$\omega_0 = \frac{1}{\sqrt{LC}} \qquad \textbf{(RLC circuit)}. \tag{9.11}$$

So long as neither L nor C is zero or ∞, the transfer function $\mathbf{H}(\omega) = \mathbf{V}_R/\mathbf{V}_s$ will exhibit a two-sided spectrum with a peak at ω_0—similar in shape to that of the bandpass filter response shown in Fig. 9-2(c).

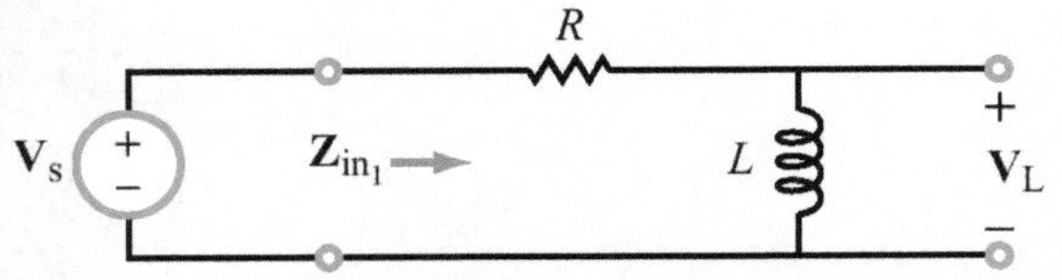

(a) First-order RL filter

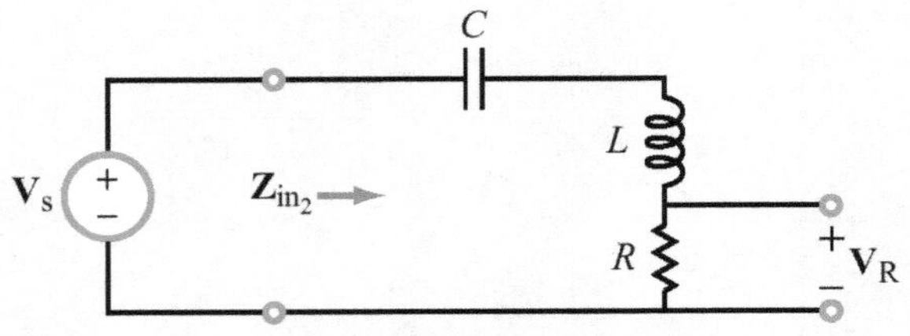

(b) Series RLC circuit

Figure 9-4: Resonance occurs when the imaginary part of the input impedance is zero. For the RL circuit, $\mathfrak{Im}\,[\mathbf{Z}_{\text{in}_1}] = 0$ when $\omega = 0$ (dc), but for the RLC circuit, $\mathfrak{Im}\,[\mathbf{Z}_{\text{in}_2}] = 0$ requires that $\mathbf{Z}_L = -\mathbf{Z}_C$ or, equivalently, $\omega^2 = 1/LC$.

Roll-off rate S_g

Outside the passband, the rectangle-shaped idealized responses shown in Fig. 9-2 have infinite slopes, but of course, the actual responses have finite slopes. The steeper the slope, the more discriminating the filter is, and the closer it approaches the idealized response. Hence, the slope S_g outside the passband (called the *gain roll-off rate*) is an important attribute of the filter response.

9-1.2 RC Circuit Example

To illustrate the transfer-function concept with a concrete example, let us consider the series RC circuit shown in Fig. 9-5(a). Voltage source $\mathbf{V}_s$ is designated as the input phasor, and on the output side, we have designated two voltage phasors, namely $\mathbf{V}_R$ and $\mathbf{V}_C$. We now examine the frequency responses of the transfer functions corresponding to each of those two output voltages.

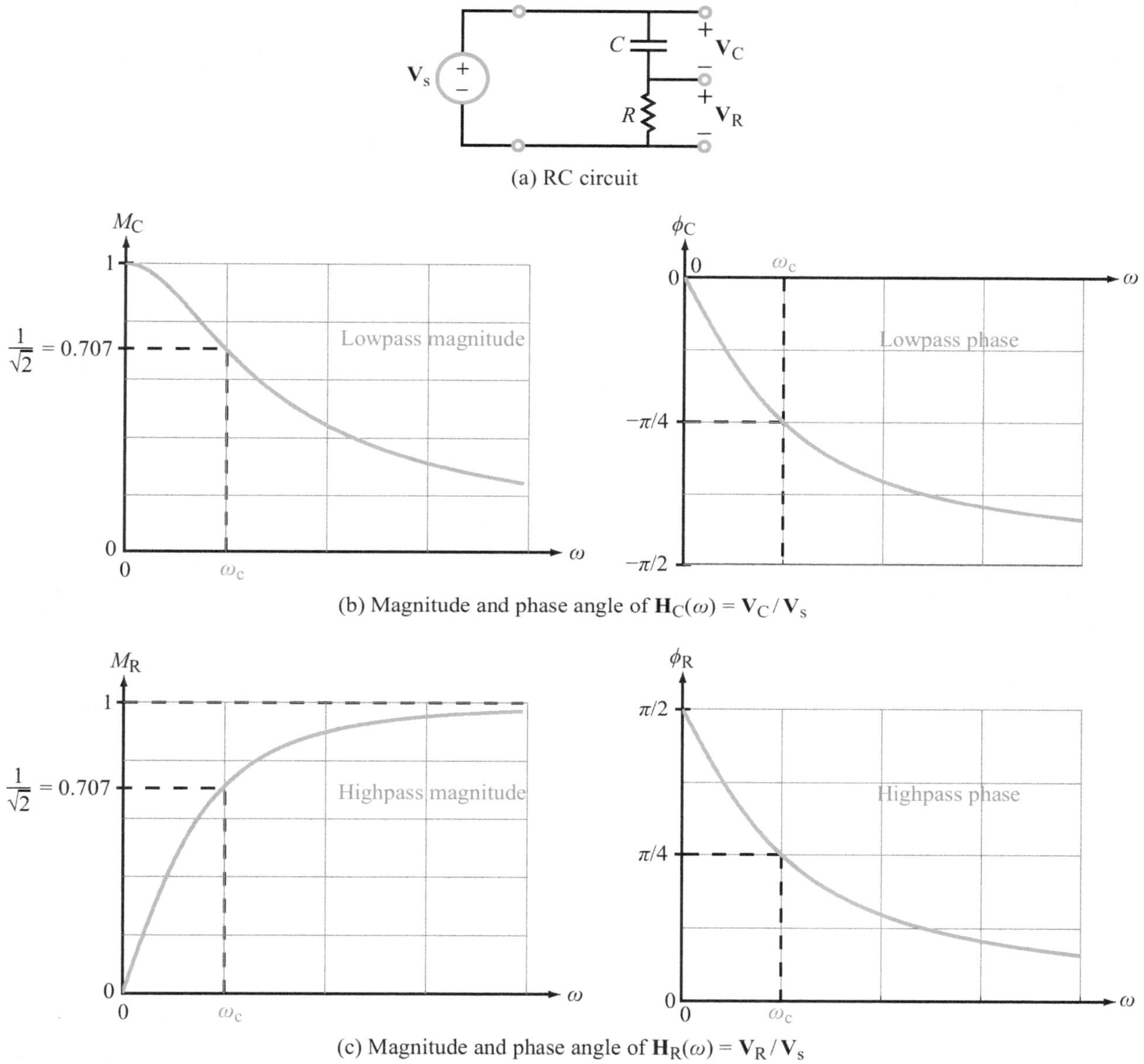

Figure 9-5: Lowpass and highpass transfer functions.

Lowpass filter

Application of voltage division gives

$$\mathbf{V}_C = \frac{\mathbf{V}_s \mathbf{Z}_C}{R + \mathbf{Z}_C} = \frac{\mathbf{V}_s / j\omega C}{R + \frac{1}{j\omega C}}. \tag{9.12}$$

The transfer function corresponding to $\mathbf{V}_C$ is

$$\mathbf{H}_C(\omega) = \frac{\mathbf{V}_C}{\mathbf{V}_s} = \frac{1}{1 + j\omega RC}, \tag{9.13}$$

where we have multiplied the numerator and denominator of Eq. (9.12) by $j\omega C$ to simplify the form of the expression. In terms of its magnitude $M_C(\omega)$ and phase angle $\phi_C(\omega)$, the transfer function is given by

$$\mathbf{H}_C(\omega) = M_C(\omega)\, e^{j\phi_C(\omega)}, \tag{9.14}$$

with

$$M_C(\omega) = |\mathbf{H}_C(\omega)| = \frac{1}{\sqrt{1+\omega^2R^2C^2}} \tag{9.15a}$$

and

$$\phi_C(\omega) = -\tan^{-1}(\omega RC). \tag{9.15b}$$

Spectral plots for $M_C(\omega)$ and $\phi_C(\omega)$ are displayed in Fig. 9-5(b). It is clear from the plot of its magnitude that the expression given by Eq. (9.13) represents the transfer function of a lowpass filter with a dc gain factor $M_0 = 1$. At dc, the capacitor acts like an open circuit—allowing no current to flow through the loop—with the obvious consequence that $\mathbf{V}_C = \mathbf{V}_s$. At very high values of ω, the capacitor acts like a short circuit, in which case the voltage across it is approximately zero.

Application of Eq. (9.6) allows us to determine the corner frequency ω_c as follows

$$M_C^2(\omega_c) = \frac{1}{1+\omega_c^2R^2C^2} = \frac{1}{2}, \tag{9.16}$$

which leads to

$$\omega_c = \frac{1}{RC}. \tag{9.17}$$

Note that ω_c is the inverse of the time constant $\tau = RC$ introduced in Chapter 5.

Highpass filter

The output across R in Fig. 9-5(a) leads to

$$\mathbf{H}_R(\omega) = \frac{\mathbf{V}_R}{\mathbf{V}_s} = \frac{j\omega RC}{1+j\omega RC}. \tag{9.18}$$

The magnitude and phase angle of $\mathbf{H}_R(\omega)$ are given by

$$M_R(\omega) = |\mathbf{H}_R(\omega)| = \frac{\omega RC}{\sqrt{1+\omega^2R^2C^2}} \tag{9.19a}$$

and

$$\phi_R(\omega) = \frac{\pi}{2} - \tan^{-1}(\omega RC). \tag{9.19b}$$

Their spectral plots are displayed in Fig. 9-5(c).

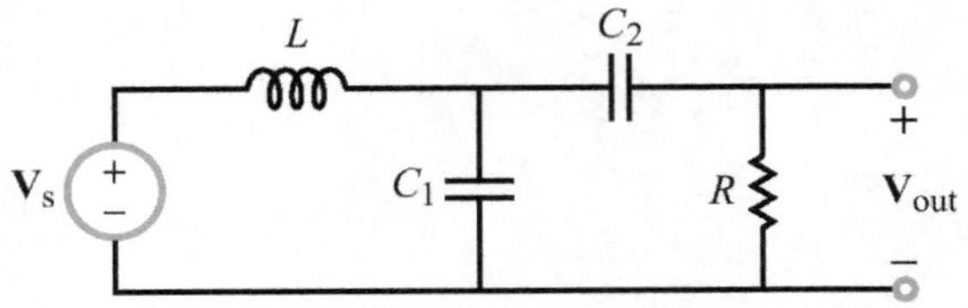

Figure 9-6: Circuit of Example 9-1.

Example 9-1: Resonant Frequency

For the circuit in Fig. 9-6, (a) obtain an expression for $\mathbf{H}(\omega) = \mathbf{V}_{out}/\mathbf{V}_s$ and (b) show that $\mathbf{H}(\omega)$ becomes purely real at $\omega_0 = 1/\sqrt{L(C_1+C_2)}$.

Solution: (a) Application of KCL and KVL leads to

$$\mathbf{H}(\omega) = \frac{\mathbf{V}_{out}}{\mathbf{V}_s} = \frac{RZ_{C_1}}{Z_{C_1}Z_{C_2} + Z_L(Z_{C_1}+Z_{C_2}) + R(Z_{C_1}+Z_L)},$$

where $Z_L = j\omega L$, $Z_{C_1} = 1/j\omega C_1$, and $Z_{C_2} = 1/j\omega C_2$. After a few steps of algebra aimed at transforming the expression into a form whose denominator is purely real, we end up with

$$\mathbf{H}(\omega) = \frac{\omega^2R^2C_2^2(1-\omega^2LC_1) + j\omega RC_2[1-\omega^2L(C_1+C_2)]}{[1-\omega^2L(C_1+C_2)]^2 + \omega^2R^2C_2^2(1-\omega^2LC_1)^2}.$$

(b) At

$$\omega = \omega_0 = \frac{1}{\sqrt{L(C_1+C_2)}},$$

the imaginary part of the expression becomes equal to zero and the expression simplifies to

$$\mathbf{H}(\omega_0) = \frac{C_1+C_2}{C_2}.$$

> **Concept Question 9-1:** Is the transfer function of a circuit always the same as its voltage gain? (See CAD)

> **Concept Question 9-2:** Is the gain factor M_0 always the peak value of $M(\omega)$? (See CAD)

> **Concept Question 9-3:** When is a circuit in a resonance condition? (See CAD)

Concept Question 9-4: Why is the corner frequency also called the half-power frequency? (See CAD)

Exercise 9-1: A series RL circuit is connected to a voltage source $\mathbf{V}_s$. Obtain an expression for $\mathbf{H}(\omega) = \mathbf{V}_R/\mathbf{V}_s$, where $\mathbf{V}_R$ is the phasor voltage across R. Also, determine the corner frequency of $\mathbf{H}(\omega)$.

Answer: $\mathbf{H}(\omega) = R/(R + j\omega L)$, $\omega_c = R/L$. (See CAD)

Exercise 9-2: Obtain an expression for the input impedance of the circuit in Fig. E9.2 and then use it to determine the resonant frequency.

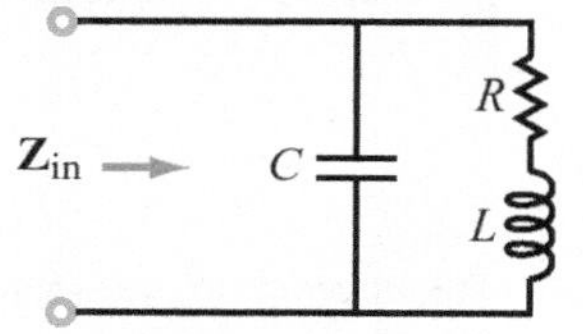

Figure E9.2

Answer: $\omega_0 = \sqrt{\dfrac{1}{LC} - \dfrac{R^2}{L^2}}$. (See CAD)

9-2 Scaling

When designing a frequency filter, it often is convenient to start by designing a *prototype model* in which the elements have values on the order of ohms, henrys, and farads and then to scale the prototype circuit into a *practical circuit* that not only contains elements with realistic values but also provides the specified frequency response. A circuit can be scaled in magnitude, in frequency, or both.

▶ *Magnitude scaling* changes the values of the elements in the circuit, but it does not modify its frequency response. *Frequency scaling* allows the designer to translate the frequency response into higher or lower frequency ranges while keeping the impedances of the circuit elements unchanged. ◀

9-2.1 Magnitude Scaling

The transfer function of a circuit is based on the impedances of its elements. If all impedances are multiplied (scaled) by the same *magnitude scaling factor* K_m, the absolute level of the transfer function may or may not change, but its relative frequency response will remain the same. To distinguish between the prototype and scaled circuits, we shall:

(a) Denote elements and impedances of the prototype circuit with unprimed symbols:

$$\mathbf{Z}_R = R, \qquad \mathbf{Z}_L = j\omega L, \qquad \text{and} \qquad \mathbf{Z}_C = \frac{1}{j\omega C}. \qquad \text{(prototype circuit)} \tag{9.20}$$

(b) Denote elements and impedances of the scaled circuit with primed symbols:

$$\mathbf{Z}'_R = R', \qquad \mathbf{Z}'_L = j\omega L', \qquad \text{and} \qquad \mathbf{Z}'_C = \frac{1}{j\omega C'}. \qquad \text{(scaled circuit)} \tag{9.21}$$

Magnitude scaling by a factor K_m implies that:

$$\mathbf{Z}'_R = K_m \mathbf{Z}_R, \quad \mathbf{Z}'_L = K_m \mathbf{Z}_L, \quad \text{and} \quad \mathbf{Z}'_C = K_m \mathbf{Z}_C, \tag{9.22}$$

which translates into the relations

$$\begin{aligned} R' &= K_m R, \\ L' &= K_m L \\ C' &= \frac{C}{K_m}, \\ \omega &= \omega'. \end{aligned} \tag{9.23}$$

(magnitude scaling only)

Thus, resistor and inductor values scale by K_m, but capacitor values scale by $1/K_m$.

To illustrate with an example, consider the transfer function given by Eq. (9.18),

$$\mathbf{H}_R(\omega) = \frac{j\omega RC}{1 + j\omega RC} \tag{9.24a}$$

and its scaled version

$$\mathbf{H}'_R(\omega) = \frac{j\omega R'C'}{1 + j\omega R'C'}. \tag{9.24b}$$

Applying the recipe given by Eq. (9.23) leads to

$$\mathbf{H}'_R(\omega) = \mathbf{H}_R(\omega),$$

which means that the frequency response remains unchanged.

9-2.2 Frequency Scaling

To shift the profile of a transfer function along the ω axis by a *frequency scaling factor* K_f while keeping its relative shape the same, we can replace ω in the transfer function of the prototype circuit with $\omega' = K_f\omega$ and scale the element values so that their impedances remain unchanged. For an inductor, $\mathbf{Z}_L = j\omega L$, so if ω is to be scaled up by K_f, L has to be scaled down by the same factor in order for $\mathbf{Z}_L$ to stay the same. Hence, the impedance condition requires that

$$R' = R,\quad L' = \frac{L}{K_f},\quad C' = \frac{C}{K_f},\quad \omega' = K_f\omega. \tag{9.25}$$

(frequency scaling only)

9-2.3 Combined Magnitude and Frequency Scaling

To transfer the prototype circuit design into a realizable circuit, we often apply magnitude and frequency scaling simultaneously, in which case the relationships between the prototype and scaled circuits become

$$R' = K_m R,\quad L' = \frac{K_m}{K_f}\,L,\quad C' = \frac{1}{K_m K_f}\,C,\quad \omega' = K_f\omega. \tag{9.26}$$

(magnitude and frequency scaling)

Example 9-2: Third-Order LP Filter

As discussed in Section 9-5, the order of a filter provides a measure of how steep its response is as a function of ω. The circuit in Fig. 9-7 is a prototype model of a third-order lowpass filter with a cutoff frequency of $\omega_c = 1$ rad/s. Develop a scaled version with a cutoff frequency of $\omega_c' = 10^6$ rad/s and a resistor value of 2 kΩ.

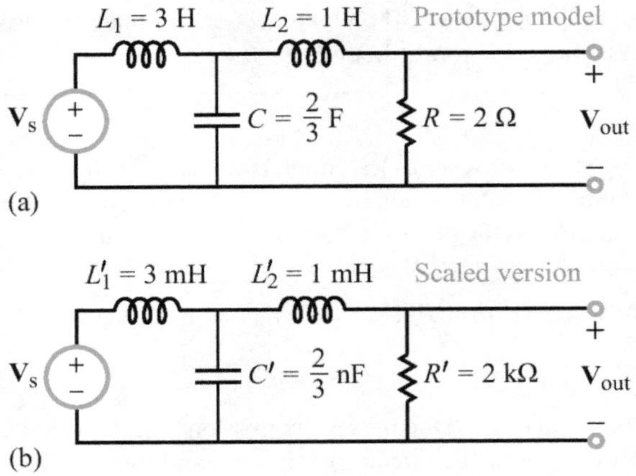

Figure 9-7: Prototype and scaled circuits of Example 9-2.

Solution: Based on the given information, the scaling factors are

$$K_m = \frac{R'}{R} = \frac{2\text{k}}{2} = 10^3 \quad\text{and}\quad K_f = \frac{\omega_c'}{\omega_c} = \frac{10^6}{1} = 10^6.$$

Application of Eq. (9.26) leads to

$$L_1' = \frac{K_m}{K_f}\,L_1 = \frac{10^3}{10^6}\times 3 = 3\text{ mH},$$

$$L_2' = \frac{K_m}{K_f}\,L_2 = 1\text{ mH},$$

and

$$C' = \frac{1}{K_m K_f}\,C = \frac{1}{10^3\times 10^6}\times\frac{2}{3} = \frac{2}{3}\text{ nF}.$$

The scaled circuit is displayed in Fig. 9-7(b).

Concept Question 9-5: How is the scaling concept used in the design of resonant circuits and filters? (See CAD)

Concept Question 9-6: What remains unchanged in (a) magnitude scaling alone and (b) frequency scaling alone? (See CAD)

Technology Brief 22
Noise-Cancellation Headphones

Noise-cancellation headphones are a class of devices that use *active noise-control* technology to reduce the level of environmental noise reaching a listener's ear. They were invented by Amar Bose in 1978, based on concepts developed in the mid-20th century (most notably by Paul Lueg, who developed a system for cancelling noise in air ducts using loudspeakers). The primary advantage of such systems is the ability to selectively reduce noise without having to use heavy and expensive sound padding. Beyond hearing aids and the commercial headphones used by airline passengers, specialized active noise-control systems have been in use by pilots and heavy equipment operators for several decades. In small enclosed environments, active noise-control systems can employ microphone and speaker arrays to lower the amount of ambient noise experienced by the listener. Examples of this include noise-cancellation systems used to dampen engine noise in cockpits, active mufflers for industrial exhaust stacks, noise reduction around large fans and, recently, systems for reducing road and traffic noise in automobile interiors.

Active Noise Control

In its most basic form, active noise control consists of measuring the sound levels at certain points in the environment and then using that data to emit noise from speakers whose frequency, phase shift, and amplitude are selected in order to cancel out the incoming environmental noise (Fig. TF22-1). In noise-cancellation headphones, small microphones outside the headphones measure the incoming ambient noise, and the measured signal is then fed to circuitry that produces output noise in the headset that cancels out the ambient sound (Fig. TF22-2). The general phenomenon whereby one waveform is added to another to cancel it out is called *destructive interference*. The basic idea is to add a replica of the environmental noise signal, but shifted in phase by 180°, which is equivalent to multiplying the added signal waveform by (-1). Consider a vibrating wave traveling along a one-dimensional string (Fig. TF22-3), which is analogous to a sound pressure wave moving through the air. If we superimpose a second traveling wave onto the string (perhaps by waving the end up and down with a second hand), the two waves will overlap and the result will be the sum of the two individual waves (superposition). If we precisely time the second wave so that it is the exact mirror of the first (i.e., it is phase-shifted by $180°$ from the first wave), the two waves will cancel out exactly, and the string will not vibrate. This (in principle) is what active noise control aims to do, even though (in reality) the technology faces a number of limitations.

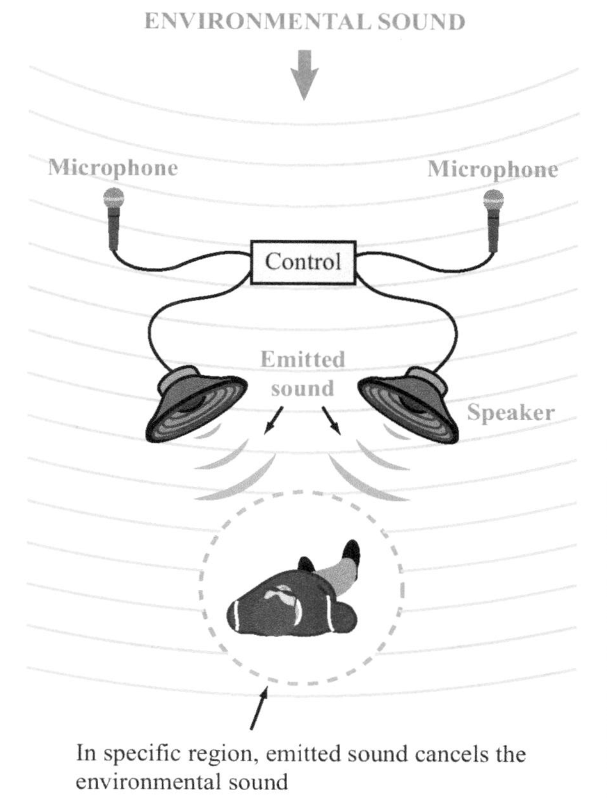

Figure TF22-1: Active noise control. A basic active noise-control system uses a set of microphones to sense incoming ambient sound; the microphone signals are fed into control circuits which drive a set of speakers. The control circuit generates exactly the signals required to cancel out the incoming ambient sound in a specific region.

Limitations

In order to truly cancel out all ambient noise, the emitted noise would have to exactly match the ambient noise in both space and time for all audible frequencies across a three-dimensional volume (such as the interior of a car or an airplane cabin). This is very difficult to accomplish in real environments. High frequencies are the hardest

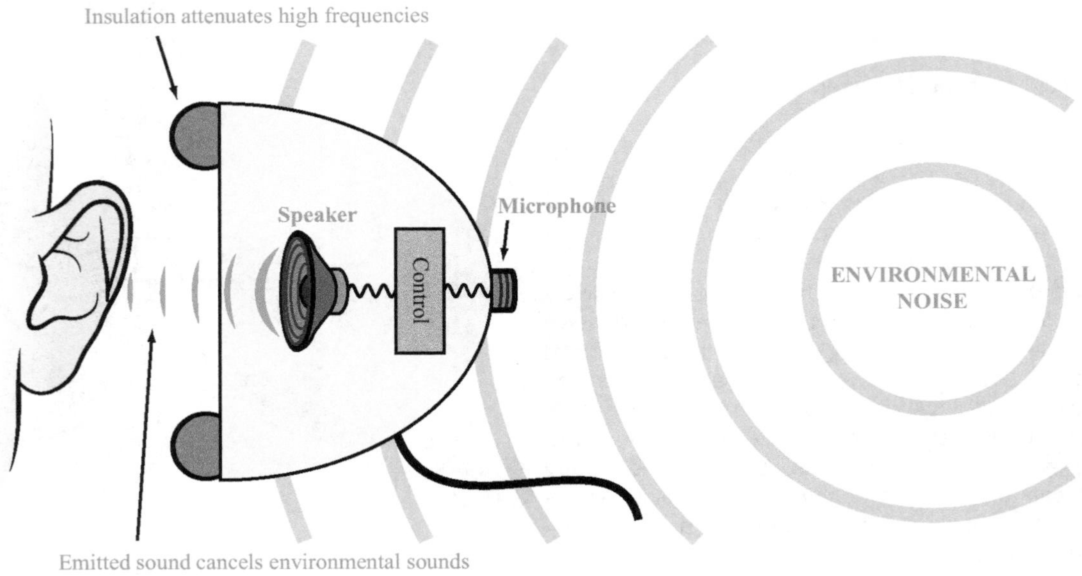

Figure TF22-2: Noise-cancellation headphones employ the active noise-control principle to eliminate noise around the ear. Sound insulation is used commonly to remove high-frequency noise signals, while the active system removes low-frequency sound.

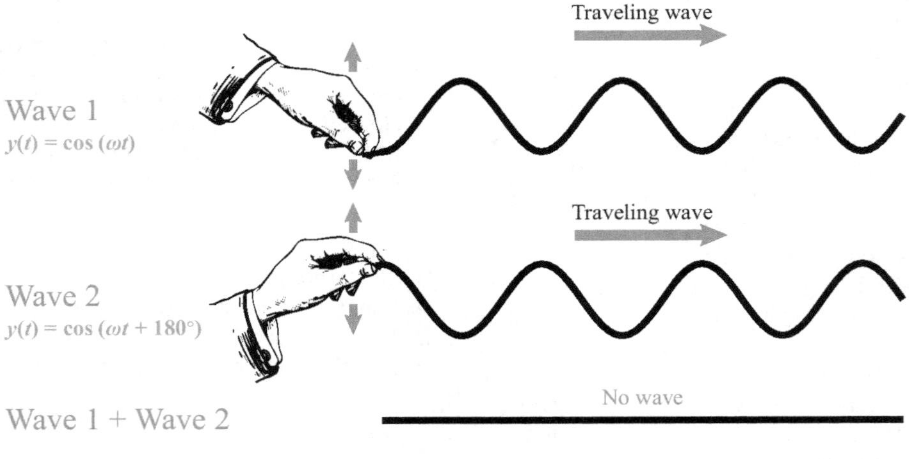

Figure TF22-3: Destructive interference via the superposition of two waves. A string can be agitated so as to generate traveling waves along its length. In order to cancel out a wave traveling along the string (Wave 1), we simultaneously can generate a second wave of the same amplitude and frequency with a 180° phase shift (Wave 2). Because the vibration of the string will be the result of the two waves superimposed, the two waves cancel out, and no vibration occurs. This is analogous to what active control systems do: Microphones sense ambient waves that are then canceled out by emitting an appropriately phase-shifted wave from the speakers.

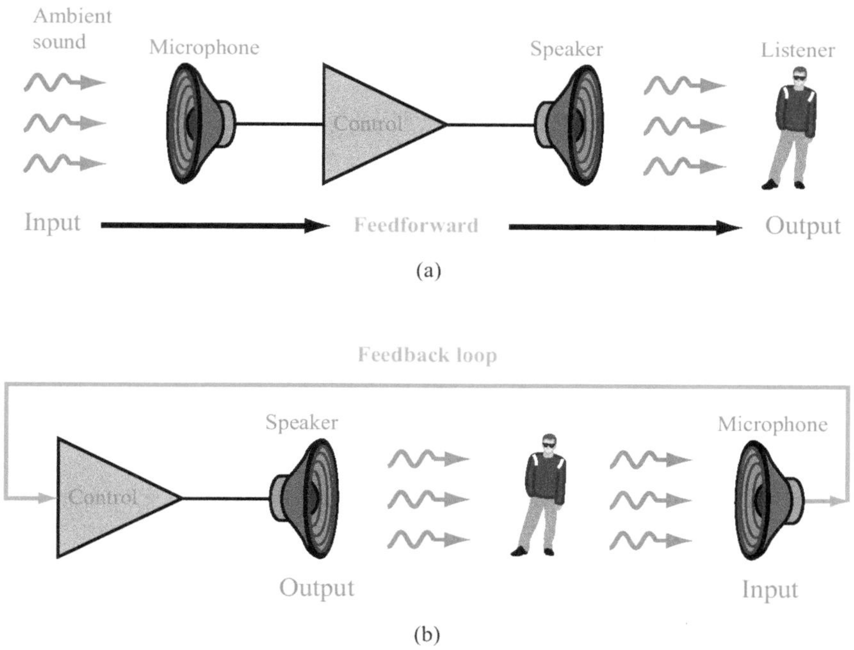

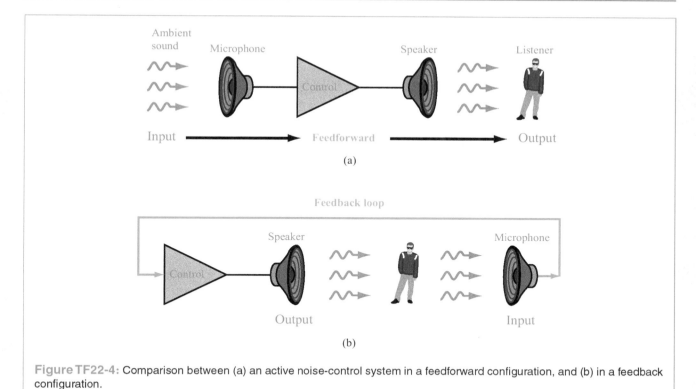

Figure TF22-4: Comparison between (a) an active noise-control system in a feedforward configuration, and (b) in a feedback configuration.

to match, because to correctly cancel them, the system would need to employ large arrays of microphones and speakers. Moreover, objects within the environment will reflect, absorb, and emit sound—further complicating the signals required to cancel the ambient sound. The situation is somewhat easier for headphones, as the area of interest is simply the user's ear (a much smaller physical region). However, most commercial noise-cancellation headphones do not attempt to cancel high frequencies. Padding and passive layers are instead used to absorb the high frequencies and the system actively cancels out only the lower frequencies (e.g., the airplane engine hum). In general, noise cancellation only works well for sound that is periodic. Noise that is random or has very fast changes is very hard to mask, because the system cannot compute what the interfering signal should be instantaneously.

Feedforward versus Feedback Control

Active noise-control systems provide an interesting comparison between feedback control (which we examined in Chapter 4) and feedforward control (Fig. TF22-4). Consider again Fig. TF22-1. If the microphones of this system are positioned relatively far away from the speakers, they sample the incoming ambient sound signal and send it ahead to the control circuit, which then drives the speakers. There is no microphone at the speaker location and thus no way to measure the "output" of the system (i.e., there is no microphone that measures how well the system is canceling the sound at the listener). This is an example of ***feedforward*** control. If we were to move the microphones very close to the speakers (or, better yet right next to the listener), the microphones would continuously report how well the speakers were canceling the sound. If the control system is doing poorly, the microphones will detect some sound, and the control circuit can attempt to correct for this. In such a configuration, the system is operating with ***feedback*** control. Figure TF22-4 illustrates both of these control configurations. Some sophisticated active noise-control systems use both modes simultaneously: They have distant microphones as well as microphones near the listener. In general, feedforward systems are less practical to implement in consumer systems, and feedback systems tend to be less stable.

Exercise 9-3: Determine (a) $\mathbf{Z}_{\text{in}}$ of the prototype circuit shown in Fig. E9.3 at $\omega = 1$ rad/s and (b) $\mathbf{Z}'_{\text{in}}$ of the same circuit after scaling it by $K_{\text{m}} = 1000$ and $K_{\text{f}} = 1000$.

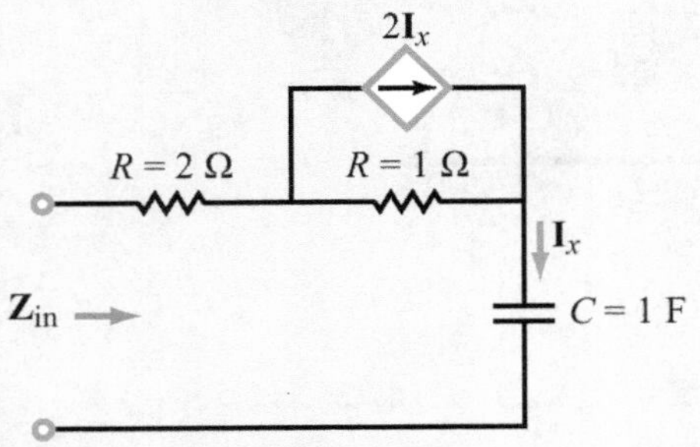

Figure E9.3

Answer: (a) $\mathbf{Z}_{\text{in}} = (1 - j)\ \Omega$, (b) $\mathbf{Z}'_{\text{in}} = (1 - j)$ kΩ. (See CAD)

Table 9-1: **Correspondence between power ratios in natural numbers and their dB values (left table) and between voltage or current ratios and their dB values (right table).**

$\dfrac{P}{P_0}$	dB
10^N	$10N$ dB
10^3	30 dB
100	20 dB
10	10 dB
4	≈ 6 dB
2	≈ 3 dB
1	0 dB
0.5	≈ -3 dB
0.25	≈ -6 dB
0.1	-10 dB
10^{-N}	$-10N$ dB

$\left\lvert\dfrac{\mathbf{V}}{\mathbf{V}_0}\right\rvert$ or $\left\lvert\dfrac{\mathbf{I}}{\mathbf{I}_0}\right\rvert$	dB
10^N	$20N$ dB
10^3	60 dB
100	40 dB
10	20 dB
4	≈ 12 dB
2	≈ 6 dB
1	0 dB
0.5	≈ -6 dB
0.25	≈ -12 dB
0.1	-20 dB
10^{-N}	$-20N$ dB

9-3 Bode Plots

In the late 1930s, inventor Hendrik Bode (pronounced Boh-dee) developed a graphical technique that has since become a standard tool for the analysis and design of resonant circuits, including filters, oscillators, and amplifiers.

▶ Bode's technique, which generates what we today call ***Bode plots*** or a ***Bode diagram***, relies on *using a logarithmic scale for ω and on expressing the magnitude of the transfer function in decibels (dB).* ◀

To make sure the reader is fully familiar with the properties of the dB operator, a quick review is in order.

9-3.1 The dB Scale

The ratio of the power P relative to a reference power level P_0—such as the output power generated by an amplifier, relative to the input power supplied by the source—is called ***relative*** or ***normalized power***. In many engineering applications, P/P_0 may vary over several orders of magnitude when plotted against a specific variable of interest, such as the frequency ω of the circuit. The dB scale originally was introduced as a logarithmic conversion tool to facilitate the generation of plots involving relative power, but its use has since been expanded to other physical quantities. The dB operator is intended as a scale converter of relative quantities, such as P/P_0, rather than of P itself, but it still can be applied to P by setting P_0 equal to a specified value, such as 1 watt or 1 mwatt, so long as P is expressed in the same units as P_0.

If G is defined as the power gain,

$$G = \frac{P}{P_0}, \tag{9.27}$$

then the corresponding gain in dB is defined as

$$G\ [\text{dB}] = 10 \log G = 10 \log \left(\frac{P}{P_0}\right) \qquad (\text{dB}). \tag{9.28}$$

The logarithm is in base 10. *The dB scale converts a power ratio to its logarithmic value and then multiplies it by 10.* Table 9-1 (left side) provides a listing of some values of G and the corresponding values of G [dB]. Note that when G varies across six orders of magnitude, from 10^{-3} to 10^3, G [dB] varies from -30 dB to $+30$ dB. Also note that the dB value of 2 is $\approx +3$ dB and the dB value of 0.5 is ≈ -3 dB.

Even though the scale originally was applied to power ratios, it now is used to express voltage and current ratios as well. If P and P_0 are the average powers absorbed by resistors of equal value and the corresponding phasor voltages across the resistors are $\mathbf{V}$ and $\mathbf{V}_0$, respectively, then

$$G\ [\text{dB}] = 10 \log \left(\frac{\frac{1}{2}|\mathbf{V}|^2/R}{\frac{1}{2}|\mathbf{V}_0|^2/R}\right) = 20 \log \left(\frac{|\mathbf{V}|}{|\mathbf{V}_0|}\right). \tag{9.29}$$

Similarly,

$$G\text{ [dB]} = 20\log\left(\frac{|\mathbf{I}|}{|\mathbf{I}_0|}\right). \tag{9.30}$$

▶ Whereas the dB definition for power ratio includes a scaling factor of 10, the scaling factor for voltage and current is 20. ◀

A useful property of the log operator is that *the log of the product of two numbers is equal to the sum of their logs.* That is if

$$G = XY \Rightarrow G\text{ [dB]} = X\text{ [dB]} + Y\text{ [dB]}. \tag{9.31}$$

This result follows from

$$G\text{ [dB]} = 10\log(XY) = 10\log X + 10\log Y = X\text{ [dB]} + Y\text{ [dB]}.$$

By the same token, if:

$$G = \frac{X}{Y} \Rightarrow G\text{ [dB]} = X\text{ [dB]} - Y\text{ [dB]}. \tag{9.32}$$

Conversion of products and ratios into sums and differences will prove to be quite useful when constructing the frequency response of a resonant circuit.

Example 9-3: RL Highpass Filter

For the series RL circuit shown in Fig. 9-8(a):

(a) Obtain an expression for the transfer function $\mathbf{H} = \mathbf{V}_{\text{out}}/\mathbf{V}_{\text{s}}$ in terms of ω/ω_c where $\omega_c = R/L$.

(b) Determine the magnitude M [dB] $= 20\log|\mathbf{H}|$ and plot it as a function of ω on a log scale with ω expressed in units of ω_c.

(c) Determine and plot the phase angle of $\mathbf{H}$.

Solution: (a) Voltage division gives

$$\mathbf{V}_{\text{out}} = \frac{j\omega L\mathbf{V}_{\text{s}}}{R + j\omega L},$$

which leads to

$$\mathbf{H} = \frac{\mathbf{V}_{\text{out}}}{\mathbf{V}_{\text{s}}} = \frac{j\omega L}{R + j\omega L} = \frac{j(\omega/\omega_c)}{1 + j(\omega/\omega_c)}, \tag{9.33}$$

with $\omega_c = R/L$.

(b) The magnitude of $\mathbf{H}$ is given by

$$M = |\mathbf{H}| = \frac{(\omega/\omega_c)}{|1 + j(\omega/\omega_c)|} = \frac{(\omega/\omega_c)}{\sqrt{1 + (\omega/\omega_c)^2}}. \tag{9.34}$$

Since H is a voltage ratio, the appropriate dB scaling factor is 20, so to find the power gain in dB,

$$\begin{aligned} M\text{ [dB]} &= 20\log M \\ &= 20\log(\omega/\omega_c) - 20\log[1 + (\omega/\omega_c)^2]^{1/2} \\ &= \underset{①}{20\log(\omega/\omega_c)} - \underset{②}{10\log[1 + (\omega/\omega_c)^2]}. \end{aligned} \tag{9.35}$$

In the Bode-diagram terminology introduced later in Section 9-3.2, the components of M [dB] are called *factors*, so in the present case, M [dB] consists of two factors with the second one having a negative coefficient. A magnitude plot is displayed on semilog graph paper with the vertical axis in dB and the horizontal axis in (rad/s). If in the expression for M [dB], ω appears in a normalized format—as in (ω/ω_c)—we may choose to express the horizontal axis in units of ω_c. Figure 9-8(b) contains individual plots for each of the two factors comprising M [dB] as well as a plot for their sum.

On semilog graph paper, the plot of $\log(\omega/\omega_c)$ is a straight line that crosses the ω axis at $(\omega/\omega_c) = 1$. This is because $\log 1 = 0$. At $(\omega/\omega_c) = 10$, $20\log 10 = 20$ dB. Hence;

① $20\log\left(\frac{\omega}{\omega_c}\right)$ ➡ straight line with slope = 20 dB/decade and ω axis crossing at $\omega/\omega_c = 1$.

▶ At $\omega/\omega_c = 10$, $20\log(10) = 20$ dB, at $\omega/\omega_c = 100$, $20\log(100) = 40$ dB, and so on. Hence, the slope is 20 dB/decade. ◀

Note that a *decade* refers to a change by a factor of 10. Thus, the range from 1 to 10 is a decade, and so are the ranges from 3 to 30 and 10 to 100.

② The second factor has a nonlinear plot, with the following properties:

Low-Frequency Asymptote

(2a) As $(\omega/\omega_c) \Rightarrow 0$, $\quad -10\log\left[1 + \left(\frac{\omega}{\omega_c}\right)^2\right] \Rightarrow 0.$

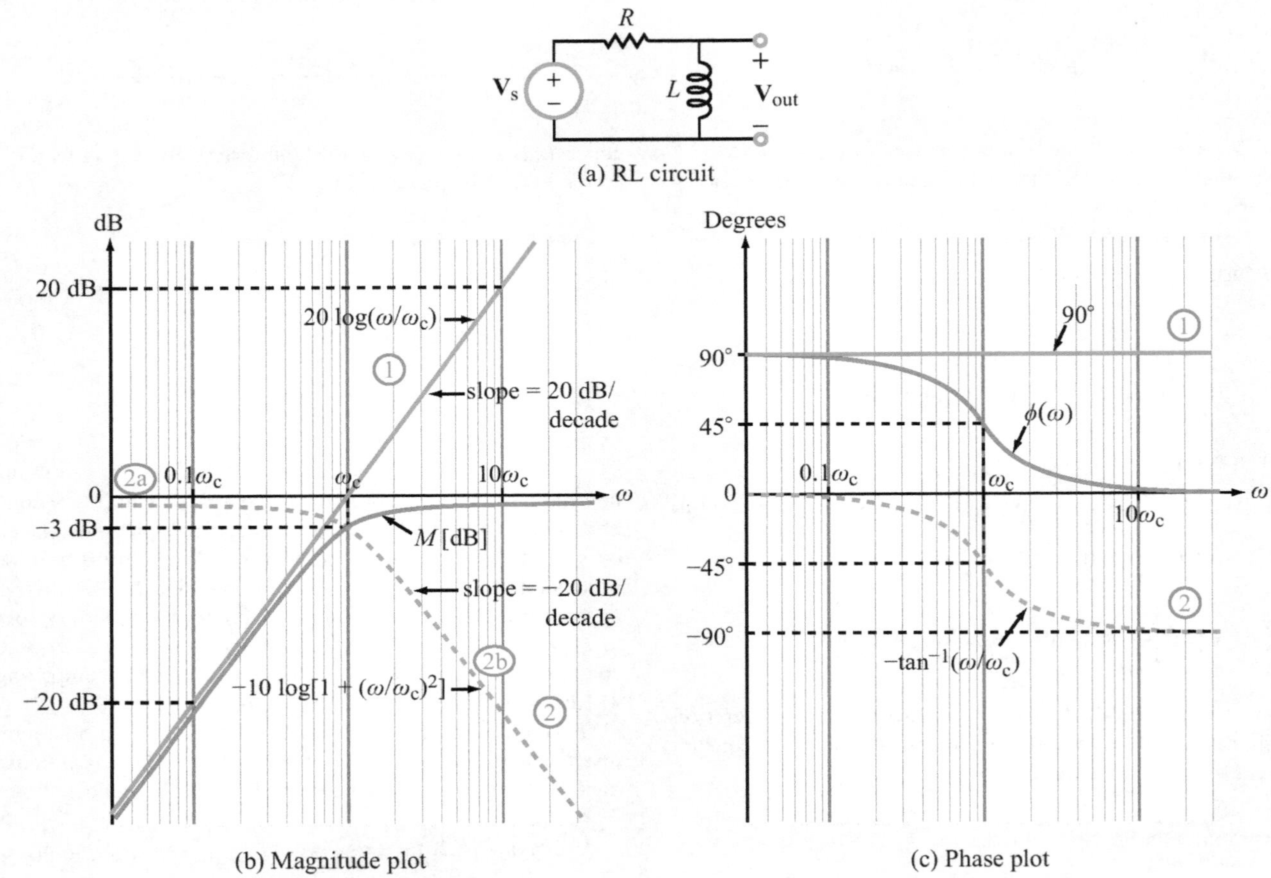

Figure 9-8: Magnitude and phase plots of $\mathbf{H} = \mathbf{V}_{\text{out}}/\mathbf{V}_{\text{s}}$.

High-Frequency Asymptote

(2b) As $(\omega/\omega_c) \Rightarrow \infty$,

$$-10\log\left[1+\left(\frac{\omega}{\omega_c}\right)^2\right] \Rightarrow -20\log\left(\frac{\omega}{\omega_c}\right).$$

The plot of M [dB] is obtained by graphically adding together the two plots of its individual factors (Fig. 9-8(b)). At low frequencies such that $(\omega/\omega_c \ll 1)$, M [dB] is dominated by its first factor; at $\omega/\omega_c = 1$, M [dB] $= -3$ dB; and at high frequencies $(\omega/\omega_c \gg 1)$, M [dB] $\to 0$, because its two factors cancel each other out. The overall profile (in red in Fig. 9-8(b)) is typical of the spectral response of a highpass filter with a cutoff frequency ω_c.

(c) From Eq. (9.33), the phase angle of **H** is

$$\phi(\omega) = \underset{(1)}{90^\circ} - \underset{(2)}{\tan^{-1}\left(\frac{\omega}{\omega_c}\right)}. \tag{9.36}$$

The 90° component is contributed by j in the numerator and the second term is the phase angle of the denominator. The phase plot is displayed in Fig. 9-8(c).

> **Concept Question 9-7:** When is it helpful to use the dB scale? (See CAD)

> **Concept Question 9-8:** What is the scaling factor for power ratio? For current ratio? (See CAD)

Exercise 9-4: Convert the following voltage ratios to dB: (a) 20, (b) 0.03, (c) 6×10^6.

Answer: (a) 26.02 dB, (b) −30.46 dB, (c) 135.56 dB. (See CAD)

Exercise 9-5: Convert the following dB values to voltage ratios: (a) 36 dB, (b) −24 dB, (c) −0.5 dB.

Answer: (a) 63.1, (b) 0.063, (c) 0.94. (See CAD)

9-3.2 Poles and Zeros

In polar coordinates, the transfer function $\mathbf{H}(\omega)$ is composed of a magnitude $M(\omega)$ and a phase angle $\phi(\omega)$,

$$\mathbf{H}(\omega) = M(\omega)\, e^{j\phi(\omega)}. \tag{9.37}$$

For any circuit, the expression for $\mathbf{H}(\omega)$ in general can be cast as the product of multiple *factors* $\mathbf{A}_1(\omega)$ to $\mathbf{A}_n(\omega)$,

$$\mathbf{H}(\omega) = \mathbf{A}_1(\omega)\, \mathbf{A}_2(\omega) \ldots \mathbf{A}_n(\omega). \tag{9.38}$$

Discussion of the functional forms of $\mathbf{A}_1$ to $\mathbf{A}_n$ will follow shortly, but to clarify what we mean by Eq. (9.38), let us consider the simple example of a transfer function given by

$$\mathbf{H}(\omega) = 10\, \frac{1 + j\omega/\omega_z}{1 + j\omega/\omega_p}. \tag{9.39}$$

In this case,

$$\mathbf{A}_1 = 10, \tag{9.40a}$$

$$\mathbf{A}_2 = 1 + j\omega/\omega_z, \tag{9.40b}$$

and

$$\mathbf{A}_3 = \frac{1}{1 + j\omega/\omega_p}. \tag{9.40c}$$

▶ The expression for $\mathbf{H}(\omega)$ was structured intentionally into a form—called the *standard form*—in which the two terms involving ω each are written such that the real part is unity and the coefficient of ω in the imaginary part is defined as the reciprocal of an angular frequency (ω_z or ω_p). ◀

For the circuit represented by the transfer function given by Eq. (9.39), ω_z and ω_p are related to the circuit architecture and the element values of the circuit. The quantity ω_z (which has the same units as ω) is called a *zero* of $\mathbf{H}(\omega)$, because it appears in a factor contained in the numerator of $\mathbf{H}(\omega)$. Similarly, ω_p is called a *pole* because it is part of a factor contained in the denominator of $\mathbf{H}(\omega)$. If the numerator is a product of multiple factors, $\mathbf{H}(\omega)$ will have multiple zeros—one associated with each factor (except for frequency independent factors, such as $\mathbf{A}_1 = 10$). A transfer function also may have multiple poles if the denominator of $\mathbf{H}(\omega)$ is the product of multiple factors. Moreover, the factors may assume functional forms different from those given by $\mathbf{A}_1$ to $\mathbf{A}_3$ in Eq. (9.40).

To analyze the frequency response of the circuit, we need to extract from Eq. (9.38) explicit expressions for the magnitude $M(\omega)$ and the phase angle $\phi(\omega)$. For any two complex numbers, the phase angle of their product is equal to the sum of their individual phase angles. Application of this multiplication principle to Eq. (9.38) gives

$$\phi(\omega) = \phi_{\mathbf{A}_1}(\omega) + \phi_{\mathbf{A}_2}(\omega) + \cdots + \phi_{\mathbf{A}_n}(\omega), \tag{9.41}$$

where $\phi_{\mathbf{A}_1}(\omega)$ to $\phi_{\mathbf{A}_n}(\omega)$ are the phase angles of factors $\mathbf{A}_1$ to $\mathbf{A}_n$, respectively. Transformation from a product form (as in Eq. (9.38)), into a sum (as in Eq. (9.41)) allows us to generate a phase plot for each factor separately and then add them together graphically—rather than having to deal with a single complicated expression all at once. The dB conversion introduced by Bode accomplishes a similar transformation for the magnitude $M(\omega)$. Application of the log property described by Eq. (9.31) leads to

$$\begin{aligned} M\ [\text{dB}] &= 20 \log |\mathbf{H}| \\ &= 20 \log |\mathbf{A}_1| + 20 \log |\mathbf{A}_2| + \cdots + 20 \log |\mathbf{A}_n| \\ &= A_1\ [\text{dB}] + A_2\ [\text{dB}] + \cdots + A_n\ [\text{dB}] \end{aligned} \tag{9.42}$$

where

$$A_1\ [\text{dB}] = 20 \log |\mathbf{A}_1|, \tag{9.43}$$

and a similar definition applies to the other factors. The transformations represented by Eqs. (9.41) and (9.42) constitute the basic framework for generating Bode diagrams. Our next step is to examine the possible functional forms that factors $\mathbf{A}_1$ to $\mathbf{A}_n$ may assume.

Standard form refers to an arrangement in which factors $\mathbf{A}_1$ to $\mathbf{A}_n$ of Eq. (9.38) each can assume any one of only seven possible functional forms. We will examine the general character of each of these factors individually (as if it were the only component of $\mathbf{H}(\omega)$) by considering two types of plots: *exact plots* based on the exact expression for $\mathbf{H}(\omega)$ and *straight-line approximations*—called *Bode plots*—that are much easier to generate and yet provide reasonable accuracy. The symbol N (which we will call the *order* of a factor) is an integer equal to or greater than 1.

9-3.3 Functional Forms

Constant factor: $\mathbf{H} = K$

This is a frequency-independent constant that may be positive or negative.

Magnitude: $M\,[\text{dB}] = 20\log|K|$
Phase: $\phi = 0°$ if $K > 0$, or $\pm 180°$ if $K < 0$
Bode plots: Same as exact plots; straight horizontal lines (Table 9-2)

Table 9-2: Bode straight-line approximations for magnitude and phase.

Factor	Bode Magnitude	Bode Phase
Constant K	$20\log K$; 0 dB	$\pm 180°$ if $K < 0$; $0°$ if $K > 0$
Zero @ Origin $(j\omega)^N$	0 dB at 1; slope = $20N$ dB/decade	$(90N)°$
Pole @ Origin $(j\omega)^{-N}$	0 dB at 1; slope = $-20N$ dB/decade	$(-90N)°$
Simple Zero $(1 + j\omega/\omega_c)^N$	0 dB at ω_c; slope = $20N$ dB/decade	$0°$ at $0.1\omega_c$; $(90N)°$ at $10\omega_c$
Simple Pole $\left(\dfrac{1}{1 + j\omega/\omega_c}\right)^N$	0 dB at ω_c; slope = $-20N$ dB/decade	$0°$ at $0.1\omega_c$; $(-90N)°$ at $10\omega_c$
Quadratic Zero $[1 + j2\xi\omega/\omega_c + (j\omega/\omega_c)^2]^N$	0 dB at ω_c; slope = $40N$ dB/decade	$0°$ at $0.1\omega_c$; $(180N)°$ at $10\omega_c$
Quadratic Pole $\dfrac{1}{[1 + j2\xi\omega/\omega_c + (j\omega/\omega_c)^2]^N}$	0 dB at ω_c; slope = $-40N$ dB/decade	$0°$ at $0.1\omega_c$; $(-180N)°$ at $10\omega_c$

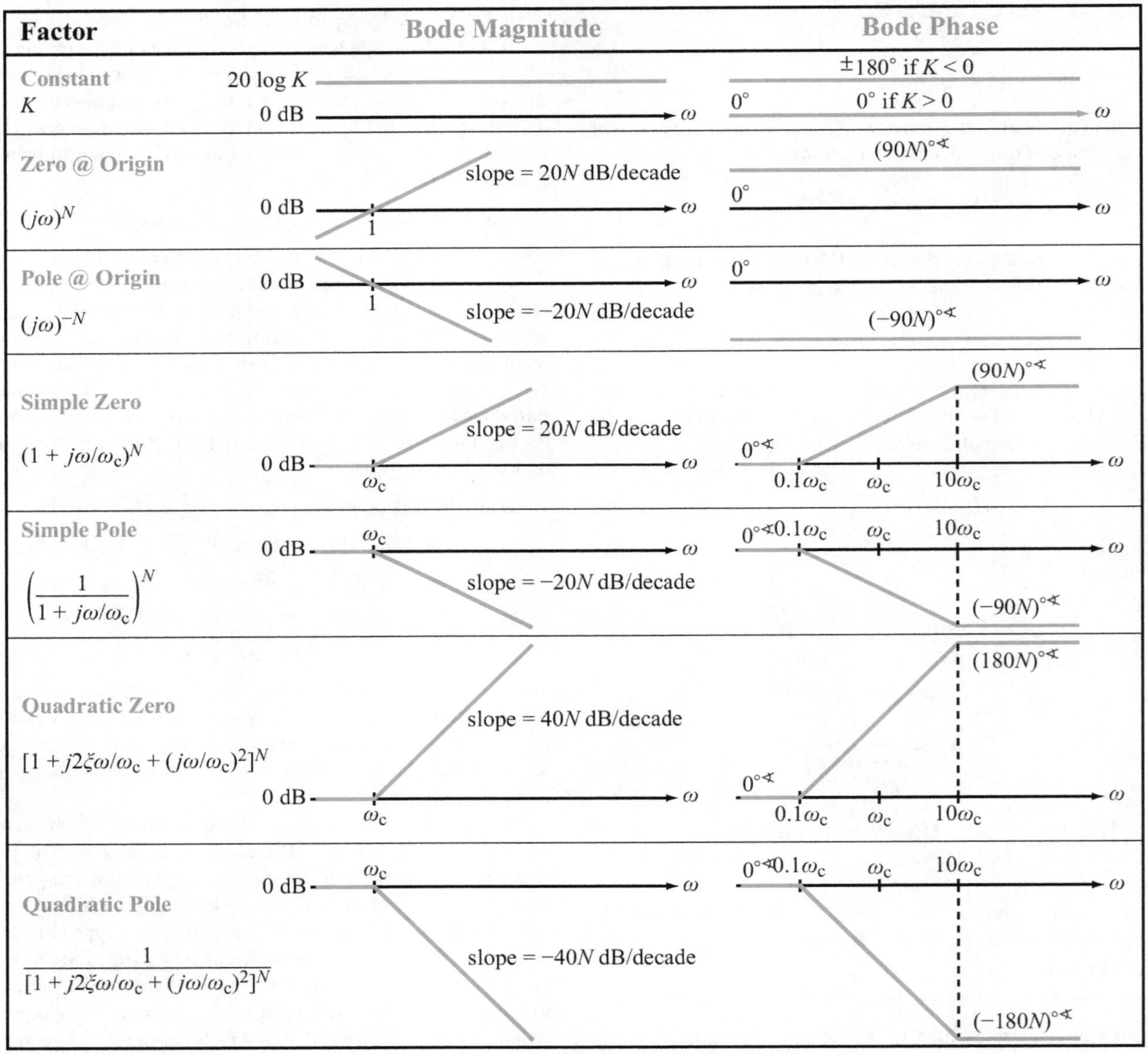

Zero @ origin: $\mathbf{H} = (j\omega)^N$, (N = positive integer)

The name of this factor reflects the fact that $\mathbf{H} \to 0$ as $\omega \to 0$. N is its order, so for example, $(j\omega)^2$ is called a second-order zero @ origin.

Magnitude:	M [dB] $= 20 \log \lvert (j\omega)^N \rvert = 20N \log \omega$
Phase:	$\phi = (90N)^\circ$
Bode plots:	Same as exact **Magnitude:** Straight line through $\omega = 1$ with slope = $20N$ dB/decade **Phase:** Constant level (Table 9-2)

Pole @ origin: $\mathbf{H} = 1/(j\omega)^N$

This factor is called a pole because $\mathbf{H} \to \infty$ when $\omega \to 0$. The function $1/(j\omega)^3$ is called a third-order pole @ origin, for example.

Magnitude:
$$M \text{ [dB]} = 20 \log \left| \frac{1}{(j\omega)^N} \right| = -20N \log \omega$$

Phase: $\phi = (-90N)^\circ$

Bode plots: Same as exact (Table 9-2)

Magnitude and phase plots are identical to those of zero @ origin except for (−) sign in both cases.

Simple zero: $\mathbf{H} = (1 + j\omega/\omega_c)^N$

Standard form requires that the real part be 1 and the imaginary part be positive. The constant ω_c is the corner frequency of the simple-zero factor, and N is its order.

Magnitude:
$$M \text{ [dB]} = 20 \log \lvert (1 + j\omega/\omega_c)^N \rvert = 10N \log[1 + (\omega/\omega_c)^2] \approx \begin{cases} 0 \text{ dB}, & \text{for } \omega/\omega_c \ll 1 \\ 20N \log(\omega/\omega_c), & \text{for } \omega/\omega_c \gg 1 \end{cases}$$

Phase:
$$\phi = N \tan^{-1}\left(\frac{\omega}{\omega_c}\right) \approx \begin{cases} 0, & \text{for } \omega/\omega_c \ll 1 \\ (90N)^\circ, & \text{for } \omega/\omega_c \gg 1 \end{cases}$$

Bode plots: Straight-line approximation is different from exact; for magnitude, the maximum difference is $3N$ dB and it occurs at $\omega/\omega_c = 1$

Magnitude: 0 dB horizontal line to $\omega = \omega_c$, followed by straight line with slope = $20N$ dB/decade

Phase: 0° horizontal line to $\omega = 0.1\omega_c$; straight line connecting coordinates $[0.1\omega_c, 0]$ to $[10\omega_c, (90N)^\circ]$; followed by horizontal line $(90N)^\circ$

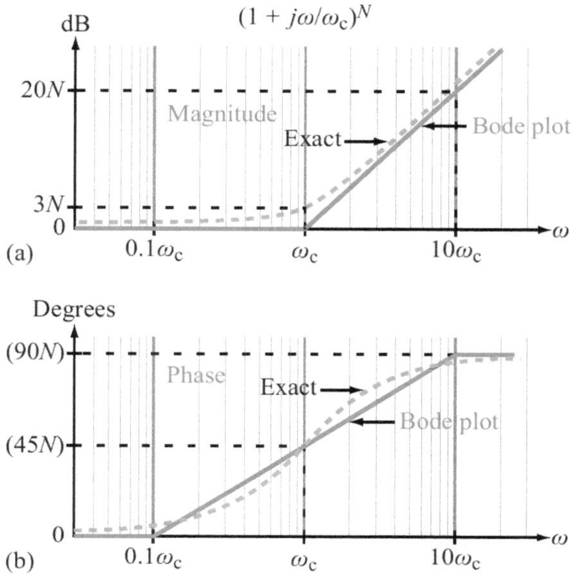

Figure 9-9: Comparison of exact plots with the Bode straight-line approximations for a simple zero with a corner frequency ω_c.

Figure 9-9 provides a comparison between the Bode straight-line approximation and the exact solution for both magnitude and phase. The corner frequency ω_c gets its name from the Bode magnitude plot, which turns the corner at ω_c.

Simple pole: $\mathbf{H} = 1/(1 + j\omega/\omega_c)^N$

Plots are mirror images (relative to ω axis) of those for the simple zero (Table 9-2).

Quadratic zero: $\mathbf{H} = [1 + j2\xi\omega/\omega_c + (j\omega/\omega_c)^2]^N$

N is the order of the quadratic zero, ω_c is its corner frequency, and ξ is its *damping factor*.

Magnitude:
$$M \text{ [dB]} = 10N \log \left\{ \left[1 - \left(\frac{\omega}{\omega_c}\right)^2 \right]^2 + 4\xi^2 \left(\frac{\omega}{\omega_c}\right)^2 \right\} \approx \begin{cases} 0 \text{ dB} & \text{for } \omega/\omega_c \ll 1, \\ 40N \log(\omega/\omega_c) & \text{for } \omega/\omega_c \gg 1 \end{cases}$$

Phase:
$$\phi = N \tan^{-1}\left[\frac{2\xi(\omega/\omega_c)}{1 - (\omega/\omega_c)^2}\right] \approx \begin{cases} 0^\circ & \text{for } \omega/\omega_c \ll 1, \\ (180N)^\circ & \text{for } \omega/\omega_c \gg 1 \end{cases}$$

Bode plots: **Magnitude:** Same as simple zero except at twice the slope
Phase: Same as simple zero except at twice the slope and twice the level at $\omega/\omega_c \gg 1$

Figure 9-10 displays plots of the magnitude and phase of the quadratic-zero factor with $N = 1$ for three different values of the damping coefficient. Whereas the value of ξ has little influence on the shape of the plots when $\omega/\omega_c \ll 1$ or $\omega/\omega_c \gg 1$, it exercises significant influence when ω is in the neighborhood of ω_c. *In terms of the Bode straight-line approximation, the Bode plots (Table 9-2) for a quadratic factor of order N are identical to those for a simple factor of order 2N.*

Quadratic pole: $\mathbf{H} = [1 + j2\xi\omega/\omega_c + (j\omega/\omega_c)^2]^{-N}$
Plots are mirror images, relative to ω axis, as those for the quadratic zero.

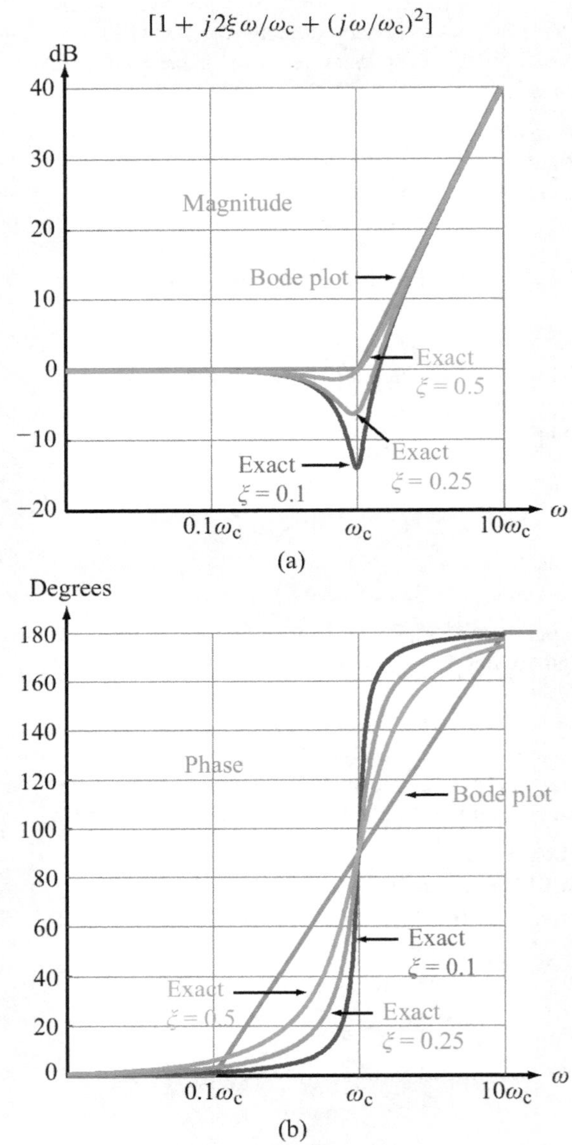

Figure 9-10: Comparison of exact plots with Bode straight-line approximations for a quadratic zero
$[1 + j2\xi\omega/\omega_c + (j\omega/\omega_c)^2]$.

9-3.4 General Observations

A few general observations about the Bode straight-line plots shown in Table 9-2 are in order.

(1) For $N = 1$, the slope of the nonhorizontal Bode magnitude line (called *gain roll-off rate*) is 20 dB/decade for both the zero @ origin and simple-zero factors. The corresponding slope for their pole counterparts is −20 dB/decade.

(2) For $N = 1$, the slopes of the nonhorizontal Bode magnitude lines for the quadratic zero and quadratic pole factors are 40 dB/decade and −40 dB/decade, respectively.

(3) The slopes of all Bode magnitude and phase lines are proportional to N. For example, the slope of the magnitude of a first-order simple-zero factor $(1 + j\omega/\omega_c)$ is 20 dB/decade, so the slope of a third-order simple-zero factor $(1 + j\omega/\omega_c)^3$ is 60 dB/decade.

Example 9-4: Bode Plots I

The voltage transfer function of a certain circuit is given by

$$\mathbf{H}(\omega) = \frac{(20 + j4\omega)^2}{j40\omega(100 + j2\omega)}.$$

(a) Rearrange the expression into standard form. (b) Generate Bode plots for the magnitude and phase of $\mathbf{H}(\omega)$.

Solution: (a) By factoring out 20^2 from the factor in the numerator and 100 from the factor in the denominator, we have

$$\mathbf{H}(\omega) = \frac{400(1 + j\omega/5)^2}{j4000\omega(1 + j\omega/50)} = \frac{-j0.1(1 + j\omega/5)^2}{\omega(1 + j\omega/50)}.$$

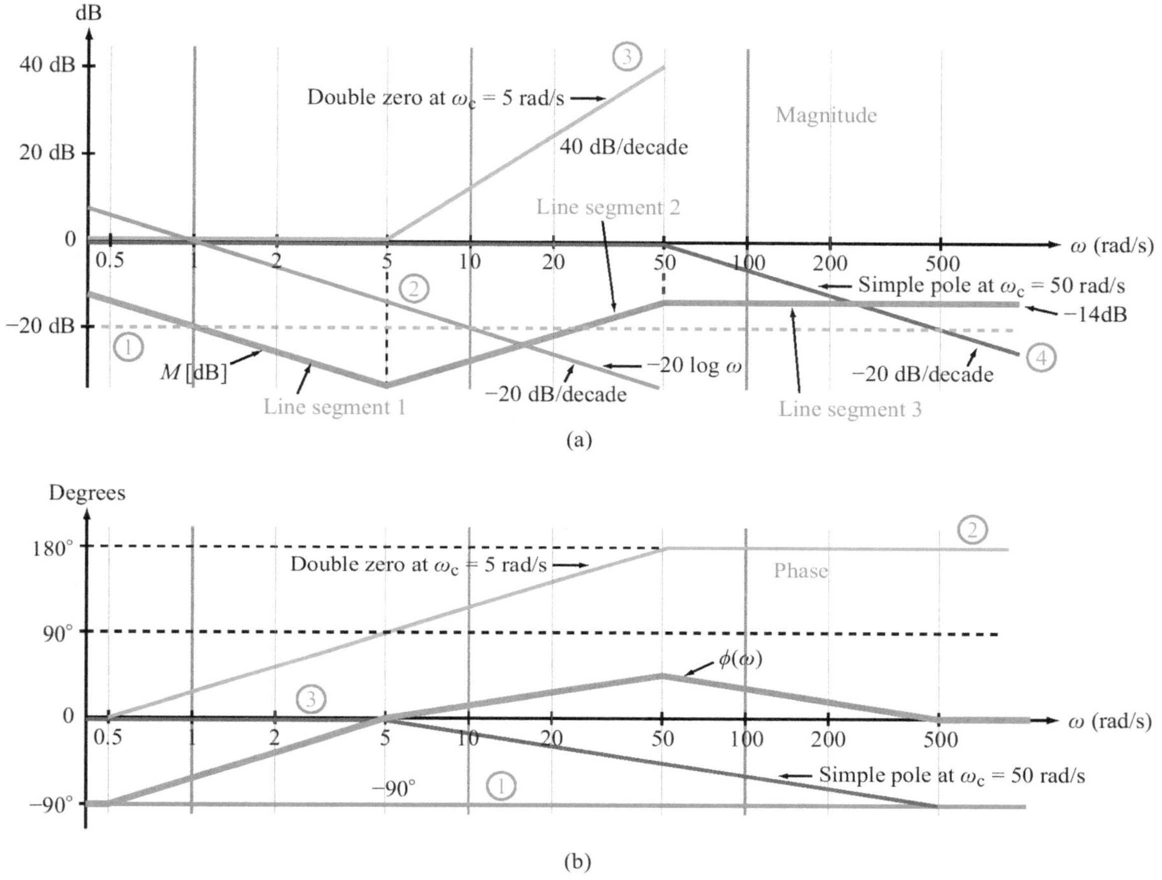

Figure 9-11: Bode amplitude and phase plots for the transfer function of Example 9-4.

The corner frequency of the double-zero factor given by $(1+j\omega/5)^2$ is $\omega_{c_1} = 5$ rad/s, and similarly, the corner frequency of the simple-pole factor given by $(1+j\omega/50)$ is $\omega_{c_2} = 50$ rad/s.

(b)

$$\begin{aligned} M\ [\text{dB}] &= 20 \log |\mathbf{H}| \\ &= 20 \log 0.1 + 40 \log |1 + j\omega/5| \\ &\quad - 20 \log \omega - 20 \log |1 + j\omega/50| \\ &= \underset{①}{-20\ \text{dB}} + \underset{③}{40 \log |1 + j\omega/5|} \\ &\quad \underset{②}{- 20 \log \omega} \underset{④}{- 20 \log |1 + j\omega/50|}. \end{aligned}$$

The Bode line-approximations for the four terms constituting M [dB] and their sum are shown in Fig. 9-11(a). The sum is obtained by graphically adding the line-approximations corresponding to the four individual terms. An alternative method for generating the Bode magnitude plot is to start by plotting the line-approximation of the term with the lowest corner frequency, and then to move forward along the ω axis while sequentially changing the slope of the line as we encounter terms with higher corner frequencies. To illustrate the procedure, we labeled the three line segments of M [dB] in Fig. 9-11(a) as line segments 1, 2, 3, and 4.

(1) The constant term is −20 dB (horizontal line with zero slope, labeled ① in Fig. 9-11(a)).

(2) The lowest-frequency term is the pole @ origin $(1/\omega)$. Its Bode line goes through 0 dB at $\omega = 1$ rad/s and has a slope of -20 dB/decade. It is labeled ② in Fig. 9-11(a)).

(3) The combination of (1) and (2) generates line segment 1, which goes through $\omega = 1$ at -20 dB.

(4) The term with the next higher corner frequency is the double zero with $\omega_c = 5$ rad/s (labeled ③ in Fig. 9-11(a)). A double zero has a Bode line with a slope of $+40$ dB/decade. Hence, at $\omega_c = 5$ rad/s, we change the slope of line segment 1 by adding 40 dB/decade to its original slope of -20 dB/decade. This step generates line segment 2, with a slope of $+20$ dB/decade.

(5) Line segment 2 continues until we encounter the corner frequency of the next term, namely the simple pole with $\omega_c = 50$ rad/s (labeled ④ in Fig. 9-11(a)). Adding a slope of -20 dB/decade leads to line segment 3, with a net slope of 0, and a constant level of -14 dB.

The phase of $\mathbf{H}(\omega)$ is given by

$$\phi = \underset{①}{-90^\circ} + \underset{②}{2\tan^{-1}\frac{\omega}{5}} - \underset{③}{\tan^{-1}\frac{\omega}{50}}.$$

Bode plots for ϕ and its three components are shown in Fig. 9-11(b).

Example 9-5: Bode Plots II

Transfer function $\mathbf{H}(\omega)$ is given by

$$\mathbf{H}(\omega) = \frac{(j10\omega + 30)^2}{(300 - 3\omega^2 + j90\omega)}.$$

(a) Rearrange $\mathbf{H}(\omega)$ into standard form. (b) Generate Bode plots for its magnitude and phase.

Solution: (a) Upon reversing the order of the real and imaginary components in the numerator, factoring out 30^2 from it, and factoring out 300 from the denominator, we get

$$\mathbf{H}(\omega) = \frac{3(1 + j\omega/3)^2}{[1 + j3\omega/10 + (j\omega/10)^2]},$$

which consists of a constant factor $K = 3$, a zero factor with a corner frequency of 3 (rad/s), and a quadratic pole with a corner frequency of 10 rad/s.

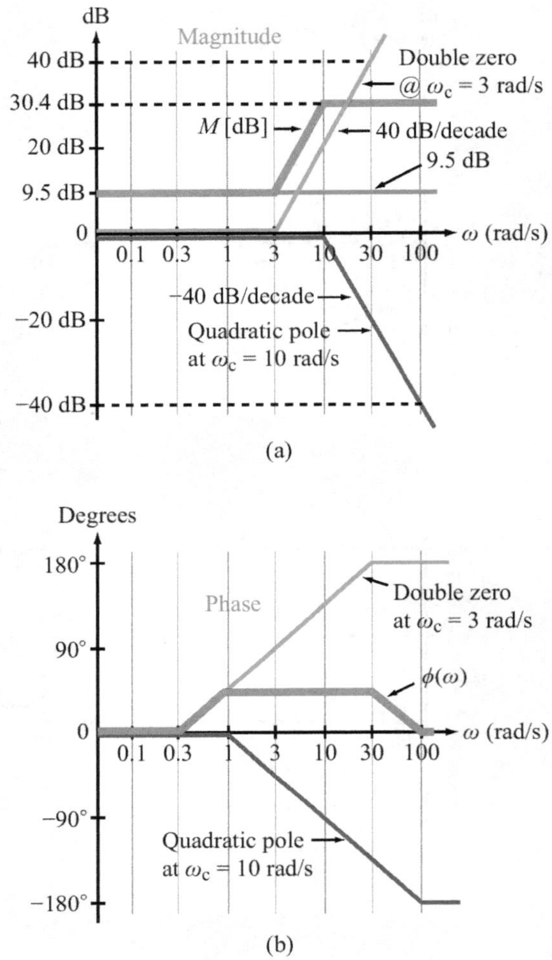

Figure 9-12: Bode magnitude and phase plots for Example 9-5.

(b)

$$\begin{aligned} M\ [\text{dB}] &= 20\log|\mathbf{H}| \\ &= 20\log 3 + 40\log|1 + j\omega/3| \\ &\quad - 20\log|1 + j3\omega/10 + (j\omega/10)^2| \\ &= 9.5\ \text{dB} + 40\log|1 + j\omega/3| \\ &\quad - 20\log|1 + j3\omega/10 + (j\omega/10)^2|, \end{aligned}$$

$$\phi = 2\tan^{-1}(\omega/3) - \tan^{-1}\left(\frac{3\omega/10}{1 - \omega^2/100}\right).$$

Bode plots of M [dB] and ϕ are shown in Fig. 9-12. We note that M [dB] exhibits a highpass filter-like response.

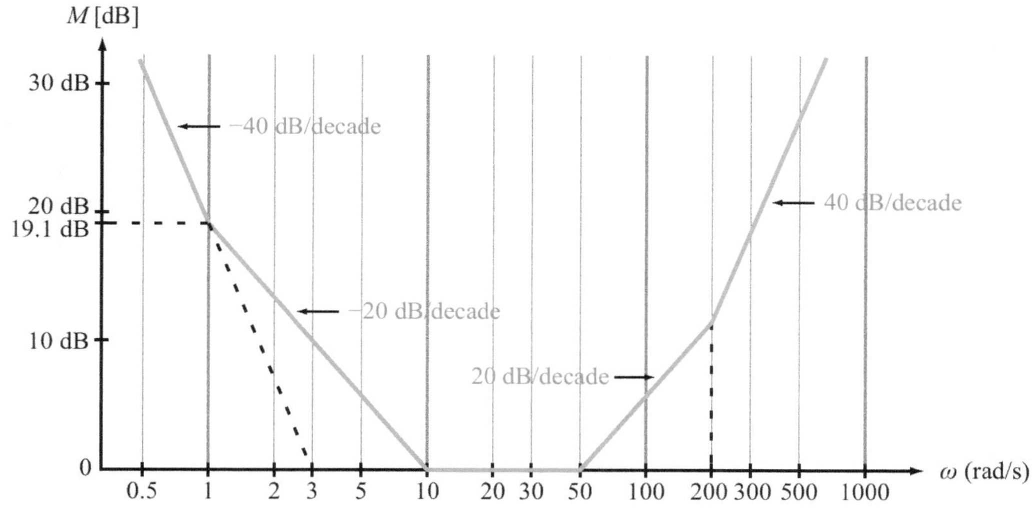

Figure 9-13: Bode plot of bandreject filter of Example 9-6.

Exercise 9-6: Generate a Bode magnitude plot for the transfer function

$$\mathbf{H} = \frac{10(100 + j\omega)(1000 + j\omega)}{(10 + j\omega)(10^4 + j\omega)}.$$

Answer:

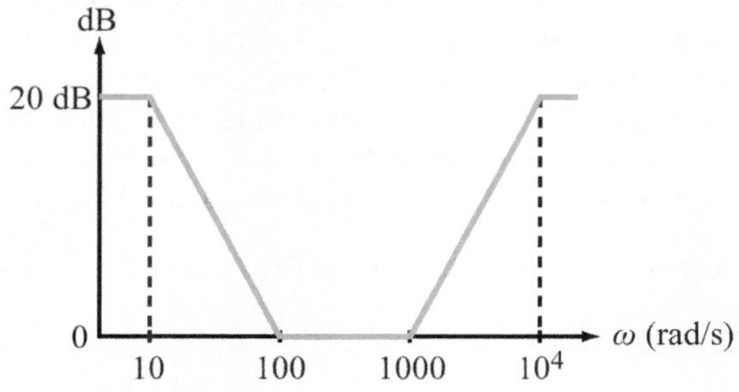

(See CAD)

Example 9-6: Bandreject Filter

The Bode magnitude plot shown in Fig. 9-13 belongs to a bandreject filter that provides significant gain at low and high frequencies, but no gain to frequencies in the 10 to 50 (rad/s) range. Obtain the transfer function $\mathbf{H}(\omega)$.

Solution: The Bode plot consists of five segments.

The first segment, corresponding to $\omega \le 1$ rad/s, is generated by a pole @ the origin that goes through $\omega = 3$ rad/s and has a slope of −40 dB/decade. The slope indicates that it is a double pole, so it must be given by

$$\mathbf{H}_1 = -\left(\frac{1}{\omega/3}\right)^2 = -\frac{9}{\omega^2}.$$

To verify the validity of our expression, let us convert it to dB as

$$M_1 \text{ [dB]} = 20 \log \frac{9}{\omega^2} = 20 \log 9 - 40 \log \omega$$
$$= 19.1 \text{ dB} - 40 \log \omega.$$

At $\omega = 1$ rad/s, M_1 [dB] = 19.1 dB, which matches the figure.

As we progress along the ω axis, the second segment has a slope of only −20 dB/decade, which means that a simple-zero factor with a corner frequency of 1 rad/s has come into play. Hence,

$$\mathbf{H}_2 = (1 + j\omega).$$

At $\omega = 10$ rad/s, the slope becomes zero, signifying the introduction of another simple-zero factor given by

$$\mathbf{H}_3 = (1 + j\omega/10).$$

Similarly,

$$\mathbf{H}_4 = (1 + j\omega/50)$$

and

$$\mathbf{H}_5 = (1 + j\omega/200).$$

Hence,

$$\mathbf{H}(\omega) = \mathbf{H}_1\mathbf{H}_2\mathbf{H}_3\mathbf{H}_4\mathbf{H}_5$$
$$= \frac{9(1 + j\omega)(1 + \frac{j\omega}{10})(1 + \frac{j\omega}{50})(1 + \frac{j\omega}{200})}{\omega^2}.$$

Since we are not given any information about the phase pattern of $\mathbf{H}(\omega)$, the solution we obtained is correct within a multiplication factor of j^N, which can accommodate j (for $N = 1$), -1 (for $N = 2$), $-j$ (for $N = 3$), and 1 (for $N = 4$). The magnitude of $\mathbf{H}(\omega)$ is the same, regardless of the value of N.

Concept Question 9-9: What does the term *standard form* of a transfer function refer to, and what purpose does it serve? (See CAD)

Concept Question 9-10: For which of the seven standard factors are the Bode plots identical to the exact plots and for which are they different? (See CAD)

Concept Question 9-11: What is the *gain roll-off rate*? (See CAD)

Exercise 9-7: Determine the functional form of the transfer function whose Bode magnitude plot is shown in Fig. E9.7, given that its phase angle at dc is 90°.

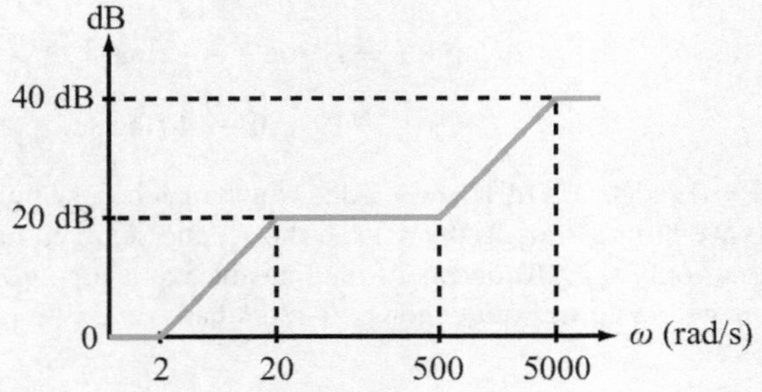

Figure E9.7

Answer: $\mathbf{H} = \dfrac{j(1 + j\omega/2)(1 + j\omega/500)}{(1 + j\omega/20)(1 + j\omega/5000)}$. (See CAD)

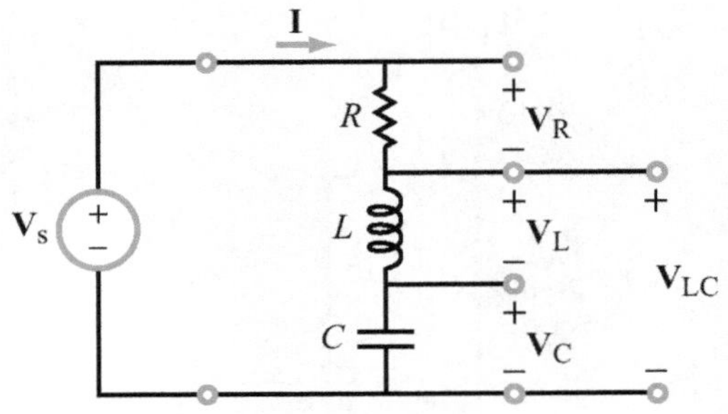

Figure 9-14: Series RLC circuit.

9-4 Passive Filters

Filters are of two types: passive and active.

► *Passive filters* are resonant circuits that contain only passive elements, namely resistors, capacitors, and inductors. In contrast, *active filters* contain op amps, transistors, and/or other active devices, in addition to the passive elements. ◄

Passive and active filters are the subject of the next four sections.

Any circuit that does not have a uniform frequency response is (by definition) a filter, simply because its output favors certain frequency ranges over others. Of particular interest to circuit designers are the four basic types of filters we introduced in Section 9-1. As we mentioned there, a filter transfer function is characterized by a number of attributes, including the following:

1. The frequency ranges of its passband(s) and stopband(s).
2. The gain factor M_0.
3. The gain roll-off rate S_g.

The objective of this section is to examine the basic properties of passive filters by analyzing their transfer functions. To that end, we use the series RLC circuit shown in Fig. 9-14, in which we have designated four voltage outputs, namely $\mathbf{V}_R$, $\mathbf{V}_L$, and $\mathbf{V}_C$ across the individual elements, and $\mathbf{V}_{LC}$ across the combination of L and C. We will examine the frequency responses of the transfer functions corresponding to all four output voltages.

9-4.1 Bandpass Filter

The current $\mathbf{I}$ flowing through the loop in Fig. 9-15(a) is given by

$$\mathbf{I} = \frac{\mathbf{V}_s}{R + j\left(\omega L - \dfrac{1}{\omega C}\right)} = \frac{j\omega C \mathbf{V}_s}{(1-\omega^2 LC) + j\omega RC}, \tag{9.44}$$

where we multiplied the numerator and denominator by $j\omega C$ to simplify the form of the expression. The transfer function corresponding to $\mathbf{V}_R$ is

$$\mathbf{H}_{BP}(\omega) = \frac{\mathbf{V}_R}{\mathbf{V}_s} = \frac{R\mathbf{I}}{\mathbf{V}_s} = \frac{j\omega RC}{(1-\omega^2 LC) + j\omega RC}, \tag{9.45}$$

where we added the subscript "BP" in anticipation of the fact that $\mathbf{H}_{BP}(\omega)$ is the transfer function of a bandpass filter. Its magnitude and phase angle are given by

$$M_{BP}(\omega) = |\mathbf{H}_{BP}(\omega)| = \frac{\omega RC}{\sqrt{(1-\omega^2 LC)^2 + \omega^2 R^2 C^2}}, \tag{9.46}$$

and

$$\phi_{BP}(\omega) = 90^\circ - \tan^{-1}\left[\frac{\omega RC}{1-\omega^2 LC}\right]. \tag{9.47}$$

According to the spectral plot displayed in Fig. 9-15(b), M_{BP} goes to zero at both extremes of the frequency spectrum and exhibits a maximum across an intermediate range centered at ω_0. Hence, the circuit functions like a bandpass (BP) filter, allowing the transmission (through it) of signals whose angular frequencies are close to ω_0 and discriminating against those with frequencies that are far away from ω_0.

The general profile of $M_{BP}(\omega)$ can be discerned by examining the circuit of Fig. 9-15(a) at specific values of ω. At $\omega = 0$, the capacitor behaves like an open circuit, allowing no current to flow and no voltage to develop across R. At $\omega = \infty$, it is the inductor that acts like an open circuit, again allowing no current to flow. In the intermediate frequency range when the value of ω is such that $\omega L = 1/\omega C$, the impedances of L and C cancel each other out, reducing the total impedance of the RLC circuit to R and the current to $\mathbf{I} = \mathbf{V}_s/R$. Consequently, $\mathbf{V}_R = \mathbf{V}_s$, and $\mathbf{H}_{BP} = 1$. To note the significance of this specific condition, we call it the *resonance condition*, and we refer to the frequency at which it occurs as the *resonant frequency* ω_0, which is given by

$$\omega_0 = \frac{1}{\sqrt{LC}}. \tag{9.48}$$

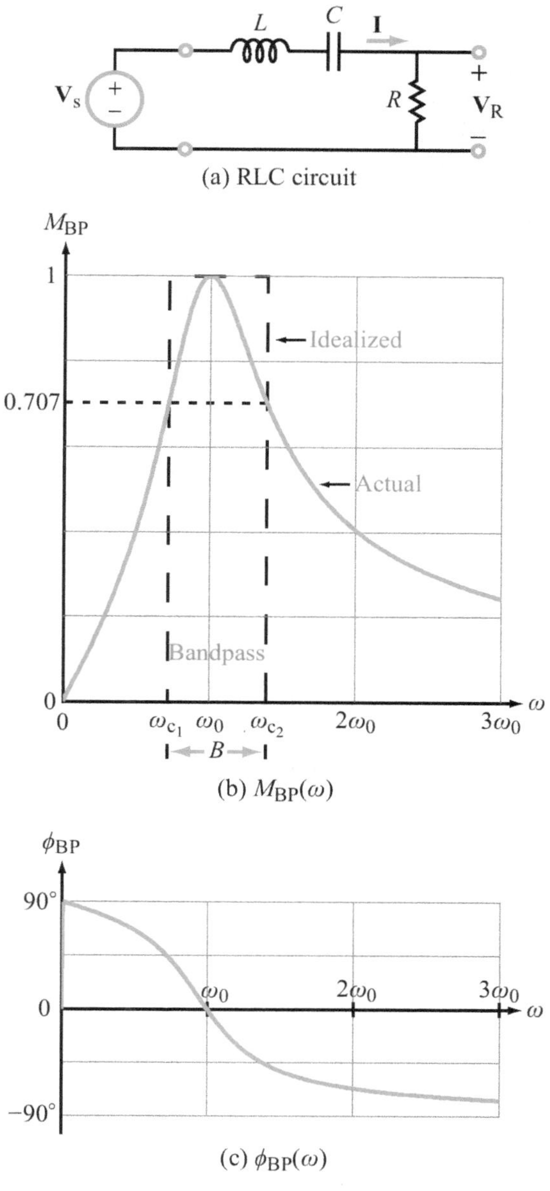

Figure 9-15: Series RLC bandpass filter.

The phase plot in Fig. 9-15(c) conveys the fact that ϕ_{BP} is dominated by the phase of C at low frequencies and by the phase of L at high frequencies, and $\phi_{BP} = 0$ at $\omega = \omega_0$.

At resonance, the energy in the circuit oscillates back and forth between L and C.

Filter bandwidth

The ***bandwidth*** of the bandpass filter is defined as the frequency range extending between ω_{c_1} and ω_{c_2}, where ω_{c_1} and ω_{c_2} are the values of ω at which $M^2_{\text{BP}}(\omega) = 0.5$ or $M_{\text{BP}}(\omega) = 1/\sqrt{2} = 0.707$. As we will see shortly, M^2_{BP} is proportional to the power delivered to the resistor in the RLC circuit. At resonance, the power is at its maximum, and at ω_{c_1} and ω_{c_2}, the power delivered to R is equal to 1/2 of the maximum possible, which explains why ω_{c_1} and ω_{c_2} also are referred to as the ***half-power frequencies*** (or the ***−3 dB frequencies*** on a dB scale). Thus,

$$M^2_{\text{BP}}(\omega) = \frac{1}{2} \qquad @\ \omega_{c_1} \text{ and } \omega_{c_2}. \tag{9.49}$$

Upon inserting the expression for $M_{\text{BP}}(\omega)$ given by Eq. (9.46) and carrying out several steps of algebra, we obtain the solutions

$$\omega_{c_1} = -\frac{R}{2L} + \sqrt{\left(\frac{R}{2L}\right)^2 + \frac{1}{LC}} \tag{9.50a}$$

and

$$\omega_{c_2} = \frac{R}{2L} + \sqrt{\left(\frac{R}{2L}\right)^2 + \frac{1}{LC}}. \tag{9.50b}$$

The bandwidth then is given by

$$B = \omega_{c_2} - \omega_{c_1} = \frac{R}{L}. \tag{9.51}$$

It is worth noting that ω_0 is equal to the geometric mean of ω_{c_1} and ω_{c_2}:

$$\omega_0 = \sqrt{\omega_{c_1}\omega_{c_2}}. \tag{9.52}$$

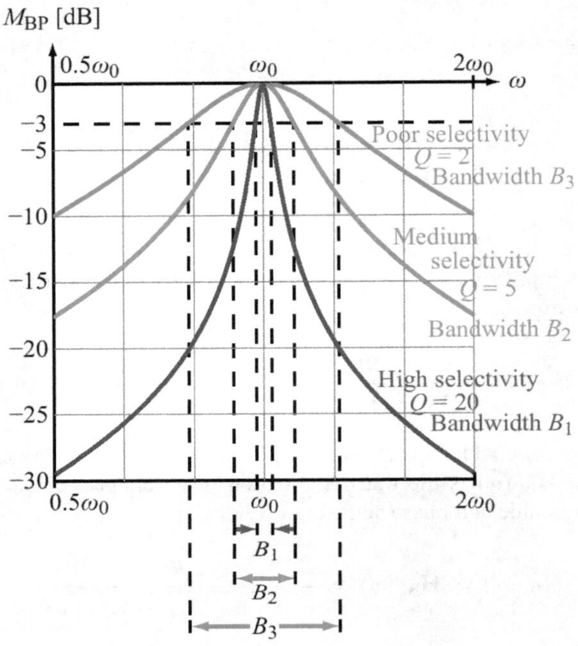

Figure 9-16: Examples of bandpass-filter responses.

Quality factor

According to the foregoing discussion, the choice of values we make for R, L, and C will specify the overall shape of the transfer function completely, as well as its center frequency ω_0 and bandwidth B.

> ▶ The ***quality factor*** of a circuit Q is an attribute commonly used to characterize the ***degree of selectivity*** of the circuit. A high Q circuit has a narrow bandwidth (relative to the center frequency) and high selectivity. ◀

Figure 9-16 displays frequency responses for three circuits, all with the same ω_0. The high-Q circuit exhibits a sharp response with a narrow bandwidth (relative to ω_0), the medium-Q circuit has a broader pattern, and the low-Q circuit has a pattern with limited selectivity.

For the bandpass-filter response, Q obviously is related to the ratio ω_0/B, but the formal definition of Q applies to any resonant circuit and is based on energy considerations, namely

$$Q = 2\pi \left(\frac{W_{\text{stor}}}{W_{\text{diss}}}\right)\bigg|_{\omega=\omega_0}, \tag{9.53}$$

where W_{stor} is the maximum energy that can be *stored* in the circuit at resonance ($\omega = \omega_0$), and W_{diss} is the *energy dissipated* by the circuit during a single period T. The stored energy is stored in L and C and the dissipated energy is dissipated in R. The factor 2π is an artificial multiplier introduced solely so that the expression for Q (that we will be deriving shortly) is simple in form and easy to remember.

In the RLC circuit of Fig. 9-15(a), the source is represented by the phasor voltage $\mathbf{V}_s$. In order to obtain expressions for the stored and dissipated energies at resonance, we need to (a) go back to the time domain and (b) specify ω as ω_0. For the sake of convenience and without loss of generality, we will assign the source the functional form

$$\upsilon_s(t) = V_0 \cos\omega_0 t \quad \longleftrightarrow \quad \mathbf{V}_s = V_0. \tag{9.54}$$

At resonance, $\omega_0 L = 1/\omega_0 C$, so the expressions for the total impedance of the circuit and the current flowing through it simplify to

$$\mathbf{Z} = R + j\omega_0 L - j/\omega_0 C = R \qquad (@\ \omega_0 = 1/\sqrt{LC}) \tag{9.55}$$

and

$$\mathbf{I} = \frac{\mathbf{V}_s}{\mathbf{Z}} = \frac{\mathbf{V}_s}{R} = \frac{V_0}{R}. \tag{9.56}$$

The time-domain current then is given by

$$i(t) = \mathfrak{Re}\left[\frac{V_0}{R}\, e^{j\omega_0 t}\right] = \frac{V_0}{R}\cos\omega_0 t. \tag{9.57}$$

At any instant in time, the instantaneous energies stored in the inductor and the capacitor are given by

$$w_{\text{L}}(t) = \frac{1}{2}\, L\, i_{\text{L}}^2(t) = \frac{V_0^2 L}{2R^2}\cos^2\omega_0 t \qquad \text{(J)} \tag{9.58a}$$

and

$$\begin{aligned} w_{\text{C}}(t) &= \frac{1}{2}\, C\, \upsilon_{\text{C}}^2(t) = \frac{1}{2}\, C\left(\frac{1}{C}\int i\; dt\right)^2 \\ &= \frac{1}{2}\, C\left(\frac{V_0}{\omega_0 RC}\sin\omega_0 t\right)^2 \\ &= \frac{V_0^2 L}{2R^2}\sin^2\omega_0 t \qquad \text{(J)}. \end{aligned} \tag{9.58b}$$

Even though both w_{L} and w_{C} vary with time, their sum is always a constant and equal to the maximum energy stored in the circuit,

$$\begin{aligned} W_{\text{stor}} &= w_{\text{L}}(t) + w_{\text{C}}(t) \\ &= \frac{V_0^2 L}{2R^2}\,[\cos^2\omega_0 t + \sin^2\omega_0 t] = \frac{V_0^2 L}{2R^2}. \end{aligned} \tag{9.59}$$

The energy dissipated by R during a single period is obtained by integrating the expression for the power p_{R} over a period $T = 1/f_0 = 2\pi/\omega_0$ so that

$$\begin{aligned} W_{\text{diss}} &= \int_0^T p_{\text{R}}\; dt \\ &= \int_0^T i^2 R\; dt = \int_0^{2\pi/\omega_0} \frac{V_0^2}{R}\cos^2\omega_0 t\; dt = \frac{\pi V_0^2}{\omega_0 R}. \end{aligned} \tag{9.60}$$

Upon substituting Eqs. (9.59) and (9.60) into Eq. (9.53), we obtain the result

$$Q = \frac{\omega_0 L}{R} \qquad \textbf{(bandpass filter)}. \tag{9.61}$$

Using the relation given by Eq. (9.51), the expression for the quality factor becomes

$$Q = \frac{\omega_0}{B} \qquad \textbf{(bandpass filter)}, \tag{9.62}$$

which is dimensionless. Thus, for a bandpass filter, Q is the inverse of the bandwidth B normalized to the center frequency ω_0.

To highlight the role of Q, we can use the expressions for Q and ω_0 to rewrite Eqs. (9.46) and (9.47) for the magnitude and phase angle of $\mathbf{H}_{\text{BP}}(\omega)$ in the forms

$$M_{\text{BP}}(\omega) = \frac{(\omega/Q\omega_0)}{\{[1-(\omega/\omega_0)^2]^2 + (\omega/Q\omega_0)^2\}^{1/2}} \tag{9.63a}$$

and

$$\phi_{\text{BP}}(\omega) = 90^\circ - \tan^{-1}\left\{\frac{(\omega/\omega_0)}{Q[1-(\omega/\omega_0)^2]}\right\}. \tag{9.63b}$$

Hence, the spectral response of the transfer function is specified completely by the combination of Q and ω_0.

Also, in view of Eq. (9.61), the expressions given by Eq. (9.50) for the half-power frequencies ω_{c_1} and ω_{c_2} can be rewritten as

$$\frac{\omega_{c_1}}{\omega_0} = -\frac{1}{2Q} + \sqrt{1 + \frac{1}{4Q^2}} \tag{9.64a}$$

Table 9-3: Attributes of series and parallel RLC bandpass circuits.

RLC Circuit	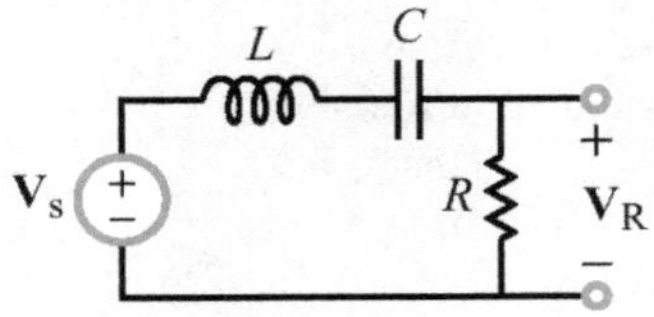	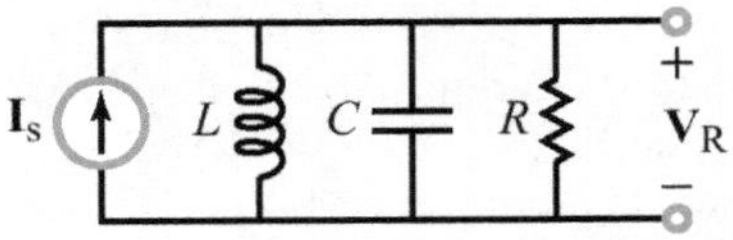
Transfer Function	$\mathbf{H} = \dfrac{\mathbf{V}_R}{\mathbf{V}_s}$	$\mathbf{H} = \dfrac{\mathbf{V}_R}{\mathbf{I}_s}$
Resonant Frequency, ω_0	$\dfrac{1}{\sqrt{LC}}$	$\dfrac{1}{\sqrt{LC}}$
Bandwidth, B	$\dfrac{R}{L}$	$\dfrac{1}{RC}$
Quality Factor, Q	$\dfrac{\omega_0}{B} = \dfrac{\omega_0 L}{R}$	$\dfrac{\omega_0}{B} = \dfrac{R}{\omega_0 L}$
Lower Half-Power Frequency, ω_{c_1}	$\left[-\dfrac{1}{2Q} + \sqrt{1 + \dfrac{1}{4Q^2}}\right]\omega_0$	$\left[-\dfrac{1}{2Q} + \sqrt{1 + \dfrac{1}{4Q^2}}\right]\omega_0$
Upper Half-Power Frequency, ω_{c_2}	$\left[\dfrac{1}{2Q} + \sqrt{1 + \dfrac{1}{4Q^2}}\right]\omega_0$	$\left[\dfrac{1}{2Q} + \sqrt{1 + \dfrac{1}{4Q^2}}\right]\omega_0$

Notes: (1) The expression for Q of the series RLC circuit is the inverse of that for Q of the parallel circuit.
(2) For $Q \geq 10$, $\omega_{c_1} \approx \omega_0 - \frac{B}{2}$, and $\omega_{c_2} \approx \omega_0 + \frac{B}{2}$.

and

$$\frac{\omega_{c_2}}{\omega_0} = \frac{1}{2Q} + \sqrt{1 + \frac{1}{4Q^2}}. \tag{9.64b}$$

For a circuit with $Q > 10$, the expressions for ω_{c_1} and ω_{c_2} simplify to

$$\omega_{c_1} \approx \omega_0 - \frac{B}{2}, \qquad \omega_{c_2} \approx \omega_0 + \frac{B}{2}, \qquad \text{(if } Q > 10) \quad (9.65)$$

thereby forming a symmetrical bandpass centered at ω_0. Table 9-3 provides a summary of the salient features of the series RLC bandpass filter. For comparison, the table also includes the corresponding list for the parallel RLC circuit.

Example 9-7: Filter Design

(a) Design a series RLC bandpass filter with a center frequency $f_0 = 1$ MHz (Fig. 9-17) and a quality factor $Q = 20$, given that $L = 0.1$ mH.

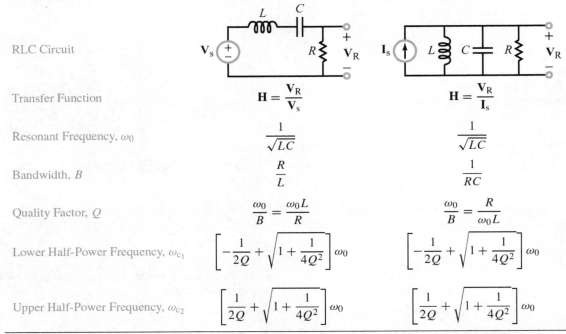

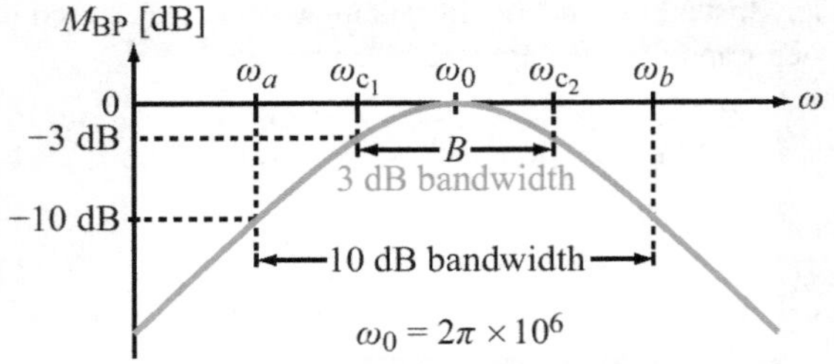

Figure 9-17: 10 dB bandwidth extends from ω_a to ω_b, corresponding to M_{BP} (dB) = −10 dB.

(b) Determine the 10 dB bandwidth of the filter, which is defined as the bandwidth between frequencies at which the power level is 10 dB below the peak value.

Solution: (a) Application of

$$\omega_0 = 2\pi f_0 = 2\pi \times 10^6 = \frac{1}{\sqrt{LC}} = \frac{1}{\sqrt{10^{-4}C}}$$

leads to $C = 0.25$ nF.

Solving Eq. (9.61) for R gives

$$R = \frac{\omega_0 L}{Q} = \frac{2\pi \times 10^6 \times 10^{-4}}{20} = 31.4\ \Omega.$$

(b) Voltage is proportional to $M_{\rm BP}$, and power is proportional to $M_{\rm BP}^2$. The definition for power in dB is

$$P\ [\text{dB}] = 10 \log P = 10 \log M_{\rm BP}^2 = 20 \log M_{\rm BP} = M_{\rm BP}\ [\text{dB}].$$

We seek to find angular frequencies ω_a and ω_b corresponding to $M_{\rm BP}$ [dB] $= -10$ dB (Fig. 9-17). If

$$20 \log M_{\rm BP} = -10 \text{ dB},$$

it follows that

$$M_{\rm BP} = 10^{-0.5} = 0.316.$$

The expression for $M_{\rm BP}$ is given by Eq. (9.46) as

$$M_{\rm BP} = \frac{\omega RC}{\sqrt{(1-\omega^2 LC)^2 + \omega^2 R^2 C^2}}.$$

With $M_{\rm BP} = 0.316$, $R = 31.4\ \Omega$, $L = 10^{-4}$ H, and $C = 0.25$ nF, solution of the expression yields

$$\frac{\omega_a}{\omega_0} = 0.93 \qquad \text{and} \qquad \frac{\omega_b}{\omega_0} = 1.08.$$

The corresponding bandwidth in Hz is

$$B_{10\ \text{dB}} = (1.08 - 0.93) \times 1 \text{ MHz} = 0.15 \text{ MHz}.$$

Example 9-8: Two-Stage Bandpass Filter

Determine $\mathbf{H}(\omega) = \mathbf{V}_o/\mathbf{V}_s$ for the two-stage BP-filter circuit shown in Fig. 9-18. If $Q_1 = \omega_0 L/R$ is the quality factor of a single stage alone, what is Q_2 for the two stages in combination, given that $R = 2\ \Omega$, $L = 10$ mH, and $C = 1\ \mu$F?

Solution: For each stage alone,

$$\omega_0 = \frac{1}{\sqrt{LC}} = \frac{1}{\sqrt{10^{-2} \times 10^{-6}}} = 10^4 \text{ rad/s}$$

and

$$Q_1 = \frac{\omega_0 L}{R} = \frac{10^4 \times 10^{-2}}{2} = 50.$$

The loop equations for mesh currents $\mathbf{I}_1$ and $\mathbf{I}_2$ are

$$-\mathbf{V}_s + \mathbf{I}_1 \left(j\omega L + \frac{1}{j\omega C} + R \right) - R\mathbf{I}_2 = 0$$

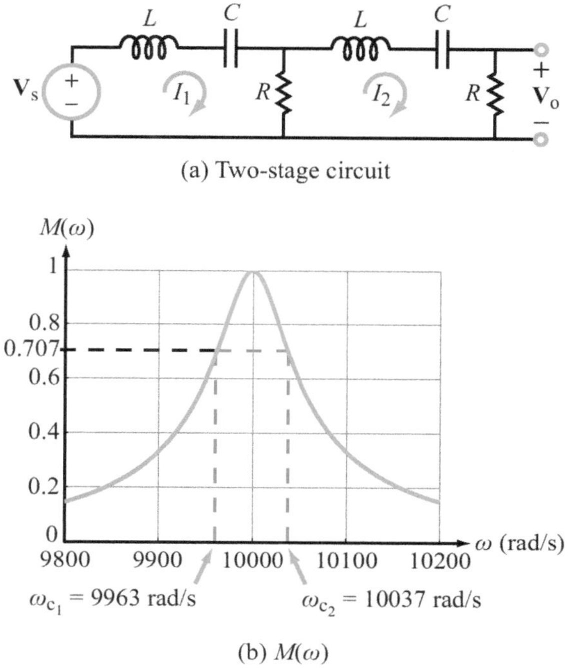

Figure 9-18: Two-stage RLC circuit of Example 9-8.

and

$$-R\mathbf{I}_1 + \mathbf{I}_2 \left(2R + j\omega L + \frac{1}{j\omega C} \right) = 0.$$

Simultaneous solution of the two equations leads to

$$\mathbf{H}(\omega) = \frac{\mathbf{V}_o}{\mathbf{V}_s}$$

$$= \frac{\omega^2 R^2 C^2}{\omega^2 R^2 C^2 - (1-\omega^2 LC)^2 - j3\omega RC(1-\omega^2 LC)}$$

$$= \frac{\omega^2 R^2 C^2 [\omega^2 R^2 C^2 - (1-\omega^2 LC)^2 + j3\omega RC(1-\omega^2 LC)]}{[\omega^2 R^2 C^2 - (1-\omega^2 LC)^2]^2 + 9\omega^2 R^2 C^2 (1-\omega^2 LC)^2}.$$

Resonance occurs when the imaginary part of $\mathbf{H}(\omega)$ is zero, which is satisfied either when $\omega = 0$ (which is a trivial resonance) or when $\omega = 1/\sqrt{LC}$. Hence, the two-stage circuit has the same resonance frequency as a single-stage circuit.

Using the specified values of R, L, and C, we can calculate the magnitude $M(\omega) = |\mathbf{H}(\omega)|$ and plot it as a function of ω.

The result is displayed in Fig. 9-18(b). From the spectral plot, we have

$$\omega_{c_1} = 9963 \text{ rad/s},$$
$$\omega_{c_2} = 10037 \text{ rad/s},$$
$$B_2 = \omega_{c_2} - \omega_{c_1} = 10037 - 9963 = 74 \text{ rad/s},$$

and

$$Q_2 = \frac{\omega_0}{B_2} = \frac{10^4}{74} = 135,$$

where B_2 is the bandwidth of the two-stage BP-filter response. The two-stage combination increases the quality factor from 50 to 135.

Exercise 9-8: Show that for the parallel RLC circuit shown in Fig. E9.8, the transfer-impedance transfer function $\mathbf{H_Z} = \mathbf{V_R}/\mathbf{I_s}$ exhibits a bandpass-filter response.

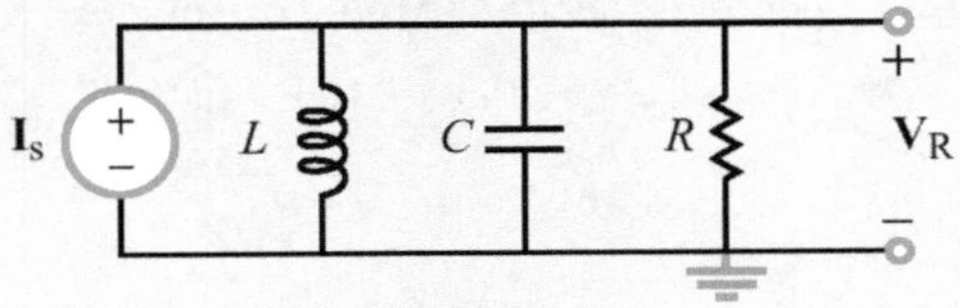

Figure E9.8

Answer:

$$\mathbf{H_Z} = \frac{\mathbf{V_R}}{\mathbf{I_s}} = \frac{j\omega L}{(1 - \omega^2 LC) + j\omega L/R}.$$

The functional form of $\mathbf{H_Z}(\omega)$ is identical to that given by Eq. (9.45) for the series RLC bandpass filter. Moreover, both circuits resonate at $\omega = 1/\sqrt{LC}$. (See CAD)

9-4.2 Highpass Filter

At low frequencies, the capacitor C in the circuit of Fig. 9-19(a) acts like an open circuit, so $\mathbf{V}_L$ across the inductor is essentially zero. Conversely, at high frequencies, the capacitor acts like a short circuit and the inductor acts like an open circuit. Consequently, $\mathbf{V}_L \approx \mathbf{V}_s$. This behavior constitutes a highpass filter.

Transfer function $\mathbf{H}_{HP}(\omega)$, corresponding to $\mathbf{V}_L$ in the circuit of Fig. 9-19(a), is given by

$$\mathbf{H}_{HP}(\omega) = \frac{\mathbf{V}_L}{\mathbf{V}_s} = \frac{j\omega L\mathbf{I}}{\mathbf{V}_s} = \frac{-\omega^2 LC}{(1 - \omega^2 LC) + j\omega RC} \qquad (9.66)$$

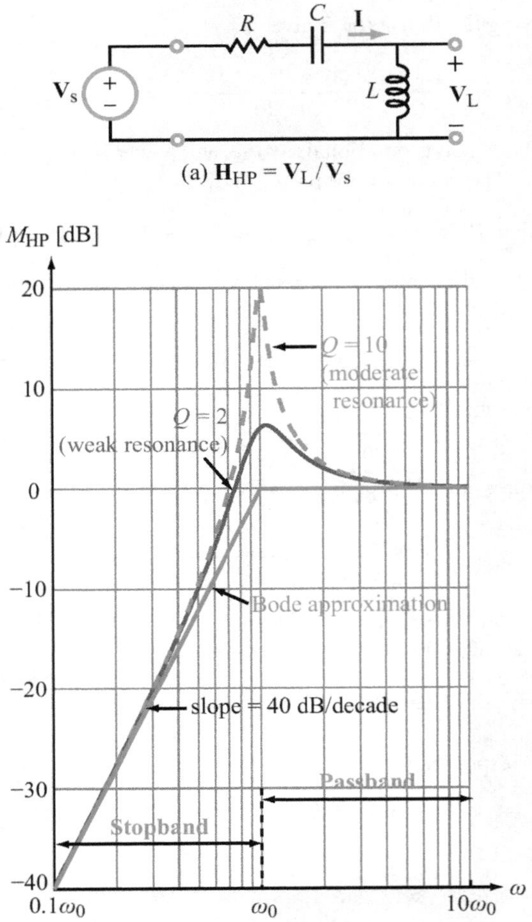

(a) $\mathbf{H}_{HP} = \mathbf{V}_L/\mathbf{V}_s$

(b) Magnitude spectrum

Figure 9-19: Plots of M_{HP} [dB] for $Q = 2$ (weak resonance) and $Q = 10$ (moderate resonance).

with magnitude and phase angle

$$M_{HP}(\omega) = \frac{\omega^2 LC}{[(1 - \omega^2 LC)^2 + \omega^2 R^2 C^2]^{1/2}}$$
$$= \frac{(\omega/\omega_0)^2}{\{[1 - (\omega/\omega_0)^2]^2 + (\omega/Q\omega_0)^2\}^{1/2}} \qquad (9.67a)$$

and

$$\phi_{HP}(\omega) = 180^\circ - \tan^{-1}\left[\frac{\omega RC}{1 - \omega^2 LC}\right]$$
$$= 180^\circ - \tan^{-1}\left\{\frac{(\omega/\omega_0)}{Q[1 - (\omega/\omega_0)^2]}\right\}, \qquad (9.67b)$$

where ω_0 and Q are defined by Eqs. (9.48) and (9.61), respectively. Figure 9-19(b) displays logarithmic plots of M_{HP} [dB] for two values of Q. Because $M_{HP}(\omega)$ has a quadratic zero, its slope in the stopband is 40 dB/decade.

Exercise 9-9: How should R be related to L and C so that the denominator of Eq. (9.66) becomes a simple pole of order 2? What will the value of Q be in that case?

Answer: $R = 2\sqrt{L/C}$, $Q = 1/2$. (See CAD)

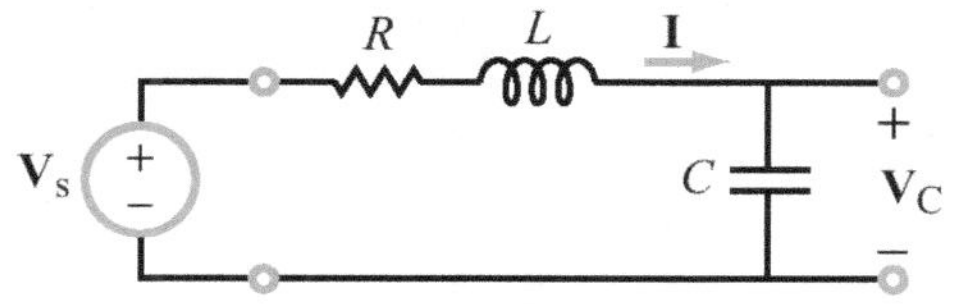

(a) $\mathbf{H}_{LP} = \mathbf{V}_C/\mathbf{V}_s$

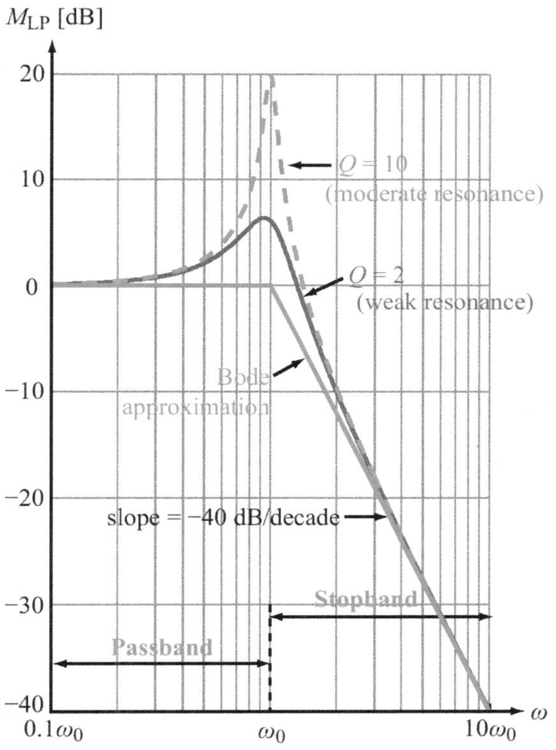

(b) Magnitude spectrum

Figure 9-20: RLC lowpass filter.

9-4.3 Lowpass Filter

The voltage across the capacitor in Fig. 9-20(a) generates a lowpass-filter transfer function given by

$$\mathbf{H}_{LP}(\omega) = \frac{\mathbf{V}_C}{\mathbf{V}_s} = \frac{(1/j\omega C)\mathbf{I}}{\mathbf{V}_s} = \frac{1}{(1-\omega^2 LC) + j\omega RC}, \tag{9.68}$$

with magnitude and phase angle given by

$$M_{LP}(\omega) = \frac{1}{[(1-\omega^2 LC)^2 + \omega^2 R^2 C^2]^{1/2}} = \frac{1}{\{[1-(\omega/\omega_0)^2]^2 + (\omega/Q\omega_0)^2\}^{1/2}}, \tag{9.69a}$$

and

$$\phi_{LP}(\omega) = -\tan^{-1}\left(\frac{\omega RC}{1-\omega^2 LC}\right) = -\tan^{-1}\left\{\frac{(\omega/\omega_0)}{Q[1-(\omega/\omega_0)^2]}\right\}. \tag{9.69b}$$

The spectral plots of M_{LP} [dB] shown in Fig. 9-20(b) are mirror images of the highpass filter plots displayed in Fig. 9-19(b).

9-4.4 Bandreject Filter

The output voltage across the combination of L and C in Fig. 9-21(a) generates a bandreject filter transfer function and is equal to $\mathbf{V}_s - \mathbf{V}_R$:

$$\mathbf{H}_{BR}(\omega) = \frac{\mathbf{V}_L + \mathbf{V}_C}{\mathbf{V}_s} = \frac{\mathbf{V}_s - \mathbf{V}_R}{\mathbf{V}_s} = 1 - \mathbf{H}_{BP}(\omega), \tag{9.70}$$

where $\mathbf{H}_{BP}(\omega)$ is the bandpass filter transfer function given by Eq. (9.45). The spectral response of $\mathbf{H}_{BP}$ passes all frequencies except for an intermediate band centered at ω_0, as shown in Fig. 9-21(b). The width of the stopband is determined by the values of ω_0 and Q.

Exercise 9-10: Is $M_{BR} = 1 - M_{BP}$?

Answer: No, because $M_{BR} = |\mathbf{H}_{BR}| = |1 - \mathbf{H}_{BP}| \neq 1 - |\mathbf{H}_{BP}| = 1 - M_{BP}$. (See CAD)

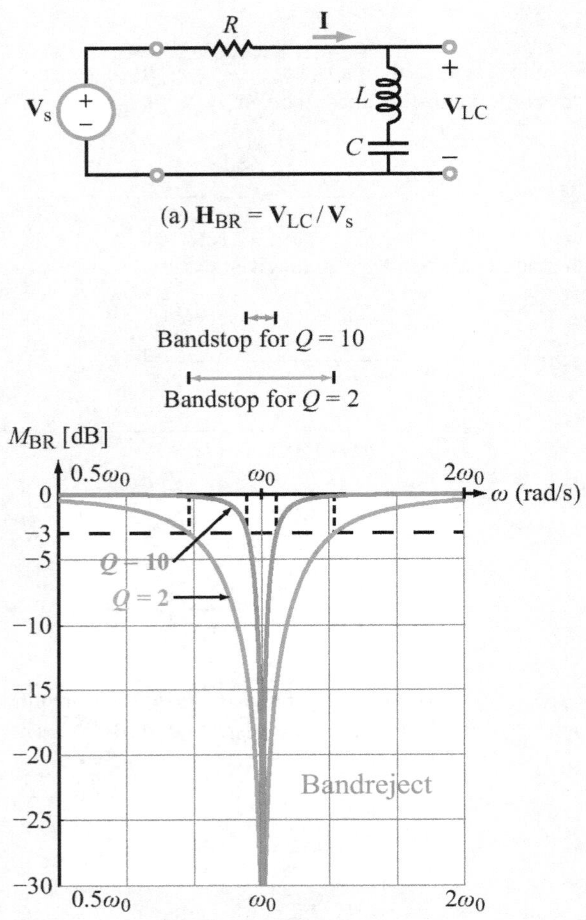

(a) $\mathbf{H}_{BR} = \mathbf{V}_{LC}/\mathbf{V}_s$

(b) Spectral response

Figure 9-21: Bandreject filter.

9-5 Filter Order

In Section 9-3, we associated the term *order* with the power of ω, so a factor given by $(1 + j\omega/\omega_c)^2$ was called a second-order zero because the highest power of ω in the expression is 2. Similarly, $(1 + j\omega/\omega_c)^{-2}$ was called a second-order pole because the highest power of ω is also 2, but the expression appears in the denominator. The term order also is used to describe the overall filter response, which may be composed of the product of several zero and pole factors—each with its own order. Multiple, different characterizations have been used over the years to define the order of a filter, so to avoid ambiguity, we adopt the following definition.

► The order of a filter is equal to the absolute value of the highest power of ω in its transfer function when ω is in the filter's stopband(s). ◄

Let us examine this definition for three circuit configurations.

9-5.1 First-Order Lowpass RC Filter

The transfer function of the RC circuit shown in Fig. 9-22(a) is given by

$$\mathbf{H}_1(\omega) = \frac{\mathbf{V}_C}{\mathbf{V}_s} = \frac{1/j\omega C}{R + 1/j\omega C} = \frac{1}{1 + j\omega RC} = \frac{1}{1 + j\omega/\omega_{c_1}} \quad \text{(first-order)}, \tag{9.71}$$

where we multiplied both the numerator and denominator by $j\omega C$ so as to rearrange the expression into the *standard form* we discussed in Section 9-3.2. The expression given by Eq. (9.71) is a simple pole with a *corner frequency* given by

$$\omega_{c_1} = \frac{1}{RC} \quad \text{(RC filter)}. \tag{9.72}$$

It is evident from the expression given by Eq. (9.71) that the highest order of ω is 1, and therefore the RC circuit is a first-order filter. Strict application of the definition for the order of a filter requires that we evaluate the power of ω when ω is in the stopband of the filter. In the present case, the stopband covers the range $\omega \geq \omega_{c_1}$. When ω is well into the stopband $(\omega/\omega_{c_1} \gg 1)$, Eq. (9.71) simplifies to

$$\mathbf{H}_1(\omega) \approx \frac{-j\omega_{c_1}}{\omega} \quad (\text{for } \omega/\omega_{c_1} \gg 1), \tag{9.73}$$

which confirms the earlier conclusion that the RC circuit is first-order.

► A circuit containing a single reactive element (capacitor or inductor) generates a first-order transfer function. Generally speaking, the order of a filter depends on the number of reactive elements contained in the circuit. The order may be smaller or equal to the number of reactive elements, but not greater. ◄

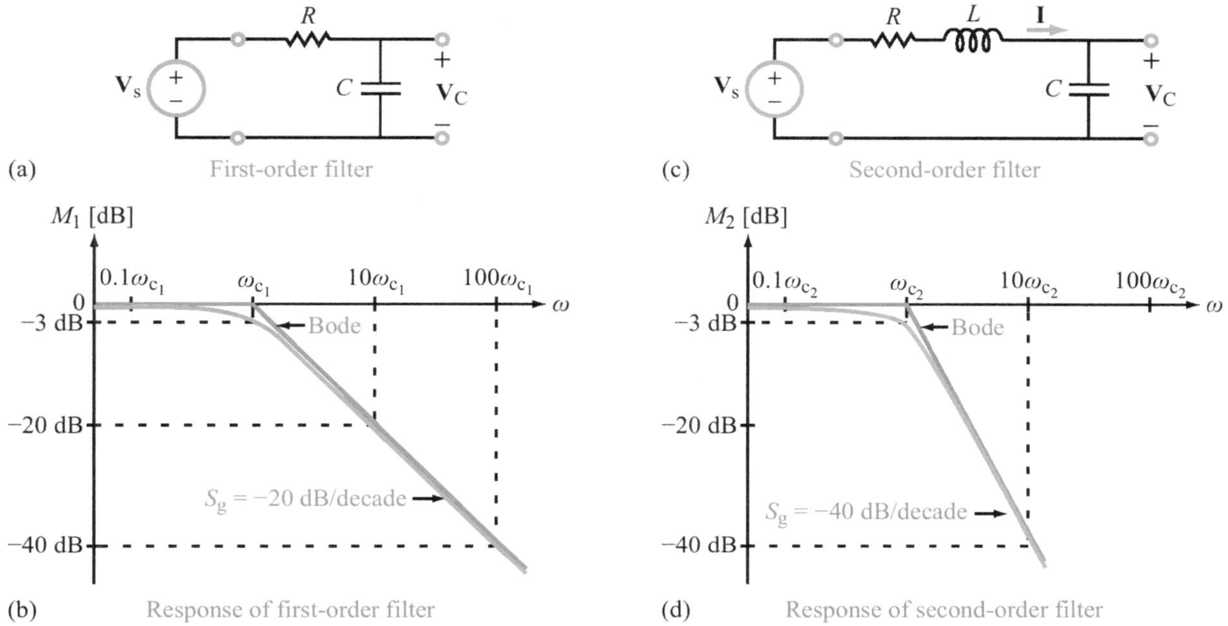

Figure 9-22: Comparison of magnitude responses of the first-order RC filter and the second-order RLC filter. The corner frequencies are given by $\omega_{c_1} = 1/RC$ and $\omega_{c_2} = 1.28/RC$.

The magnitude of $\mathbf{H}_1(\omega)$ is given by

$$M_1 = |\mathbf{H}_1(\omega)| = \frac{1}{|1 + j\omega/\omega_{c_1}|} = \frac{1}{\sqrt{1 + (\omega/\omega_{c_1})^2}}. \quad (9.74)$$

At $\omega = \omega_{c_1}$,

$$M_1(\omega_{c_1}) = \frac{1}{\sqrt{1+1}} = 0.707. \quad (9.75)$$

When expressed in dB, M_1 becomes

$$\begin{aligned} M_1 \text{ [dB]} &= 20 \log M_1 \\ &= -10 \log[1 + (\omega/\omega_{c_1})^2] \\ &= \begin{cases} 0 \text{ dB} & @\ \omega = 0, \\ -3 \text{ dB} & @\ \omega/\omega_{c_1} = 1, \\ -20 \log(\omega/\omega_{c_1}) & @\ \omega/\omega_{c_1} \gg 1. \end{cases} \end{aligned} \quad (9.76)$$

On the semilog scale of Fig. 9-22(b), M_1 [dB] starts out at 0 dB—corresponding to $M_1 = 1$ in natural units—decreases to −3 dB at $\omega = \omega_{c_1}$, and then its slope accelerates towards a steady-state value of −20 dB/decade at much greater values of ω. As noted earlier, the steepness (slope) of the transfer function after it has transitioned from its passband to its stopband is called its gain roll-off rate S_g.

► For a first-order filter, $S_g = -20$ dB/decade. To achieve a faster rate of decay, second- or higher-order filters are called for. ◄

Example 9-9: Filter Transmission Spectrum

An RC lowpass filter uses a capacitor $C = 10\ \mu$F. (a) Specify R so that $\omega_{c_1} = 1$ krad/s. (b) The filter is considered acceptably transparent to a signal if the signal's voltage amplitude is reduced by no more than 12 dB as it passes through the filter. What is the filter's transmission spectrum according to this criterion?

Solution: (a) Application of Eq. (9.72) leads to

$$R = \frac{1}{\omega_{c_1} C} = \frac{1}{10^3 \times 10^{-5}} = 100\ \Omega.$$

(b) If

$$M_1 \text{ [dB]} = 20 \log M_1 = -12 \text{ dB},$$

then

$$\log M_1 = -\frac{12}{20} = -0.6$$

and

$$M_1 = 10^{-0.6} = 0.25.$$

Equating this value of M_1 to the expression given by Eq. (9.74) leads to

$$M_1 = \frac{1}{\sqrt{1 + (\omega/\omega_{c_1})^2}} = 0.25,$$

which yields the solution

$$\frac{\omega}{\omega_{c_1}} = 3.87 \qquad \text{or} \qquad \omega = 3.87 \text{ krad/s}.$$

Hence, the transmission spectrum of the filter extends from 0 to 3.87 krad/s or equivalently from 0 to 616 Hz.

9-5.2 Second-Order Lowpass Filter

For the RLC circuit shown in Fig. 9-22(c), we determined in Section 9-4.3 that its transfer function is given by Eq. (9.68) as

$$\mathbf{H}_2(\omega) = \frac{\mathbf{V}_C}{\mathbf{V}_s} = \frac{1}{(1 - \omega^2 LC) + j\omega RC} \qquad \text{(RLC filter)}. \qquad (9.77)$$

The magnitude spectrum of the RLC lowpass filter was presented earlier in Fig. 9-20(b), where it was observed that the response may exhibit a resonance phenomenon in the neighborhood of $\omega_0 = 1/\sqrt{LC}$, and that it decays with $S_g = -40$ dB/decade in the stopband ($\omega \geq \omega_0$). This is consistent with the fact that the RLC circuit generates a second-order lowpass filter when the output voltage is taken across the capacitor. In terms of our definition for the order of a filter in the stopband ($\omega^2 \gg 1/LC$), Eq. (9.77) reduces to

$$\mathbf{H}_2(\omega) \approx \frac{-1}{\omega^2 LC} \qquad (\text{for } \omega \gg \omega_0), \qquad (9.78)$$

which assumes the form of a second-order pole.

The ripple-like effect exhibited by the RLC filter in Fig. 9-20 can be avoided through a judicious choice of the values of R, L, and C. By replacing the minus sign in Eq. (9.77) with j^2 and selecting R such that

$$R = 2\sqrt{\frac{L}{C}} \qquad \text{(no-ripple condition)}, \qquad (9.79)$$

the expression given by Eq. (9.77) can be converted into a perfect square:

$$\begin{aligned} \mathbf{H}_2(\omega) &= \frac{1}{1 + j^2\omega^2 LC + j2\sqrt{LC}} \\ &= \frac{1}{(1 + j\omega\sqrt{LC})^2} \qquad \text{(second order)}. \qquad (9.80) \end{aligned}$$

The constraint given by Eq. (9.79) allowed us to convert $\mathbf{H}_2(\omega)$ from a quadratic pole into a simple pole of second-order. Its magnitude response is displayed in Fig. 9-22(d).

The corner frequency of the RLC lowpass filter (ω_{c_2}) is determined by setting the magnitude of $\mathbf{H}_2(\omega)$ equal to $1/\sqrt{2}$. Thus,

$$|\mathbf{H}_2(\omega_{c_2})| = \frac{1}{1 + \omega_{c_2}^2 LC} = \frac{1}{\sqrt{2}},$$

which leads to

$$\omega_{c_2} = \left\{ \frac{\sqrt{2} - 1}{LC} \right\}^{1/2} = \frac{0.64}{\sqrt{LC}}. \qquad (9.81)$$

From Eq. (9.79), $L = R^2C/4$. When used in Eq. (9.81), the expression for ω_{c_2} becomes

$$\omega_{c_2} = \frac{1.28}{RC} \qquad \text{(RLC filter)}. \qquad (9.82)$$

The foregoing analysis warrants the following observations:

1. The RC lowpass filter is first-order, its corner frequency is $\omega_{c_1} = 1/RC$, and its gain roll-off rate is $S_g = -20$ dB/decade.

2. By adding a series inductor whose value is specified by $L = R^2C/4$, the filter becomes second-order, its corner frequency shifts upward to $1.28/RC$, and its slope becomes twice as steep.

Technology Brief 23
Spectral and Spatial Filtering

Filtering is applied in electrical systems, sound systems, optical systems, mechanical systems, and more. It is so ubiquitous that we often are unaware of the filtering that occurs in every day life. This Technology Brief provides an overview of how filters are applied in sight and photography, and how this affects what we see. There are two major types of filters in what we see—*spectral filters* that control color and *spatial filters* that control smoothing and edges.

Figure TF23-2: The eyes are a nonlinear bandpass filter. Greenish yellow provides high visibility in both dim and bright conditions because of the nonlinear sensitivity of our eyes to this color. (Credit: Agoora.co.uk.)

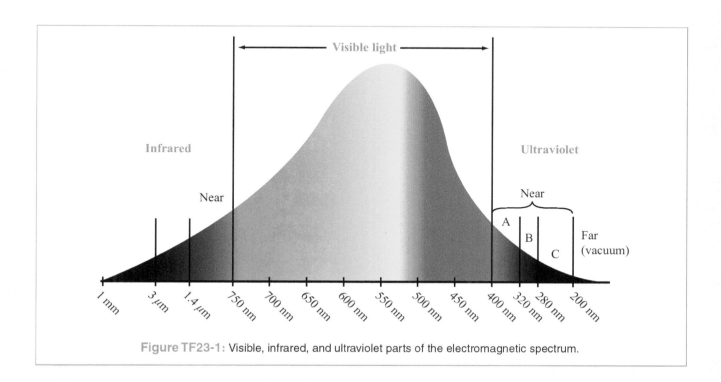

Spectral Filters

Our eye is, itself, a spectral filter. The colors we can see range in wavelength from about 750 nm (400 THz, red) to 400 nm (750 THz, violet). Other frequencies, such as the ultraviolet (UV) and infrared (IR), are not detectable by the human eye. Thus, our eye is a spectral bandpass filter with a bandwidth of about 350 THz. But we do not see all colors equally well. Our eye's frequency response is nonlinear. Figure TF23-1 shows the relative response of the eye to different colors of light. The eye is more sensitive to colors in the yellow-green region than to red and blue colors. This is why neon-yellow clothing provides high visibility under both dim and bright conditions (Fig. TF23-2).

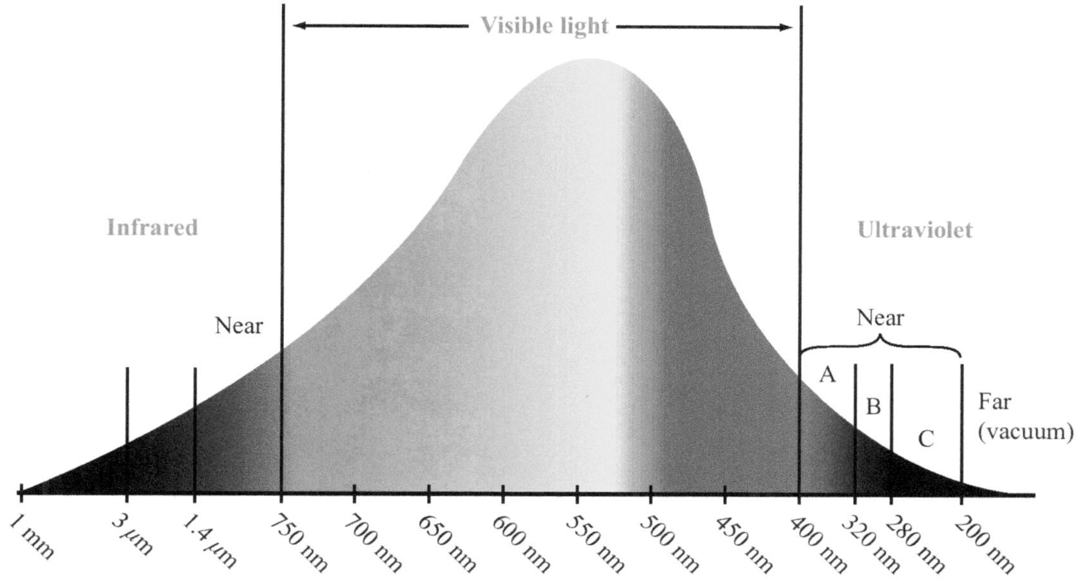

Figure TF23-1: Visible, infrared, and ultraviolet parts of the electromagnetic spectrum.

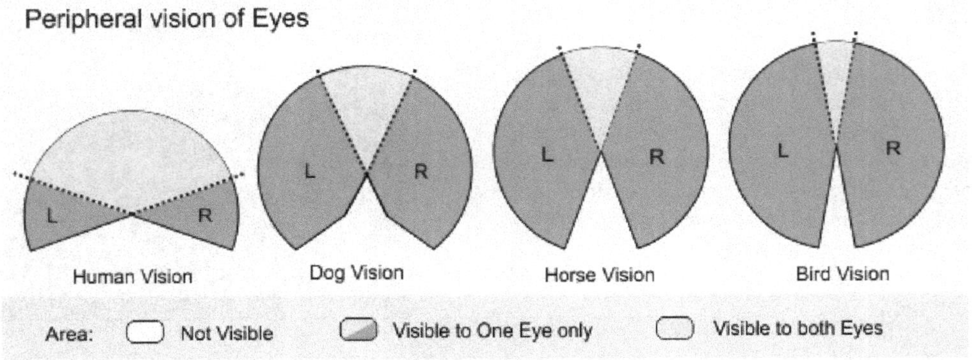

Figure TF23-3: Field of view (spatial filter) for humans, dogs, horses, and birds. (Courtesy of Wordpress.com.)

Our eyes are also a spatial filter. Humans can see across an angular range of about 180°, including a narrow range where we focus and see colors, and a much wider peripheral field of view where vision is not as clear and colors are more limited (Fig. TF23-3). The field of view is controlled mainly by the placement of the eyes on the head. Some birds have 360° fields of view. Some animals (horses, for instance) do not see directly in front of them, but have broader fields of view that let them see along both of their sides and even almost behind them. Prey animals tend to have larger fields of view than predators, whose eyes have more focus in front of them.

Tinted lenses and photo editing are spectral filters that can selectively filter out various parts of the optical spectrum (Fig. TF23-4), acting as band pass or band reject filters. Gray tinted lenses are most common,

(a) Grey filter

(a) Amber filter

(c) Rose filter

(d) Blue filter

Figure TF23-4: Color filters. (Credit: SunglassWarehouse.com and Krista Ward.)

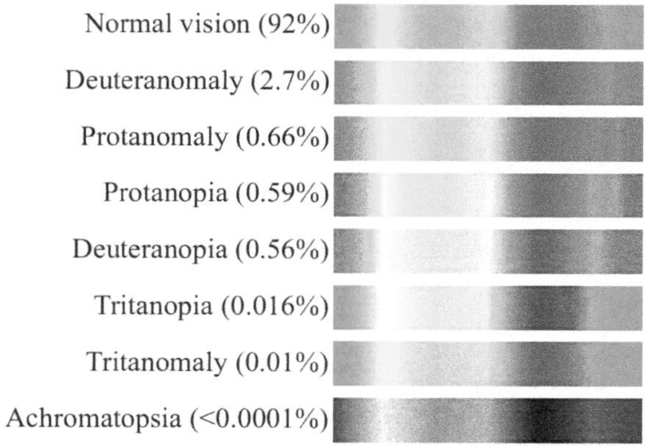

Figure TF23-5: Percentage of people with various types of color blindness.

because they absorb all colors roughly equally, thus reducing the overall brightness with minimal change to the color vision. They act like optical attenuators. Amber or brown lenses reduce brightness, but they also absorb (filter out) blue and UV light. The blue hues in an image often appear to us as hazy or blurry, so absorbing this range brings out the apparent contrast and sharpnes in an image, and absorbing UV helps reduce the risk of cataracts. Yellow lenses, often used in ski goggles, absorb almost all of the blue light, making the image appear bright and sharp, but colors are significantly distorted. Green lenses also absorb the blue range, thus enhancing contrast and sharpness, and they have the highest color contrast. Purple and Rose colored lenses give emphasis to objects outdoors against green and blue backgrounds, thus raising the contrast in an image, especially in low light conditions. These images are often perceived as more beautiful, and seeing the world "through rose colored glasses" has become synonymous with optimism.

Color blindness is a natural type of filtering that occurs when the eyes are less sensitive to color. Figure TF23-5 shows color filtering and the incidence of different types of color blindness in people.

Spatial Filters

When we use a camera to record a digital image, we want that image to reflect what we see in real life. Or do we?

(a) Lowpass filtering is a smoothing (averaging) operation

(b) Edge detection is highlighted by high-frequency spatial filtering

Figure TF23-6: (a) Lowpass filters remove small imperfections and (b) highpass filters accentuate edges.

Photoshop^{TM} and other photo editing tools allow us to control just how "real" our photos appear. One common spatial filter applied to images after the fact is a smoothing filter. This allows us to remove blemishes from the skin by smoothing or averaging the skin tones as in Fig. TF23-6. Smoothing (or blurring) is, effectively, a spatial lowpass filter. By contrast, a highpass filter emphasizes the outline and provides edge detection. It makes things less blurry. Things that change suddenly (edges) represent the high-frequency spatial content in an image.

9-5.3 RLC Bandpass Filter

We just concluded that the RLC lowpass filter is second-order. The RLC circuit also behaves like a bandpass filter when the output voltage is taken across R instead of C. Does it then follow that the RLC bandpass filter is second-order?

To answer the question, we start by examining the expression for the transfer function when ω is in the stopbands of the filter. The expression for $\mathbf{H}_{BP}(\omega)$ is given by Eq. (9.45) as

$$\mathbf{H}_{BP}(\omega) = \frac{j\omega RC}{1 - \omega^2 LC + j\omega RC}, \qquad (9.83)$$

and the spectral plot of its magnitude is shown in Fig. 9-15(b). For $\omega \ll \omega_0$ and $\omega \gg \omega_0$ (where $\omega_0 = 1/\sqrt{LC}$), the expression simplifies to

$$\mathbf{H}_{BP}(\omega) \approx \begin{cases} j\omega RC & (\text{for } \omega \ll \omega_0 \text{ and } \omega \ll 1/RC), \\ \dfrac{-jR}{\omega L} & (\text{for } \omega \gg \omega_0). \end{cases} \qquad (9.84)$$

At the low-frequency end, $\mathbf{H}_{BP}(\omega)$ reduces to a first-order zero @ origin, and at the high-frequency end, it reduces to a first-order pole @ origin.

► Hence, the RLC bandpass filter is first-order, not second-order. ◄

Concept Question 9-12: How does S_g of a third-order filter compare with that of a first-order filter? (See CAD)

Concept Question 9-13: When is a series RLC circuit a first-order circuit, and when is it a second-order circuit? (See CAD)

Exercise 9-11: What is the order of the two-stage bandpass filter circuit shown in Fig. 9-18(a)?

Answer: Second-order. (See CAD)

Exercise 9-12: Determine the order of

$$\mathbf{H}(\omega) = \mathbf{V}_{out}/\mathbf{V}_s$$

for the circuit in Fig. E9.12.

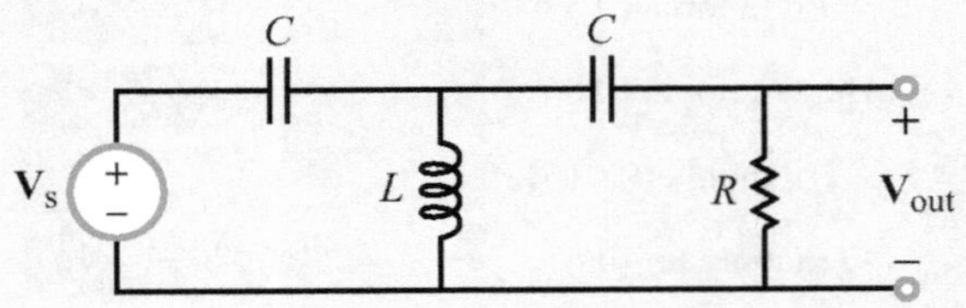

Figure E9.12

Answer:

$$\mathbf{H}(\omega) = \frac{j\omega^3 RLC^2}{\omega^2 LC - (1 - \omega^2 LC)(1 + j\omega RC)},$$

which describes a highpass filter. In the stopband (very small values of ω), $\mathbf{H}(\omega)$ varies as ω^3. Hence, it is third-order. (See CAD)

9-6 Active Filters

The four basic types of filters we examined in earlier sections (lowpass, highpass, bandpass, and bandreject) are all relatively easy to design, but they do have a number of drawbacks. Passive elements cannot generate energy, so the power gain of a passive filter cannot exceed 1. Active filters (by comparison) can be designed to provide significant gain in addition to realizing the specified filter performance. The "active" part of the name refers to the fact that op amps require external dc sources to operate. A second drawback of passive filters has to do with inductors. Whereas capacitors and resistors can be fabricated easily in planar form on machine-assembled printed circuit boards, inductors generally are more expensive to fabricate and more difficult to integrate into the rest of the circuit, because they are bulky and three-dimensional in shape. In contrast, op-amp circuits can be designed to function as filters without the use of inductors. The intended operating-frequency range is an important determining factor with regard to what type of filter is best to design and use. Except for a few specially designed op-amp models, most op amps do not perform reliably at frequencies above 1 MHz, so their use as filters is limited to lower frequencies. Fortunately, inductor size becomes less of a problem above 1 MHz (because $\mathbf{Z}_L = j\omega L$ necessitating a smaller value for L, and consequently a physically smaller

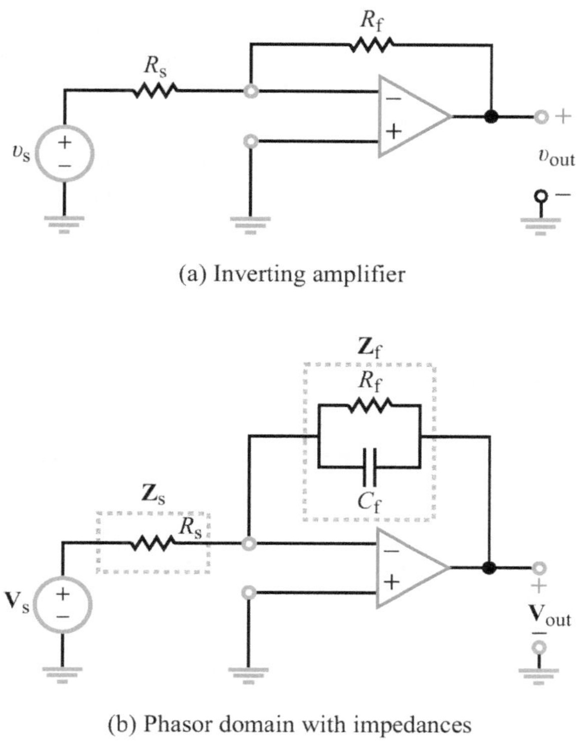

Figure 9-23: Inverting amplifier functioning like a lowpass filter.

inductor), so passive filters are the predominant type used at the higher frequencies.

One of the major assets of op-amp circuits is that they easily can be cascaded together (both in series and in parallel) to realize the intended function. Moreover, by inserting buffer circuits (see Section 4-7) between successive stages, impedance mismatch and loading problems can be minimized or avoided altogether.

9-6.1 Single-Pole Lowpass Filter

Consider the circuit shown in Fig. 9-23(a), which essentially is a replica of the inverting amplifier circuit of Fig. 4-11(a) and for which the input and output voltages are related by

$$\upsilon_{out} = -\frac{R_f}{R_s}\,\upsilon_s. \tag{9.85}$$

Let us now transform the circuit into the phasor domain and generalize it by replacing resistors R_s and R_f with impedances $\mathbf{Z}_s$ and $\mathbf{Z}_f$, respectively, as shown in Fig. 9-23(b). Further, let us retain $\mathbf{Z}_s$ as R_s, but specify $\mathbf{Z}_f$ as the parallel combination of a resistor R_f and a capacitor C_f. By analogy with Eq. (9.85), the equivalent phasor relationship for the circuit in Fig. 9-23(b) is

$$\mathbf{V}_{out} = -\frac{\mathbf{Z}_f}{\mathbf{Z}_s}\,\mathbf{V}_s, \tag{9.86}$$

with

$$\mathbf{Z}_s = R_s \tag{9.87a}$$

and

$$\mathbf{Z}_f = R_f \parallel \left(\frac{1}{j\omega C_f}\right) = \frac{R_f}{1 + j\omega R_f C_f}. \tag{9.87b}$$

The transfer function of the circuit, which we soon will recognize as that of a lowpass filter, is given by

$$\begin{aligned} \mathbf{H}_{LP}(\omega) &= \frac{\mathbf{V}_{out}}{\mathbf{V}_s} = -\frac{\mathbf{Z}_f}{\mathbf{Z}_s} = -\frac{R_f}{R_s}\left(\frac{1}{1 + j\omega R_f C_f}\right) \\ &= G_{LP}\left(\frac{1}{1 + j\omega/\omega_{LP}}\right), \end{aligned} \tag{9.88}$$

where

$$G_{LP} = -\frac{R_f}{R_s} \quad \text{and} \quad \omega_{LP} = \frac{1}{R_f C_f}. \tag{9.89}$$

The expression for G_{LP} is the same as that of the original inverting amplifier, and ω_{LP} is the cutoff frequency of the lowpass filter. Except for the gain factor, the expression given by Eq. (9.88) is identical in form to Eq. (9.71), which is the transfer function of the RC lowpass filter. A decided advantage of the active lowpass filter over its passive counterpart is that ω_{LP} is independent of both the input resistance R_s and any nonzero load resistance R_L that may be connected across the op amp's output terminals.

9-6.2 Single-Pole Highpass Filter

If in the inverting amplifier circuit we were to specify the input and feedback impedances as

$$\mathbf{Z}_s = R_s - \frac{j}{\omega C_s} \quad \text{and} \quad \mathbf{Z}_f = R_f, \tag{9.90}$$

as shown in Fig. 9-24, we would obtain the highpass-filter transfer function given by

$$\begin{aligned} \mathbf{H}_{HP}(\omega) &= \frac{\mathbf{V}_{out}}{\mathbf{V}_s} = -\frac{\mathbf{Z}_f}{\mathbf{Z}_s} = -\frac{R_f}{R_s - j/\omega C_s} \\ &= G_{HP}\left[\frac{j\omega/\omega_{HP}}{1 + j\omega/\omega_{HP}}\right], \end{aligned} \tag{9.91}$$

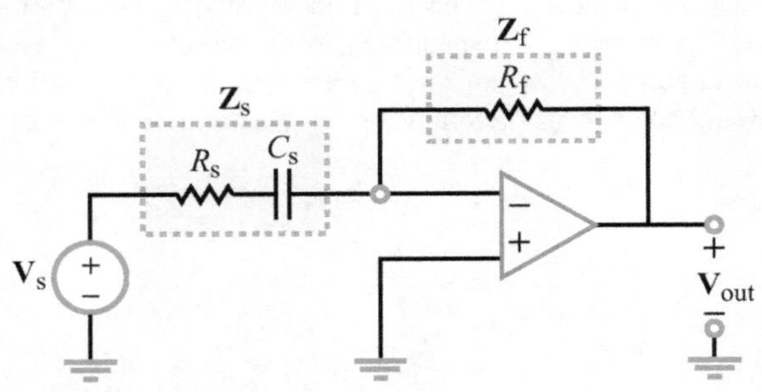

Figure 9-24: Single-pole active highpass filter.

where

$$G_{HP} = -\frac{R_f}{R_s} \quad \text{and} \quad \omega_{HP} = \frac{1}{R_s C_s}. \qquad (9.92)$$

The expression given by Eq. (9.91) represents a first-order highpass filter with a cutoff frequency ω_{HP} and a gain factor G_{HP}.

Concept Question 9-14: What are the major advantages of active filters over their passive counterparts? (See CAD)

Concept Question 9-15: Are active filters used mostly at frequencies below 1 MHz or above 1 MHz? (See CAD)

Exercise 9-13: Choose values for R_s and R_f in the circuit of Fig. 9-23(b) so that the gain magnitude is 10 and the corner frequency is 10^3 rad/s given that $C_f = 1$ μF.

Answer: $R_s = 100\ \Omega$, $R_f = 1$ kΩ. (See CAD)

9-7 Cascaded Active Filters

The active lowpass and highpass filters we examined thus far—as well as other op-amp configurations that provide these functions—can be regarded as basic building blocks that easily can be cascaded together to create second- or higher-order lowpass and highpass filters or to design bandpass and bandreject filters (Fig. 9-25).

▶ The cascading approach allows the designer to work with each stage separately and then combine all of the stages together to achieve the desired specifications. ◀

Moreover, inverting or noninverting amplifier stages can be added to the filter cascade to adjust the gain or polarity of the output signal, and buffer circuits can be inserted in between stages to provide impedance isolation, if necessary.

▶ Throughout the multistage process, it is prudent to compare the positive and negative peak values of the voltage at the output of every stage with the op amp's power-supply voltages V_{CC} and $-V_{CC}$ to make sure that the op amp will not go into saturation mode. ◀

Example 9-10: Third-Order Lowpass Filter

For the three-stage active filter shown in Fig. 9-26, generate dB plots for M_1, M_2, and M_3, where $M_1 = |\mathbf{V}_1/\mathbf{V}_s|$, $M_2 = |\mathbf{V}_2/\mathbf{V}_s|$, and $M_3 = |\mathbf{V}_3/\mathbf{V}_s|$.

Solution: Since all three stages have the same values for R_f and C_f, they have the same cutoff frequency give by

$$\omega_{LP} = \frac{1}{R_f C_f} = \frac{1}{10^4 \times 10^{-9}} = 10^5 \text{ rad/s}.$$

The input resistance of the first stage is 10 Ω, but the input resistances of the second and third stages are 10 kΩ. Hence,

$$G_1 = -\frac{10\text{k}}{10} = -10^3 \quad \text{and} \quad G_2 = G_3 = -\frac{10\text{k}}{10\text{k}} = -1.$$

Transfer function M_1 therefore is given by

$$M_1 = \left|\frac{\mathbf{V}_1}{\mathbf{V}_s}\right| = \left|\frac{G_1}{1 + j\omega/\omega_{LP}}\right| = \frac{10^3}{\sqrt{1 + (\omega/10^5)^2}}$$

and

$$M_1 \text{ [dB]} = 20 \log\left[\frac{10^3}{\sqrt{1 + (\omega/10^5)^2}}\right] = 60 \text{ dB} - 10 \log[1 + (\omega/10^5)^2].$$

The transfer function corresponding to $\mathbf{V}_2$ is

$$M_2 = \left|\frac{\mathbf{V}_2}{\mathbf{V}_1} \cdot \frac{\mathbf{V}_1}{\mathbf{V}_s}\right| = \left|\frac{G_1}{1 + j\omega/\omega_{LP}}\right| \left|\frac{G_2}{1 + j\omega/\omega_{LP}}\right| = \frac{10^3}{1 + (\omega/10^5)^2}$$

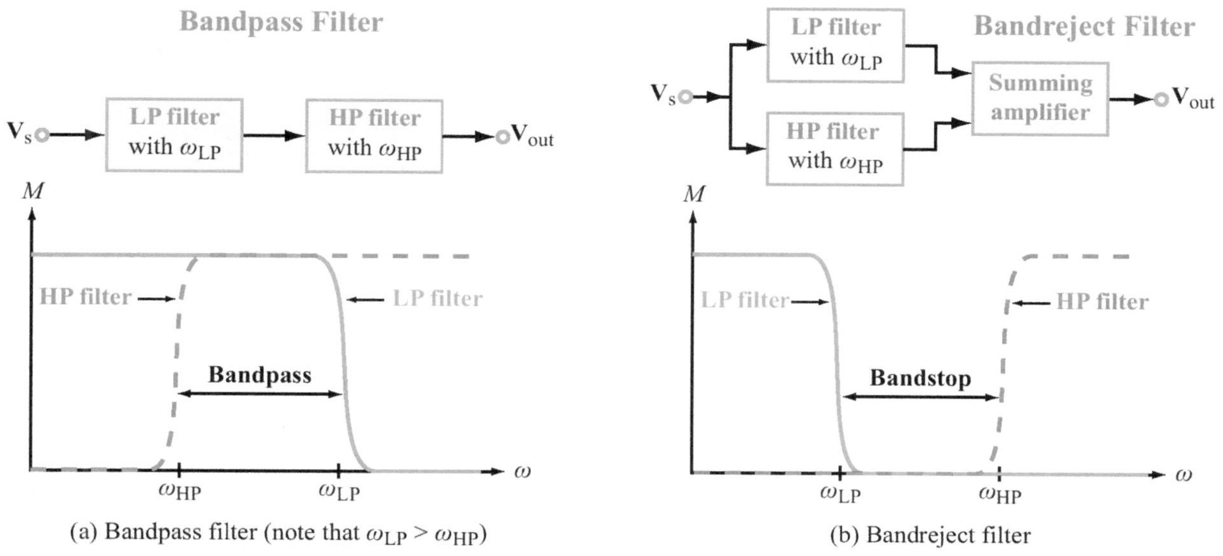

(a) Bandpass filter (note that $\omega_{LP} > \omega_{HP}$)

(b) Bandreject filter

Figure 9-25: (a) In-series cascade of a lowpass and a highpass filter generates a bandpass filter; (b) in-parallel cascading generates a bandreject filter.

and

$$M_2\ [\text{dB}] = 20 \log \left[\frac{10^3}{1 + (\omega/10^5)^2} \right]$$
$$= 60\ \text{dB} - 20 \log[1 + (\omega/10^5)^2].$$

Similarly,

$$M_3\ [\text{dB}] = 60\ \text{dB} - 30 \log[1 + (\omega/10^5)^2].$$

The three-stage process is shown in Fig. 9-26(b) in block-diagram form, and spectral plots of M_1 [dB], M_2 [dB], and M_3 [dB] are displayed in Fig. 9-26(c). We note that the gain roll-off rate S_g is -20 dB for M_1 [dB], -40 dB for M_2 [dB], and -60 dB for M_3 [dB]. We also note that the -3 dB corner frequencies are not the same for the three stages.

Concept Question 9-16: Why is it more practical to cascade multiple stages of active filters than to cascade multiple stages of passive filters? (See CAD)

Concept Question 9-17: What determines the gain factors of the highpass and lowpass op-amp filters? (See CAD)

Exercise 9-14: What are the values of the corner frequencies associated with M_1, M_2, and M_3 of Example 9-10?

Answer: $\omega_{c_1} = 10^5$ rad/s, $\omega_{c_2} = 0.64\omega_{c_1} = 6.4 \times 10^4$ rad/s, $\omega_{c_3} = 0.51\omega_{c_1} = 5.1 \times 10^4$ rad/s. (See CAD)

Analogy to AND and OR gates

An *AND logic gate* has two inputs (whose logic states can each be either 0 or 1) and one output. Its output state is 1 if and only if both input states are 1. Otherwise, its output state is zero. Hence, *the output of an AND gate is equal to the product of its input states.* The cascaded bandpass filter diagrammed in Fig. 9-25(a) is analogous to an AND gate. The filter consists of an idealized lowpass filter with cutoff frequency ω_{LP} *connected in series* with an idealized highpass filter with cutoff frequency ω_{HP}. *The frequency of the input signal has to be in the passband of both filters in order for the signal to make it to the output.* The filter passband is defined by the bandwidth extending from ω_{HP} to ω_{LP}, as illustrated in Fig. 9-25(a). The combination of the two filters when cascaded in series is equivalent to an AND gate, because the final output is proportional to the product of their transfer functions, $\mathbf{H}_{LP}\mathbf{H}_{HP}$. The process is illustrated through Example 9-11.

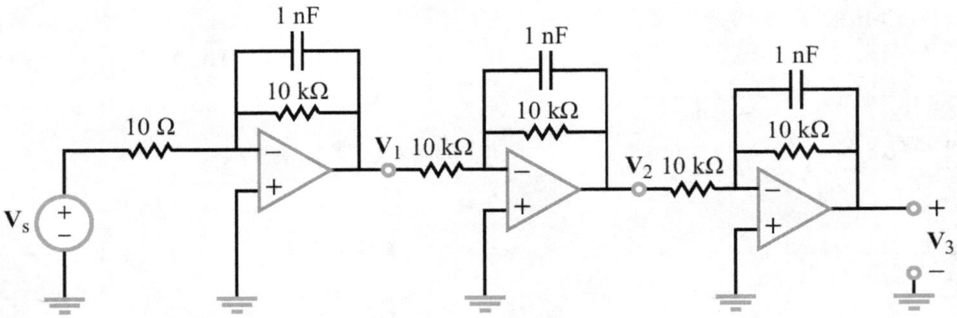

(a) Circuit diagram

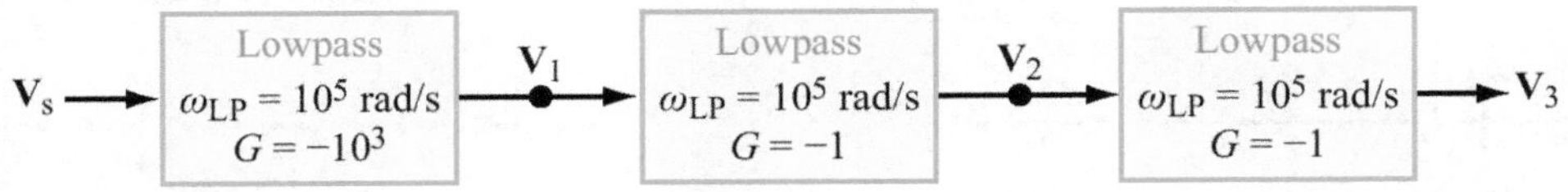

(b) Block diagram

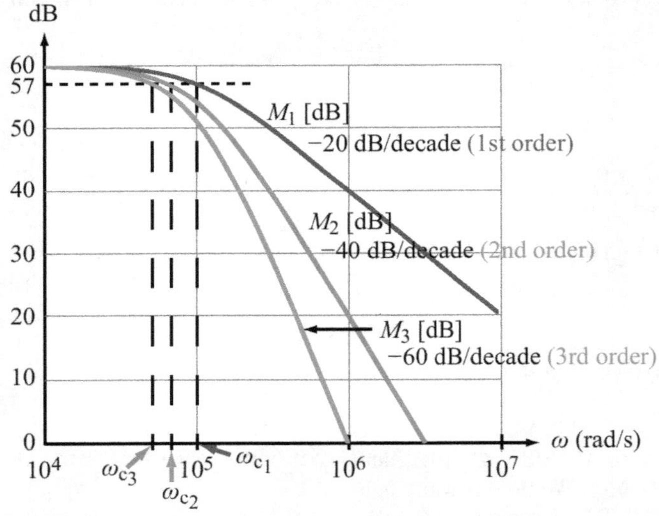

(c) Transfer function plots

Figure 9-26: Three-stage lowpass filter and corresponding transfer functions.

In contrast, the bandreject filter (Fig. 9-25(b)) is analogous to an *OR gate*, for which the state of its output is 1 if either one or both of its inputs has a state of 1. The cascade configuration of the bandreject filter consists of a lowpass filter *connected in parallel* with a highpass filter—thereby offering the input signal to pass through either or both of them—and then their outputs are added by a summing amplifier. Example 9-12 provides more details.

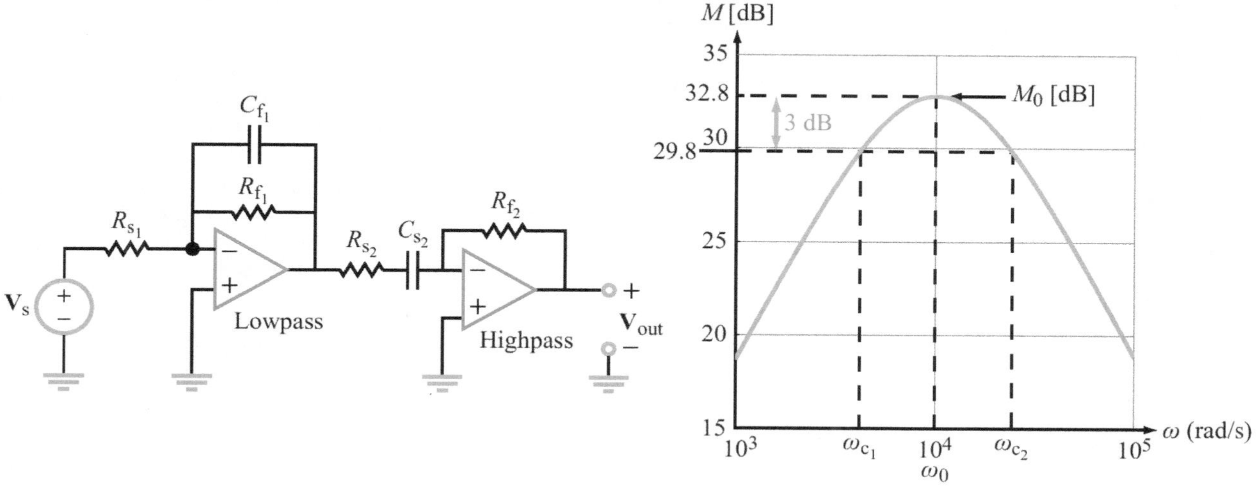

(a) Two-stage bandpass filter

(b) M [dB]

Figure 9-27: Active bandpass filter of Example 9-11.

Example 9-11: Bandpass Filter

The block diagram shown in Fig. 9-27(a) is a two-stage bandpass filter with the following elements: $R_{s_1} = 1\ \text{k}\Omega$, $R_{f_1} = 10\ \text{k}\Omega$, $C_{f_1} = 9$ nF, $R_{s_2} = 12\ \text{k}\Omega$, $C_{s_2} = 9$ nF, and $R_{f_2} = 96\ \text{k}\Omega$. Determine and plot the magnitude of the transfer function and obtain the values of ω_0, ω_{c_1}, ω_{c_2}, B, and Q.

Solution: The first stage is a lowpass filter with a transfer function given by Eq. (9.88) as

$$\mathbf{H}_{\text{LP}}(\omega) = G_{\text{LP}}\left(\frac{1}{1 + j\omega/\omega_{\text{LP}}}\right)$$

with

$$G_{\text{LP}} = -\frac{R_{f_1}}{R_{s_1}} = -\frac{10^4}{10^3} = -10$$

and

$$\omega_{\text{LP}} = \frac{1}{R_{f_1}C_{f_1}} = \frac{1}{10^4 \times 9 \times 10^{-9}} = 11.11 \text{ krad/s}.$$

The transfer function of the highpass filter in the second stage is characterized by Eq. (9.91) as

$$\mathbf{H}_{\text{HP}}(\omega) = G_{\text{HP}}\left(\frac{j\omega/\omega_{\text{HP}}}{1 + j\omega/\omega_{\text{HP}}}\right)$$

with

$$G_{\text{HP}} = -\frac{R_{f_2}}{R_{s_2}} = -\frac{96 \times 10^3}{12 \times 10^3} = -8$$

and

$$\omega_{\text{HP}} = \frac{1}{R_{s_2}C_{s_2}} = \frac{1}{12 \times 10^3 \times 9 \times 10^{-9}} = 9.26 \text{ krad/s}.$$

The combined transfer function is then given by

$$\begin{aligned}\mathbf{H}(\omega) &= \mathbf{H}_{\text{LP}}\mathbf{H}_{\text{HP}}\\ &= G_1G_2\left[\frac{j\omega/\omega_{\text{HP}}}{(1 + j\omega/\omega_{\text{LP}})(1 + j\omega/\omega_{\text{HP}})}\right],\end{aligned} \tag{9.93}$$

and its magnitude is

$$\begin{aligned}M &= |\mathbf{H}(\omega)|\\ &= \left|\frac{80\omega/(9.26 \times 10^3)}{[1 + j\omega/(11.11 \times 10^3)][1 + j\omega/(9.26 \times 10^3)]}\right|\\ &= \frac{80\omega/(9.26 \times 10^3)}{\{[1 + (\omega/11.11 \times 10^3)^2][1 + (\omega/9.26 \times 10^3)^2]\}^{1/2}}.\end{aligned} \tag{9.94}$$

The angular frequency ω_0 of a bandpass filter is defined as the frequency at which the transfer function is a maximum.

For a high-Q filter, ω_0 is approximately midway between the lower and upper cutoff frequencies ω_{c_1} and ω_{c_2}, but ω_0 may be significantly closer to ω_1 or ω_2 if Q is not very large. We do not yet know the value of Q for the present filter, but we can generate an approximate estimate from knowledge of the values of ω_{LP} and ω_{HP}. The estimated bandwidth is

$$B \text{ (est)} = \omega_{LP} - \omega_{HP} = (11.11 - 9.26)\text{k} = 1.85 \text{ krad/s},$$

and if we assume that ω_0 is midway between ω_{HP} and ω_{LP}, then

$$\omega_0 \text{ (est)} = \left(9.26 + \frac{1.85}{2}\right)\text{k} = 10.185 \text{ krad/s}$$

and

$$Q \text{ (est)} = \frac{\omega_0 \text{ (est)}}{B \text{ (est)}} = \frac{10.185}{1.85} = 5.5.$$

Since Q is not greater than 10, the estimated values of B, ω_0, and Q are not likely to be very accurate, so we should return to the expression for M given by Eq. (9.94) and use it to determine the exact values of ω_0, ω_{c_1}, and ω_{c_2}. We can do so by calculating (or plotting) M as a function of ω to identify: (a) M_0 and ω_0, the maximum value of $M(\omega)$ and the corresponding value of ω at which it occurs, respectively, and (b) ω_{c_1} and ω_{c_2}, the corner frequencies at which $M = M_0/\sqrt{2}$ (or -3 dB below the peak on a dB scale). According to the spectral plot of M [dB] shown in Fig. 9-27(b),

$$M_0 \text{ [dB]} = 32.8 \text{ dB}, \qquad \omega_0 \text{ (exact)} = 10.14 \text{ krad/s},$$

$$\omega_{c_1} = 4.19 \text{ krad/s}, \qquad \text{and} \qquad \omega_{c_2} = 24.56 \text{ krad/s}.$$

Hence,

$$B \text{ (exact)} = \omega_{c_2} - \omega_{c_1} = 20.37 \text{ krad/s}$$

and

$$Q \text{ (exact)} = \frac{\omega_0}{B} = \frac{10.14}{20.37} \approx 0.51.$$

The obvious conclusion is that our estimated values for B and Q are way off in comparison with their exact counterparts. We assumed that the corner frequencies ω_{LP} and ω_{HP} associated with functions $\mathbf{H}_{LP}$ and $\mathbf{H}_{HP}$, respectively, are good estimates of the corner frequencies ω_{c_1} and ω_{c_2} of the product of the two functions. That was obviously a poor assumption.

Example 9-12: Bandreject Filter

Design a bandreject filter with the specifications: (a) Gain $= -50$, (b) bandstop extends from 20 kHz to 40 kHz, and (c) gain roll-off rate $= -40$ dB/decade along both boundaries of the bandstop.

Solution: The specified roll-off rate requires cascading two identical lowpass filters and two identical highpass filters. To minimize performance variations among identical pairs, identical resistors will be used in all four units (Fig. 9-28(b)), which means that they all will have unity gain. The overall gain of -50 will be provided by the summing amplifier.

Somewhat arbitrarily, we select $R = 1$ kΩ, and the value of R_f is specified by the gain of the summing amplifier as

$$G = -50 = -\frac{R_f}{R} \quad \Rightarrow \quad R_f = 50 \text{ k}\Omega.$$

The transfer function of the bandreject filter is given by

$$\mathbf{H}(\omega) = G[\mathbf{H}_{LP}^2 + \mathbf{H}_{HP}^2]$$
$$= -50\left[\left(\frac{1}{1 + j\omega RC_{LP}}\right)^2 + \left(\frac{j\omega RC_{HP}}{1 + j\omega RC_{HP}}\right)^2\right].$$

Next, we need to specify values for C_{LP} and C_{HP}. As an approximation, we assume that over the passband of the lowpass filters, the highpass filters exercise minimal impact, and vice versa. This allows us to deal with the transfer functions of the lowpass and highpass filters separately. The magnitude of the transfer function of the two cascaded lowpass filters is

$$M_{LP} = |\mathbf{H}_{LP}^2| = \left|\left(\frac{1}{1 + j\omega RC_{LP}}\right)^2\right| = \frac{1}{1 + \omega^2 R^2 C_{LP}^2}.$$

The specifications call for a lower corner frequency of 20 kHz or, equivalently, $\omega_1 = 2\pi \times 2 \times 10^4 = 4\pi \times 10^4$ rad/s. Setting $M_{LP} = 1/\sqrt{2}$, $R = 1$ kΩ, and $\omega = \omega_1$ leads to

$$C_{LP} \approx 5 \text{ nF}.$$

A similar analysis for the highpass filter chain with $f_2 = 40$ kHz (or $\omega_2 = 8\pi \times 10^4$ rad/s) leads to

$$\frac{\omega^2 R^2 C_{HP}^2}{1 + \omega^2 R^2 C_{HP}^2} = \left.\frac{1}{\sqrt{2}}\right|_{\text{at } \omega=\omega_2},$$

which provides the solution

$$C_{HP} \approx 6 \text{ nF}.$$

The spectrum of M [dB] $= 20 \log |\mathbf{H}|$ is displayed in Fig. 9-28(c).

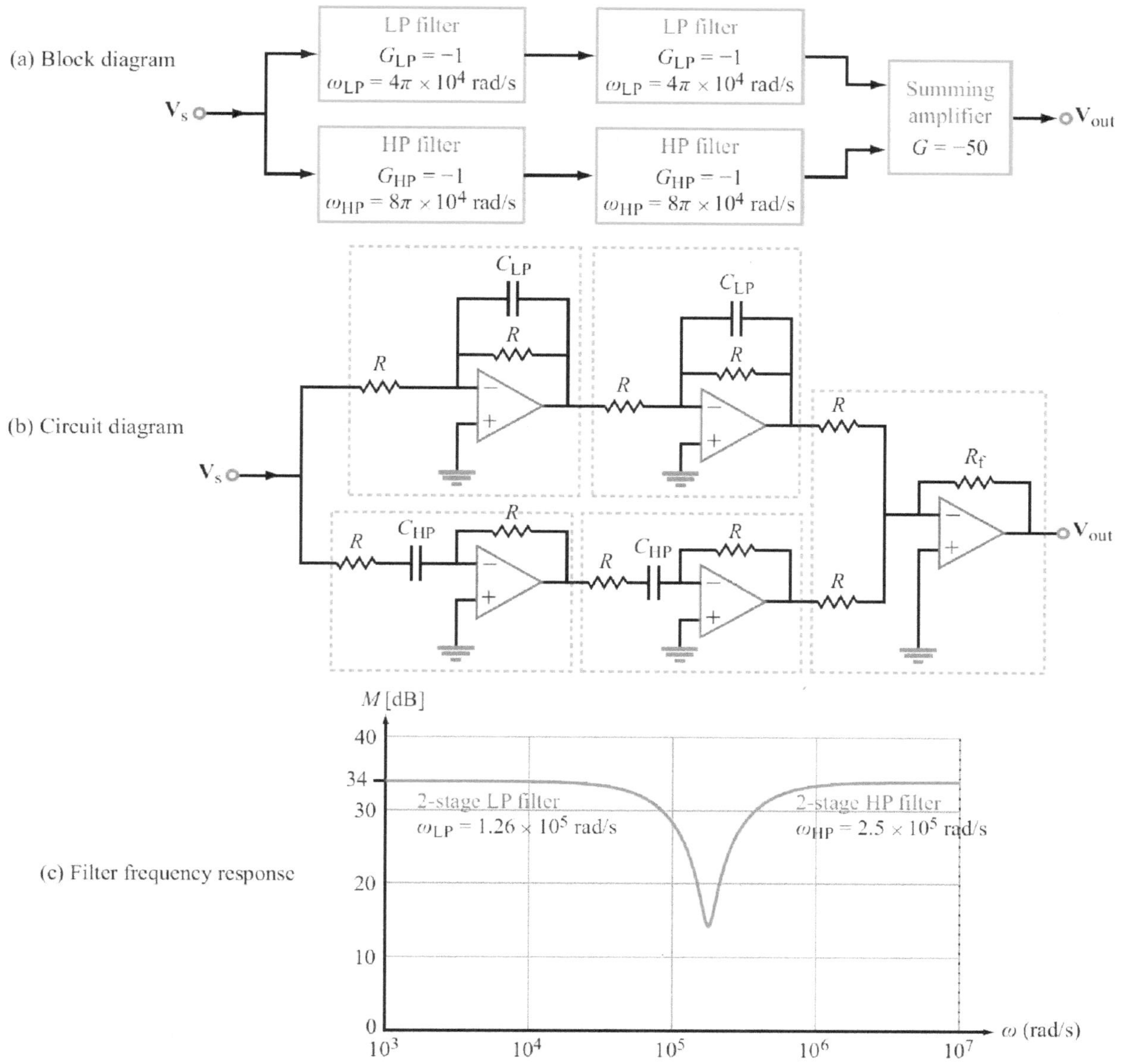

Figure 9-28 Bandreject filter of Example 9-12.

Exercise 9-15: The bandreject filter of Example 9-12 uses two lowpass-filter stages and two highpass-filter stages. If three stages of each were used instead, what would the expression for $\mathbf{H}(\omega)$ be in that case?

Answer:

$$\mathbf{H}(\omega) = 50\left[\left(\frac{1}{1 + j\omega/4\pi \times 10^4}\right)^3 + \left(\frac{j\omega/8\pi \times 10^4}{1 + j\omega/8\pi \times 10^4}\right)^3\right].$$

(See CAD)

Technology Brief 24
Electrical Engineering and the Audiophile

The reproduction of high-quality music with sufficient fidelity to sound like a live performance in one's living room was one of the technological hallmarks of the 20th century. In these days of iPods and online music distribution, good music is increasingly accessible to many people. The price of good quality tuners, amplifiers, and speakers continues to drop, and driven mostly by demand for home entertainment audio/video systems, audio equipment is increasingly "user-friendly." The reproduction of theater-quality or live-performance sound in a confined space is challenging enough to be a profession unto itself. It also can be a very rewarding technical hobby for the well-versed electrical engineer. In this Technology Brief, we will cover some of the basics of audio equipment and relate them directly to the concepts taught in this book. Several good audiophile websites exist with more in-depth treatments of these (and other) topics; beyond the audiophile community, the sub-field of audio engineering has an extensive academic and professional literature to consult.

The Basics

Reproduced sound starts out as an analog (e.g., the vinyl record) or digital (e.g., the mp3 file) recording. How that recording is made from real sound with high fidelity is beyond the scope of this Brief (and is a large component of the audio engineering profession). That recording is converted into an electrical signal that is first amplified and then transmitted via cables to speakers. Figure TF24-1 shows a schematic of the process.

The ***audible spectrum*** of the human ear extends from about 20 Hz to 20 kHz, although the frequency response may vary among different individuals depending on age and other factors. An audio signal is a superposition of many sinusoids oscillating at different frequencies—each with its own individual amplitude. When we say a sound has a lot of ***bass***, for example, we mean that the low-frequency segment of its spectrum (20 to 100 Hz) has a large amplitude when compared with higher-frequency components. Conversely, very shrill or high-pitched sounds have large-amplitude components in the high-frequency range (10 kHz to 20 kHz). When converting an electrical recording back into the original sound that generated it in the first place, the reproduction fidelity is determined by the degree of distortion that the spectrum undergoes during the playback process. In practice, minimizing ***spectral distortion*** can be quite a challenge!

Each component of the sound-reproduction system shown in Fig. TF24-1 is characterized by its own transfer function relating its output to its input, and since each of these components is equivalent to a circuit composed of resistive and reactive elements, its transfer function

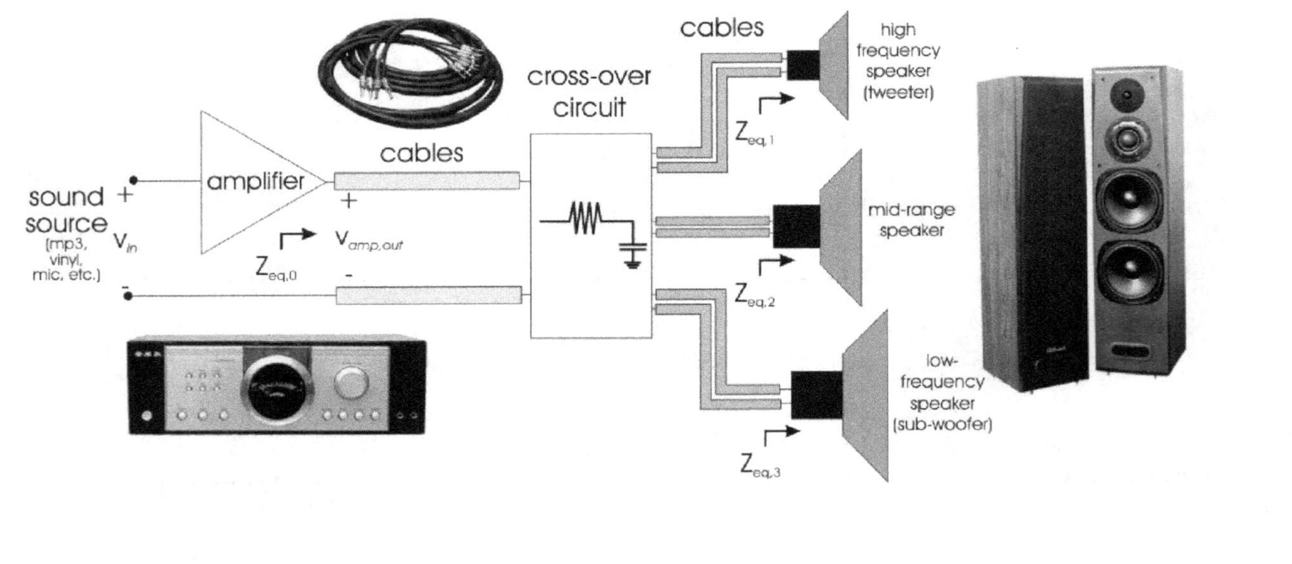

Figure TF24-1: Schematic of a basic audio-reproduction system.

is bound to exhibit a non-uniform spectral response. The amplifier, for example, may act like a filter, favoring parts of the audible spectrum over others. The cables, which behave (electrically) like the RC transmission line of Fig. 7-35, will favor low-frequency spectral components over high-frequency components. Thus, unless the components of an audio-reproduction system are well designed in order to generate a transfer function with a nearly flat spectral response over the audible range, the reproduced sound will exhibit a distorted spectrum when compared with the original spectrum. While there are many objective metrics by which to judge the fidelity of audio equipment, every listener processes a given sound differently, introducing a subjective component into the experience. A great way to appreciate the concepts introduced in this Technology Brief is to walk into a high-end audio-systems store with three favorite CDs and then to listen to them on many different amplifier-speaker combinations.

Amplifiers

It takes quite a bit of power to drive speakers to produce sound in a room. The function of an amplifier is to boost the audio signal's power high enough to drive the speakers. In doing so, the amplifier must:

- keep frequency distortion to a minimum, and
- introduce as little noise as possible into the signal.

In order to keep frequency distortion to a minimum, the amplifier's response must be as uniform as possible; in other words, signals of different frequencies and different amplitudes must be amplified with exactly the same gain. To address this, many different transistor–amplifier topologies have been developed over the years. These amplifiers are grouped into classes based on behavior and topology; the principal differences lie in circuit complexity, power consumption, and the degree of fidelity with which the circuit reproduces an input signal. Audio-amplifier circuit topologies are categorized by letter—currently from A to G. Although a description of each class lies beyond the scope of this discussion, very succinct overviews can be found in many places online. In order to reduce the noise during the amplification step, two-amp stages often are used. The first stage is called the pre-amplifier. Preamps have very good noise characteristics and amplify the signal partway (this mid-level signal is called the *line signal*). Often, this is simply an amplification of the voltage level. The power amp then boosts this signal (which does not have much noise) to a level high enough to drive speakers; this usually requires significant current amplification to provide enough overall power to the speakers.

Cables

The cables that transfer a signal between sources, amplifiers, crossovers, and speakers can themselves distort the signal. Cables behave exactly like the transmission line of Fig. 7-35; the distributed resistance and capacitance act like a filter with an associated frequency response. In general, cables should be:

- as short as possible,
- properly impedance-matched to both the output of the amplifier and the input of the speakers, and
- properly terminated so the cable connections to the equipment do not introduce capacitances.

All three of these objectives easily are accomplished using industry-standard cables and connectors. In some modern systems, transmission of audio signals between non-speaker components (e.g., from a tuner to an amp or from an amp to a TV) is often performed in digital form so as to eliminate both noise issues and frequency distortion.

Speakers

A speaker is any electro-mechanical device or transducer that converts an electrical signal into sound. Electromagnetic transducers are the most commonly used type for consumer audio applications (Fig. TF24-2),

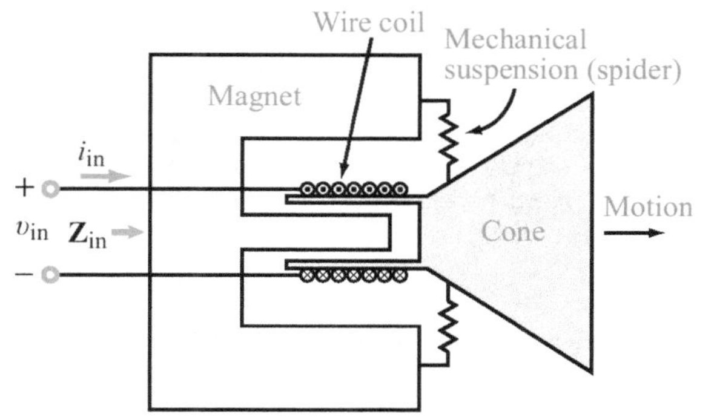

Figure TF24-2: Conceptual illustration of an electromagnetic speaker transducer. The current from the amplifier runs through a coil that induces an electromagnetic force on the cone in proportion to the amplitude of the input signal. The cone motion produces pressure waves and, hence, sound.

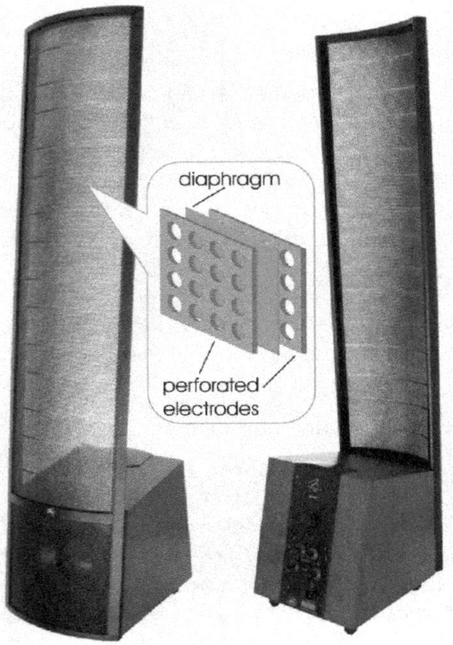

Figure TF24-3: Electrostatic speakers consist of a very thin ($\sim 20\ \mu$m) polymer membrane, called a diaphragm, which is coated with a conductor. This membrane is suspended between two perforated electrodes. The diaphragm is held at a dc potential of several kV. The voltage between the electrodes and the diaphragm is driven by the amplified audio signal so as to displace the diaphragm and move air (which produces sound). The principle behind actuation is very similar to that discussed in Technology Brief 10: Micromechanical Sensors and Actuators. Most electrostatic speakers have poor base response, so they are usually paired with subwoofers (like the one shown here). (Image courtesy of MartinLogan, Ltd.)

although several other technologies, such as electrostatic speakers (Fig. TF24-3), exist as well. The principal metric when choosing a speaker is arguably its frequency response (Fig. TF24-4). Ideally, a speaker will provide a very flat response. This means that signals at different frequencies recreated into sound all at the same audio level. Generally speaking, very small speakers have difficulty reproducing very low frequencies (i.e., bass); a deep drum or baseline may be lost entirely when listening through a small speaker.

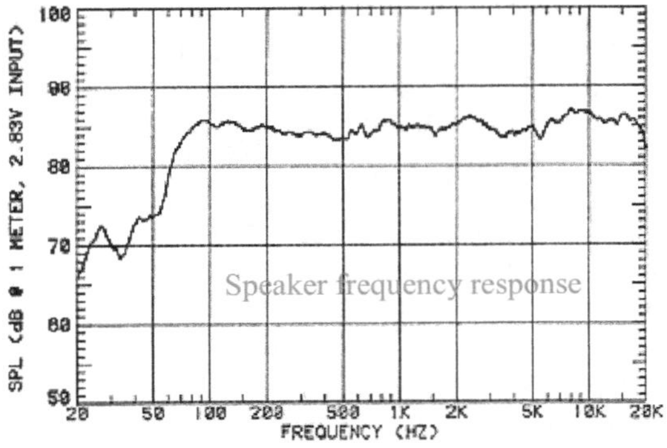

Figure TF24-4: Frequency response of a good consumer-quality speaker consisting of a tweeter and woofer.

The most common method for obtaining a nice flat frequency response is to drive several speakers together—each with a different but complementary frequency response. When listened to as a group, the frequency response is close to flat. For example, ***tweeters*** are small speakers intended for reproducing high-frequency sound, while ***woofers*** only reproduce the lowest frequencies. A common entry-level speaker consists of a tweeter, a mid-range speaker, and a woofer all housed together. With appropriate crossover circuits the ensemble can exhibit a good response.

Crossover circuits

As we noted earlier, most speakers cannot handle the entire range of frequencies in the audio range. In order to split the signal for use by the different speakers (such as a tweeter, a mid-range speaker, and a sub-woofer), passive filters are used. The signal is applied to a set of filters that produce three outputs: one output contains only low frequencies in some range, a second contains mid-range frequencies and a third output contains only high-frequency harmonics. In this way, each speaker receives a dedicated signal that contains only the frequencies it can reproduce properly. Designing crossovers can be an involved process that takes into account many variables, including the amount of current in the input signal, the input impedances of all of the speakers, and the frequency range of each speaker. Without careful design, the crossover circuit can provide too much signal power to one speaker and too little to another, thereby distorting the overall frequency response heard by the listener.

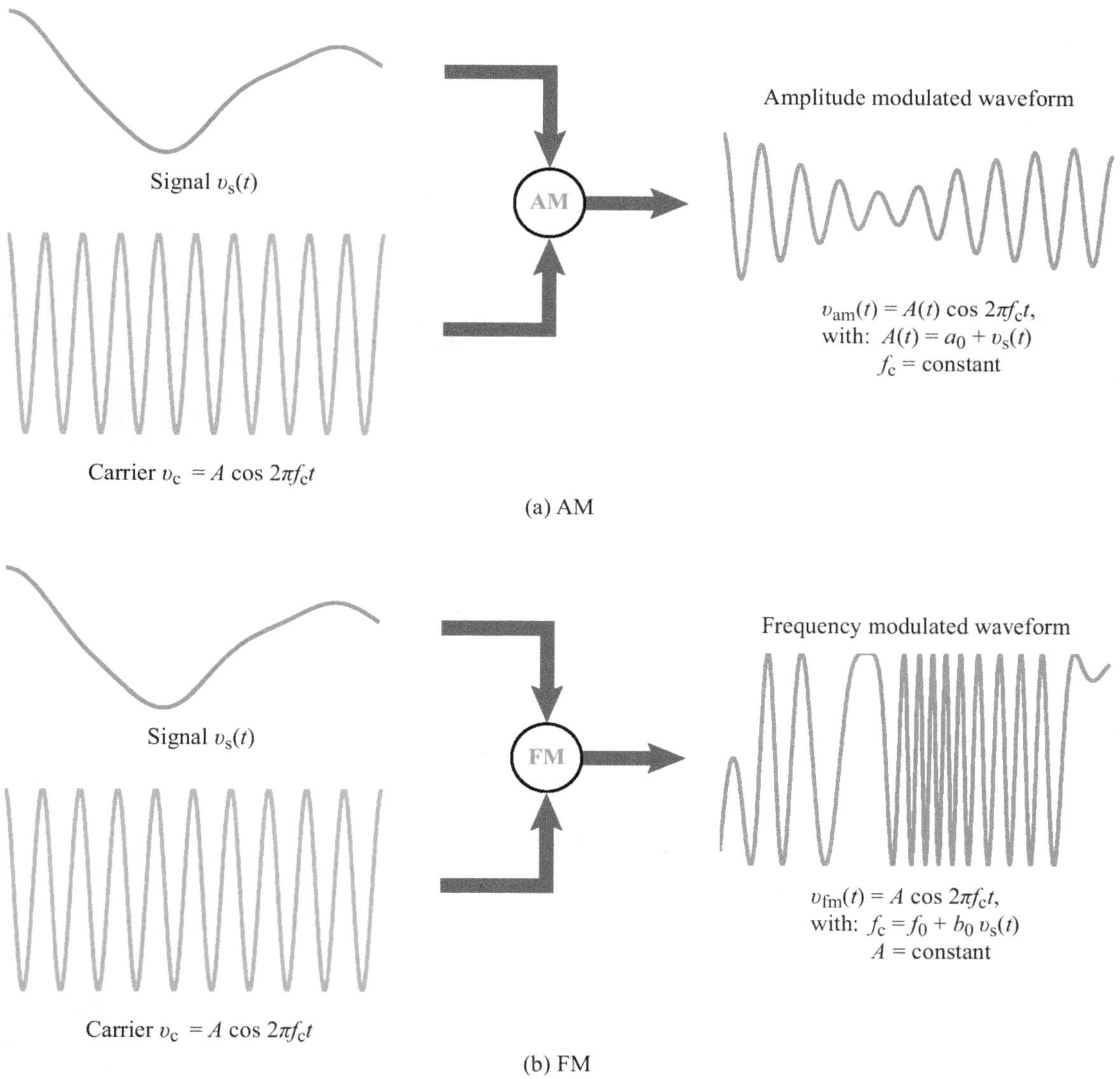

Figure 9-29: Overview of AM and FM.

9-8 Application Note: Modulation and the Superheterodyne Receiver

9-8.1 Modulation

In the language of electronic communication, the term *signal* refers to the information to be communicated between two different locations or between two different circuits, and the term *carrier* refers to the sinusoidal waveform that *carries* the information. The latter is of the form

$$v_c(t) = A\cos 2\pi f_c t, \tag{9.95}$$

where A is its amplitude and f_c is its *carrier frequency*. The sinusoid can be used to carry information by *modulating* (varying) its amplitude—in which case A becomes $A(t)$—while keeping f_c constant. In the example shown in Fig. 9-29(a),

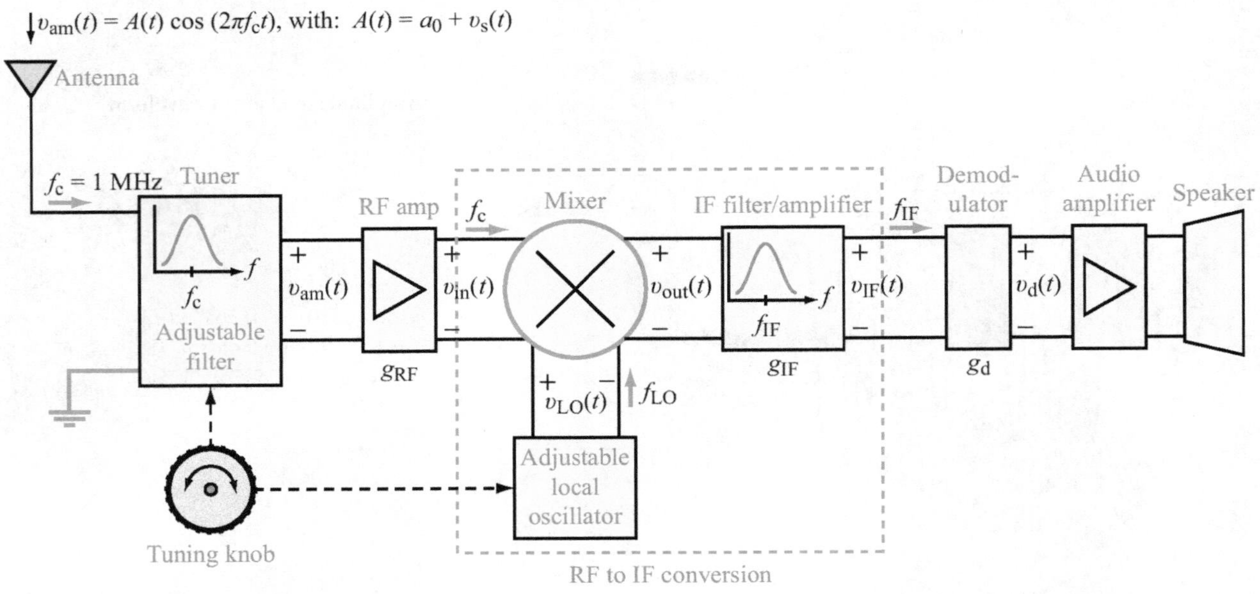

Figure 9-30: Block diagram of superheterodyne receiver.

multiplication of the signal waveform by the sinusoidal carrier generates an *amplitude-modulated (AM)* carrier whose envelope is identical to the signal waveform. Alternatively, we can apply *frequency modulation (FM)* by keeping A constant and varying f_c in a fashion that mimics the variation of the signal waveform, as illustrated by Fig. 9-29(b). FM usually requires more bandwidth than AM, but it is also more immune to noise and interference, thereby delivering a higher-quality sound than AM. Many other types of modulation techniques also are available, including phase modulation (changing the phase of the carrier) and pulse-code modulation.

9-8.2 The Superheterodyne Receiver

Let us assume the signal $v_s(t)$ in Fig. 9-29(a) is an audio signal and the carrier frequency $f_c = 1$ MHz. Let us also assume that the signal was used to generate an amplitude-modulated waveform, which was then fed into a transmit antenna. After propagating through the air along many different directions (as dictated by the antenna radiation pattern), part of the AM waveform was intercepted by a receive antenna connected to an AM receiver. Prior to 1918, the receiver would have been a *tuned-radio frequency receiver* or a *regenerative receiver*, both of which suffered from poor frequency selectivity and low immunity to noise. In either case, the receiver would have *demodulated* the AM signal by suppressing the carrier and preserving the envelope, thereby retrieving the original signal $v_s(t)$ (or more realistically, some distorted version of $v_s(t)$). To overcome the shortcomings of such receivers, *Edwin Armstrong* introduced the *heterodyne receiver* in 1918 by proposing the addition of a receiver stage to convert the carrier frequency of the AM signal f_c to a fixed lower frequency (now called the *intermediate frequency* f_{IF}) before detection (demodulation). [Armstrong also invented frequency modulation in 1935.] The superheterodyne concept proved to be one of the foundational enablers of 20th-century radio transmission. It is still in use in most AM and FM analog receivers, although it slowly is getting supplanted by digital techniques (Section 9-8.4).

Figure 9-30 shows a basic block diagram of a superheterodyne receiver. The *tuner* is a bandpass filter whose center frequency can be adjusted to allow the intended signal at $f_c = 1$ MHz (for example) to pass through, while rejecting signals at other carrier frequencies. After amplification by the *radio-frequency (RF)* amplifier, the AM signal either can be demodulated directly (which is what receivers did prior to 1918) or it can be converted into an IF signal by *mixing* (multiplying) it with another locally generated sinusoidal signal provided by a *local oscillator*. As will be explained in Section 9-8.3, the *mixer* is a device that multiplies the two signals available at its input and generates an output signal whose frequency is

$$f_{\rm IF} = f_{\rm LO} - f_{\rm c}, \tag{9.96}$$

where $f_{\rm LO}$ is the local-oscillator frequency. The frequency conversion given by Eq. (9.96) assumes that $f_{\rm LO} \geq f_{\rm c}$; otherwise, $f_{\rm IF} = f_{\rm c} - f_{\rm LO}$ if $f_{\rm LO} < f_{\rm c}$. It is important to note that frequency conversion changes the carrier frequency of the AM waveform from $f_{\rm c}$ to $f_{\rm IF}$, but the audio signal $\upsilon_{\rm s}(t)$ remains unchanged; it merely is getting carried by a different carrier frequency.

The diagram in Fig. 9-30 indicates that the tuning knob controls the center of the adjustable tuner as well as the local oscillator frequency. By *synchronizing* these two frequencies to each other, the IF frequency remains always a constant. This is an important feature of the superheterodyne receiver, because it insures that the same *IF filter/amplifier* can be used to provide high-selectivity filtering and high-gain amplification, regardless of the carrier frequency of the AM signal. In the *AM radio band*, the carrier frequency of the audio signals transmitted by an AM radio station may be at any frequency between 530 and 1610 kHz. Because of the built-in synchronization between the tuner and the local oscillator, the IF frequency of an AM receiver is always at 455 kHz, which is the standard IF for AM radio. Similarly, the standard IF for FM radio is 10 MHz, and the standard IF for television is 45 MHz.

It is impractical to design and manufacture high-performance components at every frequency in the radio spectrum. By designating certain frequencies as IF standards, industry was able to develop devices and systems that operate with very high performance at those frequencies. Consequently, frequency conversion to an IF band is very prevalent not only in radio and TV receivers but also in radar sensors, satellite communication systems and transponders, among others.

9-8.3 Frequency Conversion

Regardless of the specific type of modulation used in a modern communication system, it will employ one or more steps of frequency conversion, whereby the carrier frequency is changed from an initial frequency f_1 to a new frequency f_2. If f_2 is higher than f_1, it is called *up-conversion*, and the reverse is called *down-conversion*. In the AM example of Fig. 9-30, $f_1 = f_{\rm c} = 1$ MHz and $f_2 = f_{\rm IF} = 455$ kHz. To explain how the conversion takes place, consider the general case of two signals given by

$$\upsilon_{\rm in}(t) = A(t)\ \cos 2\pi f_{\rm c} t \tag{9.97a}$$

and

$$\upsilon_{\rm LO}(t) = A_{\rm LO} \cos 2\pi f_{\rm LO} t, \tag{9.97b}$$

where $A(t)$ represents the amplitude of the audio signal waveform, $\upsilon_{\rm s}(t)$ (Fig. 9-29(a)), and $A_{\rm LO}$ is a constant amplitude associated with the *local oscillator* (LO) signal.

A *mixer* is a diode circuit that has two inputs and one output with its output voltage $\upsilon_{\rm out}(t)$ being equal to the product of its input voltages:

$$\upsilon_{\rm out}(t) = \upsilon_{\rm in}(t) \times \upsilon_{\rm LO}(t) = A(t)\ A_{\rm LO} \cos 2\pi f_{\rm c} t \cos 2\pi f_{\rm LO} t. \tag{9.98}$$

Application of the trigonometric identity

$$\cos x \cos y = \tfrac{1}{2}[\cos(x+y) + \cos(x-y)] \tag{9.99}$$

leads to

$$\upsilon_{\rm out}(t) = \frac{A(t)\ A_{\rm LO}}{2} \cos[2\pi(f_{\rm c} + f_{\rm LO})t] + \frac{A(t)\ A_{\rm LO}}{2} \cos[2\pi(f_{\rm LO} - f_{\rm c})t]. \tag{9.100}$$

Let us consider the case where $f_{\rm c} = 1$ MHz and $f_{\rm LO} = 1.445$ MHz. The expression for $\upsilon_{\rm out}(t)$ becomes

$$\upsilon_{\rm out}(t) = A'(t)\ \cos 2\pi f_{\rm s} t + A'(t)\ \cos 2\pi f_{\rm d} t, \tag{9.101}$$

where

$$A'(t) = \frac{A(t)\ A_{\rm LO}}{2}, \tag{9.102}$$

and $f_{\rm s}$ and $f_{\rm d}$ are the sum and difference frequencies:

$$f_{\rm s} = f_{\rm c} + f_{\rm LO} = 2.445 \text{ MHz} \tag{9.103a}$$

and

$$f_{\rm d} = f_{\rm LO} - f_{\rm c} = 0.445 \text{ MHz}. \tag{9.103b}$$

Thus, $\upsilon_{\rm out}(t)$ consists of two signal components with markedly different carrier frequencies. By selecting a narrow IF filter/amplifier in Fig. 9-30 with a center frequency $f_{\rm IF} = f_{\rm d}$, only the difference-frequency component of $\upsilon_{\rm out}(t)$ will make it through the filter. Consequently, its output is given by

$$\upsilon_{\rm IF}(t) = g_{\rm IF}\ A'(t)\ \cos 2\pi f_{\rm IF} t,$$

where g_{IF} is the voltage gain factor of the IF filter/amplifier. Demodulation, which is a low-frequency filtering process, removes the IF carrier, leaving behind a detected signal given by

$$v_d(t) = g_d g_{IF} A'(t),$$

where g_d is a demodulator constant. Since $A'(t)$ is directly proportional to the original audio signal $v_s(t)$, $v_d(t)$ becomes (ideally) a replica of $v_s(t)$.

9-8.4 Software Radio

The steady increase in the speed of digital circuits has made it possible to perform all of the functions of a superheterodyne receiver directly in the digital domain. Some implementations consist of little more than an antenna connected to the input pin of a digital chip. The chip converts the input signal into digital format and then performs all of the mixing, filtering, amplifying, and demodulating functions by direct computation. This digital approach, first proposed in the 1980s, is sometimes called *software radio*. In practice, analog-to-digital converters do not usually have the specifications required to directly sample signals coming from an antenna, and low-noise amplifiers are needed at the front end of the digital system. Additionally, many software radio implementations still use mixers at the front end, and simply digitize the signal coming out of the mixer.

Concept Question 9-18: What are the advantages of FM over AM? (See CAD)

Concept Question 9-19: What is the fundamental contribution of the superheterodyne receiver, and why is it significant? (See CAD)

Concept Question 9-20: What does a mixer do? (See CAD)

9-9 Spectral Response with Multisim

The AC Analysis and Parameter Sweep tools are very useful when analyzing the frequency response of a circuit. The Network Analyzer, first introduced in Chapter 8, also provides a convenient way to evaluate the frequency response of a circuit using Multisim. These tools are illustrated in the next three examples.

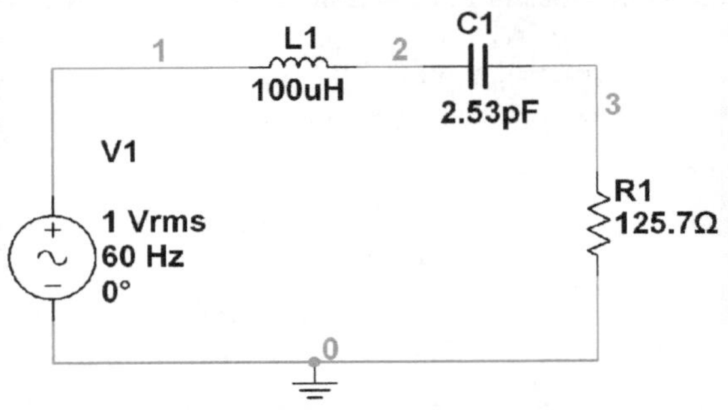

Figure 9-31: A series RLC filter implemented in Multisim.

Example 9-13: RLC Circuit

Design a series RLC bandpass filter with a center frequency of 10 MHz and $Q = 50$. Use Multisim to generate magnitude and phase plots covering the range from 8 to 12 MHz.

Solution: The specified filter can be designed with an infinite number of different combinations of R, L, and C. We will choose a realistic value for L, namely 0.1 mH, which will dictate that C be

$$C = \frac{1}{\omega_0^2 L} = \frac{1}{(2\pi \times 10^7)^2 \times 10^{-4}} = 2.53 \text{ pF}.$$

Next, we select the value of R to satisfy the requirement on Q. From Table 9-3, we obtain

$$R = \frac{\omega_0 L}{Q} = \frac{2\pi \times 10^7 \times 10^{-4}}{50} = 125.7\ \Omega.$$

With all three elements specified, we construct the Multisim circuit shown in Fig. 9-31. Before performing AC Analysis, we should double-click on the ac source and change its value to 1 V

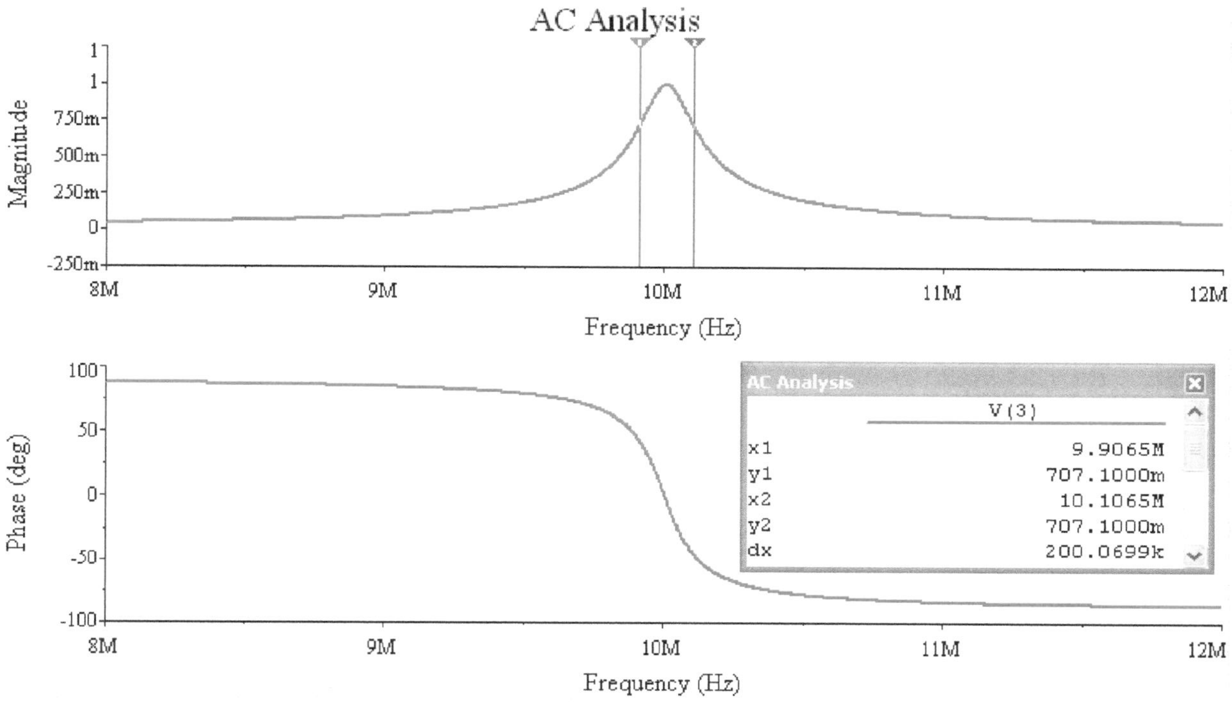

Figure 9-32: Magnitude and phase plots for the circuit of Fig. 9-31 generated by AC Analysis for Example 9-13.

(rms). Next, we select Simulate → Analyses → AC Analysis, and then we set FSTART to 8 MHz and FSTOP to 12 MHz. With both the Sweep Type and Vertical Scale set to Linear, the number of points set to 1000, and the variable selected for analysis is V(3), AC Analysis generates the plots displayed in Fig. 9-32. The magnitude plot exhibits a peak at 10 MHz, and the phase goes through 0° at that frequency. To verify that the circuit has a $Q = 50$, we use the cursors to establish the locations at which the vertical value of the curve is $1/\sqrt{2} = 0.707$ V. The separation between the two cursors (labeled "dx" in the cursor box) is 200.0699 kHz. This is the half-power bandwidth B. The quality factor is

$$\begin{aligned} Q &= \frac{\omega_0}{B} \\ &= \frac{10^7}{200.0699 \times 10^3} \\ &= 49.98, \end{aligned}$$

which is approximately equal to the specified value.

Example 9-14: Parameter Sweep

Apply Parameter Sweep to the circuit in Fig. 9-31 to generate spectral responses for C1 = 1 pF, 4 pF, 7 pF, and 10 pF.

Solution: Starting with the circuit in Fig. 9-31, we set V1 = 1 V. Next, we select Simulate → Analyses → Parameter Sweep. Upon selecting the parameter we wish to vary (capacitance C1), its minimum (1 pF), maximum (10 pF), step size (3 pF), and number of points (4), we select AC Analysis in the More Options box. This allows us to set the frequency range, the type of sweep (linear), and the number of points—just as we did previously in Example 9-13, except that the frequency range is 0 to 20 MHz. The Simulate command generates the plots shown in Fig. 9-33.

Example 9-15: Bode Plots

Reproduce the circuit of Fig. 9-26(a) in Multisim and generate Bode plots corresponding to the outputs of the three stages.

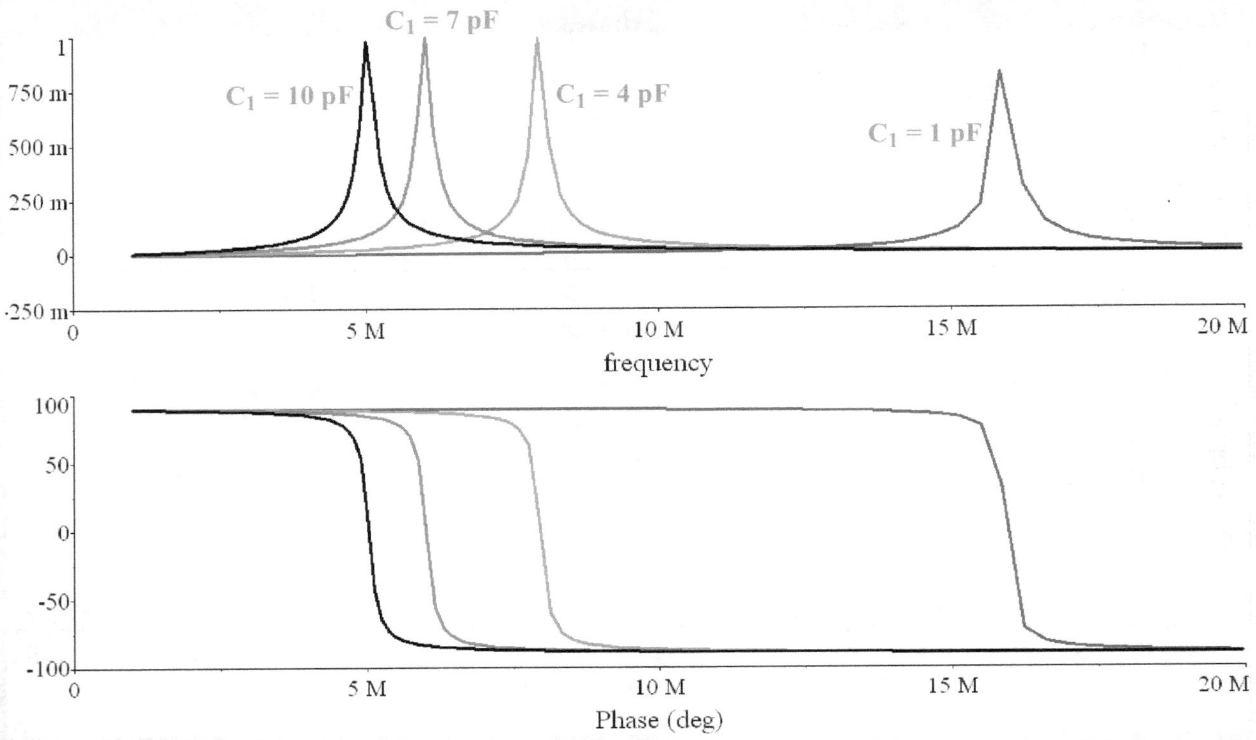

Figure 9-33: AC analysis plots for the circuit in Fig. 9-31 generated with the Parameter Sweep tool in Example 9-14. The capacitance was varied from 1 to 10 pF.

Solution: The circuit is reproduced in Fig. 9-34(a), and part (b) displays the results. In order to generate these plots, we use AC Analysis with

FSTART $= 10^4$ (rad/s)$/2\pi = 1.592$ kHz

and

FSTOP $= 10^7/2\pi = 1.592$ MHz.

The number of points was set to 200,

Sweep Type = Decade,

and

Vertical Scale = Decibel.

Example 9-16: Bode Plotter Instrument

Use the Bode Plotter Instrument to generate magnitude and phase plots of the circuit in Fig. 9-31 over the frequency range of 8 to 12 MHz.

Solution: Go to Simulate → Instruments → Bode Plotter. Connect the "IN" terminals across the V1 source and connect the "OUT" terminals across the resistor R1. Bring up the Bode Plotter Instrument window. With the Magnitude Mode selected, set the horizontal scale to Lin (for linear), set I (initial frequency) to 8 MHz, and set F (final frequency) to 12 MHz. For the vertical scale, leave it on Log, set I to −50 dB, and set F to 5 dB. Select the Phase mode and for the vertical scale set I to −100 deg and F to 100 deg. Run the Interactive Simulation by pressing F5 or the appropriate button or toggle switch on the toolbar. In the Magnitude and Phase mode, you will generate plots similar to those shown in Fig. 9-35(a) and (b), respectively.

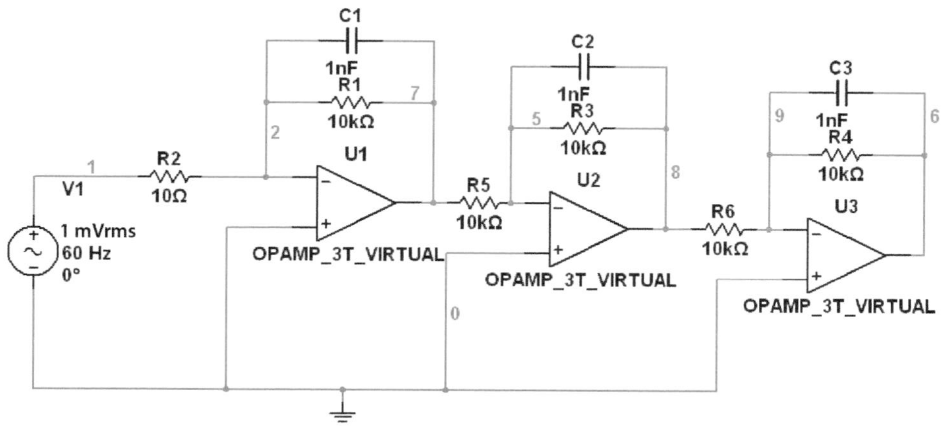

(a) Three-stage circuit of Fig. 9-26(a)

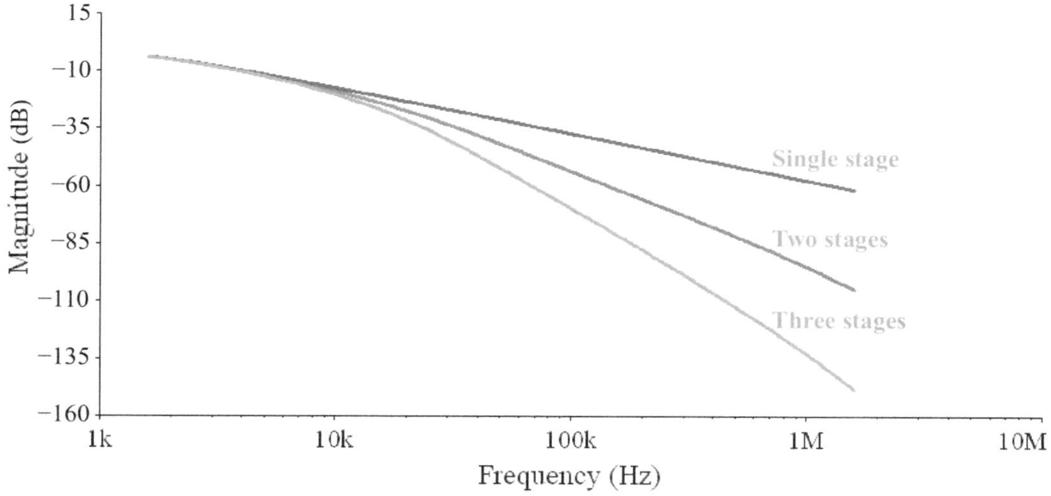

(b) Bode plots

Figure 9-34: Three-stage op-amp circuit of Fig. 9-26(a) reproduced in Multisim for Example 9-15.

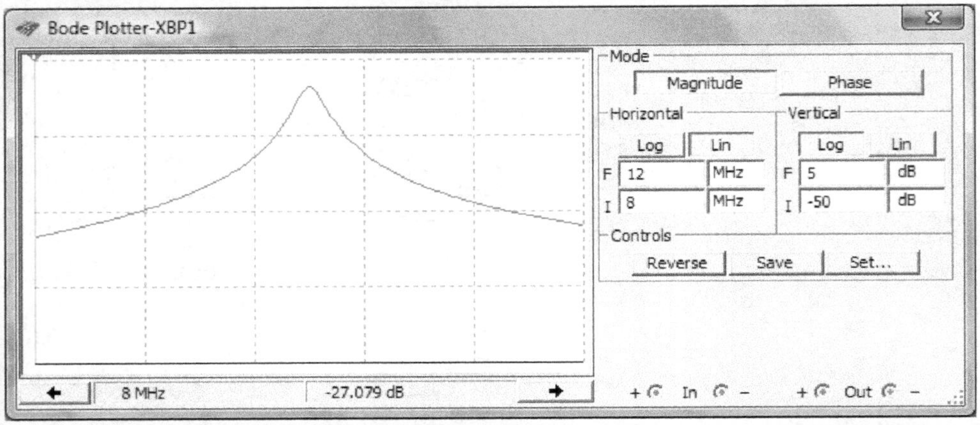

(a) Magnitude plot

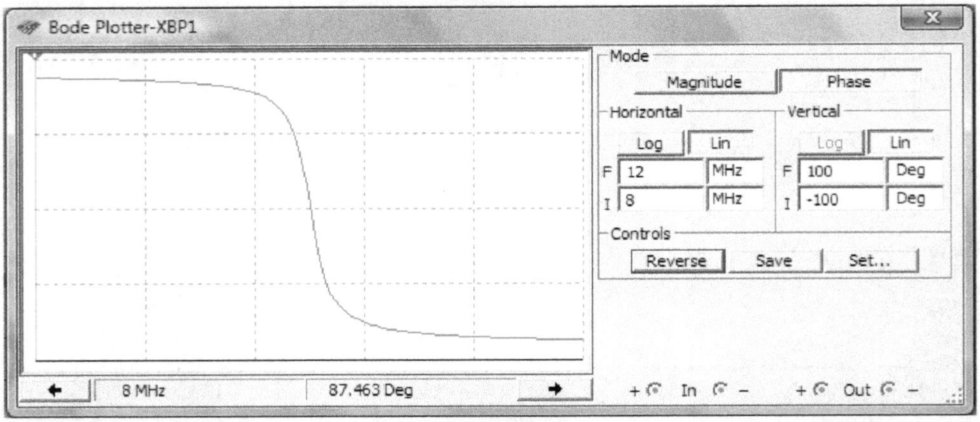

(b) Phase plot

Figure 9-35: Output of Bode Plotter Instrument for Example 9-16.

Summary

Concepts

- The transfer function of a circuit is the ratio of a phasor output voltage or current to a phasor input voltage or current.
- The transfer function is characterized by magnitude and phase plots describing the spectral response of the circuit.
- At the resonant frequency ω_0, the input impedance of the circuit is purely real.
- The Bode diagram uses straight-line approximations on a semilog-scale to display the magnitude and phase spectra of the transfer function.
- The quality factor Q of a bandpass filter defines the degree of frequency selectivity of the filter.
- The order of a filter defines the gain roll-off rate of the magnitude spectrum in the stopband.
- Active filters are used primarily at frequencies below 1 MHz, whereas passive filters are better suited at higher frequencies.
- Active filters can provide power gain, and they easily can be cascaded in series or in parallel to generate the desired frequency response.
- In a superheterodyne receiver, the RF frequency is converted into an IF frequency for amplification and filtering prior to demodulation.
- Parameter sweep can be used in Multisim to compare the circuit response for different values of a key parameter.

Mathematical and Physical Models

Resonant Frequency ω_0

$$\Im\{\mathbf{Z}_{\text{in}}(\omega)\} = 0 \qquad @\ \omega = \omega_0$$

Magnitude and Frequency Scaling

$$R' = K_{\text{m}} R, \qquad L' = \frac{K_{\text{m}}}{K_{\text{f}}}\, L$$

$$C' = \frac{1}{K_{\text{m}} K_{\text{f}}}\, C, \qquad \omega' = K_{\text{f}}\,\omega$$

dB Scale

$$\text{If } G = XY \quad \Rightarrow \quad G\ [\text{dB}] = X\ [\text{dB}] + Y\ [\text{dB}]$$

$$\text{If } G = \frac{X}{Y} \quad \Rightarrow \quad G\ [\text{dB}] = X\ [\text{dB}] - Y\ [\text{dB}]$$

Series and Parallel Bandpass RLC Filters

$$\omega_0 = \sqrt{\omega_{\text{c}_1}\omega_{\text{c}_2}} = \frac{1}{\sqrt{LC}}$$

$$Q = \frac{\omega_0 L}{R} \qquad \text{(series)}$$

$$Q = \frac{R}{\omega_0 L} \qquad \text{(parallel)}$$

Active Filters

Sections 9-6 and 9-7

Important Terms

Provide definitions or explain the meaning of the following terms:

−3 dB frequency
active filter
amplitude modulation
AM radio band
AND logic gate
bandpass filter
bandreject filter
bandwidth
Bode plot
Bode diagram
carrier
carrier frequency
connected in parallel
connected in series
corner frequency
cutoff frequency
damping factor
dc gain
degree of selectivity
demodulate
down-conversion
Edwin Armstrong
energy dissipated
exact plots

Important Terms (continued)

factor
filter order
frequency conversion
frequency modulation
frequency response
frequency scaling
frequency scaling factor
frequency-selective circuits
gain factor
gain roll-off rate
half-power frequency
heterodyne receiver
high-frequency gain
highpass filter
idealized response
IF
IF filter/amplifier
intermediate frequency
local oscillator
lowpass filter
magnitude
magnitude scaling
magnitude scaling factor
magnitude response
mix
mixer
normalized power
order
OR gate
passband
passive filter
phase angle
phase response
pole
pole @ origin factor
pole factor
practical circuit
prototype model
quadratic-pole factor
quadratic-zero factor
quality factor
radio-frequency
regenerative receiver
relative power
resonance condition
resonant frequency
RF
signal
simple-pole factor
simple-zero factor
software radio
standard form
store
straight-line approximation
stopband
transfer function
trivial resonance
tuned-radio frequency receiver
tuner
unity input
up-conversion
voltage transfer function
zero
zero factor
zero @ origin factor

PROBLEMS

Section 9-1: Transfer Function

*9.1 Determine the resonant frequency of the circuit shown in Fig. P9.1, given that $R = 100\ \Omega$, $L = 5$ mH, and $C = 1\ \mu$F.

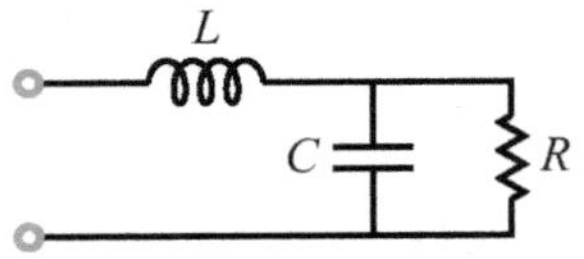

Figure P9.1: Circuit for Problem 9.1.

9.2 Determine the resonant frequency of the circuit shown in Fig. P9.2, given that $R = 100\ \Omega$, $L = 5$ mH, and $C = 1\ \mu$F.

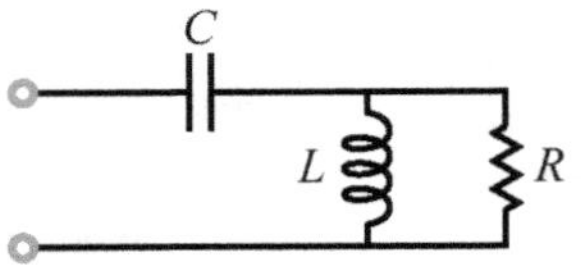

Figure P9.2: Circuit for Problem 9.2.

*Answer(s) available in Appendix G.

*9.3 Determine the resonant frequency of the circuit shown in Fig. P9.3, given that $R = 1$ kΩ, $L = 10$ mH, and $C = 10$ nF.

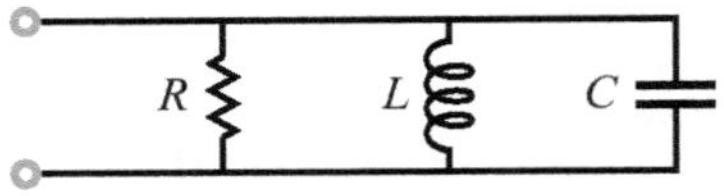

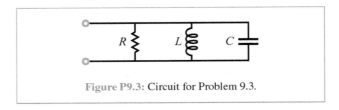

Figure P9.3: Circuit for Problem 9.3.

9.4 Determine the resonant frequency of the circuit shown in Fig. P9.4, given that $R = 1$ kΩ, $L = 10$ mH, and $C = 10$ nF.

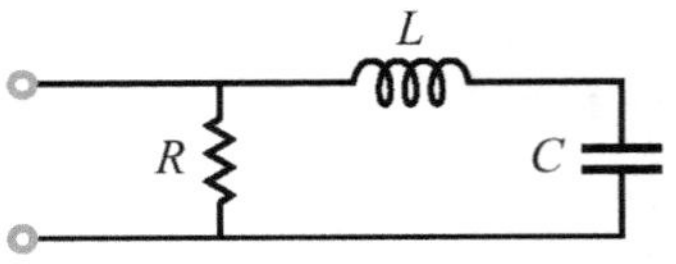

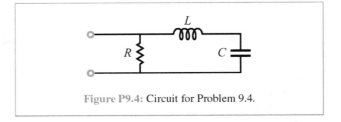

Figure P9.4: Circuit for Problem 9.4.

*9.5 Determine the resonant frequency of the circuit shown in Fig. P9.5, given that $R_1 = 10\ \Omega$, $R_2 = 100\ \Omega$, $L = 5$ mH, and $C = 0.1\ \mu$F.

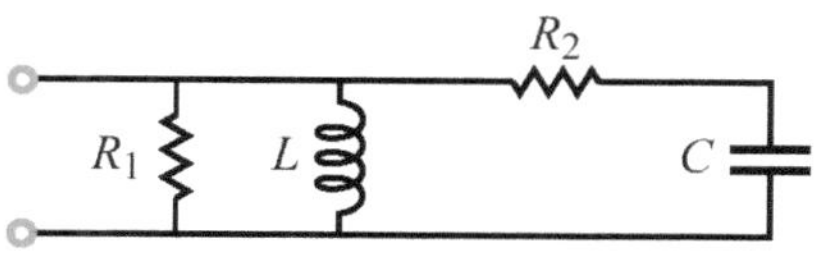

Figure P9.5: Circuit for Problem 9.5.

9.6 For the circuit shown in Fig. P9.6, determine (a) the transfer function $\mathbf{H} = \mathbf{V}_o/\mathbf{V}_i$ and (b) the frequency ω_o at which $\mathbf{H}$ is purely real.

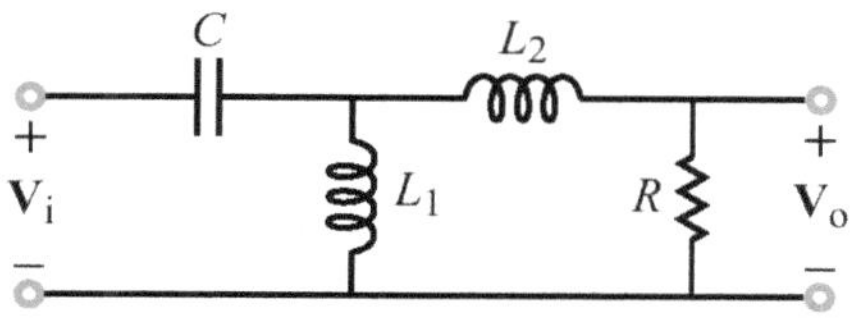

Figure P9.6: Circuit for Problem 9.6.

9.7 For the circuit shown in Fig. P9.7, determine (a) the transfer function $\mathbf{H} = \mathbf{V}_o/\mathbf{V}_i$ and (b) the frequency ω_o at which $\mathbf{H}$ is purely real.

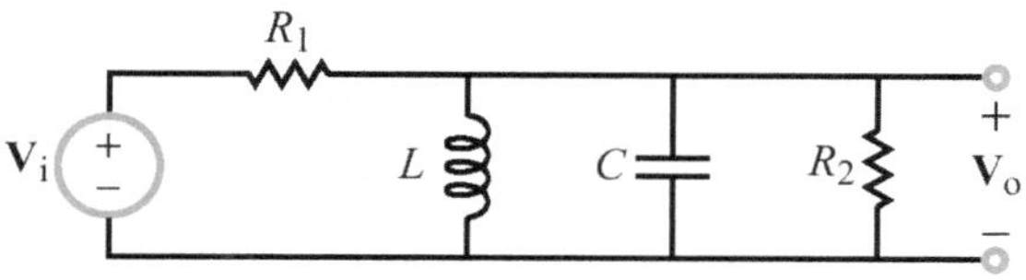

Figure P9.7: Circuit for Problem 9.7.

Section 9-2: Scaling

*__9.8__ What values of the scaling factors K_m and K_f should be applied to scale a circuit containing a 1 F capacitor and 4 H inductor into one containing 1 μF and 10 mH, respectively?

9.9 The corner frequency of the highpass-filter circuit shown in Fig. P9.9 is approximately 1 Hz. Scale the circuit up in frequency by a factor of 10^5 while keeping the values of the inductors unchanged.

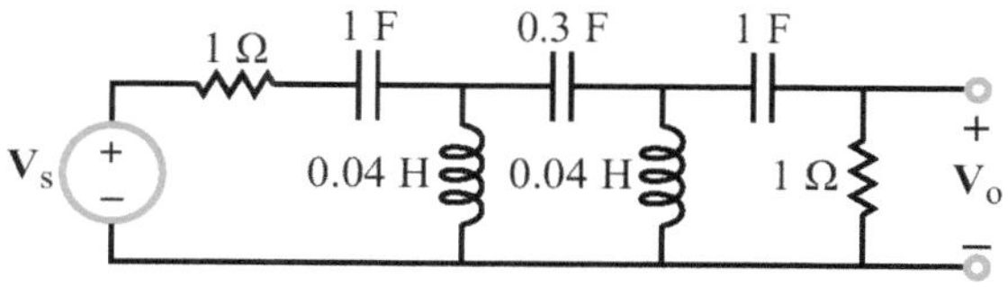

Figure P9.9: Circuit for Problem 9.9.

9.10 For the circuit shown in Fig. P9.10:

(a) Obtain an expression for the input impedance $\mathbf{Z}_{in}(\omega)$.

(b) If $R_1 = R_2 = 1\ \Omega$, $C = 1$ F, and $L = 5$ H, at what angular frequency is $\mathbf{Z}_{in}$ purely real?

(c) Scale the circuit by $K_m = 20$ and write down the new expression for the input impedance.

(d) Is the value of ω at which the input impedance of the scaled circuit is real the same or different from the answer of part (b)?

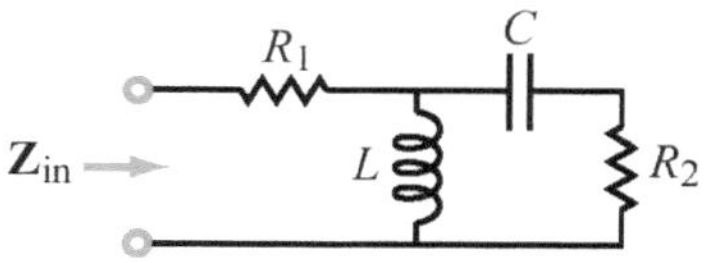

Figure P9.10: Circuit for Problem 9.10.

9.11 For the circuit shown in Fig. P9.11:

(a) Obtain an expression for the input impedance $\mathbf{Z}_{in}(\omega)$.

*(b) If $R_1 = 1\ \Omega$, $R_2 = R_3 = 2\ \Omega$, $L = 1$ H, and $C = 1$ F, at what angular frequency is $\mathbf{Z}_{in}$ purely real?

(c) Redraw the circuit after scaling it by $K_m = 10^3$ and $K_f = 10^5$. Specify the new element values.

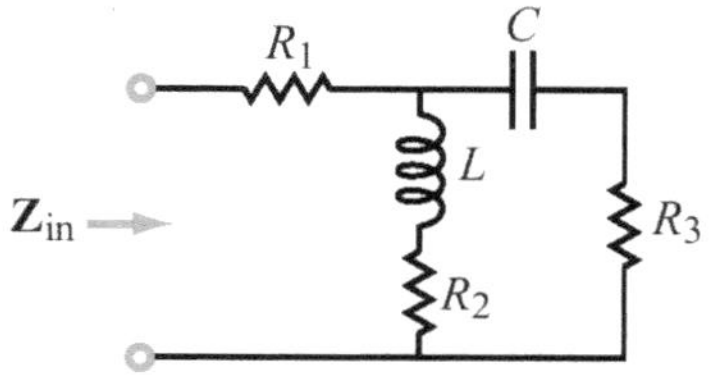

Figure P9.11: Circuit for Problem 9.11.

9.12 Circuit (b) in Fig. P9.12 is a scaled version of circuit (a). The scaling process may have involved magnitude or frequency scaling, or both simultaneously. If $R_1 = 1$ kΩ gets scaled to $R_1' = 10$ kΩ, supply the impedance values of the other elements in the scaled circuit.

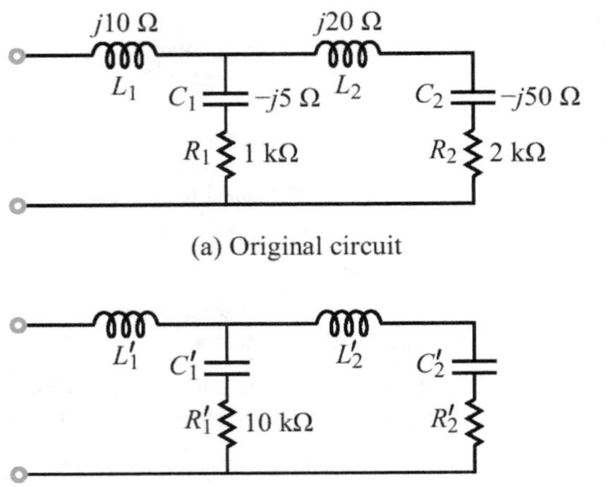

(a) Original circuit

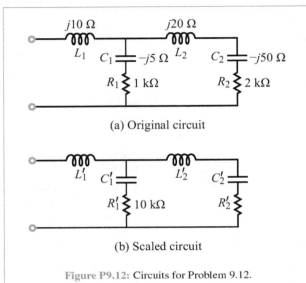

(b) Scaled circuit

Figure P9.12: Circuits for Problem 9.12.

Section 9-3: Bode Plots

9.13 Convert the following power ratios to dB.

(a) 3×10^2

*(b) 0.5×10^{-2}

(c) $\sqrt{2000}$

(d) $(360)^{1/4}$

9.14 Convert the following power ratios to dB.

*(a) $6e^3$

(b) $2.3 \times 10^3 + 60$

(c) $24(3 \times 10^7)$

(d) $4/(5 \times 10^3)$

9.15 Convert the following voltage ratios to dB.

(a) 2×10^{-4}

(b) 3000

(c) $\sqrt{30}$

(d) $6/(5 \times 10^4)$

9.16 Convert the following dB values to voltage ratios.

(a) 46 dB

(b) 0.4 dB

*(c) −12 dB

(d) −66 dB

9.17 Generate Bode magnitude and phase plots (straight-line approximations) for the following voltage transfer functions.

(a) $\mathbf{H}(\omega) = \dfrac{j100\omega}{10 + j\omega}$

(b) $\mathbf{H}(\omega) = \dfrac{0.4(50 + j\omega)^2}{(j\omega)^2}$

(c) $\mathbf{H}(\omega) = \dfrac{(40 + j80\omega)}{(10 + j50\omega)}$

(d) $\mathbf{H}(\omega) = \dfrac{(20 + j5\omega)(20 + j\omega)}{j\omega}$

9.18 Generate Bode magnitude and phase plots (straight-line approximations) for the following voltage transfer functions.

(a) $\mathbf{H}(\omega) = \dfrac{30(10 + j\omega)}{(200 + j2\omega)(1000 + j2\omega)}$

(b) $\mathbf{H}(\omega) = \dfrac{j100\omega}{(100 + j5\omega)(100 + j\omega)^2}$

(c) $\mathbf{H}(\omega) = \dfrac{(200 + j2\omega)}{(50 + j5\omega)(1000 + j\omega)}$

9.19 Generate Bode magnitude and phase plots (straight-line approximations) for the following voltage transfer functions.

(a) $\mathbf{H}(\omega) = \dfrac{4 \times 10^4(60 + j6\omega)}{(4 + j2\omega)(100 + j2\omega)(400 + j4\omega)}$

(b) $\mathbf{H}(\omega) = \dfrac{(1 + j0.2\omega)^2(100 + j2\omega)^2}{(j\omega)^3(500 + j\omega)}$

(c) $\mathbf{H}(\omega) = \dfrac{8 \times 10^{-2}(10 + j10\omega)}{j\omega(16 - \omega^2 + j4\omega)}$

(d) $\mathbf{H}(\omega) = \dfrac{4 \times 10^4\omega^2(100 - \omega^2 + j50\omega)}{(5 + j5\omega)(200 + j2\omega)^3}$

9.20 Generate Bode magnitude and phase plots (straight-line approximations) for the following voltage transfer functions.

(a) $\mathbf{H}(\omega) = \dfrac{j5 \times 10^3\omega(20 + j2\omega)}{(2500 - \omega^2 + j20\omega)}$

(b) $\mathbf{H}(\omega) = \dfrac{512(1 + j\omega)(4 + j40\omega)}{(256 - \omega^2 + j32\omega)^2}$

(c) $\mathbf{H}(\omega) = \dfrac{j(10 + j\omega) \times 10^8}{(20 + j\omega)^2(500 + j\omega)(1000 + j\omega)}$

***9.21** Determine the voltage transfer function $\mathbf{H}(\omega)$ corresponding to the Bode magnitude plot shown in Fig. P9.21. The phase of $\mathbf{H}(\omega)$ approaches 180° as ω approaches 0.

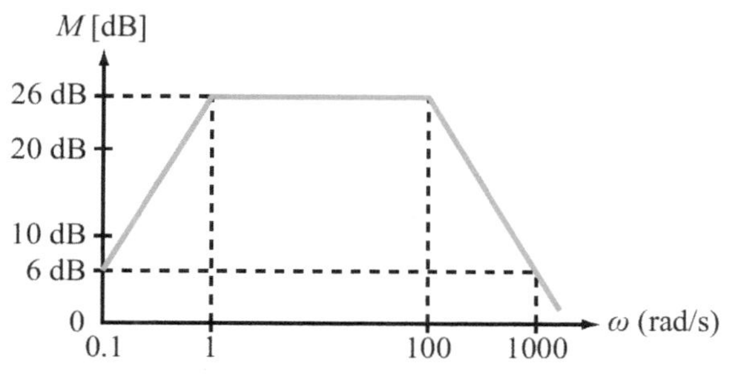

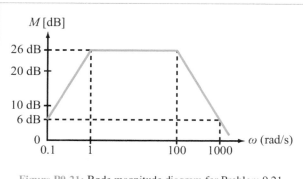

Figure P9.21: Bode magnitude diagram for Problem 9.21.

9.22 Determine the voltage transfer function $\mathbf{H}(\omega)$ corresponding to the Bode magnitude plot shown in Fig. P9.22. The phase of $\mathbf{H}(\omega)$ is 90° at $\omega = 0$.

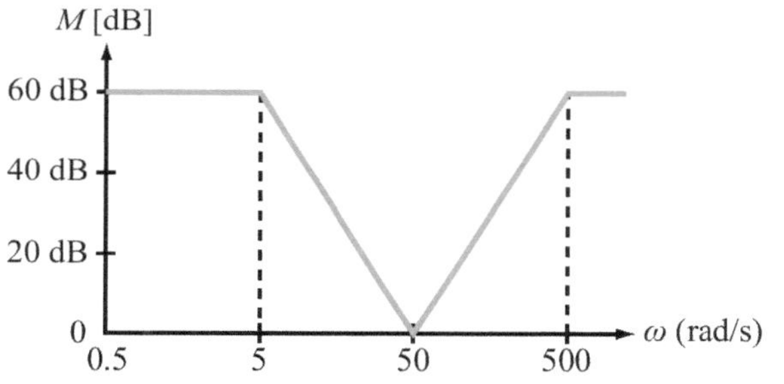

Figure P9.22: Bode magnitude plot for Problem 9.22.

***9.23** Determine the voltage transfer function $\mathbf{H}(\omega)$ corresponding to the Bode magnitude plot shown in Fig. P9.23. The phase of $\mathbf{H}(\omega)$ is 180° at $\omega = 0$.

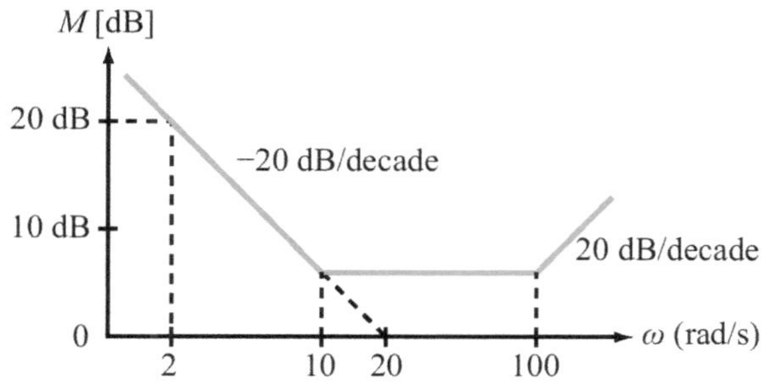

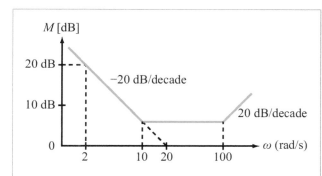

Figure P9.23: Bode magnitude plot for Problem 9.23.

9.24 Determine the voltage transfer function $\mathbf{H}(\omega)$ corresponding to the Bode magnitude plot shown in Fig. P9.24. The phase of $\mathbf{H}(\omega)$ is −90° at $\omega = 0$.

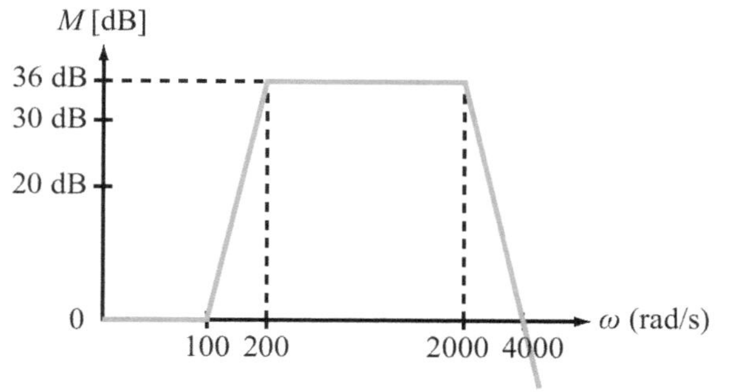

Figure P9.24: Bode magnitude plot for Problem 9.24.

9.25 Determine the voltage transfer function $\mathbf{H}(\omega)$ corresponding to the Bode magnitude plot shown in Fig. P9.25. The phase of $\mathbf{H}(\omega)$ is 0° at $\omega = 0$.

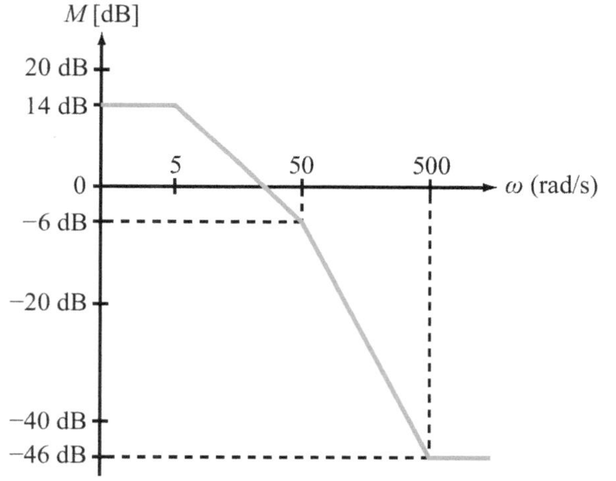

Figure P9.25: Bode magnitude plot for Problem 9.25.

Sections 9-4 and 9-5: Passive Filters

9.26 The element values of a series RLC bandpass filter are $R = 5\ \Omega$, $L = 20$ mH, and $C = 0.5\ \mu$F.

*(a) Determine ω_0, Q, B, ω_{c_1}, and ω_{c_2}.

(b) Is it possible to double the magnitude of Q by changing the values of L and/or C while keeping ω_0 and R unchanged? If yes, propose such values, and if no, why not?

9.27 A series RLC bandpass filter has half-power frequencies at 1 kHz and 10 kHz. If the input impedance at resonance is 6 Ω, what are the values of R, L, and C?

9.28 A series RLC circuit is driven by an ac source with a phasor voltage $\mathbf{V}_s = 10\angle 30^\circ$ V. If the circuit resonates at 10^3 rad/s and the average power absorbed by the resistor at resonance is 2.5 W, determine the values of R, L, and C, given that $Q = 5$.

*__9.29__ The element values of a parallel RLC circuit are $R = 100$ Ω, $= L = 10$ mH, and $C = 0.4$ mF. Determine ω_0, Q, B, ω_{c_1}, and ω_{c_2}.

9.30 Design a parallel RLC filter with $f_0 = 4$ kHz, $Q = 100$, and an input impedance of 25 kΩ at resonance.

9.31 For the circuit shown in Fig. P9.31:

(a) Obtain an expression for $\mathbf{H}(\omega) = \mathbf{V}_o/\mathbf{V}_i$ in standard form.

(b) Generate spectral plots for the magnitude and phase of $\mathbf{H}(\omega)$, given that $R_1 = 1$ Ω, $R_2 = 2$ Ω, $C_1 = 1$ μF, and $C_2 = 2$ μF.

(c) Determine the cutoff frequency ω_c and the slope of the magnitude (in dB) when $\omega/\omega_c \ll 1$ and when $\omega/\omega_c \gg 1$.

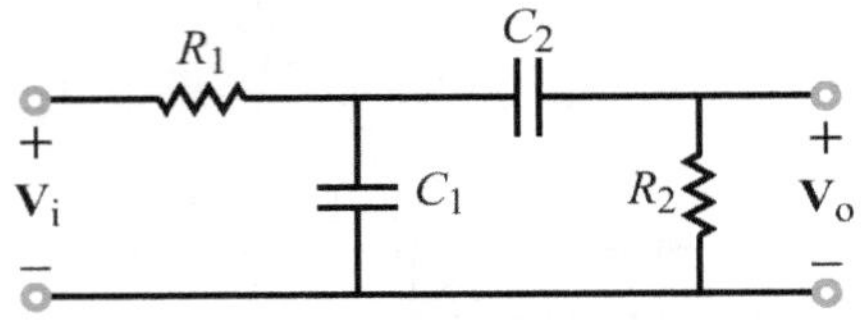

Figure P9.31: Circuit for Problem 9.31.

9.32 For the circuit shown in Fig. P9.32:

(a) Obtain an expression for $\mathbf{H}(\omega) = \mathbf{V}_o/\mathbf{V}_i$ in standard form.

*(b) Generate spectral plots for the magnitude and phase of $\mathbf{H}(\omega)$, given that $R_1 = 1$ Ω, $R_2 = 2$ Ω, $L_1 = 1$ mH, and $L_2 = 2$ mH.

(c) Determine the cutoff frequency ω_c and the slope of the magnitude (in dB) when $\omega/\omega_c \ll 1$ and when $\omega/\omega_c \gg 1$.

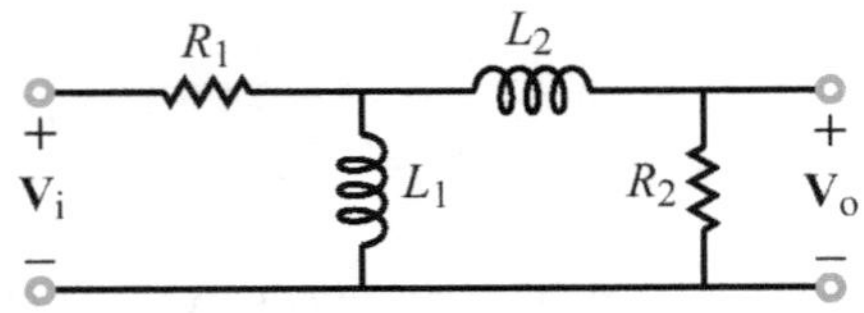

Figure P9.32: Circuit for Problem 9.32.

9.33 For the circuit shown in Fig. P9.33:

(a) Obtain an expression for $\mathbf{H}(\omega) = \mathbf{V}_o/\mathbf{V}_i$ in standard form.

(b) Generate spectral plots for the magnitude and phase of $\mathbf{H}(\omega)$, given that $R = 100$ Ω, $L = 0.1$ mH, and $C = 1$ μF.

(c) Determine the cutoff frequency ω_c and the slope of the magnitude (in dB) when $\omega/\omega_c \gg 1$.

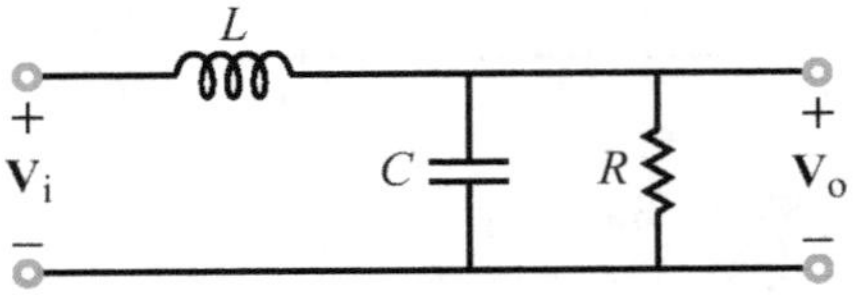

Figure P9.33: Circuit for Problem 9.33.

9.34 For the circuit shown in Fig. P9.34:

(a) Obtain an expression for $\mathbf{H}(\omega) = \mathbf{V}_o/\mathbf{V}_i$ in standard form.

(b) Generate spectral plots for the magnitude and phase of $\mathbf{H}(\omega)$, given that $R = 10$ Ω, $L = 1$ mH, and $C = 10$ μF.

(c) Determine the cutoff frequency ω_c and the slope of the magnitude (in dB) when $\omega/\omega_c \ll 1$.

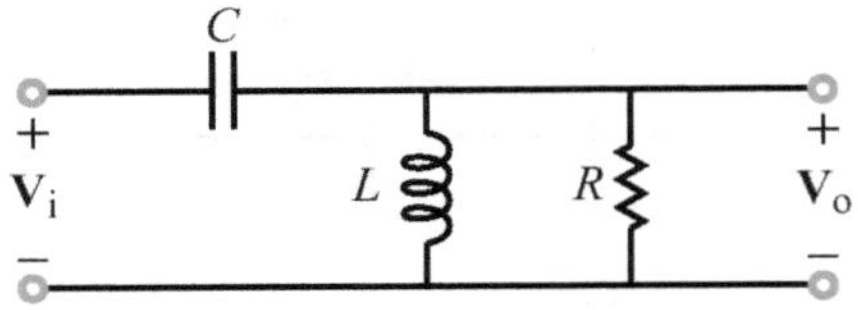

Figure P9.34: Circuit for Problem 9.34.

9.35 For the circuit shown in Fig. P9.35:

*(a) Obtain an expression for $\mathbf{H}(\omega) = \mathbf{V}_o/\mathbf{V}_i$ in standard form.

(b) Generate spectral plots for the magnitude and phase of $\mathbf{H}(\omega)$, given that $R = 50$ Ω and $L = 2$ mH.

(c) Determine the cutoff frequency ω_c and the slope of the magnitude (in dB) when $\omega/\omega_c \ll 1$.

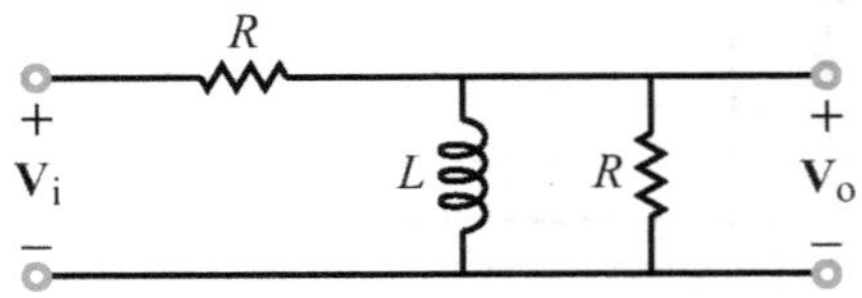

Figure P9.35: Circuit for Problem 9.35.

9.36 For the circuit shown in Fig. P9.36:

(a) Obtain an expression for $\mathbf{H}(\omega) = \mathbf{V}_o/\mathbf{V}_i$ in standard form.

(b) Generate spectral plots for the magnitude and phase of $\mathbf{H}(\omega)$, given that $R = 50\ \Omega$ and $L = 2$ mH.

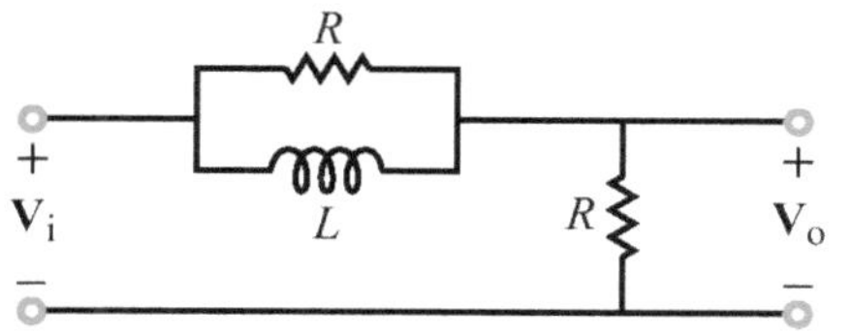

Figure P9.36: Circuit for Problem 9.36.

Sections 9-6 and 9-7: Active Filters

9.37 For the op-amp circuit of Fig. P9.37:

*(a) Obtain an expression for $\mathbf{H}(\omega) = \mathbf{V}_o/\mathbf{V}_s$ in standard form.

(b) Generate spectral plots for the magnitude and phase of $\mathbf{H}(\omega)$, given that $R_1 = 1$ kΩ, $R_2 = 4$ kΩ, and $C = 1\ \mu$F.

(c) What type of filter is it? What is its maximum gain?

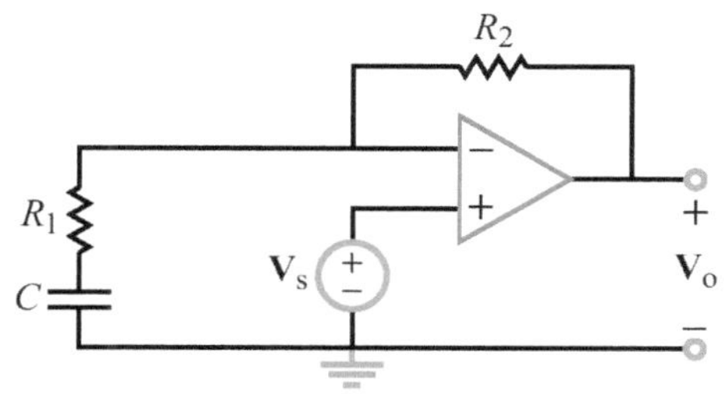

Figure P9.37: Circuit for Problem 9.37.

9.38 For the op-amp circuit of Fig. P9.38:

(a) Obtain an expression for $\mathbf{H}(\omega) = \mathbf{V}_o/\mathbf{V}_s$ in standard form.

(b) Generate spectral plots for the magnitude and phase of $\mathbf{H}(\omega)$, given that $R_1 = 99$ kΩ, $R_2 = 1$ kΩ, and $C = 0.1\ \mu$F.

(c) What type of filter is it? What is its maximum gain?

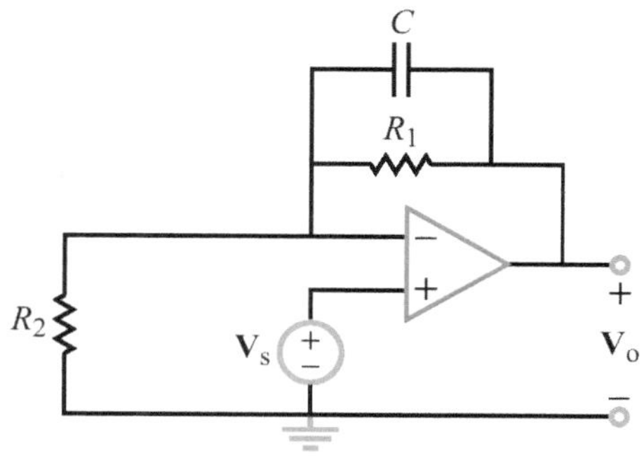

Figure P9.38: Circuit for Problem 9.38.

9.39 For the op-amp circuit of Fig. P9.39:

*(a) Obtain an expression for $\mathbf{H}(\omega) = \mathbf{V}_o/\mathbf{V}_i$ in standard form.

(b) Generate spectral plots for the magnitude and phase of $\mathbf{H}(\omega)$, given that $R_1 = R_2 = 100\ \Omega$, $C_1 = 10\ \mu$F, and $C_2 = 0.4\ \mu$F.

(c) What type of filter is it? What is its maximum gain?

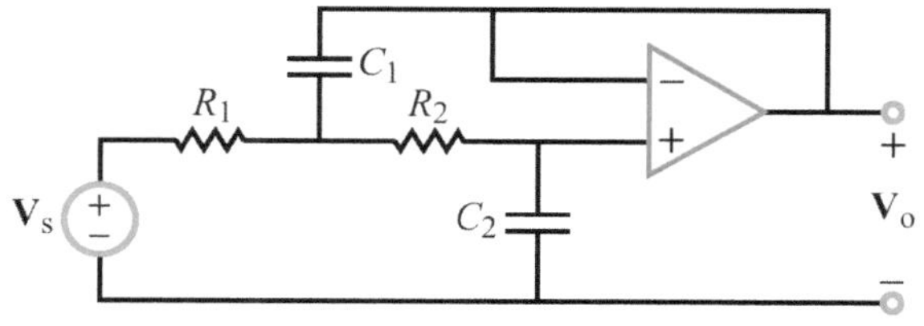

Figure P9.39: Circuit for Problems 9.39 and 9.40.

9.40 Repeat Problem 9.39 after interchanging the values of C_1 and C_2 to $C_1 = 0.4\ \mu$F and $C_2 = 10\ \mu$F.

9.41 For the op-amp circuit of Fig. P9.41:

*(a) Obtain an expression for $\mathbf{H}(\omega) = \mathbf{V}_o/\mathbf{V}_s$ in standard form.

(b) Generate spectral plots for the magnitude and phase of $\mathbf{H}(\omega)$, given that $R_1 = 1$ kΩ, $R_2 = 20\ \Omega$, $C_1 = 5\ \mu$F, and $C_2 = 25$ nF.

(c) What type of filter is it? What is its maximum gain?

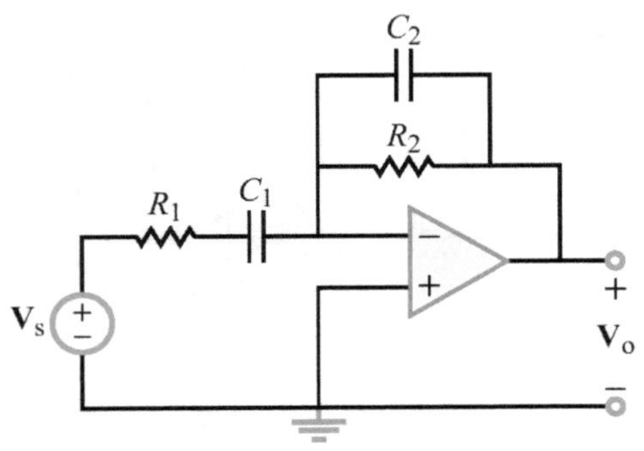

Figure P9.41: Circuit for Problem 9.41.

9.42 Design an active lowpass filter with a gain of 4, a corner frequency of 1 kHz, and a gain roll-off rate of −60 dB/decade.

9.43 Design an active highpass filter with a gain of 10, a corner frequency of 2 kHz, and a gain roll-off rate of 40 dB/decade.

9.44 Show that the transfer function of the circuit shown in Fig. P9.44 is given by

$$\mathbf{H}(\omega) = \frac{\mathbf{V}_o}{\mathbf{V}_s} = -G\left(1 + j\frac{\omega}{\omega_c}\right),$$

and relate G and ω_c to R_1, R_2, and C.

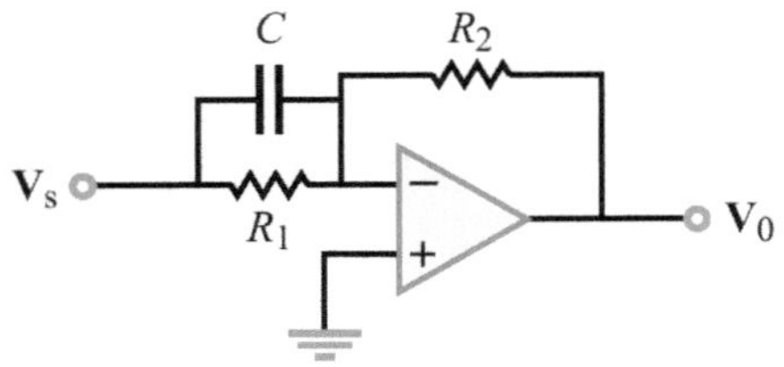

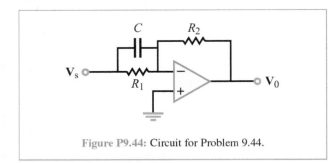

Figure P9.44: Circuit for Problem 9.44.

9.45 Repeat Problem 9.41 after replacing the series combination of R_1 and C_1 with a parallel combination.

*__9.46__ Consider the circuit shown in Fig. P9.46. Obtain its transfer function $\mathbf{H}(\omega) = \mathbf{V}_o/\mathbf{V}_s$ for $R_1 = 1$ kΩ, $R_2 = 10$ kΩ, and $C = 1$ μF. What role, if any, does the capacitor play? Explain.

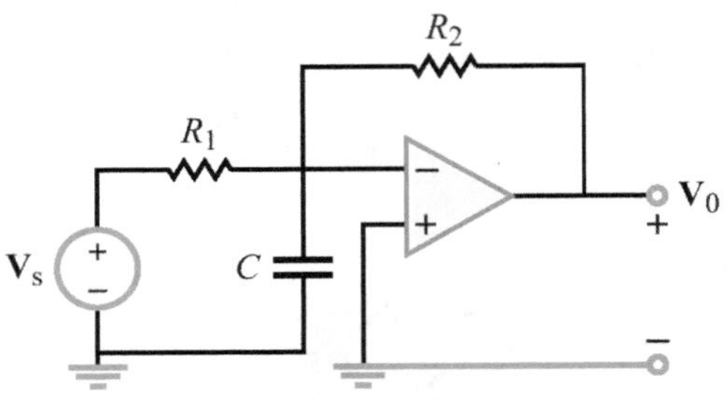

Figure P9.46: Circuit for Problem 9.46.

9.47 Use resistors, capacitors, and a single op amp to design a circuit with input voltage $\mathbf{V}_s$, output voltage $\mathbf{V}_o$, and transfer function

$$\mathbf{H}(\omega) = \frac{\mathbf{V}_o}{\mathbf{V}_s} = -10\left(1 + j\frac{\omega}{100}\right).$$

9.48 The element values in the circuit of the second-order bandpass filter shown in Fig. P9.48 are $R_{f_1} = 100$ kΩ, $R_{s_1} = 10$ kΩ, $R_{f_2} = 100$ kΩ, $R_{s_2} = 10$ kΩ, $C_{f_1} = 3.98 \times 10^{-11}$ F, and $C_{s_2} = 7.96 \times 10^{-10}$ F. Generate a spectral plot for the magnitude of $\mathbf{H}(\omega) = \mathbf{V}_o/\mathbf{V}_s$. Determine the frequency locations of the maximum value of M [dB] and its half-power points.

Section 9-8: Superheterodyne Receiver

9.49 Using the circuit layout shown in Fig. 9-15, design a tuner that uses a variable inductor, a capacitor, and a resistor. The input impedance of the tuner should be 377 Ω at 1 MHz, and its bandwidth should be 2 percent.

*__9.50__ What range of frequencies should the local oscillator be able to provide to mix the FM radio range (88 to 108 MHz) down to 10 MHz?

Section 9-9: Multisim

9.51 Generate plots in Multisim for the magnitude and phase of the transfer function for a series bandpass filter with $L = 1$ mH, $f_0 = 1$ MHz, and $Q = 10$. Choose FSTART = 100 kHz and FSTOP = 10 MHz.

9.52 Perform a Parameter Sweep in Multisim for capacitor C_{s_2} of the two-stage bandpass filter shown in Fig. 9-27. Generate response plots from 10 Hz to 100 kHz for each of five equally spaced values of C_{s_2} starting at 1 nF and ending at 15 nF.

9.53 Use Multisim to generate spectral plots for the magnitudes and phases of voltages $\mathbf{V}_C$ and $\mathbf{V}_o$ in the circuit of

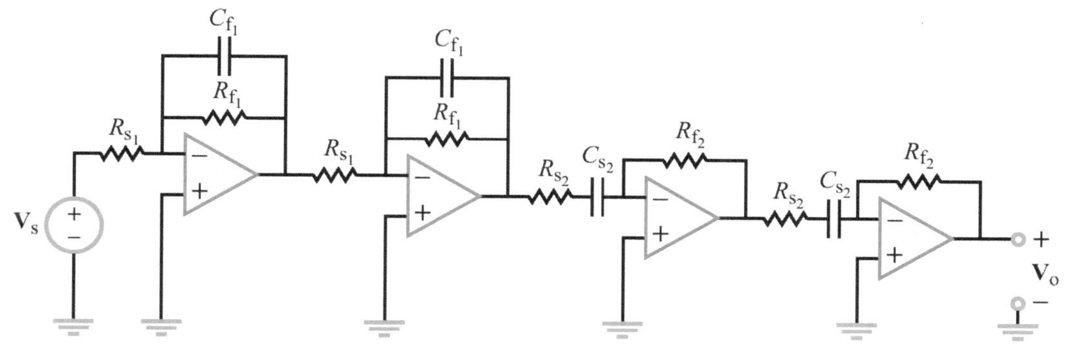

Figure P9.48: Circuit for Problem 9.48.

Fig. P9.53. The circuit is a second-order passive lowpass filter followed by an active highpass filter. Use the following element values: $R_1 = 20.3\ \Omega$, $R_2 = R_3 = 1.592\ \text{k}\Omega$, $L_1 = 100$ nH, $C_1 = C_2 = 1$ nF, and $\upsilon_s(t) = \cos 2\pi f t$ V.

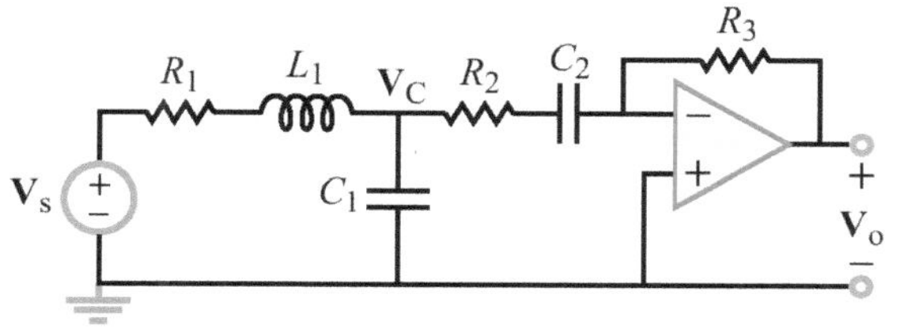

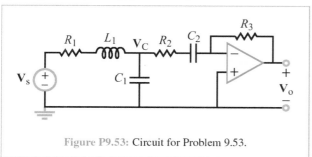

Figure P9.53: Circuit for Problem 9.53.

9.54 For the circuit in Fig. P9.54, use Multisim to generate spectral plots for the magnitude and phase of $\mathbf{H}(\omega) = \mathbf{V}_o/\mathbf{V}_i$ over the range from 100 Hz to 100 kHz. When performing the AC Analysis, use 200 points per decade. Determine the frequencies at which M [dB] is a maximum or a minimum.

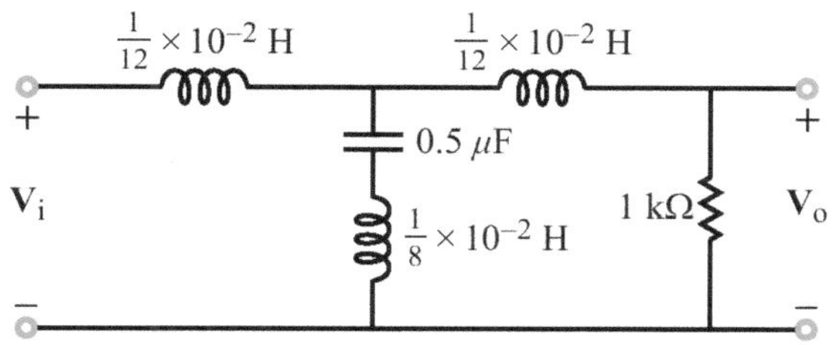

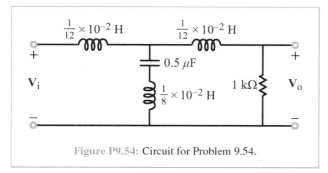

Figure P9.54: Circuit for Problem 9.54.

9.55 For the circuit in Fig. P9.55, use Multisim to generate spectral plots for the magnitude and phase of $\mathbf{H}(\omega) = \mathbf{V}_o/\mathbf{V}_i$ over the range from 1 to 15 kHz. When performing the AC Analysis, use 10^4 points in linear scan. Determine the frequencies at which M [dB] is a maximum or a minimum.

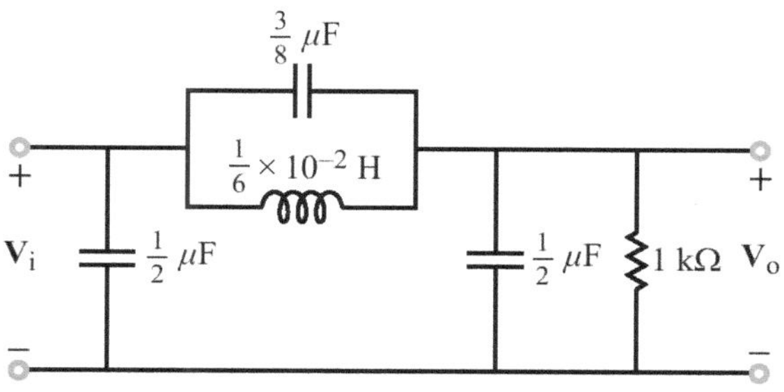

Figure P9.55: Circuit for Problem 9.55.

9.56 For the circuit in Fig. P9.56, use Multisim to generate spectral plots for the magnitude and phase of $\mathbf{H}(\omega) = \mathbf{V}_o/\mathbf{V}_i$ over the range from 100 Hz to 10 kHz. When performing the AC Analysis, use 200 points per decade. Determine the frequencies at which M [dB] is a maximum or a minimum.

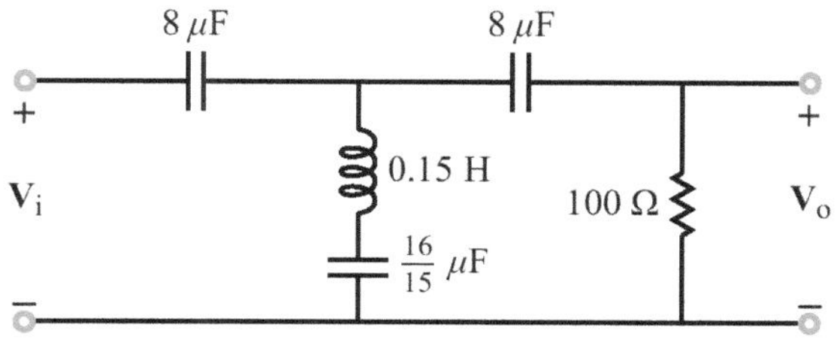

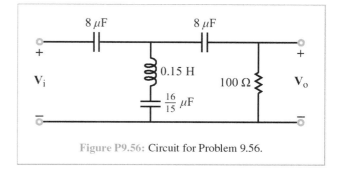

Figure P9.56: Circuit for Problem 9.56.

9.57 Figure P9.57 depicts a band-stop filter composed of a high-pass filter, a low-pass filter, and a summing amplifier. Construct it in Multisim using the values $R = 5$ kΩ, $C_{LP} = 26$ nF, and $C_{HP} = 1$ nF. Using Multisim's AC Analysis, plot the transfer function from 100 Hz to 100 kHz for the high-pass component alone, the low-pass component alone, and the overall filter all on the same graph. Find the frequency where minimum gain occurs for the overall filter.

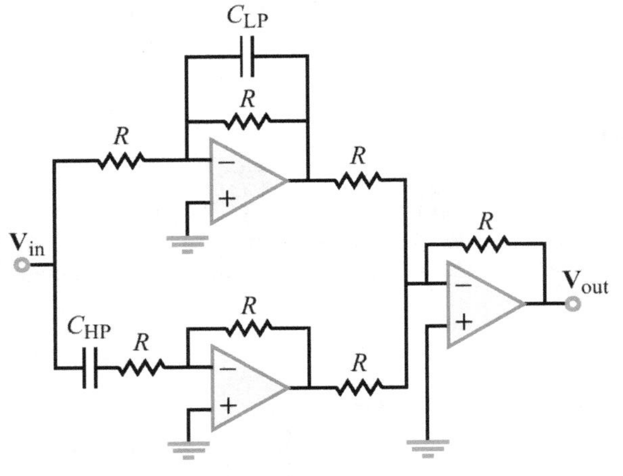

Figure P9.57: Circuit for Problem 9.57.

9.58 Build the circuit shown in Fig. 9-15 in Multisim with values $C = 1$ pF and $R = 377$ Ω. Simulate the circuit with $L = 5$ mH, 10 mH, and 15 mH. Plot the output of the filter at the three tunings on the same plot from 100 kHz to 100 MHz.

Potpourri Questions

9.59 What role does phase play in the operation of noise-cancelling headphones?

9.60 What does high-frequency filtering do to an image? What about low-frequency filtering?

9.61 What is the range of the audible spectrum?

Integrative Problems: Analytical / Multisim / myDAQ

To master the material in this chapter, solve the following problems using three complementary approaches: (a) analytically, (b) with Multisim, and (c) by constructing the circuit and using the myDAQ interface unit to measure quantities of interest via your computer. [myDAQ tutorials and videos are available on (CAD).]

m9.1 Scaling: Figure m9.1 shows a prototype bandreject filter with center frequency $\omega_0 = 1$ rad/s. The prototype component values are $R = 1$ Ω, $L = 1.817$ H, and $C = 0.5505$ F.

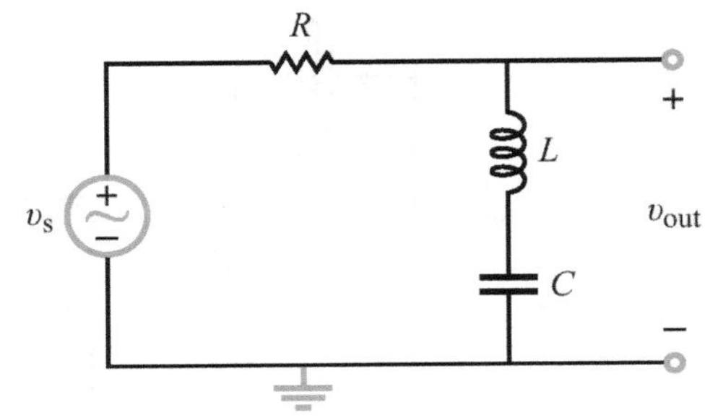

Figure m9.1 Circuit for Problem m9.1.

(a) Apply magnitude and frequency scaling to the bandreject filter so that $R' = 100$ Ω and $L' = 33$ mH. Draw the finished circuit diagram.

(b) Determine the center frequency in Hz of the scaled bandreject filter.

m9.2 Bode Plots: For the circuit in Fig. m9.2:

(a) Determine the voltage transfer function $\mathbf{H}(\omega)$ of the filter circuit. Write your finished result in standard form for creating a Bode plot.

(b) Substitute $\omega = 2\pi f$ to express the voltage transfer function in terms of oscillation frequency f in Hz.

(c) Generate Bode magnitude and phase plots for $\mathbf{H}(f)$ using oscillation frequency f as the independent variable. Use the following component values: $R_1 = 3.3$ kΩ, $R_2 = 10$ kΩ, $C_1 = 0.01$ μF, and $C_2 = 0.1$ μF.

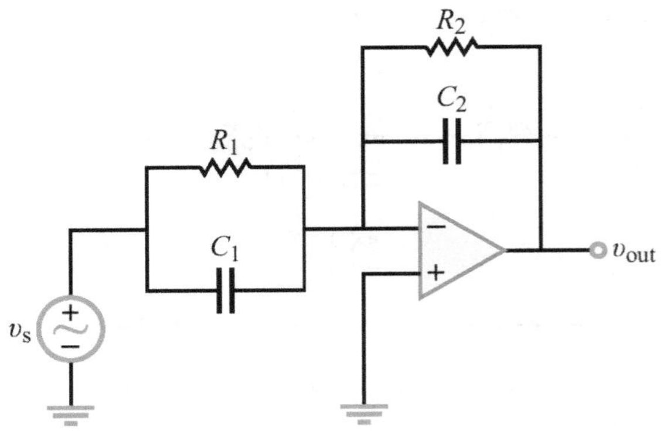

Figure m9.2 Circuit for Problem m9.2.

(d) Determine the following filter circuit properties by inspecting the Bode plot:

(1) Low-frequency asymptotes for magnitude and phase

(2) High-frequency asymptotes for magnitude and phase

(3) Corner frequencies (this filter circuit has two such frequencies)

m9.3 Filter Order: The filter circuit shown in Fig. m9.3 uses the component values $R = 1.0\ \text{k}\Omega$ and $C = 1.0\ \mu\text{F}$.

(a) Obtain an expression for $\mathbf{H}(\omega) = \mathbf{V}_o/\mathbf{V}_i$ in standard form.

(b) Substitute $\omega = 2\pi f$ to express $\mathbf{H}(\omega)$ in terms of the oscillation frequency f in Hz.

(c) Generate spectral plots for the magnitude and phase of $\mathbf{H}(f)$.

(d) Determine the cutoff frequency f_c.

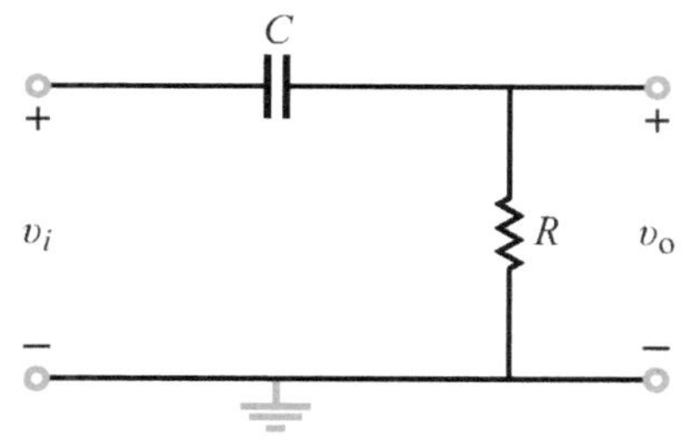

Figure m9.3 Circuit for Problem m9.3.

m9.4 Cascaded Active Filters: A telephone line provides sufficient bandwidth (3 kHz) for intelligible voice conversations, but human hearing has a much higher bandwidth, typically 20 Hz to 20,000 Hz

(a) Design an active bandpass filter to mimic the bandwidth of a telephone line subject to the following constraints:

(1) Cascade a first-order active lowpass filter and a first-order active highpass filter,

(2) Set the corner frequencies to 300 Hz and 3.0 kHz,

(3) Set the passband gain to 0 dB,

(4) Choose resistors in the range 1.0 kΩ to 100 kΩ, and

(5) Use the total of four fixed-value resistors and two fixed-value capacitors selected from the parts listed in Appendix A of the tutorial on CAD.

Draw the schematic diagram of your finished design.

(b) Predict the performance of your finished design by calculating the following values:

(1) Low-frequency passband corner in Hz,

(2) High-frequency passband corner in Hz, and

(3) Passband gain in dB.

m9.5 Bode Plot for an RLC Circuit I

(a) Determine the transfer function of the RLC circuit in Fig. m9.5. Compute the magnitude and phase of the transfer function at 100, 1,000, 5,000, and 10,000 Hz.

(b) Using Multisim and myDAQ, capture the Bode plot for the circuit. How does your answer from part (b) verify your answer from part (a)?

(c) Determine v_{out} when the input is $0.5 \cos(200\pi t)$ V.

(d) Determine v_{out} when the input is $0.5 \cos(10^3\pi t)$ V.

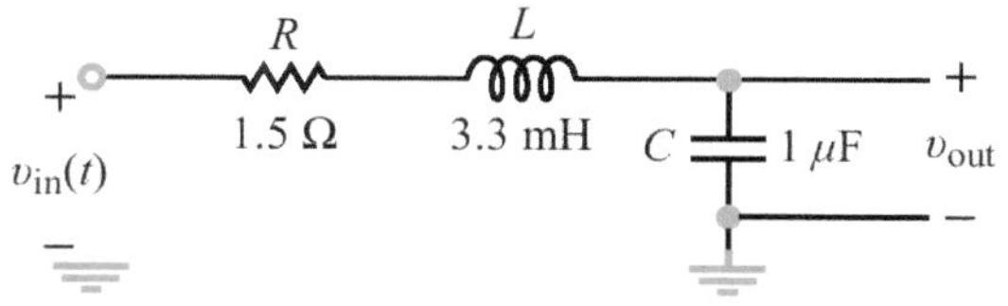

Figure m9.5 Circuit for Problem m9.5.

m9.6 Bode Plot for an RLC Circuit II The circuit in Fig. m9.6 is an RLC circuit. For this problem, there are separate directions for the handwritten, Multisim, and myDAQ portions.

(a) The transfer function for the circuit in Fig. m9.6 is given by

$$\mathbf{H}(\omega) = \frac{j\omega RC}{(1 - \omega^2 LC) + j\omega RC} .$$

Using the transfer function, compute the magnitude and phase at each of the following frequencies: 100, 500, 1,000, 2,000, 5,000, and 10,000 Hz. Plot the magnitude and phase on separate graphs.

(b) For the Multisim and myDAQ portions of this problem, capture a Bode plot for this circuit. Do your answers from part (a) and part (b) agree?

(c) Determine v_{out} when $v_{\text{in}} = 1 \cos(5000\pi t)$ V.

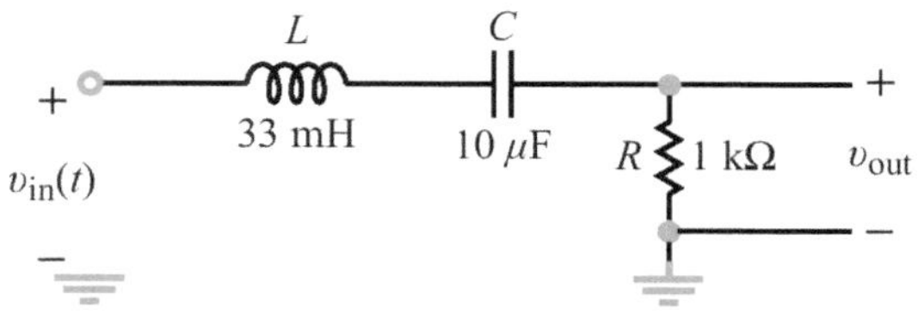

Figure m9.6 Circuit for Problem m9.6.

CHAPTER 10

Three-Phase Circuits

Contents

Objectives

Learn to:

- Analyze both balanced and unbalanced three-phase circuits.
- Convert a Y-source configuration into a Δ-source configuration, and vice versa.
- Convert a Y-load configuration into a Δ-load configuration, and vice versa.
- Compute complex power delivered by a three-phase source or extracted by a three-phase load.
- Apply power-factor compensation.
- Calculate power quantities based on wattmeter measurements.

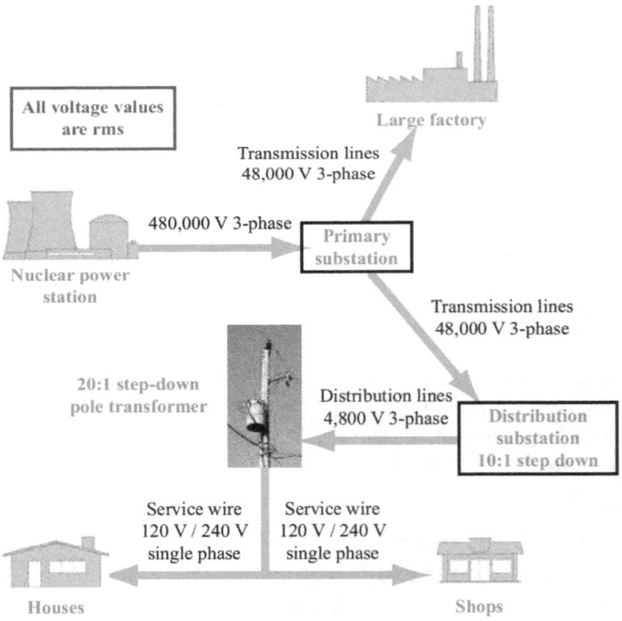

Between the power generating station and a residence or shop, power is transferred across *transmission lines* in a form known as *three-phase power*. What is three-phase power and why is it used? The intent of the present chapter is to answer these questions and to provide the tools for analyzing *three-phase* networks.

Overview

How does the electrical power generated by a power station get distributed to users with diverse voltage-level and power-consumption requirements? The requirements of a large factory are very different from those of a building or a single residence. The power distribution network—often called the *power grid*—uses above- or underground transmission lines to transfer power between different locations, and employs *step-up and step-down transformers* to change voltage levels at specific nodes in the grid. As explained later in this chapter, at a typical large electrical station, power is generated by spinning a large electromagnet past three separate stationary coils arranged evenly around a circular tube. The energy used to spin the electromagnet may come from a hydroelectric dam, diesel or gas engine, or a steam turbine driven by burning coal, oil, or natural gas, or generated by a nuclear reactor. By spinning the electromagnet at 60 revolutions per second, magnetic induction generates an ac voltage across the terminals of each of the three coils. The three induced voltages have the same amplitude and frequency (60 Hz), but their phase angles are staggered by $360°/3 = 120°$ between any two of them. Hence, the power generated by this arrangement is called *three-phase*.

In the power-distribution model shown in Fig. 10-1, the nuclear power station uses large *transformers* to step-up each of the coil voltages from its initial level to 480,000 V (rms). This is done before distributing the power across the grid. The conversion changes the voltage level, but not the amount of power made available by the transmission line. As noted in Section 7-10, when a transformer steps up the voltage by the turns ratio N_2/N_1, it simultaneously steps down the current by the same ratio. By stepping up the voltage at the power

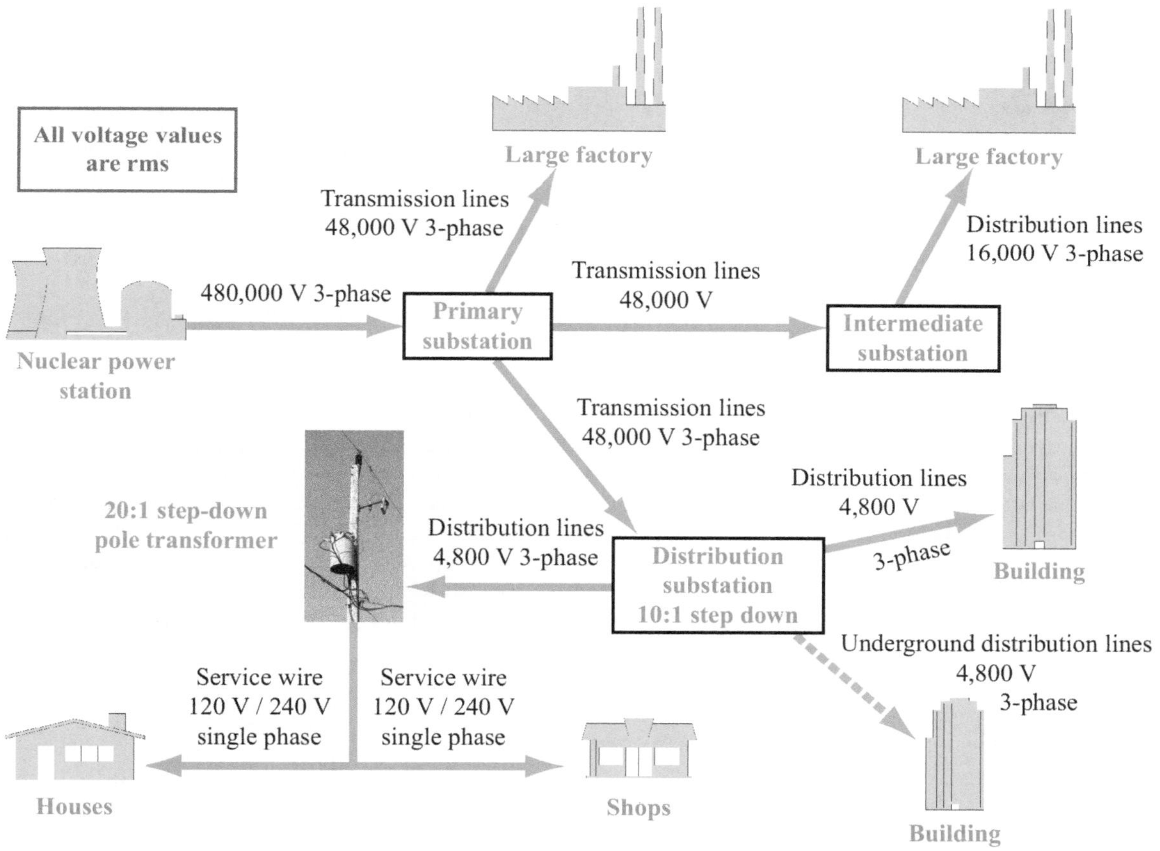

Figure 10-1: Typical electrical power grid.

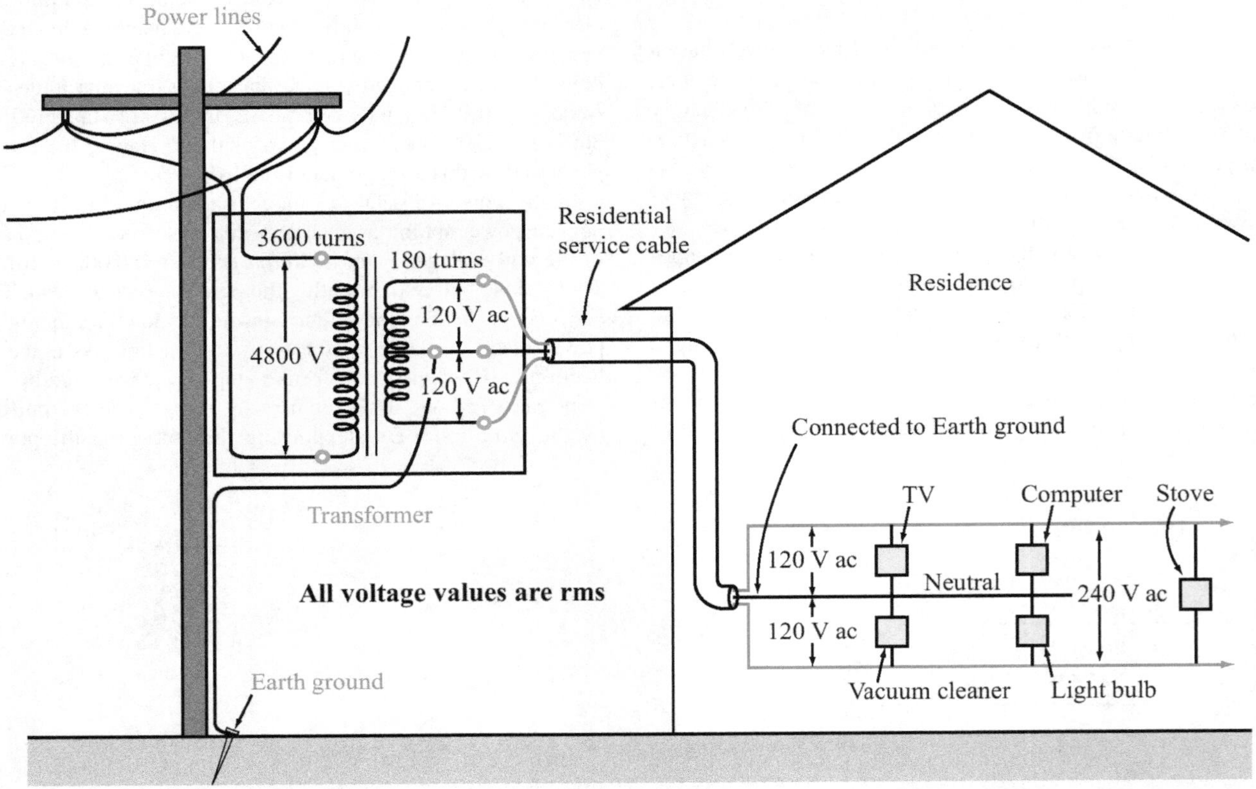

Figure 10-2: A 4800 V rms single-phase ac source connected to a residential user through a 20 : 1 step-down transformer.

station to a very high level, the current that flows along the transmission line gets reduced significantly. Since the power loss in the transmission line is $P_{\text{loss}} = I_{\text{rms}}^2 R$, where R is the total resistance of the transmission line, the voltage up-conversion step serves to reduce the power loss by several orders of magnitude.

The power grid includes several substations designed to convert power from *transmission* to *distribution*. This is accomplished through the use of step-down transformers and a *bus circuit* that can split the power into multiple directions. For a single residence, a *center-tapped pole transformer* is used to step-down one of the three-phase lines to a level manageable by household applicances (Fig. 10-2). The power carried to the house from the transformer is called *three-wire single phase*, with the middle wire assuming the role of the neutral wire. It is single phase because the two 120 V rms voltages at the secondary side of the transformer have the same frequency and phase.

Instead of providing a return path to the generating station through an actual wire, the earth ground is used to provide the feedback path for electrons. This is accomplished by using a cable to connect the middle wire at the transformer output to ground (Fig. 10-2). Use of the ground cable is a safety measure to prevent charging up machinery to dangerous levels, as well as for the discharging of high-voltage events (like lightning) and to prevent current-related heating of wires due to unbalanced loads.

10-1 Balanced Three-Phase Generators

Figure 10-3(a) is a representative cross-sectional view of a typical *three-phase ac generator*. The generator consists of a rotating electromagnet, called the *rotor*, and three separate stationary coils distributed evenly around a circular tube called

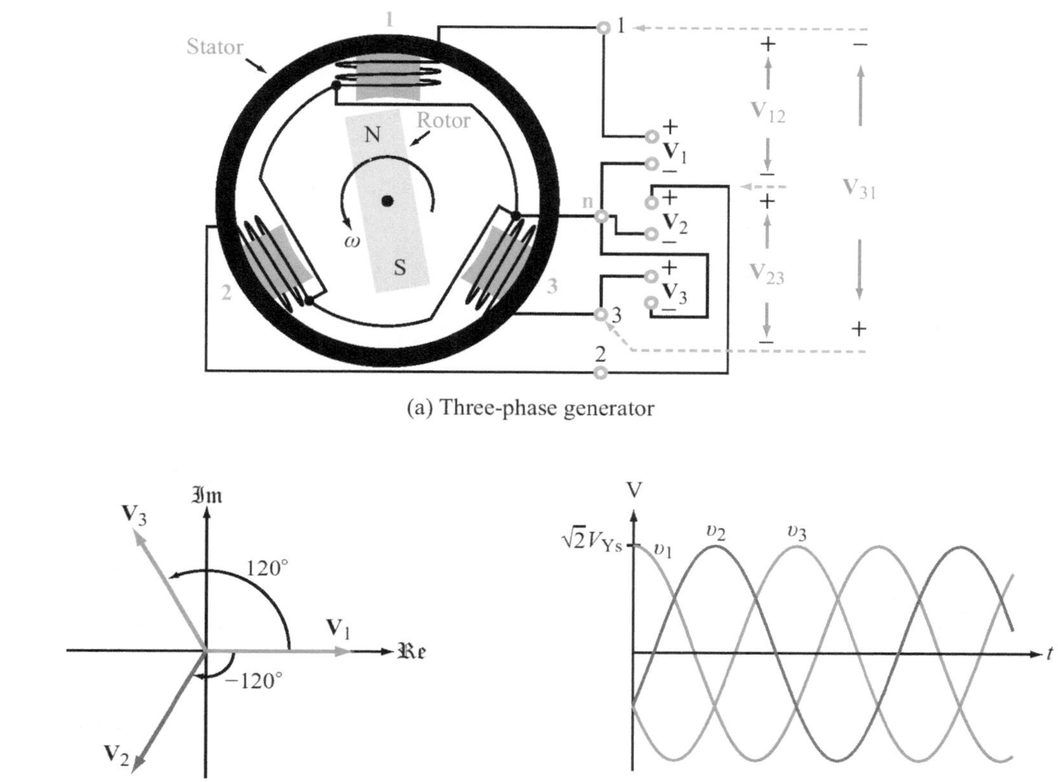

(a) Three-phase generator

(b) Phasor voltages $\mathbf{V}_1$ to $\mathbf{V}_3$ in the complex plane

(c) Voltage waveforms

Figure 10-3: Three-phase ac generator and associated voltage waveforms.

the *stator*. The rotor is spun around by a turbine or some other external force. The three coils are arranged 120° apart over the circumference of the stator. As the electromagnet rotates, its magnetic field induces a sinusoidal voltage at the terminals of each of the three coils. If the coils are identical in shape and number of turns, the three induced phasor voltages, $\mathbf{V}_1$ to $\mathbf{V}_3$, will all have the same amplitude and their time-domain counterparts, $\upsilon_1(t)$ to $\upsilon_3(t)$, will vary sinusoidally at the same frequency $f = \omega/2\pi$, where ω is the angular rotation frequency of the rotor. However, because the coils are physically distributed 120° apart, the voltages induced in adjacent coils will be delayed in time and shifted in phase by 120° relative to one another. By designating the common terminal n as the neutral (ground) terminal with $\mathbf{V}_n = 0$ and selecting $\mathbf{V}_1$ in Fig. 10-3(a) as the reference voltage with zero phase, the phase of $\mathbf{V}_2$ will be either 120° or $-120°$, relative to the phase of $\mathbf{V}_1$, depending on the relative directions of the two windings. If all windings are the same, which usually is the case, common practice is to adopt a ***positive (123) phase sequence*** in which case the phase of $\mathbf{V}_2$ follows behind that of $\mathbf{V}_1$ by 120° and the phase of $\mathbf{V}_3$ follows behind that of $\mathbf{V}_2$ by 120°. Hence, the phases of $\mathbf{V}_1$, $\mathbf{V}_2$, and $\mathbf{V}_3$ are 0, $-120°$, and $-240°$ (or equivalently, $+120°$) (Fig. 10-3(b)), and their waveforms are shifted in time accordingly (Fig. 10-3(c)). We will refer to this arrangement as a ***balanced three-phase Y-source configuration*** with a positive phase sequence:

Y-Source Configuration

$$\begin{aligned} \mathbf{V}_1 &= V_{Ys}\,\underline{/0^\circ}\,, \\ \mathbf{V}_2 &= V_{Ys}\,\underline{/-120^\circ}\,, \\ \mathbf{V}_3 &= V_{Ys}\,\underline{/-240^\circ}\,, \end{aligned} \tag{10.1}$$

with rms *magnitude* V_{Ys}. Voltages $\mathbf{V}_1$ to $\mathbf{V}_3$ are called the *phase voltages* of the Y-source configuration, and they all have the same magnitude, V_{Ys}.

▶ Throughout the remaining material in this chapter, all magnitudes of phasor voltages and currents will denote rms values. ◀

In a positive phase sequence, clockwise rotation between sources in the complex plane (Fig. 10-3(b)) entails an incremental phase shift of $-120°$. In a *negative phase sequence*, the phase-shift increment is $+120°$.

We should note that for a balanced three-phase source

$$\mathbf{V}_1 + \mathbf{V}_2 + \mathbf{V}_3 = 0, \qquad (10.2)$$

which can be verified numerically by inserting Eq. (10.1) into Eq. (10.2) or graphically by summing the three vectors in Fig. 10-3(b).

In the wiring configuration of Fig. 10-3(a), which is redrawn diagrammatically in Fig. 10-4(a), the three voltage sources share *neutral terminal* n and a common wire called the *neutral wire*. This configuration, which may or may not include the neutral wire, is the most common in North America.

▶ In reality, associated with each source is a complex source impedance, but the three source impedances usually are ignored because their values are much smaller than those of the impedances of the loads connected to the generator circuit. ◀

The time-domain counterpart of phasor voltage $\mathbf{V}_1$ of the Y-source configuration is

$$\upsilon_1(t) = \mathfrak{Re}[\sqrt{2}\,\mathbf{V}_1 e^{j\omega t}] = \sqrt{2}\,V_{Ys}\cos\omega t,$$

where we included the factor $\sqrt{2}$ because V_{Ys} was specified as an rms value. Similarly, for the other two phasor voltage sources

$$\upsilon_2(t) = \sqrt{2}\,V_{Ys}\cos(\omega t - 120°)$$

and

$$\upsilon_3(t) = \sqrt{2}\,V_{Ys}\cos(\omega t - 240°).$$

A common alternative to the Y-source configuration is the *Δ-source configuration* shown in Fig. 10-4(b). Note that the Δ-source configuration does not have a neutral wire. The relationships between voltages $\mathbf{V}_{12}$, $\mathbf{V}_{23}$, and $\mathbf{V}_{31}$ of the Δ configuration and the three voltages of the Y configuration are

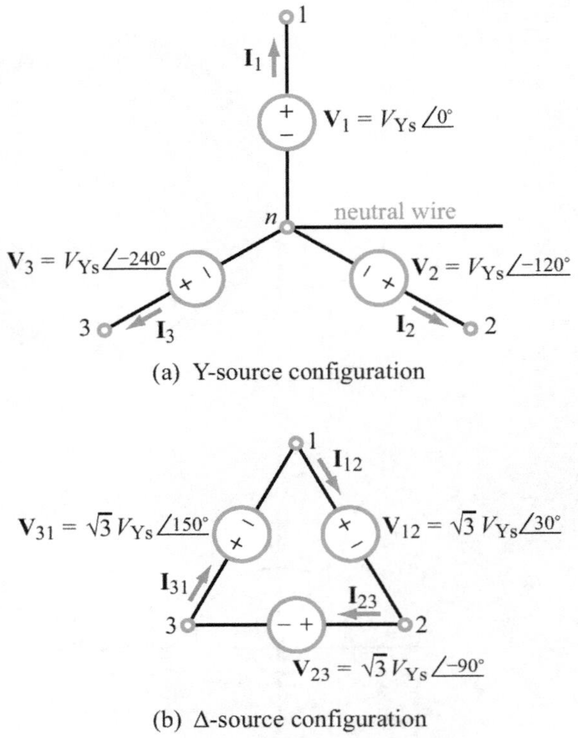

Figure 10-4: Y- and Δ-source configurations, with V_{Ys} = rms value of the phase-voltage magnitude of the Y-source. The rms magnitude of the Δ-source phase voltages is $\sqrt{3}\,V_{Ys}$.

Δ-Source Configuration

$$\begin{aligned}
\mathbf{V}_{12} &= \mathbf{V}_1 - \mathbf{V}_2 \\
&= V_{Ys}\,\underline{/0°} - V_{Ys}\,\underline{/-120°} \\
&= \sqrt{3}\,V_{Ys}\,\underline{/30°} = V_{\Delta s}\,\underline{/30°}\,, \\
\mathbf{V}_{23} &= \mathbf{V}_2 - \mathbf{V}_3 = V_{\Delta s}\,\underline{/-90°}\,, \\
\mathbf{V}_{31} &= \mathbf{V}_3 - \mathbf{V}_1 = V_{\Delta s}\,\underline{/150°}
\end{aligned} \qquad (10.3)$$

with $V_{\Delta s} = \sqrt{3}\,V_{Ys}$.

We note that the *magnitude of the Δ-source*, $V_{\Delta s}$, is $\sqrt{3}$ times larger than that of the Y-source, and the phases of the Δ sources are 30° ahead of their Y counterparts (see Fig. E10.1 of Exercise 10-1).

► It is important to remember that the phase angle of a phasor is defined relative to a reference. In Fig. 10-4, we assigned $\mathbf{V}_1$ a phase angle of zero, and all of the other voltages are defined relative to $\mathbf{V}_1$. Had its phase angle been ϕ instead, the phases of all of the other voltages would have had to be adjusted accordingly ($\mathbf{V}_2 = V_{\text{Ys}} \underline{/\phi - 120^\circ}$, etc.). ◄

Example 10-1: Y-Δ Sources

If in a balanced Δ-source configuration with a positive phase sequence, $\upsilon_{12}(t) = 440\cos(120\pi t + 45^\circ)$ V, determine $\upsilon_1(t)$, $\upsilon_2(t)$, and $\upsilon_3(t)$ of the equivalent Y-source configuration.

Solution: The rms phasor counterpart of $\upsilon_{12}(t)$ is

$$\mathbf{V}_{12} = \frac{440}{\sqrt{2}} \underline{/45^\circ} \text{ V (rms)}.$$

From Fig. 10-4, we observe that $\mathbf{V}_1$ can be obtained from $\mathbf{V}_{12}$ by reducing the magnitude of the latter by $\sqrt{3}$ and decrementing its phase by 30°:

$$\mathbf{V}_1 = \frac{\mathbf{V}_{12}}{\sqrt{3}} \underline{/-30^\circ} = \frac{440}{\sqrt{2}\sqrt{3}} \underline{/45^\circ - 30^\circ} = 179.63\underline{/15^\circ} \quad \text{(rms)}.$$

Hence,

$$\upsilon_1(t) = \sqrt{2} \times 179.63\cos(120\pi t + 15^\circ)$$
$$= 254.03\cos(120\pi t + 15^\circ) \text{ V}.$$

By extension,

$$\upsilon_2(t) = 254.03\cos(120\pi t - 105^\circ) \text{ V},$$
$$\upsilon_3(t) = 254.03\cos(120\pi t + 135^\circ) \text{ V}.$$

Concept Question 10-1: Why are the voltage sources of a three-phase source separated by 120° increments in phase? (See CAD)

Concept Question 10-2: In a Y-source configuration, each voltage source is measured across one of the generator's three coils, with one terminal serving as a common neutral terminal for all three. How are the sources of a Δ configuration measured? (See CAD)

Exercise 10-1: Superimpose onto Fig. 10-4(b) the three source voltages of the Δ configuration.

Answer:

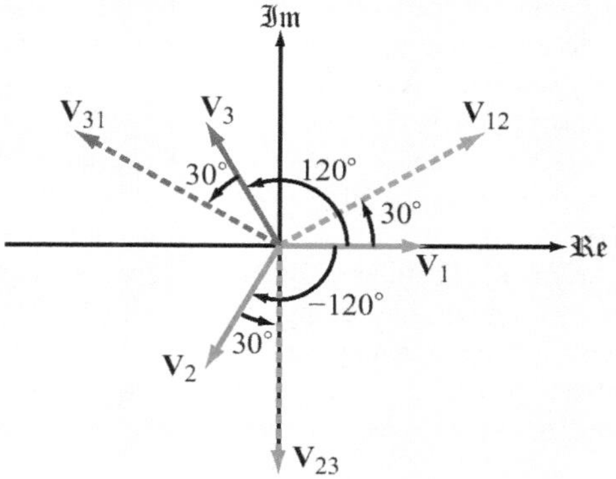

Figure E10.1

(See CAD)

Exercise 10-2: Given a balanced Δ-source configuration with a positive phase sequence and $\mathbf{V}_{12} = 208\underline{/45^\circ}$ V (rms), determine (a) phase voltages $\mathbf{V}_{23}$ and $\mathbf{V}_{31}$, and (b) $\mathbf{V}_1$, $\mathbf{V}_2$, and $\mathbf{V}_3$ of the equivalent Y-source configuration.

Answer: (a) $\mathbf{V}_{23} = 208\underline{/-75^\circ}$ V (rms), $\mathbf{V}_{31} = 208\underline{/-195^\circ}$ V (rms), (b) $\mathbf{V}_1 = 120\underline{/15^\circ}$ V (rms), $\mathbf{V}_2 = 120\underline{/-105^\circ}$ V (rms), $\mathbf{V}_3 = 120\underline{/-225^\circ}$ V (rms). (See CAD)

Exercise 10-3: Show graphically why the phase magnitude of $\mathbf{V}_{12}$ of the Δ-source is $\sqrt{3}$ times larger than the phase magnitude of the Y-source.

Answer:

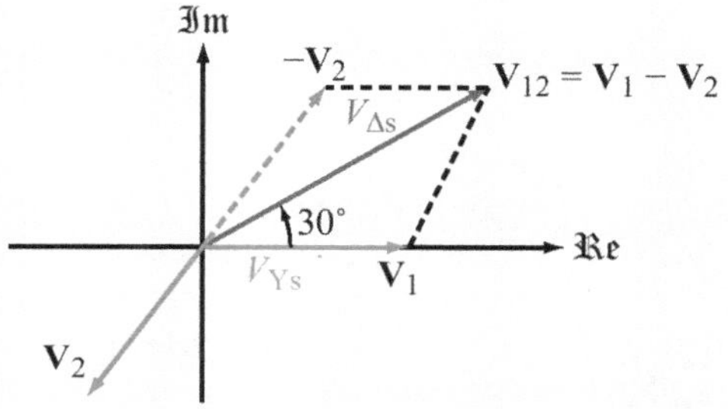

Figure E10.3

(See CAD)

10-2 Source-Load Configurations

The three-phase generator is equivalent to three separate *single-phase generators*, each one of which can be connected to a separate load. Transmission of three-phase power from the generator to loads is more efficient than three separate, single-phase transmissions. A three-phase Δ-source configuration uses 3 transmission wires and a Y-source configuration uses 3 or 4 wires (the neutral wire is not always used), whereas 6 wires are required to support three separate single-phase transmissions. The three loads connected to the three sources may be arranged in either a *Y- or a Δ-load configuration* (with the Δ configuration being more common). Hence, the source-load connections may assume any one of four possible combinations: Y-Y, Y-Δ, Δ-Y, and Δ-Δ (Fig. 10-5).

> ▶ By applying equivalent-circuit transformations, any one of the four connection configurations can be converted into any of the other three. ◀

10-2.1 Y and Δ Notation

In view of the several sources, currents, and impedances involved in the four source-load topologies, it will prove helpful to summarize our notation. We will be guided by the circuits in Fig. 10-5.

Nodes

1, 2, 3	nodes in source circuit
a, b, c	nodes in load circuit
n and N	neutral node in source and load (Y-Y configuration only)

Voltages

$\mathbf{V}_1, \mathbf{V}_2, \mathbf{V}_3$	*phase voltages* of Y-source, with rms magnitude V_{Ys}
$\mathbf{V}_{12}, \mathbf{V}_{23}, \mathbf{V}_{31}$	*phase voltages* of Δ-source, with rms magnitude $V_{\Delta\text{s}} = \sqrt{3}\, V_{\text{Ys}}$
$\mathbf{V}_n = 0$	*neutral node* in Y-source
$\mathbf{V}_{ab}, \mathbf{V}_{bc}, \mathbf{V}_{ca}$	*line voltages* (voltages between transmission-line pairs), with rms magnitude V_{L}; same as *phase voltages* of Δ-load
$\mathbf{V}_{aN}, \mathbf{V}_{bN}, \mathbf{V}_{cN}$	*phase voltages* of Y-load

Currents

$\mathbf{I}_1, \mathbf{I}_2, \mathbf{I}_3$	*phase currents* in Y-source configuration
$\mathbf{I}_{12}, \mathbf{I}_{23}, \mathbf{I}_{31}$	*phase currents* in Δ-source configuration
$\mathbf{I}_{\text{L}_1}, \mathbf{I}_{\text{L}_2}, \mathbf{I}_{\text{L}_3}$	*line currents* through transmission lines, same as *phase currents* $\mathbf{I}_a, \mathbf{I}_b, \mathbf{I}_c$ of Y-load
$\mathbf{I}_{ab}, \mathbf{I}_{bc}, \mathbf{I}_{ca}$	*phase currents* of Δ-load

Impedances

$\mathbf{Z}_{\text{TL}_1}, \mathbf{Z}_{\text{TL}_2}, \mathbf{Z}_{\text{TL}_3}$	transmission-line impedances
$\mathbf{Z}_n$	impedance of neutral line (Y-Y configuration only)
$\mathbf{Z}_a, \mathbf{Z}_b, \mathbf{Z}_c$	impedances of Y-load configuration (for balanced load $\mathbf{Z}_a = \mathbf{Z}_b = \mathbf{Z}_c = \mathbf{Z}_{\text{Y}}$)
$\mathbf{Z}_{ab}, \mathbf{Z}_{bc}, \mathbf{Z}_{ca}$	impedances of Δ-load configuration (for balanced load $\mathbf{Z}_{ab} = \mathbf{Z}_{bc} = \mathbf{Z}_{ca} = \mathbf{Z}_\Delta$)

> ▶ We should note that the term *line voltage* is short-hand for *line-to-line voltage* at the point of use (i.e., at the load). ◀

In the present context, the line voltages are the voltages between transmission-line pairs, namely the voltages between nodes a and b, b and c, and c and a in Fig. 10-5. The associated *line currents* are the currents flowing through the transmission lines. The terms *phase voltages* and *phase currents* usually refer to the voltages across and currents through the load impedances, but often they are also used in connection with the source circuit. To avoid confusion, it is best to specify whether it is the source circuit or the load circuit that the phase voltages and currents belong to.

10-2.2 Balanced Conditions

- In view of how the voltage sources are induced, it is safe to assume that the three Y or Δ sources are *balanced*, meaning that they have the same amplitude and frequency and their phases are separated by 120° increments.

- The load circuit, however, may or may not be balanced. *The load circuit is considered balanced if all of its impedances are the same.* Three-phase motors are one such example.

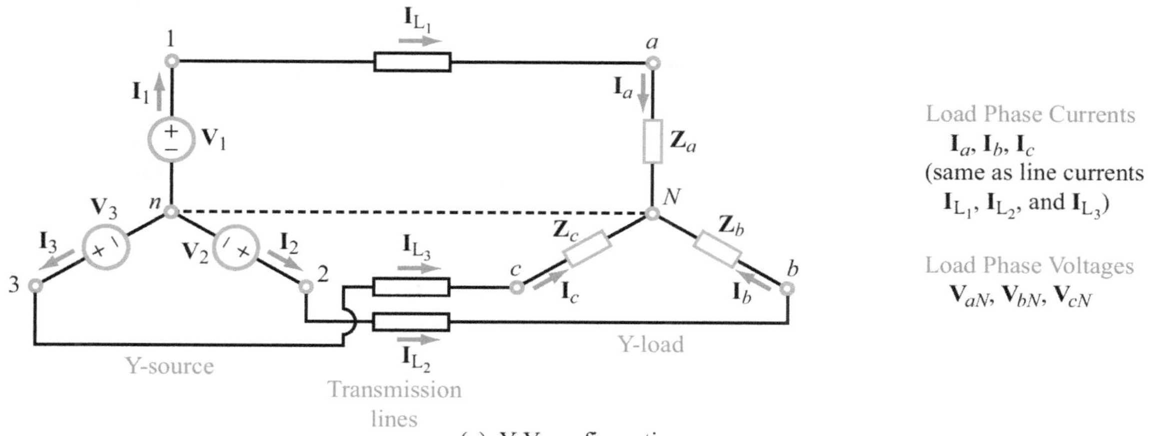

(a) Y-Y configuration

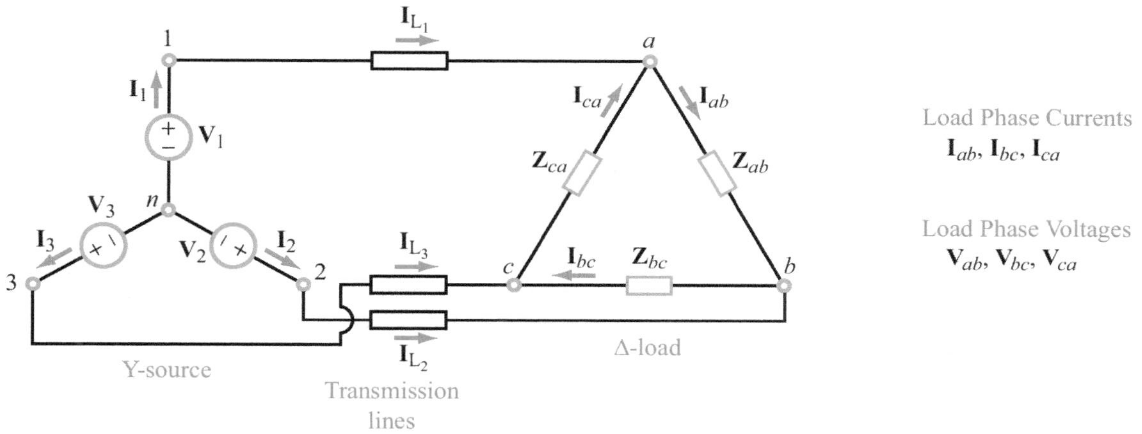

(b) Y-Δ configuration

Figure 10-5: The three-phase source and load circuits can be connected in four possible arrangements: Y-Y, Y-Δ, Δ-Y, and Δ-Δ. In each arrangement, the source and load circuits are connected via transmission lines carrying *line currents* I_{L_1}, I_{L_2}, and I_{L_3}. [Parts (c) and (d) follow on the next page.]

- A ***network*** is said to be ***balanced*** if its source voltages are balanced and if it has identical transmission lines and identical loads.

The techniques presented in Chapters 7–9 are more than sufficient to analyze any three-phase circuit, for any combination of source and load configurations.

> **Concept Question 10-3:** What is a *balanced* source? *Balanced* load? *Balanced* network? (See CAD)

> **Concept Question 10-4:** Which *line(s)* do the line voltages refer to? (See CAD)

> **Concept Question 10-5:** For which load configuration are (a) the phase voltages the same as the line voltages? and (b) the phase currents the same as the line currents? (See CAD)

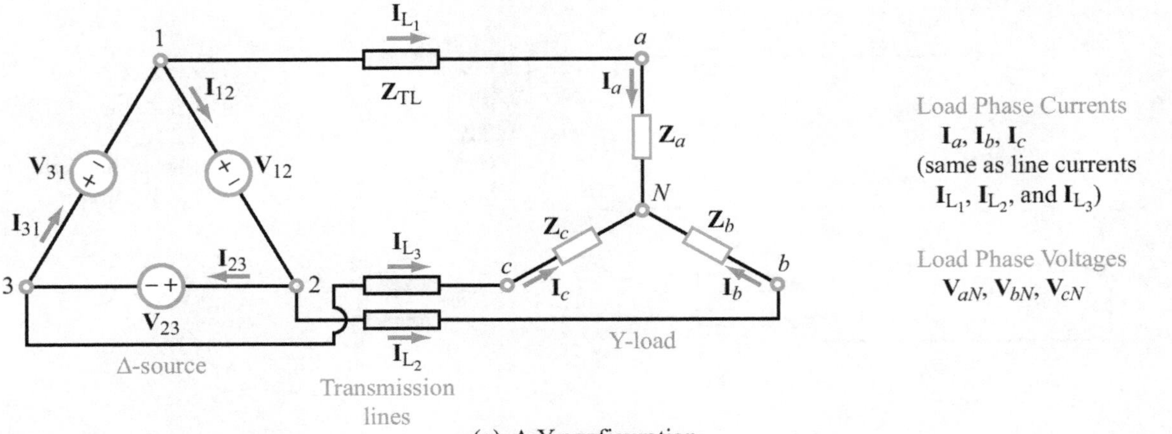

(c) Δ-Y configuration

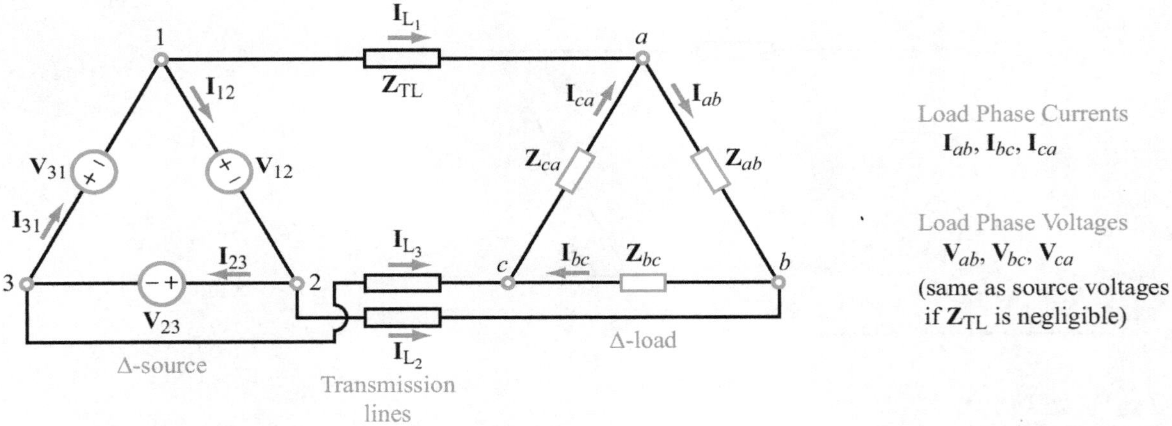

(d) Δ-Δ configuration

(Fig. 10-5 continued)

10-3 Y-Y Configuration

In the Y-Y configuration depicted in Fig. 10-6, the Y-load network is connected to the Y-source circuit through four wires (*transmission lines*). The transmission lines, which may or may not be identical, are characterized by impedances $\mathbf{Z}_{TL_1}$, $\mathbf{Z}_{TL_2}$, and $\mathbf{Z}_{TL_3}$, connecting sources $\mathbf{V}_1$ through $\mathbf{V}_3$ to loads $\mathbf{Z}_a$ to $\mathbf{Z}_c$. In addition, a fourth transmission line of impedance $\mathbf{Z}_n$ connects node n of the source configuration to node N of the load configuration.

Example 10-2: Balanced Y-Y Network

With reference to the network shown in Fig. 10-6:

(a) Develop a node voltage equation for $\mathbf{V}_N$, the voltage at node N, with node n treated as (the ground) reference.

(b) Determine *line currents* $i_{L_1}(t)$ to $i_{L_3}(t)$ and $i_n(t)$ for a balanced network, given that $V_{Ys} = 120$ V (rms), $f = 60$ Hz, $\mathbf{Z}_{TL_1} = \mathbf{Z}_{TL_2} = \mathbf{Z}_{TL_3} = (1 + j1)$ Ω, and $\mathbf{Z}_a = \mathbf{Z}_b = \mathbf{Z}_c = (29 + j9)$ Ω.

(c) Determine *line voltages* $\upsilon_{ab}(t)$, $\upsilon_{bc}(t)$, and $\upsilon_{ca}(t)$.

Solution: (a) Relative to node n (i.e., with $\mathbf{V}_n = 0$), the node equation at node N is:

$$\mathbf{I}_n - \mathbf{I}_{L_1} - \mathbf{I}_{L_2} - \mathbf{I}_{L_3} = 0, \tag{10.4}$$

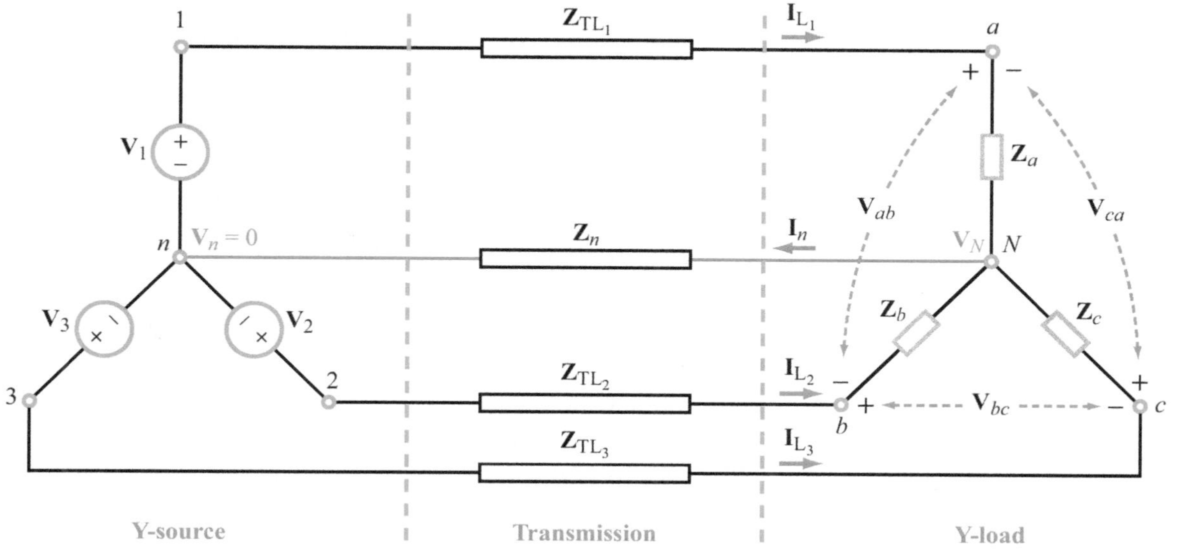

Figure 10-6: Three-phase Y source connected to a Y load circuit via transmission lines.

or equivalently

$$\frac{\mathbf{V}_N}{\mathbf{Z}_n} - \frac{(\mathbf{V}_1 - \mathbf{V}_N)}{\mathbf{Z}_{TL_1} + \mathbf{Z}_a} - \frac{(\mathbf{V}_2 - \mathbf{V}_N)}{\mathbf{Z}_{TL_2} + \mathbf{Z}_b} - \frac{(\mathbf{V}_3 - \mathbf{V}_N)}{\mathbf{Z}_{TL_3} + \mathbf{Z}_c} = 0. \quad (10.5)$$

(b) For a balanced network, the denominators of terms 2 to 4 in Eq. (10.5) become identical, which we shall denote as $\mathbf{Z}_0$:

$$\mathbf{Z}_0 = \mathbf{Z}_{TL_1} + \mathbf{Z}_a = (1 + j1) + (29 + j9) = (30 + j10)\ \Omega. \quad (10.6)$$

The node voltage equation given by Eq. (10.5) simplifies to:

$$\mathbf{V}_N \left(\frac{1}{\mathbf{Z}_n} + \frac{3}{\mathbf{Z}_0}\right) = \frac{\mathbf{V}_1 + \mathbf{V}_2 + \mathbf{V}_3}{\mathbf{Z}_0}. \quad (10.7)$$

According to Eq. (10.2), for a balanced source, $\mathbf{V}_1 + \mathbf{V}_2 + \mathbf{V}_3 = 0$. Hence,

$$\mathbf{V}_N = 0 \quad \textbf{(balanced network)}. \quad (10.8)$$

Consequently,

$$\mathbf{I}_n = \frac{\mathbf{V}_N}{\mathbf{Z}_n} = 0, \quad (10.9)$$

$$\mathbf{I}_{L_1} = \frac{\mathbf{V}_1 - \mathbf{V}_N}{\mathbf{Z}_0} = \frac{120}{30 + j10} = 3.80e^{-j18.4^\circ} \text{ A (rms)},$$

$$\mathbf{I}_{L_2} = \frac{\mathbf{V}_2 - \mathbf{V}_N}{\mathbf{Z}_0} = \frac{120e^{-j120^\circ}}{30 + j10} = 3.80e^{-j138.4^\circ} \text{ A (rms)},$$

and

$$\mathbf{I}_{L_3} = \frac{\mathbf{V}_3 - \mathbf{V}_N}{\mathbf{Z}_0} = \frac{120e^{-j240^\circ}}{30 + j10} = 3.80e^{j101.6^\circ} \text{ A (rms)}.$$

The numerical values of phasor currents $\mathbf{I}_{L_1}$ to $\mathbf{I}_{L_3}$ are in rms, so to obtain their corresponding time-domain expressions, we need to multiply them by $\sqrt{2}$:

$$i_{L_1}(t) = \mathfrak{Re}[\sqrt{2}\,\mathbf{I}_{L_1} e^{j\omega t}] = 5.37\cos(2\pi f t - 18.4^\circ) \text{ A},$$

$$i_{L_2}(t) = \mathfrak{Re}[\sqrt{2}\mathbf{I}_{L_2} e^{j\omega t}] = 5.37\cos(2\pi f t - 138.4^\circ) \text{ A},$$

and

$$i_{L_3}(t) = \mathfrak{Re}[\sqrt{2}\mathbf{I}_{L_3} e^{j\omega t}] = 5.37\cos(2\pi f t + 101.6^\circ) \text{ A},$$

with $f = 60$ Hz.

(c)

$$\begin{aligned}\mathbf{V}_{ab} &= \mathbf{I}_{L_1}\mathbf{Z}_a - \mathbf{I}_{L_2}\mathbf{Z}_b \\ &= (3.80e^{-j18.4^\circ} - 3.80e^{-j138.4^\circ})(29 + j9) \\ &= 200e^{j28.8^\circ} \text{ V (rms)}.\end{aligned}$$

Similarly,

$$\mathbf{V}_{bc} = \mathbf{I}_{L_2}\mathbf{Z}_b - \mathbf{I}_{L_3}\mathbf{Z}_c = 200e^{-j91.2^\circ} \text{ V (rms)}$$

$$\mathbf{V}_{ca} = \mathbf{I}_{L_3}\mathbf{Z}_c - \mathbf{I}_{L_1}\mathbf{Z}_a = 200e^{j148.8^\circ} \text{ V (rms)}.$$

The corresponding time-domain line voltages are

$$\upsilon_{ab}(t) = 282.2\cos(2\pi ft + 28.8°) \text{ V},$$
$$\upsilon_{bc}(t) = 282.2\cos(2\pi ft - 91.2°) \text{ V},$$

and

$$\upsilon_{ca}(t) = 282.2\cos(2\pi ft + 148.8°) \text{ V}.$$

Concept Question 10-6: What is the magnitude of $\mathbf{I}_n$ in a balanced Y-Y configuration (Fig. 10-6)? (See CAD)

Exercise 10-4: Were we to repeat Example 10-2, but with the transmission-line impedances set to zero, which of the following line-current quantities will change and which will remain the same: (a) amplitudes, (b) absolute phases, and (c) phases relative to each other?

Answer: (a) Amplitudes will change, (b) absolute phases will change, but (c) relative phases will continue to be 120° apart (between pairs). (See CAD)

10-4 Balanced Networks

A three-phase network is balanced if it has a balanced three-phase source, transmission lines with identical impedances, and a balanced three-phase load (with equal impedances). The inherent symmetry associated with a balanced network allows us to simplify the circuit analysis considerably.

10-4.1 Transformation Between Balanced Sources

A balanced, three-phase Y-source with a positive phase sequence is completely specified by two parameters: (a) V_{Ys}, the rms magnitude of the phase voltages and (b) ϕ_1, the phase of $\mathbf{V}_1$. That is,

$$\textbf{Balanced Y-source} = \begin{cases} \mathbf{V}_1 = V_{Ys} \angle \phi_1 , \\ \mathbf{V}_2 = V_{Ys} \angle \phi_1 - 120° , \\ \mathbf{V}_3 = V_{Ys} \angle \phi_1 + 120° . \end{cases} \quad (10.10)$$

According to our earlier discussion in connection with Fig. 10-4, the Y-source can be transformed into an equivalent Δ-source,

$$\textbf{Balanced Δ-source} = \begin{cases} \mathbf{V}_{12} = \mathbf{V}_1 \times \sqrt{3} \angle 30° , \\ \mathbf{V}_{23} = \mathbf{V}_2 \times \sqrt{3} \angle 30° , \\ \mathbf{V}_{31} = \mathbf{V}_3 \times \sqrt{3} \angle 30° . \end{cases} \quad (10.11)$$

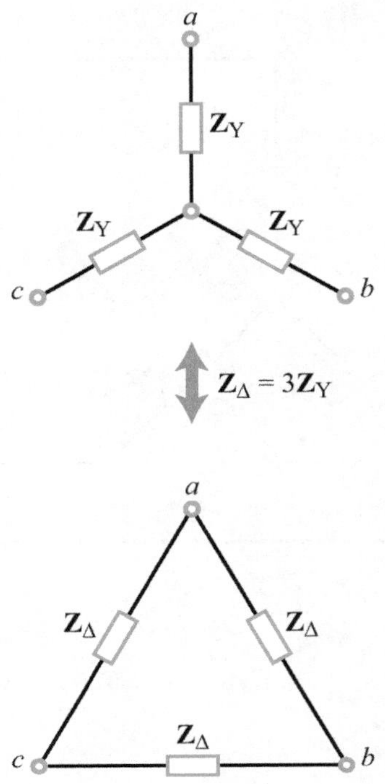

Figure 10-7: Y-Δ transformation for balanced load circuits.

► Transformation of $(\mathbf{V}_1, \mathbf{V}_2, \mathbf{V}_3)$ of the Y-source into $(\mathbf{V}_{12}, \mathbf{V}_{23}, \mathbf{V}_{31})$ of the Δ-source involves multiplication of the magnitudes by $\sqrt{3}$ and advancing the phases by 30°. ◄

Conversely, the phase voltages of the Δ-source can be transferred into those of the Y-source by dividing the magnitudes by $\sqrt{3}$ and delaying the phases by 30°.

10-4.2 Transformation Between Balanced Loads

For load circuits, Section 7-5.2 provides a comprehensive treatment of how to transform the impedances of any Y-load configuration into those of an equivalent Δ configuration, and vice versa. The relationships apply to both balanced and unbalanced load circuits. For a balanced Y-load circuit with equal impedances $\mathbf{Z}_Y$ (Fig. 10-7), its equivalent Δ-load circuit has impedances $\mathbf{Z}_\Delta$ given by

$$\mathbf{Z}_\Delta = 3\mathbf{Z}_Y. \quad (10.12)$$

Technology Brief 25
Miniaturized Energy Harvesting

Energy is present in our environment in many forms. One can think of energy as a quantity that measures the ability of a system to do some amount of work on its environment (or other systems). For example, moving objects possess kinetic energy, objects within gravity wells possess potential energy, and charged particles within an electric field possess electrical energy. Real systems often *transduce energy* from one form to another as part of normal operation: photodiodes (solar cells) convert electromagnetic waves (light) into the movement of charge particles in a potential field (thus providing a system like your phone with the ability to do work), sensors often transduce chemical energy into electrical energy to measure the state of a system, an electrostatic actuator might transduce electrical energy into mechanical energy to perform mechanical work, etc. There is useful energy present all around you, in the gentle mechanical vibrations moving through your building, in the radio frequency waves generated by radio emitters, and even in the heat that your body or your car engine emits.

As computation and communication technologies miniaturize and become ever-more pervasive, many everyday computing objects require less and less *power* to operate. A typical circa-2012 laptop might consume 50–100 W (joules per second) during normal operation, a smartphone might consume 0.5–4 W (depending on what the user is doing, whether the radio is on, etc.), and a good low power wristwatch might consume 10 μW down to ~100 nW.

As power consumption decreases for some functions, it turns out that there is, in many cases, just enough energy in the environment to power these systems. This is often known as *energy harvesting* or *energy scavenging*. Below, we'll look at some interesting devices that have been built to scavenge energy from the environment to power everyday systems. A great many scavenging systems have been built in recent years, so we'll focus on general classes of scavenging. It is also important to note that the line between power scavenging and conventional power generation can become blurred: is a normal solar cell scavenging light to produce power? Sure! Is a wind turbine scavenging wind power to produce electricity? Of course. The idea is to focus on technologies that convert very small sources of power which, in the past, were often too small to be useful or were ignored.

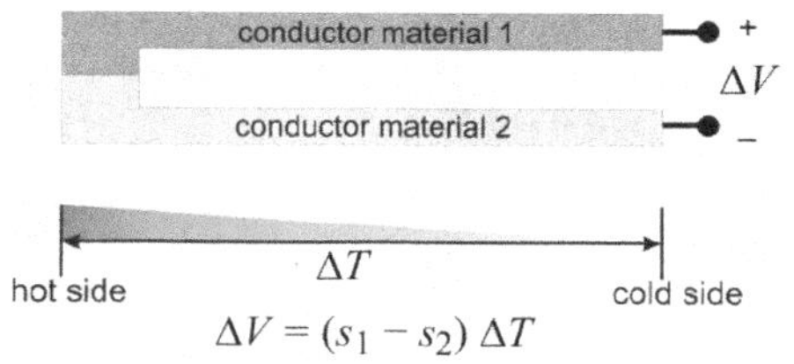

Figure TF25-1: The most common class of thermoelectric materials operate according to the *Seebeck effect*. When two conductors are joined at one end and exposed to a temperature gradient (ΔT), a potential difference (ΔV) is measurable across the free ends of the two conductors. The relationship between ΔV and ΔT depends on $(s_1 - s_2)$, where s_1 and s_2 are the Seebeck coefficients of the two materials. The Seebeck coefficient is a material-specific property thet depends on the molecular structure of the material. This potential difference can be used to drive a current through an external load and thus do work. Interestingly, the effect can be run in reverse—known as the *Peltier effect*—such that an applied voltage can be used to create a temperature difference. This is the basis of cryogenic cooling systems.

Thermoelectric

Almost every system that does useful work also produces heat. This "waste heat" is often exhausted to the environment but, since time immemorial, humans have also used heat to do work. Steam engines, internal combustion engines, the turbine systems at power plants and thousands of other *heat engines* extract useful energy from a heat source. Modern, top of the line power plants contain *combined heat and power* (CHP) systems that internally recover waste heat to increase efficiency.

A number of miniaturized technologies have been explored for scavenging tiny amounts of waste heat. Although these efforts have included making tiny heat engines, fundamental physical limitations have so far prevented successful scaling down of mechanical heat engines down to the millimeter. A different approach is to use materials that convert heat directly to electrical power; *thermoelectric* materials are in this category.

In the early 19th century Thomas Johann Seebeck observed that a voltage was induced when two dissimilar conductors were placed in a thermal gradient (Fig. TF25-1). Thermoelectric materials are used extensively to sense temperature. More recently,

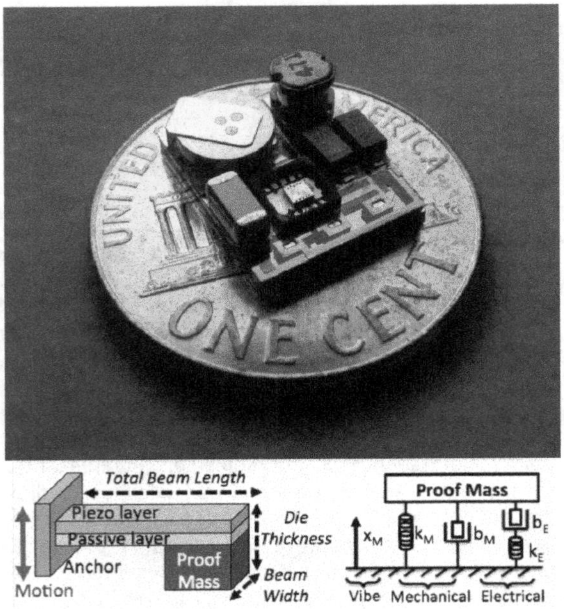

Figure TF25-2: (*top*) A tiny micromechanical harvesting system developed at the University of Michigan. It occupies 27 mm^3 and can harvest > 200 μW delivered at 1.85 V when exposed to 1.5 g's of acceleration when vibrating at or near ~155 Hz.

researchers have built tiny thermoelectric systems that scavenge power from small thermal gradients (such as is found between the surface of your skin and the environment); the small temperature difference (usually 1°–5 °C) and the low efficiency of existing thermoelectric materials has limited this to low scavenged energy densities (<1–10 mW per °C of temperature gradient per cm^2 of converter area) when compared to batteries or other conventional sources.

The efficiency of conversion (that is, how much of the heat energy is successfully converted to electrical power) hovers in the 1–6% range for implemented systems and this limitation comes from the thermomelectric material itself. A number of material-science efforts are under way to produce thermoelectric materials with higher efficiencies.

Mechanical Harvesting

A number of technologies have been developed to harvest the small motions present in everyday life. Perpetual motion watches use the regular motion of your arm to power tiny spring mass systems which either charge a battery or drive clockwork. Similar spring-mass systems (similar to those presented in Technology Brief 15: Micromechanical Sensor and Actuators) have been developed to power sensor motes deployed in areas with continuous environmental force or vibrations (Fig. TF25-2). As the environmental vibrations (which are usually very small, like those caused by the whirring of gears, the hum of an engine or the regular force applied by your heel to the rubber sole of your shoe) move the scavenger system, that force or motion can be converted to electrical power.

The conversion of mechanical work to electrical power can occur via the motion of charged conductor plates, electromagnetic windings or even a class of materials called piezoelectrics. A piezoelectric material converts mechanical deformation into a voltage or current (and vice versa as described in Technology Brief 19: Crystal Oscillators). Several technology development groups, for example, have introduced piezoelectric flexures to the soles of running shoes. The amount of energy scavenged depends on acceleration or force experienced by the system, but typical systems can range from ~1 μW/cm for normal vibrations encountered in daily life to ~ 100 μW/cm^2 in industrial or high impact settings (such as the vibrations given off by heavy machinery). Figure TF25-2 shows an example of a complete energy scavenging system that employs a ***cantilever beam*** (the diving board structure in Fig. TF25-2) that acts as a spring with a proof mass at its tip. The cantilever structure oscillates when external vibrations are applied, deforming the cantilever. A piezoelectric film on the cantilever converts the oscillating mechanical energy to oscillating electrical energy that can power a circuit or sensor.

Radio Frequency Scavenging

A more recent class of devices attempts to scavenge power from the radio frequency electromagnetic energy. One common approach involves coupling oscillating radio frequency signals between (usually flat) conductive coils placed very close to each other; this is the basis of ***radio frequency identification*** (RFID) systems. A different approach is to collect or scavenge energy form electromagnetic energy present in our environment (from radio transmitters, phones, Wifi, etc.). Many working systems have been demonstrated in this second class over the years. One popular concept is the rectifying antenna, or ***rectenna***.

10-4.3 Single-Phase Equivalent Circuits

▶ A balanced three-phase network can be subdivided into three, independent, single-phase equivalent circuits (Fig. 10-8). ◀

In the solution of Example 10-2, we discovered that $\mathbf{V}_N = 0$ and $\mathbf{I}_n = 0$, both a consequence of the balanced-network condition. Since there is no voltage drop across the transmission line between nodes n and N (Fig. 10-8(a)), *we can treat those two nodes as the same node, even when no neutral line exists between them.* Hence, the balanced Y-Y network is equivalent to the sum of three, independent, single-phase circuits as shown in Fig. 10-8(b). The equivalency allows us to analyze each single-phase circuit separately.

The equivalence afforded by the balanced condition to the Y-Y network can be extended to the three other topologies by transforming them into Y-Y networks:

- For a balanced Y-Δ network, transform its Δ-load into a Y-load using Eq. (10.12).
- For a balanced Δ-Y network, transform its Δ-source into a Y-source using the recipe of Section 10-4.1.
- For a balanced Δ-Δ network, transform both its Δ-source and Δ-load to Y-configurations.

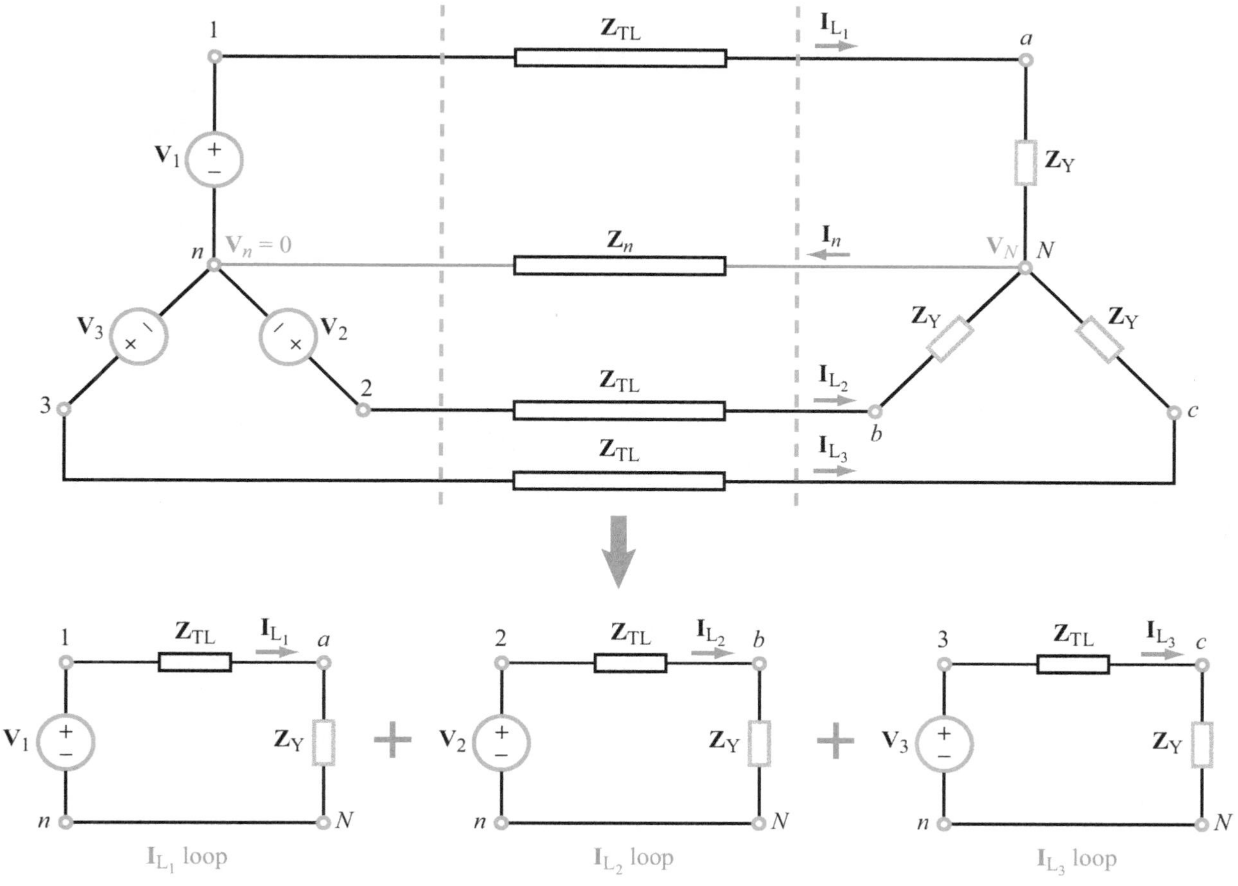

Figure 10-8: The balanced Y-Y network is equivalent to the sum of three, independent single-phase circuits.

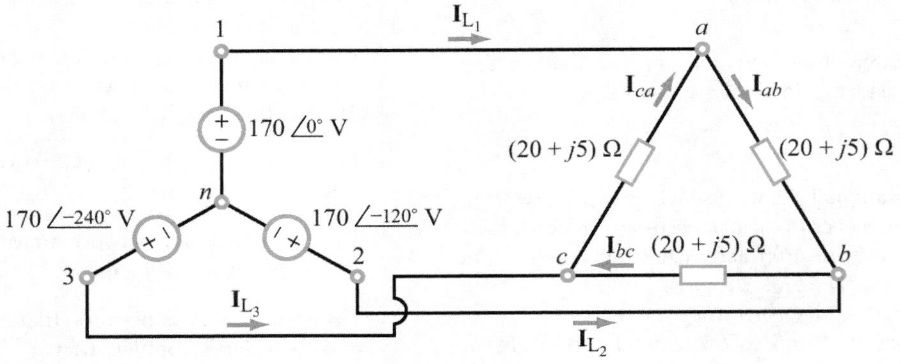

(a) Y-Δ network

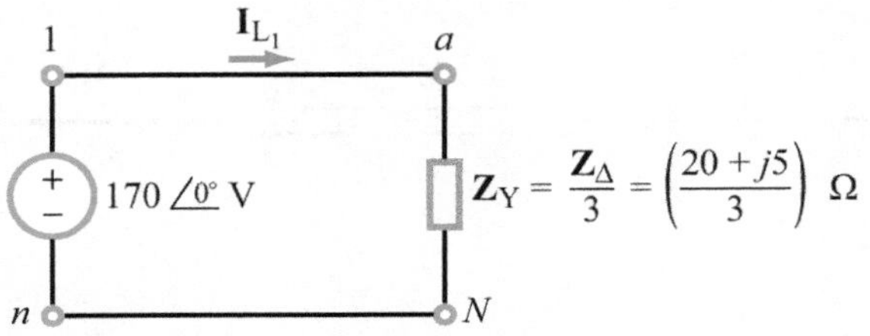

(b) Single-phase equivalent circuit

Figure 10-9: Circuit of Example 10-3 (voltage values are rms).

Example 10-3: Balanced Y-Δ Network

In the balanced Y-Δ network shown in Fig. 10-9(a), transmission-line impedances have been ignored. Determine line currents $\mathbf{I}_{L_1}$ to $\mathbf{I}_{L_3}$, and compare them with those of the phase currents in the load circuit.

Solution:

Method 1: Y-Δ Network

Application of KVL to loop $(n1ab2n)$ in Fig. 10-9 leads to

$$-170\underline{/0^\circ} + (20 + j5)\mathbf{I}_{ab} + 170\underline{/-120^\circ} = 0.$$

Solving for phase current $\mathbf{I}_{ab}$ gives

$$\mathbf{I}_{ab} = \frac{170\underline{/0^\circ} - 170\underline{/-120^\circ}}{20 + j5} = 14.28\underline{/16^\circ} \text{ A (rms)}.$$

Similarly,

$$\mathbf{I}_{bc} = \frac{170\underline{/-120^\circ} - 170\underline{/-240^\circ}}{20 + j5} = 14.28\underline{/-104^\circ} \text{ A (rms)}$$

and

$$\mathbf{I}_{ca} = \frac{170\underline{/-240^\circ} - 170\underline{/0^\circ}}{20 + j5} = 14.28\underline{/136^\circ} \text{ A (rms)}.$$

The line currents are

$$\begin{aligned}\mathbf{I}_{L_1} &= \mathbf{I}_{ab} - \mathbf{I}_{ca} \\ &= 14.28\underline{/16^\circ} - 14.28\underline{/136^\circ} \\ &= 14.28\sqrt{3}\,\underline{/-14^\circ} = 24.74\,\underline{/-14^\circ} \text{ A (rms)}, \\ \mathbf{I}_{L_2} &= \mathbf{I}_{bc} - \mathbf{I}_{ab} = 24.74\underline{/-134^\circ} \text{ A (rms)},\end{aligned}$$

and

$$\mathbf{I}_{L_3} = \mathbf{I}_{ca} - \mathbf{I}_{bc} = 24.74\underline{/106^\circ} \text{ A (rms)}.$$

We observe that the amplitudes of the line currents are $\sqrt{3}$ times the amplitudes of the phase currents in the load circuit.

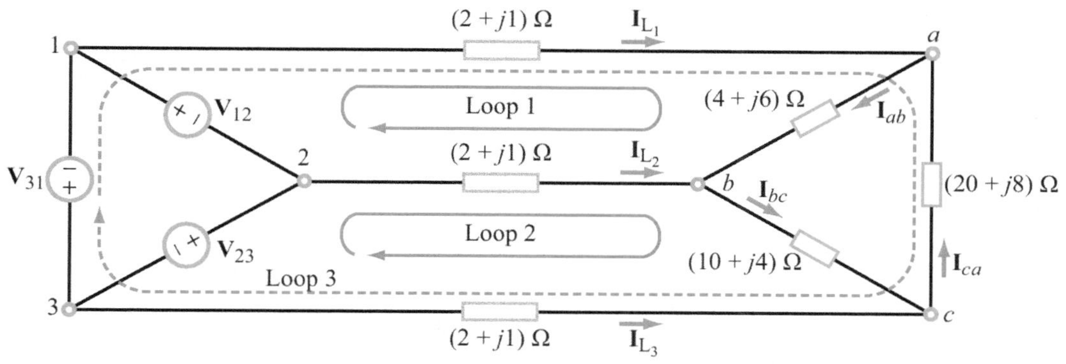

Figure 10-10: Δ-Δ network of Example 10-4.

Method 2: Equivalent Single-Phase

After transforming the Δ-load into an equivalent Y-load with $\mathbf{Z}_Y = \mathbf{Z}_\Delta/3$, we can apply the single-phase equivalency to generate three single-phase circuits. The single-phase circuit associated with line current $\mathbf{I}_{L_1}$ is shown in Fig. 10-9(b). Hence,

$$\mathbf{I}_{L_1} = \frac{170\,\angle 0^\circ}{\mathbf{Z}_Y} = \frac{3 \times 170\,\angle 0^\circ}{20 + j5} = 24.74\,\angle -14^\circ \text{ A (rms)},$$

which is identical to the result provided by Method 1. Line currents $\mathbf{I}_{L_2}$ and $\mathbf{I}_{L_3}$ can then be obtained successively by subtracting a phase angle of 120° in each step.

Example 10-4: Unbalanced Δ-Load

The unbalanced Δ-load circuit in Fig. 10-10 is connected to a balanced Δ-source with a positive phase sequence. If $\mathbf{V}_{12} = 416\angle 60^\circ$ V (rms), determine the line currents $\mathbf{I}_{L_1}$ to $\mathbf{I}_{L_3}$.

Solution: Given that the generator has a positive-sequence source, its phase voltages are

$$\begin{aligned}
\mathbf{V}_{12} &= 416\angle 60^\circ = (208 + j360) \text{ V (rms)},\\
\mathbf{V}_{23} &= \mathbf{V}_{12} \times 1\angle -120^\circ\\
&= 416\angle -60^\circ = (208 - j360) \text{ V (rms)},\\
\mathbf{V}_{31} &= \mathbf{V}_{12} \times 1\angle -240^\circ\\
&= 416\angle -180^\circ = (-416 + j0) \text{ V (rms)}.
\end{aligned}$$

The mesh-current equations for loops 1, 2, and 3 are

$$\begin{aligned}
-\mathbf{V}_{12} + (2 + j1)\mathbf{I}_{L_1} + (4 + j6)\mathbf{I}_{ab} - (2 + j1)\mathbf{I}_{L_2} &= 0,\\
-\mathbf{V}_{23} + (2 + j1)\mathbf{I}_{L_2} + (10 + j4)\mathbf{I}_{bc} - (2 + j1)\mathbf{I}_{L_3} &= 0,\\
\mathbf{V}_{31} + (2 + j1)\mathbf{I}_{L_1} - (20 + j8)\mathbf{I}_{ca} - (2 + j1)\mathbf{I}_{L_3} &= 0.
\end{aligned}$$

At nodes a, b, and c:

$$\begin{aligned}
\mathbf{I}_{L_1} &= \mathbf{I}_{ab} - \mathbf{I}_{ca},\\
\mathbf{I}_{L_2} &= \mathbf{I}_{bc} - \mathbf{I}_{ab},\\
\mathbf{I}_{L_3} &= \mathbf{I}_{ca} - \mathbf{I}_{bc}.
\end{aligned}$$

Making these substitutions in the three mesh-current equations, and then solving the three simultaneous equations, leads to

$$\begin{aligned}
\mathbf{I}_{ab} &= (33.31 + j5.95) \text{ A (rms)},\\
\mathbf{I}_{bc} &= (6.38 - j25.37) \text{ A (rms)},\\
\mathbf{I}_{ca} &= (-11.25 + j4.72) \text{ A (rms)}.
\end{aligned}$$

The line currents are then

$$\begin{aligned}
\mathbf{I}_{L_1} &= (33.31 + j5.95) - (-11.25 + j4.72)\\
&= (44.56 + j1.23) \text{ A (rms)},\\
\mathbf{I}_{L_2} &= (6.38 - j25.35) - (33.30 + j5.94)\\
&= (-26.94 - j31.32) \text{ A (rms)},\\
\mathbf{I}_{L_3} &= (-11.25 + j4.72) - (6.38 - j25.35)\\
&= (-17.63 + j30.09) \text{ A (rms)}.
\end{aligned}$$

Note that even though the load circuit is unbalanced,

$$\mathbf{I}_{L_1} + \mathbf{I}_{L_2} + \mathbf{I}_{L_3} = 0.$$

Concept Question 10-7: What conditions must apply in order to use the equivalent single-phase method? What advantage does it offer? (See CAD)

Exercise 10-5: Determine $\mathbf{I}_{L_1}$ in the balanced Y-Y network of Fig. 10-8, given that $\mathbf{V}_1 = 120\angle 0°$ V, $\mathbf{Z}_{TL} = (2 + j1)\ \Omega$ and $\mathbf{Z}_Y = (28 + j9)\ \Omega$.

Answer: $\mathbf{I}_{L_1} = 3.80\angle -18.4°$ A. (See CAD)

10-5 Power in Balanced Three-Phase Networks

Having examined the relationships for the currents through and voltages across the loads in balanced Y- and Δ-load circuits, we will now consider the power quantities associated with them.

10-5.1 Y-Load Configuration

The balanced Y-load circuit shown in Fig. 10-11 has equal impedances $\mathbf{Z}_Y$. Using $\mathbf{V}_{aN}$ as a reference, the circuit's three phase voltages can be expressed as

$$\mathbf{V}_{aN} = V_{YL}\angle 0° , \quad (10.13a)$$

$$\mathbf{V}_{bN} = V_{YL}\angle -120° , \quad (10.13b)$$

$$\mathbf{V}_{cN} = V_{YL}\angle -240° , \quad (10.13c)$$

where V_{YL} is the *rms magnitude of the phase voltages* of the Y-load. Their phase sequence is consistent with the phase sequence of the Y-source circuit (Fig. 10-6). Current $\mathbf{I}_a$ flowing through impedance $\mathbf{Z}_Y = Z_Y\angle\phi_Y$ is

$$\mathbf{I}_a = \frac{\mathbf{V}_{aN}}{\mathbf{Z}_Y} = \frac{V_{YL}\angle 0°}{Z_Y\angle\phi_Y} = \frac{V_{YL}}{Z_Y}\angle -\phi_Y = I_{YL}\angle -\phi_Y , \quad (10.14a)$$

where

$$I_{YL} = \frac{V_{YL}}{Z_Y}$$

is the *rms magnitude of the phase current*. Similarly,

$$\mathbf{I}_b = I_{YL}\angle -120° - \phi_Y , \quad (10.14b)$$

$$\mathbf{I}_c = I_{YL}\angle -240° - \phi_Y . \quad (10.14c)$$

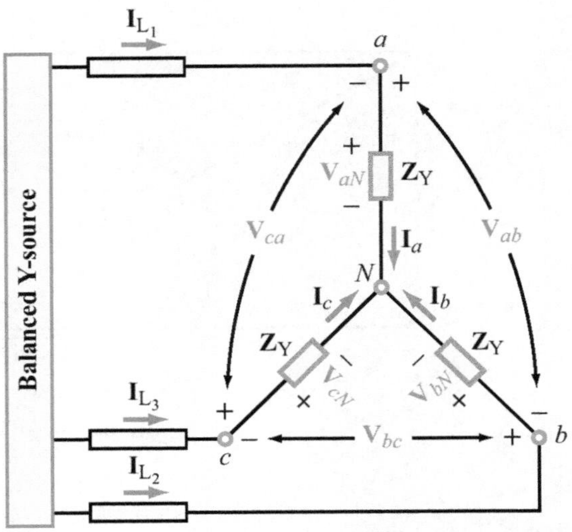

Figure 10-11: Balanced Y-load circuit with *line voltages* $\mathbf{V}_{ab}$ to $\mathbf{V}_{ca}$, *line currents* $\mathbf{I}_{L_1}$ to $\mathbf{I}_{L_3}$, *phase voltages* $\mathbf{V}_{aN}$ to $\mathbf{V}_{cN}$, and *phase currents* $\mathbf{I}_a$ to $\mathbf{I}_c$.

In accordance with Eq. (8.32), the *complex power* associated with an impedance carrying current $\mathbf{I}_a$ is

$$\mathbf{S}_a = \mathbf{V}_{aN}\mathbf{I}_a^* = V_{YL}(I_{YL}\angle -\phi_Y)^* = V_{YL}I_{YL}\angle\phi_Y . \quad (10.15)$$

The real and imaginary components of $\mathbf{S}_a$ represent P_a, the *average real power* dissipated in $\mathbf{Z}_Y$, and the *reactive power* stored in it, Q_a:

$$P_a = \mathfrak{Re}[\mathbf{S}_a] = V_{YL}I_{YL}\cos\phi_Y, \quad (10.16a)$$

$$Q_a = \mathfrak{Im}[\mathbf{S}_a] = V_{YL}I_{YL}\sin\phi_Y. \quad (10.16b)$$

It is a straightforward exercise to show that

$$P_a = P_b = P_c = V_{YL}I_{YL}\cos\phi_Y \quad (10.17a)$$

and

$$Q_a = Q_b = Q_c = V_{YL}I_{YL}\sin\phi_Y. \quad (10.17b)$$

Hence, for the overall balanced Y-load circuit, the *total average power* P_T and *total reactive power* Q_T are

$$P_T = 3V_{YL}I_{YL}\cos\phi_Y, \quad (10.18a)$$

$$Q_T = 3V_{YL}I_{YL}\sin\phi_Y. \quad (10.18b)$$

(balanced network)

The *power factor* of the overall load circuit is

$$pf = \frac{P_T}{|\mathbf{S}_T|} = \frac{P_T}{\sqrt{P_T^2 + Q_T^2}} = \cos\phi_Y, \qquad (10.18c)$$

which is also the power factor of the individual loads. From the circuit in Fig. 10-11, it is evident that the phase currents in the Y-load circuit are identical to the line currents ($\mathbf{I}_{L_1} = \mathbf{I}_a$, $\mathbf{I}_{L_2} = \mathbf{I}_b$, and $\mathbf{I}_{L_3} = \mathbf{I}_c$). Hence, the rms magnitude of the line currents, I_L, is equal to the rms magnitude of the phase currents, I_{YL}. Moreover, it is easy to show that the rms magnitude V_L of the three line voltages is related to the rms magnitude of the phase voltages by $V_L = \sqrt{3}\, V_{YL}$. Consequently, Eq. (10.18) can be expressed in terms of the line voltages and currents as

$$P_T = 3\left(\frac{V_L}{\sqrt{3}}\right) I_L \cos\phi_Y = \sqrt{3}\, V_L I_L \cos\phi_Y \qquad (10.19a)$$

and

$$Q_T = \sqrt{3}\, V_L I_L \sin\phi_Y. \qquad (10.19b)$$

Moreover, the two expressions can be combined into

$$\mathbf{S}_T = \mathbf{P}_T + j\,Q_T = \sqrt{3}\, V_L I_L \underline{/\phi_Y} \qquad (10.20)$$

(balanced Y-load).

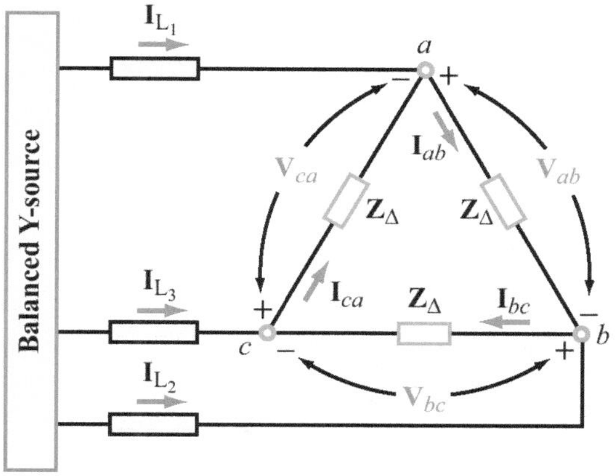

Figure 10-12: Balanced Δ-load circuit connected to a balanced Y-source.

10-5.2 Δ-Load Configuration

In the Y-load circuit of the preceding subsection, the line and phase currents were the same, and the amplitudes of the line and phase voltages were different but related by a factor of $\sqrt{3}$. The opposite is true for the Δ-load circuit shown in Fig. 10-12; the line and phase voltages are the same ($V_{\Delta L} = V_L$) and the corresponding currents are related by $I_{\Delta L} = I_L/\sqrt{3}$. Hence, with

$$|\mathbf{V}_{ab}| = |\mathbf{V}_{bc}| = |\mathbf{V}_{ca}| = V_{\Delta L}, \qquad (10.21a)$$

and

$$|\mathbf{I}_{ab}| = |\mathbf{I}_{bc}| = |\mathbf{I}_{ca}| = I_{\Delta L}, \qquad (10.21b)$$

the total complex power is

$$\mathbf{S}_T = P_T + j\,Q_T = 3V_{\Delta L} I_{\Delta L} \underline{/\phi_\Delta}\,, \qquad (10.22)$$

where ϕ_Δ is the phase angle of $\mathbf{Z}_\Delta$. In terms of the rms amplitudes of the line voltages and currents, $\mathbf{S}_T$ is

$$\mathbf{S}_T = 3V_L\left(\frac{I_L}{\sqrt{3}}\right)\underline{/\phi_\Delta} = \sqrt{3}\, V_L I_L \underline{/\phi_\Delta}\,. \qquad (10.23)$$

(balanced Δ-load)

The form of Eq. (10.23) for the Δ-load circuit is the same as that of Eq. (10.20) for the Y-load circuit. If we transform a balanced Δ-load into a balanced Y-load, the magnitude of the Y-impedances decreases by a factor of 3, but their phase angles remain the same ($\phi_\Delta = \phi_Y$). *Consequently, Eq. (10.18c) applies to both Y- and Δ-loads.* **Table 10-1** provides a summary of the voltage and power relations for balanced Y- and Δ-networks.

10-5.3 Total Instantaneous Power

For the Y-load circuit of Fig. 10-11, the instantaneous power $P_a(t)$ extracted by the load with phasor voltage $\mathbf{V}_{aN}$ across it and current $\mathbf{I}_a$ through it is

$$\begin{aligned} P_a(t) &= \upsilon_{aN}(t)\, i_a(t) \\ &= \mathfrak{Re}[\sqrt{2}\,\mathbf{V}_{aN} e^{j\omega t}]\, \mathfrak{Re}[\sqrt{2}\,\mathbf{I}_a e^{j\omega t}] \\ &= \mathfrak{Re}[\sqrt{2}\, V_{YL} e^{j\omega t}]\, \mathfrak{Re}[\sqrt{2}\, I_{YL} e^{-j\phi_Y} e^{j\omega t}] \\ &= 2V_{YL} I_{YL} \cos\omega t \cos(\omega t - \phi_Y), \end{aligned} \qquad (10.24a)$$

where we used the expressions for $\mathbf{V}_{aN}$ and $\mathbf{I}_a$ given by Eqs. (10.13a) and (10.14a), respectively , and introduced a factor of $\sqrt{2}$ in both to convert their magnitudes from rms to peak amplitude.

Table 10-1: Balanced networks.

$\mathbf{V}_1 = V_{Ys} \angle \phi_1$
$\mathbf{V}_2 = V_{Ys} \angle \phi_1 - 120°$
$\mathbf{V}_3 = V_{Ys} \angle \phi_1 + 120°$

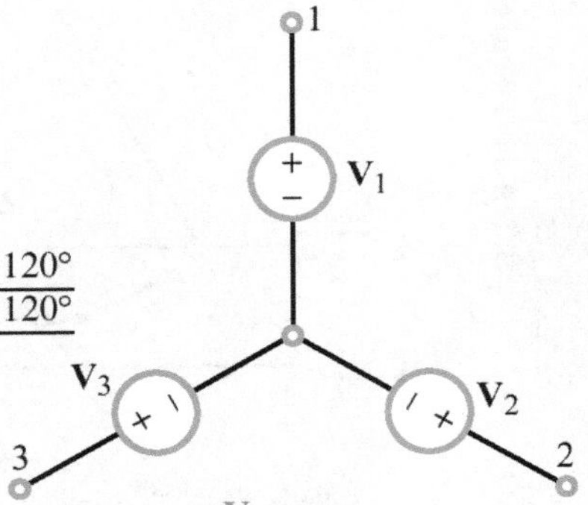

Y-source

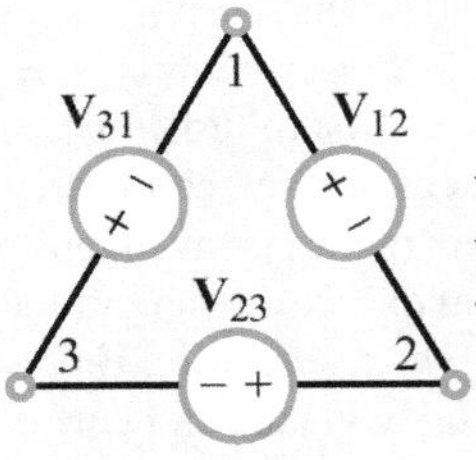

Δ-source

$\mathbf{V}_{12} = \mathbf{V}_1 \times \sqrt{3} \angle 30°$
$\mathbf{V}_{23} = \mathbf{V}_2 \times \sqrt{3} \angle 30°$
$\mathbf{V}_{31} = \mathbf{V}_3 \times \sqrt{3} \angle 30°$

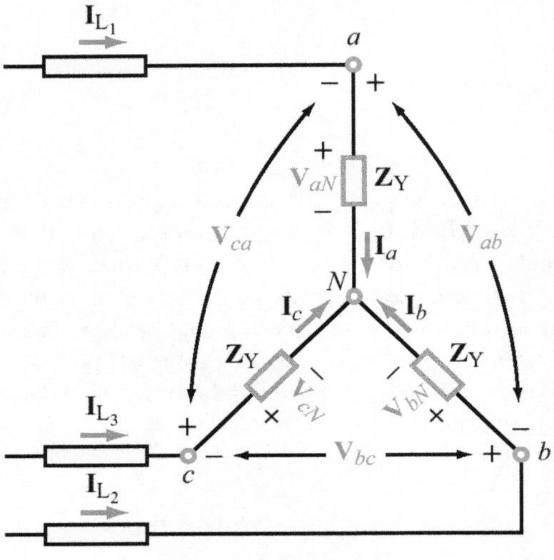

Y-load

$\mathbf{Z}_\Delta = 3\mathbf{Z}_Y$

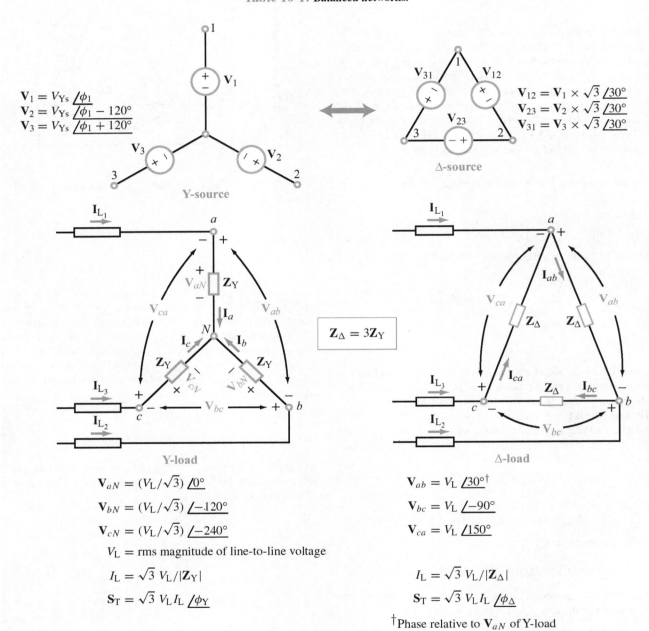

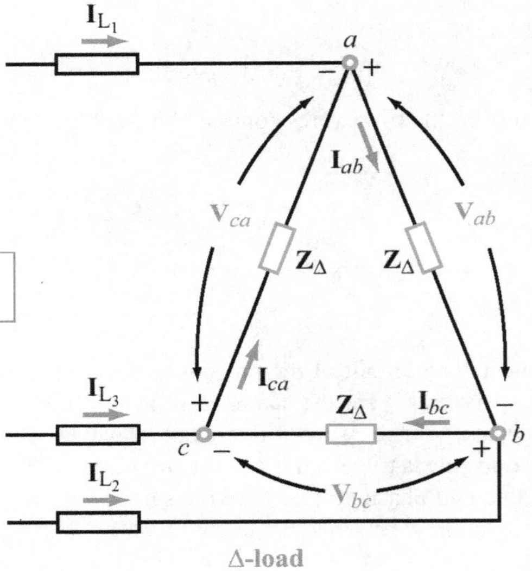

Δ-load

$\mathbf{V}_{aN} = (V_L/\sqrt{3}) \angle 0°$

$\mathbf{V}_{bN} = (V_L/\sqrt{3}) \angle -120°$

$\mathbf{V}_{cN} = (V_L/\sqrt{3}) \angle -240°$

V_L = rms magnitude of line-to-line voltage

$I_L = \sqrt{3}\, V_L/|\mathbf{Z}_Y|$

$\mathbf{S}_T = \sqrt{3}\, V_L I_L \angle \phi_Y$

$\mathbf{V}_{ab} = V_L \angle 30°^{\dagger}$

$\mathbf{V}_{bc} = V_L \angle -90°$

$\mathbf{V}_{ca} = V_L \angle 150°$

$I_L = \sqrt{3}\, V_L/|\mathbf{Z}_\Delta|$

$\mathbf{S}_T = \sqrt{3}\, V_L I_L \angle \phi_\Delta$

†Phase relative to $\mathbf{V}_{aN}$ of Y-load

Similarly, for the other two loads of the Y-load configuration

$$P_b(t) = 2V_{YL} I_{YL} \cos(\omega t - 120°)\cos(\omega t - 120° - \phi_Y), \quad (10.24b)$$

and

$$P_c(t) = 2V_{YL} I_{YL} \cos(\omega t - 240°)\cos(\omega t - 240° - \phi_Y). \quad (10.24c)$$

The total instantaneous power of the Y-load circuit, $P_T(t)$, is the sum of the three expressions given by Eqs. (10.24a to c),

$$\begin{aligned} P_T(t) = 2V_{YL} I_{YL}[&\cos\omega t \cos(\omega t - \phi_Y) \\ &+ \cos(\omega t - 120°)\cos(\omega t - 120° - \phi_Y) \\ &+ \cos(\omega t - 240°)\cos(\omega t - 240° - \phi_Y)]. \end{aligned} \quad (10.25)$$

Upon using the identities

$$\cos x \cos y = \frac{1}{2}[\cos(x + y) + \cos(x - y)], \quad (10.26a)$$

and

$$\cos\theta + \cos(\theta - 120°) + \cos(\theta - 240°) = 0 \quad (10.26b)$$

for any θ, the expression for $P_T(t)$ reduces to

$$P_T(t) = 3V_{YL} I_{YL} \cos\phi_Y. \quad (10.27)$$

This is a very significant result.

> ▶ The instantaneous power extracted by a balanced three-phase load from a balanced three-phase source is *not* a function of time. ◀

The power extracted by each load impedance individually varies sinusoidally with time at an angular frequency of 2ω, but the total power extracted by the three loads in combination does not oscillate at all. Consequently, the power generated by a balanced three-phase source is very steady, as is the power delivered to a balanced three-phase load, such as a three-phase motor. This observation is true for both Y and Δ configurations.

Example 10-5: Power in Balanced Δ-Load

A balanced Y-source connected to the Δ-load circuit of Fig. 10-12 generates line currents with rms magnitude $I_L = 5$ A. If the total power consumed by the load is 3120 W and the phase angle of the individual load impedances is 30° leading, determine the magnitude of the line voltage and the load impedances $\mathbf{Z}_\Delta$.

Solution: From Eq. (10.23), the total power P_T is

$$P_T = \mathfrak{Re}[\mathbf{S}_T] = \sqrt{3}\, V_L I_L \cos\phi_\Delta.$$

Since the impedance phase angle is leading, $\phi_\Delta = -30°$. Solving for V_L:

$$V_L = \frac{P_T}{\sqrt{3}\, I_L \cos\phi_\Delta} = \frac{3120}{\sqrt{3}\, 5\cos(-30°)} = 416 \text{ V (rms)}.$$

In the Δ-load circuit, $I_L = \sqrt{3}\, I_{YL}$ and $V_L = V_{YL}$. Hence,

$$Z_\Delta = \frac{V_{YL}}{I_{YL}} = \frac{V_L}{(I_L/\sqrt{3})} = \sqrt{3}\,\frac{416}{5} = 144.11\ \Omega,$$

and

$$\mathbf{Z}_\Delta = Z_\Delta \angle{-30°} = (124.8 - j72.1)\ \Omega.$$

Example 10-6: Power in Unbalanced Δ-Δ Network

This is a reexamination of the circuit analyzed earlier in Example 10-4, but from the standpoint of the power supplied by its source. For the sake of convenience, the circuit is reintroduced in Fig. 10-13, along with the values of the line currents and the phase currents in the load circuit. Determine the complex power supplied by the three-phase source, and the associated power factor.

Solution: The total complex power supplied by the source equals the sum of the complex powers extracted by the transmission-line impedances and by the three load impedances:

$$\begin{aligned} \mathbf{S} = (2 + j1)[|\mathbf{I}_{L_1}|^2 + |\mathbf{I}_{L_2}|^2 + |\mathbf{I}_{L_3}|^2] + (4 + j6)|\mathbf{I}_{ab}|^2 \\ + (10 + j4)|\mathbf{I}_{bc}|^2 + (20 + j8)|\mathbf{I}_{ca}|^2. \end{aligned}$$

Using the values listed in Fig. 10-13 leads to

$$\mathbf{S} = (24.2 + j15.7) \text{ kVA}$$

and

$$pf = \frac{P}{|\mathbf{S}|} = \frac{24.2}{|24.2 + j15.7|} = 0.84 \text{ lagging}.$$

Technology Brief 26
Inside a Power Generation Station

Many of the other Technology Briefs in this book are about small circuits with high component densities, such as Technology Brief 1 on Nano- and Microtechnology and Technology Brief 7 on Integrated Circuit Fabrication. In contrast, this Technology Brief is about *big* circuits used to support high-voltage and high-power systems. Household power sources provide 120—240 V. As seen in Fig. 10-1, the voltages in local distribution systems are on the order of *1–10 kV*, in transmission systems they are on the order of *10s of kV*, and in power generation stations they are on the order of *100s of kV*. As the power increases, the wires and electrical components must be physically larger to accommodate the heat generated by large currents passing through even small resistances. They must also be physically separated by greater distances, to prevent breakdown across the air gap between them.

High-power systems use many of the same electrical components as other electrical circuits, but the physical scale of high-power components is much bigger. Consider, for example, the very large ***iron-core inductor*** at the utility substation shown in Fig. TF26-1. Each of its three phases is connected via cables at the top. A bank of ***capacitors*** is shown in Fig. TF26-2. Both the capacitor and inductor are physically isolated from the ground below them in order to prevent them from arcing or shorting to the ground.

As noted in Section 10-2, 3-phase circuits are arranged in either a *Y* or a Δ ***configuration***. Figure TF26-3 shows photos of inductors connected in both configurations.

Not only are the inductors and capacitors used in the distribution substations very large in size, but so is the power generator. As described in Section 10-1, 3-phase electrical power is generated by creating a rotating magnetic field (often with rotor coils) inside three ***stator coils***. Figure TF26-4 shows one of the large stator coils for a coal-fired power ***generation station***. The sheer scale of the coils is evident from the fact that several technicians are working inside of it.

The large size theme also applies to electrical insulation, connections, and fuses used in high power applications. When connecting any electrical system, physical/mechanical connections are needed to hold the system together, and electrical connections to provide the appropriate paths for current. It is best practice to separate the mechanical connections from the electrical connections (similar to a smaller system where the electrical solder should not serve as the mechanical support). High-power systems use ceramic or glass insulator strings to electrically isolate the mechanical system that holds wires in place. These are seen and labeled in Fig. TF26-3(a), and a close-up is shown in Fig. TF26-5.

Opening a switch in a high power system requires special care, as illustrated in Fig. TF26-6. A fuse for high voltage systems is shown in Fig. TF26-7. Unlike smaller fuses where the metal component is meant to melt away, this type of fuse snaps open when the current gets too high. This is useful when maintaining long transmission lines, for instance, because a maintainer can visually observe which fuse is open, indicating the location for repair.

High-power systems provide exciting jobs for electrical engineers. Unlike commercial products, where circuits are designed to be used in thousands or millions of identical devices, most power systems are built for a single individualized application.

Figure TF26-1: A large 50 MVAR (Mega-Volts-Amps Reactive) loading inductor.

Figure TF26-2: A bank of large capacitors for three phase power. (Note the 3 lines going in and out of each capacitor.

(a) Δ configuration

(b) Y configuration

Figure TF26-3: 3-phase Y and Δ inductor configurations in a 345 kV substation. (Photos courtesy of Intermountain Power Project.)

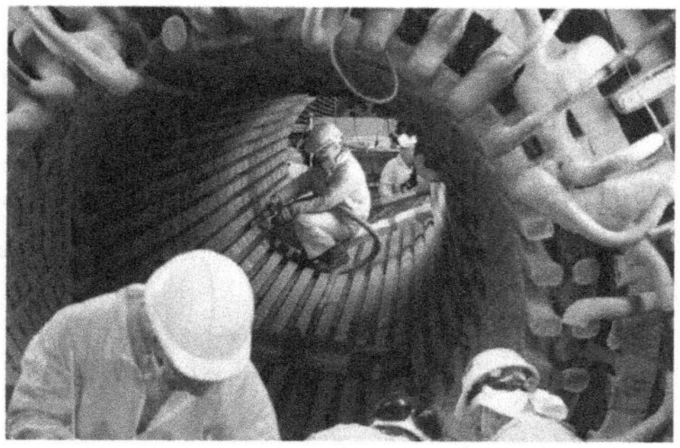

Figure TF26-4: Maintenance technicians working inside one of the generator coils at the Intermountain Power Project, a coal-fired power generation station. (Photo courtesy of Intermountain Power Project.)

Figure TF26-5: Cap and pin insulator string (the vertical string of discs) on a 275 kV suspension pylon.

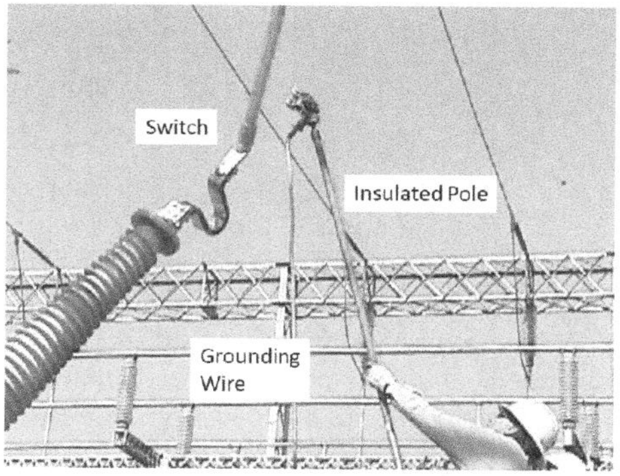

Figure TF26-6: Opening a switch using a fiberglass-insulated pole with a grounding wire. (Credit: IECACA.)

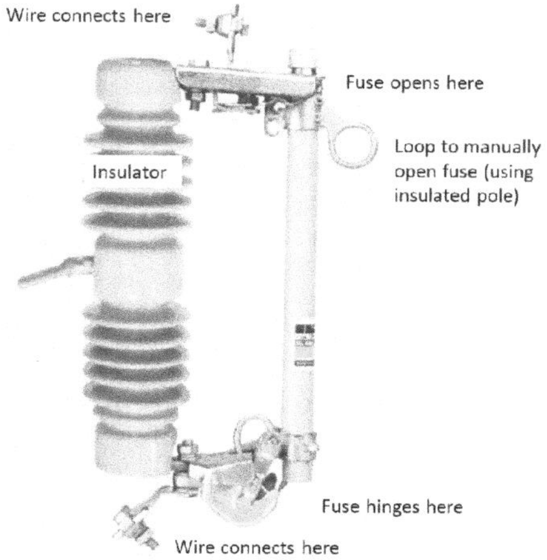

Figure TF26-7: Expulsion fuse cutout for 15 kV–27 kV.

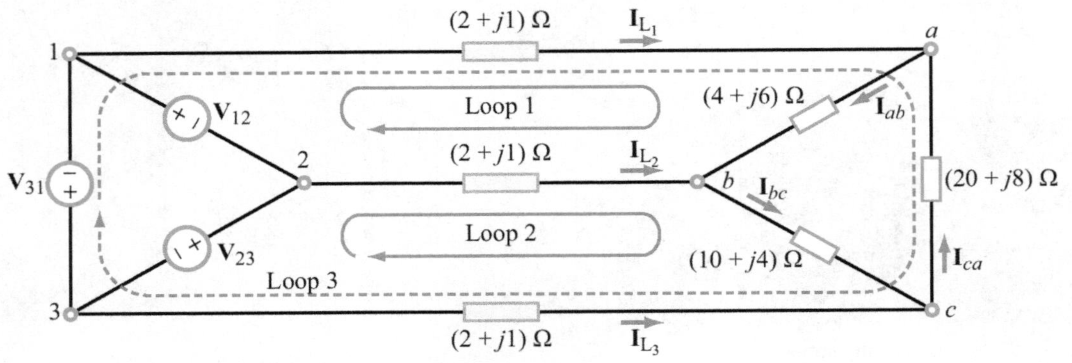

$\mathbf{I}_{L_1} = (44.56 + j1.23)$ A (rms) $\qquad \mathbf{I}_{ab} = (33.31 + j5.95)$ A (rms)

$\mathbf{I}_{L_2} = (-26.94 - j31.32)$ A (rms) $\qquad \mathbf{I}_{bc} = (6.38 - j25.37)$ A (rms)

$\mathbf{I}_{L_3} = (-17.63 - j30.09)$ A (rms) $\qquad \mathbf{I}_{ca} = (-11.25 + j4.72)$ A (rms)

Figure 10-13: Circuit of Example 10-4 with calculated values of the currents.

Concept Question 10-8: According to Eq. (10.27), the total instantaneous power supplied by a balanced three-phase source to a balanced three-phase load does not change with time. Why is such an attribute significant? (See CAD)

Concept Question 10-9: In each source of a three-phase configuration, the voltage varies sinusoidally at an angular frequency ω, and so does the current. At what angular frequency does the power vary and why? (See CAD)

Exercise 10-6: Prove Eq. (10.26b).

Answer: (See CAD)

10-6 Power-Factor Compensation

In Section 8-4.2, we examined how a shunt capacitor can be used to improve the *power factor* of single-phase inductive loads. By way of illustration, we showed in Example 8-6 that the addition of a 4.62 mF capacitor changes the power factor from 0.8 for the load alone to 0.95 for the load-capacitor combination. The circuit specifications included the rms voltage supplied by the source (220 V) and the power consumed by the load (200 kW).

The *pf*-compensation technique can be extended to balanced three-phase circuits. Figure 10-14(a) and (b) display the single-phase network and its three-phase extension. Note that the per-phase line voltage in Fig. 10-14(b) is the same (220 V rms) as the input voltage in the single-phase circuit, as is the per-phase consumed power (600 kW/3 = 200 kW). From the standpoint of *pf*-compensation, the circuit in Fig. 10-14(b) is equivalent to three identical single-phase circuits. Hence, the values of the three compensation capacitors are all the same as the value of C in the single-phase circuit.

Example 10-7: Balanced Source Connected to Multiple Loads

A balanced three-phase source supplies a 1200 V (rms) line voltage to three separate, balanced, three-phase loads, connected in parallel as shown in Fig. 10-15(a). Determine:

(a) the total complex power supplied by the source,

(b) the power factor at the source end,

(c) the line currents, and

(d) the capacitance of the three shunt capacitors added to raise the source's power factor to 0.92 lagging (Fig. 10-15(b)).

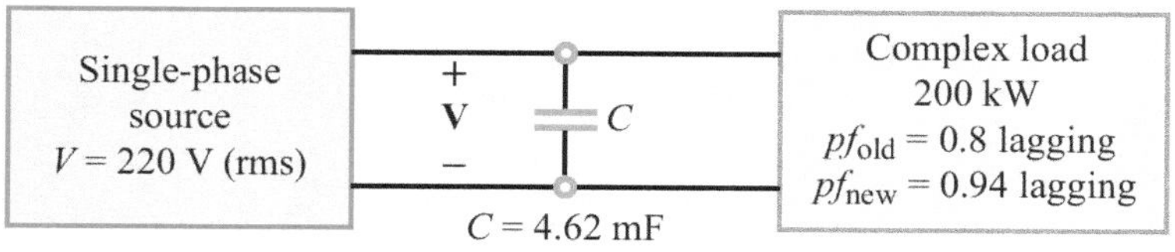

(a) Single-phase

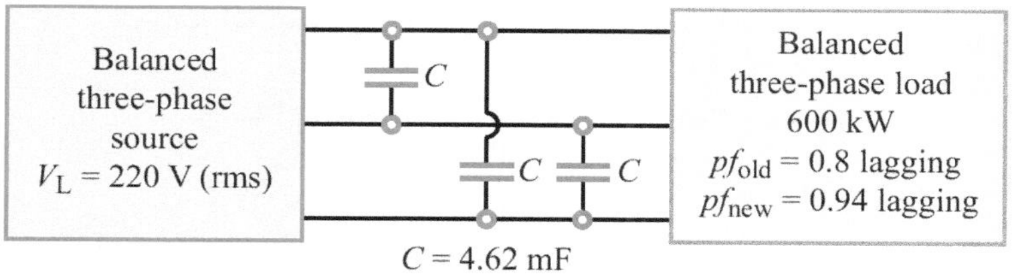

(b) Three-phase

Figure 10-14: A balanced three-phase load can be compensated by treating it as three individual circuits each consuming one-third of the total power.

Solution: (a) For load 1, $S_1 = 6,000$ VA and $pf_1 = 0.8$ lagging. Hence,

$$\phi_1 = \cos^{-1}(0.8) = 36.87^\circ$$

and

$$\mathbf{S}_1 = S_1(\cos\phi_1 + j\sin\phi_1) = (4800 + j3600) \text{ VA}.$$

Similarly, for loads 2 and 3, with $S_2 = 12,000$ VA and $pf_2 = 0.6$ lagging and $S_3 = 18,000$ VA and $pf_3 = 0.9$ lagging, we have

$$\phi_2 = \cos^{-1}(0.6) = 53.13^\circ,$$

$$\phi_3 = \cos^{-1}(0.9) = 25.84^\circ,$$

$$\mathbf{S}_2 = S_2(\cos\phi_2 + j\sin\phi_2) = (7200 + j9600) \text{ VA},$$

and

$$\mathbf{S}_3 = S_3(\cos\phi_3 + j\sin\phi_3) = (16200 + j7846) \text{ VA}.$$

The total complex power is then

$$\mathbf{S}_T = \mathbf{S}_1 + \mathbf{S}_2 + \mathbf{S}_3 = (28200 + j21046) \text{ VA},$$

of which

$$P_T = 28200 \text{ W} \quad \text{and} \quad Q_T = 21046 \text{ VAR}.$$

(b) From the standpoint of the source, the combination of the three loads is equivalent to a single, balanced, three-phase load with phase angle ϕ given by

$$\phi = \tan^{-1}\left(\frac{Q_T}{P_T}\right) = \tan^{-1}\left(\frac{21046}{28200}\right) = 36.73^\circ,$$

and a corresponding power factor

$$pf_s = \cos\phi = \cos 36.73^\circ = 0.8.$$

(c) Line current $\mathbf{I}_{L_1}$ is

$$\mathbf{I}_{L_1} = \mathbf{I}_{a_1} + \mathbf{I}_{a_2} + \mathbf{I}_{a_3},$$

where $\mathbf{I}_{a_1}$, $\mathbf{I}_{a_2}$, and $\mathbf{I}_{a_3}$ are the current branches into loads 1, 2, and 3, respectively (Fig. 10-15(a)). According to Eqs. (10.20) and (10.23), for both Y- and Δ-loads, the complex power extracted by a balanced three-phase load is

$$\mathbf{S} = \sqrt{3}\, V_L I_L \angle\phi,$$

where V_L and I_L are the magnitudes of the line voltage and current, and ϕ is the phase angle of the load impedance. For all three loads, $V_L = 1200$ V (rms), and for load 1, $\phi_1 = 36.87^\circ$. Hence,

$$S_1 = \sqrt{3}\, V_L I_{a_1},$$

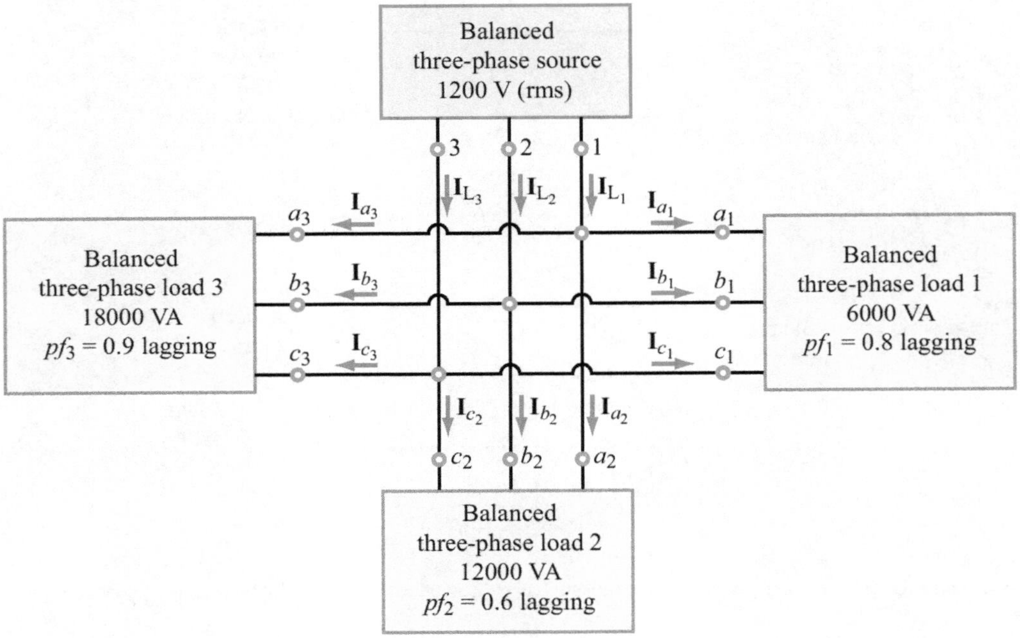

(a) Original circuit

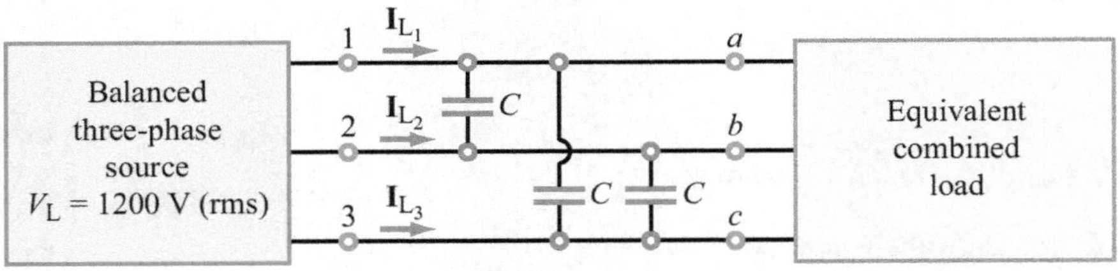

(b) Equivalent circuit with shunt capacitors added

Figure 10-15: Three-phase source connected in parallel to three loads, each a balanced three-phase load (Example 10-7).

which leads to

$$I_{a_1} = \frac{S_1}{\sqrt{3}\,V_L} = \frac{6000}{\sqrt{3}\,1200} = 2.89 \text{ A (rms)}.$$

The phase angle of I_{a_1} is $-36.87°$; this is negative because the power factor is lagging. Hence,

$$\mathbf{I}_{a_1} = 2.89\underline{/-36.87°} \text{ A (rms)}.$$

Similarly,

$$\mathbf{I}_{a_2} = \frac{S_2}{\sqrt{3}\,V_L}\underline{/-\phi_2} = 5.77\underline{/-53.13°} \text{ A (rms)},$$

$$\mathbf{I}_{a_3} = \frac{S_3}{\sqrt{3}\,V_L}\underline{/-\phi_3} = 8.66\underline{/-25.84°} \text{ A (rms)}.$$

The phasor sum of all three currents is

$$\mathbf{I}_{L_1} = \mathbf{I}_{a_1} + \mathbf{I}_{a_2} + \mathbf{I}_{a_3} = 16.93\underline{/-36.73°} \text{ A (rms)}.$$

Currents $\mathbf{I}_{L_2}$ and $\mathbf{I}_{L_3}$ follow suit by subtracting and adding 120° to their phases, respectively,

$$\mathbf{I}_{L_2} = 16.93\underline{/-156.73^\circ} \text{ A (rms)},$$
$$\mathbf{I}_{L_3} = 16.93\underline{/83.27^\circ} \text{ A (rms)}.$$

(d) Raising the power factor from $pf_s = 0.8$ to a new value $pf'_s = 0.92$ corresponds to changing ϕ of the combined load from 36.73° to

$$\phi' = \cos^{-1}(0.92) = 23.07^\circ.$$

It also means that the total complex power changes from

$$\mathbf{S}_T = (28200 + j21046) \text{ VA} \quad \text{(before)}$$

to

$$\mathbf{S}'_T = (28200 + jQ'_T) \text{ VA} \quad \text{(after adding capacitors)},$$

with P_T remaining unchanged, but the reactive power decreasing to

$$Q'_T = P_T \tan\phi' = 28200 \tan 23.07^\circ = 12013.15 \text{ VAR}.$$

The incremental reactive power contributed by the three shunt capacitors is

$$Q_C = 12013.15 - 21046 = -9032.87 \text{ VAR}.$$

Given that the voltage across each capacitor has a magnitude $V_L = 1200$ V and Q_C is related to C by

$$Q_C = -3V_L^2\omega C,$$

where the factor 3 accounts for the fact that Q_C is the sum of the reactive powers of the three capacitors, we deduce that

$$C = \frac{-Q_C}{3\omega V_L^2} = \frac{-(-9032.87)}{3 \times 2\pi \times 60 \times (1200)^2} = 5.55\ \mu\text{F}.$$

Concept Question 10-10: Can the pf-compensation method used in single-phase circuits be extended to a three-phase source connected to a balanced three-phase load? (See CAD)

Concept Question 10-11: How can one apply the pf-compensation method to a three-phase source connected in parallel to multiple, balanced, three-phase load circuits? (See CAD)

Exercise 10-7: Suppose the circuit shown in Fig. 10-15(a) contains only balanced loads 1 and 2. What value should C have in order to raise the source's power factor to 0.92 lagging?

Answer: $C = 4.97\ \mu$F. (See CAD)

10-7 Power Measurement in Three-Phase Circuits

The *wattmeter* is the standard instrument used to measure the average power consumed by a load. The classic analog wattmeter uses two coils, a *current coil* that measures the current flowing to the load, as shown in Fig. 10-16(a), and a *voltage coil* that measures a current proportional to the voltage across the load. One terminal of each coil has a double polarity mark (±) next to it.

► If the ± terminal of the current coil is on the end toward the source and the ± terminal of the voltage coil is connected to the line in which the current coil is inserted, the calibrated deflection on the wattmeter (or digital display) is equal to the average power P absorbed by the load. ◄

That is, if in Fig. 10-16(a), $\mathbf{I}$ and $\mathbf{V}$ are in rms and

$$\mathbf{I} = I\underline{/\phi_i} \text{ (rms)}, \qquad (10.28a)$$
$$\mathbf{V} = V\underline{/\phi_v} \text{ (rms)}, \qquad (10.28b)$$

then the quantity measured by the wattmeter is

$$P = \mathfrak{Re}[\mathbf{S}] = \mathfrak{Re}[\mathbf{VI}^*] = VI\cos(\phi_v - \phi_i). \qquad (10.29)$$

► The presence of the wattmeter in a circuit has negligible impact on the voltages and currents in the circuit. The advent of digital circuits that can rapidly sample voltage and current has enabled a class of *digital wattmeters* that does not require coils but performs substantially the same function. ◄

The wattmeter power-measurement technique can be extended to measure P_T, the total average power absorbed by any three-phase load (Y or Δ, balanced or unbalanced). By inserting three wattmeters, one between each pair of lines connected to the three-phase load, the total power absorbed

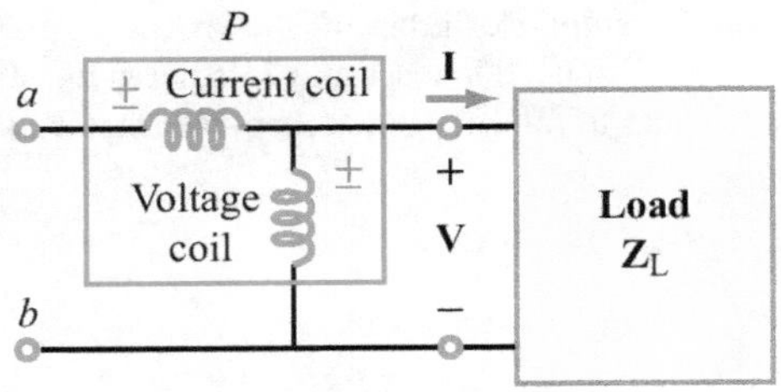

(a) Single-phase power measurement

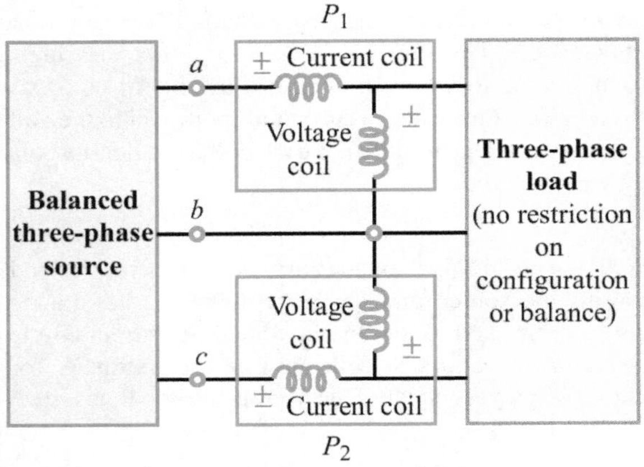

(b) Three-phase power measurement

Figure 10-16: A wattmeter uses two coils. The double polarity mark (±) of the current coil denotes the terminal that should be toward the source, and on the voltage coil, ± marks the terminal that should be connected to the line containing the current coil.

by the load, P_T, is determined by simply summing the power measurements made by the three wattmeters.

Alternatively, P_T can be determined by using only two wattmeters. The *two-wattmeter method* is not only simpler to apply, but if the load is balanced, it also provides a measurement of Q_T, and hence the total complex power $\mathbf{S}_T = P_T + jQ_T$. The arrangement is shown in Fig. 10-16(b). The wattmeters are inserted in the lines connected to nodes a and c, with the line connected to node b acting as a common reference. Actually, any of the three lines can be used as the common reference, so long as the wattmeters are inserted in the other two lines with their voltage-coil terminals bearing the ± mark connected to those lines (and not to the reference line). The wattmeters measure average powers P_1 and P_2. Our task is to demonstrate that the sum of these two powers is equal to P_T. We will do so by considering the unbalanced Δ-load configuration shown in Fig. 10-17, in which the three impedances may have different values.

At the input terminals to the Δ-load, the balanced three-phase source supplies line voltages $\mathbf{V}_{ab}$, $\mathbf{V}_{bc}$, and $\mathbf{V}_{ca}$ with a positive phase sequence. For convenience, we assign $\mathbf{V}_{ab}$ a phase angle of zero,

$$\mathbf{V}_{ab} = V_L \angle 0^\circ \text{ (rms)} \tag{10.30a}$$

$$\mathbf{V}_{bc} = V_L \angle -120^\circ \text{ (rms)} \tag{10.30b}$$

$$\mathbf{V}_{ca} = V_L \angle 120^\circ \text{ (rms)}, \tag{10.30c}$$

where V_L is the rms magnitude. Line currents $\mathbf{I}_1$ to $\mathbf{I}_3$ are hitherto unknown.

Total power absorbed by the load

The total average power absorbed by the three-phase load is the sum of the average powers absorbed by its three impedances. For load $\mathbf{Z}_{ab} = Z_{ab} \angle \phi_{ab}$,

$$P_{ab} = \Re\mathfrak{e}[\mathbf{S}_{ab}] = \Re\mathfrak{e}[\mathbf{V}_{ab}\mathbf{I}^*_{ab}] = \Re\mathfrak{e}\left[\mathbf{V}_{ab}\frac{\mathbf{V}^*_{ab}}{\mathbf{Z}^*_{ab}}\right] = \frac{V_L^2}{Z_{ab}}\cos\phi_{ab}. \tag{10.31}$$

Similar expressions apply to loads $\mathbf{Z}_{bc}$ and $\mathbf{Z}_{ca}$. The total power is

$$P_T = P_{ab} + P_{bc} + P_{ca}$$
$$= V_L^2\left[\frac{\cos\phi_{ab}}{Z_{ab}} + \frac{\cos\phi_{bc}}{Z_{bc}} + \frac{\cos\phi_{ca}}{Z_{ca}}\right]. \tag{10.32}$$

Two-wattmeter power measurement

With current $\mathbf{I}_1$ through it and voltage $\mathbf{V}_{ab}$ across it, the top wattmeter measures

$$P_1 = \Re\mathfrak{e}[\mathbf{S}_1] = \Re\mathfrak{e}[\mathbf{V}_{ab}\mathbf{I}^*_1]. \tag{10.33}$$

From the Δ-circuit,

$$\mathbf{I}_1 = \mathbf{I}_{ab} - \mathbf{I}_{ca} = \frac{\mathbf{V}_{ab}}{\mathbf{Z}_{ab}} - \frac{\mathbf{V}_{ca}}{\mathbf{Z}_{ca}} = \frac{V_L \angle 0^\circ}{Z_{ab}\angle \phi_{ab}} - \frac{V_L \angle 120^\circ}{Z_{ca}\angle \phi_{ca}}. \tag{10.34}$$

Using Eq. (10.34) in Eq. (10.33) leads to

$$P_1 = \Re\mathfrak{e}\left[V_L \angle 0^\circ\left(\frac{V_L\angle 0^\circ}{Z_{ab}\angle -\phi_{ab}} - \frac{V_L \angle -120^\circ}{Z_{ca}\angle -\phi_{ca}}\right)\right]$$
$$= V_L^2\left[\frac{\cos\phi_{ab}}{Z_{ab}} - \frac{\cos(\phi_{ca} - 120^\circ)}{Z_{ca}}\right]$$
$$= V_L^2\left[\frac{\cos\phi_{ab}}{Z_{ab}} + \frac{\cos(\phi_{ca} + 60^\circ)}{Z_{ca}}\right], \tag{10.35}$$

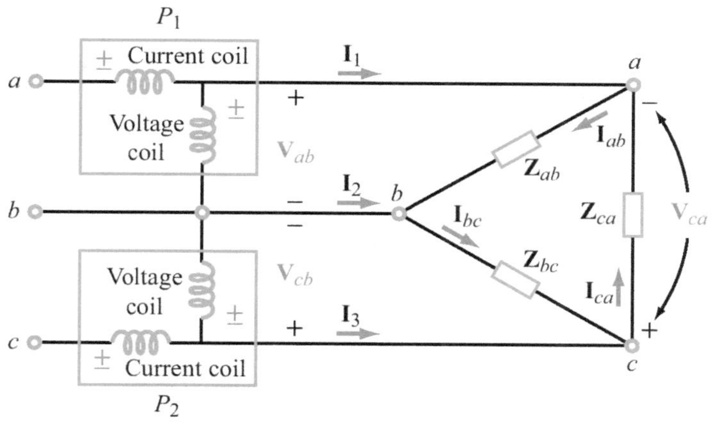

Figure 10-17: Two-wattmeter method applied to an unbalanced Δ-load.

where in the last step the negative sign (−) preceding the second term was replaced with a positive sign (+) and 180° was added to the phase.

For the lower wattmeter, the current through it is $\mathbf{I}_3$ and the voltage across it is $\mathbf{V}_{cb}$. Hence,

$$P_2 = \mathfrak{Re}[\mathbf{V}_{cb}\mathbf{I}_3^*]. \tag{10.36}$$

Using Eq. (10.30b) for $\mathbf{V}_{bc}$, as well as adding 180° in phase to convert $\mathbf{V}_{bc}$ to $\mathbf{V}_{cb}$, gives

$$\mathbf{V}_{cb} = V_L \angle 60^\circ \text{ (rms)}. \tag{10.37a}$$

Moreover, from the circuit,

$$\mathbf{I}_3 = \mathbf{I}_{ca} - \mathbf{I}_{bc} = \frac{\mathbf{V}_{ca}}{\mathbf{Z}_{ca}} - \frac{\mathbf{V}_{bc}}{\mathbf{Z}_{bc}} = \frac{V_L \angle 120^\circ}{Z_{ca} \angle \phi_{ca}} - \frac{V_L \angle -120^\circ}{Z_{bc} \angle \phi_{bc}}. \tag{10.37b}$$

Hence,

$$\begin{aligned} P_2 &= \mathfrak{Re}\left[V_L \angle 60^\circ \left(\frac{V_L \angle -120^\circ}{Z_{ca} \angle -\phi_{ca}} - \frac{V_L \angle 120^\circ}{Z_{bc} \angle -\phi_{bc}}\right)\right] \\ &= V_L^2 \left[\frac{\cos(\phi_{ca} - 60^\circ)}{Z_{ca}} - \frac{\cos(\phi_{bc} + 180^\circ)}{Z_{bc}}\right] \\ &= V_L^2 \left[\frac{\cos(\phi_{ca} - 60^\circ)}{Z_{ca}} + \frac{\cos\phi_{bc}}{Z_{bc}}\right]. \end{aligned} \tag{10.38}$$

The sum of the powers measured by the two wattmeters is

$$P_1 + P_2 = V_L^2 \left\{\frac{\cos\phi_{ab}}{Z_{ab}} + \frac{\cos\phi_{bc}}{Z_{bc}} + \frac{1}{Z_{ca}}\left[\cos(\phi_{ca} + 60^\circ) + \cos(\phi_{ca} - 60^\circ)\right]\right\}. \tag{10.39}$$

Applying the identity

$$\cos x + \cos y = 2\cos\left(\frac{x-y}{2}\right)\cos\left(\frac{x+y}{2}\right)$$

to the last term leads to

$$P_1 + P_2 = V_L^2 \left[\frac{\cos\phi_{ab}}{Z_{ab}} + \frac{\cos\phi_{bc}}{Z_{bc}} + \frac{\cos\phi_{ca}}{Z_{ca}}\right], \tag{10.40}$$

which is identical to the expression for P_T given by Eq. (10.32). Hence,

$$P_T = P_1 + P_2. \tag{10.41}$$

(any 3-phase load)

► The sum of the power measurements performed by the two wattmeters is the same as the total power absorbed by the three-phase load. ◄

This conclusion was reached for an unbalanced Δ-load configuration. It is equally true for a balanced load as well as for any Y-load configuration.

Example 10-8: Reactive Power of a Balanced Load

Show that if the Δ-load circuit in Fig. 10-17 is balanced, with equal impedances $\mathbf{Z}_{ab} = \mathbf{Z}_{bc} = \mathbf{Z}_{ca} = \mathbf{Z}_\Delta$, the total reactive power of the circuit is

$$Q_T = \sqrt{3}\,(P_2 - P_1). \tag{10.42}$$

Solution: For load $\mathbf{Z}_{ab} = \mathbf{Z}_\Delta = Z_\Delta \angle \phi_\Delta$,

$$Q_{ab} = \mathfrak{Im}\{\mathbf{S}_{ab}\} = \mathfrak{Im}\left\{\frac{V_L^2}{Z_\Delta \angle -\phi_\Delta}\right\} = \frac{V_L^2}{Z_\Delta}\sin\phi_\Delta\,.$$

For all three loads,

$$Q_T = 3\,\frac{V_L^2}{Z_\Delta}\sin\phi_\Delta. \tag{10.43}$$

Next, if we set

$$Z_{ab} = Z_{bc} = Z_{ca} = Z_\Delta$$

and

$$\phi_{ab} = \phi_{bc} = \phi_{ca} = \phi_\Delta$$

in Eqs. (10.35) and (10.38), and then subtract P_1 from P_2, we end up with

$$\begin{aligned} P_2 - P_1 &= \frac{V_L^2}{Z_\Delta}\,[\cos(\phi_\Delta - 60^\circ) + \cos\phi_\Delta \\ &\qquad - \cos\phi_\Delta - \cos(\phi_\Delta + 60^\circ)] \\ &= \frac{V_L^2}{Z_\Delta}\,[\cos(\phi_\Delta - 60^\circ) - \cos(\phi_\Delta + 60^\circ)]. \end{aligned}$$

Applying the trigonometric identity

$$\cos(x \pm y) = \cos x \cos y \mp \sin x \sin y,$$

leads to

$$P_2 - P_1 = \frac{2V_L^2}{Z_\Delta}\sin\phi_\Delta \sin 60^\circ = \sqrt{3}\,\frac{V_L^2}{Z_\Delta}\sin\phi_\Delta. \tag{10.44}$$

Upon comparing Eq. (10.44) with Eq. (10.43), we conclude that

$$Q_T = \sqrt{3}\,(P_2 - P_1) \qquad \text{(balanced load)}. \tag{10.45}$$

We note that Eq. (10.45) is equally valid for a balanced Y-load. Moreover, the combination of P_T and Q_T can be used to determine the phase angle ϕ associated with the system's power factor, namely

$$\tan\phi = \frac{Q_T}{P_T} = \frac{\sqrt{3}\,(P_2 - P_1)}{P_2 + P_1}\,, \tag{10.46}$$

from which we have

$$pf = \cos\phi. \tag{10.47}$$

Additionally, the sign of Q_T provides information about the load:

- Load is inductive if $Q_T > 0$.
- Load is capacitive if $Q_T < 0$.
- Load is resistive if $Q_T = 0$.

Concept Question 10-12: A wattmeter uses a current coil and a voltage coil. How should the two coils be connected relative to the source and load of a single-phase circuit? (See CAD)

Concept Question 10-13: The two-wattmeter method can provide a measurement of what power quantity in a three-phase network? Is the method constrained to balanced networks? (See CAD)

Exercise 10-8: When used on a balanced three-phase load, the two-wattmeter method provided measurements $P_1 = 4{,}800$ W and $P_2 = 10{,}200$ W. What is the total complex power $\mathbf{S}_T$ of the load?

Answer: $\mathbf{S}_T = (15000 + j5400)$ VA. (See CAD)

Summary

Concepts

- In a balanced three-phase source with a positive phase sequence, the three sources have identical voltage magnitudes and frequency, but their phase angles are shifted clockwise by $-120°$ increments.
- Three-phase networks can assume four configurations: Y-Y, Y-Δ, Δ-Y, and Δ-Δ, and each can be transformed into any of the other three.
- A balanced three-phase network can be subdivided into three independent, single-phase circuits.
- The total instantaneous power supplied by a balanced three-phase source is a constant (not a function of time), thereby rendering smooth power delivery to a balanced three-phase load, such as a three-phase motor.
- The two-wattmeter method provides a measurement of the total real power P_T consumed by any three-phase load, whether balanced or not. Moreover, if the load is balanced, the method also provides a measurement of the reactive power Q_T.

Mathematical and Physical Models

Balanced Y-Source

$$\mathbf{V}_1 = V_{Ys}\angle\phi_1$$
$$\mathbf{V}_2 = V_{Ys}\angle\phi_1 - 120°$$
$$\mathbf{V}_3 = V_{Ys}\angle\phi_1 + 120°$$
$$\mathbf{V}_1 + \mathbf{V}_2 + \mathbf{V}_3 = 0$$

Y $\leftrightarrow$ Δ Balanced Loads

$$\mathbf{Z}_\Delta = 3\mathbf{Z}_Y$$

Two-Wattmeter Measurement

$P_T = P_1 + P_2$ (any load)

$Q_T = \sqrt{3}\,(P_2 - P_1)$ (Balanced Y- or Δ-load)

Balanced Δ-Source

$$\mathbf{V}_{12} = V_{\Delta s}\angle\phi_1 + 30°$$
$$\mathbf{V}_{23} = V_{\Delta s}\angle\phi_1 - 90°$$
$$\mathbf{V}_{31} = V_{\Delta s}\angle\phi_1 + 150°$$
$$V_{\Delta s} = \sqrt{3}\,V_{Ys}$$
$$\mathbf{V}_{12} + \mathbf{V}_{23} + \mathbf{V}_{31} = 0$$

Total Complex Power

$\mathbf{S}_T = \sqrt{3}\,V_L I_L\angle\phi_Y$ (balanced Y-load)

$\mathbf{S}_T = \sqrt{3}\,V_L I_L\angle\phi_\Delta$ (balanced Δ-load)

Total Instantaneous Power

$P_T(t) = 3V_{YL} I_{YL} \cos\phi_Y$ (balanced Y-network)

$P_T(t) = 3V_{\Delta L} I_{\Delta L} \cos\phi_\Delta$ (balanced Δ-network)

Important Terms

Provide definitions or explain the meaning of the following terms:

Δ-load configuration
Δ-source configuration
average real power
balanced
balanced load
balanced network
balanced source
balanced three-phase Y-source configuration
bus circuit
center-tapped pole transformer
complex power
current coil
digital wattmeter
line current
line voltage
line-to-line voltage
magnitude
magnitude of the Δ-source
negative phase sequence
network
neutral node
neutral terminal
neutral wire
phase current
phase magnitude
phase voltage
positive (123) phase sequence
power factor
power grid
reactive power
real power
rms magnitude of the phase current
rms magnitude of the phase voltages
rotor
single-phase generators
stator
step-down transformer
step-up transformer
three-phase
three-phase ac generator
three-wire single phase
total average power
total reactive power
transformer
transmission line
two-wattmeter method
voltage coil
wattmeter
Y-load configuration
Y-source configuration
Y-Y, Y-Δ, Δ-Y, Δ-Δ configurations

PROBLEMS

Section 10-1: Balanced Three-Phase Generators

10.1 For each of the following groups of sources, determine if the three sources constitute a balanced source, and if it is, determine if it has a positive or negative phase sequence.

(a) $v_a(t) = 169.7\cos(377t + 15°)$ V
$v_b(t) = 169.7\cos(377t - 105°)$ V
$v_c(t) = 169.7\sin(377t - 135°)$ V

(b) $v_a(t) = 311\cos(\omega t - 12°)$ V
$v_b(t) = 311\cos(\omega t + 108°)$ V
$v_c(t) = 311\cos(\omega t + 228°)$ V

(c) $\mathbf{V}_1 = 140\angle{-140°}$ V
$\mathbf{V}_2 = 114\angle{-20°}$ V
$\mathbf{V}_3 = 124\angle{100°}$ V

10.2 In a balanced Y-source with a positive phase sequence, $\mathbf{V}_1 = (103.92 - j60)$ V (rms). Determine: (a) $\mathbf{V}_2$ and $\mathbf{V}_3$ and (b) $\mathbf{V}_{12}$, $\mathbf{V}_{23}$, and $\mathbf{V}_{31}$ of the equivalent Δ-source configuration, all in polar form.

10.3 In a balanced Δ-source with a positive phase sequence, $v_{12}(t) = 240\cos(120\pi t - 20°)$ V. Determine (a) $v_{23}(t)$, $v_{31}(t)$, and (b) $v_1(t)$, $v_2(t)$, and $v_3(t)$ of the equivalent Y-source configuration.

*__10.4__ In a balanced Y-source, $\mathbf{V}_2 = 100\angle{60°}$ V (rms) and $\mathbf{V}_3 = -100$ V (rms). Is the phase sequence positive or negative? Also, determine $\mathbf{V}_1$.

10.5 In a balanced Δ-source with a positive phase sequence, $\mathbf{V}_{23} = (56.94 + j212.5)$ V (rms). Determine $v_{12}(t)$, $v_{23}(t)$, and $v_{31}(t)$. Assume $f = 60$ Hz.

10.6 In a balanced Y-source with a positive phase sequence, $\mathbf{V}_1 = (16.93 + j46.51)$ V (rms). Determine $v_{12}(t)$, $v_{23}(t)$, and $v_{31}(t)$ of the equivalent Δ-source configuration. Assume $f = 60$ Hz.

Sections 10-2 to 10-4: Configurations and Networks

*__10.7__ In the circuit of Fig. P10.7, $\mathbf{V}_{aN} = 65\angle{47°}$ V (rms), $\mathbf{V}_{bN} = 60\angle{15°}$ V (rms), $\mathbf{V}_{cN} = 45\angle{-86°}$ V (rms), and $f = 60$ Hz. Calculate $v_{ab}(t)$, $v_{bc}(t)$, and $v_{ca}(t)$.

*Answer(s) available in Appendix G.

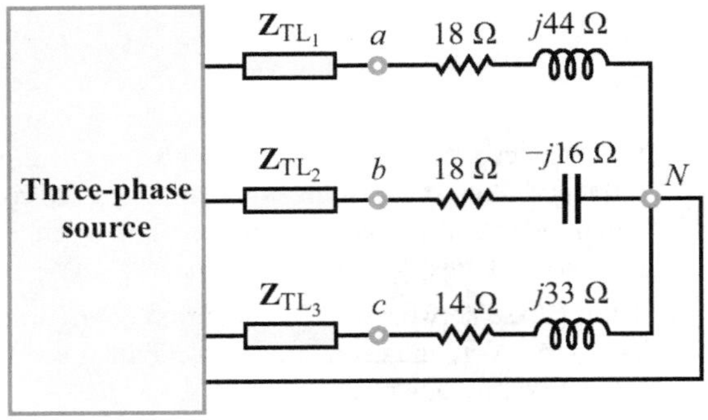

Figure P10.7: Circuit for Problem 10.7.

10.8 In the network of Fig. P10.8, $\mathbf{Z}_a = \mathbf{Z}_b = \mathbf{Z}_c = (25 + j5)$ Ω. Determine the line currents.

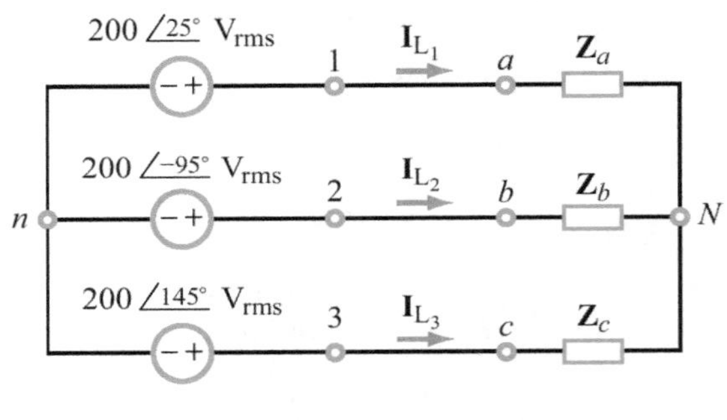

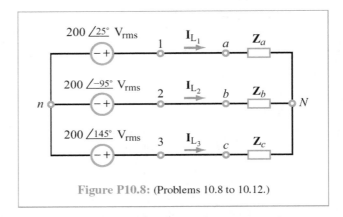

Figure P10.8: (Problems 10.8 to 10.12.)

10.9 In the network of Fig. P10.8, $\mathbf{Z}_a = (10 + j2)$ Ω, $\mathbf{Z}_b = 6$ Ω, and $\mathbf{Z}_c = (8 + j1)$ Ω. Determine the line currents.

10.10 Repeat Problem 10.8 after inserting transmission-line impedances $\mathbf{Z}_{TL} = (3 + j1)$ Ω between nodes 1 and a, 2 and b, and 3 and c.

*__10.11__ Repeat Problem 10.9 after inserting transmission-line impedances $\mathbf{Z}_{TL} = (3 + j1)$ Ω between nodes 1 and a, 2 and b, and 3 and c.

10.12 Repeat Problem 10.9 after adding a zero-impedance wire between nodes n and N. Also, determine the current through it.

10.13 Apply single-phase equivalency to determine the line currents in the Y-Δ network shown in Fig. P10.13. The load impedances are $\mathbf{Z}_{ab} = \mathbf{Z}_{bc} = \mathbf{Z}_{ca} = (25 + j5)$ Ω.

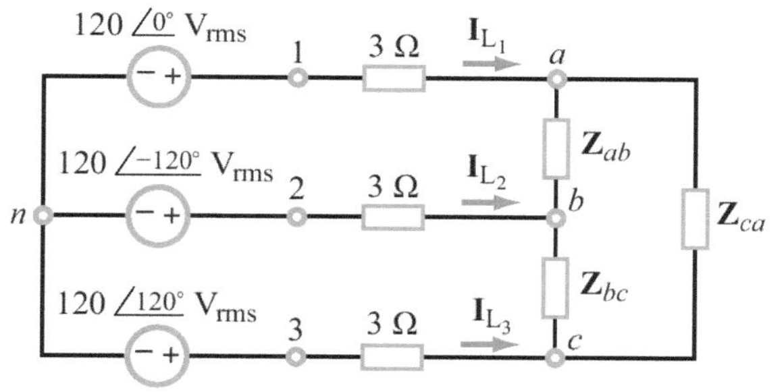

Figure P10.13: (Problems 10.13 to 10.14.)

*10.14 Determine the line currents in the network of Fig. P10.13, given that $\mathbf{Z}_{ab} = 10\ \Omega$, $\mathbf{Z}_{bc} = 5\ \Omega$, and $\mathbf{Z}_{ca} = (10 - j5)\ \Omega$.

10.15 Given a balanced Δ-Y network with $\mathbf{V}_{12} = 440\angle 0°$ V (rms) connected in a positive phase sequence, and $\mathbf{Z}_{\mathrm{Y}} = (10 - j2)\ \Omega$, apply the necessary transformation in order to use single-phase equivalency to determine the three line currents. Transmission-line impedances are ignored.

10.16 Given a balanced Δ-Δ network with $\mathbf{V}_{12} = 440\angle 0°$ V (rms) connected in a positive phase sequence, and $\mathbf{Z}_{\Delta} = (6 - j2)\ \Omega$, apply the necessary transformation in order to use single-phase equivalency to determine the three line currents. Transmission-line impedances are ignored.

*10.17 Determine $\mathbf{I}_{\mathrm{L}_1}$ in the network of Fig. P10.17. (Hint: This is a balanced network. If you apply the correct transformations, the solution simplifies considerably.)

10.18 The network shown in Fig. P10.18 consists of a balanced 120 V (rms) Y-source connected in a positive phase sequence, lossless transmission lines, and an unbalanced Δ-load with $\mathbf{Z}_{ab} = 14\ \Omega$, $\mathbf{Z}_{bc} = (6 - j2)\ \Omega$, and $\mathbf{Z}_{ca} = (24 + j6)\ \Omega$. Assign $\mathbf{V}_1$ a phase angle of 0°.

(a) Determine the phase voltages at the load: $\mathbf{V}_{ab}$, $\mathbf{V}_{bc}$, and $\mathbf{V}_{ca}$.

(b) Determine the phase currents: $\mathbf{I}_{ab}$, $\mathbf{I}_{bc}$, and $\mathbf{I}_{ca}$.

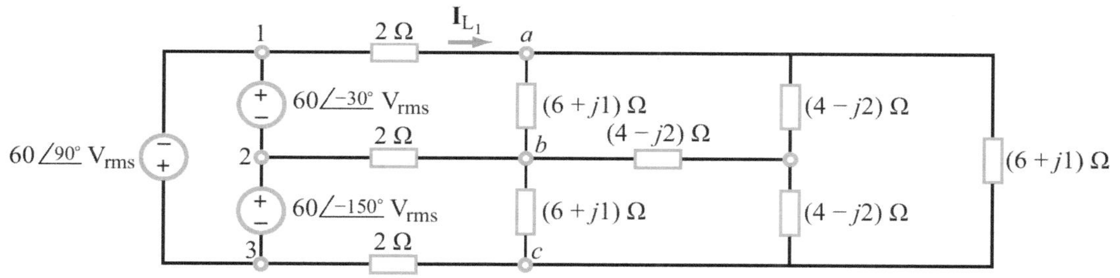

Figure P10.17: (Problem 10.17.)

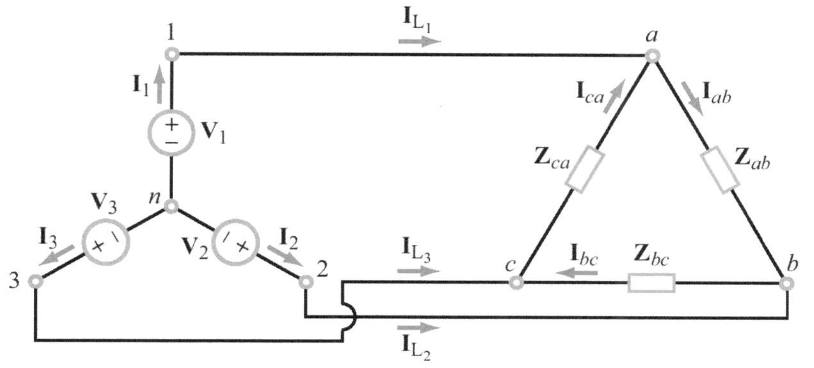

Figure P10.18: (Problem 10.18.)

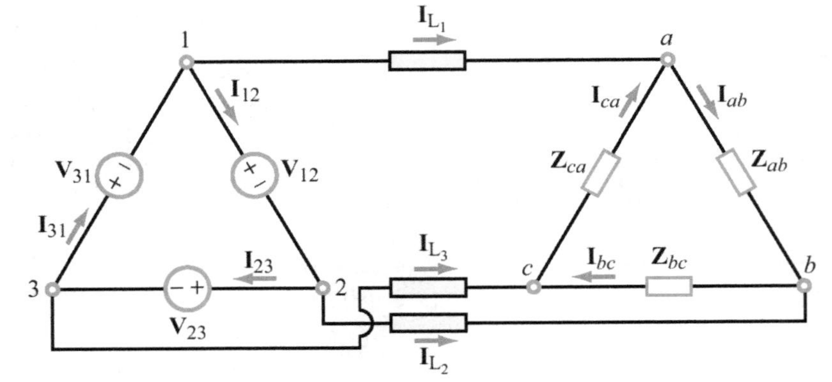

Figure P10.19: (Problem 10.19.)

(c) Determine the line currents: $\mathbf{I}_{L_1}$, $\mathbf{I}_{L_2}$, and $\mathbf{I}_{L_3}$.

(d) Determine the total power absorbed by the load, given that the power absorbed by an impedance $\mathbf{Z}$ due to the flow of current $\mathbf{I}$ through it is $P = |\mathbf{I}|^2\ \Re\mathfrak{e}[\mathbf{Z}]$, where $\mathbf{I}$ is rms.

10.19 The network shown in Fig. P10.19 consists of a balanced 120 V (rms) Δ-source connected in a positive phase sequence, lossless transmission lines, and an unbalanced Δ-load with $\mathbf{Z}_{ab} = (6 + j4)\ \Omega$, $\mathbf{Z}_{bc} = 4\ \Omega$, and $\mathbf{Z}_{ca} = 12\ \Omega$. Assign $\mathbf{V}_{12}$ a phase angle of 0°.

(a) Determine the phase voltages at the load: $\mathbf{V}_{ab}$, $\mathbf{V}_{bc}$, and $\mathbf{V}_{ca}$.

(b) Determine the phase currents: $\mathbf{I}_{ab}$, $\mathbf{I}_{bc}$, and $\mathbf{I}_{ca}$.

(c) Determine the line currents: $\mathbf{I}_{L_1}$, $\mathbf{I}_{L_2}$, and $\mathbf{I}_{L_3}$.

(d) Determine the total power absorbed by the load, given that the power absorbed by an impedance $\mathbf{Z}$ due to the flow of current $\mathbf{I}$ through it is $P = |\mathbf{I}|^2\ \Re\mathfrak{e}[\mathbf{Z}]$, where $\mathbf{I}$ is rms.

*__10.20__ Determine $\mathbf{I}_n$ in the circuit of Fig. P10.20. All sources are rms.

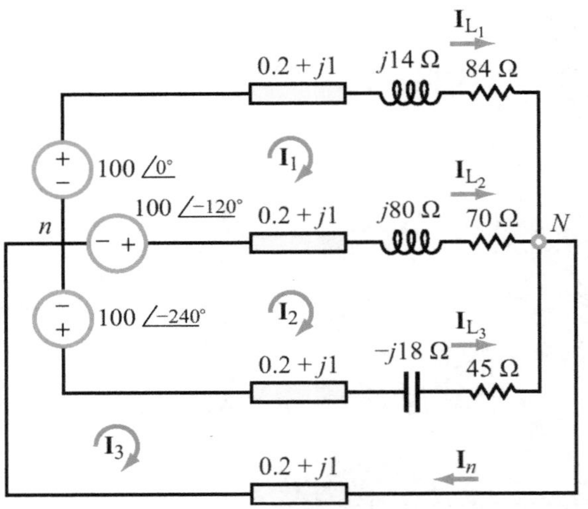

Figure P10.20: Circuit for Problem 10.20.

Sections 10-5 to 10-7: Power

10.21 For the network in Fig. P10.21, (a) generate three single-phase equivalent circuits and (b) determine the complex power supplied by the three-phase source.

10.22 For the network in Fig. P10.22, (a) generate three single-phase equivalent circuits and (b) determine the complex power supplied by the three-phase source.

*__10.23__ Determine the complex power supplied by the source in the network of Fig. P10.23.

10.24 A balanced Y-load is supplied by a three-phase generator at a line voltage of 416 V (rms). If the real power absorbed by the load is 6 kW at a power factor 0.7 lagging, determine $\mathbf{Z}_Y$ and the magnitude of the line current.

10.25 A balanced Δ-load is supplied by a three-phase generator at a line voltage of 208 V (rms). If the complex power extracted by the load is $(8 + j4)$ kVA, determine $\mathbf{Z}_\Delta$ and the magnitude of the line current.

*__10.26__ For the network in Fig. P10.23, determine (a) the average real power supplied by the three-phase source and (b) what fraction of it is absorbed by the three-phase load.

10.27 Determine the complex power extracted by the load in Fig. P10.27. Also determine the power factor of the overall load circuit as seen by the source.

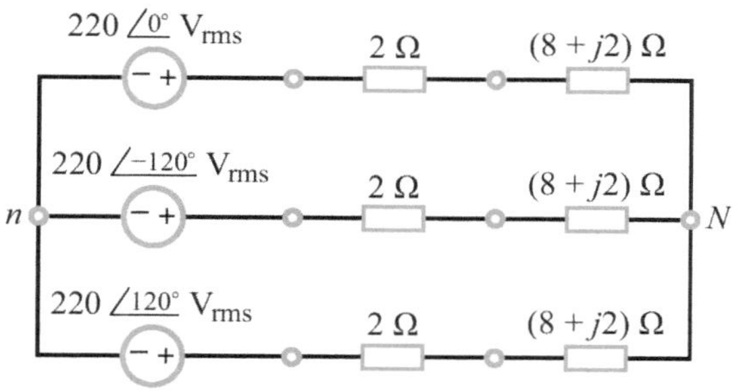

Figure P10.21: (Problem 10.21.)

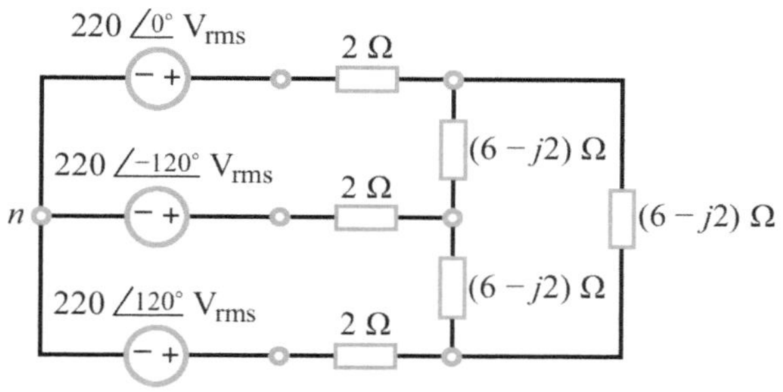

Figure P10.22: (Problem 10.22.)

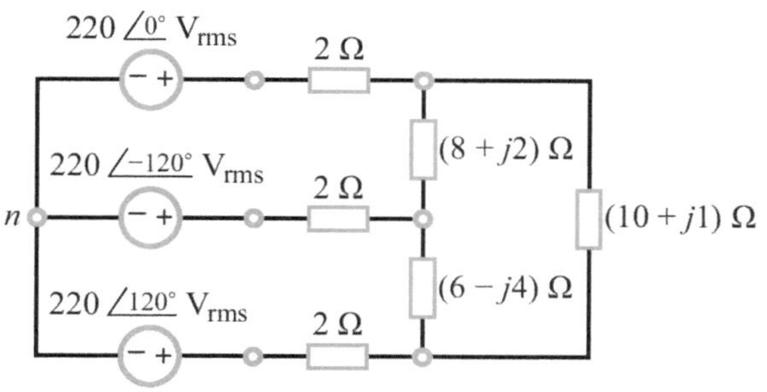

Figure P10.23: (Problems 10.23 and 10.26.)

*10.28 In the three-phase network of Fig. P10.28, $\mathbf{Z}_{TL} = 4\ \Omega$, $\mathbf{Z}_{\Delta_1} = (10 - j4)\ \Omega$, and $\mathbf{Z}_{Y_2} = (5 + j2)\ \Omega$. Determine the line currents.

10.29 A 208 V (rms) balanced three-phase source supports two loads connected in parallel. Each load is itself a balanced three-phase load. Determine the line current, given that load 1 is 12 kVA at $pf_1 = 0.7$ leading and load 2 is 18 kVA at $pf_2 = 0.9$ lagging.

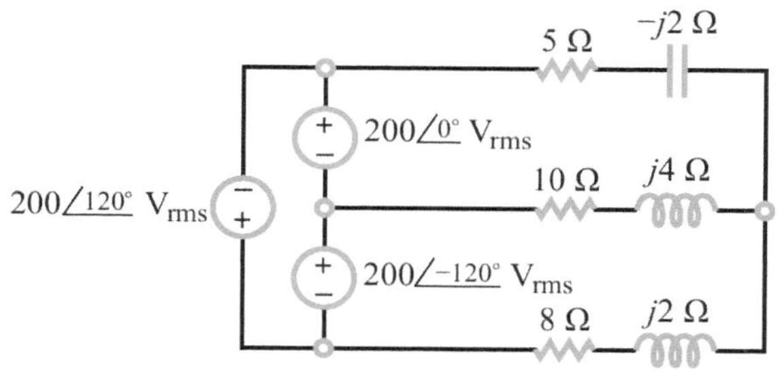

Figure P10.27: (Problem 10.27.)

*10.30 A 416 V balanced three-phase source supports four loads connected in parallel. Each load is itself a balanced three-phase load. Determine the line current and the power factor at the source, given that load 1 is 12 kVA at $pf_1 = 0.7$ leading, load 2 is 18 kVA at $pf_2 = 0.9$ lagging, load 3 is 6 kW at $pf_3 = 1$, and load 4 is 24 kVA at $pf_4 = 0.7$ lagging.

10.31 A 240 V (rms), 60 Hz Y-source is connected to a balanced three-phase Y-load by four wires, one of which is the neutral wire. If the load is 400 kVA at $pf_{\text{old}} = 0.6$ lagging, what size capacitors should be added to change the power factor to $pf_{\text{new}} = 0.95$ lagging?

*10.32 A 416 V (rms), 60 Hz Y-source is connected to a balanced three-phase Y-load by four wires, one of which is the neutral wire. If the load is 800 kW at $pf_{\text{old}} = 0.75$ lagging, what size capacitors should be added to change the power factor to $pf_{\text{new}} = 0.95$ lagging?

10.33 A balanced three-phase source with a line voltage of 208 V (rms) is connected to a three-phase motor designed as a balanced Y-load. The powers measured using the two-wattmeter method are $P_1 = 800$ W and $P_2 = 300$ W. Determine the impedances of the motor and its power factor.

*10.34 Determine the power readings of the two wattmeters shown in the circuit of Fig. P10.34 given that $\mathbf{Z}_\Delta = (10 + j6)\ \Omega$ and the amplitudes of the voltage sources are rms.

10.35 Determine the power readings of the two wattmeters shown in the circuit of Fig. P10.35 given that $\mathbf{Z}_Y = (15 - j5)\ \Omega$.

*10.36 Determine the power readings of the two wattmeters shown in the circuit of Fig. P10.36 given that $\mathbf{Z}_\Delta = (10 + j6)\ \Omega$ and the amplitudes of the voltage sources are rms.

10.37 Repeat Problem 10.36 after replacing the balanced Δ-load with a balanced Y-load with $\mathbf{Z}_Y = (10 + j6)\ \Omega$.

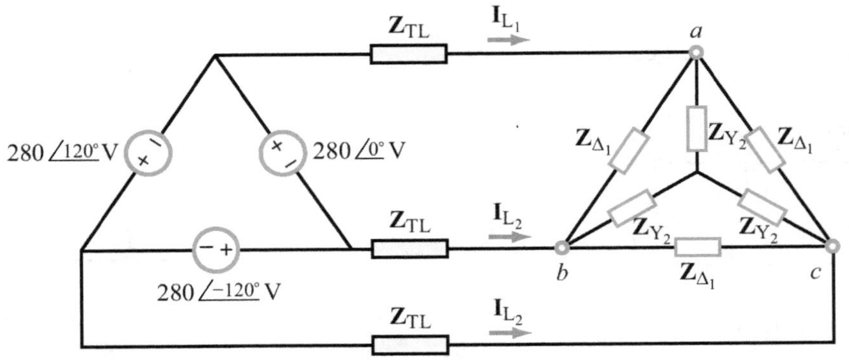

Figure P10.28: Circuit for Problem 10.28.

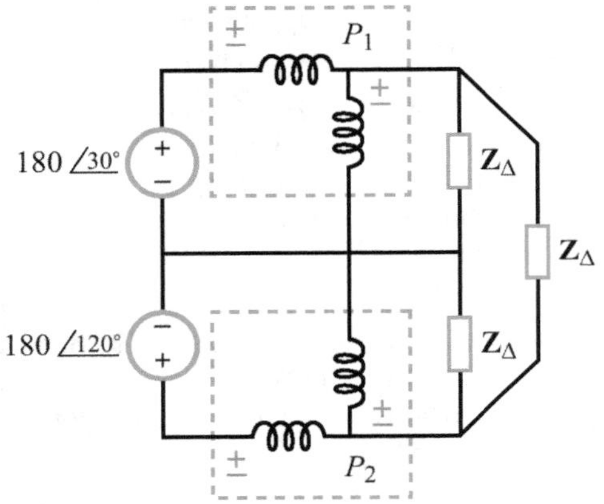

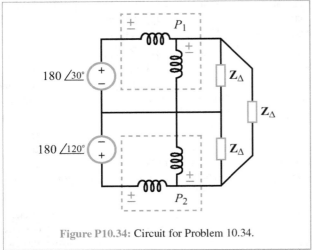

Figure P10.34: Circuit for Problem 10.34.

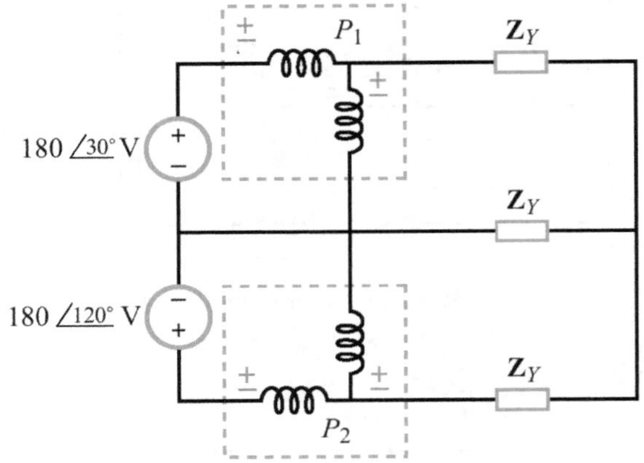

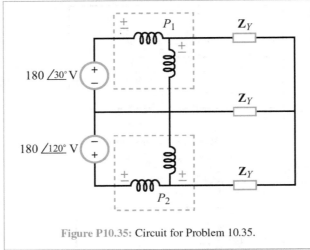

Figure P10.35: Circuit for Problem 10.35.

Potpourri Questions

10.38 Name three types of energy harvesting and describe the energy conversion process in each case.

10.39 How are the Seebeck and Peltier effects related?

10.40 Why do power stations use large-size capacitors and inductors?

10.41 What is a *stator coil*?

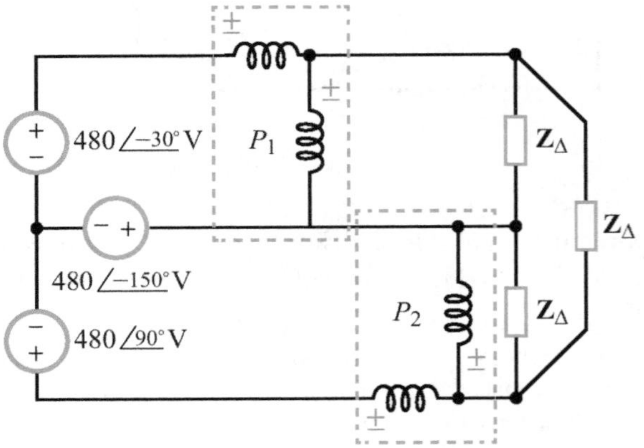

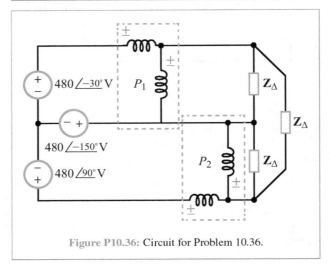

Figure P10.36: Circuit for Problem 10.36.

CHAPTER 11

Magnetically Coupled Circuits

Contents

Objectives

Learn to:

- Incorporate mutual coupling in magnetically coupled circuits.
- Analyze circuits containing magnetically coupled coils.
- Relate input to output voltages, currents, and impedances for magnetically coupled transformers, including ideal transformers and three-phase transformers.

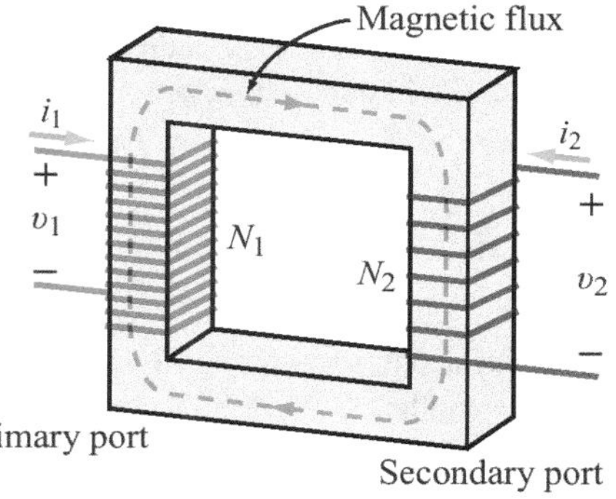

When two physically unconnected *inductors* are in close proximity to one another, current flow through one of them induces a *magnetically coupled voltage* across the other one. Magnetic coupling may be intentional or not. Highly coupled *voltage transformers* used in power distribution networks are an example of intentional coupling. If the coupling between two coils in a circuit is unintentional but significant, its effects should be incorporated into the analysis of the circuit.

Overview

Voltage transformers are used in many electrical systems, including power supply circuits (Section 7-9) and power distribution networks (Chapter 10). Whereas resistors, capacitors, and inductors are one-port, two-terminal devices, a transformer is a two-port (four-terminal) device with a *primary port* and a *secondary port*. Coupling of energy between the two ports is realized through a shared magnetic field, without the need for direct contact between them. Transformers are part of a family of devices and circuits called *magnetically coupled circuits*, whose operation relies on magnetic coupling rather than current conduction.

We begin this chapter by examining the voltage and current relationships between the primary and secondary ports of a coupled two-coil system. We did so previously in Section 7-9.1, but then our treatment was limited to the special case of the ideal transformer with perfect coupling. In this more comprehensive examination, we introduce the concepts of mutual inductance, equivalent circuits, and impedance transformations, and we learn how three-phase transformers are configured to step-up or step-down voltage levels in three-phase power circuits.

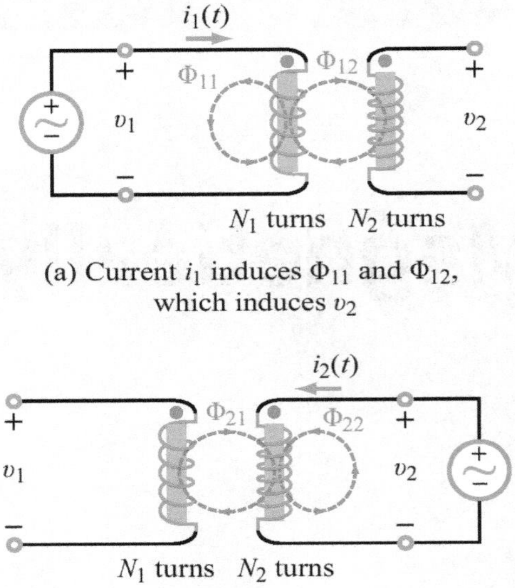

(a) Current i_1 induces Φ_{11} and Φ_{12}, which induces v_2

(b) Current i_2 induces Φ_{22} and Φ_{21}, which induces v_1

Figure 11-1: Magnetically coupled coils.

11-1 Magnetic Coupling

Magnetic coupling can occur between any two inductors in close proximity of one another. Current flow through the coils of one of the inductors induces a *mutual inductance voltage* across the other inductor, and vice versa. The induction process is described in terms of a *mutual inductance*, measured in henrys (H), that depends on the degree of magnetic coupling between the two inductors, which in turn depends on their physical shapes, orientations relative to one another, spacing between them, and the *magnetic permeability* μ of the medium between them. Mutual inductance may be intentional or not. It is key to the operation of highly coupled transformers used for stepping up and stepping down voltage levels. On the other hand, mutual inductance between two inductors, transmission lines, or wires in a certain circuit may be totally unintentional, as well as unavoidable. In that case, we should learn how to account for the voltages induced by the mutual inductance and how to incorporate them in the analysis of the circuit.

The two magnetically coupled coils in Fig. 11-1(a) have N_1 turns on *primary port 1* and N_2 turns on *primary port 2*. Port 1 is connected to a source that causes current $i_1(t)$ to flow through coil 1, which generates *magnetic flux* Φ_{11} linking coil 1 alone and flux Φ_{12} linking both coils.

▶ Magnetic fluxes form closed loops because the magnetic field lines that emerge from one end of the primary coil will also flow back in at the other end of the coil. The direction of the magnetic field is dictated by the direction of the current in the coil: if the four fingers of the right hand point in the direction of the current, the thumb will point in the direction of the magnetic field. Alternatively, if the four fingers wrap in the direction of the magnetic flux, the thumb will point in the direction of the current. This is known as the *right hand rule*. ◀

The *total magnetic flux* through coil 1 is

$$\Phi_1 = \Phi_{11} + \Phi_{12}. \tag{11.1}$$

Magnetic flux linkage Λ_1 is defined as the total flux linking all N_1 turns of coil 1,

$$\Lambda_1 = N_1 \Phi_1. \tag{11.2a}$$

For coil 2, the flux linking coil 1 to coil 2 is Φ_{12}, and the corresponding Λ_2 is

$$\Lambda_2 = N_2 \Phi_{12}. \tag{11.2b}$$

Self inductance L_1 of coil 1 is defined as the ratio of the magnetic flux linkage Λ_1 to the current i_1 responsible for inducing Λ_1,

$$L_1 = \frac{\Lambda_1}{i_1}, \tag{11.3}$$

and the voltage induced across inductor L_1 is

$$\upsilon_1 = \frac{d\Lambda_1}{dt} = L_1 \frac{di_1}{dt}. \tag{11.4a}$$

By analogy, Λ_2 in coil 2 induces voltage υ_2, with

$$\upsilon_2 = \frac{d\Lambda_2}{dt} = N_2 \frac{d\Phi_{12}}{dt}. \tag{11.4b}$$

Both υ_1 and υ_2 are induced by di_1/dt. In the case of υ_1, the link is self inductance L_1, as given by Eq. (11.4a). To establish an analogous relationship between di_1/dt and υ_2, we rewrite Eq. (11.4b) as

$$\upsilon_2 = N_2 \frac{d\Phi_{12}}{di_1} \times \frac{di_1}{dt}. \tag{11.4c}$$

Next, we define the ***mutual inductance*** M_{21}, as

$$M_{21} = N_2 \frac{d\Phi_{12}}{di_1}, \tag{11.5}$$

and the expression for υ_2 becomes

$$\upsilon_2 = \pm M_{21} \frac{di_1}{dt}. \tag{11.6}$$

► Subscripts 21 refer to the fact that M_{21} is the inductance of coil 2 due to the magnetic field induced by current i_1. ◄

The expression in Eq. (11.6) includes a (±) on the right-hand side. This is because mutual inductance M_{21} is a positive quantity measured in henrys (H), whereas υ_2 may be positive or negative, depending on the direction of the winding in coil 2 relative to the direction of the winding in coil 1. For the specific winding directions shown in Fig. 11-1(a), the appropriate sign is (+).

If we were to reverse the roles of coils 1 and 2, by connecting the source to coil 2 instead of to coil 1, thereby causing current i_2 to flow through coil 2, as depicted in Fig. 11-1(b), we would end up with the following expressions for υ_1 and υ_2:

$$\upsilon_1 = \pm M_{12} \frac{di_2}{dt} \tag{11.7a}$$

and

$$\upsilon_2 = L_2 \frac{di_2}{dt}. \tag{11.7b}$$

► Because the coupled coils constitute a linear system, energy considerations (Section 11-3) require that $M_{12} = M_{21} = M$, where M is now called the mutual inductance between the two coils. ◄

The ambiguity between the (+) and (−) signs in Eqs. (11.6) and (11.7a) is resolved through the use of a standard ***dot convention*** based on the directions of the two windings. For a specific direction of i_1 (left-hand side Fig. 11-2), the polarity of υ_2 depends on whether the dots are on the same or opposite terminals of the windings and whether i_1 enters coil 1 at its dotted or undotted terminal.

► In a two-coil magnetically coupled system, if current enters the first one at its dotted terminal, the polarity of the mutual-inductance voltage induced across the second coil is positive at its dotted terminal. The polarity of the induced voltage is reversed if the current in the first coil enters at the undotted terminal. Moreover, reciprocity applies: current in the second coil induces a mutual-inductance voltage across the first one in accordance with the same dot convention. ◄

This dot convention covers all combinations of current directions and dot locations outlined in Fig. 11-2.

Finally, if we generalize to the configuration shown in Fig. 11-3(a) in which currents flow through both coils simultaneously, voltage υ_1 will contain two components, one due to self-inductance of coil 1 and another due to the mutual inductance between the two coils. That is, υ_1 will be the sum of Eqs. (11.4a) and (11.7a), and similarly, υ_2 becomes the sum of Eqs. (11.6) and (11.7b). Specifically:

For dots on same ends and currents entering coils at dotted ends (**Fig. 11-3(a)**):

$$\upsilon_1 = L_1 \frac{di_1}{dt} + M \frac{di_2}{dt} \tag{11.8a}$$

and

$$\upsilon_2 = L_2 \frac{di_2}{dt} + M \frac{di_1}{dt}. \tag{11.8b}$$

For dots on opposite ends but current entering coils at same ends (**Fig. 11-3(b)**):

$$\upsilon_1 = L_1 \frac{di_1}{dt} - M \frac{di_2}{dt} \tag{11.9a}$$

and

$$\upsilon_2 = L_2 \frac{di_2}{dt} - M \frac{di_1}{dt}. \tag{11.9b}$$

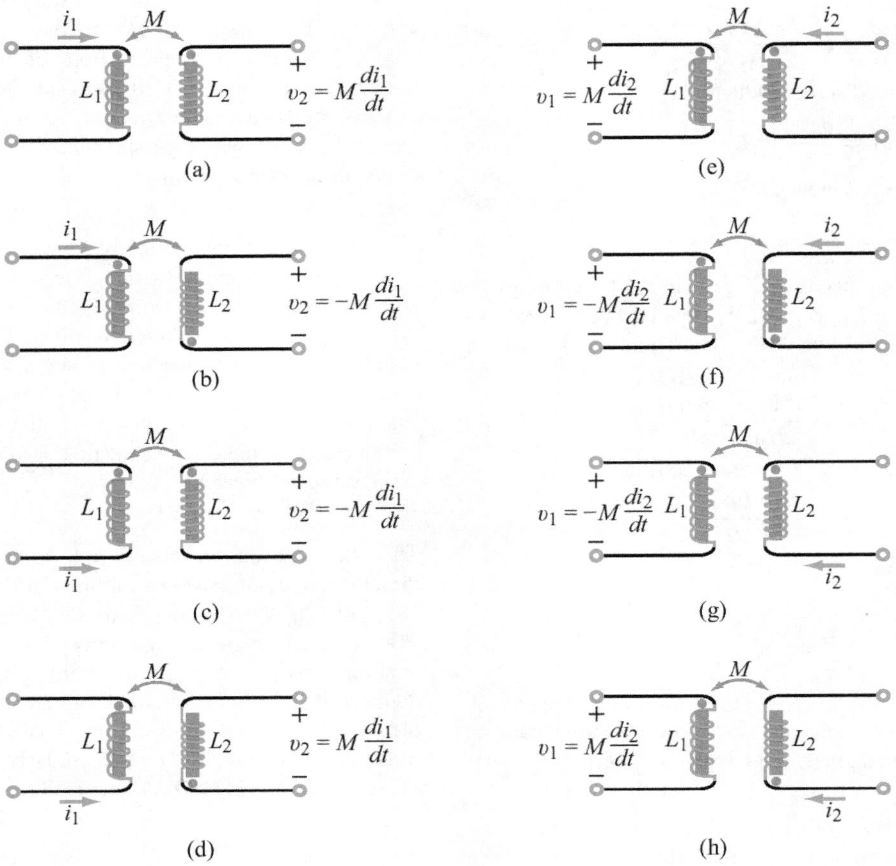

Figure 11-2: Dot convention for the mutual-inductance voltage induced in coil 2 by current i_1 in coil 1, and vice versa.

Sign Convention

Primary Side: v_1

(1) The ***self-induced*** component $L_1\, di_1/dt$ of v_1 is assigned a (+) sign if:

(a) the (+) polarity of v_1 is defined at the dotted terminal, and i_1 enters coil 1 at the dotted terminal, or

(b) the (+) polarity of v_1 is defined at the undotted terminal, and i_1 enters coil 1 at the undotted terminal.

Otherwise, $L_1\, di_1/dt$ is assigned a (−) sign.

(2) The ***mutually induced*** component $M\, di_2/dt$ of v_1 is assigned a (+) sign if:

(a) the (+) polarity of v_1 is defined at the dotted terminal, and i_2 enters coil 2 at the dotted terminal, or

(b) the (+) polarity of v_1 is defined at the undotted terminal, and i_2 enters coil 2 at the undotted terminal.

Otherwise, $M\, di_2/dt$ is assigned a (−) sign.

Secondary Side: v_2

The same sign convention applies from the perspective of the secondary side: simply replace "primary" with "secondary," subscript 1 with 2, and vice versa.

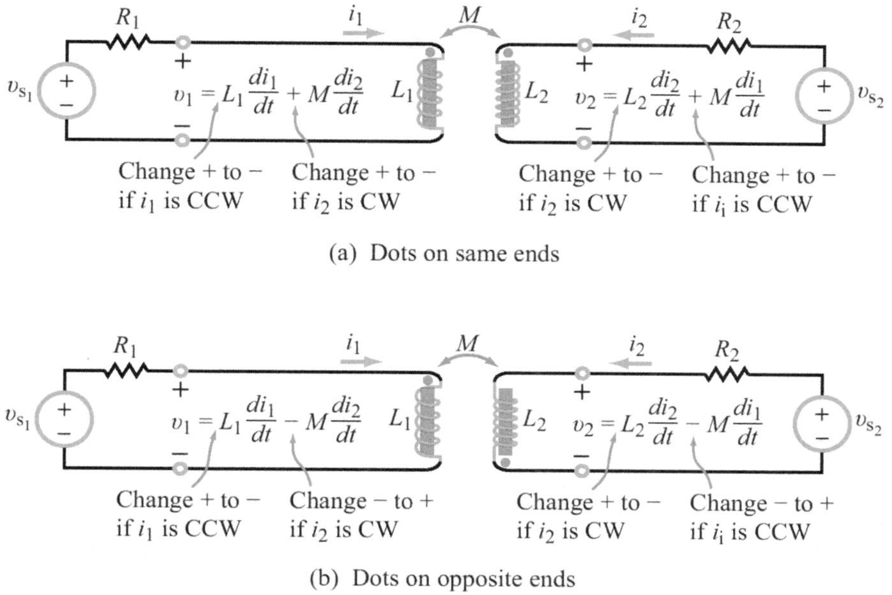

Figure 11-3: Polarities of voltage components for clockwise (CW) and counterclockwise (CCW) current directions.

Example 11-1: 1 kHz Circuit

Determine load current $i_L(t)$ in the circuit of Fig. 11-4(a), given that $v_s(t) = 10\cos(2\pi \times 10^3 t)$ (V), $R_1 = 5\ \Omega$, $C_1 = C_2 = 10\ \mu$F, $L_1 = 1$ mH, $L_2 = 3$ mH, $M = 0.5$ mH, and $R_L = 20\ \Omega$.

Solution: We start by transforming the ac circuit from the time domain to the phasor domain (Fig. 11-4(b)). Thus, $v_s(t)$ becomes phasor voltage $\mathbf{V}_s$, C gets transformed into an impedance $\mathbf{Z}_C = -j/\omega C$, and L gets transformed into an impedance $\mathbf{Z}_L = j\omega L$. The angular frequency is

$$\omega = 2\pi f = 2\pi \times 10^3 \text{ rad/s}.$$

Denoting $\mathbf{I}_1$ and $\mathbf{I}_2$ as the mesh currents in the two loops, both defined with clockwise directions, the mesh-current equations are

$$-\mathbf{V}_s + \left(R_1 - \frac{j}{\omega C} + j\omega L_1\right)\mathbf{I}_1 - j\omega M\mathbf{I}_2 = 0 \qquad (11.10a)$$

and

$$-j\omega M\mathbf{I}_1 + \left(j\omega L_2 - \frac{j}{\omega C} + R_L\right)\mathbf{I}_2 = 0, \qquad (11.10b)$$

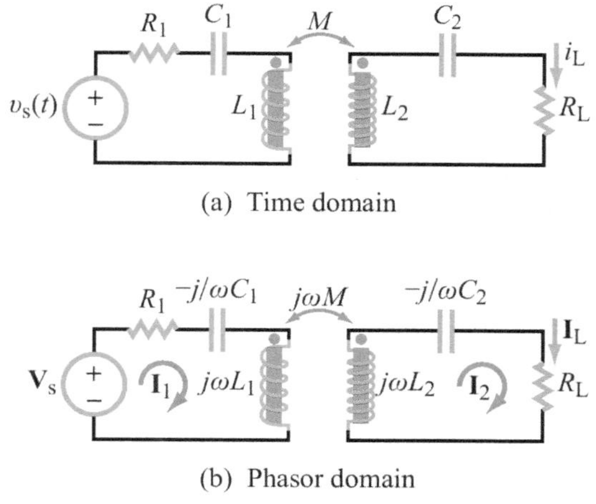

Figure 11-4: Circuit of Example 11-1.

where $C = C_1 = C_2$. Note that the polarity of the last term in Eq. (11.10a) is negative because, in accordance with the convention shown in Fig. 11-2(g), the winding dots are on the same end in Fig. 11-4(b) but $\mathbf{I}_2$ enters the undotted terminal of coil 2. Simultaneous solution of the two equations for $\mathbf{I}_L = \mathbf{I}_2$ gives

$$\mathbf{I}_L = \frac{j\omega M \mathbf{V}_s}{\left(R_1 + j\omega L_1 - \frac{j}{\omega C}\right)\left(R_L + j\omega L_2 - \frac{j}{\omega C}\right) + \omega^2 M^2}. \tag{11.11}$$

Substitution of the specified values leads to

$$\mathbf{I}_L = 139.5 e^{j142.2°} \text{ mA},$$

and its time-domain equivalent is

$$i_L(t) = \mathfrak{Re}[\mathbf{I}_L e^{j\omega t}] = 139.5 \cos(2\pi \times 10^3 t + 142.2°) \text{ mA}.$$

Example 11-2: Coupled Inductors

The circuit in Fig. 11-5 has the following element values: $\mathbf{V}_s = 30e^{j60°}$ V, $L_1 = 10$ mH, $L_2 = 30$ mH, $R_1 = 5\ \Omega$, $R_L = 10\ \Omega$, and the ac source operates at 60 Hz. The circuit layout is such that inductors L_1 and L_2 experience a relatively small mutual inductance M. Determine the average power delivered to the load R_L for (a) $M = 4$ mH, (b) $M = 1$ mH, and (c) $M = 0$.

Solution: Before we apply mesh analysis, let us determine $\mathbf{V}_1$ and $\mathbf{V}_2$ across the two inductors. Voltage $\mathbf{V}_1$ consists of two terms, $j\omega L_1 \mathbf{I}_1$ due to current $\mathbf{I}_1$ entering at the (+) terminal of $\mathbf{V}_1$, and $j\omega M(\mathbf{I}_2 - \mathbf{I}_1)$ due to current $(\mathbf{I}_2 - \mathbf{I}_1)$ through L_2. The polarity of the second term is governed by the dot convention: if current enters a coil at its dotted terminal, the polarity of the mutual-inductance voltage induced across the second coil is positive at its dotted terminal. In the present case, $(\mathbf{I}_2 - \mathbf{I}_1)$ enters L_2 at its dotted terminal, so the voltage it induces across L_1 is positive at the dotted terminal of L_1. Hence,

$$\mathbf{V}_1 = j\omega L_1 \mathbf{I}_1 + j\omega M(\mathbf{I}_2 - \mathbf{I}_1). \tag{11.12a}$$

Application of the same rule to L_2 gives

$$\mathbf{V}_2 = j\omega L_2(\mathbf{I}_2 - \mathbf{I}_1) + j\omega M \mathbf{I}_1. \tag{11.12b}$$

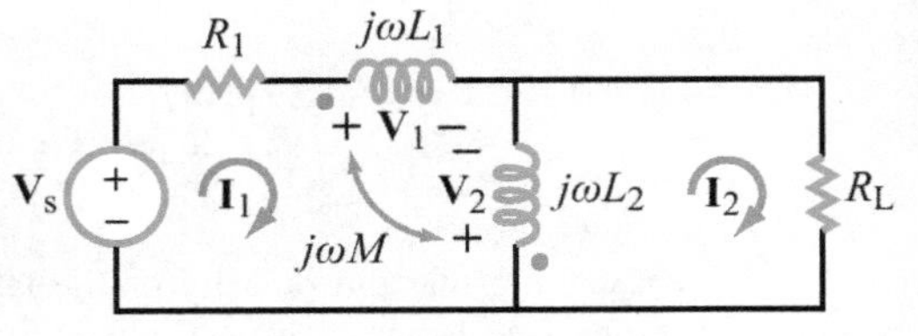

Figure 11-5: Circuit of Example 11-2.

The first mesh equation is

$$-\mathbf{V}_s + R_1 \mathbf{I}_1 + \mathbf{V}_1 - \mathbf{V}_2 = 0,$$

or equivalently,

$$(R_1 + j\omega L_1 + j\omega L_2 - j2\omega M)\mathbf{I}_1 - (j\omega L_2 - j\omega M)\mathbf{I}_2 = \mathbf{V}_s. \tag{11.13a}$$

Similarly, for the second mesh,

$$\mathbf{V}_2 + R_L \mathbf{I}_2 = 0,$$

or equivalently,

$$-(j\omega L_2 - j\omega M)\mathbf{I}_1 + (R_L + j\omega L_2)\mathbf{I}_2 = 0. \tag{11.13b}$$

(a) $M = 4$ mH

Upon replacing R_1, R_L, L_1, L_2 with their specified values, setting $M = 4$ mH, and multiplying inductances by $\omega = 2\pi f = 2\pi \times 60 = 377$ rad/s, matrix solution of the two equations gives

$$\mathbf{I}_2 = 1.657 e^{j63.1°} \text{ A}, \tag{11.14}$$

and according to Eq. (8.3), the corresponding average power absorbed by R_L is

$$P_L = \frac{1}{2}\, |\mathbf{I}_2|^2 R_L = \frac{1}{2}\, (1.657)^2 \times 10 = 13.73 \text{ W}. \tag{11.15}$$

(b) $M = 1$ mH

Repetition of the process with $M = 1$ mH gives

$$\mathbf{I}_2 = 1.64 e^{j62.15°} \text{ A} \tag{11.16}$$

and

$$P_L = 13.45 \text{ W}. \tag{11.17}$$

(c) $M = 0$

In the absence of mutual coupling between the two coils,

$$\mathbf{I}_2 = 1.635 e^{j62.03°} \text{ A} \tag{11.18}$$

and

$$P_L = 13.372 \text{ W}. \tag{11.19}$$

Error

Ignoring M altogether would incur an error in P_L of

$$\% \text{ error} = \frac{P_L(@M = 4 \text{ mH}) - P_L(@M = 0)}{P_L(@M = 4 \text{ mH})} \times 100$$
$$= 2.61\%, \qquad \text{when true } M \text{ is 4 mH,}$$

and

$$\% \text{ error} = 0.61\%, \qquad \text{when true } M \text{ is 1 mH.}$$

Example 11-3: Equivalent Inductance

For the circuit in Fig. 11-6(a), obtain an expression for the equivalent inductance, L_{eq}, defined such that it would exhibit the same i-υ characteristic at nodes (a, b) as the actual circuit.

Solution: Equivalency means that circuits in Figs. 11-6(b) and (c) will have the same current $\mathbf{I}$ flowing through both loops

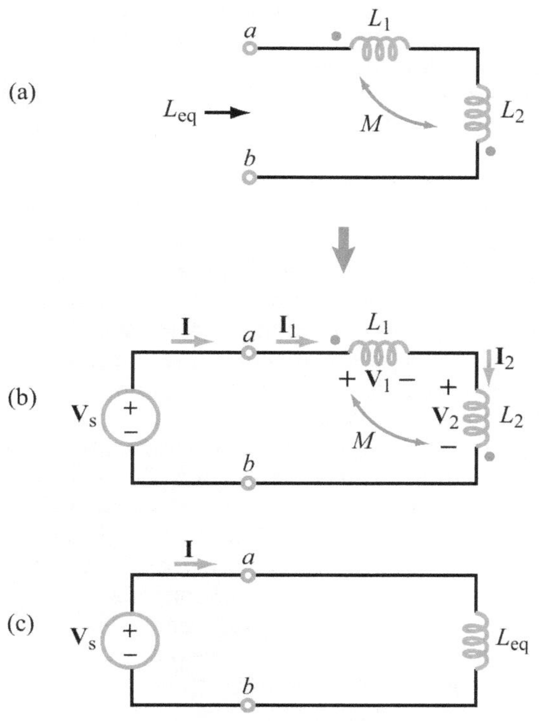

Figure 11-6: Finding L_{eq} of two series-coupled inductors (Example 11-3).

when connected to the same voltage source $\mathbf{V}_s$. For the two-inductor circuit in Fig. 11-6(b),

$$\mathbf{I}_1 = \mathbf{I}_2 = \mathbf{I},$$

and while $\mathbf{I}_1$ enters L_1 at its dotted terminal, $\mathbf{I}_2$ enters L_2 at its undotted terminal. While guided by Fig. 11-3(b), application of the dot convention to the loop in Fig. 11-6(b) gives

$$\mathbf{V}_1 = j\omega L_1\mathbf{I}_1 - j\omega M\mathbf{I}_2 = j\omega(L_1 - M)\mathbf{I}$$

and

$$\mathbf{V}_2 = j\omega L_2\mathbf{I}_2 - j\omega M\mathbf{I}_1. = j\omega(L_2 - M)\mathbf{I}.$$

At terminals (a, b),

$$\mathbf{V}_s = \mathbf{V}_1 + \mathbf{V}_2 = j\omega(L_1 + L_2 - 2M)\mathbf{I}.$$

For the circuit in circuit Fig. 11-6(c),

$$\mathbf{V}_s = j\omega L_{eq}\mathbf{I}.$$

Equivalency leads to

$$L_{eq} = L_1 + L_2 - 2M.$$

Concept Question 11-1: What determines the polarity of the mutual inductance voltage? Summarize the rules of the dot convention. (See CAD)

Concept Question 11-2: What factors determine how strong or weak the magnetic coupling is between two coils? (See CAD)

Exercise 11-1: Repeat Example 11-1 after moving the dot location on the side of L_2 from the top end of the coil to the bottom.

Answer: $i_L(t) = 139.5 \cos(2\pi \times 10^3 t - 37.8°)$ mA. (See CAD)

Exercise 11-2: Repeat Example 11-3 for the two in-series inductors in Fig. 11-6(a), but with the dot location on L_2 being on the top end.

Answer: $L_{eq} = L_1 + L_2 + 2M$. (See CAD)

Technology Brief 27
Magnetic Resonance Imaging (MRI)

Magnetic resonance imaging (MRI), also called *nuclear magnetic resonance* (NMR), is a powerful medical imaging tool that provides extremely detailed 2-D and 3-D images of the body, an example of which is shown in Fig. TF27-1. MRI is particularly useful for imaging soft tissues (organs, ligaments, spinal column, arteries and veins, etc.). Unlike X-ray, which uses very high frequency ionizing radiation, MRI uses lower frequency magnetic and radio frequency fields, which are non-ionizing and do not damage cells. Since MRI's early demonstration in the 1970s, its applications have burgeoned, and research is continually opening up new and improved MRI techniques.

MRI utilizes the fact that the bulk of the human body contains water and every water molecule has a permanent *magnetic dipole moment*, which means that a water molecule behaves like a small magnet. The hydrogen atoms in the (H_2O) water molecule have a natural spin associated with them, and because they are weakly charged, this spin (charges moving in a circle) creates a *magnetic field* as shown by the S-to-N arrow in Fig. TF27-2. Thus, the water molecule acts like a weak bar magnet with North and South poles. Normally, these spins are randomly aligned in the body, but if a strong external magnetic field is applied, they all line up with the applied magnetic field. Almost exactly half line up in the N-S direction, and almost exactly half in the opposite S-N direction. This *dc magnetic* field will just hold them in place for the rest of the MRI scan time.

But you can never have *exactly* half of the spins in each of the two directions, so a few extra spins (about 9 out of 2 million for a typical 1.5 tesla magnet) always show up in neither the N-S or S-N configuration. These extra outlier molecules are the ones the MRI scanner actually uses to create the image.

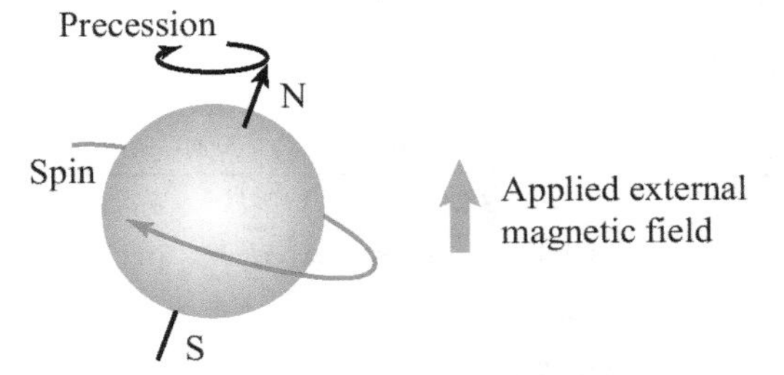

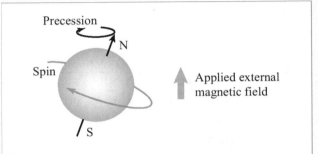

Figure TF27-2: The spin of the charged hydrogen atoms in H_2O produces a magnetic field.

A second magnetic field is then applied, but unlike the strong dc magnetic field imposed to hold the rest of the molecules in place, this magnetic field is a *radiofrequency* (RF) pulsed signal, at a particular frequency for which the outlier molecules are known to resonate strongly (this is the resonance part of MRI). This natural resonant frequency is called the *Larmor frequency*, and it depends on the chemical makeup, density, and structure of the tissue, thus allowing MRI to distinguish different tissues, identify and detect chemical composition, and even determine the status (such as inflammation) of various tissues. When the RF pulse turns on, the outlier molecules align with that magnetic field. When the pulse turns off, they relax back to their original state. As they relax, their spin precesses (becomes tipped like a toy top slowing down) and produces yet another magnetic field, which returns to and is picked up by the same coil that produced the original RF signal.

Now let's look at the MRI machine and the hardware (Fig. TF27-3) that makes this all happen. The large applied magnetic field is produced by a large *superconducting electromagnet*. A typical medical MRI scanner is 1.5 teslas (1 tesla = 10k gauss). By comparison, a strong refrigerator magnet is 100 gauss, and the Earth's magnetic field at its surface is around 0.5 gauss. The superconducting magnet is cooled with liquid nitrogen down to the point where its resistance is virtually zero (see Technology Brief 3 on superconductivity), so that the current and hence magnetic field can be maximized. These magnets weigh several tons and cost hundreds of thousands of dollars a year in electricity and liquid nitrogen to keep them running. They take weeks to cool down enough to reach superconductivity, and days to ramp up the current to produce their large magnetic field.

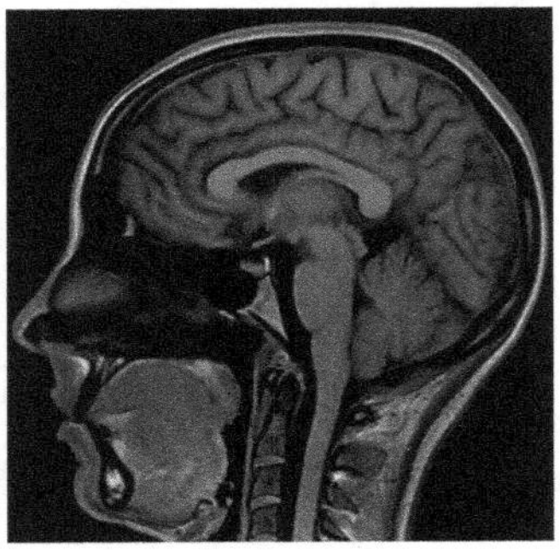

Figure TF27-1: MRI scan of the head.

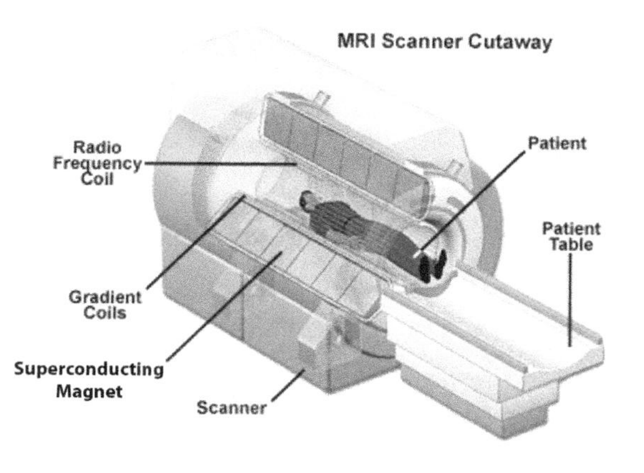

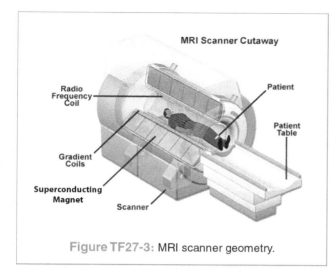

Figure TF27-3: MRI scanner geometry.

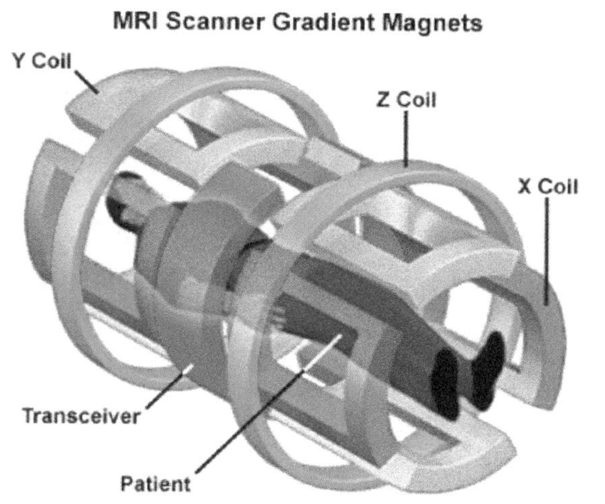

Figure TF27-4: MRI scanner gradient coils are used to adjust which slice the scanner is imaging.

The patient is slid into the bore (hole) in the center of the magnet, where the field is strong, and also quite uniform. Because the body is "nonmagnetic" ($\mu_r = 1$), this strong magnetic field does not move or hurt the person, although it does polarize (line up) the spins in his/her hydrogen atoms. If the person has any metal inside of him/her (such as implantable medical devices, artificial joints, or bone repair surgeries), this can preclude the use of MRI for that patient. It is also important to keep all metal (oxygen tanks, wheelchairs, pens, clipboards, etc.) away from the magnet, or it can be pulled irretrievably and dangerously into the bore. The magnetic field is so strong that these materials cannot be removed without slowly (days or weeks) reducing the current in the electromagnet to turn it off.

The MRI scans the body in slices, like a loaf of bread. The slice being scanned is adjusted by creating a very small gradient in the RF field using yet another set of coils, the *gradient coils* shown in Fig. TF27-4. The current in these coils is ramped up and turned on and off very quickly to move through hundreds of scan slices quickly. When the current in a coil decreases, the magnetic field decreases too. The energy stored in this magnetic field has to go somewhere—it is returned to the circuit, creating a voltage spike at the source driving the coil. The voltage spike must be controlled by managing the current decay. In addition, this decreasing magnetic field creates another current that tries to oppose the change (you may have learned about *Lenz's law* in physics), making it impossible to instantly turn off a large magnetic field. Another interesting (and often very noisy!) effect is seen in these gradient coils. The coils themselves vibrate from their strong magnetic field, which creates a constant hum and often loud thumps and even crashing noises, so patients receiving MRI scans generally wear earplugs to block the noise.

Now let's look at the RF coils that transmit and receive the RF pulses. RF coils come in many different designs, shapes, and sizes as shown in Fig. TF27-5. The coil must be large enough to surround the region being imaged (the head, torso, knee, etc.). The most common coils for whole-body or whole-head imaging are the birdcage coil shown in part (a) of the figure and the planar or surface coils shown in part (b). The *signal* of interest is determined by how much of the RF energy gets from the coil to the feature of interest, and back to the coil. Coils such as the birdcage coil are designed to have a field as constant as possible over the head, for instance, with no particular focus on the optic nerve. So, the signal (S) from the head is relatively large but the signal from the optic nerve is relatively small. The *electrical noise* (N) picked up by the receiver is generated by the entire field of view of the coil. So for imaging the head, the *signal-to-noise ratio* (SNR), given by S/N, is relatively large, but for the optic nerve, the signal is lower while the noise is the same, so SNR is smaller and not good enough to provide an accurate measurement of the signal. Much of the research on MRI coils today is therefore focused on development of coils such as the one in part (c) of Fig. TF27-5, which focus the RF energy on a specific feature of interest (in this case, the optic nerve) thus increasing the signal (S) in the SNR and often decreasing the noise (N) as well. The effect of SNR on image quality is seen in Fig. TF27-6.

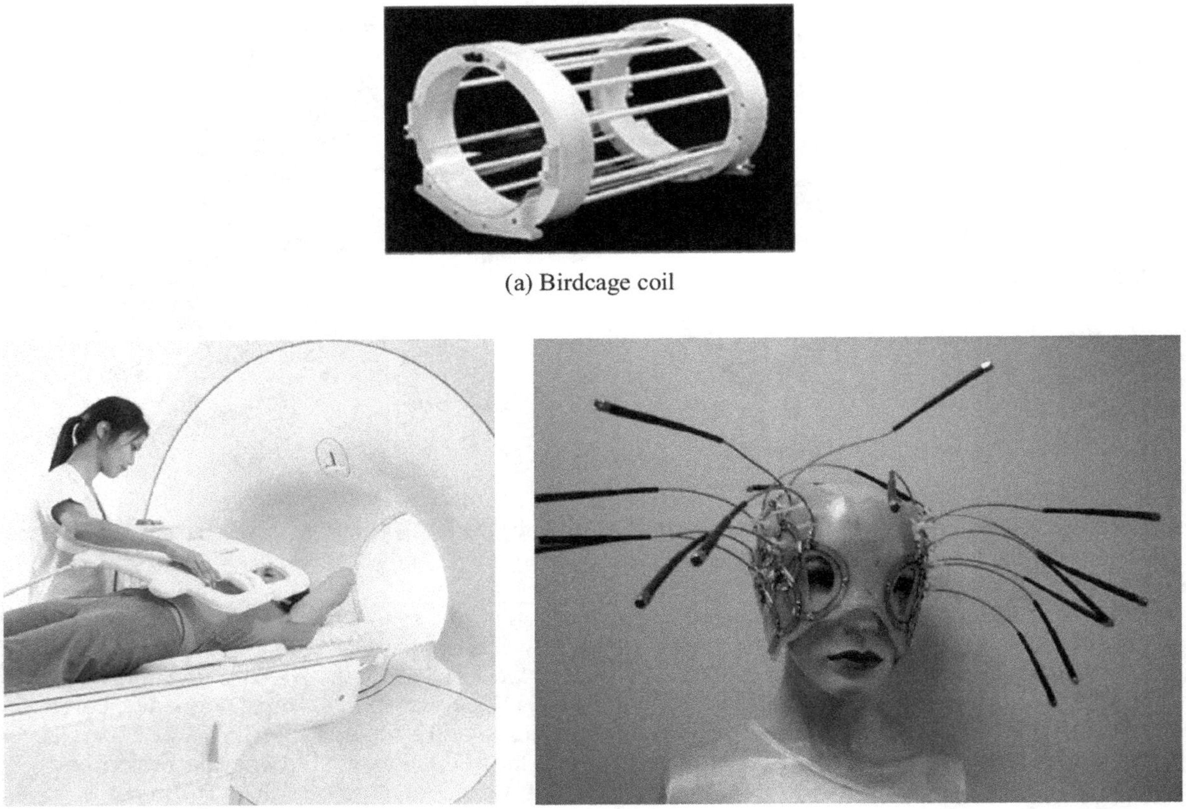

(a) Birdcage coil

(b) Surface coil

(c)

Figure TF27-5: Coils used in MRI: (a) birdcage coil; (b) surface coil (one on the front and another on the back of the body). (Credit: Emilee Minalga.)

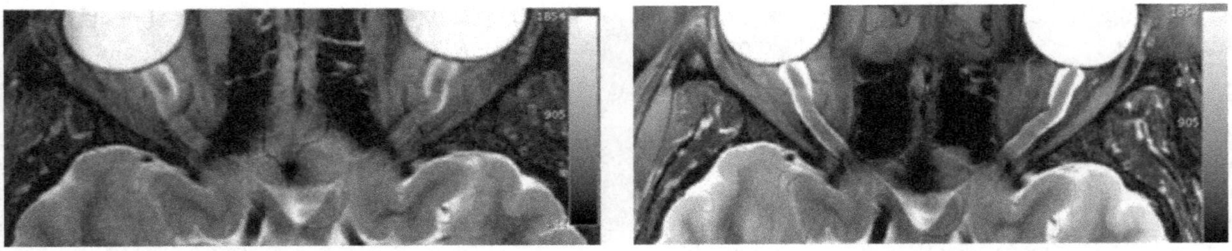

(a) Low SNR

(b) High SNR

Figure TF27-6: Imaging of the optic nerve (seen just below the eyeballs) with (a) low SNR and (b) high SNR. (Photo courtesy Robb Merrill.)

11-2 Transformers

11-2.1 Coupling Coefficient

To couple magnetic flux between two coils, the coils may be wound around a common core (Fig. 11-7(a)), on two separate arms of a rectangular core (Fig. 11-7(b)), or in any other arrangement conducive to having a significant fraction of the magnetic flux generated by each coil shared with the other. The *coupling coefficient* k defines the degree of magnetic coupling between the coils, with $0 \leq k \leq 1$. For a *loosely coupled* pair of coils, $k < 0.5$; for *tightly coupled* coils, $k > 0.5$; and for *perfectly coupled* coils, $k = 1$. The magnitude of k depends on the physical geometry of the two-coil configuration and the magnetic permeability μ of the core material.

> ► A transformer is said to be *linear* if μ of its core material is a constant, independent of the magnitude of the currents flowing through the coils (and hence, the strength of the induced magnetic field). ◄

Most core materials, including air, wood, and ceramics, are nonferromagnetic, and their μ is approximately equal to μ_0, the *permeability of free space*. When nonferromagnetic materials are used for the common core around which the coils are wound, the magnitude of k depends entirely on how tightly coupled the two windings are. Such transformers are indeed linear, but the magnitude of k is seldom greater than 0.4. Increasing k requires the use of ferromagnetic cores, but the transformer becomes heavier in weight and its behavior becomes nonlinear. The degree of nonlinearity depends on the choice of materials. *With certain types of purified iron, transformers can be designed to exhibit coupling coefficients approaching unity.*

As was noted earlier in connection with Fig. 11-1(a), current i_1, through coil 1 generates magnetic fluxes Φ_{11} through coil 1 and Φ_{12} through both coils 1 and 2. The coupling coefficient is given by

$$k = \frac{\Phi_{12}}{\Phi_{11} + \Phi_{12}} = \frac{\Phi_{12}}{\Phi_1}. \qquad (11.20a)$$

where $\Phi_1 = \Phi_{11} + \Phi_{12}$. The perfectly coupled case corresponds to when the flux coupled to coil 2, namely Φ_{12}, is equal to the self-coupled flux Φ_1. Similarly, from the standpoint of coil 2,

$$k = \frac{\Phi_{21}}{\Phi_{22} + \Phi_{21}} = \frac{\Phi_{21}}{\Phi_2}. \qquad (11.20b)$$

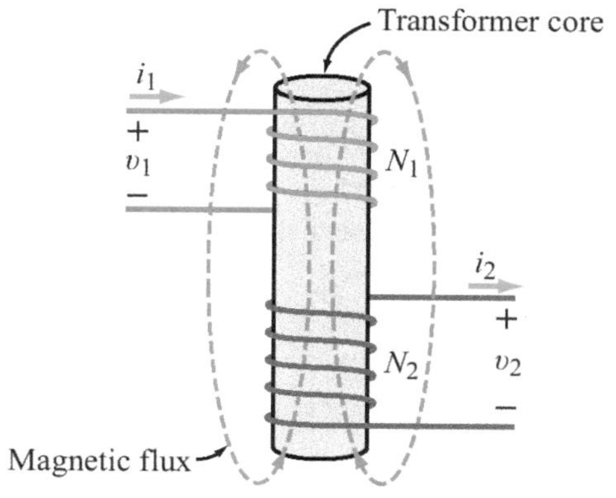

(a) Cylindrical core

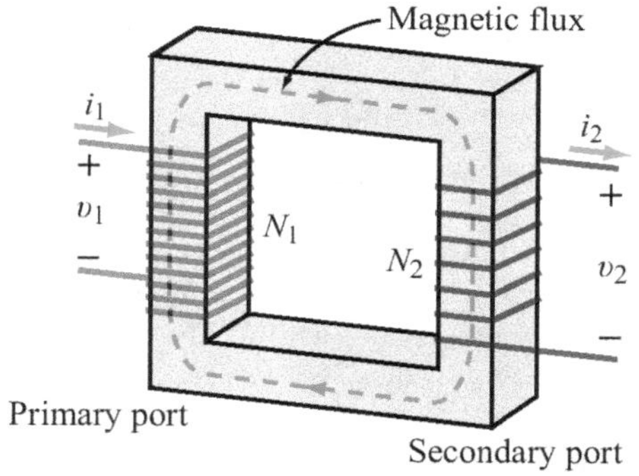

(b) Rectangular core

Figure 11-7: Magnetically coupled coils.

Through energy considerations, k can be related to L_1, L_2, and M as

$$k = \frac{M}{\sqrt{L_1 L_2}}. \qquad (11.21)$$

The mutual inductance M is a maximum when $k = 1$ (perfectly coupled transformer),

$$M(\max) = \sqrt{L_1 L_2}. \qquad (11.22)$$

(perfectly coupled transformer with $k = 1$)

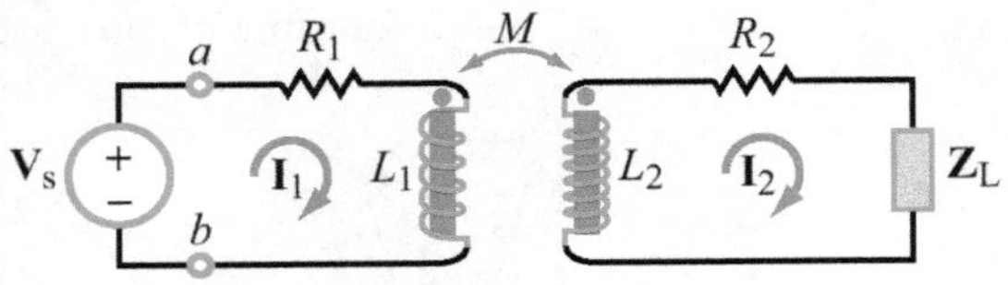

(a) Original circuit

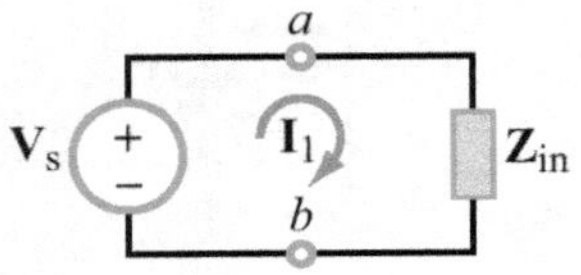

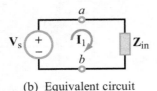

(b) Equivalent circuit

Figure 11-8: (a) Transformer circuit with coil resistors R_1 and R_2, and (b) in terms of an equivalent input impedance $\mathbf{Z}_{in}$.

11-2.2 Input Impedance

In addition to the two coupled coils, a realistic *transformer* circuit should include two series resistors, R_1 and R_2, to account for ohmic losses in the coils. The circuit shown in Fig. 11-8 reflects this reality by including resistor R_1 on the side of coil 1 and resistor R_2 on the side of coil 2. The circuit is driven by a voltage source $\mathbf{V}_s$ on the *primary side* and terminated in a complex load $\mathbf{Z}_L$ on the *secondary side*.

In terms of the designated mesh currents $\mathbf{I}_1$ and $\mathbf{I}_2$, the KVL mesh equations are

$$-\mathbf{V}_s + (R_1 + j\omega L_1)\mathbf{I}_1 - j\omega M\mathbf{I}_2 = 0 \qquad (11.23a)$$

and

$$-j\omega M\mathbf{I}_1 + (R_2 + j\omega L_2 + \mathbf{Z}_L)\mathbf{I}_2 = 0. \qquad (11.23b)$$

From the standpoint of source $\mathbf{V}_s$, the circuit to the right of terminals (a, b) can be represented by an equivalent *input impedance* $\mathbf{Z}_{in}$, as depicted in Fig. 11-8(b). By manipulating Eqs. (11.23) to eliminate $\mathbf{I}_2$, we can generate the following expression for $\mathbf{Z}_{in}$:

$$\mathbf{Z}_{in} = \frac{\mathbf{V}_s}{\mathbf{I}_1} = (R_1 + j\omega L_1) + \frac{\omega^2 M^2}{R_2 + j\omega L_2 + \mathbf{Z}_L}$$

$$= (R_1 + j\omega L_1) + \mathbf{Z}_R, \qquad (11.24)$$

where we define the second term as the *reflected impedance* $\mathbf{Z}_R$, namely

$$\mathbf{Z}_R = \frac{\omega^2 M^2}{R_2 + j\omega L_2 + \mathbf{Z}_L}. \qquad (11.25)$$

We note that

$$\mathbf{Z}_R = \frac{\omega^2 M^2}{\text{impedance of secondary loop}}.$$

In the absence of coupling between the two windings of the transformer (i.e., $M = 0$), $\mathbf{Z}_{in}$ reduces to

$$\mathbf{Z}_{in}\,(M = 0) = R_1 + j\omega L_1.$$

This is exactly what we expect for a series RL circuit connected to a source $\mathbf{V}_s$. When M is not zero, the impedance of the secondary circuit, $(R_2 + j\omega L_2 + \mathbf{Z}_L)$, becomes part of the input impedance of the primary circuit, enabled by the magnetic coupling represented by M. This dependence is akin to *reflecting* the impedance of the secondary circuit onto the primary circuit. The input and reflected impedances are related by

$$\mathbf{Z}_{in} = \mathbf{Z}_{in}\,(M = 0) + \mathbf{Z}_R. \qquad (11.26)$$

The expressions given by Eqs. (11.24) and (11.25) were derived for a transformer circuit in which the windings have dots on the same ends. Repetition of the process for windings whose dots are on opposite ends leads to the same results.

► $\mathbf{Z}_{in}$ depends on the degree of magnetic coupling, but not on whether the coupling is additive or subtractive. ◄

Example 11-4: Input Impedance

Determine current $\mathbf{I}_1$ in the circuit of Fig. 11-9.

Solution: From the given circuit, we deduce that $\omega M = 2\ \Omega$. By analogy with Eq. (11.24), $\mathbf{Z}_{in}$ is given by

$$\mathbf{Z}_{in} = (3 - j2 + j5) + \frac{2^2}{6 + 4 - j4 + j20} = 4.2e^{j42.2^\circ}\ \Omega.$$

Hence,

$$\mathbf{I}_1 = \frac{\mathbf{V}_s}{\mathbf{Z}_{in}} = \frac{120e^{j30^\circ}}{4.2e^{j42.2^\circ}} = 28.6e^{-j12.2^\circ}\ \text{A}.$$

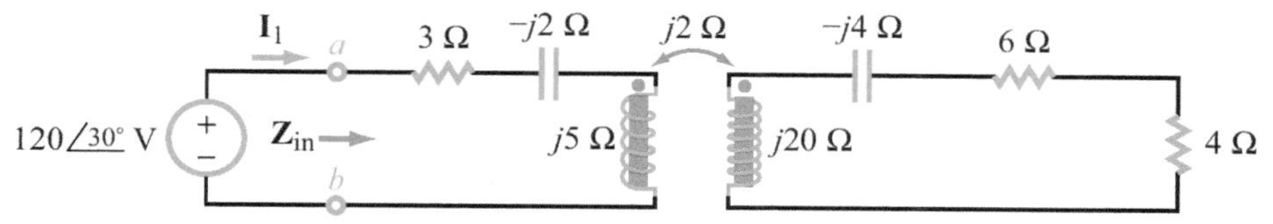

Figure 11-9: Circuit of Example 11-4.

11-2.3 Equivalent Circuits

A circuit is said to be electrically equivalent to another if both exhibit the same **I-V** relationships at a specified set of terminals. For the transformer in Fig. 11-10(a), phase voltages $\mathbf{V}_1$ and $\mathbf{V}_2$ are related to $\mathbf{I}_1$ and $\mathbf{I}_2$ by

$$\mathbf{V}_1 = j\omega L_1 \mathbf{I}_1 + j\omega M \mathbf{I}_2 \tag{11.27a}$$

and

$$\mathbf{V}_2 = j\omega L_2 \mathbf{I}_2 + j\omega M \mathbf{I}_1, \tag{11.27b}$$

which can be cast in matrix form as

$$\begin{bmatrix} \mathbf{V}_1 \\ \mathbf{V}_2 \end{bmatrix} = \begin{bmatrix} j\omega L_1 & j\omega M \\ j\omega M & j\omega L_2 \end{bmatrix} \begin{bmatrix} \mathbf{I}_1 \\ \mathbf{I}_2 \end{bmatrix}. \tag{11.27c}$$

(transformer)

T-Equivalent Circuit

In anticipation of next steps, we have joined in Fig. 11-10(a) the negative terminals of $\mathbf{V}_1$ and $\mathbf{V}_2$ together, which has no impact on the operation of the transformer. Part (b) of the figure displays a proposed *T-equivalent circuit* whose element values L_x, L_y and L_z automatically incorporate the magnetic coupling present in the transformer coils, thereby avoiding the need to account for the mutual-inductance terms when writing KVL equations. The **I-V** matrix equation for the T-circuit (also called a Y-circuit) is

$$\begin{bmatrix} \mathbf{V}_1 \\ \mathbf{V}_2 \end{bmatrix} = \begin{bmatrix} j\omega(L_x + L_z) & j\omega L_z \\ j\omega L_z & j\omega(L_y + L_z) \end{bmatrix} \begin{bmatrix} \mathbf{I}_1 \\ \mathbf{I}_2 \end{bmatrix}. \tag{11.28}$$

(T-equivalent circuit)

The transformer and its T-equivalent circuit exhibit the same **I-V** relationships if the four terms in the matrix of Eq. (11.27)

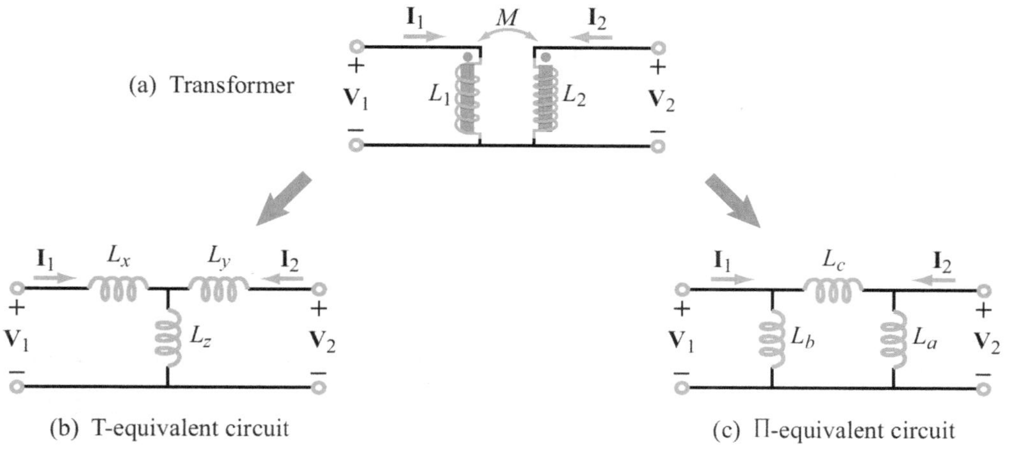

Figure 11-10: The transformer can be modeled in terms of T- or Π-equivalent circuits.

are identical with their corresponding terms in the matrix of Eq. (11.28). Equalization of the two matrices leads to

Transformer dots on same ends

$$L_x = L_1 - M, \tag{11.29a}$$

$$L_y = L_2 - M, \tag{11.29b}$$

and

$$L_z = M. \tag{11.29c}$$

Had the transformer dots been located on opposite ends, the two terms involving M in Eq. (11.27) would have been preceded by minus signs. Consequently, the element values of inductors L_x, L_y, and L_z would be

Transformer dots on opposite ends

$$L_x = L_1 + M, \tag{11.30a}$$

$$L_y = L_2 + M, \tag{11.30b}$$

and

$$L_z = -M. \tag{11.30c}$$

Even though a negative value for inductance L_z is not physically realizable, the mathematical equivalency holds nonetheless and the equivalent circuit is perfectly applicable.

Π-Equivalent Circuit

In some situations, it may be easier to analyze the larger circuit within which the transformer resides by replacing the transformer with a ***Π-equivalent circuit*** instead of the T-equivalent circuit. In such cases, we can use the model shown in Fig. 11-10(c). The expressions for L_a, L_b, and L_c can be obtained either by repeating the procedure we used for the T-equivalent circuit or by applying the Y-Δ transformation equations given in Section 7-4.2. Either route leads to:

Transformer with dots on same ends

$$L_a = \frac{L_1 L_2 - M^2}{L_1 - M}, \tag{11.31a}$$

$$L_b = \frac{L_1 L_2 - M^2}{L_2 - M}, \tag{11.31b}$$

and

$$L_c = \frac{L_1 L_2 - M^2}{M}. \tag{11.31c}$$

If the transformer dots are located on opposite ends, M in Eq. (11.31) should be replaced with $-M$.

Example 11-5: Equivalent Circuit

Use the T-equivalent circuit model to determine $\mathbf{I}_1$ in the circuit of Fig. 11-11.

Solution: Use of Eq. (11.29) gives

$$j\omega L_x = j\omega L_1 - j\omega M = j5 - j2 = j3\ \Omega,$$

$$j\omega L_y = j\omega L_2 - j\omega M = j20 - j2 = j18\ \Omega,$$

and

$$j\omega L_z = j\omega M = j2\ \Omega.$$

The T-equivalent circuit is shown in Fig. 11-11(b). Application of the mesh analysis by-inspection method leads to the matrix equation

$$\begin{bmatrix} (3 - j2 + j3 + j2) & -j2 \\ -j2 & (j2 + j18 - j4 + 6 + 4) \end{bmatrix} \begin{bmatrix} \mathbf{I}_1 \\ \mathbf{I}_2 \end{bmatrix} = \begin{bmatrix} 120e^{j30^\circ} \\ 0 \end{bmatrix}.$$

Its solution is

$$\mathbf{I}_1 = 28.6e^{-j12.2^\circ}\ \text{A},$$

which is identical with the answer obtained in Example 11-4 using the input impedance method.

Concept Question 11-3: What does the coupling coefficient represent? What is its range? (See CAD)

Concept Question 11-4: How is the mutual inductance M related to L_1 and L_2 for a perfectly coupled transformer? (See CAD)

Concept Question 11-5: Why does the reflected impedance $\mathbf{Z}_R$ bear that name? (See CAD)

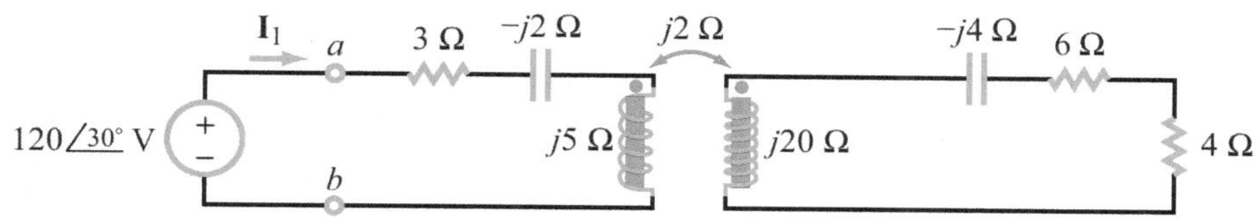

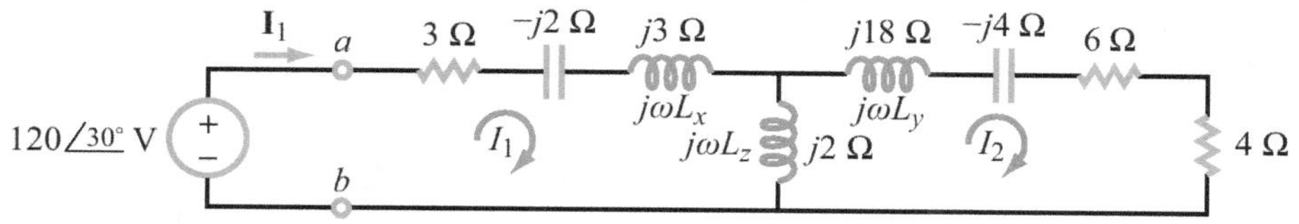

Figure 11-11: (a) Original circuit and (b) after replacing transformer with T-equivalent circuit.

Exercise 11-3: The expression for $\mathbf{Z}_{\text{in}}$ given by Eq. (11.25) was derived for the circuit in Fig. 11-8, in which both dots are on the upper end of the coils. What would the expression look like were the two dots located on opposite ends?

Answer: The expression remains the same. (See CAD)

Exercise 11-4: What are the element values of the Π-equivalent circuit of the transformer in Fig. 11-11(a)?

Answer: $j\omega L_a = j32\ \Omega$, $j\omega L_b = j5.33\ \Omega$, $j\omega L_c = j48\ \Omega$. (See CAD)

11-3 Energy Considerations

Given that the transformer in Fig. 11-12(a)—with inductance L_1 on the primary side, L_2 on the secondary side, and mutual inductance M coupling the two coils—is equivalent to the T-equivalent circuit in Fig. 11-12(b), we can use the latter to compute the total amount of energy stored in the transformer for any specified values of currents i_1 and i_2. According to Eq. (5.59), the magnetic energy stored in an inductor L due to the flow of current i through it is

$$w(t) = \frac{1}{2}\, L\, i^2(t) \qquad \text{(J)}. \tag{11.32}$$

For the circuit in Fig. 11-12(b),

$$\begin{aligned} w(t) &= \frac{1}{2}\, L_x i_1^2 + \frac{1}{2}\, L_y i_2^2 + \frac{1}{2}\, L_z (i_1 + i_2)^2 \\ &= \frac{1}{2}(L_1 - M) i_2^2 + \frac{1}{2}(L_2 - M) i_2^2 + \frac{1}{2}\, M (i_1 + i_2)^2 \\ &= \frac{1}{2}\, L_1 i_1^2 + \frac{1}{2}\, L_2 i_2^2 + M i_1 i_2. \end{aligned} \tag{11.33}$$

Equation (11.33) applies to transformers in which i_1 and i_2 both enter or both leave the dotted terminals, and both dotted terminals are on the same end (as in Fig. 11-12). Reversing the direction of either current or reversing the locations of the dots requires replacing M with $-M$.

Example 11-6: Magnetic Energy

In the circuit in Fig. 11-13, determine the magnetic energy stored in the transformer at $t = 0$, given that $v_s(t) = 12\cos(377t + 60°)$ V.

Solution: We start by replacing the transformer with its T-equivalent circuit and then transforming the new circuit to the phasor domain (Fig. 11-13(b)). Per Eq. (11.30), the values of L_x, L_y, and L_z are

$$L_x = L_1 + M = (10 + 6) \text{ mH} = 16 \text{ mH},$$

$$L_y = L_2 + M = (30 + 6) \text{ mH} = 36 \text{ mH},$$

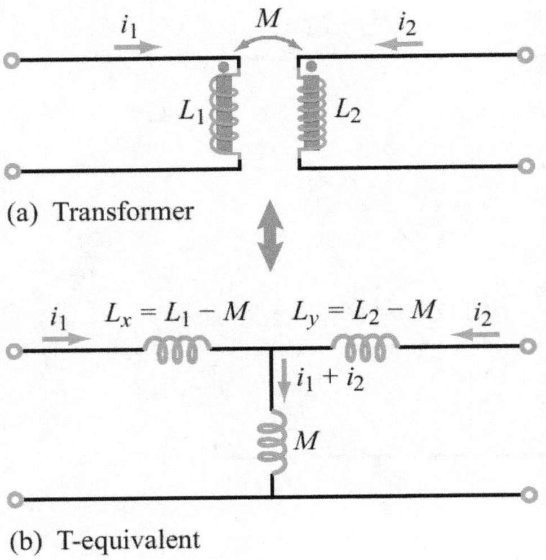

Figure 11-12: Transformer and its T-equivalent circuit. Reversing the direction of either current or if dots are on opposite ends, M should be replaced with $-M$.

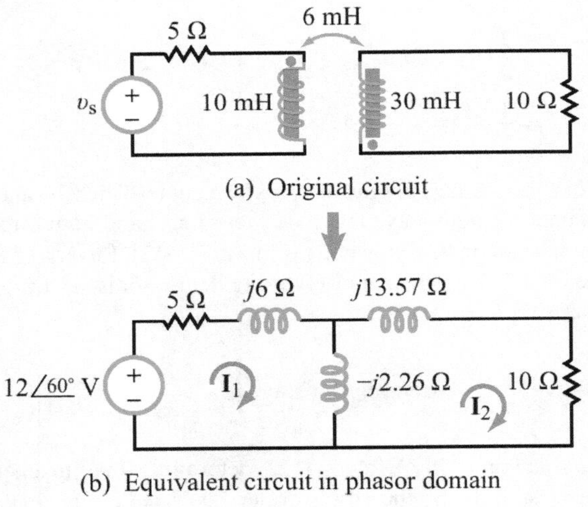

Figure 11-13: Circuits of Example 11-6.

and

$$L_z = -M = -6 \text{ mH}.$$

Transforming the inductors to the phasor domain entails multiplying the inductance values by $j\omega = j377$ rad/s, which leads to

$$j\omega L_x = j377 \times 16 \times 10^{-3} = j6\ \Omega,$$
$$j\omega L_y = j377 \times 36 \times 10^{-3} = j13.57\ \Omega,$$

and

$$j\omega L_z = j377 \times (-6 \times 10^{-3}) = -j2.26\ \Omega.$$

Mesh analysis by inspection gives

$$\begin{bmatrix} (5 + j6 - j2.26) & +j2.26 \\ +j2.26 & (10 + j13.57 - j2.26) \end{bmatrix} \begin{bmatrix} \mathbf{I}_1 \\ \mathbf{I}_2 \end{bmatrix} = \begin{bmatrix} 12e^{j60^\circ} \\ 0 \end{bmatrix}.$$

Solution of the matrix equation for $\mathbf{I}_1$ and I_2 leads to

$$\mathbf{I}_1 = 1.91e^{j26.06^\circ} \text{ A}$$

and

$$\mathbf{I}_2 = 0.29e^{-j112.5^\circ} \text{ A}.$$

The time-domain equivalents are

$$i_1(t) = 1.91\cos(377t + 26.06^\circ) \text{ A}$$

and

$$i_2(t) = 0.29\cos(377t - 112.5^\circ) \text{ A}.$$

The magnetic energy stored in the three inductors of the circuit in Fig. 11-13(b) at $t = 0$ is

$$\begin{aligned} w(0) &= \left[\frac{1}{2} L_x i_1^2 + \frac{1}{2} L_y i_2^2 + \frac{1}{2} L_z (i_1 - i_2)^2 \right] \Bigg|_{t=0} \\ &= \frac{1}{2} \times 16 \times 10^{-3} \times (1.91 \cos 26.06^\circ)^2 \\ &\quad + \frac{1}{2} \times 36 \times 10^{-3} \times [0.29 \cos(-112.5^\circ)]^2 \\ &\quad + \frac{1}{2} \times (-6 \times 10^{-3}) \\ &\quad \cdot [1.91 \cos 26.06^\circ - 0.29 \cos(-112.5^\circ)]^2 \\ &= 13.7 \text{ mJ}. \end{aligned}$$

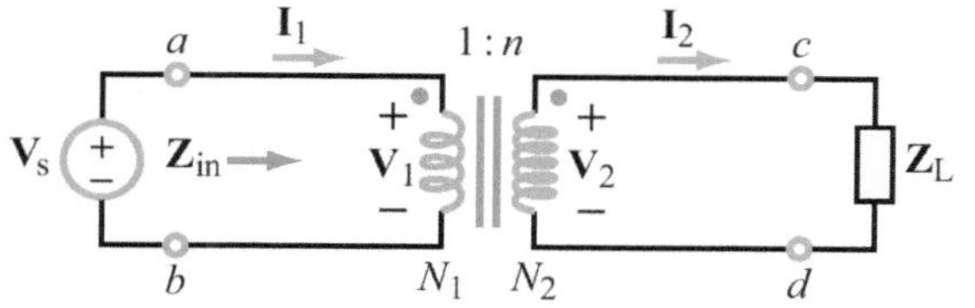

(a) Dots on same ends

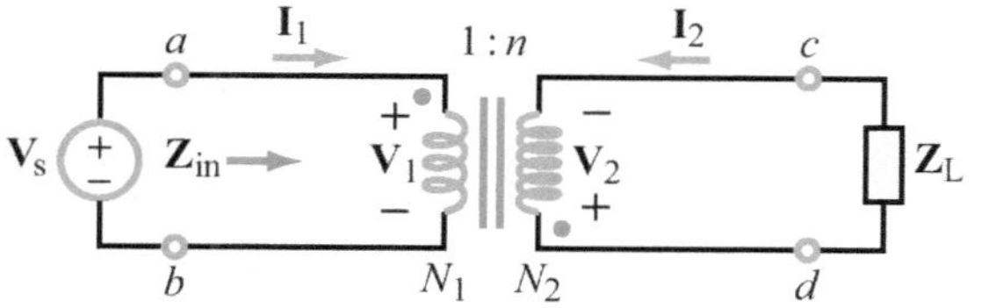

(b) Dots on opposite ends

Figure 11-14: Schematic symbol for an ideal transformer. Note the reversal of the voltage polarity and current direction when the dot location at the secondary is moved from the top end of the coil to the bottom end. For both configurations: $\mathbf{V}_2/\mathbf{V}_1 = N_2/N_1 = n$, $\mathbf{I}_2/\mathbf{I}_1 = N_1/N_2 = 1/n$.

11-4 Ideal Transformers

An ideal transformer is characterized by lossless coils ($R_1 = R_2 = 0$) and a maximum coupling coefficient $k = 1$. The mutual inductance M is a maximum and given by Eq. (11.22) as

$$M(\text{max}) = \sqrt{L_1 L_2}\,. \quad \text{(ideal transformer)} \tag{11.34}$$

According to Eq. (5.52), the inductance L of a solenoid-shaped inductor is proportional to N^2, where N is the number of turns wound around the core. Hence, for an ideal transformer with N_1 turns on the primary side and N_2 on the secondary, as depicted in Fig. 11-14,

$$\frac{L_2}{L_1} = \frac{N_2^2}{N_1^2} = n^2, \tag{11.35}$$

where $n = N_2/N_1$ is called the *turns ratio*.

▶ Note that the symbol for the ideal transformer includes *two parallel lines* between the coils, an indicator that the coils are wound around a ferromagnetic core with perfect coupling ($k = 1$). Also note that because μ of a ferromagnetic material is very large, so are L_1 and L_2. ◀

For the ideal transformer with dots on the same ends (Fig. 11-14(a)), using Eqs. (11.34) and (11.35) in Eq. (11.27a) [while noting that the direction of $\mathbf{I}_2$ in Fig. 11-14(a) is opposite of that used to derive Eq. (11.27)] leads to

$$\begin{aligned}\mathbf{V}_1 &= j\omega L_1\mathbf{I}_1 - j\omega M\mathbf{I}_2\\ &= j\omega L_1\mathbf{I}_1 - j\omega\sqrt{L_1L_2}\,\mathbf{I}_2\\ &= j\omega L_1\mathbf{I}_1 - j\omega L_1 n\mathbf{I}_2 = j\omega L_1(\mathbf{I}_1 - n\mathbf{I}_2).\end{aligned}$$

Similarly, for Eq. (11.27b),

$$\mathbf{V}_2 = j\omega L_2\mathbf{I}_2 - j\omega M\mathbf{I}_1 = [j\omega L_1(\mathbf{I}_1 - n\mathbf{I}_2)]n.$$

Hence,

$$\frac{\mathbf{V}_2}{\mathbf{V}_1} = n \quad \text{(ideal transformer with dots on same ends).} \tag{11.36}$$

▶ The transformer is called a *step-up transformer* ($V_2 > V_1$) when $n > 1$ and a *step-down transformer* ($V_2 < V_1$) when $n < 1$. ◀

For a lossless transformer, complex power $\mathbf{S}_1$ supplied by its primary side must match complex power $\mathbf{S}_2$ absorbed by its secondary:

$$\mathbf{S}_1 = \mathbf{S}_2, \tag{11.37}$$

or equivalently

$$\mathbf{V}_1\mathbf{I}_1^* = \mathbf{V}_2\mathbf{I}_2^*. \tag{11.38}$$

In view of Eq. (11.36), it follows that

$$\frac{\mathbf{I}_2}{\mathbf{I}_1} = \frac{1}{n} \quad \text{(ideal transformer with dots on same ends).} \tag{11.39}$$

The expressions for the voltage and current ratios given by Eqs. (11.36) and (11.39) apply to the current directions, voltage polarities and dot locations indicated in both configurations of Fig. 11-14.

11-4.1 Input Impedance

At the input terminals in the transformer circuits of Fig. 11-14, source $\mathbf{V}_s$ *views* an input impedance $\mathbf{Z}_{in} = \mathbf{V}_1/\mathbf{I}_1$, and at the load end, $\mathbf{Z}_L = \mathbf{V}_2/\mathbf{I}_2$. The combination leads to

$$\mathbf{Z}_{in} = \frac{\mathbf{V}_1}{\mathbf{I}_1} = \frac{1}{n^2}\frac{\mathbf{V}_2}{\mathbf{I}_2} = \frac{\mathbf{Z}_L}{n^2}. \qquad (11.40)$$

Thus, from the standpoint of the source $\mathbf{V}_s$, the entire circuit to the right of terminals (a, b) is equivalent to input impedance $\mathbf{Z}_{in}$. We note that $\mathbf{Z}_{in}$ is equal to $(1/n^2)$ of $\mathbf{Z}_L$.

▶ For the ideal transformer, the voltage, current, and impedance ratios are defined in terms of the turns ratio, independently of the values of L_1 and L_2. ◀

11-4.2 Equivalent Circuits

If the circuit connected to the transformer on its secondary side consists of only a passive impedance, the input-impedance equivalency is all we need to simplify the overall circuit, but what if the secondary side contains a voltage or current source? Also, by extension, what if we wish to view the circuit from the perspective of the secondary side, rather than that of the primary side? These questions are addressed by Example 11-7.

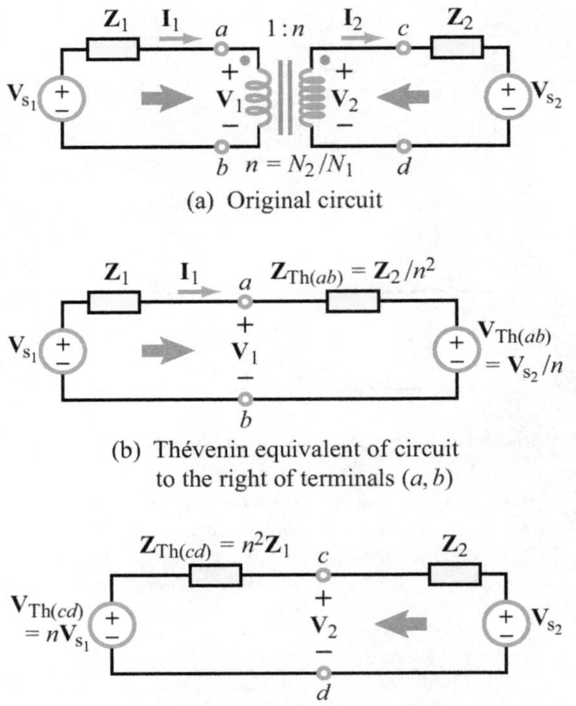

Figure 11-15: Thévenin equivalent circuit of Example 11-7.

Example 11-7: Thévenin Equivalent

For the circuit of Fig. 11-15, (a) obtain the Thévenin equivalent of the circuit segment to the right of terminals (a, b), then (b) repeat the process for the circuit segment to the left of terminals (c, d).

Solution: (a) The Thévenin equivalent circuit shown in Fig. 11-15(b) consists of a Thévenin voltage source $\mathbf{V}_{Th(ab)}$ and a Thévenin impedance $\mathbf{Z}_{Th(ab)}$. Our task is to relate them to the element values of the circuit to the right of terminals (a, b).

The Thévenin voltage is defined as the open-circuit voltage across terminals (a, b), which amounts to determining $\mathbf{V}_1$ after disconnecting the input circuit ($\mathbf{V}_{s_1}$ and $\mathbf{Z}_1$). With $\mathbf{V}_{s_1}$ absent, $\mathbf{I}_1 = 0$, and by virtue of Eq. (11.39), $\mathbf{I}_2 = 0$. Hence, with no voltage drop across $\mathbf{Z}_2$, $\mathbf{V}_2 = \mathbf{V}_{s_2}$, and

$$\mathbf{V}_{Th(ab)} = \mathbf{V}_1 = \frac{\mathbf{V}_2}{n} = \frac{\mathbf{V}_{s_2}}{n}.$$

To determine $\mathbf{Z}_{Th(ab)}$, we replace $\mathbf{V}_{s_2}$ with a short circuit and use Eq. (11.40),

$$\mathbf{Z}_{Th(ab)} = \mathbf{Z}_{in} = \frac{\mathbf{Z}_2}{n^2}. \qquad (11.41)$$

(b) Looking to the left from terminals (c, d) involves identically the same process, except that the turns ratio going from terminals (c, d) to terminals (a, b) is now $1/n$. The result is diagrammed in Fig. 11-15(c).

Example 11-8: Ideal Autotransformer

In an *autotransformer*, the primary and secondary sides share a part or all of the same winding. Develop expressions for the voltage and current ratios in the autotransformer circuits shown in Fig. 11-16(a) and (b).

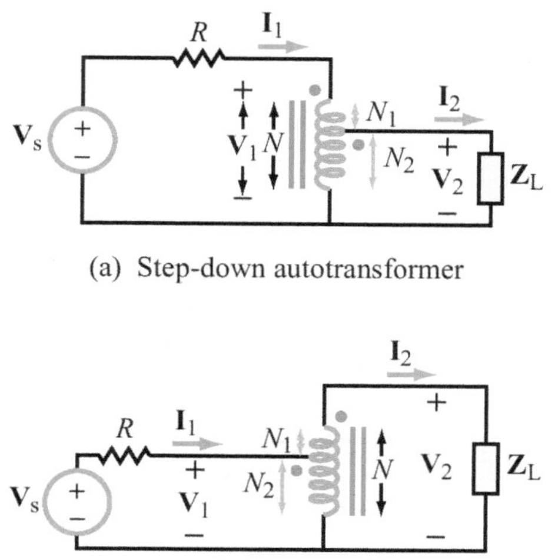

(a) Step-down autotransformer

(b) Step-up autotransformer

Figure 11-16: Autotransformer circuits.

Solution: (a) For the step-down autotransformer in Fig. 11-16(a),

$$\frac{\mathbf{V}_2}{\mathbf{V}_1} = \frac{N_2}{N} = \frac{N_2}{N_1 + N_2}$$

and (11.42)

$$\frac{\mathbf{I}_2}{\mathbf{I}_1} = \frac{\mathbf{V}_1}{\mathbf{V}_2} = \frac{N_1 + N_2}{N_2} .$$

(step-down autotransformer)

(b) The converse is true for the step-up autotransformer of Fig. 11-16(b). That is,

$$\frac{\mathbf{V}_2}{\mathbf{V}_1} = \frac{N}{N_2} = \frac{N_1 + N_2}{N_2}$$

and (11.43)

$$\frac{\mathbf{I}_2}{\mathbf{I}_1} = \frac{\mathbf{V}_1}{\mathbf{V}_2} = \frac{N_2}{N_1 + N_2} .$$

(step-up autotransformer)

Concept Question 11-6: What is the coupling coefficient of an ideal transformer? (See CAD)

Concept Question 11-7: Are the secondary-to-primary voltage and current ratios in an ideal transformer dependent on L_1 and L_2? (See CAD)

Concept Question 11-8: What is an autotransformer? (See CAD)

Exercise 11-5: Determine the Thévenin equivalent of the circuit to the right of terminals (a, b) in Fig. E11.5.

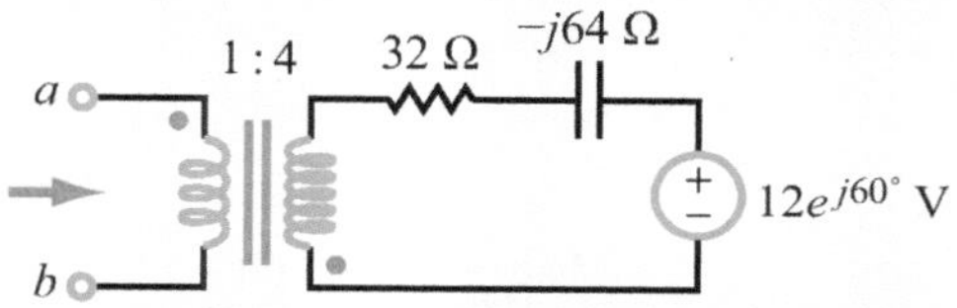

Figure E11.5

Answer:

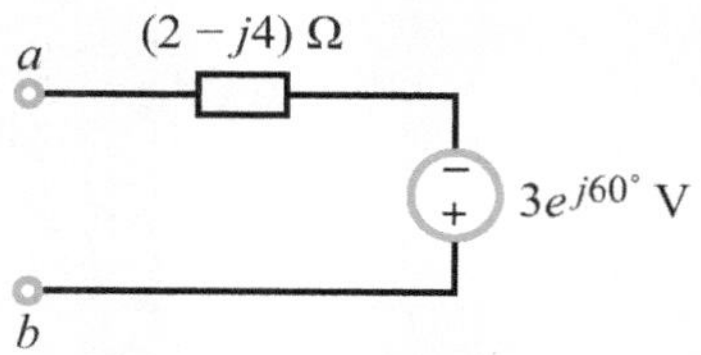

(See CAD)

Exercise 11-6: An autotransformer is used to step up the voltage by a factor of 10. If $N = 200$, what are the values of N_1 and N_2?

Answer: $N_1 = 180$ and $N_2 = 20$. (See CAD)

11-5 Three-Phase Transformers

In Chapter 10 we observed that distribution of power is cheaper (fewer wires) and more efficient (less ohmic loss) when transferred by a three-phase system than by three

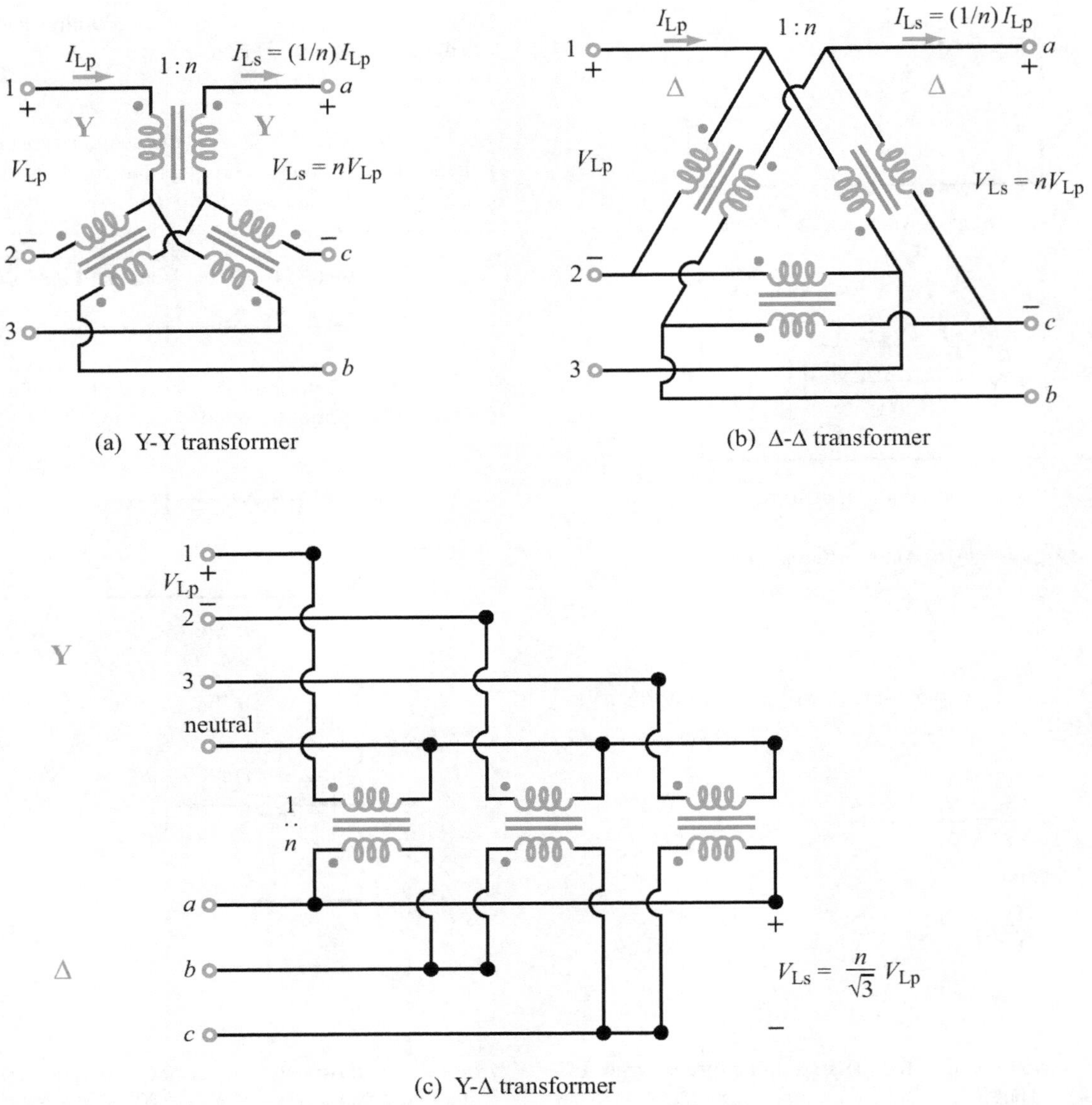

Figure 11-17: Three possible connection configurations for three-phase transformers.

independent single-phase systems. A similar argument holds for transformers. To step up or step down voltage levels in a three-phase network, it is better to use a single three-phase transformer than three single-phase transformers.

Three-phase transformers can be designed to couple any combination of Y- and Δ-configurations on the primary and secondary sides.

The three possible combinations are diagrammed in Fig. 11-17 (Y-Δ and Δ-Y transformers are mirror images of one another). In each case, the line voltages and currents in the secondary circuit are related to those on the primary side. For Y-Y and Δ-Δ connections (Fig. 11-17(a) and (b)),

$$\frac{V_{Ls}}{V_{Lp}} = \frac{I_{Lp}}{I_{Ls}} = n \qquad \text{(Y-Y and Δ-Δ)}, \tag{11.44}$$

where V_{Lp} and V_{Ls} are the *rms magnitudes of the line voltages on the primary and secondary sides of the transformer*, and a similar notation applies to the line currents I_{Lp} and I_{Ls}. The transformer is assumed to be ideal (lossless and perfectly coupled), in which case conservation of energy requires that

$$S_{Tp} = S_{Ts},$$

where S_{Tp} and S_{Ts} are the *total apparent powers* at the primary and secondary sides of the transformer. Recall from Eqs. (10.20) and (10.23) that the magnitude of the total apparent power is given by

$$S_T = \sqrt{3}\, V_L I_L \qquad \text{(Y and }\Delta\text{)}, \qquad (11.45)$$

for both Y- and Δ-configurations and on either side of the transformer. Consequently, for Y-Δ and Δ-Y transformers (Fig. 11-17(c) and (d)),

$$\frac{V_{Ls}}{V_{Lp}} = \frac{I_{Lp}}{I_{Ls}} = \frac{n}{\sqrt{3}} \qquad (\text{Y-}\Delta)$$

and

$$\frac{V_{Ls}}{V_{Lp}} = \frac{I_{Lp}}{I_{Ls}} = \sqrt{3}\, n \qquad (\Delta\text{-Y}). \qquad (11.46)$$

Example 11-9: Δ-Δ Transformer

The secondary side of a Δ-Δ three-phase, ideal transformer is connected to a 16 kVA, balanced, three-phase load. If each transformer has a turns ratio of 1 : 4 and the line voltage at the load side is 120 V (rms), determine the line voltage and line current at the primary side.

Solution: For the Δ-Δ connection, Eq. (11.44) gives

$$V_{Lp} = \frac{V_{Ls}}{n} = \frac{120}{4} = 30 \text{ V (rms)}.$$

Application of the apparent power expressions given by Eq. (11.45) to the secondary side leads to

$$I_{Ls} = \frac{S_T}{\sqrt{3}\, V_{Ls}} = \frac{16,000}{\sqrt{3} \times 120} = 77 \text{ A (rms)}.$$

The line current at the primary side is

$$I_{Lp} = n I_{Ls} = 4 \times 77 = 308 \text{ A (rms)}.$$

Example 11-10: Three-Phase Transformer Circuit

Determine $\mathbf{I}_x$ in the circuit of Fig. 11-18.

Solution: The three mesh equations are

$$-36 + (10 + 50)\mathbf{I}_1 - 50\mathbf{I}_2 + \mathbf{V}_1 = 0,$$
$$-50\mathbf{I}_1 + (30 + 40 + 50)\mathbf{I}_2 - 30\mathbf{I}_3 - \mathbf{V}_2 = 0,$$

and

$$-30\mathbf{I}_2 + (20 + 30)\mathbf{I}_3 + \mathbf{V}_2 - \mathbf{V}_1 = 0.$$

Additionally, because the transformer dots are on opposite ends [see Fig. 11-14(b) for reference],

$$\mathbf{V}_2 = -n\mathbf{V}_1 = -4\mathbf{V}_1$$

and

$$\mathbf{I}_2 = \frac{-\mathbf{I}_1}{n} = \frac{-\mathbf{I}_1}{4}.$$

After multiple substitutions, we arrive at the solution

$$\mathbf{I}_x = \mathbf{I}_2 = -0.06 \text{ A}.$$

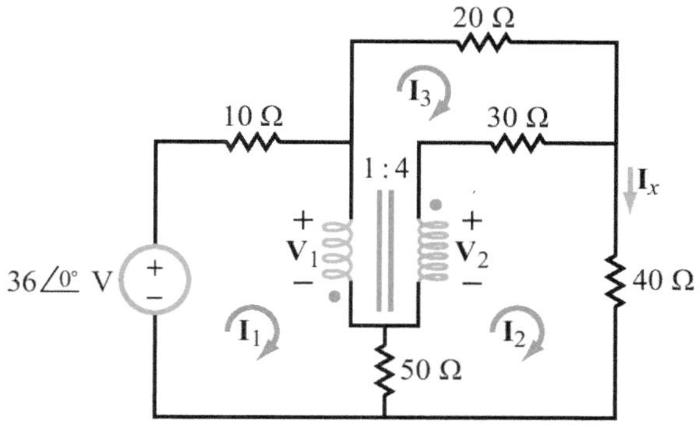

Figure 11-18: Circuit of Example 11-10.

Concept Question 11-9: What are three-phase transformers used for? (See CAD)

Concept Question 11-10: How is the secondary-to-primary voltage ratio related to the turns ratio n for Y-Y, Δ-Δ, Y-Δ, and Δ-Y configurations? (See CAD)

Exercise 11-7: Determine $\mathbf{I}_x$ in the circuit of Fig. 11-18 after replacing the 20 Ω resistor with an open circuit.

Answer: $I_x = -0.097$ A. (See CAD)

Summary

Concepts

- Current flow through a coil in close proximity of a second coil induces a mutual inductance voltage across the second coil through a shared magnetic field.
- The dot convention, which accounts for the directions of the windings in the two coupled coils, defines the polarities of the induced mutual-inductance voltages.
- A transformer can be modeled in terms of T- or Π-equivalent circuits.
- The coupling coefficient of an ideal transformer is $k = 1$. Its secondary-to-primary voltage and current ratios are defined by the turns ratio $n = N_2/N_1$.
- Three-phase transformers are used to couple any combination of Y- and Δ-configurations on the primary and secondary sides.

Mathematical and Physical Models

Magnetic Coupling

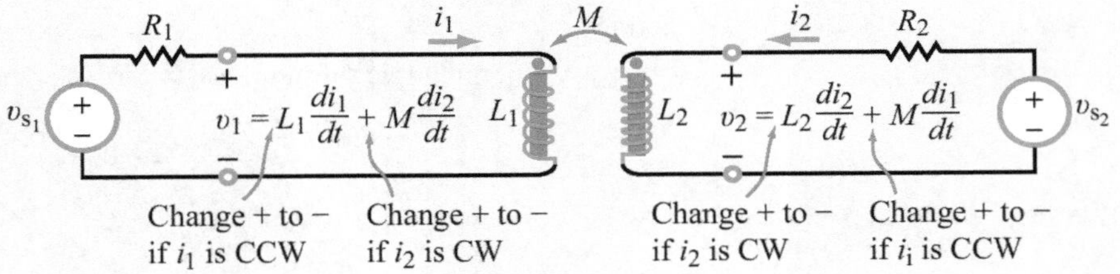

(a) Dots on same ends

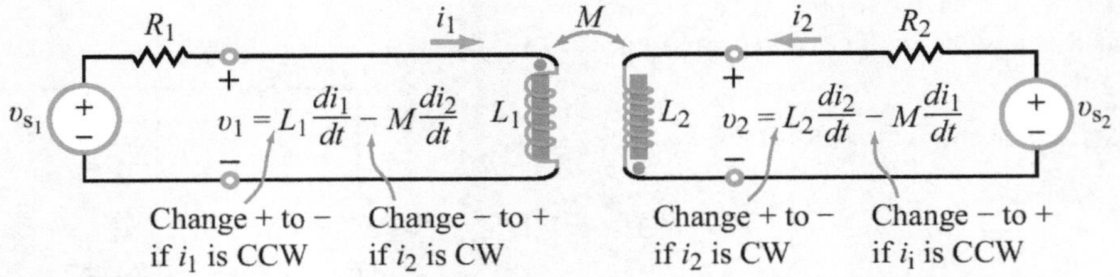

(b) Dots on opposite ends

$$k = \frac{M}{\sqrt{L_1 L_2}}$$

Mathematical and Physical Models (continued)

Equivalent Circuits

(a) Transformer

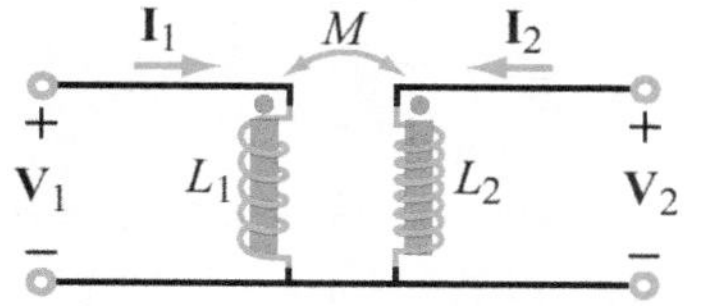

(b) T-equivalent circuit

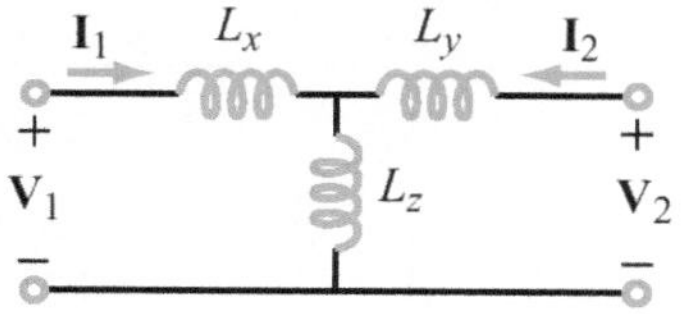

$L_x = L_1 - M$
$L_y = L_2 - M$
$L_z = M$

(c) Π-equivalent circuit

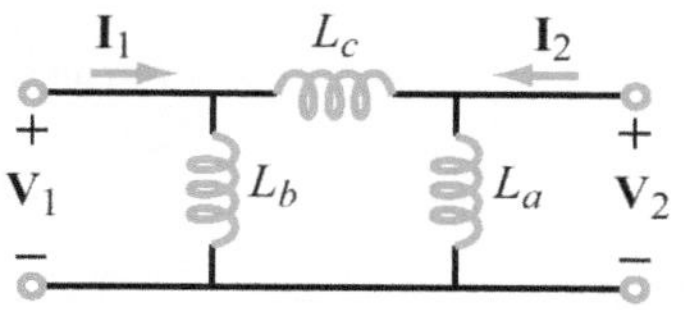

$$L_a = \frac{L_1 L_2 - M^2}{L_1 - M}$$

$$L_b = \frac{L_1 L_2 - M^2}{L_2 - M}$$

$$L_c = \frac{L_1 L_2 - M^2}{M}$$

Replace M with $-M$ if dots are on opposite ends.

Equivalent Inductance

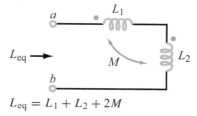

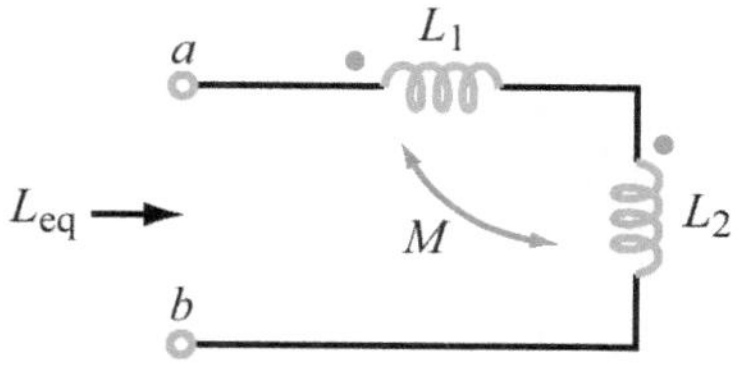

$L_{eq} = L_1 + L_2 + 2M$

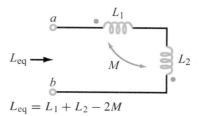

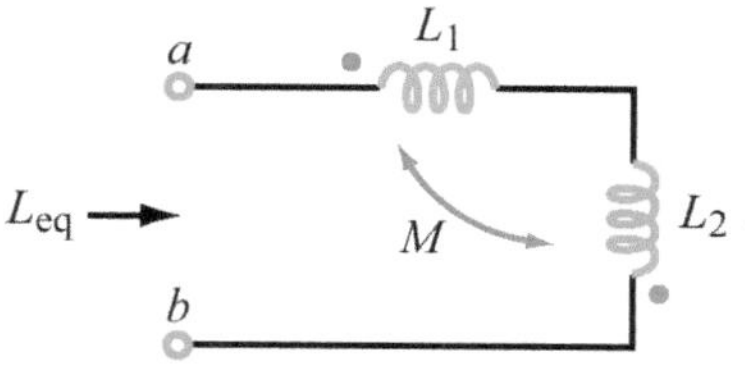

$L_{eq} = L_1 + L_2 - 2M$

Ideal Transformer

$$\frac{\mathbf{V}_2}{\mathbf{V}_1} = n$$

$$\frac{\mathbf{I}_2}{\mathbf{I}_1} = \frac{1}{n}$$

$$\mathbf{Z}_{in} = \frac{\mathbf{V}_1}{\mathbf{I}_1} = \frac{1}{n^2}\frac{\mathbf{V}_2}{\mathbf{I}_2} = \frac{\mathbf{Z}_L}{n^2}$$

Important Terms

Provide definitions or explain the meaning of the following terms:

Π-equivalent circuit
autotransformer
coupling coefficient
dot convention
input impedance
linear
loosely coupled
magnetic flux
magnetic flux linkage
magnetic permeability
magnetically coupled circuit
mutual inductance
mutual inductance voltage
mutual voltage
perfectly coupled
permeability of free space
primary port
primary side
reflected impedance
secondary port
secondary side
self inductance
step-down transformer
step-up transformer
T-equivalent circuit
three-phase transformer
tightly coupled
total flux
transformer
turns ratio
two parallel lines
voltage transformers

PROBLEMS

Section 11-1: Magnetic Coupling

*11.1 For the circuit shown in Fig. P11.1, determine (a) $i(t)$ and (b) the average power absorbed by R_L.

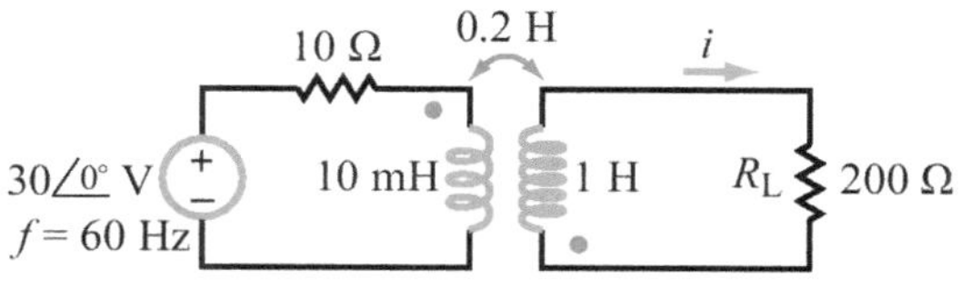

Figure P11.1: Circuit for Problem 11.1.

*Answer(s) available in Appendix G.

11.2 For the circuit in Fig. P11.2, determine (a) $i_L(t)$ and (b) the average power dissipated in R_L.

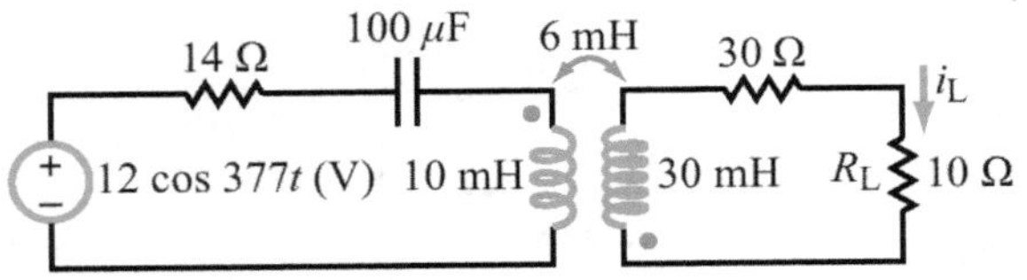

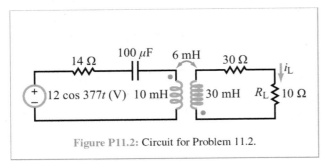

Figure P11.2: Circuit for Problem 11.2.

11.3 For the circuit in Fig. P11.3, determine $\mathbf{V}_{out}$.

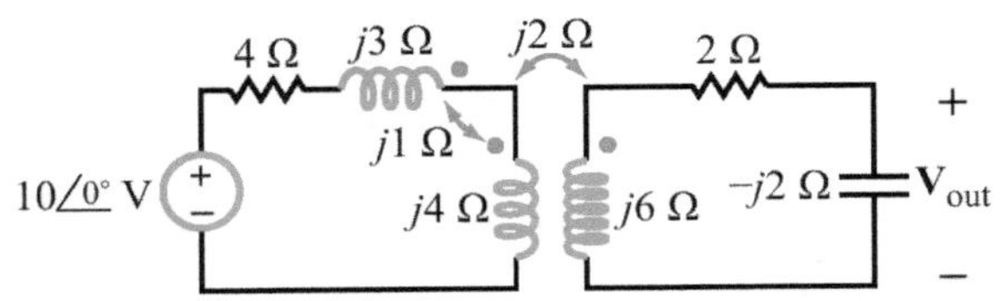

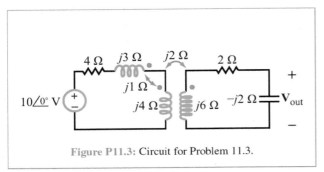

Figure P11.3: Circuit for Problem 11.3.

*__11.4__ Determine $\mathbf{V}_{out}$ in the circuit shown in Fig. P11.4.

11.5 Determine the average power dissipated in the 4 Ω resistor of the circuit in Fig. P11.5.

11.6 Determine $\mathbf{I}_x$ in the circuit of Fig. P11.6.

*__11.7__ Determine $\mathbf{I}_x$ in the circuit of Fig. P11.7.

11.8 Determine the average power dissipated in the 4 Ω resistor of the circuit in Fig. P11.8.

*__11.9__ Determine $\mathbf{V}_{out}$ in the circuit of Fig. P11.9.

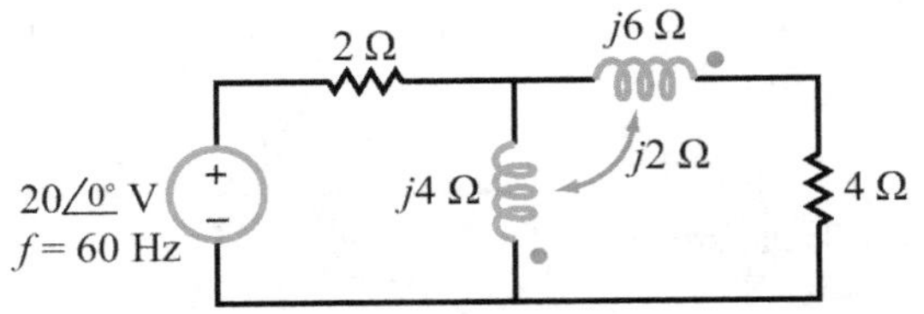

Figure P11.5: Circuit for Problem 11.5.

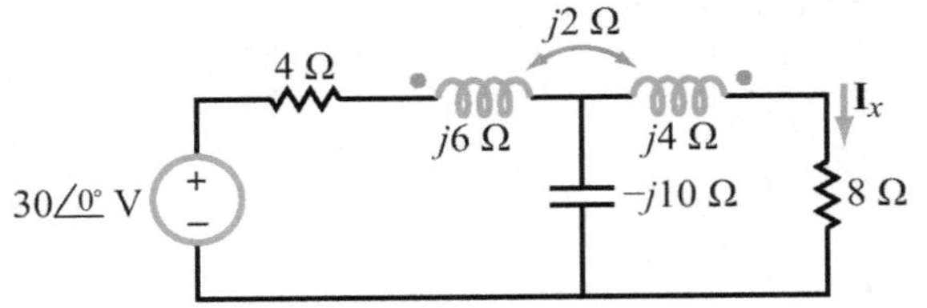

Figure P11.6: Circuit for Problem 11.6.

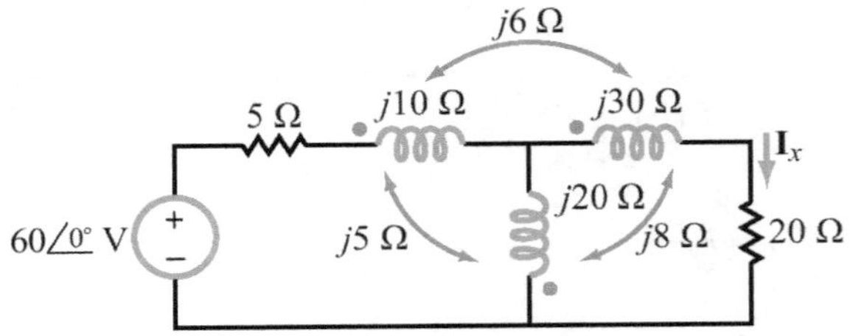

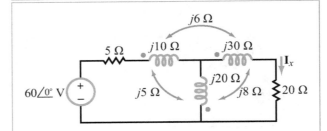

Figure P11.7: Circuit for Problem 11.7.

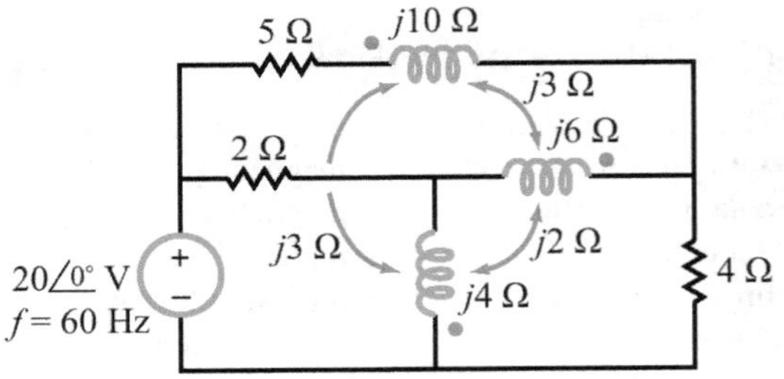

Figure P11.8: Circuit for Problem 11.8.

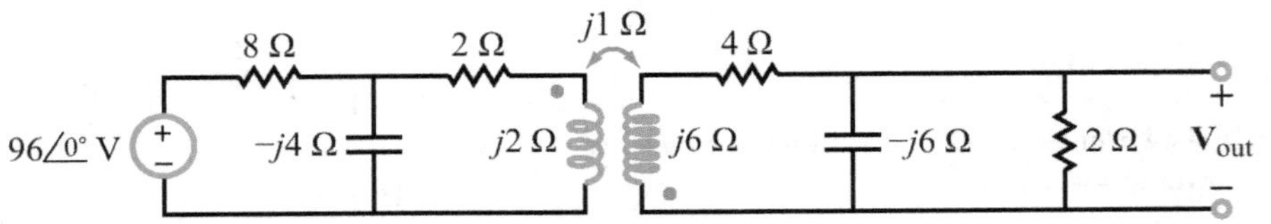

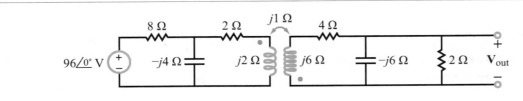

Figure P11.4: Circuit for Problem 11.4.

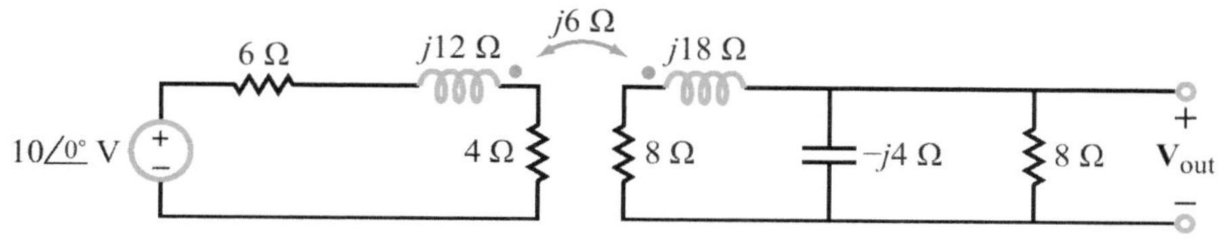

Figure P11.9: Circuit for Problem 11.9.

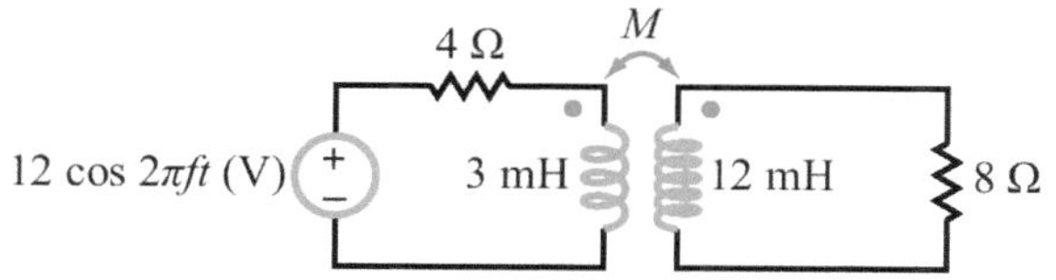

Figure P11.10: Circuit for Problem 11.10.

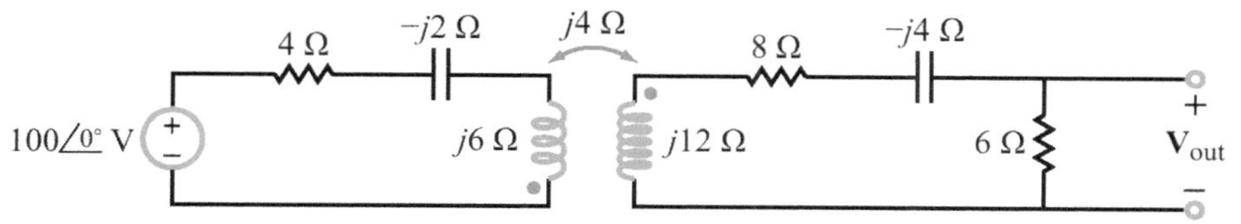

Figure P11.11: Circuit for Problem 11.11.

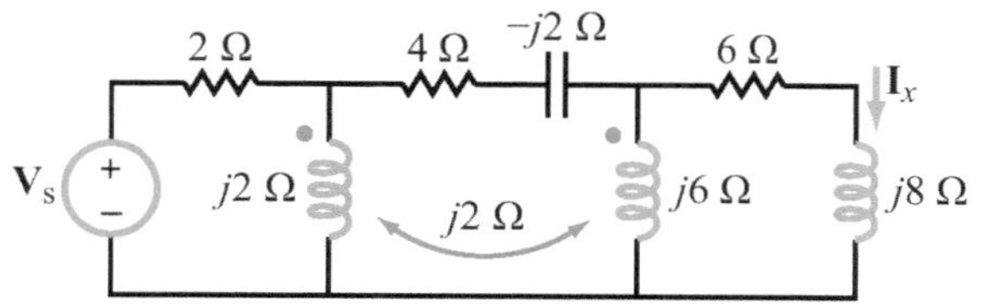

Figure P11.12: Circuit for Problem 11.12.

11.10 The circuit shown in Fig. P11.10 uses a 12 V ac source to deliver power to an 8 Ω speaker. If the average power delivered to the speaker is 1.8 W at an audio frequency $f = 1$ kHz, what is the value of the coupling coefficient k?

*__11.11__ Determine $\mathbf{V}_{\text{out}}$ in the circuit in Fig. P11.11.

11.12 Determine $\mathbf{I}_x$ in the circuit of Fig. P11.12, given that $\mathbf{V}_s = 20\angle 30^\circ$ (V).

11.13 Determine:

(a) L_{eq} at terminals (a, b) in Fig. P11.13(a).

(b) L_{eq} at terminals (a, b) in Fig. P11.13(b).

*(c) L_{eq} at terminals (a, b) in Fig. P11.13(c).

(d) L_{eq} at terminals (a, b) in Fig. P11.13(d).

(e) L_{eq} at terminals (a, b) in Fig. P11.13(e).

(f) L_{eq} at terminals (a, b) in Fig. P11.13(f).

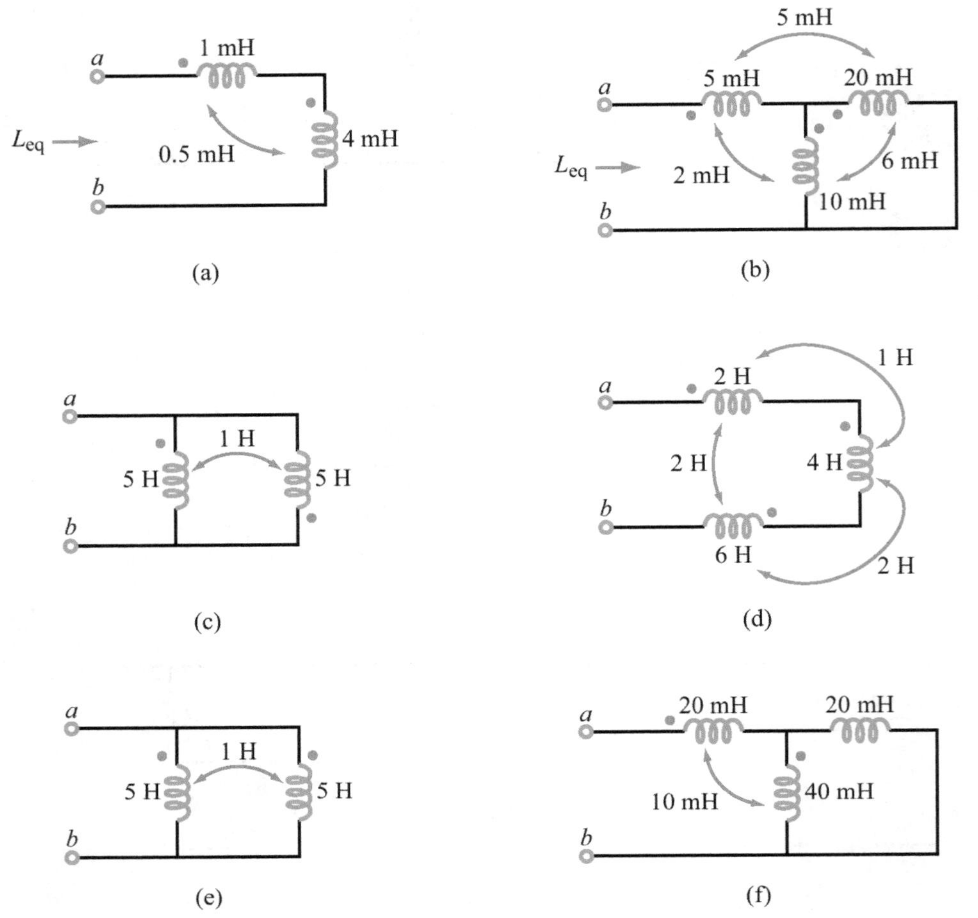

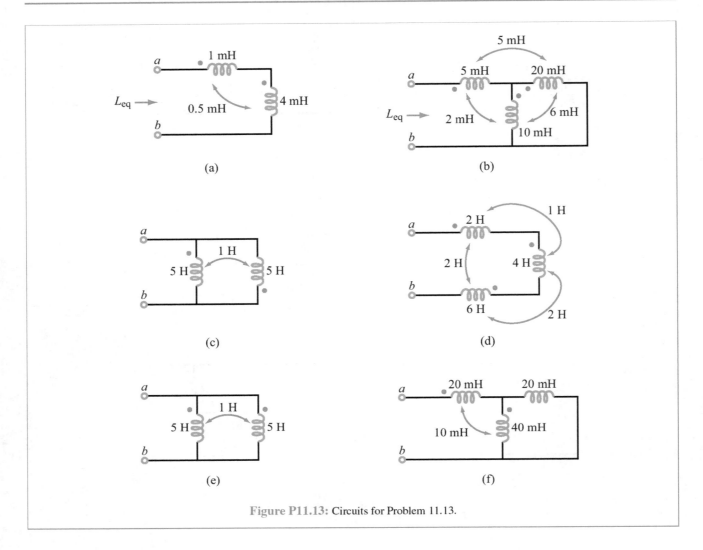

Figure P11.13: Circuits for Problem 11.13.

Sections 11-2 and 11-3: Transformers and Energy

11.14 Determine (a) the input impedance and (b) the reflected impedance, both at terminals (a, b) in the circuit of Fig. P11.14.

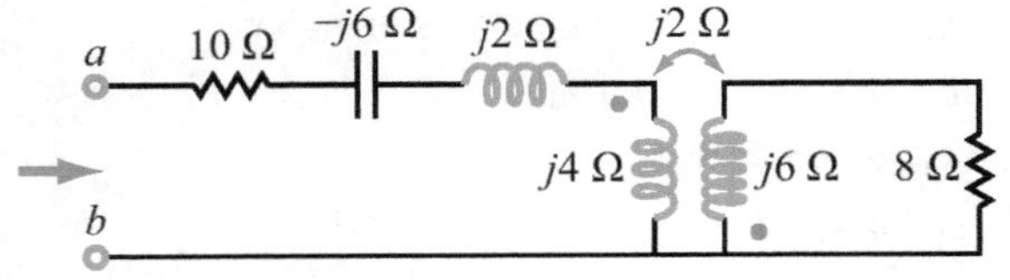

Figure P11.14: Circuit for Problem 11.14.

*__11.15__ Determine (a) the input impedance and (b) the reflected impedance, both at terminals (c, d) in the circuit of Fig. P11.15.

11.16 For the circuit in Fig. P11.16 (a) determine the Thévenin equivalent to the left of $\mathbf{Z}_L$, (b) choose $\mathbf{Z}_L$ for maximum power transfer, and (c) compute the average power absorbed by $\mathbf{Z}_L$.

*__11.17__ Determine the input impedance $\mathbf{Z}_{in}$ of the circuit in Fig. P11.17.

11.18 In the circuit of Fig. P11.18, what should the value of the coupling coefficient k be so that $\mathbf{V}_{out}/\mathbf{V}_{in} = 0.49$?

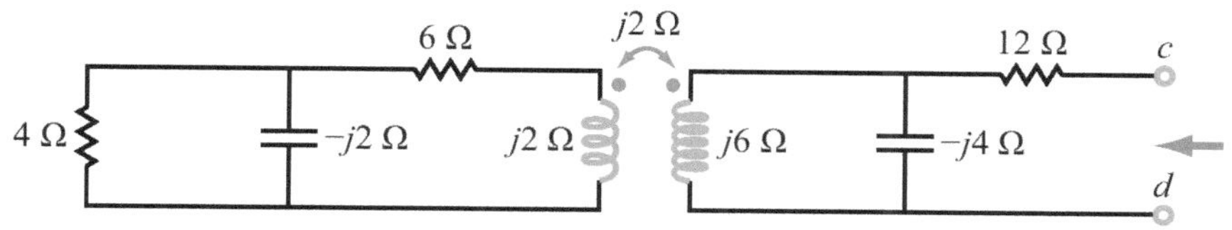

Figure P11.15: Circuit for Problem 11.15.

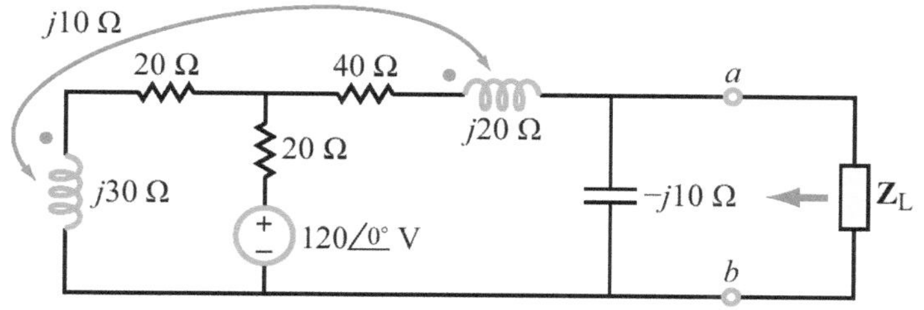

Figure P11.16: Circuit for Problem 11.16.

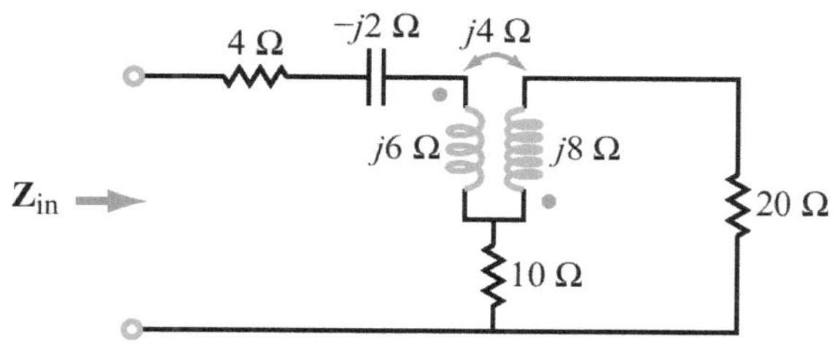

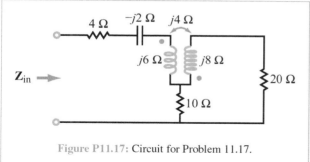

Figure P11.17: Circuit for Problem 11.17.

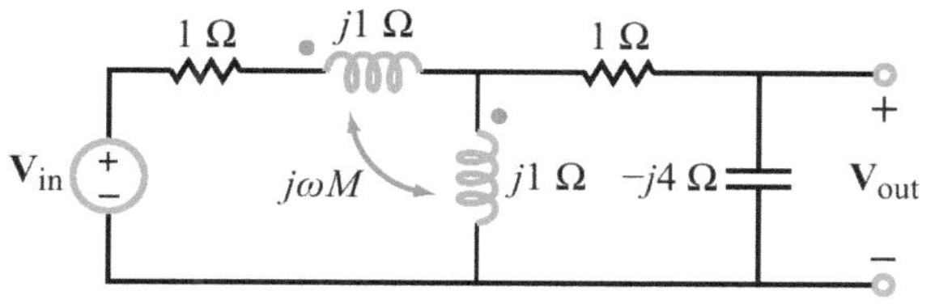

Figure P11.18: Circuit for Problem 11.18.

*11.19 Apply T- and Π-transformations to determine L_{eq} in the circuit of Fig. P11.19.

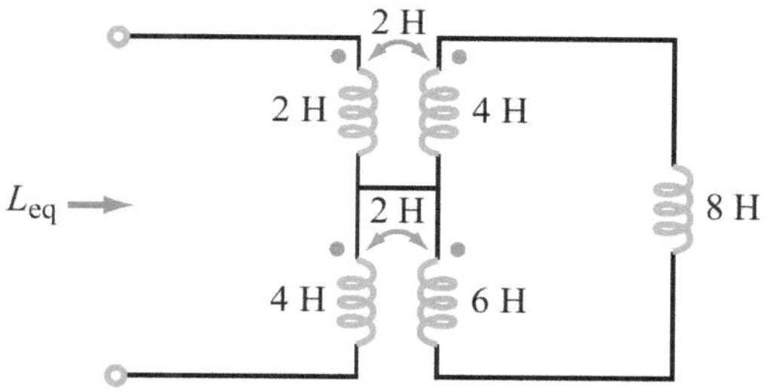

Figure P11.19: Circuit for Problem 11.19.

11.20 For the circuit in Fig. P11.20, determine the complex powers: (a) supplied by the source, (b) stored by the two inductors, and (c) dissipated by the source and load resistors.

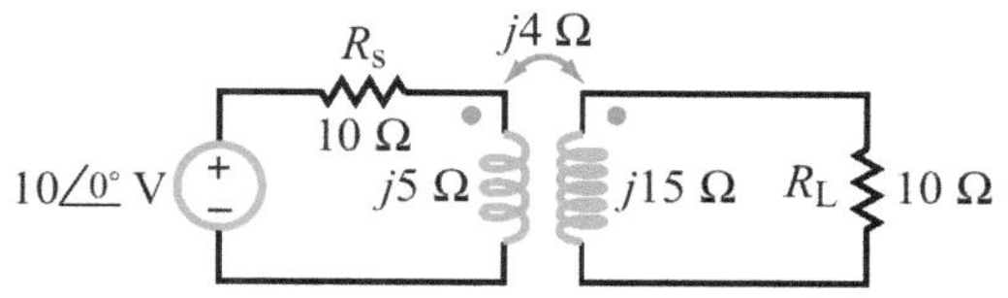

Figure P11.20: Circuit for Problem 11.20.

*11.21 Determine input impedance $\mathbf{Z}_{\text{in}}$ at terminals (a, b) for the circuit in Fig. P11.21.

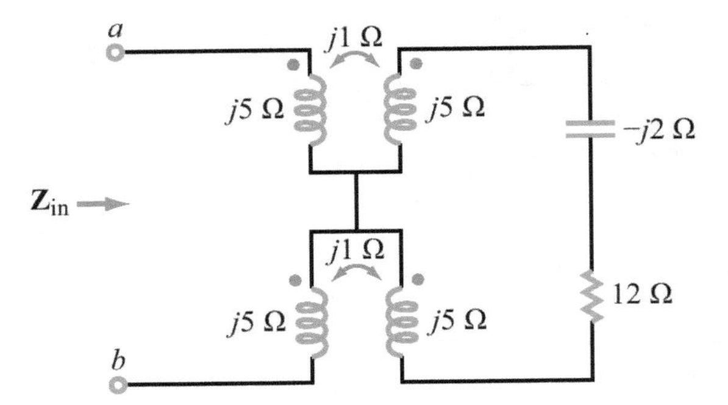

Figure P11.21: Circuit for Problem 11.21.

11.22 Determine the average power dissipated in the 10 Ω load in the circuit of Fig. P11.22, given that $\mathbf{V}_s = 10\angle 0^\circ$ V (rms).

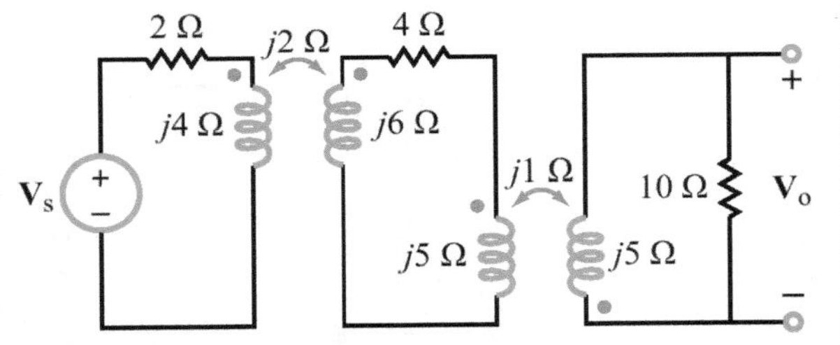

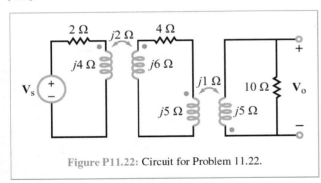

Figure P11.22: Circuit for Problem 11.22.

Section 11-4: Ideal Transformer

11.23 Determine $\mathbf{V}_{\text{out}}$ in the circuit of Fig. P11.23.

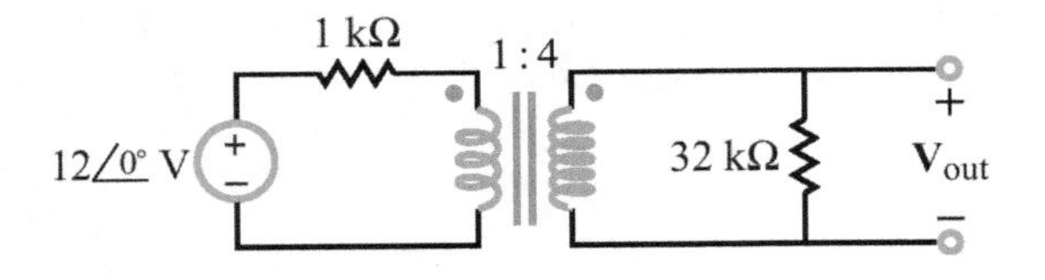

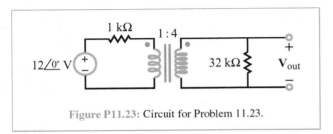

Figure P11.23: Circuit for Problem 11.23.

*11.24 Repeat Problem 11.23 after inserting a 0.5 μF capacitor in series with the 1 kΩ resistor. The angular frequency is 10^3 rad/s.

11.25 Determine the complex power supplied by the source in the circuit of Fig. P11.25.

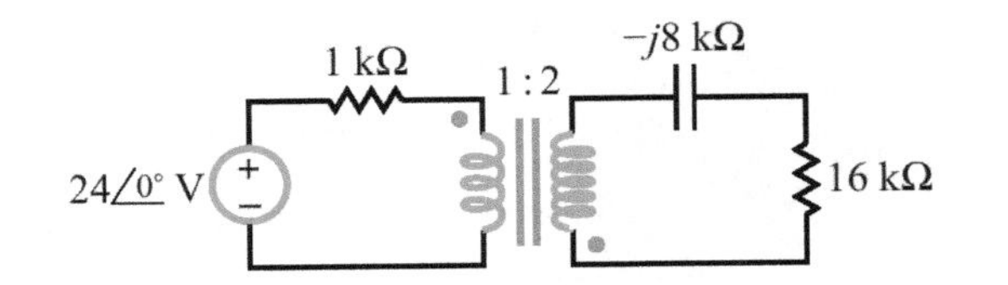

Figure P11.25: Circuit for Problem 11.25.

*11.26 Determine current $\mathbf{I}_x$ in the circuit of Fig. P11.26.

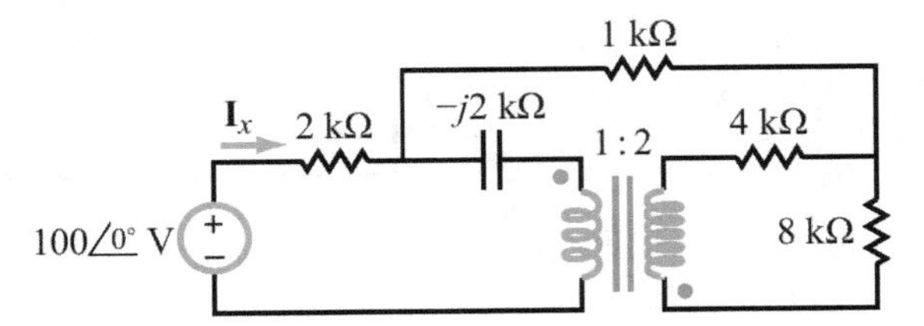

Figure P11.26: Circuit for Problem 11.26.

11.27 Determine currents $\mathbf{I}_1$ and $\mathbf{I}_2$ in the circuit of Fig. P11.27.

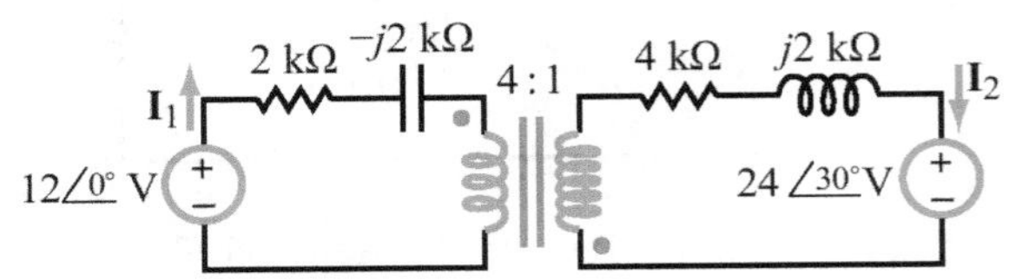

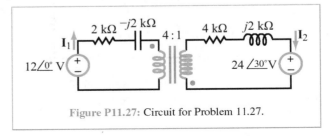

Figure P11.27: Circuit for Problem 11.27.

*11.28 Determine the average power delivered to R_{L} in the circuit of Fig. P11.28.

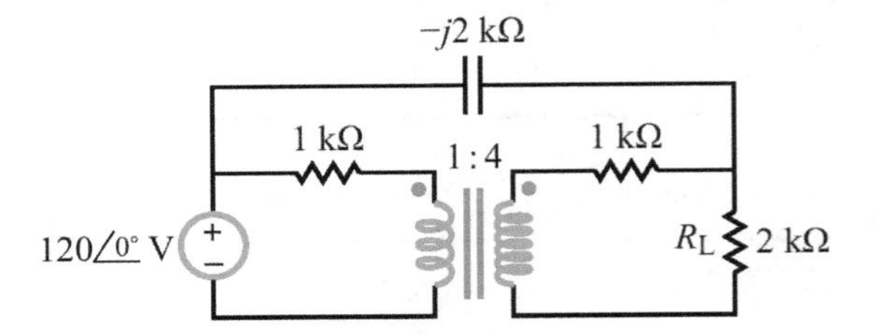

Figure P11.28: Circuit for Problem 11.28.

11.29 Determine the input impedance $\mathbf{Z}_{in}$ of the circuit in Fig. P11.29.

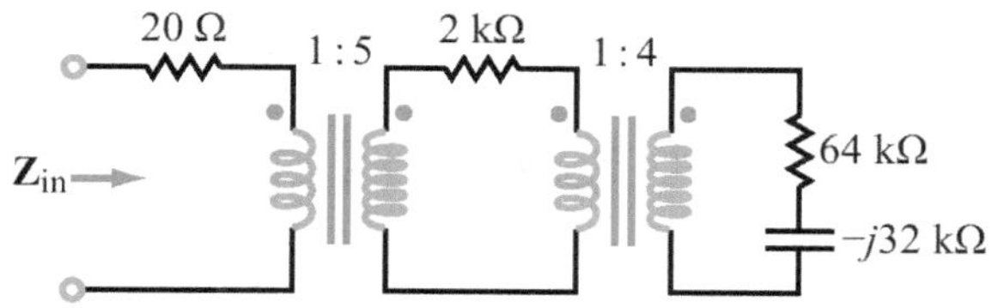

Figure P11.29: Circuit for Problem 11.29.

11.30 Determine the power absorbed by R_L in the circuit of Fig. P11.30.

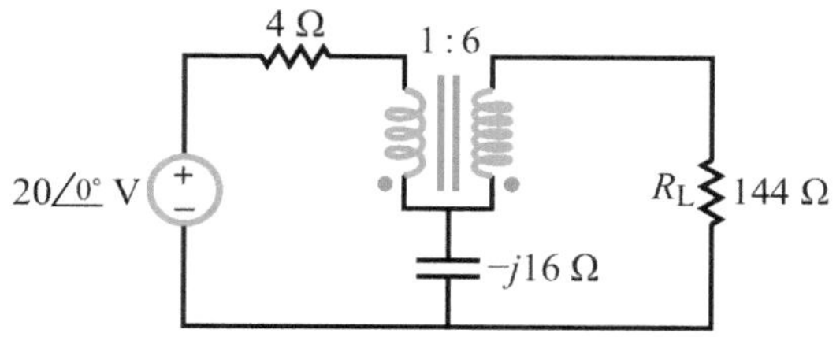

Figure P11.30: Circuit for Problem 11.30.

*11.31 Determine the average power delivered to $\mathbf{Z}_L$ in the circuit of Fig. P11.31, given that $N_1 = 50$ turns and $N_2 = 10$ turns.

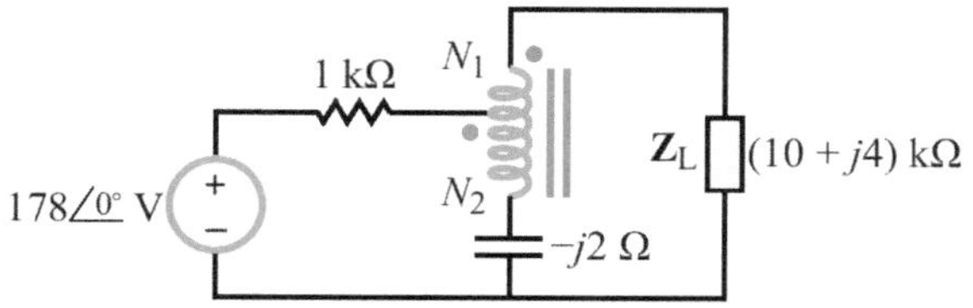

Figure P11.31: Circuit for Problem 11.31.

Section 11-5: Three-Phase Transformers

11.32 A Y-Δ ideal three-phase transformer with a turns ratio of 1 : 10 supplies a 32 kVA load at a line voltage of 208 V. Determine the line voltage and line current at the primary side.

*11.33 A Δ-Y ideal three-phase transformer supplies a 32 kVA load at a line voltage of 240 V. If the line voltage at the primary side is 51.96 V, what is the turns ratio?

Potpourri Questions

11.34 What particular features give an MRI advantages over other imaging systems?

11.35 An MRI machine uses a superconducting magnet to create a high magnetic field. For what purpose? Another magnetic field called the RF field is also used. Why?

CHAPTER 12

Circuit Analysis by Laplace Transform

Contents

Objectives

Learn to:

- Compute the Laplace transform of a time-dependent function.
- Perform partial fraction expansion.
- Analyze circuits using the Laplace transform technique.

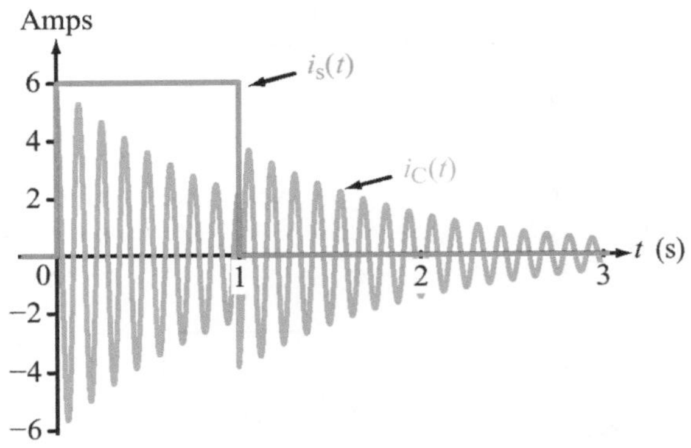

Source and capacitor currents.

Overview

In Chapters 5 and 6 we examined the transient response to a sudden change in voltage or current in circuits containing RC, RL, and RLC configurations. The excitation sources were dc voltage and current sources, in combination with SPST and SPDT switches. Voltage and current responses in such circuits are characterized by exponential functions of the form $e^{-\alpha t}$, where α is a damping coefficient that defines the rate at which the response transitions from its initial value immediately after the sudden change to its final value at $t = \infty$. The time-domain solution method employed in Chapters 5 and 6 is quite satisfactory, so long as the forcing function is a dc source and the differential equation describing the voltages and currents in the circuit is not higher than second order. Now, we introduce the Laplace transform analysis technique, which can be easily applied to a wide range of circuits and any type of realistic forcing function. In fact, the Laplace transform reduces to the phasor transform of Chapter 7 when the circuit sources are time-harmonic sinusoidal functions (and if the Laplace-transform analysis is confined to the steady state component of the overall solution). The phasor-domain technique served us well in Chapters 7–9, but it does not account for the transient component of the circuit response. Because in most ac circuits of interest, the transient component decays rapidly after connecting the sources to the circuit, the steady state solution provided by the phasor-domain technique is all that is needed. However, if we were to seek a solution that incorporates both the transient and steady state components, then the Laplace transform technique is the solution method of choice.

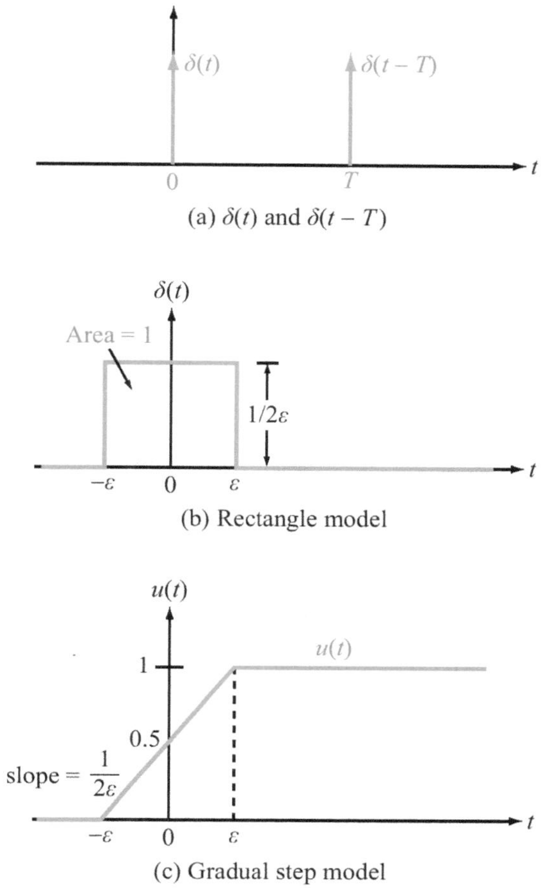

Figure 12-1: Unit impulse function.

12-1 Unit Impulse Function

The waveforms commonly encountered in electric circuits include a variety of continuous-time functions—such as the exponential and sinusoid—as well as some discontinuous functions, most notably the step and impulse functions. The *step function*, defined earlier in Section 5-3, is used to describe mathematically the instantaneous action by a switch to connect or disconnect a source to the circuit. In like manner, the *impulse function* is a useful mathematical tool for describing a sudden action of very short duration, or for sampling a continuous function at discrete points in time. An example of the latter is when a continuous visual image is stored by recording only 24 images per second, thereby sampling a continuous signal only at specific times.

Graphically, the *unit impulse function*—also known as the *delta function* $\delta(t)$—is represented by a vertical arrow, as shown in Fig. 12-1(a). If it is located at $t = T$, it is designated $\delta(t - T)$. For any fixed value T, the unit impulse function is defined through the combination of two properties:

$$\delta(t - T) = 0 \qquad \text{for } t \neq T, \tag{12.1a}$$

$$\int_{-\infty}^{\infty} \delta(t - T)\, dt = 1. \tag{12.1b}$$

► The first property states that the unit impulse function $\delta(t - T)$ is zero everywhere, except at its own location ($t = T$), but its value is infinite at that location! The second property states that the total area under the unit impulse function is equal to 1, regardless of its location. ◄

To visualize the meaning of the second property, we can represent the unit impulse function by the graphical rectangle shown in Fig. 12-1(b), with the understanding that $\delta(t)$ is defined in the limit as $\varepsilon \to 0$. The rectangle is of width $w = 2\varepsilon$ and height $h = \frac{1}{2}\,\varepsilon$, so its area is always unity, even as $\varepsilon \to 0$.

Since $\delta(t - T)$ in the integral of Eq. (12.1b) is, by definition, equal to zero everywhere except over an infinitesimally narrow range surrounding $t = T$, Eq. (12.1b) can be reexpressed as

$$\int_{T-\varepsilon}^{T+\varepsilon} \delta(t - T)\, dt = 1. \tag{12.2}$$

12-1.1 Relationship Between $u(t)$ and $\delta(t)$

According to the rectangle model displayed in Fig. 12-1(b), $\delta(t) = 1/(2\varepsilon)$ over the narrow range $-\varepsilon < t < \varepsilon$. For the gradual step model of $u(t)$ shown in Fig. 12-1(c), its slope also is $1/(2\varepsilon)$. Hence,

$$\frac{du(t)}{dt} = \delta(t). \tag{12.3}$$

Even though this relationship between the unit impulse and unit step functions was obtained on the basis of specific geometrical models for $\delta(t)$ and $u(t)$, its validity can be demonstrated to be true always. The corresponding expression for $u(t)$ is

$$u(t) = \int_{-\infty}^{t} \delta(\tau)\, d\tau, \tag{12.4}$$

where τ is a dummy integration variable introduced so as to avoid confusion between the integration variable τ and the upper limit of the integral t in Eq. (12.4). For the time-shifted case,

$$u(t - T) = \int_{-\infty}^{t} \delta(\tau - T)\, d\tau. \tag{12.5a}$$

$$\frac{d}{dt}\,[u(t - T)] = \delta(t - T), \tag{12.5b}$$

By extension, a scaled impulse $A\,\delta(t)$ has an area A and

$$\int_{-\infty}^{t} A\,\delta(\tau)\, d\tau = A\,u(t). \tag{12.6}$$

12-1.2 Sampling Property of $\delta(t)$

As was noted earlier, multiplying an impulse function by a constant A gives a scaled impulse of area A. Now we consider what happens when a continuous-time function $x(t)$ is multiplied by $\delta(t)$. Since $\delta(t)$ is zero everywhere except at $t = 0$, it follows that

$$x(t)\,\delta(t) = x(0)\,\delta(t), \tag{12.7}$$

provided that $x(t)$ is continuous at $t = 0$. By extension, multiplication of $x(t)$ by the time-shifted impulse function $\delta(t - T)$ gives

$$x(t)\,\delta(t - T) = x(T)\,\delta(t - T). \tag{12.8}$$

> ▶ Multiplication of a time-continuous function $x(t)$ by an impulse located at $t = T$ generates a scaled impulse of magnitude $x(T)$ at $t = T$, provided $x(t)$ is continuous at $t = T$. ◀

One of the most useful features of the impulse function is its *sampling property*. For any function $x(t)$ known to be continuous at $t = T$:

$$\int_{-\infty}^{\infty} x(t)\,\delta(t - T)\, dt = x(T). \tag{12.9}$$

(sampling property)

Derivation of the sampling property relies on Eqs. (12.1b) and (12.8):

$$\int_{-\infty}^{\infty} x(t)\,\delta(t - T)\, dt = \int_{-\infty}^{\infty} x(T)\,\delta(t - T)\, dt$$

$$= x(T) \int_{-\infty}^{\infty} \delta(t - T)\, dt = x(T).$$

> **Concept Question 12-1:** How is $u(t)$ related to $\delta(t)$? (See CAD)

> **Concept Question 12-2:** Why is Eq. (12.9) called the *sampling property* of the impulse function? (See CAD)

Exercise 12-1: If $x(t)$ is the rectangular pulse shown in Fig. E12.1(a), determine its time derivative $x'(t)$ and plot it.

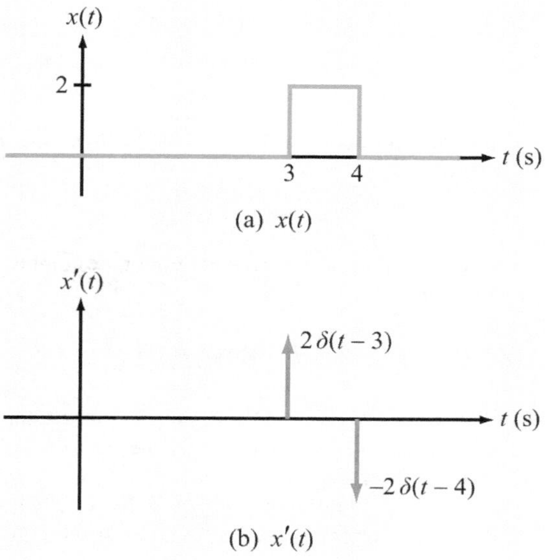

Figure E12.1

Answer: $x'(t) = 2\delta(t-3) - 2\delta(t-4)$. (See CAD)

12-2 The Laplace Transform Technique

▶ A *domain transformation* is a mathematical process that converts a set of variables from their domain into a corresponding set of variables defined in another domain. ◀

In reality, a circuit functions entirely in the time domain, with both its inputs (sources) and outputs (voltages and currents) expressed as functions of the time variable t. In the top horizontal sequence depicted in Fig. 12-2, application of KCL and KVL leads to the differential equation characterizing the output of interest, and its solution then yields the desired response. All mathematical steps are performed entirely in the time domain.

A transform is a mathematical operator that converts functions, such as $\upsilon(t)$ and $i(t)$, into counterpart functions defined in another domain. The Laplace transform is one such operator; it converts a function $\upsilon(t)$ defined in the time domain into a counterpart $\mathbf{V}(\mathbf{s})$ defined in another domain called the **s**-domain. The two domains may be thought of as "*parallel universes*" and a transformation is a "*transport*" between the two universes. By applying the Laplace transform, a circuit (or its associated differential equation) can be transformed to the **s**-domain. The transformed circuit is characterized by an algebraic equation—instead of a differential equation—with a relatively straightforward solution. By transforming

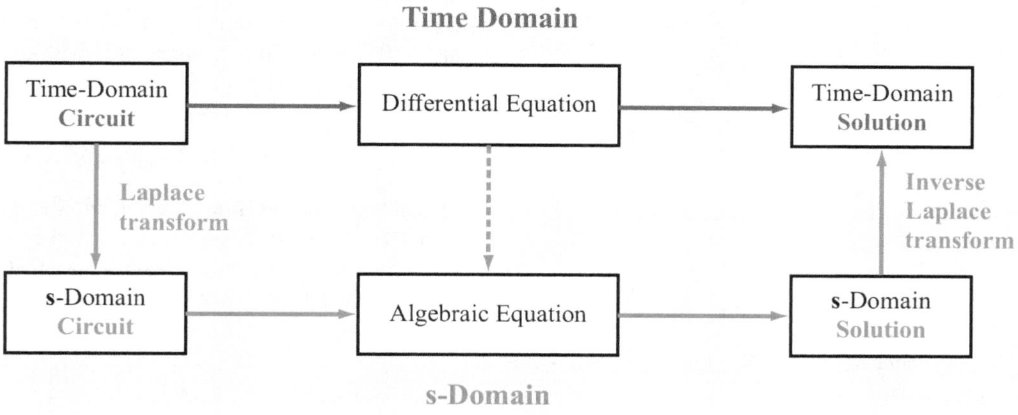

Figure 12-2: The top horizontal sequence involves solving a differential equation entirely in the time domain. The bottom horizontal sequence involves a much easier solution of a linear equation in the **s**-domain.

the solution back to the time domain, we end up with the same solution that we would have obtained had we solved the differential equation directly in the time domain. Even though the **s**-domain route involves transforming the circuit to the **s**-domain and transforming the solution to the time domain (which is called an inverse Laplace transformation), the overall solution process is considerably easier to implement than the traditional differential-equation solution method.

Solution Procedure: Laplace Transform

Step 1: The circuit is transformed to the Laplace domain—also known as the **s**-domain.

Step 2: In the **s**-domain, application of KVL and KCL yields a set of algebraic equations.

Step 3: The equations are solved for the variable of interest.

Step 4: The **s**-domain solution is transformed back to the time domain.

After introducing the Laplace transform and exploring its properties, we will demonstrate its capabilities by applying the outlined four-step procedure to analyze several types of passive and active circuits.

12-2.1 Definition of the Laplace Transform

The symbol $\mathcal{L}[f(t)]$ is a short-hand notation for "the Laplace transform of function $f(t)$." Usually denoted $\mathbf{F}(\mathbf{s})$, the Laplace transform is defined by

$$\mathbf{F}(\mathbf{s}) = \mathcal{L}[f(t)] = \int_{0^-}^{\infty} f(t)\, e^{-\mathbf{s}t}\, dt, \qquad (12.10)$$

where $\mathbf{s}$ is a complex variable with a real part σ and an imaginary part ω:

$$\mathbf{s} = \sigma + j\omega. \qquad (12.11)$$

Given that the exponent $\mathbf{s}t$ has to be dimensionless, $\mathbf{s}$ has the unit of inverse second, which is the same as Hz or rad/s. Moreover, since $\mathbf{s}$ is a complex quantity, it is often termed *complex frequency*.

In view of the definite limits on the integral in Eq. (12.10), the outcome of the integration will be an expression that depends on a single variable, $\mathbf{s}$. The transform operation converts a function $f(t)$ defined in the time domain into a function $\mathbf{F}(\mathbf{s})$ defined in the $\mathbf{s}$-domain. Functions $f(t)$ and $\mathbf{F}(\mathbf{s})$ are called a *Laplace transform pair*.

The *uniqueness property* of the Laplace transform states:

▶ A given $f(t)$ has a unique $\mathbf{F}(\mathbf{s})$, and vice versa. ◀

The uniqueness property can be expressed in symbolic form by

$$f(t) \longleftrightarrow \mathbf{F}(\mathbf{s}). \qquad (12.12a)$$

The two-way arrow is a short-hand notation for the combination of the two statements

$$\mathcal{L}[f(t)] = \mathbf{F}(\mathbf{s}), \qquad \mathcal{L}^{-1}[\mathbf{F}(\mathbf{s})] = f(t). \qquad (12.12b)$$

The first statement asserts that $\mathbf{F}(\mathbf{s})$ is the Laplace transform of $f(t)$, and the second one asserts that the *inverse Laplace transform* ($\mathcal{L}^{-1}[\ ]$) of $\mathbf{F}(\mathbf{s})$ is $f(t)$.

Because the lower limit on the integral in Eq. (12.10) is 0^-, $\mathbf{F}(\mathbf{s})$ is called a *one-sided transform*, in contrast with the *two-sided transform* for which the lower limit is $-\infty$. When we apply the Laplace transform technique to electric circuits, we select the start time for the circuit operation as $t = 0$, so the single-sided transform is plenty suitable for our intended use, and we will adhere to it exclusively in this book. Moreover, *unless noted to the contrary, it will be assumed that $f(t)$ is always multiplied by an implicit invisible step function $u(t)$*. The inquisitive reader may ask why we use 0^-, instead of simply 0, as our lower limit in the integral of Eq. (12.10). We use it as a reminder that the integration can include initial conditions at $t = 0^-$, which may be associated with the voltage across a capacitor or the current through an inductor.

12-2.2 Convergence Condition

Depending on the functional form of $f(t)$, the Laplace transform integral given by Eq. (12.10) may or may not converge to a finite value. If it does not, the Laplace transform does not exist. Convergence requires that

$$\int_{0^-}^{\infty} |f(t)\, e^{-\mathbf{s}t}|\, dt = \int_{0^-}^{\infty} |f(t)||e^{-\sigma t}||e^{-j\omega t}|\, dt$$
$$= \int_{0^-}^{\infty} |f(t)|e^{-\sigma t}\, dt < \infty, \qquad (12.13)$$

for some real value of σ. We used the fact that $|e^{-j\omega t}| = 1$ for any value of ωt and, since σ is real, $|e^{-\sigma t}| = e^{-\sigma t}$. If σ_c is the smallest value of σ for which the integral converges, then the ***region of convergence*** is $\sigma > \sigma_c$. *Fortunately, this convergence issue is somewhat esoteric to circuit analysts and designers because the waveforms of the excitation sources used in electric circuits are such that they do satisfy the convergence condition for all values of σ, and hence, their Laplace transforms do exist.*

12-2.3 Inverse Laplace Transform

Equation (12.10) allows us to obtain Laplace transform $\mathbf{F}(\mathbf{s})$ corresponding to time function $f(t)$. The inverse process, denoted $\mathcal{L}^{-1}[\mathbf{F}(\mathbf{s})]$, allows us to perform an integration on $\mathbf{F}(\mathbf{s})$ to obtain $f(t)$:

$$f(t) = \mathcal{L}^{-1}[\mathbf{F}(\mathbf{s})] = \frac{1}{2\pi j} \int_{\sigma - j\infty}^{\sigma + j\infty} \mathbf{F}(\mathbf{s})\, e^{\mathbf{s}t}\, d\mathbf{s}, \qquad (12.14)$$

where $\sigma > \sigma_c$. The integration, which has to be performed in the two-dimensional complex plane, is rather cumbersome and to be avoided if an alternative approach is available for converting $\mathbf{F}(\mathbf{s})$ into $f(t)$. Fortunately, there is an alternative approach. Instead of applying Eq. (12.14), we can generate a table of Laplace transform pairs for all of the time functions commonly encountered in electric circuits, and then use it, sort of like a look-up table, to transform the s-domain solution to the time domain. The validity of this approach is supported by the uniqueness property of the Laplace transform, which guarantees a one-to-one correspondence between every $f(t)$ and its corresponding $\mathbf{F}(\mathbf{s})$. The details of the inverse-transform process are covered in Section 12-4.

Example 12-1: Laplace Transforms of Singularity Functions

The step, rectangle, and impulse waveforms displayed in Fig. 12-3 are known as ***singularity functions***, because either they or their time derivatives exhibit discontinuities. Determine their Laplace transforms.

Solution: (a) The step function in Fig. 12-3(a) is given by

$$f_1(t) = A\, u(t - T).$$

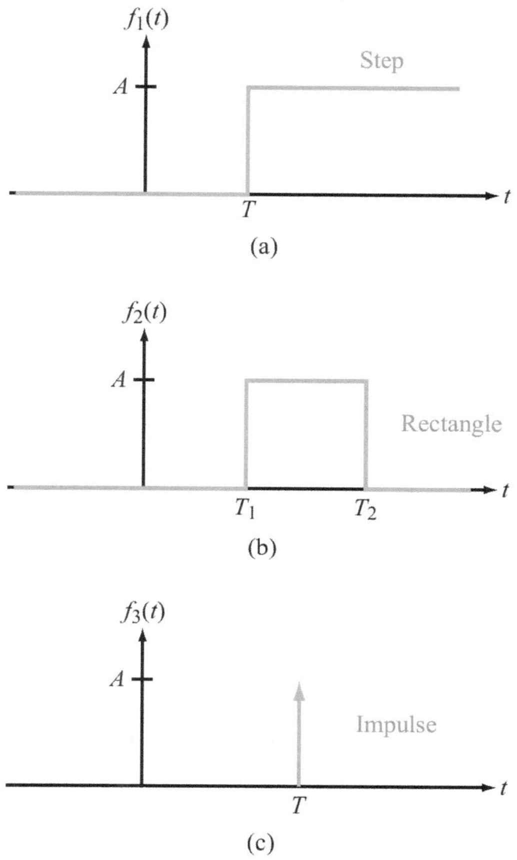

Figure 12-3: Singularity functions for Example 12-1.

Application of Eq. (12.10) gives

$$\mathbf{F}_1(\mathbf{s}) = \int_{0^-}^{\infty} f_1(t)\, e^{-\mathbf{s}t}\, dt$$

$$= \int_{0^-}^{\infty} A\, u(t - T)\, e^{-\mathbf{s}t}\, dt$$

$$= A \int_{T}^{\infty} e^{-\mathbf{s}t}\, dt = -\frac{A}{\mathbf{s}}\, e^{-\mathbf{s}t}\Big|_{T}^{\infty} = \frac{A}{\mathbf{s}}\, e^{-\mathbf{s}T}.$$

For the special case where $A = 1$ and $T = 0$ (the step occurs at $t = 0$), the transform pair becomes

$$u(t) \quad \longleftrightarrow \quad \frac{1}{\mathbf{s}} . \tag{12.15}$$

(b) The rectangle function in Fig. 12-3(b) can be constructed as the sum of two step functions:

$$f_2(t) = A[u(t - T_1) - u(t - T_2)],$$

and its Laplace transform is

$$\begin{aligned}\mathbf{F}_2(\mathbf{s}) &= \int_{0^-}^{\infty} A[u(t-T_1) - u(t-T_2)]e^{-\mathbf{s}t}\, dt \\ &= A\int_{0^-}^{\infty} u(t-T_1)\, e^{-\mathbf{s}t}\, dt - A\int_{0^-}^{\infty} u(t-T_2)\, e^{-\mathbf{s}t}\, dt \\ &= \frac{A}{\mathbf{s}}[e^{-\mathbf{s}T_1} - e^{-\mathbf{s}T_2}].\end{aligned}$$

(c) The impulse function in Fig. 12-3(c) is given by

$$f_3(t) = A\, \delta(t - T),$$

and the corresponding Laplace transform is

$$\begin{aligned}\mathbf{F}_3(\mathbf{s}) &= \int_{0^-}^{\infty} A\, \delta(t-T)\, e^{-\mathbf{s}t}\, dt \\ &= A\int_{T-\varepsilon}^{T+\varepsilon} \delta(t-T)\, e^{-\mathbf{s}t}\, dt \\ &= Ae^{-\mathbf{s}T}\int_{T-\varepsilon}^{T+\varepsilon} \delta(t-T)\, dt = Ae^{-\mathbf{s}T},\end{aligned}$$

where we have used the procedure introduced earlier in connection with Eq. (12.9). For the special case where $A = 1$ and $T = 0$, the Laplace transform pair simplifies to

$$\delta(t) \quad \longleftrightarrow \quad 1. \tag{12.16}$$

Example 12-2: Laplace Transform of $\cos\omega t$

Obtain the Laplace transform of $f(t) = [\cos(\omega t)]\, u(t)$.

Solution: Inclusion of $u(t)$ in the expression for $f(t)$ is simply a reminder of the implicit assumption common to all excitations considered in this chapter, namely that $f(t) = 0$ for $t < 0^-$.

The solution is facilitated by expressing $\cos\omega t$ in terms of complex exponentials (see Table 7-2), namely

$$\cos\omega t = \frac{1}{2}[e^{j\omega t} + e^{-j\omega t}].$$

Use of this expression in Eq. (12.10) gives

$$\begin{aligned}\mathbf{F}(\mathbf{s}) &= \int_{0^-}^{\infty} \cos\omega t\; u(t)\; e^{-\mathbf{s}t}\, dt \\ &= \frac{1}{2}\left[\int_{0}^{\infty} e^{j\omega t}e^{-\mathbf{s}t}\, dt + \int_{0}^{\infty} e^{-j\omega t}e^{-\mathbf{s}t}\, dt\right] \\ &= \frac{1}{2}\left[\frac{e^{(j\omega-\mathbf{s})t}}{j\omega-\mathbf{s}} + \frac{e^{-(j\omega+\mathbf{s})t}}{-(j\omega+\mathbf{s})}\right]\Bigg|_0^{\infty} = \frac{\mathbf{s}}{\mathbf{s}^2+\omega^2} .\end{aligned}$$

Hence,

$$[\cos(\omega t)]\, u(t) \quad \longleftrightarrow \quad \frac{\mathbf{s}}{\mathbf{s}^2+\omega^2} . \tag{12.17}$$

Concept Question 12-3: Is the uniqueness property of the Laplace transform unidirectional or bidirectional? Why is that significant? (See CAD)

Concept Question 12-4: Is convergence of the Laplace transform integral in doubt when applied to circuit analysis? If not, why not? (See CAD)

Exercise 12-2: Determine the Laplace transform of (a) $[\sin(\omega t)]\, u(t)$, (b) $e^{-at}\, u(t)$, and (c) $t\, u(t)$. Assume all waveforms are zero for $t < 0$.

Answer: (a) $[\sin(\omega t)]\, u(t) \longleftrightarrow \dfrac{\omega}{\mathbf{s}^2+\omega^2}$,
(b) $e^{-at}\, u(t) \longleftrightarrow \dfrac{1}{\mathbf{s}+a}$, (c) $t\, u(t) \longleftrightarrow \dfrac{1}{\mathbf{s}^2}$. (See CAD)

Technology Brief 28
3-D TV

Attempts to produce stereoscopic perception of depth in images have been recorded since at least the mid-19th century. *Stereopsis* is the impression of depth that arises when humans and other animals view the world using two eyes. Since each eye is at a slightly different location with respect to any object in a viewed scene, the brain can use the difference between the left and right eye images to extract information about depth (and, thus, perceived three dimensionality). Stereopsis was first described in detail by Charles Wheatstone in the 1830s (although it had been observed but not properly understood during the Italian Renaissance).

A variety of devices, usually termed ***stereoscopes***, were constructed during the Victorian era which presented viewers with slightly different images to the left and right eyes (usually by projecting them through lenses into each eye separately).

Anaglyphic 3-D

The rise of modern cinema saw several additional attempts to convey stereoscopic information to the big screen. The most popular of these, at its peak in the 1950s and 60s, was ***anaglyphic*** projection (Fig. TF28-1(a)). Viewers watching a projected movie (or television screen) wore glasses with different color filters in front of each eye (usually cyan and red). Two images were projected simultaneously on the screen, such that only one image could pass each filter and be perceived by the eye. Since each of the two images or films had been captured by cameras slightly offset from each other (to mimic the separation of human eyes), the brain stitched these images together somewhat naturally and perceived depth and "3-D" in the film. Anaglyphic technologies suffer somewhat in that they do not provide perfect color reconstruction and can often produce blurry or ghost images (depending on the quality of the filters used).

The New Rise of 3-D TV

The rapid maturation of flat-screen television, leading to very high resolution, very fast refresh times, high contrast ratios and deep color reproduction, enabled a new resurgence in 3-D technologies in the last few years. Several technologies are currently competing for market dominance (with others in development).

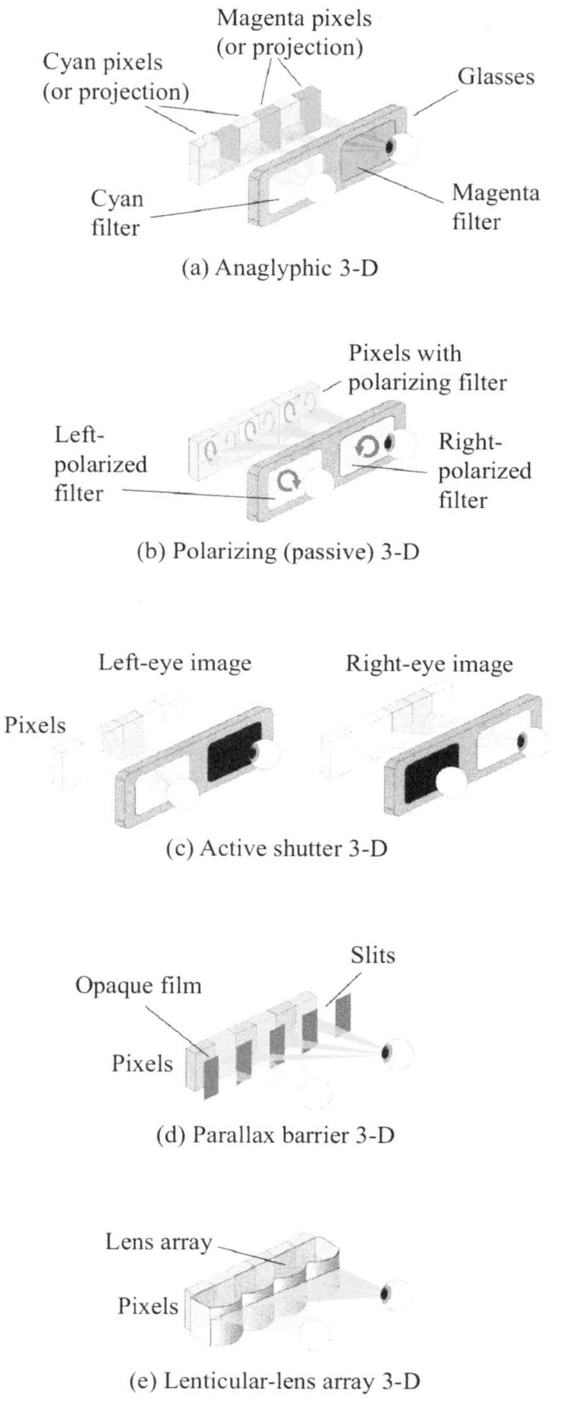

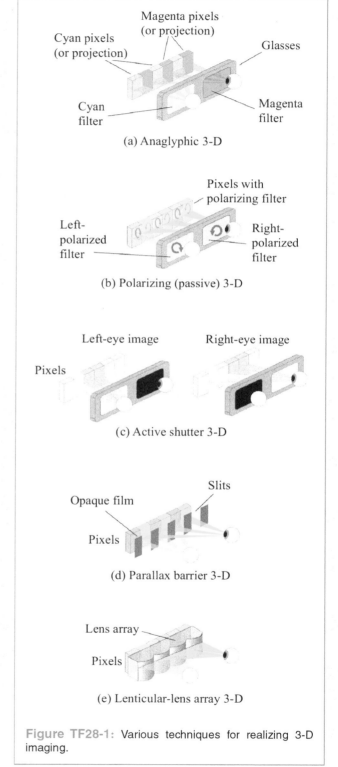

Figure TF28-1: Various techniques for realizing 3-D imaging.

Polarizing 3-D

So-called *polarizing* or *passive* 3-D TV is in some ways similar to anaglyphic systems. Instead of employing color filters, 3-D passive TV glasses use *light polarization* to deliver different images to the left and right eye (Fig. TF28-1(b)). Light is, of course, electromagnetic radiation perceived by the human eye. Traveling waves of electromagentic radiation are composed of oscillating electric and magnetic fields with specific orientations or *polarizations* (this does not refer to the direction of travel of the light, but rather, the directions in which the light's electric and magnetic fields oscillate as the wave travels through space). The details of light polarization can be complex, but of importance to us is the fact that polarizations can be complementary; for example, light can travel with clockwise polarity or counter-clockwise polarity with respect to its source. The human eye cannot tell the difference between different polarizations of light.

Passive 3-D television sets contain polarizing filters placed in front of the pixels in the display; with these filters, half of the pixels in the television emit clockwise polarized light and the other half produce counterclockwise polarized light. The viewer wears glasses which let only one polarity of light through to the left eye and the other to the right eye. In this way, each eye is presented with a different image. This type of 3-D TV has advantages in that the glasses are very lightweight. Historically, the main drawback of 3-D images was the decrease in resolution of the image (since each eye received half of the total pixels of the television). Recent advances, however, may be solving this problem. One approach, made possible by the speed of modern pixels, is to present each eye with half of each full-resolution stereoscopic image (as above) and then, *very quickly*, present the eyes with the other half of a full resolution image. This requires televisions that can refresh images at 120 Hz (120 times each second or a new image every 8.3 milliseconds); this is about double the speed at which a human eye can perceive flicker!

Active Shutter 3-D

Active shutter 3-D or *alternate frame-sequencing 3-D* sets also use glasses but they tend to be heavier and more expensive (Fig. TF28-1(c)). This type of technology uses a normal flat-screen TV (but it must be fast) to display the images intended for the left eye alternating in time with the images intended for the right eye. In other words, while watching a movie, the television displays a frame intended for the left eye followed by a frame intended for the right eye, and so on. The glasses hold LCD screens over each eye and receive a synchronization signal from the television (either infrared or radio frequency). In sync with the TV, the glasses block light to the left eye (by darkening the LCD), then block light to the right eye, and so on, repeating this sequence 24 times a second or faster. In this way, each eye receives the stereoscopic pair intended for it at full resolution. Unlike traditional passive 3-D TVs, all pixels are used for each frame of the image.

Parallax Barrier 3-D

Parallax-barrier displays are a *glasses-free* 3-D display technology that has been around for a number of years but is beginning to make it into prototype flat-panel televisions. Parallax-barrier technology was used in the Nintendo 3-DS hand-held and several 3-D smartphones. The idea behind parallax barrier technology is shown in Fig. TF28-1(d). An opaque film with precisely aligned slots is fabricated over the display pixels; the slots are intended to block light from some pixels from reaching the left eye and to block light from other pixels from reaching the right eye for a viewer standing directly in front of the display. The principal advantage of parallax-barrier displays is that no glasses are needed; anyone standing in front of the TV sees images in stereoscopic 3-D. The two principal disadvantages are the halved resolution (similar to traditional passive 3-D, as explained above), since light from only about half the pixels reach each eye, and the limited viewing angle for 3-D. Typical prototypes only work within a 20° angle on either side of center, making it less attractive for consumer use (although the technology is evolving fast).

Lenticular-Lens Arrays

Lenticular-lens arrays (Fig. TF28-1(e)) work in a similar manner to parallax viewing except that the light from a given pixel is *focused* onto the right or left eyes (as opposed to blocked) by an array of precisely fabricated lenses (a very thin plastic sheet is usually molded into a regular array of lenses) sitting over the display. Lenticular-lens displays currently suffer from similar drawbacks to parallax-barrier displays and are currently very expensive. As with parallax-barrier technology, several companies are actively pursuing this technology and prices may drop rapidly as the technology matures.

12-3 Properties of the Laplace Transform

The Laplace transform has a number of useful, *universal properties* that apply to any function $f(t)$, greatly facilitating the process of transforming a circuit from the t-domain to the s-domain. This section will conclude with a table outlining several universal properties of the Laplace transform, to which we will be making frequent reference throughout this chapter. Some of these properties are intuitively obvious, while others may require some elaboration.

12-3.1 Time Scaling

If

$$f(t) \leftrightarrow \mathbf{F}(\mathbf{s}), \qquad (12.18)$$

then the transform of the time-scaled function $f(at)$ is

$$f(at) \leftrightarrow \frac{1}{a}\,\mathbf{F}\left(\frac{\mathbf{s}}{a}\right), \qquad a > 0. \qquad (12.19)$$

(time-scaling property)

▶ The *time scaling* property states that stretching the time axis by a factor a corresponds to shrinking the $\mathbf{s}$ axis and the amplitude of $\mathbf{F}(\mathbf{s})$ by the same factor, and vice versa. ◀

To prove Eq. (12.19), we start with the standard definition of the Laplace transform given by Eq. (12.10):

$$\mathcal{L}[f(at)] = \int_{0^-}^{\infty} f(at)\, e^{-\mathbf{s}t}\, dt. \qquad (12.20)$$

In the integral, if we set $\tau = at$ and $dt = \frac{1}{a}\, d\tau$, we have

$$\mathcal{L}[f(at)] = \frac{1}{a}\int_{0^-}^{\infty} f(\tau)\, e^{-(\mathbf{s}/a)\tau}\, d\tau$$

$$= \frac{1}{a}\int_{0^-}^{\infty} f(\tau)\, e^{-\boldsymbol{\mu}\tau}\, d\tau, \qquad \text{with } \boldsymbol{\mu} = \frac{\mathbf{s}}{a}. \qquad (12.21)$$

The definite integral is identical in form to the Laplace transform definition given by Eq. (12.10), except that the dummy variable is τ instead of t, and the coefficient of the exponent is $\boldsymbol{\mu} = \mathbf{s}/a$ instead of just $\mathbf{s}$. Hence,

$$\mathcal{L}[f(at)] = \frac{1}{a}\,\mathbf{F}(\boldsymbol{\mu}) = \frac{1}{a}\,\mathbf{F}\left(\frac{\mathbf{s}}{a}\right), \qquad a > 0, \qquad (12.22)$$

which proves the time-scaling property defined by Eq. (12.19).

12-3.2 Time Shift

If t is shifted by T along the time axis, with $T \geq 0$, then

$$f(t-T)\, u(t-T) \leftrightarrow e^{-T\mathbf{s}}\, \mathbf{F}(\mathbf{s}), \quad T \geq 0. \qquad (12.23)$$

(time-shift property)

The validity of this property is demonstrated as follows:

$$\begin{aligned}
\mathcal{L}[f(t-T)\, u(t-T)] &= \int_{0^-}^{\infty} f(t-T)\, u(t-T)\, e^{-\mathbf{s}t}\, dt \\
&= \int_{T}^{\infty} f(t-T)\, e^{-\mathbf{s}t}\, dt \\
&= \int_{0}^{\infty} f(x)\, e^{-\mathbf{s}(x+T)}\, dx \\
&= e^{-T\mathbf{s}} \int_{0}^{\infty} f(x)\, e^{-\mathbf{s}x}\, dx \\
&= e^{-T\mathbf{s}}\, \mathbf{F}(\mathbf{s}),
\end{aligned} \qquad (12.24)$$

where we made the substitutions $t - T = x$ and $dt = dx$, and then applied the definition for $\mathbf{F}(\mathbf{s})$ given by Eq. (12.10).

To illustrate the utility of the timeshift property, we consider the cosine function of Example 12-2, where it was shown that

$$[\cos(\omega t)]\, u(t) \leftrightarrow \frac{\mathbf{s}}{\mathbf{s}^2 + \omega^2}. \qquad (12.25)$$

According to Eq. (12.23),

$$[\cos \omega(t-T)]\, u(t-T) \leftrightarrow e^{-T\mathbf{s}}\, \frac{\mathbf{s}}{\mathbf{s}^2 + \omega^2}. \qquad (12.26)$$

Had we analyzed a linear circuit driven by a sinusoidal voltage source that started at $t = 0$, and then we wanted to reanalyze it anew, but wanted to delay both the cosine function and the start time by T, Eq. (12.26) would provide an expedient solution to obtaining the transform of the delayed cosine function.

Exercise 12-3: Determine $\mathcal{L}[\sin \omega(t-T)\, u(t-T)]$ for $T \geq 0$.

Answer:

$$e^{-T\mathbf{s}}\, \frac{\omega}{\mathbf{s}^2 + \omega^2}.$$

(See CAD)

12-3.3 Frequency Shift

According to the time-shift property, if t is replaced with $(t - T)$ in the time domain, $\mathbf{F}(\mathbf{s})$ gets multiplied by $e^{-T\mathbf{s}}$ in the $\mathbf{s}$-domain. Within a $(-)$ sign, the converse is also true: if $\mathbf{s}$ is replaced with $(\mathbf{s} + a)$ in the $\mathbf{s}$-domain, $f(t)$, gets multiplied by e^{-at} in the time domain. Thus,

$$e^{-at} f(t) \quad \longleftrightarrow \quad \mathbf{F}(\mathbf{s}+a). \qquad (12.27)$$

(frequency shift property)

Proof of Eq. (12.27) is part of Exercise 12-4.

Concept Question 12-5: According to the time scaling property of the Laplace transform, "stretching the time axis corresponds to shrinking the $\mathbf{s}$ axis." What does that mean? (See CAD)

Concept Question 12-6: Explain the similarities and differences between the time-shift and frequency-shift properties of the Laplace transform. (See CAD)

Exercise 12-4: (a) Prove Eq. (12.27) and (b) apply it to determine $\mathcal{L}\{[e^{-at}\cos(\omega t)]\, u(t)\}$.

Answer: (a) (See C3),
(b) $[e^{-at}\cos(\omega t)]\, u(t) \longleftrightarrow \dfrac{\mathbf{s}+a}{(\mathbf{s}+a)^2+\omega^2}$. (See CAD)

12-3.4 Time Differentiation

Differentiating $f(t)$ in the time domain is equivalent to: (a) multiplying $\mathbf{F}(\mathbf{s})$ by $\mathbf{s}$ in the $\mathbf{s}$-domain, and then (b) subtracting $f(0^-)$ from $\mathbf{s}\,\mathbf{F}(\mathbf{s})$:

$$f' = \frac{df}{dt} \quad \longleftrightarrow \quad \mathbf{s}\,\mathbf{F}(\mathbf{s}) - f(0^-). \qquad (12.28)$$

(time-differentiation property)

To verify the validity of Eq. (12.28), we start with the standard definition for the Laplace transform:

$$\mathcal{L}[f'] = \int_{0^-}^{\infty} \frac{df}{dt}\, e^{-\mathbf{s}t}\, dt. \qquad (12.29)$$

Integration by parts, with

$$x = e^{-\mathbf{s}t}, \qquad dx = -\mathbf{s}e^{-\mathbf{s}t}\, dt,$$
$$dy = \left(\frac{df}{dt}\right) dt, \qquad \text{and} \qquad y = f,$$

gives

$$\begin{aligned}\mathcal{L}[f'] &= xy\Big|_{0^-}^{\infty} - \int_{0^-}^{\infty} y\, dx \\ &= e^{-\mathbf{s}t} f(t)\Big|_{0^-}^{\infty} - \int_{0^-}^{\infty} -\mathbf{s}\, f(t)\, e^{-\mathbf{s}t}\, dt \\ &= -f(0^-) + \mathbf{s}\,\mathbf{F}(\mathbf{s}), \end{aligned} \qquad (12.30)$$

which is equivalent to Eq. (12.28).

Higher derivatives can be obtained by repeating the application of Eq. (12.28). For the second derivative of $f(t)$,

$$f'' = \frac{d^2 f}{dt^2} \quad \longleftrightarrow \quad \mathbf{s}^2\,\mathbf{F}(\mathbf{s}) - \mathbf{s}\, f(0^-) - f'(0^-),$$

(second-derivative property) (12.31)

where $f'(0^-)$ is the derivative of $f(t)$, evaluated at $t = 0^-$.

12-3.5 Time Integration

Integration of $f(t)$ in the time domain is equivalent to dividing $\mathbf{F}(\mathbf{s})$ by $\mathbf{s}$ in the $\mathbf{s}$-domain:

$$\int_{0^-}^{t} f(\tau)\, d\tau \quad \longleftrightarrow \quad \frac{1}{\mathbf{s}}\,\mathbf{F}(\mathbf{s}). \qquad (12.32)$$

(time-integration property)

Application of the Laplace transform definition gives

$$\mathcal{L}\left[\int_{0^-}^{t} f(\tau)\, d\tau\right] = \int_{0^-}^{\infty}\left[\int_{0^-}^{t} f(\tau)\, d\tau\right] e^{-st}\, dt, \qquad (12.33)$$

where, for the sake of clarity, we changed the dummy variable in the inner integral from t to τ. Integration by parts with

$$x = \int_{0}^{t} f(\tau)\, d\tau \qquad dx = f(\tau)\, d\tau,$$

$$dy = e^{-st}\, dt \qquad y = -\frac{e^{-st}}{\mathbf{s}},$$

leads to

$$\mathcal{L}\left[\int_{0^-}^{t} f(\tau)\, d\tau\right]$$

$$= xy\Big|_{0^-}^{\infty} - \int_{0^-}^{\infty} y\, dx$$

$$= \left[-\frac{e^{-st}}{\mathbf{s}}\int_{0^-}^{t} f(\tau)\, d\tau\right]\Bigg|_{0^-}^{\infty} + \frac{1}{\mathbf{s}}\int_{0^-}^{\infty} f(t)\, e^{-st}\, dt = \frac{1}{\mathbf{s}}\,\mathbf{F}(\mathbf{s}). \qquad (12.34)$$

Both limits on the first term on the right-hand side yield zero values; at the upper limit,

$$e^{-st}\Big|_{t=\infty} = 0,$$

and at the lower limit,

$$\int_{0^-}^{0^-} f(\tau)\, d\tau = 0.$$

To illustrate the utility of the time-integration property given by Eq. (12.32), we consider the relationships between $\delta(t)$, $u(t)$, and $r(t)$. From Eq. (12.16), we have

$$\delta(t) \;\leftrightarrow\; 1.$$

Since $u(t)$ is equal to the time integral of $\delta(t)$, and $r(t)$ is the time integral of $u(t)$, it follows that

$$u(t) = \int_{0^-}^{t} \delta(\tau)\, d\tau \;\leftrightarrow\; \frac{1}{\mathbf{s}},$$

and

$$r(t) = \int_{0^-}^{t} u(\tau)\, d\tau \;\leftrightarrow\; \frac{1}{\mathbf{s}^2}.$$

Table 12-1 provides a summary of the major properties of the Laplace transform, and Table 12-2 provides a list of Laplace transforms of commonly encountered time functions.

Exercise 12-5: Obtain the Laplace transform of
(a) $f_1(t) = 2(2 - e^{-t})\, u(t)$ and
(b) $f_2(t) = e^{-3t}\cos(2t + 30°)\, u(t)$.

Answer: (a) $\mathbf{F}_1(\mathbf{s}) = \dfrac{2\mathbf{s}+4}{\mathbf{s}(\mathbf{s}+1)}$,
(b) $\mathbf{F}_2(\mathbf{s}) = \dfrac{0.866\mathbf{s} + 3.6}{\mathbf{s}^2 + 6\mathbf{s} + 13}$. (See CAD)

12-4 Circuit Analysis Procedure

Now that we have learned how to transform a time-domain function $f(t)$ to its Laplace counterpart $\mathbf{F}(\mathbf{s})$, we shall demonstrate the basic steps of the Laplace transform technique by analyzing a relatively simple circuit. Figure 12-4 contains a series RLC circuit, with no stored energy, connected to a dc voltage source V_o via a SPST switch that closes at $t = 0$. Hence, the source should be represented as

$$\upsilon_s(t) = V_o\, u(t). \qquad (12.35)$$

Fundamentally, the Laplace transfer technique consists of four steps:

Step 1: Apply KCL and/or KVL to obtain the integrodifferential equation(s) of the circuit for $t \geq 0$

For the circuit in Fig. 12-4(a), KVL at $t \geq 0$ gives

$$R\, i(t) + \upsilon_C(t) + L\,\frac{di}{dt} = V_o\, u(t), \qquad (12.36)$$

Table 12-1: **Properties of the Laplace transform ($f(t) = 0$ for $t < 0^-$).**

Property	$f(t)$		$\mathbf{F}(\mathbf{s}) = \mathcal{L}[f(t)]$
1. Multiplication by constant	$K\, f(t)$	$\longleftrightarrow$	$K\, \mathbf{F}(\mathbf{s})$
2. Linearity	$K_1\, f_1(t) + K_2\, f_2(t)$	$\longleftrightarrow$	$K_1\, \mathbf{F}_1(\mathbf{s}) + K_2\, \mathbf{F}_2(\mathbf{s})$
3. Time scaling	$f(at), \quad a > 0$	$\longleftrightarrow$	$\frac{1}{a}\, \mathbf{F}\left(\frac{\mathbf{s}}{a}\right)$
4. Time shift	$f(t-T)\, u(t-T)$	$\longleftrightarrow$	$e^{-T\mathbf{s}}\, \mathbf{F}(\mathbf{s}), \quad T \geq 0$
5. Frequency shift	$e^{-at}\, f(t)$	$\longleftrightarrow$	$\mathbf{F}(\mathbf{s}+a)$
6. Time 1st derivative	$f' = \frac{df}{dt}$	$\longleftrightarrow$	$\mathbf{s}\, \mathbf{F}(\mathbf{s}) - f(0^-)$
7. Time 2nd derivative	$f'' = \frac{d^2 f}{dt^2}$	$\longleftrightarrow$	$\mathbf{s}^2\mathbf{F}(\mathbf{s}) - \mathbf{s} f(0^-) - f'(0^-)$
8. Time integral	$\int_{0^-}^{t} f(\tau)\, d\tau$	$\longleftrightarrow$	$\frac{1}{\mathbf{s}}\, \mathbf{F}(\mathbf{s})$
9. Frequency derivative	$t\, f(t)$	$\longleftrightarrow$	$-\frac{d}{d\mathbf{s}}\, \mathbf{F}(\mathbf{s})$
10. Frequency integral	$\frac{f(t)}{t}$	$\longleftrightarrow$	$\int_{\mathbf{s}}^{\infty} \mathbf{F}(\mathbf{s}')\, d\mathbf{s}'$

where $i(t)$ is the current flowing through the loop and $\upsilon_C(t)$ is the voltage across C. By invoking the i-υ relationship for C, Eq. (12.36) becomes

$$Ri + \left[\frac{1}{C}\int_{0^-}^{t} i\, dt + \upsilon_C(0^-)\right] + L\,\frac{di}{dt} = V_0\, u(t), \qquad (12.37)$$

which now contains a single dependent variable, $i(t)$.

Step 2: Define Laplace transform currents and voltages corresponding to the time-domain currents and voltages and then transform the equation to the s-domain

We designate $\mathbf{I}(\mathbf{s})$ as the $\mathbf{s}$-domain counterpart of $i(t)$,

$$i(t) \longleftrightarrow \mathbf{I}(\mathbf{s}). \qquad (12.38)$$

To transform Eq. (12.37) to the $\mathbf{s}$-domain, we apply the appropriate property or *Laplace transformation* (LT) from Tables 12-1 and 12-2, as follows:

$$R\, i(t) \longleftrightarrow R\, \mathbf{I}(\mathbf{s})$$

(multiplication by constant),

$$\frac{1}{C}\int_{0^-}^{t} i\, dt \longleftrightarrow \frac{1}{C}\,\frac{\mathbf{I}(\mathbf{s})}{\mathbf{s}} \quad \textbf{(time-integral property)},$$

$$\upsilon_C(0^-) \longleftrightarrow \frac{\upsilon_C(0^-)}{\mathbf{s}} \quad \textbf{(LT of a constant)},$$

$$L\,\frac{di}{dt} \longleftrightarrow L[\mathbf{s}\,\mathbf{I}(\mathbf{s}) - i(0^-)]$$

(time derivative property),

$$V_0\, u(t) \longleftrightarrow \frac{V_0}{\mathbf{s}} \quad \textbf{(LT of a constant)}.$$

The opening paragraph of this section stated that the circuit had no stored energy prior to $t = 0$. Hence, $\upsilon_C(0^-) = 0$ and $i(0^-) = 0$. Replacing each of the terms in Eq. (12.37) with its s-domain counterpart leads to

$$R\mathbf{I} + \frac{\mathbf{I}}{C\mathbf{s}} + L\mathbf{s}\mathbf{I} = \frac{V_0}{\mathbf{s}} \quad \textbf{(s-domain)}. \qquad (12.39)$$

Table 12-2: **Examples of Laplace transform pairs for $T \geq 0$. Note that multiplication by $u(t)$ guarantees that $f(t) = 0$ for $t < 0^-$.**

Laplace Transform Pairs			
	$f(t)$		$\mathbf{F}(\mathbf{s}) = \mathcal{L}[f(t)]$
1	$\delta(t)$	$\longleftrightarrow$	1
1a	$\delta(t-T)$	$\longleftrightarrow$	$e^{-T\mathbf{s}}$
2	1 or $u(t)$	$\longleftrightarrow$	$\dfrac{1}{\mathbf{s}}$
2a	$u(t-T)$	$\longleftrightarrow$	$\dfrac{e^{-T\mathbf{s}}}{\mathbf{s}}$
3	$e^{-at}\ u(t)$	$\longleftrightarrow$	$\dfrac{1}{\mathbf{s}+a}$
3a	$e^{-a(t-T)}\ u(t-T)$	$\longleftrightarrow$	$\dfrac{e^{-T\mathbf{s}}}{\mathbf{s}+a}$
4	$t\ u(t)$	$\longleftrightarrow$	$\dfrac{1}{\mathbf{s}^2}$
4a	$(t-T)\ u(t-T)$	$\longleftrightarrow$	$\dfrac{e^{-T\mathbf{s}}}{\mathbf{s}^2}$
5	$t^2\ u(t)$	$\longleftrightarrow$	$\dfrac{2}{\mathbf{s}^3}$
6	$te^{-at}\ u(t)$	$\longleftrightarrow$	$\dfrac{1}{(\mathbf{s}+a)^2}$
7	$t^2e^{-at}\ u(t)$	$\longleftrightarrow$	$\dfrac{2}{(\mathbf{s}+a)^3}$
8	$t^{n-1}e^{-at}\ u(t)$	$\longleftrightarrow$	$\dfrac{(n-1)!}{(\mathbf{s}+a)^n}$
9	$\sin\omega t\ u(t)$	$\longleftrightarrow$	$\dfrac{\omega}{\mathbf{s}^2+\omega^2}$
10	$\sin(\omega t+\theta)\ u(t)$	$\longleftrightarrow$	$\dfrac{\mathbf{s}\sin\theta+\omega\cos\theta}{\mathbf{s}^2+\omega^2}$
11	$\cos\omega t\ u(t)$	$\longleftrightarrow$	$\dfrac{\mathbf{s}}{\mathbf{s}^2+\omega^2}$
12	$\cos(\omega t+\theta)\ u(t)$	$\longleftrightarrow$	$\dfrac{\mathbf{s}\cos\theta-\omega\sin\theta}{\mathbf{s}^2+\omega^2}$
13	$e^{-at}\sin\omega t\ u(t)$	$\longleftrightarrow$	$\dfrac{\omega}{(\mathbf{s}+a)^2+\omega^2}$
14	$e^{-at}\cos\omega t\ u(t)$	$\longleftrightarrow$	$\dfrac{\mathbf{s}+a}{(\mathbf{s}+a)^2+\omega^2}$
15	$2e^{-at}\cos(bt-\theta)\ u(t)$	$\longleftrightarrow$	$\dfrac{e^{j\theta}}{\mathbf{s}+a+jb}+\dfrac{e^{-j\theta}}{\mathbf{s}+a-jb}$
16	$\dfrac{2t^{n-1}}{(n-1)!}\ e^{-at}\cos(bt-\theta)\ u(t)$	$\longleftrightarrow$	$\dfrac{e^{j\theta}}{(\mathbf{s}+a+jb)^n}+\dfrac{e^{-j\theta}}{(\mathbf{s}+a-jb)^n}$

Note: $(n-1)! = (n-1)(n-2)\dots 1$.

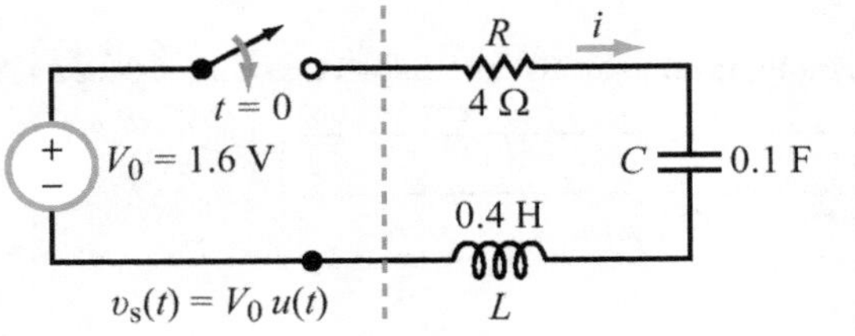

(a) RLC circuit

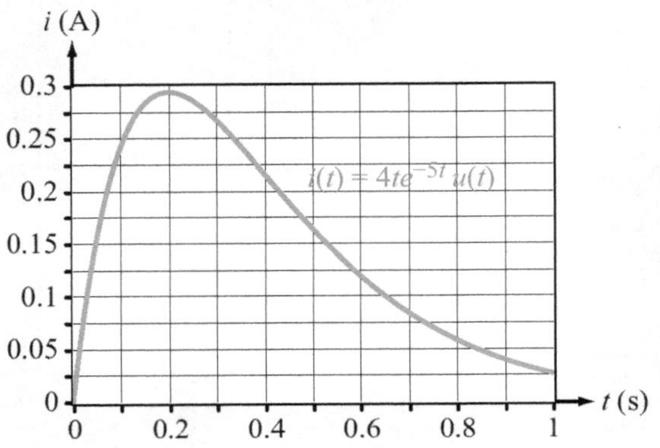

(b) $i(t)$

Figure 12-4: The dc source, in combination with the switch, constitutes an input excitation $\upsilon_s(t) = V_0\, u(t)$.

Step 3: Solve for the variable of interest in the s-domain

Solving for $\mathbf{I(s)}$, and then replacing R, L, C, and V_0 with their numerical values, leads to

$$\mathbf{I(s)} = \frac{V_0}{L\left[\mathbf{s}^2 + \dfrac{R}{L}\mathbf{s} + \dfrac{1}{LC}\right]} = \frac{4}{\mathbf{s}^2 + 10\mathbf{s} + 25} = \frac{4}{(\mathbf{s}+5)^2}. \tag{12.40}$$

Step 4: Transform the solution back to the time domain with the help of Tables 12-1 and 12-2

According to entry #6 in Table 12-2,

$$\mathcal{L}^{-1}\left[\frac{1}{(\mathbf{s}+a)^2}\right] = te^{-at}\, u(t).$$

Hence,

$$i(t) = 4te^{-5t}\, u(t), \tag{12.41}$$

and its plot is displayed in Fig. 12-4(b).

In this particular example, the expression for $\mathbf{I(s)}$ given by Eq. (12.40) matches one of the entries available in Table 12-2, but what should we do if it does not? We have two options:

(1) we can apply the inverse Laplace transform relation given by Eq. (12.14), which in general involves a rather cumbersome integration, or

(2) we can apply the *partial-fraction-expansion* method to rearrange the expression for $\mathbf{I(s)}$ into a sum of terms, each of which has an appropriate match in Table 12-2. This latter approach is the subject of the next section.

12-5 Partial Fraction Expansion

Let us assume that after transforming the integrodifferential equation associated with a circuit of interest to the s-domain, and then solving it for the voltage or current whose behavior we wish to examine, we end up with an expression $\mathbf{F(s)}$. Our next step is to inverse transform $\mathbf{F(s)}$ to the time domain, thereby completing our solution. The degree of mathematical difficulty associated with the implementation of the inverse transformation depends on the mathematical form of $\mathbf{F(s)}$.

Consider, for example, the expression

$$\mathbf{F(s)} = \frac{4}{\mathbf{s}+2} + \frac{6}{(\mathbf{s}+5)^2} + \frac{8}{\mathbf{s}^2 + 4\mathbf{s} + 5}. \tag{12.42}$$

The inverse transform, $f(t)$, is given by

$$f(t) = \mathcal{L}^{-1}[\mathbf{F(s)}] = \mathcal{L}^{-1}\left[\frac{4}{\mathbf{s}+2}\right] + \mathcal{L}^{-1}\left[\frac{6}{(\mathbf{s}+5)^2}\right] + \mathcal{L}^{-1}\left[\frac{8}{\mathbf{s}^2+4\mathbf{s}+5}\right]. \tag{12.43}$$

By comparison with the entries in Table 12-2, we note that:

(a) The first term in Eq. (12.43), $4/(\mathbf{s}+2)$, is functionally the same as entry #3 in Table 12-2, with $a = 2$. Hence,

$$\mathcal{L}^{-1}\left[\frac{4}{\mathbf{s}+2}\right] = 4e^{-2t}\, u(t). \tag{12.44a}$$

(b) The second term, $6/(\mathbf{s}+5)^2$, is functionally the same as entry #6 in Table 12-2, with $a = 5$. Thus,

$$\mathcal{L}^{-1}\left[\frac{6}{(\mathbf{s}+5)^2}\right] = 6te^{-5t}\, u(t). \tag{12.44b}$$

(c) The third term, $1/(\mathbf{s}^2+4\mathbf{s}+5)$, is similar but not identical in form to entry #13 in Table 12-2. However, it can be rearranged to assume the proper form:

$$\frac{1}{\mathbf{s}^2+4\mathbf{s}+5} = \frac{1}{(\mathbf{s}+2)^2+1}.$$

Consequently,

$$\mathcal{L}^{-1}\left[\frac{8}{(\mathbf{s}+2)^2+1}\right] = 8e^{-2t}\sin t\; u(t). \qquad (12.44c)$$

Combining the results represented by Eqs. (12.44a–c) gives:

$$f(t) = [4e^{-2t} + 6te^{-5t} + 8e^{-2t}\sin t]\; u(t). \qquad (12.45)$$

The preceding example demonstrated that the implementation of the inverse Laplace transform is a rather painless process, so long as the expression for $\mathbf{F}(\mathbf{s})$ is composed of a series of terms similar to those in Eq. (12.42). If, however, $\mathbf{F}(\mathbf{s})$ is not in the proper form, we will need to reconfigure it before we can apply the inverse transform. Specifically, $\mathbf{F}(\mathbf{s})$ should be expanded into a sum of partial fractions by applying the applicable recipe from among those outlined in the forthcoming subsections.

12-5.1 Distinct Real Poles

Consider the $\mathbf{s}$-domain function

$$\mathbf{F}(\mathbf{s}) = \frac{\mathbf{s}^2-4\mathbf{s}+3}{\mathbf{s}(\mathbf{s}+1)(\mathbf{s}+3)}. \qquad (12.46)$$

The *poles* of $\mathbf{F}(\mathbf{s})$ are the values of $\mathbf{s}$ at which its denominator is zero. In the present case, the poles of $\mathbf{F}(\mathbf{s})$ are $\mathbf{s}=0$, $\mathbf{s}=-1$, and $\mathbf{s}=-3$. All three poles are *real* and *distinct*. By distinct we mean that no two or more poles are the same; in $(\mathbf{s}+4)^2$, for example, the pole $\mathbf{s}=-4$ occurs twice, and therefore it is not distinct.

$\mathbf{F}(\mathbf{s})$ can be decomposed into partial fractions corresponding to the three factors in the denominator of $\mathbf{F}(\mathbf{s})$:

$$\mathbf{F}(\mathbf{s}) = \frac{A_1}{\mathbf{s}} + \frac{A_2}{(\mathbf{s}+1)} + \frac{A_3}{(\mathbf{s}+3)}, \qquad (12.47)$$

where A_1 to A_3 are *expansion coefficients* to be determined shortly. Equating the two functional forms of $\mathbf{F}(\mathbf{s})$, we have

$$\frac{A_1}{\mathbf{s}} + \frac{A_2}{(\mathbf{s}+1)} + \frac{A_3}{(\mathbf{s}+3)} = \frac{\mathbf{s}^2-4\mathbf{s}+3}{\mathbf{s}(\mathbf{s}+1)(\mathbf{s}+3)}. \qquad (12.48)$$

Associated with each expansion coefficient is a *pole factor*; $\mathbf{s}$, $(\mathbf{s}+1)$, and $(\mathbf{s}+3)$ are the pole factors associated with A_1, A_2, and A_3, respectively. To determine the value of any expansion coefficient we multiply both sides of Eq. (12.48) by the pole factor of that expansion coefficient, and then we evaluate them at $\mathbf{s}$ = pole value of that pole factor. The procedure is called the *residue method*.

To determine A_2, for example, we multiply both sides of Eq. (12.48) by $(\mathbf{s}+1)$, we reduce the expressions, and then we evaluate them at $\mathbf{s}=-1$:

$$\left\{(\mathbf{s}+1)\left[\frac{A_1}{\mathbf{s}} + \frac{A_2}{(\mathbf{s}+1)} + \frac{A_3}{(\mathbf{s}+3)}\right]\right\}\Bigg|_{\mathbf{s}=-1} = \left[\frac{(\mathbf{s}+1)(\mathbf{s}^2-4\mathbf{s}+3)}{\mathbf{s}(\mathbf{s}+1)(\mathbf{s}+3)}\right]\Bigg|_{\mathbf{s}=-1}. \qquad (12.49)$$

After reduction, the expression becomes

$$\left[\frac{A_1(\mathbf{s}+1)}{\mathbf{s}} + A_2 + \frac{A_3(\mathbf{s}+1)}{(\mathbf{s}+3)}\right]\Bigg|_{\mathbf{s}=-1} = \left[\frac{(\mathbf{s}^2-4\mathbf{s}+3)}{\mathbf{s}(\mathbf{s}+3)}\right]\Bigg|_{\mathbf{s}=-1}. \qquad (12.50)$$

We note that (1) the presence of $(\mathbf{s}+1)$ in the numerators of terms 1 and 3 on the left-hand side will force those terms to go to zero when evaluated at $\mathbf{s}=-1$, (2) the middle term has only A_2 in it, and (3) the reduction on the right-hand side of Eq. (12.50) eliminated the pole factor $(\mathbf{s}+1)$ from the expression. Consequently,

$$A_2 = \frac{(-1)^2+4+3}{(-1)(-1+3)} = -4.$$

Similarly,

$$A_1 = \mathbf{s}\,\mathbf{F}(\mathbf{s})|_{\mathbf{s}=0} = \frac{\mathbf{s}^2-4\mathbf{s}+3}{(\mathbf{s}+1)(\mathbf{s}+3)}\Bigg|_{\mathbf{s}=0} = 1,$$

and

$$A_3 = (\mathbf{s}+3)\,\mathbf{F}(\mathbf{s})|_{\mathbf{s}=-3} = \frac{\mathbf{s}^2-4\mathbf{s}+3}{\mathbf{s}(\mathbf{s}+1)}\Bigg|_{\mathbf{s}=-3} = 4.$$

Having established the values of A_1 to A_3, we now are ready to apply the inverse Laplace transform to Eq. (12.47):

$$\mathbf{F}(\mathbf{s}) = \frac{1}{\mathbf{s}} - \frac{4}{\mathbf{s}+1} + \frac{4}{\mathbf{s}+3},$$

and with the help of Table 12-2, we obtain the result

$$f(t) = \mathcal{L}^{-1}[\mathbf{F}(\mathbf{s})] = \mathcal{L}^{-1}\left[\frac{1}{\mathbf{s}} - \frac{4}{\mathbf{s}+1} + \frac{4}{\mathbf{s}+3}\right] = [1 - 4e^{-t} + 4e^{-3t}]\; u(t). \qquad (12.51)$$

Building on this example, we can generalize the process to:

Distinct Real Poles

Give a proper rational function defined by

$$\mathbf{F}(\mathbf{s}) = \frac{\mathbf{N}(\mathbf{s})}{\mathbf{D}(\mathbf{s})} = \frac{\mathbf{N}(\mathbf{s})}{(\mathbf{s}+p_1)(\mathbf{s}+p_2)\dots(\mathbf{s}+p_n)}, \tag{12.52}$$

with numerator $\mathbf{N}(\mathbf{s})$ and distinct real poles $-p_1$ to $-p_n$, such that $p_i \neq p_j$ for all $i \neq j$, $\mathbf{F}(\mathbf{s})$ can be expanded into the equivalent form:

$$\begin{aligned} \mathbf{F}(\mathbf{s}) &= \frac{A_1}{\mathbf{s}+p_1} + \frac{A_2}{\mathbf{s}+p_2} + \cdots + \frac{A_n}{\mathbf{s}+p_n} \\ &= \sum_{i=1}^{n} \frac{A_i}{\mathbf{s}+p_i}, \end{aligned} \tag{12.53}$$

with expansion coefficients A_1 to A_n given by

$$A_1 = (\mathbf{s}+p_i)\,\mathbf{F}(\mathbf{s})\big|_{\mathbf{s}=-p_i}, \qquad i = 1, 2, \dots, n. \tag{12.54}$$

In view of entry #3 in Table 12-2, the inverse Laplace transform or Eq. (12.53) is

$$\begin{aligned} f(t) &= \mathcal{L}^{-1}[\mathbf{F}(\mathbf{s})] \\ &= [A_1 e^{-p_1 t} + A_2 e^{-p_2 t} + \cdots + A_n e^{-p_n t}]\, u(t). \end{aligned} \tag{12.55}$$

Exercise 12-6: Apply the partial-fraction-expansion method to determine $f(t)$, given that its Laplace transform is

$$\mathbf{F}(\mathbf{s}) = \frac{10\mathbf{s}+16}{\mathbf{s}(\mathbf{s}+2)(\mathbf{s}+4)}.$$

Answer: $f(t) = [2 + e^{-2t} - 3e^{-4t}]\, u(t)$. (See CAD)

12-5.2 Repeated Real Poles

We now consider the case when $\mathbf{F}(\mathbf{s})$ contains repeated real poles or a combination of distinct and repeated real poles. The partial-fraction-expansion method is outlined by the following steps.

Step 1. We are given a function $\mathbf{F}(\mathbf{s})$ composed of the product

$$\mathbf{F}(\mathbf{s}) = \mathbf{F}_1(\mathbf{s})\,\mathbf{F}_2(\mathbf{s}), \tag{12.56}$$

with

$$\mathbf{F}_1(\mathbf{s}) = \frac{\mathbf{N}(\mathbf{s})}{(\mathbf{s}+p_1)(\mathbf{s}+p_2)\dots(\mathbf{s}+p_n)}, \tag{12.57}$$

and

$$\mathbf{F}_2(\mathbf{s}) = \frac{1}{(\mathbf{s}+p)^m}. \tag{12.58}$$

We note that $\mathbf{F}_1(\mathbf{s})$ is identical in form to Eq. (12.52) and contains only distinct real poles, $-p_1$ to $-p_n$, thereby qualifying it for representation by a series of terms as in Eq. (12.53). The second function, $\mathbf{F}_2(\mathbf{s})$, has an m-repeated pole at $\mathbf{s} = -p$, where m is a positive integer. Also, the repeated pole is not a pole of $\mathbf{F}_1(\mathbf{s})$; $p \neq p_i$ for $i = 1, 2, \dots, n$.

Step 2. Partial fraction representation for an m-repeated pole at $\mathbf{s} = -p$ consists of m terms:

$$\frac{B_1}{\mathbf{s}+p} + \frac{B_2}{(\mathbf{s}+p)^2} + \cdots + \frac{B_m}{(\mathbf{s}+p)^m}. \tag{12.59}$$

Step 3. Partial fraction expansion for the combination of the product $\mathbf{F}_1(\mathbf{s})\,\mathbf{F}_2(\mathbf{s})$ is then given by

$$\begin{aligned} \mathbf{F}(\mathbf{s}) &= \frac{A_1}{\mathbf{s}+p_1} + \frac{A_2}{\mathbf{s}+p_2} + \cdots + \frac{A_n}{\mathbf{s}+p_n} \\ &\quad + \frac{B_1}{\mathbf{s}+p} + \frac{B_2}{(\mathbf{s}+p)^2} + \cdots + \frac{B_m}{(\mathbf{s}+p)^m} \\ &= \sum_{i=1}^{n} \frac{A_i}{\mathbf{s}+p_i} + \sum_{j=1}^{m} \frac{B_j}{(\mathbf{s}+p)^j}. \end{aligned} \tag{12.60}$$

Step 4. Expansion coefficients A_1 to A_n are determined by applying Eq. (12.54):

$$A_i = (\mathbf{s}+p_i)\,\mathbf{F}(\mathbf{s})|_{\mathbf{s}=-p_i}, \qquad i = 1, 2, \dots, n. \tag{12.61}$$

Repeated Real Poles

Expansion coefficients B_1 to B_m are determined through a procedure that involves multiplication by $(\mathbf{s}+p)^m$, differentiation with respect to $\mathbf{s}$, and evaluation at $\mathbf{s}=-p$:

$$B_j = \left\{ \frac{1}{(m-j)!} \frac{d^{m-j}}{d\mathbf{s}^{m-j}} [(\mathbf{s}+p)^m \, \mathbf{F}(\mathbf{s})] \right\} \bigg|_{\mathbf{s}=-p}, \quad j = 1, 2, \ldots, m. \tag{12.62}$$

For the m, $m-1$, and $m-2$ terms, Eq. (12.62) reduces to:

$$B_m = (\mathbf{s}+p)^m \, \mathbf{F}(\mathbf{s})|_{\mathbf{s}=-p}, \tag{12.63a}$$

$$B_{m-1} = \left\{ \frac{d}{d\mathbf{s}} [(\mathbf{s}+p)^m \, \mathbf{F}(\mathbf{s})] \right\} \bigg|_{\mathbf{s}=-p}, \tag{12.63b}$$

$$B_{m-2} = \left\{ \frac{1}{2!} \frac{d^2}{d\mathbf{s}^2} [(\mathbf{s}+p)^m \, \mathbf{F}(\mathbf{s})] \right\} \bigg|_{\mathbf{s}=-p}. \tag{12.63c}$$

Thus, the evaluation of B_m does not involve any differentiation, that of B_{m-1} involves differentiation with respect to $\mathbf{s}$ only once (and division by 1!), and that of B_{m-2} involves differentiation twice and division by 2!. *In practice, it is easiest to start by evaluating B_m first and then evaluating the other expansion coefficients in descending order.*

5. Once the values of all of the expansion coefficients of Eq. (12.60) have been determined, transformation to the time domain is accomplished by applying entry #8 of Table 12-2,

$$\mathcal{L}^{-1}\left[\frac{(n-1)!}{(\mathbf{s}+a)^n}\right] = t^{n-1} e^{-at} \, u(t). \tag{12.64}$$

The result is

$$f(t) = \mathcal{L}^{-1}[\mathbf{F}(\mathbf{s})] = \left[\sum_{i=1}^{n} A_i e^{-p_i t} + \sum_{j=1}^{m} \frac{B_j t^{j-1}}{(j-1)!} e^{-pt} \right] u(t). \tag{12.65}$$

Example 12-3: Repeated Poles

Determine the inverse Laplace transform of

$$\mathbf{F}(\mathbf{s}) = \frac{\mathbf{N}(\mathbf{s})}{\mathbf{D}(\mathbf{s})} = \frac{\mathbf{s}^2+3\mathbf{s}+3}{\mathbf{s}^4+11\mathbf{s}^3+45\mathbf{s}^2+81\mathbf{s}+54}.$$

Solution: In theory, any polynomial with real coefficients can be expressed as a product of linear and quadratic factors (of the form $(s+p)$ and (s^2+as+b), respectively). The process involves long division, but it requires knowledge of the roots of the polynomial, which can be determined through the application of numerical techniques. In the present case, a random check reveals that $\mathbf{s}=-2$ and $\mathbf{s}=-3$ are roots of $\mathbf{D}(\mathbf{s})$. Given that $\mathbf{D}(\mathbf{s})$ is fourth order, it should have four roots, including possible duplicates.

Since $\mathbf{s}=-2$ is a root of $\mathbf{D}(\mathbf{s})$, we should be able to factor out $(\mathbf{s}+2)$ from it. Long division gives

$$\mathbf{D}(\mathbf{s}) = \mathbf{s}^4+11\mathbf{s}^3+45\mathbf{s}^2+81\mathbf{s}+54 = (\mathbf{s}+2)(\mathbf{s}^3+9\mathbf{s}^2+27\mathbf{s}+27).$$

Next, we factor out $(\mathbf{s}+3)$, which yields

$$\mathbf{D}(\mathbf{s}) = (\mathbf{s}+2)(\mathbf{s}+3)(\mathbf{s}^2+6\mathbf{s}+9) = (\mathbf{s}+2)(\mathbf{s}+3)^3.$$

Hence, $\mathbf{F}(\mathbf{s})$ has a distinct real pole at $\mathbf{s}=-2$ and a triply repeated pole at $\mathbf{s}=-3$, and the given expression can be rewritten as

$$\mathbf{F}(\mathbf{s}) = \frac{\mathbf{s}^2+3\mathbf{s}+3}{(\mathbf{s}+2)(\mathbf{s}+3)^3}.$$

Through partial fraction expansion, $\mathbf{F}(\mathbf{s})$ can be decomposed into

$$\mathbf{F}(\mathbf{s}) = \frac{A}{\mathbf{s}+2} + \frac{B_1}{\mathbf{s}+3} + \frac{B_2}{(\mathbf{s}+3)^2} + \frac{B_3}{(\mathbf{s}+3)^3},$$

with

$$A = (\mathbf{s}+2)\,\mathbf{F}(\mathbf{s})|_{\mathbf{s}=-2} = \frac{\mathbf{s}^2+3\mathbf{s}+3}{(\mathbf{s}+3)^3}\bigg|_{\mathbf{s}=-2} = 1,$$

$$B_3 = (\mathbf{s}+3)^3\,\mathbf{F}(\mathbf{s})|_{\mathbf{s}=-3} = \frac{\mathbf{s}^2+3\mathbf{s}+3}{\mathbf{s}+2}\bigg|_{\mathbf{s}=-3} = -3,$$

$$B_2 = \frac{d}{d\mathbf{s}}[(\mathbf{s}+3)^3\,\mathbf{F}(\mathbf{s})]\bigg|_{\mathbf{s}=-3} = 0,$$

$$B_1 = \frac{1}{2}\frac{d^2}{d\mathbf{s}^2}[(\mathbf{s}+3)^3\,\mathbf{F}(\mathbf{s})]\bigg|_{\mathbf{s}=-3} = -1.$$

Hence,

$$\mathbf{F}(\mathbf{s}) = \frac{1}{\mathbf{s}+2} - \frac{1}{\mathbf{s}+3} - \frac{3}{(\mathbf{s}+3)^3},$$

and use of Table 12-2 for the first two terms and application of Eq. (12.64) to the last term leads to

$$\mathcal{L}^{-1}[\mathbf{F}(\mathbf{s})] = \left[e^{-2t} - e^{-3t} - \frac{3}{2}t^2 e^{-3t}\right] u(t).$$

Technology Brief 29
Mapping the Entire World in 3-D

Mapping software has become increasingly indispensable in the 21st-century industrialized world. Giving verbal directions to someone's house has been replaced with directing them to Google Maps with the relevant address. Even more exciting and controversial, however, is the growing suite of 3-D *virtual globe* mapping software. Of these, arguably the most famous is the currently free Google Earth software. These packages allow the user to fly in virtual space around the world, into cities and remote areas and (in densely mapped areas) to view their own backyards, streets signs, and local landscapes; Google Sky includes virtual, navigable representations of the Moon, Mars, and the night sky. The tools are becoming an enabler for a new generation of armchair historians, archaeologists, demographers. They have already been used in search and rescue operations. How do these packages work? Where does this data come from? How is the world mapped?

Planes, Satellites, and Automobiles

Data for these packages is acquired by specialized companies (including Google itself) that make use of satellites, aircraft, and (more recently) large fleets of specially equipped vans. The majority of data comes from several satellites orbiting earth financed by either national governments or private companies. For example, the U.S. National Aeronautics and Space Administration (NASA) has the long-standing Landsat 7 program which has 30-m imaging resolution and scans the earth in about 16 days. The European Space Agency's ERS and Envisat satellites perform similar functions. All of these satellites perform functions other than visual spectrum imaging; some have infrared sensors, radar sensors, temperature sensors, etc. Several commercial satellites are now in orbit whose primary function is to map the globe in high-resolution mode; DigitalGlobe's WorldView-2 satellite, for example, provides 0.46 m spatial resolution—although not all data are publicly available. Most of these satellites maintain sun-synchronous orbits, which means that their orbits loop over or near the north and south poles and cross the equator twice on each loop. In this type of orbit, the satellite "visits" a given place at the same local time each visit, which is great for maintaining constant lighting for satellite images. Additionally, 2-D visual information is supplemented with digital elevation model (DEM) data collected by NASA's Shuttle Radar Topography Mission (SRTM). The SRTM (Fig. TF29-1) consisted of two radar antennas deployed on the space shuttle *Endeavour* during the 11-day mission of STS-99 in February 2000. A sample product is shown in Fig. TF29-2.

Figure TF29-1: The Shuttle Radar Topography Mission used an antenna located in the payload bay of the shuttle, and a second outboard antenna attached to the end of a 60-m mast. (Courtesy of NASA.)

Aircraft imaging complements the satellite data, although it is more expensive and available in limited areas. Several companies have launched fleets of specially equipped vans with multiple cameras, laser distance sensors, and on-board computation to collect, merge, and store high-resolution, three-dimensional data at street level. Additionally, Google Earth allows for user-inputted 3-D information and models. Figure TF29-3 shows one such vehicle developed by TeleAtlas. Hundreds of similar vehicles roam the earth; the cameras provide images over 360° around the vehicle, and a laser system measures important distances like bridge and building heights; GPS tracking hardware records the vehicle's position; and onboard computers synthesize everything and store it. As these vehicles visit more and more places, the 3-D map of the world continues to grow.

Imaging Software

To produce a navigable, virtual representation of our globe, all of this data is then compiled, corrected, and

Figure TF29-2: A shaded relief image of Mount St. Helens in the state of Washington. (Courtesy of NASA.)

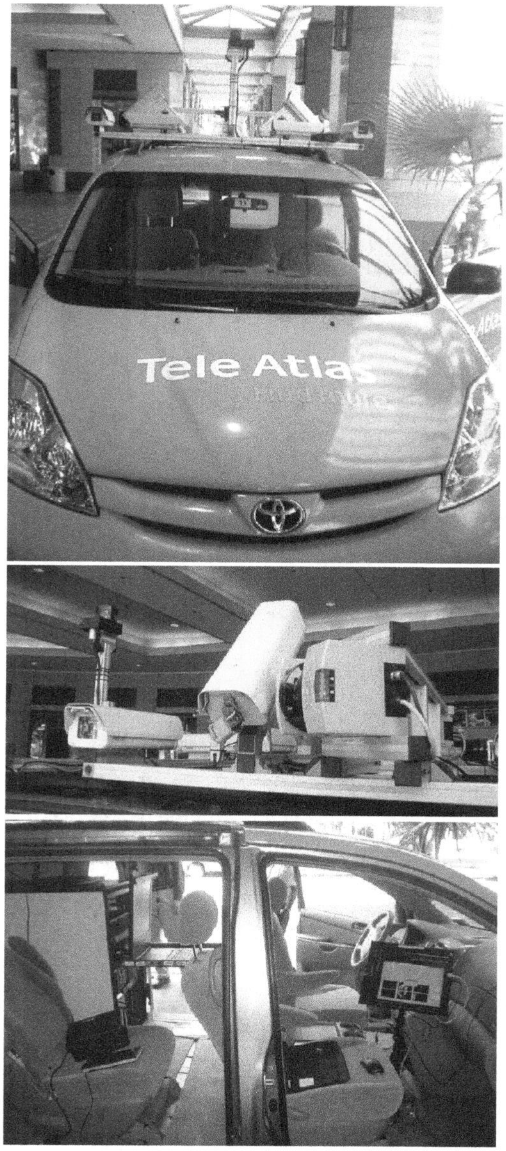

Figure TF29-3: A TeleAtlas van showing the imaging and laser equipment and the computation hardware inside the van.

merged. This is not just a massive storage operation. Often, imagery comes from multiple sources that do not match exactly, there may be gaps between images and, very commonly, the color of the images must be corrected and made consistent. Fine-scale errors often are detectable with these map programs when data is incorrectly merged or have different dates; for example, pictures of a city might incorrectly show data from adjacent areas taken before and after major events, stitched together. Problems with incorporating 3-D topographical data with the visual information are still common. For public-accessible programs, not all data is taken at the same time nor with the same frequency; for example, Google Earth guarantees that image data is no more than three years old. More expensive commercial software is often more timely.

Beyond the compilation and merging of datasets, programs like Google Earth are increasingly integrating their software with both other software and mobile hardware. For example, Google Earth interfaces with both Wikipedia and the Google search engine as well as an increasing suite of information-providing programs. In a similar manner, some versions of commercial virtual globe programs can interface with GPS position-finding devices. Such programs take waypoints and tracks from the mobile GPS devices and merge them with available topographic, imaging, and other virtual globe datasets.

Concept Question 12-7: What purpose does the partial-fraction-expansion method serve? (See CAD)

Concept Question 12-8: When evaluating the expansion coefficients of a function containing repeated poles, is it more practical to start by evaluating the coefficient of the fraction with the lowest-order pole or that with the highest-order pole? Why? (See CAD)

Exercise 12-7: Determine the inverse Laplace transform of

$$\mathbf{F}(\mathbf{s}) = \frac{4\mathbf{s}^2 - 15\mathbf{s} - 10}{(\mathbf{s}+2)^3}.$$

Answer: $f(t) = (18t^2 - 31t + 4)e^{-2t}\,u(t)$. (See CAD)

12-5.3 Distinct Complex Poles

The Laplace transform of a certain circuit is given by

$$\mathbf{F}(\mathbf{s}) = \frac{4\mathbf{s}+1}{(\mathbf{s}+1)(\mathbf{s}^2+4\mathbf{s}+13)}. \tag{12.66}$$

In addition to the simple-pole factor, the denominator includes a quadratic-pole factor with roots $\mathbf{s}_1$ and $\mathbf{s}_2$. Solution of $\mathbf{s}^2 + 4\mathbf{s} + 13 = 0$ gives

$$\mathbf{s}_1 = -2 + j3, \qquad \mathbf{s}_2 = -2 - j3. \tag{12.67}$$

The fact that the two roots are complex conjugates of one another is a consequence of the property that *for any physically realizable circuit, if it has any complex poles, those poles always appear in conjugate pairs.*

In view of Eq. (12.67), the quadratic factor is given by

$$\mathbf{s}^2 + 4\mathbf{s} + 13 = (\mathbf{s}+2-j3)(\mathbf{s}+2+j3), \tag{12.68}$$

and $\mathbf{F}(\mathbf{s})$ can now be expanded into partial fractions:

$$\mathbf{F}(\mathbf{s}) = \frac{A}{\mathbf{s}+1} + \frac{\mathbf{B}_1}{\mathbf{s}+2-j3} + \frac{\mathbf{B}_2}{\mathbf{s}+2+j3}. \tag{12.69}$$

Expansion coefficients $\mathbf{B}_1$ and $\mathbf{B}_2$ are printed in bold letters to signify the fact that they may be complex quantities. Determination of A, $\mathbf{B}_1$, and $\mathbf{B}_2$ follows the same factor-multiplication technique employed in Section 12-5.1 with

$$A = (\mathbf{s}+1)\,\mathbf{F}(\mathbf{s})|_{\mathbf{s}=-1} = \left.\frac{4\mathbf{s}+1}{\mathbf{s}^2+4\mathbf{s}+13}\right|_{\mathbf{s}=-1} = -0.3, \tag{12.70a}$$

$$\begin{aligned}\mathbf{B}_1 &= (\mathbf{s}+2-j3)\,\mathbf{F}(\mathbf{s})|_{\mathbf{s}=-2+j3}\\ &= \left.\frac{4\mathbf{s}+1}{(\mathbf{s}+1)(\mathbf{s}+2+j3)}\right|_{\mathbf{s}=-2+j3}\\ &= \frac{4(-2+j3)+1}{(-2+j3+1)(-2+j3+2+j3)}\\ &= \frac{-7+j12}{-18-j6} = 0.73e^{-j78.2^\circ},\end{aligned} \tag{12.70b}$$

and

$$\begin{aligned}\mathbf{B}_2 &= (\mathbf{s}+2+j3)\,\mathbf{F}(\mathbf{s})|_{\mathbf{s}=-2-j3}\\ &= \left.\frac{4\mathbf{s}+1}{(\mathbf{s}+1)(\mathbf{s}+2-j3)}\right|_{\mathbf{s}=-2-j3} = 0.73e^{j78.2^\circ}.\end{aligned} \tag{12.70c}$$

We observe that $\mathbf{B}_2 = \mathbf{B}_1^*$.

► In fact, the expansion coefficients associated with conjugate poles are always conjugate pairs themselves. ◄

The inverse Laplace transform of Eq. (12.69) is

$$\begin{aligned}f(t) &= \mathcal{L}^{-1}[\mathbf{F}(\mathbf{s})] = \mathcal{L}^{-1}\left(\frac{-0.3}{\mathbf{s}+1}\right) + \mathcal{L}^{-1}\left(\frac{0.73e^{-j78.2^\circ}}{\mathbf{s}+2-j3}\right)\\ &\quad + \mathcal{L}^{-1}\left(\frac{0.73e^{j78.2^\circ}}{\mathbf{s}+2+j3}\right)\\ &= [-0.3e^{-t} + 0.73e^{-j78.2^\circ}e^{-(2-j3)t}\\ &\quad + 0.73e^{j78.2^\circ}e^{-(2+j3)t}]\,u(t).\end{aligned} \tag{12.71}$$

Because complex numbers do not belong in the time domain, our initial reaction to their presence in the solution given by Eq. (12.71) is that perhaps an error was committed somewhere along the way. The truth is the solution is correct, but incomplete. Terms 2 and 3 are conjugate pairs, so by applying Euler's formula, they can be combined into a single term containing real quantities only:

$$\begin{aligned}&0.73e^{-j78.2^\circ}e^{-(2-j3)t} + 0.73e^{j78.2^\circ}e^{-(2+j3)t}\\ &\quad = 0.73e^{-2t}[e^{j(3t-78.2^\circ)} + e^{-j(3t-78.2^\circ)}]\\ &\quad = 2\times 0.73e^{-2t}\cos(3t-78.2^\circ)\\ &\quad = 1.46e^{-2t}\cos(3t-78.2^\circ).\end{aligned} \tag{12.72}$$

Hence, the final time-domain solution is

$$f(t) = [-0.3e^{-t} + 1.46e^{-2t}\cos(3t - 78.2°)]\,u(t). \quad (12.73)$$

Exercise 12-8: Determine the inverse Laplace transform of

$$\mathbf{F(s)} = \frac{2\mathbf{s} + 14}{\mathbf{s}^2 + 6\mathbf{s} + 25}.$$

Answer: $f(t) = [2\sqrt{2}\, e^{-3t}\cos(4t - 45°)]\,u(t)$. (See CAD)

12-5.4 Repeated Complex Poles

If the Laplace transform $\mathbf{F(s)}$ contains repeated complex poles, we can expand it into partial fractions by using a combination of the tools introduced in Sections 12-5.2 and 12-5.3. The process is illustrated in Example 12-4.

Example 12-4: Five-Pole Function

Determine the inverse Laplace transform of

$$\mathbf{F(s)} = \frac{108(\mathbf{s}^2 + 2)}{(\mathbf{s} + 2)(\mathbf{s}^2 + 10\mathbf{s} + 34)^2}.$$

Solution: The roots of

$$\mathbf{s}^2 + 10\mathbf{s} + 34 = 0$$

are

$$\mathbf{s}_1 = -5 - j3$$

and

$$\mathbf{s}_2' = -5 + j3.$$

Hence,

$$\mathbf{F(s)} = \frac{108(\mathbf{s}^2 + 2)}{(\mathbf{s} + 2)(\mathbf{s} + 5 + j3)^2(\mathbf{s} + 5 - j3)^2},$$

and its partial fraction expansion can be expressed as

$$\mathbf{F(s)} = \frac{A}{\mathbf{s} + 2} + \frac{\mathbf{B}_1}{\mathbf{s} + 5 + j3} + \frac{\mathbf{B}_2}{(\mathbf{s} + 5 + j3)^2} + \frac{\mathbf{B}_1^*}{\mathbf{s} + 5 - j3} + \frac{\mathbf{B}_2^*}{(\mathbf{s} + 5 - j3)^2},$$

where $\mathbf{B}_1^*$ and $\mathbf{B}_2^*$ are the complex conjugates of $\mathbf{B}_1$ and $\mathbf{B}_2$, respectively. Coefficients A, $\mathbf{B}_1$, and $\mathbf{B}_2$ are evaluated as follows:

$$A = (\mathbf{s} + 2)\,\mathbf{F(s)}|_{\mathbf{s}=-2} = \left.\frac{108(\mathbf{s}^2 + 2)}{(\mathbf{s}^2 + 10\mathbf{s} + 34)^2}\right|_{\mathbf{s}=-2} = 2,$$

$$\begin{aligned}
\mathbf{B}_2 &= (\mathbf{s} + 5 + j3)^2\,\mathbf{F(s)}|_{\mathbf{s}=-5-j3} \\
&= \left.\frac{108(\mathbf{s}^2 + 2)}{(\mathbf{s} + 2)(\mathbf{s} + 5 - j3)^2}\right|_{\mathbf{s}=-5-j3} \\
&= \frac{108[(-5 - j3)^2 + 2]}{(-5 - j3 + 2)(-5 - j3 + 5 - j3)^2} \\
&= 24 + j6 = 24.74e^{j14°},
\end{aligned}$$

and

$$\begin{aligned}
\mathbf{B}_1 &= \left.\frac{d}{d\mathbf{s}}[(\mathbf{s} + 5 + j3)^2\,\mathbf{F(s)}]\right|_{\mathbf{s}=-5-j3} \\
&= \left.\frac{d}{d\mathbf{s}}\left[\frac{108(\mathbf{s}^2 + 2)}{(\mathbf{s} + 2)(\mathbf{s} + 5 - j3)^2}\right]\right|_{\mathbf{s}=-5-j3} \\
&= \left[\frac{108(2\mathbf{s})}{(\mathbf{s} + 2)(\mathbf{s} + 5 - j3)^2} - \frac{108(\mathbf{s}^2 + 2)}{(\mathbf{s} + 2)^2(\mathbf{s} + 5 - j3)^2}\right. \\
&\quad \left.\left. - \frac{2 \times 108(\mathbf{s}^2 + 2)}{(\mathbf{s} + 2)(\mathbf{s} + 5 - j3)^3}\right]\right|_{\mathbf{s}=-5-j3} \\
&= -(1 + j9) = 9.06e^{-j96.34°}.
\end{aligned}$$

The remaining constants are

$$\mathbf{B}_1^* = 9.06e^{j96.34°},$$

and

$$\mathbf{B}_2^* = 24.74e^{-j14°},$$

and the inverse Laplace transform is

$$\begin{aligned}
f(t) &= \mathcal{L}^{-1}[\mathbf{F(s)}] \\
&= \mathcal{L}^{-1}\left[\frac{2}{\mathbf{s} + 2} + \frac{9.06e^{-j96.34°}}{\mathbf{s} + 5 + j3} + \frac{9.06e^{j96.34°}}{\mathbf{s} + 5 - j3}\right. \\
&\quad \left. + \frac{24.74e^{j14°}}{(\mathbf{s} + 5 + j3)^2} + \frac{24.74e^{-j14°}}{(\mathbf{s} + 5 - j3)^2}\right] \\
&= \Big[2e^{-2t} \\
&\quad + 9.06(e^{-j96.34°}e^{-(5+j3)t} + e^{j96.34°}e^{-(5-j3)t}) \\
&\quad + 24.74t\,(e^{j14°}e^{-(5+j3)t} + e^{-j14°}e^{-(5-j3)t})\Big]\,u(t) \\
&= [2e^{-2t} + 18.12e^{-5t}\cos(3t + 96.34°) \\
&\quad + 49.48te^{-5t}\cos(3t - 14°)]\,u(t).
\end{aligned}$$

Table 12-3: **Transform pairs for four types of poles.**

Pole	F(s)	$f(t)$
1. Distinct real	$\dfrac{A}{\mathbf{s}+a}$	$Ae^{-at}\,u(t)$
2. Repeated real	$\dfrac{A}{(\mathbf{s}+a)^n}$	$A\,\dfrac{t^{n-1}}{(n-1)!}\,e^{-at}\,u(t)$
3. Distinct complex	$\left[\dfrac{Ae^{j\theta}}{\mathbf{s}+a+jb}+\dfrac{Ae^{-j\theta}}{\mathbf{s}+a-jb}\right]$	$2Ae^{-at}\cos(bt-\theta)\,u(t)$
4. Repeated complex	$\left[\dfrac{Ae^{j\theta}}{(\mathbf{s}+a+jb)^n}+\dfrac{Ae^{-j\theta}}{(\mathbf{s}+a-jb)^n}\right]$	$\dfrac{2At^{n-1}}{(n-1)!}\,e^{-at}\cos(bt-\theta)\,u(t)$

Example 12-5: Interesting Transform!

Determine the time-domain equivalent of the Laplace transform

$$\mathbf{F}(\mathbf{s})=\frac{\mathbf{s}e^{-3\mathbf{s}}}{\mathbf{s}^2+4}.$$

Solution: We start by separating out the exponential $e^{-3\mathbf{s}}$ from the remaining polynomial fraction. We do so by defining

$$\mathbf{F}(\mathbf{s})=e^{-3\mathbf{s}}\,\mathbf{F}_1(\mathbf{s}),$$

where

$$\mathbf{F}_1(\mathbf{s})=\frac{\mathbf{s}}{\mathbf{s}^2+4}=\frac{\mathbf{s}}{(\mathbf{s}+j2)(\mathbf{s}-j2)}=\frac{\mathbf{B}_1}{\mathbf{s}+j2}+\frac{\mathbf{B}_2}{\mathbf{s}-j2},$$

with

$$\mathbf{B}_1=(\mathbf{s}+j2)\,\mathbf{F}(\mathbf{s})|_{\mathbf{s}=-j2}=\left.\frac{\mathbf{s}}{\mathbf{s}-j2}\right|_{\mathbf{s}=-j2}=\frac{-j2}{-j4}=\frac{1}{2},$$

and

$$\mathbf{B}_2=\mathbf{B}_1^*=\frac{1}{2}.$$

Hence,

$$\mathbf{F}(\mathbf{s})=e^{-3\mathbf{s}}\,\mathbf{F}_1(\mathbf{s})=\frac{e^{-3\mathbf{s}}}{2(\mathbf{s}+j2)}+\frac{e^{-3\mathbf{s}}}{2(\mathbf{s}-j2)}.$$

By invoking property #3a of Table 6-4, we obtain the inverse Laplace transform

$$\begin{aligned}f(t)=\mathcal{L}^{-1}[\mathbf{F}(\mathbf{s})]&=\mathcal{L}^{-1}\left[\frac{1}{2}\,\frac{e^{-3\mathbf{s}}}{\mathbf{s}+j2}+\frac{1}{2}\,\frac{e^{-3\mathbf{s}}}{\mathbf{s}-j2}\right]\\&=\left[\frac{1}{2}(e^{-j2(t-3)}+e^{j2(t-3)})\right]\,u(t-3)\\&=[\cos(2t-6)]\,u(t-3).\end{aligned}$$

We conclude this section with Table 12-3, which lists $\mathbf{F}(\mathbf{s})$ and its corresponding inverse transform $f(t)$ for all combinations of real versus complex, and distinct versus repeated, poles.

12-6 s-Domain Circuit Element Models

▶ The s-domain technique can be used to analyze circuits excited by sources with any type of variation—including pulse, step, ramp, sinusoid, and exponential—and provides a complete solution that incorporates both the steady state and transient components of the overall response. ◀

We can apply the technique by transforming the differential equation associated with the circuit, or, equivalently, by *transforming the circuit itself*, which entails representing R, L, and C by s-domain models.

12-6.1 Resistor in the s-Domain

Application of the Laplace transform to Ohm's law,

$$\mathcal{L}[\upsilon] = \mathcal{L}[Ri], \tag{12.74}$$

leads to

$$\mathbf{V} = R\mathbf{I}, \tag{12.75}$$

where, by definition,

$$\mathbf{V} = \mathcal{L}[\upsilon], \qquad \mathbf{I} = \mathcal{L}[i]. \tag{12.76}$$

Hence, for the resistor the correspondence between the time and **s**-domains is

$$\upsilon = Ri \quad \Longleftrightarrow \quad \mathbf{V} = R\mathbf{I}. \tag{12.77}$$

12-6.2 Inductor in the s-Domain

For R, the form of the i–υ relationship remained invariant under the transformation to the **s**-domain. That is not the case for L and C. Application of the Laplace transform to the i–υ relationship of the inductor,

$$\mathcal{L}[\upsilon] = \mathcal{L}\left[L\,\frac{di}{dt}\right], \tag{12.78}$$

gives

$$\mathbf{V} = L[\mathbf{sI} - i(0^-)], \tag{12.79}$$

where $i(0^-)$ is the current that was flowing through the inductor at $t = 0^-$. The time differentiation property (#6 in Table 12-1) was used in obtaining Eq. (12.79). The correspondence between the two domains is expressed as

$$\upsilon = L\,\frac{di}{dt} \quad \Longleftrightarrow \quad \mathbf{V} = \mathbf{s}L\mathbf{I} - L\,i(0^-). \tag{12.80}$$

In the **s**-domain, an inductor is represented by an impedance $\mathbf{Z}_L = \mathbf{s}L$, in series with a dc voltage source given by $L\,i(0^-)$ or—through source transformation—in parallel with a dc current source $i(0^-)/\mathbf{s}$, as shown in Table 12-4. Note that the current $\mathbf{I}$ flows from $(-)$ to $(+)$ through the dc voltage source (if $i(0^-)$ is positive).

12-6.3 Capacitor in the s-Domain

Similarly,

$$i = C\,\frac{d\upsilon}{dt} \quad \Longleftrightarrow \quad \mathbf{I} = \mathbf{s}C\mathbf{V} - C\,\upsilon(0^-), \tag{12.81}$$

where $\upsilon(0^-)$ is the initial voltage across the capacitor. The **s**-domain circuit models for the capacitor are available in Table 12-4.

> ▶ The **s**-domain transformation of circuit elements incorporates initial conditions associated with any energy storage that may have existed in capacitors and inductors at $t = 0^-$. ◀

12-6.4 Impedance

Impedances $\mathbf{Z}_R$, $\mathbf{Z}_L$, and $\mathbf{Z}_C$ are defined in the **s**-domain in terms of voltage to current ratios under zero initial conditions $[i(0^-) = \upsilon(0^-) = 0]$:

$$\mathbf{Z}_R = R, \quad \mathbf{Z}_L = \mathbf{s}L, \quad \text{and} \quad \mathbf{Z}_C = \frac{1}{\mathbf{s}C}. \tag{12.82}$$

Exercise 12-9: Convert the circuit in Fig. E12.9 into the **s**-domain.

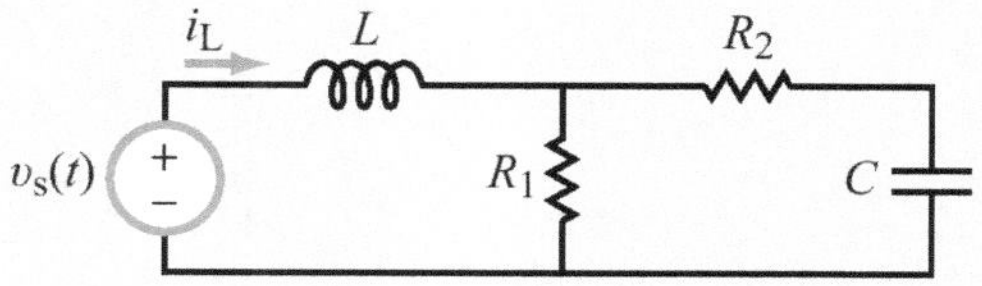

Figure E12.9

Answer:

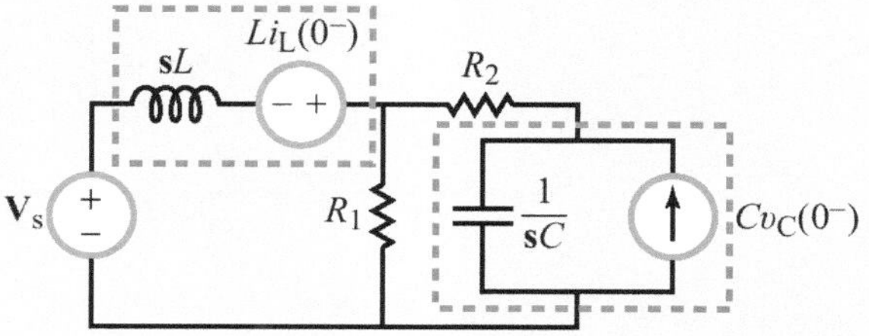

(See CAD)

Table 12-4: **Circuit models for R, L, and C in the s-domain.**

Time-Domain	s-Domain		
Resistor			
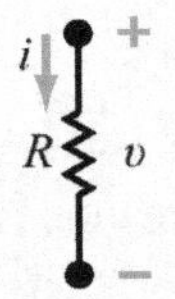	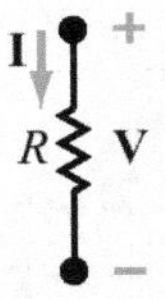		
$\upsilon = Ri$	$\mathbf{V} = R\mathbf{I}$		
Inductor			
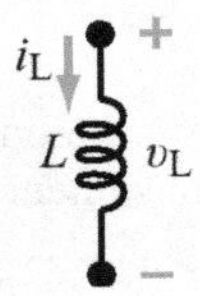	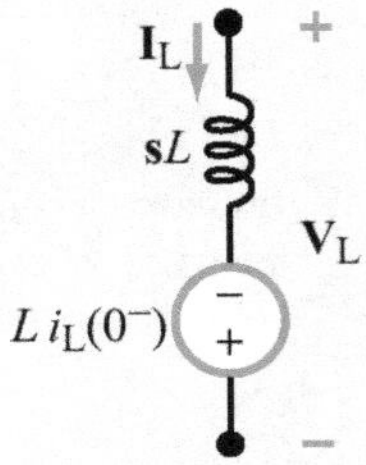	OR	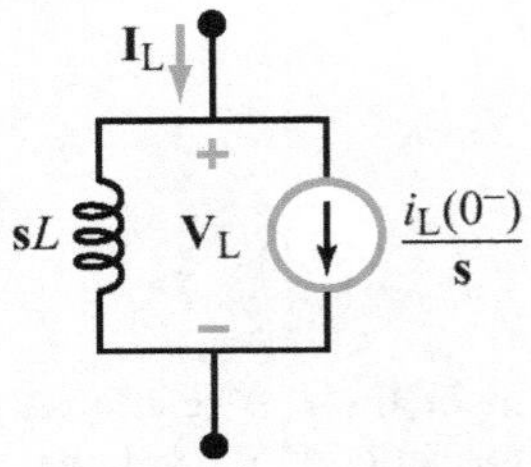
$\upsilon_L = L\dfrac{di_L}{dt}$ $i_L = \dfrac{1}{L}\displaystyle\int_{0^-}^{t} \upsilon_L\, dt + i_L(0^-)$	$\mathbf{V}_L = \mathbf{s}L\mathbf{I}_L - L\, i_L(0^-)$		$\mathbf{I}_L = \dfrac{\mathbf{V}_L}{\mathbf{s}L} + \dfrac{i_L(0^-)}{\mathbf{s}}$
Capacitor			
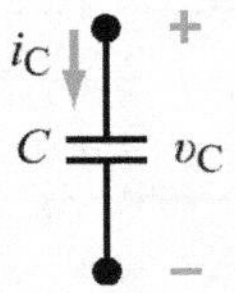	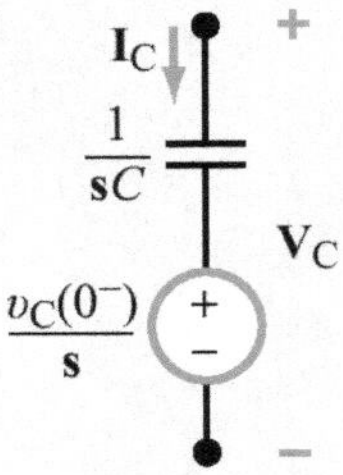	OR	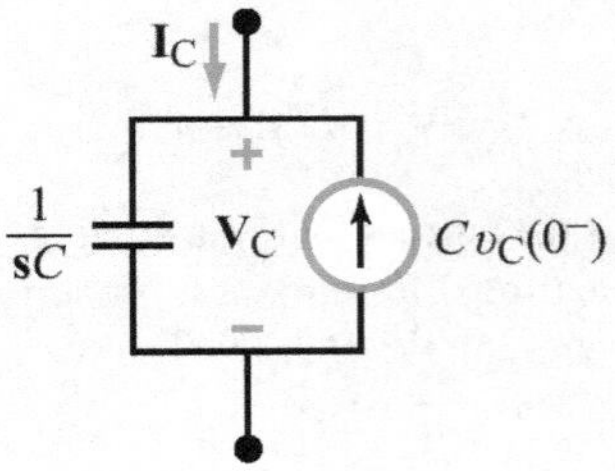
$i_C = C\dfrac{d\upsilon_C}{dt}$ $\upsilon_C = \dfrac{1}{C}\displaystyle\int_{0^-}^{t} i_C\, dt + \upsilon_C(0^-)$	$\mathbf{V}_C = \dfrac{\mathbf{I}_C}{\mathbf{s}C} + \dfrac{\upsilon_C(0^-)}{\mathbf{s}}$		$\mathbf{I}_C = \mathbf{s}C\mathbf{V}_C - C\,\upsilon_C(0^-)$

12-7 s-Domain Circuit Analysis

The circuit laws and analysis tools we used earlier in the time domain are equally applicable in the **s**-domain. They include KVL and KCL; voltage and current division; source transformation; source superposition; and Thévenin and Norton equivalent circuits. Execution of the **s**-domain analysis technique entails the following four steps:

> **Solution Procedure: s-Domain Technique**
>
> **Step 1:** Evaluate the circuit at $t = 0^-$ to determine voltages across capacitors and currents through inductors. Use this information in conjunction with Table 12-4 to transform the circuit from the time domain to the **s**-domain.
>
> **Step 2:** Apply KVL, KCL, and the other circuit tools to obtain an explicit expression for the voltage or current of interest.
>
> **Step 3:** If necessary, expand the expression into partial fractions.
>
> **Step 4:** Use the list of transform pairs given in Tables 12-2 and 12-3 and the list of properties in Table 12-1 (if needed) to transform the partial fraction to the time domain.

This process is illustrated through the next five examples.

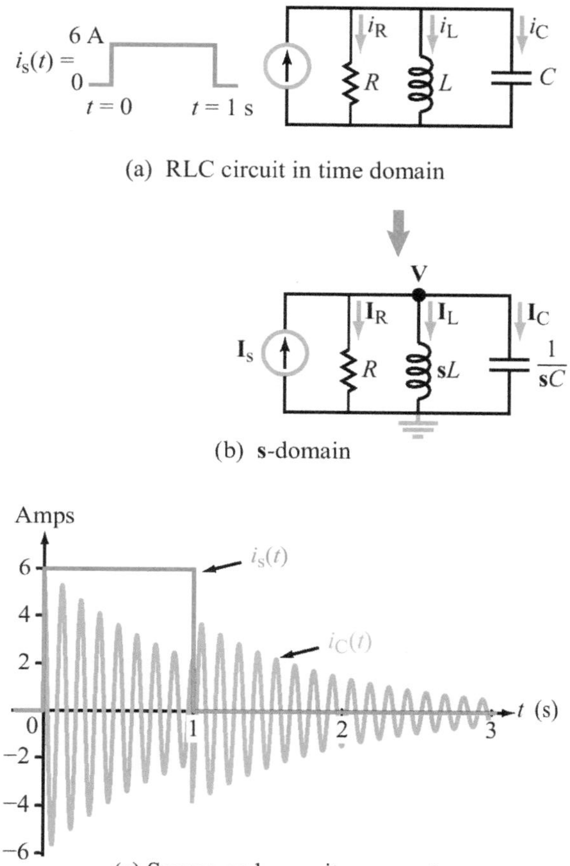

Figure 12-5: Circuit for Example 12-6.

Example 12-6: Parallel RLC Circuit

Determine the capacitor current response $i_C(t)$ to the rectangular pulse shown in Fig. 12-5(a), given that $R = 125\ \Omega$, $L = 0.1$ H, and $C = 4$ mF.

Solution: Per the proposed solution recipe, our first step should be to determine $i_L(0^-)$ and $\upsilon_C(0^-)$. In the present case, prior to activating the current source, the circuit contained no energy. Hence, $i_L(0^-) = 0$ and $\upsilon_C(0^-) = 0$, in which case transformation of the circuit elements to the **s**-domain entails replacing L with $\mathbf{s}L$ and C with $1/(\mathbf{s}C)$. The **s**-domain circuit is shown in Fig. 12-5(b).

The input source is

$$i_s(t) = 6u(t) - 6u(t-1), \tag{12.83}$$

and, according to entries #2 and #2a in Table 12-2, the **s**-domain expression for the source should be

$$\mathbf{I}_s = \frac{6}{\mathbf{s}} - \frac{6}{\mathbf{s}} e^{-\mathbf{s}}. \tag{12.84}$$

Our intermediate goal is to determine $\mathbf{I}_C$, the **s**-domain current through the capacitor in Fig. 12-5(b). Application of KCL at node **V** in Fig. 12-5(b) gives

$$\mathbf{V}\left(\frac{1}{R} + \frac{1}{\mathbf{s}L} + \mathbf{s}C\right) = \mathbf{I}_s. \tag{12.85}$$

Also, since $\upsilon_C(0) = 0$, $\mathbf{I}_C$ is related to **V** by

$$\mathbf{I}_C = \mathbf{s}C\mathbf{V}.$$

Solution for $\mathbf{I}_C$ leads to

$$\mathbf{I}_C = \left(\frac{\mathbf{s}^2}{\mathbf{s}^2 + \dfrac{\mathbf{s}}{RC} + \dfrac{1}{LC}}\right)\mathbf{I}_s, \tag{12.86}$$

where, in anticipation of applying partial-fraction expansion later on, we configured the denominator such that the coefficient of the highest-order **s**-term is 1.

Since $\mathbf{I}_s$ is the sum of two similar terms, we shall apply the superposition principle as follows:

$$\mathbf{I}_C = \left(\frac{\mathbf{s}^2}{\mathbf{s}^2 + \dfrac{\mathbf{s}}{RC} + \dfrac{1}{LC}}\right)\left(\frac{6}{\mathbf{s}} - \frac{6e^{-\mathbf{s}}}{\mathbf{s}}\right) = \mathbf{I}_{C_1} + \mathbf{I}_{C_2}, \tag{12.87}$$

where

$$\mathbf{I}_{C_1} = \left(\frac{\mathbf{s}^2}{\mathbf{s}^2 + \dfrac{\mathbf{s}}{RC} + \dfrac{1}{LC}}\right)\frac{6}{\mathbf{s}} = \frac{6\mathbf{s}}{\mathbf{s}^2 + \dfrac{\mathbf{s}}{RC} + \dfrac{1}{LC}}, \tag{12.88a}$$

and

$$\mathbf{I}_{C_2} = \left(\frac{6\mathbf{s}}{\mathbf{s}^2 + \dfrac{\mathbf{s}}{RC} + \dfrac{1}{LC}}\right)(-e^{-\mathbf{s}}). \tag{12.88b}$$

Solution for $\mathbf{I}_{C_1}$:

Inserting the specified element values, namely $R = 125\ \Omega$, $L = 0.1$ H, and $C = 4$ mF, leads to

$$\mathbf{I}_{C_1} = \frac{6\mathbf{s}}{\mathbf{s}^2 + 2\mathbf{s} + 2500}. \tag{12.89}$$

The roots of $\mathbf{s}^2 + 2\mathbf{s} + 2500 = 0$ are

$$\mathbf{s}_1 = -1 - j\sqrt{2499} \approx -1 - j50, \tag{12.90a}$$

$$\mathbf{s}_2 = -1 + j\sqrt{2499} \approx -1 + j50. \tag{12.90b}$$

Hence,

$$\mathbf{I}_{C_1} = \frac{6\mathbf{s}}{(\mathbf{s} + 1 + j50)(\mathbf{s} + 1 - j50)}. \tag{12.91}$$

Partial fraction expansion takes the form

$$\mathbf{I}_{C_1} = \frac{\mathbf{B}}{\mathbf{s} + 1 + j50} + \frac{\mathbf{B}^*}{\mathbf{s} + 1 - j50},$$

with

$$\mathbf{B} = (\mathbf{s} + 1 + j50)\left.\frac{6\mathbf{s}}{(\mathbf{s} + 1 + j50)(\mathbf{s} + 1 - j50)}\right|_{\mathbf{s}=-1-j50} = 3e^{-j1.15^\circ}. \tag{12.92}$$

Hence,

$$\mathbf{I}_{C_1} = \frac{3e^{-j1.15^\circ}}{(\mathbf{s} + 1 + j50)} + \frac{3e^{j1.15^\circ}}{(\mathbf{s} + 1 - j50)}. \tag{12.93}$$

Per entry #3 in Table 12-3, the time-domain equivalent of $\mathbf{I}_{C_1}$ is

$$i_{C_1}(t) = 6e^{-t}\cos(50t + 1.15^\circ)\ u(t)\ \text{A}. \tag{12.94}$$

Solution for $\mathbf{I}_{C_2}$:

The expression for $\mathbf{I}_{C_2}$ given by Eq. (12.88b) is identical to that for $\mathbf{I}_{C_1}$ except for a multiplication factor of $(-e^{-\mathbf{s}})$. Per the time-shift property in Table 12-1, $i_{C_2}(t)$ can be obtained from the expression for $i_{C_1}(t)$ by multiplying $i_{C_1}(t)$ by (-1) and delaying t by 1 s. Hence,

$$i_{C_2}(t) = -6e^{-(t-1)}\cos[50(t-1) + 1.15^\circ]\ u(t-1)\ \text{A}. \tag{12.95}$$

Total solution:

$$\begin{aligned} i_C(t) &= i_{C_1}(t) + i_{C_2}(t) \\ &= 6\{e^{-t}\cos(50t + 1.15^\circ)\ u(t) \\ &\quad - e^{-(t-1)}\cos[50(t-1) + 1.15^\circ]\ u(t-1)\}\ \text{A}. \end{aligned} \tag{12.96}$$

Figure 12-5(c) displays the waveforms of the source current $i_s(t)$ and the capacitor current $i_C(t)$.

Example 12-7: Two-State Power Supply

In the circuit shown in Fig. 12-6(a), the voltage source can operate at 125 V or 250 V, and when it switches between the two states, it does so gradually. Determine $i_L(t)$ for $t \geq 0$, in response to

$$\upsilon_s(t) = \begin{cases} 125\ \text{V} & \text{for } t < 0 \\ (250 - 125e^{-2t})\ u(t)\ \text{V} & \text{for } t \geq 0. \end{cases} \tag{12.97}$$

Solution: The circuit condition at $t < 0$ is depicted in Fig. 12-6(b) wherein the capacitor and inductor have been replaced with an open circuit and short circuit, respectively. KVL leads to

$$i_L(0^-) = \frac{125}{R} = \frac{125}{12.5 \times 10^3} = 10\ \text{mA},$$

$$\upsilon_C(0^-) = \upsilon_L(0^-) = 0.$$

With the help of Table 12-2, the **s**-domain expression for $\mathbf{V}_s$ at $t \geq 0$ is

$$\mathbf{V}_s = \frac{250}{\mathbf{s}} - \frac{125}{\mathbf{s} + 2}. \tag{12.98}$$

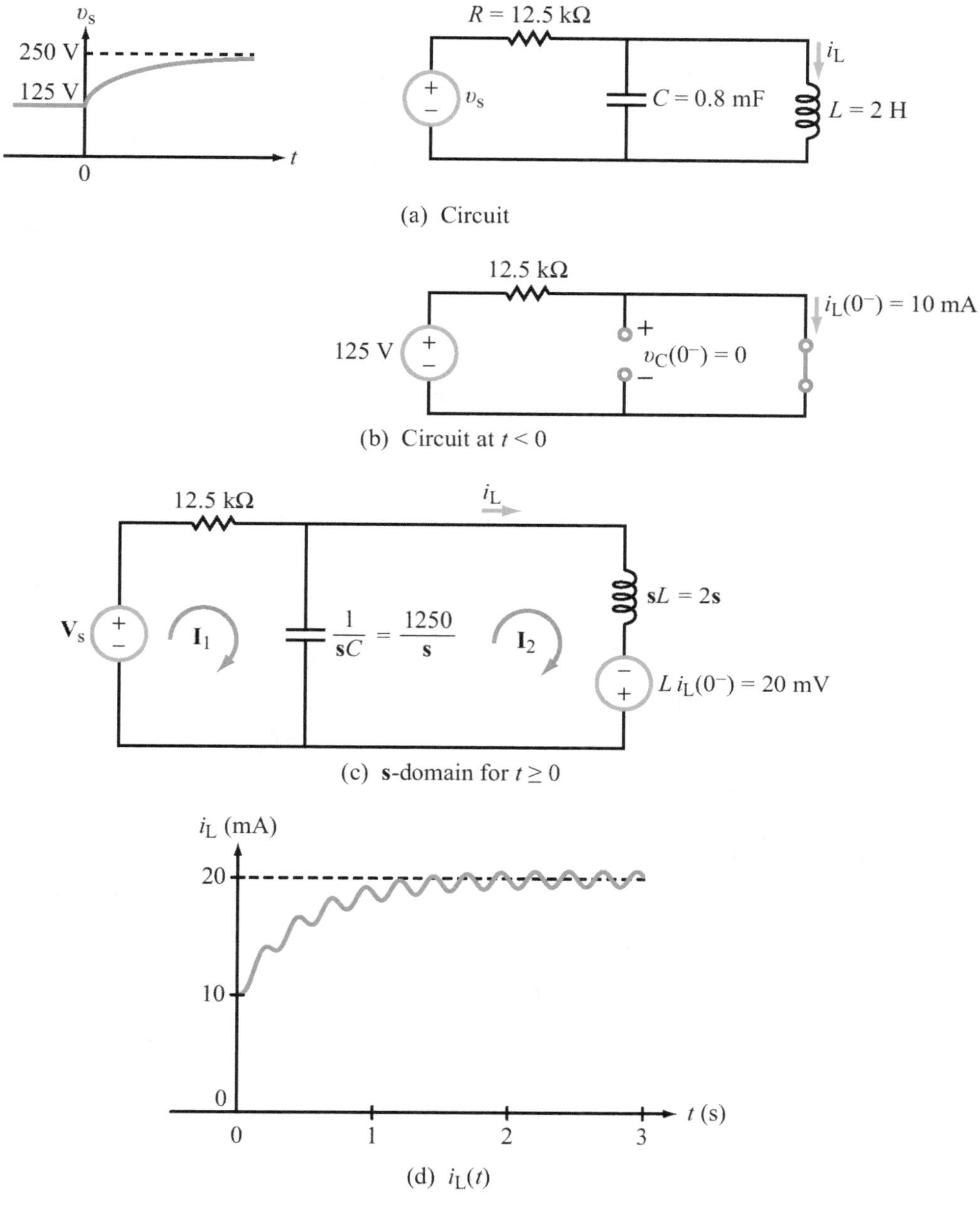

Figure 12-6: Example 12-7.

The **s**-domain circuit is shown in Fig. 12-6(c), in which L and C are represented by their **s**-domain models in accordance with Table 12-4, namely

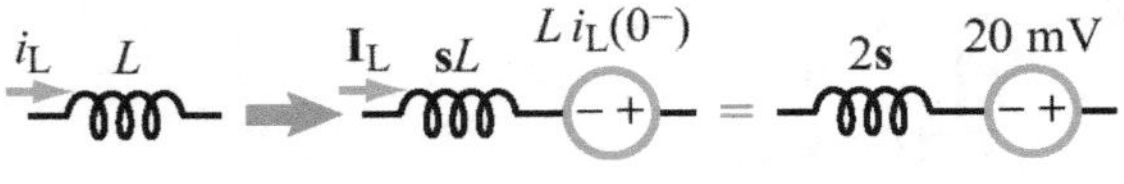

and

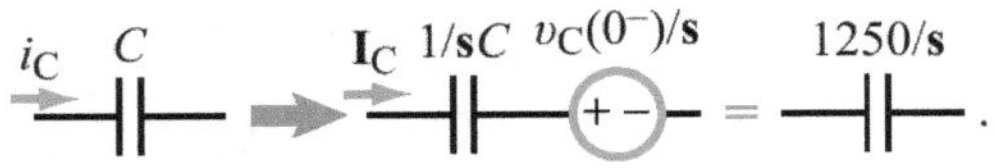

By inspection, the mesh current equations for the two loops in Fig. 12-6(c) are

$$\left(12.5\times 10^3+\frac{1250}{\mathbf{s}}\right)\mathbf{I}_1-\frac{1250}{\mathbf{s}}\,\mathbf{I}_2=\mathbf{V}_s, \tag{12.99a}$$

$$-\frac{1250}{\mathbf{s}}\,\mathbf{I}_1+\left(\frac{1250}{\mathbf{s}}+2\mathbf{s}\right)\mathbf{I}_2=20\times 10^{-3}. \tag{12.99b}$$

After replacing $\mathbf{V}_s$ with the expression given by Eq. (12.98), simultaneous solution of the two linear equations leads to

$$\mathbf{I}_2=\frac{10\mathbf{s}^3+21\mathbf{s}^2+6252\mathbf{s}+25000}{\mathbf{s}(\mathbf{s}+2)(\mathbf{s}^2+0.1\mathbf{s}+625)}\ \text{mA}. \tag{12.100}$$

The roots of the quadratic term in the denominator are

$$\mathbf{s}_1\approx -0.05-j25,$$
$$\mathbf{s}_2\approx -0.05+j25.$$

Hence, Eq. (12.100) can be rewritten as

$$\mathbf{I}_2=\frac{10\mathbf{s}^3+21\mathbf{s}^2+6252\mathbf{s}+25000}{\mathbf{s}(\mathbf{s}+2)(\mathbf{s}+0.05+j25)(\mathbf{s}+0.05-j25)}. \tag{12.101}$$

Application of the partial-fractions expansion recipes outlined earlier in Section 12-5 leads to

$$\mathbf{I}_2=\left[\frac{20}{\mathbf{s}}-\frac{10}{\mathbf{s}+2}+\frac{0.38e^{-j90^\circ}}{\mathbf{s}+0.05+j25}+\frac{0.38e^{j90^\circ}}{\mathbf{s}+0.05-j25}\right]\text{mA},$$

and transformation to the time domain gives

$$\begin{aligned} i_L(t)&=i_2(t)\\ &=\left[20-10e^{-2t}+0.76e^{-0.05t}\cos(25t+90^\circ)\right]u(t)\ \text{mA}\\ &=\left[20-10e^{-2t}-0.76e^{-0.05t}\sin 25t\right]u(t)\ \text{mA}. \end{aligned}$$

The waveform of $i_L(t)$ is displayed in Fig. 12-6(d).

Example 12-8: ac Source with dc Bias

The current source shown in the circuit of Fig. 12-7(a) is given by the displayed waveform, which consists of a 1.5 A dc source prior to $t=0$, and the combination of a cosinusoidal waveform and a 1 A dc bias after $t=0$. Determine $v_{out}(t)$ for $t\geq 0$, given that $R_1=1\ \Omega$, $R_2=0.5\ \Omega$, and $L=0.5$ H.

Solution: The current source is characterized by

$$i_s(t)=\begin{cases}1.5\ \text{A} & \text{for } t\leq 0\\ (1+0.5\cos 4t)\ \text{A} & \text{for } t\geq 0.\end{cases}$$

Since $i_s(t)$ was nonzero before $t=0$, we need to examine initial conditions by analyzing the circuit shown in Fig. 12-7(b), from which we deduce that

$$i_L(0^-)=\frac{1.5R_1}{R_1+R_2}=\frac{1.5\times 1}{1+0.5}=1\ \text{A}.$$

The **s**-domain expression for $i_s(t)$ for $t\geq 0$ is

$$\mathbf{I}_s=\frac{1}{\mathbf{s}}+\frac{0.5\mathbf{s}}{\mathbf{s}^2+16}=\frac{1.5\mathbf{s}^2+16}{\mathbf{s}(\mathbf{s}^2+16)}.$$

Figure 12-7(c) depicts the circuit in the **s**-domain, where we applied source transformation to convert $(\mathbf{I}_s, R_1)$ into a voltage source $\mathbf{V}_s=\mathbf{I}_sR_1$, in series with R_1.

At node $\mathbf{V}_{out}$ in the circuit of Fig. 12-7(c),

$$\frac{\mathbf{V}_{out}-\mathbf{V}_s-L\,i_L(0^-)}{R_1+\mathbf{s}L}+\frac{\mathbf{V}_{out}}{R_2}=0,$$

which gives

$$\begin{aligned}\mathbf{V}_{out}&=\frac{R_2[1.5\mathbf{s}^2+16+\mathbf{s}(\mathbf{s}^2+16)L\,i_L(0^-)]}{\mathbf{s}(\mathbf{s}^2+16)[(R_1+R_2)+\mathbf{s}L]}\\ &=\frac{\mathbf{s}^3+3\mathbf{s}^2+16\mathbf{s}+32}{2\mathbf{s}(\mathbf{s}+3)(\mathbf{s}^2+16)}\\ &=\frac{\mathbf{s}^3+3\mathbf{s}^2+16\mathbf{s}+32}{2\mathbf{s}(\mathbf{s}+3)(\mathbf{s}+j4)(\mathbf{s}-j4)}.\end{aligned}$$

Partial fraction expansion gives

$$\mathbf{V}_{out}=\frac{A_1}{\mathbf{s}}+\frac{A_2}{\mathbf{s}+3}+\frac{\mathbf{B}}{\mathbf{s}+j4}+\frac{\mathbf{B}^*}{\mathbf{s}-j4},$$

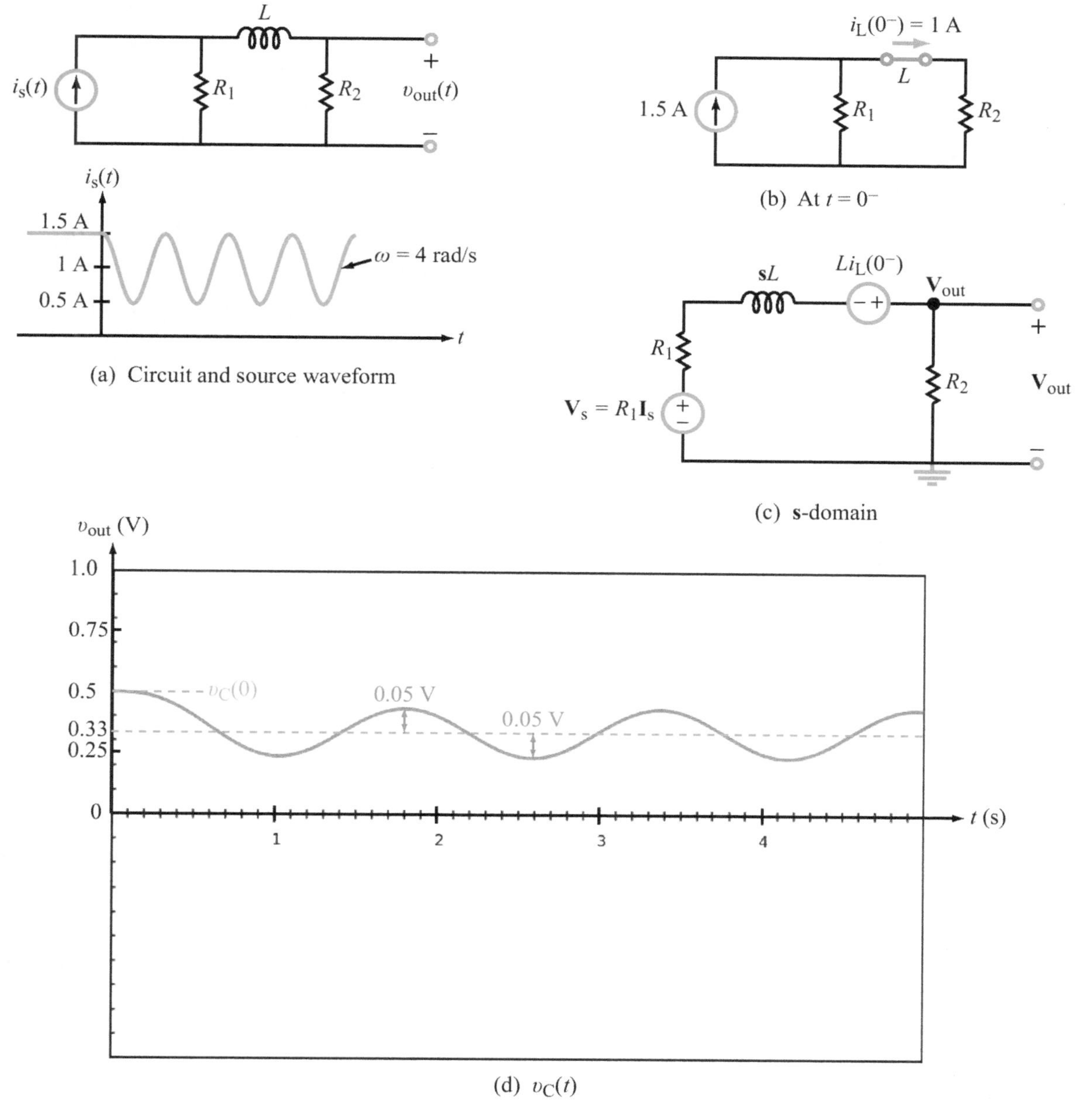

Figure 12-7: Example 12-8.

with

$$A_1 = \mathbf{s}\,\mathbf{V}_{\text{out}}|_{\mathbf{s}=0} = \left.\frac{\mathbf{s}^3+3\mathbf{s}^2+16\mathbf{s}+32}{2(\mathbf{s}+3)(\mathbf{s}^2+16)}\right|_{\mathbf{s}=0} = \frac{1}{3},$$

$$A_2 = (\mathbf{s}+3)\,\mathbf{V}_{\text{out}}|_{\mathbf{s}=-3} = \left.\frac{\mathbf{s}^3+3\mathbf{s}^2+16\mathbf{s}+32}{2\mathbf{s}(\mathbf{s}^2+16)}\right|_{\mathbf{s}=-3} = \frac{8}{75},$$

$$\mathbf{B} = (\mathbf{s}+j4)\,\mathbf{V}_{\text{out}}|_{\mathbf{s}=-j4} = \left.\frac{\mathbf{s}^3+3\mathbf{s}^2+16\mathbf{s}+32}{2\mathbf{s}(\mathbf{s}+3)(\mathbf{s}-j4)}\right|_{\mathbf{s}=-j4} = \frac{1}{20}\,e^{j53^\circ}.$$

Hence,

$$\mathbf{V}_{\text{out}} = \frac{1}{3\mathbf{s}} + \frac{8}{75(\mathbf{s}+3)} + \frac{e^{j53^\circ}}{20(\mathbf{s}+j4)} + \frac{e^{-j53^\circ}}{20(\mathbf{s}-j4)}.$$

The corresponding time-domain voltage is

$$\upsilon_{\text{out}}(t) = \left[\frac{1}{3} + \frac{8}{75}\,e^{-3t} + \frac{1}{10}\cos(4t-53^\circ)\right] u(t) \quad \text{(V)}.$$

After the decay of the negative exponential term, the output becomes the sum of a dc term (1/3 V) and an ac term with an amplitude of 0.1 V. The circuit output response is displayed in Fig. 12-7(d).

Example 12-9: Circuit with a Switch

Determine $i_{\text{L}}(t)$ in the circuit shown in Fig. 12-8(a) for $t \geq 0$.

Solution: We start by examining the state of the circuit at $t = 0^-$ (before closing the switch). Upon replacing L with a short circuit and C with an open circuit, as portrayed by the configuration in Fig. 12-8(b), we establish that

$$i_{\text{L}}(0^-) = 1\text{ A} \quad \text{and} \quad \upsilon_{\text{C}}(0^-) = 12\text{ V}. \qquad (12.102)$$

For $t \geq 0$, the s-domain equivalent of the original circuit is shown in Fig. 12-8(c), where we have replaced R_2 with a short circuit, converted the dc source into its s-domain equivalent and, in accordance with the circuit models given in Table 12-4, converted L and C into impedances—each with its own appropriate voltage source. By inspection, the two mesh current equations are given by

$$(4+12+2\mathbf{s})\mathbf{I}_1 - (12+2\mathbf{s})\mathbf{I}_2 = \frac{24}{\mathbf{s}} + 2, \qquad (12.103a)$$

$$-(12+2\mathbf{s})\mathbf{I}_1 + \left(12+2\mathbf{s}+\frac{5}{\mathbf{s}}\right)\mathbf{I}_2 = -2 - \frac{12}{\mathbf{s}}. \qquad (12.103b)$$

Simultaneous solution of the two equations leads to

$$\mathbf{I}_1 = \frac{12\mathbf{s}^2+77\mathbf{s}+60}{\mathbf{s}(4\mathbf{s}^2+29\mathbf{s}+40)} \qquad (12.104a)$$

and

$$\mathbf{I}_2 = \frac{8(\mathbf{s}+6)}{4\mathbf{s}^2+29\mathbf{s}+40}. \qquad (12.104b)$$

The associated inductor current $\mathbf{I}_{\text{L}}$ is

$$\mathbf{I}_{\text{L}} = \mathbf{I}_1 - \mathbf{I}_2 = \frac{4\mathbf{s}^2+29\mathbf{s}+60}{\mathbf{s}(4\mathbf{s}^2+29\mathbf{s}+40)} = \frac{4\mathbf{s}^2+29\mathbf{s}+60}{4\mathbf{s}(\mathbf{s}+1.85)(\mathbf{s}+5.4)}, \qquad (12.105)$$

which can be represented by the partial fraction expansion

$$\mathbf{I}_{\text{L}} = \frac{A_1}{\mathbf{s}} + \frac{A_2}{\mathbf{s}+1.85} + \frac{A_3}{\mathbf{s}+5.4}. \qquad (12.106)$$

The values of A_1 through A_3 are obtained from

$$A_1 = \mathbf{s}\mathbf{I}_{\text{L}}|_{\mathbf{s}=0} = \frac{60}{40} = 1.5, \qquad (12.107a)$$

$$A_2 = (\mathbf{s}+1.85)\mathbf{I}_{\text{L}}|_{\mathbf{s}=-1.85} = \left.\frac{4\mathbf{s}^2+29\mathbf{s}+60}{4\mathbf{s}(\mathbf{s}+5.4)}\right|_{\mathbf{s}=-1.85} = -0.76, \qquad (12.107b)$$

and

$$A_3 = (\mathbf{s}+5.4)\mathbf{I}_{\text{L}}|_{\mathbf{s}=-5.4} = 0.26. \qquad (12.107c)$$

Hence,

$$\mathbf{I}_{\text{L}} = \frac{1.5}{\mathbf{s}} - \frac{0.76}{\mathbf{s}+1.85} + \frac{0.26}{\mathbf{s}+5.4}, \qquad (12.108)$$

and the corresponding time-domain current for $t \geq 0$ is

$$i_{\text{L}}(t) = [1.5 - 0.76e^{-1.85t} + 0.26e^{-5.4t}]\,u(t)\text{ A}. \qquad (12.109)$$

The time variation of $i_{\text{L}}(t)$ is displayed in Fig. 12-8(d).

Example 12-10: Op-Amp Integrator

The op-amp integrator circuit shown in Fig. 12-9(a) was first introduced in Section 5-6.1. We now examine its behavior by applying the s-domain analysis technique to a step-function input given by $\upsilon_{\text{i}}(t) = 10u(t)$ mV. The capacitor had no charge prior to $t = 0$ and the op-amp's power supply voltage is $V_{\text{cc}} = 10$ V.

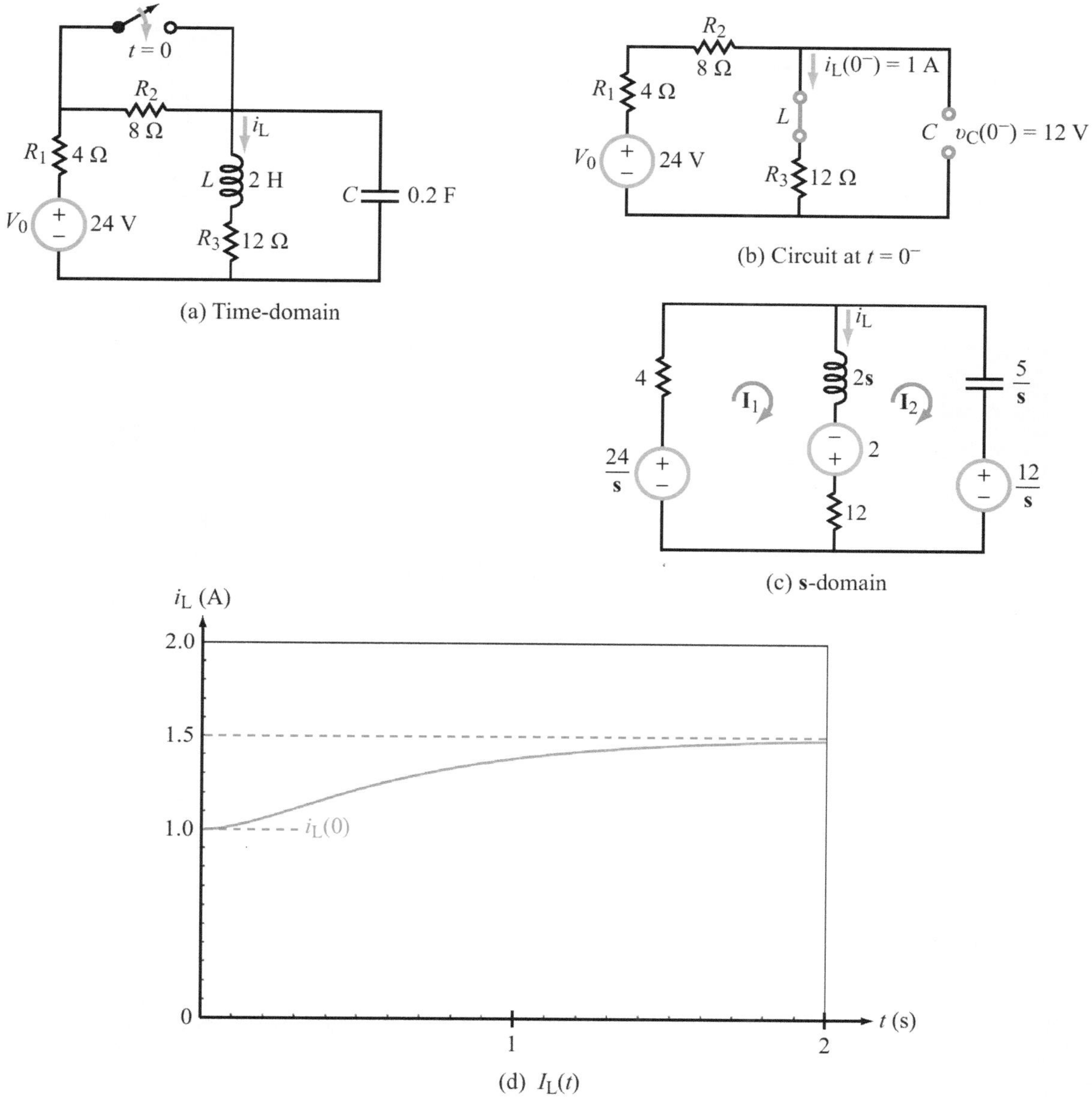

Figure 12-8: Circuit for Example 12-9.

Solution: The **s**-domain circuit is shown in Fig. 12-9(b), from which we deduce that

$$\mathbf{V}_{\text{out}} = -\left(\frac{\mathbf{Z}_C}{\mathbf{Z}_R}\right)\mathbf{V}_i = -\frac{\mathbf{V}_i}{\mathbf{s}R_1C_1} = -\frac{10^{-2}}{\mathbf{s}}\left(\frac{50}{\mathbf{s}}\right) = -\frac{0.5}{\mathbf{s}^2}.$$

Application of entry #4 of Table 12-2 leads to

$$\upsilon_{\text{out}_1}(t) = -0.5t\;u(t).$$

We observe in Fig. 12-9(c) that $\upsilon_{\text{out}}(t)$ is a negative ramp function that saturates at $V_{cc} = -10$ V when t reaches 20 s.

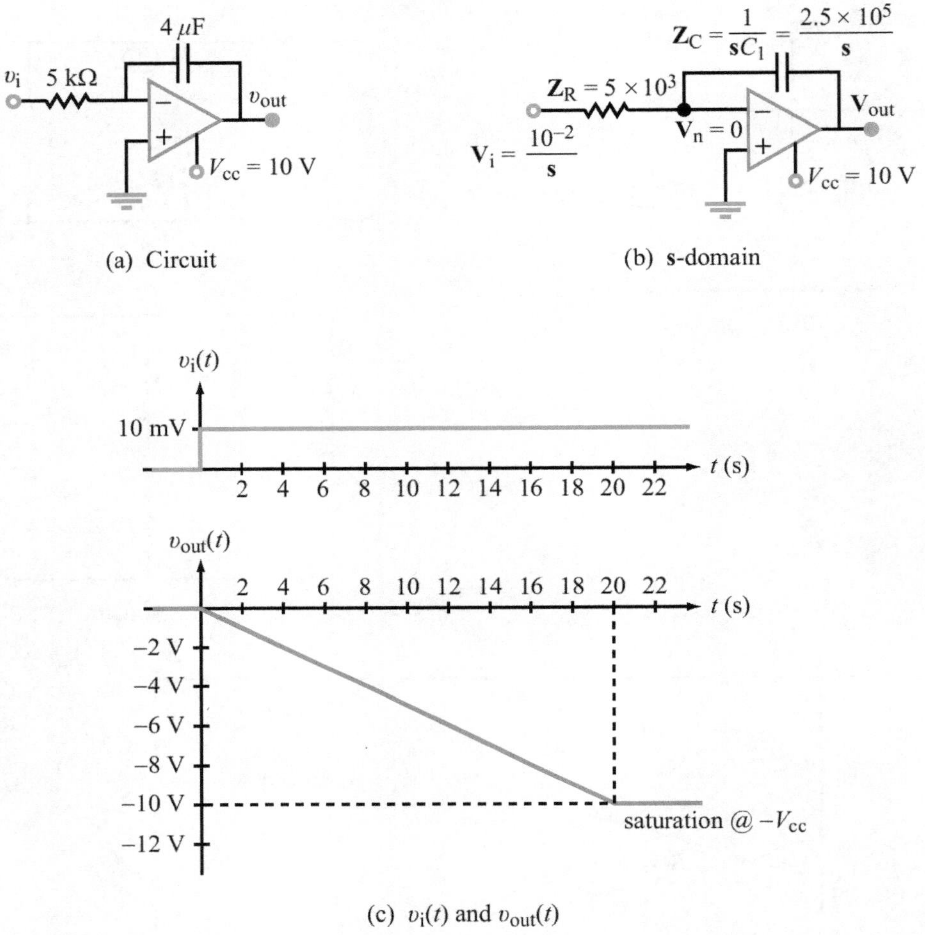

Figure 12-9: Circuit for Example 12-10.

12-8 Multisim Analysis of Circuits Driven by Nontrivial Inputs

The utility of SPICE simulators becomes most evident when trying to simulate circuits driven by nontrivial inputs—in contrast with sinusoids or dc voltages. In this section, we will revisit some of the examples we examined earlier in this chapter to demonstrate how easy it is to obtain solutions with Multisim and to compare the solutions with the analytical results based on the Laplace transform method. As a learning tool, Multisim is also very useful, in that it allows the user to test his/her understanding of core concepts by simulating circuits over a wide range of conditions and for a variety of different input waveforms.

Example 12-11: RC Circuit Response

A series RC circuit, with $R = 500$ kΩ and $C = 1$ μF, is excited by a voltage source that delivers a 1 V, 1 s rectangular pulse. Draw the circuit in Multisim and generate the output response using the Transient Analysis tool.

Solution: By now, we should be very familiar with how to create a pulse source in Multisim. The circuit is shown in

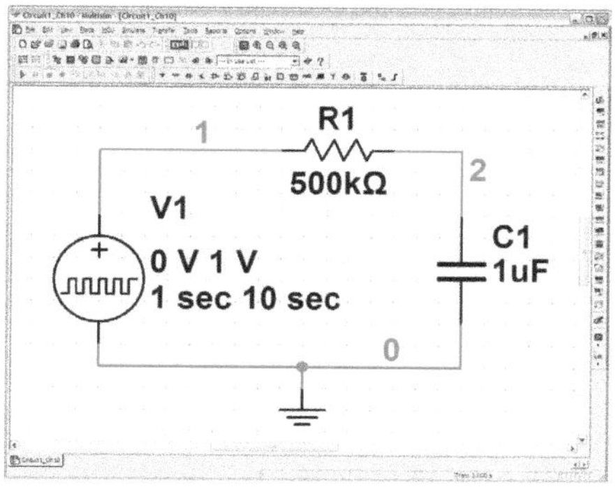

(a) Circuit in Multisim

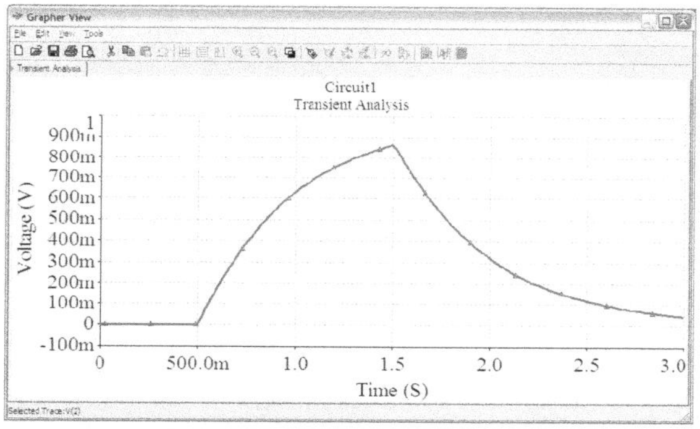

(b) Response

Figure 12-10: (a) RC circuit excited by a 1 V, 1 s rectangular pulse at 0.5 s, and (b) the corresponding response at node 2.

Fig. 12-10(a), and the output response across the capacitor is displayed in Fig. 12-10(b). Note that a *delay time* of 0.5 s was introduced in the parameter selections of the pulse source in order to generate a clearer plot.

Example 12-12: Interrupted Voltage Source in Multisim

Draw the circuit shown in Fig. 12-11(a) in Multisim and then use Transient Analysis to generate a plot of the voltage across the 3-Ω resistor in response to an input excitation given by 15 V prior to $t = 0$, and $15(1 - e^{-2t})$ V afterwards.

Solution: The circuit is reproduced in Fig. 12-11(b). To model the exponential input voltage, we use the EXPONENTIAL_VOLTAGE source which can accommodate both rising and falling exponentials. Multisim divides the exponential voltage into two segments, with the first called the *Rise* segment and the second called the *Fall* segment, and this order is independent of whether the change in level is actually a rise or a fall. In the present case we need to simulate in the first segment an instantaneous change in level from 15 V down to 0 V. We do so by setting the Initial Value to 15 V, the Pulsed Value to 0 V, the Rise Delay Time to 0 s, and the Rise Time to 1 ns (which is practically the same as instantaneous). To

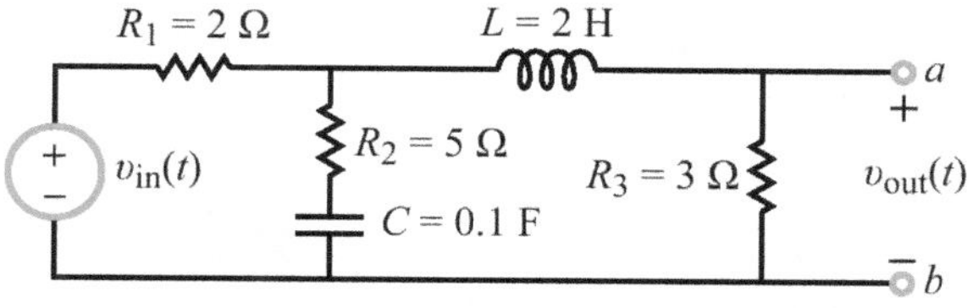

(a) Time domain

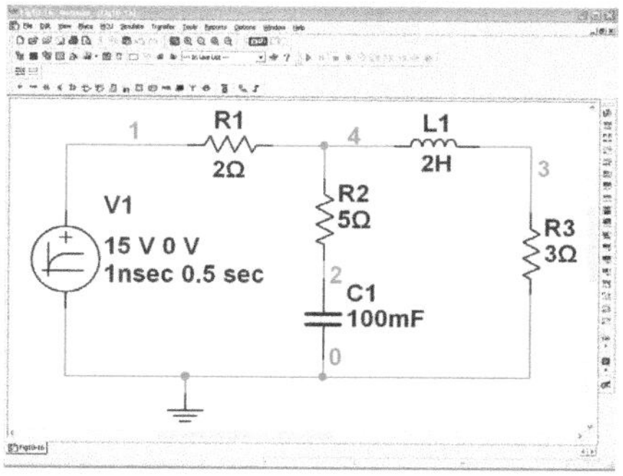

(b) Circuit in Multisim

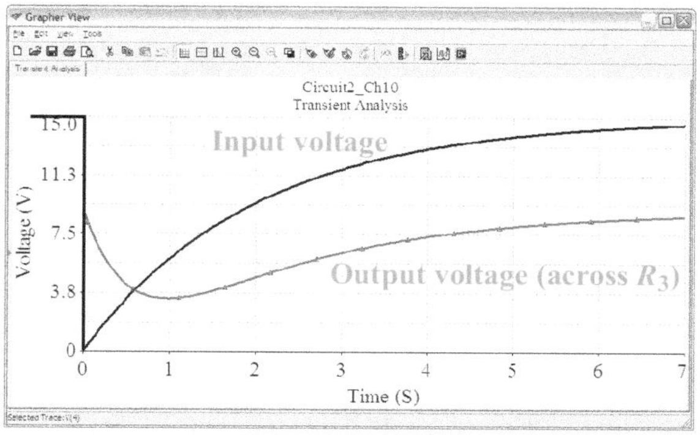

(c) Response

Figure 12-11: Multisim rendition of the circuit response to a sudden (but temporary) change in supply voltage level.

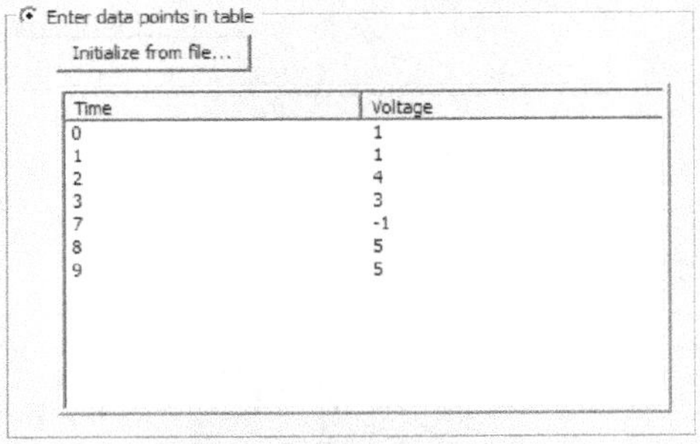

(a) Time-voltage pairs for circuit

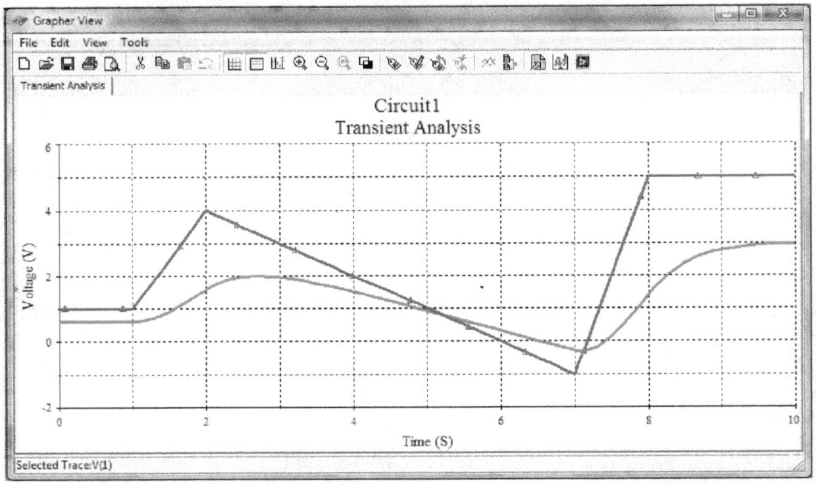

(b) Response

Figure 12-12: Multisim rendition of the circuit response to an arbitrary input signal produced by the PWL (Piecewise Linear) source.

simulate the second segment during which the voltage increases from 0 to 15 V with a time constant of 0.5 s, we set the Fall Delay at 0 and the Fall Time Constant at 0.5 s. Applying Transient Analysis results in the responses displayed in Fig. 12-11(c).

Example 12-13: Piecewise Linear Voltage Source

The PIECEWISE_LINEAR_VOLTAGE source allows you to define time-voltage pairs such that the source will be at a given voltage at the corresponding time and the source will "connect the dots" in between using a linear progression. Hence, entering the time-voltage pairs of (0,1), (1,1), and (2,4) will create a source which starts at 1 V and stays steady until 1 s, at which time it will increase with a slope of 3 V/s to reach a value of 4 V at 2 s.

Replace the Exponential voltage source in the circuit in Fig. 12-11(a) with a piecewise linear (PWL) voltage source with the time-voltage pairs: (0,1), (1,1), (2,4), (3,3), (7,−1), (8,5), and (9,5). Plot both the input and output in Transient Analysis from 0 to 10 s.

Solution: With the PIECEWISE_LINEAR_VOLTAGE source in place, double-click on it and then make sure the Value tab is selected. Click on Enter data points in table, and insert the time-voltage pairs shown in Fig. 12-12(a). Applying Transient Analysis results in the responses displayed in Fig. 12-12(b).

Summary

Concepts

- The Laplace-transform analysis technique transforms the circuit to a new domain, solves for the quantity of interest in that domain, and then transforms the solution back to the time domain. The technique can be applied to circuits with any type of excitation.
- The Laplace transform has many useful properties that can facilitate the process of finding the Laplace transform of a time function.
- Under zero initial conditions, circuit elements R, L, and C transform to R, $\mathbf{s}L$, and $1/\mathbf{s}C$, respectively, in the $\mathbf{s}$-domain.

Mathematical and Physical Models

Unit Impulse Function

$$\delta(t - T) = 0 \qquad \text{for } t \neq T$$

$$\int_{-\infty}^{\infty} \delta(t - T)\, dt = 1$$

Laplace Transform

$$\mathbf{F}(\mathbf{s}) = \mathcal{L}[f(t)] = \int_{0^-}^{\infty} f(t)\, e^{-\mathbf{s}t}\, dt$$

Properties ➡ Table 12-1

Transform Pair ➡ Tables 12-2 and 12-3

Time/s-Domain Equivalents

Resistor	$\upsilon = Ri$	⟷	$\mathbf{V} = R\mathbf{I}$
Inductor	$\upsilon = L\,\dfrac{di}{dt}$	⟷	$\mathbf{V} = \mathbf{s}L\mathbf{I} - L\,i(0^-)$
Capacitor	$i = C\,\dfrac{d\upsilon}{dt}$	⟷	$\mathbf{I} = \mathbf{s}C\mathbf{V} - C\,\upsilon(0^-)$

Important Terms

Provide definitions or explain the meaning of the following terms:

complex frequency
convergence condition
critically damped response
damped natural frequency
damping coefficient
delay time
delta function
distinct
domain transformation
expansion coefficient
Fall
improper rational function
impulse function
initial condition
initial value
inverse Laplace transform
final condition
Laplace transform
Laplace transformation
Laplace transform pair
natural response
one-sided transform
overdamped response
partial fraction expansion
pole
pole factor
proper rational function
real
region of convergence
residue method
resonant frequency
Rise
sampling property
second-order circuit
singularity function
steady state response
step function
time scaling
transient response
two-sided transform
underdamped response
uniqueness property
unit impulse function
unit step function
universal property

PROBLEMS

Sections 12-1 and 12-2: Impulse Response and Complex Algebra

12.1 Evaluate each of the following integrals.

(a) $G_1 = \displaystyle\int_{-\infty}^{\infty} (3t^3 + 2t^2 + 1)[\delta(t) + 4\delta(t-2)]\, dt$

(b) $G_2 = \displaystyle\int_{-2}^{4} 4(e^{-2t} + 1)[\delta(t) - 2\delta(t-2)]\, dt$

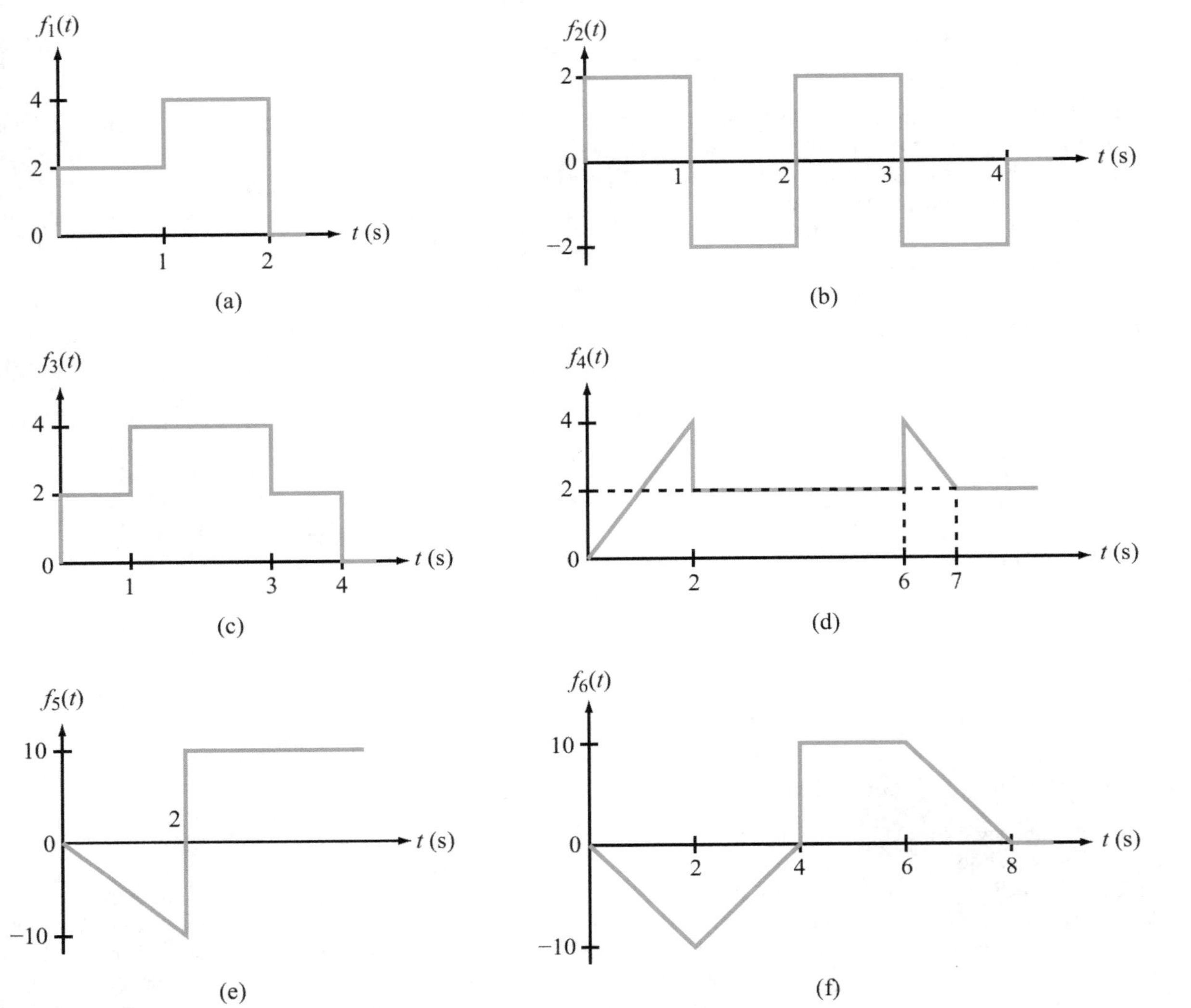

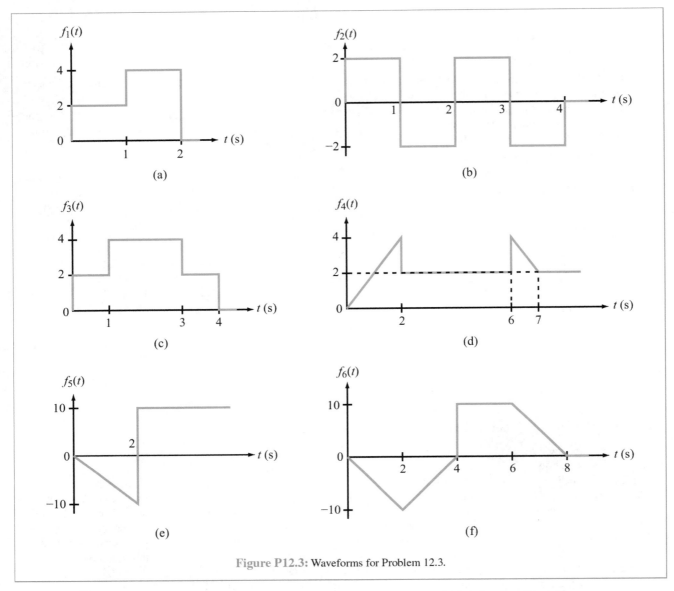

Figure P12.3: Waveforms for Problem 12.3.

(c) $G_3 = \int_{-20}^{20} 3(t \cos 2\pi t - 1)[\delta(t) + \delta(t - 10)]\, dt$

12.2 Evaluate each of the following integrals:

(a) $G_1 = \int_{-\infty}^{\infty} (3t^3 - 4t^2 + 3)[\delta(t) + 2\delta(t - 2)]\, dt.$

(b) $G_2 = \int_{-4}^{4} 2(e^{-3t} + 1)[\delta(t) - 2\delta(t - 2)]\, dt.$

(c) $G_3 = \int_{-12}^{16} 4[t \sin(2\pi t) - 1][\delta(t - 1) + \delta(t - 6)]\, dt.$

Sections 12-2 and 12-4: Laplace Transform

12.3 Express each of the waveforms in Fig. P12.3 in terms of step functions and then determine its Laplace transform. [Recall that the ramp function is related to the step function by $r(t - T) = (t - T)\, u(t - T)$.] Assume that all waveforms are zero for $t < 0$.

12.4 Determine the Laplace transform of each of the following functions by applying the properties given in Tables 12-1 and 12-2.

(a) $f_1(t) = 4te^{-2t}\, u(t)$
(b) $f_2(t) = 10\cos(12t + 60°)\, u(t)$
*(c) $f_3(t) = 12e^{-3(t-4)}\, u(t-4)$
(d) $f_4(t) = 30(e^{-3t} + e^{3t})\, u(t)$
(e) $f_5(t) = 16e^{-2t}\cos 4t\, u(t)$
(f) $f_6(t) = 20te^{-2t}\sin 4t\, u(t)$

12.5 Determine the Laplace transform of each of the following functions by applying the properties given in Tables 12-1 and 12-2.

*(a) $h_1(t) = 12te^{-3(t-4)}\, u(t-4)$
(b) $h_2(t) = 27t^2\sin(6t - 60°)\, u(t)$
*(c) $h_3(t) = 10t^3e^{-2t}\, u(t)$
(d) $h_4(t) = 5(t-6)\, u(t-3)$
(e) $h_5(t) = 10e^{-3t}\, u(t-4)$
(f) $h_6(t) = 4e^{-2(t-3)}\, u(t-4)$

12.6 Determine the Laplace transform of the following functions.

(a) $f_1(t) = 25\cos(4\pi t + 30°)\, \delta(t)$
(b) $f_2(t) = 25\cos(4\pi t + 30°)\, \delta(t-0.2)$
(c) $f_3(t) = 10\,\dfrac{\sin 3t}{t}\, u(t)$
(d) $f_4(t) = \dfrac{d^2}{dt^2}\,[e^{-4t}\, u(t)]$
(e) $f_5(t) = \dfrac{d}{dt}\,[4te^{-2t}\cos(4\pi t + 30°)\, u(t)]$
(f) $f_6(t) = e^{-3t}\cos(4t + 30°)\, u(t)$
(g) $f_7(t) = t^2[u(t) - u(t-4)]$
(h) $f_8(t) = 10\cos(6\pi t + 30°)\, \delta(t-0.2)$

12.7 Determine the Laplace transform of each of the following functions:

(a) $f_1(t) = 2t^2e^{-3t}\, u(t)$
(b) $f_2(t) = 5\sin(6t + 30°)\, u(t)$
*(c) $f_3(t) = 10e^{-4(t-3)}\, u(t-3)$
(d) $f_4(t) = 15(e^{-2t} - e^{2t})\, u(t)$
(e) $f_5(t) = 8e^{-4t}\cos(2t)\, u(t)$
(f) $f_6(t) = 10te^{-t}\sin(2t)\, u(t)$

Section 12-5: Partial Fractions

12.8 Obtain the inverse Laplace transform of each of the following functions by first applying the partial-fraction-expansion method.

(a) $\mathbf{F}_1(\mathbf{s}) = \dfrac{6}{(\mathbf{s}+2)(\mathbf{s}+4)}$
(b) $\mathbf{F}_2(\mathbf{s}) = \dfrac{4}{(\mathbf{s}+1)(\mathbf{s}+2)^2}$
(c) $\mathbf{F}_3(\mathbf{s}) = \dfrac{3\mathbf{s}^3 + 36\mathbf{s}^2 + 131\mathbf{s} + 144}{\mathbf{s}(\mathbf{s}+4)(\mathbf{s}^2 + 6\mathbf{s} + 9)}$
(d) $\mathbf{F}_4(\mathbf{s}) = \dfrac{2\mathbf{s}^2 + 4\mathbf{s} - 10}{(\mathbf{s}+6)(\mathbf{s}+2)^2}$

12.9 Obtain the inverse Laplace transform of each of the following functions.

(a) $\mathbf{F}_1(\mathbf{s}) = \dfrac{\mathbf{s}^2 + 17\mathbf{s} + 15}{\mathbf{s}(\mathbf{s}^2 + 6\mathbf{s} + 5)}$
*(b) $\mathbf{F}_2(\mathbf{s}) = \dfrac{2\mathbf{s}^2 + 10\mathbf{s} + 16}{(\mathbf{s}+2)(\mathbf{s}^2 + 6\mathbf{s} + 10)}$
(c) $\mathbf{F}_3(\mathbf{s}) = \dfrac{4}{(\mathbf{s}+2)^3}$
(d) $\mathbf{F}_4(\mathbf{s}) = \dfrac{2(\mathbf{s}^3 + 12\mathbf{s}^2 + 16)}{(\mathbf{s}+1)(\mathbf{s}+4)^3}$

12.10 Obtain the inverse Laplace transform of each of the following functions.

(a) $\mathbf{F}_1(\mathbf{s}) = \dfrac{(\mathbf{s}+2)^2}{\mathbf{s}(\mathbf{s}+1)^3}$
(b) $\mathbf{F}_2(\mathbf{s}) = \dfrac{1}{(\mathbf{s}^2 + 4\mathbf{s} + 5)^2}$
*(c) $\mathbf{F}_3(\mathbf{s}) = \dfrac{\sqrt{2}(\mathbf{s}+1)}{\mathbf{s}^2 + 6\mathbf{s} + 13}$
(d) $\mathbf{F}_4(\mathbf{s}) = \dfrac{-2(\mathbf{s}^2 + 20)}{\mathbf{s}(\mathbf{s}^2 + 8\mathbf{s} + 20)}$

12.11 Obtain the inverse Laplace transform of each of the following functions.

(a) $\mathbf{F}_1(\mathbf{s}) = 2 + \dfrac{4(\mathbf{s}-4)}{\mathbf{s}^2 + 16}$
*(b) $\mathbf{F}_2(\mathbf{s}) = \dfrac{4}{\mathbf{s}} + \dfrac{4\mathbf{s}}{\mathbf{s}^2 + 9}$
(c) $\mathbf{F}_3(\mathbf{s}) = \dfrac{(\mathbf{s}+5)e^{-2\mathbf{s}}}{(\mathbf{s}+1)(\mathbf{s}+3)}$
(d) $\mathbf{F}_4(\mathbf{s}) = \dfrac{(1 - e^{-4\mathbf{s}})(24\mathbf{s} + 40)}{(\mathbf{s}+2)(\mathbf{s}+10)}$
(e) $\mathbf{F}_5(\mathbf{s}) = \dfrac{\mathbf{s}(\mathbf{s}-8)e^{-6\mathbf{s}}}{(\mathbf{s}+2)(\mathbf{s}^2 + 16)}$
(f) $\mathbf{F}_6(\mathbf{s}) = \dfrac{4\mathbf{s}(2 - e^{-4\mathbf{s}})}{\mathbf{s}^2 + 9}$

*Answer(s) available in Appendix G.

Sections 12-6 and 12-7: s-Domain Analysis

*12.12 In the circuit of Fig. P12.12(a), $i_s(t)$ is given by the waveform shown in Fig. P12.12(b). Determine $i_L(t)$ for $t \geq 0$, given that $R_1 = R_2 = 2\ \Omega$ and $L = 4$ H.

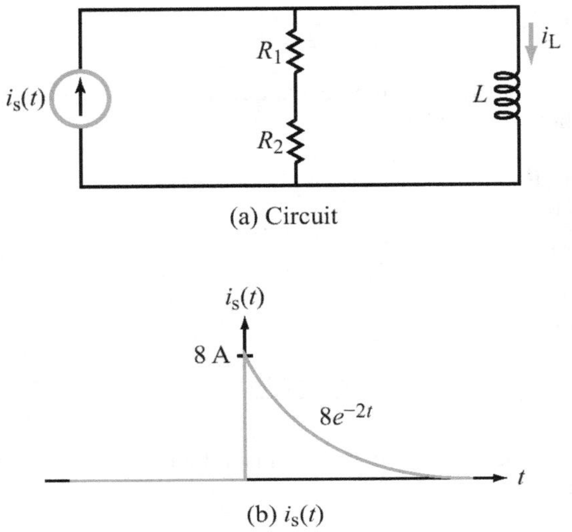

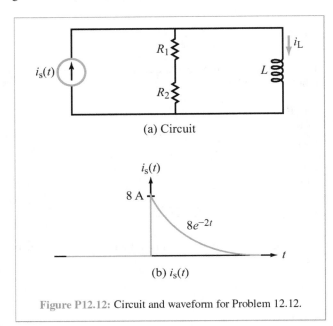

Figure P12.12: Circuit and waveform for Problem 12.12.

12.13 In the circuit of Fig. P12.13, $\upsilon_s(t)$ is given by $\upsilon_s(t) = [20u(t) - 5\delta(t)]$ V. Determine $\upsilon_C(t)$ for $t \geq 0$, given that $L = 1$ H, $C = 0.5$ F, and $R = 6\ \Omega$.

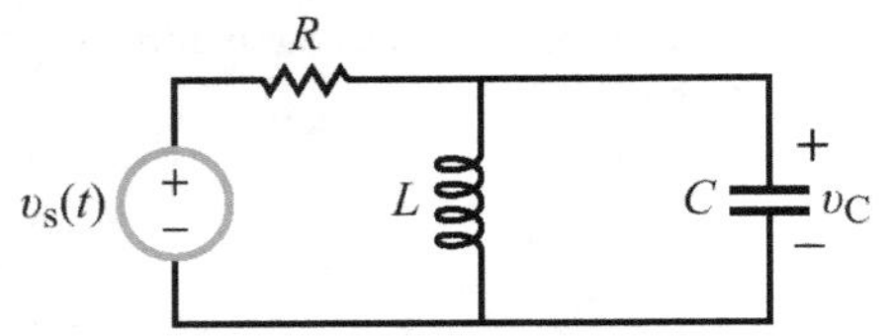

Figure P12.13: Circuit for Problem 12.13.

12.14 In the circuit of Fig. P12.14(a), $\upsilon_s(t)$ is given by the waveform displayed in Fig. P12.14(b). Determine $i(t)$ for $t \geq 0$, given that $R = 2\ \Omega$, $L = 6$ H, and $C = 2$ F.

*12.15 In the circuit of Fig. P12.15(a), $\upsilon_s(t)$ is given by the waveform displayed in Fig. P12.15(b). Determine $i_L(t)$, given that $R_1 = R_2 = 4\ \Omega$, $L = 2$ H, and $C = 0.5$ F.

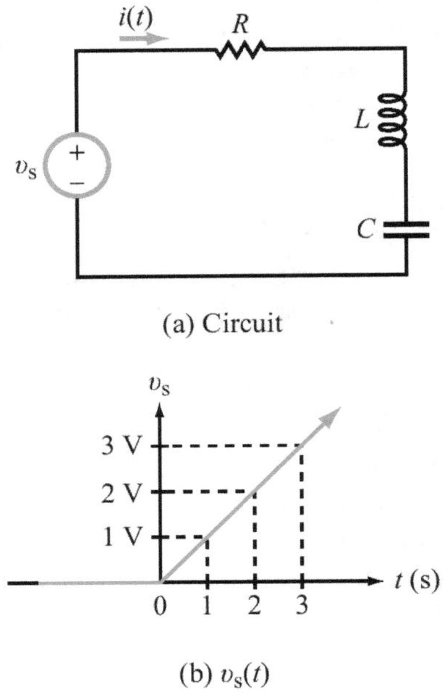

Figure P12.14: Circuit and waveform for Problem 12.14.

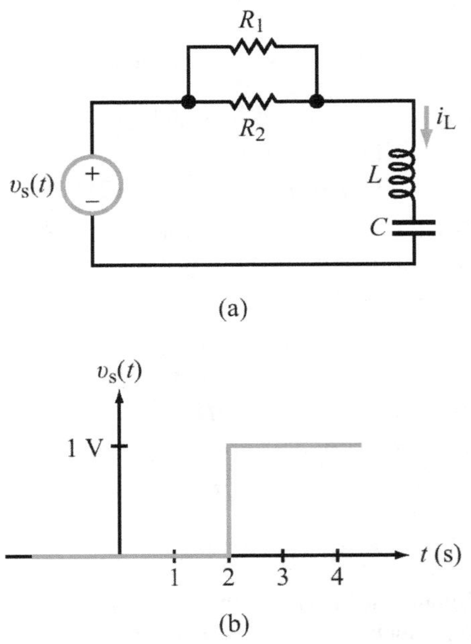

Figure P12.15: Circuit and waveform for Problem 12.15.

12.16 In the circuit of Fig. P12.16,

$$i_s(t) = 7u(t) + 4\delta(t) \text{ (A)}.$$

Initially the capacitor had 8 J of energy stored in it. Determine $v_{R_1}(t)$ for $t \geq 0$, given that $R_1 = 2\ \Omega$, $R_2 = 4\ \Omega$, and $C = 1$ F.

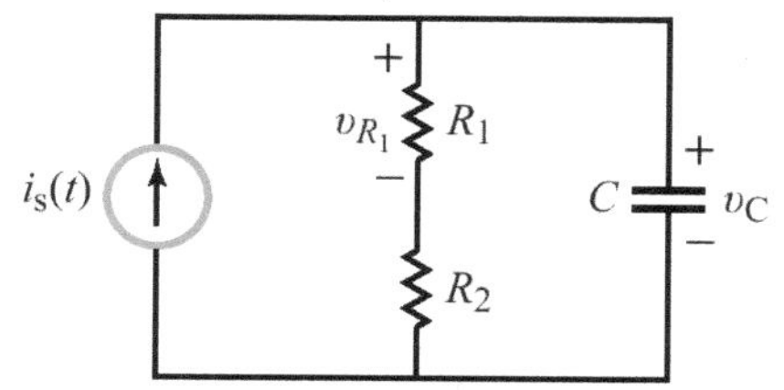

Figure P12.16: Circuit for Problem 12.16.

12.17 In the circuit of Fig. P12.17, $v_s(t)$ is a rectangular 5 V pulse of duration 3 seconds starting at $t = 0$. Determine $i_L(t)$ for $t \geq 0$, given that $R_1 = 6\ \Omega$, $L = 1$ H, and $C = \frac{1}{3}$ F. Assume that initially the capacitor had no charge stored in it.

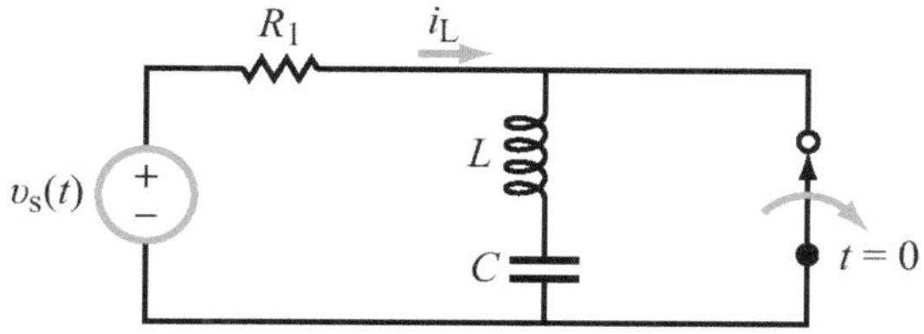

Figure P12.17: Circuit for Problem 12.17.

***12.18** Determine $v(t)$ in the circuit of Fig. P12.18, given that $v_s(t) = 2u(t)$ V, $R_1 = 1\ \Omega$, $R_2 = 3\ \Omega$, $C = 0.3689$ F, and $L = 0.2259$ H.

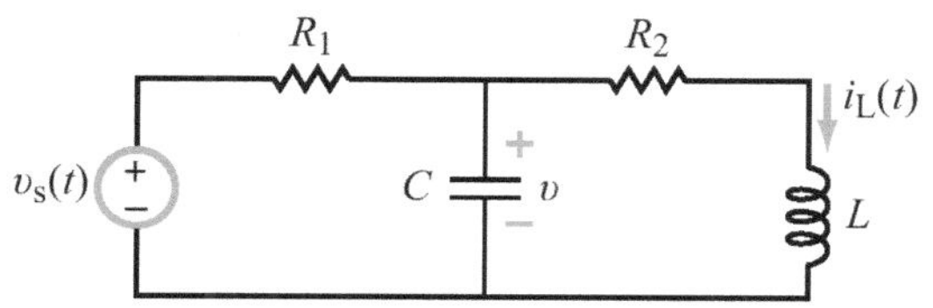

Figure P12.18: Circuit for Problems 12.18 and 12.19.

12.19 Determine $i_L(t)$ in the circuit in Fig. P12.18, given that $v_s(t) = 2u(t)$, $R_1 = 2\ \Omega$, $R_2 = 6\ \Omega$, $L = 2.215$ H, and $C = 0.0376$ F.

12.20 Determine $v_{out}(t)$ in the circuit in Fig. P12.20, given that $v_s(t) = 35u(t)$ V, $v_{C_1}(0^-) = 20$ V, $v_{C_2}(0^-) = 0$, $R_1 = 1\ \Omega$, $C_1 = 1$ F, $R_2 = 0.5\ \Omega$, and $C_2 = 2$ F.

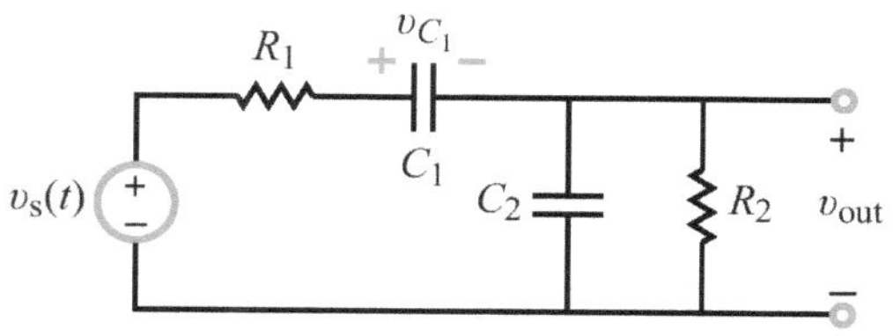

Figure P12.20: Circuit for Problem 12.20.

12.21 Determine $i_L(t)$ in the circuit of Fig. P12.21 for $t \geq 0$, given that the switch was opened at $t = 0$ after it had been closed for a long time, $v_s = 12$ mV, $R_0 = 5\ \Omega$, $R_1 = 10\ \Omega$, $R_2 = 20\ \Omega$, $L = 0.2$ H, and $C = 6$ mF.

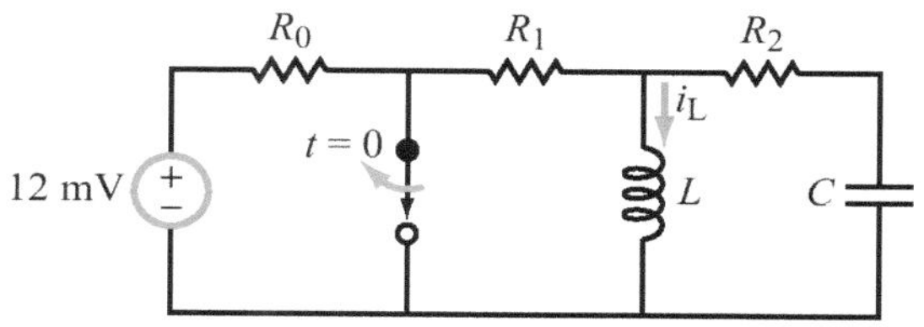

Figure P12.21: Circuit for Problems 12.21 and 12.22.

***12.22** Repeat Problem 12.21, but assume that the switch had been open for a long time and then closed at $t = 0$. Set the dc source at 12 mV and the element values at $R_0 = 5\ \Omega$, $R_1 = 10\ \Omega$, $R_2 = 20\ \Omega$, $L = 2$ H, and $C = 0.4$ F.

12.23 Determine $i_L(t)$ in the circuit of Fig. P12.23, given that $R_1 = 1\ \Omega$, $R_2 = 6\ \Omega$, $L = 1$ H, and $C = 0.5$ F. Assume no energy was stored in the circuit segment to the right of the switch prior to $t = 0$.

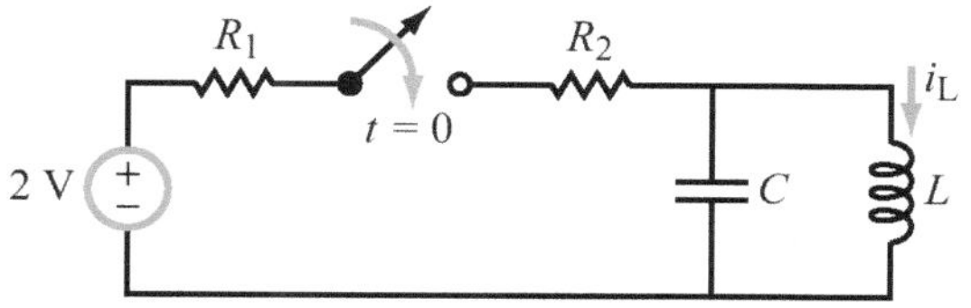

Figure P12.23: Circuit for Problem 12.23.

12.24 Determine $v_{C_2}(t)$ in the circuit of Fig. P12.24, given that $R = 200\ \Omega$, $C_1 = 1$ mF, and $C_2 = 5$ mF.

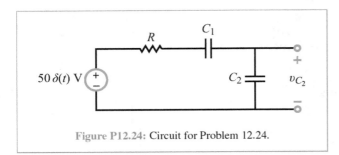

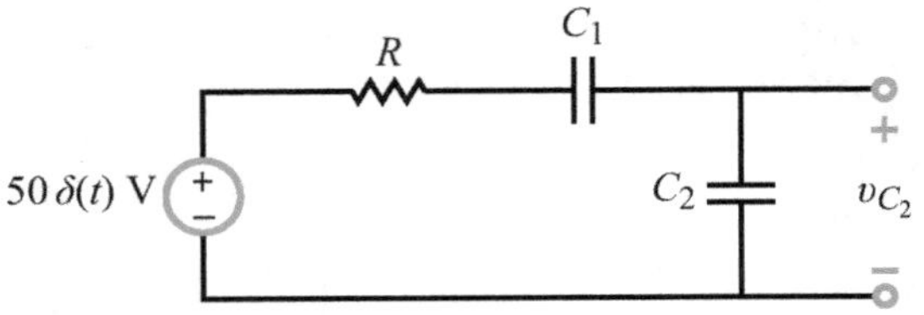

Figure P12.24: Circuit for Problem 12.24.

*12.25 Determine $i_L(t)$ in the circuit of Fig. P12.25, given that before closing the switch $v_C(0^-) = 12$ V. Also, the element values are $R = 2\ \Omega$, $L = 1.5$ H, and $C = 0.5$ F.

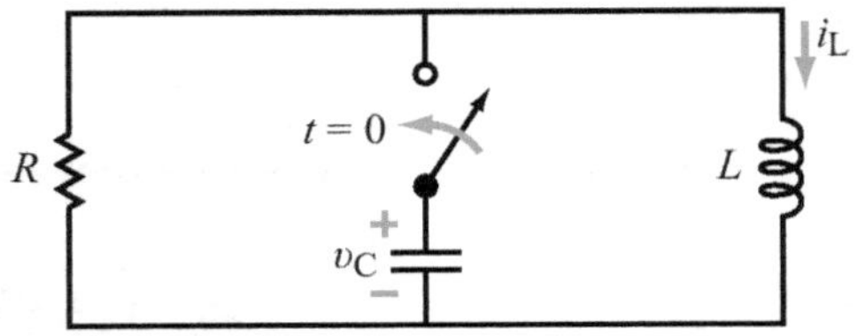

Figure P12.25: Circuit for Problem 12.25.

12.26 Determine $v_{out}(t)$ in the circuit of Fig. P12.26, given that $v_s(t) = 11u(t)$ V, $R_1 = 2\ \Omega$, $R_2 = 4\ \Omega$, $R_3 = 6\ \Omega$, $L = 1$ H, and $C = 0.5$ F.

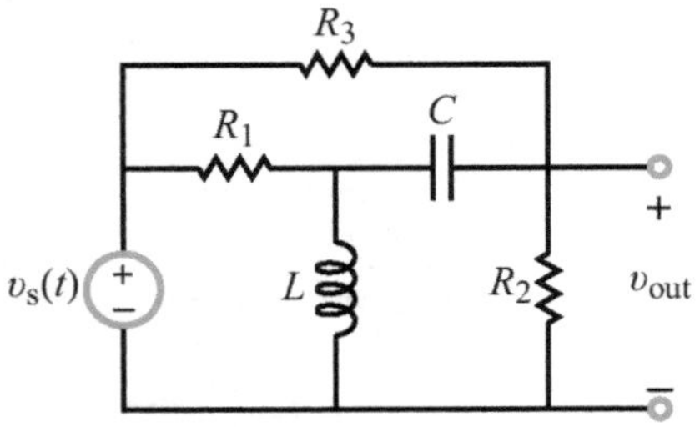

Figure P12.26: Circuit for Problem 12.26.

12.27 Determine $i_L(t)$ in the circuit of Fig. P12.27 for $t \geq 0$, given that $R = 4\ \Omega$, $L = 1$ H, and $C = 0.5$ F.

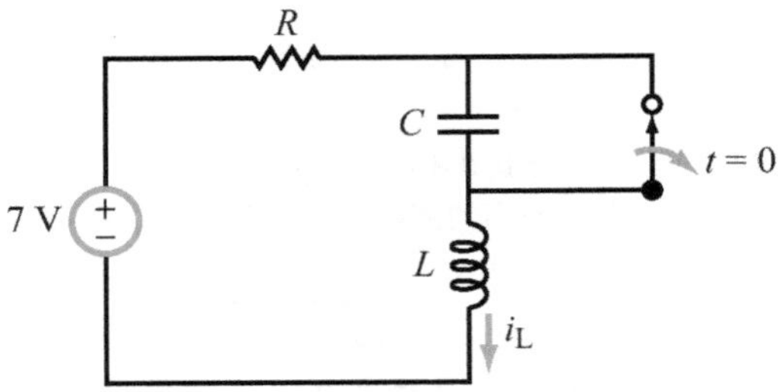

Figure P12.27: Circuit for Problem 12.27.

*12.28 Apply mesh-current analysis in the s-domain to determine $i_L(t)$ in the circuit of Fig. P12.28, given that $v_s(t) = 44u(t)$ V, $R_1 = 2\,\Omega$, $R_2 = 4\,\Omega$, $R_3 = 6\,\Omega$, $C = 0.1$ F, and $L = 4$ H.

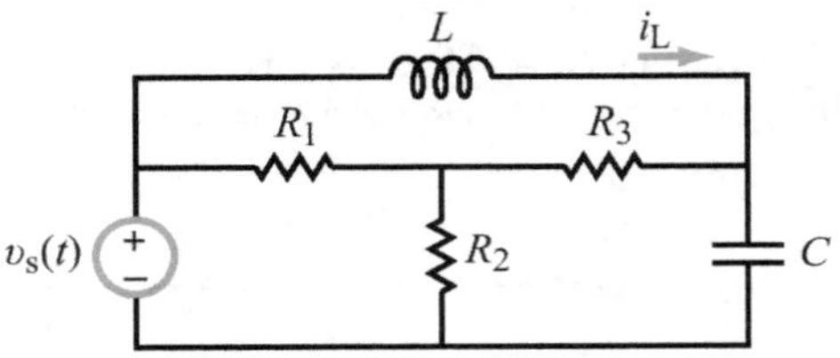

Figure P12.28: Circuit for Problem 12.28.

12.29 Determine $v_{out}(t)$ in the circuit of Fig. P12.29, given that $v_s(t) = 3u(t)$ V, $R_1 = 4\ \Omega$, $R_2 = 10\ \Omega$, and $L = 2$ H.

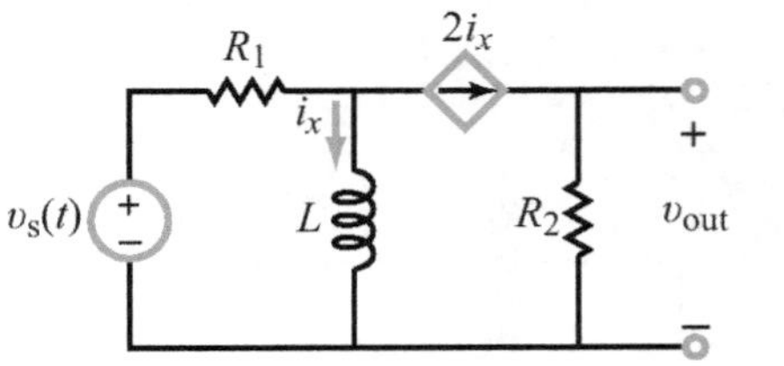

Figure P12.29: Circuit for Problems 12.29 and 12.30.

12.30 Repeat Problem 12.29 with $v_s(t) = 3\delta(t)$ V.

*12.31 The voltage source in the circuit of Fig. P12.31 is, given by $v_s(t) = [10+5u(t)]$ V. Determine $i_L(t)$ for $t \geq 0$, given that $R_1 = 1\ \Omega$, $R_2 = 1\ \Omega$, $L = 2$ H, and $C = 1$ F.

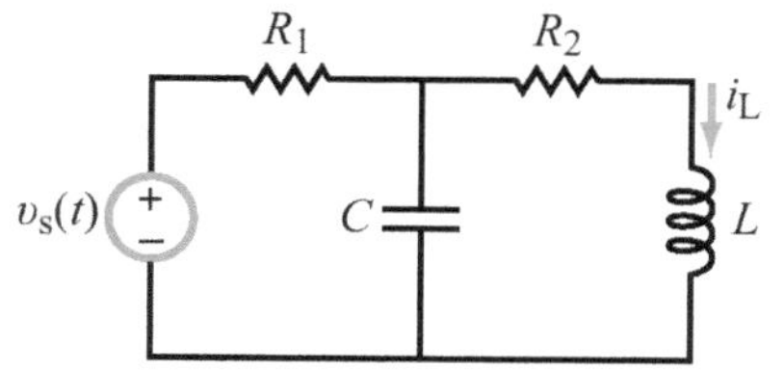

Figure P12.31: Circuit for Problems 12.31 and 12.35.

12.32 The current source in the circuit of Fig. P12.32 is given by $i_s(t) = [10u(t) + 15\delta(t)]$ mA. Determine $v_C(t)$ for $t \geq 0$, given that $R_1 = 1$ kΩ, $R_2 = 1$ kΩ, and $C = 2$ mF.

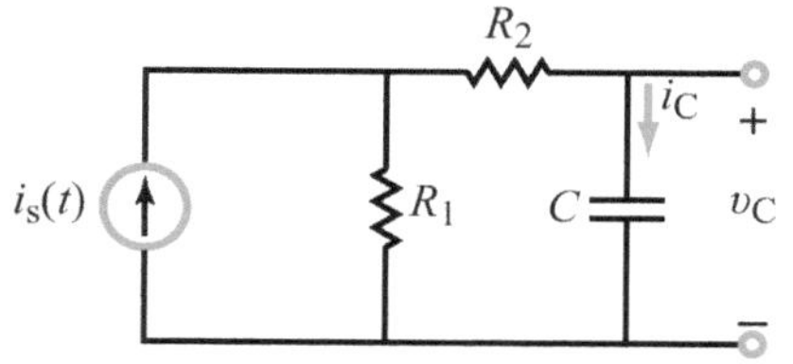

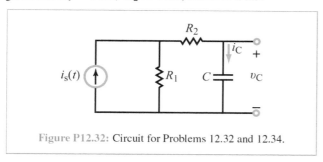

Figure P12.32: Circuit for Problems 12.32 and 12.34.

12.33 The circuit in Fig. P12.33 is excited by a 10 V, 1 s long rectangular pulse. Determine $i(t)$, given that $R_1 = 1\ \Omega$, $R_2 = 2\ \Omega$, and $L = 1/3$ H.

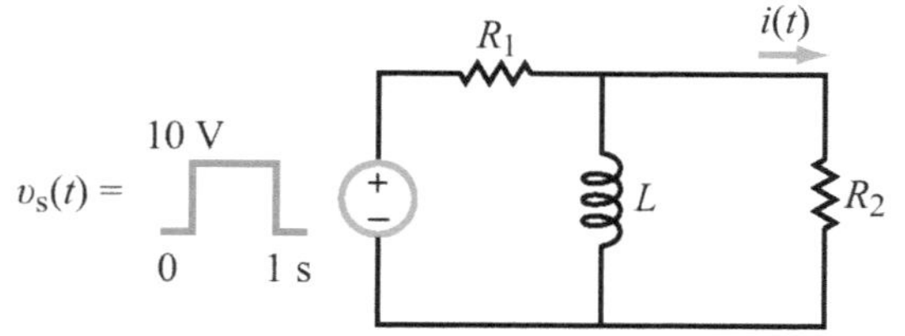

Figure P12.33: Circuit for Problem 12.33.

***12.34** Repeat Problem 12.32 after replacing the current source with a 10 mA, 2 s long rectangular pulse.

12.35 Analyze the circuit shown in Fig. P12.31 to determine $i_L(t)$ in response to a voltage excitation $v_s(t)$ in the form of a 10 V rectangular pulse that starts at $t = 0$ and ends at $t = 5$ s. The element values are $R_1 = 1\ \Omega$, $R_2 = 3\ \Omega$, $L = 2$ H, and $C = 0.5$ F.

12.36 The current source in the circuit of Fig. P12.36 is given by $i_s(t) = 6e^{-2t}\ u(t)$ A. Determine $i_L(t)$ for $t \geq 0$, given that $R_1 = 10\ \Omega$, $R_2 = 5\ \Omega$, $L = 0.6196$ H, and $LC = (1/15)$ s.

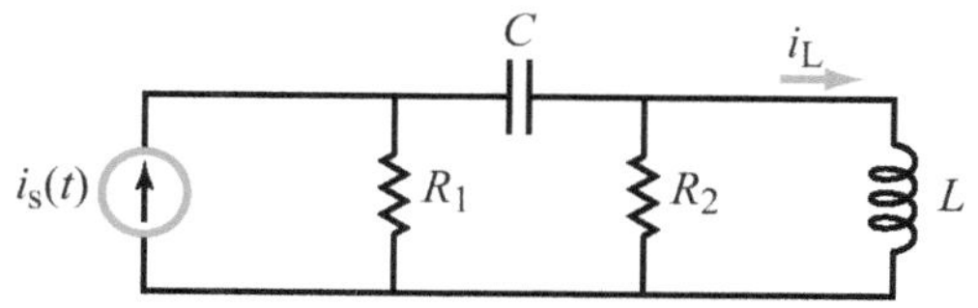

Figure P12.36: Circuit for Problems 12.36 and 12.37.

***12.37** Given the current-source waveform displayed in Fig. P12.37, determine $i_L(t)$ in the circuit of Fig. P12.36, given that $R_1 = 10\ \Omega$, $R_2 = 5\ \Omega$, $L = 0.6196$ H, and $LC = (1/15)$ s.

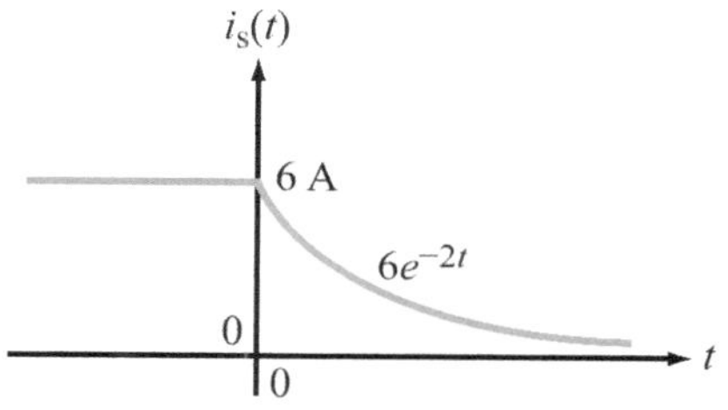

Figure P12.37: Waveform for Problem 12.37.

12.38 If the circuit shown in Fig. P12.38(a) is excited by the current waveform $i_s(t)$ shown in Fig. P12.38(b), determine $i(t)$ for $t \geq 0$, given that $R_1 = 10\ \Omega$, $R_2 = 5\ \Omega$, and $C = 0.02$ F.

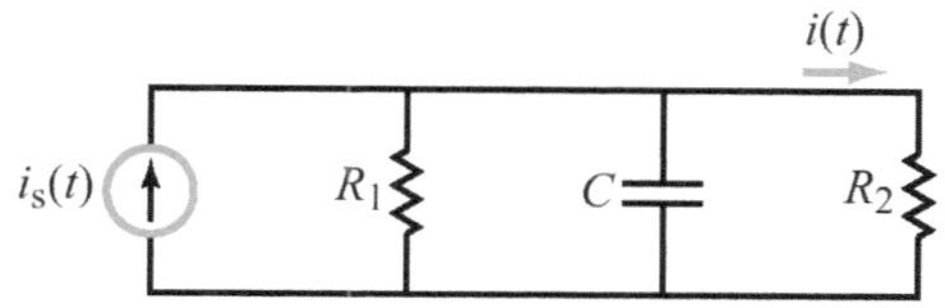

(a) Circuit

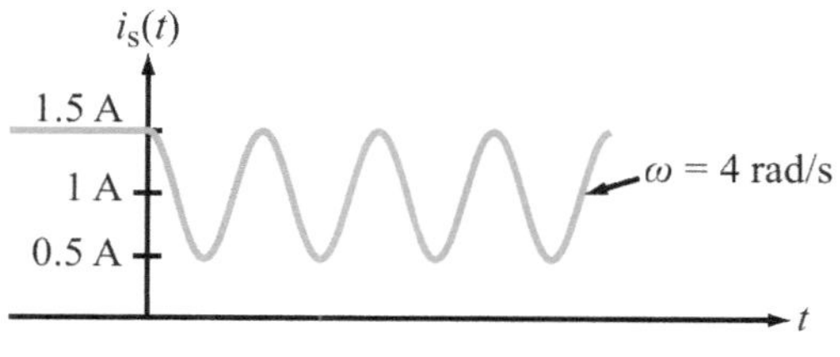

(b) Waveform

Figure P12.38: Circuit for Problems 12.38 to 12.40.

12.39 If the circuit shown in Fig. P12.38(a) is excited by current waveform $i_s(t) = 36te^{-6t}\,u(t)$ mA, determine $i(t)$ for $t \geq 0$, given that $R_1 = 2\ \Omega$, $R_2 = 4\ \Omega$, and $C = (1/8)$ F.

12.40 If the circuit shown in Fig. P12.38(a) is excited by a current waveform given by $i_s(t) = 9te^{-3t}\,u(t)$ mA, determine $i(t)$ for $t \geq 0$, given that $R_1 = 1\ \Omega$, $R_2 = 3\ \Omega$, and $C = 1/3$ F.

12.41 The circuit shown in Fig. P12.41 first was introduced in Problem 5.68. Then, a time-domain solution was sought for $\upsilon_{\text{out}_1}(t)$ and $\upsilon_{\text{out}_2}(t)$ for $t \geq 0$, given that $\upsilon_i(t) = 10u(t)$ mV, $V_{CC} = 10$ V for both op amps, and the two capacitors had no change prior to $t = 0$. Analyze the circuit and plot $\upsilon_{\text{out}_1}(t)$ and $\upsilon_{\text{out}_2}(t)$ using the Laplace transform technique.

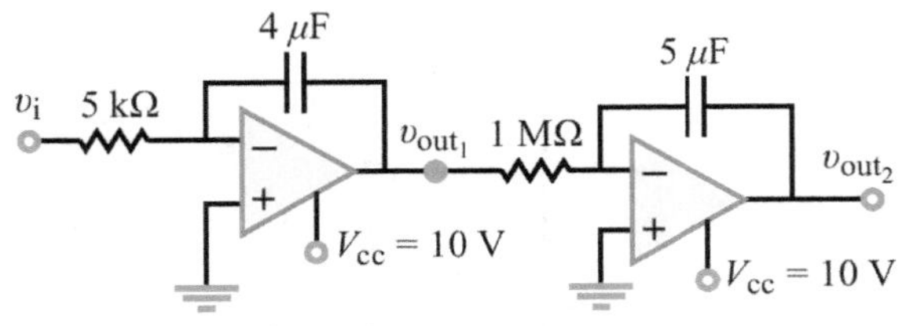

Figure P12.41: Circuit for Problems 12.41 and 12.42.

12.42 Repeat Problem 12.41 retaining all element values and conditions but changing the input voltage to $\upsilon_i(t) = 0.4te^{-2t}\,u(t)$.

12.43 For the circuit shown in Fig. P12.43, determine $\upsilon_{\text{out}}(t)$ given that $R_1 = 1$ kΩ, $R_2 = 4$ kΩ, and $C = 1\ \mu$F, and

(a) $\upsilon_s(t) = 2u(t)$ (V),

(b) $\upsilon_s(t) = 2\cos(1000t)$ (V),

(c) $\upsilon_s(t) = 2e^{-t}\,u(t)$ (V).

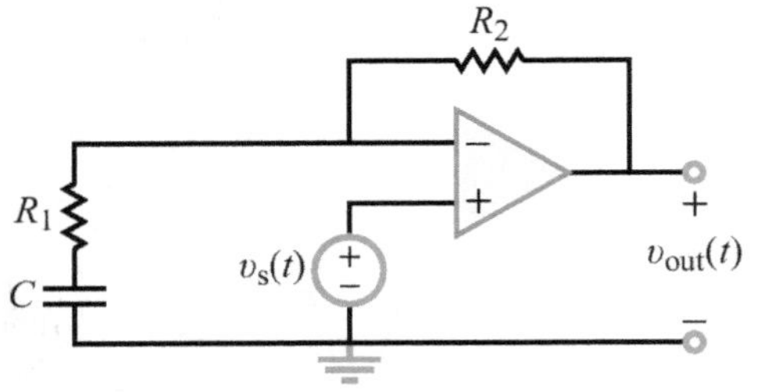

Figure P12.43: Op-amp circuit for Problem 12.43.

12.44 For the circuit shown in Fig. P12.44, determine $\upsilon_{\text{out}}(t)$ given that $R_1 = R_2 = 100\ \Omega$, $C_1 = C_2 = 1\ \mu$F, and

*(a) $\upsilon_s(t) = 2u(t)$ (V),

(b) $\upsilon_s(t) = 2\cos(1000t)\,u(t)$ (V),

(c) $\upsilon_s(t) = 2e^{-t}\,u(t)$ (V).

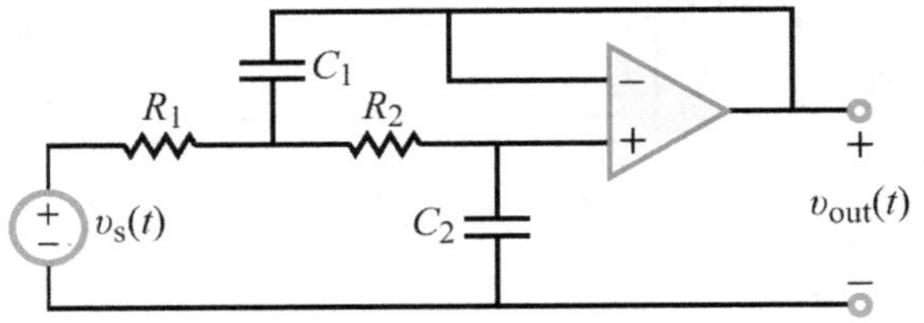

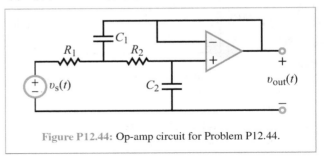

Figure P12.44: Op-amp circuit for Problem P12.44.

Section 12-8: Multisim Analysis of Circuits Driven by Nontrivial Inputs

12.45 Apply a 1 V, 1 Hz signal with a 10 V dc offset to the circuit in Fig. P12.45 and plot both $\upsilon_C(t)$ and $\upsilon_R(t)$ for $R = 1\ \Omega$ and $C = 1$ F. Why are the dc offsets different for the voltages across the two components?

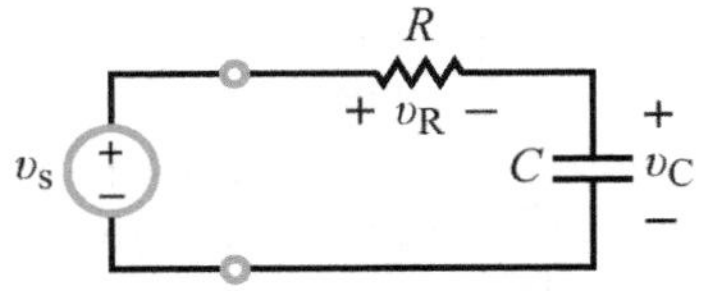

Figure P12.45: Circuit for Problems 12.45 and 12.47.

12.46 Repeat Example 12-12, but vary the time constant of the exponent from 0.1 s to 5 s (pick more than 3 points). Plot all responses on the same display.

12.47 In Multisim, apply input signal $\upsilon_s(t) = 5t$ V to the circuit shown in Fig. P12.45 and plot both $\upsilon_C(t)$ and $\upsilon_R(t)$ for 0 to 5 s. Find the point at which $\upsilon_C(t) = \upsilon_R(t)$. If we change the input signal to $\upsilon_s(t) = 10t$ V, will the point in time where $\upsilon_C(t) = \upsilon_R(t)$ change? Explain.

12.48 Using a Piecewise_Linear source in Multisim, build and simulate the circuit found in Fig. P12.33 (including the specified source). Plot $i(t)$ for 0 to 2 s. Use $R_1 = 1\ \Omega$, $R_2 = 2\ \Omega$, and $L = 1/3$ H.

Potpourri Questions

12.49 What techniques are used to generate the 3-D effect in 3-D TVs?

12.50 What type of imaging sensor is used to map the Earth surface in 3-D? How many antennas does it use?

Integrative Problems: Analytical / Multisim / myDAQ

To master the material in this chapter, solve the following problems using three complementary approaches: (a) analytically, (b) with Multisim, and (c) by constructing the circuit and using the myDAQ interface unit to measure quantities of interest via your computer. [myDAQ tutorials and videos are available on CAD.]

m12.1 Step Response: For the circuit in Fig. m12.1:

(a) Determine the transfer function $\mathbf{H(s)} = \mathbf{V_o(s)}/\mathbf{V_s(s)}$. Write the transfer function in simplified standard form with symbolic values.

(b) Determine the output response $v_o(t)$ to the input $v_s(t) = 4u(t)$ by working in the Laplace domain. Assume the capacitor is initially discharged.

(c) Plot $v_s(t)$ and $v_o(t)$ on the same graph from 0 to 5 ms using a tool such as MathScript or MATLAB for $R = 5.6$ kΩ and $C = 0.1$ μF. Include a hard copy of the script used to create the plot.

(d) Determine the following values for $v_o(t)$:

(1) Initial value $v_o(0)$,

(2) Time to reach 50% of the initial value, and

(3) Final value.

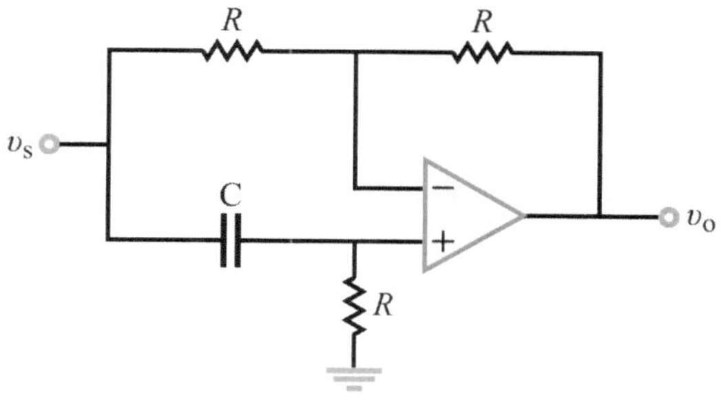

Figure m12.1 Circuit for Problem m12.1.

CHAPTER 13

Fourier Analysis Technique

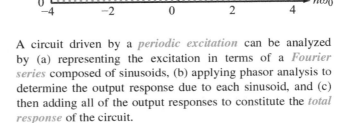

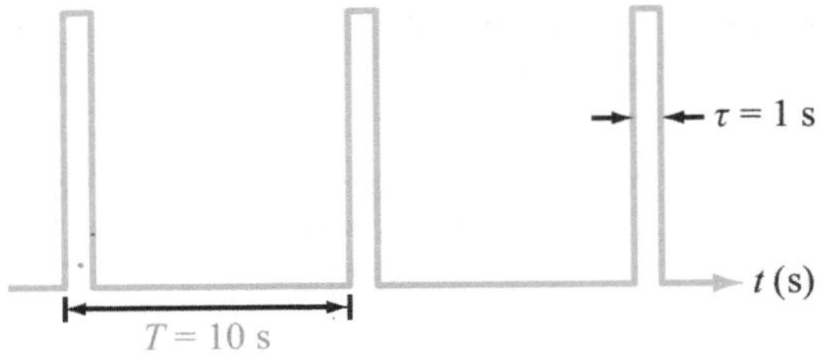

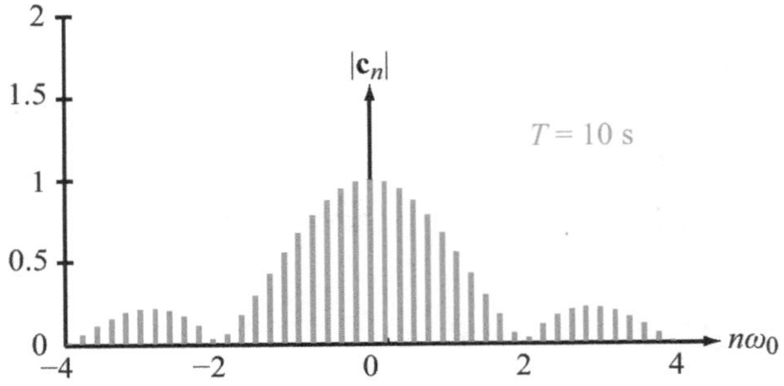

Contents

Objectives

Learn to:

- Express a periodic function in terms of a Fourier series using the cosine/sine, the amplitude/phase, and the complex exponential representations.
- Determine the line spectrum of a periodic waveform.
- Utilize symmetry consideratins in the evaluation of Fourier coefficients.
- Explain the Gibbs phenomenon.
- Analyze circuits excited by periodic waveforms.
- Calculate the average power dissipated in or delivered by a component in a circuit excited by a periodic voltage or current.
- Evaluate the Fourier transform of a nonperiodic waveform.
- Apply the Fourier transform technique to analyze circuits excited by nonperiodic waveforms.
- Use Multisim to model the behavior of the Sigma-Delta modulator.

A circuit driven by a *periodic excitation* can be analyzed by (a) representing the excitation in terms of a *Fourier series* composed of sinusoids, (b) applying phasor analysis to determine the output response due to each sinusoid, and (c) then adding all of the output responses to constitute the *total response* of the circuit.

Overview

First introduced in Chapter 7, the phasor-domain analysis technique has proven to be a potent—and easy to implement—tool for determining the steady-state response of circuits excited by sinusoidal waveforms. As a periodic function with a period T, a sinusoidal signal shares a distinctive property with all other members of the family of *periodic functions*, namely the *periodicity property* given by Eq. (8.4) as

$$x(t) = x(t + nT), \tag{13.1}$$

where n is an integer. Given this natural connection between sinusoids and other periodic functions, can we somehow extend the phasor-domain solution technique to nonsinusoidal periodic excitations such as a square wave or a train of pulses? The answer is yes, and the process for realizing it is facilitated by two enabling mechanisms: the *Fourier theorem* and the superposition principle. The Fourier theorem makes it possible to mathematically characterize any periodic excitation in the form of a sum of multiple sinusoidal harmonics. The superposition principle allows us to apply phasor analysis to calculate the circuit response due to each harmonic as if it were the only excitation in the circuit and then to add all of the responses together, thereby realizing the response to the original periodic excitation. The first half of this chapter aims to demonstrate the mechanics of the solution process as well as to explain the physics associated with the circuit response to the different harmonics.

The second half of the chapter is devoted to the *Fourier transform*, which is particularly useful for analyzing circuits excited by nonperiodic waveforms, such as single pulses or step functions. As we will see in Section 13-7, the Fourier transform is related to the Laplace transform of Chapter 12 and becomes identical to it under certain circumstances, but the two techniques are generally distinct (as are their conditions of applicability from the standpoint of circuit analysis).

13-1 Fourier Series Analysis Technique

By way of introducing the Fourier series analysis technique, let us consider the RL circuit shown in Fig. 13-1(a), which is excited by the square-wave voltage waveform shown in Fig. 13-1(b). The waveform amplitude is 3 V and its period $T = 2$ s. Our goal is to determine the output voltage response, $\upsilon_{\text{out}}(t)$. The solution procedure consists of three basic steps.

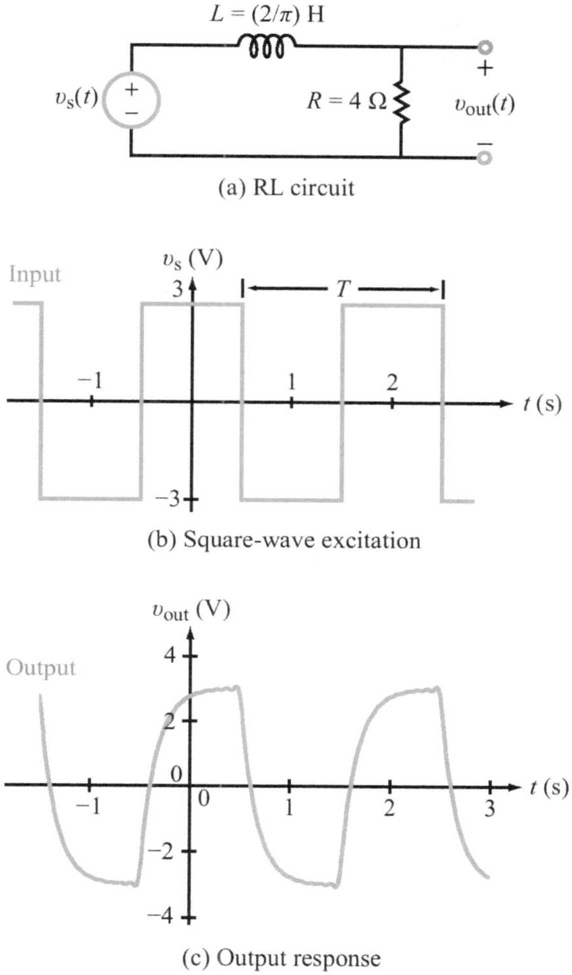

Figure 13-1: RL circuit excited by a square wave and corresponding output response.

Step 1: Express the periodic excitation in terms of Fourier harmonics

According to the Fourier theorem (which we will introduce and examine in detail in Section 11-2), the waveform shown in Fig. 13-1(b) can be represented by the series

$$\upsilon_s(t) = \frac{12}{\pi}\left(\cos\omega_0 t - \frac{1}{3}\cos 3\omega_0 t + \frac{1}{5}\cos 5\omega_0 t - \cdots\right) \text{ V}, \tag{13.2}$$

where $\omega_0 = 2\pi/T = 2\pi/2 = \pi$ (rad/s) is the *fundamental angular frequency* of the waveform. Since our present objective is to outline the solution procedure, we will accept it as a given that the infinite-series representation given by Eq. (13.2) is indeed equivalent to the square wave of Fig. 13-1(b). The series consists of cosine functions of the form $\cos m\omega_0 t$ with m assuming only odd values (1, 3, 5, etc.). Thus, the series contains only odd *harmonics* of ω_0. Other periodic waveforms may include both odd and even harmonics. The coefficient of the mth harmonic is equal to $1/m$ (relative to the coefficient of the fundamental), and its polarity is positive if $m = 3, 7, \ldots$ and negative if $m = 5, 9, \ldots$. In view of these properties, we can replace m with $(2n-1)$ and cast $\upsilon_s(t)$ in the form

$$\upsilon_s(t) = \frac{12}{\pi}\sum_{n=1}^{\infty}(-1)^{n+1}\frac{1}{2n-1}\cos[(2n-1)\pi t] \text{ V}. \tag{13.3}$$

In terms of its components, $\upsilon_s(t)$ is given by

$$\upsilon_s(t) = \upsilon_{s_1}(t) + \upsilon_{s_2}(t) + \upsilon_{s_3}(t) + \cdots \tag{13.4}$$

with

$$\upsilon_{s_1}(t) = \frac{12}{\pi}\cos\omega_0 t \text{ V}, \tag{13.5a}$$

$$\upsilon_{s_2}(t) = -\frac{12}{3\pi}\cos 3\omega_0 t \text{ V}, \tag{13.5b}$$

and

$$\upsilon_{s_3}(t) = \frac{12}{5\pi}\cos 5\omega_0 t \text{ V}, \quad \text{etc.} \tag{13.5c}$$

In the phasor domain, the counterpart of $\upsilon_s(t)$ is given by:

$$\mathbf{V}_s(t) = \mathbf{V}_{s_1}(t) + \mathbf{V}_{s_2}(t) + \mathbf{V}_{s_3}(t) + \cdots \tag{13.6}$$

with

$$\mathbf{V}_{s_1} = \frac{12}{\pi} \text{ V} \quad (\text{for } \omega = \omega_0), \tag{13.7a}$$

$$\mathbf{V}_{s_2} = -\frac{12}{3\pi} \text{ V} \quad (\text{for } \omega = 3\omega_0), \tag{13.7b}$$

and

$$\mathbf{V}_{s_3} = \frac{12}{5\pi} \text{ V} \quad (\text{for } \omega = 5\omega_0), \quad \text{etc.} \tag{13.7c}$$

Phasor voltages $\mathbf{V}_{s_1}$, $\mathbf{V}_{s_2}$, $\mathbf{V}_{s_3}$, etc., are the counterparts of $\upsilon_{s_1}(t)$, $\upsilon_{s_2}(t)$, $\upsilon_{s_3}(t)$, etc., respectively.

Step 2: Determine output responses to input harmonics

For the circuit in Fig. 13-1(a), input voltage $\mathbf{V}_{s_1}$ acting alone would generate a corresponding output voltage $\mathbf{V}_{\text{out}_1}$. Keeping in mind that $\mathbf{V}_{s_1}$ corresponds to $\upsilon_{s_1}(t)$ at $\omega = \omega_0 = \pi$ rad/s, voltage division gives

$$\begin{aligned}\mathbf{V}_{\text{out}_1} &= \left.\left(\frac{R}{R+j\omega L}\right)\right|_{\omega=\omega_0=\pi} \mathbf{V}_{s_1} \\ &= \frac{4}{4+j\pi\times\frac{2}{\pi}}\cdot\frac{12}{\pi} = 3.42\angle{-26.56^\circ}\end{aligned} \tag{13.8}$$

with a corresponding time-domain voltage

$$\upsilon_{\text{out}_1}(t) = \mathfrak{Re}[\mathbf{V}_{\text{out}_1}e^{j\omega_0 t}] = 3.42\cos(\omega_0 t - 26.56^\circ) \text{ V}. \tag{13.9}$$

Similarly, at $\omega = 3\omega_0 = 3\pi$ rad/s,

$$\begin{aligned}\mathbf{V}_{\text{out}_2} &= \left.\frac{R}{R+j\omega L}\right|_{\omega=3\omega_0=3\pi} \mathbf{V}_{s_2} \\ &= \frac{4}{4+j3\pi\times\frac{2}{\pi}}\cdot\left(-\frac{12}{3\pi}\right) = -0.71\angle{-56.31^\circ} \text{ V}\end{aligned} \tag{13.10}$$

and

$$\upsilon_{\text{out}_2}(t) = \mathfrak{Re}[\mathbf{V}_{\text{out}_2}e^{j3\omega_0 t}] = -0.71\cos(3\omega_0 t - 56.31^\circ) \text{ V}. \tag{13.11}$$

In view of the harmonic pattern expressed in the form of Eq. (13.3), for the harmonic at angular frequency $\omega = (2n-1)\omega_0 = (2n-1)\pi$ rad/s,

$$\begin{aligned}\mathbf{V}_{\text{out}_n} &= \frac{4}{4+j(2n-1)\pi\times\frac{2}{\pi}}\cdot(-1)^{n+1}\frac{12}{\pi(2n-1)} \\ &= (-1)^{n+1}\frac{24}{\pi(2n-1)\sqrt{4+(2n-1)^2}} \\ &\quad \cdot \angle{-\tan^{-1}[(2n-1)/2]} \text{ V}.\end{aligned} \tag{13.12}$$

The corresponding time domain voltage is

$$\begin{aligned}\upsilon_{\text{out}_n}(t) &= \mathfrak{Re}[\mathbf{V}_{\text{out}_n}e^{j(2n-1)\omega_0 t}] \\ &= (-1)^{n+1}\frac{24}{\pi(2n-1)\sqrt{4+(2n-1)^2}} \\ &\quad \cdot\cos\left[(2n-1)\omega_0 t - \tan^{-1}\left(\frac{2n-1}{2}\right)\right] \text{ V},\end{aligned} \tag{13.13}$$

with $\omega_0 = \pi$ rad/s.

Step 3: Apply the superposition principle to determine $v_{\text{out}}(t)$

According to the superposition principle, if v_{out_1} is the output generated by a linear circuit when excited by an input voltage v_{s_1} acting alone and if similarly v_{out_2} is the output due to v_{s_2} acting alone, then the output due to the combination of v_{s_1} and v_{s_2} acting simultaneously is simply the sum of v_{out_1} and v_{out_2}. Moreover, the principle is extendable to any number of sources. In the present case, the square-wave excitation is equivalent to a series of sinusoidal sources $v_{s_1}, v_{s_2}, \ldots$ generating corresponding output voltages $v_{\text{out}_1}, v_{\text{out}_2}, \ldots$. Consequently,

$$\begin{aligned} v_{\text{out}}(t) &= \sum_{n=1}^{\infty} v_{\text{out}_n}(t) \\ &= \sum_{n=1}^{\infty} (-1)^{n+1} \frac{24}{\pi(2n-1)\sqrt{4+(2n-1)^2}} \\ &\quad \cdot \cos\left[(2n-1)\omega_0 t - \tan^{-1}\left(\frac{2n-1}{2}\right)\right] \\ &= 3.42\cos(\omega_0 t - 26.56^\circ) \\ &\quad - 0.71\cos(3\omega_0 t - 56.31^\circ) \\ &\quad + 0.28\cos(5\omega_0 t - 68.2^\circ) + \cdots \text{ V}, \end{aligned} \tag{13.14}$$

with $\omega_0 = \pi$ rad/s.

We note that the fundamental component of $v_{\text{out}}(t)$ has the dominant amplitude and that the higher the harmonic is, the smaller is its amplitude. This allows us to approximate $v_{\text{out}}(t)$ by retaining only a few terms, such as up to $n = 10$, depending on the level of desired accuracy. The plot of $v_{\text{out}}(t)$ displayed in Fig. 13-1(c) (which is based on only the first 10 terms) is sufficiently accurate for most practical applications.

The foregoing three-step procedure (which is equally applicable to any linear circuit excited by any realistic periodic function) relied on the use of the Fourier theorem to express the square-wave pattern in terms of sinusoids. In the next section, we examine the attributes of the Fourier theorem and how we may apply it to any periodic function.

> **Concept Question 13-1:** The Fourier series technique is applied to analyze circuits excited by what type of functions? (See CAD)

> **Concept Question 13-2:** How is the angular frequency of the nth harmonic related to that of the fundamental ω_0? How is ω_0 related to the period T of the periodic function? (See CAD)

> **Concept Question 13-3:** What steps constitute the Fourier series solution procedure? (See CAD)

13-2 Fourier Series Representation

In 1822, the French mathematician Jean-Baptiste Joseph Fourier developed an elegant formulation for representing periodic functions in terms of a series of sinusoidal harmonics. The representation is known today as the *Fourier series*, and the formulation is called the *Fourier theorem*. To guarantee that a periodic function $f(t)$ has a realizable Fourier series, it should satisfy a set of conditions known as the *Dirichlet conditions*.

> ▶ Fortunately, any periodic function generated by a real circuit will meet these conditions automatically and therefore, we are assured that its Fourier series does indeed exist. ◀

The *Fourier theorem* states that a periodic function $f(t)$ of period T can be cast in the form

$$f(t) = a_0 + \sum_{n=1}^{\infty} (a_n \cos n\omega_0 t + b_n \sin n\omega_0 t),$$

(sine/cosine representation) (13.15)

where ω_0, the *fundamental angular frequency* of $f(t)$, is related to T by

$$\omega_0 = \frac{2\pi}{T}. \tag{13.16}$$

The summation is an infinite series whose first pair of terms (for $n = 1$) involve $\cos\omega_0 t$ and $\sin\omega_0 t$. Higher values of n involve sine and cosine functions at harmonic multiples of ω_0, namely $2\omega_0$, $3\omega_0$, etc. The constants a_0, a_n, and b_n (for $n = 1$ to ∞) are collectively called the *Fourier coefficients* of $f(t)$. Their values are determined by evaluating integral expressions involving $f(t)$, namely,

$$a_0 = \frac{1}{T}\int_0^T f(t)\, dt, \tag{13.17a}$$

$$a_n = \frac{2}{T}\int_0^T f(t) \cos n\omega_0 t \, dt, \tag{13.17b}$$

and

$$b_n = \frac{2}{T} \int_0^T f(t) \sin n\omega_0 t \, dt. \tag{13.17c}$$

We will derive these expressions shortly.

▶ Even though the indicated limits of integration are from 0 to T, the expressions are equally valid if the lower limit is changed to t_0 and the upper limit to (t_0+T) for any value of t_0. In some cases, the evaluation is easier to perform by integrating from $-T/2$ to $T/2$. ◀

Coefficient a_0 is equal to the time-average value of $f(t)$. It is called the *dc component* of $f(t)$, because the average values of the ac components are all zero.

13-2.1 Sine/Cosine Representation

To verify the validity of the expressions given by Eq. (13.17), we make use of the trigonometric integral properties listed in Table 13-1.

dc Fourier component a_0

Equation (8.5) in Chapter 8 states that the average value of a periodic function is obtained by integrating it over a complete period T and then dividing the integral by T. Applying the definition to Eq. (13.15) gives

$$\begin{aligned} \frac{1}{T}\int_0^T f(t)\,dt &= \frac{1}{T}\int_0^T a_0\,dt \\ &\quad + \frac{1}{T}\int_0^T \left[\sum_{n=1}^{\infty} a_n \cos n\omega_0 t + b_n \sin n\omega_0 t\right] dt \\ &= a_0 + \frac{1}{T}\int_0^T a_1 \cos \omega_0 t\,dt + \frac{1}{T}\int_0^T a_2 \cos 2\omega_0 t\,dt + \cdots \\ &\quad + \frac{1}{T}\int_0^T b_1 \sin \omega_0 t\,dt + \frac{1}{T}\int_0^T b_2 \sin 2\omega_0 t\,dt + \cdots . \end{aligned} \tag{13.18}$$

According to Property 1 in Table 13-1, the average value of a sine function is zero, and the same is true for a cosine function (Property 2). Hence, all of the terms in Eq. (13.18) containing $\cos n\omega_0 t$ or $\sin n\omega_0 t$ will vanish, leaving behind

$$\frac{1}{T}\int_0^T f(t)\,dt = a_0, \tag{13.19}$$

which is identical to the definition given by Eq. (13.17a).

a_n Fourier coefficients

Multiplication of both sides of Eq. (13.15) by $\cos m\omega_0 t$ (with m being any integer value equal to or greater than 1), followed

Table 13-1: Trigonometric integral properties for any integers m and n. The integration period $T = 2\pi/\omega_0$.

Property	Integral
1	$\int_0^T \sin n\omega_0 t\,dt = 0$
2	$\int_0^T \cos n\omega_0 t\,dt = 0$
3	$\int_0^T \sin n\omega_0 t \sin m\omega_0 t\,dt = 0, \quad n \neq m$
4	$\int_0^T \cos n\omega_0 t \cos m\omega_0 t\,dt = 0, \quad n \neq m$
5	$\int_0^T \sin n\omega_0 t \cos m\omega_0 t\,dt = 0$
6	$\int_0^T \sin^2 n\omega_0 t\,dt = T/2$
7	$\int_0^T \cos^2 n\omega_0 t\,dt = T/2$

Note: All integral properties remain valid when the arguments $n\omega_0 t$ and $m\omega_0 t$ are phase shifted by a constant angle ϕ_0. Thus, Property 1, for example, becomes $\int_0^T \sin(n\omega_0 t + \phi_0)\,dt = 0$, and Property 5 becomes $\int_0^T \sin(n\omega_0 t + \phi_0)\cos(m\omega_0 t + \phi_0)\,dt = 0$.

with integration over $[0, T]$ yields

$$\int_0^T f(t)\cos m\omega_0 t\,dt = \int_0^T a_0 \cos m\omega_0 t\,dt$$
$$+ \int_0^T \sum_{n=1}^{\infty} a_n \cos n\omega_0 t \cos m\omega_0 t\,dt$$
$$+ \int_0^T \sum_{n=1}^{\infty} b_n \sin n\omega_0 t \cos m\omega_0 t\,dt. \quad (13.20)$$

On the right-hand side of Eq. (13.20):

(1) The term containing a_0 is equal to zero (Property 2 in Table 13-1).

(2) All terms containing b_n are equal to zero (Property 5).

(3) All terms containing a_n are equal to zero (Property 4), except when $m = n$, in which case Property 7 applies.

Hence, after eliminating all of the zero-valued terms and then setting $m = n$ in the two remaining terms, we have

$$\int_0^T f(t)\cos n\omega_0 t\,dt = a_n \frac{T}{2}, \quad (13.21)$$

which proves Eq. (13.17b).

b_n Fourier coefficients

Similarly, if we were to repeat the preceding process, after multiplication of Eq. (13.15) by $\sin m\omega_0 t$ (instead of $\cos m\omega_0 t$), we would conclude with a result affirming the validity of Eq. (13.17c).

To develop an appreciation for how the components of the Fourier series add up to represent the periodic waveform, let us consider the square-wave voltage waveform shown in Fig. 13-2(a). Over the period extending from $-T/2$ to $T/2$, $\upsilon(t)$ is given by

$$\upsilon(t) = \begin{cases} -A, & \text{for } -T/2 < t < -T/4, \\ A, & \text{for } -T/4 < t < T/4, \\ -A, & \text{for } T/4 < t < -T/2. \end{cases}$$

If we apply Eq. (13.17)—with integration limits $[-T/2, T/2]$—to evaluate the Fourier coefficients and then use them in Eq. (13.15), we end up with the series

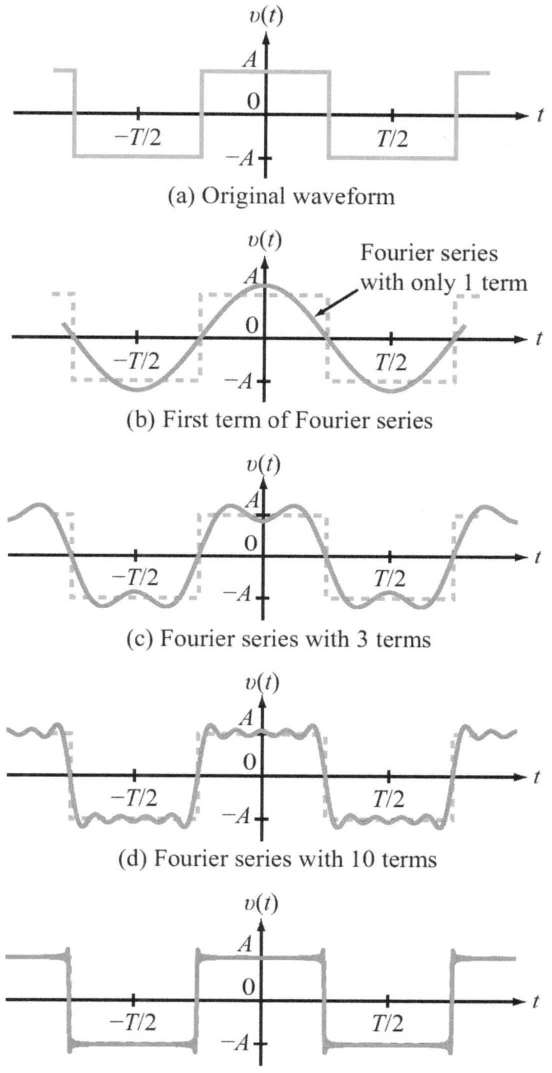

Figure 13-2: Comparison of the square-wave waveform with its Fourier series representation using only the first term (b), the sum of the first three (c), ten (d), and 100 terms (e).

$$\upsilon(t) = \sum_{n=1}^{\infty} \frac{4A}{n\pi} \sin\left(\frac{n\pi}{2}\right) \cos\left(\frac{2n\pi t}{T}\right)$$
$$= \frac{4A}{\pi}\cos\left(\frac{2\pi t}{T}\right) - \frac{4A}{3\pi}\cos\left(\frac{6\pi t}{T}\right)$$
$$+ \frac{4A}{5\pi}\cos\left(\frac{10\pi t}{T}\right) - \cdots .$$

Alone, the first term of the series provides a crude approximation of the square wave (Fig. 13-2(b)), but as we add more and more terms, the sum starts to better resemble the general shape of the square wave, as demonstrated by the waveforms in Figs. 13-2(c) to (e).

Example 13-1: Sawtooth Waveform

Express the sawtooth waveform shown in Fig. 13-3(a) in terms of a Fourier series, and then evaluate how well the original waveform is represented by a truncated series in which the summation stops when n reaches a specified truncation number $n_{\max}$. Generate plots for $n_{\max} = 1, 2, 10$, and 100.

Solution: The sawtooth waveform is characterized by a period $T = 4$ s and $\omega_0 = 2\pi/T = \pi/2$ (rad/s). Over the waveform's first cycle ($t = 0$ to $t = 4$ s), its amplitude variation is given by

$$f(t) = 5t \qquad (\text{for } 0 \leq t \leq 4 \text{ s}).$$

Application of Eq. (13.17) yields:

$$a_0 = \frac{1}{T}\int_0^T f(t)\,dt = \frac{1}{4}\int_0^4 5t\,dt = 10,$$

$$a_n = \frac{2}{T}\int_0^T f(t)\,\cos(n\omega_0 t)\,dt$$

$$= \frac{2}{4}\int_0^4 5t\cos\left(\frac{n\pi}{2}\,t\right)\,dt = 0,$$

and

$$b_n = \frac{2}{T}\int_0^T f(t)\,\sin(n\omega_0 t)\,dt$$

$$= \frac{2}{4}\int_0^4 5t\sin\left(\frac{n\pi}{2}\,t\right)\,dt = -\frac{20}{n\pi}.$$

Upon inserting these results into Eq. (13.15), we obtain the following *complete* Fourier series representation for the sawtooth waveform:

$$f(t) = 10 - \frac{20}{\pi}\sum_{n=1}^{\infty}\frac{1}{n}\sin\left(\frac{n\pi}{2}\,t\right).$$

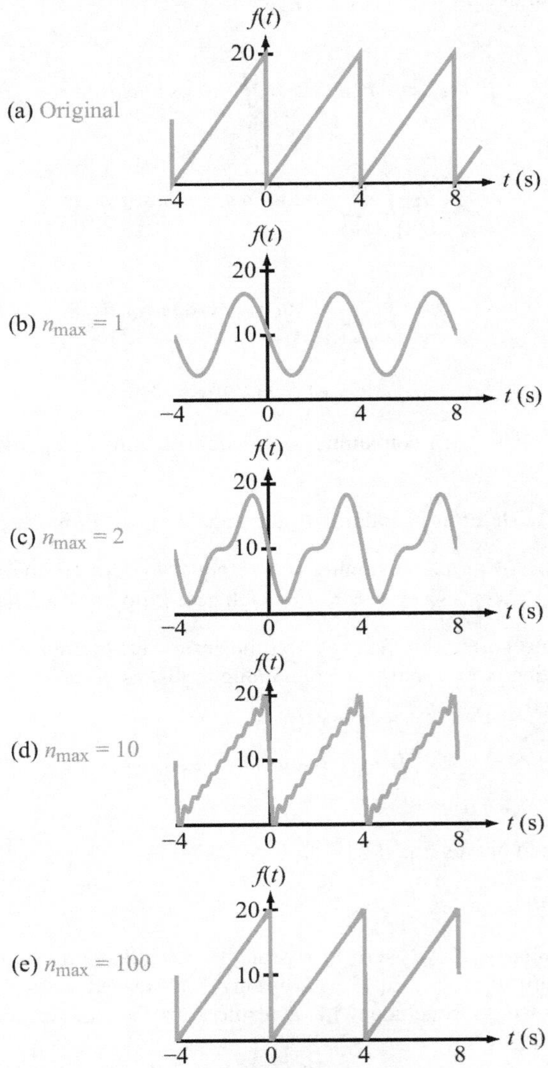

Figure 13-3: Sawtooth waveform: (a) original waveform, (b)–(e) representation by a truncated Fourier series with $n_{\max} = 1, 2, 10$, and 100, respectively.

▶ The $n_{\max}$-*truncated series* is identical in form to the complete series, except that the summation is terminated after the index n reaches $n_{\max}$. ◀

Figures 13-3(b) through (e) display the waveforms calculated using the truncated series with $n_{\max} = 1, 2, 10$, and 100. As expected, the addition of more terms improves the accuracy of the Fourier series representation, but even with only 10 terms (in addition to the dc component), the truncated series appears to provide a reasonable approximation of the original waveform.

Concept Question 13-4: Is the Fourier series representation given by Eq. (13.15) applicable to a periodic function that starts at $t = 0$ (and is zero for $t < 0$)? (See CAD)

Concept Question 13-5: What is a *truncated* series? (See CAD)

Exercise 13-1: Obtain the Fourier series representation for the waveform shown in Fig. E13.1.

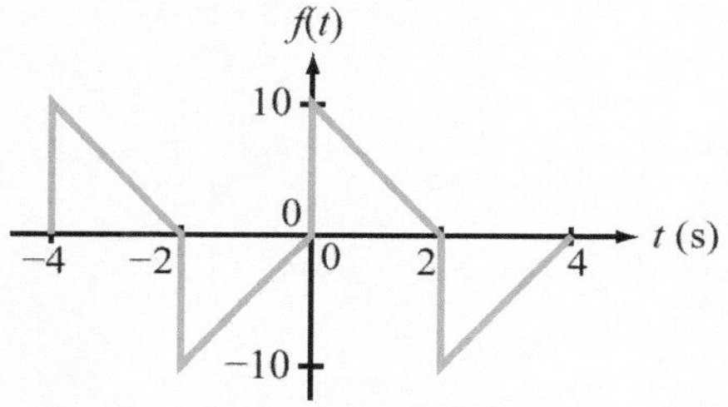

Figure E13.1

Answer:

$$f(t) = \sum_{n=1}^{\infty} \left[\frac{20}{n^2\pi^2}(1 - \cos n\pi)\cos\frac{n\pi t}{2} + \frac{10}{n\pi}(1 - \cos n\pi)\sin\frac{n\pi t}{2} \right].$$

(See CAD)

13-2.2 Amplitude and Phase Representation

In the sine/cosine Fourier series representation given by Eq. (13.15), at each value of the integer index n, the summation contains the sum of a sine term and a cosine term, with both at angular frequency $n\omega_0$. The sum can be converted into a single sinusoid as follows. For $n \geq 0$,

$$a_n \cos n\omega_0 t + b_n \sin n\omega_0 t = A_n \cos(n\omega_0 t + \phi_n), \qquad (13.22)$$

where A_n is called the *amplitude of the nth harmonic* and ϕ_n is its associated *phase*. The relationships between (A_n, ϕ_n) and (a_n, b_n) are obtained by expanding the right-hand side of Eq. (13.22) in accordance with the trigonometric identity

$$\cos(x + y) = \cos x \cos y - \sin x \sin y. \qquad (13.23)$$

Thus,

$$a_n \cos n\omega_0 t + b_n \sin n\omega_0 t = A_n \cos\phi_n \cos n\omega_0 t - A_n \sin\phi_n \sin n\omega_0 t. \qquad (13.24)$$

Upon equating the coefficients of $\cos n\omega_0 t$ and $\sin n\omega_0 t$ on one side of the equation to their respective counterparts on the other side, we have

$$a_n = A_n \cos\phi_n \quad \text{and} \quad b_n = -A_n \sin\phi_n, \qquad (13.25)$$

which can be combined to yield the relationships

$$A_n = \sqrt{a_n^2 + b_n^2}$$

and

$$\phi_n = \begin{cases} -\tan^{-1}\left(\dfrac{b_n}{a_n}\right) & a_n > 0, \\ \pi - \tan^{-1}\left(\dfrac{b_n}{a_n}\right) & a_n < 0. \end{cases} \qquad (13.26)$$

In complex form,

$$A_n \angle \phi_n = a_n - jb_n. \qquad (13.27)$$

In view of Eq. (13.22), the cosine/sine Fourier series representation of $f(t)$ can be rewritten in the alternative *amplitude/phase* format

$$f(t) = a_0 + \sum_{n=1}^{\infty} A_n \cos(n\omega_0 t + \phi_n). \qquad (13.28)$$

(amplitude/phase representation)

Associated with each discrete frequency harmonic $n\omega_0$ is an amplitude A_n and a phase ϕ_n. A *line spectrum* of a periodic signal $f(t)$ is a visual depiction of its Fourier coefficients, A_n and ϕ_n. Its *amplitude spectrum* consists of vertical lines located at discrete values along the ω axis, with a line of height a_0 located at dc ($\omega = 0$), another of height A_1 at $\omega = \omega_0$, a third of height A_2 at $\omega = 2\omega_0$, and so on. Similarly, the *phase spectrum* of $f(t)$ consists of lines of lengths proportional to the values of ϕ_n with each located at its corresponding harmonic $n\omega_0$. Line spectra show at a glance which frequencies in the spectrum of $f(t)$ are most significant and which are not. Example 13-2 provides an illustration.

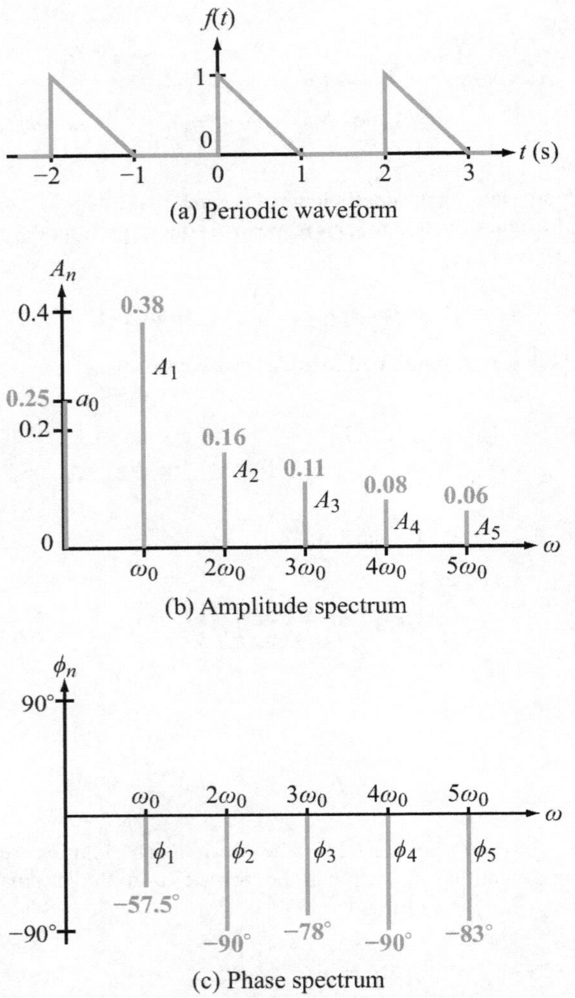

Figure 13-4: Periodic waveform of Example 13-2 with its associated line spectra.

Example 13-2: Line Spectra

Generate and plot the amplitude and phase spectra of the periodic waveform displayed in Fig. 13-4(a).

Solution: The periodic waveform has a period $T = 2$ s. Hence, $\omega_0 = 2\pi/T = 2\pi/2 = \pi$ rad/s, and the functional expression for $f(t)$ over its first cycle along the positive t axis is

$$f(t) = \begin{cases} 1 - t & \text{for } 0 < t \leq 1 \text{ s}, \\ 0 & \text{for } 1 \leq t \leq 2 \text{ s}. \end{cases}$$

The dc component of $f(t)$ is given by

$$a_0 = \frac{1}{T}\int_0^T f(t)\,dt = \frac{1}{2}\int_0^1 (1-t)\,dt = 0.25,$$

which is equal to the area under a single triangle, divided by the period $T = 2$ s.

For the other Fourier coefficients, evaluation of the expressions given by Eqs. (13.17b and c) leads to

$$\begin{aligned} a_n &= \frac{2}{T}\int_0^T f(t)\cos n\omega_0 t\,dt \\ &= \frac{2}{2}\int_0^1 (1-t)\cos n\pi t\,dt \\ &= \frac{1}{n\pi}\sin n\pi t\Big|_0^1 - \left(\frac{1}{n^2\pi^2}\cos n\pi t + \frac{t}{n\pi}\sin n\pi t\right)\Big|_0^1 \\ &= \frac{1}{n^2\pi^2}[1 - \cos n\pi] \end{aligned}$$

and

$$\begin{aligned} b_n &= \frac{2}{T}\int_0^T f(t)\sin n\omega_0 t\,dt = \frac{2}{2}\int_0^1 (1-t)\sin n\pi t\,dt \\ &= -\frac{1}{n\pi}\cos n\pi t\Big|_0^1 - \left(\frac{1}{n^2\pi^2}\sin n\pi t - \frac{t}{n\pi}\cos n\pi t\right)\Big|_0^1 \\ &= \frac{1}{n\pi}. \end{aligned}$$

By Eq. (13.26), the harmonic amplitudes and phases are given by

$$\begin{aligned} A_n &= \sqrt[+]{a_n^2 + b_n^2} = \left[\left(\frac{1}{n^2\pi^2}[1 - \cos n\pi]\right)^2 + \left(\frac{1}{n\pi}\right)^2\right]^{1/2} \\ &= \begin{cases} \left(\dfrac{4}{n^4\pi^4} + \dfrac{1}{n^2\pi^2}\right)^{1/2} & \text{for } n = \text{odd} \\ \dfrac{1}{n\pi} & \text{for } n = \text{even} \end{cases} \end{aligned}$$

and

$$\begin{aligned} \phi_n &= -\tan^{-1}\frac{b_n}{a_n} = -\tan^{-1}\left(\frac{n\pi}{[1 - \cos n\pi]}\right) \\ &= \begin{cases} -\tan^{-1}\left(\dfrac{n\pi}{2}\right) & \text{for } n = \text{odd}, \\ -90^\circ & \text{for } n = \text{even}. \end{cases} \end{aligned}$$

The values of A_n and ϕ_n for the first three terms are

$$A_1 = 0.38, \qquad \phi_1 = -57.5^\circ,$$
$$A_2 = 0.16, \qquad \phi_2 = -90^\circ,$$
$$A_3 = 0.11, \qquad \phi_3 = -78^\circ.$$

Spectral plots of A_n and ϕ_n are shown in Figs. 13-4(b) and (c), respectively.

Exercise 13-2: Obtain the line spectra associated with the periodic function of Exercise 13-1.

Answer:

$$A_n = [1 - \cos(n\pi)]\,\frac{20}{n^2\pi^2}\sqrt{1 + \frac{n^2\pi^2}{4}}$$

and

$$\phi_n = -\tan^{-1}\left(\frac{n\pi}{2}\right).$$

(See CAD)

13-2.3 Symmetry Considerations

According to Eq. (13.17), determination of the Fourier coefficients involves evaluation of three definite integrals involving $f(t)$. If $f(t)$ exhibits symmetry properties, the evaluation process can be simplified significantly.

dc Symmetry

Since a_0 is equal to the average value of $f(t)$ which is proportional to the net area under the waveform over the span of a complete cycle, it follows that $a_0 = 0$ if the waveform is such that the area above the zero horizontal axis is equal to the area below it. The waveforms in Figs. 13-5(a), (b), (d), and (e) are examples of dc-symmetrical functions.

Even and odd symmetry

The waveform of a function $f(t)$ possesses *even symmetry* if it is symmetrical with respect to the vertical axis; the shape of the waveform on the left-hand side of the vertical axis is the mirror image of the waveform on the right-hand side. Mathematically, an *even function* satisfies the condition

$$f(t) = f(-t) \qquad \textbf{(even symmetry)}. \qquad (13.29)$$

The waveforms displayed in Figs. 13-5(b) and (c) exhibit even symmetry, as do the waveforms of $\sin^2 \omega t$ and $|\sin \omega t|$, among many others.

In contrast, the sine and square waves shown in Fig. 13-5(d) and (e) exhibit *odd symmetry*; in each case, the shape of the waveform on the left-hand side of the vertical axis is the *inverted* mirror image of the waveform on the right-hand side. Thus, for an *odd function*:

$$f(t) = -f(-t) \qquad \textbf{(odd symmetry)}. \qquad (13.30)$$

In the case of the square wave, were we to shift the waveform by $T/4$ to the left, it would switch from an odd function into an even function.

13-2.4 Even-Function Fourier Coefficients

Even symmetry allows us to simplify Eq. (13.17) to the following expressions:

Even Symmetry: $f(t) = f(-t)$

$$a_0 = \frac{2}{T}\int_0^{T/2} f(t)\,dt,$$
$$a_n = \frac{4}{T}\int_0^{T/2} f(t)\cos(n\omega_0 t)\,dt, \qquad (13.31)$$
$$b_n = 0,$$
$$A_n = |a_n|, \quad \text{and} \quad \phi_n = \begin{cases} 0 & \text{if } a_n > 0, \\ 180^\circ & \text{if } a_n < 0. \end{cases}$$

The expressions for a_0 and a_n are the same as given earlier by Eqs. (13.17a and b), except that the integration limits are now over half of a period and the integral has been multiplied by a factor of 2. The simplification is justified by the even symmetry of $f(t)$. As was stated in connection with Eq. (13.17), the only restriction associated with the integration limits is that the upper limit has to be greater than the lower limit by exactly T. Hence, by choosing the limits to be $[-T/2, T/2]$, and then recognizing that the integral of $f(t)$ over $[-T/2, 0]$ is equal to the integral over $[0, T/2]$, we justify the changes reflected in the expression for a_0. A similar argument applies to the expression for a_n based on the fact that multiplication of an even function $f(t)$ by $\cos n\omega_0 t$ (which itself is an even function) yields an even function. The rationale for setting $b_n = 0$ for all n

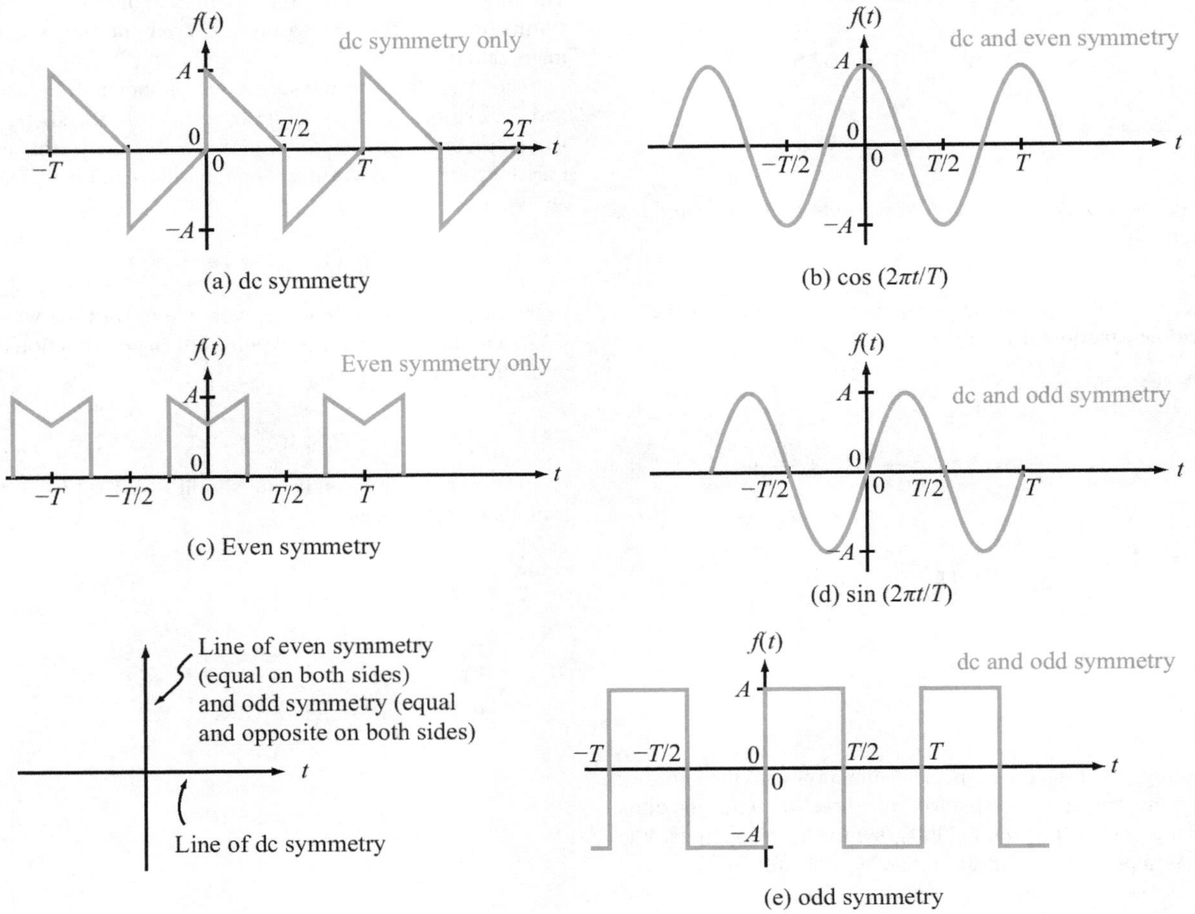

Figure 13-5: Waveforms with (a) dc symmetry, (b and c) even symmetry, and (d and e) odd symmetry.

relies on the fact that multiplication of an even function $f(t)$ by $\sin n\omega_0 t$ (which is an odd function) yields an odd function, and integration of an odd function over $[-T/2, T/2]$ is always equal to zero. This is because the integral of an odd function over $[-T/2, 0]$ is equal in magnitude but opposite in sign to the integral over $[0, T/2]$.

13-2.5 Odd-Function Fourier Coefficients

In view of the preceding discussion, it follows that for a function with odd symmetry we have the following equations:

Odd Symmetry: $f(t) = -f(-t)$

$$a_0 = 0, \qquad a_n = 0,$$

$$b_n = \frac{4}{T}\int_0^{T/2} f(t)\,\sin(n\omega_0 t)\,dt, \tag{13.32}$$

$$A_n = |b_n| \quad \text{and} \quad \phi_n = \begin{cases} -90^\circ & \text{if } b_n > 0, \\ 90^\circ & \text{if } b_n < 0. \end{cases}$$

Selected waveforms are displayed in Table 13-2, together with their corresponding Fourier series expressions.

Table 13-2: **Fourier series expressions for a select set of periodic waveforms.**

	Waveform	Fourier Series
1. Square Wave	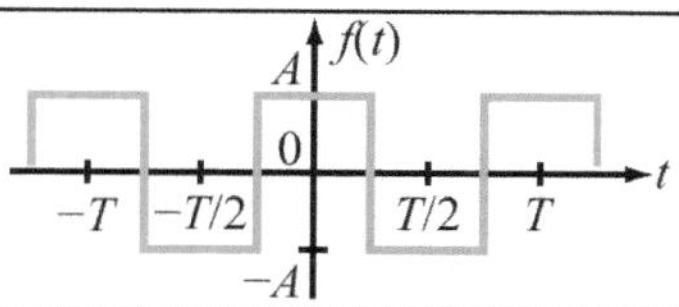	$f(t)=\sum_{n=1}^{\infty}\frac{4A}{n\pi}\sin\left(\frac{n\pi}{2}\right)\cos\left(\frac{2n\pi t}{T}\right)$
2. Time-Shifted Square Wave	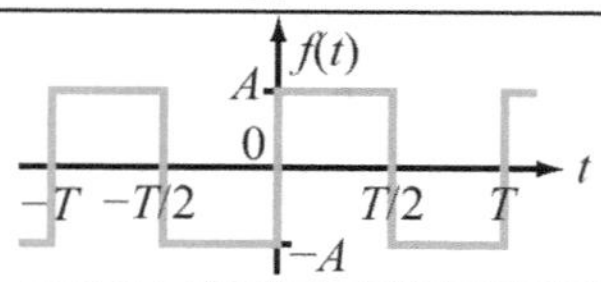	$f(t)=\sum_{\substack{n=1\\ n=\text{odd}}}^{\infty}\frac{4A}{n\pi}\sin\left(\frac{2n\pi t}{T}\right)$
3. Pulse Train	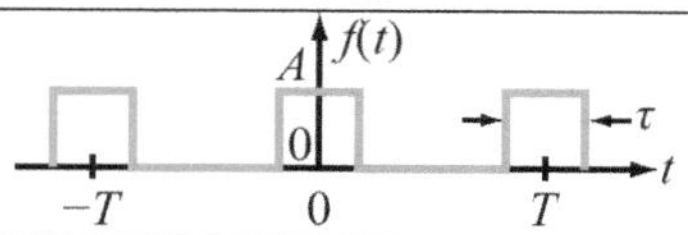	$f(t)=\frac{A\tau}{T}+\sum_{n=1}^{\infty}\frac{2A}{n\pi}\sin\left(\frac{n\pi\tau}{T}\right)\cos\left(\frac{2n\pi t}{T}\right)$
4. Triangular Wave	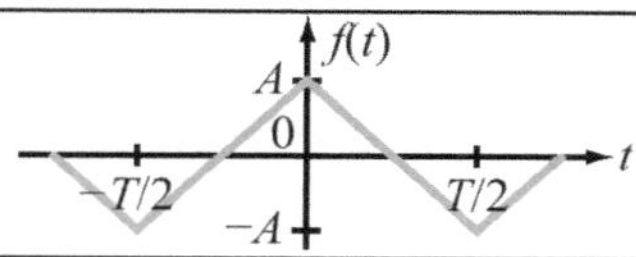	$f(t)=\sum_{\substack{n=1\\ n=\text{odd}}}^{\infty}\frac{8A}{n^2\pi^2}\cos\left(\frac{2n\pi t}{T}\right)$
5. Shifted Triangular Wave	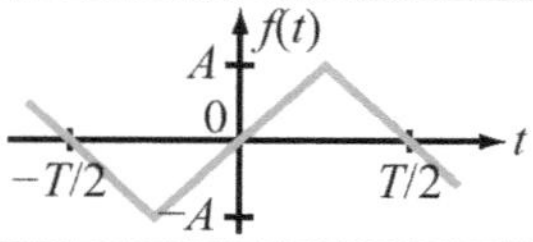	$f(t)=\sum_{\substack{n=1\\ n=\text{odd}}}^{\infty}\frac{8A}{n^2\pi^2}\sin\left(\frac{n\pi}{2}\right)\sin\left(\frac{2n\pi t}{T}\right)$
6. Sawtooth	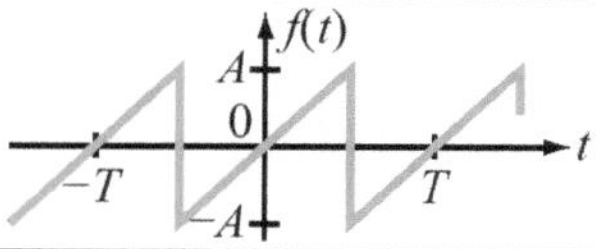	$f(t)=\sum_{n=1}^{\infty}(-1)^{n+1}\frac{2A}{n\pi}\sin\left(\frac{2n\pi t}{T}\right)$
7. Backward Sawtooth	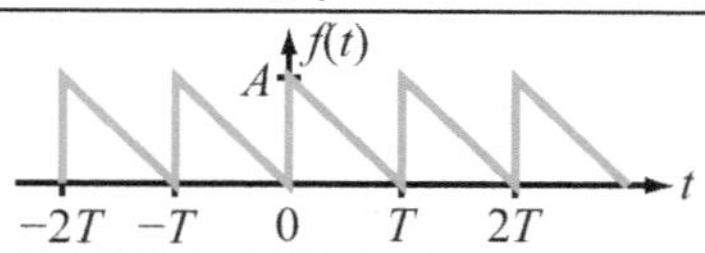	$f(t)=\frac{A}{2}+\sum_{n=1}^{\infty}\frac{A}{n\pi}\sin\left(\frac{2n\pi t}{T}\right)$
8. Full-Wave Rectified Sinusoid	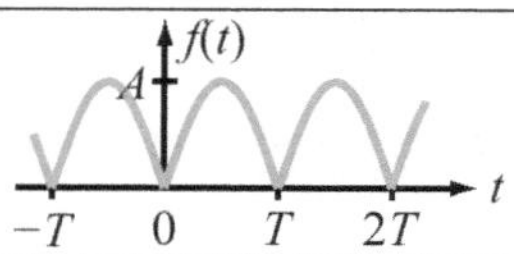	$f(t)=\frac{2A}{\pi}+\sum_{n=1}^{\infty}\frac{4A}{\pi(1-4n^2)}\cos\left(\frac{2n\pi t}{T}\right)$
9. Half-Wave Rectified Sinusoid	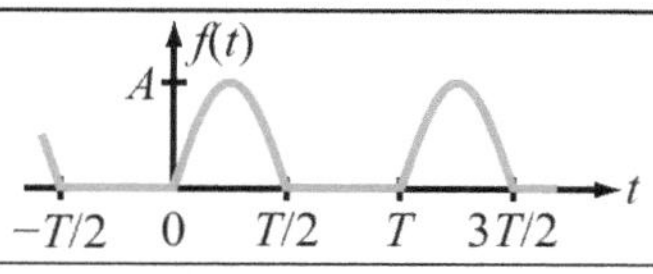	$f(t)=\frac{A}{\pi}+\frac{A}{2}\sin\left(\frac{2\pi t}{T}\right)+\sum_{\substack{n=2\\ n=\text{even}}}^{\infty}\frac{2A}{\pi(1-n^2)}\cos\left(\frac{2n\pi t}{T}\right)$

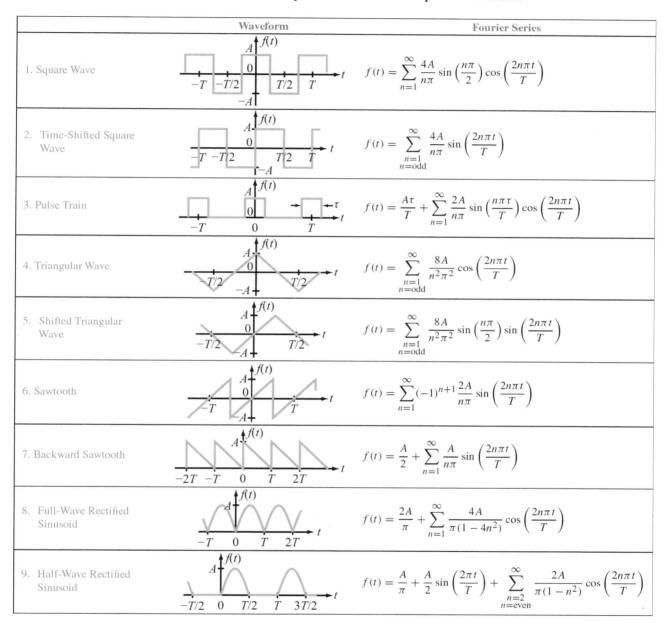

Example 13-3: M-Periodic Waveform

Evaluate the Fourier coefficients of the M-periodic waveform shown in Fig. 13-6(a),

Solution: The M waveform is even-symmetrical, its period is $T = 4$ s, $\omega_0 = 2\pi/T = \pi/2$ rad/s, and its functional form over the positive half period is:

$$f(t)=\begin{cases}\frac{1}{2}(1+t) & 0\le t\le 1 \text{ s},\\ 0 & 1\le t\le 2 \text{ s}.\end{cases}$$

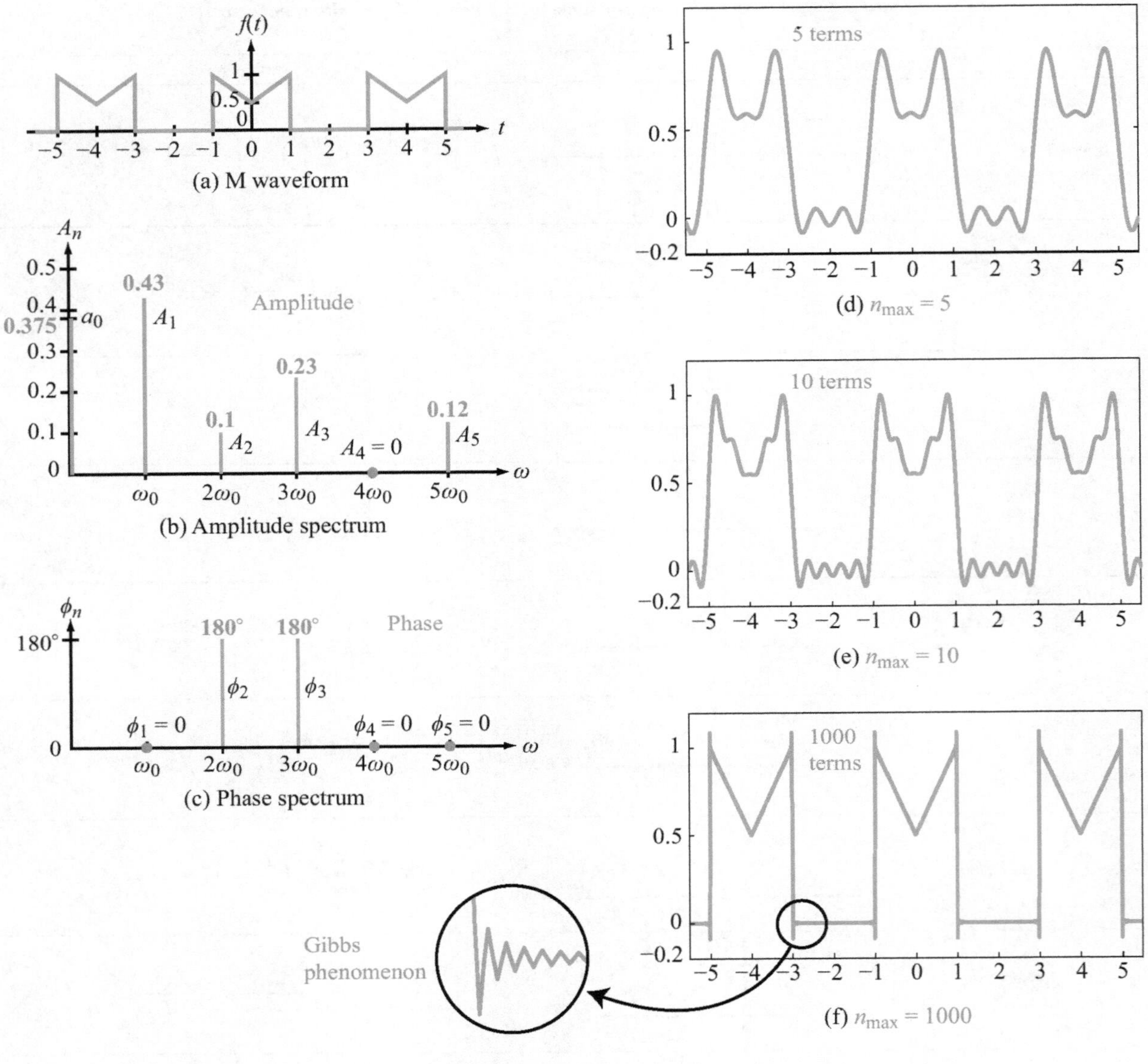

Figure 13-6: Plots for Example 13-3.

Application of Eq. (13.31) yields:

$$a_0 = \frac{2}{T} \int_0^{T/2} f(t)\, dt = \frac{2}{4} \int_0^1 \frac{1}{2}(1+t)\, dt = 0.375,$$

$$a_n = \frac{4}{T} \int_0^{T/2} f(t) \cos n\omega_0 t\, dt$$

$$= \frac{4}{4} \int_0^1 \frac{1}{2}(1+t) \cos n\omega_0 t\, dt$$

$$= \frac{2}{n\pi} \sin \frac{n\pi}{2} + \frac{2}{n^2\pi^2} \left(\cos \frac{n\pi}{2} - 1 \right),$$

and

$$b_n = 0.$$

Since $b_n = 0$, we have for $n \neq 0$

$$A_n = |a_n| \qquad \text{and} \qquad \phi_n = \begin{cases} 0 & \text{if } a_n > 0, \\ 180^\circ & \text{if } a_n < 0. \end{cases}$$

Figures 13-6(b) and (c) display the amplitude and phase line spectra of the M-periodic waveform, and parts (d) through (f) display the waveforms based on the first five terms, the first ten terms, and the first 1000 terms of the Fourier series, respectively.

> ▶ As expected, the addition of more terms in the Fourier series improves the overall fidelity of the reproduced waveform. However, no matter how many terms are included in the series representation, the reproduction cannot duplicate the original M-waveform at points of discontinuity, such as when the waveform jumps from zero to 1. *Discontinuities generate oscillations.* ◀

Increasing the number of terms (adding more harmonics) reduces the period of the oscillation. Ultimately, the oscillations fuse into a solid line, except at the discontinuities (see expanded view of the discontinuity at $t = -3$ s in Fig. 13-6(f)).

> ▶ As n approaches ∞, the Fourier series representation will reproduce the original waveform with perfect fidelity at all nondiscontinuous points, but at a point where the waveform jumps discontinuously between two different levels, the Fourier series will converge to a level half-way between them. ◀

At $t = 1$ s, 3 s, 5 s, ..., the Fourier series will converge to 0.5. This oscillatory behavior of the Fourier series in the neighborhood of discontinuous points is called the ***Gibbs phenomenon***.

Example 13-4: Waveform Synthesis

Given that waveform $f_1(t)$ in Fig. 13-7(a) is represented by the Fourier series

$$f_1(t) = \sum_{n=1}^{\infty} \frac{4A}{n\pi} \sin\left(\frac{n\pi}{2}\right) \cos\left(\frac{2n\pi t}{T}\right),$$

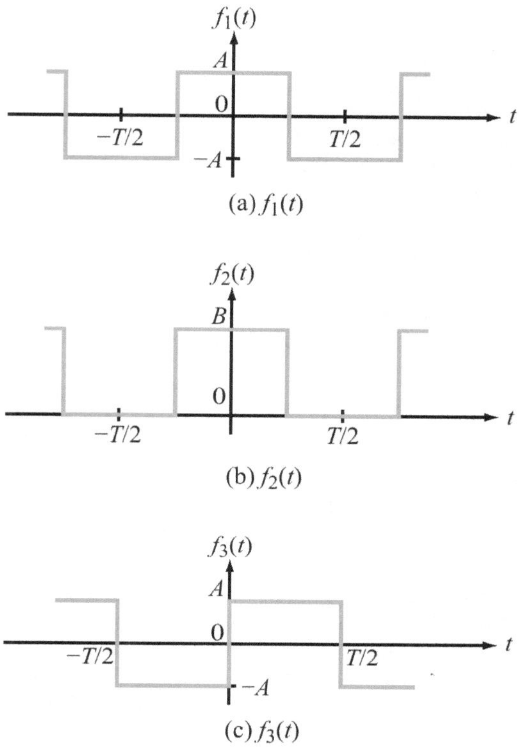

Figure 13-7: Waveforms for Example 13-4.

generate the Fourier series corresponding to the waveforms displayed in Figs. 13-7(b) and (c).

Solution:

(1) Waveform $f_2(t)$

Waveforms $f_1(t)$ and $f_2(t)$ are similar in shape and have the same period, but they also exhibit two differences: (1) the dc value of $f_1(t)$ is zero because it has dc symmetry, whereas the dc value of $f_2(t)$ is $B/2$, and (2) the peak-to-peak value of $f_1(t)$ is $2A$, compared with only B for $f_2(t)$. Mathematically, $f_2(t)$ is related to $f_1(t)$ by

$$\begin{aligned} f_2(t) &= \frac{B}{2} + \left(\frac{B}{2A}\right) f_1(t) \\ &= \frac{B}{2} + \sum_{n=1}^{\infty} \frac{2B}{n\pi} \sin\left(\frac{n\pi}{2}\right) \cos\left(\frac{2n\pi t}{T}\right). \end{aligned}$$

Technology Brief 30
Bandwidth, Data Rate, and Communication

In Section 9-4.1, we defined the ***bandwidth*** B of a resonant circuit as the frequency span over which power transfer through the circuit is greater than half of the maximum level possible. This common half-power (or -3 dB) definition for B can be extended to many devices, circuits, and transmission channels. But how does the everyday use of the word ***bandwidth*** refer to the data rate of a transmission channel, such as the rate at which your internet connection can download data?

Signal and Noise in Communication Channels

Every circuit (including switches, amplifiers, filters, phase shifters, rectifiers, etc.) and every transmission medium (air, wires, and optical fibers) operates with acceptable performance over some specific range of frequencies, outside of which ac signals are severely damped. The actual span of this operational frequency range is dictated by the physical characteristics of the circuit or transmission medium. One such example is the ***coaxial cable*** commonly used to connect a TV to a "cable network" or to an outside antenna. The coaxial cable is a high-fidelity transmission medium—causing negligible distortion or attenuation of the signal passing through it—so long as the carrier frequency of the signal is not much higher than about 10 GHz. The "cutoff" frequency of a typical coaxial cable is determined by the cable's distributed capacitance, inductance, and resistance, which are governed in turn by the geometry of the cable, the conductivity of its inner and outer conductors, and the permittivity of the insulator that separates them.

The ***MOSFET*** offers another example; in Section 5-7 we noted that the switching speed of a MOSFET circuit is limited by parasitic capacitances, setting an upper limit on the switching frequency that a given MOSFET circuit can handle. A circuit with a maximum switching speed of 100 ps, for example, cannot respond to frequencies greater than 1/100 ps (or 10 GHz) without distorting the output waveform in some significant way. Our third example is ***Earth's atmosphere***. According to **Fig. TF20-1** (in Technology Brief 20: The Electromagnetic Spectrum), the transmission spectrum for the atmosphere is characterized by a limited set of ***transmission windows***, with each window extending over a specific range of frequencies.

The overall effective bandwidth B of a communication system is determined by the operational bandwidths of its constituent circuits and the transmission spectra of the cables or other transmission media it uses. As we will see shortly, the channel capacity (or data rate) of the system is directly proportional to B, but it also is influenced by the intensity and character of the ***noise*** in the system. Noise is random power self-generated by all real devices, circuits, and transmission media. In fact, *any material at a temperature greater than 0 K (which includes all physical materials, since no material can exist at exactly 0 K) emits noise power all of the time.* **Figure TF30-1** illustrates how the noise generated by a circuit modifies the waveform of the signal passing through it. The input is an ideal sine wave, whereas the output consists of the same sine wave but with a fluctuating component added to it. The fluctuating component represents the noise generated by the circuit, which is random in polarity because the voltage associated with the noise fluctuates randomly between positive and negative values.

If we know both the power P_S carried by the signal and the average power P_N associated with the noise, we then can determine the ***signal-to-noise ratio*** (SNR):

$$\text{SNR} = \frac{P_S}{P_N}.$$

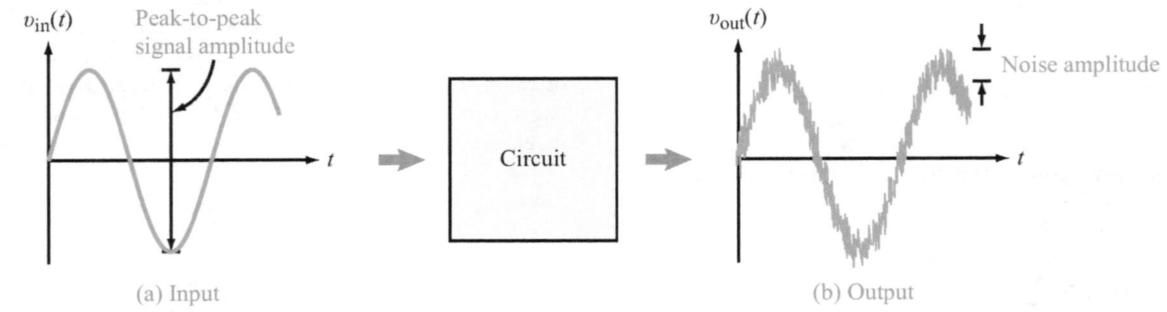

Figure TF30-1: The noise generated by the circuit adds a fluctuating component to the input signal.

Table TT30-1: Theoretical limits for data transmission capacity given typical SNR and channel bandwidth number for several common digital communication layers.

Physical Channel	Channel Capacity (C)	Typical SNR (P_S/P_N)	Bandwidth of Channel (B)
Analog phone line	~30 kbps	10^3	3000 Hz
802.11g Wifi	<60 Mbps	<10	20 MHz
100-BaseT Ethernet	100 Mbps	10^3	100 MHz
10 Gb/s Ethernet	10 Gbps	10^2	1600 MHz

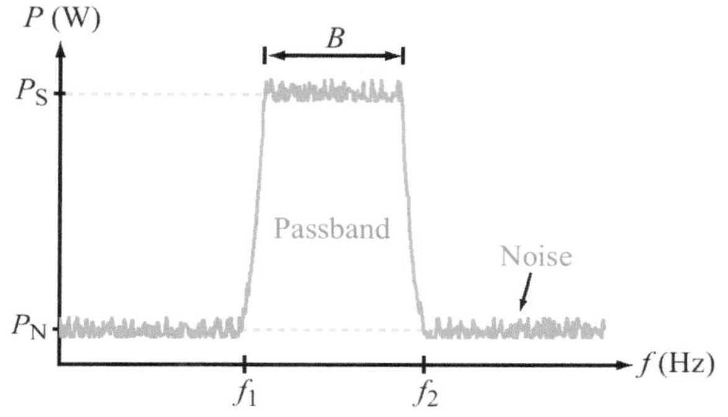

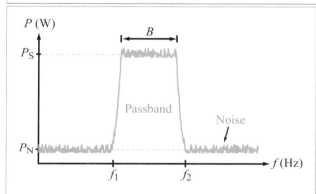

Figure TF30-2: Typical spectral response of a communication system with bandwidth B. The signal-to-noise ratio is given by P_S/P_N, where P_S and P_N are the average power levels of the signal and noise, respectively.

The bandwidth B and the SNR (Fig. TF30-2), jointly determine the highest data rate that can be transmitted reliably through a circuit or a communication system.

Shannon-Hartley Theorem

In the late 1940s and early 1950s, ***Claude Shannon***, building on earlier work by ***Harry Nyquist*** and ***Ralph Hartley***, developed a complete theory that established the limits of information transfer in a communication system. This seminal work represents the foundation of ***information theory*** and underlies all of the subsequent developments that shaped today's ***information revolution***, including the Internet, cell phones, satellite communications, and much more. Shannon's foundational work also has impacted the development of many related disciplines, including encryption, encoding, jamming, efficient use of frequency space, and even quantum-level information manipulations.

The ***Shannon-Hartley theorem*** defines how much data can be transferred through a channel (with no error) in terms of the bandwidth B and the SNR. It states that

$$C = B \log_2 \left(1 + \frac{P_S}{P_N}\right),$$

where C is the ***channel capacity*** (or ***data rate***) in bits/second (bps), B is the bandwidth in Hz, and P_S/P_N is the SNR. As an example, let us consider a communication channel with $B = 100$ MHz, $P_S = 1$ mW, and $P_N = 1\ \mu$W. The corresponding SNR is 1000, and the corresponding value of C is 996×10^6 bits/s or ~ 996 Mb/s. By way of comparison, a 100GbE Ethernet connection can operate at 100 Gbps (or approximately two orders of magnitude faster), a 100 Base-T Ethernet connection can manage a maximum rate of only 100 Mbps, 802.11 Wifi networks are rated at 54 Mbps, and the Bluetooth 2.0 protocol used by many portable devices is limited to 2.1 Mbps. The channel capacity of a conventional telephone used to support audio transmissions is only $\sim$33 kbps. Table TT30-1 provides some theoretical limits for data transmission capacity given typical SNR and channel bandwidth number for several common digital communication layers. We should note that when people use the term "bandwidth" in everyday speech, they really mean channel capacity; B and C are directly proportional to one another, but they obviously are not the same quantity.

According to the expression for C, for a sufficiently high bandwidth, it is possible to achieve reasonably high data-transfer rates even when SNR < 1 ! The implication of this statement is that information can be transmitted reliably on channels whose noise levels exceed that of the signal, provided the signal is spread across a wide frequency spectrum. For example, with a bandwidth of 1 GHz, it is possible to transfer error-free data at a rate of 95 Mbps, even when SNR is only 0.1 (that is, with the signal power an order of magnitude smaller than the noise power). This is (in part) the basis for ***ultra-wideband*** communication schemes used in cell phones and GPS.

(2) Waveform $f_3(t)$

Comparison of waveform $f_1(t)$ with waveform $f_3(t)$ reveals that the latter is shifted by $T/4$ along the t axis relative to $f_1(t)$. That is,

$$f_3(t) = f_1\left(t - \frac{T}{4}\right) = \sum_{n=1}^{\infty} \frac{4A}{n\pi} \sin\left(\frac{n\pi}{2}\right) \cos\left[\frac{2n\pi}{T}\left(t - \frac{T}{4}\right)\right].$$

Examination of the first few terms of $f_3(t)$ demonstrates that $f_3(t)$ can be rewritten in the simpler form

$$f_3(t) = \sum_{\substack{n=1 \\ n=\text{odd}}}^{\infty} \frac{4A}{n\pi} \sin\left(\frac{2n\pi t}{T}\right).$$

Concept Question 13-6: What purpose is served by the symmetry properties of a periodic function? (See CAD)

Concept Question 13-7: What distinguishes the phase angles ϕ_n of an even-symmetrical function from those of an odd-symmetrical function? (See CAD)

Concept Question 13-8: What is the Gibbs phenomenon? (See CAD)

Exercise 13-3: (a) Does the waveform $f(t)$ shown in Fig. E13.3 exhibit either even or odd symmetry? (b) What is the value of a_0? (c) Does the function $g(t) = f(t) - a_0$ exhibit either even or odd symmetry?

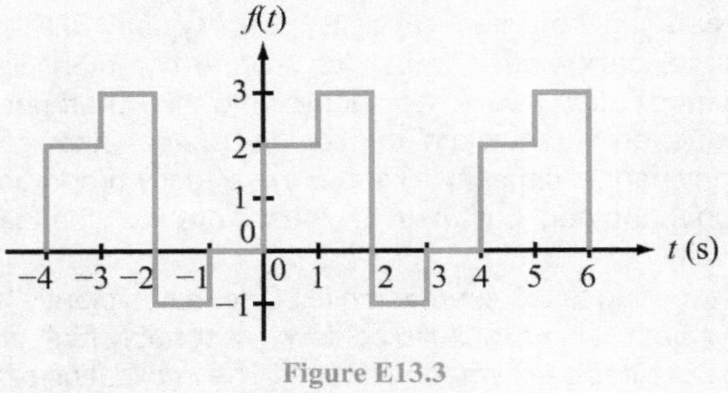

Figure E13.3

Answer: (a) Neither even nor odd symmetry, (b) $a_0 = 1$, (c) odd symmetry. (See CAD)

13-3 Circuit Applications

Given the tools we developed in the preceding section for how to express a periodic function in terms of a Fourier series, we now examine how to analyze linear circuits when excited by periodic voltage or current sources. The method of solution relies on the application of the phasor-domain technique that we introduced in Chapter 7 for analyzing circuits with sinusoidal signals. A periodic function can be expressed as the sum of cosine and sine functions with coefficients a_n and b_n, and zero phase angles, or as the sum of only cosine functions with amplitudes A_n and phase angles ϕ_n. The latter form is amenable to direct application of the phasor-domain technique, whereas the former will require converting all $\sin n\omega_0 t$ terms into $\cos(n\omega_0 t - 90°)$ before implementation of the phasor-domain technique.

Even though the basic solution procedure was outlined earlier in Section 13-1, it is worth repeating it in a form that incorporates the concepts and terminology introduced in Section 13-2. To that end, we shall use $\upsilon_s(t)$ (or $i_s(t)$ if it is a current source) to denote the input excitation and $\upsilon_{out}(t)$ (or $i_{out}(t)$) to denote the output response for which we seek a solution.

Solution Procedure: Fourier Series Analysis Procedure

Step 1: Use the expression for $\upsilon_s(t)$ over one cycle to compute the Fourier coefficients a_0, a_n, and b_n (Eq. (13.17)).

Step 2: Express $\upsilon_s(t)$ in terms of an amplitude/phase Fourier series as

$$\upsilon_s(t) = a_0 + \sum_{n=1}^{\infty} A_n \cos(n\omega_0 t + \phi_n) \qquad (13.33)$$

with $A_n \angle \phi_n = a_n - jb_n$.

Step 3: Establish the generic transfer function of the circuit at frequency ω as

$$\mathbf{H}(\omega) = \mathbf{V}_{out} \qquad \text{when } \upsilon_s = 1 \cos \omega t. \qquad (13.34)$$

Step 4: Write down the time-domain solution as

$$\upsilon_{out}(t) = a_0\, \mathbf{H}(\omega = 0) + \sum_{n=1}^{\infty} A_n \mathfrak{Re}\{\mathbf{H}(\omega = n\omega_0)\, e^{j(n\omega_0 t + \phi_n)}\}. \qquad (13.35)$$

For each value of n, coefficient $A_n e^{j\phi_n}$ is associated with frequency harmonic $n\omega_0$. Hence, in Step 4, each harmonic amplitude is multiplied by its corresponding $e^{jn\omega_0 t}$ before application of the $\mathfrak{Re}\{\ \}$ operator.

Example 13-5: RC Circuit

Determine $\upsilon_{\text{out}}(t)$ when the circuit in Fig. 13-8(a) is excited by the voltage waveform shown in Fig. 13-8(b). The element values are $R = 20$ kΩ and $C = 0.1$ mF.

Solution:

Step 1: The period of $\upsilon_s(t)$ is 4 s. Hence, $\omega_0 = 2\pi/4 = \pi/2$ rad/s, and by Eq. (13.17),

$$a_0 = \frac{1}{T}\int_0^T f(t)\,dt = \frac{1}{4}\int_0^1 10\,dt = 2.5 \text{ V},$$

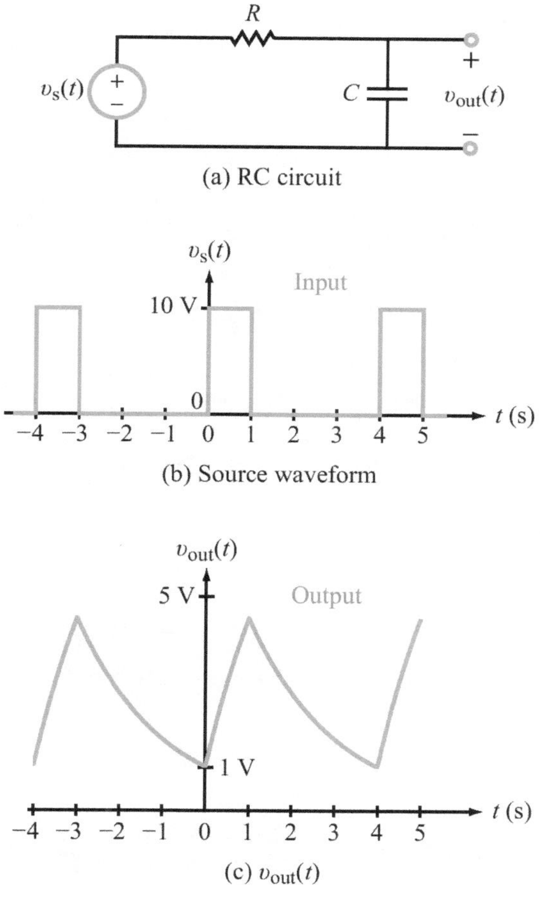

Figure 13-8: Circuit response to periodic pulses (Example 13-5).

$$a_n = \frac{2}{4}\int_0^1 10\cos\frac{n\pi}{2}t\,dt = \frac{10}{n\pi}\sin\frac{n\pi}{2} \text{ V},$$

$$b_n = \frac{2}{4}\int_0^1 10\sin\frac{n\pi}{2}t\,dt = \frac{10}{n\pi}\left(1-\cos\frac{n\pi}{2}\right) \text{ V}.$$

Step 2: For $n \geq 1$,

$$A_n\angle\phi_n = a_n - jb_n = \frac{10}{n\pi}\left[\sin\frac{n\pi}{2} - j\left(1-\cos\frac{n\pi}{2}\right)\right].$$

The values of $A_n\angle\phi_n$ for the first four terms are

$$A_1\angle\phi_1 = \frac{10\sqrt{2}}{\pi}\angle{-45°},$$

$$A_2\angle\phi_2 = \frac{10}{\pi}\angle{-90°},$$

$$A_3\angle\phi_3 = \frac{10\sqrt{2}}{3\pi}\angle{-135°},$$

and

$$A_4\angle\phi_4 = 0.$$

Step 3: With $RC = 2\times10^4 \times 10^{-4} = 2$ s, the generic phasor-domain transfer function of the circuit is

$$\begin{aligned}\mathbf{H}(\omega) &= \mathbf{V}_{\text{out}} \qquad (\text{with } \mathbf{V}_s = 1)\\ &= \frac{1}{1+j\omega RC}\\ &= \frac{1}{\sqrt{1+\omega^2R^2C^2}}\,e^{-j\tan^{-1}(\omega RC)}\\ &= \frac{1}{\sqrt{1+4\omega^2}}\,e^{-j\tan^{-1}(2\omega)}.\end{aligned} \tag{13.36}$$

Step 4: The time-domain output voltage is

$$\upsilon_{\text{out}}(t) = 2.5 + \sum_{n=1}^{\infty}\mathfrak{Re}\left\{A_n\,\frac{1}{\sqrt{1+4n^2\omega_0^2}}\,e^{j[n\omega_0 t+\phi_n-\tan^{-1}(2n\omega_0)]}\right\}. \tag{13.37}$$

Using the values of $A_n\angle\phi_n$ determined earlier for the first four terms and replacing ω_0 with its numerical value of $\pi/2$ rad/s, the expression becomes

$$\begin{aligned}\upsilon_{\text{out}}(t) &= 2.5 \\ &+ \frac{10\sqrt{2}}{\pi\sqrt{1+\pi^2}}\cos\left[\frac{\pi t}{2} - 45^\circ - \tan^{-1}(\pi)\right] \\ &+ \frac{10}{\pi\sqrt{1+4\pi^2}}\cos[\pi t - 90^\circ - \tan^{-1}(2\pi)] \\ &+ \frac{10\sqrt{2}}{3\pi\sqrt{1+9\pi^2}}\cos\left[\frac{3\pi t}{2} - 135^\circ - \tan^{-1}(3\pi)\right]\cdots \\ &= 2.5 + 1.37\cos\left(\frac{\pi t}{2} - 117^\circ\right) + 0.5\cos(\pi t - 171^\circ) \\ &+ 0.16\cos\left(\frac{3\pi t}{2} + 141^\circ\right)\cdots \text{ V.}\end{aligned}$$

The voltage response $\upsilon_{\text{out}}(t)$ is displayed in Fig. 13-8(c) and is computed using the series solution given by the preceding expression with $n_{\max} = 1000$.

Example 13-6: Three-Stage Phase Shifter

In Chapter 7, we showed that the phasor-domain transfer function of the three-stage phase shifter shown in Fig. 13-9(a) is given by Eq. (7.99) as

$$\mathbf{H}(\omega) = \frac{\mathbf{V}_{\text{out}}}{\mathbf{V}_{\text{s}}} = \frac{x^3}{(x^3 - 5x) + j(1 - 6x^2)},$$

where $x = \omega RC$. Determine the output response to the periodic waveform shown in Fig. 13-9(b) given that $RC = 1$ s.

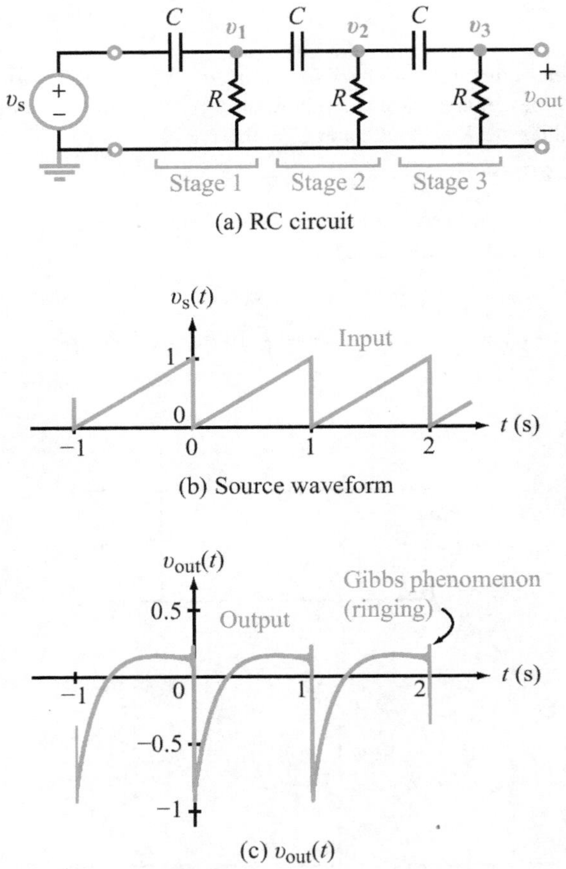

Figure 13-9: Circuit and plots for Example 13-6.

Solution:

Step 1: With $T = 1$ s, $\omega_0 = 2\pi/T = 2\pi$ rad/s, and $\upsilon_s(t) = t$ over [0, 1],

$$a_0 = \frac{1}{T}\int_0^T \upsilon_s(t)\,dt = \int_0^1 t\,dt = 0.5,$$

$$\begin{aligned}a_n &= \frac{2}{1}\int_0^1 t\cos 2n\pi t\,dt \\ &= 2\left[\frac{1}{(2n\pi)^2}\cos 2n\pi t + \frac{t}{2n\pi}\sin 2n\pi t\right]\Bigg|_0^1 = 0,\end{aligned}$$

$$\begin{aligned}b_n &= \frac{2}{1}\int_0^1 t\sin 2n\pi t\,dt \\ &= 2\left[\frac{1}{(2n\pi)^2}\sin 2n\pi t - \frac{t}{2n\pi}\cos 2n\pi t\right]\Bigg|_0^1 \\ &= -\frac{1}{n\pi}.\end{aligned}$$

Step 2: For $n \geq 1$,

$$A_n\angle\phi_n = 0 - jb_n = 0 + j\,\frac{1}{n\pi} = \frac{1}{n\pi}\angle 90^\circ \text{ V.}$$

Step 3: With $RC = 1$ and $x = \omega RC = \omega$, $\mathbf{H}(\omega)$ becomes

$$\mathbf{H}(\omega) = \frac{\omega^3}{(\omega^3 - 5\omega) + j(1 - 6\omega^2)}.$$

Step 4: With $\omega_0 = 2\pi$ rad/s, $\mathbf{H}(\omega = 0) = 0$, and $A_n = (1/n\pi)e^{j90°}$, the time-domain voltage is obtained by multiplying each term in the summation by its corresponding $e^{jn\omega_0 t} = e^{j2n\pi t}$, and then taking the real part of the entire expression:

$$\upsilon_{\text{out}}(t) =$$

$$\sum_{n=1}^{\infty} \Re\mathfrak{e}\left\{\frac{8n^2\pi^2}{[(2n\pi)((2n\pi)^2 - 5) + j(1 - 24n^2\pi^2)]} e^{j(2n\pi t + 90°)}\right\}.$$

Evaluating the first few terms of $\upsilon_{\text{out}}(t)$ leads to

$$\upsilon_{\text{out}}(t) = 0.25\cos(2\pi t + 137°) + 0.15\cos(4\pi t + 116°) + 0.10\cos(6\pi t + 108°) + \cdots.$$

A plot of $\upsilon_{\text{out}}(t)$ with 100 terms is displayed in Fig. 13-9(c).

Concept Question 13-9: What is the connection between the Fourier series solution method and the phasor-domain solution technique? (See CAD)

Concept Question 13-10: Application of the Fourier series method in circuit analysis relies on which fundamental property of the circuit? (See CAD)

Exercise 13-4: The RL circuit shown in Fig. E13.4(a) is excited by the square-wave voltage waveform of Fig. E13.4(b). Determine $\upsilon_{\text{out}}(t)$.

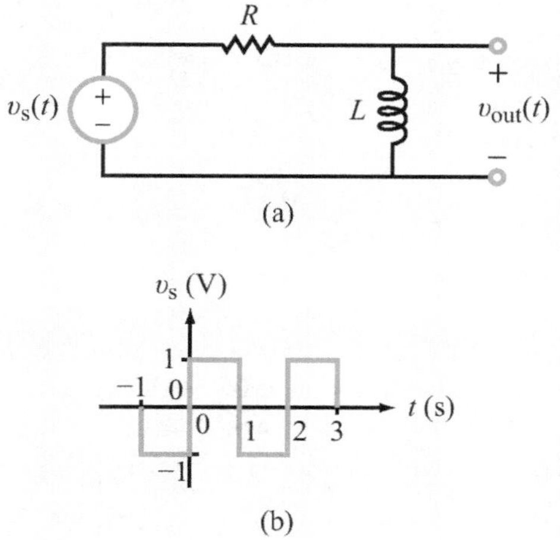

Figure E13.4

Answer:

$$\upsilon_{\text{out}}(t) = \sum_{\substack{n=1\\ n=\text{odd}}}^{\infty} \frac{4L}{\sqrt{R^2 + n^2\pi^2L^2}} \cos(n\pi t + \theta_n);$$

$$\theta_n = -\tan^{-1}\left(\frac{n\pi L}{R}\right).$$

(See CAD)

13-4 Average Power

If a circuit is excited by a periodic voltage or current of period T and associated fundamental angular frequency $\omega_0 = 2\pi/T$, then every segment of the circuit will exhibit a voltage across it (Fig. 13-10) characterized by a Fourier series of the form

$$\upsilon(t) = V_{\text{dc}} + \sum_{n=1}^{\infty} V_n \cos(n\omega_0 t + \phi_{\upsilon_n}), \tag{13.38}$$

where V_{dc} is the average value of $\upsilon(t)$, V_n is the amplitude of the nth harmonic, and ϕ_{υ_n} is the associated phase angle. Similarly, the current flowing into the circuit segment is also given by a Fourier series as

$$i(t) = I_{\text{dc}} + \sum_{m=1}^{\infty} I_m \cos(m\omega_0 t + \phi_{i_m}), \tag{13.39}$$

where similar definitions apply to I_{dc}, I_m, and ϕ_{i_m}. In Eq. (13.39) we used the integer index m instead of n in order to keep the two summations distinguishable from one another.

By denoting the current direction such that it is flowing into the (+) voltage terminal (Fig. 13-10), the passive sign

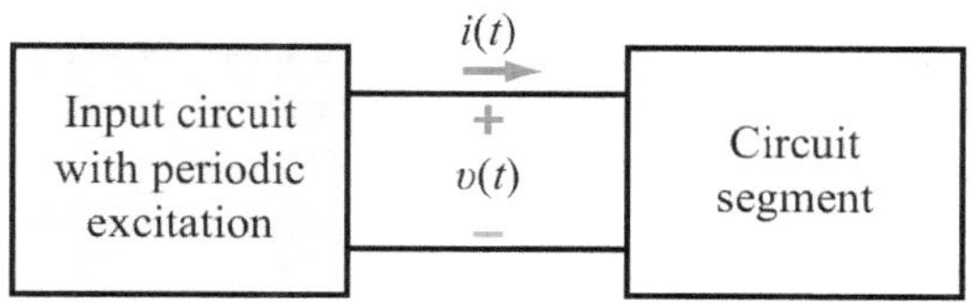

Figure 13-10: Voltage across and current into a circuit segment.

convention stipulates that the product υi represents power flow into the circuit segment. Hence, the average power is given by

$$\begin{aligned} P_{\text{av}} &= \frac{1}{T}\int_0^T \upsilon i\,dt = \frac{1}{T}\int_0^T V_{\text{dc}} I_{\text{dc}}\,dt \\ &+ \sum_{n=1}^{\infty} \frac{1}{T}\int_0^T V_n I_{\text{dc}} \cos(n\omega_0 t + \phi_{\upsilon_n})\,dt \\ &+ \sum_{m=1}^{\infty} \frac{1}{T}\int_0^T V_{\text{dc}} I_m \cos(m\omega_0 t + \phi_{i_m})\,dt \\ &+ \sum_{m=1}^{\infty}\sum_{n=1}^{\infty} \frac{1}{T}\int_0^T V_n I_m \cos(n\omega_0 t + \phi_{\upsilon_n}) \\ &\cdot \cos(m\omega_0 t + \phi_{i_m})\,dt. \end{aligned} \quad (13.40)$$

From Table 13-1, integral Property 2 states that

$$\int_0^T \cos(n\omega_0 t + \phi_0)\,dt = 0 \quad (13.41)$$

for any integer $n \geq 1$ and any constant phase angle ϕ_0. Consequently, terms 2 and 3 in Eq. (13.40) vanish. Moreover, the product of the two cosine functions inside the last term is expandable into

$$\begin{aligned} &\cos(n\omega_0 t + \phi_{\upsilon_n}) \cos(m\omega_0 t + \phi_{i_m}) \\ &= \frac{1}{2}\cos(\text{sum of arguments}) \\ &+ \frac{1}{2}\cos(\text{difference between arguments}) \\ &= \frac{1}{2}\cos[(n+m)\omega_0 t + \phi_{\upsilon_n} + \phi_{i_m}] \\ &+ \frac{1}{2}\cos[(n-m)\omega_0 t + \phi_{\upsilon_n} - \phi_{i_m}]. \end{aligned} \quad (13.42)$$

Performing the integration over the two terms of Eq. (13.42) yields zero values, except when $n = m$. Implementing these considerations in Eq. (13.40) leads to

$$P_{\text{av}} = V_{\text{dc}} I_{\text{dc}} + \frac{1}{2}\sum_{n=1}^{\infty} V_n I_n \cos(\phi_{\upsilon_n} - \phi_{i_n}). \quad (13.43)$$

We note that $\frac{1}{2} V_n I_n \cos(\phi_{\upsilon_n} - \phi_{i_n})$ represents the average power associated with frequency harmonic $n\omega_0$. Hence:

► The total average power is equal to the dc power ($V_{\text{dc}} I_{\text{dc}}$) plus the sum of the average ac powers associated with the fundamental frequency ω_0 and its harmonic multiples. ◄

Example 13-7: ac Power Fraction

The periodic voltage across a certain circuit truncated to the first three ac terms of its Fourier series is given by

$$\begin{aligned} \upsilon(t) &= 2 + 3\cos(4t + 30^\circ) + 1.5\cos(8t - 30^\circ) \\ &+ 0.5\cos(12t - 135^\circ)\ \text{V}, \end{aligned}$$

and the associated current flowing into the (+) voltage terminal of the circuit is

$$\begin{aligned} i(t) &= 60 + 10\cos(4t - 30^\circ) \\ &+ 5\cos(8t + 15^\circ) + 2\cos 12t\ \text{mA}. \end{aligned}$$

Determine the ac fraction of the average power.

Solution: Application of Eq. (13.43) yields

$$\begin{aligned} P_{\text{av}} &= 2 \times 60 + \frac{3 \times 10}{2}\cos(30^\circ + 30^\circ) \\ &+ \frac{1.5 \times 5}{2}\cos(-30^\circ - 15^\circ) + \frac{0.5 \times 2}{2}\cos(-135^\circ) \\ &= 120 + 7.5 + 2.65 - 0.353 = 129.80\ \text{W}. \end{aligned}$$

The ac fraction is

$$\frac{7.5 + 2.65 - 0.353}{129.8} = 7.55\%.$$

Exercise 13-5: What will the expression given by Eq. (13.43) simplify to if the associated circuit segment is (a) purely resistive or (b) purely reactive?

Answer: (a) $P_{\text{av}} = V_{\text{dc}} I_{\text{dc}} + \frac{1}{2}\sum_{n=1}^{\infty} V_n I_n$; because $\phi_{\upsilon_n} = \phi_{i_n}$. (b) $P_{\text{av}} = 0$, because for a capacitor, $I_{\text{dc}} = 0$ and $\phi_{\upsilon_n} - \phi_{i_n} = -90^\circ$; and for an inductor, $V_{\text{dc}} = 0$ and $\phi_{\upsilon_n} - \phi_{i_n} = 90^\circ$. (See CAD)

Technology Brief 31
Synthetic Biology

Whether amplifying, sensing, computing, or communicating, all of the circuits discussed in this book manipulate electric charge to process information. Voltage levels and current intensities all represent the collective properties of charges inside metals, insulators and semiconductors. The processing of information, however, can be accomplished in other media as well. Mechanical circuits (like the Babbage Engine, described in Technology Brief 1), optical circuits (where computation is accomplished by manipulating light), and chemical circuits (where the operators are the reactants and products of chemical reactions) all have been demonstrated or in use for many years. Recently, engineers have begun to make synthetic information processing circuits inside biological cells. This new branch of engineering grew out of biochemical engineering and is called **synthetic biology**. It promises to revolutionize the way we interact with biological systems.

In order to understand why synthetic biology is so powerful, and why it is so closely related to electrical engineering, we need to understand how biological cells process information. A cell, whether a free swimming bacterium or a human liver cell, is constantly transducing, storing and processing information from its environment. Cells produce molecules called **proteins**, each of which can perform a specific function on a specific molecule. They can be thought of as little molecular robots. For example, certain proteins on a cell's membrane act as sensors, detecting the presence of molecules in the liquid around the cell. These surface proteins can change the state of other proteins inside the cell, which in turn affect other proteins, and so on. These chains of chemical reactions are called **biochemical pathways**, and in this way the cell can adjust what molecules it produces based on what molecules are in its environment. If, say, the environment contains glucose, the cell's sensors can

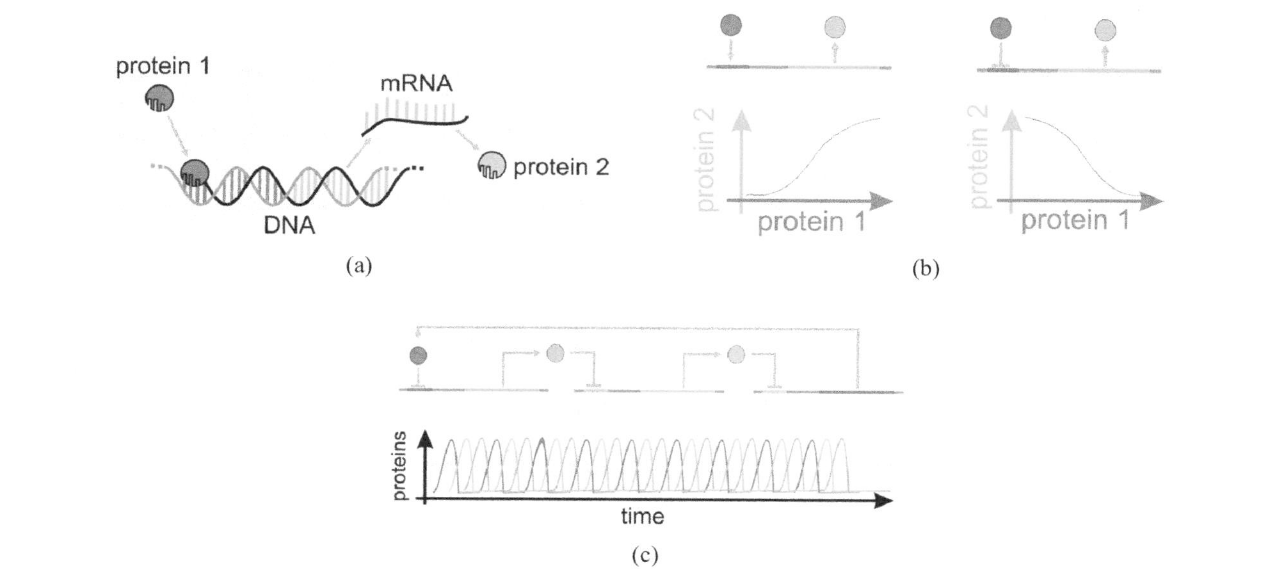

Figure TF31-1: *Building functions with genes and proteins.* (a) Consider a protein (red circle) which can bind to a certain, specific section of DNA within a cell. Upon binding to this sequence (shown in red), the production of another protein (Protein 2, in yellow) is affected; the gene which encodes for Protein 2 is shown as yellow base pairs in the DNA. In this example, increasing the amount of Protein 1 increases the production (and thus the amount) of Protein 2 in the cell. This motif is encountered often in cells. Cells ranging from bacteria to mammals contain relationships like these between proteins and genes. (b) The example on the left shows a simplified representation of (a) along with a plot showing how the amount of Protein 1 affects the amount of Protein 2 in the cell. The example on the right shows a similar cartoon, except in this case Protein 1 *inhibits* the production of Protein 2 (more Protein 1 in the cell causes less Protein 2). Notice how this right example behaves as a rudimentary protein-protein **inverter**. (c) An example of a synthetic gene circuit in which three of the inverters in (b) are linked so that a protein produced by one gene inhibits the production of the next gene. This produces a chemical oscillator known as a **repressilator**.

Figure TF31-2: By introducing new genes into *E. coli*, the bacteria were made sensitive to light. In this example, by a team from the University of Texas at Austin and the University of California, San Francisco, a thin film of bacteria was grown in a Petri dish and exposed to patterns of light with the message "Hello World." Each bacterium in the film responded by changing color depending on the light level it was exposed to.

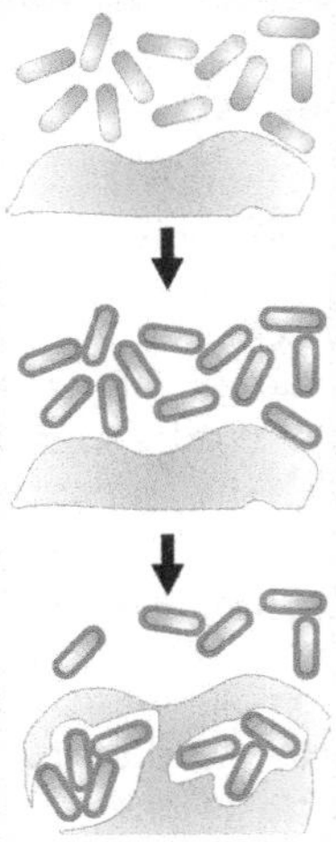

Figure TF31-3: Researchers at the University of California, Berkeley, and the University of California, San Francisco have devised synthetic pathways that may one day allow engineered bacteria to invade and kill tumors. (*top*) Modified *E. coli* bacteria at normal cell densities or oxygen conditions behave normally. (*middle*) Upon encountering low oxygen environments (associated with tumors) or high cell densities, the modified bacteria express invasin, a molecule that adheres to tumor cells and tricks the cells into absorbing the bacteria. (*bottom*) Once inside the tumor, the bacteria begin to invade and destroy it.

detect this and begin producing proteins that enable the cell to use glucose as fuel.

A key component of this regulation process depends on the *genes* the cell possesses. Although a discussion of genes is well beyond the scope of this book, for our purposes a cell contains a set of molecules called genes, which store descriptions of all the proteins it can make; a given gene will usually *encode* a single type of protein. In many ways, you can think of the genes as the cell's software. When we say a cell *expresses* a gene, we mean it has used the information in that gene to make the gene's protein. What is important here is that the biochemical pathways determine which genes are expressed. Thus, the surface proteins that detect glucose, for example, cause the cell to express the genes that encode the proteins that consume glucose. A single cell can make thousands of different proteins, which can interact with each other as well as with the cell's genes.

With the advent of modern molecular biology, our knowledge of the cell's pathways has expanded dramatically. In the latter half of the 20th century, biochemical engineers put this information to work by growing cells from many different species and modifying them to perform many useful functions. Waste water treatment, drug production, food additive production, and many other useful chemical processes are now carried out using cells.

Even more recently, synthetic biologists have begun to build *information processing* circuits into cells (Fig. TF31-1). These engineers hope to design components and circuits that perform many of the same functions you have studied in this book, using the cell's biochemical pathways! Amplification, logical functions, clocks, memory, multi-channel communication, sensing, and even rudimentary "software" programs are all being developed using cells, proteins, and genes instead of circuit boards and solid-state materials. If these efforts succeed, our ability to interact and guide the behavior of existing cells and to build entirely new types of cells with human-made programs will have a profound impact on the world of science and technology (Fig. TF31-2 and Fig. TF31-3). Along with these challenges comes a great responsibility to understand how our inventions can affect the natural world. What is very exciting is that synthetic biologists are realizing that many of the concepts electrical engineers developed for their electric circuits—noise, bandwidth, linear analysis, circuit diagrams, etc.—are proving useful for designing biological circuits!

13-5 Fourier Transform

The Fourier series is a perfectly suitable construct for representing periodic functions, but what about nonperiodic functions? The pulse-train waveform shown in Fig. 13-11(a) consists of a sequence of rectangular pulses—with a width of $\tau = 2$ s. The period $T = 4$ s. In part (b) of the same figure, the individual pulses have the same shape as before, but T has been increased to 7 s. So long as T is finite, both waveforms are amenable to representation by the Fourier series, but what would happen if we let $T \to \infty$, ending up with the single pulse shown in Fig. 13-11(c)? Can we then represent the no-longer periodic pulse by a Fourier series? We will discover shortly that as $T \to \infty$, the summation $\sum_{n=1}^{\infty}$ in the Fourier series evolves into a continuous integral, which we call the *Fourier transform*.

> ▶ When analyzing electric circuits, we apply the Fourier series representation if the excitation is periodic in character, and we use the Fourier transform representation if the excitation is nonperiodic. ◀

Does that mean that we can use both the Laplace transform (Chapter 12) and the Fourier transform techniques to analyze circuits containing nonperiodic sources? If so, which of the two transforms should we use, and why? We address these questions later (Section 13-7), after formally introducing the Fourier transform and discussing some of its salient features.

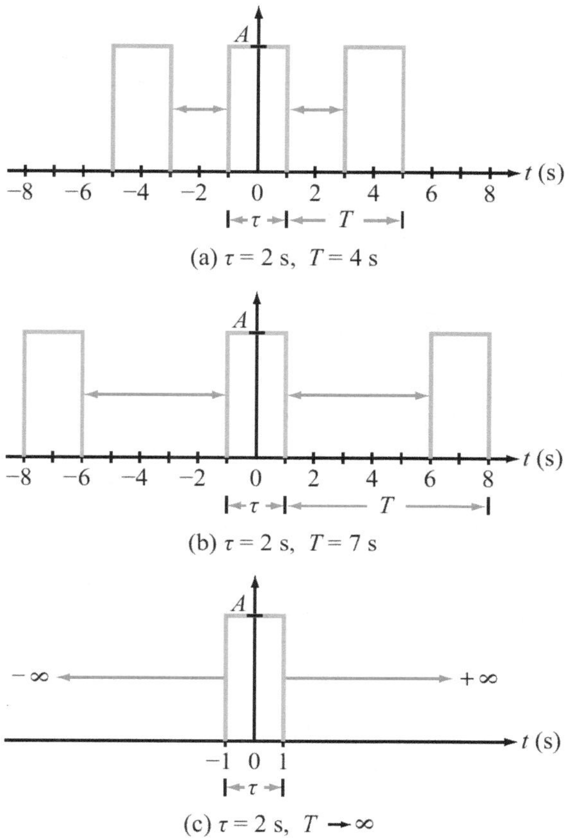

Figure 13-11: The single pulse in (c) is equivalent to a periodic pulse train with $T = \infty$.

13-5.1 Exponential Fourier Series

According to Eq. (13.15), a periodic function of period T and corresponding fundamental frequency $\omega_0 = 2\pi/T$ can be represented by the series

$$f(t) = a_0 + \sum_{n=1}^{\infty} a_n \cos n\omega_0 t + b_n \sin n\omega_0 t. \qquad (13.44)$$

Sine and cosine functions can be converted into complex exponentials via Euler's identity:

$$\cos n\omega_0 t = \frac{1}{2}(e^{jn\omega_0 t} + e^{-jn\omega_0 t}), \qquad (13.45a)$$

$$\sin n\omega_0 t = \frac{1}{j2}(e^{jn\omega_0 t} - e^{-jn\omega_0 t}). \qquad (13.45b)$$

Upon inserting Eqs. (13.45a and b) into Eq. (13.44), we have

$$f(t) = a_0 + \sum_{n=1}^{\infty}\left[\frac{a_n}{2}(e^{jn\omega_0 t} + e^{-jn\omega_0 t}) + \frac{b_n}{j2}(e^{jn\omega_0 t} - e^{-jn\omega_0 t})\right]$$

$$= a_0 + \sum_{n=1}^{\infty}\left[\left(\frac{a_n - jb_n}{2}\right)e^{jn\omega_0 t} + \left(\frac{a_n + jb_n}{2}\right)e^{-jn\omega_0 t}\right]$$

$$= a_0 + \sum_{n=1}^{\infty}[\mathbf{c}_n e^{jn\omega_0 t} + \mathbf{c}_{-n} e^{-jn\omega_0 t}], \qquad (13.46)$$

Table 13-3: **Fourier series representations for a periodic function $f(t)$.**

Cosine/Sine	Amplitude/Phase	Complex Exponential
$f(t) = a_0 + \sum_{n=1}^{\infty}(a_n \cos n\omega_0 t + b_n \sin n\omega_0 t)$	$f(t) = a_0 + \sum_{n=1}^{\infty} A_n \cos(n\omega_0 t + \phi_n)$	$f(t) = \sum_{n=-\infty}^{\infty} \mathbf{c}_n e^{jn\omega_0 t}$
$a_0 = \frac{1}{T}\int_0^T f(t)\, dt$	$A_n e^{j\phi_n} = a_n - jb_n$	$\mathbf{c}_n = \lvert\mathbf{c}_n\rvert e^{j\phi_n};\ \mathbf{c}_{-n} = \mathbf{c}_n^*$
$a_n = \frac{2}{T}\int_0^T f(t) \cos n\omega_0 t\, dt$	$A_n = \sqrt{a_n^2 + b_n^2}$	$\lvert\mathbf{c}_n\rvert = A_n/2;\ c_0 = a_0$
$b_n = \frac{2}{T}\int_0^T f(t) \sin n\omega_0 t\, dt$	$\phi_n = \begin{cases} -\tan^{-1}\left(\frac{b_n}{a_n}\right), & a_n > 0 \\ \pi - \tan^{-1}\left(\frac{b_n}{a_n}\right), & a_n < 0 \end{cases}$	$\mathbf{c}_n = \frac{1}{T}\int_0^T f(t)\, e^{-jn\omega_0 t}\, dt$
$a_0 = c_0;\ a_n = A_n \cos\phi_n;\ b_n = -A_n \sin\phi_n;\ \mathbf{c}_n = \frac{1}{2}(a_n - jb_n)$		

where we introduced the complex coefficients

$$\mathbf{c}_n = \frac{a_n - jb_n}{2}$$

and

$$\mathbf{c}_{-n} = \frac{a_n + jb_n}{2} = \mathbf{c}_n^*. \tag{13.47}$$

As the index n is incremented from 1 to ∞, the second term in Eq. (13.46) generates the series

$$\mathbf{c}_{-1}e^{-j\omega_0 t} + \mathbf{c}_{-2}e^{-j2\omega_0 t} + \cdots,$$

which also can be generated by $\mathbf{c}_n e^{jn\omega_0 t}$ with n incremented from -1 to $-\infty$. This equivalence allows us to express $f(t)$ in the compact *exponential form* as

$$f(t) = \sum_{n=-\infty}^{\infty} \mathbf{c}_n e^{jn\omega_0 t}, \tag{13.48}$$

(exponential representation)

where

$$c_0 = a_0, \tag{13.49}$$

and the range of n has been expanded to $[-\infty, \infty]$. For all coefficients $\mathbf{c}_n$ including c_0, it is easy to show that

$$\mathbf{c}_n = \frac{1}{T}\int_{-T/2}^{T/2} f(t)\, e^{-jn\omega_0 t}\, dt. \tag{13.50}$$

► Even though the integration limits indicated in Eq. (13.50) are from $-T/2$ to $T/2$, they can be chosen arbitrarily so long as the upper limit exceeds the lower limit by exactly T. ◄

For easy reference, Table 13-3 provides a summary of the relationships associated with all three Fourier series representations introduced in this chapter, namely the cosine/sine, amplitude/phase, and complex exponential.

Example 13-8: Pulse Train

Obtain the Fourier series exponential representation for the pulse-train waveform displayed in Fig. 13-11(a) in terms of the pulse width τ and the period T. Evaluate and plot the line spectrum of $\lvert\mathbf{c}_n\rvert$ for $A = 10$ and $\tau = 1$ s for each of the following values of T: 5 s, 10 s, and 20 s.

Solution: Over a period T extending from $-T/2$ to $T/2$,

$$f(t) = \begin{cases} A & \text{for } -\tau/2 \le t \le \tau/2, \\ 0 & \text{otherwise.} \end{cases}$$

With the integration domain chosen to be from $-T/2$ to $T/2$, Eq. (13.50) gives

$$\begin{aligned} \mathbf{c}_n &= \frac{1}{T} \int_{-T/2}^{T/2} f(t)\, e^{-jn\omega_0 t}\, dt \\ &= \frac{1}{T} \int_{-\tau/2}^{\tau/2} A e^{-jn\omega_0 t}\, dt \\ &= \frac{A}{-jn\omega_0 T} e^{-jn\omega_0 t} \Big|_{-\tau/2}^{\tau/2} \\ &= \frac{2A}{n\omega_0 T} \left[\frac{e^{jn\omega_0 \tau/2} - e^{-jn\omega_0 \tau/2}}{2j} \right]. \end{aligned} \tag{13.51}$$

The quantity inside the square bracket matches the form of one of Euler's formulas, namely

$$\sin x = \frac{e^{jx} - e^{-jx}}{2j}. \tag{13.52}$$

Hence, Eq. (13.51) can be rewritten in the form

$$\begin{aligned} \mathbf{c}_n &= \frac{2A}{n\omega_0 T} \sin(n\omega_0 \tau/2) \\ &= \frac{A\tau}{T} \frac{\sin(n\omega_0 \tau/2)}{(n\omega_0 \tau/2)} = \frac{A\tau}{T} \operatorname{sinc}(n\omega_0 \tau/2), \end{aligned} \tag{13.53}$$

where in the last step we introduced the ***sinc function***, which is defined as*

$$\operatorname{sinc}(x) = \frac{\sin x}{x}. \tag{13.54}$$

Among the important properties of the sinc function are the following:

(a) When its argument is zero, the sinc function is equal to 1,

$$\operatorname{sinc}(0) = \frac{\sin(x)}{x} \Big|_{x=0} = 1. \tag{13.55}$$

Verification of this property can be established by applying l'Hôpital's rule to Eq. (13.54) and then setting $x = 0$.

(b) Since $\sin(m\pi) = 0$ for any integer value of m, the same is true for the sinc function,

$$\operatorname{sinc}(m\pi) = 0, \qquad m \neq 0. \tag{13.56}$$

(c) Because both $\sin x$ and x are odd functions, their ratio is an even function. Hence, the sinc function possesses even symmetry relative to the vertical axis. Consequently,

$$\mathbf{c}_n = \mathbf{c}_{-n}. \tag{13.57}$$

Evaluation of Eq. (13.53) with $A = 10$ leads to the line spectra displayed in Fig. 13-12. The general shape of the envelope is dictated by the sinc function, exhibiting a symmetrical pattern with a peak at $n = 0$, a major lobe extending between $n = -T/\tau$ and $n = T/\tau$, and progressively smaller amplitude lobes on both sides. The density of spectral lines depends on the ratio of T/τ, so in the limit as $T \to \infty$, the line spectrum becomes a continuum.

13-5.2 Nonperiodic Waveforms

In Example 13-8, we noted that as the period $T \to \infty$ the periodic function becomes nonperiodic and the associated line spectrum evolves from one containing discrete lines into a continuum. We now explore this evolution in mathematical terms, culminating in a definition for the Fourier transform of a nonperiodic function. To that end, we begin with the pair of expressions given by Eqs. (13.48) and (13.50), namely

$$f(t) = \sum_{n=-\infty}^{\infty} \mathbf{c}_n e^{jn\omega_0 t} \tag{13.58a}$$

and

$$\mathbf{c}_n = \frac{1}{T} \int_{-T/2}^{T/2} f(t)\, e^{-jn\omega_0 t}\, dt. \tag{13.58b}$$

These two quantities form a complementary pair with $f(t)$ defined in the continuous time domain and $\mathbf{c}_n$ defined in the discrete frequency domain as $n\omega_0$, with $\omega_0 = 2\pi/T$. For a given value of T, the nth frequency harmonic is at $n\omega_0$ and the next harmonic after that is at $(n+1)\omega_0$. Hence, the ***spacing between adjacent harmonics*** is

$$\Delta\omega = (n+1)\omega_0 - n\omega_0 = \omega_0 = \frac{2\pi}{T}. \tag{13.59}$$

*An alternative definition for the sinc function is $\operatorname{sinc}(x) = \sin(\pi x)/(\pi x)$, and it is used in MATLAB and MathScript. Both definitions are used in the literature, and in this book we adhere to the one given by Eq. (13.54).

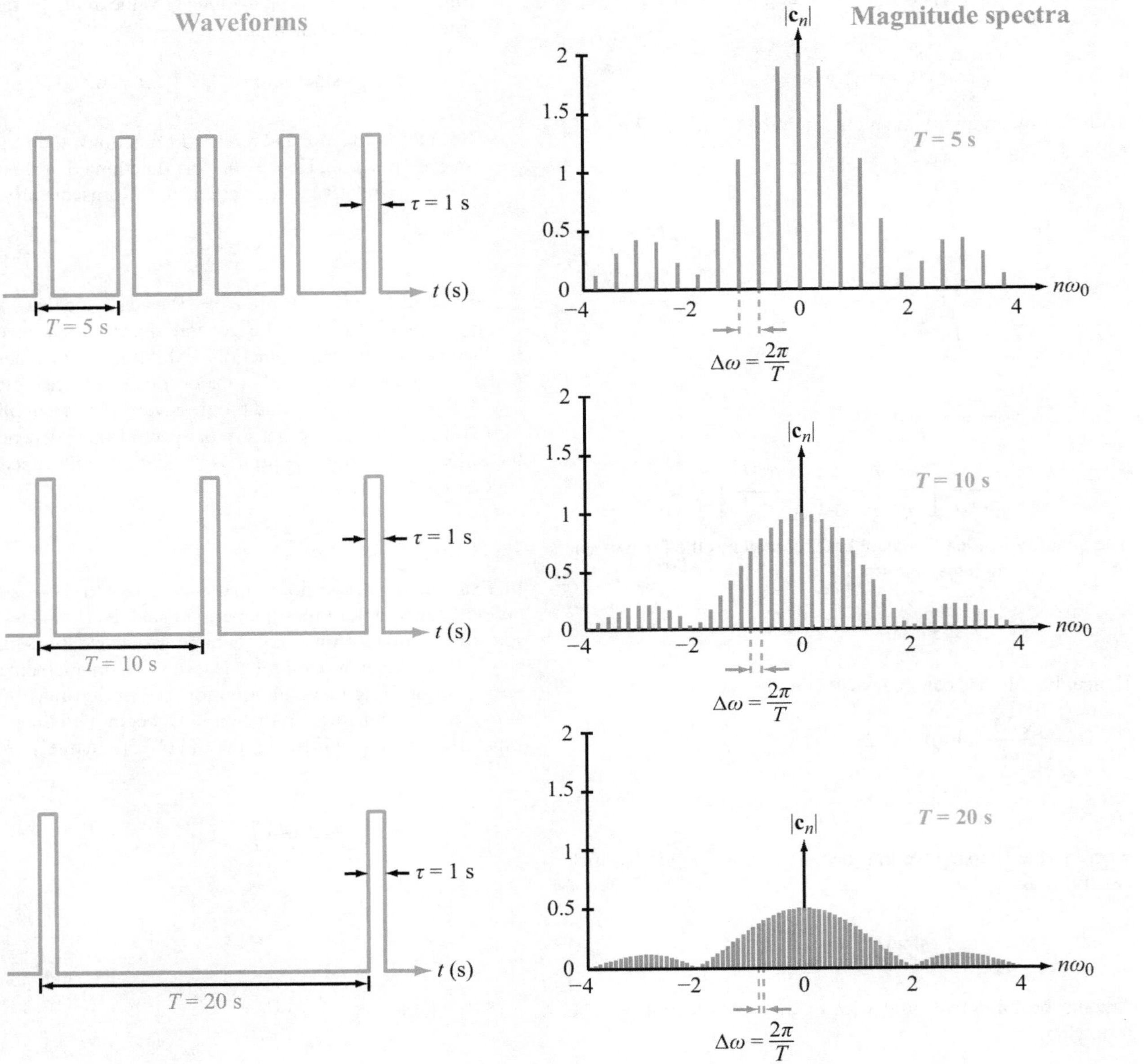

Figure 13-12: Line spectra for pulse trains with $T/\tau = 5$, 10, and 20.

If we insert Eq. (13.58b) into Eq. (13.58a) and replace $1/T$ with $\Delta\omega/2\pi$, we get

$$f(t) = \sum_{n=-\infty}^{\infty} \left[\frac{1}{2\pi} \int_{-T/2}^{T/2} f(t)\, e^{-jn\omega_0 t}\, dt \right] e^{jn\omega_0 t}\, \Delta\omega. \tag{13.60}$$

As $T \to \infty$, $\Delta\omega \to d\omega$ and $n\omega_0 \to \omega$, and the sum becomes a continuous integral:

$$f(t) = \frac{1}{2\pi} \int_{-\infty}^{\infty} \left[\int_{-\infty}^{\infty} f(t)\, e^{-j\omega t}\, dt \right] e^{j\omega t}\, d\omega. \tag{13.61}$$

Given this new arrangement, we now are ready to offer formal definitions for the Fourier transform $\mathbf{F}(\omega)$ and its inverse transform $f(t)$:

$$\mathbf{F}(\omega) = \mathcal{F}[f(t)] = \int_{-\infty}^{\infty} f(t)\, e^{-j\omega t}\, dt \qquad (13.62a)$$

and

$$f(t) = \mathcal{F}^{-1}[\mathbf{F}(\omega)] = \frac{1}{2\pi} \int_{-\infty}^{\infty} \mathbf{F}(\omega)\, e^{j\omega t}\, d\omega, \qquad (13.62b)$$

where $\mathcal{F}[f(t)]$ is a shorthand notation for "the Fourier transform of $f(t)$," and similarly, $\mathcal{F}^{-1}[\mathbf{F}(\omega)]$ represents the inverse operation. Occasionally, we also may use the symbolic form

$$f(t) \longleftrightarrow \mathbf{F}(\omega).$$

Example 13-9: Rectangular Pulse

Determine the Fourier transform of the solitary rectangular pulse shown in Fig. 13-13(a) and then plot its *magnitude spectrum* $|\mathbf{F}(\omega)|$ for $A = 5$ and $\tau = 1$ s.

Solution: Application of Eq. (13.62a) with

$$f(t) = \text{rect}(t/\tau) = A$$

over the integration interval $[-\tau/2, \tau/2]$ leads to

$$\mathbf{F}(\omega) = \int_{-\tau/2}^{\tau/2} Ae^{-j\omega t}\, dt = \frac{A}{-j\omega}\, e^{-j\omega t}\Big|_{-\tau/2}^{\tau/2}$$

$$= A\tau\, \frac{\sin \omega\tau/2}{(\omega\tau/2)} = A\tau \text{ sinc}\left(\frac{\omega\tau}{2}\right). \qquad (13.63)$$

The *frequency spectrum* of $|\mathbf{F}(\omega)|$ is displayed in Fig. 13-13(b) for the specified values of $A = 5$ and $\tau = 1$ s. The *nulls* in the spectrum occur when the argument of the sinc function is a multiple of $\pm\pi$ (rad/s), which in this specific case correspond to ω equal to multiples of 2π (rad/s).

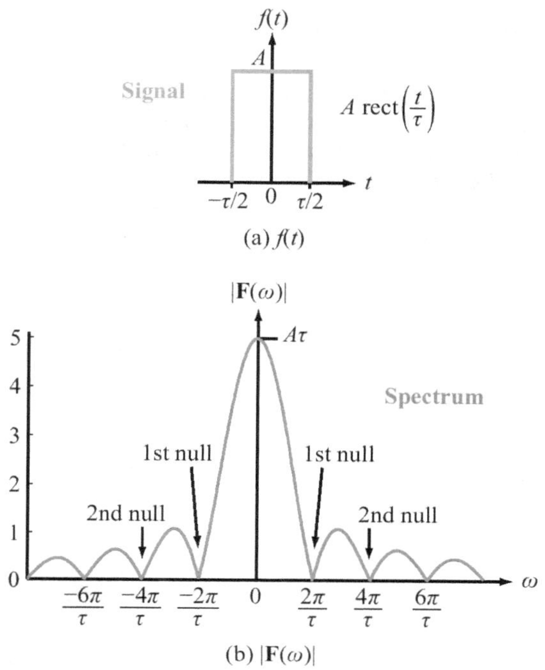

Figure 13-13: (a) Rectangular pulse of amplitude A and width τ; (b) frequency spectrum of $|\mathbf{F}(\omega)|$ for $A = 5$ and $\tau = 1$ s.

Concept Question 13-11: For the cosine/sine and amplitude/phase Fourier series representations, the summation extends from $n = 1$ to $n = \infty$. What are the limits on the summation for the complex exponential representation? (See CAD)

Concept Question 13-12: What is a sinc function, and what are its primary properties? Why is sinc(0) = 1? (See CAD)

Concept Question 13-13: What is the functional form for the Fourier transform $\mathbf{F}(\omega)$ of a rectangular pulse of amplitude 1 and duration τ? (See CAD)

Exercise 13-6: For a single rectangular pulse of width τ, what is the spacing $\Delta\omega$ between first nulls? If τ is very wide, will its frequency spectrum be narrow and peaked or wide and gentle?

Answer: $\Delta\omega = 4\pi/\tau$. Wide τ leads to narrow spectrum. (See CAD)

Technology Brief 32
Brain-Machine Interfaces (BMI)

In most vertebrates (like you), nerves extend from the brain (the *central nervous system*), through the spinal cord and out to your many organs (the nerves that lie outside the brain are collectively called the *peripheral nervous system*). Peripheral nerves carry information in both directions. On the one hand, peripheral neurons can fire at the behest of neurons in the brain and trigger muscle contraction or chemical release (via certain glands); on the other hand, sensor cells in the periphery can cause peripheral neurons to fire, sending signals to the brain to indicate pain, temperature, pressure, etc. Most of the peripheral neurons pass through the spinal cord; injuries to the spinal cord can be very dangerous, as trauma and inflammation can sever these connections, leading to paralysis, lack of sensory function, etc.

There is, of course, a long history of medical and scientific approaches to helping individuals afflicted with *motor dysfunctions* (whether due to trauma or congenital effects). Among these is the use of *prosthetic devices* that can supplement or replace lost function: prosthetic arms, prosthetic legs, advanced wheelchairs and exoskeletons have all been developed to aid those with motor problems (Fig. TF32-1). Historically, the way to drive these prosthetics is either by making use of motor functions that a patient still has (using hands to drive a wheelchair or sucking on a straw to drive a keyboard, for example) or reading signals from non-damaged peripheral nerves (recording from a pectoral muscle nerve, for example, to drive a robot elbow).

In the last decade or so, a slightly different paradigm has arisen that—while still in its infancy—promises a radical new way to communicate with prosthetics. The basic idea is to *directly record from neurons in the brain*, use those signals to drive a controller and communicate the

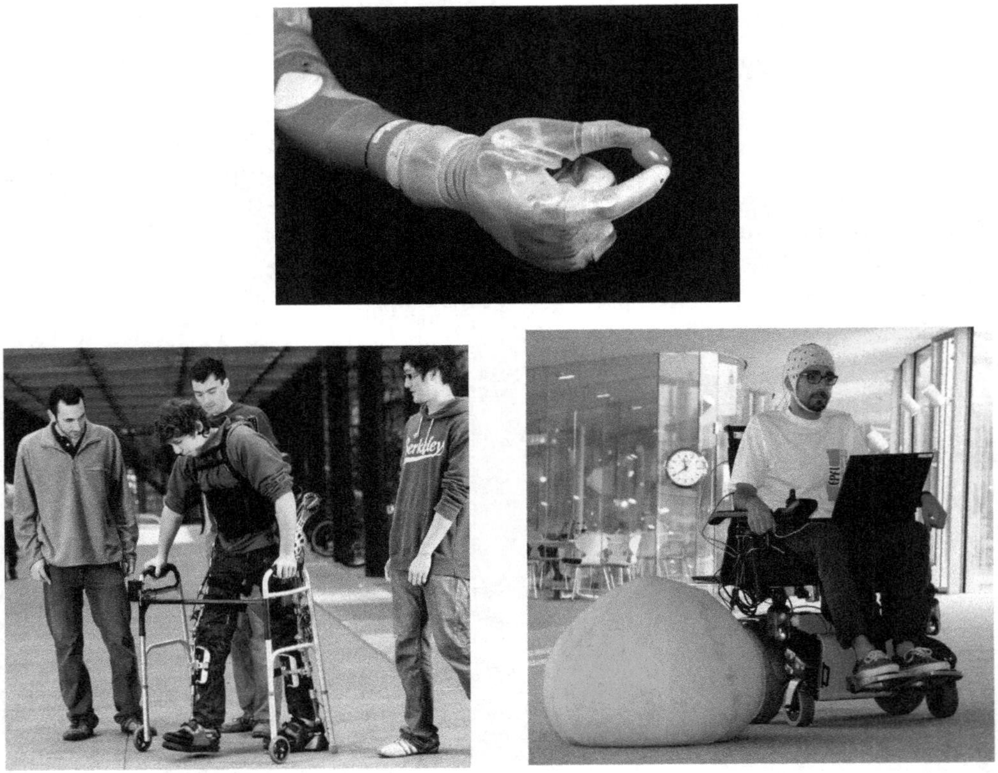

Figure TF32-1: A variety of existing, functional prosthetics. (a) The "Luke arm" built by DEKA Corp.; (b) an exoskeleton built at the Kazerooni Lab, University of California, Berkeley; (c) an EEG-controlled wheelchair, developed by José del R. Millán's group at the École Polytechnique Fédérale de Lausanne (EPFL, Switzerland).

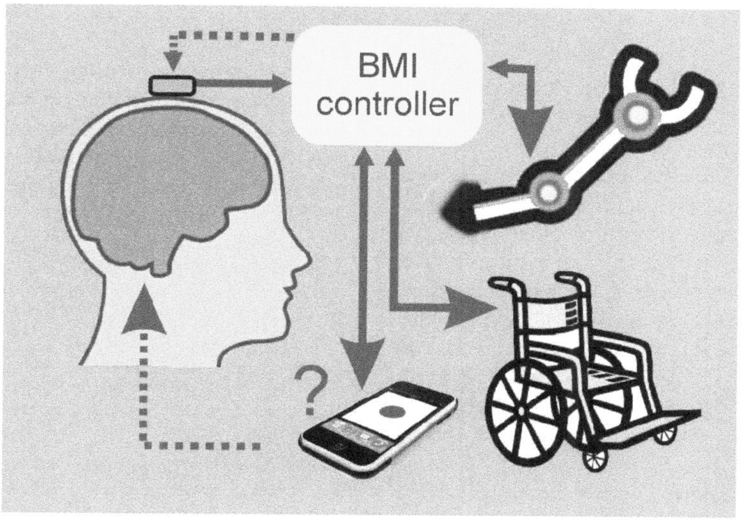

Figure TF32-2: The basic BMI loop depends on (a) neural recording, (b) a computational device or controller that maps neural signals to control signals to the prosthetics, and (c) feedback to the user or patient. Perhaps one day BMIs will even drive the use of portable consumer devices!

control signals to a prosthetic (Fig. TF32-2). In a sense, a computer—via a neural recording interface—records signals directly from the brain and uses them, without a spinal cord, to directly drive a robot prosthetic. This arena is currently seeing something of a gold rush as scientific results over the last 10 years and a medical trial (*BrainGate*) have encouraged researchers to explore and develop new technologies. Most central nervous system recording for BMI applications involves implanting arrays of ***recording electrodes*** (see Section 4-12) into the motor cortex of the subject. This is a very dangerous procedure (which involves a craniotomy), so research in humans has been limited to individuals with severe dysfunctions for whom the risk is appropriate. Once inserted, the subject trains over days and weeks to drive the prosthetic via the electrode array. Among the many remarkable findings in the recent BMI literature is that the neurons into which the recording array is inserted themselves learn to modify their firing behavior as the subject learns to use the BMI! That is, although scientists initially focused on what control algorithm would best decode the neural signals to drive say, a robot arm, they soon found—to their surprise—that the brain itself would learn to use whatever algorithm the controller employed. Researchers could even change algorithms and the subject could relearn the task, eventually able to switch between controllers.

Many challenges remain and it is an area of heavy overlap between electrical engineering, computer science, and neuroscience. Making electrode arrays that last an appreciable fraction of a patient's lifetime is still an unsolved problem: recording arrays typically fail after a few years. It is not at all clear what signals are optimal to drive a prosthetic nor which technology (or energy modality) is optimal for a long-lasting implant; many approaches are currently being explored. It is not known what the limits of such control are: could a subject be trained to operate a complex, multi-parameter non-motor task, for example (like imagining speech or interfacing in complex ways to a tablet)? Ultra-low power electrical recording front ends and ultra-low power radios are another area of intense study, as these systems must ultimately be miniaturized and implanted into a person as unobtrusively as possible. Lastly, ethical issues abound, ranging from the acceptability of animal testing to the possible enhancement possibilities. Some of this remains squarely in the realm of science fiction, but there is no doubt these approaches are a possible route to helping people with severe motor dysfunctions, making it a worthwhile endeavor in which EE's are making very big contributions.

13-5.3 Convergence of the Fourier Integral

Not every function $f(t)$ has a Fourier transform. The Fourier transform $\mathbf{F}(\omega)$ exists if the Fourier integral given by Eq. (13.62a) converges to a finite number or to an equivalent expression, but as we shall discuss shortly, it also may exist even if the Fourier integral does not converge. Convergence depends on the character of $f(t)$ over the integration range $[-\infty, \infty]$. By character, we mean (1) whether or not $f(t)$ exhibits infinite discontinuities and (2) how $f(t)$ behaves as t approaches $\pm\infty$. As a general rule, the Fourier integral does converge if $f(t)$ has no discontinuities (i.e., it is single-valued) and the integral of its absolute magnitude is finite (***absolutely integrable***). That is,

$$\int_{-\infty}^{\infty} |f(t)|\, dt < \infty. \qquad (13.64)$$

A function $f(t)$ still can have a Fourier transform—even if it has discontinuities so long as those discontinuities are bounded. The step function $A\ u(t)$ exhibits a finite discontinuity at $t = 0$ if A is finite.

The stated conditions for the existence of the Fourier transform are sufficient but not necessary conditions. In other words, some functions may still have transforms even though their Fourier integrals do not converge. Among such functions are the constant $f(t) = A$ and the unit step function $f(t) = A\ u(t)$, both of which represent important excitation waveforms in linear circuits. To realize the Fourier transform of a function whose transform exists but its Fourier integral does not converge, we need to employ an indirect approach. The approach entails the following ingredients:

(a) If $f(t)$ is a function whose Fourier integral does not converge, we select a related second function $f_\varepsilon(t)$ whose functional form includes a parameter ε. When allowed to approach a certain limit, $f_\varepsilon(t)$ becomes identical to $f(t)$.

(b) The choice of function $f_\varepsilon(t)$ should be such that its Fourier integral does converge, and therefore, $f_\varepsilon(t)$ has a definable Fourier transform $\mathbf{F}_\varepsilon(\omega)$.

(c) By taking parameter ε in the expression for $\mathbf{F}_\varepsilon(\omega)$ to its limit, $\mathbf{F}_\varepsilon(\omega)$ reduces to the transform $\mathbf{F}(\omega)$ corresponding to the original function $f(t)$.

This procedure is illustrated through some of the examples presented in the next section.

13-6 Fourier Transform Pairs

In this section, we develop fluency in how to move back and forth between the time domain and the ω domain. We will learn how to circumvent the convergence issues we noted earlier in Section 13-5.3, and in the process, we will identify a number of useful properties of the Fourier transform.

13-6.1 Linearity Property

If

$$f_1(t) \leftrightarrow \mathbf{F}_1(\omega)$$

and

$$f_2(t) \leftrightarrow \mathbf{F}_2(\omega),$$

then

$$K_1\, f_1(t) + K_2\, f_2(t) \leftrightarrow K_1\, \mathbf{F}_1(\omega) + K_2\, \mathbf{F}_2(\omega), \qquad (13.65)$$

(linearity property)

(superposition)

where K_1 and K_2 are constants. Proof of Eq. (13.65) is easily ascertained through the application of Eq. (13.62a).

13-6.2 Fourier Transform of $\delta(t - t_0)$

By Eq. (13.62a), the Fourier transform of $\delta(t - t_0)$ is given by

$$\mathbf{F}(\omega) = \mathcal{F}[\delta(t - t_0)] = \int_{-\infty}^{\infty} \delta(t - t_0) e^{-j\omega t}\, dt$$

$$= e^{-j\omega t}\Big|_{t=t_0} = e^{-j\omega t_0}. \qquad (13.66)$$

Hence,

$$\delta(t - t_0) \leftrightarrow e^{-j\omega t_0} \qquad (13.67a)$$

and

$$\delta(t) \leftrightarrow 1. \qquad (13.67b)$$

Thus, a unit impulse function $\delta(t)$ generates a constant of unit amplitude that extends over $[-\infty, \infty]$ in the ω domain, as shown in Fig. 13-14(a), and vice versa.

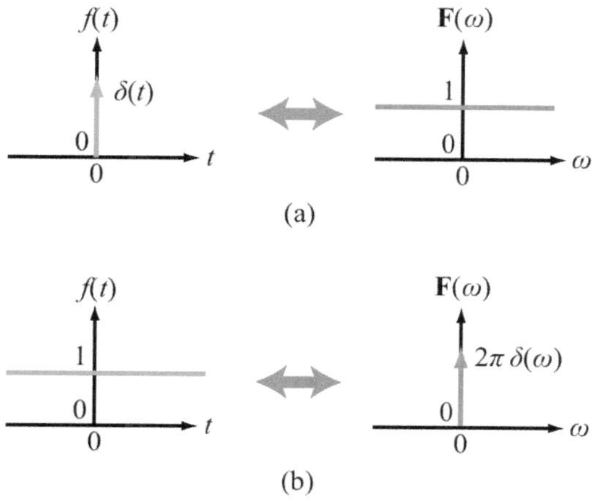

Figure 13-14: (a) The Fourier transform of $\delta(t)$ is 1 and (b) the Fourier transform of 1 is $2\pi\ \delta(\omega)$.

13-6.3 Shift Properties

By Eq. (13.62b), the inverse Fourier transform of $\mathbf{F}(\omega) = \delta(\omega - \omega_0)$ is

$$f(t) = \mathcal{F}^{-1}[\delta(\omega - \omega_0)] = \frac{1}{2\pi} \int_{-\infty}^{\infty} \delta(\omega - \omega_0)\, e^{j\omega t}\, d\omega = \frac{e^{j\omega_0 t}}{2\pi}.$$

Hence,

$$e^{j\omega_0 t} \quad \longleftrightarrow \quad 2\pi\ \delta(\omega - \omega_0) \qquad (13.68a)$$

and

$$1 \quad \longleftrightarrow \quad 2\pi\ \delta(\omega). \qquad (13.68b)$$

Comparison of the plots in Fig. 13-14(a) and (b) demonstrates the correspondence between the time domain and the ω domain: an impulse $\delta(t)$ in the time domain generates a uniform spectrum in the frequency domain; conversely, a uniform (constant) waveform in the time domain generates an impulse $\delta(\omega)$ in the frequency domain. By the same token, a rectangular pulse in the time domain generates a sinc pattern in the frequency domain, and a sinc pulse in the time domain generates a rectangular spectrum in the frequency domain, as illustrated by Fig. 13-15.

It is straightforward to show that the result given by Eq. (13.68a) can be generalized to

$$e^{j\omega_0 t}\ f(t) \quad \longleftrightarrow \quad \mathbf{F}(\omega - \omega_0), \qquad (13.69)$$

(frequency-shift property)

which is known as the *frequency-shift property* of the Fourier transform. It states that multiplication of a function $f(t)$ by $e^{j\omega_0 t}$ in the time domain corresponds to shifting the Fourier transform of $f(t)$ by ω_0 along the ω axis. The converse of the frequency-shift property is the *time-shift property* given by

$$f(t - t_0) \quad \longleftrightarrow \quad e^{-j\omega t_0}\ \mathbf{F}(\omega). \qquad (13.70)$$

(time-shift property)

13-6.4 Fourier Transform of $\cos \omega_0 t$

By Euler's identity,

$$\cos \omega_0 t = \frac{e^{j\omega_0 t} + e^{-j\omega_0 t}}{2}.$$

In view of Eq. (13.68a),

$$\mathbf{F}(\omega) = \mathcal{F}\left[\frac{e^{j\omega_0 t}}{2} + \frac{e^{-j\omega_0 t}}{2}\right] = \pi\ \delta(\omega - \omega_0) + \pi\ \delta(\omega + \omega_0).$$

Hence,

$$\cos \omega_0 t \quad \longleftrightarrow \quad \pi[\delta(\omega - \omega_0) + \delta(\omega + \omega_0)], \qquad (13.71)$$

and similarly,

$$\sin \omega_0 t \quad \longleftrightarrow \quad j\pi[\delta(\omega + \omega_0) - \delta(\omega - \omega_0)]. \qquad (13.72)$$

As shown in Fig. 13-16, the Fourier transform of $\cos \omega_0 t$ consists of impulse functions at $\pm\omega_0$.

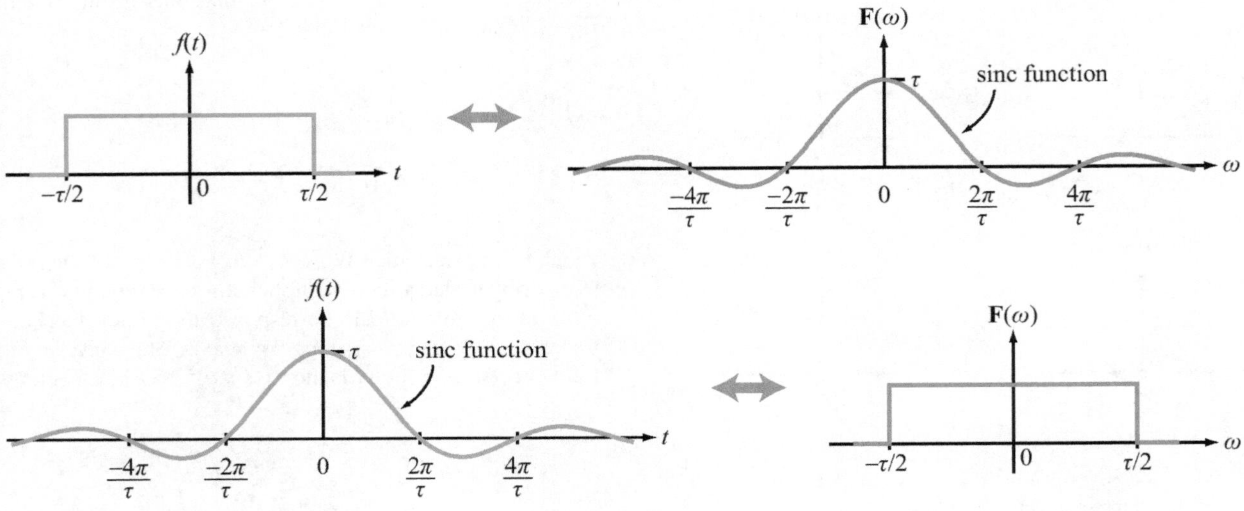

Figure 13-15: Time-frequency duality: a rectangular pulse generates a sinc spectrum and, conversely, a sinc-pulse generates a rectangular spectrum.

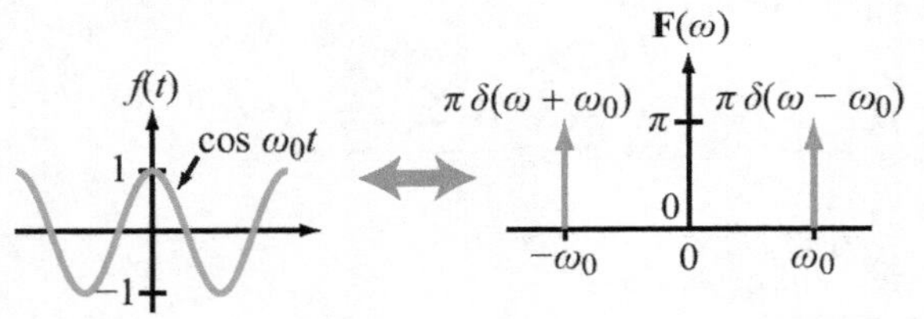

Figure 13-16: The Fourier transform of $\cos \omega_0 t$ is equal to two impulse functions—one at ω_0 and another at $-\omega_0$.

13-6.5 Fourier Transform of $Ae^{-at}\,u(t)$ with $a > 0$

The Fourier transform of an exponentially decaying function that starts at $t = 0$ is

$$\mathbf{F}(\omega) = \mathcal{F}[Ae^{-at}\,u(t)]$$

$$= \int_0^\infty Ae^{-at}e^{-j\omega t}\,dt = A\left.\frac{e^{-(a+j\omega)t}}{-(a+j\omega)}\right|_0^\infty = \frac{A}{a+j\omega}.$$

Hence,

$$Ae^{-at}\,u(t) \quad \longleftrightarrow \quad \frac{A}{a+j\omega}, \qquad \text{for } a > 0. \tag{13.73}$$

13-6.6 Fourier Transform of $u(t)$

The direct approach to finding $\mathbf{F}(\omega)$ for the unit step function leads to

$$\mathbf{F}(\omega) = \mathcal{F}[u(t)] = \int_{-\infty}^{\infty} u(t)\,e^{-j\omega t}\,dt$$

$$= \int_0^\infty e^{-j\omega t}\,dt = \left.\frac{e^{-j\omega t}}{-j\omega}\right|_0^\infty = \frac{j}{\omega}(e^{-j\infty} - 1),$$

which is problematic because $e^{-j\infty}$ does not converge. To avoid the convergence problem, we can pursue an alternative approach that involves the ***signum function***, which is defined by

$$\text{sgn}(t) = u(t) - u(-t). \tag{13.74}$$

Shown graphically in Fig. 13-17(a), the signum function resembles a step-function waveform (with an amplitude of 2 units) that has been slid downward by 1 unit. Looking at the waveform, it is easy to see that one can generate a step function from the signum function as follows:

$$u(t) = \frac{1}{2} + \frac{1}{2}\,\text{sgn}(t). \tag{13.75}$$

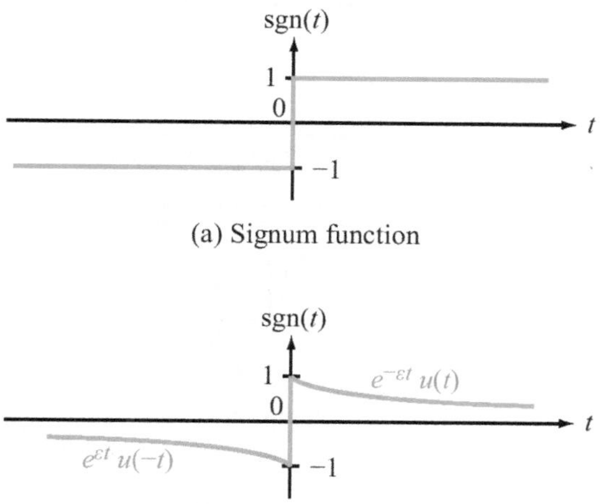

(a) Signum function

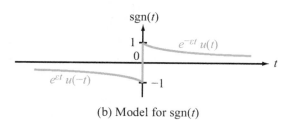

(b) Model for sgn(t)

Figure 13-17: The model shown in (b) approaches the exact definition of sgn(t) as $\varepsilon \rightarrow 0$.

The corresponding Fourier transform is given by

$$\mathcal{F}[u(t)] = \mathcal{F}\left[\frac{1}{2}\right] + \frac{1}{2}\,\mathcal{F}[\text{sgn}(t)] = \pi\,\delta(\omega) + \frac{1}{2}\,\mathcal{F}[\text{sgn}(t)], \tag{13.76}$$

where in the first term we used the relationship given by Eq. (13.68b). Next, we will obtain $\mathcal{F}[\text{sgn}(t)]$ by modeling the signum function as

$$\text{sgn}(t) = \lim_{\varepsilon\to 0}[e^{-\varepsilon t}\,u(t) - e^{\varepsilon t}\,u(-t)] \tag{13.77}$$

with $\varepsilon > 0$. The shape of the modeled waveform is shown in Fig. 13-17(b) for a small value of ε.

Now we are ready to apply the formal definition of the Fourier transform given by Eq. (13.62a):

$$\begin{aligned}\mathcal{F}[\text{sgn}(t)] &= \int_{-\infty}^{\infty} \lim_{\varepsilon\to 0}[e^{-\varepsilon t}\,u(t) - e^{\varepsilon t}\,u(-t)]e^{-j\omega t}\,dt \\ &= \lim_{\varepsilon\to 0}\left[\int_0^{\infty} e^{-(\varepsilon+j\omega)t}\,dt - \int_{-\infty}^{0} e^{(\varepsilon-j\omega)t}\,dt\right] \\ &= \lim_{\varepsilon\to 0}\left[\left.\frac{e^{-(\varepsilon+j\omega)t}}{-(\varepsilon+j\omega)}\right|_0^{\infty} - \left.\frac{e^{(\varepsilon-j\omega)t}}{\varepsilon-j\omega}\right|_{-\infty}^{0}\right] \\ &= \lim_{\varepsilon\to 0}\left[\frac{1}{\varepsilon+j\omega} - \frac{1}{\varepsilon-j\omega}\right] = \frac{2}{j\omega}. \end{aligned} \tag{13.78}$$

Use of Eq. (13.78) in Eq. (13.76) gives

$$\mathcal{F}[u(t)] = \pi\,\delta(\omega) + \frac{1}{j\omega}.$$

Equivalently, the preceding result can be expressed in the form

$$u(t) \quad \longleftrightarrow \quad \pi\,\delta(\omega) + \frac{1}{j\omega}. \tag{13.79}$$

Table 13-4 provides a list of commonly used time functions together with their corresponding Fourier transforms, and Table 13-5 offers a summary of the major properties of the Fourier transform—many of which resemble those we encountered earlier in Chapter 12 in connection with the Laplace transform.

Example 13-10: Fourier Transform Properties

Establish the validity of the time derivative and modulation properties of the Fourier transform (Properties 6 and 10 in Table 13-5).

Solution:

Time derivative property: From Eq. (13.62b),

$$f(t) = \frac{1}{2\pi}\int_{-\infty}^{\infty} \mathbf{F}(\omega)\,e^{j\omega t}\,d\omega. \tag{13.80}$$

Differentiating both sides with respect to t gives

$$\begin{aligned} f'(t) = \frac{df}{dt} &= \frac{1}{2\pi}\int_{-\infty}^{\infty} j\omega\,\mathbf{F}(\omega)\,e^{j\omega t}\,d\omega \\ &= j\omega\left[\frac{1}{2\pi}\int_{-\infty}^{\infty}\mathbf{F}(\omega)\,e^{j\omega t}\,d\omega\right]. \end{aligned}$$

Hence, differentiating $f(t)$ in the time domain is equivalent to multiplying $\mathbf{F}(\omega)$ by $j\omega$ in the frequency domain:

$$f'(t) \quad \longleftrightarrow \quad j\omega\,\mathbf{F}(\omega). \tag{13.81}$$

Table 13-4: Examples of Fourier transform pairs. Note that constant $a \geq 0$.

	$f(t)$	$\mathbf{F}(\omega) = \mathcal{F}[f(t)]$	$\lvert\mathbf{F}(\omega)\rvert$
	BASIC FUNCTIONS		
1.	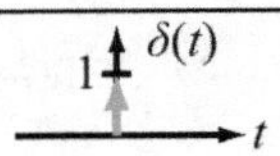	$\delta(t) \quad \longleftrightarrow \quad 1$	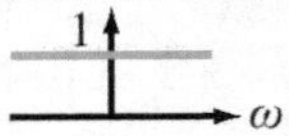
1a.	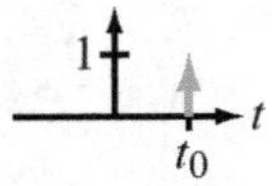	$\delta(t - t_0) \quad \longleftrightarrow \quad e^{-j\omega t_0}$	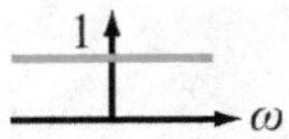
2.	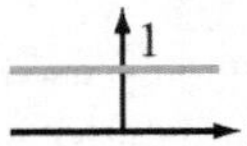	$1 \quad \longleftrightarrow \quad 2\pi\ \delta(\omega)$	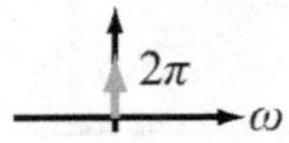
3.	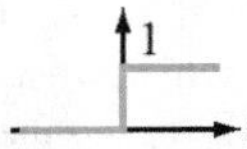	$u(t) \quad \longleftrightarrow \quad \pi\ \delta(\omega) + 1/j\omega$	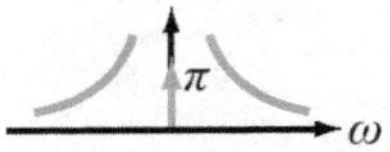
4.	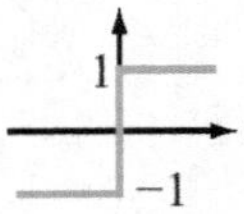	$\text{sgn}(t) \quad \longleftrightarrow \quad 2/j\omega$	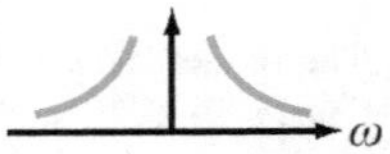
5.	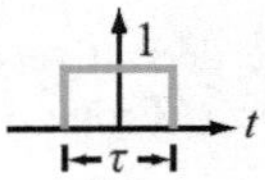	$\text{rect}(t/\tau) \quad \longleftrightarrow \quad \tau\ \text{sinc}(\omega\tau/2)$	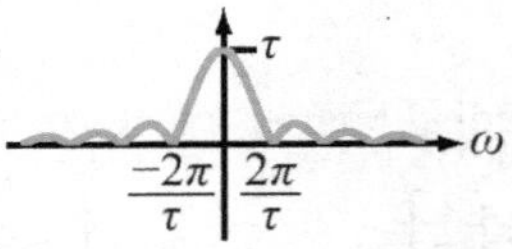
6.	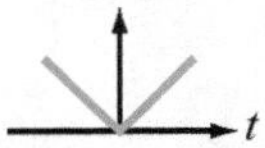	$\lvert t\rvert \quad \longleftrightarrow \quad -2/\omega^2$	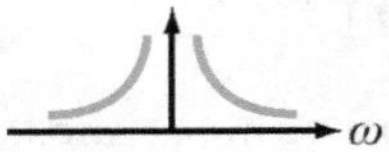
7.	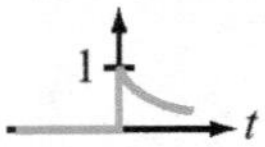	$e^{-at}\ u(t) \quad \longleftrightarrow \quad 1/(a + j\omega)$	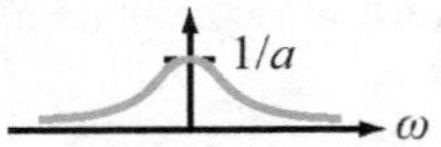
8.	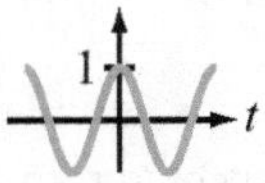	$\cos\omega_0 t \quad \longleftrightarrow \quad \pi[\delta(\omega - \omega_0) + \delta(\omega + \omega_0)]$	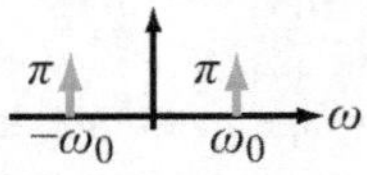
9.	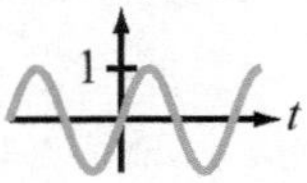	$\sin\omega_0 t \quad \longleftrightarrow \quad j\pi[\delta(\omega + \omega_0) - \delta(\omega - \omega_0)]$	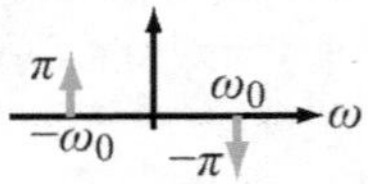
	ADDITIONAL FUNCTIONS		
10.		$e^{j\omega_0 t} \quad \longleftrightarrow \quad 2\pi\ \delta(\omega - \omega_0)$	
11.		$te^{-at}\ u(t) \quad \longleftrightarrow \quad 1/(a + j\omega)^2$	
12.		$[e^{-at}\sin\omega_0 t]\ u(t) \quad \longleftrightarrow \quad \omega_0/[(a + j\omega)^2 + \omega_0^2]$	
13.		$[e^{-at}\cos\omega_0 t]\ u(t) \quad \longleftrightarrow \quad (a + j\omega)/[(a + j\omega)^2 + \omega_0^2]$	

Time modulation property: We start by multiplying both sides of Eq. (13.80) by $\cos\omega_0 t$, and for convenience, we change the dummy variable ω to ω':

$$\cos\omega_0 t\ f(t) = \frac{1}{2\pi} \int_{-\infty}^{\infty} \cos\omega_0 t\ \mathbf{F}(\omega')\ e^{j\omega' t}\ d\omega'.$$

Table 13-5: **Major properties of the Fourier transform.**

Property	$f(t)$		$\mathbf{F}(\omega) = \mathcal{F}[f(t)]$
1. Multiplication by a constant	$K\ f(t)$	$\longleftrightarrow$	$K\ \mathbf{F}(\omega)$
2. Linearity	$K_1\ f_1(t) + K_2\ f_2(t)$	$\longleftrightarrow$	$K_1\ \mathbf{F}_1(\omega) + K_2\ \mathbf{F}_2(\omega)$
3. Time scaling	$f(at)$	$\longleftrightarrow$	$\frac{1}{\lvert a\rvert}\ \mathbf{F}\left(\frac{\omega}{a}\right)$
4. Time shift	$f(t - t_0)$	$\longleftrightarrow$	$e^{-j\omega t_0}\ \mathbf{F}(\omega)$
5. Frequency shift	$e^{j\omega_0 t}\ f(t)$	$\longleftrightarrow$	$\mathbf{F}(\omega - \omega_0)$
6. Time 1st derivative	$f' = \frac{df}{dt}$	$\longleftrightarrow$	$j\omega\ \mathbf{F}(\omega)$
7. Time nth derivative	$\frac{d^n f}{dt^n}$	$\longleftrightarrow$	$(j\omega)^n\ \mathbf{F}(\omega)$
8. Time integral	$\int_{-\infty}^{t} f(t)\ dt$	$\longleftrightarrow$	$\frac{\mathbf{F}(\omega)}{j\omega} + \pi\ \mathbf{F}(0)\ \delta(\omega)$
9. Frequency derivative	$t^n\ f(t)$	$\longleftrightarrow$	$(j)^n \frac{d^n \mathbf{F}(\omega)}{d\omega^n}$
10. Modulation	$\cos\omega_0 t\ f(t)$	$\longleftrightarrow$	$\frac{1}{2}[\mathbf{F}(\omega - \omega_0) + \mathbf{F}(\omega + \omega_0)]$
11. Convolution in t	$f_1(t) * f_2(t)$	$\longleftrightarrow$	$\mathbf{F}_1(\omega)\ \mathbf{F}_2(\omega)$
12. Convolution in ω	$f_1(t)\ f_2(t)$	$\longleftrightarrow$	$\frac{1}{2\pi}\ \mathbf{F}_1(\omega) * \mathbf{F}_2(\omega)$

Applying Euler's identity to $\cos\omega_0 t$ on the right-hand side leads to

$$\cos\omega_0 t\ f(t) = \frac{1}{2\pi} \int_{-\infty}^{\infty} \left(\frac{e^{j\omega_0 t} + e^{-j\omega_0 t}}{2}\right) \mathbf{F}(\omega')\ e^{j\omega' t}\ d\omega'$$
$$= \frac{1}{4\pi}\left[\int_{-\infty}^{\infty} \mathbf{F}(\omega')\ e^{j(\omega'+\omega_0)t}\ d\omega' + \int_{-\infty}^{\infty} \mathbf{F}(\omega')\ e^{j(\omega'-\omega_0)t}\ d\omega'\right].$$

Upon making the substitution $(\omega = \omega' + \omega_0)$ in the first integral and independently making the substitution $(\omega = \omega' - \omega_0)$ in the second integral, we have

$$\cos\omega_0 t\ f(t) = \frac{1}{2}\left[\frac{1}{2\pi} \int_{-\infty}^{\infty} \mathbf{F}(\omega - \omega_0)\ e^{j\omega t}\ d\omega + \frac{1}{2\pi} \int_{-\infty}^{\infty} \mathbf{F}(\omega + \omega_0)\ e^{j\omega t}\ d\omega\right],$$

which can be cast in the abbreviated form

$$\cos\omega_0 t\ f(t) \longleftrightarrow \frac{1}{2}[\mathbf{F}(\omega - \omega_0) + \mathbf{F}(\omega + \omega_0)]. \tag{13.82}$$

Concept Question 13-14: What is the Fourier transform of a dc voltage? (See CAD)

Concept Question 13-15: "An impulse in the time domain is equivalent to an infinite number of sinusoids, all with equal amplitude." Is this a true statement? Can one construct an ideal impulse function? (See CAD)

Exercise 13-7: Use the entries in Table 13-4 to determine the Fourier transform of $u(-t)$.

Answer: $\mathbf{F}(\omega) = \pi\,\delta(\omega) - 1/j\omega$. (See CAD)

Exercise 13-8: Verify the Fourier transform expression for entry #10 in Table 13-4.

Answer: (See CAD)

13-6.7 Parseval's Theorem

If $f(t)$ represents the voltage across a $1\ \Omega$ resistor, then $f^2(t)$ represents the power dissipated in the resistor, and the integrated value of $f^2(t)$ over $[-\infty, \infty]$ represents the cumulative energy W expended in the resistor. Thus,

$$W = \int_{-\infty}^{\infty} f^2(t)\,dt = \int_{-\infty}^{\infty} f(t)\left[\frac{1}{2\pi}\int_{-\infty}^{\infty} \mathbf{F}(\omega)\,e^{j\omega t}\,d\omega\right] dt, \tag{13.83}$$

where one $f(t)$ was replaced with the inverse Fourier transform relationship given by Eq. (13.62b). By reversing the order of $f(t)$ and $\mathbf{F}(\omega)$, and reversing the order of integration, we have

$$\begin{aligned} W &= \frac{1}{2\pi}\int_{-\infty}^{\infty} \mathbf{F}(\omega)\left[\int_{-\infty}^{\infty} f(t)\,e^{j\omega t}\,dt\right] d\omega \\ &= \frac{1}{2\pi}\int_{-\infty}^{\infty} \mathbf{F}(\omega)\left[\int_{-\infty}^{\infty} f(t)\,e^{-j(-\omega)t}\,dt\right] d\omega \\ &= \frac{1}{2\pi}\int_{-\infty}^{\infty} \mathbf{F}(\omega)\,\mathbf{F}(-\omega)\,d\omega \\ &= \frac{1}{2\pi}\int_{-\infty}^{\infty} \mathbf{F}(\omega)\,\mathbf{F}^*(\omega)\,d\omega, \end{aligned} \tag{13.84}$$

where we used the *reversal property* (also known as the *conjugate symmetry property*) of the Fourier transform (see Exercise 13-9), which is given by

$$\mathbf{F}(-\omega) = \mathbf{F}^*(\omega). \tag{13.85}$$

(reversal property)

The combination of Eqs. (13.83) and (13.84) can be written as

$$\int_{-\infty}^{\infty} f^2(t)\,dt = \frac{1}{2\pi}\int_{-\infty}^{\infty} |\mathbf{F}(\omega)|^2\,d\omega. \tag{13.86}$$

(Parseval's theorem)

▶ ***Parseval's theorem*** states that the total energy in the time domain is equal to the total energy in the ω domain. ◀

Exercise 13-9: Verify the reversal property given by Eq. (13.85).

Answer: (See CAD)

13-7 Phasor vs. Laplace vs. Fourier

Consider a linear circuit with input excitation $x(t)$ and output response $y(t)$, where $x(t)$ is a voltage or current source and $y(t)$ is a voltage between two nodes in the circuit or a current through one of its branches. Our goal is to analyze the circuit to determine the desired output $y(t)$. Beyond the time-domain differential equation solution method, which in practice can accommodate only first- and second-order circuits, we have available to us three techniques by which to determine $y(t)$.

(a) The phasor-domain technique (Chapter 7).

(b) The Laplace transform technique (Chapter 12).

(c) The Fourier series and transform techniques (Chapter 13).

The applicability conditions for the three techniques, summarized in Table 13-6, are governed by the duration and shape of the waveform of the input excitation. Based on its duration, an input excitation $x(t)$ is said to be:

(1) *everlasting* (two-sided): if it exists over all time $(-\infty, \infty)$,

(2) *one-sided*: if it starts at or after $t = 0$; i.e., $x(t) = 0$ for $t < 0^-$.

In real life, there is no such thing as an everlasting excitation. When we deal with real circuits, there is always a starting point in time for both the input excitation and output response. In general, an output signal consists of two components, a *transient component* associated with the initial onset of the input signal, and a *steady-state component* that alone remains after the transient component decays to zero. If the input excitation is sinusoidal and we are interested in only the steady-state component of the output response, it is often convenient to regard the input excitation as everlasting, even though, strictly speaking, it cannot be so. We regard it as such because we can then apply the phasor-domain technique, which is easier to implement than the other two techniques.

According to the summary provided in Table 13-6:

- If $x(t)$ is an *everlasting sinusoid*, the phasor-domain technique is the solution method of choice.

- If $x(t)$ is an *everlasting periodic excitation*, such as a square wave or any repetitive waveform that can be represented by a Fourier series, then by virtue of the superposition principle, the phasor-domain technique can be used to compute the output responses corresponding to the individual Fourier components of the input excitation, and then all of the output components can be added up to generate the total output.

- If $x(t)$ is a *one-sided excitation*, the Laplace transform technique is the preferred solution method. An important feature of the technique is that it can accommodate nonzero initial conditions of the circuit, if they exist.

- If $x(t)$ is everlasting and its waveform is nonperiodic, we can obtain $y(t)$ by applying either a bilateral form of the Laplace transform or the Fourier transform. For input excitations $x(t)$ whose Laplace transforms do not exist but their Fourier transforms do, the Fourier transform approach becomes the only viable option, and the converse is true for excitations whose Fourier transforms do not exist but their Laplace transforms do.

The Laplace transform operates in the $\mathbf{s}$ domain, where $\mathbf{s} = \sigma + j\omega$ is represented by a complex plane with real and imaginary axes along σ and $j\omega$. The Fourier transform operates along the $j\omega$ axis of the $\mathbf{s}$ plane, corresponding to $\sigma = 0$. Thus, for circuits excited by everlasting sinusoidal waveforms,

$$\mathbf{F}(\omega) = \mathbf{F}(\mathbf{s})\big|_{\sigma=0}.$$

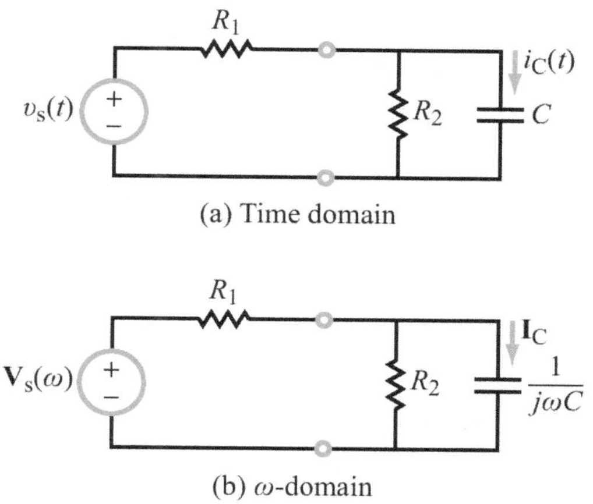

Figure 13-18: Circuits for Example 13-11.

13-8 Circuit Analysis with Fourier Transform

As was mentioned earlier, the Fourier transform technique can be used to analyze circuits excited by either one-sided or two-sided nonperiodic waveforms—as long as the circuit has no initial conditions. The procedure (which is analogous to the Laplace transform technique) with $\mathbf{s}$ replaced by $j\omega$ is demonstrated through Example 13-11.

Example 13-11: RC Circuit

The RC circuit shown in Fig. 13-18(a) is excited by a voltage source $\upsilon_s(t)$. Apply Fourier analysis to determine $i_C(t)$ if: (a) $\upsilon_s = 10u(t)$, (b) $\upsilon_s(t) = 10e^{-2t}\,u(t)$, and (c) $\upsilon_s(t) = 10 + 5\cos 4t$, all measured in volts. The element values are $R_1 = 2$ kΩ, $R_2 = 4$ kΩ, and $C = 0.25$ mF.

Solution:

Step 1: Transfer circuit to ω domain

In the frequency-domain circuit shown in Fig. 13-18(b), $\mathbf{V}_s(\omega)$ is the Fourier transform of $\upsilon_s(t)$.

Step 2: Determine $\mathbf{H}(\omega) = \mathbf{I}_C(\omega)/\mathbf{V}_s(\omega)$

Application of source transformation to the circuit in Fig. 13-18(b) followed with current division leads to

$$\mathbf{H}(\omega) = \frac{\mathbf{I}_C(\omega)}{\mathbf{V}_s(\omega)} = \frac{j\omega/R_1}{\dfrac{R_1 + R_2}{R_1 R_2 C} + j\omega} = \frac{j0.5\omega \times 10^{-3}}{3 + j\omega}. \tag{13.87}$$

Table 13-6: **Methods of solution.**

Input $x(t)$			
Duration	**Waveform**	**Solution Method**	**Output** $y(t)$
Everlasting	Sinusoid	Phasor Domain	Steady-State Component (no transient exists)
Everlasting	Periodic	Phasor Domain and Fourier Series	Steady-State Component (no transient exists)
One-sided, $x(t) = 0$, for $t < 0^-$	Any	Laplace Transform (unilateral) (can accommodate nonzero initial conditions)	Complete Solution (transient + steady-state)
Everlasting	Any	Bilateral Laplace Transform or Fourier Transform	Complete Solution (transient + steady-state)

Step 3: Solve for $\mathbf{I}_C(\omega)$ and $i_C(t)$

(a) $\upsilon_s(t) = 10u(t)$: The corresponding Fourier transform per entry #3 in Table 13-4 is

$$\mathbf{V}_s(\omega) = 10\pi\ \delta(\omega) + \frac{10}{j\omega}.$$

The corresponding current is

$$\mathbf{I}_C(\omega) = \mathbf{H}(\omega)\ \mathbf{V}_s(\omega) = \frac{j5\pi\omega\ \delta(\omega) \times 10^{-3}}{3 + j\omega} + \frac{5 \times 10^{-3}}{3 + j\omega}. \tag{13.88}$$

The inverse Fourier transform of $\mathbf{I}_C(\omega)$ is given by

$$i_C(t) = \frac{1}{2\pi} \int_{-\infty}^{\infty} \frac{j5\pi\omega\ \delta(\omega) \times 10^{-3}}{3 + j\omega}\ e^{j\omega t}\ d\omega + \mathcal{F}^{-1}\left[\frac{5 \times 10^{-3}}{3 + j\omega}\right],$$

where we applied the formal definition of the inverse Fourier transform to the first term—because it includes a delta function—and the functional form to the second term—because we intend to use look-up entry #7 in Table 13-4. Accordingly,

$$i_C(t) = 0 + 5e^{-3t}\ u(t) \text{ mA}. \tag{13.89}$$

(b) $\upsilon_s(t) = 10e^{-2t}\ u(t)$: By entry #7 in Table 13-4,

$$\mathbf{V}_s(\omega) = \frac{10}{2 + j\omega}.$$

The corresponding current $\mathbf{I}_C(\omega)$ is given by

$$\mathbf{I}_C(\omega) = \mathbf{H}(\omega)\ \mathbf{V}_s(\omega) = \frac{j5\omega \times 10^{-3}}{(2 + j\omega)(3 + j\omega)}. \tag{13.90}$$

Application of partial fraction expansion (Section 12-5) gives

$$\mathbf{I}_C(\omega) = \frac{A_1}{2 + j\omega} + \frac{A_2}{3 + j\omega},$$

with

$$A_1 = (2 + j\omega)\ \mathbf{I}_C(\omega)|_{j\omega=-2} = \left.\frac{j5\omega \times 10^{-3}}{3 + j\omega}\right|_{j\omega=-2} = -10 \times 10^{-3}$$

and

$$A_2 = (3 + j\omega)\ \mathbf{I}_C(\omega)|_{j\omega=-3} = \left.\frac{j5\omega \times 10^{-3}}{2 + j\omega}\right|_{j\omega=-3} = 15 \times 10^{-3}.$$

Hence,

$$\mathbf{I}_C(\omega) = \left(\frac{-10}{2 + j\omega} + \frac{15}{3 + j\omega}\right) \times 10^{-3}$$

and

$$i_C(t) = (15e^{-3t} - 10e^{-2t})\ u(t) \text{ mA}. \tag{13.91}$$

(c) $\upsilon_s(t) = 10 + 5\cos 4t$:

By entries #2 and #8 in Table 13-4,

$$\mathbf{V}_s(\omega) = 20\pi\ \delta(\omega) + 5\pi[\delta(\omega - 4) + \delta(\omega + 4)],$$

and the capacitor current is

$$\mathbf{I}_C(\omega) = \mathbf{H}(\omega)\ \mathbf{V}_s(\omega) = \frac{j10\pi\omega\ \delta(\omega) \times 10^{-3}}{3 + j\omega} + j2.5\pi \times 10^{-3}\left[\frac{\omega\ \delta(\omega - 4)}{3 + j\omega} + \frac{\omega\ \delta(\omega + 4)}{3 + j\omega}\right].$$

The corresponding time-domain current is obtained by applying Eq. (13.62b) as

$$
\begin{aligned}
i_C(t) &= \frac{1}{2\pi}\int_{-\infty}^{\infty} \frac{j10\pi\omega\,\delta(\omega)\times 10^{-3}e^{j\omega t}\,d\omega}{3+j\omega} \\
&\quad + \frac{1}{2\pi}\int_{-\infty}^{\infty} \frac{j2.5\pi\omega\times 10^{-3}}{3+j\omega}\,\delta(\omega-4)\,e^{j\omega t}\,d\omega \\
&\quad + \frac{1}{2\pi}\int_{-\infty}^{\infty} \frac{j2.5\pi\omega\times 10^{-3}}{3+j\omega}\,\delta(\omega+4)\,e^{j\omega t}\,d\omega \\
&= 0 + \frac{j5\times 10^{-3}e^{j4t}}{3+j4} - \frac{j5\times 10^{-3}e^{-j4t}}{3-j4} \\
&= 5\times 10^{-3}\left(\frac{e^{j4t}e^{j36.9^\circ}}{5} + \frac{e^{-j4t}e^{-j36.9^\circ}}{5}\right) \\
&= 2\cos(4t+36.9^\circ)\text{ mA.}
\end{aligned}
\tag{13.92}
$$

13-9 Multisim: Mixed-Signal Circuits and the Sigma-Delta Modulator

Historically, circuit designers tended to fall into two broad classes: those who designed digital circuits and those who designed analog circuits. As a broad generalization, digital-circuit designers built logic gates, computational elements, memories, and so on, whereas analog designers tended to work on circuits that interfaced with the noncircuit world: amplifiers, drivers, radio frequency circuits, analog-to-digital converters, among others. Moreover, digital designers tended to have more comprehensive and powerful software design tools, mainly because digital circuits could be abstracted into modules that made hierarchical analysis possible. For example, a transistor could be modeled as a simple switch, several of these switches could be wired together as a simple logic gate, many logic gates could be wired together to make a counter or a memory, and so on, all of which can be readily modeled as "black boxes" in software. Analog circuits, by contrast, defied this type of compartmentalization due to feedback loops, nonlinear behavior, and complex topologies; this made analog design almost an art form.

Advances in silicon fabrication technologies have now blurred the line between these two worlds considerably. A new generation of circuits, known as *mixed-signal circuits*, contains elements of both worlds (Fig. 13-19). This exciting area combines the power of analog designs with the scalability, modularity and computational power of digital circuits. Modern analog-to-digital conversion (ADC) and digital-to-analog conversion (DAC) circuits, cell phone communication circuits, software radio, internet routers, and audio synthesis circuits are all examples of mixed signal circuits. The advantages are numerous. Consider software radio, for example. We saw in Section 9-8 how a superheterodyne receiver works, which is a perfect example of a multistage analog circuit. But what if many of the functions of the superheterodyne receiver could be performed by digital circuits instead? What advantages might there be? One obvious advantage is the introduction of computational "intelligence" into the radio itself. If the receiver is designed, in part or in whole, with digital circuits, these circuits can be built around computation and memory. Programs can be loaded that allow the radio to change its power consumption, transmission patterns and protocols based on user or environmental parameters; this is often known as *cognitive radio*. The integration of analog and digital circuits comes with certain drawbacks, however. Design remains a challenging, and highly paid, exercise. Design and testing software for mixed signal circuits is nowhere near as advanced as that for digital circuits. The fabrication of these circuits is often confined to specialized processes not compatible with standard, digital-processor fabrication methodologies (although this is rapidly changing).

13-9.1 The Sigma-Delta ($\Sigma\Delta$) Modulator and Analog-to-Digital Converters

So far in this book we have used Multisim to model amplifiers, digital circuits, filters, resonators, and circuits that employ feedback. In this section, we will put it all together and show you how to design a very useful circuit, the ***Sigma-Delta*** ***(*** $\Sigma\Delta$ ***) Modulator***, which since its early development by Inose and Yasuda in 1962 has now become a standard tool for the design of inexpensive analog-to-digital converters (ADCs).

There are many ways to convert an analog waveform into a digital sequence of pulses. The classic ADC circuit takes a time-varying analog voltage, $\upsilon_{\text{in}}(t)$, and produces a corresponding time-varying digital output consisting of a number of bits ($V_{\text{out}0}$, $V_{\text{out}1}$, etc.). Figure 13-20(a) shows the process schematically. Here, a linearly increasing voltage is fed into a 4-bit ADC; as the input voltage changes with time, the four digital output bits change their state (either "0" or "1"). All of the pulses have the same duration and can change states instantaneously. With 4 binary bits, we can construct 16 different values (i.e., 0000, 0001, ..., 1111), so this 4-bit ADC converts any input voltage to one of 2^4 or, equivalently, 16 different digital values. Modern ADCs commonly have 12, 16 or even 24 output bits, giving them very high resolution (e.g., $2^{24} = 16{,}777{,}216$ different values!).

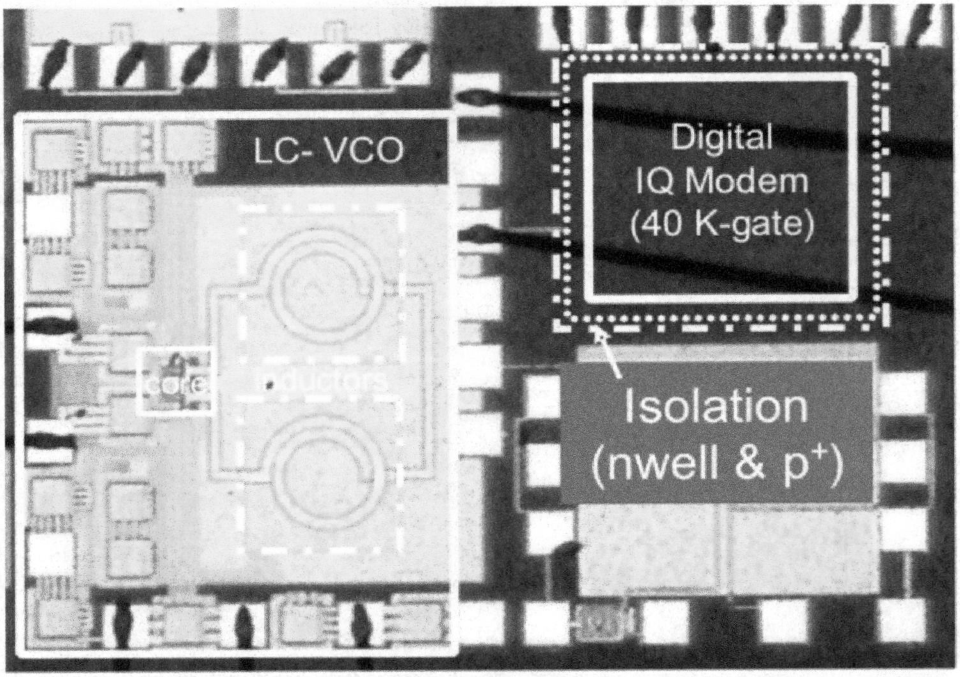

Figure 13-19: This mixed-signal chip implements a highly reconfigurable RF receiver based on a down-converting Sigma-Delta A/D (courtesy Renaldi Winoto and Prof. Borivoje Nikolic, U.C. Berkeley)

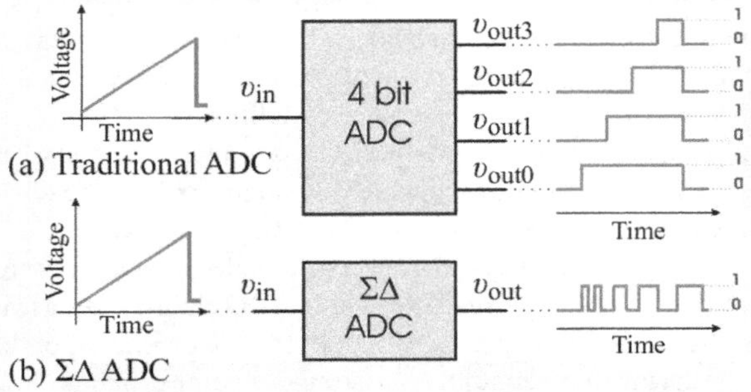

Figure 13-20: (a) A traditional 4-bit ADC converts an analog input voltage and produces 4 digital output bits; (b) a $\Sigma\Delta$ ADC generates a pulse train where the pulse duration is governed by the magnitude of the input voltage.

One usually trades off speed for resolution; the bits in a fast 12-bit ADC integrated circuit can change states once every 2 microseconds, which means that the ADC can measure the input voltage approximately 500,000 times per second.

Unlike conventional ADCs, the *ΣΔ ADC* generates an output consisting of a single digital bit. The *duration* of the voltage pulse, however, depends on the value of the input voltage (Fig. 13-20(b)), thereby encoding the magnitude of the input voltage into the duration of a single pulse, instead of encoding it into the binary states of several pulses (bits) of equal duration. The $\Sigma\Delta$ modulator is particularly attractive for designing and building inexpensive ADC circuits because it can (a) be made using digital components, which are less expensive to build and easier to test than analog components, and (b) the digital components can be reprogrammed and modified by the user using firmware. Although invented in the 1960s, the $\Sigma\Delta$ modulator was not used commercially until digital CMOS processes became sufficiently fast (to produce time-varying output pulses faster than the changes exhibited by the input signal), small (enabling the mixing of digital with analog circuit components) and inexpensive to fabricate.

The entire circuit (Fig. 13-21) can be built from analog components introduced in this book, namely a subtractor,

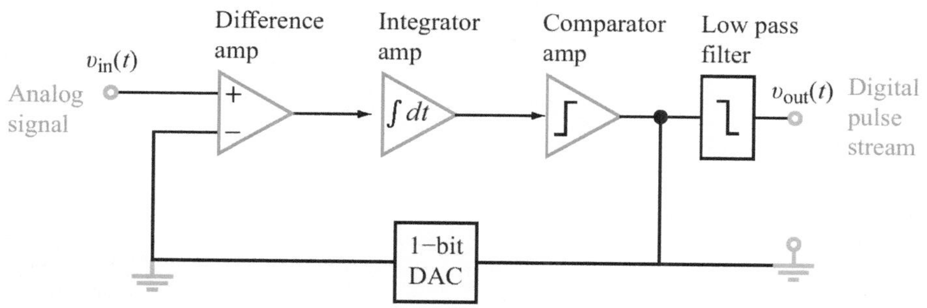

Figure 13-21: Block diagram of a ΣΔ modulator.

an integrator, a comparator, a 1-bit DAC and a low-pass filter. In real implementations, most of these components are replaced with digital substitutes, often reprogrammable during operation. Thus, one could replace the analog filter with an adjustable digital filter, the integrator with a digital integrator, and so on. The only analog component in modern ΣΔs is usually the DAC.

13-9.2 How the ΣΔ Works

Table 13-6 shows the individual subcircuits of the ΣΔ modulator and Fig. 13-22 displays the complete circuit, all drawn in Multisim. Our basic ΣΔ circuit takes an analog input, $\upsilon_{in}(t)$, subtracts from it a feedback signal, $\upsilon_{bit}(t)$, then integrates this signal, producing $\upsilon_{int}(t)$. The integrated signal is then compared with a reference voltage (in our case, 0 V)

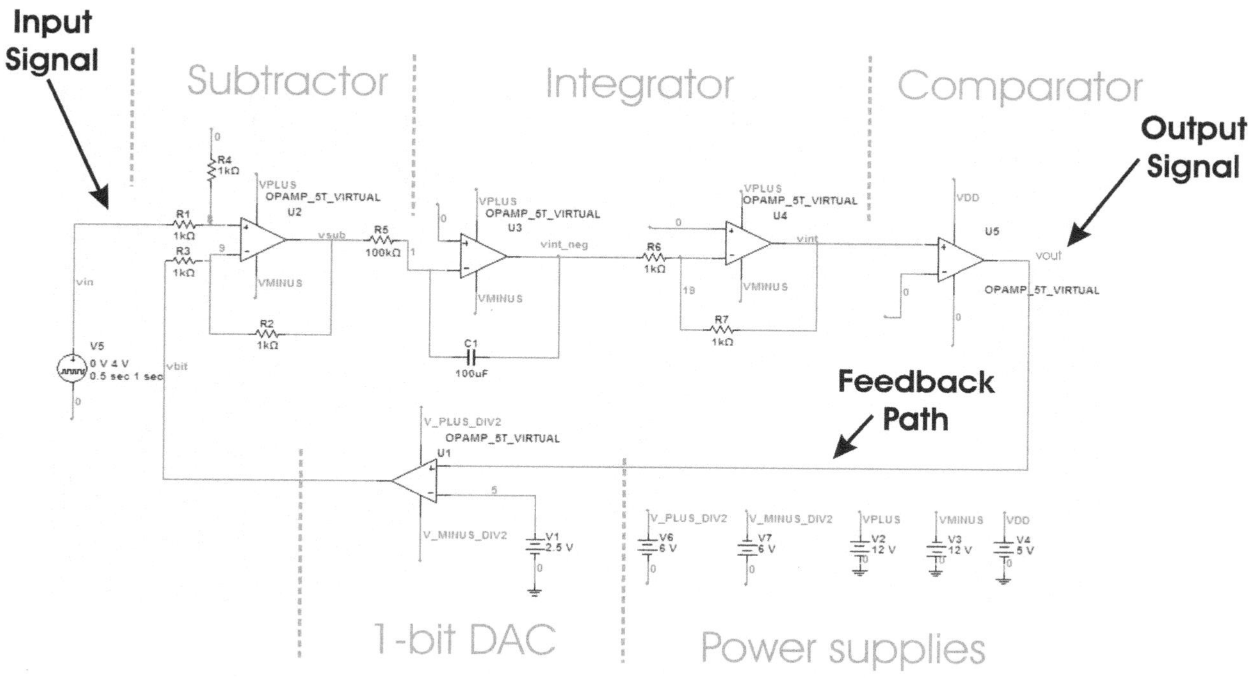

Figure 13-22: Complete Multisim circuit of the ΣΔ modulator.

Table 13-7: Multisim circuits of the ΣΔ modulator.

Multisim Circuit	Description and Notes
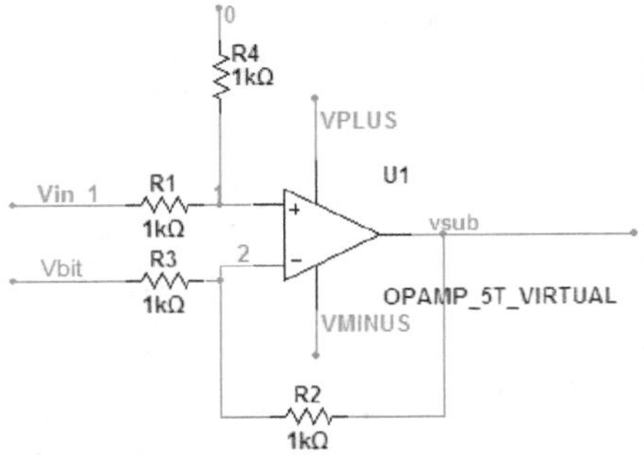	**Subtractor:** This is a difference amplifier (following Table 4-3) with a voltage gain of 1. VPLUS and VMINUS are the extremes of the analog input (in the complete circuit, they are set to ±12 V).
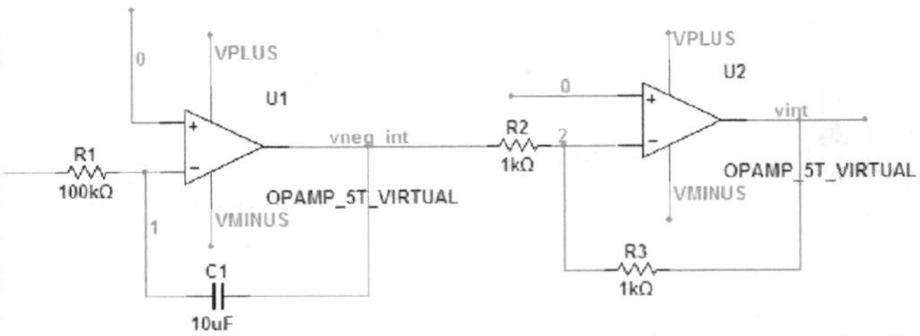	**Integrator:** This circuit consists of an inverting integrator amplifier (Section 5-6.1) and an inverting amplifier (following Table 4-3) with a voltage gain of 1 (to remove the integrator's negative sign).
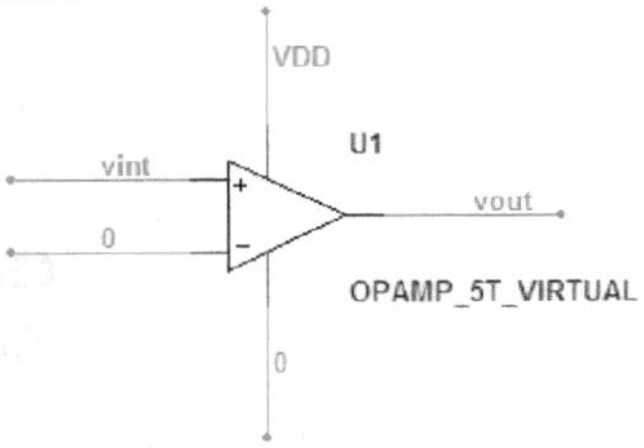	**Comparator:** The comparator is a simple op amp with no feedback (open loop). Since the internal voltage gain A of the op amp is so high (Section 4-1.2), any positive difference between the noninverting and the inverting inputs immediately drives the amplifier output to V_{DD}; a negative difference drives the amplifier output to 0 V. V_{DD} is set to the desired digital voltage level (5 V, in the case of the complete circuit in Fig. 13-22).
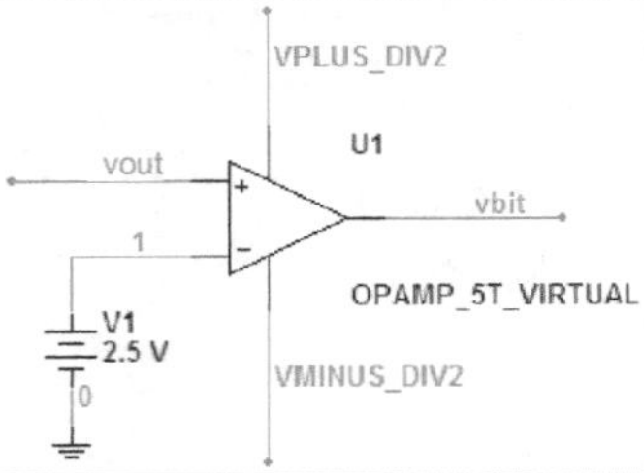	**1-Bit Digital-to-Analog Converter (DAC):** The DAC is very similar to the comparator. The input voltage is compared to a voltage level halfway between 0 and V_{DD}; this has the effect of transforming an input of V_{DD} into an output voltage of VPLUS/2 (+6 V in Fig. 13-22) and an input voltage of 0 V into an output voltage of VMINUS/2 (−6 V in Fig. 13-22).

and produces $\upsilon_{\text{out}}(t)$. The output of the comparator, $\upsilon_{\text{out}}(t)$, can only have two values: V_{DD} or 0, where V_{DD} is the dc power supply voltage of the comparator. Hence, $\upsilon_{\text{out}}(t)$ is a time-varying digital signal. Note that $\upsilon_{\text{out}}(t)$ is also sent to a 1-bit DAC that converts the digital signal to an analog signal, $\upsilon_{\text{bit}}(t)$, which is fed back to the subtractor.

The overall functionality of the $\Sigma\Delta$ modulator is illustrated by Fig. 13-23 for an input signal composed of a 1 Hz sinusoid with an amplitude of 4 V. We observe that the corresponding output, $\upsilon_{\text{out}}(t)$, consists of a sequence of pulses whose durations are proportional to the instantaneous level of the input voltage, $\upsilon_{\text{in}}(t)$. Thus, the $\Sigma\Delta$ circuit encodes amplitude information contained in an analog signal into pulse-duration information in a digital sequence. After transmission of $\upsilon_{\text{out}}(t)$ through downstream digital circuits, the original information can be retrieved by measuring the durations of the pulses. This can be accomplished by a digital counter, either in hardware (using a counter circuit) or in software on a microcontroller. In hardware, transitions can be detected with *Schmidt triggers* or similar edge detectors, and the counter is made to "count" the duration between transitions.

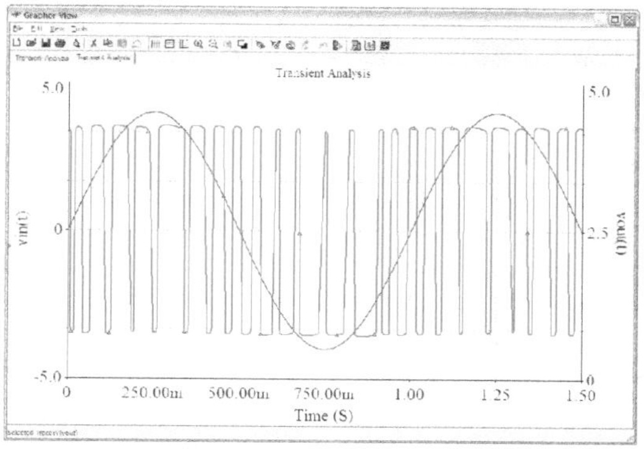

Figure 13-23: A 1 Hz sinusoidal ac signal, $\upsilon_{\text{in}}(t)$, blue trace, is converted to a series of pulses at the output, $\upsilon_{\text{out}}(t)$, red trace, by the Sigma Delta modulator. Note that the duration of the pulses is related to the instantaneous level of voltage $\upsilon_{\text{in}}(t)$.

Summary

Concepts

- A periodic waveform of period T can be represented by a Fourier series consisting of a dc term and sinusoidal terms that are harmonic multiples of $\omega_0 = 2\pi/T$.
- The Fourier series can be represented in terms of a cosine/sine form, amplitude/phase form, and a complex exponential form.
- Circuits excited by a periodic waveform can be analyzed by applying the superposition theorem to the individual terms of the harmonic series.
- Nonperiodic waveforms can be represented by a Fourier transform.
- Upon transforming the circuit to the frequency domain, the circuit can be analyzed for the desired voltage or current of interest and then the result can be inverse transformed to the time domain.
- The Sigma-Delta modulator is an example of a mixed-signal circuit. It converts an analog waveform into a single-bit digital pulse whose duration is proportional to the instantaneous magnitude of the waveform.

Mathematical and Physical Models

Fourier Series Table 13-3

Average Power

$$P_{\text{av}} = V_{\text{dc}} I_{\text{dc}} + \frac{1}{2}\sum_{n=1}^{\infty} V_n I_n \cos(\phi_{\upsilon_n} - \phi_{i_n})$$

Fourier Transform

$$\mathbf{F}(\omega) = \mathcal{F}[f(t)] = \int_{-\infty}^{\infty} f(t)\, e^{-j\omega t}\, dt$$

$$f(t) = \mathcal{F}^{-1}[\mathbf{F}(\omega)] = \frac{1}{2\pi}\int_{-\infty}^{\infty} \mathbf{F}(\omega)\, e^{j\omega t}\, d\omega$$

sinc Function $\text{sinc}(x) = \dfrac{\sin x}{x}$

Properties of Fourier Transform Table 13-5

2-D Fourier Transform

$$\mathbf{F}(\omega_x, \omega_y) = \mathcal{F}[f(x, y)] = \int_{-\infty}^{\infty} f(x, y)\, e^{-j\omega_x x} e^{-j\omega_y y}\, dx\, dy$$

$$f(x, y) = \frac{1}{(2\pi)^2} \int_{-\infty}^{\infty} \mathbf{F}(\omega_x, \omega_y)\, e^{j\omega_x x} e^{j\omega_y y}\, d\omega_x\, d\omega_y$$

Important Terms Provide definitions or explain the meaning of the following terms:

ΣΔ ADC
absolutely integrable
amplitude of the nth harmonic
amplitude/phase
amplitude spectrum
cognitive radio
complete
conjugate symmetry property
dc component
Dirichlet conditions
duration
even function
even symmetry
everlasting
everlasting periodic excitation
everlasting sinusoid
exponential form
Fourier coefficient
Fourier series
Fourier theorem
Fourier transform
frequency-shift property
frequency spectrum
fundamental angular frequency
Gibbs phenomenon
harmonic
inverted
line spectrum
magnitude spectrum
mixed signal circuit
n_{max}-truncated series
null
odd function
odd symmetry
one-sided
one-sided excitation
Parseval's theorem
periodic function
periodicity property
periodic waveform
phase
phase spectrum
reversal property
Schmidt triggers
Sigma-Delta modulator
signum function
sinc function
spacing between adjacent harmonics
steady-state component
time-shift property
transient component
truncated series

PROBLEMS

Sections 13-1 and 13-2: Fourier Series

For each of the waveforms in Problems 13.1 through 13.10:

(a) Determine if the waveform has dc, even, or odd symmetry.

(b) Obtain its cosine/sine Fourier series representation.

(c) Convert the representation to amplitude/phase format and plot the line spectra for the first five nonzero terms.

(d) Use MATLAB software to plot the waveform using a truncated Fourier series representation with $n_{max} = 100$.

13.1 Waveform in Fig. P13.1 with $A = 10$.

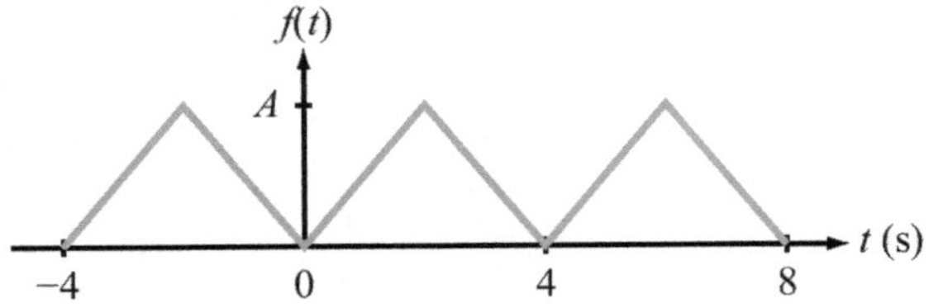

Figure P13.1: Waveform for Problem 13.1.

13.2 Waveform in Fig. P13.2 with $A = 4$.

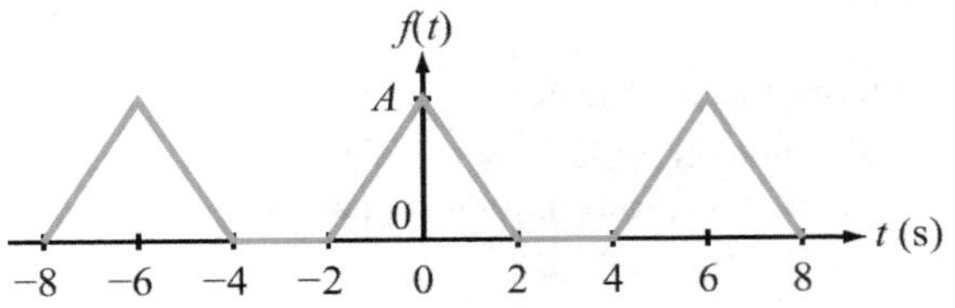

Figure P13.2: Waveform for Problem 13.2.

13.3 Waveform in Fig. P13.3 with $A = 6$.

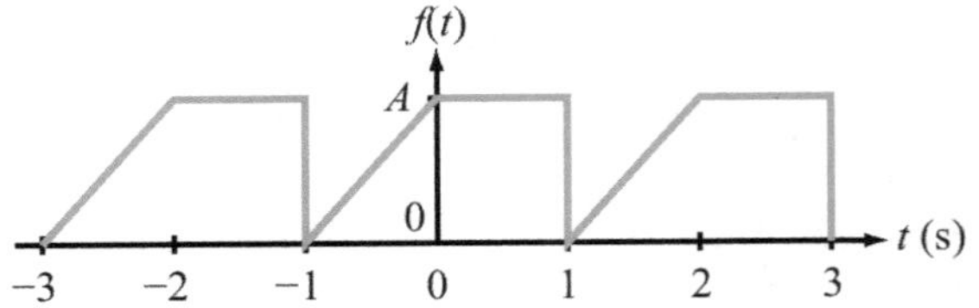

Figure P13.3: Waveform for Problem 13.3.

13.4 Waveform in Fig. P13.4 with $A = 10$.

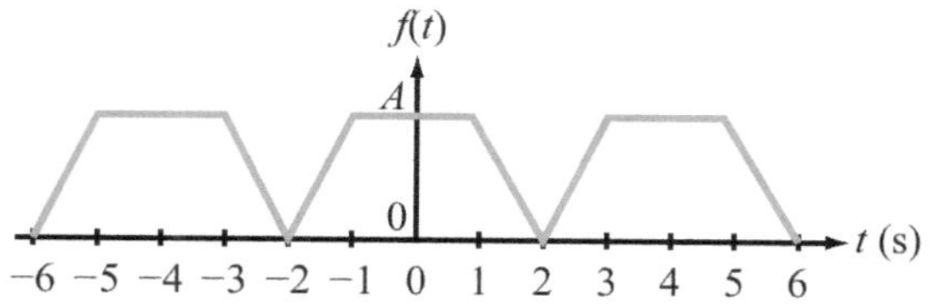

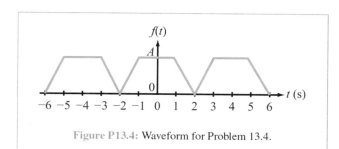

Figure P13.4: Waveform for Problem 13.4.

*13.5 Waveform in Fig. P13.5 with $A = 20$.

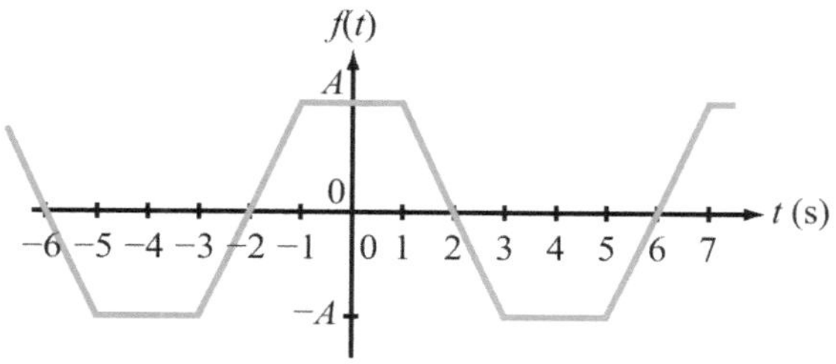

Figure P13.5: Waveform for Problem 13.5.

13.6 Waveform in Fig. P13.6 with $A = 100$.

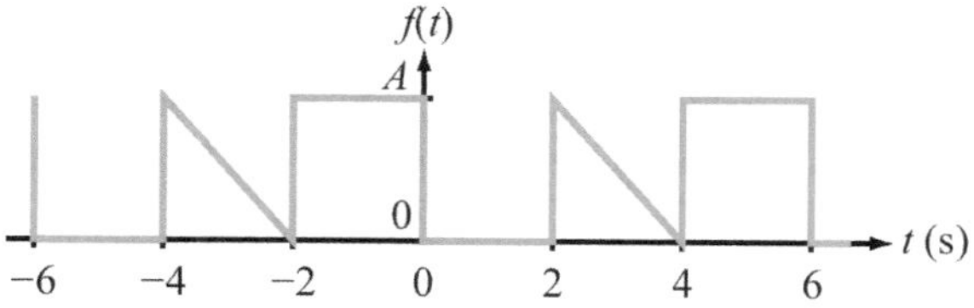

Figure P13.6: Waveform for Problem 13.6.

13.7 Waveform in Fig. P13.7 with $A = 4$.

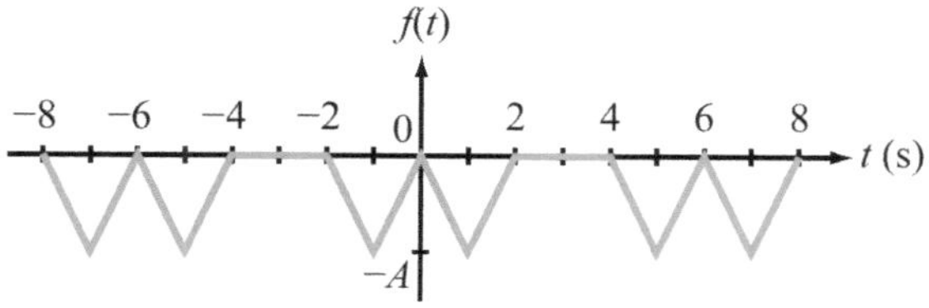

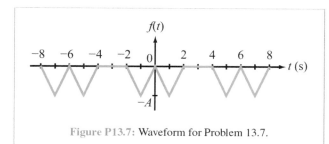

Figure P13.7: Waveform for Problem 13.7.

13.8 Waveform in Fig. P13.8 with $A = 10$.

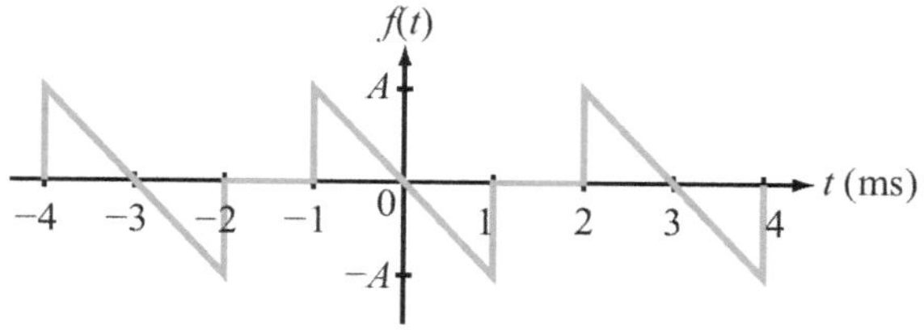

Figure P13.8: Waveform for Problem 13.8.

*13.9 Waveform in Fig. P13.9 with $A = 10$.

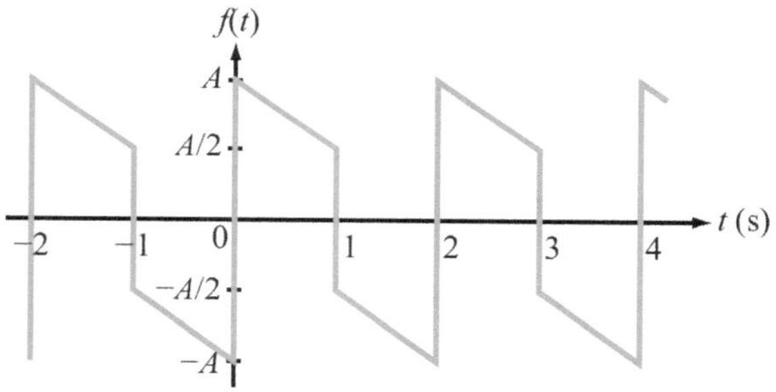

Figure P13.9: Waveform for Problem 13.9.

13.10 Waveform in Fig. P13.10 with $A = 20$.

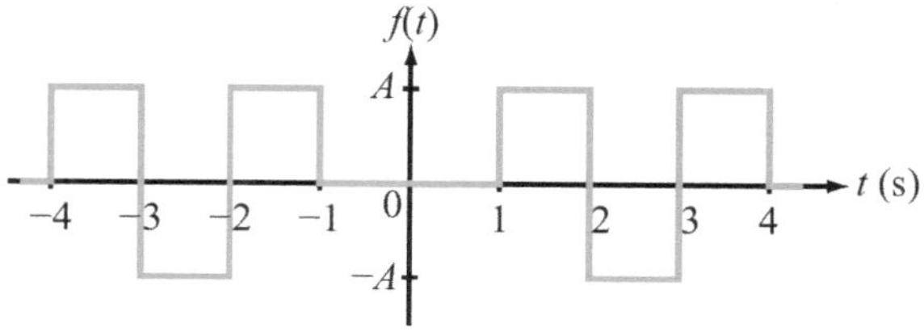

Figure P13.10: Waveform for Problem 13.10.

13.11 Obtain the cosine/sine Fourier series representation for $f(t) = \cos^2(4\pi t)$, and use MATLAB software to plot it with $n_{\max} = 2$, 10, and 100.

*13.12 Repeat Problem 13.11 for $f(t) = \sin^2(4\pi t)$.

13.13 Repeat Problem 13.11 for $f(t) = |\sin(4\pi t)|$.

*Answer(s) available in Appendix G.

13.14 Which of the six waveforms shown in Figs. P13.1 through P13.6 will exhibit the Gibbs oscillation phenomenon when represented by a Fourier series? Why?

13.15 Consider the sawtooth waveform shown in Fig. 13-3(a). Evaluate the Gibbs phenomenon in the neighborhood of $t = 4$ s by plotting the Fourier series representation with $n_{\max} = 100$ over the range between 3.99 s and 4.01 s (using expanded scales if necessary).

*__13.16__ The Fourier series of the periodic waveform shown in Fig. P13.16(a) is given by

$$f_1(t) = 10 - \frac{20}{\pi} \sum_{n=1}^{\infty} \frac{1}{n} \sin\left(\frac{n\pi t}{2}\right).$$

Determine the Fourier series of waveform $f_2(t)$ in Fig. P13.16(b).

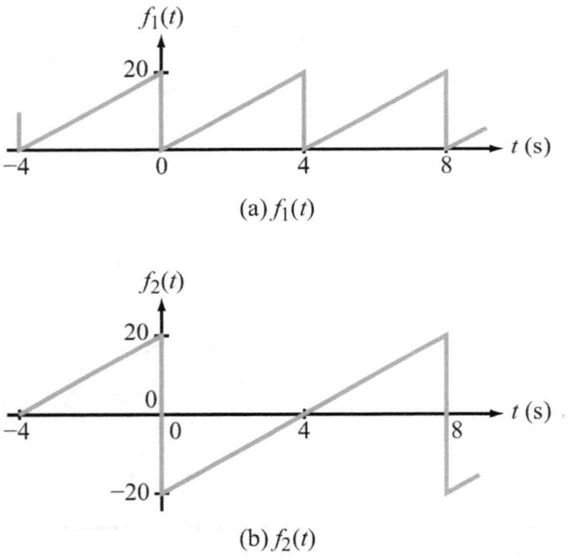

Figure P13.16: Waveforms for Problem 13.16.

Section 13-3: Circuit Applications

13.17 The voltage source $v_s(t)$ in the circuit of Fig. P13.17 generates a square wave (waveform #1 in Table 13-2) with $A = 10$ V and $T = 1$ ms.

(a) Derive the Fourier series representation of $v_{\text{out}}(t)$.

(b) Calculate the first five terms of $v_{\text{out}}(t)$ using $R_1 = R_2 = 2$ kΩ and $C = 1$ μF.

(c) Plot $v_{\text{out}}(t)$ using $n_{\max} = 100$.

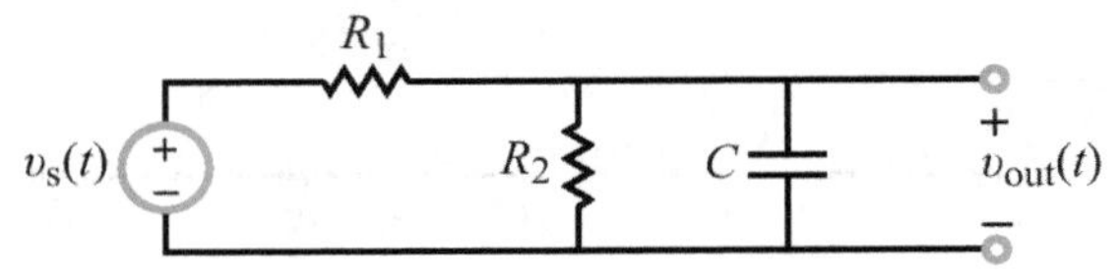

Figure P13.17: Circuit for Problem 13.17.

13.18 The current source $i_s(t)$ in the circuit of Fig. P13.18 generates a sawtooth wave (waveform in Fig. 13-3(a)) with a peak amplitude of 20 mA and a period $T = 5$ ms.

(a) Derive the Fourier series representation of $v_{\text{out}}(t)$.

(b) Calculate the first five terms of $v_{\text{out}}(t)$ using $R_1 = 500$ Ω, $R_2 = 2$ kΩ, and $C = 0.33$ μF.

(c) Plot $v_{\text{out}}(t)$ and $i_s(t)$ using $n_{\max} = 100$.

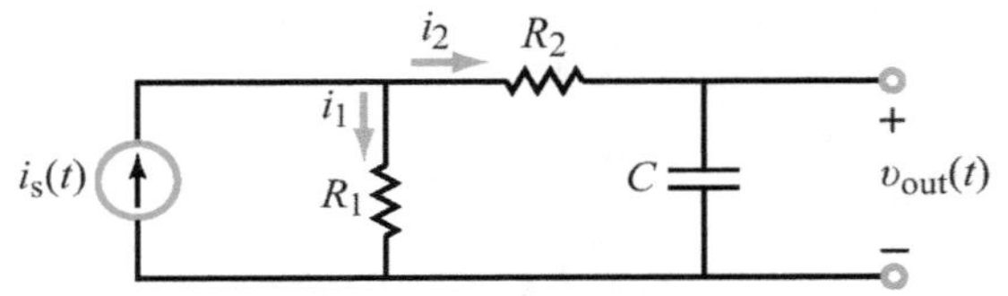

Figure P13.18: Circuit for Problem 13.18.

13.19 The current source $i_s(t)$ in the circuit of Fig. P13.19 generates a train of pulses (waveform #3 in Table 13-2) with $A = 6$ mA, $\tau = 1$ μs, and $T = 10$ μs.

(a) Derive the Fourier series representation of $i(t)$.

(b) Calculate the first five terms of $i(t)$ using $R = 1$ kΩ, $L = 1$ mH, and $C = 1$ μF.

(c) Plot $i(t)$ and $i_s(t)$ using $n_{\max} = 100$.

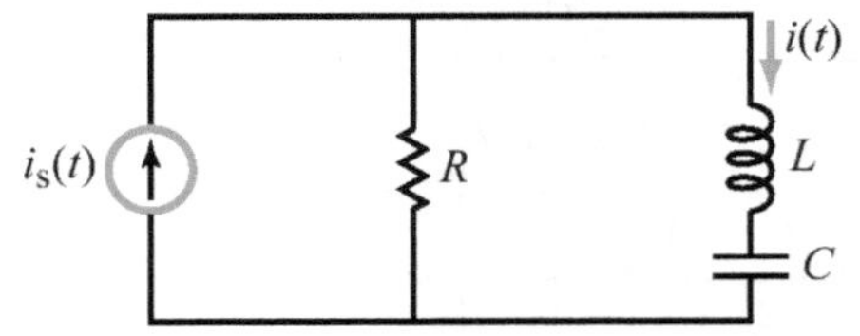

Figure P13.19: Circuit for Problem 13.19.

13.20 Voltage source $v_s(t)$ in the circuit of Fig. P13.20(a) has the waveform displayed in Fig. P13.20(b).

(a) Derive the Fourier series representation of $i(t)$.

(b) Calculate the first five terms of $i(t)$ using $R_1 = R_2 = 10\ \Omega$ and $L_1 = L_2 = 10$ mH.

(c) Plot $i(t)$ and $\upsilon_s(t)$ using $n_{max} = 100$.

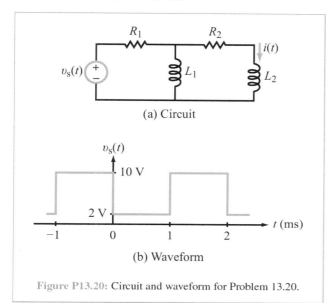

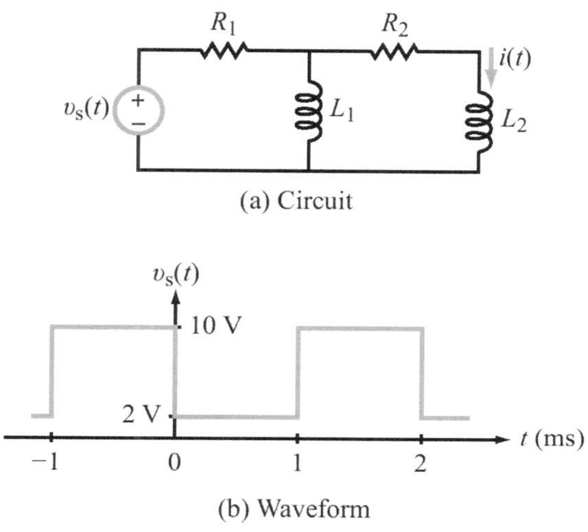

Figure P13.20: Circuit and waveform for Problem 13.20.

13.21 Determine the output voltage $\upsilon_{out}(t)$ in the circuit of Fig. P13.21, given that the input voltage $\upsilon_{in}(t)$ is a full-wave rectified sinusoid (waveform #8 in Table 13-2) with $A = 120$ V and $T = 1\ \mu$s.

(a) Derive the Fourier series representation of $\upsilon_{out}(t)$.

(b) Calculate the first five terms of $\upsilon_{out}(t)$ using $R = 1$ kΩ, $L = 1$ mH, and $C = 1$ nF.

(c) Plot $\upsilon_{out}(t)$ and $\upsilon_{in}(t)$ using $n_{max} = 100$.

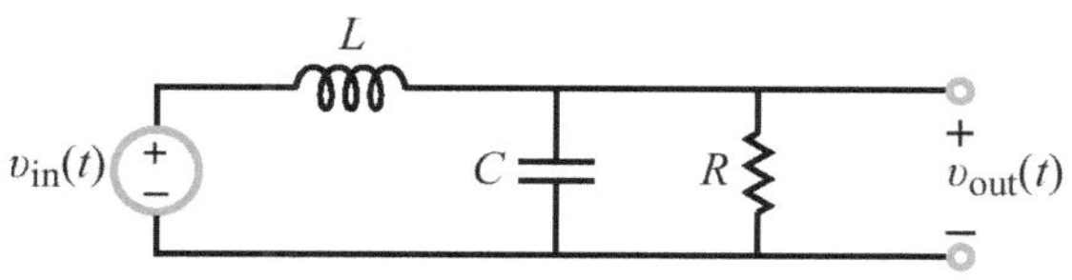

Figure P13.21: Circuit for Problem 13.21.

13.22

(a) Repeat Example 13-5, after replacing the capacitor with an inductor $L = 0.1$ H and reducing the value of R to 1 Ω.

(b) Calculate the first five terms of $\upsilon_{out}(t)$.

(c) Plot $\upsilon_{out}(t)$ and $\upsilon_s(t)$ using $n_{max} = 100$.

13.23 Determine $\upsilon_{out}(t)$ in the circuit of Fig. P13.23, given that the input excitation is characterized by a triangular waveform (#4 in Table 13-2) with $A = 24$ V and $T = 20$ ms.

(a) Derive Fourier series representation of $\upsilon_{out}(t)$.

(b) Calculate first five terms of $\upsilon_{out}(t)$ using $R = 470\ \Omega$, $L = 10$ mH, and $C = 10\ \mu$F.

(c) Plot $\upsilon_{out}(t)$ and $\upsilon_s(t)$ using $n_{max} = 100$.

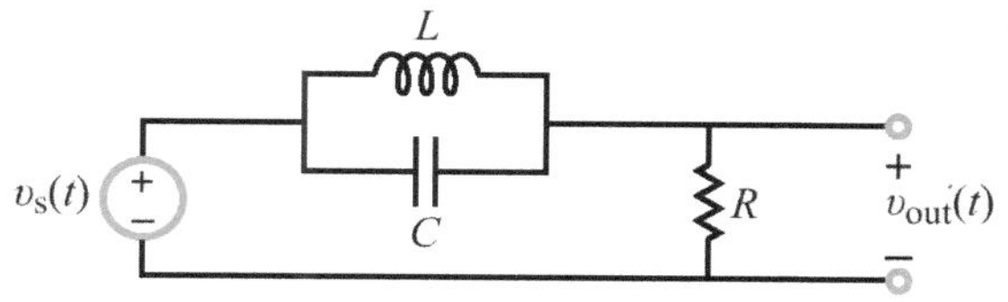

Figure P13.23: Circuit for Problem 13.23.

13.24 A backward-sawtooth waveform (#7 in Table 13-2) with $A = 100$ V and $T = 1$ ms is used to excite the circuit in Fig. P13.24.

(a) Derive Fourier series representation of $\upsilon_{out}(t)$.

(b) Calculate the first five terms of $\upsilon_{out}(t)$ using $R_1 = 1$ kΩ, $R_2 = 100\ \Omega$, $L = 1$ mH, and $C = 1\ \mu$F.

(c) Plot $\upsilon_{out}(t)$ and $\upsilon_s(t)$ using $n_{max} = 100$.

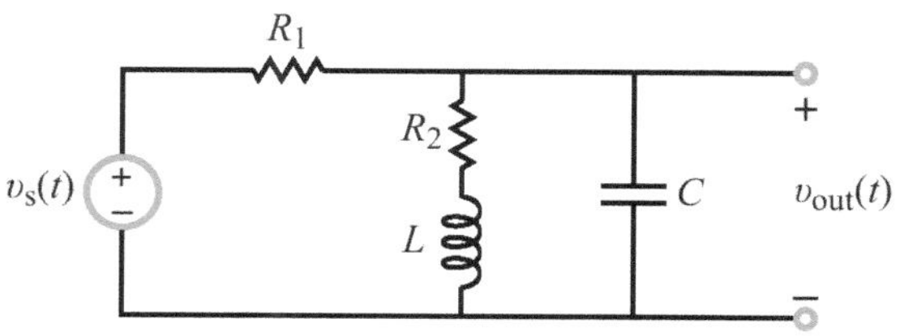

Figure P13.24: Circuit for Problem 13.24.

13.25 The circuit in Fig. P13.25 is excited by the source waveform shown in Fig. P13.20(b).

(a) Derive Fourier series representation of $i(t)$.

(b) Calculate the first five terms of $\upsilon_{out}(t)$ using $R_1 = R_2 = 100\ \Omega$, $L = 1$ mH, and $C = 1\ \mu$F.

(c) Plot $i(t)$ and $\upsilon_s(t)$ using $n_{max} = 100$.

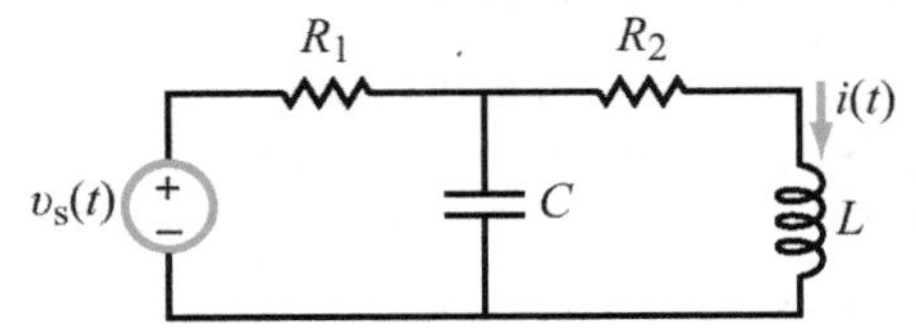

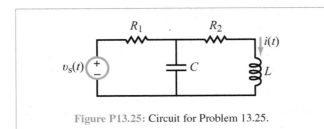

Figure P13.25: Circuit for Problem 13.25.

*13.26 The RC op-amp integrator circuit of Fig. P13.26 is excited by a square wave (waveform #1 in Table 13-2) with $A = 4$ V and $T = 2$ s.

(a) Derive Fourier series representation of $\upsilon_{\text{out}}(t)$.

(b) Calculate the first five terms of $\upsilon_{\text{out}}(t)$ using $R_1 = 1$ kΩ, $R_1 = 10$ kΩ, and $C = 10$ μF.

(c) Plot $\upsilon_{\text{out}}(t)$ using $n_{\max} = 100$.

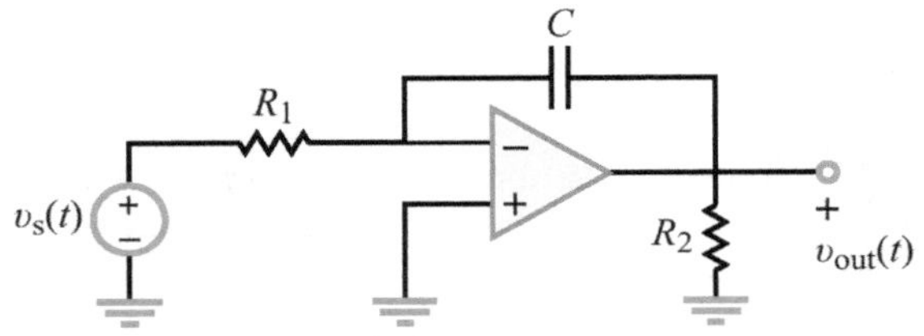

Figure P13.26: Circuit for Problem 13.26.

13.27 Repeat Problem 13.26 after interchanging the locations of the 1 kΩ resistor and the 10 μF capacitor.

Section 13-4: Average Power

13.28 The voltage across the terminals of a certain circuit and the current entering into its (+) voltage terminal are given by

$$\upsilon(t) = [4 + 12\cos(377t + 60^\circ) - 6\cos(754t - 30^\circ)] \text{ V},$$

$$i(t) = [5 + 10\cos(377t + 45^\circ) + 2\cos(754t + 15^\circ)] \text{ mA}.$$

Determine the average power consumed by the circuit, and the ac power fraction.

*13.29 The current flowing through a 2 kΩ resistor is given by

$$i(t) = [5 + 2\cos(400t + 30^\circ) + 0.5\cos(800t - 45^\circ)] \text{ mA}.$$

Determine the average power consumed by the resistor, as well as the ac power fraction.

13.30 The current flowing through a 10 kΩ resistor is given by a triangular waveform (#4 in Table 13-2) with $A = 4$ mA and $T = 0.2$ s.

(a) Determine the exact value of the average power consumed by the resistor.

(b) Using a truncated Fourier series representation of the waveform with only the first four terms, obtain an approximate value for the average power consumed by the resistor.

(c) What is the percentage of error in the value given in (b)?

13.31 The current source in the parallel RLC circuit of Fig. P13.31 is given by

$$i_s(t) = [10 + 5\cos(100t + 30^\circ) - \cos(200t - 30^\circ)] \text{ mA}.$$

Determine the average power dissipated in the resistor given that $R = 1$ kΩ, $L = 1$ H, and $C = 1$ μF.

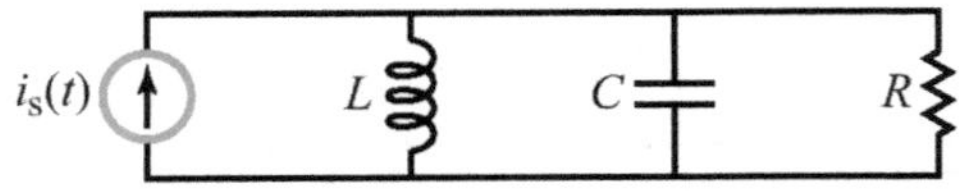

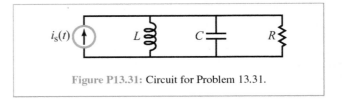

Figure P13.31: Circuit for Problem 13.31.

13.32 A series RC circuit is connected to a voltage source whose waveform is given by waveform #5 in Table 13-2, with $A = 12$ V and $T = 1$ ms. Using a truncated Fourier series representation composed of only the first three nonzero terms, determine the average power dissipated in the resistor, given that $R = 2$ kΩ and $C = 1$ μF.

Sections 13-5 and 13-6: Fourier Transform

For each of the waveforms in Problems 13.33 through 13.42, determine the Fourier transform.

*13.33 Waveform in Fig. P13.33 with $A = 5$ and $T = 3$ s.

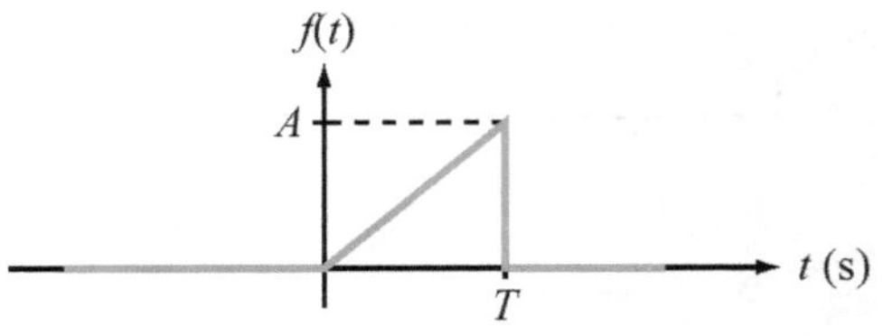

Figure P13.33: Waveform for Problem 13.33.

13.34 Waveform in Fig. P13.34 with $A = 10$ and $T = 6$ s.

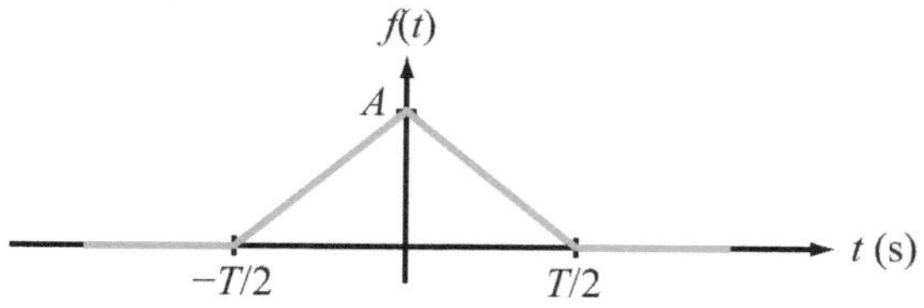

Figure P13.34: Waveform for Problem 13.34.

13.35 Waveform in Fig. P13.35 with $A = 12$ and $T = 3$ s.

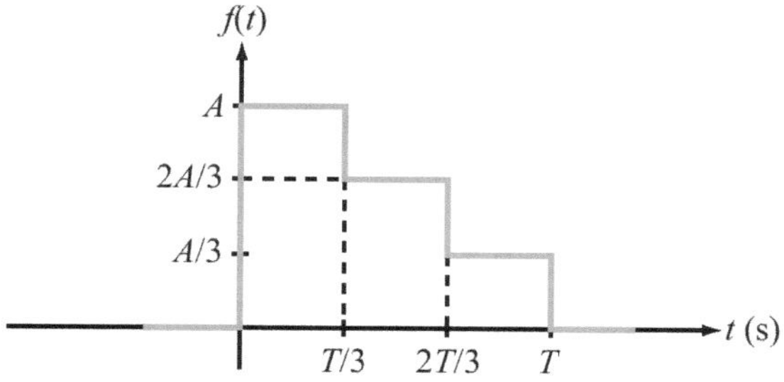

Figure P13.35: Waveform for Problem 13.35.

13.36 Waveform in Fig. P13.36 with $A = 2$ and $T = 12$ s.

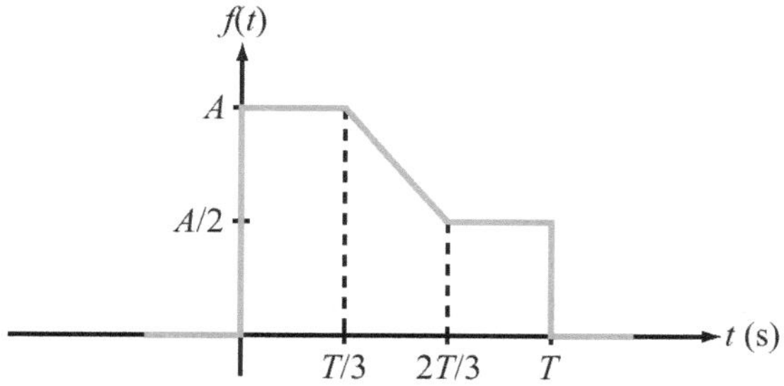

Figure P13.36: Waveform for Problem 13.36.

13.37 Waveform in Fig. P13.37 with $A = 1$ and $T = 3$ s.

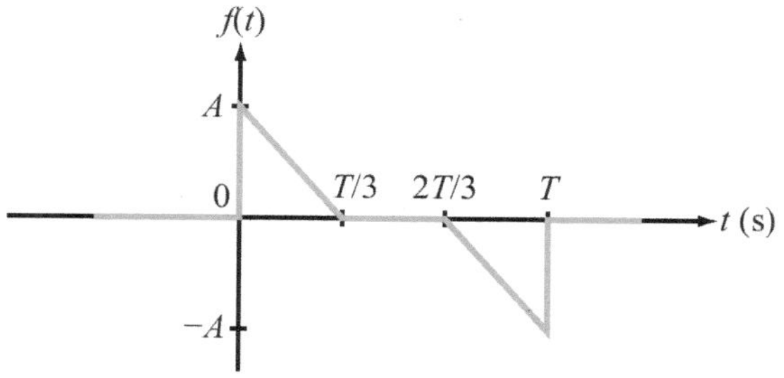

Figure P13.37: Waveform for Problem 13.37.

13.38 Waveform in Fig. P13.38 with $A = 1$ and $T = 2$ s.

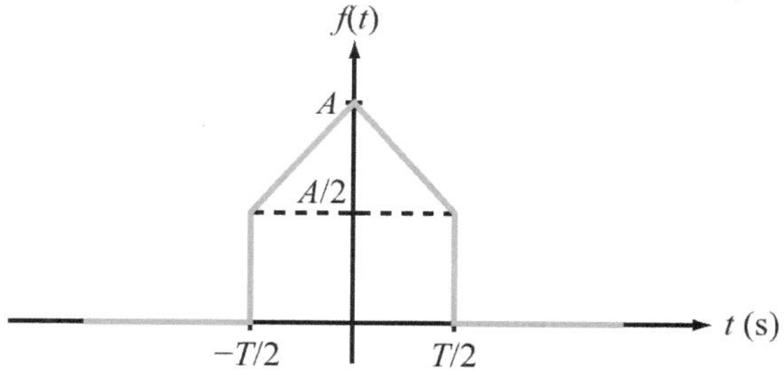

Figure P13.38: Waveform for Problem 13.38.

***13.39** Waveform in Fig. P13.39 with $A = 3$ and $T = 1$ s.

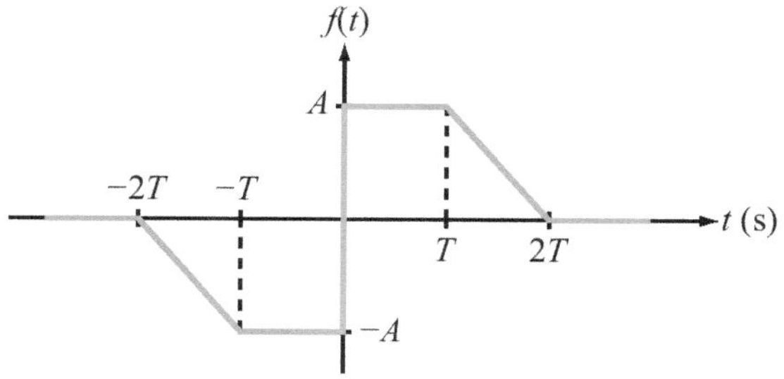

Figure P13.39: Waveform for Problem 13.39.

13.40 Waveform in Fig. P13.40 with $A = 5$, $T = 1$ s, and $\alpha = 10\ \text{s}^{-1}$.

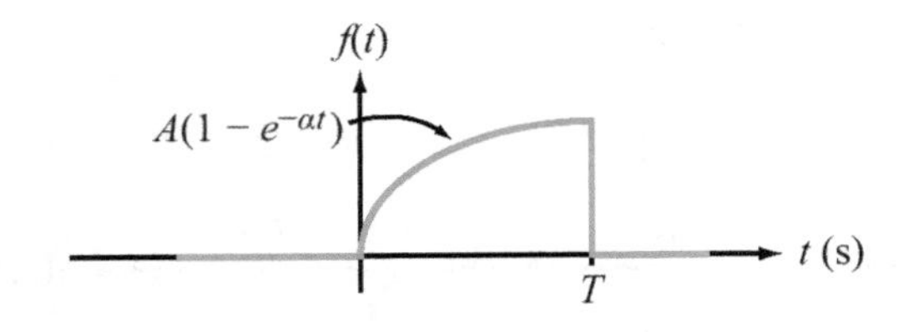

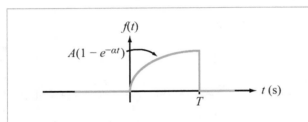

Figure P13.40: Waveform for Problem 13.40.

13.41 Waveform in Fig. P13.41 with $A = 10$ and $T = 2$ s.

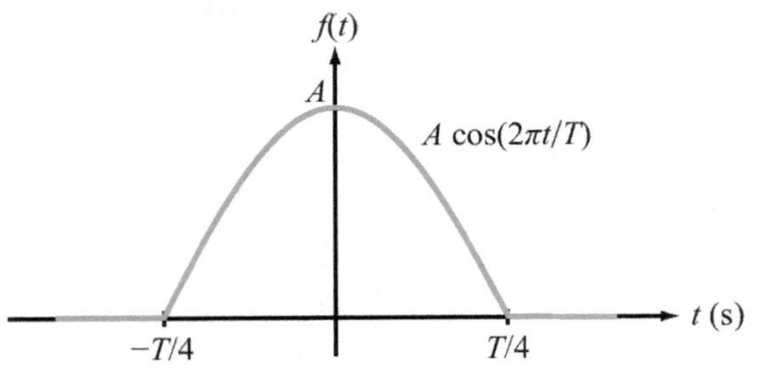

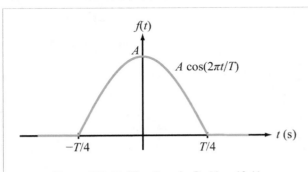

Figure P13.41: Waveform for Problem 13.41.

13.42 Find the Fourier transform of the following signals with $A = 2$, $\omega_0 = 5$ rad/s, $\alpha = 0.5\ \text{s}^{-1}$, and $\phi_0 = \pi/5$.

(a) $f(t) = A\cos(\omega_0 t - \phi_0), \quad -\infty \le t \le \infty$

(b) $g(t) = e^{-\alpha t}\cos(\omega_0 t)\, u(t)$

13.43 Find the Fourier transform of the following signals with $A = 3$, $B = 2$, $\omega_1 = 4$ rad/s, and $\omega_2 = 2$ rad/s.

(a) $f(t) = [A + B\sin(\omega_1 t)]\sin(\omega_2 t)$

(b) $g(t) = A|t|, \quad |t| < (2\pi/\omega_1)$

13.44 Find the Fourier transform of the following signals with $\alpha = 0.5\ \text{s}^{-1}$, $\omega_1 = 4$ rad/s, and $\omega_2 = 2$ rad/s.

(a) $f(t) = e^{-\alpha t}\sin(\omega_1 t)\cos(\omega_2 t)\, u(t)$

(b) $g(t) = te^{-\alpha t}, \quad 0 \le t \le 10\alpha$

13.45 Using the definition of Fourier transform, prove that

$$\mathcal{F}[t\, f(t)] = j\,\frac{d}{d\omega}\,\mathcal{F}(\omega).$$

13.46 Let the Fourier transform of $f(t)$ be

$$F(\omega) = \frac{A}{(B + j\omega)}.$$

Determine the transforms of the following signals (using $A = 5$ and $B = 2$).

(a) $\mathcal{F}(3t - 2)$

*(b) $t\, f(t)$

(c) $d\, f(t)/dt$

13.47 Let the Fourier transform of $f(t)$ be

$$F(\omega) = \frac{1}{(A + j\omega)}\, e^{-j\omega} + B.$$

Determine the Fourier transforms of the following signals (set $A = 2$ and $B = 1$).

(a) $f\left(\frac{5}{8}\, t\right)$

(b) $f(t)\cos(At)$

(c) $d^3 f/dt^3$

13.48 Prove the following two Fourier transform pairs.

(a) $\cos(\omega T)\ F(\omega) \quad \leftrightarrow \quad \frac{1}{2}[f(t - T) + f(t + T)]$

(b) $\sin(\omega T)\ F(\omega) \quad \leftrightarrow \quad \frac{1}{2j}[f(t + T) - f(t - T)]$

Section 13-8: Circuit Analysis with Fourier Transform

13.49 The circuit in Fig. P13.19 is excited by the source waveform shown in Fig. P13.33.

(a) Derive the expression for $\upsilon_{\text{out}}(t)$ using Fourier analysis.

(b) Plot $\upsilon_{\text{out}}(t)$ using $A = 5$, $T = 3$ ms, $R_1 = 500\ \Omega$, $R_2 = 2\ \text{k}\Omega$, and $C = 0.33\ \mu\text{F}$.

(c) Repeat (b) with $C = 0.33$ mF and comment on the results.

13.50 The circuit in Fig. P13.19 is excited by the source waveform shown in Fig. P13.34.

(a) Derive the expression for $\upsilon_{\text{out}}(t)$ using Fourier analysis.

(b) Plot $\upsilon_{\text{out}}(t)$ using $A = 5$, $T = 3$ s, $R_1 = 500\ \Omega$, $R_2 = 2\ \text{k}\Omega$, and $C = 0.33$ mF.

Section 13-10: Multisim

13.51 Design a Sigma-Delta converter that converts a sinusoidal voltage input with a magnitude always $\le |1\ \text{V}|$ and generates a digital signal with 0–5 V range. No voltage into any op amp can exceed ± 20 V.

13.52 Design a Sigma-Delta converter that converts a sinusoidal current input with a magnitude always $\le |1\ \text{mA}|$ and generates a digital signal with 0–5 V range. (Hint: The easiest way to do this is to add an additional op-amp buffer ahead of the subtractor input to convert the current signal into a voltage signal.) No voltage into any op amp can exceed ± 20 V.

Potpourri Questions

13.53 How is data rate related to bandwidth and SNR?

13.54 In brain-machine interfaces, what are the electrodes connected to?

Integrative Problems: Analytical / Multisim / myDAQ

To master the material in this chapter, solve the following problems using three complementary approaches: (a) analytically, (b) with Multisim, and (c) by constructing the circuit and using the myDAQ interface unit to measure quantities of interest via your computer. [myDAQ tutorials and videos are available on CAD.]

†**m13.1 Fourier Series Representation:** Consider the periodic voltage waveform $v(t)$ shown in Fig. m13.1.

(a) Determine if the waveform has dc, even, or odd symmetry.

(b) Obtain its cosine/sine Fourier series representation.

(c) Convert the representation to amplitude/phase format and plot the amplitude line spectrum for $n = 0$ using $A = 10$ V and $T = 4$ ms.

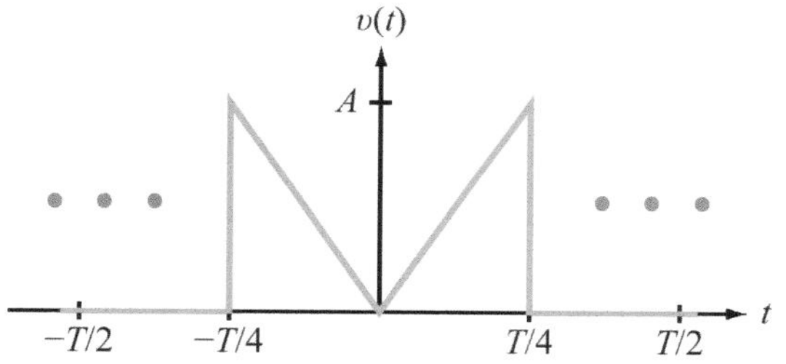

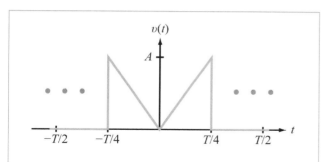

Figure m13.1 Voltage waveform for Problem m13.1.

m13.2 Circuit Applications: The sawtooth voltage waveform $v_s(t)$ shown in Fig. m13.2(a) with $A = 5$ V and $T = 2$ ms serves as the input to the circuit of Fig. m13.2(b).

(a) Determine the Fourier series representation of $v_o(t)$.

(b) Plot $v_o(t)$ and $v_s(t)$ with MathScript or MATLAB as follows:

(1) Time $0 \le t \le 5$ ms,

(2) Sum of $n_{max} = 100$ terms, and

(3) Circuit components $R = 5.6$ kΩ and $C = 0.1$ μF.

Use sufficient time resolution to display Gibbs phenomenon ringing.

(c) Measure the maximum value of $v_o(t)$ from the plot, and also measure the first time at which the maximum value occurs after $t = 0$.

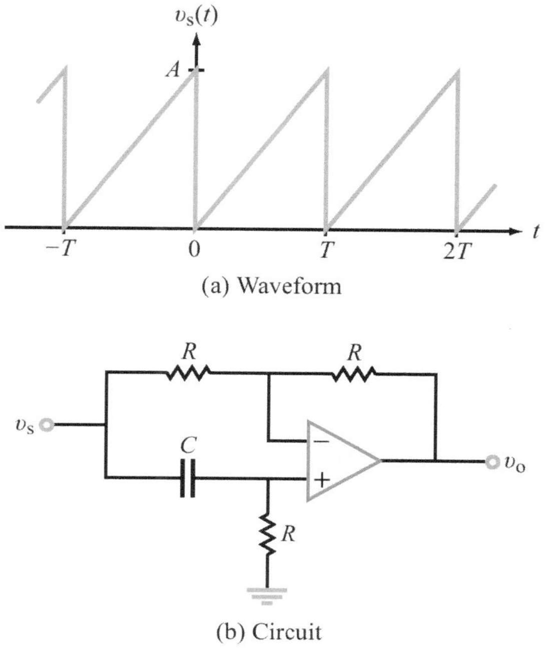

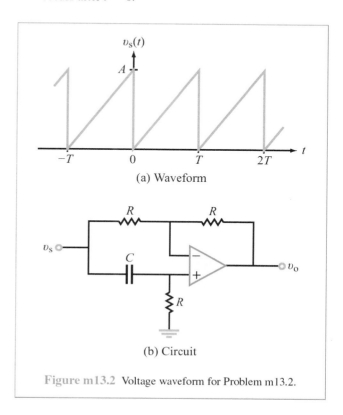

Figure m13.2 Voltage waveform for Problem m13.2.

m13.3 Fourier Transform: For the waveform shown in Fig. m13.3:

(a) Determine its Fourier transform.

(b) Plot the amplitude $|\mathbf{F}(\omega)|$ with MathScript or MATLAB as follows:

(1) Frequency $0 \le f \le 4000$ Hz (remember to convert angular frequency ω to oscillation frequency f),

(2) $A = 10$, and

(3) $\tau = 1$, 2, and 4 ms (create three distinct plots)

(c) Determine the frequency at which the first null occurs in each of the three plots.

(d) Discuss the relationship between the rectangular pulse width and the width of the main lobe of the amplitude spectrum.

† Complete solution available on CAD.

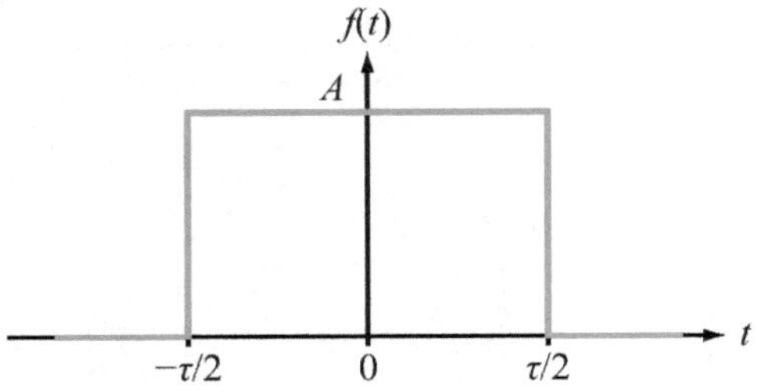

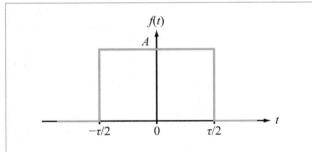

Figure m13.3 Rectangular pulse waveform for Problem m13.3.

m13.4 Circuit Analysis with Fourier Transform: The circuit of Fig. m13.4(a) is excited by the double-pulse waveform shown in Fig. m13.4(b).

(a) Derive the expression for $\upsilon_o(t)$ using Fourier analysis.

(b) Plot $\upsilon_s(t)$ and $\upsilon_o(t)$ on the same graph over the time span $0 \leq t \leq 5$ ms with MathScript or MATLAB for the following values: $A = 5$ V, $T = 1$ ms, $R = 5.6$ kΩ, and $C = 0.1$ μF.

(c) Determine the value of $\upsilon_o(t)$ at time $t = 2$ ms and $t = 3$ ms.

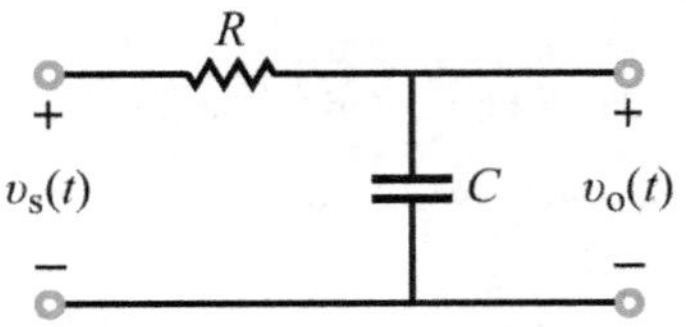

(a) Circuit

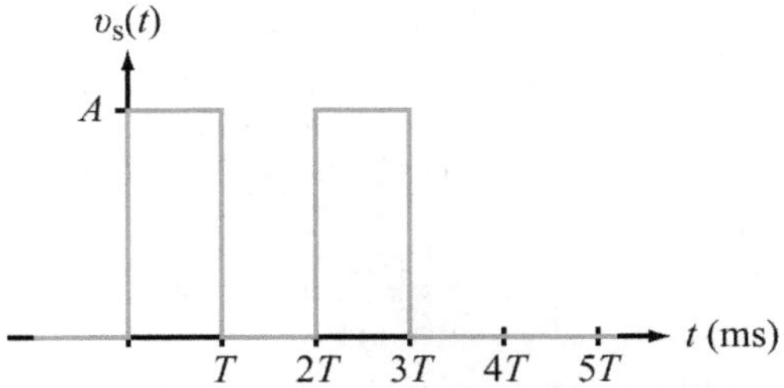

(b) Waveform

Figure m13.4 Double-plus waveform and circuit for Problem m13.4.

APPENDIX A

Symbols, Quantities, and Units

Symbol	Quantity	SI Unit	Abbreviation
A	Cross-sectional area	meter^2	m^2
A	Op-amp gain	dimensionless	—
B	Bandwidth	radians/second	rad/s
C	Capacitance	farad	F
d	Distance or spacing	meter	m
E	Electric field	volt/meter	V/m
F	Force	newton	N
$\mathcal{F}$	Fourier transform	(variable)	(variable)
f	Frequency	hertz	Hz
G	Conductance	siemen	S
G	Closed-loop gain	dimensionless	—
G	Power gain	dimensionless	—
g	MOSFET gain constant	amperes/volt	A/V
$\mathbf{H}$	Transfer function	(variable)	(variable)
I, i	Current	ampere	A
k	Spring constant	newtons/meter	N/m
$\mathcal{L}$	Laplace transform	(variable)	(variable)

Symbol	Quantity	SI Unit	Abbreviation
L	Inductance	henry	H
L, ℓ	Length	meter	m
N	Number of turns	dimensionless	—
P, p	Power	watt	W
P	Mechanical stress	newtons/meter^2	N/m^2
pf	Power factor	dimensionless	—
Q, q	Charge	coulomb	C
Q	Reactive power	volt·ampere reactive	VAR
Q	Quality factor	dimensionless	—
R	Resistance	ohm	Ω
S	Cross-sectional area	meter^2	m^2
$\mathbf{S}$	Complex power	volt·ampere	VA
$\mathbf{s}$	Complex frequency	radians/second	rad/s
T, t	Time	second	s
u	Velocity	meters/second	m/s
V, v	Voltage (potential difference)	volt	V
W	Width	meter	m
W, w	Energy	joule	J
X	Reactance part of impedance	ohm	Ω
$\mathbf{Y}$	Admittance	siemen	S
$\mathbf{Z}$	Impedance	ohm	Ω
α	Piezoresistive coefficient	meters^2/newton	m^2/N
α	Damping coefficient	nepers/second	Np/s
β	Common-emitter current gain	dimensionless	—
β	Air resistance constant	newtons·second/meter	N·s/m
ε	Permittivity	farads/meter	F/m
Λ	Magnetic flux linkage	weber	Wb
λ	Time shift	second	s
λ	Wavelength	meters	m
μ	Magnetic permeability	henrys/meter	H/m
ρ	Resistivity	ohms/meter	Ω/m
σ	Conductivity	siemens/meter	S/m
τ	Time constant or duration	second	s
ϕ	Phase	radians	rad
χ_e	Electrical susceptibility	dimensionless	—
ξ	Damping factor	dimensionless	—
ω	Angular frequency	radians/second	rad/s

APPENDIX B

Solving Simultaneous Equations

Electric circuit analysis methods help us solve for the unknown voltages and currents in the circuits:

- Kirchhoff's circuit law solves for unknown branch currents.
- Node-voltage method solves for unknown node voltages.
- Mesh-current method solves for unknown mesh currents.

In order to solve for n of these unknown currents or voltages, we need n independent equations that relate known values (such as the resistors, voltage and current sources, and other elements in the circuit) to the unknown voltages or currents. Once we have these n equations, we can write them in standard matrix form and solve them using either Cramer's rule (by hand) or matrix inversion (using MATLAB, MathScript RF Module, similar software solvers, or your engineering calculator). This Appendix provides a brief overview of these approaches.

B-1 Review of Cramer's Rule

Let us assume that the application of Kirchhoff's current and voltage laws to a certain circuit led to the following set of equations:

$$2(i_1 + i_2) - 10 + (3i_2 - i_1 - 4i_3) = 0 \qquad \text{(B.1a)}$$

$$-3(i_1 + i_2) + 2(i_1 + 3i_3) = 0 \qquad \text{(B.1b)}$$

$$i_1 - 5 - i_2 = 0 \qquad \text{(B.1c)}$$

Our task is to solve the three independent, simultaneous, linear equations to determine the values of the three unknowns, i_1 to i_3. [Recall that *independence* means that none of the three equations can be generated through a linear combination of the other two.] One way to accomplish the specified task is to apply the method of elimination of variables. If we solve for i_1 in Eq. (B.1c), for example, and then use the expression $i_1 = (i_2 + 5)$ to replace i_1 in Eqs. (B.1a and b), we end up with two new equations containing only two variables, i_2 and i_3. Repeat of the substitution procedure leads to a single equation in only one unknown, which can be solved directly. Once that unknown has been determined, it is a straightforward process to solve for the values of the other two variables.

Such a solution method might prove effective for solving a simple set of three simultaneous equations, but what if the circuit we wish to analyze happens to contain a large number of variables? In that case, the more expeditious approach is to take advantage of Cramer's rule, whose implementation procedure is both systematic and straightforward.

Our review of Cramer's rule will initally use the set of three simultaneous equations given by Eq. (B.1) to demonstrate the mechanics of the solution procedure for a system of order 3. Afterwards, we will treat the general case of a *system of order n* (consisting of n independent equations in n unknowns).

B-1.1 System of Order 3

Step 1: Cast Equations in Standard Form

Before we can apply Cramer's rule, we need to regularize the simultaneous equations into a *standard system* of the following form:

$$a_{11}i_1 + a_{12}i_2 + a_{13}i_3 = b_1, \qquad \text{(B.2a)}$$

$$a_{21}i_1 + a_{22}i_2 + a_{23}i_3 = b_2, \qquad \text{(B.2b)}$$

$$a_{31}i_1 + a_{32}i_2 + a_{33}i_3 = b_3, \qquad \text{(B.2c)}$$

where the a's are the coefficients of the variables, i_1 to i_3, and the b's are the unaffiliated constants. By expanding the bracketed quantities in Eq. (B.1) and collecting terms, we can convert the equations into the standard form defined by Eq. (B.2). Such a process leads to:

$$i_1 + 5i_2 - 4i_3 = 10, \tag{B.3a}$$

$$-i_1 - 3i_2 + 6i_3 = 0, \tag{B.3b}$$

$$i_1 - i_2 = 5. \tag{B.3c}$$

This generates a matrix equation of the form **AI** = **B**, where **I** is the vector of n unknowns (currents i_1 through i_3 in this case):

$$\begin{bmatrix} 1 & 5 & -4 \\ -1 & -3 & 6 \\ 1 & -1 & 0 \end{bmatrix} \begin{bmatrix} i_1 \\ i_2 \\ i_3 \end{bmatrix} = \begin{bmatrix} 10 \\ 0 \\ 5 \end{bmatrix}. \tag{B.4}$$

Note that $a_{11} = 1$, $a_{21} = -1$, and $a_{33} = 0$. The regularized set of three linear, simultaneous equations given by Eq. (B.4) is a system of order 3.

Step 2: General Solution

According to Cramer's rule, the solutions for i_1 to i_3 are given by

$$i_1 = \frac{\Delta_1}{\Delta}, \tag{B.5a}$$

$$i_2 = \frac{\Delta_2}{\Delta}, \tag{B.5b}$$

$$i_3 = \frac{\Delta_3}{\Delta}, \tag{B.5c}$$

where Δ is the value of the *characteristic determinant* of the system represented by Eq. (B.4), and Δ_1 to Δ_3 are the *affiliated determinants* for variables i_1 to i_3. The procedure for evaluating these determinants is covered in Steps 3 and 4. Before we do so, however, we should note that in view of the fact that Δ appears in the denominator in Eq. (B.5), *Cramer's rule cannot provide solutions for the unknown variables when* $\Delta = 0$. This is not surprising, because for any system of n unknowns, the condition $\Delta = 0$ occurs when one, or more, of the equations is not independent. This means that the system contains more unknowns than the available number of independent equations, in which case it has no unique solution.

Step 3: Evaluating the Characteristic Determinant

The characteristic determinant is composed of the a-coefficients of the 3×3 system of equations:

$$\Delta = \begin{vmatrix} a_{11} & a_{12} & a_{13} \\ a_{21} & a_{22} & a_{23} \\ a_{31} & a_{32} & a_{33} \end{vmatrix}. \tag{B.6}$$

Each element in the determinant has an *address* jk specified by its row number j and column number k. Thus, a_{12} is in the first row ($j = 1$) and second column ($k = 2$). For the system given by Eq. (B.4),

$$\Delta = \begin{vmatrix} 1 & 5 & -4 \\ -1 & -3 & 6 \\ 1 & -1 & 0 \end{vmatrix}. \tag{B.7}$$

To evaluate Δ, we *expand* it in terms of the elements of one of its rows. For simplicity, we will always perform the expansion using the top row. The expansion process converts Δ from a determinant of order 3 into the sum of 3 terms, each containing a determinant of order 2. Expanding Eq. (B.7) by its top row gives

$$\Delta = a_{11}C_{11} + a_{12}C_{12} + a_{13}C_{13} = C_{11} + 5C_{12} - 4C_{13}, \tag{B.8}$$

where C_{11}, C_{12}, and C_{13} are the *cofactors* of elements a_{11}, a_{12}, and a_{13}, respectively. The cofactor of any element a_{jk} located at the intersection of row j and column k is related to the minor determinant of that element by

$$C_{jk} = (-1)^{j+k} M_{jk}, \tag{B.9}$$

and the minor determinant M_{jk} is obtained by deleting from the parent determinant all elements contained in row j and column k. Hence, M_{11} is given by Δ after removal of the top row and the left column,

$$M_{11} = \begin{vmatrix} a_{22} & a_{23} \\ a_{32} & a_{33} \end{vmatrix} = \begin{vmatrix} -3 & 6 \\ -1 & 0 \end{vmatrix}. \tag{B.10}$$

For a determinant of order 2, expansion by the top row gives

$$M_{11} = a_{22}M_{22} - a_{23}M_{23} = a_{22}a_{33} - a_{23}a_{32}, \tag{B.11}$$

which is equivalent to diagonal multiplication of the upper-left and lower-right corners to get $a_{22}a_{33}$, followed with multiplication of the other two corners to get $a_{23}a_{32}$, and then subtracting the latter term from the former. Substituting the values of the coefficients we have

$$M_{11} = (-3) \times 0 - 6 \times (-1) = 6. \tag{B.12a}$$

Similarly, M_{12} is obtained by removing from Δ the elements in row 1 and the elements in column 2,

$$\begin{aligned} M_{12} = \begin{vmatrix} a_{21} & a_{23} \\ a_{31} & a_{33} \end{vmatrix} &= a_{21}a_{33} - a_{23}a_{31} \\ &= (-1) \times 0 - 6 \times 1 = -6. \end{aligned} \tag{B.12b}$$

Finally, M_{13} is obtained by removing from Δ row 1 and column 3,

$$M_{13} = \begin{vmatrix} -1 & -3 \\ 1 & -1 \end{vmatrix} = (-1)\times(-1) - (-3)\times 1 = 4. \quad \text{(B.12c)}$$

Inserting the values of the three minor determinants in Eq. (B.8) gives

$$\begin{aligned} \Delta &= C_{11} + 5C_{12} - 4C_{13} \\ &= M_{11} - 5M_{12} - 4M_{13} \\ &= 6 - 5\times(-6) - 4\times 4 = 20. \end{aligned} \quad \text{(B.13)}$$

Step 4: Evaluating the Affiliated Determinants

The affiliated determinant Δ_1 for variable i_1 is obtained by replacing column 1 in the characteristic determinant Δ with a column comprised of the b's in Eq. (B.2). That is,

$$\Delta_1 = \begin{vmatrix} b_1 & a_{12} & a_{13} \\ b_2 & a_{22} & a_{23} \\ b_3 & a_{32} & a_{33} \end{vmatrix} = \begin{vmatrix} 10 & 5 & -4 \\ 0 & -3 & 6 \\ 5 & -1 & 0 \end{vmatrix}. \quad \text{(B.14)}$$

Evaluation of Δ_1 follows the same rules of expansion discussed earlier in Step 3 in connection with the evaluation of Δ. Hence

$$\begin{aligned} \Delta_1 &= 10\begin{vmatrix} -3 & 6 \\ -1 & 0 \end{vmatrix} - 5\begin{vmatrix} 0 & 6 \\ 5 & 0 \end{vmatrix} - 4\begin{vmatrix} 0 & -3 \\ 5 & -1 \end{vmatrix} \\ &= 10\times 6 - 5\times(-30) - 4\times 15 = 150. \end{aligned}$$

Application of Eq. (B.5a) gives

$$i_1 = \frac{\Delta_1}{\Delta} = \frac{150}{20} = 7.5.$$

Similarly, Δ_2 is obtained from Δ upon replacing column 2 with the b-column, and Δ_3 is obtained from Δ upon replacing column 3 with the b-column. The procedure leads to $\Delta_2 = 50$, $\Delta_3 = 50$, $i_2 = \Delta_2/\Delta = 50/20 = 2.5$, and $i_3 = \Delta_3/\Delta = 2.5$.

B-1.2 System of Order n

For a regularized system of linear simultaneous equations given by

$$a_{11}i_1 + a_{12}i_2 + a_{13}i_3 + \cdots + a_{1n}i_n = b_1, \quad \text{(B.15a)}$$

$$a_{21}i_1 + a_{22}i_2 + a_{23}i_3 + \cdots + a_{2n}i_n = b_2, \quad \text{(B.15b)}$$

$$\vdots \quad \vdots \quad \vdots \quad \vdots \quad \vdots \quad \vdots$$

$$a_{n1}i_1 + a_{n2}i_2 + a_{n3}i_3 + \cdots + a_{nn}i_n = b_n, \quad \text{(B.15n)}$$

the solution for any variable i_k of the system is

$$i_k = \frac{\Delta_k}{\Delta}, \quad \text{(B.16)}$$

where Δ is the characteristic determinant and Δ_k is the affiliated determinant for variable i_k. Analogous with the 3×3 system of Section 3-1, Δ is composed of the a-coefficients:

$$\Delta = \begin{vmatrix} a_{11} & a_{12} & a_{13} & \cdots & a_{1n} \\ a_{21} & a_{22} & a_{23} & \cdots & a_{2n} \\ \vdots & \vdots & \vdots & & \vdots \\ a_{n1} & a_{n2} & a_{n3} & \cdots & a_{nn} \end{vmatrix}, \quad \text{(B.17)}$$

and Δ_k is obtained from Δ by replacing column k with the b-column. For example, Δ_2 is given by

$$\Delta_2 = \begin{vmatrix} a_{11} & b_1 & a_{13} & \cdots & a_{1n} \\ a_{21} & b_2 & a_{23} & \cdots & a_{2n} \\ \vdots & \vdots & \vdots & & \vdots \\ a_{n1} & b_n & a_{n3} & \cdots & a_{nn} \end{vmatrix}. \quad \text{(B.18)}$$

To determine the value of a determinant of order n, we can carry out a process of successive expansion, analogous with that outlined in Step 3 of Section 3-1. The first step in the expansion process converts Δ from a determinant of order n into a sum of n terms, each containing a determinant of order $(n-1)$. Each of those new determinants can then be expanded into the sum of determinants of order $(n-2)$, and the process can be contined until it reaches a determinant of order 1, which consists of a single element.

B-2 Matrix Solution Method

The system of three simultaneous equations given by Eq. (B.2) can be cast in matrix form as

$$\begin{bmatrix} a_{11} & a_{12} & a_{13} \\ a_{21} & a_{22} & a_{23} \\ a_{31} & a_{32} & a_{33} \end{bmatrix} \begin{bmatrix} i_1 \\ i_2 \\ i_3 \end{bmatrix} = \begin{bmatrix} b_1 \\ b_2 \\ b_3 \end{bmatrix}, \quad \text{(B.19)}$$

or in symbolic form as

$$\mathbf{AI} = \mathbf{B}, \quad \text{(B.20)}$$

where

$$\mathbf{A} = \begin{bmatrix} a_{11} & a_{12} & a_{13} \\ a_{21} & a_{22} & a_{23} \\ a_{31} & a_{32} & a_{33} \end{bmatrix}, \quad \text{(B.21a)}$$

$$\mathbf{I} = \begin{bmatrix} i_1 \\ i_2 \\ i_3 \end{bmatrix}, \quad \text{(B.21b)}$$

$$\mathbf{B} = \begin{bmatrix} b_1 \\ b_2 \\ b_3 \end{bmatrix}. \quad \text{(B.21c)}$$

Matrix **A** is always a square matrix (same number of rows and columns), so long as the system of simultaneous equations contains the same number of independent equations as the number of unknowns. The solution for the unknown vector **I** is given by

$$\mathbf{I} = \mathbf{A}^{-1}\mathbf{B}, \quad \text{(B.22)}$$

where $\mathbf{A}^{-1}$ is the *inverse* of matrix **A**. The inverse of a square matrix is given by

$$\mathbf{A}^{-1} = \frac{\text{adj}\,\mathbf{A}}{\Delta}, \quad \text{(B.23)}$$

where adj **A** is the *adjoint* of **A** and Δ is the determinant of **A**. The adjoint of **A** is obtained from **A** by replacing each element a_{jk} with its cofactor C_{jk}, and then *transposing* the resultant matrix, wherein the rows and columns are interchanged. Thus,

$$\text{adj}\,\mathbf{A} = [C_{jk}]^T. \quad \text{(B.24)}$$

To illustrate the matrix solution method, let us return to the three simultaneous equations given by Eq. (B.3). Matrices **A** and **B** are given by

$$\mathbf{A} = \begin{bmatrix} 1 & 5 & -4 \\ -1 & -3 & 6 \\ 1 & -1 & 0 \end{bmatrix}, \quad \text{(B.25a)}$$

$$\mathbf{B} = \begin{bmatrix} 10 \\ 0 \\ 5 \end{bmatrix}. \quad \text{(B.25b)}$$

According to Eq. (B.24), adj **A** is given by

$$\text{adj}\,\mathbf{A} = \begin{bmatrix} C_{11} & C_{12} & C_{13} \\ C_{21} & C_{22} & C_{23} \\ C_{31} & C_{32} & C_{33} \end{bmatrix}^T = \begin{bmatrix} C_{11} & C_{21} & C_{31} \\ C_{12} & C_{22} & C_{32} \\ C_{13} & C_{23} & C_{33} \end{bmatrix}. \quad \text{(B.26)}$$

Each cofactor is a 2×2 determinant. Application of the definition given by Eq. (B.9) leads to

$$\text{adj}\,\mathbf{A} = \begin{bmatrix} 6 & 4 & 18 \\ 6 & 4 & -2 \\ 4 & 6 & 2 \end{bmatrix}. \quad \text{(B.27)}$$

Upon incorporating Eqs. (B.22) and (B.23) and using the value of Δ obtained in Eq. (B.13), we have

$$\mathbf{I} = \begin{bmatrix} i_1 \\ i_2 \\ i_3 \end{bmatrix} = \frac{1}{20}\begin{bmatrix} 6 & 4 & 18 \\ 6 & 4 & -2 \\ 4 & 6 & 2 \end{bmatrix}\begin{bmatrix} 10 \\ 0 \\ 5 \end{bmatrix}. \quad \text{(B.28)}$$

Standard matrix multiplication leads to

$$i_1 = \frac{1}{20}\begin{bmatrix} 6 & 4 & 18 \end{bmatrix}\begin{bmatrix} 10 \\ 0 \\ 5 \end{bmatrix} = \frac{1}{20}(6\times 10 + 4\times 0 + 18\times 5) = 7.5. \quad \text{(B.29)}$$

Similarly, multiplication using the second and third rows of adj **A** leads to $i_2 = i_3 = 2.5$.

B-3 MATLAB or MathScript Solution

In MATLAB or MathScript software, matrices **A** and **B** of Eq. (B.25) are entered as:

```
A = [1 5 -4; -1 -3 6; 1 -1 0];
B = [10; 0 ; 5];
```

The solution of **AI** = **B** is obtained by entering the statement

```
I = A \ B;
```

The MATLAB or MathScript response would be

```
I =
    7.5000
    2.5000
    2.5000.
```

More information on using MATLAB and MathScript is available in Appendix E.

APPENDIX C

Overview of Multisim

by Joe Steinmeier

For information on the acquisition of the Multisim software, see the book's website (http://CAD.eecs.umich.edu).

Included on the book's website is a brief tutorial for getting started with Multisim. Multisim is a useful program for simulating and analyzing circuits. While the textbook introduces many of Multisim's fundamental concepts, in the interest of space, many others are left out. The tutorial strives to review as well as continue the textbook's coverage of Multisim. Importantly, the demos help clear up common stumbling blocks and the sometimes strange idiosyncrasies of the Multisim software.

The tutorial consists of 43 basic "Demos," which are mixtures of problems, solutions, investigations, and experiments in Multisim. The Demos are distributed among Chapters 2–9 and also Chapters 12 and 13. Each demo attempts to focus on introducing one main concept of Multisim, although it is unavoidable that other concepts are introduced throughout. An index is included to allow for the quick referencing of the tutorial.

The demos are intended to help you become proficient in Multisim via simple but powerful examples. As you study each chapter in class, it is a good idea to at least skim over them and do some of the sample problems on the book website. Multisim can be an invaluable tool when trying to understand how any circuit works!

The demos, grouped by chapter, cover the following material:

Chapter 2

2.1 Introduction to Multisim/The Three-way Switch

Reviews *layout basics* with a simple, yet elegant, circuit involving three-way switches and light bulbs.

2.2 Resistor Network Analysis

Introduces the Multimeter tool and resistor circuits with many resistors.

2.3 Thermal Sensing Wheatstone Bridge

Demonstrates a variable resistor sensor in a *Wheatstone bridge circuit*.

2.4 The Wattmeter in Multisim

Introduces power measurement in resistive circuits using the Wattmeter tool.

2.5 A Study of Dependent Sources

Discusses the various *dependent sources* in Multisim, their limitations and uses.

2.6 An Introduction to ABM Sources

Discusses the *Analog Behavioral Modeling (ABM) sources* in Multisim, which allow for the creation of formula-based dependent sources.

Chapter 3

3.1 DC Circuit Analysis I

Discusses how to use the Measurement Probes and the Interactive Simulation mode to solve for the voltages and currents in simple circuits.

3.2 DC Circuit Analysis II

Discusses how to use the DC Operating Point Analysis tool to solve for the voltages and currents in simple circuits.

3.3 Multisim and Thévenin and Norton Circuits

Discusses a general technique, using the Measurement Probe, for determining *Thévenin/Norton equivalents* of any circuit.

3.4 Maximum Power Transfer

Uses the Wattmeter and the Interactive Simulation mode to examine power transfer in resistive circuits.

3.5 Plotting Power Transfer in Multisim

Demonstrates how to make an *ABM source* behave like a time-varying resistor and uses this component in a Parameter Sweep analysis to plot power transfer as a function of a varying resistance. This demo presents a very useful technique for plotting how a DC output changes as a function of a changing device parameter.

Chapter 4

4.1 Operational Amplifiers in Multisim

Introduces the DC Operating Point Analysis tool and uses it to analyze an *inverting op-amp circuit*. Also provides very nice tips on making attractive, easy-to-read *plots* in Multisim.

4.2 Introducing the Function Generator and the Oscilloscope

Introduces the Function Generator and Oscilloscope instruments and uses them to measure the voltage gain of an op-amp circuit. Also discusses the Interactive Simulation tool in more detail.

4.3 Introduction to Signal Sources and the Transient Analysis

Introduces *time-varying sources* and the Transient Analysis tool in the context of a simple op-amp circuit.

4.4 Using an Operational Amplifier in a Simple Audio Mixer

Combines the lessons of Demos 4.1, 4.2, and 4.3 to build a three-channel op-amp *audio mixer* and analyze its time-dependent behavior.

Chapter 5

5.1 Introduction to Transient Circuits

Discusses how to build *interactive switch-based transients* in circuits with the Interactive Simulation tool.

5.2 Transient Analysis and First-Order Circuits

Discusses how to build *voltage-controlled switches* to generate transients with the Transient Analysis tool.

5.3 Inductors in Multisim

Extends Demos 5.1 and 5.2 to *switch-driven transients* in circuits with *inductors*.

5.4 Time Constants in RC Circuits

Uses the Parameter Sweep tool to plot the response of an *RC circuit* as you vary the value of R.

Chapter 6

6.1 Parallel RLC Circuit Analysis

Applies the Oscilloscope and both types of switches discussed in Demos 5.1 and 5.2 to analyze the *transient behavior of RLC circuits*.

6.2 An Over-, Under-, and Critically Damped Circuit

Applies *ABM sources* to modeling the three fundamental types of transient responses in an RLC circuit.

Chapter 7

7.1 Measuring Impedance with the Network Analyzer

A good introduction into the Network Analyzer tool.

7.2 Introduction to AC Analysis

Discusses how to produce *frequency response plots* using the AC Analysis tool. The demo uses an RLC circuit as an example, but the technique is useful for any type of circuit.

7.3 AC Thévenin Circuit Determination

Uses the AC Analysis tool and the Network Analyzer instrument to determine the open circuit voltage and complex impedance of an RLC circuit at a specific frequency. Using this data, the demo shows how to calculate the *Thévenin equivalent* circuit and demonstrates that the *transient response* of the original circuit and its Thévenin equivalent are the same.

7.4 Making an Impedance Purely Real

Uses the AC Analysis tool and the Network Analyzer instrument to adjust a circuit's frequency response.

7.5 Modeling an AC-to-DC Power Supply

Builds, tests, and analyzes a *rectifier circuit*, very similar to that in Section 7-9, using Multisim.

7.6 Phase Shift Circuits in Multisim

A good companion demo to Example 7-19 in Chapter 7.

7.7 The Logic Analyzer: An Introduction

Describes the Logic Analyzer used in Example 7-19 of Section 7-10 in more detail.

Chapter 8

8.1 Introduction to RMS Values in Multisim

Discusses how to use the Multimeter instrument to obtain *root-mean-square (rms)* values for voltage, current, and power in a circuit.

8.2 AC Power Using AC Analysis

8.3 Power Factor in Multisim

These two demos (8.2 and 8.3) discuss the somewhat tricky business of plotting *complex power* and *power factors* as a function of frequency using the AC Analysis tool. Because of Multisim's variable and equation nomenclature, a user can easily spend a long time trying to enter the right equations in the analysis tools. This demo clarifies the jargon!

8.4 Three-Phase in Multisim

Demonstrates the *three-phase source* component in Multisim using the Measurement Probe and the Transient Analysis tool.

8.5 Maximizing Power Delivered to a Load in a Complex Circuit

Uses the AC Analysis tool to calculate the power delivered to a load as a function of frequency. A good companion demo to Section 8-6.

Chapter 9

9.1 Introduction to Filters in Multisim

Demonstrates how to use the Bode Plotter instrument and the AC Analysis tool to generate *Bode plots* of any circuit.

9.2 Modeling a (Very)-Low-Pass Filter in Multisim

Models a real-life application of a *low-pass filter*. Frequency response and transient response are shown with the various Multisim analysis tools.

9.3 Speaker Crossover Circuit (Plotting Multiple Filters at Once)

This is a great companion demo to Technology Brief 18: Electrical Engineering and the Audiophile. It demonstrates how to design and test a basic *audio crossover* circuit with the AC Analysis tool.

9.4 AC Parameter Sweep in a Radio Tuner Circuit

Uses the Parameter Sweep tool and the AC Analysis tool to demonstrate how varying a capacitor's value adjusts a filter's response, thereby acting as a *tuner* (or "station selector") for a radio receiver.

9.5 60-Hz Active Notch Filter

Offers an analysis of a *multi-stage op-amp bandstop* (or notch) filter. A good companion demo to Sections 9-7 and 9-9.

Chapter 12

12.1 Piecewise Linear Sources

Provides more detail on the *piecewise linear source*.

12.2 Exponential Sources

Provides more detail on the *exponential source*.

Chapter 13

13.1 Introduction to the Spectrum Analyzer

13.2 Fourier Analysis in Multisim

These two demos (13.1 and 13.2) describe how to measure a signal's various frequency components using either the Spectrum Analyzer instrument and/or the Fourier Analysis tool.

13.3 Analysis of a Square Wave

This demo uses the Spectrum Analyzer and Oscilloscope to demonstrate the construction of a square wave from superimposed sinusoidal components at different frequencies (i.e., the sum of the Fourier components).

APPENDIX D

Mathematical Formulas

D-1 Trigonometric Relations

$$\sin x = \pm \cos(x \mp 90^\circ)$$

$$\cos x = \pm \sin(x \pm 90^\circ)$$

$$\sin x = -\sin(x \pm 180^\circ)$$

$$\cos x = -\cos(x \pm 180^\circ)$$

$$\sin(-x) = -\sin x$$

$$\cos(-x) = \cos x$$

$$\sin^2 x = \frac{1}{2}(1 - \cos 2x)$$

$$\cos^2 x = \frac{1}{2}(1 + \cos 2x)$$

$$\sin(x \pm y) = \sin x \cos y \pm \cos x \sin y$$

$$\cos(x \pm y) = \cos x \cos y \mp \sin x \sin y$$

$$2 \sin x \sin y = \cos(x - y) - \cos(x + y)$$

$$2 \sin x \cos y = \sin(x + y) + \sin(x - y)$$

$$2 \cos x \cos y = \cos(x + y) + \cos(x - y)$$

$$\sin 2x = 2 \sin x \cos x$$

$$\cos 2x = 1 - 2 \sin^2 x$$

$$\sin x + \sin y = 2 \sin\left(\frac{x + y}{2}\right) \cos\left(\frac{x - y}{2}\right)$$

$$\sin x - \sin y = 2 \cos\left(\frac{x + y}{2}\right) \sin\left(\frac{x - y}{2}\right)$$

$$\cos x + \cos y = 2 \cos\left(\frac{x + y}{2}\right) \cos\left(\frac{x - y}{2}\right)$$

$$\cos x - \cos y = -2 \sin\left(\frac{x + y}{2}\right) \sin\left(\frac{x - y}{2}\right)$$

$$e^{jx} = \cos x + j \sin x \qquad \text{(Euler's identity)}$$

$$\sin x = \frac{e^{jx} - e^{-jx}}{2j}$$

$$\cos x = \frac{e^{jx} + e^{-jx}}{2}$$

$$\cos^2 x + \sin^2 x = 1$$

$$2\pi \text{ rad} = 360^\circ$$

$$1 \text{ rad} = 57.30^\circ$$

D-2 Indefinite Integrals (*a* and *b* are constants)

$$\int \sin ax \, dx = -\frac{1}{a} \cos ax$$

$$\int \cos ax \, dx = \frac{1}{a} \sin ax$$

$$\int e^{ax} \, dx = \frac{1}{a} e^{ax}$$

$$\int \ln x \, dx = x \ln x - x$$

$$\int x e^{ax} \, dx = \frac{e^{ax}}{a^2} (ax - 1)$$

$$\int x^2 e^{ax}\,dx = \frac{e^{ax}}{a^3}\,(a^2x^2 - 2ax + 2)$$

$$\int x \sin ax\,dx = \frac{1}{a^2}\,\sin ax - \frac{x}{a}\,\cos ax$$

$$\int x \cos ax\,dx = \frac{1}{a^2}\,\cos ax + \frac{x}{a}\,\sin ax$$

$$\int x^2 \sin ax\,dx = \frac{2x}{a^2}\,\sin ax - \frac{a^2x^2 - 2}{a^3}\,\cos ax$$

$$\int x^2 \cos ax\,dx = \frac{2x}{a^2}\,\cos ax + \frac{a^2x^2 - 2}{a^3}\,\sin ax$$

$$\int e^{ax} \sin bx\,dx = \frac{e^{ax}}{a^2 + b^2}\,(a \sin bx - b \cos bx)$$

$$\int e^{ax} \cos bx\,dx = \frac{e^{ax}}{a^2 + b^2}\,(a \cos bx + b \sin bx)$$

$$\int e^{ax} \sin^2 bx\,dx = \frac{e^{ax}}{a^2 + 4b^2}\left[(a \sin bx - 2b \cos bx) \sin bx + \frac{2b^2}{a}\right]$$

$$\int e^{ax} \cos^2 bx\,dx = \frac{e^{ax}}{a^2 + 4b^2}\left[(a \cos bx + 2b \sin bx) \cos bx + \frac{2b^2}{a}\right]$$

$$\int \sin ax \sin bx\,dx = \frac{\sin(a-b)x}{2(a-b)} - \frac{\sin(a+b)x}{2(a+b)}, \quad a^2 \neq b^2$$

$$\int \cos ax \cos bx\,dx = \frac{\sin(a-b)x}{2(a-b)} + \frac{\sin(a+b)x}{2(a+b)}, \quad a^2 \neq b^2$$

$$\int \sin ax \cos bx\,dx = -\frac{\cos(a-b)x}{2(a-b)} - \frac{\cos(a+b)x}{2(a+b)}, \quad a^2 \neq b^2$$

$$\int \sin^2 ax\,dx = \frac{x}{2} - \frac{\sin 2ax}{4a}$$

$$\int \cos^2 ax\,dx = \frac{x}{2} + \frac{\sin 2ax}{4a}$$

$$\int \frac{dx}{x^2 + a^2} = \frac{1}{a}\,\tan^{-1}\frac{x}{a}$$

$$\int \frac{dx}{(x^2 + a^2)^2} = \frac{1}{2a^2}\left(\frac{x}{x^2 + a^2} + \frac{1}{a}\,\tan^{-1}\frac{x}{a}\right)$$

$$\int \frac{x^2\,dx}{a^2 + x^2} = x - a \tan^{-1}\frac{x}{a}$$

D-3 Definite Integrals (*m* and *n* are integers)

$$\int_0^{2\pi} \sin nx\,dx = \int_0^{2\pi} \cos nx\,dx = 0$$

$$\int_0^{\pi} \sin^2 nx\,dx = \int_0^{\pi} \cos^2 nx\,dx = \frac{\pi}{2}$$

$$\int_0^{\pi} \sin nx \sin mx\,dx = 0, \quad n \neq m$$

$$\int_0^{\pi} \cos nx \cos mx\,dx = 0, \quad n \neq m$$

$$\int_0^{\pi} \sin nx \cos nx\,dx = 0$$

$$\int_0^{\pi} \sin nx \cos mx\,dx = \begin{cases} 0, & \text{if } m + n = \text{even and } m \neq n \\ \dfrac{2n}{n^2 - m^2}, & \text{if } m + n = \text{odd and } m \neq n \end{cases}$$

$$\int_0^{2\pi} \sin nx \cos mx\,dx = 0$$

$$\int_0^{\infty} \frac{\sin ax}{ax}\,dx = \frac{\pi}{2a}$$

APPENDIX E

MATLAB and MathScript

E-1 Background

MATLAB

MATLAB is a computer program developed and sold by the Mathworks, Inc.

"MATLAB" is an abbreviation for MATrix LABoratory. It was originally based on a set of numerical linear algebra programs, written in FORTRAN, called LINPACK. So MATLAB tends to formulate problems in terms of vectors and arrays of numbers, and often solves problems by formulating them as linear algebra problems.

MathScript: **For information on the acquisition of MathScript, see the book's website (CAD.eecs.umich.edu).**

MathScript is a computer program developed and sold by National Instruments, as a module in LabView. The basic commands used by MATLAB also work in MathScript, but higher-level MATLAB commands may not work in MathScript. For the purposes of this *Circuits* book, all necessary commands work equally well in both MATLAB and MathScript.

A student version of MathScript is available for free for students using this book (see http://c3.eecs.umich.edu/ for instructions). *Access to MATLAB is not required to use this book. In this sequel, we use "M/M" to designate "MATLAB or MathScript."*

Getting Started

To use the student version of MathScript, download LabView and then select MathScript under "Tools."

We will use `this font` to represent typed commands and generated output. You can get help for any command, such as `plot`, by typing at the prompt `help plot`.

Some basic things to know about M/M:

- Inserting a **semicolon** ";" at the end of a command suppresses the output; without it M/M will type the results of the computation. This is harmless, but it is irritating to have numbers flying by on your screen.
- Inserting **ellipses** "`...`" at the end of a command means it is continued on the next line. This is useful for long commands.
- Inserting "`%`" at the beginning of a line makes the line a **comment**; it will not be executed. Comments are used to explain what the program is doing at that point.
- `clear` eliminates all present variables. Programs should start with a `clear`.
- `whos` shows all variables and their sizes.
- M/M variables are case-sensitive: `t` and `T` are different variables.
- `save myfile X,Y` saves the variables `X` and `Y` in the file myfile.mat for use in another session of M/M at another time.
- `load myfile` loads all variables saved in myfile.mat, so they can now be used in the present session of M/M.
- `quit` ends the present session of M/M.

.m Files

An M/M program is a list of commands executed in succession. Programs are called "m-files" since their extension is "`.m`," or "scripts."

To write an .m file, at the upper left, click:
File → New → m-file
This opens a window with a text editor.
Type in your commands and then type:
File → Save as → myname.m
Make sure you save it with an .m extension. Then you can run the file by typing its name at the prompt: `>> myname`. Make sure the file name is not the same as a MATLAB command! Using your own name is a good idea.

You can access previously-typed commands using uparrow and downarrow on your keyboard.

To *download a file* from a web site, right-click on it, select **save target as**, and use the menu to select the proper file type (specified by its file extension).

E-2 Basic Computation

E-2.1 Basic Arithmetic

- Addition: `3+2` gives `ans=5`
- Subtraction: `3-2` gives `ans=1`
- Multiplication: `2*3` gives `ans=6`
- Division: `6/2` gives `ans=3`
- Powers: `2^3` gives `ans=8`
- Others: `sin,cos,tan,exp,log,log10`
- Square root: `sqrt(49)` gives `ans=7`
- Conjugate: `conj(3+2j)` gives `ans=3-2i`

Both `i` or `j` represent $\sqrt{-1}$; answers use `i`. `pi` represents π. `e` does not represent 2.71828.

E-2.2 Entering Vectors and Arrays

To enter *row vector* [1 2 3] and store it in `A` type at the prompt `A=[1 2 3];` or `A=[1,2,3];`

To enter the same numbers as a *column vector* and store it in `A`, type at the prompt *either* `A=[1;2;3];` *or* `A=[1 2 3];A=A';` Note `A=A'` replaces `A` with its transpose. "Transpose" means "convert rows to columns, and vice-versa."

To enter a vector of consecutive or equally-spaced numbers, follow these examples:

- `[2:6]` gives `ans=2 3 4 5 6`
- `[3:2:9]` gives `ans=3 5 7 9`
- `[4:-1:1]` gives `ans=4 3 2 1`

To enter an *array* or matrix of numbers, type, for example, `B=[3 1 4;1 5 9;2 6 5];` This gives the array `B` and its transpose `B'`

$$B = \begin{bmatrix} 3 & 1 & 4 \\ 1 & 5 & 9 \\ 2 & 6 & 5 \end{bmatrix} \qquad B' = \begin{bmatrix} 3 & 1 & 2 \\ 1 & 5 & 6 \\ 4 & 9 & 5 \end{bmatrix}$$

Other basics of arrays:

- `ones(M,N)` is an $M \times N$ array of "1"
- `zeros(M,N)` is an $M \times N$ array of "0"
- `length(X)` gives the length of vector `X`
- `size(X)` gives the size of array `X`
  For `B` above, `size(B)` gives `ans=3 3`
- `A(I,J)` gives the (I,J)th element of `A`. For `B` above, `B(2,3)` gives `ans=9`

E-2.3 Array Operations

Arrays add and subtract point-by-point:
`X=[3 1 4];Y=[2 7 3];X+Y` gives `ans=5 8 7`
But `X*Y` generates an error message.
To compute various types of vector products:

- To multiply element-by-element, use `X.*Y` This gives `ans=6 7 12`. To divide element-by-element, type `X./Y`
- To find the inner product of `X` and `Y` $(3)(2)+(1)(7)+(4)(3)=25$, use `X*Y'` This gives `ans=25`
- To find the outer product of `X` and `Y`

$$\begin{bmatrix} (3)(2) & (3)(7) & (3)(3) \\ (1)(2) & (1)(7) & (1)(3) \\ (4)(2) & (4)(7) & (4)(3) \end{bmatrix} \quad \text{use } X'*Y$$

This gives the above matrix.

A common problem is when you think you have a row vector when in fact you have a column vector. Check by using `size(X)`; in the present example, the command gives `ans=1,3` which tells you that `X` is a 1×3 (row) vector.

- The following functions operate on each element of an array separately, giving another array: `sin,cos,tan,exp,log,log10,sqrt` `cos([0:3]*pi)` gives `ans=1 -1 1 -1`
- To compute n^2 for $n = 0, 1 \ldots 5$, use `[0:5].^2` which gives `ans=0 1 4 9 16 25`
- To compute 2^n for $n = 0, 1 \ldots 5$, use `2.^[0:5]` which gives `ans=1 2 4 8 16 32`

Other array operations include:

- `A=[1 2 3;4 5 6];(A(:))'` Stacks A by columns into a column vector and transposes the result to a row vector. In the present example, the command gives `ans=1 4 2 5 3 6`
- `reshape(A(:),2,3)` Unstacks the column vector to a 2×3 array which, in this case, is the original array `A`.
- `X=[1 4 1 5 9 2 6 5];C=X(2:8)-X(1:7)` Takes differences of successive values of `X`. In the present example, the command gives `C=3 -3 4 4 -7 4 -1`
- `D=[1 2 3]; E=[4 5 6]; F=[D E]` This *concatenates* the vectors `D` and `E` (i.e., it appends `E` after `D` to get vector `F`) In the present example, the command gives `F=1 2 3 4 5 6`
- `I=find(A>2)` stores in `I` locations (indices) elements of vector `A` that exceed 2. `find([3 1 4 1 5]<2)` gives `ans=2 4`
- `A(A>2)=0` sets to 0 all values of elements of vector `A` exceeding 2. `A=[3 1 4 1 5]; A(A<2)=0` gives `A=3 0 4 0 5`

M/M indexing of arrays starts with 1, while signals and systems indexing starts with 0. For example, the DFT is defined using index $n = 0, 1 \ldots N-1$, for $k = 0, 1 \ldots N-1$. `fft(X)`, which computes the DFT of `X`, performs

```
fft(X)=X*exp(-j*2*pi*[0:N-1]'*[0:N-1]/N);
```

E-2.4 Solving Systems of Equations

To solve the linear system of equations

$$\begin{bmatrix} 1 & 2 \\ 3 & 4 \end{bmatrix} \begin{bmatrix} x \\ y \end{bmatrix} = \begin{bmatrix} 17 \\ 39 \end{bmatrix}$$

using

```
A=[1 2;3 4];Y=[17;39];X=A\Y;X'
```

gives `ans=5.000 6.000`, which is the solution $[x\ y]'$.

To solve the complex system of equations

$$\begin{bmatrix} 1+2j & 3+4j \\ 5+6j & 7+8j \end{bmatrix} \begin{bmatrix} x \\ y \end{bmatrix} = \begin{bmatrix} 16+32j \\ 48+64j \end{bmatrix}$$

`[1+2j 3+4j;5+6j 7+8j]\[16+32j;48+64j]` gives

$$\texttt{ans=}\begin{matrix} 2-2i \\ 6+2i \end{matrix},$$

which is the solution.

These systems can also be solved using `inv(A)*Y`, but we do not recommend it because computing the matrix inverse of `A` takes much more computation than just solving the system of equations. Computing a matrix inverse can lead to numerical difficulties for large matrices.

E-3 Partial Fractions

E-3.1 Rectangular-to-Polar Complex Conversion

If an M/M result is a complex number, then it is presented in its rectangular form `a+bj`. M/M recognizes both `i` and `j` as $\sqrt{-1}$, so that complex numbers can be entered as `3+2j` or `3+2i`.

To convert a complex number `X` to polar form, use `abs(X),angle(X)` to get its magnitude and phase (in radians), respectively. To get its phase in degrees, use `angle(X)*180/pi`

Note `atan(imag(X)/real(X))` will **not** give the correct phase, since this formula is only valid if the real part is positive. `angle` corrects this.

The real and imaginary parts of `X` are found using `real(X)` and `imag(X)`, respectively.

E-3.2 Polynomial Zeros

To compute the zeros of a polynomial, enter its coefficients as a *row* vector `P` and use `R=roots(P)`. For example, to find the zeros of $3x^3 - 21x+18$ (the roots of $3x^3 - 21x+18=0$) use `P=[3 0 -21 18];R=roots(P);R'`, giving `ans= -3.0000 2.0000 1.0000`, which are the roots.

To find the monic (leading coefficient is one) polynomial from the values of its zeros, enter the numbers as a *column* vector `R` and use `P=poly(R)`. For example, to find the polynomial having $\{1, 3, 5\}$ as its zeros, use `R=[1;3;5];P=poly(R)`, giving `P=1 -9 23 -15`. The polynomial is therefore $x^3 - 9x^2 + 23x$-15.

Note that polynomial are stored as row vectors, and roots are stored as column vectors.

Pole-zero diagrams are made using `zplane`. To produce the pole-zero diagram of

$$\mathbf{H}(\mathbf{z}) = \frac{\mathbf{z}^2 + 3\mathbf{z} + 2}{\mathbf{z}^2 + 5\mathbf{z} + 6},$$

type `zplane([1 3 2],[1 5 6])`. The unit circle $|\mathbf{z}|=1$ is also plotted, as a dotted line.

E-3.3 Partial Fraction Expansions

Partial fraction expansions are a vital part of signals and systems, and their computation is onerous (see Chapter 3). M/M computes partial fraction expansions using `residue`. Specifically,

$$\mathbf{H}(\mathbf{s}) = \frac{b_0\mathbf{s}^M + b_1\mathbf{s}^{M-1} + \cdots + b_M}{a_0\mathbf{s}^N + a_1\mathbf{s}^{N-1} + \cdots + a_N}$$

has the partial fraction expansion (if $M \leq N$)

$$\mathbf{H}(\mathbf{s}) = K + \frac{\mathbf{R_1}}{\mathbf{s} - \mathbf{p_1}} + \cdots + \frac{\mathbf{R_N}}{\mathbf{s} - \mathbf{p_N}}$$

The poles $\{\mathbf{p_i}\}$ and residues $\{\mathbf{R_i}\}$ can be computed from coefficients $\{a_i\}$ and $\{b_i\}$ using

`B=`$[b_0\ b_1 \ldots b_M]$;`A=`$[a_0\ a_1 \ldots a_N]$
```
[R P]=residue(B,A);[R P]
```

The residues $\{R_i\}$ are given in column vector `R`, and poles $\{p_i\}$ are given in column vector `P`.

To compute the partial fraction expansion of

$$\mathbf{H}(\mathbf{s}) = \frac{3\mathbf{s} + 6}{\mathbf{s}^2 + 5\mathbf{s} + 4},$$

use the command

```
[R P]=residue([3 6],[1 5 4]);[R P]
```

This gives $\begin{bmatrix} 2 & -4 \\ 1 & -1 \end{bmatrix}$, so `R` $= \begin{bmatrix} 2 \\ 1 \end{bmatrix}$ and `P` $= \begin{bmatrix} -4 \\ -1 \end{bmatrix}$, from which we read off

$$\mathbf{H}(\mathbf{s}) = \frac{2}{\mathbf{s} + 4} + \frac{1}{\mathbf{s} + 1}.$$

In practice, the poles and residues both often occur in complex conjugate pairs. Then use

$$\mathbf{R}e^{\mathbf{p}t} + \mathbf{R}^*e^{\mathbf{p}*t} = 2|\mathbf{R}|e^{at}\cos(\omega t + \theta),$$

$\mathbf{R} = |\mathbf{R}|e^{j\theta}$ and $\mathbf{p} = a + j\omega$, to simplify the result.

To compute the partial fraction expansion of

$$\mathbf{H}(\mathbf{s}) = \frac{\mathbf{s} + 7}{\mathbf{s}^2 + 8\mathbf{s} + 25},$$

use the command

```
[R P]=residue([1 7],[1 8 25]);[R P]
```

This gives

$$\begin{bmatrix} 0.5000 - 0.5000i & -4.000 + 3.000i \\ 0.5000 + 0.5000i & -4.000 - 3.000i \end{bmatrix}$$

from which we have

$$\mathbf{H}(\mathbf{s}) = \frac{0.5 - j0.5}{\mathbf{s} + 4 - j3} + \frac{0.5 + j0.5}{\mathbf{s} + 4 + j3},$$

which has the inverse Laplace transform

$$h(t) = (0.5 - j0.5)e^{(-4+j3)t} + (0.5 + j0.5)e^{(-4-j3)t}$$

From `abs(0.5-0.5j),angle(0.5-0.5j)`,

$$h(t) = 2\frac{\sqrt{2}}{2}e^{-4t}\cos(3t - \frac{\pi}{4}) = \sqrt{2}e^{-4t}\cos(3t - \frac{\pi}{4}).$$

Both $h(t)$ expressions are valid for $t > 0$.

If $\mathbf{H(s)}$ is proper but not strictly proper, the constant K is nonzero. It is computed using

```
[R P K]=residue(B,A);[R P],K
```

since `K` has size different from `R` and `P`.

To find the partial fraction expansion of

$$\mathbf{H(s)} = \frac{\mathbf{s}^2 + 8\mathbf{s} + 9}{\mathbf{s}^2 + 3\mathbf{s} + 2},$$

use the command

```
[R P K]=residue([1 8 9],[1 3 2]);[R P] K
```

gives $\begin{bmatrix} 3 & -2 \\ 2 & -1 \end{bmatrix}$, `K=1` so `R=` $\begin{bmatrix} 3 \\ 2 \end{bmatrix}$, `P=` $\begin{bmatrix} -2 \\ -1 \end{bmatrix}$, from which we read off

$$\mathbf{H(s)} = 1 + \frac{3}{\mathbf{s}+2} + \frac{2}{\mathbf{s}+1}.$$

Double poles are handled as follows:

To find the partial fraction expansion of

$$\mathbf{H(s)} = \frac{8\mathbf{s}^2 + 33\mathbf{s} + 30}{\mathbf{s}^3 + 5\mathbf{s}^2 + 8\mathbf{s} + 4},$$

use the command

```
[R P]=residue([8 33 30],[1 5 8 4]);
```

`[R P]` gives $\begin{bmatrix} 3 & -2 \\ 4 & -2 \\ 5 & -1 \end{bmatrix}$, so `R=` $\begin{bmatrix} 3 \\ 4 \\ 5 \end{bmatrix}$, `P=` $\begin{bmatrix} -2 \\ -2 \\ -1 \end{bmatrix}$. We then read off

$$\mathbf{H(s)} = \frac{3}{\mathbf{s}+2} + \frac{4}{(\mathbf{s}+2)^2} + \frac{5}{\mathbf{s}+1}.$$

In practice, we are interested not in an analytic expression for $h(t)$, but in computing $h(t)$ sampled every T_s seconds. These samples can be computed directly from `R` and `P`, for $0 \leq t \leq T$:

```
t=[0:Ts:T];H=real(R.'*exp(P*t));
```

Since `R` and `P` are column vectors, and `t` is a row vector, `H` is the inner products of `R` with each column of the array `exp(P*t)`. `R.'` transposes `R` without also taking complex conjugates of its elements. `real` is necessary since roundoff error creates a tiny (incorrect) imaginary part in `H`.

APPENDIX F

myDAQ Quick Reference Guide

by Nathan Sawicki*

Note: The myDAQ board, which does not come with the book, can be purchased directly from National Instruments.

NI myDAQ is a student instrumentation and data acquisition device that allows students to make electronic measurements from their PC (Fig. F-1). Usually the circuit is built on a breadboard, which allows for making electrical connections between components without permanently soldering the connections. myDAQ converts the PC into an electronic instrumentation laboratory containing a variety of voltage sources (DC, AC, pulse, etc.) with which to excite the circuit, and standard test equipment (multimeter, oscilloscope, etc.) for measuring the voltage and current responses of the circuit. The physical wire connections made between the circuit and the myDAQ are essentially the same as those one would make to a real source or test equipment, and the graphical displays on the PC are visually very much like those one would see on the real test equipment.

This appendix is a quick reference guide for the myDAQ and how to use it. Additional material including video tutorials, are available on the book website http://c3.eecs.umich.edu/.

F-1 Getting Started with the myDAQ

Follow the directions that came with your myDAQ to (a) install the NI ELVISmx software (if you have a MAC instead of a PC, you will need to install a Windows emulator before installing the myDAQ software), (b) install the 20-pin connector in the myDAQ, and (c) connect the myDAQ to the PC.

F-1.1 USB Port

Connect the myDAQ to the PC using the USB cable that came with the MyDAQ. The light next to the USB port will indicate that the myDAQ is functional (Fig. F-2).

*This appendix was originally written by Mr. Sawicki, a fourth-year student at the University of Michigan, and later edited by the book authors.

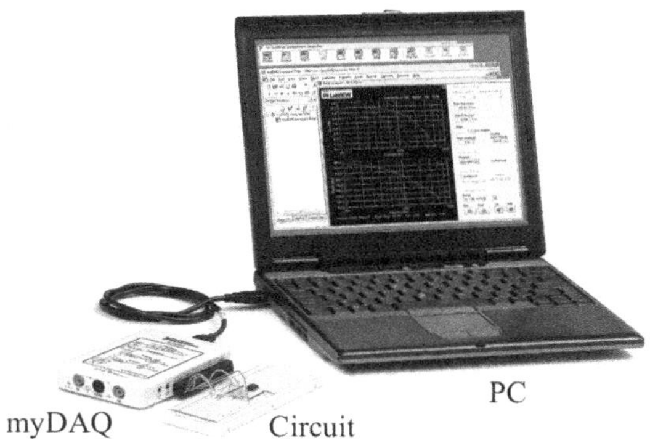

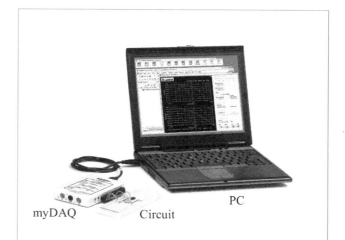

Figure F-1: myDAQ connected to a PC and an electric circuit. The PC is used to provide the voltage inputs to the circuit, as well as to measure its output voltages and currents.

F-1.2 Source Ports

Install the 20-Screw Terminal Connector into the myDAQ (you may need to press hard to insert the connector). This will give you access to:

(1) analog voltage sources ($\pm$ 15 V (left) and +5 V (far right pin)), which are always "on,"

(2) two grounds, AGND and DGND, which are connected internally in the myDAQ,

(3) two analog output ports, A0 and A1, which can provide variable DC, AC, and other waveforms up to $\pm$10 V peak,

(4) a digital I/O port DIO (which we will not be using in this book), and

(5) Audio In and Audio Out ports.

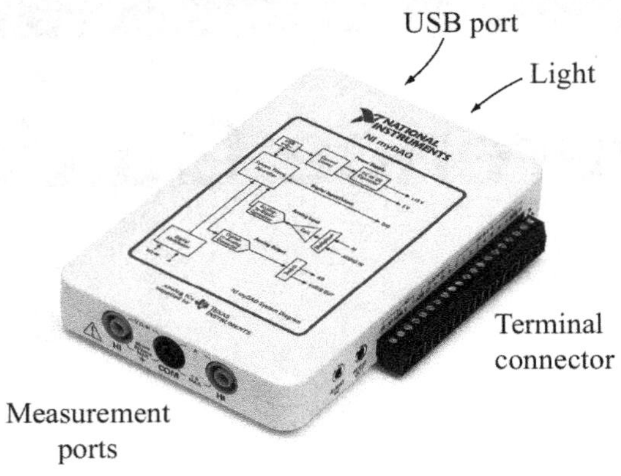

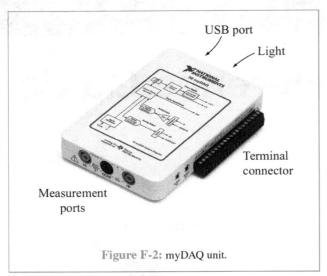

Figure F-2: myDAQ unit.

The myDAQ does not provide a current source, so we will demonstrate how to build one later on in this appendix. To access any of these ports, use the screwdriver that came with the myDAQ to loosen the screw at the top of the terminal, install the (stripped) end of the wire, and retighten the screw.

▶ The ±15 V and 5 V sources are always "on" when the myDAQ is operational. The other sources are "on" or "off," depending on the choices selected in the control panel on the PC display. Be careful not to short (touch) these sources to each other, to ground, to parts of your circuit where you do not intend them to be, or to you. Be careful when measuring within your circuit that metal probe tips do not accidentally touch (short) multiple points in the circuit. Although the current produced by the myDAQ is low, do not touch these sources (you can touch the insulated parts of wires, etc., but avoid touching the stripped ends and other bare metal parts attached to these sources). It is always a good idea when building or modifying circuits to disconnect the source from the circuit while making changes. ◀

F-1.3 Measurement Ports

The three measurement ports (Fig. F-2) provide access to the digital multimeter (DMM: voltmeter, ohmmeter, ammeter, as well as diode test). Depending on the type of measurement you wish to make, connect the probes that came with the myDAQ to these ports. One probe (black) should always be connected to the COM port, which is the common/reference/ground node. Then, connect the other (red) probe to one of the HI ports:

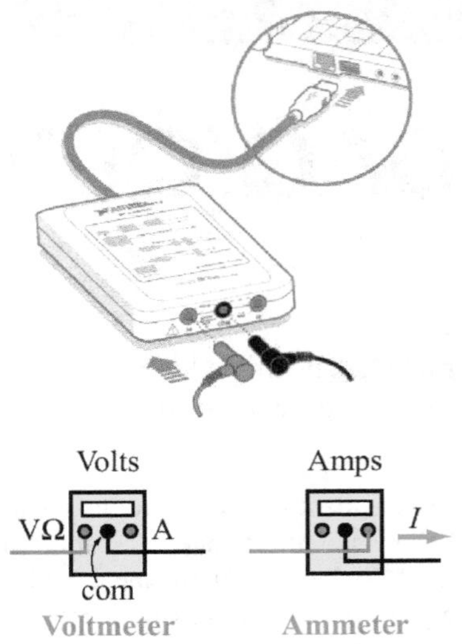

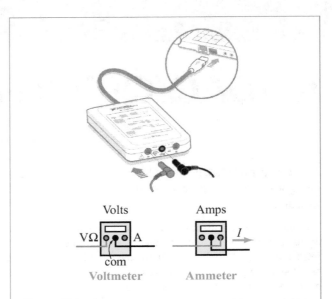

Figure F-3: DMM ports. Note that these three connections represent the same three connections on the DMM graphic shown throughout the book.

(a) The left port is for voltage, resistance, and diode measurements, and

(b) the right port is used for current measurements. Figure F-3 shows the proper connection for voltage and resistance measurements. The connection for current uses the right red port instead of the left one.

▶ WARNING: The maximum allowable myDAQ voltage is 60 VDC/Vrms. DO NOT plug the myDAQ into circuits with hazardous voltages, such as wall outlets and car batteries; doing so could damage the myDAQ or cause injury. READ the *NI myDAQ User Guide and Specifications* prior to NI myDAQ use. Be careful when using the ammeter to connect it in SERIES with the current being measured, NOT in parallel, or you could blow out the fuse on the myDAQ. ◀

F-1.4 NI ELVISmx Instrument Launcher

Run the NI ELVISmx software. If the instrument launcher is not already open, go to **Start≫All Programs≫National Instruments≫NI ELVISmx for NI ELVISMX & NI myDAQ≫NI ELVISmx Instrument Launcher**.

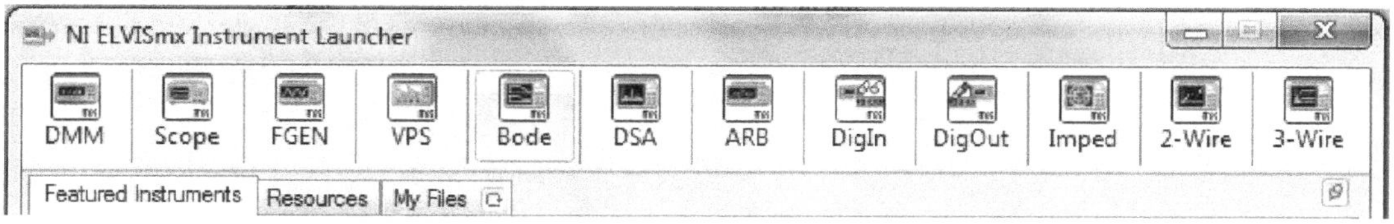

Figure F-4: Instrument launcher.

The instruments used in this book include (Fig. F-4):

DMM: **(Digital Multimeter):** Measures magnitudes of voltage, current, and resistance.

Scope: Measures time-varying voltages.

FGEN **(Function Generator):** Source for time varying (oscillating) voltages including sine waves, square waves, rectangular functions, and pulse trains.

Bode: Measures the frequency response of a circuit.

ARB **(Arbitrary Waveform Generator):** Source for arbitrary waveforms including variable DC voltages.

F-2 Measuring Resistance

1. Software: Run the NI ELVISmx software. Select the digital multimeter (DMM) from the instrument launcher. Select the ohmmeter, and its setup as shown in Fig. F-5. Select either Auto Range (a good start) or a range slightly above the value you expect to measure. Follow the instructions in step 3 below to connect the resistor you want to test to myDAQ, then press RUN. It will take a few seconds from when you connect the resistor to when the correct value is shown on the PC display.

2. Connection: Connect probe cables to the left HI and COM ports on the myDAQ as shown.

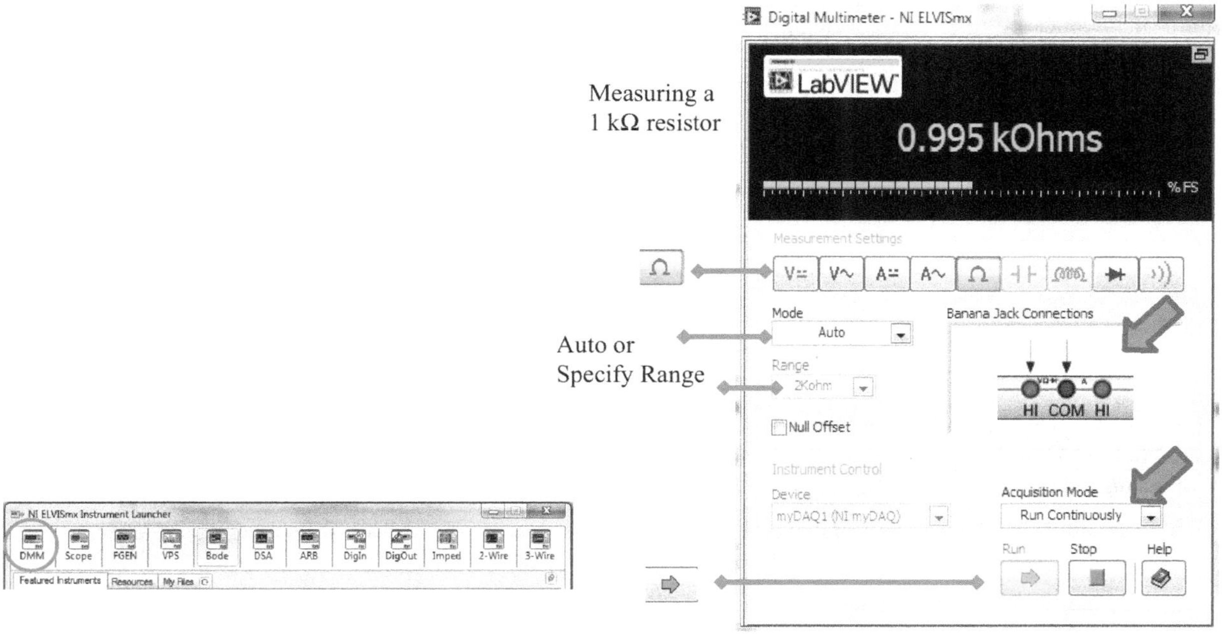

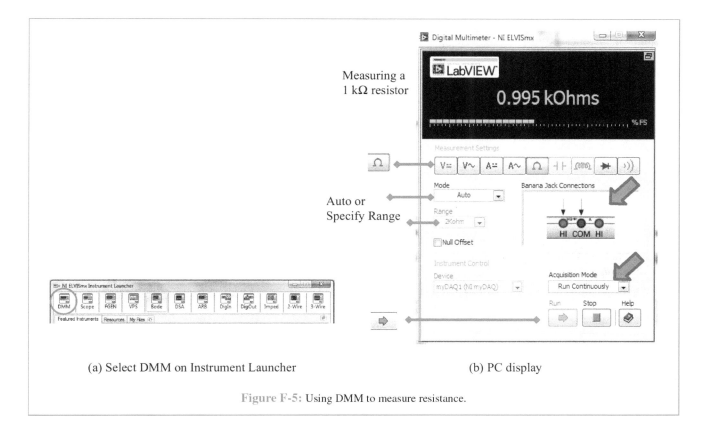

(a) Select DMM on Instrument Launcher

(b) PC display

Figure F-5: Using DMM to measure resistance.

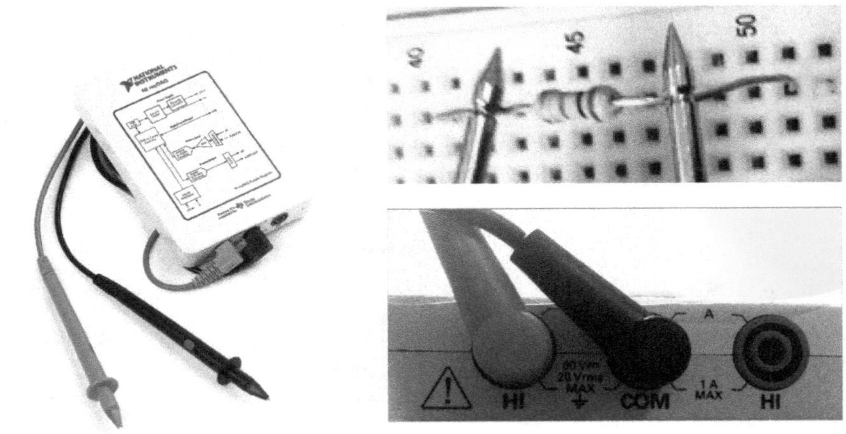

Figure F-6: Connecting probes to a resistor.

3. Measure: Touch the pointy ends of the probes to either end of a resistor (red/black polarity does not matter). It is safe to hold the resistor and probes in your fingers. The probes are also pointy enough to poke directly into the protoboard holes, plus they have small gouges to help steady them against wires, as shown in Fig. F-6. For easier connection, you can use additional clips to connect the probes to your wires. It is important to have good quality connections between the measurement probes and your circuit, which requires steady pressure or additional clips. The resistor in the figure (Brown-Black-Red) is a 1 kΩ resistor.

NOTE: If you are measuring a resistor in a circuit, your measurement will include the effect of other connected circuit elements. Remove the resistor from the circuit if you are trying to measure its resistance alone.

F-3 Measuring Voltage

Specification: **The maximum voltage the myDAQ can be connected to is 60 V DC/Vrms. Do not connect the myDAQ to a voltage higher than this.**

Software: Run the NI ELVISMX software. Select the DMM option (Fig. F-7).

Connection: Connect probe cables to the *leftr* HI and COM ports on the myDAQ as shown in Fig. F-6.

When connecting the voltmeter to the circuit, the red lead should be connected to the (+) terminal of the voltage being measured, and the black lead should be connected to the (−) terminal, as shown in Fig. F-8(b). Figure 1-18(b) of the text shows the connections for measurement of differential voltages and node voltages.

Measure AA Battery: Touch the red DMM probe to the positive battery terminal while touching the black DMM probe to the negative battery terminal. Press hard enough to make a good electrical connection. You should see the DMM reading change to a value close to the battery's voltage rating. AAA, AA, C, D cells should all give a reading of around 1.5 V. Now see what happens if you switch the leads and put the red lead on the − terminal and the black lead on the + terminal.

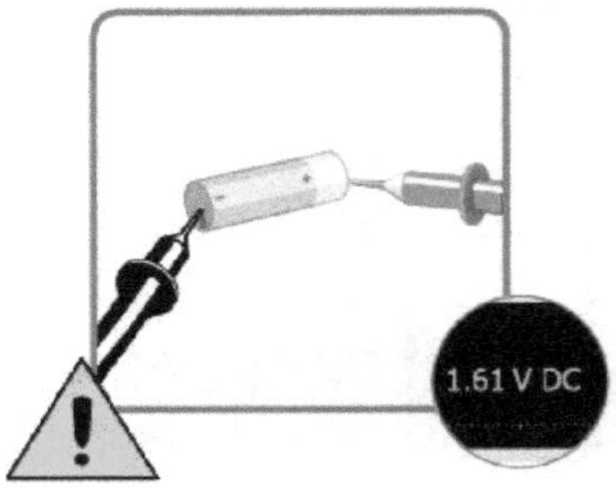

Measure myDAQ Voltages: Now let's measure the voltages that come out of the myDAQ. Press RUN on the DMM, so that it is measuring voltage. You should see 0.0 V on the PC display until you connect the probes across an electrical device or a circuit. Press the red myDAQ probe to the screw on the +15 *V*

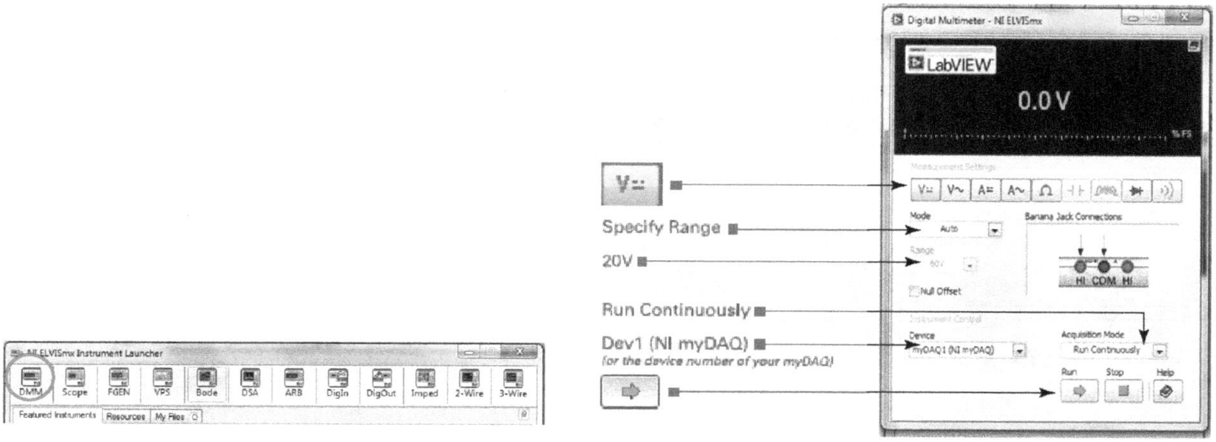

(a) Select DMM on Instrument Launcher (b) PC display

Figure F-7: Using DMM to measure voltage.

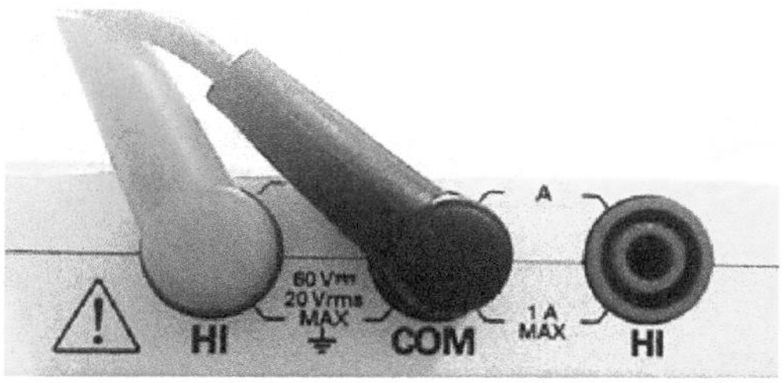

(a) Probe leads

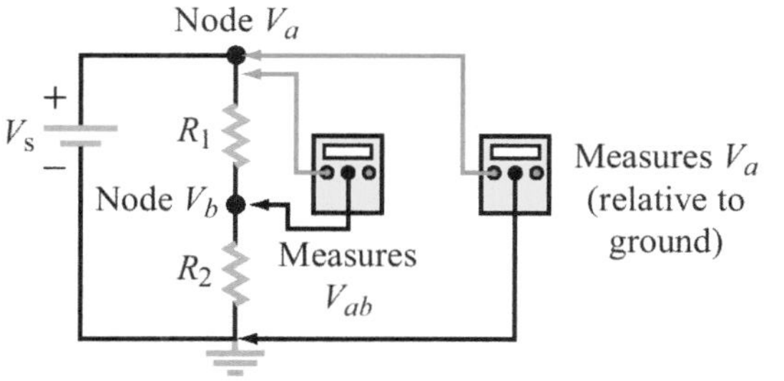

(b) Voltmeters connected to measure voltage difference V_{ab} and node voltage V_a (relative to ground)

Figure F-8: Connecting probes to measure voltage.

connection on the 20-pin terminal and the black probe to the AGND pin. You should see a voltage close to +15 V. Repeat for the −15 V and +5 V measurements. What happens if you switch the + (red) and − (black) DMM leads in your measurements? Press the *Stop button* when you have finished your readings.

Evaluate: Was the voltage you measured from your battery lower than its voltage rating? If so, why? What happened when you switched the leads?

F-4 Measuring Current

Specification: **The maximum current the myDAQ can be connected to is 1A DC/Arms. Do not connect the myDAQ to a current higher than this.**

Software: Run the NI ELVISMX software. Select the DMM software. Select the DC ammeter (A⎓), and its setup as shown in Fig. F-9. Select either Auto Range (a good start) or a range slightly above the value you expect to measure (20 mA is a good range for this test). Press RUN. The DMM is now measuring current (which will be 0.0 until you connect the probes in series with a current you want to measure).

Connection: Connect the Probes to the myDAQ as shown in Fig. F-10. When connecting the ammeter to the circuit, the red lead is where the current comes in (the "tail" of the current arrow), and the black lead is where the current leaves (the "tip/head" of the arrow). Figure F-10(b) shows the connections. If

(a) Select DMM

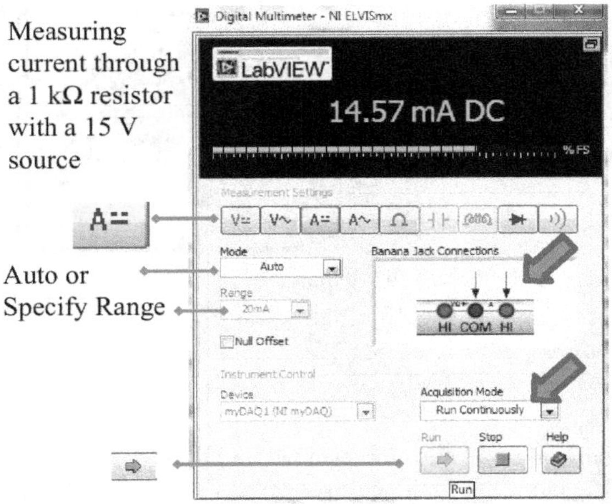

(b) PC display

Figure F-9: Using DMM to measure current.

(a) Probe connection

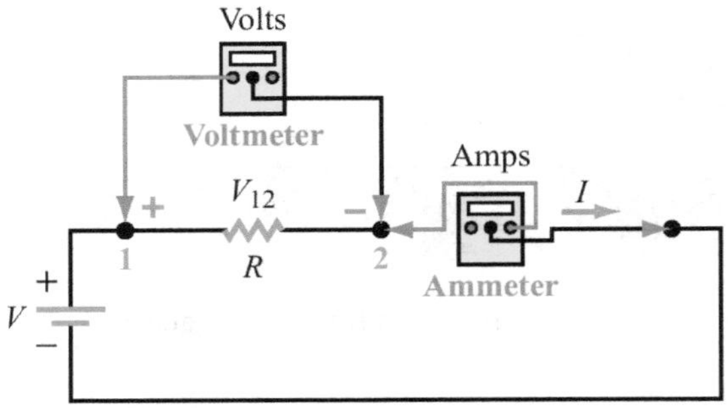

(b) DMM connections for measuring current

Figure F-10: Probe connections for measuring current.

the display result is negative, it means the current is flowing in the opposite direction.

Measuring I: Since the myDAQ does not have a built-in current source (current sources are not standard sources), we need to build a simple circuit and measure the current flowing through it.

To that end, build the circuit in Fig. F-11(a) with $V = +15$ V (from the myDAQ) and $R = 1$ kΩ:

- Connect a red wire (with both ends stripped) to the myDAQ +15 V and a similar black wire to the myDAQ AGND (Fig. F-11(b)). Tighten the screws in the 20-terminal connector to hold them securely in place.

> ▶ **The myDAQ is on, and the voltages are live, so don't let the ends of the red and black wires touch each other.** ◀

- Insert a 1 kΩ resistor (a resistor with brown-black-red lines) in your protoboard as shown in Fig. F-11(b). In the protoboard, Row 2 corresponds to node 1 in Fig. F-11(a). The other end of the resistor (row 8) is node 2.
- Connect the red wire from the myDAQ +15 V output to another point on row 2 on the protoboard. This is also node 1 in Fig. F-11(a).
- Connect the red probe from the myDAQ to the other end of the resistor, using an alligator clip to make a good connection. This is node 2.
- Connect the black wire from AGND to another row in the protoboard. This will be node 3. Connect a small piece of wire onto which you can clip the black alligator lead in this row also.
- Connect the black probe from the myDAQ ammeter (COM) to the black wire alligator clip connected to AGND (node 3). Connect the red probe from the myDAQ ammeter (right-HI) to the other end of the red alligator clip (node 2).
- You have now completed the circuit, and (ideally) the display is $I = 15$ V/1 kΩ $= 15$ mA.

> ▶ NOTE: Unclip the leads when you finish your measurement, as prolonged current will heat up the resistor. ◀

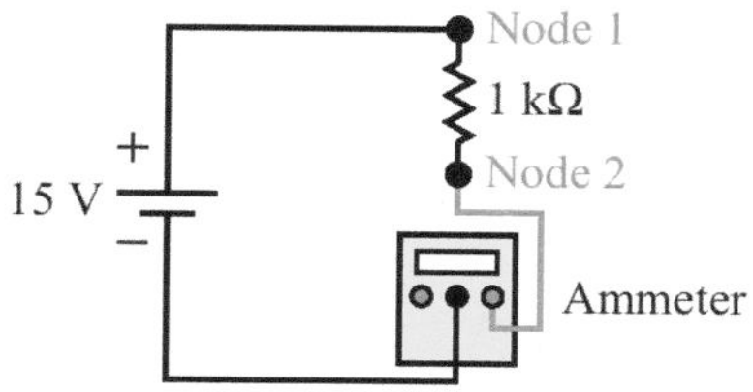

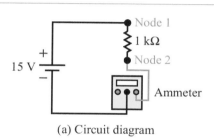

(a) Circuit diagram

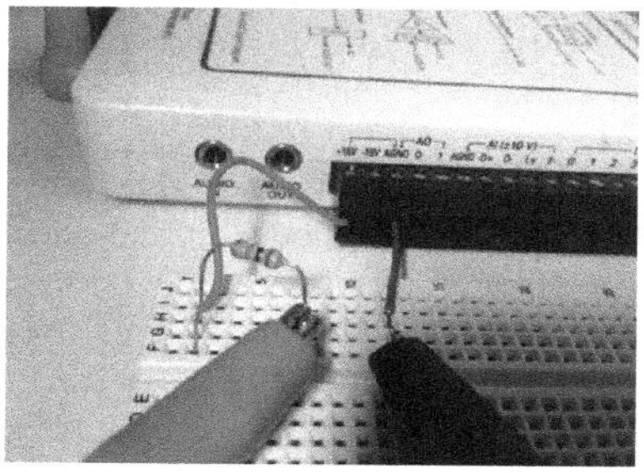

(b) Circuit connections

Figure F-11: Measuring current through a 1 kΩ resistor.

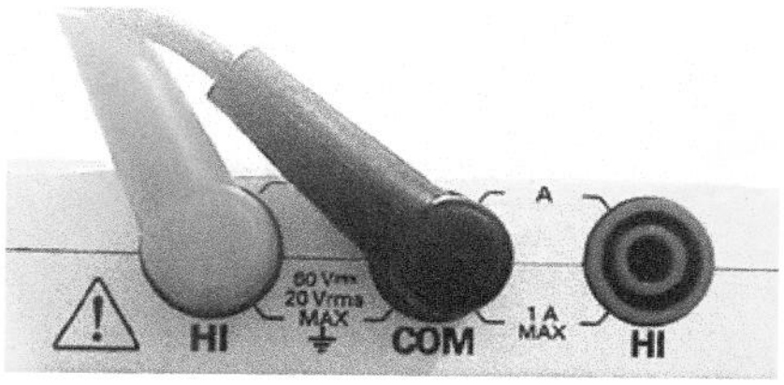

(a) DMM connections

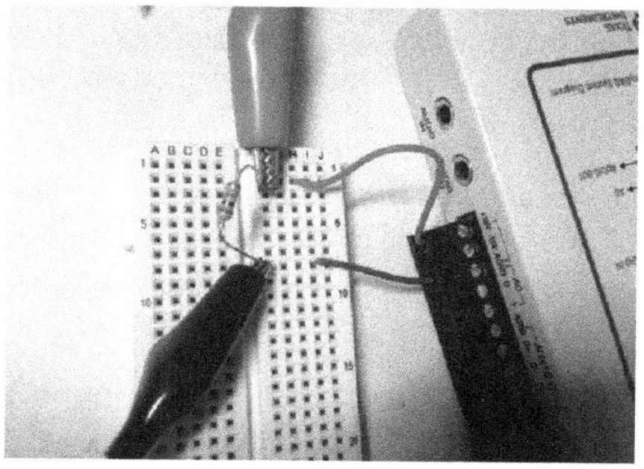

(b) Measuring voltage

Figure F-12: Measuring current by applying Ohm's law.

Measuring current with ammeter: On the DMM panel on the PC, press Run. You should see the DMM reading change to a value close to the expected current. Repeat using 5 V from the myDAQ. Now see what happens if you switch the ammeter probes. Once you have taken your readings, hit the Stop button *and* unclip the leads. Current flows so long as the leads are connected, and prolonged current will heat up the resistor.

> ▶ **If you accidentally try to make a voltage or resistance reading with the ammeter probe connections instead of the voltmeter connections, you can blow out the fuse on the myDAQ. Plan ahead. If you are not planning to make additional current measurements, return the probes to their voltmeter connections.** ◀

Measuring current with voltmeter: A very common way to "measure" the current is to measure the voltage and use Ohm's law to calculate the current. This is shown in Fig. F-12.

- Return the DMM probes to the voltmeter position.
- Build the circuit in Fig. F-11(a) with a +15 V myDAQ source and 1 kΩ resistor as shown in Fig. F-12. As soon as you connect the red and black wires and insert the resistor, current starts to flow, so don't leave it this way too long or the resistor will heat up.
  - Connect one end of the resistor to +15 V in row 2, which corresponds to node 1 in Fig. F-12(a).
  - Connect the black wire from the myDAQ AGND to the other end of the resistor. (this will be node 2 in Fig. F-12(a)).
- Measure the voltage from node 1 (red voltmeter probe) to node 2 (black voltmeter probe).
- Remove the resistor and then measure the resistance between nodes 1 and 2.
- Determine the current $I = V/R$.

Once you have taken your readings, hit the Stop button and disconnect the voltage (+15 V) from the circuit.

Evaluate: Were the currents you measured higher or lower than expected? If so, why? What happened when you switched the DMM leads? How similar were the current readings using the ammeter with those using the voltmeter/Ohm's law?

Debug: If current = 0

There are several reasons you might have read zero current:

- Did you make solid connections between power/ground and the circuit? Are the connections in the 20-pin socket tight, and did they make good connection to the resistor. If the voltage measured across the resistor is 0, the answer is "probably not."
- Did you connect to the circuit properly? Look very carefully at the picture of the connections to the resistor. Note that the connections on the protoboard need to be in the same row as the resistor. Read about the protoboard in the next section.
- Did you make solid connections to the measurement probes? Are they in the correct DMM locations, depending on if you are measuring volts, ohms, or amps? Did you make a solid connection to the circuit?
- Is the myDAQ plugged in and running (probably, or you would have noticed this by now).
- Is something broken . . . possibly the fuse?

Debug: If you think you might have blown the fuse

Exceeding the NI myDAQ ammeter maximum current rating of 1 amp will most likely blow the protection fuse, in which case the ammeter will always read zero and appear as an open circuit. To check the fuse, follow the video tutorial: https://decibel.ni.com/content/docs/DOC-12879.

F-5 Building with the Protoboard / Breadboard

A *breadboard* (also known as a *protoboard* or *plugboard*) is a base on which circuits are hand-built or prototyped. Breadboards allow connections between components and wires without permanently soldering the connection. They come in different shapes and sizes, but their basic connections are all the same. See the picture in Fig. F-13, showing the front (plug side) of the breadboard and the back side, which shows the metal clips that make up the nodes on the breadboard. (The back side is normally protected by a paper or plastic backing. If you remove this backing, the clips fall out, so just leave it in place.) The metal clips pinch onto the wires stuck into the plug side to make electrical connections.

Each horizontal clip on the breadboard creates a node, so each ROW of the board is one node, and any wires plugged in on that row are connected to the same node. The vertical clips on the sides of the board create nodes that extend the full length of each side of the board. They are marked with red and blue lines and are often called *rails*. These rails are commonly used for power and ground by plugging a voltage (such as from the myDAQ)

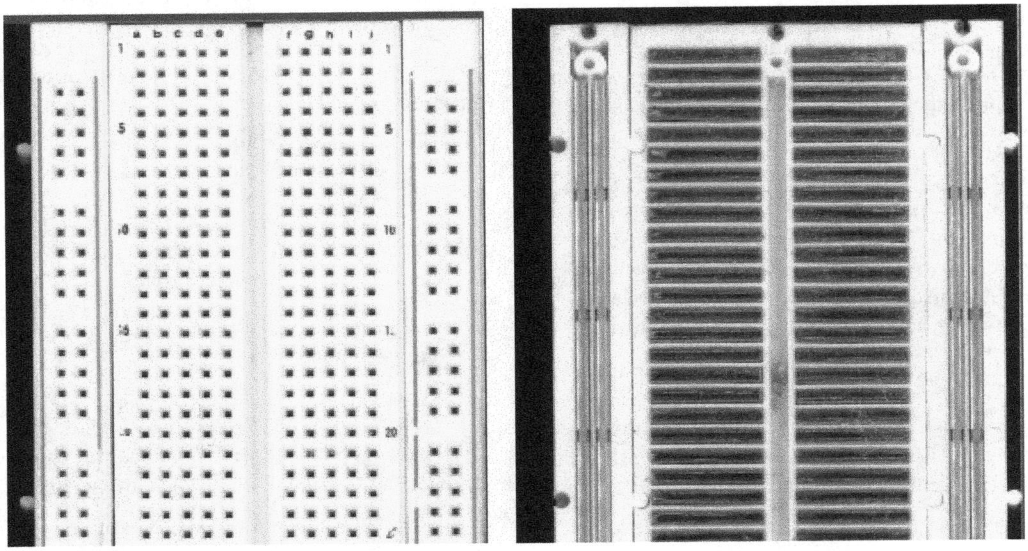

Figure F-13: Breadboard front (plug side) and back (metal clips create nodes).

Figure F-14: 8-pin dual-in-line package (DIP) chip plugged in across the divider on the breadboard. Each of the 8 pins is in a separate row (node). A small dot on the upper left corner of the chip, or a small dibet in the top of the chip, indicates which side is up.

on the red rail and ground (such as AGND from the myDAQ) on the blue rail. This makes power and ground convenient to the many places it is needed throughout the breadboard.

The plastic center divider is just the right width to allow a dual-in-line package (DIP) chip to be plugged in with its legs in the holes on either side of this divider. We use this for op-amp chips (see Fig. F-14).

Build example: Series and parallel resistors

Figure F-15 shows light bulbs connected in series and in parallel, and Fig. F-16 shows an example of two 1 kΩ resistors connected in series (between rows 40 and 50) and in parallel (between rows 30 and 35). Note that for the series combination, one end of each of the two resistors is plugged into row 45.

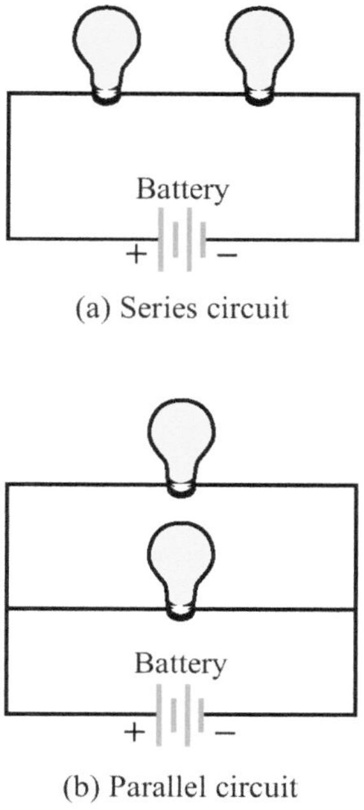

Figure F-15: Two light bulbs connected (a) in series and (b) in parallel.

Evaluate: Measure the resistance of the series and parallel pair of resistors. Build other combinations of these four resistors as well. How close were the measurements to what you expected?

Connect +5 V across the series and parallel resistors. For the parallel combination, plug +5 V into row 30, and AGND into row 35. For the series combination, plug +5 V into row 40 and AGND into row 50. Measure the voltage across each resistor. Calculate the current through each resistor using Ohm's law. Verify that series resistors have the same current through them, and parallel resistors have the same voltage across them.

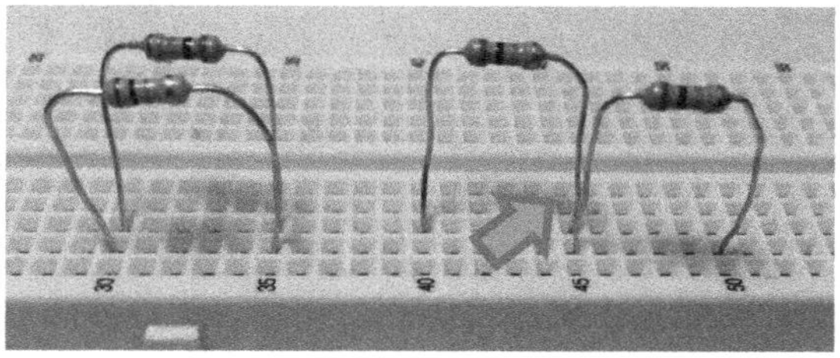

Figure F-16: Protoboard connections for two resistors in series and in parallel.

F-6 Using the NI myDAQ as a Current Source

The NI myDAQ cannot act as a stand-alone current source. However, we can use circuit elements to convert an input voltage into a steady current, thereby acting like a current source. One such circuit element is the *LM317 regulator*. This tutorial explains how to use the LM317 to generate a current source.

Using the LM317 voltage regulator

The LM317 has three wires: V_{in}, V_{out}, and V_{Adjust}. There are several different packages for the LM317, three of which are shown in Fig. F-17.

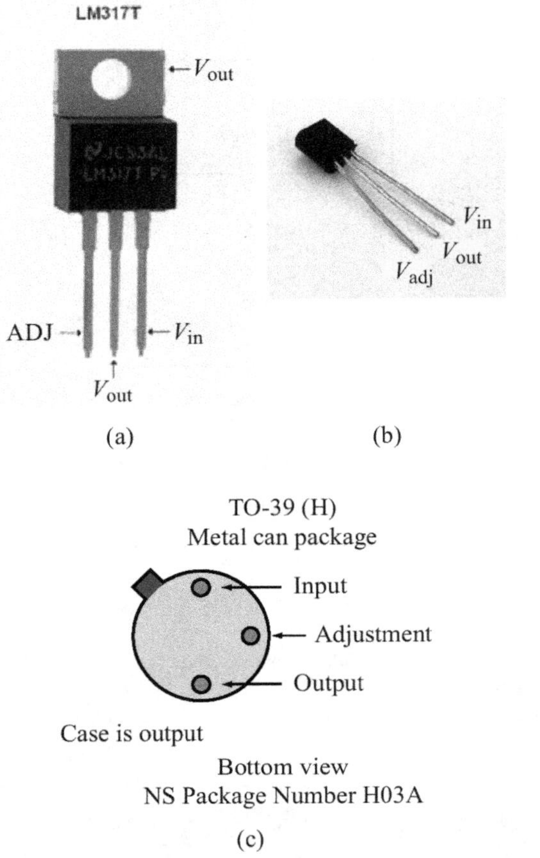

Figure F-17: Three types of LM317 packages. Note that (b) has the flat side up. For the circular configuration in (c), the pin arrangement corresponds to a bottom view of the metal can.

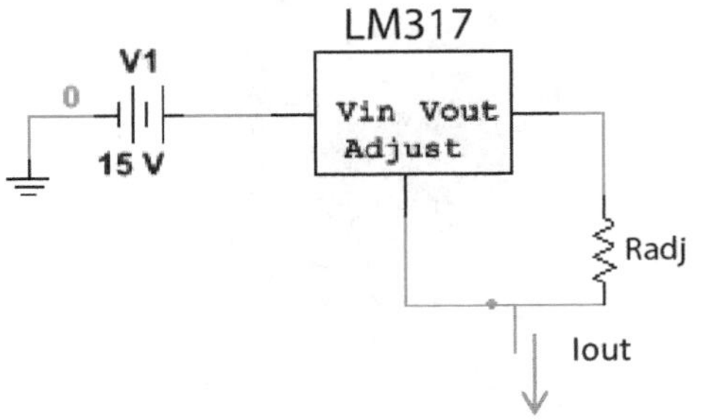

Figure F-18: Circuit configuration for building a current source using the LM317 regulator.

Building a current source

Figure F-18 shows how to connect the LM317 to build a current source.

- Connect +15 V to V_{in}.
- Calculate the adjustment resistor based on your desired current current output:

$$I_{out} = \frac{1.25}{R_{adj}}.$$

- Connect R_{adj} between V_{out} and V_{Adjust}.
- The current output comes from the V_{Adjust} port, so connect one end of a wire there and the other end to the + node of your load.
- Connect AGND to the ground node of your circuit.

Example 1: Build the circuit in Fig. F-19(a).

- Build the LM317 current source circuit shown in Fig. F-19(b). The circuit requires a 1.84 mA current source, so select

$$R_{adj} = \frac{1.25}{0.00184} = 679.3\ \Omega \text{ (round to } 680\ \Omega).$$

Connect I_{out} to the 1 kΩ load resistor.

- Measure the voltage across the 1 kΩ resistor. You should get $V = (1.84A)(1\text{ k}\Omega) = 1.84$ V.

Example 2: Easily adjustable current source.

The current source of Example 1 is fine, except that it requires us to recalculate and purchase a new resistor every time we change the magnitude of the current source. If you want to use your myDAQ at home, this may not always be practical. Instead, replace the 680 Ω resistor with a potentiometer, which you can adjust with a screw or dial. However, the potentiometer can be turned down to $R = 0$, which conceptually would make the current go to infinity. In reality, the myDAQ (and most sources) has an inherent current limit. The myDAQ can provide 32 mA from the +15 V source. Although the myDAQ will limit the current to about 32 mA, even if you try to draw more, it is bad practice and can potentially damage the source if you try

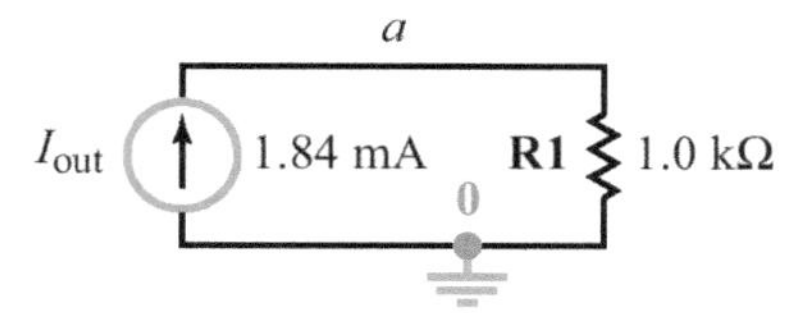

(a) Desired circuit

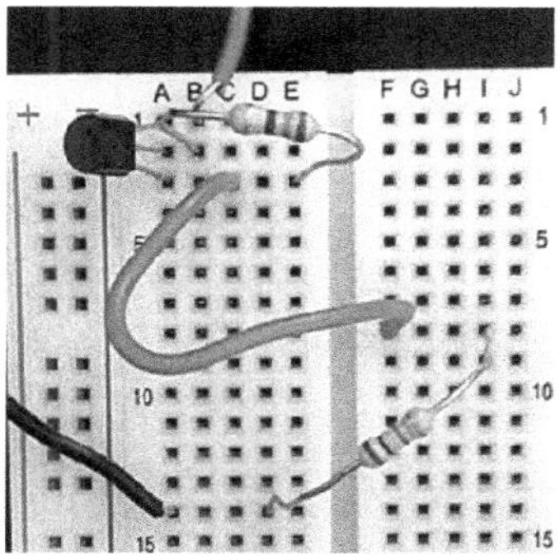

(b) Wiring configuration

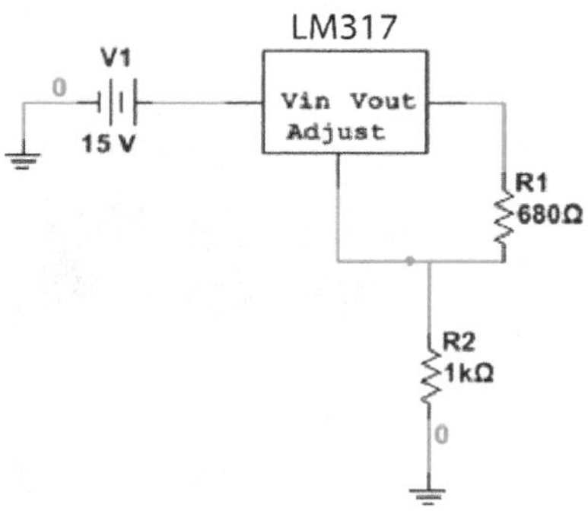

(c) Equivalent current source circuit

Figure F-19: Current source circuit example.

to drive it beyond its limit. Therefore, the minimum R_{adj} you should use is:

$$R_{adj}(\text{min}) = \frac{1.25}{32 \text{ mA}} = 39\ \Omega.$$

To be safe, insert a 39 Ω *current limiting resistor* in series with the potentiometer. Now let's consider what size potentiometer to use. The larger the potentiometer, the smaller the current. Also the larger the potentiometer, the more sensitive it is when you turn the dial (the harder it is to dial the current you want). So, we should use the smallest potentiometer that gives us the minimum current required. For $R_{adj}(max) = 1\ \text{k}\Omega$, the minimum current would be

$$I_{min} = \frac{1.25}{1\ \text{k}\Omega} = 1.25 \text{ mA},$$

which is small enough for any of the examples in this book. A 1 kΩ potentiometer is therefore a good choice. Our final circuit should look like that shown in Fig. F-20.

Storing the current source

The current source configuration is small enough that it can be stored on a breadboard without being disassembled. It is recommended that you build the LM317 circuit once, neatly, perhaps trimming the leads of the LM317 to make it fit snugly against the board. This way, you can use it whenever the need arises to construct a circuit with a current source.

F-7 Creating Waveforms with the Function Generator (FGEN)

A function generator is used for creating AC and periodic voltage waveforms. The NI ELVISMX function generator is capable of creating sinusoidal, ramp, and square wave sources. It can also perform a sweep over a range of frequencies.

Specifications: The A0 output is limited to ±10 VDC or V_{rms} and a current maximum of 2 mA.

Software: Run the NI ELVISmx software. Select the function generator FGEN from the instrument launcher. Press RUN to start a continual waveform or SWEEP to sweep sequentially through a range of frequencies.

Select the settings for your waveform:

- Produces a sinusoidal waveform
- Produces a ramp waveform
- Produces a square wave

Frequency Range: From 200 mHz to 20 kHz. Note: Period T (s) = 1/Frequency (Hz).

Amplitude: Peak-to-peak (V_{pp}) up to 10 V. V_{rms} (for a sine wave) = $\sqrt{2}\ V_{pp}$.

DC Offset: Adds V_{DC} to the waveform, from −5 V to +5 V.

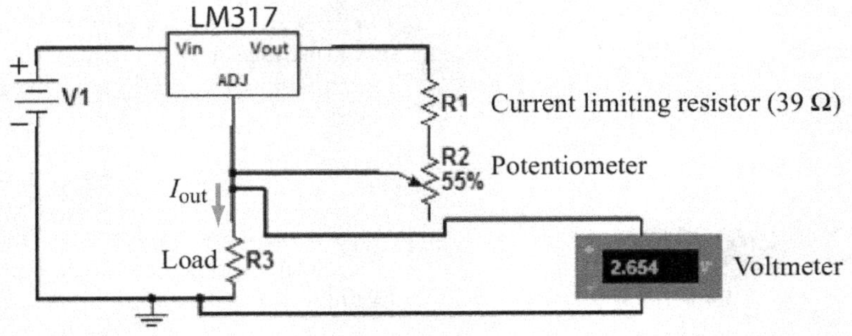

Figure F-20: Current source with LM317 regulator and a potentiometer to adjust I_{out}.

(a) Select FGEN

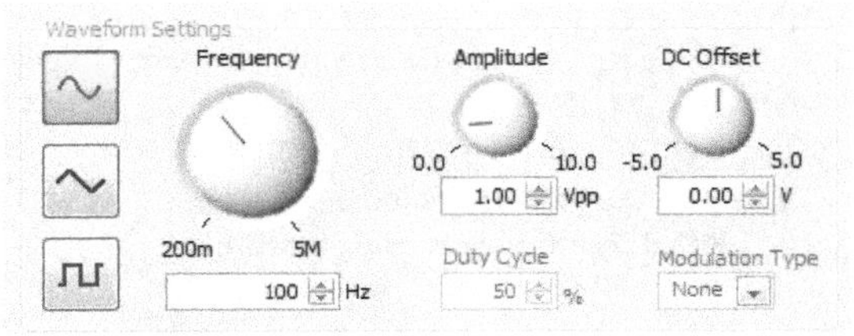

(b) Waveform selection

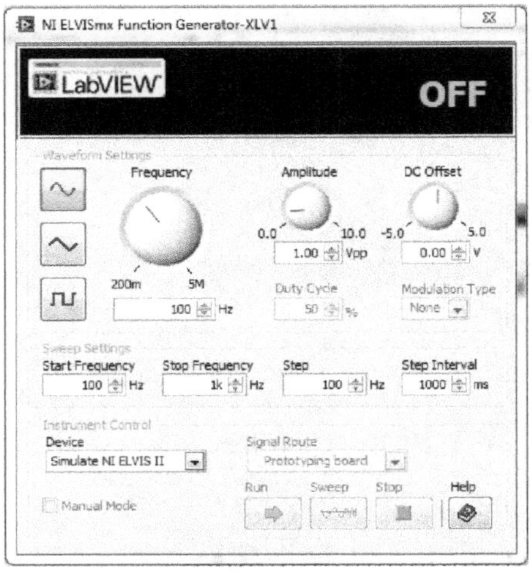

(c) PC display

Figure F-21: Using the function generator.

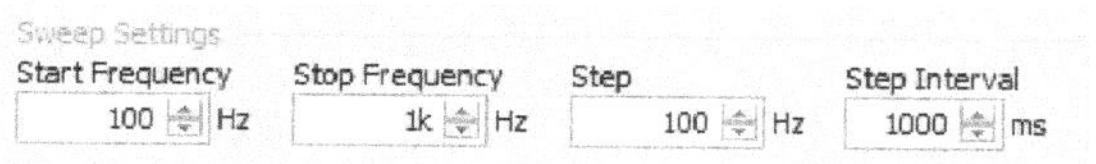

Figure F-22: Sweep settings for function generator.

Figure F-23: Analog output port for voltage output for function generator.

Duty Cycle (only available for square wave): % of time the waveform is HI.

Sweep Settings: Sweep (step) through frequencies from *Start Frequency* through *Stop Frequency* in steps of size *Step*. The *Step Interval* (sometimes called *dwell time*) is the amount of time the function generator stays at each frequency (Fig. F-22).

Signal Route: Select which Analog Output (AO = 0 or 1) channel the waveform will appear on.

Connection: The voltage waveform you create is accessible from the Analog Output ports (AO = 0 or AO = 1) on the 20-pin connector. Connect a red wire to the AO port (0 or 1), as shown in Fig. F-23. This will be the (+) side of your voltage source. Connect a black wire to the AGND port, and this will be the (−) side of your voltage source. Tighten the screws to hold the wires in place. Press RUN when you are ready to turn on the function generator.

F-8 Measure a Time-Varying Voltage with the Oscilloscope

The oscilloscope is used to measure time-varying voltages (Fig. F-24). You can think of it as a voltmeter that measures signals as a function of time. The scope can measure either one or up to two individual channels simultaneously, which is often useful when comparing input and output voltages.

Software: Run the NI ELVISmx software. Select the oscilloscope from the instrument launcher. If you are not sure what settings to use, the Autoscale button is a good start. Press RUN.

Channel Settings: Decide which Analog Input (AI = 0 and/or AI = 1) to use, and check the Enable box(es) for channels to be viewed.

Scale Volts/Div (y axis): Select how large each division is appropriate along the y axis (volts/division).

Vertical Position (DC offset): This feature introduces a vertical offset to create a better view of the waveform, if needed. Set to 0 initially.

Timebase (x axis): Select how large each division is appropriate along the x axis (seconds/division).

Trigger Type: Toggle for when the waveform appears on screen.

(a) *Immediate*, displays the waveform instantly.

(b) *Edge*, the waveform is displayed only if its magnitude is higher or lower than the *Level* (V) setting. This is used to stabilize your view of the waveform.

Acquisition Mode: It is recommended that you *Run Continuously*. The *Once* setting will display one waveform without re-acquiring a new signal.

Connection: Connect the analog inputs to the circuit to be measured. The scope connection works essentially the same as the voltmeter connection. There are two channels (AI = 0 and AI = 1). Each channel has a (+) connection that should be connected to the (+) node of the voltage to be measured, and a (−) connection to be connected to the (−) node of the voltage to be measured. Figure F-25 shows a red lead connected to 0+ and a black lead connected to 0−. These should then be connected across the voltage to be measured. Channels 0 and 1 can be used at the same time or separately. Click Run on the user interface and you are ready to view your oscillating voltage.

Example: Measuring the AC voltage from the function generator

Use the myDAQ Function Generator to create a 4 V_{pp} 100 Hz AC sine wave. Connect it to the Scope by matching the red and black wires in the two photos of Fig. F-26. You should obtain the waveform seen in the scope window in Fig. F-22(b).

Evaluate: Experiment with all of the settings on the function generator and verify their magnitudes and time periods on the scope. Remember that the *period* = 1/*frequency*.

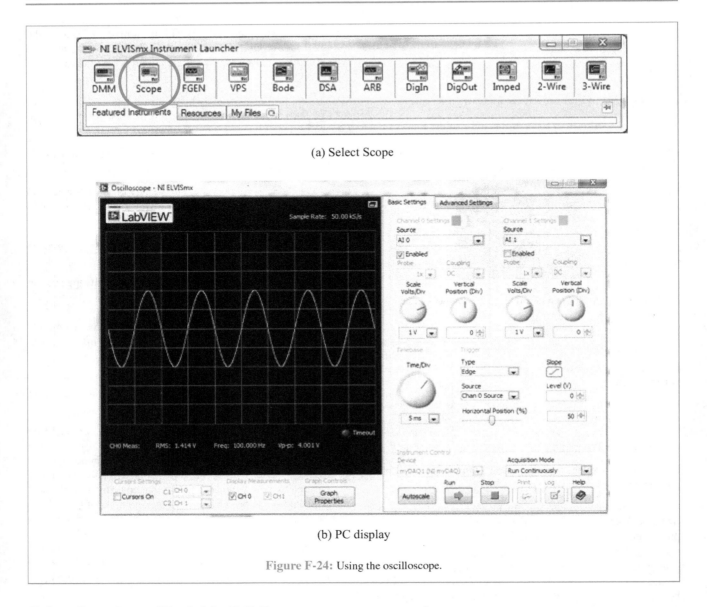

(a) Select Scope

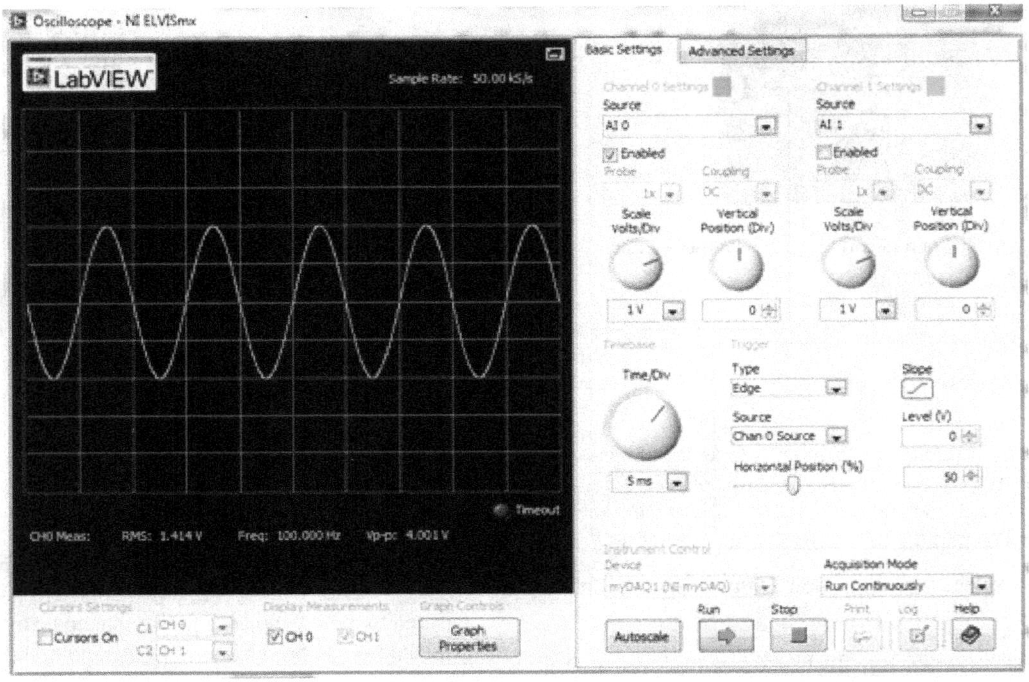

(b) PC display

Figure F-24: Using the oscilloscope.

F-9 Creating a Variable DC Source with the Arbitrary Waveform Generator

The arbitrary waveform generator is used to create variable DC voltage sources or user-defined arbitrary time domain waveforms. We will use the ARB generator to create a variable voltage source. The generator can output one or two distinct waveforms.[†] This tutorial covers only the basics of the ARB generator. See the online information for more details.[‡]

Specifications: The A0 output is limited to ±10 VDC or V_{rms} and a current maximum of 2 mA.

Software: Run the NI ELVISmx software. Select the arbitrary waveform generator (ARB) from the instrument launcher

[†]If two waveforms are to run continuously, they must be the same length in time and number of samples.

[‡]Video tutorial: https://decibel.ni.com/content/docs/DOC-12941

Figure F-25: Scope mode cable connections.

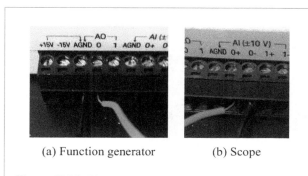

(a) Function generator (b) Scope

Figure F-26: Wire connections for displaying the waveform generator output shown in Fig. F-22(b).

(Fig. F-27(a)). The waveform window (Fig. F-27(b)) provides a view of available functions, just like the oscilloscope option. Select and enable the output channel(s). These can also be the A0 or audio output channels. Select a waveform for each channel you enable. (You will create a waveform file using the editor described below if you want to use a custom output.) Select the *Update Rate* on the waveform window to be the same as the *Sample Rate* on the editor window. Select Run Continuously or Run Once. Select the Gain. The system creates a variable DC voltage source by creating and saving a DC waveform with magnitude 1.0 and then adjusting the gain to create the desired DC voltage. Press RUN to start the generator.

Use the Waveform Editor to create a custom waveform as described below. Start this waveform by selecting it from the Waveform Name folder, and then set the Gain. In this example, you create a 1 V DC voltage with a Sampling/Update Rate of 1 kHz named 1VDC 1kHz.wdt. To change its voltage level, set the Gain to whatever voltage is needed (2 V in this case). The voltage waveform (2 V DC in this example) will appear in the User Interface window, and you can then use it for other applications.

 Click the Waveform Editor icon to begin creating your custom waveform (Fig. F-28 will appear).

Create a waveform:

- Set the Sample Rate. Remember this setting, or add it to the filename, because you will need to specify this as the Update Rate in the ARB user interface (Fig. F-27).
- Add as many segments as desired in your waveform. For the Variable DC Voltage example, only one segment is needed.
- Set the time duration of each segment (10 ms for this example).
- For each segment add a New Component. Specify the component from the Function Library, an expression, or sketch. For this example, use Function Library ≫ DC Level (Offset = 1.0) to create a 1 V DC voltage.
- You can add additional components to each segment. Specify if these are to be added (+), subtracted (−), multiplied (×), divided (/), or frequency modulated (FM). For the variable DC voltage, no additional components are needed.
- Save your file as a *.wdt file. Choose a name that describes the waveform and include in it its sample rate. Use 1 VDC 1 kHz as the filename for this example.

Connection: The waveform voltage you create is accessible from the Analog Output ports (AO = 0 or AO = 1) on the 20-pin connector (Fig. F-29). Connect a red wire to the AO port (0 or 1). This will be the (+) side of your voltage source. Connect a black wire to the AGND port; this will be the (−) side of your voltage source. Tighten the screws to hold the wires in place.

F-10 Measuring Frequency Response with the Bode Analyzer

A *Bode analyzer* is used to plot the frequency response of a circuit or system (known as a Bode plot).[§] The frequency and phase of voltages in a circuit become very important when capacitors and inductors are used. When these or other frequency-dependent components are used, the circuit may act differently at different frequencies. A Bode plot yields two important pieces of information for any measurable voltage in a circuit:

[§] http://www.ni.com/white-paper/11504/en/.

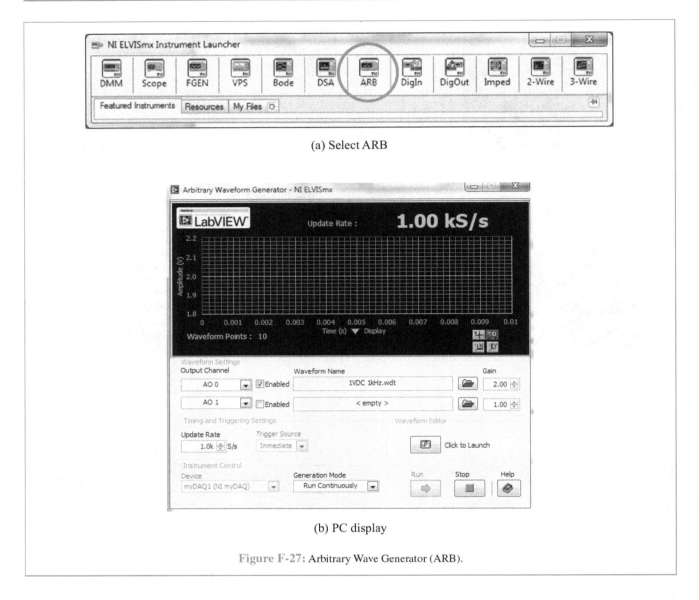

(a) Select ARB

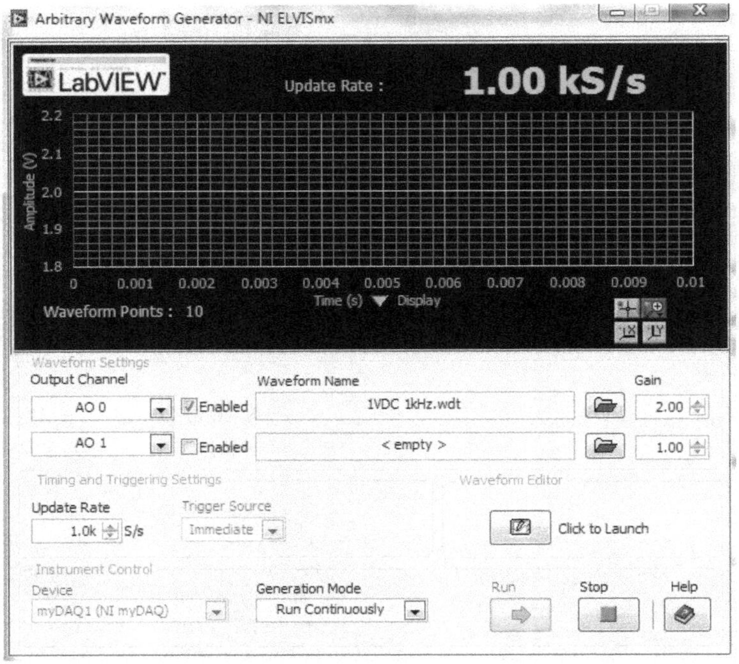

(b) PC display

Figure F-27: Arbitrary Wave Generator (ARB).

(1) The magnitude of a specific voltage in the circuit over a specified range of frequencies.

(2) The associated phase of said voltage over the same range of frequencies.

Software: Run the NI ELVISmx software. Select the Bode Analyzer from the instrument launcher (Fig. F-30). The Bode analyzer effectively drives the circuit with a V_{in} frequency sweep from the function generator (AO = 0). It reads V_{in} in the stimulus channel (AO = 0 by default) and V_{out} in the response channel (AI = 1). It then compares them and plots the Gain $= V_{out}/V_{in}$ in either linear or log scale. This is the frequency response of the system.

To demonstrate how this is done, let's measure the frequency response of the voltage across the capacitor in the series RC circuit of Fig. F-31(a). The Bode Analyzer User Interface is shown in Fig. F-31(b).

(1) Connect V_{in} and set up its frequency sweep.

V_{in} will be generated at the AO = 0 output.

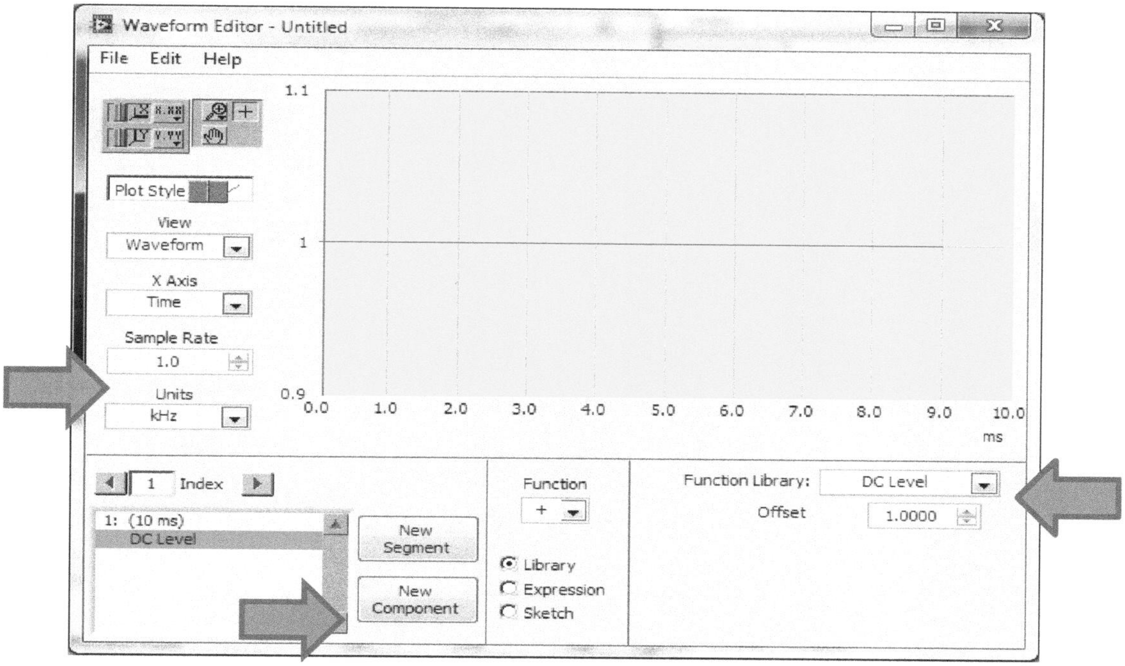

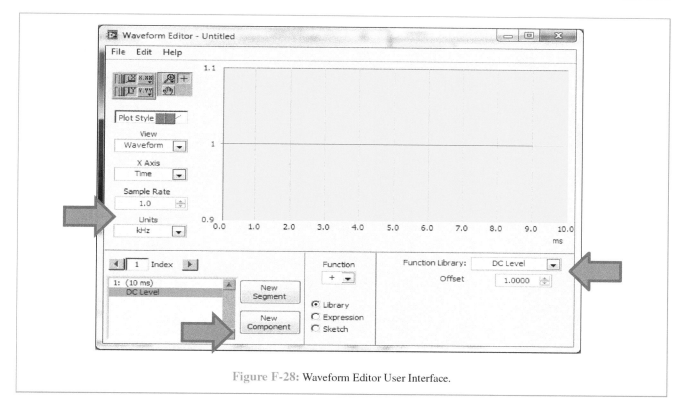

Figure F-28: Waveform Editor User Interface.

Figure F-29: Cable connections for ARB.

- Connect AO = 0 to the $V_{in}+$ node in the circuit (node a).
- Connect AGND to the ground node in the circuit.

Start/Stop Frequency: Set the range of frequencies to be measured.

Steps per decade: This specifies how many data points are measured per frequency decade (10–100 Hz is a decade, for example).

Peak Amplitude: The Bode analyzer sends out a signal at AO = 0. *Peak amplitude* selects the amplitude of this signal.

Mapping: Choose whether the x axis (frequency) is graphed logarithmically or linearly.

(2) Connect the stimulus (measurement) channel.

Stimulus Channel: This channel measures the stimulus. The myDAQ defaults to AI = 0 as the stimulus channel. Connect AI = 0+ in the same place as $V_{in}+$ (node a) and AI = 0− at the ground of your circuit.

(3) Connect the response (measurement) channel.

Response Channel: This channel measures the frequency response of V_{out}. The myDAQ defaults to AI = 1 as the response channel. Connect the AI = 1+ port to the (+) node of V_{out} (node b) and the AI = 1− port to the (−) node of V_{out} (ground, in this example).

(4) Press Run. You should see plots of the magnitude and phase of the frequency response of V_{out} on the PC display.

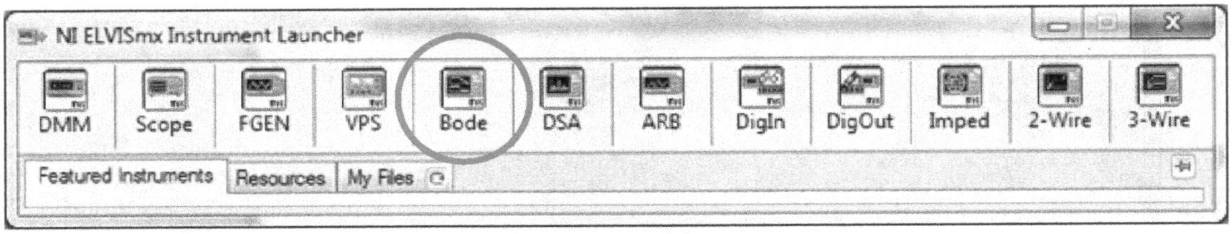

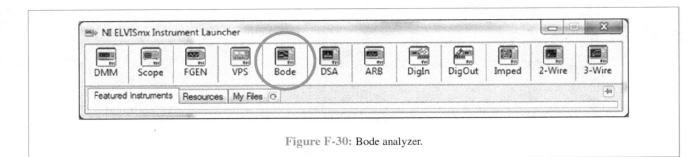

Figure F-30: Bode analyzer.

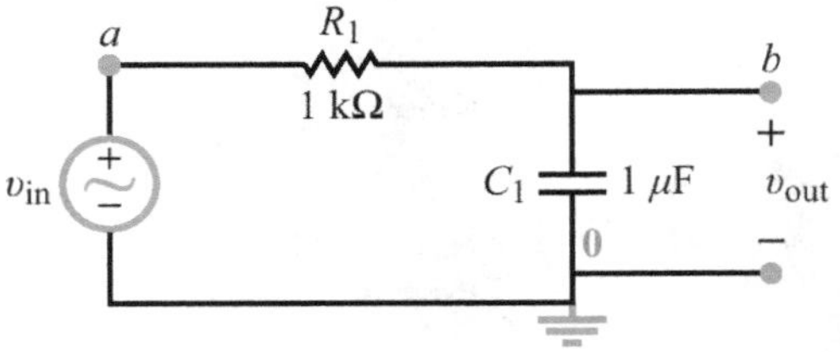

(a) RC circuit

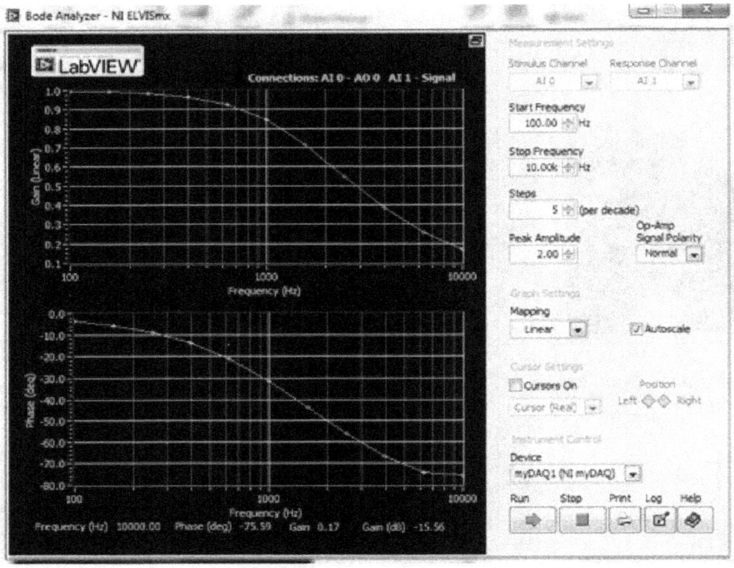

(b) Bode analyzer display and user interface

Figure F-31: (a) RC circuit and (b) its gain magnitude and phase plots.

APPENDIX G

Answers to Selected Problems

Chapter 1

1.1 (b) 4 μA
(d) 390 GV

1.3 (c) 4 pF

1.13 $I = 18$ A along $-z$

1.14 (c) $i(t) = 0.12e^{-0.4t}$ (pA)

1.15 (d) $i(t) = 1.7(1 - e^{-1.2t} + 1.2te^{-1.2t})$ nA

1.16 (b) $\Delta Q(1, 12) = 2.948$ (C)

1.17 (b) $i = 0$ @ $t = 3$ s.

1.20 $i = 4.8$ A

1.22 (b) $i = 0$

1.23 (a) $i = 0$

1.25 (a) $V_2 = 48$ V

1.26 (a) -4 V

1.29 $I = 5$ A, $W = 432$ kJ

1.31 (a) $p(0) = 0.5$ W; $p(0.25\text{ s}) = 0.5$ W

1.36 $P_{\max} = 6$ W

1.38 $V_y = 1.2$ V

1.40 $V_z = 2.5$ V

Chapter 2

2.1 $\ell \approx 2$ km

2.3 (b) $R \approx 1,174\ \Omega$

2.7 $R = 6.41\ \Omega$; $i = 17.2$ A

2.11 $R = 2500\ \Omega$

2.15 $I_x = 2.43$ A

2.17 $I_1 = 2$ A, $I_2 = 1$ A, $I_3 = 2$ A, $I_4 = 1$ A

2.19 $I_x = 3.57$ A; $I_y = 2.86$ A

2.23 $P = 0.32$ W

2.25 $V_1 = -6$ V, $V_2 = 0$, $V_3 = 6$ V

2.29 $I_0 = 31$ A

2.34 $R = 3$ kΩ

2.36 $V_x = 8$ V

2.38 $R_{\text{eq}} = 9\ \Omega$

2.41 (a) $R_{\text{eq}} = 5.5\ \Omega$

2.44 $I = 2$ A

2.49 $I = 1.97$ A

2.51 $I = 3.8$ mA

2.55 $P = 40$ W

2.59 $P = 4$ W

2.61 $R_{eq} = 10\ \Omega$

2.63 (a) $R_3 = 1.5\ \Omega$

2.66 $H = 0.6$ mm

2.67 $I_1 = 0,\ I_2 = 0.1$ A

Chapter 3

3.1 $I = 3$ A

3.3 $P = -120$ W

3.5 $V_R = 7$ V

3.7 $I_x = -0.1$ A

3.11 $P = -24$ W

3.13 $I_x = 0.77$ A

3.15 $I_x = 8$ A

3.17 $V_x = 1.41$ V

3.20 $I_x = 0.151$ A

3.23 $V_x = -2.85$ V

3.26 $V = 10$ V

3.28 $V = 5$ V

3.30 $I_x = -0.1$ A

3.32 $V_x = 4$ V

3.34 $V_x = -3$ V

3.36 $I_x = 6.25$ A

3.39 $I_x = 8$ A

3.42 $V_x = 1.67$ V

3.45 $I_0 = 0.6$ A

3.48 $I_x = 2$ A

3.52 $V_1 = 25.5$ V; $V_2 = 4.5$ V

3.56 $V_x = 1.5$ V

3.58 $I = 0.05$ A

3.62 $V_x = -2.094$ V

3.64 $V_{Th} = 4$ V; $R_{Th} = 5.2\ \Omega$

3.68 $V_{Th} = -7.6$ V; $R_{Th} = 1.6\ \Omega$

3.72 $I_N = 7.71$ A, $R_N = 4.54\ \Omega$

3.74 $I_N = 0.217$ A; $R_{Th} = 9.2\ \Omega$

3.77 $V_{Th} = 1$ V, $R_{Th} = 2.4\ \Omega$

3.80 $V_{Th} = 3$ V, $R_{Th} = 1.5\ \Omega$

3.83 $P_{max} = 2.09$ mW

3.85 $P_{max} = 10$ mW

3.89 $I_0 \approx I_{REF}$

3.93 $V_{out} \approx (R_L/R_E)V_{in}$

Chapter 4

4.1 $\upsilon_o = -10$ V

4.3 $\upsilon_o = -10$ V

4.6 (d) $G = -100$

4.9 $R_f = 16$ kΩ

4.13 $G = R_L(R_1 + R_2)/[R_1(R_3 + R_L)]$

4.14 (b) $R_f = 180$ kΩ

4.16 $R_L = 4$ kΩ

4.19 $G = 0.33$; $-21\text{ V} \le \upsilon_s \le 21\text{ V}$

4.21 $-2\text{ V} \le \upsilon_s \le 2\text{ V}$

4.24 $\upsilon_o = 6.5$ V

4.30 $\upsilon_o = (38 - 4\upsilon_s)$ V; $5.5\text{ V} \le \upsilon_s \le 13.5\text{ V}$

4.35 $\upsilon_o = -3.23$ V

4.38 $\upsilon_o = 0.826$ V, $P = 0.34$ mW

4.40 (a) $G_1 = -6,\ G_2 = -20$

4.46 $\upsilon_o = -[(R_3/R_2)(R_1 + R_2)/(R_1 + R_s)]\upsilon_s$

4.50 $\upsilon_o = 8.5\upsilon_s$

4.52 $\upsilon_o = -5.19$ V

4.54 $\upsilon_s = -0.1$ V

4.56 $G = 2 \times 10^5$ to 2×10^6

4.60 $\upsilon_o = 2.5 - 10^4\upsilon_s$

Chapter 5

5.3 $v(t) = 10u(t - 2\ \mu s) - 10u(t - 7\ \mu s)$ V

5.7 $v_o = 12$ V; $\tau = 2$ s.

5.9 $v(t) = 2 + 8e^{-0.5t}$ V

5.15 $v_1 = -12$ V; $v_2 = 6$ V; $v_3 = 2$ V

5.17 $C_{eq} = 4\ \mu$F

5.19 $C_{eq} = 2.95$ F

5.21 8 μF: 60 V; 3 μF: 60 V; 6 μF: 20 V; 6 μF: 30 V; 12 μF: 10 V; 10 μF: 30 V

5.25 $i(t) = (0.25 - e^{-0.2t})$ A

5.29 $v_{C_1} = 20$ V; $v_{C_2} = 12$ V; $i_{L_1} = 0$; $i_{L_2} = 2$ A

5.31 $L_{eq} = 5$ mH

5.33 (c) $i_C(\infty) = 0$; $v_C(\infty) = 8$ V

5.36 $v(t) = 14e^{-11.67t}$ V

5.39 $v(t) = [-18 + 24e^{-1.25t}]$ V

5.43 $v(t) = 5.35e^{-1000(t-0.1)/45}$ V

5.45 $v_C(t) = 4e^{-100t/1.5}$ V

5.48 $i(t) = 2e^{-100t/3}$ A

5.50 $i_1(t) = 2.88e^{-10t}$ mA; $i_2(t) = 0.72e^{-20t}$ mA

5.53 $i(t) = 0.3(1 - e^{-8t})$ A

5.56 $i(t) = [20 - 30e^{-500t}]$ mA

5.60 $v_{out}(t) = v_{out}(0) + v_i(t) + \dfrac{1}{RC}\displaystyle\int_0^t v_i\, dt$

5.63 $i_{out}(t) = 0.6e^{-2t}$ (mA)

5.65 (b) $v_{out}(t) = -24(1 - e^{-2t})$ V

5.68 (a) $v_{out_1}(t) = -0.5t$ V

5.73 $W = 0.2\ \mu$m

Chapter 6

6.1 $v_C(0) = 12$ V, $i_L(0) = 0$, $i_C(0) = -3$ A, $v_L(0) = 12$ V, $v_C(\infty) = 0$, $i_L(\infty) = 4$ A

6.3 (a) $v_C(0) = -12$ V, $i_L(0) = 3$ mA

6.7 $i_1(0) = 1$ A; $i_2(0) = 2$ A

6.10 $i_C(0) = -3$ A, $v_C(0) = -24$ V, $i_R(0) = -3$ A, $v_R(0) = -24$ V, $i_L(0) = 0$, $v_L(0) = -24$ V, $v_L(\infty) = 0$, $i_R(\infty) = 0$, $v_C(\infty) = 0$, $i_L(\infty) = 0$

6.12 $v_C(t) = (12.64e^{-2.68t} - 3.64e^{-9.32t})$

6.14 $i_L(t) = -90e^{-6t}$ A

6.16 $v_C(t) = V_0 \cos(\omega_d t)$; $\omega_d = 1/\sqrt{LC}$

6.18 $i_C(t) = -[(40/3)e^{-0.5t} \sin 0.375t]$ A

6.22 $i_C(t) = (12 \sin t - 6 \cos t)e^{-2t}$ A

6.25 $v_C(t) = 0.4 + (2.44t - 0.2)e^{-5.81t}$ mV

6.28 (a) $v_C(\infty) = 10$ V

6.29 $v_C(t) = (12 + 3e^{-60t})$ V

6.32 $v_C(t) = 4 - e^{-6000t}[10\cos(11431t) - 5.25\sin(11431t)]$V

6.35 $v_C(t) =$
$\frac{4}{3} + \left[\frac{20}{3}\cos(745.4t) + 5.96\sin(745.4t)\right]e^{-666.7t}$ V

6.38 $i_L(t) = 1.5$ mA

6.40 $i_L(t) = [32 - (32 + 6400t)e^{-400t}]$ mA

6.43 (a) $i_L(t) =$
$-2 + [4.5\cos(526.8t) + 7.26\sin(526.8t)]e^{-850t}$ A,
(b) $w_C(\infty) = 0$

6.46 (a) $i_L(t) = 2$ A
(b) No, the solution method is applicable to dc sources only.

6.48 $i_C(t) =$
$$\begin{cases} (-0.0045e^{-83t} + 0.1045e^{-1917t})\ \text{A} & \text{for } 0 \le t \le 1\ \text{ms} \\ (2.43 \times 10^{-3}\cos 400t - 0.017\sin 400t)\ \text{A} & \text{for } t \ge 1\ \text{ms} \end{cases}$$

6.50 $v(t) = (24 - 14.4e^{-2t} + 6.4e^{-3t})$ V

6.52 (c) $[3\cos 0.433t + 5.2\sin 0.433t]e^{-0.75t}$ A

6.54 $v_C(t) = 2.4 - (0.4\cos 3.74t - 0.428\cos 3.74t)e^{-6t}$ V

6.56 $i_2(t) = (2.3e^{0.89t} + 97.7e^{-5.61t})\ \mu$A

6.58 $v_{out}(t) = -8e^{-20t} \sin 74.83t$ V

Chapter 7

7.1 $A = 4$ V, $f = 4$ kHz, $t = 0.25$ ms, $\phi = 45°$

7.3 $\upsilon(t) = 12\cos(8\pi \times 10^3 t + 60°)$ V

7.6 $\Delta t = 1/9$ μs, waveform shifts backwards

7.9 $\upsilon(t) = 12\cos(100\pi t - 45°)$ V

7.10 (c) $\mathbf{z}_3 = 7.21e^{-j146.3°}$
(e) $\mathbf{z}_5 = 5e^{-j53.13°}$

7.11 (c) $\mathbf{z}_3 = -1.22(1 + j)$
(f) $\mathbf{z}_6 = 3 + j4$

7.14 (b) $\mathbf{z}^2 = 100e^{-j73.74°}$
(f) $\mathfrak{Im}(\mathbf{z}^*) = -6$

7.15 (c) $\mathbf{z}_1\mathbf{z}_2^* = 10e^{-j105°}$

7.16 (b) $e^{\mathbf{z}} = -2.45 - j2.24$

7.18 (b) $\mathbf{B} = 42.5e^{-j161.99°}$

7.20 (b) $62.2e^{-j4.61°}$

7.22 (b) $\mathbf{V}_2 = 2e^{j108°}$ V

7.23 (c) $i_2(t) = 2\cos(2\pi \times 10^3 t + 150°)$ A

7.25 (c) $\mathbf{Z}_C = -j3.18$ Ω

7.27 (b) $\upsilon_2(t) = 5.49\cos(1000t - 18°)$ V

7.29 $i_C(t) = 9.4\cos(2\pi \times 10^4 t - 21.48°)$ mA

7.32 $\upsilon_{ab}(t) = 0.42\cos(300t - 186.35°)$ V

7.34 (b) $\mathbf{Z}_2 = (98.5 + j1524)$ Ω

7.36 $\mathbf{Z} = (5.32 - j1.69)$ Ω

7.38 $\mathbf{I}_R = 2.528e^{j35.56°}$ A

7.40 (a) $\mathbf{Z} = (5 + j5)$ Ω
(b) $\mathbf{I} = 3.54$ A

7.44 $L = 0$ or 2.5 mH

7.48 $\mathbf{V}_{Th} = -12$ V; $\mathbf{Z}_{Th} = 0$

7.50 $\mathbf{Z}_L = (6 + j2)$ Ω

7.51 $\mathbf{Z}_L = (2 - j1)$ kΩ

7.56 $f = 795.8$ Hz

7.58 $i_C(t) = 1.25\cos(400t - 6.35°)$ A

7.60 $\mathbf{V}_1 = (20.7 + j16.1)$ V,
$\mathbf{V}_2 = -(20.7 + j42.1)$ V

7.62 $\mathbf{I}_C = 1.93e^{j4.9°}$ A

7.65 $\mathbf{V}_{out} = \left(\dfrac{3 + j1}{5}\right)\mathbf{V}_s$

7.68 $i_x(t) = 24.72\cos(5 \times 10^5 t - 74.06°)$ A

7.71 $\mathbf{V}_{Th} = 10e^{-j30.5°}$ V, $\mathbf{Z}_{Th} = (2.9 - j3)$ Ω

7.74 $i_a(t) = 2.06\cos(35t + 152.75°)$ A

7.77 $\mathbf{I}_a = 4.4e^{-j21.45°}$ A

7.80 $\upsilon_{out}(t) = V_0 \sin \omega t$

7.83 $\upsilon_{out}(t) = 11.32\cos(377t + 152.05°)$ V

Chapter 8

8.1 (a) $V_{av} = 2$ V
(b) $V_{rms} = 2.31$ V

8.4 (a) $I_{av} = 3$ A
(b) $I_{rms} = 3.46$ A

8.7 (a) $V_{av} = 2$ V
(b) $V_{rms} = 3.79$ V

8.10 (a) $V_{av} = 2$ V
(b) $V_{rms} = 2.38$ V

8.12 (a) $V_{av} = 0.432$ V
(b) $V_{rms} = 0.5136$ V

8.15 (c) $V_{av} = 12$ V, $V_{rms} = 12.32$ V

8.17 (c) $\mathbf{S} = 330e^{j15°}$ VA, $P_{av} = 318.76$ W,
$Q = 85.41$ VAR, $pf = 0.97$ (lagging)

8.18 (c) $\mathbf{S} = 2.5e^{-j75°}$ VA, $P_{av} = 0.65$ W,
$Q = -2.415$ VAR, $pf = 0.26$ (leading)

8.21 $\mathbf{S} = 0.665e^{-j79.35°}$ VA, $pf = 0.185$ (leading)

8.23 $P_{av}(200\ \Omega) = 5.52$ W, $P_{av}(100\ \Omega) = 1.38$ W,
$P_{av}(\text{source}) = 6.9$ W

8.26 $P_{av} = 496.4$ mW

8.29 $P_L = 0.186$ W

8.31 $\mathbf{S}_{\text{load}} = (0.38 + j0.26)$ VA

8.33 $P_L = 3.933$ mW

8.36 $P_L = 0.713$ W

8.39 $P_{av} = 0.2$ mW

8.43 $\mathbf{Z}_{eq} = (29 + j8.93)\ \Omega$, inductive
$\mathbf{Z}_1$ = inductive, $\mathbf{Z}_2$ = capacitive, $\mathbf{Z}_3$ = inductive

8.45 $C = L/(R^2 + \omega^2 L^2)$

8.48 (a) $\mathbf{Z}_1 = (10.5 - j5.2)\ \Omega$; $\mathbf{Z}_2 = (7.23 + j5.05)\ \Omega$

8.50 (a) $pf = 0.673$ lagging

8.52 $\mathbf{Z}_L = (0.6 - j0.2)\ \Omega$, $P_{max} = 6.78$ W

8.55 $\mathbf{Z}_L = (1.33 - j4)\ \Omega$, $P_{max} = 4.18$ W

8.58 $\mathbf{Z}_L = (5.1568 + j0.72)\ \Omega$, $P_{max} = 0.8$ W

8.60 $\mathbf{Z}_L = (22.9 + j39.8)\ \Omega$, $P_{max} = 38.9$ mW

Chapter 9

9.1 $\omega_0 = 10^4$ rad/s

9.3 $\omega_0 = 10^5$ rad/s

9.5 $\omega_0 = 5 \times 10^4$ rad/s

9.8 $K_m = 50$; $K_f = 2 \times 10^4$

9.11 (b) $\omega_0 = 1$ rad/s

9.13 (b) −23 dB

9.14 (a) 20.81 dB

9.16 (c) 0.25

9.21 $\mathbf{H}(\omega) = \dfrac{-2000\omega}{(1 + j\omega)(100 + j\omega)}$

9.23 $\mathbf{H}(\omega) = -(10 + j\omega)(100 + j\omega)/50\omega$

9.26 (a) $\omega_0 = 10^4$ rad/s, $Q = 40$, $B = 250$ rad/s,
$\omega_{c_1} = 9875$ rad/s, $\omega_{c_2} = 10125$ rad/s

9.29 $\omega_0 = 5$ rad/s; $Q = 20$; $B = 25$ rad/s; $\omega_{c_1} = 487.5$ rad/s; $\omega_{c_2} = 512.5$ rad/s

9.32 (b) $\mathbf{H}(\omega) = \dfrac{j10^{-3}\omega}{1 + j2.5\omega/1000 + (j\omega/1000)^2}$

9.35 (a) $\mathbf{H}(\omega) = \dfrac{1}{2}\left(\dfrac{j\omega/\omega_c}{1 + j\omega/\omega_c}\right)$, with $\omega_c = 1.25 \times 10^4$ rad/s

9.37 (a) $\mathbf{H}(\omega) = \dfrac{1 + j\omega C(R_1 + R_2)}{1 + j\omega R_1 C}$

9.39 (a) $\mathbf{H}(\omega) = \dfrac{1}{1 + j0.4\omega/5000 + (j\omega/5000)^2}$

9.41 (a) $\mathbf{H}(\omega) = \left[\dfrac{-j(\omega/\omega_{c_1})}{(1 + j\omega/\omega_{c_2})(1 + j\omega/\omega_{c_3})}\right]$, with
$\omega_{c_1} = 10^4$ rad/s, $\omega_{c_2} = 200$ rad/s, $\omega_{c_3} = 2 \times 10^6$ rad/s

9.46 $\mathbf{H}(\omega) = -10$

9.50 78 to 98 MHz, or 98 to 118 MHz

Chapter 10

10.4 $\mathbf{V}_1 = 100\angle{-60^\circ}$ V (rms), negative phase sequence

10.7 $\upsilon_{ab}(t) = 34.79\cos(377t + 113.06^\circ)$ V
$\upsilon_{bc}(t) = 81.58\cos(377t + 47.78^\circ)$ V
$\upsilon_{ca}(t) = 101.2\cos(377t - 114.02^\circ)$ V

10.11 $i_{L_1}(t) = 22.73\cos(2\pi ft + 16.9^\circ)$ A
$i_{L_2}(t) = 27.71\cos(2\pi ft - 100.5^\circ)$ A
$i_{L_3}(t) = 26.56\cos(2\pi ft + 129.0^\circ)$ A

10.14 $i_{L_1}(t) = 27.31\cos(2\pi ft + 7.62^\circ)$ A
$i_{L_2}(t) = 32.71\cos(2\pi ft - 114.2^\circ)$ A
$i_{L_3}(t) = 29.54\cos(2\pi ft + 117.5^\circ)$ A

10.17 $\mathbf{I}_{L_1} = 10.02\angle{-59.3^\circ}$ A (rms)

10.20 $\mathbf{I}_n = 1.64\angle{143.2^\circ}$ A (rms)

10.23 $\mathbf{S}_T = 29.68\angle{-3.56^\circ}$ kVA

10.26 $\mathbf{S}_T = 29.68\angle{-3.56^\circ}$ kVA, 58.2%

10.28 $\mathbf{I}_{L_1} = 25.61\angle{-28.37^\circ}$ A (rms)
$\mathbf{I}_{L_2} = 25.61\angle{-148.37^\circ}$ A (rms)
$\mathbf{I}_{L_3} = 25.61\angle{91.69^\circ}$ A (rms)

10.30 $\mathbf{I}_{L_1} = 69.62\angle{-19.1^\circ}$ A (rms)
$\mathbf{I}_{L_2} = 69.62\angle{-139.1^\circ}$ A (rms)
$\mathbf{I}_{L_3} = 69.62\angle{100.9^\circ}$ A (rms)
$pf = 0.9449$

10.32 $C = 2.261$ mF

10.34 $P_1 = 3.35$ kW, $P_2 = 6.194$ kW

10.36 $P_1 = P_2 = 50.82$ W

Chapter 11

11.1 (a) $i(t) = 0.293\cos(120\pi t - 125.8°)$ A
(b) $P = 8.57$ W

11.4 $\mathbf{V}_{\text{out}} = 2.629\underline{/164.4°}$ V

11.7 $\mathbf{I}_x = 0.8622\underline{/-51.9°}$ A

11.9 $\mathbf{V}_{\text{out}} = 0.8\underline{/-73.3°}$ V

11.11 $\mathbf{V}_{\text{out}} = 24.83\underline{/-155.6°}$ V

11.13 (c) $L_{\text{eq}} = 2$ H

11.15 $\mathbf{Z}_{\text{in}} = 18.28\underline{/-38.9°}$ Ω

11.17 $\mathbf{Z}_{\text{in}} = 14.03\underline{/30.8°}$ Ω

11.19 $L_{\text{eq}} = 5.1$ H

11.21 $\mathbf{Z}_{\text{in}} = (0.3 + j9.9)$ Ω

11.24 $\mathbf{V}_{\text{out}} = 26.63\underline{/33.7°}$ V

11.26 $\mathbf{I}_x = 24.25\underline{/-129°}$ mA

11.28 $P_{\text{L}} = 0.83$ W

11.31 $P_{\text{L}} = 2.676$ W

11.33 $n = 8/3$

Chapter 12

12.4 (c) $\mathbf{F}_3(\mathbf{s}) = \dfrac{12e^{-4\mathbf{s}}}{\mathbf{s}+3}$

12.5 (a) $\mathbf{H}_1(\mathbf{s}) = \left[\dfrac{48}{\mathbf{s}+3} + \dfrac{12}{(\mathbf{s}+3)^2}\right]e^{-4\mathbf{s}}$
(c) $\mathbf{H}_3(\mathbf{s}) = \dfrac{60}{(\mathbf{s}+2)^4}$

12.7 (c) $\mathbf{F}_3(\mathbf{s}) = \dfrac{10e^{-3\mathbf{s}}}{\mathbf{s}+4}$

12.9 (b) $f_2(t) = [2e^{-2t} - 4e^{-3t}\cos t]\, u(t)$

12.10 (c) $f_3(t) = 2e^{-3t}\cos(2t + 45°)\, u(t)$

12.11 (b) $f_2(t) = 4(1 + \cos 3t)\, u(t)$

12.12 $i_{\text{L}}(t) = 8(e^{-t} - e^{-2t})\, u(t)$ A

12.15 $i_{\text{L}}(t) = 1.14e^{-0.5(t-2)}\sin(0.87(t-2))\, u(t-2)$ A

12.18 $\upsilon(t) = [1.5 - 1.56e^{-4t} + 0.072e^{-12t}]\, u(t)$ V

12.22 $i_{\text{L}}(t) = (0.012e^{-0.13t} - 0.81e^{-3.31t})\, u(t)$ mA

12.25 $i_{\text{L}}(t) = 30e^{-2t}\sin t\; u(t)$

12.28 $i_{\text{L}}(t) = [2 + 7e^{-0.67t}\cos(1.43t - 1012.5°)]\, u(t)$ A

12.31 $i_{\text{L}}(t) = [1.25 + 2.58e^{-1.75t}\cos(0.97t - 61°)]\, u(t)$ A

12.34 $\upsilon_{\text{C}}(t) =$
$10(1 - e^{-0.25t})\, u(t) - 10[1 - e^{-0.25(t-2)}]\, u(t-2)$ V

12.37 $i_{\text{L}}(t) = (21.52e^{-2t} - 112.14e^{-t} - 5.38e^{-5t})\, u(t)$ A

12.44 (a) $\upsilon_{\text{out}}(t) = 2[1 - e^{-10^4 t}(1 + 10^4 t)]\, u(t)$ V

Chapter 13

13.5 (a) even and dc symmetry
(b) $$f(t) = \sum_{n=1}^{\infty} \frac{160}{(n\pi)^2}\left[\cos\left(\frac{n\pi}{4}\right) - \cos\left(\frac{3n\pi}{4}\right)\right] \cdot \cos\left(\frac{n\pi t}{4}\right)$$

13.5 (a) Odd and dc symmetry
(b) $$f(t) = \sum_{n=1}^{\infty} \frac{10}{n\pi}\,[2 - \cos(n\pi)]\sin(n\pi t)$$

13.12 $f(t) = \frac{1}{2} - \frac{1}{2}\cos(8\pi t)$

13.16 $$f_2(t) = -\frac{40}{\pi}\sum_{n=1}^{\infty}\frac{1}{n}\sin\left(\frac{n\pi t}{4}\right)$$

13.26 (a) $\upsilon_{\text{out}}(t) =$
$$\sum_{n=1}^{\infty} 100 \times \left(\frac{4}{n\pi}\right)^2 \sin\left(\frac{n\pi}{2}\right)\cos(n\pi t + 90°)\ \text{V}$$

13.29 $P_{\text{av}} = 54.5$ mW; ac power fraction = 8.26%

13.33 $\mathbf{F}(\omega) = \dfrac{5}{3\omega^2}\,[(1 + j3\omega)e^{-j3\omega} - 1]$

13.39 $\mathbf{F}(\omega) = \dfrac{3}{\omega}\left[-2j - \dfrac{2j}{\omega}(\sin\omega - \sin 2\omega)\right]$

13.46 (b) $t\, f(t) \;\longleftrightarrow\; \dfrac{5}{(2 + j\omega)^2}$

Index

A

B

C

D

E

F

G

H

I

J

K

L

M

N

O

Q

R

S

T

W

X

Y

Z

CPSIA information can be obtained
at www.ICGtesting.com
Printed in the USA
LVHW062257070721
692124LV00003B/21

9 781607 854838